P9-DEL-368

www.mhhe.com/seeley10

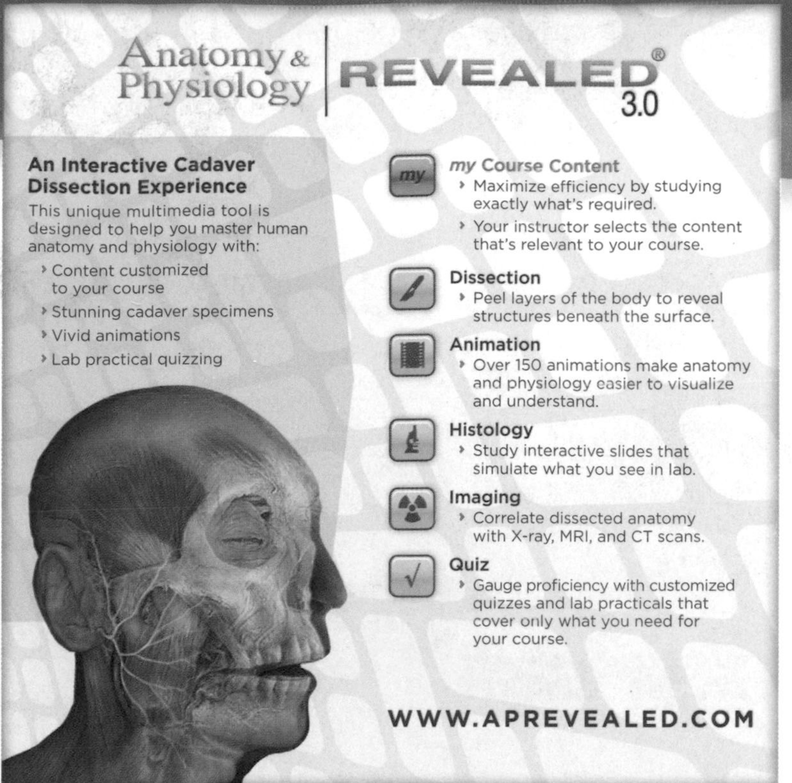
Anatomy & Physiology
REVEALED® 3.0
An Interactive Cadaver Dissection Experience
This unique multimedia tool is designed to help you master human anatomy and physiology with:
Content customized to your course
Stunning cadaver specimens
Vivid animations
Lab practical quizzing
my Course Content
Maximize efficiency by studying exactly what's required.
Your instructor selects the content that's relevant to your course.
Dissection
Peel layers of the body to reveal structures beneath the surface.
Animation
Over 150 animations make anatomy and physiology easier to visualize and understand.
Histology
Study interactive slides that simulate what you see in lab.
Imaging
Correlate dissected anatomy with X-ray, MRI, and CT scans.
Quiz
Gauge proficiency with customized quizzes and lab practicals that cover only what you need for your course.
WWW.APREVEALED.COM

Full Textbook Integration! APR

Icons throughout the book indicate specific McGraw-Hill Anatomy & Physiology | REVEALED® content that corresponds to the text and figures.

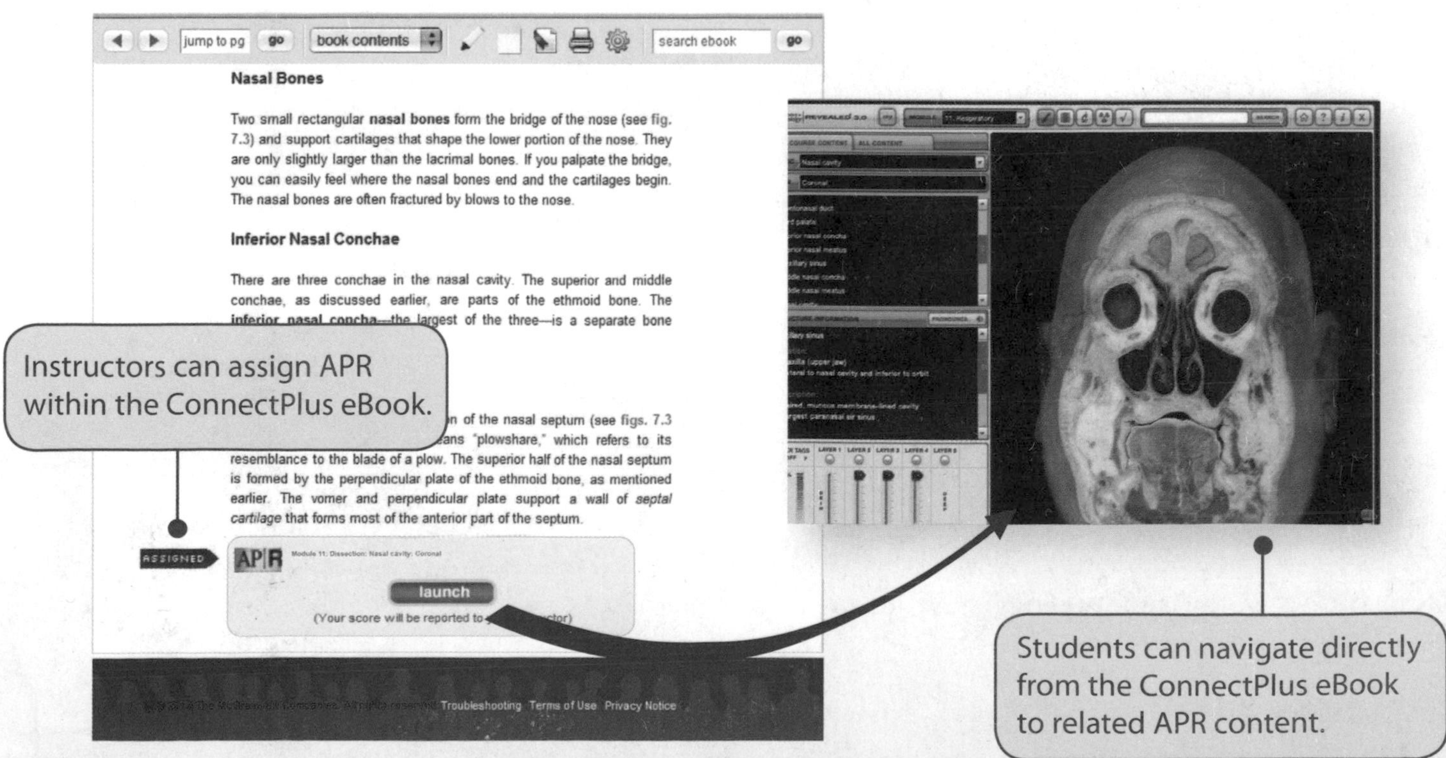
jump to pg
go
book contents
search ebook
go
Nasal Bones
Two small rectangular nasal bones form the bridge of the nose (see fig. 7.3) and support cartilages that shape the lower portion of the nose. They are only slightly larger than the lacrimal bones. If you palpate the bridge, you can easily feel where the nasal bones end and the cartilages begin. The nasal bones are often fractured by blows to the nose.
Inferior Nasal Conchae
There are three conchae in the nasal cavity. The superior and middle conchae, as discussed earlier, are parts of the ethmoid bone. The inferior nasal concha—the largest of the three—is a separate bone
Instructors can assign APR within the ConnectPlus eBook.
of the nasal septum (see figs. 7.3
ans "plowshare," which refers to its resemblance to the blade of a plow. The superior half of the nasal septum is formed by the perpendicular plate of the ethmoid bone, as mentioned earlier. The vomer and perpendicular plate support a wall of septal cartilage that forms most of the anterior part of the septum.
ASSIGNED
APR
launch
(Your score will be reported to
Troubleshooting Terms of Use Privacy Notice
Students can navigate directly from the ConnectPlus eBook to related APR content.

Brief Contents

SEELEY'S ANATOMY & PHYSIOLOGY

10TH EDITION

Cinnamon VanPutte
SOUTHWESTERN ILLINOIS COLLEGE

Jennifer Regan
UNIVERSITY OF SOUTHERN MISSISSIPPI

Andrew Russo
UNIVERSITY OF IOWA

Rod Seeley
IDAHO STATE UNIVERSITY

Trent Stephens
IDAHO STATE UNIVERSITY

Philip Tate
STEVENS COLLEGE

Mc Graw Hill
Connect Learn Succeed™

The McGraw·Hill Companies

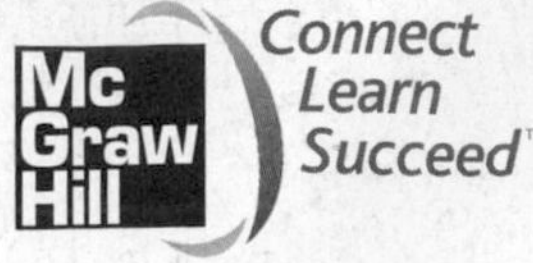
Mc
Graw
Hill
Connect
Learn
Succeed™

SEELEY'S ANATOMY & PHYSIOLOGY, TENTH EDITION

This book is printed on acid-free paper.

1 2 3 4 5 6 7 8 9 0 DOW/DOW 1 0 9 8 7 6 5 4 3

ISBN 978-0-07-340363-2
MHID 0-07-340363-6

Senior Vice President, Products & Markets: *Kurt L. Strand*
Vice President, General Manager, Products & Markets: *Marty Lange*
Vice President, Content Production & Technology Services: *Kimberly Meriwether David*
Director of Development: *Rose Koos*
Managing Director: *Michael S. Hackett*
Director: *James F. Connely*
Developmental Editor: *Mandy C. Clark*
Senior Project Manager: *Jayne L. Klein*
Senior Buyer: *Laura Fuller*
Designer: *Tara McDermott*
Cover/Interior Design: *Elise Lansdon*
Cover Illustration: © *The McGraw-Hill Companies*
Cover Image: © *Koji Aoki/Corbis/RF*
Content Licensing Specialist: *John Leland*
Photo Research: *Jerry Marshall*
Senior Media Project Manager: *Tammy Juran*
Compositor: *ArtPlus Ltd.*
Typeface: *10/12 Minion*
Printer: *RR Donnelley*

Library of Congress Cataloging-in-Publication Data

Seeley, Rod R.
[Anatomy & physiology]
Seeley's anatomy & physiology. — 10th ed. / Rod Seeley, Cinnamon VanPutte, Jennifer Regan, Andrew Russo.
p. cm.
Includes index.
ISBN 978-0-07-340363-2 — ISBN 0-07-340363-6 (hard copy : alk. paper) 1. Human physiology. 2. Human anatomy.
I. VanPutte, Cinnamon L. II. Regan, Jennifer. III. Russo, Andrew. IV. Title. V. Title: Seeley's anatomy and physiology.
QP34.5.S4 2014
612–dc23

2012028548

www.mhhe.com

ABOUT THE **Authors**

Cinnamon L. VanPutte
Associate Professor of Biology
Southwestern Illinois College

Cinnamon has been teaching biology and human anatomy and physiology for almost two decades. At Southwestern Illinois College she is a full-time faculty member and the coordinator for the anatomy and physiology courses. Cinnamon is an active member of several professional societies, including the Human Anatomy & Physiology Society (HAPS). Her Ph.D. in zoology, with an emphasis in endocrinology, is from Texas A&M University. She worked in Dr. Duncan MacKenzie's lab, where she was indoctrinated in the major principles of physiology and the importance of critical thinking. The critical thinking component of *Seeley's Essentials of Human Anatomy & Physiology* epitomizes Cinnamon's passion for the field of human anatomy and physiology; she is committed to maintaining this tradition of excellence. Cinnamon and her husband, Robb, have two children: a daughter, Savannah, and a son, Ethan. Savannah is very creative and artistic; she loves to sing, write novels and do art projects. Robb and Ethan have their black belts in karate and Ethan is one of the youngest black belts at his martial arts school. Cinnamon is also active in martial arts and is a competitive Brazilian Jiu-Jitsu practitioner. She has competed at both the Pan Jiu-Jitsu Championship and the World Jiu-Jitsu Championship.

GRAY'S
ANATOMY

Jennifer L. Regan
Instructor
University of Southern Mississippi

For over ten years, Jennifer has taught introductory biology, human anatomy and physiology, and genetics at the university and community college level. She has received the Instructor of the Year Award at both the departmental and college level while teaching at USM. In addition, she has been recognized for her dedication to teaching by student organizations such as the Alliance for Graduate Education in Mississippi and Increasing Minority Access to Graduate Education. Jennifer has dedicated much of her career to improving lecture and laboratory instruction at her institutions. Critical thinking and lifelong learning are two characteristics Jennifer hopes to instill in her students. She appreciates the Seeley approach to learning and is excited about contributing to further development of the textbook. She received her Ph.D. in biology at the University of Houston, under the direction of Edwin H. Bryant and Lisa M. Meffert. She is an active member of several professional organizations, including the Human Anatomy and Physiology Society. During her free time, Jennifer enjoys spending time with her husband, Hobbie, and two sons, Patrick and Nicholas.

Andrew F. Russo
Professor of Molecular
Physiology and Biophysics
University of Iowa

Andrew has over 20 years of classroom experience with human physiology, neurobiology, molecular biology, and cell biology courses at the University of Iowa. He is a recipient of the Collegiate Teaching Award and is currently the course director for Medical Cell Biology and Director of the Biosciences Graduate Program. He is also a member of several professional societies, including the American Physiological Society and the Society for Neuroscience. Andrew received his Ph.D. in biochemistry from the University of California at Berkeley. His research interests are focused on the molecular neurobiology of migraine. His decision to join the author team for *Seeley's Essentials of Human Anatomy & Physiology* is the culmination of a passion for teaching that began in graduate school. He is excited about the opportunity to hook students' interest in learning by presenting cutting-edge clinical and scientific advances. Andy is married to Maureen, a physical therapist, and has three daughters Erilynn, Becky, and Colleen, now in college and graduate school. He enjoys all types of outdoor sports, especially bicycling, skiing, ultimate Frisbee and, before moving to Iowa, bodyboard surfing.

This text is dedicated to the students of human anatomy and physiology. Helping students develop a working knowledge of anatomy and physiology is a satisfying challenge, and we have a great appreciation for the effort and enthusiasm of so many who want to know more. It is difficult to imagine anything more exciting, or more important, than being involved in the process of helping people learn about the subject we love so much.

WHAT SETS **Seeley** APART?

Seeley's Anatomy & Physiology is written for the two-semester anatomy and physiology course. The writing is comprehensive enough to provide the depth necessary for those courses not requiring prerequisites, and yet is presented with such clarity that it nicely balances the thorough coverage. Clear descriptions and exceptional illustrations combine to help students develop a firm understanding of the concepts of anatomy and physiology and to teach them how to use that information.

What Makes this Text a Market Leader?

Seeley Learning System—*Emphasis on Critical Thinking*

An emphasis on critical thinking is integrated throughout this textbook. This approach can be found in questions starting each chapter and embedded within the narrative; in clinical material that is designed to bridge concepts explained in the text with real-life applications and scenarios; in end-of-chapter questions that go beyond rote memorization; and in a visual program that presents material in understandable, relevant images.

- Problem-solving perspective from the book's inception
- Pedagogy builds student comprehension from knowledge to application (Predict questions, Critical Thinking questions, and Learn To Predict Answer)

Predict Questions challenge students to use their understanding of new concepts to solve a problem. Answers to the questions are provided at the end of the book, allowing students to evaluate their responses and to understand the logic used to arrive at the correct answer. All Predict question answers have been rewritten in teaching style format to model the answer for the student. Helps students learn how to think critically.

CRITICAL THINKING

1. The hypothalamohypophysial portal system connects the hypothalamus with the anterior pituitary. Why is such a special circulatory system advantageous?
2. A patient exhibits polydipsia (thirst), polyuria (excess urine production), and urine with a low specific gravity (contains few ions and no glucose). If you wanted to reverse the symptoms, would you administer insulin, glucagon, ADH, or aldosterone? Explain.
3. A patient complains of headaches and visual disturbances. A casual glance reveals enlarged finger bones, a heavy deposition of bone over the eyes, and a prominent jaw. The doctor determines that the headaches and visual disturbances result from increased pressure within the skull and that the presence of a pituitary tumor is affecting hormone secretion. Name the hormone causing the problem, and explain why increased pressure exists within the skull.
4. Most laboratories are able to determine blood levels of TSH, T_3, and T_4. Given that ability, design a method of determining whether hyperthyroidism in a patient results from a pituitary abnormality or from the production of a nonpituitary thyroid stimulatory substance.
5. Over the past year, Julie has gradually gained weight. The increase in adipose tissue is distributed over her trunk, face, and neck, and her muscle mass appears to be decreased. Julie also feels weak and bruises easily. Her physician suspects Cushing syndrome and orders a series
6. An anatomy and physiology instructor asks two students to predict a patient's response to chronic vitamin D deficiency. One student claims the person would suffer from hypocalcemia. The other student claims the calcium levels would remain within their normal range, although at the low end, and that bone reabsorption would occur to the point that advanced osteomalacia might occur. With whom do you agree, and why?
7. A patient arrives at the emergency room in an unconscious condition. A medical emergency bracelet reveals that he has diabetes. The patient is in either diabetic coma or insulin shock. How can you tell which, and what treatment do you recommend for each condition?
8. Predict some of the consequences of exposure to intense and prolonged stress.
9. Katie was getting nervous. At 16, she was the only one in her group of friends who had not started menstruating. Katie had always dreamed of having three beautiful children someday and she was worried. Her mother took her to see Dr. Josephine, who ordered several blood tests. When the results came back, Dr. Josephine gently explained to Katie and her mother that Katie would never be able to have children and would never menstruate. Dr. Josephine then asked Katie to wait in the outer room while she spoke privately to her mother. She explained to Katie's mom that Katie has androgen insensitivity syndrome. Though Katie is genetically male and her gonads produce more of the male

Predict 4

Explain the advantages of having elastic ligaments that extend from vertebra to vertebra in the vertebral column and why it would be a disadvantage if tendons, which connect skeletal muscles to bone, were elastic.

Critical Thinking These innovative exercises encourage students to apply chapter concepts to solve a problem. These questions help build student's knowledge of anatomy & physiology while developing reasoning and critical thinking skills.

Clinical IMPACT | Acquired Immunodeficiency Syndrome

Acquired immunodeficiency syndrome (AIDS) is a life-threatening disease caused by the **human immunodeficiency virus (HIV).** HIV is transmitted from an infected person to a noninfected person in body fluids, such as blood, semen, or vaginal secretions. The major methods of transmission are through unprotected sexual contact, through contaminated needles used by intravenous drug users, through tainted blood products, and from a pregnant woman to her fetus. Evidence indicates that household, school, and work contacts do not result in transmission. Reduced exposure to HIV is the best prevention for its transmission. Practices such as abstinence, the use of latex condoms, monogamy, and avoiding sharing needles are effective ways to reduce exposure to HIV. Medical professionals should also use care when handling body fluids, such as wearing latex gloves.

HIV infection begins when a protein on the surface of the virus, called gp120, binds to a CD4 molecule on the surface of a cell. The CD4 molecule is found primarily on helper T cells, and it normally enables helper T cells to adhere to other lymphocytes—for example, during antigen presentation. Certain monocytes, macrophages, neurons, and neuroglia also have CD4 molecules. Once attached to the CD4 molecules, the virus injects its genetic material (RNA) and enzymes into the cell and begins to replicate. Copies of the virus are manufactured using the organelles and materials within the cell. Replicated viruses escape from the cell and infect other cells.

Following infection by HIV, within 3 weeks to 3 months, many patients develop mononucleosis-like symptoms, such as fever, sweats, fatigue, muscle and joint aches, headache, sore throat, diarrhea, rash, and swollen lymph nodes. Within 1–3 weeks, these symptoms disappear as the immune system responds to the virus by producing antibodies and activating cytotoxic T cells that kill HIV-infected cells. However, the immune system is not able to eliminate HIV completely, and by about 6 months a kind of "set point" is achieved in which the virus continues to replicate at a low but steady rate. This chronic stage of infection lasts, on average, 8–10 years, and the infected person feels good and exhibits few, if any, symptoms.

Although helper T cells are infected and destroyed during the chronic stage of HIV infection, the body responds by producing large numbers of helper T cells. Nonetheless, over a period of years the HIV numbers gradually increase, and helper T cell numbers decrease. Normally, approximately 1200 helper T cells are present per cubic millimeter of blood. An HIV-infected person is diagnosed with AIDS when one or more of the following conditions appear: The helper T cell count falls below 200 cells/mm^3, an opportunistic infection occurs, or Kaposi sarcoma develops.

Opportunistic infections involve organisms that normally do not cause disease but do so when the immune system is depressed. Without helper T cells, cytotoxic T and B cell activation is impaired, and adaptive resistance is suppressed. Examples of opportunistic infections include pneumocystis (noo-mō-sis′tis) pneumonia (caused by an intracellular fungus, *Pneumocystis carinii*), tuberculosis (caused by an intracellular bacterium, *Mycobacterium tuberculosis*), syphilis (caused by a sexually transmitted bacterium, *Treponema pallidum*), candidiasis (kan-di-dī′ă-sis; a yeast infection of the mouth or vagina caused by *Candida albicans*), and protozoans that cause severe, persistent diarrhea. Kaposi sarcoma is a type of cancer that produces lesions in the skin, lymph nodes, and visceral organs. AIDS symptoms resulting from the effects of HIV on the nervous system include motor retardation, behavioral changes, progressive dementia, and possibly psychosis.

A cure for AIDS has yet to be discovered. Management of AIDS can be divided into two categories: (1) management of secondary infections or malignancies associated with AIDS and (2) control of HIV replication. In order for HIV to replicate, the viral RNA is used to make viral DNA, which is inserted into the host cell's DNA. The inserted viral DNA directs the production of new viral RNA and proteins, which are assembled to form new HIV. Key steps in the replication of HIV require viral enzymes. The enzyme **reverse transcriptase** promotes the formation of viral DNA from viral RNA, and **integrase** (in′te-grās) inserts the viral DNA into the host cell's DNA. A viral **protease** (prō′tē-ās) breaks large viral proteins into smaller proteins, which are incorporated into the new HIV.

Blocking the activity of HIV enzymes can inhibit the replication of HIV. The first effective treatment of AIDS was the drug azidothymidine (AZT; az′i-dō-thī′mi-dēn), also called zidovudine (zī-dō′voo-dēn). AZT is a **reverse transcriptase inhibitor,** which prevents HIV RNA from producing viral DNA. AZT can delay the onset of AIDS but does not appear to increase the survival time of AIDS patients. However, the number of babies who contract AIDS from their HIV-infected mothers can be dramatically reduced by giving AZT to the mothers during pregnancy and to the babies following birth.

Protease inhibitors are drugs that interfere with viral proteases. The current treatment for suppressing HIV replication is **highly active antiretroviral therapy (HAART).** This therapy uses drugs from at least two classes of antivirals. Treatment may involve combining three drugs, such as two reverse transcriptase inhibitors and one protease inhibitor, because HIV is unlikely to develop resistance to all three drugs. This strategy has proven very effective in reducing the death rate from AIDS and partially restoring health in some individuals.

Still in the research stage are **integrase inhibitors,** which prevent the insertion of viral DNA into the host cell's DNA. Another advance in AIDS treatment is a test for measuring **viral load,** which measures the number of viral RNA molecules in a milliliter of blood. The actual level of HIV is one-half the RNA count because each HIV has two RNA strands. Viral load is a good predictor of how soon a person will develop AIDS. If viral load is high, the onset of AIDS is likely to occur sooner than if the viral load is low. It is also possible to detect developing viral resistance by an increase in viral load. In response, a change in drug dose or type may slow viral replication. Current treatment goals are to keep viral load below 500 RNA molecules per milliliter of blood.

Effective treatment for AIDS is not the same as a cure. Even if viral load decreases to the point that the virus is undetected in the blood, the virus still remains in cells throughout the body. The virus may eventually mutate and escape drug suppression. The long-term goal for deterring AIDS is to develop a vaccine that prevents HIV infection.

Because of improved treatment, people with HIV/AIDS can now live for many years. Thus, HIV/AIDS is being viewed increasingly as a chronic disease, not a death sentence. Working together, a multidisciplinary team of occupational therapists, physical therapists, nutritionists/dieticians, psychologists, infectious disease physicians, and others can help patients with HIV/AIDS have a better quality of life.

Clinical Impact boxes These in-depth boxed essays explore relevant topics of clinical interest. Subjects covered include pathologies, current research, sports medicine, exercise physiology, and pharmacology.

Clinical Emphasis—*Case Studies*

Bring Relevance to the Reader

- **NEW!** Chapter opening photos and scenarios have been correlated to provide a more complete story and begin critical thinking from the start of the chapter
- **UPDATED!** Learn to Predict and chapter Predict questions with unique Learn to Predict Answers
- Clinical Impact boxes (placed at key points in the text)
- Case Studies
- **UPDATED!** Clinical Genetics essays have been updated and streamlined for accuracy and impact
- **UPDATED!** Diseases and Disorders tables
- **UPDATED!** Systems Pathologies with System Interactions

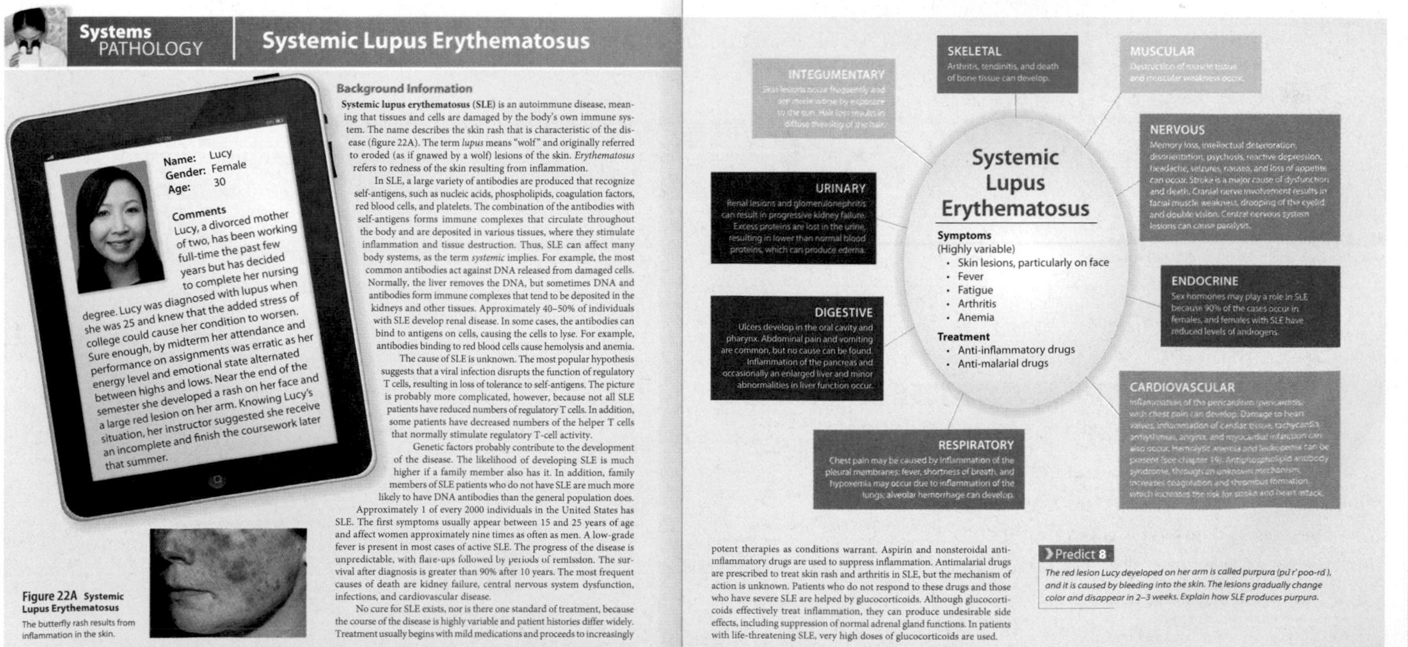

Systems PATHOLOGY | Systemic Lupus Erythematosus

Name: Lucy
Gender: Female
Age: 30

Comments
Lucy, a divorced mother of two, has been working full-time the past few years but has decided to complete her nursing degree. Lucy was diagnosed with lupus when she was 25 and knew that the added stress of college could cause her condition to worsen. Sure enough, by midterm her attendance and performance on assignments was erratic as her energy level and emotional state alternated between highs and lows. Near the end of the semester she developed a rash on her face and a large red lesion on her arm. Knowing Lucy's situation, her instructor suggested she receive an incomplete and finish the coursework later that summer.

Background Information

Systemic lupus erythematosus (SLE) is an autoimmune disease, meaning that tissues and cells are damaged by the body's own immune system. The name describes the skin rash that is characteristic of the disease (figure 22A). The term *lupus* means "wolf" and originally referred to eroded (as if gnawed by a wolf) lesions of the skin. *Erythematosus* refers to redness of the skin resulting from inflammation.

In SLE, a large variety of antibodies are produced that recognize self-antigens, such as nucleic acids, phospholipids, coagulation factors, red blood cells, and platelets. The combination of the antibodies with self-antigens forms immune complexes that circulate throughout the body and are deposited in various tissues, where they stimulate inflammation and tissue destruction. Thus, SLE can affect many body systems, as the term *systemic* implies. For example, the most common antibodies act against DNA released from damaged cells. Normally, the liver removes the DNA, but sometimes DNA and antibodies form immune complexes that tend to be deposited in the kidneys and other tissues. Approximately 40–50% of individuals with SLE develop renal disease. In some cases, the antibodies can bind to antigens on cells, causing the cells to lyse. For example, antibodies binding to red blood cells cause hemolysis and anemia.

The cause of SLE is unknown. The most popular hypothesis suggests that a viral infection disrupts the function of regulatory T cells, resulting in loss of tolerance to self-antigens. The picture is probably more complicated, however, because not all SLE patients have reduced numbers of regulatory T cells. In addition, some patients have decreased numbers of the helper T cells that normally stimulate regulatory T-cell activity.

Genetic factors probably contribute to the development of the disease. The likelihood of developing SLE is much higher if a family member also has it. In addition, family members of SLE patients who do not have SLE are much more likely to have DNA antibodies than the general population does.

Approximately 1 of every 2000 individuals in the United States has SLE. The first symptoms usually appear between 15 and 25 years of age and affect women approximately nine times as often as men. A low-grade fever is present in most cases of active SLE. The progress of the disease is unpredictable, with flare-ups followed by periods of remission. The survival after diagnosis is greater than 90% after 10 years. The most frequent causes of death are kidney failure, central nervous system dysfunction, infections, and cardiovascular disease.

No cure for SLE exists, nor is there one standard of treatment, because the course of the disease is highly variable and patient histories differ widely. Treatment usually begins with mild medications and proceeds to increasingly potent therapies as conditions warrant. Aspirin and nonsteroidal anti-inflammatory drugs are used to suppress inflammation. Antimalarial drugs are prescribed to treat skin rash and arthritis in SLE, but the mechanism of action is unknown. Patients who do not respond to these drugs and those who have severe SLE are helped by glucocorticoids. Although glucocorticoids effectively treat inflammation, they can produce undesirable side effects, including suppression of normal adrenal gland functions. In patients with life-threatening SLE, very high doses of glucocorticoids are used.

Figure 22A Systemic Lupus Erythematosus
The butterfly rash results from inflammation in the skin.

Systemic Lupus Erythematosus

Symptoms
(Highly variable)
- Skin lesions, particularly on face
- Fever
- Fatigue
- Arthritis
- Anemia

Treatment
- Anti-inflammatory drugs
- Anti-malarial drugs

INTEGUMENTARY
Skin lesions occur frequently and are made worse by exposure to the sun. Hair loss results in diffuse thinning of the hair.

SKELETAL
Arthritis, tendinitis, and death of bone tissue can develop.

MUSCULAR
Destruction of muscle tissue and muscular weakness occur.

NERVOUS
Memory loss, intellectual deterioration, disorientation, psychosis, reactive depression, headache, seizures, nausea, and loss of appetite can occur. Stroke is a major cause of dysfunction and death. Cranial nerve involvement results in facial muscle weakness, drooping of the eyelid, and double vision. Central nervous system lesions can cause paralysis.

URINARY
Renal lesions and glomerulonephritis can result in progressive kidney failure. Excess proteins are lost in the urine, resulting in lower than normal blood proteins, which can produce edema.

ENDOCRINE
Sex hormones may play a role in SLE because 90% of the cases occur in females, and females with SLE have reduced levels of androgens.

DIGESTIVE
Ulcers develop in the oral cavity and pharynx. Abdominal pain and vomiting are common, but no cause can be found. Inflammation of the pancreas and occasionally an enlarged liver and minor abnormalities in liver function occur.

CARDIOVASCULAR
Inflammation of the pericardium (pericarditis) with chest pain can develop. Damage to heart valves, inflammation of cardiac tissue, tachycardia, arrhythmias, angina, and myocardial infarction can also occur. Hemolytic anemia and leukopenia can be present (see chapter 19). Antiphospholipid antibody syndrome, through an unknown mechanism, increases coagulation and thrombus formation, which increases the risk for stroke and heart attack.

RESPIRATORY
Chest pain may be caused by inflammation of the pleural membranes; fever, shortness of breath, and hypoxemia may occur due to inflammation of the lungs; alveolar hemorrhage can develop.

Predict 8
The red lesion Lucy developed on her arm is called purpura (pŭr′poo-ră), and it is caused by bleeding into the skin. The lesions gradually change color and disappear in 2–3 weeks. Explain how SLE produces purpura.

Systems Pathologies boxes These spreads explore a specific condition or disorder related to a particular body system. Presented in a simplified case study format, each Systems Pathology vignette begins with a patient history followed by background information about the featured topic.

Exceptional Art—*Always created from the student perspective*

A picture is worth a thousand words—especially when you're learning anatomy and physiology. Because words alone cannot convey the nuances of anatomy or the intricacies of physiology, *Seeley's Anatomy & Physiology* employs a dynamic program of full-color illustrations and photographs that support and further clarify the textual explanations:

- **UPDATED!** Fundamental Figures teamed with special online support and now linked to APR
- **UPDATED!** Homeostasis figures were revised to draw a correlation from the text description of feedback system components to the figure. Maintains consistency throughout each organ system
- **NEW!** All figures were visually linked to create consistency throughout the text. The same colors are always used for the same type of arrow, cytoplasm in a cell, symbols for ions, and molecules, etc.
- Step-by-step Process figures
- Atlas-quality cadaver images
- Illustrated tables
- Photos side-by-side with illustrations
- **NEW!** Color saturation of art makes the art more engaging
- Macro-to-micro art

Renal capsule
Cortex
Medulla
Segmental artery
Hilum (indentation)
Renal artery
Renal vein
Renal pelvis
Ureter
Artery and vein in the renal sinus
Renal sinus (space)
Renal pyramid
Renal papilla
Minor calyx
Major calyx
Renal column
Medullary rays
(a)
Renal capsule
Cortex
Medulla
Renal sinus (space)
Renal pyramid
Renal papilla
Renal column
Major calyx
Minor calyx
Hilum (indentation)
Renal artery
Renal vein
Renal pelvis
Ureter
(b)

FIGURE 26.3 APR **Frontal Section of the Kidney and Ureter**
(*a*) A frontal kidney section shows that the cortex forms the outer part of the kidney, and the medulla forms the inner part. A central cavity called the renal sinus contains the renal pelvis. The renal columns of the kidney project from the cortex into the medulla and separate the pyramids. (*b*) Photograph of a longitudinal section of a human kidney and ureter.

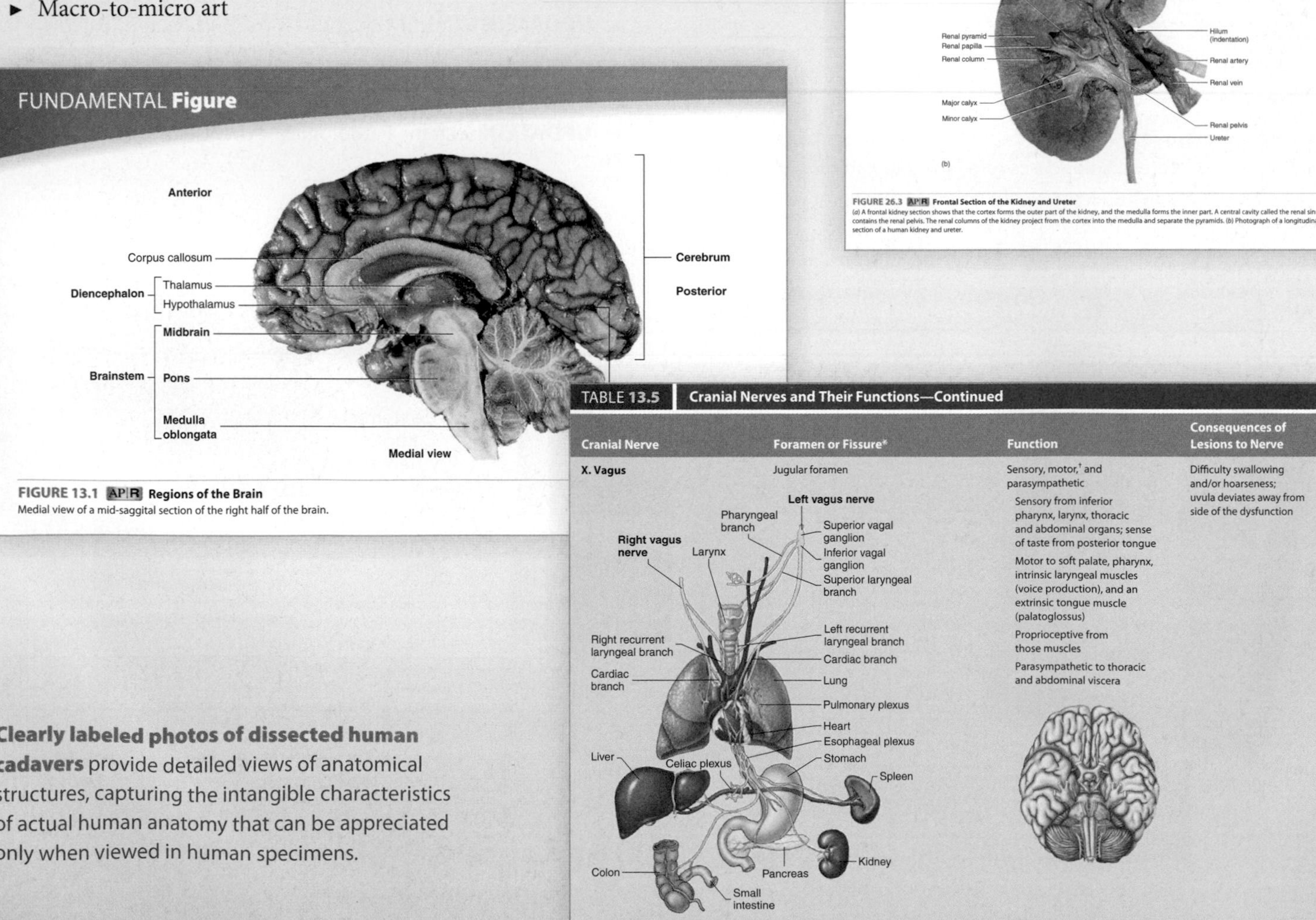
FUNDAMENTAL Figure
Anterior
Corpus callosum
Diencephalon
Thalamus
Hypothalamus
Midbrain
Brainstem
Pons
Medulla oblongata
Cerebrum
Posterior
Medial view
TABLE 13.5 Cranial Nerves and Their Functions—Continued
Cranial Nerve
Foramen or Fissure*
Function
Consequences of Lesions to Nerve
X. Vagus
Jugular foramen
Sensory, motor,† and parasympathetic
Sensory from inferior pharynx, larynx, thoracic and abdominal organs; sense of taste from posterior tongue
Motor to soft palate, pharynx, intrinsic laryngeal muscles (voice production), and an extrinsic tongue muscle (palatoglossus)
Proprioceptive from those muscles
Parasympathetic to thoracic and abdominal viscera
Difficulty swallowing and/or hoarseness; uvula deviates away from side of the dysfunction
Left vagus nerve
Pharyngeal branch
Superior vagal ganglion
Inferior vagal ganglion
Superior laryngeal branch
Right vagus nerve
Larynx
Left recurrent laryngeal branch
Right recurrent laryngeal branch
Cardiac branch
Cardiac branch
Lung
Pulmonary plexus
Heart
Esophageal plexus
Stomach
Liver
Celiac plexus
Spleen
Kidney
Pancreas
Colon
Small intestine

FIGURE 13.1 APR **Regions of the Brain**
Medial view of a mid-saggital section of the right half of the brain.

Clearly labeled photos of dissected human cadavers provide detailed views of anatomical structures, capturing the intangible characteristics of actual human anatomy that can be appreciated only when viewed in human specimens.

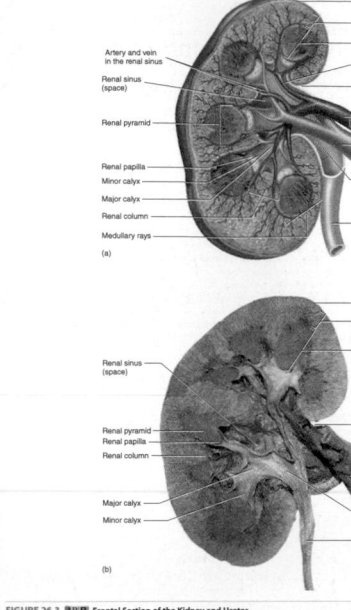

Incomparable Instructor and Student Resources—*Making teaching easier and learning smarter*

- **NEW!** Chapter opener rewritten with a focus on maintenance of homeostasis, a major underlying theme of the book
- **NEW!** In-text Learning Outcomes and Assessment Questions
- **NEW!** Learning Outcomes Correlation guide between Predict, Learn to Predict, Review and Comprehension, and Critical Thinking Questions
- Anatomy and Physiology | REVEALED® (APR) features "melt-away" dissection of real cadavers
- **NEW!** McGraw-Hill Anatomy & Physiology REVEALED® (APR) links to figures for eBook and is now also available for mobile devices
- Enhanced Lecture PowerPoints with APR cadaver images
- **NEW!** All figures are visually linked to create consistency throughout the text and art coloration has been saturated to help make the art more engaging
- Lecture PowerPoints with embedded animations
- **NEW!** Author Revised Testbank
- ConnectPlus® Course Management system
- **NEW!** Access to media-rich eBooks directly linked to APR
- **NEW!** LearnSmart™ tailors study time and identifies at-risk students and is now available for mobile devices
- Flex Art lets you take it apart and build it back during lecture
- **NEW!** Based on the same world-class super-adaptive technology as LearnSmart™, McGraw-Hill LabSmart™ is a must-see, outcomes-based lab simulation

FAQ
Give feedback
Open eBook
LearnSmart Debriefing
Debriefing HOME
The 5 most challenging learning objectives
Incorrect answers
Meta-cognitive skills
Status
Tree of wisdom
Learning plan
Metacognitive skills
Metacognition is knowledge about your own knowledge and learning. Speed up your learning with this report.
Aware that you were correct: 80%
You know these questions and know that you do. When all questions are in this category, you truly master the subject.
Aware that you were wrong: 4%
You are wrong, but you know it. That is the first step towards improving. Don't worry if you have many questions in this category. Just keep learning!
Unaware that you were correct: 1%
If you have many questions here, you actually know more than you think you know. Maybe think twice before you conclude that you don't know the answer.
You are a true master - and you know it. Good job!
Unaware
Aware
Correct
1%
80%
Wrong
12%
4%
< Back
connect
Charles Xavier
FAQ
Give feedback
Open eBook
Cellular form and function
The organelle marked with a pointer is the:
Easy!
Think I've got it
Unsure...
No idea
Skip
< Home

TENTH EDITION Changes

What's New?

The tenth edition of *Seeley's Anatomy & Physiology* is the result of extensive analysis of the text and evaluation of input from instructors who have thoroughly reviewed chapters. The outcome is a retention of the beloved features which foster student understanding, with an emphasis on a sharper focus within many sections, affording an even more logical flow within the text. Throughout every chapter the writing style is clean and more accessible to students.

Learning Outcomes and Assessment—*Helping instructors track student progress*

- **NEW!** Learning Outcomes are carefully written and labeled to outline expectations for each section
- **NEW!** Author Correlation of Review and Comprehension, Predict, and Critical Thinking Questions to Learning Outcomes are provided online to assist with linking course measuring standards and student comprehension
- **NEW!** Online student questions and test bank questions are correlated with Learning Outcomes to further scaffold and measure student progress and understanding
- The Clinical Genetics feature has been updated and streamlined to provide the newest and most accurate information available

14.2 Control of Skeletal Muscles

LEARNING OUTCOMES

After reading this section, you should be able to

A. Describe the primary motor area of the cerebra and discuss how it interacts with other parts of frontal lobe.

B. Distinguish between upper and lower motor n and between direct and indirect tracts.

ASSESS YOUR PROGRESS

12. *Compare upper motor neurons with lower motor neurons.*
13. *Where are the primary motor, premotor, and prefrontal areas of the cerebral cortex located? Explain the sequential nature of their functions.*
14. *Why are some areas of the body represented as larger than oth areas on the topographic map of the primary motor cortex?*

Clinical GENETICS | Skin Cancer

Skin cancer is the most common type of cancer. Most skin cancers result from damage caused by the ultraviolet (UV) radiation in sunlight. Some skin cancers are induced by chemicals, x-rays, depression of the immune system, or inflammation, whereas others are inherited.

UV radiation damages the genes (DNA) in epidermal cells, producing mutations. If a mutation is not repaired, the mutation is passed to one of the two daughter cells when a cell divides by mitosis. If mutations affecting oncogenes and tumor suppressor genes in epidermal cells accumulate, uncontrolled cell division and skin cancer can result (see Clinical Genetics, "Genetic Changes in Cancer Cells," in chapter 3).

The amount of protective melanin in the skin affects the likelihood of developing skin cancer. Fair-skinned individuals, who have less melanin, are at an increased risk of developing skin cancer compared with dark-skinned individuals, who have more melanin. Long-term or intense exposure to UV radiation also increases the risk. Thus, individuals who are older than 50, who have engaged in repeated recreational or occupational exposure to the sun, or who have experienced sunburn are at increased risk. Most skin cancers develop on the parts of the body that are frequently exposed to sunlight, such as the face, neck, ears, and dorsum of the forearm and hand. A physician should be consulted if skin cancer is suspected.

There are three types of skin cancer: basal cell carcinoma, squamous cell carcinoma, and melanoma (figure 5A). **Basal cell carcinoma,** the most common type, affects cells in the stratum basale. Basal cell carcinomas have a varied appearance. Some are open sores that bleed, ooze, or crust for several weeks. Others are reddish patches; shiny, pearly, or translucent bumps; or scarlike areas of shiny, taut skin. Removal or destruction of the tumor cures most cases.

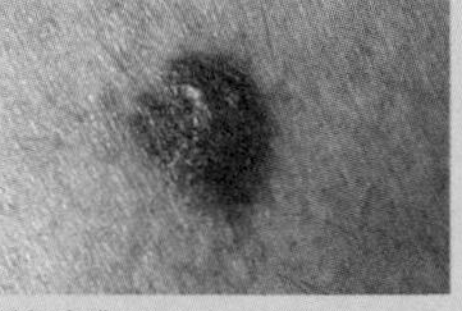

(a) Basal cell carcinoma

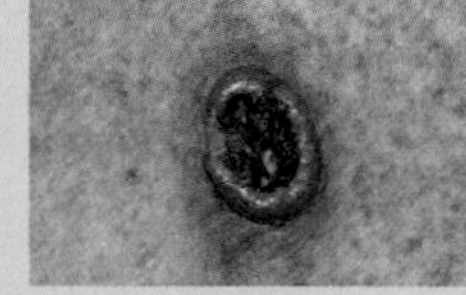

(b) Squamous cell carcinoma

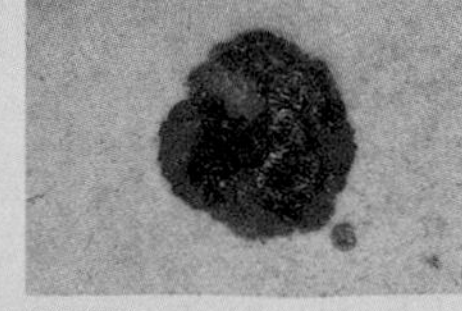

(c) Melanoma

FIGURE 5A Cancer of the Skin

ConnectPlus® with LearnSmart™—
Making teaching easier and learning smarter

- A personalized cognitive mapping learning system, LearnSmart™, tailors study time and identifies at-risk students
- Deliver and track assignments, quizzes, and tests easily online
- Use various ready-to-go Lecture PowerPoints including embedded animations—or make your own using McGraw-Hill art
- "Flex" art—take it apart and build it back during lecture
- 24/7 eBook access with animations, videos, and practice quizzing

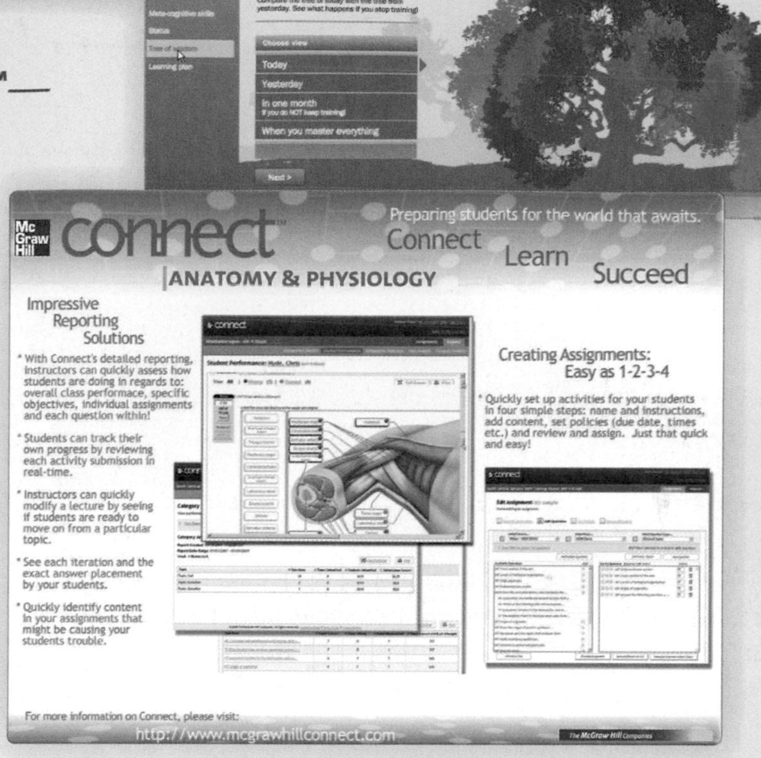

LearnSmart Debriefing
Your tree
Today
Yesterday
In one month
When you master everything
Next >
Mc Graw Hill
connect
ANATOMY & PHYSIOLOGY
Preparing students for the world that awaits.
Connect
Learn
Succeed
Impressive Reporting Solutions
With Connect's detailed reporting, instructors can quickly assess how students are doing in regards to: overall class performace, specific objectives, individual assignments and each question within!
Students can track their own progress by reviewing each activity submission in real-time.
Instructors can quickly modify a lecture by seeing if students are ready to move on from a particular topic.
See each iteration and the exact answer placement by your students.
Quickly identify content in your assignments that might be causing your students trouble.
Creating Assignments: Easy as 1-2-3-4
Quickly set up activities for your students in four simple steps: name and instructions, add content, set policies (due date, times etc.) and review and assign. Just that quick and easy!
For more information on Connect, please visit:
http://www.mcgrawhillconnect.com

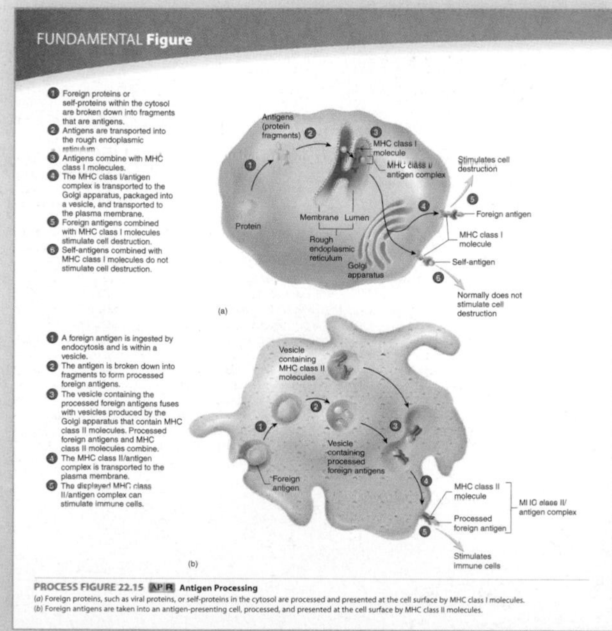

FUNDAMENTAL Figure
Antigens (protein fragments)
MHC class I molecule
MHC class I/antigen complex
Stimulates cell destruction
Foreign antigen
MHC class I molecule
Self-antigen
Normally does not stimulate cell destruction
Protein
Membrane
Lumen
Rough endoplasmic reticulum
Golgi apparatus
(a)
Vesicle containing MHC class II molecules
Vesicle containing processed foreign antigens
Foreign antigen
MHC class II molecule
Processed foreign antigen
MHC class II/antigen complex
Stimulates immune cells
(b)
PROCESS FIGURE 22.15 APR Antigen Processing
(a) Foreign proteins, such as viral proteins, or self-proteins in the cytosol are processed and presented at the cell surface by MHC class I molecules.
(b) Foreign antigens are taken into an antigen-presenting cell, processed, and presented at the cell surface by MHC class II molecules.

Fundamental Figures—
Integrated with special ConnectPlus® assets!

- **NEW!** Special icons now link Fundamental Figures with corresponding modules within APR
- Additional online ConnectPlus® resources support these important figures
- Grouped together, the Fundamental Figures represent an excellent summary and study tool

[Lea]rn to Predict and Learn to Predict Answer—*Helping students learn how to think*

❯ Learn to Predict

While weight training, Pedro strained his back and damaged a vertebral disk. The bulged disk placed pressure on the left side of the spinal cord, compressing the third lumbar spinal nerve, which innervates the following muscles: psoas major, iliacus, pectineus, sartorius, vastus lateralis, vastus medius, vastus intermedius, and rectus femoris. As a result, action potential conduction to these muscles was reduced. Using your new knowledge about the histology and physiology of the muscular system from chapter 9 and combining it with the information about gross muscle anatomy in this chapter, predict Pedro's symptoms and which movements of his lower limb were affected, other than walking on a flat surface. What types of daily tasks would be difficult for Pedro to perform?

- ► Part of the overall critical thinking Predict questions that appear throughout each chapter, a special Learn to Predict question now opens every chapter. This specifically written scenario takes knowledge acquired from previous chapters, and ties it into content in the current chapter.

Answer

Learn to Predict ❮ From page 309

The description of Pedro's injury provided specific information about the regions of the body affected: the left hip and thigh. In addition, we are told that the injury affected action potential conduction to the muscles of these regions. These facts will help us determine Pedro's symptoms and predict the movements that may be affected by his injury.

Chapter 9 described the relationship between action potential conduction and the force of muscle contractions. The reduction in action potential conduction to the muscles of the hip and thigh reduced the stimulation of these muscles, reducing the contraction force. As a result of his injury, we can predict that Pedro experienced weakness in his left hip and thigh, limiting his activity level.

We read in chapter 10 that the muscles affected by Pedro's injury (psoas major, iliacus, pectineus, sartorius, vastus lateralis, vastus medius, vastus intermedius, and rectus femoris) are involved in flexing the hip, the knee, or both. Therefore, we can conclude that movements involving hip and knee flexion, such as walking up and down stairs, would be affected. Any tasks that require Pedro to walk up and down stairs would be more difficult for him. Sitting and standing may also be affected, but the weakness in Pedro's left hip and thigh may be compensated for by increased muscle strength on his right side.

Answers to the rest of this chapter's Predict questions are in Appendix G.

- ► The Learn to Predict Answer box at the end of each chapter teaches students step-by-step how to answer the chapter-opening critical thinking question. This is foundational to real learning and is a crucial part of helping students put facts together to reach that "Aha" moment of true comprehension.

Specialized Figures Clarify Tough Concepts

Studying anatomy and physiology does not have to be an intimidating task mired in memorization. *Seeley's Anatomy & Physiology* uses two special types of illustrations to help students not only learn the steps involved in specific processes, but also apply the knowledge as they predict outcomes in similar situations. Process Figures organize the key occurrences of physiological processes in an easy-to-follow format. Homeostasis figures summarize the mechanisms of homeostasis by diagramming how a given system regulates a parameter within a narrow range of values.

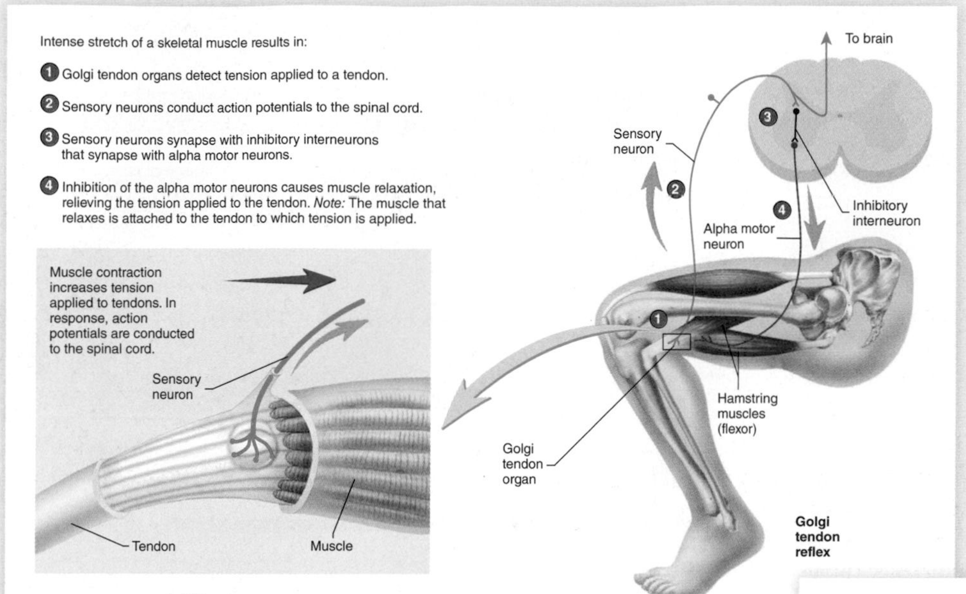

Intense stretch of a skeletal muscle results in:
1 Golgi tendon organs detect tension applied to a tendon.
2 Sensory neurons conduct action potentials to the spinal cord.
3 Sensory neurons synapse with inhibitory interneurons that synapse with alpha motor neurons.
4 Inhibition of the alpha motor neurons causes muscle relaxation, relieving the tension applied to the tendon. Note: The muscle that relaxes is attached to the tendon to which tension is applied.
Muscle contraction increases tension applied to tendons. In response, action potentials are conducted to the spinal cord.
Sensory neuron
Tendon
Muscle
Golgi tendon organ
To brain
Sensory neuron
Alpha motor neuron
Inhibitory interneuron
Golgi tendon organ
Hamstring muscles (flexor)
Golgi tendon reflex

PROCESS FIGURE 12.7 Golgi Tendon Reflex

Step-by-Step Process Figures

Process Figures break down physiological processes into a series of smaller steps, allowing readers to build their understanding by learning each important phase. Numbers are placed carefully in the art, permitting students to zero right in to where the action described in each step takes place.

NEW Correlated With APR! Homeostasis Figures with in-art explanations and organ icons

- These specialized flowcharts illustrating the mechanisms that body systems employ to maintain homeostasis have been refined and improved in the tenth edition.
- More succinct explanations
- Small icon illustrations included in boxes depict the organ or structure being discussed.
- All homeostasis figures were revised to draw a correlation from the text description of feedback system components to the figure. Maintains consistency throughout each organ system.

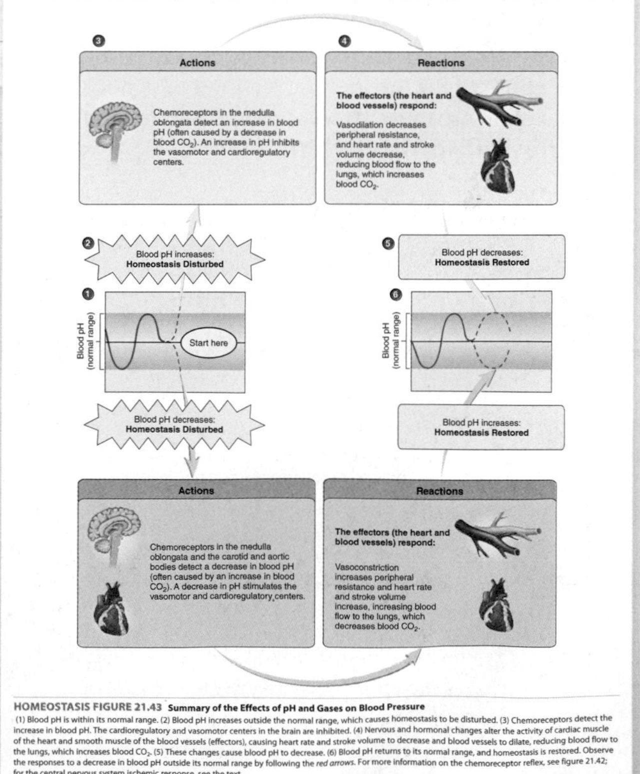

3
Actions
Chemoreceptors in the medulla oblongata detect an increase in blood pH (often caused by a decrease in blood CO_2). An increase in pH inhibits the vasomotor and cardioregulatory centers.
4
Reactions
The effectors (the heart and blood vessels) respond:
Vasodilation decreases peripheral resistance, and heart rate and stroke volume decrease, reducing blood flow to the lungs, which increases blood CO_2.
2
Blood pH increases: Homeostasis Disturbed
5
Blood pH decreases: Homeostasis Restored
1
Blood pH (normal range)
Start here
6
Blood pH (normal range)
Blood pH decreases: Homeostasis Disturbed
Blood pH increases: Homeostasis Restored
Actions
Chemoreceptors in the medulla oblongata and the carotid and aortic bodies detect a decrease in blood pH (often caused by an increase in blood CO_2). A decrease in pH stimulates the vasomotor and cardioregulatory centers.
Reactions
The effectors (the heart and blood vessels) respond:
Vasoconstriction increases peripheral resistance and heart rate and stroke volume increase, increasing blood flow to the lungs, which decreases blood CO_2.

HOMEOSTASIS FIGURE 21.43 Summary of the Effects of pH and Gases on Blood Pressure
(1) Blood pH is within its normal range. (2) Blood pH increases outside the normal range, which causes homeostasis to be disturbed. (3) Chemoreceptors detect the increase in blood pH. The cardioregulatory and vasomotor centers in the brain are inhibited. (4) Nervous and hormonal changes alter the activity of cardiac muscle of the heart and smooth muscle of the blood vessels (effectors), causing heart rate and stroke volume to decrease and blood vessels to dilate, reducing blood flow to the lungs, which increases blood CO_2. (5) These changes cause blood pH to decrease. (6) Blood pH returns to its normal range, and homeostasis is restored. Observe the responses to a decrease in blood pH outside its normal range by following the *red arrows*. For more information on the chemoreceptor reflex, see figure 21.42; for the central nervous system ischemic response, see the text.

TEACHING AND Learning Supplements

McGraw-Hill Connect® Anatomy & Physiology—
www.mcgrawhillconnect.com

Mc Graw Hill
connect
plus+
ANATOMY & PHYSIOLOGY

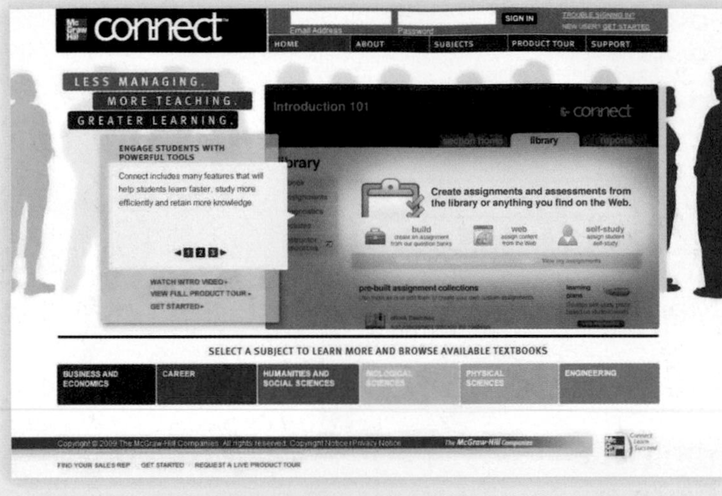
connect
SIGN IN
HOME
ABOUT
SUBJECTS
PRODUCT TOUR
SUPPORT
LESS MANAGING.
MORE TEACHING.
GREATER LEARNING.
Introduction 101
library
Create assignments and assessments from the library or anything you find on the Web.
build
web
self-study
pre-built assignment collections
SELECT A SUBJECT TO LEARN MORE AND BROWSE AVAILABLE TEXTBOOKS
BUSINESS AND ECONOMICS
CAREER
HUMANITIES AND SOCIAL SCIENCES
PHYSICAL SCIENCES
ENGINEERING

McGraw-Hill Connect® Anatomy & Physiology provides online presentation, assignment, and assessment solutions. It connects your students with the tools and resources they'll need to achieve success.

With **Connect Anatomy & Physiology** you can deliver assignments, quizzes, and tests online. A robust set of questions and activities are presented and aligned with the textbook's learning outcomes. As an instructor, you can edit existing questions and author entirely new problems. Track individual student performance—by question, assignment, or in relation to the class overall—with detailed grade reports. Integrate grade reports easily with Learning Management Systems (LMS), such as WebCT and Blackboard.

McGraw-Hill ConnectPlus® Anatomy & Physiology provides students with all the advantages of **Connect Anatomy & Physiology**, plus 24/7 online access to an eBook. This media-rich version of the book is available through the Connect platform and allows seamless integration of text, media, and assessments.
www.mcgrawhillconnect.com

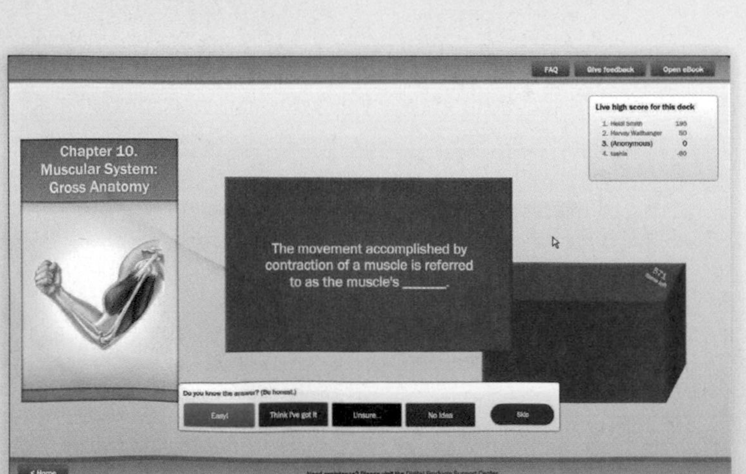
FAQ
Give feedback
Open eBook
Live high score for this deck
Chapter 10.
Muscular System:
Gross Anatomy
The movement accomplished by contraction of a muscle is referred to as the muscle's ______.
Easy!
Think I've got it
Unsure
No Idea
Skip
< Home

Mc Graw Hill LearnSmart™

www.mhlearnsmart.com

McGraw-Hill LearnSmart™ is available as an integrated feature of **McGraw-Hill Connect® Anatomy & Physiology**. It is an adaptive learning system designed to help students learn faster, study more efficiently, and retain more knowledge for greater success. LearnSmart assesses a student's knowledge of course content through a series of adaptive questions. It pinpoints concepts the student does not understand and maps out a personalized study plan for success. This innovative study tool also has features that allow instructors to see exactly what students have accomplished and a built-in assessment tool for graded assignments. Visit the following site for a demonstration.
www.mhlearnsmart.com

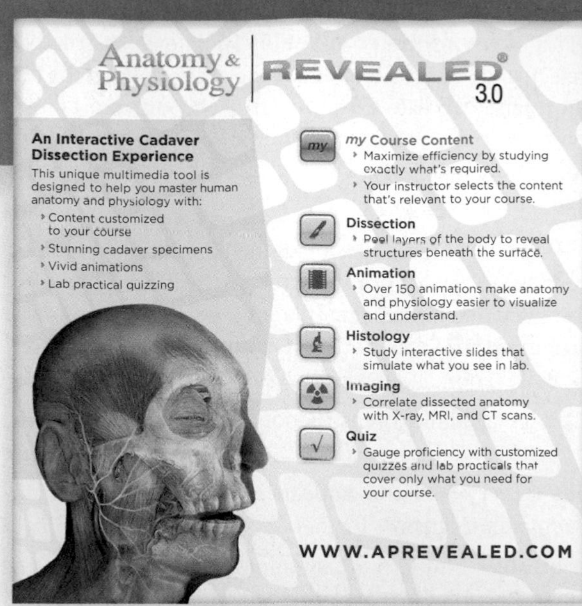
Anatomy & Physiology
REVEALED® 3.0
An Interactive Cadaver Dissection Experience
This unique multimedia tool is designed to help you master human anatomy and physiology with:
Content customized to your course
Stunning cadaver specimens
Vivid animations
Lab practical quizzing
my Course Content
Maximize efficiency by studying exactly what's required.
Your instructor selects the content that's relevant to your course.
Dissection
Peel layers of the body to reveal structures beneath the surface.
Animation
Over 150 animations make anatomy and physiology easier to visualize and understand.
Histology
Study interactive slides that simulate what you see in lab.
Imaging
Correlate dissected anatomy with X-ray, MRI, and CT scans.
Quiz
Gauge proficiency with customized quizzes and lab practicals that cover only what you need for your course.
WWW.APREVEALED.COM

Complete Access

Anatomy & Physiology REVEALED® provides simple yet complete access to the entire human body. The app is organized into systems. For each, you will find dissection tools, histological and radiological imagery, videos, and a helpful quiz.

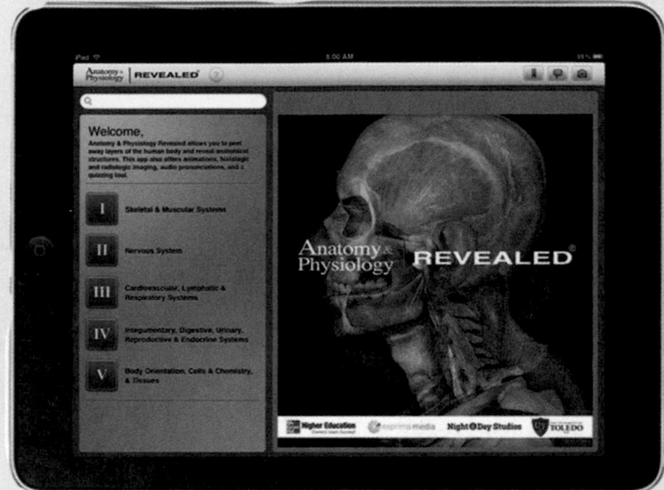
REVEALED
Welcome,
Skeletal & Muscular Systems
Nervous System
Cardiovascular, Lymphatic & Respiratory Systems
Integumentary, Digestive, Urinary, Reproductive & Endocrine Systems
Body Orientation, Cells & Chemistry, & Tissues
Anatomy & Physiology REVEALED

Additional Mobile Apps Available

CourseSmart

COURSESMART LEARNSMART TEGRITY VITALSOURCE

Performance-Enhancing Study Tools and Low-Cost Textbook Alternatives

Electronic Books

If you or your students are ready for an alternative version of the traditional textbook, McGraw-Hill offers a cheaper and ecofriendly alternative to traditional textbooks. By purchasing eBooks from McGraw-Hill, students can save as much as 50% on selected titles delivered on the most advanced eBook platform available.

Contact your McGraw-Hill sales representative to discuss eBook packaging options.

Mc Graw Hill create

McGraw-Hill Create™

With McGraw-Hill Create™, you can easily rearrange chapters, combine material from other content sources, and quickly upload content you have written, like your course syllabus or teaching notes. Find the content you need in Create by searching through thousands of leading McGraw-Hill textbooks. Arrange your book to fit your teaching style. Create even allows you to personalize your book's appearance by selecting the cover and adding your name, school, and course information. Order a Create book and you'll receive a complimentary print review copy in 3–5 business days or a complimentary electronic review copy (eComp) via e-mail in minutes. Go to www.mcgrawhillcreate.com today and register to experience how McGraw-Hill Create empowers you to teach your students your way. **www.mcgrawhillcreate.com**

Other Resources Available

Mc Graw Hill
tegrity

My Lectures–McGraw-Hill Tegrity
www.tegrity.com

McGraw-Hill Tegrity® records and distributes your class lecture, with just a click of a button. Students can view anytime/anywhere via computer, iPod, or mobile device. It indexes as it records your PowerPoint® presentations and anything shown on your computer so students can use keywords to find exactly what they want to study. Tegrity is available as an integrated feature of McGraw-Hill Connect Anatomy & Physiology or as standalone.

Physiology Interactive Lab Simulations (Ph.I.L.S) 4.0

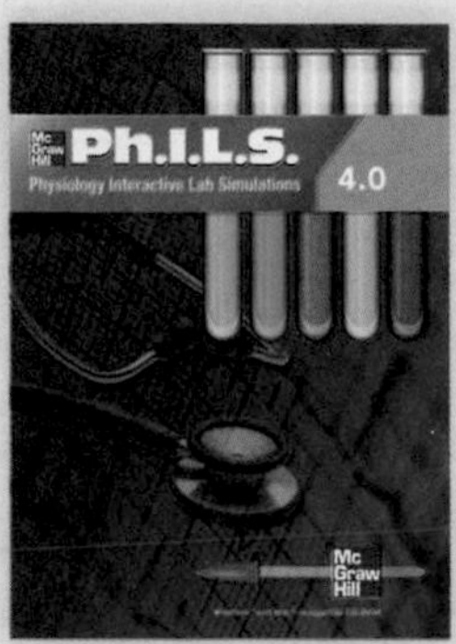
Ph.I.L.S.
Physiology Interactive Lab Simulations
4.0

Ph.I.L.S. 4.0 is the perfect way to reinforce key physiology concepts with powerful lab experiments. Created by Dr. Phil Stephens at Villanova University, this program offers ***42 laboratory simulations*** that may be used to supplement or substitute for wet labs. All 42 labs are self-contained experiments—no lengthy instruction manual required. Users can adjust variables, view outcomes, make predictions, draw conclusions, and print lab reports. This easy-to-use software offers the flexibility to change the parameters of the lab experiment. There are no limits!

Laboratory Manual

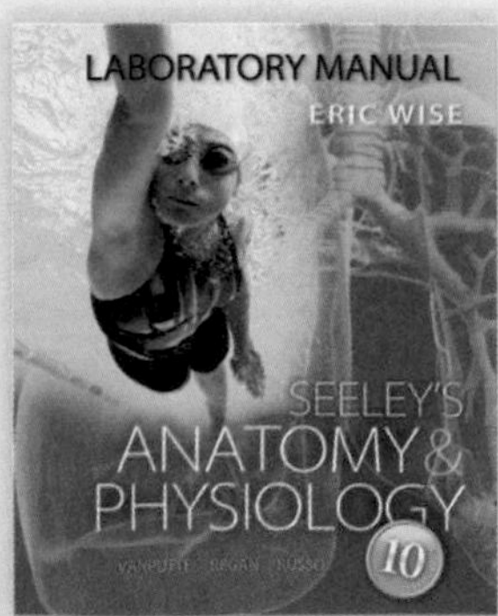
LABORATORY MANUAL
ERIC WISE
SEELEY'S
ANATOMY & PHYSIOLOGY
10

The Laboratory Manual to accompany *Seeley's Anatomy & Physiology*, authored by Eric Wise of Santa Barbara City College, contains 43 laboratory exercises that are integrated closely with the textbook. Each exercise demonstrates the anatomical and physiological facts and principles presented in the textbook by investigating specific concepts in greater detail. Key features of the lab manual include over 12 new cat dissection photos and many new human cadaver images, step-by-step explanations and a complete materials list for each experiment, precisely labeled, full-color drawings and photographs, self-contained presentations with the essentials background needed to complete each exercise, and extensive lab reports at the end of every exercise.

Correlated Website

The website that accompanies *Seeley's Essentials of Anatomy & Physiology* at **www.mhhe.com/seeley10** allows instructors to browse, select, and export files containing artwork from the text in multiple formats to create customized classroom presentations, visually based tests and quizzes, dynamic course website content, or printed support materials. The digital assets on the website are all available for teaching presentations:

- **Art** Full-color digital files of all the illustrations in the book and unlabeled versions of the same artwork can be readily incorporated into lecture presentations, exams, or custom-made classroom materials. In addition, all files are pre-inserted into blank PowerPoint slides for easy lecture presentations.
- **Photos** Digital files of instructionally significant photographs from the text can be reproduced for multiple classroom uses.
- **Tables** Every table that appears in the text is available to instructors in electronic form.
- **Animations** Numerous full-color animations illustrating physiological processes are provided. Harness the visual impact of processes in motion by importing these files into classroom presentations or online course materials.

Students will benefit from **practice quizzing, animation quizzing,** and other **study tools,** all correlated by chapter. Help with difficult concepts is only a click away!

Mc Graw Hill LabSmart™

McGraw-Hill LabSmart™
THE Virtual Lab Experience

Based on the same world-class super-adaptive technology as McGraw-Hill LearnSmart™, McGraw-Hill LabSmart™ is a must-see, outcomes-based lab simulation. It assesses a student's knowledge and adaptively corrects deficiencies, allowing the student to learn faster and retain more knowledge with greater success.

First, a student's knowledge is adaptively leveled on core learning outcomes: Questioning reveals knowledge deficiencies that are corrected by the delivery of content that is conditional on a student's response. Then, a simulated lab experience requires the student to think and act like a scientist: Recording, interpreting, and analyzing data using simulated equipment found in labs and clinics. The student is allowed to make mistakes—a powerful part of the learning experience! A virtual coach provides subtle hints when needed; asks questions about the student's choices; and allows the student to reflect upon and correct those mistakes. Whether your need is to overcome the logistical challenges of a traditional lab, provide better lab prep, improve student performance, or make your online experience one that rivals the real world, LabSmart accomplishes it all.

Learn more at www.mhlabsmart.com

Acknowledgments

A great deal of effort is required to produce a heavily illustrated textbook like *Seeley's Anatomy & Physiology*. Many hours of work are required to organize and develop the components of the textbook while also creating and designing illustrations, but no text is solely the work of the authors. It is not possible to adequately acknowledge the support and encouragement provided by our loved ones. They have had the patience and understanding to tolerate our absences and our frustrations. They have also been willing to provide assistance and unwavering support.

Many hands besides our own have touched this text, guiding it through various stages of development and production. We wish to express our gratitude to the staff of McGraw-Hill for their help and encouragement. We appreciate the guidance and tutelage of Director James Connely. We are sincerely grateful to Developmental Editor Mandy Clark for her careful scrutiny of the manuscript, her creative ideas and suggestions, and her tremendous patience and encouragement. Special thanks are also offered to Copyeditor Deb DeBord for her attention to detail and for carefully polishing our words. A special acknowledgement of gratitude is owed to Project Manager Jayne Klein for her patience and detail-tracking abilities. Content Licensing Specialist John Leland, Production Supervisor Sandy Ludovissy, Designer Tara McDermott, and Media Project Manager Tammy Juran, we thank you for your time spent turning our manuscript into a book and its accompanying website. The McGraw-Hill employees with whom we have worked are excellent professionals. They have been consistently helpful and their efforts are truly appreciated. Their commitment to this project has clearly been more than a job to them.

Finally, we sincerely thank the reviewers and instructors who've provided us time and time again with remarkable feedback. We wish we could pay you what you're really worth to us! To conscientiously review a textbook requires a true commitment and dedication to excellence in teaching. Your helpful criticisms and suggestions for improvement were significant in revising the ninth edition. Our advisory board was a special group of exceptional reviewers to whom we could turn to at any time during the development of this text for almost immediate valuable input. To those of you who've participated in focus groups, we'd like to recognize the time you spent away from family and students in order to provide us with significant information about the future of anatomy and physiology at your institution. We gratefully acknowledge all of you who played a part in this edition by name in the next section.

Cinnamon VanPutte
Jennifer Regan
Andy Russo
Rod Seeley

Reviewers

Jerry M. Allen
KCTCS Somerset Community College

Timothy A. Ballard
University of North Carolina Wilmington

David M. Bastedo
San Bernadino Valley College

Mary L. Bonine
Northeast Iowa Community College–Peosta

Nishi Sood Bryska
University of North Carolina Charlotte

Ronald A. Canterbury
University of Cincinnati at Cincinnati

Claire Michelle Carpenter
Yakima Valley Community College

James Davis
University of Southern Maine

Kathryn A. Durham
Lorain County Community College

Clair B. Eckersell
Brigham Young University Idaho

Angela M. Edwards
Trident Technical College

Jeffrey G. Edwards
Brigham Young University

Maria Florez
Lone Start College–CyFair

Purti Gadkari
Wharton County Junior College

Ewa Gorski
Community College of Baltimore Cantonsville

Edwin Griff
University of Cincinnati at Cincinnati

Richard S. Groover
J. Sargeant Reynolds Community College

Robert F. Halliwell
TJL School of Pharmacy & Health Sciences

Clare Hays
Metro State College of Denver

Christopher W. Herman
Eastern Michigan University

William Huber
St. Louis Community College Forest Park

Jason Hunt
Brigham Young University Idaho

Susanne Kalup
Westmoreland County Community College

Kamal Kamal
Valencia Community College West

L. Henry Kermott
St. Olaf College

April Kilgore
KCTCS Somerset Community College

Tyjuanna R. S. LaBennett
North Carolina Central University

Steven D. Leidich
Cuyahoga Community College

Jerri K. Lindsey
Tarrant County College NE

Mary Katherine Lockwood
University of New Hampshire

Karen K. McLellan
Indiana University/Purdue University Fort Wayne

Kenneth Michalis
San Bernadino Valley College

Claire A. Miller
Community College of Denver

William R. Millington
Albany College of Pharmacy

Dan Miska
Wright State University

J. Jean Mitchell
Okaloosa-Walton College; University of Florida

Ramzi A. Ockaili
J. Sargeant Reynolds Community College

Sidney L. Palmer
Brigham Young University

Betsy Peitz
California State University, Los Angeles

Raffaelle Pernice
Hudson County Community College

Terri Pope
Cuyahoga Community College

Helen M. Rarick
Wilbur Wright College

Marilyn Shannon
Indiana University/Purdue University Fort Wayne

Jeff Simpson
Metropolitan State College of Denver

Jason Tasch
Cuyahoga Community College Metro

Janis G. Thompson
Lorain County Community College

Corinne Ulbright
Indiana University/Purdue University Indianapolis

Anthony J. Uzwiak
The College of New Jersey

Charles Wright III
Community College of Baltimore County–Essex

Accuracy Checkers

Nishi Bryska
UNC Charlotte

Lois Brewer Borek
Georgia State University

Ronald A. Canterbury
University of Cincinnati

Ethel Cornforth
San Jacinto College South

Emily Y. F. González
Northern Essex Community College

Clare Hays
Metropolitan State College of Denver

Chris Herman
Eastern Michigan University

Kristopher Kelley
University of Louisiana at Monroe

Allart Kok
Community College of Baltimore County

Cynthia Littlejohn
The University of Southern Mississippi

Karen McLellan
Indiana University/Purdue University Fort Wayne

Margaret Ott
Tyler Junior College

Josephine Rogers
University of Cincinnati

Tim Roye
San Jacinto College

Fadi N. Salloum
J. Sargeant Reynolds Community College

Marilyn M. Shannon
Indiana University/Purdue University Fort Wayne

Denise Slayback-Barry
Indiana University/Purdue University Indianapolis

Eric L. Sun
Macon State College

Corinne Ulbright
Indiana University/Purdue University Indianapolis

MaryJo A. Witz
Monroe Community College

Dwight Wray
Brigham Young University–Idaho

Martin Zahn
Thomas Nelson Community College

Contents

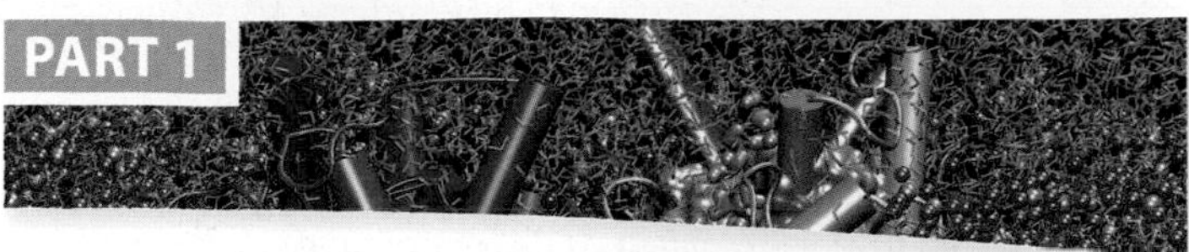
PART 1

Organization of the Human Body

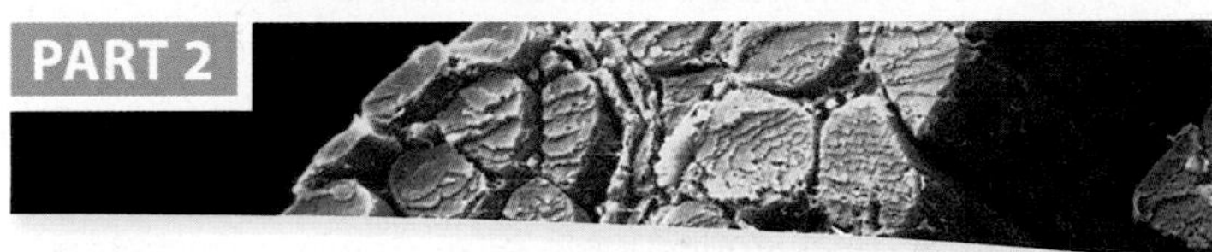
PART 2

Support and Movement

PART 3 Integration and Control Systems

PART 5 Reproduction and Development

Chapter-by-Chapter Changes

Chapter 1

- Chapter opener rewritten with a focus on maintenance of homeostasis, a major underlying theme of the book.
- Chapter opener revised to link opening photo with Learn to Predict and chapter introduction. Provides a cohesive theme for better student learning and engagement.
- Learning outcomes goals at the beginning of the chapter were numbered to correlate with Predict questions and end-of-chapter questions.
- Clinical Impact "Anatomical Imaging" was converted to an illustrated table, table 1.1, which increases the perceived importance to students and makes the information easier to interpret.
- The homeostasis section was revised per reviewer feedback for a more accurate description of negative and positive feedback.

Chapter 2

- Redesigned and combined former figures 2.9 and 2.10 on synthesis and decomposition reactions into new figure 2.9. Eliminated redundant information and made information less daunting by showing simple schematics adjacent to more complex representations of protein and carbohydrate molecules.
- New figures 2.10 and 2.11 provide more intuitive presentations of energy in chemical reactions and concept of activation energy.
- New figure on buffers (figure 2.13) illustrates an important physiological concept previously described only with text.
- Hydrogen bonding and water sections have been rewritten to emphasize importance of H bonds in the structure and unique functions of water.
- Legend for covalent bonding figure 2.5 has been rewritten to increase clarity.
- Descriptions of both the conservation of energy and the release of energy during ATP hydrolysis have been rewritten to more clearly describe these fundamental points.
- Tertiary folding of proteins has been rewritten to clearly distinguish secondary from tertiary folding.
- New electron micrograph (figure 2.15c) has been added that better illustrates glycogen granules in a cell.
- Chapter opening material has been tied into the cover figure and the Learn to Predict question.
- Background coloring on several figures has been changed to make them more visually striking.

Chapter 3

- New chapter opener figure of aquaporin to tie in to Learn to Predict question.
- Clinical Impact "Microscopic Imaging" has been updated.
- Table 3.2 is now illustrated to better represent membrane protein function.
- Section 3.6 is reorganized into Passive Membrane Transport and Active Membrane Transport mechanisms.
- Table 3.3 has been reorganized to reflect revision of section 3.6.
- Section 3.12 Genetics has been moved to chapter 29.
- Clinical Genetics "Genetic Changes in Cancer Cells" updated.
- All figures illustrating the plasma membrane have been updated so that the cytoplasmic side is yellow. This provides consistency throughout the text and is more visually appealing.

Chapter 4

- Relationship between structure and function in A&P has been emphasized with a new paragraph and examples in section 4.1.
- Microscopy Clinical Impact has been moved to chapter 3. In the process, we removed extraneous technical information (such as fixation methods) and updated it with addition of atomic force microscopy (AFM). We have included new images of nuclear pores seen by light, TEM, SEM, and AFM to illustrate the different types of microscopy.
- Embryological terms in section 4.2 have been updated (epiblast and hypoblast).
- Figure 4.5 on matrix proteins has been greatly simplified. The figure had acquired too many unnecessary details, especially on collagen biosynthesis. The revised figure emphasizes the concepts that collagen, elastin, and proteoglycans have different properties. Corresponding changes in the text emphasizing the rope-like nature of collagen fibers and rubber-band like nature of elastin fibers have been made.

- Description of the basement membrane has been modified and now also included that its porous substance that allows diffusion of substances to and from the epithelium.
- Description of endocrine glands, including their different ontogenies, has been removed since this concept is not needed until later in the textbook.
- Ground substance of the matrix has been emphasized with a new heading.
- Based on increasing and solid evidence that brown fat plays important roles in the human adult, and not just infants, the statement that brown fat is primarily in infants has been removed.
- New cover image showing microvilli. This image matches the Learn to Predict question and the intense fluorescent signal will help grab student's attention.
- More vibrant color and contrast in several histology images to better display cell types in tissues (figures in tables 4.2, 4.3, 4.10c, 4.14).
- Eliminated neuroglia image since this topic is not emphasized in this chapter and glia are indicated in table 4.15 figure.
- Clinical Impact on Marfan syndrome has been streamlined by removing unimportant genetic details (chromosome number, types and number of allelic variants, protein name, etc).
- Clinical Impact on cancer has been updated and rewritten to focus on types of cancer arising from different tissues.
- Clinical Genetics on cancer has been moved to Chapter 3 and has been streamlined and updated. The relevant critical thinking question also moved to chapter 3.

Chapter 5

- Clinical Genetics "Skin Cancer" has been updated.
- New Systems Pathology presentation.

Chapter 6

- Chapter opener rewritten with a focus on maintenance of homeostasis, a major underlying theme of the book.
- Added osteoclast figure to fill in an important gap in information for bone growth and development and calcium homeostasis.
- Updated information on osteoclast function.
- Clinical Genetics box "Osteogenesis Imperfecta" updated for accuracy and currency.
- Figures 6.13 and 6.14 were combined so students can see the "big picture" and better correlate ideas.
- Added actual x-ray images to figure 6.20 for real world correlation.
- Figure 6.21 revised for better link between physiological process components.

Chapter 7

- Chapter opener rewritten with a focus on maintenance of homeostasis, a major underlying theme of the book.
- Clinical Impact "Herniated Discs" was revised and updated to include stem cell techniques for treatment and surgical methods.
- All figures were visually linked to create consistency throughout the chapter.

Chapter 8

- New Learn to Predict question that ties into the accompanying chapter opener figure of a knee MRI.
- Clarification that joints are where bones move in close contact with each other, but are not bone on bone.
- Clarification of difference between sutures and synostosis.
- Clarification of different fates of synchondrosis joints (convert to synostosis, synovial joints, or persist as synchondrosis joints).
- New presentation of types of synovial joints from six separate figures and one table into one figure (figure 8.8) to allow a more concise and organized presentation with better visualization and comparison between the different joints with respect to their structure, connecting bones, and movements.
- Revision of major knee ligament information. Text now emphasizes the two clinically important sets of ligaments (cruciate and collateral) and uses the more common terms of medial and lateral collateral ligaments. Role of the popliteal ligament has been deemphasized.
- New Predict question focused on PCL tears and posterior drawer test.
- Clinical Impact on joint changes in pregnancy has been updated and information added describing the importance and effectiveness of early diagnosis of congenital hip dislocation.
- Clinical Impact on TMJ disorders has been updated and rewritten to emphasize the symptoms of common chronic cases and successful treatment paradigms.
- Description of bunions has been corrected to indicate that they are deformations of the great toe that may have associated bursitis, but are distinct from bursitis.
- New Critical Thinking question brings information on inflammation and bones from chapter 7 with vertebral joints from chapter 8.
- Aging section has been clarified to describe how protein cross-linking causes loss of joint flexibility by changes in fibrous connective tissue of tendons and ligaments.
- Arrow colors in the figures that indicate movement have been changed to dark blue for consistency.

Chapter 9

- Figures 9.3, 9.4, 9.15, 9.17 and any other figure with myosin myofilaments were revised to more accurately reflect relative sizes of thick and thin filaments.
- Sections 9.4 and 9.5 were combined and reorganized to follow a more logical sequence; new information is built upon previous information.
- A new figure 9.16 was added per reviewer feedback to have information culminate in a "big picture" summary figure of skeletal muscle contraction.
- Figure 9.6 was revised so that a photomicrograph, which shows the actual process, was added.
- Throughout the chapter, the membrane potential figure scale was modified to more accurately reflect the level for skeletal muscle.
- Figure 9.21 was revised for clarity based on reviewer feedback.
- Based on reviewer feedback, new information on sarcopenia was added to the section on aging.
- Table 9.3 was revised for clarity and information on type of work supported by each path was added.
- Updated information on fiber types and distribution.

Chapter 10

- Added new table for muscle shapes (figures 10.2 and 10.3 were reorganized into an illustrated table) and the terminology was updated.
- Updated information on aging in Clinical Impact "Bodybuilding" per reviewer feedback.
- In all figures with a background screen, the color of the screen was changed to yellow, which looks more modern and increases student engagement.

Chapter 11

- Revised figure 11.2 into a flow chart so students may conceptually follow the organization of the nervous system.
- Reorganized glial cells into a single illustrated table to give a "big picture" among these cells.
- Section 11.5 was reorganized and revised for clarity.
- Combined old figures 11.12 and 11.13 into a new figure (11.7) to create a "big picture" figure to give students a greater connectivity.
- Revised figure 11.20 (new figure 11.14) for accuracy and clarity of concept.
- Revised figure 11.22 (new figure 11.16) for clarity.
- Revised section 11.7 to update terminology.
- Revised the Learn to Predict answer for accuracy.

Chapter 12

- Cervical rib syndrome case study has been renamed Thoracic Outlet Syndrome to reflect the more commonly used medical term and has been extensively modified and updated with new information, including treatments.
- Median nerve damage Clinical Impact has been rewritten and updated to include causes of carpal tunnel syndrome and that typing at a keyboard is no longer a recognized cause.
- Diseases and Disorders table has been updated and modified. Have added Marie-Charcot-Tooth syndrome, one of the most common inherited neurological disorders, and diabetic neuropathy, an increasingly common, but poorly understood disorder. Myotonic dystrophy has been removed since current research is still not clear whether this is a primary neuropathy. Grouping in infection categories has also been eliminated since the role of infection is not clear in some diseases.
- Multiple figures have been modified to improve presentation of information.
  - Consistent colors for sensory (green) and motor (purple) tracts in the spinal cord (figures 12.3, 12.11) and changed arrow colors in other figures for consistency.
  - Figure 12.9 process figure better describes the action of inhibitory neurons (dashed line) in the withdrawal reflex.
- The clinical connection of a lung tumor potentially compressing the phrenic nerve has been updated as the second most common and most lethal cancer among men.
- Minor wording changes to improve clarity—e.g. superficial and deep to describe white and gray matter of spinal cord instead of peripheral and central to avoid confusion with terms used to describe divisions of the nervous system (CNS, PNS). Consistent use of term motor when describing autonomic motor neurons to emphasize their motor functions. Revised coat/sleeve analogy to describe the dura and epineurium relationship.

Chapter 13

- New chapter opener photo (MRI) and introductory paragraph to better illustrate theme of chapter and match topic of the Learn to Predict question.
- Rewritten brainstem section to describe overall function, followed by anatomy.
- Revised reticular formation section to clarify that it is not an anatomical division of the brainstem, it spans all divisions of brainstem, and is involved in many functions in addition to the reticular activating system.
- Included description of the solitary nucleus and nucleus ambiguous serving as nuclei for multiple cranial nerves and clarified that several cranial nerves have more than one nucleus in the brainstem.
- Included general description of diencephalon in table 13.1.
- Thalamic nuclei have been highlighted with colors in figure 13.7 to allow better visualization.

- Added that the hypothalamus is the major coordinating center of the autonomic nervous system.
- Added prefrontal cortex and its functions to the description of the frontal lobe.
- Added that taste information is received and processed by the insula.
- Added arachnoid villi to the description of recirculation of cerebrospinal fluid by arachnoid granulations.
- Added general functions and comparison to spinal nerves to introduction of cranial nerves.
- Added that trigeminal sensory nerves also innervate meninges and their role in migraine. Description of migraine was also added to the Diseases and Disorders table.
- Added traumatic brain injury as the signature wound of the Iraq/Afghanistan wars.
- Rewritten facial palsy section of the Disease and Disorder table, including likely role of viral infections in Bell Palsy.
- Added the more commonly used clinical term torticollis for wry neck in Predict question.
- Removed Clinical Genetics box on neurofibromatosis since this is a rare disease and did not illustrate any pertinent contribution of genetics to A+P.
- More saturated colors in 5 figures, modified 4 other figures for better clarity.
- Added new schematic that better illustrates the layers and cell types in the cortex (figure 13.8c).

Chapter 14

- Evoked potentials have been added to the section on brain waves as a diagnostic tool for neurological disorders.
- Clarified the difference between sensation and perception, with sensation as the stimulus and perception as how our brain interprets the stimulus.
- The section on pain has been modified. Definition of pain receptors has been clarified and peripheral-acting analgesics have been included.
- Have clarified the origin of indirect motor pathways in the brainstem. Included the tectospinal tract as one of the major indirect pathways.
- Clinical Genetics material on Tay-Sachs has been shortened and rewritten to emphasize how this disorder exemplifies the application and power of genetic testing and counseling.
- Added that the reasoning behind clinical lesions of the corpus callosum is to treat intractable epilepsy.
- Added the sensation of tickle to table 14.2.
- Have removed statement that secondary receptor cells do not generate action potentials since taste receptor cells are exceptions that can generate both graded and action potentials.
- Figures 14.15 and 14.18 have been redrawn to include anatomical schematics of brain and other tissues to aid conceptualization of descending pathways and the cerebellar comparator function, respectively. In addition, the comparator pathways have been simplified with removal of the red nucleus.
- Updated image of an EEG net on a patient shown in figure 14.21.
- Direction of the action potential has been added to figure 14.23 to help students place LTP in the context of signal transmission.
- Updated and expanded Clinical Impact on headaches includes common triggers and a more complete description of symptoms.
- Updated Systems Pathology on stroke includes comparison of the two types of stroke with differences in diagnosis and treatments.
- New chapter opener figure shows a colorful and diverse image of labeled hippocampal neurons from transgenic mice.

Chapter 15

- New Learn to Predict question added.
- Function of conjunctiva has been added.
- Clinical Impact "Color Blindness" has been updated as a Clinical Genetics reading.
- In all figures with a background screen, the color of the screen was changed to yellow, which looks more modern and increases student engagement.

Chapter 16

- Overview of the Autonomic Nervous System added.
- Clarified differences between neural pathways presented in Sympathetic Division and Parasympathetic Division, and the means by which postganglionic fibers reach target organs in Autonomic Nerve Plexuses and Distribution of Autonomic Nerve Fibers.
- Dual innervation introduced at the beginning of the Physiology of the Autonomic Nervous System section.
- Comparison of sympathetic and parasympathetic activities moved to the beginning of the Physiology of the Autonomic Nervous System section.
- Definitions of agonist and antagonist drugs added to Neurotransmitter section.

Chapter 17

- Revised figure 17.3 for clarity.
- Revised figure 17.5 for clarity.
- Revised figure 17.9 for cohesion with other sections of the text.
- Revised figure 17.11 for accuracy.
- Revised figure 17.16 and 17.14 (combined two) and reordered for a more logical presentation of the information (old figures 17.14 and 17.16).
- Section 17.4 was reorganized for a more logical flow of information.

Chapter 18

- Figures 18.7, 18.9, 18.10, 18.12, 18.13 (hormone names added for each layer), and 18.17 revised for clarity.

- Figure 18.19 was revised into an illustrated table to help students make better connections.
- Added a new Critical Thinking question to enhance student learning and problem solving.

Chapter 19

- Production of Formed Elements revised to include intermediate stem cells: myeloid stem cell and lymphoid stem cell.
- Figure 19.2 revised to include myeloid stem cell and lymphoid stem cell.
- Figure 19.12 now includes a reference figure to illustrate the factors inside and outside the blood involved in coagulation.
- Figure 19.15 revised to better represent the interactions between maternal and fetal circulation.

Chapter 20

- Figure 20.2, revised making the inset figure larger and easier to see reference points for heart location.
- Section 20.7 Cardiac Cycle revised so that the discussion of the cardiac cycle begins with Atrial Systole. This correlates better with the discussion of EEG and the normal events associated with heart contraction and relaxation.
- Figure 20.18 and table 20.2 also revised to correlate with new organization of the cardiac cycle discussion.
- Systems Pathology "Myocardial Infarction" presented in new format.

Chapter 21

- Figure 21.6 revised moving the diagram of valves in veins to a separate figure.
- Figure 21.9 now illustrates the splenic and renal arteries more accurately.
- Figure 21.37 revised so blood flow is more obvious.

Chapter 22

- New chapter opener photo to correlate with Learn to Predict question.
- Figure 22.1 revised so components of lymphatic system are clear.
- Function of thymic corpuscles updated.
- Eosinophil function updated.
- Suppressor T cells are introduced as regulatory T cells.
- Genetic relationship of MHC molecules discussed to assist reader in understanding the need for genetic matches in tissue transplants.
- Systems Pathology "Systemic Lupus Erythematosus" presented in new format.

Chapter 23

- Reorganized the layout of section 23.2 on a functional basis to help students make connections between the anatomy and physiology.
- Corrected an error in section 23.3, "Airflow Into and Out of Alveoli" per reviewer feedback.
- Corrected figure 23.15 per reviewer feedback.
- Revised figure 23.21.

Chapter 24

- New chapter opener figure showing a gallstone in a colorful abdominal CT scan. Matches the Learn to Predict.
- Introduction revised to incorporate the points made by the Learn to Predict question.
- Clarified that ENS is a division of the ANS.
- Rewrote the section on stomach filling to clarify the rugae actions and regulation.
- New Predict question for the Case Study on spinal cord injury.
- Removed unnecessary information from the Clinical Genetics box.
- New Systems Pathology organization and new art to highlight the story.
- Reduced the number of learning outcomes for Oral Cavity section from six to three to better emphasize the important points.
- Corrected misstatements referring to Giardia and a bolus of food.
- Clarified summary statements on pancreatic secretions and regulation of pancreatic secretions.
- Added the kidney to the view of retroperitoneal organs in figure 24.5.
- Improved visualization of swallowing phases by highlighting movement of the larynx and epiglottis in figure 24.10.
- Changes in six figures to provide color consistency for arrows indicating functions, ion channels, and other molecules.

Chapter 25

- New Learn to Predict question.
- MyPlate replaces the MyPyramid discussion.
- Metabolism figures updated so that background color represents the cellular location (cytosol or mitochondrion) of each process.

Chapter 26

- Revised table 26.1 for accuracy.
- Added an introductory paragraph to Section 26.3—Urine Concentration Mechanism to help students make connections.
- Learning outcomes goals at the beginning of the chapter were numbered to correlate with Predict questions and end of chapter questions.

Chapter 27

- Moved table 27.3 to appear after the introductory text to make the information flow more logical.
- In all figures with a background screen, the color of the screen was changed to yellow, which looks more modern and increases student engagement.
- Chapter opener rewritten with a focus on maintenance of homeostasis, a major underlying theme of the book.

Chapter 28

- Estradiol introduced as a specific type of estrogen.
- Figures 28.8 and 28.18 revised so the hypothalamohypophysial portal system is more accurately represented.
- Atresia introduced in Oogenesis and Fertilization section.
- New figure 28.13 presents the process of oogenesis in context of ovarian follicle development.
- Clinical Impact—Cervical Cancer updated with new recommendations for HPV vaccination for males.
- Systems Pathology "Benign Uterine Tumors" presented in new format.

Chapter 29

- New chapter opener figure to correlate with Learn to Predict question.
- New chapter introduction discusses changes in perception of age over generations.
- Figure 29.21 revised so the hypothalamohypophysial portal system is more accurately represented.
- Genetics is now presented in this chapter instead of chapter 3.

1

The Human Organism

> **Learn to Predict**
>
> Renzo, the dancer in the photo, is perfectly balanced, yet a slight movement in any direction would cause him to adjust his position. The human body adjusts its balance among all its parts through a process called homeostasis.
>
> Let's imagine that Renzo is suffering from a blood sugar disorder. Earlier, just before this photo was taken, he'd eaten an energy bar. As an energy bar is digested, blood sugar rises. Normally, tiny collections of cells embedded in the pancreas respond to the rise in blood sugar by secreting the chemical insulin. Insulin increases the movement of sugar from the blood into his cells. However, Renzo did not feel satisfied from his energy bar. He felt dizzy and was still hungry, all symptoms he worried could be due to a family history of diabetes. Fortunately, the on-site trainer tested his blood sugar and noted that it was much higher than normal. After a visit to his regular physician, Renzo was outfitted with an insulin pump, and his blood sugar levels are more consistent.
>
> After reading about homeostasis in this chapter, create an explanation for Renzo's blood sugar levels before and after his visit to the doctor.

What lies ahead is an astounding adventure—learning about the structure and function of the human body and the intricate checks and balances that regulate it. Renzo's response to eating the energy bar is a good example of how important this system of checks and balances is in the body. Perhaps you have had a similar experience, but with a different outcome. You have overslept, rushed to your 8 a.m. class, and missed breakfast. Afterwards, on the way to Anatomy & Physiology class, you bought an energy bar from the vending machine. Eating the energy bar helped you feel better. The explanation for these experiences is the process of homeostasis; for you, homeostasis was maintained, but for Renzo, there was a disruption in homeostasis. Throughout this book, the major underlying theme is homeostasis. As you think about Renzo's case, you will come to realize just how capable the human body is of an incredible coordination of thousands upon thousands of processes. Knowing human anatomy and physiology is also the basis for understanding disease. The study of human anatomy and physiology is important for students who plan a career in the health sciences because health professionals need a sound knowledge of structure and function in order to perform their duties. In addition, understanding anatomy and physiology prepares all of us to evaluate recommended treatments, critically review advertisements and reports in the popular literature, and rationally discuss the human body with health professionals and nonprofessionals.

Module 1
Body Orientation

Anatomy & Physiology
REVEALED
aprevealed.com

1.1 Anatomy and Physiology

LEARNING OUTCOMES

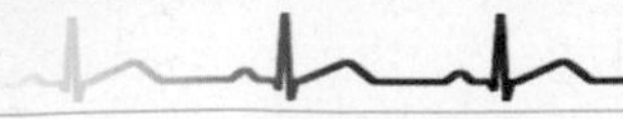

After reading this section, you should be able to

A. Define *anatomy* and describe the levels at which anatomy can be studied.

B. Define *physiology* and describe the levels at which physiology can be studied.

C. Explain the importance of the relationship between structure and function.

Anatomy is the scientific discipline that investigates the body's structure—for example, the shape and size of bones. In addition, anatomy examines the relationship between the structure of a body part and its function. Thus, the fact that bone cells are surrounded by a hard, mineralized substance enables the bones to provide strength and support. Understanding the relationship between structure and function makes it easier to understand and appreciate anatomy. Anatomy can be considered at different levels. **Developmental anatomy** studies the structural changes that occur between conception and adulthood. **Embryology** (em-brē-ol′ō-jē), a subspecialty of developmental anatomy, considers changes from conception to the end of the eighth week of development.

Some structures, such as cells, are so small that they must be studied using a microscope. **Cytology** (sī-tol′ō-jē) examines the structural features of cells, and **histology** (his-tol′ō-jē) examines tissues, which are composed of cells and the materials surrounding them.

Gross anatomy, the study of structures that can be examined without the aid of a microscope, can be approached from either a systemic or a regional perspective. In systemic anatomy, the body is studied system by system. A system is a group of structures that have one or more common functions, such as the cardiovascular, nervous, respiratory, skeletal, or muscular system. The systemic approach is taken in this and most other introductory textbooks. In regional anatomy, the body is studied area by area. Within each region, such as the head, abdomen, or arm, all systems are studied simultaneously. The regional approach is taken in most graduate programs at medical and dental schools.

Surface anatomy is the study of the external form of the body and its relation to deeper structures. For example, the sternum (breastbone) and parts of the ribs can be seen and palpated (felt) on the front of the chest. Health professionals use these structures as anatomical landmarks to identify regions of the heart and points on the chest where certain heart sounds can best be heard. **Anatomical imaging** uses radiographs (x-rays), ultrasound, magnetic resonance imaging (MRI), and other technologies to create pictures of internal structures (table 1.1). Anatomical imaging has revolutionized medical science. Some scientists estimate that the past 20 years have seen as much progress in clinical medicine as occurred in all of medicine's previous history. Anatomical imaging has made a major contribution to that progress. Anatomical imaging allows medical personnel to look inside the body with amazing accuracy and without the trauma and risk of exploratory surgery. Although most of the technology used in anatomical imaging is very new, the concept and earliest technology are quite old. In 1895, Wilhelm Roentgen (1845–1923) became the first medical scientist to use **x-rays** to see inside the body. The rays were called x-rays because no one knew what they were. Whenever the human body is exposed to x-rays, ultrasound, electromagnetic fields, or radioactively labeled substances, a potential risk exists. This risk must be weighed against the medical benefit. Numerous studies have been conducted and are still being done to determine the effects of diagnostic and therapeutic exposure to x-rays. The risk of anatomical imaging is minimized by using the lowest possible doses providing the necessary information. No known risks exist from ultrasound or electromagnetic fields at the levels used for diagnosis. Both surface anatomy and anatomical imaging provide important information for diagnosing disease.

However, no two humans are structurally identical. **Anatomical anomalies** are physical characteristics that differ from the normal pattern. Anatomical anomalies can vary in severity from relatively harmless to life-threatening. For example, each kidney is normally supplied by one blood vessel, but in some individuals a kidney is supplied by two blood vessels. Either way, the kidney receives adequate blood. On the other hand, in the condition called "blue baby" syndrome, certain blood vessels arising from an infant's heart are not attached in their correct locations; blood is not effectively pumped to the lungs, and so the tissues do not receive adequate oxygen.

Physiology is the scientific investigation of the processes or functions of living things. The major goals when studying human physiology are to understand and predict the body's responses to stimuli and to understand how the body maintains conditions within a narrow range of values in a constantly changing environment.

Like anatomy, physiology can be considered at many levels. **Cell physiology** examines the processes occurring in cells, and **systemic physiology** considers the functions of organ systems. **Neurophysiology** focuses on the nervous system, and **cardiovascular physiology** deals with the heart and blood vessels. Physiology often examines systems rather than regions because a particular function can involve portions of a system in more than one region.

Studies of the human body must encompass both anatomy and physiology because structures, functions, and processes are interwoven. **Pathology** (pa-thol′ō-jē) is the medical science dealing with all aspects of disease, with an emphasis on the cause and development of abnormal conditions, as well as the structural and functional changes resulting from disease. **Exercise physiology** focuses on the changes in function and structure caused by exercise.

ASSESS YOUR PROGRESS

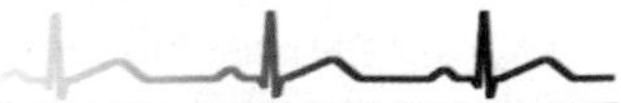

1. *How does the study of anatomy differ from the study of physiology?*
2. *What is studied in gross anatomy? In surface anatomy?*
3. *What type of physiology is employed when studying the endocrine system?*
4. *Why are anatomy and physiology normally studied together?*

TABLE 1.1 Anatomical Imaging

Imaging Technique	Image	Clinical Examples
X-ray	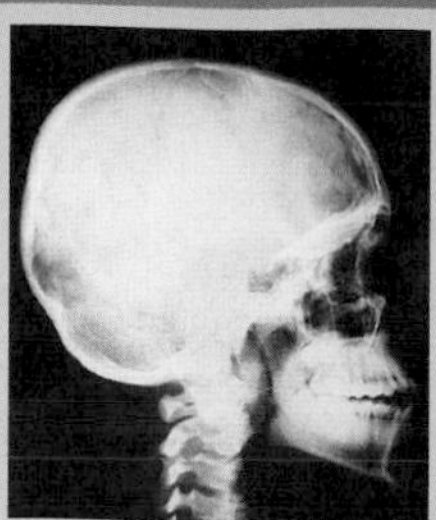	This extremely shortwave electromagnetic radiation (see chapter 2) moves through the body, exposing a photographic plate to form a **radiograph** (rā′dē-ō-graf). Bones and radiopaque dyes absorb the rays and create underexposed areas that appear white on the photographic film. Almost everyone has had a radiograph taken, either to visualize a broken bone or to check for a cavity in a tooth. However, a major limitation of radiographs is that they give only flat, two-dimensional (2-D) images of the body.
Ultrasound	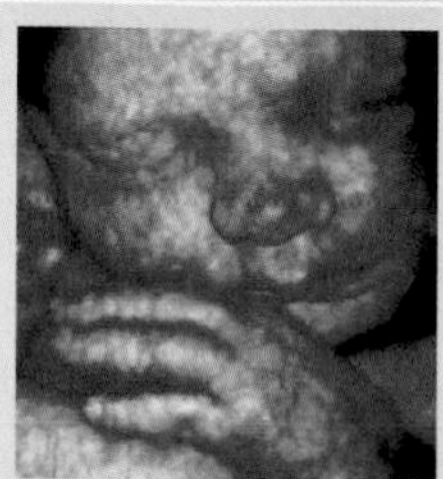	**Ultrasound,** the second oldest imaging technique, was first developed in the early 1950s as an extension of World War II sonar technology. It uses high-frequency sound waves, which are emitted from a transmitter-receiver placed on the skin over the area to be scanned. The sound waves strike internal organs and bounce back to the receiver on the skin. Even though the basic technology is fairly old, the most important advances in the field occurred only after it became possible to analyze the reflected sound waves by computer. Once a computer analyzes the pattern of sound waves, the information is transferred to a monitor and visualized as a **sonogram** (son′ō-gram) image. One of the more recent advances in ultrasound technology is the ability of more advanced computers to analyze changes in position through "real-time" movements. Among other medical applications, ultrasound is commonly used to evaluate the condition of the fetus during pregnancy.
Computed Tomography (CT)	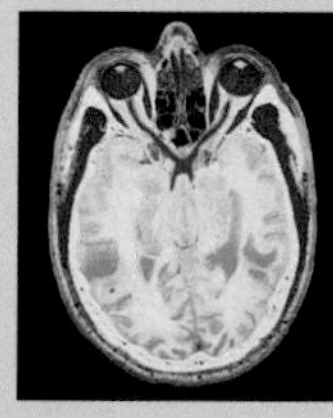 (a) 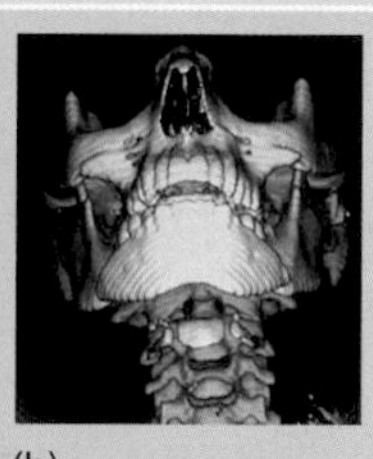 (b)	**Computed tomographic** (tō′mō-graf′ik) **(CT) scans,** developed in 1972 and originally called *computerized axial tomographic (CAT) scans,* are computer-analyzed x-ray images. A low-intensity x-ray tube is rotated through a 360-degree arc around the patient, and the images are fed into a computer. The computer then constructs the image of a "slice" through the body at the point where the x-ray beam was focused and rotated (*a*). Some computers are able to take several scans short distances apart and stack the slices to produce a 3-D image of a body part (*b*).
Dynamic Subtraction Angiography (DSA)	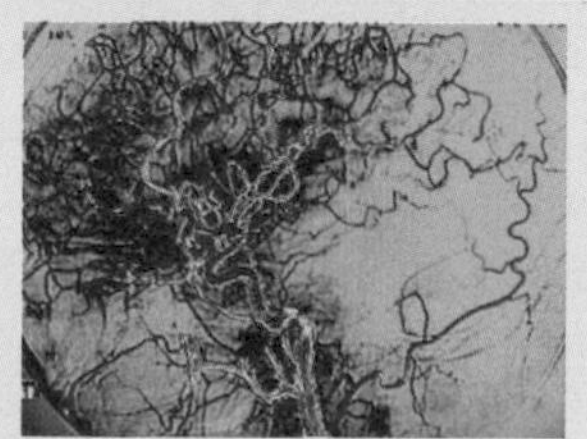	**Digital subtraction angiography** (an-jē-og′ră-fē) **(DSA)** is one step beyond CT scanning. A 3-D radiographic image of an organ, such as the brain, is made and stored in a computer. Then a radiopaque dye is injected into the blood, and a second radiographic computer image is made. The first image is subtracted from the second one, greatly enhancing the differences revealed by the injected dye. These dynamic computer images can be used, for example, to guide a catheter into a carotid artery during angioplasty, a procedure by which a tiny balloon compresses the material clogging the artery.
Magnetic Resonance Imaging (MRI)	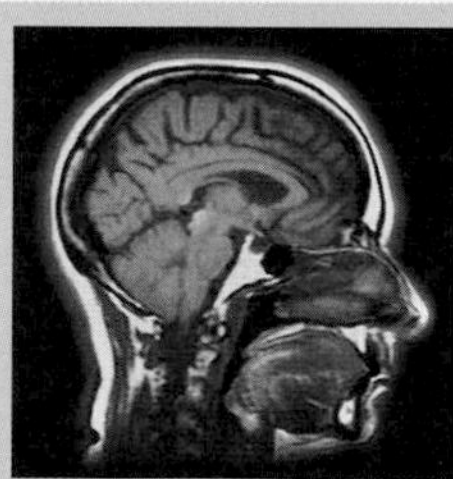	**Magnetic resonance imaging (MRI)** directs radio waves at a person lying inside a large electromagnetic field. The magnetic field causes the protons of various atoms to align (see chapter 2). Because of the large amounts of water in the body, the alignment of hydrogen atom protons is most important in this imaging system. Radio waves of certain frequencies, which change the alignment of the hydrogen atoms, then are directed at the patient. When the radio waves are turned off, the hydrogen atoms realign in accordance with the magnetic field. The time it takes the hydrogen atoms to realign is different for various body tissues. These differences can be analyzed by computer to produce very clear sections through the body. The technique is also very sensitive in detecting some forms of cancer far more readily than can a CT scan.
Positron Emission Tomography (PET)	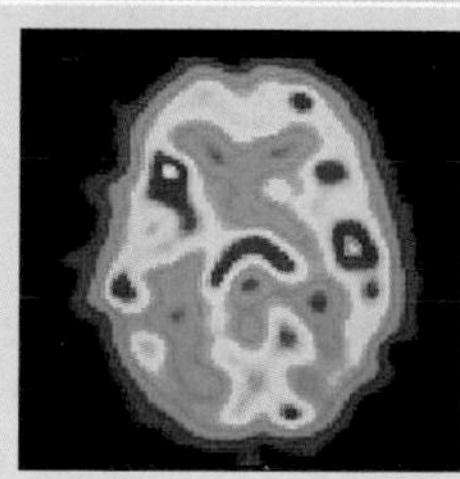	**Positron emission tomographic (PET) scans** can identify the metabolic states of various tissues. This technique is particularly useful in analyzing the brain. When cells are active, they are using energy. The energy they need is supplied by the breakdown of glucose (blood sugar). If radioactively treated ("labeled") glucose is given to a patient, the active cells take up the labeled glucose. As the radioactivity in the glucose decays, positively charged subatomic particles called positrons are emitted. When the positrons collide with electrons, the two particles annihilate each other and gamma rays are given off. The gamma rays can be detected, pinpointing the cells that are metabolically active.

1.2 Structural and Functional Organization of the Human Body

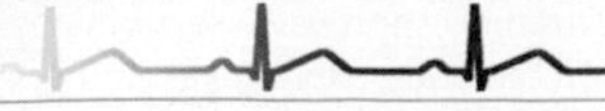

LEARNING OUTCOMES

After reading this section, you should be able to

A. Name the six levels of organization of the body, and describe the major characteristics of each level.

B. List the 11 organ systems, identify their components, and describe the major functions of each system.

The body can be studied at six levels of organization: the chemical, cell, tissue, organ, organ system, and whole organism levels (figure 1.1).

1. *Chemical level.* The chemical level involves interactions between atoms, which are tiny building blocks of matter. Atoms combine to form molecules, such as water, sugar, fats, and proteins. The function of a molecule is intimately related to its structure. For example, collagen molecules are ropelike protein fibers that give skin structural strength and flexibility. With old age, the structure of collagen changes, and the skin becomes fragile and more easily torn. We present a brief overview of chemistry in chapter 2.
2. *Cell level.* **Cells** are the basic structural and functional units of plants and animals. Molecules combine to form **organelles** (or′gă-nelz; little organs), which are the small structures that make up cells. For example, the nucleus is an organelle that contains the cell's hereditary information, and mitochondria are organelles that manufacture adenosine triphosphate (ATP), a molecule cells use for energy. Although cell types differ in their structure and function, they have many characteristics in common. Knowledge of these characteristics, as well as their variations, is essential to understanding anatomy and physiology. We discuss the cell in chapter 3.
3. *Tissue level.* A **tissue** is composed of a group of similar cells and the materials surrounding them. The characteristics of the cells and surrounding materials determine the functions of the tissue. The numerous tissues that make up the body are classified into four basic types: epithelial, connective, muscle, and nervous. We discuss tissues in chapter 4.
4. *Organ level.* An **organ** is composed of two or more tissue types that perform one or more common functions. The urinary bladder, heart, stomach, and lung are examples of organs (figure 1.2).
5. *Organ system level.* An **organ system** is a group of organs that together perform a common function or set of functions and are therefore viewed as a unit. For example, the urinary system consists of the kidneys, ureter, urinary bladder, and urethra. The kidneys produce urine, which the ureters transport to the urinary bladder, where it is stored until being eliminated from the body through the urethra. In this text, we consider 11 major organ systems: the integumentary, skeletal, muscular, nervous, endocrine, cardiovascular, lymphatic, respiratory, digestive, urinary, and reproductive systems. Figure 1.3 presents a brief summary of these organ systems and their functions.
6. *Organism level.* An **organism** is any living thing considered as a whole—whether composed of one cell, such as a bacterium, or of trillions of cells, such as a human. The human organism is a complex of organ systems, all mutually dependent on one another.

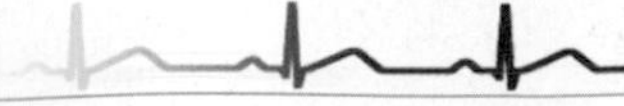

ASSESS YOUR PROGRESS

5. *From simplest to complex, list and define the body's six levels of organization.*
6. *What are the four basic types of tissues?*
7. *Referring to figure 1.3, which two organ systems are responsible for regulating the other organ systems? Which two are responsible for support and movement?*

Predict 2

In one type of diabetes, the pancreas fails to produce insulin, a chemical normally made by pancreatic cells and released into the blood. List as many levels of organization as you can at which this disorder could be corrected.

1.3 Characteristics of Life

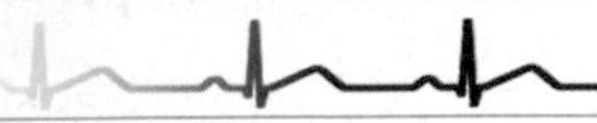

LEARNING OUTCOME

After reading this section, you should be able to

A. List and define the six characteristics of life.

Humans are organisms, sharing characteristics with other organisms. The most important common feature of all organisms is life. This text recognizes six essential characteristics of life:

1. **Organization** refers to the specific interrelationships among the parts of an organism and how those parts interact to perform specific functions. Living things are highly organized. All organisms are composed of one or more cells. Cells in turn are composed of highly specialized organelles, which depend on the precise organization of large molecules. Disruption of this organized state can result in loss of functions, or even death.
2. **Metabolism** (mĕ-tab′ō-lizm) refers to all of the chemical reactions taking place in an organism. It includes an organism's ability to break down food molecules, which the organism uses as a source of energy and raw materials to synthesize its own molecules. Energy is also used when one part of a molecule moves relative to another part, changing the shape of the molecule. Changes in molecular shape can lead to changes in cellular shape, which can produce movement of the organism. Metabolism is necessary for vital functions, such as responsiveness, growth, development, and reproduction.
3. **Responsiveness** is an organism's ability to sense changes in its external or internal environment and adjust to those changes. Responses include such actions as moving toward food or water and moving away from danger or poor environmental conditions. Organisms can also make adjustments that maintain their internal environment. For example, if the

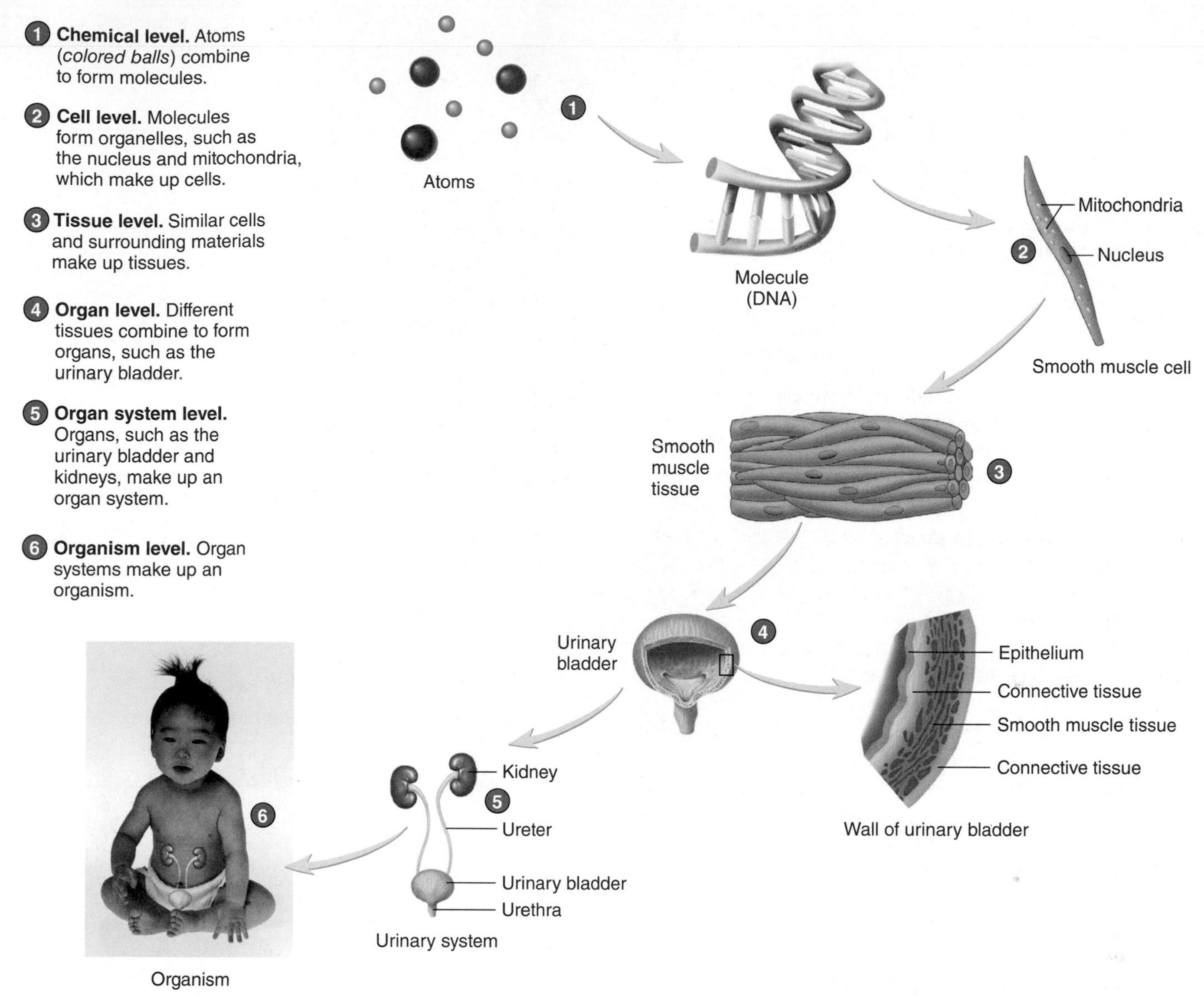

1 Chemical level. Atoms (colored balls) combine to form molecules.
2 Cell level. Molecules form organelles, such as the nucleus and mitochondria, which make up cells.
3 Tissue level. Similar cells and surrounding materials make up tissues.
4 Organ level. Different tissues combine to form organs, such as the urinary bladder.
5 Organ system level. Organs, such as the urinary bladder and kidneys, make up an organ system.
6 Organism level. Organ systems make up an organism.
Atoms
Molecule (DNA)
Mitochondria
Nucleus
Smooth muscle cell
Smooth muscle tissue
Urinary bladder
Epithelium
Connective tissue
Smooth muscle tissue
Connective tissue
Wall of urinary bladder
Kidney
Ureter
Urinary bladder
Urethra
Urinary system
Organism

PROCESS FIGURE 1.1 Levels of Organization for the Human Body

external environment causes the body temperature to rise, sweat glands produce sweat, which can lower body temperature back toward its normal range.

4. **Growth** refers to an increase in the size or number of cells, which produces an overall enlargement of all or part of an organism. For example, a muscle enlarged by exercise is composed of larger muscle cells than those of an untrained muscle, and the skin of an adult has more cells than the skin of an infant. An increase in the materials surrounding cells can also contribute to growth. For instance, bone grows because of an increase in cell number and the deposition of mineralized materials around the cells.
5. **Development** includes the changes an organism undergoes through time, beginning with fertilization and ending at death. The greatest developmental changes occur before birth, but many changes continue after birth, and some go on throughout life. Development usually involves growth, but it also involves differentiation and morphogenesis. **Differentiation** is change in cell structure and function from generalized to specialized, and **morphogenesis** (mōr-fō-jen′ĕ-sis) is change in the shape of tissues, organs, and the entire organism. For example, following fertilization, generalized cells specialize to become specific cell types, such as skin, bone, muscle, or nerve cells. These differentiated cells form the tissues and organs.
6. **Reproduction** is the formation of new cells or new organisms. Without reproduction of cells, growth and development are not possible. Without reproduction of organisms, species become extinct.

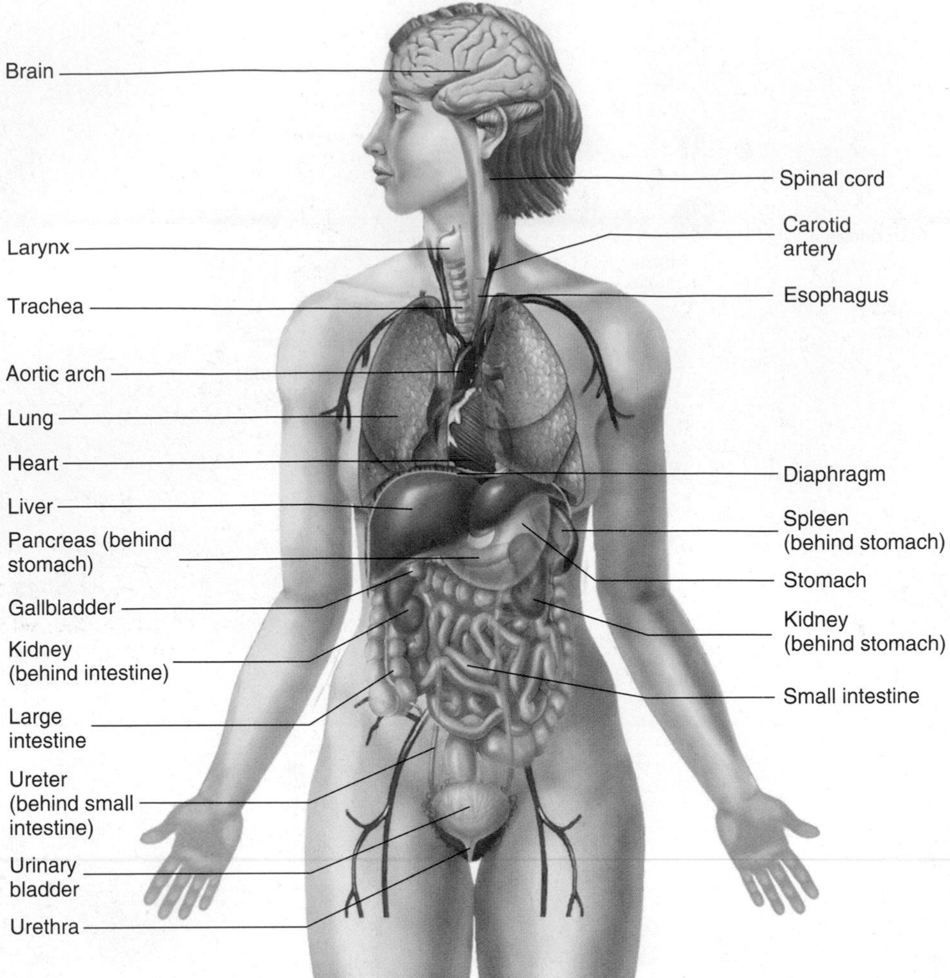
Brain
Spinal cord
Larynx
Carotid artery
Trachea
Esophagus
Aortic arch
Lung
Heart
Diaphragm
Liver
Spleen (behind stomach)
Pancreas (behind stomach)
Stomach
Gallbladder
Kidney (behind stomach)
Kidney (behind intestine)
Small intestine
Large intestine
Ureter (behind small intestine)
Urinary bladder
Urethra

FIGURE 1.2 APR **Major Organs of the Body**

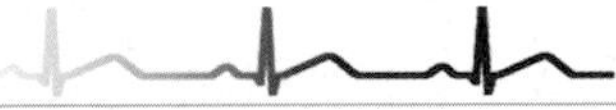

ASSESS YOUR PROGRESS

8. *What are the six characteristics of living things? Briefly explain each.*
9. *How does differentiation differ from morphogenesis?*

1.4 Biomedical Research

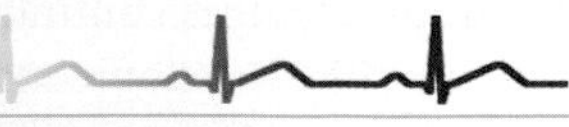

LEARNING OUTCOME

After reading this section, you should be able to

A. Explain why it is important to study other organisms along with humans.

Studying other organisms has increased our knowledge about humans because humans share many characteristics with other organisms. For example, studying single-celled bacteria provides much information about human cells. However, some biomedical research cannot be accomplished using single-celled organisms or isolated cells. Sometimes other mammals must be studied, as evidenced by the great progress in open heart surgery and kidney transplantation made possible by perfecting surgical techniques on other mammals before attempting them on humans. Strict laws govern the use of animals in biomedical research; these laws are designed to ensure minimal suffering on the part of the animal and to discourage unnecessary experimentation.

Although much can be learned from studying other organisms, the ultimate answers to questions about humans can be obtained only from humans because other organisms differ from humans in significant ways. A failure to appreciate the differences between humans and other animals led to many misconceptions by early scientists. One of the first great anatomists was a Greek physician, Claudius Galen (ca. 130–201). Galen described a large number of anatomical structures supposedly present in humans but observed only in other animals. For example, he described the liver as having five lobes. This is true for rats, but not for humans, who have four-lobed livers. The errors introduced by Galen persisted for more than 1300 years until a Flemish anatomist, Andreas Vesalius (1514–1564), who is considered the first modern anatomist, carefully examined human cadavers and began

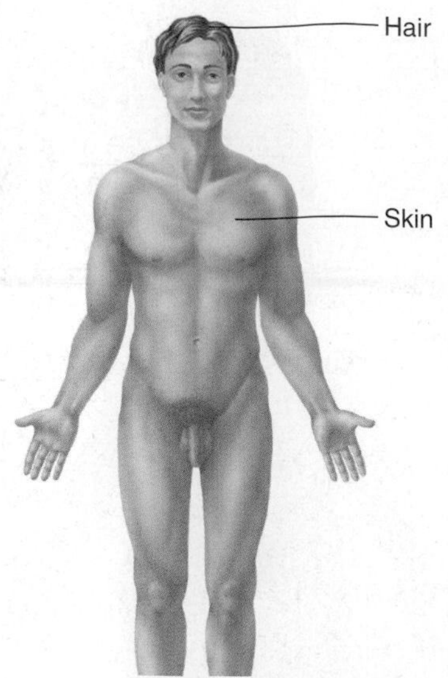
Hair
Skin

Integumentary System

Provides protection, regulates temperature, prevents water loss, and helps produce vitamin D. Consists of skin, hair, nails, and sweat glands.

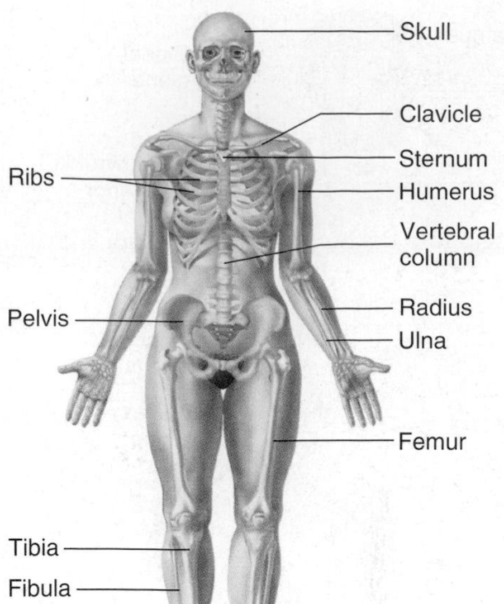
Skull
Clavicle
Sternum
Ribs
Humerus
Vertebral column
Pelvis
Radius
Ulna
Femur
Tibia
Fibula

Skeletal System

Provides protection and support, allows body movements, produces blood cells, and stores minerals and fat. Consists of bones, associated cartilages, ligaments, and joints.

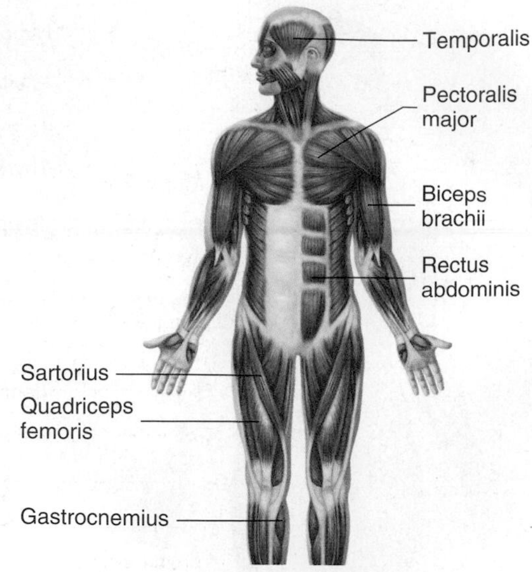
Temporalis
Pectoralis major
Biceps brachii
Rectus abdominis
Sartorius
Quadriceps femoris
Gastrocnemius

Muscular System

Produces body movements, maintains posture, and produces body heat. Consists of muscles attached to the skeleton by tendons.

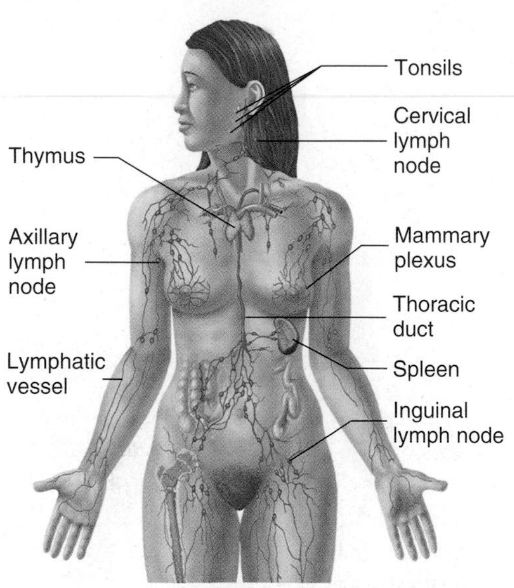
Tonsils
Cervical lymph node
Thymus
Axillary lymph node
Mammary plexus
Thoracic duct
Lymphatic vessel
Spleen
Inguinal lymph node

Lymphatic System

Removes foreign substances from the blood and lymph, combats disease, maintains tissue fluid balance, and absorbs fats from the digestive tract. Consists of the lymphatic vessels, lymph nodes, and other lymphatic organs.

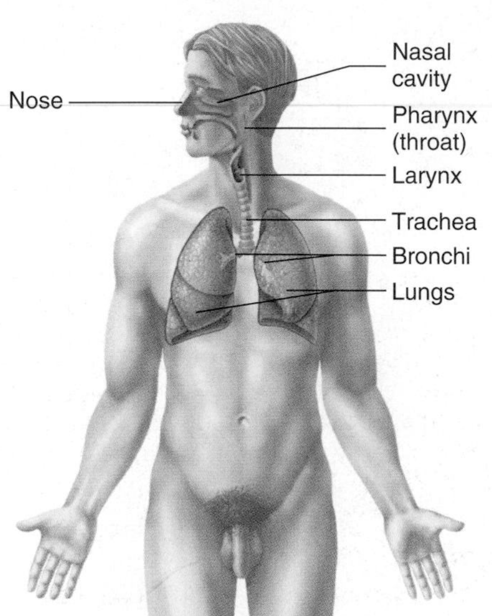
Nasal cavity
Nose
Pharynx (throat)
Larynx
Trachea
Bronchi
Lungs

Respiratory System

Exchanges oxygen and carbon dioxide between the blood and air and regulates blood pH. Consists of the lungs and respiratory passages.

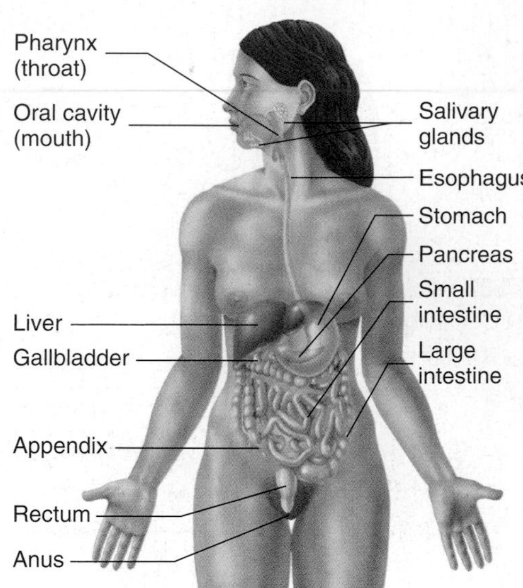
Pharynx (throat)
Oral cavity (mouth)
Salivary glands
Esophagus
Stomach
Pancreas
Small intestine
Liver
Gallbladder
Large intestine
Appendix
Rectum
Anus

Digestive System

Performs the mechanical and chemical processes of digestion, absorption of nutrients, and elimination of wastes. Consists of the mouth, esophagus, stomach, intestines, and accessory organs.

FIGURE 1.3 Organ Systems of the Body

to correct the textbooks. This example should serve as a word of caution: Some current knowledge in molecular biology and physiology has not been confirmed in humans.

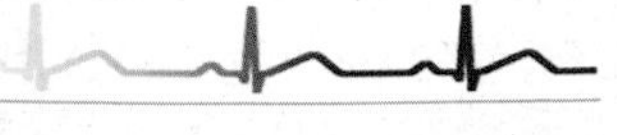
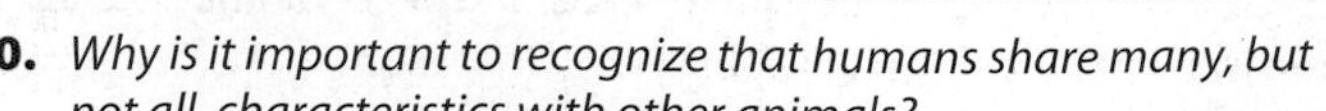

ASSESS YOUR PROGRESS

10. *Why is it important to recognize that humans share many, but not all, characteristics with other animals?*

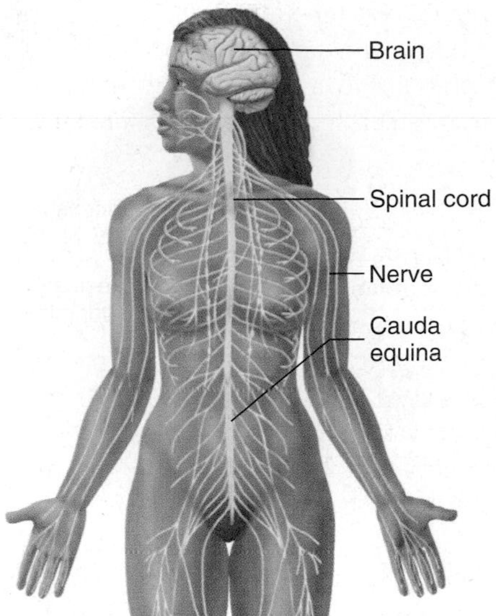
Brain
Spinal cord
Nerve
Cauda equina

Nervous System

A major regulatory system that detects sensations and controls movements, physiological processes, and intellectual functions. Consists of the brain, spinal cord, nerves, and sensory receptors.

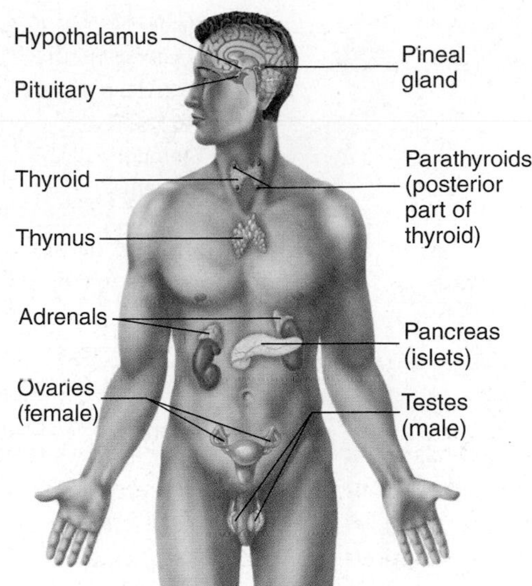
Hypothalamus
Pituitary
Pineal gland
Thyroid
Parathyroids (posterior part of thyroid)
Thymus
Adrenals
Pancreas (islets)
Ovaries (female)
Testes (male)

Endocrine System

A major regulatory system that influences metabolism, growth, reproduction, and many other functions. Consists of glands, such as the pituitary, that secrete hormones.

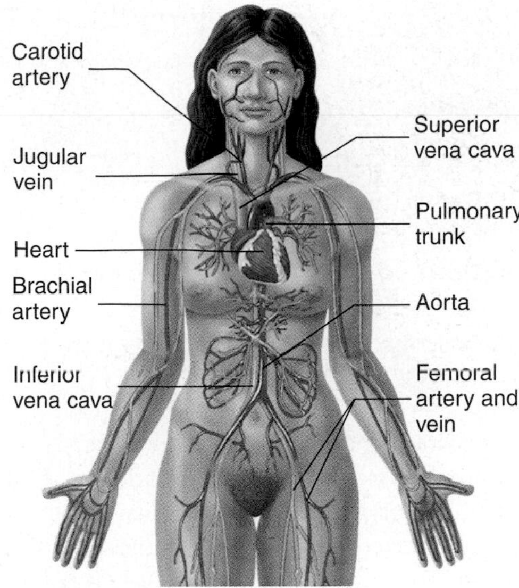
Carotid artery
Superior vena cava
Jugular vein
Pulmonary trunk
Heart
Brachial artery
Aorta
Inferior vena cava
Femoral artery and vein

Cardiovascular System

Transports nutrients, waste products, gases, and hormones throughout the body; plays a role in the immune response and the regulation of body temperature. Consists of the heart, blood vessels, and blood.

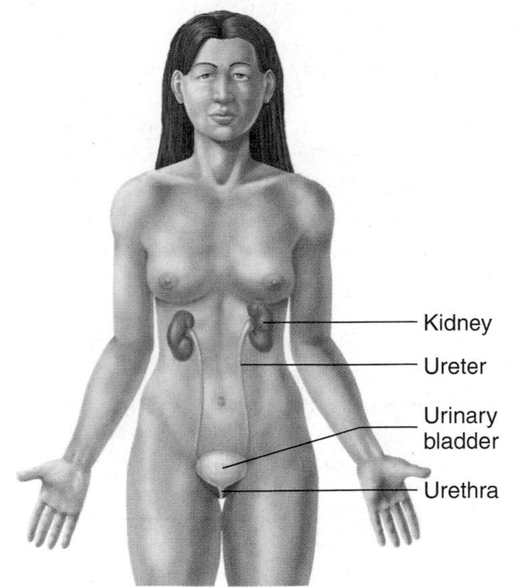
Kidney
Ureter
Urinary bladder
Urethra

Urinary System

Removes waste products from the blood and regulates blood pH, ion balance, and water balance. Consists of the kidneys, urinary bladder, and ducts that carry urine.

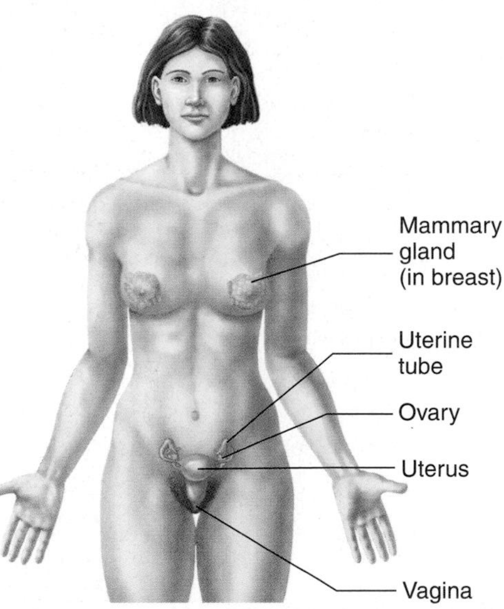
Mammary gland (in breast)
Uterine tube
Ovary
Uterus
Vagina

Female Reproductive System

Produces oocytes and is the site of fertilization and fetal development; produces milk for the newborn; produces hormones that influence sexual function and behaviors. Consists of the ovaries, vagina, uterus, mammary glands, and associated structures.

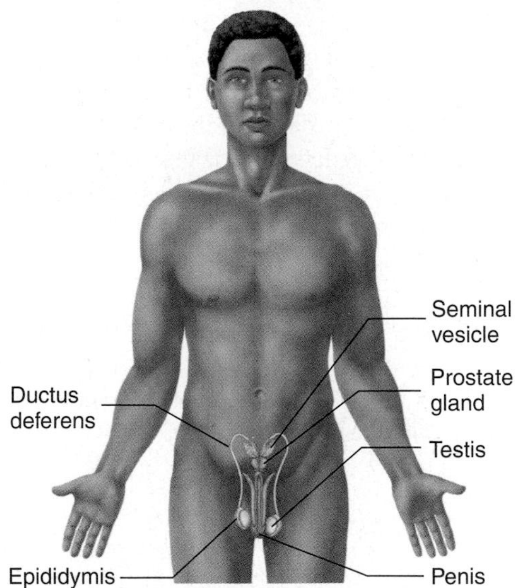
Seminal vesicle
Prostate gland
Ductus deferens
Testis
Epididymis
Penis

Male Reproductive System

Produces and transfers sperm cells to the female and produces hormones that influence sexual functions and behaviors. Consists of the testes, accessory structures, ducts, and penis.

FIGURE 1.3 (continued)

1.5 Homeostasis

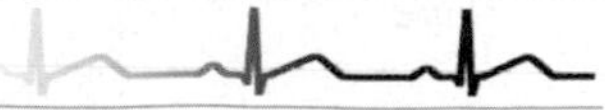

LEARNING OUTCOMES

After reading this section, you should be able to

A. Define *homeostasis* and explain why it is important for proper body function.

B. Describe a negative-feedback mechanism and give an example.

C. Describe a positive-feedback mechanism and give an example.

Homeostasis (hō′mē-ō-stā′sis) is the existence and maintenance of a relatively constant environment within the body. A small amount of fluid surrounds each body cell. For cells to function normally, the volume, temperature, and chemical content of this fluid—conditions known as **variables** because their values can change—must remain within a narrow range. Body temperature is a variable that can increase in a hot environment or decrease in a cold one.

Homeostatic mechanisms, such as sweating or shivering, normally maintain body temperature near an ideal normal value, or **set point** (figure 1.4). Note that these mechanisms are not able to maintain body temperature *precisely* at the set point. Instead, body temperature increases and decreases slightly around the set point to produce a **normal range** of values. As long as body temperature remains within this normal range, homeostasis is maintained. Keep in mind that the fluctuations are minimal, however. Note in figure 1.4 that the normal body temperature range is no more than 1 degree Fahrenheit above or below normal. Our *average* body temperature is 98.6 degrees Fahrenheit. Just as your home's thermostat does not keep the air temperature exactly at 75 degrees Fahrenheit at all times, your body's temperature does not stay perfectly stable.

The organ systems help keep the body's internal environment relatively constant. For example, the digestive, respiratory, cardiovascular, and urinary systems work together, so that each cell in the body receives adequate oxygen and nutrients and waste products do not accumulate to a toxic level. If body fluids deviate from homeostasis, body cells do not function normally and can even die. Disease disrupts homeostasis and sometimes results in death. Modern medicine attempts to understand disturbances in homeostasis and works to reestablish a normal range of values.

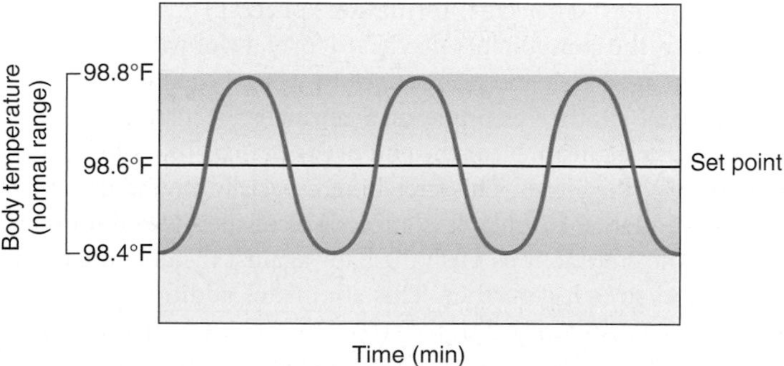
Body temperature (normal range)
98.8°F
98.6°F
98.4°F
Set point
Time (min)

FIGURE 1.4 Homeostasis
Homeostasis is the maintenance of a variable around an ideal normal value, or set point. The value of the variable fluctuates around the set point to establish a normal range of values.

Negative Feedback

Most systems of the body are regulated by **negative-feedback** mechanisms, which maintain homeostasis. *Negative* means that any deviation from the set point is made smaller or is resisted; therefore, in a negative-feedback mechanism, the response to the original stimulus results in deviation from the set point, becoming smaller. An example of important negative-feedback mechanisms in the body are those maintaining normal blood pressure. Normal blood pressure is critical to our health because blood pressure helps move blood from the heart to tissues. The blood transports essential materials to and from the tissues. Because a disruption of normal blood pressure could result in a disease state, maintaining homeostasis through negative feedback is a critical activity. Most negative-feedback mechanisms have three components: (1) a **receptor,** which monitors the value of a variable; (2) a **control center,** which receives

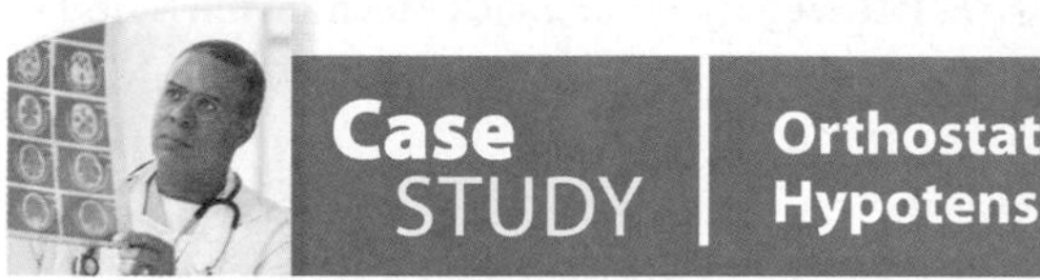

Case STUDY | Orthostatic Hypotension

Molly is a 75-year-old widow who lives alone. For 2 days, she had a fever and chills and mainly stayed in bed. On rising to go to the bathroom, she felt dizzy, fainted, and fell to the floor. Molly quickly regained consciousness and managed to call her son, who took her to the emergency room, where a physician diagnosed orthostatic hypotension.

Orthostasis literally means "to stand," and *hypotension* refers to low blood pressure; thus, **orthostatic hypotension** is a significant drop in blood pressure upon standing. When a person moves from lying down to standing, blood "pools" within the veins below the heart because of gravity, and less blood returns to the heart. Consequently, blood pressure drops because the heart has less blood to pump.

Predict 3

Although orthostatic hypotension has many causes, in the elderly it can be due to age-related decreases in neural and cardiovascular responses. Decreased fluid intake while feeling ill and sweating due to a fever can result in dehydration. Dehydration can decrease blood volume and lower blood pressure, increasing the likelihood of orthostatic hypotension. Use figure 1.6 to answer the following:

a. Describe the normal response to a decrease in blood pressure on standing.

b. What happened to Molly's heart rate just before she fainted? Why did Molly faint?

c. How did Molly's fainting and falling to the floor help establish homeostasis (assuming she was not injured)?

FUNDAMENTAL Figure

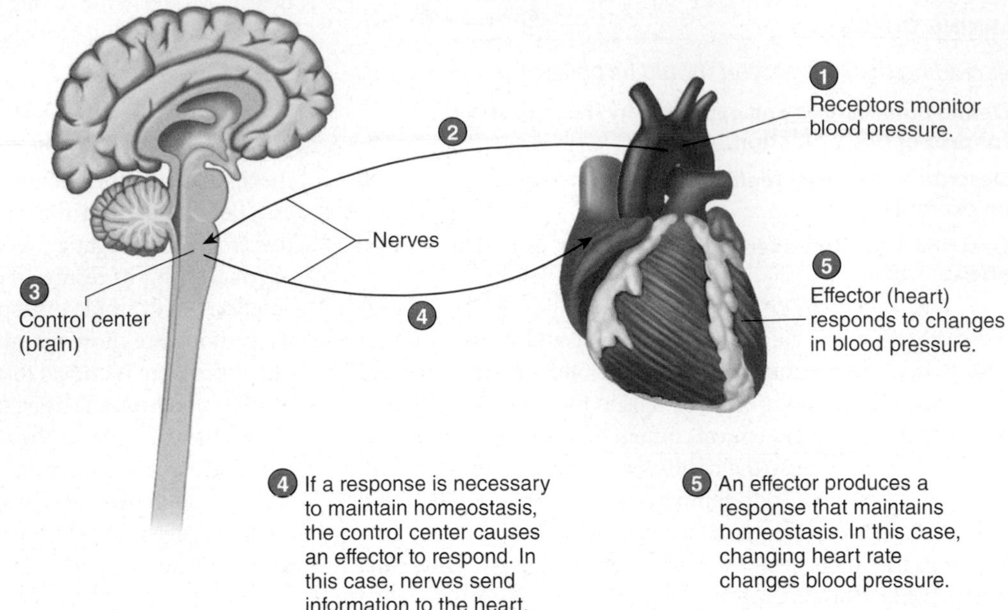

1 Receptors monitor the value of a variable. In this case, receptors in the wall of a blood vessel monitor blood pressure.
2 Information about the value of the variable is sent to a control center. In this case, nerves send information to the part of the brain responsible for regulating blood pressure.
3 The control center compares the value of the variable against the set point.
3 Control center (brain)
2
Nerves
4
1 Receptors monitor blood pressure.
5 Effector (heart) responds to changes in blood pressure.
4 If a response is necessary to maintain homeostasis, the control center causes an effector to respond. In this case, nerves send information to the heart.
5 An effector produces a response that maintains homeostasis. In this case, changing heart rate changes blood pressure.

PROCESS FIGURE 1.5 Negative-Feedback Mechanism: Blood Pressure

information about the variable from the receptor, establishes the set point, and controls the effector; and (3) an **effector,** which produces responses that change the value of the variable. A changed variable is a **stimulus** because it initiates a homeostatic mechanism. Several negative-feedback mechanisms regulate blood pressure, and they are described more fully in chapters 20 and 21. Here we describe one of them: Receptors that monitor blood pressure are located within large blood vessels near the heart and the head. A control center in the brain receives signals sent through nerves from the receptors. The control center evaluates the information and sends signals through nerves to the heart. The heart is the effector, and the heart rate increases or decreases in response to signals from the brain (figure 1.5).

If blood pressure increases slightly, receptors detect that change and send the information to the control center in the brain. The control center causes the heart rate to decrease, lowering blood pressure. If blood pressure goes down slightly, the receptors inform the control center, which elevates the heart rate, thereby producing an increase in blood pressure (figure 1.6). As a result, blood pressure constantly rises and falls within a normal range of values.

Although homeostasis is the maintenance of a normal range of values, this does not mean that all variables remain within the same narrow range of values at all times. Sometimes a deviation from the usual range of values can be beneficial. For example, during exercise the normal range for blood pressure differs from the range under resting conditions and the blood pressure is significantly elevated (figure 1.7). Muscle cells require increased oxygen and nutrients and increased removal of waste products to support their heightened level of activity during exercise. Elevated blood pressure increases delivery of blood to muscles during exercise, thereby increasing the delivery of oxygen and nutrients and the removal of waste products—ultimately maintaining muscle cell homeostasis.

Positive Feedback

Positive-feedback mechanisms occur when a response to the original stimulus results in the deviation from the set point becoming even greater. At times, this type of response is required to re-achieve homeostasis. For example, during blood loss, a chemical responsible for blood clot formation, called thrombin, stimulates production of even more thrombin (figure 1.8). In this way, a disruption in homeostasis is resolved through a positive-feedback mechanism. What prevents the entire vascular system from clotting? The clot formation process is self-limiting. Eventually, the components needed to form a clot will be depleted in the damaged area and no more clot material can be formed.

Birth is another example of a normally occurring positive-feedback mechanism. Near the end of pregnancy, the baby's larger size stretches the uterus. This stretching, especially around the opening of the uterus, stimulates contractions of the uterine muscles. The uterine contractions push the baby against the opening of the uterus and stretch it further. This stimulates additional contractions, which result in additional stretching. This positive-feedback sequence ends only when the baby is delivered from the uterus and the stretching stimulus is eliminated.

Two basic principles to remember are that (1) many disease states result from the failure of negative-feedback mechanisms to maintain homeostasis and (2) some positive-feedback mechanisms

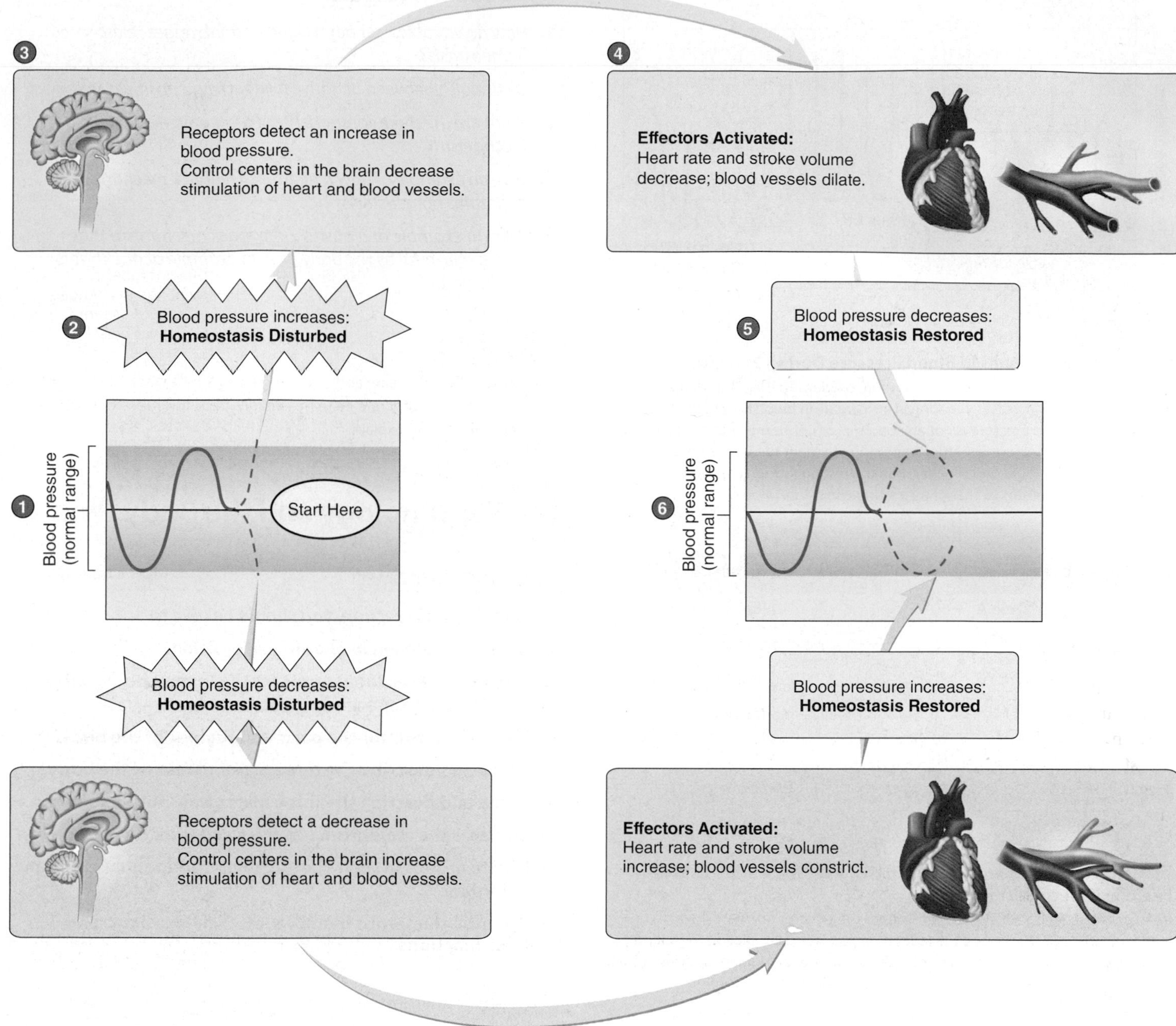
3
Receptors detect an increase in blood pressure.
Control centers in the brain decrease stimulation of heart and blood vessels.
4
Effectors Activated:
Heart rate and stroke volume decrease; blood vessels dilate.
2
Blood pressure increases:
Homeostasis Disturbed
5
Blood pressure decreases:
Homeostasis Restored
1
Blood pressure (normal range)
Start Here
6
Blood pressure (normal range)
Blood pressure decreases:
Homeostasis Disturbed
Blood pressure increases:
Homeostasis Restored
Receptors detect a decrease in blood pressure.
Control centers in the brain increase stimulation of heart and blood vessels.
Effectors Activated:
Heart rate and stroke volume increase; blood vessels constrict.

HOMEOSTASIS PROCESS FIGURE 1.6 Negative-Feedback Control of Blood Pressure

Throughout this book, all homeostasis figures have the same format as shown here. The changes caused by the increase of a variable outside the normal range are shown in the *green boxes,* and the changes caused by a decrease are shown in the *red boxes.* To help you learn how to interpret homeostasis figures, some of the steps in this figure are numbered: (1) Blood pressure is within its normal range. (2) Blood pressure increases outside the normal range; this change is detected by receptors and causes homeostasis to be disturbed. (3) The blood pressure control center in the brain responds to the change in blood pressure. (4) The control center activates the effectors: heart rate and stroke volume decrease and blood vessels dilate. (5) These changes cause blood pressure to decrease. (6) Blood pressure returns to its normal range, and homeostasis is restored. Observe the responses to a decrease in blood pressure outside its normal range by following the *red arrows.*

can be detrimental instead of helpful. One example of a detrimental positive-feedback mechanism is inadequate delivery of blood to cardiac (heart) muscle. Contraction of cardiac muscle generates blood pressure and the heart pumps blood to itself through a system of blood vessels on the outside of the heart. Just as with other tissues, blood pressure must be maintained to ensure adequate delivery of blood to the cardiac muscle. Following extreme blood loss, blood pressure decreases to the point that the delivery of blood to cardiac muscle is inadequate. As a result, cardiac muscle does not function normally. The heart pumps less blood, which

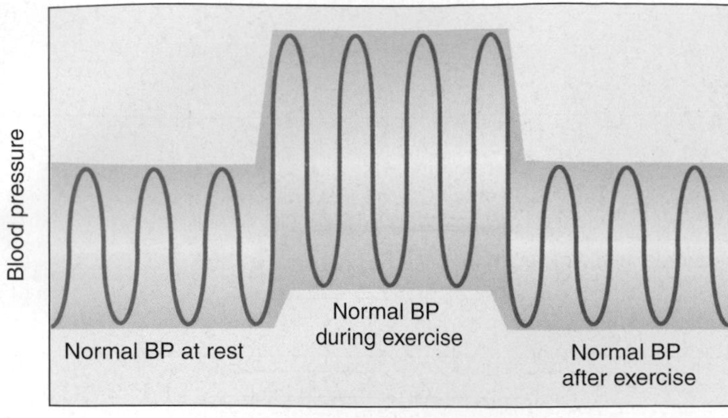
Blood pressure
Normal BP at rest
Normal BP during exercise
Normal BP after exercise
Time

FIGURE 1.7 Changes in Blood Pressure During Exercise
During exercise, muscle tissue demands more oxygen. To meet this demand, blood pressure (BP) increases, resulting in an increase in blood flow to the tissues. The increased blood pressure is not an abnormal or nonhomeostatic condition but a resetting of the normal homeostatic range to meet the increased demand. The reset range is higher and broader than the resting range. After exercise ceases, the range returns to that of the resting condition.

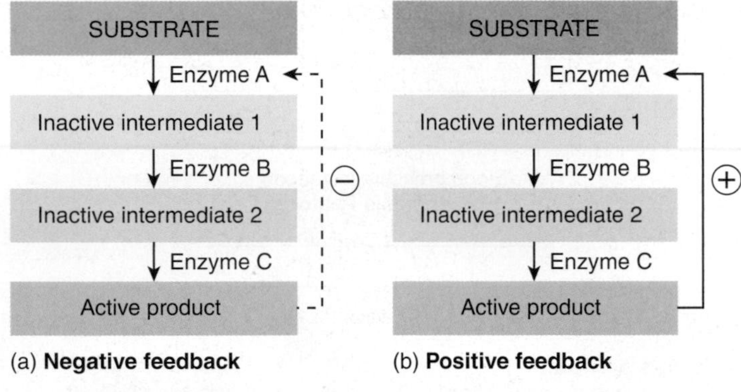
SUBSTRATE
Enzyme A
Inactive intermediate 1
Enzyme B
Inactive intermediate 2
Enzyme C
Active product
(a) Negative feedback
SUBSTRATE
Enzyme A
Inactive intermediate 1
Enzyme B
Inactive intermediate 2
Enzyme C
Active product
(b) Positive feedback

FIGURE 1.8 Comparison of Negative-Feedback and Positive-Feedback Mechanisms
(*a*) In negative feedback, the response *stops* the effector. (*b*) In positive feedback, the response keeps the reaction going. For example, during blood clotting, the "active product" represents thrombin, which triggers "enzyme A," the first step in the cascade that leads to the production of thrombin.

causes the blood pressure to drop even further—a deviation further from the setpoint. The additional decrease in blood pressure further reduces blood delivery to cardiac muscle, and the heart pumps even less blood, which again decreases the blood pressure. The process self-propagates until the blood pressure is too low to sustain the cardiac muscle, the heart stops beating, and death results. In this example, we see the deviation from the heart rate set point becoming larger and larger—this is a positive-feedback mechanism. Thus, if blood loss is severe, negative-feedback mechanisms may not be able to maintain homeostasis, and the postive feedback of ever-decreasing blood pressure can develop. On the other hand, following a moderate amount of blood loss (e.g., after donating a pint of blood), *negative-feedback mechanisms* result in an *increase* in heart rate, which restores blood pressure.

ASSESS YOUR PROGRESS

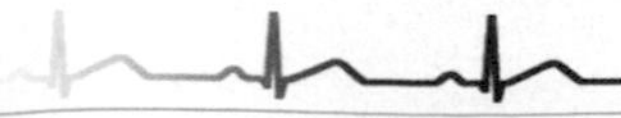

11. *How do variables, set points, and normal ranges relate to homeostasis?*
12. *Distinguish between negative feedback and positive feedback.*
13. *What are the three components of a negative-feedback mechanism?*
14. *Give an example of how a negative-feedback mechanism maintains homeostasis.*
15. *Give an example of a positive-feedback mechanism that may be harmful to the body and an example of one that is not harmful.*

Predict 4

Ashley is on the track team and is running an 800-meter race. Throughout the race, her respiratory rate increases rapidly. Does this represent negative or positive feedback? Explain.

1.6 Terminology and the Body Plan

LEARNING OUTCOMES

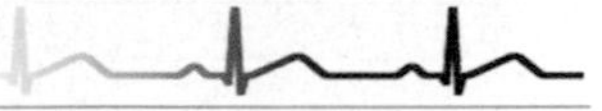

After reading this section, you should be able to

A. Describe a person in anatomical position.
B. Define the directional terms for the human body, and use them to locate specific body structures.
C. Know the terms for the parts and regions of the body.
D. Name and describe the three major planes of the body.
E. Name and describe the three major ways to cut an organ.
F. Describe the major trunk cavities and their divisions.
G. Locate organs in their specific cavity, abdominal quadrant, or region.
H. Describe the serous membranes, their locations, and their functions.

As you study anatomy and physiology, you will be learning many new words. Knowing the derivation, or **etymology** (et′uh-mol′ŏ-jē), of these words can make learning them easy and fun. Most anatomical terms are derived from Latin or Greek. For example, *foramen* is a Latin word for "hole," and *magnum* means "large." The foramen magnum is therefore a large hole in the skull through which the spinal cord attaches to the brain.

Prefixes and suffixes can be added to words to expand their meaning. For example, the suffix *-itis* means an inflammation, so appendicitis is an inflammation of the appendix. As new terms are introduced in this text, their meanings are often explained. The glossary and the list of word roots, prefixes, and suffixes on the inside back cover provide additional information about the new terms.

It is very helpful to learn these new words, so that your message is clear and correct when you speak to colleagues or write reports.

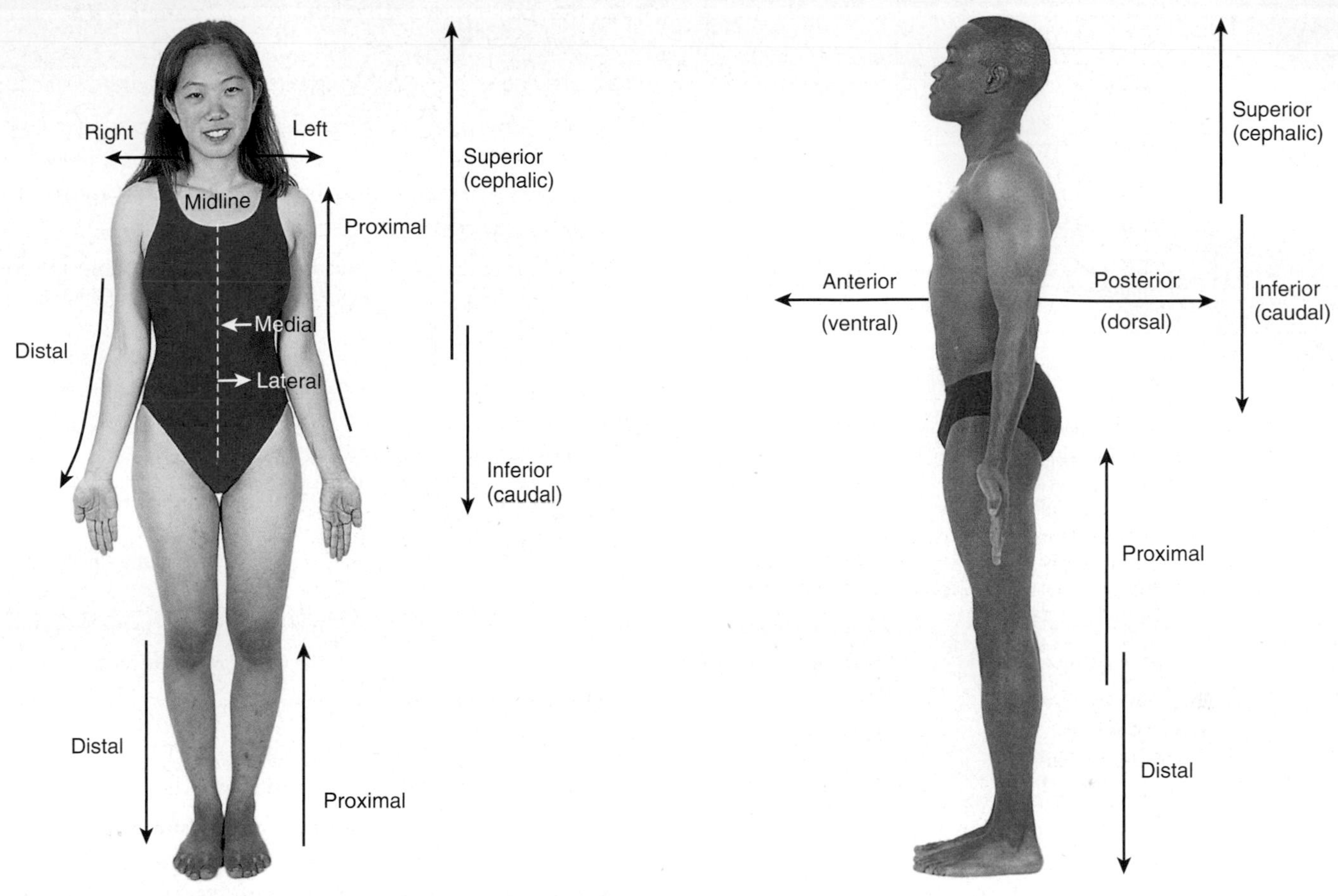
Right
Left
Midline
Proximal
Superior
(cephalic)
Medial
Lateral
Distal
Inferior
(caudal)
Distal
Proximal
Superior
(cephalic)
Anterior
(ventral)
Posterior
(dorsal)
Inferior
(caudal)
Proximal
Distal

FIGURE 1.9 APIR **Directional Terms**

All directional terms are in relation to the body in the anatomical position: a person standing erect with the face directed forward, the arms hanging to the sides, and the palms of the hands facing forward.

Body Positions

Anatomical position refers to a person standing erect with the face directed forward, the upper limbs hanging to the sides, and the palms of the hands facing forward (figure 1.9). A person is **supine** when lying face upward and **prone** when lying face downward.

The position of the body can affect the description of body parts relative to each other. In the anatomical position, the elbow is above the hand but, in the supine or prone position, the elbow and hand are at the same level. To avoid confusion, relational descriptions are always based on the anatomical position, no matter the actual position of the body. Thus, the elbow is always described as being "above" (superior to) the wrist, whether the person is lying down or is even upside down.

Directional Terms

Directional terms describe parts of the body relative to each other. Important directional terms are illustrated in figure 1.9 and summarized in table 1.2. It is important to become familiar with these directional terms as soon as possible because you will see them repeatedly throughout this text. **Right** and **left** are retained as directional terms in anatomical terminology. *Up* is replaced by **superior,** *down* by **inferior,** *front* by **anterior,** and *back* by **posterior.**

In humans, *superior* is synonymous with **cephalic** (se-fal′ik), which means toward the head, because, when we are in the anatomical position, the head is the highest point. In humans, the term *inferior* is synonymous with **caudal** (kaw′dăl), which means toward the tail, which would be located at the end of the vertebral column if humans had tails. The terms *cephalic* and *caudal* can be used to describe directional movements on the trunk, but they are not used to describe directional movements on the limbs.

The word *anterior* means "that which goes before," and **ventral** means "belly." The anterior surface of the human body is therefore the ventral surface, or belly, because the belly "goes first" when we are walking. The word *posterior* means "that which follows," and **dorsal** means "back." The posterior surface of the body is the dorsal surface, or back, which follows as we are walking.

TABLE 1.2 Directional Terms for Humans

Terms	Etymology*	Definition	Examples
Right		Toward the right side of the body	Right ear
Left		Toward the left side of the body	Left eye
Superior	L. higher	A structure above another	The chin is superior to the navel.
Inferior	L. lower	A structure below another	The navel is inferior to the chin.
Cephalic	G. *kephale,* head	Closer to the head than another structure (usually synonymous with *superior*)	The chin is cephalic to the navel.
Caudal	L. *cauda,* a tail	Closer to the tail than another structure (usually synonymous with *inferior*)	The navel is caudal to the chin.
Anterior	L. before	The front of the body	The navel is anterior to the spine.
Posterior	L. *posterus,* following	The back of the body	The spine is posterior to the breastbone.
Ventral	L. *ventr-,* belly	Toward the belly (synonymous with *anterior*)	The navel is ventral to the spine.
Dorsal	L. *dorsum,* back	Toward the back (synonymous with *posterior*)	The spine is dorsal to the breastbone.
Proximal	L. *proximus,* nearest	Closer to the point of attachment to the body than another structure	The elbow is proximal to the wrist.
Distal	L. *di-* plus *sto,* to stand apart or be distant	Farther from the point of attachment to the body than another structure	The wrist is distal to the elbow.
Lateral	L. *latus,* side	Away from the midline of the body	The nipple is lateral to the breastbone.
Medial	L. *medialis,* middle	Toward the midline of the body	The nose is medial to the eye.
Superficial	L. *superficialis,* toward the surface	Toward or on the surface (not shown in figure 1.10)	The skin is superficial to muscle.
Deep	O.E. *deop,* deep	Away from the surface, internal (not shown in figure 1.10)	The lungs are deep to the ribs.

*Origin and meaning of the word: L., Latin; G., Greek; O.E., Old English.

Predict 5

The anatomical position of a cat refers to the animal standing erect on all four limbs and facing forward. On the basis of the etymology of the directional terms, which two terms indicate movement toward the cat's head? What two terms mean movement toward the cat's back? Compare these terms with those referring to a human in the anatomical position.

Proximal means "nearest," whereas **distal** means "distant." These terms are used to refer to linear structures, such as the limbs, in which one end is near another structure and the other end is farther away. Each limb is attached at its proximal end to the body, and the distal end, such as the hand, is farther away.

Medial means "toward the midline," and **lateral** means "away from the midline." The nose is in a medial position in the face, and the eyes are lateral to the nose. **Superficial** describes a structure close to the surface of the body, and **deep** is toward the interior of the body. The skin is superficial to muscle and bone.

ASSESS YOUR PROGRESS

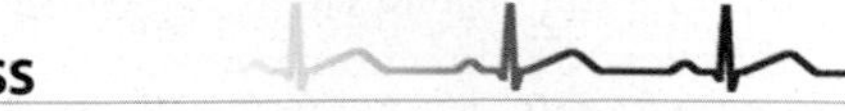

16. *What is* anatomical position *in humans? Why is it important?*

17. *What two directional terms indicate "toward the head" in humans? What are the opposite terms?*

18. *What two directional terms indicate "the back" in humans? What are the opposite terms?*

19. *Define the following directional terms and give the term that means the opposite:* proximal, lateral, *and* superficial.

Predict 6

Use as many directional terms as you can to describe the relationship between your kneecap and your heel.

Body Parts and Regions

Health professionals use a number of terms when referring to different parts or regions of the body. Figure 1.10 shows the anatomical terms, with the common terms in parentheses.

The central region of the body consists of the **head, neck,** and **trunk.** The trunk can be divided into the **thorax** (chest), **abdomen** (region between the thorax and pelvis), and **pelvis** (the inferior end of the trunk associated with the hips). The upper limb is divided into the arm, forearm, wrist, and hand. The **arm** extends from the shoulder to the elbow, and the **forearm** extends from the elbow to the wrist. The lower limb is divided into the thigh, leg, ankle, and foot. The **thigh** extends from the hip to the knee, and the **leg** extends from the knee to the ankle. Note that, contrary to popular usage, the terms *arm* and *leg* refer to only a part of the respective limb.

The abdomen is often subdivided superficially into **quadrants** by two imaginary lines—one horizontal and one vertical—that intersect at the navel (figure 1.11*a*). The quadrants formed are the right-upper, left-upper, right-lower, and left-lower quadrants. In addition to these quadrants, the abdomen is sometimes subdivided into **regions** by four imaginary lines: two horizontal and two vertical. These four lines create a "virtual" tic-tac-toe grid on the abdomen, resulting in nine regions: epigastric, right and

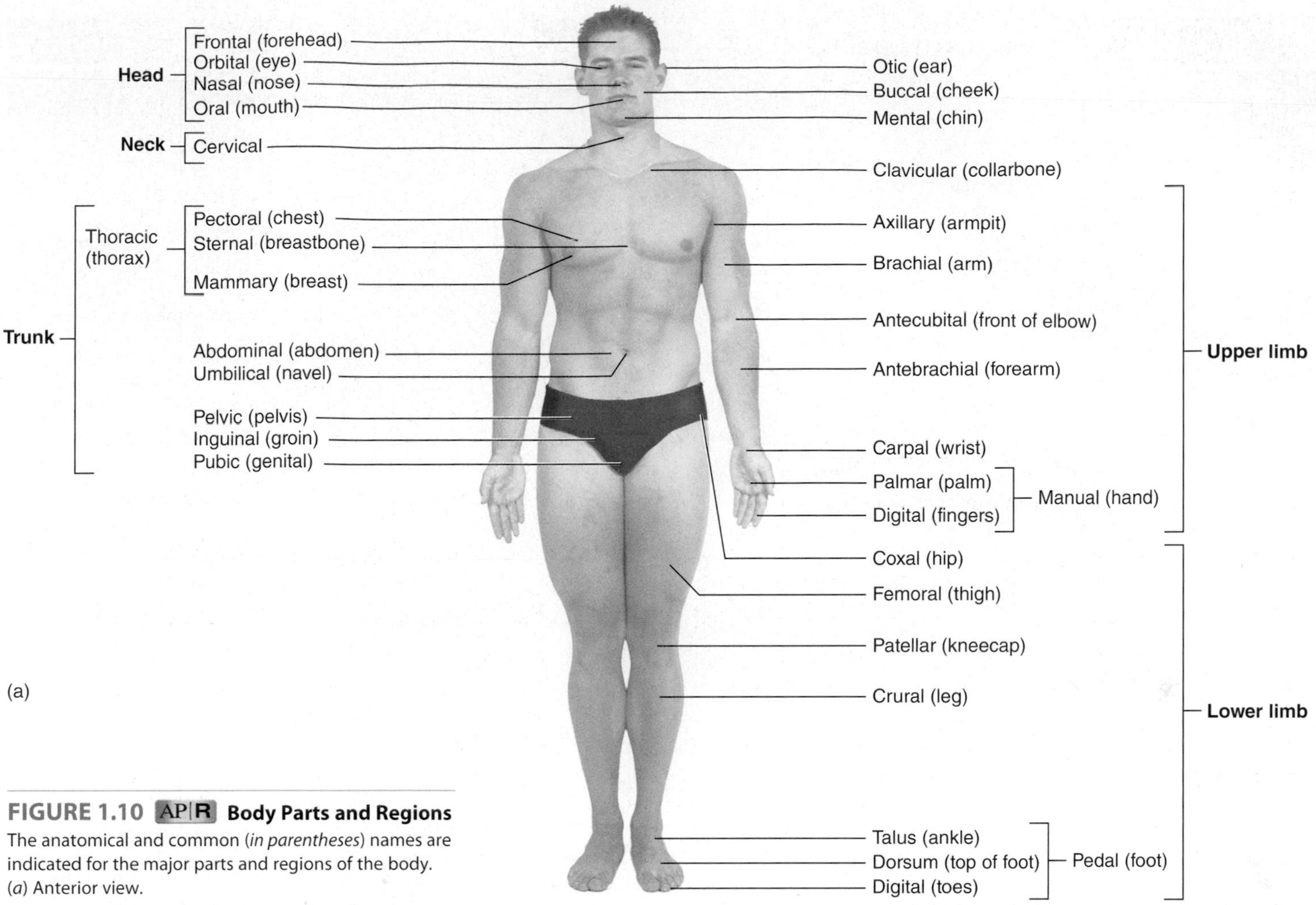

Head
Frontal (forehead)
Orbital (eye)
Nasal (nose)
Oral (mouth)
Neck
Cervical
Trunk
Thoracic (thorax)
Pectoral (chest)
Sternal (breastbone)
Mammary (breast)
Abdominal (abdomen)
Umbilical (navel)
Pelvic (pelvis)
Inguinal (groin)
Pubic (genital)
Otic (ear)
Buccal (cheek)
Mental (chin)
Clavicular (collarbone)
Axillary (armpit)
Brachial (arm)
Antecubital (front of elbow)
Antebrachial (forearm)
Carpal (wrist)
Palmar (palm)
Digital (fingers)
Manual (hand)
Upper limb
Coxal (hip)
Femoral (thigh)
Patellar (kneecap)
Crural (leg)
Talus (ankle)
Dorsum (top of foot)
Digital (toes)
Pedal (foot)
Lower limb
(a)

FIGURE 1.10 APIR **Body Parts and Regions**
The anatomical and common (*in parentheses*) names are indicated for the major parts and regions of the body. (*a*) Anterior view.

left hypochondriac, umbilical, right and left lumbar, hypogastric, and right and left iliac (figure 1.11*b*). Health professionals use the quadrants and regions as reference points for locating underlying organs. For example, the appendix is in the right-lower quadrant, and the pain of an acute appendicitis is usually felt there.

Planes

At times, it is useful to describe the body as having imaginary flat surfaces, called **planes,** passing through it (figure 1.12). A plane divides, or sections, the body, making it possible to "look inside" and observe the body's structures. A **sagittal** (saj′i-tăl) **plane** runs vertically through the body, separating it into right and left portions. The word *sagittal* literally means "the flight of an arrow" and refers to the way the body would be split by an arrow passing anteriorly to posteriorly. A **median plane** is a sagittal plane that passes through the midline of the body, dividing it into equal right and left halves. A **transverse** (**horizontal**) **plane** runs parallel to the ground, dividing the body into superior and inferior portions. A **frontal** (**coronal**) (kōr′ŏ-năl, kō-rō′năl; crown) **plane** runs vertically from right to left and divides the body into anterior and posterior parts.

Organs are often sectioned to reveal their internal structure (figure 1.13). A cut through the long axis of the organ is a **longitudinal section,** and a cut at right angles to the long axis is a **transverse** (**cross**) **section.** If a cut is made across the long axis at other than a right angle, it is called an **oblique section.**

ASSESS **YOUR PROGRESS**

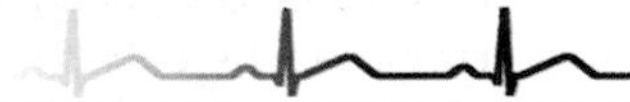

20. *What makes up the central region of the body?*
21. *What is the difference between the arm and the upper limb? Between the leg and the lower limb?*
22. *What are the anatomical terms for the following common body terms—neck, mouth, hand, front of elbow, calf, sole?*
23. *In what quadrant would the majority of the stomach be located? In which region(s)?*
24. *List and describe the three planes of the body.*
25. *In what three ways can you cut an organ?*

Body Cavities

The body contains many cavities. Some of these cavities, such as the nasal cavity, open to the outside of the body, and some do not. The trunk contains three large cavities that do not open to the outside of the body: the thoracic, the abdominal, and the pelvic

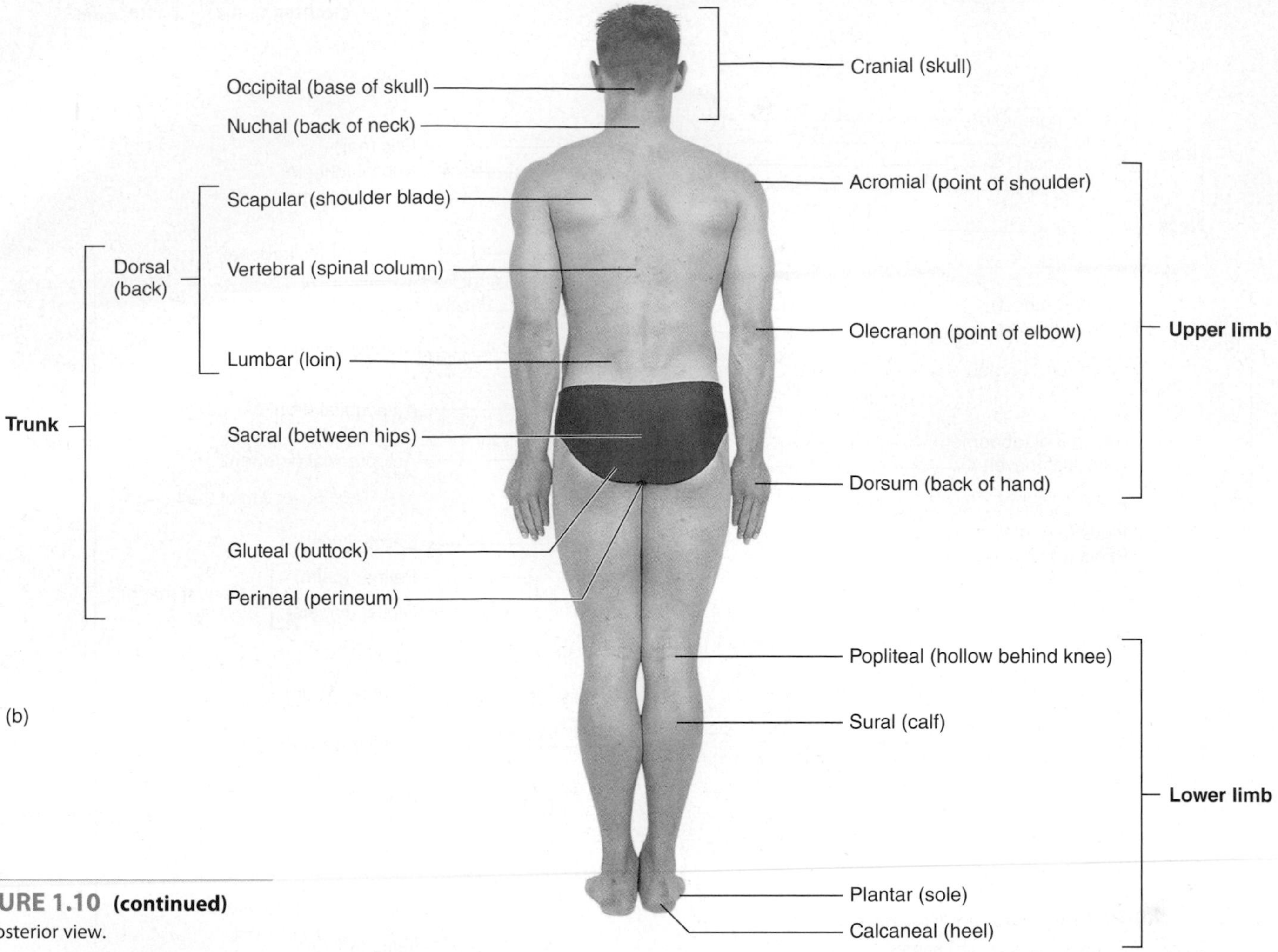

Cranial (skull)
Occipital (base of skull)
Nuchal (back of neck)
Acromial (point of shoulder)
Scapular (shoulder blade)
Dorsal (back)
Vertebral (spinal column)
Olecranon (point of elbow)
Upper limb
Lumbar (loin)
Trunk
Sacral (between hips)
Dorsum (back of hand)
Gluteal (buttock)
Perineal (perineum)
Popliteal (hollow behind knee)
(b)
Sural (calf)
Lower limb
Plantar (sole)
Calcaneal (heel)

FIGURE 1.10 (continued)
(*b*) Posterior view.

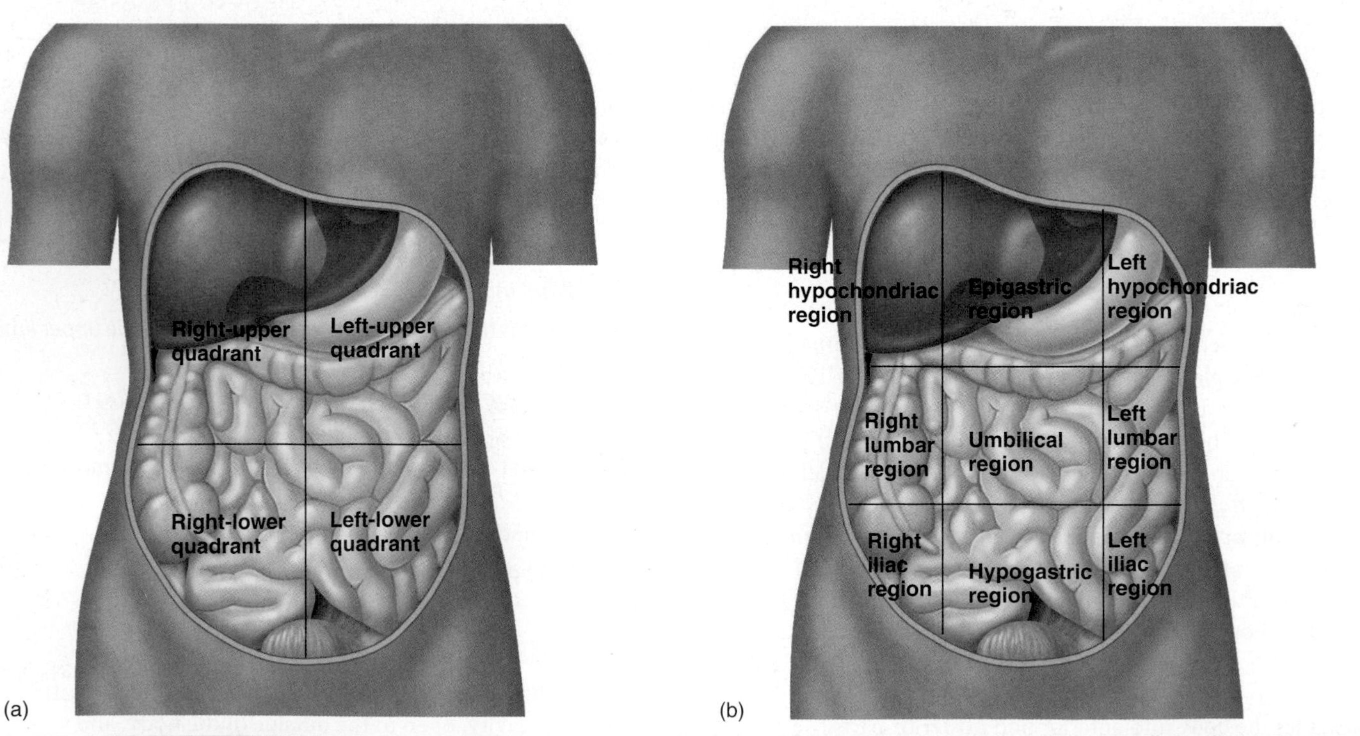
Right-upper quadrant
Left-upper quadrant
Right-lower quadrant
Left-lower quadrant
(a)
Right hypochondriac region
Epigastric region
Left hypochondriac region
Right lumbar region
Umbilical region
Left lumbar region
Right iliac region
Hypogastric region
Left iliac region
(b)

FIGURE 1.11 AP|R **Subdivisions of the Abdomen**
Lines are superimposed over internal organs to demonstrate the subdivisions they lie in. (*a*) Abdominal quadrants. (*b*) There are nine abdominal regions.

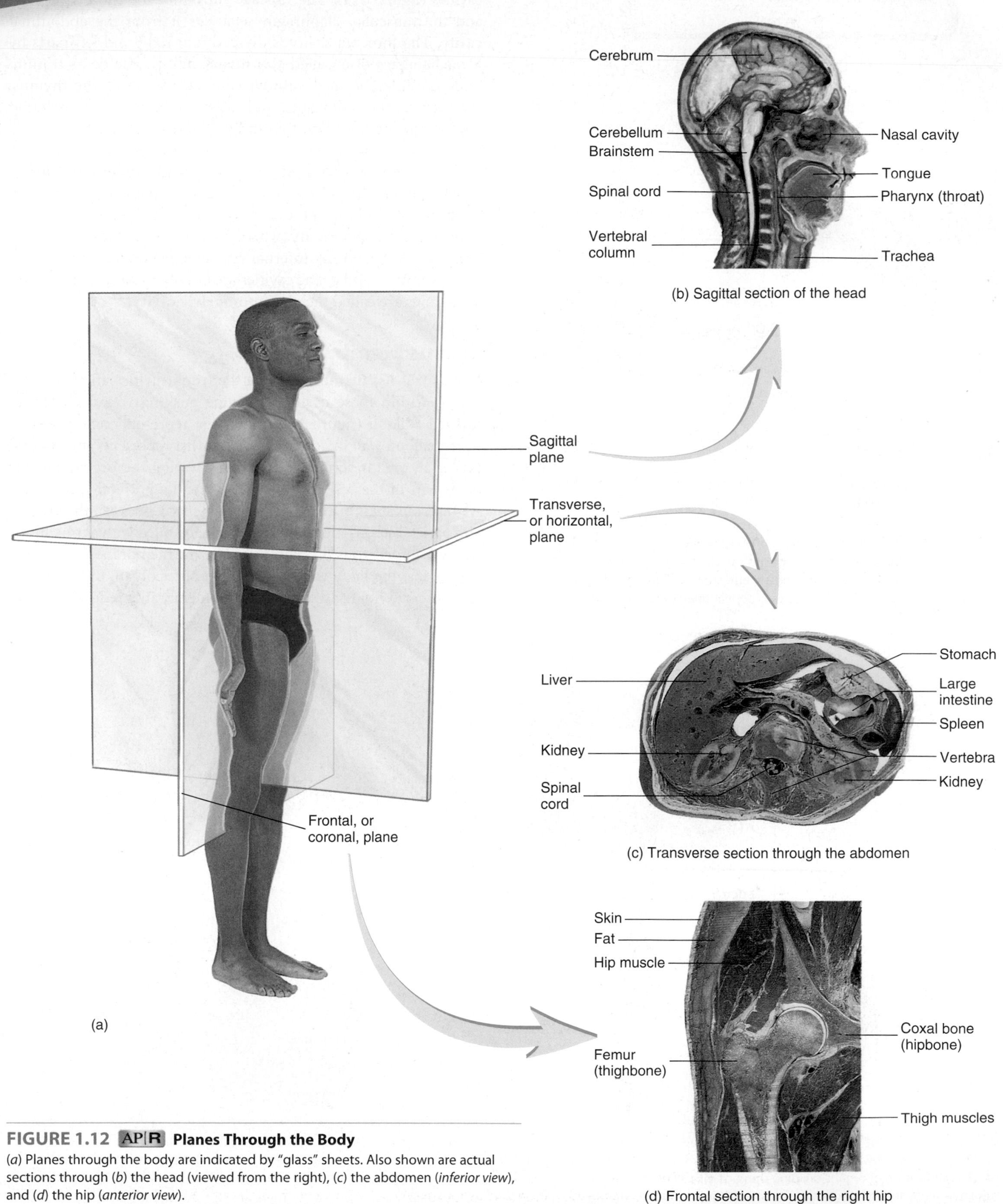

Cerebrum
Cerebellum
Brainstem
Spinal cord
Vertebral column
Nasal cavity
Tongue
Pharynx (throat)
Trachea
(b) Sagittal section of the head
Sagittal plane
Transverse, or horizontal, plane
Liver
Kidney
Spinal cord
Stomach
Large intestine
Spleen
Vertebra
Kidney
(c) Transverse section through the abdomen
Frontal, or coronal, plane
(a)
Skin
Fat
Hip muscle
Femur (thighbone)
Coxal bone (hipbone)
Thigh muscles
(d) Frontal section through the right hip

FIGURE 1.12 APR **Planes Through the Body**

(*a*) Planes through the body are indicated by "glass" sheets. Also shown are actual sections through (*b*) the head (viewed from the right), (*c*) the abdomen (*inferior view*), and (*d*) the hip (*anterior view*).

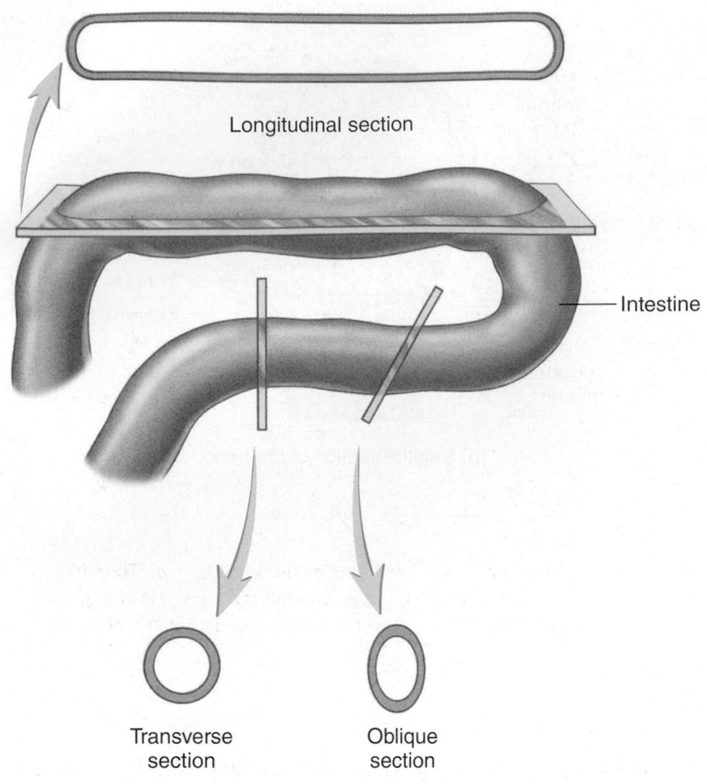
Longitudinal section
Intestine
Transverse section
Oblique section

FIGURE 1.13 Planes Through an Organ
Planes through the small intestine are indicated by "glass" sheets. The views of the small intestine after sectioning are also shown. Although the small intestine is basically a tube, the sections appear quite different in shape.

cavities (figure 1.14). The rib cage surrounds the **thoracic cavity,** and the muscular diaphragm separates it from the abdominal cavity. The thoracic cavity is divided into right and left parts by a median partition called the **mediastinum** (mē′dē-as-tī′nŭm; middle wall). The mediastinum contains the heart, the thymus, the trachea, the esophagus, and other structures, such as blood vessels and nerves. The two lungs are located on each side of the mediastinum.

Abdominal muscles primarily enclose the **abdominal cavity,** which contains the stomach, the intestines, the liver, the spleen, the pancreas, and the kidneys. Pelvic bones encase the small space known as the **pelvic cavity,** where the urinary bladder, part of the large intestine, and the internal reproductive organs are housed. The abdominal and pelvic cavities are not physically separated and sometimes are called the **abdominopelvic cavity.**

Serous Membranes

Serous (sēr′ŭs) **membranes** line the trunk cavities and cover the organs within these cavities. Imagine pushing your fist into an inflated balloon (figure 1.15). Your fist represents an organ; the inner balloon wall in contact with your fist represents the **visceral** (vis′er-ăl; organ) **serous membrane** covering the organ; and the outer part of the balloon wall represents the **parietal** (pă-rī′ĕ-tăl; wall) **serous membrane.** The cavity, or space, between the visceral and parietal serous membranes is normally filled with a thin, lubricating film of serous fluid produced by the membranes. As organs rub against the body wall or against another organ, the combination of serous fluid and smooth serous membranes reduces friction.

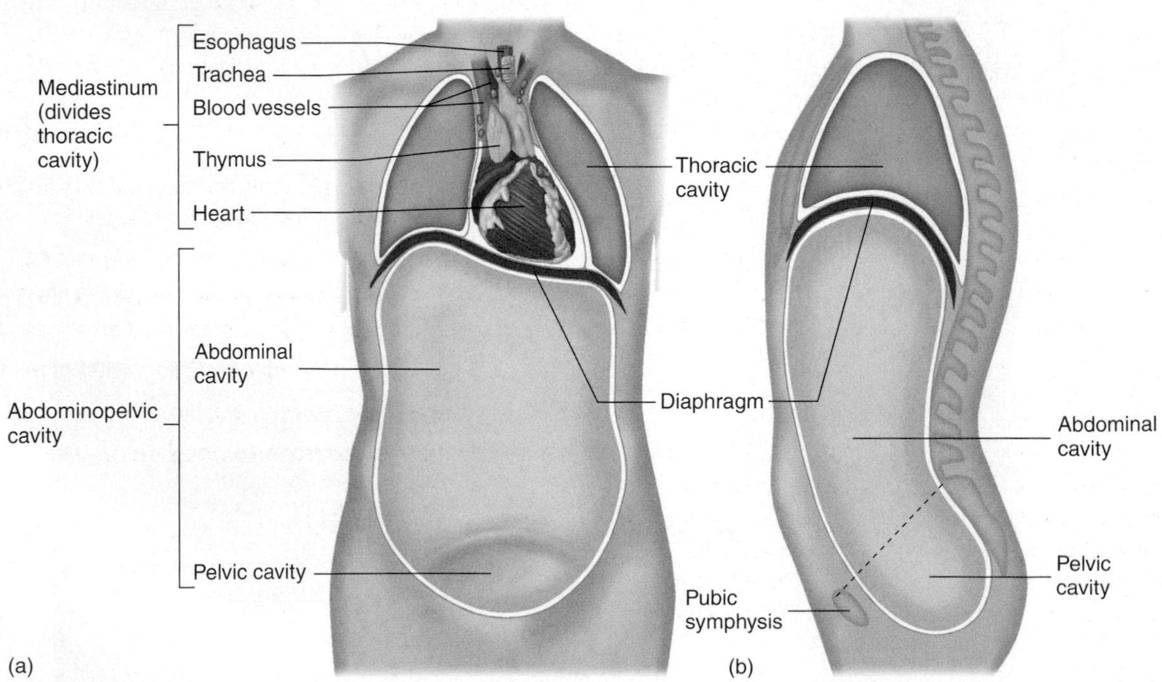
Mediastinum (divides thoracic cavity)
Esophagus
Trachea
Blood vessels
Thymus
Heart
Abdominopelvic cavity
Abdominal cavity
Pelvic cavity
Thoracic cavity
Diaphragm
Pubic symphysis
Abdominal cavity
Pelvic cavity
(a)
(b)

FIGURE 1.14 AP|R Trunk Cavities
(*a*) Anterior view showing the major trunk cavities. The diaphragm separates the thoracic cavity from the abdominal cavity. The mediastinum, which includes the heart, is a partition of organs dividing the thoracic cavity. (*b*) Sagittal section of the trunk cavities viewed from the left. The dashed line shows the division between the abdominal and pelvic cavities. The mediastinum has been removed to show the thoracic cavity.

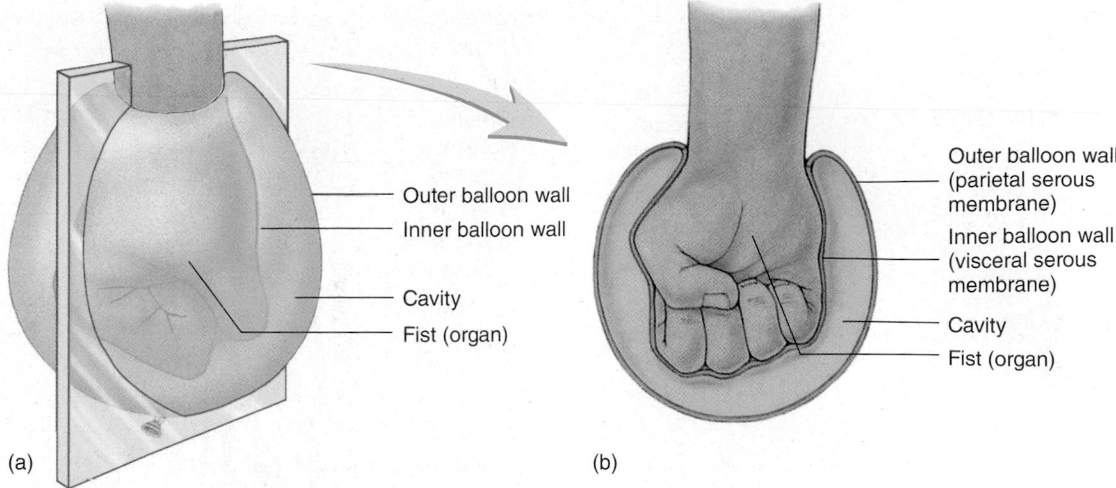
Outer balloon wall
Inner balloon wall
Cavity
Fist (organ)
(a)
Outer balloon wall
(parietal serous
membrane)
Inner balloon wall
(visceral serous
membrane)
Cavity
Fist (organ)
(b)

FIGURE 1.15 Serous Membranes

(*a*) A fist pushing into a balloon. A "glass" sheet indicates the location of a section through the balloon. (*b*) Interior view produced by the section in (*a*). The fist represents an organ, and the walls of the balloon represent the serous membranes. The inner wall of the balloon represents a visceral serous membrane in contact with the fist (organ). The outer wall of the balloon represents a parietal serous membrane. Figure 1.16 shows the relationship of the parietal and visceral membranes to the heart.

The thoracic cavity contains three serous membrane–lined cavities: a pericardial cavity and two pleural cavities. The **pericardial** (per-i-kar′dē-ăl; around the heart) **cavity** surrounds the heart (figure 1.16*a*). The visceral pericardium covers the heart, which is contained within a connective tissue sac lined with the parietal pericardium. The pericardial cavity, which contains pericardial fluid, is located between the visceral pericardium and the parietal pericardium.

Each lung is covered by visceral pleura and surrounded by a **pleural** (ploor′ ăl; associated with the ribs) **cavity** (figure 1.16*b*). Parietal pleura line the inner surface of the thoracic wall, the outer surface of the parietal pericardium, and the superior surface of the diaphragm. The pleural cavity lies between the visceral pleura and the parietal pleura and contains pleural fluid.

The abdominopelvic cavity contains a serous membrane–lined cavity called the **peritoneal** (per′i-tō-nē′ăl; to stretch over) **cavity** (figure 1.16*c*). Visceral peritoneum covers many of the organs of the abdominopelvic cavity. Parietal peritoneum lines the wall of the abdominopelvic cavity and the inferior surface of the diaphragm. The peritoneal cavity is located between the visceral peritoneum and the parietal peritoneum and contains peritoneal fluid.

The serous membranes can become inflamed, usually as a result of an infection. **Pericarditis** (per′i-kar-dī′tis) is inflammation of the pericardium, **pleurisy** (ploor′i-sē) is inflammation of the pleura, and **peritonitis** (per′i-tō-nī′tis) is inflammation of the peritoneum.

Mesenteries (mes′en-ter-ēz), which consist of two layers of peritoneum fused together (figure 1.16*c*), connect the visceral peritoneum of some abdominopelvic organs to the parietal peritoneum on the body wall or to the visceral peritoneum of other abdominopelvic organs. The mesenteries anchor the organs to the body wall and provide a pathway for nerves and blood vessels to reach the organs. Other abdominopelvic organs are more closely attached to the body wall and do not have mesenteries. Parietal peritoneum covers these other organs, which are said to be **retroperitoneal** (re′trō-per′i-tō-nē′ăl; behind the peritoneum). The retroperitoneal organs include the kidneys, the adrenal glands, the pancreas, parts of the intestines, and the urinary bladder (figure 1.16*c*).

ASSESS YOUR PROGRESS

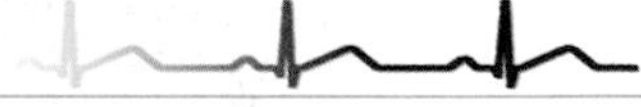

26. *What structure separates the thoracic cavity from the abdominal cavity? The abdominal cavity from the pelvic cavity?*
27. *What structure divides the thoracic cavity into right and left parts?*
28. *What is a serous membrane and its function? Differentiate between the parietal and visceral portions of a serous membrane.*
29. *Name the serous membrane–lined cavities of the trunk.*
30. *What are mesenteries? Explain their function.*
31. *What are retroperitoneal organs? List five examples.*

Predict 7

Explain how an organ can be located within the abdominopelvic cavity but not be within the peritoneal cavity.

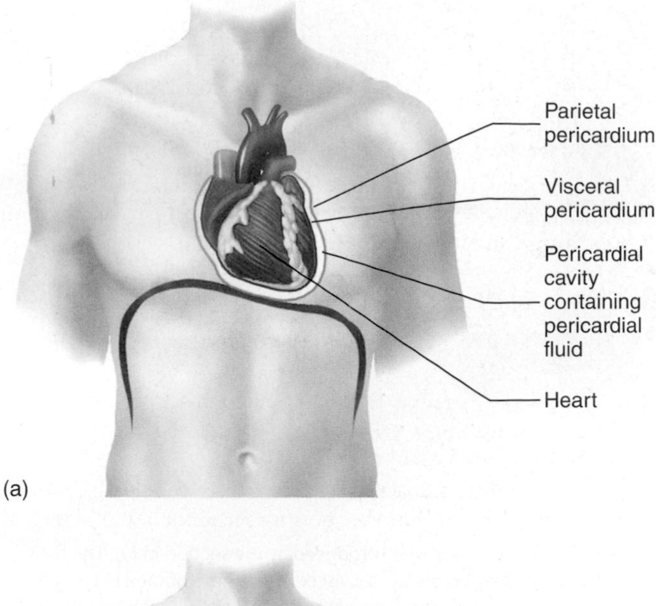
Parietal pericardium
Visceral pericardium
Pericardial cavity containing pericardial fluid
Heart
(a)

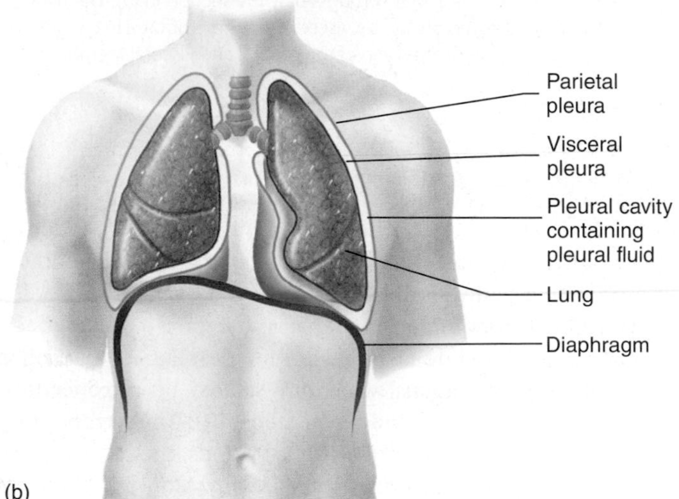
Parietal pleura
Visceral pleura
Pleural cavity containing pleural fluid
Lung
Diaphragm
(b)

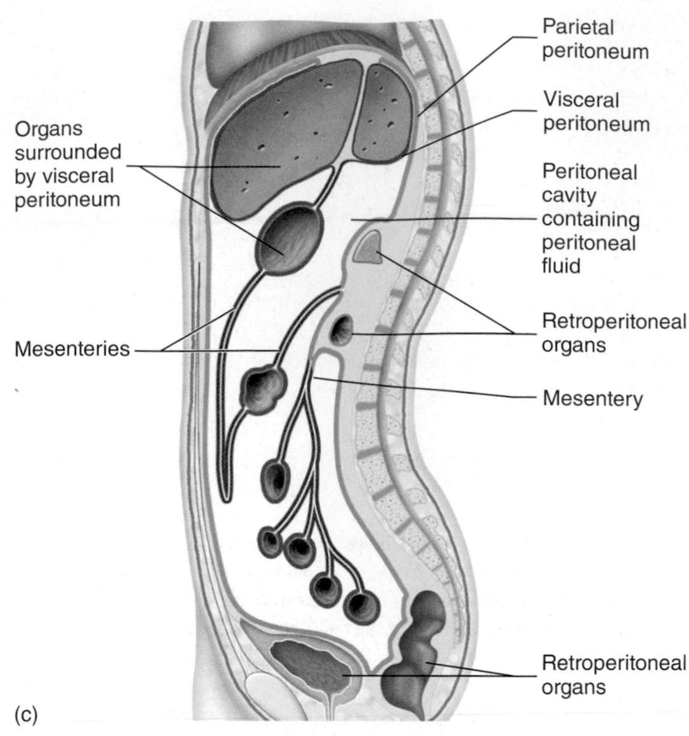
Parietal peritoneum
Visceral peritoneum
Organs surrounded by visceral peritoneum
Peritoneal cavity containing peritoneal fluid
Retroperitoneal organs
Mesenteries
Mesentery
Retroperitoneal organs
(c)

FIGURE 1.16 Location of Serous Membranes

(*a*) Frontal section showing the parietal pericardium (*blue*), the visceral pericardium (*red*), and the pericardial cavity. (*b*) Frontal section showing the parietal pleura (*blue*), the visceral pleura (*red*), and the pleural cavities. (*c*) Sagittal section through the abdominopelvic cavity showing the parietal peritoneum (*blue*), the visceral peritoneum (*red*), the peritoneal cavity, mesenteries (*purple*), and parts of the retroperitoneal organs.

Learn to Predict ❮ Answer

From page 1

The first Predict in every chapter of this text is designed to help you develop the skills to successfully answer critical thinking questions. The first step in the process is always to analyze the question itself. In this case, the question asks you to evaluate the mechanisms governing Renzo's blood sugar levels, and it provides the clue that there's a homeostatic mechanism involved. In addition, the question describes a series of events that help create an explanation: Renzo doesn't feel satisfied after eating, has elevated blood sugar, and then is prescribed an insulin pump. In chapter 1, we learn that homeostasis is the existence and maintenance of a relatively constant internal environment. Renzo experienced hunger despite eating, and his blood sugar levels were higher than normal. In this situation, we saw a disruption in homeostasis because his blood sugar stayed too high after eating. Normally, increased blood sugar after a meal would return to the normal range by the activity of insulin secreted by the pancreas. When blood sugar returns to normal, insulin secretion stops. In Renzo's case, his pancreas has stopped making insulin. Thus, the doctor prescribed an insulin pump to take over for his pancreas. Now when Renzo eats, the insulin pump puts insulin into his blood and his blood sugar levels are maintained near the set point.

Answers to the rest of this chapter's Predict questions are in Appendix G.

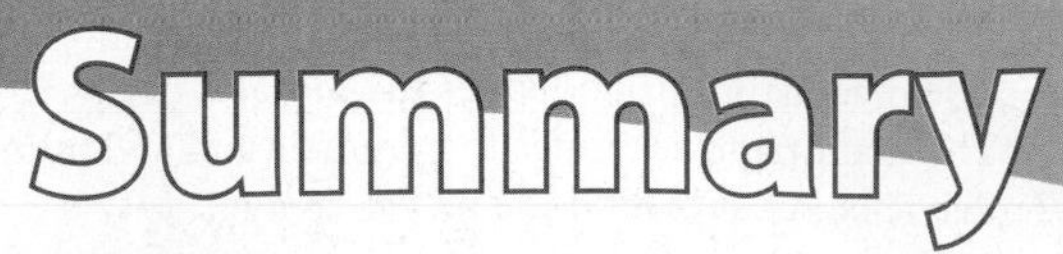
Summary

Knowledge of anatomy and physiology can be used to solve problems concerning the body when healthy or diseased.

1.1 Anatomy and Physiology (p. 2)

1. Anatomy is the study of the body's structures.
   - Developmental anatomy considers anatomical changes from conception to adulthood. Embryology focuses on the first 8 weeks of development.
   - Cytology examines cells, and histology examines tissues.
   - Gross anatomy studies organs from either a systemic or a regional perspective.
2. Surface anatomy uses superficial structures to locate deeper structures, and anatomical imaging is a noninvasive technique for identifying deep structures.
3. Physiology is the study of the body's functions. It can be approached from a cellular or a systems point of view.
4. Pathology deals with all aspects of disease. Exercise physiology examines changes caused by exercise.

1.2 Structural and Functional Organization of the Human Body (p. 4)

1. Basic chemical characteristics are responsible for the structure and functions of life.
2. Cells are the basic structural and functional units of organisms, such as plants and animals. Organelles are small structures within cells that perform specific functions.
3. Tissues are composed of groups of cells of similar structure and function and the materials surrounding them. The four primary tissue types are epithelial, connective, muscle, and nervous tissues.
4. Organs are structures composed of two or more tissues that perform specific functions.
5. Organs are arranged into the 11 organ systems of the human body (see figure 1.3).
6. Organ systems interact to form a whole, functioning organism.

1.3 Characteristics of Life (p. 4)

Humans share many characteristics with other organisms, such as organization, metabolism, responsiveness, growth, development, and reproduction.

1.4 Biomedical Research (p. 6)

Much of our knowledge about humans is derived from research on other organisms.

1.5 Homeostasis (p. 9)

Homeostasis is the condition in which body functions, body fluids, and other factors of the internal environment are maintained at levels suitable to support life.

Negative Feedback

1. Negative-feedback mechanisms maintain homeostasis.
2. Many negative-feedback mechanisms consist of a receptor, a control center, and an effector.

Positive Feedback

1. Positive-feedback mechanisms usually result in deviations further from the set point.
2. Although a few positive-feedback mechanisms are normal for maintaining homeostasis in the body, some positive-feedback mechanisms can be harmful.
3. Normal postive-feedback mechanisms include blood clotting and childbirth labor. Harmful positive-feedback examples include decreased blood flow to the heart.

1.6 Terminology and the Body Plan (p. 12)

Body Positions

1. A human standing erect with the face directed forward, the arms hanging to the sides, and the palms facing forward is in the anatomical position.
2. A person lying face upward is supine; a person lying face downward is prone.

Directional Terms

Directional terms always refer to the anatomical position, no matter what the actual position of the body (see table 1.2).

Body Parts and Regions

1. The body can be divided into a central region, consisting of the head, neck, and trunk, and the upper limbs and lower limbs.
2. Superficially, the abdomen can be divided into quadrants or into nine regions. These divisions are useful for locating internal organs or describing the location of a pain or a tumor.

Planes

1. Planes of the body
   - A sagittal plane divides the body into right and left parts. A median plane divides the body into equal right and left halves.
   - A transverse (horizontal) plane divides the body into superior and inferior portions.
   - A frontal (coronal) plane divides the body into anterior and posterior parts.
2. Sections of an organ
   - A longitudinal section of an organ divides it along the long axis.
   - A transverse (cross) section cuts at a right angle to the long axis of an organ.
   - An oblique section cuts across the long axis of an organ at an angle other than a right angle.

Body Cavities

1. The mediastinum subdivides the thoracic cavity.
2. The diaphragm separates the thoracic and abdominal cavities.
3. Pelvic bones surround the pelvic cavity.

Serous Membranes

1. Serous membranes line the trunk cavities. The parietal portion of a serous membrane lines the wall of the cavity, and the visceral portion is in contact with the internal organs.
   - The serous membranes secrete fluid, which fills the space between the visceral and parietal membranes. The serous membranes protect organs from friction.
   - The pericardial cavity surrounds the heart, the pleural cavities surround the lungs, and the peritoneal cavity surrounds certain abdominal and pelvic organs.
2. Mesenteries are parts of the peritoneum that hold the abdominal organs in place and provide a passageway for blood vessels and nerves to the organs.
3. Retroperitoneal organs are located "behind" the parietal peritoneum.

REVIEW AND COMPREHENSION

1. Physiology
   a. deals with the processes or functions of living things.
   b. is the scientific discipline that investigates the body's structures.
   c. is concerned with organisms and does not deal with levels of organization, such as cells and systems.
   d. recognizes the static (as opposed to the dynamic) nature of living things.
   e. can be used to study the human body without considering anatomy.
2. The following are organizational levels for considering the body.
   (1) cell
   (2) chemical
   (3) organ
   (4) organ system
   (5) organism
   (6) tissue

   Choose the correct order for these organizational levels, from simplest to most complex.
   a. 1,2,3,6,4,5
   b. 2,1,6,3,4,5
   c. 3,1,6,4,5,2
   d. 4,6,1,3,5,2
   e. 1,6,5,3,4,2

For questions 3–8, match each organ system with one of the following functions.

   a. regulates other organ systems
   b. removes waste products from the blood; maintains water balance
   c. regulates temperature; reduces water loss; provides protection
   d. removes foreign substances from the blood; combats disease; maintains tissue fluid balance
   e. produces movement; maintains posture; produces body heat

3. Endocrine system
4. Integumentary system
5. Muscular system
6. Nervous system
7. Urinary system
8. The characteristic of life that is defined as "all the chemical reactions taking place in an organism" is
   a. development.
   b. growth.
   c. metabolism.
   d. organization.
   e. responsiveness.
9. The following events are part of a negative-feedback mechanism.
   (1) Blood pressure increases.
   (2) The control center compares actual blood pressure to the blood pressure set point.
   (3) The heart beats faster.
   (4) Receptors detect a decrease in blood pressure.

   Choose the arrangement that lists the events in the order they occur.
   a. 1,2,3,4
   b. 1,3,2,4
   c. 3,1,4,2
   d. 4,2,3,1
   e. 4,3,2,1
10. Which of these statements concerning positive feedback is correct?
    a. Positive-feedback responses maintain homeostasis.
    b. Positive-feedback responses occur continuously in healthy individuals.
    c. Birth is an example of a normally occurring positive-feedback mechanism.
    d. When cardiac muscle receives an inadequate supply of blood, positive-feedback mechanisms increase blood flow to the heart.
    e. Medical therapy seeks to overcome illness by aiding positive-feedback mechanisms.
11. A term that means nearer the attached end of a limb is
    a. distal.
    b. lateral.
    c. medial.
    d. proximal.
    e. superficial.
12. Which of these directional terms are paired most appropriately as opposites?
    a. superficial and deep
    b. medial and proximal
    c. distal and lateral
    d. superior and posterior
    e. anterior and inferior
13. The part of the upper limb between the elbow and the wrist is called the
    a. arm.
    b. forearm.
    c. hand.
    d. inferior arm.
    e. lower arm.
14. A patient with appendicitis usually has pain in the ____________ quadrant of the abdomen.
    a. left-lower
    b. right-lower
    c. left-upper
    d. right-upper

15. A plane that divides the body into anterior and posterior parts is a
    a. frontal (coronal) plane.
    b. sagittal plane.
    c. transverse plane.
16. The lungs are
    a. part of the mediastinum.
    b. surrounded by the pericardial cavity.
    c. found within the thoracic cavity.
    d. separated from each other by the diaphragm.
    e. surrounded by mucous membranes.
17. Given the following organ and cavity combinations:
    (1) heart and pericardial cavity
    (2) lungs and pleural cavity
    (3) stomach and peritoneal cavity
    (4) kidney and peritoneal cavity

    Which of the organs is correctly paired with a space that surrounds that organ?
    a. 1,2 b. 1,2,3 c. 1,2,4 d. 2,3,4 e. 1,2,3,4
18. Which of the following membrane combinations are found on the superior and inferior surface of the diaphragm?
    a. parietal pleura—parietal peritoneum
    b. parietal pleura—visceral peritoneum
    c. visceral pleura—parietal peritoneum
    d. visceral pleura—visceral peritoneum
19. Which of the following organs is *not* retroperitoneal?
    a. adrenal glands
    b. urinary bladder
    c. kidneys
    d. pancreas
    e. stomach

Answers in Appendix E

CRITICAL THINKING

1. Exposure to a hot environment causes the body to sweat. The hotter the environment, the greater the sweating. Two anatomy and physiology students are arguing about the mechanisms involved. Student A claims that they are positive feedback, and student B claims they are negative feedback. Do you agree with student A or student B, and why?
2. A male has lost blood as a result of a gunshot wound. Even though the bleeding has been stopped, his blood pressure is low and dropping and his heart rate is elevated. Following a blood transfusion, his blood pressure increases and his heart rate decreases. Which of the following statement(s) is (are) consistent with these observations?
    a. Negative-feedback mechanisms can be inadequate without medical intervention.
    b. The transfusion interrupted a positive-feedback mechanism.
    c. The increased heart rate after the gunshot wound and before the transfusion is a result of a positive-feedback mechanism.
    d. a and b
    e. a, b, and c
3. Provide the correct directional term for the following statement: When a boy is standing on his head, his nose is ____________ to his mouth.
4. During pregnancy, which of the mother's body cavities increases most in size?
5. A woman falls while skiing and is accidentally impaled by her ski pole. The pole passes through the abdominal body wall and into and through the stomach, pierces the diaphragm, and finally stops in the left lung. List, in order, the serous membranes the pole pierces.

Answers in Appendix F

Visit this book's website at **www.mhhe.com/seeley10** for chapter quizzes, interactive learning exercises, and other study tools.

Mc Graw Hill
connect plus+
ANATOMY & PHYSIOLOGY

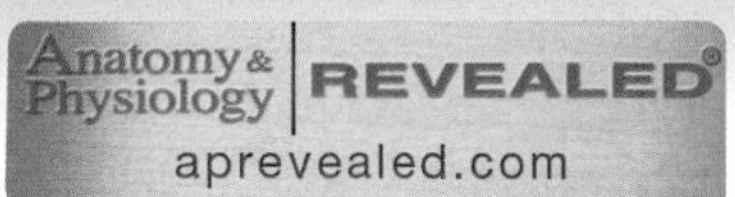
Anatomy & Physiology
REVEALED
aprevealed.com

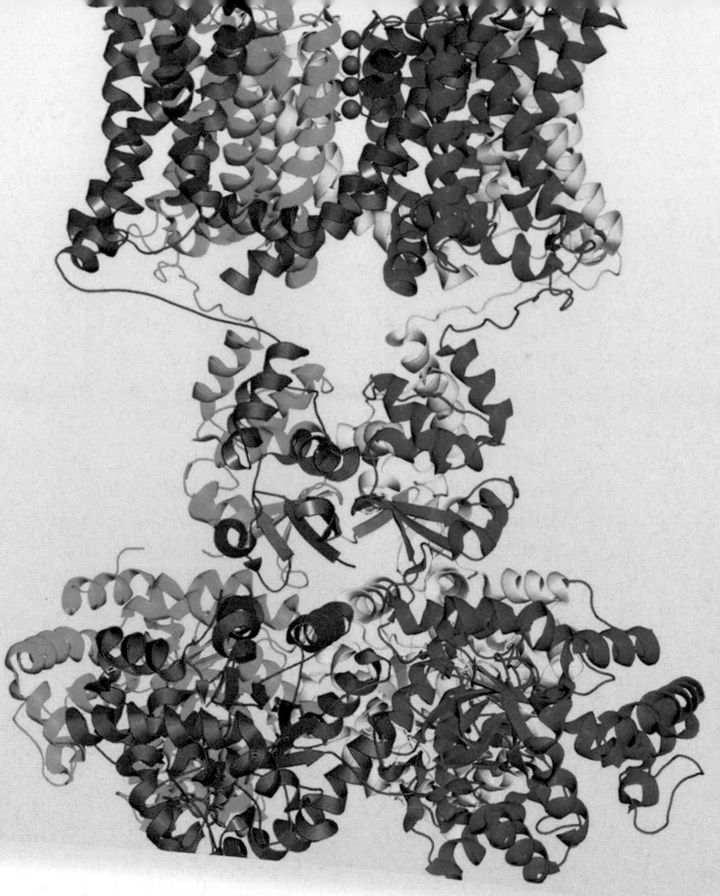

2

Learn to Predict

Brad and his lab partner, Angie, mixed a small amount of starch into a test tube of water. Then they added iodine, which stained the starch molecules blue. Next they added saliva to the test tube. After 30 minutes, the blue color disappeared. This exercise demonstrates the consequences of metabolism in the absence of homeostasis (described in chapter 1). Homeostasis often involves a balance of chemical reactions that will make and break molecules, such as starch. In the test tube, only one reaction occurred. After reading the chapter, you will have learned that the rate of a chemical reaction can be greatly increased by enzymes in cells and body fluids, and you will understand the roles of two chemical phenomena—activation energy and decomposition reactions. Considering the properties of enzymes and chemical reactions, can you explain why the blue color in the test tube disappeared?

The Chemical Basis of Life

Life is chemistry. Chemicals compose the structures of the body, and the interactions of chemicals with one another are responsible for the body's functions. Nerve impulse generation, digestion, muscle contraction, and metabolism can be described in chemical terms, and so can many abnormal conditions and illnesses, as well as their treatments. These diverse functions all involve intricate interactions between molecules. For example, chemicals in saliva interact with food to aid in digestion, and membrane proteins assemble to form a pore for ions to pass into and out of our cells to aid in nerve impulse generation and muscle contraction. To understand anatomy and physiology, it is essential to have a basic knowledge of chemistry—the scientific discipline concerned with the atomic composition of substances and the reactions they undergo. This chapter is not a comprehensive treatment of chemistry, but it does review some of the basic chemical concepts related to living systems. When necessary, refer back to this chapter when chemical processes are discussed later in the book.

Photo: The chemical composition of the body's structures determines their function. This ribbon diagram of a potassium channel protein (Kv1.2) shows the four subunits (in different colors) assembled together to create a pore for the passage of potassium ions (purple balls) across the plasma membrane. A view from the side (right) and a top-down view (left) of the channel are shown.

Module 2
Cells and Chemistry

Anatomy & Physiology
REVEALED
aprevealed.com

2.1 Basic Chemistry

LEARNING OUTCOMES

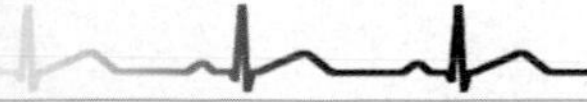

After reading this section, you should be able to

A. Define *matter, mass,* and *weight*.

B. Distinguish between elements and atoms, and state the four most abundant elements in the body.

C. Name the subatomic particles of an atom, and indicate their mass, charge, and location in an atom.

D. Define *atomic number, mass number, isotope, atomic mass,* and *mole*.

E. Compare and contrast ionic and covalent bonds.

F. Differentiate between a molecule and a compound.

G. Explain what creates a hydrogen bond, and relate its importance.

H. Describe solubility and the process of dissociation, and predict if a compound or molecule is an electrolyte or a nonelectrolyte.

Matter, Mass, and Weight

All living and nonliving things are composed of **matter,** which is anything that occupies space and has mass. **Mass** is the amount of matter in an object, and **weight** is the gravitational force acting on an object of a given mass. For example, the weight of an apple results from the force of gravity "pulling" on the apple's mass.

❯ Predict 2

The difference between mass and weight can be illustrated by considering an astronaut. How do an astronaut's mass and weight in outer space compare with the astronaut's mass and weight on the earth's surface?

The international unit for mass is the **kilogram (kg),** which is the mass of a platinum-iridium cylinder kept at the International Bureau of Weights and Measurements in France. The mass of all other objects is compared with this cylinder. For example, a 2.2-pound lead weight and 1 liter (L) (1.06 qt) of water each have a mass of approximately 1 kg. An object with 1/1000 the mass of a kilogram has a mass of 1 **gram (g).**

Elements and Atoms

An **element** is the simplest type of matter, having unique chemical properties. A list of the elements commonly found in the human body appears in table 2.1. About 96% of the body's weight results from the elements oxygen, carbon, hydrogen, and nitrogen. The majority of the body's weight is from oxygen. Oxygen is also the most abundant element in the earth's crust. Carbon plays an especially important role in the chemistry of the body, due in part to its propensity to form covalent bonds with itself and other molecules. Many elements are present in only trace amounts, but still play essential roles in the body. Elements can have multiple roles and exist in different states in the body. For example, mineralized calcium contributes to the solid matrix of bones, while dissolved calcium helps regulate enzyme activities and nervous system signaling.

An **atom** (*atomos,* indivisible) is the smallest particle of an element that has the chemical characteristics of that element. An element is composed of atoms of only one kind. For example, the element carbon is composed of only carbon atoms, and the element oxygen is composed of only oxygen atoms.

An element, or an atom of that element, is often represented by a symbol. Usually, the symbol is the first letter or letters of the element's name—for example, C for carbon, H for hydrogen, and

TABLE 2.1 Common Elements in the Human Body

Element	Symbol	Atomic Number	Mass Number	Atomic Mass	Percent in Human Body by Weight (%)	Percent in Human Body by Number of Atoms (%)
Hydrogen	H	1	1	1.008	9.5	63.0
Carbon	C	6	12	12.01	18.5	9.5
Nitrogen	N	7	14	14.01	3.3	1.4
Oxygen	O	8	16	16.00	65.0	25.5
Fluorine	F	9	19	19.00	Trace	Trace
Sodium	Na	11	23	22.99	0.2	0.3
Magnesium	Mg	12	24	24.31	0.1	0.1
Phosphorus	P	15	31	30.97	1.0	0.22
Sulfur	S	16	32	32.07	0.3	0.05
Chlorine	Cl	17	35	35.45	0.2	0.03
Potassium	K	19	39	39.10	0.4	0.06
Calcium	Ca	20	40	40.08	1.5	0.31
Iron	Fe	26	56	55.85	Trace	Trace
Iodine	I	53	127	126.9	Trace	Trace

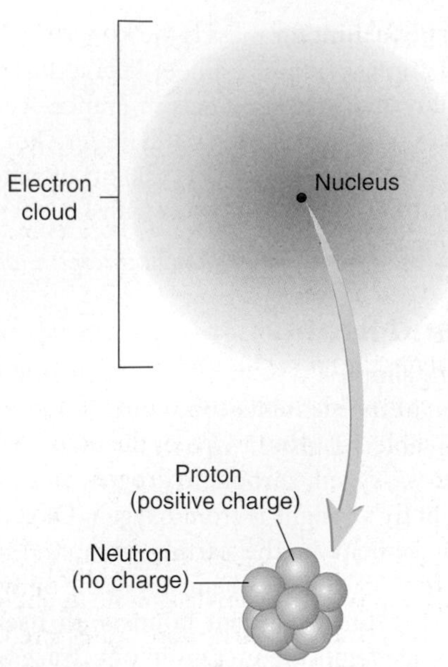
Electron cloud
Nucleus
Proton (positive charge)
Neutron (no charge)

FIGURE 2.1 APR **Model of an Atom**
The tiny, dense nucleus consists of positively charged protons and uncharged neutrons. Most of the volume of an atom is occupied by rapidly moving, negatively charged electrons, which can be represented as an electron cloud. The probable location of an electron is indicated by the color of the electron cloud. The darker the color in each small part of the electron cloud, the more likely the electron is located there.

Ca for calcium. Occasionally, the symbol is taken from the Latin, Greek, or Arabic name for the element—for example, the symbol for sodium is Na, from the Latin word *natrium.*

Atomic Structure

The characteristics of matter result from the structure, organization, and behavior of atoms. Atoms are composed of subatomic particles, some of which have an electrical charge. The three major types of subatomic particles are neutrons, protons, and electrons (figure 2.1). A **neutron** has no electrical charge, a **proton** has one positive charge, and an **electron** has one negative charge. The positive charge of a proton is equal in magnitude to the negative charge of an electron. The number of protons and the number of electrons in each atom are equal, and the individual charges cancel each other. Therefore, each atom is electrically neutral.

Protons and neutrons form the **nucleus** at the center of an atom, and electrons move around the nucleus (figure 2.1). The nucleus accounts for 99.97% of an atom's mass but only 1 ten-trillionth of its volume. Most of the volume of an atom is occupied by the electrons. Because electrons are always moving around the nucleus, the region where they are most likely to be found can be represented by an **electron cloud.**

Atomic Number and Mass Number

Each element is uniquely defined by the number of protons in the atoms of that element. For example, only hydrogen atoms have 1 proton, only carbon atoms have 6 protons, and only oxygen atoms have 8 protons (figure 2.2; see table 2.1). The **atomic number** of an element is equal to the number of protons in each atom and, because the number of electrons is equal to the number of

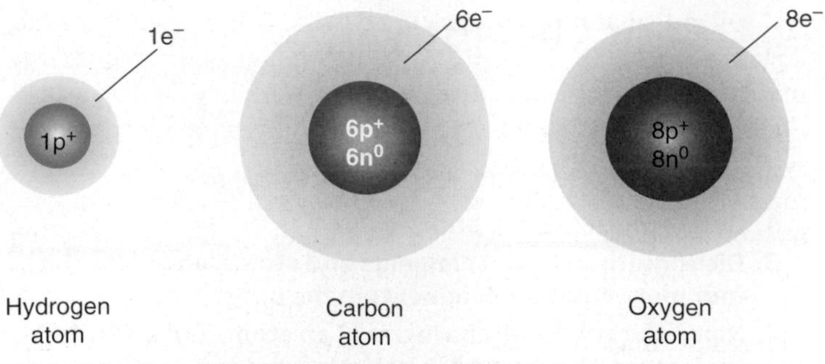
1e⁻
1p⁺
Hydrogen atom
6e⁻
6p⁺
6n⁰
Carbon atom
8e⁻
8p⁺
8n⁰
Oxygen atom

FIGURE 2.2 Hydrogen, Carbon, and Oxygen Atoms
Within the nucleus, the number of positively charged protons (p^+) and uncharged neutrons (n^0) is indicated. The negatively charged electrons (e^-) are around the nucleus. Atoms are electrically neutral because the number of protons and the number of electrons within an atom are equal.

protons, the atomic number is also the number of electrons. There are 90 naturally occurring elements, but additional elements have been synthesized by altering atomic nuclei. See the periodic table in appendix A for additional information about the elements.

Protons and neutrons have about the same mass, and they are responsible for most of the mass of atoms. Electrons, on the other hand, have very little mass. The **mass number** of an element is the number of protons plus the number of neutrons in each atom. For example, the mass number for carbon is 12 because it has 6 protons and 6 neutrons.

Predict 3

The atomic number of potassium is 19, and the mass number is 39. How many protons, neutrons, and electrons are in an atom of potassium?

Isotopes and Atomic Mass

Isotopes (ī′sō-tōpz) are two or more forms of the same element that have the same number of protons and electrons but a different number of neutrons. Thus, isotopes have the *same* atomic number but *different* mass numbers. There are three isotopes of hydrogen: hydrogen, deuterium, and tritium. All three isotopes have 1 proton and 1 electron, but hydrogen has no neutrons in its nucleus, deuterium has 1 neutron, and tritium has 2 neutrons (figure 2.3). Isotopes can be denoted using the symbol of the element preceded by the mass number (number of protons and neutrons) of the isotope. Thus, hydrogen is ^{1}H, deuterium is ^{2}H, and tritium is ^{3}H.

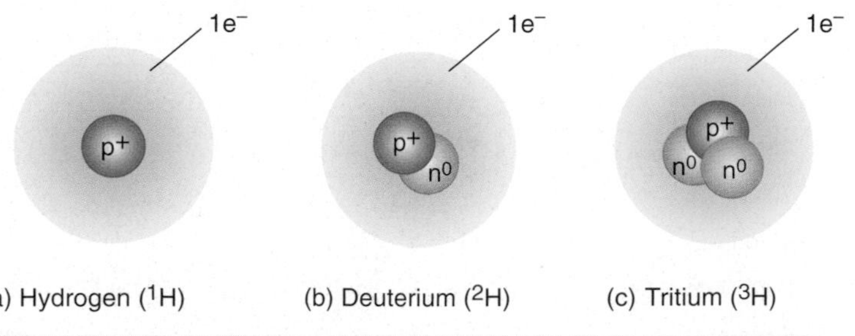
1e⁻
p⁺
1e⁻
p⁺
n⁰
1e⁻
p⁺
n⁰
n⁰

(a) Hydrogen (^{1}H) (b) Deuterium (^{2}H) (c) Tritium (^{3}H)

FIGURE 2.3 Isotopes of Hydrogen
(*a*) Hydrogen has 1 proton and no neutrons in its nucleus. (*b*) Deuterium has 1 proton and 1 neutron in its nucleus. (*c*) Tritium has 1 proton and 2 neutrons in its nucleus.

Individual atoms have very little mass. A hydrogen atom has a mass of 1.67×10^{-24} g (see appendix B for an explanation of the scientific notation of numbers). To avoid working with such small numbers, scientists use a system of relative atomic mass. In this system, a **dalton (Da),** or *unified atomic mass unit* (*u*), is 1/12 the mass of ^{12}C, a carbon atom with 6 protons and 6 neutrons. Thus, ^{12}C has an atomic mass of exactly 12 Da. However, a naturally occurring sample of carbon contains mostly ^{12}C and a small quantity of other carbon isotopes, such as ^{13}C, which has 6 protons and 7 neutrons. The **atomic mass** of an element is the *average* mass of its naturally occurring isotopes, taking into account the relative abundance of each isotope. For example, the atomic mass of the element carbon is 12.01 Da (see table 2.1), which is slightly more than 12 Da because of the additional mass of the small amount of other carbon isotopes. Because the atomic mass is an average, a sample of carbon can be treated as if all the carbon atoms had an atomic mass of 12.01 Da.

The Mole and Molar Mass

Just as a grocer sells eggs in lots of a dozen, a chemist groups atoms in lots of 6.022×10^{23}, which is called **Avogadro's number,** or 1 **mole** (abbreviated **mol**). Stated another way, a mole of a substance contains Avogadro's number of entities, such as atoms, ions, or molecules. The mass of 1 mole of a substance expressed in grams is called the **molar mass.** Molar mass is a convenient way to determine the number of atoms in a sample of an element. Because 12 g of ^{12}C is used as the standard, the atomic mass of an entity expressed in unified atomic mass units is the same as the molar mass expressed in grams. Thus, carbon atoms have an atomic mass of 12.01 Da, and 12.01 g of carbon has Avogadro's number (1 mol) of carbon atoms. By the same token, 1.008 g of hydrogen (1 mol) has the same number of atoms as 12.01 g of carbon (1 mol).

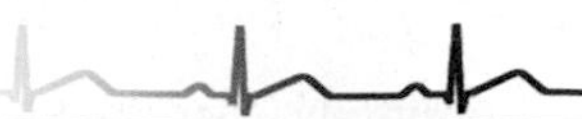

ASSESS **YOUR PROGRESS**

1. *Define* matter. *How are the mass and the weight of an object different?*
2. *Differentiate between* element *and* atom. *What four elements are found in the greatest abundance in the human body?*
3. *For each subatomic particle of an atom, state its charge and location. Which region of an atom is most responsible for the mass of the atom? Its volume?*
4. *Which subatomic particle determines the atomic number? What determines the mass number?*
5. *What is an* isotope? *How are isotopes denoted?*
6. *What is Avogadro's number? How is it related to a mole and molar mass?*

Electrons and Chemical Bonding

The outermost electrons of an atom determine its chemical behavior. When these outermost electrons are transferred, or shared, between atoms, **chemical bonding** occurs. Two major types of chemical bonds are ionic and covalent bonds.

Ionic Bonds

Recall that an atom is electrically neutral because it has equal numbers of protons and electrons. However, an atom can donate or lose electrons to other atoms. When this occurs, the numbers of protons and electrons are no longer equal, and a charged particle, called an **ion** (ī′on), is formed. After an atom loses an electron, it has 1 more proton than it has electrons and is positively charged. A sodium atom (Na) can lose an electron to become a positively charged sodium ion (Na^+) (figure 2.4*a*). After an atom gains an electron, it has 1 more electron than it has protons and is negatively charged. A chlorine atom (Cl) can accept an electron to become a negatively charged chloride ion (Cl^-).

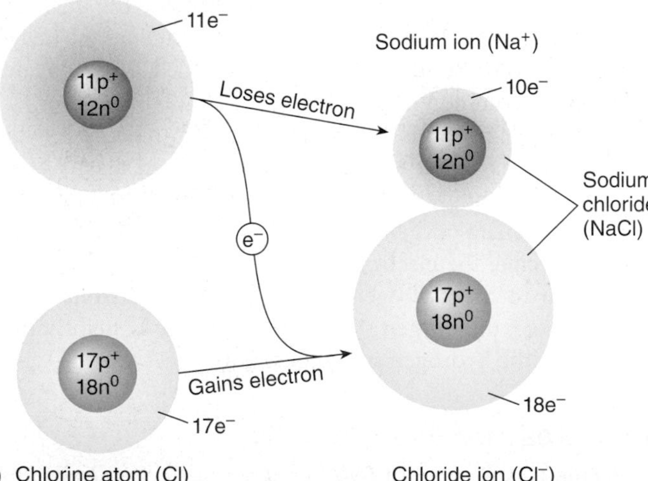
Sodium atom (Na)
11e⁻
11p⁺
12n⁰
Loses electron
Sodium ion (Na⁺)
10e⁻
11p⁺
12n⁰
Sodium
chloride
(NaCl)
e⁻
17p⁺
18n⁰
17p⁺
18n⁰
Gains electron
17e⁻
18e⁻
(a) Chlorine atom (Cl)
Chloride ion (Cl⁻)

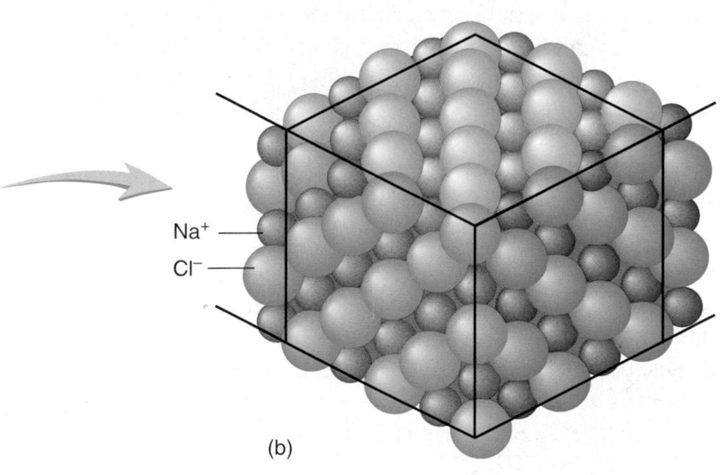
Na⁺
Cl⁻
(b)

(c)

FIGURE 2.4 AP|R **Ionic Bonds**

(*a*) A sodium atom (Na) loses an electron to become a smaller, positively charged ion, and a chlorine atom (Cl) gains an electron to become a larger, negatively charged ion. The attraction between the oppositely charged ions results in ionic bonding and the formation of sodium chloride. (*b*) The sodium ions (Na^+) and the chlorine ions (Cl^-) are organized to form a cube-shaped array. (*c*) A photomicrograph of salt crystals reflects the cubic arrangement of the ions.

Positively charged ions are called **cations** (kat′ī-onz), and negatively charged ions are called **anions** (an′ī-onz). Because oppositely charged ions are attracted to each other, cations and anions tend to remain close together, forming an **ionic** (ī-on′ik) **bond.** For example, Na^+ and Cl^- are held together by ionic bonding to form an array of ions called sodium chloride (NaCl), or table salt (figure 2.4*b,c*). Some ions commonly found in the body are listed in table 2.2.

Covalent Bonds

A **covalent bond** forms when atoms share one or more pairs of electrons. The resulting combination of atoms is called a molecule. An example is the covalent bond between two hydrogen atoms to form a hydrogen molecule (figure 2.5). Each hydrogen atom has 1 electron. As the two hydrogen atoms get closer together, the positively charged nucleus of each atom begins to attract the electron of the other atom. At an optimal distance, the 2 nuclei mutually attract the 2 electrons, and each electron is shared by both nuclei. The two hydrogen atoms are now held together by a covalent bond.

The sharing of one pair of electrons by two atoms results in a **single covalent bond.** A single line between the symbols of the atoms involved (e.g., H—H) represents a single covalent bond. A **double covalent bond** results when two atoms share 4 electrons, 2 from each atom. When a carbon atom combines with two oxygen atoms to form carbon dioxide, two double covalent bonds form. Double covalent bonds are indicated by a double line between the atoms (O═C═O).

TABLE 2.2 Important Ions in the Human Body

Common Ions	Symbols	Significance*
Calcium	Ca^{2+}	Part of bones and teeth; functions in blood clotting, muscle contraction, release of neurotransmitters
Sodium	Na^+	Membrane potentials, water balance
Potassium	K^+	Membrane potentials
Hydrogen	H^+	Acid-base balance
Hydroxide	OH^-	Acid-base balance
Chloride	Cl^-	Water balance
Bicarbonate	HCO_3^-	Acid-base balance
Ammonium	NH_4^+	Acid-base balance
Phosphate	PO_4^{3-}	Part of bones and teeth; functions in energy exchange, acid-base balance
Iron	Fe^{2+}	Red blood cell formation
Magnesium	Mg^{2+}	Necessary for enzymes
Iodide	I^-	Present in thyroid hormones

*The ions are part of the structures or play important roles in the processes listed.

e−
e−
p+
p+

1 There is no interaction between the two hydrogen atoms because they are too far apart.

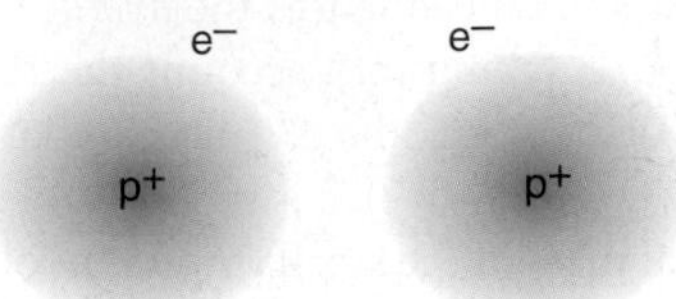

e−
e−
p+
p+

2 The positively charged nucleus of each hydrogen atom begins to attract the electron of the other.

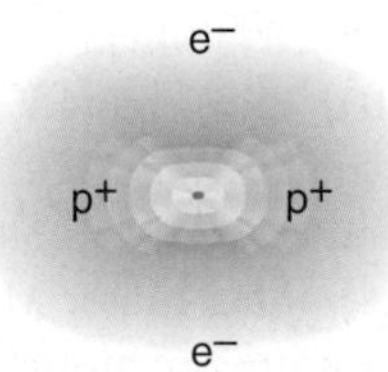

e−
p+
p+
e−

3 A covalent bond is formed when the electrons are shared between the nuclei because the electrons are equally attracted to each nucleus.

PROCESS FIGURE 2.5 Covalent Bonding

When electrons are shared equally between atoms, as in a hydrogen molecule, the bonds are called **nonpolar covalent bonds.** However, atoms bound to one another by a covalent bond do not always share their electrons equally because the nucleus of one atom attracts the electrons more strongly than does the nucleus of the other atom. Bonds of this type are called **polar covalent bonds** and are common in both living and nonliving matter.

Polar covalent bonds can result in polar molecules, which are electrically asymmetric. For example, oxygen atoms attract electrons more strongly than do hydrogen atoms. When covalent bonding between an oxygen atom and two hydrogen atoms forms a water molecule, the electrons are located closer to the oxygen nucleus than to the hydrogen nuclei. Because electrons have a negative charge, the oxygen side of the molecule is slightly more negative than the hydrogen side (figure 2.6).

Molecules and Compounds

A **molecule** is composed of two or more atoms chemically combined to form a structure that behaves as an independent unit. Sometimes the atoms that combine are of the same type, such as two hydrogen atoms combining to form a hydrogen molecule. However, more typically, a molecule consists of two or more different types of atoms, such as two hydrogen atoms and an oxygen atom combining to form water. Thus, a glass of water consists of a collection of individual water molecules positioned next to one another.

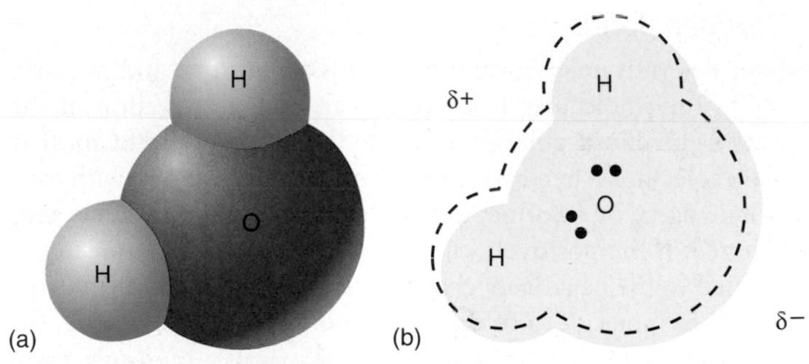
H
O
H
(a)
δ+
H
O
H
δ−
(b)

FIGURE 2.6 Polar Covalent Bonds

(*a*) A water molecule forms when two hydrogen atoms form covalent bonds with an oxygen atom. (*b*) The hydrogen atoms and oxygen atoms are sharing electron pairs (indicated by the black dots), but the sharing is unequal. The dashed outline shows the expected location of the electron cloud if the electrons are shared equally. But the actual electron cloud (yellow) is shifted toward the oxygen. Consequently, the oxygen side of the molecule has a slightly negative charge (indicated by δ^-), and the hydrogen side of the molecule has a slightly positive charge (indicated by δ^+).

A **compound** is a substance resulting from the chemical combination of two or more *different types* of atoms. Water is a molecule that is also a compound because it is a combination of two different atoms, hydrogen and oxygen. But not all molecules are compounds. For example, a hydrogen molecule is not a compound because it does not consist of different types of atoms.

Some compounds are molecules and some are not. (Remember that, to be a molecule, a structure must be an independent unit.) Covalent compounds, in which different types of atoms are held together by covalent bonds, are molecules because the sharing of electrons results in distinct units. On the other hand, ionic compounds, in which ions are held together by the force of attraction between opposite charges, are not molecules because they do not consist of distinct units. Table salt (NaCl) is an example of a substance that is a compound but not a molecule. A piece of NaCl does not consist of individual sodium chloride molecules positioned next to one another. Instead, NaCl is an organized array of individual Na^+ and individual Cl^- in which each charged ion is surrounded by several ions of the opposite charge (see figure 2.4*b*).

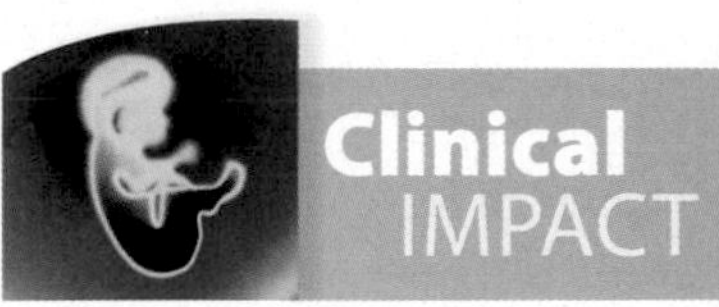
Clinical
IMPACT

Applications of Atomic Particles

Protons, neutrons, and electrons are responsible for the chemical properties of atoms. They also have other properties that can be useful in a clinical setting. For example, they have been used to develop methods for examining the inside of the body. Health professionals and researchers commonly use **radioactive isotopes** because sensitive measuring devices can detect their radioactivity, even when they are present in very small amounts.

Radioactive isotopes have unstable nuclei, which spontaneously change to form more stable nuclei. As a result, either new isotopes or new elements are produced. In this process of nuclear change, the nuclei of radioactive isotopes emit alpha particles, beta particles, and gamma rays. Alpha (α) particles are positively charged helium ions (He^{2+}), which consist of 2 protons and 2 neutrons. Beta (β) particles are electrons formed as neutrons change into protons. An electron is ejected from the neutron, and the proton that is produced remains in the nucleus. Gamma (γ) rays are a form of electromagnetic radiation (high-energy photons) released from nuclei as they lose energy.

All isotopes of an element have the same atomic number, and their chemical behavior is very similar. For example, ^3H (tritium) can substitute for ^1H (hydrogen), and either ^{125}iodine or ^{131}iodine can substitute for ^{126}iodine in chemical reactions.

Several procedures that are used to determine the concentration of substances, such as hormones, depend on the incorporation of small amounts of radioactive isotopes, such as ^{125}iodine, into the substances being measured. These procedures enable clinicians to more accurately diagnose disorders of the thyroid gland, the adrenal gland, and the reproductive organs.

Radioactive isotopes are also used to treat cancer. Some of the particles released from isotopes have a very high energy content and can penetrate and destroy tissues. Rapidly growing tissues, such as tumors, are more sensitive to radiation than are healthy cells. Thus, radioactive isotopes can be used to destroy tumors. Medical facilities also use radiation to sterilize materials that cannot be exposed to high temperatures (e.g., some fabric and plastic items used during surgical procedures). In addition, radioactive emissions can be used to sterilize food and other items.

X-rays are electromagnetic radiations with a much shorter wavelength than visible light. When electric current is used to heat a filament to very high temperatures, the energy of the electrons becomes so great that some electrons are emitted from the hot filament. When these electrons strike a positive electrode at high speeds, they release some of their energy in the form of x-rays.

X-rays do not penetrate dense material as readily as they penetrate less dense material, and x-rays can expose photographic film. Consequently, an x-ray beam can pass through a person and onto photographic film. Dense tissues of the body absorb the x-rays; on the film, these areas are underexposed and, so, appear white or light in color. By contrast, the x-rays readily pass through less dense tissue, so the film in these areas is overexposed and they appear black or dark in color. In an x-ray film of the skeletal system, the dense bones are white, and the less dense soft tissues are dark, often so dark that no details can be seen. Health professionals use x-rays to determine whether bones are broken or have other abnormalities.

Soft tissues can be photographed by using low-energy x-rays. Mammograms are low-energy x-rays of the breast that can reveal tumors because tumors are slightly denser than normal tissue.

Radiopaque substances are dense materials that absorb x-rays. If a radiopaque liquid is given to a patient, the liquid assumes the shape of the organ into which it is placed. For example, if a patient swallows a barium solution, the outline of the upper digestive tract can be photographed using x-rays to detect any abnormality, such as an ulcer.

The properties of elements can change when they are combined to form compounds. For example, the element hydrogen is extremely flammable, and oxygen, although not flammable on its own, promotes fire. However, when combined, hydrogen and oxygen form the very nonflammable compound water. Likewise, elements that are dangerous or toxic to humans can become useful as compounds. For example, sodium is very explosive when placed in water, and chlorine is a strong disinfectant in solutions, such as bleach and swimming pool water. Chlorine is so toxic that it was used as a poison gas in World War I, yet, when combined, sodium and chloride form the relatively safe and nonexplosive compound table salt.

The kinds and numbers of atoms (or ions) in a molecule or compound are typically represented by a formula consisting of the symbols of the atoms (or ions) plus subscripts denoting the quantity of each type of atom (or ion). The formula for glucose (a sugar) is $C_6H_{12}O_6$, indicating that glucose has 6 carbon, 12 hydrogen, and 6 oxygen atoms (table 2.3).

The **molecular mass** of a molecule or compound can be determined by adding up the atomic masses of its atoms (or ions). The term *molecular mass* is used for convenience for ionic compounds, even though they are not molecules. For example, the atomic mass of sodium is 22.99 and that of chloride is 35.45. The molecular mass of NaCl is therefore 58.44 (22.99 + 35.45).

Intermolecular Forces

The weak electrostatic attractions that exist between the oppositely charged parts of molecules, or between ions and molecules, are called **intermolecular forces.** These forces, which are much weaker than the forces producing chemical bonding, include hydrogen bonds and the properties of solubility and dissociation.

Hydrogen Bonds

Molecules with polar covalent bonds have positive and negative "ends." Intermolecular force results from the attraction of the positive end of one polar molecule to the negative end of another polar molecule. When hydrogen forms a covalent bond with oxygen, nitrogen, or fluorine, the resulting molecule becomes very polarized. If the positively charged hydrogen of one molecule is attracted to the negatively charged oxygen, nitrogen, or fluorine of another molecule, a **hydrogen bond** forms. For example, the positively charged hydrogen atoms of a water molecule form hydrogen bonds with the negatively charged oxygen atoms of other water molecules (figure 2.7). These hydrogen bonds are essential for the unique properties of water (see section 2.3).

Hydrogen bonds play an important role in determining the shape of complex molecules. The bonds can occur between different polar parts of a molecule to hold the molecule in its normal three-dimensional shape (see "Proteins" and "Nucleic Acids: DNA and RNA" in section 2.4).

Table 2.4 summarizes the important characteristics of chemical bonds (ionic and covalent) and intermolecular forces (hydrogen bonds).

Solubility and Dissociation

Solubility is the ability of one substance to dissolve in another—for example, sugar dissolving in water. Charged substances, such as sodium chloride, and polar substances, such as glucose, readily dissolve in water, whereas nonpolar substances, such as oils, do not. We all have seen how oil floats on water. Substances dissolve in water when they become surrounded by water molecules. If the positive and negative ends of the water molecules are more attracted to the charged ends of other molecules than to each other, the

TABLE 2.3 Picturing Molecules

Representation	Hydrogen	Carbon Dioxide	Glucose
Chemical Formula The formula shows the kind and number of atoms present.	H_2	CO_2	$C_6H_{12}O_6$
Electron-Dot Formula The bonding electrons are shown as dots between the symbols of the atoms.	H:H Single covalent bond	O::C::O Double covalent bond	Not used for complex molecules
Bond-Line Formula The bonding electrons are shown as lines between the symbols of the atoms.	H—H Single covalent bond	O=C=O Double covalent bond	CH_2OH, O, OH, HO, OH, OH
Models Atoms are shown as different-sized and different-colored spheres.	Hydrogen atom	Oxygen atom, Carbon atom	

FUNDAMENTAL **Figure**

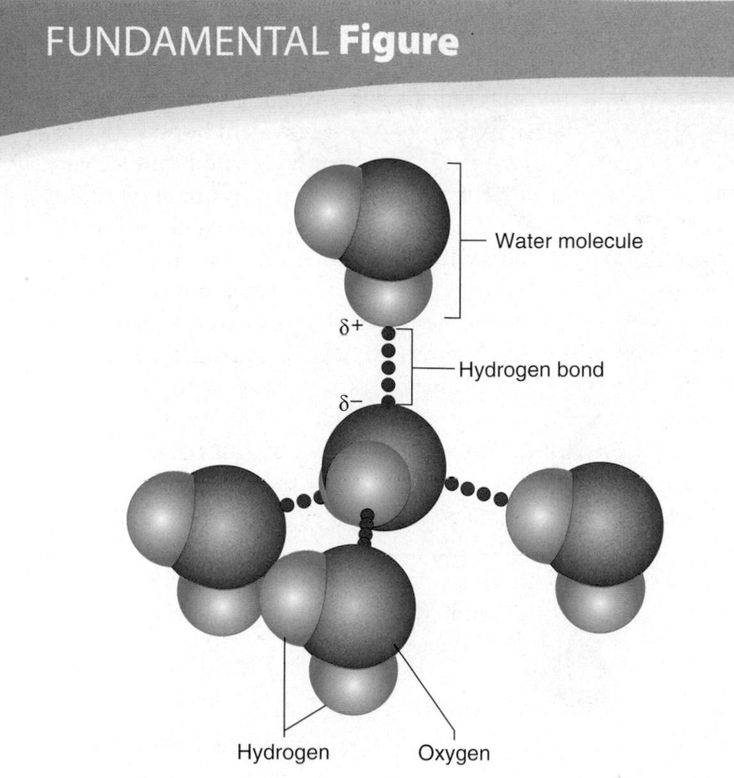
Water molecule
δ+
Hydrogen bond
δ−
Hydrogen
Oxygen

FIGURE 2.7 Hydrogen Bonds
The positive (δ^+) hydrogen part of one water molecule (blue) forms a hydrogen bond (red dotted line) with the negative (δ^-) oxygen part of another water molecule (red). As a result, hydrogen bonds hold the water molecules together.

TABLE 2.4 Comparison of Bonds

Bond	Example
Ionic Bond A complete transfer of electrons between two atoms results in separate positively charged and negatively charged ions.	Na^+Cl^- Sodium chloride
Polar Covalent Bond An unequal sharing of electrons between two atoms results in a slightly positive charge (δ^+) on one side of the molecule and a slightly negative charge (δ^-) on the other side of the molecule.	H, δ^+, O δ^-, H Water
Nonpolar Covalent Bond An equal sharing of electrons between two atoms results in an even charge distribution among the atoms of the molecule.	H, H—C—H, H Methane
Hydrogen Bond The attraction of oppositely charged ends of one polar molecule to another polar molecule holds molecules or parts of molecules together.	H, O···H—O, H H Water molecules

hydrogen bonds between the ends of the water molecules break, and water molecules surround the other molecules, which become dissolved in water.

When ionic compounds dissolve in water, their ions **dissociate,** or separate, from one another because cations are attracted to the negative ends of water molecules and anions are attracted to the positive ends of water molecules. When NaCl dissociates in water, sodium and chloride ions separate, and water molecules surround and isolate the ions, thereby keeping them in solution (figure 2.8).

When molecules (covalent compounds) dissolve in water, they usually remain intact, even though they are surrounded by water molecules. Thus, in a glucose solution, glucose molecules are surrounded by water molecules.

Cations and anions that dissociate in water are sometimes called **electrolytes** (ē-lek′trō-lītz) because they have the capacity to conduct an electric current, which is the flow of charged particles. An electrocardiogram (ECG) is a recording of electric currents produced by the heart. These currents can be detected by electrodes on the surface of the body because the ions in the body fluids conduct electric currents. Molecules that do not dissociate form solutions that do not conduct electricity and are called **nonelectrolytes.** Pure water is a nonelectrolyte.

Maintaining the proper balance of electrolytes is important for keeping the body hydrated, controlling blood pH, and ensuring the proper function of muscles and nerves. Under most conditions, including moderate exercise, the body's usual regulatory mechanisms are sufficient to maintain electrolyte homeostasis. However, people engaging in prolonged exercise, such as competing in a triathlon, are advised to consume sports drinks containing electrolytes. In an emergency, administering intravenous solutions can restore electrolyte and fluid balance.

ASSESS **YOUR PROGRESS**

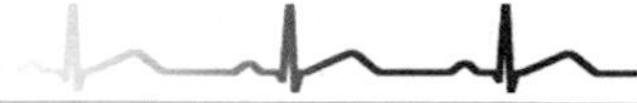

7. *Describe how an ionic bond is formed. What are a cation and an anion?*
8. *What occurs in the formation of a covalent bond? What is the difference between polar and nonpolar covalent bonds?*
9. *Distinguish between a molecule and a compound. Give an example of each. Are all molecules compounds? Are all compounds molecules?*
10. *What are intermolecular forces, and how do they create a hydrogen bond?*
11. *What is meant by the statement "table sugar is soluble in water?"*
12. *Describe what occurs during the dissociation of NaCl in water. What occurs when glucose ($C_6H_{12}O_6$) dissolves in water?*
13. *Explain the difference between electrolytes and nonelectrolytes. Classify each of the following water solutions as an electrolyte or a nonelectrolyte: potassium iodide (KCl), sucrose ($C_{12}H_{22}O_{11}$), magnesium bromide ($MgBr_2$), lactose ($C_{12}H_{22}O_{11}$).*

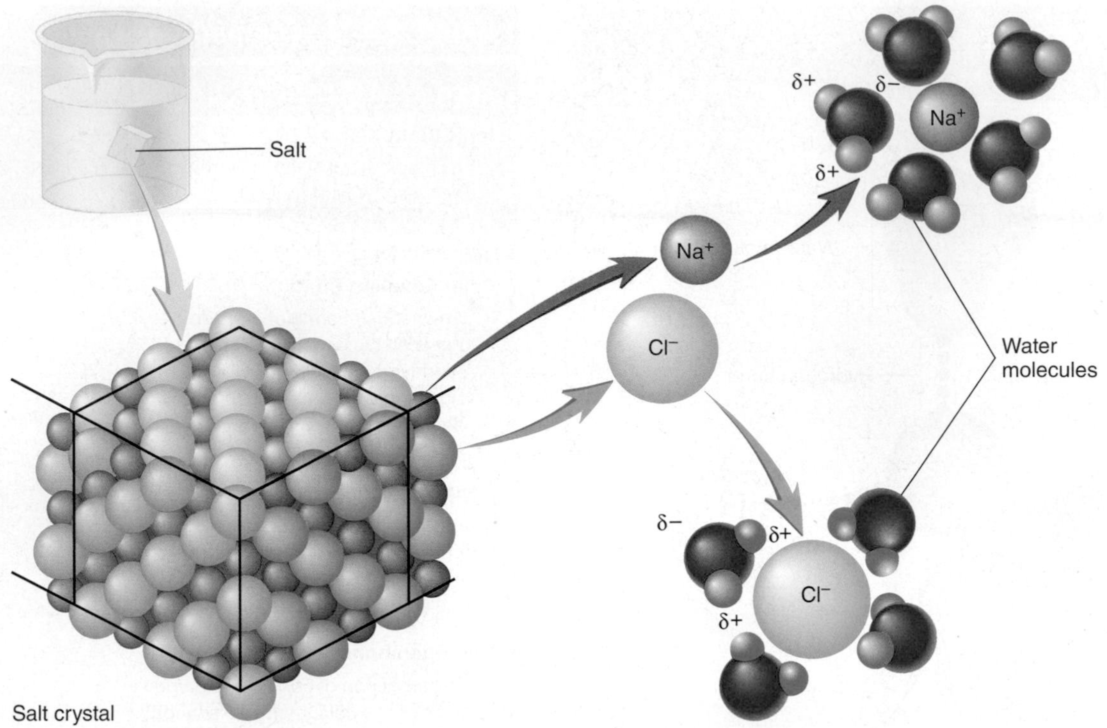
Salt
δ+
δ−
Na+
δ+
Na+
Cl−
Water
molecules
δ−
δ+
Cl−
δ+
Salt crystal

FIGURE 2.8 AP|R **Dissociation**
Sodium chloride (table salt) dissociates in water. The positively charged Na^+ are attracted to the negatively charged (δ^-) oxygen (red) end of the water molecule, and the negatively charged Cl^- are attracted to the positively charged (δ^+) hydrogen (blue) end of the water molecule.

2.2 Chemical Reactions and Energy

LEARNING OUTCOMES

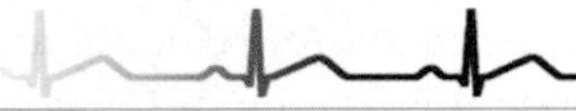

After reading this section, you should be able to

A. Summarize the characteristics of synthesis, decomposition, reversible reactions, and oxidation-reduction reactions.

B. Illustrate what occurs in dehydration and hydrolysis reactions.

C. Explain how reversible reactions produce chemical equilibrium.

D. Contrast potential and kinetic energy.

E. Distinguish between chemical reactions that release energy and those that take in energy.

F. Describe the factors that can affect the rate of chemical reactions.

In a **chemical reaction,** atoms, ions, molecules, or compounds interact either to form or to break chemical bonds. The substances that enter into a chemical reaction are called **reactants,** and the substances that result from the chemical reaction are called **products.**

For our purposes, three important points can be made about chemical reactions. First, in some reactions, less complex reactants are combined to form a larger, more complex product. An example is the synthesis of the complex proteins of the human body from amino acid "building blocks" obtained from food (figure 2.9*a*). Second, in other reactions, a reactant can be broken down, or decomposed, into simpler, less complex products. An example is the breakdown of carbohydrate molecules into glucose molecules (figure 2.9*b*). Third, atoms are generally associated with other atoms through chemical bonding or intermolecular forces; therefore, to synthesize new products or break down reactants, it is necessary to change the relationship between atoms.

Synthesis Reactions

When two or more reactants chemically combine to form a new and larger product, the process is called a **synthesis reaction.** An example of a synthesis reaction is the combination of two amino acids to form a dipeptide (figure 2.9*a*). As the amino acids are bound together, water results. Synthesis reactions in which water is a product are called **dehydration** (water out) **reactions.** As the atoms rearrange as a result of a synthesis reaction, old chemical bonds are broken and new chemical bonds are formed.

Another example of a synthesis reaction in the body is the formation of adenosine triphosphate (ATP; see section 2.4 for the details of ATP structure). ATP, which is composed of adenosine and three phosphate groups, is synthesized from adenosine diphosphate (ADP), which has two phosphate groups, and an inorganic phosphate (H_2PO^4) that is often symbolized as P_i:

A-P-P	+	P_i	→	A-P-P-P
(ADP)		(Inorganic phosphate)		(ATP)

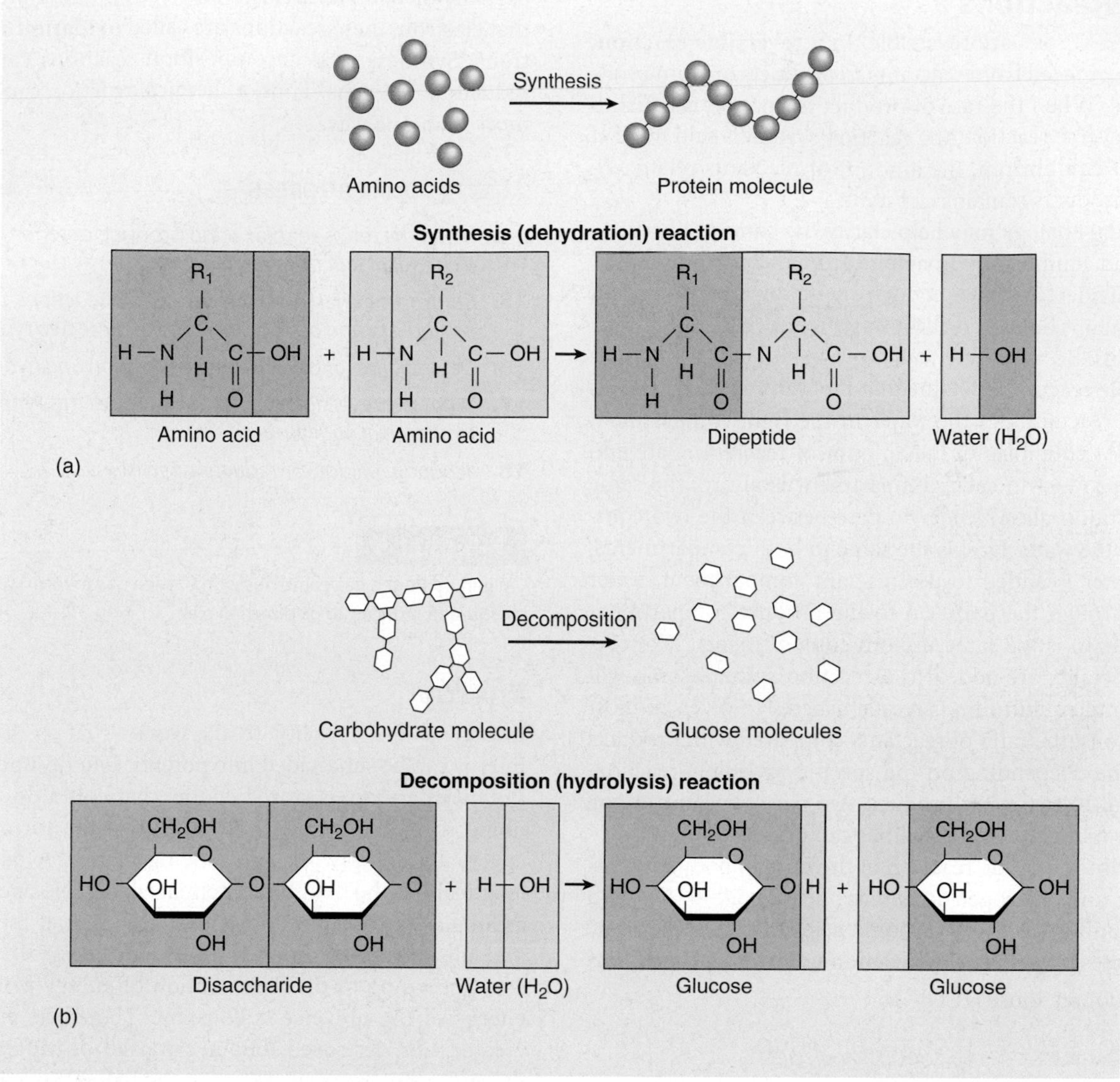
Synthesis
Amino acids
Protein molecule
Synthesis (dehydration) reaction
Amino acid
Amino acid
Dipeptide
Water (H_2O)
(a)
Decomposition
Carbohydrate molecule
Glucose molecules
Decomposition (hydrolysis) reaction
Disaccharide
Water (H_2O)
Glucose
Glucose
(b)

FIGURE 2.9 Synthesis and Decomposition Reactions

(*a*) A synthesis reaction in which amino acids combine to form a protein molecule. The combination of two amino acids is shown in detail. This reaction is also a dehydration reaction because it results in the removal of a water molecule from the amino acids. (*b*) A decomposition reaction in which a carbohydrate breaks apart into individual glucose molecules. The breakdown of a disaccharide is shown in detail. This reaction is also a hydrolysis reaction because it involves the splitting of a water molecule.

Synthesis reactions produce the molecules characteristic of life, such as ATP, proteins, carbohydrates, lipids, and nucleic acids. All of the synthesis reactions that occur within the body are collectively referred to as **anabolism** (ă-nab′ō-lizm). The growth, maintenance, and repair of the body could not take place without anabolic reactions.

Decomposition Reactions

A **decomposition reaction** is the reverse of a synthesis reaction—a larger reactant is chemically broken down into two or more smaller products. The breakdown of a disaccharide (a type of carbohydrate) into glucose molecules (figure 2.9*b*) is an example. Note that this reaction requires that water be split into two parts and that each part be contributed to one of the new glucose molecules. Reactions that use water in this manner are called **hydrolysis** (hī-drol′i-sis; water dissolution) **reactions.**

The breakdown of ATP to ADP and an inorganic phosphate is another example of a decomposition reaction:

A-P-P-P	→	A-P-P	+	P_i
(ATP)		(ADP)		(Inorganic phosphate)

The decomposition reactions occurring in the body are collectively called **catabolism** (kă-tab′-ō-lizm). They include the digestion of food molecules in the intestine and within cells, the breakdown of fat stores, and the breakdown of foreign matter and microorganisms in certain blood cells that protect the body. All of the anabolic and catabolic reactions in the body are collectively defined as **metabolism.**

Reversible Reactions

Some chemical reactions are reversible. In a **reversible reaction,** the reaction can proceed from reactants to products or from products to reactants. When the rate of product formation is equal to the rate of the reverse reaction, the reaction system is said to be at **equilibrium.** At equilibrium, the amount of reactants relative to the amount of products remains constant.

The following analogy may help clarify the concept of reversible reactions and equilibrium. Imagine a trough containing water. The trough is divided into two compartments by a partition, but the partition contains holes that allow water to move freely between the compartments. Because water can move in either direction, this is like a reversible reaction. Imagine that the water in the left compartment is the reactant and the water in the right compartment is the product. At equilibrium, the amount of reactant relative to the amount of product in each compartment is always the same because the partition allows water to pass between the two compartments until the water level is the same in both compartments. If additional water is added to the reactant compartment, water flows from it through the partition to the product compartment until the water level is the same in both compartments. Likewise, if additional reactants are added to a reaction system, some will form product until equilibrium is reestablished. However, in most reversible reactions, the ratio of reactants compared with products is not one to one. Depending on the specific reversible reaction, one part reactant to two parts product, two parts reactant to one part product, or many other possibilities can occur.

An important reversible reaction in the human body involves carbon dioxide and hydrogen ions. Carbon dioxide (CO_2) and water (H_2O) combine to form carbonic acid (H_2CO_3). Carbonic acid then separates by a reversible reaction to form hydrogen ions (H^+) and bicarbonate ions (HCO_3^-):

$$CO_2 + H_2O \rightleftarrows H_2CO_3 \rightleftarrows H^+ + HCO_3^-$$

If CO_2 is added to H_2O, additional H_2CO_3 forms, which causes more H^+ and HCO_3^- to form. The amount of H^+ and HCO_3^- relative to CO_2 therefore remains constant. Maintaining a constant level of H^+ is necessary for proper functioning of the nervous system. This can be achieved, in part, by regulating blood CO_2 levels. For example, slowing down the respiration rate causes blood carbon dioxide levels to increase.

Predict 4

If the respiration rate increases, CO_2 is eliminated from the blood. What effect does this change have on blood H^+ levels?

Oxidation-Reduction Reactions

Chemical reactions that result from the exchange of electrons between the reactants are called oxidation-reduction reactions. When sodium and chlorine react to form sodium chloride, the sodium atom loses an electron and the chlorine atom gains an electron. The loss of an electron by an atom is called **oxidation,** and the gain of an electron is called **reduction.** The transfer of the electron can be complete, resulting in an ionic bond, or it can be partial, resulting in a covalent bond. Because one atom partially or completely loses an electron and another atom gains that electron, these reactions are called **oxidation-reduction reactions.** Synthesis and decomposition reactions can be oxidation-reduction reactions. Thus, a chemical reaction can be described in more than one way.

ASSESS YOUR PROGRESS

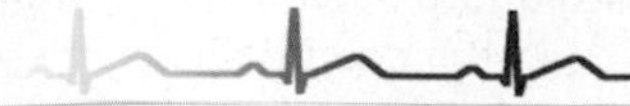

14. *Using the terms* reactant *and* product, *describe what occurs in a chemical reaction.*

15. *Contrast synthesis and decomposition reactions, and explain how catabolism and anabolism relate to these two types of reactions.*

16. *Describe the role of water in dehydration and hydrolysis reactions.*

17. *What is a reversible reaction? How does this type of reaction lead to chemical equilibrium?*

18. *What are oxidation-reduction reactions?*

Predict 5

When hydrogen gas combines with oxygen gas to form water, is the hydrogen reduced or oxidized? Explain.

Energy

Energy is the capacity to do **work**—that is, to move matter. Energy can be subdivided into potential energy and kinetic energy. **Potential energy** is stored energy that could do work but is not doing so. **Kinetic** (ki-net′ik) **energy** is the form of energy that is actually doing work and moving matter. A ball held at arm's length above the floor has potential energy. No energy is expended as long as the ball does not move. However, if the ball is released and falls toward the floor, it has kinetic energy.

According to the conservation of energy principle, the total energy of the universe is constant. Therefore, energy is neither created nor destroyed, but it can take on different forms. For example, the potential energy in the ball is converted into kinetic energy as the ball falls toward the floor. Conversely, the kinetic energy required to raise the ball from the floor is converted back into potential energy.

Potential and kinetic energy exist in many different forms. Here we consider mechanical, chemical, and heat energy. **Mechanical energy** results from the position or movement of objects. Many of the activities of the human body, such as moving a limb, breathing, and circulating blood, involve mechanical energy.

Chemical Energy

The **chemical energy** of a substance is the potential energy stored within its chemical bonds. In any chemical reaction, the potential energy in the chemical bonds of the reactants can be compared with the potential energy in the chemical bonds of the products. If the potential energy in the reactants is less than that in the products, energy must be supplied for the reaction to occur. An example is the synthesis of ATP from ADP:

$$\underset{\text{(Less potential energy in reactants)}}{ADP + H_2PO_4^-} + \text{Energy} \rightarrow \underset{\text{(More potential energy in products)}}{ATP + H_2O}$$

For simplicity, the H_2O is often not shown in this reaction, and P_i is used to represent inorganic phosphate ($H_2PO_4^-$). For this reaction to occur, bonds in $H_2PO_4^-$ are broken, and bonds are formed in ATP and H_2O. As a result of the breaking of existing bonds, the formation of new bonds, and the input of energy, these products have more potential energy than the reactants (figure 2.10*a*).

If the potential energy in the chemical bonds of the reactants is greater than that of the products, the reaction releases energy. For example, the chemical bonds of food molecules contain more potential energy than the waste products that are produced when food molecules are decomposed. The energy released from the chemical bonds of food molecules is used by living systems to synthesize ATP. Once ATP is produced, the breakdown of ATP to ADP results in the release of energy:

$ATP + H_2O$	$\rightarrow$	$ADP + H_2PO_4^-$	+	Energy
(More potential energy in reactants)		(Less potential energy in products)		

For this reaction to occur, the bonds in ATP and H_2O are broken and bonds in $H_2PO_4^-$ are formed. As a result of breaking the existing bonds and forming new bonds, these products have less potential energy than the reactants, and energy is released (figure 2.10*b*). Note that there are two quantities of energy in this reaction. The first is energy required to break the reactant chemical bonds. The second is energy released from those chemical bonds, which yields the net release of energy in the reaction. Thus, breakdown of ATP results in the net release of energy when the overall reaction is considered. The energy released when ATP is broken down can be used to synthesize other molecules, to do work (such as muscle contraction), or to produce heat.

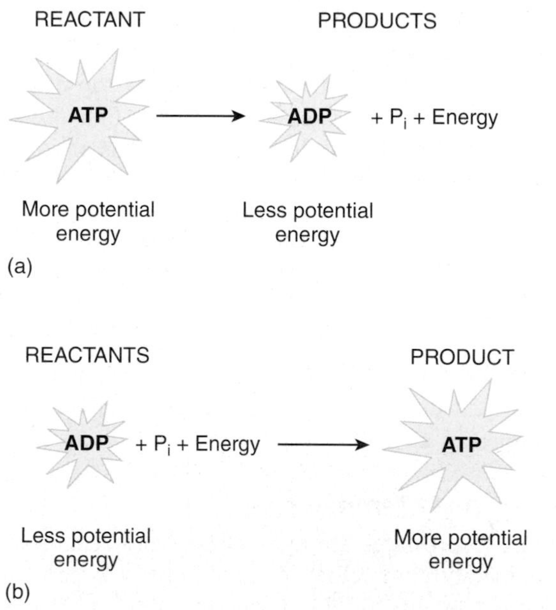
REACTANT
PRODUCTS
ATP
ADP
+ P$_i$ + Energy
More potential energy
Less potential energy
(a)
REACTANTS
PRODUCT
ADP
+ P$_i$ + Energy
ATP
Less potential energy
More potential energy
(b)

FIGURE 2.10 APIR **Energy and Chemical Reactions**
In the two reactions shown here, the larger "sunburst" represents greater potential energy and the smaller "sunburst" represents less potential energy. (*a*) Energy is released as a result of the breakdown of ATP. (*b*) The input of energy is required for the synthesis of ATP.

Heat Energy

Heat energy is the energy that flows between objects that are at different temperatures. Temperature is a measure of how hot or cold a substance is relative to another substance. Heat is always transferred from a hotter object to a cooler object, such as from a hot stove top to a finger.

All other forms of energy can be converted into heat energy. For example, when a moving object comes to rest, its kinetic energy is converted into heat energy by friction. Some of the potential energy of chemical bonds is released as heat energy during chemical reactions. Human body temperature is maintained by heat produced as a by-product of chemical reactions.

> **Predict 6**
>
> Energy from the breakdown of ATP provides the kinetic energy for muscle movement. Why does body temperature increase during exercise?

Speed of Chemical Reactions

Molecules are constantly in motion and therefore have kinetic energy. A chemical reaction occurs only when molecules with sufficient kinetic energy collide with each other. As two molecules move closer together, the negatively charged electron cloud of one molecule repels the negatively charged electron cloud of the other molecule. If the molecules have sufficient kinetic energy, they overcome this repulsion and come together. The nuclei in some atoms attract the electrons of other atoms, resulting in the breaking and formation of new chemical bonds. **Activation energy** is the minimum amount of energy that the reactants must have to start a chemical reaction (figure 2.11). Even reactions that release energy must overcome the activation energy barrier for the reaction to proceed. For example, heat in the form of a spark is required to start the reaction between oxygen and gasoline vapor. Once some oxygen molecules react with gasoline, the energy released can start additional reactions.

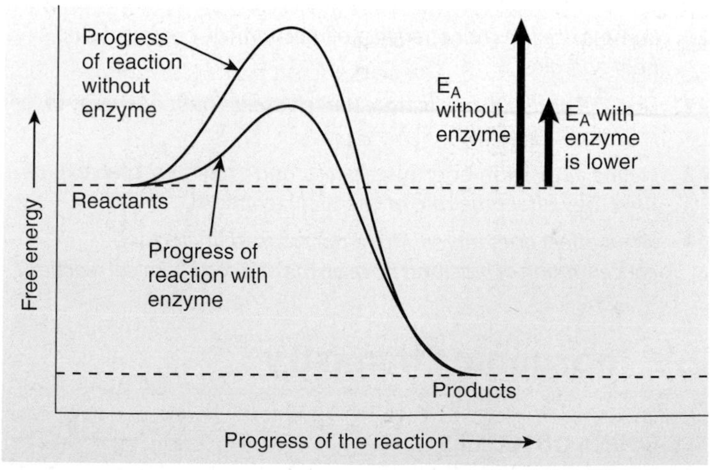
Progress of reaction without enzyme
E$_A$ without enzyme
E$_A$ with enzyme is lower
Reactants
Free energy
Progress of reaction with enzyme
Products
Progress of the reaction

FIGURE 2.11 Activation Energy and Enzymes
Activation energy is required to initiate chemical reactions. Without an enzyme, a chemical reaction can proceed, but it needs more energy input. Enzymes lower the activation energy, making it easier for the reaction to proceed.

Given any population of molecules, some of them have more kinetic energy and move about faster than others. Even so, at normal body temperatures, most of the chemical reactions necessary for life proceed too slowly to support life because few molecules have enough energy to start a chemical reaction. **Catalysts** (kat′ă-listz) are substances that increase the rate of chemical reactions without being permanently changed or depleted themselves. **Enzymes** (en′zīmz), which are discussed in greater detail later in the chapter, are proteins that act as catalysts. Enzymes increase the rate of chemical reactions by lowering the activation energy necessary for the reaction to begin (figure 2.11). As a result, more molecules have sufficient energy to undergo chemical reactions. An enzyme allows the rate of a chemical reaction to take place more than a million times faster than it would without the enzyme.

Temperature can also affect the speed of chemical reactions. As temperature increases, reactants have more kinetic energy, move at faster speeds, and collide with one another more frequently and with greater force, thereby increasing the likelihood of a chemical reaction. For example, when a person has a fever of only a few degrees, reactions occur throughout the body at an accelerated rate, increasing activity in the organ systems, such as the heart and respiratory rates. When body temperature drops, various metabolic processes slow. For example, in cold weather, the fingers are less agile, largely because of the reduced rate of chemical reactions in cold muscle tissue.

Within limits, the greater the concentration of the reactants, the greater the rate at which a given chemical reaction proceeds. This is true because, as the concentration of reactants increases, they are more likely to come into contact with one another. For example, the normal concentration of oxygen inside cells enables oxygen to come into contact with other molecules and produce the chemical reactions necessary for life. If the oxygen concentration decreases, the rate of chemical reactions decreases. A decrease in oxygen in cells can impair cell function and even result in death.

ASSESS YOUR PROGRESS

19. *Define* energy. *How are potential and kinetic energies different from each other?*
20. *Summarize the characteristics of mechanical, chemical, and heat energies.*
21. *Use ATP and ADP to illustrate the release or input of energy in chemical reactions.*
22. *Define* activation energy, catalyst, *and* enzymes; *then explain how they affect the rate of chemical reactions.*
23. *What effect does increasing temperature or increasing concentration of reactants have on the rate of a chemical reaction?*

2.3 Inorganic Chemistry

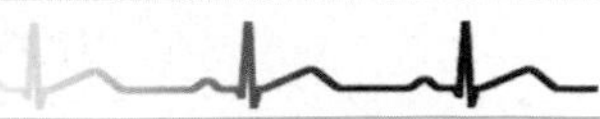

LEARNING OUTCOMES

After reading this section, you should be able to

A. Distinguish between inorganic and organic compounds.

B. Describe how the properties of water contribute to its physiological functions.

C. Describe the pH scale and its relationship to acidic, basic, and neutral solutions.

D. Explain the importance of buffers in organisms.

E. Compare the roles of oxygen and carbon dioxide in the body.

Inorganic chemistry generally deals with substances that do not contain carbon, although a more rigorous definition is the lack of carbon-hydrogen bonds. **Organic chemistry** is the study of carbon-containing substances, with a few exceptions. For example, carbon monoxide (CO), carbon dioxide (CO_2), and bicarbonate ions (HCO_3^-), which lack C—H bonds, are classified as inorganic molecules.

Inorganic substances play many vital roles in human anatomy and physiology. Examples include the oxygen we breathe, the calcium phosphate that makes up our bones, and the many metals required for protein functions, ranging from iron in blood gas transport to zinc in alcohol detoxification. In this section, we discuss the important roles of oxygen, carbon dioxide, and water—all inorganic molecules—in the body.

Water

Water has remarkable properties due to its polar nature. A molecule of **water** is formed when an atom of oxygen forms polar covalent bonds with two atoms of hydrogen. This gives a partial positive charge to the hydrogen atoms and a partial negative charge to the oxygen atom. Because of water's polarity, hydrogen bonds form between the positively charged hydrogen atoms of one water molecule and the negatively charged oxygen atoms of another water molecule. These hydrogen bonds organize the water molecules into a lattice, which holds the water molecules together and are responsible for many unique properties of water (see figures 2.6 and 2.7). The attraction of water to another water molecule is called **cohesion.** An example of cohesion is the surface tension exhibited when water bulges over the top of a full glass without spilling over. The same attractive force of hydrogen bonds with water will also attract other molecules. This process is called **adhesion.** The combination of cohesion and adhesion helps hold cells together and move fluids through the body.

Water accounts for approximately 50% of the weight of a young adult female and 60% of a young adult male. Females have a lower percentage of water than males because they typically have more body fat, which is relatively free of water. Plasma, the liquid portion of blood, is 92% water. Water has physical and chemical properties well suited for its many functions in living organisms. These properties are outlined in the following discussion.

Stabilizing Body Temperature

Water can absorb large amounts of heat and remain at a fairly stable temperature; therefore, it tends to resist large temperature fluctuations. Because of this property, blood, which is mostly water, can transfer heat from deep in the body to the surface, where the heat is released. In addition, when water evaporates, it changes from a liquid to a gas; because heat is required for that process, the evaporation of water from the surface of the body rids the body of excess heat.

Protection

Water is an effective lubricant that provides protection against damage resulting from friction. For example, tears protect the surface of the eye from rubbing of the eyelids. Water also forms a fluid cushion around organs, helping protect them from trauma. The cerebrospinal fluid that surrounds the brain is an example.

Chemical Reactions

Many of the chemical reactions necessary for life do not take place unless the reacting molecules are dissolved in water. For example, sodium chloride must dissociate in water into Na^+ and Cl^-, which can then react with other ions. Water also directly participates in many chemical reactions. As previously mentioned, a dehydration reaction is a synthesis reaction that produces water, and a hydrolysis reaction is a decomposition reaction that requires water (see figure 2.9).

Mixing Medium

A **mixture** is a combination of two or more substances physically blended together, but not chemically combined. A **solution** is any mixture of liquids, gases, or solids in which the substances are uniformly distributed with no clear boundary between them. For example, a salt solution consists of salt dissolved in water, air is a solution containing a variety of gases, and wax is a solid solution composed of several fatty substances. Solutions are often described in terms of one substance dissolving in another: The **solute** (sol′ūt) dissolves in the **solvent.** In a salt solution, water is the solvent and the dissolved salt is the solute. Sweat is a salt solution in which sodium chloride and other solutes are dissolved in water.

A **suspension** is a mixture containing materials that separate from each other unless they are continually, physically blended together. Blood is a suspension—that is, red blood cells are suspended in a liquid called plasma. As long as the red blood cells and plasma are mixed together as they pass through blood vessels, the red blood cells remain suspended in the plasma. However, if the blood is allowed to sit in a container, the red blood cells and plasma separate from each other.

A **colloid** (kol′oyd) is a mixture in which a dispersed (solutelike) substance is distributed throughout a dispersing (solventlike) substance. The dispersed particles are larger than a simple molecule but small enough that they remain dispersed and do not settle out. Proteins, which are large molecules, are common dispersed particles; proteins and water form colloids. For instance, the plasma portion of blood and the liquid interior of cells are colloids containing many important proteins.

In living organisms, the complex fluids inside and outside cells consist of solutions, suspensions, and colloids. Blood is an example of all of these mixtures. It is a solution containing dissolved nutrients, such as sugar; a suspension holding red blood cells; and a colloid containing proteins.

Water's ability to mix with other substances enables it to act as a medium for transport, moving substances from one part of the body to another. Body fluids, such as plasma, transport nutrients, gases, waste products, and a variety of molecules involved in regulating body functions.

Solution Concentrations

The concentration of solute particles dissolved in solvents can be expressed in several ways. One common way is to indicate the percent of solute by weight per volume of solution. A 10% solution of sodium chloride can be made by dissolving 10 g of sodium chloride into enough water to make 100 mL of solution.

Physiologists often determine concentrations in **osmoles** (os′mōlz), which express the number of particles in a solution. A particle can be an atom, an ion, or a molecule. An osmole (Osm) is Avogadro's number of particles of a substance in 1 kilogram (kg) of water. The **osmolality** (os-mō-lal′i-tē) of a solution reflects the number, not the type, of particles in a solution. For example, a 1 Osm glucose solution and a 1 Osm NaCl solution both contain Avogadro's number of particles per kg of water. The glucose solution has 1.0 Osm of glucose molecules, whereas the NaCl solution has 0.5 Osm of Na^+ and 0.5 Osm of Cl^- because NaCl dissociates into Na^+ and Cl^- in water.

Because the concentration of particles in body fluids is so low, physiologists use the measurement **milliosmole** (mOsm), 1/1000 of an osmole. Most body fluids have a concentration of about 300 mOsm and contain many different ions and molecules. The concentration of body fluids is important because it influences the movement of water into or out of cells (see chapter 3). Appendix C contains more information on calculating concentrations.

ASSESS YOUR PROGRESS 

24. *What is the difference between inorganic and organic chemistry?*

25. *What two properties of water are the result of hydrogen bonding, and how are these two properties different?*

26. *List and briefly describe the four functions that water performs in living organisms.*

27. *Using the terms* solute *and* solvent*, summarize the properties of solutions, suspensions, and colloids.*

28. *How is the osmolality of a solution determined? What is a milliosmole?*

Acids and Bases

The body contains many molecules and compounds, called acids and bases, that can alter body functions by releasing and binding protons. A normal balance of acids and bases is maintained by homeostatic mechanisms involving buffers, the respiratory system, and the kidneys (see chapter 27).

For most purposes, an **acid** is defined as a proton donor. A hydrogen ion (H^+) is a proton because it results when an electron is lost from a hydrogen atom, which consists of a proton and an electron. Therefore, a molecule or compound that releases H^+ is an acid. Hydrochloric acid (HCl) forms hydrogen ions (H^+) and chloride ions (Cl^-) in solution and therefore is an acid:

$$HCl \rightarrow H^+ + Cl^-$$

A **base** is defined as a proton acceptor, and any substance that binds to (accepts) H^+ is a base. Many bases function as proton acceptors by releasing hydroxide ions (OH^-) when they dissociate. The base sodium hydroxide (NaOH) dissociates to form Na^+ and OH^-:

$$NaOH \rightarrow Na^+ + OH^-$$

The OH^- are proton acceptors that combine with H^+ to form water:

$$OH^- + H^+ \rightarrow H_2O$$

Acids and bases are classified as strong or weak. Strong acids or bases dissociate almost completely when dissolved in water. Consequently, they release almost all of their H^+ or OH^-. The more completely the acid or base dissociates, the stronger it is. For example, HCl is a strong acid because it completely dissociates in water:

$$HCl \rightarrow H^+ + Cl^-$$
Not easily reversible

Weak acids or bases only partially dissociate in water. Consequently, they release only some of their H^+ or OH^-. For example, when acetic acid (CH_3COOH) is dissolved in water, some of it dissociates, but some of it remains in the undissociated form. An equilibrium is established between the ions and the undissociated weak acid:

$$CH_3COOH \rightleftarrows CH_3COO^- + H^+$$
Easily reversible

For a given weak acid or base, the amount of the dissociated ions relative to the weak acid or base is a constant.

The pH Scale

The **pH scale** is a means of referring to the H^+ concentration in a solution (figure 2.12). The scale ranges from 0 to 14. A **neutral solution** has equal concentrations of H^+ and OH^-; pure water is considered a neutral solution and has a pH of 7. Solutions with a pH less than 7 are **acidic** and have a greater concentration of H^+ than OH^-. Solutions with a pH greater than 7 are **alkaline** (al′kă-līn), or **basic,** and have fewer H^+ than OH^-.

A change in the pH of a solution by 1 pH unit represents a 10-fold change in the H^+ concentration. For example, a solution of pH 6 has a H^+ concentration 10 times greater than a solution of pH 7 and 100 times greater than a solution of pH 8. As the pH value becomes smaller, the solution has more H^+ and is more acidic; as the pH value becomes larger, the solution has fewer H^+ and is more basic. Appendix D considers pH in greater detail.

The normal pH range for human blood is 7.35 to 7.45. **Acidosis** results if blood pH drops below 7.35, in which case the nervous system becomes depressed and the individual may become disoriented and possibly comatose. **Alkalosis** results if blood pH rises above 7.45. Then the nervous system becomes overexcitable, and the individual may become extremely nervous or have convulsions. Both acidosis and alkalosis can be fatal.

Salts

A **salt** is a compound consisting of a cation other than H^+ and an anion other than OH^-. Salts are formed by the interaction of an acid and a base in which the H^+ of the acid are replaced by the positive ions of the base. For example, in a solution in which hydrochloric acid (HCl) reacts with the base sodium hydroxide (NaOH), the salt sodium chloride (NaCl) is formed:

$$\underset{\text{(Acid)}}{HCl} + \underset{\text{(Base)}}{NaOH} \rightarrow \underset{\text{(Salt)}}{NaCl} + \underset{\text{(Water)}}{H_2O}$$

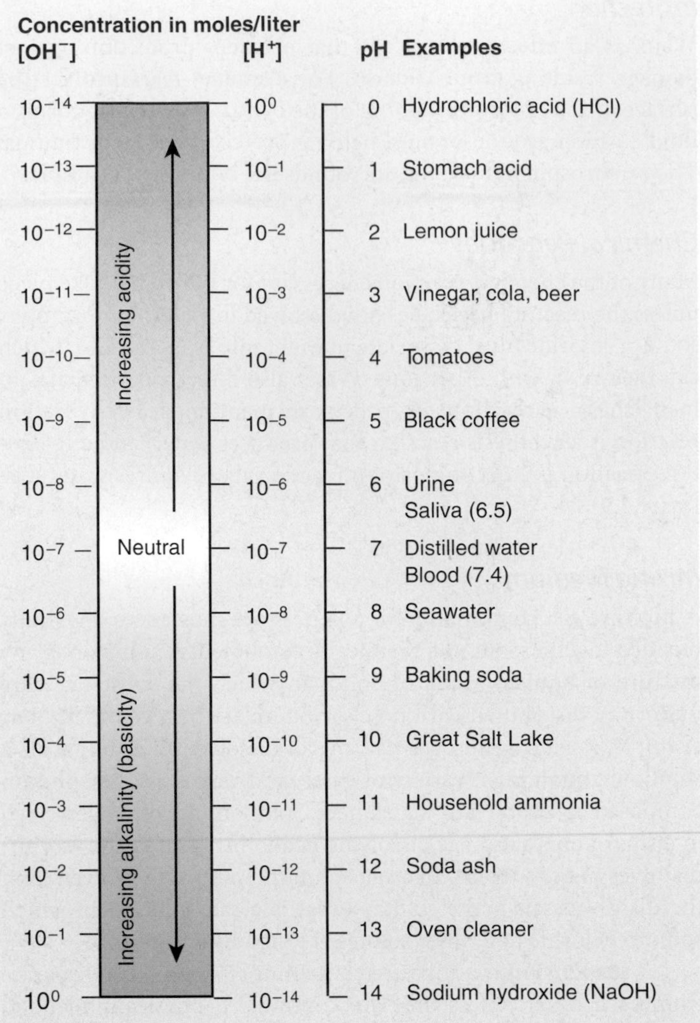
Concentration in moles/liter
[OH⁻]
[H⁺]
pH
Examples
10^-14 10^0 0 Hydrochloric acid (HCl)
10^-13 10^-1 1 Stomach acid
10^-12 10^-2 2 Lemon juice
10^-11 10^-3 3 Vinegar, cola, beer
10^-10 10^-4 4 Tomatoes
10^-9 10^-5 5 Black coffee
10^-8 10^-6 6 Urine
Saliva (6.5)
10^-7 10^-7 7 Distilled water
Blood (7.4)
10^-6 10^-8 8 Seawater
10^-5 10^-9 9 Baking soda
10^-4 10^-10 10 Great Salt Lake
10^-3 10^-11 11 Household ammonia
10^-2 10^-12 12 Soda ash
10^-1 10^-13 13 Oven cleaner
10^0 10^-14 14 Sodium hydroxide (NaOH)
Increasing acidity
Neutral
Increasing alkalinity (basicity)

FIGURE 2.12 The pH Scale

A pH of 7 is considered neutral. Values less than 7 are acidic (the lower the number, the more acidic). Values greater than 7 are basic (the higher the number, the more basic). Representative fluids and their approximate pH values are listed.

Typically, when salts such as sodium chloride dissociate in water, they form positively and negatively charged ions (see figure 2.8).

Buffers

The chemical behavior of many molecules changes as the pH of the solution in which they are dissolved changes. For example, many enzymes work best within narrow ranges of pH. The survival of an organism depends on its ability to maintain homeostasis by keeping body fluid pH within a narrow range. Deviations from the normal pH range for human blood are life-threatening.

One way body fluid pH is regulated involves the use of buffers. **Buffers** are chemicals that resist changes in pH when either acids or bases are added to a solution. For example, when an acid is added to a buffered solution, the buffer binds to the H^+, preventing these ions from causing a decrease in the pH of the solution (figure 2.13).

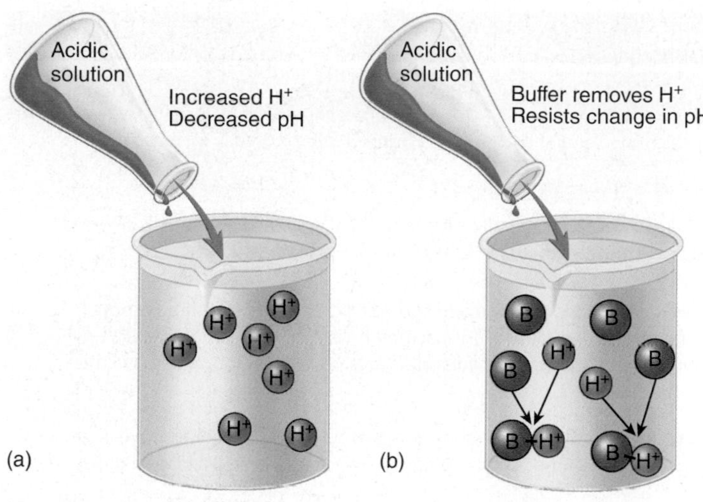

Acidic solution
Increased H+
Decreased pH
Acidic solution
Buffer removes H+
Resists change in pH
H+
B
(a)
(b)

FIGURE 2.13 Buffers
(*a*) The addition of an acid to a nonbuffered solution results in an increase of H^+ and a decrease in pH. (*b*) The addition of an acid to a buffered solution results in a much smaller change in pH. The added H^+ bind to the buffer (symbolized by the letter *B*).

Important buffers in living systems are composed of bicarbonate, phosphates, amino acids, and proteins. Buffers prevent large changes in pH values by acting as **conjugate acid-base pairs.** A conjugate base is what remains of an acid after the H^+ (proton) is lost. A conjugate acid is formed when a H^+ is transferred to the conjugate base. Two substances related in this way are a conjugate acid-base pair.

A major buffer in our body fluids is the bicarbonate system. A bicarbonate ion (HCO_3^-) is formed by the dissociation of carbonic acid (H_2CO_3).

$$H_2CO_3 \rightleftarrows H^+ + HCO_3^-$$

Carbonic acid and bicarbonate are a conjugate acid-base pair. The sodium salt of bicarbonate, also known as baking soda, is an active ingredient in some antacids taken to reduce stomach acidity. In the forward reaction, H_2CO_3 loses a H^+ to produce HCO_3^-, which is a conjugate base. In the reverse reaction, a H^+ is transferred to the HCO_3^- (conjugate base) to produce H_2CO_3, which is a conjugate acid.

For a given condition, this reversible reaction results in an equilibrium, in which the amounts of H_2CO_3 relative to the amounts of H^+ and HCO_3^- remain constant. The conjugate acid-base pair can resist changes in pH because of this equilibrium. If an acid is added to a buffer, the H^+ from the added acid can combine with the base component of the conjugate acid-base pair. As a result, the concentration of H^+ does not increase as much as it would without this reaction. If H^+ is added to a H_2CO_3 solution, many of the H^+ combine with HCO_3^- to form H_2CO_3.

On the other hand, if a base is added to a buffered solution, the conjugate acid can release H^+ to counteract the effects of the added base. For example, if OH^- are added to a H_2CO_3 solution, the OH^- combine with H^+ to form water. As the H^+ are incorporated into water, H_2CO_3 dissociates to form H^+ and HCO_3^-, thereby maintaining the H^+ concentration (pH) within a normal range.

The greater the buffer concentration, the more effectively it can resist a change in pH; however, buffers cannot entirely prevent some change in the pH of a solution. For example, when an acid is added to a buffered solution, the pH decreases, but not to the extent it would have without the buffer.

Predict 7

Dihydrogen phosphate ion ($H_2PO_4^-$) and monohydrogen phosphate ion (HPO_4^{2-}) form the phosphate buffer system. Identify the conjugate acid and the conjugate base in the phosphate buffer system:

$$H_2PO_4^- \rightleftarrows H^+ + HPO_4^{2-}$$

Explain how they function as a buffer when either H^+ or OH^- are added to the solution.

Oxygen and Carbon Dioxide

Oxygen (O_2) is an inorganic molecule consisting of two oxygen atoms bound together by a double covalent bond. About 21% of the gas in the atmosphere is oxygen, and it is essential for most living organisms. Humans require oxygen in the final step of a series of reactions that extract energy from food molecules (see chapter 25).

Carbon dioxide (CO_2) consists of one carbon atom bound to two oxygen atoms. Each oxygen atom is bound to the carbon atom by a double covalent bond. Carbon dioxide is produced when organic molecules, such as glucose, are metabolized within the cells of the body (see chapter 25). Much of the energy stored in the covalent bonds of glucose is transferred to other organic molecules when glucose is broken down and carbon dioxide is released. Once carbon dioxide is produced, it is eliminated from the cell as a metabolic by-product, transferred to the lungs by the blood, and exhaled during respiration. If carbon dioxide is allowed to accumulate within cells, it becomes toxic.

ASSESS YOUR PROGRESS

29. *Define* acid *and* base, *and describe the pH scale.*
30. *What is the difference between a strong acid or base and a weak acid or base?*
31. *The blood pH of a patient is 7.30. What condition does this patient have, and what are the symptoms?*
32. *How are salts related to acids and bases?*
33. *What is a buffer, and why are buffers important in the body?*
34. *What is a conjugate acid-base pair?*
35. *What are the functions of oxygen and carbon dioxide in living systems?*

2.4 Organic Chemistry

LEARNING OUTCOMES

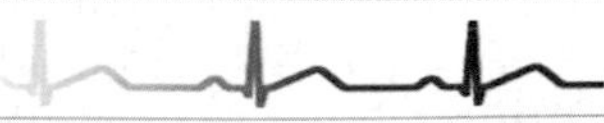

After reading this section, you should be able to

A. Describe the structural organization and major functions of carbohydrates, lipids, proteins, and nucleic acids.

TABLE 2.5 Major Functional Groups of Organic Compounds

Name and Structural Formula*	Functional Significance
Hydroxyl R—O—H	*Alcohols* contain a hydroxyl group, which is polar and hydrophilic. Hydroxyl groups greatly increase the solubility of molecules in water.
Sulfhydryl R—S—H	*Thiols* have a sulfhydryl group, which is polar and hydrophilic.The amino acid cysteine contains a sulfhydryl group that can form a disulfide bond with another cysteine to help stabilize protein structure.
Carbonyl R—C(=O)—R	*Ketones* and *aldehydes* have a carbonyl group, which is polar and hydrophilic. Ketones contain a carbonyl group within the carbon chain. Ketones are formed during normal metabolism, but they can be elevated in the blood during starvation or certain diabetic states. Aldehydes are similar to ketones, but they have the carbonyl group at the end of the carbon chain.
Carboxyl R—C(=O)—OH	*Carboxylic acids* have a carboxyl group, which is hydrophilic and can act as an acid by donating a hydrogen ion. All amino acids have a carboxyl group at one end. At physiological pH, the amino acid carboxyl group is predominantly negatively charged.
Ester R—C(=O)—O—R	*Esters* are structures with an ester group, which is less hydrophilic than hydroxyl or carboxyl groups. Triglycerides and dietary fats are esters. Other types of esters include the volatile compounds in perfumes.
Amino R—N(H)(H)	*Amines* have an amino group, which is less hydrophilic than carboxyl groups. Amines can act as a base by accepting a hydrogen ion. All amino acids have an amine group at one end. At physiological pH, the amino acid amine group is predominantly positively charged.
Phosphate R—O—P(=O)(—O⁻)—O⁻	*Phosphates* have a phosphate group, which is very hydrophilic due to the double negative charge. Phosphates are used as an energy source (adenosine triphosphate), in biological membranes (phospholipids), and as intracellular signaling molecules (protein phosphorylation).

*R = variable group.

B. Explain how enzymes work.

C. Describe the roles of nucleotides in the structures and functions of DNA, RNA, and ATP.

Carbon's ability to form covalent bonds with other atoms makes possible the formation of the large, diverse, complicated molecules necessary for life. Carbon atoms bound together by covalent bonds constitute the "backbone" of many large molecules. Two mechanisms that allow the formation of a wide variety of molecules are (1) variation in the length of the carbon chains and (2) the combination of the atoms involved. For example, some protein molecules have thousands of carbon atoms bound by covalent bonds to one another or to other atoms, such as nitrogen, sulfur, hydrogen, and oxygen. Selected major functional groups of organic compounds are listed in table 2.5.

The four major groups of organic molecules essential to living organisms are carbohydrates, lipids, proteins, and nucleic acids. In addition, a high energy form of a nucleic acid building block, called ATP, is an important organic molecule in cellular processes. Each of these groups and ATP have specific structural and functional characteristics.

Carbohydrates

Carbohydrates are composed primarily of carbon, hydrogen, and oxygen atoms and range in size from small to very large. In most carbohydrates, there are approximately two hydrogen atoms and one oxygen atom for each carbon atom. Note that this two-to-one ratio is the same as in water (H_2O). The molecules are called carbohydrates because carbon (*carbo*) atoms are combined with the same atoms that form water (*hydrated*). The large number of oxygen atoms in carbohydrates makes them relatively polar molecules. Consequently, they are soluble in polar solvents, such as water.

Carbohydrates are important parts of other organic molecules, and they can be broken down to provide the energy necessary for life. Undigested carbohydrates also provide bulk in feces, which helps maintain the normal function and health of the digestive tract. Table 2.6 summarizes the role of carbohydrates in the body.

Monosaccharides

Large carbohydrates are composed of numerous, relatively simple building blocks called **monosaccharides** (mon-ō-sak′ă-rīdz; *mono-*, one + *saccharide,* sugar), or simple sugars. Monosaccharides

TABLE 2.6 Role of Carbohydrates in the Body

Role	Example
Structure	Ribose forms part of RNA and ATP molecules, and deoxyribose forms part of DNA.
Energy	Monosaccharides (glucose, fructose, galactose) can be used as energy sources. Disaccharides (sucrose, lactose, maltose) and polysaccharides (starch, glycogen) must be broken down to monosaccharides before they can be used for energy. Glycogen is an important energy-storage molecule in muscles and in the liver.
Bulk	Cellulose forms bulk in the feces.

commonly contain 3 carbons (trioses), 4 carbons (tetroses), 5 carbons (pentoses), or 6 carbons (hexoses).

The monosaccharides most important to humans include both 5- and 6-carbon sugars. Common 6-carbon sugars, such as glucose, fructose, and galactose, are **isomers** (ī′sō-merz), which are molecules that have the same number and types of atoms but differ in their three-dimensional arrangement (figure 2.14). Glucose, or blood sugar, is the major carbohydrate in the blood and a major nutrient for most cells of the body. Blood glucose levels are tightly regulated by insulin and other hormones. In people with diabetes, the body is unable to regulate glucose levels properly. Diabetics need to monitor their blood glucose carefully to minimize the deleterious effects of this disease. Fructose and galactose are also important dietary nutrients. Important 5-carbon sugars include ribose and deoxyribose (see figure 2.24), which are components of ribonucleic acid (RNA) and deoxyribonucleic acid (DNA), respectively.

Disaccharides

Disaccharides (dī-sak′ă-rīdz; *di-*, two) are composed of two simple sugars bound together through a dehydration reaction. Glucose and fructose, for example, combine to form a disaccharide called **sucrose** (table sugar) plus a molecule of water (figure 2.15*a*). Several disaccharides are important to humans, including sucrose, lactose, and maltose. Lactose, or milk sugar, is glucose combined with galactose; maltose, or malt sugar, is two glucose molecules joined together.

Polysaccharides

Polysaccharides (pol-ē-sak′ă-rīdz; *poly-*, many) consist of many monosaccharides bound together to form long chains that are either straight or branched. **Glycogen,** or animal starch, is a polysaccharide composed of many glucose molecules (figure 2.15*b*). Because glucose can be metabolized rapidly and the resulting

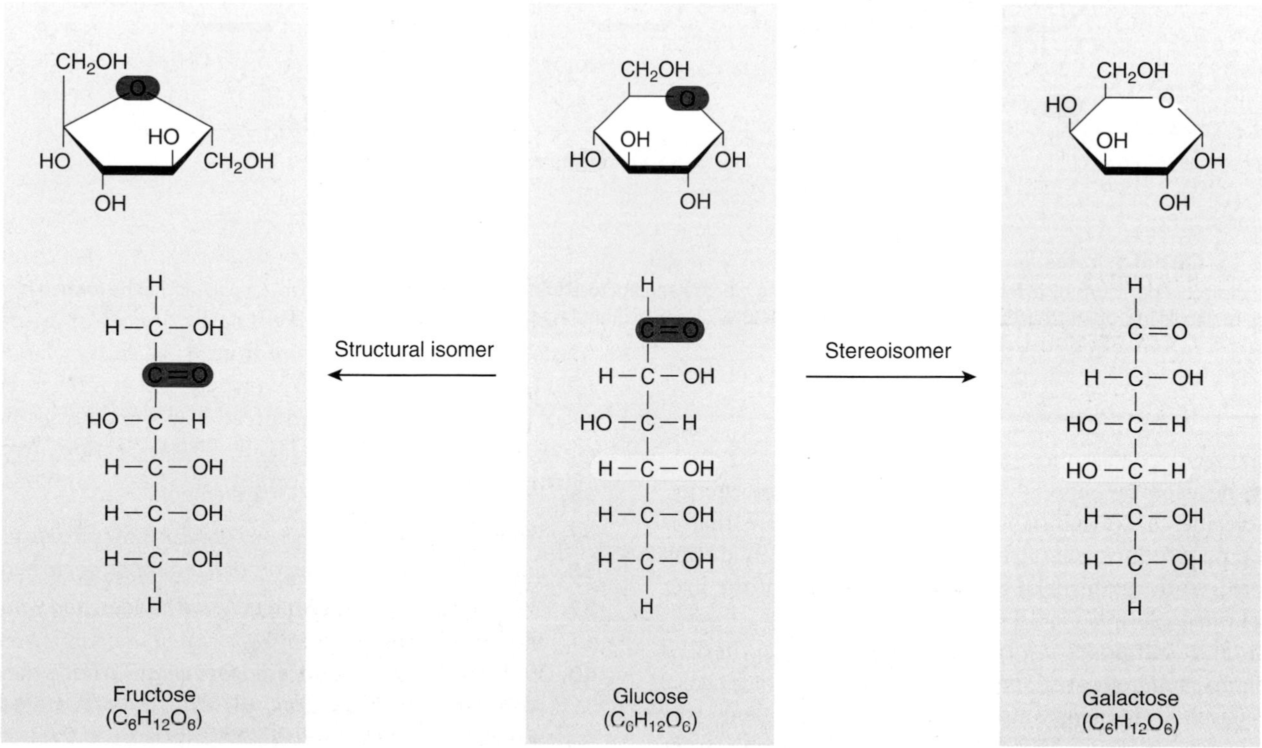

FIGURE 2.14 Monosaccharides

These monosaccharides almost always form a ring-shaped molecule. Although not labeled with a C, carbon atoms are located at the corners of the ring-shaped molecules. Linear models readily illustrate the relationships between the atoms of the molecules. Fructose is a structural isomer of glucose because it has identical chemical groups bonded in a different arrangement in the molecule (red shading). Galactose is a stereoisomer of glucose because it has exactly the same groups bonded to each carbon atom but located in a different three-dimensional orientation (yellow shading).

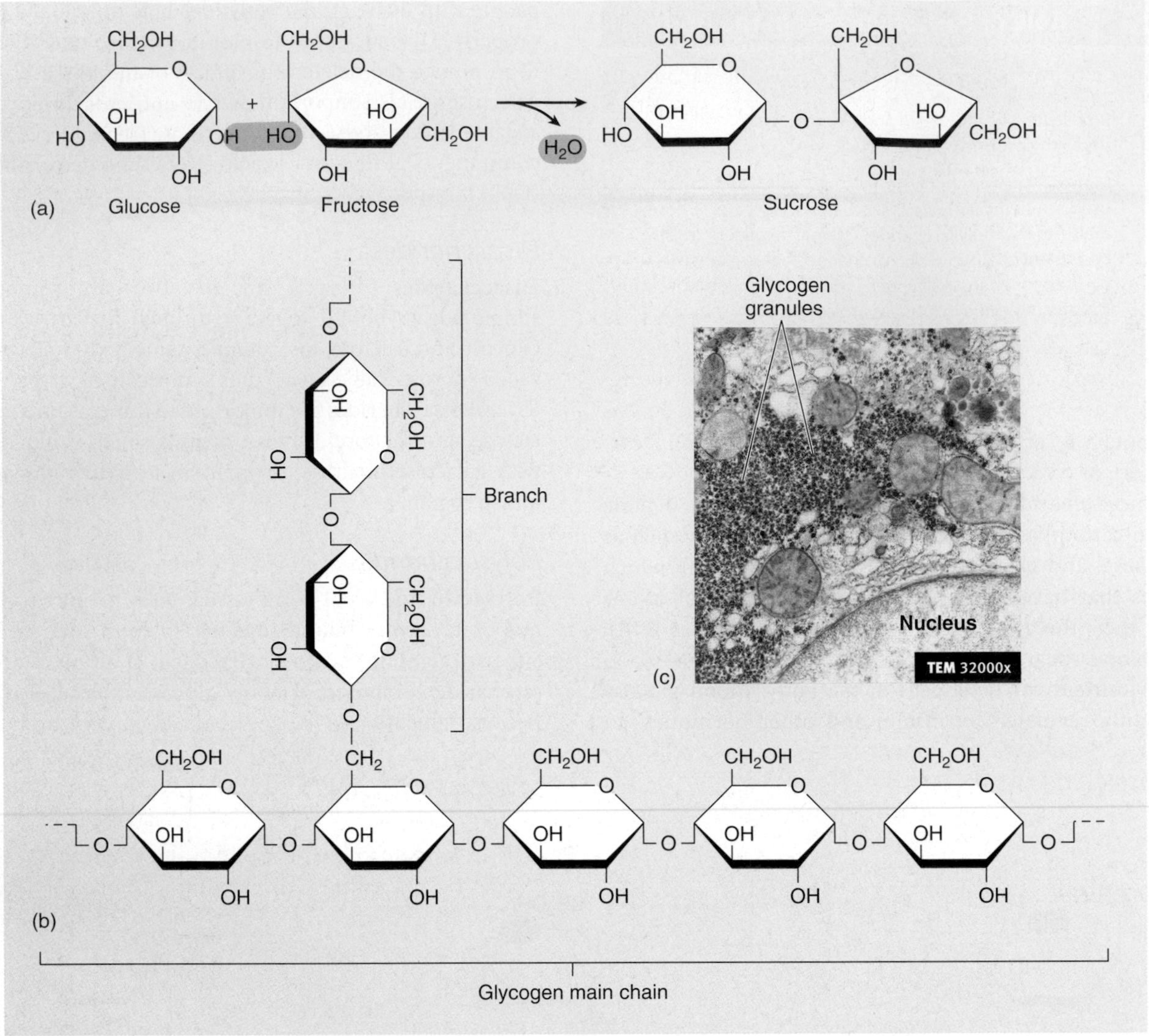

(a) Glucose + Fructose → Sucrose + H_2O

(b) Branch; Glycogen main chain

(c) Glycogen granules; Nucleus; TEM 32000x

FIGURE 2.15 Carbohydrates
(*a*) Sucrose, a disaccharide, forms by a dehydration reaction involving glucose and fructose (monosaccharides). (*b*) Glycogen is a polysaccharide formed by combining many glucose molecules. (*c*) The transmission electron micrograph shows glycogen granules in a liver cell.

energy can be used by cells, glycogen is an important energy-storage molecule. A substantial amount of the glucose that is metabolized to produce energy for muscle contraction during exercise is stored in the form of glycogen in the cells of the liver and skeletal muscles.

Starch and **cellulose** are two important polysaccharides found in plants, and both are composed of long chains of glucose. Plants use starch as an energy-storage molecule in the same way that animals use glycogen, and cellulose is an important structural component of plant cell walls. When humans ingest plants, the starch can be broken down and used as an energy source. Humans, however, do not have the digestive enzymes necessary to break down cellulose. Cellulose is eliminated in the feces, where it provides bulk.

ASSESS YOUR PROGRESS

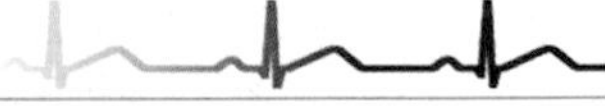

36. *Why is carbon such a versatile element?*

37. *What is the building block of carbohydrates? What are isomers?*

38. *List the 5- and 6-carbon sugars that are important to humans.*

39. *What are disaccharides and polysaccharides, and what type of reaction is used to make them?*

40. *Which carbohydrates are used for energy? What is the function of starch and cellulose in plants? What is the function of glycogen and cellulose in animals?*

Lipids

Lipids are a second major group of organic molecules common to living systems. Like carbohydrates, they are composed principally of carbon, hydrogen, and oxygen, but some lipids contain small amounts of other elements, such as phosphorus and nitrogen. Lipids have a lower ratio of oxygen to carbon than do carbohydrates, which makes them less polar. Consequently, lipids can be dissolved in nonpolar organic solvents, such as alcohol or acetone, but they are relatively insoluble in water.

Lipids have many important functions in the body (table 2.7). They provide protection and insulation, help regulate many physiological processes, and form plasma membranes. In addition, lipids are major energy-storage molecules and can be broken down and used as a source of energy. Several kinds of molecules, such as fats, phospholipids, eicosanoids, steroids, and fat-soluble vitamins, are classified as lipids.

Fats are a major type of lipid. Like carbohydrates, the fats humans ingest are broken down by hydrolysis reactions in cells to release energy for use by those cells. Conversely, if fat intake exceeds need, excess chemical energy from any source can be stored in the body as fat for later use. Fats also provide protection by surrounding and padding organs, and under-the-skin fats act as an insulator to prevent heat loss.

Triglycerides (trī-glis′er-īdz) constitute 95% of the fats in the human body. Triglycerides consist of two different types of building blocks: one glycerol and three fatty acids. **Glycerol** is a 3-carbon molecule with a hydroxyl group attached to each carbon atom, and each **fatty acid** consists of a straight chain of carbon atoms with a carboxyl group attached at one end (figure 2.16). A **carboxyl** (kar-bok′sil) **group** (—COOH) consists of both an oxygen atom and a hydroxyl group attached to a carbon atom:

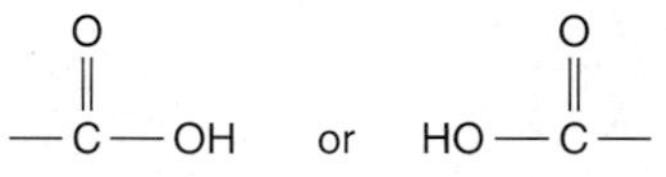

$$-\overset{\displaystyle O}{\overset{\|}{C}}-OH \quad \text{or} \quad HO-\overset{\displaystyle O}{\overset{\|}{C}}-$$

TABLE 2.7 Role of Lipids in the Body

Role	Example
Protection	Fat surrounds and pads organs.
Insulation	Fat under the skin prevents heat loss. Myelin surrounds nerve cells and electrically insulates the cells from one another.
Regulation	Steroid hormones regulate many physiological processes. For example, estrogen and testosterone are the reproductive hormones responsible for many of the differences between males and females. Prostaglandins help regulate tissue inflammation and repair.
Vitamins	Fat-soluble vitamins perform a variety of functions. Vitamin A forms retinol, which is necessary for seeing in the dark; active vitamin D promotes calcium uptake by the small intestine; vitamin E promotes wound healing; and vitamin K is necessary for the synthesis of proteins responsible for blood clotting.
Structure	Phospholipids and cholesterol are important components of the membranes of cells.
Energy	Lipids can be stored and broken down later for energy; per unit of weight, they yield more energy than carbohydrates or proteins.

The carboxyl group is responsible for the acidic nature of the molecule because it releases hydrogen ions into solution. Glycerides can be described according to the number and kinds of fatty acids that combine with glycerol through dehydration reactions. Monoglycerides have one fatty acid, diglycerides have two fatty acids, and triglycerides have three fatty acids bound to glycerol.

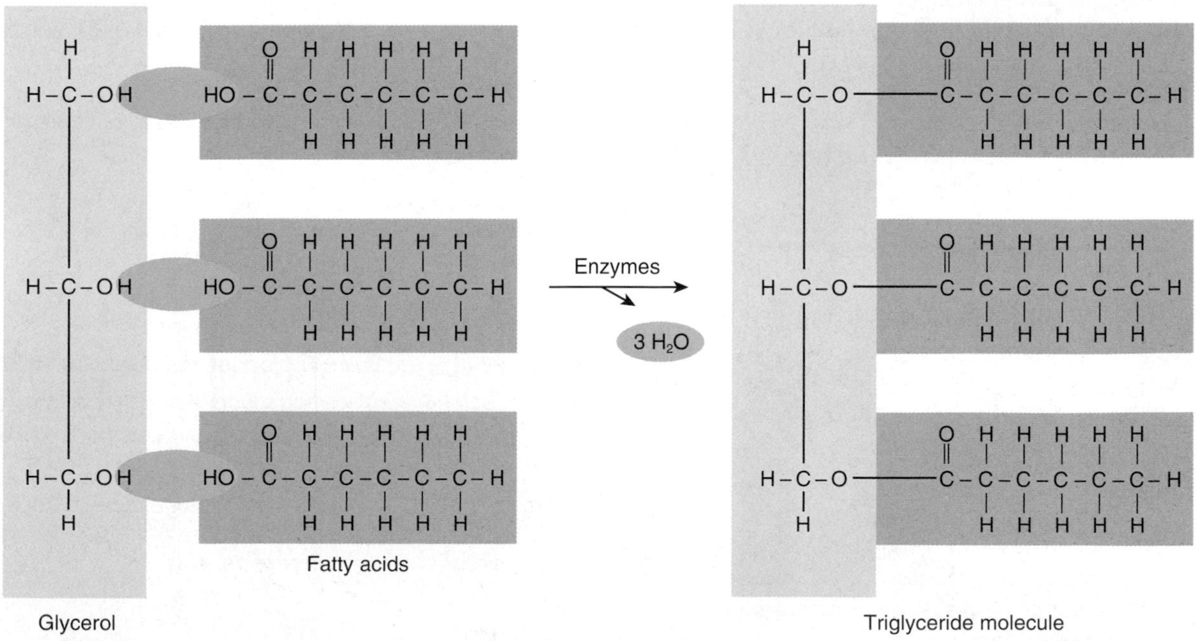

Glycerol · Fatty acids · Enzymes · $3\ H_2O$ · Triglyceride molecule

FIGURE 2.16 Triglyceride

A triglyceride is produced from one glycerol molecule and three fatty acid molecules. One water molecule (H_2O) is given off for each covalent bond formed between a fatty acid molecule and glycerol.

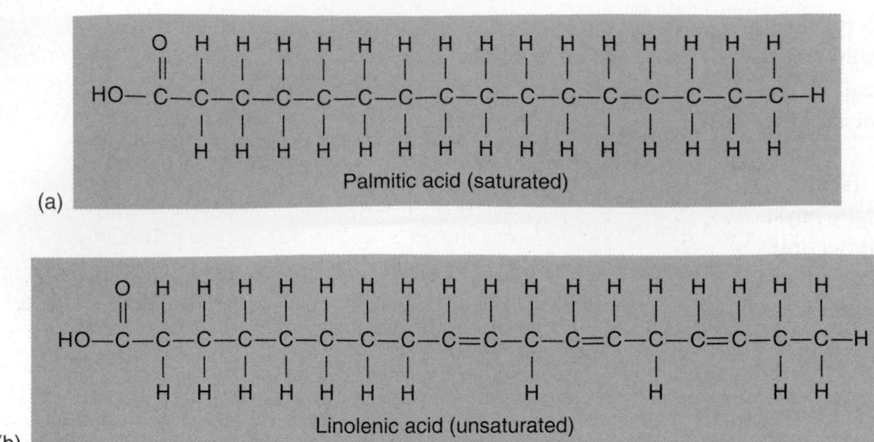
Palmitic acid (saturated)
(a)
Linolenic acid (unsaturated)
(b)

FIGURE 2.17 Fatty Acids
(*a*) Palmitic acid is saturated (having no double bonds between the carbons). (*b*) Linolenic acid is unsaturated (having three double bonds between the carbons). For clarity, the kinks at the double covalent bonds are not shown.

Fatty acids differ from one another according to the length and the degree of saturation of their carbon chains. Most naturally occurring fatty acids contain an even number of carbon atoms, with 14- to 18-carbon chains the most common. *Saturation* refers to the number of hydrogen atoms in the carbon chain. A fatty acid is **saturated** if it contains only single covalent bonds between the carbon atoms (figure 2.17*a*). Sources of saturated fats include beef, pork, whole milk, cheese, butter, eggs, coconut oil, and palm oil. The carbon chain is **unsaturated** if it has one or more double covalent bonds between carbon atoms (figure 2.17*b*). The double covalent bond introduces a kink into the carbon chain, which tends to keep them liquid at room temperature. **Monounsaturated fats,** such as olive and peanut oils, have one double covalent bond between carbon atoms. **Polyunsaturated fats,** such as safflower, sunflower, corn, and fish oils, have two or more double covalent bonds between carbon atoms. Unsaturated fats are the best type of fats in the diet because, unlike saturated fats, they do not contribute to the development of cardiovascular disease.

***Trans* fats** are unsaturated fats that have been chemically altered by the addition of H atoms. The process makes the fats more saturated and hence more solid and stable (longer shelf-life). However, the double covalent bonds that do not become saturated are changed from the usual *cis* configuration (H on the same side of the double bond) to a *trans* configuration (H on different sides.) This change in structure makes the consumption of *trans* fats an even greater factor than saturated fats in the risk for cardiovascular disease.

Phospholipids are similar to triglycerides, except that one of the fatty acids bound to the glycerol is replaced by a molecule containing phosphate and, usually, nitrogen (figure 2.18). A phospholipid is polar at the end of the molecule to which the phosphate is bound and nonpolar at the other end. The polar end of the molecule is attracted to water and is said to be **hydrophilic** (water-loving). The nonpolar end is repelled by water and is said to be **hydrophobic** (water-fearing). Phospholipids are important structural components of the membranes of cells (see figure 3.2).

The **eicosanoids** (ī′kō-să-noydz) are a group of important chemicals derived from fatty acids. They include **prostaglandins** (pros′tă-glan′dinz), **thromboxanes** (throm′bok-zānz), and **leukotrienes** (loo-kō-trī′ēnz). Eicosanoids are made in most cells and are important regulatory molecules. Among their numerous effects is their role in the response of tissues to injuries.

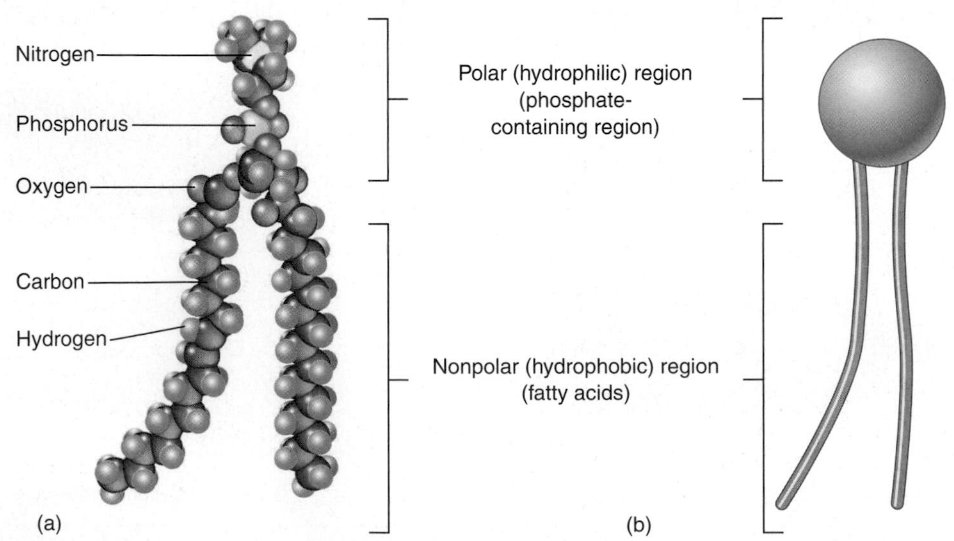
Nitrogen
Phosphorus
Oxygen
Carbon
Hydrogen
(a)
Polar (hydrophilic) region
(phosphate-
containing region)
Nonpolar (hydrophobic) region
(fatty acids)
(b)

FIGURE 2.18 Phospholipids
(*a*) A molecular model of a phospholipid. The bent carbon chain indicates a kink from a double covalent bond. (*b*) A simplified depiction of a phospholipid.

Prostaglandins have been implicated in regulating the secretion of certain hormones, blood clotting, some reproductive functions, and many other processes. Many of the therapeutic effects of aspirin and other anti-inflammatory drugs result from their ability to inhibit prostaglandin synthesis.

Steroids differ in chemical structure from other lipid molecules, but their solubility characteristics are similar. All steroid molecules are composed of carbon atoms bound together into four ringlike structures (figure 2.19). Important steroid molecules include cholesterol, bile salts, estrogen, progesterone, and testosterone. Cholesterol is an especially important steroid because other steroid molecules are synthesized from it. For example, bile salts, which increase fat absorption in the intestines, are derived from cholesterol, as are the reproductive hormones estrogen, progesterone, and testosterone. In addition, cholesterol is an important component of plasma membranes. Although high levels of cholesterol in the blood increase the risk for cardiovascular disease, a certain amount of cholesterol is vital for normal function.

Another class of lipids is the **fat-soluble vitamins.** Their structures are not closely related to one another, but they are nonpolar molecules essential for many normal body functions.

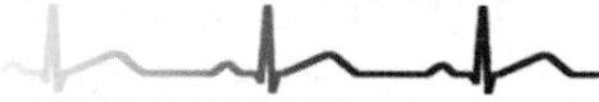

ASSESS **YOUR PROGRESS**

41. *State six roles of lipids in the body, and give an example of each.*

42. *What is the most common fat in the body, and what are its basic building blocks?*

43. *What is the difference between a saturated fat and an unsaturated fat? What is a trans fat?*

44. *Describe the structure of a phospholipid. Which end of the molecule is hydrophilic? Explain why.*

45. *What are three examples of eicosanoids and their general functions?*

46. *Why is cholesterol an important steroid?*

Proteins

All **proteins** contain carbon, hydrogen, oxygen, and nitrogen bound together by covalent bonds, and most proteins contain some sulfur. In addition, some proteins contain small amounts of phosphorus, iron, and iodine. The molecular mass of proteins can be very large. For the purpose of comparison, the molecular mass of water is approximately 18, sodium chloride 58, and glucose 180, but the molecular mass of proteins ranges from approximately 1000 to several million.

Proteins regulate body processes, act as a transportation system, provide protection, help muscles contract, and provide structure and energy. Table 2.8 summarizes the role of proteins in the body.

Protein Structure

The basic building blocks for proteins are the 20 **amino** (ă-mē′nō) **acid** molecules. Each amino acid has an amine (ă-mēn′) group (—NH_2), a carboxyl group (—COOH), a hydrogen atom, and a side chain designated by the symbol R attached to the same carbon atom. The side chain can be a variety of chemical structures, and the differences in the side chains make the amino acids different from one another (figure 2.20).

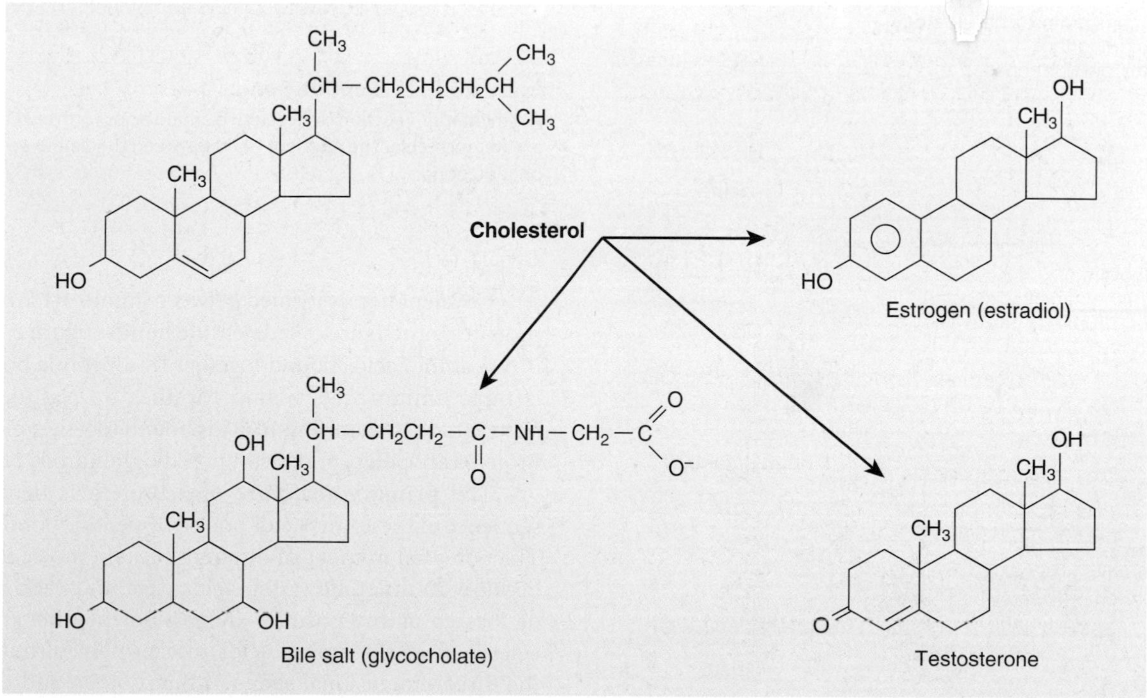

FIGURE 2.19 Steroids
Steroids are four-ringed molecules that differ from one another according to the groups attached to the rings. Cholesterol, the most common steroid, can be modified to produce other steroids.

TABLE 2.8 Role of Proteins in the Body

Role	Example
Regulation	Enzymes control chemical reactions. Hormones regulate many physiological processes; for example, insulin affects glucose transport into cells.
Transport	Hemoglobin transports oxygen and carbon dioxide in the blood. Plasma proteins transport many substances in the blood. Proteins in plasma membranes control the movement of materials into and out of the cell.
Protection	Antibodies protect against microorganisms and other foreign substances.
Contraction	Actin and myosin in muscle are responsible for muscle contraction.
Structure	Collagen fibers form a structural framework in many parts of the body. Keratin adds strength to skin, hair, and nails.
Energy	Proteins can be broken down for energy; per unit of weight, they yield as much energy as carbohydrates do.

The general structure of an amino acid showing the amine group ($-NH_2$), carboxyl group ($-COOH$), and hydrogen atom highlighted in yellow. The R side chain is the part of an amino acid that makes it different from other amino acids.

Glycine is the simplest amino acid. The side chain is a hydrogen atom.

Tyrosine, which has a more complicated side chain, is an important component of thyroid hormones.

Improper metabolism of phenylalanine in the genetic disease phenylketonuria (PKU) can cause mental retardation.

Aspartic acid combined with phenylalanine forms the artificial sweetener aspartame (Nutrasweet™ and Equal™).

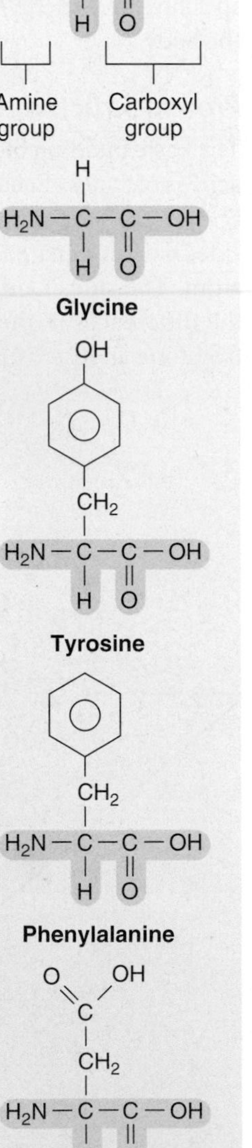
Amine group
Carboxyl group
Glycine
Tyrosine
Phenylalanine
Aspartic acid

FIGURE 2.20 **Amino Acids**

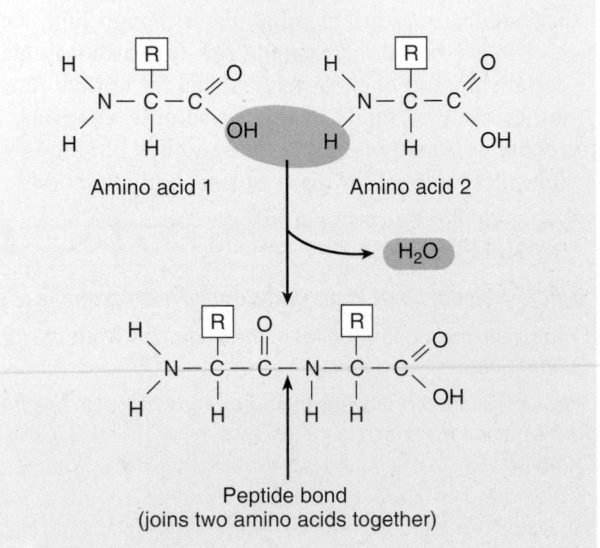
Amino acid 1
Amino acid 2
H_2O
Peptide bond
(joins two amino acids together)

FIGURE 2.21 **Peptide Bond**
A dehydration reaction between two amino acids forms a dipeptide and a water molecule. The covalent bond between the amino acids is called a peptide bond.

Covalent bonds formed between amino acid molecules during protein synthesis are called **peptide bonds** (figure 2.21). A dipeptide is two amino acids bound together by a peptide bond, a tripeptide is three amino acids bound together by peptide bonds, and a polypeptide is many amino acids bound together by peptide bonds. Proteins are polypeptides composed of hundreds of amino acids.

The **primary structure** of a protein is determined by the sequence of the amino acids bound by peptide bonds (figure 2.22*a*). The potential number of different protein molecules is enormous because 20 different amino acids exist and each amino acid can be located at any position along a polypeptide chain. The characteristics of the amino acids in a protein ultimately determine the three-dimensional shape of the protein, and the shape of the protein determines its function. A change in one or a few amino acids in the primary structure can alter protein function, usually making the protein less functional or even nonfunctional.

(a) Primary structure—the amino acid sequence. A protein consists of a chain of different amino acids (represented by different colored spheres).

(b) Secondary structure results from hydrogen bonding (*dotted red lines*). The hydrogen bonds cause the amino acid chain to form pleated (folded) sheets or helices (coils).

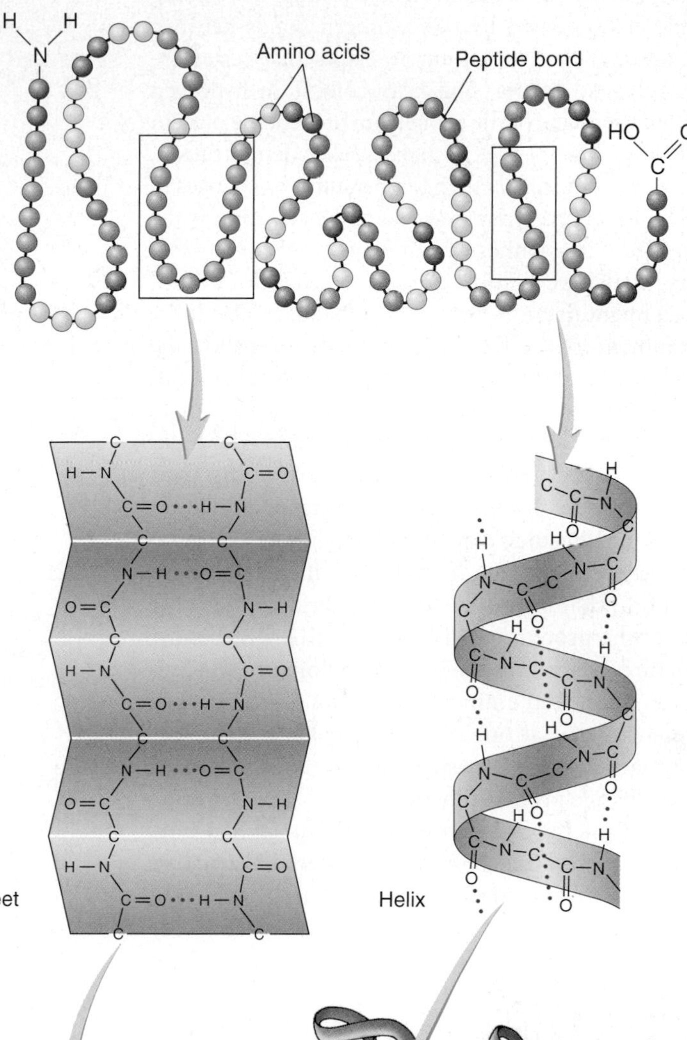
Amino acids
Peptide bond
Pleated sheet
Helix

(c) Tertiary structure with secondary folding caused by interactions within the polypeptide and its immediate environment

(d) Quaternary structure—the relationships between individual subunits

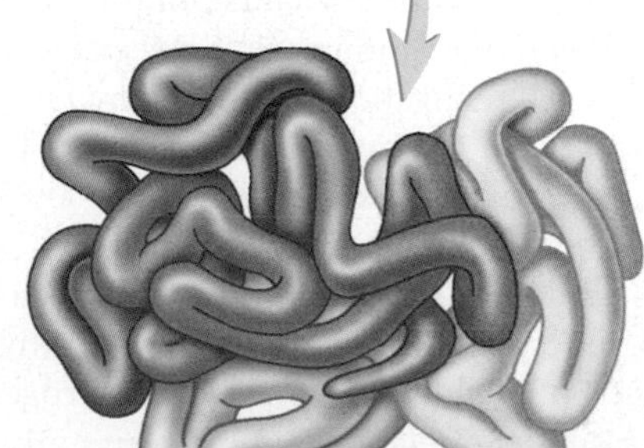

FIGURE 2.22 Protein Structure

The **secondary structure** results from the folding or bending of the polypeptide chain caused by the hydrogen bonds between amino acids (figure 2.22*b*). Two common shapes that result are pleated (folded) sheets and helices (sing. *helix,* coil). If the hydrogen bonds that maintain the shape of the protein are broken, the protein becomes nonfunctional. This change in shape is called **denaturation,** and it can be caused by abnormally high temperatures or changes in the pH of body fluids. An everyday example of denaturation is the change in the proteins of egg whites when they are cooked.

The **tertiary structure** results from large-scale folding of the protein driven by interactions within the protein and with the immediate environment (figure 2.22*c*). These interactions allow the pleated sheets and helices of the secondary structure to be arranged and organized relative to each other. Some amino acids are quite polar and therefore form hydrogen bonds with water. The polar portions of proteins tend to remain unfolded, maximizing their contact with water, whereas the less polar regions tend to fold into a globular shape, minimizing their contact with water. The formation of covalent bonds between sulfur atoms located at different locations along the polypeptide chain produces disulfide bridges that hold different regions of the protein together in the tertiary structure. The tertiary structure determines the shape of a **domain,** which is a folded sequence of 100–200 amino acids within a protein. The functions of proteins occur at one or more domains. Therefore, changes in the primary or secondary structure that affect the shape of the domain can change protein function.

If two or more proteins associate to form a functional unit, the individual proteins are called subunits. The **quaternary structure** results from the spatial relationships between the individual subunits (figure 2.22*d*).

Enzymes

Proteins perform many roles in the body, including acting as enzymes. An **enzyme** is a protein catalyst that increases the rate at which a chemical reaction proceeds without the enzyme being permanently changed. The three-dimensional shape of enzymes is critical for their normal function because it determines the structure of the enzyme's **active site.** According to the **lock-and-key model** of enzyme action, a reaction occurs when the reactants (key) bind to the active site (lock) on the enzyme. This view of enzymes and reactants as rigid structures fitting together has been modified by the **induced fit model,** in which the enzyme is able to change shape slightly and better fit the reactants. The enzyme is like a glove that does not achieve its functional shape until the hand (reactants) moves into place.

At the active site, reactants are brought into close proximity and the reaction occurs (figure 2.23). After the reactants combine, they are released from the active site, and the enzyme is capable of catalyzing additional reactions. The activation energy required for a chemical reaction to occur is lowered by enzymes (see figure 2.11) because they orient the reactants toward each other in such a way that a chemical reaction is more likely to occur.

Slight changes in the structure of an enzyme can destroy the active site's ability to function. Enzymes are very sensitive to changes in temperature or pH, which can break the hydrogen bonds within them. As a result, the relationship between amino acids changes, thereby producing a change in shape that prevents the enzyme from functioning normally.

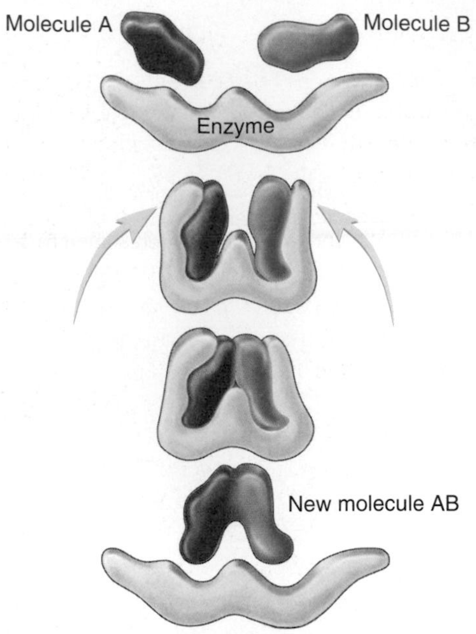
Molecule A
Molecule B
Enzyme
New molecule AB

FIGURE 2.23 Enzyme Action
The enzyme brings the two reacting molecules together. After the reaction, the unaltered enzyme can be used again.

To be functional, some enzymes require additional, nonprotein substances called **cofactors.** A cofactor can be an ion, such as magnesium or zinc, or an organic molecule. Cofactors that are organic molecules, such as certain vitamins, may be referred to as **coenzymes.** Cofactors normally form part of the enzyme's active site and are required to make the enzyme functional.

Because an enzyme's active site can bind only to certain reactants, each enzyme catalyzes a specific chemical reaction and no others. Therefore, many different enzymes are needed to catalyze the many chemical reactions of the body. Enzymes are often named by adding the suffix *-ase* to the name of the molecules on which they act. For example, an enzyme that catalyzes the breakdown of lipids is a **lipase** (lip′ās, lī′pās), and an enzyme that breaks down proteins is a **protease** (prō′tē-ās).

Enzymes control the rate at which most chemical reactions proceed in living systems. Consequently, they control essentially all cellular activities. At the same time, the activity of enzymes themselves is regulated by several mechanisms within the cells. Some mechanisms control the enzyme concentration by influencing the rate at which the enzymes are synthesized; others alter the activity of existing enzymes. Much of our knowledge about the regulation of cellular activity involves understanding how enzyme activity is controlled.

ASSESS YOUR PROGRESS

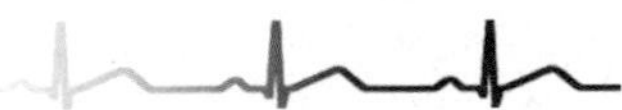

47. *What are the building blocks of proteins? What type of bond chemically connects these building blocks? What is the importance of the R group?*
48. *What determines the primary, secondary, tertiary, and quaternary structures of a protein?*
49. *What is denaturation? Name two factors that can cause it.*
50. *Compare the lock-and-key and the induced fit models of enzyme activity. What determines the active site of an enzyme? State the difference between a cofactor and a coenzyme.*

Nucleic Acids: DNA and RNA

The **nucleic** (noo-klē′ik, noo-klā′ik) **acids** are large molecules composed of carbon, hydrogen, oxygen, nitrogen, and phosphorus. **Deoxyribonucleic** (dē-oks′ē-rī′bō-noo-klē′ik) **acid** (**DNA**) is the genetic material of cells, and copies of DNA are transferred from one generation of cells to the next generation. DNA contains the information that determines the structure of proteins. **Ribonucleic** (rī′bō-noo-klē′ik) **acid** (**RNA**) is structurally related to DNA, and three types of RNA also play important roles in protein synthesis. Chapter 3 describes the means by which DNA and RNA direct the functions of the cell.

Both DNA and RNA consist of basic building blocks called **nucleotides** (noo′klē-ō-tīdz). Each nucleotide is composed of a monosaccharide to which a nitrogenous base and a phosphate group are attached (figure 2.24). The 5-carbon monosaccharide is **deoxyribose** for DNA; it is **ribose** for RNA. The nitrogenous bases consist of carbon and nitrogen atoms organized into rings. They are bases because the nitrogen atoms tend to take up H^+ from solution. The nitrogenous bases are cytosine (sī′tō-sēn), thymine (thī′mēn, thī′min), and uracil (ūr′ă-sil), which have a single ring, and guanine (gwahn′ēn) and adenine (ad′ĕ-nēn), which have two rings each. Single-ringed bases are called pyrimidines (pī-rim′i-dēnz), and double-ringed bases are called purines (pūr′ēnz; figure 2.25).

DNA has two strands of nucleotides joined together to form a twisted, ladderlike structure called a double helix (figure 2.26). The sides of the ladder are formed by covalent bonds between the deoxyribose molecules and phosphate groups of adjacent nucleotides. The rungs of the ladder are formed by the bases of the nucleotides of one side connected to the bases of the other side by hydrogen bonds. Each nucleotide of DNA contains one of the organic bases: adenine, thymine, cytosine, or guanine. **Complementary base pairs** are bases held together by hydrogen bonds. Adenine and thymine are complementary base pairs because the structure of these bases allows two hydrogen bonds to form between them. Cytosine and guanine are complementary base pairs because the structure of these bases allows three hydrogen bonds to form between them. The two strands of a DNA molecule are said to be complementary. If the sequence of bases in one DNA strand is known, the sequence of bases in the other strand can be predicted because of complementary base pairing.

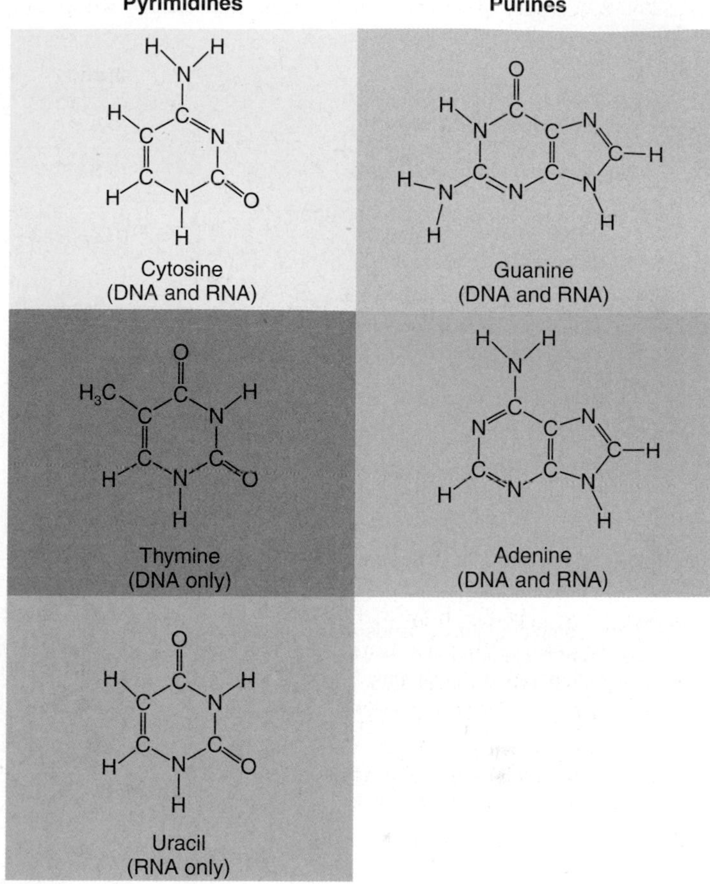
Pyrimidines
Purines
Cytosine
(DNA and RNA)
Guanine
(DNA and RNA)
Thymine
(DNA only)
Adenine
(DNA and RNA)
Uracil
(RNA only)

FIGURE 2.25 Nitrogenous Bases
The organic bases found in nucleic acids are separated into two groups. Pyrimidines are single-ringed molecules, and purines are double-ringed molecules.

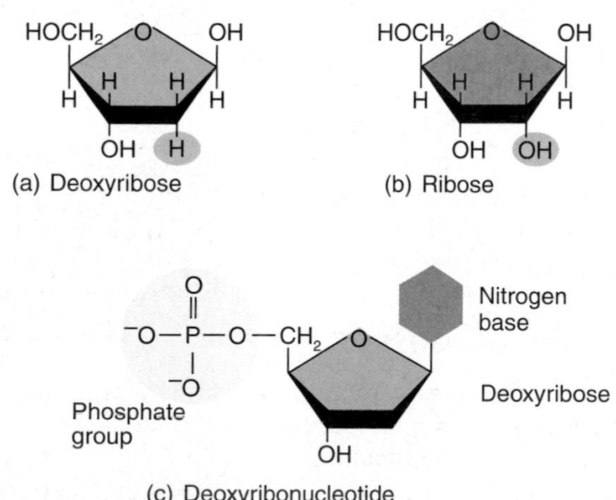
(a) Deoxyribose
(b) Ribose
Nitrogen base
Deoxyribose
Phosphate group
(c) Deoxyribonucleotide

FIGURE 2.24 Components of Nucleotides
(a) Deoxyribose sugar, which forms nucleotides used in DNA production. *(b)* Ribose sugar, which forms nucleotides used in RNA production. Note that deoxyribose is ribose minus an oxygen atom. *(c)* A deoxyribonucleotide consists of deoxyribose, a nitrogen base, and a phosphate group.

The two nucleotide strands of a DNA molecule are **antiparallel,** meaning that the two strands lie side by side but their sugar-phosphate "backbones" extend in opposite directions because of the orientation of their nucleotides (figure 2.26). A nucleotide has a 5′ end and a 3′ end. The prime sign is used to indicate the carbon atoms of the deoxyribose sugar, which are numbered 1′ to 5′.

The sequence of nitrogenous bases in DNA is a "code" that stores information used to determine the structures and functions of cells. A sequence of DNA bases that directs the synthesis of proteins or RNA molecules is called a **gene** (see chapter 3 for more information on genes). Genes determine the type and sequence of amino acids in protein molecules. Because enzymes are proteins, DNA structure determines the rate and type of chemical reactions

1 The building blocks of nucleic acids are nucleotides, which consist of a phosphate group, a sugar, and a nitrogenous base.

2 The phosphate groups connect the sugars to form two strands of nucleotides (*purple columns*).

3 Hydrogen bonds (*dotted red lines*) between the nucleotides join the two nucleotide strands together. Adenine binds to thymine and cytosine binds to guanine.

4 Deoxyribose carbon atoms are numbered. One end of a DNA strand has a 3′ end because of the orientation of its nucleotides.

5 The other end of a DNA strand has a 5′ end.

6 The complementary strands are antiparallel in that the 5′→ 3′ direction of one strand runs counter to the 5′→ 3′ direction of the other strand.

7 The nucleotide strands coil to form a double-stranded helix.

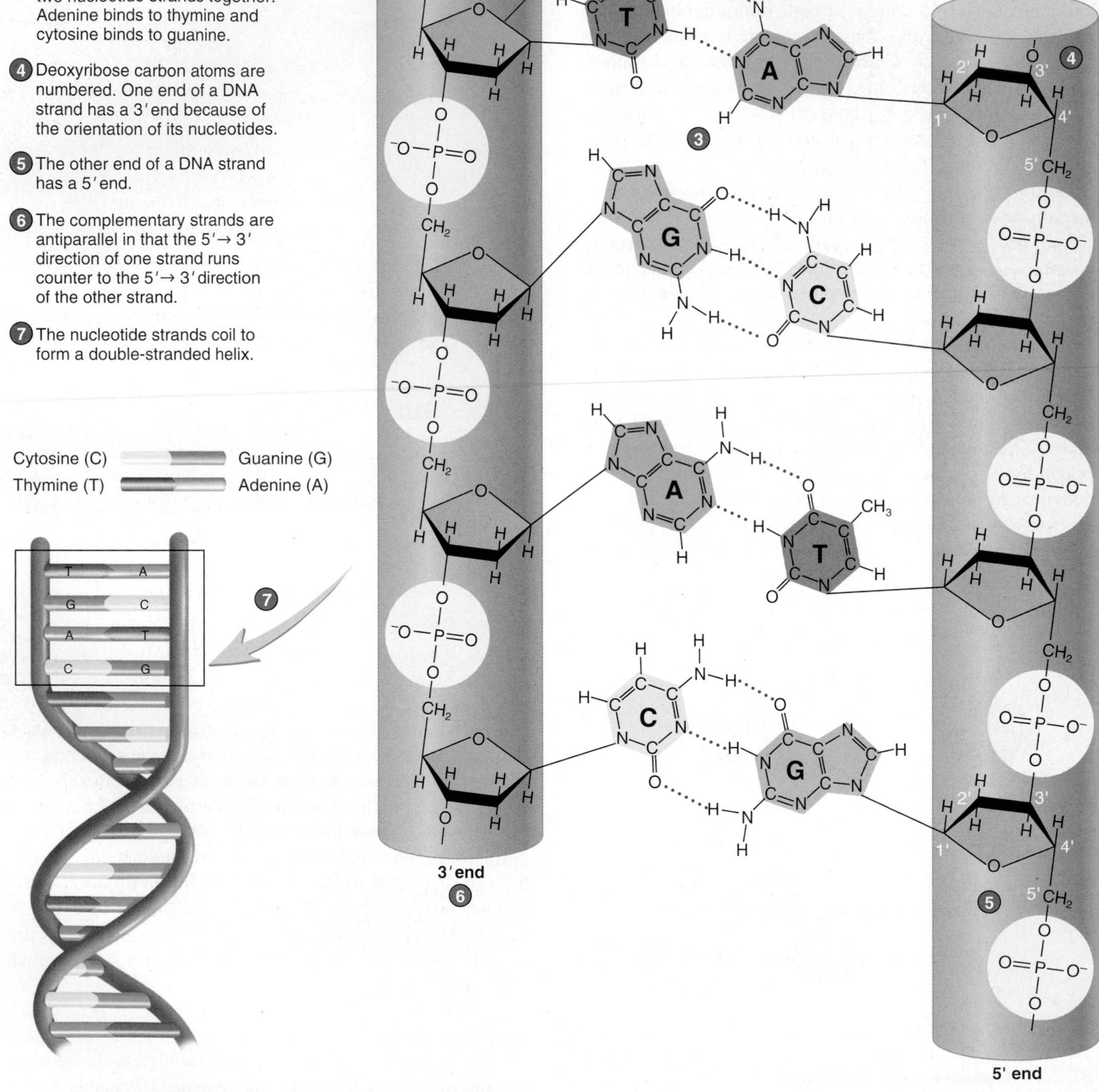

Cytosine (C)
Guanine (G)
Thymine (T)
Adenine (A)
5′ end
3′ end
Phosphate group
Sugar (deoxyribose)
Nitrogen base (thymine)
Nucleotide
3′ end
5′ end

PROCESS FIGURE 2.26 APR **Structure of DNA**

that occur in cells by controlling enzyme structure. Therefore, the information contained in DNA ultimately defines all cellular activities. Other proteins that are coded by DNA, such as collagen, determine many of the structural features of humans.

RNA's structure is similar to a single strand of DNA. Like DNA, four different nucleotides make up the RNA molecule, and the nitrogenous bases are the same, except that thymine is replaced with uracil (see figure 2.25). Uracil can bind only to adenine.

Adenosine Triphosphate

Adenosine triphosphate (ă-den′ō-sēn trī-fos′fāt; **ATP**) is an especially important organic molecule in all living organisms. It consists of adenosine (the sugar ribose with the nitrogenous base adenine) and three phosphate groups (figure 2.27). The potential energy stored in the covalent bond between the second and third phosphate groups of ATP is important to living organisms because it provides the energy used in nearly all of the chemical reactions within cells. Removal of the third phosphate generates adenosine diphosphate (ADP), which has only two phosphate groups.

The catabolism of glucose and other nutrient molecules results in chemical reactions that release energy. Some of that energy is used to synthesize ATP from ADP and an inorganic phosphate group (P_i):

$$\text{ADP} + P_i + \text{Energy (from catabolism)} \rightarrow \text{ATP}$$

The transfer of energy from nutrient molecules to ATP involves a series of oxidation-reduction reactions in which a high-energy electron is transferred from one molecule to the next molecule in the series. In chapter 25, the oxidation-reduction reactions of metabolism are considered in greater detail.

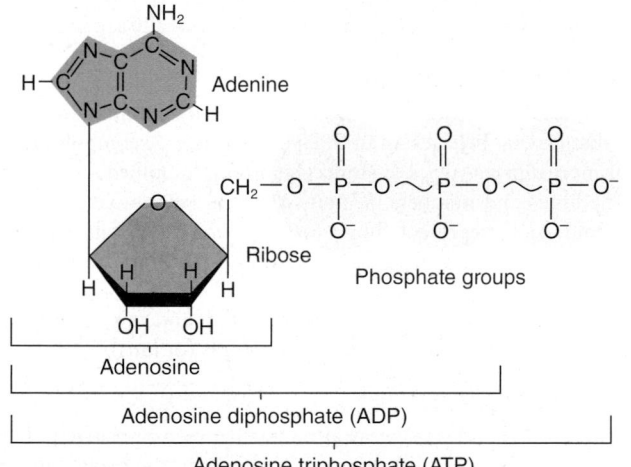

NH_2
Adenine
$CH_2-O-P-O\sim P-O\sim P-O^-$
Ribose
Phosphate groups
OH OH
Adenosine
Adenosine diphosphate (ADP)
Adenosine triphosphate (ATP)

FIGURE 2.27 AP|R **Structure of an Adenosine Triphosphate (ATP) Molecule**

Once produced, ATP is used to provide energy for other chemical reactions (anabolism) or to drive cell processes, such as muscle contraction. In the process, ATP is converted back to ADP and an inorganic phosphate group:

$$\text{ATP} \rightarrow \text{ADP} + P_i + \text{Energy (for anabolism and other cell processes)}$$

ATP is often called the energy currency of cells because it is capable of both storing and providing energy. The concentration of ATP is maintained within a narrow range of values, and essentially all energy-requiring chemical reactions stop when the ATP levels become inadequate.

ASSESS **YOUR PROGRESS**

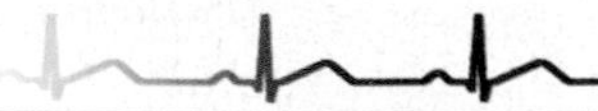

51. *Name two types of nucleic acids, and state their functions.*
52. *What are the basic building blocks of nucleic acids? What kinds of sugars and bases are found in DNA? In RNA?*
53. *DNA is like a twisted ladder. What forms the sides of the ladder? The rungs?*
54. *Name the complementary base pairs in DNA and RNA.*
55. *What is meant by the statement "DNA strands are antiparallel"?*
56. *Define* gene, *and explain how genes determine the structures and functions of cells.*
57. *Describe the structure of ATP. Where does the energy to synthesize ATP come from? What is the energy stored in ATP used for?*

Clinical IMPACT

Cyanide Poisoning

Cyanide compounds can be lethal to humans because they interfere with the production of ATP in mitochondria (see chapter 25). Without sufficient ATP, cells die because there is inadequate energy for anabolic chemical reactions, active transport, and other energy-requiring cell processes. The heart and brain are especially susceptible to cyanide poisoning. The most common cause of cyanide poisoning is inhalation of smoke released by the burning of rubber and plastic in household fires. Cyanide poisoning by inhalation or absorption through the skin can also occur in certain manufacturing processes, and cyanide gas was used to kill people during the Holocaust. Deliberate suicide by ingesting cyanide is rare but was made famous by suicide capsules in spy movies. In 1982, seven people in the Chicago area died after taking Tylenol that someone had laced with cyanide. Subsequent copycat tamperings occurred and led to the widespread use of tamper-proof capsules and packaging.

Answer

Learn to Predict ❮ From page 24

To understand the reactions occurring in the test tube, you will need to ask yourself the following questions: (1) What was in the saliva that changed the starch? (2) How do enzymes speed up chemical reactions in the body? To answer these questions, let us determine what important information was provided in the question. First, we know that starch, which is a polysaccharide, is the primary material in the test tube and that iodine stains starch a blue color. Also, saliva is added to the test tube. Saliva contains digestive enzymes that break down carbohydrates, such as starch. Finally, we are told that after 30 minutes the blue color disappeared. This indicates that starch is no longer present in the test tube. We can therefore conclude that digestive enzymes in the saliva catalyzed a decomposition reaction, breaking down the starch to a different material.

Chapter 2 teaches us that enzymes are protein catalysts that speed up chemical reactions by lowering the activation energy. Activation energy is the minimum energy that the reactants must have to start the chemical reaction. Therefore, the digestive enzymes in the saliva lowered the activation energy needed to break the bonds in the starch molecules.

Answers to the rest of this chapter's Predict questions are in Appendix G.

Summary

Chemistry is the study of the composition, structure, and properties of substances and the reactions they undergo. Much of the structure and function of healthy or diseased organisms can be understood at the chemical level.

2.1 Basic Chemistry (p. 25)

Matter, Mass, and Weight

1. Matter is anything that occupies space and has mass.
2. Mass is the amount of matter in an object.
3. Weight results from the force exerted by earth's gravity on matter.

Elements and Atoms

1. An element is the simplest type of matter having unique chemical and physical properties.
2. An atom is the smallest particle of an element that has the chemical characteristics of that element. An element is composed of only one kind of atom.
3. Atoms consist of protons, neutrons, and electrons.
   - Protons are positively charged, electrons are negatively charged, and neutrons have no charge.
   - Protons and neutrons are in the nucleus; electrons are located around the nucleus and can be represented by an electron cloud.
4. The atomic number is the unique number of protons in an atom. The mass number is the sum of the protons and the neutrons.
5. Isotopes are atoms that have the same atomic number but different mass numbers.
6. The atomic mass of an element is the average mass of its naturally occurring isotopes weighted according to their abundance.
7. A mole of a substance contains Avogadro's number (6.022×10^{23}) of atoms, ions, or molecules. The molar mass of a substance is the mass of 1 mole of the substance expressed in grams.

Electrons and Chemical Bonding

1. The chemical behavior of atoms is determined mainly by their outermost electrons. A chemical bond occurs when atoms share or transfer electrons.
2. Ions are atoms that have gained or lost electrons.
   - An atom that loses 1 or more electrons becomes positively charged and is called a cation. An anion is an atom that becomes negatively charged after accepting 1 or more electrons.
   - An ionic bond results from the attraction of the oppositely charged cation and anion to each other.
3. A covalent bond forms when electron pairs are shared between atoms. A polar covalent bond results when the sharing of electrons is unequal and can produce a polar molecule that is electrically asymmetric.

Molecules and Compounds

1. A molecule is two or more atoms chemically combined to form a structure that behaves as an independent unit. A compound is two or more *different* types of atoms chemically combined.
2. The kinds and numbers of atoms (or ions) in a molecule or compound can be represented by a formula consisting of the symbols of the atoms (or ions) plus subscripts denoting the number of each type of atom (or ion).
3. The molecular mass of a molecule or compound can be determined by adding up the atomic masses of its atoms (or ions).

Intermolecular Forces

1. A hydrogen bond is the weak attraction between a positively charged hydrogen and negatively charged oxygen or other polar molecule. Hydrogen bonds are important in determining properties of water and the three-dimensional structure of large molecules.
2. Solubility is the ability of one substance to dissolve in another. Ionic substances that dissolve in water by dissociation are electrolytes. Molecules that do not dissociate are nonelectrolytes.

2.2 Chemical Reactions and Energy (p. 32)

Synthesis Reactions

1. A synthesis reaction is the chemical combination of two or more substances to form a new or larger substance.

2. A dehydration reaction is a synthesis reaction in which water is produced.
3. The sum of all the synthesis reactions in the body is called anabolism.

Decomposition Reactions

1. A decomposition reaction is the chemical breakdown of a larger substance to two or more different and smaller substances.
2. A hydrolysis reaction is a decomposition reaction in which water is depleted.
3. The sum of all the decomposition reactions in the body is called catabolism.

Reversible Reactions

Reversible reactions produce an equilibrium condition in which the amount of reactants relative to the amount of products remains constant.

Oxidation-Reduction Reactions

Oxidation-reduction reactions involve the complete or partial transfer of electrons between atoms.

Energy

1. Energy is the ability to do work. Potential energy is stored energy, and kinetic energy is energy resulting from the movement of an object.
2. Chemical energy
   - Chemical bonds are a form of potential energy.
   - Chemical reactions in which the products contain more potential energy than the reactants require the input of energy.
   - Chemical reactions in which the products have less potential energy than the reactants release energy.
3. Heat energy
   - Heat energy is energy that flows between objects that are at different temperatures.
   - Heat energy is released in chemical reactions and is responsible for body temperature.

Speed of Chemical Reactions

1. Activation energy is the minimum energy that the reactants must have to start a chemical reaction.
2. Enzymes are specialized protein catalysts that lower the activation energy for chemical reactions. Enzymes speed up chemical reactions but are not consumed or altered in the process.
3. Increased temperature and concentration of reactants can increase the rate of chemical reactions.

2.3 Inorganic Chemistry (p. 36)

Inorganic chemistry is mostly concerned with non-carbon-containing substances but does include some carbon-containing substances, such as carbon dioxide and carbon monoxide that lack carbon-hydrogen bonds. Some inorganic chemicals play important roles in the body.

Water

1. Water is a polar molecule composed of one atom of oxygen and two atoms of hydrogen.
2. Because water molecules form hydrogen bonds with each other, water is good at stabilizing body temperature, protecting against friction and trauma, making chemical reactions possible, directly participating in chemical reactions (e.g., dehydration and hydrolysis reactions), and serving as a mixing medium (e.g., solutions, suspensions, and colloids).
3. A mixture is a combination of two or more substances physically blended together, but not chemically combined.
4. A solution is any liquid, gas, or solid in which the substances are uniformly distributed, with no clear boundary between the substances.
5. A solute dissolves in a solvent.
6. A suspension is a mixture containing materials that separate from each other unless they are continually, physically blended together.
7. A colloid is a mixture in which a dispersed (solutelike) substance is distributed throughout a dispersing (solventlike) substance. Particles do not settle out of a colloid.

Solution Concentrations

1. One measurement of solution concentration is the osmole, which contains Avogadro's number (6.022×10^{23}) of particles (i.e., atoms, ions, or molecules) in 1 kilogram of water.
2. A milliosmole is 1/1000 of an osmole.

Acids and Bases

1. Acids are proton (H^+) donors, and bases (e.g., OH^-) are proton acceptors.
2. A strong acid or base almost completely dissociates in water. A weak acid or base partially dissociates.
3. The pH scale shows the H^+ concentrations of various solutions.
   - A neutral solution has an equal number of $H^+$and OH^- and is assigned a pH of 7.
   - Acidic solutions, in which the number of H^+ is greater than the number of OH^-, have pH values less than 7.
   - Basic, or alkaline, solutions have more OH^- than H^+ and a pH greater than 7.
4. A salt is a molecule consisting of a cation other than H^+ and an anion other than OH^-. Salts form when acids react with bases.
5. A buffer is a solution of a conjugate acid-base pair that resists changes in pH when acids or bases are added to the solution.

Oxygen and Carbon Dioxide

Oxygen is necessary for the reactions that extract energy from food molecules in living organisms. When the organic molecules are broken down during metabolism, carbon dioxide and energy are released.

2.4 Organic Chemistry (p. 39)

Organic molecules contain carbon and hydrogen atoms bound together by covalent bonds.

Carbohydrates

1. Monosaccharides are the basic building blocks of other carbohydrates. Examples are ribose, deoxyribose, glucose, fructose, and galactose. Glucose is an especially important source of energy.
2. Disaccharide molecules are formed by dehydration reactions between two monosaccharides. They are broken apart into monosaccharides by hydrolysis reactions. Examples of disaccharides are sucrose, lactose, and maltose.
3. A polysaccharide is composed of many monosaccharides bound together to form a long chain. Examples include cellulose, starch, and glycogen.

Lipids

1. Triglycerides are composed of glycerol and fatty acids. One, two, or three fatty acids can attach to the glycerol molecule.
   - Fatty acids are straight chains of carbon molecules with a carboxyl group. Fatty acids can be saturated (having only single covalent bonds between carbon atoms) or unsaturated (having one or more double covalent bonds between carbon atoms).
   - Energy is stored in fats.
2. Phospholipids are lipids in which a fatty acid is replaced by a phosphate-containing molecule. Phospholipids are a major structural component of plasma membranes.
3. Steroids are lipids composed of four interconnected ring molecules. Examples are cholesterol, bile salts, and sex hormones.
4. Other lipids include fat-soluble vitamins, prostaglandins, thromboxanes, and leukotrienes.

Proteins

1. The building blocks of a protein are amino acids, which are joined by peptide bonds.
2. The number, kind, and arrangement of amino acids determine the primary structure of a protein. Hydrogen bonds between amino acids determine secondary structure, and hydrogen bonds between amino acids and water determine tertiary structure. Interactions between different protein subunits determine quaternary structure.
3. Enzymes are protein catalysts that speed up chemical reactions by lowering their activation energy.
4. The active sites of enzymes bind only to specific reactants.
5. Cofactors are ions or organic molecules, such as vitamins, that are required for some enzymes to function.

Nucleic Acids: DNA and RNA

1. The basic unit of nucleic acids is the nucleotide, which is a monosaccharide with an attached phosphate and a nitrogenous base.
2. DNA nucleotides contain the monosaccharide deoxyribose and the nitrogenous base adenine, thymine, guanine, or cytosine. DNA occurs as a double strand of joined nucleotides. Each strand is complementary and antiparallel to the other strand.
3. A gene is a sequence of DNA nucleotides that determines the structure of a protein or RNA.
4. RNA nucleotides are composed of the monosaccharide ribose. The nitrogenous bases are the same as for DNA, except that thymine is replaced with uracil.

Adenosine Triphosphate

Adenosine triphosphate (ATP) stores energy derived from catabolism. The energy released from ATP is used in anabolism and other cell processes.

REVIEW AND COMPREHENSION

1. The smallest particle of an element that still has the chemical characteristics of that element is a(n)
   a. electron. c. neutron. e. atom.
   b. molecule. d. proton.
2. ^{12}C and ^{14}C are
   a. atoms of different elements.
   b. isotopes.
   c. atoms with different atomic numbers.
   d. atoms with different numbers of protons.
   e. compounds.
3. A cation is a(n)
   a. uncharged atom.
   b. positively charged atom.
   c. negatively charged atom.
   d. atom that has gained an electron.
4. A polar covalent bond between two atoms occurs when
   a. one atom attracts shared electrons more strongly than another atom.
   b. atoms attract electrons equally.
   c. an electron from one atom is completely transferred to another atom.
   d. the molecule becomes ionized.
   e. a hydrogen atom is shared between two different atoms.
5. Table salt (NaCl) is
   a. an atom. c. a molecule. e. a cation.
   b. organic. d. a compound.
6. The weak attractive force between two water molecules forms a(n)
   a. covalent bond. c. ionic bond. e. isotope.
   b. hydrogen bond. d. compound.
7. Electrolytes are
   a. nonpolar molecules.
   b. covalent compounds.
   c. substances that usually don't dissolve in water.
   d. found in solutions that do not conduct electricity.
   e. cations and anions that dissociate in water.
8. In a decomposition reaction,
   a. anabolism occurs.
   b. proteins are formed from amino acids.
   c. large molecules are broken down to form small molecules.
   d. a dehydration reaction may occur.
   e. All of these are correct.
9. Oxidation-reduction reactions
   a. can be synthesis or decomposition reactions.
   b. have one reactant gaining electrons.
   c. have one reactant losing electrons.
   d. can create ionic or covalent bonds.
   e. All of these are correct.
10. Potential energy
    a. is energy caused by movement of an object.
    b. is the form of energy that is actually doing work.
    c. includes energy within chemical bonds.
    d. can never be converted to kinetic energy.
    e. All of these are correct.
11. Which of these descriptions of heat energy is *not* correct?
    a. Heat energy flows between objects that are at different temperatures.
    b. Heat energy can be produced from all other forms of energy.
    c. Heat energy can be released during chemical reactions.
    d. Heat energy must be added to break apart ATP molecules.
    e. Heat energy is always transferred from a hotter object to a cooler object.
12. Which of these statements concerning enzymes is correct?
    a. Enzymes increase the rate of reactions but are permanently changed as a result.
    b. Enzymes are proteins that function as catalysts.
    c. Enzymes increase the activation energy requirement for a reaction to occur.
    d. Enzymes usually can only double the rate of a chemical reaction.
    e. Enzymes increase the kinetic energy of the reactants.
13. Water
    a. is composed of two oxygen atoms and one hydrogen atom.
    b. has a low specific heat.
    c. is composed of polar molecules into which ionic substances dissociate.
    d. is produced in a hydrolysis reaction.
    e. is a very small organic molecule.
14. When sugar is dissolved in water, the water is called the
    a. solute. b. solution. c. solvent.
15. Which of these is an example of a suspension?
    a. sweat c. sugar dissolved in water
    b. water and proteins inside cells d. red blood cells in plasma

16. A solution with a pH of 5 is ________ and contains ________ H^+ than (as) a neutral solution.
   a. a base, more
   b. a base, fewer
   c. an acid, more
   d. an acid, fewer
   e. neutral, the same number of
17. A buffer
   a. slows down chemical reactions.
   b. speeds up chemical reactions.
   c. increases the pH of a solution.
   d. maintains a relatively constant pH.
   e. works by forming salts.
18. A conjugate acid-base pair
   a. acts as a buffer.
   b. can combine with H^+ in a solution.
   c. can release H^+ to combine with OH^-.
   d. describes carbonic acid (H_2CO_3) and bicarbonate ions (HCO_3^-).
   e. All of these are correct.
19. The polysaccharide used for energy storage in the human body is
   a. cellulose.
   b. glycogen.
   c. lactose.
   d. sucrose.
   e. starch.
20. The basic units or building blocks of triglycerides are
   a. simple sugars (monosaccharides).
   b. double sugars (disaccharides).
   c. amino acids.
   d. glycerol and fatty acids.
   e. nucleotides.
21. A ________ fatty acid has one double covalent bond between carbon atoms.
   a. cholesterol
   b. monounsaturated
   c. phospholipid
   d. polyunsaturated
   e. saturated
22. The ________ structure of a protein results from the folding of the pleated sheets or helices.
   a. primary
   b. secondary
   c. tertiary
   d. quaternary
23. According to the lock-and-key model of enzyme action,
   a. reactants must first be heated.
   b. enzyme shape is not important.
   c. each enzyme can catalyze many types of reactions.
   d. reactants must bind to an active site on the enzyme.
   e. enzymes control only a small number of reactions in the cell.
24. DNA molecules
   a. contain genes.
   b. contain a single strand of nucleotides.
   c. contain the nucleotide uracil.
   d. are of three different types that have roles in protein synthesis.
   e. contain up to 100 nitrogenous bases.
25. ATP
   a. is formed by the addition of a phosphate group to ADP.
   b. is formed with energy released during catabolic reactions.
   c. provides the energy for anabolic reactions.
   d. contains three phosphate groups.
   e. All of these are correct.

Answers in Appendix E

CRITICAL THINKING

1. Iron has an atomic number of 26 and a mass number of 56. How many protons, neutrons, and electrons are in an atom of iron? If an atom of iron lost 3 electrons, what would be the charge of the resulting ion? Write the correct symbol for this ion.
2. Why is the conversion of a triglyceride molecule to fatty acids and glycerol a catabolic hydrolysis reaction? Would the reverse anabolic reaction generate water?
3. A mixture of chemicals is warmed slightly. As a consequence, although no more heat is added, the solution becomes very hot. Explain what has occurred to make the solution so hot.
4. Two solutions, when mixed together at room temperature, produce a chemical reaction. However, when the solutions are boiled and allowed to cool to room temperature before mixing, no chemical reaction takes place. Explain.
5. In terms of the potential energy in food, explain why eating food is necessary for increasing muscle mass.
6. Solution A is a strong acid of pH 2, and solution B is an equally strong base of pH 8. Each chemical can donate or receive a single proton. If equal amounts of solutions A and B are mixed, is the resulting solution acidic or basic?
7. Carbon dioxide that accumulates in the blood can become toxic, in part because it alters the blood pH. Some of the carbon dioxide molecules react with water to form carbonic acid ($CO_2 + H_2O \rightleftarrows H_2CO_3$). Ned can swim across the swimming pool under water. Before diving into the water, he breathes rapidly for a few seconds, and while he is under the water he does not breathe at all. Explain how the pH of his blood changes while breathing rapidly and while swimming under water. Also explain why the pH of his blood does not change dramatically.
8. An enzyme (E) catalyzes the following reaction:

$$A + B \xrightarrow{E} C$$

However, the product (C) binds to the active site of the enzyme in a reversible fashion and keeps the enzyme from functioning. What happens if A and B are continually added to a solution that contains a fixed amount of the enzyme?
9. Using the materials commonly found in a kitchen, explain how to distinguish between a protein and a lipid.

Answers in Appendix F

Visit this book's website at **www.mhhe.com/seeley10** for chapter quizzes, interactive learning exercises, and other study tools.

Mc Graw Hill
connect plus+
ANATOMY & PHYSIOLOGY

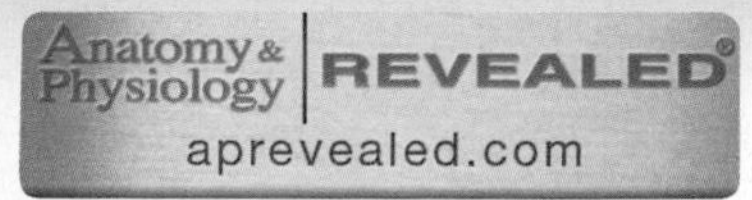
Anatomy & Physiology
REVEALED
aprevealed.com

3

Learn to Predict

Carlos always carries a water bottle, and he never likes to be too far from a restroom. Carlos has diabetes insipidus, an incurable disease that causes his kidneys to produce an unusually large volume of dilute urine. To keep his body fluids in a state of homeostasis (see chapter 1), Carlos has to drink enough water and other solutes to replace what he loses as urine. Diabetes insipidus results from a gene mutation, a change in the DNA, that prevents the kidneys from responding normally to an important hormone, called ADH, that regulates water loss from the kidneys. After reading about cell structure and gene expression in chapter 3, explain how Carlos's condition developed at the cellular level.

Photo: Model of an aquaporin, or water channel, in the plasma membrane of a cell. The image also represents water molecules (blue shapes) moving through the channel.

Module 2
Cells & Chemistry

Anatomy & Physiology
REVEALED
aprevealed.com

Cell Biology

The human body is composed of trillions of cells. If each of these cells were the size of a standard brick, the colossal human statue erected from those bricks would be 6 miles high! In reality, an average-sized cell is only one-fifth the size of the smallest dot you can make on a sheet of paper with a sharp pencil. Although they are minute, cells act as complex factories to carry out the functions of life.

All of the cells of an individual originate from a single fertilized cell. During development, cell division and specialization give rise to a wide variety of cell types, such as nerve, muscle, bone, and blood cells. Each cell type has important characteristics that are critical to normal body function, including cell metabolism and energy use; synthesis of molecules, such as proteins and nucleic acids; communication between cells; reproduction; and inheritance. One of the important reasons for maintaining homeostasis is to keep the trillions of cells that form the body functioning normally.

FUNDAMENTAL **Figure**

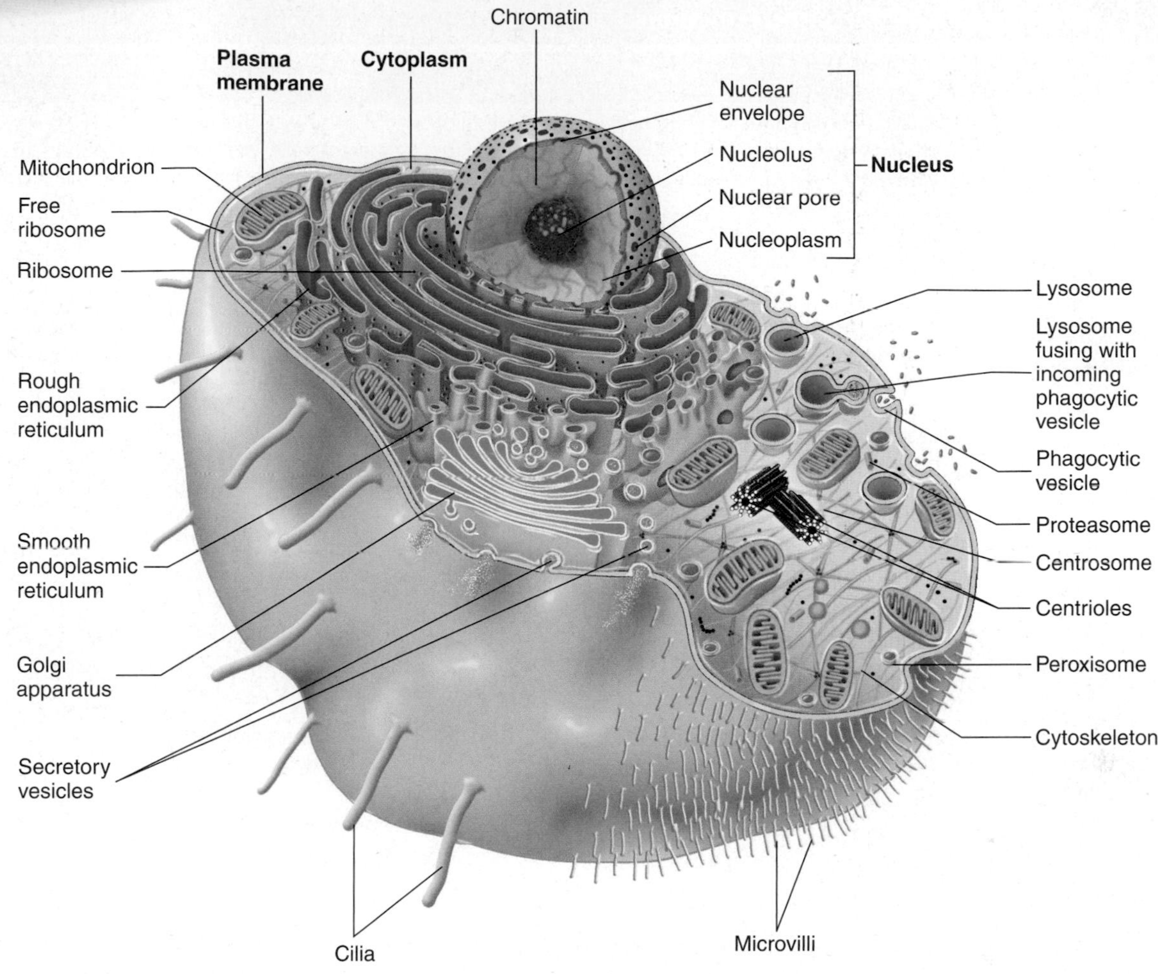

Chromatin
Plasma membrane
Cytoplasm
Nuclear envelope
Nucleolus
Nuclear pore
Nucleoplasm
Nucleus
Mitochondrion
Free ribosome
Ribosome
Rough endoplasmic reticulum
Smooth endoplasmic reticulum
Golgi apparatus
Secretory vesicles
Lysosome
Lysosome fusing with incoming phagocytic vesicle
Phagocytic vesicle
Proteasome
Centrosome
Centrioles
Peroxisome
Cytoskeleton
Cilia
Microvilli

FIGURE 3.1 AP|R **A Human Cell**

A generalized human cell showing the plasma membrane, nucleus, and cytoplasm with its organelles. Although no single cell contains all these organelles, many cells contain a large number of them.

3.1 Functions of the Cell

LEARNING OUTCOMES

After reading this section, you should be able to

A. List the general parts of a cell.

B. Relate and explain the four main functions of cells.

Cells are the basic units of all living things, including humans. Cells within the body may have quite different structures and functions, yet they share several common characteristics (figure 3.1; table 3.1). The **plasma** (plaz′mă) **membrane,** or *cell membrane*, forms the outer boundary of the cell, through which the cell interacts with its external environment. The nucleus (noo′klē-ŭs) is usually located centrally; it directs cell activities, most of which take place in the **cytoplasm** (sī′tō-plazm), located between the plasma membrane and the nucleus. Within cells, specialized structures called **organelles** (or′gă-nelz) perform specific functions.

The characteristic functions of the cell include the following:

1. *Cell metabolism and energy use.* The chemical reactions that occur within cells are referred to as metabolic reactions and are collectively known as cell metabolism. The energy released from some metabolic reactions fuels cellular activities, such as the synthesis of molecules and muscle contraction. During some metabolic reactions, energy is released as heat, which helps maintain body temperature.
2. *Synthesis of molecules.* The different cells of the body synthesize various types of molecules, including proteins, nucleic acids, and lipids. The structural and functional characteristics of cells are determined by the types of molecules they produce.
3. *Communication.* Cells produce and respond to chemical and electrical signals that allow them to communicate with one another. For example, nerve cells produce chemical signals by which they communicate with muscle cells, and muscle cells respond by contracting or relaxing.

TABLE 3.1 Summary of Cell Parts and Functions

Cell Parts	Structure	Function
Plasma membrane	Lipid bilayer composed of phospholipids and cholesterol; proteins extend across or are embedded in either surface of the lipid bilayer	Functions as the outer boundary of cells; controls the entry and exit of substances; receptor proteins function in intercellular communication; marker molecules enable cells to recognize one another
Nucleus (Nuclear pores, Nuclear envelope, Nucleolus)	Enclosed by nuclear envelope, a double membrane with nuclear pores; contains chromatin (dispersed, thin strands of DNA and associated proteins), which condenses to become visible mitotic chromosomes during cell division; also contains one or more nucleoli, dense bodies consisting of ribosomal RNA and proteins	Is the control center of the cell; DNA within the nucleus regulates protein (e.g., enzyme) synthesis and therefore the chemical reactions of the cell
Cytoplasmic Organelles		
Ribosome	Ribosomal RNA and proteins form large and small subunits; some are attached to endoplasmic reticulum, whereas others (free ribosomes) are distributed throughout the cytoplasm	Serves as site of protein synthesis
Rough endoplasmic reticulum	Membranous tubules and flattened sacs with attached ribosomes	Synthesizes proteins and transports them to Golgi apparatus
Smooth endoplasmic reticulum	Membranous tubules and flattened sacs with no attached ribosomes	Manufactures lipids and carbohydrates; detoxifies harmful chemicals; stores calcium
Golgi apparatus	Flattened membrane sacs stacked on each other	Modifies, packages, and distributes proteins and lipids for secretion or internal use
Lysosome	Membrane-bound vesicle pinched off Golgi apparatus	Contains digestive enzymes
Peroxisome	Membrane-bound vesicle	Serves as one site of lipid and amino acid degradation; breaks down hydrogen peroxide
Proteasomes	Tubelike protein complexes in the cytoplasm	Break down proteins in the cytoplasm
Mitochondria	Spherical, rod-shaped, or threadlike structures; enclosed by double membrane; inner membrane forms projections called cristae	Are major sites of ATP synthesis when oxygen is available

TABLE 3.1 Summary of Cell Parts and Functions

Cell Parts	Structure	Function
Centrioles	Pair of cylindrical organelles in the centrosome, consisting of triplets of parallel microtubules	Serve as centers for microtubule formation; determine cell polarity during cell division; form the basal bodies of cilia and flagella
Cilia	Extensions of the plasma membrane containing doublets of parallel microtubules; 10 μm in length	Move materials over the surface of cells
Flagellum	Extension of the plasma membrane containing doublets of parallel microtubules; 55 μm in length	In humans, propels spermatozoa
Microvilli	Extension of the plasma membrane containing microfilaments	Increase surface area of the plasma membrane for absorption and secretion; modified to form sensory receptors

4. *Reproduction and inheritance.* Most cells contain a complete copy of all the genetic information of the individual. This genetic information ultimately determines the structural and functional characteristics of the cell. As a person grows, cells divide to produce new cells, each containing the same genetic information. Specialized body cells called gametes are responsible for transmitting genetic information to the next generation.

ASSESS YOUR PROGRESS

1. *What parts are common to most cells?*
2. *Explain the four characteristic functions of the cell.*

3.2 How We See Cells

LEARNING OUTCOME

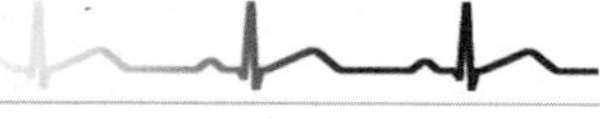

After reading this section, you should be able to

A. Relate the kinds of microscopes used to study cells.

Most cells are too small to be seen with the unaided eye, so we must use microscopes to study them. **Light microscopes** allow us to visualize the general features of cells, whereas **electron microscopes** enable us to study the fine structure of cells. A **scanning electron microscope (SEM)** can reveal features of the cell surface and the surfaces of internal structures. A **transmission electron microscope (TEM)** allows us to see "through" parts of the cell and thus to discover detailed aspects of cell structure. A more detailed description of microscopes and their use appears in Clinical Impact, "Microscopic Imaging," in this chapter.

ASSESS YOUR PROGRESS

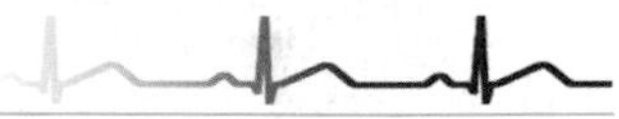

3. *Which cell features can be seen with a light microscope? With electron microscopes?*

3.3 Plasma Membrane

LEARNING OUTCOMES

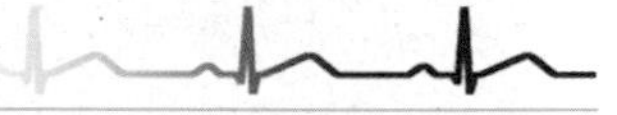

After reading this section, you should be able to

A. Describe the functions and general structure of the plasma membrane.

B. Relate why a membrane potential is formed.

The plasma membrane is the outermost component of a cell. It functions as a boundary separating the substances inside the cell, which are **intracellular,** from substances outside the cell, which are **extracellular.** The plasma membrane encloses and supports the cell contents. It attaches cells to the extracellular environment or to other cells. The cells' ability to recognize and communicate with each other takes place through the plasma membrane. In addition, the plasma membrane determines what moves into and out of cells. As a result, the intracellular contents of cells are different from the extracellular environment.

An electrical charge difference across the plasma membrane called the **membrane potential** is a result of the cell's regulation of ion movement into and out of the cell. Because there are more positively charged ions immediately on the outside of the plasma membrane and more negatively charged ions and proteins inside, the outside of the plasma membrane is positively charged, compared with the inside of the plasma membrane. The membrane potential, an important feature of a living cell's normal function, is considered in greater detail in chapters 9 and 11.

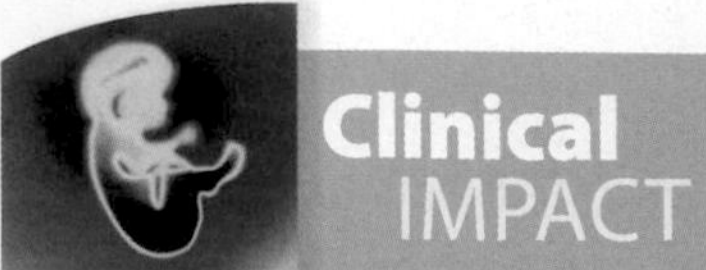

Clinical IMPACT — Microscopic Imaging

We see objects because light either passes through them or is reflected off them and enters our eyes (see chapter 15). However, vision with the unaided eye has limitations. The smallest objects we can resolve, or identify as separate objects, are approximately 100 μm, or 0.1 mm, in diameter, which is approximately the size of a fine pencil dot. The details of cells and tissues, which are much smaller than 100 μm, cannot be examined without the aid of a microscope.

Two basic types of microscopes have been developed: light microscopes and electron microscopes. As their names imply, light microscopes use light to produce an image, and electron microscopes use beams of electrons to produce an image. The resolution of light microscopes is limited by the wavelength of light, the lower limit being approximately 0.1 μm—about the size of a small bacterium.

Light microscopy is regularly used to examine biopsy specimens because samples can be quickly prepared and the resolution is adequate to diagnose most conditions that cause changes in tissue structure. Because most tissues are colorless and transparent when thinly sectioned, the tissue must be stained with a dye, so that the structural details can be seen, or must be stained with antibodies that recognize specific molecules. The bound antibodies are detected by fluorescent tags.

To see objects much smaller than a cell requires an electron microscope, which has a limit of resolution of approximately 0.1 nm, about the size of some molecules. In a transmission electron microscope (TEM), a beam of electrons is passed through the object to be viewed; using a scanning electron microscope (SEM), the beam of electrons is reflected off the surface of the object. For TEM, the specimen must be embedded in plastic and thinly sectioned (0.01–0.15 μm thick). The magnification ability of SEM is not as great as that of TEM; however, the depth of focus of SEM is much greater and produces a clearer three-dimensional image of tissue structure.

A newer type of microscopy, which expands on the advantages of SEM, is the atomic force microscope (AFM). This type of microscope scans the sample using a tiny mechanical probe that can be deflected by small forces between the probe and sample. This generates a three-dimensional surface map of the sample. AFM combines the high resolution of TEM with the topographical visualization of SEM and added benefit that samples can generally be viewed under more physiological conditions. Because TEM, SEM, and AFM do not transmit color information, the micrographs are black and white unless assigned false colors.

Examples of various microscopic images of a nuclear pore (see "The Nucleus and Cytoplasmic Organelles," later in this chapter) are shown in figure 3A.

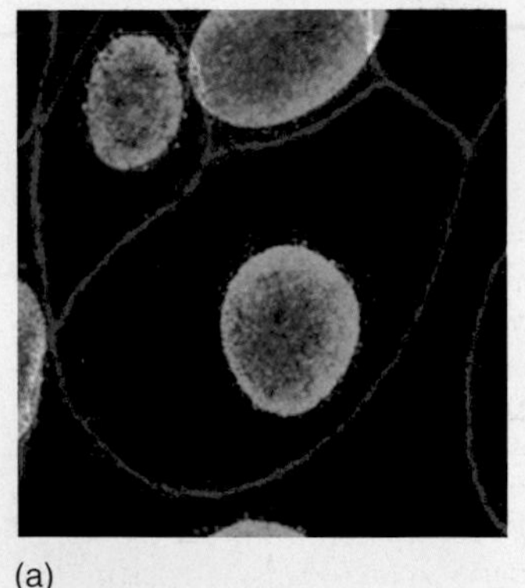
(a)

(b)

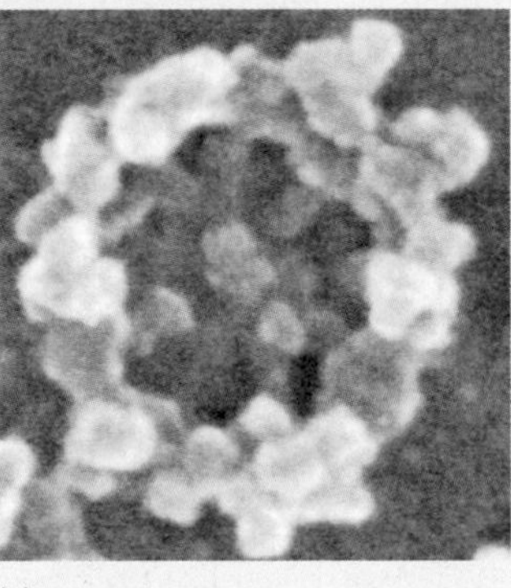
(c)

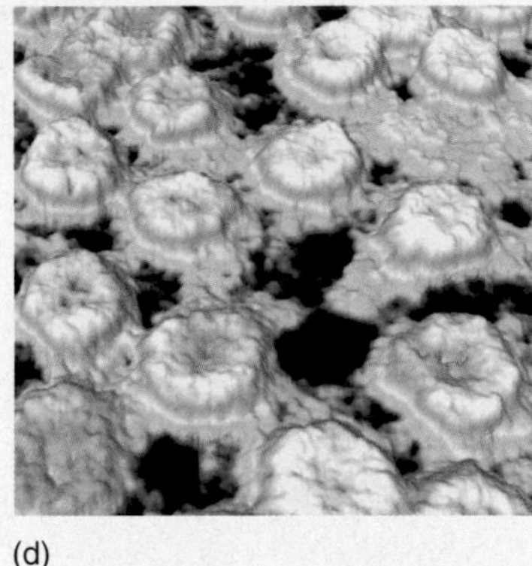
(d)

FIGURE 3A
(*a*) Light microscopy image of a cell. The nucleus is labeled with antibodies that bind nuclear pores (green) and proteins on the plasma membrane (red). (*b*) Transmission electron micrograph of a single nuclear pore. (*c*) Scanning electron micrograph of a single nuclear pore. View is from the cytoplasm of the cell into the pore. (*d*) Color-enhanced atomic force microscopy image of nuclear pores.

The plasma membrane consists of 45–50% lipids, 45–50% proteins, and 4–8% carbohydrates (figure 3.2). The carbohydrates combine with lipids to form glycolipids and with proteins to form glycoproteins. The glycocalyx (glī′kō-kā′liks) is the collection of glycolipids, glycoproteins, and carbohydrates on the outer surface of the plasma membrane. The glycocalyx also contains molecules absorbed from the extracellular environment, so there is often no precise boundary between the plasma membrane and the extracellular environment.

ASSESS YOUR PROGRESS

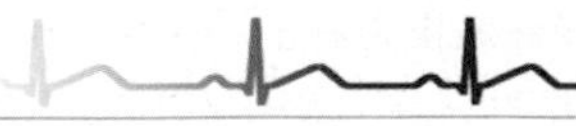

4. *Explain five functions of the plasma membrane.*
5. *Differentiate between intracellular and extracellular.*
6. *What is the membrane potential? Is the outside of the plasma membrane positively or negatively charged compared with the inside?*
7. *What are the main chemical components of the plasma membrane?*

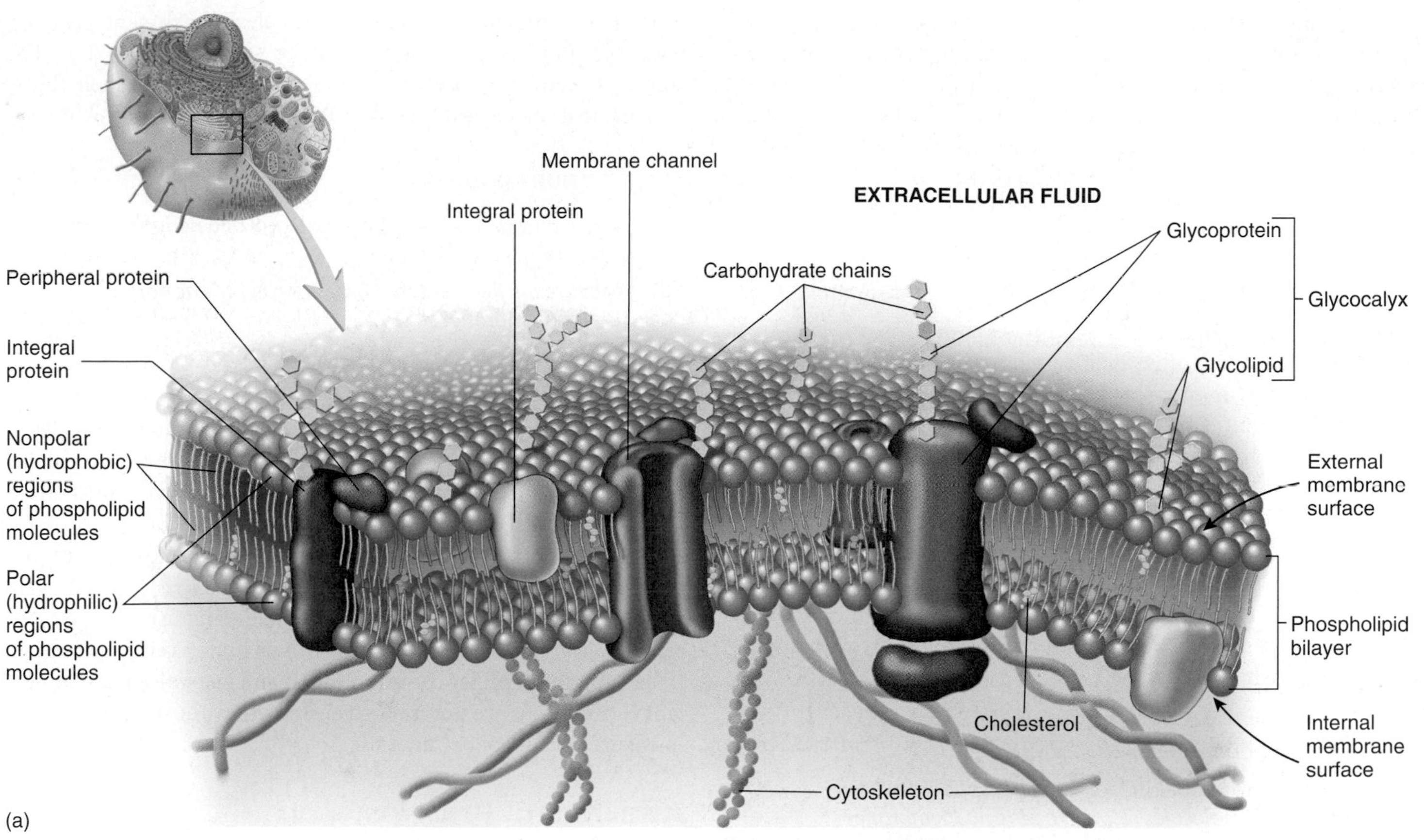

Membrane channel
Integral protein
EXTRACELLULAR FLUID
Glycoprotein
Glycocalyx
Peripheral protein
Carbohydrate chains
Integral protein
Glycolipid
Nonpolar (hydrophobic) regions of phospholipid molecules
External membrane surface
Polar (hydrophilic) regions of phospholipid molecules
Phospholipid bilayer
Cholesterol
Internal membrane surface
Cytoskeleton
(a)
CYTOPLASM (Intracellular Fluid)

(b)
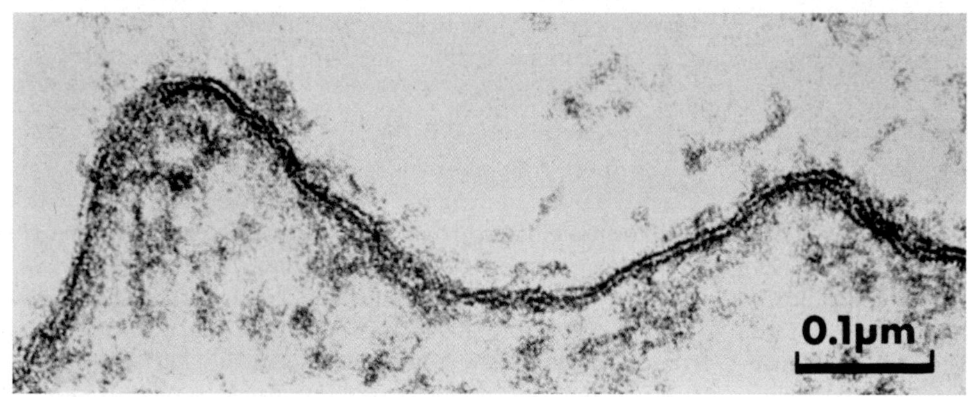
0.1µm

FIGURE 3.2 AP|R Plasma Membrane

(*a*) Fluid-mosaic model of the plasma membrane. The membrane is composed of a bilayer of phospholipids and cholesterol with proteins "floating" in the membrane. The nonpolar hydrophobic region of each phospholipid molecule is directed toward the center of the membrane, and the polar hydrophilic region is directed toward either the extracellular fluid or the cytoplasm. (*b*) Transmission electron micrograph showing the plasma membrane of a single cell. Proteins at either surface of the lipid bilayer stain more readily than the lipid bilayer does and give the membrane the appearance of having three parts: The two outer parts consist of proteins and the phospholipid heads, and the central part is composed of the phospholipid tails and cholesterol.

3.4 Membrane Lipids

LEARNING OUTCOMES

After reading this section, you should be able to

A. List and describe the functions of membrane lipids.

B. Explain the nature of the fluid-mosaic model of membrane structure.

The predominant lipids of the plasma membrane are phospholipids and cholesterol. **Phospholipids** readily assemble to form a **lipid bilayer,** a double layer of phospholipid molecules, because they have a polar (charged) head and a nonpolar (uncharged) tail (see chapter 2). The polar, **hydrophilic** (water-loving) heads are exposed to the aqueous extracellular and intracellular fluids of the cell, whereas the nonpolar, **hydrophobic** (water-fearing) tails face one another in the interior of the plasma membrane (figure 3.2).

The **fluid-mosaic model** concept of the plasma membrane suggests that the plasma membrane is neither rigid nor static in structure but is highly flexible and can change its shape and composition through time. The lipid bilayer functions as a dense liquid in which other molecules, such as proteins, are suspended. The fluid nature of the lipid bilayer has several important consequences: It provides an important means of distributing molecules within the plasma membrane. In addition, slight damage to the membrane can be repaired because the phospholipids tend to reassemble around damaged sites and close them. The fluid nature of the lipid bilayer also enables membranes to fuse with one another.

Cholesterol is the other major lipid in the plasma membrane (see chapter 2). It is interspersed among the phospholipids and accounts for about one-third of the total lipids in the plasma membrane. The hydrophilic hydroxyl (—OH) group of cholesterol extends between the phospholipid heads to the hydrophilic surface of the membrane, whereas the hydrophobic part of the cholesterol molecule lies within the hydrophobic region of the phospholipids. The amount of cholesterol in a particular plasma membrane is a major factor in determining the fluid nature of the membrane. Cholesterol limits the movement of phospholipids, providing stability to the plasma membrane.

ASSESS YOUR PROGRESS

8. *How do the hydrophilic heads and hydrophobic tails of phospholipid molecules result in a plasma membrane?*
9. *Summarize the characteristics of the fluid-mosaic model of membrane structure.*
10. *What is the function of cholesterol in plasma membranes?*

3.5 Membrane Proteins

LEARNING OUTCOMES

After reading this section, you should be able to

A. List and explain the functions of membrane proteins.

B. Describe the characteristics of specificity, competition, and saturation of transport proteins.

Although the basic structure of the plasma membrane and some of its functions are determined by its lipids, many of its other functions are determined by its proteins. Some protein molecules, called **integral membrane proteins,** penetrate deeply into the lipid bilayer, in many cases extending from one surface to the other (figure 3.2), whereas other proteins, called **peripheral membrane proteins,** are attached to either the inner or the outer surfaces of the lipid bilayer. Integral membrane proteins consist of regions made up of amino acids with hydrophobic R groups and other regions of amino acids with hydrophilic R groups (see chapter 2). The hydrophobic regions are located within the hydrophobic part of the membrane, and the hydrophilic regions lie at the inner or outer surface of the membrane or line channels through the membrane. Some peripheral proteins may be bound to integral membrane proteins, whereas others are bound to the polar heads of the phospholipid molecules. Membrane proteins can function as marker molecules, attachment proteins, transport proteins, receptor proteins, or enzymes (table 3.2). The ability of membrane proteins to function depends on their three-dimensional shapes and their chemical characteristics.

ASSESS YOUR PROGRESS

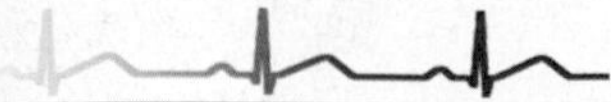

11. *Describe the difference between integral and peripheral proteins in the plasma membrane.*
12. *What are the five roles that proteins can play as part of the plasma membrane?*

Marker Molecules

Marker molecules are cell surface molecules that allow cells to identify other cells or other molecules. They are mostly **glycoproteins** (proteins with attached carbohydrates; table 3.2) or **glycolipids** (lipids with attached carbohydrates). The protein portions of glycoproteins may be either integral or peripheral membrane proteins. Examples of marker protein function include a sperm cell's recognition of an oocyte and the immune system's ability to distinguish between self-cells and foreign cells, such as bacteria or donor cells in an organ transplant. Intercellular communication and recognition are important because cells are not isolated entities; they must work together to ensure normal body function.

Attachment Proteins

Integral proteins may function as **attachment proteins,** which allow cells to attach to other cells or to extracellular molecules (table 3.2). Many attachment proteins also attach to intracellular molecules. **Cadherins** are proteins that attach cells to other cells; **integrins** are proteins that attach cells to extracellular molecules. Integrins function in pairs of integral membrane proteins, which interact with both intracellular and extracellular molecules. Because of the interaction with intracellular molecules, integrins also function in cellular communication.

Transport Proteins

Transport proteins are integral proteins that allow ions or molecules to move from one side of the plasma membrane to the other. These proteins have three characteristics: specificity, competition, and saturation. **Specificity** means that each transport protein binds to and transports only a certain type of molecule or ion (figure 3.3*a*). For example, the transport protein that moves glucose does not move amino acids. The chemical structure of the binding site determines the specificity of the transport protein (figure 3.3*a*). **Competition** is the result of similar molecules binding to the transport protein (figure 3.3*b*). Although the binding sites of transport proteins exhibit specificity, closely related substances, in which regions of two different molecules have the same shape, may bind to the same binding site. The substance in the greater concentration or the substance that binds to the binding site more readily is moved across the plasma membrane at the greater rate. **Saturation** means that the rate of movement of molecules across the membrane is limited by the number of available transport proteins (figure 3.4). As the concentration of a transported substance increases, more

TABLE 3.2 Functions of Membrane Proteins

Membrane Protein	Protein Function
Marker molecules 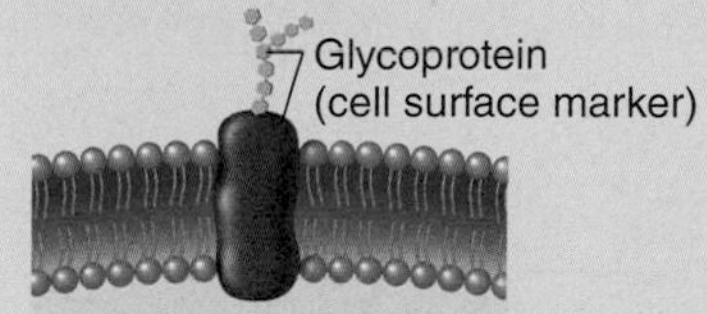 Glycoprotein (cell surface marker)	Allow cells to identify other cells or other molecules
Attachment proteins 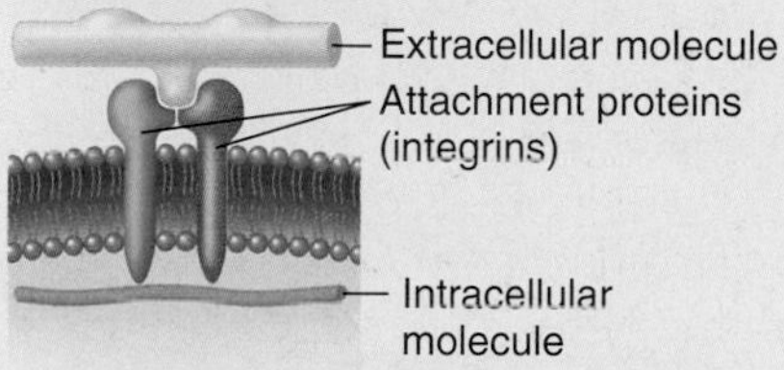 Extracellular molecule Attachment proteins (integrins) Intracellular molecule	Anchor cells to other cells (cadherins) or to extracellular molecules (integrins)
Transport proteins Channel proteins 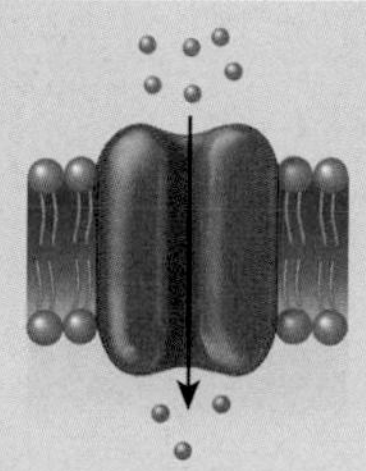	Form passageways through the plasma membrane, allowing specific ions or molecules to enter or exit the cell; may be gated or nongated
Carrier proteins (transporters) 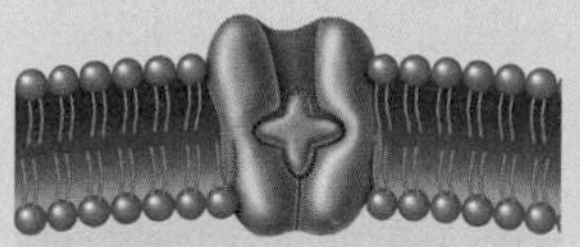	Move ions or molecules across the membrane; binding of specific chemical to carrier proteins causes changes in the shape of the carrier proteins; the carrier proteins then move the specific chemical across the membrane
ATP-powered pumps 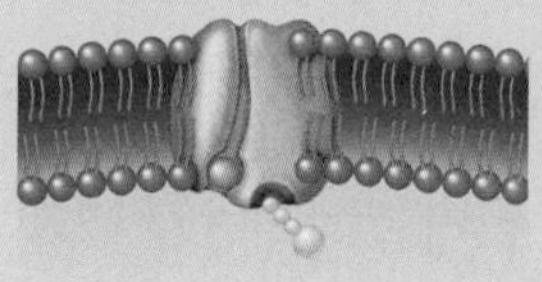	Move specific ions or molecules across the membrane; require ATP molecules to function
Receptor proteins 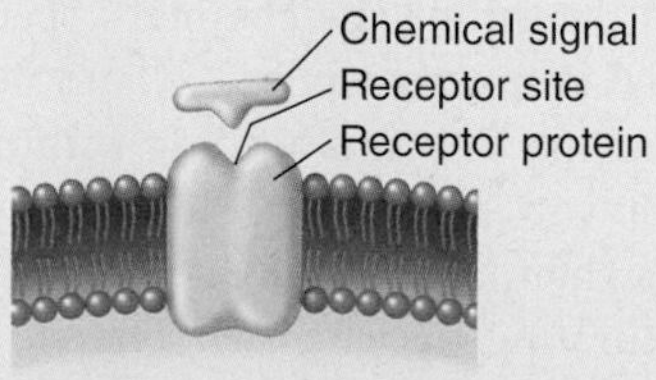 Chemical signal Receptor site Receptor protein	Function as binding sites for chemical signals in the extracellular fluid; binding of chemical signals to receptors triggers cellular responses
Enzymes 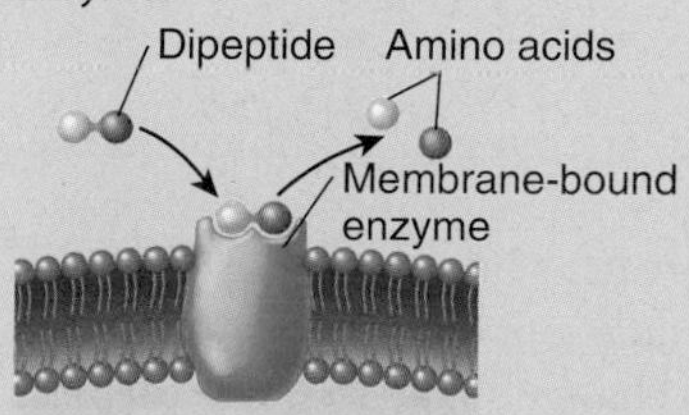 Dipeptide Amino acids Membrane-bound enzyme	Catalyze chemical reactions either inside or outside cells

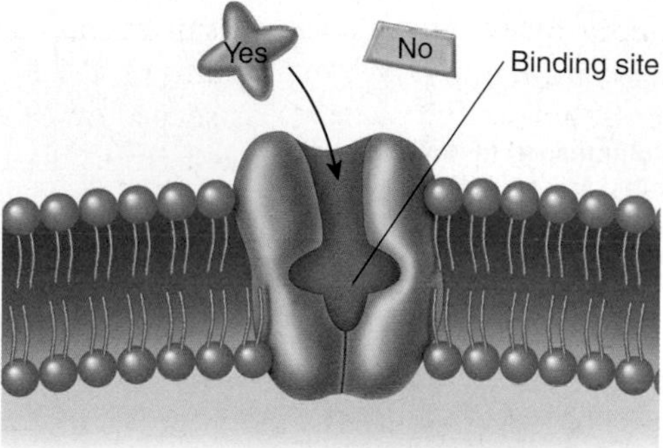
Yes
No
Binding site

(a) Specificity. Only molecules that are the right shape to bind to the binding site are transported.

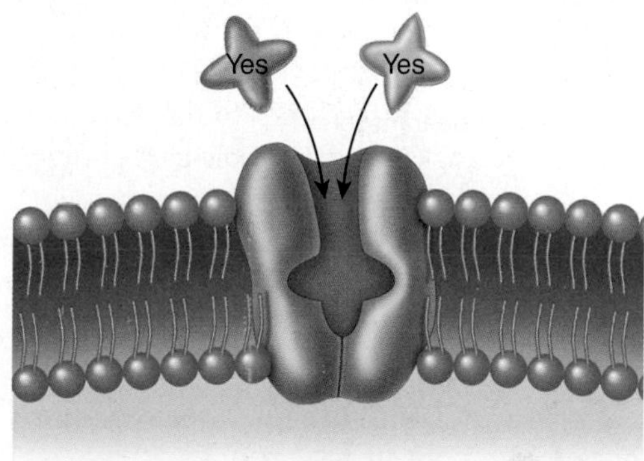
Yes
Yes

(b) Competition. Similarly shaped molecules can compete for the same binding site.

FIGURE 3.3 Transport Proteins: Specificity and Competition

transport proteins have their binding sites occupied, so the rate at which the substance is moved across the plasma membrane increases. However, once the concentration of the substance is increased so that all the binding sites are occupied, the rate of movement remains constant, even though the concentration of the substance increases further. Transport proteins include channel proteins, carrier proteins, and ATP-powered pumps.

Channel Proteins

Channel proteins are one or more integral membrane proteins arranged so that they form a tiny channel through the plasma membrane (figure 3.5; table 3.2). The hydrophobic regions of the proteins face outward toward the hydrophobic part of the plasma membrane, and the hydrophilic regions of the protein face inward and line the channel. Ions or small molecules of the right size, charge, and shape can pass through the channel. The charges in the hydrophilic part of the channel proteins determine which types of ions can pass through the channel.

Some channel proteins, called **leak ion channels,** or *nongated ion channels*, are always open and are responsible for the plasma membrane's permeability to ions when the plasma membrane is at rest. Other channels, called **gated ion channels,** can be open or closed. Some gated ion channels open or close in response to chemical signals, or **ligands** (lig′andz, lī′gandz), which are small molecules that bind to the proteins or glycoproteins. These are called **ligand-gated ion channels.** Other gated ion channels open or close when there is a change in the membrane potential. These are called **voltage-gated ion channels.**

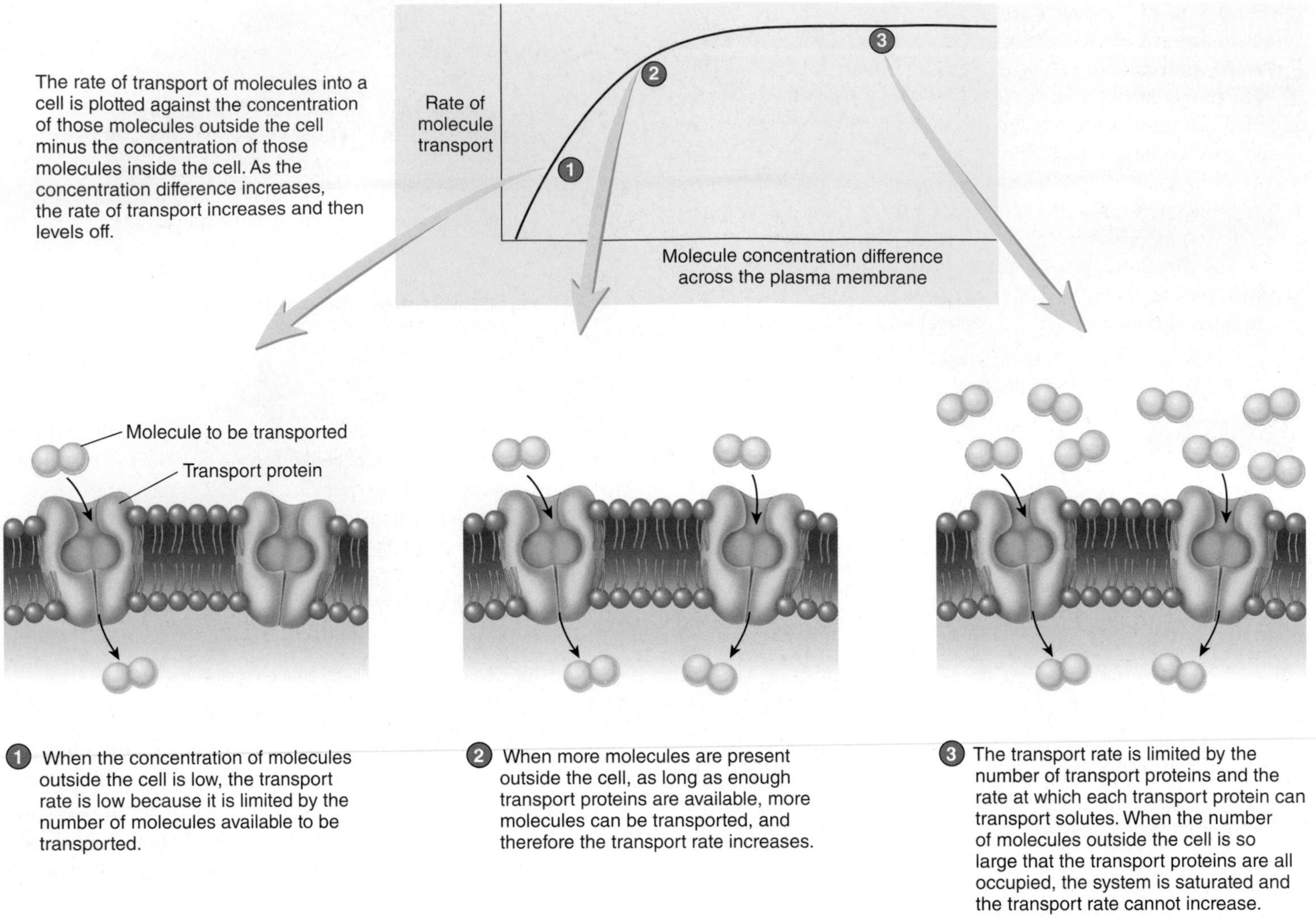

The rate of transport of molecules into a cell is plotted against the concentration of those molecules outside the cell minus the concentration of those molecules inside the cell. As the concentration difference increases, the rate of transport increases and then levels off.

Rate of molecule transport

Molecule concentration difference across the plasma membrane

Molecule to be transported

Transport protein

1 When the concentration of molecules outside the cell is low, the transport rate is low because it is limited by the number of molecules available to be transported.

2 When more molecules are present outside the cell, as long as enough transport proteins are available, more molecules can be transported, and therefore the transport rate increases.

3 The transport rate is limited by the number of transport proteins and the rate at which each transport protein can transport solutes. When the number of molecules outside the cell is so large that the transport proteins are all occupied, the system is saturated and the transport rate cannot increase.

PROCESS FIGURE 3.4 Saturation of a Transport Protein

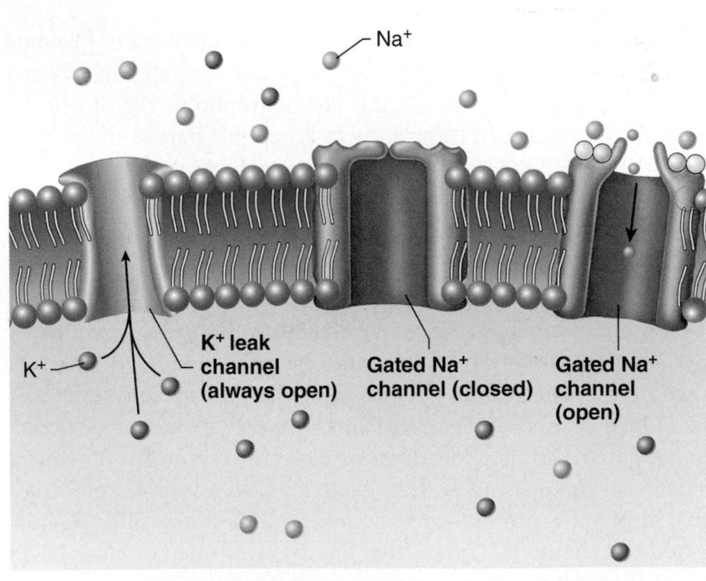
Na^+
K^+
K^+ leak channel (always open)
Gated Na^+ channel (closed)
Gated Na^+ channel (open)

FIGURE 3.5 Leak and Gated Membrane Channels

Channel proteins were once thought of as simple tubes, with or without gates, through which ions pass. Many channel proteins, however, are more complex than that. It now appears that ions briefly bind to specific sites inside channels and that the shapes of those channels change as ions are transported through them. The size and charge within a channel determine the channel's specificity. For example, Na^+ channels do not transport K^+, and vice versa. In addition, similar ions moving into and binding within a channel protein are in competition with each other. Furthermore, the number of ions moving into a channel protein can exceed the capacity of the channel, thus saturating the channel. Therefore, channel proteins exhibit specificity, competition, and saturation.

Carrier Proteins

Carrier proteins, or *transporters*, are integral membrane proteins that move ions or molecules from one side of the plasma membrane to the other. The carrier proteins have specific binding sites to which ions or molecules attach on one side of the plasma membrane. The carrier proteins change shape to move the bound ions or molecules to the other side of the plasma membrane, where the ions or molecules are released (figure 3.6). The carrier protein then resumes its original shape and is available to transport more molecules.

The movement of ions or molecules by carrier proteins can be classified as uniport, symport, or antiport. **Uniport** is the movement of one specific ion or molecule across the membrane. **Symport** (*cotransport*) is the movement of two different ions or molecules in the same direction across the plasma membrane, whereas **antiport** (*countertransport*) is the movement of two different ions or molecules in opposite directions across the plasma membrane. Carrier proteins involved in these types of movement are called **uniporters, symporters,** and **antiporters,** respectively.

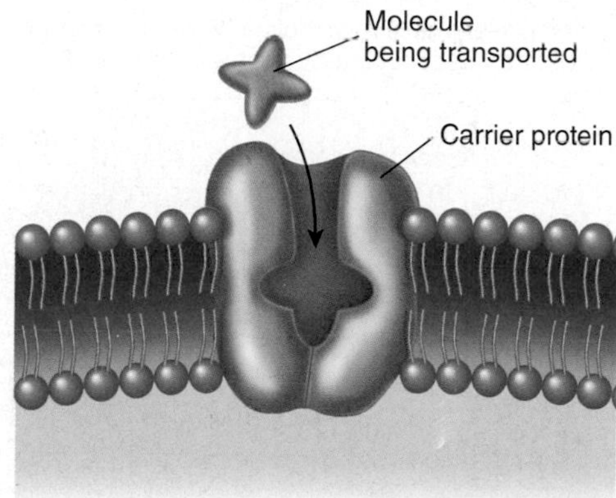
Molecule being transported
Carrier protein

1 A molecule enters the carrier protein from one side of the plasma membrane.

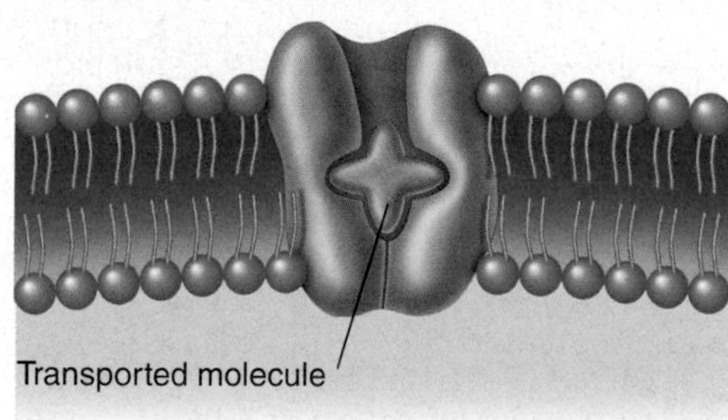
Transported molecule

2 The carrier protein briefly binds the transported molecule.

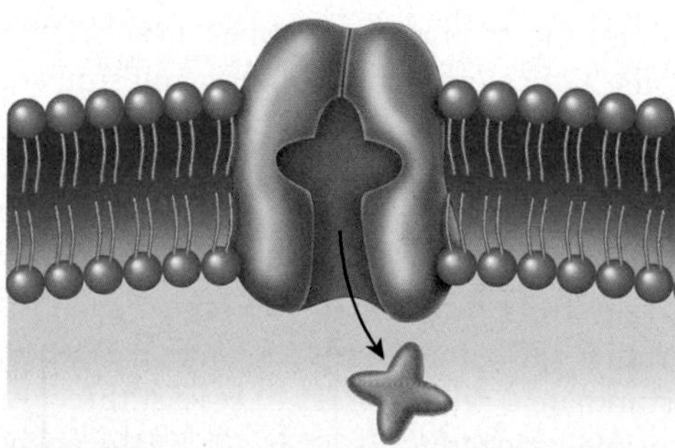

3 The carrier protein changes shape and releases the transported molecule on the other side of the plasma membrane. The carrier protein then changes back to its original shape (go to step 1).

PROCESS FIGURE 3.6 Transport by a Carrier Protein

ATP-Powered Pumps

ATP-powered pumps are transport proteins that move specific ions or molecules from one side of the plasma membrane to the other. Unlike movement by carrier proteins, however, these movements are fueled by the breakdown of adenosine triphosphate (ATP). Recall from chapter 2 that energy stored in ATP molecules is used to power many cellular activities. ATP-powered pumps have binding sites, to which specific ions or molecules can bind, as well as a binding site for ATP. The breakdown of ATP to adenosine diphosphate (ADP) releases energy, changing the shape of the protein, which moves the ion or molecule across the membrane (figure 3.7).

Receptor Proteins

Receptor proteins are proteins or glycoproteins in the plasma membrane that have an exposed **receptor site** on the outer cell surface, which can attach to specific chemical signals. Many receptors and the chemical signals they bind are part of an intercellular communication system that coordinates cell activities. One cell can release a chemical signal that diffuses to another cell and binds to its receptor. The binding acts as a signal that triggers a response. The same chemical signal would have no effect on other cells that lacked the specific receptor molecule.

Receptors Linked to Channel Proteins

Some membrane-bound receptors also help form ligand-gated ion channels. The ion channels are composed of proteins that span the plasma membrane. Parts of one or more of the channel proteins form receptors on the cell surface. When chemical signals, or ligands, bind to these receptors, the combination alters the

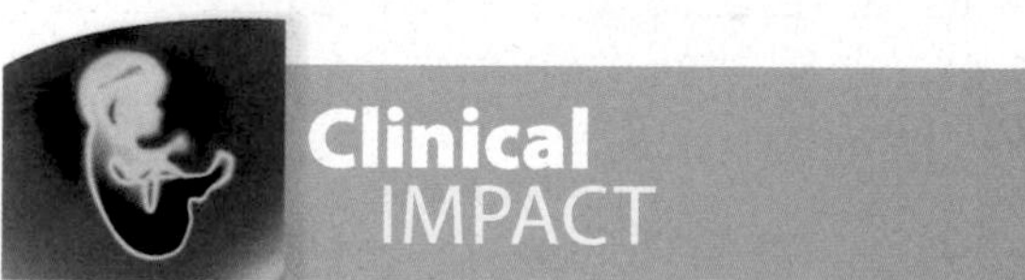
Clinical
IMPACT

Cystic Fibrosis

Cystic fibrosis is a genetic disorder that affects chloride ion channels. There are three types of cystic fibrosis. In about 70% of cases, a defective channel protein fails to reach the plasma membrane from its site of production inside the cell. In the remaining cases, the channel protein is incorporated into the plasma membrane but does not function normally. In some cases, the channel protein fails to bind ATP. In others, ATP binds to the channel protein but the channel does not open. The failure of these ion channels to function causes the affected cells to produce thick, viscous secretions. Although cystic fibrosis affects many cell types, its most profound effects are in the pancreas and the lungs. In the pancreas, the thick secretions block the release of digestive enzymes, resulting in an inability to digest certain types of food and sometimes leading to serious cases of pancreatitis (inflammation of the pancreas). In the lungs, the thick secretions block airways and make breathing difficult. A more detailed description of cystic fibrosis and its consequences appears in chapter 23.

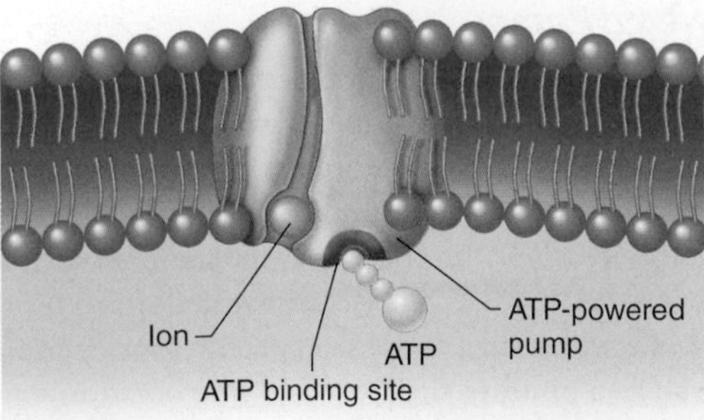
Ion
ATP
ATP binding site
ATP-powered pump

1 ATP and ion bind to the ATP-powered pump.

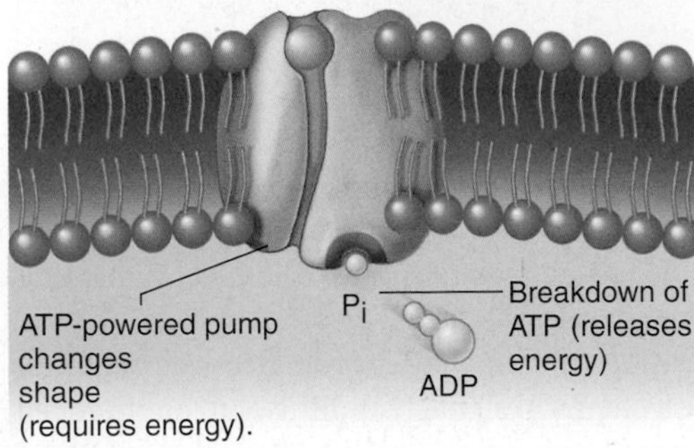
ATP-powered pump changes shape (requires energy).
P_i
Breakdown of ATP (releases energy)
ADP

2 The ATP breaks down to adenosine diphosphate (ADP) and a phosphate (P_i) and releases energy, which powers the shape change in the ATP-powered pump. As a result, the ion moves across the membrane.

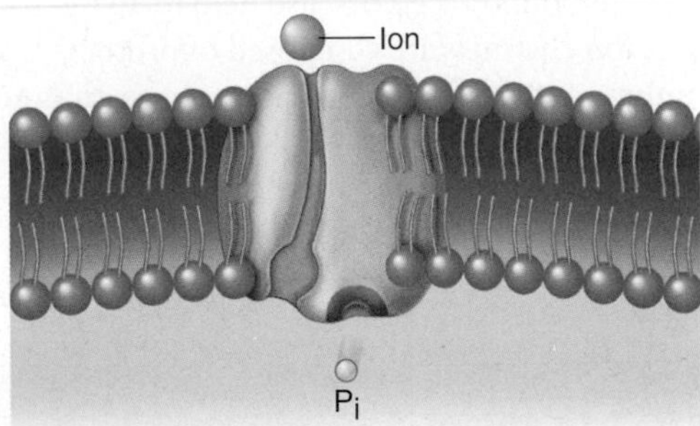
Ion
P_i

3 The ion and phosphate are released from the ATP-powered pump. The pump resumes its original shape (go to step 1).

PROCESS FIGURE 3.7 Transport by an ATP-Powered Pump

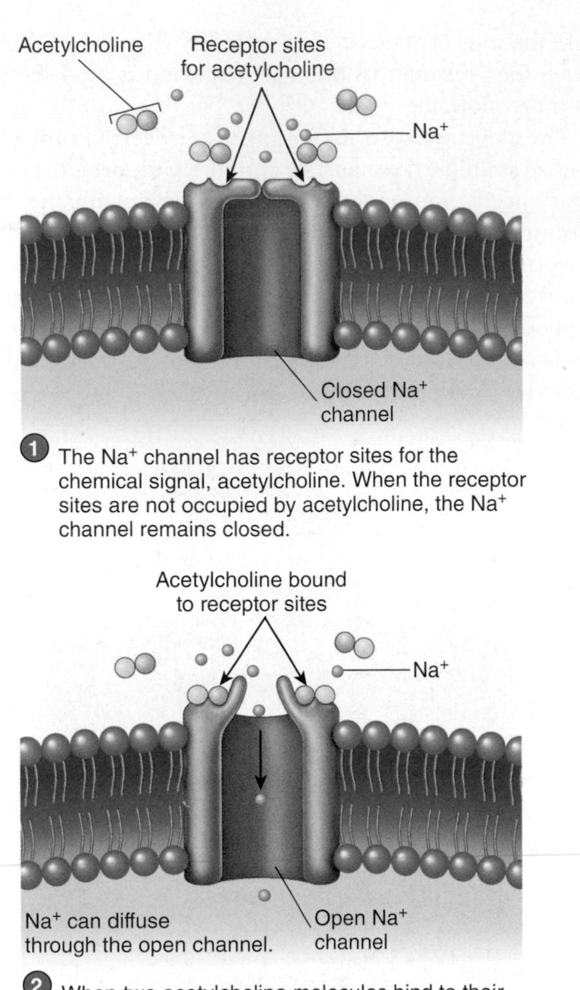
Acetylcholine
Receptor sites for acetylcholine
Na^+
Closed Na^+ channel
Acetylcholine bound to receptor sites
Na^+
Na^+ can diffuse through the open channel.
Open Na^+ channel

1 The Na^+ channel has receptor sites for the chemical signal, acetylcholine. When the receptor sites are not occupied by acetylcholine, the Na^+ channel remains closed.

2 When two acetylcholine molecules bind to their receptor sites on the Na^+ channel, the channel opens to allow Na^+ to diffuse through the channel into the cell.

PROCESS FIGURE 3.8 Opening and Closing of a Channel Protein

three-dimensional structure of the proteins of the ion channels, causing the channels either to open or to close. The result is a change in the permeability of the plasma membrane to the specific ions passing through the ion channels (figure 3.8). For example, acetylcholine released from nerve cells is a chemical signal that combines with membrane-bound receptors of skeletal muscle cells. The combination of acetylcholine molecules with the receptor sites opens Na^+ channels in the plasma membrane. Consequently, the Na^+ diffuse into the skeletal muscle cells and trigger events that cause the cells to contract.

Receptors Linked to G Protein Complexes

Some membrane-bound receptor molecules function by altering the activity of a **G protein complex** located on the inner surface of the plasma membrane (figure 3.9), which acts as an intermediary between a receptor and other cellular proteins. The G protein complex consists of three proteins: alpha (α), beta (β), and gamma (γ) proteins. The G protein complex will only associate with a receptor that has a chemical signal bound to it. In its unassociated state, the α subunit of the G protein complex has guanosine diphosphate (GDP) attached to it (figure 3.9, *step 1*). When a chemical signal binds to the receptor, the receptor becomes associated with the G protein complex. The α subunit releases the GDP and attaches to guanosine triphosphate (GTP; figure 3.9, *step 2*). The G protein complex separates from the receptor, and the α subunit separates from the β and γ subunits (figure 3.9, *step 3*). The activated α subunit can stimulate a cell response in at least three ways: (1) by means of intracellular chemical signals, (2) by the opening of ion channels in the plasma membrane, and (3) by the activation of enzymes associated with the plasma membrane.

Drugs with structures similar to those of specific chemical signals may compete with those chemical signals for their receptor sites. Depending on the exact characteristics of a drug, it binds

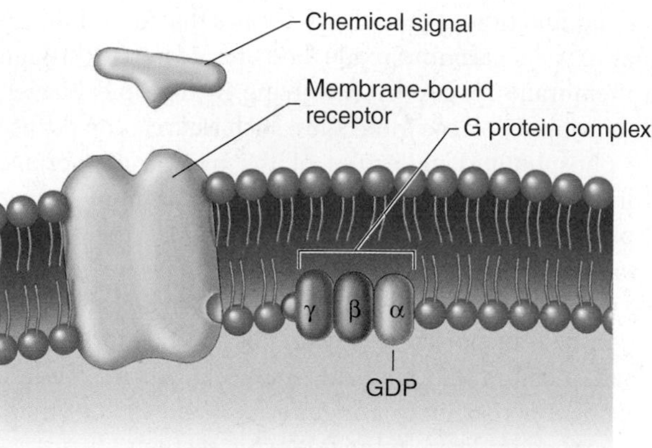
Chemical signal
Membrane-bound receptor
G protein complex
γ
β
α
GDP

1 A G protein complex will only associate with a receptor that has a chemical signal bound to it. In its unassociated state, the α subunit of the G protein complex has guanosine diphosphate (GDP) attached to it.

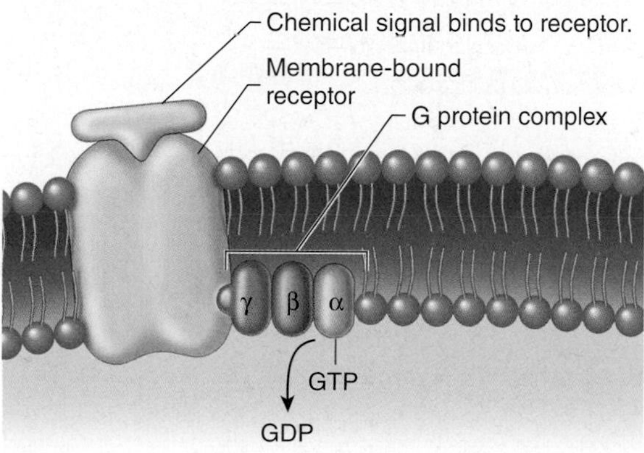
Chemical signal binds to receptor.
Membrane-bound receptor
G protein complex
γ
β
α
GTP
GDP

2 When a chemical signal binds to the receptor, the receptor becomes associated with the G protein complex. GDP is released from the α subunit and a guanosine triphosphate (GTP) is attached to it.

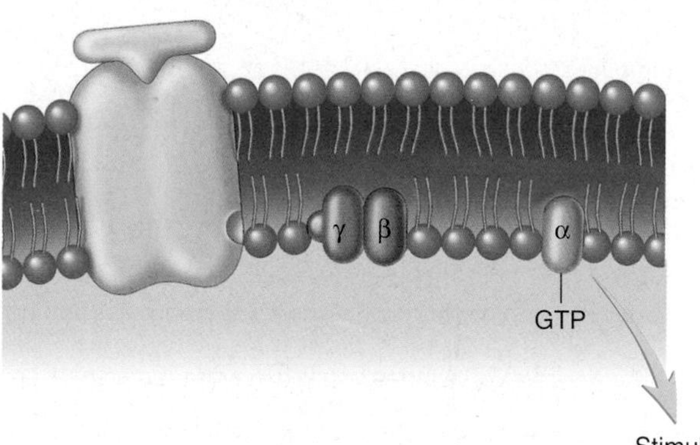
γ
β
α
GTP
Stimulates a cell response

3 The G protein complex separates from the receptor and the α subunit separates from the other subunits. The α subunit stimulates a cell response.

PROCESS FIGURE 3.9 Receptor Linked to a G Protein Complex

to a receptor site and either activates or inhibits the action of the receptor. For example, some drugs compete with the chemical signal epinephrine for its receptor sites. Some of these drugs activate epinephrine receptors; others inhibit them.

Enzymes

Some membrane proteins function as **enzymes,** which can catalyze chemical reactions on either the inner or the outer surface of the plasma membrane. For example, some enzymes on the surface of cells in the small intestine break the peptide bonds of dipeptides to form two single amino acids (table 3.2). Some membrane-associated enzymes are always active; others are activated by membrane-bound receptors or G protein complexes.

ASSESS YOUR PROGRESS

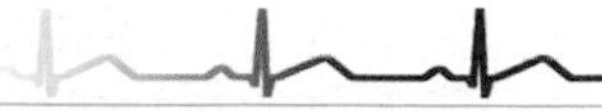

13. *What are the three classes of transport proteins?*
14. *Describe specificity, competiton, and saturation as characteristics of transport proteins.*
15. *What are the three types of channel proteins, and what signal causes each to open or close?*
16. *Define* uniport, symport, *and* antiport.
17. *Compare and contrast how carrier proteins and ATP-powered pumps move ions or molecules across the plasma membrane.*
18. *To what part of a receptor molecule does a chemical signal attach? Explain how a chemical signal can bind to a receptor on a channel protein and cause a change in membrane permeability.*
19. *Describe how receptors alter the activity of G protein complexes. List three ways in which activated α subunits can stimulate a cell response.*
20. *Give an example of the action of a plasma membrane enzyme.*

3.6 Movement Through the Plasma Membrane

LEARNING OUTCOMES

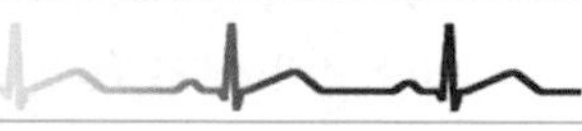

After reading this section, you should be able to

A. Describe the nature of the plasma membrane in reference to the passage of materials through it.
B. List and explain the three ways that molecules and ions can pass through the plasma membrane.
C. Discuss the process of diffusion, and relate it to a concentration gradient.
D. Explain the role of osmosis and osmotic pressure in controlling the movement of water across the plasma membrane. Illustrate the differences among hypotonic, isotonic, and hypertonic solutions in terms of water movement.
E. Describe mediated transport.
F. Compare and contrast facilitated diffusion, active transport, and secondary active transport.
G. Describe the processes of endocytosis and exocytosis.

The plasma membrane separates extracellular material from intracellular material and is **selectively permeable**—that is, it allows only certain substances to pass through it. The intracellular material has a different composition than the extracellular material, and the cell's survival depends on the maintenance of these differences. Enzymes, other proteins, glycogen, and potassium ions are present in higher concentrations intracellularly; sodium, calcium, and chloride ions exist in greater concentrations extracellularly. In addition, even though nutrients must continually enter the cell and waste products must exit, the cell's volume remains unchanged. Because of the plasma membrane's permeability and its ability to transport molecules selectively, the cell is able to maintain homeostasis. Rupture of the membrane, alteration of its permeability characteristics, or inhibition of transport processes can disrupt the normal concentration differences across the plasma membrane and lead to cell death.

Molecules and ions can pass through the plasma membrane in different ways, depending on their chemical characteristics and the structure and function of the cell. Molecules that are soluble in lipids, such as oxygen, carbon dioxide, and steroids, pass through the plasma membrane readily by dissolving in the lipid bilayer. Some small, non-lipid-soluble molecules, such as urea, can diffuse between the phospholipid molecules of the plasma membrane. Large, non-lipid-soluble molecules and ions that cannot diffuse across the phospholipid bilayer may move across the plasma membrane with the help of transport proteins. Finally, large, non-lipid-soluble molecules, as well as small pieces of matter and even whole cells, can be transported across the plasma membrane in a **vesicle,** a small, membrane-bound sac. Table 3.3 lists the specific types of movement across plasma membranes, and each of these methods is described in detail in the following sections.

TABLE 3.3 Comparison of Membrane Transport Mechanisms

Transport Mechanism	Description	Substances Transported	Example
Passive Transport Mechanisms			
Diffusion	Random movement of molecules results in net movement from areas of higher to lower concentration.	Lipid-soluble molecules dissolve in the lipid bilayer and diffuse through it; ions and small molecules diffuse through membrane channels.	Oxygen, carbon dioxide, and lipids, such as steroid hormones, dissolve in the lipid bilayer; Cl^- and urea move through membrane channels.
Osmosis	Water diffuses across a selectively permeable membrane.	Water diffuses through the lipid bilayer.	Water moves from the intestines into the blood.
Facilitated diffusion	Carrier proteins combine with substances and move them across the plasma membrane; no ATP is used; substances are always moved from areas of higher to lower concentration; it exhibits the characteristics of specificity, saturation, and competition.	Some substances too large to pass through membrane channels and too polar to dissolve in the lipid bilayer are transported.	Glucose moves by facilitated diffusion into muscle cells and adipocytes.
Active Transport Mechanisms			
Active transport	ATP-powered pumps combine with substances and move them across the plasma membrane; ATP is used; substances can be moved from areas of lower to higher concentration; it exhibits the characteristics of specificity, saturation, and competition.	Substances too large to pass through channels and too polar to dissolve in the lipid bilayer are transported; substances that are accumulated in concentrations higher on one side of the membrane than on the other are transported.	Ions, such as Na^+, K^+, and Ca^{2+}, are actively transported.
Secondary active transport	Ions are moved across the plasma membrane by active transport, which establishes an ion concentration gradient; ATP is required; ions then move back down their concentration gradient by facilitated diffusion, and another ion or molecule moves with the diffusion ion (symport) or in the opposite direction (antiport).	Some sugars, amino acids, and ions are transported.	There is a concentration gradient for Na^+ into intestinal epithelial cells. This gradient provides the energy for the symport of glucose. As Na^+ enter the cell, down their concentration gradient, glucose also enters the cell. In many cells, H^+ are moved in the opposite direction of Na^+ (antiport).
Vesicular Transport			
Endocytosis	The plasma membrane forms a vesicle around the substances to be transported, and the vesicle is taken into the cell; this requires ATP; in receptor-mediated endocytosis, specific substances are ingested.	Phagocytosis takes in cells and solid particles; pinocytosis takes in molecules dissolved in liquid.	Immune system cells called phagocytes ingest bacteria and cellular debris; most cells take in substances through pinocytosis.
Exocytosis	Materials manufactured by the cell are packaged in secretory vesicles that fuse with the plasma membrane and release their contents to the outside of the cell; this requires ATP.	Proteins and other water-soluble molecules are transported out of cells.	Digestive enzymes, hormones, neurotransmitters, and glandular secretions are transported, and cell waste products are eliminated.

Passive Membrane Transport

Membrane transport mechanisms can be classified based on whether or not the cell expends metabolic energy during the transport process. During passive membrane transport, the cell does not expend metabolic energy; however, active membrane transport, discussed later in this section, does require the cell to expend metabolic energy. Passive membrane transport includes diffusion, osmosis, and facilitated diffusion.

Diffusion

A solution consists of one or more substances, called **solutes,** dissolved in the predominant liquid or gas, which is called the **solvent. Diffusion** is the movement of solutes from an area of higher solute concentration to an area of lower solute concentration (figure 3.10). Diffusion is a product of the constant random motion of all atoms, molecules, or ions in a solution. Because there are more solute particles in an area of higher concentration than in an area of lower concentration and because the particles move randomly, the chances are greater that solute particles will move from the higher to the lower concentration than in the opposite direction. Thus, the overall, or net, movement is from the area of higher solute concentration to that of lower solute concentration. At equilibrium, the net movement of solutes stops, although the random molecular motion continues, and the movement of solutes in any one direction is balanced by an equal movement in the opposite direction. Examples of diffusion include the movement and distribution of smoke or perfume throughout a room without air currents and the dispersion of a dye throughout a beaker of still water.

A concentration difference occurs when the solute concentration in a solvent is greater at one point than at another point. The concentration difference between two points, divided by the distance between the two points, is called the **concentration gradient.** Solutes diffuse *down* their concentration gradients (from a higher to a lower solute concentration) until an equilibrium is achieved. The greater the concentration gradient, the greater the rate of diffusion of a solute down that gradient. Adjusting the concentration difference or distance between the two points changes the concentration gradient. Increasing the concentration difference between the two points or decreasing the distance between the two points causes the concentration gradient to increase, whereas decreasing the concentration difference between the two points or increasing the distance between the two points causes the concentration gradient to decrease.

The rate of diffusion is influenced by several factors, including the magnitude of the concentration gradient, the temperature of the solution, the size of the diffusing molecules, and the viscosity of the solvent. The greater the concentration gradient, the greater the number of solute particles moving from a higher to a lower solute concentration. As the temperature of a solution increases, the speed at which all molecules move increases, resulting in a greater diffusion rate. Small molecules diffuse through a solution more readily than do large ones. **Viscosity** is a measure of a fluid's resistance to flow. A fluid with a low viscosity flows more easily and a fluid with a high viscosity flows less easily. Thick solutions, such as syrup, are more viscous than water. Diffusion occurs more slowly in viscous solvents than in thin, watery solvents.

Diffusion of molecules is an important means by which substances move between the extracellular and intracellular fluids in the body. Substances that can diffuse through either the lipid bilayer or the membrane channels can pass through the plasma membrane (figure 3.11). Some nutrients enter and some waste products leave the cell by diffusion, and maintenance of the appropriate intracellular concentration of these substances depends to a large degree on diffusion. For example, if the extracellular concentration of oxygen is reduced the concentration gradient decreases, inadequate oxygen diffuses into the cell, and the cell cannot function normally.

ASSESS YOUR PROGRESS

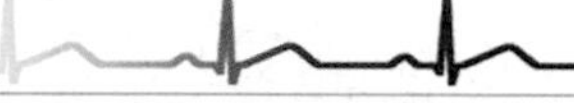

21. *Explain why the plasma membrane is selectively permeable.*
22. *List three ways that substances can move across the plasma membrane.*
23. *Describe how the amount of solute in a solvent creates a concentration gradient. Do solutes diffuse with (down) or against their concentration gradient?*
24. *How is the rate of diffusion affected by an increased concentration gradient? By increased temperature of a solution? By increased viscosity of the solvent?*

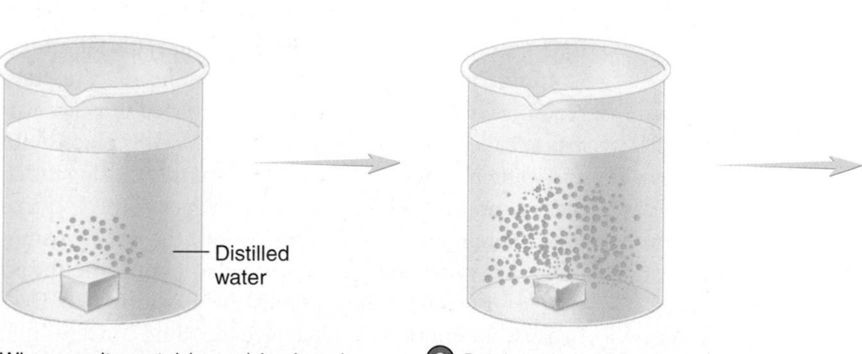
Distilled water

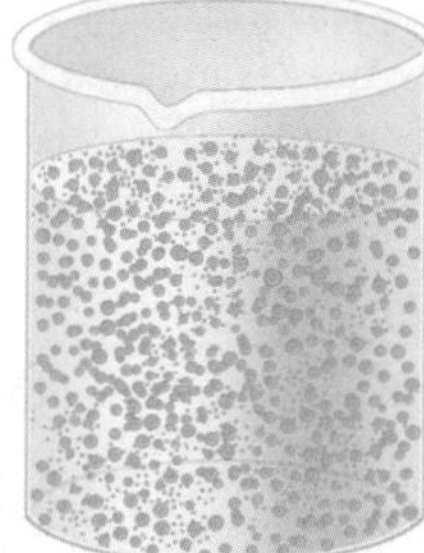

1 When a salt crystal (*green*) is placed into a beaker of water, a concentration gradient exists between the salt from the salt crystal and the water that surrounds it.

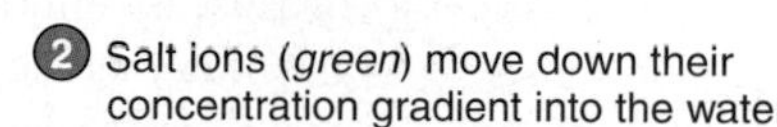

2 Salt ions (*green*) move down their concentration gradient into the water.

3 Salt ions and water molecules are distributed evenly throughout the solution. Even though the salt ions and water molecules continue to move randomly, an equilibrium exists, and no net movement occurs because no concentration gradient exists.

PROCESS FIGURE 3.10 APR **Diffusion**

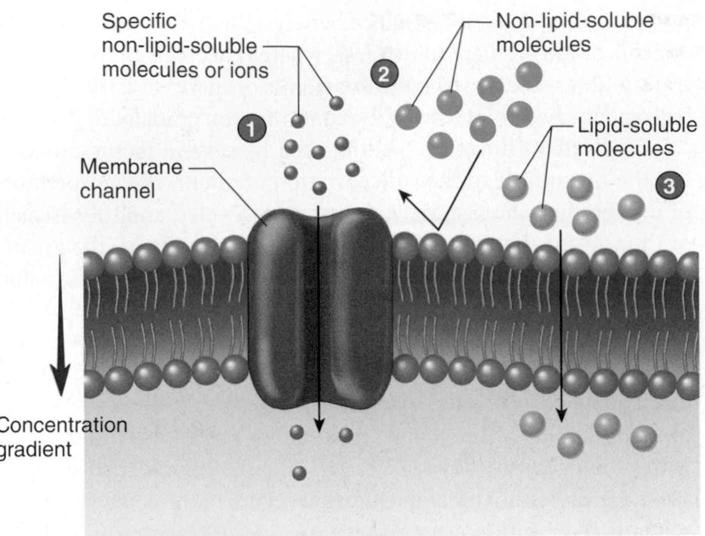
Specific non-lipid-soluble molecules or ions
Non-lipid-soluble molecules
Lipid-soluble molecules
Membrane channel
Concentration gradient
1
2
3

1 Certain specific non-lipid-soluble molecules or ions diffuse through membrane channels.

2 Other non-lipid-soluble molecules or ions, for which membrane channels are not present in the cell, cannot enter the cell.

3 Lipid-soluble molecules diffuse directly through the plasma membrane.

PROCESS FIGURE 3.11 Diffusion Through the Plasma Membrane

Predict 2

Mr. Smith is suffering from chronic renal failure, characterized by a gradual decrease in his kidneys' ability to perform their normal functions. Over the past few months, blood tests have indicated an increasing concentration of urea in his blood, which is a sign that the concentration of urea in the extracellular fluid is also increasing. Urea, a toxic waste produced inside cells, diffuses across the plasma membrane into the extracellular fluid. The kidneys eliminate excess urea in the urine. Explain why the extracellular concentration of urea is increasing, and predict how the intracellular concentration of urea is changing.

Osmosis

Osmosis (os-mō′sis) is the diffusion of water (solvent) across a selectively permeable membrane, such as a plasma membrane (figure 3.12). *Selectively permeable* means that the membrane allows water but not all the solutes dissolved in the water to diffuse through it. **Aquaporins,** or water channel proteins, increase membrane permeability to water in some cell types, such as kidney cells. Water diffuses from a solution with proportionately more water, across a selectively permeable membrane, and into a solution with proportionately less water. Because solution concentrations are defined in terms of solute concentrations, not in terms of water content (see chapter 2), water diffuses from the less concentrated solution (fewer solutes, more water) into the more concentrated solution (more solutes, less water). Osmosis is important to cells because large volume changes caused by water movement disrupt normal cell function.

Osmotic pressure is the force required to prevent water from moving by osmosis across a selectively permeable membrane. The osmotic pressure of a solution can be determined by placing the solution into a tube that is closed at one end by a selectively permeable membrane (figure 3.12). The tube is then immersed in distilled water. Water molecules move by osmosis through the membrane into the tube, forcing the solution to move up the tube. As the solution rises into the tube, its weight produces hydrostatic pressure, which moves water out of the tube back into the distilled water surrounding the tube. At equilibrium, net movement of water stops, which means that the movement of water into the tube by osmosis is equal to the movement of water out of the tube caused by hydrostatic pressure. The osmotic pressure of the solution in the tube is equal to the hydrostatic pressure that prevents net movement of water into the tube.

The osmotic pressure of a solution provides information about the tendency for water to move by osmosis across a selectively permeable membrane. Because water moves from less concentrated solutions (fewer solutes, more water) into more concentrated solutions (more solutes, less water), the greater the concentration of a solution (the less water it has), the greater the tendency for water to move into the solution, and the greater the osmotic pressure to prevent that movement.

Predict 3

Given the demonstration in figure 3.12, what happens to osmotic pressure if the membrane is not selectively permeable but instead allows all solutes and water to pass through it?

Three terms describe the osmotic pressure of solutions. Solutions with the same concentration of solute particles (see chapter 2) have the same osmotic pressure and are referred to as **isosmotic** (ī′sos-mot′ik). The solutions are isosmotic even if the types of solute particles in the two solutions differ from each other. If one solution has a greater concentration of solute particles, and therefore a greater osmotic pressure than another solution, the first solution is said to be **hyperosmotic** (hī′per-oz-mot′ik) compared with the more dilute solution. The more dilute solution, with the lower osmotic pressure, is **hyposmotic** (hī-pos-mot′ik) compared with the more concentrated solution.

Three additional terms describe the tendency of cells to shrink or swell when placed into a solution (figure 3.13). If a cell placed into a solution neither shrinks nor swells, the solution is said to be **isotonic** (ī-sō-ton′ik). In an isotonic solution, the shape of the cell remains constant, maintaining its internal tension or tone, a condition called **tonicity** (tō-nis′i-tē). If a cell is placed into a solution and water moves out of the cell by osmosis, causing the cell to shrink, the solution is called **hypertonic** (hī-per-ton′ik). If a cell is placed into a solution and water moves into the cell by osmosis, causing the cell to swell, the solution is called **hypotonic** (hī-pō-ton′ik).

An isotonic solution may be isosmotic to the cytoplasm. Because isosmotic solutions have the same concentration of solutes and water as the cytoplasm of the cell, no net movement of water occurs, and the cell neither swells nor shrinks (figure 3.13*b*). Hypertonic solutions can be hyperosmotic and have a greater concentration of solute molecules and a lower concentration of water than the cytoplasm of the cell. Therefore, water moves by osmosis from the cell into the hypertonic solution, causing the cell

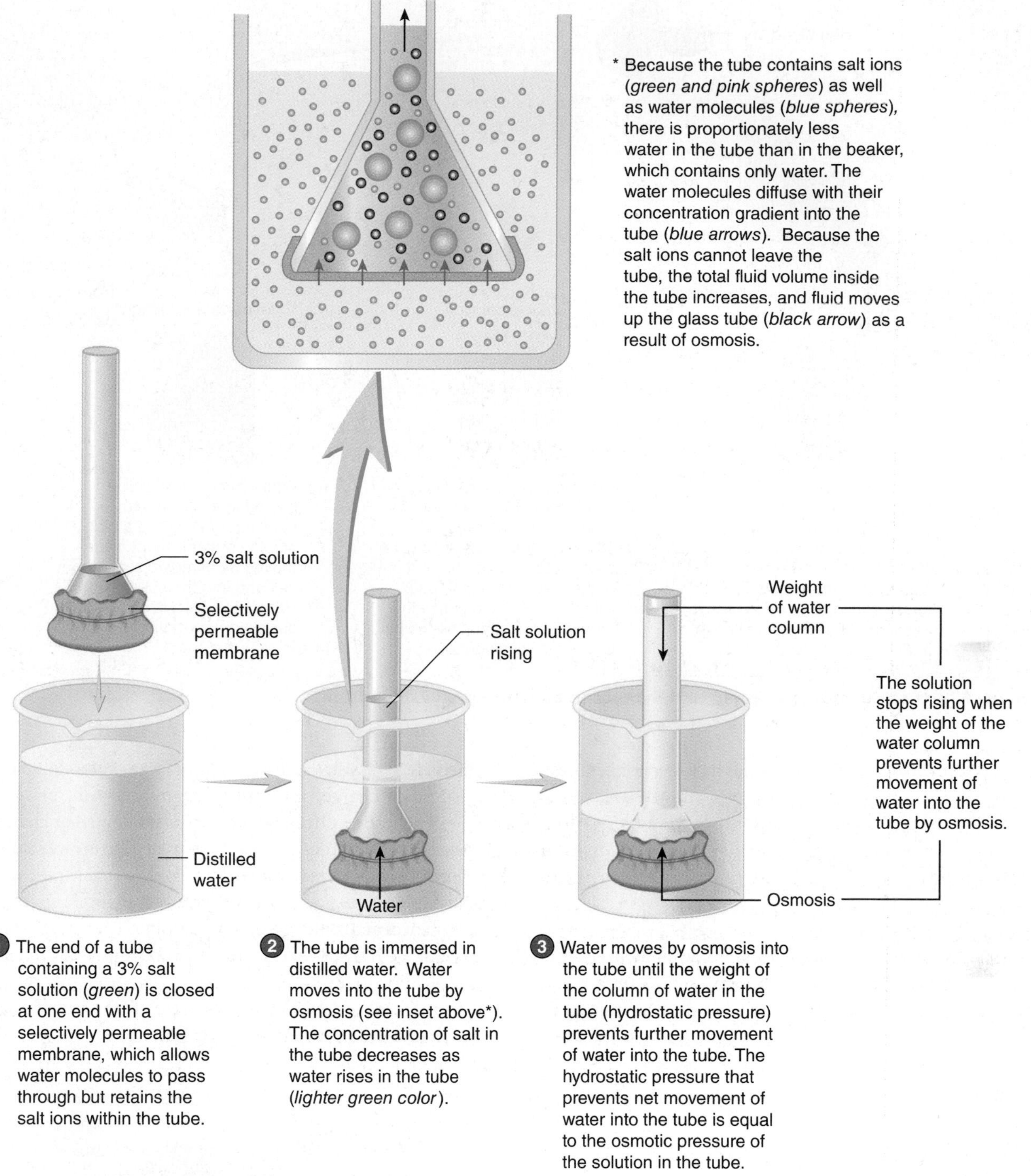
* Because the tube contains salt ions (*green and pink spheres*) as well as water molecules (*blue spheres*), there is proportionately less water in the tube than in the beaker, which contains only water. The water molecules diffuse with their concentration gradient into the tube (*blue arrows*). Because the salt ions cannot leave the tube, the total fluid volume inside the tube increases, and fluid moves up the glass tube (*black arrow*) as a result of osmosis.
3% salt solution
Selectively permeable membrane
Distilled water
Salt solution rising
Water
Weight of water column
The solution stops rising when the weight of the water column prevents further movement of water into the tube by osmosis.
Osmosis
1 The end of a tube containing a 3% salt solution (*green*) is closed at one end with a selectively permeable membrane, which allows water molecules to pass through but retains the salt ions within the tube.
2 The tube is immersed in distilled water. Water moves into the tube by osmosis (see inset above*). The concentration of salt in the tube decreases as water rises in the tube (*lighter green color*).
3 Water moves by osmosis into the tube until the weight of the column of water in the tube (hydrostatic pressure) prevents further movement of water into the tube. The hydrostatic pressure that prevents net movement of water into the tube is equal to the osmotic pressure of the solution in the tube.

PROCESS FIGURE 3.12 AP|R **Osmosis**

to shrink, a process called **crenation** (krē-nā′shŭn) in red blood cells (figure 3.13*c*). Hypotonic solutions can be hyposmotic and have a smaller concentration of solute molecules and a greater concentration of water than the cytoplasm of the cell. Therefore, water moves by osmosis into the cell, causing it to swell. If the cell swells enough, it can rupture, a process called **lysis** (lī′sis; figure 3.13*a*). Solutions injected into the bloodstream or the tissues must be isotonic because shrinkage or swelling of cells disrupts their normal function and can lead to cell death.

The *-osmotic* terms refer to the concentration of the solutions, and the *-tonic* terms refer to the tendency of cells to swell or shrink. These terms should not be used interchangeably. Not all isosmotic solutions are isotonic. For example, it is possible to prepare a solution of glycerol and a solution of mannitol that are isosmotic to the cytoplasm of the cell. Because the solutions are isosmotic, they have the same concentration of solutes and water as the cytoplasm. However, glycerol can diffuse across the plasma membrane, whereas mannitol cannot. When glycerol diffuses into the cell, the solute

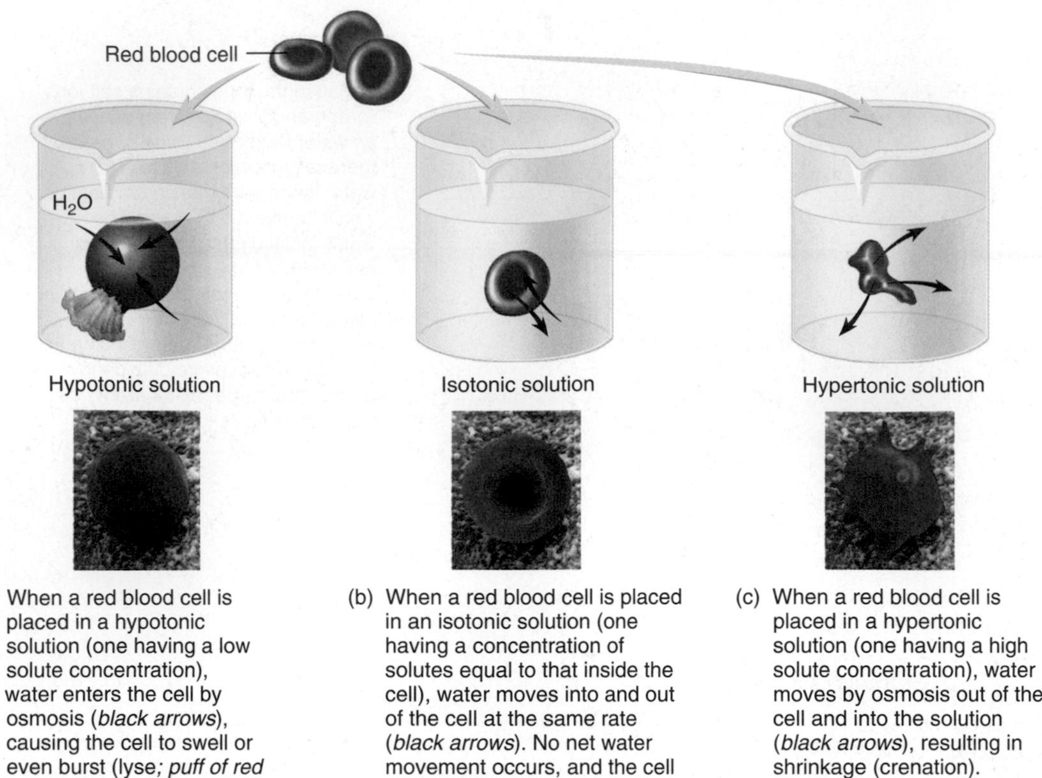
Red blood cell
H_2O
Hypotonic solution
Isotonic solution
Hypertonic solution

(a) When a red blood cell is placed in a hypotonic solution (one having a low solute concentration), water enters the cell by osmosis (*black arrows*), causing the cell to swell or even burst (lyse; *puff of red in lower part of cell*).

(b) When a red blood cell is placed in an isotonic solution (one having a concentration of solutes equal to that inside the cell), water moves into and out of the cell at the same rate (*black arrows*). No net water movement occurs, and the cell shape remains normal.

(c) When a red blood cell is placed in a hypertonic solution (one having a high solute concentration), water moves by osmosis out of the cell and into the solution (*black arrows*), resulting in shrinkage (crenation).

FIGURE 3.13 Effects of Hypotonic, Isotonic, and Hypertonic Solutions on Red Blood Cells

concentration of the cytoplasm increases, and its water concentration decreases. Therefore, water moves by osmosis into the cell, causing it to swell, and the glycerol solution is both isosmotic and hypotonic. In contrast, mannitol cannot enter the cell, and the isosmotic mannitol solution is also isotonic.

Facilitated Diffusion

Many essential molecules, such as amino acids and glucose, cannot enter the cell by diffusion, and many products, such as urea, cannot exit the cell by diffusion directly through the plasma membrane. **Mediated transport** is the process by which transport proteins mediate, or assist, the movement of large, water-soluble molecules or electrically charged molecules or ions across the plasma membrane.

Facilitated diffusion is a carrier-mediated or channel-mediated passive membrane transport process that moves substances into or out of cells from a higher to a lower concentration (figure 3.14). Facilitated diffusion does not require metabolic energy to transport substances across the plasma membrane. The rate at which molecules or ions are transported is directly proportional to their concentration gradient up to the point of saturation, when all the carrier proteins or channels are occupied. Then the rate of transport remains constant at its maximum rate (see figure 3.4).

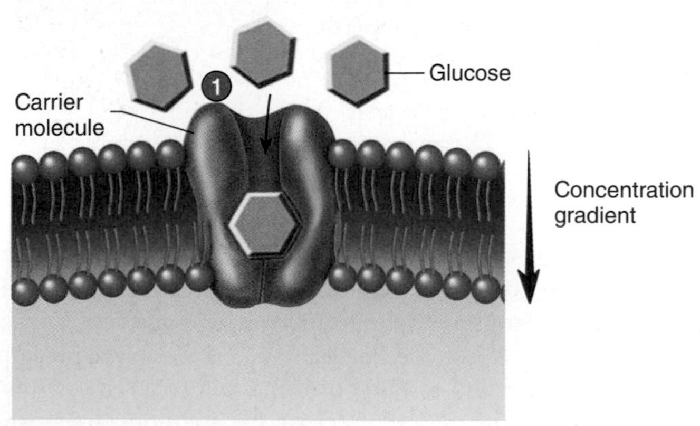
Carrier molecule
1
Glucose
Concentration gradient

1 The carrier molecule binds with a molecule, such as glucose, on the outside of the plasma membrane.

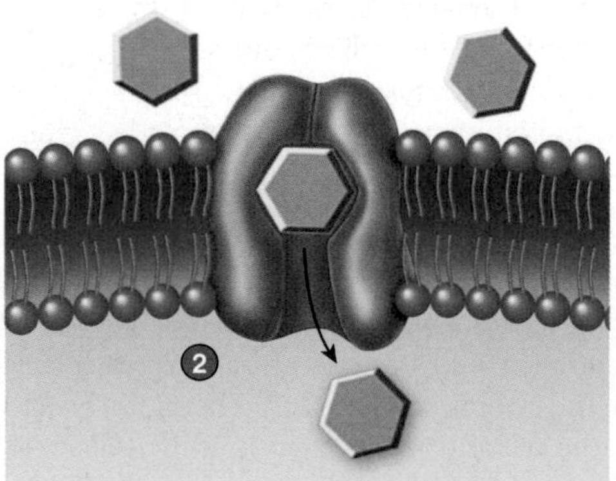
2

2 The carrier molecule changes shape and releases the molecule on the inside of the plasma membrane.

PROCESS FIGURE 3.14 Facilitated Diffusion

> **Predict 4**
>
> The transport of glucose into and out of most cells, such as muscle cells and adipocytes, occurs by facilitated diffusion. Once glucose enters a cell, it is rapidly converted to another molecule, such as glucose-6-phosphate or glycogen. What effect does this conversion have on the cell's ability to acquire glucose? Explain.

Active Membrane Transport

Active Transport

Active transport is a mediated transport process that requires energy provided by ATP (figure 3.15). Movement of the transported substance to the opposite side of the membrane and its subsequent release from the ATP-powered pump are fueled by the breakdown of ATP. The maximum rate at which active transport proceeds depends on the number of ATP-powered pumps in the plasma membrane and the availability of adequate ATP. Active transport is important because it can move substances against their concentration gradients—that is, from lower concentrations to higher concentrations. Consequently, it can accumulate substances on one side of the plasma membrane at concentrations many times greater than those on the other side. Active transport can also move substances from higher to lower concentrations.

Some active-transport mechanisms exchange one substance for another. For example, the **sodium-potassium (Na^+–K^+) pump** moves Na^+ out of cells and K^+ into cells (figure 3.15). The result is a higher concentration of Na^+ outside the cell and a higher concentration of K^+ inside the cell. Because ATP is broken down during the transport of Na^+ and K^+, the pump is also called **sodium-potassium ATP-ase.** The Na^+–K^+ pump is very important to a number of cell functions, as discussed in chapters 9 and 11.

Secondary Active Transport

Secondary active transport involves the active transport of an ion, such as sodium, out of a cell, establishing a concentration gradient, with a higher concentration of the ions outside the cell. The tendency for the ions to move back into the cell (down their concentration gradient) provides the energy necessary to move a different ion or some other molecule into the cell. For example, glucose moves from the lumen of the intestine into epithelial cells by secondary active transport (figure 3.16). This process requires two transport proteins: (1) A Na^+–K^+ pump actively moves Na^+ out of the cell, and (2) a carrier protein facilitates the movement of Na^+ and glucose into the cell. Both Na^+ and glucose are necessary for the carrier protein to function.

The movement of Na^+ down its concentration gradient provides the energy to move glucose molecules into the cell against their concentration gradient. Thus, glucose can accumulate at concentrations higher inside the cell than outside. Because the movement of glucose molecules against their concentration gradient results from the formation of a concentration gradient of Na^+ by an active-transport mechanism, the process is called secondary active transport.

1 Three sodium ions (Na^+) and adenosine triphosphate (ATP) bind to the sodium–potassium (Na^+–K^+) pump.

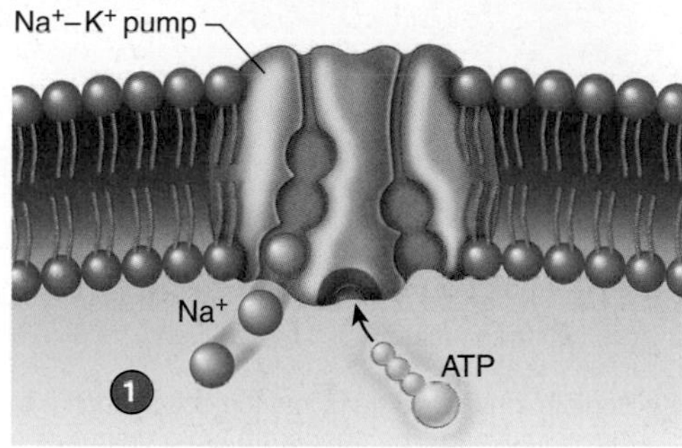
Na⁺–K⁺ pump
Na⁺
ATP
1

2 The ATP breaks down to adenosine diphosphate (ADP) and a phosphate (P) and releases energy. That energy is used to power the shape change in the Na^+–K^+ pump.

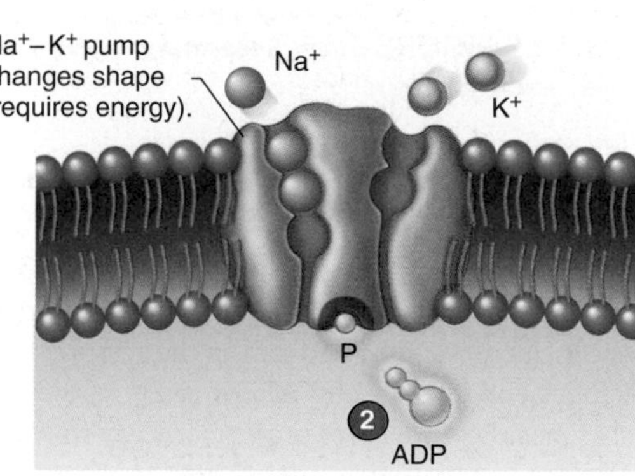
Na⁺–K⁺ pump changes shape (requires energy).
Na⁺
K⁺
P
2
ADP

3 The Na^+–K^+ pump changes shape, and the Na^+ are transported across the membrane and into the extracellular fluid.

4 Two potassium ions (K^+) bind to the Na^+–K^+ pump.

5 The phosphate is released from the Na^+–K^+ pump binding site.

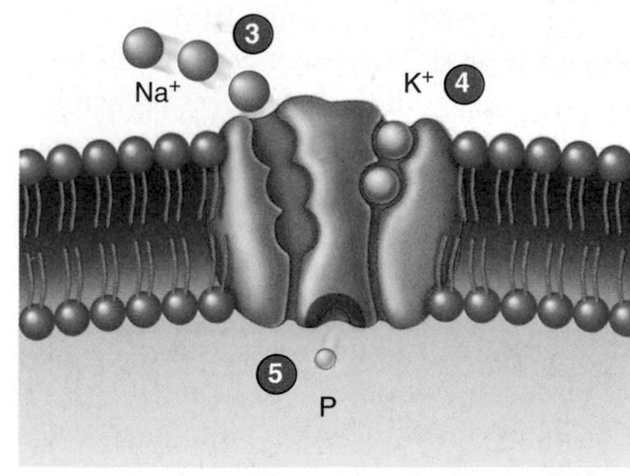
3
Na⁺
K⁺ 4
5
P

6 The Na^+–K^+ pump changes shape, transporting K^+ across the membrane and into the cytoplasm. The Na^+–K^+ pump can again bind to Na^+ and ATP.

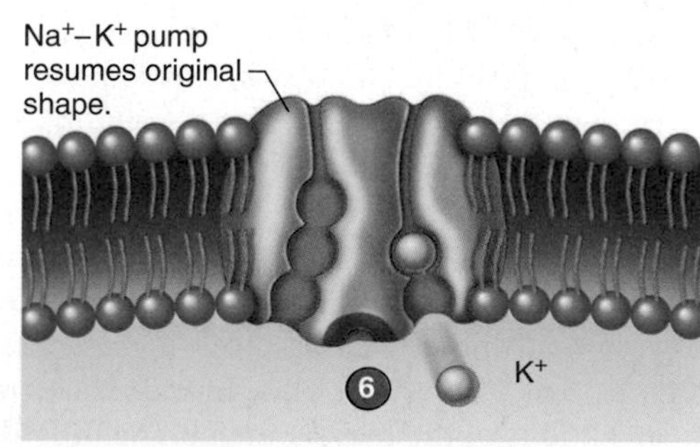
Na⁺–K⁺ pump resumes original shape.
6
K⁺

PROCESS FIGURE 3.15 APR **Active Transport: Sodium-Potassium Pump**

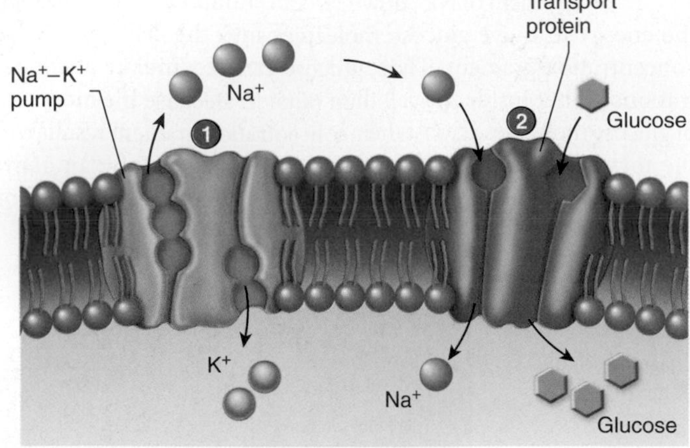
Transport protein
$Na^+–K^+$ pump
Na^+
1
2
Glucose
K^+
Na^+
Glucose

1 A $Na^+–K^+$ pump maintains a concentration of Na^+ that is higher outside the cell than inside.

2 Sodium ions move back into the cell through a transport protein that also moves glucose. The concentration gradient for Na^+ provides energy required to move glucose against its concentration gradient.

PROCESS FIGURE 3.16 Secondary Active Transport (Symport) of Na^+ and Glucose

Vesicular Transport

Vesicular transport is the movement of larger volumes of substances across the plasma membrane through the formation or release of vesicles, membrane-bound sacs, in the cytoplasm. Vesicular transport includes endocytosis and exocytosis.

Vesicular transport requires energy in the form of ATP and therefore is an active membrane transport process. However, because it involves the bulk movement of material into the cell, vesicular transport does not demonstrate the degree of specificity or saturation that other forms of active membrane transport exhibit.

Endocytosis (en′dō-sī-tō′sis) is the uptake of material through the plasma membrane by the formation of a vesicle. A portion of the plasma membrane wraps around a particle or droplet and fuses, so that the particle or droplet is surrounded by a membrane. That portion of the membrane then "pinches off," so that the enclosed particle or droplet is within the cytoplasm of the cell, and the plasma membrane is left intact.

Endocytosis includes both phagocytosis and pinocytosis. In **phagocytosis** (făg-ō-sī-tō′sis), which means "cell-eating," solid particles are ingested and phagocytic vesicles are formed (figure 3.17). White blood cells and some other cell types phagocytize bacteria, cell debris, and foreign particles. Phagocytosis is therefore important in eliminating harmful substances from the body.

Pinocytosis (pin′ō-sī-tō′sis), which means "cell-drinking," is distinguished from phagocytosis in that smaller vesicles form and they contain molecules dissolved in liquid rather than particles (figure 3.18). Pinocytosis often forms vesicles near the tips of deep invaginations of the plasma membrane. It is a common transport phenomenon in a variety of cell types, occurring in certain cells of the kidneys, epithelial cells of the intestines, cells of the liver, and cells that line capillaries.

Endocytosis can exhibit specificity. For example, cells that phagocytize bacteria and necrotic tissue do not phagocytize healthy cells. The plasma membrane may contain specific receptor molecules that recognize certain substances and allow them to be transported into the cell by phagocytosis or pinocytosis. This is called **receptor-mediated endocytosis,** and the receptor sites combine only with certain molecules (figure 3.19). This mechanism increases the rate at which the cells take up specific substances. Cholesterol and growth factors are examples of molecules that can be taken into a cell by receptor-mediated endocytosis.

Hypercholesterolemia is a common genetic disorder characterized by the reduction in or absence of low-density lipoprotein (LDL) receptors on cell surfaces, which interferes with the receptor-mediated endocytosis of LDL cholesterol. As a result

(a)

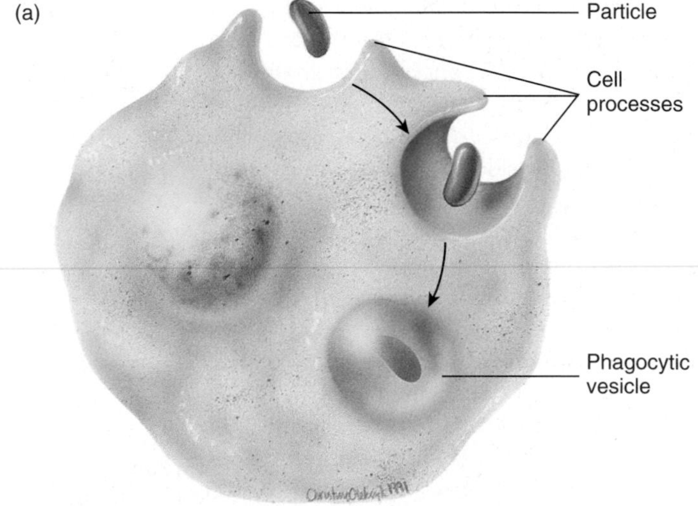
Particle
Cell processes
Phagocytic vesicle

(b)

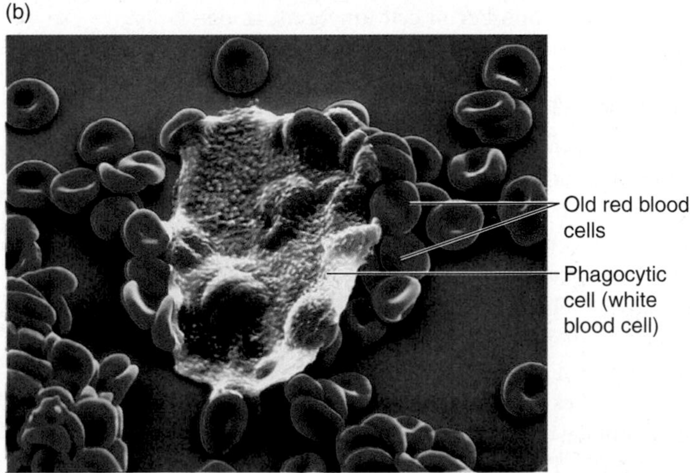
Old red blood cells
Phagocytic cell (white blood cell)

FIGURE 3.17 Phagocytosis

(*a*) In this type of endocytosis, a solid particle is ingested, and a phagocytic vesicle forms around it. (*b*) Scanning electron micrograph of phagocytosis of red blood cells.

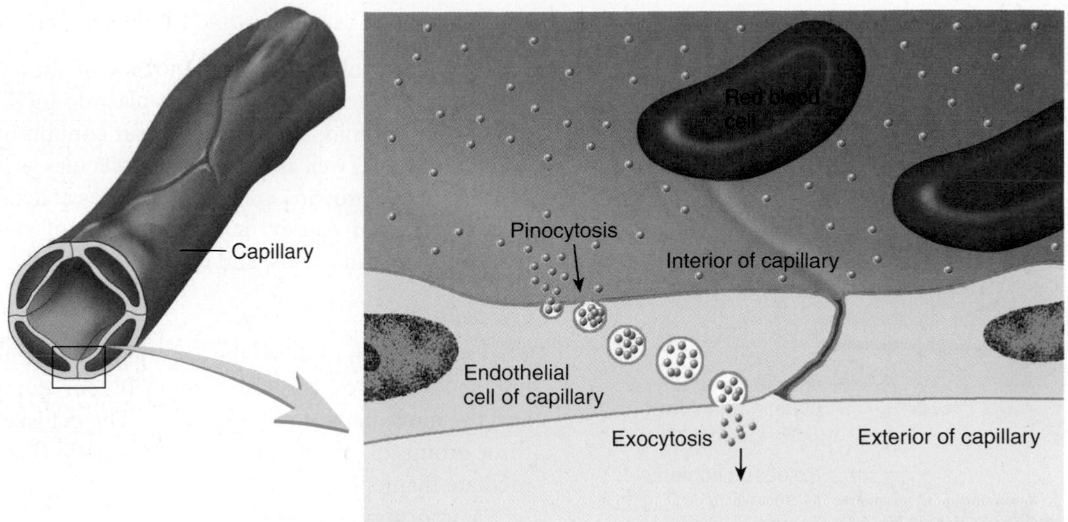
Capillary
Red blood cell
Pinocytosis
Interior of capillary
Endothelial cell of capillary
Exocytosis
Exterior of capillary

FIGURE 3.18 Pinocytosis
Pinocytosis is much like phagocytosis, except that the cell processes—and therefore the vesicles formed—are much smaller and the material inside the vesicle is liquid rather than particulate. Pinocytotic vesicles form on the internal side of a capillary, are transported across the cell, and open by exocytosis outside the capillary.

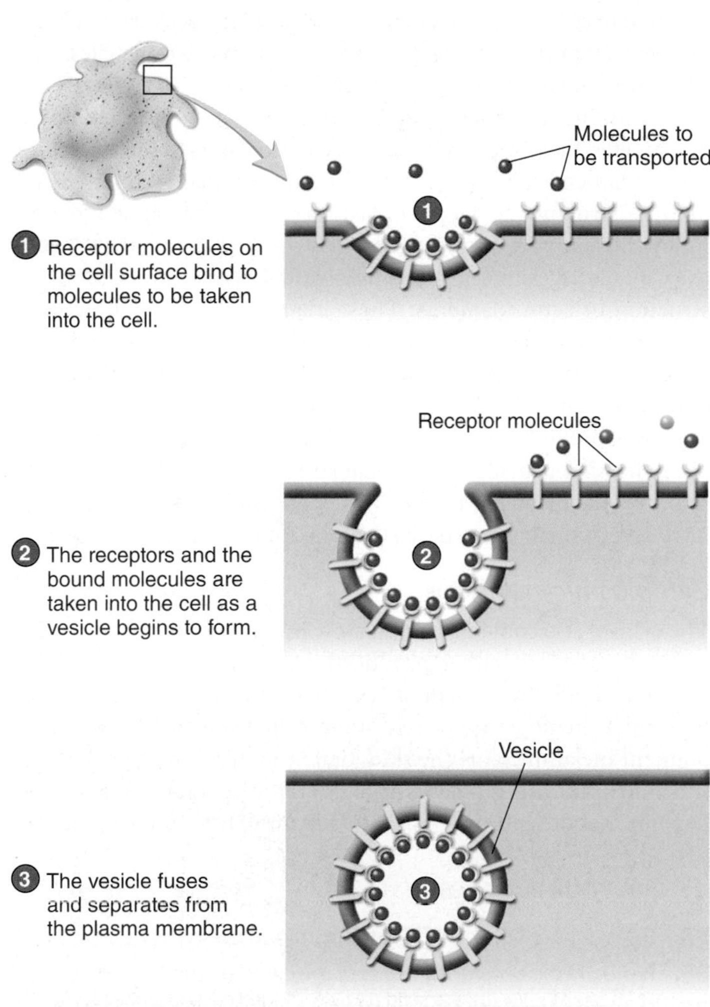
Molecules to be transported
1
1 Receptor molecules on the cell surface bind to molecules to be taken into the cell.
Receptor molecules
2
2 The receptors and the bound molecules are taken into the cell as a vesicle begins to form.
Vesicle
3
3 The vesicle fuses and separates from the plasma membrane.

PROCESS FIGURE 3.19 Receptor-Mediated Endocytosis

of inadequate cholesterol uptake, cholesterol synthesis within these cells is not regulated, and too much cholesterol is produced. The excess cholesterol accumulates in blood vessels, resulting in atherosclerosis. Atherosclerosis can cause heart attacks or strokes. A more detailed description of hypercholesterolemia can be found in chapter 24.

In some cells, secretions accumulate within vesicles. These secretory vesicles then move to the plasma membrane, where the vesicle membrane fuses with the plasma membrane and the vesicle contents are expelled from the cell. This process is called **exocytosis** (ek′sō-sī-tō′sis; figure 3.20). The secretion of digestive enzymes by the pancreas and the secretion of mucus by the salivary glands are examples of exocytosis.

ASSESS YOUR PROGRESS

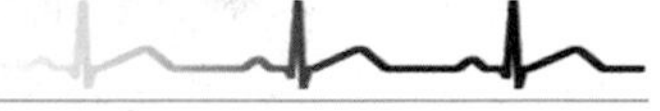

25. *Define* osmosis, *and describe how osmotic pressure is created. As the concentration of a solution increases, what happens to its osmotic pressure and to the tendency for water to move into the solution?*

26. *Compare isosmotic, hyperosmotic, and hyposmotic solutions with isotonic, hypertonic, and hypotonic solutions.*

27. *What is mediated transport? What types of particles move through the plasma membrane by mediated transport?*

28. *Contrast facilitated diffusion and active transport in relation to energy expenditure and direction of movement with respect to the concentration gradient.*

29. *What is secondary active transport? Describe how it functions.*

30. *What occurs during endocytosis? What role do vesicles play?*

31. *How do phagocytosis and pinocytosis differ from each other?*

32. *What is receptor-mediated endocytosis?*

33. *Describe and give examples of exocytosis.*

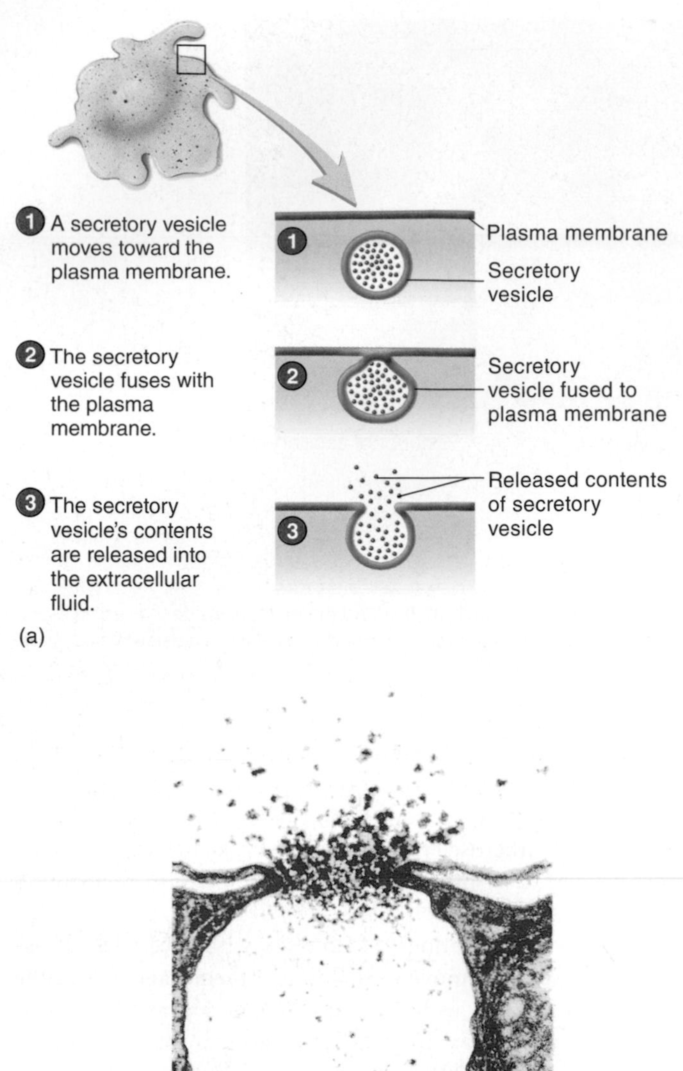
1 A secretory vesicle moves toward the plasma membrane.
Plasma membrane
Secretory vesicle
2 The secretory vesicle fuses with the plasma membrane.
Secretory vesicle fused to plasma membrane
Released contents of secretory vesicle
3 The secretory vesicle's contents are released into the extracellular fluid.
(a)
TEM 30,000x
(b)

PROCESS FIGURE 3.20 Exocytosis
(*a*) Diagram of exocytosis. (*b*) Transmission electron micrograph of exocytosis.

3.7 Cytoplasm

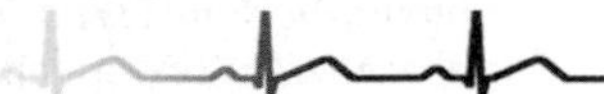

LEARNING OUTCOMES

After reading this section, you should be able to

A. Describe the composition and functions of the cytoplasm.

B. Describe the composition and function of the cytoskeleton.

Cytoplasm, the cellular material outside the nucleus but inside the plasma membrane, is about half cytosol and half organelles.

Cytosol

Cytosol (sī′tō-sol) is the fluid portion of the cytoplasm, which contains the cytoskeleton and cytoplasmic inclusions. The fluid portion is a colloid, a viscous solution containing dissolved ions and molecules as well as suspended molecules, especially proteins. Many of these proteins are enzymes that catalyze the breakdown of molecules for energy or the synthesis of sugars, fatty acids, nucleotides, amino acids, and other molecules.

Cytoskeleton

The **cytoskeleton** supports the cell and holds the nucleus and other organelles in place. It is also responsible for changes in cell shape and the movement of cell organelles. The cytoskeleton consists of three groups of proteins: microtubules, actin filaments, and intermediate filaments (figure 3.21).

Microtubules are hollow tubes composed primarily of protein units called **tubulin.** The microtubules are about 25 nanometers (nm) in diameter, with walls about 5 nm thick. They vary in length but are normally several micrometers (μm) long. Microtubule length can change as tubulin subunits are added or removed. Microtubules play a variety of roles within cells. They help provide support and structure to the cytoplasm of the cell, much like an internal scaffolding; they are involved in cell division and in the transport of intracellular materials; and they form essential components of certain cell organelles, such as centrioles, spindle fibers, cilia, and flagella.

Actin filaments, or *microfilaments*, are small fibrils, about 8 nm in diameter, that form bundles, sheets, or networks in the cytoplasm. These filaments have a spiderweb-like appearance. Actin filaments provide structure to the cytoplasm and mechanical support for microvilli, support the plasma membrane, and define the shape of the cell. Changes in cell shape involve the breakdown and reconstruction of actin filaments. These changes in shape allow some cells to move about. Muscle cells contain a large number of highly organized actin filaments, which are responsible for the muscle's contractile capabilities (see chapter 9).

Intermediate filaments are protein fibers about 10 nm in diameter that provide mechanical strength to cells. For example, intermediate filaments support the extensions of nerve cells, which have a very small diameter but can be up to a meter in length.

Cytoplasmic Inclusions

The cytosol also contains **cytoplasmic inclusions,** which are aggregates of chemicals either produced or taken in by the cell. For example, lipid droplets or glycogen granules store energy-rich molecules; hemoglobin in red blood cells transports oxygen; the pigment melanin colors the skin, hair, and eyes; and **lipochromes** (lip′ō-krōmz) are pigments that increase in amount with age. Dust, minerals, and dyes can also accumulate in the cytoplasm.

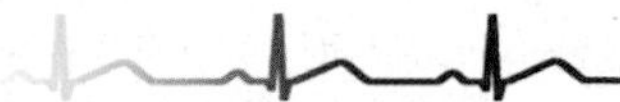

ASSESS YOUR PROGRESS

34. *Differentiate between the cytoplasm and the cytosol.*
35. *What are the general functions of the cytoskeleton?*
36. *List and describe the functions of microtubules, actin filaments, and intermediate filaments.*
37. *What are cytoplasmic inclusions? Give several examples.*

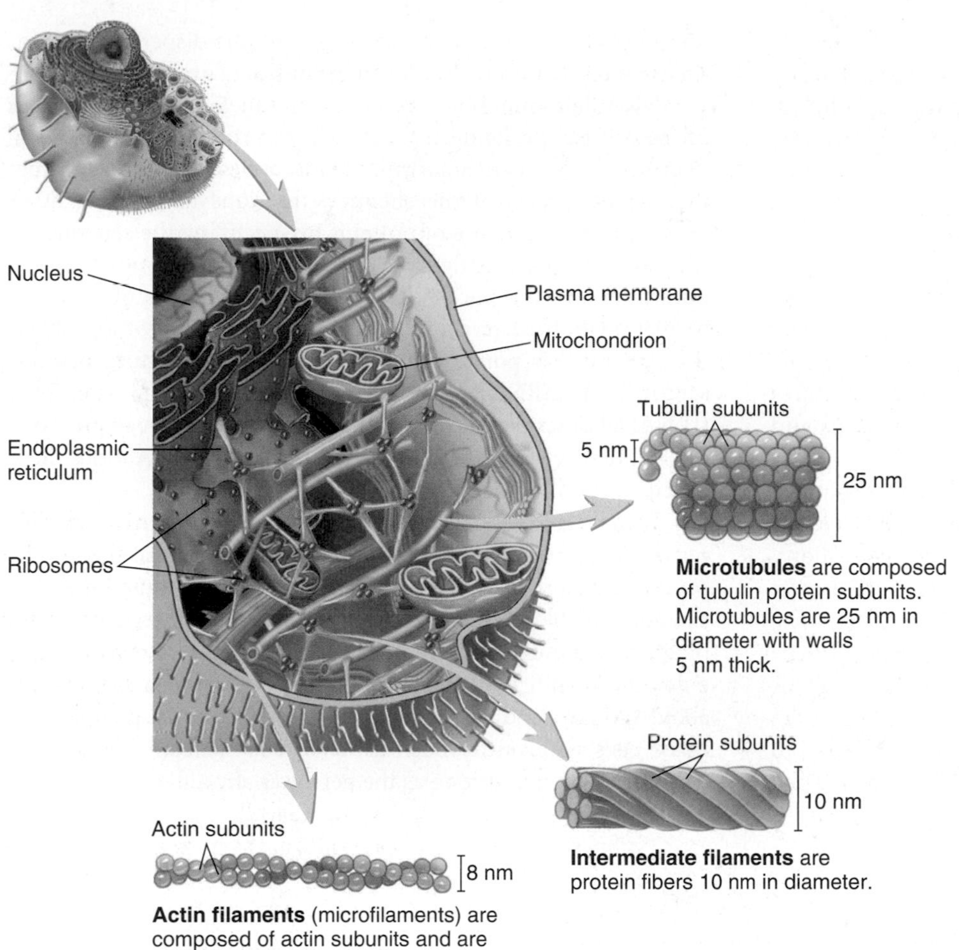
Nucleus
Plasma membrane
Mitochondrion
Endoplasmic reticulum
Ribosomes
Tubulin subunits
5 nm
25 nm
Microtubules are composed of tubulin protein subunits. Microtubules are 25 nm in diameter with walls 5 nm thick.
Protein subunits
10 nm
Intermediate filaments are protein fibers 10 nm in diameter.
Actin subunits
8 nm
Actin filaments (microfilaments) are composed of actin subunits and are about 8 nm in diameter.
(a)

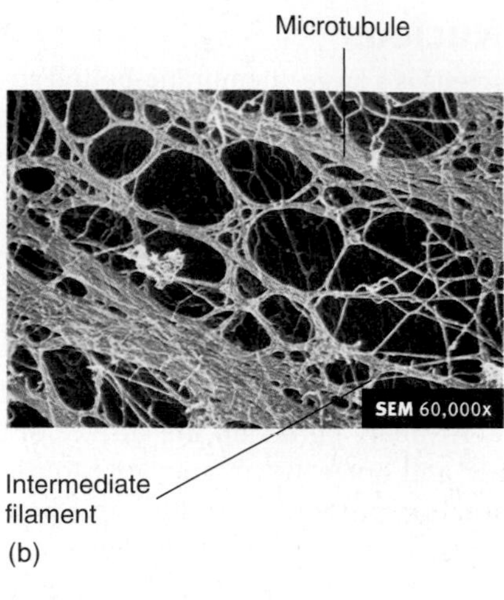
Microtubule
SEM 60,000x
Intermediate filament
(b)

FIGURE 3.21 Cytoskeleton
(*a*) Diagram of the cytoskeleton. (*b*) Scanning electron micrograph of the cytoskeleton.

3.8 The Nucleus and Cytoplasmic Organelles

LEARNING OUTCOMES

After reading this section, you should be able to

A. Define *organelle*.
B. Describe the structure and function of the nucleus and nucleoli.
C. Explain the structure and function of ribosomes.
D. Compare the structure and functions of rough and smooth endoplasmic reticula.
E. Discuss the structure and function of the Golgi apparatus.
F. Describe the role of secretory vesicles in the cell.
G. Compare the structure and roles of lysosomes and peroxisomes in digesting material within the cell.
H. Relate the structure and function of proteasomes.
I. Describe the structure and function of mitochondria.
J. Explain the structure and function of the centrosome.
K. Compare the structure and function of cilia, flagella, and microvill.

Organelles are structures within cells that are specialized for particular functions, such as manufacturing proteins or producing ATP. Organelles can be thought of as individual workstations within the cell, each responsible for performing specific tasks. One class of organelles has membranes that are similar to the plasma membrane, whereas other organelles are clusters of proteins and other molecules not surrounded by a membrane. The interior of the membrane-bound organelles is separated from the cytoplasm, creating subcellular compartments having their own enzymes capable of carrying out unique chemical reactions. The nucleus is the largest organelle of the cell. The remaining organelles are considered cytoplasmic organelles (see table 3.1).

The number and type of cytoplasmic organelles within each cell are related to the specific structure and function of the cell. Cells secreting large amounts of protein contain well-developed organelles that synthesize and secrete protein, whereas cells actively transporting substances, such as sodium ions, across their plasma membrane contain highly developed organelles that produce ATP. The following sections describe the structure and main functions of the nucleus and major cytoplasmic organelles in cells.

The Nucleus

The **nucleus** is a large, membrane-bound structure usually located near the center of the cell. It may be spherical, elongated, or lobed, depending on the cell type. All body cells have a nucleus at some point in their life cycle (see section 3.10), although some cells, such as red blood cells, lose their nuclei as they develop. Other cells, such as skeletal muscle cells and certain bone cells called osteoclasts, contain more than one nucleus.

The nucleus consists of **nucleoplasm** surrounded by a **nuclear envelope** (figure 3.22) composed of two membranes separated by a space. At many points on the surface of the nuclear envelope, the inner and outer membranes fuse to form porelike structures called **nuclear pores.** Molecules move between the nucleus and the cytoplasm through these openings.

Deoxyribonucleic acid (DNA) is mostly found within the nucleus (see figure 2.26), although small amounts of DNA are also found within mitochondria (described later in this section). Nuclear DNA and associated proteins are organized into discrete structures called **chromosomes** (krō′mō-sōmz; figure 3.23). The proteins include **histones** (his′tōnz), which are important for the structural organization of DNA, and other proteins that regulate DNA function. During most of the cell's life cycle, the chromosomes are dispersed throughout the nucleus as delicate filaments referred to as **chromatin** (krō′ma-tin; see figures 3.22 and 3.23). During cell division (see "Cell Division," section 3.10), the dispersed chromatin becomes densely coiled, forming compact chromosomes.

DNA determines the structural and functional characteristics of the cell by specifying the structure of proteins. Proteins form many of a cell's structural components, as well as all the enzymes that regulate most of the chemical reactions in the cell. DNA establishes the structure of proteins by specifying the sequence of their amino acids (see figure 2.22*a*). DNA is a large molecule that does not leave the nucleus but functions by means of an intermediate, **ribonucleic acid (RNA),** which can leave the nucleus through nuclear pores. DNA determines the structure of messenger RNA (mRNA), ribosomal RNA (rRNA), and transfer RNA (tRNA; all described in more detail in section 3.9). A sequence of nucleotides in a DNA molecule that specifies an RNA molecule is called a **gene.**

Because mRNA synthesis occurs within the nucleus, cells without nuclei accomplish protein synthesis only as long as the mRNA produced before the nucleus degenerates remains functional. For example, red blood cells lack nuclei—the nuclei of developing red blood cells are expelled from the cells before the red blood cells enter the blood. Red blood cells survive without a nucleus for about 120 days and must be continually replaced. In comparison, many cells with nuclei, such as nerve and skeletal muscle cells, potentially survive as long as the person is alive.

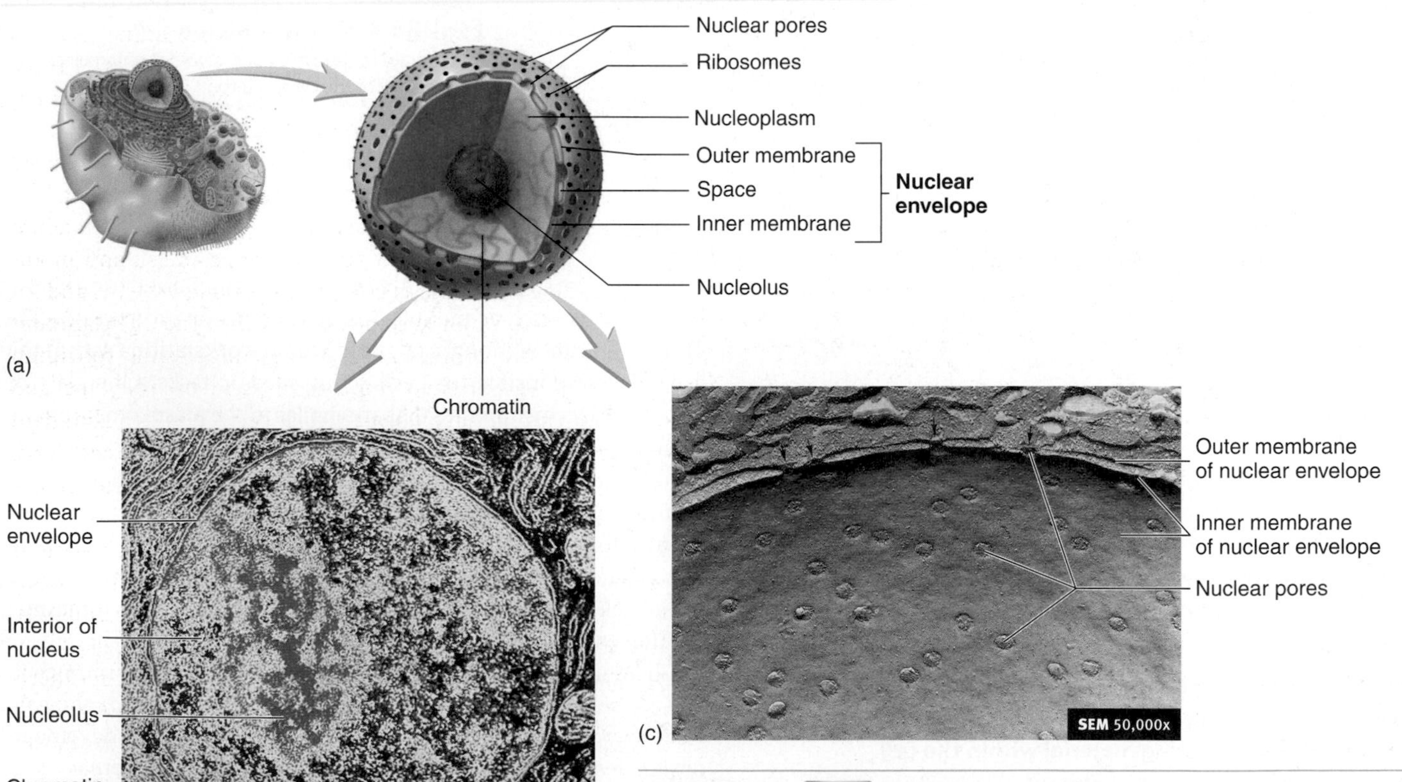
Nuclear pores
Ribosomes
Nucleoplasm
Outer membrane
Space
Inner membrane
Nuclear envelope
Nucleolus
(a)
Chromatin
Nuclear envelope
Interior of nucleus
Nucleolus
Chromatin
TEM 20,000x
(b)
Outer membrane of nuclear envelope
Inner membrane of nuclear envelope
Nuclear pores
SEM 50,000x
(c)

FIGURE 3.22 APR **Nucleus**

(*a*) The nuclear envelope consists of inner and outer membranes that become fused at the nuclear pores. The nucleolus is a condensed region of the nucleus not bound by a membrane and consisting mostly of RNA and protein. (*b*) Transmission electron micrograph of the nucleus. (*c*) Scanning electron micrograph showing the inner surface of the nuclear envelope and the nuclear pores.

A **nucleolus** (noo-klē′ō-lŭs) is a dense region within the nucleus that lacks a surrounding membrane (see figure 3.1). Usually, one nucleolus exists per nucleus, but several nucleoli may be seen in the nuclei of rapidly dividing cells. The nucleolus incorporates portions of chromosomes that contain DNA from which rRNA is produced. Within the nucleolus, the subunits of ribosomes are manufactured.

Ribosomes

Ribosomes (rī′bō-sōmz) are the sites of protein synthesis. Each ribosome is composed of a large subunit and a small subunit. The ribosomal subunits consist of **ribosomal RNA (rRNA)** produced in the nucleus and proteins produced in the cytoplasm. The ribosomal subunits are assembled separately in the nucleolus of the nucleus (figure 3.24). The ribosomal subunits then move through the nuclear pores into the cytoplasm, where the ribosomal subunits assemble with mRNA to form the functional ribosome during protein synthesis. Ribosomes can be found free in the cytoplasm or associated with an intracellular membrane complex called the endoplasmic reticulum. **Free ribosomes** primarily synthesize proteins used inside the cell, whereas ribosomes attached to the endoplasmic reticulum produce integral membrane proteins and proteins that are secreted from the cell.

ASSESS YOUR PROGRESS

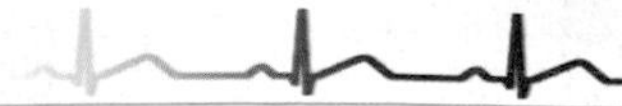

38. *Define* organelles. *Are all organelles found in all cells?*
39. *Describe the structure of the nucleus and the nuclear envelope. What is the function of the nuclear pores?*
40. *Distinguish between chromatin and a chromosome. What molecule is found in chromatin? What are histones?*
41. *How can DNA control the structural and functional characteristics of the cell without leaving the nucleus? List the types of RNA.*
42. *What is the function of the nucleolus?*
43. *What molecules combine to form ribosomes? Where are ribosomal subunits formed? And assembled?*
44. *Compare the functions of free ribosomes and ribosomes attached to the endoplasmic reticulum.*

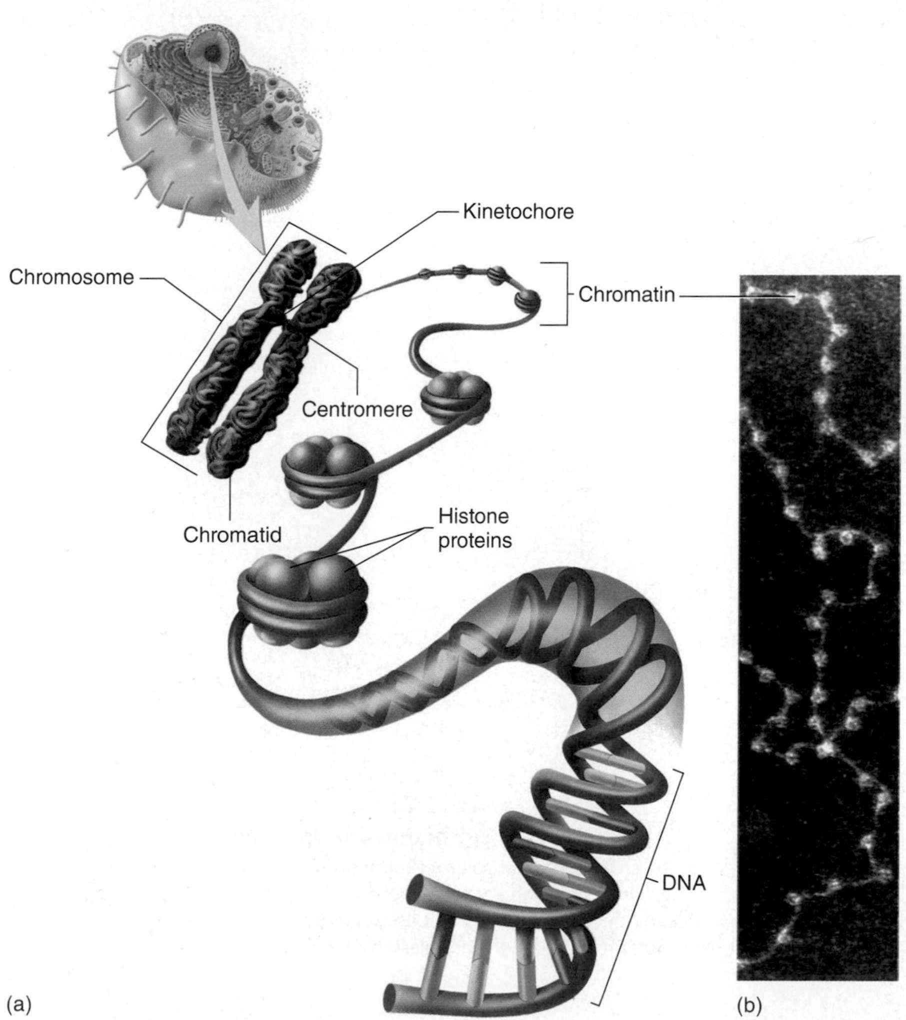

Kinetochore
Chromosome
Chromatin
Centromere
Chromatid
Histone proteins
DNA
(a)
(b)

FIGURE 3.23 Chromosome Structure
(*a*) DNA is associated with globular histone proteins and other DNA-binding proteins. DNA molecules and bound proteins are called chromatin. During cell division, the chromatin condenses, so that individual structures, called chromosomes, become visible. (*b*) Transmission electron micrograph of chromatin.

Endoplasmic Reticulum

The outer membrane of the nuclear envelope is continuous with a series of membranes distributed throughout the cytoplasm of the cell (see figure 3.1), collectively referred to as the **endoplasmic reticulum** (en′dō-plas′mik re-tik′ū-lŭm; network inside the cytoplasm). The endoplasmic reticulum consists of broad, flattened, interconnecting sacs and tubules (figure 3.25). The interior spaces of those sacs and tubules are called **cisternae** (sis-ter′nē) and are isolated from the rest of the cytoplasm.

The **rough endoplasmic reticulum** is called "rough" because ribosomes are attached to it. The ribosomes of the rough endoplasmic reticulum are sites where proteins are produced and modified for use as integral membrane proteins and for secretion into the extracellular space. The amount and configuration of the endoplasmic reticulum within the cytoplasm depend on the type and function of the particular cell. Cells with abundant rough endoplasmic reticulum synthesize large amounts of protein, which are secreted for use outside the cell.

Smooth endoplasmic reticulum, which is endoplasmic reticulum without attached ribosomes, manufactures lipids, such as phospholipids, cholesterol, and steroid hormones, as well as carbohydrates. Enzymes required for lipid synthesis are associated with the membranes of the smooth endoplasmic reticulum, and cells that synthesize large amounts of lipids contain dense accumulations of smooth endoplasmic reticulum. Many phospholipids produced in the smooth endoplasmic reticulum help form vesicles within the cell and contribute to the plasma membrane. Smooth endoplasmic reticulum also participates

1. Ribosomal proteins, produced in the cytoplasm, are transported through nuclear pores into the nucleolus.
2. rRNA, most of which is produced in the nucleolus, is assembled with ribosomal proteins to form small and large ribosomal subunits.
3. The small and large ribosomal subunits leave the nucleolus and the nucleus through nuclear pores.
4. The small and large subunits, now in the cytoplasm, combine with each other and with mRNA during protein synthesis.

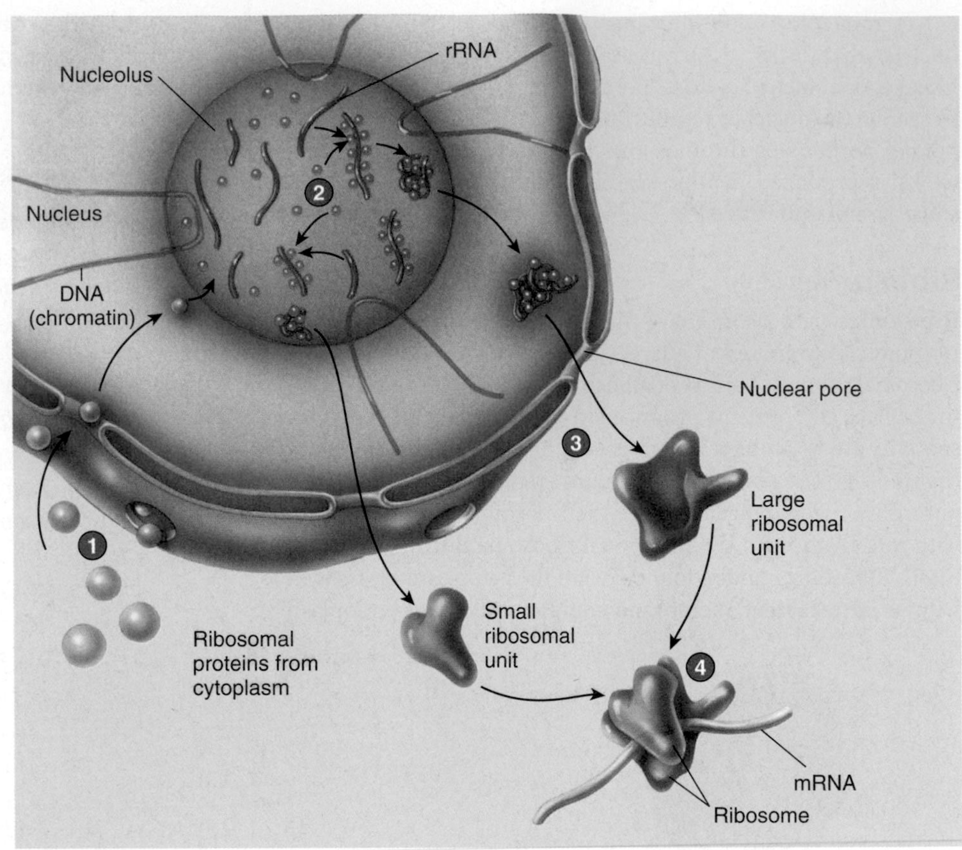
Nucleolus
rRNA
Nucleus
DNA (chromatin)
Nuclear pore
Large ribosomal unit
Small ribosomal unit
Ribosomal proteins from cytoplasm
mRNA
Ribosome
1
2
3
4

PROCESS FIGURE 3.24 Production of Ribosomes

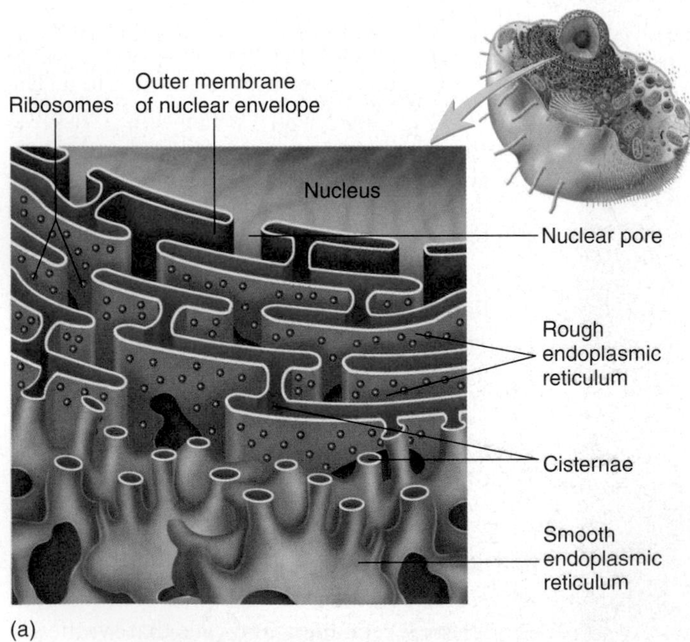
Ribosomes
Outer membrane of nuclear envelope
Nucleus
Nuclear pore
Rough endoplasmic reticulum
Cisternae
Smooth endoplasmic reticulum
(a)

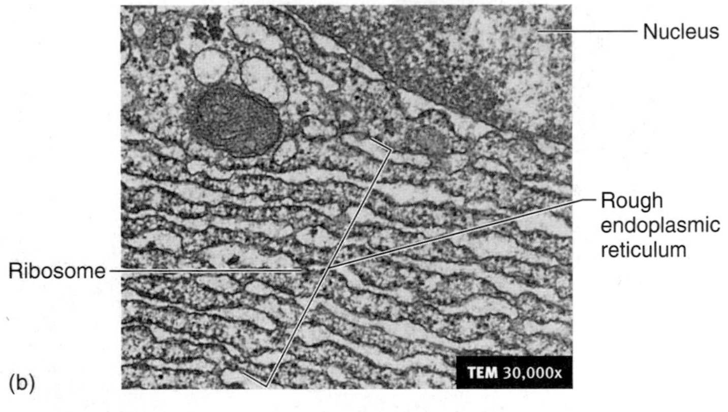
Nucleus
Rough endoplasmic reticulum
Ribosome
(b)
TEM 30,000x

FIGURE 3.25 AP|R Endoplasmic Reticulum

(*a*) The endoplasmic reticulum is continuous with the nuclear envelope and occurs either as rough endoplasmic reticulum (with ribosomes) or as smooth endoplasmic reticulum (without ribosomes). (*b*) Transmission electron micrograph of the rough endoplasmic reticulum.

in **detoxification**, the processes by which enzymes act on chemicals and drugs to change their structure and reduce their toxicity. The smooth endoplasmic reticulum of skeletal muscle stores the calcium ions that function in muscle contraction.

Golgi Apparatus

The **Golgi** (gōl′jē) **apparatus** is composed of flattened, membranous sacs, containing cisternae, stacked on each other like dinner plates (figure 3.26). The Golgi apparatus can be thought of as a

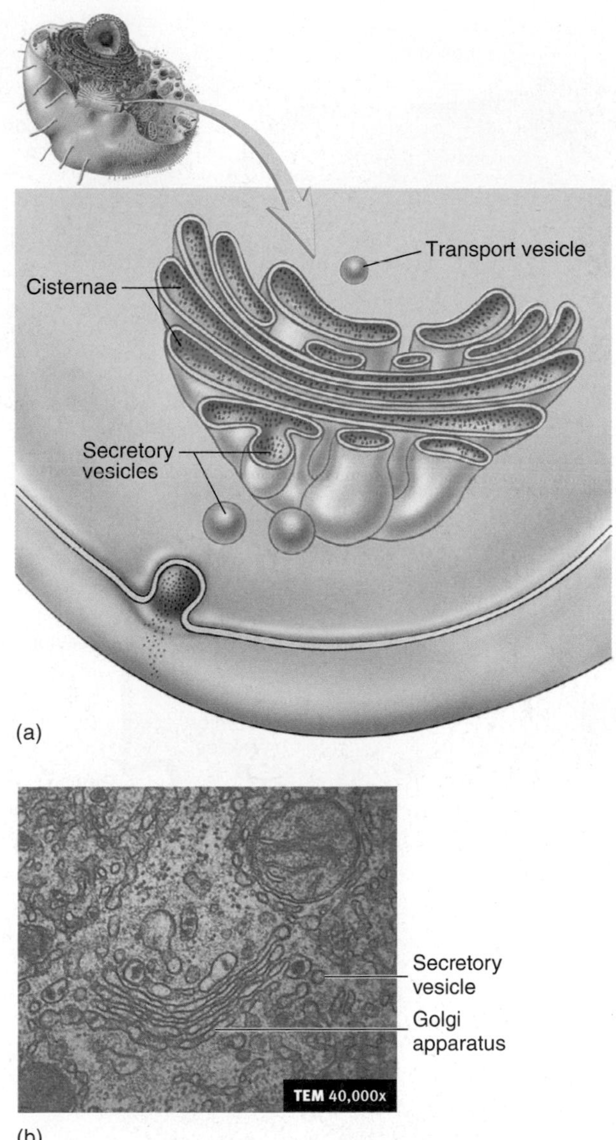
Transport vesicle
Cisternae
Secretory
vesicles
(a)
Secretory
vesicle
Golgi
apparatus
TEM 40,000x
(b)

FIGURE 3.26 APR **Golgi Apparatus**
(*a*) The Golgi apparatus is composed of flattened, membranous sacs containing cisternae. It resembles a stack of dinner plates or pancakes.
(*b*) Transmission electron micrograph of the Golgi apparatus.

packaging and distribution center because it modifies, packages, and distributes proteins and lipids manufactured by the rough and smooth endoplasmic reticula (figure 3.27). Proteins produced at the ribosomes of the rough endoplasmic reticulum enter the endoplasmic reticulum. These proteins are later packaged into **transport vesicles** that move to the Golgi apparatus, fuse with the Golgi apparatus membrane, and release the protein into the Golgi apparatus cisterna. The Golgi apparatus concentrates and, in some cases, chemically modifies the proteins by synthesizing and attaching carbohydrate molecules to the proteins to form glycoproteins or by attaching lipids to proteins to form lipoproteins. The proteins are then packaged into vesicles that pinch off from the margins of the Golgi apparatus and are distributed to various locations. Some vesicles carry proteins to the plasma membrane, where the proteins are secreted from the cell by exocytosis; other vesicles contain proteins that become part of the plasma membrane; and still other vesicles contain enzymes that are used within the cell.

The Golgi apparatus is most highly developed in cells that secrete large amounts of protein or glycoproteins, such as cells in the salivary glands and the pancreas.

ASSESS **YOUR PROGRESS**

45. *Describe the structure and location of the endoplasmic reticula.*

46. *What are the functions of the rough endoplasmic reticulum?*

47. *Explain the functions of the smooth endoplasmic reticulum.*

48. *Relate the structure and function of the Golgi apparatus.*

49. *Name three ways in which proteins are distributed from the Golgi apparatus.*

Secretory Vesicles

The membrane-bound **secretory vesicles** (see figure 3.26) that pinch off from the Golgi apparatus move to the surface of the cell, their membranes fuse with the plasma membrane, and the contents of the vesicles are released to the exterior by exocytosis. The membranes of the vesicles are then incorporated into the plasma membrane.

Secretory vesicles accumulate in some cells, but their contents frequently are not released to the exterior until the cell receives a signal. For example, secretory vesicles that contain the hormone insulin do not release it until the concentration of glucose in the blood increases and acts as a signal for the secretion of insulin from the cells.

Lysosomes

Lysosomes (lī′sō-sōmz) are membrane-bound vesicles that form at the Golgi apparatus (figure 3.27). They contain a variety of hydrolytic enzymes that function as intracellular digestive systems. Vesicles taken into the cell fuse with the lysosomes to form one vesicle and to expose the endocytized materials to hydrolytic enzymes (figure 3.28). Various enzymes within lysosomes digest nucleic acids, proteins, polysaccharides, and lipids. Certain white blood cells have large numbers of lysosomes that contain enzymes to digest phagocytized bacteria. Lysosomes also digest the organelles of the cell that are no longer functional, a process called **autophagy** (aw-tō-fā′jē; self-eating). In other cells, the lysosomes move to the plasma membrane, and the enzymes are secreted by exocytosis. For example, the normal process of bone remodeling involves the breakdown of bone tissue by specialized bone cells. Lysosomes produced by those cells release the enzymes responsible for that degradation into the extracellular fluid.

Peroxisomes

Peroxisomes (per-ok′si-sōmz) are membrane-bound vesicles that are smaller than lysosomes. Peroxisomes contain enzymes that break down fatty acids and amino acids. Hydrogen peroxide (H_2O_2), which can be toxic to the cell, is a by-product of that breakdown. Peroxisomes also contain the enzyme **catalase,** which breaks down hydrogen peroxide to water and oxygen. Cells that are active in detoxification, such as liver and kidney cells, have many peroxisomes.

1. Some proteins are produced at ribosomes on the surface of the rough endoplasmic reticulum and are transferred into the cisterna of the endoplasmic reticulum as they are produced.
2. The proteins are surrounded by a vesicle that forms from the membrane of the endoplasmic reticulum.
3. This transport vesicle moves from the endoplasmic reticulum to the Golgi apparatus, fuses with its membrane, and releases the proteins into its cisterna.
4. The Golgi apparatus concentrates and, in some cases, modifies the proteins into glycoproteins or lipoproteins.
5. The proteins are packaged into vesicles that form from the membrane of the Golgi apparatus.
6. Some vesicles, such as lysosomes, contain enzymes that are used within the cell.
7. Secretory vesicles carry proteins to the plasma membrane, where the proteins are secreted from the cell by exocytosis.
8. Some vesicles contain proteins that become part of the plasma membrane.

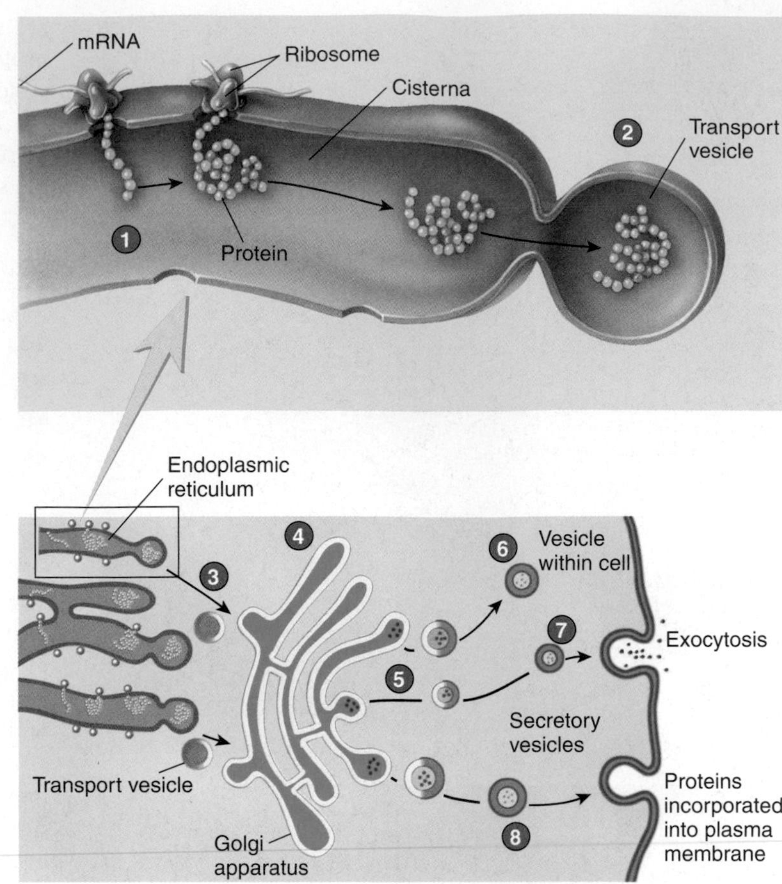
mRNA
Ribosome
Cisterna
Transport vesicle
Protein
Endoplasmic reticulum
Vesicle within cell
Exocytosis
Secretory vesicles
Transport vesicle
Proteins incorporated into plasma membrane
Golgi apparatus

PROCESS FIGURE 3.27 Function of the Golgi Apparatus

1. A vesicle forms around material outside the cell.
2. The vesicle is pinched off from the plasma membrane and becomes a separate vesicle inside the cell.
3. A lysosome is pinched off the Golgi apparatus.
4. The lysosome fuses with the vesicle.
5. The enzymes from the lysosome mix with the material in the vesicle, and the enzymes digest the material.

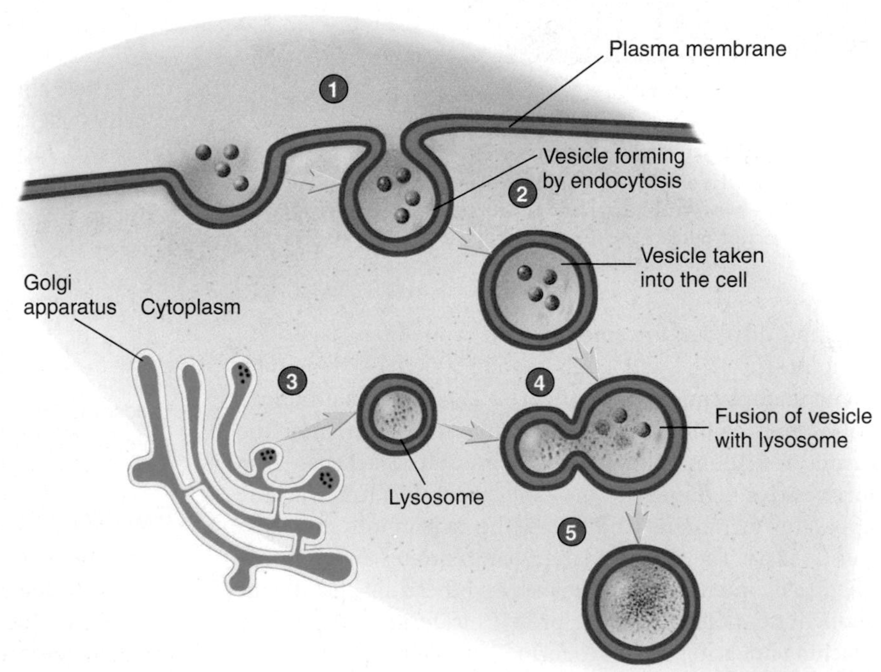
Plasma membrane
Vesicle forming by endocytosis
Vesicle taken into the cell
Golgi apparatus
Cytoplasm
Fusion of vesicle with lysosome
Lysosome

PROCESS FIGURE 3.28 Action of Lysosomes

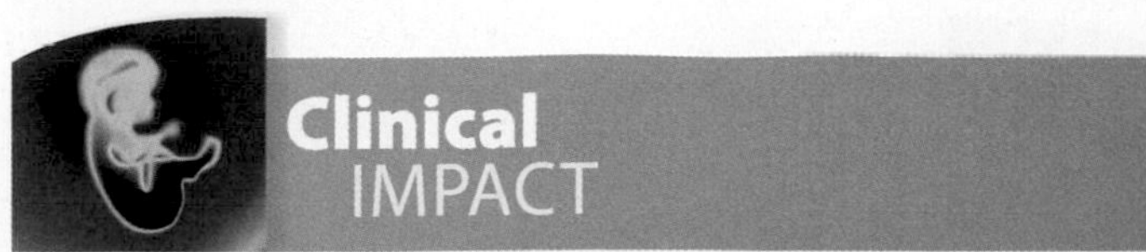
Clinical
IMPACT

Diseases Involving Lysosomal Enzymes

Some diseases result from nonfunctional lysosomal enzymes. For example, **Pompe disease** is a rare genetic disorder caused by the inability of lysosomal enzymes to break down glycogen. The glycogen accumulates in large amounts in the heart, liver, and skeletal muscles, and this accumulation often leads to heart failure. **Familial hyperlipoproteinemia** is a group of genetic disorders that occur when large amounts of lipids accumulate in phagocytic cells that lack the normal enzymes required to break down the lipid droplets. Symptoms include abdominal pain, enlarged spleen and liver, and yellow nodules on the skin, which are filled with the affected phagocytic cells. **Mucopolysaccharidoses,** such as **Hurler syndrome,** are genetic diseases in which lysosomal enzymes are unable to break down mucopolysaccharides (glycosaminoglycans), so these molecules accumulate in the lysosomes of connective tissue cells and nerve cells. People affected by these diseases suffer mental retardation and skeletal deformities.

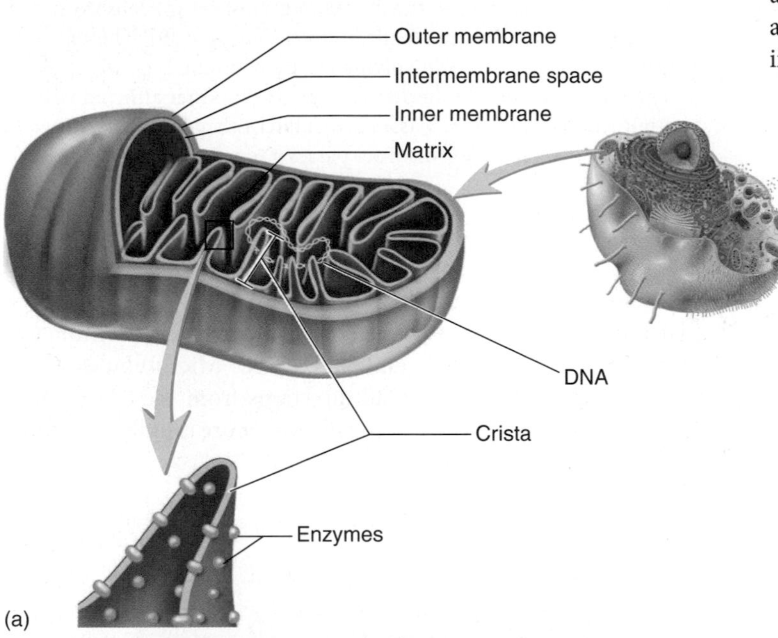
Outer membrane
Intermembrane space
Inner membrane
Matrix
DNA
Crista
Enzymes
(a)

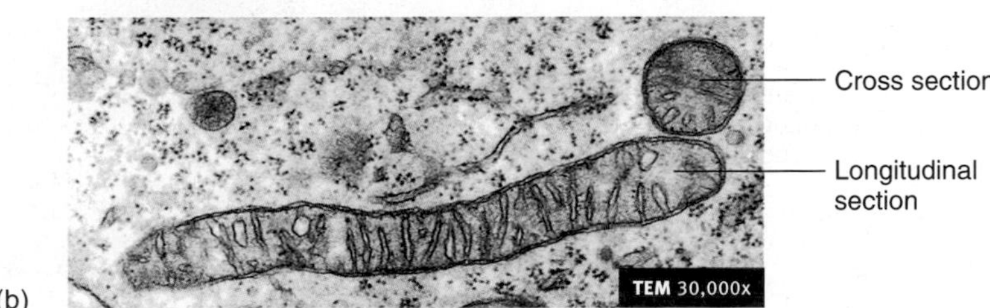
Cross section
Longitudinal section
TEM 30,000x
(b)

FIGURE 3.29 APR **Mitochondrion**

(*a*) Typical mitochondrion structure. (*b*) Transmission electron micrograph showing mitochondria in longitudinal and cross section.

Proteasomes

Proteasomes (prō′tē-ă-sōmz) are large protein complexes containing enzymes that break down and recycle other proteins within the cell. Proteasomes are not surrounded by membranes. They are barrel-like structures; the inner surfaces of the barrel have enzymatic regions that break down the proteins. Proteins at the ends of the barrel regulate which proteins are taken in for digestion.

Mitochondria

Mitochondria (mī-tō-kon′drē-ă) provide energy for the cell. Consequently, they are often called the cell's power plants. Mitochondria are usually depicted as small, rod-shaped structures (figure 3.29). However, in living cells mitochondria are very dynamic and constantly change shape and number as they split and fuse with each other. Mitochondria are the major sites for the production of ATP, which is the primary energy source for most energy-requiring chemical reactions within the cell. Each mitochondrion has an inner and an outer membrane, separated by an intermembrane space. The outer membrane has a smooth contour, but the inner membrane has numerous infoldings called **cristae** (kris′tē; sing. crista) that project like shelves into the interior of the mitochondrion.

A complex series of mitochondrial enzymes forms two major enzyme systems, which are responsible for oxidative metabolism and most ATP synthesis (see chapter 25). The enzymes of the citric acid (Krebs) cycle are in the **matrix,** which is the substance located in the space formed by the inner membrane. The enzymes of the electron-transport chain are embedded within the inner membrane. Cells with a greater energy requirement have more mitochondria with more cristae than do cells with lower energy requirements. Within the cytoplasm of a cell, the mitochondria are more numerous in areas where ATP is used. For example, mitochondria are numerous in cells that perform active transport and are packed near the membrane where active transport occurs.

Increases in the number of mitochondria result from the division of preexisting mitochondria. When muscles enlarge as a result of exercise, the number of mitochondria within the muscle cells increases to provide the additional ATP required for muscle contraction.

The information for making some mitochondrial proteins is stored in DNA contained within the mitochondria themselves, and those proteins are synthesized on ribosomes within the mitochondria. However, the structure of most mitochondrial proteins is determined by nuclear DNA, and these proteins are synthesized on ribosomes within the cytoplasm and then transported into the mitochondria. Both mitochondrial DNA and mitochondrial ribosomes are very different from those within the cell's nucleus and cytoplasm, respectively. Mitochondrial DNA is a closed circle of about 16,500 base pairs (bp) coding for 37 genes, compared with the open strands of nuclear DNA, which is composed of 3 billion bp coding for about 20,000 genes. In addition, unlike nuclear DNA, mitochondrial DNA does not have associated histone proteins. Mitochondrial ribosomes are more similar in size and structure to bacterial ribosomes than to cytoplasmic ribosomes.

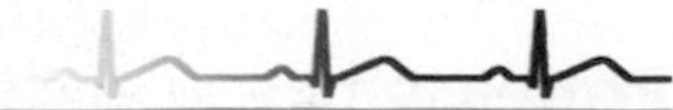

ASSESS YOUR PROGRESS

50. *How are secretory vesicles formed?*

51. *Describe the process by which lysosomal enzymes digest phagocytized materials. What is autophagy?*

52. *What is the function of peroxisomes? How does catalase protect cells?*

53. *What are the structure and function of proteosomes?*

54. *Describe the structure of a mitochondrion. How does its nickname, the cell's power plant, relate to its function?*

55. *What enzymes are found on the cristae? In the matrix? How can the number of mitochondria in a cell increase?*

Predict 5

Describe the structural characteristics of cells that are highly specialized to do the following: (a) synthesize and secrete proteins, (b) actively transport substances into the cell, (c) synthesize lipids, and (d) phagocytize foreign substances.

Centrioles and Spindle Fibers

The **centrosome** (sen′trō-sōm), a specialized zone of cytoplasm close to the nucleus, is the center of microtubule formation in the cell. It contains two **centrioles** (sen′trē-ōlz). Each centriole is a small, cylindrical organelle about 0.3–0.5 μm in length and 0.15 μm in diameter, and the two centrioles are normally oriented perpendicular to each other within the centrosome (see figure 3.1). The wall of the centriole is composed of nine evenly spaced, longitudinally oriented, parallel units, or triplets. Each unit consists of three parallel microtubules joined together (figure 3.30).

Microtubules appear to influence the distribution of actin and intermediate filaments. Through its control of microtubule formation, the centrosome is closely involved in determining cell shape and movement. The microtubules extending from the centrosomes are very dynamic—constantly growing and shrinking.

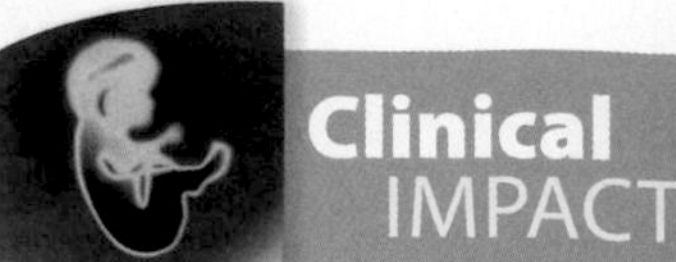

Clinical IMPACT

Mitochondrial Diseases

Mitochondria play a major role in the synthesis of ATP. Each mitochondrion contains a single DNA molecule with at least 37 genes, 13 of which code for proteins that are important for ATP synthesis. The other 24 genes are important for the expression of mitochondrial genes. Therefore, mutations, or changes, in mitochondrial genes can lead to disruptions in normal ATP synthesis, reducing the amount of ATP produced by the cells. Disorders that result from such mutations are collectively called **mitochondrial diseases.** The effects of these diseases are most obvious in tissues that require large amounts of ATP, particularly nervous and muscle tissues. As a consequence, the common symptoms of mitochondrial diseases are loss of neurological function and defects in muscular activity. For example, **Leber hereditary optic neuropathy** results in sudden vision loss due to optic nerve degeneration. Mutations associated with this disorder have been found in the genes that function in ATP synthesis, especially in the cells of the optic nerve. Because of their high energy demands, the cells of the optic nerve are damaged or die due to the lack of ATP.

In humans, mitochondria are passed only from the mother to her offspring because the mitochondria of sperm cells do not enter the oocyte during fertilization (see chapter 29). Therefore, mitochondrial diseases, involving the mitochondrial genes, show a pattern of maternal inheritance—that is, a mother afflicted with a mitochondrial disease passes it to all her offspring, but a father suffering from the same disorder passes it to none of his offspring.

Before cell division, the two centrioles double in number; the centrosome divides into two; and one centrosome, containing two centrioles, moves to each end of the cell. Microtubules called **spindle fibers** extend out in all directions from the centrosome. These microtubules grow and shrink even more rapidly than those

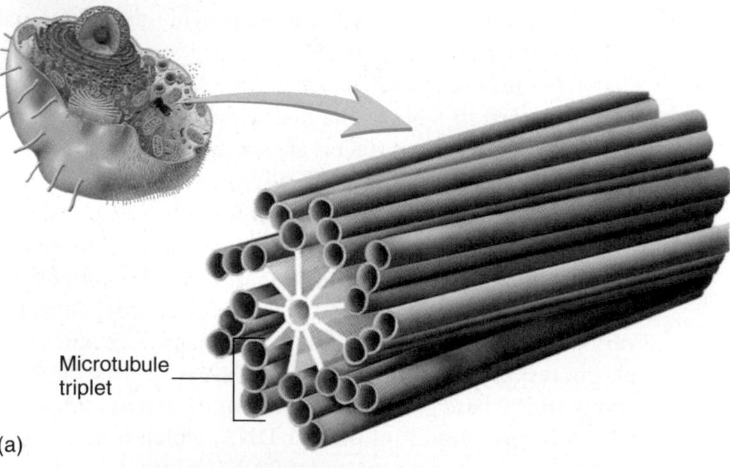
Microtubule triplet
(a)

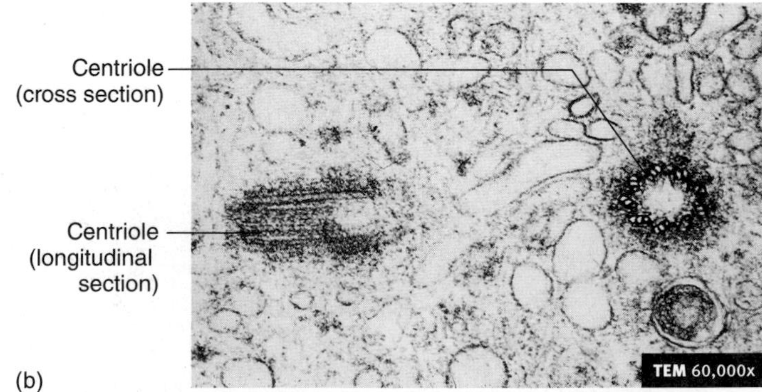
Centriole (cross section)
Centriole (longitudinal section)
TEM 60,000x
(b)

FIGURE 3.30 AP|R **Centriole**

(*a*) Structure of a centriole, which is composed of nine triplets of microtubules. Each triplet contains one complete microtubule fused to two incomplete microtubules. (*b*) Transmission electron micrograph of a pair of centrioles, which are normally located together near the nucleus. One is shown in cross section and one in longitudinal section.

of nondividing cells. If the extended end of a spindle fiber comes in contact with a chromosome, the spindle fiber attaches to the chromosome and stops growing or shrinking. Eventually, spindle fibers from each centrosome bind to all the chromosomes. During cell division, the spindle microtubules facilitate the movement of chromosomes toward the two centrosomes (see "Cell Division" in section 3.10).

Cilia and Flagella

Cilia (sil′ē-ă) are structures that project from the surface of cells and are capable of movement. They vary in number from one to thousands per cell. Cilia are cylindrical in shape, measuring about 10 μm in length and 0.2 μm in diameter. The shaft of each cilium is enclosed by the plasma membrane. Each cilium contains two centrally located microtubules and nine peripheral pairs of fused microtubules (the so-called 9 + 2 arrangement) that extend from the base to the tip of the cilium (figure 3.31). Movement of the cilium results when the microtubules move past each other, a process that requires energy from ATP. **Dynein arms,** proteins connecting adjacent pairs of microtubules, push the microtubules past each other. A **basal body** (a modified centriole) is located in the cytoplasm at the base of the cilium.

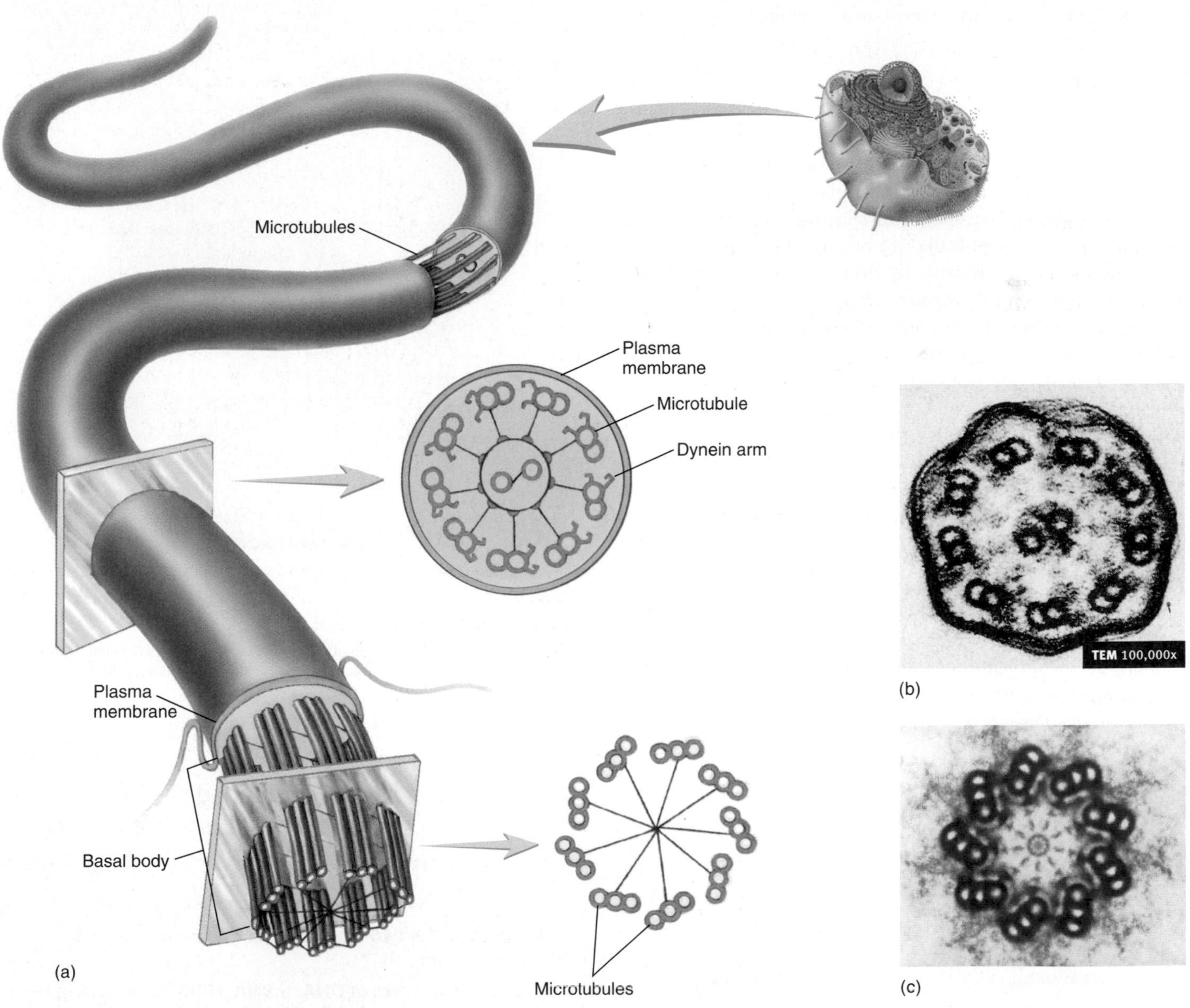
Microtubules
Plasma membrane
Microtubule
Dynein arm
Plasma membrane
Basal body
(a)
Microtubules
TEM 100,000x
(b)
(c)

FIGURE 3.31 Structure of Cilia and Flagella
(*a*) The shaft of a cilium or flagellum has nine microtubule doublets around its periphery and two in the center. Dynein arms are proteins that connect one pair of microtubules to another pair. Dynein arm movement, which requires ATP, causes the microtubules to slide past each other, resulting in bending or movement of the cilium or flagellum. A basal body attaches the cilium or flagellum to the plasma membrane. (*b*) Transmission electron micrograph through a cilium. (*c*) Transmission electron micrograph through the basal body of a cilium.

Cilia are numerous on surface cells that line the respiratory tract and the female reproductive tract. In these regions, cilia move in a coordinated fashion, with a power stroke in one direction and a recovery stroke in the opposite direction (figure 3.32). Their motion moves materials over the surface of the cells. For example, cilia in the trachea move mucus containing trapped dust particles upward and away from the lungs, thus helping keep the lungs clear of debris.

Flagella (flă-jel′ă) have a structure similar to that of cilia, but they are longer (45 μm). Sperm cells are the only human cells that possess flagella, and usually only one flagellum exists per cell. Furthermore, whereas cilia move small particles across the cell surface, flagella move the entire cell. For example, each sperm cell is propelled by a single flagellum. In contrast to cilia, which have a power stroke and a recovery stroke, flagella move in a wavelike fashion.

Microvilli

Microvilli (mī-krō-vil′ī; figure 3.33) are cylindrically shaped extensions of the plasma membrane about 0.5–1.0 μm in length and 90 nm in diameter. Normally, each cell has many microvilli, and they increase the cell surface area. A student looking at photographs may confuse microvilli with cilia, but microvilli are only one-tenth to one-twentieth the size of cilia. Individual microvilli can usually be seen only with an electron microscope, whereas cilia can be seen with a light microscope. Microvilli do not move, and they are supported with actin filaments, not microtubules. Microvilli are found on the cells of the intestine, kidney, and other areas where absorption is an important function. In certain locations of the body, microvilli are highly modified to function as sensory receptors. For example, elongated microvilli in hair cells of the inner ear respond to sound.

ASSESS YOUR PROGRESS

56. *What is the centrosome? Relate the structure of centrioles.*

57. *What are spindle fibers? Explain the relationship among centrosomes, spindle fibers, and chromosomes during cell division.*

58. *Contrast the structure and function of cilia with those of flagella.*

59. *Describe the structure and function of microvilli. How are microvilli different from cilia?*

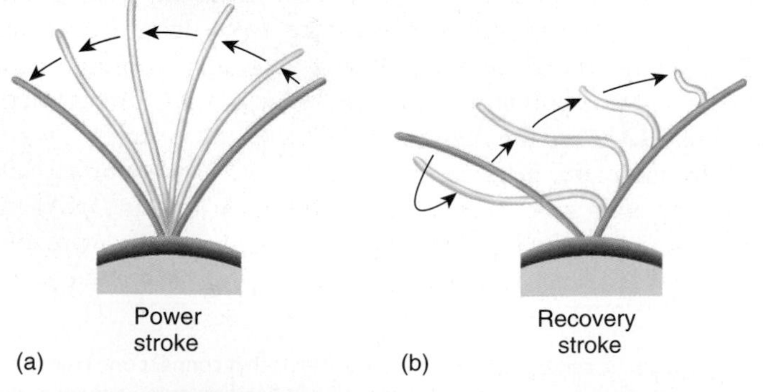
Power
stroke
(a)
Recovery
stroke
(b)

FIGURE 3.32 Ciliary Movement
(*a*) Power and (*b*) recovery strokes.

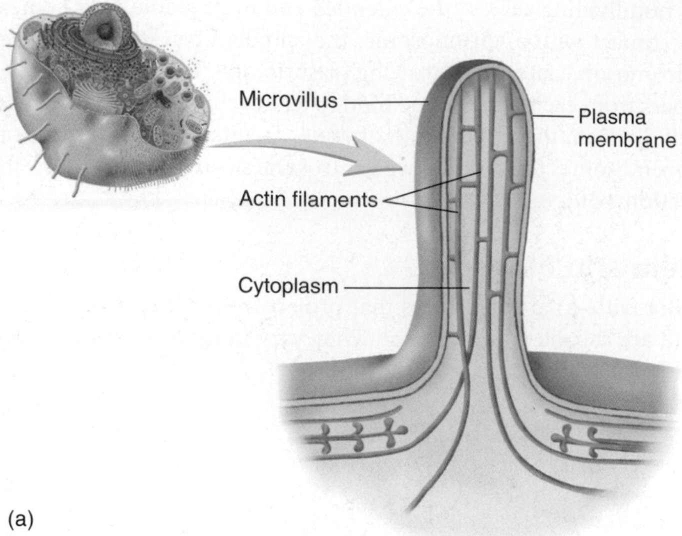
Microvillus
Plasma
membrane
Actin filaments
Cytoplasm
(a)

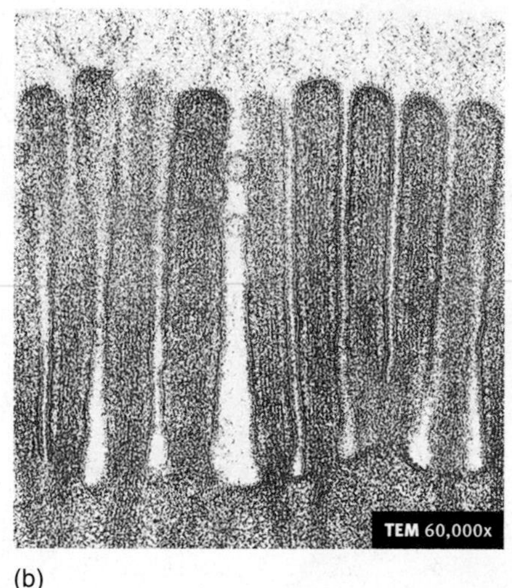
TEM 60,000x
(b)

FIGURE 3.33 Microvillus
(*a*) A microvillus is a tiny, tubular extension of the cell; it contains cytoplasm and some actin filaments (microfilaments). (*b*) Transmission electron micrograph of microvilli.

3.9 Genes and Gene Expression

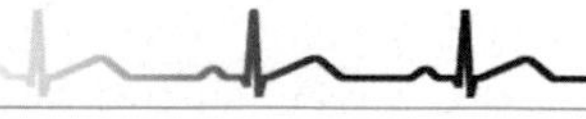

LEARNING OUTCOMES

After reading this section, you should be able to

A. **Describe the two-step process that results in gene expression.**

B. **Explain the roles of DNA, mRNA, tRNA, and rRNA in the production of a protein.**

C. **Explain what the genetic code is and what it is coding for.**

D. **Describe what occurs during posttranscriptional processing and posttranslational processing.**

E. **Describe the regulation of gene expression.**

FUNDAMENTAL **Figure**

1. DNA contains the information necessary to produce proteins.
2. Transcription of one DNA strand results in mRNA, which is a complementary copy of the information in the DNA strand needed to make a protein.
3. The mRNA leaves the nucleus and goes to a ribosome.
4. Amino acids, the building blocks of proteins, are carried to the ribosome by tRNAs.
5. In the process of translation, the information contained in mRNA is used to determine the number, kinds, and arrangement of amino acids in the polypeptide chain.

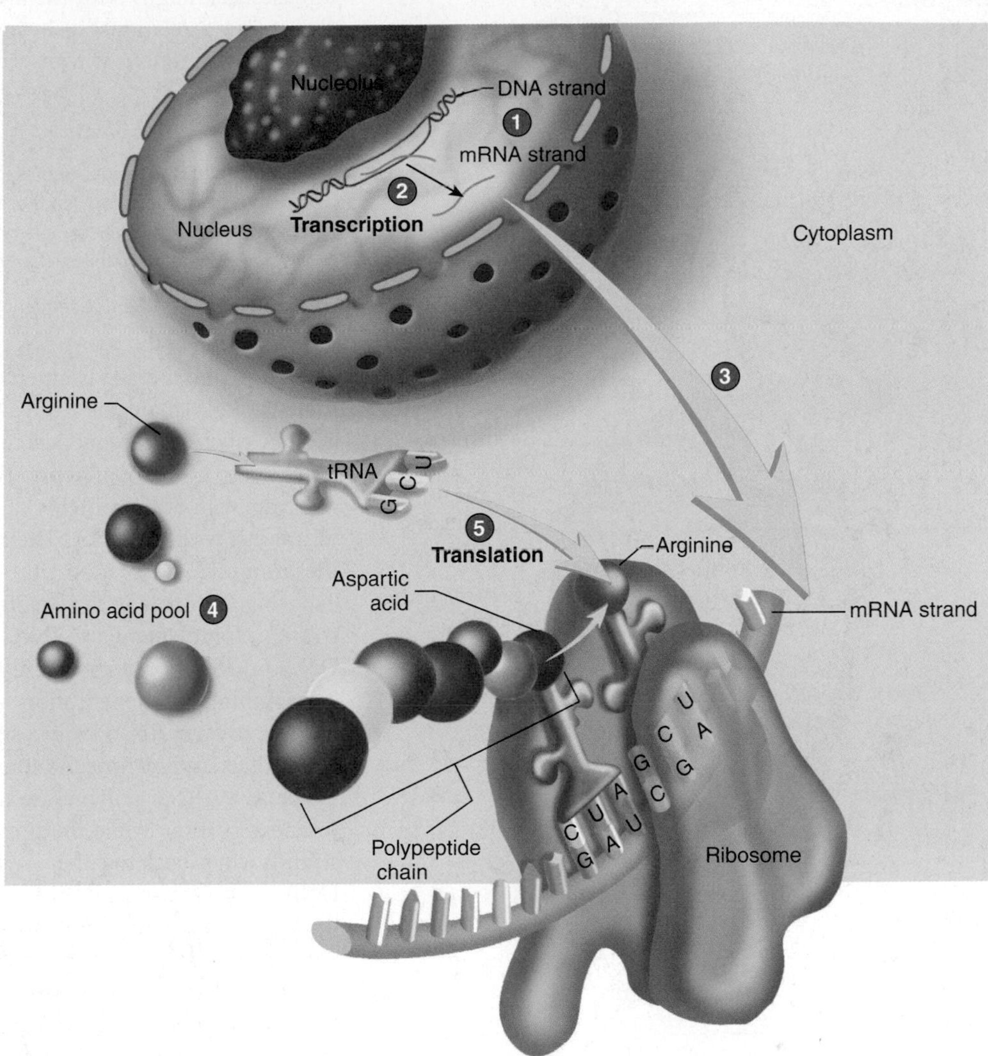
Nucleolus
DNA strand
1
mRNA strand
2
Nucleus
Transcription
Cytoplasm
3
Arginine
tRNA
G C U
5
Translation
Arginine
Aspartic acid
Amino acid pool 4
mRNA strand
U
C A
G
G
A
C
C U
U
G A
Polypeptide chain
Ribosome

PROCESS FIGURE 3.34 **Overview of Gene Expression**

Genes are the functional units of **heredity,** the transmission of genetic traits from parent to offspring. Each gene is a segment of a DNA molecule that specifies the structure of an RNA molecule. This RNA can be functional on its own, or it can produce a protein. The production of RNA and/or proteins from the information stored in DNA is called **gene expression** (figure 3.34). Gene expression that yields proteins involves two steps: transcription and translation. This process can be illustrated with an analogy. Suppose a cook wants a cake recipe that is found only in a reference book in the library. Because the book cannot be checked out, the cook makes a copy, or **transcription,** of the recipe. Later, in the kitchen, the information contained in the copied recipe is used to make the cake. The changing of something from one form to another (from recipe to cake) is called **translation.** In this analogy, DNA is the reference book that contains many recipes (genes) for making different proteins. DNA, however, is too large a molecule to pass through the nuclear envelope to go to the cytoplasm (the kitchen), where the proteins are synthesized. Just as the reference book stays in the library, DNA remains in the nucleus. Therefore, through transcription, the cell makes a copy of the gene (the recipe) necessary to make a particular protein (the cake). The copy, which is called mRNA, travels from the nucleus to ribosomes (the kitchen) in the cytoplasm, where the information in the copy is used to construct a protein (i.e., translation). Of course, to turn a recipe into a cake, ingredients are needed. The ingredients necessary to synthesize a protein are amino acids. Specialized transport molecules, called **transfer RNA (tRNA),** carry the amino acids to the ribosomes (figure 3.34).

In summary, gene expression involves transcription (making a copy of a small part of the stored information in DNA) and translation (converting that copied information into a protein). The details of transcription and translation are considered next.

Transcription

Transcription is the synthesis of mRNA, tRNA, and rRNA based on the nucleotide sequence in DNA (figure 3.35). Transcription

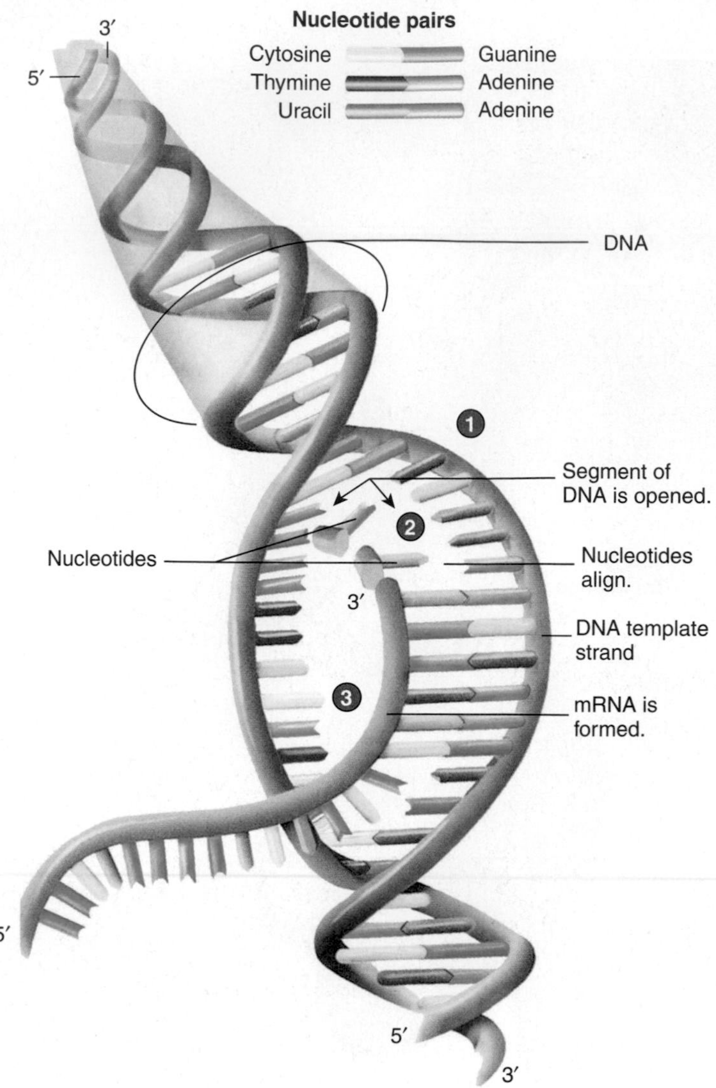
Nucleotide pairs
Cytosine
Guanine
Thymine
Adenine
Uracil
Adenine
3′
5′
DNA
Segment of DNA is opened.
Nucleotides
Nucleotides align.
3′
DNA template strand
mRNA is formed.
5′
5′
3′

1. The strands of the DNA molecule separate from each other. One DNA strand serves as a template for mRNA synthesis.
2. Nucleotides that will form mRNA pair with DNA nucleotides according to the base-pair combinations shown in the key at the top of the figure. Thus, the sequence of nucleotides in the template DNA strand (*purple*) determines the sequence of nucleotides in mRNA (*gray*). RNA polymerase (the enzyme is not shown) joins the nucleotides of mRNA together.
3. As nucleotides are added, an mRNA molecule is formed.

PROCESS FIGURE 3.35 APR **Formation of mRNA by Transcription of DNA**

occurs when a section of a DNA molecule unwinds and its complementary strands separate. One of the DNA strands serves as the template strand for the process of transcription. Nucleotides that form RNA align with the DNA nucleotides in the template strand by complementary base pairing. For example, suppose the DNA base sequence TGCA is to be transcribed. An adenine aligns with the thymine of DNA, cytosine aligns with guanine, and guanine aligns with cytosine. Instead of thymine, uracil of RNA (see figure 2.25) aligns with adenine of DNA. Thus, the sequence of bases that aligns with the TGCA sequence of DNA is ACGU. This pairing relationship between nucleotides ensures that the information in DNA is transcribed correctly into RNA.

Predict 6

Given the following sequence of nucleotides of a template DNA strand, predict the sequence of mRNA that is transcribed from it. What is the nucleotide sequence of the complementary strand of the DNA molecule? How does it differ from the nucleotide sequence of RNA?

DNA nucleotide sequence: CGTACGCCGAGACGTCAAC

RNA polymerase is an enzyme that synthesizes the complementary RNA molecule from DNA. RNA polymerase attaches to a DNA nucleotide sequence called a **promoter.** However, RNA polymerase does not attach to the promoter by itself. It must first associate with other proteins called **transcription factors** in order to interact with the DNA. The attachment of RNA polymerase to the promoter causes a portion of the DNA molecule to unwind, exposing the nucleotide sequence for that region of the template strand. Complementary RNA nucleotides then align with the DNA nucleotides of the template strand. The RNA nucleotides are combined by dehydration reactions, catalyzed by RNA polymerase, to form RNA. Only a small portion of the DNA molecule unwinds at any one time. As complementary nucleotides are added to the RNA, RNA polymerase moves along the DNA, unwinding the next portion, while the previously unwound section of DNA strands winds back together. When RNA polymerase encounters a DNA nucleotide sequence called the **terminator,** it detaches from the DNA, releasing the newly formed RNA. The type of RNA that will make proteins is called messenger RNA (mRNA).

The region of a DNA molecule between the promoter and termination of transcription is a gene. The structure of a gene is more complex than just the nucleotides that code for a protein; some regions that are transcribed to form mRNA do not code for parts of a protein. Regions of the mRNA that code for proteins are called **exons,** whereas the non-protein-coding regions are called **introns.** An mRNA that contains introns is called a **pre-mRNA** (figure 3.36). The introns are removed from the pre-mRNA, and the exons are spliced together. The functional mRNA consists only of exons.

Before a pre-mRNA leaves the nucleus, it undergoes several modifications called **posttranscriptional processing,** which produces the functional mRNA that is used in translation to produce a protein (figure 3.36). A **7-methylguanosine cap** is added to one end of mRNA, and a series of adenine nucleotides, called a **poly-A tail,** is added to the other end. These modifications to the ends of the mRNA ensure that mRNA travels from the nucleus to the cytoplasm and interacts with ribosomes during translation.

In a process called **alternative splicing,** various combinations of exons are incorporated into mRNA. Which exons—and how many—are used to make mRNA can vary between cells of different tissues, resulting in different mRNAs transcribed from the same gene. Alternative splicing allows a single gene to produce more than one specific protein; however, the various proteins usually have similar functions in different tissues. In humans, nearly all mRNAs undergo alternative RNA splicing.

1. The two DNA strands separate, and one strand is transcribed to produce a pre-mRNA strand.
2. An intron is cleaved from between two exons and is discarded.
3. The exons are spliced together to make the functional mRNA.
4. A 7-methyl guanosine cap and a poly-A tail are added to the mRNA.

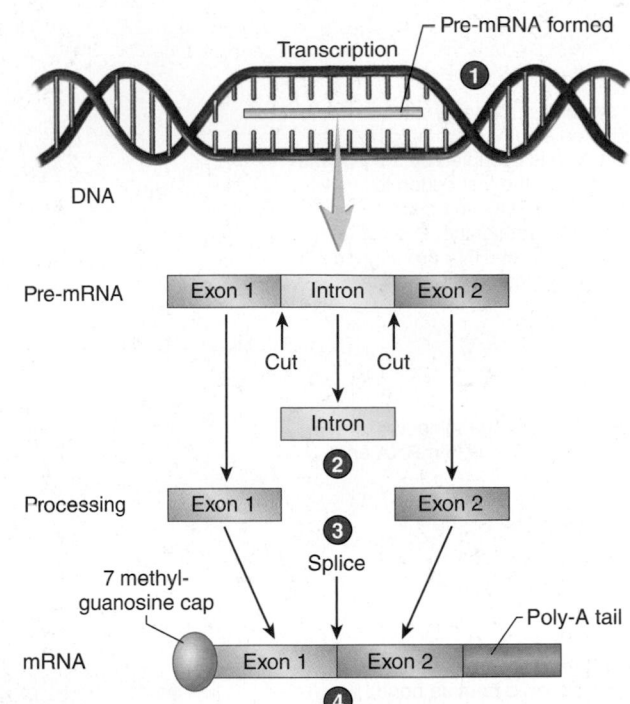

Transcription
Pre-mRNA formed
DNA
Pre-mRNA
Exon 1 | Intron | Exon 2
Cut
Cut
Intron
Processing
Exon 1
Exon 2
Splice
7 methyl-guanosine cap
Poly-A tail
mRNA
Exon 1 | Exon 2

PROCESS FIGURE 3.36 Posttranscriptional Change in mRNA

Genetic Code

The information contained in mRNA, called the **genetic code,** is carried in sets of three nucleotide units called **codons.** A codon specifies an amino acid during translation. For example, the codon GAU specifies the amino acid aspartic acid, and the codon CGA specifies arginine. Although there are only 20 different amino acids commonly found in proteins, 64 possible codons exist. Therefore, an amino acid can have more than one codon. The codons for arginine include CGA, CGG, CGU, and CGC. Furthermore, some codons act as signals during translation. AUG, which specifies methionine, also acts as a **start codon,** which signals the beginning of translation. UAA, UGA, and UAG act as **stop codons,** which signal the end of translation. Unlike the start codon, those codons do not specify amino acids. Therefore, the protein-coding region of an mRNA begins at the start codon and ends at a stop codon.

Translation

Translation is the synthesis of a protein at the ribosome in response to the codons of mRNA (figure 3.37). In addition to mRNA, translation requires ribosomes and tRNA. Ribosomes consist of **ribosomal RNA (rRNA)** and proteins. Like mRNA, tRNA and rRNA are produced in the nucleus by transcription.

The function of tRNA is to match a specific amino acid to a specific codon of mRNA. To do this, one end of each kind of tRNA combines with a specific amino acid. Another part of the tRNA, called the **anticodon,** consists of three nucleotides and is complementary to a particular codon of mRNA. On the basis of the pairing relationships between nucleotides, the anticodon can combine only with its matched codon. For example, the tRNA that binds to aspartic acid has the anticodon CUA, which combines with the codon GAU of mRNA. Therefore, the codon GAU codes for aspartic acid.

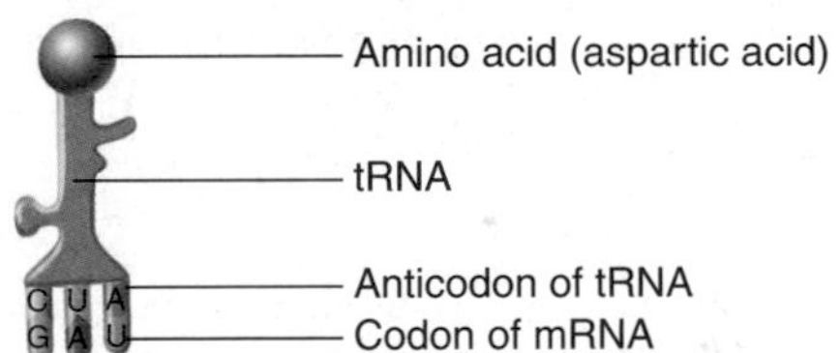
Amino acid (aspartic acid)
tRNA
Anticodon of tRNA
Codon of mRNA
C U A
G A U

Ribosomes align the codons of the mRNA with the anticodons of tRNA and then enzymatically join the amino acids of adjacent tRNA molecules. The mRNA moves through the ribosome one codon at a time. With each move, a new tRNA enters the ribosome and the amino acid is linked to the growing chain, forming a polypeptide. The step-by-step process of translation at the ribosome is described in detail in figure 3.37.

After a ribosome uses the initial part of mRNA, another ribosome can attach to the mRNA and begin to make a protein. The resulting cluster of ribosomes attached to the mRNA is called a **polyribosome** (figure 3.37). Each ribosome in a polyribosome produces an identical protein, and polyribosomes are an efficient way to produce many copies of the same protein using a single mRNA molecule.

Many proteins are longer when they are first made than in their final, functional state. These proteins are called **proproteins,** and the extra piece of the molecule is cleaved off by enzymes to make the proprotein into a functional protein. Many proteins are enzymes, and the proproteins of those enzymes are called **proenzymes.**

1 To start protein synthesis, a ribosome binds to mRNA. The ribosome also has two binding sites for tRNA, one of which is occupied by a tRNA with its amino acid. Note that the first codon to associate with a tRNA is AUG, the start codon,which codes for methionine. The codon of mRNA and the anticodon of tRNA are aligned and joined. The other tRNA binding site is open.

2 By occupying the open tRNA binding site, the next tRNA is properly aligned with mRNA and with the other tRNA.

3 An enzyme within the ribosome catalyzes a synthesis reaction to form a peptide bond between the amino acids. Note that the amino acids are now associated with only one of the tRNAs.

4 The ribosome shifts position by three nucleotides. The tRNA without the amino acid is released from the ribosome, and the tRNA with the amino acids takes its position. A tRNA binding site is left open by the shift. Additional amino acids can be added by repeating steps 2 through 4.

5 Eventually, a stop codon in the mRNA, such as UAA, ends the process of translation. At this point, the mRNA and polypeptide chain are released from the ribosome.

6 Multiple ribosomes attach to a single mRNA to form a polyribosome. As the ribosomes move down the mRNA, proteins attached to the ribosomes lengthen and eventually detach from the mRNA.

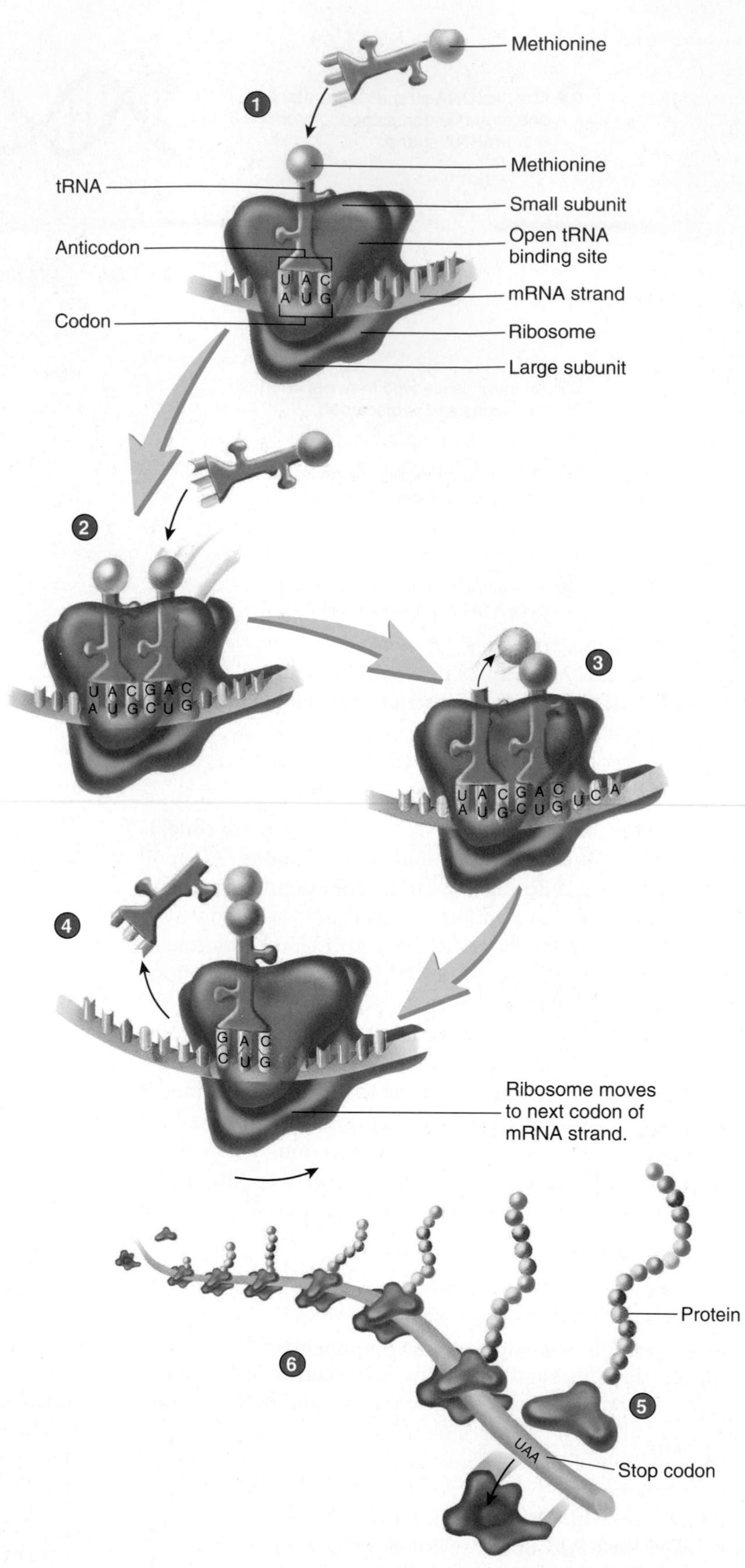
Methionine
1
Methionine
tRNA
Small subunit
Open tRNA binding site
Anticodon
UAC
AUG
mRNA strand
Codon
Ribosome
Large subunit
2
UACGAC
AUGCUG
3
UACGAC
AUGCUGUCA
4
GAC
CUG
Ribosome moves to next codon of mRNA strand.
Protein
6
5
UAA
Stop codon

PROCESS FIGURE 3.37 APIR **Translation of mRNA to Produce a Protein**

If many proenzymes were made within cells as functional enzymes, they could digest the cell that made them. Instead, they are made as proenzymes and are not converted to active enzymes until they reach a protected region of the body, such as inside the small intestine, where they are functional. Many proteins have side chains, such as polysaccharides, added to them following translation. Some proteins are composed of two or more amino acid chains that are joined after each chain is produced on separate ribosomes. These various modifications to proteins are referred to as **posttranslational processing.**

Regulation of Gene Expression

Most of the cells in the body have the same DNA. However, the transcription of mRNA in cells is regulated so that all portions of all DNA molecules are not continually transcribed. The proteins associated with DNA in the nucleus play a role in regulating transcription. As cells differentiate and acquire specialized functions during development, part of the DNA is no longer transcribed, whereas other segments of DNA become more active. For example, the DNA coding for hemoglobin is not expressed in most cells, and little if any hemoglobin is synthesized. But in developing red blood cells the DNA coding for hemoglobin is transcribed, and hemoglobin synthesis occurs rapidly.

Gene expression in a single cell is not normally constant but fluctuates in response to changes in signals from within and outside the cell. Regulatory molecules that interact with nuclear proteins can either increase or decrease the transcription rate of specific DNA segments. For example, triiodothyronine (T_3), a hormone released by cells of the thyroid gland, enters cells, such as skeletal muscle cells; interacts with specific nuclear proteins; and increases transcription of mRNAs from specific genes. Consequently, the production of certain proteins increases. As a further result, an increase in the number of mitochondria and an increase in metabolism occur in these cells.

ASSESS YOUR PROGRESS

60. *What is gene expression, and what two processes result in gene expression?*
61. *What type of molecule results from transcription? Where do the events of transcription occur?*
62. *Place the events of transcription in sequence.*
63. *What are exons and introns? How do they relate to pre-mRNA and posttranscriptional processing?*
64. *What is the role of alternative splicing in variation?*
65. *What is the genetic code?*
66. *What are start and stop codons? How are they different from promoters and terminators?*
67. *Place the steps of translation in sequence. In what molecules are codons and anticodons found? What is a polyribosome?*
68. *What occurs in posttranslational processing? How does it relate to proproteins and proenzymes?*
69. *State two ways the cell controls what part of DNA is transcribed.*

3.10 Cell Life Cycle

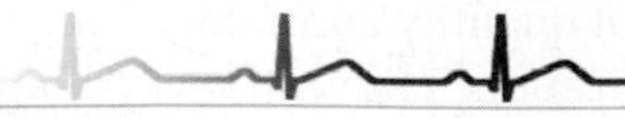

LEARNING OUTCOMES

After reading this section, you should be able to

A. Describe the stages of the cell life cycle.
B. Give the details of DNA replication.
C. Explain what occurs during mitosis and cytokinesis.

The **cell life cycle** includes the changes a cell undergoes from the time it is formed until it divides to produce two new cells. The life cycle of a cell has two stages: interphase and cell division (mitosis and cytokinesis; figure 3.38).

Interphase

Interphase is the phase between cell divisions; nearly all of the life cycle of a typical cell is spent in interphase. During this time, the cell carries out the metabolic activities necessary for life and performs its specialized functions—for example, secreting digestive

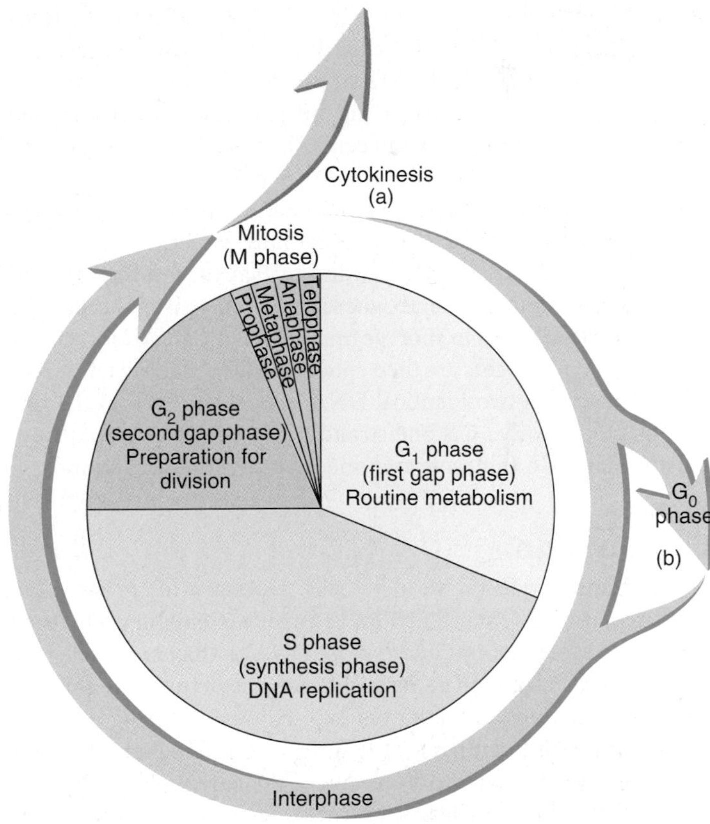

Cytokinesis
(a)
Mitosis
(M phase)
Telophase
Anaphase
Metaphase
Prophase
G2 phase
(second gap phase)
Preparation for
division
G1 phase
(first gap phase)
Routine metabolism
G0
phase
(b)
S phase
(synthesis phase)
DNA replication
Interphase

FIGURE 3.38 Cell Cycle

The cell cycle is divided into interphase (blue) and cell division (mitosis and cytokinesis). Interphase is divided into G_1, S, and G_2 subphases. During G_1, the cell carries out routine metabolic activities. During the S phase, DNA is replicated. During the G_2 phase, the cell prepares for division. (*a*) Following mitosis, two cells are formed by the process of cytokinesis. Each new cell begins a new cell cycle. (*b*) Many cells exit the cell cycle and enter the G_0 phase, where they remain until stimulated to divide, at which point they reenter the cell cycle.

enzymes. In addition, the cell prepares to divide. This preparation includes an increase in cell size as many cell components double in quantity and a doubling of the DNA content during DNA replication. The centrioles within the centrosome are also duplicated. Consequently, when the cell divides, each new cell receives the organelles and DNA necessary for continued functioning.

Interphase can be divided into three subphases, called G_1, S, and G_2. During G_1 (the first *gap* phase), the cell carries out routine metabolic activities. During the S phase (the *synthesis* phase), the DNA is replicated (new DNA is synthesized). During the G_2 phase (the second *gap* phase), the cell prepares for cell division. Many body cells do not divide for days, months, or even years. These "resting" cells exit the cell cycle and enter what is called the G_0 phase, in which they remain unless stimulated to divide.

DNA Replication

DNA **replication** is the process by which two new strands of DNA are made, using the two existing strands as templates. During interphase, DNA and its associated proteins appear as dispersed chromatin threads within the nucleus. When DNA replication begins, the two strands of each DNA molecule separate from each other for some distance (figure 3.39). Each strand then functions as a template, or pattern, for the production of a new, complementary strand of DNA. Each new strand forms as complementary nucleotides pair with the existing nucleotides of each template strand of the original DNA molecule. The production of the new nucleotide strands is catalyzed by **DNA polymerase,** an enzyme that adds new nucleotides to the 3′ end of the growing strands. Because of the antiparallel orientation of the two DNA strands, the strands form differently. One strand, called the **leading strand,** forms as a continuous strand, whereas the other strand, called the **lagging strand,** forms in short segments called *Okazaki fragments.* The Okazaki fragments are then spliced by **DNA ligase.** DNA replication results in two identical DNA molecules. Each of the two new DNA molecules has one strand of nucleotides derived from the original DNA molecule and one newly synthesized strand.

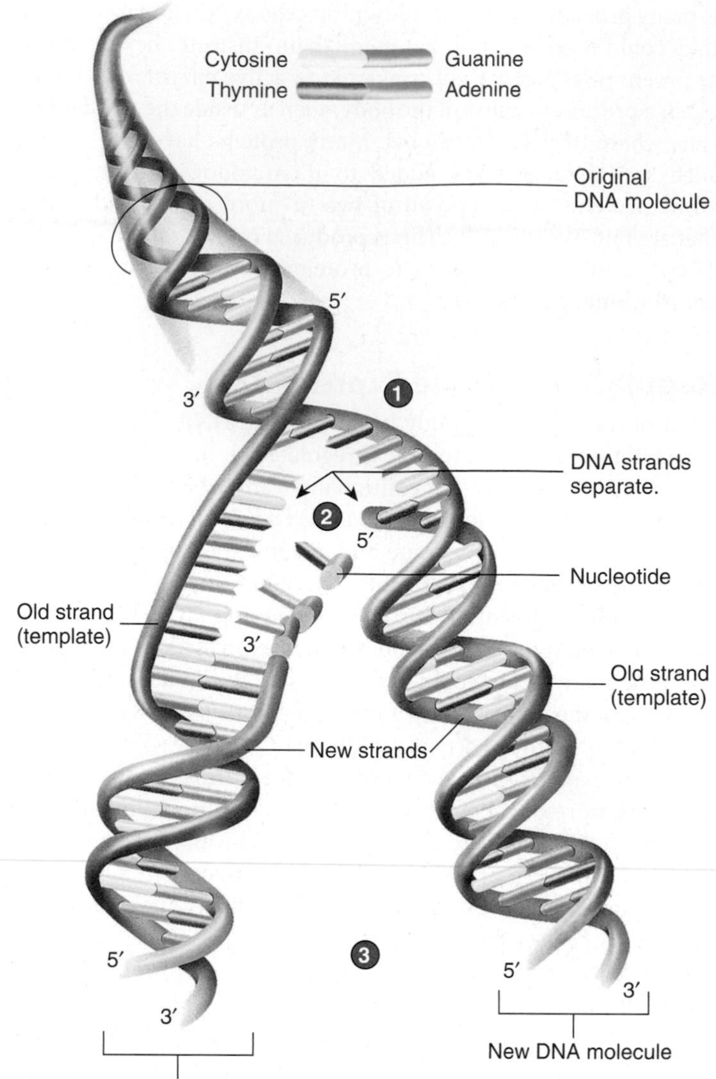
Cytosine
Guanine
Thymine
Adenine
Original DNA molecule
5′
3′
1
DNA strands separate.
2
5′
Nucleotide
Old strand (template)
3′
Old strand (template)
New strands
5′
3
5′
3′
3′
New DNA molecule
New DNA molecule

1 The strands of the DNA molecule separate from each other.

2 Each old strand (dark purple) functions as a template on which a new, complementary strand (light purple) is formed. The base-pairing relationship between nucleotides determines the sequence of nucleotides in the newly formed strands.

3 Two identical DNA molecules are produced.

PROCESS FIGURE 3.39 APR **Replication of DNA**

Replication during the S phase of interphase produces two identical molecules of DNA.

Cell Division

Cell division produces the new cells necessary for growth and tissue repair. A parent cell divides to form two daughter cells, each having the same amount and type of DNA as the parent cell. The daughter cells also tend to have the same structure and perform the same functions as the parent cell. However, during development and cell differentiation, the functions of daughter cells may differ from each other and from that of the parent cell.

Cell division involves two major events: division of the chromosomes into two new nuclei and division of the cytoplasm to form two new cells, each of which contains one of the newly formed nuclei. The nuclear events are called mitosis, and the cytoplasmic division is called cytokinesis.

Mitosis

Mitosis (mī-to′sis) is the division of a cell's chromosomes into two new nuclei, each of which has the same amount and type of DNA as the original nucleus. During mitosis, the chromatin becomes very densely coiled to form compact chromosomes called **mitotic chromosomes.** Mitotic chromosomes are discrete bodies that can be stained and easily seen with a light microscope. Because the DNA has been replicated, each mitotic chromosome consists of two copies of the original chromosome, which are individually called **chromatids** (krō′ma-tids). The chromatids are attached at the **centromere** (sen′trō-mēr) (figure 3.40; see figure 3.23). The **kinetochore** (ki-nē′tō-kōr, ki-net′ō-kōr) is a protein structure that

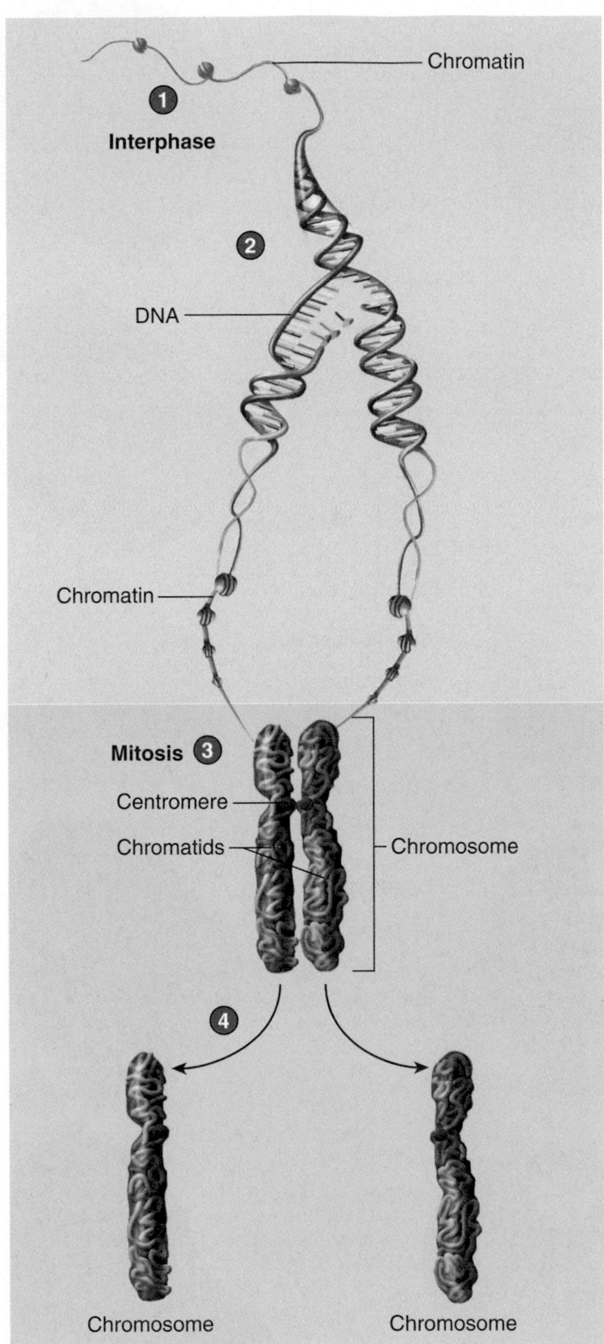
Chromatin
1
Interphase
2
DNA
Chromatin
Mitosis 3
Centromere
Chromatids
Chromosome
4
Chromosome
Chromosome

1 The DNA of a chromosome is dispersed as chromatin.

2 The DNA molecule unwinds, and each strand of the molecule is replicated.

3 During mitosis, the chromatin from each replicated DNA strand condenses to form a chromatid. The chromatids are joined at the centromere to form a single chromosome.

4 The chromatids separate to form two new, identical chromosomes. The chromosomes will unwind to form chromatin in the nuclei of the two daughter cells.

PROCESS FIGURE 3.40 Replication of a Chromosome

binds the centromere and provides a point of attachment for microtubules that will separate and move the chromatids during mitosis. As two daughter cells form, a nucleus is reformed around the chromatids. After the chromatids have separated, each is considered a chromosome. Each daughter cell receives a copy of the chromosomes. Thus, the daughter cells receive the same complement of chromosomes and are genetically identical.

Mitosis is divided into four phases: prophase, metaphase, anaphase, and telophase (tel′ō-fāz):

- During **prophase** (figure 3.41, *step 2*), the chromatin condenses to form mitotic chromosomes. The chromosomes are visible with a light microscope and it is evident that each has replicated. Also the centrioles divide and migrate to each pole of the cell and microtubules called spindle fibers extend from the centrioles to the centromeres of the chromosomes. In late prophase, the nucleolus and nuclear envelope disappear.
- In **metaphase** (figure 3.41, *step 3*), the chromosomes align near the center of the cell.
- At the beginning of **anaphase** (figure 3.41, *step 4*), the chromatids separate. At this point, one of the two identical sets of chromosomes are moved by the spindle fibers toward the centrioles at each of the poles of the cell. At the end of anaphase, each set of chromosomes has reached an opposite pole of the cell, and the cytoplasm begins to divide.
- During **telophase** (figure 3.41, *step 5*), nuclear envelopes form around each set of chromosomes to form two separate nuclei. The chromosomes begin to uncoil and resemble the genetic material characteristic of interphase.

Cytokinesis

Cytokinesis (sī′tō-ki-nē′sis) is the division of the cell's cytoplasm to produce two new cells. Cytokinesis begins in anaphase and continues through telophase (figure 3.41). The first sign of cytokinesis is the formation of a **cleavage furrow,** an indentation of the plasma membrane that forms midway between the centrioles. A contractile ring composed primarily of actin filaments pulls the plasma membrane inward, dividing the cell into halves. Cytokinesis is complete when the membranes of the halves separate at the cleavage furrow to form two separate cells.

ASSESS YOUR PROGRESS

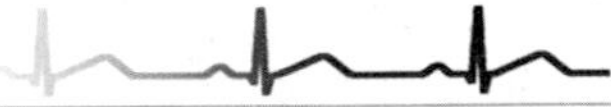

70. *What are the two stages of the cell life cycle? Which stage is the longest?*

71. *Describe the cell's activities during the G_1, S, and G_2 phases of interphase.*

72. *Describe the process of DNA replication. What are the functions of DNA polymerase and DNA ligase?*

73. *What are the two major events of cell division? What happens in each?*

74. *Differentiate among chromatin, chromatids, and chromosomes.*

75. *List the events that occur during prophase, metaphase, anaphase, and telophase of mitosis.*

76. *What is the end result of mitosis and cytokinesis?*

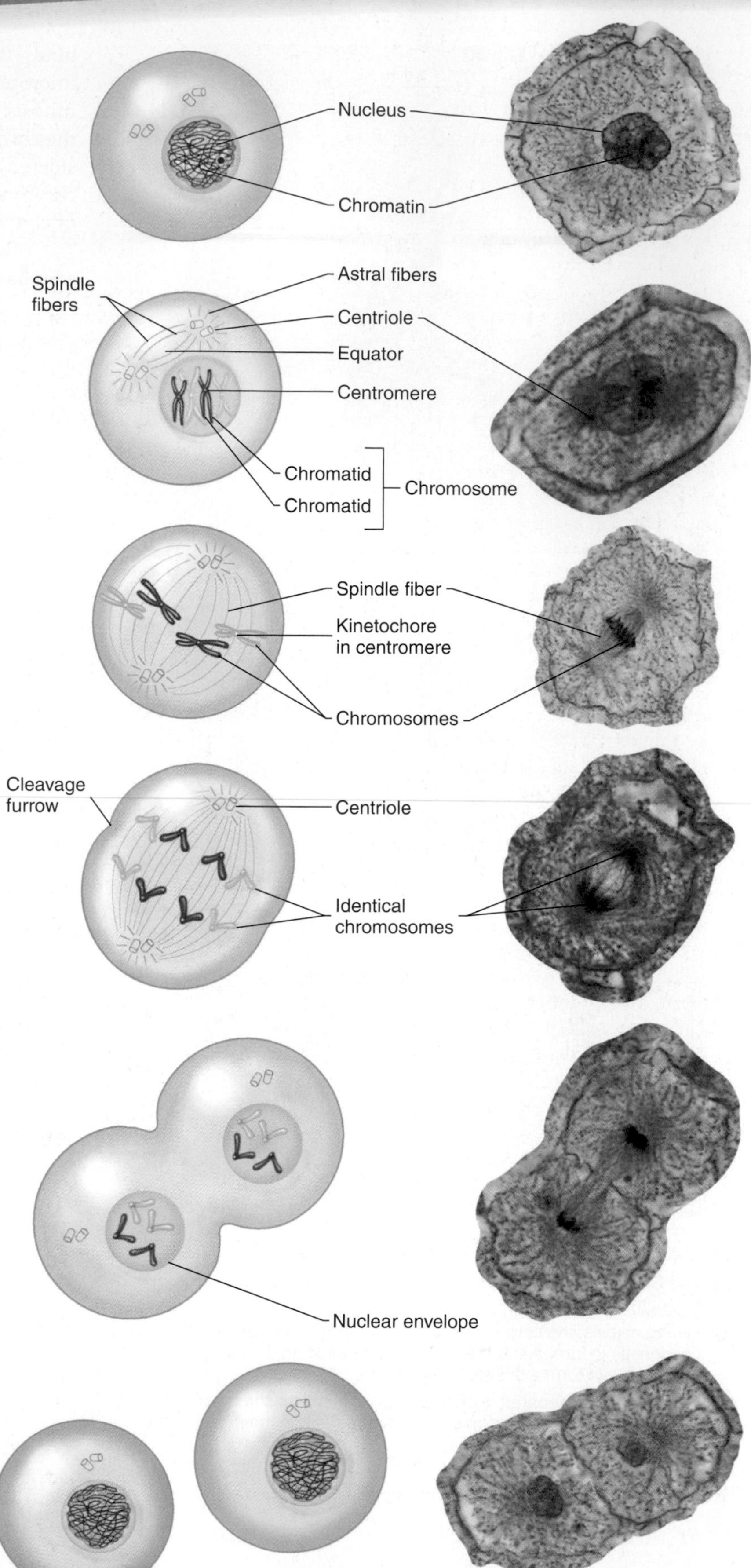
Nucleus
Chromatin
Spindle fibers
Astral fibers
Centriole
Equator
Centromere
Chromatid
Chromatid
Chromosome
Spindle fiber
Kinetochore in centromere
Chromosomes
Cleavage furrow
Centriole
Identical chromosomes
Nuclear envelope

1. **Interphase** is the time between cell divisions. DNA is present as thin threads of chromatin in the nucleus. DNA replication occurs during the S phase of interphase. Organelles, other than the nucleus, and centrioles duplicate during interphase.

2. In **prophase,** the chromatin condenses into chromosomes. Each chromosome consists of two chromatids joined at the centromere. The centrioles move to the opposite ends of the cell, and the nucleolus and the nuclear envelope disappear. Microtubules form near the centrioles and project in all directions. Some of the microtubules end blindly and are called astral fibers. Others, known as spindle fibers, project toward an invisible line called the equator and overlap with fibers from opposite centrioles.

3. In **metaphase,** the chromosomes align in the center of the cell in association with the spindle fibers. Some spindle fibers are attached to kinetochores in the centromere of each chromosome.

4. In **anaphase,** the chromatids separate, and each chromatid is then referred to as a chromosome. Thus, when the centromeres divide, the chromosome number is double, and there are two identical sets of chromosomes. The chromosomes, assisted by the spindle fibers, move toward the centrioles at each end of the cell. Separation of the chromatids signals the beginning of anaphase, and by the time anaphase has ended, the chromosomes have reached the poles of the cell. Cytokinesis begins during anaphase as a cleavage furrow forms around the cell.

5. In **telophase,** migration of each set of chromosomes is complete. The chromosomes unravel to become less distinct chromatin threads. The nuclear envelope forms from the endoplasmic reticulum. The nucleoli form, and cytokinesis continues to produce two cells.

6. Mitosis is complete, and a new interphase begins. The chromosomes have unraveled to become chromatin. Cell division has produced two daughter cells, each with DNA that is identical to the DNA of the parent cell.

PROCESS FIGURE 3.41 APR **Cell Division: Mitosis and Cytokinesis**

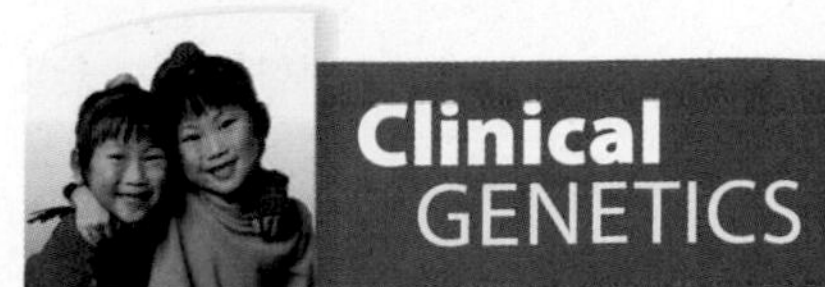

Clinical GENETICS | Genetic Changes in Cancer Cells

Cancer (kan′ser) refers to a malignant, spreading tumor. A tumor (too′mŏr) is a growth of neoplastic tissue. The term **neoplasm** (nē′ō-plazm; new growth) refers to abnormal tissue growth from rapid cell proliferation that continues after normal growth would have stopped. **Oncology** (ong-kol′ō-jē) is the study of tumors and the problems they cause. A neoplasm can be either **benign** (bē-nīn′; L. kind), unlikely to spread or to become worse, or **malignant** (ma-lig′nănt; with malice or intent to cause harm), able to spread and become worse. Although benign tumors are usually less dangerous than malignant tumors, they can cause problems if they compress surrounding tissues and impair their functions.

Cells of malignant neoplasms, or cancer cells, differ from cells of normal tissues in two fundamental ways:

1. Cancer cells have unregulated growth and altered morphology. They tend to be more spherical than normal cells. They appear more embryonic, or less mature, than the normal tissue from which they arise. For example, a skin cancer cell is more spherical and softer than the stratified squamous epithelial cells of the skin.
2. Cancer cells are invasive. They have the ability to squeeze into spaces and enter surrounding tissues. They secrete enzymes that cut paths through healthy tissue. Cancer cells can dislodge; enter blood vessels, lymphatic vessels, or body cavities; and travel to distant sites, where they invade tissues. The process by which cancer spreads to distant sites is called **metastasis** (me-tas′ta-sis).

Most cancers are caused by mutations of genes within somatic cells. It is estimated that less than 10% of cancers are inherited. Cancer develops in somatic cells because of mutations that occur during cell division. When DNA is replicated prior to cell division, a small number of replication errors occur. A DNA sequence with replication errors in it is a mutation. Other factors that cause mutations by damaging or altering DNA include radiation, certain chemicals and toxins, and some viruses. Because mutations are most likely to occur during DNA replication, cancer usually develops in tissues that are undergoing frequent cell divisions, such as epithelial cells (see chapter 4). Cancer requires multiple mutations that accumulate over many cell generations. This is one reason that cancer becomes more common in older people.

Two major mechanisms help prevent the development of cancer in cells: (1) DNA repair enzymes detect and correct errors that occur during replication. The likelihood that cancer will develop is increased if the genes controlling DNA repair enzymes undergo mutations and become defective, so that mutated genes persist in cells. (2) A self-destruction mechanism called **apoptosis** (ap′op-to′sis) destroys cells with abnormal DNA. Apoptosis is a normal process involved in the self-destruction of cells that have a limited life span, but it can also cause self-destruction in cells with damaged DNA. Thus, apoptosis can cause mutated cells to self-destruct and thereby remove cells with mutations before cancer develops. Mutation of the genes responsible for apoptosis can also result in the persistence of cells with mutations, and these cells can continue to divide.

Some genes promote cell division, whereas others suppress it. Genes that promote cell division are called **proto-oncogenes.** Mutations in proto-oncogenes can give rise to abnormal regulatory genes, called **oncogenes** (ong′ko-jēnz), which increase the rate of cell division. Oncogenes often code for proteins that control cell division. **Tumor suppressor genes** are normal genes that slow or stop cell division. Mutations that delete or inactivate tumor suppressor genes can also increase the rate of cell division by taking off the brakes, so to speak, of the processes that promote cell division. Many types of oncogenes and altered tumor suppressor genes have been identified in human cancer cells.

Additional mutations cause the structure and functions of the cancer cells to differ from those of normal cells. For example, these mutations increase the ability of cancer cells to invade and destroy surrounding tissues and to metastasize. The continued accumulation of mutations in cancer cells is also responsible for changes in the characteristics of the cancer cells over time. These changes can cause the cancer cells in a tumor to become less sensitive to treatments designed to kill them, such as chemotherapeutic drugs.

Cancer therapy concentrates primarily on confining and then killing the malignant cells. This goal is currently accomplished by killing the tissue with x-rays or lasers, by removing the tumor surgically, or by treating the patient with drugs that kill rapidly dividing cells. The major problem with current therapy is that many cancers cannot be completely eliminated. Also, these treatments often kill normal cells in rapidly growing tissues, such as bone marrow, where new blood cells are produced, and the lining of the intestinal tract. Loss of these tissues can result in anemia, caused by the lack of red blood cells, and nausea, caused by the loss of the intestinal lining.

Answer

Learn to Predict

From page 56

Consider first the important points made in the question itself. First, Carlos suffers from a genetic disease (diabetes insipidus) and, second, this disease results in excessive water loss at the kidneys. In this chapter, we learned that genes determine the structure of cellular proteins. But the question is, what type of protein? There are two possibilities. We also learned that cellular proteins have many functions, including membrane receptors and transport proteins. Mutations in the gene for either type can lead to disruption in water homeostasis. ADH is a hormone, or chemical signal, that regulates water loss from the kidneys. A mutation in the membrane receptor gene specific for ADH can, therefore, disrupt ADH regulation, since the kidney cells cannot respond to ADH. On the other hand, aquaporins are membrane proteins that regulate osmosis or diffusion of water across the membrane. A mutation in an aquaporin gene can also lead to the disruption of water homeostasis.

Answers to the rest of this chapter's Predict questions are in Appendix G.

Summary

3.1 Functions of the Cell (p. 57)

1. The plasma membrane forms the outer boundary of the cell.
2. The nucleus directs the cell's activities.
3. The cytoplasm, between the nucleus and the plasma membrane, is where most cell activities take place.
4. Cells perform the following functions:
   - Cells metabolize and release energy.
   - Cells synthesize molecules.
   - Cells provide a means of communication.
   - Cells reproduce and provide for inheritance.

3.2 How We See Cells (p. 59)

1. Light microscopes allow us to visualize the general features of cells.
2. Electron microscopes allow us to visualize the fine structure of cells.

3.3 Plasma Membrane (p. 59)

1. The plasma membrane passively or actively regulates what enters or leaves the cell.
2. The plasma membrane is composed of a phospholipid bilayer, in which proteins are suspended (commonly depicted by the fluid-mosaic model).

3.4 Membrane Lipids (p. 61)

Lipids give the plasma membrane most of its structure and some of its function.

3.5 Membrane Proteins (p. 62)

1. Membrane proteins function as marker molecules, attachment proteins, transport proteins, receptor proteins, and enzymes.
2. Transport proteins include channel proteins, carrier proteins, and ATP-powered pumps.
3. Some receptor proteins are linked to and control channel proteins.
4. Some receptor molecules are linked to G protein complexes, which control numerous cellular activities.

3.6 Movement Through the Plasma Membrane (p. 67)

1. Lipid-soluble molecules pass through the plasma membrane readily by dissolving in the lipid bilayer. Small molecules diffuse between the phospholipid molecules of the plasma membrane.
2. Large non-lipid-soluble molecules and ions (e.g., glucose and amino acids) are transported through the membrane by transport proteins.
3. Large non-lipid-soluble molecules, as well as very large molecules and even whole cells, can be transported across the membrane in vesicles.

Passive Membrane Transport

1. Diffusion is the movement of a substance from an area of higher solute concentration to one of lower solute concentration (down a concentration gradient).
2. The concentration gradient is the difference in solute concentration between two points divided by the distance separating the points.
3. The rate of diffusion increases with an increase in the concentration gradient, an increase in temperature, a decrease in molecular size, and a decrease in viscosity.
4. The end result of diffusion is uniform distribution of molecules.
5. Diffusion requires no expenditure of energy.
6. Osmosis is the diffusion of water (solvent) across a selectively permeable membrane.
7. Osmotic pressure is the force required to prevent the movement of water across a selectively permeable membrane.
8. Isosmotic solutions have the same concentration of solute particles, hyperosmotic solutions have a greater concentration of solute particles, and hyposmotic solutions have a lower concentration of solute particles.
9. Cells placed in an isotonic solution neither swell nor shrink. In a hypertonic solution, they shrink (crenate); in a hypotonic solution, they swell and may burst (lyse).

10. Mediated transport is the movement of a substance across a membrane by means of a transport protein. The substances transported tend to be large, water-soluble molecules.
11. Facilitated diffusion moves substances down their concentration gradient and does not require energy (ATP).

Active Membrane Transport

1. Active transport can move substances against their concentration gradient and requires ATP. An exchange pump is an active-transport mechanism that simultaneously moves two substances in opposite directions across the plasma membrane.
2. In secondary active transport, an ion is moved across the plasma membrane by active transport, and the energy produced by the ion diffusing back down its concentration gradient can transport another molecule, such as glucose, against its concentration gradient.
3. Vesicular transport is the movement of large volumes of substances across the plasma membrane through the formation or release of vesicles.
4. Endocytosis is the bulk movement of materials into cells.
   - Phagocytosis is the bulk movement of solid material into cells by the formation of a vesicle.
   - Pinocytosis is similar to phagocytosis, except that the ingested material is much smaller and is in solution.
5. Receptor-mediated endocytosis allows for endocytosis of specific molecules.
6. Exocytosis is the secretion of materials from cells by vesicle formation.
7. Both endocytosis and exocytosis require energy.

3.7 Cytoplasm (p. 76)

The cytoplasm is the material outside the nucleus and inside the plasma membrane.

Cytosol

1. Cytosol consists of a fluid part (the site of chemical reactions), the cytoskeleton, and cytoplasmic inclusions.
2. The cytoskeleton supports the cell and is responsible for cell movements. It consists of protein fibers.
   - Microtubules are hollow tubes composed of the protein tubulin. They form spindle fibers and are components of centrioles, cilia, and flagella.
   - Actin filaments are small protein fibrils that provide structure to the cytoplasm or cause cell movements.
   - Intermediate filaments are protein fibers that provide structural strength to cells.
3. Cytoplasmic inclusions, such as lipochromes, are not surrounded by membranes.

3.8 The Nucleus and Cytoplasmic Organelles (p. 77)

Organelles are subcellular structures specialized for specific functions.

The Nucleus

1. The nuclear envelope consists of a double membrane with nuclear pores.
2. DNA and associated proteins are found inside the nucleus as chromatin.
3. DNA is the hereditary material of the cell. It controls cell activities by producing proteins through RNA.
4. A gene is a portion of a DNA molecule. Genes determine the proteins in a cell.
5. Nucleoli consist of RNA and proteins and are the sites of ribosomal subunit assembly.

Ribosomes

1. Ribosomes consist of small and large subunits manufactured in the nucleolus and assembled in the cytoplasm.
2. Ribosomes are the sites of protein synthesis.
3. Ribosomes can be free or associated with the endoplasmic reticulum.

Endoplasmic Reticulum

1. The endoplasmic reticulum is an extension of the outer membrane of the nuclear envelope; it forms tubules or sacs (cisternae) throughout the cell.
2. The rough endoplasmic reticulum has ribosomes and is a site of protein synthesis and modification.
3. The smooth endoplasmic reticulum lacks ribosomes and is involved in lipid production, detoxification, and calcium storage.

Golgi Apparatus

The Golgi apparatus is a series of closely packed, modified cisternae that modify, package, and distribute lipids and proteins produced by the endoplasmic reticulum.

Secretory Vesicles

Secretory vesicles are membrane-bound sacs that carry substances from the Golgi apparatus to the plasma membrane, where the contents of the vesicles are released by exocytosis.

Lysosomes

1. Lysosomes are membrane-bound sacs containing hydrolytic enzymes. Within the cell, the enzymes break down phagocytized material and nonfunctional organelles (autophagy).
2. Enzymes released from the cell by lysis or enzymes secreted from the cell can digest extracellular material.

Peroxisomes

Peroxisomes are membrane-bound sacs containing enzymes that digest fatty acids and amino acids, as well as enzymes that catalyze the breakdown of hydrogen peroxide.

Proteasomes

Proteasomes are large, multienzyme complexes, not bound by membranes, that digest selected proteins within the cell.

Mitochondria

1. Mitochondria are the major sites for the production of ATP, which cells use as an energy source.
2. The mitochondria have a smooth outer membrane and an inner membrane that is infolded to form cristae.
3. Mitochondria contain their own DNA, can produce some of their own proteins, and can replicate independently of the cell.

Centrioles and Spindle Fibers

1. Centrioles are cylindrical organelles located in the centrosome, a specialized zone of the cytoplasm that serves as the site of microtubule formation.
2. Spindle fibers are involved in the separation of chromosomes during cell division.

Cilia and Flagella

1. Cilia facilitate the movement of materials over the surface of the cell.
2. Flagella, which are much longer than cilia, propel sperm cells.

Microvilli

Microvilli increase the surface area of the plasma membrane for absorption or secretion.

3.9 Genes and Gene Expression (p. 86)

1. During transcription, information stored in DNA is copied to form mRNA.
2. During translation, the mRNA goes to ribosomes, where it directs the synthesis of proteins.

Transcription

1. DNA unwinds and, through nucleotide pairing, produces pre-mRNA (transcription).
2. Introns are removed and exons are spliced together during posttranscriptional processing.
3. Modifications to the ends of mRNA also occur during posttranscriptional processing.

Genetic Code

The genetic code specifies amino acids and consists of codons, which are sequences of three nucleotides in mRNA.

Translation

1. mRNA moves through the nuclear pores to ribosomes.
2. Transfer RNA (tRNA), which carries amino acids, interacts at the ribosome with mRNA. The anticodons of tRNA bind to the codons of mRNA, and the amino acids join to form a protein (translation).
3. During posttranslational processing, proproteins, some of which are proenzymes, are modified into proteins, some of which are enzymes.

Regulation of Gene Expression

1. Cells become specialized because certain parts of the DNA molecule are activated but other parts are not.
2. The level of DNA activity and thus protein production can be controlled internally or can be affected by regulatory substances secreted by other cells.

3.10 Cell Life Cycle (p. 91)

The cell life cycle has two stages: interphase and cell division.

Interphase

1. Interphase, the period between cell divisions, is the time of DNA replication.
2. During replication, DNA unwinds, and each strand produces a new DNA molecule.

Cell Division

1. Cell division includes nuclear division and cytoplasmic division.
2. Mitosis is the replication of the cell's nucleus, and cytokinesis is division of the cell's cytoplasm.
3. Mitosis is a continuous process divided into four phases.
   - *Prophase.* Chromatin condenses to become visible as chromosomes. Each chromosome consists of two chromatids joined at the centromere. Centrioles move to opposite poles of the cell, and astral fibers and spindle fibers form. Nucleoli disappear, and the nuclear envelope degenerates.
   - *Metaphase.* Chromosomes align at the center of the cell.
   - *Anaphase.* The chromatids of each chromosome separate at the centromere. Each chromatid is then called a chromosome. The chromosomes migrate to opposite poles.
   - *Telophase.* Chromosomes unravel to become chromatin. The nuclear envelope and nucleoli reappear.
4. Cytokinesis begins with the formation of the cleavage furrow during anaphase. It is complete when the plasma membrane comes together at the equator, producing two new daughter cells.

REVIEW AND COMPREHENSION

1. In the plasma membrane, ____________ form(s) the lipid bilayer, ____________ determine(s) the fluid nature of the membrane, and ____________ mainly determine(s) the function of the membrane.
   a. phospholipids, cholesterol, proteins
   b. phospholipids, proteins, cholesterol
   c. proteins, cholesterol, phospholipids
   d. cholesterol, phospholipids, proteins
   e. cholesterol, proteins, phospholipids
2. Which of the following functioning proteins are found in the plasma membrane?
   a. channel proteins
   b. marker molecules
   c. receptor molecules
   d. enzymes
   e. All of these are correct.
3. In general, lipid-soluble molecules diffuse through the ____________; small, water-soluble molecules diffuse through the ____________.
   a. membrane channels, membrane channels
   b. membrane channels, lipid bilayer
   c. lipid bilayer, carrier proteins
   d. membrane channels, carrier proteins
   e. carrier proteins, membrane channels
4. Small pieces of matter, and even whole cells, can be transported across the plasma membrane in
   a. membrane channels.
   b. carrier molecules.
   c. receptor molecules.
   d. marker molecules.
   e. vesicles.
5. The rate of diffusion increases if the
   a. concentration gradient decreases.
   b. temperature of a solution decreases.
   c. viscosity of a solution decreases.
   d. All of these are correct.
6. Concerning the process of diffusion, at equilibrium
   a. the net movement of solutes stops.
   b. random molecular motion continues.
   c. there is an equal movement of solute in opposite directions.
   d. the concentration of solute is equal throughout the solution.
   e. All of these are correct.
7. If a cell is placed in a(n) ____________ solution, lysis of the cell may occur.
   a. hypertonic
   b. hypotonic
   c. isotonic
   d. isosmotic

8. Suppose that a woman runs a long-distance race in the summer. During the race, she loses a large amount of hyposmotic sweat. You would expect her cells to
   a. shrink.
   b. swell.
   c. stay the same.
9. Which of these statements about facilitated diffusion is true?
   a. In facilitated diffusion, net movement is down the concentration gradient.
   b. Facilitated diffusion requires the expenditure of energy.
   c. Facilitated diffusion does not require a carrier protein.
   d. Facilitated diffusion moves materials through membrane channels.
   e. Facilitated diffusion moves materials in vesicles.
10. Which of these statements concerning the symport of glucose into cells is true?
    a. The sodium-potassium exchange pump moves Na^+ into cells.
    b. The concentration of Na^+ outside cells is less than inside cells.
    c. A carrier protein moves Na^+ into cells and glucose out of cells.
    d. The concentration of glucose can be greater inside cells than outside cells.
    e. As Na^+ are actively transported into the cell, glucose is carried along.
11. A white blood cell ingests solid particles by forming vesicles. This describes the process of
    a. exocytosis.
    b. facilitated diffusion.
    c. secondary active transport.
    d. phagocytosis.
    e. pinocytosis.
12. Given these characteristics:
    (1) requires energy
    (2) requires carrier proteins
    (3) requires membrane channels
    (4) requires vesicles

    Choose the characteristics that apply to exocytosis.
    a. 1,2
    b. 1,4
    c. 1,3,4
    d. 1,2,3
    e. 1,2,3,4
13. Cytoplasm is found
    a. in the nucleus.
    b. outside the nucleus and inside the plasma membrane.
    c. outside the plasma membrane.
    d. inside mitochondria.
    e. everywhere in the cell.
14. Which of these elements of the cytoskeleton is composed of tubulin and forms essential components of centrioles, spindle fibers, cilia, and flagella?
    a. actin filaments
    b. intermediate filaments
    c. microtubules
15. A large structure, normally visible in the nucleus of a cell, where ribosomal subunits are produced is called a(n)
    a. endoplasmic reticulum.
    b. mitochondrion.
    c. nucleolus.
    d. lysosome.
16. A cell that synthesizes large amounts of protein for use outside the cell has a large
    a. number of cytoplasmic inclusions.
    b. number of mitochondria.
    c. amount of rough endoplasmic reticulum.
    d. amount of smooth endoplasmic reticulum.
    e. number of lysosomes.
17. Which of these organelles produces large amounts of ATP?
    a. nucleus
    b. mitochondria
    c. ribosomes
    d. endoplasmic reticulum
    e. lysosomes
18. Cylindrically shaped extensions of the plasma membrane that do not move, are supported by actin filaments, and may function in absorption or as sensory receptors are
    a. centrioles.
    b. spindle fibers.
    c. cilia.
    d. flagella.
    e. microvilli.
19. A portion of an mRNA molecule that determines one amino acid in a polypeptide chain is called a(n)
    a. nucleotide.
    b. gene.
    c. codon.
    d. exon.
    e. intron.
20. In which of these organelles is mRNA synthesized?
    a. nucleus
    b. ribosome
    c. endoplasmic reticulum
    d. nuclear envelope
    e. peroxisome
21. During the cell life cycle, DNA replication occurs during the
    a. G_1 phase.
    b. G_2 phase.
    c. M phase.
    d. S phase.
22. Given the following activities:
    (1) repair
    (2) growth
    (3) gamete production
    (4) differentiation

    Which of the activities is (are) the result of mitosis?
    a. 2
    b. 3
    c. 1,2
    d. 3,4
    e. 1,2,4

Answers in Appendix E

CRITICAL THINKING

1. Why does a surgeon use sterile distilled water rather than sterile isotonic saline to irrigate a surgical wound from which a tumor has been removed?
2. Solution A is hyperosmotic to solution B. If solution A is separated from solution B by a selectively permeable membrane, does water move from solution A into solution B, or vice versa? Explain.
3. A researcher wants to determine the nature of the transport mechanism that moved substance X into a cell. She could measure the concentration of substance X in the extracellular fluid and within the cell, as well as the rate of movement of substance X into the cell. She does a series of experiments and gathers the data shown in the accompanying graph. Choose the transport process that is consistent with the data.
   a. diffusion
   b. active transport
   c. facilitated diffusion
   d. There is not enough information to make a judgment.

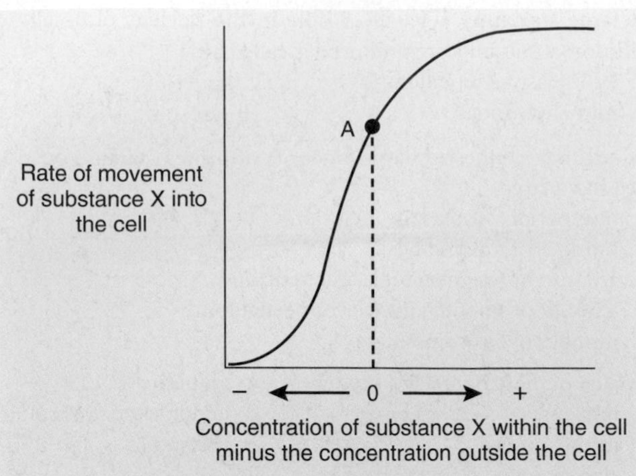
A
Rate of movement of substance X into the cell
– 0 +
Concentration of substance X within the cell minus the concentration outside the cell

Graph depicting the rate of movement of substance X from a fluid into a cell (*y* axis) versus the concentration of substance X within the cell minus the concentration outside the cell (*x* axis). At point *A*, the extracellular concentration of substance X is equal to the intracellular concentration of substance X (designated *0* on the *x* axis).

4. A dialysis membrane is selectively permeable, and substances smaller than proteins are able to pass through it. If you wanted to use a dialysis machine to remove only urea (a small molecule) from blood, what could you use for the dialysis fluid?
   a. a solution that is isotonic and contains only large molecules, such as protein
   b. a solution that is isotonic and contains the same concentration of all substances except that it has no urea
   c. distilled water, which contains no ions or dissolved molecules
   d. blood, which is isotonic and contains the same concentration of all substances, including urea
5. Luke started a training program at the beginning of the summer. For the first week, he jogged 1 mile. Each week after that, he increased the distance he jogged by about 0.5 mile and increased his speed. At the end of 2 months, he was able to jog 4.5 miles each day at a faster pace than he had been able to jog at the beginning of the summer. During the 2-month period, the muscle cells in his heart and his legs increased in size. Identify a critical membrane-bound organelle that increased in number and made it possible for Luke to run the longer distance. Also, explain how these organelles increased in number, and describe the location of the genetic code for the proteins in the organelles.
6. If you had the ability to inhibit mRNA synthesis with a drug, explain how you could distinguish between proteins released from secretory vesicles in which they had been stored and proteins released from cells in which they had been newly synthesized.

Answers in Appendix F

4

Tissues

In some ways, the human body is like a car. Both consist of many parts that are made of materials consistent with their specialized functions. For example, car tires are made of synthetic rubber reinforced with a variety of fibers, the engine is composed of various metal parts, and the windows are transparent glass. Similarly, the many parts of the human body are made of groups of specialized cells and the materials surrounding them. Muscle cells, which contract to produce body movements, differ both structurally and functionally from epithelial cells, which protect, secrete, or absorb. Conversely, cells in the retina of the eye, which are specialized to detect light and enable us to see, do not contract as muscle cells do—nor do they exhibit the functions of epithelial cells.

The structure and the function of tissues are so closely related that we can often predict the function of a tissue when given its structure, and vice versa. Knowledge of tissue structure and function is important in understanding organs, organ systems, and the complete organism.

Learn to Predict

It is Matt's birthday, but he will not be eating any cake. Matt has gluten enteropathy, also called celiac disease, which results from an inappropriate immune response to gluten, a group of proteins found in wheat and various other grains. After eating food containing gluten, such as most breads and cereals, Matt has bouts of diarrhea because his intestinal lining is unable to properly absorb water and nutrients. The poor absorption is due to a reduced number of villi, or fingerlike protrusions of the intestinal lining, and reduced transport capacity of the remaining cells within the villi. In chapter 3 we learned that water and nutrients enter and exit the body's cells by osmosis and other transport processes. Chapter 4 describes how tissues are specialized to allow this flow of water and nutrients. After reading this chapter, identify the type of tissue affected by Matt's disease and which parts of the cells in this tissue are damaged, thus reducing their ability to absorb water and nutrients. Then explain why Matt has diarrhea after eating food containing gluten.

Photo: Fluorescent image of intestinal tissue. A section of small intestine has been immunolabeled to detect actin (red) and laminin (green), structural proteins that help support the villi, which protrude into the lumen of the intestine. Nuclei are shown in blue.

Module 3
Tissues

Anatomy & Physiology
REVEALED
aprevealed.com

4.1 Tissues and Histology

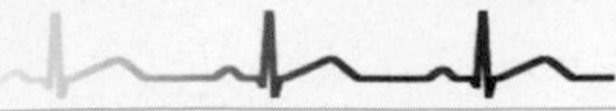

LEARNING OUTCOMES

After reading this section, you should be able to

A. Describe the general make-up of a tissue.

B. List the four primary tissue types.

C. Explain how histology relates to biopsies and autopsies.

Tissues (tish′ūz) are collections of specialized cells and the extracellular substances surrounding them. Body tissues are classified into four types, based on the structure of the cells, the composition of the noncellular substances surrounding the cells (called the **extracellular matrix**), and the functions of the cells. The four primary tissue types, from which all organs of the body are formed, are epithelial tissue, connective tissue, muscle tissue, and nervous tissue.

Epithelial and connective tissues are the most diverse in form. The different types of epithelial and connective tissues are classified by structure, including cell shape, relationship of cells to one another, and composition of the extracellular matrix. Muscle and nervous tissues are also classified by structure as well as by functional characteristics and location.

The relationship between the function and the structure of body parts is an important concept in anatomy and physiology. The function of a body structure can often be determined by its specific type of tissue. For example, the lungs have a thin layer of specialized epithelial tissue that enhances the exchange of gases between air and blood. In contrast, the outer layer of the skin is composed of a different type of epithelial tissue, which provides protection to underlying tissues.

The tissues of the body are interdependent. For example, muscle tissue cannot produce movement unless it receives oxygen carried by red blood cells, and new bone tissue cannot form unless epithelial tissue absorbs calcium and other nutrients from the digestive tract. Also, the loss of one vital tissue through disease or injury can lead to organ failure and death.

Histology (his-tol′ō-jē) is the microscopic study of tissues. Much information about a person's health can be gained by examining tissues. A **biopsy** (bī′op-sē) is the process of removing tissue samples from patients surgically or with a needle for diagnostic purposes. Examining tissue samples can distinguish various disorders. For example, some red blood cells have an abnormal shape in people suffering from sickle-cell disease, and red blood cells are smaller than normal in people with iron-deficiency anemia. Cancer is identified and classified based on characteristic changes in tissues. For example, changes in the structure of epithelial cells can indicate cancer of the uterine cervix, and changes in white blood cells identify people who have leukemia. Also, a greatly increased number of white blood cells can be a sign of infection. Epithelial cells from respiratory passages have an abnormal structure in people with chronic bronchitis, as well as in people with lung cancer.

Tissue samples can be sent to a laboratory for examination. In some cases, tissues are removed surgically and examined quickly, so that the results can be reported while the patient is still anesthetized. These results help determine the appropriate therapy—for example, the amount of tissue removed as part of breast or other types of cancer treatment.

An **autopsy** (aw′top-sē) is an examination of the organs of a dead body to determine the cause of death or to study the changes caused by a disease. Microscopic examination of tissue is often part of an autopsy.

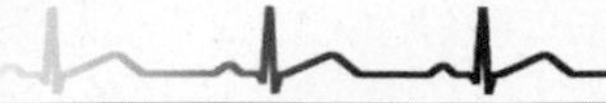

ASSESS YOUR PROGRESS

1. *What components are found in a tissue?*
2. *Name the four primary tissue types and the characteristics that are used to classify them.*
3. *Define* histology. *Explain how the histology of tissues taken by biopsy or autopsy can be used to diagnose some diseases.*

4.2 Embryonic Tissue

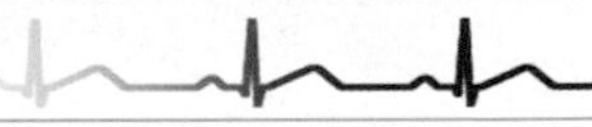

LEARNING OUTCOME

After reading this section, you should be able to

A. Identify the three embryonic germ layers and name the adult structures that are derived from each.

Approximately 13 or 14 days after fertilization, the embryonic stem cells that give rise to a new individual form a slightly elongated disk consisting of two layers, the epiblast and the hypoblast. Cells of the epiblast then migrate between the two layers to form the three embryonic germ layers: the ectoderm, the mesoderm, and the endoderm. The ectoderm, mesoderm, and endoderm are called germ layers because the beginning of all adult structures can be traced back to one of them and they give rise to all the tissues of the body (see chapter 29).

The **endoderm** (en′dō-derm), the inner layer, forms the lining of the digestive tract and its derivatives. The **mesoderm** (mez′ō-derm), the middle layer, forms tissues such as muscle, bone, and blood vessels. The **ectoderm** (ek′tō-derm), the outer layer, forms the skin; a portion of the ectoderm called **neuroectoderm** (noor-ō-ek′tō-derm) becomes the nervous system (see chapter 29). Groups of cells that break away from the neuroectoderm during development, called **neural crest cells,** give rise to parts of the peripheral nerves (see chapter 29), skin pigment (see chapter 5), the medulla of the adrenal gland (see chapter 18), and many tissues of the face.

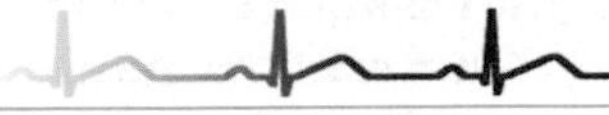

ASSESS YOUR PROGRESS

4. *Name the three embryonic germ layers.*
5. *What adult structures are derived from each layer?*
6. *What is formed from neural crest cells?*

FUNDAMENTAL **Figure**

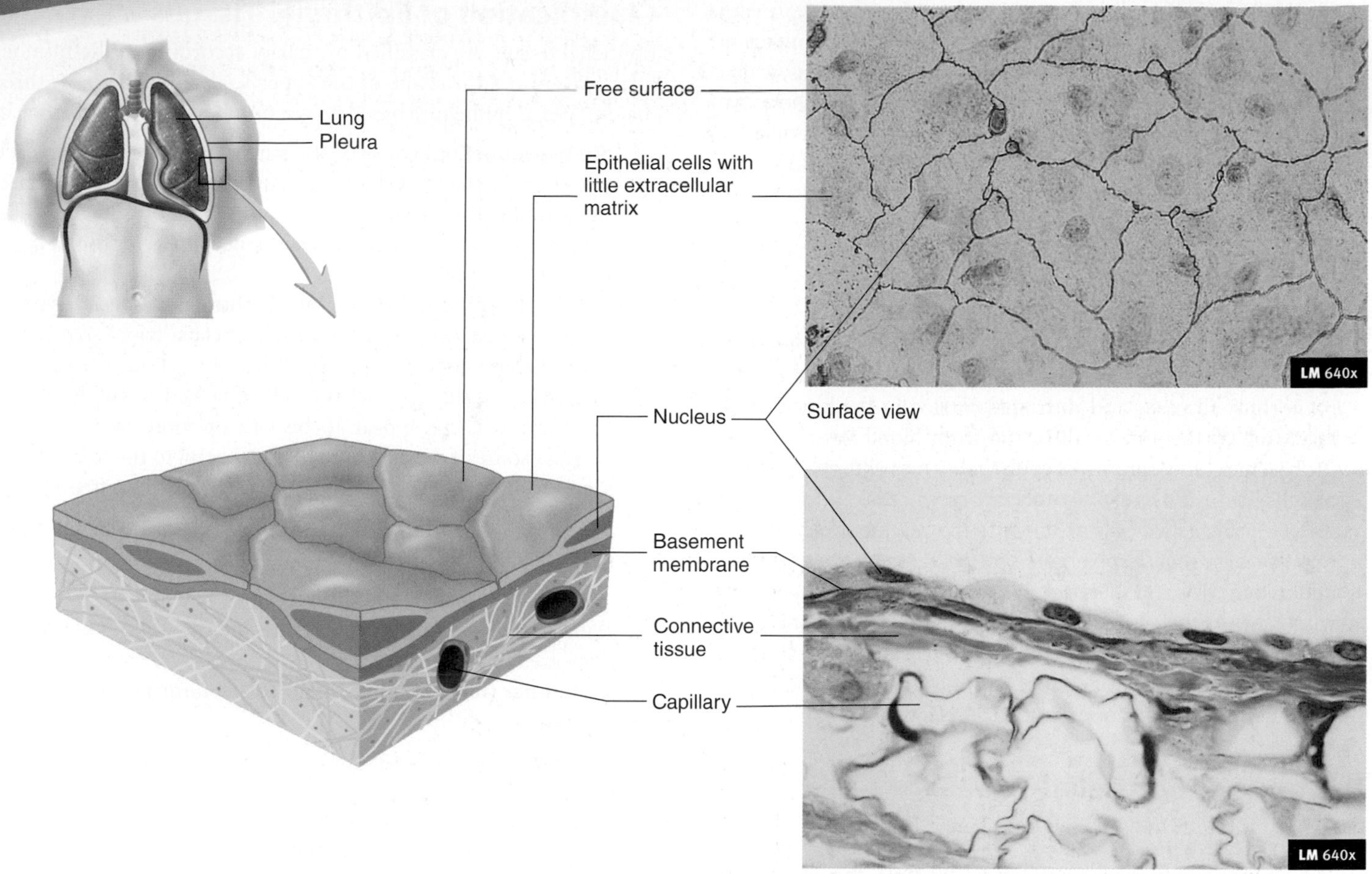

Lung
Pleura
Free surface
Epithelial cells with little extracellular matrix
Nucleus
Basement membrane
Connective tissue
Capillary
LM 640x
Surface view
LM 640x
Cross-sectional view

FIGURE 4.1 Characteristics of Epithelial Tissue
Surface and cross-sectional views of epithelial tissue illustrate the following characteristics: little extracellular material between cells, a free surface, and a basement membrane attaching epithelial cells to underlying tissues. Capillaries in connective tissue do not penetrate the basement membrane. Nutrients, oxygen, and waste products must diffuse across the basement membrane between the capillaries and the epithelial cells.

4.3 Epithelial Tissue

LEARNING OUTCOMES

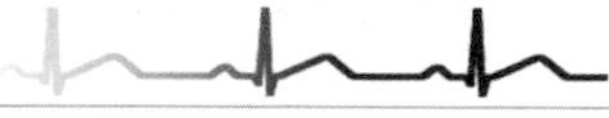

After reading this section, you should be able to

A. List and explain the general characteristics of epithelial tissue.

B. Describe the major functions of epithelial tissue.

C. Classify epithelial tissues based on the number of cell layers and the shape of the cells.

D. Name and describe the various types of epithelial tissue, including their chief functions and locations.

E. Relate the structural specializations of epithelial tissue with the functions they perform.

F. Differentiate between exocrine and endocrine glands, and unicellular and multicellular glands.

G. Categorize glands based on their structure and function.

Epithelial (ep-i-thē′lē-ăl) **tissue,** or *epithelium* (ep-i-the′lē-ŭm), covers and protects surfaces, both outside and inside the body. The characteristics common to most types of epithelial tissue are shown in figure 4.1 and listed here:

1. *Mostly composed of cells.* Epithelial tissue consists almost entirely of cells, with very little extracellular matrix between them.
2. *Covers body surfaces.* Epithelial tissue covers body surfaces and forms glands that are derived developmentally from body surfaces. The body surfaces include the exterior surface, the lining of the digestive and respiratory tracts, the heart and blood vessels, and the linings of many body cavities.
3. *Distinct cell surfaces.* Most epithelial tissues have cells with one **free,** or *apical* (ap′i-kăl), **surface** not attached to other cells; a **lateral surface** attached to other epithelial cells; and a **basal surface** attached to a basement membrane. The free surface often lines the lumen of ducts, vessels, and cavities. The **basement membrane** is a specialized type of extracellular material secreted by epithelial and connective tissue cells. Like the adhesive on Scotch™ tape, the basement membrane helps

attach the epithelial cells to the underlying tissues, and it plays an important role in supporting and guiding cell migration during tissue repair. The basement membrane is typically porous, which allows substances to move to and from the epithelial tissue above it. A few epithelial tissues, such as those in lymphatic capillaries and liver sinusoids, do not have basement membranes, and some epithelial tissues, such as those in some endocrine glands, do not have a free surface or a basal surface with a basement membrane.

4. *Cell and matrix connections.* Specialized cell contacts bind adjacent epithelial cells together and to the extracellular matrix of the basement membrane.
5. *Nonvascular.* Blood vessels in the underlying connective tissue do not penetrate the basement membrane to reach the epithelium; thus, all gases and nutrients carried in the blood must reach the epithelium by diffusing from blood vessels across the basement membrane. In epithelial tissues with many layers of cells, diffusion must also occur across cells, and the most metabolically active cells are close to the basement membrane.
6. *Capable of regeneration.* Epithelial cells retain the ability to undergo mitosis and therefore are able to replace damaged cells with new epithelial cells. Undifferentiated cells (stem cells) continuously divide and produce new cells. In some types of epithelial tissues, such as those in the skin and the digestive tract, new cells continuously replace cells that die.

Functions of Epithelial Tissues

The major functions of epithelial tissue are

1. *Protecting underlying structures.* For example, the outer layer of the skin and the epithelium of the oral cavity protect the underlying structures from abrasion.
2. *Acting as a barrier.* Epithelium prevents many substances from moving through it. For example, the skin acts as a barrier to water and reduces water loss from the body. The skin also prevents many toxic molecules and microorganisms from entering the body.
3. *Permitting the passage of substances.* Epithelium allows many substances to move through it. For example, oxygen and carbon dioxide are exchanged between the air and blood by diffusion through the epithelium in the lungs. Epithelium acts as a filter in the kidney, allowing many substances to pass from the blood into the urine but retaining other substances, such as blood cells and proteins, in the blood.
4. *Secreting substances.* Mucous glands, sweat glands, and the enzyme-secreting portions of the pancreas are all composed of epithelial cells that secrete their products onto surfaces or into ducts that carry them to other areas of the body.
5. *Absorbing substances.* The plasma membranes of certain epithelial tissues contain carrier proteins (see chapter 3), which regulate the absorption of materials.

ASSESS YOUR PROGRESS

7. *List six characteristics common to most types of epithelial tissue.*
8. *What are the distinct cell surfaces found in epithelial tissue? Describe them.*
9. *List and describe the major functions of epithelial tissue.*

Classification of Epithelial Tissues

Epithelial tissues are classified primarily according to the number of cell layers and the shape of the superficial cells. There are three major types of epithelium based on the number of cell layers in each:

1. **Simple epithelium** consists of a single layer of cells, with each cell extending from the basement membrane to the free surface.
2. **Stratified epithelium** consists of more than one layer of cells, but only the basal layer attaches the deepest layer to the basement membrane.
3. **Pseudostratified columnar epithelium** is a special type of simple epithelium. The prefix *pseudo-* means "false," so this type of epithelium appears to be stratified but is not. It consists of one layer of cells, with all the cells attached to the basement membrane. There appear to be two or more layers of cells because some of the cells are tall and extend to the free surface, whereas others are shorter and do not extend to the free surface.

There are three types of epithelium based on idealized shapes of the epithelial cells:

1. **Squamous** (skwā′mŭs) cells are flat or scalelike.
2. **Cuboidal** (cubelike) cells are cube-shaped—about as wide as they are tall.
3. **Columnar** (tall and thin, similar to a column) cells tend to be taller than they are wide.

In most cases, an epithelium is given two names, such as simple squamous, stratified squamous, simple columnar, or pseudostratified columnar. The first name indicates the number of layers, and the second indicates the shape of the cells at the free surface (table 4.1). Tables 4.2–4.4 provide an overview of the major types of epithelial tissues and their distribution.

Simple squamous epithelium consists of one layer of flat, or scalelike, cells that rest on a basement membrane (table 4.2*a*). Stratified squamous epithelium consists of several layers of cells. Near the basement membrane, the cells are more cube-shaped, but at the free surface the cells are flat or scalelike (table 4.3*a*). Pseudostratified columnar epithelial cells are columnar in shape (taller than they are wide) and, although they appear to consist of more than one layer, all the cells rest on the basement membrane (table 4.4*a*).

TABLE 4.1 Classification of Epithelium

Number of Layers or Category	Shape of Cells
Simple (single layer of cells)	Squamous Cuboidal Columnar
Stratified (more than one layer of cells)	Squamous Nonkeratinized (moist) Keratinized Cuboidal (very rare) Columnar (very rare)
Pseudostratified (modification of simple epithelium)	Columnar
Transitional (modification of stratified epithelium)	Roughly cuboidal to columnar when not stretched and squamouslike when stretched

TABLE 4.2 Simple Epithelium

(a) Simple Squamous Epithelium APR

Structure: Single layer of flat, often hexagonal cells; the nuclei appear as bumps when viewed as a cross section because the cells are so flat

Function: Diffusion, filtration, some secretion, and some protection against friction

Location: Lining of blood vessels and the heart, lymphatic vessels (endothelium) and small ducts, alveoli of the lungs, portions of the kidney tubules, lining of serous membranes (mesothelium) of the body cavities (pleural, pericardial, peritoneal), and inner surface of the tympanic membranes

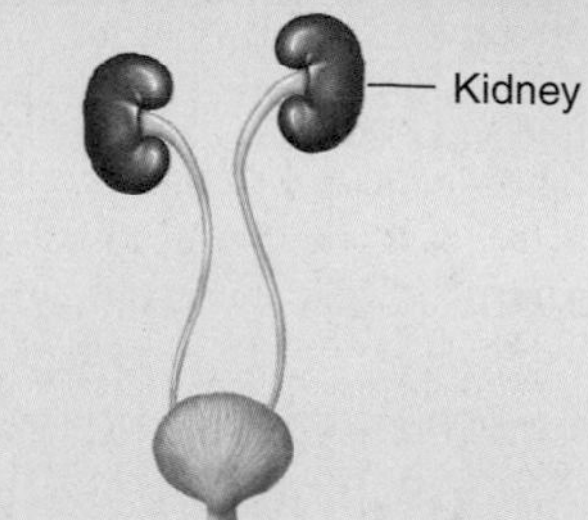
Kidney

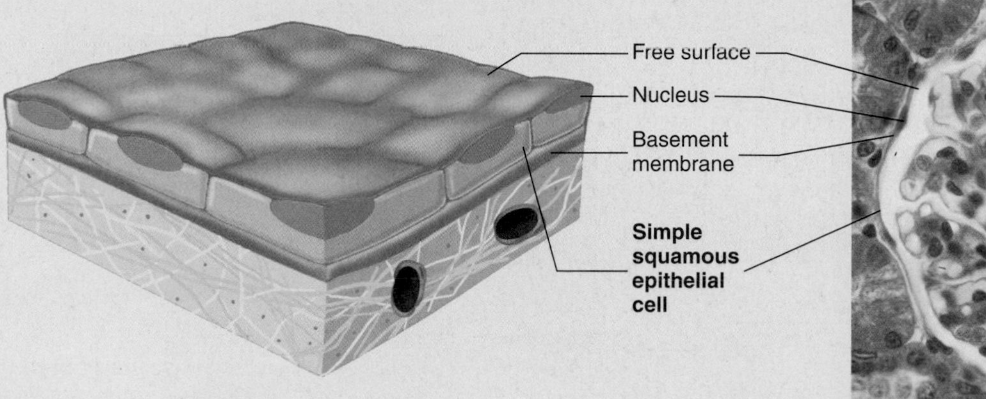
Free surface
Nucleus
Basement membrane
Simple squamous epithelial cell
LM 640x

(b) Simple Cuboidal Epithelium APR

Structure: Single layer of cube-shaped cells; some cells have microvilli (kidney tubules) or cilia (terminal bronchioles of the lungs)

Function: Secretion and absorption by cells of the kidney tubules; secretion by cells of glands and choroid plexuses; movement of particles embedded in mucus out of the terminal bronchioles by ciliated cells

Location: Kidney tubules, glands and their ducts, choroid plexuses of the brain, lining of terminal bronchioles of the lungs, surfaces of the ovaries

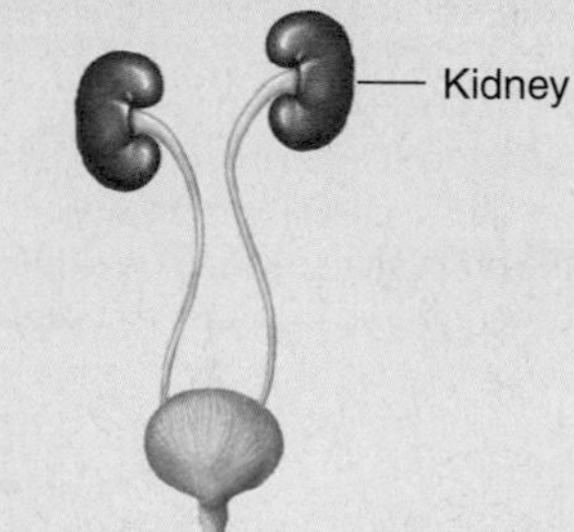
Kidney

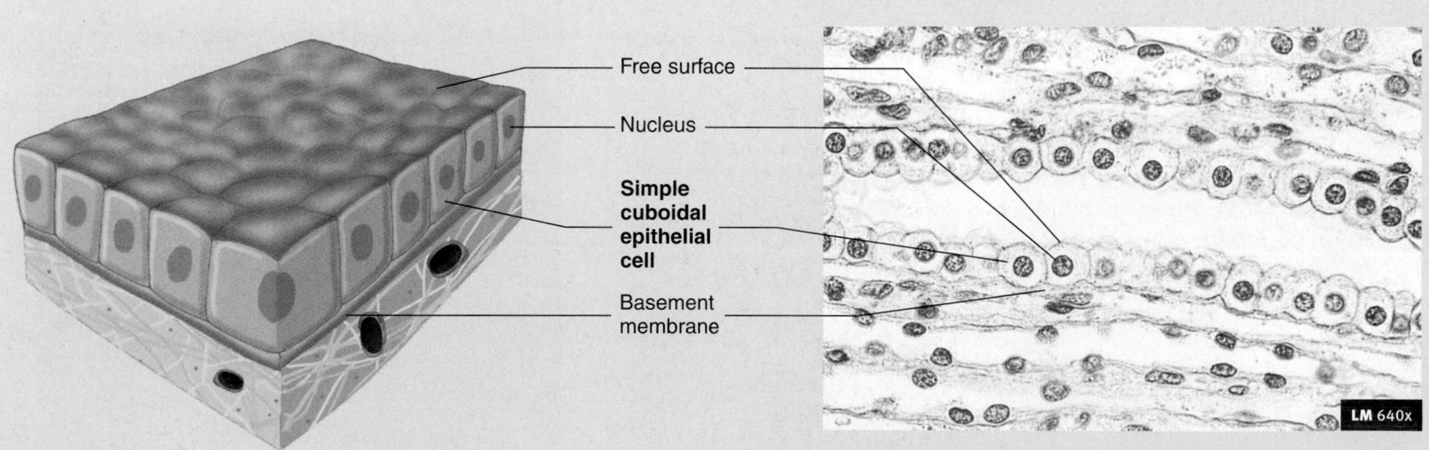
Free surface
Nucleus
Simple cuboidal epithelial cell
Basement membrane
LM 640x

TABLE 4.2 Simple Epithelium—Continued

(c) Simple Columnar Epithelium APR

Structure: Single layer of tall, narrow cells; some cells have cilia (bronchioles of lungs, auditory tubes, uterine tubes, and uterus) or microvilli (intestines)

Function: Movement of particles out of the bronchioles of the lungs by ciliated cells; partially responsible for the movement of oocytes through the uterine tubes by ciliated cells; secretion by cells of the glands, the stomach, and the intestines; absorption by cells of the small and large intestines

Location: Glands and some ducts, bronchioles of the lungs, auditory tubes, uterus, uterine tubes, stomach, intestines, gallbladder, bile ducts, ventricles of the brain

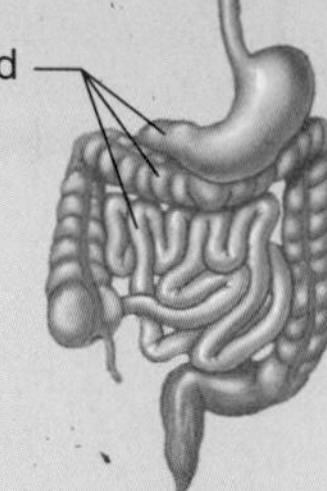
Lining of stomach and intestines

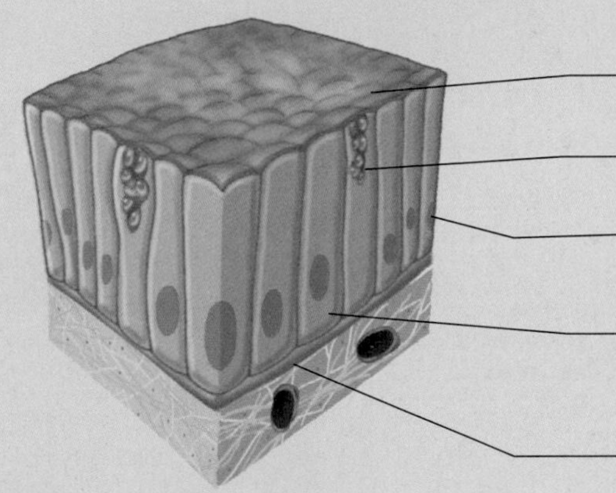
Free surface
Goblet cell containing mucus
Nucleus
Simple columnar epithelial cell
Basement membrane

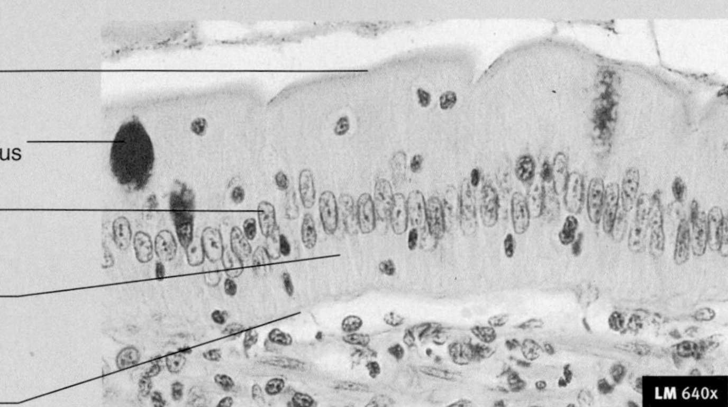
LM 640x

TABLE 4.3 Stratified Epithelium

(a) Stratified Squamous Epithelium APR

Structure: Multiple layers of cells that are cube-shaped in the basal layer and progressively flattened toward the surface; the epithelium can be nonkeratinized (moist) or keratinized; in nonkeratinized stratified squamous epithelium, the surface cells retain a nucleus and cytoplasm; in keratinized stratified epithelium, the cytoplasm of cells at the surface is replaced by a protein called keratin, and the cells are dead

Function: Protection against abrasion, a barrier against infection, reduction of water loss from the body

Location: Keratinized—primarily In the skin; nonkeratinized—mouth, throat, larynx, esophagus, anus, vagina, inferior urethra, cornea

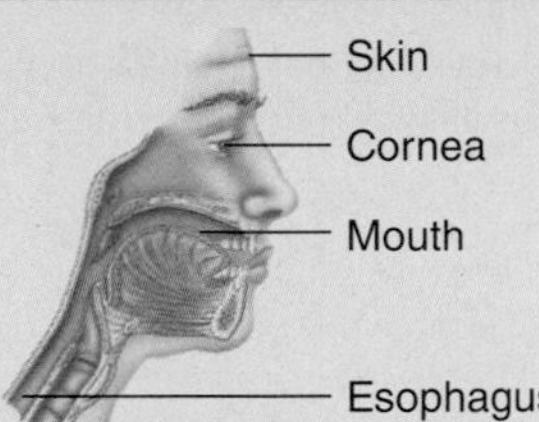
Skin
Cornea
Mouth
Esophagus

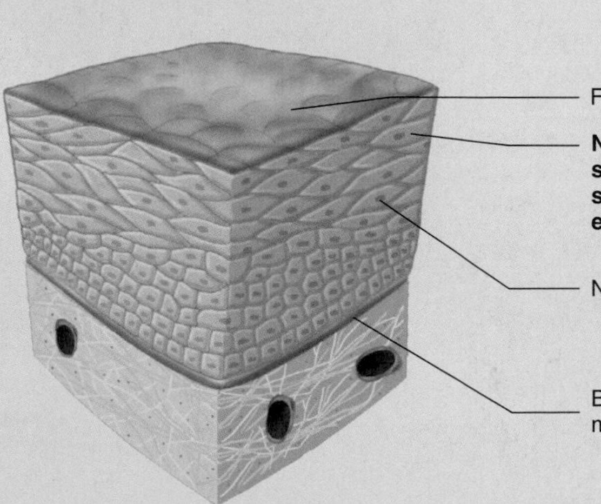
Free surface
Nonkeratinized stratified squamous epithelial cell
Nuclei
Basement membrane

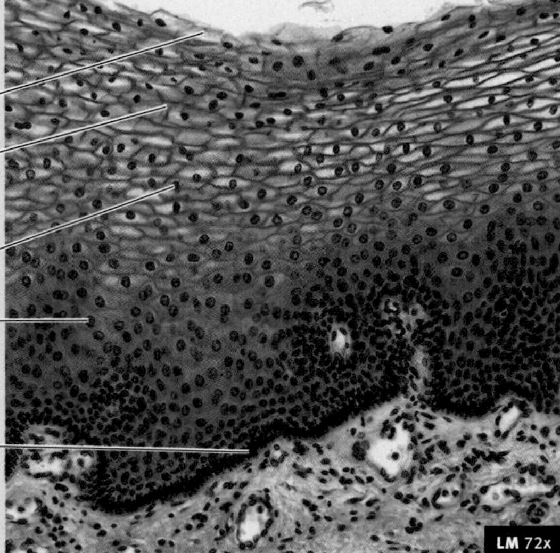
LM 72x

TABLE 4.3 Stratified Epithelium—Continued

(b) Stratified Cuboidal Epithelium

Structure: Multiple layers of somewhat cube-shaped cells

Function: Secretion, absorption, protection against infection

Location: Sweat gland ducts, ovarian follicular cells, salivary gland ducts

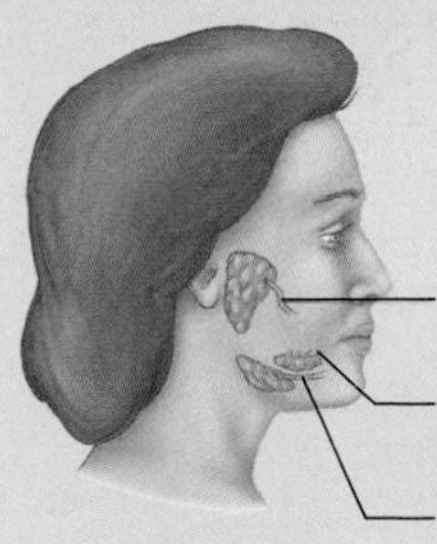
Parotid gland duct
Sublingual gland duct
Submandibular gland duct

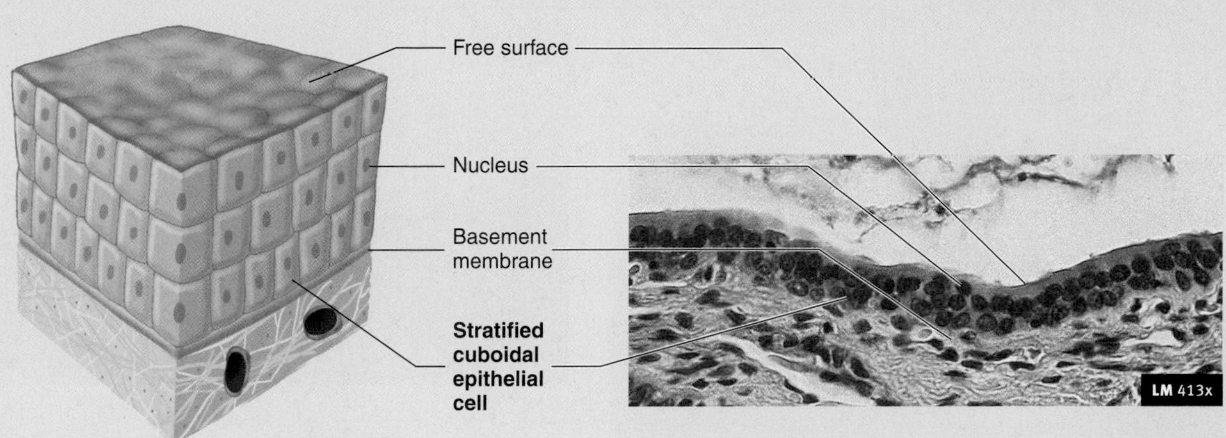
Free surface
Nucleus
Basement membrane
Stratified cuboidal epithelial cell
LM 413x

(c) Stratified Columnar Epithelium

Structure: Multiple layers of cells with tall, thin cells resting on layers of more cube-shaped cells; the cells are ciliated in the larynx

Function: Protection, secretion

Location: Mammary gland ducts, larynx, a portion of the male urethra

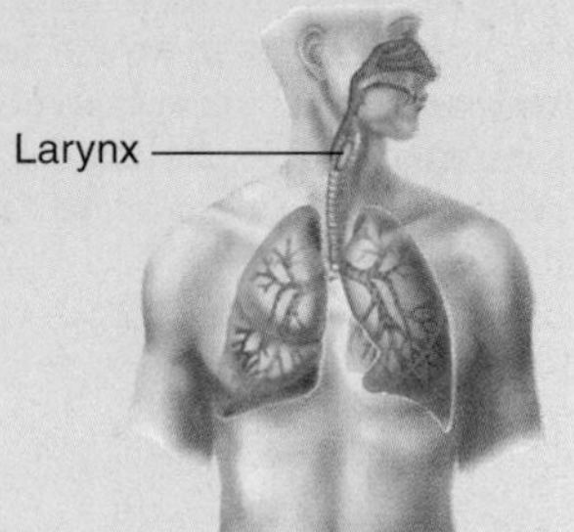
Larynx

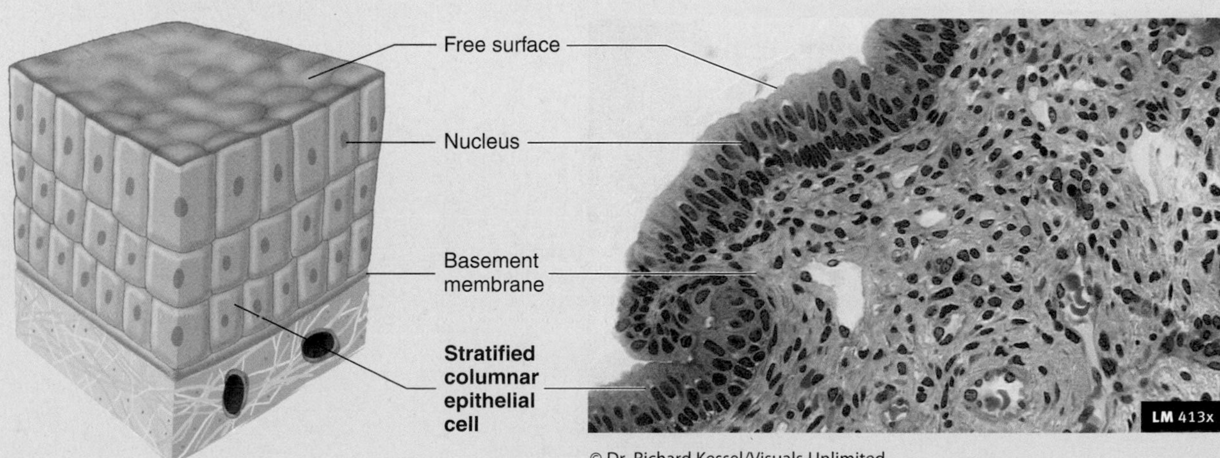
Free surface
Nucleus
Basement membrane
Stratified columnar epithelial cell
LM 413x

TABLE 4.4 Pseudostratified Columnar Epithelium and Transitional Epithelium

(a) Pseudostratified Columnar Epithelium APR

Structure: Single layer of cells; some cells are tall and thin and reach the free surface, and others do not; the nuclei of these cells are at different levels and appear stratified; the cells are almost always ciliated and are associated with goblet cells that secrete mucus onto the free surface

Function: Synthesize and secrete mucus onto the free surface; move mucus (or fluid) that contains foreign particles over the surface of the free surface and from passages

Location: Lining of the nasal cavity, nasal sinuses, auditory tubes, pharynx, trachea, bronchi of the lungs

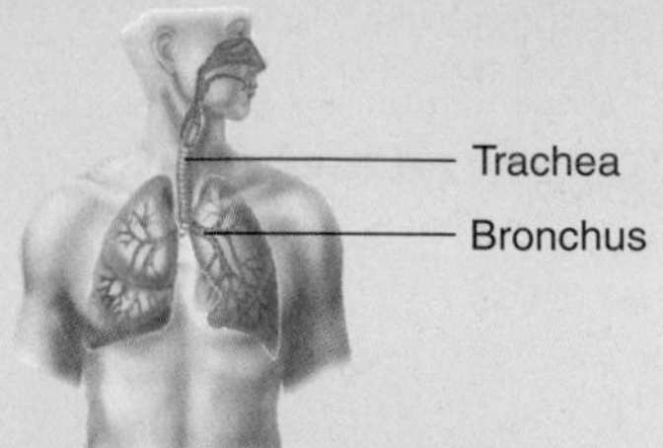
Trachea
Bronchus

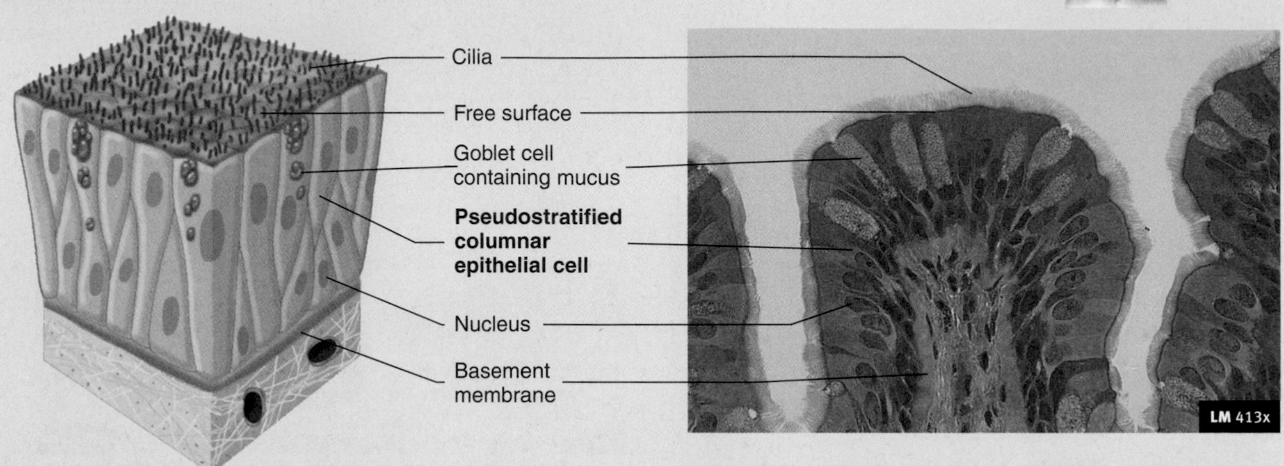
Cilia
Free surface
Goblet cell containing mucus
Pseudostratified columnar epithelial cell
Nucleus
Basement membrane
LM 413x

(b) Transitional Epithelium APR

Structure: Stratified cells that appear cube-shaped when the organ or tube is not stretched and squamous when the organ or tube is stretched by fluid; the number of layers also decreases on stretch

Function: Accommodate fluctuations in the volume of fluid in organs or tubes; protect against the caustic effects of urine

Location: Lining of the urinary bladder, ureters, superior urethra

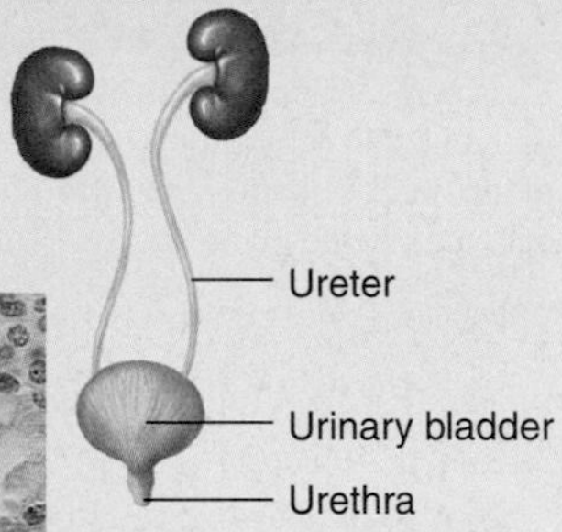

Ureter
Urinary bladder
Urethra

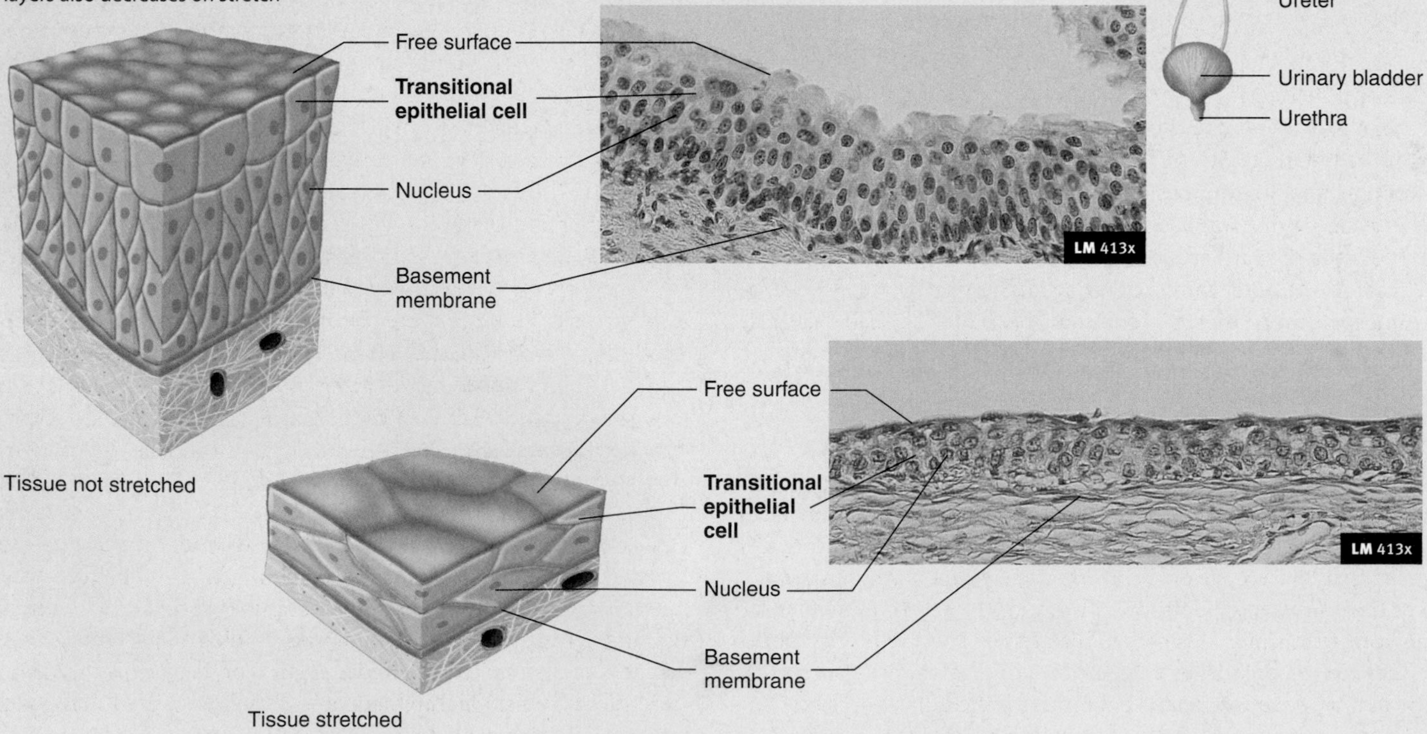

Free surface
Transitional epithelial cell
Nucleus
Basement membrane
LM 413x

Tissue not stretched

Free surface
Transitional epithelial cell
Nucleus
Basement membrane
LM 413x

Tissue stretched

Stratified squamous epithelium can be classified further as either nonkeratinized or keratinized, according to the condition of the outermost layer of cells. **Nonkeratinized (moist) stratified squamous epithelium** (table 4.3*a*), found in areas such as the mouth, esophagus, rectum, and vagina, consists of living cells in the deepest and outermost layers. A layer of fluid covers the outermost layers of cells, which makes them moist. In contrast, **keratinized** (ker′ă-ti-nīzd) **stratified squamous epithelium,** found in the skin (see chapter 5), consists of living cells in the deepest layers, and the outer layers are composed of dead cells containing the protein keratin. The dead, keratinized cells give the tissue a dry, durable, moisture-resistant character. In addition to the skin, keratinized stratified squamous epithelium is also found in the gums and hard palate of the mouth.

A unique type of stratified epithelium called **transitional epithelium** (table 4.4*b*) lines the urinary bladder, ureters, pelvis of the kidney (including the major and minor calyces; kal′-i-sēz), and superior part of the urethra (see chapter 26). These are structures where considerable expansion can occur. The shape of the cells and the number of cell layers vary, depending on the degree to which transitional epithelium is stretched. The surface cells and the underlying cells are roughly cuboidal or columnar when the epithelium is not stretched, and they become more flattened or squamouslike as the epithelium is stretched. Also, as the epithelium is stretched, the epithelial cells can shift on one another, so that the number of layers decreases from five or six to two or three.

Functional Characteristics

Epithelial tissues have many functions, including forming a barrier between a free surface and the underlying tissues and secreting, transporting, and absorbing selected molecules (table 4.5). The structure and organization of cells within each epithelial type reflect these functions.

Cell Layers and Cell Shapes

Simple epithelium, with its single layer of cells, covers surfaces. In the lungs it facilitates the diffusion of gases; in the kidneys it filters blood; in glands it secretes cellular products; and in the intestines it absorbs nutrients. Stratified epithelium is found in areas where protection is a major function because it is able to hinder the selective movement of materials through the epithelium. The multiple layers of cells in stratified epithelium are well adapted for a protective role. As the outer cells are damaged, they are replaced by cells from deeper layers; thus, a continuous barrier of epithelial cells is maintained in the tissue. Stratified squamous epithelium is found in areas of the body where abrasion can occur, such as the skin, mouth, throat, esophagus, anus, and vagina.

Differing functions are also reflected in cell shape. Cells that filter substances and allow diffusion are normally flat and thin. For example, simple squamous epithelium forms blood and lymphatic capillaries, alveoli (air sacs) of the lungs, and parts of the kidney tubules. Cells that secrete or absorb are usually cuboidal or columnar. They have greater cytoplasmic volume relative to surface area than seen with squamous cells. This greater volume results from the presence of organelles responsible for the tissues' functions. For example, pseudostratified columnar epithelium, which secretes large amounts of mucus, lines the respiratory tract (see chapter 23) and contains large **goblet cells,** which are specialized columnar epithelial cells. The goblet cells contain abundant organelles, such as ribosomes, endoplasmic reticulum, Golgi apparatuses, and secretory vesicles, that are responsible for synthesizing and secreting mucus.

ASSESS YOUR PROGRESS

10. *Describe simple, stratified, and pseudostratified epithelial tissues. Distinguish among squamous, cuboidal, and columnar epithelial cells.*
11. *How do nonkeratinized stratified squamous epithelium and keratinized stratified squamous epithelium differ? Where is each type found?*
12. *Describe the changes in cell shape and number of cell layers in transitional epithelium as it is stretched. Where is transitional epithelium found?*
13. *List the types of epithelial tissue, giving the structure, functions, and major locations of each.*
14. *What functions would a single layer of epithelial cells be expected to perform? A stratified layer?*
15. *Why are cuboidal or columnar cells found where secretion or absorption is occurring?*

Predict 2

Explain the consequences of having (a) nonkeratinized stratified epithelium rather than simple columnar epithelium lining the digestive tract, (b) nonkeratinized stratified squamous epithelium rather than keratinized stratified squamous epithelium in the skin, and (c) simple columnar epithelium rather than stratified squamous epithelium lining the mouth.

Free Surfaces

The free surfaces of epithelial tissues can be smooth or folded; they may have microvilli or cilia. Smooth surfaces reduce friction. For example, the lining of blood vessels is a simple squamous epithelium that reduces friction as blood flows through the vessels (see chapter 21).

Microvilli and cilia were described in chapter 3. Microvilli are nonmotile and contain microfilaments. They greatly increase free surface area and occur in cells that absorb or secrete, such as serous membranes and the lining of the small intestine (see chapter 24). Stereocilia are elongated microvilli found in sensory structures, such as the inner ear (see chapter 15), and they play a role in sound detection. They are also found in some places where absorption is important, such as in the epithelium of the epididymis. Motile cilia, which contain microtubules, move materials across the free surface of the cell (see chapter 3). Three types of ciliated epithelium line the respiratory tract (see chapter 23), where cilia move mucus that contains foreign particles out of the respiratory passages. Cilia are also found on the apical surface of the simple columnar epithelial cells of the uterus and uterine tubes, where the cilia help move mucus and oocytes.

Transitional epithelium has a rather unusual plasma membrane specialization: rigid sections of membrane separated by very flexible regions in which the plasma membrane is folded. When transitional epithelium is stretched, the folded regions of the plasma membrane can unfold. Transitional epithelium is specialized to expand in tissues such as the urinary bladder.

TABLE 4.5 Function and Location of Epithelial Tissue

	LOCATION				
Function	**Simple Squamous Epithelium**	**Simple Cuboidal Epithelium**	**Simple Columnar Epithelium**	**Stratified Squamous Epithelium**	**Stratified Cuboidal Epithelium**
Diffusion	Blood and lymphatic capillaries, alveoli of lungs, thin segments of loops of Henle				
Filtration	Bowman capsules of kidneys				
Secretion or absorption	Mesothelium (serous fluid)	Choroid plexus (produces cerebrospinal fluid), part of kidney tubules, many glands and their ducts	Stomach, small intestine, large intestine, uterus, many glands		
Protection (against friction and abrasion)	Endothelium (e.g., epithelium of blood vessels) Mesothelium (e.g., epithelium of body cavities)			Skin (epidermis), corneas, mouth and throat, epiglottis, larynx, esophagus, anus, vagina	
Movement of mucus (ciliated)		Terminal bronchioles of lungs	Bronchioles of lungs, auditory tubes, uterine tubes, uterus		
Capable of great stretching					
Miscellaneous	Inner part of tympanic membranes, smallest ducts of glands	Surface of ovaries, inside lining of eyes (pigmented epithelium of retina), ducts of glands	Bile duct, gallbladder, ependyma (lining of brain ventricles and central canal of spinal cord), ducts of glands	Lower part of urethra, sebaceous gland ducts	Sweat gland ducts

Cell Connections

Cells have structures that hold them to one another or to the basement membrane. These structures do three things: (1) mechanically bind the cells together, (2) help form a permeability barrier, and (3) provide a mechanism for intercellular communication.

Epithelial cells secrete glycoproteins that attach the cells to the basement membrane and to one another. This relatively weak binding between cells is reinforced by **desmosomes** (dez′mō-sōmz), disk-shaped structures with especially adhesive glycoproteins that bind cells to one another and intermediate filaments that extend into the cytoplasm of the cells (figure 4.2). Many desmosomes are found in epithelial tissues that are subjected to stress, such as the stratified squamous epithelium of the skin. **Hemidesmosomes,** similar to one-half of a desmosome, attach epithelial cells to the basement membrane.

Tight junctions hold cells together and form a permeability barrier (figure 4.2). Tight junctions are formed by plasma membranes of adjacent cells that join one another in a jigsaw fashion to make a tight seal. Near the free surface of simple epithelial cells, the tight junctions form a ring that completely surrounds each cell and binds adjacent cells together to prevent the passage of materials between cells. For example, in the stomach and the urinary bladder, chemicals cannot pass between cells. Thus, water and other substances must pass through the epithelial cells, which can actively regulate what is absorbed or secreted. Tight junctions are found in areas where a layer of simple epithelium forms a permeability barrier. For example, water can diffuse through epithelial cells, and active transport, symport, and facilitated diffusion move most nutrients through the epithelial cells of the small intestine.

An **adhesion belt** of glycoproteins is found just below the tight junction. It is located between the plasma membranes of adjacent cells and acts as a weak glue that holds cells together. These connections are not as strong as those of desmosomes.

A **gap junction** is a small, specialized contact region between cells containing protein channels that aid intercellular communication by allowing ions and small molecules to pass from one cell to another (figure 4.2). In epithelium, the function of gap junctions

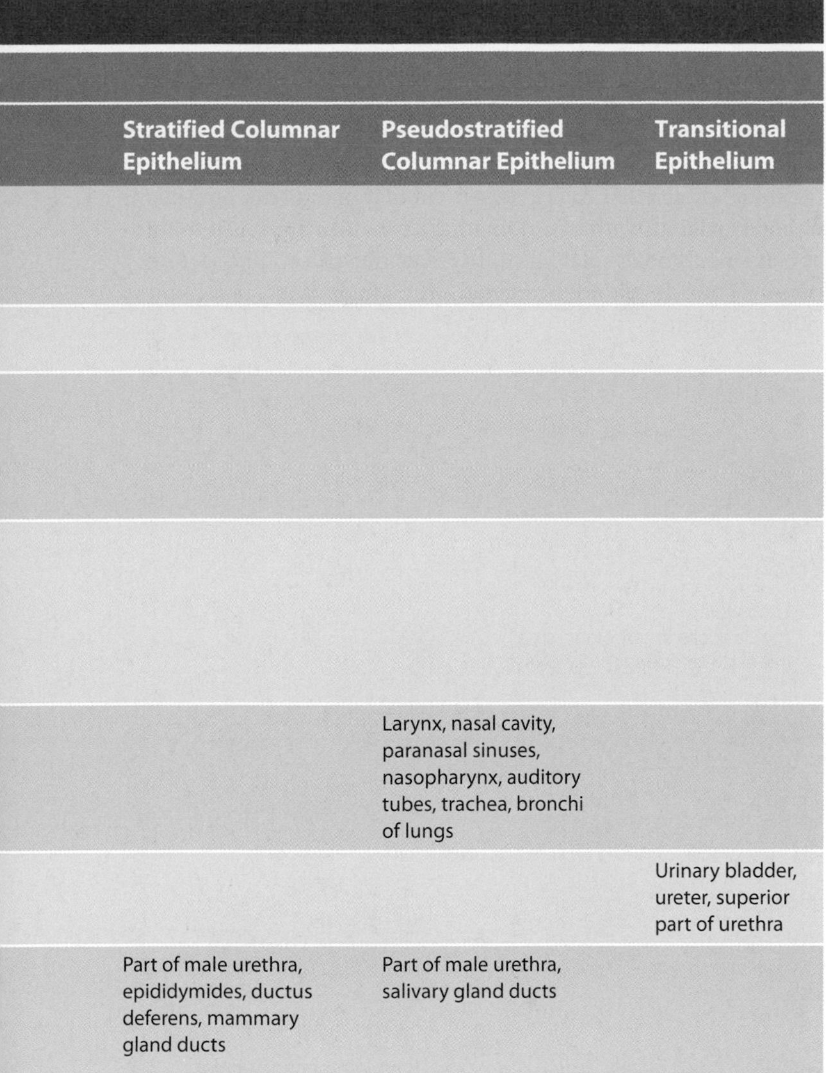

Stratified Columnar Epithelium	Pseudostratified Columnar Epithelium	Transitional Epithelium
	Larynx, nasal cavity, paranasal sinuses, nasopharynx, auditory tubes, trachea, bronchi of lungs	
		Urinary bladder, ureter, superior part of urethra
Part of male urethra, epididymides, ductus deferens, mammary gland ducts	Part of male urethra, salivary gland ducts	

FUNDAMENTAL **Figure**

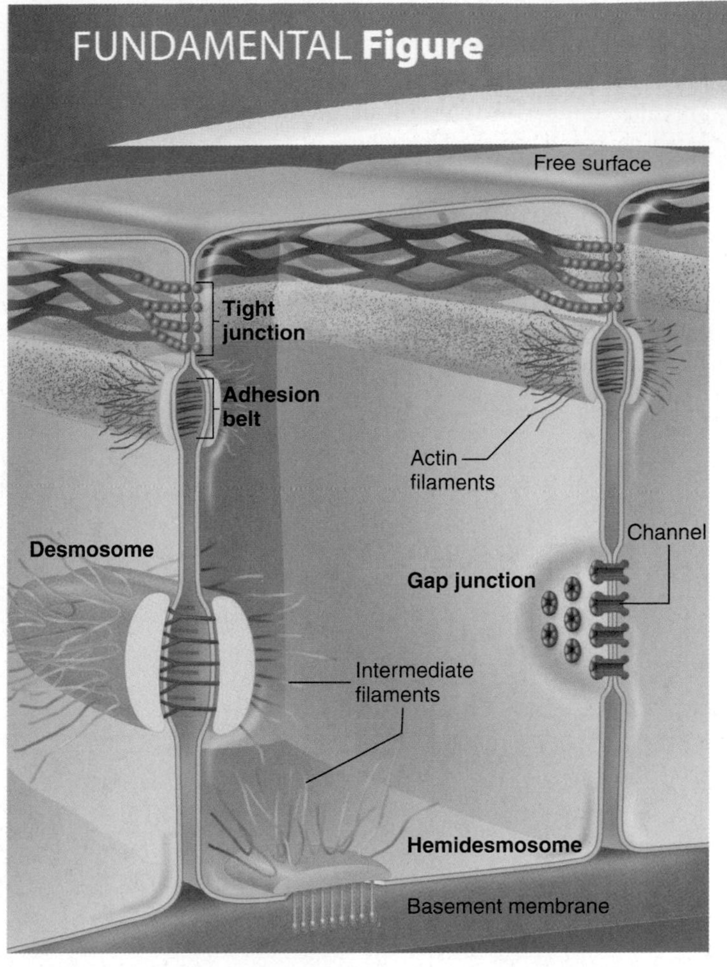
Free surface
Tight junction
Adhesion belt
Actin filaments
Channel
Desmosome
Gap junction
Intermediate filaments
Hemidesmosome
Basement membrane

FIGURE 4.2 Cell Connections
Desmosomes, adhesion belts, and tight junctions anchor cells to one another, and hemidesmosomes anchor cells to the basement membrane. Gap junctions allow adjacent cells to communicate with each other. Few cells have all of these different connections.

is not entirely clear; gap junctions between ciliated epithelial cells may coordinate the movements of cilia. In cardiac and smooth muscle tissues, gap junctions are important in coordinating important functions. Because ions can pass through the gap junctions from one cell to the next, electrical signals can pass from cell to cell to coordinate the contraction of cardiac and smooth muscle cells. Thus, electrical signals that originate in one cell of the heart can spread from cell to cell and cause the entire heart to contract. The gap junctions between cardiac muscle cells are found in specialized cell-to-cell connections called **intercalated disks** (see chapter 20). In addition to containing gap junctions, the intercalated disks have desmosomes that help hold adjacent cells in close contact.

ASSESS **YOUR PROGRESS**

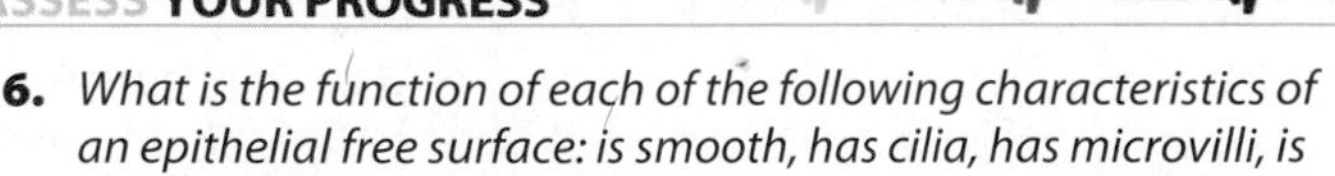

16. *What is the function of each of the following characteristics of an epithelial free surface: is smooth, has cilia, has microvilli, is folded? Give an example of where each surface type is found in the body.*

17. *Name the possible ways by which epithelial cells are bound to one another and to the basement membrane.*

18. *What role do desmosomes play in the skin?*

19. *What is the general function of gap junctions?*

Predict 3

If a simple epithelium has well-developed tight junctions, explain how NaCl can move from one side of the epithelial layer to the other, what type of epithelium it is likely to be, and how the movement of NaCl causes water to move in the same direction.

Glands

Glands are secretory organs. Many glands are composed primarily of epithelium, with a supporting network of connective tissue. These glands develop from an infolding or outfolding of epithelium in the embryo. If the gland maintains an open contact with the epithelium from which it developed, a duct is present. Glands with ducts are called **exocrine** (ek′sō-krin) **glands,** and their ducts are lined with epithelium. Alternatively, some glands become separated from the epithelium of their origin and have no ducts; these are called **endocrine** (en′dō-krin) **glands.** Endocrine glands have extensive blood vessels. The cellular products of endocrine glands, which are called **hormones** (hōr′mōnz), are secreted into the bloodstream and carried throughout the body.

Most exocrine glands are composed of many cells and are called **multicellular glands,** but some exocrine glands are composed of a single cell and are called **unicellular glands** (figure 4.3*a*). **Goblet cells** (see table 4.2*c*) are unicellular glands that secrete mucus.

Multicellular glands can be classified according to the structure of their ducts and secretory regions (figure 4.3). Glands that have a single duct are called **simple,** and glands with ducts that branch are called **compound.** Glands with secretory regions shaped as tubules (small tubes) are called **tubular,** whereas those shaped in saclike structures are called **acinar** (as′i-nar) or **alveolar** (al-vē′ō-lar). Tubular glands can be straight or coiled. Glands with a combination of the two are called tubuloacinar or tubuloalveolar. If multiple acinar or tubular secretory regions (not ducts) are branched off a single duct, the gland is called branched.

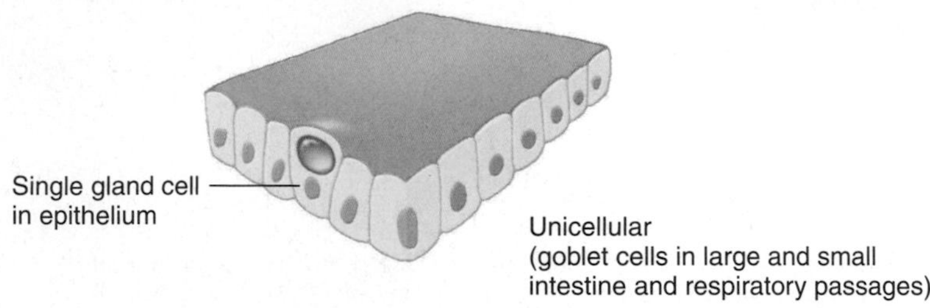
Single gland cell in epithelium
Unicellular
(goblet cells in large and small intestine and respiratory passages)

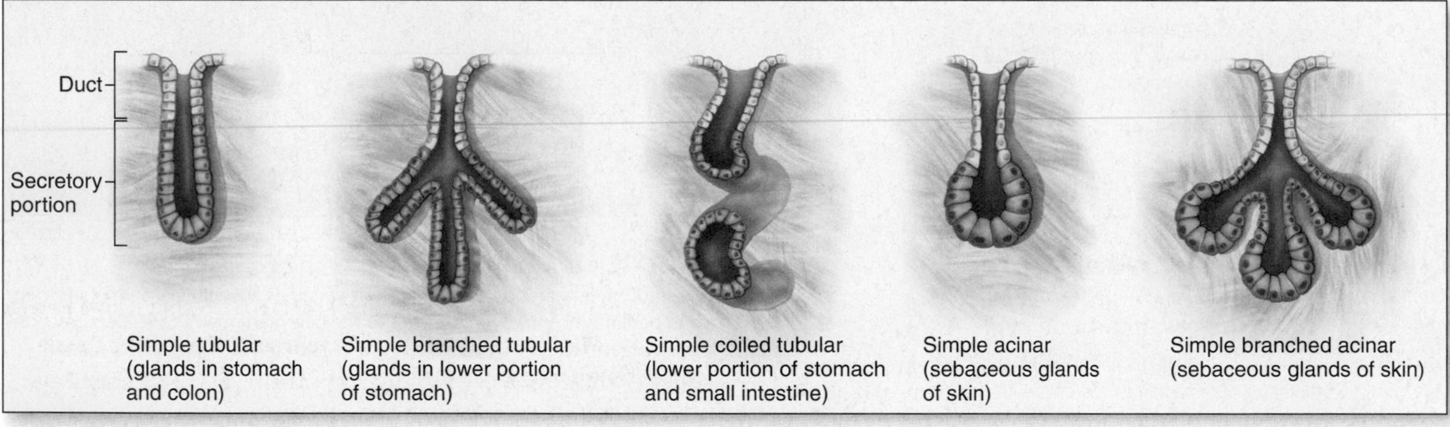
Duct
Secretory portion
Simple tubular (glands in stomach and colon)
Simple branched tubular (glands in lower portion of stomach)
Simple coiled tubular (lower portion of stomach and small intestine)
Simple acinar (sebaceous glands of skin)
Simple branched acinar (sebaceous glands of skin)

(a) Simple glands

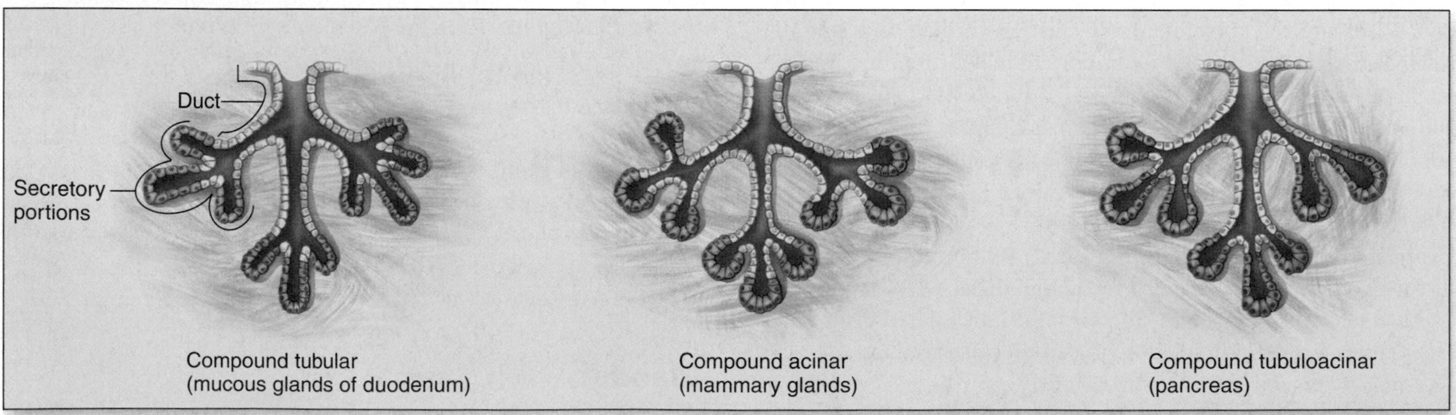
Duct
Secretory portions
Compound tubular (mucous glands of duodenum)
Compound acinar (mammary glands)
Compound tubuloacinar (pancreas)

(b) Compound glands

FIGURE 4.3 Structure of Exocrine Glands

The names of exocrine glands are based on the shapes of their secretory units and their ducts.

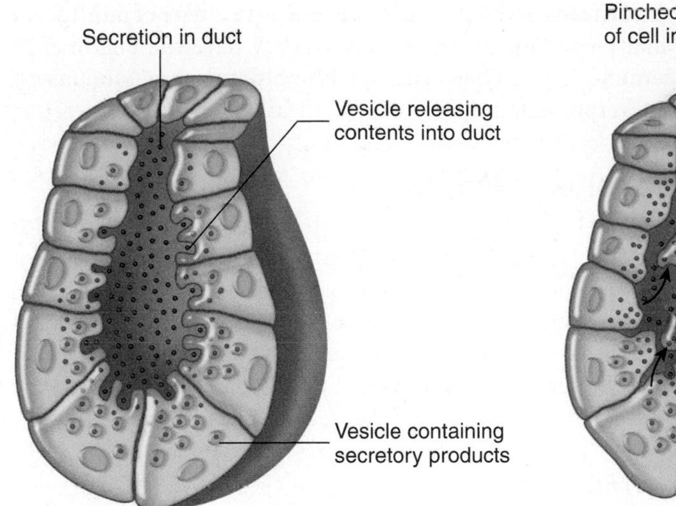
Secretion in duct
Vesicle releasing contents into duct
Vesicle containing secretory products

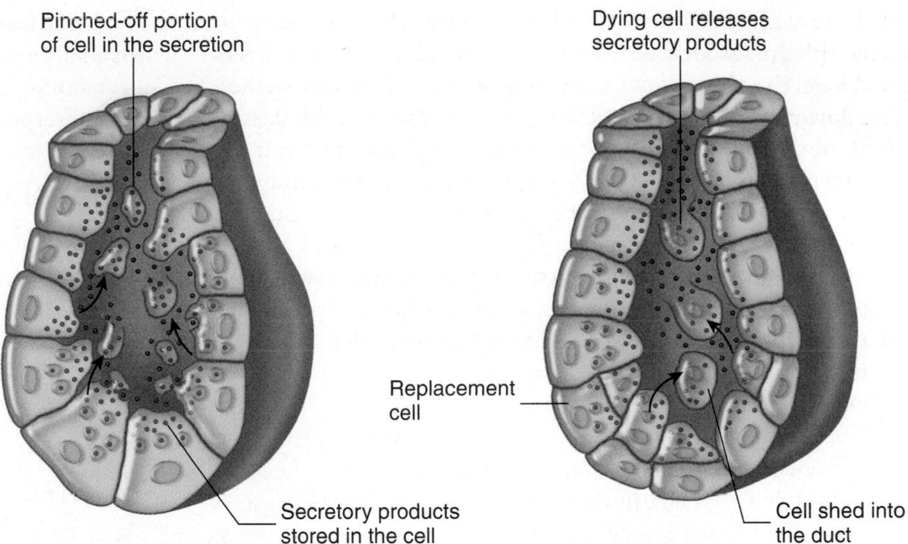
Pinched-off portion of cell in the secretion
Secretory products stored in the cell
Dying cell releases secretory products
Replacement cell
Cell shed into the duct

(a) **Merocrine gland**
Cells of the gland produce secretions by active transport or produce vesicles that contain secretory products, and the vesicles empty their contents into the duct through exocytosis.

(b) **Apocrine gland**
Secretory products are stored in the cell near the lumen of the duct. A portion of the cell near the lumen containing secretory products is pinched off the cell and joins secretions produced by a merocine process.

(c) **Holocrine gland**
Secretory products are stored in the cells of the gland. Entire cells are shed by the gland and become part of the secretion. The lost cells are replaced by other cells deeper in the gland.

FIGURE 4.4 Exocrine Glands and Secretion Types
Exocrine glands are classified by type of secretion.

Exocrine glands can also be classified according to how products leave the cell. The most common type of secretion is **merocrine** (mer′ō-krin) **secretion.** In merocrine secretion, products are released, but no actual cellular material is lost (figure 4.4*a*). Secretions are either actively transported or packaged in vesicles and then released by the process of exocytosis at the free surface of the cell. Merocrine secretion is used by water-producing sweat glands and the exocrine portion of the pancreas. In **apocrine** (ap′ō-krin) **secretion,** the secretory products are released as fragments of the gland cells (figure 4.4*b*). Secretory products are retained within the cell, and portions of the cell are pinched off to become part of the secretion. The milk-producing mammary glands release milk by a combination of apocrine and mostly merocrine secretion. **Holocrine** (hol′ō-krin) **secretion** involves the shedding of entire cells (figure 4.4*c*). Products accumulate in the cytoplasm of each epithelial cell, the cell ruptures and dies, and the entire cell becomes part of the secretion. Holocrine secretion is used by the sebaceous (oil) glands of the skin.

ASSESS YOUR PROGRESS

20. *Distinguish between exocrine and endocrine glands. How are multicellular exocrine glands classified on the basis of their duct system? Their secretory portion shape?*

21. *Give an example of a unicellular exocrine gland. What does it secrete?*

22. *Describe three ways in which exocrine glands release secretions. Give an example of each method.*

4.4 Connective Tissue

LEARNING OUTCOMES

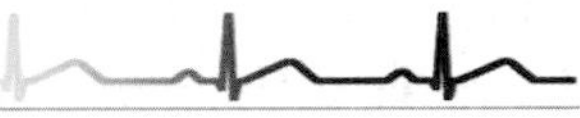

After reading this section, you should be able to

A. List and describe the major functions of connective tissue.

B. Identify the specialized cells found in connective tissue.

C. Describe the three main components of the extracellular matrix of connective tissue.

D. Discuss the types and functions of embryonic connective tissue.

E. Explain how adult connective tissue is classified.

F. Give an example of each type of connective tissue, describe its characteristic functions, and state its location in the body.

Connective tissue is abundant—it makes up part of every organ in the body. Connective tissue differs from the other three tissue types in that it consists of cells separated from each other by abundant extracellular matrix. Connective tissue is diverse in both structure and function.

Functions of Connective Tissue

Connective tissue performs the following major functions:

1. *Enclosing and separating other tissues.* Sheets of connective tissue form capsules around organs, such as the liver and kidneys. Connective tissue also forms layers that separate tissues and organs. For example, connective tissues separate muscles, arteries, veins, and nerves from one another.

2. *Connecting tissues to one another.* Strong cables, or bands, of connective tissue called tendons attach muscles to bone, whereas connective tissue bands called ligaments hold bones together.
3. *Supporting and moving parts of the body.* Bones of the skeletal system provide rigid support for the body, and the semirigid cartilage supports structures such as the nose, ears, and joint surfaces. Joints between bones allow one part of the body to move relative to other parts.
4. *Storing compounds.* Adipose tissue (fat) stores high-energy molecules, and bones store minerals, such as calcium and phosphate.
5. *Cushioning and insulating.* Adipose tissue cushions and protects the tissue it surrounds and provides an insulating layer beneath the skin that helps conserve heat.
6. *Transporting.* Blood transports the gases, nutrients, enzymes, hormones, and cells of the immune system throughout the body.
7. *Protecting.* Cells of the immune system and blood protect against toxins and tissue injury, as well as against microorganisms. Bones protect underlying structures from injury.

Cells of Connective Tissue

The specialized cells of the various connective tissues produce the extracellular matrix. The name of the cell identifies the cell functions by means of one of the following suffixes: *-blast, -cyte,* or *-clast.* **Blasts** create the matrix, **cytes** maintain it, and **clasts** break it down for remodeling. For example, **fibroblasts** are cells that form fibrous connective tissue, and **fibrocytes** maintain it. **Chondroblasts** form cartilage (*chondro-*, cartilage), and **chondrocytes** maintain it. **Osteoblasts** form bone (*osteo-*, bone), **osteocytes** maintain it, and **osteoclasts** break it down (see chapter 6).

Adipose (ad′i-pōs; fat) **cells,** or *fat cells,* also called **adipocytes** (ad′i-pō-sītz), contain large amounts of lipid. The lipid pushes the rest of the cell contents to the periphery, so that each cell appears to contain a large, centrally located lipid droplet with a thin layer of cytoplasm around it. Adipose cells are rare in some connective tissue types, such as cartilage; abundant in others, such as loose connective tissue.

Mast cells commonly lie beneath membranes in loose connective tissue and along small blood vessels of organs. They contain chemicals, such as heparin, histamine, and proteolytic enzymes, that are released in response to injury, such as trauma and infection, and play important roles in inflammation.

White blood cells, or *leukocytes* (see chapter 19), continuously move from blood vessels into connective tissues. The rate of movement increases dramatically in response to injury or infection. In addition, accumulations of lymphocytes, a type of white blood cell, are common in some connective tissues, such as that beneath the epithelial lining of certain parts of the digestive system.

Macrophages are found in some connective tissue types. They are derived from monocytes, a type of white blood cell. Macrophages are either **fixed,** meaning that they do not move through the connective tissue in which they are found, or **wandering,** moving in ameboid fashion through the connective tissue. Macrophages phagocytize foreign and injured cells, and they play a major role in protecting against infections. **Platelets** are fragments of hemopoetic cells containing enzymes and special proteins that function in the clotting process to reduce bleeding from a wound.

Undifferentiated mesenchymal cells are a type of **adult stem cell** that persist in connective tissue. They have the potential to form multiple cell types, such as fibroblasts or smooth muscle cells, in response to injury.

Extracellular Matrix

The extracellular matrix of connective tissue has three major components: (1) protein fibers, (2) ground substance consisting of nonfibrous protein and other molecules, and (3) fluid. The structure of the matrix gives connective tissue types most of their functional characteristics—for example, they enable bones and cartilage to bear weight, tendons and ligaments to withstand tension, and the skin's dermis to withstand punctures, abrasions, and other abuse.

Protein Fibers of the Matrix

Three types of protein fibers—collagen, reticular, and elastic—help form connective tissue.

Collagen (kol′a-jen) **fibers** consist of collagen, which is the most abundant protein in the body. Collagen accounts for one-fourth to one-third of total body protein, or 6% of total body weight. Collagen is synthesized within fibroblasts and secreted into the extracellular space, After collagen molecules are secreted, they are linked together to make long collagen fibrils. The collagen fibrils are then joined together in bundles to form collagen fibers (figure 4.5*a*). Collagen fibers are very strong and flexible, like microscopic ropes, but quite inelastic. There are at least 20 types of collagen fibers, many of which are specific to certain tissues.Type I collagen is the most abundant in the body. The flexible, ropelike strength of type I collagen fibers makes them well suited for tendons, ligaments, skin, and bone. These body structures need to resist being pulled yet have some flexibility. Cartilage is mainly type II collagen, and reticular fibers are mainly type III collagen.

Reticular (re-tik′ū-lār; netlike) **fibers** are very fine collagen fibers and therefore not a chemically distinct category of fibers. They are very short, thin fibers that branch to form a network and appear different microscopically from other collagen fibers. Reticular fibers are not as strong as most collagen fibers, but networks of reticular fibers fill spaces between tissues and organs.

Elastic fibers consist of a protein called **elastin** (e-las′tin). As the name suggests, this protein has the ability to return to its original shape after being stretched or compressed, giving tissue an elastic quality. Fibroblasts secrete polypeptide chains, which are linked together to form a network. The elastin network stretches like a rubber band in response to force and recoils when relaxed (figure 4.5*b*).

Ground Substance of the Matrix

Two types of large, nonfibrous molecules, called hyaluronic acid and proteoglycans, are part of the extracellular matrix. These molecules constitute most of the **ground substance** of the matrix, the "shapeless" background against which the collagen fibers are seen through the microscope. However, the molecules themselves are not shapeless but highly structured. **Hyaluronic** (hī′ă-loo-ron′ik; glassy appearance) **acid** is a long, unbranched polysaccharide chain composed of repeating disaccharide units. It gives a very slippery quality to the fluids that contain it; for that reason, it is a good

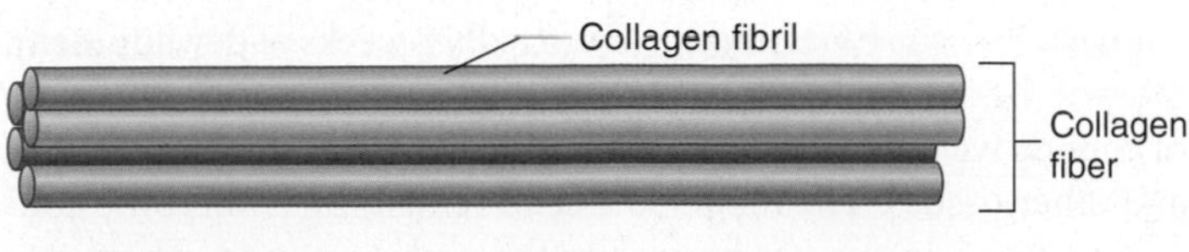
Collagen fibril
Collagen fiber

(a) Collagen fibers

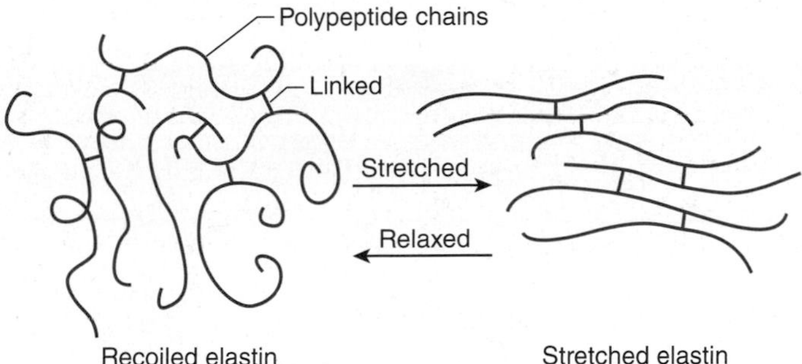
Polypeptide chains
Linked
Stretched
Relaxed
Recoiled elastin
Stretched elastin

(b) Elastic fibers

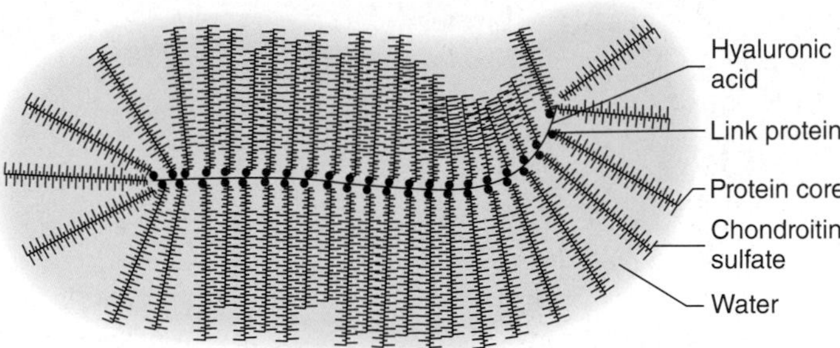
Hyaluronic acid
Link protein
Protein core
Chondroitin sulfate
Water

(c) Proteoglycan aggregates

FIGURE 4.5 Molecules of the Connective Tissue Matrix

lubricant for joint cavities (see chapter 8). Hyaluronic acid is also present in large quantities in connective tissue and is the major component of the vitreous humor of the eye (see chapter 15).

A **proteoglycan** (prō′tē-ō-glī′kan; formed from proteins and polysaccharides) **monomer** is a large molecule that consists of 80 to 100 polysaccharides, called **glycosaminoglycans** (glī-kōs-am-i-nō-glī′kanz), such as **chondroitin** (kon-drō′i-tin) **sulfate,** each attached by one end to a protein core. The protein cores of many proteoglycan monomers can attach through link proteins to a long molecule of hyaluronic acid to form a **proteoglycan aggregate** (figure 4.5*c*). Proteoglycan aggregates trap large quantities of water, which allows them to return to their original shape when compressed or deformed. There are several types of glycosaminoglycans, and their abundance varies with each connective tissue type.

Several **adhesive molecules** are found in ground substance. These adhesive molecules hold the proteoglycan aggregates together and to structures such as plasma membranes. Specific adhesive molecules predominate in certain types of ground substance. For example, **chondronectin** is in the ground substance of cartilage, **osteonectin** is in the ground substance of bone, and **fibronectin** is in the ground substance of fibrous connective tissue.

ASSESS YOUR PROGRESS

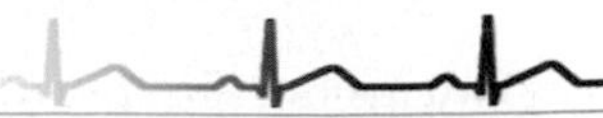

23. *What is the main characteristic that distinguishes connective tissue from other tissues?*

24. *List the major functions of connective tissue, and give an example of a type of connective tissue that performs each function.*

25. *Explain the differences among -blast, -cyte, and -clast cells of connective tissue.*

26. *What are the three components of the extracellular matrix of connective tissue?*

27. *Contrast the structure and characteristics of collagen fibers, reticular fibers, and elastic fibers.*

28. *Describe the structure and functions of hyaluronic acid and proteoglycan aggregates.*

29. *What is the function of adhesive molecules? Give some specific examples.*

Connective Tissue Classifications

Connective tissue types blend into one another, and the transition points cannot be identified precisely. As a result, connective tissue is somewhat arbitrarily classified by the type and proportions of cells and extracellular matrix components. Table 4.6 presents the classification of connective tissue used in this text.

TABLE 4.6 Classification of Connective Tissue
Embryonic Connective Tissue
Mesenchyme
Mucous connective tissue
Adult Connective Tissue
Connective Tissue Proper
Loose (fewer fibers, more ground substance)
Areolar
Adipose
Reticular
Dense (more fibers, less ground substance)
Dense, regular collagenous
Dense, regular elastic
Dense, irregular collagenous
Dense, irregular elastic
Supporting Connective Tissue
Cartilage (semisolid matrix)
Hyaline
Fibrocartilage
Elastic
Bone (solid matrix)
Spongy
Compact
Fluid Connective Tissue
Blood
Red blood cells
White blood cells
Platelets
Hemopoietic tissue
Red marrow
Yellow marrow

Two major categories of connective tissue are embryonic and adult. Embryonic connective tissue is called **mesenchyme** (mez′en-kīm). It is composed of irregularly shaped fibroblasts surrounded by abundant, semifluid extracellular matrix in which delicate collagen fibers are distributed (table 4.7*a*). It forms in the embryo during the third and fourth weeks of development from mesoderm and neural crest cells (see chapter 29), and all adult connective tissue types develop from it. By 8 weeks of development, most of the mesenchyme has become specialized to form the types of connective tissue seen in adults, as well as muscle, blood vessels, and other tissues. The major source of remaining embryonic connective tissue in the newborn is in the umbilical cord, where it is called **mucous connective tissue,** or *Wharton's jelly* (table 4.7*b*). The structure of mucous connective tissue is similar to that of

TABLE 4.7 Embryonic Connective Tissue

(a) Mesenchyme

Structure: The mesenchymal cells are irregularly shaped; the extracellular matrix is abundant and contains scattered reticular fibers

Location: Mesenchyme is the embryonic tissue from which connective tissues, as well as other tissues, arise

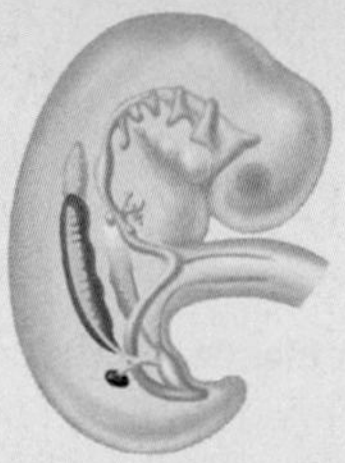

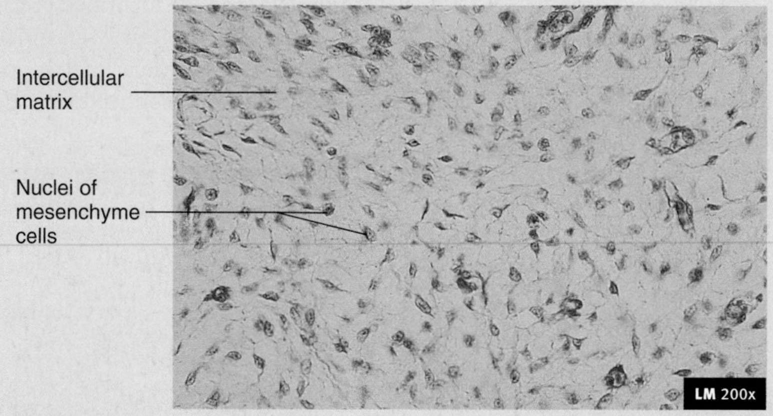

Intercellular matrix

Nuclei of mesenchyme cells

LM 200x

(b) Mucous Connective Tissue

Structure: Mucous tissue is mesenchymal tissue that remains unspecialized; the cells are irregularly shaped; the extracellular matrix is abundant and contains scattered reticular fibers

Location: Umbilical cord of newborn

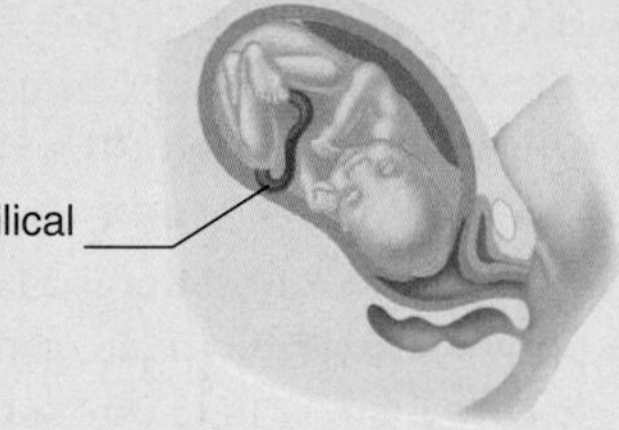
Umbilical cord

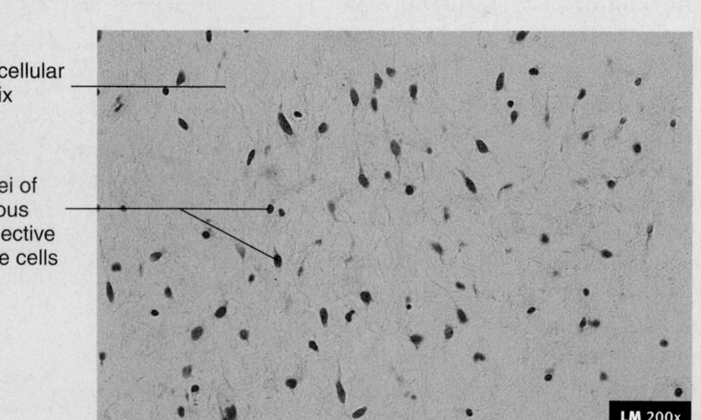
Intercellular matrix
Nuclei of mucous connective tissue cells
LM 200x

mesenchyme. The mucous connective tissue helps support the umbilical cord blood vessels between the mother and the child. After birth, the mucous connective tissue can also be a rich source of stem cells.

Adult connective tissue consists of three types: connective tissue proper (loose and dense), supporting connective tissue (cartilage and bone), and fluid connective tissue (blood).

Connective Tissue Proper

Loose Connective Tissue

Loose connective tissue (table 4.8) consists of relatively few protein fibers that form a lacy network, with numerous spaces filled with ground substance and fluid. Three subdivisions of loose connective tissue are areolar, adipose, and reticular. **Areolar** (ă-r ē′ō-lăr) **tissue** is the "loose packing" material of most organs and other tissues; it attaches the skin to underlying tissues (table 4.8*a*). It contains collagen, reticular, and elastic fibers and a variety of cells. For example, fibroblasts produce the fibrous matrix; macrophages move through the tissue, engulfing bacteria and cell debris; mast cells contain chemicals that help mediate inflammation; and lymphocytes are involved in immunity. The loose packing of areolar tissue is often associated with the other loose connective tissue types, adipose and reticular tissue.

Adipose tissue and reticular tissue are connective tissues with special properties. **Adipose tissue** (table 4.8*b*) consists of adipocytes, which contain large amounts of lipid. Unlike other connective tissue types, adipose tissue is composed of large cells and a small amount of extracellular matrix, which consists of loosely arranged collagen and reticular fibers with some scattered elastic fibers. Blood vessels form a network in the extracellular matrix. The adipocytes are usually arranged in clusters, or lobules, separated from one another by loose connective tissue. Adipose tissue functions as an insulator, a protective tissue, and a site of energy storage. Lipids take up less space per calorie than either carbohydrates or proteins and therefore are well adapted for energy storage.

Adipose tissue exists in both yellow and brown forms. **Yellow adipose** tissue is by far the most abundant. Yellow adipose tissue appears white at birth, but it turns yellow with age because of the accumulation of pigments, such as carotene, a plant pigment that humans can metabolize as a source of vitamin A. In humans, **brown adipose** tissue is found in specific areas of the body, such as the axillae (armpits), the neck, and near the kidneys. The brown color results from the cytochrome pigments in the tissue's numerous mitochondria and its abundant blood supply. It is difficult to distinguish brown adipose from yellow adipose in babies because the color difference is not great. Brown adipose fat is specialized to generate heat as a result of oxidative metabolism of lipid molecules in mitochondria. It can play a significant role in regulating body temperature in newborns and may also play a role in adult metabolism (see chapter 25).

Reticular tissue forms the framework of lymphatic tissue (table 4.8*c*), such as in the spleen and lymph nodes, as well as in bone marrow and the liver. It is characterized by a network of reticular fibers and reticular cells. **Reticular cells** produce the reticular fibers and remain closely attached to them. The spaces between the reticular fibers can contain a wide variety of other cells, such as dendritic cells, which look very much like reticular cells but are cells of the immune system; macrophages; and blood cells (see chapter 22).

Dense Connective Tissue

Dense connective tissue has a relatively large number of protein fibers, which form thick bundles and fill nearly all of the extracellular space. Most of the cells of developing dense connective tissue are spindle-shaped fibroblasts. Once the fibroblasts become completely surrounded by matrix, they are fibrocytes. Dense connective tissue can be subdivided into two major groups: regular and irregular.

Dense regular connective tissue has protein fibers in the extracellular matrix that are oriented predominantly in one direction. **Dense regular collagenous connective tissue** (table 4.9*a*) has abundant collagen fibers, which give this tissue a white appearance. Dense regular collagenous connective tissue forms structures such as tendons, which connect muscles to bones (see chapter 9), and most ligaments, which connect bones to bones (see chapter 8). The collagen fibers of dense connective tissue resist stretching and give the tissue considerable strength in the direction of the fiber orientation. Tendons and most ligaments consist almost entirely of thick bundles of densely packed parallel collagen fibers with the orientation of the collagen fibers in one direction, which makes the tendons and ligaments very strong, cablelike structures.

The general structures of tendons and ligaments are similar, but they differ in the following respects: (1) The collagen fibers of ligaments are often less compact, (2) some fibers of many ligaments are not parallel, and (3) ligaments are usually more flattened than tendons and form sheets or bands of tissues.

Dense regular elastic connective tissue (table 4.9*b*) consists of parallel bundles of collagen fibers and abundant elastic fibers. The elastin in elastic ligaments gives them a slightly yellow color. Dense regular elastic connective tissue forms some elastic ligaments, such as those in the vocal folds and the **nuchal** (noo′kăl; back of the neck) **ligament,** which lies along the posterior of the neck, helping hold the head upright. When elastic ligaments are stretched, they tend to shorten to their original length, much as an elastic band does.

Predict 4

Explain the advantages of having elastic ligaments that extend from vertebra to vertebra in the vertebral column and why it would be a disadvantage if tendons, which connect skeletal muscles to bone, were elastic.

Dense irregular connective tissue contains protein fibers arranged as a meshwork of randomly oriented fibers. Alternatively, the fibers within a given layer of dense irregular connective tissue can be oriented in one direction, whereas the fibers of adjacent layers are oriented at nearly right angles to that layer. Dense irregular connective tissue forms sheets of connective tissue that have strength in many directions but less strength in any single direction than does regular connective tissue.

Predict 5

Scars consist of dense irregular connective tissue made of collagen fibers. Vitamin C is required for collagen synthesis. Predict the effect of scurvy, which is a nutritional disease caused by vitamin C deficiency, on wound healing.

TABLE 4.8 Connective Tissue Proper: Loose Connective Tissue

(a) Areolar Connective Tissue APR

Structure: Cells (e.g., fibroblasts,macrophages, and lymphocytes) within a fine network of mostly collagen fibers; often merges with denser connective tissue

Function: Loose packing, support, and nourishment for the structures with which it is associated

Location: Widely distributed throughout the body; substance on which epithelial basement membranes rest; packing between glands, muscles, and nerves; attaches the skin to underlying tissues

Nucleus
Elastic fiber
Collagen fiber
LM 400x

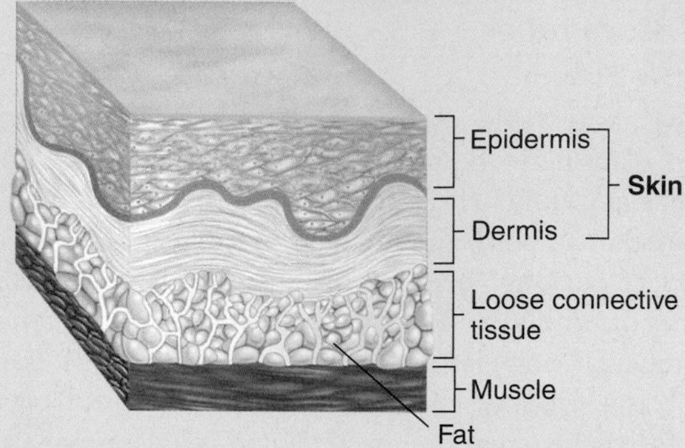
Epidermis
Dermis
Skin
Loose connective tissue
Muscle
Fat

(b) Adipose Tissue APR

Structure: Little extracellular matrix surrounding cells; the adipocytes are so full of lipid that the cytoplasm is pushed to the periphery of the cell

Function: Packing material, thermal insulation, energy storage, and protection of organs against injury from being bumped or jarred

Location: Predominantly in subcutaneous areas, in mesenteries, in renal pelvis, around kidneys, attached to the surface of the colon, in mammary glands, in loose connective tissue that penetrates spaces and crevices

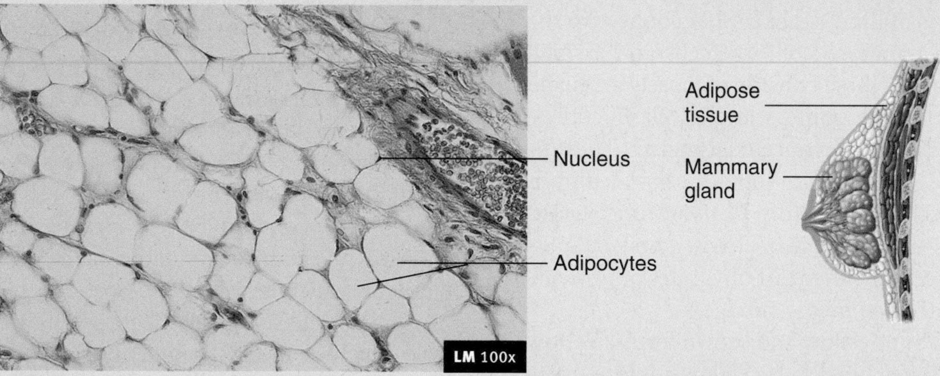
Nucleus
Adipocytes
LM 100x
Adipose tissue
Mammary gland

(c) Reticular Tissue

Structure: Fine network of reticular fibers irregularly arranged

Function: Provides a superstructure for lymphatic and hemopoietic tissues

Location: Within the lymph nodes, spleen, bone marrow

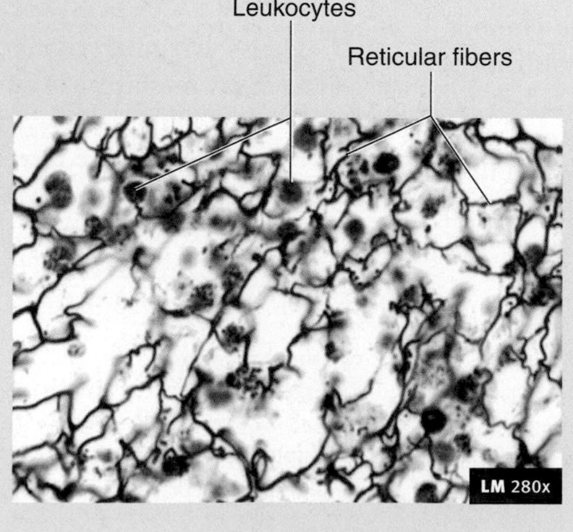
Leukocytes
Reticular fibers
LM 280x

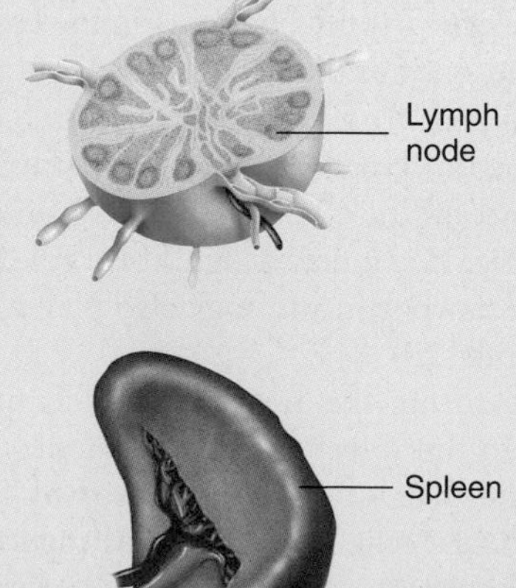
Lymph node
Spleen

TABLE 4.9 Connective Tissue Proper: Dense Connective Tissue

(a) Dense Regular Collagenous Connective Tissue APR

Structure: Matrix composed of collagen fibers running in somewhat the same direction

Function: Able to withstand great pulling forces exerted in the direction of fiber orientation; great tensile strength and stretch resistance

Location: Tendons (attach muscle to bone) and ligaments (attach bones to each other)

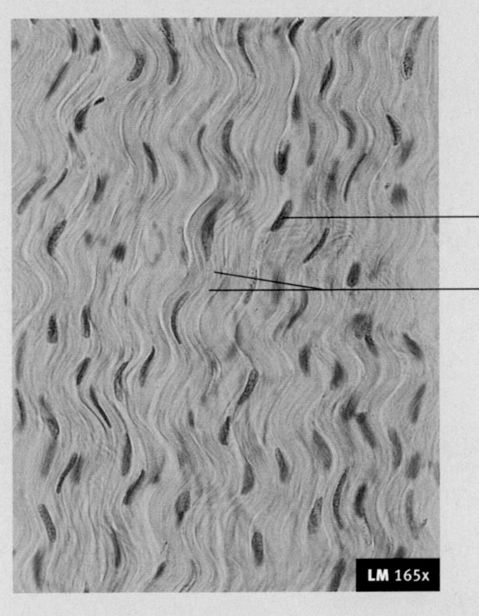
LM 165x

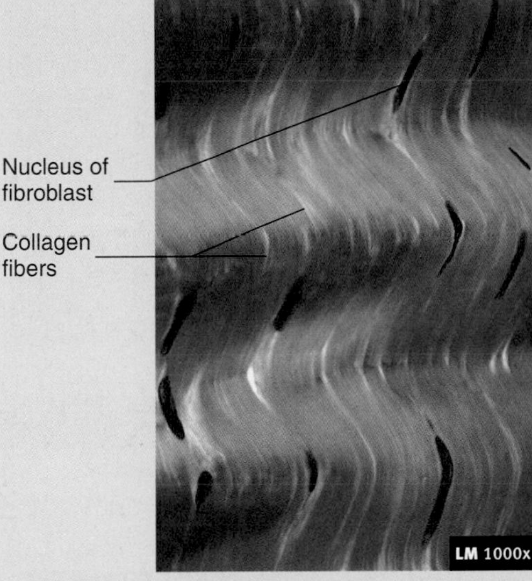
Nucleus of fibroblast
Collagen fibers
LM 1000x

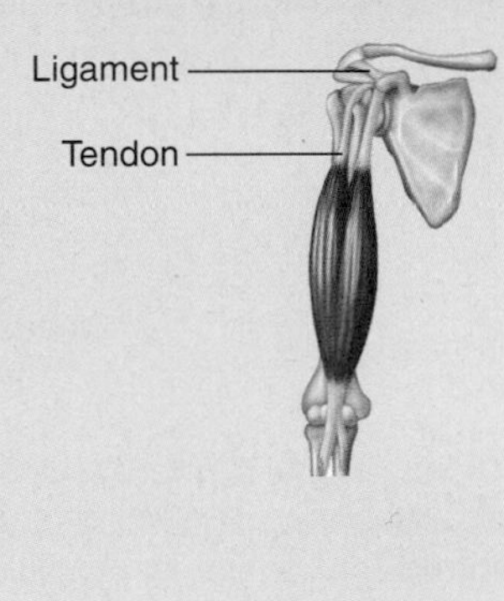
Ligament
Tendon

(b) Dense Regular Elastic Connective Tissue APR

Structure: Matrix composed of regularly arranged collagen fibers and elastic fibers

Function: Able to stretch and recoil like a rubber band, with strength in the direction of fiber orientation

Location: Vocal folds and elastic ligaments between the vertebrae and along the dorsal aspect of the neck

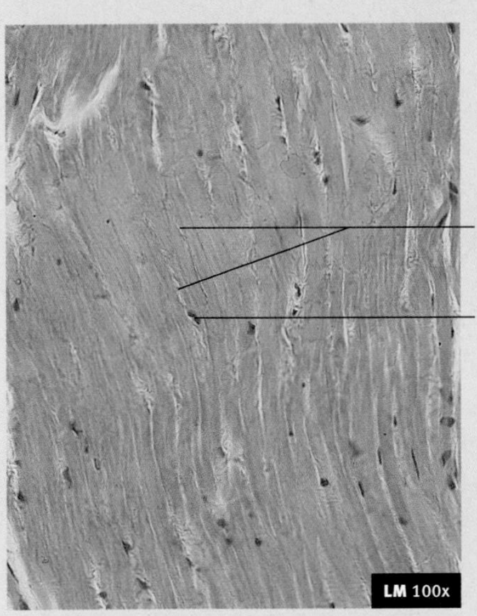
LM 100x

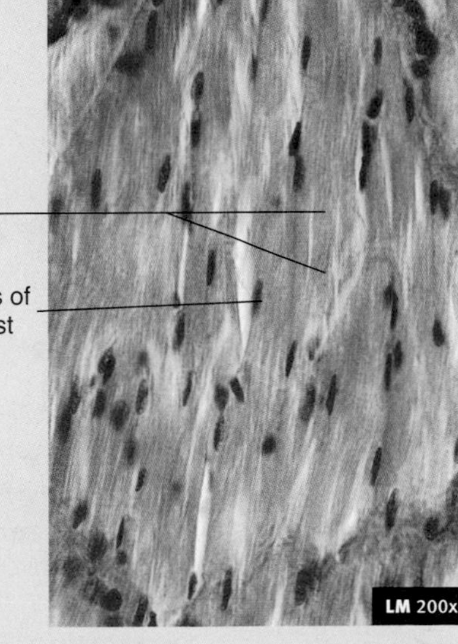
Elastic fibers
Nucleus of fibroblast
LM 200x

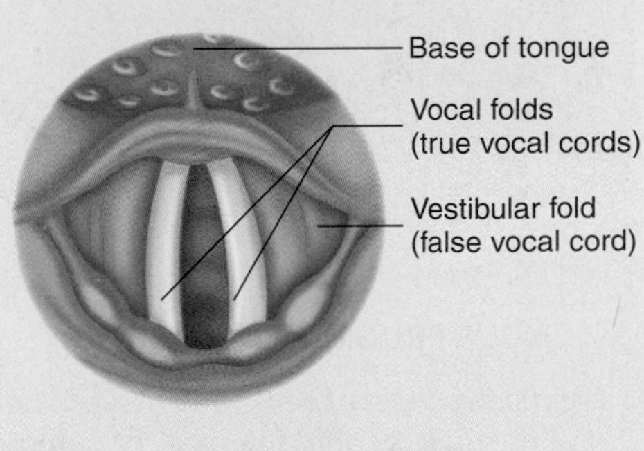
Base of tongue
Vocal folds (true vocal cords)
Vestibular fold (false vocal cord)

Dense irregular collagenous connective tissue (table 4.9*c*) forms most of the dermis, which is the tough, inner portion of the skin (see chapter 5), as well as the connective tissue capsules that surround organs such as the kidney and spleen.

Dense irregular elastic connective tissue (table 4.9*d*) is found in the walls of elastic arteries. In addition to collagen fibers, oriented in many directions, the layers of this tissue contain abundant elastic fibers.

TABLE 4.9 Connective Tissue Proper: Dense Connective Tissue—Continued

(c) Dense Irregular Collagenous Connective Tissue

Structure: Matrix composed of collagen fibers that run in all directions or in alternating planes of fibers oriented in a somewhat single direction

Function: Tensile strength capable of withstanding stretching in all directions

Location: Sheaths; most of the dermis of the skin; organ capsules and septa; outer covering of body tubes

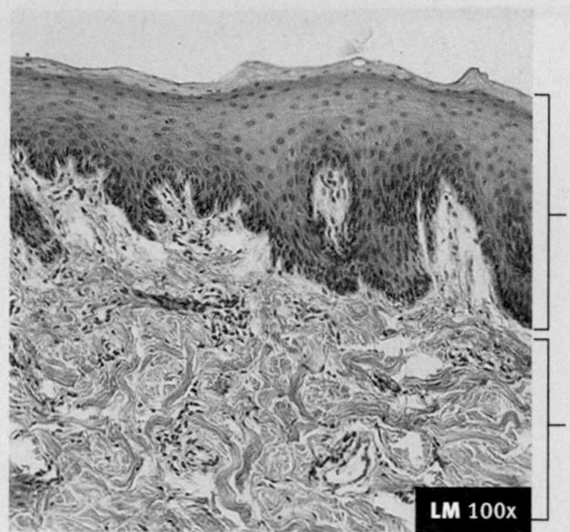
LM 100x

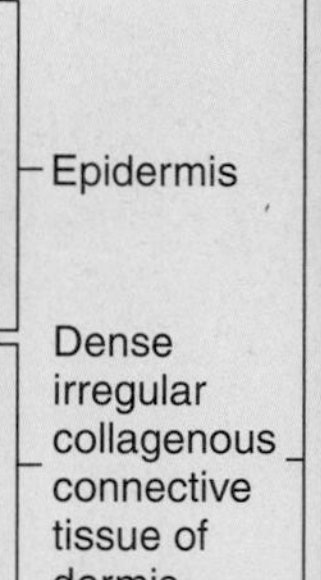
Epidermis
Dense irregular collagenous connective tissue of dermis

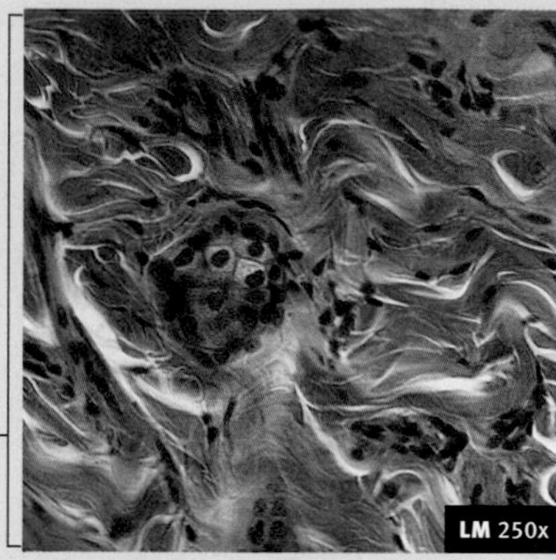
LM 250x

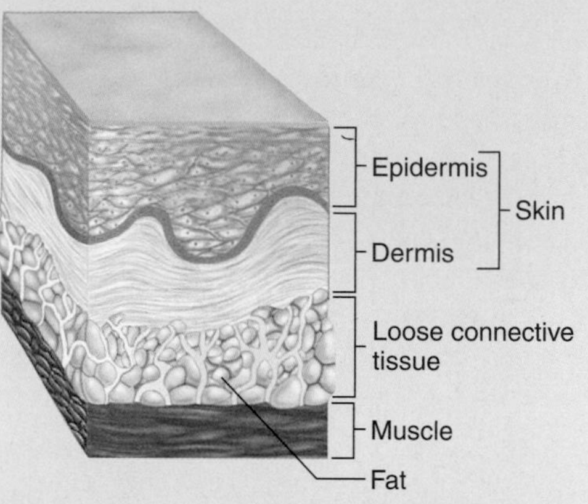
Epidermis
Dermis
Skin
Loose connective tissue
Muscle
Fat

(d) Dense Irregular Elastic Connective Tissue

Structure: Matrix composed of bundles and sheets of collagenous and elastic fibers oriented in multiple directions

Function: Capable of strength, with stretching and recoil in several directions

Location: Elastic arteries

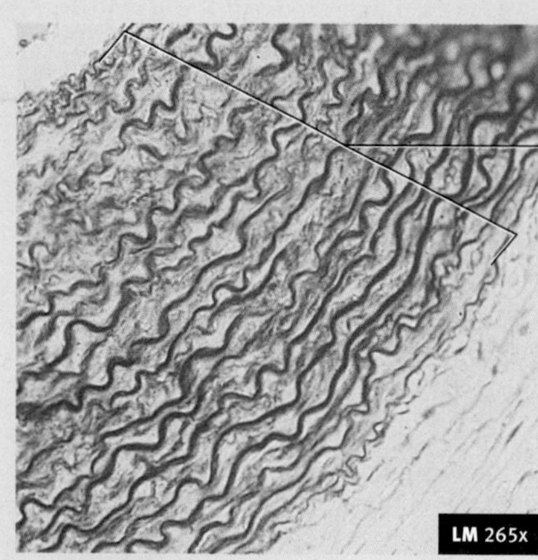
LM 265x
Dense irregular elastic connective tissue

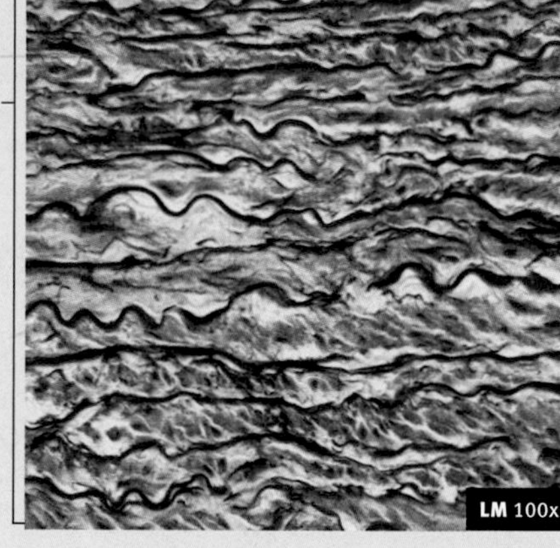
LM 100x

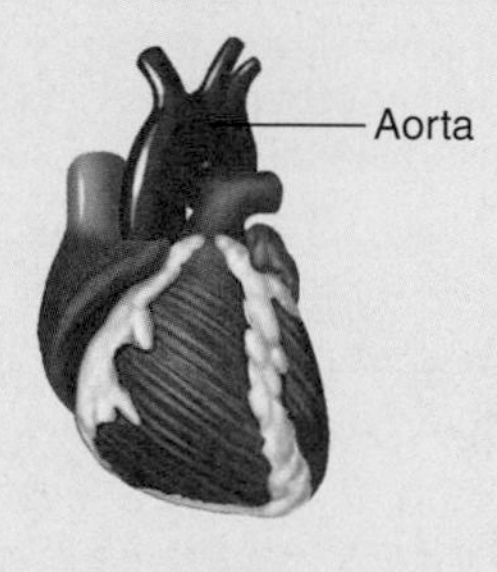
Aorta

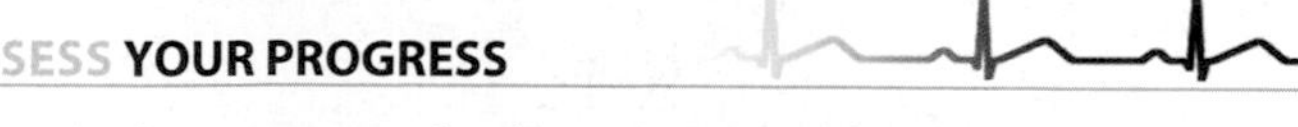

ASSESS YOUR PROGRESS

30. *List the two types of embryonic tissue. What does mesenchyme give rise to in the adult? What is the purpose of mucous connective tissue?*
31. *What are the three classifications of adult connective tissue, and what tissue types are included in each?*
32. *Describe the fiber arrangement in loose (areolar) connective tissue. What are the functions of this tissue type, and where it is found in the body?*
33. *What structural features distinguish adipose tissue from other types of connective tissue? What is an adipocyte?*
34. *Name the two types of adipose tissue, and give the functions of each. Which type is primarily found in infants?*
35. *What is the function of reticular tissue? Where is it found?*
36. *Structurally and functionally, what is the difference between dense regular connective tissue and dense irregular connective tissue?*
37. *Name the two kinds of dense regular connective tissue, and give an example of each. Do the same for dense irregular connective tissue.*

Supporting Connective Tissue

Cartilage

Cartilage (kar′ti-lij) is composed of cartilage cells within an extensive and relatively rigid matrix. The surface of nearly all cartilage is surrounded by a layer of dense irregular connective tissue called the **perichondrium** (per-i-kon′drē-ŭm), described in more detail in chapter 6. Cartilage cells arise from the perichondrium and secrete cartilage matrix. Once completely surrounded by matrix, the cartilage cells are called **chondrocytes** (kon′ drō-sītz), and the spaces in which they are located are called **lacunae** (lă-koo′nē). The matrix contains protein fibers, ground substance, and fluid. The protein fibers are collagen fibers or a mixture of collagen and elastic fibers. The ground substance consists of proteoglycans and other organic molecules. Most of the proteoglycans in the matrix form aggregates with hyaluronic acid. Within the cartilage matrix, proteoglycan aggregates function as minute sponges capable of trapping large quantities of water. This trapped water allows cartilage to spring back after being compressed. The collagen fibers give cartilage considerable strength. Next to bone, cartilage is the firmest structure in the body.

Cartilage has no blood vessels or nerves, except those of the perichondrium; it therefore heals very slowly after an injury because the cells and nutrients necessary for tissue repair cannot reach the damaged area easily.

There are three types of cartilage:

1. **Hyaline** (hī′ă-lin) **cartilage** has large amounts of both collagen fibers and proteoglycans (table 4.10*a*). Collagen fibers are evenly dispersed throughout the ground substance, and hyaline cartilage in joints has a very smooth surface. Specimens appear to have a glassy, translucent matrix when viewed through a microscope. Hyaline cartilage is found where strong support and some flexibility are needed, such as in the rib cage and within the trachea and bronchi (see chapter 23). It also covers the surfaces of bones that move smoothly against each other in joints. Hyaline cartilage forms most of the skeleton before it is replaced by bone in the embryo, and it is involved in growth that increases the length of bones (see chapter 6).
2. **Fibrocartilage** has more collagen fibers than proteoglycans (table 4.10*b*). Compared with hyaline cartilage, fibrocartilage has much thicker bundles of collagen fibers dispersed through its matrix. Fibrocartilage is slightly compressible and very tough. It is found in areas of the body where a great deal of pressure is applied to joints, such as in the knee, in the jaw, and between vertebrae. Some joints, such as the knee, have both hyaline and fibrocartilage connective tissue. In these joints, pads of fibrocartilage help absorb shocks and prevent bone-to-bone abrasion. Fibrocartilage injuries of the knee joint (meniscus tears) are common sports-related injuries.
3. **Elastic cartilage** has numerous elastic fibers in addition to collagen and proteoglycans dispersed throughout its matrix (table 4.10*c*). It is found in areas that have rigid but elastic properties, such as the external ears.

Predict 6

One of several changes caused by rheumatoid arthritis in joints is the replacement of hyaline cartilage with dense irregular collagenous connective tissue. Predict the effect of replacing hyaline cartilage with fibrous connective tissue.

Bone

Bone is a hard connective tissue that consists of living cells and mineralized matrix. Bone matrix has organic and inorganic portions. The organic portion consists of protein fibers, primarily collagen, and other organic molecules. The mineral, or inorganic, portion consists of specialized crystals called **hydroxyapatite** (hī-drok′sē-ap-ă-tīt), which contain calcium and phosphate. The strength and rigidity of the mineralized matrix allow bones to support and protect

TABLE 4.10 Supporting Connective Tissue: Cartilage

(a) Hyaline Cartilage APR

Structure: Collagen fibers are small and evenly dispersed in the matrix, making the matrix appear transparent; the cartilage cells, or chondrocytes, are found in spaces, or lacunae, within the firm but flexible matrix

Function: Allows the growth of long bones; provides rigidity with some flexibility in the trachea, bronchi, ribs, and nose; forms rugged, smooth, yet somewhat flexible articulating surfaces; forms the embryonic skeleton

Location: Growing long bones, cartilage rings of the respiratory system, costal cartilage of ribs, nasal cartilages, articulating surface of bones, embryonic skeleton

Chondrocyte in a lacuna
Nucleus
Matrix
LM 240x

Bone
Hyaline cartilage

TABLE 4.10 Supporting Connective Tissue: Cartilage—Continued

(b) Fibrocartilage APR

Structure: Collagen fibers similar to those in hyaline cartilage; the fibers are more numerous than in other cartilages and are arranged in thick bundles

Function: Somewhat flexible and capable of withstanding considerable pressure; connects structures subjected to great pressure

Location: Intervertebral disks, symphysis pubis articular disks (e.g., knee and temporomandibular [jaw] joints)

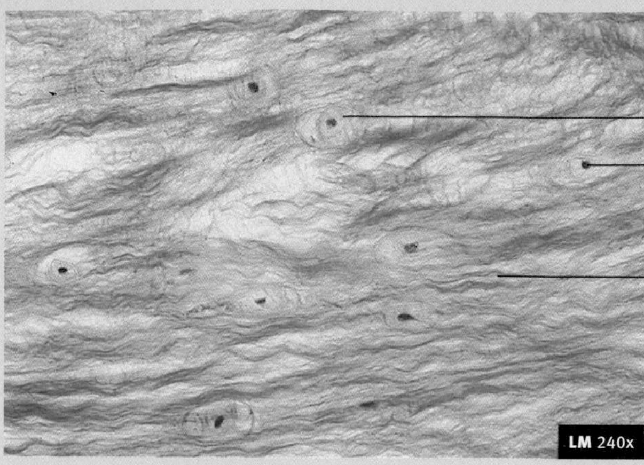
Chondrocyte in lacuna
Nucleus
Collagen fibers in matrix
LM 240x

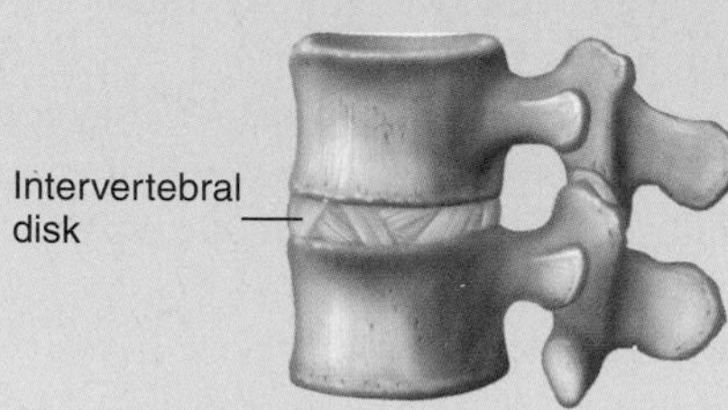
Intervertebral disk

(c) Elastic Cartilage APR

Structure: Similar to hyaline cartilage, but matrix also contains elastic fibers

Function: Provides rigidity with even more flexibility than hyaline cartilage because elastic fibers return to their original shape after being stretched

Location: External ears, epiglottis, auditory tubes

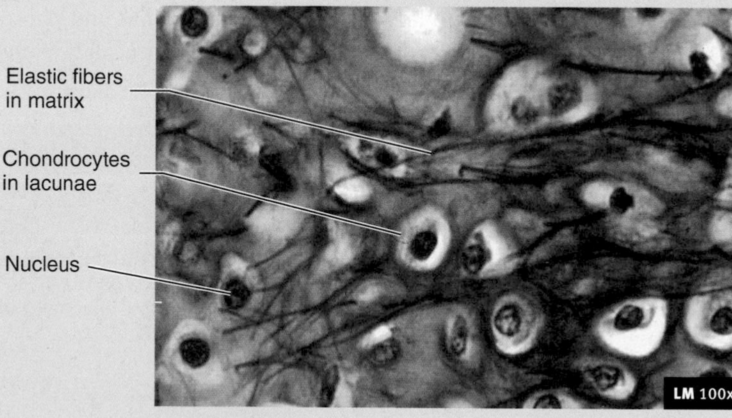
Elastic fibers in matrix
Chondrocytes in lacunae
Nucleus
LM 100x

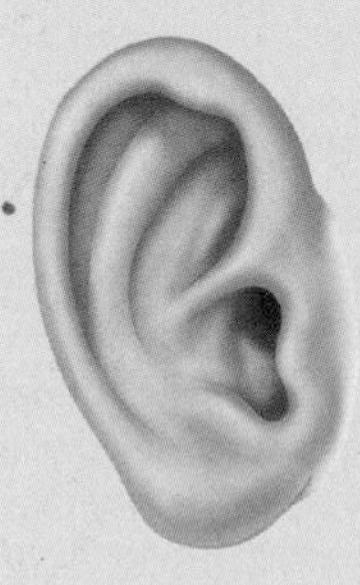

other tissues and organs. Bone cells, or **osteocytes** (os′tē-ō-sītz), are located within holes in the matrix, which are called lacunae and are similar to the lacunae of cartilage.

Two types of bone exist:

1. **Spongy bone** has spaces between **trabeculae** (tră-bek′ū-lē; beams), or plates, of bone and therefore resembles a sponge (table 4.11*a*).
2. **Compact bone** is more solid, with almost no space between many thin layers, or **lamellae** (lă-mel′ē; sing. lă-mel′ă) of bone (table 4.11*b*).

Bone, unlike cartilage, has a rich blood supply. For this reason, bone can repair itself much more readily than can cartilage. Bone and bone cells are described more fully in chapter 6.

Fluid Connective Tissue

Blood

Blood is unusual among the connective tissues because the matrix between the cells is liquid (table 4.12*a*). The cells of most other connective tissues are more or less stationary within a relatively rigid matrix, but blood cells move freely within a fluid matrix.

TABLE 4.11 Supporting Connective Tissue: Bones

(a) Spongy Bone

Structure: Latticelike network of scaffolding characterized by trabeculae with large spaces between them filled with hemopoietic tissue; the osteocytes, or bone cells, are located within lacunae in the trabeculae

Function: Acts as scaffolding to provide strength and support without the greater weight of compact bone

Location: In the interior of the bones of the skull, vertebrae, sternum, and pelvis; in the ends of the long bones

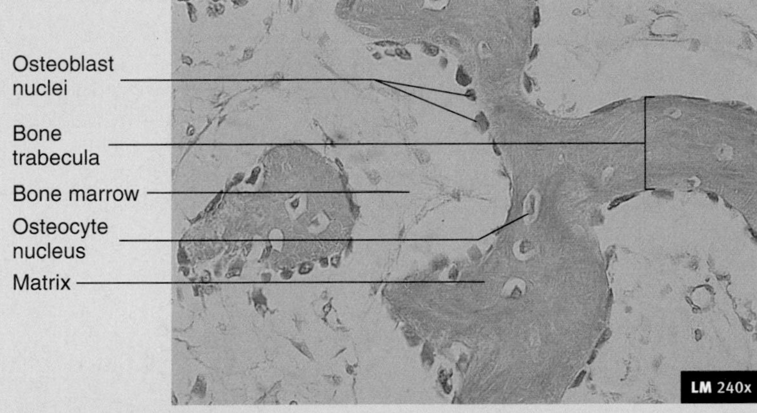
Osteoblast nuclei
Bone trabecula
Bone marrow
Osteocyte nucleus
Matrix
LM 240x

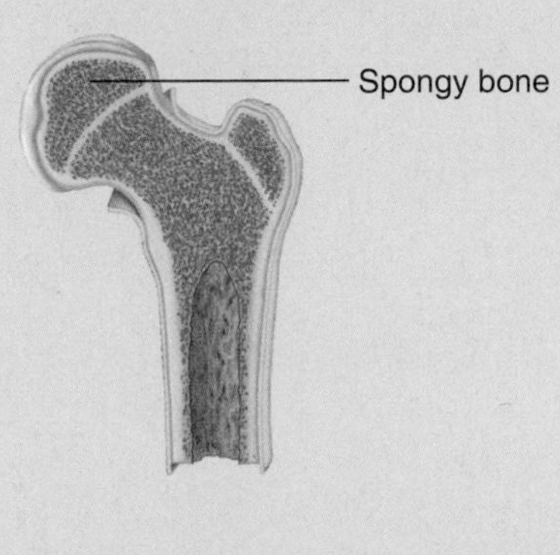
Spongy bone

(b) Compact Bone

Structure: Hard, bony matrix predominates; many osteocytes (not seen in this bone preparation) are located within lacunae that are distributed in a circular fashion around the central canals; small passageways connect adjacent lacunae

Function: Provides great strength and support; forms a solid outer shell on bones that keeps them from being easily broken or punctured

Location: Outer portions of all bones, the shafts of long bones

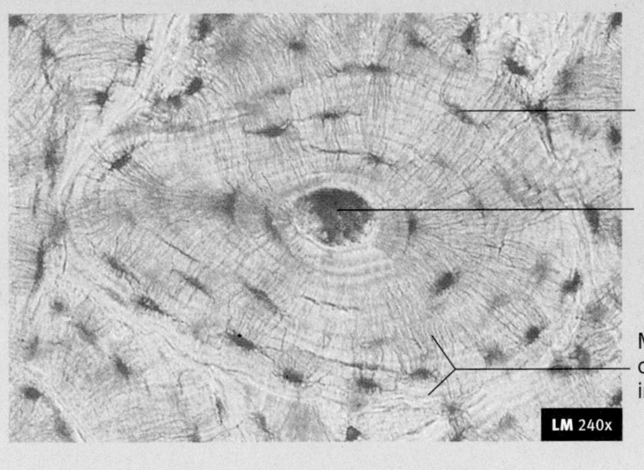
Lacuna
Central canal
Matrix organized into lamellae
LM 240x

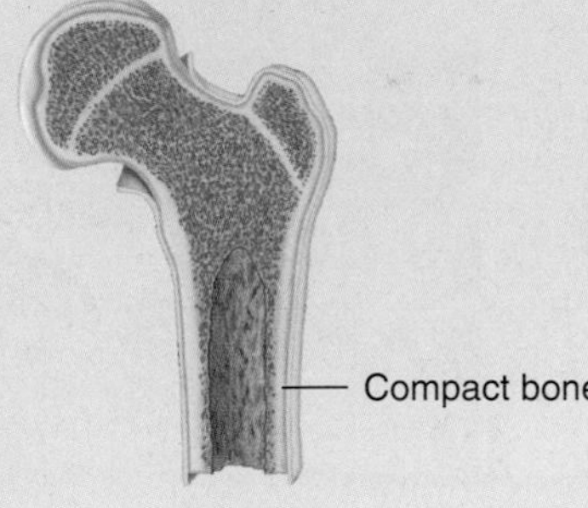
Compact bone

Blood's liquid matrix allows it to flow rapidly through the body, carrying nutrients, oxygen, waste products, and other materials. The matrix of blood is also unusual in that most of it is produced by cells contained in other tissues, rather than by blood cells. There are three types of cellular structures: red blood cells, white blood cells, and cell fragments called platelets. White blood cells sometimes leave the bloodstream and wander through other tissues. Blood is discussed more fully in chapter 19.

Hemopoietic (hē′mō-poy-et′ik) **tissue** forms blood cells. In adults, hemopoietic tissue is found in **bone marrow** (mar′ō; table 4.12*b*), which is the soft connective tissue in the cavities of bones. There are two types of bone marrow: **red marrow** and **yellow marrow** (see chapter 6). Red marrow is hemopoietic tissue surrounded by a framework of reticular fibers. Hemopoietic tissue produces red and white blood cells; it is described in detail in chapter 19. In children, the marrow of most bones is red marrow. Yellow marrow consists of yellow adipose tissue and does not produce blood cells. As children grow, yellow marrow replaces much of the red marrow in bones.

TABLE 4.12 Fluid Connective Tissue: Blood and Hemopoietic Tissue

(a) Blood APR

Structure: Blood cells and a fluid matrix

Function: Transports oxygen, carbon dioxide, hormones, nutrients, waste products, and other substances; protects the body from infections and is involved in temperature regulation

Location: Within the blood vessels; white blood cells frequently leave the blood vessels and enter the interstitial spaces

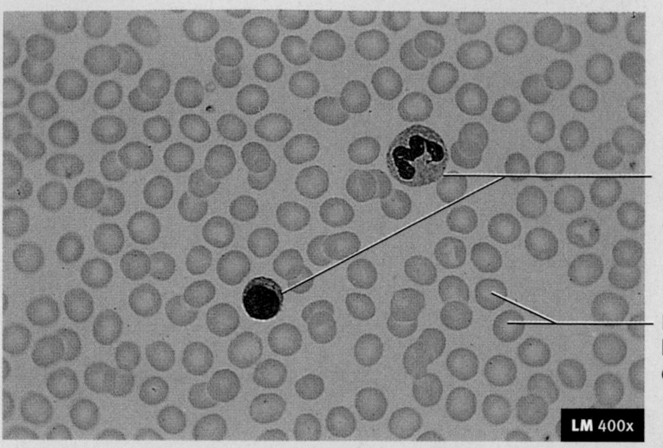
White blood cells
Red blood cells
LM 400x

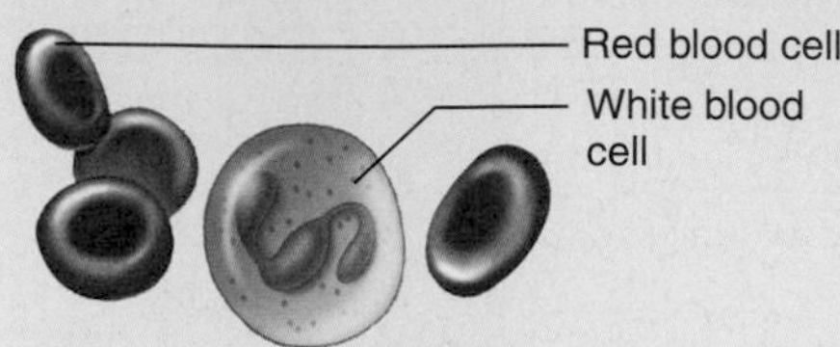
Red blood cell
White blood cell

(b) Bone Marrow APR

Structure: Reticular framework with numerous blood-forming cells (red marrow)

Function: Produces new blood cells (red marrow); stores lipids (yellow marrow)

Location: Within marrow cavities of bone; two types: (1) red marrow (hemopoietic, or blood-forming, tissue) in the ends of long bones and in short, flat, and irregularly shaped bones and (2) yellow marrow, mostly adipose tissue, in the shafts of long bones

Cells destined to become red blood cells

Fat

Nuclei

LM 600x

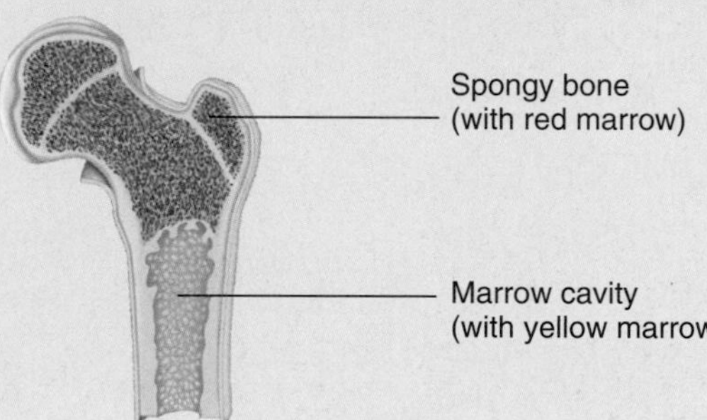
Spongy bone (with red marrow)
Marrow cavity (with yellow marrow)

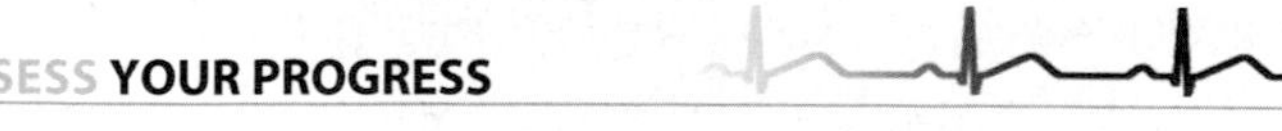

ASSESS YOUR PROGRESS

38. *Describe the cells and matrix of cartilage. What are lacunae? What is the perichondrium? Why does cartilage heal slowly?*

39. *What are the three types of cartilage? How do they differ in structure and function? Where would each type be found in the body?*

40. *Describe the cells and matrix of bone. Differentiate between spongy and compact bone.*

41. *What characteristic separates blood from other connective tissues? What are the three formed elements in blood?*

42. *Describe the function of hemopoietic tissue. Explain the difference between red marrow and yellow marrow.*

4.5 Muscle Tissue

LEARNING OUTCOME

After reading this section, you should be able to

A. Discuss the three types of muscle tissue by describing their general structures, locations in the body, and functions.

The main characteristic of **muscle tissue** is that it contracts, or shortens, with a force and therefore is responsible for movement. Muscle contraction is accomplished by the interaction of contractile proteins, which are described in chapter 9. Muscles contract

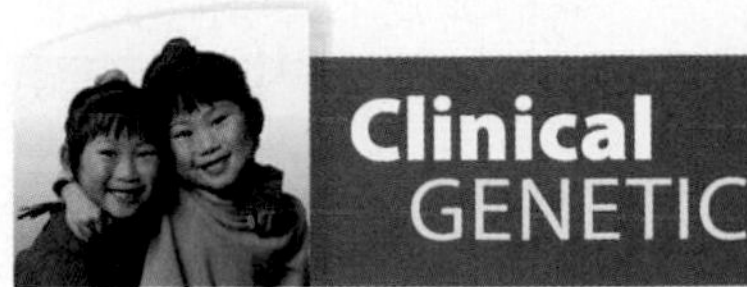

Clinical GENETICS | Marfan Syndrome

Marfan syndrome is an autosomal dominant disorder that affects approximately 1 in 5000 people. The gene for Marfan syndrome codes for a protein called fibrillin-1, which is necessary for the normal structure of the elastic fibers of connective tissue. Children of a person with Marfan syndrome have a 50% chance of inheriting the disorder because it is an autosomal dominant trait (see chapter 29). However, about 25% of the cases of Marfan syndrome occur in children whose parents do not have the disorder. In these cases, a mutation of the gene occurs during the formation of sperm cells or oocytes.

Many people with Marfan syndrome have limbs, fingers, and toes that are disproportionately long in relation to the rest of the body. Connective tissues are weakened; as a consequence, the heart valves, which are composed largely of connective tissue, do not function normally, resulting in heart murmurs (abnormal heart sounds). Poor vision is common because the lenses of the eyes, which are normally held in place by elastic fibers, are positioned abnormally. The lungs are prone to collapse, and dilation of large arteries, such as the aorta, can occur. A common cause of death in people with Marfan syndrome is rupture of the aorta. There is no cure for the condition, but treatments can reduce the danger of the symptoms. For example, drugs that lower blood pressure reduce the vascular risks.

It has been speculated that President Lincoln may have had Marfan syndrome, but some geneticists now think it more likely he had a rare inherited form of endocrine cancer that includes physical features of Marfan syndrome.

to move the entire body, to pump blood through the heart and blood vessels, and to decrease the size of hollow organs, such as the stomach and urinary bladder. The three types of muscle tissue—skeletal, cardiac, and smooth muscle—are grouped according to both structure and function (table 4.13).

Skeletal muscle is what we normally think of as "muscle" (table 4.14*a*). It is the meat of animals and constitutes about 40% of a person's body weight. As the name implies, skeletal muscle attaches to the skeleton and enables the body to move. Skeletal muscle is under voluntary (conscious) control because a person can purposefully cause skeletal muscle contraction to achieve specific body movements. However, the nervous system can cause skeletal muscles to contract without conscious involvement, as occurs during reflex movements and the maintenance of muscle tone. Skeletal muscle cells are long, cylindrical cells, each containing many nuclei located at the periphery of the cell. Some skeletal muscle cells extend the entire length of a muscle. Skeletal muscle cells are **striated** (strī′āt-ed), or banded, because of the arrangement of contractile proteins within the cells (see chapter 9).

Cardiac muscle is the muscle of the heart; it is responsible for pumping blood (table 4.14*b*). Cardiac muscle is under involuntary

TABLE 4.13 Comparison of Muscle Types

	Skeletal Muscle	Cardiac Muscle	Smooth Muscle
Location	Attached to bones	In the heart	In the walls of hollow organs, blood vessels, eyes, glands, skin
Cell Shape	Very long, cylindrical cells (1–4 cm and may extend the entire length of the muscle, 10–100 μm in diameter)	Cylindrical cells that branch (100–500 μm in length, 12–20 μm in diameter)	Spindle-shaped cells (15–200 μm in length, 5–8 μm in diameter)
Nucleus	Multinucleated, peripherally located	Single, centrally located	Single, centrally located
Striations	Yes	Yes	No
Control	Voluntary (conscious)	Involuntary (unconscious)	Involuntary (unconscious)
Ability to Contract Spontaneously	No	Yes	Yes
Function	Moves the body	Provides the major force for moving blood through the blood vessels	Moves food through the digestive tract, empties the urinary bladder, regulates blood vessel diameter, changes pupil size, contracts many gland ducts, moves hair, performs many other functions
Special Features	None	Branching fibers, intercalated disks containing gap junctions joining the cells to each other	Gap junctions

TABLE 4.14 Muscle Tissue

(a) Skeletal Muscle APR

Structure: Skeletal muscle cells or fibers appear striated (banded); cells are large, long, and cylindrical, with many nuclei located at the periphery

Function: Moves the body; is under voluntary (conscious) control

Location: Attached to bone or other connective tissue

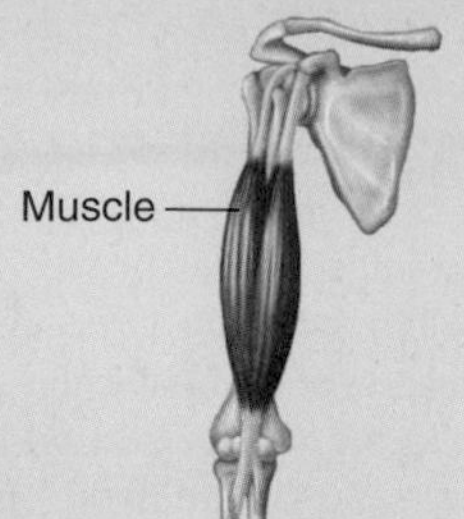
Muscle

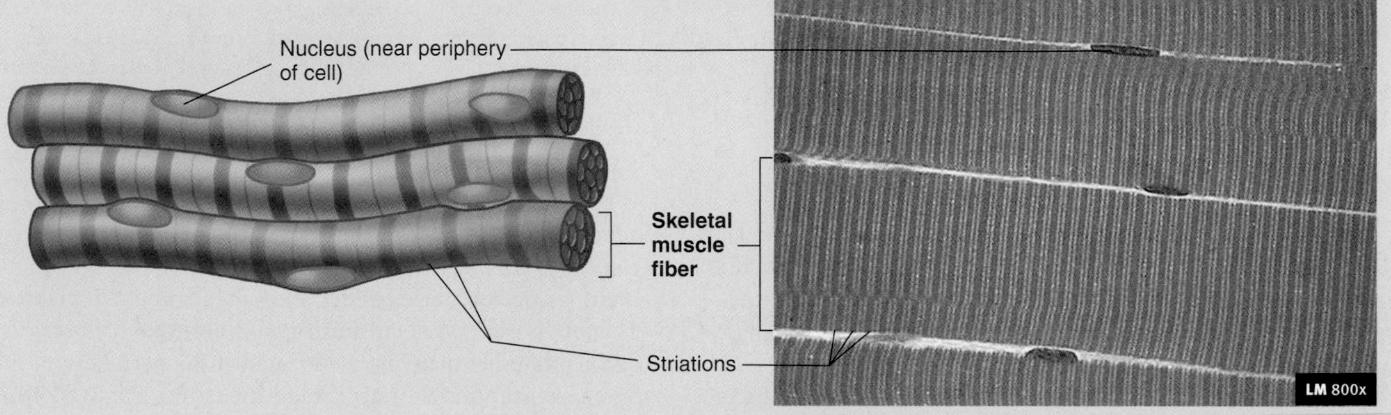
Nucleus (near periphery of cell)
Skeletal muscle fiber
Striations
LM 800x

(b) Cardiac Muscle APR

Structure: Cardiac muscle cells are cylindrical and striated and have a single, centrally located nucleus; they are branched and connected to one another by intercalated disks, which contain gap junctions

Function: Pumps the blood; is under involuntary (unconscious) control

Location: In the heart

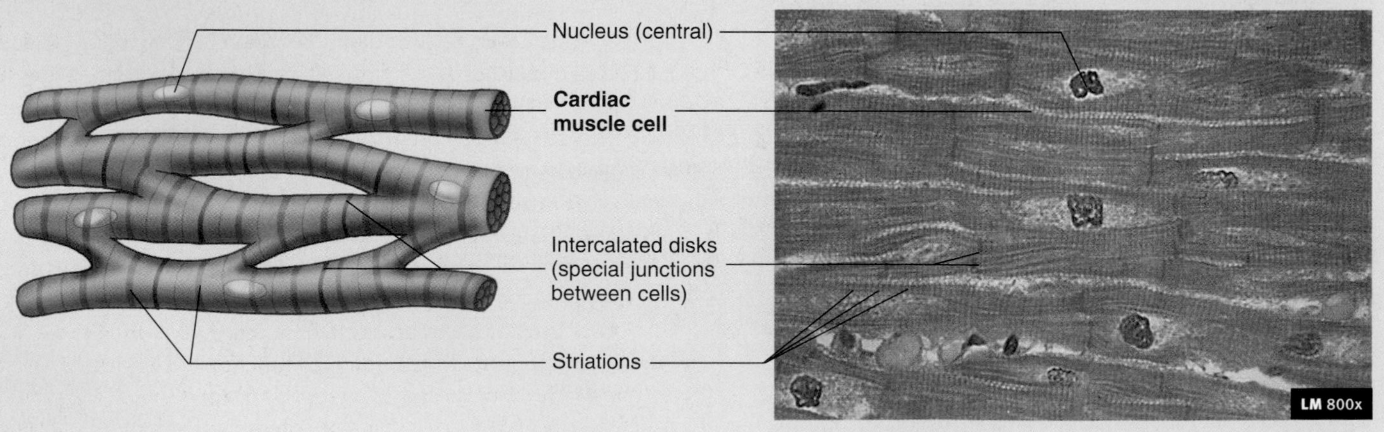
Nucleus (central)
Cardiac muscle cell
Intercalated disks (special junctions between cells)
Striations
LM 800x

TABLE 4.14 Muscle Tissue—Continued

(c) Smooth Muscle APR

Structure: Smooth muscle cells are tapered at each end, are not striated, and have a single nucleus

Function: Regulates the size of organs, forces fluid through tubes, controls the amount of light entering the eye, and produces "goose flesh" in the skin; is under involuntary (unconscious) control

Location: In hollow organs, such as the stomach and small and large intestines

Wall of stomach
Wall of colon
Wall of small intestine

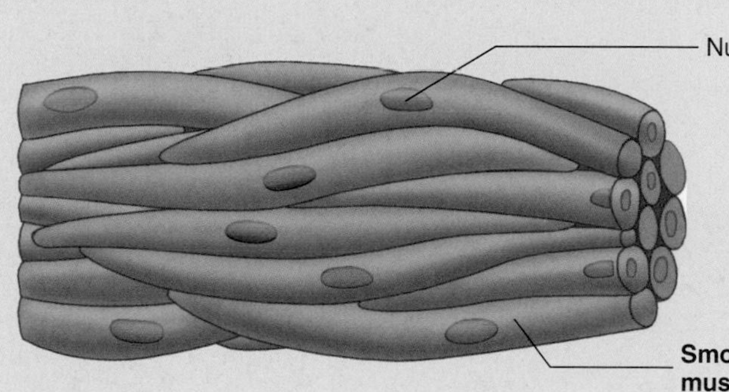
Nucleus
Smooth muscle cell

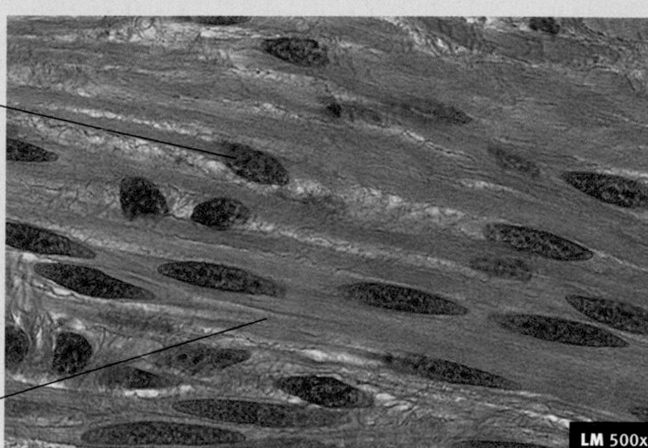
LM 500x

(unconscious) control, although a person can learn to influence the heart rate by using techniques such as meditation and biofeedback. Cardiac muscle cells are cylindrical but much shorter than skeletal muscle cells. Cardiac muscle cells are striated and usually have one nucleus per cell. They are often branched and connected to one another by **intercalated** (in-ter′kă-lā-ted; inserted between) **disks,** which contain specialized gap junctions and are important in coordinating cardiac muscle cell contractions (see chapter 20).

Smooth muscle forms the walls of hollow organs (except the heart); it is also found in the skin and eyes (table 4.14*c*). Smooth muscle is responsible for a number of functions, such as moving food through the digestive tract and emptying the urinary bladder. Like cardiac muscle, smooth muscle is controlled involuntarily. Smooth muscle cells are tapered at each end, have a single nucleus, and are not striated.

ASSESS YOUR PROGRESS

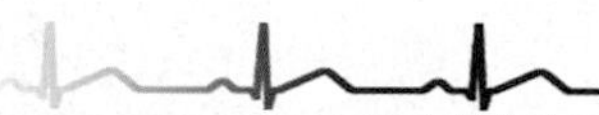

43. *Functionally, what is unique about muscle tissue?*
44. *Compare the structure of skeletal, cardiac, and smooth muscle cells.*
45. *Which type of muscle is under voluntary control?*
46. *Where is each type of muscle tissue found, and what tasks does each perform?*

4.6 Nervous Tissue

LEARNING OUTCOME

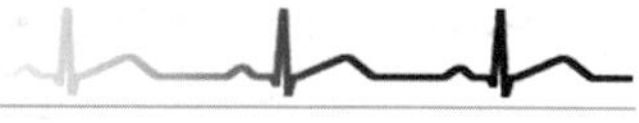

After reading this section, you should be able to

A. Describe the structural and functional roles of neurons and neuroglia in the nervous tissue.

Nervous tissue is found in the brain, spinal cord, and nerves and is characterized by the ability to conduct electrical signals called **action potentials.** Nervous tissue consists of neurons, which are responsible for its conductive ability, and support cells called neuroglia.

Neurons, or *nerve cells,* are the conducting cells of nervous tissue. Just as an electrical wiring system transports electricity throughout a house, neurons transport electrical signals throughout the body. A neuron is composed of three major parts: a cell body, dendrites, and an axon. The **cell body** contains the nucleus and is the site of general cell functions. Dendrites and axons consist of projections of cytoplasm surrounded by membrane. **Dendrites** (den′drītz) usually receive action potentials. They are much shorter than axons and have multiple branches at their ends. The **axon** (ak′son) usually conducts action potentials away from the cell body. Axons can be much longer than dendrites, and they have a constant diameter along their entire length.

Neurons that possess several dendrites and one axon are called **multipolar neurons** (table 4.15*a*). Neurons that possess a single dendrite and an axon are called **bipolar neurons. Pseudo-unipolar neurons** have only a single, short process that extends from the cell body (table 4.15*b*). This process then divides into two branches, which extend to the periphery and to the central nervous system. The two branches function as a single axon, although there are dendritelike receptors on the peripheral branch. Within each subgroup are many shapes and sizes of neurons, especially in the brain and spinal cord.

Neuroglia (noo-rog′lē-ă; nerve glue) are the support cells of the brain, spinal cord, and peripheral nerves (table 4.15). Neuroglia nourish, protect, and insulate neurons. Neurons and neuroglial cells are described in greater detail in chapter 11.

TABLE 4.15 Types of Neurons

(a) Multipolar Neuron APR

Structure: The neuron consists of dendrites, a cell body, and a long axon; neuroglia, or support cells, surround the neurons

Function: Neurons transmit information in the form of action potentials, store "information," and integrate and evaluate data; neuroglia support, protect, and form specialized sheaths around axons

Location: In the brain, spinal cord, ganglia

(b) Pseudo-Unipolar Neuron

Structure: The neuron consists of a cell body with one axon

Function: Conducts action potentials from the periphery to the brain or spinal cord

Location: In ganglia outside the brain and spinal cord

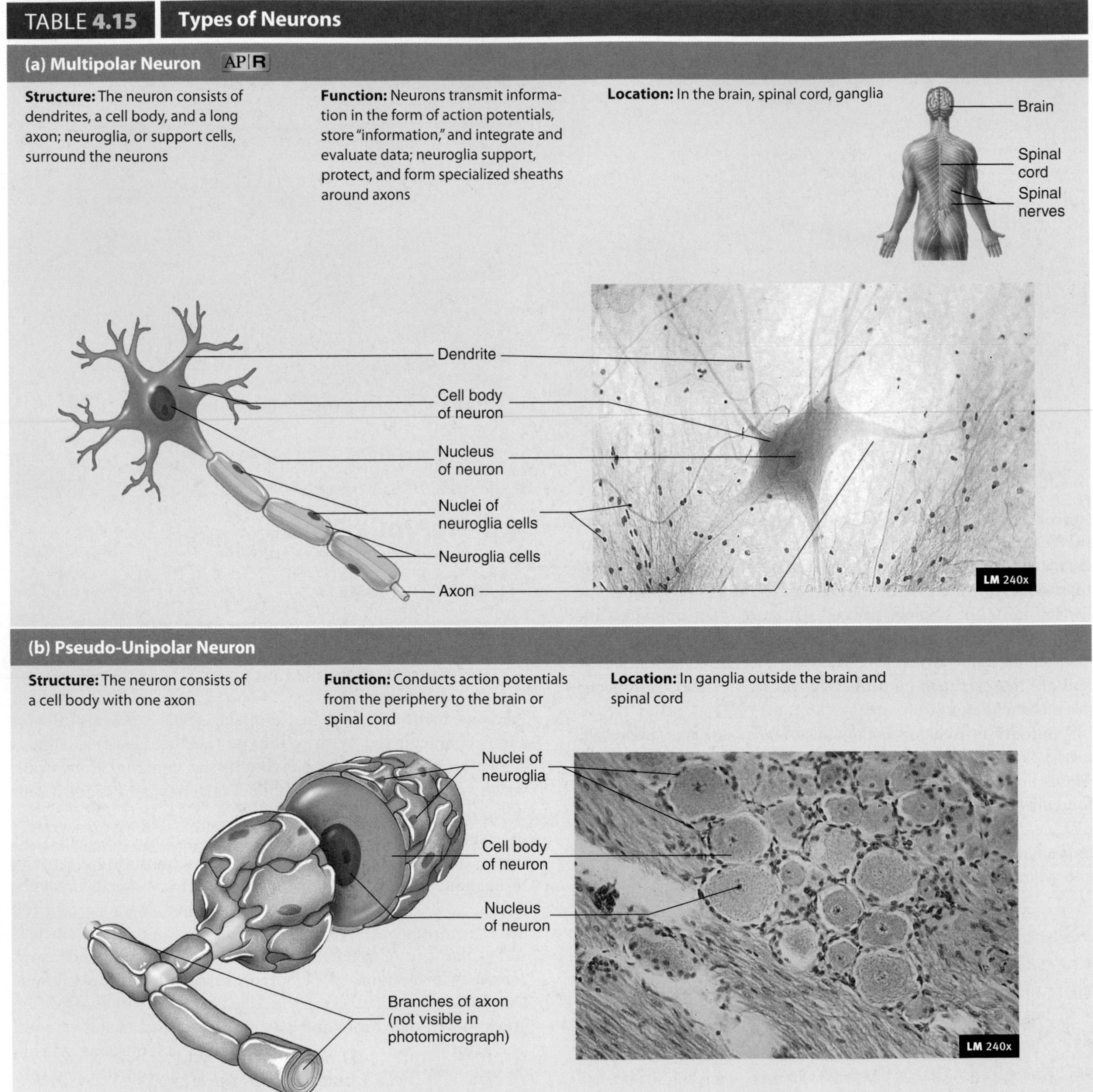
Brain
Spinal cord
Spinal nerves
Dendrite
Cell body of neuron
Nucleus of neuron
Nuclei of neuroglia cells
Neuroglia cells
Axon
LM 240x
Nuclei of neuroglia
Cell body of neuron
Nucleus of neuron
Branches of axon (not visible in photomicrograph)
LM 240x

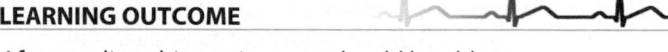

ASSESS YOUR PROGRESS

47. *What is the characteristic function of nervous tissue?*
48. *Define and list the functions of the cell body, dendrites, and axon of a neuron.*
49. *Differentiate among multipolar, bipolar, and pseudo-unipolar neurons.*
50. *What are the functions of neuroglia?*

4.7 Tissue Membranes

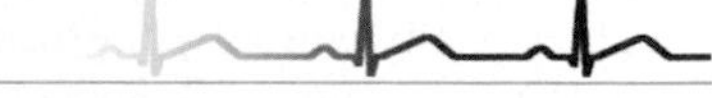

LEARNING OUTCOME

After reading this section, you should be able to

A. List the structural and functional characteristics of mucous, serous, and synovial membranes.

A membrane is a thin sheet of tissue that covers a structure or lines a cavity. Most membranes are formed from a superficial epithelial tissue and the connective tissue on which it rests. The skin, or cutaneous membrane (see chapter 5), is the external membrane. The three major categories of internal membranes are mucous, serous, and synovial membranes.

Mucous Membranes

A **mucous** (mū′kŭs) **membrane** consists of epithelial cells, their basement membrane, a thick layer of loose connective tissue called the **lamina propria** (lam′i-nă prō′prē-ă), and sometimes a layer of smooth muscle cells. Mucous membranes line cavities and canals that open to the outside of the body, such as the digestive, respiratory, excretory, and reproductive passages (figure 4.6*a*). Many, but not all, mucous membranes contain goblet cells or multicellular mucous glands that secrete a viscous substance called **mucus** (mū′kŭs). The functions of the mucous membranes vary, depending on their location, and include protection, absorption, and secretion. For example, the stratified squamous epithelium of the oral cavity performs a protective function, whereas the simple columnar epithelium of the small intestine absorbs nutrients and secretes digestive enzymes and mucus. Mucous membrane also lines the nasal passages. When that membrane becomes inflamed, we experience the "runny nose" characteristic of the common cold or an allergy.

Serous Membranes

A **serous** (sēr′ŭs) **membrane** consists of three components: a layer of simple squamous epithelium called **mesothelium** (mez-ō-thē′lē-ŭm), its basement membrane, and a delicate layer of loose connective tissue. Serous membranes line cavities, such as the pericardial, pleural, and peritoneal cavities, that do not open to the exterior (figure 4.6*b*). Serous membranes do not contain glands, but they secrete a small amount of fluid called **serous fluid,** which lubricates the serous membranes, making their surfaces slippery. Serous membranes protect the internal organs from friction, help hold them in place, and act as selectively permeable barriers to prevent large amounts of fluid from accumulating within the serous cavities.

FUNDAMENTAL **Figure**

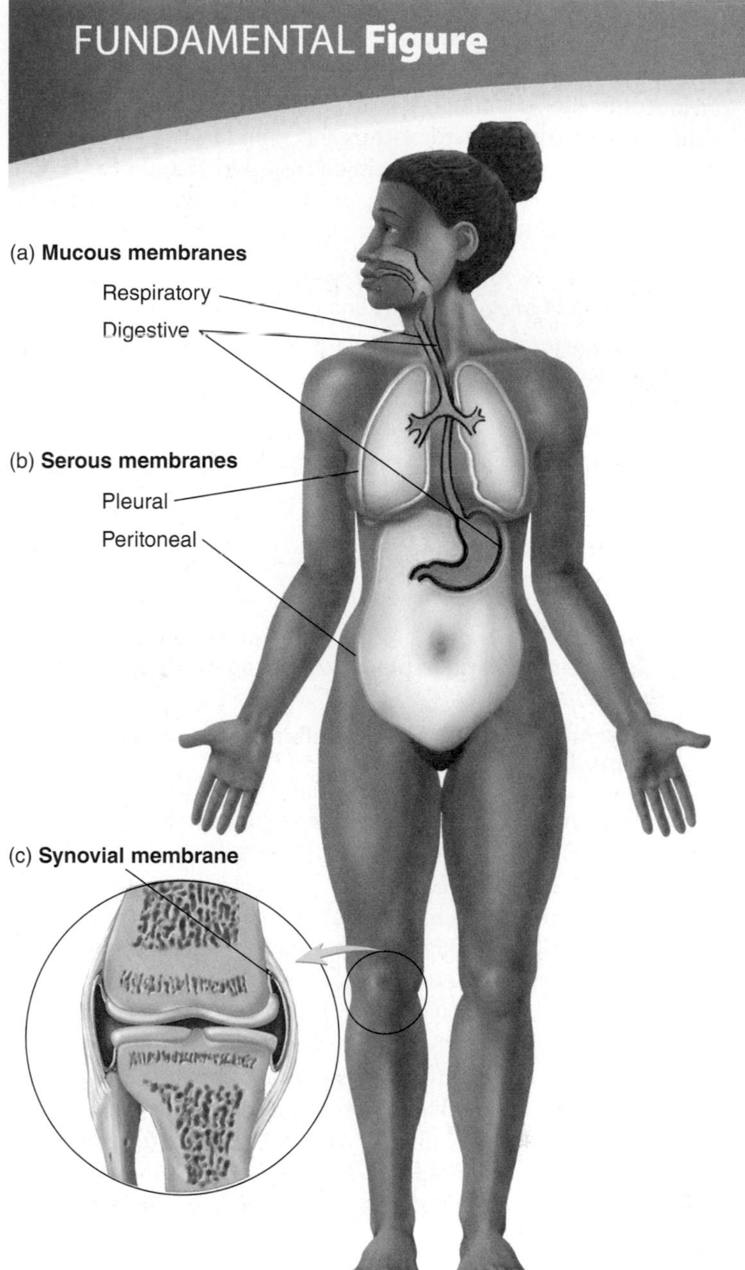
(a) Mucous membranes
Respiratory
Digestive
(b) Serous membranes
Pleural
Peritoneal
(c) Synovial membrane

FIGURE 4.6 Internal Membranes
(*a*) Mucous membranes line cavities that open to the outside and often contain mucous glands, which secrete mucus. (*b*) Serous membranes line cavities that do not open to the exterior; they do not contain glands but do secrete serous fluid. (*c*) Synovial membranes line cavities that surround synovial joints.

Synovial Membranes

A **synovial** (si-nō′vē-ăl) **membrane** is made up of only connective tissue. Synovial membranes consist of modified connective tissue cells, either intermixed with part of the dense connective tissue of the joint capsule or separated from the capsule by areolar or adipose tissue. Synovial membranes line freely movable joints (figure 4.6*c*). They produce **synovial fluid,** which is rich in hyaluronic acid, making the joint fluid very slippery, thus facilitating smooth movement within the joint. Synovial and other connective tissue membranes are discussed in chapter 8.

ASSESS YOUR PROGRESS 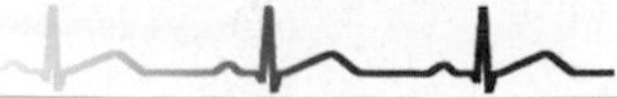

51. *Compare mucous, serous, and synovial membranes according to the types of cavities they line and their secretions.*
52. *What are the functions of mucous, serous, and synovial membranes?*

4.8 Tissue Damage and Inflammation

LEARNING OUTCOMES 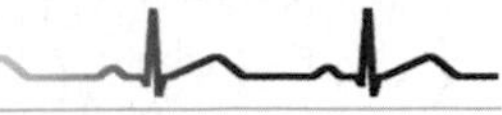

After reading this section, you should be able to

A. **Describe the process of inflammation in response to tissue damage and explain how inflammation protects the body.**
B. **Relate the five major signs of inflammation and how they are produced.**

Inflammation (*flamma*, flame) is the response that occurs when tissues are damaged. Although many agents cause injury, such as microorganisms, cold, heat, radiant energy, chemicals, electricity, and mechanical trauma, the inflammatory response to all of them is similar. The inflammatory response mobilizes the body's defenses, isolates and destroys microorganisms and other injurious agents, and removes foreign materials and damaged cells, so that tissue repair can proceed (see chapter 22). Figure 4.7 illustrates the stages of the inflammatory response.

Inflammation has five major manifestations: redness, heat, swelling, pain, and disturbed function. Although unpleasant, these processes usually aid recovery, and each of the symptoms can be understood in terms of events that occur during the inflammatory response.

After a person is injured, chemical substances called **chemical mediators** are released or activated in the tissues and the adjacent blood vessels. The mediators are histamine, kinins, prostaglandins, leukotrienes, and others. Some mediators induce dilation of blood vessels and produce redness and heat. Dilation of blood vessels is beneficial because it speeds the arrival of white blood cells and other substances important for fighting infections and repairing the injury.

Chemical mediators also stimulate pain receptors and increase the permeability of blood vessels. The increased permeability allows materials such as clotting proteins and white blood cells to move out of the blood vessels and into the tissue, where they can deal directly with the injury. As proteins from the blood move into the tissue, they change the osmotic relationship between the blood and the tissue. Water follows the proteins by osmosis, and the tissue swells, producing **edema** (e-dē′mă). Edema increases the pressure in the tissue, which can also stimulate neurons and cause pain.

Clotting proteins present in blood diffuse into the interstitial spaces and form a clot. Clotting also occurs by platelet aggregation in the injured blood vessels. Clotting isolates the injurious agent and separates it from the rest of the body. Foreign particles and microorganisms at the site of injury are "walled off" from tissues by the clotting process. Pain, limitation of movement resulting from edema, and tissue destruction all contribute to the disturbed function. This disturbance can be valuable because it warns the person to protect the injured structure from further damage. Sometimes the inflammatory response lasts longer or is more intense than is desirable, and the patient is given drugs to suppress the symptoms. Antihistamines block the effects of histamine, aspirin prevents the synthesis of prostaglandins, and cortisone reduces the release of several chemical mediators that cause inflammation. Still, the inflammatory response by itself may not be enough to combat the effects of injury or fight off an infection, and the patient may require antibiotics.

ASSESS YOUR PROGRESS 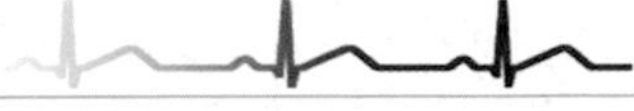

53. *What is the function of the inflammatory response?*
54. *Name the five manifestations of inflammation; explain how each is produced and the benefits of each.*

Predict 7

In some injuries, tissues are so severely damaged that some cells die and blood vessels are destroyed. For such injuries, where do the signs of inflammation, such as redness, heat, edema, and pain, occur?

4.9 Tissue Repair

LEARNING OUTCOMES 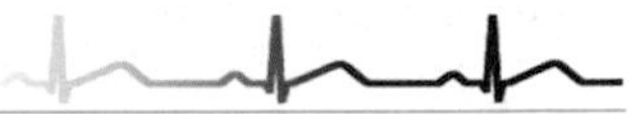

After reading this section, you should be able to

A. **Describe the three groups of cells based on their ability to regenerate.**
B. **Explain the major events involved in tissue repair.**

Tissue repair is the substitution of viable cells for dead cells. Tissue repair can occur by regeneration or replacement. In **regeneration** (rē′jen-er-ā ′shŭn), the new cells are the same type as those that were destroyed, and normal function is usually restored. In **replacement,** a new type of tissue develops, which eventually produces a scar and causes the loss of some tissue function. Most wounds heal through regeneration and replacement; which process dominates depends on the tissues involved and the nature and extent of the wound.

Cells are classified into three groups—labile, stable, or permanent cells—according to their ability to regenerate. **Labile cells** continue to divide throughout life. Labile cells include adult stem cells and other cells of the skin, mucous membranes, and hemopoietic and lymphatic tissues. Damage to these cells can be repaired completely by regeneration. **Stable cells,** such as those of connective tissues and glands, including the liver, pancreas, and

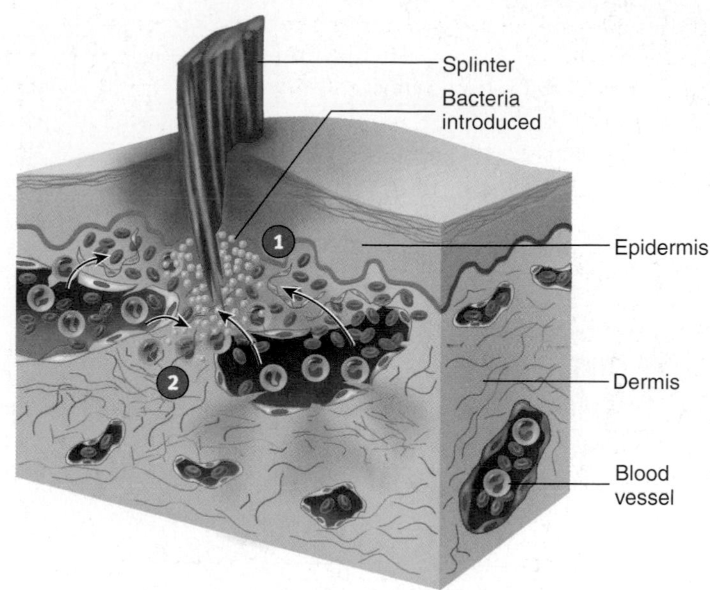
Splinter
Bacteria introduced
Epidermis
Dermis
Blood vessel
1
2

1 A splinter in the skin causes damage and introduces bacteria. Chemical mediators of inflammation are released or activated in injured tissues and adjacent blood vessels. Some blood vessels rupture, causing bleeding.

2 Chemical mediators cause capillaries to dilate and the skin to become red. Chemical mediators also increase capillary permeability, and fluid leaves the capillaries, producing swelling *(arrows)*.

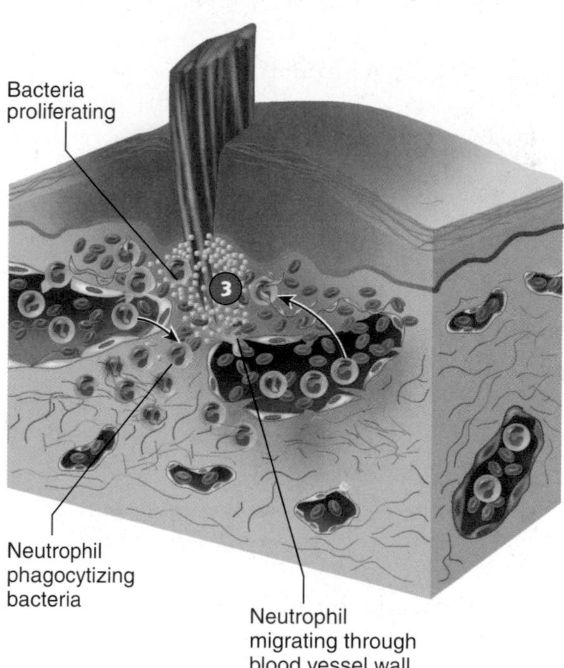
Bacteria proliferating
3
Neutrophil phagocytizing bacteria
Neutrophil migrating through blood vessel wall

3 White blood cells (e.g., neutrophils) leave the dilated blood vessels and move to the site of bacterial infection, where they begin to phagocytize bacteria and other debris.

PROCESS FIGURE 4.7 Stages of the Inflammatory Response

endocrine glands, do not divide after growth ceases, but they retain the ability to divide and are capable of regeneration in response to injury. **Permanent cells** have a very limited ability to replicate and, if killed, are usually replaced by a different type of cell. Some permanent cells, such as neurons, are postmitotic. If damaged, neurons may recover if the cell body is not destroyed; however, if the neuron cell body is destroyed, the remainder of the neuron dies. Some undifferentiated cells of the central nervous system are stem cells that can undergo mitosis and form functional neurons in the adult. This has raised hope that damaged areas of the brain may be regenerated. Undifferentiated cells of skeletal and cardiac muscle also have a very limited ability to regenerate in response to injury, although individual skeletal and cardiac muscle cells can repair themselves. In contrast, smooth muscle readily regenerates following injury.

Skin repair is a good example of tissue repair (figure 4.8). The basic pattern of repair is the same as for other tissues, especially those covered by epithelium. If the edges of the wound are close together, as in a surgical incision, the wound heals by a process called **primary union,** or *primary intention.* If the edges are not close together, or if tissue loss has been extensive, the process is called **secondary union,** or *secondary intention.*

FUNDAMENTAL **Figure**

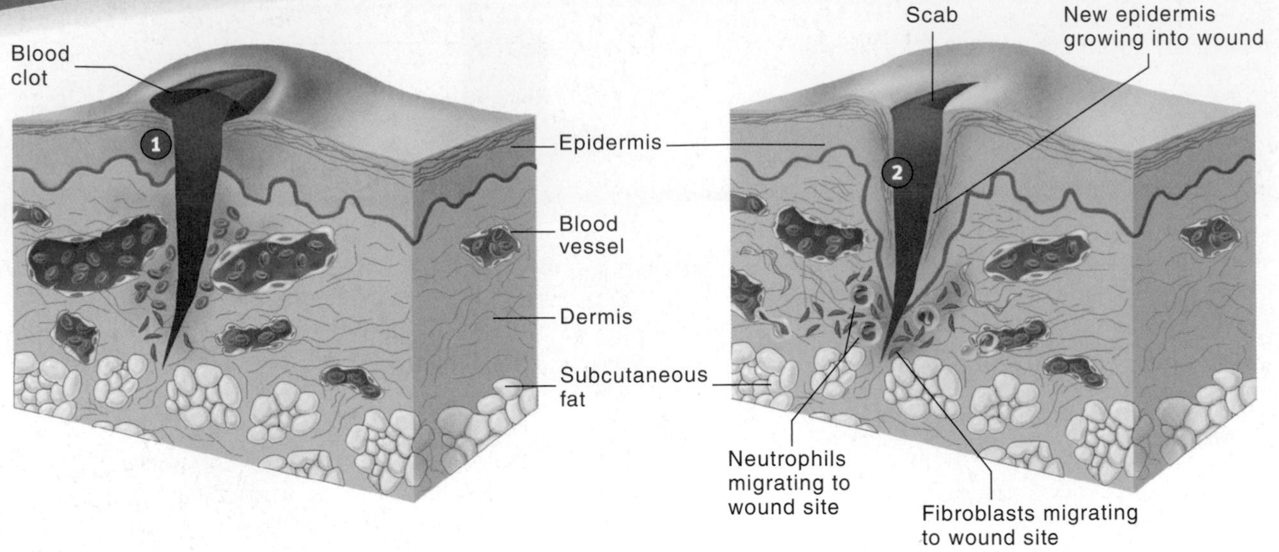
Blood clot
Epidermis
Blood vessel
Dermis
Subcutaneous fat
Scab
New epidermis growing into wound
Neutrophils migrating to wound site
Fibroblasts migrating to wound site

❶ Fresh wound cuts through the epithelium (epidermis) and underlying connective tissue (dermis), and a clot forms.

❷ Approximately 1 week after the injury, a scab is present, and epithelium (new epidermis) is growing into the wound.

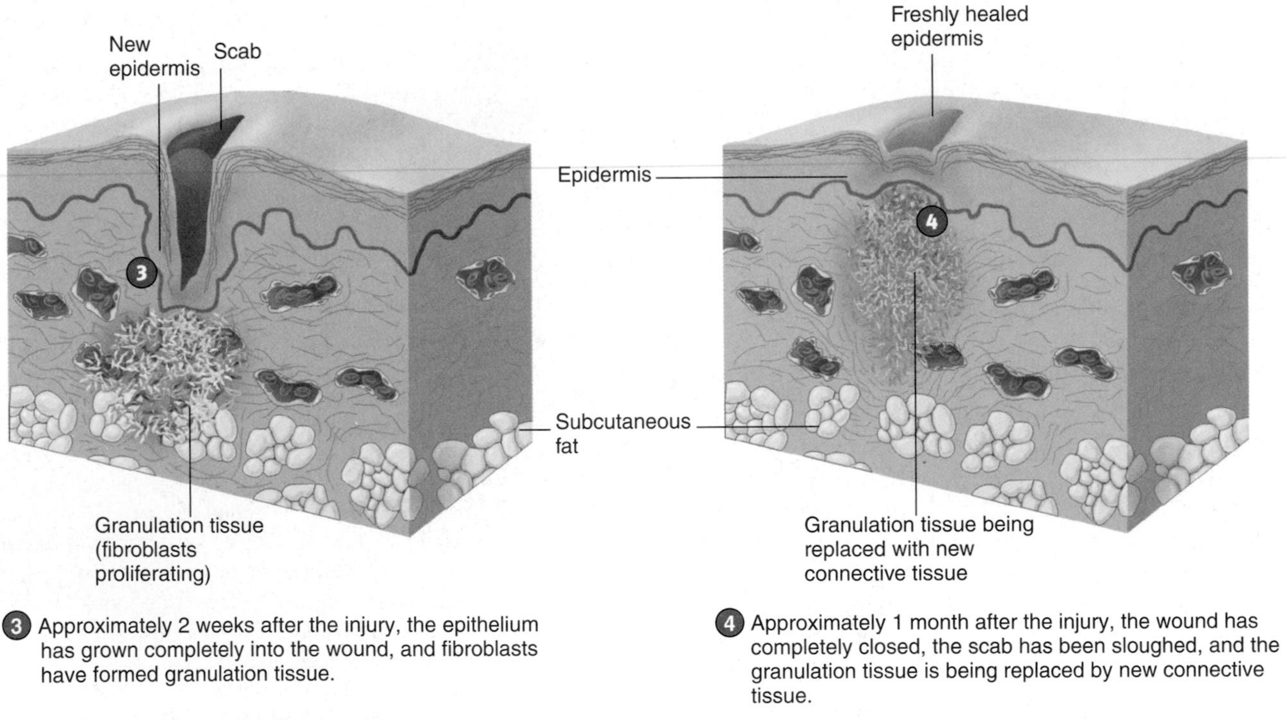
New epidermis
Scab
Epidermis
Subcutaneous fat
Granulation tissue (fibroblasts proliferating)
Freshly healed epidermis
Granulation tissue being replaced with new connective tissue

❸ Approximately 2 weeks after the injury, the epithelium has grown completely into the wound, and fibroblasts have formed granulation tissue.

❹ Approximately 1 month after the injury, the wound has completely closed, the scab has been sloughed, and the granulation tissue is being replaced by new connective tissue.

PROCESS FIGURE 4.8 Tissue Repair

In primary union, the wound fills with blood and a clot forms (see chapter 19). The clot contains the threadlike protein **fibrin** (fī′brin), which binds the edges of the wound together. The surface of the clot dries to form a **scab,** which seals the wound and helps prevent infection. An inflammatory response induces vasodilation and takes more blood cells and other substances to the area. Blood vessel permeability increases, resulting in edema (swelling). Fibrin and blood cells move into the wounded tissues because of the increased vascular permeability. Fibrin isolates and walls off microorganisms and other foreign matter. Some of the white blood cells that move into the tissue are phagocytic cells called **neutrophils** (noo′trō-filz; figure 4.8). They ingest bacteria, thus helping fight infection, and they ingest tissue debris and clear the area for repair. Neutrophils are killed in this process and can accumulate as a mixture of dead cells and fluid called **pus** (pŭs).

Fibroblasts from surrounding connective tissue migrate into the clot and produce collagen and other extracellular matrix

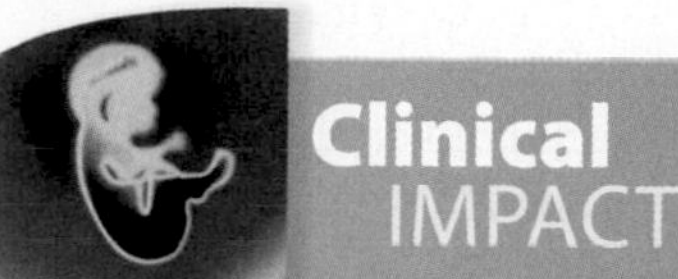
Clinical
IMPACT

Chronic Inflammation

If the agent responsible for an injury is not removed or if the healing process is disrupted, the inflammatory response persists and is called **chronic inflammation.** For example, some lung infections are brief and end in repair, but a prolonged infection can cause chronic inflammation that results in tissue destruction and permanent damage to the lung. Chronic inflammation of the stomach or small intestine may cause an ulcer. Chronic inflammation can also result from prolonged infections; prolonged exposure to irritants, such as silica, in the lung; or abnormal immune responses. White blood cells invade areas of chronic inflammation, and healthy tissues are ultimately destroyed and replaced by fibrous connective tissue, which is an important cause of loss of organ function. Chronic inflammation of the lungs, the liver, the kidneys, or other vital organs can lead to death.

components. Capillaries grow from blood vessels at the edge of the wound and revascularize the area, and fibrin in the clot is broken down and removed. The result is the replacement of the clot by **granulation tissue,** a delicate, granular-appearing connective tissue that consists of fibroblasts, collagen, and capillaries. A large amount of granulation tissue is converted to a **scar,** which consists of dense irregular collagenous connective tissue. At first, a scar is bright red because numerous blood vessels are present. Later, the scar becomes white as collagen accumulates and the vascular channels are compressed.

Repair by secondary union proceeds in a similar fashion, but with some differences. Because the wound edges are far apart, the clot may not close the gap completely, and it takes the epithelial cells much longer to regenerate and cover the wound. Also, the increased tissue damage means that both the degree of inflammation and the risk of infection are greater and there is more cell debris for the phagocytes to remove. Much more granulation tissue forms, and the contraction of fibroblasts in the granulation tissue leads to **wound contracture,** resulting in disfiguring and debilitating scars. Thus, it is advisable to suture a large wound, so that it can heal by primary rather than secondary union. Healing is faster, with a lowered risk of infection and a reduced degree of scarring.

ASSESS YOUR PROGRESS

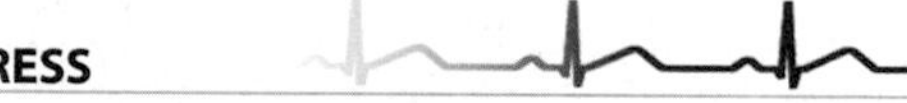

55. *Define* tissue repair. *Differentiate between repair by regeneration and repair by replacement.*

56. *Compare labile, stable, and permanent cells according to their ability to regenerate. Give examples of each type.*

57. *Describe the process of wound repair. Contrast healing by primary union and healing by secondary healing. Which process is better, and why?*

58. *What is granulation tissue? How does granulation tissue contribute to scars and wound contracture?*

4.10 Effects of Aging on Tissues

LEARNING OUTCOME

After reading this section, you should be able to

A. Describe the age-related changes that occur in cells and in extracellular matrix.

Age-related changes—for example, reduced visual acuity and reduced smell, taste, and touch sensations—are well documented. A clear decline in many types of athletic performance can be measured after approximately age 30–35. With advanced age, the number of neurons and muscle cells decreases substantially. The functional capacity of body systems, such as the respiratory and cardiovascular systems, declines. The rates of healing and scarring in the elderly are very different from those in the very young, and major changes in skin structure develop. Characteristic alterations in brain function also develop in the elderly. All these changes result in the differences among young, middle-aged, and older people.

At the tissue level, age-related changes affect cells and the extracellular materials they produce. In general, cells divide more slowly in older people. Collagen fibers become more irregular in

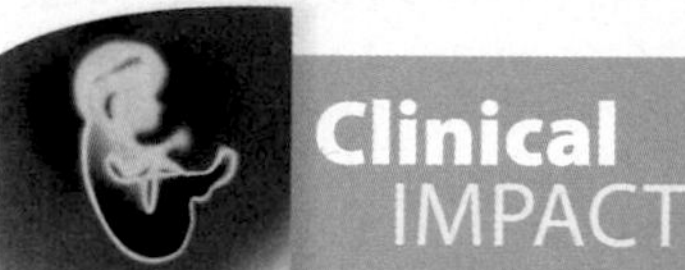
Clinical
IMPACT

Molecular Tissue Profiles of Cancer Tissue

There are many types of cancer and special names for them based on their tissue of origin. The most common types of cancer are those from epithelial tissue. A **carcinoma** (kar-si-nō′ma) is a cancer derived from epithelial tissue. Carcinomas include nearly all lung, breast, colon, prostate, and skin cancers. Basal cell and squamous cell carcinomas are types of skin cancer derived from epithelial tissue. **Adenocarcinomas** (ad′ĕ-nō-kar-si-nō′maz) are types of carcinomas derived from glandular epithelium. Most breast cancers are adenocarcinomas. A **sarcoma** (sar-kō′mă) is a relatively rare type of cancer derived from mesodermal tissue (muscle and connective tissue). For example, an osteosarcoma (os′tē-ō-sar-kō′mă) is cancer of bone, and a chondrosarcoma (kon′drō-sar-kō′mă) is cancer of cartilage.

Identifying the tissue of origin is useful for the diagnosis and treatment of cancer. Since tumor cells have altered shapes compared with their morphology in tissues (see Clinical Genetics, "Genetic Changes in Cancer Cells," in chapter 3), molecular markers are commonly used to identify the type of tumor. For example, specific types of carcinomas express keratin filaments that are characteristic of different types of epithelial tissue. Other intermediate filaments are diagnostic of sarcomas and other types of cancers. Advances in nucleic acid technologies have opened the door for even more extensive gene expression profiling of cancers. In the future, it is likely that distinct molecular profiles of cancer will allow a more definitive diagnosis and prognosis, which may lead to targeted therapies tailored for individual patients.

structure, even though their number may increase. As a consequence, connective tissues with abundant collagen, such as tendons and ligaments, become less flexible and more fragile. Elastic fibers fragment, bind to calcium ions, and become less elastic. Consequently, elastic connective tissues become less elastic. Reduced flexibility and elasticity of connective tissue are responsible for increased wrinkling of the skin, as well as the increased tendency for bones to break in older people.

Arterial walls also become less elastic due to changes in the structure of elastic and collagen fibers. Atherosclerosis results as plaques form in the walls of blood vessels, which contain collagen fibers, lipids, and calcium deposits (see chapter 21). These changes result in a reduced blood supply to tissues and an increased susceptibility to blockage and rupture. The rate of red blood cell synthesis declines in the elderly as well.

Injuries in the very young heal more rapidly and more completely than in older people. A fracture in an infant's femur is likely to heal quickly and eventually leave no evidence in the bone. A similar fracture in an adult heals more slowly, and a scar, seen in x-rays of the bone, is likely to persist throughout life.

However, there is good news. It is increasingly evident that many of the cell losses and functional declines of aging can be slowed by physical and mental exercise. Staying active, both physically and mentally, is often a good prescription for better health.

ASSESS YOUR PROGRESS

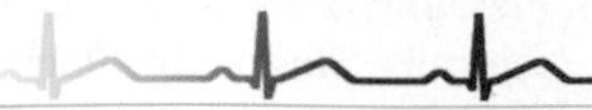

59. *How do cells respond to the effects of aging?*

60. *Describe the age-related changes in tissues with abundant collagen and elastic fibers.*

61. *How does healing of injuries change in older people?*

Answer

Learn to Predict ❮ From page 101

The question tells us that gluten enteropathy affects the intestinal lining, reducing its ability to absorb nutrients and water. It also reminds us that nutrient and water absorption occurs at the cellular level via several different transport processes.

Let us first identify the tissue type affected by gluten enteropathy. In chapter 4 we learned that epithelial tissue covers body surfaces, including the lining of the intestines. Further reading showed that the intestinal lining is composed of simple columnar epithelial tissue. Therefore, the tissue type affected by Matt's gluten enteropathy is simple columnar epithelium.

We are then asked to identify the specific cell parts affected by this disease. As stated in the question, the intestinal lining is organized into fingerlike projections called villi, which are covered by the simple columnar epithelium. Chapter 4 stated that the epithelial cells of this tissue have microvilli. In chapter 3, we learned that microvilli are extensions of the plasma membrane that increase the surface area for absorption. Matt's gluten enteropathy reduced his ability to absorb nutrients and water, so we can conclude that the cell parts affected by the disease are the microvilli.

Finally, the question asks us to explain why Matt suffers from bouts of diarrhea after eating gluten. We know that gluten damages the intestinal lining by decreasing the number of villi and microvilli. This reduces the surface area for absorption. If the surface area decreases, fewer nutrients are absorbed. Chapter 3 showed us that water moves by osmosis to areas of higher solute concentration. The nutrient molecules are solutes in the intestines. Since the solutes are not being absorbed, the solute concentration remains high in the intestines, and water absorption decreases. As a result, the nutrients and water accumulate in the intestines, resulting in the watery feces of diarrhea.

Answers to the rest of this chapter's Predict questions are in Appendix G.

Summary

4.1 Tissues and Histology (p. 102)

1. Tissues are collections of similar cells and the extracellular substances surrounding them.
2. The four primary tissue types are epithelial, connective, muscle, and nervous tissues.
3. Histology is the microscopic study of tissues.

4.2 Embryonic Tissue (p. 102)

All four of the primary tissue types are derived from each of the three germ layers (mesoderm, ectoderm, and endoderm).

4.3 Epithelial Tissue (p. 103)

1. Epithelium consists of cells with little extracellular matrix. It covers surfaces, usually has a basement membrane, and does not have blood vessels.
2. The basement membrane is secreted by the epithelial cells and attaches the epithelium to the underlying tissues.

Functions of Epithelial Tissues

Epithelial tissues protect underlying structures, act as barriers, permit some substances to pass through epithelial layers, secrete substances, and absorb substances.

Classification of Epithelial Tissues

1. Simple epithelium has a single layer of cells, stratified epithelium has two or more layers, and pseudostratified epithelium has a single layer that appears stratified.
2. Cells can be squamous (flat), cuboidal, or columnar.
3. Stratified squamous epithelium can be nonkeratinized or keratinized.
4. Transitional epithelium is stratified, with cells that can change shape from cuboidal to flattened.

Functional Characteristics

1. Simple epithelium is usually involved in diffusion, filtration, secretion, or absorption. Stratified epithelium serves a protective role. Squamous cells function in diffusion and filtration. Cuboidal or columnar cells, with a larger cell volume that contains many organelles, secrete or absorb.
2. A smooth free surface reduces friction (mesothelium and endothelium), microvilli increase absorption (intestines), and cilia move materials across the free surface (respiratory tract and uterine tubes). Transitional epithelium has a folded surface that allows the cell to change shape, and the number of cells making up the epithelial layers changes.
3. Cells are bound together mechanically by glycoproteins, desmosomes, and adhesion belts and to the basement membrane by hemidesmosomes. Tight junctions form a permeability barrier, and gap junctions allow intercellular communication.

Glands

1. Glands are organs that secrete. Exocrine glands secrete through ducts, and endocrine glands release hormones that are absorbed directly into the blood.
2. Glands are classified as unicellular or multicellular. Goblet cells are unicellular glands. Multicellular exocrine glands have ducts, which are simple or compound. The ducts can be tubular or end in small sacs (acini or alveoli). Tubular glands can be straight or coiled.
3. Glands are classified according to their mode of secretion. Merocrine glands (pancreas) secrete substances as they are produced, apocrine glands (mammary glands) accumulate secretions that are released when a portion of the cell pinches off, and holocrine glands (sebaceous glands) accumulate secretions that are released when the cell ruptures and dies.

4.4 Connective Tissue (p. 113)

Connective tissue is distinguished by its extracellular matrix.

Functions of Connective Tissue

Connective tissues enclose and separate organs and tissues; connect tissues to one another; help support and move body parts; store compounds; cushion and insulate the body; transport substances; and protect against toxins and injury.

Cells of Connective Tissue

1. The extracellular matrix results from the activity of specialized connective tissue cells; in general, blast cells form the matrix, cyte cells maintain it, and clast cells break it down. Fibroblasts form protein fibers of many connective tissues, osteoblasts form bone, and chondroblasts form cartilage.
2. Connective tissue commonly contains adipose cells, mast cells, white blood cells, macrophages, and mesenchymal cells (stem cells).

Extracellular Matrix

1. The major components of the extracellular matrix of connective tissue are protein fibers, ground substance, and fluid.
2. Protein fibers of the matrix have the following characteristics:
   - Tropocollagens are linked together to form collagen fibrils, which are joined to form collagen fibers. The collagen fibers resemble ropes. They are strong and flexible but resist stretching.
   - Reticular fibers are fine collagen fibers that form a branching network that supports other cells and tissues.
   - Elastic fibers have a structure similar to that of a spring. After being stretched, they tend to return to their original shape.
3. Ground substance has the following major components:
   - Hyaluronic acid makes fluids slippery.
   - Proteoglycan aggregates trap water, which gives tissues the capacity to return to their original shape when compressed or deformed.
   - Adhesive molecules hold proteoglycans together and to plasma membranes.

Connective Tissue Classifications

Connective tissue is classified according to the type and proportions of cells and extracellular matrix fibers, ground substance, and fluid.

1. Embryonic connective tissue is called mesenchyme, consists of irregularly shaped cells and abundant matrix, and gives rise to adult connective tissue.
2. Adult connective tissue consists of connective tissue proper, supporting connective tissue, and fluid connective tissue.

Connective Tissue Proper

1. Loose connective tissue
   - Areolar connective tissue has many different cell types and a random arrangement of protein fibers with space between the fibers. This tissue fills spaces around the organs and attaches the skin to underlying tissues.
   - Adipose tissue has adipocytes filled with lipid and very little extracellular matrix (a few reticular fibers). It functions in energy storage, insulation, and protection. Adipose tissue can be yellow or brown. Brown adipose is specialized for generating heat.
   - Reticular tissue is a network of reticular fibers; it forms the framework of lymphatic tissue, bone marrow, and the liver.
2. Dense connective tissue
   - Dense regular connective tissue is composed of fibers arranged in one direction, which provides strength in a direction parallel to the fiber orientation. Two types of dense regular connective tissue exist: collagenous (tendons and most ligaments) and elastic (ligaments of vertebrae).
   - Dense irregular connective tissue has fibers organized in many directions, which produces strength in different directions. Two types of dense irregular connective tissue exist: collagenous (capsules of organs and dermis of skin) and elastic (large arteries).

Supporting Connective Tissue

1. Cartilage
   - Cartilage has a relatively rigid matrix composed of protein fibers and proteoglycan aggregates. The major cell type is the chondrocyte, which is located within lacunae.
   - Hyaline cartilage has evenly dispersed collagen fibers that provide rigidity with some flexibility. Examples include the costal cartilage, the covering over the ends of bones in joints, the growing portion of long bones, and the embryonic skeleton.
   - Fibrocartilage has collagen fibers arranged in thick bundles; it can withstand great pressure, and it is found between vertebrae, in the jaw, and in the knee.
   - Elastic cartilage is similar to hyaline cartilage, but it contains elastin. It is more flexible than hyaline cartilage and is found in the external ear.
2. Bone
   - Bone cells, or osteocytes, are located in lacunae surrounded by a mineralized matrix (hydroxyapatite) that makes bone very hard. Spongy bone has spaces between bony trabeculae; compact bone is more solid.

Fluid Connective Tissue

1. Blood
   - Blood cells are suspended in a fluid matrix.
2. Hemopoietic tissue
   - Hemopoietic tissue forms blood cells.

4.5 Muscle Tissue (p. 124)

1. Muscle tissue has the ability to contract.
2. Skeletal (striated voluntary) muscle attaches to bone and is responsible for body movement. Skeletal muscle cells are long and cylindrically shaped with many peripherally located nuclei.
3. Cardiac (striated involuntary) muscle cells are cylindrical, branching cells with a single, central nucleus. Cardiac muscle is found in the heart and is responsible for pumping blood through the circulatory system.
4. Smooth (nonstriated involuntary) muscle forms the walls of hollow organs, the iris of the eye, and other structures. Its cells are spindle-shaped with a single, central nucleus.

4.6 Nervous Tissue (p. 127)

1. Nervous tissue is able to conduct electrical impulses and is composed of neurons (conductive cells) and neuroglia (support cells).
2. Neurons have cell processes called dendrites and axons. Dendrites receive electrical impulses, and axons conduct them. Neurons can be multipolar (several dendrites and an axon), bipolar (one dendrite and one axon), or pseudo-unipolar (one axon).

4.7 Tissue Membranes (p. 129)

1. Mucous membranes consist of epithelial cells, their basement membrane, the lamina propria, and sometimes smooth muscle cells; they line cavities that open to the outside and often contain mucous glands, which secrete mucus.
2. Serous membranes line cavities that do not open to the exterior and do not contain glands but do secrete serous fluid.
3. Synovial membranes are formed by connective tissue, line joint cavities, and secrete a lubricating fluid.

4.8 Tissue Damage and Inflammation (p. 130)

1. Inflammation involves a response that isolates injurious agents from the rest of the body and destroys the injurious agent.
2. Inflammation produces five symptoms: redness, heat, swelling, pain, and disturbed function.

4.9 Tissue Repair (p. 130)

1. Tissue repair is the substitution of viable cells for dead ones. Tissue repair occurs by regeneration or replacement.
   - Labile cells divide throughout life and can undergo regeneration.
   - Stable cells do not ordinarily divide after growth is complete but can regenerate if necessary.
   - Permanent cells cannot replicate. If killed, permanent tissue is repaired by replacement.
2. Tissue repair by primary union occurs when the edges of the wound are close together. Secondary union occurs when the edges are far apart.

4.10 Effects of Aging on Tissues (p. 133)

1. Age-related changes in tissues result from reduced rates of cell division and changes in the extracellular fibers.
2. Collagen fibers become less flexible and have reduced strength.
3. Elastic fibers become fragmented and less elastic.

REVIEW AND COMPREHENSION

1. Given these characteristics:
   (1) capable of contraction
   (2) covers free body surfaces
   (3) lacks blood vessels
   (4) composes various glands
   (5) anchored to connective tissue by a basement membrane

   Which of these are characteristics of epithelial tissue?
   a. 1,2,3 c. 3,4,5 e. 2,3,4,5
   b. 2,3,5 d. 1,2,3,4
2. Which of these embryonic germ layers gives rise to muscle, bone, and blood vessels?
   a. ectoderm b. endoderm c. mesoderm
3. A tissue that covers a surface, is one cell layer thick, and is composed of flat cells is
   a. simple squamous epithelium.
   b. simple cuboidal epithelium.
   c. simple columnar epithelium.
   d. stratified squamous epithelium.
   e. transitional epithelium.

4. Stratified epithelium is usually found in areas of the body where the principal activity is
   a. filtration.
   b. protection.
   c. absorption.
   d. diffusion.
   e. secretion.
5. Which of these characteristics do *not* describe nonkeratinized stratified squamous epithelium?
   a. many layers of cells
   b. flat surface cells
   c. living surface cells
   d. found in the skin
   e. outer layers covered by fluid
6. In parts of the body where considerable expansion occurs, such as the urinary bladder, which type of epithelium would you expect to find?
   a. cuboidal
   b. pseudostratified
   c. transitional
   d. squamous
   e. columnar
7. Epithelial cells with microvilli are most likely found
   a. lining blood vessels.
   b. lining the lungs.
   c. lining the uterine tube.
   d. lining the small intestine.
   e. in the skin.
8. Pseudostratified ciliated columnar epithelium can be found lining the
   a. digestive tract.
   b. trachea.
   c. thyroid gland.
   d. kidney tubules.
   e. urinary bladder.
9. A type of cell connection whose *only* function is to prevent the cells from coming apart is a
   a. desmosome.
   b. gap junction.
   c. tight junction.
10. The glands that lose their connection with epithelium during embryonic development and secrete their cellular products into the bloodstream are called ____________ glands.
    a. apocrine
    b. endocrine
    c. exocrine
    d. holocrine
    e. merocrine
11. A ____________ gland has a duct that branches repeatedly, and the ducts end in saclike structures.
    a. simple tubular
    b. compound tubular
    c. simple coiled tubular
    d. simple acinar
    e. compound acinar
12. The fibers in dense connective tissue are produced by
    a. fibroblasts.
    b. adipocytes.
    c. osteoblasts.
    d. osteoclasts.
    e. macrophages.
13. Mesenchymal cells
    a. form embryonic connective tissue.
    b. give rise to all adult connective tissues.
    c. in adults produce new connective tissue cells in response to injury.
    d. All of these are correct.
14. A tissue with a large number of collagen fibers organized parallel to each other would most likely be found in
    a. a muscle.
    b. a tendon.
    c. adipose tissue.
    d. a bone.
    e. cartilage.
15. Extremely delicate fibers that make up the framework for organs such as the liver, spleen, and lymph nodes are
    a. elastic fibers.
    b. reticular fibers.
    c. microvilli.
    d. cilia.
    e. collagen fibers.
16. In which of these locations is dense irregular elastic connective tissue found?
    a. ligaments
    b. nuchal ligament
    c. dermis of the skin
    d. large arteries
    e. adipose tissue
17. Which of these is *not* true of adipose tissue?
    a. It is the site of energy storage.
    b. It is a type of connective tissue.
    c. It acts as a protective cushion.
    d. Brown adipose is found only in older adults.
    e. It functions as a heat insulator.
18. Which of these types of connective tissue has the smallest amount of extracellular matrix?
    a. adipose
    b. bone
    c. cartilage
    d. loose connective tissue
    e. blood
19. Fibrocartilage is found
    a. in the cartilage of the trachea.
    b. in the rib cage.
    c. in the external ear.
    d. on the surface of bones in movable joints.
    e. between vertebrae.
20. A tissue composed of cells located in lacunae surrounded by a hard matrix of hydroxyapatite is
    a. hyaline cartilage.
    b. bone.
    c. nervous tissue.
    d. dense regular collagenous connective tissue.
    e. fibrocartilage.
21. Which of these characteristics apply to smooth muscle?
    a. striated, involuntary
    b. striated, voluntary
    c. unstriated, involuntary
    d. unstriated, voluntary
22. Which of these statements about nervous tissue is *not* true?
    a. Neurons have cytoplasmic extensions called axons.
    b. Electrical signals (action potentials) are conducted along axons.
    c. Bipolar neurons have two axons.
    d. Neurons are nourished and protected by neuroglia.
    e. Dendrites receive electrical signals and conduct them toward the cell body.
23. The linings of the digestive, respiratory, excretory, and reproductive passages are composed of
    a. serous membranes.
    b. mucous membranes.
    c. mesothelium.
    d. synovial membranes.
    e. endothelium.
24. Chemical mediators
    a. cause blood vessels to constrict.
    b. decrease the permeability of blood vessels.
    c. initiate processes that lead to edema.
    d. help prevent clotting.
    e. decrease pain.
25. Which of these types of cells are labile?
    a. neurons
    b. skin
    c. liver
    d. pancreas
26. Permanent cells
    a. divide and replace damaged cells in replacement tissue repair.
    b. form granulation tissue.
    c. are responsible for removing scar tissue.
    d. are usually replaced by a different cell type if they are destroyed.
    e. are replaced during regeneration tissue repair.

Answers in Appendix E

CRITICAL THINKING

1. Given the observation that a tissue has more than one layer of cells lining a free surface, (1) list the possible tissue types that exhibit those characteristics, and (2) explain what additional observations are needed to identify the tissue type.
2. A patient suffered from kidney failure a few days after being exposed to a toxic chemical. A biopsy of his kidney indicated that many of the thousands of epithelium-lined tubules making up the kidney had lost their simple cuboidal epithelial cells, although the basement membranes appeared mostly intact. How likely is a full recovery for this person?
3. Willie B. Coffin has smoked for years. In the past few months, mucus has accumulated in his lungs and he coughs often. A tissue sample (biopsy) taken from the lower portion of his trachea indicated that stratified squamous epithelium has replaced the normal pseudostratified columnar epithelium lining the trachea. Willie's physician explained that he has bronchitis, inflammation of the respiratory passages, caused by smoking. As a result, some of the normal epithelium of the large respiratory passageways has been converted to stratified squamous epithelium. Explain why mucus has accumulated in Willie's lungs to a greater degree than normal.
4. How can you distinguish between a gland that produces a merocrine secretion and a gland that produces a holocrine secretion? Assume that you have the ability to chemically analyze the composition of the secretions.
5. Name a tissue that has the following characteristics: abundant extracellular matrix consisting almost entirely of collagen fibers that are parallel to each other. Then state which of the following injuries results from damage to this kind of tissue: dislocated neck vertebrae, torn tendon, or ruptured intervertebral disk.
6. Antihistamines block the effect of a chemical mediator called histamine, which is released during the inflammatory response. What effect does administering antihistamines have on the inflammatory response, and is the use of an antihistamine beneficial?

Answers in Appendix F

Visit this book's website at **www.mhhe.com/seeley10** for chapter quizzes, interactive learning exercises, and other study tools.

Mc Graw Hill
connect plus+
ANATOMY & PHYSIOLOGY

Anatomy & Physiology
REVEALED
aprevealed.com

5

Integumentary System

Learn to Predict

It was a dream job—summer days spent poolside, soaking up the sun. Following her freshman year, Laura worked as a lifeguard at the country club in her hometown. By the end of summer, she had a golden tan and was anxious to show off her hot new look on campus. However, after returning to school in the fall, she was disappointed to see that her skin kept getting lighter, and in only a few weeks it had returned to its normal pale tone. By combining your understanding of epithelial tissue from chapter 4 with further information about skin structure and pigmentation in this chapter, explain how and why Laura's tan faded in the fall.

Photo: The number of skin care products is amazing. The woman in this photo is shopping at a retail store that specializes in products primarily dedicated to improving the appearance of the integumentary system.

It is the morning of "the big day." You look in the mirror and, much to your dismay, there is a big red bump on your chin. Just when you needed to look your best, this had to happen! For most people, blemish-free skin is highly desirable, and any sign of acne is cause for embarrassment. Hair loss and crows' feet also cause consternation in some people. It goes without saying that much time, effort, and money are spent on changing the appearance of the integumentary system. Think about the amount of counter space dedicated to skin care products, hair care products, and cosmetics in a typical discount store. People apply lotion to their skin, color their hair, and trim their nails. They try to prevent sweating by using antiperspirants and reduce or even mask body odor by washing or using deodorants and perfumes.

The **integumentary** (in-teg-ū-men′tă-rē) **system** consists of the skin and accessory structures, such as hair, glands, and nails. Although people are concerned about the appearance of their integumentary system for vanity's sake, its appearance can also indicate physiological imbalances. Some disorders, such as acne or warts, affect only the integumentary system. Other disorders affect different body parts but are reflected in the integumentary system, which provides useful signs for diagnosis. For example, reduced blood flow through the skin during a heart attack can cause a person to look pale, whereas increased blood flow as a result of fever can cause a flushed appearance. Also, some diseases cause skin rashes, such as those characteristic of measles, chickenpox, and allergic reactions.

Module 4
Integumentary System

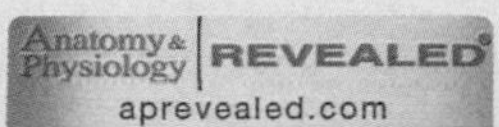
Anatomy & Physiology
REVEALED
aprevealed.com

5.1 Functions of the Integumentary System

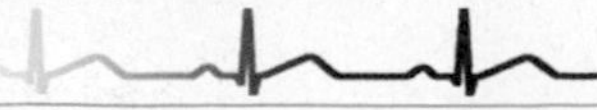

LEARNING OUTCOME

After reading this section, you should be able to

A. Describe the general functions of the integumentary system.

Although we are often concerned with how the integumentary system looks, it has many important functions that go beyond appearance. The integumentary system forms the boundary between the body and the external environment, thereby separating us from the external environment while allowing us to interact with it. Following are the major functions of the integumentary system:

1. *Protection.* The skin protects against abrasion and the harmful effects of ultraviolet light. It also keeps microorganisms from entering the body and prevents dehydration by reducing water loss from the body.
2. *Sensation.* The integumentary system has sensory receptors that can detect heat, cold, touch, pressure, and pain.
3. *Temperature regulation.* The amount of blood flow through the skin and the activity of sweat glands help regulate body temperature.
4. *Vitamin D production.* When exposed to ultraviolet light, the skin produces a molecule that can be transformed into vitamin D, an important regulator of calcium homeostasis.
5. *Excretion.* Small amounts of waste products are excreted through the skin and glands.

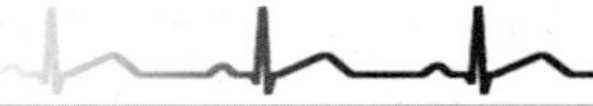

ASSESS YOUR PROGRESS

1. *Provide an example for each function of the integumentary system.*

5.2 Skin

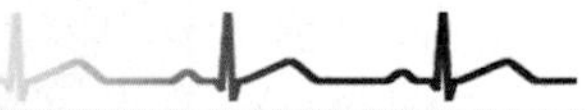

LEARNING OUTCOMES

After reading this section, you should be able to

A. Describe the structure and function of the epidermis.
B. Discuss the epidermal strata and relate them to the process of keratinization.
C. Differentiate between thick and thin skin as to the layers present and their locations.
D. Explain the major factors affecting skin color.
E. Describe the structure and functions of the dermis.

The skin is made up of two major tissue layers, the epidermis and the dermis (figure 5.1). The **epidermis** (ep-i-der′mis; upon the dermis) is the superficial layer of the skin, consisting of epithelial tissue. The epidermis resists abrasion on the skin's surface and reduces water loss through the skin. The epidermis rests on the **dermis** (der′mis; skin), a layer of connective tissue. The dermis is responsible for most of the structural strength of the skin. The strength of the dermis is seen in leather, which is produced from the hide (skin) of an animal by removing the epidermis and preserving the dermis in a process called tanning.

The skin rests on the **subcutaneous tissue,** or *hypodermis* (hi-pō-der′mis), a layer of loose connective tissue (figure 5.1). The subcutaneous tissue is not part of the skin or the integumentary system, but it does connect the skin to underlying muscle or bone. Table 5.1 summarizes the structures and functions of the skin and subcutaneous tissue.

Epidermis

The epidermis is stratified squamous epithelium. It is separated from the underlying dermis by a basement membrane. The epidermis is not as thick as the dermis and contains no blood vessels. The living cells of the epidermis receive nutrients and excrete waste products by the diffusion of substances between the epidermis and the capillaries of the dermis (figure 5.2).

The epidermis is composed of several types of cells. Most cells of the epidermis are called **keratinocytes** (ke-rat′i-nō-sītz) because they produce a protein mixture called **keratin** (ker′ă-tin), which makes the cells more durable. Keratinocytes give the epidermis its ability to resist abrasion and reduce water loss. Other cells of the epidermis include **melanocytes** (mel′ă-nō-sītz), which contribute to skin color; **Langerhans cells,** which are part of the immune system (see chapter 22); and **Merkel cells,** which are specialized epidermal cells associated with the nerve endings responsible for detecting light touch and superficial pressure (see chapter 14).

New keratinocytes are produced when keratinocyte stem cells undergo mitosis in the deepest layer of the epidermis. As new cells form, they push older cells to the surface, where they slough off. The outermost cells in this stratified arrangement protect the cells underneath, and the deeper replicating cells replace cells lost from the surface. As they move from the deeper epidermal layers to the surface, the keratinocytes change shape and chemical composition, through the process called **keratinization** (ker′ă-tin-i-zā′shŭn), where the cells accumulate keratin. During keratinization, the cells eventually die and produce an outer layer of dead, hard cells that resists abrasion and forms a permeability barrier. The study of keratinization is important because many skin diseases result from malfunctions in this process. For example, large scales of epidermal tissue are sloughed off in **psoriasis** (sō-rī′ă-sis). By comparing normal and abnormal keratinization, scientists may be able to develop effective therapies for psoriasis.

Although keratinization is a continual process, distinct transitional stages can be recognized as the cells change. On the basis of these stages, the many layers of cells in the epidermis are divided into regions, or **strata** (sing. *stratum*; see figures 5.2*b* and 5.3). From the deepest to the most superficial, the five strata are the stratum basale, stratum spinosum, stratum granulosum, stratum lucidum, and stratum corneum. The number of cell layers in each stratum and even the number of strata in the skin vary, depending on their location in the body.

Stratum Basale

The deepest portion of the epidermis is a single layer of cuboidal or columnar cells called the **stratum basale** (bā′să-lē), or *stratum germinativum* (jer′mi-nă-tīv′um; figure 5.3, *stratum 1*). Structural

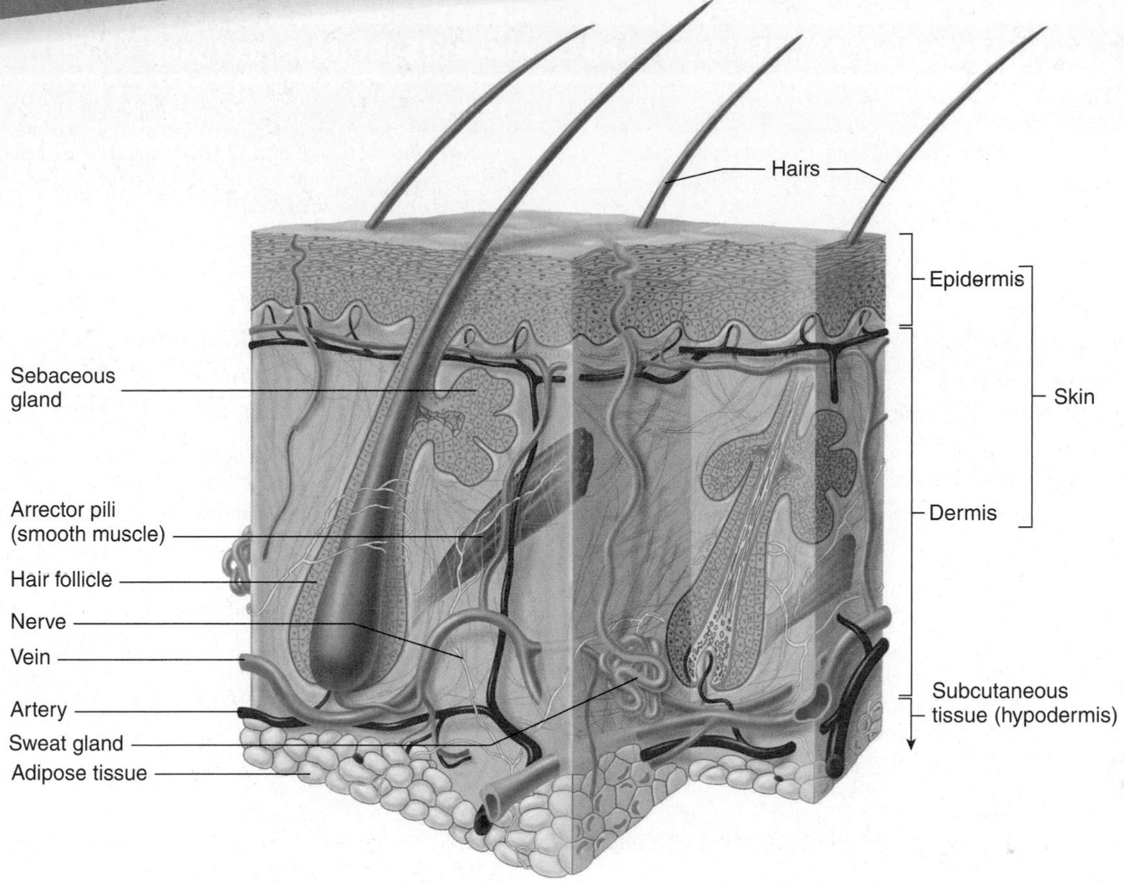
Hairs
Epidermis
Skin
Dermis
Subcutaneous tissue (hypodermis)
Sebaceous gland
Arrector pili (smooth muscle)
Hair follicle
Nerve
Vein
Artery
Sweat gland
Adipose tissue

FIGURE 5.1 AP|R **Skin and Subcutaneous Tissue**

The skin, consisting of the epidermis and the dermis, is connected by the subcutaneous tissue to underlying structures. Note the accessory structures (hairs, glands, and arrector pili), some of which project into the subcutaneous tissue, as well as the large amount of adipose tissue in the subcutaneous tissue.

strength is provided by hemidesmosomes, which anchor the epidermis to the basement membrane, and by desmosomes, which hold the keratinocytes together (see chapter 4). Keratinocytes are strengthened internally by keratin fibers (intermediate filaments) that insert into the desmosomes. Keratinocyte stem cells of the stratum basale undergo mitotic divisions approximately every 19 days. One daughter cell remains a stem cell in the stratum basale and divides again, but the other daughter cell is pushed toward the surface and becomes keratinized. It takes approximately 40–56 days for the cell to reach the epidermal surface and slough off.

Stratum Spinosum

Superficial to the stratum basale is the **stratum spinosum** (spī-nō′sŭm), consisting of 8–10 layers of many-sided cells (figure 5.3, *stratum 2*). As the cells in this stratum are pushed to the surface, they flatten; desmosomes break apart, and new desmosomes form. During preparation for microscopic observation, the cells usually shrink from one another, except where they are attached by desmosomes, causing the cells to appear spiny—hence the name stratum spinosum. Additional keratin fibers and lipid-filled, membrane-bound organelles called **lamellar** (lam′ĕ-lăr, lă-mel′ar) **bodies** form inside the keratinocytes.

Stratum Granulosum

The **stratum granulosum** (gran-ū-lō′sŭm) consists of two to five layers of somewhat flattened, diamond-shaped cells with long axes that are oriented parallel to the surface of the skin (figure 5.3, *stratum 3*). This stratum derives its name from the presence of nonmembrane-bound protein granules of **keratohyalin** (ker′ă-tō-hī′ă-lin), which accumulate in the cytoplasm of the cell. The lamellar bodies, formed as the cells pass through the stratum spinosum, move to the plasma membrane and release their lipid contents into the extracellular space. Inside the cell, a protein envelope forms beneath the plasma membrane. In the most superficial layers of the stratum granulosum, the nucleus and other organelles degenerate, and the cell dies. Unlike the other organelles and the nucleus, however, the keratin fibers and keratohyalin granules within the cytoplasm do not degenerate.

Stratum Lucidum

The **stratum lucidum** (loo′si-dŭm) appears as a thin, clear zone above the stratum granulosum (figure 5.3, *stratum 4*); it consists of several layers of dead cells with indistinct boundaries. Keratin fibers are present, but the keratohyalin, which was evident as granules in the stratum granulosum, has dispersed around the keratin fibers, and the cells appear somewhat transparent. The stratum lucidum is present in only a few areas of the body (see the section "Thick and Thin Skin").

TABLE 5.1 Comparison of the Skin (Epidermis and Dermis) and Subcutaneous Tissue

Part	Structure	Function
Epidermis	Superficial part of skin; stratified squamous epithelium; composed of four or five strata	Prevents water loss and the entry of chemicals and microorganisms; protects against abrasion and harmful effects of ultraviolet light; produces vitamin D; gives rise to hair, nails, and glands
Stratum corneum	Most superficial stratum of the epidermis; 25 or more layers of dead squamous cells	Provides structural strength due to keratin within cells; prevents water loss due to lipids surrounding cells; sloughing off of most superficial cells resists abrasion
Stratum lucidum	Three to five layers of dead cells; appears transparent; present in thick skin, absent in most thin skin	Disperses keratohyalin around keratin fibers
Stratum granulosum	Two to five layers of flattened, diamond-shaped cells	Produces keratohyalin granules; lamellar bodies release lipids from cells; cells die
Stratum spinosum	A total of 8–10 layers of many-sided cells	Produces keratin fibers; lamellar bodies form inside keratinocytes
Stratum basale	Deepest stratum of the epidermis; single layer of cuboidal or columnar cells; basement membrane of the epidermis attaches to the dermis	Produces cells of the most superficial strata; melanocytes produce and contribute melanin, which protects against ultraviolet light
Dermis	Deep part of skin; connective tissue composed of two layers	Is responsible for the structural strength and flexibility of the skin; the epidermis exchanges gases, nutrients, and waste products with blood vessels in the dermis
Papillary layer	Papillae project toward the epidermis; loose connective tissue	Brings blood vessels close to the epidermis; dermal papillae form fingerprints and footprints
Reticular layer	Mat of collagen and elastic fibers; dense irregular connective tissue	Is the main fibrous layer of the dermis; strong in many directions; forms cleavage lines
Subcutaneous tissue	Not part of the skin; loose connective tissue with abundant deposits of adipose tissue	Attaches the dermis to underlying structures; adipose tissue provides energy storage, insulation, and padding; blood vessels and nerves from the subcutaneous tissue supply the dermis

Stratum Corneum

The last, and most superficial, stratum of the epidermis is the **stratum corneum** (kōr′nē-ŭm; figure 5.3, *stratum 5*). This stratum is composed of 25 or more layers of dead, overlapping squamous cells joined by desmosomes. Eventually, the desmosomes break apart, and the cells are shed from the surface of the skin. Excessive shedding of the stratum corneum of the scalp results in dandruff. Less noticeably, skin cells are continually shed from other areas as clothes rub against the body or as the skin is washed.

The stratum corneum consists of **cornified cells,** which are dead cells, with a hard protein envelope, filled with the protein keratin. Keratin is a mixture of keratin fibers and keratohyalin. The envelope and the keratin are responsible for the structural strength of the stratum corneum. The type of keratin found in the skin is soft keratin. Another type of keratin, hard keratin, is found in nails and the external parts of hair. Cells containing hard keratin are more durable than cells with soft keratin, and they are not shed.

Lipids are released from lamellar bodies surrounding the skin cells. The lipids are responsible for many of the skin's permeability characteristics.

Predict 2

Some drugs are administered by applying them to the skin (e.g., a nicotine skin patch to help a person stop smoking). The drug diffuses through the epidermis to blood vessels in the dermis. What kind of substances can pass easily through the skin by diffusion? What kinds of substances have difficulty diffusing through the skin?

Thick and Thin Skin

When we say a person has thick or thin skin, we are usually referring metaphorically to the person's ability to take criticism. However, in a literal sense all of us have both thick and thin skin. Skin is classified as thick or thin based on the structure of the epidermis. **Thick skin** has all five epithelial strata, and the stratum corneum has many layers of cells. Thick skin is found in areas subject to pressure or friction, such as the palms of the hands, the soles of the feet, and the fingertips.

Thin skin covers the rest of the body and is more flexible than thick skin. Each of its strata contains fewer layers of cells than are found in thick skin; the stratum granulosum frequently consists of only one or two layers of cells, and the stratum lucidum is generally absent. Hair is found only in thin skin.

The entire skin, including both the epidermis and the dermis, varies in thickness from 0.5 mm in the eyelids to 5.0 mm on the back and shoulders. The terms *thin* and *thick*, which refer to the epidermis only, should not be used when total skin thickness is considered. Most of the difference in total skin thickness results from variation in the thickness of the dermis. For example, the skin of the back is thin skin, whereas that of the palm of the hand is thick skin; however, because the dermis of the skin of the back is thicker, the total skin thickness of the back is greater than that of the palm.

In skin subjected to friction or pressure, the number of layers in the stratum corneum greatly increases to produce a thickened area called a **callus** (kal′ŭs). The skin over bony prominences may

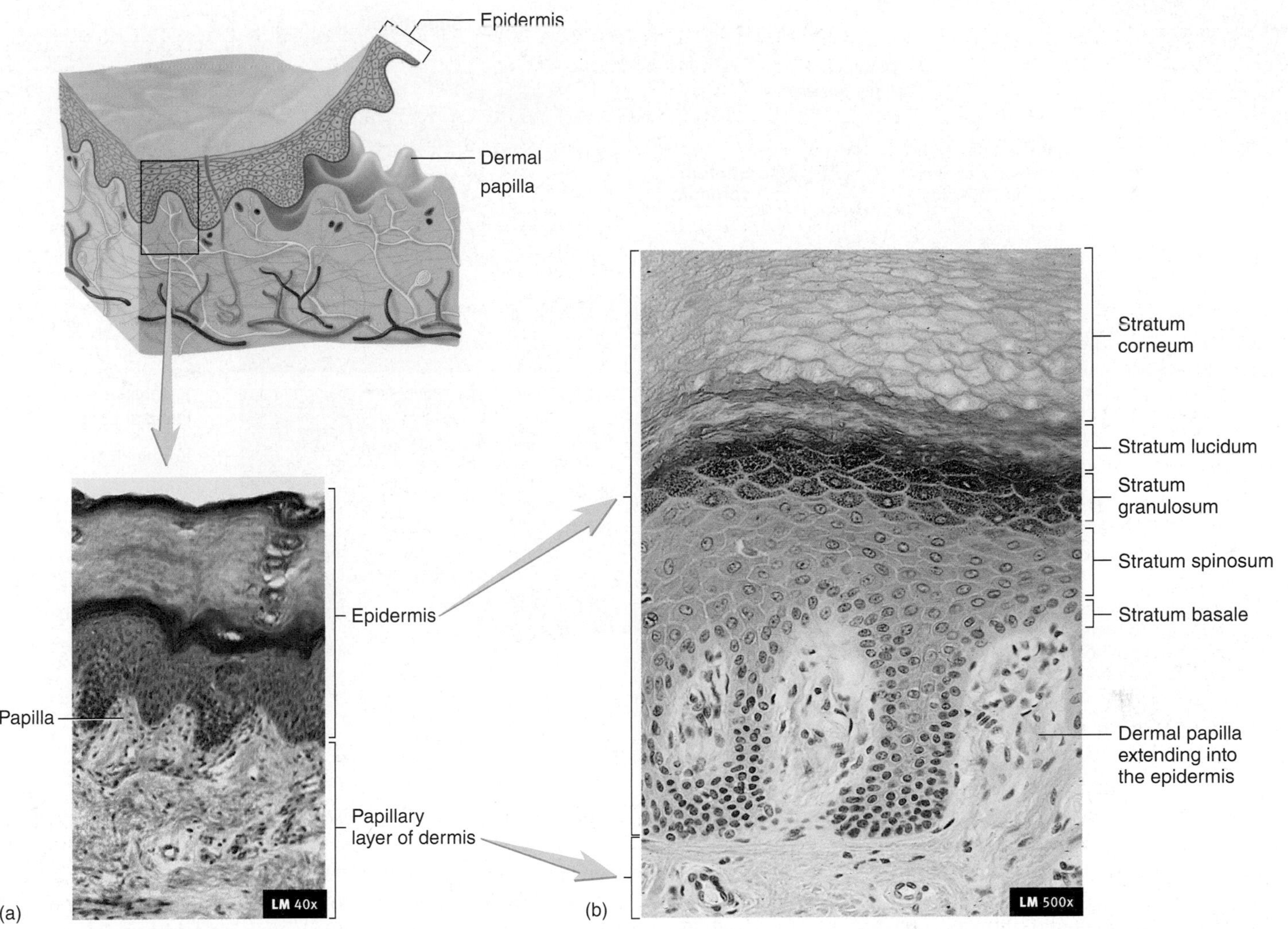
Epidermis
Dermal papilla
Papilla
Epidermis
Papillary layer of dermis
LM 40x
(a)
(b)
Stratum corneum
Stratum lucidum
Stratum granulosum
Stratum spinosum
Stratum basale
Dermal papilla extending into the epidermis
LM 500x

FIGURE 5.2 APR **Dermis and Epidermis**

(*a*) Photomicrograph of the dermis covered by the epidermis. The dermis consists of the papillary and reticular layers. The papillary layer has projections, called papillae, that extend into the epidermis. (*b*) Higher-magnification photomicrograph of the epidermis resting on the papillary layer of the dermis. Note the strata of the epidermis.

develop a cone-shaped structure called a **corn.** The base of the cone is at the surface, but the apex extends deep into the epidermis, and pressure on the corn may be quite painful. Calluses and corns can develop in both thin and thick skin.

Skin Color

The factors that determine skin color include pigments in the skin, blood circulating through the skin, and the thickness of the stratum corneum. **Melanin** (mel′ă-nin) is the group of pigments primarily responsible for skin, hair, and eye color. Melanin also provides protection against ultraviolet light from the sun. Large amounts of melanin are found in certain regions of the skin, such as freckles, moles, the nipples, the areolae of the breasts, the axillae, and the genitalia. Other areas of the body, such as the lips, palms of the hands, and soles of the feet, contain less melanin.

Melanin is produced by **melanocytes** (mel′ă-nō-sītz), irregularly shaped cells with many long processes that extend between the keratinocytes of the stratum basale and the stratum spinosum (figure 5.4). The Golgi apparatuses of the melanocytes package melanin into vesicles called **melanosomes** (mel′ă-nō-sōmz), which move into the cell processes of the melanocytes. Keratinocytes phagocytize (see chapter 3) the tips of the melanocyte cell processes, thereby acquiring melanosomes. Although all keratinocytes can contain melanin, only the melanocytes produce it.

To produce melanin, the enzyme tyrosinase (tī′rō-si-nās, tir′ō-si-nās) converts the amino acid tyrosine to dopaquinone (dō′pă-kwin′ōn, dō′pă-kwī-nōn). Dopaquinone can be converted to a variety of related molecules, most of which are brown to black pigments but some of which are yellowish or reddish.

Melanin production is determined by genetic factors, exposure to light, and hormones. Genetic factors are primarily responsible for the variations in skin color among different races and among people of the same race. Since all races have about the same number of melanocytes, racial variations in skin color are determined by the amount and types of melanin produced by the melanocytes, as well as by the size, number, and distribution of the melanosomes. Although many genes are responsible for skin color, a single mutation

FUNDAMENTAL **Figure**

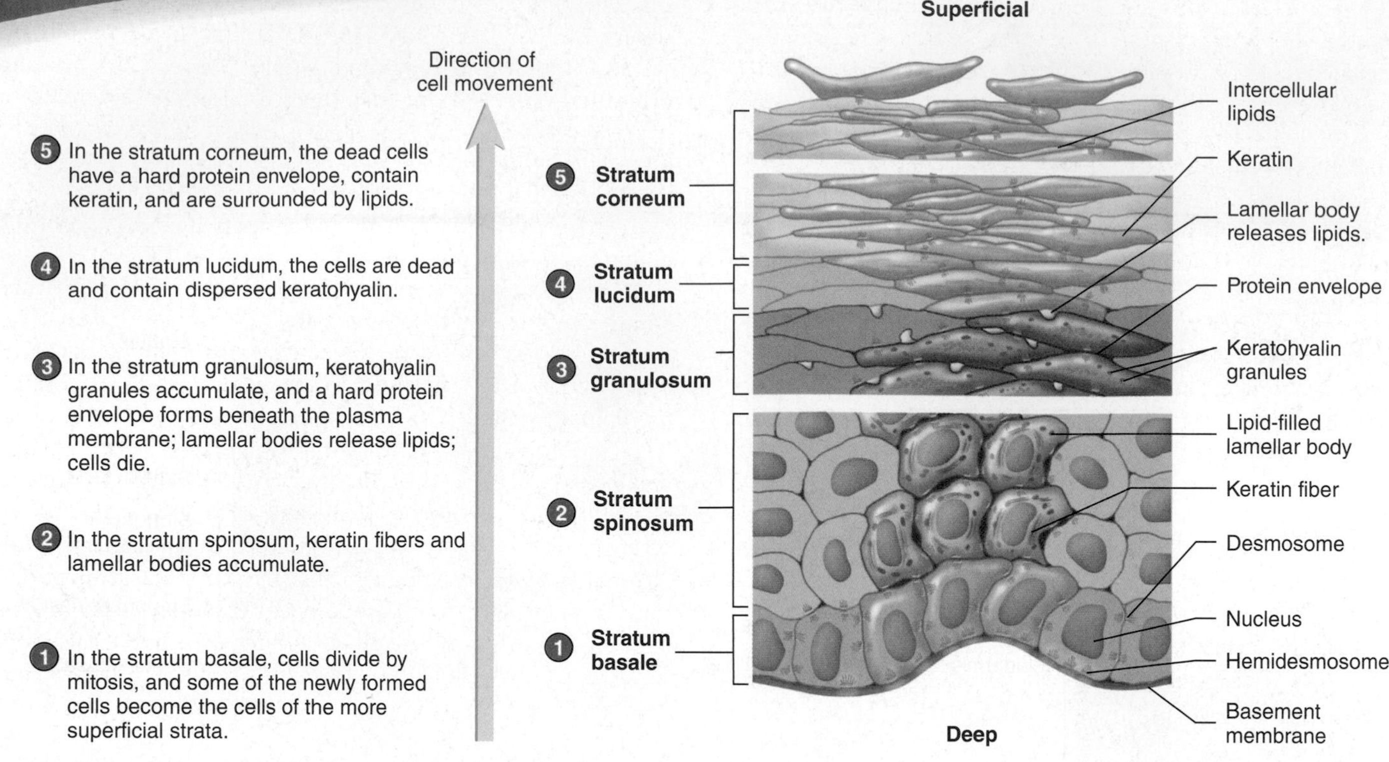

Superficial
Direction of cell movement
5 In the stratum corneum, the dead cells have a hard protein envelope, contain keratin, and are surrounded by lipids.
4 In the stratum lucidum, the cells are dead and contain dispersed keratohyalin.
3 In the stratum granulosum, keratohyalin granules accumulate, and a hard protein envelope forms beneath the plasma membrane; lamellar bodies release lipids; cells die.
2 In the stratum spinosum, keratin fibers and lamellar bodies accumulate.
1 In the stratum basale, cells divide by mitosis, and some of the newly formed cells become the cells of the more superficial strata.
5 Stratum corneum
4 Stratum lucidum
3 Stratum granulosum
2 Stratum spinosum
1 Stratum basale
Intercellular lipids
Keratin
Lamellar body releases lipids.
Protein envelope
Keratohyalin granules
Lipid-filled lamellar body
Keratin fiber
Desmosome
Nucleus
Hemidesmosome
Basement membrane
Deep

PROCESS FIGURE 5.3 **Epidermal Layers and Keratinization**

(see chapter 29) can prevent the manufacture of melanin. **Albinism** (al′bi-nizm) is usually a recessive genetic trait that results from an inability to produce tyrosinase. The result is a deficiency or an absence of pigment in the skin, the hair, and the irises of the eyes. Exposure to ultraviolet light darkens the melanin already present in the skin and stimulates melanin production, resulting in tanning.

During pregnancy, certain hormones, such as estrogen and melanocyte-stimulating hormone, cause the mother's body to increased melanin production, which causes darkening of the nipples, areolae, and genitalia. The cheekbones, forehead, and chest also may darken, resulting in the "mask of pregnancy," and a dark line of pigmentation may appear on the midline of the abdomen. Diseases that cause increased secretion of adrenocorticotropic hormone and melanocyte-stimulating hormone, such as Addison disease, also cause increased pigmentation.

Blood flowing through the skin imparts a reddish hue, a condition called **erythema** (er-ĭ-thē′mă). An inflammatory response (see chapter 4) stimulated by infection, sunburn, allergic reactions,

1. Melanosomes are produced by the Golgi apparatus of the melanocyte.
2. Melanosomes move into melanocyte cell processes.
3. Epithelial cells phagocytize the tips of the melanocyte cell processes.
4. The melanosomes, which were produced inside the melanocytes, have been transferred to epithelial cells and are now inside them.

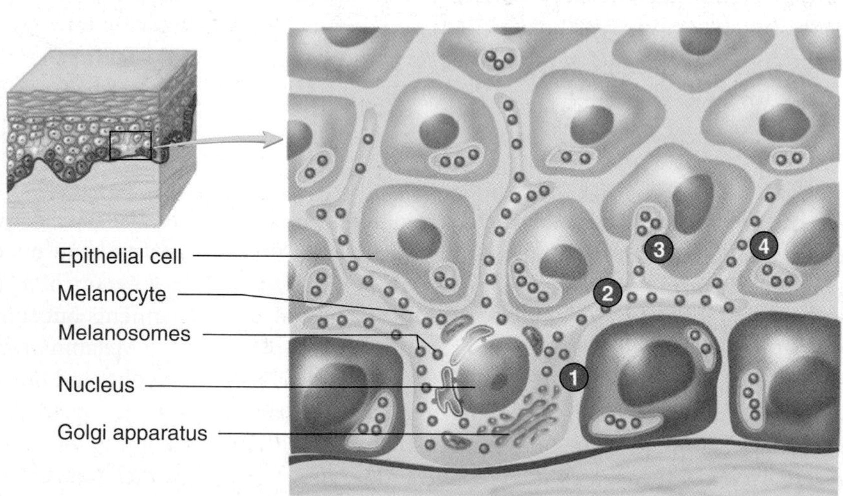

Epithelial cell
Melanocyte
Melanosomes
Nucleus
Golgi apparatus
1
2
3
4

PROCESS FIGURE 5.4 AP|R **Melanin Transfer to Keratinocytes**

Melanocytes make melanin, which is packaged into melanosomes and transferred to many keratinocytes.

insect bites, or other causes can produce erythema, as can exposure to the cold and blushing or flushing when angry or hot. A decrease in blood flow, as occurs in shock, can make the skin appear pale, and a decrease in the blood oxygen content produces **cyanosis** (sī-ă-nō′sis), a bluish skin color (see Clinical Impact, "The Integumentary System as a Diagnostic Aid," later in this chapter).

Carotene (kar′ō-tēn) is a yellow pigment found in plants, such as carrots and corn. Humans normally ingest carotene and use it as a source of vitamin A. Carotene is lipid-soluble and, when large amounts of carotene are consumed, the excess accumulates in the stratum corneum and in adipocytes of the dermis and subcutaneous tissue, causing the skin to develop a yellowish tint. The yellowish tint slowly disappears once carotene intake is reduced.

The location of pigments and other substances in the skin affects the color produced. For example, light reflected off dark pigment in the dermis or subcutaneous tissue can be scattered by collagen fibers of the dermis to produce a blue color. The same effect produces the blue color of the sky as light is reflected from dust particles in the air. The deeper within the dermis or subcutaneous tissue any dark pigment is located, the bluer the pigment appears because of the light-scattering effect of the overlying tissue. This effect causes the blue color of tattoos, bruises, and some superficial blood vessels.

ASSESS **YOUR PROGRESS**

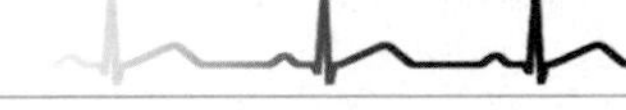

2. *From deepest to most superficial, name and describe the five strata of the epidermis. In which stratum are new cells formed by mitosis? Which strata have live cells, and which strata have dead cells?*
3. *Describe the structural features resulting from keratinization that make the epidermis structurally strong and resistant to water loss.*
4. *Compare the structure and location of thick and thin skin. Is hair in thick or thin skin?*
5. *Which cells of the epidermis produce melanin? What happens to the melanin once it is produced?*
6. *How do genetic factors, exposure to sunlight, and hormones determine the amount of melanin in the skin?*
7. *How do carotene, blood flow, oxygen content, and collagen affect skin color?*

Dermis

The dermis is connective tissue containing fibroblasts, a few adipocytes, and macrophages. Collagen is the main connective tissue fiber, but elastic and reticular fibers are also present. Adipocytes and blood vessels are scarce in the dermis compared with the subcutaneous tissue. The dermis contains nerve endings, hair follicles, smooth muscles, glands, and lymphatic vessels (see figure 5.1). The nerve endings are varied in structure and function. They include free nerve endings for pain, itch, tickle, and temperature sensations; hair follicle receptors for light touch; Pacinian corpuscles for deep pressure; Meissner corpuscles for detecting simultaneous stimulation at two points on the skin; and Ruffini end organs for sensing continuous touch or pressure (see figure 14.1).

The dermis is divided into two layers (see figure 5.2*a*): the superficial **papillary** (pap′i-lār-ē) **layer** and the deeper **reticular** (re-tik′ū-lăr) **layer.** The papillary layer derives its name from projections, called **dermal papillae** (pă-pil′ē), that extend toward the epidermis. The papillary layer is loose connective tissue with thin fibers that are somewhat loosely arranged. The papillary layer also contains blood vessels that supply the overlying epidermis with oxygen and nutrients, remove waste products, and aid in regulating body temperature.

The dermal papillae under the thick skin of the palms of the hands and soles of the feet lie in parallel, curving ridges that shape the overlying epidermis into fingerprints and footprints. The ridges increase friction and improve the grip of the hands and feet. Everyone has unique fingerprints and footprints, even identical twins.

The reticular layer, which is composed of dense irregular connective tissue, is the main layer of the dermis. It is continuous with the subcutaneous tissue and forms a mat of irregularly arranged fibers that are resistant to stretching in many directions. The elastic and collagen fibers are oriented more in some directions than in others and produce **cleavage lines,** or *tension lines,* in the skin (figure 5.5). It is important for health professionals to understand cleavage line directions because an incision made

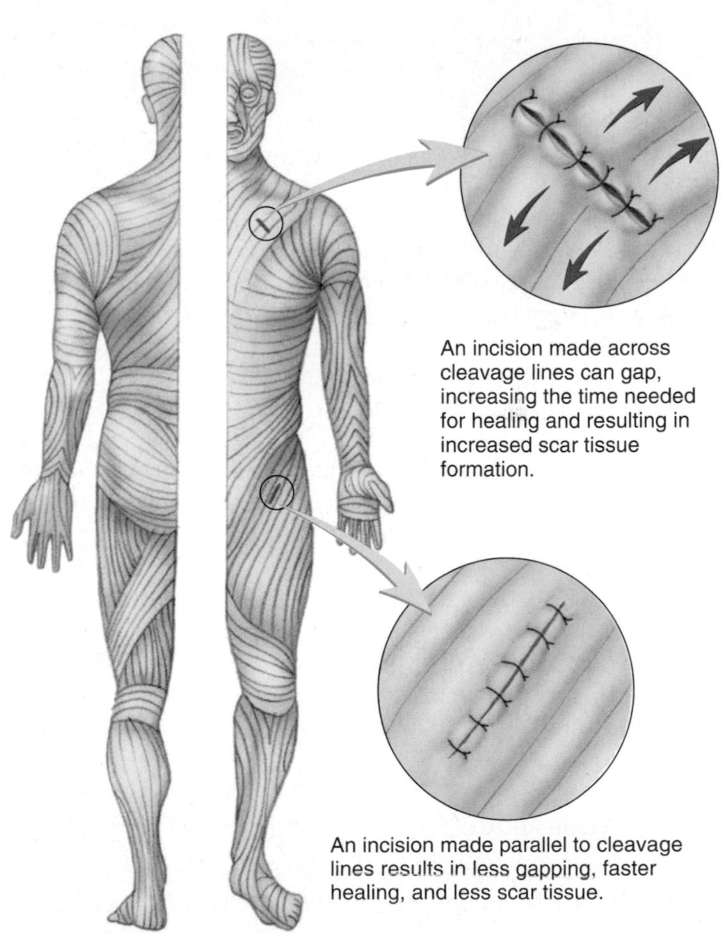
An incision made across cleavage lines can gap, increasing the time needed for healing and resulting in increased scar tissue formation.
An incision made parallel to cleavage lines results in less gapping, faster healing, and less scar tissue.

FIGURE 5.5 Cleavage Lines
The orientation of collagen fibers produces cleavage lines, or tension lines, in the skin.

Clinical GENETICS | Skin Cancer

Skin cancer is the most common type of cancer. Most skin cancers result from damage caused by the ultraviolet (UV) radiation in sunlight. Some skin cancers are induced by chemicals, x-rays, depression of the immune system, or inflammation, whereas others are inherited.

UV radiation damages the genes (DNA) in epidermal cells, producing mutations. If a mutation is not repaired, the mutation is passed to one of the two daughter cells when a cell divides by mitosis. If mutations affecting oncogenes and tumor suppressor genes in epidermal cells accumulate, uncontrolled cell division and skin cancer can result (see Clinical Genetics, "Genetic Changes in Cancer Cells," in chapter 3).

The amount of protective melanin in the skin affects the likelihood of developing skin cancer. Fair-skinned individuals, who have less melanin, are at an increased risk of developing skin cancer compared with dark-skinned individuals, who have more melanin. Long-term or intense exposure to UV radiation also increases the risk. Thus, individuals who are older than 50, who have engaged in repeated recreational or occupational exposure to the sun, or who have experienced sunburn are at increased risk. Most skin cancers develop on the parts of the body that are frequently exposed to sunlight, such as the face, neck, ears, and dorsum of the forearm and hand. A physician should be consulted if skin cancer is suspected.

There are three types of skin cancer: basal cell carcinoma, squamous cell carcinoma, and melanoma (figure 5A). **Basal cell carcinoma,** the most common type, affects cells in the stratum basale. Basal cell carcinomas have a varied appearance. Some are open sores that bleed, ooze, or crust for several weeks. Others are reddish patches; shiny, pearly, or translucent bumps; or scarlike areas of shiny, taut skin. Removal or destruction of the tumor cures most cases.

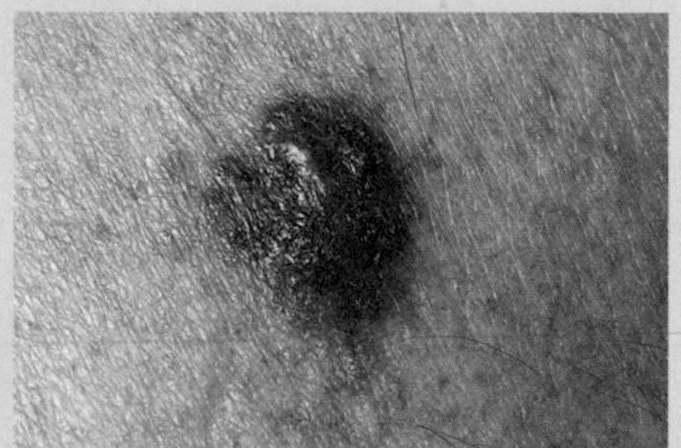

(a) Basal cell carcinoma

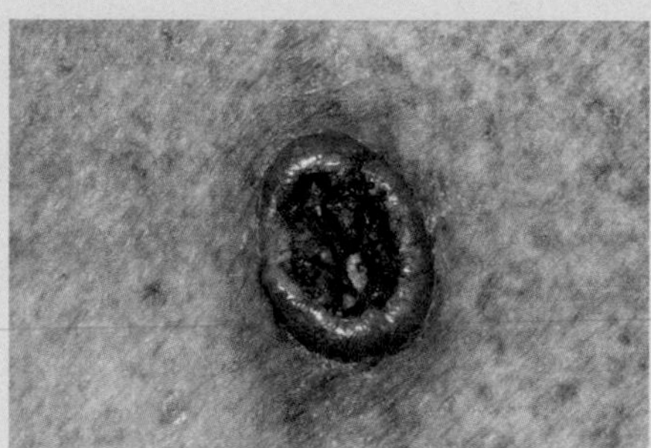

(b) Squamous cell carcinoma

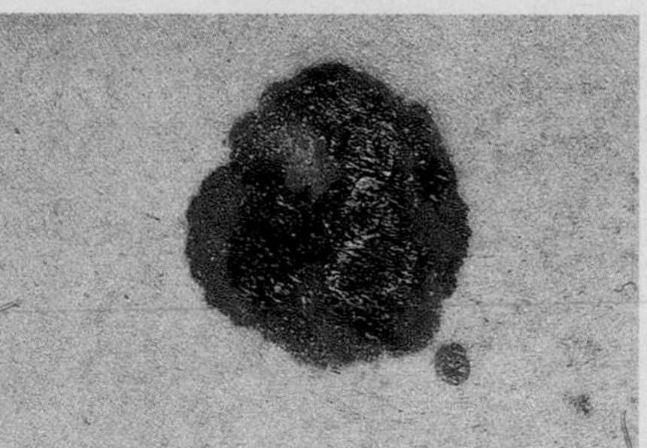

(c) Melanoma

FIGURE 5A Cancer of the Skin

parallel to the cleavage lines is less likely to gap than an incision made across them. The development of infections and the formation of scar tissue are reduced in wounds where the edges are closer together.

If the skin is overstretched, the dermis may rupture and leave lines that are visible through the epidermis. These lines of scar tissue, called **stretch marks,** can develop on the abdomen and breasts of a woman during pregnancy or on the skin of athletes who have quickly increased muscle size by intense weight training.

ASSESS YOUR PROGRESS

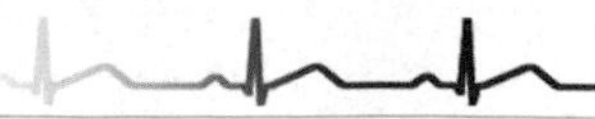

8. *Name and compare the two layers of the dermis. Which layer is responsible for most of the structural strength of the skin?*
9. *What are formed by the dermal papillae in thick skin? What roles do they have?*
10. *What are cleavage lines, and how are they related to the healing of a cut?*

5.3 Subcutaneous Tissue

LEARNING OUTCOME

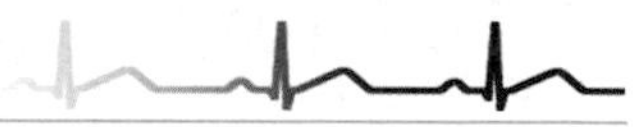

After reading this section, you should be able to

A. Describe the structure and functions of the subcutaneous tissue underlying the skin.

Just as a house rests on a foundation, the skin rests on subcutaneous tissue, which attaches it to underlying bone and muscle and supplies it with blood vessels and nerves (see figure 5.1). The subcutaneous tissue consists of loose connective tissue with collagen and elastic fibers. The main types of cells within the subcutaneous tissue are fibroblasts, adipocytes, and macrophages. The subcutaneous tissue, which is not part of the skin, is sometimes called the *hypodermis.*

Approximately half the body's stored lipids are in the subcutaneous tissue, where they function in insulation and padding and as a source of energy. The subcutaneous tissue can be used to estimate

Squamous cell carcinoma is the second most common type of skin cancer. Squamous cell carcinoma affects cells in the stratum spinosum and can appear as a wartlike growth; a persistent, scaly red patch; an open sore; or an elevated growth with a central depression. These lesions may bleed. Removal or destruction of the tumor cures most cases.

Melanoma (mel′ă-nō′mă) is the least common, but most deadly, type of skin cancer, accounting for over 77% of the skin cancer deaths in the United States. Because they arise from melanocytes, most melanomas are black or brown, but occasionally a melanoma stops producing melanin and appears skin-colored, pink, red, or purple. About 40% of melanomas develop in preexisting moles. Treatment of melanomas when they are confined to the epidermis is almost always successful. However, if a melanoma invades the dermis and metastasizes to other parts of the body, it is difficult to treat and can be deadly.

Early detection and treatment of melanoma before it metastasizes can prevent death. Melanoma can be detected by routine examination of the skin and application of the **ABCDE rule,** which states the signs of melanoma: *A* stands for asymmetry (one side of the lesion does not match the other side), *B* is for border irregularity (the edges are ragged, notched, or blurred), *C* is for color (pigmentation is not uniform), *D* is for diameter (greater than 6 mm), and *E* is for evolving (lesion changes over time). Evolving lesions change size, shape, elevation, or color; they may bleed, crust, itch, or become tender.

In order for cancer cells to metastasize, they must leave their site of origin, enter the circulation, and become established in a new location. For example, melanoma cells first spread within the epidermis. Some of those cells may then break through the basement membrane and invade the dermis; from there, they may enter lymphatic or blood vessels and spread to other parts of the body. The ability of cancer cells to metastasize requires an accumulation of mutations that enables the cells to detach from similar cells, recognize and digest the basement membrane, and become established elsewhere when surrounded by different cell types.

Basal cell carcinomas very rarely metastasize, and only 2–6% of squamous cell carcinomas metastasize. Compared with keratinocytes, melanocytes are more likely to give rise to tumors that metastasize because, in their developmental past, they had the ability to migrate and become established in new locations. In the embryo, melanocytes are derived from a population of cells called neural crest cells (see chapter 13). A gene called *Slug* regulates neural crest cell migration. In normal melanocytes, the *Slug* gene is inactive, but in metastasizing melanoma cells it is reactivated. The reactivation of embryonic genes, such as *Slug,* may also play a role in other metastasizing cancers.

Most skin cancers result from a series of genetic changes in somatic cells. Some people, however, have a genetic susceptibility to developing skin cancer. **Xeroderma pigmentosum** (zēr′ō-der′mă pig′men-tō′sŭm) is a rare inherited disorder in which a DNA repair gene is defective. Because damage to genes by UV radiation is not repaired, exposure to UV radiation results in the development of fatal skin cancers in childhood. Limiting exposure to the sun and using sunscreens can reduce everyone's likelihood of developing skin cancer, especially those who have a genetic susceptibility. Two types of UV radiation play a role. Ultraviolet-B (UVB; 290–320 nm) radiation is the most potent for causing sunburn; it is also the main cause of basal and squamous cell carcinomas and a significant cause of melanoma. Ultraviolet-A (UVA; 320–400 nm) also contributes to skin cancer development, especially melanoma. It, too, penetrates the dermis, causing wrinkling and leathering of the skin. The Skin Cancer Foundation recommends using a broad-spectrum sunscreen that protects against both UVB and UVA, with a sun protection factor (SPF) of at least 15.

total body fat by pinching the skin at selected locations and measuring the thickness of the skin fold and underlying subcutaneous tissue. The thicker the fold, the greater the amount of total body fat.

The amount of adipose tissue in the subcutaneous tissue varies with age, sex, and diet. Most babies have a chubby appearance because they have proportionately more adipose tissue than adults. Women have proportionately more adipose tissue than men, especially over the thighs, buttocks, and breasts, which accounts for some of the differences in body shape between women and men. The amount of adipose tissue in the subcutaneous tissue is also responsible for some of the differences in body shape between individuals of the same sex.

ASSESS YOUR PROGRESS

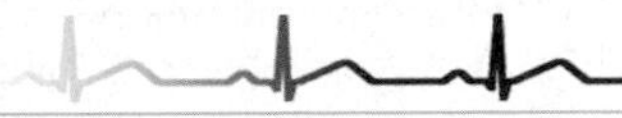

11. *Name the types of tissue forming the subcutaneous tissue layer.*
12. *How is the subcutaneous tissue related to the skin?*
13. *List the functions of the adipose tissue within the subcutaneous tissue.*

Clinical
IMPACT

Injections

Injections are used to introduce certain substances, such as medication and vaccines, into the body. There are three types of injections. An **intradermal injection,** as is used for the tuberculin skin test, goes into the dermis. It is administered by drawing the skin taut and inserting a small needle at a shallow angle into the skin. A **subcutaneous injection** extends into the subcutaneous tissue; an insulin injection is one example. A subcutaneous injection is achieved by pinching the skin to form a "tent" into which a short needle is inserted. An **intramuscular injection** reaches a muscle deep to the subcutaneous tissue. It is accomplished by inserting a long needle at a 90-degree angle to the skin. Intramuscular injections are used for injecting most vaccines and certain antibiotics.

5.4 Accessory Skin Structures

LEARNING OUTCOMES

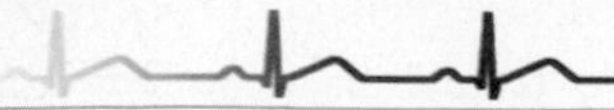

After reading this section, you should be able to

A. Describe the structure of a hair and discuss the phases of hair growth.

B. Explain the function of the arrector pili muscle.

C. Name the glands of the skin and describe the secretions they produce.

D. Describe the parts of a nail and explain how nails grow.

The accessory skin structures include the hair, glands, and nails.

Hair

The presence of **hair** is one of the characteristics of all mammals; if the hair is dense and covers most of the body surface, it is called fur. In humans, hair is found everywhere on the skin except the palms, the soles, the lips, the nipples, parts of the external genitalia, and the distal segments of the fingers and toes.

By the fifth or sixth month of fetal development, delicate, unpigmented hair called **lanugo** (lă-noo′gō) has developed and covered the fetus. Near the time of birth, **terminal hairs,** which are long, coarse, and pigmented, replace the lanugo of the scalp, eyelids, and eyebrows. **Vellus** (vel′ŭs) **hairs,** which are short, fine, and usually unpigmented, replace the lanugo on the rest of the body. At puberty, terminal hair, especially in the pubic and axillary regions, replaces much of the vellus hair. The hair of the chest, legs, and arms is approximately 90% terminal hair in males and approximately 35% in females. In males, terminal hairs replace the vellus hairs of the face to form the beard. The beard, pubic, and axillary hair are signs of sexual maturity. In addition, pubic and axillary hair may function as wicks for dispersing odors produced by secretions from specialized glands in the pubic and axillary regions. It also has been suggested that pubic hair protects against abrasion during intercourse and axillary hair reduces friction when the arms move.

Hair Structure

A hair is divided into the **shaft,** which protrudes above the surface of the skin, and the **root,** located below the surface (figure 5.6*a*). The base of the root is expanded to form the **hair bulb.** Most of the root and the shaft are composed of columns of dead, keratinized epithelial cells arranged in three concentric layers: the medulla, the cortex, and the cuticle (figure 5.6*c*). The **medulla** (me-dool′ă) is the central axis of the hair; it consists of two or three layers of cells containing soft keratin. The **cortex** forms the bulk of the hair; it consists of cells containing hard keratin. The cortex is covered by the **cuticle** (kū′ti-kl), a single layer of cells that contain hard keratin. The edges of the cuticle cells overlap like shingles on a roof.

The **hair follicle** is a tubelike invagination of the epidermis that extends into the dermis from which hair develops. A hair follicle consists of a **dermal root sheath** and an **epithelial root sheath.** The dermal root sheath is the portion of the dermis that surrounds the epithelial root sheath. The epithelial root sheath is divided into external and internal parts (figure 5.6*b*). At the opening of the follicle, the external epithelial root sheath has all the strata found in thin skin. Deeper in the hair follicle, the number of cells decreases until at the hair bulb only the stratum basale is present. This has important consequences for skin repair. If the epidermis and the superficial part of the dermis are damaged, the undamaged part of the hair follicle that lies deep in the dermis can be a source of new epithelium. The internal epithelial root sheath has raised edges that mesh closely with the raised edges of the hair cuticle and hold the hair in place. When a hair is pulled out, the internal epithelial root sheath usually comes out as well and is plainly visible as whitish tissue around the root of the hair.

The hair bulb is an expanded knob at the base of the hair root (figure 5.6*a,b*). Inside the hair bulb is a mass of undifferentiated epithelial cells, the **matrix,** which produces the hair and the internal epithelial root sheath. The dermis of the skin projects into the hair bulb as a **hair papilla;** it contains blood vessels that provide nourishment to the cells of the matrix.

Clinical IMPACT

The Integumentary System as a Diagnostic Aid

The integumentary system can be used in diagnosis because it is easily observed and often reflects events occurring in other parts of the body. For example, **cyanosis** (sī-ă-nō′sis), a bluish color to the skin that results from decreased blood oxygen content, is an indication of impaired cardiovascular or respiratory function. When red blood cells wear out, they are broken down, and the liver excretes part of their contents as bile pigments into the small intestine. **Jaundice** (jawn′dis), a yellowish skin color, occurs when excess bile pigments accumulate in the blood. If a disease, such as viral hepatitis, damages the liver, bile pigments are not excreted and accumulate in the blood.

Rashes and lesions in the skin can be symptomatic of problems elsewhere in the body. For example, scarlet fever results from a bacterial infection in the throat. The bacteria release a toxin into the blood that causes the pink-red rash for which this disease was named. In allergic reactions (see chapter 22), histamine released into the tissues produces swelling and reddening. Development of the skin rash called hives can indicate an allergy to foods or drugs, such as penicillin.

The condition of the skin, hair, and nails is affected by nutritional status. Vitamin A deficiency causes the skin to produce excess keratin and assume a characteristic sandpaper texture, whereas iron-deficiency anemia causes the nails to lose their normal contour and become flat or concave (spoon-shaped).

Hair concentrates many substances, which can be detected by laboratory analysis, and comparing a patient's hair with "normal" hair can be useful in certain diagnoses. For example, lead poisoning results in high levels of lead in the hair. However, hair analysis as a screening test to determine a person's general health or nutritional status is unreliable.

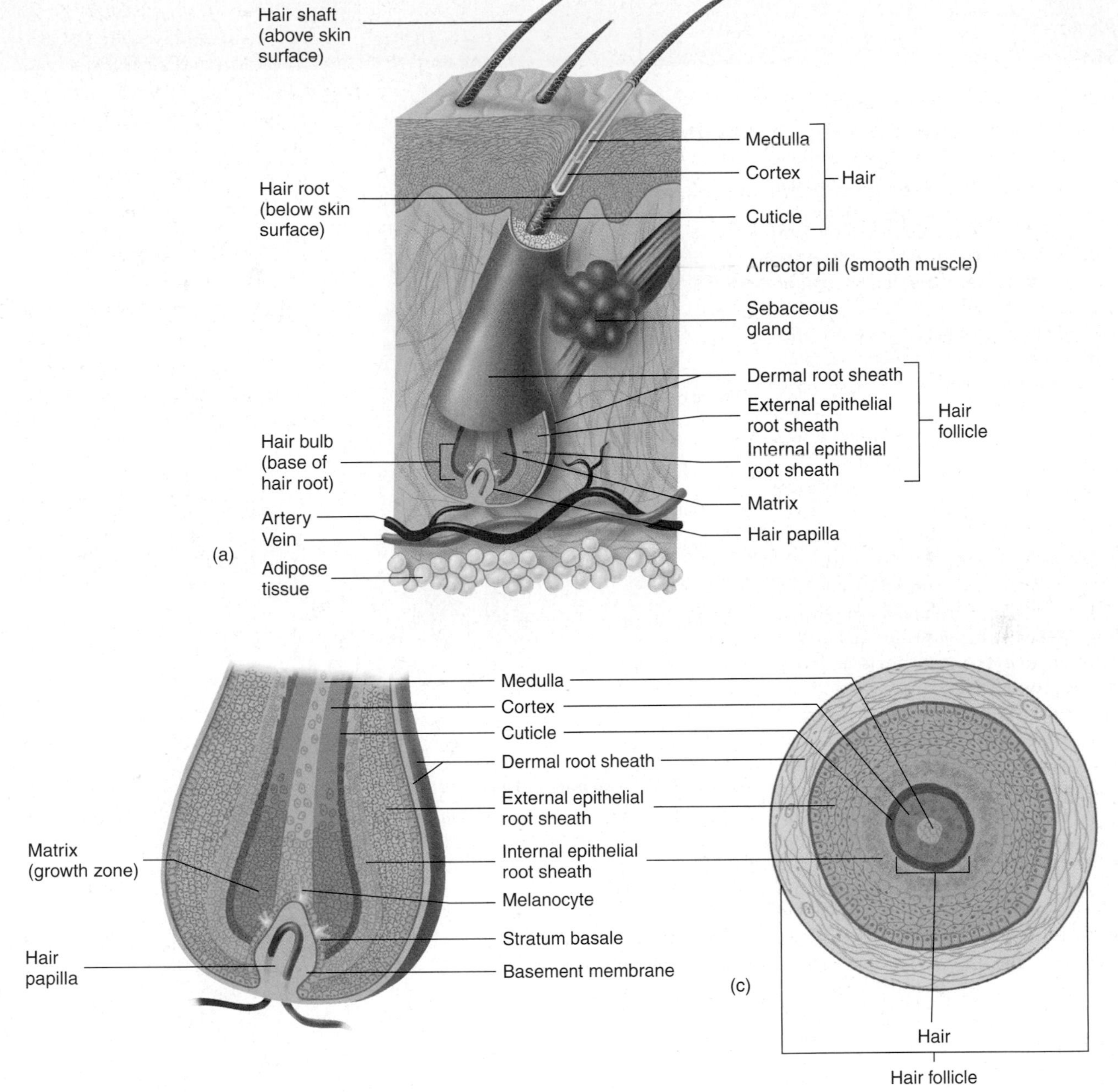
Hair shaft
(above skin
surface)
Hair root
(below skin
surface)
Medulla
Cortex
Cuticle
Hair
Arrector pili (smooth muscle)
Sebaceous
gland
Dermal root sheath
External epithelial
root sheath
Internal epithelial
root sheath
Hair
follicle
Hair bulb
(base of
hair root)
Matrix
Hair papilla
Artery
Vein
Adipose
tissue
(a)
Medulla
Cortex
Cuticle
Dermal root sheath
External epithelial
root sheath
Internal epithelial
root sheath
Melanocyte
Stratum basale
Basement membrane
Matrix
(growth zone)
Hair
papilla
(b)
(c)
Hair
Hair follicle

FIGURE 5.6 APR **Hair Follicle**

(*a*) The hair follicle contains the hair and consists of a dermal and an epithelial root sheath. (*b*) Enlargement of the hair follicle wall and hair bulb. (*c*) Cross section of a hair within a hair follicle.

Hair Growth

Hair is produced in cycles that involve a **growth stage** and a **resting stage.** During the growth stage, hair is formed by matrix cells that differentiate, become keratinized, and die. The hair grows longer as cells are added at the base of the hair root. Eventually, hair growth stops; the hair follicle shortens and holds the hair in place. A resting period follows, after which a new cycle begins and a new hair replaces the old hair, which falls out of the hair follicle. Thus, losing a hair normally means that the hair is being replaced. The length of each stage depends on the hair—eyelashes grow for approximately 30 days and rest for 105 days, whereas scalp hairs grow for 3 years and rest for 1–2 years. At any given time, an estimated 90% of the scalp hairs are in the growing stage, and loss of approximately 100 scalp hairs per day is normal.

The most common kind of permanent hair loss is "pattern baldness." Hair follicles shrink and revert to producing vellus hair,

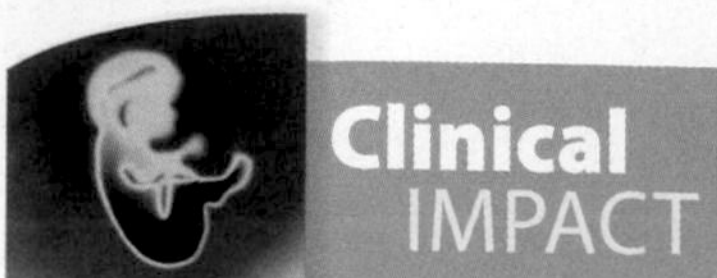

Clinical IMPACT — Burns

A **burn** is injury to a tissue caused by heat, cold, friction, chemicals, electricity, or radiation. Burns are classified according to the extent of surface area involved and the depth of the burn. For an adult, the surface area that is burned can be conveniently estimated by "the rule of nines," which divides the body into areas that are approximately 9%, or multiples of 9%, of the body surface area (BSA; figure 5B). For younger patients, surface area relationships are different. For example, in an infant, the head and neck are 21% of BSA, whereas in an adult they are 9%. For burn victims younger than age 15, a table specifically developed for them should be consulted.

On the basis of depth, burns are classified as either partial-thickness or full-thickness burns (figure 5C). **Partial-thickness burns** are subdivided into first- and second-degree burns. **First-degree burns** involve only the epidermis and may result in redness, pain, and slight edema (swelling). They can be caused by sunburn or brief exposure to hot or cold objects, and they heal in a week or so without scarring.

Second-degree burns damage the epidermis and dermis. Minimal dermal damage causes redness, pain, edema, and blisters. Healing takes approximately 2 weeks, and no scarring results. However, if the burn goes deep into the dermis, the wound appears red, tan, or white; may take several months to heal; and might scar. In all second-degree burns, the epidermis regenerates from epithelial tissue in hair follicles and sweat glands, as well as from the edges of the wound.

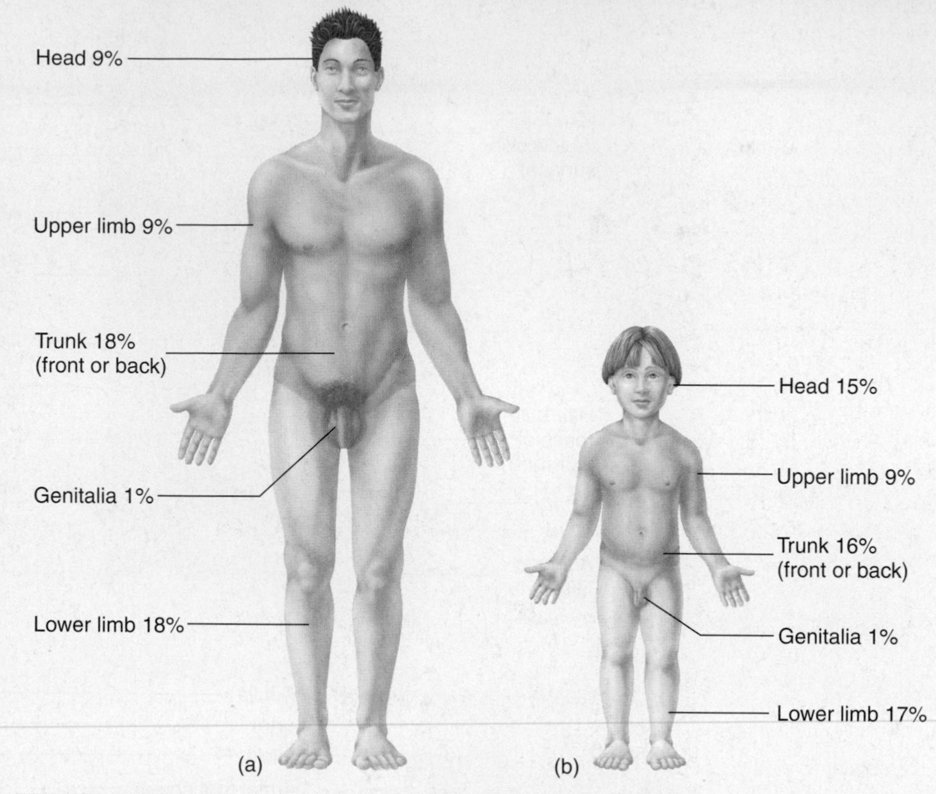

Head 9%
Upper limb 9%
Trunk 18% (front or back)
Genitalia 1%
Lower limb 18%
(a)
Head 15%
Upper limb 9%
Trunk 16% (front or back)
Genitalia 1%
Lower limb 17%
(b)

FIGURE 5B The Rule of Nines
(*a*) In an adult, surface areas can be estimated using the rule of nines: Each major area of the body is 9%, or a multiple of 9%, of the total body surface area. (*b*) In infants and children, the head represents a larger proportion of surface area, so the rule of nines is not as accurate, as can be seen in this depiction of a 5-year-old child.

Full-thickness burns are also called **third-degree burns.** The epidermis and dermis are completely destroyed, and deeper tissue may be involved. Third-degree burns are often sur-

which is very short, transparent, and for practical purposes invisible. Eventually, hair production in these smaller follicles may completely cease. Although baldness is more common and more pronounced in certain men, it can also occur in women. Genetic factors and the hormone testosterone are involved in causing pattern baldness.

The average rate of hair growth is approximately 0.3 mm per day, although hairs grow at different rates, even in the same approximate location. Cutting, shaving, or plucking hair does not alter the growth rate or the character of the hair, but hair can feel coarse and bristly shortly after shaving because the short hairs are less flexible. Maximum hair length is determined by the rate of hair growth and the length of the growing phase. For example, scalp hair can become very long, but eyelashes are short.

Hair Color

Melanocytes within the hair bulb matrix produce melanin and pass it to keratinocytes in the hair cortex and medulla. As with the skin, varying amounts and types of melanin cause different shades of hair color. Blonde hair has little black-brown melanin, whereas jet black hair has the most. Intermediate amounts of melanin account for different shades of brown. Red hair is caused by varying amounts of a red type of melanin. Hair sometimes contains both black-brown and red melanin. With age, the amount of melanin in hair can decrease, causing hair color to fade or become white (i.e., no melanin). Gray hair is usually a mixture of unfaded, faded, and white hairs. Hair color is controlled by several genes, and dark hair color is not necessarily dominant over light.

rounded by first- and second-degree burns. Although the areas that have first- and second-degree burns are painful, the region of third-degree burn is usually painless because the sensory receptors have been destroyed. Third-degree burns appear white, tan, brown, black, or deep cherry-red. Skin can regenerate only from the edges, and skin grafts are often necessary.

The depth and percentage of BSA affected can be combined with other criteria to classify the seriousness of a burn. The following criteria define a **major burn:** a third-degree burn over 10% or more of the BSA; a second-degree burn over 25% or more of the BSA; or a second- or third-degree burn of the hands, feet, face, genitals, or anal region. Facial burns are often associated with damage to the respiratory tract, and burns of joints often heal with scar tissue formation that limits movement. A **moderate burn** is a third-degree burn over 2–10% of the BSA or a second-degree burn over 15–25% of the BSA. A **minor burn** is a third-degree burn over less than 2% or a second-degree burn over less than 15% of the BSA.

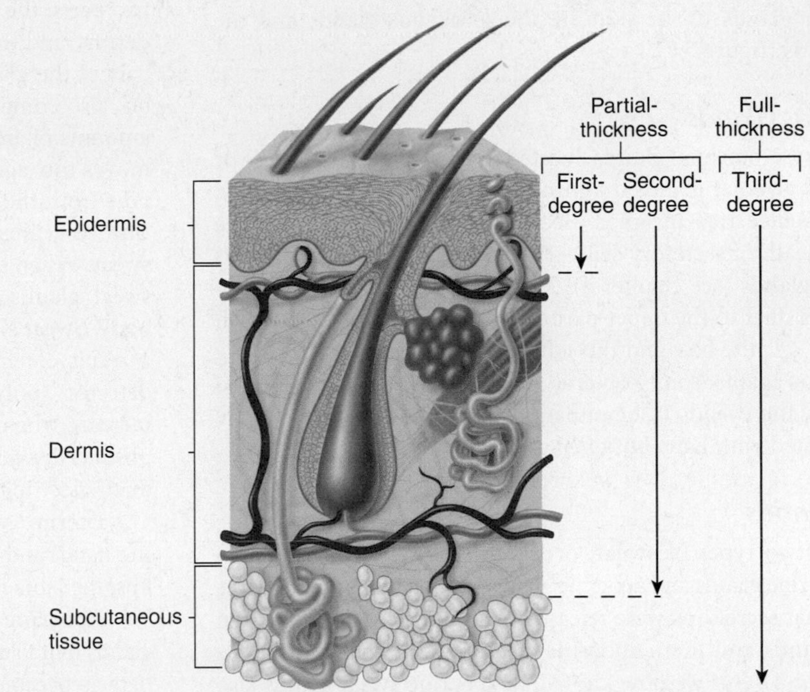
Partial-thickness
Full-thickness
First-degree
Second-degree
Third-degree
Epidermis
Dermis
Subcutaneous tissue

FIGURE 5C Burns
Partial-thickness burns are subdivided into first-degree burns (damage to only the epidermis) and second-degree burns (damage to the epidermis and part of the dermis). Full-thickness, or third-degree, burns destroy the epidermis, the dermis, and sometimes deeper tissues.

Deep partial-thickness and full-thickness burns take a long time to heal, and they tend to form scar tissue with disfiguring and debilitating wound contracture. To prevent these complications and speed healing, skin grafts are performed. In a split skin graft, the epidermis and part of the dermis are removed from another part of the body and placed over the burn. Interstitial fluid from the burned area nourishes the graft until its dermis becomes vascularized. At the graft donation site, part of the dermis is still present. The deep parts of hair follicles and sweat gland ducts remain in this dermis, where they serve as a source of epithelial cells that form a new epidermis. This is the same process of epidermis formation that occurs following superficial second-degree burns.

When it is not possible or practical to move skin from one part of the body to a burn site, artificial skin or grafts from human cadavers or pigs are used. These techniques are often unsatisfactory because the body's immune system recognizes the graft as a foreign substance and rejects it. A solution to this problem is laboratory-grown skin. A piece of healthy skin from the burn victim is removed and placed into a flask with nutrients and hormones that stimulate rapid growth. The skin that is produced consists only of epidermis and does not contain glands or hair.

Muscles

Associated with each hair follicle are smooth muscle cells called the **arrector pili** (ă-rek′tōr pī′lī), which extend from the dermal root sheath of the hair follicle to the papillary layer of the dermis (figure 5.6*a*). Normally, the hair follicle and the hair inside it are at an oblique angle to the surface of the skin. When the arrector pili muscles contract, however, they pull the follicle into a more perpendicular position, causing the hair to "stand on end." Movement of the hair follicles produces raised areas called "goose bumps."

ASSESS YOUR PROGRESS

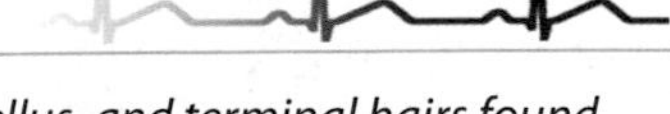

14. *When and where are lanugo, vellus, and terminal hairs found in the skin?*
15. *What are the regions of a hair? What type of cells make up most of a hair?*
16. *Describe the three layers of a hair seen in cross section.*
17. *Describe the parts of a hair follicle. How is the epithelial root sheath important in skin repair?*
18. *In what part of a hair does growth take place? What are the stages of hair growth? Do all hairs grow at the same rate?*
19. *What determines the different shades of hair color?*
20. *Explain the location and action of arrector pili muscles.*

Glands

The major glands of the skin are the sebaceous glands and the sweat glands (figure 5.7).

Sebaceous Glands

Sebaceous (sē-bā′shŭs) **glands,** located in the dermis, are simple or compound alveolar glands that produce **sebum** (sē′bŭm), an oily, white substance rich in lipids. Because sebum is released by the lysis and death of secretory cells, sebaceous glands are classified as holocrine glands (see chapter 4). Most sebaceous glands are connected by a duct to the upper part of the hair follicles, from which the sebum oils the hair and the skin surface. This prevents drying and protects against some bacteria. A few sebaceous glands located in the lips, the eyelids (meibomian glands), and the genitalia are not associated with hairs but open directly onto the skin surface.

Sweat Glands

There are two types of **sweat,** or *sudoriferous* (soo-dō-rif′er-ŭs) **glands:** eccrine glands and apocrine glands. At one time, physiologists believed that secretions were released in a merocrine fashion from eccrine glands and in an apocrine fashion from apocrine glands (see chapter 4). But we now know that apocrine sweat glands also release some of their secretions in a merocrine fashion, and possibly some in a holocrine fashion. Traditionally, they are still referred to as apocrine sweat glands.

Eccrine (ek′rin) **sweat glands** (sometimes called *merocrine* [mer′ō-krin, mer′ō-krīn, mer′ō-krēn] *sweat glands*) are the most common type of sweat gland. They are simple, coiled, tubular glands that open directly onto the surface of the skin through sweat pores (figure 5.7). Eccrine sweat glands can be divided into two parts: the deep, coiled portion, which is located mostly in the dermis, and the duct, which passes to the skin surface. The coiled part of the gland produces an isotonic fluid that is mostly water but also contains some salts (mainly sodium chloride) and small amounts of ammonia, urea, uric acid, and lactic acid. As this fluid moves through the duct, sodium chloride moves by active transport from the duct back into the body, thereby conserving salts. The resulting hyposmotic fluid that leaves the duct is called **sweat.** When the body temperature starts to rise above normal, the sweat glands produce sweat, which evaporates and cools the body. Sweat also can be released in the palms, soles, and axillae as a result of emotional stress. Emotional sweating is used in lie detector (polygraph) tests because sweat gland activity may increase when a person tells a lie. Such tests can detect the sweat produced, even in small amounts, because the salt solution conducts electricity and lowers the electrical resistance of the skin.

Eccrine sweat glands are most numerous in the palms of the hands and soles of the feet but are absent from the margin of the lips, the labia minora, and the tips of the penis and clitoris.

Apocrine (ap′ō-krin) **sweat glands** are simple, coiled, tubular glands that usually open into hair follicles superficial to the opening of the sebaceous glands (figure 5.7). Apocrine sweat glands are found in the axillae and genitalia (scrotum and labia majora) and around the anus. They do not help regulate temperature in humans. Apocrine sweat glands become active at puberty as a result of sex hormones. Their secretions contain organic substances, such as 3-methyl-2-hexenoic acid, that are essentially odorless when first released but are quickly metabolized by bacteria to cause what is commonly known as body odor. Many mammals use scent as a means of communication, and physiologists have suggested that the activity of apocrine sweat glands may signal sexual maturity.

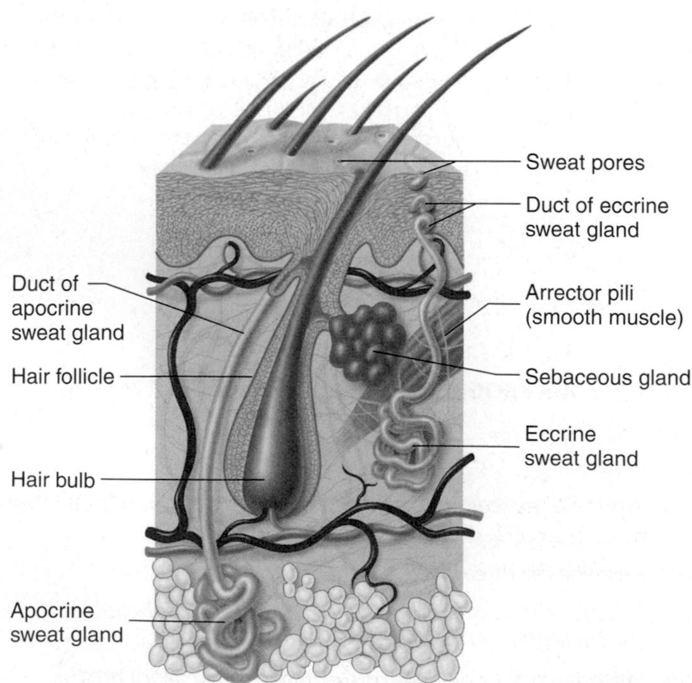
Sweat pores
Duct of eccrine sweat gland
Duct of apocrine sweat gland
Arrector pili (smooth muscle)
Hair follicle
Sebaceous gland
Eccrine sweat gland
Hair bulb
Apocrine sweat gland

FIGURE 5.7 Glands of the Skin
Sebaceous glands and apocrine sweat glands empty into a hair follicle. Eccrine sweat glands empty onto the surface of the skin.

Other Glands

Other skin glands are the ceruminous glands and the mammary glands. The **ceruminous** (sĕ-roo′mi-nŭs) **glands** are modified eccrine sweat glands located in the ear canal (external auditory canal). **Cerumen,** or earwax, is composed of the combined secretions of ceruminous glands and sebaceous glands. Cerumen and hairs in the ear canal protect the tympanic membrane by preventing the entry of dirt and small insects. However, an accumulation of cerumen can block the ear canal and make hearing more difficult.

The **mammary glands** are modified apocrine sweat glands located in the breasts. They produce milk. The structure and regulation of mammary glands are discussed in chapters 28 and 29.

Nails

A **nail** is a thin plate consisting of layers of dead stratum corneum cells that contain a very hard type of keratin. Nails are located on the distal ends of the digits (fingers and toes). A nail consists of the proximal **nail root** and the distal **nail body** (figure 5.8*a*). The nail root is covered by skin, and the nail body is the visible portion of the nail. The lateral and proximal edges of the nail are covered by skin called the **nail fold,** and the edges are held in place by the **nail groove** (figure 5.8*b*). The stratum corneum of the nail fold

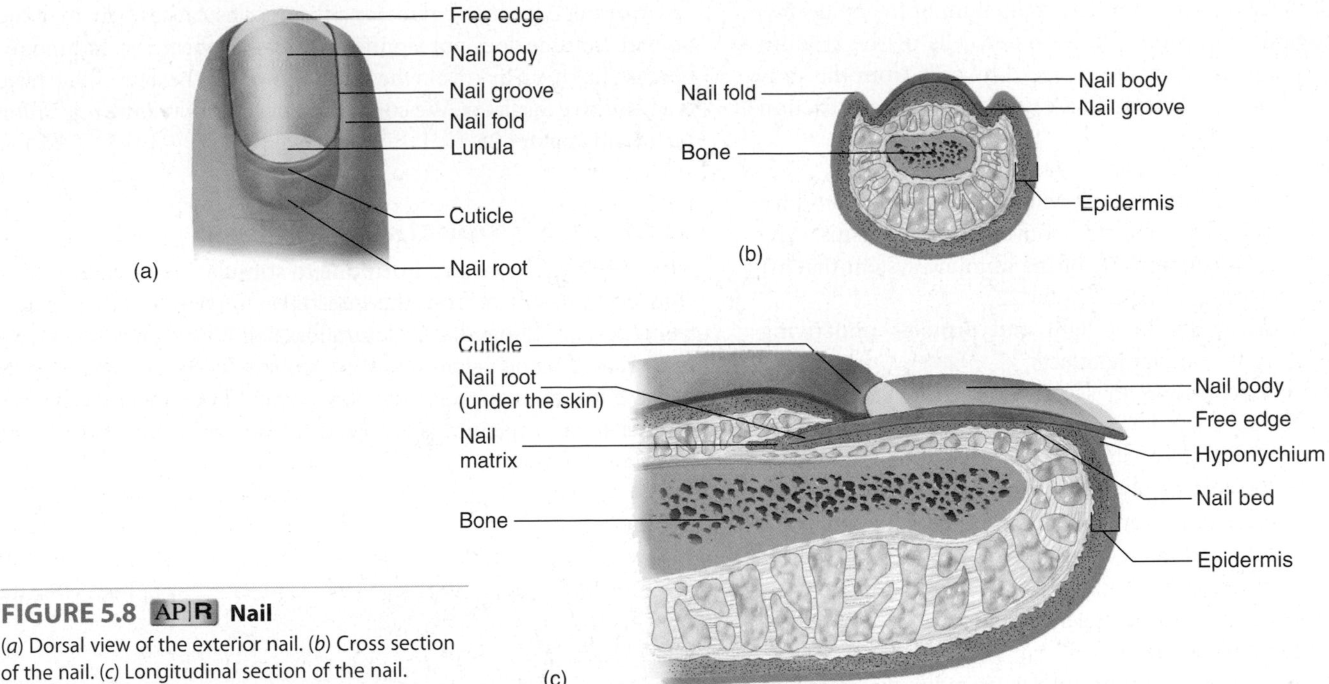
Free edge
Nail body
Nail groove
Nail fold
Lunula
Cuticle
Nail root
(a)
Nail fold
Bone
Nail body
Nail groove
Epidermis
(b)
Cuticle
Nail root
(under the skin)
Nail
matrix
Bone
Nail body
Free edge
Hyponychium
Nail bed
Epidermis
(c)

FIGURE 5.8 APIR **Nail**
(*a*) Dorsal view of the exterior nail. (*b*) Cross section of the nail. (*c*) Longitudinal section of the nail.

grows onto the nail body as the **cuticle,** or *eponychium* (ep-ō-nik′ē-ŭm). Beneath the free edge of the nail body is the **hyponychium** (hī′pō-nik′ē-ŭm), a thickened region of the stratum corneum (figure 5.8*c*).

The nail root extends distally from the **nail matrix.** The nail also attaches to the underlying **nail bed,** which is located between the nail matrix and the hyponychium. The nail matrix and bed are composed of epithelial tissue, with a stratum basale that gives rise to the cells that form the nail. The nail matrix is thicker than the nail bed and produces nearly all of the nail. The nail bed is visible through the clear nail and appears pink because of blood vessels in the dermis. A small part of the nail matrix, the **lunula** (loo′noo-lă), is seen through the nail body as a whitish, crescent-shaped area at the base of the nail. The lunula, seen best on the thumb, appears white because the blood vessels do not show through the thicker nail matrix.

As the nail forms in the nail matrix and bed, it slides over the nail bed toward the distal end of the digit. Nails grow at an average rate of 0.5–1.2 mm per day, and fingernails grow more rapidly than toenails. Unlike hair, they grow continuously throughout life and do not have a resting phase.

Predict 3

While trying to fix some loose boards on his deck, Bob hit his left thumb with his hammer. The hammer struck his thumbnail distal to the lunula and proximal to the hyponychium. After a short period, a dark area appeared in the area of the nail bed. The injury was very painful until a physician drilled a small hole through Bob's nail, releasing bloody fluid. After nearly 2 months, the dark area moved to the free edge of the nail. Explain why a dark area developed in the nail. What caused Bob's pain, and why did drilling a hole in his nail relieve it? Why did the dark area move distally over time?

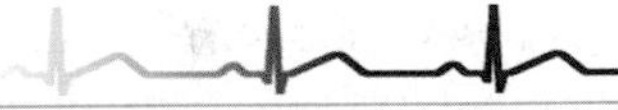

ASSESS YOUR PROGRESS

21. *What do sebaceous glands secrete? What is the function of the secretion?*
22. *Which glands of the skin are responsible for cooling the body? Where are they located? Which glands are involved with producing body odor? Where are they located?*
23. *Name the parts of a nail. Which part produces most of the nail? What is the lunula?*
24. *What makes a nail hard? Do nails have growth stages?*

5.5 Physiology of the Integumentary System

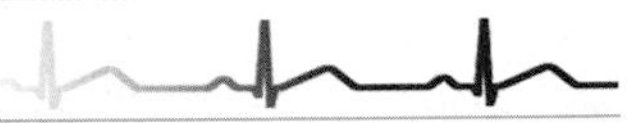

LEARNING OUTCOMES

After reading this section, you should be able to

A. Relate the protective functions of the skin, hair, glands, and nails.
B. Explain how the skin acts as a sense organ.
C. Discuss the importance of the skin in temperature regulation.
D. Describe the involvement of the skin in vitamin D production and in excretion.

Protection

The integumentary system is the body's fortress, defending it from harm:

1. The skin protects underlying structures from mechanical damage. The dermis provides structural strength, preventing

tearing of the skin. The stratified epithelium of the epidermis protects against abrasion. As the outer cells of the stratum corneum slough off, they are replaced by cells from the stratum basale. Calluses develop in areas subject to heavy friction or pressure.

2. The skin prevents microorganisms and other foreign substances from entering the body. Secretions from skin glands produce an environment unsuitable for some microorganisms. The skin also contains components of the immune system that act against microorganisms (see chapter 22).
3. Melanin absorbs ultraviolet light and protects underlying structures from its damaging effects.
4. Hair provides protection in several ways. The hair on the head acts as a heat insulator and protects against ultraviolet light and abrasion. The eyebrows keep sweat out of the eyes, the eyelashes protect the eyes from foreign objects, and hair in the nose and ears prevents dust and other materials from entering. Axillary and pubic hair are a sign of sexual maturity and protect against abrasion.
5. Nails protect the ends of the fingers and toes from damage and can be used in defense.
6. The intact skin plays an important role in reducing water loss because its lipids act as a barrier to the diffusion of water.

Some lipid-soluble substances readily pass through the epidermis. Lipid-soluble medications can be administered by applying them to the skin, after which the medication slowly diffuses through the skin into the blood. For example, nicotine patches are applied to help reduce withdrawal symptoms in people attempting to quit smoking.

Sensation

Receptors in the skin can detect pain, heat, cold, and pressure. For example, the epidermis and dermal papillae are well supplied with touch receptors. The dermis and deeper tissues contain pain, heat, cold, touch, and pressure receptors. Hair follicles (but not the hair) are well innervated, and sensory receptors surrounding the base of hair follicles can detect hair movement. Sensory receptors are discussed in more detail in chapter 14.

Temperature Regulation

Body temperature is affected by blood flow through the skin. When blood vessels (arterioles) in the dermis dilate, more warm blood flows from deeper structures to the skin, and heat loss increases (figure 5.9, *steps 1 and 2*). Body temperature tends to increase as a result of exercise, fever, or a rise in environmental temperature. In order to maintain homeostasis, this excess heat must be lost. The body accomplishes this by producing sweat. The sweat spreads over the surface of the skin; as it evaporates, the body loses heat.

When blood vessels in the dermis constrict, less warm blood flows from deeper structures to the skin, and heat loss decreases (figure 5.9, *steps 3 and 4*). If body temperature begins to drop below normal, heat can be conserved by a decrease in the diameter of dermal blood vessels.

Contraction of the arrector pili muscles causes hair to stand on end, but this does not significantly reduce heat loss in humans because so little hair covers the body. However, the hair on the head is an effective insulator. We consider general temperature regulation further in chapter 25.

Vitamin D Production

Vitamin D functions as a hormone to stimulate the uptake of calcium and phosphate from the intestines, to promote their release from bones, and to reduce calcium loss from the kidneys, resulting in increased blood calcium and phosphate levels. Adequate levels of these minerals are necessary for normal bone metabolism (see chapter 6), and calcium is required for normal nerve and muscle function (see chapter 9).

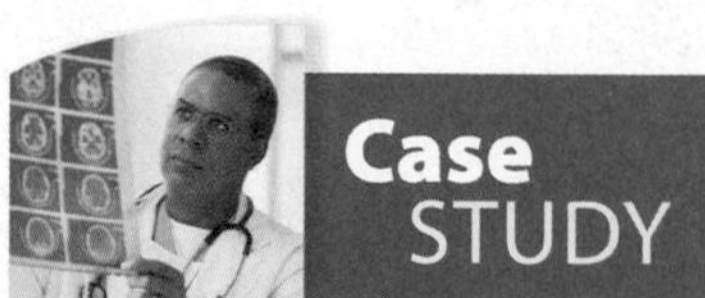

Case STUDY | Frostbite

Billy was hiking in the mountains one autumn day. Unexpectedly, a cold front moved in and the temperature dropped to well below freezing. Billy was unprepared for the temperature change, and he did not have a hat or earmuffs. As the temperature dropped, his ears and nose became pale in color. After continued exposure to the dropping temperatures, every 15–20 minutes, his ears and nose turned red for 5–10 minutes and then became pale again. After several hours, Billy managed to hike back to the trail head. By then, he was very chilled and had no sensation in his ears or nose. As he looked in the rearview mirror of his car, he could see that the skin of his ears and nose had turned white. It took Billy 2 hours to drive to the nearest emergency room, where he learned that the white skin meant he had frostbite of his ears and nose. About 2 weeks later, the frostbitten skin peeled. Despite treatment with an antibiotic, the skin of his right ear became infected. Eventually, Billy recovered, but he lost part of his right ear.

Predict 4

Frostbite is the most common type of freezing injury. When skin temperature drops below 0°C (32°F), the skin freezes and ice crystal formation damages tissues.

a. Using figure 5.9, describe the mechanism that caused Billy's ears and nose to become pale. How is this mechanism beneficial when the ambient temperature is decreasing?
b. Explain what happened when Billy's ears and nose periodically turned red. How is this beneficial when the ambient temperature is very cold?
c. What is the significance of Billy's ears and nose turning and staying white?
d. Why is a person with frostbite likely to develop an infection of the affected part of the body?

FUNDAMENTAL **Figure**

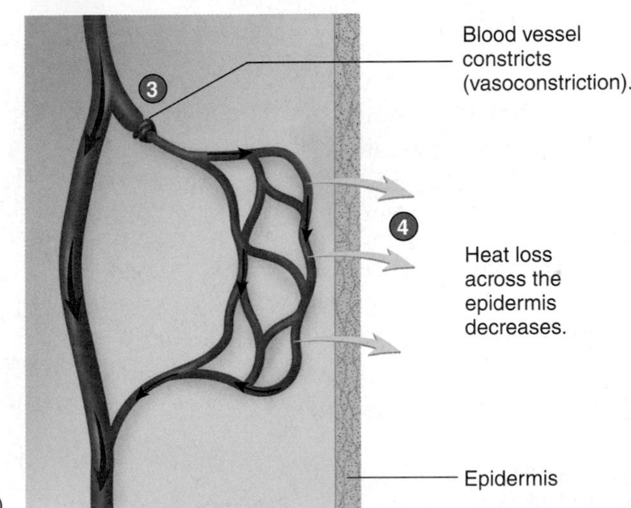

1 Blood vessel dilation results in increased blood flow toward the surface of the skin.
Blood vessel dilates (vasodilation).
2 Increased blood flow beneath the epidermis results in increased heat loss (*gold arrows*).
Heat loss across the epidermis increases.
Epidermis
(a)
3 Blood vessel constriction results in decreased blood flow toward the surface of the skin.
Blood vessel constricts (vasoconstriction).
4 Decreased blood flow beneath the epidermis results in decreased heat loss.
Heat loss across the epidermis decreases.
Epidermis
(b)

PROCESS FIGURE 5.9 AP|R **Heat Exchange in the Skin**

Vitamin D synthesis begins in skin exposed to ultraviolet light, and people can produce all the vitamin D they require by this process if enough ultraviolet light is available. However, because people live indoors and wear clothing, their exposure to ultraviolet light may not be adequate for the manufacture of sufficient vitamin D. This is especially likely for people living in cold climates because they remain indoors or are covered by warm clothing when outdoors. Fortunately, vitamin D can also be ingested and absorbed in the small intestine. Natural sources of vitamin D are liver (especially fish liver), egg yolks, and dairy products (e.g., butter, cheese, and milk). In addition, the diet can be supplemented with vitamin D in fortified milk or vitamin pills.

Vitamin D synthesis begins when the precursor molecule 7-dehydrocholesterol (7-dē-hī′drō-kō-les′ter-ol) is exposed to ultraviolet light and converted into cholecalciferol (kō′lē-kal-sif′er-ol). Cholecalciferol is released into the blood and modified by hydroxylation (hydroxide ions are added) in the liver and kidneys to form active vitamin D, calcitriol (kal-si-trī′ol).

Excretion

Excretion is the removal of waste products from the body. In addition to water and salts, sweat contains a small amount of waste products, including urea, uric acid, and ammonia. However, even when large amounts of sweat are lost, the quantity of waste products eliminated in the sweat is insignificant because the urinary system excretes most of the body's waste products.

ASSESS YOUR PROGRESS

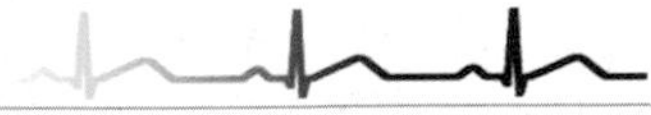

25. *In what ways does the skin provide protection?*

26. *What kinds of sensory receptors are in the skin, and why are they important?*

27. *How does the skin help regulate body temperature?*

28. *Name the locations where cholecalciferol is produced and then modified into vitamin D. What are the functions of vitamin D?*

29. *What substances are excreted in sweat? Is the skin an important site of excretion?*

Systems PATHOLOGY | Burns

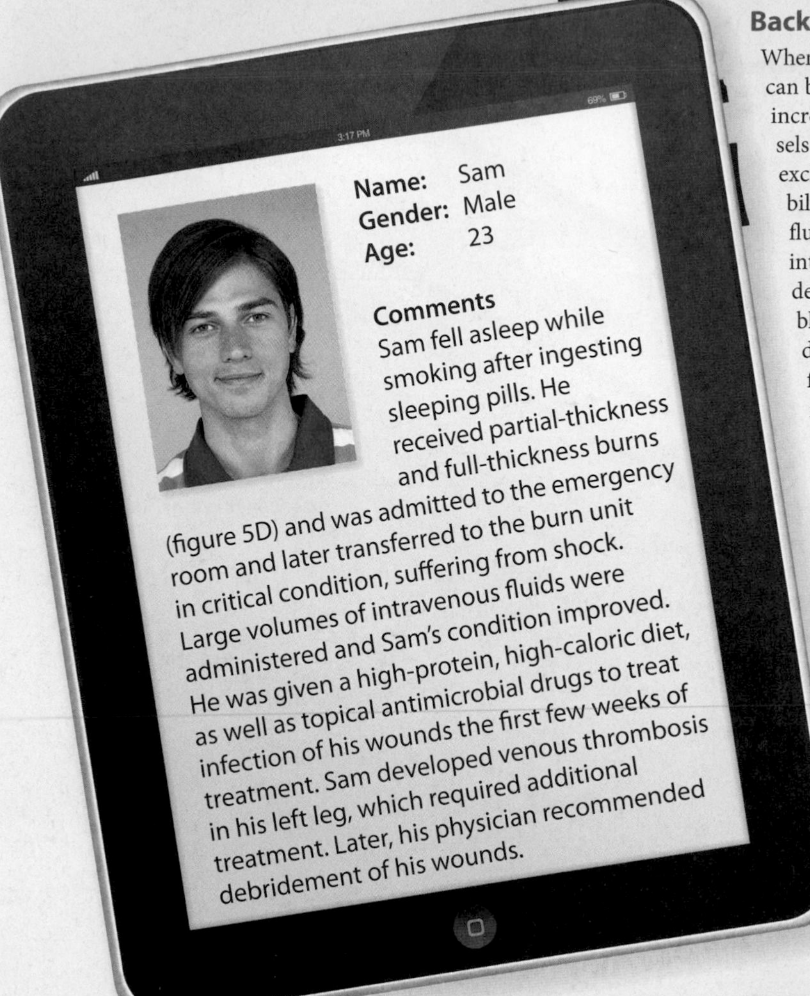

Name: Sam
Gender: Male
Age: 23
Comments
Sam fell asleep while smoking after ingesting sleeping pills. He received partial-thickness and full-thickness burns (figure 5D) and was admitted to the emergency room and later transferred to the burn unit in critical condition, suffering from shock. Large volumes of intravenous fluids were administered and Sam's condition improved. He was given a high-protein, high-caloric diet, as well as topical antimicrobial drugs to treat infection of his wounds the first few weeks of treatment. Sam developed venous thrombosis in his left leg, which required additional treatment. Later, his physician recommended debridement of his wounds.

Background Information

When large areas of skin are severely burned, the resulting systemic effects can be life-threatening. Within minutes of a major burn injury, there is increased permeability of the capillaries, which are the small blood vessels in which fluid, gases, nutrients, and waste products are normally exchanged between the blood and the tissues. This increased permeability occurs at the burn site and throughout the body. As a result, fluid and electrolytes (see chapter 2) are lost from the burn wound and into tissue spaces. The loss of fluid decreases blood volume, which decreases the heart's ability to pump blood. The resulting decrease in blood delivery to tissues can cause tissue damage, shock, and even death. Treatment consists of administering intravenous fluid at a faster rate than it leaks out of the capillaries. Although this fluid replacement can reverse the shock and prevent death, fluid continues to leak into tissue spaces, causing pronounced **edema** (swelling).

Typically, after 24 hours, capillary permeability returns to normal, and the amount of intravenous fluid administered can be greatly decreased. How burns cause capillary permeability to change is not well understood; however, it is clear that, following a burn, immunological and metabolic changes affect not only capillaries but the rest of the body as well. For example, chemical mediators (see chapter 4), which are released in response to tissue damage, contribute to changes in capillary permeability throughout the body.

Substances released from the burn may also cause cells to function abnormally. Burn injuries result in an almost immediate hypermetabolic state, which persists until wound closure. Two other factors contributing to the increased metabolism are (1) a resetting of the temperature control center in the brain to a higher temperature and (2) hormones released by the endocrine system (e.g., epinephrine and norepinephrine from the adrenal glands, which can increase cell metabolism).

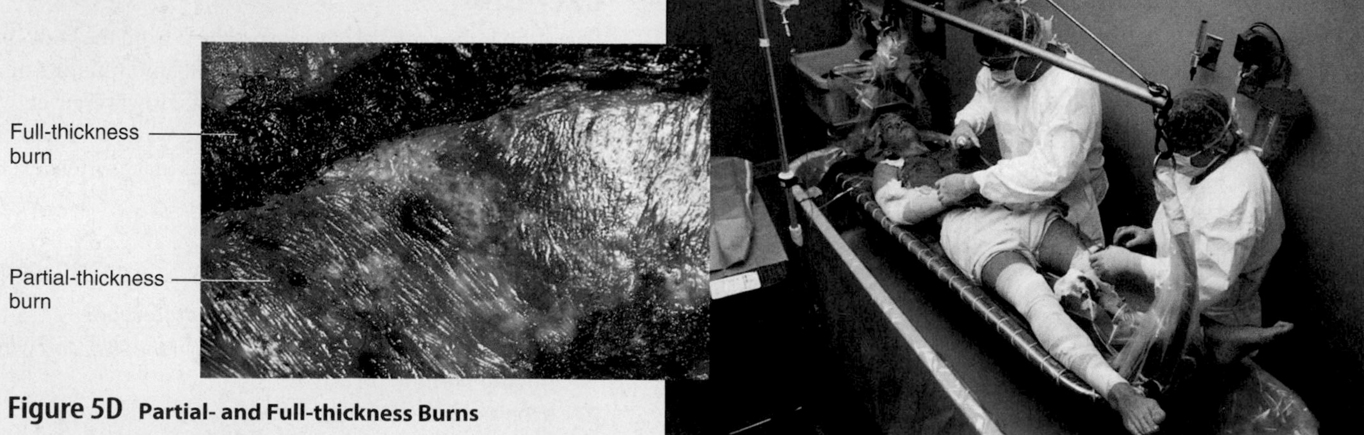

Full-thickness burn
Partial-thickness burn

Figure 5D Partial- and Full-thickness Burns

Figure 5E Patient in a Burn Unit

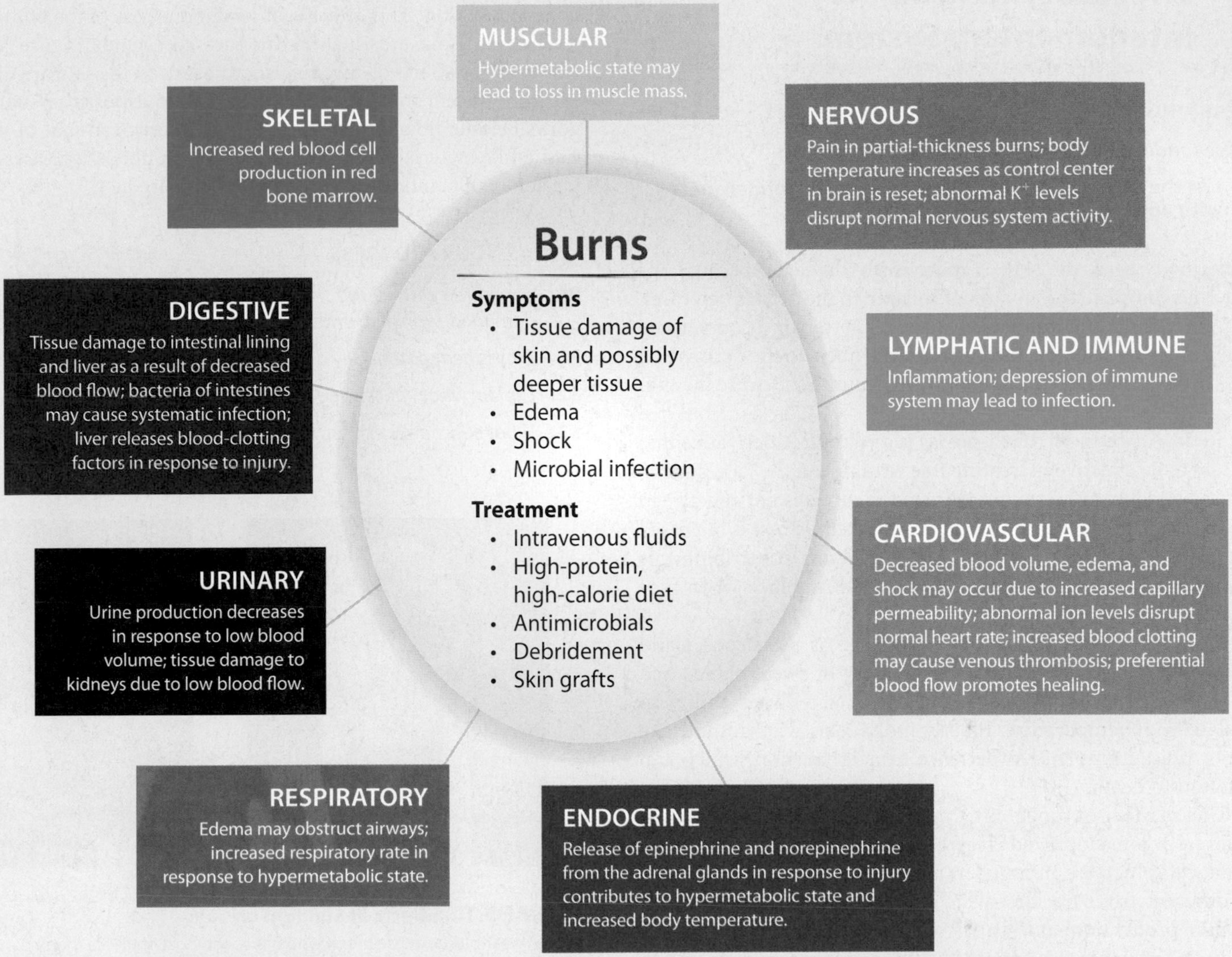

MUSCULAR
Hypermetabolic state may lead to loss in muscle mass.

SKELETAL
Increased red blood cell production in red bone marrow.

NERVOUS
Pain in partial-thickness burns; body temperature increases as control center in brain is reset; abnormal K^+ levels disrupt normal nervous system activity.

Burns

Symptoms
- Tissue damage of skin and possibly deeper tissue
- Edema
- Shock
- Microbial infection

Treatment
- Intravenous fluids
- High-protein, high-calorie diet
- Antimicrobials
- Debridement
- Skin grafts

DIGESTIVE
Tissue damage to intestinal lining and liver as a result of decreased blood flow; bacteria of intestines may cause systematic infection; liver releases blood-clotting factors in response to injury.

LYMPHATIC AND IMMUNE
Inflammation; depression of immune system may lead to infection.

URINARY
Urine production decreases in response to low blood volume; tissue damage to kidneys due to low blood flow.

CARDIOVASCULAR
Decreased blood volume, edema, and shock may occur due to increased capillary permeability; abnormal ion levels disrupt normal heart rate; increased blood clotting may cause venous thrombosis; preferential blood flow promotes healing.

RESPIRATORY
Edema may obstruct airways; increased respiratory rate in response to hypermetabolic state.

ENDOCRINE
Release of epinephrine and norepinephrine from the adrenal glands in response to injury contributes to hypermetabolic state and increased body temperature.

Compared with a normal body temperature of approximately 37°C (98.6°F), a typical burn patient may have a body temperature of 38.5°C (101.3°F) despite the higher loss of water by evaporation from the burn.

In severe burns, the increased metabolic rate can result in a loss of as much as 30–40% of the patient's preburn weight. To help compensate, treatment may include doubling or tripling the patient's caloric intake. In addition, the need for protein, which is necessary for tissue repair, is greater.

Normal skin maintains homeostasis by preventing microorganisms from entering the body. Because burns damage and sometimes completely destroy the skin, microorganisms can cause infections. For this reason, burn patients are maintained in an aseptic (sterile) environment in an attempt to prevent the entry of microorganisms into the wound. They are also given antimicrobial drugs, which kill microorganisms or suppress their growth. **Debridement** (dā-brēd-mont′), the removal of dead tissue from the burn (figure 5E), helps prevent infections by cleaning the wound and removing tissue in which infections could develop. Skin grafts, performed within a week of the injury, also help close the wound and prevent the entry of microorganisms.

Despite these efforts, however, infections are still the major cause of death for burn victims. Depression of the immune system during the first or second week after the injury contributes to the high infection rate. First, the thermally altered tissue is recognized as a foreign substance, which stimulates the immune system. Then, the immune system is overwhelmed as immune system cells become less effective and production of the chemicals that normally provide resistance to infections decreases (see chapter 22). The greater the magnitude of the burn, the greater the depression of the immune system and the greater the risk for infection.

Venous thrombosis, the development of a clot in a vein, is another complication of burns. Blood normally forms a clot when exposed to damaged tissue, such as at a burn site, but clotting can also occur elsewhere, such as in veins, where clots can block blood flow, resulting in tissue destruction. The concentration of chemicals that cause clotting (called clotting factors) increases for two reasons: Loss of fluid from the burn patient concentrates the chemicals, and the liver releases an increased amount of clotting factors.

Predict 5

When Sam was first admitted to the burn unit, the nurses carefully monitored his urine output. Why does that make sense in light of his injuries?

5.6 Effects of Aging on the Integumentary System

LEARNING OUTCOME

After reading this section, you should be able to

A. List the changes the integumentary system undergoes with age.

As the body ages, the skin is more easily damaged because the epidermis thins and the amount of collagen in the dermis decreases. Skin infections are more likely, and skin repair occurs more slowly. A decrease in the number of elastic fibers in the dermis and a loss of adipose tissue from the subcutaneous tissue cause the skin to sag and wrinkle.

Retin-A (tretinoin; tret′i-nō-in) is a vitamin A derivative that appears to be effective in treating fine wrinkles on the face, such as those caused by long-term exposure to the sun; it is not effective in treating deep lines. One ironic side effect of Retin-A use is increased sensitivity to the sun's ultraviolet rays. Doctors prescribing this cream caution their patients to always use a sunblock when they are going to be outdoors.

The skin also becomes drier with age as sebaceous gland activity decreases. Decreases in the activity of sweat glands and the blood supply to the dermis result in a reduced ability to regulate body temperature. Elderly individuals who do not take proper precautions may experience heat exhaustion, which can even lead to death.

The number of functioning melanocytes generally decreases; however, in some localized areas, especially on the hands and face, melanocytes increase in number, producing age spots. (Age spots are different from freckles, which are caused by an increase in melanin production, not an increase in melanocyte numbers.) White or gray hairs also appear because of a decrease in or lack of melanin production.

Skin that is exposed to sunlight appears to age more rapidly than nonexposed skin. This effect is observed on areas of the body that receive sun exposure, such as the face and hands (figure 5.10). However, the effects of chronic sun exposure on the skin are different from the effects of normal aging. In skin exposed to sunlight, normal elastic fibers are replaced by an interwoven mat of thick, elastic-like material, the number of collagen fibers decreases, and the ability of keratinocytes to divide is impaired.

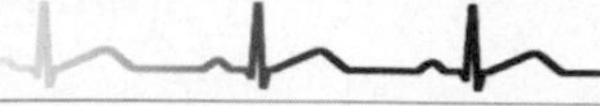

ASSESS YOUR PROGRESS

30. *Compared with young skin, why is aged skin more likely to be damaged, wrinkled, and dry?*

31. *Why is heat potentially dangerous to the elderly?*

32. *Explain what causes age spots and white hair.*

33. *What effect does exposure to sunlight have on skin?*

(a) 

(b) 

FIGURE 5.10 Effects of Sunlight on Skin
(*a*) A 91-year-old Japanese monk who has spent most of his life indoors. (*b*) A 62-year-old Native American woman who has spent most of her life outdoors.

Diseases and Disorders

TABLE 5.2 Skin

Condition	Description
Birthmarks	Congenital (present at birth) disorders of the dermal capillaries
Ringworm	Fungal infection that produces patchy scaling and inflammatory response in the skin
Eczema and dermatitis	Inflammatory conditions of the skin caused by allergy, infection, poor circulation, or exposure to chemical or environmental factors
Psoriasis	Chronic skin disease characterized by thicker than normal epidermal layer (stratum corneum) that sloughs to produce large, silvery scales; bleeding may occur if the scales are scraped away
Vitiligo	Development of patches of white skin where melanocytes are destroyed, apparently by an autoimmune response
Bacterial Infections	
Impetigo	Small blisters containing pus; easily rupture to form a thick, yellowish crust; usually affects children
Erysipelas	Swollen patches in the skin caused by the bacterium *Streptococcus pyogenes*
Decubitus ulcers (bedsores, pressure sores)	Develop in people who are bedridden or confined to a wheelchair; compression of tissue and reduced circulation result in destruction of the skin and subcutaneous tissue, which later become infected by bacteria, forming ulcers
Acne	Disorder of sebaceous glands and hair follicles that occurs when sloughed cells block the hair follicle, resulting in the formation of a lesion or pimple; the lesion may become infected and result in scarring; acne appears to be affected by hormones, sebum, abnormal keratinization within hair follicles, and the bacterium *Propionibacterium acnes*
Viral Infections	
Rubeola (measles)	Skin lesions; caused by a virus contracted through the respiratory tract; may develop into pneumonia or infect the brain, causing damage
Rubella (German measles)	Skin lesions; usually mild viral disease contracted through the respiratory tract; may be dangerous if contracted during pregnancy because the virus can cross the placenta and damage the fetus
Chickenpox	Skin lesions; usually mild viral disease contracted through the respiratory tract
Shingles	Painful skin lesions; caused by the chickenpox virus after childhood infection; can recur when the dormant virus is activated by trauma, stress, or another illness
Cold sores (fever blisters)	Skin lesions; caused by herpes simplex I virus; transmitted by oral or respiratory routes; lesions recur
Genital herpes	Genital lesions; caused by herpes simplex II virus; transmitted by sexual contact

Go to www.mhhe.com/seeley10 for additional information on these pathologies.

Answer

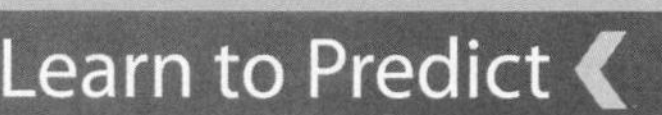
Learn to Predict

From page 139

To begin, we must identify the important information provided in the question. First, we are told that Laura is spending her summer at the pool, constantly exposed to the sun, and develops a "golden tan." Second, she returns to school in the fall, thus spending less time outside, and her tan fades.

We know that sun exposure leads to tanning of the skin, but to answer this question fully we must consider what specifically causes the skin to darken and why tanning is not permanent. We learned in chapter 5 that melanocytes in the epidermis produce melanin in response to sun exposure and package it in melanosomes. Keratinocytes phagocytize the tips of melanocyte processes containing the melanosomes, and the result is pigmented keratinocytes, and thus darker skin. We can therefore conclude that Laura's skin is darker due to increased sun exposure and increased melanin production during the summer.

We also learned in this chapter that the epidermis is a stratified squamous epithelium (as described in chapter 4) composed of many layers of cells. Mitosis of cells in the deepest layer (stratum basale) of the epidermis produces new cells, which gradually undergo keratinization and eventually die. At the surface of the skin, keratinized (dead) keratinocytes are sloughed off and replaced by new ones daily.

So why did Laura's tan fade? It faded because the older, heavily pigmented keratinocytes were sloughed off each day and replaced with new cells. These new cells, produced *after* she returned to school, were less pigmented and her skin became a lighter color.

Answers to the rest of this chapter's Predict questions are in Appendix G.

Summary

The integumentary system consists of the skin, hair, glands, and nails.

5.1 Functions of the Integumentary System (p. 140)

The integumentary system separates and protects us from the external environment. Other functions include sensation, temperature regulation, vitamin D production, and excretion of small amounts of waste products.

5.2 Skin (p. 140)

Epidermis

1. The epidermis is stratified squamous epithelium divided into five strata.
2. The stratum basale consists of keratinocytes, which produce the cells of the more superficial strata.
3. The stratum spinosum consists of several layers of cells held together by many desmosomes.
4. The stratum granulosum consists of cells filled with granules of keratohyalin. Cell death occurs in this stratum.
5. The stratum lucidum consists of a layer of dead, transparent cells.
6. The stratum corneum consists of many layers of dead squamous cells. The most superficial cells slough off.
7. Keratinization is the transformation of the living cells of the stratum basale into the dead squamous cells of the stratum corneum.
   - Keratinized cells are filled with keratin and have a protein envelope, both of which contribute to structural strength. The cells are also held together by many desmosomes.
   - Intercellular spaces are filled with lipids from the lamellae that help make the epidermis impermeable to water.
8. Soft keratin is present in skin and the inside of hairs, whereas hard keratin occurs in nails and the outside of hairs. Hard keratin makes cells more durable, and these cells are not shed.

Thick and Thin Skin

1. Thick skin has all five epithelial strata.
2. Thin skin contains fewer cell layers per stratum, and the stratum lucidum is usually absent. Hair is found only in thin skin.

Skin Color

1. Melanocytes produce melanin inside melanosomes and then transfer the melanin to keratinocytes. The size and distribution of melanosomes determine skin color. Melanin production is determined genetically but can be influenced by ultraviolet light (tanning) and hormones.
2. Carotene, an ingested plant pigment, can cause the skin to appear yellowish.
3. Increased blood flow produces a red skin color, whereas decreased blood flow causes pale skin. Decreased oxygen content in the blood results in a bluish color, a condition called cyanosis.

Dermis

1. The dermis is connective tissue divided into two layers.
2. The papillary layer has projections called dermal papillae and is composed of loose connective tissue that is well supplied with capillaries.
3. The reticular layer is the main layer. It is dense irregular connective tissue consisting mostly of collagen.

5.3 Subcutaneous Tissue (p. 146)

1. Located beneath the dermis, the subcutaneous tissue is loose connective tissue that contains collagen and elastic fibers.
2. The subcutaneous tissue attaches the skin to underlying structures and is a site of lipid storage.

5.4 Accessory Skin Structures (p. 148)

Hair

1. Lanugo (fetal hair) is replaced near the time of birth by terminal hairs (scalp, eyelids, and eyebrows) and vellus hairs. At puberty, vellus hairs can be replaced with terminal hairs.
2. A hair has three parts: shaft, root, and hair bulb.
3. The root and shaft of a hair are composed of dead keratinized epithelial cells. In the center, a cortex of cells containing hard keratin surrounds a medulla composed of cells containing soft keratin. The cortex is covered by the cuticle, a single layer of cells filled with hard keratin.
4. The hair bulb produces the hair in cycles, with a growth stage and a resting stage.
5. Hair color is determined by the amount and kind of melanin present.
6. Contraction of the arrector pili muscles, which are smooth muscles, causes hair to "stand on end" and produces "goose bumps."

Glands

1. Sebaceous glands produce sebum, which oils the hair and the surface of the skin.
2. Eccrine sweat glands produce sweat, which cools the body. Apocrine sweat glands produce an organic secretion that can be broken down by bacteria to cause body odor.
3. Other skin glands are ceruminous glands, which make cerumen (earwax), and mammary glands, which produce milk.

Nails

1. The nail root is covered by skin, and the nail body is the visible part of the nail.
2. Nearly all of the nail is formed by the nail matrix, but the nail bed contributes.
3. The lunula is the part of the nail matrix visible through the nail body.
4. The nail is stratum corneum containing hard keratin.

5.5 Physiology of the Integumentary System (p. 153)

Protection

1. The skin protects against abrasion and ultraviolet light, prevents the entry of microorganisms, helps regulate body temperature, and prevents water loss.
2. Hair protects against abrasion and ultraviolet light and is a heat insulator.
3. Nails protect the ends of the digits.

Sensation

The skin contains sensory receptors for pain, touch, hot, cold, and pressure, which allow for proper responses to the environment.

Temperature Regulation

1. Through dilation and constriction of blood vessels, the skin controls heat loss from the body.
2. Sweat glands produce sweat, which evaporates and lowers body temperature.

Vitamin D Production

1. Skin exposed to ultraviolet light produces cholecalciferol, which is modified in the liver and then in the kidneys to form active vitamin D.
2. Vitamin D increases blood calcium levels by promoting calcium uptake from the small intestine, calcium release from bone, and the reduction of calcium loss from the kidneys.

Excretion

Skin glands remove small amounts of waste products (e.g., urea, uric acid, and ammonia) but are not important in excretion.

5.6 Effects of Aging on the Integumentary System (p. 158)

1. As the body ages, blood flow to the skin declines, the skin becomes thinner, and elasticity is lost.
2. Sebaceous and sweat glands become less active, and the number of melanocytes decreases.

REVIEW AND COMPREHENSION

1. If a splinter penetrates the skin of the palm of the hand to the second epidermal layer from the surface, the last layer damaged is the
   a. stratum granulosum.
   b. stratum basale.
   c. stratum corneum.
   d. stratum lucidum.
   e. stratum spinosum.

For questions 2–6, match the layer of the epidermis with the correct description or function:

a. stratum basale
b. stratum corneum
c. stratum granulosum
d. stratum lucidum
e. stratum spinosum

2. Production of keratin fibers; formation of lamellar bodies; limited amount of cell division
3. Sloughing occurs; 25 or more layers of dead squamous cells
4. Production of cells; melanocytes produce and contribute melanin; hemidesmosomes present
5. Production of keratohyalin granules; lamellar bodies release lipids; cells die
6. Dispersion of keratohyalin around keratin fibers; layer appears transparent; cells dead
7. The function of melanin in the skin is to
   a. lubricate the skin.
   b. prevent skin infections.
   c. protect the skin from ultraviolet light.
   d. reduce water loss.
   e. help regulate body temperature.
8. Concerning skin color, which pair of statements is *not* correctly matched?
   a. skin appears yellow—carotene present
   b. no skin pigmentation (albinism)—genetic disorder
   c. skin tans—increased melanin production
   d. skin appears blue (cyanosis)—oxygenated blood
   e. African-Americans darker than Caucasians—more melanin in African-American skin

For questions 9–11, match the layer of the dermis with the correct description or function:

a. papillary layer
b. reticular layer

9. Layer of dermis responsible for most of the structural strength of the skin
10. Layer of dermis responsible for fingerprints and footprints
11. Layer of dermis responsible for cleavage lines and stretch marks
12. After birth, the type of hair on the scalp, eyelids, and eyebrows is
    a. lanugo.
    b. terminal hair.
    c. vellus hair.
13. Hair
    a. is produced by the dermal root sheath.
    b. consists of living, keratinized epithelial cells.
    c. is colored by melanin.
    d. contains mostly soft keratin.
    e. grows from the tip.
14. Given these parts of a hair and hair follicle:
    (1) cortex
    (2) cuticle
    (3) dermal root sheath
    (4) epithelial root sheath
    (5) medulla

Arrange the structures in the correct order from the outside of the hair follicle to the center of the hair.

a. 1,4,3,5,2
b. 2,1,5,3,4
c. 3,4,2,1,5
d. 4,3,1,2,5
e. 5,4,3,2,1

15. Concerning hair growth,
    a. hair falls out of the hair follicle at the end of the growth stage.
    b. most of the hair on the body grows continuously.
    c. cutting or plucking the hair increases its growth rate and thickness.
    d. genetic factors and the hormone testosterone are involved in "pattern baldness."
    e. eyebrows have a longer growth stage and resting stage than scalp hair.
16. Smooth muscles that produce "goose bumps" when they contract and are attached to hair follicles are called
    a. external root sheaths.
    b. arrector pili.
    c. dermal papillae.
    d. internal root sheaths.
    e. hair bulbs.

For questions 17–19, match the type of gland with the correct description or function:

a. apocrine sweat gland
b. eccrine sweat gland
c. sebaceous gland

17. Alveolar glands that produce a white, oily substance; usually open into hair follicles
18. Coiled, tubular glands that secrete a hyposmotic fluid that cools the body; most numerous in the palms of the hands and soles of the feet
19. Secretions from these coiled, tubular glands are broken down by bacteria to produce body odor; found in the axillae, in the genitalia, and around the anus
20. The lunula of the nail appears white because
    a. it lacks melanin.
    b. blood vessels cannot be seen through the thick nail matrix.
    c. the cuticle decreases blood flow to the area.
    d. the nail root is much thicker than the nail body.
    e. the hyponychium is thicker than the cuticle.
21. Most of the nail is produced by the
    a. cuticle.
    b. hyponychium.
    c. nail bed.
    d. nail matrix.
    e. dermis.
22. The skin helps maintain optimum calcium and phosphate levels in the body by participating in the production of
    a. vitamin A.
    b. vitamin B.
    c. vitamin D.
    d. melanin.
    e. keratin.
23. Which of these processes increase(s) heat loss from the body?
    a. dilation of dermal arterioles
    b. constriction of dermal arterioles
    c. increased sweating
    d. Both a and c are correct.
    e. Both b and c are correct.
24. In third-degree (full-thickness) burns, both the epidermis and the dermis are destroyed. Which of the following conditions does *not* occur as a result of a third-degree burn?
    a. dehydration (increased water loss)
    b. increased likelihood of infection
    c. increased sweating
    d. loss of sensation in the burned area
    e. poor temperature regulation in the burned area

Answers in Appendix E

CRITICAL THINKING

1. The skin of infants is more easily penetrated and injured by abrasion than is the skin of adults. Based on this fact, which stratum of the epidermis is probably much thinner in infants than in adults?
2. Melanocytes are found primarily in the stratum basale of the epidermis. In reference to their function, why does this location make sense?
3. The rate of water loss from the skin of a hand was measured. Following the measurement, the hand was soaked in alcohol for 15 minutes. After all the alcohol had been removed from the hand, the rate of water loss was again measured. Compared with the rate of water loss before soaking, what difference, if any, would you expect in the rate of water loss after soaking the hand in alcohol?
4. It has been several weeks since Goodboy Player has competed in a tennis match. After a match, he discovers that a blister has formed beneath an old callus on his foot, and the callus has fallen off. When he examines the callus, it appears yellow. Can you explain why?
5. A woman has stretch marks on her abdomen, yet she states that she has never been pregnant. Is this possible?
6. Why are your eyelashes not a foot long? Your fingernails?
7. Pulling on hair can be quite painful, yet cutting hair is not painful. Explain.
8. A patient has an ingrown toenail, in which the nail grows into the nail fold. Would cutting the nail away from the nail fold permanently correct this condition? Why or why not?
9. Defend or refute the following statement: Dark-skinned children are more susceptible to rickets (insufficient calcium in the bones) than fair-skinned children.
10. Harry, age 55, went to a health fair and had a PSA test. The test results and subsequent examinations indicated prostate cancer. Harry was given radiation treatments and chemotherapeutic drugs. These drugs adversely affect cancer cells by interrupting mitosis, but they also interrupt mitosis in normal cells. Describe the probable effect of chemotherapy on Harry's epidermis, hair, nails, skin pigmentation, and sebaceous glands.

Answers in Appendix F

Visit this book's website at **www.mhhe.com/seeley10** for chapter quizzes, interactive learning exercises, and other study tools.

Mc Graw Hill
connect plus+
ANATOMY & PHYSIOLOGY

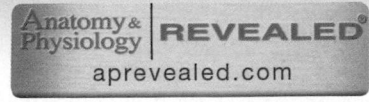

Anatomy & Physiology REVEALED
aprevealed.com

6

Skeletal System

BONES AND BONE TISSUE

"Break a leg!" your friend might say just before your first job interview, but if you really broke your leg, your chances of completing the interview would be slim to none. One of the most iconic symbols of the human form, the skeleton, is essential for our day-to-day activities. Sitting, standing, walking, picking up a pencil, and taking a breath all involve the skeletal system. Besides helping the body move and breathe, the skeleton is the structural framework that gives the body its shape and protects the internal organs and soft tissues. Although the skeleton consists of the mineralized material left after the flesh and organs have been removed and is often associated with death, it is composed of dynamic, living tissues that are able to grow, adapt to stress, and undergo repair after injury.

Learn to Predict

Amir lived for football. He had played football since second grade and regularly watched every televised college and professional football game. He dreamed of the day he would be on TV, playing football. One memorable game, Amir's favorite player broke his leg and Amir wanted to do everything he could to prevent that from happening to him. He asked his doctor to help him create a bone-healthy diet. Chapter 5 explained the anatomy and physiology of the skin, and chapter 6 explores the structure of bone tissue—how it forms and grows, as well as the nutritional requirements for normal bone growth. After reading this chapter, explain the specific nutritional requirements of the diet Amir's doctor advised him to consume for injury prevention.

Photo: University of Alabama wide receiver Tyrone Prothro suffered a career-ending injury in 2005 when he broke both his tibia and his fibula while attempting a touchdown reception. Unfortunately, the healed bone would never be as strong as it was before.

Module 5
Skeletal System

Anatomy & Physiology REVEALED
aprevealed.com

6.1 Functions of the Skeletal System

LEARNING OUTCOMES

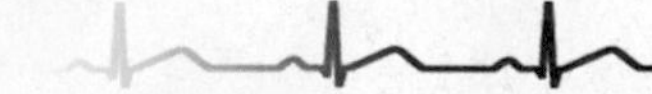

After reading this section, you should be able to

A. List the components of the skeletal system.

B. Explain the functions of the skeletal system.

The skeletal system has four components: bones, cartilage, tendons, and ligaments. The skeleton is usually thought of as the framework of the body, but the skeletal system has many other functions as well, including the following:

1. *Support.* Rigid, strong bone is well suited for bearing weight and is the major supporting tissue of the body. Cartilage provides a firm yet flexible support within certain structures, such as the nose, external ear, thoracic cage, and trachea. **Ligaments** are strong bands of fibrous connective tissue that attach to bones and hold them together.
2. *Protection.* Bone is hard and protects the organs it surrounds. For example, the skull encloses and protects the brain, and the vertebrae surround the spinal cord. The rib cage protects the heart, lungs, and other organs of the thorax.
3. *Movement.* Skeletal muscles attach to bones by **tendons,** which are strong bands of connective tissue. Contraction of the skeletal muscles moves the bones, producing body movements. Joints, which are formed where two or more bones come together, allow movement between bones. Smooth cartilage covers the ends of bones within some joints, allowing the bones to move freely. Ligaments allow some movement between bones but prevent excessive movements.
4. *Storage.* Some minerals in the blood are taken into bone and stored. Should blood levels of these minerals decrease, the minerals are released from bone into the blood. The principal minerals stored are calcium and phosphorus, two minerals essential for many physiological processes. Adipose tissue is also stored within bone cavities. If needed, the lipids are released into the blood and used by other tissues as a source of energy.
5. *Blood cell production.* Many bones contain cavities filled with red bone marrow, which gives rise to blood cells and platelets (see chapter 19).

ASSESS YOUR PROGRESS

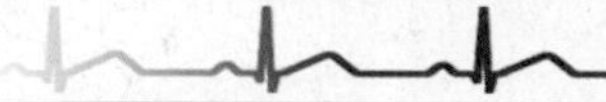

1. *Name the four components of the skeletal system.*
2. *Describe the five major functions of the skeletal system.*

6.2 Cartilage

LEARNING OUTCOMES

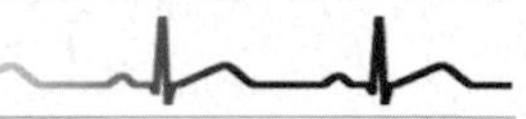

After reading this section, you should be able to

A. Relate the importance of cartilage to the structure of the skeletal system.

B. Describe the structure of hyaline cartilage.

C. Explain the types of cartilage growth.

Cartilage comes in three types: hyaline cartilage, fibrocartilage, and elastic cartilage (see chapter 4). Although each type of cartilage can provide support, hyaline cartilage is most intimately associated with bone. An understanding of hyaline cartilage structure is important because it is the precursor for most bones in the body. In addition, bone lengthening and bone repair often involve the production of hyaline cartilage, followed by its replacement with bone.

Hyaline cartilage consists of specialized cells called **chondroblasts** (kon′drō-blasts; cartilage) that produce a matrix surrounding themselves (figure 6.1). When matrix surrounds a chondroblast, it becomes a **chondrocyte** (kon′drō-sīt), a rounded cell that occupies a space

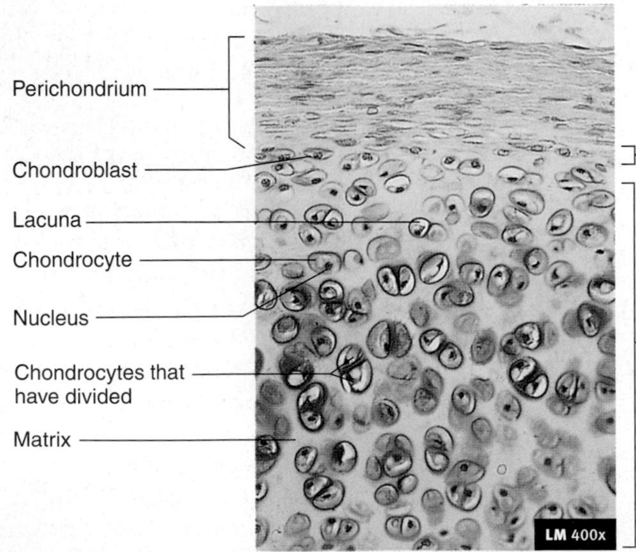
Perichondrium
Chondroblast
Lacuna
Chondrocyte
Nucleus
Chondrocytes that have divided
Matrix
LM 400x
Appositional growth
(new cartilage is added to the surface of the cartilage by chondroblasts from the inner layer of the perichondrium)
Interstitial growth
(new cartilage is formed within the cartilage by chondrocytes that divide and produce additional matrix)

FIGURE 6.1 Hyaline Cartilage
Photomicrograph of hyaline cartilage covered by perichondrium. Chondrocytes within lacunae are surrounded by a cartilage matrix.

called a **lacuna** (lă-koo′nă) within the matrix. The matrix contains collagen, which provides strength, and proteoglycans, which make cartilage resilient by trapping water.

The **perichondrium** (per-i-kon′drē-ŭm) is a double-layered connective tissue sheath covering most cartilage (figure 6.1). The outer layer of the perichondrium is dense irregular connective tissue containing fibroblasts. The inner, more delicate layer has fewer fibers and contains chondroblasts. Blood vessels and nerves penetrate the outer layer of the perichondrium but do not enter the cartilage matrix, so nutrients must diffuse through the cartilage matrix to reach the chondrocytes. **Articular** (ar-tik′ū-lăr) **cartilage,** which is hyaline cartilage that covers the ends of bones where they come together to form joints, has no perichondrium, blood vessels, or nerves.

Predict 2

Explain why damaged cartilage takes a long time to heal. What are the advantages of articular cartilage having no perichondrium, blood vessels, or nerves?

Cartilage grows in two ways. In **appositional growth,** chondroblasts in the perichondrium add new cartilage to the outside edge of the existing cartilage. The chondroblasts lay down new matrix and add new chondrocytes to the outside of the tissue. In **interstitial growth,** chondrocytes within the tissue divide and add more matrix between the existing cells (figure 6.1).

ASSESS YOUR PROGRESS

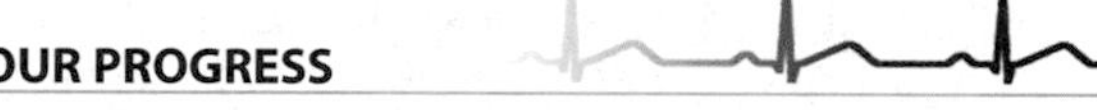

3. *What are the three types of cartilage? Which type is more closely associated with bone?*
4. *Describe the structure of hyaline cartilage. Name the two types of cartilage cells. What is a lacuna?*
5. *Differentiate between appositional and interstitial growth of cartilage.*

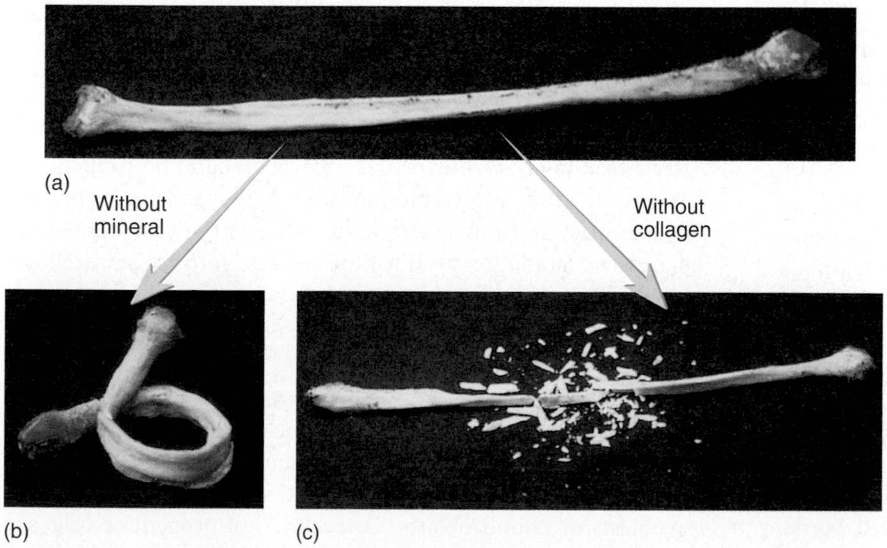

6.3 Bone Histology

LEARNING OUTCOMES

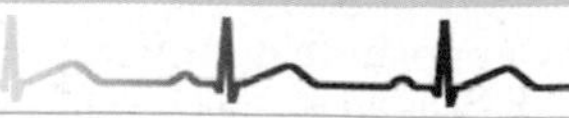

After reading this section, you should be able to

A. Describe the components of the extracellular bone matrix and state the function of each.

B. List each type of bone cell and give the function and origin of each.

C. Describe the structure of woven and lamellar bone.

D. Explain the structural differences between compact and spongy bone.

Bone consists of extracellular bone matrix and bone cells. The bone cells produce the bone matrix, become entrapped within it, and break it down so that new matrix can replace the old matrix. The composition of the bone matrix is responsible for the characteristics of bone.

Bone Matrix

By weight, mature bone matrix is normally about 35% organic and 65% inorganic material. The organic material consists primarily of collagen and proteoglycans. The inorganic material consists primarily of a calcium phosphate crystal called **hydroxyapatite** (hī-drok′sē-ap-ă-tīt), which has the molecular formula $Ca_{10}(PO_4)_6(OH)_2$.

The collagen and mineral components are responsible for the major functional characteristics of bone. Bone matrix can be compared to reinforced concrete. Like reinforcing steel bars, the collagen fibers lend flexible strength to the matrix; like concrete, the mineral components give the matrix weight-bearing strength.

If all the mineral is removed from a long bone, collagen becomes the primary constituent and the bone is overly flexible. On the other hand, if the collagen is removed from the bone, the mineral component becomes the primary constituent and the bone is very brittle (figure 6.2).

(a) Without mineral — Without collagen (b) (c)

FIGURE 6.2 Effects of Changing the Bone Matrix
(*a*) Normal bone. (*b*) Demineralized bone, in which collagen is the primary remaining component, can be bent without breaking. (*c*) When collagen is removed, mineral is the primary remaining component, making the bone so brittle that it is easily shattered.

Bone Cells

Bone cells are categorized as osteoblasts, osteocytes, and osteoclasts. Each cell type has different functions and a different origin.

Osteoblasts

Osteoblasts (os′tē-ō-blastz), which are bone-forming cells, have an extensive endoplasmic reticulum and numerous ribosomes. They produce collagen and proteoglycans, which are packaged into vesicles by the Golgi apparatus and released from the cell by exocytosis. Osteoblasts also release **matrix vesicles,** membrane-bound sacs formed when the plasma membrane buds, or protrudes outward, and pinches off. The matrix vesicles concentrate Ca^{2+} and PO_4^{3-} and form needlelike hydroxyapatite crystals. When the hydroxyapatite crystals are released from the matrix vesicles, they act as templates, or "seeds," which stimulate further hydroxyapatite formation and mineralization of the matrix.

Ossification (os′i-fi-kā′shŭn), or *osteogenesis* (os′tē-ō-jen′ĕ-sis), is the formation of bone by osteoblasts. Ossification occurs by appositional growth on the surface of previously existing bone or cartilage. For example, osteoblasts beneath the periosteum cover the surface of preexisting bone (figure 6.3*a*). Elongated cell processes from osteoblasts connect to the cell processes of other osteoblasts through gap junctions (see chapter 4). Bone matrix produced by the osteoblasts covers the older bone surface and surrounds the osteoblast cell bodies and processes. The result is a new layer of bone.

Osteocytes

Once an osteoblast becomes surrounded by bone matrix, it is referred to as an **osteocyte** (os′tē-ō-sīt; figure 6.3*b*). Osteocytes become relatively inactive, compared with most osteoblasts, but it is possible for them to produce the components needed to maintain the bone matrix.

The spaces occupied by the osteocyte cell bodies are called **lacunae** (lă-koo′nē), and the spaces occupied by the osteocyte cell processes are called **canaliculi** (kan-ă-lik′ū-lī; little canals; figure 6.3*c*). In a sense, the cells and their processes form a "mold" around which the matrix is formed. Bone differs from cartilage in that the processes of bone cells are in contact with one another through the canaliculi. Instead of diffusing through the mineralized matrix, nutrients and gases can pass through the small amount of fluid surrounding the cells in the canaliculi and lacunae or pass from cell to cell through the gap junctions connecting the cell processes.

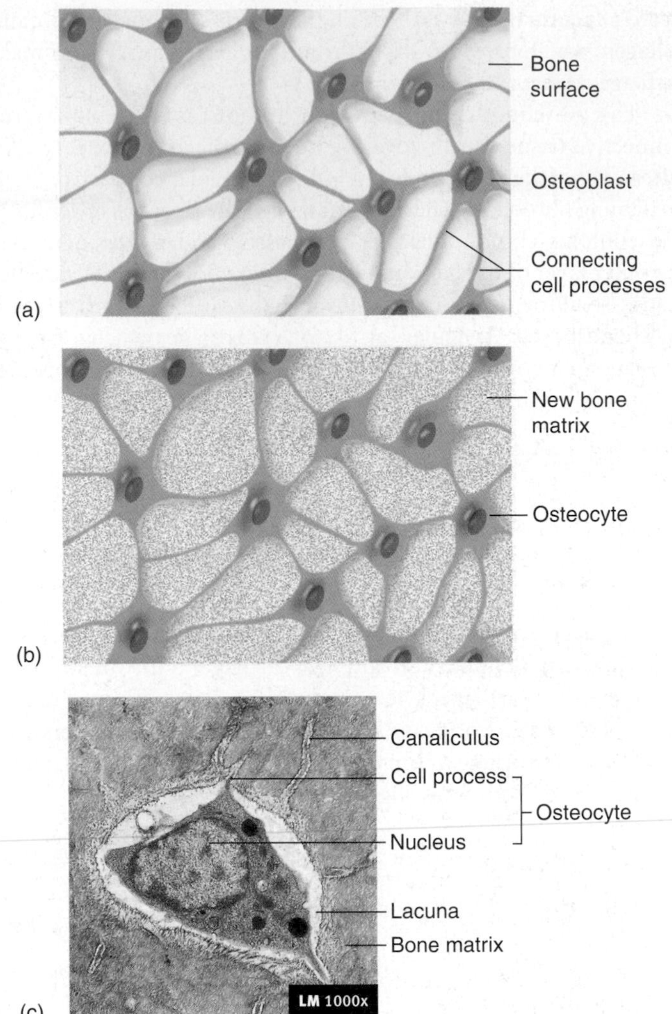
(a)
Bone surface
Osteoblast
Connecting cell processes
(b)
New bone matrix
Osteocyte
(c)
Canaliculus
Cell process
Nucleus
Osteocyte
Lacuna
Bone matrix
LM 1000x

FIGURE 6.3 Ossification
(*a*) On a preexisting surface, such as cartilage or bone, the cell processes of different osteoblasts join together. (*b*) Osteoblasts have produced bone matrix and are now osteocytes. (*c*) Photomicrograph of an osteocyte in a lacuna with cell processes in the canaliculi.

Osteoclasts

Osteoclasts (os′tē-ō-klastz) are bone-destroying cells (figure 6.4). These cells perform reabsorption, or breakdown, of bone that mobilizes crucial Ca^{2+} and phosphate ions for use in many metabolic processes. These cells are massive, multinucleated cells whose differentiation follows a complex pathway. They originate within the red bone marrow monocyte/macrophage lineage. These precursors attach to the bone matrix where direct contact with osteoblasts is required to allow eventual maturation into functional osteosclasts.

Mature osteoclasts carry out bone reabsorption through a multi-step process. First, the osteoclasts must access the bone matrix. The current model proposes that osteoblasts lining the connective tissue around bone regulate the movement of mature osteoclasts into a bone remodeling area. Once the osteoclasts have come in contact with the bone surface, they form attachment structures via interactions with cell-surface proteins called integrins (see chapter 3). Soon after, structures called podosomes develop and form a sealed compartment under the osteoclast. The osteoclast cell membrane then further differentiates into a highly folded form called the **ruffled border** (see figure 6.4).

The ruffled border is a specialized reabsorption-specific area of the membrane. Acidic vesicles fuse with the membrane of the ruffled border, while ATP-powered H^+ pumps, and protein-digesting enzymes are inserted into the membrane of the ruffled border. Secretion of H^+ creates an acidic environment within the sealed compartment, which causes decalcification of the bone matrix. The protein-digestion enzymes are secreted into the sealed compartment and digest the organic, protein component of the bone matrix. After breakdown of the matrix, the degradation products are removed by a transcytosis mechanism, whereby the products enter the osteoclast and move across the cytoplasm to the other side. There, the degradation products are secreted into the extracellular space, enter the blood, and are used elsewhere in the body.

Origin of Bone Cells

Connective tissue develops embryologically from mesenchymal cells (see chapter 4). Some of the mesenchymal cells become **stem cells,** which can replicate and give rise to more specialized cell types. **Osteochondral progenitor cells** are stem cells that can become osteoblasts or chondroblasts. Osteochondral progenitor cells are located in the inner layer of the perichondrium and in layers of connective tissue that cover bone (periosteum and endosteum). From these locations, they are a potential source of new osteoblasts or chondroblasts.

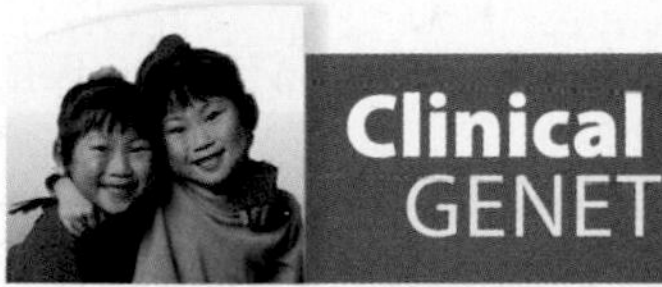

Clinical GENETICS | Osteogenesis Imperfecta

Osteogenesis imperfecta (OI; os′tē-ō-jen′ĕ-sis im-per-fek′tă) is also known as brittle bone disorder. This connective tissue disease is caused by mutations that yield reduced or defective type I collagen. Type I collagen is the major collagen of bone, tendon, and skin. There is considerable variability in the appearance and severity of OI symptoms, which is partially explained by different types of mutations.

The mildest and most common form of OI is called type I. It is caused by too little formation of normal type I collagen. The majority of patients have a mutation in a type I collagen gene that creates a stop codon, so that the gene no longer encodes a functional protein (see chapter 3). The patient produces approximately half as much type I collagen as usual. Thus, type I OI is a collagen-deficiency disorder. Patients may exhibit any of the following characteristics: bones predisposed to fracture, especially before puberty; a tendency to develop spinal curvature; loose joints; brittle teeth; hearing loss; and a blue tint to the whites of the eyes. Unlike people who have more severe forms of OI, these patients have normal or near-normal stature and minimal or no bone deformities. The number of fractures over a lifetime can vary from a few to more than 100. Children with mild type I OI may exhibit few obvious clinical features, except for a history of broken bones. It is important for children with OI to be diagnosed properly because broken bones can be associated with child abuse.

The more severe types of OI occur when mutated collagen genes are transcribed and the resulting mRNA codes for a defective protein. The more defective the protein, the weaker the collagen fiber and the more severe the disorder. In addition, depending on the mutation, bone cells may produce less bone matrix because they are inefficient at making matrix containing defective collagen. Thus, the most severe forms of OI are defective-collagen and collagen-deficiency disorders. Type II, the most severe OI, is usually lethal within the first week of life because of breathing failure due to rib fractures and underdeveloped lungs. Type III OI is characterized by bones that fracture very easily, even before and during birth. Fractures occurring before birth often heal in poor alignment, leaving the limbs short and bent (figure 6A). The symptoms of type III are the same as those of type I OI. Type IV OI has symptoms between types I and III in severity. Types V–VIII are more infrequent.

In the United States, 20,000 to 50,000 people may have OI. Almost all cases are caused by autosomal dominant mutations (see chapter 29). *Autosomal dominant* means that a mutation in only one copy of the gene will cause OI, even if the other gene is normal. Because OI is an autosomal dominant trait, there is a 50% chance that a child will inherit OI from an affected parent. However, the disorder's great variability may cause the child to be affected in different ways than the parent. For example, the tendency for fractures and bone deformity may be different. Approximately 25% of children with OI have parents who do not have the disorder. In these cases, a new mutation, which occurs during the formation of sperm cells or oocytes, is responsible. Unfortunately, there is no cure for OI, and treatments are primarily directed at reducing the risk for fractures.

Arm
Forearm
Hand

FIGURE 6A Osteogenesis Imperfecta

Osteoblasts are derived from osteochondral progenitor cells, and osteocytes are derived from osteoblasts. Whether or not osteocytes freed from their surrounding bone matrix by reabsorption can revert to become active osteoblasts is a debated issue. As discussed in the previous section, osteoclasts are not derived from osteochondral progenitor cells but from stem cells in red bone marrow (see chapter 19).

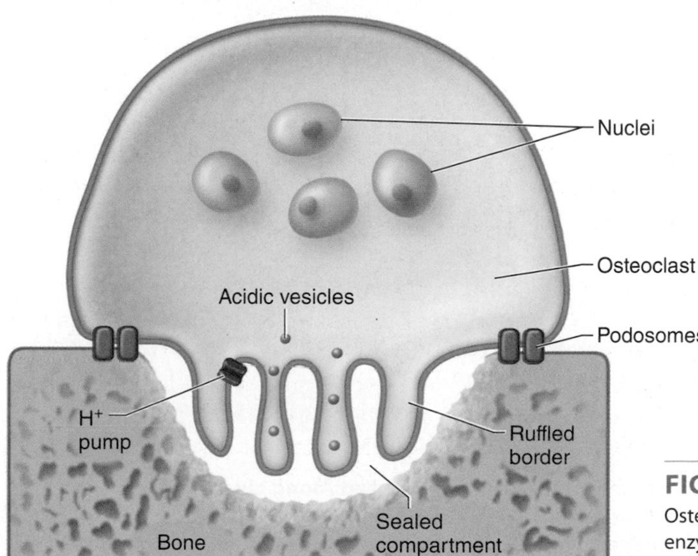
Nuclei
Osteoclast
Acidic vesicles
Podosomes
H+ pump
Ruffled border
Sealed compartment
Bone

FIGURE 6.4 Osteoclast Structure
Osteoclasts are massive, multinucleated cells that secrete acid and protein-digesting enzymes, which degrade bone. These cells then transport the digested matrix from the bone into the extracellular fluid.

ASSESS YOUR PROGRESS 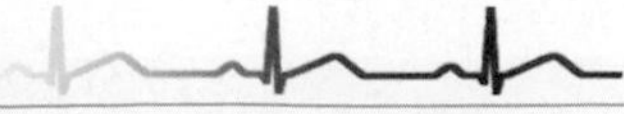

6. *Name the components of bone matrix, and explain their contribution to bone flexibility and bones' ability to bear weight.*
7. *Differentiate among the characteristics and functions of osteoblasts, osteocytes, and osteoclasts.*
8. *Describe the formation of new bone by appositional growth. Name the spaces that are occupied by osteocyte cell bodies and cell processes.*
9. *What cells give rise to osteochondral progenitor cells? What kinds of cells are derived from osteochondral progenitor cells? What types of cells give rise to osteoclasts?*

Woven and Lamellar Bone

Bone tissue is classified as either woven or lamellar, according to the organization of collagen fibers within the bone matrix.

In **woven bone,** the collagen fibers are randomly oriented in many directions. Woven bone is first formed during fetal development or during the repair of a fracture. After its formation, osteoclasts break down the woven bone and osteoblasts build new matrix. The process of removing old bone and adding new bone is called **bone remodeling** and is discussed in section 6.7. Woven bone is remodeled to form lamellar bone.

Lamellar bone is mature bone that is organized into thin sheets or layers approximately 3–7 micrometers (μm) thick called **lamellae** (lă-mel′ē). In general, the collagen fibers of one lamella lie parallel to one another, but at an angle to the collagen fibers in the adjacent lamellae. Osteocytes, within their lacunae, are arranged in layers sandwiched between lamellae.

FIGURE 6.5 AP|R **Spongy Bone**
(*a*) Beams of bone, the trabeculae, surround spaces in the bone. In life, the spaces are filled with red or yellow bone marrow and blood vessels. (*b*) Transverse section of a trabecula.

Spongy and Compact Bone

Bone, whether woven or lamellar, can be classified according to the amount of bone matrix relative to the amount of space within the bone. Spongy bone, which appears porous, has less bone matrix and more space than compact bone. Compact bone, by contrast, has more bone matrix and less space than spongy bone.

Spongy bone consists of interconnecting rods or plates of bone called **trabeculae** (tră-bek′ū-lē; beam; figure 6.5*a*). Between the trabeculae are spaces, which in life are filled with bone marrow and blood vessels.

Most trabeculae are thin (50–400 μm) and consist of several lamellae with osteocytes located in lacunae between the lamellae (figure 6.5*b*). Each osteocyte is associated with other osteocytes through canaliculi. Usually, no blood vessels penetrate the trabeculae, so osteocytes must obtain nutrients through their canaliculi. The surfaces of trabeculae are covered with a single layer of cells consisting mostly of osteoblasts with a few osteoclasts.

Trabeculae are oriented along the lines of stress within a bone (figure 6.6). If the stress on a bone is changed slightly (e.g., because of a fracture that heals improperly), the trabecular pattern realigns with the new lines of stress.

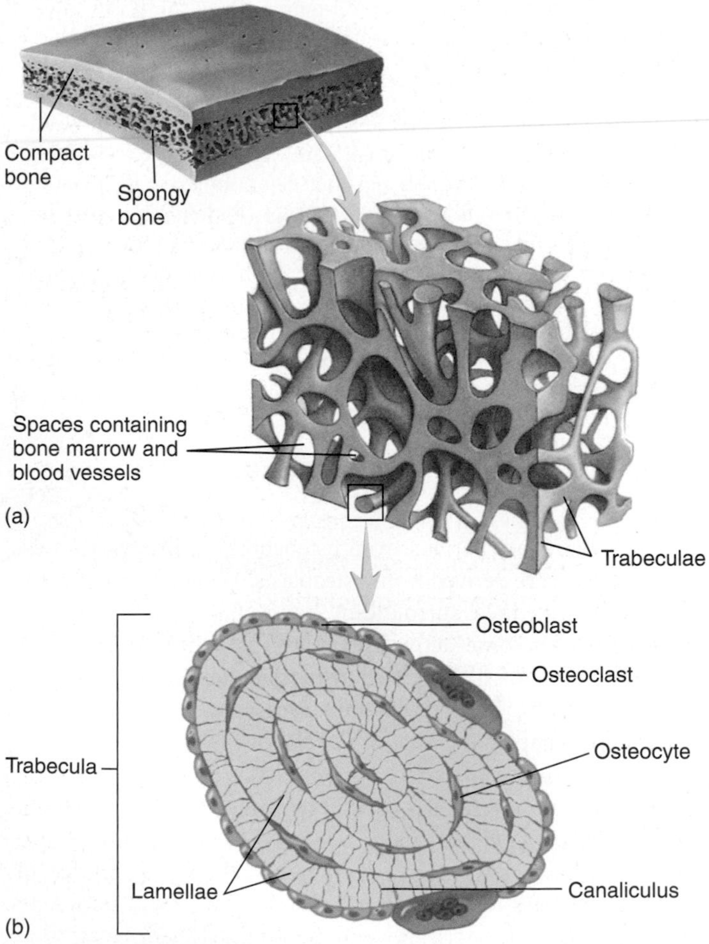

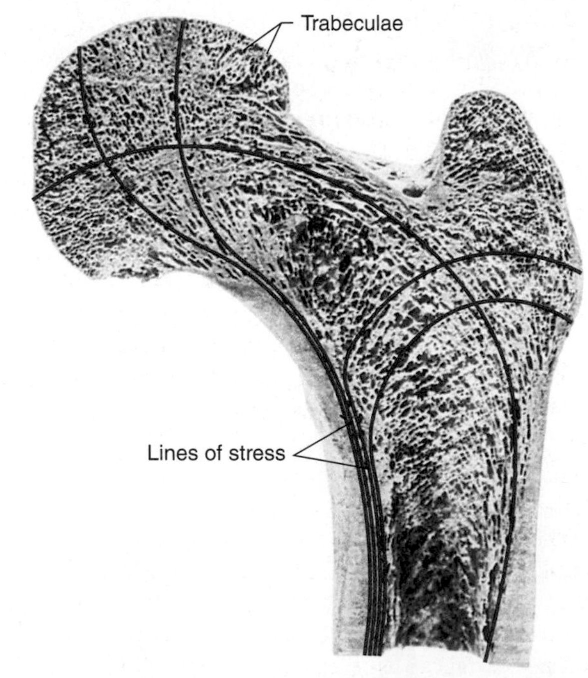
Trabeculae
Lines of stress

FIGURE 6.6 Trabeculae Oriented Along Lines of Stress
The proximal end of a long bone (femur) showing trabeculae oriented along lines of stress (red lines). The trabeculae bear weight and help bones resist bending and twisting.

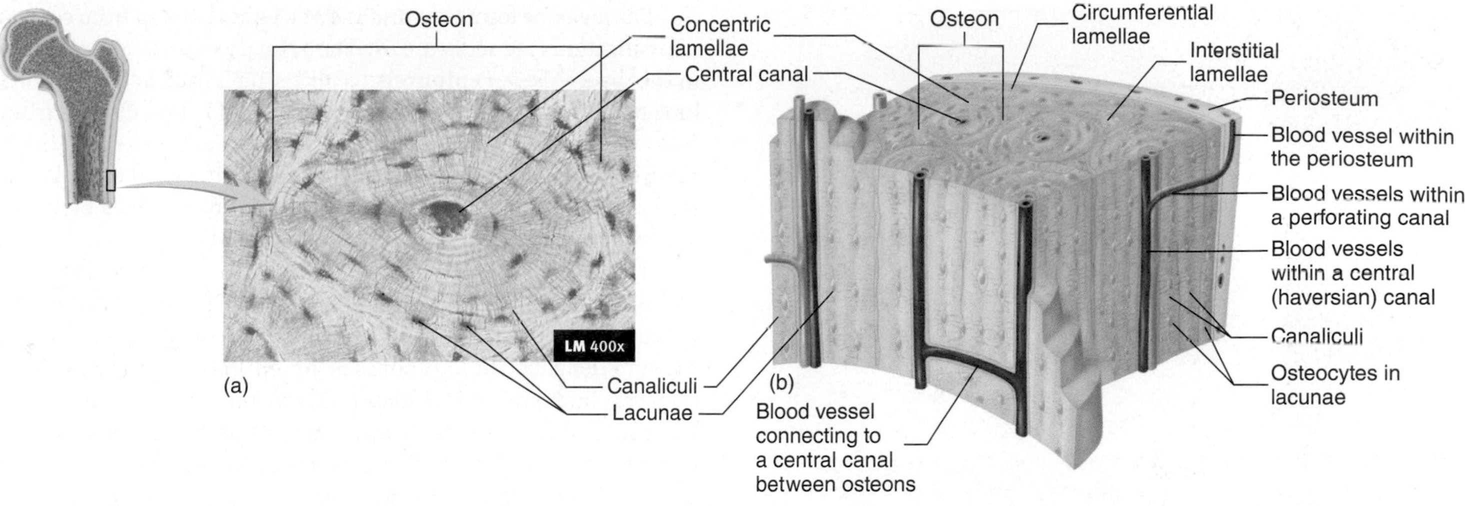

Osteon
Concentric lamellae
Central canal
Osteon
Circumferential lamellae
Interstitial lamellae
Periosteum
Blood vessel within the periosteum
Blood vessels within a perforating canal
Blood vessels within a central (haversian) canal
Canaliculi
Osteocytes in lacunae
LM 400x
(a)
Canaliculi
Lacunae
(b)
Blood vessel connecting to a central canal between osteons

FIGURE 6.7 AP|R **Compact Bone**

(*a*) Photomicrograph of an osteon. (*b*) Compact bone consists mainly of osteons, which are concentric lamellae surrounding blood vessels within central canals. The outer surface of the bone is formed by circumferential lamellae, and bone between the osteons consists of interstitial lamellae.

Compact bone is denser and has fewer spaces than spongy bone (figure 6.7). Blood vessels enter the substance of the bone itself, and the lamellae of compact bone are primarily oriented around those blood vessels. Vessels that run parallel to the long axis of the bone are contained within **central** (*haversian*; ha-ver′shan) **canals.** Central canals are lined with endosteum and contain blood vessels, nerves, and loose connective tissue. **Concentric lamellae** are circular layers of bone matrix that surround a common center, the central canal. An **osteon** (os′tē-on), or *haversian system*, consists of a single central canal, its contents, and associated concentric lamellae and osteocytes. In cross section, an osteon resembles a circular target; the "bull's-eye" of the target is the central canal, and 4–20 concentric lamellae form the rings. Osteocytes are located in lacunae between the lamellar rings, and canaliculi radiate between lacunae across the lamellae, looking like minute cracks across the rings of the target.

The outer surfaces of compact bone are formed by **circumferential lamellae,** which are thin plates that extend around the bone (figure 6.7*b*). In some bones, such as certain bones of the face, the layer of compact bone can be so thin that no osteons exist and the compact bone is composed of only circumferential lamellae. Between the osteons are **interstitial lamellae,** which are remnants of concentric or circumferential lamellae that were partially removed during bone remodeling.

Osteocytes receive nutrients and eliminate waste products through the canal system within compact bone. Blood vessels from the periosteum or medullary cavity enter the bone through **perforating canals** (*Volkmann canals*), which run perpendicular to the long axis of the bone (figure 6.7*b*). Perforating canals are not surrounded by concentric lamellae but pass through the concentric lamellae of osteons. The central canals receive blood vessels from perforating canals. Nutrients in the blood vessels enter the central canals, pass into the canaliculi, and move through the cytoplasm of the osteocytes that occupy the canaliculi and lacunae to the most peripheral cells within each osteon. Waste products are removed in the reverse direction.

ASSESS **YOUR PROGRESS**

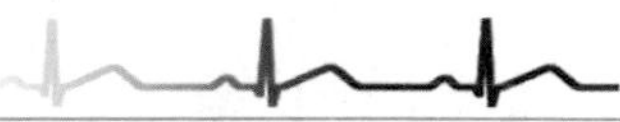

10. *How is the organization of collagen fibers different in woven and lamellar bone? What process produces woven bone?*
11. *Describe the structure of spongy bone. What are trabeculae, and what is their function? How do osteocytes within trabeculae obtain nutrients?*
12. *Describe the structure of compact bone. What is an osteon? Name three types of lamellae found in compact bone.*
13. *Trace the pathway nutrients must follow from blood vessels in the periosteum to osteocytes within lacunae in osteons.*

6.4 Bone Anatomy

LEARNING OUTCOMES

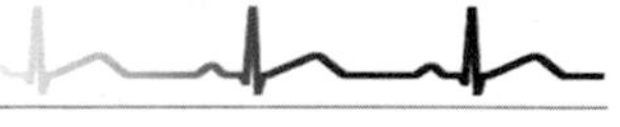

After reading this section, you should be able to

A. Classify bones according to their shape.

B. Label the parts of a typical long bone.

C. Explain the differences in structure between long bones and flat, short, and irregular bones.

Bone Shapes

Individual bones are classified according to shape: long, flat, short, or irregular (figure 6.8). **Long bones** are longer than they are wide. Most of the bones of the upper and lower limbs are long bones. **Flat bones** have a relatively thin, flattened shape and are usually curved. Examples of flat bones include certain skull bones, the ribs, the breastbone (sternum), and the shoulder blades (scapulae). **Short bones** are round or nearly cube-shaped, as exemplified by the bones of the wrist (carpal bones) and ankle (tarsal bones). **Irregular bones,** such as the vertebrae and facial bones, have shapes that do not fit readily into the other three categories.

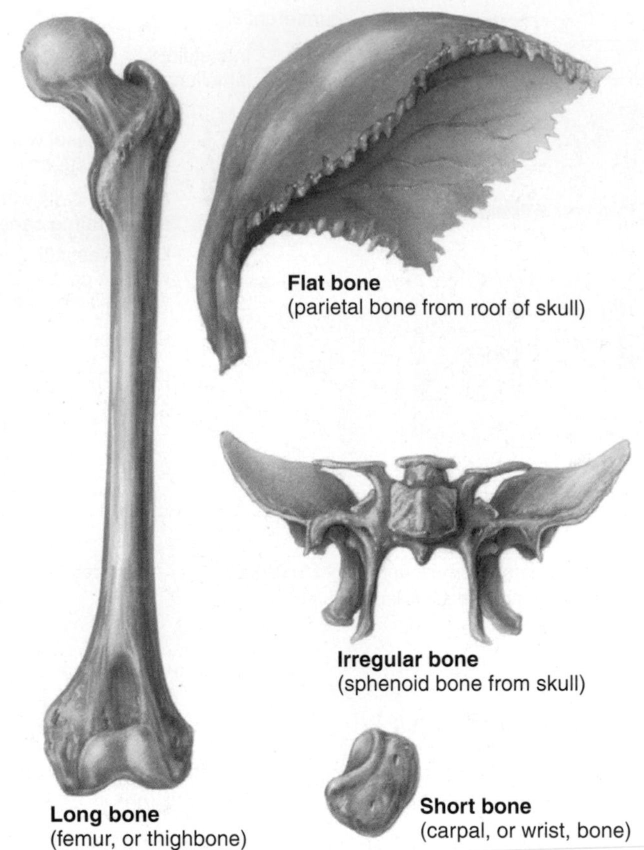
Flat bone
(parietal bone from roof of skull)
Irregular bone
(sphenoid bone from skull)
Long bone
(femur, or thighbone)
Short bone
(carpal, or wrist, bone)

FIGURE 6.8 **Bone Shapes**

Structure of a Long Bone

A long bone serves as a convenient model for overall bone structure (figure 6.9). Table 6.1 lists the major parts of a long bone. The **diaphysis** (dī-af′i-sis), or shaft, is composed primarily of compact bone, but it can also contain some spongy bone. The end of a long bone is mostly spongy bone, with an outer layer of compact bone. Within joints, the end of a long bone is covered with hyaline cartilage called **articular cartilage** (figure 6.9*a*, *b*).

During bone formation and growth, bones develop from centers of ossification (see section 6.5). The primary ossification center is in the diaphysis. An **epiphysis** (e-pif′i-sis; pl. -sēz) is the part of a long bone that develops from a center of ossification distinct from that of the diaphysis. Each long bone of the arm, forearm, thigh (figure 6.9*a*,*b*), and leg has one or more epiphyses on each end of the bone. Each long bone of the hand and foot has one epiphysis, which is located on the proximal or distal end of the bone.

The **epiphyseal** (ep-i-fiz′ē-ăl) **plate,** or *growth plate*, separates the epiphysis from the diaphysis (figure 6.9*a*). Growth in bone length (discussed in section 6.6) occurs at the epiphyseal plate. Consequently, growth in length of the long bones of the arm, forearm, thigh, and leg occurs at both ends of the diaphysis, whereas growth in length of the hand and foot bones occurs at one end of the diaphysis. When bone stops growing in length, the epiphyseal plate becomes ossified and is called the **epiphyseal line** (figure 6.9*b*).

In addition to the small spaces within spongy bone and compact bone, the diaphysis of a long bone can have a large internal space called the **medullary cavity.** The cavities of spongy bone and the medullary cavity are filled with marrow. **Red marrow** is the site of blood cell formation, and **yellow marrow** is mostly adipose tissue. In the fetus, the spaces within bones are filled with red marrow. The conversion of red marrow to yellow marrow begins just before birth and continues well into adulthood. Yellow marrow completely replaces the red marrow in the long bones of the limbs, except for some red marrow in the proximal part of the arm bones and thighbones. Elsewhere, varying proportions of yellow and red marrow are found. In some locations, red marrow is completely replaced by yellow marrow; in others, there is a mixture of red and yellow marrow. For example, part of the hipbone (ilium) may contain 50% red marrow and 50% yellow marrow. The hipbone is used as a source of donated red bone marrow because it is a large bone with more marrow than smaller bones and it can be accessed relatively easily.

The **periosteum** (per-ē-os′tē-ŭm) is a connective tissue membrane that covers the outer surface of a bone (figure 6.9*c*). The outer fibrous layer is dense irregular collagenous connective tissue that contains blood vessels and nerves. The inner layer is a single layer of bone cells, including osteoblasts, osteoclasts, and osteochondral

TABLE 6.1 **Gross Anatomy of a Long Bone**

Part	Description
Diaphysis	Shaft of the bone
Epiphysis	Part of the bone that develops from a center of ossification distinct from the diaphysis
Periosteum	Double-layered connective tissue membrane covering the outer surface of bone except where articular cartilage is present; ligaments and tendons attach to bone through the periosteum; blood vessels and nerves from the periosteum supply the bone; the periosteum is where bone grows in diameter
Endosteum	Thin connective tissue membrane lining the inner cavities of bone
Articular cartilage	Thin layer of hyaline cartilage covering a bone where it forms a joint (articulation) with another bone
Epiphyseal plate	Area of hyaline cartilage between the diaphysis and epiphysis; cartilage growth followed by endochondral ossification results in growth in bone length
Spongy bone	Bone having many small spaces; found mainly in the epiphysis; arranged into trabeculae
Compact bone	Dense bone with few internal spaces organized into osteons; forms the diaphysis and covers the spongy bone of the epiphyses
Medullary cavity	Large cavity within the diaphysis
Red marrow	Connective tissue in the spaces of spongy bone or in the medullary cavity; the site of blood cell production
Yellow marrow	Fat stored within the medullary cavity or in the spaces of spongy bone

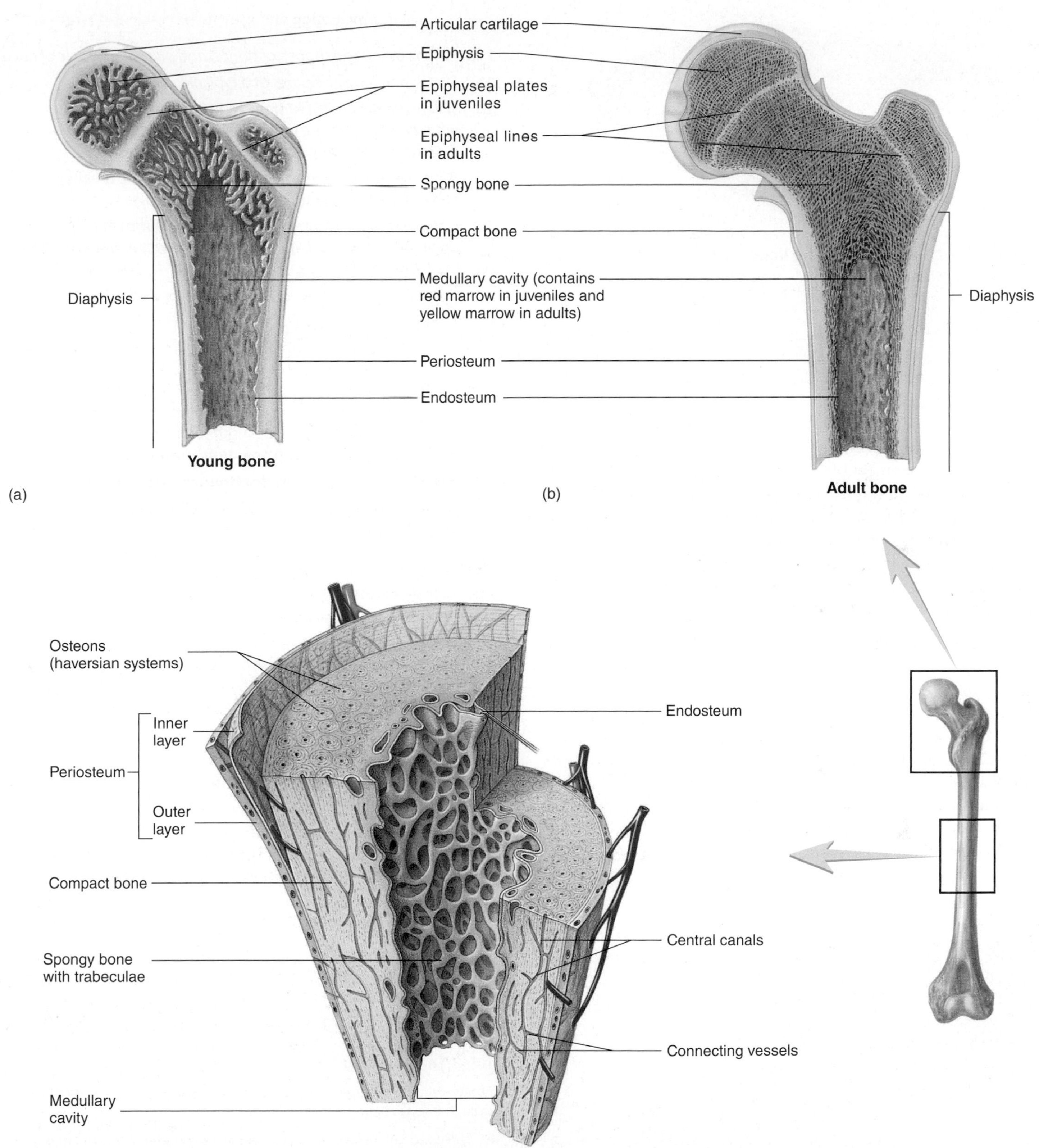
Articular cartilage
Epiphysis
Epiphyseal plates
in juveniles
Epiphyseal lines
in adults
Spongy bone
Compact bone
Diaphysis
Medullary cavity (contains
red marrow in juveniles and
yellow marrow in adults)
Periosteum
Endosteum
Young bone
(a)
(b)
Diaphysis
Adult bone
Osteons
(haversian systems)
Inner
layer
Periosteum
Outer
layer
Compact bone
Spongy bone
with trabeculae
Medullary
cavity
Endosteum
Central canals
Connecting vessels
(c)
Adult bone

FIGURE 6.9 Long Bone

The femur (thighbone) serves as a model of the parts of a long bone. (*a*) The presence of epiphyseal plates, as well as the condition of the diaphysis and epiphyses, shows that this is a young bone. The femur is unusual in that it has two epiphyses at its proximal end. (*b*) Adult long bone with epiphyseal lines. (*c*) Internal features of a portion of the diaphysis in (*a*).

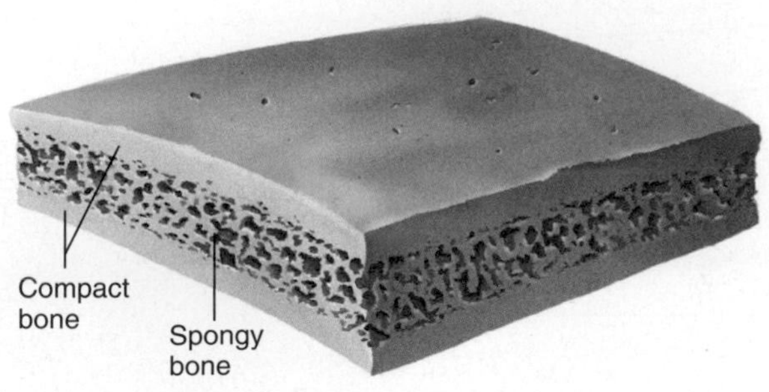
Compact bone
Spongy bone

FIGURE 6.10 Structure of a Flat Bone
Outer layers of compact bone surround spongy bone.

progenitor cells (see "Bone Cells" in section 6.3). Where tendons and ligaments attach to bone, the collagen fibers of the tendon or ligament become continuous with those of the periosteum. In addition, some of the collagen fibers of the tendons or ligaments penetrate the periosteum into the outer part of the bone. These bundles of collagen fibers are called **perforating fibers,** or *Sharpey fibers*, and they strengthen the attachment of the tendons or ligaments to the bone.

The **endosteum** (en-dos′tē-ŭm) is a single layer of cells that lines the internal surfaces of all cavities within bones, such as the medullary cavity of the diaphysis and the smaller cavities in spongy and compact bone (figure 6.9*c*). The endosteum includes osteoblasts, osteoclasts, and osteochondral progenitor cells.

Structure of Flat, Short, and Irregular Bones

Flat bones contain an interior framework of spongy bone sandwiched between two layers of compact bone (figure 6.10). Short and irregular bones have a composition similar to the epiphyses of long bones—compact bone surfaces surrounding a spongy bone center with small spaces that are usually filled with marrow. Short and irregular bones are not elongated and have no diaphyses. However, certain regions of these bones, such as the processes (projections), have epiphyseal growth plates and therefore small epiphyses.

Some of the flat and irregular bones of the skull have air-filled spaces called **sinuses** (sī′nŭs-ĕz; see chapter 7), which are lined by mucous membranes.

ASSESS YOUR PROGRESS

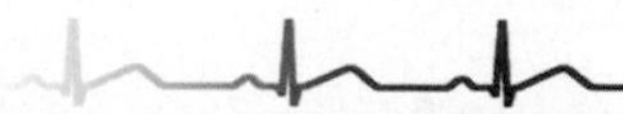

14. *List the four basic shapes of bones, and give an example of each.*
15. *Sketch and label the parts of a typical long bone.*
16. *Where are the periosteum and endosteum located, and what types of cells are found in each? What is the function of perforating (Sharpey) fibers?*
17. *What are red and yellow bone marrows? Where are they located in a child and in an adult?*
18. *Compare the structure of a long bone with those of flat, short, and irregular bones. Explain where compact and spongy bones are found in each type.*

6.5 Bone Development

LEARNING OUTCOMES

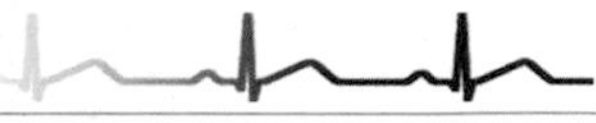

After reading this section, you should be able to

A. Outline the process of intramembranous ossification.
B. Describe the steps of endochondral ossification.
C. List the bones, or parts of bones, that develop from each type of ossification.

During fetal development, bone forms in two patterns—**intramembranous ossification** and **endochondral ossification.** Intramembranous ossification takes place in connective tissue membranes, and endochondral ossification takes place in cartilage. Both methods initially produce woven bone, which is then remodeled. After remodeling, bone formed by intramembranous ossification cannot be distinguished from bone formed by endochondral ossification. Table 6.2 compares the two types of ossification.

Intramembranous Ossification

At approximately the fifth week of development in an embryo, embryonic mesenchyme condenses around the developing brain to form a membrane of connective tissue with delicate, randomly oriented collagen fibers. Intramembranous ossification of the membrane begins at approximately the eighth week of embryonic

TABLE 6.2 Comparison of Intramembranous and Endochondral Ossification

Intramembranous Ossification	Endochondral Ossification
Embryonic mesenchyme forms a collagen membrane containing osteochondral progenitor cells.	Embryonic mesenchymal cells become chondroblasts, which produce a cartilage template surrounded by the perichondrium.
No stage is comparable.	Chondrocytes hypertrophy, the cartilage matrix becomes calcified, and the chondrocytes die.
Embryonic mesenchyme forms the periosteum, which contains osteoblasts.	The perichondrium becomes the periosteum when osteochondral progenitor cells within the periosteum become osteoblasts.
Osteochondral progenitor cells become osteoblasts at centers of ossification; internally, the osteoblasts form spongy bone; externally, the periosteal osteoblasts form compact bone.	Blood vessels and osteoblasts from the periosteum invade the calcified cartilage template; internally, these osteoblasts form spongy bone at primary ossification centers (and later at secondary ossification centers); externally, the periosteal osteoblasts form compact bone.
Intramembranous bone is remodeled and becomes indistinguishable from endochondral bone.	Endochondral bone is remodeled and becomes indistinguishable from intramembranous bone.

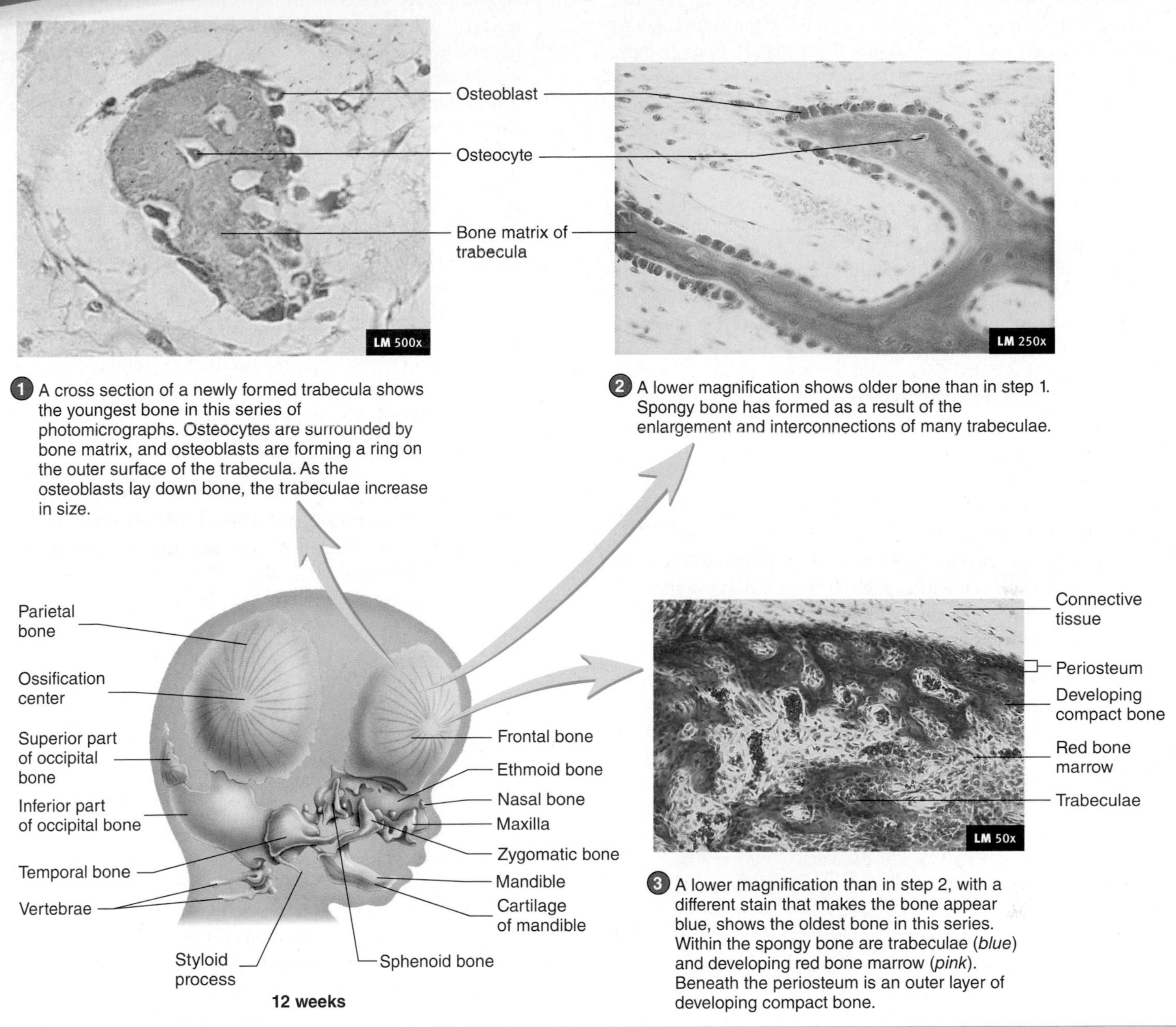
Osteoblast
Osteocyte
Bone matrix of trabecula
LM 500x
LM 250x
1 A cross section of a newly formed trabecula shows the youngest bone in this series of photomicrographs. Osteocytes are surrounded by bone matrix, and osteoblasts are forming a ring on the outer surface of the trabecula. As the osteoblasts lay down bone, the trabeculae increase in size.
2 A lower magnification shows older bone than in step 1. Spongy bone has formed as a result of the enlargement and interconnections of many trabeculae.
Parietal bone
Ossification center
Superior part of occipital bone
Inferior part of occipital bone
Temporal bone
Vertebrae
Styloid process
Sphenoid bone
Frontal bone
Ethmoid bone
Nasal bone
Maxilla
Zygomatic bone
Mandible
Cartilage of mandible
12 weeks
Connective tissue
Periosteum
Developing compact bone
Red bone marrow
Trabeculae
LM 50x
3 A lower magnification than in step 2, with a different stain that makes the bone appear blue, shows the oldest bone in this series. Within the spongy bone are trabeculae (blue) and developing red bone marrow (pink). Beneath the periosteum is an outer layer of developing compact bone.

PROCESS FIGURE 6.11 Intramembranous Ossification

The inset (lower left) shows a 12-week-old fetus. Bones formed by intramembranous ossification are yellow, and bones formed by endochondral ossification are blue. Intramembranous ossification starts at a center of ossification and expands outward. Therefore, the youngest bone is at the edge of the expanding bone and the oldest bone is at the center of ossification.

Source: Photo 2 © Dr. Richard Kessel/Visuals Unlimited.

development and is completed by approximately 2 years of age. Many skull bones, part of the mandible (lower jaw), and the diaphyses of the clavicles (collarbones) develop by intramembranous ossification (figure 6.11).

The locations in the membrane where ossification begins are called **centers of ossification.** The centers of ossification expand to form a bone by gradually ossifying the membrane. Thus, the centers have the oldest bone, and the expanding edges the youngest bone. The larger, membrane-covered spaces between the developing skull bones that have not yet been ossified are called **fontanels,** or soft spots (figure 6.12; see chapter 8). The bones eventually grow together, and all the fontanels have usually closed by 2 years of age. The steps in intramembranous ossification are as follows:

1. Intramembranous ossification begins when some of the mesenchymal cells in the membrane become osteochondral progenitor cells, which specialize to become osteoblasts. The osteoblasts produce bone matrix that surrounds the collagen fibers of the connective tissue membrane, and the osteoblasts become osteocytes. As a result of this process, many tiny trabeculae of woven bone develop (figure 6.11, *step 1*).

2. Additional osteoblasts gather on the surfaces of the trabeculae and produce more bone, thereby causing the trabeculae to become larger and longer (figure 6.11, *step 2*). Spongy bone forms as the trabeculae join together, resulting in an interconnected network of trabeculae separated by spaces.
3. Cells within the spaces of the spongy bone specialize to form red bone marrow, and cells surrounding the developing bone specialize to form the periosteum. Osteoblasts from the periosteum lay down bone matrix to form an outer surface of compact bone (figure 6.11, *step 3*).

Thus, the end products of intramembranous bone formation are bones with outer compact bone surfaces and spongy centers (see figure 6.10). Remodeling converts woven bone to lamellar bone and contributes to the final shape of the bone.

Endochondral Ossification

The formation of cartilage begins at approximately the end of the fourth week of embryonic development. Endochondral ossification of some of this cartilage starts at approximately the eighth week of embryonic development, but this process might not begin in other cartilage until as late as 18–20 years of age. Bones of the base of the skull, part of the mandible, the epiphyses of the clavicles, and most of the remaining skeletal system develop through endochondral ossification (figures 6.11 and 6.12). The steps in endochondral ossification are as follows:

1. Endochondral ossification begins as mesenchymal cells aggregate in regions of future bone formation. The mesenchymal cells become osteochondral progenitor cells that become chondroblasts. The chondroblasts produce a hyaline **cartilage model** having the approximate shape of the bone that will later be formed (figure 6.13, *step 1*). As the chondroblasts are surrounded by cartilage matrix, they become chondrocytes. The cartilage model is surrounded by perichondrium, except where a joint will form connecting one bone to another bone. The perichondrium is continuous with tissue that will become the joint capsule later in development (see chapter 8).
2. When blood vessels invade the perichondrium surrounding the cartilage model (figure 6.13, *step 2*), osteochondral progenitor cells within the perichondrium become osteoblasts. The perichondrium becomes the periosteum when the osteoblasts begin to produce bone. The osteoblasts produce compact bone on the surface of the cartilage model, forming a **bone collar.** Two other events occur at the same time that the bone collar is forming. First, the cartilage model increases in size as a result of interstitial and appositional cartilage growth. Second, the chondrocytes in the center of the cartilage model absorb some of the cartilage matrix and **hypertrophy** (hī-per′trō-fē), or enlarge. The chondrocytes also release matrix vesicles, which initiate the formation of hydroxyapatite crystals in the cartilage matrix. At this point, the cartilage is called **calcified cartilage.** The chondrocytes in this calcified area eventually die, leaving enlarged lacunae with thin walls of calcified matrix.
3. Blood vessels grow into the enlarged lacunae of the calcified cartilage (figure 6.13, *step 3*). Osteoblasts and osteoclasts migrate into the calcified cartilage area from the periosteum by way of the connective tissue surrounding the outside of the blood vessels. The osteoblasts produce bone on the surface of the calcified cartilage, forming bone trabeculae, which changes the calcified cartilage of the diaphysis into spongy bone. This area of bone formation is called the **primary ossification center.**
4. As bone development proceeds, the cartilage model continues to grow, more perichondrium becomes periosteum, and the bone collar thickens and extends farther along the diaphysis. Additional cartilage within both the diaphysis and the epiphysis is calcified (figure 6.13, *step 4*). Remodeling converts woven bone to lamellar bone and contributes to the final shape of the bone. Osteoclasts remove bone from the center of the diaphysis to form the medullary cavity, and cells within the medullary cavity specialize to form red bone marrow.
5. In long bones, the diaphysis is the primary ossification center, and additional sites of ossification, called **secondary ossification centers,** appear in the epiphyses (figure 6.13, *step 5*). The events occurring at the secondary ossification centers are the same as those at the primary ossification centers, except that the spaces in the epiphyses do not enlarge to form a medullary cavity as in the diaphysis. Primary ossification centers appear during early fetal development, whereas secondary ossification centers appear in the proximal epiphysis of the femur, humerus, and tibia about 1 month before birth. A baby is considered full-term if one of these three ossification centers can be seen on radiographs at the time of birth. At about

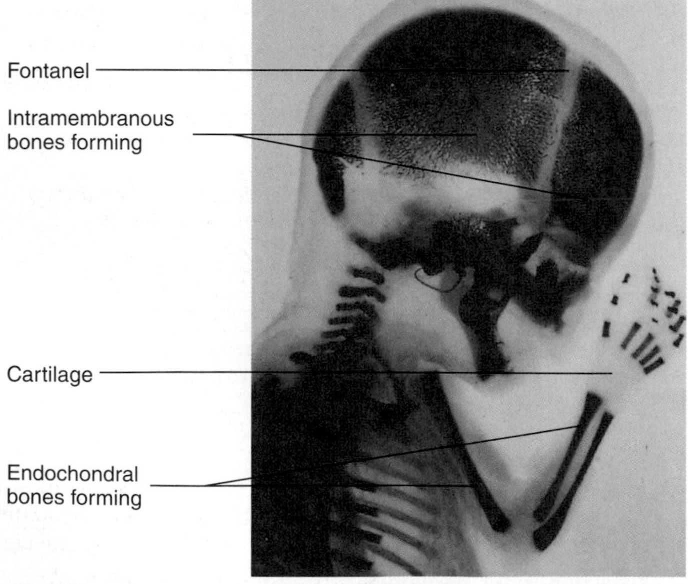
Fontanel
Intramembranous bones forming
Cartilage
Endochondral bones forming

FIGURE 6.12 Bone Formation in a Fetus
Eighteen-week-old fetus showing intramembranous and endochondral ossification. Intramembranous ossification occurs at ossification centers in the flat bones of the skull. Endochondral ossification has formed bones in the diaphyses of long bones. The ends of the long bones are still cartilage at this stage of development.

FUNDAMENTAL **Figure**

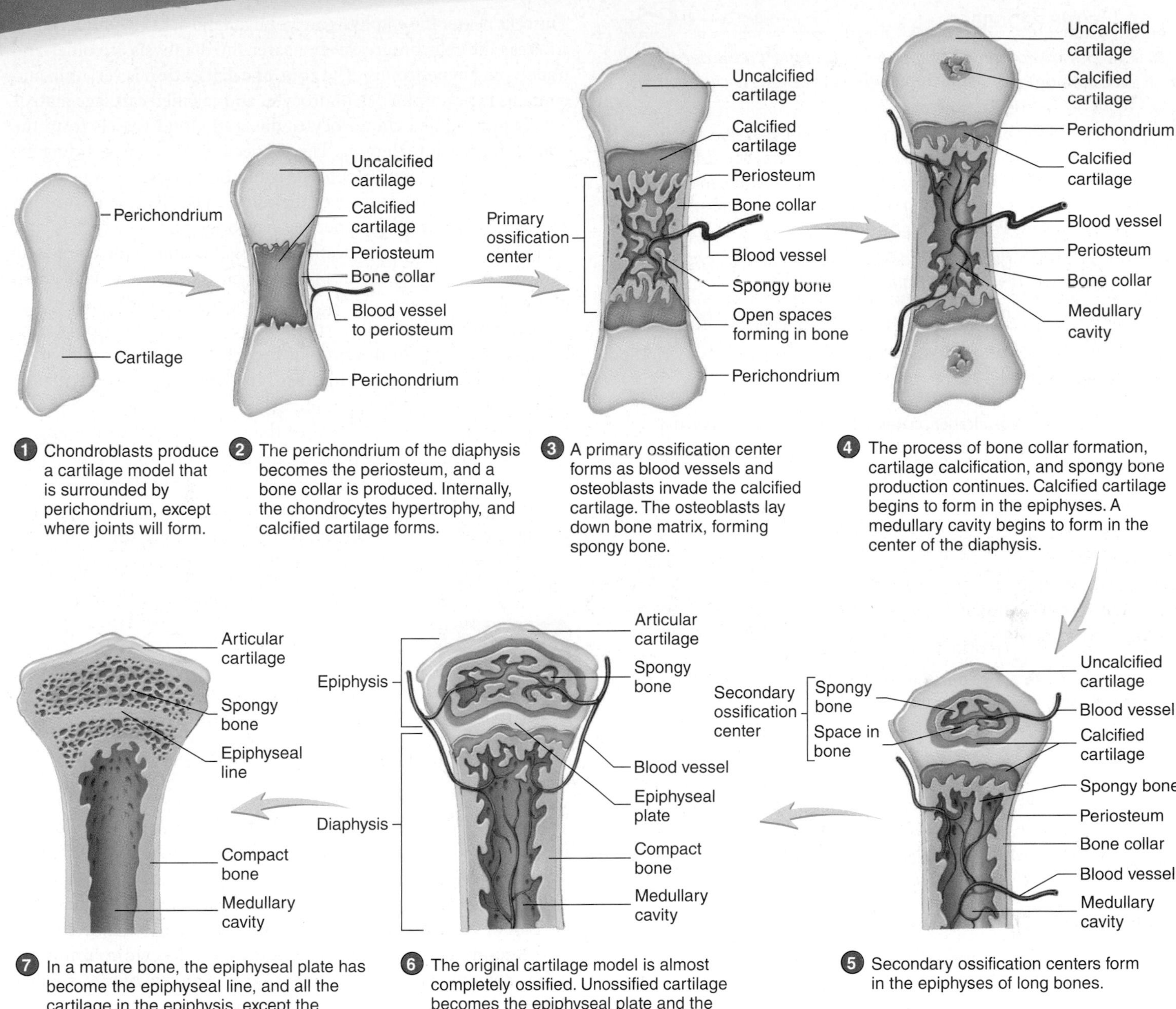

Perichondrium
Cartilage
Uncalcified cartilage
Calcified cartilage
Periosteum
Bone collar
Blood vessel to periosteum
Perichondrium
Primary ossification center
Uncalcified cartilage
Calcified cartilage
Periosteum
Bone collar
Blood vessel
Spongy bone
Open spaces forming in bone
Perichondrium
Uncalcified cartilage
Calcified cartilage
Perichondrium
Calcified cartilage
Blood vessel
Periosteum
Bone collar
Medullary cavity
1 Chondroblasts produce a cartilage model that is surrounded by perichondrium, except where joints will form.
2 The perichondrium of the diaphysis becomes the periosteum, and a bone collar is produced. Internally, the chondrocytes hypertrophy, and calcified cartilage forms.
3 A primary ossification center forms as blood vessels and osteoblasts invade the calcified cartilage. The osteoblasts lay down bone matrix, forming spongy bone.
4 The process of bone collar formation, cartilage calcification, and spongy bone production continues. Calcified cartilage begins to form in the epiphyses. A medullary cavity begins to form in the center of the diaphysis.
Articular cartilage
Spongy bone
Epiphyseal line
Compact bone
Medullary cavity
Epiphysis
Diaphysis
Articular cartilage
Spongy bone
Blood vessel
Epiphyseal plate
Compact bone
Medullary cavity
Secondary ossification center
Spongy bone
Space in bone
Uncalcified cartilage
Blood vessel
Calcified cartilage
Spongy bone
Periosteum
Bone collar
Blood vessel
Medullary cavity
7 In a mature bone, the epiphyseal plate has become the epiphyseal line, and all the cartilage in the epiphysis, except the articular cartilage, has become bone.
6 The original cartilage model is almost completely ossified. Unossified cartilage becomes the epiphyseal plate and the articular cartilage.
5 Secondary ossification centers form in the epiphyses of long bones.

PROCESS FIGURE 6.13 AP|R **Endochondral Ossification**

Endochondral ossification begins with the formation of a cartilage model. See successive steps as indicated by the blue arrows.

18–20 years of age, the last secondary ossification center appears in the medial epiphysis of the clavicle.

6. Replacement of cartilage by bone continues in the cartilage model until all the cartilage, except that in the epiphyseal plate and on articular surfaces, has been replaced by bone (figure 6.13, *step 6*). The epiphyseal plate, which exists during the time a person's bones are actively growing, and the articular cartilage, which is a permanent structure, are derived from the original embryonic cartilage model. After a person's bones have stopped growing, the epiphyseal plate regresses into a "scar," called the epiphyseal line (see "Growth in Bone Length" in section 6.6).
7. In mature bone, spongy and compact bone are fully developed, and the epiphyseal plate has become the epiphyseal line. The only cartilage present is the articular cartilage at the ends of the bone (figure 6.13, *step 7*). All the original perichondrium that surrounded the cartilage model has become periosteum.

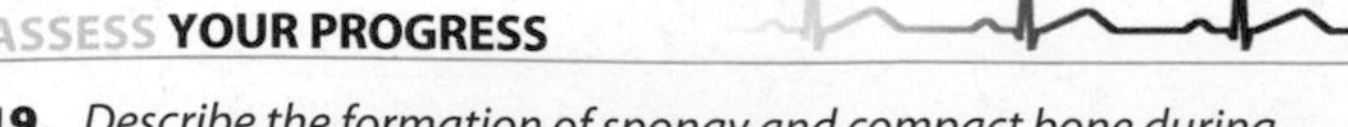

ASSESS YOUR PROGRESS

19. *Describe the formation of spongy and compact bone during intramembranous ossification. What are centers of ossification? What are fontanels?*
20. *For the process of endochondral ossification, describe the formation of the following structures: cartilage model, bone collar, calcified cartilage, primary ossification center, medullary cavity, secondary ossification center, epiphyseal plate, epiphyseal line, and articular cartilage.*
21. *When do primary and secondary ossification centers appear during endochondral ossification?*
22. *What bones, or parts of bones, are formed from each type of ossification?*

Predict 3

During endochondral ossification, calcification of cartilage results in the death of chondrocytes. However, ossification of the bone matrix does not result in the death of osteocytes. Explain.

6.6 Bone Growth

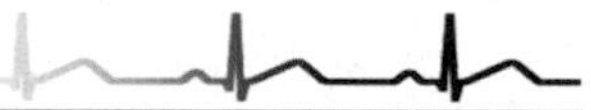

LEARNING OUTCOMES

After reading this section, you should be able to

A. Demonstrate an understanding of bone growth in length and width, as well as at the articular cartilage.

B. Describe the factors that affect bone growth.

Unlike cartilage, bones cannot grow by interstitial growth. Bones increase in size only by appositional growth, the formation of new bone on the surface of older bone or cartilage. For example, trabeculae grow in size when osteoblasts deposit new bone matrix onto the surface of the trabeculae (see figure 6.11).

Predict 4

Explain why bones cannot undergo interstitial growth as cartilage does.

Growth in Bone Length

Long bones and bony projections increase in length because of growth at the epiphyseal plate. In a long bone, the epiphyseal plate separates the epiphysis from the diaphysis (figure 6.14*a*). Long projections of bones, such as the processes of vertebrae (see chapter 7), also have epiphyseal plates.

Growth at the epiphyseal plate involves the formation of new cartilage by interstitial cartilage growth followed by appositional bone growth on the surface of the cartilage. The epiphyseal plate is organized into four zones (figure 6.14*b*). The **zone of resting cartilage** is nearest the epiphysis and contains randomly arranged chondrocytes that do not divide rapidly. The chondrocytes in the **zone of proliferation** produce new cartilage through interstitial cartilage growth. The chondrocytes divide and form columns resembling stacks of plates or coins. In the **zone of hypertrophy,** the chondrocytes produced in the zone of proliferation mature and enlarge. Thus, a maturation gradient exists in each column: The cells nearer the epiphysis are younger and actively proliferating, whereas the cells progressively nearer the diaphysis are older and undergoing hypertrophy. The **zone of calcification** is very thin and contains hypertrophied chondrocytes and calcified cartilage matrix. The hypertrophied chondrocytes die, and blood vessels from the diaphysis grow into the area. The connective tissue surrounding the blood vessels contains osteoblasts from the endosteum. The osteoblasts line up on the surface of the calcified cartilage and, through appositional bone growth, deposit new bone matrix, which is later remodeled. Cartilage calcification and ossification in the epiphyseal plate occur by the same basic process as calcification and ossification of the cartilage model during endochondral bone formation.

As new cartilage cells form in the zone of proliferation, and as these cells enlarge in the zone of hypertrophy, the overall length of the diaphysis increases (figure 6.14*c*). However, the thickness of the epiphyseal plate does not increase because the rate of cartilage growth on the epiphyseal side of the plate is equal to the rate of cartilage replacement by bone on the diaphyseal side of the plate.

As the bones achieve normal adult size, they stop growing in length because the epiphyseal plate has ossified and become the epiphyseal line. This event, called closure of the epiphyseal plate, occurs between approximately 12 and 25 years of age, depending on the bone and the individual.

Predict 5

An x-ray revealed that Jill had suffered a fracture of her left femur (figure 6.15) while playing soccer in junior high school. The bone was set, without surgery, and she had to have a cast on her leg for longer than the 6 weeks normally required for a fracture. A year after the accident, her left femur is shorter than her right femur. Explain how this occurred.

Growth at Articular Cartilage

Epiphyses increase in size because of growth at the articular cartilage. In addition, growth at the articular cartilage increases the size of bones that do not have an epiphysis, such as short bones. The process of growth in articular cartilage is similar to that occurring in the epiphyseal plate, except that the chondrocyte columns are not as obvious. The chondrocytes near the surface of the articular cartilage are similar to those in the zone of resting cartilage of the epiphyseal plate. In the deepest part of the articular cartilage, nearer bone tissue, the cartilage is calcified and ossified to form new bone.

When the epiphyses reach their full size, the growth of cartilage and its replacement by bone cease. The articular cartilage, however, persists throughout life and does not become ossified as the epiphyseal plate does.

Predict 6

Explain why it is advantageous for the articular cartilage never to become ossified.

Growth in Bone Width

Long bones increase in width (diameter) and other bones increase in size or thickness because of appositional bone growth beneath the periosteum. When a bone rapidly grows in width, as

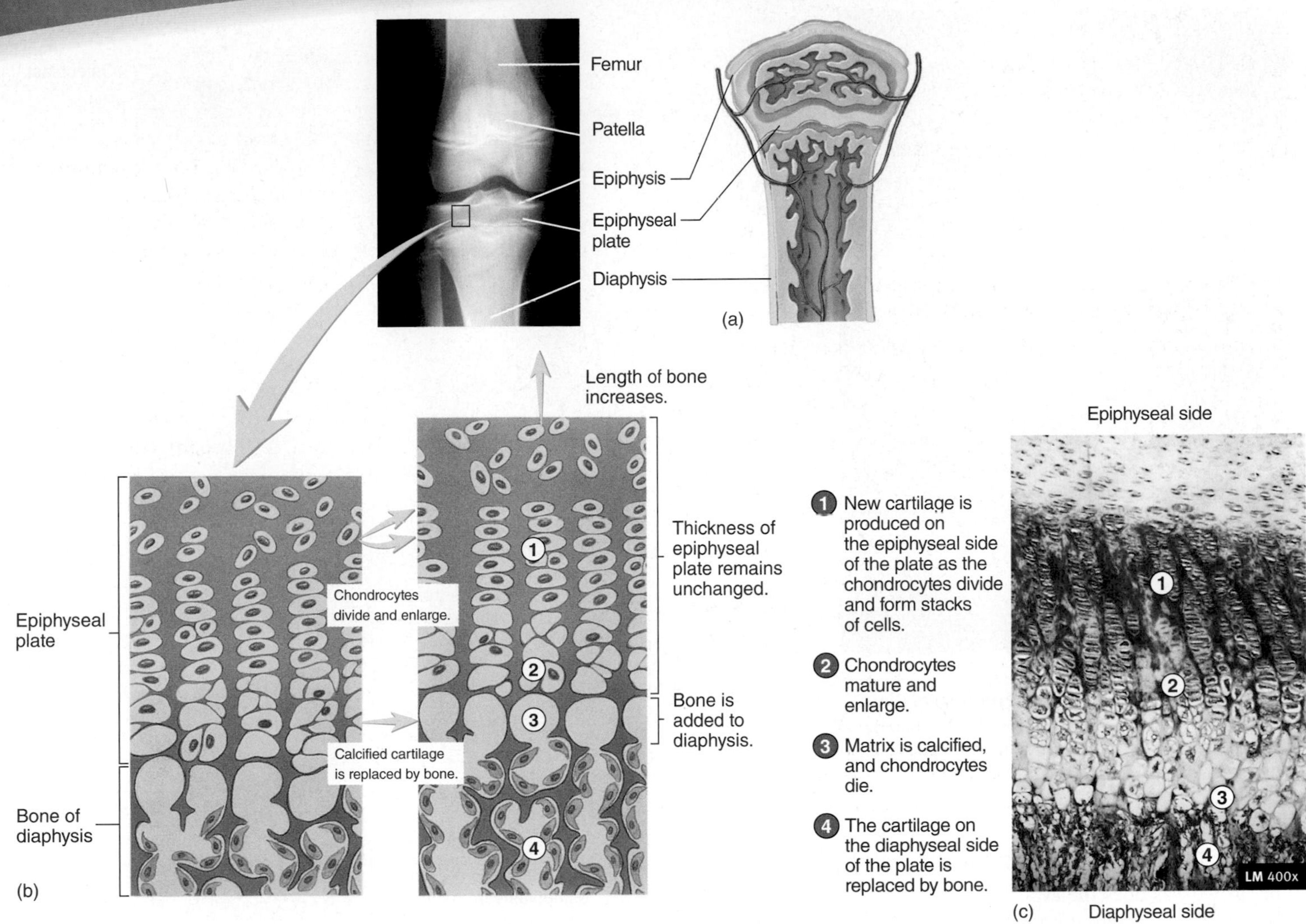
Femur
Patella
Epiphysis
Epiphyseal plate
Diaphysis
(a)
Length of bone increases.
Epiphyseal plate
Bone of diaphysis
(b)
Chondrocytes divide and enlarge.
Calcified cartilage is replaced by bone.
Thickness of epiphyseal plate remains unchanged.
Bone is added to diaphysis.
1 New cartilage is produced on the epiphyseal side of the plate as the chondrocytes divide and form stacks of cells.
2 Chondrocytes mature and enlarge.
3 Matrix is calcified, and chondrocytes die.
4 The cartilage on the diaphyseal side of the plate is replaced by bone.
Epiphyseal side
LM 400x
(c)
Diaphyseal side

PROCESS FIGURE 6.14 Epiphyseal Plate
(*a*) Radiograph and drawing of the knee, showing the epiphyseal plate of the tibia (shinbone). Because cartilage does not appear readily on x-ray film, the epiphyseal plate appears as a black area between the white diaphysis and the epiphyses. (*b*) Zones of the epiphyseal plate, including newly ossified bone. (*c*) New cartilage forms on the epiphyseal side of the plate at the same rate that new bone forms on the diaphyseal side of the plate. Consequently, the epiphyseal plate remains the same thickness but the diaphysis increases in length.

occurs in young bones or during puberty, osteoblasts from the periosteum lay down bone to form a series of ridges with grooves between them (figure 6.16, *step 1*). The periosteum covers the bone ridges and extends down into the bottom of the grooves, and one or more blood vessels of the periosteum lie within each groove. As the osteoblasts continue to produce bone, the ridges increase in size, extend toward each other, and meet to change the groove into a tunnel (figure 6.16, *step 2*). The name of the periosteum in the tunnel changes to *endosteum* because the membrane now lines an internal bone surface. Osteoblasts from the endosteum lay down bone to form a concentric lamella (figure 6.16, *step 3*). The production of additional lamellae fills in the tunnel, encloses the blood vessel, and produces an osteon (figure 6.16, *step 4*).

When a bone grows in width slowly, the surface of the bone becomes smooth as osteoblasts from the periosteum lay down even layers of bone to form circumferential lamellae. The circumferential lamellae break down during remodeling to form osteons (see section 6.7).

Factors Affecting Bone Growth

The bones of an individual's skeleton usually reach a certain length, thickness, and shape through the processes described in the previous sections. The potential shape and size of a bone and an individual's final adult height are determined genetically, but factors such as nutrition and hormones can greatly modify the expression of those genetic factors.

Nutrition

Because bone growth requires chondroblast and osteoblast proliferation, any metabolic disorder that affects the rate of cell proliferation

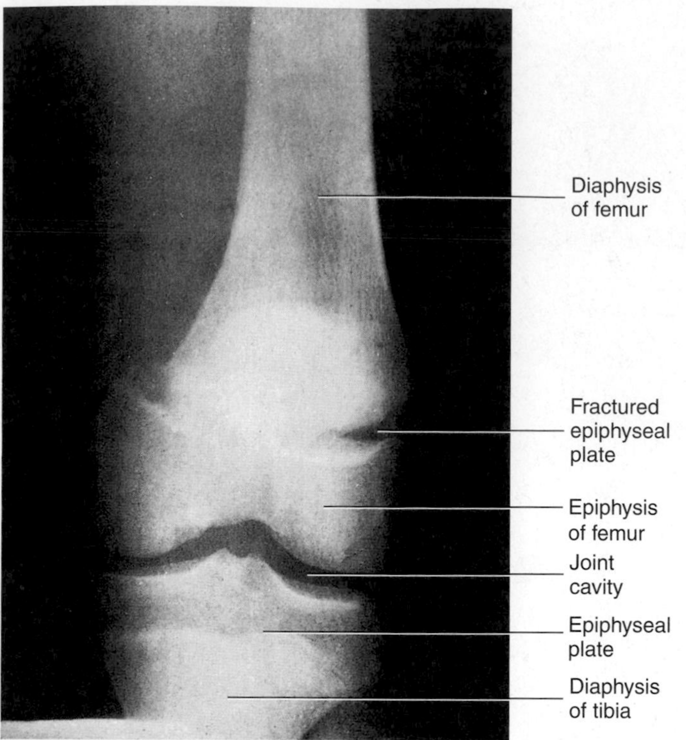
Diaphysis of femur
Fractured epiphyseal plate
Epiphysis of femur
Joint cavity
Epiphyseal plate
Diaphysis of tibia

FIGURE 6.15 Fracture of the Epiphyseal Plate
Radiograph of an adolescent's knee. The femur (thighbone) is separated from the tibia (leg bone) by a joint cavity. The epiphyseal plate of the femur is fractured, thereby separating the diaphysis from the epiphysis.

or the production of collagen and other matrix components affects bone growth, as does the availability of calcium or other minerals needed in the mineralization process.

The long bones of a child sometimes exhibit lines of arrested growth, which are transverse regions of greater bone density crossing an otherwise normal bone (figure 6.17). These lines are caused by greater calcification below the epiphyseal plate of a bone, where it has grown at a slower rate during an illness or severe nutritional deprivation. They demonstrate that illness or malnutrition during the time of bone growth can cause a person to be shorter than he or she would have been otherwise.

Certain vitamins are important to bone growth in very specific ways. **Vitamin D** is necessary for the normal absorption of calcium from the intestines (see chapters 5 and 24). The body can either synthesize or ingest vitamin D. Its rate of synthesis increases when the skin is exposed to sunlight. Insufficient vitamin D in children causes **rickets,** a disease resulting from reduced mineralization of the bone matrix. Children with rickets may have bowed bones and inflamed joints. During the winter in northern climates, children who are not exposed to sufficient sunlight can take supplementary vitamin D to prevent rickets. The body's inability to absorb lipids in which vitamin D is soluble can also result in vitamin D deficiency. This condition sometimes occurs in adults who have digestive disorders and can be one cause of "adult rickets," or **osteomalacia** (os′tē-ō-mă-lā′shē-ă), which is softening of the bones as a result of calcium depletion.

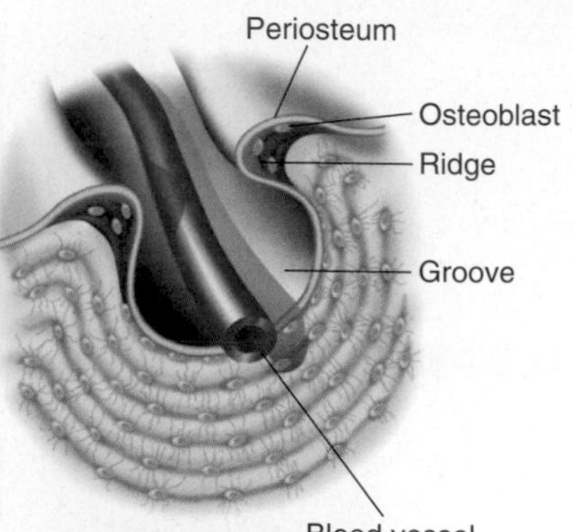
1 Osteoblasts beneath the periosteum lay down bone (*dark brown*) to form ridges separated by grooves. Blood vessels of the periosteum lie in the grooves.
Periosteum
Osteoblast
Ridge
Groove
Blood vessel

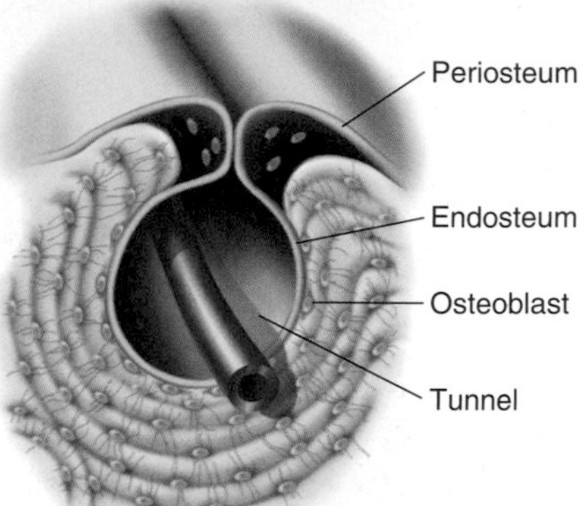
2 The groove is transformed into a tunnel when the bone built on adjacent ridges meets. The periosteum of the groove becomes the endosteum of the tunnel.
Periosteum
Endosteum
Osteoblast
Tunnel

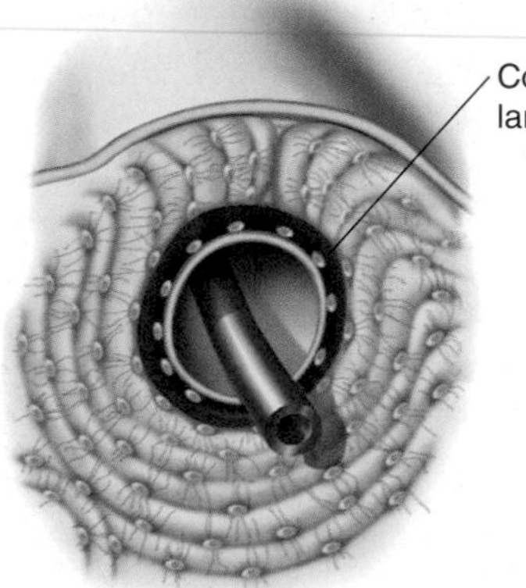
3 Appositional growth by osteoblasts from the endosteum results in the formation of a new concentric lamella.
Concentric lamella

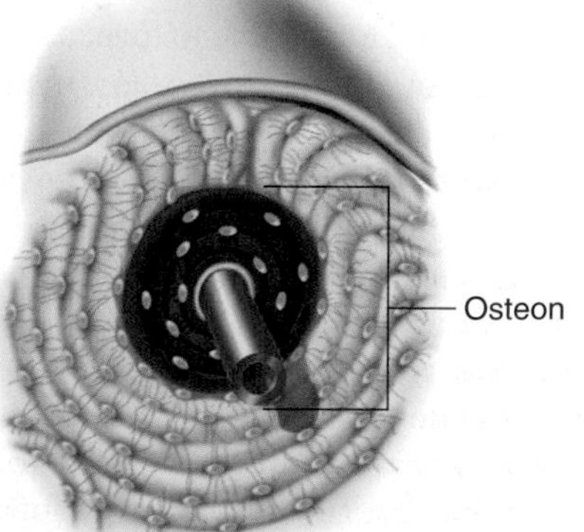
4 The production of additional concentric lamellae fills in the tunnel and completes the formation of the osteon.
Osteon

PROCESS FIGURE 6.16 APR Bone Growth in Width
Bones can increase in width by the formation of new osteons beneath the periosteum.

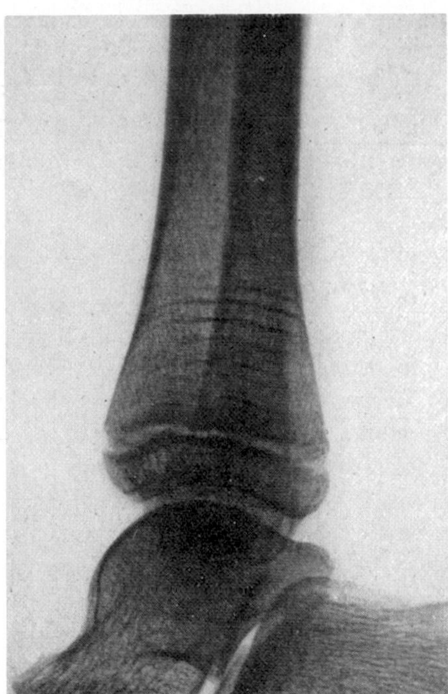

FIGURE 6.17 **Arrested Growth Lines**
In this radiograph, the horizontal dark lines are lines of arrested growth.

Vitamin C is necessary for collagen synthesis by osteoblasts. Normally, as old collagen breaks down, new collagen is synthesized to replace it. Vitamin C deficiency results in bones and cartilage that are deficient in collagen because collagen synthesis is impaired. In children, vitamin C deficiency can retard growth. In both children and adults, vitamin C deficiency can result in **scurvy,** which is marked by ulceration and hemorrhage in almost any area of the body because normal collagen synthesis is not occurring in connective tissues. Wound healing, which requires collagen synthesis, is hindered in patients with vitamin C deficiency. In extreme cases, the teeth fall out because the ligaments that hold them in place break down.

Hormones

Hormones are very important in bone growth. **Growth hormone** from the anterior pituitary increases general tissue growth (see chapters 17 and 18), including overall bone growth, by stimulating interstitial cartilage growth and appositional bone growth. Disruptions in normal growth hormone can cause dramatic changes in an individual's height. Excessive growth hormone secretion results in pituitary gigantism, whereas insufficient growth hormone secretion results in pituitary dwarfism (figure 6.18). **Thyroid hormone** is also required for normal growth of all tissues, including cartilage; therefore, a decrease in this hormone can result in a smaller individual (see table 6.3). **Sex hormones** also influence bone growth. Estrogen (a class of female sex hormones) and testosterone (a male sex hormone) initially stimulate bone growth, which accounts for the burst of growth at puberty when production of these hormones increases. However, both hormones also stimulate ossification of epiphyseal plates, and thus the cessation of growth. Females usually stop growing earlier than males because estrogens cause quicker closure of the epiphyseal plate than testosterone does. Because their entire growth period is somewhat shorter, females usually do not reach the same height as males. Decreased levels of testosterone or estrogen can prolong the growth phase of the epiphyseal plates, even though the bones grow more slowly. Overall, growth is very complex and is influenced by many factors besides sex hormones, such as other hormones, genetics, and nutrition.

FIGURE 6.18 **Effect of Growth Hormone on Stature**
The taller man (giant) has excessive growth hormone secretion, whereas the shorter man (dwarf) has insufficient growth hormone secretion.

ASSESS YOUR PROGRESS

23. *Name and describe the events occurring in the four zones of the epiphyseal plate. Explain how the epiphyseal plate remains the same thickness while the bone increases in length.*

24. *Explain the process of growth at the articular cartilage. What happens to the epiphyseal plate and the articular cartilage when bone growth ceases?*

25. *Describe how new osteons are produced as a bone increases in width.*

26. *Explain how illness or malnutrition can affect bone growth. How do vitamins D and C affect bone growth?*

27. *Bone growth is greatly affected by growth hormone and thyroid hormone. Explain these effects.*

28. *How do estrogen and testosterone affect bone growth? How do these effects account for the average height difference observed in men and women?*

> **Predict 7**
>
> Nellie is a 12-year-old female who has an adrenal tumor that is producing a large amount of estrogen. If untreated, what effect will this condition have on her growth for approximately the next 6 months? How will her height have been affected by the time she is 18?

6.7 Bone Remodeling

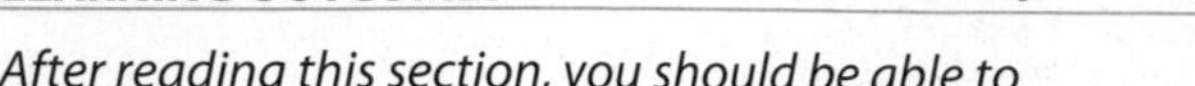

LEARNING OUTCOMES

After reading this section, you should be able to

A. Explain the need for bone remodeling, particularly in long bones.

B. Describe the role of a basic multicellular unit (BMU) in the remodeling process.

C. Discuss how mechanical stress affects bone remodeling and bone strength.

Just as our homes must be remodeled when they become outdated, bone that becomes old is replaced with new bone in a process called **bone remodeling.** In this process, osteoclasts remove old bone and osteoblasts deposit new bone. Bone remodeling converts woven bone into lamellar bone and is involved in several important functions, including bone growth, changes in bone shape, adjustment of the bone to stress, bone repair, and calcium ion (Ca^{2+}) regulation in the body.

The structure of a long bone—a hollow cylinder with a medullary cavity in the center—has two mechanical advantages: (1) A hollow cylinder is lighter in weight than a solid rod, and (2) a hollow cylinder with the same height, weight, and composition as a solid rod, but with a greater diameter, can support much more weight without bending. As a long bone increases in length and diameter, the size of the medullary cavity also increases (figure 6.19), keeping the bone from becoming very heavy. In addition, as the bone grows in diameter, the relative thickness of compact bone is maintained as osteoclasts remove bone on the inside and osteoblasts add bone to the outside.

Bone remodeling involves a **basic multicellular unit (BMU),** a temporary assembly of osteoclasts and osteoblasts that travels through or across the surface of bone, removing old bone matrix and replacing it with new bone matrix. The average life span of a BMU is approximately 6 months, and BMU activity renews the entire skeleton every 10 years. In compact bone, the osteoclasts of a BMU break down bone matrix, forming a tunnel. Interstitial lamellae (see figure 6.7*b*) are remnants of osteons that were not completely removed when a BMU formed a tunnel. Blood vessels grow into the tunnel, and osteoblasts of the BMU move in and lay down a layer of bone on the tunnel wall, forming a concentric lamella. Additional concentric lamellae are produced, filling in the tunnel from the outside to the inside, until an osteon is formed, with the center of the tunnel becoming a central canal containing blood vessels. In spongy bone, the BMU removes bone matrix from the surface of a trabecula, forming a cavity, which the BMU then fills in with new bone matrix.

Mechanical Stress and Bone Strength

The amount of stress applied to a bone can modify the bone's strength through remodeling, the formation of additional bone, alteration in trabecular alignment to reinforce the scaffolding, or other changes. Mechanical stress applied to bone increases osteoblast activity in bone tissue, and the removal of mechanical stress decreases osteoblast activity. Under conditions of reduced stress, as when a person is bedridden or paralyzed, osteoclast activity continues at a nearly normal rate but osteoblast activity decreases, resulting in less bone density. In addition, pressure in bone causes an electrical change that increases the activity of osteoblasts; therefore, applying weight (pressure) to a broken bone can speed the healing process. Weak pulses of electric current are sometimes applied to a broken bone to speed healing.

ASSESS YOUR PROGRESS

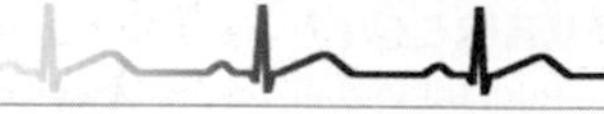

29. *Why is it important for bone remodeling to occur?*

30. *What is a basic multicellular unit (BMU)? Explain how a BMU directs remodeling in compact bone and in spongy bone.*

31. *How does bone adjust to mechanical stress? Describe the roles of osteoblasts and osteoclasts in this process. What happens to bone that is not subject to mechanical stress?*

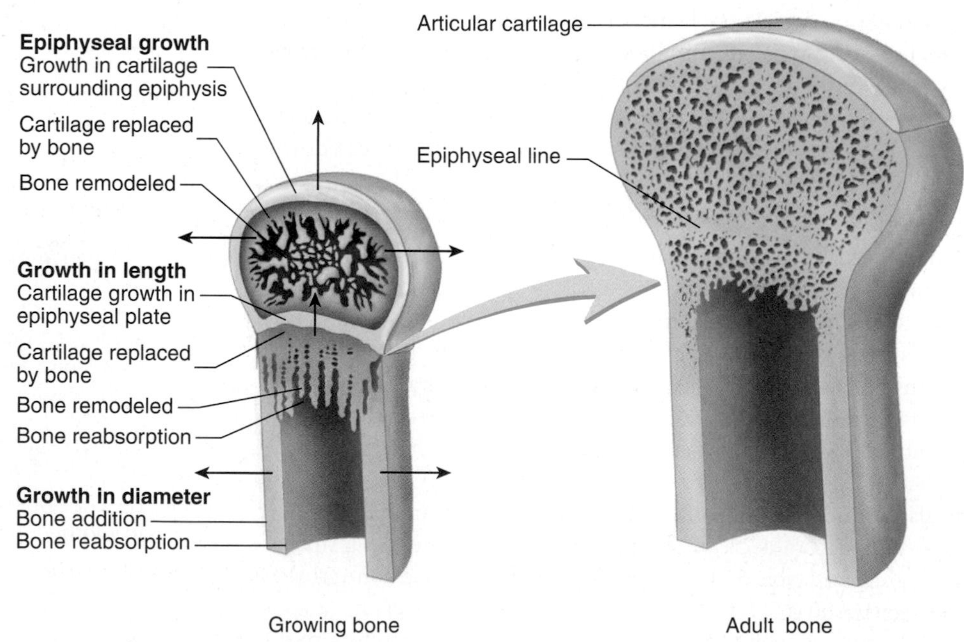

Epiphyseal growth
Growth in cartilage surrounding epiphysis
Cartilage replaced by bone
Bone remodeled
Growth in length
Cartilage growth in epiphyseal plate
Cartilage replaced by bone
Bone remodeled
Bone reabsorption
Growth in diameter
Bone addition
Bone reabsorption
Articular cartilage
Epiphyseal line
Growing bone
Adult bone

FIGURE 6.19 Remodeling of a Long Bone
The epiphysis enlarges and the diaphysis increases in length as new cartilage forms and is replaced by bone during remodeling. The diameter of the bone increases as a result of bone growth on the outside of the bone, and the size of the medullary cavity increases because of bone reabsorption.

6.8 Bone Repair

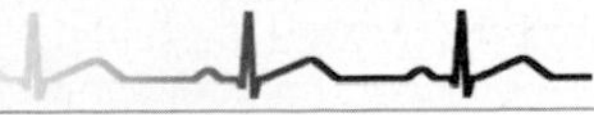

LEARNING OUTCOME

After reading this section, you should be able to

A. Outline and explain the steps in bone repair.

Bone is a living tissue that can undergo repair if it is damaged (figure 6.20*a*). This process has four major steps:

1. *Hematoma formation* (figure 6.20*b*, *step 1*). When a bone is fractured, the blood vessels in the bone and surrounding periosteum are damaged and a hematoma forms. A **hematoma** (hē-mă-tō′mă) is a localized mass of blood released from blood vessels but confined within an organ or a space. Usually, the blood in a hematoma forms a clot, which consists of fibrous proteins that stop the bleeding. Disruption of blood vessels in the central canals results in inadequate blood delivery to osteocytes, and bone tissue adjacent to the fracture site dies. Tissues around the bone often become inflamed and swollen following the injury.
2. *Callus formation* (figure 6.20*b*, *step 2*). A **callus** (kal′ŭs) is a mass of tissue that forms at a fracture site and connects the broken ends of the bone. An **internal callus** forms *between* the ends of the broken bone, as well as in the marrow cavity if the fracture occurs in the diaphysis of a long bone. Several days after the fracture, blood vessels grow into the clot. As the clot dissolves, macrophages clean up cell debris, osteoclasts break down dead bone tissue, and fibroblasts produce collagen and other extracellular materials to form granulation tissue (see chapter 4). As the fibroblasts continue to produce collagen fibers, a denser fibrous network, which helps hold the bone together, is produced. Chondroblasts derived from osteochondral progenitor cells of the periosteum and endosteum begin to produce cartilage in the fibrous network. As these events are occurring, osteochondral progenitor cells in the endosteum become osteoblasts and produce new bone, which contributes to the internal callus. If formation of the internal callus is prevented by infection, bone movements, or the nature of the injury, the two ends of the bone do not rejoin, a condition called nonunion of the bone. This condition can be treated surgically by implanting an appropriate substrate, such as living bone from another site in the body or dead bone from a cadaver. Other substrates have also been used. For example, a specific marine coral calcium phosphate is converted into a predominantly hydroxyapatite biomatrix that is very much like spongy bone.

   The **external callus** forms a collar *around* the opposing ends of the bone fragments. Osteochondral progenitor cells from the periosteum become osteoblasts, which produce bone, and chondroblasts, which produce cartilage. Cartilage production is more rapid than bone production, and the cartilage from each side of the break eventually grows together. The external callus is a bone-cartilage collar that stabilizes the ends of the broken bone. In modern medical practice, applying a cast or surgically implanting metal supports can help stabilize the bone.
3. *Callus ossification* (figure 6.20*b*, *step 3*). Like the cartilage models formed during fetal development, the cartilage in the external callus is replaced by woven spongy bone through endochondral ossification. The result is a stronger external callus. Even as the internal callus is forming and replacing the hematoma, osteoblasts from the periosteum and endosteum enter the internal callus and begin to produce bone. Eventually, the fibers and cartilage of the internal callus are replaced by woven spongy bone, which further stabilizes the broken bone.
4. *Bone remodeling* (figure 6.20*b*, *step 4*). Filling the gap between bone fragments with an internal callus of woven bone is not the end of the repair process because woven bone is not as structurally strong as the original lamellar bone. Repair is complete only when the woven bone of the internal callus and the dead bone

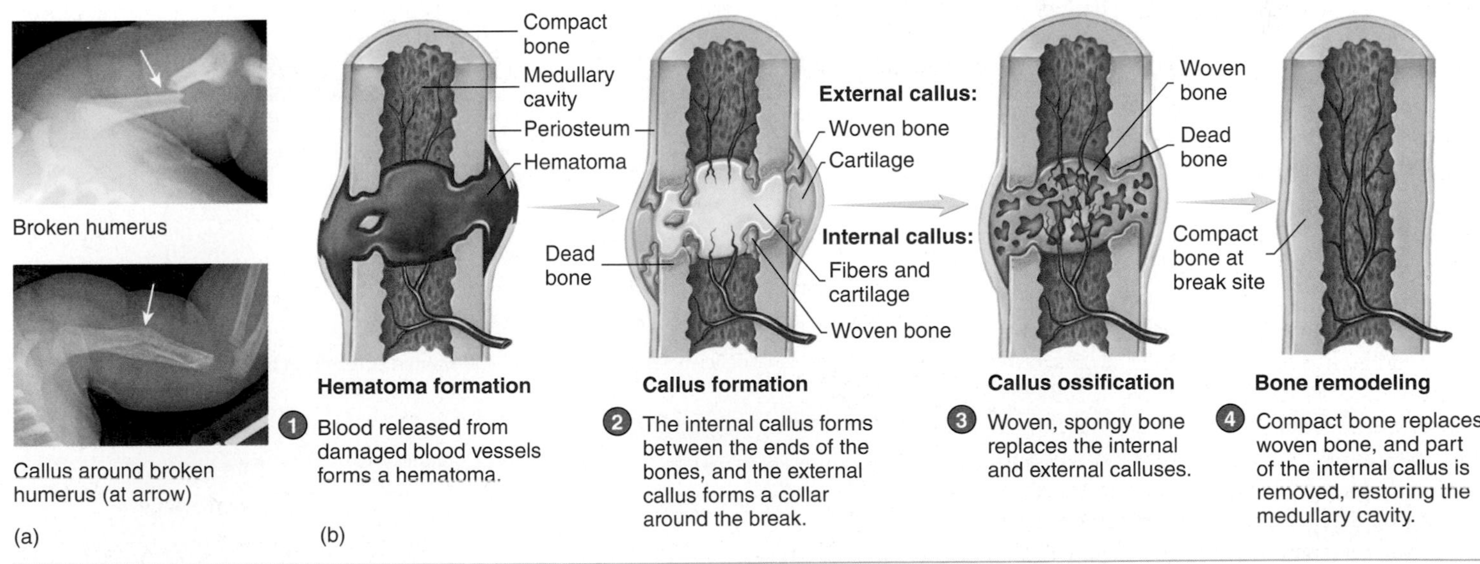
Broken humerus
Callus around broken humerus (at arrow)
(a)
Compact bone
Medullary cavity
Periosteum
Hematoma
Dead bone
(b)
Hematoma formation
1 Blood released from damaged blood vessels forms a hematoma.
External callus:
Woven bone
Cartilage
Internal callus:
Fibers and cartilage
Woven bone
Callus formation
2 The internal callus forms between the ends of the bones, and the external callus forms a collar around the break.
Woven bone
Dead bone
Callus ossification
3 Woven, spongy bone replaces the internal and external calluses.
Compact bone at break site
Bone remodeling
4 Compact bone replaces woven bone, and part of the internal callus is removed, restoring the medullary cavity.

PROCESS FIGURE 6.20 Bone Repair
(*a*) On the top is a radiograph of the broken humerus of author A. Russo's granddaughter, Viviana Russo. On the bottom is the same humerus a few weeks later, with a callus now formed around the break. (*b*) The steps in bone repair.

Clinical IMPACT — Classification of Bone Fractures

Bone fractures are classified in several ways. The most commonly used classification involves the severity of the injury to the soft tissues surrounding the bone. An **open fracture** (formerly called a *compound fracture*) occurs when an open wound extends to the site of the fracture or when a fragment of bone protrudes through the skin. If the skin is not perforated, the fracture is called a **closed fracture** (formerly called a *simple fracture*). If the soft tissues around a closed fracture are damaged, it is called a **complicated fracture.**

Figure 6B illustrates several other types of bone fractures. An **incomplete fracture** does not extend completely across the bone, whereas in a **complete fracture** the bone is broken into at least two fragments. An incomplete fracture on the convex side of the curve of the bone is a **greenstick fracture. Hairline fractures** are incomplete fractures in which the two sections of bone do not separate; hairline fractures are common in the skull.

A **comminuted** (kom′i-noo-ted) **fracture** is a complete fracture in which the bone breaks into more than two pieces—usually two major fragments and a smaller fragment. In an **impacted fracture,** one fragment is driven into the spongy portion of the other fragment.

Fractures are also classified according to the direction of the fracture within the bone. **Linear fractures** run parallel to the long axis of the bone, and **transverse fractures** are at right angles to the long axis. **Spiral fractures** take a helical course around the bone, and **oblique fractures** run obliquely in relation to the long axis. **Dentate fractures** have rough, toothed, broken ends, and **stellate fractures** have breakage lines radiating from a central point.

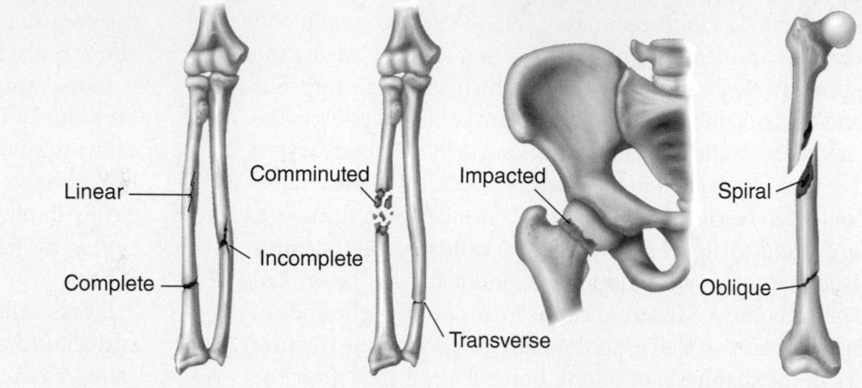
Linear
Complete
Comminuted
Incomplete
Transverse
Impacted
Spiral
Oblique

FIGURE 6B Types of Bone Fractures

adjacent to the fracture site have been replaced by compact bone. In this compact bone, osteons from both sides of the break extend across the fracture line to "peg" the bone fragments together. This remodeling process takes time—as much as a year or more. As the internal callus is remodeled and becomes stronger, the external callus is reduced in size by osteoclast activity. Eventually, repair may be so complete that no evidence of the break remains; however, the repaired zone usually remains slightly thicker than the adjacent bone. If the fracture has occurred in the diaphysis of a long bone, remodeling also restores the medullary cavity.

Diseases and Disorders

TABLE 6.3 Skeletal System

Condition	Description
Tumors	May be malignant or benign and cause a range of bone defects
Growth and Developmental Disorders	
Gigantism	Abnormally increased body size due to excessive growth at the epiphyseal plates
Dwarfism	Abnormally small body size due to improper growth at the epiphyseal plates
Osteogenesis imperfecta	Brittle bones that fracture easily due to insufficient or abnormal collagen
Rickets	Growth retardation due to nutritional deficiencies in minerals (Ca^{2+}) or vitamin D; results in bones that are soft, weak, and easily broken
Bacterial Infections	
Osteomyelitis	Bone inflammation often due to a bacterial infection that may lead to complete destruction of the bone
Tuberculosis	Typically, a lung bacterium that can also affect bone
Decalcification	
Osteomalacia	Softening of adult bones due to calcium depletion; often caused by vitamin D deficiency
Osteoporosis	Reduction in overall quantity of bone tissue; see Systems Pathology

Go to www.mhhe.com/seeley10 for additional information on these pathologies.

FUNDAMENTAL **Figure**

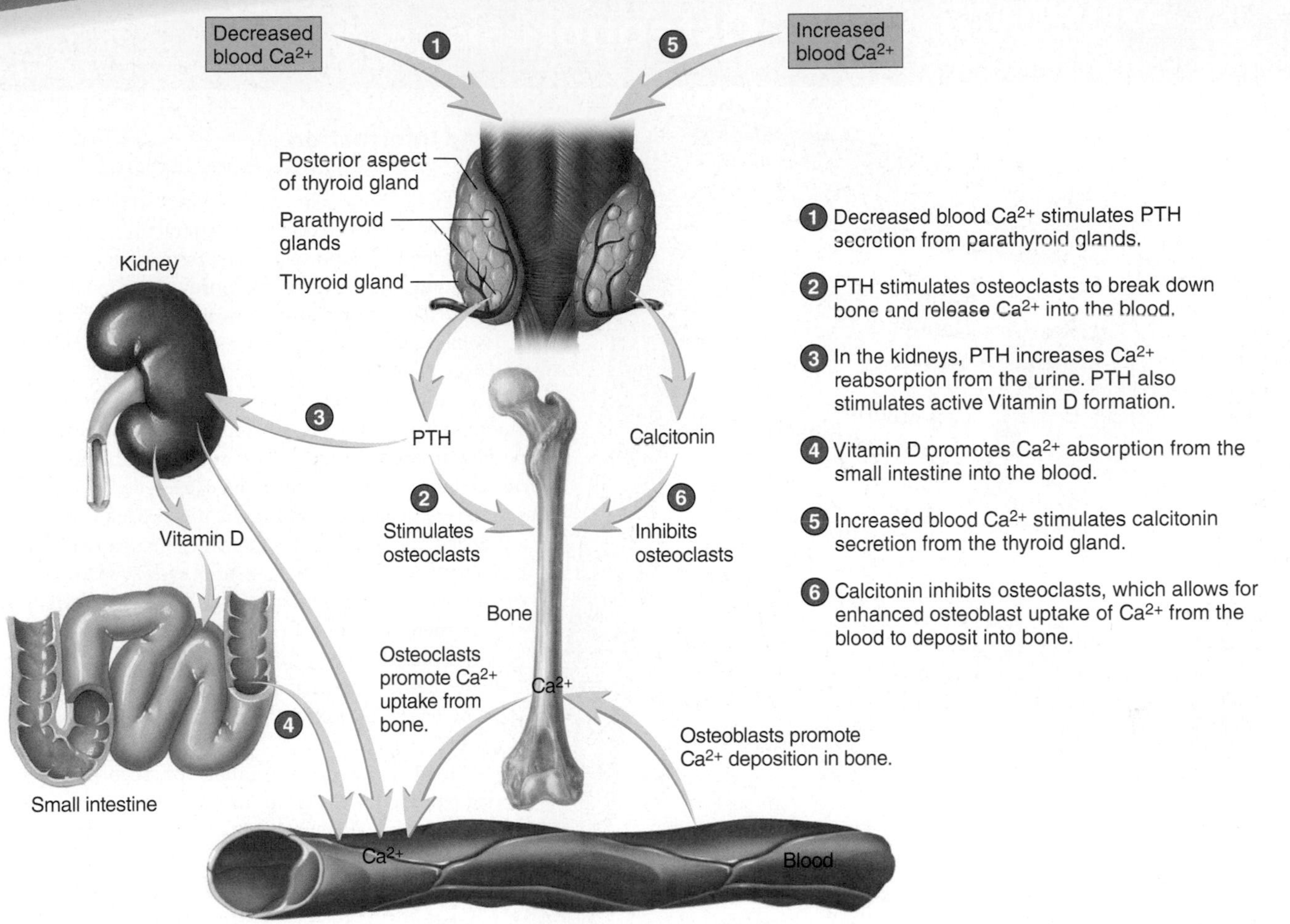

Decreased blood Ca^{2+}

Increased blood Ca^{2+}

Posterior aspect of thyroid gland

Parathyroid glands

Thyroid gland

Kidney

PTH

Calcitonin

Stimulates osteoclasts

Inhibits osteoclasts

Vitamin D

Bone

Osteoclasts promote Ca^{2+} uptake from bone.

Ca^{2+}

Osteoblasts promote Ca^{2+} deposition in bone.

Small intestine

Ca^{2+}

Blood

1. Decreased blood Ca^{2+} stimulates PTH secretion from parathyroid glands.
2. PTH stimulates osteoclasts to break down bone and release Ca^{2+} into the blood.
3. In the kidneys, PTH increases Ca^{2+} reabsorption from the urine. PTH also stimulates active Vitamin D formation.
4. Vitamin D promotes Ca^{2+} absorption from the small intestine into the blood.
5. Increased blood Ca^{2+} stimulates calcitonin secretion from the thyroid gland.
6. Calcitonin inhibits osteoclasts, which allows for enhanced osteoblast uptake of Ca^{2+} from the blood to deposit into bone.

PROCESS FIGURE 6.21 Calcium Homeostasis

ASSESS YOUR PROGRESS

32. *What are the four steps of bone repair?*
33. *How does breaking a bone result in hematoma formation?*
34. *Distinguish between the location and composition of the internal callus and those of the external callus.*
35. *Why is remodeling of the ossified callus necessary?*

6.9 Calcium Homeostasis

LEARNING OUTCOMES

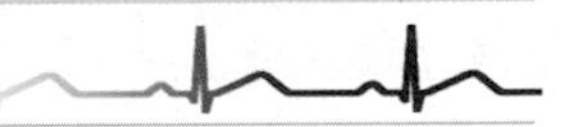

After reading this section, you should be able to

A. Explain the role of bone in calcium homeostasis.

B. Describe how parathyroid hormone and calcitonin influence bone health and calcium homeostasis.

Bones play an important role in regulating blood Ca^{2+} levels, which must be maintained within narrow limits for functions such as muscle contraction and membrane potentials to occur normally (see chapters 9 and 11). Bone is the major storage site for calcium in the body, and movement of Ca^{2+} into and out of bone helps determine blood Ca^{2+} levels. Calcium ions move into bone as osteoblasts build new bone, and they move out of bone as osteoclasts break down bone (figure 6.21; see section 6.3). When osteoblast and osteoclast activities are balanced, the movements of Ca^{2+} into and out of a bone are equal.

When blood Ca^{2+} levels are too low, osteoclast activity increases. Osteoclasts release more Ca^{2+} from bone into the blood than osteoblasts remove from the blood to make new bone. Consequently, a net movement of Ca^{2+} occurs from bone into blood, and blood Ca^{2+} levels increase. Conversely, if blood Ca^{2+} levels are too high, osteoclast activity decreases. Osteoclasts release fewer Ca^{2+} from bone into the blood than osteoblasts remove from the blood to produce new bone. As a result, a net movement of Ca^{2+} occurs from the blood to bone, and blood Ca^{2+} levels decrease.

Calcium homeostasis is maintained by two hormones: parathyroid hormone and calcitonin. **Parathyroid hormone (PTH)** is the major regulator of blood Ca^{2+} levels. PTH, secreted from the parathyroid glands (see figure 18.11) when blood Ca^{2+} levels are too low, stimulates an increase in the number of osteoclasts, which break down bone and elevate blood Ca^{2+} levels (see figure 6.20).

Systems PATHOLOGY | Osteoporosis

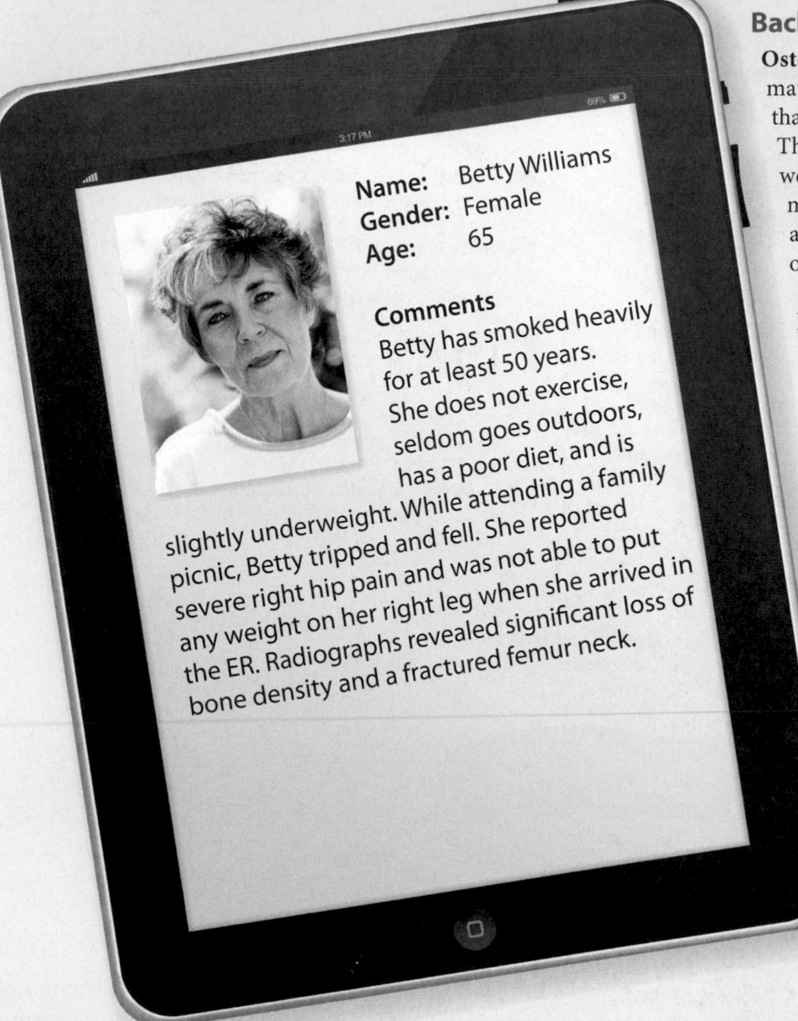
Name: Betty Williams
Gender: Female
Age: 65
Comments
Betty has smoked heavily for at least 50 years. She does not exercise, seldom goes outdoors, has a poor diet, and is slightly underweight. While attending a family picnic, Betty tripped and fell. She reported severe right hip pain and was not able to put any weight on her right leg when she arrived in the ER. Radiographs revealed significant loss of bone density and a fractured femur neck.

Background Information

Osteoporosis (os′tē-ō-pō-rō′sis), or porous bone, is a loss of bone matarix. The loss of bone mass makes bones so porous and weakened that they become deformed and prone to fracture (figures 6C and 6D). The occurrence of osteoporosis increases with age. In both men and women (although it is 2.5 times more common in women), bone mass starts to decrease at about age 40 and continually decreases thereafter. Women can eventually lose approximately one-half, and men one-quarter, of their spongy bone.

In women, decreased production of the female reproductive hormone estrogen can cause osteoporosis, mostly in spongy bone, especially in the vertebrae of the spine and the bones of the forearm. Collapse of the vertebrae can cause a decrease in height or, in more severe cases, kyphosis in the upper back (figure 6E).

Estrogen levels decrease after menopause, removal of the ovaries, amenorrhea (lack of a menstrual cycle) due to extreme exercise or anorexia nervosa (self-starvation), and cigarette smoking.

In men, reduction in testosterone levels can cause a loss of bone tissue. However, this is less of a problem in men than in women because men have denser bones than women, and testosterone levels generally don't decrease significantly until after age 65.

Inadequate dietary intake or absorption of calcium, sometimes due to certain medications, can also contribute to osteoporosis. Absorption of calcium from the small intestine decreases with age.

Finally, too little exercise or disuse from injury can also cause osteoporosis. Significant amounts of bone are lost after 8 weeks of immobilization.

Early diagnosis of osteoporosis can lead to more preventive treatments. Instruments that measure the absorption of photons (particles of light) by bone are used; of these, dual-energy x-ray absorptiometry (DEXA) is considered the

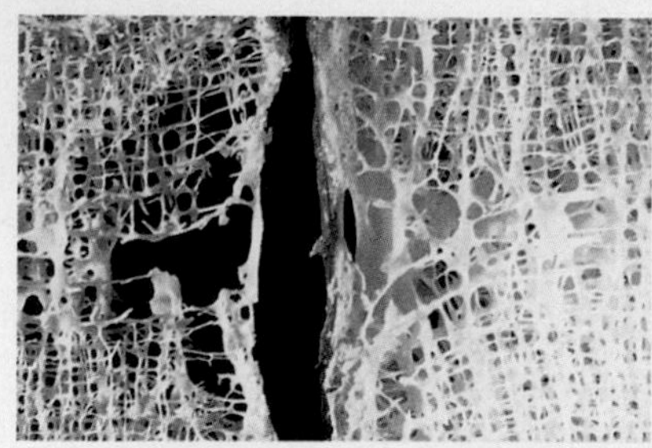
Osteoporotic bone
Normal bone

Figure 6C Photomicrograph of Osteoporotic Bone and Normal Bone

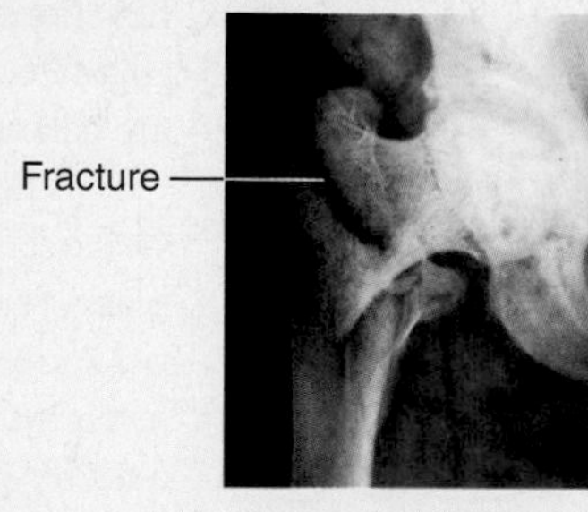
Fracture

Figure 6D Radiograph

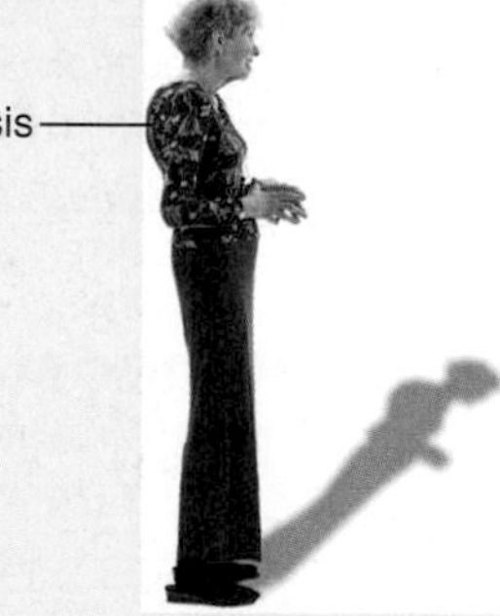
Kyphosis

Figure 6E

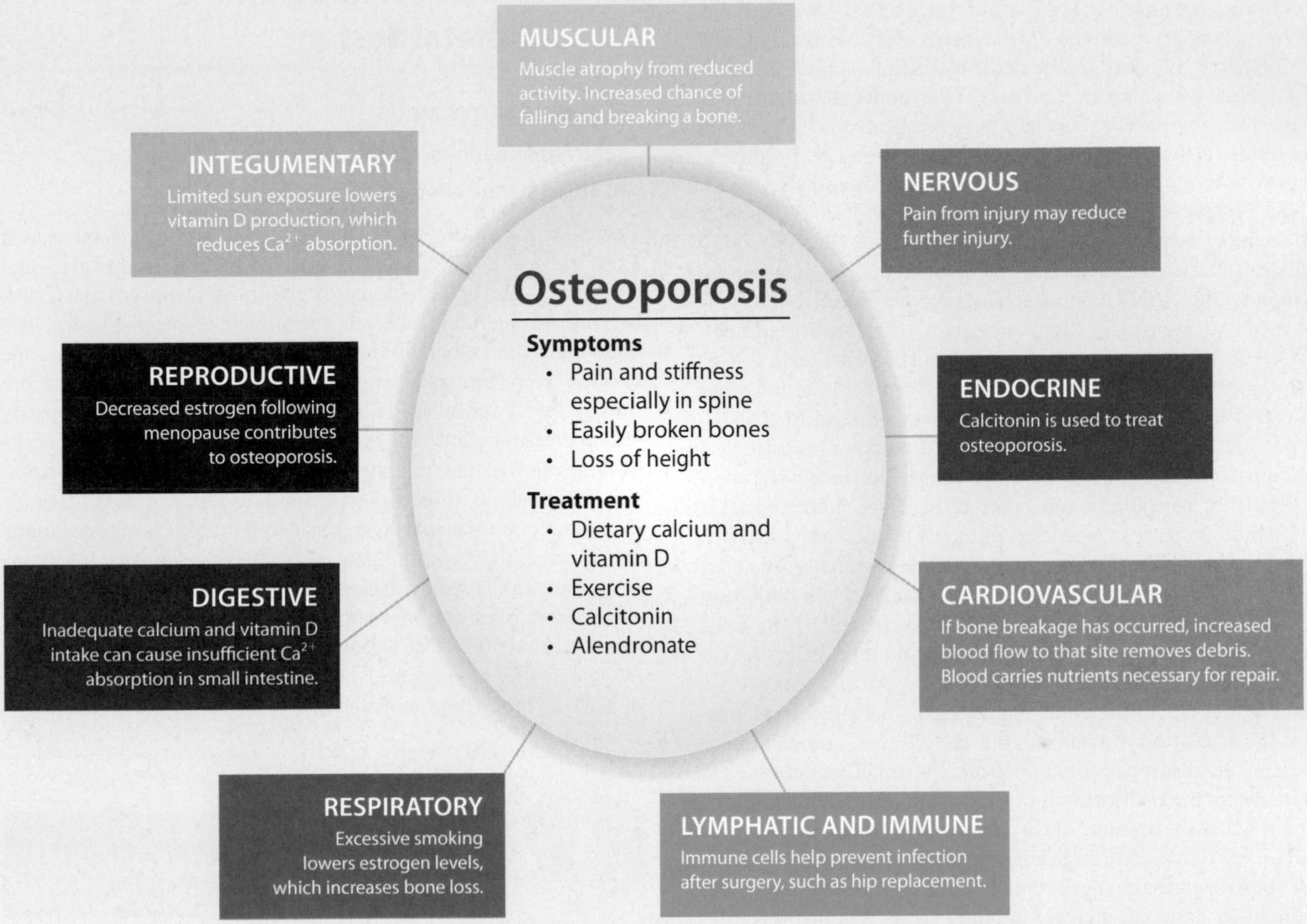

MUSCULAR
Muscle atrophy from reduced activity. Increased chance of falling and breaking a bone.
INTEGUMENTARY
Limited sun exposure lowers vitamin D production, which reduces Ca^{2+} absorption.
NERVOUS
Pain from injury may reduce further injury.
Osteoporosis
Symptoms
• Pain and stiffness especially in spine
• Easily broken bones
• Loss of height
Treatment
• Dietary calcium and vitamin D
• Exercise
• Calcitonin
• Alendronate
REPRODUCTIVE
Decreased estrogen following menopause contributes to osteoporosis.
ENDOCRINE
Calcitonin is used to treat osteoporosis.
DIGESTIVE
Inadequate calcium and vitamin D intake can cause insufficient Ca^{2+} absorption in small intestine.
CARDIOVASCULAR
If bone breakage has occurred, increased blood flow to that site removes debris. Blood carries nutrients necessary for repair.
RESPIRATORY
Excessive smoking lowers estrogen levels, which increases bone loss.
LYMPHATIC AND IMMUNE
Immune cells help prevent infection after surgery, such as hip replacement.

best. Supplementation with dietary calcium and vitamin D and exercise are the best preventive and rehabilitory measures to prevent bone loss or regain mild bone loss. Calcitonin (Miacalcin) inhibits osteoclast activity, as does alendronate (Fosamax), which binds to hydroxyapatite. Although osteoporosis is linked to estrogen loss, estrogen therapy has been associated with many side effects, including breast cancer, and is no longer recommended as a treatment.

Predict 8

What advice should Betty give her granddaughter, so that the granddaughter will be less likely to develop osteoporosis when she is Betty's age?

In addition, PTH stimulates osteoblasts to release enzymes that break down the layer of unmineralized organic bone matrix covering bone, thereby making the mineralized bone matrix available to osteoclasts.

Osteoclast numbers are regulated by the interactions of osteoblasts and red bone marrow stem cells of the monocyte/macrophage lineage. Osteoblasts and stem cells have receptors for PTH. When PTH binds to these receptors, osteoblasts respond by producing **receptor activator of nuclear factor kappaB ligand (RANKL).** RANKL is expressed on the surface of the osteoblasts and combines with **receptor activator of nuclear factor kappaB (RANK)** found on the cell surfaces of osteoclast precursor stem cells. In a cell-to-cell interaction, RANKL on osteoblasts binds to RANK on osteoclast precursor stem cells, stimulating them to become osteoclasts.

Osteoclast production is inhibited by **osteoprotegerin** (os'tē-ō-prō-teg'er-in) **(OPG),** which is secreted by osteoblasts and other cells. OPG inhibits osteoclast production by binding to RANKL and preventing RANKL from binding to its receptor on osteoclast precursor stem cells. Increased PTH causes decreased secretion of OPG from osteoblasts and other cells. Thus, increased PTH promotes an increase in osteoclast numbers by increasing RANKL and decreasing OPG. The increased RANKL stimulates osteoclast precursor cells, and the decreased OPG results in less inhibition of osteoclast precursor cells. Conversely, decreased PTH results in fewer osteoclasts by decreasing RANKL and increasing OPG.

PTH also regulates blood Ca^{2+} levels by increasing Ca^{2+} uptake in the small intestine (figure 6.21). Increased PTH promotes the activation of vitamin D in the kidneys, and vitamin D increases the absorption of Ca^{2+} from the small intestine. PTH also increases the reabsorption of Ca^{2+} from urine in the kidneys, which reduces the amount of Ca^{2+} lost in the urine.

Tumors that secrete large amounts of PTH can cause so much bone breakdown that bones become weakened and fracture easily. On the other hand, an increase in blood Ca^{2+} results in less PTH secretion, decreased osteoclast activity, reduced Ca^{2+} release from bone, and eventually decreased blood Ca^{2+} levels.

Calcitonin (kal-si-tō'nin), secreted from the thyroid gland when blood Ca^{2+} levels are too high (see figure 18.8), decreases osteoclast activity (see figure 6.20) by binding to receptors on the osteoclasts. An increase in blood Ca^{2+} stimulates the thyroid gland to secrete calcitonin, which inhibits osteoclast activity. PTH and calcitonin are described more fully in chapters 18 and 27.

ASSESS YOUR PROGRESS

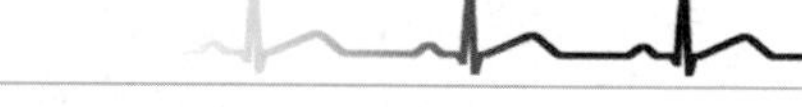

36. *How is calcium moved into and out of bone? What happens in bone when blood calcium levels decrease? When blood calcium levels increase?*
37. *Name the hormone that is the major regulator of Ca^{2+} levels in the body. What stimulates the secretion of this hormone?*
38. *Describe how PTH controls the number of osteoclasts. What are the effects of PTH on the formation of vitamin D, Ca^{2+} uptake in the small intestine, and reabsorption of Ca^{2+} from the urine?*
39. *What stimulates calcitonin secretion? How does calcitonin affect osteoclast activity?*

6.10 Effects of Aging on the Skeletal System

LEARNING OUTCOME

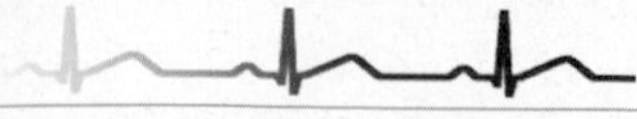

After reading this section, you should be able to

A. Describe the effects of aging on bones.

The most significant age-related changes in the skeletal system affect the quality and quantity of bone matrix. Recall that a mineral (hydroxyapatite) in the bone matrix gives bone compression (weight-bearing) strength, but collagen fibers make the bone flexible. The bone matrix in an older bone is more brittle than a younger bone because decreased collagen production results in relatively more mineral and fewer collagen fibers. With aging, the amount of matrix also decreases because the rate of matrix formation by osteoblasts becomes slower than the rate of matrix breakdown by osteoclasts.

Bone mass is at its highest around age 30, and men generally have denser bones than women because of the effects of testosterone and greater body weight. Race also affects bone mass. African-Americans and Hispanics have higher bone masses than Caucasians and Asians. After age 35, both men and women experience bone loss at a rate of 0.3–0.5% a year. This loss can increase by 10 times in women after menopause, when they can lose bone mass at a rate of 3–5% a year for approximately 5–7 years (see Systems Pathology).

Case STUDY Bone Density

Henry is a 65-year-old man who was admitted to the emergency room after a fall. A radiograph confirmed that he had fractured the proximal part of his arm bone (surgical neck of the humerus). The radiograph also revealed that his bone matrix was not as dense as it should be for a man his age. A test for blood Ca^{2+} was normal. On questioning, Henry confessed that he is a junk food addict who eats few vegetables and never consumes dairy products. In addition, Henry never exercises and seldom goes outdoors except at night.

Predict 9

Use your knowledge of bone physiology and figure 6.21 to answer the following questions.

A. Why is Henry more likely to break a bone than are most men his age?
B. How have Henry's eating habits contributed to his low bone density?
C. Would Henry's PTH levels be lower than normal, normal, or higher than normal?
D. What effect has Henry's nocturnal lifestyle had on his bone density?
E. How has lack of exercise affected his bone density?

At first, spongy bone is lost as the trabeculae become thinner and weaker. The ability of the trabeculae to provide support also decreases as they become disconnected from each other. Eventually, some of the trabeculae completely disappear. Trabecular bone loss is greatest in the trabeculae that are under the least stress. In other words, more sedentary individuals experience greater bone loss.

A slow loss of compact bone begins about age 40 and increases after age 45. However, the rate of compact bone loss is approximately half the rate of trabecular bone loss. Bones become thinner, but their outer dimensions change little, because most compact bone is lost under the endosteum on the inner surfaces of bones. In addition, the remaining compact bone weakens as a result of incomplete bone remodeling. In a young bone, when osteons are removed, the resulting spaces are filled with new osteons. With aging, the new osteons fail to completely fill in the spaces produced when the older osteons are removed.

Significant bone loss increases the likelihood of bone fractures. For example, the loss of trabeculae greatly increases the risk for compression fractures of the vertebrae (backbones) because the weight-bearing body of a vertebra consists mostly of spongy bone. In addition, bone loss can lead to deformity, loss of height, pain, and stiffness. For example, compression fractures of the vertebrae can cause an exaggerated curvature of the spine, resulting in a bent-forward, stooped posture. Loss of bone from the jaws can lead to tooth loss.

The most effective preventive measure against the effects of aging on the skeletal system is the combination of increasing physical activity and taking dietary calcium and vitamin D supplements. Intensive exercise, especially weight-bearing exercise, can even reverse the loss of bone matrix.

ASSESS YOUR PROGRESS

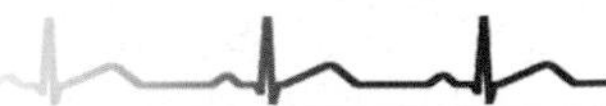

40. *What effect does aging have on the quality and quantity of bone matrix?*

41. *How can a person be protected from the effects of aging on the skeletal system?*

Answer

Learn to Predict ❮ **From page 163**

The question suggests that Amir is interested in eating a healthy diet, especially one that supports strong bones and joints. In chapter 6, we learned that children whose diets lack adequate vitamin D develop rickets, a condition characterized by inflamed joints and bowed legs. Chapter 5 explained that vitamin D is necessary for normal Ca^{2+} absorption in the small intestine. Although we learned that the body can manufacture vitamin D in sun-exposed skin, chapter 6 informed us that vitamin D deficiency can also result from a lack of lipids in the diet. Growth of strong bones depends on adequate Ca^{2+} absorption. We also read in chapter 6 that vitamin C is important for normal bone growth. Vitamin C is necessary for the synthesis of collagen, a major organic component of the bone matrix that is responsible for some of the major functional characteristics of bone. Collagen is also a common matrix protein in other connective tissues, such as ligaments and the dermis of the skin. Inadequate collagen synthesis due to vitamin C deficiency can result in short stature, poor wound healing, and loose permanent teeth. Thus, to help prevent any serious injuries, Amir should be sure to consume plenty of vitamin D– and vitamin C–rich foods in addition to healthy lipid-rich foods.

Answers to the rest of this chapter's Predict questions are in Appendix G.

Summary

6.1 Functions of the Skeletal System (p. 164)

1. The skeletal system consists of bones, cartilage, tendons, and ligaments.
2. The skeletal system supports the body, protects the organs it surrounds, allows body movements, stores minerals and lipids, and is the site of blood cell production.

6.2 Cartilage (p. 164)

1. Chondroblasts produce cartilage and become chondrocytes. Chondrocytes are located in lacunae surrounded by matrix.
2. The matrix of cartilage contains collagen fibers (for strength) and proteoglycans (to trap water).
3. The perichondrium surrounds cartilage.
   - The outer layer contains fibroblasts.
   - The inner layer contains chondroblasts.
4. Cartilage grows by both appositional and interstitial growth.

6.3 Bone Histology (p. 165)

Bone Matrix

1. Collagen provides flexible strength.
2. Hydroxyapatite provides compressional strength.

Bone Cells

1. Osteoblasts produce bone matrix and become osteocytes.
   - Osteoblasts connect to one another through cell processes and surround themselves with bone matrix to become osteocytes.
   - Osteocytes are located in lacunae and are connected to one another through canaliculi.
2. Osteoclasts break down bone (with assistance from osteoblasts).
3. Osteoblasts originate from osteochondral progenitor cells, whereas osteoclasts originate from monocyte/macrophage lineage stem cells in red bone marrow.
4. Ossification, the formation of bone, occurs through appositional growth.

Woven and Lamellar Bone

1. Woven bone has collagen fibers oriented in many directions. It is remodeled to form lamellar bone.
2. Lamellar bone is arranged in thin layers, called lamellae, which have collagen fibers oriented parallel to one another.

Spongy and Compact Bone

1. Spongy bone has many spaces.
   - Lamellae combine to form trabeculae, beams of bone that interconnect to form a latticelike structure with spaces filled with bone marrow and blood vessels.
   - The trabeculae are oriented along lines of stress and provide structural strength.
2. Compact bone is dense with few spaces.
   - Compact bone consists of organized lamellae: Circumferential lamellae form the outer surface of compact bones; concentric lamellae surround central canals, forming osteons; interstitial lamellae are remnants of lamellae left after bone remodeling.
   - Canals within compact bone provide a means for exchanging gases, nutrients, and waste products. From the periosteum or endosteum, perforating canals carry blood vessels to central canals, and canaliculi connect central canals to osteocytes.

6.4 Bone Anatomy (p. 169)

Bone Shapes

Individual bones can be classified as long, flat, short, or irregular.

Structure of a Long Bone

1. The diaphysis is the shaft of a long bone, and the epiphyses are distinct from the diaphysis and house the epiphyseal plate.
2. The epiphyseal plate is the site of lengthwise bone growth.
3. The medullary cavity is a space within the diaphysis.
4. Red marrow is the site of blood cell production, and yellow marrow consists of fat.
5. The periosteum covers the outer surface of bone.
   - The outer layer contains blood vessels and nerves.
   - The inner layer contains osteoblasts, osteoclasts, and osteochondral progenitor cells.
   - Perforating fibers hold the periosteum, ligaments, and tendons in place.
6. The endosteum lines cavities inside bone and contains osteoblasts, osteoclasts, and osteochondral progenitor cells.

Structure of Flat, Short, and Irregular Bones

Flat, short, and irregular bones have an outer covering of compact bone surrounding spongy bone.

6.5 Bone Development (p. 172)

Intramembranous Ossification

1. Some skull bones, part of the mandible, and the diaphyses of the clavicles develop from membranes.
2. Within the membrane at centers of ossification, osteoblasts produce bone along the membrane fibers to form spongy bone.
3. Beneath the periosteum, osteoblasts lay down compact bone to form the outer surface of the bone.
4. Fontanels are areas of membrane that are not ossified at birth.

Endochondral Ossification

1. Most bones develop from a cartilage model.
2. The cartilage matrix is calcified, and chondrocytes die. Osteoblasts form bone on the calcified cartilage matrix, producing spongy bone.
3. Osteoblasts build an outer surface of compact bone beneath the periosteum.
4. Primary ossification centers form in the diaphysis during fetal development. Secondary ossification centers form in the epiphyses.
5. Articular cartilage on the ends of bones and the epiphyseal plate does not ossify.

6.6 Bone Growth (p. 176)

1. Bones increase in size only by appositional growth, the addition of new bone to the surface of older bone or cartilage.
2. Trabeculae grow by appositional growth.

Growth in Bone Length

1. Epiphyseal plate growth involves the interstitial growth of cartilage followed by appositional bone growth on the cartilage.
2. Epiphyseal plate growth results in increased length of the diaphysis and bony processes. Bone growth in length ceases when the epiphyseal plate becomes ossified and forms the epiphyseal line.

Growth at Articular Cartilage

1. Articular cartilage growth involves the interstitial growth of cartilage followed by appositional bone growth on the cartilage.
2. Articular cartilage growth results in larger epiphyses and an increase in the size of bones that do not have epiphyseal plates.

Growth in Bone Width

1. Appositional bone growth beneath the periosteum increases the diameter of long bones and the size of other bones.
2. Osteoblasts from the periosteum form ridges with grooves between them. The ridges grow together, converting the grooves into tunnels filled with concentric lamellae to form osteons.
3. Osteoblasts from the periosteum lay down circumferential lamellae, which can be remodeled.

Factors Affecting Bone Growth

1. Genetic factors determine bone shape and size. The expression of genetic factors can be modified.
2. Factors that alter the mineralization process or the production of organic matrix, such as deficiencies in vitamins D and C, can affect bone growth.
3. Growth hormone, thyroid hormone, estrogen, and testosterone stimulate bone growth.
4. Estrogen and testosterone cause increased bone growth and closure of the epiphyseal plate.

6.7 Bone Remodeling (p. 180)

1. Remodeling converts woven bone to lamellar bone and allows bone to change shape, adjust to stress, repair itself, and regulate body calcium levels.
2. Bone adjusts to stress by adding new bone and by realigning bone through remodeling.

6.8 Bone Repair (p. 181)

1. Fracture repair begins with the formation of a hematoma.
2. The hematoma is replaced by an internal callus consisting of fibers and cartilage.
3. The external callus is a bone-cartilage collar that stabilizes the ends of the broken bone.
4. The internal and external calluses are ossified to become woven bone.
5. The woven bone is replaced by compact bone.

6.9 Calcium Homeostasis (p. 183)

PTH increases blood Ca^{2+} by increasing bone breakdown, Ca^{2+} absorption from the small intestine, and the reabsorption of Ca^{2+} from the urine. Calcitonin decreases blood Ca^{2+} by decreasing bone breakdown.

6.10 Effects of Aging on the Skeletal System (p. 186)

1. With aging, bone matrix is lost and the matrix becomes more brittle.
2. Spongy bone loss results from thinning and loss of trabeculae. Compact bone loss mainly occurs from the inner surface of bones and involves formation of fewer osteons.
3. Loss of bone increases the risk for fractures and causes deformity, loss of height, pain, stiffness, and loss of teeth.
4. Exercise and dietary supplements are effective at preventing bone loss.

REVIEW AND COMPREHENSION

1. Which of these is *not* a function of bone?
   a. internal support and protection
   b. attachment for the muscles
   c. calcium and phosphate storage
   d. blood cell production
   e. vitamin D storage
2. Chondrocytes are mature cartilage cells within the __________, and they are derived from __________.
   a. perichondrium, fibroblasts
   b. perichondrium, chondroblasts
   c. lacunae, fibroblasts
   d. lacunae, chondroblasts
3. Which of these statements concerning cartilage is correct?
   a. Cartilage often occurs in thin plates or sheets.
   b. Chondrocytes receive nutrients and oxygen from blood vessels in the matrix.
   c. Articular cartilage has a thick perichondrium layer.
   d. The perichondrium contains both chondrocytes and osteocytes.
   e. Appositional growth of cartilage occurs when chondrocytes within the tissue add more matrix from the inside.
4. Which of these substances makes up the major portion of bone?
   a. collagen
   b. hydroxyapatite
   c. proteoglycan aggregates
   d. osteocytes
   e. osteoblasts
5. The flexible strength of bone occurs because of
   a. osteoclasts.
   b. ligaments.
   c. hydroxyapatite.
   d. collagen fibers.
   e. periosteum.
6. The primary function of osteoclasts is to
   a. prevent osteoblasts from forming.
   b. become osteocytes.
   c. break down bone.
   d. secrete calcium salts and collagen fibers.
   e. form the periosteum.
7. Central canals
   a. connect perforating canals to canaliculi.
   b. connect spongy bone to compact bone.
   c. are where blood cells are produced.
   d. are found only in spongy bone.
   e. are lined with periosteum.
8. The lamellae found in osteons are __________ lamellae.
   a. circumferential
   b. concentric
   c. interstitial
9. Spongy bone consists of interconnecting rods or plates of bone called
   a. osteons.
   b. canaliculi.
   c. circumferential lamellae.
   d. a haversian system.
   e. trabeculae.
10. A fracture in the shaft of a bone is a break in the
    a. epiphysis.
    b. perichondrium.
    c. diaphysis.
    d. articular cartilage.
11. Yellow marrow is
    a. found mostly in children's bones.
    b. found in the epiphyseal plate.
    c. important for blood cell production.
    d. mostly adipose tissue.
12. The periosteum
    a. is an epithelial tissue membrane.
    b. covers the outer and internal surfaces of bone.
    c. contains only osteoblasts.
    d. becomes continuous with collagen fibers of tendons or ligaments.
    e. has a single fibrous layer.
13. Given these events:
    (1) Osteochondral progenitor cells become osteoblasts.
    (2) Connective tissue membrane is formed.
    (3) Osteoblasts produce woven bone.

    Which sequence best describes intramembranous bone formation?
    a. 1,2,3 b. 1,3,2 c. 2,1,3 d. 2,3,1 e. 3,2,1
14. Given these processes:
    (1) Chondrocytes die.
    (2) Cartilage matrix calcifies.
    (3) Chondrocytes hypertrophy.
    (4) Osteoblasts deposit bone.
    (5) Blood vessels grow into lacunae.

    Which sequence best represents the order in which these processes occur during endochondral bone formation?
    a. 3,2,1,4,5
    b. 3,2,1,5,4
    c. 5,2,3,4,1
    d. 3,2,5,1,4
    e. 3,5,2,4,1

15. Intramembranous bone formation
    a. occurs at the epiphyseal plate.
    b. is responsible for growth in diameter of a bone.
    c. gives rise to the flat bones of the skull.
    d. occurs within a hyaline cartilage model.
    e. produces articular cartilage in the long bones.
16. The ossification regions formed during early fetal development
    a. are secondary ossification centers.
    b. become articular cartilage.
    c. become medullary cavities.
    d. become the epiphyses.
    e. are primary ossification centers.
17. Growth in the length of a long bone occurs
    a. at the primary ossification center.
    b. beneath the periosteum.
    c. at the center of the diaphysis.
    d. at the epiphyseal plate.
    e. at the epiphyseal line.
18. During growth in length of a long bone, cartilage forms and then ossifies. The location of the ossification is the zone of
    a. calcification.
    b. hypertrophy.
    c. proliferation.
    d. resting cartilage.
19. Given these processes:
    (1) An osteon is produced.
    (2) Osteoblasts from the periosteum form a series of ridges.
    (3) The periosteum becomes the endosteum.
    (4) Osteoblasts lay down bone to produce a concentric lamella.
    (5) Grooves are changed into tunnels.

    Which sequence best represents the order in which these processes occur during growth in the width of a long bone?
    a. 1,4,2,3,5
    b. 2,5,3,4,1
    c. 3,4,2,1,5
    d. 4,2,1,5,3
    e. 5,4,2,1,3
20. Chronic vitamin D deficiency results in which of these consequences?
    a. Bones become brittle.
    b. The percentage of bone composed of hydroxyapatite increases.
    c. Bones become soft and pliable.
    d. Scurvy occurs.
    e. Both a and b are correct.
21. Estrogen
    a. stimulates a burst of growth at puberty.
    b. causes a later closure of the epiphyseal plate than testosterone does.
    c. causes a longer growth period in females than testosterone causes in males.
    d. tends to prolong the growth phase of the epiphyseal plates.
    e. All of these are correct.
22. Bone remodeling can occur
    a. when woven bone is converted into lamellar bone.
    b. as bones are subjected to varying patterns of stress.
    c. as a long bone increases in diameter.
    d. when new osteons form in compact bone.
    e. All of these are correct.
23. Given these processes:
    (1) cartilage ossification
    (2) external callus formation
    (3) hematoma formation
    (4) internal callus formation
    (5) remodeling of woven bone into compact bone

    Which sequence best represents the order in which the processes occur during repair of a fracture?
    a. 1,2,3,4,5
    b. 2,4,3,1,5
    c. 3,4,2,1,5
    d. 4,1,5,2,3
    e. 5,3,4,2,1
24. Which of these processes during bone repair requires the longest period of time?
    a. cartilage ossification
    b. external callus formation
    c. hematoma formation
    d. internal callus formation
    e. remodeling of woven bone into compact bone
25. If the secretion of parathyroid hormone (PTH) increases, osteoclast activity ____________ and blood Ca^{2+} levels ____________.
    a. decreases, decrease
    b. decreases, increase
    c. increases, decrease
    d. increases, increase

Answers in Appendix E

CRITICAL THINKING

1. When a person develops Paget disease, for unknown reasons the collagen fibers in the bone matrix run randomly in all directions. In addition, the amount of trabecular bone decreases. What symptoms would you expect to observe?
2. Explain why running helps prevent osteoporosis in the elderly. Does the benefit include all bones or mainly those of the lower limbs and spine?
3. Astronauts can experience a dramatic decrease in bone density while in a weightless environment. Explain how this happens, and suggest a way to slow the loss of bone tissue.
4. In some cultures, eunuchs were responsible for guarding harems, which are the collective wives of one male. Eunuchs are males who were castrated as boys. Castration removes the testes, the major site of testosterone production in males. Because testosterone is responsible for the sex drive in males, the reason for castration is obvious. As a side effect of this procedure, the eunuchs grew to above normal heights. Can you explain why?
5. When a long bone breaks, blood vessels at the fracture line are severed. The formation of blood clots stops the bleeding. Within a few days, bone tissue on both sides of the fracture site dies. However, the bone dies back only a certain distance from the fracture line. Explain.
6. A patient has hyperparathyroidism because a tumor in the parathyroid gland is producing excessive amounts of PTH. How does this hormone affect bone? Would the administration of large doses of vitamin D help the situation? Explain.

Answers in Appendix F

Visit this book's website at **www.mhhe.com/seeley10** for chapter quizzes, interactive learning exercises, and other study tools.

connect plus+
ANATOMY & PHYSIOLOGY

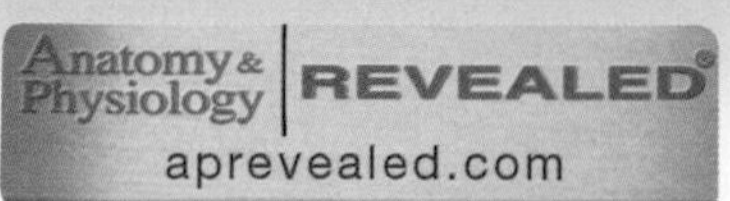
Anatomy & Physiology
REVEALED
aprevealed.com

7

Skeletal System

GROSS ANATOMY

If the body had no skeleton, it would look like a poorly stuffed rag doll. The skeletal system is the framework that helps maintain the body's shape and enables us to move normally. Most muscles act on bones to produce movement, often pulling on the bones with considerable force. Human bones are very strong and can resist tremendous bending and compression forces without breaking. Nonetheless, each year approximately 6.8 million Americans break a bone.

The skeletal system includes bones, cartilage, and ligaments. However, to study skeletal gross anatomy, anatomists use dried, prepared bones, so that they can view the major features of individual bones unobstructed by associated soft tissues. As you study the bones depicted in this chapter, keep in mind that living bones not only contain soft tissue, such as the periosteum (see chapter 6), but also have important relationships with many soft tissues, including muscles, tendons, ligaments, cartilage, nerves, and blood vessels.

Learn to Predict

Dave Plummer loves telling people around the pool that he has a pig valve in his heart. At the age of 70, Dave required replacement of a heart valve. The surgeon opened Dave's thoracic cavity by making a longitudinal midline incision from the superior to the inferior margin of his sternum through the skin and underlying soft tissue. Then he cut the sternum with a bone saw along the same line, so that the right and left halves of the sternum could be spread apart enough to expose the heart. After the defective valve had been replaced, the surgeon wired the two halves of the sternum back together. For several days after the surgery, Dave experienced significant discomfort in his back, and although he started walking within a few days, he could not resume his normal swimming routine until 2 months later. After learning about bone repair and anatomy in chapter 6, and studying the structure of individual bones and their relationships to each other in chapter 7, name the specific parts of the skeletal system and the tissue layers of the bone that the surgeon cut. Also, explain Dave's back discomfort and why he could not resume swimming sooner.

Photo: Swiming is an effective form of exercise, but after surgery it may take a while before a patient can return to normal activities.

Module 5
Skeletal System

Anatomy & Physiology
REVEALED
aprevealed.com

7.1 Skeletal Anatomy Overview

LEARNING OUTCOMES

After reading this section, you should be able to

A. Define the anatomical terms for bone features.

B. List the two anatomical portions of the skeleton.

The average adult skeleton has 206 bones (table 7.1; figure 7.1). Although this is the traditional number, the actual number of bones varies from person to person and decreases with age as some bones become fused.

Anatomists use several common terms to describe the features of bones (table 7.2). Most of these features involve the relationship between the bones and associated soft tissues. If a bone possesses a **tubercle** (too′ber-kl; lump) or a **process** (projection), most likely a ligament or tendon was attached to that tubercle or process during life.

TABLE 7.1 Number of Named Bones Listed by Category

Bones		Number
Axial Skeleton		
Skull (Cranium)		
Braincase (neurocranium)		
Paired (left and right)	Parietal	2
	Temporal	2
Unpaired (single)	Frontal	1
	Sphenoid	1
	Occipital	1
	Ethmoid	1
Face (viscerocranium)		
Paired	Maxilla	2
	Zygomatic	2
	Palatine	2
	Lacrimal	2
	Nasal	2
	Inferior nasal concha	2
Unpaired	Mandible	1
	Vomer	1
	Total skull bones	22
Bones Associated with the Skull		
Auditory ossicles		
Malleus		2
Incus		2
Stapes		2
Hyoid		1
	Total associated bones	7
Vertebral Column		
Cervical vertebrae		7
Thoracic vertebrae		12
Lumbar vertebrae		5
Sacrum		1
Coccyx		1
	Total vertebral column bones	26
Rib Cage (Thoracic Cage)		
Ribs		24
Sternum		1
	Total rib cage bones	25
	Total axial skeleton bones	80

Bones		Number
Appendicular Skeleton		
Pectoral Girdle		
Scapula		2
Clavicle		2
Upper Limb		
Humerus		2
Ulna		2
Radius		2
Carpal bones		16
Metacarpal bones		10
Phalanges		28
	Total girdle and upper limb bones	64
Pelvic Girdle		
Coxal bone		2
Lower Limb		
Femur		2
Tibia		2
Fibula		2
Patella		2
Tarsal bones		14
Metatarsal bones		10
Phalanges		28
	Total girdle and lower limb bones	62
	Total appendicular skeleton bones	126
	Total axial skeleton bones	80
	Total appendicular skeleton bones	126
	Total bones	206

FUNDAMENTAL **Figure**

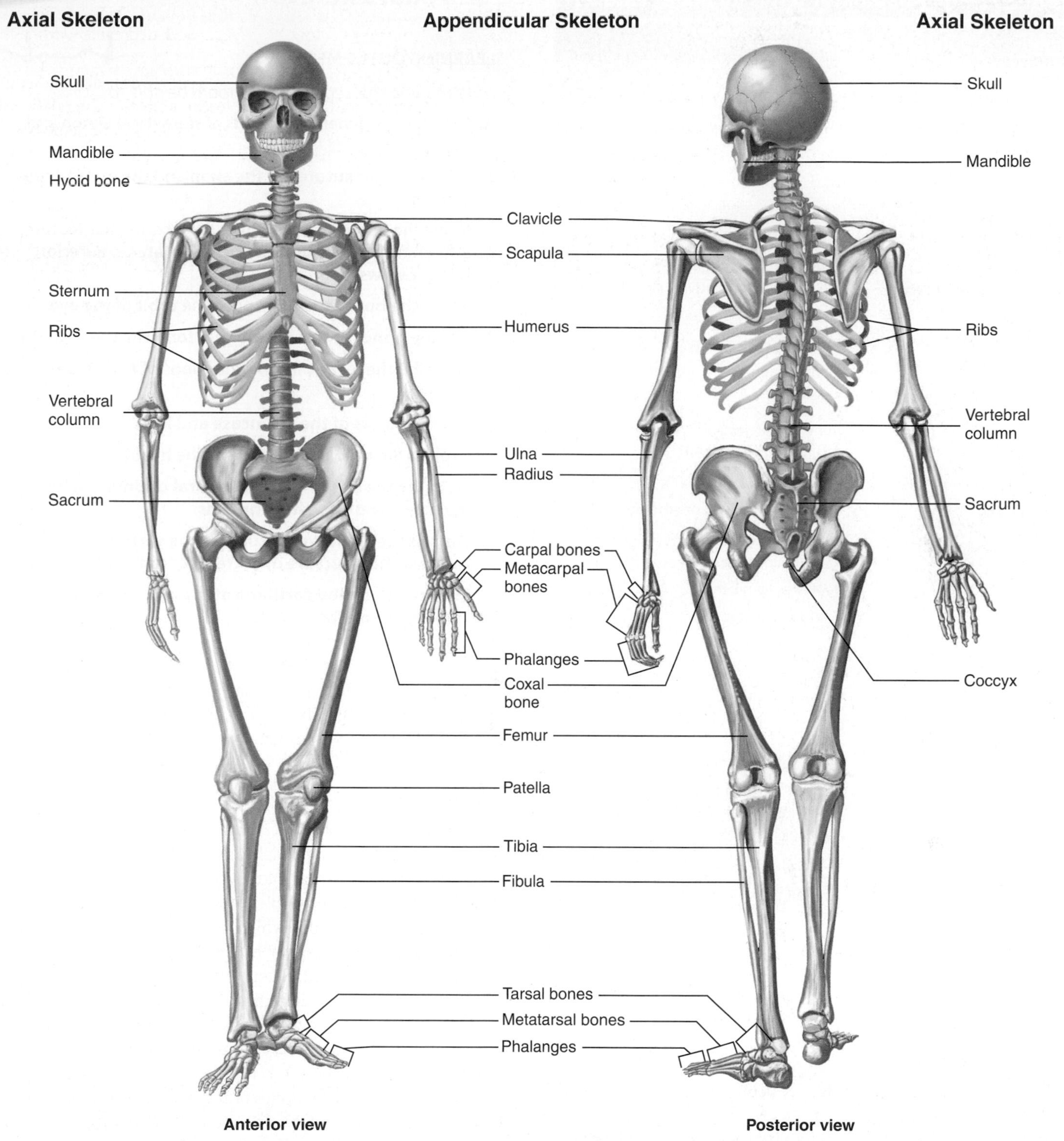
Axial Skeleton
Appendicular Skeleton
Axial Skeleton
Skull
Mandible
Hyoid bone
Sternum
Ribs
Vertebral column
Sacrum
Clavicle
Scapula
Humerus
Ulna
Radius
Carpal bones
Metacarpal bones
Phalanges
Coxal bone
Femur
Patella
Tibia
Fibula
Tarsal bones
Metatarsal bones
Phalanges
Skull
Mandible
Ribs
Vertebral column
Sacrum
Coccyx
Anterior view
Posterior view

FIGURE 7.1 AP|R **Complete Skeleton**

Bones of the axial skeleton are listed in the far left- and right hand columns; bones of the appendicular skeleton are listed in the center. (The skeleton is not shown in the anatomical position.)

TABLE 7.2 Anatomical Terms for Bone Features

Term	Description
Body	Main part
Head	Enlarged, often rounded end
Neck	Constriction between head and body
Margin, border	Edge
Angle	Bend
Ramus	Branch off the body beyond the angle
Condyle	Smooth, rounded articular surface
Facet	Small, flattened articular surface
Ridges	
Line, linea	Low ridge
Crest, crista	Prominent ridge
Spine	Very high ridge
Projections	
Process	Prominent projection
Tubercle	Small, rounded bump
Tuberosity, tuber	Knob; larger than a tubercle
Trochanter	Tuberosity on the proximal femur
Epicondyle	Upon a condyle
Lingula	Flat, tongue-shaped process
Hamulus	Hook-shaped process
Cornu	Horn-shaped process
Openings	
Foramen	Hole
Canal, meatus	Tunnel
Fissure	Cleft
Sinus, labyrinth	Cavity
Depressions	
Fossa	General term for a depression
Notch	Depression in the margin of a bone
Fovea	Little pit
Groove, ulcus	Deep, narrow depression

If a bone has a smooth, articular surface, that surface was part of a joint and was covered with articular cartilage. If the bone has a **foramen** (fō-rā′men; pl. *foramina;* fō-ram′i-nă; hole) in it, that foramen was the opening through which a nerve or blood vessel passed. Some skull bones contain mucous membrane–lined air spaces called **sinuses.**

The bones of the skeleton are divided into axial and appendicular portions (figure 7.1).

ASSESS YOUR PROGRESS

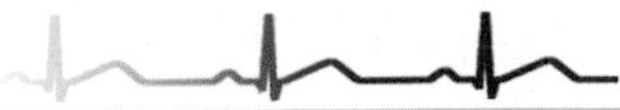

1. *How are lumps, projections, and openings in bones related to soft tissues?*
2. *What do each of the following terms mean: tubercle, condyle, spine, foramen, fossa?*
3. *What are the two anatomical portions of the skeleton?*

7.2 Axial Skeleton

LEARNING OUTCOMES

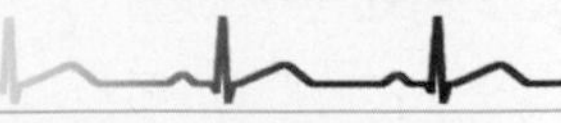

After reading this section, you should be able to

A. Describe the general functions of the axial skeleton and list its parts.
B. List the major sutures of the skull and the bones they connect.
C. Name the bones of the skull and describe their features as seen from the superior, posterior, lateral, anterior, and inferior views.
D. Name the bones that compose the orbit of the eye.
E. List the bones and cartilage that form the nasal septum.
F. Describe the locations and functions of the paranasal sinuses.
G. List the bones of the braincase and face.
H. Explain the unique structure of the hyoid bone.
I. Describe the shape of the vertebral column, list its divisions, and state its functions.
J. Discuss the common features of the vertebrae and contrast the structure of vertebrae from each region.
K. List the bones and cartilage of the rib cage, including the three types of ribs.

The **axial skeleton** is composed of the skull, auditory ossicles, hyoid bone, vertebral column, and rib cage (thoracic cage). The axial skeleton forms the upright axis of the body. It protects the brain, the spinal cord, and the vital organs housed within the thorax.

Skull

The **skull,** or *cranium* (krā′nē-ŭm), protects the brain; supports the organs of vision, hearing, smell, and taste; and provides a foundation for the structures that take air, food, and water into the body. When the skull is disassembled, the mandible is easily separated from the rest of the skull, which remains intact. Special effort is needed to separate the other bones. For this reason, it is convenient to think of the skull, except for the mandible, as a single unit. The top of the skull is called the **calvaria** (kal-vā′rē-ă), or *skullcap.* It is usually cut off to reveal the skull's interior. The exterior and interior of the skull have ridges, lines, processes, and plates. These structures are important for the attachment of muscles or for articulations between the bones of the skull. Selected features of the intact skull are listed in table 7.3 and are visible in figures 7.2–7.12. The fetal skull is considered in chapter 8.

Superior View of the Skull

The skull appears quite simple when viewed from above (figure 7.2). Only four bones are seen from this view: the frontal bone, two parietal bones, and a small part of the occipital bone. The paired **parietal bones** are joined at the midline by the **sagittal suture,** and the parietal bones are connected to the **frontal bone** by the **coronal suture.**

TABLE 7.3 Processes and Other Features of the Skull

Feature	Bone on Which Feature Is Found	Description
External Features		
Alveolar process	Mandible, maxilla	Ridges on the mandible and maxilla containing the teeth (shown in figure 7.6)
Angle	Mandible	Posterior, inferior corner of the mandible (shown in figure 7.4)
Coronoid process	Mandible	Attachment point for the temporalis muscle (shown in figure 7.4)
Mental protuberance	Mandible	Chin (resembles a bent knee; shown in figures 7.4 and 7.6)
Horizontal plate	Palatine	Posterior third of the hard palate (shown in figure 7.9)
Mandibular condyle	Mandible	Region where the mandible articulates with the skull (shown in figure 7.4)
Mandibular fossa	Temporal	Depression where the mandible articulates with the skull (shown in figure 7.12)
Mastoid process	Temporal	Enlargement posterior to the ear; attachment site for several muscles that move the head (shown in figures 7.3, 7.4, and 7.12)
Nuchal lines	Occipital	Attachment points for several posterior neck muscles (shown in figures 7.3 and 7.12)
Occipital condyle	Occipital	Point of articulation between the skull and the vertebral column (shown in figures 7.3 and 7.12)
Palatine process	Maxilla	Anterior two-thirds of the hard palate (shown in figures 7.9 and 7.12)
Pterygoid hamulus	Sphenoid	Hooked process on the inferior end of the medial pterygoid plate, around which the tendon of one palatine muscle passes; an important dental landmark (shown in table 7.7*d*)
Pterygoid plates (medial and lateral)	Sphenoid	Bony plates on the inferior aspect of the sphenoid bone; the lateral pterygoid plate is the site of attachment for two muscles of mastication (chewing; shown in figures 7.9 and 7.12)
Ramus	Mandible	Portion of the mandible superior to the angle (shown in figure 7.4)
Styloid process	Temporal	Attachment site for three muscles (to the tongue, pharynx, and hyoid bone) and some ligaments (shown in figure 7.4)
Temporal lines	Parietal	Attachment site for the temporalis muscle, which closes the jaw (shown in figure 7.4)
Internal Features		
Crista galli	Ethmoid	Process in the anterior part of the braincase to which one of the connective tissue coverings of the brain (dura mater) connects (shown in figures 7.9 and 7.11)
Petrous portion	Temporal	Thick, interior part of temporal bone containing the middle and inner ears and the auditory ossicles (shown in figure 7.11)
Sella turcica	Sphenoid	Bony structure, resembling a saddle, in which the pituitary gland is located (shown in figure 7.11)

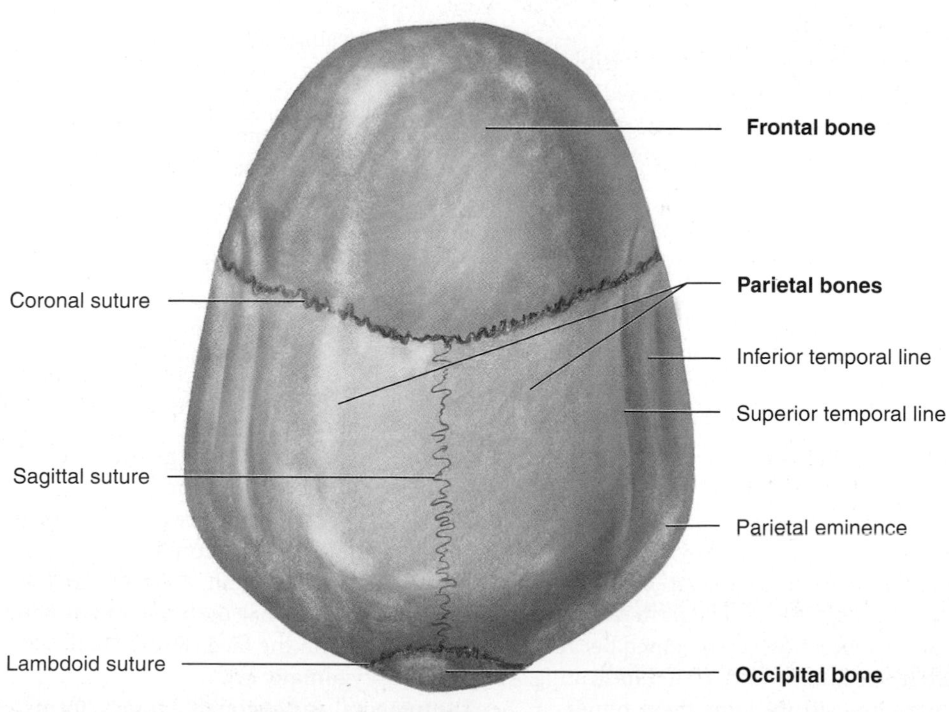
Frontal bone
Coronal suture
Parietal bones
Inferior temporal line
Superior temporal line
Sagittal suture
Parietal eminence
Lambdoid suture
Occipital bone
Superior view

FIGURE 7.2 Superior View of the Skull
(The names of the bones are in bold.)

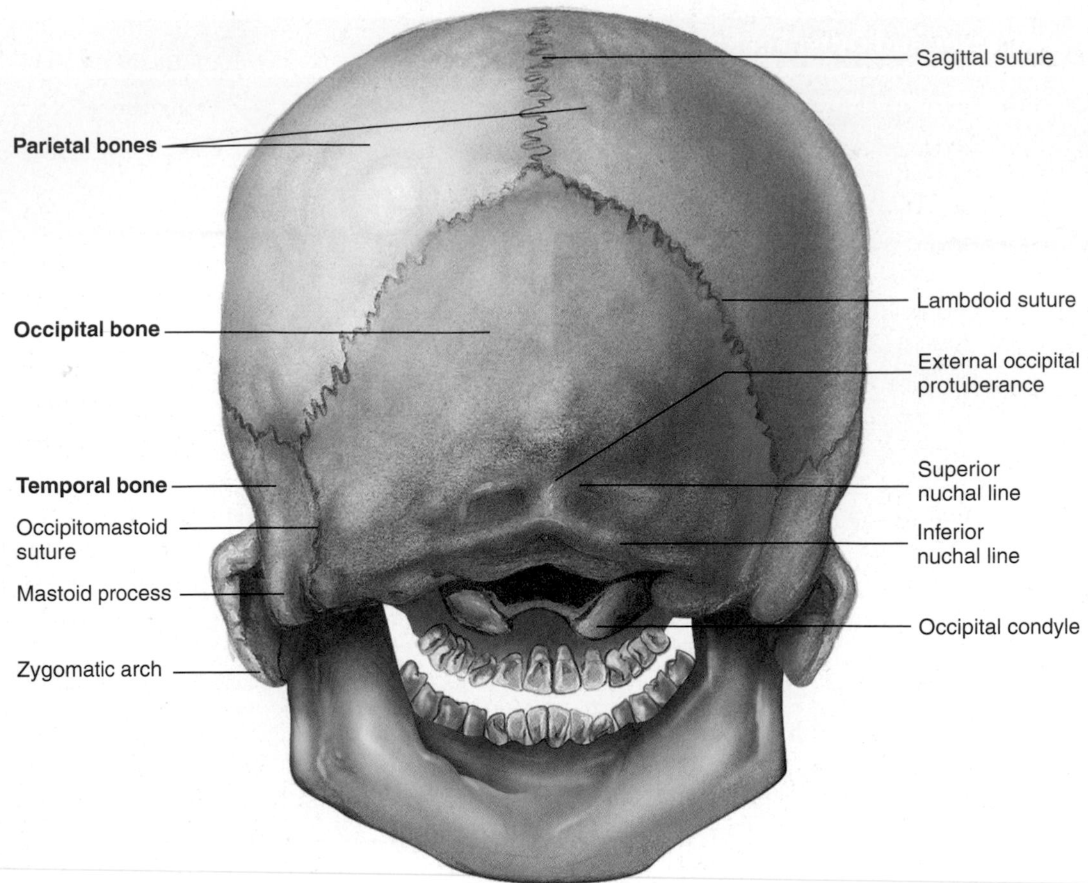
Sagittal suture
Parietal bones
Lambdoid suture
Occipital bone
External occipital protuberance
Superior nuchal line
Temporal bone
Occipitomastoid suture
Inferior nuchal line
Mastoid process
Occipital condyle
Zygomatic arch
Posterior view

FIGURE 7.3 Posterior View of the Skull
(The names of the bones are in bold.)

Posterior View of the Skull

The parietal and occipital bones are the major structures visible in the posterior view (figure 7.3). The parietal bones are joined to the **occipital bone** by the **lambdoid** (lam′doyd) **suture.** Occasionally, extra small bones called **sutural** (soo′choor-ăl) **bones** (*wormian bones*) form along the lambdoid suture.

An **external occipital protuberance** is present on the posterior surface of the occipital bone (figure 7.3). It can be felt through the scalp at the base of the head and varies considerably in size from person to person. The external occipital protuberance is the site of attachment of the **ligamentum nuchae** (noo′kē; nape of neck), an elastic ligament that extends down the neck and helps keep the head erect by pulling on the occipital region of the skull. **Nuchal lines,** a set of small ridges that extend laterally from the protuberance, are the points of attachment for several neck muscles.

Lateral View of the Skull

The parietal bone and the squamous part of the temporal bone form a large part of the side of the head (figure 7.4). The term *temporal* means related to time, and the temporal bone is so named because the hair of the temples is often the first to turn white, indicating the passage of time. The **squamous suture** joins these bones. A prominent feature of the temporal bone is a large hole, the **external auditory canal** (*external acoustic meatus*), which transmits sound waves toward the tympanic membrane. The external ear, or auricle, surrounds the canal. Just posterior and inferior to the external auditory canal is a large inferior projection, the **mastoid** (mas′toyd) **process.** The process can be seen and felt as a prominent lump just posterior to the ear. The process is not solid bone but is filled with cavities called **mastoid air cells,** which are connected to the middle ear. Important neck muscles involved in rotating the head attach to the mastoid process. The superior and inferior **temporal lines,** which are attachment points of the temporalis muscle, one of the major muscles of mastication, arch across the lateral surface of the parietal bone.

The lateral surface of the **greater wing** of the **sphenoid** (sfē′noyd) **bone** is immediately anterior to the temporal bone (figure 7.4). Although appearing to be two bones, one on each side of the skull, the sphenoid bone is actually a single bone that extends completely across the skull. Anterior to the sphenoid bone is the **zygomatic** (zī′gō-mat′ik) **bone,** or cheekbone, which can be easily seen and felt on the face (figure 7.5).

The **zygomatic arch,** which consists of joined processes from the temporal and zygomatic bones, forms a bridge across the side

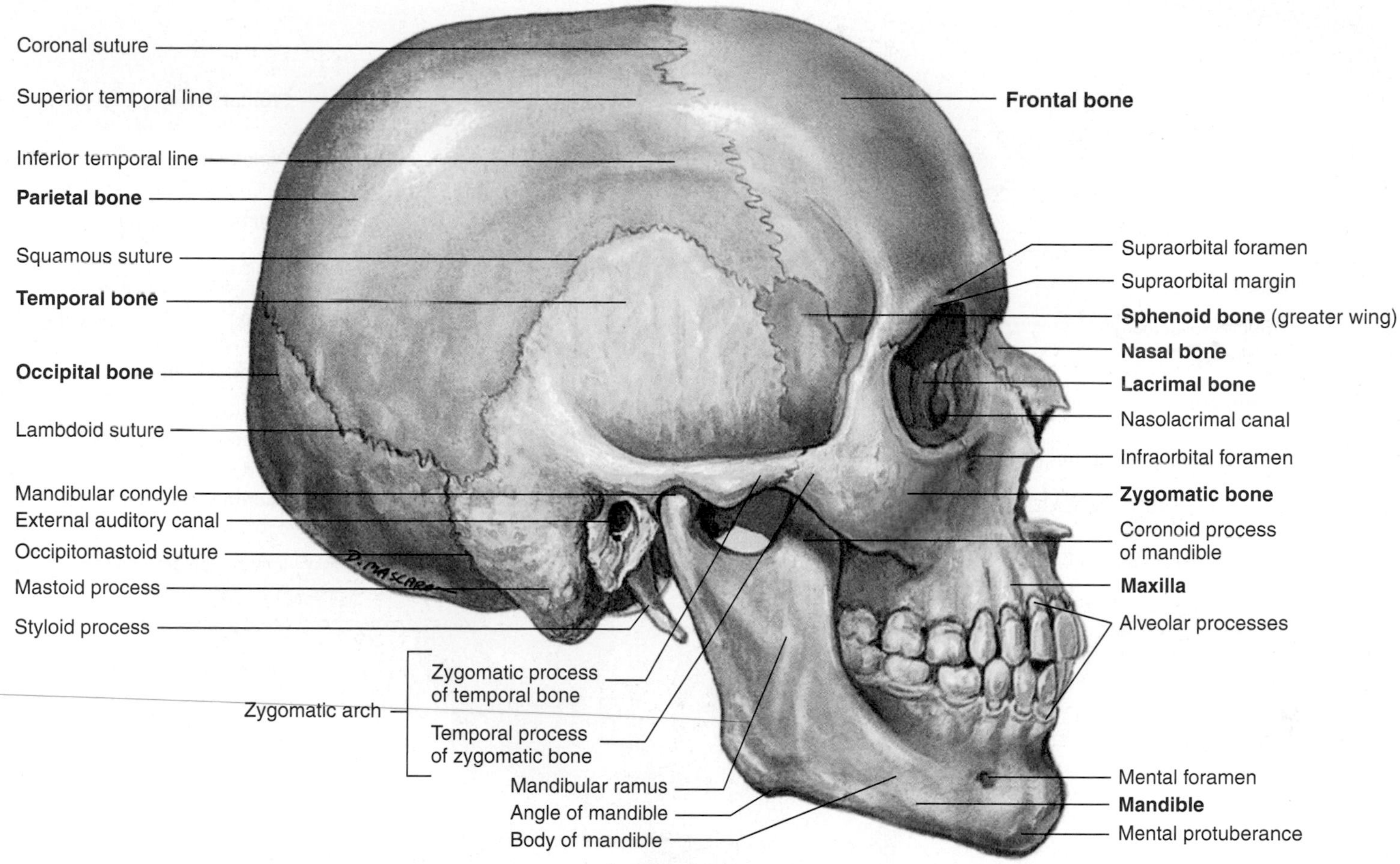

Coronal suture
Superior temporal line
Inferior temporal line
Parietal bone
Squamous suture
Temporal bone
Occipital bone
Lambdoid suture
Mandibular condyle
External auditory canal
Occipitomastoid suture
Mastoid process
Styloid process
Zygomatic arch
Zygomatic process of temporal bone
Temporal process of zygomatic bone
Mandibular ramus
Angle of mandible
Body of mandible
Frontal bone
Supraorbital foramen
Supraorbital margin
Sphenoid bone (greater wing)
Nasal bone
Lacrimal bone
Nasolacrimal canal
Infraorbital foramen
Zygomatic bone
Coronoid process of mandible
Maxilla
Alveolar processes
Mental foramen
Mandible
Mental protuberance
Lateral view

FIGURE 7.4 APR **Right Lateral View of the Skull**
(The names of the bones are in bold.)

of the skull (see figure 7.4). The zygomatic arch is easily felt on the side of the face, and the muscles on each side of the arch can be felt as the mandible opens and closes (figure 7.5).

The **maxilla** (mak-sil′ă; upper jaw) is anterior and inferior to the zygomatic bone to which it is joined. The **mandible** (lower jaw) is inferior to the maxilla and articulates posteriorly with the temporal bone (see figure 7.4). The mandible consists of two main portions: the **body,** which extends anteroposteriorly, and the **ramus** (branch), which extends superiorly from the body toward the temporal bone. The superior end of the ramus has a mandibular **condyle,** which articulates with the mandibular fossa of the temporal bone, and the **coronoid** (kōr′ŏ-noyd) **process,** to which the powerful temporalis muscle, one of the chewing muscles, attaches. The alveolar process of the maxilla contains the superior set of teeth, and the alveolar process of the mandible contains the inferior teeth.

Anterior View of the Skull

The major structures seen from the anterior view are the frontal bone (forehead), the zygomatic bones (cheekbones), the maxillae, and the mandible (figure 7.6). The teeth, which are very prominent in this view, are discussed in chapter 24. Many bones of the face can be easily felt through the skin (figure 7.7).

From this view, the most prominent openings into the skull are the orbits and the nasal cavity. Each of the two **orbits** is a cone-shaped fossa with its apex directed posteriorly (figure 7.8; see figure 7.6).

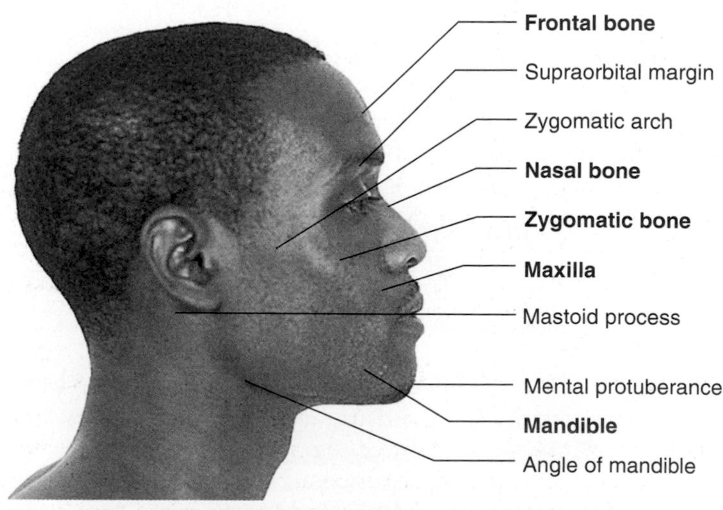
Frontal bone
Supraorbital margin
Zygomatic arch
Nasal bone
Zygomatic bone
Maxilla
Mastoid process
Mental protuberance
Mandible
Angle of mandible

FIGURE 7.5 Lateral View of Bony Landmarks on the Face

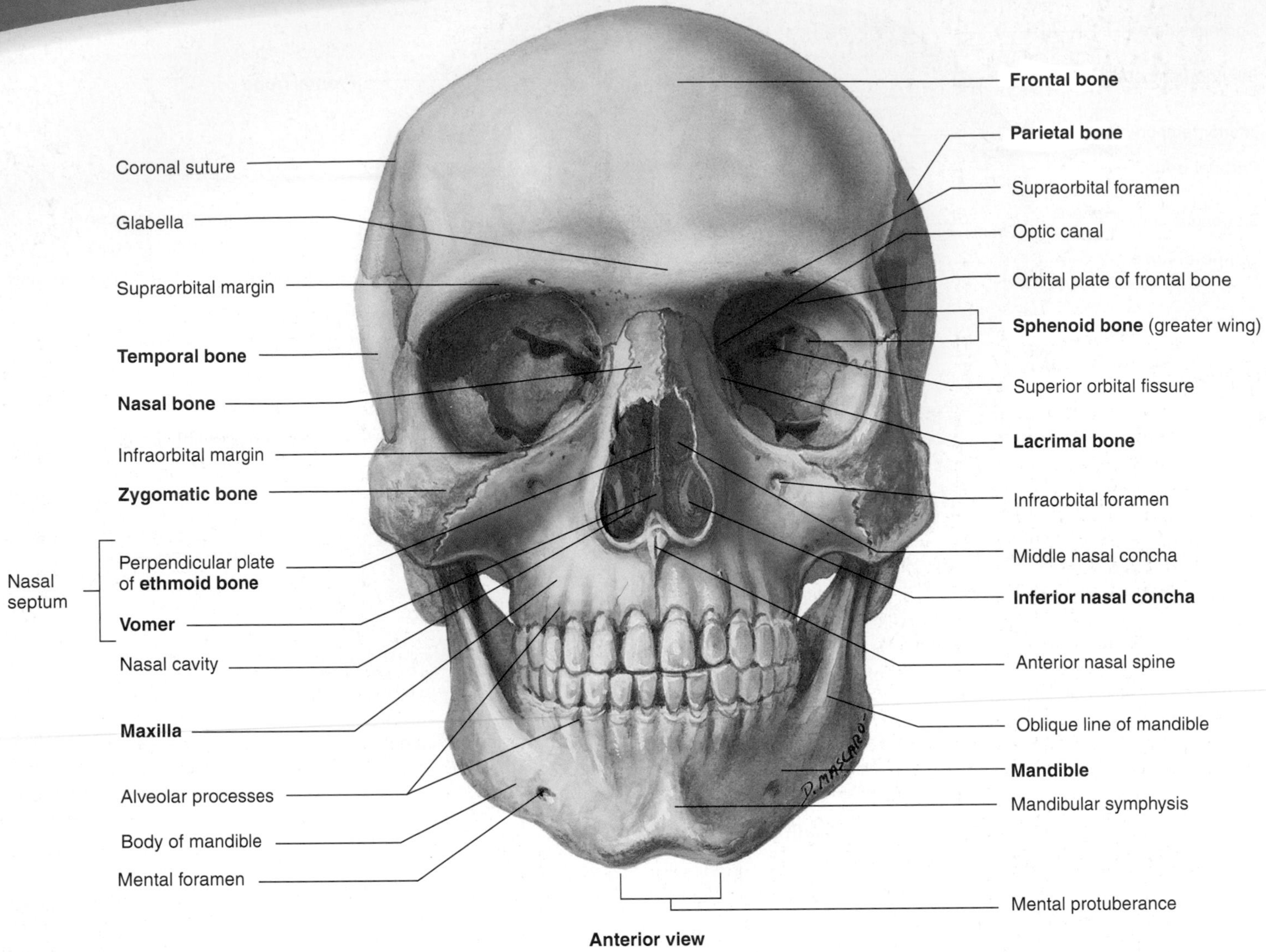

FUNDAMENTAL **Figure**

Frontal bone
Parietal bone
Coronal suture
Supraorbital foramen
Glabella
Optic canal
Supraorbital margin
Orbital plate of frontal bone
Sphenoid bone (greater wing)
Temporal bone
Superior orbital fissure
Nasal bone
Lacrimal bone
Infraorbital margin
Infraorbital foramen
Zygomatic bone
Middle nasal concha
Nasal septum
Perpendicular plate of ethmoid bone
Inferior nasal concha
Vomer
Nasal cavity
Anterior nasal spine
Maxilla
Oblique line of mandible
Mandible
Alveolar processes
Mandibular symphysis
Body of mandible
Mental foramen
Mental protuberance
D. MASCARO
Anterior view

FIGURE 7.6 APR **Anterior View of the Skull**
(The names of the bones are in bold.)

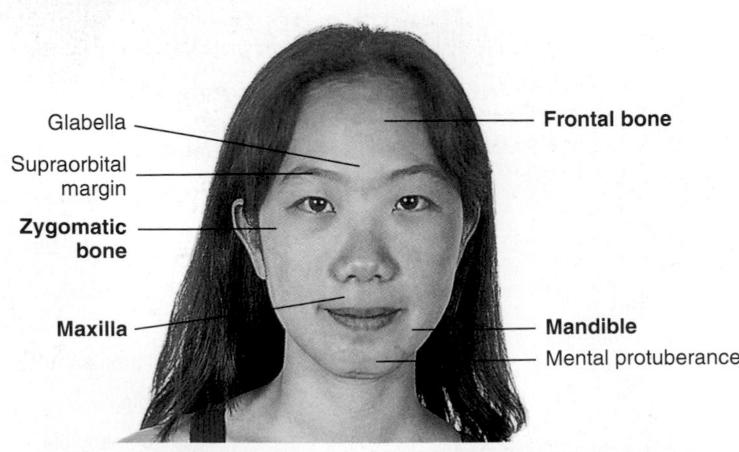
Glabella
Frontal bone
Supraorbital margin
Zygomatic bone
Maxilla
Mandible
Mental protuberance

FIGURE 7.7 Anterior View of Bony Landmarks on the Face
(The names of the bones are in bold.)

They are called orbits because the eyes rotate within the fossae. The bones of the orbits provide both protection for the eyes and attachment points for the muscles that move the eyes. The major portion of each eyeball is within the orbit, and the portion of the eye visible from the outside is relatively small. Each orbit contains blood vessels, nerves, and adipose tissue, as well as the eyeball and the muscles that move it. The bones forming the orbit are listed in table 7.4.

The orbit has several openings through which structures communicate between the orbit and other cavities. The nasolacrimal duct passes from the orbit into the nasal cavity through the **nasolacrimal canal,** carrying tears from the eyes to the nasal cavity. The optic nerve for vision passes from the eye through the **optic canal** at the posterior apex of the orbit and enters the cranial cavity. Superior and inferior fissures in the posterior region of the orbit provide openings through which nerves and blood vessels communicate with structures in the orbit or pass to the face.

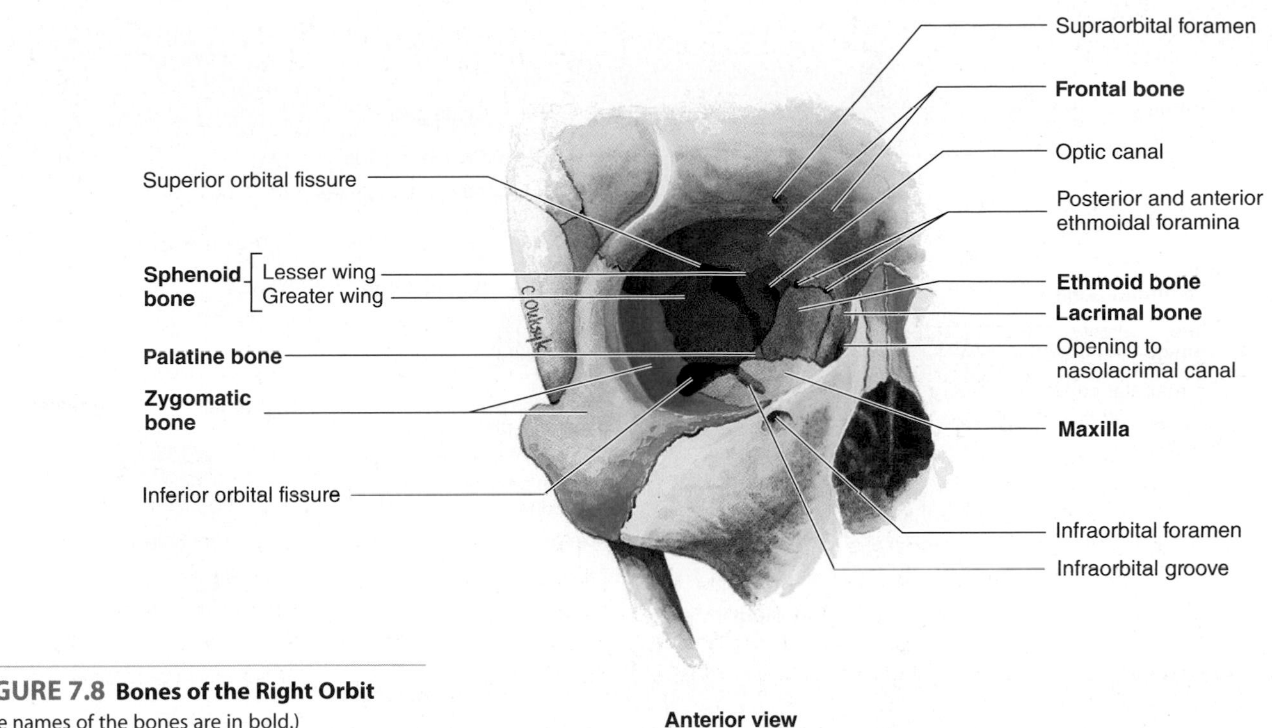

Supraorbital foramen
Frontal bone
Optic canal
Posterior and anterior ethmoidal foramina
Superior orbital fissure
Sphenoid bone
Lesser wing
Greater wing
Ethmoid bone
Lacrimal bone
Palatine bone
Opening to nasolacrimal canal
Zygomatic bone
Maxilla
Inferior orbital fissure
Infraorbital foramen
Infraorbital groove
Anterior view

FIGURE 7.8 Bones of the Right Orbit
(The names of the bones are in bold.)

The **nasal cavity** (table 7.5 and figure 7.9; see figure 7.6) has a pear-shaped opening anteriorly and is divided into right and left halves by a **nasal septum** (sep′tŭm; wall). The bony part of the nasal septum consists primarily of the vomer and the perpendicular plate of the ethmoid bone. Hyaline cartilage forms the anterior part of the nasal septum.

The external part of the nose, formed mostly of hyaline cartilage, is almost absent in the dried skeleton and is represented by the nasal bones and the frontal processes of the maxillary bones, which form the bridge of the nose.

Predict 2

A direct blow to the nose may result in a "broken nose." List at least three bones that may be broken.

TABLE 7.4 Bones Forming the Orbit (see figures 7.6 and 7.8)

Bone	Part of Orbit
Frontal	Roof
Sphenoid	Roof and posterolateral wall
Zygomatic	Lateral wall
Maxilla	Floor
Lacrimal	Medial wall
Ethmoid	Medial wall
Palatine	Medial wall

The lateral wall of the nasal cavity has three bony shelves, called the **nasal conchae** (kon′kē), which are directed inferiorly (figure 7.9). The inferior nasal concha is a separate bone, and the middle and superior nasal conchae are projections from the ethmoid bone. The conchae increase the surface area in the nasal cavity, thereby facilitating moistening of, removal of particles from, and warming of the air inhaled through the nose.

Several of the bones associated with the nasal cavity have large cavities within them called the **paranasal sinuses,** which open into the nasal cavity (figure 7.10). The sinuses decrease the weight of the skull and act as resonating chambers during voice production. Compare a normal voice with the voice of a person who has a cold and whose sinuses are "stopped up." The sinuses, which are

TABLE 7.5 Bones Forming the Nasal Cavity (see figures 7.6 and 7.9)

Bone	Part of Nasal Cavity
Frontal	Roof
Nasal	Roof
Sphenoid	Roof
Ethmoid	Roof, septum, lateral wall
Inferior nasal concha	Lateral wall
Lacrimal	Lateral wall
Maxilla	Floor
Palatine	Floor and lateral wall
Vomer	Septum

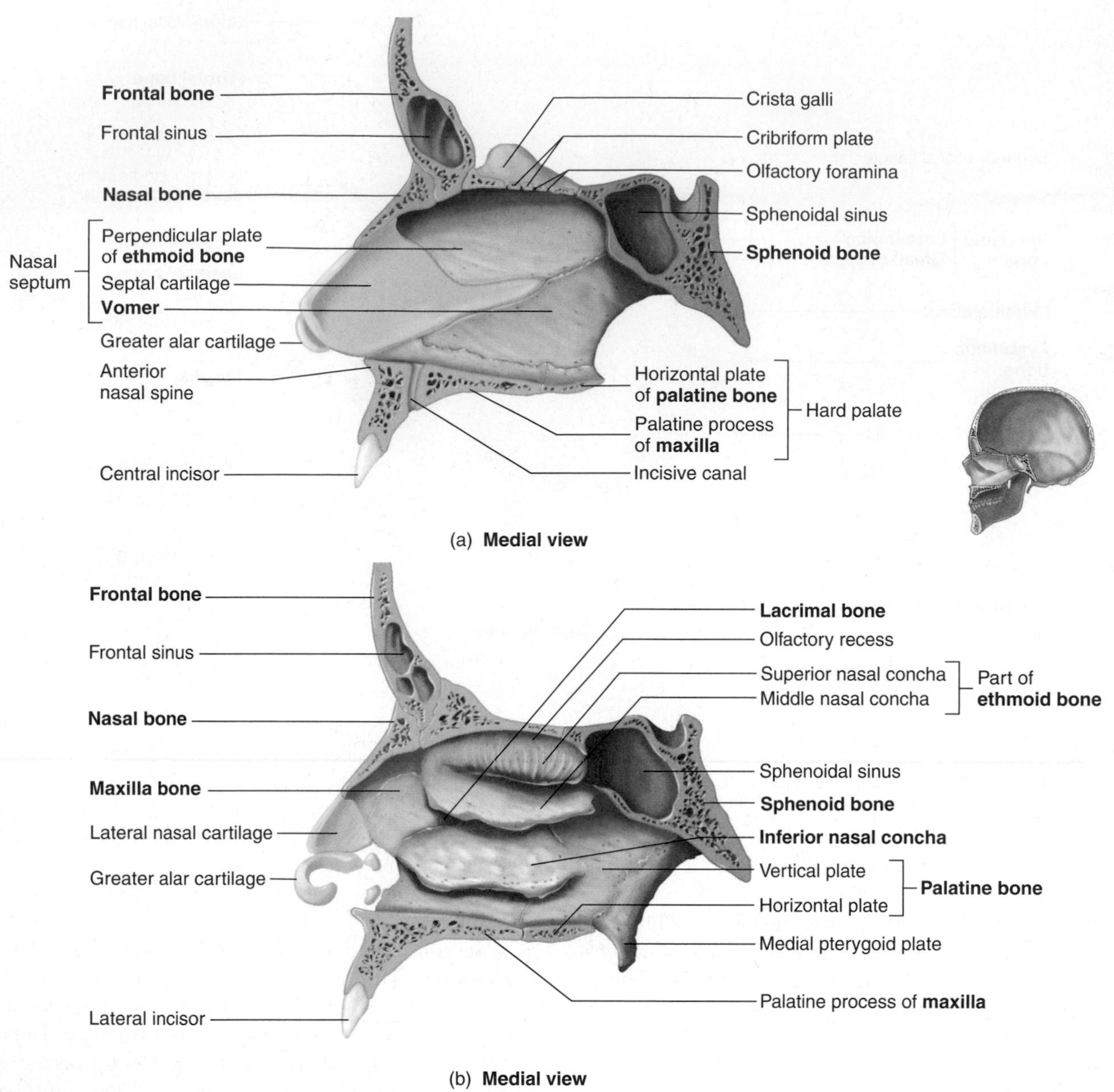
Frontal bone
Frontal sinus
Nasal bone
Nasal septum
Perpendicular plate of ethmoid bone
Septal cartilage
Vomer
Greater alar cartilage
Anterior nasal spine
Central incisor
Crista galli
Cribriform plate
Olfactory foramina
Sphenoidal sinus
Sphenoid bone
Horizontal plate of palatine bone
Palatine process of maxilla
Hard palate
Incisive canal
(a) Medial view
Frontal bone
Frontal sinus
Nasal bone
Maxilla bone
Lateral nasal cartilage
Greater alar cartilage
Lateral incisor
Lacrimal bone
Olfactory recess
Superior nasal concha
Middle nasal concha
Part of ethmoid bone
Sphenoidal sinus
Sphenoid bone
Inferior nasal concha
Vertical plate
Horizontal plate
Palatine bone
Medial pterygoid plate
Palatine process of maxilla
(b) Medial view

FIGURE 7.9 Bones of the Nasal Cavity

(*a*) Nasal septum as seen from the left nasal cavity. (*b*) Right lateral nasal wall as seen from inside the nasal cavity with the nasal septum removed. (The names of the bones are in bold.)

named for the bones in which they are located, include the frontal, maxillary, and sphenoidal sinuses. The cavities within each ethmoid bone form a maze of interconnected ethmoidal air cells collectively called the ethmoidal labyrinth, or the ethmoidal sinuses.

Interior of the Cranial Cavity

The **cranial cavity** is the skull cavity occupied by the brain. The cranial cavity can be exposed by cutting away the calvaria, the upper, domelike portion of the skull. Removing the calvaria reveals the floor of the cranial cavity (figure 7.11). That floor can be divided roughly into anterior, middle, and posterior cranial fossae, which are formed as the developing braincase conforms to the shape of the brain.

A prominent ridge, the **crista galli** (kris′tă găl′ē; rooster's comb), is located in the center of the anterior fossa. The crista galli is a point of attachment for one of the **meninges** (mĕ-nin′jēz), the dura mater, a thick connective tissue membrane that supports and protects the brain (see chapter 13). On each side of the crista galli is an olfactory fossa. An olfactory bulb rests in each fossa and receives the olfactory nerves for the sense of smell. The **cribriform** (krib′ri-fōrm; sievelike) **plate** of the ethmoid bone forms the floor of each olfactory fossa. The olfactory nerves extend from the cranial cavity into the roof of the nasal cavity through sievelike perforations in the cribriform plate called **olfactory foramina** (see figure 7.9*a* and chapter 15).

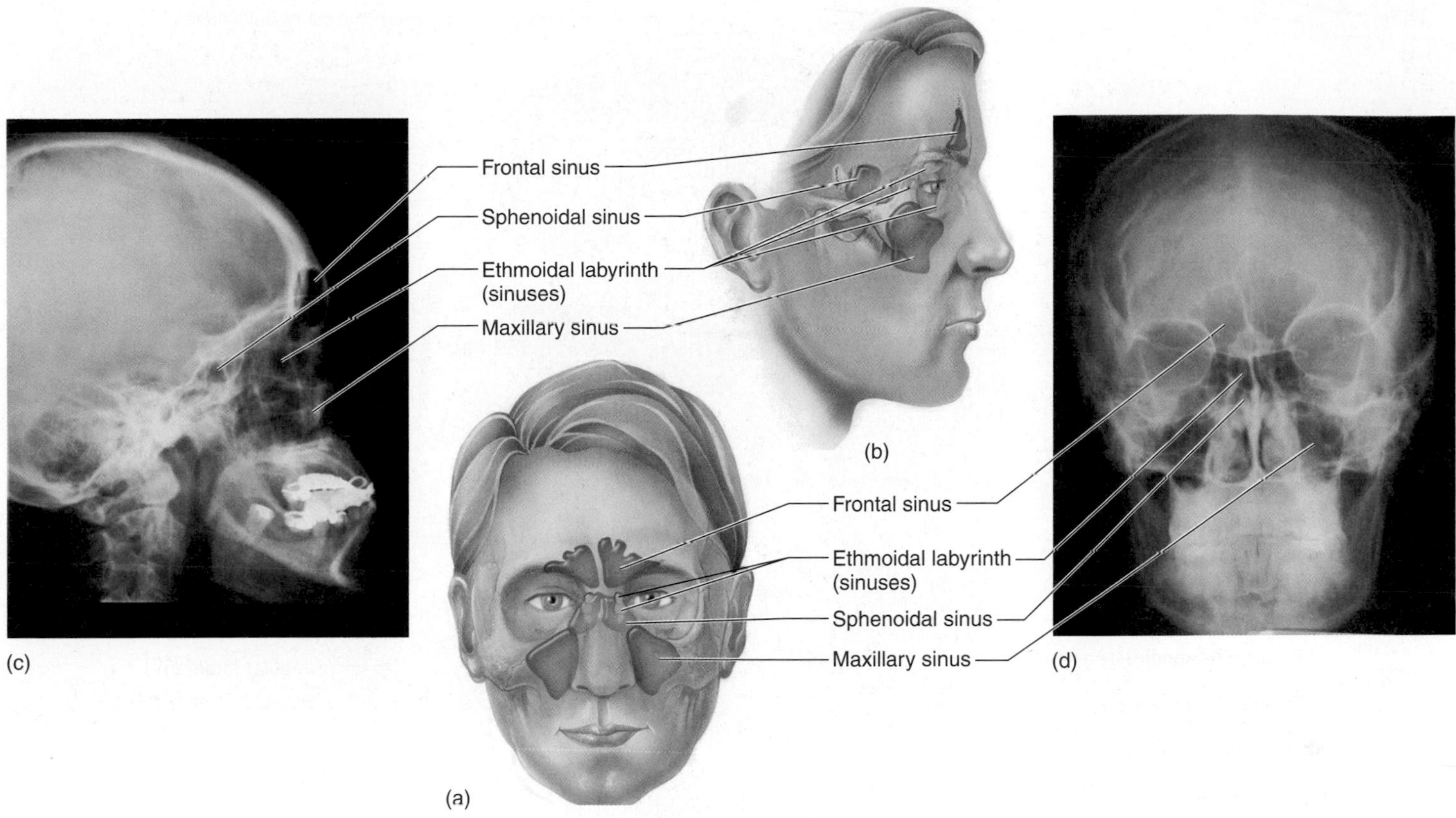
Frontal sinus
Sphenoidal sinus
Ethmoidal labyrinth (sinuses)
Maxillary sinus
(b)
Frontal sinus
Ethmoidal labyrinth (sinuses)
Sphenoidal sinus
Maxillary sinus
(c)
(d)
(a)

FIGURE 7.10 Paranasal Sinuses
(*a*) Anterior view. (*b*) Lateral view. (*c*) X-ray of the sinuses, lateral view. (*d*) X-ray of the sinuses, anterior view.

The body of the sphenoid bone forms a central prominence within the floor of the cranial cavity. This prominence is modified into a structure resembling a saddle, the **sella turcica** (sel'ă tŭr'si-kă; Turkish saddle), which is occupied by the pituitary gland. An optic canal is located on each side just anterior to the sella turcica. The lesser wings of the sphenoid bone form a ridge to each side of the optic canals. This ridge separates the **anterior cranial fossa** from the **middle cranial fossa.**

The **petrous portion** of the temporal bone extends posterolaterally from each side of the sella turcica. This thick, bony ridge (*petrous,* rocky) is hollow and contains the middle and inner ears. The petrous portion separates the middle cranial fossa from the **posterior cranial fossa.**

Several important openings in the floor of the middle cranial fossa are the **superior orbital fissure,** the **foramen rotundum,** the **foramen ovale,** the **foramen spinosum,** and the internal opening of the carotid canal (table 7.6). The **foramen lacerum** (lă-ser'um), in the floor of the carotid canal, is an artifact of the dried skull. In life, it is filled with cartilage.

The prominent **foramen magnum,** through which the brain is connected to the spinal cord, is in the posterior fossa. A **hypoglossal canal** is located on the anterolateral sides of the foramen magnum. **Jugular** (jŭg'ū-lar; throat) **foramina** lie on each side of the foramen magnum near the base of the petrous portion. An **internal auditory canal** is located about midway up the face of each petrous portion (table 7.6).

Inferior View of the Skull

Viewed from below with the mandible removed, the base of the skull is complex, with a number of foramina and specialized surfaces (figure 7.12). The foramen magnum passes through the occipital bone just slightly posterior to the center of the skull base. **Occipital condyles,** smooth points of articulation between the skull and the vertebral column, lie on the lateral and anterior margins of the foramen magnum.

The major entry and exit points for blood vessels that supply the brain can be seen from this view. Blood reaches the brain through the internal carotid arteries, which pass through the **carotid** (ka-rot'id; put to sleep) **canals,** and the vertebral arteries, which pass through the foramen magnum. An internal carotid artery enters the inferior opening of each carotid canal (see figure 7.11) and passes through the carotid canal, which runs anteromedially through the temporal bone. A thin plate of bone separates the carotid canal from the middle ear, allowing a person to hear his or her own heartbeat—for example, when frightened or after running. Most blood leaves the brain through the internal jugular veins, which exit through the jugular foramina located lateral to the occipital condyles.

Two long, pointed **styloid** (stī'loyd; stylus- or pen-shaped) **processes** project from the floor of the temporal bone (see figures 7.4 and 7.12). Three muscles involved in moving the tongue, hyoid bone, and pharynx attach to each process. The **mandibular fossa,** where the mandible articulates with the rest of the skull, is anterior to the mastoid process at the base of the zygomatic arch.

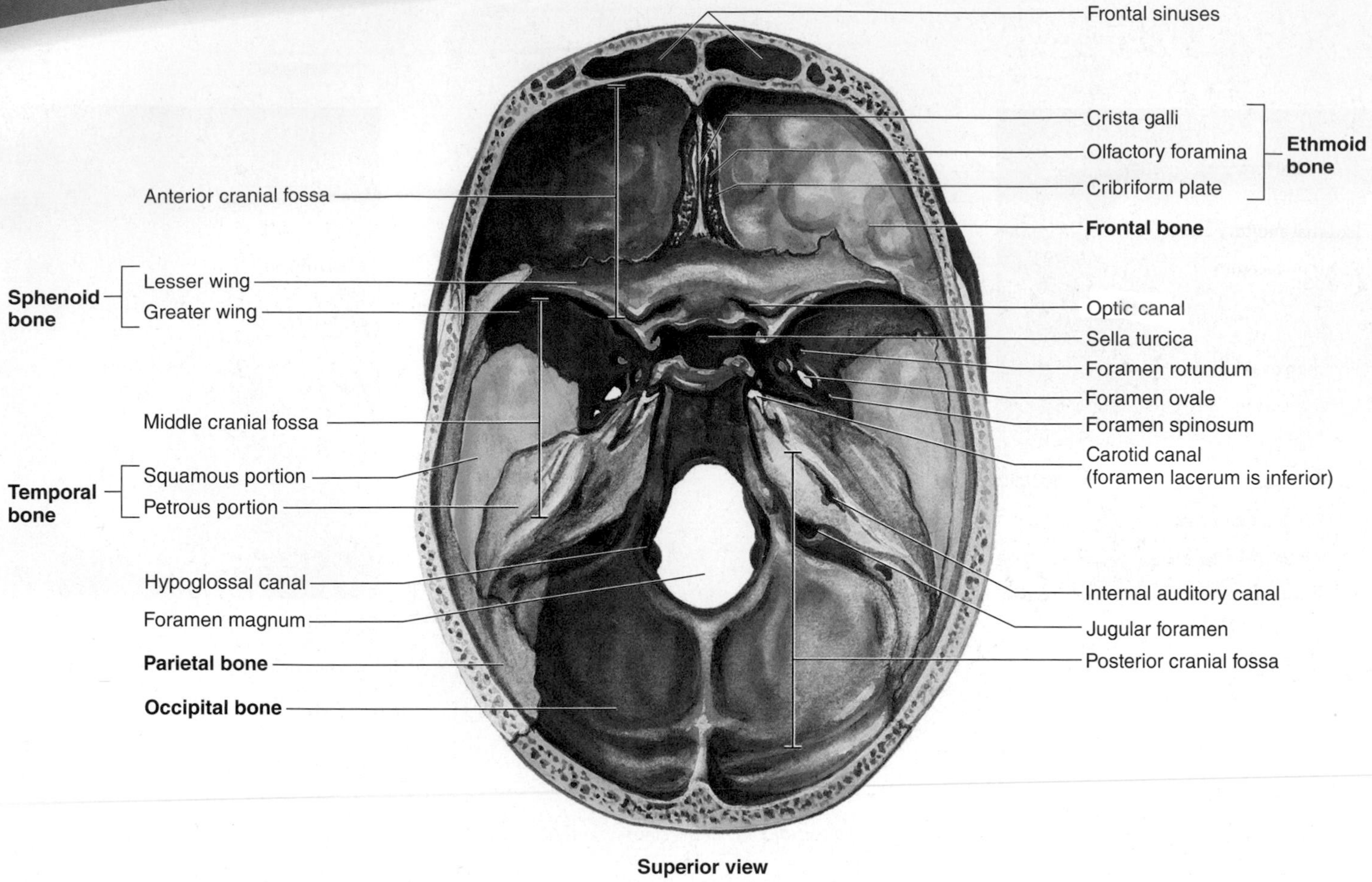

Frontal sinuses
Crista galli
Olfactory foramina
Cribriform plate
Ethmoid bone
Frontal bone
Anterior cranial fossa
Sphenoid bone
Lesser wing
Greater wing
Optic canal
Sella turcica
Foramen rotundum
Foramen ovale
Foramen spinosum
Carotid canal (foramen lacerum is inferior)
Middle cranial fossa
Temporal bone
Squamous portion
Petrous portion
Hypoglossal canal
Foramen magnum
Parietal bone
Occipital bone
Internal auditory canal
Jugular foramen
Posterior cranial fossa

Superior view

FIGURE 7.11 APIR **Floor of the Cranial Cavity**
The roof of the skull has been removed, and the floor is seen from a superior view. (The names of the bones are in bold.)

The posterior opening of the nasal cavity is bounded on each side by the vertical bony plates of the sphenoid bone: the **medial pterygoid** (ter′i-goyd; wing-shaped) **plate** and the **lateral pterygoid plate.** The medial and lateral pterygoid muscles, which help move the mandible, attach to the lateral plate (see chapter 10). The **vomer** forms most of the posterior portion of the nasal septum and can be seen between the medial pterygoid plates in the center of the nasal cavity.

The **hard palate,** or *bony palate,* forms the floor of the nasal cavity. Sutures join four bones to form the hard palate: The palatine processes of the two maxillary bones form the anterior two-thirds of the palate, and the horizontal plates of the two palatine bones form the posterior one-third of the palate. The tissues of the soft palate extend posteriorly from the hard palate. The hard and soft palates separate the nasal cavity from the mouth, enabling humans to chew and breathe at the same time.

Individual Bones of the Skull

The skull, or cranium, is composed of 22 separate bones (table 7.7; see table 7.1). In addition, the skull contains six **auditory ossicles,**

Clinical
IMPACT

Cleft Lip or Palate

During fetal development, the facial bones sometimes fail to fuse with one another. A **cleft lip** results if the maxillae do not form normally, and a **cleft palate** occurs when the palatine processes of the maxillae do not fuse with one another. A cleft palate produces an opening between the nasal and oral cavities, making it difficult to eat or drink or to speak distinctly. An artificial palate may be inserted into a newborn's mouth until the palate can be repaired. A cleft lip alone, or both cleft lip and palate, occurs approximately once in every 1000 births and is more common in males. A cleft palate alone occurs approximately once in every 2000 births and is more common in females.

TABLE 7.6 **Skull Foramina, Fissures, and Canals (see figures 7.11 and 7.12)**

Opening	Bone Containing the Opening	Structures Passing Through Openings
Carotid canal	Temporal	Carotid artery and carotid sympathetic nerve plexus
Ethmoidal foramina, anterior and posterior	Between frontal and ethmoid	Anterior and posterior ethmoidal nerves
External auditory canal	Temporal	Sound waves en route to the tympanic membrane
Foramen lacerum	Between temporal, occipital, and sphenoid	The foramen is filled with cartilage during life; the carotid canal and pterygoid canal cross its superior part but do not actually pass through it.
Foramen magnum	Occipital	Spinal cord, accessory nerves, and vertebral arteries
Foramen ovale	Sphenoid	Mandibular division of trigeminal nerve
Foramen rotundum	Sphenoid	Maxillary division of trigeminal nerve
Foramen spinosum	Sphenoid	Middle meningeal artery
Hypoglossal canal	Occipital	Hypoglossal nerve
Incisive foramen (canal)	Between maxillae	Incisive nerve
Inferior orbital fissure	Between sphenoid and maxilla	Infraorbital nerve and blood vessels and zygomatic nerve
Infraorbital foramen	Maxilla	Infraorbital nerve
Internal auditory canal	Temporal	Facial nerve and vestibulocochlear nerve
Jugular foramen	Between temporal and occipital	Internal jugular vein, glossopharyngeal nerve, vagus nerve, and accessory nerve
Mandibular foramen	Mandible	Inferior alveolar nerve to the mandibular teeth
Mental foramen	Mandible	Mental nerve
Nasolacrimal canal	Between lacrimal and maxilla	Nasolacrimal (tear) duct
Olfactory foramina	Ethmoid	Olfactory nerves
Optic canal	Sphenoid	Optic nerve and ophthalmic artery
Palatine foramina, anterior and posterior	Palatine	Palatine nerves
Pterygoid canal	Sphenoid	Sympathetic and parasympathetic nerves to the face
Sphenopalatine foramen	Between palatine and sphenoid	Nasopalatine nerve and sphenopalatine blood vessels
Stylomastoid foramen	Temporal	Facial nerve
Superior orbital fissures	Sphenoid	Oculomotor nerve, trochlear nerve, ophthalmic division of trigeminal nerve, abducens nerve, and ophthalmic veins
Supraorbital foramen or notch	Frontal	Supraorbital nerve and vessels
Zygomaticofacial foramen	Zygomatic	Zygomaticofacial nerve
Zygomaticotemporal foramen	Zygomatic	Zygomaticotemporal nerve

which function in hearing (see chapter 15). Each temporal bone holds one set of auditory ossicles, which consists of the malleus, incus, and stapes. These bones cannot be observed unless the temporal bones are cut open.

The 22 bones of the skull are divided into two portions: the braincase and the facial bones. The **braincase,** or *neurocranium,* consists of 8 bones that immediately surround and protect the brain. They include the paired parietal and temporal bones and the unpaired frontal, occipital, sphenoid, and ethmoid bones.

The 14 **facial bones,** or *viscerocranium,* form the structure of the face in the anterior skull. They are the maxilla (2), zygomatic (2), palatine (2), lacrimal (2), nasal (2), inferior nasal concha (2), mandible (1), and vomer (1) bones. The frontal and ethmoid bones, which are part of the braincase, also contribute to the face. The mandible is often listed as a facial bone, even though it is not part of the intact skull.

The facial bones protect the major sensory organs located in the face: the eyes, nose, and tongue. The bones of the face also provide attachment points for the muscles involved in **mastication** (mas-ti-kā′shŭn; chewing), facial expression, and eye movement. The jaws (mandible and maxillae) possess **alveolar** (al-vē′ō-lăr) **processes** with sockets for the attachment of the teeth. The bones of the face and their associated soft tissues determine the unique facial features of each individual.

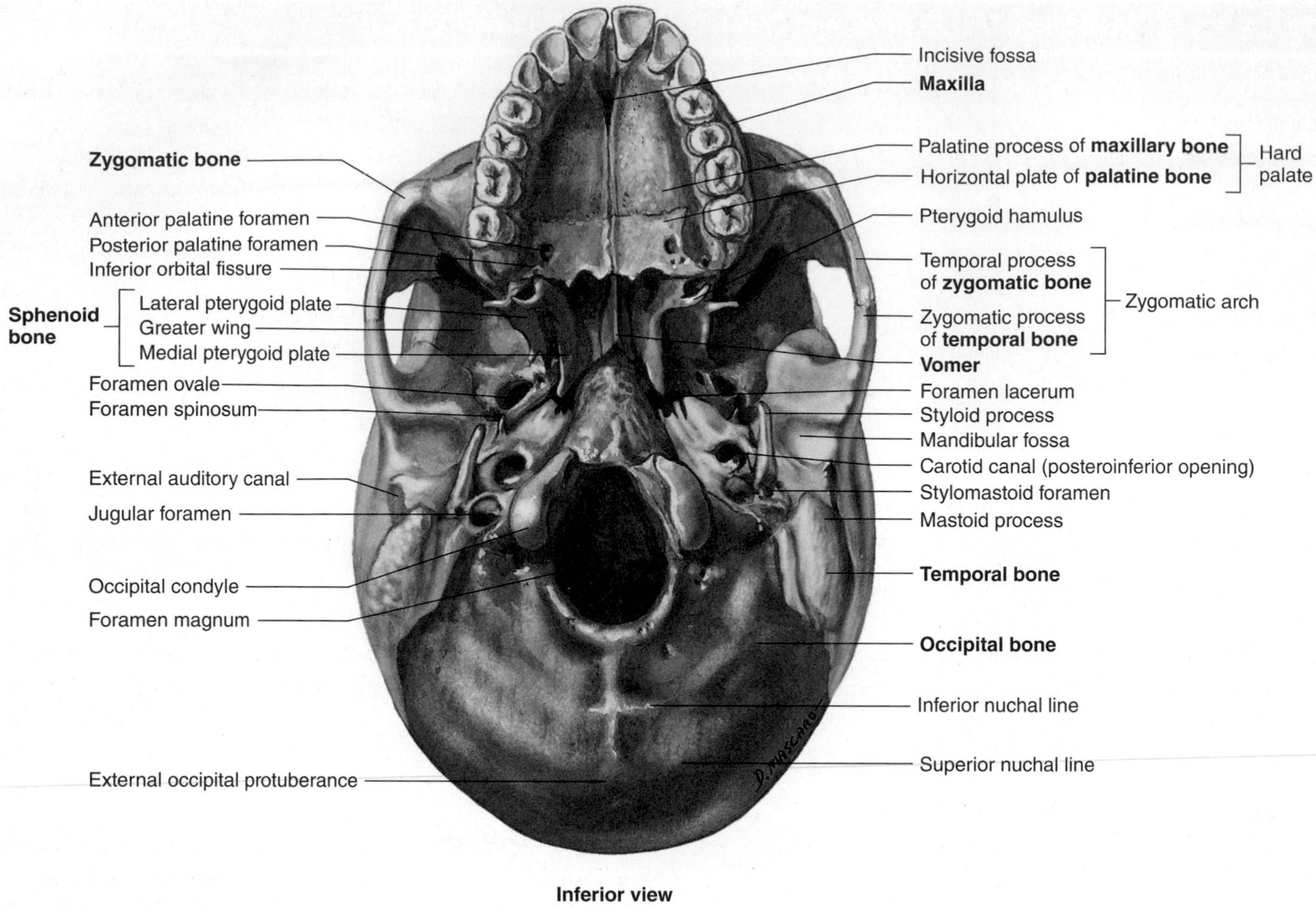

Incisive fossa
Maxilla
Zygomatic bone
Palatine process of **maxillary bone**
Horizontal plate of **palatine bone**
Hard palate
Anterior palatine foramen
Posterior palatine foramen
Inferior orbital fissure
Pterygoid hamulus
Temporal process of **zygomatic bone**
Zygomatic process of **temporal bone**
Zygomatic arch
Sphenoid bone
Lateral pterygoid plate
Greater wing
Medial pterygoid plate
Vomer
Foramen ovale
Foramen spinosum
Foramen lacerum
Styloid process
Mandibular fossa
Carotid canal (posteroinferior opening)
External auditory canal
Stylomastoid foramen
Jugular foramen
Mastoid process
Temporal bone
Occipital condyle
Foramen magnum
Occipital bone
Inferior nuchal line
Superior nuchal line
External occipital protuberance
D. Mascaro

Inferior view

FIGURE 7.12 Inferior View of the Skull
The mandible has been removed. (The names of the bones are in bold.)

ASSESS YOUR PROGRESS

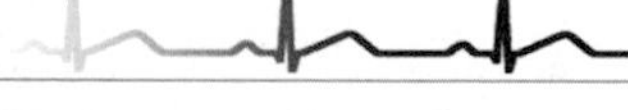

4. *What are the parts and general functions of the axial skeleton?*
5. *Name the four major sutures of the skull and the bones they connect.*
6. *List the seven bones that form the orbit of the eye.*
7. *What is a sinus? What are the functions of sinuses? Give the locations of the paranasal sinuses.*
8. *Name the bones and cartilage that compose the nasal septum.*
9. *What bones form the hard palate, and what is the function of the hard palate?*
10. *What structure allows the brainstem to connect to the spinal cord?*
11. *Name the foramina that allow the passage of the following nerves and blood vessels: optic nerve, olfactory nerve, vestibulocochlear nerve, incisive nerve, facial nerve, carotid artery, and internal jugular vein.*
12. *What structure allows sound waves to reach the tympanic membrane?*
13. *List the bones that make up the floor of the braincase.*
14. *State the bone features where the following muscles attach to the skull: neck muscles, throat muscles, muscles of mastication, muscles of facial expression, and muscles that move the eyeballs.*
15. *Name the bones of the braincase and face. What are the functions accomplished by each group?*

Hyoid Bone

The **hyoid bone** (table 7.8), which is unpaired, is often listed as part of the facial bones because it has a common developmental origin with them. It is not, however, part of the adult skull (see table 7.1). The hyoid bone has no direct bony attachment to the skull. Instead, muscles and ligaments attach it to the skull, so the hyoid "floats" in the superior aspect of the neck just below the mandible. The hyoid bone provides an attachment point for some tongue muscles and for important neck muscles that elevate the larynx during speech or swallowing.

TABLE 7.7 Skull Bones

(a) Parietal Bone (Right)—Lateral View

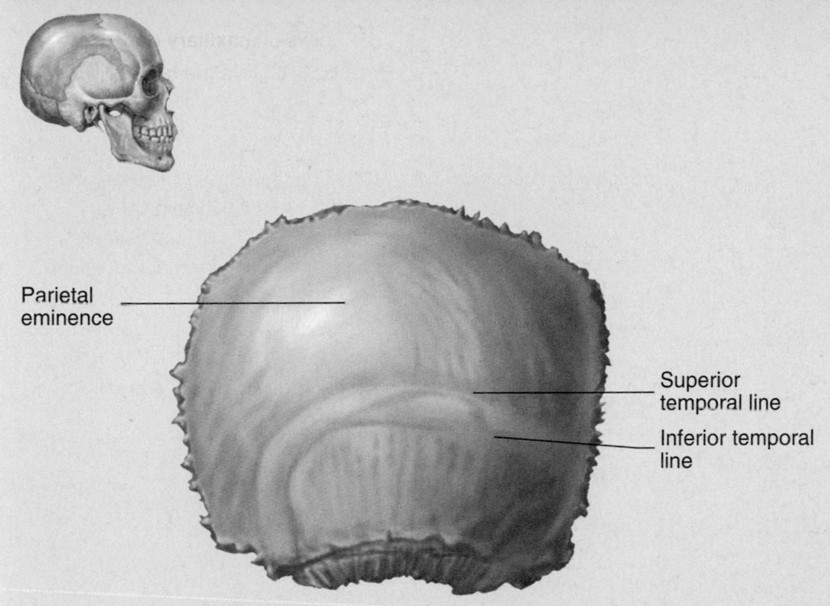
Parietal eminence
Superior temporal line
Inferior temporal line

Landmark	Description
Parietal eminence	The widest part of the head is from one parietal eminence to the other.
Superior and inferior temporal lines	Attachment point for temporalis muscle

Special Feature

Forms lateral wall of skull

(b) Temporal Bone (Right)—Lateral and Medial Views

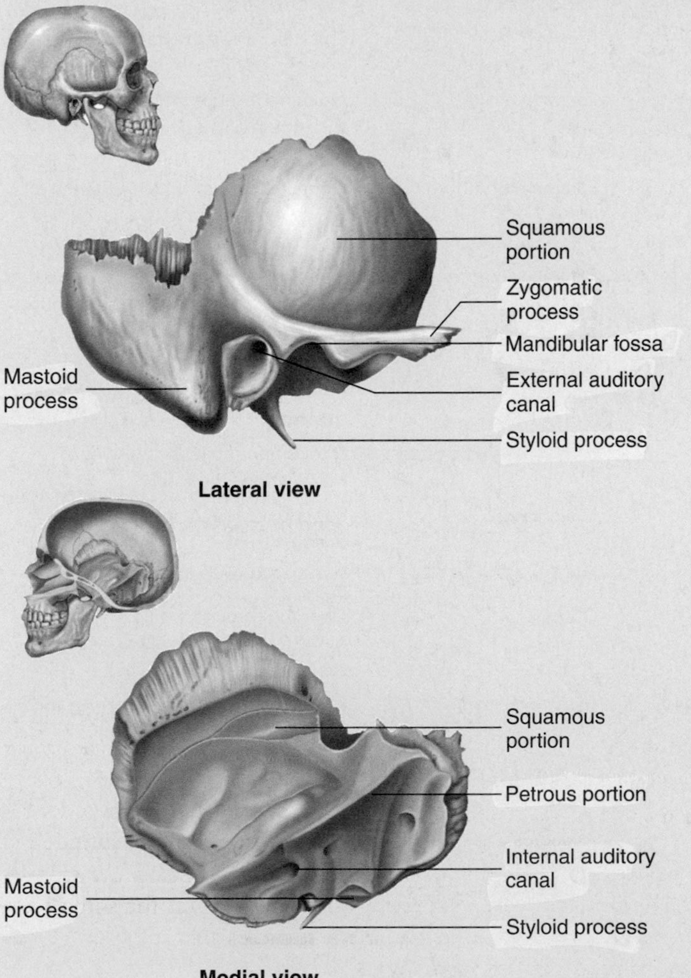
Squamous portion
Zygomatic process
Mandibular fossa
Mastoid process
External auditory canal
Styloid process
Lateral view
Squamous portion
Petrous portion
Internal auditory canal
Mastoid process
Styloid process
Medial view

Landmark	Description
Carotid canal (shown in figures 7.11 and 7.12)	Canal through which the internal carotid artery enters the cranial cavity
External auditory canal	External canal of the ear; carries sound to the ear
Internal auditory canal (shown in figure 7.11)	Opening through which the facial (cranial nerve VII) and vestibulocochlear (cranial nerve VIII) nerves enter the petrous portion of the temporal bone
Forms one side of jugular foramen (shown in figures 7.11 and 7.12)	Foramen through which the internal jugular vein exits the cranial cavity
Mandibular fossa	Articulation point between the mandible and skull
Mastoid process	Attachment point for muscles moving the head and for a hyoid muscle
Middle cranial fossa (shown in figure 7.11)	Depression in the floor of the cranial cavity formed by the temporal lobes of the brain
Petrous portion (shown in figure 7.11)	Thick portion of the temporal bone
Squamous portion (shown in figure 7.11)	Flat, lateral portion of the temporal bone
Styloid process	Attachment for muscles of the tongue, throat, and hyoid bone
Stylomastoid foramen (shown in figure 7.12)	Foramen through which the facial nerve (cranial nerve VII) exits the skull
Zygomatic process	Helps form the bony bridge extending from the cheek to just anterior to the ear; attachment for a muscle that moves the mandible

Special Features

Contains the middle and inner ear and the mastoid air cells

Place where the mandible articulates with the rest of the skull

TABLE 7.7 Skull Bones—Continued

(c) Frontal Bone—Anterior View

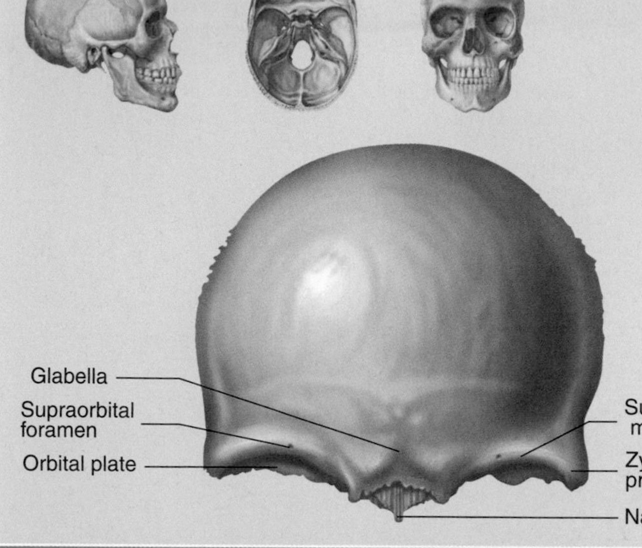
Glabella
Supraorbital foramen
Orbital plate
Supraorbital margin
Zygomatic process
Nasal spine

Landmark	Description
Glabella	Area between the supraorbital margins
Nasal spine	Superior part of the nasal bridge
Orbital plate	Roof of the orbit
Supraorbital foramen	Opening through which nerves and vessels exit the skull to the skin of the forehead
Supraorbital margin	Ridge forming the anterior superior border of the orbit
Zygomatic process	Connects to the zygomatic bone; helps form the lateral margin of the orbit

Special Features

Forms the forehead and roof of the orbit

Contains the frontal sinus

(d) Sphenoid Bone—Superior and Posterior Views

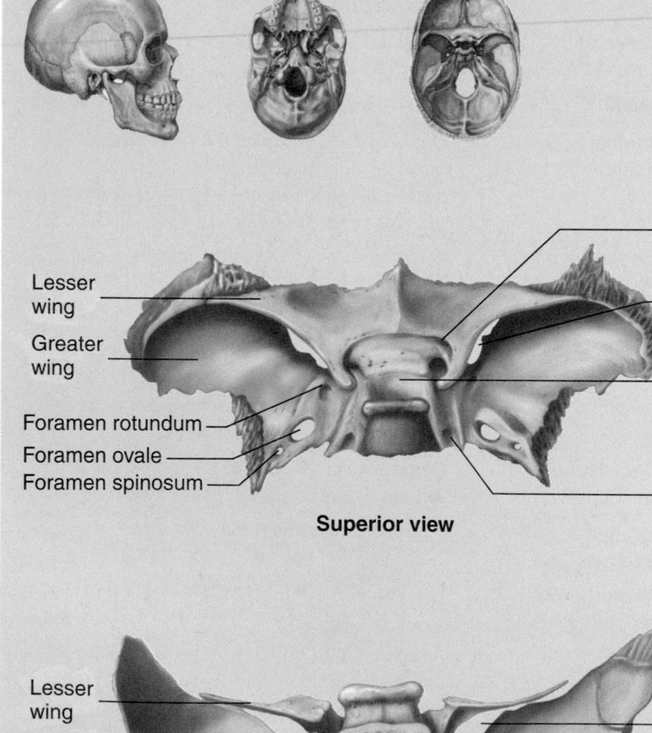
Optic canal
Lesser wing
Superior orbital fissure
Greater wing
Sella turcica
Foramen rotundum
Foramen ovale
Foramen spinosum
Groove of carotid canal
Superior view
Lesser wing
Greater wing
Superior orbital fissure
Body
Foramen rotundum
Pterygoid canal
Lateral pterygoid plate
Pterygoid hamulus
Medial pterygoid plate
Posterior view

Landmark	Description
Body	Thickest part of the bone; articulates with the occipital bone
Foramen ovale	Opening through which a branch of the trigeminal nerve (cranial nerve V) exits the cranial cavity
Foramen rotundum	Opening through which a branch of the trigeminal nerve (cranial nerve V) exits the cranial cavity
Foramen spinosum	Opening through which a major artery to the meninges (membranes around the brain) enters the cranial cavity
Greater wing	Forms the floor of the middle cranial fossa; several foramina pass through this wing
Lateral pterygoid plate	Attachment point for muscles of mastication (chewing)
Lesser wing	Superior border of the superior orbital fissure
Medial pterygoid plate	Posterolateral walls of the nasal cavity
Optic canal	Opening through which the optic nerve (cranial nerve II) passes from the orbit to the cranial cavity
Pterygoid canal	Opening through which nerves and vessels exit the cranial cavity
Pterygoid hamulus	Process around which the tendon passes from a muscle to the soft palate
Sella turcica	Fossa containing the pituitary gland
Superior orbital fissure	Opening through which nerves and vessels enter the orbit from the cranial cavity

Special Feature

Contains the sphenoidal sinus

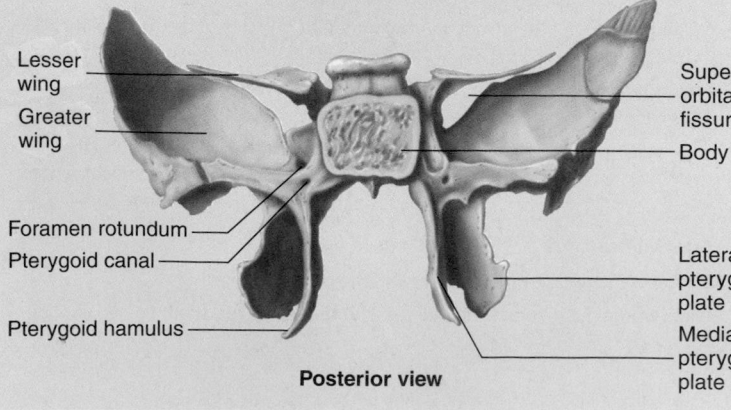

TABLE 7.7 Skull Bones—Continued

(e) Occipital Bone—Inferior View

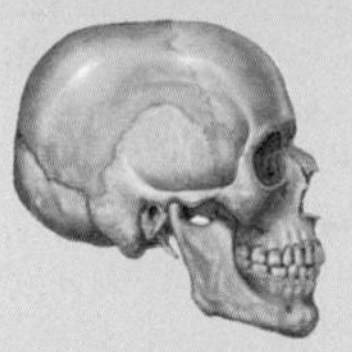
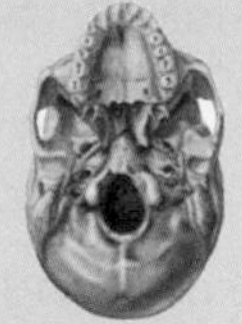
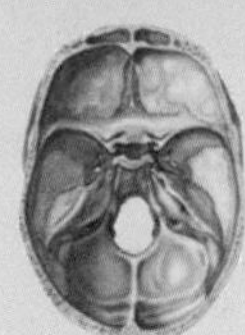
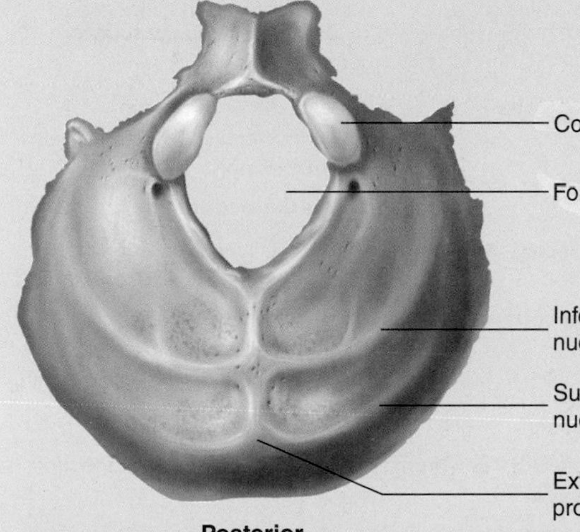

Anterior
Condyle
Foramen magnum
Inferior nuchal line
Superior nuchal line
External occipital protuberance
Posterior

Landmark	Description
Condyle	Articulation point between the skull and first vertebra
External occipital protuberance	Attachment point for a strong ligament (nuchal ligament) in the back of the neck
Foramen magnum	Opening around the point where the brain and spinal cord connect
Hypoglossal canal (shown in figure 7.11)	Opening through which the hypoglossal nerve (cranial nerve XII) passes
Inferior nuchal line	Attachment point for neck muscles
Posterior cranial fossa (shown in figure 7.11)	Depression in the posterior of the cranial cavity formed by the cerebellum
Superior nuchal line	Attachment point for neck muscles

Special Feature

Forms the base of the skull

(f) Zygomatic Bone (Right)—Lateral View

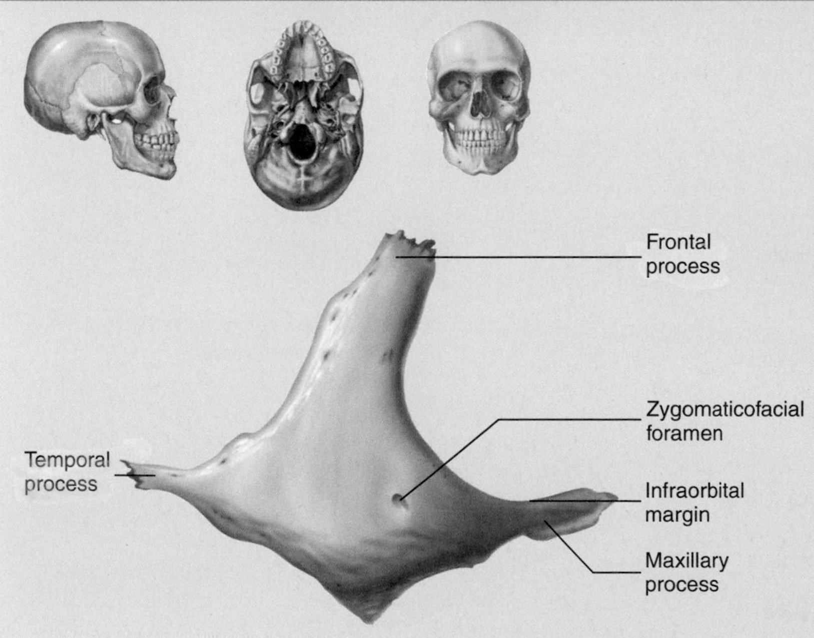

Frontal process
Zygomaticofacial foramen
Temporal process
Infraorbital margin
Maxillary process

Landmark	Description
Frontal process	Connection to the frontal bone; helps form the lateral margin of the orbit
Infraorbital margin	Ridge forming the inferior border of the orbit
Temporal process	Helps form the bony bridge from the cheek to just anterior to the ear
Zygomaticofacial foramen	Opening through which a nerve and vessels exit the orbit to the face

Special Features

Forms the prominence of the cheek

Forms the anterolateral wall of the orbit

TABLE 7.7 Skull Bones—Continued

(g) Ethmoid Bone—Superior, Lateral, and Anterior Views

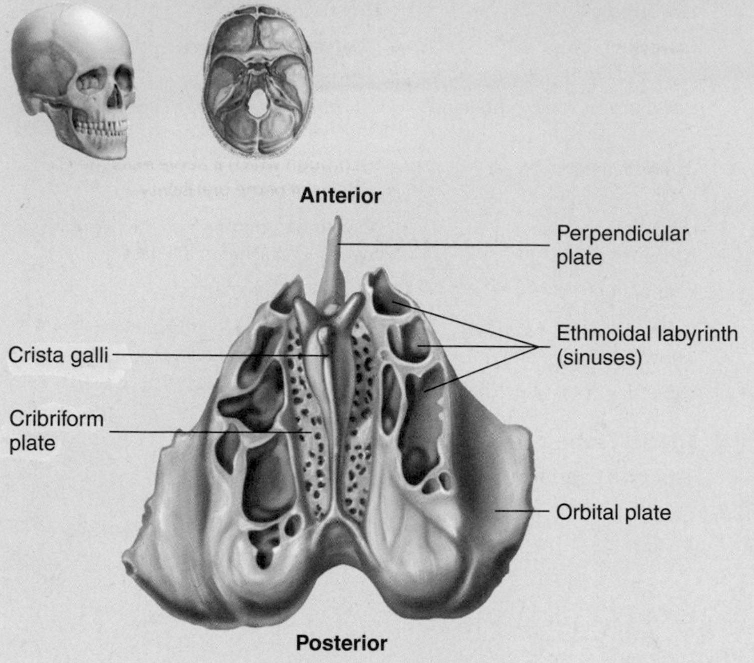
Anterior
Perpendicular plate
Ethmoidal labyrinth (sinuses)
Crista galli
Cribriform plate
Orbital plate
Posterior

Superior view

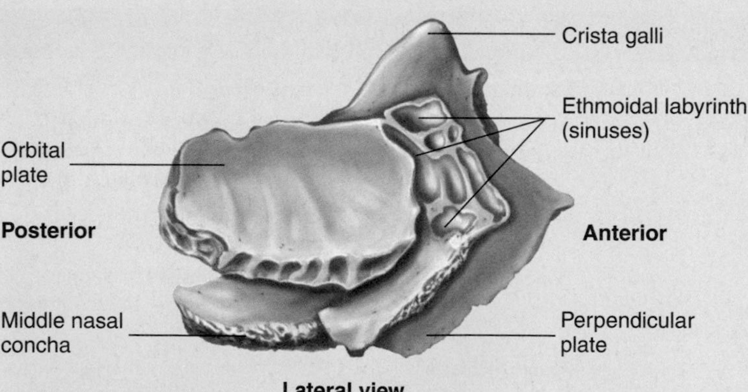
Crista galli
Ethmoidal labyrinth (sinuses)
Orbital plate
Posterior
Anterior
Middle nasal concha
Perpendicular plate

Lateral view

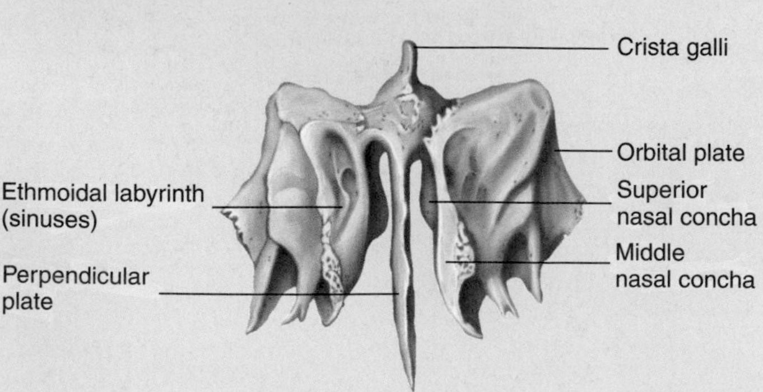
Crista galli
Orbital plate
Ethmoidal labyrinth (sinuses)
Superior nasal concha
Middle nasal concha
Perpendicular plate

Anterior view

Landmark	Description
Cribriform plate	Contains numerous olfactory foramina through which branches of the olfactory nerve (cranial nerve I) enter the cranial cavity from the nasal cavity
Crista galli	Attachment for meninges (membranes around brain)
Ethmoidal foramina (shown in figure 7.8)	Openings through which nerves and vessels pass from the orbit to the nasal cavity
Middle nasal concha	Ridge extending into the nasal cavity; increases surface area, helps warm and moisten air in the cavity
Orbital plate	Forms the medial wall of the orbit
Perpendicular plate	Forms the superior portion of the nasal septum
Superior nasal concha	Ridge extending into the nasal cavity; increases surface area, helps warm and moisten air in the cavity

Special Features

Forms part of the nasal septum and part of the lateral walls and roof of the nasal cavity

Contains the ethmoidal labyrinth, or ethmoidal sinuses; the labyrinth is divided into anterior, middle, and posterior ethmoidal cells

TABLE 7.7 Skull Bones—Continued

(h) Maxilla (Right)—Medial and Lateral Views

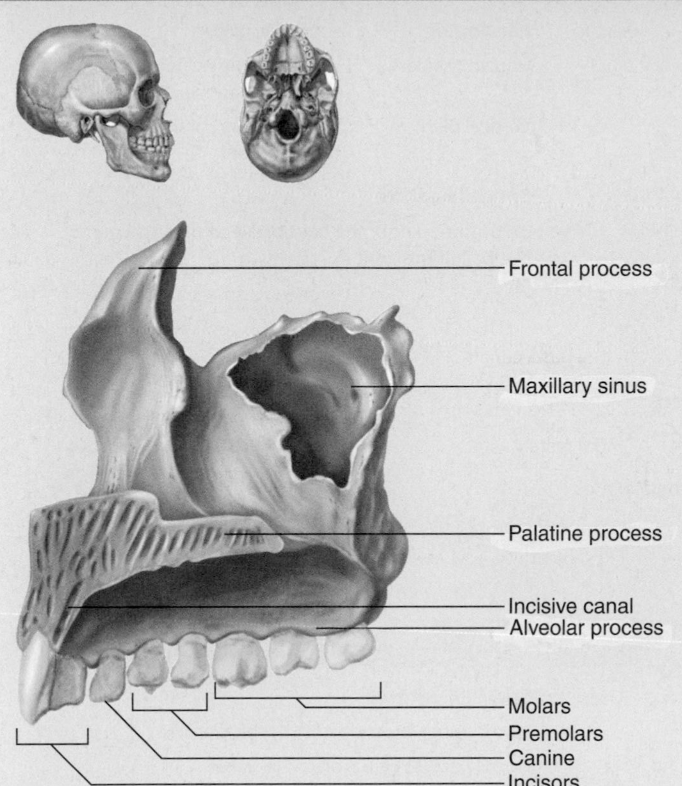
Frontal process
Maxillary sinus
Palatine process
Incisive canal
Alveolar process
Molars
Premolars
Canine
Incisors

Medial view

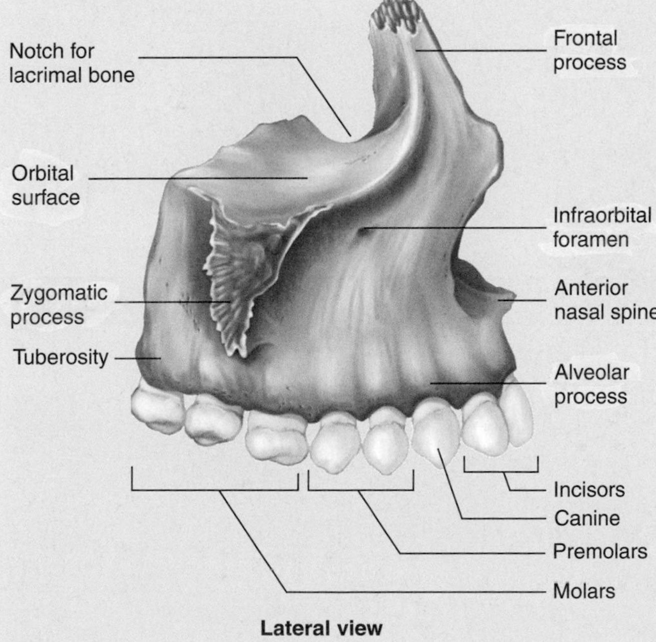
Notch for lacrimal bone
Frontal process
Orbital surface
Infraorbital foramen
Zygomatic process
Anterior nasal spine
Tuberosity
Alveolar process
Incisors
Canine
Premolars
Molars

Lateral view

Landmark	Description
Alveolar process	Ridge containing the teeth
Anterior nasal spine	Forms part of the nasal septum
Frontal process	Forms the sides of the nasal bridge
Incisive canal	Opening through which a nerve exits the nasal cavity to the roof of the oral cavity
Infraorbital foramen	Opening through which a nerve and vessels exit the orbit to the face
Orbital surface	Forms the floor of the orbit
Palatine process	Forms the anterior two-thirds of the hard palate
Maxillary tuberosity	Lump posterior to the last maxillary molar tooth
Zygomatic process	Connection to the zygomatic bone; helps form the interior margin of the orbit

Special Features

Contains the maxillary sinus and maxillary teeth

Forms part of nasolacrimal canal

TABLE 7.7 Skull Bones—Continued

(i) Palatine Bone (Right)—Medial and Anterior Views

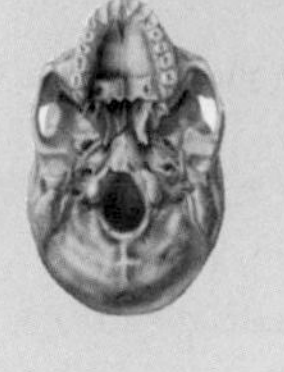

Landmark	Description
Horizontal plate	Forms the posterior one-third of the hard palate
Vertical plate	Forms part of the lateral nasal wall

Special Feature

Helps form part of the hard palate and a small part of the wall of the orbit

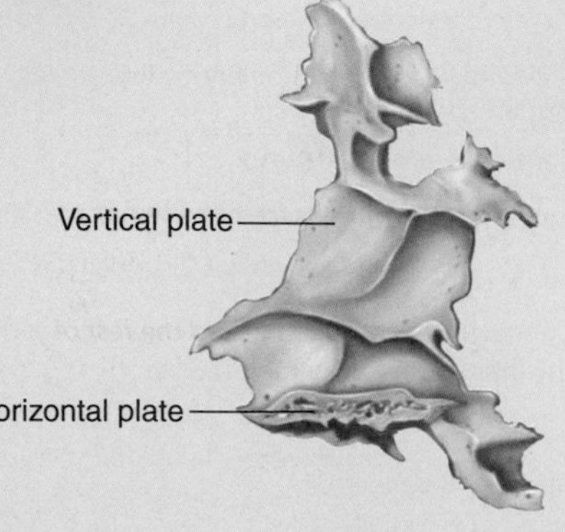
Vertical plate
Horizontal plate

Medial view

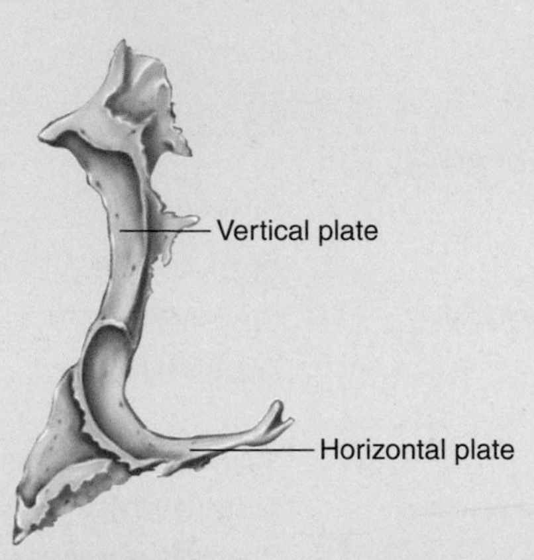
Vertical plate
Horizontal plate

Anterior view

(j) Lacrimal Bone (Right)—Anterolateral View

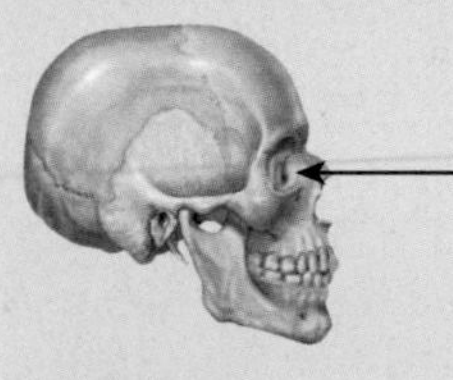

Special Features

Forms a small portion of the orbital wall

Forms part of the nasolacrimal canal

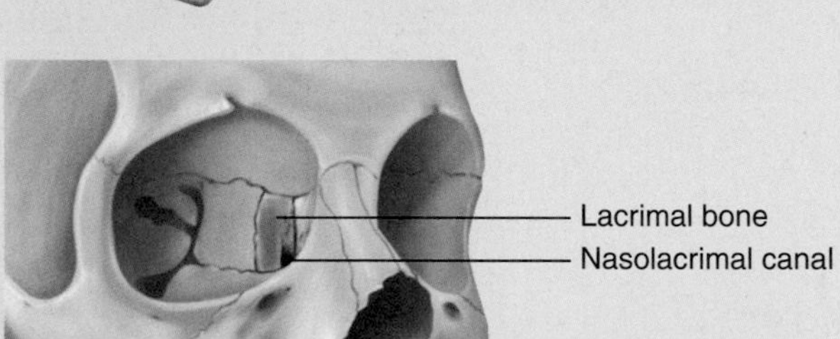
Lacrimal bone
Nasolacrimal canal

(k) Nasal Bone (Right)—Anterolateral View

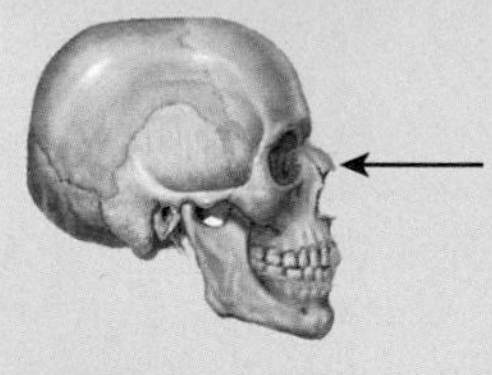

Special Feature

Forms the bridge of the nose

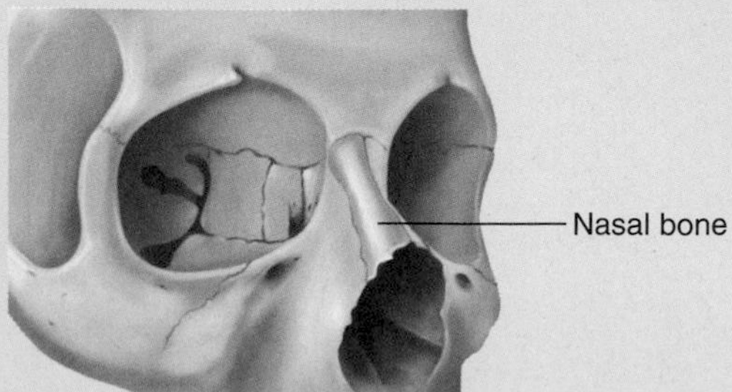
Nasal bone

TABLE 7.7 Skull Bones—Continued

(l) Mandible (Right Half)—Medial and Lateral Views

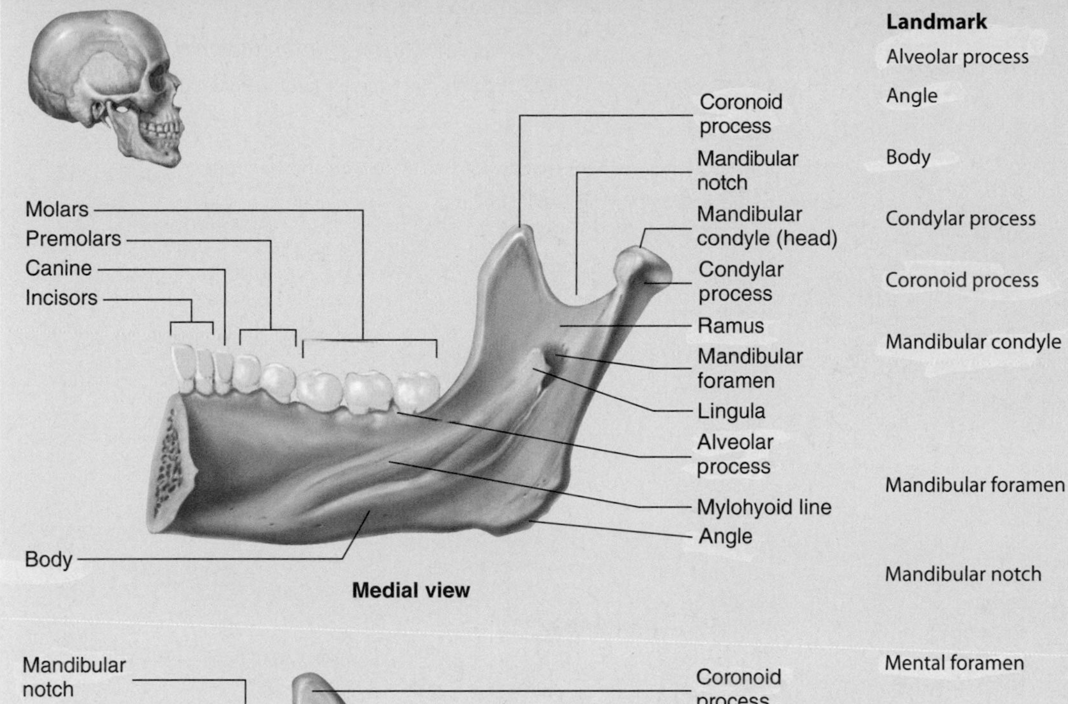
Coronoid process
Mandibular notch
Mandibular condyle (head)
Condylar process
Ramus
Mandibular foramen
Lingula
Alveolar process
Mylohyoid line
Angle
Molars
Premolars
Canine
Incisors
Body

Medial view

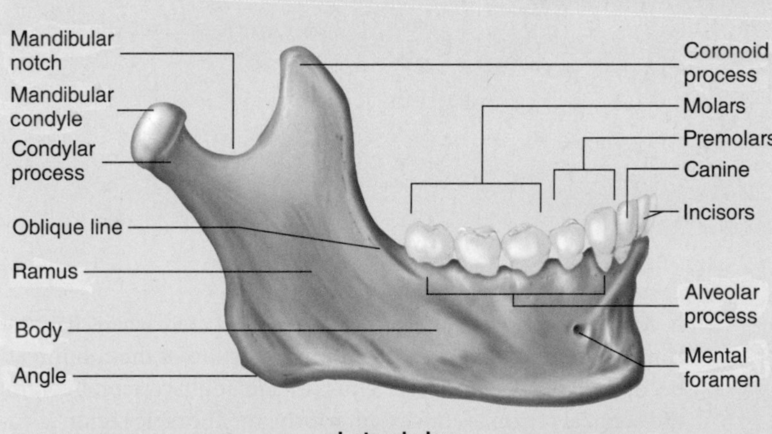
Mandibular notch
Mandibular condyle
Condylar process
Oblique line
Ramus
Body
Angle
Coronoid process
Molars
Premolars
Canine
Incisors
Alveolar process
Mental foramen

Lateral view

Landmark	Description
Alveolar process	Ridge containing the teeth
Angle	Corner between the body and ramus
Body	Major, horizontal portion of the bone
Condylar process	Extension containing the mandibular condyle
Coronoid process	Attachment for a muscle of mastication
Mandibular condyle	Helps form the temporo-mandibular joint (the point of articulation between the mandible and the rest of the skull)
Mandibular foramen	Opening through which nerves and vessels to the mandibular teeth enter the bone
Mandibular notch	Depression between the condylar process and the coronoid process
Mental foramen	Opening through which a nerve and vessels exit the mandible to the skin of the chin
Mylohyoid line	Attachment point of the mylohyoid muscle
Oblique line	Ridge from the anterior edge of the ramus onto the body of the mandible
Ramus	Major, nearly vertical portion of the bone

Special Features

The only bone in the skull that is freely movable relative to the rest of the skull bones

Holds the lower teeth

(m) Vomer—Anterior and Lateral Views

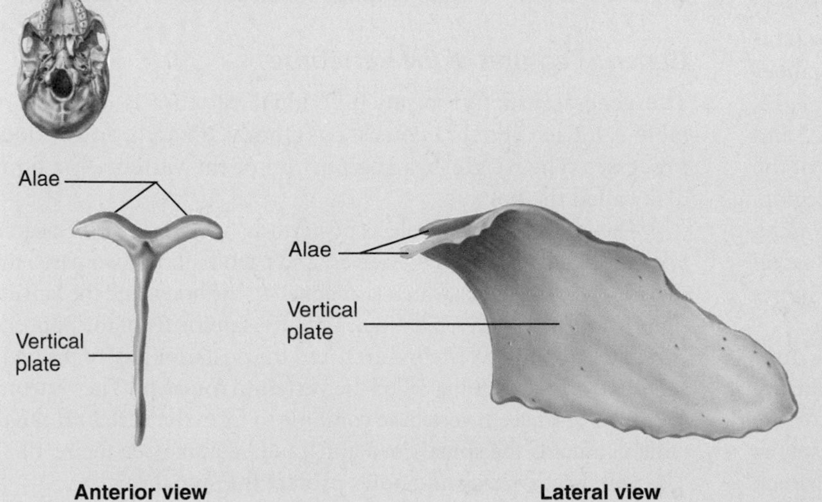
Alae
Vertical plate
Alae
Vertical plate

Anterior view

Lateral view

Landmark	Description
Alae	Attachment points between the vomer and sphenoid
Vertical plate	Forms part of the nasal septum

Special Feature

Forms most of the posterior and inferior portions of the nasal septum

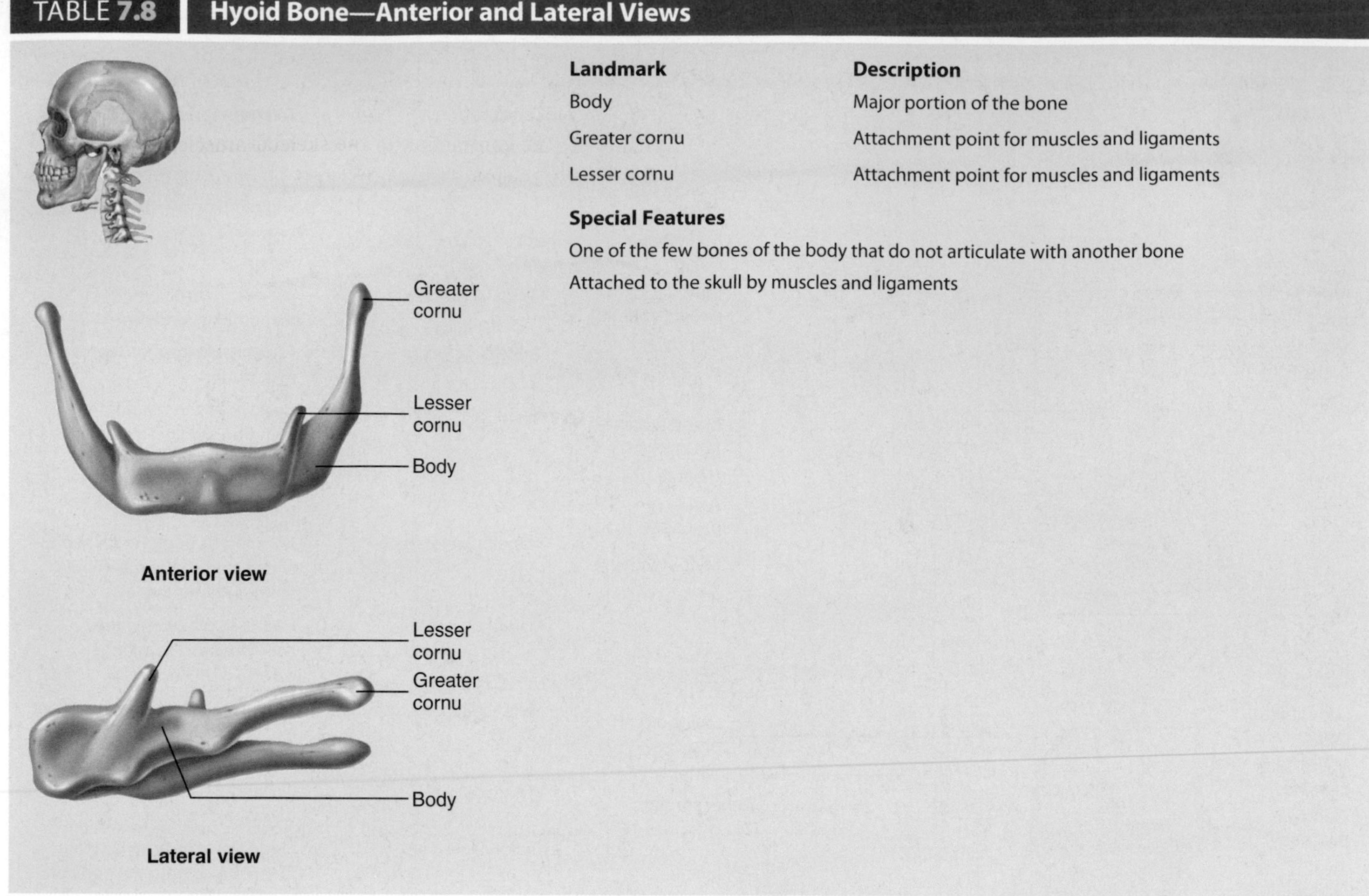
Greater
cornu
Lesser
cornu
Body
Anterior view
Lesser
cornu
Greater
cornu
Body
Lateral view

TABLE 7.8 Hyoid Bone—Anterior and Lateral Views

Landmark	Description
Body	Major portion of the bone
Greater cornu	Attachment point for muscles and ligaments
Lesser cornu	Attachment point for muscles and ligaments

Special Features

One of the few bones of the body that do not articulate with another bone

Attached to the skull by muscles and ligaments

Vertebral Column

The **vertebral column** performs five major functions: (1) It supports the weight of the head and trunk, (2) it protects the spinal cord, (3) it allows spinal nerves to exit the spinal cord, (4) it provides a site for muscle attachment, and (5) it permits movement of the head and trunk.

The vertebral column usually consists of 26 bones, called **vertebrae,** which can be divided into five regions: 7 **cervical vertebrae** (ver′tĕ-brē), 12 **thoracic vertebrae,** 5 **lumbar vertebrae,** 1 **sacral bone,** and 1 **coccygeal** (kok-sij′ē-ăl) **bone** (figure 7.13). To remember how many vertebrae are in each region, think of mealtimes: 7, 12, and 5. The cervical vertebrae are designated "C," thoracic "T," and lumbar "L." A number after the letter indicates the number of the vertebra, from superior to inferior, within each vertebral region. For example, "C1" refers to the first cervical vertebra. The developing embryo has about 33 or 34 vertebrae, but by adulthood the 5 sacral vertebrae have fused to form 1 bone, and the 4 or 5 coccygeal bones usually have fused to form 1 bone.

The five regions of the adult vertebral column have four major curvatures (figure 7.13). Two of the curves appear during embryonic development and reflect the C-shaped curve of the embryo and fetus within the uterus. When the infant raises its head in the first few months after birth, a secondary curve, which is convex anteriorly, develops in the neck. Later, when the infant learns to sit and then walk, the lumbar portion of the column also becomes convex anteriorly. Thus, in the adult vertebral column, the cervical region is convex anteriorly, the thoracic region is concave anteriorly, the lumbar region is convex anteriorly, and the sacral and coccygeal regions together are concave anteriorly. These spinal curvatures help accommodate our upright posture by aligning our body weight with our pelvis and lower limbs.

General Features of the Vertebrae

The general structure of an individual vertebra is outlined in table 7.9. Each vertebra consists of a body, an arch, and various processes. The weight-bearing portion of the vertebra is a bony disk called the **body.**

The **vertebral arch** projects posteriorly from the body. The arch is divided into left and right halves, and each half has two parts: the **pedicle** (ped′i-kl; foot), which is attached to the body, and the **lamina** (lam′i-na; thin plate), which joins with the lamina from the opposite half of the arch. The vertebral arch and the posterior part of the body surround a large opening called the **vertebral foramen.** The vertebral foramina of adjacent vertebrae combine to form the **vertebral canal,** which contains the spinal cord and cauda equina (see figure 12.1). The vertebral arches and bodies protect the spinal cord.

FUNDAMENTAL **Figure**

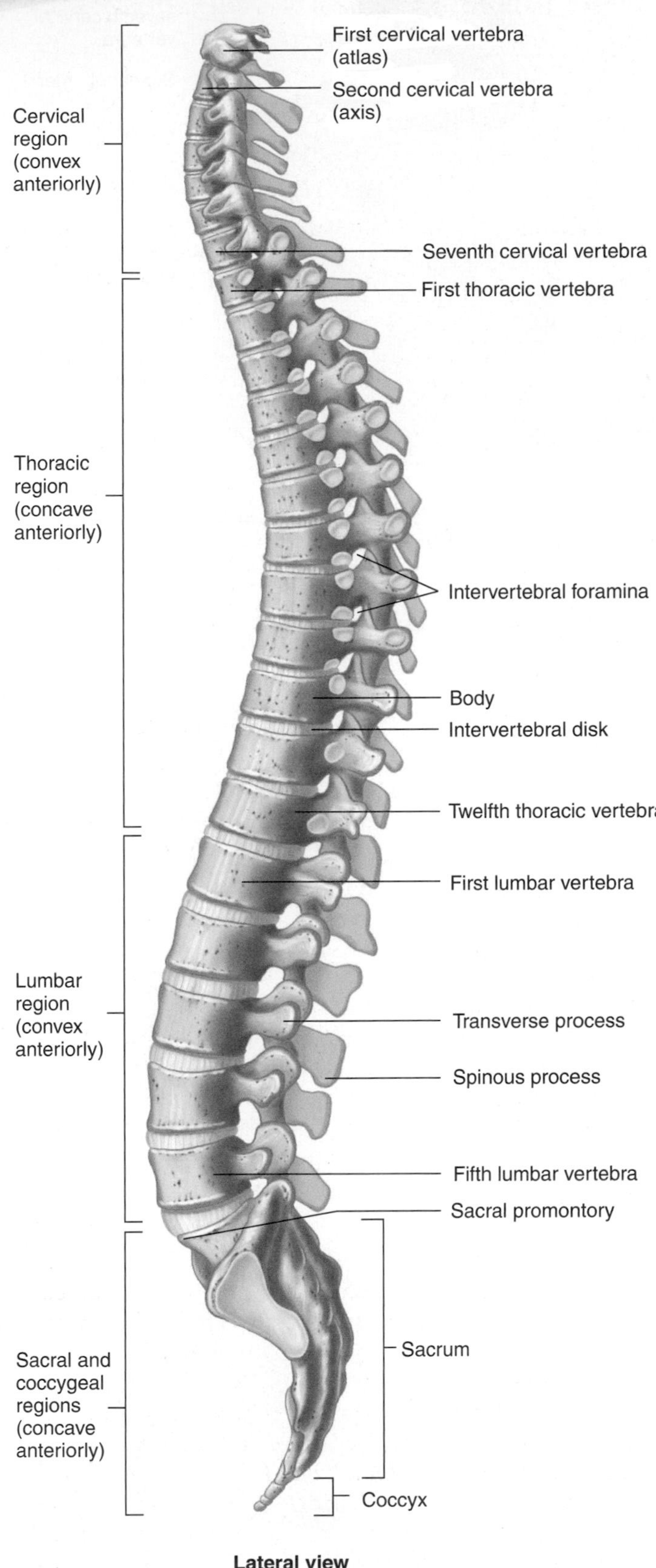

First cervical vertebra (atlas)
Second cervical vertebra (axis)
Cervical region (convex anteriorly)
Seventh cervical vertebra
First thoracic vertebra
Thoracic region (concave anteriorly)
Intervertebral foramina
Body
Intervertebral disk
Twelfth thoracic vertebra
First lumbar vertebra
Lumbar region (convex anteriorly)
Transverse process
Spinous process
Fifth lumbar vertebra
Sacral promontory
Sacrum
Sacral and coccygeal regions (concave anteriorly)
Coccyx
Lateral view

FIGURE 7.13 AP|R **Complete Vertebral Column Viewed from the Left Side**

A **transverse process** extends laterally from each side of the arch between the lamina and the pedicle, and a single **spinous process** lies at the junction between the two laminae. The spinous processes can be seen and felt as a series of lumps down the midline of the back (figure 7.14). Much vertebral movement is accomplished by the contraction of the skeletal muscles attached to the transverse and spinous processes (see chapter 10).

Clinical IMPACT

Abnormal Spinal Curvatures

In some people, the normal spinal curvature becomes distorted due to disease or a congenital defect. The three most common spinal curvatures are lordosis, kyphosis, and scoliosis. **Lordosis** (lōr-dō′sis; hollow back) is an exaggeration of the convex curve of the lumbar region. **Kyphosis** (kī-fō′sis; hump back) is an exaggeration of the concave curve of the thoracic region. It is most common in postmenopausal women but can also occur in men and becomes more prevalent as people age. **Scoliosis** (skō′lē-ō′sis) is an abnormal lateral and rotational curvature of the vertebral column, which is often accompanied by secondary abnormal curvatures, such as kyphosis (figure 7A). Contrary to popular belief, scoliosis in school-age children is not associated with carrying overly heavy backpacks. Studies have shown that, although back pain is common in backpack-bearing school kids, structural changes in the vertebral column are not. Treatments for abnormal spinal curvature depend on the age and overall medical condition of the person. However, most treatments include repeated examinations to monitor the status of the curvature, a back brace, and surgery when the curving is not slowed by bracing.

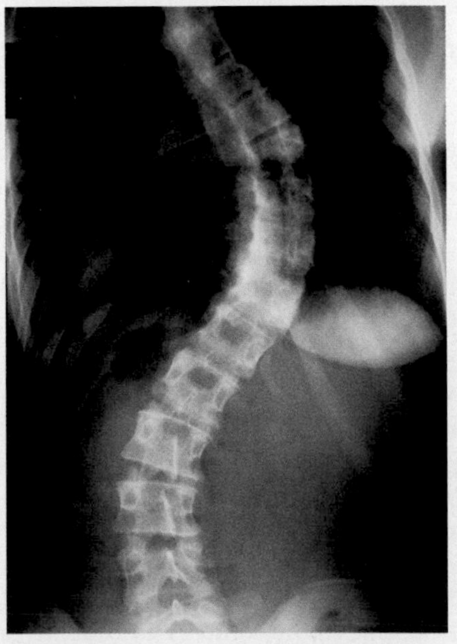

FIGURE 7A Scoliosis

Clinical IMPACT

Spina Bifida

Sometimes vertebral laminae partly or completely fail to fuse (or even fail to form) during fetal development, resulting in a condition called **spina bifida** (spī′nă bif′i-dă; split spine). This defect is most common in the lumbar region. If the defect is severe and involves the spinal cord (figure 7B), it may interfere with normal nerve function below the point of the defect.

Posterior
Dura mater
Skin of back
Enlarged fluid-filled space
Back muscles
Spinal cord
Cauda equina
Incomplete vertebral arch
Body of first lumbar vertebra
Superior view

FIGURE 7B Spina Bifida

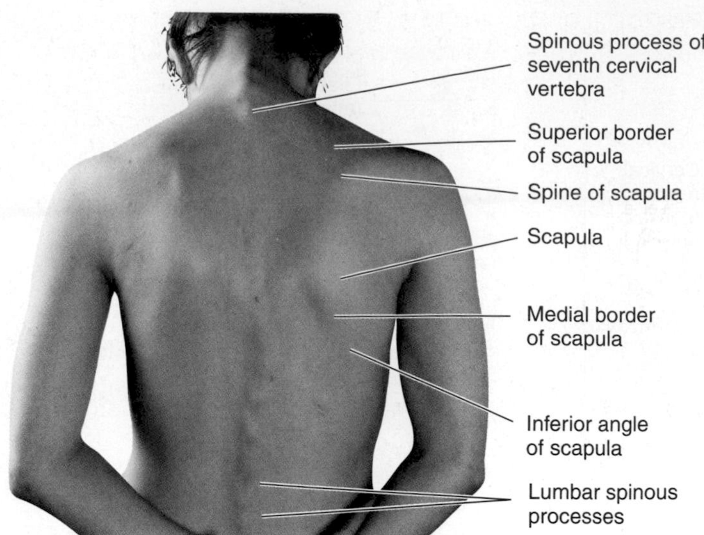
Spinous process of seventh cervical vertebra
Superior border of scapula
Spine of scapula
Scapula
Medial border of scapula
Inferior angle of scapula
Lumbar spinous processes

FIGURE 7.14 Surface View of the Back, Showing the Scapula and Vertebral Spinous Processes

Spinal nerves exit the spinal cord through the **intervertebral foramina** (table 7.9*d*; see figure 7.13). Each intervertebral foramen is formed by **intervertebral notches** in the pedicles of adjacent vertebrae.

Movement and additional support of the vertebral column are made possible by the vertebral processes. Each vertebra has two **superior** and two **inferior articular processes,** with the superior processes of one vertebra articulating with the inferior processes of the next superior vertebra (table 7.9*c,d*). Overlap of these processes increases the rigidity of the vertebral column. The region of overlap and articulation between the superior and inferior articular processes creates a smooth **articular facet** (fas′et; little face) on each articular process.

Intervertebral Disks

During life, **intervertebral disks** of fibrocartilage, which are located between the bodies of adjacent vertebrae (figure 7.15; see table 7.9 and figure 7.13), provide additional support and prevent the vertebral bodies from rubbing against each other. The intervertebral disks consist of an external **annulus fibrosus** (an′ū-lŭs fī-brō′sŭs; fibrous ring) and an internal, gelatinous **nucleus pulposus** (pŭl-pō′sŭs; pulp). The disk becomes more compressed with increasing age, so that the distance between vertebrae—and therefore the overall height of the individual—decreases. The annulus fibrosus also becomes weaker with age and more susceptible to herniation.

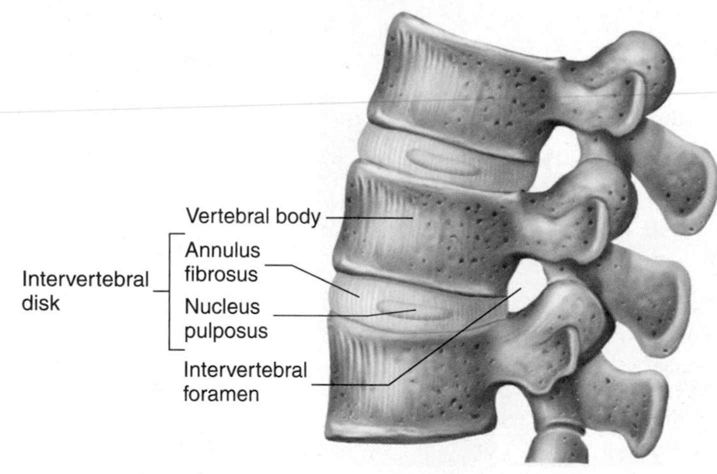
Vertebral body
Intervertebral disk
Annulus fibrosus
Nucleus pulposus
Intervertebral foramen

(a) **Lateral view**

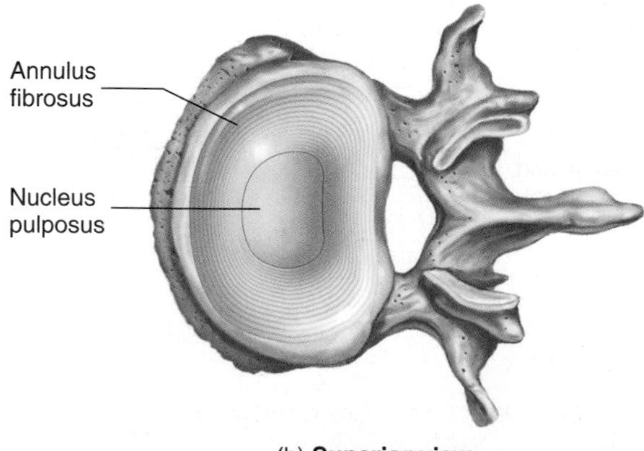
Annulus fibrosus
Nucleus pulposus

(b) **Superior view**

FIGURE 7.15 AP|R Intervertebral Disk

TABLE 7.9 General Structure of a Vertebra

Feature	Description
Body	Disk-shaped; usually the largest part with flat surfaces directed superiorly and inferiorly; forms the anterior wall of the vertebral foramen; intervertebral disks are located between the bodies
Vertebral foramen	Hole in each vertebra through which the spinal cord passes; adjacent vertebral foramina form the vertebral canal
Vertebral arch	Forms the lateral and posterior walls of the vertebral foramen; possesses several processes and articular surfaces
Pedicle	Foot of the arch with one on each side; forms the lateral walls of the vertebral foramen
Lamina	Posterior part of the arch; forms the posterior wall of the vertebral foramen
Transverse process	Process projecting laterally from the junction of the lamina and pedicle; a site of muscle attachment
Spinous process	Process projecting posteriorly at the point where the two laminae join; a site of muscle attachment; strengthens the vertebral column and allows for movement
Articular processes	Superior and inferior projections containing articular facets where vertebrae articulate with each other; strengthen the vertebral column and allow for movement
Intervertebral notches	Form intervertebral foramina between two adjacent vertebrae through which spinal nerves exit the vertebral canal

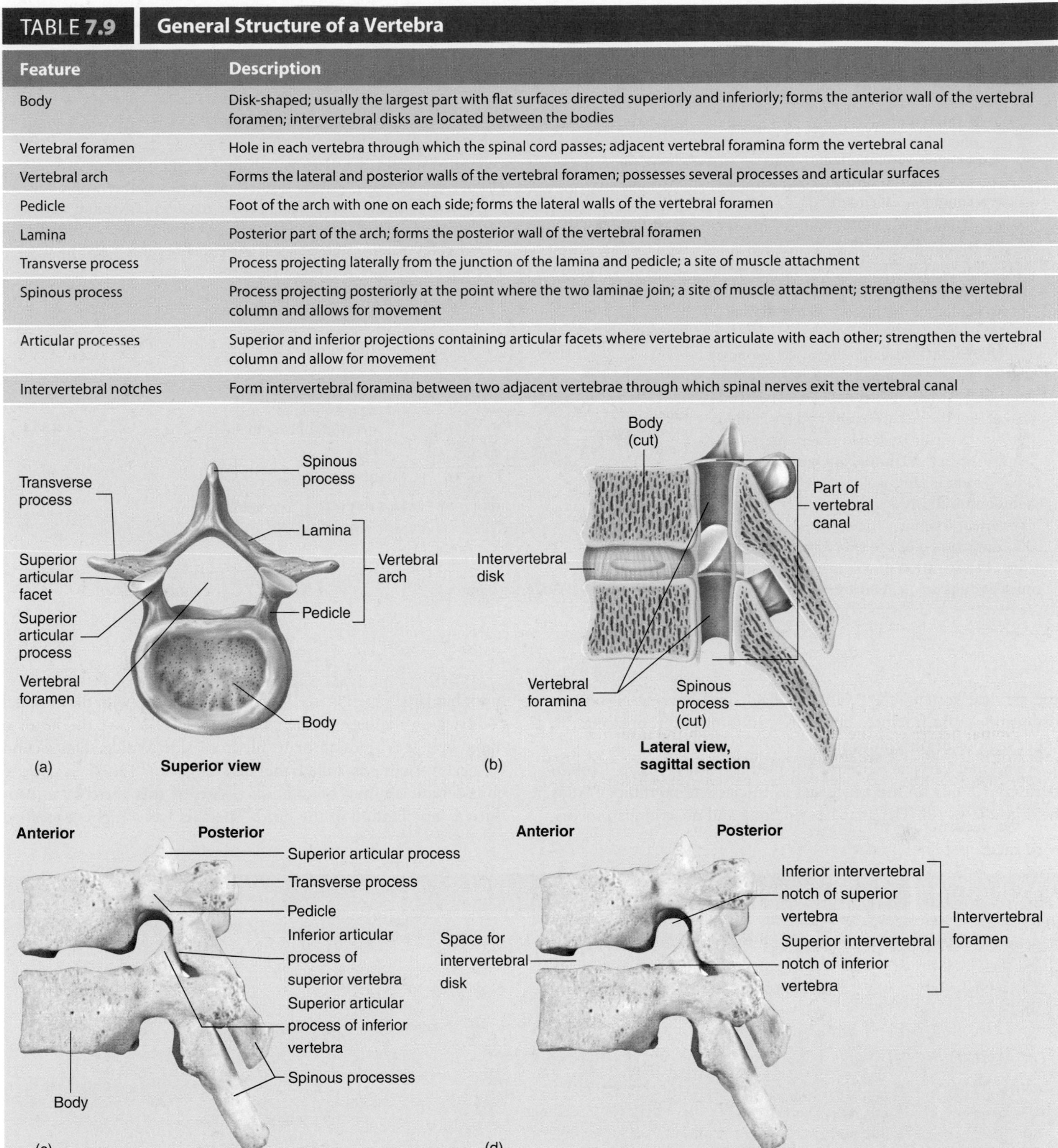
Transverse process
Spinous process
Lamina
Vertebral arch
Superior articular facet
Superior articular process
Pedicle
Vertebral foramen
Body
(a)
Superior view
Body (cut)
Part of vertebral canal
Intervertebral disk
Vertebral foramina
Spinous process (cut)
(b)
Lateral view, sagittal section
Anterior
Posterior
Superior articular process
Transverse process
Pedicle
Inferior articular process of superior vertebra
Superior articular process of inferior vertebra
Spinous processes
Body
(c)
Anterior
Posterior
Space for intervertebral disk
Inferior intervertebral notch of superior vertebra
Superior intervertebral notch of inferior vertebra
Intervertebral foramen
(d)

Regional Differences in Vertebrae

The vertebrae of each region of the vertebral column have specific characteristics that tend to blend at the boundaries between regions (table 7.10). The **cervical vertebrae** (figure 7.16; see figure 7.13) have very small bodies; most have **bifid** (bī′fid; split) **spinous processes** and a **transverse foramen** in each transverse process through which the vertebral arteries extend toward the head. Only cervical vertebrae have transverse foramina. Because

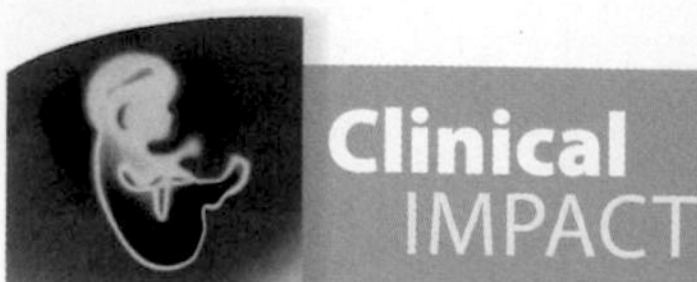

Clinical IMPACT — Herniated Intervertebral Disk

A **herniated disk** (*ruptured disk*) results when the annulus fibrosus breaks or balloons, releasing all or part of the nucleus pulposus (figure 7C). The herniated part of the disk may push against and compress the spinal cord, cauda equina, or spinal nerves, compromising their normal function and producing pain. Herniation of the inferior lumbar intervertebral disks is most common, but herniation of the inferior cervical disks is also frequent.

Herniated disks can be repaired in one of several ways. One procedure is prolonged bed rest, based on the tendency for the herniated part of the disk to recede and the annulus fibrosus to repair itself. However, many cases require surgery. A **laminectomy** is the removal of a vertebral lamina, or vertebral arch. A **hemilaminectomy** is the removal of a portion of a vertebral lamina. These procedures reduce the compression of the spinal nerve or spinal cord. **Fenestration** involves removal of the nucleus pulposus, leaving the annulus fibrosus intact. In extreme cases, the entire damaged disk is removed and a metal cage is inserted into the space previously occupied by the disk. Red bone marrow stem cells harvested from the hip are then injected into the space to allow for new bone growth. The bone marrow technique is the newest form of vertebral fusion surgery. Previously, a piece of hip bone from either the patient or a donor was inserted into the space vacated by the damaged disk. The vertebrae adjacent to the removed disk are usually further anchored together with a titanium plate held in place with titanium screws inserted into the vertebral bodies. Eventually, the adjacent vertebrae become fused by new bone growth across the gap.

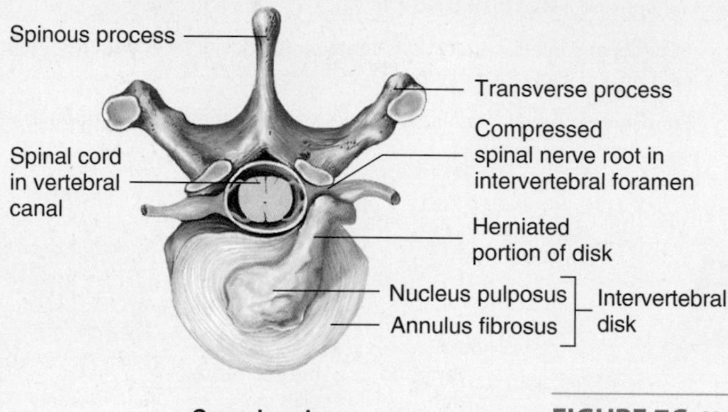
Spinous process
Transverse process
Spinal cord in vertebral canal
Compressed spinal nerve root in intervertebral foramen
Herniated portion of disk
Nucleus pulposus
Annulus fibrosus
Intervertebral disk
Superior view

FIGURE 7C Herniated Disk

the cervical vertebrae are rather delicate and have small bodies, dislocations and fractures are more common in this area than in other regions of the column.

The first cervical vertebra is called the **atlas** (figure 7.16*a,b*) because it holds up the head, just as in classical mythology Atlas held up the world. The atlas has no body and no spinous process, but it has large superior facets, where it articulates with the occipital condyles on the base of the skull. This joint allows the head to move in a "yes" motion or to tilt from side to side. The second cervical vertebra is called the **axis** (figure 7.16*c,d*) because a considerable amount of rotation occurs at this vertebra to produce a "no" motion of the head. The axis has a highly modified

TABLE 7.10 **Comparison of Vertebral Regions**

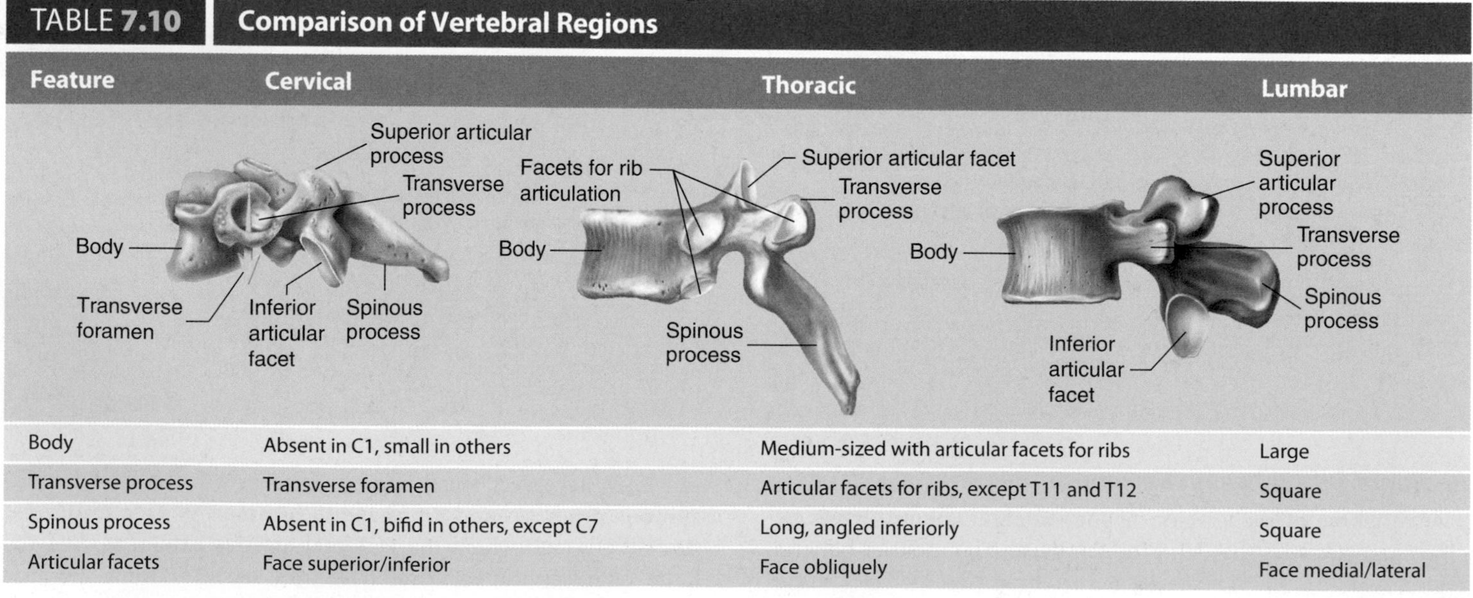
Superior articular process
Transverse process
Body
Transverse foramen
Inferior articular facet
Spinous process
Facets for rib articulation
Superior articular facet
Transverse process
Body
Spinous process
Superior articular process
Transverse process
Body
Spinous process
Inferior articular facet

Feature	Cervical	Thoracic	Lumbar
Body	Absent in C1, small in others	Medium-sized with articular facets for ribs	Large
Transverse process	Transverse foramen	Articular facets for ribs, except T11 and T12	Square
Spinous process	Absent in C1, bifid in others, except C7	Long, angled inferiorly	Square
Articular facets	Face superior/inferior	Face obliquely	Face medial/lateral

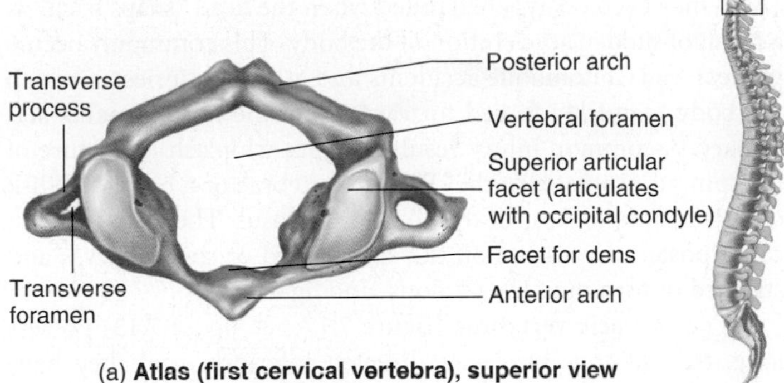
Transverse process
Posterior arch
Vertebral foramen
Superior articular facet (articulates with occipital condyle)
Facet for dens
Anterior arch
Transverse foramen

(a) **Atlas (first cervical vertebra), superior view**

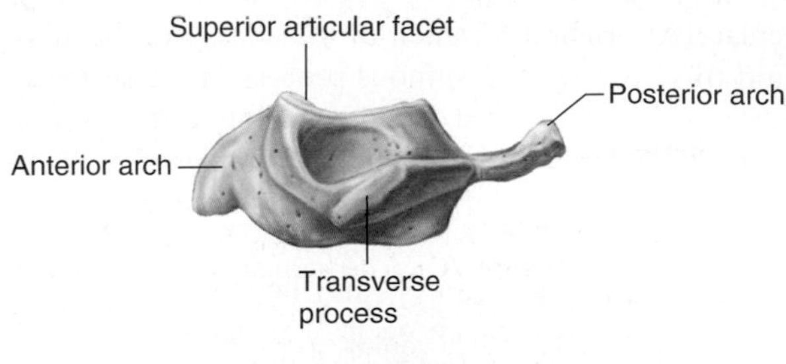
Superior articular facet
Posterior arch
Anterior arch
Transverse process

(b) **Atlas, lateral view**

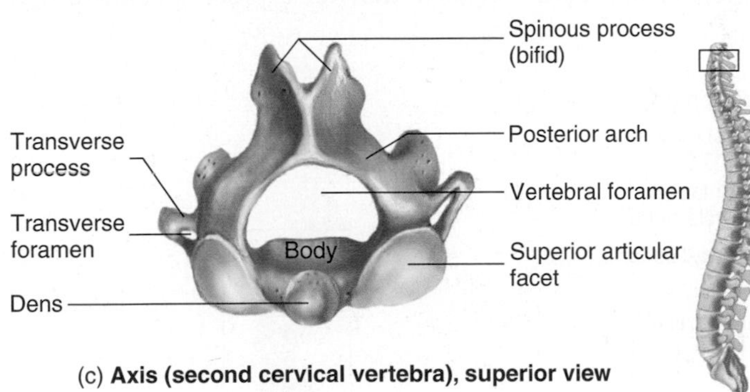
Spinous process (bifid)
Transverse process
Posterior arch
Vertebral foramen
Transverse foramen
Body
Superior articular facet
Dens

(c) **Axis (second cervical vertebra), superior view**

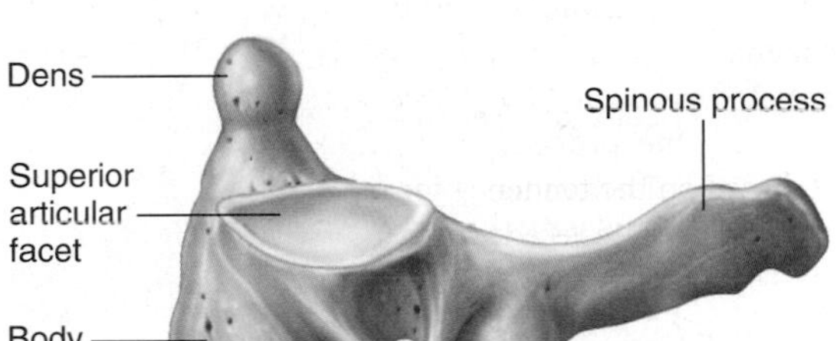
Dens
Spinous process
Superior articular facet
Body
Transverse foramen

(d) **Axis, lateral view**

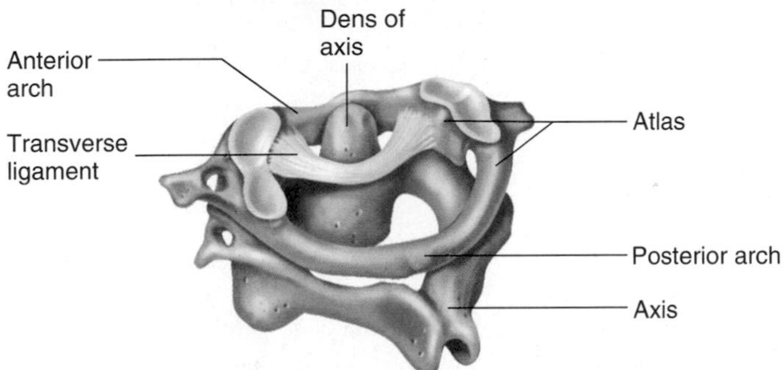
Dens of axis
Anterior arch
Atlas
Transverse ligament
Posterior arch
Axis

(e) **Atlas and axis articulated, superior view**

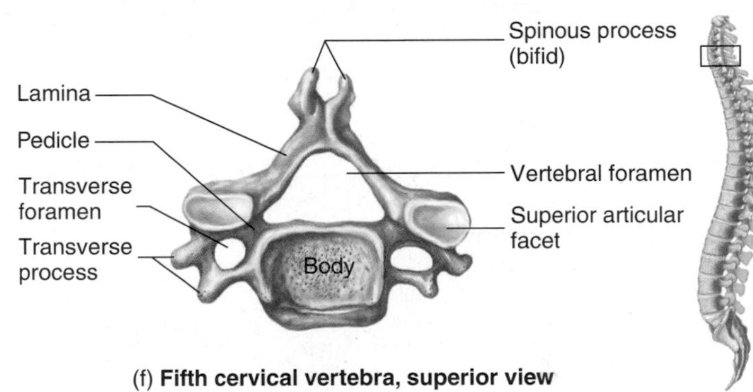
Spinous process (bifid)
Lamina
Pedicle
Vertebral foramen
Transverse foramen
Superior articular facet
Transverse process
Body

(f) **Fifth cervical vertebra, superior view**

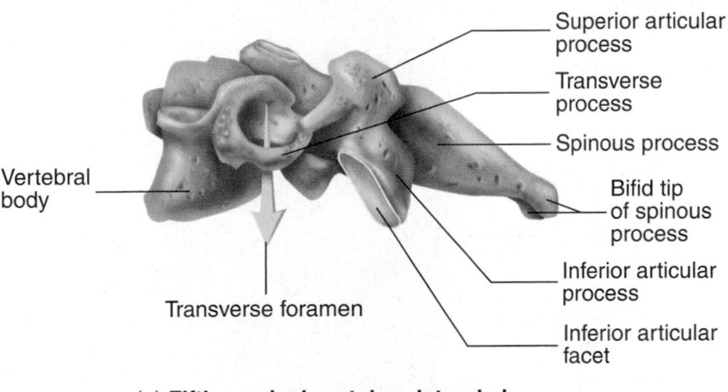
Superior articular process
Transverse process
Spinous process
Vertebral body
Bifid tip of spinous process
Inferior articular process
Transverse foramen
Inferior articular facet

(g) **Fifth cervical vertebra, lateral view**

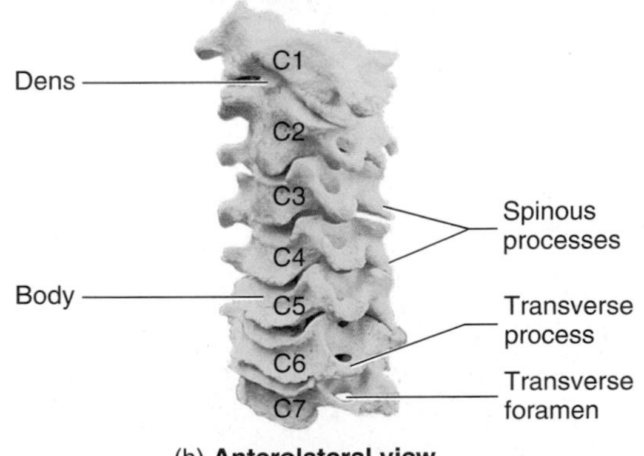
Dens
C1
C2
C3
C4
C5
C6
C7
Spinous processes
Body
Transverse process
Transverse foramen

(h) **Anterolateral view**

FIGURE 7.16 Cervical Vertebrae

(*a*) Atlas (first cervical vertebra), superior view. (*b*) Atlas, lateral view. (*c*) Axis (second cervical vertebra), superior view. (*d*) Axis, lateral view. (*e*) Atlas and axis articulated, superior view. (*f*) Fifth cervical vertebra, superior view. (*g*) Fifth cervical vertebra, lateral view. (*h*) Cervical vertebrae together from an anterolateral view.

process on the superior side of its small body called the **dens,** or *odontoid* (ō-don′toyd; tooth-shaped) *process.* The dens fits into the enlarged vertebral foramen of the atlas, and the atlas rotates around this process. The spinous process of the seventh cervical vertebra, which is not bifid, is quite pronounced and often can be seen and felt as a lump between the shoulders (see figure 7.14). The most prominent spinous process in this area is called the **vertebral prominens.** This is usually the spinous process of the seventh cervical vertebra, but it may be that of the sixth cervical vertebra or even the first thoracic. The superior articular facets face superiorly, and the inferior articular facets face inferiorly.

Whiplash is a traumatic hyperextension of the cervical vertebrae. The head is a heavy object at the end of a flexible column, and it may become hyperextended when the head "snaps back" as a result of sudden acceleration of the body. This commonly occurs in "rear-end" automobile accidents and athletic injuries, in which the body is quickly forced forward while the head remains stationary. A common injury resulting from whiplash is fracture of the spinous processes of the cervical vertebrae or a herniated disk due to an anterior tear of the annulus fibrosus. These injuries can cause posterior pressure on the spinal cord or spinal nerves and strained or torn muscles, tendons, and ligaments.

The **thoracic vertebrae** (figure 7.17; see figure 7.13) possess long, thin spinous processes directed inferiorly, and they have relatively long transverse processes. The first 10 thoracic vertebrae have articular facets on their transverse processes, where they

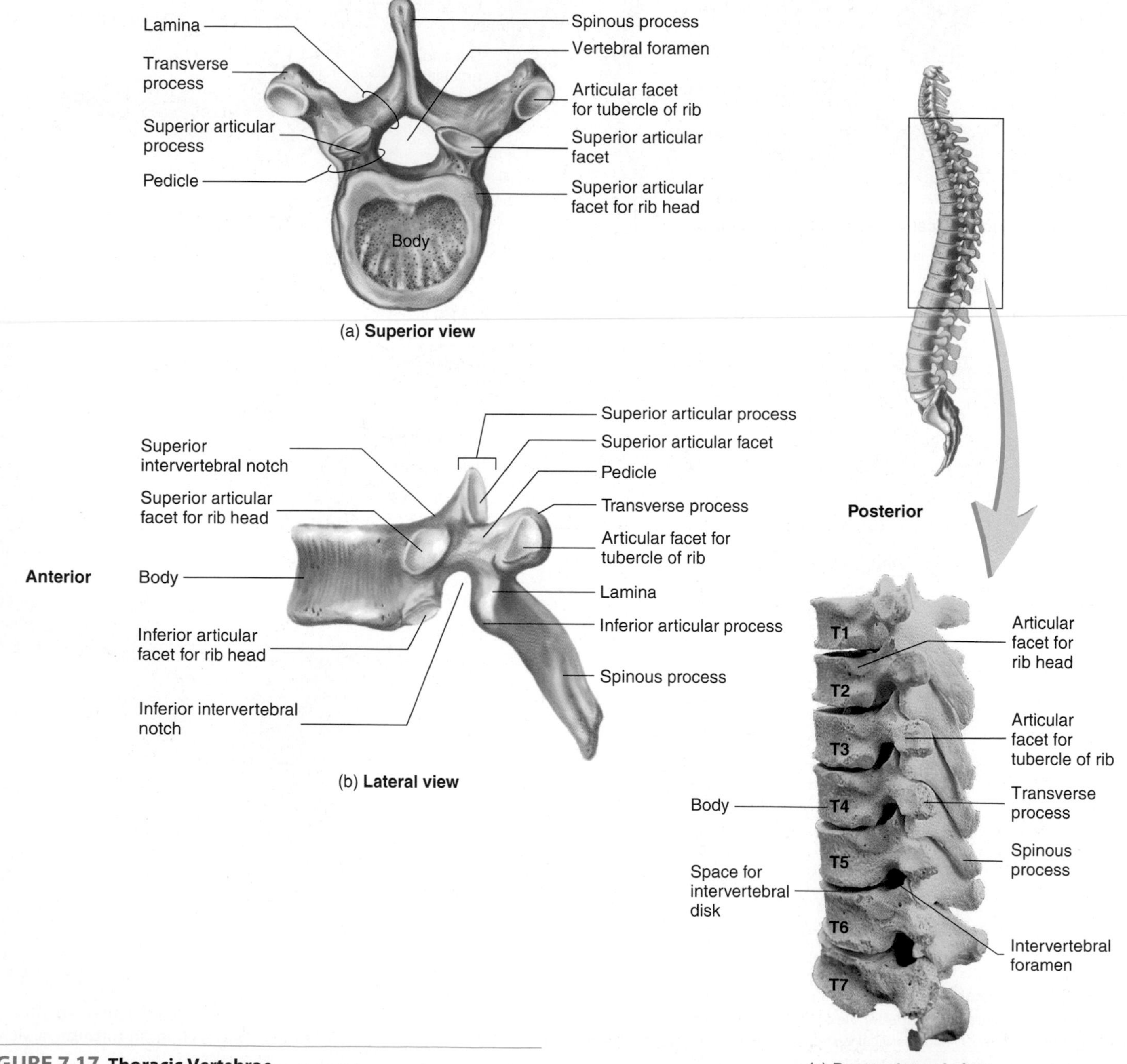
Lamina
Transverse process
Superior articular process
Pedicle
Spinous process
Vertebral foramen
Articular facet for tubercle of rib
Superior articular facet
Superior articular facet for rib head
Body
(a) Superior view
Superior intervertebral notch
Superior articular facet for rib head
Anterior
Body
Inferior articular facet for rib head
Inferior intervertebral notch
Superior articular process
Superior articular facet
Pedicle
Transverse process
Articular facet for tubercle of rib
Lamina
Inferior articular process
Spinous process
(b) Lateral view
Posterior
T1
T2
T3
T4
T5
T6
T7
Body
Space for intervertebral disk
Articular facet for rib head
Articular facet for tubercle of rib
Transverse process
Spinous process
Intervertebral foramen
(c) Posterolateral view

FIGURE 7.17 Thoracic Vertebrae

articulate with the tubercles of the ribs. Additional articular facets are on the superior and inferior margins of the body where the heads of the ribs articulate. The head of most ribs articulates with the inferior articular facet of one vertebra and with the superior articular facet for the rib head on the next vertebra down.

The **lumbar vertebrae** (figure 7.18; see figure 7.13) have large, thick bodies and heavy, rectangular transverse and spinous processes. The fifth lumbar vertebra or first coccygeal vertebra may become fused into the sacrum. Conversely, the first sacral vertebra may fail to fuse with the rest of the sacrum, resulting in six lumbar vertebrae. The superior articular facets face medially, and the inferior articular facets face laterally. When the superior articular surface of one lumbar vertebra joins the inferior articulating surface of another lumbar vertebra, the resulting arrangement adds

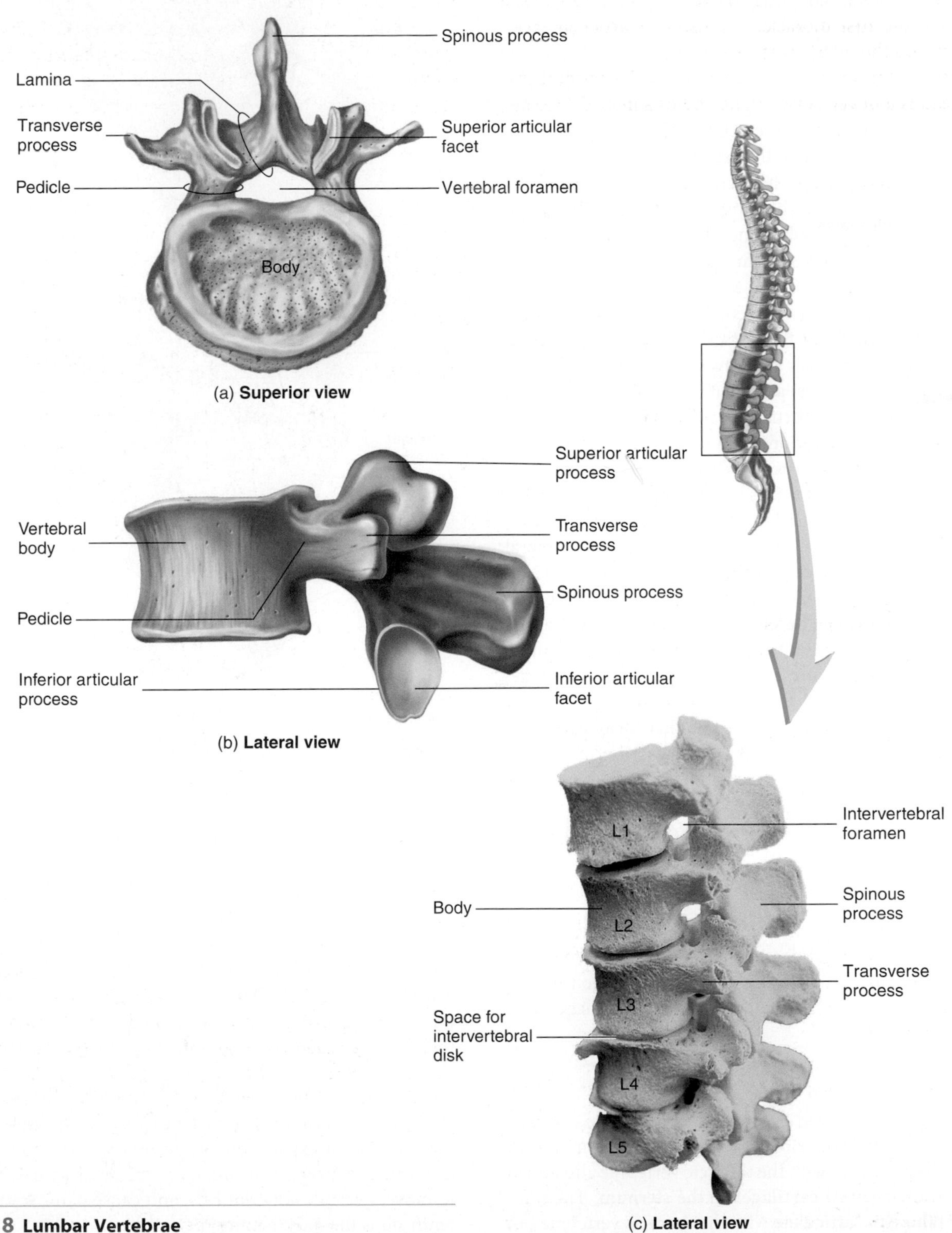
Spinous process
Lamina
Transverse process
Superior articular facet
Pedicle
Vertebral foramen
Body
(a) Superior view
Superior articular process
Vertebral body
Transverse process
Spinous process
Pedicle
Inferior articular process
Inferior articular facet
(b) Lateral view
Intervertebral foramen
L1
Body
L2
Spinous process
Transverse process
L3
Space for intervertebral disk
L4
L5
(c) Lateral view

FIGURE 7.18 Lumbar Vertebrae

strength to the inferior portion of the vertebral column and limits rotation of the lumbar vertebrae. Because the lumbar vertebrae have massive bodies and carry a large amount of weight, fractures are less common, but ruptured intervertebral disks are more common in this area than in other regions of the column.

> **Predict 3**
>
> Why are the lumbar vertebrae more massive than the cervical vertebrae? Describe some expected differences between the vertebrae of a person who engages in regular vigorous physical exercise and those of a person who never exercises.

The **sacral** (sā′krăl) **vertebrae** (figure 7.19; see figure 7.13) are highly modified compared with the others. These five vertebrae are fused into a single bone called the **sacrum** (sā′krŭm).

The transverse processes of the sacral vertebrae fuse to form the lateral parts of the sacrum. The superior surfaces of the lateral parts are wing-shaped areas called the **alae** (ā′lē; wings). Much of the lateral surfaces of the sacrum are ear-shaped **auricular surfaces,** which join the sacrum to the pelvic bones. The spinous processes of the first four sacral vertebrae partially fuse to form the **median sacral crest** along the dorsal surface of the sacrum. The spinous process of the fifth vertebra does not form, thereby leaving a **sacral hiatus** (hī-ā′tŭs) at the inferior end of the sacrum, which is often the site of anesthetic injections. The intervertebral foramina are divided into anterior and posterior foramina, called the **sacral foramina,** which are lateral to the midline. Anterior and posterior branches of the spinal nerves pass through these foramina. **Transverse lines** are where the individual sacral vertebrae fuse (see figure 7.19*a*). The anterior edge of the body of the first sacral vertebra bulges to form the **sacral promontory** (see figure 7.13), a landmark that separates the abdominal cavity from the pelvic cavity. The sacral promontory can be felt during a vaginal examination, and it is used as a reference point when measuring the pelvic inlet.

The **coccyx** (kok′siks; figure 7.19; see figure 7.13), or tailbone, is the most inferior portion of the vertebral column and usually consists of three to five semifused vertebrae that form a triangle, with the apex directed inferiorly. The coccygeal vertebrae are much smaller than the other vertebrae and have neither vertebral foramina nor well-developed processes. The coccyx is easily broken when a person falls by sitting down hard on a solid surface.

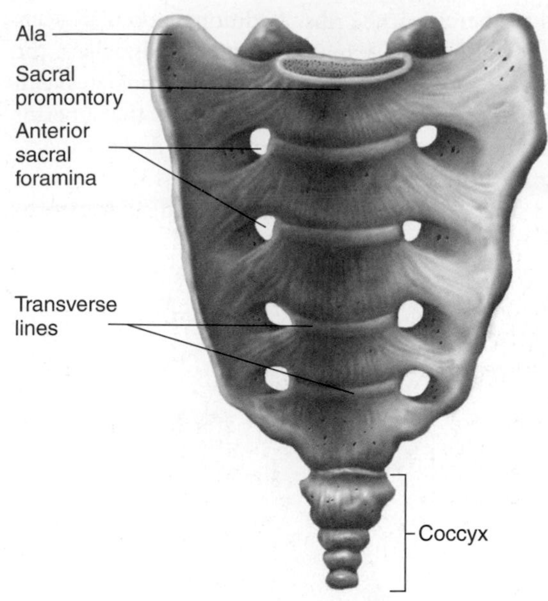
Ala
Sacral promontory
Anterior sacral foramina
Transverse lines
Coccyx

(a) **Anterior view**

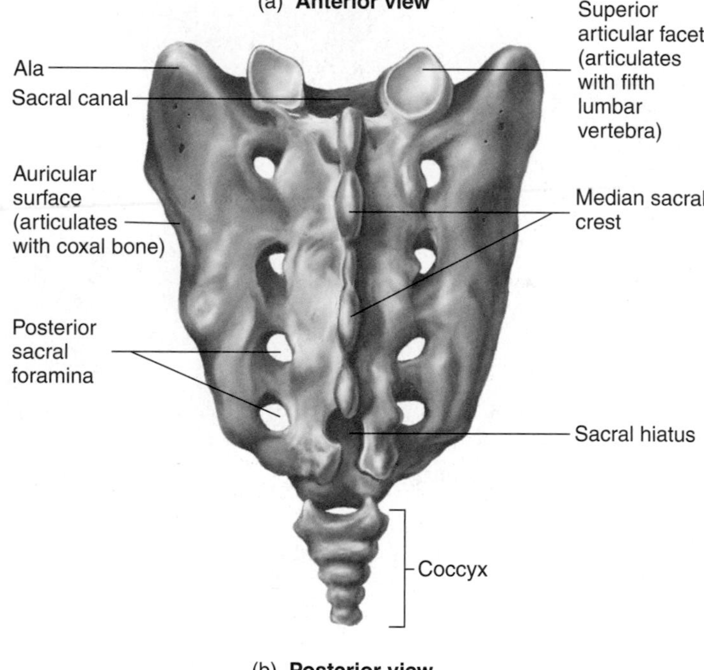
Ala
Sacral canal
Auricular surface (articulates with coxal bone)
Posterior sacral foramina
Superior articular facet (articulates with fifth lumbar vertebra)
Median sacral crest
Sacral hiatus
Coccyx

(b) **Posterior view**

FIGURE 7.19 AP|R **Sacrum**

Rib Cage

The **rib cage,** or *thoracic cage,* protects the vital organs within the thorax and forms a semirigid chamber, which can increase and decrease in volume during respiration. It consists of the thoracic vertebrae, the ribs with their associated costal (rib) cartilages, and the sternum (figure 7.20*a*).

Ribs and Costal Cartilages

The 12 pairs of **ribs** are classified as either true ribs or false ribs. The superior 7 pairs are called **true ribs,** or *vertebrosternal* (ver′tĕ′brō-ster′năl) *ribs;* they articulate with the thoracic vertebrae and attach directly through their **costal cartilages** to the sternum. The inferior 5 pairs, or **false ribs,** articulate with the thoracic vertebrae but do not attach directly to the sternum. The false ribs consist of two groups. The eighth, ninth, and tenth ribs, the **vertebrochondral** (ver′tĕ-brō-kon′drăl) **ribs,** are joined by a common cartilage to the costal cartilage of the seventh rib, which in turn is attached to the sternum. Two of the false ribs, the eleventh and twelfth ribs, are also called **floating ribs,** or *vertebral ribs,* because they do not attach to the sternum. The costal cartilages are flexible and permit the rib cage to expand during respiration. A **separated rib** is a dislocation between a rib and its costal cartilage that allows the rib to move, override adjacent ribs, and cause pain. Separation of the tenth rib is the most common.

Most ribs have two points of articulation with the thoracic vertebrae (figure 7.20*b,c*). First, the **head** articulates with the bodies of two adjacent vertebrae and the intervertebral disk between them. The head of each rib articulates with the inferior articular facet of the superior vertebra and the superior articular facet of the inferior vertebra. Second, the **tubercle** articulates with the transverse process of the inferior vertebra. The **neck** is between the head and tubercle, and the **body,** or shaft, is the main part of the rib. The **angle** of the rib is located just lateral to the tubercle and is the point of greatest curvature. The angle is the weakest part of the rib and can be fractured in a crushing injury, as may occur in an automobile accident. Sometimes the transverse processes of the seventh cervical vertebra form separate bones called **cervical ribs.** These ribs may be tiny pieces of bone or may be long enough to reach the sternum. In addition, the first lumbar vertebra may develop lumbar ribs.

Sternum

The **sternum,** or breastbone, has been described as sword-shaped and has three parts (figure 7.20*a*). The **manubrium** (mă-noo′brē-ŭm; handle) is the sword handle; the **body,** or gladiolus (sword), is the blade; and the **xiphoid** (zi′foyd; sword) **process** is the sword tip. The superior margin of the manubrium has a **jugular notch** (neck), or *suprasternal notch,* in the midline, which can be easily felt at the anterior base of the neck (figure 7.21). The first rib and

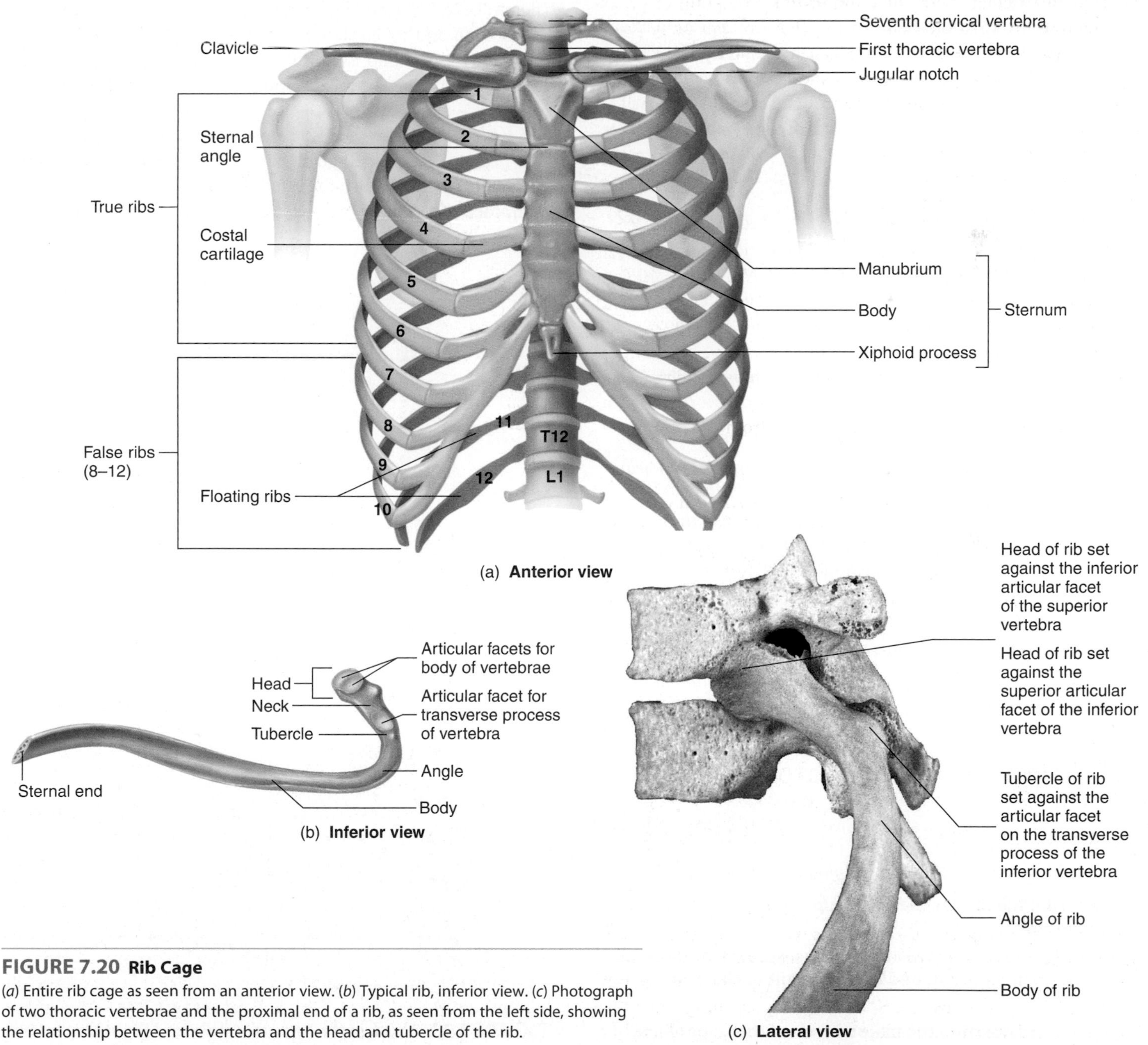
Seventh cervical vertebra
First thoracic vertebra
Jugular notch
Clavicle
Sternal angle
True ribs
Costal cartilage
False ribs (8–12)
Floating ribs
1
2
3
4
5
6
7
8
9
10
11
12
T12
L1
Manubrium
Body
Xiphoid process
Sternum
(a) Anterior view
Head
Neck
Tubercle
Sternal end
Articular facets for body of vertebrae
Articular facet for transverse process of vertebra
Angle
Body
(b) Inferior view
Head of rib set against the inferior articular facet of the superior vertebra
Head of rib set against the superior articular facet of the inferior vertebra
Tubercle of rib set against the articular facet on the transverse process of the inferior vertebra
Angle of rib
Body of rib
(c) Lateral view

FIGURE 7.20 Rib Cage
(*a*) Entire rib cage as seen from an anterior view. (*b*) Typical rib, inferior view. (*c*) Photograph of two thoracic vertebrae and the proximal end of a rib, as seen from the left side, showing the relationship between the vertebra and the head and tubercle of the rib.

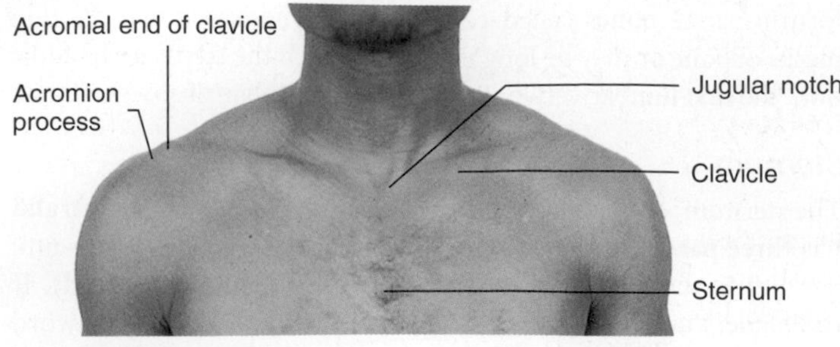
Acromial end of clavicle
Acromion process
Jugular notch
Clavicle
Sternum

FIGURE 7.21 Surface Anatomy Showing Bones of the Upper Thorax

the clavicle articulate with the manubrium. The point at which the manubrium joins the body of the sternum can be felt as a prominence on the anterior thorax called the **sternal angle** (see figure 7.20*a*). The cartilage of the second rib attaches to the sternum at the sternal angle, the third through seventh ribs attach to the body of the sternum, and no ribs attach to the xiphoid process.

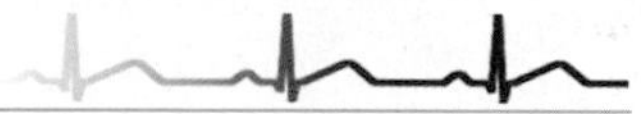

ASSESS YOUR PROGRESS

16. *Where is the hyoid bone located? Why is it a unique bone? What are its functions?*
17. *What are the functions of the vertebral column?*
18. *Name the four major curvatures of the adult vertebral column, and explain what causes them. Describe scoliosis, kyphosis, and lordosis.*
19. *Describe the structures that are common to most vertebrae.*
20. *Where do spinal nerves exit the vertebral column?*
21. *Describe the structure and function of the intervertebral disks.*
22. *Explain how the superior and inferior articular processes help support and allow movement of the vertebral column.*
23. *Name and give the number of vertebrae in each of the five regions of the vertebral column. Describe the characteristics that distinguish the different regions of vertebrae.*
24. *What is the function of the rib cage? Distinguish among true, false, and floating ribs, and give the number of each type.*
25. *Describe the articulation of the ribs with thoracic vertebrae.*
26. *What are the parts of the sternum? Name the structures that attach to the sternum.*

7.3 Appendicular Skeleton

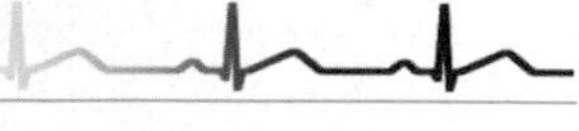

LEARNING OUTCOMES

After reading this section, you should be able to

A. Describe the girdles that make up the appendicular skeleton.

B. Identify the bones that make up the pectoral girdle and relate their structure and arrangement to the function of the girdle.

C. Name and describe the major bones of the upper limb.

D. List the bones that make up the pelvic girdle and explain why the pelvic girdle is more stable than the pectoral girdle.

E. Name the bones that make up the coxal bone. Distinguish between the male and the female pelvis.

F. Identify and describe the bones of the lower limb.

The appendicular skeleton (see figure 7.1) consists of the bones of the **upper limbs,** the **lower limbs,** and the **girdles.** The term *girdle,* which means a belt or a zone, refers to the two zones, pectoral and pelvic, where the limbs are attached to the body.

Pectoral Girdle and Upper Limb

Picture a baseball pitcher winding up to throw a fastball and you have a great demonstration of the mobility of your upper limb (figure 7.22). This mobility is possible because muscles attach the upper limb and its girdle rather loosely to the rest of the body. Thus, the upper limb is capable of a wide range of movements, including throwing, lifting, grasping, pulling, and touching.

Pectoral Girdle

The **pectoral** (pek′tō-răl) **girdle,** or *shoulder girdle,* consists of two pairs of bones that attach the upper limb to the body: Each pair is composed of a **scapula** (skap′ū-lă), or *shoulder blade* (figure 7.23),

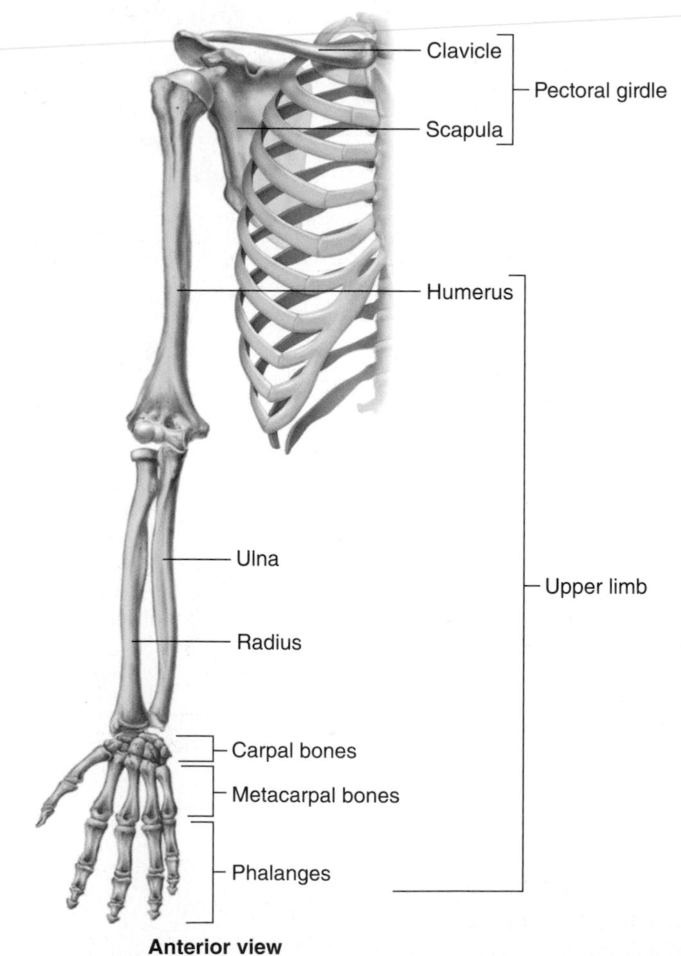
Clavicle
Scapula
Pectoral girdle
Humerus
Ulna
Radius
Carpal bones
Metacarpal bones
Phalanges
Upper limb
Anterior view

FIGURE 7.22 Bones of the Pectoral Girdle and Right Upper Limb

and a **clavicle** (klav′i-kl), or *collarbone* (see figures 7.20, 7.22, and 7.23). The scapula is a flat, triangular bone that can easily be seen and felt in a living person (see figure 7.14). The base of the triangle, the superior border, faces superiorly; the apex, the inferior angle, is directed inferiorly.

The large **acromion** (ă-krō′mē-on; shoulder tip) **process** of the scapula, which can be felt at the tip of the shoulder, has three functions: (1) to form a protective cover for the shoulder joint, (2) to form the attachment site for the clavicle, and (3) to provide attachment points for some of the shoulder muscles. The **scapular spine** extends from the acromion process across the posterior surface of the scapula and divides that surface into a small **supraspinous fossa** superior to the spine and a larger **infraspinous fossa** inferior to the spine. The deep, anterior surface of the scapula constitutes the **subscapular fossa.** The smaller **coracoid** (meaning "shaped like a crow's beak") **process** provides attachments for

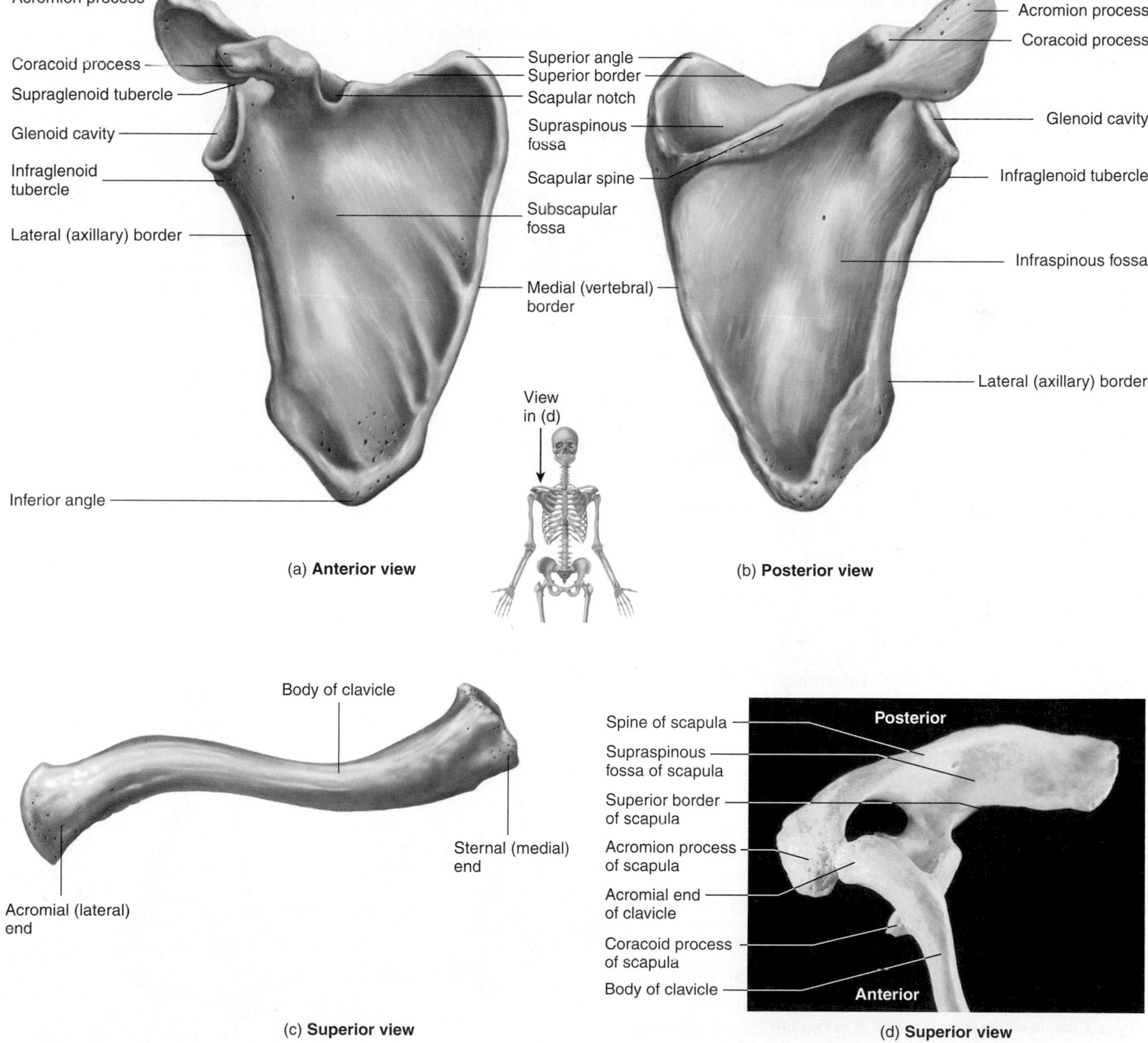
Acromion process
Coracoid process
Supraglenoid tubercle
Glenoid cavity
Infraglenoid tubercle
Lateral (axillary) border
Inferior angle
Superior angle
Superior border
Scapular notch
Supraspinous fossa
Scapular spine
Subscapular fossa
Medial (vertebral) border
View in (d)
(a) Anterior view
Acromion process
Coracoid process
Glenoid cavity
Infraglenoid tubercle
Infraspinous fossa
Lateral (axillary) border
(b) Posterior view
Body of clavicle
Sternal (medial) end
Acromial (lateral) end
(c) Superior view
Posterior
Spine of scapula
Supraspinous fossa of scapula
Superior border of scapula
Acromion process of scapula
Acromial end of clavicle
Coracoid process of scapula
Body of clavicle
Anterior
(d) Superior view

FIGURE 7.23 APR **Right Scapula and Clavicle**

(*a*) Right scapula, anterior view. (*b*) Right scapula, posterior view. (*c*) Right clavicle, superior view. (*d*) Photograph of the right scapula and clavicle from a superior view, showing the relationship between the distal end of the clavicle and the acromion process of the scapula.

some shoulder and arm muscles. A **glenoid** (glē′noyd, glen′oyd) **cavity,** located in the superior lateral portion of the bone, articulates with the head of the humerus.

The clavicle (see figures 7.20, 7.22, and 7.23*c*) is a long bone with a slight sigmoid (S-shaped) curve. It is easily seen and felt in the living human (see figure 7.21). The lateral end of the clavicle articulates with the acromion process, and its medial end articulates with the manubrium of the sternum. These articulations form the only bony connections between the pectoral girdle and the axial skeleton. Because the clavicle holds the upper limb away from the body, it facilitates the limb's mobility.

Predict 4

Sarah fell off the trampoline in her backyard. She was crying and holding her right shoulder, so her mother took her to the emergency room. Dr. Smart diagnosed a broken collarbone (clavicle), based on the position of Sarah's right upper limb. Explain.

Arm

The arm, the part of the upper limb from the shoulder to the elbow, contains only one bone, the **humerus** (figure 7.24). The humeral **head** articulates with the glenoid cavity of the scapula. The **anatomical neck,** immediately distal to the head, is almost nonexistent; thus, a surgical neck has been designated. The **surgical neck** is so named because it is a common fracture site that often requires surgical repair. Removal of the humeral head due to disease or injury occurs down to the level of the surgical neck. The **greater tubercle** is on the lateral surface, and the **lesser tubercle** is on the anterior surface of the proximal end of the humerus, where both are sites of muscle attachment. The groove between the two tubercles contains one tendon of the biceps brachii muscle and is called the **intertubercular groove,** or *bicipital* (bī-sip′i-tăl) *groove.* The **deltoid tuberosity** is located on the lateral surface of the humerus a little more than a third of the way along its length and is the attachment site for the deltoid muscle.

The articular surfaces of the distal end of the humerus exhibit unusual features where the humerus articulates with the two forearm bones. The lateral portion of the articular surface is very rounded, articulates with the radius, and is called the **capitulum** (kă-pit′ū-lŭm; head-shaped). The medial portion somewhat resembles a spool or pulley, articulates with the ulna, and is called the **trochlea** (trok′lē-ă; spool). Proximal to the capitulum and the trochlea are the **medial** and **lateral epicondyles,** which are points of attachment for the muscles of the forearm.

Forearm

The forearm has two bones. The **ulna** is on the medial side of the forearm, the side with the little finger. The **radius** is on the lateral, or thumb, side of the forearm (figure 7.25).

The proximal end of the ulna has a C-shaped articular surface, called the **trochlear notch,** or *semilunar notch,* that fits over the trochlea of the humerus. The trochlear notch is bounded by two processes. The larger, posterior process is the **olecranon** (ō-lek′ră-non; the point of the elbow) **process.** It can easily be felt and is commonly referred to as "the elbow" (figure 7.26). Posterior arm muscles attach to the olecranon process. The smaller, anterior process is the **coronoid** (kōr′ŏ-noyd; crow's beak) **process.**

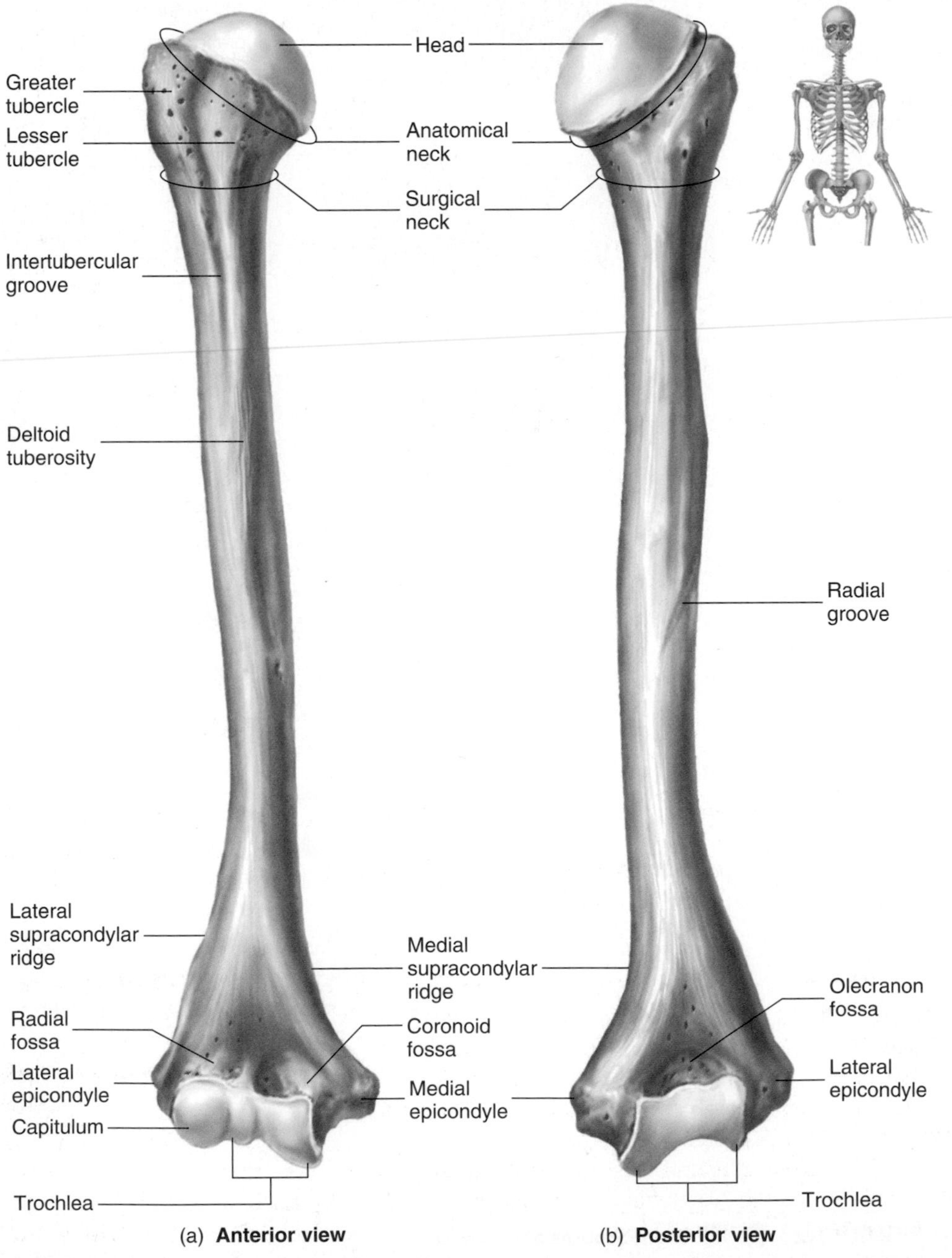
Greater tubercle
Lesser tubercle
Head
Anatomical neck
Surgical neck
Intertubercular groove
Deltoid tuberosity
Radial groove
Lateral supracondylar ridge
Medial supracondylar ridge
Radial fossa
Coronoid fossa
Olecranon fossa
Lateral epicondyle
Medial epicondyle
Lateral epicondyle
Capitulum
Trochlea
Trochlea
(a) Anterior view
(b) Posterior view

FIGURE 7.24 APR **Right Humerus**

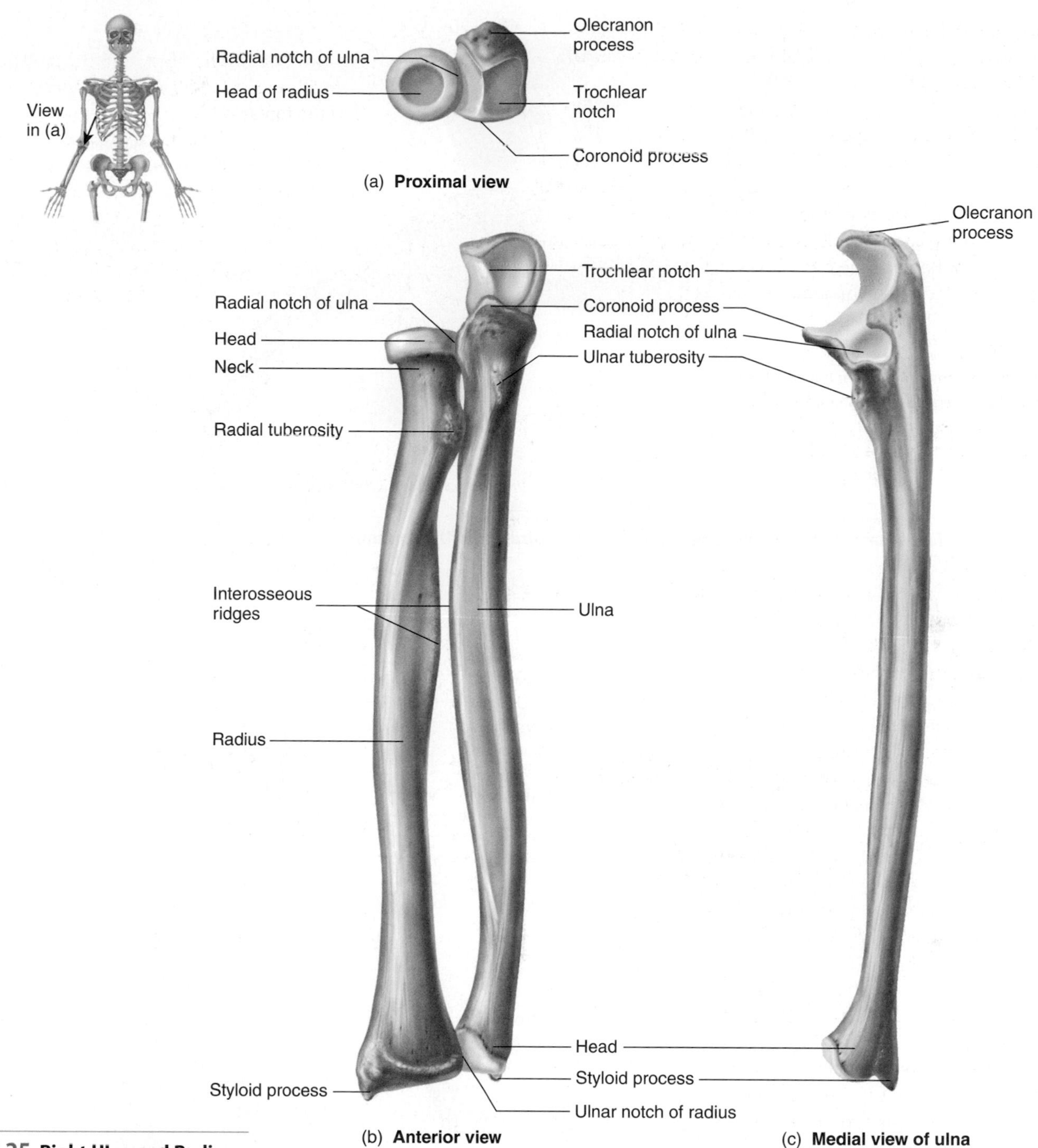
View in (a)
Radial notch of ulna
Head of radius
Olecranon process
Trochlear notch
Coronoid process
(a) Proximal view
Radial notch of ulna
Head
Neck
Radial tuberosity
Interosseous ridges
Radius
Styloid process
Trochlear notch
Coronoid process
Radial notch of ulna
Ulnar tuberosity
Ulna
Head
Styloid process
Ulnar notch of radius
(b) Anterior view
Olecranon process
(c) Medial view of ulna

FIGURE 7.25 Right Ulna and Radius

The distal end of the ulna has a small **head,** which articulates with both the radius and the wrist bones (see figures 7.25 and 7.26). The head can be seen on the posterior, medial (ulnar) side of the distal forearm. The posteromedial side of the head has a small **styloid** (stī′loyd; shaped like a stylus or writing instrument) **process,** to which ligaments of the wrist are attached.

The proximal end of the radius is the **head.** It is concave and articulates with the capitulum of the humerus. The lateral surfaces of the head constitute a smooth cylinder, where the radius rotates against the **radial notch** of the ulna. As the forearm rotates (supination and pronation; see chapter 8), the proximal end of the ulna stays in place and the radius rotates. The **radial tuberosity** is the point at which a major anterior arm muscle, the biceps brachii, attaches.

The distal end of the radius, which articulates with the ulna and the carpal bones, is somewhat broadened, and a **styloid process** to which wrist ligaments are attached is located on the lateral side of the distal radius.

Wrist

The wrist is a relatively short region between the forearm and the hand; it is composed of eight **carpal** (kar′păl) **bones** arranged into two rows of four each (figure 7.27). The proximal row of carpal

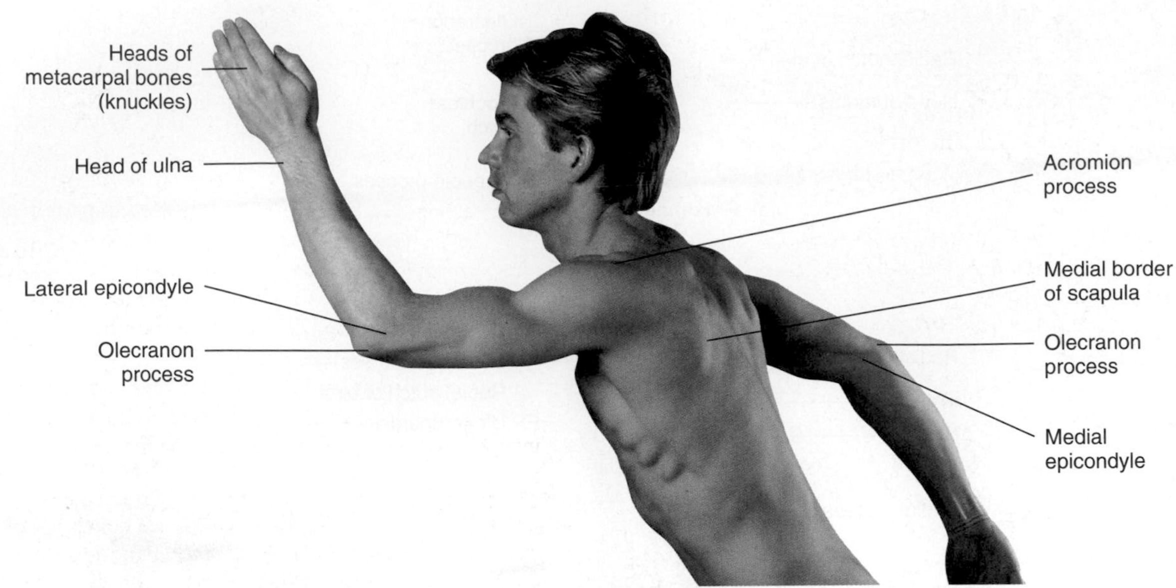
Heads of metacarpal bones (knuckles)
Head of ulna
Lateral epicondyle
Olecranon process
Acromion process
Medial border of scapula
Olecranon process
Medial epicondyle

FIGURE 7.26 Surface Anatomy Showing Bones of the Pectoral Girdle and Upper Limb

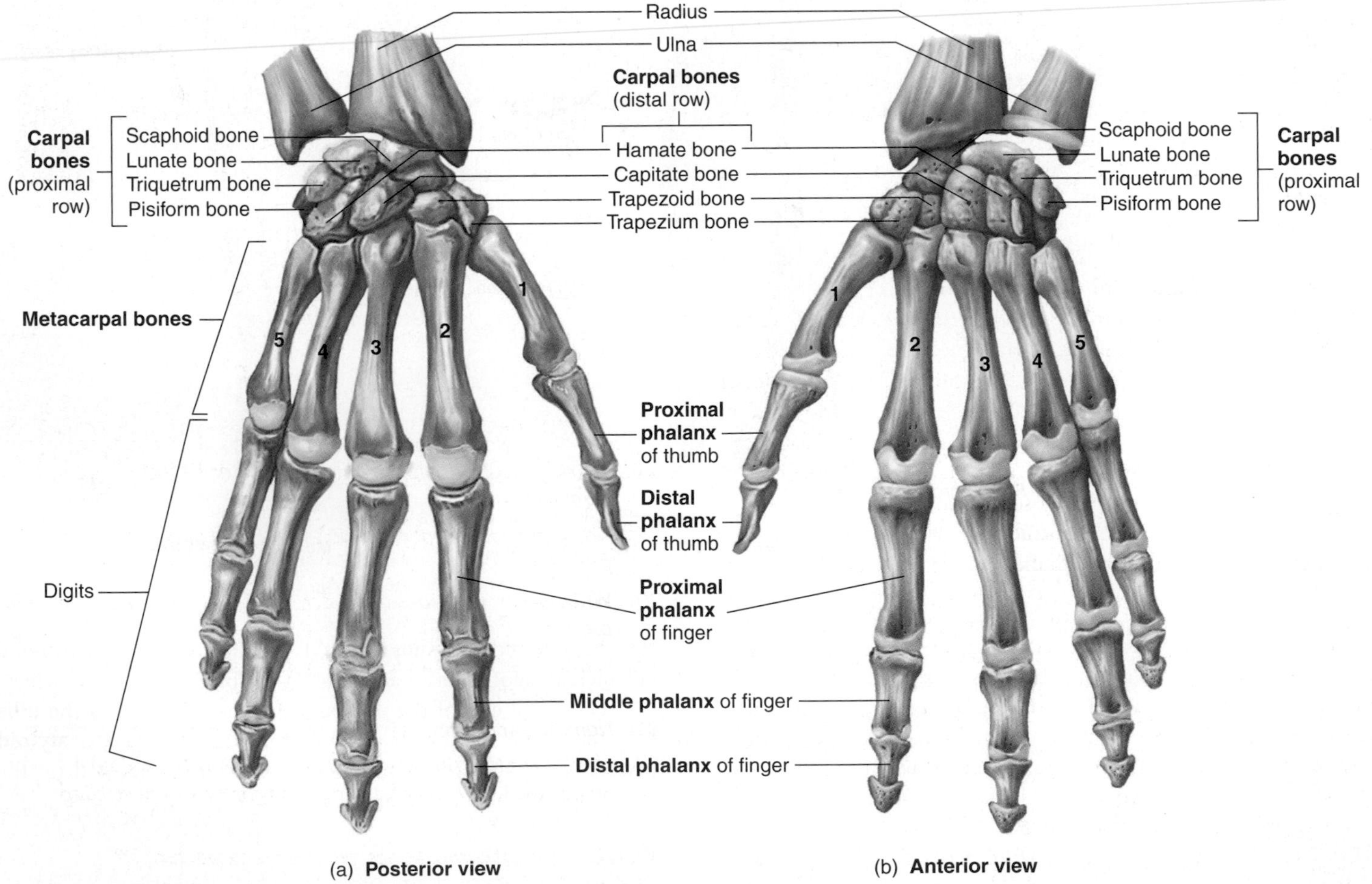

Radius
Ulna
Carpal bones (distal row)
Hamate bone
Capitate bone
Trapezoid bone
Trapezium bone
Carpal bones (proximal row)
Scaphoid bone
Lunate bone
Triquetrum bone
Pisiform bone
Metacarpal bones
1 2 3 4 5
Digits
Proximal phalanx of thumb
Distal phalanx of thumb
Proximal phalanx of finger
Middle phalanx of finger
Distal phalanx of finger

(a) **Posterior view**

(b) **Anterior view**

FIGURE 7.27 APR **Bones of the Right Wrist and Hand**

bones, lateral to medial, includes the **scaphoid** (skaf′oyd), which is boat-shaped; the **lunate** (loo′nāt), which is moon-shaped; the three-cornered **triquetrum** (trī-kwē′trŭm, trī-kwet′rŭm); and the pea-shaped **pisiform** (pis′i-fōrm), which is located on the palmar surface of the triquetrum. The distal row of carpal bones, from medial to lateral, includes the **hamate** (ha′māt), which has a hooked process on its palmar side, called the hook of the hamate; the head-shaped **capitate** (kap′i-tāt); the **trapezoid** (trap′ē-zoyd), which is named for its resemblance to a four-sided geometric form with two parallel sides; and the **trapezium** (tra-pē′zē-ŭm), which is named after a four-sided geometric form with no two sides parallel. A number of mnemonics have been developed to help students remember the carpal bones. The following one represents them in order from lateral to medial for the proximal row (top) and from medial to lateral (by the thumb) for the distal row: **S**o **L**ong **T**op **P**art, **H**ere **C**omes **T**he **T**humb—that is, **S**caphoid, **L**unate, **T**riquetrum, **P**isiform, **H**amate, **C**apitate, **T**rapezoid, and **T**rapezium.

The eight carpal bones, taken together, are convex posteriorly and concave anteriorly. The anterior concavity of the carpal bones is accentuated by the tubercle of the trapezium at the base of the thumb and the hook of the hamate at the base of the little finger. A ligament stretches across the wrist from the tubercle of the trapezium to the hook of the hamate to form a tunnel on the anterior surface of the wrist called the **carpal tunnel.** Tendons, nerves, and blood vessels pass through this tunnel to enter the hand (see Clinical Impact, "Carpal Tunnel Syndrome").

Treatments for carpal tunnel syndrome vary, depending on the severity of the condition. Mild cases can be treated nonsurgically with either anti-inflammatory medications or stretching exercises. However, if symptoms have lasted for more than 6 months, surgery is recommended. Surgical techniques involve cutting the carpal ligament to enlarge the carpal tunnel and ease pressure on the nerve.

Hand

Five **metacarpal bones** are attached to the carpal bones and constitute the bony framework of the hand (figure 7.27). They are numbered one to five, starting with the most lateral metacarpal bone, at the base of the thumb. The metacarpal bones form a curve so that, in the resting position, the palm of the hand is concave. The distal ends of the metacarpal bones help form the knuckles of the hand (see figure 7.26). The spaces between the metacarpal bones are occupied by soft tissue.

The five **digits** of each hand include one thumb and four fingers. Each digit consists of small long bones called **phalanges** (fă-lan′jēz; sing. phalanx). The thumb has two phalanges, called proximal and distal. Each finger has three phalanges, designated proximal, middle, and distal. One or two **sesamoid** (ses′ă-moyd) **bones** (not shown in figure 7.27) often form near the junction between the proximal phalanx and the metacarpal bone of the thumb. Sesamoid bones are small bones located within some tendons that increase the mechanical advantage of tendons where they cross joints.

Predict 5

Explain why the "fingers" appear much longer in a dried, articulated skeleton than in a hand with the soft tissue intact.

Clinical IMPACT

Carpal Tunnel Syndrome

The bones and ligaments that form the walls of the carpal tunnel do not stretch. Edema (fluid buildup) or connective tissue deposition may occur within the carpal tunnel as a result of trauma or some other problem. The edema or connective tissue may apply pressure against the nerve and vessels passing through the tunnel, causing carpal tunnel syndrome, which is characterized by tingling, burning, and numbness in the hand (figure 7D). Carpal tunnel syndrome occurs more frequently in people whose work involves extending the wrist and flexing the fingers.

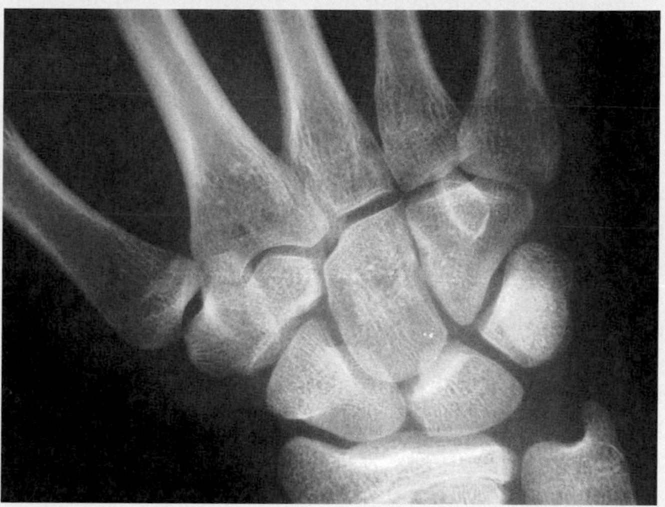

FIGURE 7D Carpal Tunnel
This x-ray shows a superior view of the carpals and metacarpals. The carpal tunnel is bordered by carpals on the superior side.

ASSESS YOUR PROGRESS

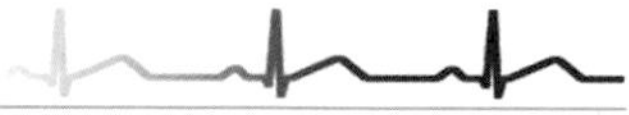

27. *Describe how the upper and lower limbs are attached to the axial skeleton.*
28. *Name the bones that make up the pectoral girdle. Describe their functions.*
29. *What are the functions of the acromion process and the coracoid process of the scapula?*
30. *Identify the bones of the upper limb, and describe their arrangement.*
31. *Name the important sites of muscle attachment on the humerus.*
32. *What is the function of the radial tuberosity? The styloid processes? Name the part of the ulna commonly referred to as "the elbow."*
33. *List the eight carpal bones. What is the carpal tunnel?*
34. *What bones form the hand? How many phalanges are in each finger and in the thumb?*

Pelvic Girdle and Lower Limb

The lower limbs support the body and are essential for normal standing, walking, and running. The general pattern of the lower limb (figure 7.28) is very similar to that of the upper limb, except that the pelvic girdle is attached much more firmly to the body than the pectoral girdle is and the bones in general are thicker, heavier, and longer than those of the upper limb. These structures reflect the function of the lower limb in supporting and moving the body.

Pelvic Girdle

The right and left **coxal** (kok′sul) **bones** (*ossa coxae* or *hipbones*) join each other anteriorly and the **sacrum** posteriorly to form a ring of bone called the **pelvic girdle.** The **pelvis** (pel′vis; basin) includes the pelvic girdle and the coccyx (figure 7.29). Each coxal bone consists of a large, concave bony plate superiorly, a slightly narrower region in the center, and an expanded bony ring inferiorly, which surrounds a large **obturator** (ob′too-rā-tŏr) **foramen.** A fossa called the **acetabulum** (as-ĕ-tab′ū-lŭm) is located on the lateral surface of each coxal bone and is the point where the lower limb articulates with the girdle. The articular surface of the acetabulum is crescent-shaped and occupies only the superior and lateral aspects of the fossa. The pelvic girdle serves as the place of attachment for the lower limbs, supports the weight of the body, and protects internal organs. Because the pelvic girdle is a complete bony ring, it provides more stable support but less mobility than the incomplete ring of the pectoral girdle. In addition, the pelvis in a woman protects a developing fetus and forms a passageway through which the fetus passes during delivery.

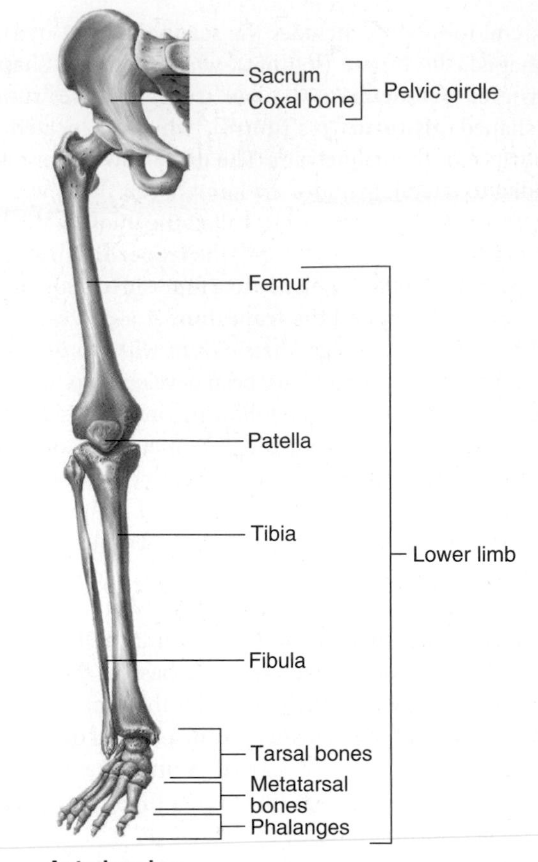
Sacrum
Coxal bone
Pelvic girdle
Femur
Patella
Tibia
Lower limb
Fibula
Tarsal bones
Metatarsal bones
Phalanges
Anterior view

FIGURE 7.28 Bones of the Right Half of the Pelvic Girdle and the Right Lower Limb

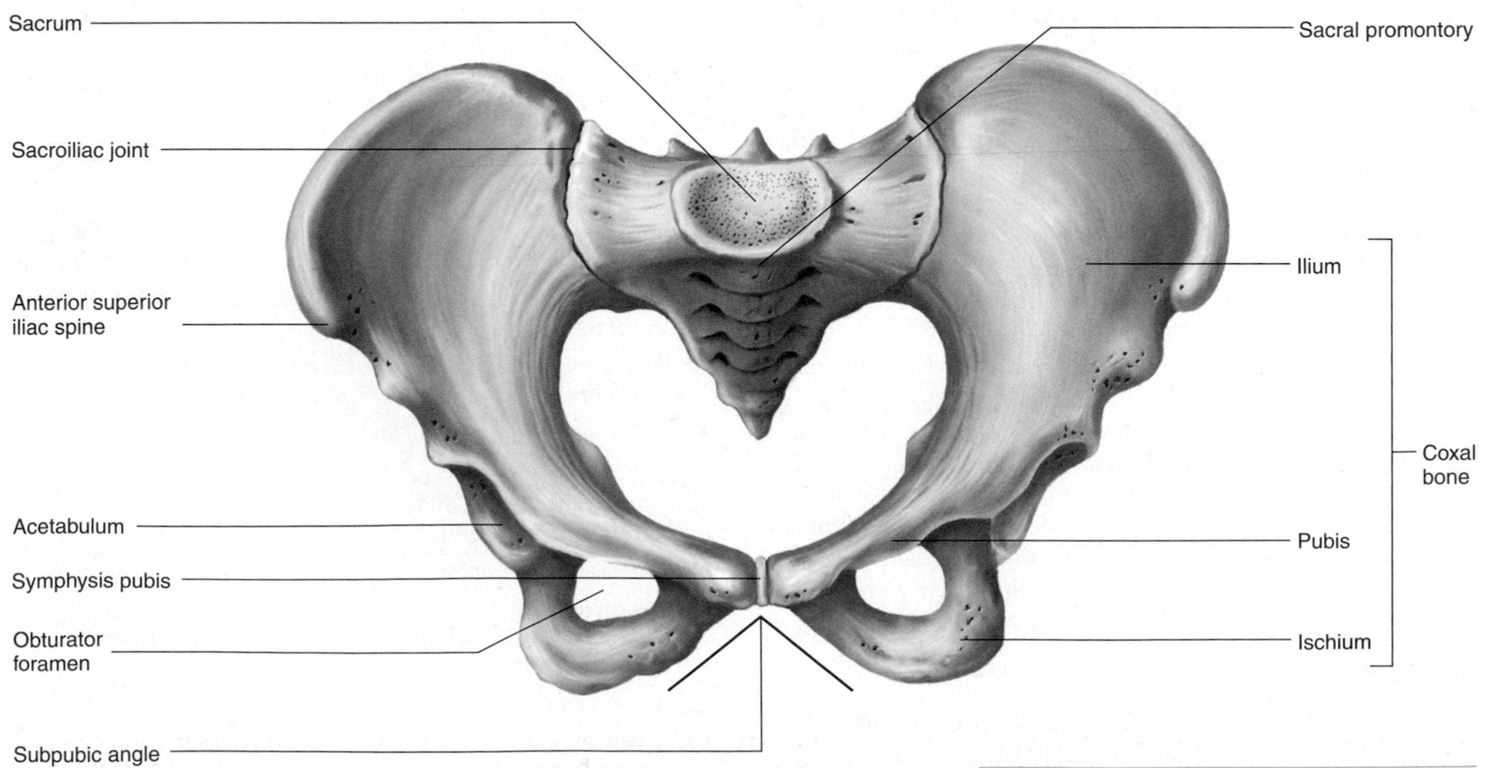
Sacrum
Sacral promontory
Sacroiliac joint
Ilium
Anterior superior iliac spine
Coxal bone
Acetabulum
Pubis
Symphysis pubis
Obturator foramen
Ischium
Subpubic angle
Anterosuperior view

FIGURE 7.29 Pelvis

Each coxal bone is formed by the fusion of three bones during development: the **ilium** (il′ē-ŭm; groin), the **ischium** (is′kē-ŭm; hip), and the **pubis** (pū′bis; genital hair). All three bones join near the center of the acetabulum (figure 7.30*a*). The superior portion of the ilium is called the **iliac crest** (figure 7.30*b,c*). The crest ends anteriorly as the **anterior superior iliac spine** and posteriorly as the **posterior superior iliac spine.** The crest and anterior spine can be felt and even seen in thin individuals (figure 7.31). The anterior superior iliac spine is an important anatomical landmark used, for example, to find the correct location for giving gluteal injections into the hip (see Clinical Impact, "Gluteal Injections"). A dimple overlies the posterior superior iliac spine just superior to the buttocks. The **greater sciatic** (sī-at′ik) **notch** is on the posterior side of the ilium, just inferior to the posterior inferior iliac spine. The sciatic nerve passes through the greater sciatic notch. The **auricular surface** of the ilium joins the sacrum to form the **sacroiliac joint** (see figure 7.29). The medial side of the ilium consists of a large depression called the **iliac fossa.**

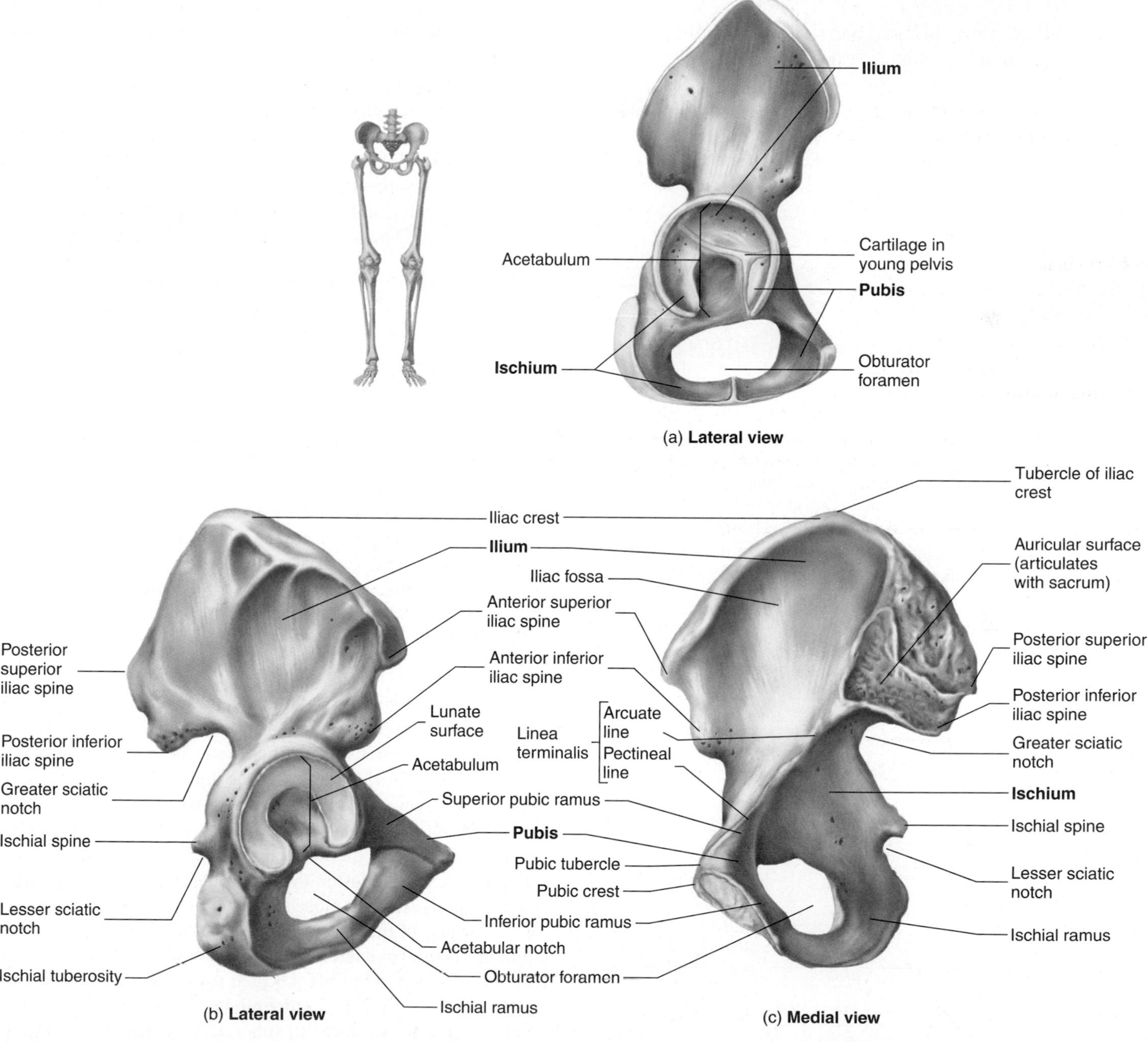
Ilium
Acetabulum
Cartilage in young pelvis
Pubis
Ischium
Obturator foramen
(a) Lateral view
Iliac crest
Ilium
Iliac fossa
Anterior superior iliac spine
Anterior inferior iliac spine
Posterior superior iliac spine
Posterior inferior iliac spine
Greater sciatic notch
Ischial spine
Lesser sciatic notch
Ischial tuberosity
Lunate surface
Acetabulum
Linea terminalis
Arcuate line
Pectineal line
Superior pubic ramus
Pubis
Pubic tubercle
Pubic crest
Inferior pubic ramus
Acetabular notch
Obturator foramen
Ischial ramus
(b) Lateral view
Tubercle of iliac crest
Auricular surface (articulates with sacrum)
Posterior superior iliac spine
Posterior inferior iliac spine
Greater sciatic notch
Ischium
Ischial spine
Lesser sciatic notch
Ischial ramus
(c) Medial view

FIGURE 7.30 AP|R **Coxal Bone**

(*a*) Right coxal bone of a growing child. Each coxal bone is formed by fusion of the ilium, ischium, and pubis. The three bones can be seen joining near the center of the acetabulum, separated by lines of cartilage. (*b*) Right coxal bone, lateral view. (*c*) Right coxal bone, medial view. (The names of the three bones forming the coxal bone are in bold.)

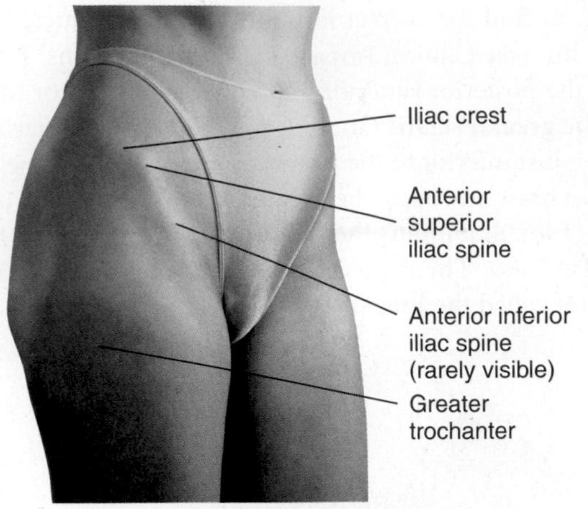
Iliac crest
Anterior superior iliac spine
Anterior inferior iliac spine (rarely visible)
Greater trochanter

FIGURE 7.31 Surface Anatomy Showing an Anterolateral View of the Coxal Bone and Femur

Clinical IMPACT

Gluteal Injections

The large gluteal muscles (hip muscles; see chapter 10) are a common site for intramuscular injections. Gluteal injections are made in the superolateral region of the hip (figure 7E) because a large nerve (the sciatic nerve; see chapter 12) lies deep to the other gluteal regions. The landmarks for such an injection are the anterior superior iliac spine and the tubercle of the iliac crest, which lies about one-third of the way along the iliac crest from anterior to posterior.

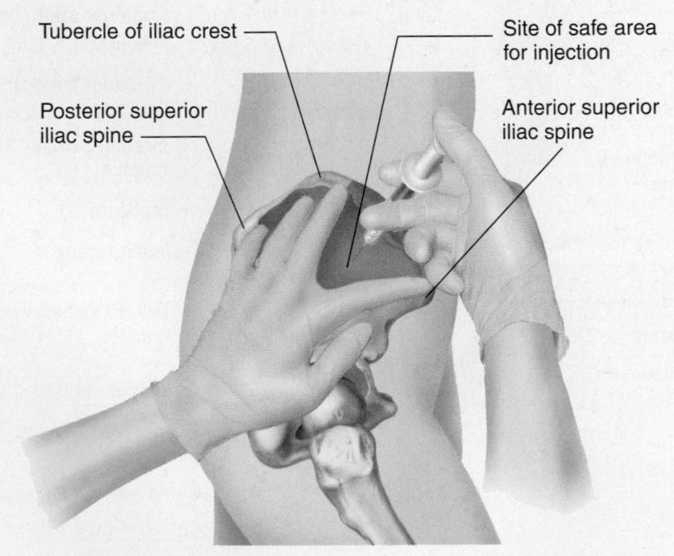
Tubercle of iliac crest
Site of safe area for injection
Posterior superior iliac spine
Anterior superior iliac spine

FIGURE 7E Gluteal Injection Site

Each sacroiliac joint is formed by the junction of the auricular surface of a coxal bone and one articular surface of the sacrum. The sacroiliac joints receive most of the weight of the upper body and are strongly supported by ligaments. However, excessive strain on the joints can allow slight joint movement and stretch connective tissue and associated nerve endings in the area, causing pain. Thus is derived the expression "My aching sacroiliac!" This problem sometimes develops in pregnant women because of the forward weight distribution of the fetus.

The ischium possesses a heavy **ischial** (is′kē-ăl) **tuberosity,** where posterior thigh muscles attach and on which a person sits (see figure 7.30*b*). The pubis has a **pubic crest** medially where abdominal muscles attach and a **pubic tubercle** laterally where the inguinal ligament attaches (see figure 7.30*c*). The pubic crest can be felt anteriorly. Just inferior to the pubic crest is the point of junction, the **symphysis** (sim′fi-sis; a coming together) **pubis,** or *pubic symphysis,* between the two coxal bones (see figure 7.29).

The pelvis is subdivided into a **true pelvis** and a **false pelvis** (figure 7.32). The opening to the true pelvis is the **pelvic inlet,** and the inferior opening of the true pelvis is the **pelvic outlet.** The false pelvis is formed by muscle overlying bone of the true pelvis.

Comparison of the Male Pelvis and the Female Pelvis

The male pelvis is usually more massive than the female pelvis as a result of the greater weight and size of the male body. The female pelvis is broader and has a larger, more rounded pelvic inlet and outlet (figure 7.32*a,b*), consistent with the need to allow the fetus to pass through these openings in the female pelvis during childbirth. A wide, circular pelvic inlet and a pelvic outlet with widely spaced ischial spines can facilitate delivery of the newborn. A smaller pelvic outlet can cause problems during delivery; thus, the size of the pelvic inlet and outlet is routinely measured during prenatal pelvic examinations. If the pelvic outlet is too small for normal delivery, the physician may perform a **cesarean section,** which is the surgical removal of the fetus through the abdominal wall. Table 7.11 lists additional differences between the male pelvis and the female pelvis.

Thigh

The thigh, like the arm, contains a single bone, the **femur.** The femur has a prominent, rounded **head,** where it articulates with the acetabulum, and a well-defined **neck;** both are located at an oblique angle to the shaft of the femur (figure 7.33). The proximal shaft exhibits two projections: a **greater trochanter** (trō-kan′ter; runner) lateral to the neck and a smaller, **lesser trochanter** inferior and posterior to the neck. Both trochanters are attachment sites for muscles that fasten the hip to the thigh. The greater trochanter and its attached muscles form a bulge that can be seen as the widest part of the hips (see figure 7.31). The distal end of the femur has **medial** and **lateral condyles,** smooth, rounded surfaces that articulate with the tibia. Located proximally to the condyles are the **medial** and **lateral epicondyles,** important sites of ligament attachment. An **adductor tubercle,** to which muscles attach, is located just proximal to the medial epicondyle.

The **patella,** or kneecap, is a large sesamoid bone located within the tendon of the quadriceps femoris muscle group, which is the major muscle group of the anterior thigh (figure 7.34). The patella

FUNDAMENTAL **Figure**

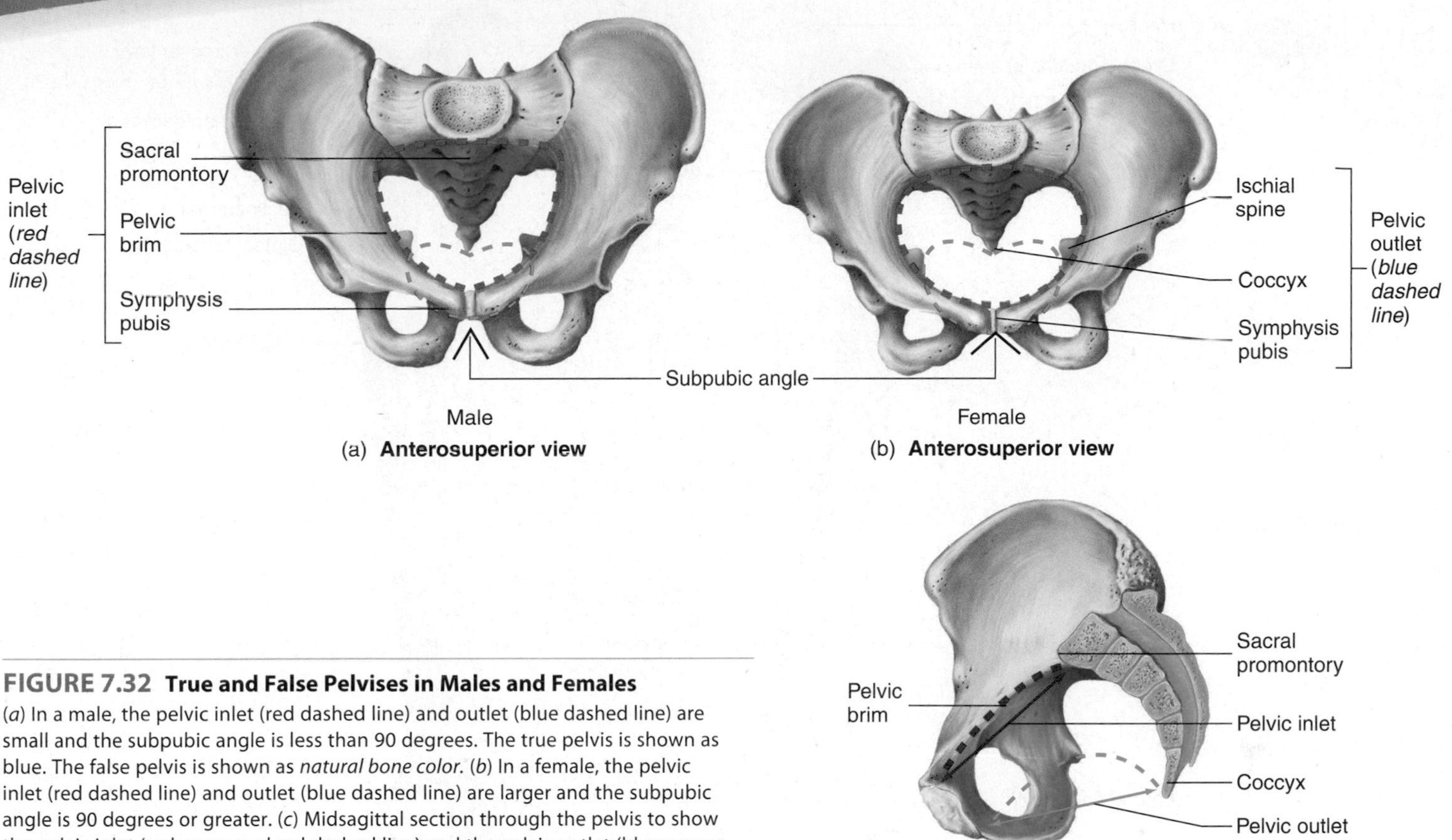

Pelvic inlet (red dashed line)
Sacral promontory
Pelvic brim
Symphysis pubis
Subpubic angle
Ischial spine
Coccyx
Symphysis pubis
Pelvic outlet (blue dashed line)
Male
(a) Anterosuperior view
Female
(b) Anterosuperior view
Pelvic brim
Sacral promontory
Pelvic inlet
Coccyx
Pelvic outlet
(c) Medial view

FIGURE 7.32 True and False Pelvises in Males and Females
(*a*) In a male, the pelvic inlet (red dashed line) and outlet (blue dashed line) are small and the subpubic angle is less than 90 degrees. The true pelvis is shown as blue. The false pelvis is shown as *natural bone color.* (*b*) In a female, the pelvic inlet (red dashed line) and outlet (blue dashed line) are larger and the subpubic angle is 90 degrees or greater. (*c*) Midsagittal section through the pelvis to show the pelvic inlet (red arrow and red dashed line) and the pelvic outlet (blue arrow and blue dashed line).

TABLE 7.11 Differences Between the Male Pelvis and the Female Pelvis (see figure 7.32)

Area	Description
General	In females, somewhat lighter in weight and wider laterally but shorter superiorly to inferiorly and less funnel-shaped; less obvious muscle attachment points in females
Sacrum	Broader in females, with the inferior part directed more posteriorly; the sacral promontory does not project as far anteriorly in females
Pelvic inlet	Heart-shaped in males; oval in females
Pelvic outlet	Broader and more shallow in females
Subpubic angle	Less than 90 degrees in males; 90 degrees or more in females
Ilium	More shallow and flared laterally in females
Ischial spines	Farther apart in females
Ischial tuberosities	Turned laterally in females and medially in males

articulates with the patellar groove of the femur to create a smooth articular surface over the anterior distal end of the femur. The patella holds the tendon away from the distal end of the femur and thereby changes the angle of the tendon between the quadriceps femoris muscle and the tibia, where the tendon attaches. This change in angle increases the force that can be applied from the muscle to the tibia. As a result, less muscle contraction is required to move the tibia.

Leg

The leg is the part of the lower limb between the knee and the ankle. Like the forearm, it consists of two bones: the **tibia** (tib′ē-ă), or shinbone, and the **fibula** (fib′ū-lă; figure 7.35). The tibia, by far the larger of the two, supports most of the weight of the leg. A **tibial tuberosity,** which is the attachment point for the quadriceps femoris muscle group, can easily be seen and felt just inferior to the patella (figure 7.36). The **anterior crest** forms the shin. The proximal end of the tibia has flat **medial** and **lateral condyles,** which articulate with the condyles of the femur. Located between the condyles is the **intercondylar eminence,** a ridge between the two articular surfaces of the proximal tibia. The distal end of the tibia is enlarged to form the **medial malleolus** (ma-lē′ō-lŭs; mallet-shaped), which helps form the medial side of the ankle joint.

The fibula does not articulate with the femur but has a small proximal head where it articulates with the tibia. The distal end of the fibula is also slightly enlarged as the **lateral malleolus** to create the lateral wall of the ankle joint. The lateral and medial malleoli can be felt and seen as prominent lumps on both sides of the ankle (figure 7.36). The thinnest, weakest portion of the fibula is just proximal to the lateral malleolus.

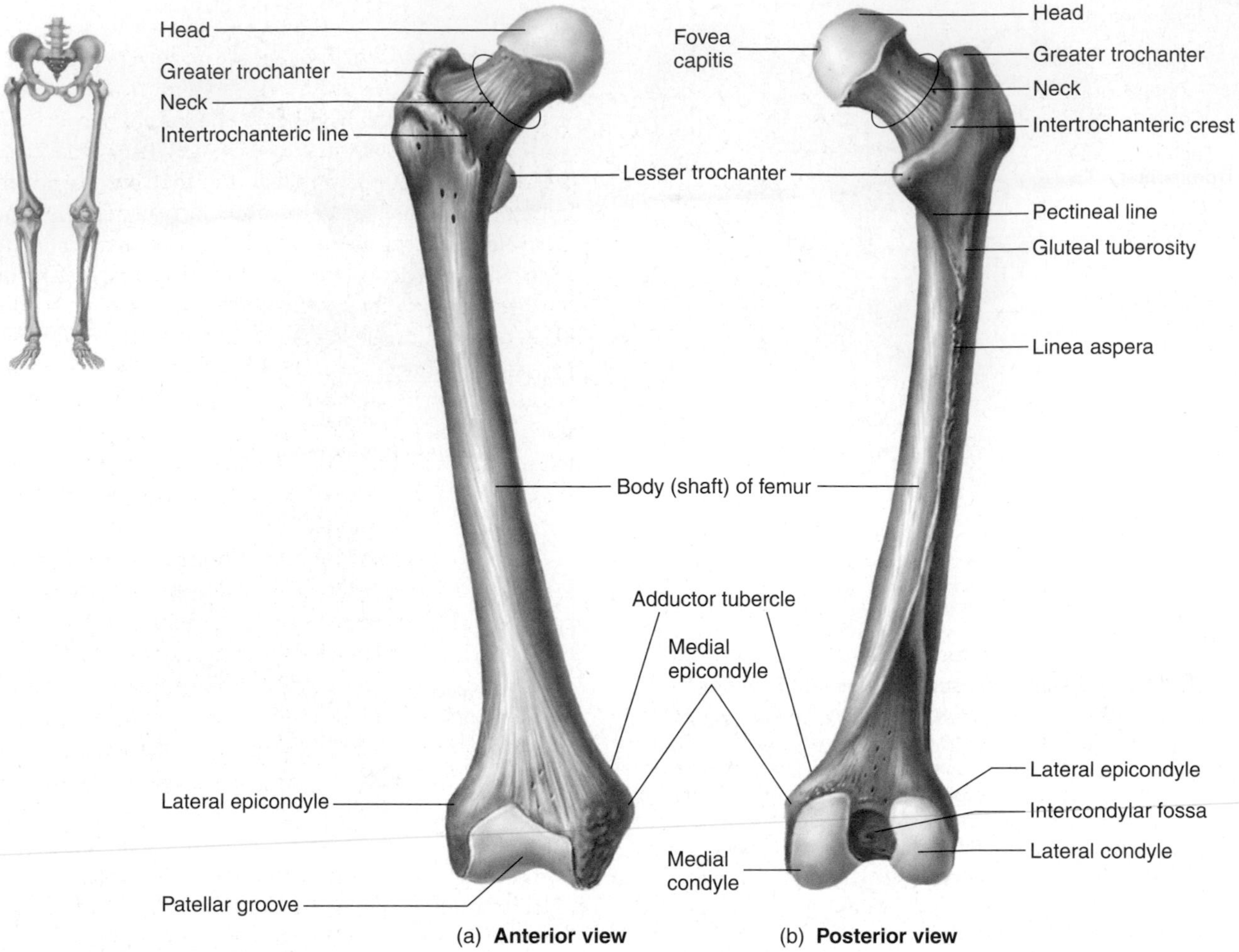

Head
Greater trochanter
Neck
Intertrochanteric line
Fovea capitis
Lesser trochanter
Head
Greater trochanter
Neck
Intertrochanteric crest
Pectineal line
Gluteal tuberosity
Linea aspera
Body (shaft) of femur
Adductor tubercle
Medial epicondyle
Lateral epicondyle
Patellar groove
Medial condyle
Lateral epicondyle
Intercondylar fossa
Lateral condyle
(a) **Anterior view**
(b) **Posterior view**

FIGURE 7.33 Right Femur

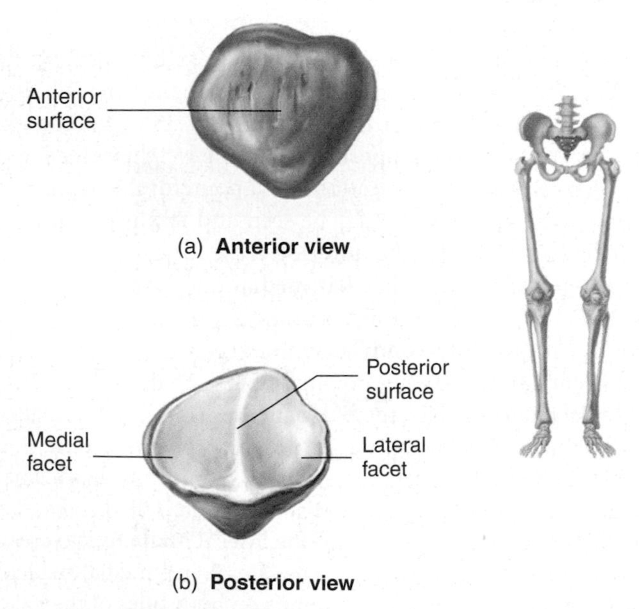
Anterior surface
(a) **Anterior view**
Posterior surface
Medial facet
Lateral facet
(b) **Posterior view**

FIGURE 7.34 Right Patella

Clinical
IMPACT

Patellar Defects

If the patella is severely fractured, the tendon from the quadriceps femoris muscle group may be torn, severely reducing muscle function. In extreme cases, it may be necessary to remove the patella to repair the tendon. Patella removal decreases the amount of power the quadriceps femoris muscle can generate at the tibia.

The patella normally tracks in the patellar groove on the anterodistal end of the femur. Abnormal tracking of the patella can become a problem in some teenagers, especially females. As a young woman's hips widen during puberty, the angles at the joints between the hips and the tibia may change considerably. As the knee becomes located more medially relative to the hip, the patella may be forced to track more laterally than normal. This lateral tracking may result in pain in the knees of physically active women.

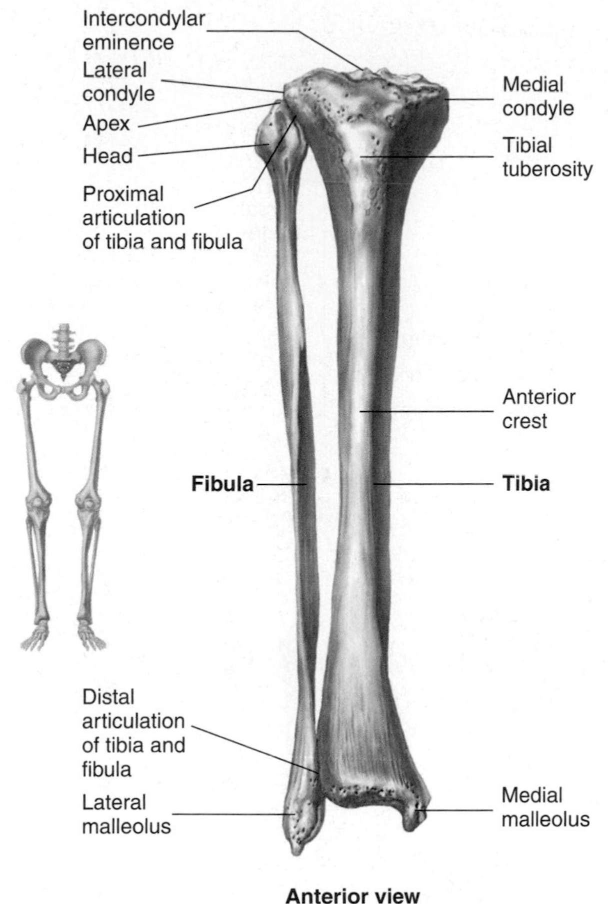
Intercondylar eminence
Lateral condyle
Apex
Head
Proximal articulation of tibia and fibula
Medial condyle
Tibial tuberosity
Anterior crest
Fibula
Tibia
Distal articulation of tibia and fibula
Lateral malleolus
Medial malleolus
Anterior view

FIGURE 7.35 Right Tibia and Fibula

> **Predict 6**
>
> Explain why modern ski boots are designed with high tops that extend partway up the leg.

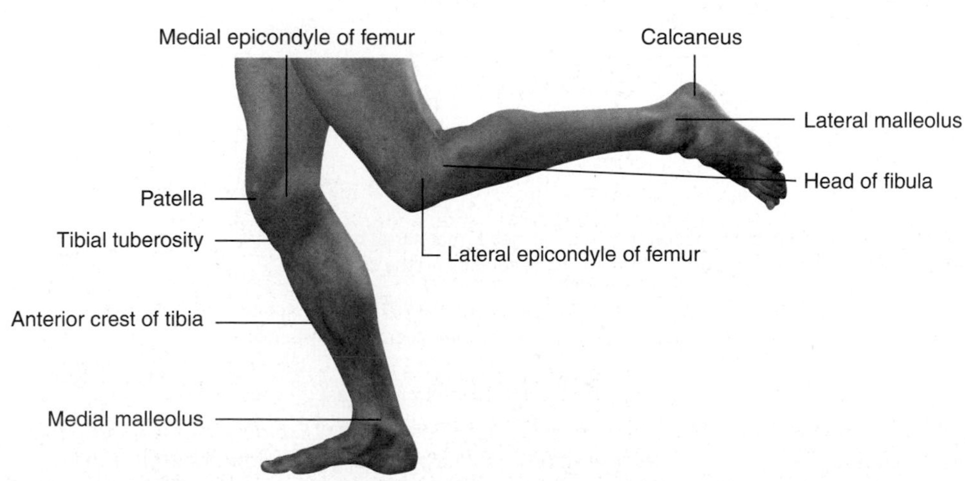
Medial epicondyle of femur
Calcaneus
Lateral malleolus
Head of fibula
Patella
Tibial tuberosity
Lateral epicondyle of femur
Anterior crest of tibia
Medial malleolus

FIGURE 7.36 Surface Anatomy Showing Bones of the Lower Limb

Foot

To continue our comparison of the lower limb with the upper limb, the proximal foot is relatively much larger than the wrist. It consists of seven **tarsal** (tar′săl; sole of the foot) **bones** (figure 7.37). Three of these bones—the talus, the calcaneus, and the navicular—are proximal in the foot but do not form a row. The **talus** (tā′lŭs), or ankle bone, articulates with the tibia and the fibula to form the ankle joint. It also articulates with the calcaneus and navicular bones. The **calcaneus** (kal-kā′nē-us; heel), the heel bone, is the largest and strongest bone in the foot. It is located inferior to the talus and supports that bone. The calcaneus protrudes posteriorly, is the important attachment point for the large calf muscles, and can be easily felt as the heel of the foot. The **navicular** (nă-vik′ū-lar), which is boat-shaped, lies between the talus posteriorly and the cuneiforms anteriorly. A mnemonic for the proximal three bones is No Thanks Cow—that is, Navicular, Talus, and Calcaneus.

Unlike the proximal tarsal bones, the distal four bones do form a row. The three medial, wedge-shaped bones are the **medial cuneiform,** the **intermediate cuneiform,** and the **lateral cuneiform,** collectively called the **cuneiforms** (kū′nē-i-fōrmz). The **cuboid** (kū′boyd), which is cube-shaped, is the most lateral of the distal row. A mnemonic for the distal row is MILC—that is, the Medial, Intermediate, and Lateral cuneiforms and the Cuboid.

The **metatarsal bones** and **phalanges** of the foot are arranged in a manner very similar to that of the metacarpal bones and phalanges of the hand, with the great toe analogous to the thumb (figure 7.37). Small sesamoid bones often form in the tendons of muscles attached to the great toe. The ball of the foot is the junction between the metatarsal bones and the phalanges.

> **Predict 7**
>
> A decubitus ulcer is a chronic ulcer that appears in pressure areas of skin overlying a bony prominence in bedridden or otherwise immobilized patients. Where are decubitus ulcers likely to occur?

The foot as a unit is convex dorsally and concave ventrally to form three major **arches:** the medial longitudinal arch, the lateral longitudiual arch, and the transverse arch (figure 7.37*b*). This system of arches distributes the weight of the body between the heel and the ball of the foot during standing and walking. As the foot is placed on the ground, weight is transferred from the tibia and fibula to the talus. From there, the weight is distributed first to the heel (calcaneus) and then through the arch system along the lateral side of the foot to the ball of the foot (head of the metatarsal bones). This effect can be observed when a person with wet, bare feet walks across a dry surface; the heel, the lateral border of the foot, and the ball of the foot make an imprint, but the middle of the plantar surface and the medial border leave

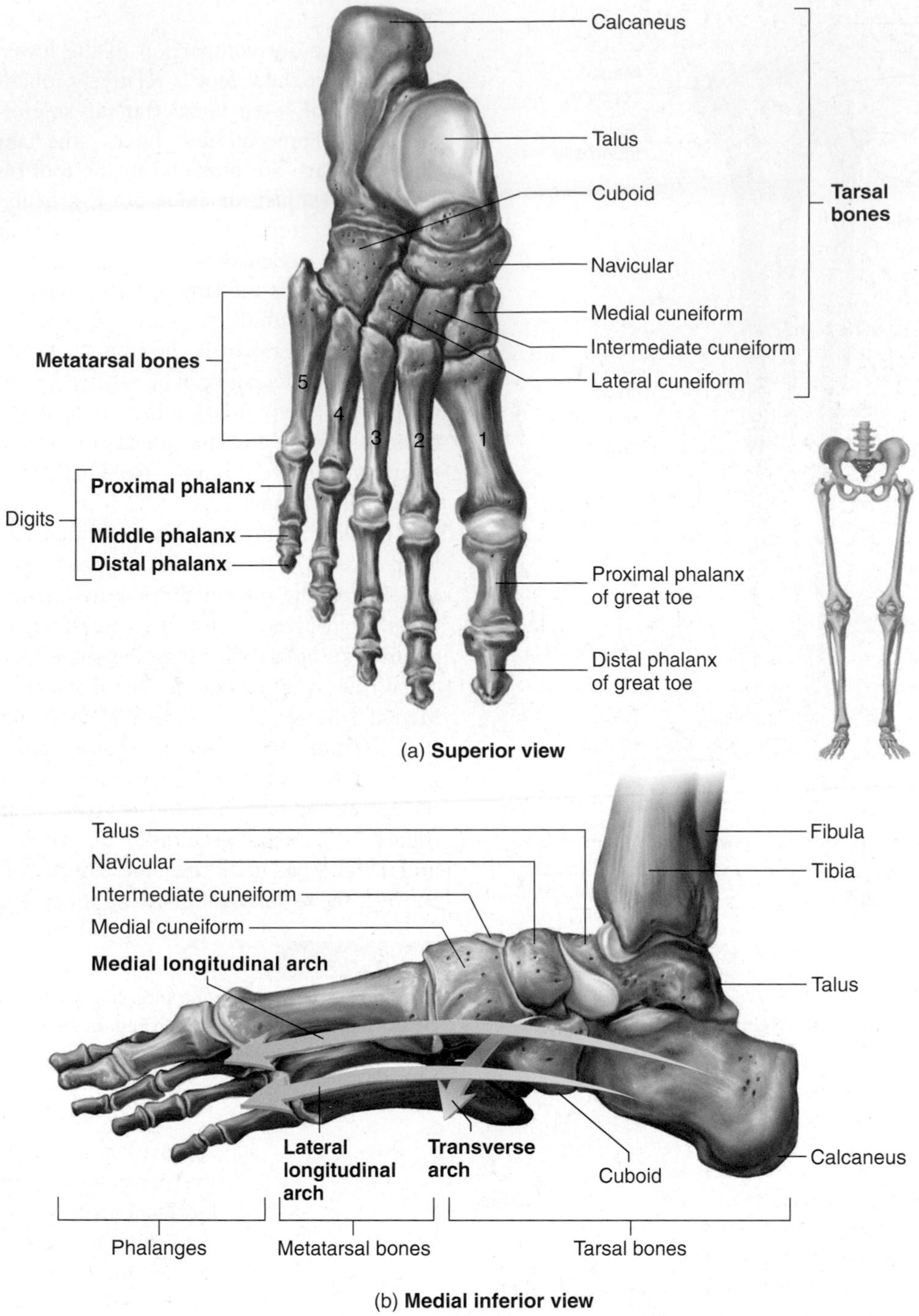

Calcaneus
Talus
Cuboid
Navicular
Medial cuneiform
Intermediate cuneiform
Lateral cuneiform
Tarsal bones
Metatarsal bones
5 4 3 2 1
Digits
Proximal phalanx
Middle phalanx
Distal phalanx
Proximal phalanx of great toe
Distal phalanx of great toe

(a) **Superior view**

Talus
Navicular
Intermediate cuneiform
Medial cuneiform
Medial longitudinal arch
Fibula
Tibia
Talus
Lateral longitudinal arch
Transverse arch
Cuboid
Calcaneus
Phalanges
Metatarsal bones
Tarsal bones

(b) **Medial inferior view**

FIGURE 7.37 APR **Bones of the Right Foot**

The medial longitudinal arch is formed by the calcaneus, the talus, the navicular, the cuneiforms, and three medial metatarsal bones. The lateral longitudinal arch is formed by the calcaneus, the cuboid, and two lateral metatarsal bones. The transverse arch is formed by the cuboid and the cuneiforms.

no impression because the arches on this side of the foot are higher than those on the lateral side. The shape of the arches is maintained by the configuration of the bones, the ligaments connecting them, and the muscles acting on the foot. The ligaments of the arches and some arch disorders are described in chapter 8.

ASSESS YOUR PROGRESS

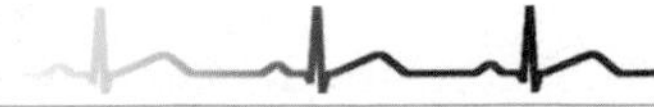

35. *What bones form the pelvic girdle? Explain why the pelvic girdle is more stable than the pectoral girdle. How does this stability affect movement?*

36. *Describe the structure of the coxal bone. What articulations does the coxal bone make?*

Case STUDY A "Broken Hip"

An 85-year-old woman who lived alone was found lying on her kitchen floor by her daughter, who had gone to check on her. The woman could not rise, even with help; when she tried, she experienced extreme pain in her right hip. Her daughter immediately dialed 911, and paramedics took the mother to the hospital.

The elderly woman's hip was x-rayed in the emergency room, and physicians determined that she had a fracture of the right femoral neck. A femoral neck fracture is commonly, but incorrectly, called a broken hip. Two days later, she received a partial hip replacement in which the head and neck of the femur were replaced, but not the acetabulum.

In the case of falls involving femoral neck fracture, it is not always clear whether the fall caused the femoral neck to fracture or a fracture of the femoral neck caused the fall.

Femoral neck fractures are among the most common injuries resulting in morbidity (disease) and mortality (death) in older adults. Four percent of women over age 85 experience femoral neck fractures each year. Despite treatment with anticoagulants and antibiotics, about 5% of patients with femoral neck fractures develop deep vein thrombosis (blood clot), and about 5% develop wound infections; either condition can be life-threatening. Hospital mortality is 1–7% among patients with femoral neck fractures, and nearly 20% of patients die within 3 months of the fracture. Only about 25% of victims ever fully recover from the injury.

Predict 8

The incidence of fracture of the femoral neck increases dramatically with age, and 81% of patients are women. The average age at such injury is 82. Why is the femoral neck so commonly injured (hint: see figure 7.1), and why are elderly women most commonly affected?

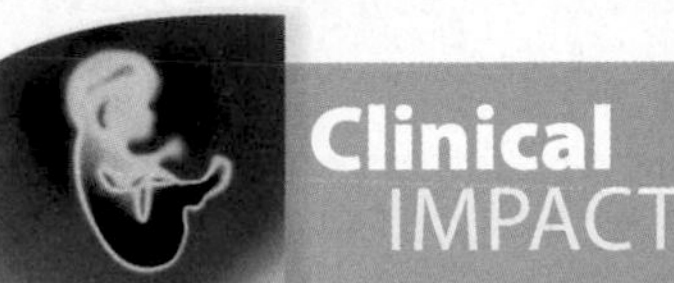

Clinical IMPACT Fractures of the Malleoli

Turning the plantar surface of the foot (the sole) outward so that it faces laterally is called eversion. Forceful eversion of the foot, as occurs when a person slips and twists the ankle or jumps and lands incorrectly on the foot, may cause the distal ends of the tibia and/or fibula to fracture (figure 7F*a*). Such fractures are common in soccer, football, and basketball players. When the foot is forcefully everted, the medial malleolus moves inferiorly toward the ground or the floor and the talus slides laterally, forcing the medial and lateral malleoli to separate. The ligament holding the medial malleolus to the tarsal bones is stronger than the bones it connects, and often it does not tear as the bones separate because of the eversion. Instead, the medial malleolus breaks. Also, as the talus slides laterally, the force can shear off the lateral malleolus or, more commonly, cause the fibula to break superior to the lateral malleolus. This type of injury to the tibia and fibula is often called a Pott fracture.

Turning the plantar surface of the foot inward so that it faces medially is called inversion. Forceful inversion of the foot can fracture the fibula just proximal to the lateral malleolus (figure 7F*b*). More often, because the ligament holding the medial malleolus to the tarsal bones is weaker than the bones it connects, inversion of the foot causes a sprain in which ligaments are damaged.

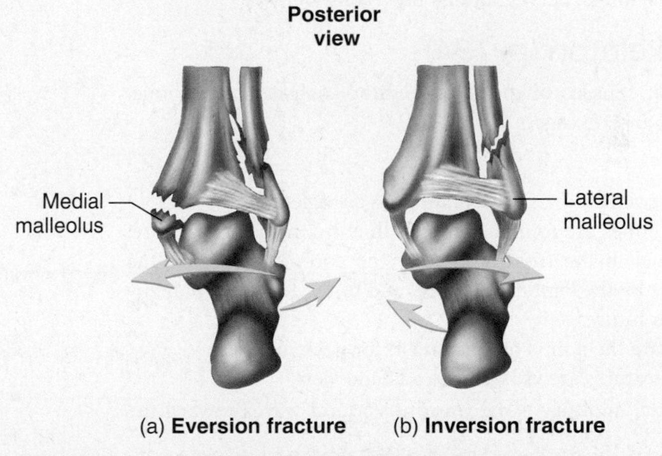
Posterior view
Medial malleolus
Lateral malleolus
(a) Eversion fracture
(b) Inversion fracture

FIGURE 7F Fracture of the Medial or Lateral Malleolus

37. *Name the important sites of muscle attachment on the pelvis.*
38. *Describe the differences between a male and a female pelvis.*
39. *Distinguish between the lower limb and the leg.*
40. *What is the function of the greater trochanter? The lesser trochanter?*
41. *Describe the function of the patella.*
42. *What is the function of the tibial tuberosity?*
43. *Name the seven tarsal bones. Which bones form the ankle joint? What bone forms the heel?*
44. *Describe the bones of the foot. How many phalanges are in each toe?*
45. *List the three arches of the foot, and describe their function.*

Answer

Learn to Predict ❮ From page 191

In this question we are asked to address four different topics; two are identification of anatomy, and two relate to the function of the skeletal system. The description of the surgery provided useful information for addressing these topics.

The anatomy questions ask us to identify (1) the specific parts of the skeletal system and (2) the layers of bone tissue the surgeon cut. We are told exactly where the surgeon made the cut: a longitudinal midline incision through the sternum. We learned in chapter 7 that the sternum is composed of the manubrium, the body, and the xiphoid process. All three of these parts would have to be cut to allow the surgeon to spread the sternum halves apart to expose the heart. We learned in chapter 6 that a bone, such as the sternum, is surrounded by a periosteum and is composed of an outer layer of compact bone and internal spongy bone. Thus, the layers of bone tissue cut during the surgery were the periosteum, compact bone, and spongy bone.

Next we must explain why Dave experienced back discomfort following the surgery, and why he needed to wait 2 months before resuming his normal swimming routine. Chapter 6 explained that movements between bones occur at joints, and chapter 7 outlined the relationships among the sternum, the ribs, and the vertebrae. We can assume that, when the two halves of the sternum were spread apart, stress was applied to other structures of the thoracic cavity. Therefore, we can attribute Dave's discomfort in his back to stress on the joints between the ribs and the vertebrae when the two halves of the sternum were separated.

The last question we need to address is why Dave needed to wait 2 months before resuming his normal swimming routine. Chapter 6 stated that the skeletal system plays a major role in body movement, specifically as attachment sites for skeletal muscles. Movement occurs when muscles contract and pull on the bone. Just like the repair of a bone fracture, healing of the sternum requires time. Contraction of the muscles involved in swimming would apply undue stress to the sternum; therefore, Dave had to delay swimming until the sternum had healed properly.

Answers to the rest of this chapter's Predict questions are in Appendix G.

Summary

The gross anatomy of the skeletal system considers the features of bone, cartilage, and ligaments that can be seen without the use of a microscope. Dried, prepared bones display the major features of bone but obscure the relationship between bone and soft tissue.

7.1 Skeletal Anatomy Overview (p. 192)

Bones have processes, smooth surfaces, and holes that are associated with ligaments, muscles, joints, nerves, and blood vessels.

7.2 Axial Skeleton (p. 194)

The axial skeleton consists of the skull, auditory ossicles, hyoid bone, vertebral column, and rib cage.

Skull

1. The skull, or cranium, can be thought of as a single unit.
2. The parietal bones are joined at the midline by the sagittal suture; they are joined to the frontal bone by the coronal suture, to the occipital bone by the lambdoid suture, and to the temporal bone by the squamous suture.
3. Nuchal lines are the points of attachment for neck muscles.
4. Several skull features are visible from a lateral view.
   - The external auditory canal transmits sound waves toward the eardrum.
   - Important neck muscles attach to the mastoid process.
   - The temporal lines are attachment points of the temporalis muscle.
   - The zygomatic arch, from the temporal and zygomatic bones, forms a bridge across the side of the skull.
5. Several skull features are visible from an anterior view.
   - The orbits contain the eyes.
   - The nasal cavity is divided by the nasal septum, and the hard palate separates the nasal cavity from the oral cavity.
   - Sinuses within bone are air-filled cavities. The paranasal sinuses, which connect to the nasal cavity, are the frontal, sphenoidal, and maxillary sinuses and the ethmoidal labyrinth.
   - The mandible articulates with the temporal bone.
6. Several skull features are inside the cranial cavity.
   - The crista galli is a point of attachment for one of the meninges.
   - The olfactory nerves extend into the roof of the nasal cavity through the cribriform plate.
   - The sella turcica is occupied by the pituitary gland.
   - The spinal cord and brain are connected through the foramen magnum.
7. Several features are on the inferior surface of the skull.
   - Occipital condyles are points of articulation between the skull and the vertebral column.
   - Blood reaches the brain through the internal carotid arteries, which pass through the carotid canals, and through the vertebral arteries, which pass through the foramen magnum.
   - Most blood leaves the brain through the internal jugular veins, which exit through the jugular foramina.
   - Styloid processes provide attachment points for three muscles involved in moving the tongue, hyoid bone, and pharynx.
   - The hard palate forms the floor of the nasal cavity.
8. The skull is composed of 22 bones.
   - The auditory ossicles, which function in hearing, are located inside the temporal bones.
   - The braincase protects the brain.
   - The facial bones protect the sensory organs of the head and are muscle attachment sites (mastication, facial expression, and eye muscles).
   - The mandible and maxillae possess alveolar processes with sockets for the attachment of the teeth.

Hyoid Bone

The hyoid bone, which "floats" in the neck, is the attachment site for the throat and tongue muscles.

Vertebral Column

1. The vertebral column provides flexible support and protects the spinal cord.
2. The vertebral column has four major curvatures: cervical, thoracic, lumbar, and sacral/coccygeal. Abnormal curvatures include lordosis in the lumbar region, kyphosis in the thoracic region, and scoliosis, an abnormal lateral curvature.
3. A typical vertebra consists of a body, a vertebral arch, and various processes.
   - Part of the body and the vertebral arch (pedicle and lamina) form the vertebral foramen, which contains and protects the spinal cord.
   - Spinal nerves exit through the intervertebral foramina.
   - The transverse and spinous processes are points of muscle and ligament attachment.
   - Vertebrae articulate with one another through the superior and inferior articular processes.
4. Adjacent bodies are separated by intervertebral disks. Each disk has a fibrous outer covering (annulus fibrosus) surrounding a gelatinous interior (nucleus pulposus).
5. Vertebrae can be distinguished by region.
   - All seven cervical vertebrae have transverse foramina, and most have bifid spinous processes.
   - The 12 thoracic vertebrae are characterized by long, downward-pointing spinous processes and demifacets.
   - The five lumbar vertebrae have thick, heavy bodies and processes.
   - The sacrum consists of five fused vertebrae and attaches to the coxal bones to form the pelvis.
   - The coccyx consists of four fused vertebrae attached to the sacrum.

Rib Cage

1. The rib cage (consisting of the ribs, their associated costal cartilages, and the sternum) protects the thoracic organs and changes volume during respiration.
2. Twelve pairs of ribs attach to the thoracic vertebrae. They are divided into seven pairs of true ribs and five pairs of false ribs. Two pairs of false ribs are floating ribs.
3. The sternum is composed of the manubrium, the body, and the xiphoid process.

7.3 Appendicular Skeleton (p. 222)

The appendicular skeleton consists of the upper and lower limbs and the girdles that attach the limbs to the body.

Pectoral Girdle and Upper Limb

1. The upper limb is attached loosely and functions in grasping and manipulation.
2. The pectoral girdle consists of the scapulae and clavicles.
   - The scapula articulates with the humerus and the clavicle. It is an attachment site for shoulder, back, and arm muscles.
   - The clavicle holds the shoulder away from the body, permitting the arm to move freely.
3. The arm bone is the humerus.
   - The humerus articulates with the scapula (head), the radius (capitulum), and the ulna (trochlea).
   - Sites of muscle attachment are the greater and lesser tubercles, the deltoid tuberosity, and the epicondyles.
4. The forearm contains the ulna and the radius.
   - The ulna and the radius articulate with each other and with the humerus and the wrist bones.
   - The wrist ligaments attach to the styloid processes of the radius and the ulna.
5. Eight carpal bones, or wrist bones, are arranged in two rows.
6. The hand consists of five metacarpal bones.
7. The phalanges are digital bones. Each finger has three phalanges, and the thumb has two phalanges.

Pelvic Girdle and Lower Limb

1. The lower limb is attached solidly to the coxal bone and functions in support and movement.
2. The pelvic girdle consists of the right and left coxal bones and the sacrum. Each coxal bone is formed by the fusion of the ilium, the ischium, and the pubis.
   - The coxal bones articulate with each other (symphysis pubis) and with the sacrum (sacroiliac joint) and the femur (acetabulum).
   - Important sites of muscle attachment are the iliac crest, the iliac spines, and the ischial tuberosity.
   - The female pelvis has a larger pelvic inlet and outlet than the male pelvis.
3. The thighbone is the femur.
   - The femur articulates with the coxal bone (head), the tibia (medial and lateral condyles), and the patella (patellar groove).
   - Sites of muscle attachment are the greater and lesser trochanters and the adductor tubercle.
   - Sites of ligament attachment are the lateral and medial epicondyles.
4. The leg consists of the tibia and the fibula.
   - The tibia articulates with the femur, the fibula, and the talus. The fibula articulates with the tibia and the talus.
   - Tendons from the thigh muscles attach to the tibial tuberosity.
5. Seven tarsal bones form the proximal portion of the foot.
6. The foot consists of five metatarsal bones.
7. The toes have three phalanges each, except for the big toe, which has two.
8. The bony arches transfer weight from the heels to the toes and allow the foot to conform to many different positions.

REVIEW AND COMPREHENSION

1. Which of these is part of the appendicular skeleton?
   a. cranium c. clavicle e. vertebra
   b. ribs d. sternum
2. A knoblike lump on a bone is called a
   a. spine. c. tuberosity. e. ramus.
   b. facet. d. sulcus.
3. The perpendicular plate of the ethmoid and the __________ form the nasal septum.
   a. palatine process of the maxilla c. vomer e. lacrimal bone
   b. horizontal plate of the palatine d. nasal bone
4. Which of these bones does *not* contain a paranasal sinus?
   a. ethmoid c. frontal e. maxilla
   b. sphenoid d. temporal
5. The mandible articulates with the skull at the
   a. styloid process. c. mandibular fossa. e. medial pterygoid.
   b. occipital condyle. d. zygomatic arch.
6. The nerves for the sense of smell pass through the
   a. cribriform plate. d. optic canal.
   b. nasolacrimal canal. e. orbital fissure.
   c. internal auditory canal.

7. The major blood supply to the brain enters through the
   a. foramen magnum.
   b. carotid canals.
   c. jugular foramina.
   d. Both a and b are correct.
   e. All of these are correct.
8. The site of the sella turcica is the
   a. sphenoid bone.
   b. maxillae.
   c. frontal bone.
   d. ethmoid bone.
   e. temporal bone.
9. Which of these bones is *not* in contact with the sphenoid bone?
   a. maxilla
   b. inferior nasal concha
   c. ethmoid
   d. parietal
   e. vomer
10. A herniated disk occurs when
    a. the annulus fibrosus ruptures.
    b. the intervertebral disk slips out of place.
    c. the spinal cord ruptures.
    d. too much fluid builds up in the nucleus pulposus.
    e. All of these are correct.
11. The weight-bearing portion of a vertebra is the
    a. vertebral arch.
    b. articular process.
    c. body.
    d. transverse process.
    e. spinous process.
12. Transverse foramina are found only in
    a. cervical vertebrae.
    b. thoracic vertebrae.
    c. lumbar vertebrae.
    d. the sacrum.
    e. the coccyx.
13. Which of these statements concerning ribs is correct?
    a. The true ribs attach directly to the sternum with costal cartilage.
    b. There are five pairs of floating ribs.
    c. The head of the rib attaches to the transverse process of the vertebra.
    d. Vertebrochondral ribs are classified as true ribs.
    e. Floating ribs do not attach to vertebrae.
14. The point where the scapula and clavicle connect is the
    a. coracoid process.
    b. styloid process.
    c. glenoid cavity.
    d. acromion process.
    e. capitulum.
15. The distal medial process of the humerus to which the ulna joins is the
    a. epicondyle.
    b. deltoid tuberosity.
    c. malleolus.
    d. capitulum.
    e. trochlea.
16. Which of these is *not* a point of muscle attachment on the pectoral girdle or upper limb?
    a. epicondyles
    b. mastoid process
    c. radial tuberosity
    d. spine of scapula
    e. greater tubercle
17. The bone(s) of the foot on which the tibia rests is (are) the
    a. talus.
    b. calcaneus.
    c. metatarsal bones.
    d. navicular.
    e. phalanges.
18. The projection on the coxal bone of the pelvic girdle that is used as a landmark for finding an injection site is the
    a. ischial tuberosity.
    b. iliac crest.
    c. anterior superior iliac spine.
    d. posterior inferior iliac spine.
    e. ischial spine.
19. When comparing the pectoral girdle with the pelvic girdle, which of these statements is correct?
    a. The pectoral girdle has greater mass than the pelvic girdle.
    b. The pelvic girdle is more firmly attached to the body than the pectoral girdle.
    c. The pectoral girdle has the limbs more securely attached than the pelvic girdle.
    d. The pelvic girdle allows greater mobility than the pectoral girdle.
20. When comparing a male pelvis with a female pelvis, which of these statements is correct?
    a. The pelvic inlet in males is larger and more circular.
    b. The subpubic angle in females is less than 90 degrees.
    c. The ischial spines in males are closer together.
    d. The sacrum in males is broader and less curved.
21. A site of muscle attachment on the proximal end of the femur is the
    a. greater trochanter.
    b. epicondyle.
    c. greater tubercle.
    d. intercondylar eminence.
    e. condyle.

Answers in Appendix E

CRITICAL THINKING

1. A patient has an infection in the nasal cavity. Name seven adjacent structures to which the infection could spread.
2. A patient is unconscious. X-rays reveal that the superior articular facet of the atlas has been fractured. Would this condition result from falling on the top of the head or being hit in the jaw with an uppercut? Explain.
3. If the vertebral column is forcefully rotated, what part of a vertebra is most likely to be damaged? In what area of the vertebral column is such damage most likely?
4. What might be the consequences of breaking both the ulna and the radius if the two bones fuse to each other during repair of the fracture?
5. A paraplegic person develops decubitus ulcers (pressure sores) on the buttocks from sitting in a wheelchair for extended periods. Name the bony protuberance responsible.
6. Why do women tend to suffer more knee pain and injuries than men?
7. On the basis of the bone structure of the lower limb, explain why it is easier to turn the foot medially (sole of the foot facing toward the midline of the body) than laterally. Why is it easier to bend the wrist medially than laterally?
8. Justin Time leaped from his second-floor hotel room to avoid burning to death in a fire. If he landed on his heels, what bone was likely fractured? Unfortunately for Justin, a 240-pound firefighter ran by and stepped heavily on the proximal part of Justin's foot (not the toes). Which bones could have been broken?

Answers in Appendix F

Visit this book's website at **www.mhhe.com/seeley10** for chapter quizzes, interactive learning exercises, and other study tools.

Mc Graw Hill
connect plus+
ANATOMY & PHYSIOLOGY

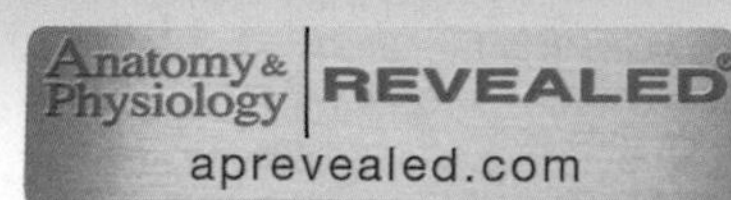
Anatomy & Physiology
REVEALED
aprevealed.com

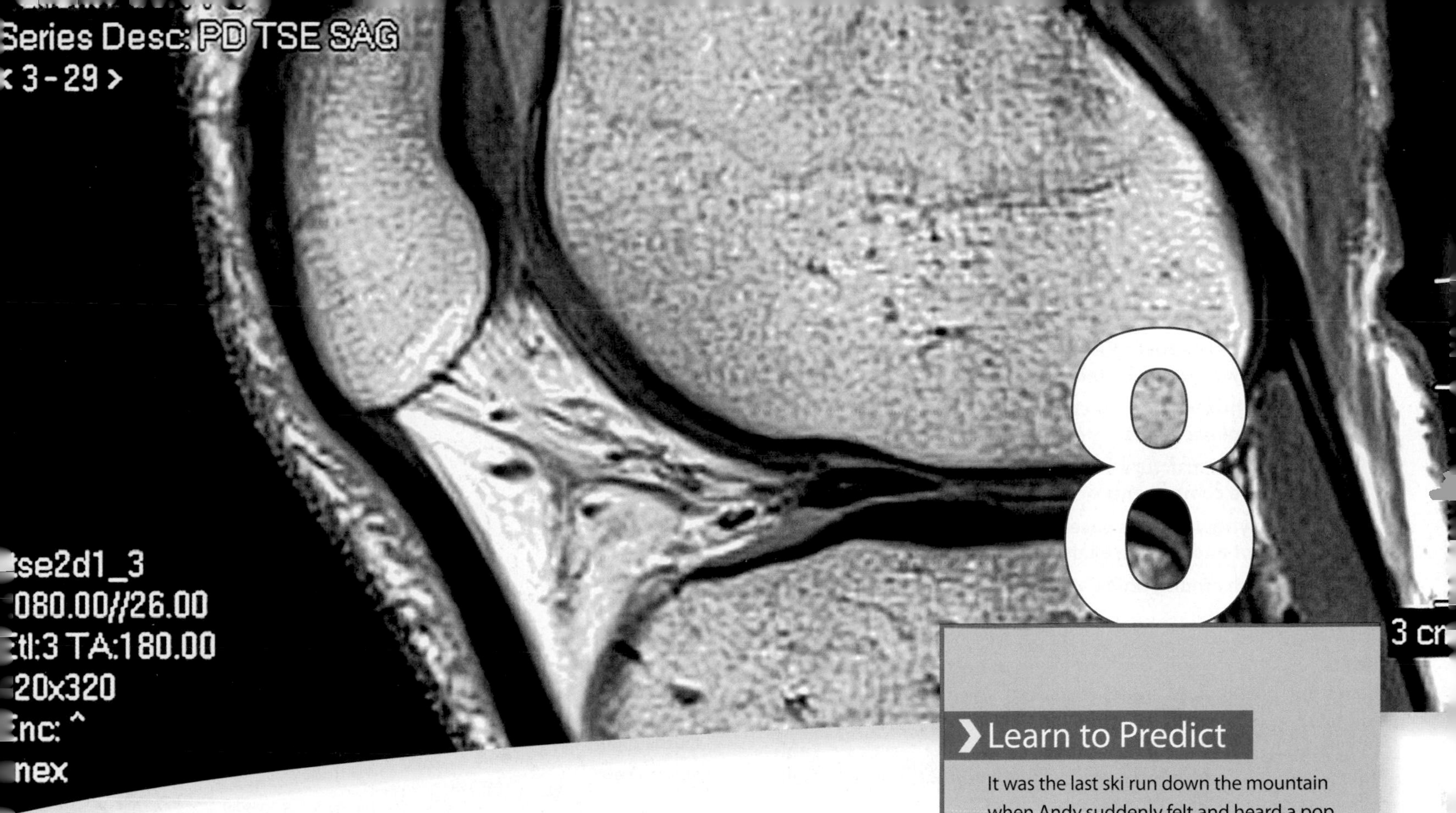

8

Joints and Movement

Watch a skier attack moguls on a mountain, or a player dribble a ball past defenders, and you are watching joints in action. Muscles pull on bones to make them move, but movement would not be possible without the joints between the bones. A joint is a place where two or more bones come together. Although we usually think of joints as movable, that is not always the case. Many joints allow only limited movement, and others allow no apparent movement at all. The structure of a given joint is directly correlated with its degree of movement.

Movable joints are places in the body where the bones move in close contact with each other. When working with machines, we know that the parts that make contact require the most maintenance. But in our bodies we tend to pay little attention to the movable joints until disease or damage makes movement very difficult. If movement is restricted—even in a highly movable joint—at any time during a person's life, the joint may be transformed into a nonmovable joint.

❯ Learn to Predict

It was the last ski run down the mountain when Andy suddenly felt and heard a pop in his left knee as his ski caught the edge of a mogul. When he tried to stand up, the pain immediately convinced him to wait for the ski patrol. In the emergency clinic, x-rays showed that bones were not broken. However, ligaments of the knee joint were clearly damaged, based on a clinical examination of mobility called the anterior drawer test. In this test, there was greater mobility than normal when the lower leg was moved in an anterior direction relative to the thigh while holding the knee flexed at a 90-degree angle. A subsequent magnetic resonance image confirmed that ligaments were torn. Fortunately, after reconstructive surgery and physical therapy to maintain range of motion and rebuild strength, Andy was soon able to play sports and ski again. In chapters 6 and 7 you learned about bone anatomy, growth, and repair. After focusing on the structure and function of joints in chapter 8, describe the likely injury to Andy's knee and how this ligament damage would affect mobility of the tibia relative to the femur.

Module 5
Skeletal System

Anatomy & Physiology
REVEALED
aprevealed.com

Photo: Magnetic resonance image (MRI) of the left knee of one of the authors (A. Russo) after a ski accident. A sagittal view of the joint is shown, with the kneecap on the left.

8.1 Classes of Joints

LEARNING OUTCOMES

After reading this section, you should be able to

A. Describe the two systems for classifying joints.

B. Explain the structure of a fibrous joint, list the three types, and give an example of each type.

C. Contrast the two types of cartilaginous joints and give examples of each type.

D. Illustrate the structure of a synovial joint and explain the roles of the components of a synovial joint.

E. Classify synovial joints based on the shape of the bones in the joint and give an example of each type.

F. Distinguish among uniaxial, biaxial, and multiaxial synovial joints.

Joints, or *articulations,* are commonly named according to the bones or portions of bones that join together; for example, the temporomandibular joint is between the temporal bone and the mandible. Some joints are simply given the Greek or Latin equivalent of the common name, such as **cubital** (kū′bi-tăl; cubit, elbow or forearm) **joint** for the elbow joint.

Joints are classified structurally as fibrous, cartilaginous, or synovial, according to the major connective tissue type that binds the bones together and whether a fluid-filled joint capsule is present. Joints are also classified according to their degree of motion as *synarthroses* (nonmovable joints), *amphiarthroses* (slightly movable joints), or *diarthroses* (freely movable joints). Because this functional classification is somewhat limited, it is not used in this book; rather, our discussions are based on the structural classification scheme, which is more precise.

Fibrous Joints

Fibrous joints consist of two bones that are united by fibrous connective tissue, have no joint cavity, and exhibit little or no movement. Joints in this group are further subdivided on the basis of structure as sutures, syndesmoses, or gomphoses (table 8.1).

Sutures

Sutures (soo′choorz) are seams between the bones of the skull (figure 8.1). Some sutures may become completely immovable in older adults. Few sutures are smooth, and the opposing bones often interdigitate (have interlocking, fingerlike processes). This interdigitation adds considerable stability to sutures. The tissue between the bones is dense regular collagenous connective tissue, and the periosteum on the inner and outer surfaces of the adjacent bones continues over the joint. The two layers of periosteum plus the dense fibrous connective tissue in between form a **sutural ligament.**

In a newborn, some of the sutures have a membranous area called a **fontanel** (fon′tă-nel′; little fountain, so named because the membrane can be seen to move with the pulse; soft spot). The fontanels make the skull flexible during the birth process and allow for growth of the head after birth (figure 8.2).

TABLE 8.1 Fibrous and Cartilaginous Joints

Class and Example of Joint	Bones or Structures Joined	Movement
Fibrous Joints		
Sutures		
Coronal	Frontal and parietal	None
Lambdoid	Occipital and parietal	None
Sagittal	The two parietal bones	None
Squamous	Parietal and temporal	Slight
Syndesmoses		
Radioulnar	Ulna and radius	Slight
Stylohyoid	Styloid process and hyoid bone	Slight
Stylomandibular	Styloid process and mandible	Slight
Tibiofibular	Tibia and fibula	Slight
Gomphoses		
Dentoalveolar	Tooth and alveolar process	Slight
Cartilaginous Joints		
Synchondroses		
Epiphyseal plate	Diaphysis and epiphysis of a long bone	None
Sternocostal	Anterior cartilaginous part of first rib; between rib and sternum	Slight
Sphenooccipital	Sphenoid and occipital	None
Symphyses		
Intervertebral	Bodies of adjacent vertebrae	Slight
Manubriosternal	Manubrium and body of sternum	None
Symphysis pubis	The two coxal bones	None except during childbirth
Xiphisternal	Xiphoid process and body of sternum	None

The margins of bones within sutures are sites of continuous intramembranous bone growth, and many sutures eventually become ossified. For example, ossification of the suture between the two frontal bones occurs shortly after birth, so that they usually form a single frontal bone in the adult skull. In most normal adults, the coronal, sagittal, and lambdoid sutures are not fused. However, in some very old adults, even these sutures ossify. When a suture becomes fully ossified, it becomes a **synostosis** (sin-os-tō′sis). A synostosis results when two bones grow together across a joint to form a single bone.

Predict 2

Predict the result of a sutural synostosis that occurs prematurely in a child's skull before the brain has reached its full size.

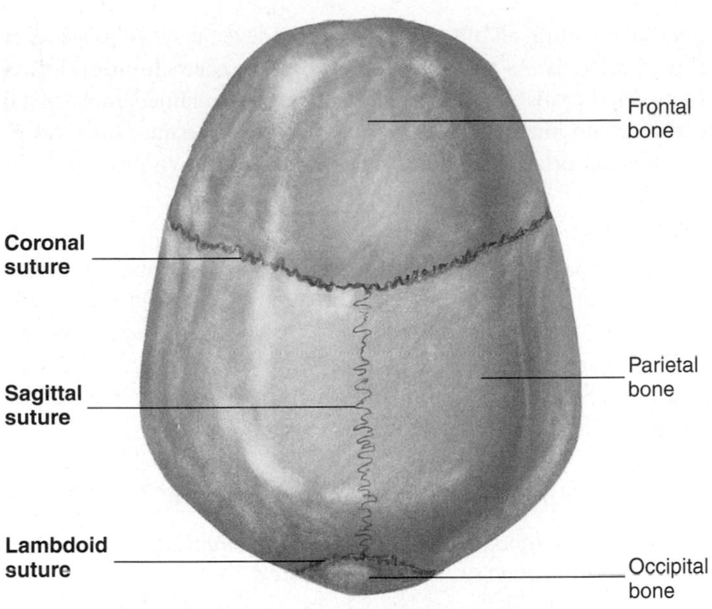
Frontal bone
Coronal suture
Sagittal suture
Parietal bone
Lambdoid suture
Occipital bone

FIGURE 8.1 **Sutures, Superior View of the Skull**

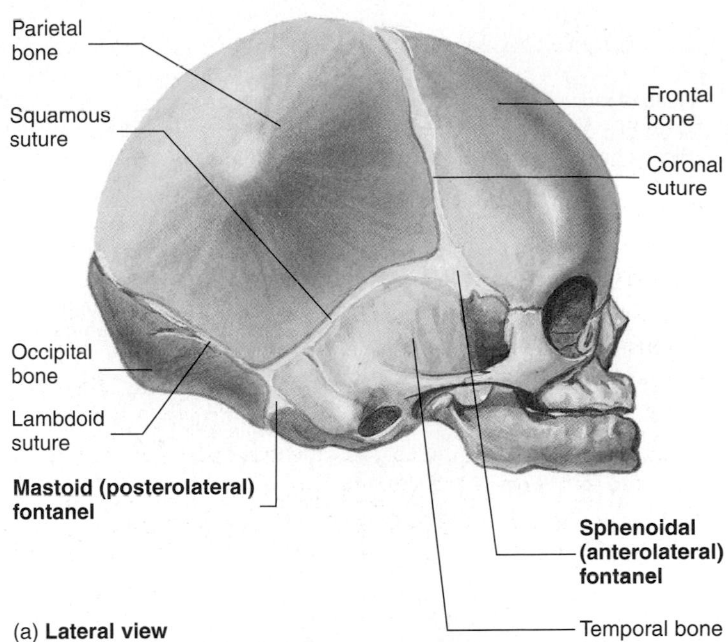
Parietal bone
Squamous suture
Frontal bone
Coronal suture
Occipital bone
Lambdoid suture
Mastoid (posterolateral) fontanel
Sphenoidal (anterolateral) fontanel
Temporal bone
(a) Lateral view

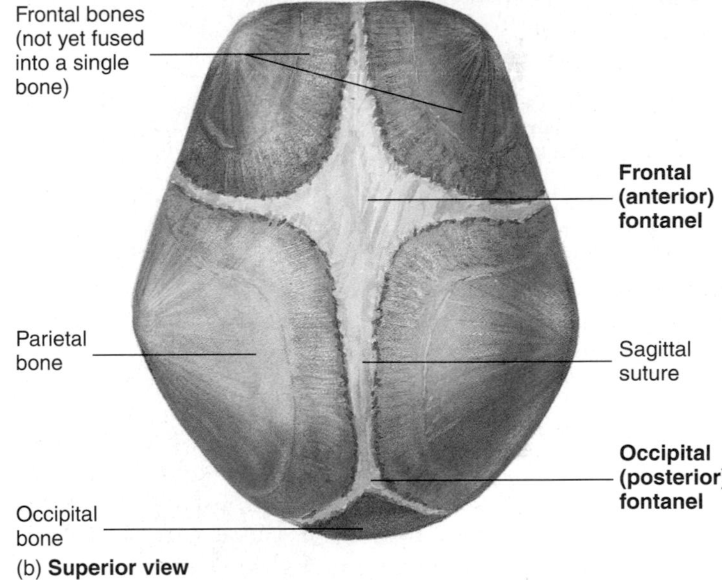
Frontal bones (not yet fused into a single bone)
Frontal (anterior) fontanel
Parietal bone
Sagittal suture
Occipital (posterior) fontanel
Occipital bone
(b) Superior view

FIGURE 8.2 **Fetal Skull Showing Fontanels and Sutures**

Syndesmoses

A **syndesmosis** (sin′dez-mō′sis) is a fibrous joint in which the bones are farther apart than in a suture and are joined by ligaments. Some movement may occur at syndesmoses because the ligaments are flexible; this occurs in the radioulnar syndesmosis, for example, which binds the radius and ulna together (figure 8.3).

Gomphoses

Gomphoses (gom-fō′sēz) are specialized joints consisting of pegs that fit into sockets and are held in place by fine bundles of regular collagenous connective tissue. The only gomphoses in the human body are the joints between the teeth and the sockets (alveoli) of the mandible and maxillae (figure 8.4). The connective tissue bundles between the teeth and their sockets are called **periodontal** (per′ē-ō-don′tăl) **ligaments;** they allow a slight amount of "give" to the teeth during mastication.

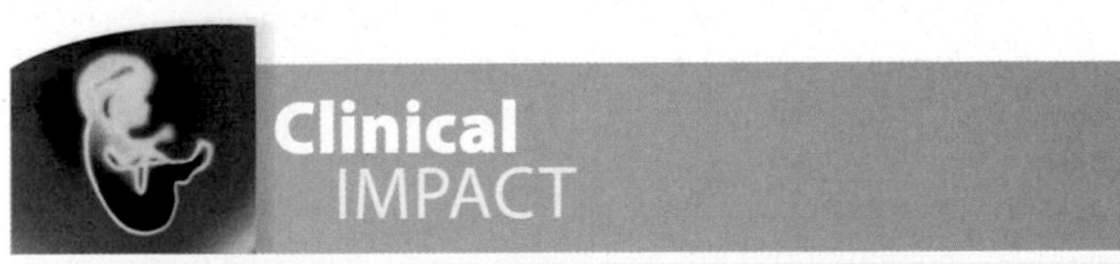
Clinical IMPACT

Gingivitis

The gingiva, or gums, are the soft tissues covering the alveolar process. Neglect of the teeth can result in **gingivitis,** an inflammation of the gingiva that is often caused by bacterial infection. Left untreated, gingivitis can spread to the tooth socket, resulting in periodontal disease, the leading cause of tooth loss in the United States. In periodontal disease, plaque and bacteria accumulate, resulting in inflammation that gradually destroys the periodontal ligaments and the bone. Teeth may become so loose that they come out of their sockets. Proper brushing, flossing, and professional cleaning to remove plaque can usually prevent gingivitis and periodontal disease.

Cartilaginous Joints

Cartilaginous joints unite two bones by means of either hyaline cartilage or fibrocartilage (table 8.1). Joints containing hyaline cartilage are called synchondroses; joints containing fibrocartilage are called symphyses.

Synchondroses

A **synchondrosis** (sin′kon-drō′sis) consists of two bones joined by hyaline cartilage where little or no movement occurs (figure 8.5*a*).

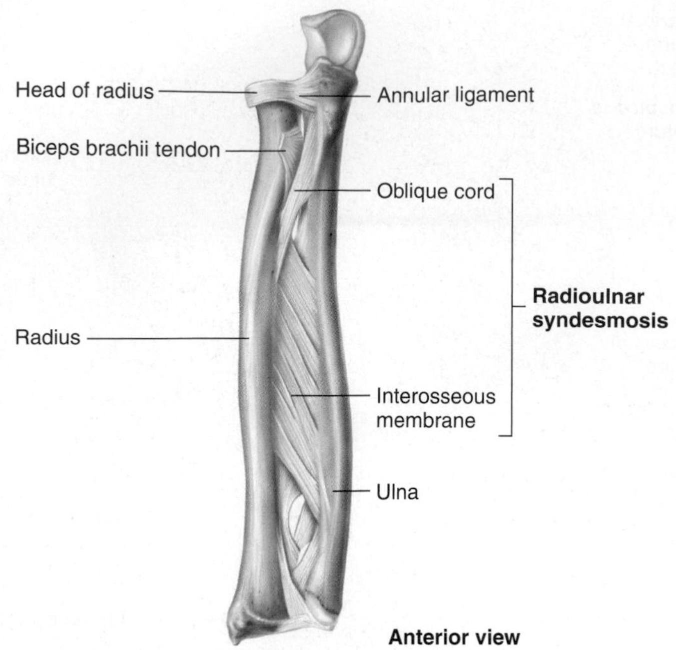
Head of radius
Annular ligament
Biceps brachii tendon
Oblique cord
Radioulnar syndesmosis
Radius
Interosseous membrane
Ulna
Anterior view

FIGURE 8.3 Right Radioulnar Syndesmosis

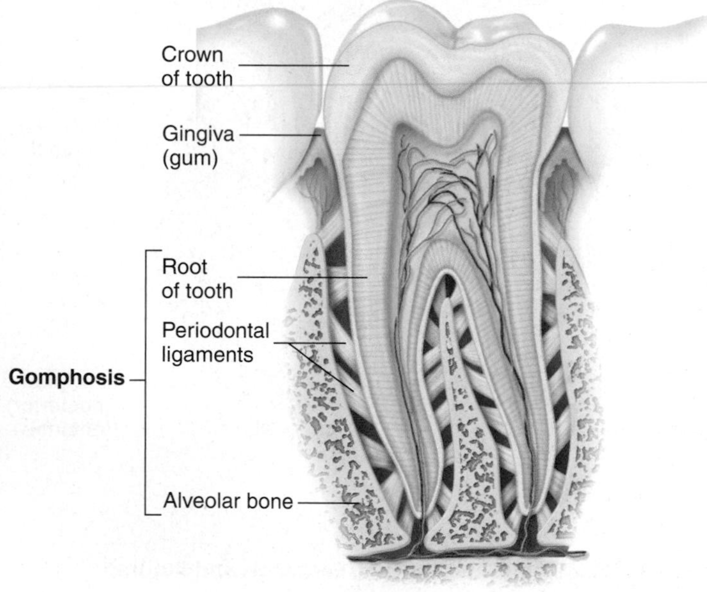
Crown of tooth
Gingiva (gum)
Root of tooth
Periodontal ligaments
Gomphosis
Alveolar bone

FIGURE 8.4 Gomphosis Between a Tooth and Alveolar Bone of the Mandible

The epiphyseal plates of growing bones are synchondroses (figure 8.5*b*). Most synchondroses are temporary. In the case of epiphyseal plates, the synchondrosis is converted to a synostosis as bone replaces the existing cartilage (see "Bone Growth" in chapter 6). Other synchondroses are converted to synovial joints, whereas others persist throughout life. An example is the sternocostal synchondrosis between the first rib and the sternum by way of the first costal cartilage (figure 8.5*c*). All the costal cartilages begin as synchondroses but, because movement occurs between them and the sternum, all but the first usually develop synovial joints at those junctions. As a result, even though the **costochondral joints** (between the ribs and the costal cartilages) are retained, most costal cartilages no longer qualify as synchondroses because one end of the cartilage attaches to bone (the sternum) by a synovial joint.

Symphyses

A **symphysis** (sim′fi-sis) consists of fibrocartilage uniting two bones. Examples of symphyses include the junction between the manubrium and the body of the sternum (figure 8.5*c*), the symphysis pubis (figure 8.6), and the intervertebral disks (see figures 7.13 and 7.15). Some of these joints are slightly movable because of the somewhat flexible nature of fibrocartilage.

ASSESS YOUR PROGRESS

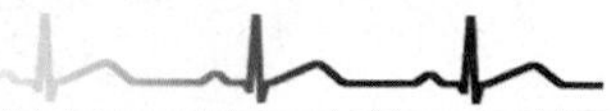

1. *What two standards are used to classify joints? List and describe the classification system used in this text.*
2. *What are the characteristics of a fibrous joint? Name the three types, and give an example of each.*
3. *What is a synostosis? Where are periodontal ligaments found?*
4. *Name the two types of cartilaginous joints, tell the type of cartilage present, and give an example of each.*

Synovial Joints

Synovial (si-nō′vē-ăl) **joints** contain synovial fluid and allow considerable movement between articulating bones (figure 8.7). These joints are anatomically more complex than fibrous and

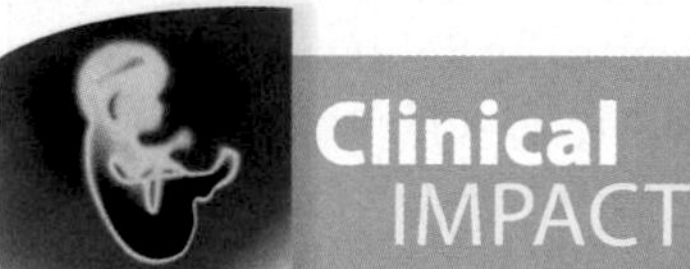

Clinical IMPACT

Joint Changes During Pregnancy

During pregnancy, certain hormones, such as estrogen, progesterone, and relaxin, act on the connective tissue of joints, particularly the symphysis pubis, making them more stretchable and allowing the joints to loosen. This change allows the pelvic opening to enlarge at the time of delivery. After delivery, the connective tissue of the symphysis pubis returns to its original condition. However, the enlarged pelvic opening may not return completely to its original size, and the woman may have slightly wider hips after the birth of the child.

The same hormones may act on the connective tissue of other joints in the body, such as the arches of the feet, causing them to relax, which may result in fallen arches (see Clinical Impact, "Arch Problems," in section 8.4). They may also act on some of a baby's joints, such as the hip, causing them to become more mobile than normal. Increased mobility of the hip can result in congenital (appearing at birth) partial or complete dislocation of the hip. Congenital hip dislocation occurs approximately once in every 1000 births. Fortunately, if detected early the condition can be corrected using a specialized harness or traction to realign the joint.

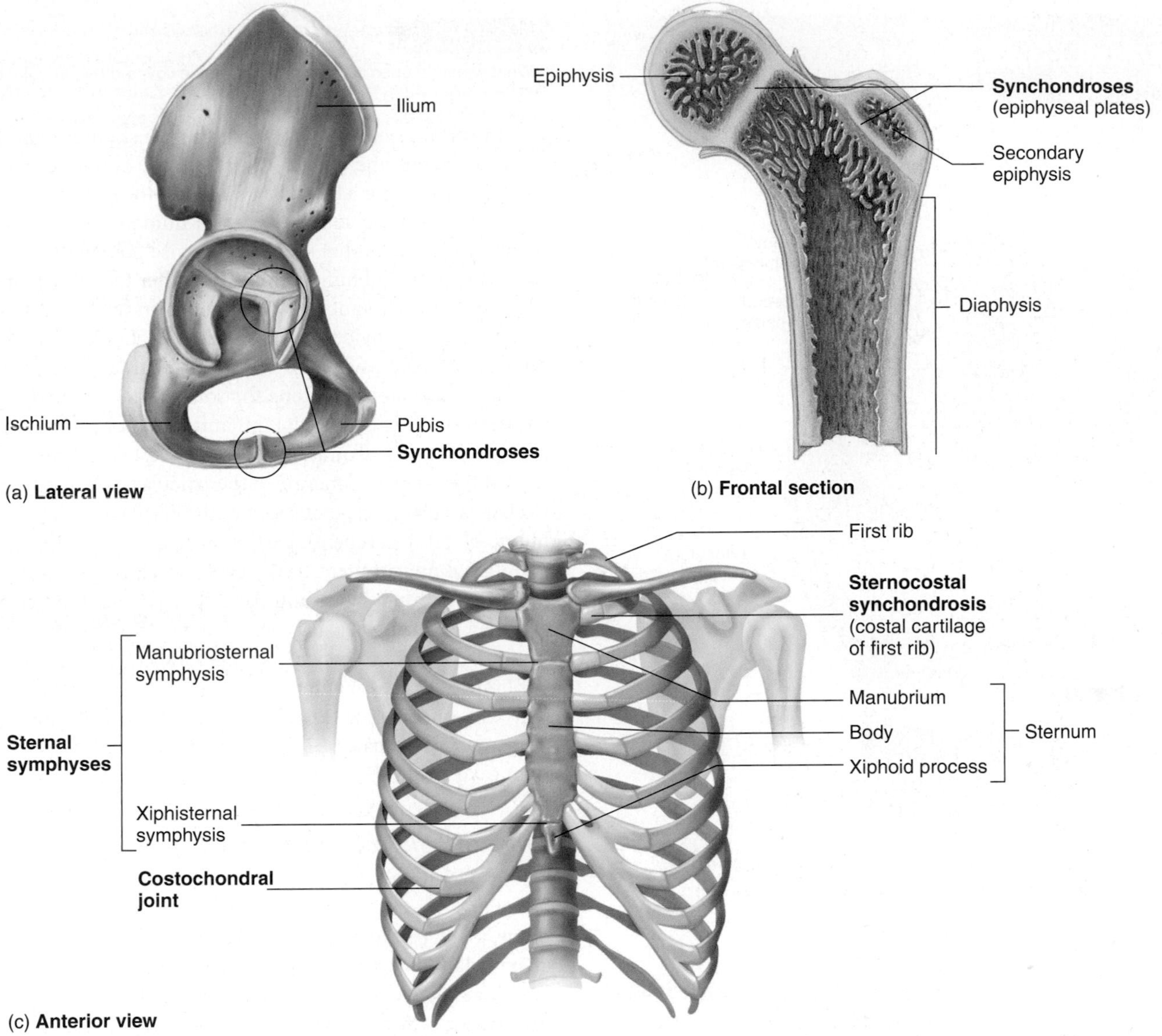

FIGURE 8.5 Synchondroses

(*a*) Synchondroses (epiphyseal plates) between the developing bones of the coxal bone of a fetus. (*b*) Epiphyseal plates (frontal section of proximal femur of a child). (*c*) Sternocostal synchondroses in a mature individual.

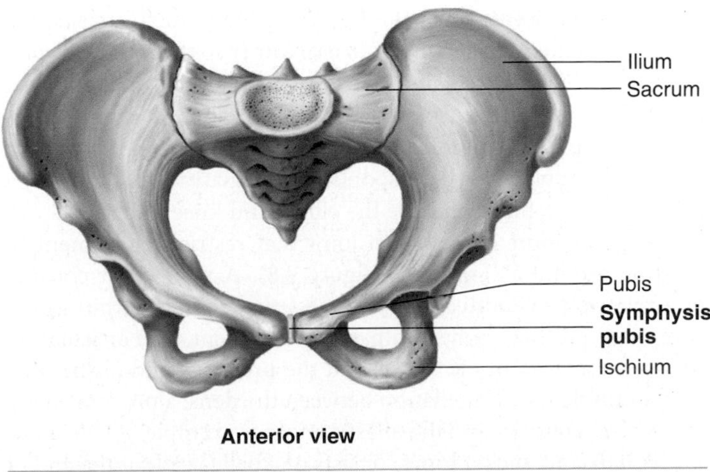
Ilium
Sacrum
Pubis
Symphysis pubis
Ischium
Anterior view

FIGURE 8.6 Symphysis Pubis

cartilaginous joints. Most joints that unite the bones of the appendicular skeleton are synovial joints, reflecting the far greater mobility of the appendicular skeleton compared with the axial skeleton.

The articular surfaces of bones within synovial joints are covered with a thin layer of hyaline cartilage called **articular cartilage,** which provides a smooth surface where the bones meet. In some synovial joints, a flat plate or pad of fibrocartilage, called an **articular disk,** lies between the articular cartilages of bones. The circumference of the disk is attached to the fibrous capsule. Joints with articular disks include the temporomandibular, sternoclavicular, and acromioclavicular joints. A **meniscus** (mĕ-nis′kus; pl. menisci, crescent-shaped) is a fibrocartilage pad found in joints such as the knee and wrist. A meniscus is much like an articular disk with a hole in the center. The circumference of the meniscus is attached to the fibrous joint capsule.

FUNDAMENTAL Figure

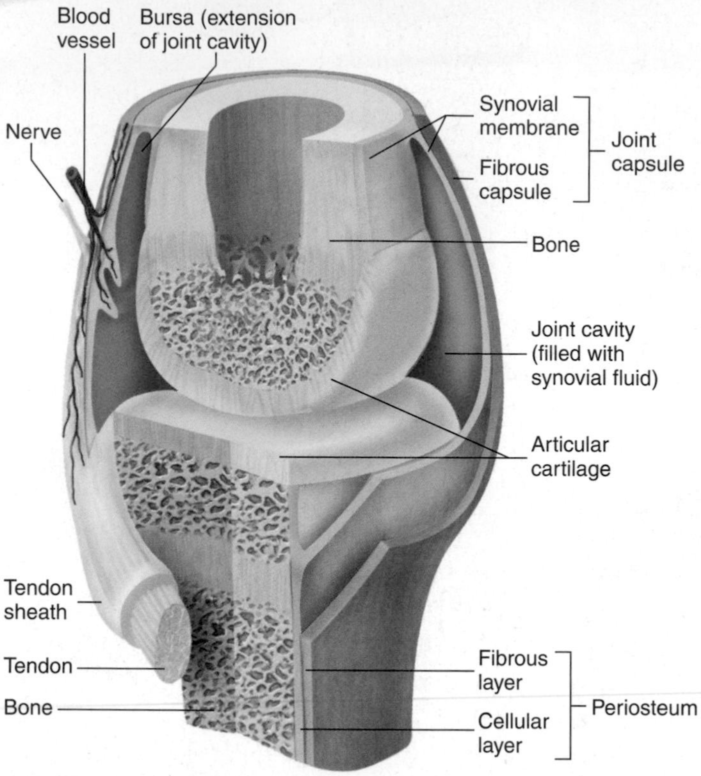

Blood vessel
Bursa (extension of joint cavity)
Nerve
Synovial membrane
Fibrous capsule
Joint capsule
Bone
Joint cavity (filled with synovial fluid)
Articular cartilage
Tendon sheath
Tendon
Bone
Fibrous layer
Cellular layer
Periosteum

FIGURE 8.7 APR **General Structure of a Synovial Joint**

The articular surfaces of the bones that meet at a synovial joint are enclosed within a synovial **joint cavity,** which is surrounded by a **joint capsule.** This capsule helps hold the bones together while still allowing for movement. The joint capsule consists of two layers: an outer **fibrous capsule** and an inner **synovial membrane** (figure 8.7). The fibrous capsule consists of dense irregular connective tissue and is continuous with the fibrous layer of the periosteum that covers the bones united at the joint. Portions of the fibrous capsule may thicken, and the collagen fibers may become regularly arranged to form ligaments. In addition, ligaments and tendons may be present outside the fibrous capsule, thereby contributing to the strength and stability of the joint while limiting movement in some directions.

The synovial membrane lines the joint cavity, except over the articular cartilage and articular disks. This thin, delicate membrane consists of a collection of modified connective tissue cells either intermixed with part of the fibrous capsule or separated from it by a layer of areolar tissue or adipose tissue. The membrane produces synovial fluid, a thin, lubricating film that covers the surfaces of a joint. Synovial fluid consists of a serum (blood fluid) filtrate and secretions from the synovial cells. It is a complex mixture of polysaccharides, proteins, lipids, and cells. The major polysaccharide, hyaluronic acid, provides much of the slippery consistency and lubricating qualities of synovial fluid.

Predict 3

What would happen if a synovial membrane covered the articular cartilage?

In certain synovial joints, such as the shoulder and knee, the synovial membrane extends as a pocket, or sac, called a **bursa** (ber′să; pocket), for a distance away from the rest of the joint cavity (figure 8.7). Bursae contain synovial fluid and provide a cushion between structures that would otherwise rub against each other, such as tendons rubbing on bones or other tendons. Some bursae, such as the subcutaneous olecranon bursae, are not associated with joints but provide a cushion between the skin and underlying bony prominences, where friction could damage the tissues. Other bursae extend along tendons for some distance, forming **tendon sheaths. Bursitis** (ber-sī′tis), inflammation of a bursa, may cause considerable pain around the joint and restrict movement.

At the peripheral margin of the articular cartilage, blood vessels form a vascular circle that supplies the cartilage with nourishment, but no blood vessels penetrate the cartilage or enter the joint cavity. Additional nourishment to the articular cartilage comes from the underlying spongy bone and from the synovial fluid covering the articular cartilage. Sensory nerves enter the fibrous capsule and, to a lesser extent, the synovial membrane. They not only supply the brain with information about pain in the joint but also furnish constant information about the joint's position and degree of movement (see chapter 14). Nerves do not enter the cartilage or joint cavity.

Types of Synovial Joints

Synovial joints are classified according to the shape of the adjoining articular surfaces. The six types of synovial joints are plane, saddle, hinge, pivot, ball-and-socket, and ellipsoid (figure 8.8). Movements at synovial joints are described as **uniaxial,** occurring around one axis; **biaxial,** occurring around two axes situated at right angles to each other; or **multiaxial,** occurring around several axes.

A **plane joint,** or *gliding joint,* consists of two flat bone surfaces of about equal size between which a slight gliding motion can occur (figure 8.8). These joints are considered uniaxial because some rotation is also possible but is limited by ligaments and adjacent bone. Examples are the articular processes between vertebrae.

A **saddle joint** consists of two saddle-shaped articulating surfaces oriented at right angles to each other so that their complementary surfaces articulate (figure 8.8). Saddle joints are biaxial joints. The carpometacarpal joint of the thumb is an example.

A **hinge joint** is a uniaxial joint in which a convex cylinder in one bone is applied to a corresponding concavity in the other bone (figure 8.8). Examples include the elbow and knee joints.

A **pivot joint** is a uniaxial joint that restricts movement to rotation around a single axis (figure 8.8). A pivot joint consists of a relatively cylindrical bony process that rotates within a ring composed partly of bone and partly of ligament. The articulation between the head of the radius and the proximal end of the ulna is an example. The articulation between the dens, a process on the axis (see chapter 7), and the atlas is another example.

A **ball-and-socket joint** consists of a ball (head) at the end of one bone and a socket in an adjacent bone into which a portion

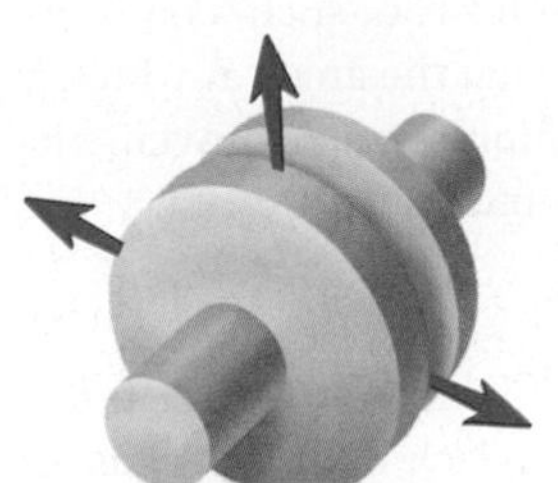

Plane

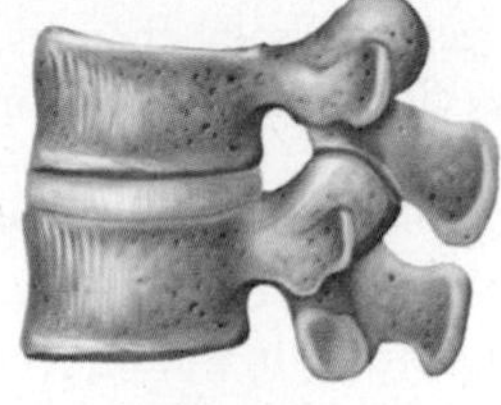
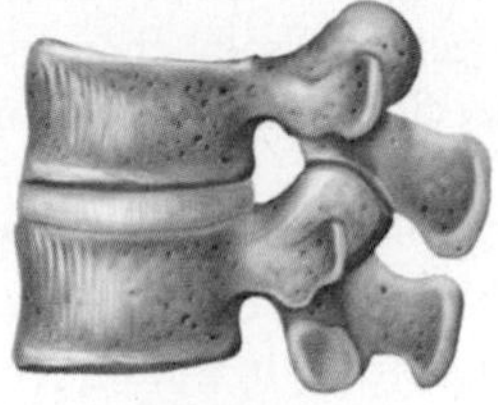
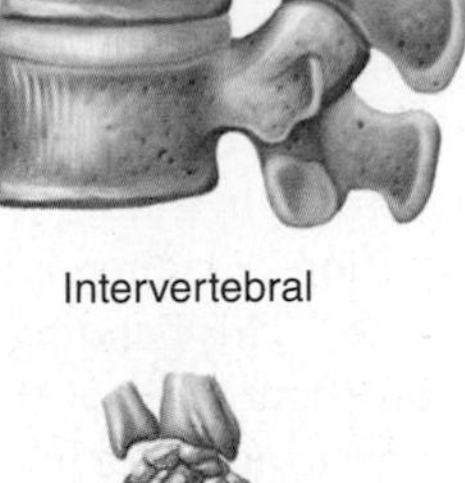

Intervertebral

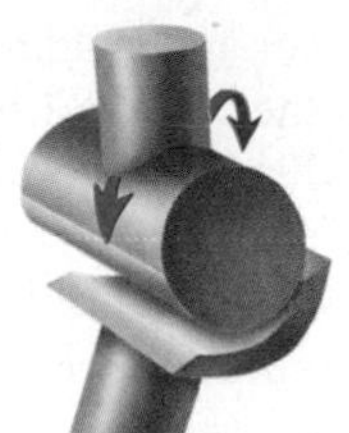

Saddle

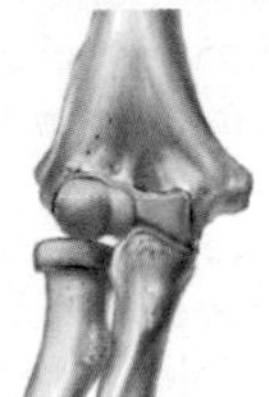

Carpometacarpal

Hinge

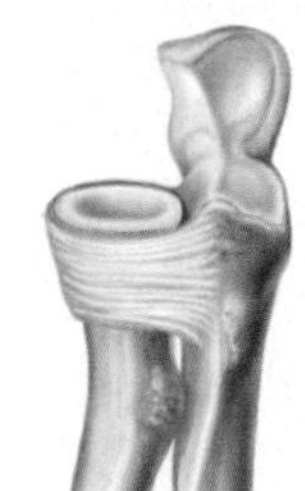

Cubital

Pivot

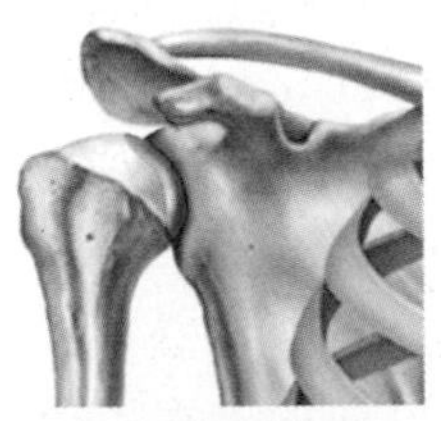

Proximal radioulnar

Ball-and-socket

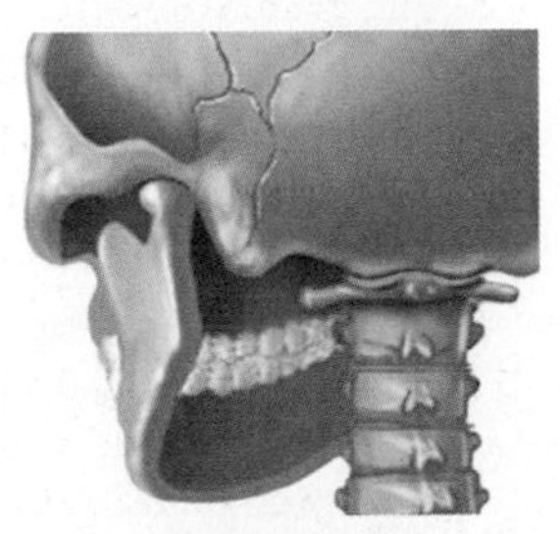

Glenohumeral

Ellipsoid

Atlantooccipital

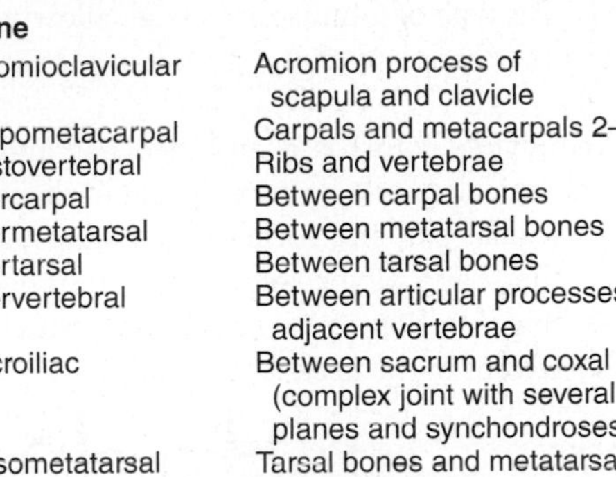
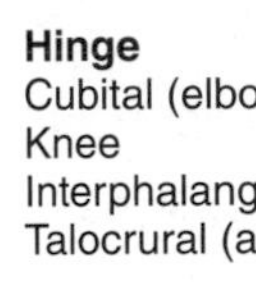
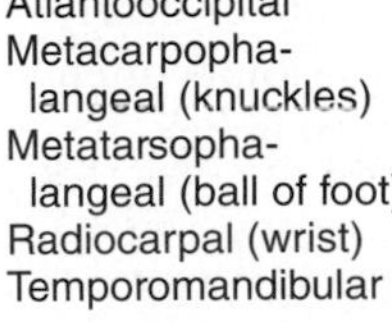

Class and Example of Joint	Structures Joined	Movement
Plane		
Acromioclavicular	Acromion process of scapula and clavicle	Slight
Carpometacarpal	Carpals and metacarpals 2–5	Slight
Costovertebral	Ribs and vertebrae	Slight
Intercarpal	Between carpal bones	Slight
Intermetatarsal	Between metatarsal bones	Slight
Intertarsal	Between tarsal bones	Slight
Intervertebral	Between articular processes of adjacent vertebrae	Slight
Sacroiliac	Between sacrum and coxal bone (complex joint with several planes and synchondroses)	Slight
Tarsometatarsal	Tarsal bones and metatarsal bones	Slight
Saddle		
Carpometacarpal pollicis	Carpal and metacarpal of thumb	Two axes
Intercarpal	Between carpal bones	Slight
Sternoclavicular	Manubrium of sternum and clavicle	Slight
Hinge		
Cubital (elbow)	Humerus, ulna, and radius	One axis
Knee	Femur and tibia	One axis
Interphalangeal	Between phalanges	One axis
Talocrural (ankle)	Talus, tibia, and fibula	Multiple axes; one predominates
Pivot		
Atlantoaxial	Atlas and axis	Rotation
Proximal radioulnar	Radius and ulna	Rotation
Distal radioulnar	Radius and ulna	Rotation
Ball-and-Socket		
Coxal (hip)	Coxal bone and femur	Multiple axes
Glenohumeral (shoulder)	Scapula and humerus	Multiple axes
Ellipsoid		
Atlantooccipital	Atlas and occipital bone	Two axes
Metacarpophalangeal (knuckles)	Metacarpal bones and phalanges	Two axes
Metatarsophalangeal (ball of foot)	Metatarsal bones and phalanges	Two axes
Radiocarpal (wrist)	Radius and carpal bones	Multiple axes
Temporomandibular	Mandible and temporal bone	Multiple axes; one predominates

FIGURE 8.8 **Types of Synovial Joints**

of the ball fits (figure 8.8). This type of joint is multiaxial, allowing a wide range of movement in almost any direction. Examples are the shoulder and hip joints.

An **ellipsoid joint** (condyloid joint) is a modified ball-and-socket joint (figure 8.8). The articular surfaces are ellipsoid in shape, rather than spherical as in regular ball-and-socket joints. Ellipsoid joints are biaxial, because the shape of the joint limits its range of movement almost to a hinge motion in two axes and restricts rotation. The atlantooccipital joint of the neck is an example.

ASSESS YOUR PROGRESS

5. *Describe the structure of a synovial joint. How do the different parts of the joint contribute to joint movement?*
6. *What are articular disks, and where are they found?*
7. *What are bursae and tendon sheaths? What is the function of each?*
8. *On what basis are synovial joints classified? List and describe the six types of synovial joints, and give an example of each.*
9. *What directional movements are permitted at each type of synovial joint?*

8.2 Types of Movement

LEARNING OUTCOMES

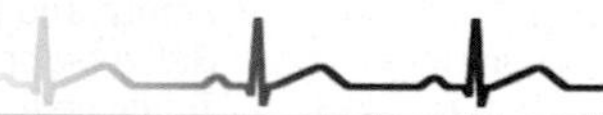

After reading this section, you should be able to

A. **Categorize movements as gliding, angular, circular, special, or a combination of types.**
B. **Demonstrate the difference between the following pairs of movements: flexion and extension; plantar flexion and dorsiflexion; abduction and adduction; supination and pronation; elevation and depression; protraction and retraction; opposition and reposition; inversion and eversion.**
C. **Distinguish between rotation and circumduction. What is excursion?**

A joint's structure relates to the movements that occur at that joint. Some joints are limited to only one type of movement; others can move in several directions. Here we break the types of movement into the following categories: gliding, angular, circular, special, and combination movements. With few exceptions, movement is best described in relation to the anatomical position—either away from it or toward it. Most movements are also possible in the opposite direction and are therefore grouped in pairs.

Gliding Movements

Gliding movements are the simplest of all the types of movement. These movements occur in plane joints between two flat or nearly flat surfaces that slide or glide over each other. These joints often allow only slight movement, as occurs between carpal bones.

Angular Movements

In angular movements, one part of a linear structure, such as the trunk or a limb, bends relative to another part of the structure, thereby changing the angle between the two parts. Angular movements also involve the movement of a solid rod, such as a limb, that is attached at one end to the body so that the angle at which it meets the body changes. The most common angular movements are flexion and extension and abduction and adduction.

Flexion and Extension

Flexion and extension are common opposing movements that may be defined in a number of ways, although each definition has an exception. The literal definitions are to bend (flex) and straighten (extend). This bending and straightening can easily be seen in the elbow (figure 8.9) or the knee (see figure 8.11). However, we have chosen to use a definition that relates to the coronal plane and that has more utility and fewer exceptions. **Flexion** is movement of a body part anterior to the coronal plane, or in the anterior direction. **Extension** is movement of a body part posterior to the coronal plane, or in the posterior direction (figure 8.10).

The exceptions to defining flexion and extension in relation to the coronal plane are the knee and foot. At the knee, flexion moves (bends) the leg in a posterior direction, and extension moves (straightens) it in an anterior direction (figure 8.11). For example, the knee flexes when you sit in a chair and extends when you stand up. Movement of the foot toward the plantar surface, as when standing on the toes, is commonly called **plantar flexion;** movement of the foot toward the shin, as when walking on the heels, is called **dorsiflexion** (figure 8.12).

Hyperextension is usually defined as abnormal, forced extension of a joint beyond its normal range of motion. For example, if a person attempts to break a fall by putting out a hand, the force of the fall directed into the hand and wrist may cause hyperextension of the wrist and result in a sprained joint or broken bone. Some health professionals, however, define hyperextension as the normal movement of structures, except the leg, into the space posterior to the anatomical position.

Abduction and Adduction

Abduction (to take away) is movement away from the midline; **adduction** (to bring together) is movement toward the midline. Moving the upper limbs away from the body, as is done in the

Flexion
Extension

FIGURE 8.9 Flexion and Extension of the Elbow

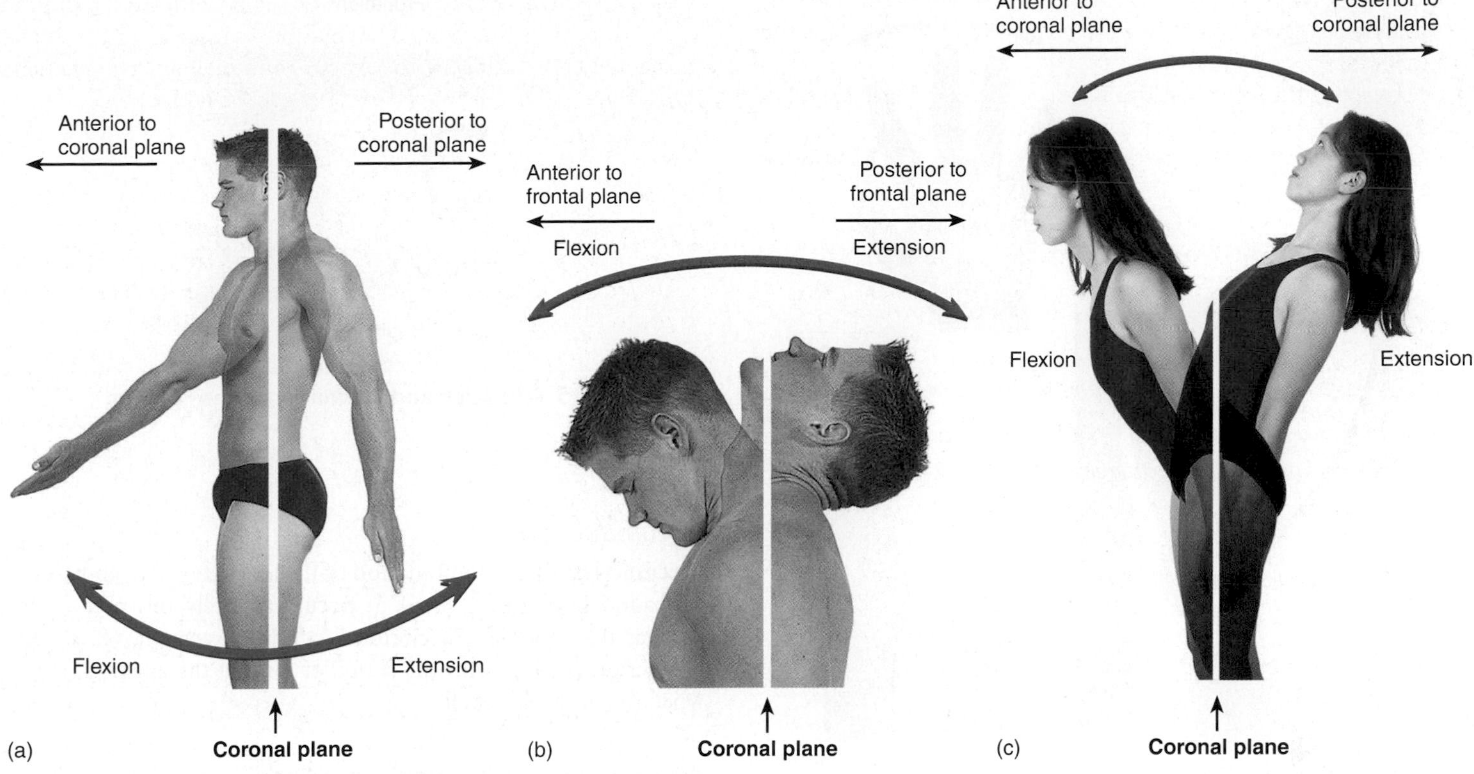

Anterior to coronal plane
Posterior to coronal plane
Flexion
Extension
(a)
Coronal plane
Anterior to frontal plane
Posterior to frontal plane
Flexion
Extension
(b)
Coronal plane
Anterior to coronal plane
Posterior to coronal plane
Flexion
Extension
(c)
Coronal plane

FIGURE 8.10 Flexion and Extension in Relation to the Coronal Plane
Flexion and extension of (*a*) the shoulder, (*b*) the neck, and (*c*) the trunk.

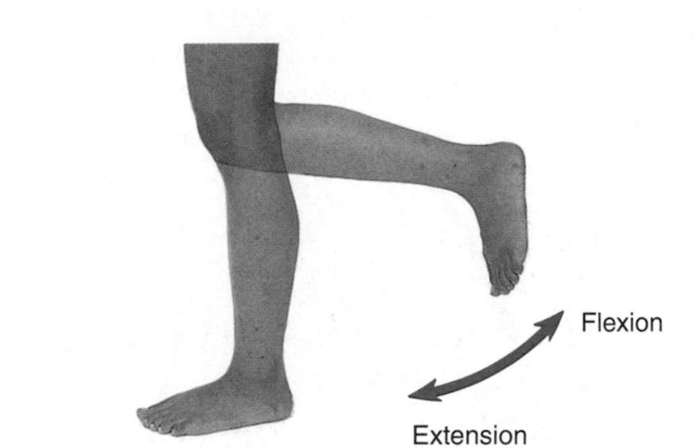

Flexion
Extension

FIGURE 8.11 Flexion and Extension of the Knee

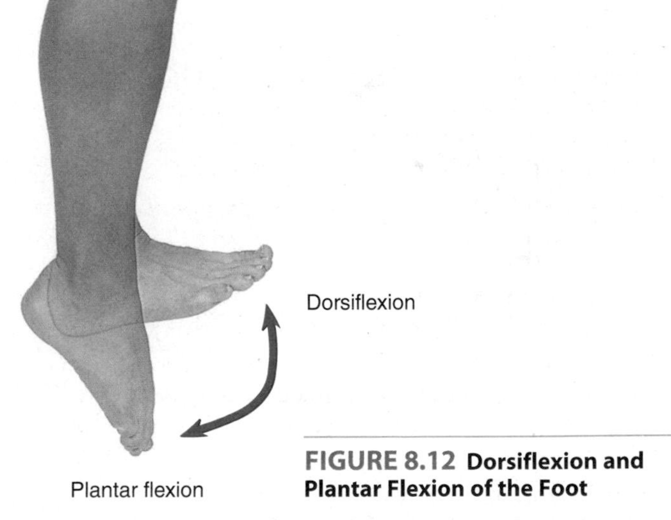

Dorsiflexion
Plantar flexion

FIGURE 8.12 Dorsiflexion and Plantar Flexion of the Foot

outward step of jumping jacks, is abduction, and bringing the upper limbs back toward the body is adduction (figure 8.13*a*). In the hand, abduction spreads the fingers apart, away from the midline of the hand, and adduction brings them back together (figure 8.13*b*). Abduction of the thumb moves it anteriorly, away from the palm. Abduction of the wrist, which is sometimes called radial deviation, moves the hand away from the midline of the body, and adduction of the wrist, sometimes called ulnar deviation, moves the hand toward the midline. Abduction of the head, which involves tilting the head to one side, is commonly called lateral flexion of the neck. Bending at the waist to one side is usually called **lateral flexion** of the vertebral column, rather than abduction.

Circular Movements

Circular movements involve rotating a structure around an axis or moving the structure in an arc.

Rotation

Rotation is the turning of a structure around its long axis, as in rotating the head, the humerus, or the entire body (figure 8.14). Medial rotation of the humerus with the forearm flexed brings the hand toward the body. Lateral rotation of the humerus moves the hand away from the body.

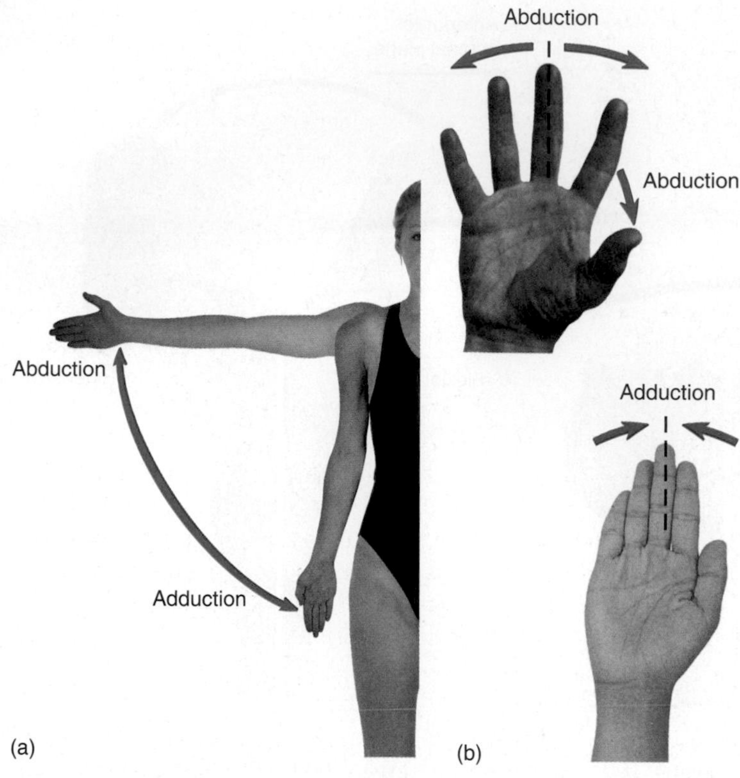
Abduction
Abduction
Abduction
Adduction
Adduction
(a)
(b)

FIGURE 8.13 Abduction and Adduction
Abduction and adduction of *(a)* the upper limb and *(b)* the fingers.

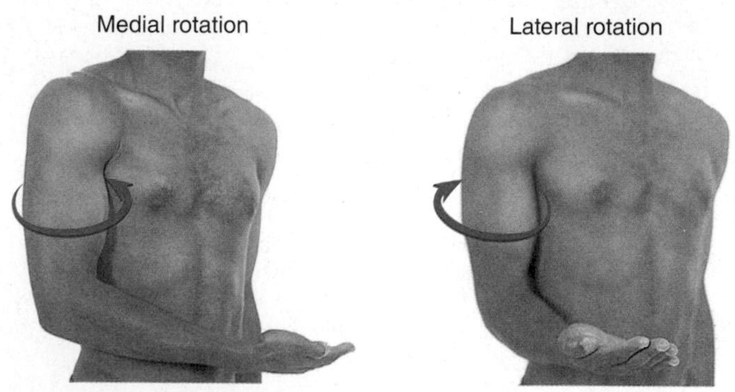
Medial rotation
Lateral rotation

FIGURE 8.14 Medial and Lateral Rotation of the Arm

Pronation and Supination

Pronation (prō-nā′shŭn) and **supination** (soo′pi-nā′shŭn) refer to the unique rotation of the forearm (figure 8.15). The word *prone* means lying facedown; the word *supine* means lying faceup. Pronation is rotation of the forearm so that the palm faces posteriorly in relation to the anatomical position. The palm of the hand faces inferiorly if the elbow is flexed to 90 degrees. Supination is rotation of the forearm so that the palm faces anteriorly in relation to the anatomical position. The palm of the hand faces superiorly if the elbow is flexed to 90 degrees. In pronation, the radius and ulna cross; in supination, they are parallel. The head of the radius rotates against the radial notch of the ulna during supination and pronation.

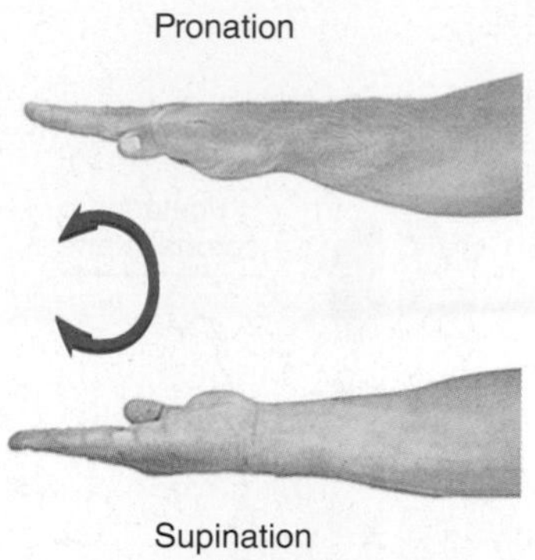
Pronation
Supination

FIGURE 8.15 Pronation and Supination of the Forearm

Circumduction

Circumduction is a combination of flexion, extension, abduction, and adduction (figure 8.16). It occurs at freely movable joints, such as the shoulder. In circumduction, the arm moves so that it describes a cone, with the shoulder joint at the apex, as occurs when pitching a baseball.

Special Movements

Special movements are those movements that are unique to only one or two joints and do not fit neatly into any of the other categories.

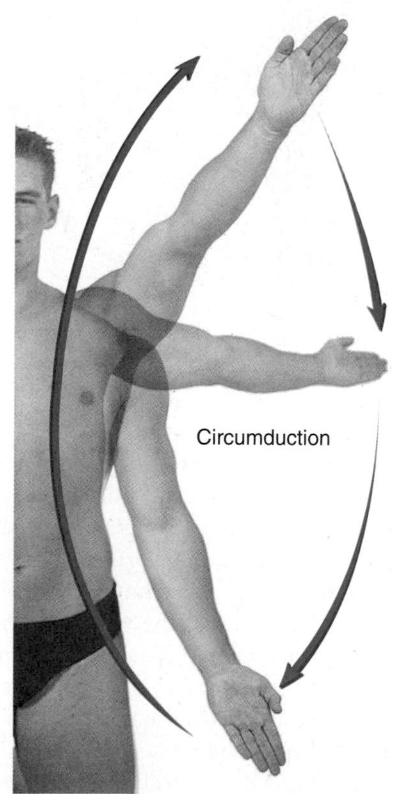
Circumduction

FIGURE 8.16 Circumduction

Elevation and Depression

Elevation moves a structure superiorly; **depression** moves it inferiorly (figure 8.17). Shrugging the shoulders is an example of scapular elevation. Depression of the mandible opens the mouth, and elevation closes it.

Protraction and Retraction

Protraction is a gliding motion that moves a structure in an anterior direction (figure 8.18). **Retraction** moves the structure back to the anatomical position or even more posteriorly. Jutting out the jaw and hunching the shoulders are examples of protraction. Pulling the jaw back and pinching or pulling the scapulae back toward the vertebral column illustrate retraction.

Excursion

Lateral excursion is moving the mandible to either the right or the left of the midline (figure 8.19), as occurs when grinding the teeth or chewing. **Medial excursion** returns the mandible to the midline position.

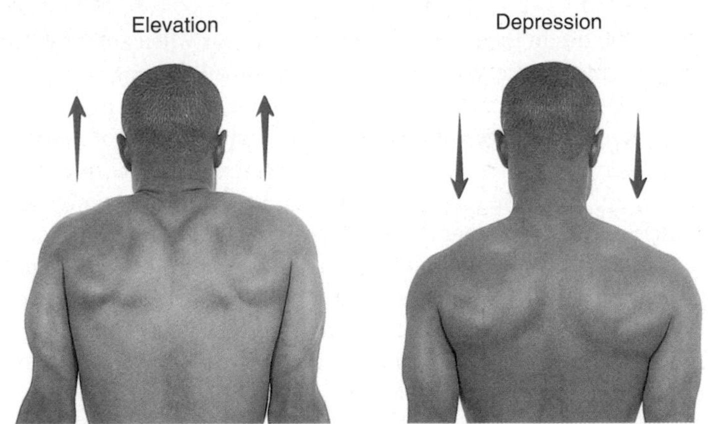
Elevation
Depression

FIGURE 8.17 Elevation and Depression of the Scapula

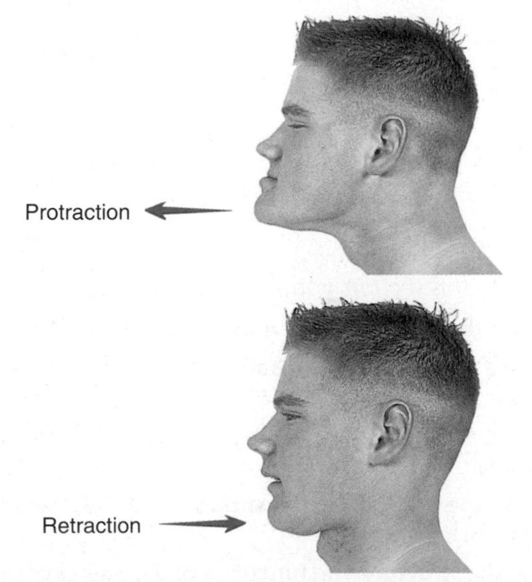
Protraction
Retraction

FIGURE 8.18 Protraction and Retraction of the Mandible

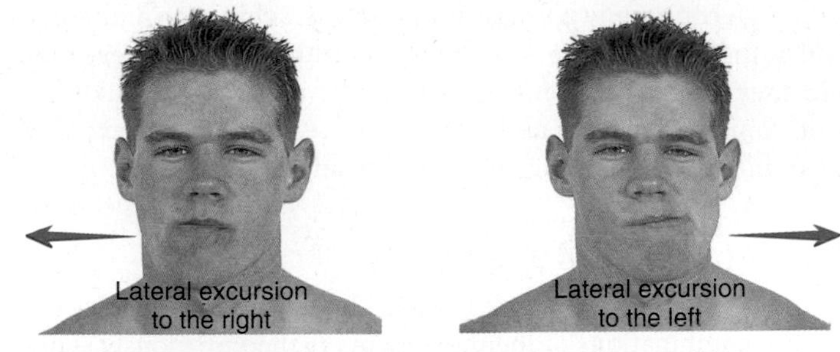
Lateral excursion to the right
Lateral excursion to the left

FIGURE 8.19 Excursion of the Mandible

Opposition and Reposition

Opposition is movement of the thumb and little finger (figure 8.20). It occurs when these two digits are brought toward each other across the palm of the hand. The thumb can also oppose the other digits, but those digits flex to touch the tip of the opposing thumb. **Reposition** returns the thumb and little finger to the neutral, anatomical position.

Inversion and Eversion

Inversion turns the ankle so that the plantar surface of the foot faces medially, toward the opposite foot, with the weight on the outside edge of the foot (rolling out). **Eversion** turns the ankle so that the plantar surface faces laterally, with the weight on the inside edge of the foot (rolling in; figure 8.21). Sometimes inversion of the foot is called supination and eversion is called pronation.

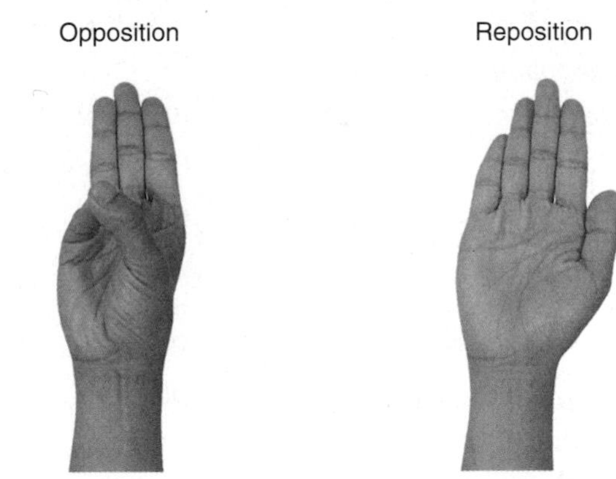
Opposition
Reposition

FIGURE 8.20 Opposition and Reposition of the Thumb and Little Finger

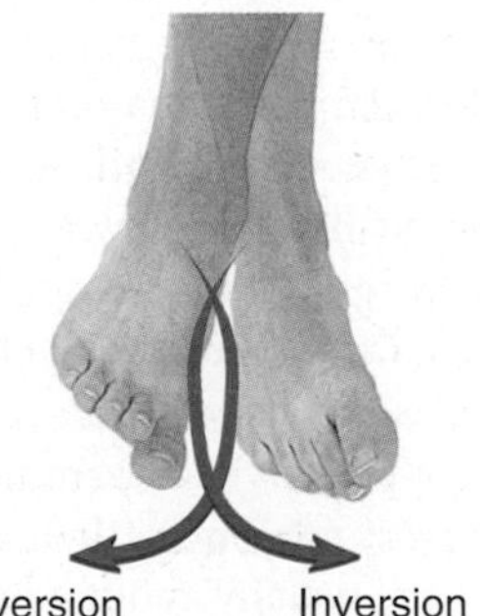
Eversion
Inversion

FIGURE 8.21 Inversion and Eversion of the Right Foot

Although commonly used as clinical terms, *supination* and *pronation* of the feet are more complex than just inversion and eversion, and they involve movements at multiple joints of the ankle and foot. Some supination and pronation are normal, but excessive pronation is a common cause of injury among runners.

Combination Movements

Most movements that we perform in the course of normal activities are combinations of the movements named previously. These combined movements are described by naming the individual movements involved. For example, when a person steps forward and to the side at a 45-degree angle, the movement at the hip is a combination of flexion and abduction.

Predict 4

What combination of movements at the shoulder and elbow joints allows a person to move the right upper limb from the anatomical position to touch the right side of the head with the fingertips?

ASSESS YOUR PROGRESS

10. *Describe flexion and extension. How are they different for the upper and lower limbs? What is hyperextension?*
11. *Contrast abduction and adduction. Describe these movements for the head, upper limbs, wrist, fingers, waist, lower limbs, and toes. For what part of the body is the term lateral flexion used?*
12. *Distinguish among rotation, circumduction, pronation, and supination. Give an example of each.*
13. *Explain the following jaw movements: protraction, retraction, lateral excursion, medial excursion, elevation, and depression.*
14. *Describe opposition and reposition.*
15. *What terms are used for turning the side of the foot medially or laterally?*

8.3 Range of Motion

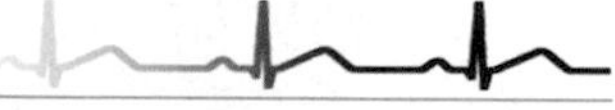

LEARNING OUTCOMES

After reading this section, you should be able to

A. **Explain the difference between active and passive range of motion.**
B. **Describe the consequences of movement beyond the normal range.**
C. **List the factors that affect normal range of motion.**

Range of motion describes the amount of mobility that can be demonstrated in a given joint. **Active range of motion** is the amount of movement that can be accomplished by contracting the muscles that normally act across a joint. **Passive range of motion** is the amount of movement that can be accomplished when the structures that meet at the joint are moved by an outside force, as when a therapist holds on to a patient's forearm and moves it toward the arm, flexing the elbow joint. The active and passive ranges of motion for normal joints are usually about equal.

Movement of joints beyond the normal range of motion can cause dislocations and sprains. A **dislocation,** or *luxation,* of a joint occurs when the articulating surfaces of the bones are moved out of proper alignment. A *subluxation* is a partial dislocation. Dislocations are often accompanied by painful damage to the supporting ligaments and articular cartilage. A **sprain** occurs when ligaments are damaged. The degree of damage can range from stretched to completely torn ligaments. Sprains often result in inflammation, swelling, and pain. Dislocations and sprains are common sports injuries.

The range of motion for a given joint is influenced by a number of factors:

1. Shape of the articular surfaces of the bones forming the joint
2. Amount and shape of cartilage covering those articular surfaces
3. Strength and location of ligaments and tendons surrounding the joint
4. Strength and location of the muscles associated with the joint
5. Amount of fluid in and around the joint
6. Amount of pain in and around the joint
7. Amount of use or disuse the joint has received over time

Abnormalities in the range of motion can occur when any of those components change. For example, damage to a ligament associated with a given joint may increase that joint's range of motion. A torn piece of cartilage within a joint can limit its range of motion. If the nerve supply to a muscle is damaged so that the muscle is weakened, the active range of motion for the joint acted on by that muscle may decrease, but the joint's passive range of motion should remain unchanged. Fluid buildup and/or pain in or around a joint (as occurs when the soft tissues around the joint develop edema following an injury) can severely limit both the active and passive ranges of motion for that joint. With disuse, both the active and passive ranges of motion for a given joint decrease.

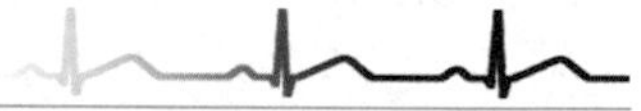

ASSESS YOUR PROGRESS

16. *What is range of motion? Contrast active and passive range of motion.*
17. *Discuss some examples of the changes that may occur with movement beyond the normal range.*

8.4 Description of Selected Joints

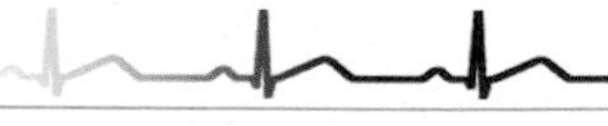

LEARNING OUTCOMES

After reading this section, you should be able to

A. **Describe the structure and movements of the TMJ.**
B. **Compare and contrast the ball-and-socket joints of the shoulder and hip.**
C. **Compare and contrast the hinge joints of the elbow, knee, and ankle.**
D. **Describe the ligaments that support the complex ellipsoid joint of the knee.**
E. **Explain the structure and functions of the arches of the foot.**
F. **Discuss the common disorders that affect these major joints.**

It is impossible in a limited space to describe all the joints of the body; therefore, we have chosen to describe only selected joints, based on their representative structure, important function, or clinical significance.

Temporomandibular Joint

The mandible articulates with the temporal bone to form the **temporomandibular joint (TMJ).** The mandibular condyle fits into the mandibular fossa of the temporal bone. A fibrocartilage articular disk is located between the mandible and the temporal bone, dividing the joint into superior and inferior joint cavities (figure 8.22). The joint is surrounded by a fibrous capsule, to which the articular disk is attached at its margin, and is strengthened by lateral and accessory ligaments.

The temporomandibular joint is a combination plane and ellipsoid joint, with the ellipsoid portion predominating. Depression of the mandible to open the mouth involves an anterior gliding motion of the mandibular condyle and articular disk relative to the temporal bone, which is about the same motion that occurs in protraction of the mandible; it is followed by a hinge motion between the articular disk and the mandibular head. The mandibular condyle is also capable of slight mediolateral movement, allowing excursion of the mandible.

Clinical IMPACT

TMJ Disorders

TMJ disorders are the most common cause of chronic orofacial pain. The primary symptom is pain in the jaw muscles and/or joint. Other symptoms include radiating pain in the face, head, and neck; reduced range of motion or locking of the jaw; and painful clicking or grating when moving the jaw. Ear pain is another symptom, which often leads patients to their physicians, who then refer them to a dentist. It is estimated that 5–12% of the population experience TMJ pain. It is at least twice as prevalent among women.

TMJ disorders often have no apparent cause or trigger. They are not easily treated, and chronic TMJ pain is often associated with other types of poorly understood chronic pain, such as fibromyalgia. Many TMJ cases can improve with treatment, although for some patients the pain is persistent or continues to recur. Therapy includes avoiding jaw movements that aggravate the problem, such as chewing gum or hard foods, and reducing stress and anxiety. Physical therapy may help relax the muscles and restore function. Analgesic and anti-inflammatory drugs are sometimes prescribed, and oral splints may be helpful, especially at night.

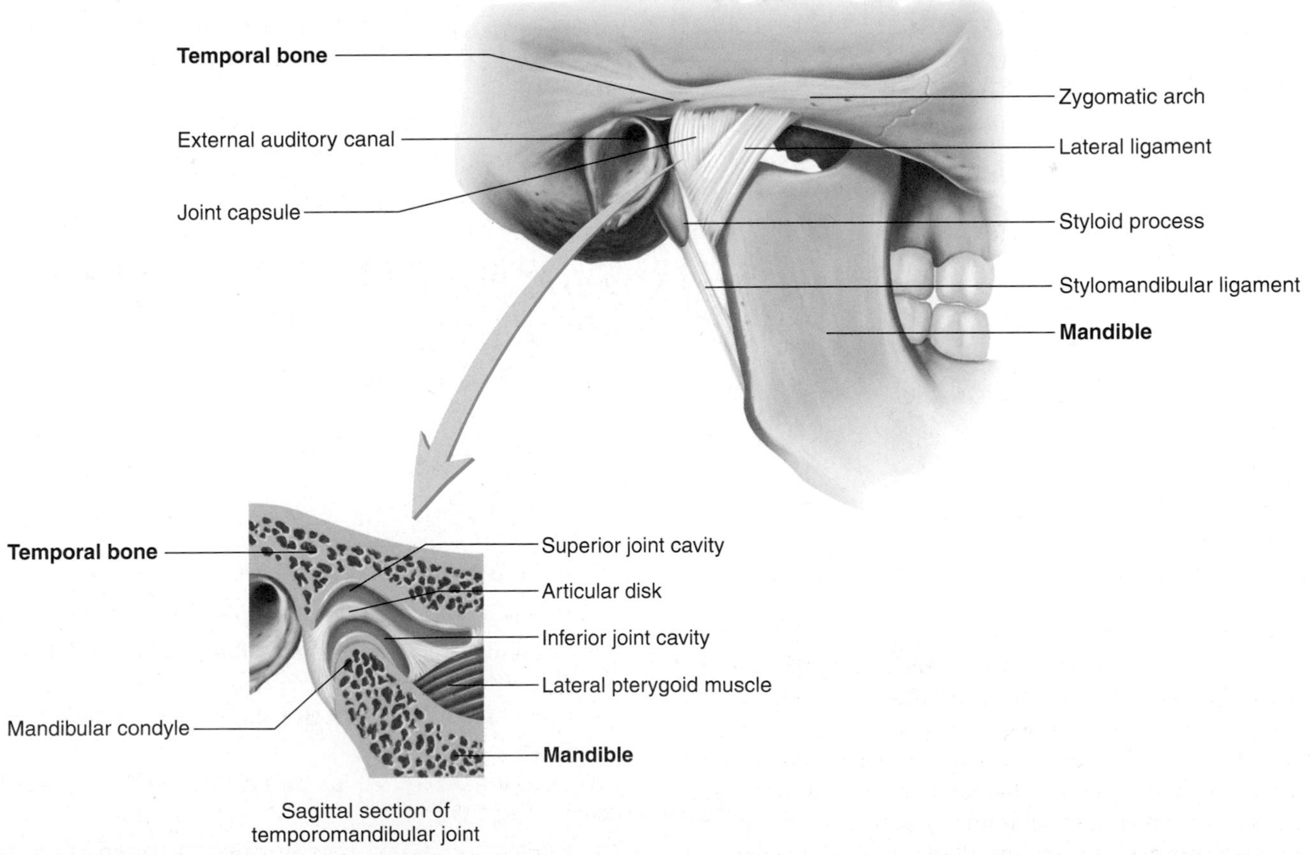
Lateral view
Temporal bone
External auditory canal
Joint capsule
Zygomatic arch
Lateral ligament
Styloid process
Stylomandibular ligament
Mandible
Temporal bone
Mandibular condyle
Superior joint cavity
Articular disk
Inferior joint cavity
Lateral pterygoid muscle
Mandible
Sagittal section of temporomandibular joint

FIGURE 8.22 APR **Right Temporomandibular Joint, Lateral View**

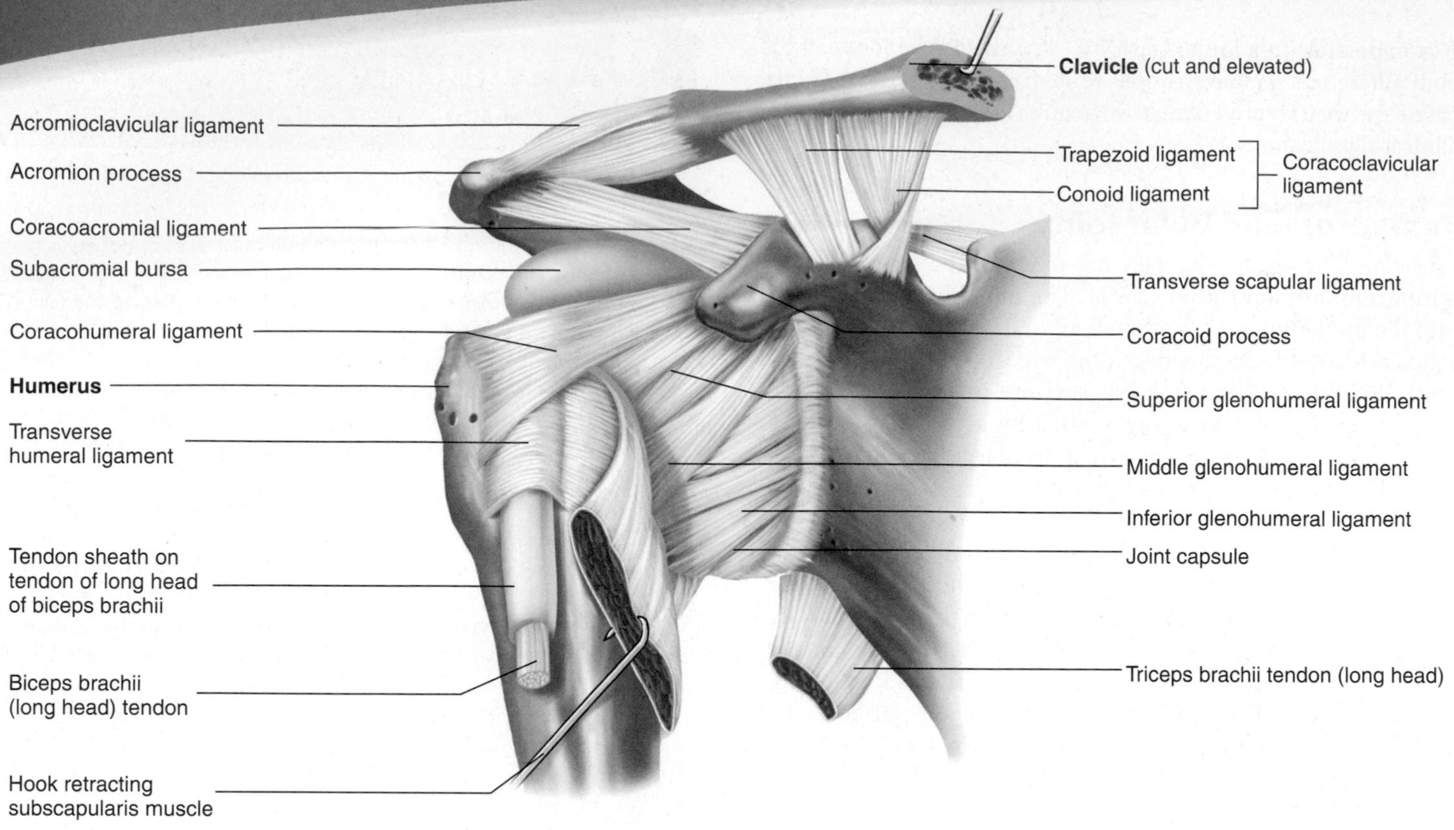
Clavicle (cut and elevated)
Acromioclavicular ligament
Acromion process
Trapezoid ligament
Conoid ligament
Coracoclavicular ligament
Coracoacromial ligament
Subacromial bursa
Transverse scapular ligament
Coracohumeral ligament
Coracoid process
Humerus
Superior glenohumeral ligament
Transverse humeral ligament
Middle glenohumeral ligament
Inferior glenohumeral ligament
Joint capsule
Tendon sheath on tendon of long head of biceps brachii
Biceps brachii (long head) tendon
Triceps brachii tendon (long head)
Hook retracting subscapularis muscle

Anterior view

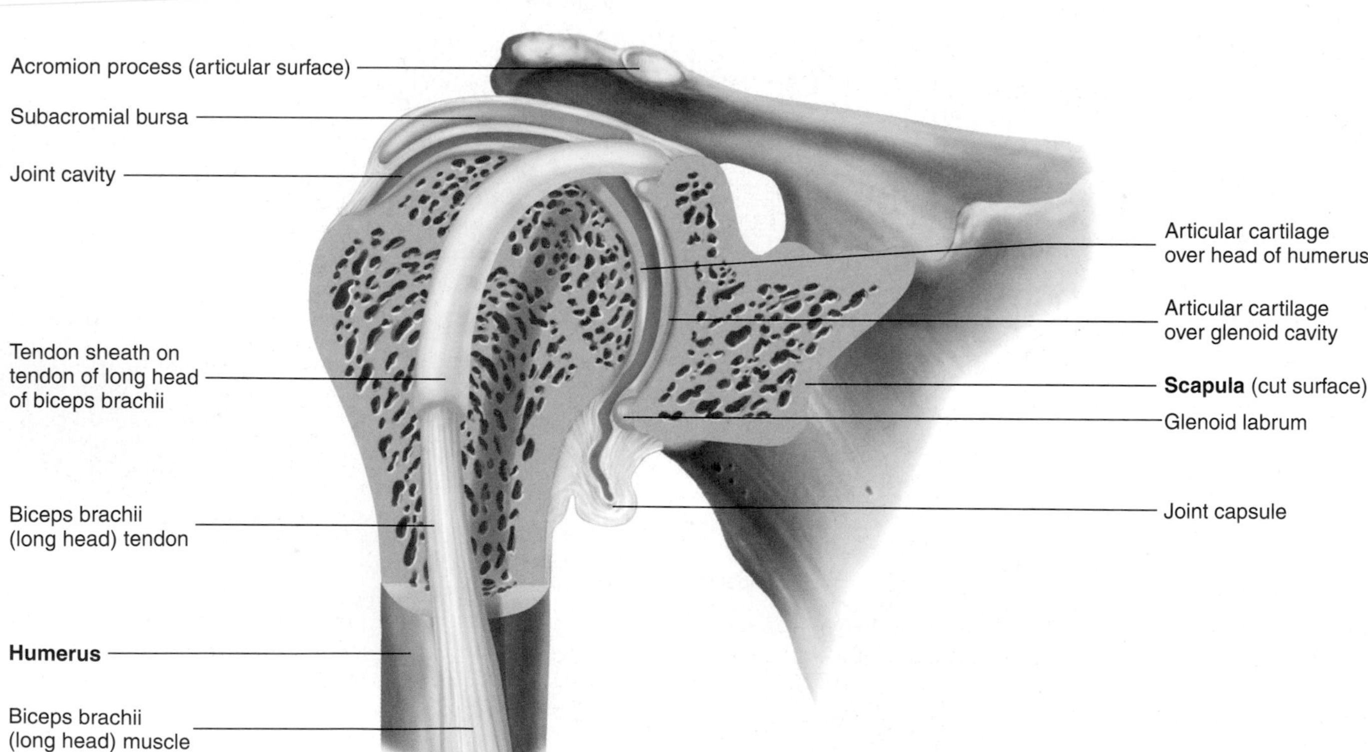
Acromion process (articular surface)
Subacromial bursa
Joint cavity
Articular cartilage over head of humerus
Articular cartilage over glenoid cavity
Tendon sheath on tendon of long head of biceps brachii
Scapula (cut surface)
Glenoid labrum
Joint capsule
Biceps brachii (long head) tendon
Humerus
Biceps brachii (long head) muscle

Frontal section

FIGURE 8.23 APR **Right Shoulder Joint**

TABLE 8.2	Ligaments of the Shoulder Joint (see figure 8.23)
Ligament	**Description**
Glenohumeral (superior, middle, and inferior)	Three slightly thickened longitudinal sets of fibers on the anterior side of the capsule; extend from the humerus to the margin of the glenoid cavity
Transverse humeral	Lateral, transverse, fibrous thickening of the joint capsule; crosses between the greater and lesser tubercles and holds down the tendon from the long head of the biceps brachii muscle
Coracohumeral	Crosses from the root of the coracoid process to the humeral neck
Coracoacromial	Crosses above the joint between the coracoid process and the acromion process; an accessory, protective ligament

Shoulder Joint

The **shoulder joint,** or *glenohumeral joint,* is a ball-and-socket joint (figure 8.23) that has less stability but more mobility than the other ball-and-socket joint, the hip. Flexion, extension, abduction, adduction, rotation, and circumduction can all occur at the shoulder joint. The rounded head of the humerus articulates with the shallow glenoid cavity of the scapula. The rim of the glenoid cavity is built up slightly by a fibrocartilage ring, the **glenoid labrum,** to which the joint capsule is attached. A **subscapular bursa** (not shown in figure 8.23) opens into the joint cavity. A **subacromial bursa** is located near the joint cavity but separated from the cavity by the joint capsule (figure 8.23).

Clinical
IMPACT

Shoulder Disorders

The most common traumatic shoulder disorders are dislocation of bones and tears in muscles or tendons. The shoulder is the most commonly dislocated joint in the body. Major ligaments cross the superior part of the shoulder joint, and no major ligaments or muscles are associated with the inferior side. As a result, the humerus is most likely to become dislocated inferiorly into the axilla. Because the axilla contains very important nerves and arteries, severe and permanent damage may occur when the humeral head dislocates inferiorly. The axillary nerve is the most commonly damaged (see chapter 12).

Chronic shoulder disorders include tendinitis (inflammation of tendons), bursitis (inflammation of bursae), and arthritis (inflammation of joints). Bursitis of the subacromial bursa can become very painful when the large shoulder muscle, called the deltoid muscle, compresses the bursa during shoulder movement.

The stability of the shoulder joint is maintained primarily by four sets of ligaments and four muscles. The ligaments are listed in table 8.2. The four muscles, referred to collectively as the **rotator cuff,** hold the humeral head tightly within the glenoid cavity (see chapter 10). The head of the humerus is also supported against the glenoid cavity by the tendon from the biceps brachii muscle in the anterior part of the arm. This tendon is unusual in that it passes through the articular capsule of the shoulder joint before crossing the head of the humerus and attaching to the scapula at the supraglenoid tubercle (see figure 7.23*a*).

Predict 5

Separation of the shoulder consists of stretching or tearing the ligaments of the acromioclavicular joint, a condition called acromioclavicular, or AC, separation. Using figure 8.23*a* and your knowledge of the articulated skeleton, explain the nature of a shoulder separation and predict the problems that may follow a separation.

Elbow Joint

The **elbow joint** (figure 8.24) is a compound hinge joint consisting of the **humeroulnar joint,** between the humerus and ulna, and the **humeroradial joint,** between the humerus and radius. The **proximal radioulnar joint,** between the proximal radius and ulna, is also closely related. The shape of the trochlear notch and its association with the trochlea of the humerus (figure 8.24*a*) limit movement at the elbow joint to flexion and extension. However, the rounded radial head rotates in the radial notch of the ulna and against the capitulum of the humerus (figure 8.24*b*), allowing pronation and supination of the hand.

The elbow joint is surrounded by a joint capsule. The humeroulnar joint is reinforced by the **ulnar collateral ligament** (figure 8.24*c*). The humeroradial and proximal radioulnar joints are reinforced by the **radial collateral ligament** and the **radial annular ligament** (figure 8.24*d*). A subcutaneous **olecranon bursa** covers the proximal and posterior surfaces of the olecranon process.

Clinical
IMPACT

Elbow Problems

Olecranon bursitis is inflammation of the olecranon bursa; it can be caused by excessive pressure of the elbow against a hard surface and is sometimes referred to as *student's elbow.* This condition typically develops over a period of months, but it can also result from a hard blow to the elbow.

Sometimes the radial head becomes subluxated (partially separated) from the annular ligament of the radius, a common condition called *nursemaid's elbow.* If a young child (usually under age 5) is lifted by one hand or swung by the arms, the action may subluxate the radial head.

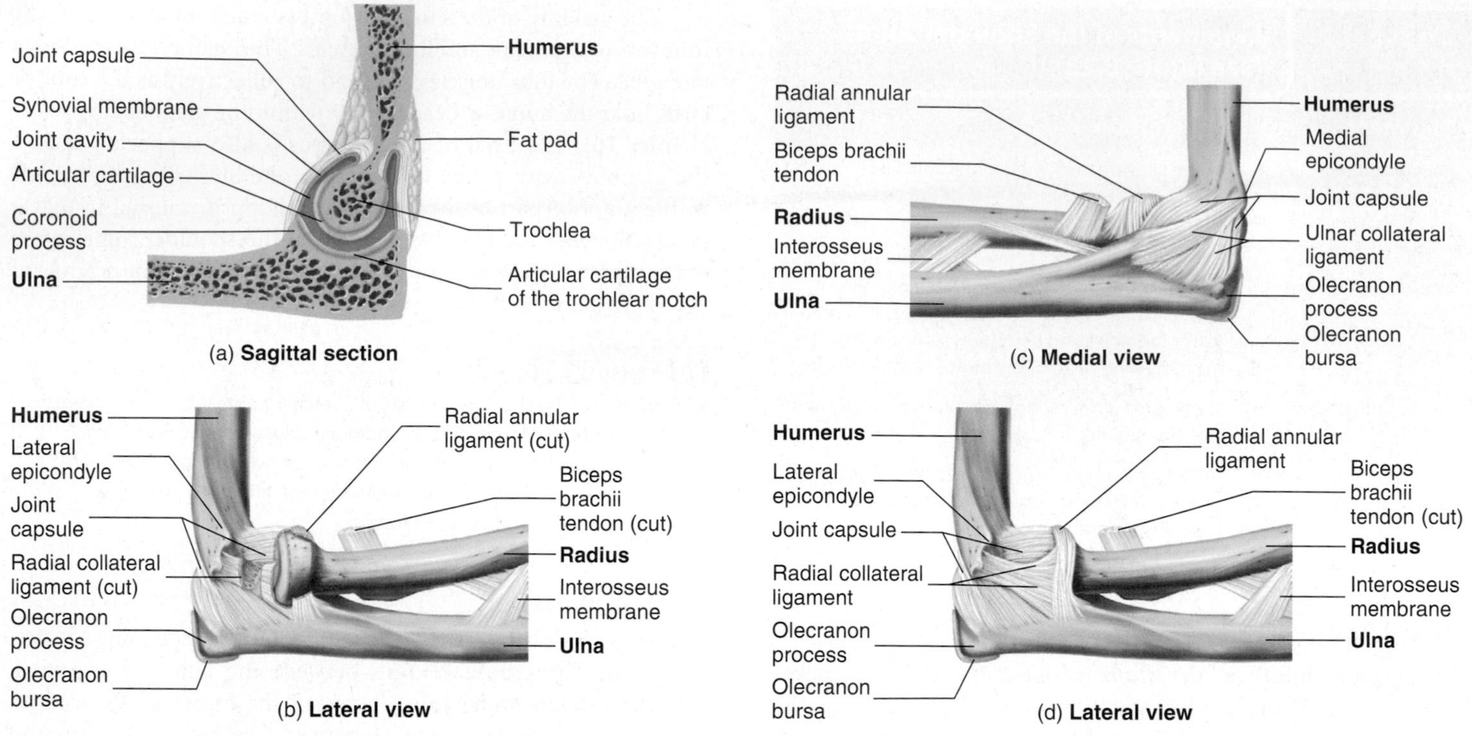
Joint capsule
Synovial membrane
Joint cavity
Articular cartilage
Coronoid process
Ulna
Humerus
Fat pad
Trochlea
Articular cartilage of the trochlear notch
(a) Sagittal section
Radial annular ligament
Biceps brachii tendon
Radius
Interosseus membrane
Ulna
Humerus
Medial epicondyle
Joint capsule
Ulnar collateral ligament
Olecranon process
Olecranon bursa
(c) Medial view
Humerus
Lateral epicondyle
Joint capsule
Radial collateral ligament (cut)
Olecranon process
Olecranon bursa
Radial annular ligament (cut)
Biceps brachii tendon (cut)
Radius
Interosseus membrane
Ulna
(b) Lateral view
Humerus
Lateral epicondyle
Joint capsule
Radial collateral ligament
Olecranon process
Olecranon bursa
Radial annular ligament
Biceps brachii tendon (cut)
Radius
Interosseus membrane
Ulna
(d) Lateral view

FIGURE 8.24 Right Elbow Joint

(*a*) Sagittal section showing the relationship between the ulna and the humerus. (*b*) Lateral view with ligaments cut to show the relationships among the radial head, ulna, and humerus. (*c*) Medial view. (*d*) Lateral view.

Hip Joint

The femoral head articulates with the relatively deep, concave acetabulum of the coxal bone to form the **hip joint,** or *coxal joint* (figure 8.25). The head of the femur is more nearly a complete ball than the articulating surface of any other bone of the body. The acetabulum is deepened and strengthened by a lip of fibrocartilage called the **acetabular labrum,** which is incomplete inferiorly, and by a **transverse acetabular ligament,** which crosses the acetabular notch on the inferior edge of the acetabulum. The hip is capable of a wide range of movement, including flexion, extension, abduction, adduction, rotation, and circumduction. Dislocation of the hip may occur when the femur is driven posteriorly while the hip is flexed, as when a person sitting in an automobile is involved in an accident. The head of the femur usually dislocates posterior to the acetabulum, tearing the acetabular labrum, the fibrous capsule, and the ligaments. Fracture of the femur and the coxal bone often accompanies hip dislocation.

An extremely strong joint capsule, reinforced by several ligaments, extends from the rim of the acetabulum to the neck of the femur (table 8.3). The **iliofemoral ligament** is especially strong. When standing, most people tend to thrust the hips anteriorly. This position is relaxing because the iliofemoral ligament supports much of the body's weight. The **ligament of the head of the femur** (round ligament of the femur) is located inside the hip joint between the femoral head and the acetabulum. This ligament does not contribute much toward strengthening the hip joint; however, it does carry a small nutrient artery to the head of the femur in about 80% of the population. The acetabular labrum, ligaments of the hip, and surrounding muscles make the hip joint much more stable but less mobile than the shoulder joint.

TABLE 8.3 Ligaments of the Hip Joint (see figure 8.25)

Ligament	Description
Transverse acetabular	Bridges gap in the inferior margin of the fibrocartilaginous acetabular labrum
Iliofemoral	Strong, thick band between the anterior inferior iliac spine and the intertrochanteric line of the femur
Pubofemoral	Extends from the pubic portion of the acetabular rim to the inferior portion of the femoral neck
Ischiofemoral	Bridges the ischial acetabular rim and the superior portion of the femoral neck; less well defined
Ligament of the head of the femur	Weak, flat band from the margin of the acetabular notch and the transverse ligament to a fovea in the center of the femoral head

Knee Joint

The **knee joint** is traditionally classified as a modified hinge joint located between the femur and the tibia (figure 8.26). Actually, it is a complex ellipsoid joint that allows flexion, extension, and a small

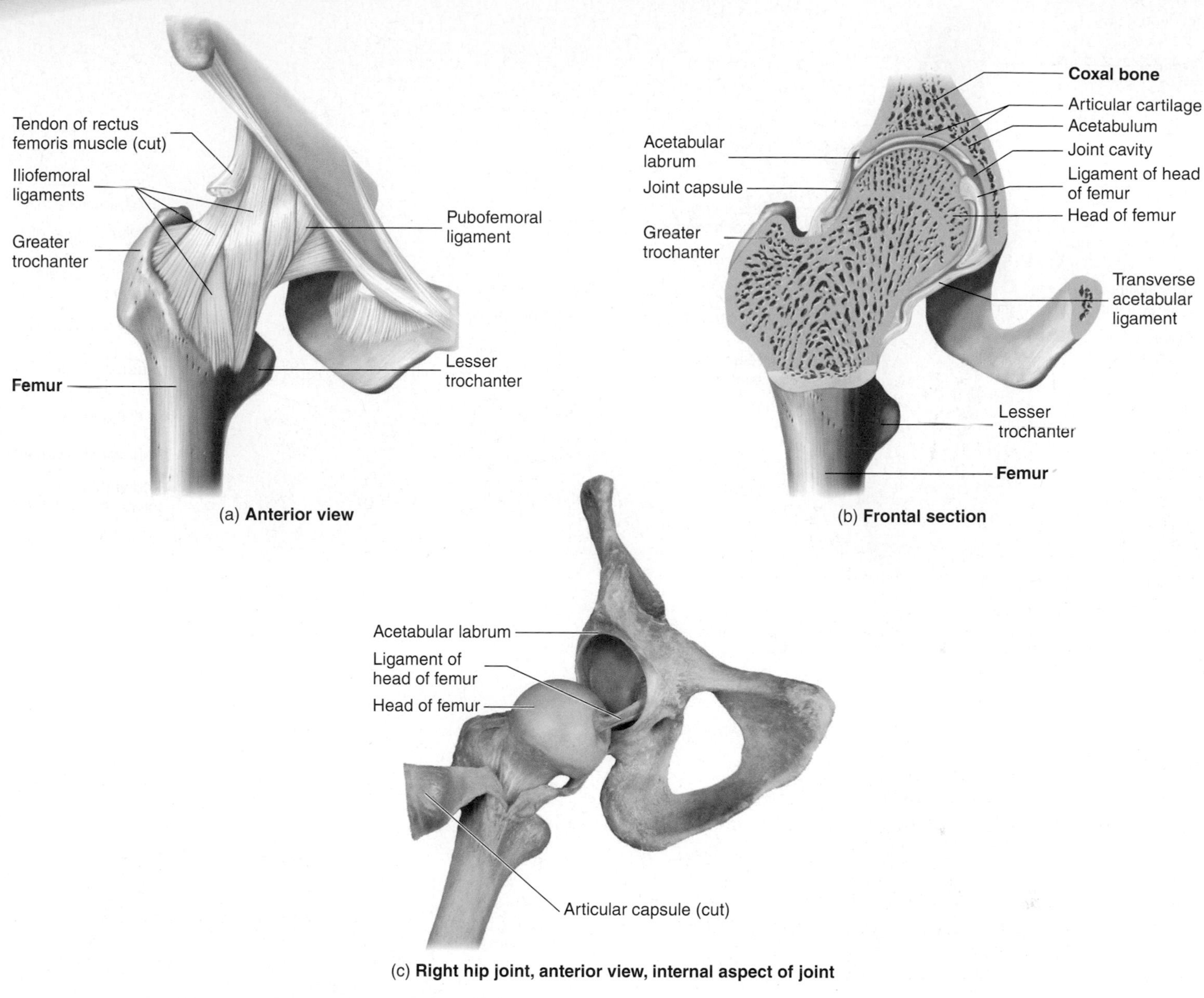
Tendon of rectus femoris muscle (cut)
Iliofemoral ligaments
Greater trochanter
Femur
Pubofemoral ligament
Lesser trochanter
(a) Anterior view
Coxal bone
Articular cartilage
Acetabulum
Joint cavity
Ligament of head of femur
Head of femur
Acetabular labrum
Joint capsule
Greater trochanter
Transverse acetabular ligament
Lesser trochanter
Femur
(b) Frontal section
Acetabular labrum
Ligament of head of femur
Head of femur
Articular capsule (cut)
(c) Right hip joint, anterior view, internal aspect of joint

FIGURE 8.25 APR **Right Hip Joint (Coxal Joint)**

amount of rotation of the leg. The distal end of the femur has two large ellipsoid surfaces with a deep fossa between them. The femur articulates with the proximal end of the tibia, which is flattened and smooth laterally, with a crest called the intercondylar eminence in the center (see figure 7.35). The margins of the tibia are built up by menisci, thick articular disks of fibrocartilage (figure 8.26*b,d*), which deepen the articular surface. The fibula articulates only with the lateral side of the tibia, not with the femur.

The major ligaments that provide knee joint stability are the cruciate and collateral ligaments. Two **cruciate** (kroo′shē-āt; crossed) **ligaments** extend between the intercondylar eminence of the tibia and the fossa of the femur (figure 8.26*b,d,e*). The **anterior cruciate ligament** prevents anterior displacement of the tibia relative to the femur, and the **posterior cruciate ligament** prevents posterior displacement of the tibia. The **medial** (tibial) and **lateral** (fibular) **collateral ligaments** stabilize the medial and lateral sides, respectively, of the knee. Joint strength is also provided by popliteal ligaments and tendons of the thigh muscles that extend around the knee (table 8.4).

A number of bursae surround the knee (figure 8.26*f*). The largest is the **suprapatellar bursa,** a superior extension of the joint capsule that allows the anterior thigh muscles to move over the distal end of the femur. Other knee bursae include the subcutaneous prepatellar bursa and the deep infrapatellar bursa, as well as the popliteal bursa, the gastrocnemius bursa, and the subcutaneous infrapatellar bursa (not shown in figure 8.26).

FUNDAMENTAL **Figure**

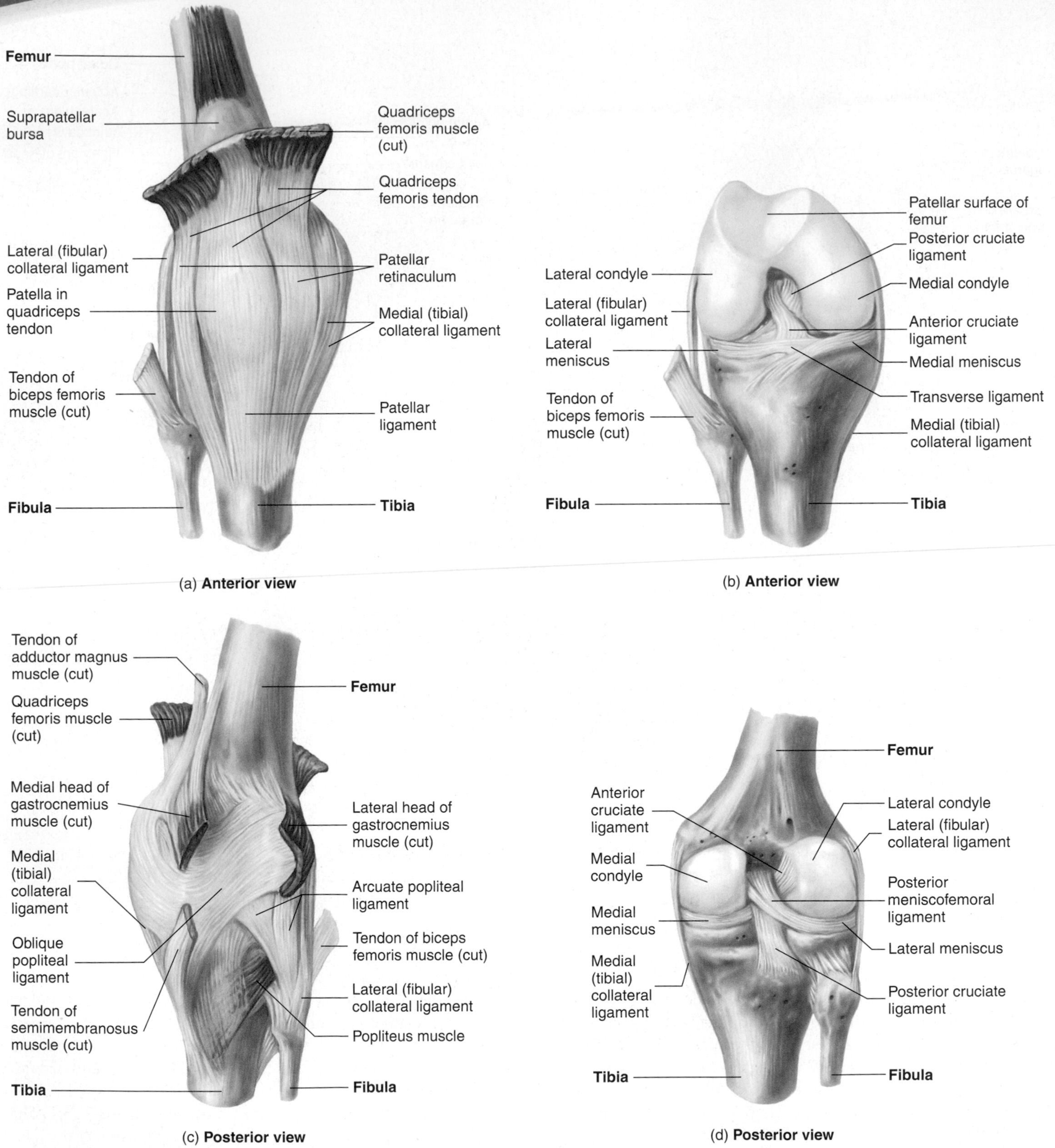

Femur
Suprapatellar bursa
Quadriceps femoris muscle (cut)
Quadriceps femoris tendon
Lateral (fibular) collateral ligament
Patellar retinaculum
Patella in quadriceps tendon
Medial (tibial) collateral ligament
Tendon of biceps femoris muscle (cut)
Patellar ligament
Fibula
Tibia

(a) **Anterior view**

Patellar surface of femur
Posterior cruciate ligament
Lateral condyle
Medial condyle
Lateral (fibular) collateral ligament
Anterior cruciate ligament
Lateral meniscus
Medial meniscus
Transverse ligament
Tendon of biceps femoris muscle (cut)
Medial (tibial) collateral ligament
Fibula
Tibia

(b) **Anterior view**

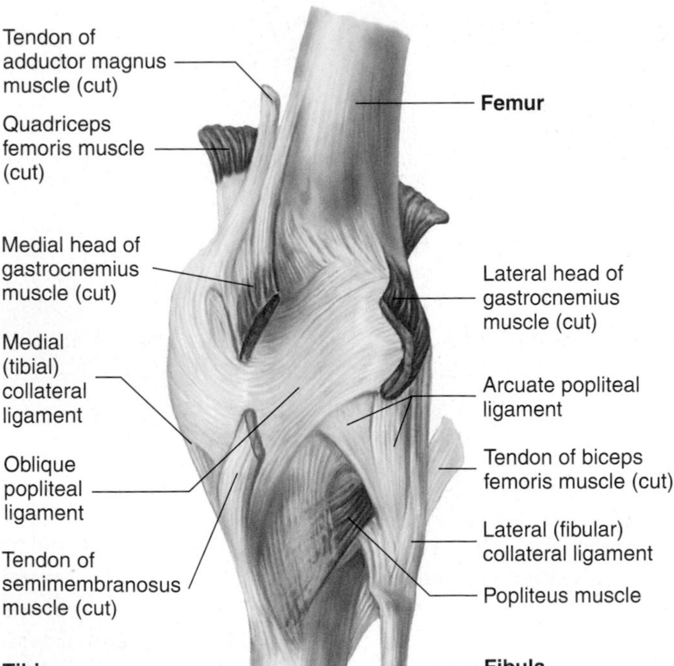
Tendon of adductor magnus muscle (cut)
Femur
Quadriceps femoris muscle (cut)
Medial head of gastrocnemius muscle (cut)
Lateral head of gastrocnemius muscle (cut)
Medial (tibial) collateral ligament
Arcuate popliteal ligament
Tendon of biceps femoris muscle (cut)
Oblique popliteal ligament
Lateral (fibular) collateral ligament
Tendon of semimembranosus muscle (cut)
Popliteus muscle
Tibia
Fibula

(c) **Posterior view**

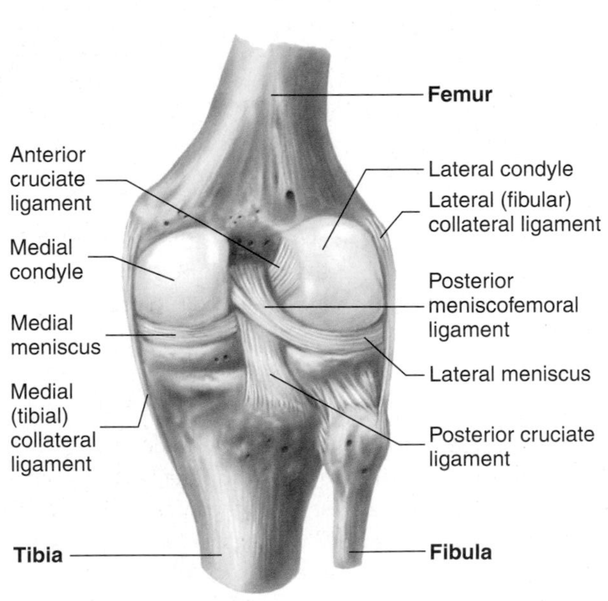
Femur
Anterior cruciate ligament
Lateral condyle
Lateral (fibular) collateral ligament
Medial condyle
Posterior meniscofemoral ligament
Medial meniscus
Lateral meniscus
Medial (tibial) collateral ligament
Posterior cruciate ligament
Tibia
Fibula

(d) **Posterior view**

FIGURE 8.26 **AP|R** **Right Knee Joint**
(*a*) Anterior superficial view. (*b*) Anterior deep view (knee flexed). (*c*) Posterior superficial view. (*d*) Posterior deep view.

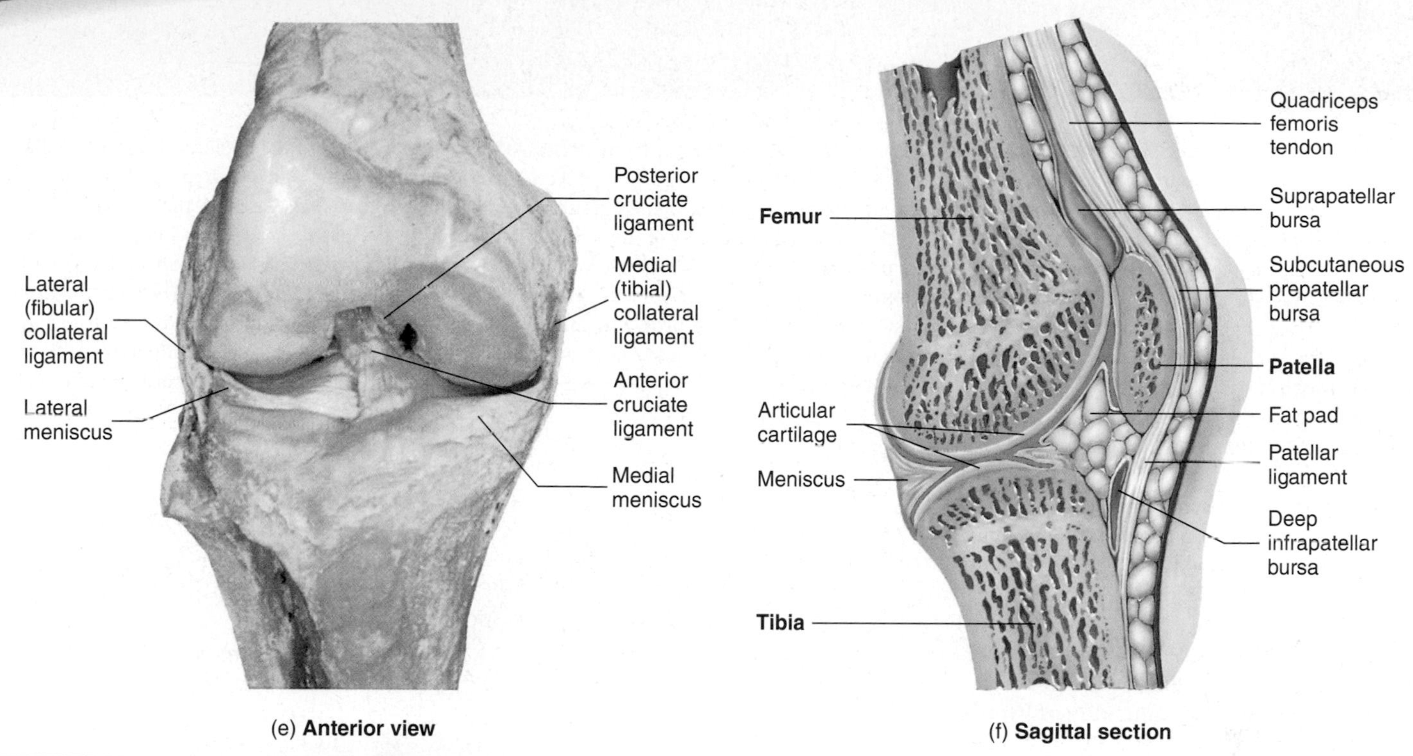

Lateral (fibular) collateral ligament
Lateral meniscus
Posterior cruciate ligament
Medial (tibial) collateral ligament
Anterior cruciate ligament
Medial meniscus
(e) Anterior view
Femur
Articular cartilage
Meniscus
Tibia
Quadriceps femoris tendon
Suprapatellar bursa
Subcutaneous prepatellar bursa
Patella
Fat pad
Patellar ligament
Deep infrapatellar bursa
(f) Sagittal section

FIGURE 8.26 Right Knee Joint ***(continued)***
(*e*) Photograph of anterior deep view. (*f*) Sagittal section.

TABLE 8.4 Ligaments of the Knee Joint (see figure 8.26)

Ligament	Description
Patellar	Thick, heavy, fibrous band between the patella and the tibial tuberosity; actually part of the quadriceps femoris tendon
Patellar retinaculum	Thin band from the margins of the patella to the sides of the tibial condyles
Oblique popliteal	Thickening of the posterior capsule; extension of the semimembranous tendon
Arcuate popliteal	Extends from the posterior fibular head to the posterior fibrous capsule
Medial (tibial) collateral	Thickening of the lateral capsule from the medial epicondyle of the femur to the medial surface of the tibia; also called the medial collateral ligament
Lateral (fibular) collateral	Round ligament extending from the lateral femoral epicondyle to the head of the fibula; also called the lateral collateral ligament
Anterior cruciate	Extends obliquely, superiorly, and posteriorly from the anterior intercondylar eminence of the tibia to the medial side of the lateral femoral condyle
Posterior cruciate	Extends superiorly and anteriorly from the posterior intercondylar eminence to the lateral side of the medial condyle
Coronary (medial and lateral)	Attaches the menisci to the tibial condyles (not illustrated)
Transverse	Connects the anterior portions of the medial and lateral menisci
Meniscofemoral (anterior and posterior)	Joins the posterior part of the lateral menisci to the medial condyle of the femur, passing anterior and posterior to the posterior cruciate ligament

Ankle Joint and Arches of the Foot

The distal tibia and fibula form a highly modified hinge joint with the talus called the **ankle joint,** or *talocrural* (tā′lō-kroo′răl) *joint* (figure 8.27). The medial and lateral malleoli of the tibia and fibula, which form the medial and lateral margins of the ankle, are rather extensive, whereas the anterior and posterior margins are almost nonexistent. As a result, a hinge joint is created. A fibrous capsule surrounds the joint, with the medial and lateral parts thickened to form ligaments. Other ligaments also help stabilize the joint (table 8.5). Dorsiflexion, plantar flexion, and limited inversion and eversion can occur at this joint.

Clinical IMPACT Knee Injuries and Disorders

Injuries to the medial side of the knee are much more common than injuries to the lateral side for several reasons. First, the **lateral** (fibular) **collateral ligament** strengthens the joint laterally and is stronger than the **medial** (tibial) **collateral ligament.** Second, severe blows to the medial side of the knee are far less common than blows to the lateral side of the knee. Finally, the **medial meniscus** is fairly tightly attached to the medial collateral ligament and is damaged 20 times more often in knee injuries than the lateral meniscus, which is thinner and not attached to the lateral collateral ligament.

A **torn meniscus** may cause a "clicking" sound during extension of the leg; if the damage is more severe, the torn piece of cartilage may move between the articulating surfaces of the tibia and femur, causing the knee to "lock" in a partially flexed position.

If the knee is driven anteriorly or hyperextended, the **anterior cruciate ligament** may be torn, which makes the knee joint very unstable. If the knee is flexed and the tibia is driven posteriorly, the **posterior cruciate ligament** may be torn. Surgical replacement of a cruciate ligament with a transplanted or an artificial ligament repairs the damage.

A common football injury results from a block or tackle to the lateral side of the knee, which can cause the knee to bend inward, tearing the medial collateral ligament and opening the medial side of the joint. The medial meniscus is often torn as well. In severe medial knee injuries, the anterior cruciate ligament is also damaged (figure 8A). Tearing of the medial collateral ligament, medial meniscus, and anterior cruciate ligament is often referred to as "the unhappy triad of injuries."

Bursitis in the **subcutaneous prepatellar bursa** (see figure 8.26*f*), commonly called *housemaid's knee,* may result from prolonged work while on the hands and knees. Another form of bursitis, *clergyman's knee,* results from excessive kneeling and affects the **subcutaneous infrapatellar bursa.** This type of bursitis is common among carpet layers and roofers.

Other common knee problems are **chondromalacia** (kon′drō-mă-lā′shē-ă), or softening of the cartilage, which results from abnormal movement of the patella within the patellar groove, and **fat pad syndrome,** which occurs when fluid accumulates in the fat pad posterior to the patella. Acute swelling in the knee appearing immediately after an injury is usually a sign of a **hemarthrosis** (hē′mar-thrō′sis, hem′ar-thrō′sis), blood accumulation within the joint cavity. A slower accumulation of fluid, "water on the knee," may be caused by bursitis.

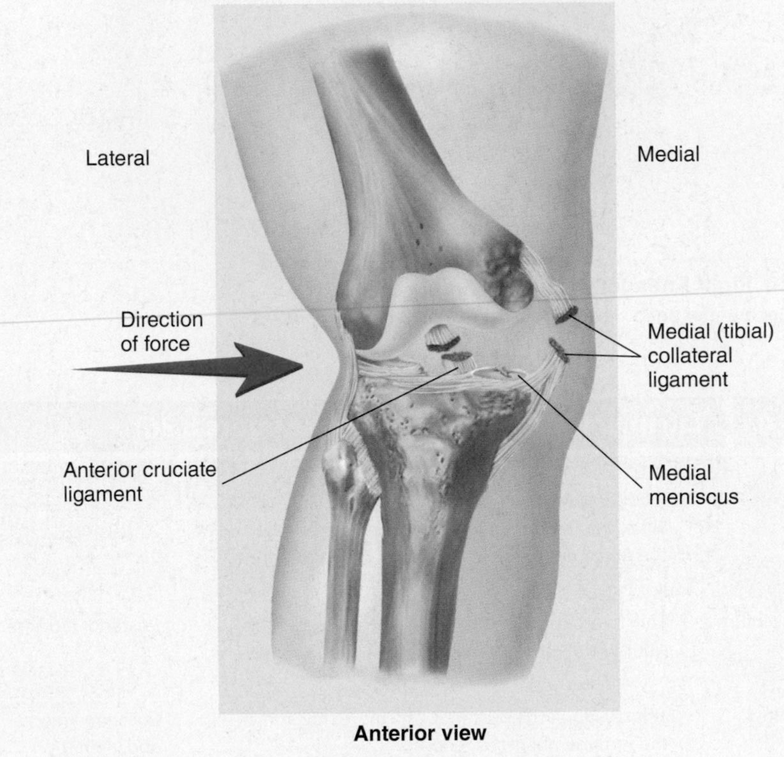
Lateral
Medial
Direction of force
Medial (tibial) collateral ligament
Anterior cruciate ligament
Medial meniscus
Anterior view

FIGURE 8A Injury to the Right Knee

The arches (see figure 7.37) have ligaments that serve two major functions: to hold the bones in their proper relationship as segments of the arch and to provide ties across the arch somewhat like a bowstring. As weight is transferred through the arch system, some of the ligaments are stretched, giving the foot more flexibility and allowing it to adjust to uneven surfaces. When weight is removed from the foot, the ligaments recoil and restore the arches to their unstressed shape.

Predict 6

Ford Dent hurt his knee in an auto accident when his knee was rammed into the dashboard. The doctor tested the knee for ligament damage by having Ford sit on the edge of a table with his knee flexed at a 90-degree angle. The doctor attempted to pull the tibia in an anterior direction (anterior drawer test) and then tried to push the tibia in a posterior direction (posterior drawer test). Results of the anterior drawer test were normal, but unusual movement did occur during the posterior drawer test. Explain the purpose of each test, and describe which ligament was damaged.

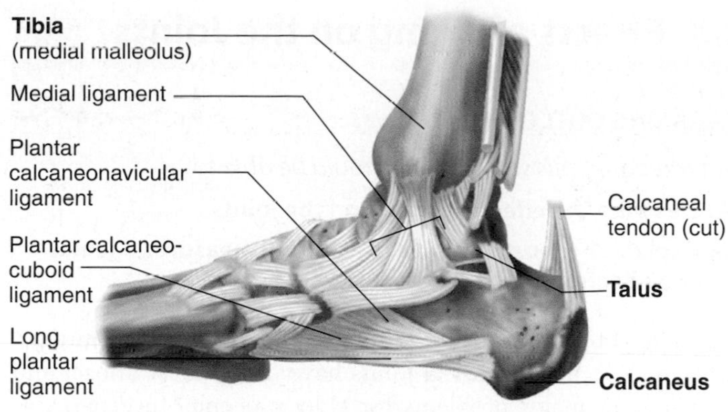
Tibia
(medial malleolus)
Medial ligament
Plantar calcaneonavicular ligament
Plantar calcaneo-cuboid ligament
Long plantar ligament
Calcaneal tendon (cut)
Talus
Calcaneus

(a) **Medial view**

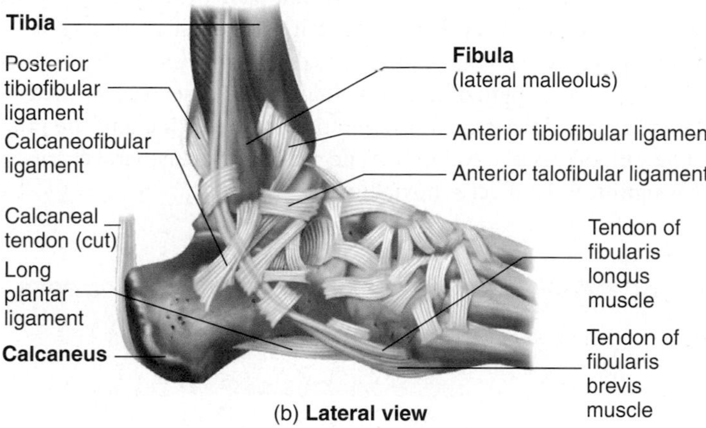
Tibia
Posterior tibiofibular ligament
Calcaneofibular ligament
Calcaneal tendon (cut)
Long plantar ligament
Calcaneus
Fibula
(lateral malleolus)
Anterior tibiofibular ligament
Anterior talofibular ligament
Tendon of fibularis longus muscle
Tendon of fibularis brevis muscle

(b) **Lateral view**

FIGURE 8.27 APIR **Ligaments of the Right Ankle Joint**

TABLE 8.5	Ligaments of the Ankle and Arch (see figure 8.27)
Ligament	**Description**
Medial	Thickening of the medial fibrous capsule that attaches the medial malleolus to the calcaneus, navicular, and talus; also called the deltoid ligament
Calcaneofibular	Extends from the lateral malleolus to the lateral surface of the calcaneus; separate from the capsule
Anterior talofibular	Extends from the lateral malleolus to the neck of the talus; fused with the joint capsule
Long plantar	Extends from the calcaneus to the cuboid and bases of metatarsal bones 2–5
Plantar calcaneocuboid	Extends from the calcaneus to the cuboid
Plantar calcaneonavicular (short plantar)	Extends from the calcaneus to the navicular

Clinical IMPACT

Ankle Injury

The ankle is the most frequently injured major joint in the body, and the most common ankle injuries are caused by forceful inversion of the foot. A **sprained ankle** results when the ligaments of the ankle are torn partially or completely. The calcaneofibular ligament tears most often (figure 8B), followed in frequency by the anterior talofibular ligament. A fibular fracture can occur with severe inversion because the talus can slide against the lateral malleolus and break it (see chapter 7).

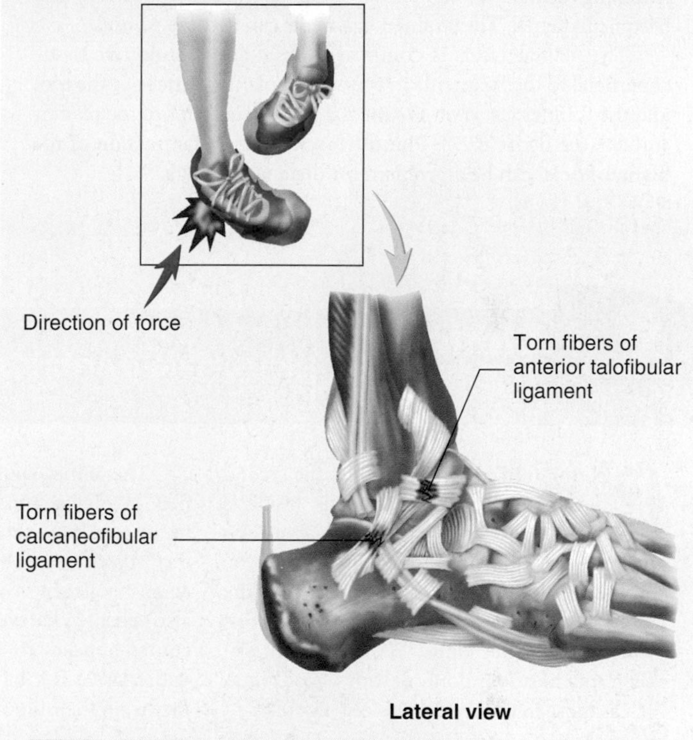
Direction of force
Torn fibers of anterior talofibular ligament
Torn fibers of calcaneofibular ligament
Lateral view

FIGURE 8B Injury to the Right Ankle

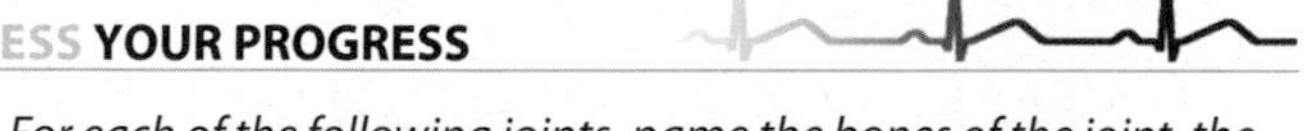

ASSESS **YOUR PROGRESS**

18. *For each of the following joints, name the bones of the joint, the specific parts of the bones that form the joint, the type of joint, and the possible movement(s) at the joint: temporomandibular, shoulder, elbow, hip, knee, and ankle.*
19. *Describe dislocations of the shoulder and hip. What conditions are most likely to cause each type?*
20. *Explain the differences in stability and movement between the shoulder and the hip joints.*
21. *List the common knee injuries, and tell which part of the knee is most often damaged in each type.*
22. *Describe a sprain, and identify which portions of the ankle joint are most commonly damaged when it is sprained.*

Clinical IMPACT

Arch Problems

The arches of the foot normally form early in fetal life. Failure to form results in congenital **flat feet,** or fallen arches, in which the arches, primarily the medial longitudinal arch, are depressed or collapsed (see figure 7.37). This condition is sometimes, but not always, painful. Flat feet may also occur when the muscles and ligaments supporting the arch fatigue and allow the arch, usually the medial longitudinal arch, to collapse. During prolonged standing, the plantar calcaneonavicular ligament may stretch, flattening the medial longitudinal arch. The transverse arch may also become flattened. The strained ligaments can become painful.

The plantar fascia is composed of the deep connective tissue superficial to the ligaments in the central plantar surface of the foot and the thinner fascia on the medial and lateral sides of the plantar surface (see figure 8.27). **Plantar fasciitis,** an inflammation of the plantar fascia, can be a problem for distance runners.

8.5 Effects of Aging on the Joints

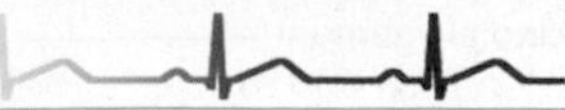

LEARNING OUTCOMES

After reading this section, you should be able to

A. Describe the effects of aging on the joints.

B. Explain the most effective preventive measures against the effects of aging on the joints.

A number of changes occur within the joints as a person ages. Those that occur in synovial joints have the greatest impact and often present major problems for elderly people. In general, as a person ages, the tissues of the body become less flexible and less elastic, in part due to changes in protein structure caused by age-related modifications that cross-link proteins together. These changes are most prevalent in long-lived proteins, such as collagen, which is abundant in fibrous connective tissue. Hence, the flexibilty and strength of tendons and ligaments decrease with age. Tissue repair slows as cell proliferation rates decline and the rate of new blood vessel development decreases. These general changes can significantly affect synovial joints.

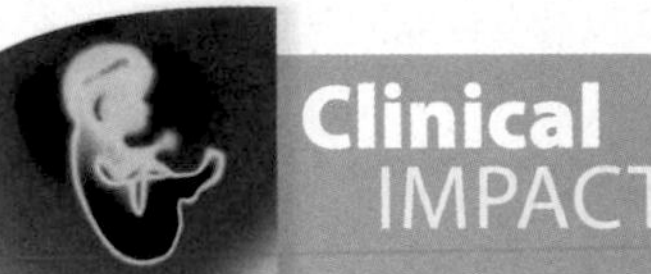

Clinical IMPACT — Rheumatoid Arthritis

Rheumatoid arthritis (RA) is the second most common type of arthritis. It affects about 3% of all women and about 1% of all men in the United States. RA is a general connective tissue disorder that affects the skin, vessels, lungs, and other organs, but it is most pronounced in the joints. RA is severely disabling and most commonly destroys small joints, such as those in the hands and feet (figure 8C).

The initial cause of RA is unknown but may involve a transient infection or an autoimmune disease (an immune reaction to one's own tissues; see chapter 22) that develops against collagen. A genetic predisposition may also exist. Whatever the cause, the ultimate course appears to be immunological. People with classic RA have a protein, **rheumatoid factor,** in their blood. In RA, the synovial fluid and associated connective tissue cells proliferate, forming a *pannus* (clothlike layer), which causes the joint capsule to become thickened and destroys the articular cartilage. In advanced stages, opposing joint surfaces can become fused. **Juvenile rheumatoid arthritis** is similar to the adult type in many ways, but no rheumatoid factor is found in the serum.

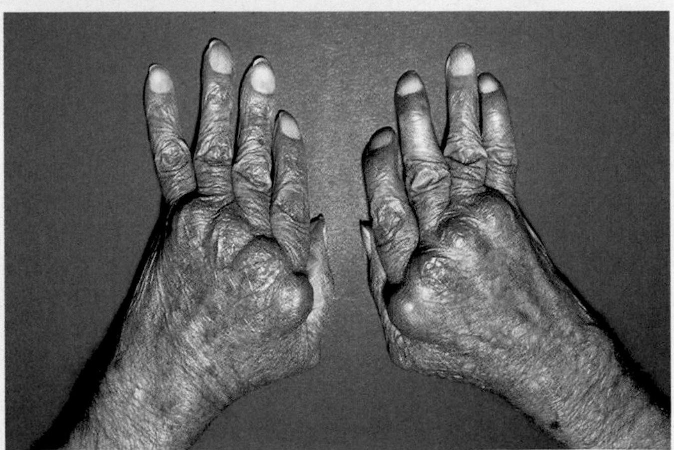

(a)

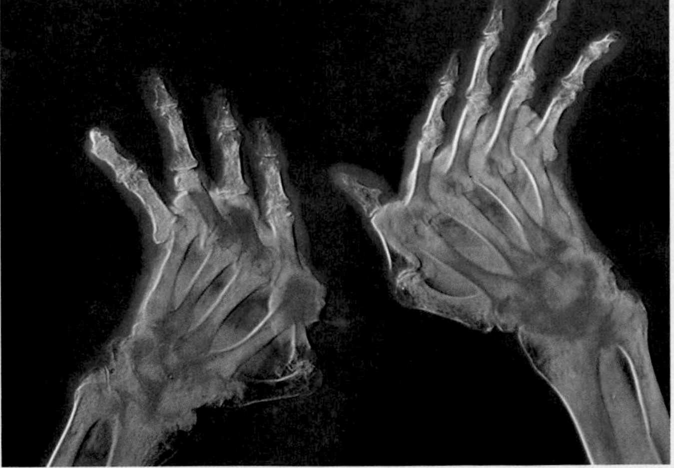

(b)

FIGURE 8C Rheumatoid Arthritis

(*a*) Photograph of hands with rheumatoid arthritis. (*b*) Radiographs of the same hands shown in (*a*).

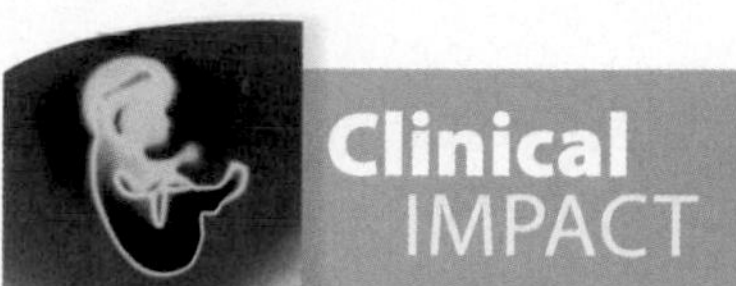
Clinical
IMPACT

Joint Replacement

As a result of recent advances in biomedical technology, many joints of the body can now be replaced by artificial ones. Joint replacement, called **arthroplasty,** was first developed in the late 1950s. One of its major purposes is to eliminate unbearable pain in patients in the age range of 55–60 who have joint disorders. Osteoarthritis is the leading disease requiring joint replacement and accounts for two-thirds of the patients. Rheumatoid arthritis accounts for more than half of the remaining cases.

The major design objectives for joint prostheses (artificial replacements) include developing stable articulations, low friction, solid fixation to the bone, and normal range of motion. Biomedical engineers are designing synthetic replacement materials to accomplish these objectives. Prosthetic joints are usually composed of metal, such as stainless steel, titanium alloys, or cobalt-chrome alloys, in combination with modern plastics, such as high-density polyethylene, silastic, or elastomer.

The bone of the articular area is removed on one side (a procedure called **hemireplacement**) or on both sides (**total replacement**) of the joint, and the artificial articular areas are glued to the bone with a synthetic adhesive, such as methylmethacrylate. The smooth metal surface rubbing against the smooth plastic surface provides low-friction contact, with a range of movement that depends on the design.

The success of joint replacement depends on the joint being replaced, the age and condition of the patient, the state of the technology, and the definition of success. Success is usually defined as minimizing pain while maintaining movement. Most reports are based on examinations of patients 2–10 years after joint replacement. The technology is improving constantly, so current reports do not adequately reflect the effect of the most recent improvements. Still, reports indicate a success rate of 80–90% in hip replacements and 60% or more in ankle and elbow replacements. The major reason for the failure of prosthetic joints is loosening of the artificial joint from the bone to which it is attached. New prostheses with porous surfaces help overcome this problem.

In addition, with use, the cartilage covering articular surfaces can wear down. When a person is young, the production of new, resilient matrix compensates for the wear. However, as a person ages, the rate of replacement declines and the matrix becomes more rigid, thus increasing its rate of wear. The production rate of lubricating synovial fluid also declines with age, further contributing to the wear of the articular cartilage. Furthermore, the ligaments and tendons surrounding a joint shorten and become less flexible with age, resulting in decreased range of motion. With age, muscles, which strengthen the joints, also tend to weaken. Finally, many older people are less physically active, which causes the joints to become less flexible and decreases their range of motion.

The most effective preventive measure against the effects of aging on the joints is to strengthen the bones and muscles and maintain flexibility. This can be accomplished through a combination of regular physical activity, stretching, and a healthy diet.

ASSESS YOUR PROGRESS

23. *List the age-related factors that contribute to cartilage wear in synovial joints.*

24. *Describe the age-related factors that cause loss of flexibility and loss of range of motion in synovial joints.*

25. *Discuss preventive measures to reduce age-related changes to joints.*

Diseases and Disorders

Joints

Condition	Description
Arthritis	Inflammation of a joint, leading to pain and stiffness of the joint; over 100 causes, including infectious agents, metabolic disorders, trauma, and immune disorders
Degenerative joint disease (osteoarthritis)	Most common type of arthritis; affects 85% of Americans over age 70; characterized by gradual degeneration of a joint with advancing age; can be delayed with exercise
Rheumatoid arthritis	General connective tissue autoimmune disorder that predominantly affects joints
Gout	Group of metabolic disorders that lead to increased production and accumulation of uric acid crystals in tissues, including joint capsules; can lead to arthritis
Lyme disease	Caused by a bacterial infection that affects multiple organs, including the joints; can lead to chronic arthritis
Bursitis	Inflammation of a bursa, often due to forceful contact or prolonged contact, such as *student's elbow* from leaning on a desk
Bunion	Most bunions are deformations of the first metatarsal (the great toe); bursitis may accompany this deformity; irritated by tight shoes
Tendinitis	Inflammation of tendon sheaths, often from overuse, such as tennis elbow
Dislocation	Movement of bones out of their correct alignment at a joint; a partial dislocation is a *subluxation*
Sprain	Stretching or tearing of ligaments supporting a joint

Go to: www.mhhe.com/seeley10 for additional information on these pathologies.

Answer

Learn to Predict ❮ From page 239

To answer this question, we first need to review the functions of the major knee ligaments that stabilize the knee joint: the cruciate and collateral ligaments. The anterior cruciate ligament (ACL) prevents extreme anterior movement of the tibia relative to the femur, which can occur from twisting the leg, such as when playing sports. The posterior cruciate ligament (PCL) prevents extreme posterior movement of the tibia relative to the femur, which can occur from a hard blow, such as in a car crash. The medial collateral ligament (MCL) prevents excessive abduction of the knee, and can be damaged from a lateral blow to the knee, such as from a tackle in football. The lateral collateral ligament (LCL) prevents excessive adduction of the knee, and can be damaged from a blow to the inside of the knee, such as from a tackle in soccer, but is less common than MCL injuries. Among these, the function of the ACL most closely fits the description of Andy's injury. The ACL and MCL are the most commonly injured ligaments in ski accidents. The ACL stabilizes the knee joint by stretching diagonally from the femur at the back of the joint to the tibia in the front, which normally prevents forward or anterior movement of the tibia from underneath the femur. It also resists medial rotation of the tibia. Thus, loss of the ACL increases mobility of the tibia in the anterior direction. This increased mobility can be detected using the anterior drawer test. In this test, the patient lies on his or her back with the hips flexed at a 45-degree angle, knees bent at a 90-degree angle, and feet flat on the examining table. A torn ACL yields increased mobility when the physician pulls the tibia forward (anterior direction) relative to the femur. Other physical tests can detect damage to the PCL, MCL, and LCL. In this case, the diagnosis of a torn ACL was confirmed by an MRI that revealed a complete tear of the ligament.

Answers to the rest of this chapter's Predict questions are in Appendix G.

Summary

8.1 Classes of Joints (p. 240)

1. A joint, or an articulation, is a place where two bones come together.
2. Joints are named according to the bones or parts of bones involved.
3. Joints are classified structurally according to the type of connective tissue that binds them together and whether fluid is present between the bones.

Fibrous Joints

1. Fibrous joints, in which bones are connected by fibrous tissue with no joint cavity, are capable of little or no movement.
2. Sutures involve interdigitating bones held together by dense fibrous connective tissue. They occur between most skull bones.
3. Syndesmoses are joints consisting of fibrous ligaments.
4. Gomphoses are joints in which pegs fit into sockets and are held in place by periodontal ligaments (teeth in the jaws).
5. Some sutures and other joints can become ossified (synostoses).

Cartilaginous Joints

1. Synchondroses are immovable joints in which bones are joined by hyaline cartilage. Epiphyseal plates are examples.
2. Symphyses are slightly movable joints made of fibrocartilage.

Synovial Joints

1. Synovial joints are capable of considerable movement. They consist of the following:
   - Articular cartilage on the ends of bones that provides a smooth surface for articulation. Articular disks can provide additional support.
   - A joint cavity is surrounded by a joint capsule of fibrous connective tissue, which holds the bones together while permitting flexibility. A synovial membrane produces synovial fluid, which lubricates the joint.
2. Bursae are extensions of synovial joints that protect skin, tendons, or bone from structures that could rub against them.
3. Synovial joints are classified according to the shape of the adjoining articular surfaces: plane (two flat surfaces), saddle (two saddle-shaped surfaces), hinge (concave and convex surfaces), pivot (cylindrical projection inside a ring), ball-and-socket (rounded surface into a socket), and ellipsoid (ellipsoid concave and convex surfaces).

8.2 Types of Movement (p. 246)

1. Gliding movements occur when two flat surfaces glide over one another.
2. Angular movements include flexion and extension, plantar flexion and dorsiflexion, and abduction and adduction.
3. Circular movements include rotation, pronation and supination, and circumduction.
4. Special movements include elevation and depression, protraction and retraction, excursion, opposition and reposition, and inversion and eversion.
5. Combination movements involve two or more of the previously mentioned movements.

8.3 Range of Motion (p. 250)

Range of motion is the amount of movement, active or passive, that can occur at a joint.

8.4 Description of Selected Joints (p. 250)

1. The temporomandibular joint is a complex hinge and gliding joint between the temporal and mandibular bones. It is capable of elevation and depression, protraction and retraction, and lateral and medial excursion.

2. The shoulder joint is a ball-and-socket joint between the head of the humerus and the glenoid cavity of the scapula that permits a wide range of motion. It is strengthened by ligaments and the muscles of the rotator cuff. The tendon of the biceps brachii passes through the joint capsule. The shoulder joint is capable of flexion and extension, abduction and adduction, rotation, and circumduction.
3. The elbow joint is a compound hinge joint between the humerus, the ulna, and the radius. Movement at this joint is limited to flexion and extension.
4. The hip joint is a ball-and-socket joint between the head of the femur and the acetabulum of the coxal bone. It is greatly strengthened by ligaments and is capable of a wide range of motion, including flexion, extension, abduction, adduction, rotation, and circumduction.
5. The knee joint is a complex ellipsoid joint between the femur and the tibia that is supported by many ligaments. The joint allows flexion and extension and slight rotation of the leg.
6. The ankle joint is a special hinge joint of the tibia, the fibula, and the talus that allows dorsiflexion and plantar flexion and inversion and eversion.
7. Ligaments hold the bones of the foot arches and transfer weight in the foot.

8.5 Effects of Aging on the Joints (p. 260)

1. With age, the connective tissue of the joints becomes less flexible and less elastic. The resulting joint rigidity increases the rate of wear in the articulating surfaces. The changes in connective tissue also reduce the range of motion.
2. The effects of aging on the joints can be slowed by exercising regularly and consuming a healthy diet.

REVIEW AND COMPREHENSION

1. Which of these joints is *not* matched with the correct joint type?
   a. parietal bone to occipital bone—suture
   b. between the coxal bones—symphysis
   c. humerus and scapula—synovial
   d. shafts of the radius and ulna—synchondrosis
   e. teeth in alveolar process—gomphosis
2. Which type of joint is the most movable?
   a. sutures c. symphyses e. gomphoses
   b. syndesmoses d. synovial
3. The intervertebral disks are examples of
   a. sutures. c. symphyses. e. gomphoses.
   b. syndesmoses. d. synovial joints.
4. Joints containing hyaline cartilage are called __________, and joints containing fibrocartilage are called __________.
   a. sutures, synchondroses
   b. syndesmoses, symphyses
   c. symphyses, syndesmoses
   d. synchondroses, symphyses
   e. gomphoses, synchondroses
5. The inability to produce the fluid that keeps most joints moist would likely be caused by a disorder of the
   a. cruciate ligaments.
   b. synovial membrane.
   c. articular cartilage.
   d. bursae.
   e. tendon sheath.
6. Assume that a sharp object penetrated a synovial joint. Given these structures:
   (1) tendon or muscle
   (2) ligament
   (3) articular cartilage
   (4) fibrous capsule (of joint capsule)
   (5) skin
   (6) synovial membrane (of joint capsule)

   Choose the order in which they would most likely be penetrated.
   a. 5,1,2,6,4,3 d. 5,1,2,4,3,6
   b. 5,2,1,4,3,6 e. 5,1,2,4,6,3
   c. 5,1,2,6,3,4
7. Which of these joints is correctly matched with the type of joint?
   a. atlas to occipital condyle—pivot
   b. tarsal bones to metatarsal bones—saddle
   c. femur to coxal bone—ellipsoid
   d. tibia to talus—hinge
   e. scapula to humerus—plane
8. When you grasp a doorknob, what movement of your forearm is necessary to unlatch the door—that is, to turn the knob in a clockwise direction? (Assume using the right hand.)
   a. pronation c. supination e. extension
   b. rotation d. flexion
9. After the door is unlatched, what movement of the elbow is necessary to open it? (Assume the door opens in and you are on the inside.)
   a. pronation c. supination e. extension
   b. rotation d. flexion
10. After the door is unlatched, what movement of the shoulder is necessary to open it? (Assume the door opens in and you are on the inside.)
    a. pronation c. supination e. extension
    b. rotation d. flexion
11. When grasping a doorknob, the thumb and little finger undergo
    a. opposition. c. lateral excursion. e. dorsiflexion.
    b. reposition. d. medial excursion.
12. A runner notices that the lateral side of her right shoe is wearing much more than the lateral side of her left shoe. This could mean that her right foot undergoes more __________ than her left foot.
    a. eversion c. plantar flexion e. lateral excursion
    b. inversion d. dorsiflexion
13. For a ballet dancer to stand on her toes, her feet must
    a. evert. c. plantar flex. e. abduct.
    b. invert. d. dorsiflex.
14. A meniscus is found in the
    a. shoulder joint. c. hip joint. e. ankle joint.
    b. elbow joint. d. knee joint.
15. A lip (labrum) of fibrocartilage deepens the joint cavity of the
    a. temporomandibular joint. d. knee joint.
    b. shoulder joint. e. ankle joint.
    c. elbow joint.

16. Which of these joints has a tendon inside the joint cavity?
    a. temporomandibular joint
    b. shoulder joint
    c. elbow joint
    d. knee joint
    e. ankle joint
17. Which of these structures help stabilize the shoulder joint?
    a. rotator cuff muscles
    b. cruciate ligaments
    c. medial and lateral collateral ligaments
    d. articular disks
    e. All of these are correct.
18. Bursitis of the subacromial bursa could result from
    a. flexing the wrist.
    b. kneeling.
    c. overusing the shoulder joint.
    d. running a long distance.
    e. extending the elbow.
19. Which of these events does *not* occur with the aging of joints?
    a. decrease in production of new cartilage matrix
    b. decline in synovial fluid production
    c. stretching of ligaments and tendons and increase in range of motion
    d. weakening of muscles
    e. increase in protein cross-linking in tissues

Answers in Appendix E

CRITICAL THINKING

1. How would body function be affected if the sternal synchondroses and the sternocostal synchondrosis of the first rib were to become synostoses?
2. Using an articulated skeleton, describe the type of joint and the movement(s) possible for each of the following joints:
    a. joint between the zygomatic bone and the maxilla
    b. ligamentous connection between the coccyx and the sacrum
    c. elbow joint
3. For each of the following muscles, describe the motion(s) produced when the muscle contracts. It may be helpful to use an articulated skeleton.
    a. The biceps brachii muscle attaches to the coracoid process of the scapula (one head) and to the radial tuberosity of the radius. Name two movements that the muscle accomplishes in the forearm.
    b. The rectus femoris muscle attaches to the anterior inferior iliac spine and the tibial tuberosity. How does contraction move the thigh? The leg?
    c. The supraspinatus muscle is located in and attached to the supraspinatus fossa of the scapula. Its tendon runs over the head of the humerus to the greater tubercle. When it contracts, what movement occurs at the glenohumeral (shoulder) joint?
    d. The gastrocnemius muscle attaches to the medial and lateral condyles of the femur and to the calcaneus. What movement of the leg results when this muscle contracts? Of the foot?
4. At first, Donnie's wife accused her once active 25-year-old husband of trying to get out of housework by constantly complaining about pain and stiffness in his lower back. But over the next 5 months, the pain and stiffness increased and seemed to be spreading up his vertebral column. The family doctor referred Donnie to a rheumatologist, who diagnosed ankylosing spondylitis (AS). AS, a chronic inflammation of joints at points where ligaments, tendons, and joint capsule insert into bone, causes fibrosis (the development of scar tissue), ossification, and fusion of joints. Combine your knowledge about bone growth, repair, and anatomy from chapters 6 and 7 and joint structure and function from this chapter to identify the category of joints primarily affected by AS, and explain how chronic inflammation of Donnie's joints led to their fusion.

Answers in Appendix F

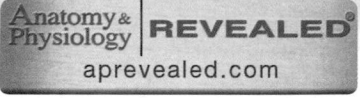

9

Muscular System

HISTOLOGY AND PHYSIOLOGY

In order to turn the pages of this chapter, move your eyes across the page, and continue breathing, electrical impulses must travel to millions of tiny motors throughout your body. These "motors" are really your muscle cells, which, on receiving a nerve impulse, convert chemical ATP energy into the mechanical energy of moving cell parts. The body has three types of muscle tissue, each with a different purpose. The muscles you voluntarily control work with the skeletal system to produce coordinated movements of your limbs. The digestive, cardiovascular, urinary, and reproductive systems all use smooth muscle to propel materials through the body. The heart contains specialized cardiac muscle tissue to pump blood. No matter where muscle tissues are in the body, they all share the same feature: contraction.

Learn to Predict

Winning a one-month membership to a health club changed Bob's life. He gave up junk food and now works out daily. In one of his aerobic exercises, he slowly flexes his elbow and supinates his right hand while lifting a 35-pound weight; then he lowers the weight back to its starting position. He repeats this process several times. After reading chapter 9 and building on your previous knowledge about bone anatomy and joints in chapters 7 and 8, explain how Bob's muscles are able to lift and lower the weight slowly.

Photo: Skeletal muscle cells lie parallel to each other, forming powerful units such as the one shown in this scanning electron micrograph. Tiny levers rhythmically pull one protein strand past another, shortening (contracting) the cell.

Module 6
Muscular System

Anatomy & Physiology
REVEALED
aprevealed.com

9.1 Functions of the Muscular System

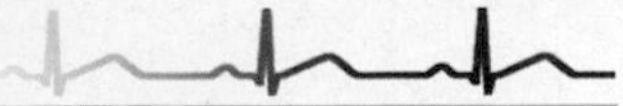

LEARNING OUTCOME

After reading this section, you should be able to

A. Summarize the functions of the muscular system.

Most of the body's movements, from the beating of the heart to the running of a marathon, result from muscle contractions. As described in chapter 4, there are three types of muscle tissue: skeletal, smooth, and cardiac. Because skeletal muscle is the most abundant and most studied type, this chapter examines the physiology of skeletal muscle in greatest detail. Chapter 10 focuses on the anatomy of the skeletal muscle system. The following list summarizes the major functions of all three types of muscle:

1. *Movement of the body.* Most skeletal muscles are attached to bones and are responsible for the majority of body movements, including walking, running, chewing, and manipulating objects with the hands.
2. *Maintenance of posture.* Skeletal muscles constantly maintain tone, which keeps us sitting or standing erect.
3. *Respiration.* Skeletal muscles of the thorax carry out the movements necessary for respiration.
4. *Production of body heat.* When skeletal muscles contract, heat is given off as a by-product. This released heat is critical for maintaining body temperature.
5. *Communication.* Skeletal muscles are involved in all aspects of communication, including speaking, writing, typing, gesturing, and smiling or frowning.
6. *Constriction of organs and vessels.* The contraction of smooth muscle within the walls of internal organs and vessels causes those structures to constrict. This constriction can help propel and mix food and water in the digestive tract; remove materials from organs, such as the urinary bladder or sweat glands; and regulate blood flow through vessels.
7. *Contraction of the heart.* The contraction of cardiac muscle causes the heart to beat, propelling blood to all parts of the body.

ASSESS YOUR PROGRESS

1. *List and describe the functions performed by skeletal muscle tissue.*
2. *State the functions of smooth and cardiac muscle tissues.*

9.2 General Properties of Muscle

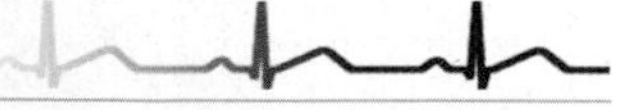

LEARNING OUTCOMES

After reading this section, you should be able to

A. Explain the four functional properties of muscle tissue.

B. Summarize the major characteristics of skeletal, smooth, and cardiac muscle.

Muscle tissue is highly specialized. It has four major functional properties: contractility, excitability, extensibility, and elasticity.

1. **Contractility** is the ability of muscle to shorten forcefully. For example, lifting this textbook requires certain muscles to contract. When muscle contracts, it either causes the structures to which it is attached to move or increases pressure inside a hollow organ or vessel. Although muscle shortens forcefully during contraction, it lengthens passively; that is, other forces cause it to lengthen, such as gravity, contraction of an opposing muscle, or the pressure of fluid in a hollow organ or vessel.
2. **Excitability** is the capacity of muscle to respond to a stimulus. Normally, the stimulus is from nerves that we consciously control. For instance, if you decide to wave to a friend, the conscious decision to lift your arm is sent via nerves. Smooth muscle and cardiac muscle can contract without outside stimuli, but they also respond to stimulation by nerves and hormones.
3. **Extensibility** means a muscle can be stretched beyond its normal resting length and still be able to contract. If you stretch to reach a dropped pencil, your muscles are longer than they are normally but you can still retrieve the pencil.
4. **Elasticity** is the ability of muscle to recoil to its original resting length after it has been stretched. Taking a deep breath demonstrates elasticity because exhalation is simply the recoil of your respiratory muscles back to the resting position, similar to releasing a stretched rubberband.

Types of Muscle Tissue

Table 9.1 compares the major characteristics of skeletal, smooth, and cardiac muscle. **Skeletal muscle,** with its associated connective tissue, constitutes about 40% of the body's weight and is responsible for locomotion, facial expressions, posture, respiratory functions, and many other body movements. The nervous system voluntarily, or consciously, controls the functions of the skeletal muscles.

Smooth muscle is the most widely distributed type of muscle in the body. It is found in the walls of hollow organs and tubes, in the interior of the eye, and in the walls of blood vessels, among other areas. Smooth muscle performs a variety of functions, including propelling urine through the urinary tract, mixing food in the stomach and the small intestine, dilating and constricting the pupil of the eye, and regulating the flow of blood through blood vessels.

Cardiac muscle is found only in the heart, and its contractions provide the major force for moving blood through the circulatory system. Unlike skeletal muscle, cardiac muscle and many smooth muscles are autorhythmic; that is, they contract spontaneously at somewhat regular intervals, and nervous or hormonal stimulation is not always required for them to contract. Furthermore, unlike skeletal muscle, smooth muscle and cardiac muscle are not consciously controlled by the nervous system. Rather, they are controlled involuntarily, or unconsciously, by the autonomic nervous and endocrine systems (see chapters 16 and 18).

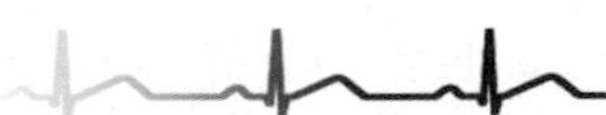

ASSESS YOUR PROGRESS

3. *Identify the four specialized functional properties of muscle tissue, and give an example of each.*
4. *Using table 9.1, distinguish among skeletal, smooth, and cardiac muscle tissues as to their locations, appearance, cell shape, and cell-to-cell attachments.*
5. *Outline the differences in control and function for skeletal, smooth, and cardiac muscle.*

TABLE 9.1 Comparison of Muscle Types

	Skeletal Muscle	Smooth Muscle	Cardiac Muscle
Location	Attached to bones	Walls of hollow organs, blood vessels, eyes, glands, and skin	Heart
Appearance			
Cell Shape	Very long and cylindrical (1 mm–4 cm, or as much as 30 cm, in length, 10 μm–100 μm in diameter)	Spindle-shaped (15–200 μm in length, 5–8 μm in diameter)	Cylindrical and branched (100–500 μm in length, 12–20 μm in diameter)
Nucleus	Multiple nuclei: peripherally located	Single, centrally located	Single, centrally located
Special Cell-to-Cell Attachments	None	Gap junctions join some visceral smooth muscle cells together	Intercalated disks join cells to one another
Striations	Yes	No	Yes
Control	Voluntary and involuntary (reflexes)	Involuntary	Involuntary
Capable of Spontaneous Contraction	No	Yes (some smooth muscle)	Yes
Function	Body movement	Moving food through the digestive tract, emptying the urinary bladder, regulating blood vessel diameter, changing pupil size, contracting many gland ducts, moving hair, and having many other functions	Pumping blood; contractions provide the major force for propelling blood through blood vessels

9.3 Skeletal Muscle Structure

LEARNING OUTCOMES

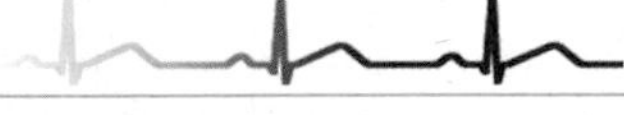

After reading this section, you should be able to

A. Describe the connective tissue components of skeletal muscle.

B. Explain the blood supply and innervation of skeletal muscle.

C. Discuss the origin of muscle fibers and explain how muscle hypertrophy occurs.

D. Describe the components of a muscle fiber.

E. Relate the types of myofilaments and describe their structures.

F. Produce diagrams that illustrate the arrangement of myofilaments in a sarcomere.

Each skeletal muscle is a complete organ consisting of cells, called **skeletal muscle fibers,** associated with smaller amounts of connective tissue, blood vessels, and nerves. The connective tissue fibers that surround a muscle and its internal components extend beyond the center of the muscle to become tendons, which connect muscles to bones or to the dermis of the skin (figure 9.1).

A muscle is composed of numerous visible bundles called muscle **fasciculi** (fă-sik′u-lī; sing. fă-sik′u-lus). Each fasciculus is surrounded by another, heavier connective tissue layer called the **perimyseum** (per′i-miz′ē-ŭm, per′i-mis′ē-ŭm). The entire muscle is surrounded by a layer of connective tissue called the **epimysium** (ep-ĭ-mis′ē-ŭm). The epimysium is composed of dense collagenous connective tissue. **Fascia** (fash′ē-ă) is a general term for connective tissue sheets within the body. **Muscular fascia** (formerly *deep fascia*), located superficial to the epimysium, separates and compartmentalizes individual muscles or groups of muscles. It consists of dense irregular collagenous connective tissue.

Nerves and Blood Vessels

Abundant nerves and blood vessels extend to skeletal muscles (figure 9.1). **Motor neurons** are specialized nerve cells that stimulate muscles to contract. Their cell bodies are located in the brain and spinal cord, and their axons extend to skeletal muscle fibers through nerves. At the level of the perimysium, the axons of motor neurons branch repeatedly, each branch projecting toward the center of one muscle fiber. The contact points between the axons and the muscle fibers, called synapses or neuromuscular junctions, are described later in the chapter (see section 9.4). Each motor neuron innervates more than one muscle fiber, and every muscle fiber receives a branch of an axon. However, most whole muscles are innervated by more than one neuron.

An artery and either one or two veins extend together with a nerve through the connective tissue layers of skeletal muscles. Numerous branches of the arteries supply the extensive capillary beds surrounding the muscle fibers, and blood is carried away from the capillary beds by branches of the veins.

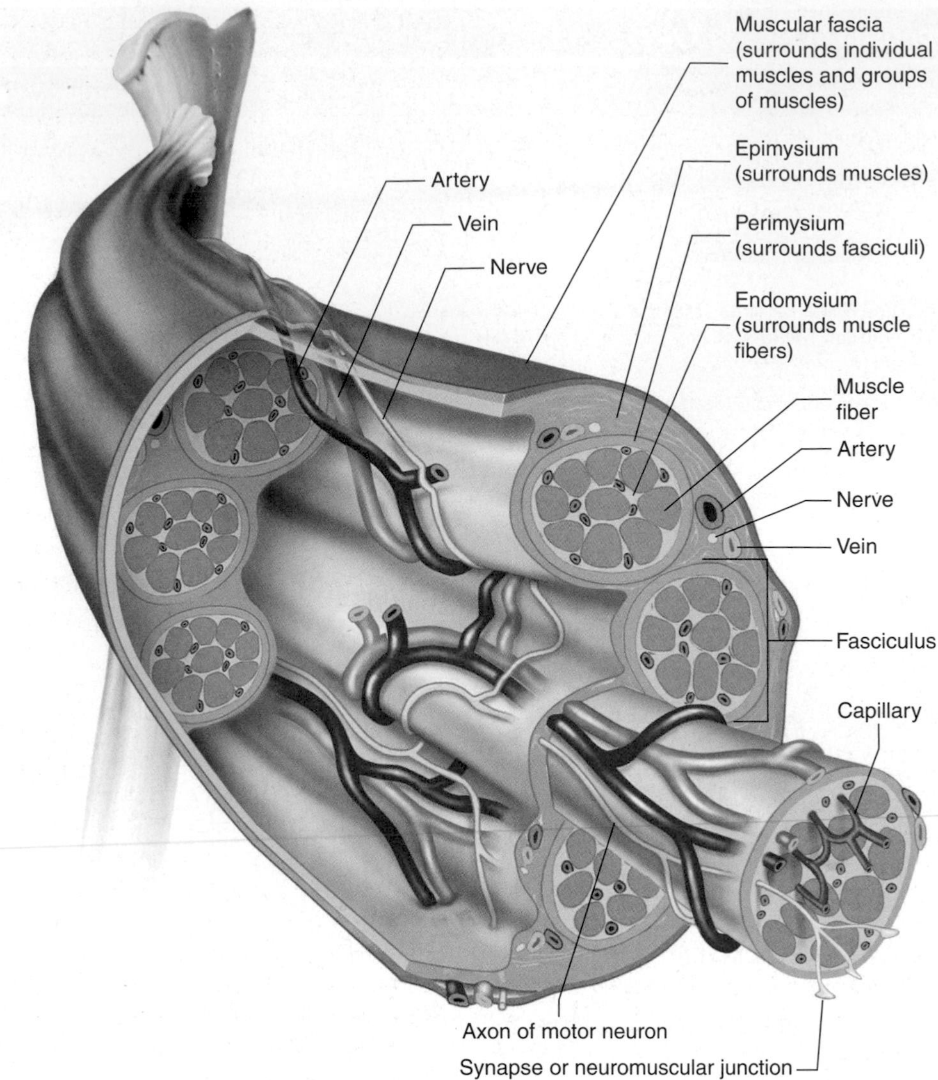
Muscular fascia (surrounds individual muscles and groups of muscles)
Epimysium (surrounds muscles)
Perimysium (surrounds fasciculi)
Endomysium (surrounds muscle fibers)
Artery
Vein
Nerve
Muscle fiber
Artery
Nerve
Vein
Fasciculus
Capillary
Axon of motor neuron
Synapse or neuromuscular junction

FIGURE 9.1 AP|R **Skeletal Muscle Structure: Connective Tissue, Innervation, and Blood Supply**

A muscle is composed of muscle fasciculi, each surrounded by perimysium. The fasciculi are composed of bundles of individual muscle fibers (muscle cells), each surrounded by endomysium. This figure shows the relationship among muscle fibers, fasciculi, and associated connective tissue layers: the epimysium, perimysium, and endomysium. Arteries, veins, and nerves course together through the connective tissue of muscles. They branch frequently as they approach individual muscle fibers. At the level of the perimysium, axons of neurons branch, and each branch extends to a muscle fiber.

Skeletal Muscle Fibers

Each skeletal muscle fiber is a single, long, cylindrical cell containing several nuclei, which are located around its periphery, near the plasma membrane. A single fiber can extend from one end of a muscle to the other. In most muscles, the fibers range from approximately 1 mm to about 4 cm in length and from 10 μm to 100 μm in diameter. Large muscles contain a large percentage of large-diameter fibers, whereas small, delicate muscles contain a large percentage of small-diameter fibers. However, any given muscle contains a mixture of small- and large-diameter fibers. As seen in a longitudinal section, alternating light and dark bands give the muscle fiber a **striated** (strī′at-ed; banded), or striped, appearance (figure 9.2).

Origin of Muscle Fibers

Muscle fibers develop from less mature, multinucleated cells called **myoblasts** (mī′ō-blasts). The multiple nuclei result from the fusion of myoblast precursor cells, not from the division of nuclei within myoblasts. Myoblasts are converted to muscle fibers as contractile proteins accumulate within their cytoplasm. Shortly after the myoblasts form, nerves grow into the area and innervate the developing muscle fibers.

The number of skeletal muscle fibers remains relatively constant after birth. Enlargement, or **hypertrophy,** of muscles after birth in children and adults results from an increase in the *size* of each muscle fiber, not from a substantial increase in the *number* of muscle fibers. Similarly, hypertrophy of muscles in response to exercise is due mainly to an increase in muscle fiber size, rather than an increase in number.

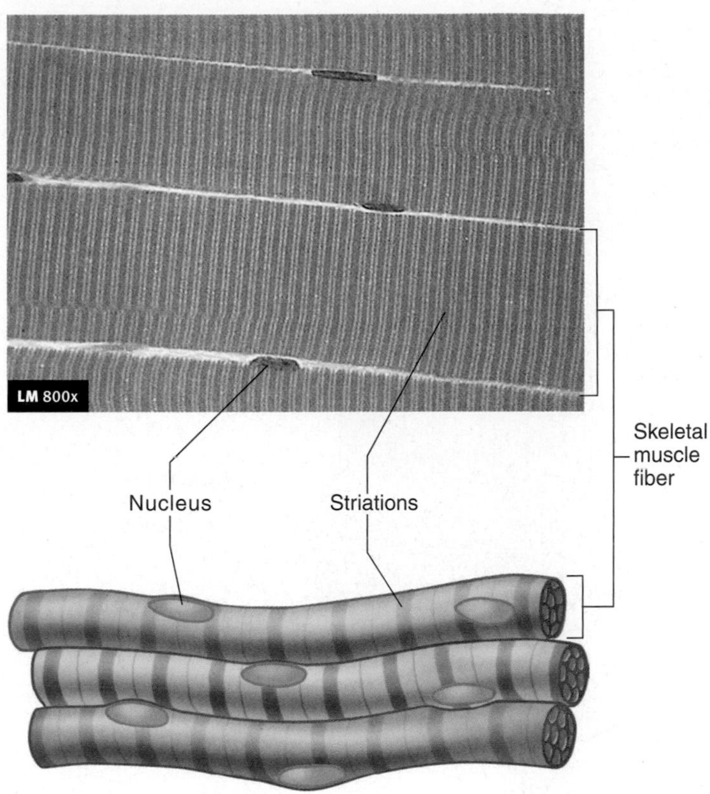
LM 800x
Nucleus
Striations
Skeletal
muscle
fiber

FIGURE 9.2 AP|R **Skeletal Muscle Fibers**
Skeletal muscle fibers in longitudinal section.

Histology of Muscle Fibers

Muscle contraction is much easier to understand when we consider the structure of a muscle fiber (figure 9.3). The plasma membrane of a muscle fiber is called the **sarcolemma** (sar′kō-lem′ă). Two delicate connective tissue layers are located just outside the sarcolemma. The deeper and thinner of the two is the **external lamina.** It consists mostly of reticular (collagen) fibers and is so thin that it cannot be distinguished from the sarcolemma when viewed under a light microscope. The second layer also consists mostly of reticular fibers, but it is a much thicker layer, called the **endomysium** (en′dō-miz′ē-ŭm; en′dō-mis′ē-ŭm; G. *mys*, muscle).

Along the surface of the sarcolemma are many tubelike invaginations of the sarcolemma, called **transverse tubules,** or **T tubules.** They occur at regular intervals along the muscle fiber and extend inward, connecting the extracellular environment with the interior of the muscle fiber (figure 9.3*b*). The T tubules are also associated with the highly organized smooth endoplasmic reticulum called the **sarcoplasmic reticulum** (sar-kō-plaz′mik re-tik′ū-lŭm) in skeletal muscle fibers. Other organelles, such as the numerous mitochondria and energy-storing glycogen granules, are packed into the cell and constitute the cytoplasm, which in muscles is called the **sarcoplasm** (sar′kō-plazm).

The sarcoplasm also contains numerous **myofibrils** (mī-ō-fī′brilz), which are bundles of protein filaments. Each myofibril is a threadlike structure, approximately 1–3 μm in diameter, that extends from one end of the muscle fiber to the other. A myofibril contains two kinds of protein filaments, called **myofilaments** (mī-ō-fil′ă-ments; figure 9.3*c*). **Actin** (ak′tin) **myofilaments,** or thin myofilaments, are approximately 8 nanometers (nm) in diameter and 1000 nm in length, whereas **myosin** (mī′ō-sin) **myofilaments,** or thick myofilaments, are approximately 12 nm in diameter and 1800 nm in length. The actin and myosin myofilaments form highly ordered units called **sarcomeres** (sar′kō-mērz), which are joined end to end to form the myofibrils (figure 9.4*a*).

ASSESS **YOUR PROGRESS** 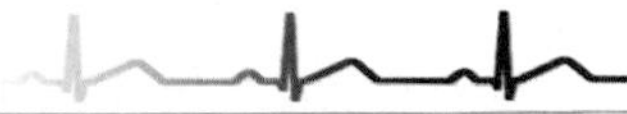

6. *Name the connective tissue layers that surround muscle fibers, muscle fasciculi, and whole muscles. Distinguish between a sarcolemma and muscular fascia.*
7. *What are motor neurons? How do the axons of motor neurons and blood vessels extend to muscle fibers?*
8. *What is the origin of muscle fibers? How do you explain the enlargement of muscle fibers after birth?*
9. *What are T tubules and the sarcoplasmic reticulum?*
10. *Describe myofibrils and myofilaments.*

Actin and Myosin Myofilaments

Each actin myofilament is composed of two strands of **fibrous actin (F actin),** a series of **tropomyosin** (trō-pō-mī′ō-sin) **molecules,** and a series of **troponin** (trō′pō-nin) **molecules** (figure 9.4*b*,*c*). The two strands of F actin are coiled to form a double helix, which extends the length of the actin myofilament. Each F actin strand is a polymer of approximately 200 small, globular units called **globular actin (G actin)** monomers. Each G actin monomer has an active site, to which myosin molecules can bind during muscle contraction. Tropomyosin is an elongated protein that winds along the groove of the F actin double helix. Each tropomyosin molecule is sufficiently long to cover seven G actin active sites. Troponin is composed of three subunits: one that binds to actin, a second that binds to tropomyosin, and a third that has a binding site for Ca^{2+}. The troponin molecules are spaced between the ends of the tropomyosin molecules in the groove between the F actin strands. The complex of tropomyosin and troponin regulates the interaction between active sites on G actin and myosin.

Myosin myofilaments are composed of many elongated **myosin molecules** shaped like golf clubs (figure 9.4*b*,*c*). Each myosin molecule consists of two **myosin heavy chains** wound together to form a **rod portion** lying parallel to the myosin myofilament and two **myosin heads** that extend laterally (figure 9.4*b*; see figure 9.3*d*). Four light myosin chains are attached to the heads of each myosin molecule. Each myosin myofilament consists of about 300 myosin molecules arranged so that about 150 of them have their heads projecting toward each end. The centers of the myosin myofilaments consist of only the rod portions of the myosin molecules. The myosin heads have three important properties: (1) The heads can bind to active sites on the actin molecules to form **cross-bridges;** (2) the heads are attached to the rod portion by a hinge region that can bend and straighten during contraction; and (3) the heads are ATPase enzymes, which break down adenosine triphosphate (ATP), releasing energy. Part of the energy is used to bend the hinge region of the myosin molecule during contraction.

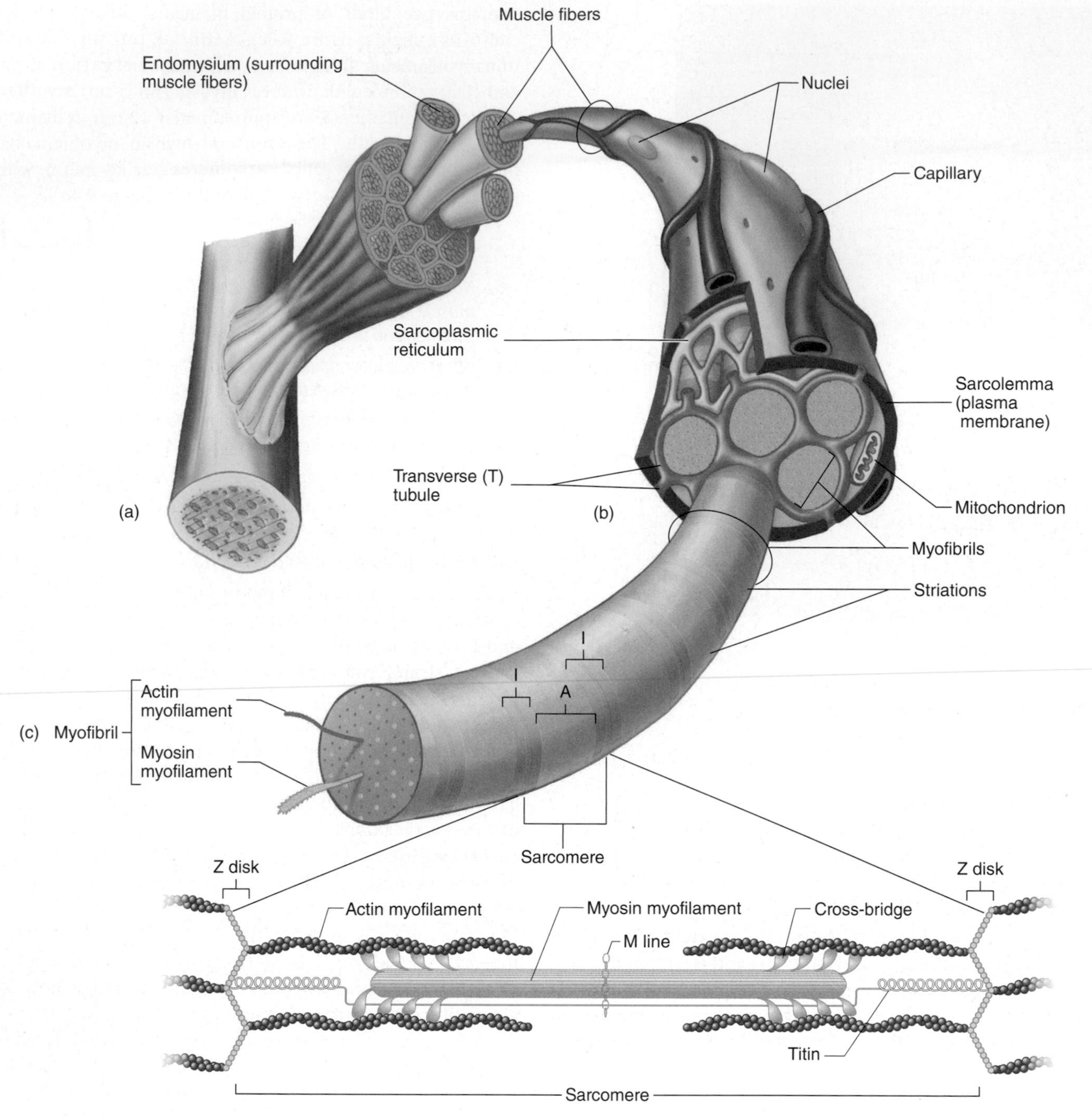
Muscle fibers
Endomysium (surrounding muscle fibers)
Nuclei
Capillary
Sarcoplasmic reticulum
Sarcolemma (plasma membrane)
Transverse (T) tubule
Mitochondrion
(a)
(b)
Myofibrils
Striations
I
I
A
(c) Myofibril
Actin myofilament
Myosin myofilament
Sarcomere
Z disk
Z disk
Actin myofilament
Myosin myofilament
Cross-bridge
M line
Titin
Sarcomere

FIGURE 9.3 Parts of a Muscle

(*a*) Part of a muscle attached by a tendon to a bone. (*b*) Enlargement of one muscle fiber. The muscle fiber contains several myofibrils. (*c*) A myofibril extended out the end of the muscle fiber. The banding patterns of the sarcomeres are shown in the myofibril. (*d*) A single sarcomere of a myofibril is composed of actin myofilaments and myosin myofilaments. The Z disk anchors the actin myofilaments, and the myosin myofilaments are held in place by titin molecules and the M line.

Sarcomeres

The sarcomere is the basic structural and functional unit of skeletal muscle because it is the smallest portion of skeletal muscle capable of contracting. Each sarcomere extends from one Z disk to an adjacent Z disk (figures 9.4 and 9.5). A **Z disk** is a filamentous network of protein forming a disklike structure for the attachment of actin myofilaments. The arrangement of the actin myofilaments and myosin myofilaments gives the myofibril a banded, or striated, appearance when viewed longitudinally. Each **isotropic** (ī-sō-trop′ik) **band,** or **I band,** includes a Z disk and extends from each side of the Z disk to the ends of the myosin myofilaments. When seen in longitudinal and cross sections, the I band on each side of the Z disk consists only of actin myofilaments.

FUNDAMENTAL **Figure**

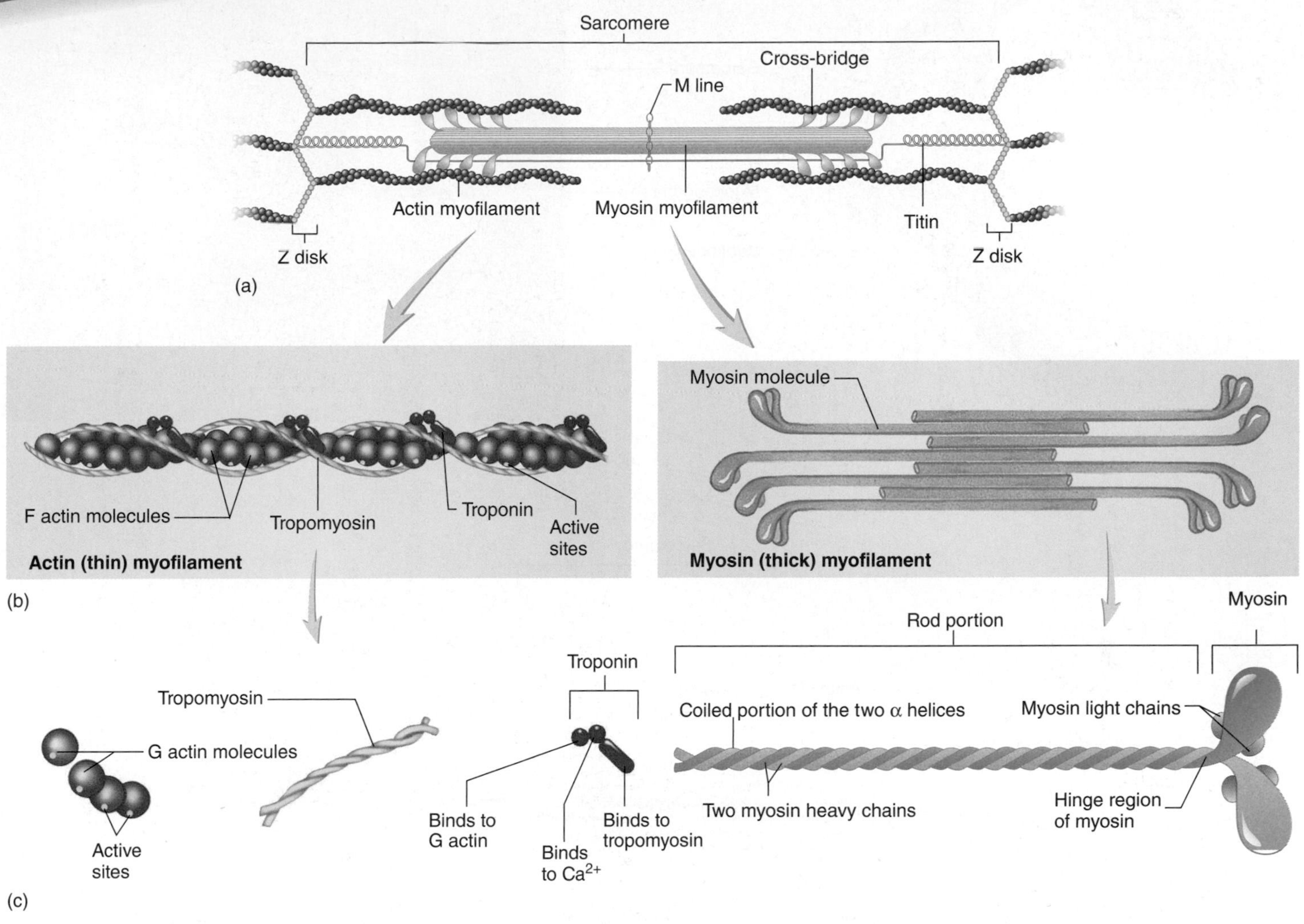

Sarcomere
Cross-bridge
M line
Actin myofilament
Myosin myofilament
Titin
Z disk
Z disk
(a)
Myosin molecule
F actin molecules
Tropomyosin
Troponin
Active sites
Actin (thin) myofilament
Myosin (thick) myofilament
(b)
Myosin
Rod portion
Troponin
Tropomyosin
G actin molecules
Coiled portion of the two α helices
Myosin light chains
Two myosin heavy chains
Hinge region of myosin
Binds to G actin
Binds to tropomyosin
Binds to Ca2+
Active sites
(c)

FIGURE 9.4 AP|R **Structure of Actin and Myosin**

(*a*) The sarcomere consists of actin (thin) myofilaments, attached to the Z disks, and myosin (thick) myofilaments, suspended between the actin myofilaments. (*b*) Actin myofilaments are composed of F actin (*chains of purple spheres*), tropomyosin (*blue strands*), and troponin (*red spheres and rod*). Myosin myofilaments are made up of many golf-club-shaped myosin molecules, with all the heads pointing in one direction at one end and the opposite direction at the other end. (*c*) G actin molecules (*purple spheres*), with their active sites (*yellow*), tropomyosin, and troponin, make up actin myofilaments. Myosin molecules (*green*) are golf-club-shaped structures composed of two molecules of heavy myosin wound together to form the rod portion and double globular heads. Four small light myosin molecules are located on the heads of each of the myosin molecules.

Each **anisotropic** (an-ī-sō-trop′ik) **band,** or **A band,** extends the length of the myosin myofilaments within a sarcomere. The actin and myosin myofilaments overlap for some distance at both ends of the A band. In a cross section of the A band where actin and myosin myofilaments overlap, each myosin myofilament is visibly surrounded by six actin myofilaments. In the center of each A band is a smaller band called the **H zone,** where the actin and myosin myofilaments do not overlap and only myosin myofilaments are present. A dark line, called the **M line,** is in the middle of the H zone and consists of delicate filaments that attach to the center of the myosin myofilaments. The M line helps hold the myosin myofilaments in place, similar to the way the Z disk holds actin myofilaments in place (figure 9.5*b*,*c*). The numerous myofibrils are oriented within each muscle fiber so that A bands and I bands of parallel myofibrils are aligned and thus produce the striated pattern seen through a microscope.

In addition to actin and myosin, there are other, less visible proteins within sarcomeres. These proteins help hold actin and myosin in place, and one of them accounts for muscle's ability to stretch (extensibility) and to recoil (elasticity). **Titin** (ti′tin; see figure 9.3) is one of the largest known proteins, consisting of a single chain of nearly 27,000 amino acids. It attaches to Z disks and extends along myosin myofilaments to the M line. The myosin myofilaments are attached to the titin molecules, which help hold them in position. Part of the titin molecule in the I band functions as a spring, allowing the sarcomere to stretch and recoil.

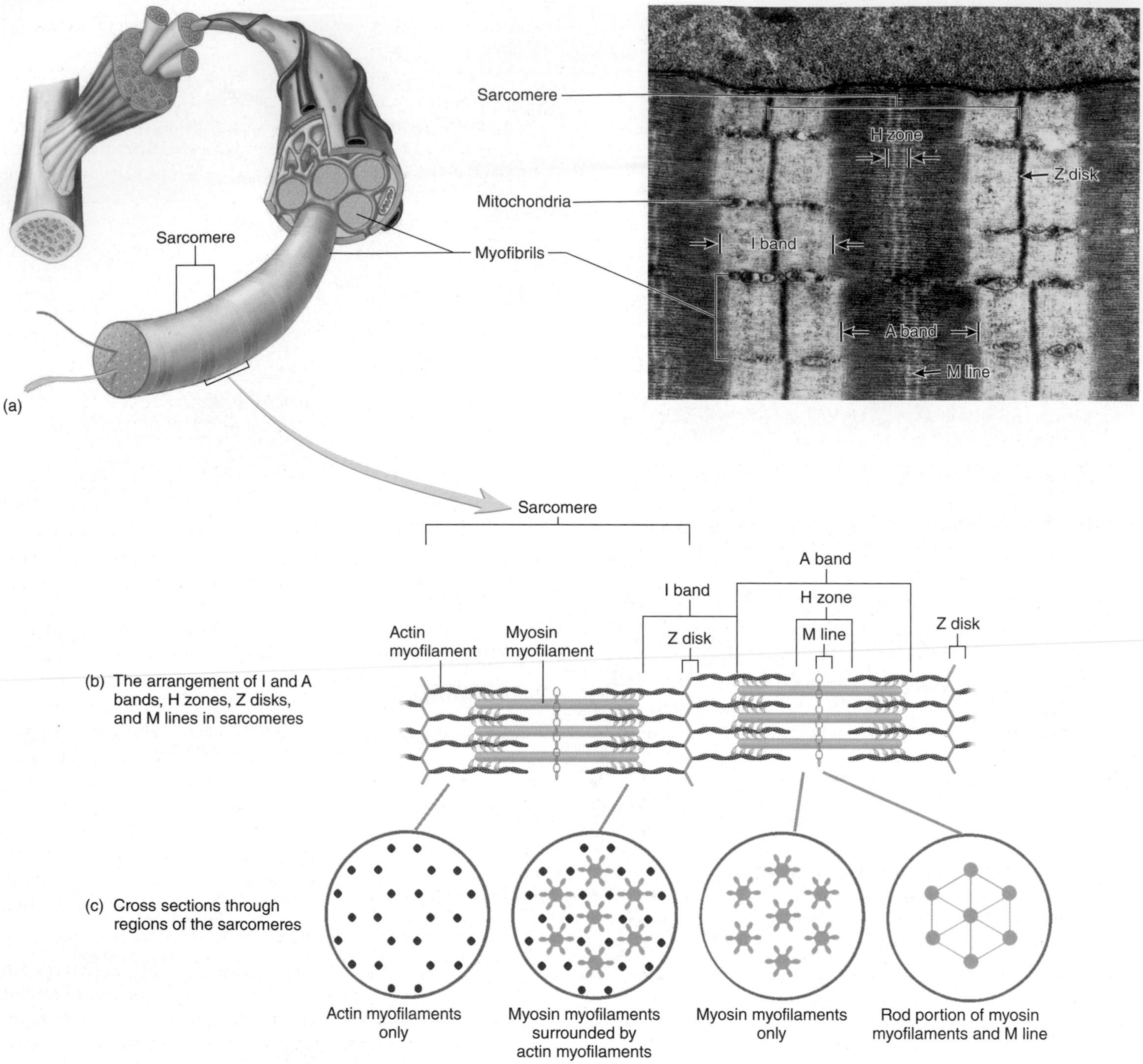

Sarcomere

Mitochondria

Myofibrils

Sarcomere

H zone

Z disk

I band

A band

M line

(a)

Sarcomere

A band

I band

H zone

Z disk

M line

Z disk

Actin myofilament

Myosin myofilament

(b) The arrangement of I and A bands, H zones, Z disks, and M lines in sarcomeres

(c) Cross sections through regions of the sarcomeres

Actin myofilaments only

Myosin myofilaments surrounded by actin myofilaments

Myosin myofilaments only

Rod portion of myosin myofilaments and M line

FIGURE 9.5 Organization of Sarcomeres

(*a*) Electron micrograph of a skeletal muscle. Several sarcomeres are shown in the myofibrils of a muscle fiber. (*b*) Diagram of two adjacent sarcomeres, depicting structures responsible for the banding pattern. The I band is between the ends of myosin myofilaments on each side of a Z disk. The A band is formed by the myosin myofilaments within a sarcomere. The H zone is between the ends of the actin myofilaments within a sarcomere. Myosin myofilaments are attached to the M line. (*c*) Cross sections through regions of the sarcomeres show the arrangement of proteins in three dimensions.

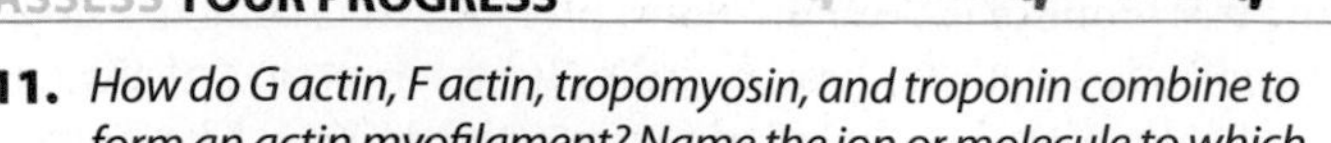
ASSESS YOUR PROGRESS

11. *How do G actin, F actin, tropomyosin, and troponin combine to form an actin myofilament? Name the ion or molecule to which each of the three subunits of troponin binds.*

12. *Describe the structure of myosin molecules and how they combine to form a myosin myofilament.*

13. *List the three important properties of a myosin head. What is a cross-bridge?*

14. *What is a sarcomere? Illustrate how Z disks, actin myofilaments, myosin filaments, the M line, and titin form a sarcomere. Describe how this arrangement produces the A band, the I band, and the H zone. Label these areas on your illustration.*

9.4 Physiology of Skeletal Muscle Fibers

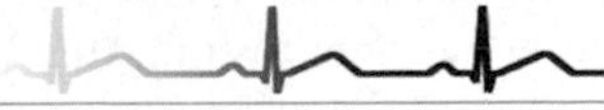

LEARNING OUTCOMES

After reading this section, you should be able to

A. Describe how the sliding filament model explains the contraction of muscle fibers.

B. Explain what happens to the length of the A band, I band, and H zone during contraction.

C. Describe the resting membrane potential and how it is generated and maintained.

D. Explain the role of ion channels in the production of an action potential.

E. Discuss the production of an action potential, including depolarization and repolarization.

F. State the all-or-none principle as it pertains to action potentials.

G. Describe the structure of a neuromuscular junction and explain how an action potential is transmitted across the junction.

H. Explain the events of excitation-contraction coupling.

I. Summarize the events of cross-bridge movement and relate them to muscle contraction.

J. State the conditions needed for muscle relaxation.

The primary function of skeletal muscle cells is to generate force by contracting, or shortening. A parallel arrangement of myofilaments in a sarcomere allows them to interact, which causes muscle contraction. This interaction is described by the sliding filament model (figure 9.6).

Sliding Filament Model

It is the shortening of sarcomeres that is responsible for the contraction of skeletal muscles. In a sarcomere, the actin and myosin myofilaments slide past one another but remain the same length as when the muscle is at rest. When the myofilaments slide past each other and the sarcomeres shorten, the myofibrils also shorten because the myofibrils consist of sarcomeres joined end to end. The myofibrils extend the length of the muscle fibers, and when they shorten the muscle fibers shorten. Groups of muscle fibers make up a muscle fascicle, and several muscle fascicles make up a whole muscle. Therefore, when sarcomeres shorten, myofibrils, muscle fibers, muscle fascicles, and muscles shorten to produce muscle contraction.

During muscle relaxation, the sarcomeres lengthen. For this to happen, an external force must be applied to a muscle by other muscles or by gravity. For example, muscle contraction causes a joint, such as the elbow or knee, to flex. Extension of the joint and lengthening of the muscle result from the contraction of muscles that produce the opposite movement (see chapter 10).

However, knowing the mechanism by which muscles shorten is a small step toward understanding muscle contraction. What stimulates the sarcomeres to shorten in the first place? In order to fully understand muscle contraction, we must first consider the electrical properties of skeletal muscle fibers.

Muscle fibers, like other cells of the body, are electrically excitable. A description of the electrical properties of skeletal muscle fibers is presented next, and later sections illustrate the role of these properties in contraction. Axons of neurons extend from the brain and spinal cord to skeletal muscle fibers. The nervous system controls the contraction of skeletal muscles through these axons. Electrical signals, called **action potentials,** travel from the brain or spinal cord along the axons to muscle fibers and cause them to contract.

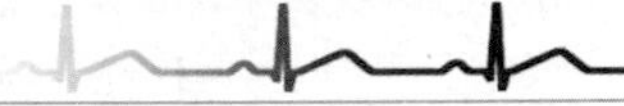

ASSESS YOUR PROGRESS

15. *How does the shortening of sarcomeres explain muscle contraction?*

16. *What must occur for a muscle to relax?*

17. *Referring to figure 9.6, explain why the I band and H zone shorten during muscle contraction, while the length of the A band remains unchanged.*

Ion Channels

In order to understand the electrical properties of skeletal muscle fibers, a review of the permeability characteristics of cell membranes and the role membrane transport proteins play in their permeability will be helpful. The phospholipid bilayer interior is a hydrophobic environment, which inhibits the movement of charged particles, particularly ions, across the membrane; however, the basis of the electrical properties of skeletal muscle cells is the movement of ions across the membrane. Recall from chapter 3 that ions can move across the membrane through ion channels. There are two major types of ion channels: nongated, or leak, and gated. These two channel types contribute to the electrical properties of both a resting cell and a stimulated cell. In resting cells, the nongated ion channels allow for the slow leak of ions down their concentration gradient. Like all membrane transport proteins, leak channels are specific for a particular ion. The gated ion channels are most important in stimulated cells. It is their presence that governs the production of action potentials. There are two major gated ion channels:

1. *Ligand-gated ion channels.* **Ligand-gated ion channels** open when a ligand (lī′gand), a chemical signal, binds to a receptor that is part of the ion channel (see figure 3.8). For example, the axons of neurons supplying skeletal muscle fibers release ligands, called **neurotransmitters** (noor′ō-trans-mit′erz), which bind to ligand-gated Na^+ channels in the membranes of the muscle fibers. As a result, the Na^+ channels open, allowing Na^+ to enter the cell.
2. *Voltage-gated ion channels.* These channels are gated membrane channels that open and close in response to a particular membrane potential. When a neuron or muscle fiber is stimulated, the charge difference changes, and a particular change causes certain voltage-gated ion channels to open or close. The voltage-gated channels that play major roles in an action potential are voltage-gated Na^+, K^+, and Ca^{2+} channels. For example, opening voltage-gated Na^+ channels allows Na^+ to cross the plasma membrane, whereas opening voltage-gated K^+ channels allows K^+ to cross and opening Ca^{2+} channels allows Ca^{2+} to cross.

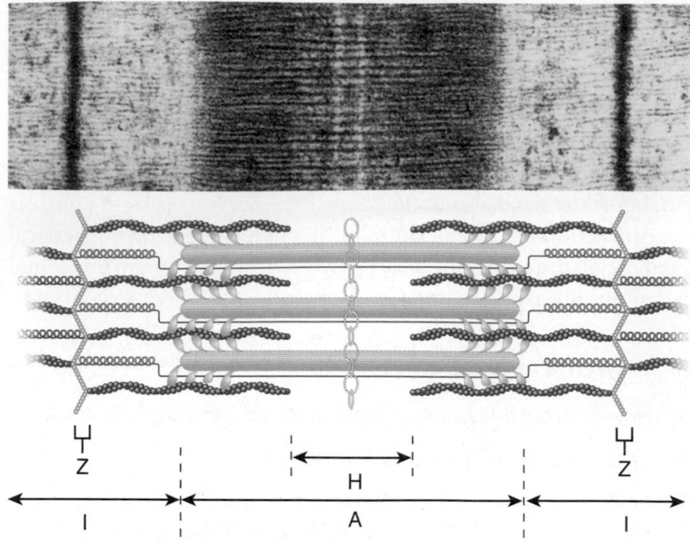
Z
H
Z
I
A
I

(a) **Relaxed sarcomere**

In a relaxed muscle, the actin and myosin myofilaments overlap slightly, and the H zone is visible. The sarcomere length is at its normal resting length. As a muscle contraction is initiated, actin myofilaments slide past the myosin myofilaments, the z disks are brought closer together, and the sarcomere begins to shorten.

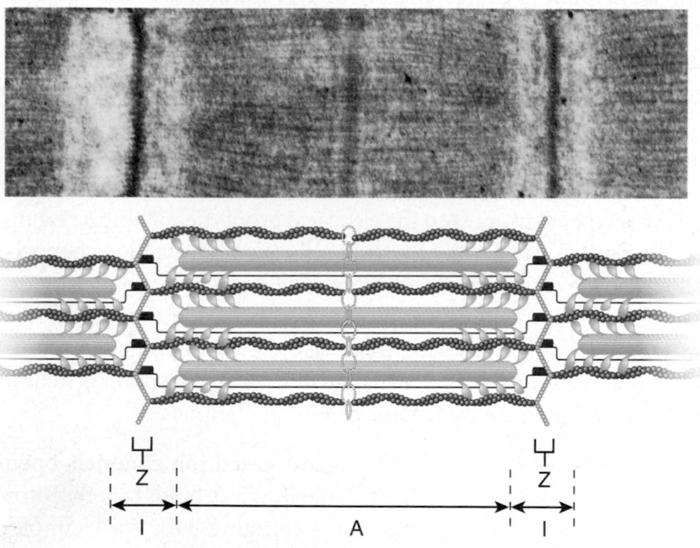
Z
Z
I
A
I

(b) **Fully contracted sarcomere**

In a contracted muscle, the A bands, which are equal to the length of the myosin myofilaments, do not narrow because the length of the myosin myofilaments does not change, nor does the length of the actin myofilaments. In addition, the ends of the actin myofilaments are pulled to and overlap in the center of the sarcomere, shortening it and the H zone disappears.

FIGURE 9.6 APIR **Sarcomere Shortening**

Resting Membrane Potential

Electrically excitable cells, like most cells, are **polarized.** That is, the inside of most plasma membranes is negatively charged compared with the outside. Thus, a voltage difference, or electrical charge difference, exists across each plasma membrane. This charge difference across the plasma membrane of an unstimulated cell is called the **resting membrane potential** (figure 9.7; see figure 11.14). Action potentials cannot be produced without a resting membrane potential. The resting membrane potential is the result of three factors: (1) The concentration of K^+ inside the plasma membrane is higher than that outside the plasma membrane; (2) the concentration of Na^+ outside the plasma membrane is higher than that inside the plasma membrane; and (3) the plasma membrane is more permeable to K^+ than to Na^+. Because the concentration gradient for an ion determines whether that ion enters or leaves the cell after its ion channel opens, when voltage-gated Na^+ channels open, Na^+ moves through the channels into the cell. In a similar fashion, when gated K^+ channels open, K^+ moves out of the cell. Since excitable cells have many K^+ leak ion channels, K^+ moves out of the cell faster than Na^+ moves into the cell. In addition, negatively charged molecules, such as proteins, are "trapped" inside the cell because the plasma membrane is impermeable to them. For these reasons, the inside of the plasma membrane is more negatively charged than the outside.

Some K^+ is able to diffuse down the concentration gradient from inside to just outside the plasma membrane. Because K^+ is positively charged, its movement from inside the cell to outside causes the inside of the plasma membrane to become even more negatively charged compared with the outside.

Potassium ions diffuse down their concentration gradient only until the charge difference across the plasma membrane is great enough to prevent any additional diffusion of K^+ out of the cell. The resting membrane potential is an equilibrium in which the tendency for K^+ to diffuse out of the cell is opposed by the negative charges inside the cell, which tend to attract the positively charged K^+ into the cell. At rest, the sodium-potassium pump transports K^+ from outside the cell to the inside and transports Na^+ from inside the cell to the outside. The active transport of Na^+ and K^+ by the sodium-potassium pump maintains the uneven distribution of Na^+ and K^+ across the plasma membrane (see chapter 3). The details of the resting membrane potential are described more fully in chapter 11.

The resting membrane potential can be measured in units called **millivolts** (mV; mV = 1/1000 volt). The potential differences across the plasma membranes of neurons and muscle fibers are between −70 and −90 mV. The potential difference is reported as a negative number because the inner surface of the plasma membrane is negative compared with the outside.

ASSESS YOUR PROGRESS

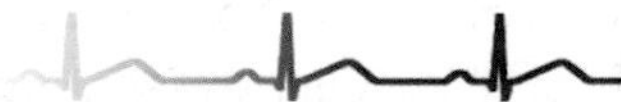

18. *What type of ion channel contributes to the resting membrane potential? Describe the permeability characteristics of the plasma membrane.*

19. *What are the two types of gated ion channels in the plasma membrane? Explain what causes each type to open and close.*

20. *What is the resting membrane potential? What three factors create the resting membrane potential?*

21. *How does the sodium-potassium pump help maintain the polarized nature of the resting membrane?*

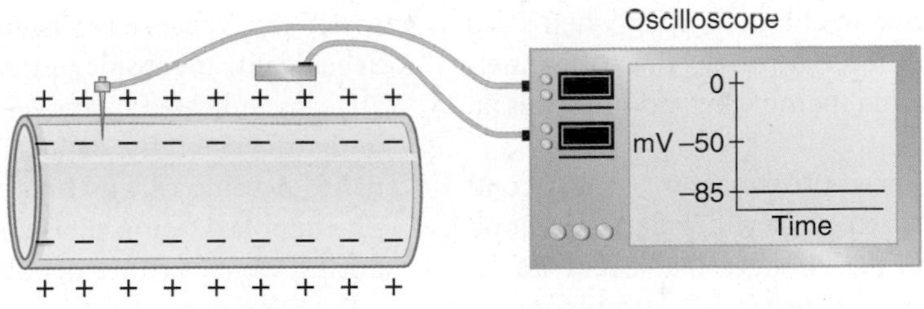
Oscilloscope
0
mV −50
−85
Time
Nerve or muscle cell

(a) **Measuring the resting membrane potential**

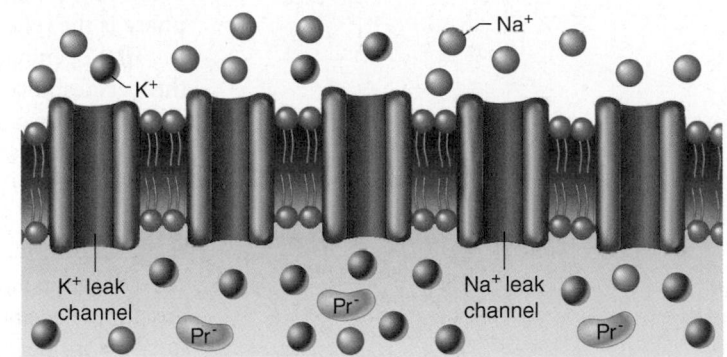

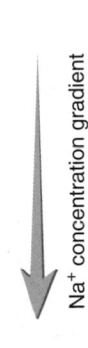

1 In a resting cell, there is a higher concentration of K^+ (*purple circles*) inside the cell membrane and a higher concentration of Na^+ (*pink circles*) outside the cell membrane. Because the membrane is not permeable to negatively charged proteins (*green*) they are isolated to inside of the cell membrane.

2 There are more K^+ leak channels than Na^+ leak channels. In the resting cell, only the leak channels are opened; the gated channels (not shown) are closed. Because of the ion concentration differences across the membrane, K^+ diffuses out of the cell down its concentration gradient and Na^+ diffuses into the cell down its concentration gradient. The tendency for K^+ to diffuse out of the cell is opposed by the tendency of the positively charged K^+ to be attracted back into the cell by the negatively charged proteins.

3 The sodium-potassium pump helps maintain the differential levels of Na^+ and K^+ by pumping three Na^+ out of the cell in exchange for two K^+ into the cell. The pump is driven by ATP hydrolysis. The resting membrane potential is established when the movement of K^+ out of the cell is equal to the movement of K^+ into the cell.

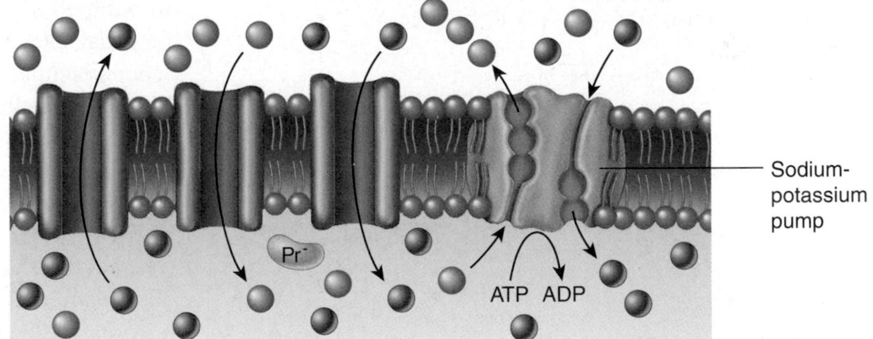
Pr⁻
ATP ADP
Sodium-potassium pump

(b) **Generation of the resting membrane potential**

PROCESS FIGURE 9.7 APR **Measuring the Resting Membrane Potential**

A device called an oscilloscope is able to measure the resting membrane potential in skeletal muscle. The recording electrode is inside the membrane, and the reference electrode is outside. Here, a potential difference of about −85 mV is recorded, with the inside of the membrane negative with respect to the outside of the membrane.

Predict 2

If ligand-gated K^+ channels were to open in an unstimulated muscle fiber, how would this affect the resting membrane potential?

Action Potentials

An action potential is a reversal of the resting membrane potential such that the inside of the plasma membrane becomes positively charged compared with the outside. The permeability characteristics of the plasma membrane change because ion channels open when a cell is stimulated. The diffusion of ions through these channels changes the charge across the plasma membrane and produces an action potential.

An action potential lasts from approximately 1 millisecond to a few milliseconds, and it has two phases, depolarization and repolarization. Figure 9.8 depicts the changes that occur in the membrane potential during an action potential. Stimulation of a cell can cause its plasma membrane to become depolarized. Specifically, the inside of the plasma membrane becomes less negative, as indicated by movement of the curve upward toward zero in figure 9.8*a*. If the depolarization changes the membrane potential to a value called **threshold,** an action potential is triggered. The **depolarization phase** of the action potential is a brief period during which further depolarization occurs and the inside of the cell becomes positively charged (figure 9.8*b*). The charge difference across the plasma membrane is said to be reversed when the membrane potential becomes a positive value. The **repolarization phase** is the return of the membrane potential to its resting value.

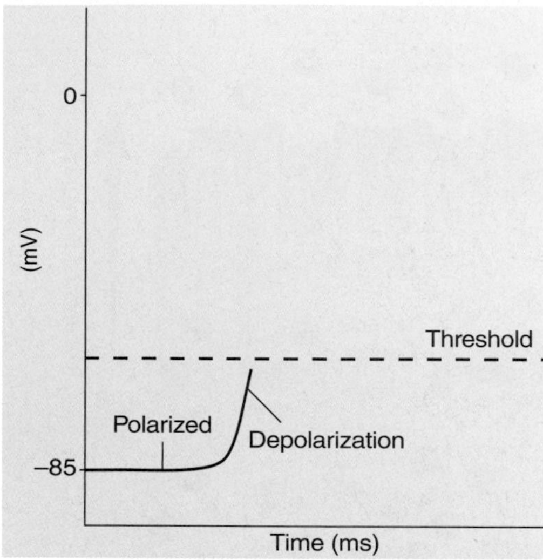
0
(mV)
Threshold
Polarized
Depolarization
–85
Time (ms)

(a) Depolarization is a change of the charge difference across the plasma membrane, making the charge inside the cell less negative and the charge outside the plasma membrane less positive. Once threshold is reached, an action potential is produced.

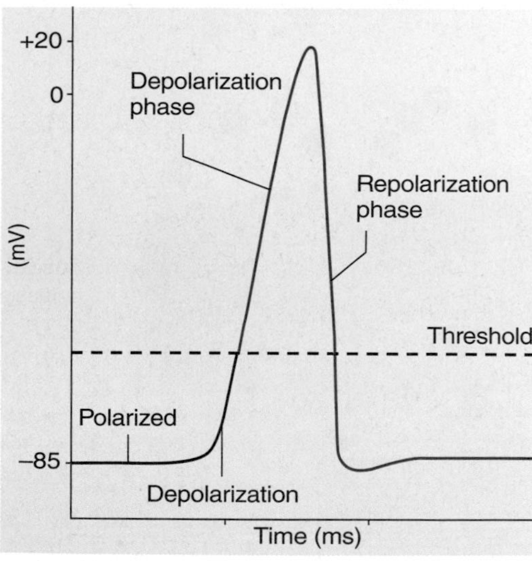
+20
0
Depolarization phase
Repolarization phase
(mV)
Threshold
Polarized
–85
Depolarization
Time (ms)

(b) During the depolarization phase of the action potential, the membrane potential changes from approximately –85 mV to approximately +20 mV. During the repolarization phase, the inside of the plasma membrane changes from approximately +20 mV back to –85 mV.

FIGURE 9.8 Depolarization and the Action Potential in Skeletal Muscle

The opening and closing of voltage-gated ion channels change the permeability of the plasma membrane to ions, resulting in depolarization and repolarization. Before a neuron or a muscle fiber is stimulated, these voltage-gated ion channels are closed (figure 9.9, *step 1*). When the cell is stimulated, gated Na^+ channels open, and Na^+ diffuses into the cell. The positively charged Na^+ makes the inside of the plasma membrane less negative. If this depolarization reaches threshold, many voltage-gated Na^+ channels are stimulated and open rapidly, causing Na^+ to diffuse into the cell until the inside of the membrane becomes positive for a brief time (figure 9.9, *step 2*). As the inside of the cell becomes positive, this voltage change causes additional permeability changes in the plasma membrane, which stop depolarization and start repolarization. Repolarization results from the closing of voltage-gated Na^+ channels and the opening of voltage-gated K^+ channels (figure 9.9, *step 3*). Thus, Na^+ stops moving into the cell, and more K^+ move out of the cell. These changes cause the inside of the plasma membrane to become more negative and the outside to become more positive. The action potential ends, and the resting membrane potential is reestablished when the voltage-gated K^+ channels close.

Action potentials occur according to the **all-or-none principle.** If a stimulus is strong enough to produce a depolarization that reaches threshold, or even if it exceeds threshold by a substantial amount, all of the permeability changes responsible for an action potential proceed without stopping. Consequently, all of the action potentials in a given cell are alike (the "all" part). If a stimulus is so weak that the depolarization does not reach threshold, few of the permeability changes occur. The membrane potential returns to its resting level after a brief period without producing an action potential (the "none" part). An action potential can be compared to the flash system of a camera. Once the shutter is triggered (reaches threshold), the camera flashes (an action potential is produced) and each flash is the same brightness (the "all" part) as the previous flashes. If the shutter is depressed but not triggered (does not reach threshold), no flash results (the "none" part).

An action potential occurs in a very small area of the plasma membrane and does not affect the entire plasma membrane at one time. Action potentials can travel, or **propagate,** across the plasma membrane because an action potential produced at one location in the plasma membrane can stimulate the production of an action potential in an adjacent location (figure 9.10). Note that an action potential does not actually move along the plasma membrane. Rather, an action potential at one location stimulates the production of another action potential in an adjacent location, which in turn stimulates the production of another, and so on. It is much

Charge Difference Across the Cell Membrane

1 Resting membrane potential. Na^+ channels (*pink*) and some, but not all, K^+ channels (*purple*) are closed. K^+ diffuses down its concentration gradient through the open K^+ channels, making the inside of the cell membrane negatively charged compared to the outside.

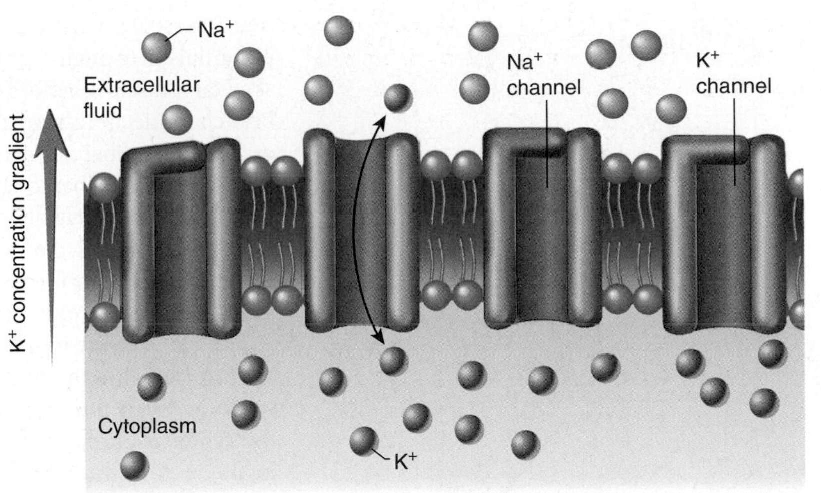
Na+
Extracellular fluid
Na+ channel
K+ channel
K+ concentration gradient
Cytoplasm
K+
1

2 Depolarization. Na^+ channels are open. Na^+ diffuses down its concentration gradient through the open Na^+ channels, making the inside of the cell membrane positively charged compared to the outside.

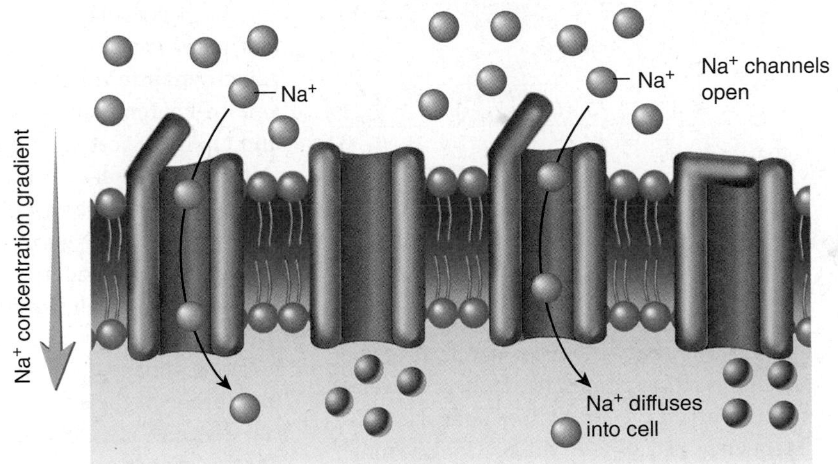
Na+
Na+
Na+ channels open
Na+ concentration gradient
Na+ diffuses into cell

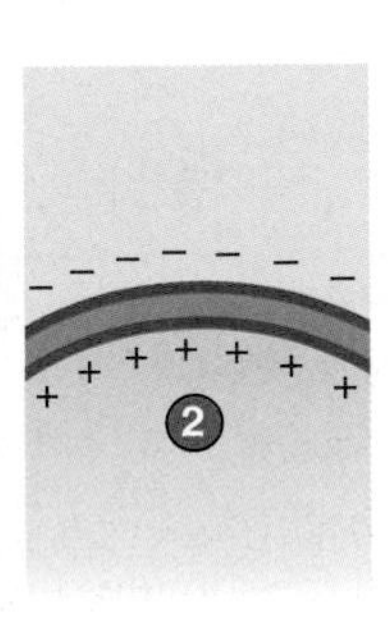
2

3 Repolarization. Na^+ channels are closed, and Na^+ movement into the cells stops. More K^+ channels open. K^+ movement out of the cell increases, making the inside of the cell membrane negatively charged compared to the outside, once again.

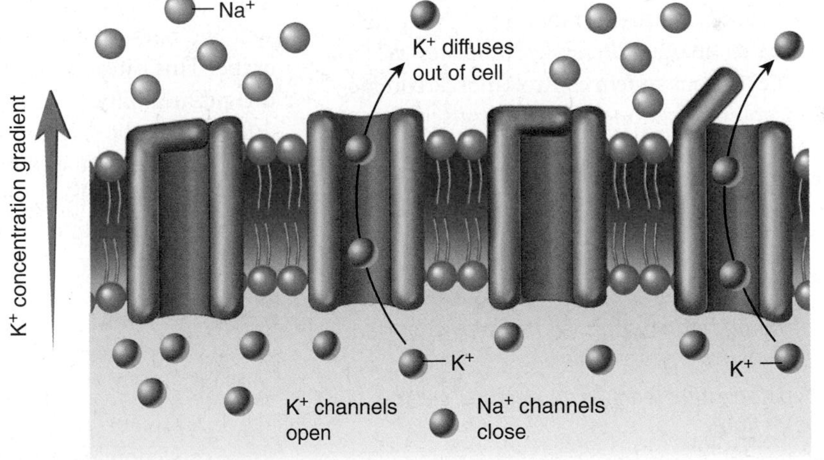
Na+
K+ diffuses out of cell
K+ concentration gradient
K+
K+
K+ channels open
Na+ channels close

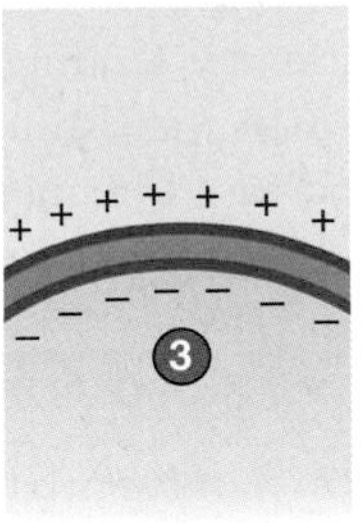
3

PROCESS FIGURE 9.9 APR **Gated Ion Channels and the Action Potential**

Step 1 illustrates the status of gated Na^+ and K^+ channels in a resting cell. Steps 2 and 3 show how the channels open and close to produce an action potential. At the far right, the charge difference across the plasma membrane at each stage is illustrated.

like a long row of toppling dominos in which each domino knocks down the next. Each domino falls, but no single domino actually travels the length of the row.

The **action potential frequency** is the number of action potentials produced per unit of time. As the strength of the stimulus applied to a neuron or a muscle fiber increases (once threshold is

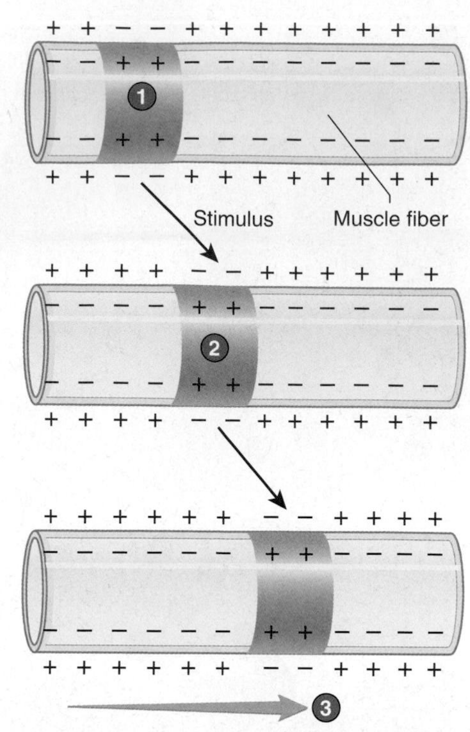

1 An action potential in a local area of the plasma membrane is indicated by the *orange band.* Note the reversal of charge across the membrane.
Stimulus
Muscle fiber
2 The action potential is a stimulus that causes another action potential to be produced in the adjacent plasma membrane.
3 The action potential propagates along the plasma membrane (orange arrow).

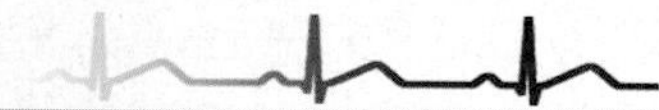

PROCESS FIGURE 9.10 APR **Action Potential Propagation in a Muscle Fiber**

reached), the action potential frequency increases as the strength of the stimulus increases. All the action potentials are identical. The action potential frequency can affect the strength of a muscle contraction (see "Stimulus Frequency and Whole Muscle Contraction," in section 9.5).

In summary, the resting membrane potential results from a charge difference across the plasma membrane. An action potential, which is a reversal of that charge difference, stimulates cells to respond by contracting. The nervous system controls muscle contractions by sending action potentials along axons to muscle fibers, which then stimulate action potentials in the muscle fibers. An increased frequency of action potentials sent to the muscle fibers can result in stronger muscle contraction.

ASSESS YOUR PROGRESS

22. *List the two types of voltage-gated channels that play important roles in the production of action potentials.*
23. *What value must depolarization reach in an electrical cell to trigger an action potential?*
24. *Describe the changes that occur during the depolarization and repolarization phases of an action potential.*
25. *Describe the propagation of an action potential.*
26. *How does the frequency of action potentials affect muscle contractions?*

Neuromuscular Junction

Action potentials carried by motor neurons stimulate action potentials in muscle fibers because of events that occur in the neuromuscular junction. Axons of motor neurons carry action potentials at a high velocity from the brain and spinal cord to skeletal muscle fibers. The axons branch repeatedly, and each branch projects toward one muscle fiber to innervate it. Thus, each muscle fiber receives a branch of an axon, and each axon innervates more than one muscle fiber (see figure 9.1).

Near the muscle fiber it innervates, each axon branch forms a cluster of enlarged axon terminals that rests in an invagination of the sarcolemma to form a **neuromuscular junction,** or **synapse.** Therefore, a neuromuscular junction consists of the axon terminals and the area of the muscle fiber sarcolemma they innervate. Each axon terminal is the **presynaptic** (prē′si-nap′tik) **terminal.** The space between the presynaptic terminal and the muscle fiber is the **synaptic** (si-nap′tik) **cleft,** and the muscle plasma membrane in the area of the junction is the **postsynaptic** (pōst-si-nap′tik) **membrane,** or **motor end-plate** (figure 9.11).

Each presynaptic terminal contains numerous mitochondria and many small, spherical sacs approximately 45 μm in diameter, called **synaptic vesicles.** The vesicles contain **acetylcholine** (**ACh;** as-e-til-kō′lēn), an organic molecule composed of acetic acid and choline. Acetylcholine is a **neurotransmitter** (noor′ō-trans-mit′er), a substance released from a presynaptic membrane that diffuses across the synaptic cleft and alters the activity of the postsynaptic cell. Neurotransmitters can stimulate (or inhibit) the production of an action potential in the postsynaptic membrane (the sarcolemma) by binding to ligand-gated ion channels.

When an action potential reaches the presynaptic terminal, it causes voltage-gated calcium ion (Ca^{2+}) channels in the plasma membrane of the axon to open; as a result, Ca^{2+} diffuses into the cell (figure 9.12). Once inside the cell, the Ca^{2+} causes the contents of a few synaptic vesicles to be secreted by exocytosis from the presynaptic terminal into the synaptic cleft. The acetylcholine molecules released from the synaptic vesicles then diffuse across the cleft and bind to receptor molecules within the postsynaptic membrane. This causes ligand-gated Na^+ channels to open, increasing the permeability of the membrane to Na^+. Sodium ions then diffuse into the cell, causing depolarization. In skeletal muscle, each action potential in the motor neuron causes a depolarization that exceeds threshold, which causes changes in voltage-gated ion channels that produce an action potential in the muscle fiber.

Predict 3

Predict the consequence if presynaptic action potentials in an axon release insufficient acetylcholine to depolarize a skeletal muscle fiber to threshold.

Acetylcholine released into the synaptic cleft is rapidly broken down to acetic acid and choline by the enzyme **acetylcholinesterase** (as′e-til-kō-lin-es′ter-ās; figure 9.12). Acetylcholinesterase keeps acetylcholine from accumulating within the synaptic cleft, where it would act as a constant stimulus at the postsynaptic terminal, producing many action potentials and continuous contraction in the muscle fiber. The release of acetylcholine and its rapid degradation in the synaptic cleft ensure that one presynaptic action potential yields only one postsynaptic action potential. Choline molecules are actively reabsorbed by the presynaptic terminal and then combined

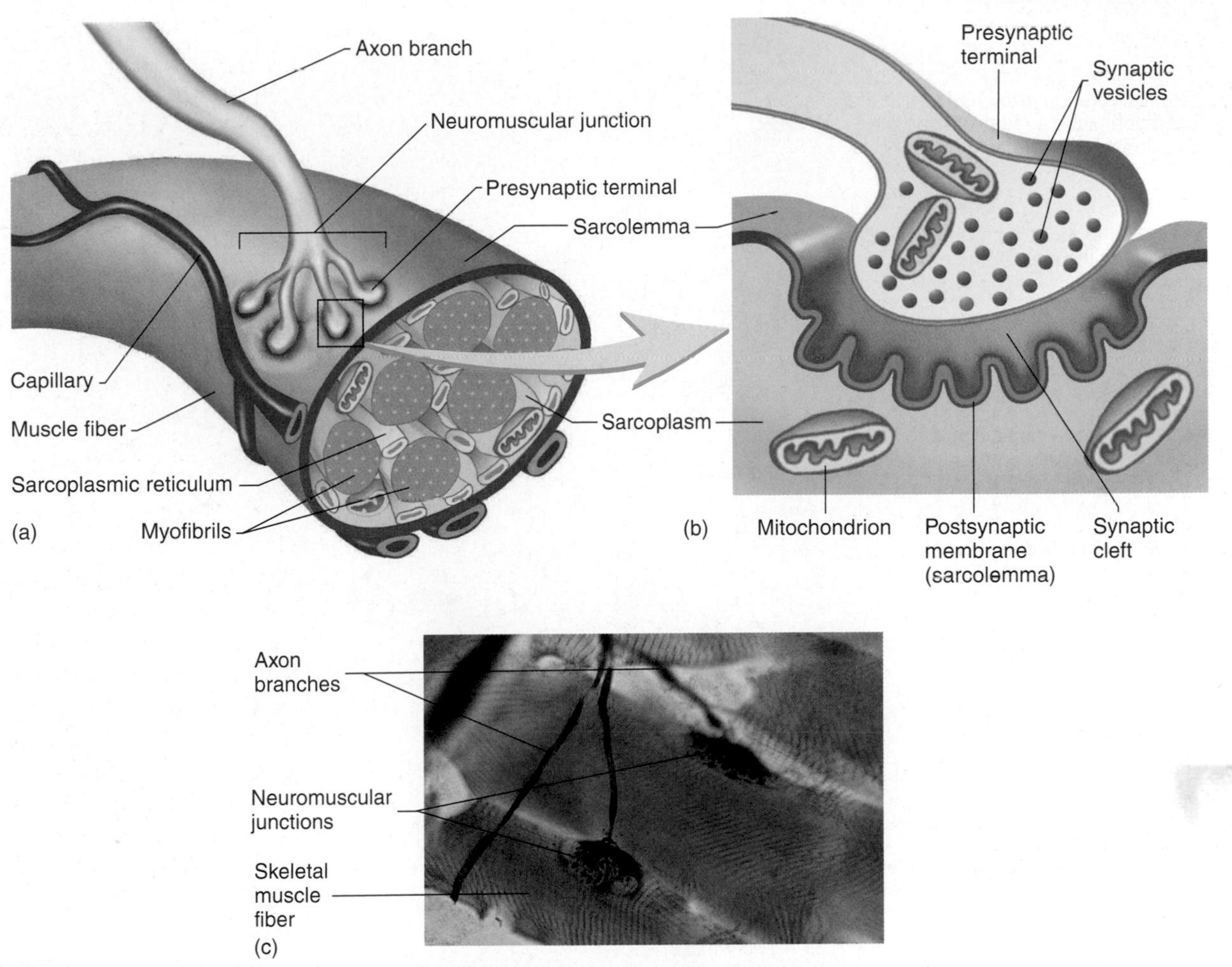
Axon branch
Neuromuscular junction
Presynaptic terminal
Sarcolemma
Capillary
Muscle fiber
Sarcoplasmic reticulum
(a)
Myofibrils
Sarcoplasm
Presynaptic terminal
Synaptic vesicles
(b)
Mitochondrion
Postsynaptic membrane (sarcolemma)
Synaptic cleft
Axon branches
Neuromuscular junctions
Skeletal muscle fiber
(c)

FIGURE 9.11 APR **Neuromuscular Junction**
(*a*) Several branches of an axon form the neuromuscular junction with a single muscle fiber. (*b*) The presynaptic terminal containing synaptic vesicles is separated from the postsynaptic membrane (sarcolemma) by the synaptic cleft. This region of the postsynaptic membrane is specialized and is shown in blue. (*c*) Photomicrograph of neuromuscular junctions.

with the acetic acid produced within the cell to form acetylcholine. Recycling choline molecules requires less energy and is more rapid than completely synthesizing new acetylcholine molecules each time they are released from the presynaptic terminal. Acetic acid is an intermediate in the process of glucose metabolism (see chapter 25). A variety of cells can take it up and use it after it diffuses from the area of the neuromuscular junction.

ASSESS YOUR PROGRESS

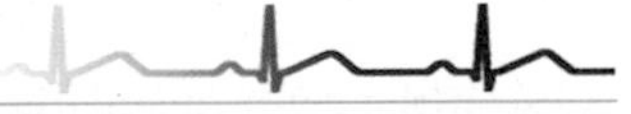

27. *Describe the structure of a neuromuscular junction, or synapse.*
28. *Outline the process of transferring the action potential in the presynaptic terminal to the postsynaptic membrane, or motor end-plate.*
29. *What ion is needed to release neurotransmitters from the synaptic vesicle? What neurotransmitter is released to skeletal muscle fibers?*
30. *What is the importance of acetylcholinesterase in the synaptic cleft? What would occur if acetylcholinesterase were not present?*

Excitation-Contraction Coupling

Action potentials produced in the sarcolemma of a skeletal muscle fiber can lead to contraction of the fiber. The mechanism by which an action potential causes contraction of a muscle fiber is called **excitation-contraction coupling,** and it involves the sarcolemma, T tubules, the sarcoplasmic reticulum, Ca^{2+}, and troponin. T tubules project into the muscle fiber and wrap around sarcomeres in the region where actin and myosin myofilaments overlap (figure 9.13; see figure 9.3). The lumen of each T tubule is filled with extracellular fluid and is continuous with the exterior of the muscle fiber. Near the T tubules, the sarcoplasmic reticulum is enlarged to form **terminal cisternae** (sis-ter′nē). A T tubule and the two adjacent terminal cisternae together are called a **triad** (tri′ad; figure 9.13). The sarcoplasmic reticulum actively transports Ca^{2+} into its lumen; thus, the concentration of Ca^{2+} is approximately 2000 times higher within the sarcoplasmic reticulum than in the sarcoplasm of a resting muscle fiber.

Excitation-contraction coupling begins at the neuromuscular junction with the production of an action potential in the sarcolemma.

FUNDAMENTAL **Figure**

1. An action potential (orange arrow) arrives at the presynaptic terminal and causes voltage-gated Ca^{2+} channels in the presynaptic membrane to open.
2. Calcium ions enter the presynaptic terminal and initiate the release of the neurotransmitter acetylcholine (ACh) from synaptic vesicles.
3. ACh is released into the synaptic cleft by exocytosis.
4. ACh diffuses across the synaptic cleft and binds to ligand-gated Na^+ channels on the postsynaptic membrane.
5. Ligand-gated Na^+ channels open and Na^+ enters the postsynaptic cell, causing the postsynaptic membrane to depolarize. If depolarization passes threshold, an action potential is generated along the postsynaptic membrane.
6. ACh unbinds from the ligand-gated Na^+ channels, which then close.
7. The enzyme acetylcholinesterase, which is attached to the postsynaptic membrane, removes acetylcholine from the synaptic cleft by breaking it down into acetic acid and choline.
8. Choline is symported with Na^+ into the presynaptic terminal, where it can be recycled to make ACh. Acetic acid diffuses away from the synaptic cleft.
9. ACh is reformed within the presynaptic terminal using acetic acid generated from metabolism and from choline recycled from the synaptic cleft. ACh is then taken up by synaptic vesicles.

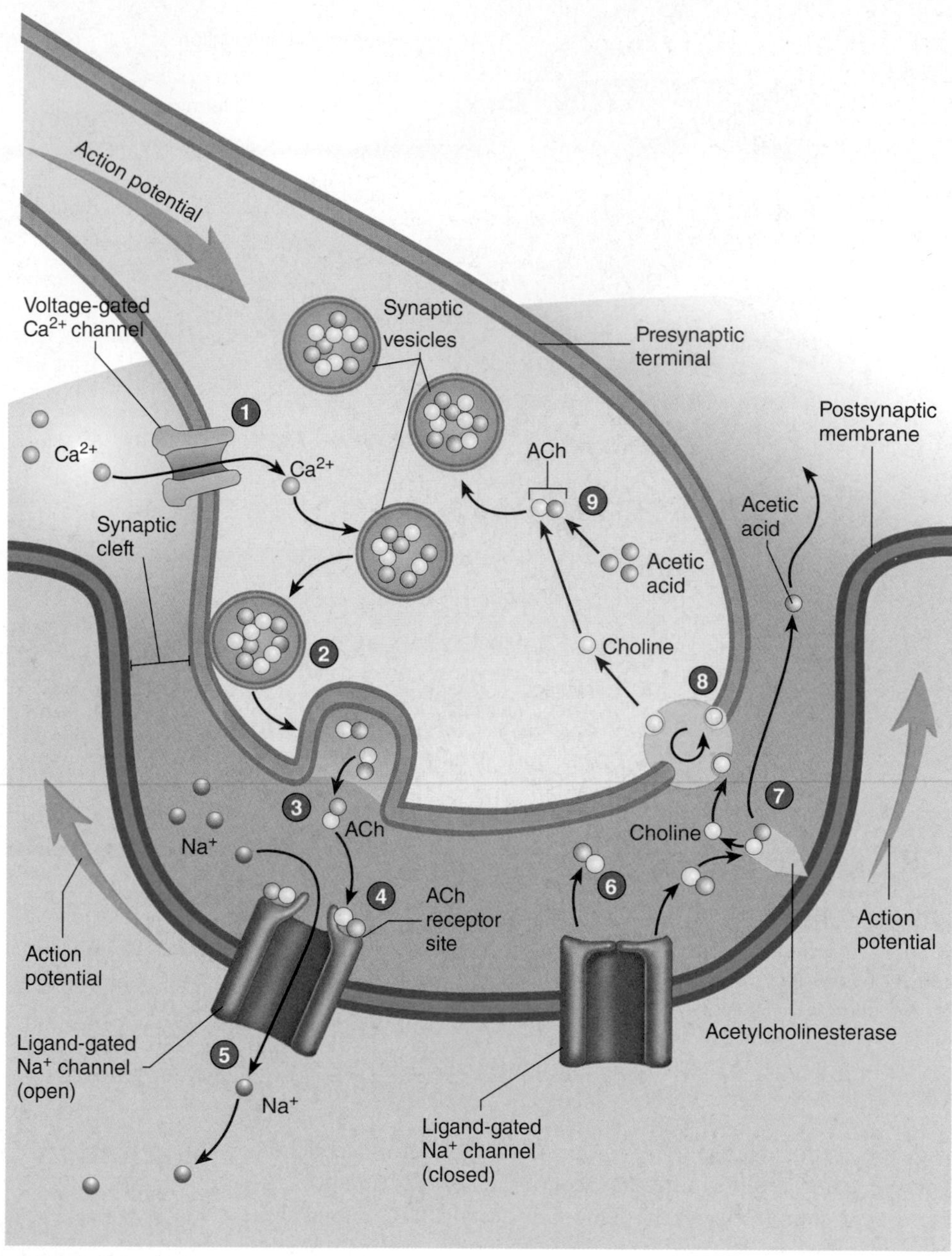

Action potential
Voltage-gated
Ca2+ channel
Synaptic
vesicles
Presynaptic
terminal
Postsynaptic
membrane
Ca2+
Ca2+
ACh
Acetic
acid
Acetic
acid
Synaptic
cleft
Choline
ACh
Na+
Choline
ACh
receptor
site
Action
potential
Action
potential
Ligand-gated
Na+ channel
(open)
Na+
Ligand-gated
Na+ channel
(closed)
Acetylcholinesterase

PROCESS FIGURE 9.12 Function of the Neuromuscular Junction

The action potential is propagated along the entire sarcolemma of the muscle fiber and into the T tubules. The T tubules carry action potentials into the interior of the muscle fiber, where they cause voltage-gated Ca^{2+} channels in the terminal cisternae of the sarcoplasmic reticulum to open. When the Ca^{2+} channels open, Ca^{2+} rapidly diffuses into the sarcoplasm surrounding the myofibrils (figure 9.14).

Calcium ions bind to the Ca^{2+} binding sites on the troponin molecules of the actin myofilaments. The combination of Ca^{2+} with troponin causes the troponin-tropomyosin complex to move deeper into the groove between the two F actin strands, which exposes active sites on the actin myofilaments. The heads of the myosin molecules then bind to the exposed active sites to form cross-bridges (figure 9.14). Movement of the cross-bridges results in contraction.

Cross-Bridge Movement

A cycle of events resulting in contraction proceeds very rapidly (figure 9.15). The heads of myosin molecules move at their hinged region, resulting in cross-bridge movement. This movement forces the actin myofilament, to which the heads of the myosin molecules

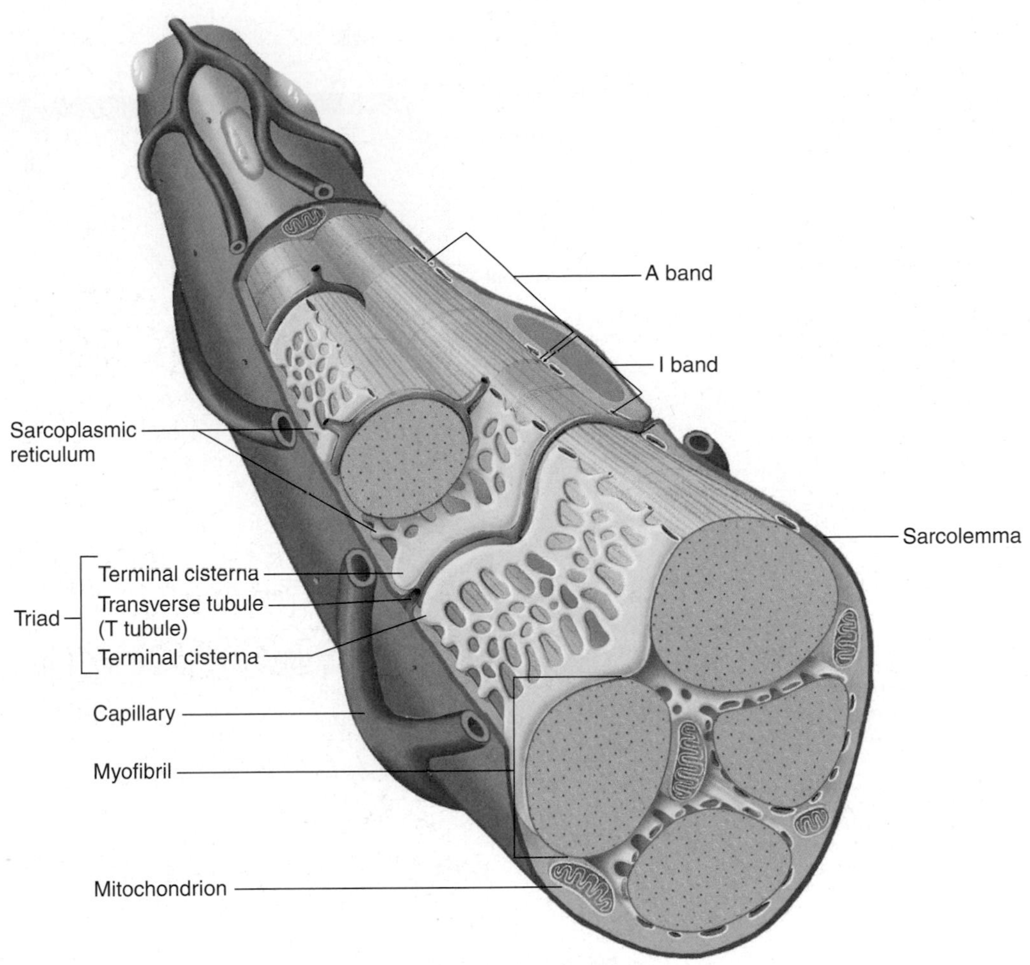
A band
I band
Sarcoplasmic reticulum
Sarcolemma
Triad
Terminal cisterna
Transverse tubule (T tubule)
Terminal cisterna
Capillary
Myofibril
Mitochondrion

FIGURE 9.13 T Tubules and Sarcoplasmic Reticulum
A T tubule and the sarcoplasmic reticulum on each side of the T tubule form a triad.

are attached, to slide over the surface of the myosin myofilament. After cross-bridge movement, each myosin head releases from the actin and returns to its original position. It can then form another cross-bridge at a different site on the actin myofilament, followed by movement, release of the cross-bridge, and return to its original position. During a single contraction, each myosin molecule undergoes the cycle of cross-bridge formation, movement, release, and return to its original position many times. This process is called **cross-bridge cycling.**

The energy from one ATP molecule is required for each cycle of cross-bridge formation, movement, and release. Before a myosin head binds to the active site on an actin myofilament, the head of the myosin molecule is in its resting position, and ADP and phosphate are bound to the head of the myosin molecule (figure 9.15). Once Ca^{2+} binds to troponin and the tropomyosin moves, the active sites on actin myofilaments are exposed. The head of a myosin molecule can then bind to an exposed active site, and the phosphate is released from the head of the myosin molecule. Energy stored in the head of the myosin molecule causes the head of the myosin molecule to move. Movement of the head causes the actin myofilament to slide past the myosin myofilament, and ADP is released from the myosin head. ATP must then bind to the head of the myosin before the cross-bridge can release. As the ATP molecule binds to the head of the myosin molecule, the myosin head breaks ATP apart into ADP and phosphate, and the myosin head releases from the active site on actin. The ADP and phosphate remain attached to the head of the myosin molecule, and the head of the myosin returns to its resting position. Energy released from the breakdown of ATP is stored in the head of the myosin molecule.

Movement of the myosin molecule while the cross-bridge is attached is called the **power stroke,** whereas return of the myosin head to its original position after cross-bridge release is called the **recovery stroke.** Many cycles of power and recovery strokes occur during each muscle contraction. While muscle is relaxed, energy stored in the heads of the myosin molecules is held in reserve until the next contraction. When Ca^{2+} is released from the sarcoplasmic reticulum in response to an action potential, the cycle of cross-bridge formation, movement, and release, which results in contraction, begins (see figures 9.14 and 9.15). Figure 9.16 summarizes the overall sequence of events that occurs as a stimulus causes a skeletal muscle to contract.

1. An action potential that was produced at the neuromuscular junction is propagated along the sarcolemma of the skeletal muscle. The depolarization also spreads along the membrane of the T tubules.

2. The depolarization of the T tubule causes voltage-gated Ca^{2+} channels in the sarcoplasmic reticulum to open, resulting in an increase in the permeability of the sarcoplasmic reticulum to Ca^{2+}, especially in the terminal cisternae. Calcium ions then diffuse from the sarcoplasmic reticulum into the sarcoplasm.

3. Calcium ions released from the sarcoplasmic reticulum bind to troponin molecules. The troponin molecules bound to G actin molecules are released, causing tropomyosin to move, and to expose the active sites on G actin.

4. Once active sites on G actin molecules are exposed, the heads of the myosin myofilaments bind to them to form cross-bridges.

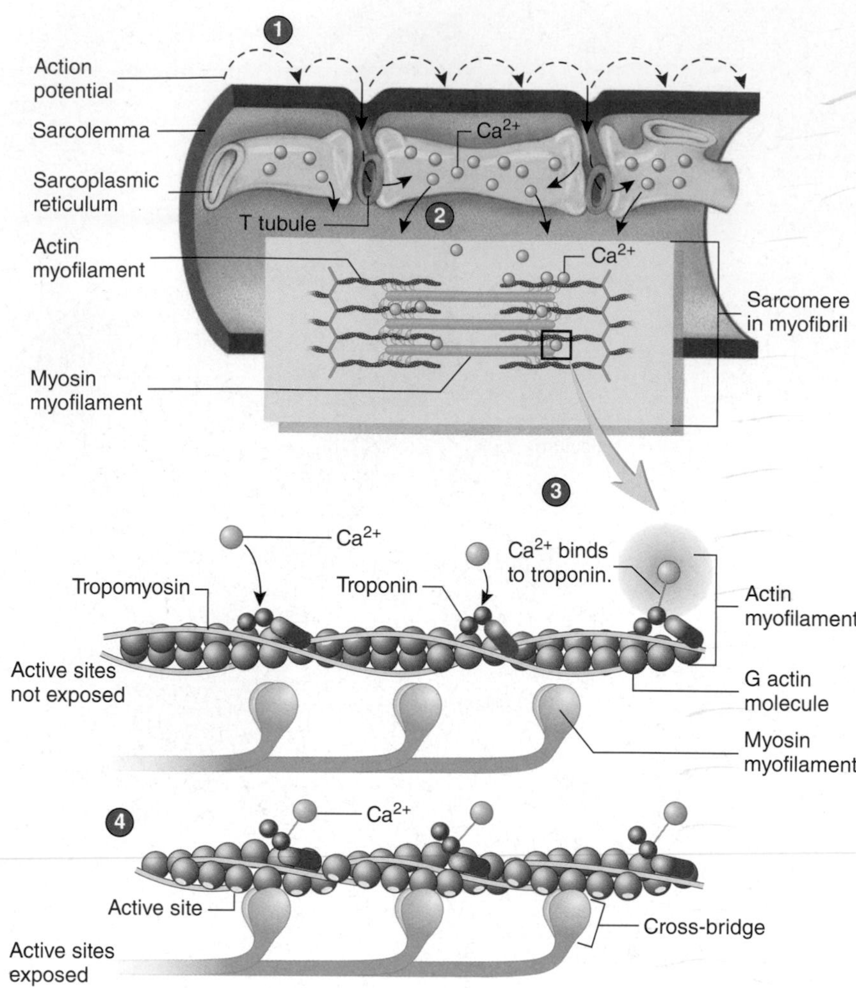
1
Action potential
Sarcolemma
Sarcoplasmic reticulum
Ca2+
T tubule
2
Actin myofilament
Ca2+
Sarcomere in myofibril
Myosin myofilament
3
Ca2+
Ca2+ binds to troponin.
Tropomyosin
Troponin
Actin myofilament
Active sites not exposed
G actin molecule
Myosin myofilament
4
Ca2+
Active site
Cross-bridge
Active sites exposed

PROCESS FIGURE 9.14 APR **Action Potentials and Muscle Contraction**

Clinical IMPACT — The Effect of Blocking Acetylcholine Receptors and Acetylcholinesterase

Any factor that affects the production, release, and degradation of acetylcholine or its ability to bind to its receptor molecule also affects the transmission of action potentials across the neuromuscular junction. For example, some insecticides contain organophosphates that bind to and inhibit the function of acetylcholinesterase. As a result, acetylcholine is not degraded but accumulates in the synaptic cleft, where it acts as a constant stimulus to the muscle fiber. Insects exposed to such insecticides die, partly because their muscles contract and cannot relax—a condition called **spastic paralysis** (spas′tik pă-ral′i-sis), which is followed by muscle fatigue.

Humans respond similarly to these insecticides. The skeletal muscles responsible for respiration cannot undergo their normal cycle of contraction and relaxation. Instead, they remain in a state of spastic paralysis until they become fatigued. Patients die of respiratory failure. Other organic poisons, such as curare, the substance originally used by South American Indians in poison arrows, bind to the acetylcholine receptors, preventing acetylcholine from binding to them. Curare does not allow the activation of the receptors, and therefore the muscle is incapable of contracting in response to nervous stimulation—a condition called **flaccid** (flak′sid, flas′id) **paralysis.** Curare is not a poison to which people are commonly exposed, but it has been used to investigate the role of acetylcholine in the neuromuscular synapse and is sometimes administered in small doses to relax muscles during certain kinds of surgery.

Myasthenia gravis (mī-as-thē′nē-ă grăv′is) results from the production of antibodies that bind to acetylcholine receptors, eventually destroying the receptor and thus reducing the number of receptors. As a consequence, muscles exhibit a degree of flaccid paralysis or are extremely weak (see table 11.5). A class of drugs that includes neostigmine partially blocks the action of acetylcholinesterase and is sometimes used to treat myasthenia gravis. The drugs increase the level of acetylcholine in the synaptic cleft, so that it can bind more effectively to the remaining acetylcholine receptors.

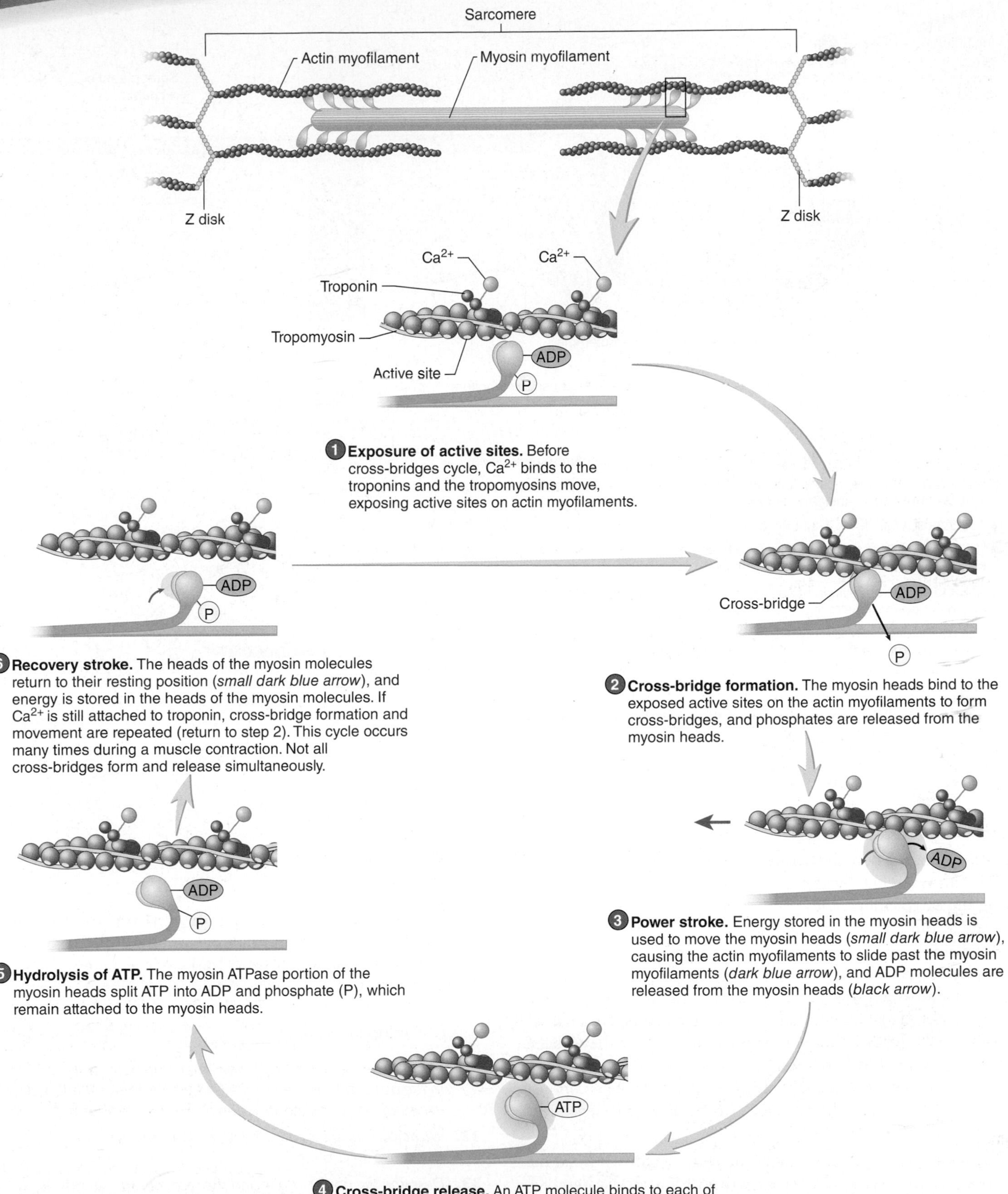
Sarcomere
Actin myofilament
Myosin myofilament
Z disk
Z disk
Ca2+
Ca2+
Troponin
Tropomyosin
Active site
ADP
P
Cross-bridge
ADP
P
ADP
ATP
ADP
P
ADP
P

1 Exposure of active sites. Before cross-bridges cycle, Ca^{2+} binds to the troponins and the tropomyosins move, exposing active sites on actin myofilaments.

2 Cross-bridge formation. The myosin heads bind to the exposed active sites on the actin myofilaments to form cross-bridges, and phosphates are released from the myosin heads.

3 Power stroke. Energy stored in the myosin heads is used to move the myosin heads (*small dark blue arrow*), causing the actin myofilaments to slide past the myosin myofilaments (*dark blue arrow*), and ADP molecules are released from the myosin heads (*black arrow*).

4 Cross-bridge release. An ATP molecule binds to each of the myosin heads, causing them to detach from the actin.

5 Hydrolysis of ATP. The myosin ATPase portion of the myosin heads split ATP into ADP and phosphate (P), which remain attached to the myosin heads.

6 Recovery stroke. The heads of the myosin molecules return to their resting position (*small dark blue arrow*), and energy is stored in the heads of the myosin molecules. If Ca^{2+} is still attached to troponin, cross-bridge formation and movement are repeated (return to step 2). This cycle occurs many times during a muscle contraction. Not all cross-bridges form and release simultaneously.

PROCESS FIGURE 9.15 AP|R **Breakdown of ATP and Cross-Bridge Movement During Muscle Contraction**

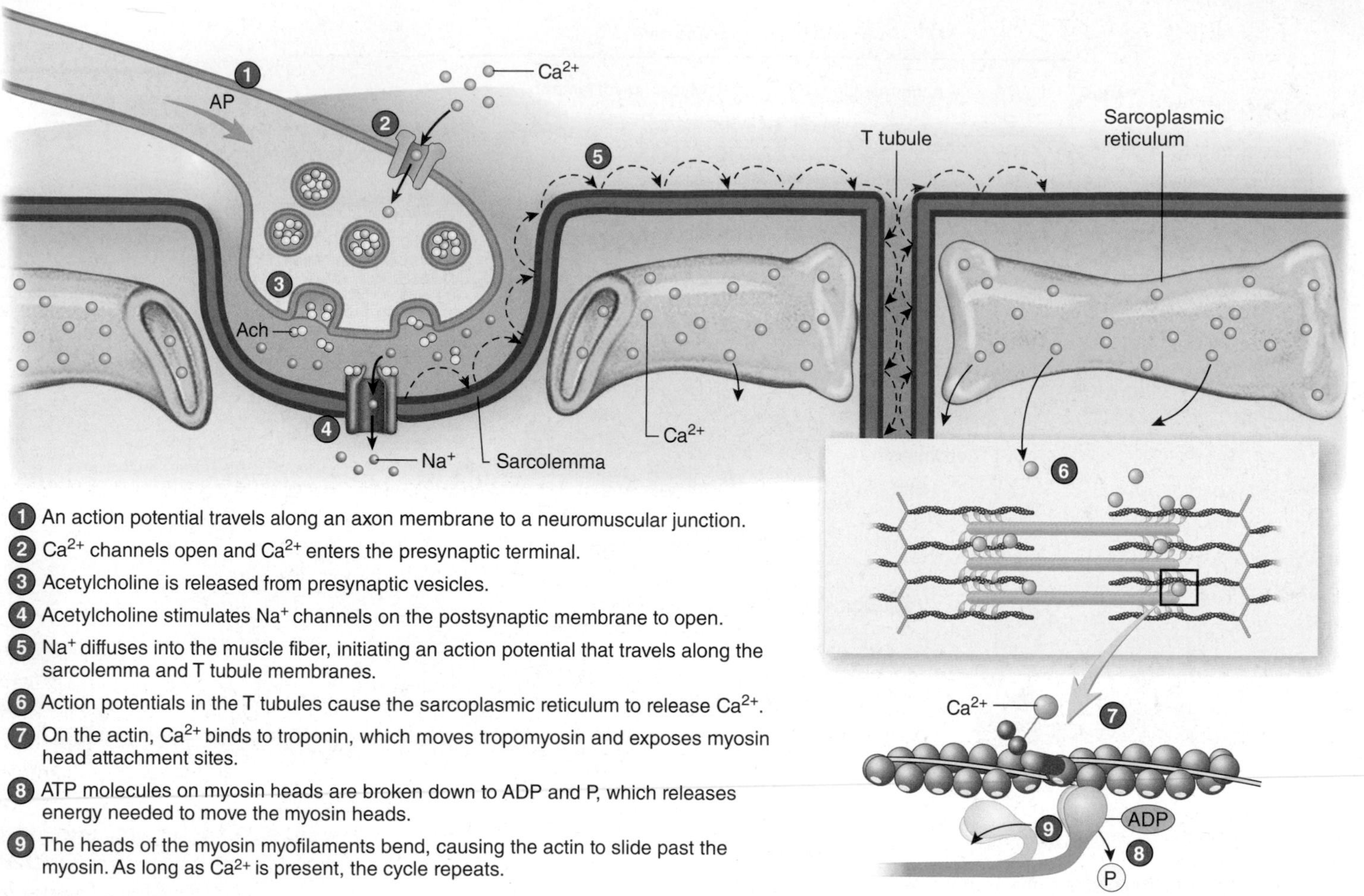

1. An action potential travels along an axon membrane to a neuromuscular junction.
2. Ca^{2+} channels open and Ca^{2+} enters the presynaptic terminal.
3. Acetylcholine is released from presynaptic vesicles.
4. Acetylcholine stimulates Na^+ channels on the postsynaptic membrane to open.
5. Na^+ diffuses into the muscle fiber, initiating an action potential that travels along the sarcolemma and T tubule membranes.
6. Action potentials in the T tubules cause the sarcoplasmic reticulum to release Ca^{2+}.
7. On the actin, Ca^{2+} binds to troponin, which moves tropomyosin and exposes myosin head attachment sites.
8. ATP molecules on myosin heads are broken down to ADP and P, which releases energy needed to move the myosin heads.
9. The heads of the myosin myofilaments bend, causing the actin to slide past the myosin. As long as Ca^{2+} is present, the cycle repeats.

PROCESS FIGURE 9.16 Summary of Skeletal Muscle Contraction

Muscle Relaxation

Muscle relaxation occurs when acetylcholine is no longer released at the neuromuscular junction. The cessation of action potentials along the sarcolemma stops Ca^{2+} release from the sarcoplasmic reticulum and Ca^{2+} is actively transported back into the sarcoplasmic reticulum. As the Ca^{2+} concentration decreases in the sarcoplasm, the Ca^{2+} diffuses away from the troponin molecules. The troponin-tropomyosin complex then reestablishes its position, which blocks the active sites on the actin molecules. As a consequence, cross-bridges cannot re-form once they have been released, and the muscle relaxes.

Thus, energy is needed not only to make muscle fibers contract but also to make muscle fibers relax. Three major ATP-dependent events are required for muscle relaxation:

1. After an action potential has occurred in the muscle fiber, the sodium-potassium pump must actively transport Na^+ and K^+ to return to and maintain resting membrane potential.
2. ATP is required to detach the myosin heads from the actin and return them to their resting position.
3. ATP is needed for the active transport of Ca^{2+} into the sarcoplasmic reticulum.

Because the reuptake of Ca^{2+} into the sarcoplasmic reticulum is much slower than the diffusion of Ca^{2+} out of the sarcoplasmic reticulum, a muscle fiber takes at least twice as long to relax as it does to contract.

ASSESS YOUR PROGRESS

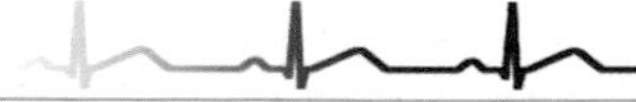

31. *Identify the steps that show how an action potential produced in the postsynaptic membrane of the neuromuscular junction eventually results in contraction of the muscle fiber.*

32. *What ion is necessary for movement of the troponin-tropomyosin complex?*

33. *Describe the power stroke and the recovery stroke. How is ATP used?*

34. *What events occur during the relaxation of the muscle fiber? What is the role of ATP?*

9.5 Physiology of Skeletal Muscle

LEARNING OUTCOMES

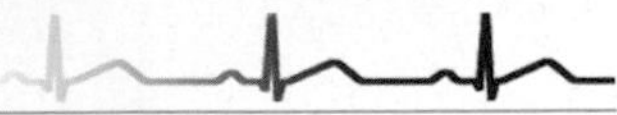

After reading this section, you should be able to

A. Describe a muscle twitch and the events that occur in each phase of a twitch.

B. Describe a motor unit and how motor unit number affects muscle control.

C. Explain how whole muscles respond in a graded fashion and how the force of contraction can be increased.

D. Summarize what occurs in treppe.

E. Relate recruitment to multiple-motor-unit summation.

F. Describe incomplete tetanus and complete tetanus.

G. Explain the connection between the initial length of a muscle and the amount of tension produced.

H. Distinguish between isometric and isotonic contractions.

I. Relate how muscle tone is maintained.

Muscle Twitch

A single, brief contraction and relaxation cycle in a muscle fiber is called a **muscle twitch.** A twitch does not last long enough or generate enough tension to perform any work. Even though the normal function of muscles is more complex, a muscle twitch can serve as an example of how muscles function in living organisms.

Figure 9.17 shows a hypothetical contraction of a single muscle fiber in response to a single action potential. The time between the application of the stimulus to the motor neuron and the beginning of contraction is the **lag phase** (*latent phase*); the time during which contraction occurs is the **contraction phase;** and the time during which relaxation occurs is the **relaxation phase.**

An action potential is an electrochemical event, but contraction is a mechanical event. An action potential is measured in millivolts and is completed in less than 2 milliseconds. Muscle contraction is measured as a force, also called tension. It is reported as the number of grams lifted, or the distance the muscle shortens, and requires up to 1 second to occur.

ASSESS YOUR PROGRESS

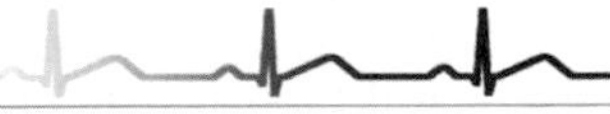

35. *List the phases of a muscle twitch, and describe the events that occur in each phase.*

Motor Units

A **motor unit** consists of a single motor neuron and all the muscle fibers it innervates (figure 9.18). An action potential in the motor neuron generates an action potential in each of the muscle fibers of its motor unit. However, not all motor units are identical. Motor units vary in terms of the number of muscle fibers they contain, and they vary in terms of their sensitivity to stimuli for contraction; some motor units respond readily to weak stimuli, whereas others respond only to strong stimuli. That is why human skeletal muscles are capable of fluid movements.

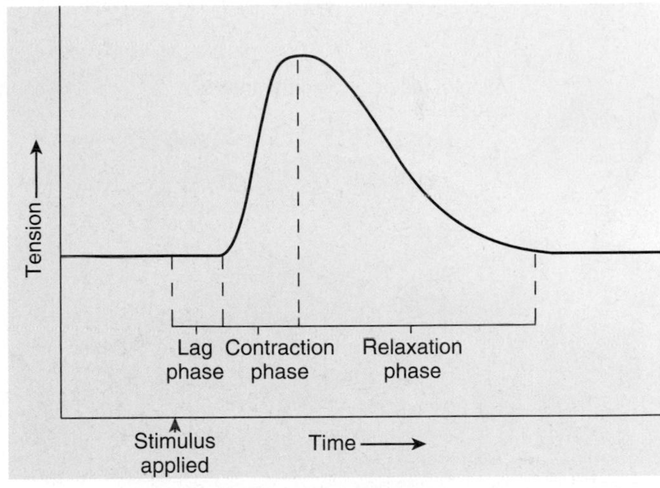
Tension
Lag phase
Contraction phase
Relaxation phase
Stimulus applied
Time

FIGURE 9.17 Phases of a Muscle Twitch in a Single Muscle Fiber
After the stimulus is applied, a short lag phase occurs, followed by a contraction phase and a relaxation phase.

Motor Unit Number

Motor units in different muscles do not always contain the same number of muscle fibers. Muscles performing delicate and precise movements have motor units with a small number of muscle fibers, whereas muscles performing more powerful but less precise contractions have motor units with many muscle fibers. For example, in very delicate muscles, such as those that move the eye, the number of muscle fibers per motor unit can be less than 10, whereas in the heavy muscles of the thigh the number can be several hundred. Therefore, the fewer fibers there are in the motor units of a muscle, the greater control a person has over that muscle.

Predict 4

The disease poliomyelitis (pō′lē-ō-mī′ĕ-lī′tis) destroys motor neurons, causing loss of muscle function and even flaccid paralysis. Some patients recover because axon branches form from the remaining motor neurons. These branches innervate the paralyzed muscle fibers to produce motor units with many more muscle fibers than usual. How does this reinnervation of muscle fibers affect the degree of muscle control in a person who has recovered from poliomyelitis?

Stimulus Strength and Motor Unit Response

The strength of muscle contractions varies from weak to strong. In other words, whole muscles respond to stimuli in a **graded** fashion. For example, the force muscles generate to lift a feather is much less than the force required to lift a 25-pound weight. The force of a contraction is increased in two ways: (1) **Summation** involves increasing the force of contraction of the muscle fibers within the muscle, and (2) **recruitment** involves increasing the number of muscle fibers contracting.

When a muscle fiber demonstrates summation, it is usually because conditions within the muscle fiber have changed. A muscle fiber, when stimulated in rapid succession, contracts with greater force with each subsequent stimulus, a phenomenon called **treppe** (trep′ē; staircase process; figure 9.19). Treppe occurs in a muscle fiber that has rested for a prolonged period. If the muscle fiber is

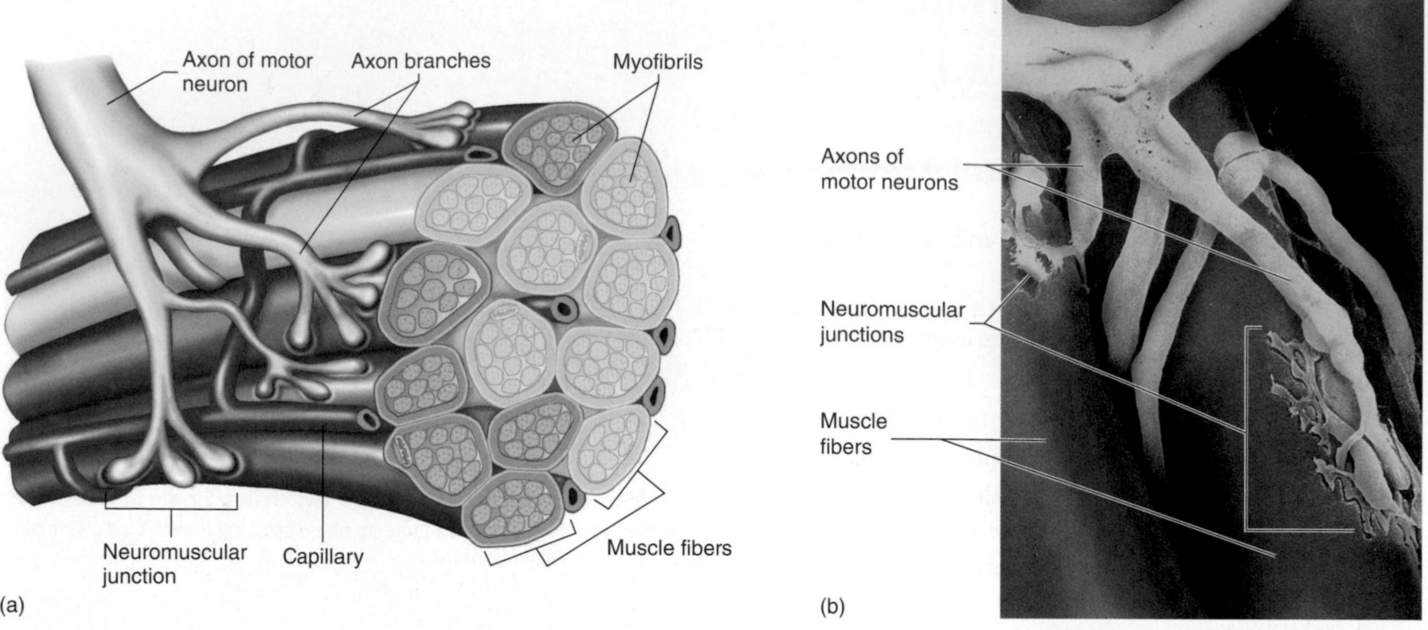

Axon of motor neuron
Axon branches
Myofibrils
Neuromuscular junction
Capillary
Muscle fibers
(a)
Axons of motor neurons
Neuromuscular junctions
Muscle fibers
(b)

FIGURE 9.18 Motor Unit

(*a*) A motor unit consists of a single motor neuron and all the muscle fibers its branches innervate. The muscle fibers shown in dark pink are part of one motor unit, and the muscle fibers shown in light pink are part of a different motor unit. In this figure the motor neuron is innervating the dark pink muscle fibers. (*b*) Photomicrograph of motor units.

maximally stimulated at a *low frequency*, which allows complete relaxation between the stimuli, the successive contractions are stronger and stronger.

A possible explanation for treppe is an increase in Ca^{2+} levels around the myofibrils. The Ca^{2+} released in response to the first stimulus is not taken up completely by the sarcoplasmic reticulum before a second stimulus causes the release of additional Ca^{2+}, even though the muscle relaxes completely between the muscle twitches.

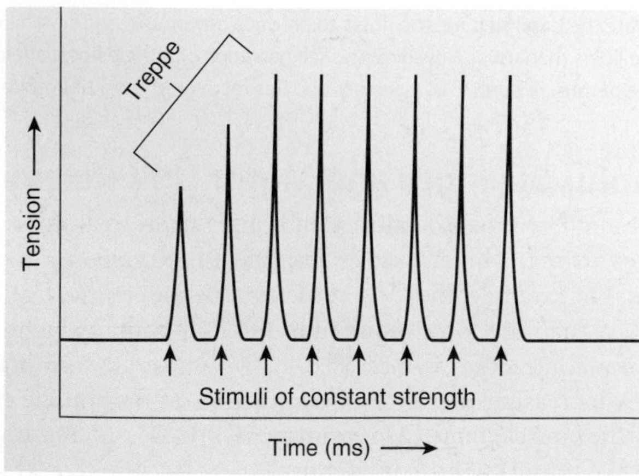

Treppe
Tension ⟶
Stimuli of constant strength
Time (ms) ⟶

FIGURE 9.19 Treppe

When a rested muscle is stimulated repeatedly with maximal stimuli at a low frequency, which allows complete relaxation between stimuli, the second contraction produces a slightly greater tension than the first, and the third contraction produces greater tension than the second. After a few contractions, the levels of tension produced by all the contractions are equal.

As a consequence, during the first few contractions of the muscle, the Ca^{2+} concentration in the sarcoplasm increases slightly, making contraction more efficient because of the increased number of ions available to bind to troponin. For athletes, treppe achieved during warm-up exercises can contribute to improved muscle efficiency. Factors such as increased blood flow to the muscle and increased muscle temperature are probably involved, and because higher temperature causes the enzymes to function more rapidly. Thus, whole muscle fibers do not display a rigid all-or-none response. Intracellular conditions can increase the fiber's response to repeated stimuli.

When a whole muscle undergoes recruitment, more and more motor units contract as the stimulus *strength* increases. The relationship between increased stimulus strength and an increased number of contracting motor units is called **multiple-motor-unit summation** because the force of contraction increases as more and more motor units are stimulated. Multiple-motor-unit summation resulting in graded responses can be demonstrated by applying brief electrical stimuli of increasing strength to the nerve supplying a muscle (figure 9.20). Various results are possible, depending on the strength of the stimulus:

- A **subthreshold stimulus** is not strong enough to cause an action potential in any of the axons in a nerve and does not cause a contraction.
- As the stimulus strength increases, it eventually becomes a **threshold stimulus,** which is strong enough to produce an action potential in a single motor unit axon, causing all the muscle fibers of the motor unit to contract.
- Progressively stronger stimuli, called **submaximal stimuli,** produce action potentials in axons of additional motor units.

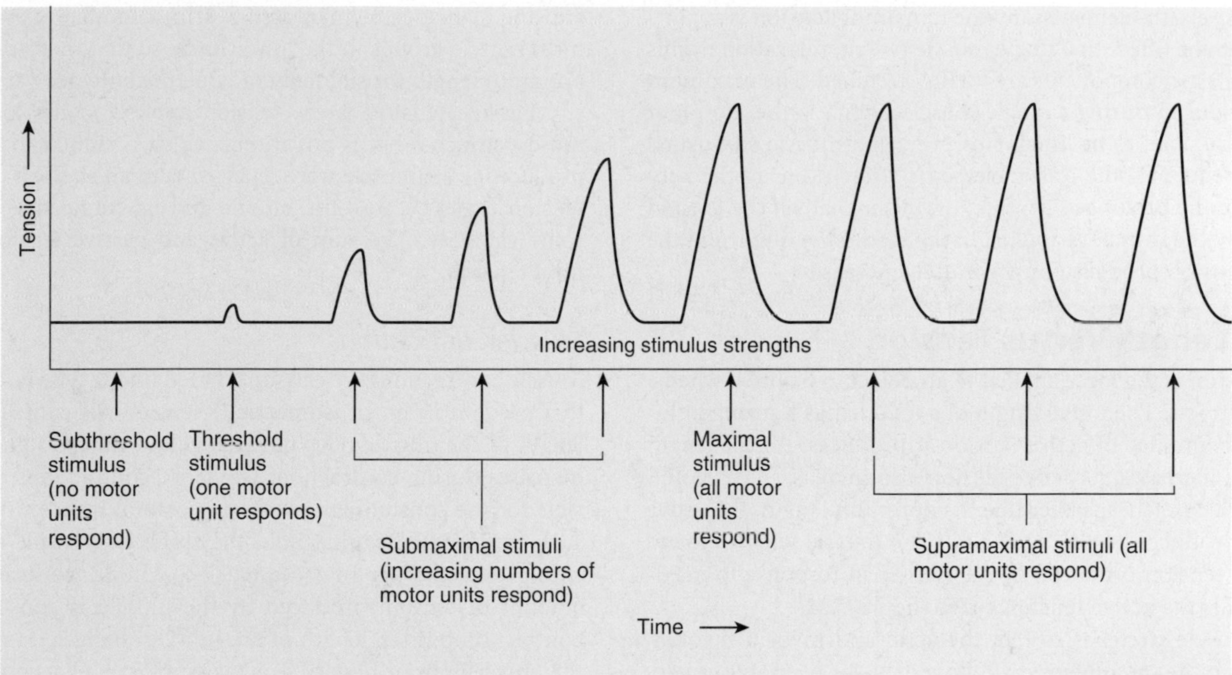
Tension
Increasing stimulus strengths
Subthreshold stimulus (no motor units respond)
Threshold stimulus (one motor unit responds)
Submaximal stimuli (increasing numbers of motor units respond)
Maximal stimulus (all motor units respond)
Supramaximal stimuli (all motor units respond)
Time

FIGURE 9.20 Multiple-Motor-Unit Summation in a Muscle
Multiple-motor-unit summation occurs as stimuli of increasing strength are applied to a nerve that innervates a muscle. The amount of tension (height of peaks) is influenced by the number of motor units responding.

- A **maximal stimulus** produces action potentials in the axons of all the motor units of that muscle. Consequently, even greater stimulus strengths (called *supramaximal stimuli*) have no additional effect.

Stimulus Frequency and Whole Muscle Contraction

An action potential in a single muscle fiber causes it to contract, but the action potential is completed long before the contraction phase is completed. In addition, the contractile mechanism in a muscle fiber exhibits no unresponsive period. That is, relaxation of a muscle fiber is not required before a second action potential can stimulate a second contraction. As the *frequency* of action potentials in a skeletal muscle fiber increases, the frequency of contraction also increases until a period of sustained contraction, or **tetanus** (tet′ă-nŭs), is achieved. In **incomplete tetanus,** muscle fibers partially relax between the contractions; in **complete tetanus,** muscle fibers produce action potentials so rapidly that no relaxation occurs between them. As the frequency of contractions increases, the increased tension produced is called **multiple-wave summation** (figure 9.21).

Tetanus of a muscle caused by stimuli of increasing frequency can be explained at the chemical level. As with treppe, a muscle fiber that has been stimulated at a *high frequency* accumulates more Ca^{2+} in the sarcoplasm, and thus the number of cross-bridges formed increases. Therefore, in comparing treppe and tetanus, we see that the mechanism is very similar—increased sarcoplasmic Ca^{2+}—but the delivery of stimulus to the muscle is different. For treppe the stimulus is maximal but delivered at a low frequency, whereas for tetanus the stimulus is at threshold but delivered at a high frequency.

A significant factor in multiple-wave summation is the fact that the sarcoplasm and the connective tissue components of muscle have some elasticity. During each separate muscle twitch, some of the tension produced by the contracting muscle fibers is used to

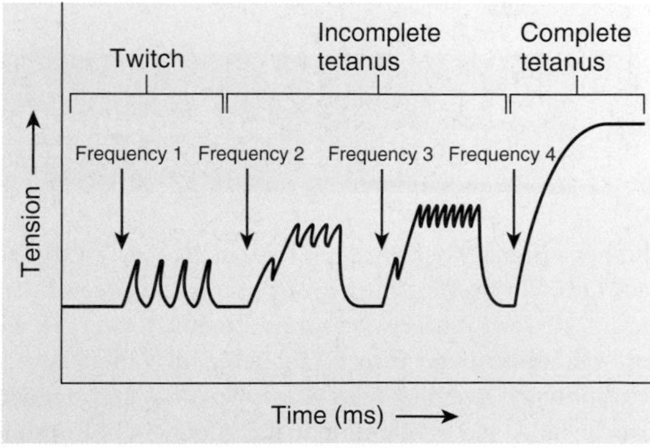
Twitch
Incomplete tetanus
Complete tetanus
Frequency 1
Frequency 2
Frequency 3
Frequency 4
Tension
Time (ms)

FIGURE 9.21 Multiple-Wave Summation
Stimuli 1–4 increase in frequency. For each stimulus, the arrow indicates the start of stimulation. Stimulus frequency 1: a single action potential arriving at a muscle fiber causes twitches that completely relax before the next action potential arrives. Stimulus frequencies 2–3: as the action potential frequency increases, muscle fibers only partially relax before the next action potential arrives and the fiber contracts again; this results in incomplete tetanus. Stimulus frequency 4: action potential frequency can increase to the point where the muscle fiber does not relax at all before the next action potential arrives, causing the muscle fiber to contract continuously; this results in complete tetanus.

stretch those elastic elements, and the remaining tension is applied to the load to be lifted. In a single muscle twitch, relaxation begins before the elastic components are totally stretched. The maximum tension produced during a single muscle twitch is therefore not applied to the load to be lifted. However, in a muscle stimulated at a high frequency, the elastic elements stretch during the very early part of the prolonged contraction. After that, all the tension produced by the muscle is applied to the load to be lifted, and the observed tension produced by the muscle increases.

Muscle Length Versus Tension

Active tension is the force applied to an object to be lifted when a muscle contracts. The initial length of a muscle has a strong influence on the amount of active tension it produces. As the length of a muscle increases, its active tension also increases, to a point. If the muscle stretches farther than its optimum length, the active tension it produces begins to decline. The muscle length plotted against the tension produced by the muscle in response to maximal stimuli is the **active tension curve** (figure 9.22).

If a muscle stretches so that the actin and myosin myofilaments within the sarcomeres do not overlap—or overlap to a very small extent—the muscle produces very little active tension when it is stimulated. Also, if the muscle does not stretch at all, the myosin myofilaments touch each of the Z disks in each sarcomere, and very little contraction of the sarcomeres can occur. If the muscle stretches to its optimum length, optimal overlap of the actin and myosin myofilaments takes place. When the muscle is stimulated, cross-bridge formation results in maximal contraction.

Before lifting heavy objects, weight lifters and others usually assume positions in which their muscles are stretched close to their optimum length. For example, the position a weight lifter assumes before power lifting stretches the upper limb and lower limb muscles to a near-optimum length for muscle contraction, and the stance a lineman assumes in a football game stretches most muscle groups in the lower limbs so that they are near their optimum length for suddenly moving the body forward.

Passive tension is the tension applied to the load when a muscle stretches but is not stimulated. It is similar to the tension produced if the muscle were replaced with an elastic band. Passive tension exists because the muscle and its connective tissue have some elasticity. The sum of active and passive tension is called **total tension.**

Muscle Contractions

Muscle contractions are classified based on the type of contraction that predominates. In **isometric** (ī-sō-met′rik) **contractions,** the length of the muscle does not change but the amount of tension increases during contraction. Isometric contractions are responsible for the constant length of the postural muscles of the body, such as the muscles that hold the spine erect while a person is sitting or standing. In **isotonic** (ī-sō-ton′ik) **contractions,** the amount of tension produced by the muscle is constant during contraction but the length of the muscle changes. Movements of the upper limbs or fingers, as in waving or using a computer keyboard, are predominantly isotonic contractions. Although some mechanical differences exist, both types of contraction result from the same contractile process within muscle fibers. Also, most muscle contractions are not strictly isometric or isotonic. For example, both the length and the tension of muscles change when a person walks or opens a heavy door.

Concentric (kon-sen′trik) **contractions** are isotonic contractions in which tension in the muscle is great enough to overcome the opposing resistance, and the muscle shortens. Concentric contractions result in an increasing tension as the muscle shortens. Many of the movements performed by muscles require concentric contractions—for example, lifting a loaded backpack from the floor to a

Case STUDY | Organophosphate Poisoning

John has a number of prize apple trees in his backyard. To prevent them from becoming infested with insects, he sprayed them with an organophosphate insecticide. Being in a rush to spray the trees before leaving town on vacation, he failed to pay attention to the safety precautions on the packaging and sprayed the trees without using any skin or respiratory protection. Soon he experienced severe stomach cramps, double vision, difficulty breathing, and spastic contractions of his skeletal muscles. John's wife took him to the emergency room, where he was diagnosed with organophosphate poisoning and given medication. Soon many of John's symptoms subsided.

Organophosphate insecticides exert their effects by binding to the enzyme acetylcholinesterase within synaptic clefts, rendering it ineffective. Thus, the organophosphate poison and acetylcholine "compete" for the acetylcholinesterase and, as the organophosphate poison increases in concentration, the enzyme is less effective in degrading acetylcholine. Organophosphate poisons affect synapses in which acetylcholine is the neurotransmitter, including skeletal muscle synapses and smooth muscle synapses, such as those in the walls of the stomach, intestines, and air passageways.

Predict 5

Organophosphate insecticides exert their effects by binding to the enzyme acetylcholinesterase within synaptic clefts, rendering it ineffective. Use figures 9.12 and 9.21 to help answer the following questions.

a. Explain the spastic contractions that occurred in John's skeletal muscles.
b. Propose as many mechanisms as you can by which a drug could counteract the effects of organophosphate poisoning.

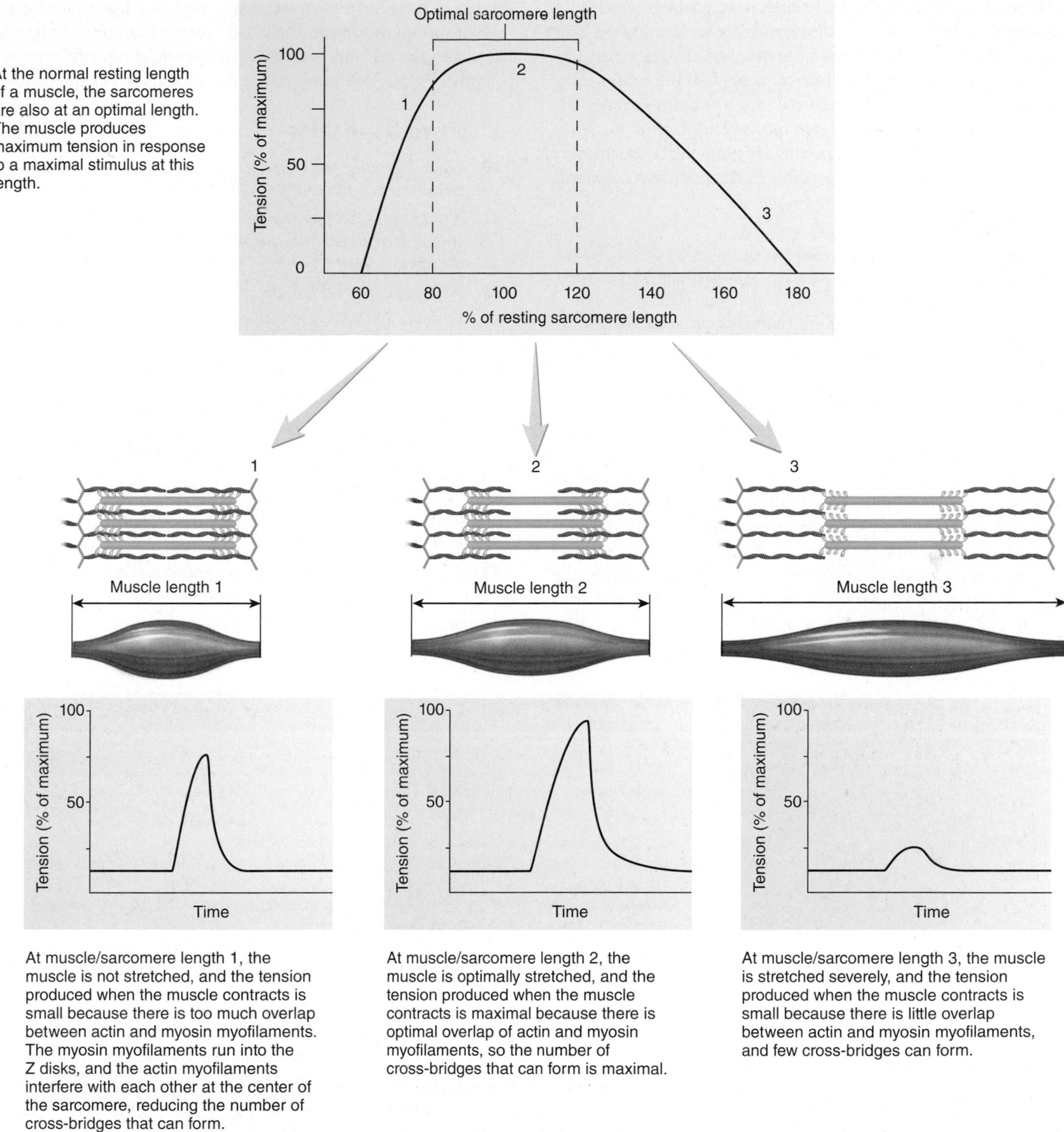
At the normal resting length of a muscle, the sarcomeres are also at an optimal length. The muscle produces maximum tension in response to a maximal stimulus at this length.
Optimal sarcomere length
Tension (% of maximum)
100
50
0
1
2
3
60
80
100
120
140
160
180
% of resting sarcomere length
Muscle length 1
Muscle length 2
Muscle length 3
Time
At muscle/sarcomere length 1, the muscle is not stretched, and the tension produced when the muscle contracts is small because there is too much overlap between actin and myosin myofilaments. The myosin myofilaments run into the Z disks, and the actin myofilaments interfere with each other at the center of the sarcomere, reducing the number of cross-bridges that can form.
At muscle/sarcomere length 2, the muscle is optimally stretched, and the tension produced when the muscle contracts is maximal because there is optimal overlap of actin and myosin myofilaments, so the number of cross-bridges that can form is maximal.
At muscle/sarcomere length 3, the muscle is stretched severely, and the tension produced when the muscle contracts is small because there is little overlap between actin and myosin myofilaments, and few cross-bridges can form.

FIGURE 9.22 Muscle Length and Tension
The length of a muscle before it is stimulated influences the muscle's force of contraction. As the muscle changes length, the sarcomeres also change length.

table top. **Eccentric** (ek-sen′trik) **contractions** are isotonic contractions in which tension is maintained in a muscle, but the opposing resistance is great enough to cause the muscle to increase in length. For example, eccentric contractions occur when a person slowly lowers a heavy weight. Eccentric contractions produce substantial force—in fact, eccentric contractions during exercise often produce greater tension than concentric contractions do. Eccentric contractions are of clinical interest because repetitive eccentric contractions, as occur in the lower limbs of people who run downhill for long distances, tend to injure muscle fibers and muscle connective tissue.

Muscle tone is the constant tension produced by muscles for long periods of time. Muscle tone is responsible for keeping the back and lower limbs straight, the head upright, and the abdomen flat. Muscle tone depends on a small percentage of all the motor units contracting out of phase with one another at any point in time. The frequency of nerve impulses causes incomplete tetanus for short periods, but the contracting motor units are stimulated in such a way that the tension produced by the whole muscle remains constant.

Predict 6

Mary Myosin overheard an argument between two students who could not decide if a weight lifter who lifts a weight above the head and then holds it there before lowering it is using isometric, concentric, or eccentric muscle contractions. Mary is an expert on muscle contractions, so she settles the debate. What is her explanation?

Movements of the body are usually smooth and occur at widely differing rates—some very slowly and others quite rapidly. Very few body movements resemble the rapid contractions of individual muscle twitches. Rather, smooth, slow contractions result from an increasing number of motor units contracting out of phase as the muscles shorten, as well as from a decreasing number of motor units contracting out of phase as muscles lengthen. Each motor unit exhibits either incomplete or complete tetanus but, because the contractions are out of phase and because the number of motor units activated varies at each point in time, a smooth contraction results. Consequently, muscles are capable of contracting either slowly or rapidly, depending on the number of motor units stimulated and the rate at which that number increases or decreases. A summary of physiological muscle responses is presented in table 9.2.

ASSESS YOUR PROGRESS

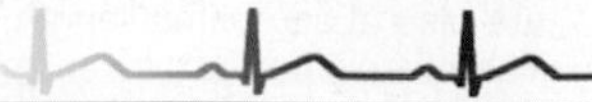

36. *What is a motor unit? Explain why the size of motor units can be different in different muscles.*
37. *What does it mean to say that a whole muscle responds to stimuli in a graded fashion? What are the two ways to increase the force of contraction?*
38. *What is treppe? Explain the physiological reason for it.*
39. *What is multiple-motor-unit summation? Explain the five possible results of multiple-motor-unit summation.*
40. *How does the lack of an unresponsive period in skeletal muscle fiber contraction explain multiple-wave summation? What is the relationship to incomplete tetanus and complete tetanus?*
41. *Distinguish between active tension and passive tension of a muscle.*
42. *Explain how the initial length of the muscle affects actin and myosin overlap, and therefore the amount of contraction that occurs.*
43. *Describe isometric, isotonic, concentric, and eccentric contractions, and give an example of each.*
44. *What is muscle tone, and how is it maintained?*

TABLE 9.2 Types of Physiological Muscle Responses

Physiological Response	Characteristics
Multiple-motor-unit summation	Each motor unit responds in an all-or-none fashion. A whole muscle is capable of producing an increasing amount of tension as the number of motor units stimulated increases.
Multiple-wave summation	Summation results when many action potentials are produced in a muscle fiber. ■ Contraction occurs in response to the first action potential, but there is not enough time for relaxation to occur between action potentials. ■ Because each action potential causes the release of Ca^{2+} from the sarcoplasmic reticulum, the ion levels remain elevated in the sarcoplasm to produce a tetanic contraction. ■ The tension produced as a result of multiple-wave summation is greater than the tension produced by a single muscle twitch. The increased tension results from the greater concentration of Ca^{2+} in the sarcoplasm and the stretch of the elastic components of the muscle early in contraction.
Treppe	Tension produced increases for the first few contractions in response to a maximal stimulus at a *low frequency* in a muscle that has been at rest for some time. Increased tension may result from the accumulation of small amounts of Ca^{2+} in the sarcoplasm for the first few contractions or from an increasing rate of enzyme activity.
Tetanus of muscles	Tetanus of muscles results from multiple-wave summation; frequency of stimulus is higher than for treppe. ■ Incomplete tetanus occurs when the action potential frequency is low enough to allow partial relaxation of the muscle fibers. ■ Complete tetanus occurs when the action potential frequency is high enough that no relaxation of the muscle fibers occurs.
Isometric contractions	A muscle produces increasing tension as it remains at a constant length; this is characteristic of postural muscles that maintain a constant tension without changing their length.
Isotonic contractions	A muscle produces a constant tension and shortens during contraction; this is characteristic of finger and hand movements. ■ In concentric contractions, a muscle produces tension as it shortens; this is characteristic of biceps brachii curl exercises. ■ In eccentric contractions, a muscle produces tension as it resists lengthening; this is characteristic of slowly descending a flight of stairs.

9.6 Muscle Fatigue

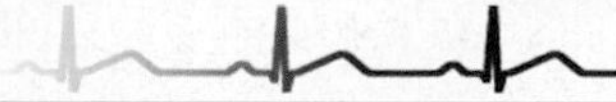

LEARNING OUTCOMES

After reading this section, you should be able to

A. Compare the mechanisms involved in the major types of muscle fatigue.

B. Contrast physiological contracture and rigor mortis.

Fatigue (fă-tēg′) is the decreased capacity to do work and the reduced efficiency of performance that normally follows a period of activity. The rate at which individuals develop fatigue is highly variable, but it is a phenomenon that everyone has experienced. Fatigue can develop at three possible sites: the nervous system, the muscles, and the neuromuscular junction.

Psychological fatigue, the most common type, involves the central nervous system. The muscles are capable of functioning, but the individual "perceives" that additional muscular work is not possible. A burst of activity in a tired athlete in response to encouragement from spectators shows how psychological fatigue can be overcome. The onset and duration of psychological fatigue vary greatly and depend on the emotional state of the individual.

The second most common type of fatigue occurs in the muscle fiber. **Muscular fatigue** results from calcium ion imbalances as ATP levels drop. Without adequate ATP levels in muscle fibers, cross-bridges and ion transport cannot function normally. As a consequence, the tension that a muscle is capable of producing declines. Fatigue in the lower limbs of marathon runners and in the upper and lower limbs of swimmers are examples.

The least common type of fatigue, called **synaptic fatigue,** occurs in the neuromuscular junction. If the action potential frequency in motor neurons is great enough, the amount of acetylcholine released from the presynaptic terminals is greater than the amount synthesized. As a result, the synaptic vesicles become depleted, and insufficient acetylcholine is released to stimulate the muscle fibers. Under normal physiological conditions, synaptic fatigue is rare, but it can occur under conditions of extreme exertion.

As a result of extreme muscular fatigue, muscles occasionally become incapable of either contracting or relaxing—a condition called **physiological contracture** (kon-trak′choor), which is caused by a lack of ATP within the muscle fibers. ATP can decline to very low levels when a muscle is stimulated strongly, as occurs during extreme exercise. When ATP levels are very low, active transport of Ca^{2+} into the sarcoplasmic reticulum slows, Ca^{2+} accumulate within the sarcoplasm, and ATP is unavailable to bind to the myosin molecules that have formed cross-bridges with the actin myofilaments. As a consequence, the previously formed cross-bridges cannot release, resulting in physiological contracture.

Rigor mortis (rig′er mōr′tĭs), the development of rigid muscles several hours after death, is similar to physiological contracture. Shortly after death, ATP production stops, and ATP levels within muscle fibers decline. Because of low ATP levels, active transport of Ca^{2+} into the sarcoplasmic reticulum stops, and Ca^{2+} leaks from the sarcoplasmic reticulum into the sarcoplasm. Calcium ions can also leak from the sarcoplasmic reticulum as its membrane breaks down after cell death. As Ca^{2+} levels increase in the sarcoplasm, cross-bridges form. However, too little ATP is available to bind to the myosin molecules, so the cross-bridges are unable to release and re-form in a cyclic fashion to produce contractions. As a consequence, the muscles remain stiff until tissues begin to degenerate.

Clinical
IMPACT

Muscle Soreness

Following vigorous exercise, people sometimes experience muscle pain, which can last for several days. The pain is more common in untrained people. In addition, highly repetitive eccentric muscle contractions produce pain more readily than concentric contractions do. The pain is related to damage to skeletal muscle fibers and the connective tissue surrounding them. In people with exercise-induced muscle soreness, enzymes that are normally found inside muscle fibers can be detected in the extracellular fluid. These enzymes are able to leave the muscle fibers because injury has increased the permeability of plasma membranes, or even ruptured them. Also found in the extracellular fluid of muscles are fragments of collagen, indicating that both muscle fibers and the connective tissue of muscles have been injured. The pain appears to result from inflammation caused by damage to muscle fibers and the connective tissue. Exercise schedules that alternate exercise with periods of rest, such as lifting weights every other day, provide time for the repair of muscle tissue.

ASSESS YOUR PROGRESS

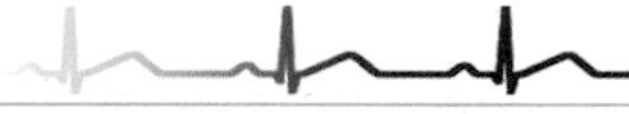

45. *What is fatigue? List the three locations where fatigue can develop.*

46. *Describe what occurs to produce each type of fatigue.*

47. *Explain the causes of physiological contracture and rigor mortis.*

9.7 Energy Sources

LEARNING OUTCOMES

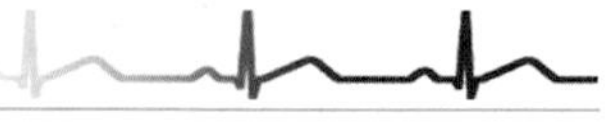

After reading this section, you should be able to

A. Describe the three sources of energy to produce ATP for muscles.

B. Distinguish between oxygen deficit and recovery oxygen consumption.

ATP is the immediate source of energy for muscle contraction. As long as adequate amounts of ATP are present, muscles can contract repeatedly for a long time. ATP must be synthesized continuously to sustain muscle contractions, and ATP synthesis must be equal to ATP breakdown because the small amount of ATP stored in muscle fibers is sufficient to support vigorous muscle contractions for only a few seconds. The energy required to produce ATP comes from three sources: (1) creatine phosphate, (2) anaerobic respiration, and (3) aerobic respiration (table 9.3). Only the main points of anaerobic respiration and aerobic respiration are considered here (a more detailed discussion can be found in chapter 25).

Creatine Phosphate

During resting conditions, energy from aerobic respiration is used to synthesize **creatine** (krē′ă-tēn, krē′ă-tin) **phosphate.** Creatine phosphate accumulates in muscle fibers, where it stores energy that can be used to synthesize ATP. As ATP levels begin to fall, ADP reacts with creatine phosphate to produce ATP and creatine.

$$\text{ADP} + \text{Creatine phosphate} \xrightarrow{\text{Creatine kinase}} \text{Creatine} + \text{ATP}$$

The reaction, catalyzed by the enzyme **creatine kinase,** occurs very rapidly and is able to maintain ATP levels as long as creatine phosphate is available in the fiber. However, during intense muscular contraction, creatine phosphate levels are quickly exhausted. ATP and creatine phosphate present in the cell provide enough energy to sustain maximum contractions for about 8–10 seconds.

Anaerobic Respiration

Anaerobic (an-ār-ō′bik) **respiration** does not require oxygen and results in the breakdown of glucose to yield ATP and lactic acid. For each molecule of glucose metabolized, two ATP molecules and two molecules of lactic acid are produced. The first stages of anaerobic respiration and aerobic respiration share an enzymatic pathway called **glycolysis** (glī-kol′-ī-sis). In glycolysis, a glucose molecule is broken down into two molecules of pyruvic acid. Two molecules of ATP are used in this process, but four molecules of ATP are produced, resulting in a net gain of two ATP molecules for each glucose molecule metabolized. In anaerobic respiration, the pyruvic acid is then converted to lactic acid. Unlike pyruvic acid, much of the lactic acid diffuses out of the muscle fibers into the bloodstream.

Anaerobic respiration is less efficient than aerobic respiration, but it is much faster, especially when insufficient oxygen is available for aerobic respiration. By using many glucose molecules, anaerobic respiration can rapidly produce ATP for a short time. During short periods of intense exercise, such as sprinting, anaerobic respiration combined with the breakdown of creatine phosphate provides enough ATP to support intense muscle contraction for up to 3 minutes. However, ATP formation from creatine phosphate and anaerobic respiration is limited by the depletion of creatine phosphate and glucose and by the brief time anaerobic respiration can produce ATP.

Aerobic Respiration

Aerobic (ār-ō′bik) **respiration** requires oxygen and breaks down glucose to produce ATP, carbon dioxide, and water. Aerobic respiration is much more efficient than anaerobic respiration. Anaerobic respiration results in a net gain of 2 ATP molecules for each glucose molecule, whereas aerobic respiration can produce up to 36 ATP molecules for each glucose molecule. In addition, aerobic respiration uses a greater variety of molecules as energy sources, such as fatty acids and amino acids. Some glucose serves as an energy source in skeletal muscles, but fatty acids provide a more important source of energy during both sustained exercise and resting conditions.

In aerobic respiration, pyruvic acid is metabolized by chemical reactions within mitochondria. Two closely coupled sequences of reactions in mitochondria, called the citric acid cycle and the electron-transport chain, produce many ATP molecules. Carbon dioxide molecules are produced and, in the last step, oxygen atoms combine with hydrogen atoms to form water. Thus, carbon dioxide, water, and ATP are major end products of aerobic respiration. The following equation represents the aerobic respiration of one molecule of glucose:

$$\text{Glucose} + 6\,O_2 + 36\,\text{ADP} + 36\,P_i \rightarrow 6\,CO_2 + 6\,H_2O + \text{About } 36\,\text{ATP}$$

Although aerobic respiration produces many more ATP molecules than anaerobic respiration does, the rate at which the ATP molecules are produced is slower. Resting muscles or muscles involved in long-term exercise, such as long-distance running or other endurance activities, depend primarily on aerobic respiration for ATP synthesis.

Clinical IMPACT

Anaerobic Exercise and Oxygen Deficit

During brief but intense exercise, such as sprinting or lifting a heavy weight, much of the ATP used by exercising muscles comes from the conversion of creatine phosphate and from anaerobic respiration. Glycogen is broken down to glucose in the skeletal muscle fibers and in the liver. The liver releases glucose into the bloodstream, where it can be taken up by skeletal muscle fibers. Anaerobic respiration within the skeletal muscle fibers converts the glucose molecules to ATP and lactic acid. Increased breathing and elevated aerobic respiration after exercise partially result from the oxygen deficit that occurs during such exercise. The increased aerobic respiration pays back a portion of the oxygen deficit by converting creatine to creatine phosphate and converting the excess lactic acid to glucose, which is once again stored as glycogen in muscles and in the liver. The magnitude of the oxygen deficit depends on the intensity of the exercise, the length of time it was sustained, and the individual's physical condition. People who are in poor physical condition do not have as great a capacity to perform aerobic respiration as well-trained athletes do.

Predict 7

A condition called McArdle disease is due to a deficiency of an enzyme necessary for the breakdown of the stored form of glucose, called glycogen. Predict how the disease affects a person's ability to exercise.

Oxygen Deficit and Recovery Oxygen Consumption

There is a lag time between when a person begins to exercise and when he or she begins to breathe more heavily because of the exercise. After exercise stops, there is another lag time before breathing returns to its preexercise rate. These changes in breathing patterns reflect muscles' need for more oxygen to produce ATP through aerobic respiration. The insufficient oxygen consumption relative

TABLE 9.3 Sources of ATP in Muscles

	Creatine Phosphate	Anaerobic Respiration	Aerobic Respiration
Metabolic Process	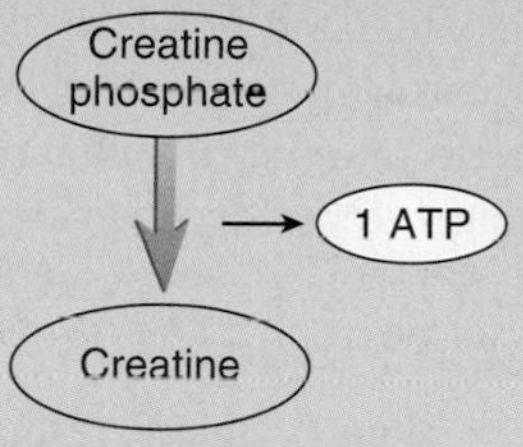 Creatine phosphate → 1 ATP; Creatine	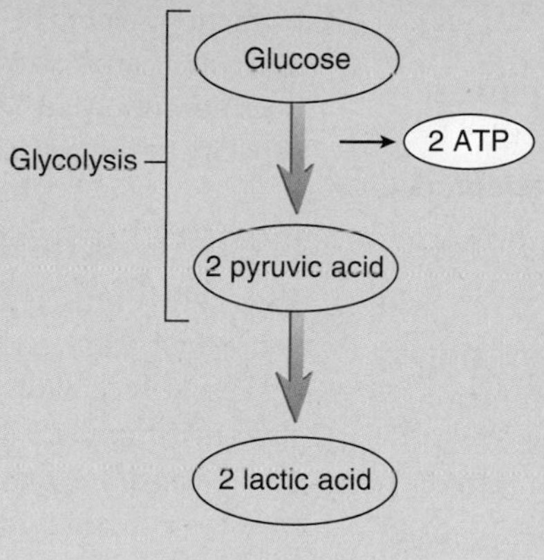 Glycolysis: Glucose → 2 ATP; 2 pyruvic acid; 2 lactic acid	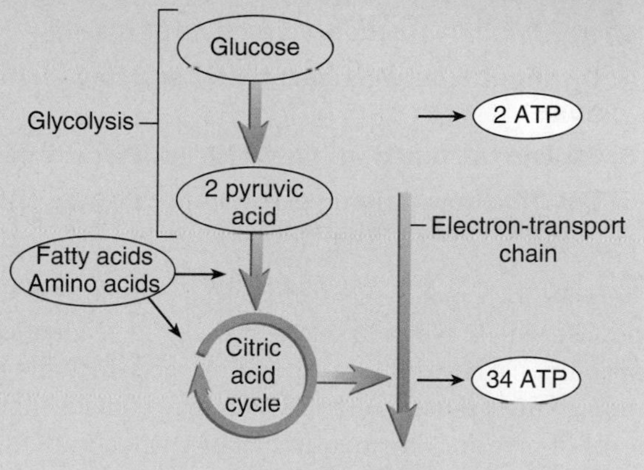 Glycolysis: Glucose → 2 ATP; 2 pyruvic acid; Fatty acids, Amino acids; Citric acid cycle; Electron-transport chain → 34 ATP
Energy Source	Creatine phosphate	Glucose	Glucose, fatty acids, amino acids
Oxygen Required	No	No	Yes
ATP Yield	1 per creatine phosphate	2 per glucose molecule	Up to 36 per glucose molecule
Duration of Energy Supply	Up to 10 seconds	Up to 3 minutes	Hours
Type of Work Supported	Moderate exercise and extreme exercise	Extreme exercise	Resting and all exercise

to increased activity at the onset of exercise creates an **oxygen deficit,** or *oxygen debt.* This deficit must be repaid during and after exercise once oxygen consumption catches up with the increased activity level. At the onset of exercise, muscles primarily acquire the ATP they need from the creatine phosphate system and anaerobic respiration—two systems that can supply ATP relatively quickly and without requiring oxygen (table 9.3). The ability of aerobic respiration to supply ATP at the onset of exercise lags behind that of the creatine phosphate system and anaerobic respiration. This explains the lag between the onset of exercise and the need for increased oxygen.

The elevated oxygen consumption that occurs after exercise has ended is called **recovery oxygen consumption.** A portion of the recovery oxygen is used to "repay" the oxygen deficit incurred at the onset of exercise, but most of the recovery oxygen is used to support metabolic processes that restore homeostasis after it was disturbed during exercise. Such disturbances include exercise-related increases in body temperature, changes in intra- and extracellular ion concentrations, and changes in metabolite and hormone levels. Recovery oxygen consumption generally lasts minutes to hours, depending on the individual's physical conditioning and on the length and intensity of the exercise session. In extreme cases, such as following a marathon, recovery oxygen consumption can last as long as 15 hours.

ASSESS YOUR PROGRESS

48. *List the energy sources used to synthesize ATP for muscle contraction.*

49. *What is the function of creatine phosphate, and when is it used?*

50. *Contrast the efficiency of anaerobic and aerobic respiration. When is each type used by cells?*

51. *When does lactic acid increase in a muscle fiber?*

52. *What is the difference between oxygen deficit and recovery oxygen consumption? Explain the factors that contribute to an oxygen deficit.*

Predict 8

Eric is a highly trained cross-country runner, and his brother John is a computer programmer who almost never exercises. While the two brothers were working on a remodeling project in the basement of their house, the doorbell rang upstairs: A package they were both very excited about was being delivered. They raced each other up the stairs to the front door to see who could get the package first. When they reached the door, both were breathing heavily. However, John continued to breathe heavily for several minutes while Eric was opening the package. Why did John breathe heavily longer than Eric, even though they had both run the same distance?

9.8 Slow-Twitch and Fast-Twitch Fibers

LEARNING OUTCOMES

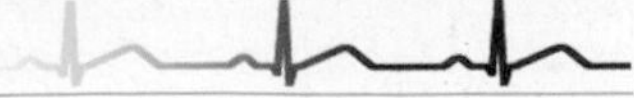

After reading this section, you should be able to

A. Distinguish between fast-twitch and slow-twitch muscle fibers.

B. Explain the functions for which each type is best adapted.

C. Describe how training can increase the size and efficiency of both types of muscle fibers.

There are two major types of skeletal muscle fibers: slow-twitch and fast-twitch. Not all skeletal muscles have identical functional capabilities. They differ in several respects, including the composition of their muscle fibers, which may contain slightly different forms of myosin. The myosin of slow-twitch muscle fibers causes the fibers to contract more slowly and to be more resistant to fatigue, whereas the myosin of fast-twitch muscle fibers causes the fibers to contract quickly and to fatigue quickly (table 9.4). The proportion of muscle fiber types differs within individual muscles.

Slow-Twitch Oxidative Muscle Fibers

Slow-twitch oxidative (SO) muscle fibers (**type I fibers**) contract more slowly, have a better-developed blood supply, have more mitochondria, and are more fatigue-resistant than fast-twitch muscle fibers. In women, type I alone, or in conjunction with type IIa are the largest-diameter fibers. Slow-twitch muscle fibers respond relatively slowly to nervous stimulation. The enzymes on the myosin heads responsible for the breakdown of ATP are called **myosin ATPase.** Slow-twitch fibers break down ATP slowly because their myosin heads have a slow form of myosin ATPase. The relatively slow breakdown of ATP means that cross-bridge movement occurs slowly, which causes the muscle to contract slowly. Aerobic respiration is the primary source for ATP synthesis in slow-twitch muscles, and their capacity to perform aerobic respiration is enhanced by a plentiful blood supply and the presence of numerous mitochondria. They are called oxidative muscle fibers because of their enhanced capacity to carry out aerobic respiration. Slow-twitch fibers also contain large amounts of **myoglobin** (mī-ō-glō′bin), a dark pigment similar to hemoglobin in red blood cells, which binds oxygen and acts as an oxygen reservoir in the muscle fiber when the blood does not supply an adequate amount. Myoglobin thus enhances the capacity of the muscle fibers to perform aerobic respiration.

Fast-Twitch Muscle Fibers

Fast-twitch muscle fibers (**type II fibers**) respond rapidly to nervous stimulation, and their myosin heads have a fast form of myosin ATPase, which allows them to break down ATP more rapidly than slow-twitch muscle fibers. This allows their cross-bridges to release and form more rapidly than those in slow-twitch muscle fibers. Muscles containing a high percentage of fast-twitch fibers have a less-well-developed blood supply than muscles containing a high percentage of slow-twitch fibers. In addition, fast-twitch muscle fibers have very little myoglobin and fewer and smaller mitochondria. Fast-twitch muscle fibers have large deposits of glycogen and are well adapted to perform anaerobic respiration. However, the anaerobic respiration of fast-twitch muscle fibers is not adapted for supplying a large amount of ATP for a prolonged period. The muscle fibers tend to contract rapidly for a shorter time and to fatigue relatively quickly. Fast-twitch muscle fibers come in two forms: type IIa, or fast-twitch oxidative glycolytic (FOG) fibers, and type IIb, or fast-twitch glycolytic (FG) fibers (table 9.4). Type IIa fibers rely on both anaerobic and aerobic ATP production, whereas type IIb fibers rely almost exclusively on anaerobic glycolysis for ATP production. In men, type IIa are the largest-diameter fibers.

Distribution of Fast-Twitch and Slow-Twitch Muscle Fibers

The muscles of many animals are composed primarily of either fast-twitch or slow-twitch muscle fibers. A chicken or pheasant

TABLE 9.4 Characteristics of Skeletal Muscle Fiber Types

	Slow-Twitch Oxidative (SO) (Type I)	Fast-Twitch Oxidative Glycolytic (FOG) (Type IIa)	Fast-Twitch Glycolytic (FG) (Type IIb)
Myoglobin Content	High	High	Low
Mitochondria	Many	Many	Few
Capillaries	Many	Many	Few
Metabolism	High aerobic capacity, low anaerobic capacity	Intermediate aerobic capacity, high anaerobic capacity	Low aerobic capacity, highest anaerobic capacity
Fatigue Resistance	High	Intermediate	Low
Myosin ATPase Activity	Slow	Fast	Fast
Glycogen Concentration	Low	High	High
Location Where Fibers Are Most Abundant	Generally in postural muscles and more in lower limbs than upper limbs	Generally in lower limbs	Generally in upper limbs
Functions	Maintenance of posture and performance of endurance activities	Endurance activities in endurance-trained muscles	Rapid, intense movements of short duration

breast, which is composed mainly of fast-twitch fibers, appears whitish because of its relatively poor blood supply and lack of myoglobin. The muscles are adapted to contract rapidly for a short time, but they fatigue quickly. By contrast, the meat of a chicken leg or a duck breast is composed of slow-twitch fibers and appears reddish or darker because of the relatively well-developed blood supply and large amount of myoglobin. These muscles are adapted to contract slowly for a longer time and to fatigue slowly. The distribution of slow-twitch and fast-twitch muscle fibers is consistent with the behavior of these animals. For example, pheasants can fly relatively fast for short distances, whereas ducks fly more slowly for long distances.

Human muscles exhibit no clear separation of slow-twitch and fast-twitch muscle fibers. Most muscles have both types of fibers, although the number of each varies for each muscle. The large postural muscles contain more slow-twitch fibers, whereas the muscles of the upper limbs contain more fast-twitch fibers.

The distribution of slow-twitch and fast-twitch muscle fibers in a given muscle is fairly constant for each individual and apparently is established during early development. Sprinters have a greater percentage of fast-twitch muscle fibers, whereas long-distance runners have a higher percentage of slow-twitch muscle fibers in their lower limb muscles. Athletes who perform a variety of anaerobic and aerobic exercises tend to have a more balanced mixture of fast-twitch and slow-twitch muscle fibers.

Effects of Exercise

Neither fast-twitch nor slow-twitch muscle fibers can be easily converted to muscle fibers of the other type without specialized training. Training can increase the size and capacity of both types of muscle fibers so that they perform more efficiently. Intense exercise that requires anaerobic respiration, such as weight lifting, increases muscular strength and mass and causes fast-twitch muscle fibers to enlarge more than slow-twitch muscle fibers. Conversely, aerobic exercise increases the vascularity of muscle and causes slow-twitch muscle fibers to enlarge more. Aerobic exercise training can convert some fast-twitch muscle fibers that fatigue readily (type IIb) to fast-twitch muscle fibers that resist fatigue (type IIa). In addition to changes in myosin, increases occur in both the number of mitochondria in the muscle fibers and their blood supply. Weight training followed by periods of rest can convert some muscle fibers from type IIa to type IIb. However, a type I muscle fiber cannot be converted to a type II fiber, and vice versa. Through specific training, a person with more fast-twitch muscle fibers can run long distances, and a person with more slow-twitch muscle fibers can increase the speed at which he or she runs.

Predict 9

Susan recently began racing her bicycle. Her training consists entirely of long rides at a steady pace. When she entered her first race, she was excited that she was able to keep pace with the rest of the riders. However, during the final sprint to the finish line, the other riders left her in their dust, and she finished in last place. Why was she unable to keep pace during the finishing sprint? As her coach, what advice would you give Susan about training for her next race?

In response to exercise, a muscle increases in size, or **hypertrophies** (hī-per′trō-fēz), and increases in strength and endurance. Conversely, a muscle that is not used decreases in size, or **atrophies** (at′rō-fēz). For example, muscular atrophy occurs in an arm or a leg that is placed in a cast for several weeks. Because muscle fiber numbers do not change appreciably during most of a person's life, atrophy and hypertrophy result from changes in the size of individual muscle fibers. As a fiber increases in size, the number of myofibrils and sarcomeres increases. The number of nuclei in each muscle fiber increases in response to exercise, but the nuclei of muscle fibers cannot divide. New nuclei are added to muscle fibers because small satellite cells near skeletal muscle fibers increase in number in response to exercise and then fuse with the skeletal muscle fibers. Other elements, such as blood vessels, connective tissue, and mitochondria, also increase in number. Atrophy due to lack of exercise results from a decrease in all these elements without a decrease in muscle fiber number. However, severe atrophy, as occurs in elderly people who cannot readily move their limbs, does involve an irreversible decrease in the number of muscle fibers and can lead to paralysis.

The increased strength of trained muscle is greater than would be expected if that strength were based only on the change in muscle size. In a trained person, part of the increase in strength results from the nervous system's ability to recruit a large number of motor units simultaneously to perform movements with better

Clinical
IMPACT

Anabolic Steroids and Growth Hormone

Some people take synthetic hormones called **anabolic steroids** (an-ă-bol′ik stēr′oydz, ster′oydz) to increase the size and strength of their muscles. Anabolic steroids are related to testosterone, a reproductive hormone secreted by the testes, except that the steroids have been altered so that their reproductive effects are minimized but their effect on skeletal muscles is maintained. Testosterone and anabolic steroids cause skeletal muscle tissue to hypertrophy. People who take large doses of anabolic steroid exhibit increases in body weight and total skeletal muscle mass, and many athletes believe that anabolic steroids improve performance that depends on strength. Unfortunately, harmful side effects are associated with taking anabolic steroids, including periods of irritability, testicular atrophy and sterility, cardiovascular diseases (such as heart attack or stroke), and abnormal liver function. Most athletic organizations prohibit the use of anabolic steroids; some analyze urine samples either randomly or periodically and have possible penalties in place for athletes whose urine shows evidence of anabolic steroid metabolites.

Some individuals use growth hormone inappropriately to increase muscle size. Growth hormone increases protein synthesis in muscle tissue, although it does not produce the same kinds of side effects as those attributed to anabolic steroids. Nevertheless, large doses of growth hormone can cause harmful side effects if taken over a long period (see chapter 18).

neuromuscular coordination. In addition, trained muscles are usually less restricted by excess adipose tissue. Metabolic enzymes increase in hypertrophied muscle fibers, resulting in a greater capacity to take in nutrients and produce ATP. Improved endurance in trained muscles is in part due to improved metabolism, increased circulation to the exercising muscles, increased numbers of capillaries, more efficient respiration, and a greater capacity of the heart to pump blood.

ASSESS YOUR PROGRESS

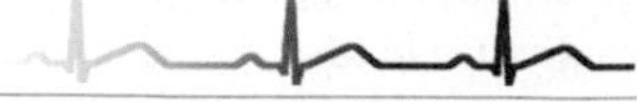

53. *Contrast the structural and physiological differences between slow-twitch and fast-twitch muscle fibers.*
54. *Explain the functions for which each type of muscle fiber is best adapted and how slow-twitch and fast-twitch fibers are distributed.*
55. *How does anaerobic versus aerobic exercise affect muscles?*
56. *What factors contribute to increases in muscle size, strength, and endurance?*

9.9 Heat Production

LEARNING OUTCOMES

After reading this section, you should be able to

A. Explain how muscle metabolism causes normal body temperature.

B. Describe how muscles respond to changes from normal body temperature.

The rate of metabolism in skeletal muscle differs before, during, and after exercise. As chemical reactions occur within cells, some energy is released in the form of heat. Normal body temperature results primarily from this heat. Because the rate of chemical reactions increases in muscle fibers during contraction, the rate of heat production also increases, causing a rise in body temperature. After exercise, increased metabolism resulting from recovery oxygen consumption helps keep the body temperature elevated, but sweating and vasodilation of blood vessels in the skin speed heat loss and keep body temperature within its normal range (see chapter 25).

When body temperature declines below a certain level, the nervous system responds by inducing **shivering,** rapid skeletal muscle contractions that produce shaking rather than coordinated movements. During shivering, the muscle movement increases heat production up to 18 times that of resting levels, and the heat produced can exceed the amount produced during moderate exercise. Thus, shivering helps raise body temperature to its normal range.

ASSESS YOUR PROGRESS

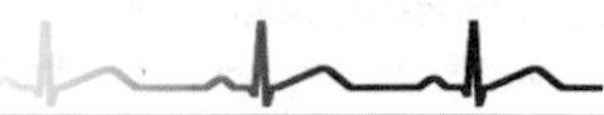

57. *How do muscles contribute to body temperature before, during, and after exercise?*
58. *What is accomplished by shivering?*

9.10 Smooth Muscle

LEARNING OUTCOMES

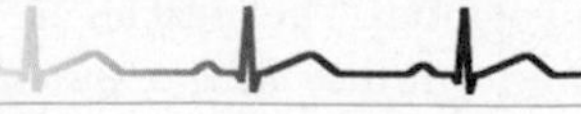

After reading this section, you should be able to

A. Describe the structural features of smooth muscle cells and contrast them with skeletal muscle fibers.

B. Explain the steps of smooth muscle contraction and how the contraction differs from skeletal muscle contraction.

C. Compare the two types of smooth muscle as to their action and locations.

D. Describe the electrical and functional properties of smooth muscle.

E. Explain how smooth muscle activities are regulated.

Smooth muscle is distributed widely throughout the body and is more variable in function than other muscle types. Smooth muscle cells (figure 9.23) are smaller than skeletal muscle fibers, ranging from 15 to 200 μm in length and from 5 to 8 μm in diameter. They are spindle-shaped, with a single nucleus in the middle of the cell. Compared with skeletal muscle, fewer actin and myosin myofilaments are present, and there are more actin than myosin myofilaments.

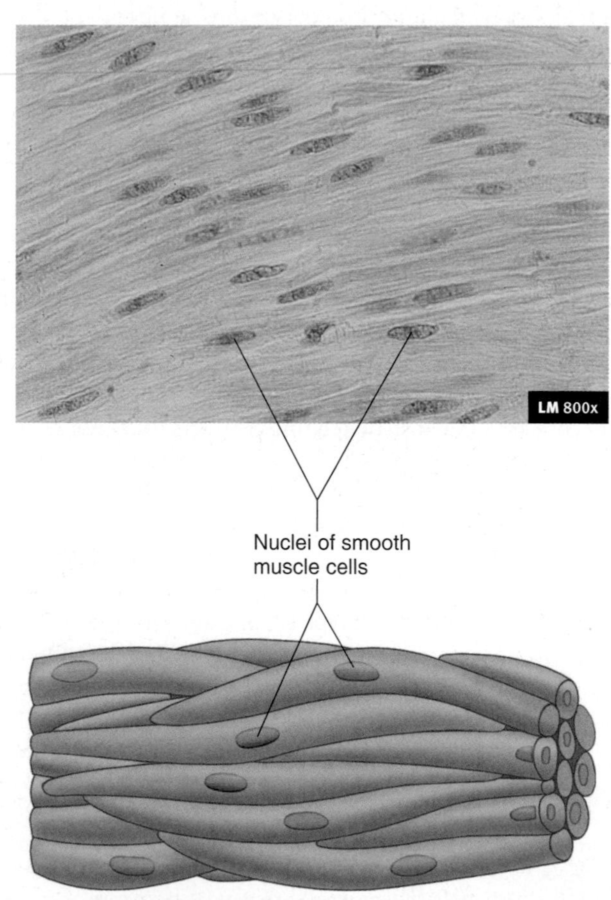
LM 800x
Nuclei of smooth muscle cells

FIGURE 9.23 APR **Smooth Muscle Histology**
Smooth muscle tissue is made up of sheets or bundles of spindle-shaped cells, with a single nucleus in the middle of each cell.

The actin and myosin myofilaments overlap, but they are organized as loose bundles. Consequently, smooth muscle does not have a striated appearance. Actin myofilaments are attached to **dense bodies,** which are scattered through the cell cytoplasm, and to **dense areas,** which are in the plasma membrane. Dense bodies and dense areas are considered equivalent to the Z disks in skeletal muscle. Noncontractile **intermediate filaments** also attach to the dense bodies. The intermediate filaments and dense bodies form an intracellular cytoskeleton, which has a longitudinal or spiral organization. The smooth muscle cells shorten when the actin and myosin slide over one another during contraction (figure 9.24).

Sarcoplasmic reticulum is present in smooth muscle cells, but no T tubule system exists. Some shallow, invaginated areas called **caveolae** (kav-ē-ō′lē) lie along the surface of the plasma membrane. The function of caveolae is not well known, but it may be similar to that of both the T tubules and the sarcoplasmic reticulum of skeletal muscle.

Other differences exist between smooth muscle and skeletal muscle. In particular, smooth muscle has a slower contraction speed than skeletal muscle. This difference is due to several factors. In smooth muscle, some of the Ca^{2+} required to initiate contractions enters the cell from the extracellular fluid and from the sarcoplasmic reticulum. Therefore, it is the greater distance that Ca^{2+} must diffuse, the slower rate at which action potentials are propagated between smooth muscle cells, and the slower rate of cross-bridge formation between actin and myosin myofilaments that are all responsible for the slower contraction of smooth muscle.

Smooth muscle contraction is stimulated both neurally and hormonally. Regardless of the stimulus source, however, calcium ions are the key to smooth muscle contraction.

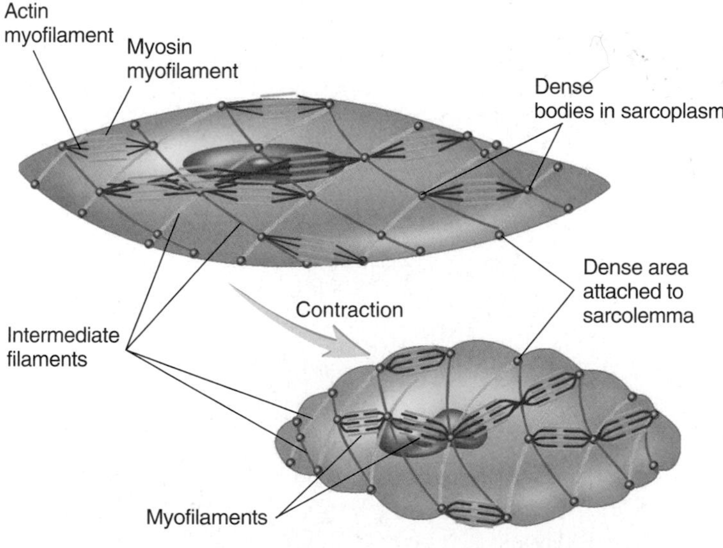
Actin myofilament
Myosin myofilament
Dense bodies in sarcoplasm
Dense area attached to sarcolemma
Contraction
Intermediate filaments
Myofilaments

FIGURE 9.24 Actin and Myosin Proteins in a Smooth Muscle Cell
Bundles of contractile myofilaments containing actin and myosin are anchored at one end to dense areas in the plasma membrane and at the other end, through dense bodies, to intermediate filaments. The contractile myofilaments are oriented with the long axis of the cell; when actin and myosin slide over one another during contraction, the cell shortens.

Although calcium ions regulate contraction in smooth muscle cells (figure 9.25), the role of Ca^{2+} in smooth muscle differs from that in skeletal muscle cells because there are no troponin molecules associated with actin fibers of smooth muscle cells. Calcium ions that enter the cytoplasm bind to a protein called **calmodulin** (kal-mod′ū-lin). Calmodulin molecules with Ca^{2+} bound to them activate an enzyme called **myosin kinase** (kī′nās), which transfers a phosphate group from ATP to light myosin molecules on the heads of myosin molecules. Cross-bridge formation occurs when myosin myofilaments have phosphate groups bound to them. The enzymes responsible for cross-bridge cycling are slower than the enzymes in skeletal muscle, resulting in slower cross-bridge formation. Once activated, cross-bridge formation has energy requirements very similar to those of cross-bridge formation in skeletal muscle fibers.

Relaxation of smooth muscle results because of the activity of another enzyme, called **myosin phosphatase** (fos′fă-tās). This enzyme removes the phosphate group from the myosin molecules (figure 9.25). If the phosphate is removed from myosin while the cross-bridges are attached to actin, the cross-bridges release very slowly. This explains how smooth muscle is able to sustain tension for long periods and without extensive energy expenditure. This period of sustained tension is often called the **latch state** of smooth muscle contraction. If myosin phosphatase removes the phosphate from myosin molecules while the cross-bridges are not attached, relaxation occurs much more rapidly.

Elevated Ca^{2+} levels in the sarcoplasm of smooth muscle cells result in the activation of myosin molecules and the formation of cross-bridge. Also, the action of myosin phosphatase results in a high percentage of myosin molecules having their phosphates removed while bound to actin. This process favors sustained contractions, or the latch state, and a low rate of energy consumption because of the slow release of cross-bridges. As long as Ca^{2+} is present, cross-bridges re-form quickly after being released. Consequently, many cross-bridges are intact at any given time in contracted smooth muscle.

Calcium ion levels in the sarcoplasm of smooth muscle are reduced as Ca^{2+} is actively transported across the plasma membrane, including the plasma membrane of caveolae, and into the sarcoplasmic reticulum. Muscles relax in response to lower intracellular levels of Ca^{2+}.

Types of Smooth Muscle

Smooth muscle can be either visceral or multiunit. **Visceral** (vis′er-ăl) **smooth muscle** (*unitary smooth muscle*) is the more common of the two types. It occurs in sheets and includes the smooth muscle of the digestive, reproductive, and urinary tracts. Visceral smooth muscle has numerous gap junctions (see chapter 4), which allow action potentials to pass directly from one cell to another. As a consequence, sheets of smooth muscle cells function as a unit, and a wave of contraction traverses the entire smooth muscle sheet. Visceral smooth muscle is often autorhythmic but in some areas it contracts only when stimulated. For example, visceral smooth muscle in the digestive tract contracts spontaneously and at relatively regular intervals, whereas visceral smooth muscle in the urinary bladder contracts when stimulated by the nervous system.

FUNDAMENTAL **Figure**

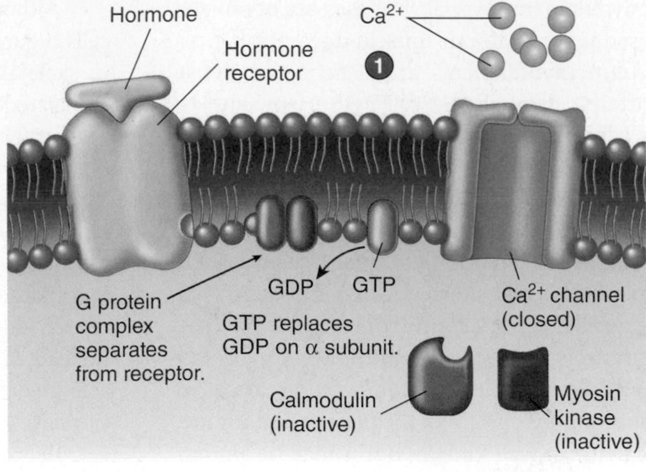

Hormone
Hormone receptor
Ca^{2+}
1
G protein complex separates from receptor.
GDP
GTP
GTP replaces GDP on α subunit.
Ca^{2+} channel (closed)
Calmodulin (inactive)
Myosin kinase (inactive)

1 A hormone combines with a hormone receptor and activates a G protein mechanism, or depolarization of the plasma membrane occurs.

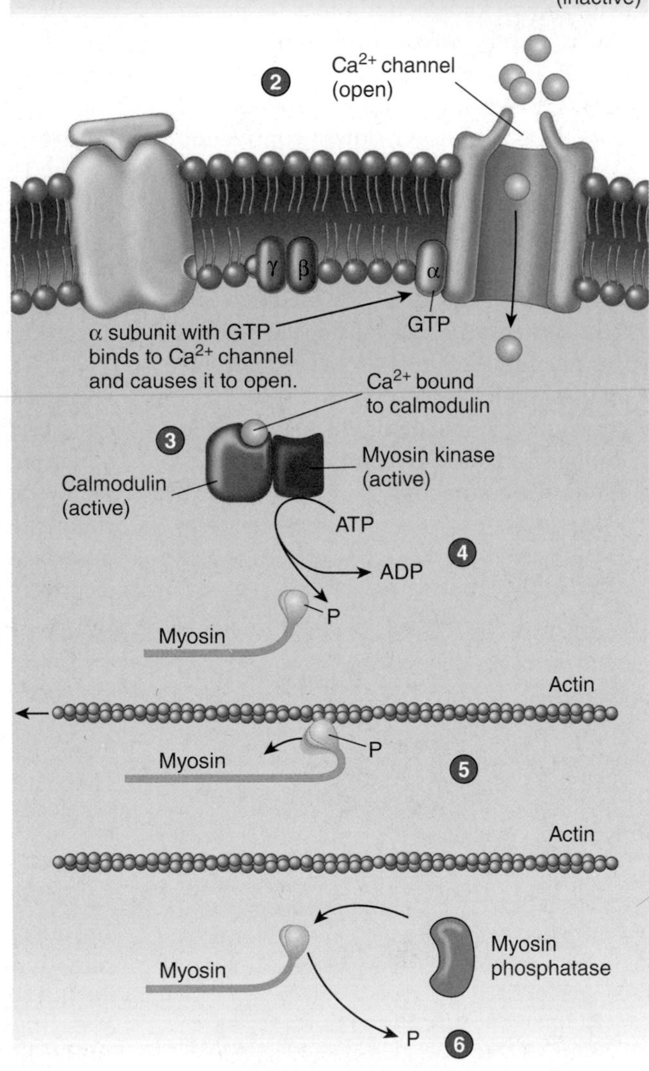

2
Ca^{2+} channel (open)
γ
β
α
α subunit with GTP binds to Ca^{2+} channel and causes it to open.
GTP
Ca^{2+} bound to calmodulin
3
Calmodulin (active)
Myosin kinase (active)
ATP
ADP
4
P
Myosin
Actin
P
Myosin
5
Actin
Myosin
Myosin phosphatase
P
6

2 An α subunit opens the Ca^{2+} channel in the plasma membrane, or depolarization opens Ca^{2+} channels. Calcium ions diffuse through the Ca^{2+} channels and combine with calmodulin.

3 Calmodulin with a Ca^{2+} bound to it binds with myosin kinase and activates it.

4 Activated myosin kinase attaches phosphate from ATP to myosin heads to activate the contractile process.

5 A cycle of cross-bridge formation, movement, detachment, and cross-bridge formation occurs.

6 Relaxation occurs when myosin phosphatase removes phosphate from myosin.

PROCESS FIGURE 9.25 Smooth Muscle Contraction

Multiunit smooth muscle occurs in various configurations: sheets, as in the walls of blood vessels; small bundles, as in the arrector pili muscles and the iris of the eye; and single cells, as in the capsule of the spleen. Multiunit smooth muscle has fewer gap junctions than visceral smooth muscle, and cells or groups of cells act as independent units. It normally contracts only when stimulated by nerves or hormones.

In visceral smooth muscle tissue, the arrangement between neurons and smooth muscle fibers differs from that in skeletal muscle tissue. The synapses are more diffuse than in skeletal muscle. Axons

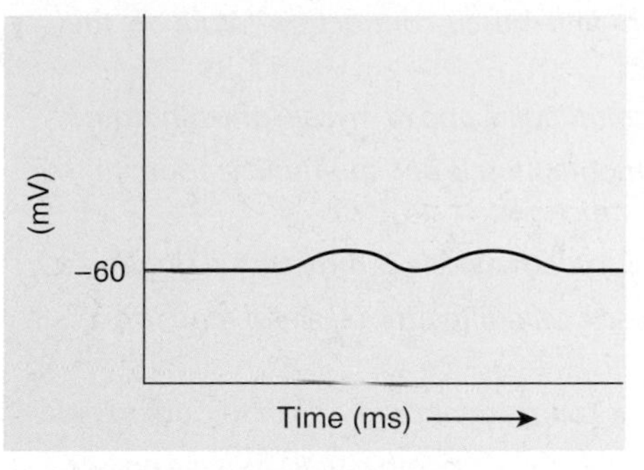
(mV)
–60
Time (ms)

(a) **Slow waves of depolarization**

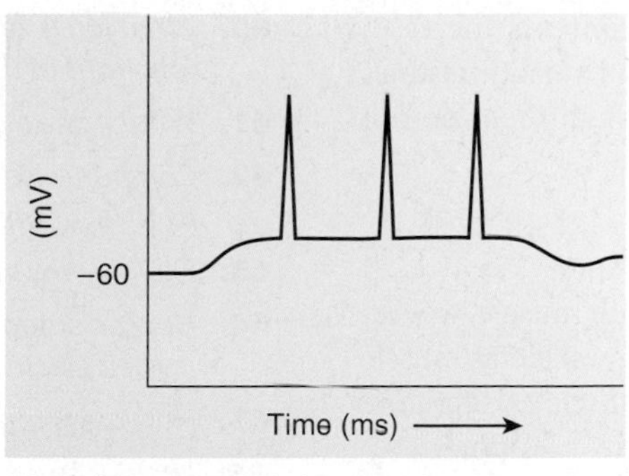
(mV)
–60
Time (ms)

(b) **Action potentials in smooth muscle superimposed on a slow wave of depolarization**

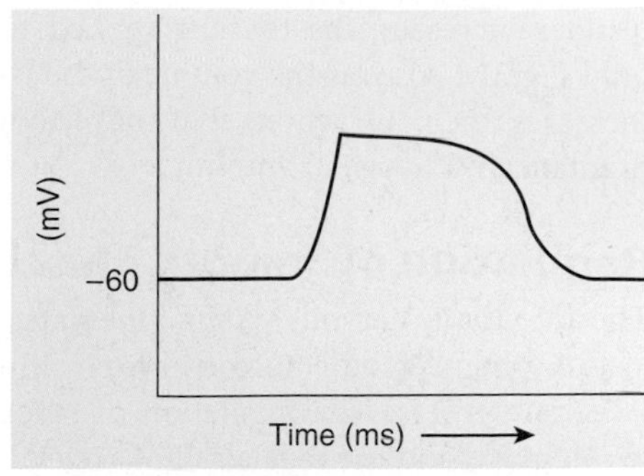
(mV)
–60
Time (ms)

(c) **Action potential with prolonged depolarization (plateau)**

FIGURE 9.26 Membrane Potentials in Smooth Muscle

of neurons terminate in a series of dilations along the branching axons within the connective tissue among the smooth muscle cells. These dilations have vesicles containing neurotransmitter molecules that, once released, diffuse among the smooth muscle cells and bind to receptors on their surfaces. Multiunit smooth muscle has synapses more like those found in skeletal muscle tissue.

Electrical Properties of Smooth Muscle

The resting membrane potential of smooth muscle cells is usually not as negative as that of skeletal muscle fibers. It normally ranges between −55 and −60 mV, compared with approximately −85 mV in skeletal muscle fibers. Furthermore, the resting membrane potential of many visceral smooth muscle cells fluctuates, with slow depolarization and repolarization phases. These slow waves of depolarization and repolarization are propagated from cell to cell for short distances (figure 9.26*a*). More "classic" action potentials can be triggered by the slow waves of depolarization and usually are propagated for longer distances (figure 9.26*b*). In addition, some smooth muscle types have action potentials with a plateau, or prolonged depolarization (figure 9.26*c*). The slow waves in the resting membrane potential may result from a spontaneous and progressive increase in the permeability of the plasma membrane to Na^+ and Ca^{2+}, or they may be controlled by neurons. Sodium ions and Ca^{2+} diffuse into the cell through their respective channels and produce the depolarization.

Smooth muscle does not respond in an all-or-none fashion to action potentials. A series of action potentials in smooth muscle can result in a single, slow contraction followed by slow relaxation instead of individual contractions in response to each action potential, as occurs in skeletal muscle. A slow wave of depolarization that has one to several more classic-appearing action potentials superimposed on it is common in many types of smooth muscle. After the wave of depolarization, the smooth muscle contracts. Action potentials with plateaus are common in smooth muscle that exhibits periods of sustained contraction.

Spontaneously generated action potentials that lead to contractions are characteristic of visceral smooth muscle in the uterus, the ureter, and the digestive tract. Certain smooth muscle cells in these organs function as **pacemaker cells,** which tend to develop action potentials more rapidly than other cells.

The nervous system can regulate smooth muscle contractions by increasing or decreasing action potentials carried by neuron axons to smooth muscle. Responses of smooth muscle cells result in depolarization and increased contraction or hyperpolarization and decreased contraction. The nervous system can also regulate the pacemaker cells.

Hormones and ligands produced locally in tissues can bind to receptors on some smooth muscle plasma membranes. The combination of a hormone or other ligands with a receptor causes ligand-gated Ca^{2+} channels in the plasma membrane to open (see figure 9.25, *step 2*). Calcium ions then enter the cell and cause smooth muscle contractions to occur without a major change in the membrane potential. For example, some smooth muscles contract when exposed to the hormone epinephrine, which combines with epinephrine receptors to activate G proteins in the plasma membrane (see figure 9.25, *step 2*). The α subunit of the G complex can produce intracellular mediator molecules, which open the ligand-gated Ca^{2+} channels in the plasma membrane or sarcoplasmic reticulum.

Predict 10

Explain how a ligand can bind to a membrane-bound receptor in a smooth muscle cell and cause sustained contraction for a prolonged period without a large increase in ATP breakdown.

Functional Properties of Smooth Muscle

Smooth muscle has four functional properties not seen in skeletal muscle: (1) Some visceral smooth muscle exhibits autorhythmic contractions; (2) smooth muscle tends to contract in response to being stretched, but a slow increase in length produces less response than a more rapid increase in length; (3) smooth muscle exhibits a relatively constant tension, called **smooth muscle tone,** over a long period and maintains that tension in response to a gradual increase in the smooth muscle length; (4) the amplitude of contraction produced by smooth muscle also remains constant, although the muscle length varies. Smooth muscle is therefore well adapted for lining the walls of hollow organs, such as the stomach and the urinary bladder. As the volume of the stomach or urinary bladder increases, the tension applied to its contents increases only slightly. Also, as the volume of the large and small intestines increases, the contractions that move food through them do not dramatically change in amplitude.

Regulation of Smooth Muscle

The autonomic nervous system innervates smooth muscle, whereas the somatic motor nervous system innervates skeletal muscle (see chapter 11). The regulation of smooth muscle is therefore involuntary, and the regulation of skeletal muscle is voluntary.

The most important neurotransmitters released from the nerves that innervate smooth muscle cells are acetylcholine and norepinephrine. Acetylcholine stimulates some smooth muscle types to contract and inhibits others.

Hormones are also important in regulating smooth muscle. Epinephrine, a hormone from the adrenal medulla, stimulates some smooth muscles, such as those in the blood vessels of the small intestine, and inhibits other smooth muscles, such as those in the intestinal wall. Oxytocin stimulates contractions of uterine smooth muscle, especially during childbirth. These and other hormones are discussed more thoroughly in chapters 17 and 18. Other chemical substances produced locally by surrounding tissues—such as histamine, prostaglandins, and by-products of metabolism—also influence smooth muscle function. For example, blood flow through capillaries is dramatically influenced by these substances (see chapter 21).

The type of receptors present on the plasma membrane to which the neurotransmitters or hormones bind determines the response of the smooth muscle. Some smooth muscle types have receptors to which acetylcholine binds, and the receptor responds by stimulating contractions; other smooth muscle types have receptors to which acetylcholine binds, and the receptor responds by inhibiting contractions. A similar relationship exists for smooth muscle receptors for norepinephrine and certain hormones.

The receptor molecules that stimulate smooth muscle contractions often open either Na^+ or Ca^{2+} channels. When these channels open, Na^+ and Ca^{2+} pass through their respective channels into the cell and cause depolarization of the plasma membrane. It is also possible for the receptor to open Ca^{2+} channels in the plasma membrane and sarcoplasmic reticulum. As a result, Ca^{2+} can diffuse into the cytoplasm of the smooth muscle cells without depolarization of the membrane potential to its threshold level and therefore not produce action potentials.

The receptor molecules that inhibit smooth muscle contractions often close Na^+ and Ca^{2+} channels or open K^+ channels. The result is hyperpolarization of the smooth muscle cells and inhibition. It is also possible for the receptors to increase the activity of the Ca^{2+} pump that transports Ca^{2+} out of the cell or into the sarcoplasmic reticulum. As a result, relaxation may occur without a change in the resting membrane potential.

The response of specific smooth muscle types to either neurotransmitters or hormones is presented in the chapters dealing with the smooth muscle types.

ASSESS YOUR PROGRESS 

59. *Describe a typical smooth muscle cell. How do its structure and its contraction process differ from those of skeletal muscle fibers?*
60. *What ion is key to smooth muscle contraction? What are the functions of this ion?*
61. *What is the role of calmodulin, and of myosin phosphatase?*
62. *Compare visceral smooth muscle and multiunit smooth muscle as to locations and structure.*
63. *Explain why visceral smooth muscle contracts as a single unit.*
64. *How do smooth muscle cells differ from skeletal muscle fibers in their electrical properties?*
65. *How are spontaneous contractions produced in smooth muscle?*
66. *List four functional properties of smooth muscle that are not seen in skeletal muscle. Can smooth muscle develop an oxygen deficit?*
67. *How do the nervous system and hormones regulate smooth muscle contraction?*
68. *How are ion channels affected by receptors that stimulate smooth muscle contractions? That inhibit smooth muscle contractions?*

9.11 Cardiac Muscle

LEARNING OUTCOME 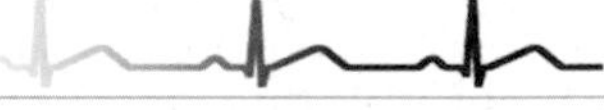

After reading this section, you should be able to

A. Discuss the structural and functional characteristics of cardiac muscle.

Cardiac muscle, which is found only in the heart, is discussed in detail in chapter 20. Like skeletal muscle tissue, cardiac muscle tissue is striated, but each cell usually contains one nucleus located near the center. Adjacent cells join to form branching fibers by specialized cell-to-cell attachments called **intercalated** (in-ter′kă-lā-ted) **disks,** which have gap junctions that allow action potentials to pass from cell to cell. Some cardiac muscle cells are autorhythmic, and one part of the heart normally acts as the pacemaker. The action potentials of cardiac muscle are similar to those of nerve and skeletal muscle but have a much longer duration and refractory (unresponsive) period. The depolarization of cardiac muscle results from the influx of both Na^+ and Ca^{2+} across the plasma membrane. The regulation of contraction in cardiac muscle by Ca^{2+} is similar to that of skeletal muscle.

ASSESS YOUR PROGRESS 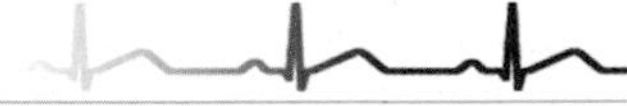

69. *Compare the structural and functional characteristics of cardiac muscle with those of skeletal muscle.*
70. *How is cardiac muscle similar to smooth muscle?*

9.12 Effects of Aging on Skeletal Muscle

LEARNING OUTCOME 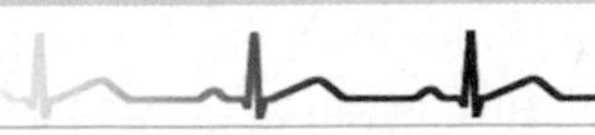

After reading this section, you should be able to

A. Describe the changes that occur in aging skeletal muscle.

Diseases and Disorders

Muscular System

Condition	Description
Cramps	Painful, spastic contractions of skeletal muscle; usually due to a buildup of lactic acid
Fibromyalgia (fī-brō-mī-al′ja)	Non-life-threatening, chronic, widespread pain in skeletal muscles with no known cure; also known as chronic muscle pain syndrome
Hypertrophy	Enlargement of skeletal muscle due to an increased number of myofibrils, as occurs with increased muscle use; in cardiac muscle, usually a result of other dieases, commonly hypertension
Atrophy	Decrease in muscle size due to a decreased number of myofilaments; can occur due to disuse of a muscle, as in paralysis; can also occur in cardiac muscle due to certain pathologies such as chronic heart failure
Muscular dystrophy	Group of genetic disorders in which muscles degenerate and atrophy; usually affects skeletal muscle and sometimes cardiac muscle
Duchenne muscular dystrophy	See Systems Pathology
Myotonic muscular dystrophy	Skeletal muscles are weak and fail to relax following forceful contractions; affects the hands most severely; dominant trait in 1/20,000 births
Myasthenia gravis	See Clinical Impact, "The Effect of Blocking Acetylcholine Receptors and Acetylcholinesterase," earlier in this chapter
Tendinitis (ten-di-nī′tis)	Inflammation of a tendon or its attachment point due to overuse of a skeletal muscle
Fibrosis	Scarring of damaged cardiac or skeletal muscle due to deposition of connective tissue
Fibrositis	Inflammation of fibrous connective tissue resulting in soreness after prolonged skeletal muscle tension; not progressive

Go to: www.mhhe.com/seeley10 for additional information on these pathologies.

A primary consequence of aging is **sarcopenia**, or muscle atrophy, the age-related reduction in muscle mass and regulation of muscle function. Aging skeletal muscle undergoes several changes that reduce muscle mass, increase the time muscle takes to contract in response to nervous stimuli, reduce stamina, and increase recovery time. Loss of muscle fibers begins as early as 25 years of age and, by age 80, the muscle mass has been reduced by approximately 50%, due primarily to the loss of muscle fibers. An important component of age-related loss of muscle mass is maintenance of independence in elderly people. In order to help delay sarcopenia, weight-lifting

Clinical GENETICS — Duchenne Muscular Dystrophy

Duchenne muscular dystrophy (DMD) is caused by mutations in the *dystrophin* (dis-trō′-fin) gene on the X chromosome. The dystrophin gene is responsible for producing a protein called **dystrophin.** Dystrophin plays a role in attaching actin myofilaments of myofibrils to, and regulating the activity of, other proteins in the sarcolemma (figure 9D). Dystrophin is thought to protect muscle fibers against mechanical stress in a normal individual. In DMD, part of the dystrophin gene is missing, and the protein it produces is nonfunctional. The lack of dystrophin results in progressive muscular weakness and muscle contractures. DMD is an X-linked recessive disorder. Thus, although females carry the gene, DMD affects males almost exclusively.

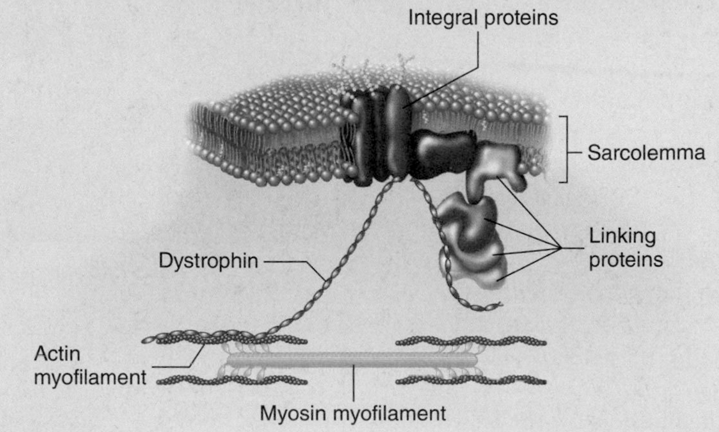
Integral proteins
Sarcolemma
Linking proteins
Dystrophin
Actin myofilament
Myosin myofilament

FIGURE 9D Role of Dystrophin
The protein dystrophin links actin myofilaments within myofibrils to integral proteins in the sarcolemma of the muscle fiber and to linking proteins.

Systems PATHOLOGY | Duchenne Muscular Dystrophy

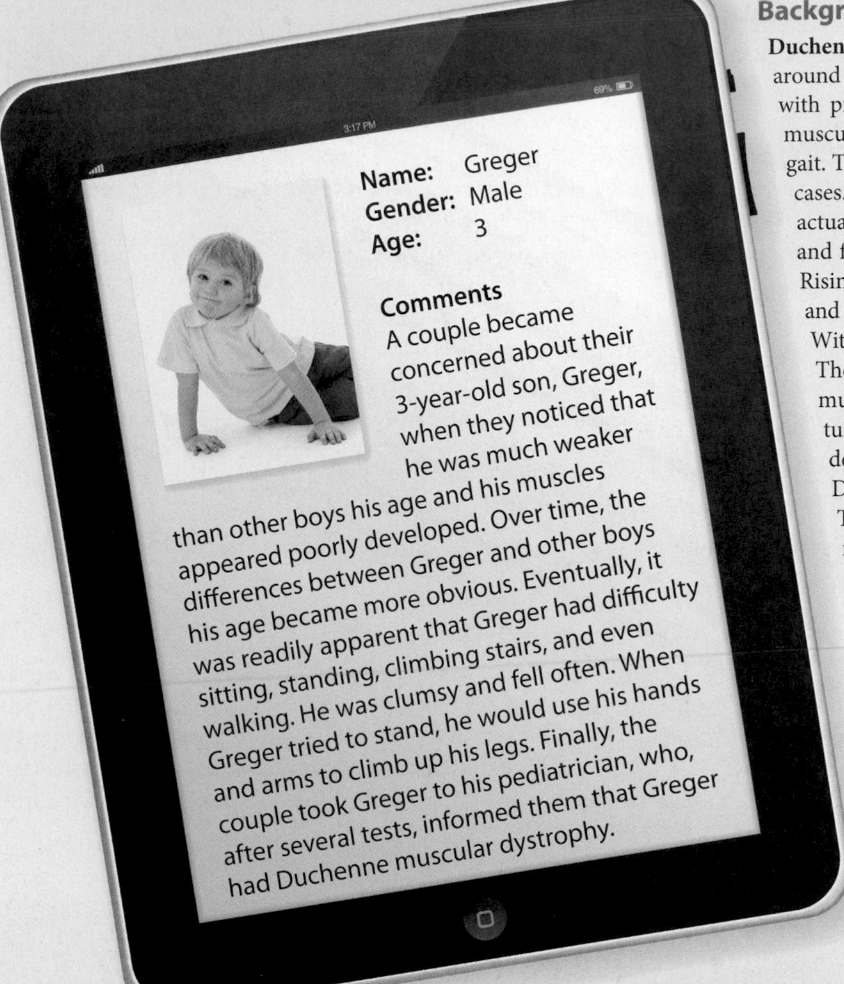
3:17 PM
69%
Name: Greger
Gender: Male
Age: 3
Comments
A couple became concerned about their 3-year-old son, Greger, when they noticed that he was much weaker than other boys his age and his muscles appeared poorly developed. Over time, the differences between Greger and other boys his age became more obvious. Eventually, it was readily apparent that Greger had difficulty sitting, standing, climbing stairs, and even walking. He was clumsy and fell often. When Greger tried to stand, he would use his hands and arms to climb up his legs. Finally, the couple took Greger to his pediatrician, who, after several tests, informed them that Greger had Duchenne muscular dystrophy.

Background Information

Duchenne muscular dystrophy (DMD) is usually identified in children around 3 years of age when their parents notice slow motor development with progressive weakness and muscle wasting (atrophy). Typically, muscular weakness begins in the hip muscles, which causes a waddling gait. Temporary enlargement of the calf muscles is apparent in 80% of cases. The enlargement is paradoxical because the muscle fibers are actually getting smaller, but the amount of fibrous connective tissue and fat between the muscle fibers is increasing (figures 9A and 9B). Rising from the floor by using the hands and arms is characteristic and is caused by weakness of the lumbar and hip muscles (figure 9C). Within 3 to 5 years, the muscles of the shoulder girdle become involved. The replacement of muscle with connective tissue contributes to muscular atrophy and shortened, inflexible muscles called contractures. The contractures limit movements and can cause severe deformities of the skeleton. By 10 to 12 years of age, people with DMD are usually unable to walk, and few live beyond age 20. There is no effective treatment to prevent the progressive deterioration of muscles in DMD. Therapy primarily involves exercises to help strengthen muscles and prevent contractures .

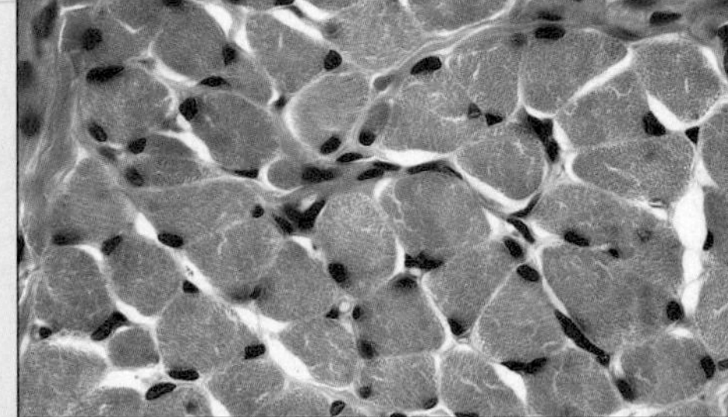

Figure 9A Normal Muscle Tissue

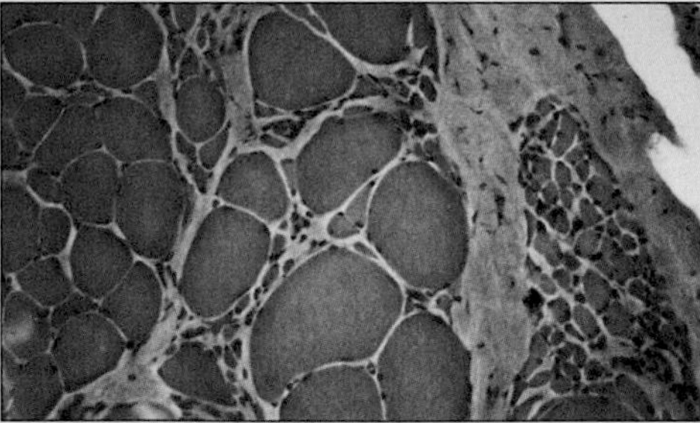

Figure 9B DMD Muscle Tissue

Figure 9C A DMD Patient

exercises are helpful to slow the loss of muscle mass but do not prevent the loss of muscle fibers. In addition, fast-twitch muscle fibers decrease in number more rapidly than slow-twitch fibers. Most of the lost strength and speed is due to the loss of fast-twitch muscle fibers.

In addition, the surface area of the neuromuscular junction decreases; as a result, action potentials in neurons stimulate action potentials in muscle fibers more slowly, so fewer action potentials are produced in muscle fibers. The number of motor neurons also decreases. Some of the muscle fibers that lose their innervation when a neuron dies are reinnervated by a branch of another motor neuron. This makes motor units in skeletal muscle fewer in number, with a greater number of muscle fibers for each neuron, which may result in less precise muscle control. Aging is also associated

Duchenne Muscular Dystrophy

Symptoms

- Muscle weakness
- Muscle atrophy
- Contractures

Treatment

- Physical therapy to prevent contractures
- No effective treatment to prevent atrophy

SKELETAL

Shortened, inflexible muscles (contractures) cause severe skeletal deformities. Curvature of the spinal column laterally and anteriorly (kyphoscoliosis) can be so severe that normal respiratory movements are impaired. Surgery is sometimes required to prevent contractures from making it impossible for the individual to sit in a wheelchair.

DIGESTIVE

Smooth muscle tissue is affected by DMD, and the reduced ability of smooth muscle to contract can result in disorders of the digestive system, including enlarged colon diameter and twisting of the small intestine that leads to intestinal obstruction, cramping, and reduced absorption of nutrients.

NERVOUS

Some degree of mental retardation occurs in a large percentage of people with DMD, although the specific cause is unknown. Research suggests that dystrophin is important in the formation of synapses between neurons.

LYMPHATIC AND IMMUNE

Although the lymphatic system is not directly affected, damaged muscle fibers are phagocytized by macrophages.

URINARY

Reduced smooth muscle function and wheelchair dependency increase the frequency of urinary tract infections.

CARDIOVASCULAR

Cardiac muscle is affected by DMD; consequently, heart failure occurs in many patients with advanced DMD. Cardiac involvement becomes serious in as many as 95% of cases and is one of the leading causes of death for DMD patients.

RESPIRATORY

Deformity of the thorax and increasing weakness of the respiratory muscles result in inadequate respiratory movements, which cause an increase in respiratory infections, such as pneumonia. Insufficient movement of air into and out of the lungs due to weak respiratory muscles is a major contributing factor in many deaths.

Predict 11

A boy with advanced Duchenne muscular dystrophy developed pulmonary edema (accumulation of fluid in the lungs) and pneumonia caused by a bacterial infection. His physician diagnosed the condition in the following way: The pulmonary edema was the result of heart failure, and the increased fluid in the lungs provided a site where bacteria could invade and grow. The fact that the boy could not breathe deeply or cough effectively made the condition worse. How would the muscle tissues in a boy with advanced DMD differ from the muscle tissues in a boy with less advanced DMD?

with decreased density of capillaries in skeletal muscles, so that a longer recovery period is required after exercise.

Many of the age-related changes in skeletal muscle can be slowed dramatically if people remain physically active instead of assuming a sedentary lifestyle. Studies show that elderly people who are sedentary can become stronger and more mobile in response to exercise. The effects of aging on people with large muscle mass are further discussed in Chapter 10 (see Clinical Impact, "Bodybuilding").

ASSESS YOUR PROGRESS

71. *Describe the changes in muscle mass and response time that occur in aging skeletal muscle.*

Learn to Predict ❮ From page 265

Answer

We must address three issues: (1) Identify the bones involved when Bob is performing the exercise, (2) describe the joint movements involved, and (3) explain how Bob's muscles lift and lower the weight slowly. We know that Bob's movement engages the bones and muscles of the arm and forearm and that Bob is actively controlling the movement of the weight as he lifts the weight and lowers it.

Chapter 7 explained that the bones of the arm and forearm are the humerus, the radius, and the ulna. These are the bones involved in the exercise. In chapter 8 we learned that these bones articulate at the elbow. Chapter 8 also described the different types of joint movements, including flexion and extension. Flexion of his elbow causes Bob to lift the weight, and extension of his elbow allows him to lower it.

We learned in chapter 9 that muscle tension can vary depending on the number of motor units stimulated. Lifting the weight requires recruitment of motor units. As the number of stimulated motor units increases, the amount of tension produced by the muscle also increases until sufficient force is produced to lift the weight (multiple-motor-unit summation). The speed of the contraction depends on the rate of motor unit recruitment. Because the rate of recruitment is slow, Bob lifts the weight slowly. If the rate of motor unit recruitment were increased, Bob would lift the weight faster (multiple-wave summation). As Bob lowers the weight, the total tension in the arm is reduced as fewer motor units contract.

Answers to the rest of this chapter's Predict questions are in Appendix G.

Summary

9.1 Functions of the Muscular System (p. 266)

Muscles are responsible for movement of the arms, legs, heart, and other parts of the body; maintenance of posture; respiration; production of body heat; communication; constriction of organs and vessels; and heartbeat.

9.2 General Properties of Muscle (p. 266)

1. Muscle exhibits contractility (shortens forcefully), excitability (responds to stimuli), extensibility (can be stretched and still contract), and elasticity (recoils to resting length).
2. Muscle tissue shortens forcefully but lengthens passively.

Types of Muscle Tissue

1. The three types of muscle are skeletal, smooth, and cardiac.
2. Skeletal muscle is responsible for most body movements; smooth muscle is found in the walls of hollow organs and tubes and moves substances through them; and cardiac muscle is in the heart and pumps blood.

9.3 Skeletal Muscle Structure (p. 267)

1. Skeletal muscle fibers are associated with connective tissue, blood vessels, and nerves.
2. Muscle fasciculi, bundles of muscle fibers, are covered by the connective tissue layer called the perimysium.
3. The entire muscle is surrounded by a connective tissue layer called the epimysium.

Nerves and Blood Vessels

1. Motor neurons extend together with arteries and veins through the connective tissue of skeletal muscles.
2. At the level of the perimysium, axons of motor neurons branch, and each branch projects to a muscle fiber to form a neuromuscular junction.

Skeletal Muscle Fibers

1. A muscle fiber is a single cell consisting of a plasma membrane (sarcolemma), cytoplasm (sarcoplasm), several nuclei, and myofibrils.
2. Myofibrils are composed of two major protein fibers: actin and myosin.
   - Actin myofilaments consist of a double helix of F actin (composed of G actin monomers), tropomyosin, and troponin.
   - Myosin molecules, consisting of two globular heads and a rodlike portion, constitute myosin myofilaments.
   - A cross-bridge forms when the myosin binds to the actin.
3. Actin and myosin are organized to form sarcomeres.
   - Sarcomeres are bound by Z disks that hold actin myofilaments.
   - Six actin myofilaments (thin filaments) surround a myosin myofilament (thick filament).
   - Myofibrils appear striated because of A bands and I bands.

9.4 Physiology of Skeletal Muscle Fibers (p. 273)

Sliding Filament Model

1. Actin and myosin myofilaments do not change in length during contraction.
2. Actin and myosin myofilaments slide past one another in a way that causes sarcomeres to shorten.
3. The I band and H zones become narrower during contraction, and the A band remains constant in length.

Ion Channels

1. Ion channels are responsible for membrane permeability and the resting membrane potential.
2. Two types of membrane channels produce action potentials: ligand-gated and voltage-gated channels.
3. Ion channels are responsible for producing action potentials.

Membrane Potentials

Plasma membranes are polarized, which means that a charge difference, called the resting membrane potential, exists across the plasma membrane. The membrane becomes polarized because the tendency for K^+ to diffuse out of the cell is resisted by the negative charges of ions and molecules inside the cell.

Action Potentials

1. The charge difference across the plasma membrane of cells is the resting membrane potential.
2. Depolarization results from an increase in the permeability of the plasma membrane to Na^+.
3. An all-or-none action potential is produced if depolarization reaches threshold.
4. The depolarization phase of the action potential results when many Na^+ channels open in an all-or-none fashion.
5. The repolarization phase of the action potential occurs when the Na^+ channels close and the K^+ channels open briefly.
6. Action potentials propagate along the plasma membranes of neurons and skeletal muscle fibers in an all-or-none fashion.

Neuromuscular Junction

1. A synaptic cleft separates the presynaptic terminal of the axon from the postsynaptic membrane of the muscle fiber.
2. Acetylcholine released from the presynaptic terminal binds to receptors of the postsynaptic membrane, thereby changing membrane permeability and producing an action potential.
3. After an action potential occurs, acetylcholinesterase splits acetylcholine into acetic acid and choline. Choline is reabsorbed into the presynaptic terminal to re-form acetylcholine.

Excitation-Contraction Coupling

1. Invaginations of the sarcolemma form T tubules, which wrap around the sarcomeres.
2. A triad is a T tubule and two terminal cisternae (an enlarged area of sarcoplasmic reticulum).
3. Action potentials move into the T tubule system, causing Ca^{2+} channels to open and release Ca^{2+} from the sarcoplasmic reticulum.
4. Calcium ions diffuse from the sarcoplasmic reticulum to the myofilaments and bind to troponin, causing tropomyosin to move and expose active sites on actin to myosin.
5. Contraction occurs when myosin heads bind to active sites on actin, myosin changes shape, and actin is pulled past the myosin.
6. Relaxation occurs when calcium is taken up by the sarcoplasmic reticulum, ATP binds to myosin, and tropomyosin moves back so that active sites on actin are no longer exposed to myosin.

Cross-Bridge Movement

1. ATP is required for the cycle of cross-bridge formation, movement, and release.
2. ATP is also required to transport Ca^{2+} into the sarcoplasmic reticulum and to maintain normal concentration gradients across the plasma membrane.

Muscle Relaxation

1. Calcium ions are transported into the sarcoplasmic reticulum.
2. Calcium ions diffuse away from troponin, preventing further cross-bridge formation.

9.5 Physiology of Skeletal Muscle (p. 285)

Muscle Twitch

1. A muscle twitch is the contraction of a single muscle fiber or a whole muscle in response to a stimulus.
2. A muscle twitch has lag, contraction, and relaxation phases.

Motor Units

1. A motor unit is one motor neuron and all the muscle fibers it controls.
2. Precise movements use small motor units. Gross movements use large motor units.

Stimulus Strength and Motor Unit Response

1. A stimulus of increasing magnitude results in graded contractions of increased force through either summation or recruitment.
2. Treppe is an increase in the force of contraction during the first few contractions of a rested muscle.
3. The force of contraction of a whole muscle increases with increased frequency of stimulation because of an increasing concentration of Ca^{2+} around the myofibrils and because of complete stretching of muscle elastic elements.

Stimulus Frequency and Whole Muscle Contraction

1. Incomplete tetanus is partial relaxation between contractions; complete tetanus is no relaxation between contractions.
2. A stimulus of increasing frequency increases the force of contraction (multiple-wave summation).

Muscle Length Versus Tension

1. Muscle contracts with less than maximum force if its initial length is shorter or longer than optimum.
2. Isometric contractions cause a change in muscle tension but no change in muscle length.
3. Isotonic contractions cause a change in muscle length but no change in muscle tension.
4. Concentric contractions cause muscles to shorten and tension to increase.
5. Eccentric contractions cause muscle to lengthen and tension to decrease gradually.
6. Muscle tone is the maintenance of steady tension for long periods.
7. Asynchronous contractions of motor units produce smooth, steady muscle contractions.

9.6 Muscle Fatigue (p. 291)

1. Fatigue, the decreased ability to do work, can be caused by the central nervous system, depletion of ATP in muscles, or depletion of acetylcholine in the neuromuscular junction.
2. Physiological contracture (the inability of muscles to contract or relax) and rigor mortis (stiff muscles after death) result from inadequate amounts of ATP.

9.7 Energy Sources (p. 291)

Energy for muscle contraction comes from ATP.

Creatine Phosphate

ATP synthesized when ADP reacts with creatine phosphate provides energy for a short time during intense exercise.

Anaerobic Respiration

The ATP synthesized by anaerobic respiration provides energy for a short time during intense exercise. Anaerobic respiration produces ATP less efficiently but more rapidly than aerobic respiration. Lactic acid levels increase because of anaerobic respiration.

Aerobic Respiration

The ATP synthesized by aerobic respiration produces energy for muscle contractions under resting conditions or during exercises such as long-distance running. Although ATP is produced more efficiently, it is produced more slowly.

Oxygen Deficit and Recovery Oxygen Consumption

After anaerobic respiration, aerobic respiration is higher than normal, as the imbalances of homeostasis that occurred during exercise become rectified.

9.8 Slow-Twitch and Fast-Twitch Fibers (p. 294)

Slow-Twitch Oxidative Muscle Fibers

Slow-twitch muscle fibers split ATP slowly and have a well-developed blood supply, many mitochondria, and myoglobin.

Fast-Twitch Muscle Fibers

Fast-twitch muscle fibers split ATP rapidly.

1. Type IIa fibers have a well-developed blood supply, more mitochondria, and more myoglobin.
2. Type IIb fibers have large amounts of glycogen, a poor blood supply, fewer mitochondria, and little myoglobin.

Distribution of Fast-Twitch and Slow-Twitch Muscle Fibers

People who are good sprinters have a greater percentage of fast-twitch muscle fibers in their leg muscles, and people who are good long-distance runners have a higher percentage of slow-twitch muscle fibers.

Effects of Exercise

1. Muscles increase (hypertrophy) or decrease (atrophy) in size because of a change in the size of muscle fibers.
2. Anaerobic exercise develops type IIb fibers. Aerobic exercise develops type I fibers and changes type IIb fibers into type IIa fast-twitch fibers.

9.9 Heat Production (p. 296)

1. Heat is a by-product of chemical reactions in muscles.
2. Shivering produces heat to maintain body temperature.

9.10 Smooth Muscle (p. 296)

1. Smooth muscle cells are spindle-shaped with a single nucleus. They have actin myofilaments and myosin myofilaments but are not striated.
2. The sarcoplasmic reticulum is poorly developed, and caveolae may function as a T tubule system.
3. Calcium ions enter the cell to initiate contraction; calmodulin binds to Ca^{2+} and activates an enzyme that transfers a phosphate group from ATP to myosin. When phosphate groups are attached to myosin, cross-bridges form.
4. Relaxation results when myosin phosphatase removes a phosphate group from the myosin molecule.
   - If phosphate is removed while the cross-bridges are attached, relaxation occurs very slowly, and this is referred to as the latch state.
   - If phosphate is removed while the cross-bridges are not attached, relaxation occurs rapidly.

Types of Smooth Muscle

1. Visceral smooth muscle fibers contract slowly, have gap junctions (and thus function as a single unit), and can be autorhythmic.
2. Multiunit smooth muscle fibers contract rapidly in response to stimulation by neurons and function independently.

Electrical Properties of Smooth Muscle

1. Spontaneous contractions result from Na^+ and Ca^{2+} leakage into cells; Na^+ and Ca^{2+} movement into the cell is involved in depolarization.
2. The autonomic nervous system, hormones, and chemicals produced locally can inhibit or stimulate action potentials (and thus contractions). Hormones can also stimulate or inhibit contractions without affecting membrane potentials.

Functional Properties of Smooth Muscle

1. Smooth muscle can contract autorhythmically in response to stretch or when stimulated by the autonomic nervous system or hormones.
2. Smooth muscle maintains a steady tension for long periods.
3. The force of smooth muscle contraction remains nearly constant, despite changes in muscle length.
4. Smooth muscle does not develop an oxygen deficit.

Regulation of Smooth Muscle

1. Smooth muscle is innervated by the autonomic nervous system and is involuntary.
2. Hormones are important in regulating smooth muscle. Certain hormones can increase the Ca^{2+} permeability of some smooth muscle membranes and therefore cause contraction without a change in the resting membrane potential.

9.11 Cardiac Muscle (p. 300)

Cardiac muscle fibers are striated, have a single nucleus, are connected by intercalated disks (and thus function as a single unit), and are capable of autorhythmicity.

9.12 Effects of Aging on Skeletal Muscle (p. 300)

Aging skeletal muscle is associated with reduced muscle mass, increased time that muscle takes to contract in response to nervous stimuli, less precise muscle control, and a longer recovery period.

REVIEW AND COMPREHENSION

1. Which of these is true of skeletal muscle?
   a. spindle-shaped cells
   b. under involuntary control
   c. many peripherally located nuclei per muscle fiber
   d. forms the walls of hollow internal organs
   e. may be autorhythmic
2. Which of these is *not* a major property of muscle?
   a. contractility
   b. elasticity
   c. excitability
   d. extensibility
   e. secretability
3. Given these structures:
   (1) whole muscle
   (2) muscle fiber (cell)
   (3) myofilament
   (4) myofibril
   (5) muscle fasciculus

   Choose the arrangement that lists the structures in the correct order from largest to smallest.
   a. 1,2,5,3,4
   b. 1,2,5,4,3
   c. 1,5,2,3,4
   d. 1,5,2,4,3
   e. 1,5,4,2,3

4. Each myofibril
   a. is made up of many muscle fibers.
   b. contains sarcoplasmic reticulum.
   c. is made up of many sarcomeres.
   d. contains T tubules.
   e. is the same thing as a muscle fiber.
5. Myosin myofilaments are
   a. attached to the Z disk.
   b. found primarily in the I band.
   c. thinner than actin myofilaments.
   d. absent from the H zone.
   e. attached to filaments that form the M line.
6. Which of these statements about the molecular structure of myofilaments is true?
   a. Tropomyosin has a binding site for Ca^{2+}.
   b. The head of the myosin molecule binds to an active site on G actin.
   c. ATPase is found on troponin.
   d. Troponin binds to the rodlike portion of myosin.
   e. Actin molecules have a hingelike portion, which bends and straightens during contraction.
7. The part of the sarcolemma that invaginates into the interior of skeletal muscle fibers is the
   a. T tubule system.
   b. sarcoplasmic reticulum.
   c. myofibrils.
   d. terminal cisternae.
   e. mitochondria.
8. During the depolarization phase of an action potential, the permeability of the plasma membrane to
   a. Ca^{2+} increases.
   b. Na^+ increases.
   c. K^+ increases.
   d. Ca^{2+} decreases.
   e. Na^+ decreases.
9. During repolarization of the plasma membrane,
   a. Na^+ move to the inside of the cell.
   b. Na^+ move to the outside of the cell.
   c. K^+ move to the inside of the cell.
   d. K^+ move to the outside of the cell.
10. Given these events:
    (1) Acetylcholine is broken down into acetic acid and choline.
    (2) Acetylcholine diffuses across the synaptic cleft.
    (3) An action potential reaches the terminal branch of the motor neuron.
    (4) Acetylcholine combines with a ligand-gated ion channel.
    (5) An action potential is produced on the muscle fiber's plasma membrane.

    Choose the arrangement that lists the events in the order they occur at a neuromuscular junction.
    a. 2,3,4,1,5
    b. 3,2,4,5,1
    c. 3,4,2,1,5
    d. 4,5,2,1,3
    e. 5,1,2,4,3
11. Acetylcholinesterase is an important molecule in the neuromuscular junction because it
    a. stimulates receptors on the presynaptic terminal.
    b. synthesizes acetylcholine from acetic acid and choline.
    c. stimulates receptors within the postsynaptic membrane.
    d. breaks down acetylcholine.
    e. causes the release of Ca^{2+} from the sarcoplasmic reticulum.
12. Given these events:
    (1) The sarcoplasmic reticulum releases Ca^{2+}.
    (2) The sarcoplasmic reticulum takes up Ca^{2+}.
    (3) Calcium ions diffuse into the sarcoplasm.
    (4) An action potential moves down the T tubule.
    (5) The sarcomere shortens.
    (6) The muscle relaxes.

    Choose the arrangement that lists the events in the order they occur following a single stimulation of a skeletal muscle fiber.
    a. 1,3,4,5,2,6
    b. 2,3,5,4,6,1
    c. 4,1,3,5,2,6
    d. 4,2,3,5,1,6
    e. 5,1,4,3,2,6
13. Given these events:
    (1) Calcium ions combine with tropomyosin.
    (2) Calcium ions combine with troponin.
    (3) Tropomyosin pulls away from actin.
    (4) Troponin pulls away from actin.
    (5) Tropomyosin pulls away from myosin.
    (6) Troponin pulls away from myosin.
    (7) Myosin binds to actin.

    Choose the arrangement that lists the events in the order they occur during muscle contraction.
    a. 1,4,7
    b. 2,5,6
    c. 1,3,7
    d. 2,4,7
    e. 2,3,7
14. With stimuli of *increasing strength*, which of these is capable of a graded response?
    a. neuron axon
    b. muscle fiber
    c. motor unit
    d. whole muscle
15. Considering the force of contraction of a skeletal muscle fiber, multiple-wave summation occurs because of
    a. increased strength of action potentials on the plasma membrane.
    b. a decreased number of cross-bridges formed.
    c. an increase in Ca^{2+} concentration around the myofibrils.
    d. an increased number of motor units recruited.
    e. increased permeability of the sarcolemma to Ca^{2+}.
16. Which of these events occurs during the lag (latent) phase of muscle contraction?
    a. cross-bridge movement
    b. active transport of Ca^{2+} into the sarcoplasmic reticulum
    c. Ca^{2+} binding to troponin
    d. sarcomere shortening
    e. breakdown of ATP to ADP
17. A weight lifter attempts to lift a weight from the floor, but the weight is so heavy that he is unable to move it. The type of muscle contraction the weight lifter is using is mostly
    a. isometric.
    b. isotonic.
    c. isokinetic.
    d. concentric.
    e. eccentric.
18. Which of these types of fatigue is the most common?
    a. muscular fatigue
    b. psychological fatigue
    c. synaptic fatigue
    d. army fatigue
19. Given these conditions:
    (1) low ATP levels
    (2) little or no transport of Ca^{2+} into the sarcoplasmic reticulum
    (3) release of cross-bridges
    (4) Na^+ accumulation in the sarcoplasm
    (5) formation of cross-bridges

    Choose the conditions that occur in both physiological contracture and rigor mortis.
    a. 1,2,3
    b. 1,2,5
    c. 1,2,3,4
    d. 1,2,4,5
    e. 1,2,3,4,5
20. Jerry Jogger's 3-mile run every morning takes about 30 minutes. Which of these sources provides most of the energy for his run?
    a. aerobic respiration
    b. anaerobic respiration
    c. creatine phosphate
    d. stored ATP

21. Which of these conditions would you expect to find within the leg muscle fibers of a world-class marathon runner?
    a. myoglobin-poor
    b. contract very quickly
    c. primarily anaerobic
    d. numerous mitochondria
22. Which of these increases the least as a result of muscle hypertrophy?
    a. number of sarcomeres
    b. number of myofibrils
    c. number of fibers
    d. blood vessels and mitochondria
    e. connective tissue
23. Relaxation in smooth muscle occurs when
    a. myosin kinase attaches phosphate to the myosin head.
    b. Ca^{2+} bind to calmodulin.
    c. myosin phosphatase removes phosphate from myosin.
    d. Ca^{2+} channels open.
    e. Ca^{2+} are released from the sarcoplasmic reticulum.
24. Compared with skeletal muscle, visceral smooth muscle
    a. has the same ability to be stretched.
    b. loses the ability to contract forcefully when stretched.
    c. maintains about the same tension, even when stretched.
    d. cannot maintain long, steady contractions.
    e. can accumulate a substantial oxygen deficit.
25. Which of these statements concerning aging and skeletal muscle is correct?
    a. A loss of muscle fibers occurs with aging.
    b. Slow-twitch fibers decrease in number faster than fast-twitch fibers.
    c. The loss of strength and speed is due mainly to loss of neuromuscular junctions.
    d. The density of capillaries in skeletal muscle increases.
    e. The number of motor neurons remains constant.

Answers in Appendix E

CRITICAL THINKING

1. Bob Canner improperly canned some homegrown vegetables. After eating the vegetables, he contracted botulism poisoning with symptoms that included difficulty swallowing and breathing. Eventually, he died of respiratory failure (his respiratory muscles relaxed and would not contract). Assuming that botulism toxin affects the neuromuscular junction, propose the ways that the toxin produces the observed symptoms.
2. A patient is thought to be suffering from either muscular dystrophy or myasthenia gravis. How would you distinguish between the two conditions?
3. Design an experiment to test the following hypothesis: Muscle A has the same number of motor units as muscle B. (Assume that you can stimulate the nerves that innervate skeletal muscles with an electronic stimulator and monitor the tension produced by the muscles.)
4. Explain what is happening at the level of individual sarcomeres when a person is using his or her biceps brachii muscle to hold a weight in a constant position. Contrast this with what is happening at the level of individual sarcomeres when a person lowers the weight, as well as when he or she raises the weight.
5. Predict the shape of an active tension curve for visceral smooth muscle. How does it differ from the active tension curve for skeletal muscle?
6. A researcher is investigating the composition of muscle tissue in the gastrocnemius muscles (in the calf of the leg) of athletes. She takes a needle biopsy from the muscle and determines the concentration (or enzyme activity) of several substances. Describe the major differences this researcher sees when comparing the muscles of athletes who perform in the following events: 100-meter dash, weight lifting, and 10,000-meter run.
7. Shorty McFleet noticed that his rate of respiration was elevated after running a 100-meter race but was not as elevated after running slowly for a much longer distance. How would you explain this?
8. High blood K^+ concentrations cause depolarization of the resting membrane potential. Predict and explain the effect of high blood K^+ levels on smooth muscle function.
9. Predict and explain the response if the ATP concentration in a muscle that was exhibiting rigor mortis could be instantly increased.
10. A hormone stimulates the smooth muscle of a blood vessel to contract. Although the hormone causes a small change in membrane potential, the smooth muscle contracts substantially. Explain.
11. Experiments were performed in an anatomy and physiology laboratory. First, the rate and depth of respiration for a resting student were determined. In experiment A, students ran in place for 30 seconds, immediately sat down and relaxed, and then had their respiration rate and depth measured. Experiment B was conducted in the same manner as experiment A, except that the students held their breath while running in place. What differences in respiration would you expect for the two experiments? Explain the basis for your predictions.
12. After learning about muscle fiber types in his anatomy and physiology class, Alex started to notice differences in the color of the turkey meat he ate for lunch. Some of the meat was very white and some of it was much darker. From the color of the meat, Alex guessed which muscles the bird used for maintenance of posture and/or slow movements, such as walking, and which muscles it used for quicker movements, such as running or flying. What type of muscle fiber predominates in white meat? In dark meat? Explain how the color of the meat relates to the function of the muscle.

Answers in Appendix F

Visit this book's website at **www.mhhe.com/seeley10** for chapter quizzes, interactive learning exercises, and other study tools.

Mc Graw Hill
connect plus+
ANATOMY & PHYSIOLOGY

Anatomy & Physiology
REVEALED
aprevealed.com

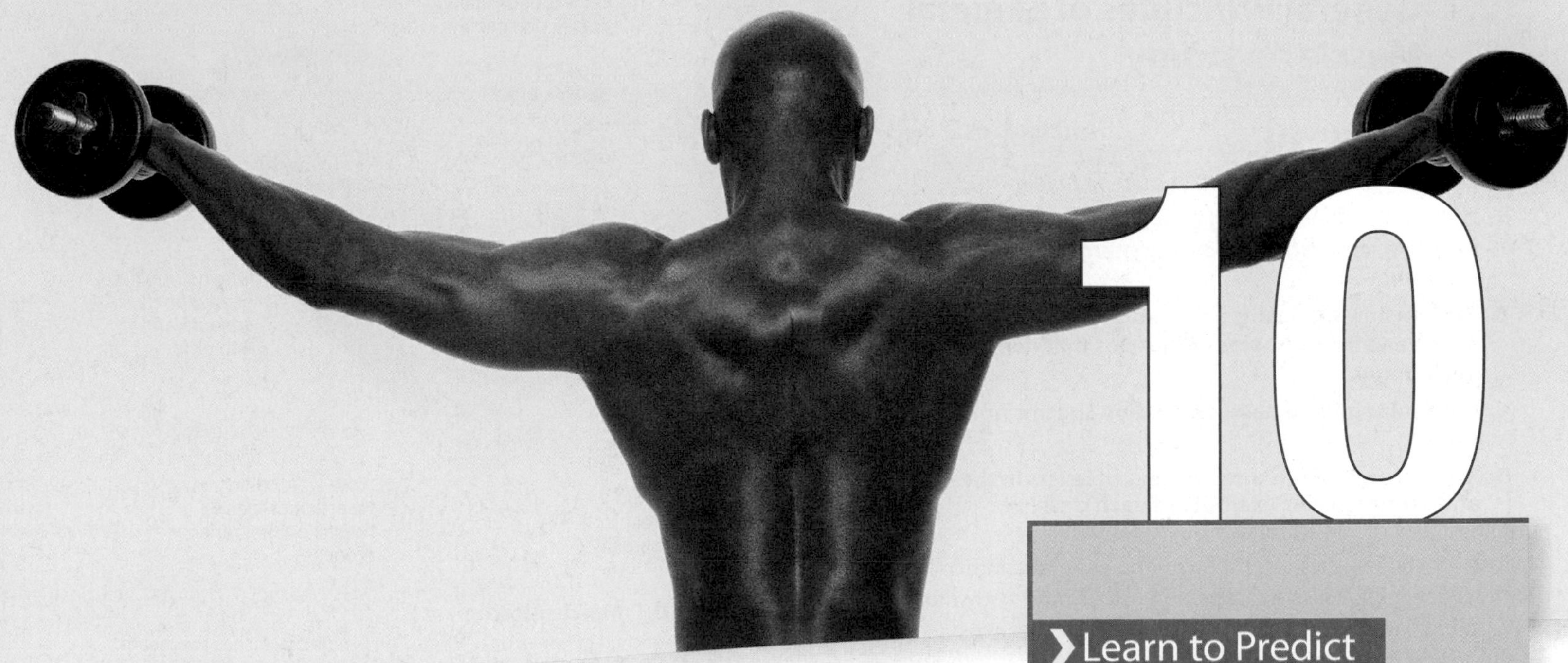
10

Muscular System

GROSS ANATOMY

Without muscles, we humans would be little more than department store mannequins—unable to walk, talk, blink our eyes, or even hold this book. But none of these inconveniences would bother us for long because we would also not be able to breathe.

One of the major characteristics of living human beings is our ability to move about. But we also use our skeletal muscles when we are not "moving." Postural muscles are constantly contracting to keep us sitting or standing upright. Respiratory muscles are constantly functioning to keep us breathing, even while we are asleep. Communication of all kinds requires skeletal muscles, whether for writing, typing, or speaking. Even silent communication using hand signals or facial expressions requires skeletal muscle function.

This chapter focuses on the anatomy of the major named skeletal muscles; cardiac muscle is considered in more depth in later chapters. The physiology of skeletal and smooth muscle was described in chapter 9, including the effects of aging on skeletal muscle.

Learn to Predict

While weight training, Pedro strained his back and damaged a vertebral disk. The bulged disk placed pressure on the left side of the spinal cord, compressing the third lumbar spinal nerve, which innervates the following muscles: psoas major, iliacus, pectineus, sartorius, vastus lateralis, vastus medius, vastus intermedius, and rectus femoris. As a result, action potential conduction to these muscles was reduced. Using your new knowledge about the histology and physiology of the muscular system from chapter 9 and combining it with the information about gross muscle anatomy in this chapter, predict Pedro's symptoms and which movements of his lower limb were affected, other than walking on a flat surface. What types of daily tasks would be difficult for Pedro to perform?

Photo: The man in this photo has clearly defined muscles. Which muscles can you identify?

Module 6
Muscular System

Anatomy & Physiology
REVEALED
aprevealed.com

10.1 General Principles of Skeletal Muscle Anatomy

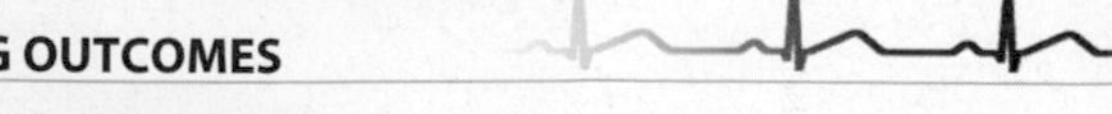

LEARNING OUTCOMES

After reading this section, you should be able to

A. Define the following and give an example of each: *origin, insertion, agonist, antagonist, synergist, fixator,* and *prime mover.*

B. Explain how fasciculus orientation determines muscle shape and list examples of muscles that demonstrate each shape.

C. Recognize muscle names based on specific nomenclature rules.

D. Explain each of the three classes of levers in the body and give a specific example of each class.

Most skeletal muscles extend from one bone to another and cross at least one joint. Muscle contraction causes most body movements by pulling one of the bones toward the other across a movable joint. Some muscles are not attached to bone at both ends. For example, some facial muscles attach to the skin, which moves as the muscles contract.

The two points of attachment of each muscle are its origin and its insertion. The **origin,** also called the *fixed end,* is usually the most stationary, proximal end of the muscle. Some muscles have multiple origins. For example, the triceps brachii has three origins that converge to form one muscle. In the case of multiple origins, each origin is also called a **head.** The **insertion,** also called the *mobile end,* is usually the distal end of the muscle attached to the bone undergoing the greatest movement. The part of the muscle between the origin and the insertion is the **belly** (figure 10.1). At the attachment point, each muscle is connected to bone by **tendons.** Tendons may be long and cablelike; broad and sheetlike (called **aponeuroses;** ap′ō-noo-rō′sēz); or short and almost nonexistent.

The **action** of a muscle is the movement accomplished when it contracts. Muscles are typically grouped so that the action of one muscle or group of muscles is opposed by that of another muscle or group of muscles. For example, the biceps brachii flexes (bends) the elbow, and the triceps brachii extends the elbow. A muscle that accomplishes a certain movement, such as flexion, is called the **agonist** (ag′ō-nist). A muscle acting in opposition to an agonist is called an **antagonist** (an-tag′ō-nist). For example, when flexing the elbow, the biceps brachii is the agonist, whereas the triceps brachii, which relaxes and stretches to allow the elbow to bend, is the antagonist. When extending the elbow, the muscles' roles are reversed; the triceps brachii is the agonist and the biceps brachii is the antagonist. Most joints in the body have agonist and antagonist groups or pairs.

Muscles also tend to function in groups to accomplish specific movements. For example, the deltoid, biceps brachii, and pectoralis major all help flex the shoulder. Furthermore, many muscles are members of more than one group, depending on the type of movement being produced. For example, the anterior part of the deltoid muscle functions with the flexors of the shoulder, whereas the posterior part functions with the extensors of the shoulder. Members

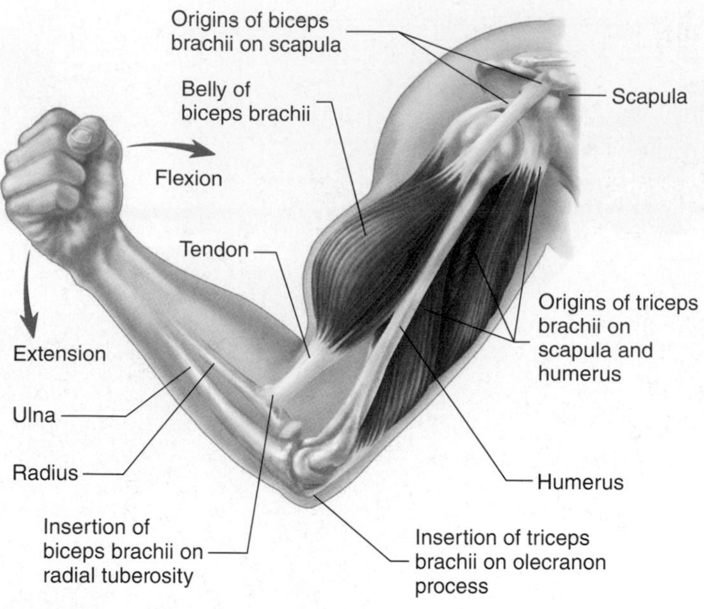
Origins of biceps brachii on scapula
Belly of biceps brachii
Scapula
Flexion
Tendon
Origins of triceps brachii on scapula and humerus
Extension
Ulna
Radius
Humerus
Insertion of biceps brachii on radial tuberosity
Insertion of triceps brachii on olecranon process

FIGURE 10.1 Muscle Attachment

Muscles are attached to bones by tendons. The biceps brachii has two heads, which originate on the scapula. The triceps brachii has three heads, which originate on the scapula and the humerus. The biceps brachii inserts onto the radial tuberosity and onto nearby connective tissue. The triceps brachii inserts onto the olecranon process of the ulna.

of a group of muscles working together to produce a movement are called **synergists** (sin′er-jistz): the biceps brachii and the brachialis are synergists in elbow flexion. Among a group of synergists, if one muscle plays the major role in accomplishing the movement, it is called the **prime mover.** The brachialis is the prime mover in flexing the elbow. **Fixators** are muscles that hold one bone in place relative to the body while a usually more distal bone is moved. The origin of a prime mover is often stabilized by fixators, so that its action occurs at its point of insertion. For example, the muscles of the scapula act as fixators to hold the scapula in place while other muscles contract to move the humerus.

Muscle Shapes

The shape and size of any given muscle greatly influence the degree to which it can contract and the amount of force it can generate. Muscles come in a wide variety of shapes, which can be grouped into five classes based on arrangement of the fasciculi (bundles of muscle fibers that can be distinguished by the unaided eye; see section 9.3): circular, convergent, parallel, pennate, and fusiform. Muscles can also have specific shapes, such as quadrate, rhomboidal, trapezium, or triangular (table 10.1).

Circular muscles, such as the orbicularis oris and orbicularis oculi, have their fasciculi arranged in a circle around an opening and act as sphincters to close the opening. Examples of circular muscles are those that surround the eyes, called the orbicularis oculi, and those that surround the mouth, called the orbicularis oris.

Convergent muscles have fascicles that arrive at one common tendon from a wide area, creating muscles that are triangular in shape. Having fibers that lie side by side can result in muscles with

TABLE 10.1 Fascicle Arrangement

Pattern of Fascicle Arrangement	Shape of Muscle	Examples
Circular Fascicles arranged in a circle around an opening; act as sphincters to close the opening		Orbicularis oris Orbicularis oculi
Convergent Broadly distributed fascicles converge at a single tendon	 Triangular	Pectoralis major Pectoralis minor
Parallel Fascicles lie parallel to one another and to the long axis of the muscle	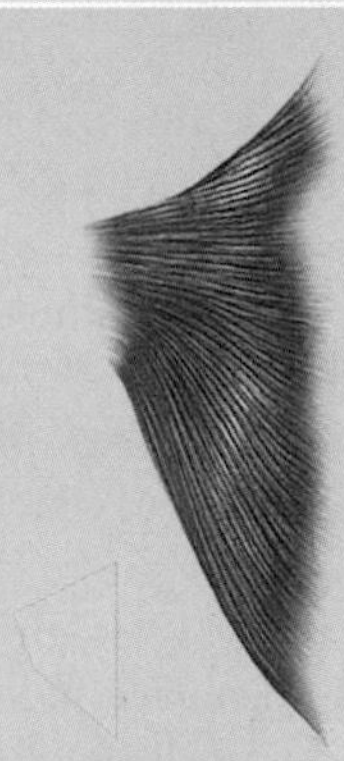 Trapezium	Trapezius
	 Rhomboidal	Rhomboideus
	 Quadrate	Rectus abdominis
Pennate Fascicles originate from a tendon that runs the length of the entire muscle. Three different patterns.		
Unipennate Fascicles on only one side of the tendon		Palmar interosseus Semimembranosus
Bipennate Fascicles on both sides of the tendon		Rectus femoris
Multipennate Fascicles arranged at many places around the central tendon. Spread out at angles to many smaller tendons.	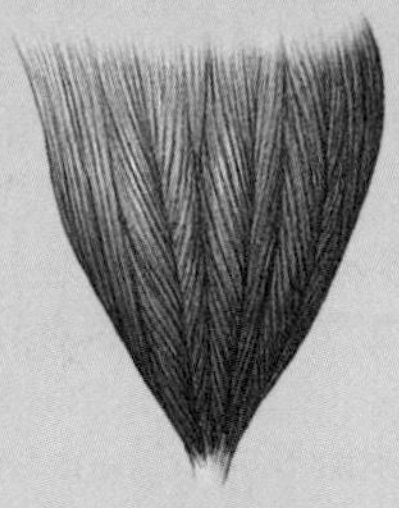	Deltoid
Fusiform Fascicles lie parallel to long axis of muscle. Belly of muscle is larger in diameter than ends.		Biceps brachii (two-headed; shown) Triceps brachii (three-headed)

less strength if the total number of fibers is low. However, if the fibers are long, these muscles can have a large range of motion. One example of convergent muscles with many long fibers is the pectoralis muscles. Similarly, in **parallel** muscles, fasciculi are organized parallel to the long axis of the muscle, but they terminate on a flat tendon that spans the width of the entire muscle. As a consequence, parallel muscles can shorten to a large degree because the fasciculi are in a direct line with the tendon; however, they contract with less force because fewer total fasciculi are attached to the tendon. The hyoid muscles are an example of parallel muscles.

The fasciculi of some muscles emerge like the barbs on a feather from a common tendon that runs the length of the entire muscle and therefore are called **pennate** (pen′āt; *pennatus*, feather) muscles. Muscles with all fasciculi on one side of the tendon are called **unipennate,** muscles with fibers arranged on two sides of the tendon are **bipennate,** and muscles with fasciculi arranged at many places around the central tendon are **multipennate.** The long tendons of pennate muscles can extend for some distance between a muscle belly and its insertion. The pennate arrangement allows a large number of fasciculi to attach to a single tendon, with the force of contraction concentrated at the tendon. The muscles that extend the knee are multipennate muscles.

Muscles whose fibers run the length of the entire muscle and taper at each end to terminate at tendons, creating a wider belly than the ends, are called **fusiform**. Because their fibers are long, but are commonly numerous, these muscles generally tend to be stronger than other muscles with parallel fascicle arrangements. The muscle that flexes the forearm is an example of a fusiform muscle.

In summary, muscle strength is primarily related to the total number of fibers in the muscle, whereas range of motion is more correlated to fascicle arrangement, with parallel fibers having the largest range of motion.

Nomenclature

Muscles are named according to several characteristics, including location, size, shape, orientation of fasciculi, origin and insertion, number of heads, and function. Recognizing the descriptive nature of muscle names makes learning those names much easier.

1. *Location.* A pectoralis (chest) muscle is located in the chest, a gluteus (buttock) muscle is in the buttock, and a brachial (arm) muscle is in the arm.
2. *Size.* The gluteus maximus (large) is the largest muscle of the buttock, and the gluteus minimus (small) is the smallest. A longus (long) muscle is longer than a brevis (short) muscle. In addition, a second part to the name immediately tells us there is more than one related muscle. For example, if there is a brevis muscle, most likely a longus muscle is present in the same area.
3. *Shape.* The deltoid (triangular) muscle is triangular in shape, a quadratus (quadrate) muscle is rectangular, and a teres (round) muscle is round.
4. *Orientation of fasciculi.* A rectus (straight, parallel) muscle has muscle fasciculi running straight with the axis of the structure to which the muscle is associated, whereas the fasciculi of an oblique muscle lie oblique to the longitudinal axis of the structure.
5. *Origin and insertion.* The sternocleidomastoid originates on the sternum and clavicle and inserts onto the mastoid process of the temporal bone. The brachioradialis originates in the arm (brachium) and inserts onto the radius.
6. *Number of heads.* A biceps muscle has two heads, and a triceps muscle has three heads. Each head has a separate origin.
7. *Function.* An abductor moves a structure away from the midline, and an adductor moves a structure toward the midline. The masseter (a chewer) is a chewing muscle.

Movements Accomplished by Muscles

Muscle movements can be explained in terms of the action of levers. A **lever** is a rigid shaft capable of turning about a hinge, or pivot point, called a **fulcrum (F)** and transferring a force applied at one point along the lever to a **weight (W),** or resistance, placed at another point along the lever. In the body, the joints function as fulcrums, and the bones function as levers. When muscles contract, the **pull (P),** or force, of muscle contraction is applied to the levers (bones), causing them to move. Three classes of levers exist, based on the relative positions of the levers, fulcrums, weights, and forces (figure 10.2): classes I, II, and III.

Class I Lever

In a **class I lever system,** the fulcrum is located between the pull and the weight (figure 10.2*a*). A child's seesaw is this type of lever. The children on the seesaw alternate between being the weight and being the pull across a fulcrum in the center of the board. In the body, the head is this type of lever; the atlantooccipital joint is the fulcrum, the posterior neck muscles provide the pull depressing the back of the head, and the face, which is elevated, is the weight. With the weight balanced over the fulcrum, only a small amount of pull is required to lift the weight. For example, only a very small shift in weight is needed for one child to lift the other on a seesaw. However, a class I lever is quite limited as to how much weight can be lifted and how high it can be lifted. For example, consider what happens when the child on one end of the seesaw is much larger than the child on the other end.

Class II Lever

In a **class II lever system,** the weight is located between the fulcrum and the pull (figure 10.2*b*). An example is a wheelbarrow; the wheel is the fulcrum, and the person lifting on the handles provides the pull. The weight, or load, carried in the wheelbarrow is placed between the wheel and the operator. In the body, a class II lever operates to depress the mandible, as in opening the mouth. (However, to compare this movement to the wheelbarrow example, the human head must be considered upside down.)

Class III Lever

In a **class III lever system,** the most common type in the body, the pull is located between the fulcrum and the weight (figure 10.2*c*). An example is a person using a shovel. The hand placed on the part of the handle closest to the blade provides the pull to lift the weight, such as a shovelful of dirt, and the hand placed near the end of the handle acts as the fulcrum. In the body, the action of the

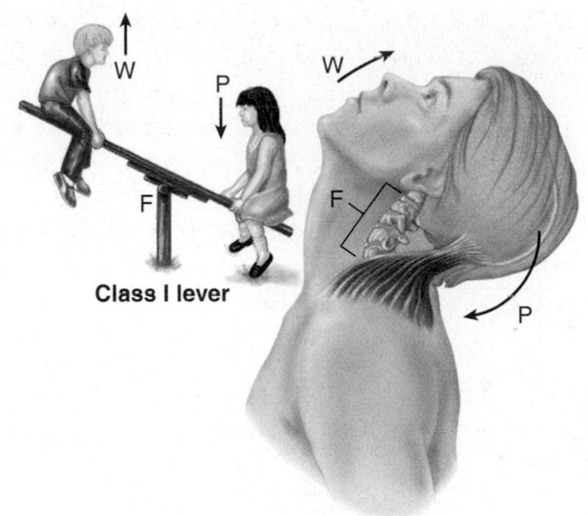
W
P
W
F
F
Class I lever
P

(a) **Class I:** The fulcrum (*F*) is located between the weight (*W*) and the pull (*P*), or force. The pull is directed downward, and the weight, on the opposite side of the fulcrum, is lifted. In the body, the fulcrum extends through several cervical vertebrae.

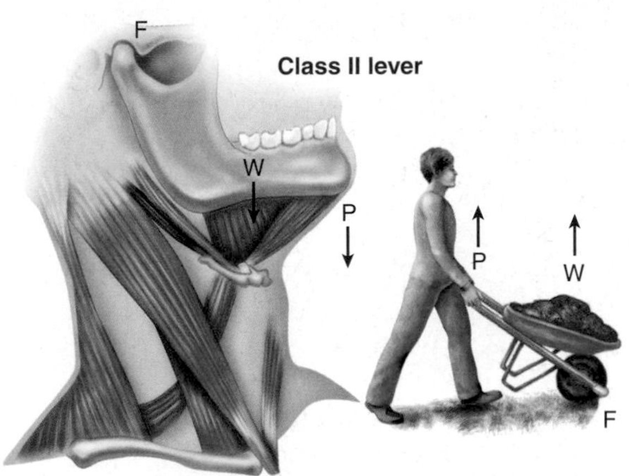
F
Class II lever
W
P
P
W
F

(b) **Class II:** The weight (*W*) is located between the fulcrum (*F*) and the pull (*P*), or force. The upward pull lifts the weight. The movement of the mandible is easier to compare to a wheelbarrow if the head is considered upside down.

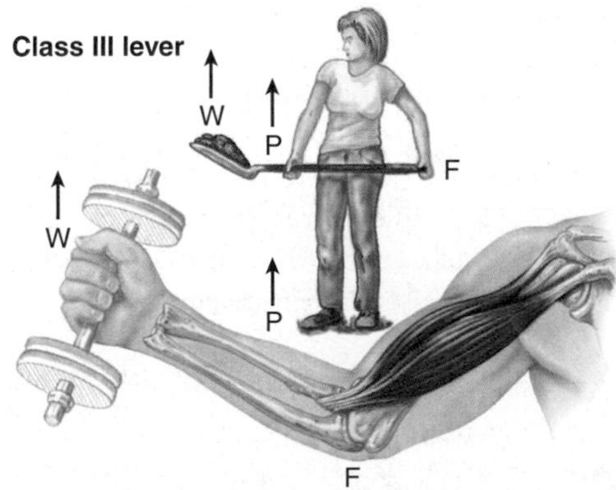
Class III lever
W
P
F
W
P
F

(c) **Class III:** The pull (*P*), or force, is located between the fulcrum (*F*) and the weight (*W*). The upward pull lifts the weight.

FIGURE 10.2 Classes of Levers

biceps brachii muscle (force) pulling on the radius (lever) to flex the elbow (fulcrum) and elevate the hand (weight) is a class III lever. This type of lever system does not allow as great a weight to be lifted, but it can be lifted a greater distance.

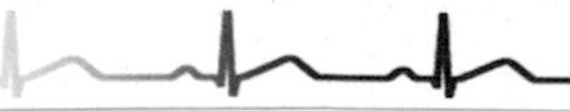

ASSESS YOUR PROGRESS

1. *Distinguish between the origin and the insertion of a muscle. In which direction is movement?*
2. *Describe the roles of the following in muscle action: agonist, antagonist, synergist, fixator, and prime mover.*
3. *Describe the different orientations of muscle fascicles, give an example of each, and explain how a muscle's shape is related to its force of contractions and the range of movement the contraction produces.*
4. *What geometric shapes can muscles have?*
5. *List the criteria used to name muscles, and give an example of each.*
6. *Using the terms fulcrum, lever, and force, explain how contraction of a muscle results in movement.*
7. *Describe the three classes of levers, and give an example of each type in the body.*

Muscle Anatomy

An overview of the superficial skeletal muscles appears in figure 10.3. Muscles of the head and neck, trunk, and limbs are described in the following sections.

10.2 Head and Neck Muscles

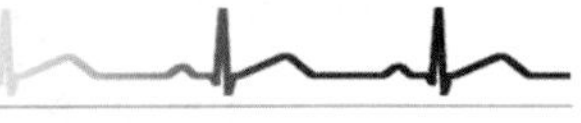

LEARNING OUTCOMES

After reading this section, you should be able to

A. Name the muscles found in the neck and list the origin, insertion, and action of each.

B. Describe movements of the head and give the muscles responsible for each movement.

C. List the muscles used to create various facial expressions.

D. Describe mastication, tongue movement, and swallowing and list the muscles or groups of muscles involved in each.

E. List the hyoid muscles and define the action of each.

F. Name the muscles responsible for movement of the eyeball and describe each movement.

Neck Muscles

The muscles that move the head and neck are listed in table 10.2. The anterior neck muscles are illustrated in figure 10.4. Most of the flexors of the head and neck lie deep within the neck along the anterior margins of the vertebral bodies (not illustrated). Extension of the neck is accomplished by the posterior neck muscles that attach to the occipital bone and mastoid process of the temporal bone (figures 10.5 and 10.6), functioning as a class I lever system. These muscles also rotate and laterally flex the neck.

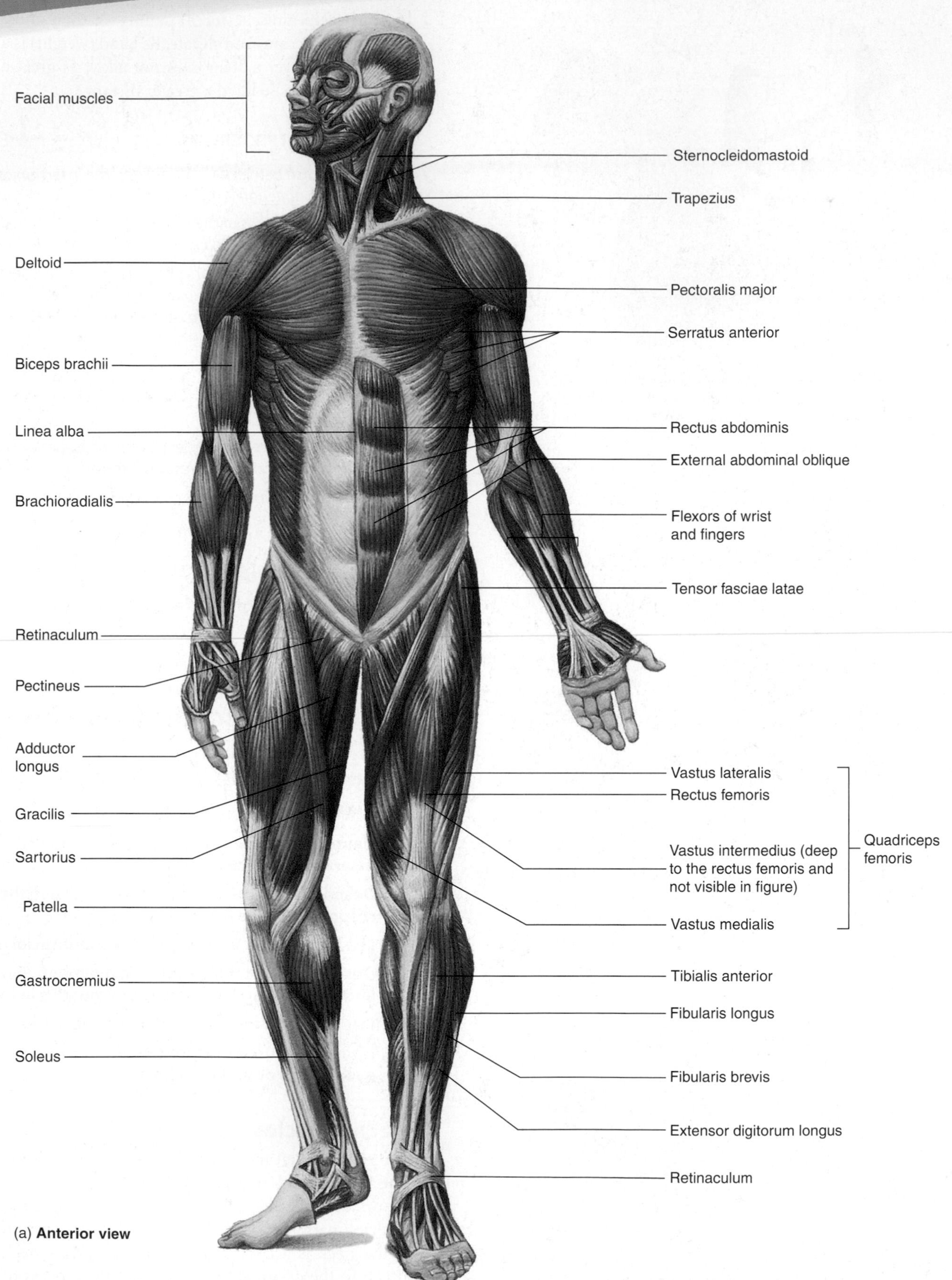
Facial muscles
Sternocleidomastoid
Trapezius
Deltoid
Pectoralis major
Serratus anterior
Biceps brachii
Linea alba
Rectus abdominis
External abdominal oblique
Brachioradialis
Flexors of wrist and fingers
Tensor fasciae latae
Retinaculum
Pectineus
Adductor longus
Vastus lateralis
Rectus femoris
Gracilis
Sartorius
Vastus intermedius (deep to the rectus femoris and not visible in figure)
Quadriceps femoris
Patella
Vastus medialis
Gastrocnemius
Tibialis anterior
Fibularis longus
Soleus
Fibularis brevis
Extensor digitorum longus
Retinaculum
(a) **Anterior view**

FIGURE 10.3 Overview of the Superficial Body Musculature

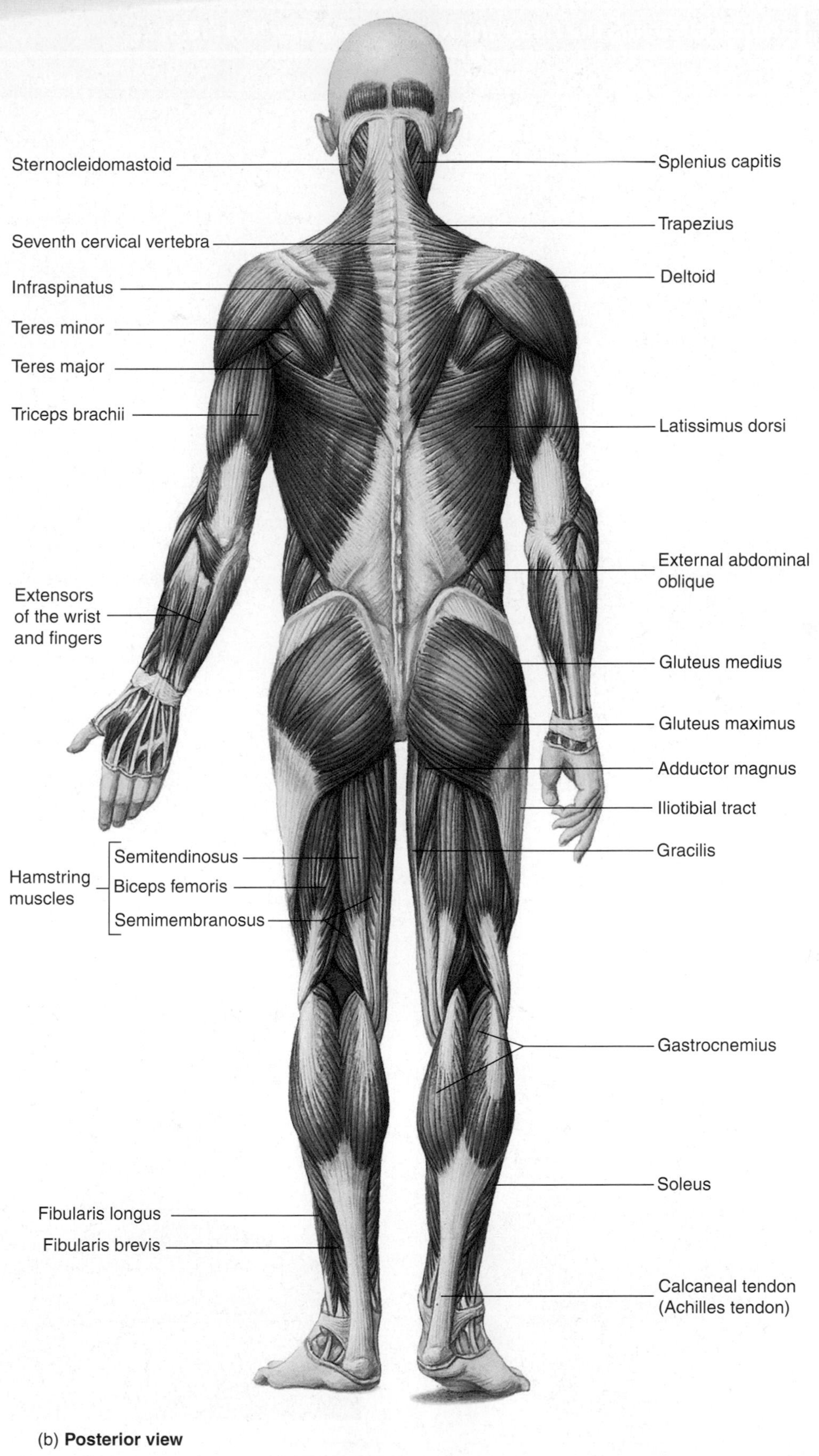
Sternocleidomastoid
Seventh cervical vertebra
Infraspinatus
Teres minor
Teres major
Triceps brachii
Extensors of the wrist and fingers
Hamstring muscles
Semitendinosus
Biceps femoris
Semimembranosus
Fibularis longus
Fibularis brevis
Splenius capitis
Trapezius
Deltoid
Latissimus dorsi
External abdominal oblique
Gluteus medius
Gluteus maximus
Adductor magnus
Iliotibial tract
Gracilis
Gastrocnemius
Soleus
Calcaneal tendon (Achilles tendon)

(b) **Posterior view**

FIGURE 10.3 (continued)

TABLE 10.2* **Muscles Moving the Head and Neck (see figures 10.4–10.6)**

Muscle	Origin	Insertion	Nerve	Action
Anterior				
Longus capitis (lon′gŭs ka′pi-tis; not illustrated)	C3–C6	Occipital bone	C1–C3	Flexes neck
Rectus capitis anterior (rek′tŭs ka′pi-tis; not illustrated)	Atlas	Occipital bone	C1–C2	Flexes neck
Posterior				
Longissimus capitis (lon-gis′ĭ-mŭs ka′pi-tis)	Upper thoracic and lower cervical vertebrae	Mastoid process	Dorsal rami of cervical nerves	Extends, rotates, and laterally flexes neck
Oblique capitis superior (ka′pi-tis)	Atlas	Occipital bone (inferior nuchal line)	Dorsal ramus of C1	Extends and laterally flexes neck
Rectus capitis posterior (rek′tŭs ka′pi-tis)	Axis, atlas	Occipital bone	Dorsal ramus of C1	Extends and rotates neck
Semispinalis capitis	C4–T6	Occipital bone	Dorsal rami of cervical nerves	Extends and rotates neck
Splenius capitis	C4–T6	Superior nuchal line and mastoid process	Dorsal rami of cervical nerves	Extends, rotates, and laterally flexes neck
Trapezius	Occipital protuberance, nuchal ligament, spinous processes of C7–T12	Clavicle, acromion process, and scapular spine	Accessory (cranial nerve XI)	Extends and laterally flexes neck
Lateral				
Rectus capitis lateralis (not illustrated)	Atlas	Occipital bone	C1	Laterally flexes neck
Sternocleidomastoid (ster′nō-klī′dō-mas′toyd)	Manubrium and medial clavicle	Mastoid process and superior nuchal line	Accessory (cranial nerve XI)	One contracting alone: laterally flexes head and neck to same side and rotates head and neck to opposite side Both contracting together: flex neck
Scalene (skā′lēn) muscles	C2–C6	First and second ribs	Cervical and brachial plexuses	Flex, laterally flex, and rotate neck

*The tables in this chapter are to be used as references. As you study the muscular system, first locate the muscle on the figure, and then find its description in the corresponding table.

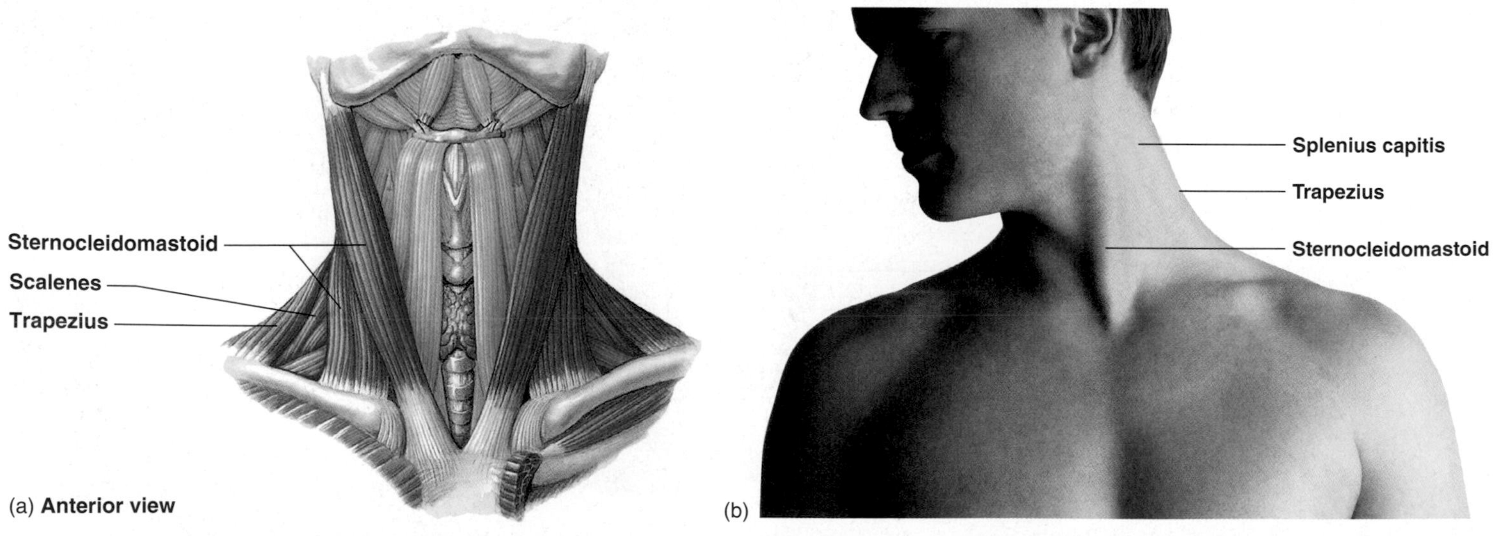
Sternocleidomastoid
Scalenes
Trapezius
(a) Anterior view
(b)
Splenius capitis
Trapezius
Sternocleidomastoid

FIGURE 10.4 APIR **Anterior Neck Muscles**
(*a*) Anterior neck muscles. (*b*) Surface anatomy of anterior neck muscles. (Muscle names are in bold.)

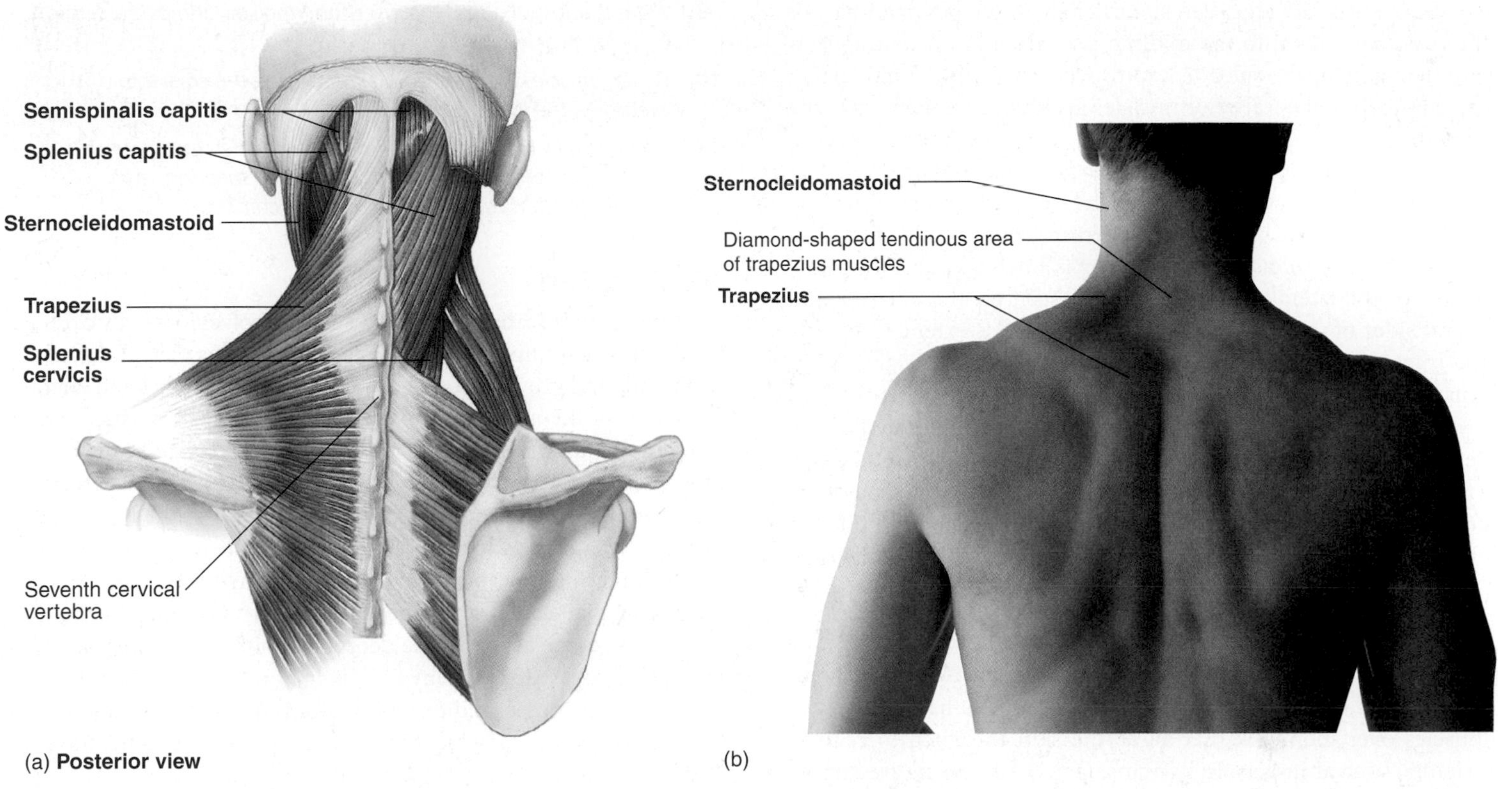
Semispinalis capitis
Splenius capitis
Sternocleidomastoid
Trapezius
Splenius cervicis
Seventh cervical vertebra
(a) Posterior view
Sternocleidomastoid
Diamond-shaped tendinous area of trapezius muscles
Trapezius
(b)

FIGURE 10.5 AP|R **Posterior Neck Muscles**
(*a*) Posterior neck muscles. (*b*) Surface anatomy of posterior neck muscles. (Muscle names are in bold.)

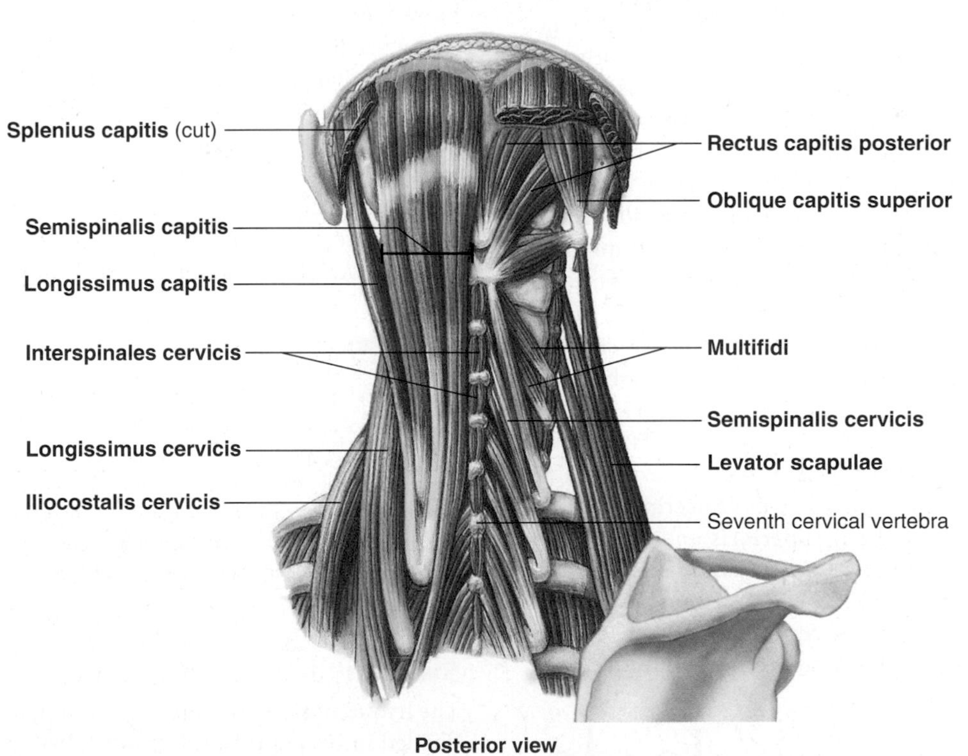
Splenius capitis (cut)
Semispinalis capitis
Longissimus capitis
Interspinales cervicis
Longissimus cervicis
Iliocostalis cervicis
Rectus capitis posterior
Oblique capitis superior
Multifidi
Semispinalis cervicis
Levator scapulae
Seventh cervical vertebra
Posterior view

FIGURE 10.6 Posterior Deep Neck Muscles
(Muscle names are in bold.)

The muscular ridge seen superficially in the posterior part of the neck and lateral to the midline is composed of the trapezius muscle overlying the splenius capitis (see figures 10.4*b* and 10.5*b*). The fasciculi of the trapezius muscles are shorter at the base of the neck and leave a diamond-shaped area over the inferior cervical and superior thoracic vertebral spines (see figure 10.5*b*).

Rotation and lateral flexion of the neck are accomplished by muscles of both the lateral and posterior groups (table 10.2). The **sternocleidomastoid** (ster′nō-klī′dō-mas′toyd) muscle is the prime mover of the lateral group. It is easily seen on the anterior and lateral sides of the neck, especially if the head is extended slightly and rotated to one side (see figure 10.4*b*). If the sternocleidomastoid muscle on only one side of the neck contracts, the neck is rotated toward the opposite side. If both contract together, they flex the neck. The scalene muscles, which are deep and lateral on the neck, assist the sternocleidomastoid in neck flexion. Lateral flexion of the neck (moving the head back to the midline after it has been tilted to one side) is accomplished by the lateral flexors of the opposite side.

Facial Expression

The skeletal muscles of the face (table 10.3; figure 10.7) are cutaneous muscles attached to the skin. Many animals have cutaneous muscles over the trunk that allow the skin to twitch to remove irritants, such as insects. In humans, facial expressions are important components of nonverbal communication, and the cutaneous muscles are confined primarily to the face and neck.

Several muscles act on the skin around the eyes and eyebrows (see figures 10.7 and 10.8). The **occipitofrontalis** (ok-sip′i-tō-frŭn-tă′lis) raises the eyebrows and furrows the skin of the forehead. The **orbicularis oculi** (ōr-bik′ū-lā′ris ok′ū-lī) closes the eyelids and causes "crow's-feet" wrinkles in the skin at the lateral corners of the eyes. The **levator palpebrae** (le-vā′ter, lē-vā′tōr pal-pē′brē) **superioris** raises the upper lids (figure 10.8*a*). A droopy eyelid on one side, called **ptosis** (tō′sis), usually indicates that the nerve to the levator palpebrae superioris, or the part of the brain controlling that nerve, has been damaged. The **corrugator supercilii** (kōr′ŭ-gā′ter, kōr′ŭ-gā′tōr soo′per-sil′ē-ī) draws the eyebrows inferiorly and medially, producing vertical corrugations (furrows) in the skin between the eyes (figure 10.8*c*; see figure 10.7).

Several muscles function in moving the lips and the skin surrounding the mouth (figure 10.8; see figure 10.7). The **orbicularis oris** (ōr-bik′ū-lā′ris ōr′is) and **buccinator** (buk′si-nā-tōr), the kissing muscles, pucker the mouth. Smiling is accomplished by the **zygomaticus** (zī′gō-mat′i-kŭs) **major** and **minor,** the **levator anguli** (ang′gū-lī) **oris,** and the **risorius** (rī-sōr′ē-ŭs). Sneering is accomplished by the **levator labii** (lā′bē-ī) **superioris** and frowning or pouting by the **depressor anguli oris,** the **depressor labii inferioris,** and the **mentalis** (men-tā′lis). If the mentalis muscles are well developed on each side of the chin, a chin dimple, where the skin is tightly attached to the underlying bone or other connective tissue, may appear between the two muscles.

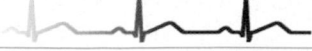

ASSESS YOUR PROGRESS 

8. *Name the major movements of the head caused by contraction of the anterior, posterior, and lateral neck muscles.*
9. *What is unusual about the insertion (and sometimes the origin) of facial muscles?*
10. *Which muscles are responsible for moving the ears, the eyebrows, the eyelids, and the nose?*
11. *What usually causes ptosis on one side? Which muscles are responsible for puckering the lips, smiling, sneering, and frowning? What causes a dimple of the chin?*

Mastication

Chewing, or **mastication** (mas-ti-kā′shŭn), involves forcefully closing the mouth (elevating the mandible: temporalis, masseter, and medial pterygoid) and grinding food between the teeth (medial and lateral excursion of the mandible; involving all muscles of mastication). The **muscles of mastication** and the **hyoid muscles** move the mandible (tables 10.4 and 10.5; figures 10.9 and 10.10). The elevators of the mandible are some of the strongest muscles of the body; they bring the mandibular teeth forcefully against the maxillary teeth to crush food. Slight mandibular depression involves relaxation of the mandibular elevators and the pull of gravity. Opening the mouth wide requires the action of the depressors of the mandible (lateral pterygoid, digastric, geniohyoid, mylohyoid). Even though the muscles of the tongue and the buccinator (table 10.6; see table 10.3) are not involved in chewing, they help move the food in the mouth and hold it in place between the teeth.

Tongue Movements

The tongue is important in mastication and speech in several ways: (1) It moves food around in the mouth; (2) with the buccinator, it holds food in place while the teeth grind it; (3) it pushes food up to the palate and back toward the pharynx to initiate swallowing; and (4) it changes shape to modify sound during speech. The tongue consists of a mass of **intrinsic muscles** (entirely within the tongue), which are involved in changing the shape of the tongue, and **extrinsic muscles** (outside of the tongue but attached to it), which help change the shape and move the tongue (figure 10.11; table 10.6). The intrinsic muscles are named for their fiber orientation in the tongue. The extrinsic muscles are named for their origin and insertion.

Predict 2

While driving to school on slick roads and talking on her cell phone, Rachel lost control of her car. The car left the road and hit a tree. Rachel was not wearing a seat belt, and her head slammed into the steering wheel, causing a fractured left mandible, as well as nerve damage. On examination, Rachel's tongue deviated toward the injured side of her face when she tried to stick out her tongue, and the left side of her tongue was paralyzed. The nerve damage affected which muscles of her tongue?

Swallowing and the Larynx

The hyoid muscles (see table 10.5 and figures 10.10 and 10.11) are divided into a **suprahyoid group** superior to the hyoid bone and an **infrahyoid group** inferior to it. When the hyoid bone is fixed by the infrahyoid muscles so that the bone is stabilized from below, the suprahyoid muscles can help depress the mandible. If the suprahyoid

TABLE 10.3 Muscles of Facial Expression (see figures 10.7 and 10.8)

Muscle	Origin	Insertion	Nerve	Action
Auricularis (aw-rik′ū-lăr′is)				
Anterior	Aponeurosis over head	Cartilage of auricle	Facial	Draws auricle superiorly and anteriorly
Posterior	Mastoid process	Posterior root of auricle	Facial	Draws auricle posteriorly
Superior	Aponeurosis over head	Cartilage of auricle	Facial	Draws auricle superiorly and posteriorly
Buccinator (buk′sĭ-nā′tōr)	Mandible and maxilla	Orbicularis oris at angle of mouth	Facial	Retracts angle of mouth; flattens cheek
Corrugator supercilii (kōr′ŭ′gā′ter soo′per-sil′ē-ī)	Nasal bridge and orbicularis oculi	Skin of eyebrow	Facial	Depresses medial portion of eyebrow; draws eyebrows together, as in frowning
Depressor anguli oris (dē-pres′ŏr ang′gū-lī ōr′is)	Lower border of mandible	Skin of lip near angle of mouth	Facial	Depresses angle of mouth
Depressor labii inferioris (dē-pres′ŏr lā′bē-ī in-fēr′ē-ōr-is)	Lower border of mandible	Skin of lower lip and orbicularis oris	Facial	Depresses lower lip
Levator anguli oris (lē-vā′tor, le-vā′ter ang′gū-lī ōr′is)	Maxilla	Skin at angle of mouth and orbicularis oris	Facial	Elevates angle of mouth
Levator labii superioris (lē-vā′tor, le-vā′ter lā′bē-ī sū-pēr′ē-ōr-is)	Maxilla	Skin and orbicularis oris of upper lip	Facial	Elevates upper lip
Levator labii superioris alaeque nasi (lē-vā′tor, le-vā′ter lā′bē-ī sū-pēr′ē-ōr-is ă-lak′ă nā′zī)	Maxilla	Ala at nose and upper lip	Facial	Elevates ala of nose and upper lip
Levator palpebrae superioris (lē-vā′tor, le-vā′ter pal-pē′brē sū-pēr′ē-ōr-is)	Lesser wing of sphenoid	Skin of eyelid	Oculomotor	Elevates upper eyelid
Mentalis (men-tā′lis)	Mandible	Skin of chin	Facial	Elevates and wrinkles skin over chin; protrudes lower lip
Nasalis (nā′ză-lis)	Maxilla	Bridge and ala of nose	Facial	Dilates nostril
Occipitofrontalis (ok-sip′i-tō-frŭn′tā′lis)	Occipital bone	Skin of eyebrow and nose	Facial	Moves scalp; elevates eyebrows
Orbicularis oculi (ōr-bik′ū-lā′ris ok′ū-lī)	Maxilla and frontal bones	Circles orbit and inserts near origin	Facial	Closes eye
Orbicularis oris (ōr-bik′ū-lā′ris ōr′is)	Nasal septum, maxilla, and mandible	Fascia and other muscles of lips	Facial	Closes lips
Platysma (plă-tiz′mă)	Fascia of deltoid and pectoralis major	Skin over inferior border of mandible	Facial	Depresses lower lip; wrinkles skin of neck and upper chest
Procerus (prō-sē′rŭs)	Bridge of nose	Frontalis	Facial	Creates horizontal wrinkles between eyes, as in frowning
Risorius (ri-sōr′ē-ŭs)	Platysma and masseter fascia	Orbicularis oris and skin at corner of mouth	Facial	Abducts angle of mouth
Zygomaticus major (zī′gō-mat′i-kŭs)	Zygomatic bone	Angle of mouth	Facial	Elevates and abducts upper lip
Zygomaticus minor (zī′gō-mat′i-kŭs)	Zygomatic bone	Orbicularis oris of upper lip	Facial	Elevates and abducts upper lip

muscles fix the hyoid and thus stabilize it from above, the thyrohyoid muscle (an infrahyoid muscle) can elevate the larynx. To observe this effect, place your hand on your larynx (Adam's apple) and swallow.

The soft palate, pharynx, and larynx contain several muscles involved in swallowing and speech (table 10.7; figure 10.12). The muscles of the soft palate close the posterior opening to the nasal cavity during swallowing.

When we swallow, muscles elevate the pharynx and larynx and then constrict the pharynx (see chapter 24). Specifically, the **palatopharyngeus** (pal′ă-tō-far-in-jē′ŭs) elevates the pharynx and the

FUNDAMENTAL **Figure**

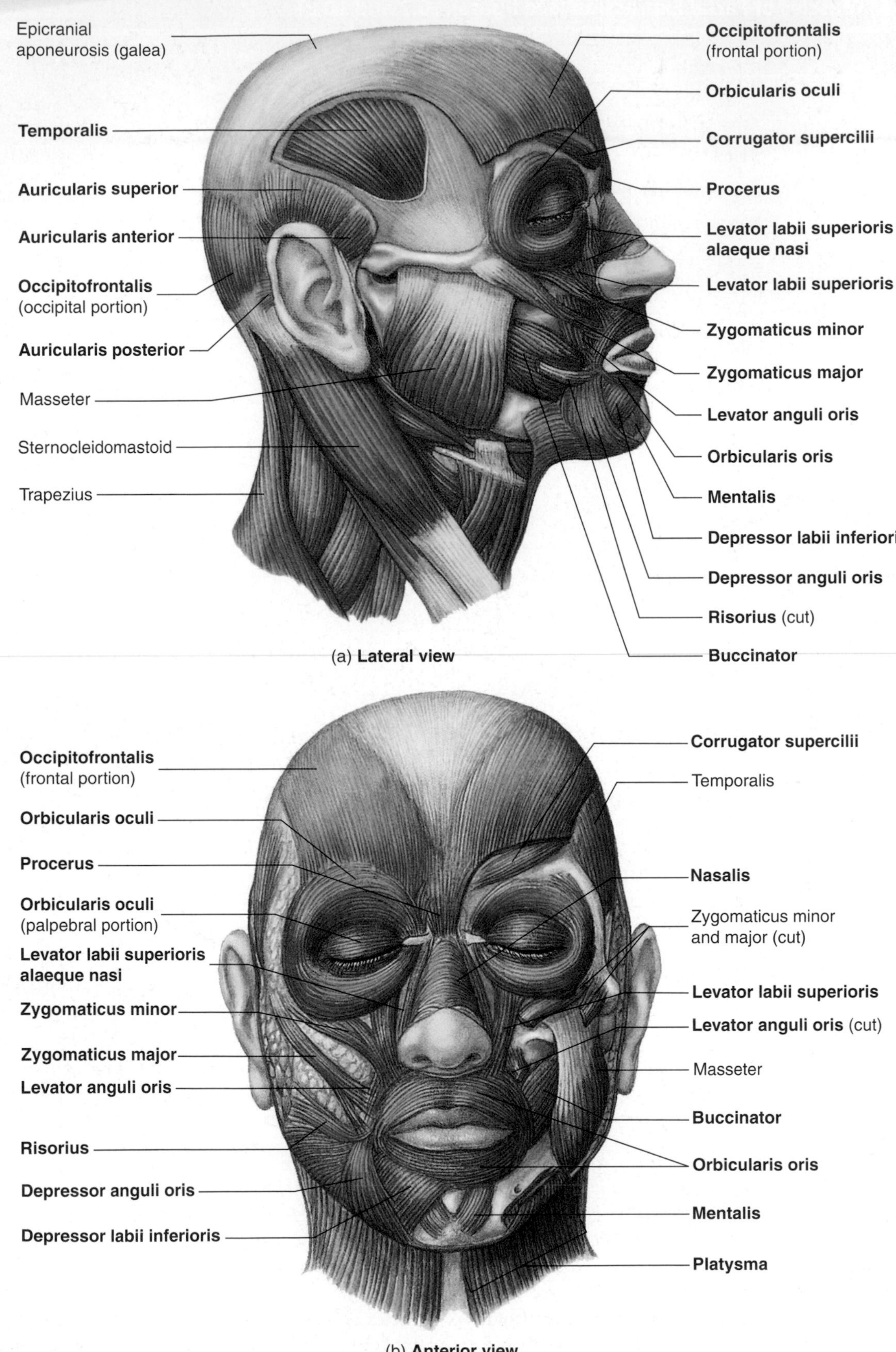

Epicranial aponeurosis (galea)
Temporalis
Auricularis superior
Auricularis anterior
Occipitofrontalis (occipital portion)
Auricularis posterior
Masseter
Sternocleidomastoid
Trapezius
Occipitofrontalis (frontal portion)
Orbicularis oculi
Corrugator supercilii
Procerus
Levator labii superioris alaeque nasi
Levator labii superioris
Zygomaticus minor
Zygomaticus major
Levator anguli oris
Orbicularis oris
Mentalis
Depressor labii inferioris
Depressor anguli oris
Risorius (cut)
Buccinator
(a) Lateral view
Occipitofrontalis (frontal portion)
Orbicularis oculi
Procerus
Orbicularis oculi (palpebral portion)
Levator labii superioris alaeque nasi
Zygomaticus minor
Zygomaticus major
Levator anguli oris
Risorius
Depressor anguli oris
Depressor labii inferioris
Corrugator supercilii
Temporalis
Nasalis
Zygomaticus minor and major (cut)
Levator labii superioris
Levator anguli oris (cut)
Masseter
Buccinator
Orbicularis oris
Mentalis
Platysma
(b) Anterior view

FIGURE 10.7 APR **Muscles of Facial Expression**
(Bold terms denote the muscles involved in facial expression.)

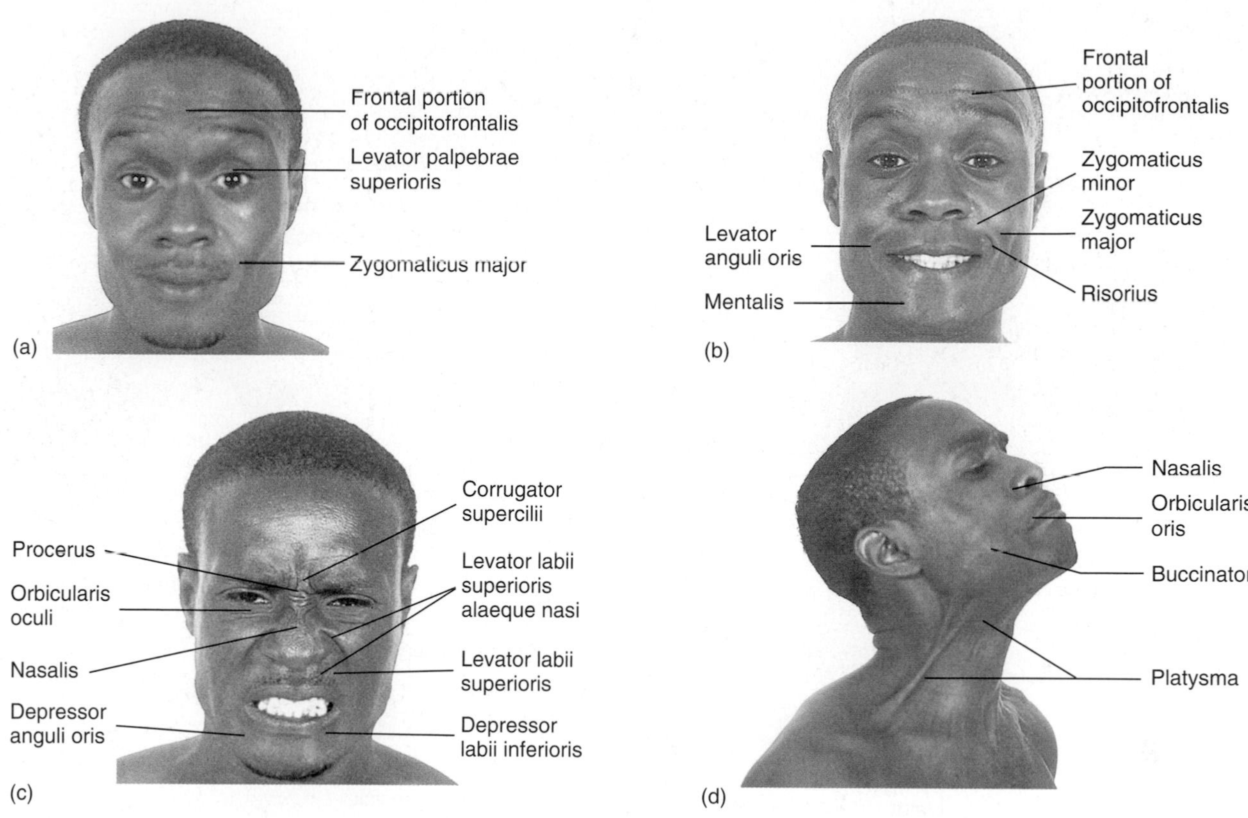
Frontal portion of occipitofrontalis
Levator palpebrae superioris
Zygomaticus major
(a)
Frontal portion of occipitofrontalis
Zygomaticus minor
Zygomaticus major
Levator anguli oris
Risorius
Mentalis
(b)
Procerus
Orbicularis oculi
Nasalis
Depressor anguli oris
Corrugator supercilii
Levator labii superioris alaeque nasi
Levator labii superioris
Depressor labii inferioris
(c)
Nasalis
Orbicularis oris
Buccinator
Platysma
(d)

FIGURE 10.8 **Surface Anatomy, Muscles of Facial Expression**

TABLE 10.4 Muscles of Mastication (see figures 10.7 and 10.9)

Muscle	Origin	Insertion	Nerve	Action
Temporalis (tem-pŏ-rā′lis)	Temporal fossa	Anterior portion of mandibular ramus and coronoid process	Mandibular division of trigeminal	Elevates and retracts mandible; involved in excursion
Masseter (ma′se-ter)	Zygomatic arch	Lateral side of mandibular ramus	Mandibular division of trigeminal	Elevates and protracts mandible; involved in excursion
Pterygoids (ter′i-goydz)				
Lateral	Lateral side of lateral pterygoid plate and greater wing of sphenoid	Condylar process of mandible and articular disk	Mandibular division of trigeminal	Protracts and depresses mandible; involved in excursion
Medial	Medial side of lateral pterygoid plate and tuberosity of maxilla	Medial surface of mandible	Mandibular division of trigeminal	Protracts and elevates mandible; involved in excursion

salpingopharyngeus (sal-pin′gō-far-in-jē′ŭs) muscles then constrict the pharynx from superior to inferior, forcing food into the esophagus.

The salpingopharyngeus also opens the auditory tube, which connects the middle ear to the pharynx. Opening the auditory tube equalizes the pressure between the middle ear and the atmosphere; this is why it is sometimes helpful to chew gum or swallow when ascending or descending a mountain in a car or when changing altitudes in an airplane.

The muscles of the larynx are listed in table 10.7 and illustrated in figure 10.12*b*. Most of the laryngeal muscles help narrow or close the laryngeal opening, so that food does not enter the larynx when a person swallows. The remaining muscles shorten (relax) the vocal cords to lower the pitch of the voice or lengthen (tense) the vocal cords to raise the pitch of the voice.

Movements of the Eyeball

The eyeball rotates within the orbit to allow vision in a wide range of directions. The movements of each eye are accomplished by six muscles, which are named for the orientation of their fasciculi relative to the spherical eye (table 10.8; figure 10.13).

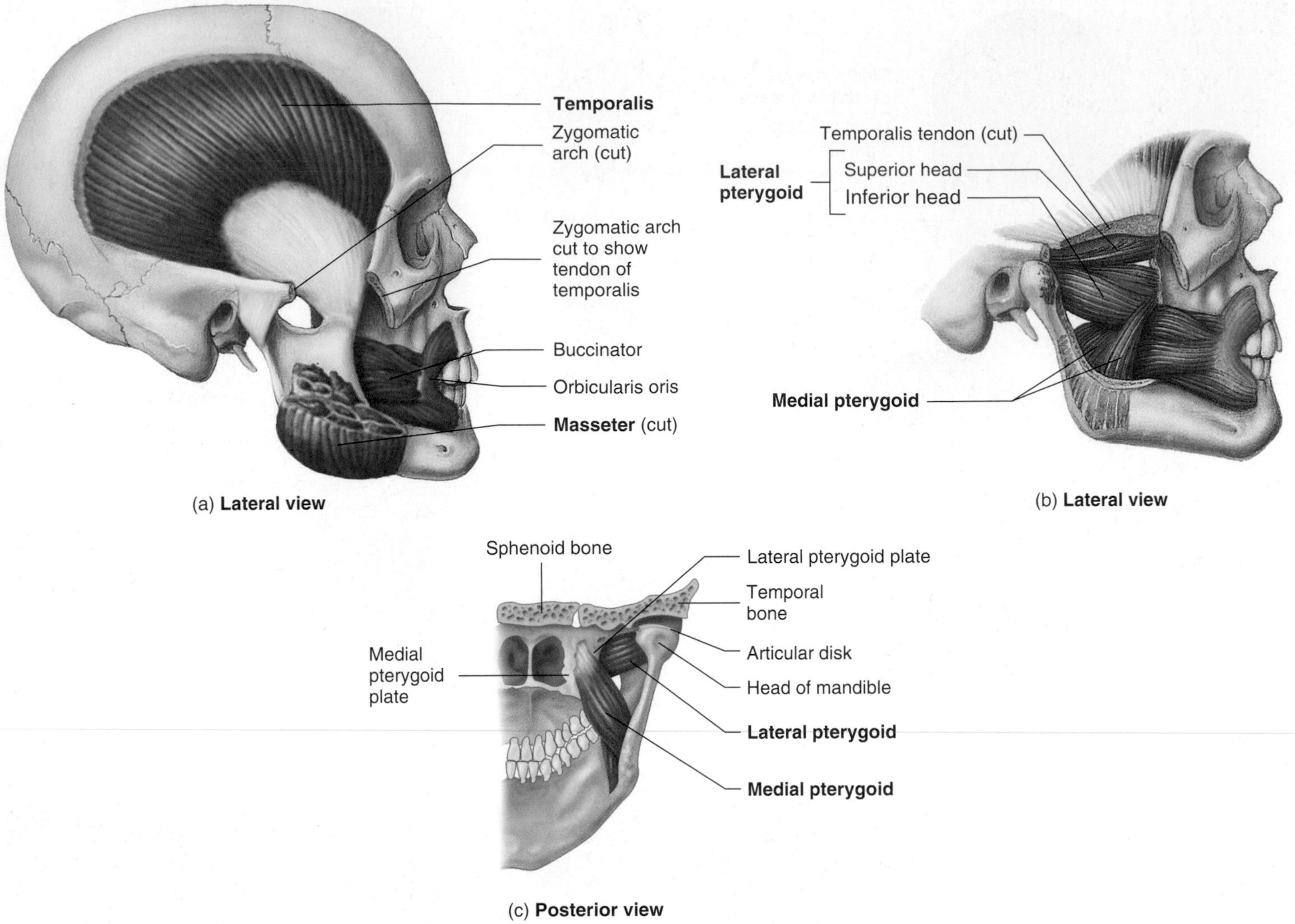

Temporalis
Zygomatic arch (cut)
Zygomatic arch cut to show tendon of temporalis
Buccinator
Orbicularis oris
Masseter (cut)

(a) Lateral view

Temporalis tendon (cut)
Lateral pterygoid
Superior head
Inferior head
Medial pterygoid

(b) Lateral view

Sphenoid bone
Medial pterygoid plate
Lateral pterygoid plate
Temporal bone
Articular disk
Head of mandible
Lateral pterygoid
Medial pterygoid

(c) Posterior view

FIGURE 10.9 APR **Muscles of Mastication**

(*a*) The masseter and zygomatic arch are cut away to expose the temporalis. (*b*) The masseter and temporalis muscles are removed, and the zygomatic arch and part of the mandible are cut away to reveal the deeper muscles. (*c*) Frontal section of the skull, showing the pterygoid muscles. (Muscle names in bold are those involved in mastication.)

Each rectus muscle (so named because the fibers are nearly straight with the axis of the eye) attaches to the eyeball anterior to the center of the sphere. The superior rectus rotates the anterior portion of the eyeball superiorly, so that the pupil, and thus the gaze, is directed superiorly (looking up). The inferior rectus depresses the gaze, the lateral rectus laterally deviates (abducts) the gaze (looking to the side), and the medial rectus medially deviates (adducts) the gaze (looking toward the nose). The superior rectus and inferior rectus are not completely straight in their orientation to the eye; thus, they also medially deviate the gaze as they contract.

The oblique muscles (so named because their fibers are oriented obliquely to the axis of the eye) insert onto the posterolateral margin of the eyeball, so that both muscles laterally deviate the gaze as they contract (see chapter 15, figure 15.11). The superior oblique elevates the posterior part of the eye, thus directing the pupil inferiorly and depressing the gaze. The inferior oblique elevates the gaze.

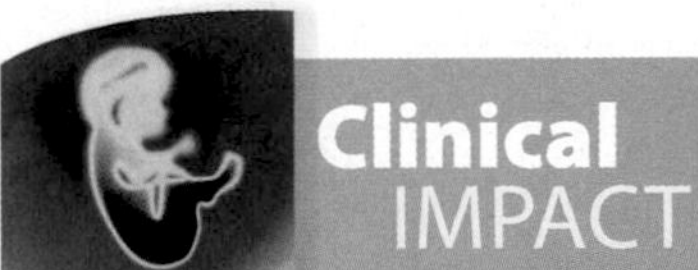
Clinical
IMPACT

Laryngospasm

Laryngospasm is a tetanic contraction of the muscles that narrows the opening of the larynx (arytenoids, lateral cricoarytenoids) and affects speech and breathing. A typical episode lasts 30–60 seconds but, in severe cases, the opening is closed completely, air can no longer pass through the larynx into the lungs, and the victim may die of asphyxiation. Laryngospasm can develop as a result of severe allergic reactions, tetanus infections, or hypocalcemia. More commonly, when food or liquid "goes down the wrong pipe," a laryngospasm episode can occur. However, some individuals may have suffered an injury to the laryngeal nerves and experience recurrent laryngospasm. For them, an injection of botulinum toxin is an effective treatment.

TABLE 10.5 Hyoid Muscles (see figure 10.10)

Muscle	Origin	Insertion	Nerve	Action
Suprahyoid Muscles				
Digastric (dī-gas′trik)	Mastoid process (posterior belly)	Mandible near midline (anterior belly)	Posterior belly—facial; anterior belly—mandibular division of trigeminal	Depresses and retracts mandible; elevates hyoid
Geniohyoid (jĕ-nī-ō-hī′oyd)	Mental protuberance of mandible	Body of hyoid	Fibers of C1 and C2 with hypoglossal	Protracts hyoid; depresses mandible
Mylohyoid (mī′lō-hī′oyd)	Body of mandible	Hyoid	Mandibular division of trigeminal	Elevates floor of mouth and tongue; depresses mandible when hyoid is fixed
Stylohyoid (stī-lō-hī′oyd)	Styloid process	Hyoid	Facial	Elevates hyoid
Infrahyoid Muscles				
Omohyoid (ō-mō-hī′oyd)	Superior border of scapula	Hyoid	Upper cervical through ansa cervicalis	Depresses hyoid; fixes hyoid in mandibular depression
Sternohyoid (ster′nō-hī′oyd)	Manubrium and first costal cartilage	Hyoid	Upper cervical through ansa cervicalis	Depresses hyoid; fixes hyoid in mandibular depression
Sternothyroid (ster′nō-thī′royd)	Manubrium and first or second costal cartilage	Thyroid cartilage	Upper cervical through ansa cervicalis	Depresses larynx; fixes hyoid in mandibular depression
Thyrohyoid (thī-rō-hī′oyd)	Thyroid cartilage	Hyoid	Upper cervical, passing with hypoglossal	Depresses hyoid and elevates thyroid cartilage of larynx; fixes hyoid in mandibular depression

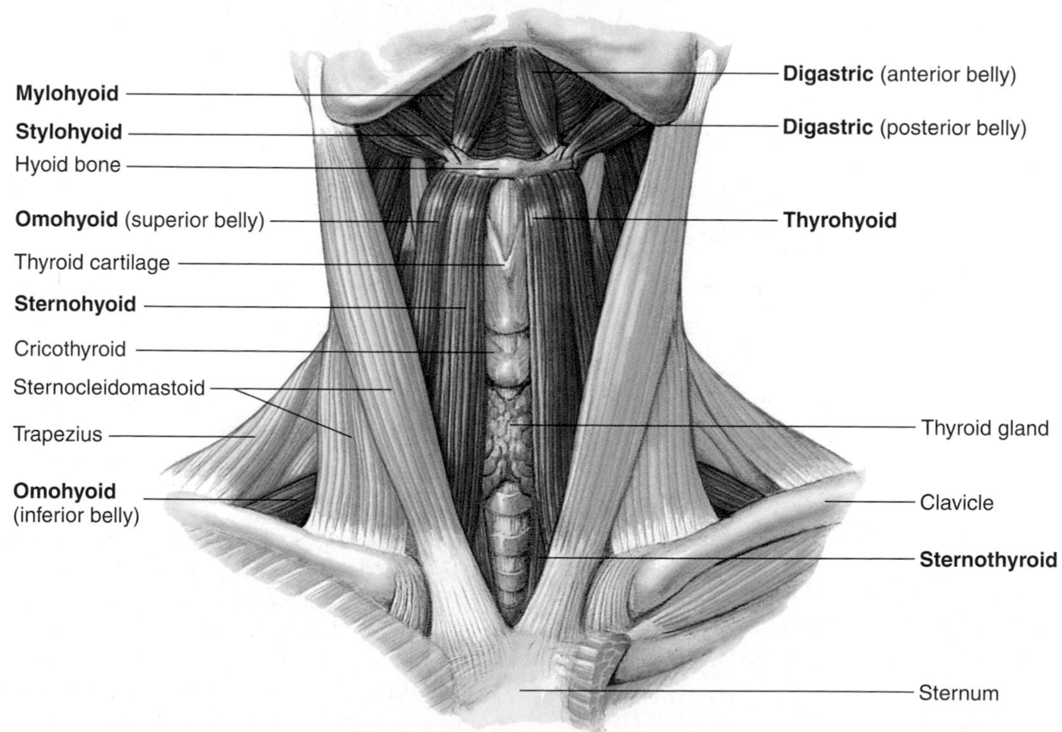

Mylohyoid
Stylohyoid
Hyoid bone
Omohyoid (superior belly)
Thyroid cartilage
Sternohyoid
Cricothyroid
Sternocleidomastoid
Trapezius
Omohyoid (inferior belly)
Digastric (anterior belly)
Digastric (posterior belly)
Thyrohyoid
Thyroid gland
Clavicle
Sternothyroid
Sternum
(a) Anterior superficial view

FIGURE 10.10 Hyoid Muscles
Hyoid muscles are shown in dark red, and the muscle names are in bold.

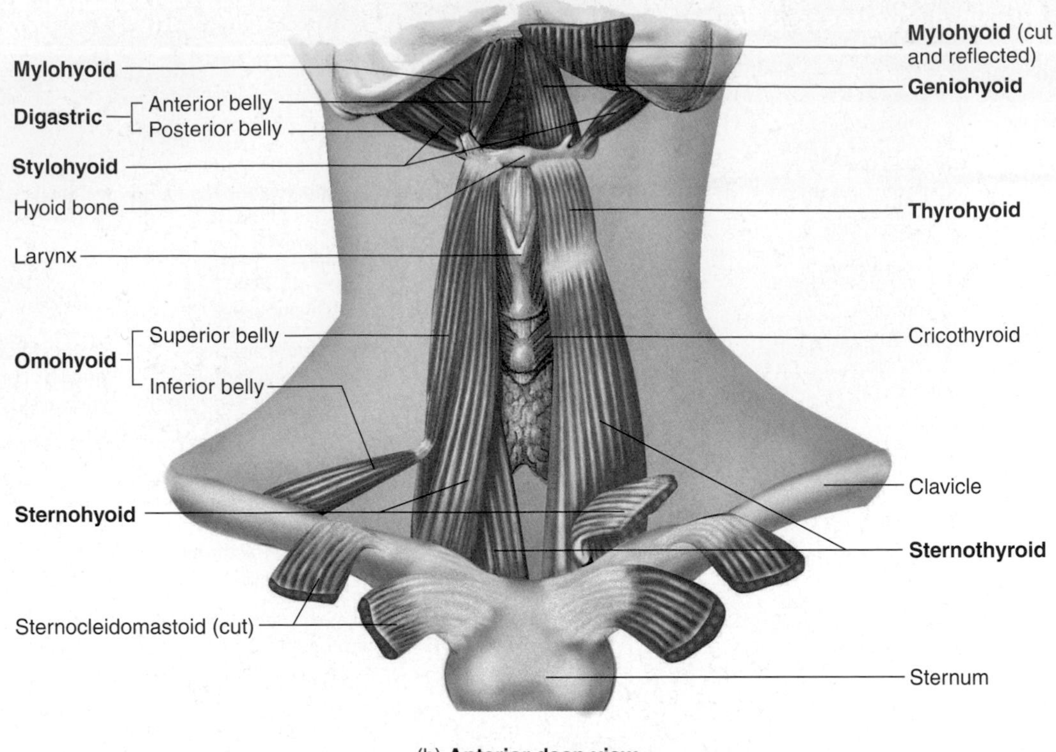

Mylohyoid
Digastric
Anterior belly
Posterior belly
Stylohyoid
Hyoid bone
Larynx
Omohyoid
Superior belly
Inferior belly
Sternohyoid
Sternocleidomastoid (cut)
Mylohyoid (cut and reflected)
Geniohyoid
Thyrohyoid
Cricothyroid
Clavicle
Sternothyroid
Sternum

(b) **Anterior deep view**

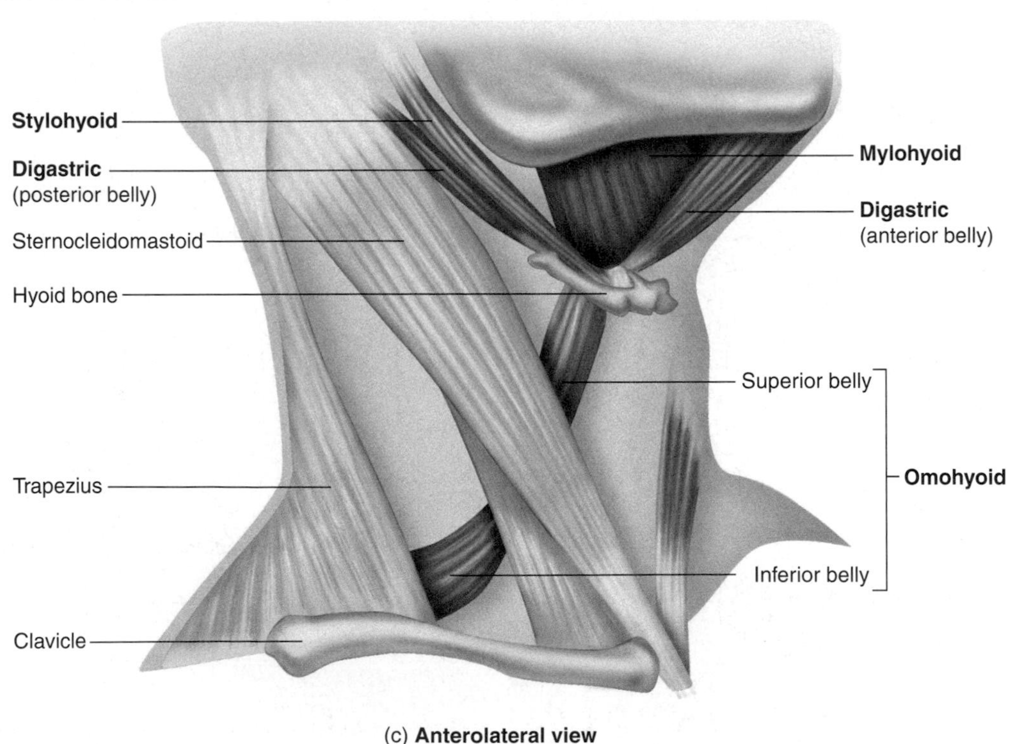

Stylohyoid
Digastric (posterior belly)
Sternocleidomastoid
Hyoid bone
Trapezius
Clavicle
Mylohyoid
Digastric (anterior belly)
Superior belly
Omohyoid
Inferior belly

(c) **Anterolateral view**

FIGURE 10.10 (continued)

TABLE 10.6 Tongue Muscles (see figure 10.11)

Muscle	Origin	Insertion	Nerve	Action
Intrinsic Muscles				
Longitudinal, transverse, and vertical (not illustrated)	Within tongue	Within tongue	Hypoglossal	Change tongue shape
Extrinsic Muscles				
Genioglossus (jĕ′nī-ō-glos′ŭs)	Mental protuberance of mandible	Tongue	Hypoglossal	Depresses and protrudes tongue
Hyoglossus (hī′ō-glos′ŭs)	Hyoid	Side of tongue	Hypoglossal	Retracts and depresses side of tongue
Styloglossus (stī′lō-glos′ŭs)	Styloid process of temporal bone	Tongue (lateral and inferior)	Hypoglossal	Retracts tongue
Palatoglossus (pal-ă-tō-glos′ŭs)	Soft palate	Tongue	Pharyngeal plexus	Elevates posterior tongue

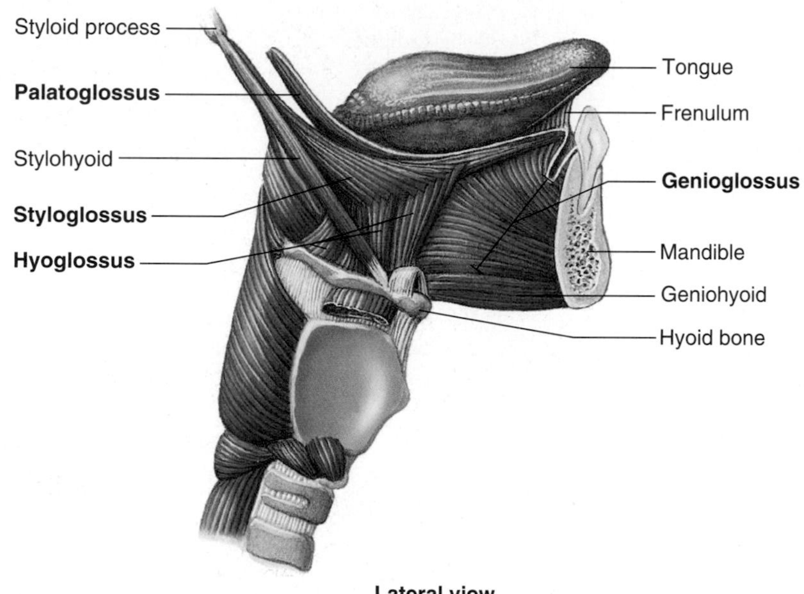

Styloid process
Palatoglossus
Stylohyoid
Styloglossus
Hyoglossus
Tongue
Frenulum
Genioglossus
Mandible
Geniohyoid
Hyoid bone
Lateral view

FIGURE 10.11 Tongue Muscles
Right lateral view. (Muscle names in bold are tongue muscles.)

ASSESS YOUR PROGRESS

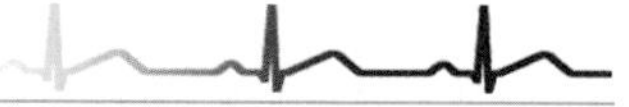

12. *Name the muscles responsible for opening and closing the jaw.*
13. *What muscles are used to cause lateral and medial excursion of the jaw?*
14. *Contrast the movements produced by the extrinsic and intrinsic tongue muscles.*
15. *Explain the interaction of the suprahyoid and infrahyoid muscles in swallowing.*
16. *Which muscles open and close the openings to the auditory tube and to the larynx?*
17. *Describe the muscles of the eye and the movements they produce.*

Predict 3

Strabismus (stra-biz′mŭs) is a condition in which one or both eyes deviate in a medial or lateral direction. In some cases, strabismus is caused by a weakness in either the medial or the lateral rectus muscle. If the lateral rectus of the right eye is weak, in which direction does the eye deviate?

TABLE 10.7 Muscles of Swallowing and the Larynx (see figures 10.11 and 10.12)

Muscle	Origin	Insertion	Nerve	Action
Larynx				
Arytenoids (ar-i-tē′noydz)				
Oblique (not illustrated)	Arytenoid cartilage	Opposite arytenoid cartilage	Recurrent laryngeal	Narrows opening to larynx
Transverse (not illustrated)	Arytenoid cartilage	Opposite arytenoid cartilage	Recurrent laryngeal	Narrows opening to larynx
Cricoarytenoids (krī′kō-ar-i-tē′noydz)				
Lateral (not illustrated)	Lateral side of cricoid cartilage	Arytenoid cartilage	Recurrent laryngeal	Narrows opening to larynx
Posterior (not illustrated)	Posterior side of cricoid cartilage	Arytenoid cartilage	Recurrent laryngeal	Widens opening of larynx
Cricothyroid (krī-kō-thī′royd)	Anterior cricoid cartilage	Thyroid cartilage	Superior laryngeal	Lengthens (tenses) vocal cords
Thyroarytenoid (thī′rō-ar′i-tē′noyd; not illustrated)	Thyroid cartilage	Arytenoid cartilage	Recurrent laryngeal	Shortens (relaxes) vocal cords
Vocalis (vō-kal′ĭs; not illustrated)	Thyroid cartilage	Arytenoid cartilage	Recurrent laryngeal	Shortens (relaxes) vocal cords
Soft Palate				
Levator veli palatini (lē-vā′tor, le-vā′ter vel′ī pal′ă-tē′nī)	Temporal bone and pharyngotympanic	Soft palate	Pharyngeal plexus	Elevates soft palate
Palatoglossus (pal-ă-tō-glos′ŭs)	Soft palate	Tongue	Pharyngeal plexus	Narrows fauces; elevates posterior tongue
Palatopharyngeus (pal′ă-tō-far-in-jē′ŭs)	Soft palate	Pharynx	Pharyngeal plexus	Narrows fauces; depresses palate; elevates pharynx
Tensor veli palatini (ten′sōr vel′ī pal′ă-tē′nī)	Sphenoid and auditory tube	Soft palate division of auditory tube	Mandibular, division of trigeminal	Tenses soft palate; opens auditory tube
Uvulae (ū′vū-lē)	Posterior nasal spine	Uvula	Pharyngeal plexus	Elevates uvula
Pharynx				
Pharyngeal constrictors (fă-rin′jē-ăl)				
Inferior	Thyroid and cricoid cartilages	Pharyngeal raphe	Pharyngeal plexus and external laryngeal nerve	Narrows inferior portion of pharynx in swallowing
Middle	Stylohyoid ligament and hyoid	Pharyngeal raphe	Pharyngeal plexus	Narrows pharynx in swallowing
Superior	Medial pterygoid plate, mandible, floor of mouth, and side of tongue	Pharyngeal raphe	Pharyngeal plexus	Narrows superior portion of pharynx in swallowing
Salpingopharyngeus (sal-ping′gō-far-in-jē′ŭs)	Auditory tube	Pharynx	Pharyngeal plexus	Elevates pharynx; opens auditory tube in swallowing
Stylopharyngeus (stī′lō-far-in-jē′ŭs)	Styloid process	Pharynx	Glossopharyngeus	Elevates pharynx

10.3 Trunk Muscles

LEARNING OUTCOMES

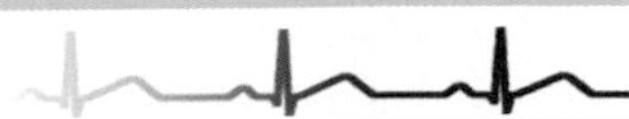

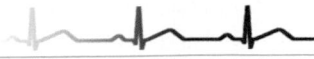

After reading this section, you should be able to

A. Describe the muscles of the vertebral column and the actions they accomplish.

B. List the muscles of the thorax and give each of their actions.

C. Describe the muscles of the abdominal wall and explain their actions.

D. List and describe the muscles of the pelvic floor and perineum.

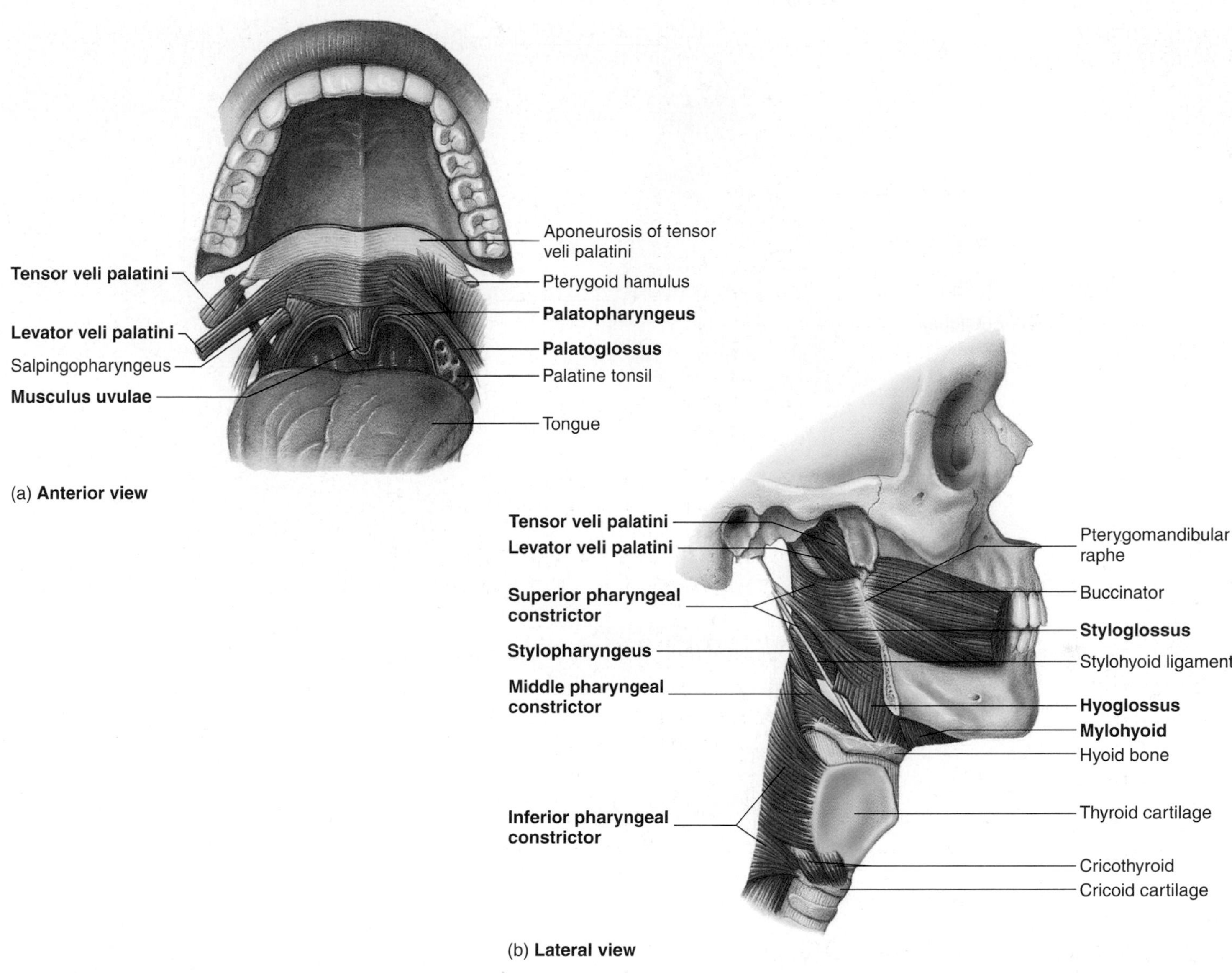
Tensor veli palatini
Levator veli palatini
Salpingopharyngeus
Musculus uvulae
Aponeurosis of tensor veli palatini
Pterygoid hamulus
Palatopharyngeus
Palatoglossus
Palatine tonsil
Tongue
(a) Anterior view
Tensor veli palatini
Levator veli palatini
Superior pharyngeal constrictor
Stylopharyngeus
Middle pharyngeal constrictor
Inferior pharyngeal constrictor
Pterygomandibular raphe
Buccinator
Styloglossus
Stylohyoid ligament
Hyoglossus
Mylohyoid
Hyoid bone
Thyroid cartilage
Cricothyroid
Cricoid cartilage
(b) Lateral view

FIGURE 10.12 Muscles of the Palate, Pharynx, and Larynx
(*a*) Anterior-inferior view of the palate. The palatoglossus and part of the palatopharyngeus muscles are cut on one side to reveal the deeper muscles. (*b*) Lateral view of the palate, pharynx, and larynx. Part of the mandible has been removed to reveal the deeper structures. (Muscle names in bold are muscles of swallowing and tongue movement.)

TABLE 10.8 Muscles Moving the Eye (see figure 10.13)

Muscle	Origin	Insertion	Nerve	Action
Oblique				
Inferior	Orbital plate of maxilla	Sclera of eye	Oculomotor	Elevates and laterally deviates gaze
Superior	Common tendinous ring	Sclera of eye	Trochlear	Depresses and laterally deviates gaze
Rectus				
Inferior	Common tendinous ring	Sclera of eye	Oculomotor	Depresses and medially deviates gaze
Lateral	Common tendinous ring	Sclera of eye	Abducens	Laterally deviates gaze
Medial	Common tendinous ring	Sclera of eye	Oculomotor	Medially deviates gaze
Superior	Common tendinous ring	Sclera of eye	Oculomotor	Elevates and medially deviates gaze

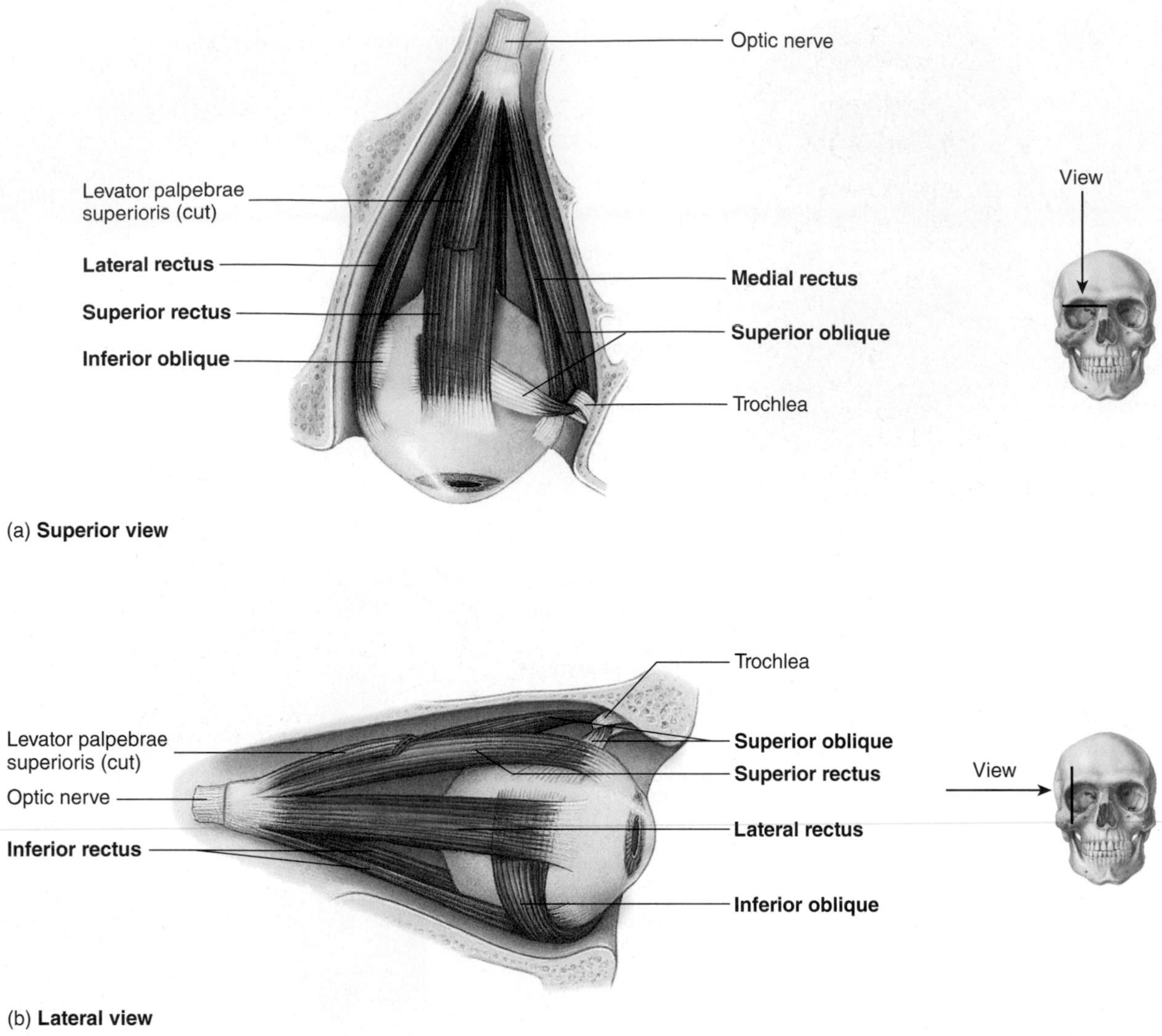

Optic nerve
Levator palpebrae superioris (cut)
Lateral rectus
Superior rectus
Inferior oblique
Medial rectus
Superior oblique
Trochlea
View

(a) **Superior view**

Trochlea
Levator palpebrae superioris (cut)
Superior oblique
Superior rectus
Optic nerve
View
Lateral rectus
Inferior rectus
Inferior oblique

(b) **Lateral view**

FIGURE 10.13 Muscles That Move the Right Eyeball
(Names of muscles of eye movement are in bold.)

Muscles Moving the Vertebral Column

The muscles that extend, laterally flex, and rotate the vertebral column are divided into superficial and deep groups (table 10.9). In general, the muscles of the deep group extend from vertebra to vertebra, whereas the muscles of the superficial group extend from the vertebrae to the ribs. These back muscles are very strong to maintain erect posture. The **erector spinae** (spī′nē) group of muscles on each side of the back consists of three subgroups: the **iliocostalis** (il′ē-ō-kos-tā′lis), the **longissimus** (lon-gis′i-mŭs), and the **spinalis** (spī-nā′lis). The longissimus group accounts for most of the muscle mass in the lower back (figure 10.14). The deepest muscles of the back attach between the spinous and transverse processes of individual vertebrae (figure 10.15).

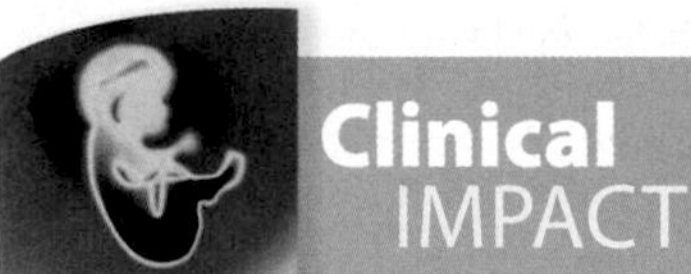
Clinical
IMPACT

Back Pain

Low back pain can result from injury, poor posture, being overweight, or lack of fitness; it is the primary cause of missed work and the second most common neurological affliction in the United States. In addition to chronic pain, a low back injury is often accompanied by muscle spasms, which are spontaneous, painful, uncontrolled muscle contractions. A few changes may help prevent more spasms and reduce pain. Patients should sit and stand up straight; use a low back support when sitting; lose weight; exercise, especially the back and abdominal muscles; and try to sleep on their side on a firm mattress. If lifestyle changes are not sufficient, treatment with muscle relaxants, anti-inflammatory drugs, or pain medication may be necessary.

FUNDAMENTAL **Figure**

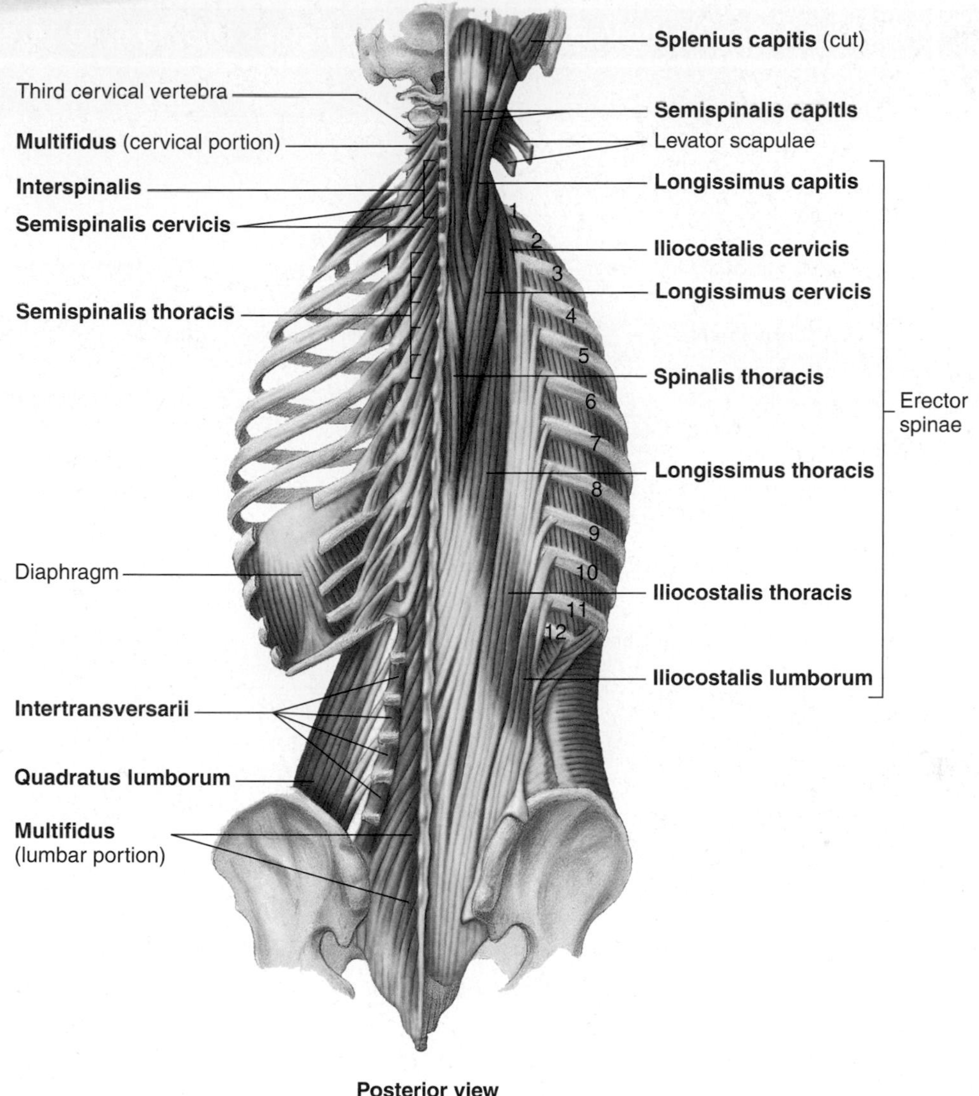
Third cervical vertebra
Multifidus (cervical portion)
Interspinalis
Semispinalis cervicis
Semispinalis thoracis
Diaphragm
Intertransversarii
Quadratus lumborum
Multifidus
(lumbar portion)
Splenius capitis (cut)
Semispinalis capitis
Levator scapulae
Longissimus capitis
Iliocostalis cervicis
Longissimus cervicis
Spinalis thoracis
Longissimus thoracis
Iliocostalis thoracis
Iliocostalis lumborum
Erector spinae
1
2
3
4
5
6
7
8
9
10
11
12
Posterior view

FIGURE 10.14 APR **Deep Neck and Back Muscles**

On the right, the erector spinae group of muscles is shown. On the left, these muscles are removed to reveal the deeper back muscles. (Names of muscles of the neck and back are in bold.)

Thoracic Muscles

The muscles of the thorax are mainly involved in the process of breathing (see chapter 23). Four major groups of thoracic muscles are associated with the rib cage (table 10.10; figure 10.16). The **scalene** (skā′lēn) muscles elevate the first two ribs during more forceful inspiration. The **external intercostals** (in-ter-kos′tălz) elevate the ribs during quiet, resting inspiration. The **internal intercostals** and **transversus thoracis** (thō-ra′sis) muscles depress the ribs during forced expiration.

The **diaphragm** (dī′ă-fram; figure 10.16*a*) causes the major movement produced during quiet breathing. It is a dome-shaped muscle; when it contracts, the dome flattens slightly, causing the volume of the thoracic cavity to increase and resulting in inspiration. If this dome of skeletal muscle or the phrenic nerve supplying it is severely damaged, the amount of air moving into and out of the lungs may be so small that the individual cannot survive without the aid of an artificial respirator.

Abdominal Wall

The muscles of the anterior abdominal wall (table 10.11; figures 10.17 and 10.18) flex and rotate the vertebral column. Contraction of the abdominal muscles when the vertebral column is fixed decreases the volume of the abdominal cavity and the thoracic cavity and can aid in such functions as forced expiration, vomiting, defecation, urination, and childbirth. The crossing pattern of the abdominal muscles creates a strong anterior wall, which holds in and protects the abdominal viscera.

In a relatively muscular person with little fat, a vertical line called the **linea alba** (lin′ē-ă al′bă), or *white line*, is visible. It extends from the area of the xiphoid process of the sternum through the navel to the pubis. The linea alba is devoid of muscle and consists of white connective tissue (figure 10.17). On each side of the linea alba is the **rectus abdominis** (figures 10.17 and 10.18), surrounded by a **rectus sheath. Tendinous intersections** (tendinous

TABLE 10.9 Muscles Acting on the Vertebral Column (see figures 10.5, 10.6, 10.14, and 10.15)

Muscle	Origin	Insertion	Nerve	Action
Superficial				
Erector spinae (ē-rek′tŏr, ē-rek′tōr spī′nē; divides into three columns)				
Iliocostalis (il′ē-ō-kos-tā′lis)	Sacrum, ilium, and lumbar spines	Ribs and vertebrae	Dorsal rami of spinal nerves	Extends vertebral column
Cervicis (ser-vī′sis)	Superior six ribs	Transverse processes of middle cervical vertebrae	Dorsal rami of thoracic nerves	Extends, laterally flexes, and rotates vertebral column
Thoracis (thō-ra′sis)	Inferior six ribs	Superior six ribs	Dorsal rami of thoracic nerves	Extends, laterally flexes, and rotates vertebral column
Lumborum (lum-bōr′ŭm)	Sacrum, ilium, and lumbar vertebrae	Inferior six ribs	Dorsal rami of thoracic and lumbar nerves	Extends, laterally flexes, and rotates vertebral column
Longissimus (lon-gis′i-mŭs)				
Capitis (ka′pĭ-tis)	Upper thoracic and lower cervical vertebrae	Mastoid process	Dorsal rami of cervical nerves	Extends head
Cervicis (ser-vī′sis)	Upper thoracic vertebrae	Transverse processes of upper cervical vertebrae	Dorsal rami of cervical nerves	Extends neck
Thoracis (thō-ra′sis)	Ribs and lower thoracic vertebrae	Transverse processes of upper lumbar vertebrae and ribs	Dorsal rami of thoracic and lumbar nerves	Extends vertebral column
Spinalis (spī-nā′lis)				
Cervicis (ser-vī′sis; not illustrated)	C6–C7	Spinous processes of C2–C3	Dorsal rami of cervical nerves	Extends neck
Thoracis (thō-ra′sis)	T11–L2	Spinous processes of middle and upper thoracic vertebrae	Dorsal rami of thoracic nerves	Extends vertebral column
Semispinalis (sem′ē-spī-nā′lis)				
Cervicis (ser-vī′sis)	Transverse processes of T2–T5	Spinous processes of C2–C5	Dorsal rami of cervical nerves	Extends neck
Thoracis (thō-ra′sis)	Transverse processes of T5–T11	Spinous processes of C5–T4	Dorsal rami of thoracic nerves	Extends vertebral column
Splenius cervicis (splē′nē-ŭs ser-vī′sis)	Spinous processes of C3–C5	Transverse processes of C1–C3	Dorsal rami of cervical nerves	Rotates and extends neck
Longus colli (lon′gŭs kō′lī; not illustrated)	Bodies of C3–T3	Bodies of C1–C6	Ventral rami of cervical nerves	Flexes neck
Deep				
Interspinales (in-ter-spī-nā′lēz)	Spinous processes of all vertebrae	Next superior spinous process	Dorsal rami of spinal nerves	Extends back and neck
Intertransversarii (in-ter-trans′ver-săr′ē-ī)	Transverse processes of all vertebrae	Next superior transverse process	Dorsal rami of spinal nerves	Laterally flexes vertebral column
Multifidus (mŭl-tif′i-dŭs)	Transverse processes of vertebrae; posterior surface of sacrum and ilium	Spinous processes of superior vertebrae	Dorsal rami of spinal nerves	Extends and rotates vertebral column
Psoas minor (sō′as mī′ner)	T12–L1	Pectineal line near pubic crest	L1	Flexes vertebral column
Rotatores (rō-tā′tōrz)	Transverse processes of all vertebrae	Base of spinous process of superior vertebrae	Dorsal rami of spinal nerves	Extends and rotates vertebral column

inscriptions) transect the rectus abdominis at three, or sometimes more, locations, causing the abdominal wall of a lean, well-muscled person to appear segmented (a "six-pack"). Lateral to the rectus abdominis is the **linea semilunaris** (sem-ē-loo-nar′is; a crescent- or half-moon-shaped line); lateral to it are three layers of muscle (figures 10.17 and 10.18). From superficial to deep, these muscles are the **external abdominal oblique, internal abdominal oblique,** and **transversus abdominis.**

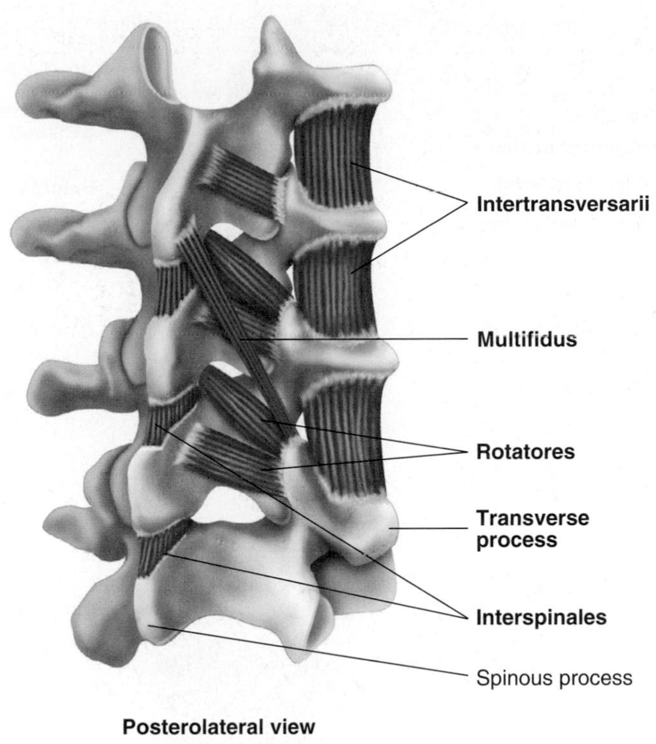

Intertransversarii
Multifidus
Rotatores
Transverse process
Interspinales
Spinous process
Posterolateral view

FIGURE 10.15 **Deep Muscles Associated with the Vertebrae** (Muscle names are in bold.)

Pelvic Floor and Perineum

The pelvis is a ring of bone (see chapter 7) with an inferior opening that is closed by a muscular wall, through which the anus and the urogenital openings penetrate (table 10.12). Most of the pelvic floor is formed by the **coccygeus** (kok-si′jē-ŭs) muscle and the **levator ani** (a′nī) muscle, referred to jointly as the **pelvic diaphragm.** The area inferior to the pelvic floor is the **perineum** (per′i-nē′ŭm), which is somewhat diamond-shaped (figure 10.19). The anterior half of the diamond is the urogenital triangle, and the posterior half is the anal triangle (see chapter 28). During pregnancy, the muscles of the pelvic diaphragm and urogenital triangle may be stretched by the extra weight of the fetus, and specific exercises are designed to strengthen them.

ASSESS YOUR PROGRESS

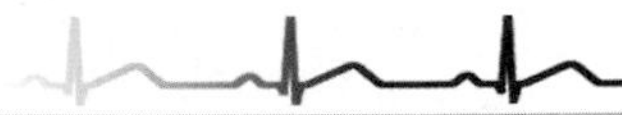

18. *List the actions of the group of back muscles that attaches to the vertebrae or ribs (or both). What is the name of the superficial subgroup?*

19. *Name the muscle that is mainly responsible for respiratory movements. What other muscles aid this movement?*

20. *Explain the anatomical basis for the segments ("cuts") seen on a well-muscled individual's abdomen. What are the functions of the abdominal muscles? List the muscles of the anterior abdominal wall.*

21. *What openings penetrate the pelvic floor muscles? Name the area inferior to the pelvic floor.*

TABLE 10.10 **Muscles of the Thorax (see figure 10.16)**

Muscle	Origin	Insertion	Nerve	Action
Diaphragm	Interior of ribs, sternum, and lumbar vertebrae	Central tendon of diaphragm	Phrenic	Inspiration; depresses floor of thorax
Intercostalis (in′ter-kos-ta′lis)				
External	Inferior margin of each rib	Superior border of next rib below	Intercostal	Quiet inspiration; elevates ribs
Internal	Superior margin of each rib	Inferior border of next rib above	Intercostal	Forced expiration; depresses ribs
Scalenus (skā-lē′nŭs)				
Anterior	Transverse processes of C3–C6	First rib	Cervical plexus	Elevates first rib
Medial	Transverse processes of C2–C6	First rib	Cervical plexus	Elevates first rib
Posterior	Transverse processes of C4–C6	Second rib	Cervical and brachial plexuses	Elevates second rib
Serratus posterior (sĕr-ā′tŭs)				
Inferior (not illustrated)	Spinous processes of T11–L2	Inferior four ribs	Ninth to eleventh intercostals and subcostal	Depresses inferior ribs and extends vertebral column
Superior (not illustrated)	Spinous processes of C6–T2	Second to fifth ribs	First to fourth intercostals	Elevates superior ribs
Transversus thoracis (trans-ver′sus thō-ra′sis; not illustrated)	Sternum and xiphoid process	Second to sixth costal cartilages	Intercostal	Decreases diameter of thorax

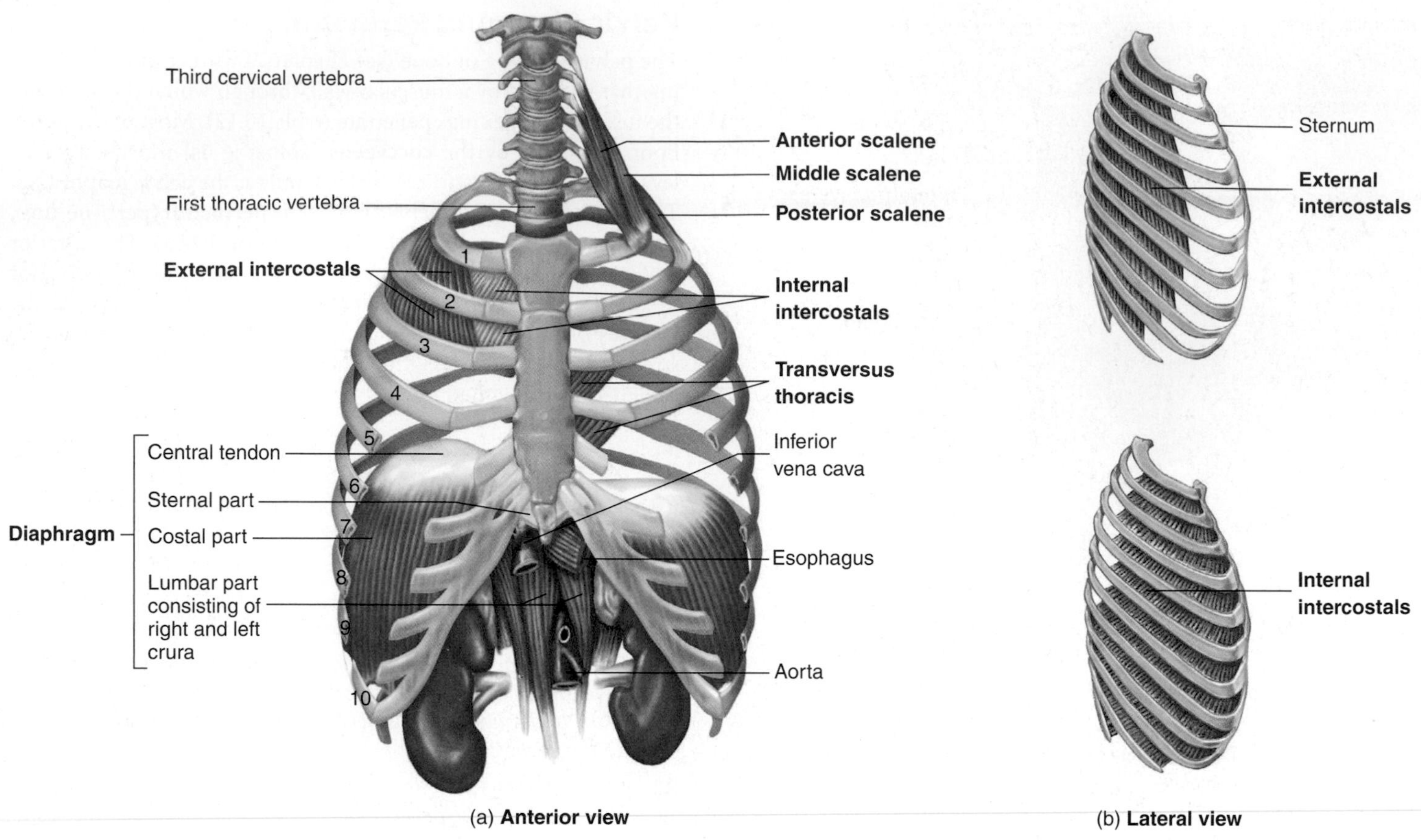
Third cervical vertebra
First thoracic vertebra
External intercostals
1
2
3
4
5
6
7
8
9
10
Central tendon
Sternal part
Costal part
Lumbar part consisting of right and left crura
Diaphragm
Anterior scalene
Middle scalene
Posterior scalene
Internal intercostals
Transversus thoracis
Inferior vena cava
Esophagus
Aorta
(a) Anterior view
Sternum
External intercostals
Internal intercostals
(b) Lateral view

FIGURE 10.16 APR **Muscles of the Thorax**
(Muscle names are in bold.)

TABLE 10.11 Muscles of the Abdominal Wall (see figures 10.3a, 10.17, and 10.18)

Muscle	Origin	Insertion	Nerve	Action
Anterior				
Rectus abdominis (rek′tŭs ab-dom′i-nis)	Pubic crest and symphysis pubis	Xiphoid process and inferior ribs	Branches of lower thoracic	Flexes vertebral column; compresses abdomen
External abdominal oblique	Fifth to twelfth ribs	Iliac crest, inguinal ligament, and rectus sheath	Branches of lower thoracic	Flexes and rotates vertebral column; compresses abdomen; depresses thorax
Internal abdominal oblique	Iliac crest, inguinal ligament, and lumbar fascia	Tenth to twelfth ribs and rectus sheath	Lower thoracic	Flexes and rotates vertebral column; compresses abdomen; depresses thorax
Transversus abdominis (trans-ver′sŭs ab-dom′i-nis)	Seventh to twelfth costal cartilages, lumbar fascia, iliac crest, and inguinal ligament	Xiphoid process, linea alba, and pubic tubercle	Lower thoracic	Compresses abdomen
Posterior				
Quadratus lumborum (kwah-drā′tŭs lŭm-bōr′ŭm)	Iliac crest and lower lumbar vertebrae	Twelfth rib and upper lumbar vertebrae	Upper lumbar	Laterally flexes vertebral column and depresses twelfth rib

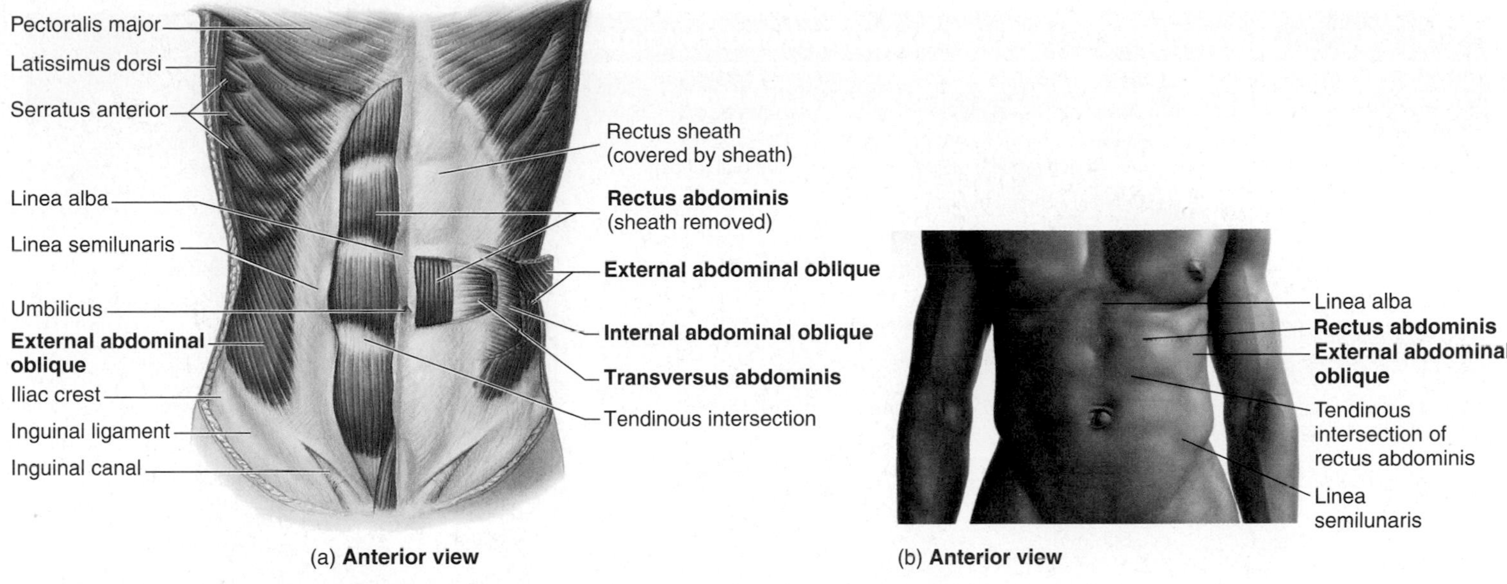
Pectoralis major
Latissimus dorsi
Serratus anterior
Linea alba
Linea semilunaris
Umbilicus
External abdominal oblique
Iliac crest
Inguinal ligament
Inguinal canal
Rectus sheath (covered by sheath)
Rectus abdominis (sheath removed)
External abdominal oblique
Internal abdominal oblique
Transversus abdominis
Tendinous intersection
(a) Anterior view
Linea alba
Rectus abdominis
External abdominal oblique
Tendinous intersection of rectus abdominis
Linea semilunaris
(b) Anterior view

FIGURE 10.17 Anterior Abdominal Wall Muscles
(*a*) Windows in the side reveal the various muscle layers. (*b*) Surface anatomy of anterior abdominal muscle. (Muscles of the abdominal wall are in bold.)

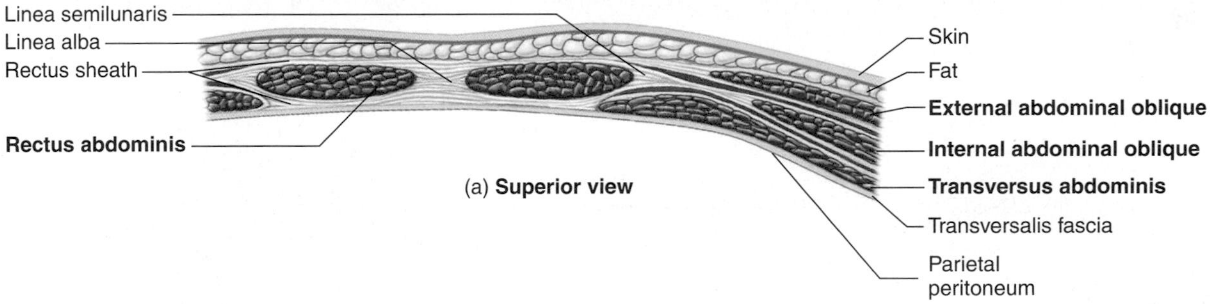

Linea semilunaris
Linea alba
Rectus sheath
Rectus abdominis
Skin
Fat
External abdominal oblique
Internal abdominal oblique
Transversus abdominis
Transversalis fascia
Parietal peritoneum

(a) **Superior view**

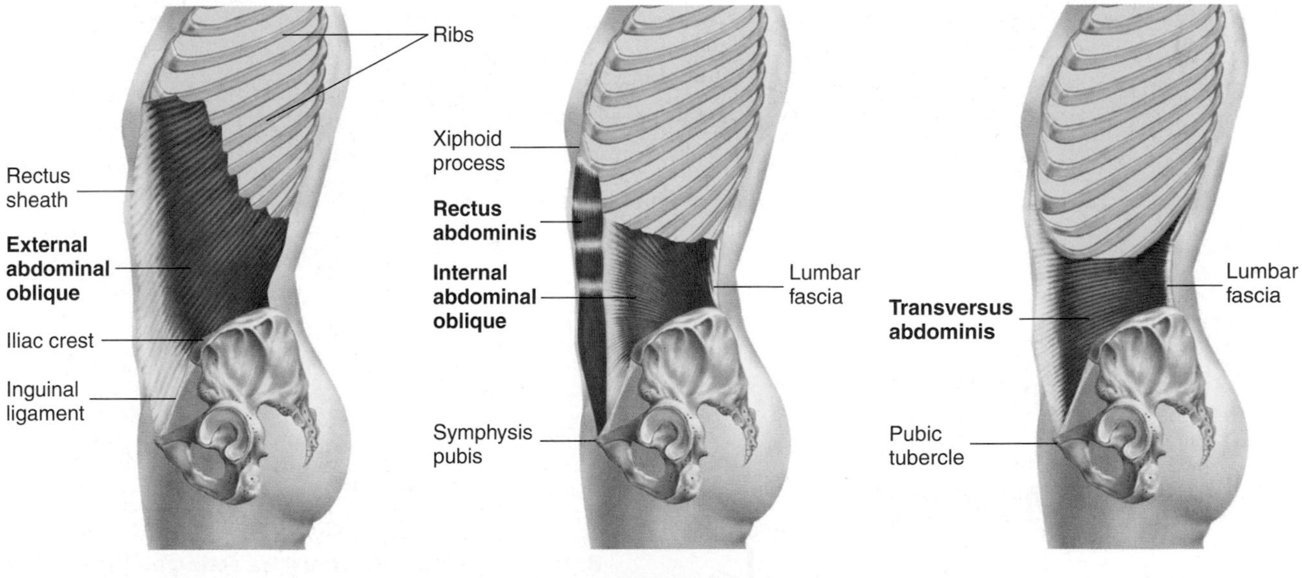

Ribs
Rectus sheath
External abdominal oblique
Iliac crest
Inguinal ligament
Xiphoid process
Rectus abdominis
Internal abdominal oblique
Symphysis pubis
Lumbar fascia
Transversus abdominis
Pubic tubercle
Lumbar fascia

(b) **Lateral views**

FIGURE 10.18 APR Anterior Abdominal Wall Muscles
(*a*) Cross section superior to the umbilicus. (*b*) Abdominal muscles shown individually. (Muscle names are in bold.)

TABLE 10.12 Muscles of the Pelvic Floor and Perineum (see figure 10.19)

Muscle	Origin	Insertion	Nerve	Action
Bulbospongiosus (bul′bō-spŭn′jē-ō′sŭs)	Male—central tendon of perineum and median raphe of penis	Dorsal surface of penis and bulb of penis	Pudendal	Constricts urethra; erects penis
	Female—central tendon of perineum	Base of clitoris	Pudendal	Erects clitoris
Coccygeus (kok-si′jē-ŭs; not illustrated)	Ischial spine	Coccyx	S3 and S4	Elevates and supports pelvic floor
Ischiocavernosus (ish′ē-ō-kav′er-nō′sŭs)	Ischial ramus	Corpus cavernosum	Perineal	Compresses base of penis or clitoris
Levator ani (lē-vā′tor, le-vā′ter ā′nī)	Posterior pubis and ischial spine	Sacrum and coccyx	Fourth sacral	Elevates anus; supports pelvic viscera
External anal sphincter (ā′năl sfingk′ter)	Coccyx	Central tendon of perineum	Fourth sacral and pudendal	Keeps orifice of anal canal closed
External urethral sphincter (ū-rē′thrăl sfingk′ter; not illustrated)	Pubic ramus	Median raphe	Pudendal	Constricts urethra
Transverse perinei (pĕr′i-nē′ī)				
Deep	Ischial ramus	Median raphe	Pudendal	Supports pelvic floor
Superficial	Ischial ramus	Central perineal	Pudendal	Fixes central tendon

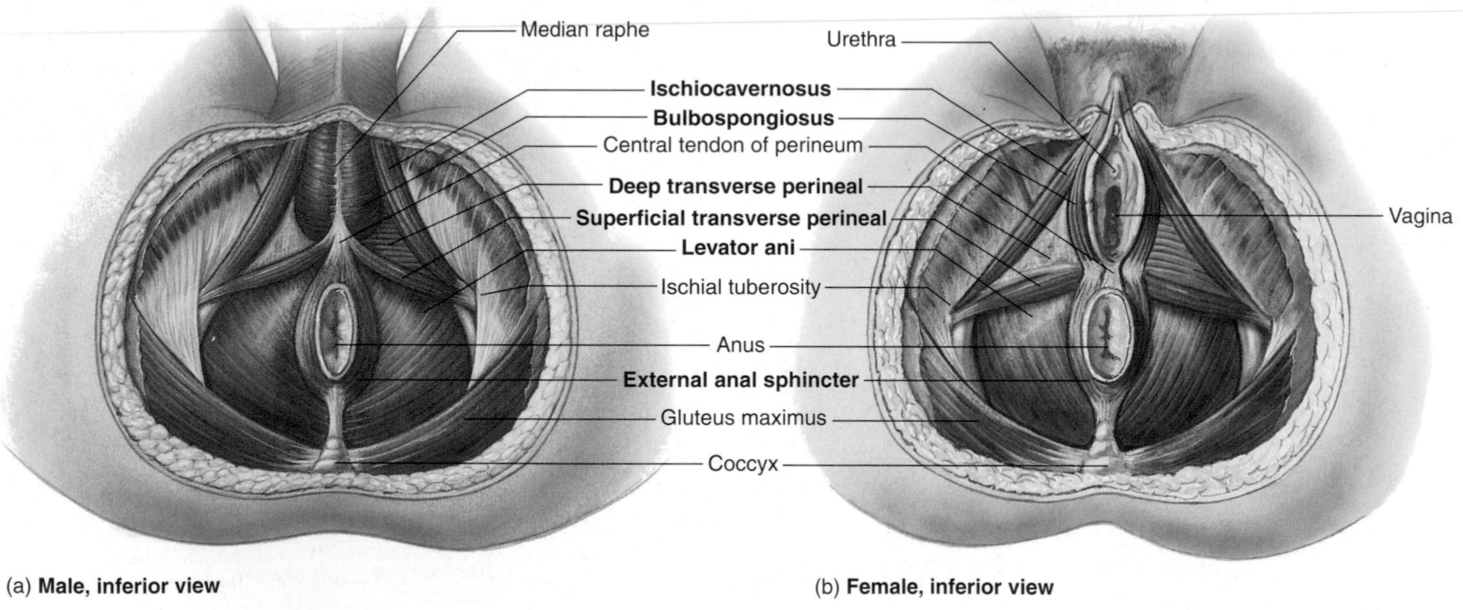
Median raphe
Urethra
Ischiocavernosus
Bulbospongiosus
Central tendon of perineum
Deep transverse perineal
Superficial transverse perineal
Levator ani
Ischial tuberosity
Anus
External anal sphincter
Gluteus maximus
Coccyx
Vagina
(a) Male, inferior view
(b) Female, inferior view

FIGURE 10.19 Muscles of the Pelvic Floor and Perineum
(Muscle names are in bold.)

10.4 Upper Limb Muscles

LEARNING OUTCOMES

After reading this section, you should be able to

A. Describe the movements of the scapula and list the muscles associated with it.

B. Name and locate the muscles acting on the shoulder and arm and explain their movements.

C. List and describe the muscles and movements of the forearm, wrist, hand, and fingers.

D. Distinguish between extrinsic and intrinsic hand muscles.

TABLE 10.13 Muscles Acting on the Scapula (see figure 10.20)

Muscle	Origin	Insertion	Nerve	Action
Levator scapulae (lē-vā′tor, le-vā′ter skap′ū-lē)	Transverse processes of C1–C4	Superior angle of scapula	Dorsal scapular	Elevates, retracts, and rotates scapula; laterally flexes neck
Pectoralis minor (pek′tō-ra′lis)	Third to fifth ribs	Coracoid process of scapula	Medial pectoral	Depresses scapula or elevates ribs
Rhomboideus (rom-bō-id′ē-ŭs)				
Major	Spinous processes of T1–T4	Medial border of scapula	Dorsal scapular	Retracts, rotates, and fixes scapula
Minor	Spinous processes of C6–C7	Medial border of scapula	Dorsal scapular	Retracts, slightly elevates, rotates, and fixes scapula
Serratus anterior (ser-ā′tŭs)	First to eighth or ninth ribs	Medial border of scapula	Long thoracic	Rotates and protracts scapula; elevates ribs
Subclavius (sŭb-klā′vē-ŭs)	First rib	Clavicle	Subclavian	Fixes clavicle or elevates first rib
Trapezius (tra-pē′zē-ŭs)	External occipital protuberance, ligamentum nuchae, and spinous processes of C7–T12	Clavicle, acromion process, and scapular spine	Accessory and cervical plexus	Elevates, depresses, retracts, rotates, and fixes scapula; extends neck

The major connection of the upper limb to the body is accomplished by muscles (table 10.13; figure 10.20). The muscles of the upper limb include those that move the scapula and those that move the arm, forearm, and hand.

Scapular Movements

The muscles attaching the scapula to the thorax include the **trapezius, levator scapulae** (skap′ū-lē), **rhomboideus** (rom-bō-id′ē-ŭs) **major** and **rhomboideus minor, serratus** (sĕr-ā′tŭs) **anterior,** and **pectoralis** (pek′tō-ra′lis) **minor.** These muscles move the scapula, permitting a wide range of movements of the upper limb, or they act as fixators to hold the scapula firmly in position when the arm muscles contract. The superficial muscles that act on the scapula can easily be seen on a living person (see figure 10.5*b*): The trapezius forms the upper line from each shoulder to the neck, and the origin of the serratus anterior from the first eight or nine ribs can be seen along the lateral thorax. The serratus anterior inserts onto the medial border of the scapula (figure 10.20*c*).

Arm Movements

The arm is attached to the thorax by several muscles, including the **pectoralis major** and the **latissimus dorsi** (lă-tis′i-mŭs dōr′sī; table 10.14; see figures 10.20*b*, 10.21, and 10.22). Notice that the pectoralis major is listed in table 10.14 as both a flexor and an extensor. The muscle flexes the extended shoulder and extends the flexed shoulder. Try these movements and notice the position and action of the muscle. The **deltoid** (deltoideus) muscle (figure 10.21) is also listed in table 10.14 as a flexor and an extensor. The deltoid muscle is like three muscles in one: The anterior fibers flex the shoulder, the lateral fibers abduct the arm, and the posterior fibers extend the shoulder. The deltoid muscle is part of the group of muscles that binds the humerus to the scapula. However, the primary muscles holding the head of the humerus in the glenoid cavity are called the **rotator cuff** muscles (listed separately in table 10.14) because they form a cuff or cap over the proximal humerus (figure 10.23). A rotator cuff injury involves damage to one or more of these muscles or their tendons, usually the supraspinatus muscle. The muscles moving the arm are involved in flexion, extension, abduction, adduction, rotation, and circumduction (table 10.15).

To visualize how these muscle groups work together, imagine that you want to raise your arm so that your hand is high above your head. First, you must abduct your arm from the anatomical position through 90 degrees (to the point at which the hand is level with the shoulder); this involves moving the humerus and is accomplished by the deltoid muscle assisted by the rotator cuff muscles, which hold the head of the humerus tightly in place. In the initial phase of abduction (first 15 degrees), the deltoid is assisted by the supraspinatus. Place your hand on your deltoid, and feel it contract as you abduct 90 degrees. Next, move your arm from 90 degrees to 180 degrees, so that your hand is high above your head; this movement primarily involves rotation of the scapula, which is accomplished by the trapezius and serratus anterior muscles. Feel the inferior angle of your scapula as you abduct your arm to 90 degrees and then rotate to 180 degrees. Do you notice a big difference? Bear in mind that your arm cannot move from 90 degrees to 180 degrees unless the head of the humerus is held tightly in the glenoid cavity by the rotator cuff muscles, especially the supraspinatus. Damage to the supraspinatus muscle can prevent abduction past 90 degrees.

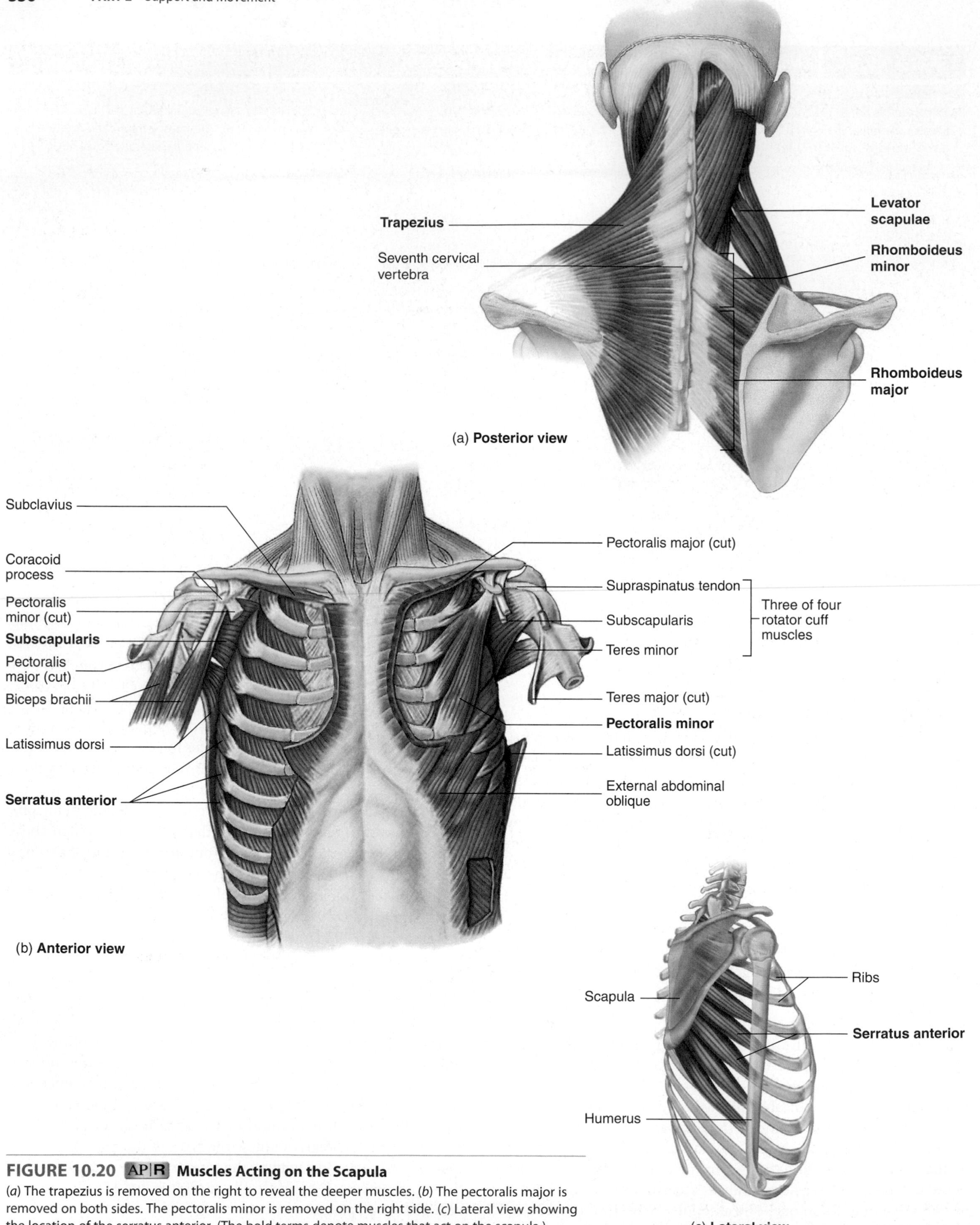
Trapezius
Seventh cervical vertebra
Levator scapulae
Rhomboideus minor
Rhomboideus major
(a) Posterior view
Subclavius
Coracoid process
Pectoralis minor (cut)
Subscapularis
Pectoralis major (cut)
Biceps brachii
Latissimus dorsi
Serratus anterior
Pectoralis major (cut)
Supraspinatus tendon
Subscapularis
Teres minor
Three of four rotator cuff muscles
Teres major (cut)
Pectoralis minor
Latissimus dorsi (cut)
External abdominal oblique
(b) Anterior view
Scapula
Ribs
Serratus anterior
Humerus
(c) Lateral view

FIGURE 10.20 APIR **Muscles Acting on the Scapula**
(*a*) The trapezius is removed on the right to reveal the deeper muscles. (*b*) The pectoralis major is removed on both sides. The pectoralis minor is removed on the right side. (*c*) Lateral view showing the location of the serratus anterior. (The bold terms denote muscles that act on the scapula.)

TABLE 10.14 Muscles Acting on the Arm (see figures 10.20–10.23)

Muscle	Origin	Insertion	Nerve	Action
Coracobrachialis (kōr′ă-kō-brā-kā-ā′lis)	Coracoid process of scapula	Midshaft of humerus	Musculocutaneous	Adducts arm and flexes shoulder
Deltoid (del′toyd)	Clavicle, acromion process, and scapular spine	Deltoid tuberosity	Axillary	Flexes and extends shoulder; abducts and medially and laterally rotates arm
Latissimus dorsi (lă-tis′i-mŭs dōr′sī)	Spinous processes of T7–L5; sacrum and iliac crest; inferior angle of scapula in some people	Medial crest of intertubercular groove	Thoracodorsal	Adducts and medially rotates arm; extends shoulder
Pectoralis major (pek′tō-rā′lis)	Clavicle, sternum, superior six costal cartilages, and abdominal aponeurosis	Lateral crest of intertubercular groove	Medial and lateral pectoral	Flexes shoulder; adducts and medially rotates arm; extends shoulder from flexed position
Teres major (ter′ēz, tēr′ēz)	Lateral border of scapula	Medial crest of intertubercular groove	Lower subscapular C5 and C6	Extends shoulder; adducts and medially rotates arm
Rotator Cuff				
Infraspinatus (in-fră-spī-nā′tŭs)	Infraspinous fossa of scapula	Greater tubercle of humerus	Suprascapular C5 and C6	Laterally rotates arm; holds head of humerus in place
Subscapularis (sŭb-skap-ū-lā′ris)	Subscapular fossa	Lesser tubercle of humerus	Upper and lower sub-scapular C5 and C6	Medially rotates arm; holds head of humerus in place
Supraspinatus (soo-pră-spī-nā′tŭs)	Supraspinous fossa	Greater tubercle of humerus	Suprascapular C5 and C6	Abducts arm; holds head of humerus in place
Teres minor (ter′ēz, tēr′ēz)	Lateral border of scapula	Greater tubercle of humerus	Axillary C5 and C6	Laterally rotates and adducts arm; holds head of humerus in place

Predict 4

A tennis player complains of pain in the shoulder when she abducts her arm while serving or reaching for an overhead volley (extreme abduction). In extreme abduction, the supraspinatus muscle rises superiorly and may be damaged by compression against what bony structure?

Several muscles that act on the arm can be seen very clearly in the living individual (see figures 10.21*c* and 10.22*c*). The pectoralis major forms the upper chest, and the deltoids are prominent over the shoulders. The deltoid is a common site for administering injections.

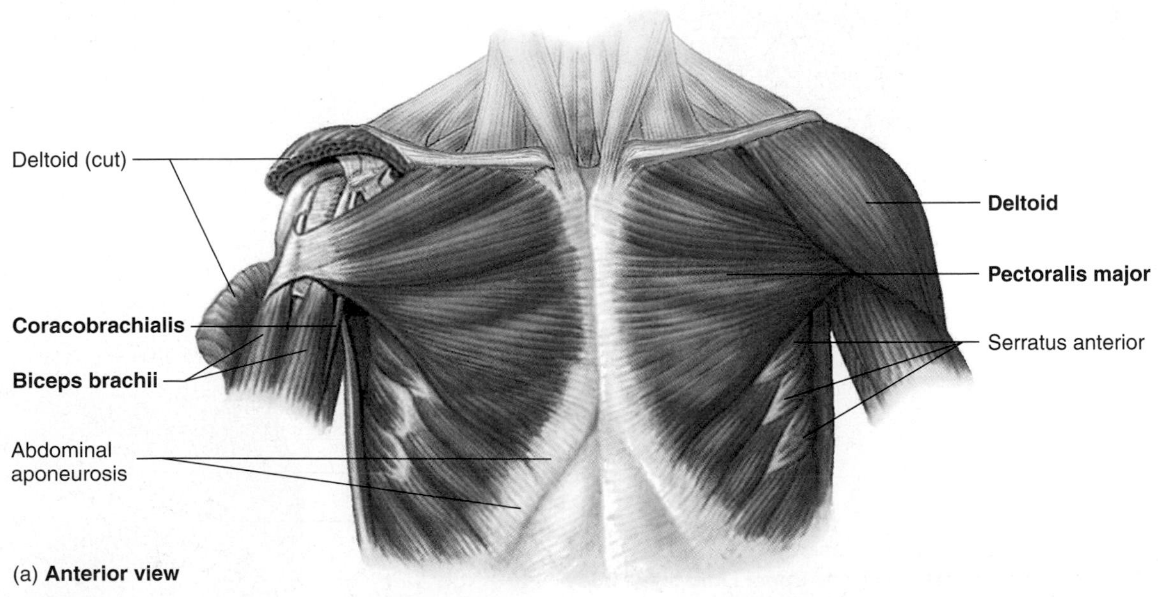
Deltoid (cut)
Coracobrachialis
Biceps brachii
Abdominal aponeurosis
(a) Anterior view
Deltoid
Pectoralis major
Serratus anterior

FIGURE 10.21 APR **Anterior Muscles Attaching the Upper Limb to the Body**
(*a*) Anterior pectoral muscles. (Names of muscles attaching the upper limb to the body are in bold.)

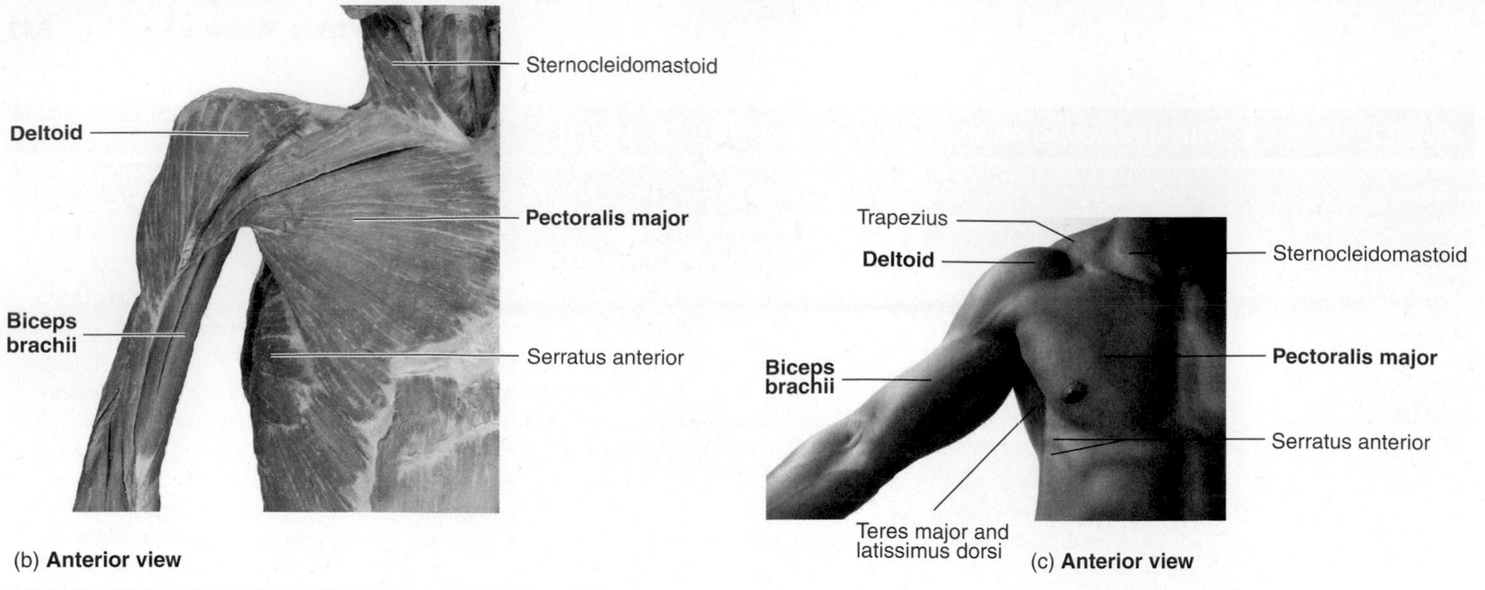
Sternocleidomastoid
Deltoid
Pectoralis major
Biceps brachii
Serratus anterior
(b) Anterior view
Trapezius
Deltoid
Sternocleidomastoid
Biceps brachii
Pectoralis major
Serratus anterior
Teres major and latissimus dorsi
(c) Anterior view

FIGURE 10.21 (continued)
(*b*) Right pectoral region of a cadaver. (*c*) Surface anatomy of the right anterior pectoral region. (Names of the upper limb muscles are in bold.)

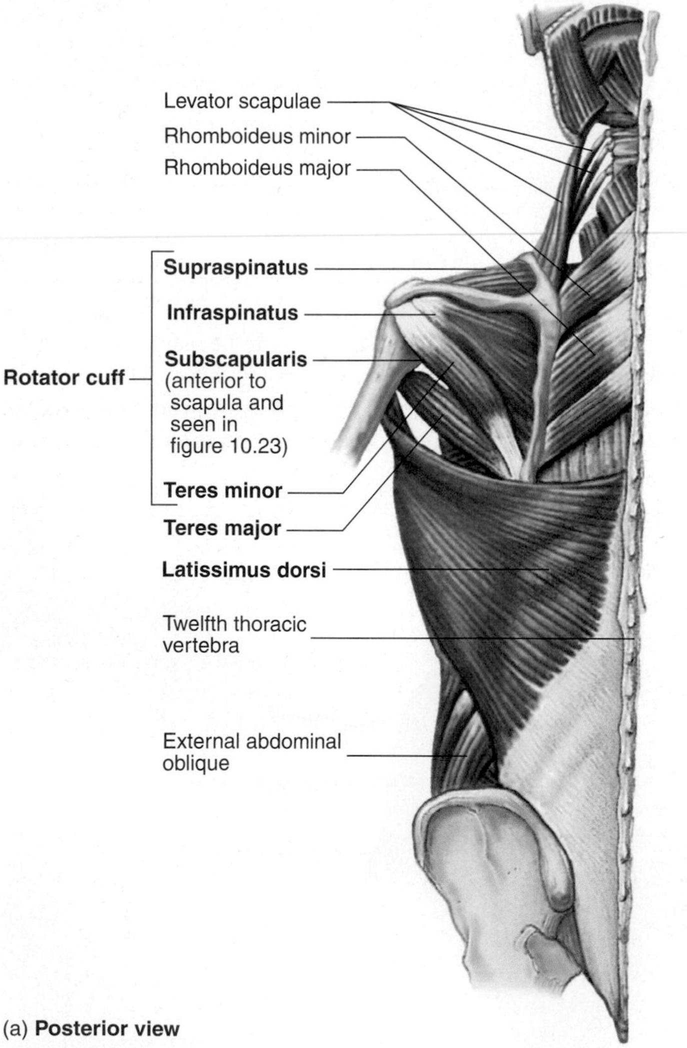

Levator scapulae
Rhomboideus minor
Rhomboideus major
Rotator cuff
Supraspinatus
Infraspinatus
Subscapularis (anterior to scapula and seen in figure 10.23)
Teres minor
Teres major
Latissimus dorsi
Twelfth thoracic vertebra
External abdominal oblique

(a) **Posterior view**

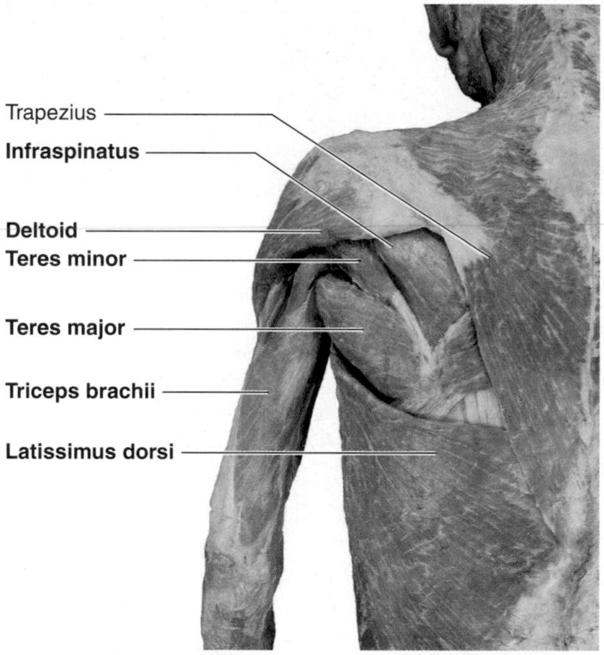

Trapezius
Infraspinatus
Deltoid
Teres minor
Teres major
Triceps brachii
Latissimus dorsi

(b) **Posterior view**

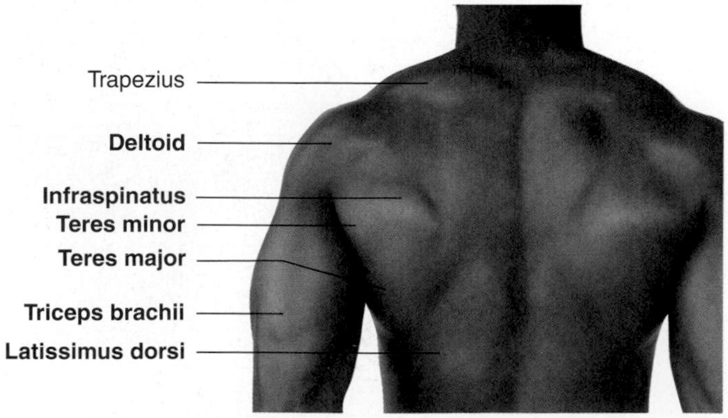

Trapezius
Deltoid
Infraspinatus
Teres minor
Teres major
Triceps brachii
Latissimus dorsi

(c) **Posterior view**

FIGURE 10.22 Posterior Muscles Attaching the Upper Limb to the Body
(*a*) Posterior view of muscles of the left posterior pectoral region. (*b*) Posterior view of a cadaver. (*c*) Surface anatomy. (Names of muscles for upper limb attachment are in bold.)

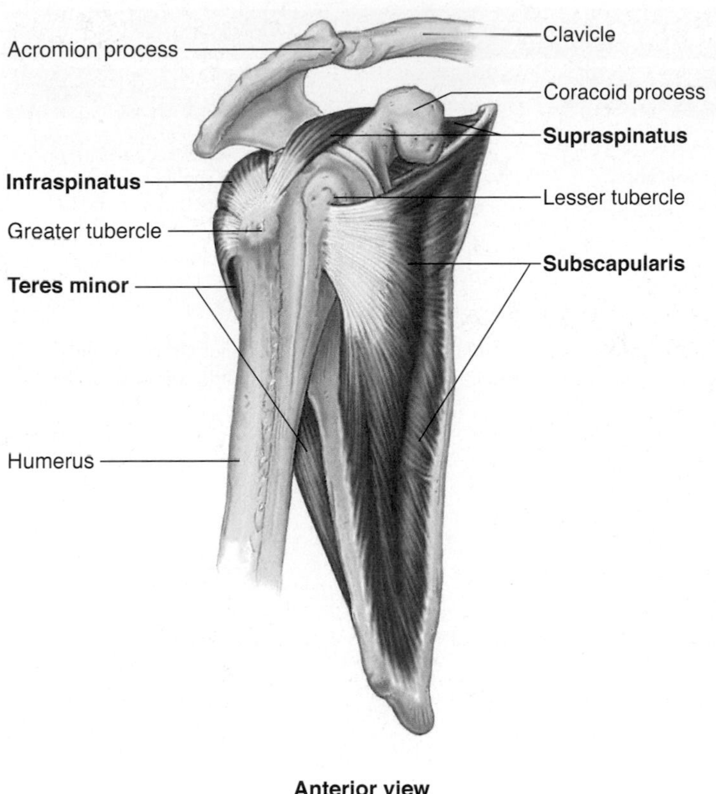

Acromion process
Clavicle
Coracoid process
Supraspinatus
Infraspinatus
Lesser tubercle
Greater tubercle
Subscapularis
Teres minor
Humerus
Anterior view

FIGURE 10.23 Right Rotator Cuff Muscles
(Muscle names are in bold.)

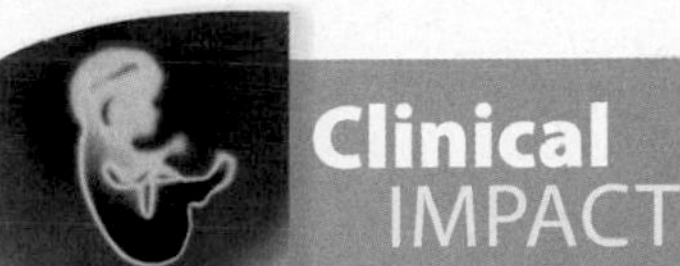

Clinical
IMPACT

Shoulder Pain and Torn Rotator Cuff

Baseball pitchers, because they throw very hard, may tear their rotator cuffs. Such tears result in pain in the anterosuperior part of the shoulder. Older people may also develop such pain because of **degenerative tendinitis** of the rotator cuff. The supraspinatus tendon is the most commonly affected part of the rotator cuff in either trauma or degeneration, probably because it has a relatively poor blood supply. If the damage is severe, surgery is required to repair the area. During surgery, the loose tissue debris is removed, and the scapula is shaved or smoothed to make more room for the supraspinatus tendon. Finally, the torn edges of the supraspinatus tendon are sewn together and to the top of the humerus.

Pain in the shoulder can also result from **subacromial bursitis,** which is inflammation of the subacromial bursa. **Biceps tendinitis,** inflammation of the biceps brachii long head tendon, can also cause shoulder pain. This inflammation is also commonly caused by throwing a baseball or football.

Forearm Movements

Extension and Flexion of the Elbow

Extension of the elbow is accomplished by the **triceps brachii** (brā′kē-ī) and the **anconeus** (ang-kō′nē-ŭs); flexion of the elbow is accomplished by the **brachialis** (brā′-kē-al′is), the **biceps brachii,** and the **brachioradialis** (brā′kē-ō-rā′dē-al′is; table 10.16; figure 10.24; see figure10.26*a*). The triceps brachii constitutes the main mass visible on the posterior aspect of the arm (see figures 10.22*c* and 10.24*c*). The biceps brachii is readily visible on the anterior aspect of the arm (see figures 10.21*c* and 10.24*d*). The brachialis lies deep to the biceps brachii and can be seen only as a mass on the medial and lateral sides of the arm. The brachioradialis forms a bulge on the anterolateral side of the forearm just distal to the elbow (figures 10.24*b* and 10.25*b*,*d*). If the elbow is forcefully flexed in the midprone position (midway between pronation and supination), the brachioradialis stands out clearly on the forearm (figure 10.25*d*).

Supination and Pronation

Supination of the forearm and hand is accomplished by muscles acting on the forearm, the **supinator** and the **biceps brachii** (see figures 10.24, 10.25*c*, and 10.26*b*). Pronation is a function of the **pronator quadratus** (kwah-drā′tŭs) and the **pronator teres** (ter′ēz, tēr′-ēz; see figures 10.24*a* and 10.25*a*,*c*).

Predict 5

Explain the difference between doing chin-ups with the forearm supinated and doing them with it pronated. The action of which muscle predominates in each type of chin-up? Which type is easier? Why?

TABLE 10.15 Summary of Muscle Actions on the Shoulder and Arm

Flexion	Extension	Abduction	Adduction	Medial Rotation	Lateral Rotation
Deltoid	Deltoid	Deltoid	Pectoralis major	Pectoralis major	Deltoid
Pectoralis major	Teres major	Supraspinatus	Latissimus dorsi	Teres major	Infraspinatus
Coracobrachialis	Latissimus dorsi		Teres major	Latissimus dorsi	Teres minor
Biceps brachii	Pectoralis major		Teres minor	Deltoid	
	Triceps brachii		Triceps brachii	Subscapularis	
			Coracobrachialis		

TABLE 10.16 Muscles Acting on the Forearm (see figure 10.24)

Muscle	Origin	Insertion	Nerve	Action
Arm				
Biceps brachii (bī′seps brā′kē-ī)	Long head—supraglenoid tubercle	Radial tuberosity and aponeurosis of biceps brachii	Musculocutaneous	Flexes shoulder and elbow; supinates forearm and hand
	Short head—coracoid process			
Brachialis (brā′kē-al′is)	Anterior surface of humerus	Ulnar tuberosity and coronoid process of ulna	Musculocutaneous and radial	Flexes elbow
Triceps brachii (trī′seps brā′kē-ī)	Long head—infraglenoid tubercle on lateral border of scapula	Olecranon process of ulna	Radial	Extends elbow; extends shoulder and adducts arm
	Lateral head—lateral and posterior surface of humerus			
	Medial head—posterior humerus			
Forearm				
Anconeus (ang-kō′nē-ŭs)	Lateral epicondyle of humerus	Olecranon process and posterior ulna	Radial	Extends elbow
Brachioradialis (brā′kē-ō-rā′dē-al′is)	Lateral supracondylar ridge of humerus	Styloid process of radius	Radial	Flexes elbow
Pronator quadratus (prō-nā′ter, prō-nā′tōr kwah-drā′tŭs)	Distal ulna	Distal radius	Anterior interosseous	Pronates forearm (and hand)
Pronator teres (prō-nā-tōr ter′ēz, tēr′ēz)	Medial epicondyle of humerus and coronoid process of ulna	Radius	Median	Pronates forearm (and hand)
Supinator (soo′pi-nā-ter, soo′pi-nā-tōr)	Lateral epicondyle of humerus and ulna	Radius	Radial	Supinates forearm (and hand)

Wrist, Hand, and Finger Movements

The forearm muscles are divided into anterior and posterior groups (table 10.17; see figures 10.25 and 10.26). Most of the anterior forearm muscles are responsible for flexion of the wrist and fingers. Most of the posterior forearm muscles cause extension of the wrist and fingers.

Extrinsic Hand Muscles

The **extrinsic hand muscles** are in the forearm but have tendons that extend into the hand. A strong band of fibrous connective tissue, the **extensor retinaculum** (ret-i-nak′ū-lŭm; bracelet), covers the flexor and extensor tendons and holds them in place around the wrist, so that they do not "bowstring" (pull away from the bone) during muscle contraction (figure 10.26*a,c*).

Two major anterior muscles, the **flexor carpi radialis** (kar′pī rā-dē-ā′lis) and the **flexor carpi ulnaris** (ŭl-nā′ris), flex the wrist; and three posterior muscles, the **extensor carpi radialis longus,** the **extensor carpi radialis brevis,** and the **extensor carpi ulnaris,** extend the wrist. The tendon of the flexor carpi radialis serves as a landmark for locating the radial pulse, which is lateral to the tendon (see figure 10.25*d*). The wrist flexors and extensors are visible on the anterior and posterior surfaces of the forearm (see figures 10.25*d* and 10.26*d*).

Flexion of the four medial digits is a function of the **flexor digitorum** (dij′i-tor′ŭm) **superficialis** and the **flexor digitorum profundus** (prō-fŭn′dŭs; deep). Extension is accomplished by the **extensor digitorum.** The tendons of this muscle are very visible on the dorsum of the hand (figure 10.26*d*). The little finger has an additional

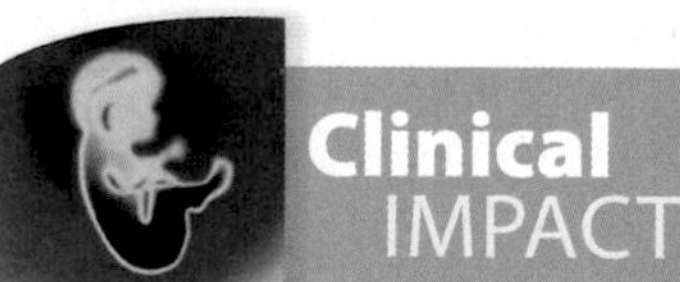

Clinical IMPACT

Tennis Elbow

Forceful, repetitive use of the forearm extensor muscles can damage them where they attach to the lateral epicondyle. This condition is often called **tennis elbow** because it can result from playing tennis. It is also called **lateral epicondylitis** because it can result from other sports and activities, including shoveling snow! Treatment rarely requires surgery; rather, noninvasive practices, such as rest, ice, compression, and elevation (RICE) and anti-inflammatory medications, are usually effective. It is also possible to experience "golfer's elbow," or **medial epicondylitis,** where the medial forearm tendons attach to the medial epicondyle of the humerus.

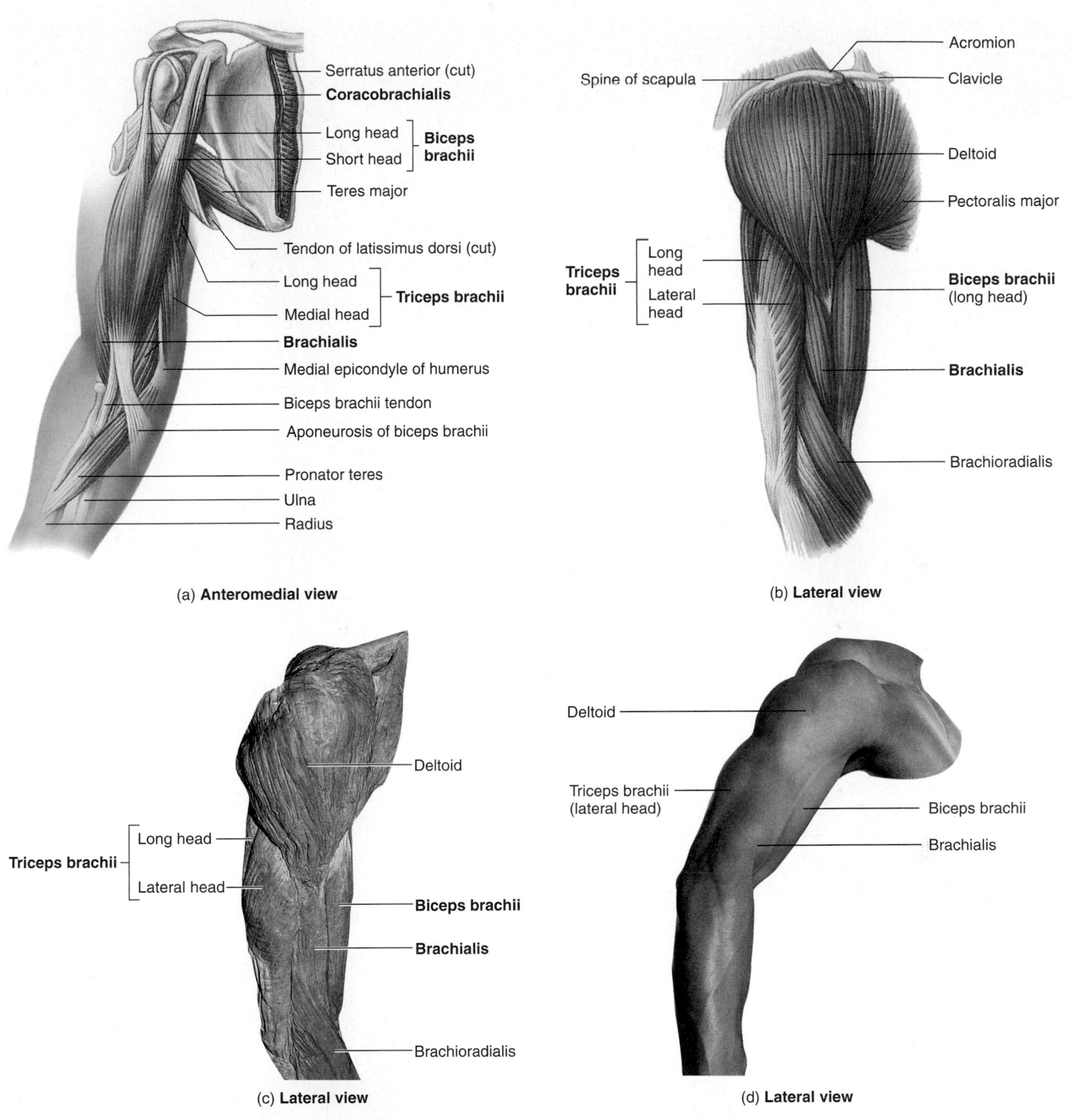
Serratus anterior (cut)
Coracobrachialis
Long head
Short head
Biceps brachii
Teres major
Tendon of latissimus dorsi (cut)
Long head
Medial head
Triceps brachii
Brachialis
Medial epicondyle of humerus
Biceps brachii tendon
Aponeurosis of biceps brachii
Pronator teres
Ulna
Radius
(a) **Anteromedial view**
Acromion
Spine of scapula
Clavicle
Deltoid
Pectoralis major
Triceps brachii
Long head
Lateral head
Biceps brachii (long head)
Brachialis
Brachioradialis
(b) **Lateral view**
Deltoid
Triceps brachii
Long head
Lateral head
Biceps brachii
Brachialis
Brachioradialis
(c) **Lateral view**
Deltoid
Triceps brachii (lateral head)
Biceps brachii
Brachialis
(d) **Lateral view**

FIGURE 10.24 APR **Lateral Right Arm Muscles**

(*a, b*) The right shoulder and arm. (*c*) The right shoulder and arm muscles of a cadaver. (*d*) Surface anatomy of the right shoulder and arm. (Names of arm muscles are in bold.)

extensor, the **extensor digiti minimi** (dij′i-tī min′i-mī). The index finger also has an additional extensor, the **extensor indicis** (in′di-sis).

Movement of the thumb is caused in part by the **abductor pollicis** (pol′i-sis) **longus,** the **extensor pollicis longus,** and the **extensor pollicis brevis.** These tendons form the sides of a depression on the posterolateral side of the wrist called the "anatomical snuffbox" (figure 10.26*d*). When snuff was in use, a small pinch could be placed into the anatomical snuffbox and inhaled through the nose.

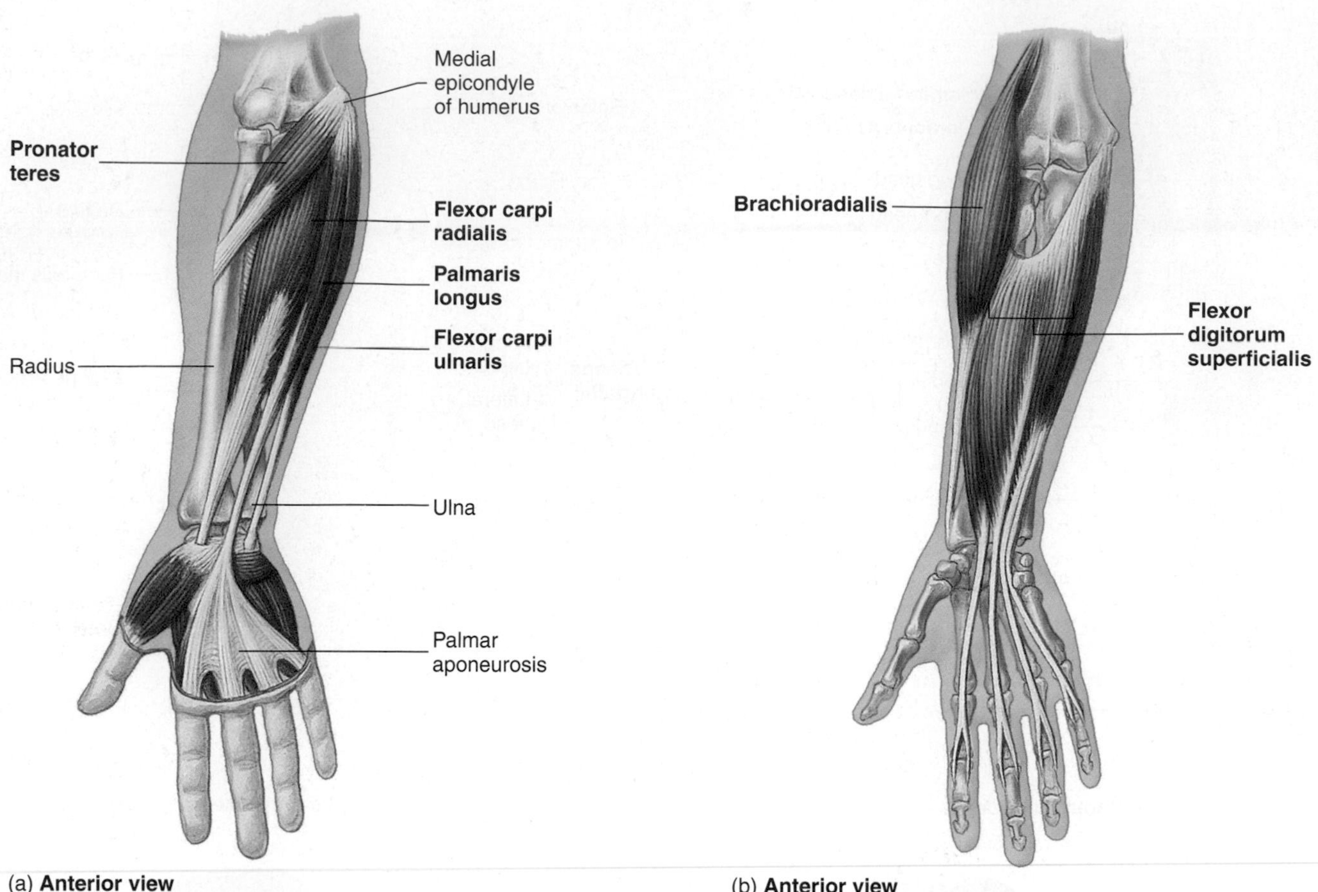
Medial
epicondyle
of humerus
**Pronator
teres**
**Flexor carpi
radialis**
**Palmaris
longus**
**Flexor carpi
ulnaris**
Radius
Ulna
Palmar
aponeurosis
(a) **Anterior view**
Brachioradialis
**Flexor
digitorum
superficialis**
(b) **Anterior view**

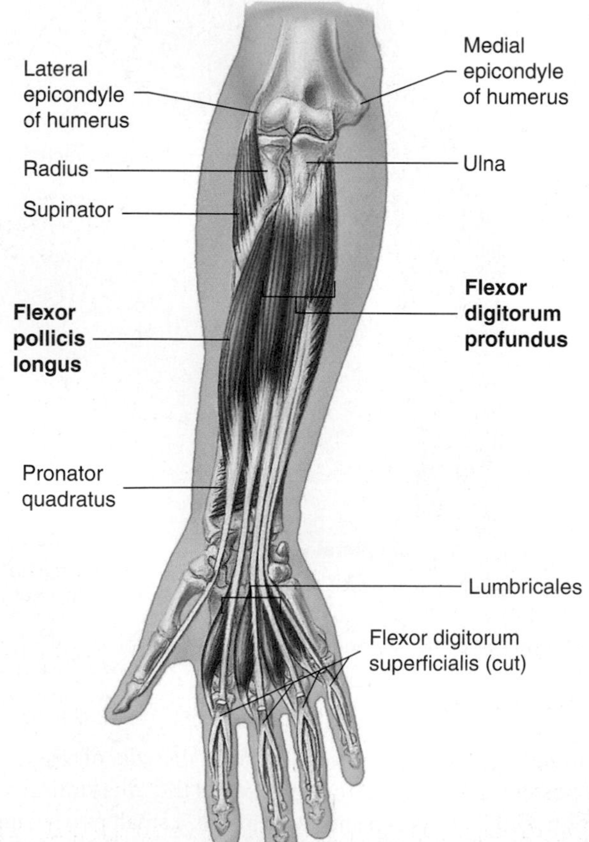
Lateral
epicondyle
of humerus
Medial
epicondyle
of humerus
Radius
Ulna
Supinator
**Flexor
digitorum
profundus**
**Flexor
pollicis
longus**
Pronator
quadratus
Lumbricales
Flexor digitorum
superficialis (cut)
(c) Anterior view

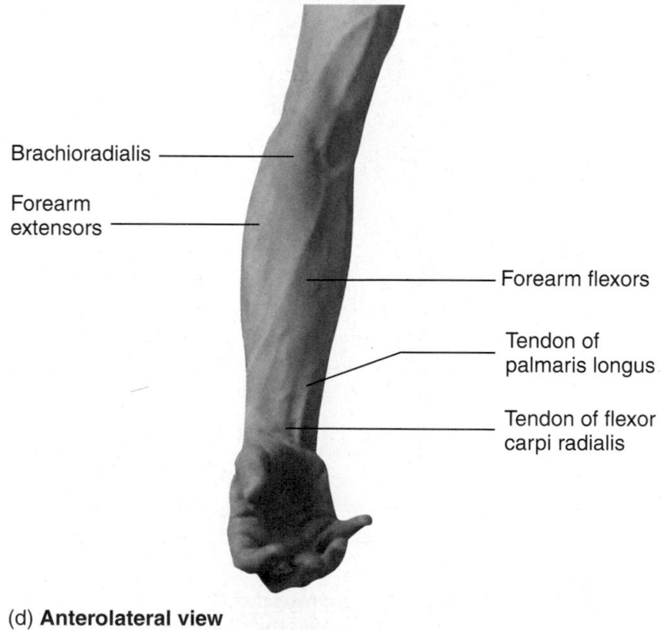
Brachioradialis
Forearm
extensors
Forearm flexors
Tendon of
palmaris longus
Tendon of flexor
carpi radialis
(d) **Anterolateral view**

FIGURE 10.25 Anterior Right Forearm Muscles
(*a*) Right forearm (superficial). The brachioradialis muscle is removed. (*b*) Right forearm (deeper than *a*). The pronator teres, flexor carpi radialis and ulnaris, and palmaris longus muscles are removed. (*c*) Right forearm (deeper than *a* or *b*). The brachioradialis, pronator teres, flexor carpi radialis and ulnaris, palmaris longus, and flexor digitorum superficialis muscles are removed. (*d*) Surface anatomy of anterior forearm muscles. (Muscle names are in bold.)

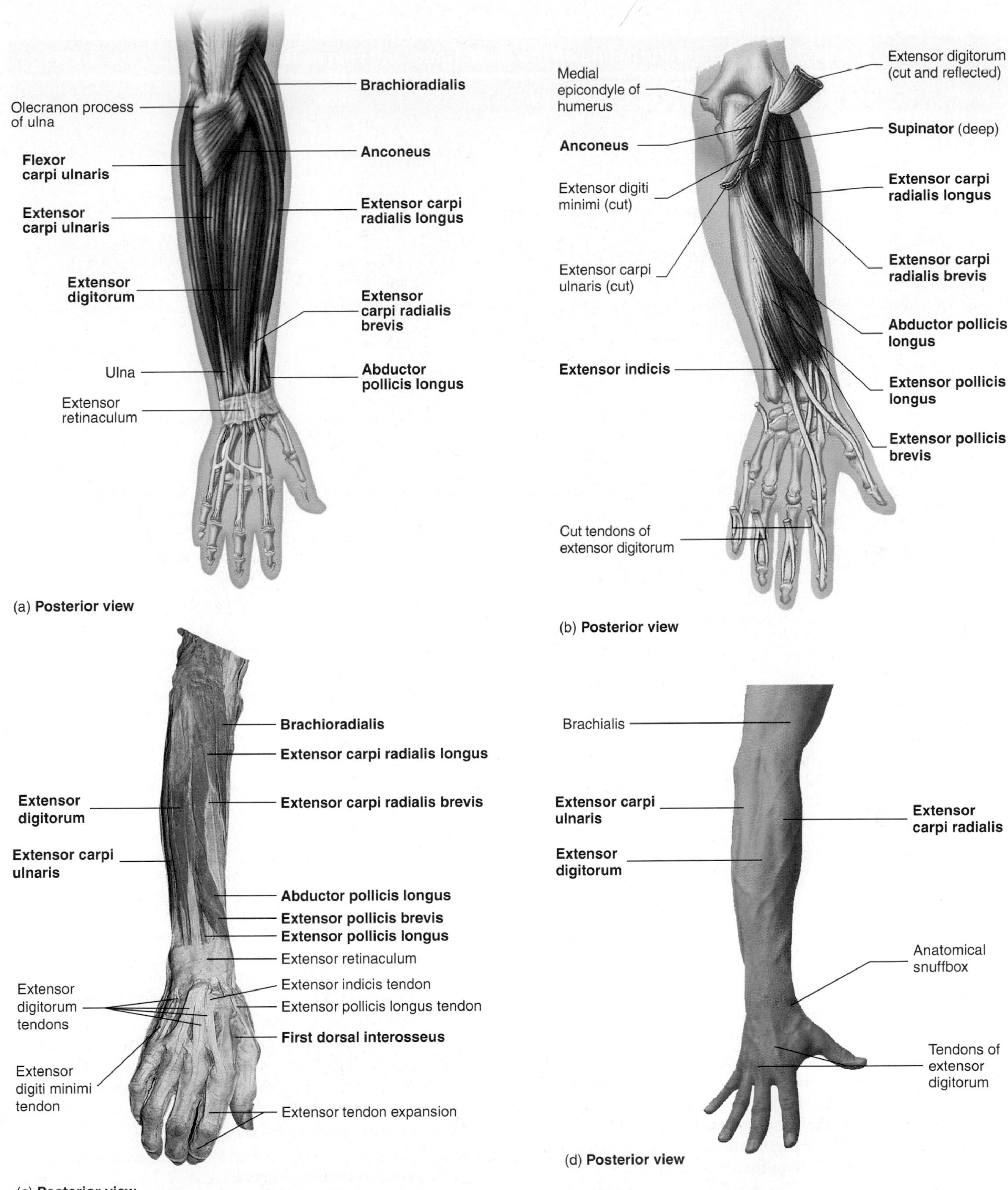
Brachioradialis
Olecranon process of ulna
Flexor carpi ulnaris
Anconeus
Extensor carpi ulnaris
Extensor carpi radialis longus
Extensor digitorum
Extensor carpi radialis brevis
Ulna
Abductor pollicis longus
Extensor retinaculum
(a) Posterior view
Medial epicondyle of humerus
Extensor digitorum (cut and reflected)
Anconeus
Supinator (deep)
Extensor digiti minimi (cut)
Extensor carpi radialis longus
Extensor carpi ulnaris (cut)
Extensor carpi radialis brevis
Abductor pollicis longus
Extensor indicis
Extensor pollicis longus
Extensor pollicis brevis
Cut tendons of extensor digitorum
(b) Posterior view
Brachioradialis
Extensor carpi radialis longus
Extensor digitorum
Extensor carpi radialis brevis
Extensor carpi ulnaris
Abductor pollicis longus
Extensor pollicis brevis
Extensor pollicis longus
Extensor retinaculum
Extensor digitorum tendons
Extensor indicis tendon
Extensor pollicis longus tendon
First dorsal interosseus
Extensor digiti minimi tendon
Extensor tendon expansion
(c) Posterior view
Brachialis
Extensor carpi ulnaris
Extensor carpi radialis
Extensor digitorum
Anatomical snuffbox
Tendons of extensor digitorum
(d) Posterior view

FIGURE 10.26 Posterior Right Forearm Muscles

(*a*) Right forearm (superficial). (*b*) Deep muscles of the right posterior forearm. The extensor digitorum, extensor digiti minimi, and extensor carpi ulnaris muscles are cut to reveal deeper muscles. (*c*) Photograph showing dissection of the posterior right forearm and hand. (*d*) Surface anatomy of posterior forearm. (Muscle names are in bold.)

TABLE 10.17 Muscles of the Forearm Acting on the Wrist, Hand, and Fingers (see figures 10.25 and 10.26)

Muscle	Origin	Insertion	Nerve	Action
Anterior Forearm				
Flexor carpi radialis (kar′pī rā-dē-ā′lis)	Medial epicondyle of humerus	Second and third metacarpal bones	Median	Flexes and abducts wrist
Flexor carpi ulnaris (kar′pī ŭl-nā′ris)	Medial epicondyle of humerus and ulna	Pisiform, hamate, and fifth metacarpal bones	Ulnar	Flexes and adducts wrist
Flexor digitorum profundus (dij′i-tōr′ŭm prō-fŭn′dŭs)	Ulna	Distal phalanges of digits 2–5	Ulnar and median	Flexes fingers at metacarpophalangeal joints and interphalangeal joints and wrist
Flexor digitorum superficialis (dij′i-tōr′ŭm soo′per-fish-ē-ā′lis)	Medial epicondyle of humerus, coronoid process, and radius	Middle phalanges of digits 2–5	Median	Flexes fingers at interphalangeal joints and wrist
Flexor pollicis longus (pol′i-sis lon′gŭs)	Radius	Distal phalanx of thumb	Median	Flexes thumb
Palmaris longus (pawl-mār′is lon′gŭs)	Medial epicondyle of humerus	Palmar fascia	Median	Tenses palmar fascia; flexes wrist
Posterior Forearm				
Abductor pollicis longus (pol′i-sis lon′gŭs)	Posterior ulna and radius and interosseous membrane	Base of first metacarpal bone	Radial	Abducts and extends thumb; abducts wrist
Extensor carpi radialis brevis (kar′pī rā-dē-ā′lis brev′is)	Lateral epicondyle of humerus	Base of third metacarpal bone	Radial	Extends and abducts wrist
Extensor carpi radialis longus (kar′pī rā-dē-ā′lis lon′gus)	Lateral supracondylar ridge of humerus	Base of second metacarpal bone	Radial	Extends and abducts wrist
Extensor carpi ulnaris (kar′pī ŭl-nā′ris)	Lateral epicondyle of humerus and ulna	Base of fifth metacarpal bone	Radial	Extends and adducts wrist
Extensor digiti minimi (dij′i-tī min′i-mī)	Lateral epicondyle of humerus	Phalanges of digit 5	Radial	Extends little finger and wrist
Extensor digitorum (dij′i-tōr′ŭm)	Lateral epicondyle of humerus	Extensor tendon expansion over phalanges of digits 2–5	Radial	Extends fingers and wrist
Extensor indicis (in′di-sis)	Ulna	Extensor tendon expansion over digit 2	Radial	Extends forefinger and wrist
Extensor pollicis brevis (pol′i-sis brev′is)	Radius	Proximal phalanx of thumb	Radial	Extends and abducts thumb; abducts wrist
Extensor pollicis longus (pol′i-sis lon′gŭs)	Ulna	Distal phalanx of thumb	Radial	Extends thumb

Intrinsic Hand Muscles

The **intrinsic hand muscles** are entirely within the hand (table 10.18; figure 10.27). Abduction of the fingers is accomplished by the **dorsal interossei** (in′ter-os′e-ī) and the **abductor digiti minimi,** whereas adduction is a function of the **palmar interossei.**

The **flexor pollicis brevis,** the **abductor pollicis brevis,** and the **opponens pollicis** form a fleshy prominence at the base of the thumb called the **thenar** (thē′nar) **eminence** (figure 10.27*a*). The **abductor digiti minimi, flexor digiti minimi brevis,** and **opponens digiti minimi** constitute the **hypothenar eminence** on the ulnar side of the hand (figure 10.27*c*). The thenar and hypothenar muscles are involved in controlling the thumb and little finger.

ASSESS YOUR PROGRESS

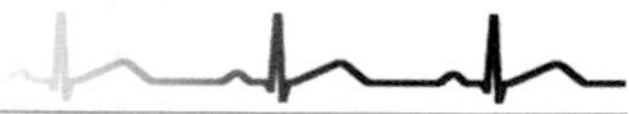

22. *Name the seven muscles that attach the scapula to the thorax. What muscles attach the arm to the thorax?*
23. *List the muscles forming the rotator cuff, and describe their function.*
24. *What muscles cause flexion and extension of the shoulder? Adduction and abduction of the arm? What muscle abducts the arm to 90 degrees? Above 90 degrees?*
25. *What muscles cause rotation of the arm?*
26. *List the muscles that cause flexion and extension of the elbow. Where are these muscles located?*
27. *What muscles produce supination and pronation of the forearm? Where are these muscles located?*
28. *Describe the muscle groups that cause flexion and extension of the wrist.*

TABLE 10.18 Intrinsic Hand Muscles (see figure 10.27)

Muscle	Origin	Insertion	Nerve	Action
Midpalmar Muscles				
Interossei (in′ter-os′e-ī)				
Dorsal	Sides of metacarpal bones	Proximal phalanges of digits 2, 3, and 4	Ulnar	Abducts second, third, and fourth digits
Palmar	Second, fourth, and fifth metacarpal bones	Digits 2, 4, and 5	Ulnar	Adducts second, fourth, and fifth digits
Lumbricals (lum′bră-kălz)	Tendons of flexor digitorum profundus	Digits 2–5	Two on radial side—median; two on ulnar side—ulnar	Flexes proximal and extends middle and distal phalanges
Thenar Muscles				
Abductor pollicis brevis (ab-dŭk-ter, ab-dŭk-tōr pol′i-sis brev′is)	Flexor retinaculum, trapezium, and scaphoid	Proximal phalanx of thumb	Median	Abducts thumb
Adductor pollicis (ă-dŭk′ter, ă-dŭk-tōr pol′i-sis)	Third metacarpal bone, second metacarpal bone, trapezoid, and capitate	Proximal phalanx of thumb	Ulnar	Adducts thumb
Flexor pollicis brevis (pol′i-sis brev′is)	Flexor retinaculum and first metacarpal bone	Proximal phalanx of thumb	Median and ulnar	Flexes thumb
Opponens pollicis (ŏ-pō′nens pol′i-sis)	Trapezium and flexor retinaculum	First metacarpal bone	Median	Opposes thumb
Hypothenar Muscles				
Abductor digiti minimi (ab-dŭk-ter, ab-dŭk-tōr dij′i-tī min′i-mī)	Pisiform	Base of digit 5	Ulnar	Abducts and flexes little finger
Flexor digiti minimi brevis (dij′i-tī min′i-mī brev′is)	Hamate	Base of proximal phalanx of digit 5	Ulnar	Flexes little finger
Opponens digiti minimi (ŏ-pō′nens dij′i-tī min′i-mī)	Hamate and flexor retinaculum	Fifth metacarpal bone	Ulnar	Opposes little finger

29. *Contrast the location and actions of the extrinsic and intrinsic hand muscles. What is the retinaculum?*

30. *Describe the muscles that move the thumb. The tendons of what muscles form the anatomical snuffbox?*

10.5 Lower Limb Muscles

LEARNING OUTCOMES

After reading this section, you should be able to

A. Summarize the muscles of the hip and thigh and explain their actions.

B. List and describe the muscles and movements of the ankle, foot, and toes.

Hip and Thigh Movements

Several hip muscles originate on the coxal bone and insert onto the femur (table 10.19; figures 10.28–10.31). These muscles are divided into three groups: anterior, posterolateral, and deep.

The anterior muscles, the **iliacus** (il-ī′ă-kŭs) and the **psoas** (sō′as) **major,** flex the hip (figure 10.28). Because these muscles share an insertion and produce the same movement, they are often referred to collectively as the **iliopsoas** (il′ē-ō-sō′as). When the thigh is fixed, the iliopsoas flexes the trunk on the thigh. For example, the iliopsoas does most of the work when a person does sit-ups.

The posterolateral hip muscles consist of the gluteal muscles and the **tensor fasciae latae** (fash′ē-ē lā′tē). The **gluteus** (gloo-tē′ŭs) **maximus** contributes most of the mass that can be seen as the buttocks (figure 10.29*c*); the **gluteus medius,** a common site for injections, creates a smaller mass just superior and lateral to the gluteus maximus. The gluteus maximus functions at its maximum force in extension of the thigh when the hip is flexed at a 45-degree angle, so that the muscle is optimally stretched, which accounts for both the sprinter's stance and the bicycle racing posture.

The deep hip muscles, as well as the gluteus maximus, function as lateral thigh rotators. The gluteus medius, gluteus minimus, and tensor fasciae latae are medial hip rotators (table 10.19; figure 10.29*b*). The gluteus medius and minimus muscles help tilt

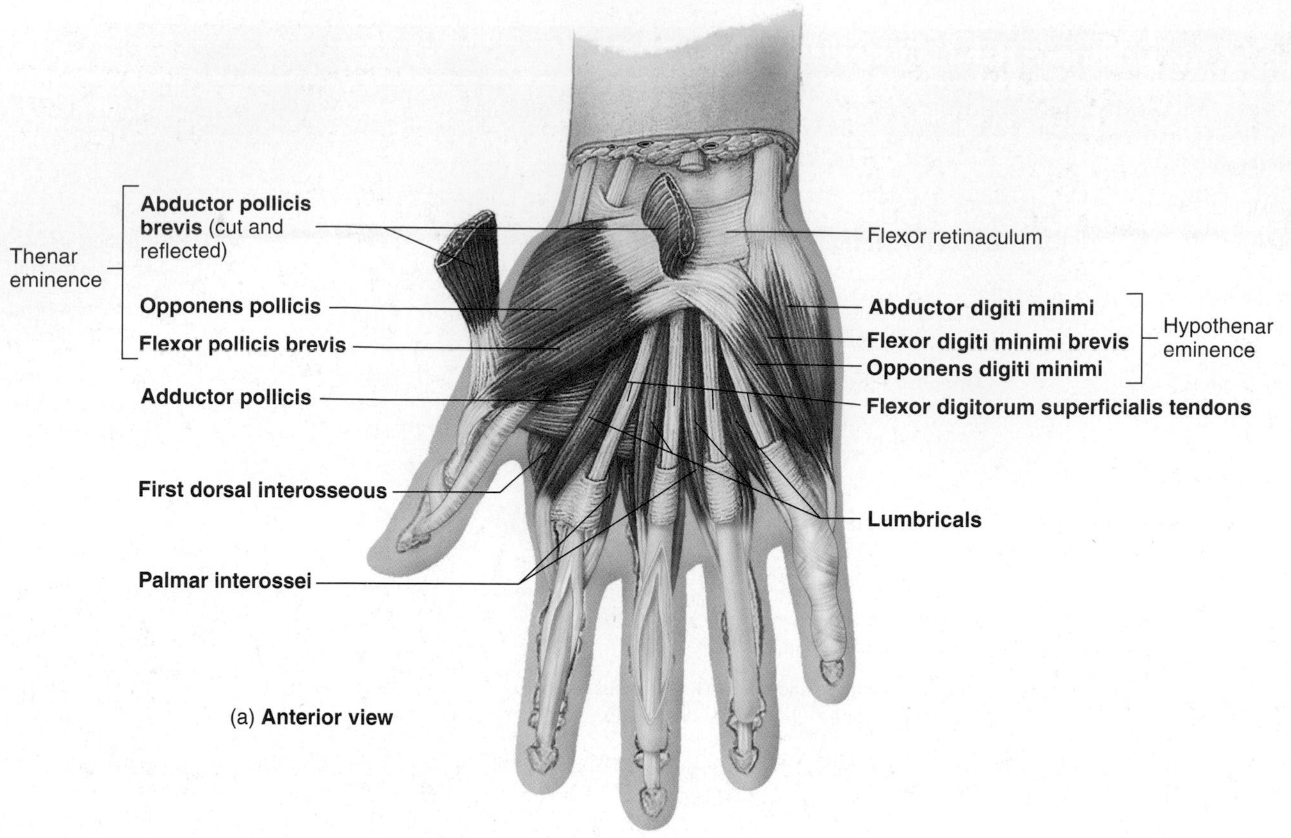
Thenar eminence
Abductor pollicis brevis (cut and reflected)
Opponens pollicis
Flexor pollicis brevis
Adductor pollicis
First dorsal interosseous
Palmar interossei
Flexor retinaculum
Abductor digiti minimi
Flexor digiti minimi brevis
Opponens digiti minimi
Hypothenar eminence
Flexor digitorum superficialis tendons
Lumbricals

(a) **Anterior view**

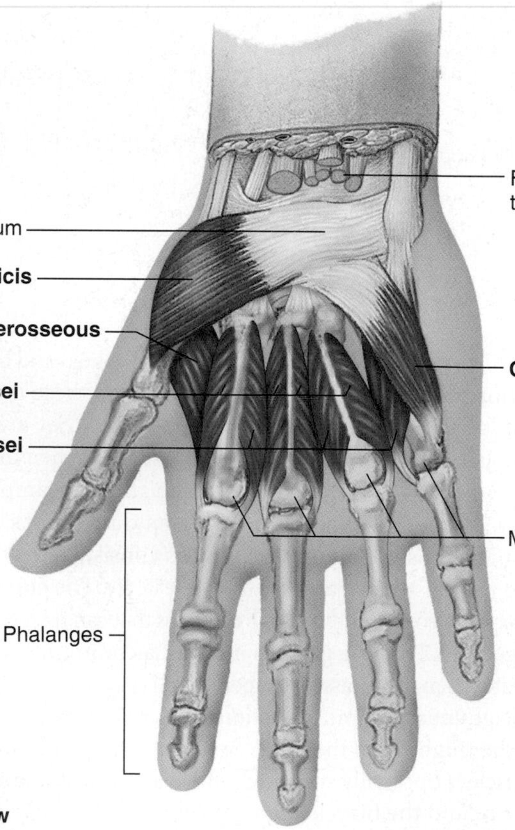

Flexor digitorum tendons (cut)
Flexor retinaculum
Opponens pollicis
First dorsal interosseous
Dorsal interossei
Palmar interossei
Opponens digiti minimi
Metacarpals
Phalanges

(b) **Anterior view**

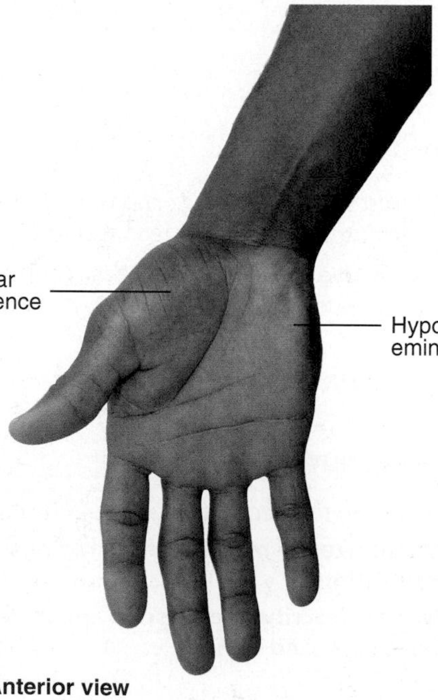

Thenar eminence
Hypothenar eminence

(c) **Anterior view**

FIGURE 10.27 Right Hand Muscles
(a) Superficial muscles of the right hand. The abductor pollicis brevis is cut. *(b)* Deep muscles of the right hand. The flexor digitorum tendons are cut. *(c)* Surface anatomy of the palmar surface of the hand. (Muscle names are in bold.)

the pelvis and maintain the trunk in an upright posture during walking, as the foot of the opposite limb is raised from the ground. Without the action of these muscles, the pelvis tends to sag downward on the unsupported side.

Leg Movements

In addition to the hip muscles, some of the muscles located in the thigh originate on the coxal bone and can cause movement of the thigh (tables 10.20 and 10.21). Three groups of thigh muscles have

TABLE 10.19 Muscles Acting on the Hip and Thigh (see figures 10.28–10.31)

Muscle	Origin	Insertion	Nerve	Action
Anterior				
Iliopsoas (il′ē-ō-sō′as)				
Iliacus (il-ī′ă-kus)	Iliac fossa	Lesser trochanter of femur and capsule of hip joint	Lumbar plexus	Flexes hip
Psoas major (sō′as)	T12–L5	Lesser trochanter of femur	Lumbar plexus	Flexes hip
Posterior and Lateral				
Gluteus maximus (gloo-tē′ŭs mak′si-mŭs)	Posterior surface of ilium, sacrum, and coccyx	Gluteal tuberosity of femur and iliotibial tract	Inferior gluteal	Extends hip; abducts and laterally rotates thigh
Gluteus medius (gloo-tē′ŭs mē′dē-ŭs)	Posterior surface of ilium	Greater trochanter of femur	Superior gluteal	Abducts and medially rotates thigh; tilts pelvis toward supported side
Gluteus minimus (gloo-tē′ŭs min′i-mŭs)	Posterior surface of ilium	Greater trochanter of femur	Superior gluteal	Abducts and medially rotates thigh; tilts pelvis toward supported side
Tensor fasciae latae (ten′sōr fash′ē-ē lā′tē)	Anterior superior iliac spine	Through iliotibial tract to lateral condyle of tibia	Superior gluteal	Tenses lateral fascia and stabilizes femur on tibia when standing; flexes hip; abducts and medially rotates thigh; tilts pelvis
Deep Thigh Rotators				
Gemellus (jĕ-mel′ŭs)				
Inferior	Ischial tuberosity	Obturator internus tendon	L5 and S1	Laterally rotates and abducts thigh
Superior	Ischial spine	Obturator internus tendon	L5 and S1	Laterally rotates and abducts thigh
Obturator (ob′too-rā-tŏr)				
Externus (eks-ter′nŭs)	Inferior margin of obturator foramen	Greater trochanter of femur	Obturator	Laterally rotates thigh
Internus (in-ter′nŭs)	Interior margin of obturator foramen	Greater trochanter of femur	L5 and S1	Laterally rotates thigh
Piriformis (pir′i-fōr′mis)	Sacrum and ilium	Greater trochanter of femur	S1 and S2	Laterally rotates and abducts thigh
Quadratus femoris (kwah′-drā′tŭs fem′ŏ-ris)	Ischial tuberosity	Intertrochanteric ridge of femur	L5 and S1	Laterally rotates thigh

TABLE 10.20 Summary of Muscle Actions on the Hip and Thigh

Flexion	Extension	Abduction	Adduction	Medial Rotation	Lateral Rotation
Iliopsoas	Gluteus maximus	Gluteus maximus	Adductor magnus	Tensor fasciae latae	Gluteus maximus
Tensor fasciae latae	Semitendinosus	Gluteus medius	Adductor longus	Gluteus medius	Obturator internus
Rectus femoris	Semimembranosus	Gluteus minimus	Adductor brevis	Gluteus minimus	Obturator externus
Sartorius	Biceps femoris	Tensor fasciae latae	Pectineus		Superior gemellus
Adductor longus	Adductor magnus	Obturator internus	Gracilis		Inferior gemellus
Adductor brevis		Gemellus superior and inferior			Quadratus femoris
Pectineus		Piriformis			Piriformis
					Adductor magnus
					Adductor longus
					Adductor brevis

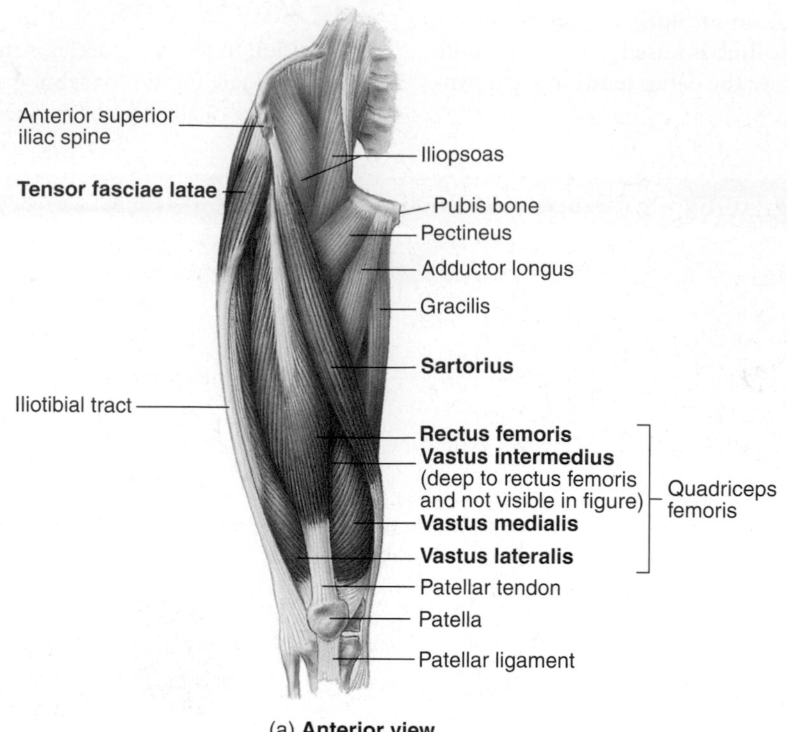
Anterior superior iliac spine
Tensor fasciae latae
Iliopsoas
Pubis bone
Pectineus
Adductor longus
Gracilis
Sartorius
Iliotibial tract
Rectus femoris
Vastus intermedius (deep to rectus femoris and not visible in figure)
Vastus medialis
Vastus lateralis
Quadriceps femoris
Patellar tendon
Patella
Patellar ligament

(a) **Anterior view**

Tensor fasciae latae
Iliopsoas
Pectineus
Adductor longus
Gracilis
Sartorius
Rectus femoris
Vastus medialis
Vastus lateralis
Quadriceps femoris

(b) **Anterior view**

Tensor fasciae latae
Sartorius
Rectus femoris (quadriceps)
Adductors
Vastus lateralis (quadriceps)
Vastus medialis (quadriceps)

(c) **Anterior view**

FIGURE 10.28 APR **Right Anterior Hip and Thigh Muscles**
(*a*) Right anterior hip and thigh muscles. (*b*) Photograph of the right anterior thigh muscles in a cadaver. (*c*) Surface anatomy of the right anterior thigh. (Muscle names are in bold.)

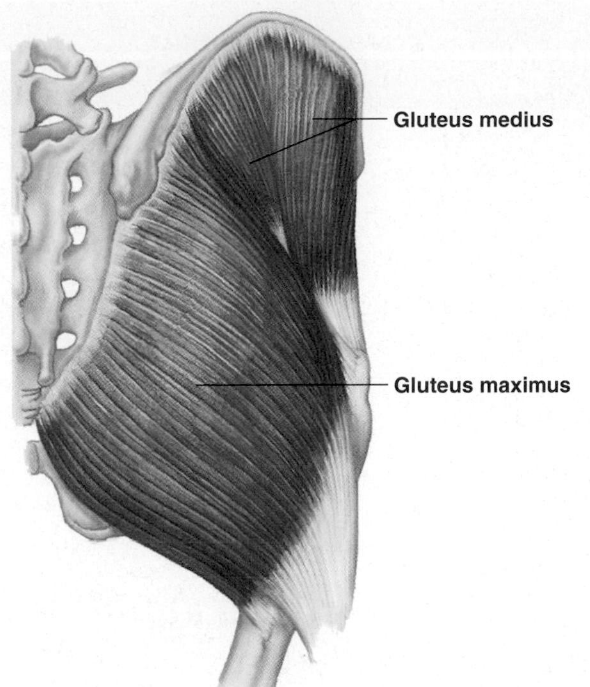

Gluteus medius
Gluteus maximus

(a) **Posterior view**

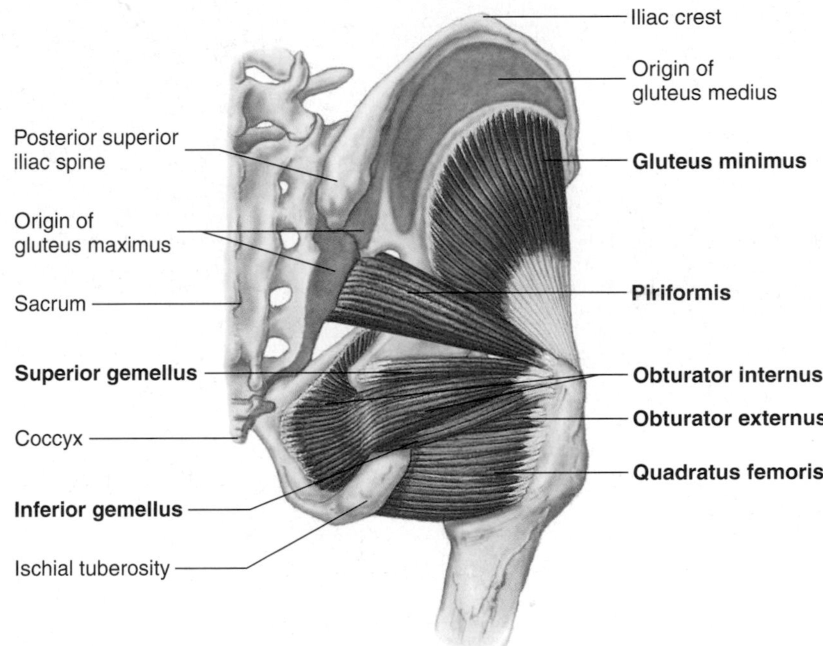

Iliac crest
Origin of gluteus medius
Posterior superior iliac spine
Gluteus minimus
Origin of gluteus maximus
Sacrum
Piriformis
Superior gemellus
Obturator internus
Obturator externus
Coccyx
Quadratus femoris
Inferior gemellus
Ischial tuberosity

(b) **Posterior view**

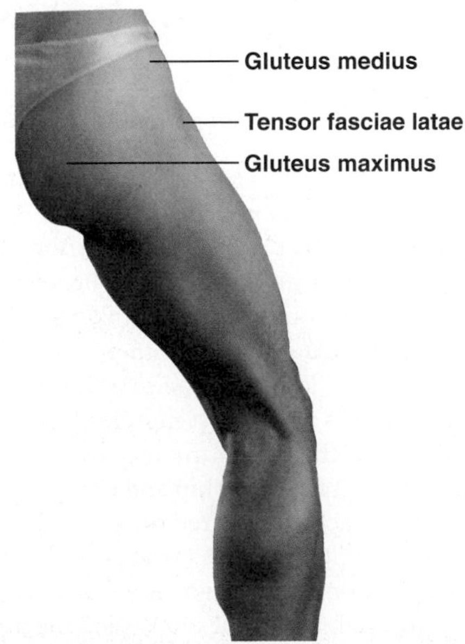

Gluteus medius
Tensor fasciae latae
Gluteus maximus

(c) **Posterior view**

FIGURE 10.29 AP|R **Right Posterior Hip Muscles**

(*a*) Right hip, superficial muscles. (*b*) Right hip, deep muscles. The gluteus maximus and medius are removed to reveal deeper muscles. (*c*) Surface anatomy of the right posterior hip muscles. (Muscle names are in bold.)

TABLE 10.21 Muscles of the Thigh (see figures 10.30 and 10.32)

Muscle	Origin	Insertion	Nerve	Action
Anterior Compartment				
Quadriceps femoris (kwah′dri-seps fem′ŏ-ris)	Rectus femoris—anterior inferior iliac spine	Patella and onto tibial tuberosity through patellar ligament	Femoral	Extends knee; rectus femoris also flexes hip
	Vastus lateralis—greater trochanter and linea aspera of femur			
	Vastus intermedius—body of femur			
	Vastus medialis—linea aspera of femur			
Sartorius (sar-tōr′ē-ŭs)	Anterior superior iliac spine	Medial side of tibial tuberosity	Femoral	Flexes hip and knee; rotates thigh laterally and leg medially
Medial Compartment				
Adductor brevis (a-dŭk′ter, a-dŭk′tōr brev′is)	Pubis	Pectineal line and linea aspera of femur	Obturator	Adducts and laterally rotates thigh; flexes hip
Adductor longus (a-dŭk′ter, a-dŭk′tōr lon′gŭs)	Pubis	Linea aspera of femur	Obturator	Adducts and laterally rotates thigh; flexes hip
Adductor magnus (a-dŭk′ter, a-dŭk′tōr mag′nŭs)	Adductor part: pubis and ischium	Adductor part: linea aspera of femur	Adductor part: obturator	Adductor part: adducts thigh and flexes hip
	Hamstring part: ischial tuberosity	Hamstring part: adductor tubercle of femur	Hamstring part: tibial	Hamstring part: extends hip and adducts thigh
Gracilis (gras′i-lis)	Pubis near symphysis	Tibia	Obturator	Adducts thigh; flexes knee
Pectineus (pek′ti-nē′ŭs)	Pubic crest	Pectineal line of femur	Femoral and obturator	Adducts thigh; flexes hip
Posterior Compartment				
Biceps femoris (bī′seps fem′ŏ-ris)	Long head—ischial tuberosity	Head of fibula	Long head—tibial	Flexes knee; laterally rotates leg; extends hip
	Short head—femur		Short head—common fibular	
Semimembranosus (sem′ē-mem-bră-nō′sŭs)	Ischial tuberosity	Medial condyle of tibia and collateral ligament	Tibial	Flexes knee; medially rotates leg; tenses capsule of knee joint; extends hip
Semitendinosus (sem′ē-ten-di-nō′sŭs)	Ischial tuberosity	Tibia	Tibial	Flexes knee; medially rotates leg; extends hip

been identified based on their location in the thigh and are organized into **compartments:** The muscles of the anterior compartment flex the hip and/or extend the knee (see figure 10.28*a*); the muscles of the medial compartment adduct the thigh (figure 10.30); and the muscles of the posterior compartment extend the hip and flex the knee (figure 10.31).

The anterior thigh muscles are the **quadriceps femoris** (fem′ŏ-ris) and the **sartorius** (sar-tōr′ē-ŭs; see table 10.20 and figure 10.28*a*). The quadriceps femoris is actually four muscles: the **rectus femoris,** the **vastus lateralis,** the **vastus medialis,** and the **vastus intermedius.** The quadriceps group extends the knee. The rectus femoris also flexes the hip because it crosses both the hip and knee joints.

The quadriceps femoris makes up the large mass on the anterior thigh (see figure 10.28*c*). The vastus lateralis is sometimes used as an injection site, especially in infants who do not have well-developed deltoid or gluteal muscles. The muscles of the quadriceps femoris have a common insertion, the patellar tendon, on and around the patella. The patellar ligament is an extension of the patellar tendon onto the tibial tuberosity. The patellar ligament is the point that is tapped with a rubber hammer when testing the knee-jerk reflex in a physical examination.

The sartorius is the longest muscle of the body, crossing from the lateral side of the hip to the medial side of the knee. As the muscle contracts, it flexes the hip and knee and laterally rotates the thigh. This is the action required for crossing the legs.

The medial thigh muscles (see figure 10.30) are involved primarily in adduction of the thigh. Some of these muscles also laterally rotate the thigh and/or flex or extend the hip. The gracilis also flexes the knee.

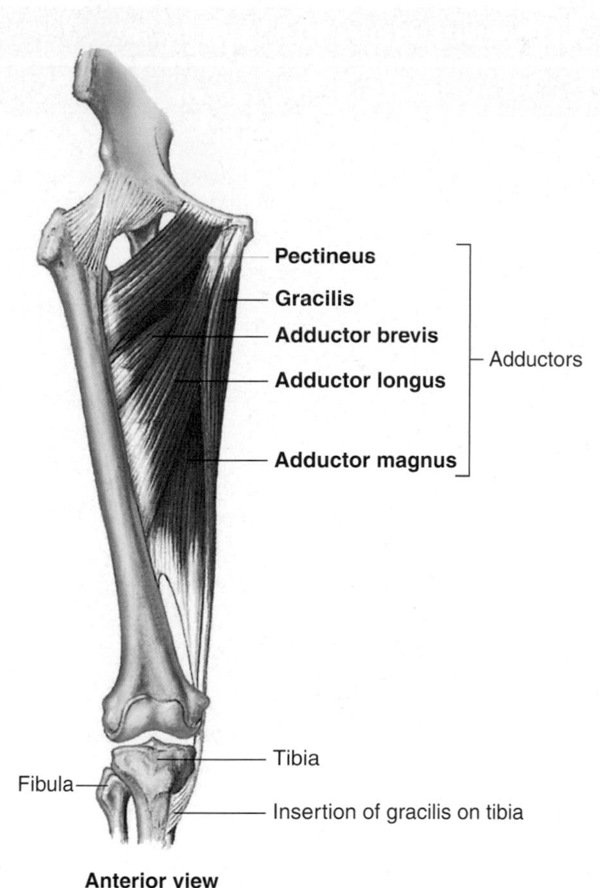
Pectineus
Gracilis
Adductor brevis
Adductor longus
Adductor magnus
Adductors
Fibula
Tibia
Insertion of gracilis on tibia
Anterior view

FIGURE 10.30 Right Medial Thigh Muscles
(Muscle names are in bold.)

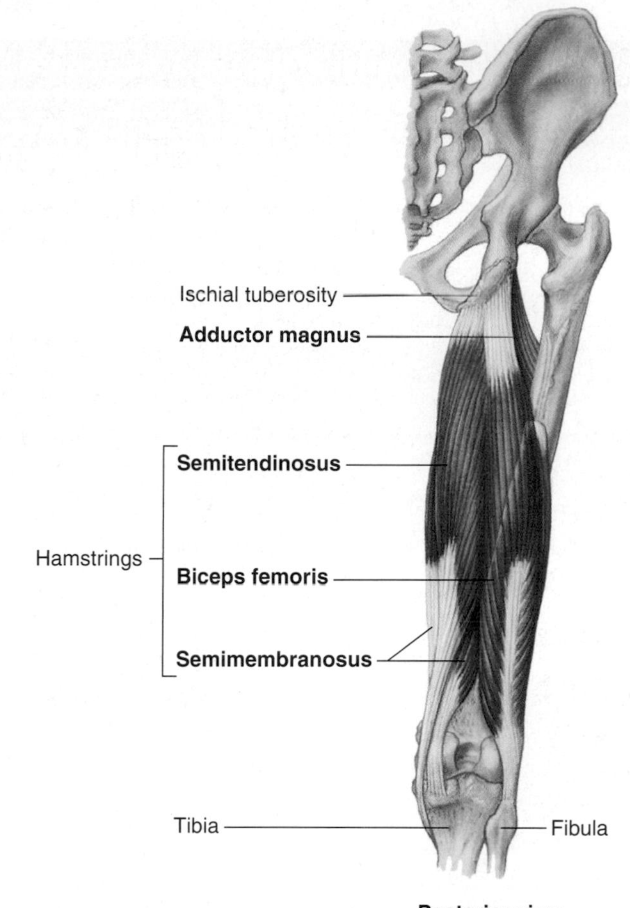
Ischial tuberosity
Adductor magnus
Hamstrings
Semitendinosus
Biceps femoris
Semimembranosus
Tibia
Fibula
Posterior view

FIGURE 10.31 Right Posterior Thigh Muscles
Hip muscles are removed. (Muscle names are in bold.)

The posterior thigh muscles (figure 10.31), collectively called the hamstring muscles, consist of the **biceps femoris,** the **semimembranosus** (sem′ē-mem-bră-nō′sŭs), and the **semitendinosus** (sem′ē-ten-di-nō′sŭs; table 10.21). Their tendons are easily seen or felt on the medial and lateral posterior aspect of a slightly bent knee (figure 10.32).

Ankle, Foot, and Toe Movements

The muscles of the leg that move the ankle and the foot are listed in table 10.22 and illustrated in figures 10.33–10.35. These **extrinsic foot muscles** are divided into three groups, each located within a separate compartment of the leg: anterior, posterior, and lateral (figure 10.33). The anterior leg muscles (figure 10.34*a*) are extensor muscles involved in dorsiflexion and eversion or inversion of the foot and extension of the toes.

The lateral muscles (figure 10.34*b*) are primarily everters of the foot, but they also aid plantar flexion. The **fibularis brevis** inserts onto the fifth metatarsal bone and everts and plantar flexes the foot. The **fibularis longus** crosses under the lateral four metatarsal bones to insert onto the first metatarsal bone and medial cuneiform. The fibular muscle was formerly referred to as *peroneal,* which means relating to the fibula; the term has now been simplified to *fibular*. The tendons of the fibularis muscles can be seen on the lateral side of the ankle (figure 10.34*d*).

Clinical
IMPACT

Shinsplints

Shinsplints is a general term involving any one of the following conditions associated with pain in the anterior portion of the leg:

1. Excessive stress on the tibialis anterior, resulting in pain along the origin of the muscle
2. Tibial periostitis, an inflammation of the tibial periosteum
3. Anterior compartment syndrome. During hard exercise, the anterior compartment muscles may swell with blood. The overlying fascia is very tough and does not expand; thus, the nerves and vessels are compressed, causing pain.
4. Stress fracture of the tibia 2–5 cm distal to the knee

Shinsplints can occur for several reasons: running with unsupportive shoes, running on a hard surface (such as concrete), or simply increasing your activity level too quickly. This injury can be treated by employing RICE and taking anti-inflammatory medicines. Runners might consider occasionally substituting a low-impact exercise, such as swimming or cycling.

TABLE 10.22 Muscles of the Leg Acting on the Leg, Ankle, and Foot (see figures 10.34 and 10.36)

Muscle	Origin	Insertion	Nerve	Action
Anterior Compartment				
Extensor digitorum longus (dij′i-tōr-ŭm lon′gŭs)	Lateral condyle of tibia and fibula	Four tendons to phalanges of four lateral toes	Deep fibular*	Extends four lateral toes; dorsiflexes and everts foot
Extensor hallucis longus (hal′i-sis lon′gŭs)	Middle fibula and interosseous membrane	Distal phalanx of great toe	Deep fibular*	Extends great toe; dorsiflexes and inverts foot
Tibialis anterior (tib-ē-a′lis)	Proximal, lateral tibia and interosseous membrane	Medial cuneiform and first metatarsal bone	Deep fibular*	Dorsiflexes and inverts foot
Fibularis tertius (peroneus tertius; fib-ū-lā′ris ter′shē-ŭs)	Fibula and interosseous membrane	Fifth metatarsal bone	Deep fibular*	Dorsiflexes and everts foot
Lateral Compartment				
Fibularis brevis (peroneus brevis; fib-ū-lā′ris brev′is)	Inferior two-thirds of lateral fibula	Fifth metatarsal bone	Superficial fibular*	Everts and plantar flexes foot
Fibularis longus (peroneus longus; fib-ū-lā′ris lon′gŭs)	Superior two-thirds of lateral fibula	First metatarsal bone and medial cuneiform	Superficial fibular*	Everts and plantar flexes foot
Posterior Compartment				
Superficial				
Gastrocnemius (gas-trok-nē′mē-ŭs)	Medial and lateral condyles of femur	Through calcaneal (Achilles) tendon to calcaneus	Tibial	Plantar flexes foot; flexes knee
Plantaris (plan-tăr′is)	Femur	Through calcaneal tendon to calcaneus	Tibial	Plantar flexes foot; flexes knee
Soleus (sō-lē′ŭs)	Fibula and tibia	Through calcaneal tendon to calcaneus	Tibial	Plantar flexes foot
Deep				
Flexor digitorum longus (dij′i-tōr′ŭm lon′gŭs)	Tibia	Four tendons to distal phalanges of four lateral toes	Tibial	Flexes four lateral toes; plantar flexes and inverts foot
Flexor hallucis longus (hal′i-sis lon′gŭs)	Fibula	Distal phalanx of great toe	Tibial	Flexes great toe; plantar flexes and inverts foot
Popliteus (pop-li-tē′ŭs)	Lateral femoral condyle	Posterior tibia	Tibial	Flexes knee; medially rotates leg
Tibialis posterior (tib-ē-a′lis)	Tibia, interosseous membrane, and fibula	Navicular, cuneiforms, cuboid, and second through fourth metatarsal bones	Tibial	Plantar flexes and inverts foot

*Also referred to as the peroneal nerve.

The superficial muscles of the posterior compartment, the **gastrocnemius** (gas-trok-nē′mē-ŭs) and the **soleus,** form the bulge of the calf (posterior leg; figure 10.35*a*,*b*,*d*). They join with the small **plantaris** muscle to form the common **calcaneal** (kal-kā′nē-al) **tendon,** or *Achilles tendon*. These muscles are involved in plantar flexion of the foot. The deep muscles of the posterior compartment plantar flex and invert the foot and flex the toes.

Intrinsic foot muscles, located within the foot itself (table 10.23; figure 10.36), flex, extend, abduct, and adduct the toes. They are arranged in a manner similar to that of the intrinsic muscles of the hand.

ASSESS **YOUR PROGRESS**

31. *Name the anterior hip muscle that flexes the hip. What muscles act as synergists to this muscle?*

32. *How is it possible for thigh muscles to move both the thigh and the leg? Name the six muscles that can do this, and give their actions.*

33. *What movements are produced by the three muscle compartments of the leg? Name the muscles of each compartment, and describe the movements for which each muscle is responsible.*

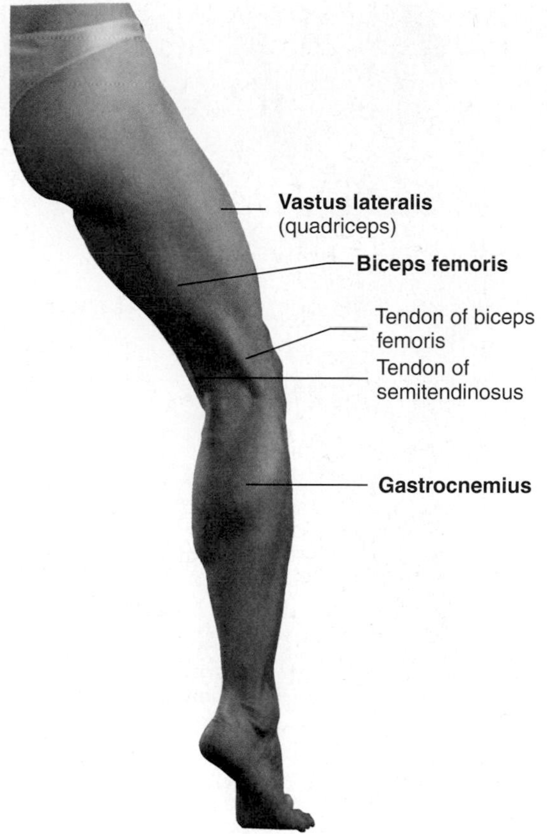
Vastus lateralis
(quadriceps)
Biceps femoris
Tendon of biceps femoris
Tendon of semitendinosus
Gastrocnemius

FIGURE 10.32 Surface Anatomy of the Posterior Lower Limb
(Muscle names are in bold.)

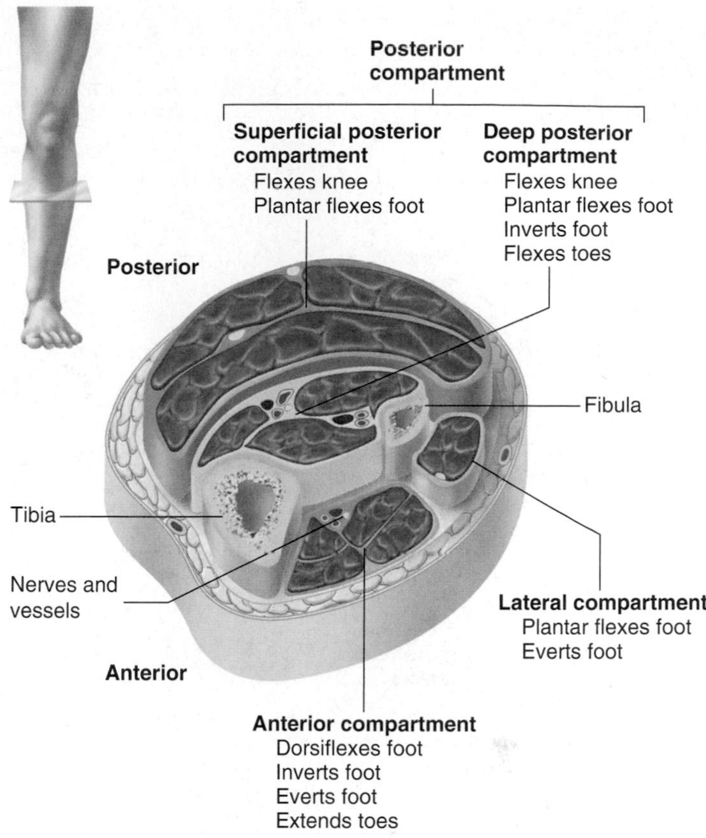
Posterior compartment
Superficial posterior compartment
Flexes knee
Plantar flexes foot
Deep posterior compartment
Flexes knee
Plantar flexes foot
Inverts foot
Flexes toes
Posterior
Fibula
Tibia
Nerves and vessels
Lateral compartment
Plantar flexes foot
Everts foot
Anterior
Anterior compartment
Dorsiflexes foot
Inverts foot
Everts foot
Extends toes

FIGURE 10.33 Cross Section Through the Left Leg
The anterior, posterior, and lateral compartments are labeled.

34. *What movement do the fibularis (peroneus) muscles have in common? The tibialis muscles?*

35. *Name the leg muscles that flex the knee. Which of them can also plantar flex the foot?*

36. *List the general actions performed by the intrinsic foot muscles.*

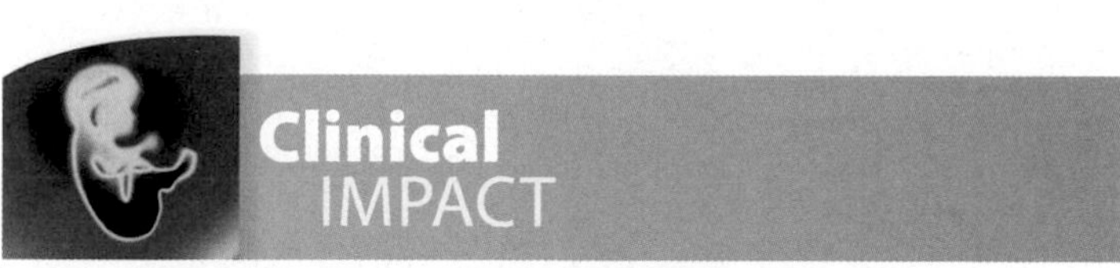
Clinical
IMPACT

Achilles Tendon

The Achilles tendon derives its name from a hero of Greek mythology. When Achilles was a baby, his mother dipped him into magic water, which made him invulnerable to harm everywhere the water touched his skin. However, his mother held him by the heel and failed to submerge this part of his body under the water. Consequently, his heel was vulnerable and proved to be his undoing; at the battle of Troy, he was shot in the heel with an arrow and died. Thus, saying that someone has an "Achilles heel" means that the person has a weak spot that can be attacked.

Achilles tendon injuries are often due to overexertion by doing too much too quickly or too soon after a break from exercise. The main keys to preventing an injury are wearing the appropriate footwear and performing proper warm-up and stretching exercises.

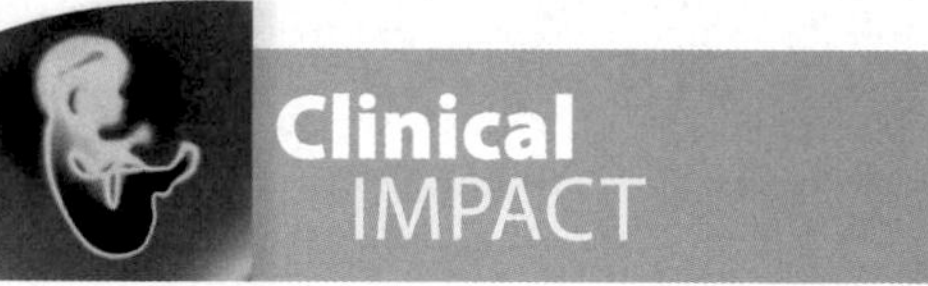
Clinical
IMPACT

Plantar Fasciitis

The muscles in the plantar region of the foot are covered with thick fascia and the plantar aponeurosis. Running on a hard surface wearing poorly fitting or worn-out shoes can result in inflammation of the plantar aponeurosis, called **plantar fasciitis.** Patients experience pain in the fascia over the heel and along the medial-inferior side of the foot. Wearing supportive shoes with heel inserts is a good first step toward treating plantar fasciitis. Studies show that wearing nonsupportive shoes, such as flip-flops, for extended periods of time leads to increased pain in the feet, heels, and ankles. RICE and anti-inflammatory medicines may relieve the pain and inflammation. About 80% of people recover fully within a year.

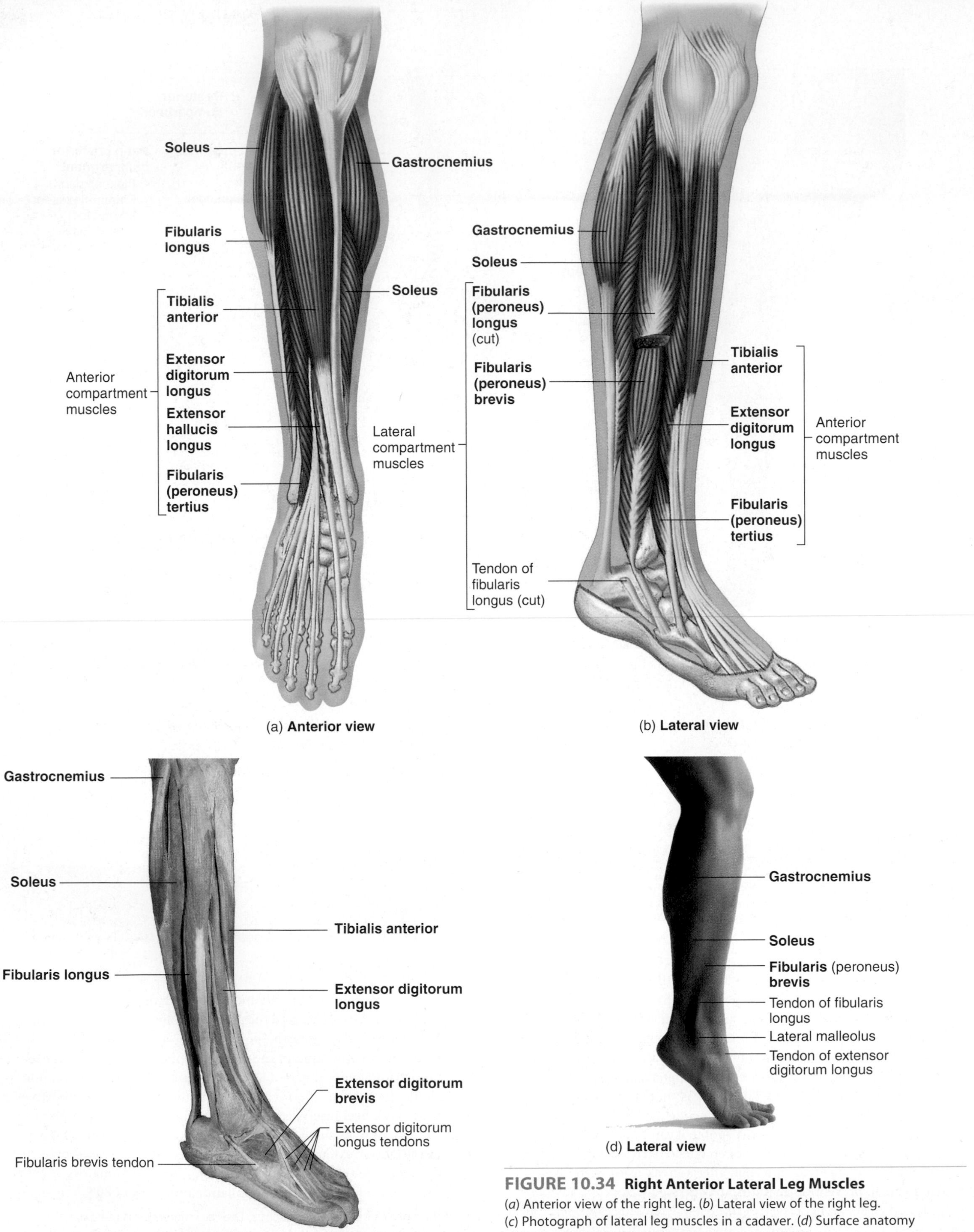
Soleus
Gastrocnemius
Fibularis
longus
Soleus
Tibialis
anterior
Extensor
digitorum
longus
Anterior
compartment
muscles
Extensor
hallucis
longus
Fibularis
(peroneus)
tertius
(a) Anterior view
Gastrocnemius
Soleus
Fibularis
(peroneus)
longus
(cut)
Fibularis
(peroneus)
brevis
Lateral
compartment
muscles
Tendon of
fibularis
longus (cut)
Tibialis
anterior
Extensor
digitorum
longus
Anterior
compartment
muscles
Fibularis
(peroneus)
tertius
(b) Lateral view
Gastrocnemius
Soleus
Tibialis anterior
Fibularis longus
Extensor digitorum
longus
Extensor digitorum
brevis
Extensor digitorum
longus tendons
Fibularis brevis tendon
(c) Lateral view
Gastrocnemius
Soleus
Fibularis (peroneus)
brevis
Tendon of fibularis
longus
Lateral malleolus
Tendon of extensor
digitorum longus
(d) Lateral view

FIGURE 10.34 Right Anterior Lateral Leg Muscles
(*a*) Anterior view of the right leg. (*b*) Lateral view of the right leg. (*c*) Photograph of lateral leg muscles in a cadaver. (*d*) Surface anatomy of the posterolateral leg. (Muscle names are in bold.)

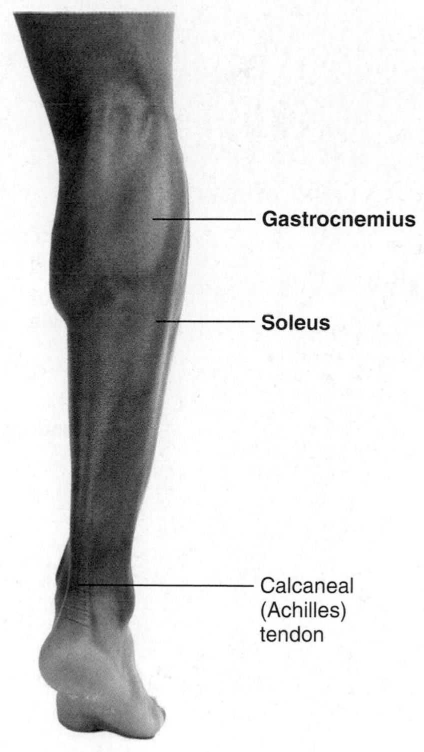

Gastrocnemius
Soleus
Calcaneal (Achilles) tendon

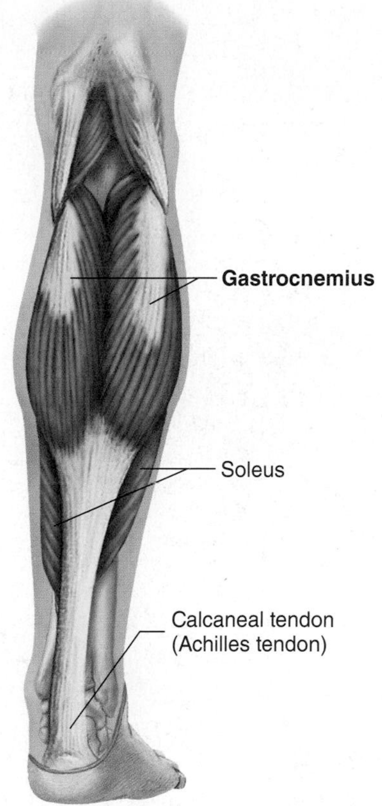

Gastrocnemius
Soleus
Calcaneal tendon (Achilles tendon)

(a) and (b) **Posterior views**

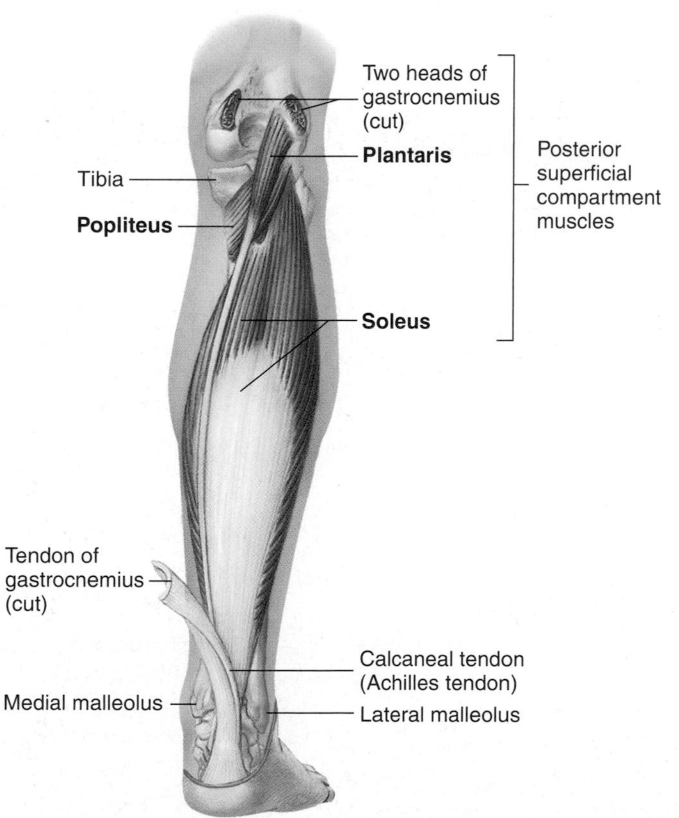

Two heads of gastrocnemius (cut)
Plantaris
Posterior superficial compartment muscles
Tibia
Popliteus
Soleus
Tendon of gastrocnemius (cut)
Calcaneal tendon (Achilles tendon)
Medial malleolus
Lateral malleolus

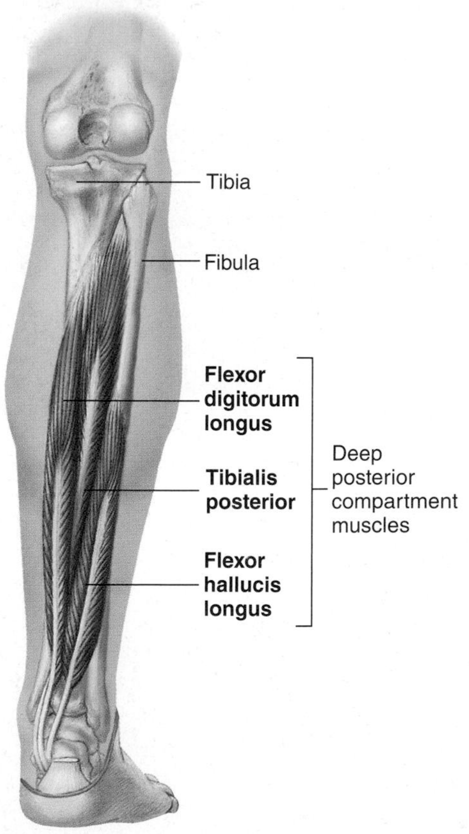

Tibia
Fibula
Flexor digitorum longus
Tibialis posterior
Flexor hallucis longus
Deep posterior compartment muscles

(c) and (d) **Posterior views**

FIGURE 10.35 APIR **Right Posterior Leg Muscles**

(*a*) Surface anatomy of the posterior right leg. (*b*) Superficial muscles. (*c*) Posterior view of the right calf, superficial muscles. The gastrocnemius is removed. (*d*) Posterior view of the right calf, deep muscles. The gastrocnemius, plantaris, and soleus muscles are removed. (Muscle names are in bold.)

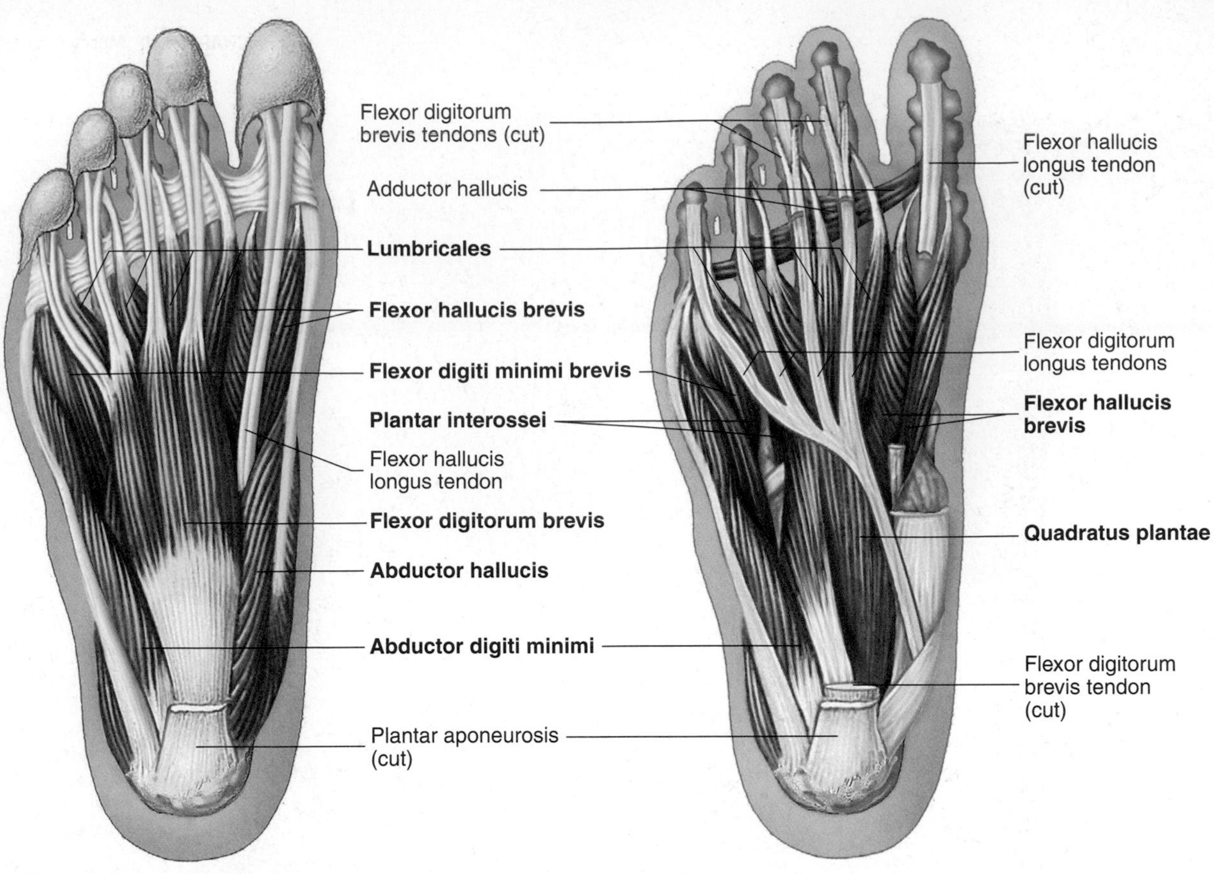
Flexor digitorum
brevis tendons (cut)
Adductor hallucis
Lumbricales
Flexor hallucis brevis
Flexor digiti minimi brevis
Plantar interossei
Flexor hallucis
longus tendon
Flexor digitorum brevis
Abductor hallucis
Abductor digiti minimi
Plantar aponeurosis
(cut)
Flexor hallucis
longus tendon
(cut)
Flexor digitorum
longus tendons
Flexor hallucis
brevis
Quadratus plantae
Flexor digitorum
brevis tendon
(cut)

FIGURE 10.36 Right Foot Muscles
(*a*) Superficial muscles of the right foot. The plantar aponeurosis is cut. (*b*) Deep muscles of the right foot. The flexor digitorum brevis and flexor hallucis longus are cut. (Muscle names are in bold.)

TABLE 10.23 Intrinsic Muscles of the Foot (see figure 10.36)

Muscle	Origin	Insertion	Nerve	Action
Abductor digiti minimi (ab-dŭk′ter, ab-dŭk′tōr dij′i-tī min′ĭ-mī)	Calcaneus	Proximal phalanx of fifth toe	Lateral plantar	Abducts and flexes little toe
Abductor hallucis (ab-dŭk′ter, ab-dŭk′tōr hal′i-sis)	Calcaneus	Base of proximal phalanx of great toe	Medial plantar	Abducts great toe
Adductor hallucis (a-dŭk′ter, a-dŭk′tōr hal′i-sis; not illustrated)	Lateral four metatarsal bones	Proximal phalanx of great toe	Lateral plantar	Adducts great toe
Extensor digitorum brevis (dij′i-tōr′ŭm brev′is; not illustrated)	Calcaneus	Four tendons fused with tendons of extensor digitorum longus	Deep fibular*	Extends toes
Flexor digiti minimi brevis (dij′i-tī min′ĭ-mī brev′is)	Fifth metatarsal bone	Proximal phalanx of digit 5	Lateral plantar	Flexes little toe (proximal phalanx)
Flexor digitorum brevis (dij′i-tōr′ŭm brev′is)	Calcaneus and plantar fascia	Four tendons to middle phalanges of four lateral toes	Medial plantar	Flexes lateral four toes
Flexor hallucis brevis (hal′i-sis brev′is)	Cuboid; medial and lateral cuneiforms	Two tendons to proximal phalanx of great toe	Medial and lateral plantar	Flexes great toe
Dorsal interossei (in′ter-os′e-ī; not illustrated)	Metatarsal bones	Proximal phalanges of digits 2, 3, and 4	Lateral plantar	Abduct second, third, and fourth toes
Plantar interossei (plan′tăr in′ter-os′e-ī)	Third, fourth, and fifth metatarsal bones	Proximal phalanges of digits 3, 4, and 5	Lateral plantar	Adduct third, fourth, and fifth toes
Lumbricales (lum′bri-kā-lēz)	Tendons of flexor digitorum longus	Extensor expansion of digits 2–5	Lateral and medial plantar	Flex proximal and extend middle and distal phalanges
Quadratus plantae (kwah′drā′tŭs plan′tē)	Calcaneus	Tendons of flexor digitorum longus	Lateral plantar	Assists flexor digitorum longus in flexing lateral four toes

*Also referred to as the peroneal nerve.

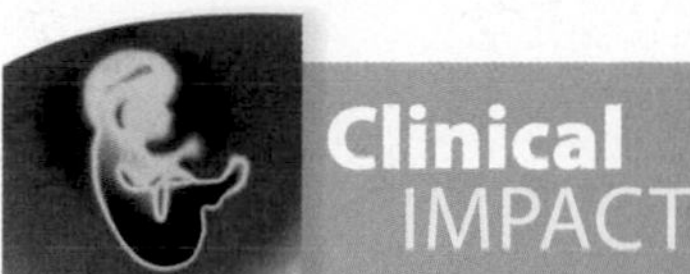

Clinical IMPACT | Bodybuilding

Bodybuilding is a popular sport worldwide. Its participants combine diet and specific weight training to develop maximum muscle mass and minimum body fat, with the goal of achieving a complete, well-balanced physique. Skill, training, and concentration are required to build a well-proportioned, muscular body and to know which exercises develop a large number of muscles and which are specialized to build up certain parts of the body. An uninformed, untrained muscle builder can build some muscles and ignore others; the result is a disproportioned body.

Is the old adage "no pain, no gain" correct? Not really. Overexercising can cause soreness and small tears in muscles. Torn muscles are weaker, and it may take up to 3 weeks to repair the damage, even though the soreness may last only 5–10 days.

Historically, although bodybuilders had a lot of muscle mass, they were not "in shape." However, today bodybuilders exercise aerobically in addition to "pumping iron."

FIGURE 10A Bodybuilders

A current topic of discussion for modern bodybuilders is whether bodybuilding shortens their life span. For instance, scientific evidence has shown that restricted-calorie diets increase life span, yet some bodybuilders consume at least 4500 calories a day when in the "bulking" phase of training. Others claim that the training process of lifting extremely heavy weights, such as squat-lifting 500 pounds in series of repetitions, and carrying the extra poundage of their acquired muscle mass causes their heart to work harder. However the overwhelming evidence at this time shows that the life span of active people is longer than that of sedentary people, even when the activity is extreme. As bodybuilders age and reduce the intensity of their workouts, their muscle mass decreases, but not at a porportionally higher rate than other people with a lower activity level. In chapter 9, see the section "Effects of Aging on Skeletal Muscle" for more information on the effects of reduced muscle mass as people age.

Bodybuilders also have their own language. They refer to "lats," "traps," and "delts" rather than latissimus dorsi, trapezius, and deltoids. The exercises have special names, such as "lat pulldowns," "preacher curls," and "triceps extensions."

Photographs of bodybuilders are very useful in the study of anatomy because they allow us to identify the surface anatomy of muscles that cannot usually be seen in untrained people (figure 10A).

Answer

Learn to Predict ❮ From page 309

The description of Pedro's injury provided specific information about the regions of the body affected: the left hip and thigh. In addition, we are told that the injury affected action potential conduction to the muscles of these regions. These facts will help us determine Pedro's symptoms and predict the movements that may be affected by his injury.

Chapter 9 described the relationship between action potential conduction and the force of muscle contractions. The reduction in action potential conduction to the muscles of the hip and thigh reduced the stimulation of these muscles, reducing the contraction force. As a result of his injury, we can predict that Pedro experienced weakness in his left hip and thigh, limiting his activity level.

We read in chapter 10 that the muscles affected by Pedro's injury (psoas major, iliacus, pectineus, sartorius, vastus lateralis, vastus medius, vastus intermedius, and rectus femoris) are involved in flexing the hip, the knee, or both. Therefore, we can conclude that movements involving hip and knee flexion, such as walking up and down stairs, would be affected. Any tasks that require Pedro to walk up and down stairs would be more difficult for him. Sitting and standing may also be affected, but the weakness in Pedro's left hip and thigh may be compensated for by increased muscle strength on his right side.

Answers to the rest of this chapter's Predict questions are in Appendix G.

Summary

Body movements result from the contraction of skeletal muscles.

10.1 General Principles of Skeletal Muscle Anatomy (p. 310)

1. The less movable end of a muscle attachment is the origin; the more movable end is the insertion.
2. An agonist causes a certain movement, and an antagonist acts in opposition to the agonist.
3. Synergists are muscles that function together to produce movement.
4. Prime movers are mainly responsible for a movement. Fixators stabilize the action of prime movers.

Muscle Shapes

Muscle shape is determined primarily by the orientation of muscle fasciculi.

Nomenclature

Muscles are named according to their location, size, shape, orientation of fasciculi, origin and insertion, number of heads, or function.

Movements Accomplished by Muscles

Contracting muscles generate a force that acts on bones (levers) across joints (fulcrums) to create movement. Three classes of levers have been identified.

Muscle Anatomy

The study of muscle anatomy is usually broken down into body regions: head and neck, trunk, upper limbs, and lower limbs.

10.2 Head and Neck Muscles (p. 313)

Neck Muscles

The origins of these muscles are mainly on the cervical vertebrae (except for the sternocleidomastoid); the insertions are on the occipital bone or mastoid process. They cause flexion, extension, rotation, and lateral flexion of the head and neck.

Facial Expression

The origins of facial muscles are on skull bones or fascia; the insertions are into the skin, causing movement of the facial skin, lips, and eyelids.

Mastication

Three pairs of muscles close the jaw; gravity opens the jaw. Forced opening is caused by the lateral pterygoids and the hyoid muscles.

Tongue Movements

Intrinsic tongue muscles change the shape of the tongue; extrinsic tongue muscles move the tongue.

Swallowing and the Larynx

1. Hyoid muscles can depress the jaw and assist in swallowing.
2. Muscles open and close the openings to the nasal cavity, auditory tubes, and larynx.

Movements of the Eyeball

Six muscles with their origins on the orbital bones insert on the eyeball and cause it to move within the orbit.

10.3 Trunk Muscles (p. 326)

Muscles Moving the Vertebral Column

1. These muscles extend, laterally flex, rotate, or flex the vertebral column.
2. A more superficial group of muscles runs from the pelvis to the skull, extending from the vertebrae to the ribs.
3. A deep group of muscles connects adjacent vertebrae.

Thoracic Muscles

1. Most respiratory movement is caused by the diaphragm.
2. Muscles attached to the ribs aid in respiration.

Abdominal Wall

Abdominal wall muscles hold and protect abdominal organs and cause flexion, rotation, and lateral flexion of the vertebral column.

Pelvic Floor and Perineum

These muscles support the abdominal organs inferiorly.

10.4 Upper Limb Muscles (p. 334)

Scapular Movements

Six muscles attach the scapula to the trunk and enable the scapula to function as an anchor point for the muscles and bones of the arm.

Arm Movements

Seven muscles attach the humerus to the scapula. Two additional muscles attach the humerus to the trunk. These muscles cause flexion and extension of the shoulder and abduction, adduction, rotation, and circumduction of the arm.

Forearm Movements

1. Flexion and extension of the elbow are accomplished by three muscles in the arm and two in the forearm.
2. Supination and pronation are accomplished primarily by forearm muscles.

Wrist, Hand, and Finger Movements

1. Forearm muscles that originate on the medial epicondyle are responsible for flexion of the wrist and fingers. Muscles extending the wrist and fingers originate on the lateral epicondyle.
2. Extrinsic hand muscles are in the forearm. Intrinsic hand muscles are in the hand.

10.5 Lower Limb Muscles (p. 345)

Hip and Thigh Movements

1. Anterior pelvic muscles cause flexion of the hip.
2. Muscles of the buttocks are responsible for extension of the hip and abduction and rotation of the thigh.

Leg Movements

1. Some muscles of the thigh also act on the leg. The anterior thigh muscles extend the leg, and the posterior thigh muscles flex the leg.

2. The thigh can be divided into three compartments.
   - The anterior compartment muscles flex the hip and extend the knee.
   - The medial compartment muscles adduct the thigh.
   - The posterior compartment muscles extend the hip and flex the knee.

Ankle, Foot, and Toe Movements

1. The leg is divided into three compartments.
   - Muscles in the anterior compartment cause dorsiflexion, inversion, or eversion of the foot and extension of the toes.
   - Muscles of the lateral compartment plantar flex and evert the foot.
   - Muscles of the posterior compartment flex the leg, plantar flex and invert the foot, and flex the toes.
2. Intrinsic foot muscles flex or extend, and abduct or adduct, the toes.

REVIEW AND COMPREHENSION

1. Muscles that oppose one another are
   a. synergists. c. hateful. e. fixators.
   b. levers. d. antagonists.
2. The most movable attachment of a muscle is its
   a. origin. c. fascia. e. belly.
   b. insertion. d. fulcrum.
3. The muscle whose name means it is to the side of midline is the
   a. gluteus maximus.
   b. vastus lateralis.
   c. teres major.
   d. latissimus dorsi.
   e. adductor magnus.
4. In a class III lever system, the
   a. fulcrum is located between the pull and the weight.
   b. weight is located between the fulcrum and the pull.
   c. pull is located between the fulcrum and the weight.
5. A prominent lateral muscle of the neck that can cause flexion of the neck or rotate the head is the
   a. digastric.
   b. mylohyoid.
   c. sternocleidomastoid.
   d. buccinator.
   e. platysma.
6. An aerial circus performer who supports her body only with her teeth while spinning around should have strong
   a. temporalis muscles.
   b. masseter muscles.
   c. buccinator muscles.
   d. Both a and b are correct.
   e. All of these are correct.
7. The tongue's shape changes *primarily* because of the action of the
   a. extrinsic tongue muscles.
   b. intrinsic tongue muscles.
8. The infrahyoid muscles
   a. elevate the mandible.
   b. move the mandible from side to side.
   c. fix (prevent movement of) the hyoid.
   d. Both a and b are correct.
   e. All of these are correct.
9. The soft palate muscles
   a. prevent food from entering the nasal cavity.
   b. close the auditory tube.
   c. force food into the esophagus.
   d. prevent food from entering the larynx.
   e. elevate the mandible.
10. Which of these movements is *not* caused by contraction of the erector spinae muscles?
    a. flexion of the vertebral column
    b. lateral flexion of the vertebral column
    c. extension of the vertebral column
    d. rotation of the vertebral column
11. Which of these muscles is *not* involved with the inspiration of air?
    a. diaphragm
    b. external intercostals
    c. scalenes
    d. transversus thoracis
12. Given these muscles:
    (1) external abdominal oblique
    (2) internal abdominal oblique
    (3) transversus abdominis

    Choose the arrangement that lists the muscles from most superficial to deepest.
    a. 1,2,3 c. 2,1,3 e. 3,1,2
    b. 1,3,2 d. 2,3,1
13. Tendinous intersections
    a. attach the rectus abdominis muscles to the xiphoid process.
    b. divide the rectus abdominis muscles into segments.
    c. separate the abdominal wall from the thigh.
    d. are the sites where blood vessels exit the abdomen into the thigh.
    e. are the central point of attachment for all the abdominal muscles.
14. Which of these muscles can both elevate and depress the scapula?
    a. rhomboideus major and minor d. trapezius
    b. levator scapulae e. pectoralis minor
    c. serratus anterior
15. Which of these muscles does *not* adduct the arm (humerus)?
    a. latissimus dorsi c. teres major e. coracobrachialis
    b. deltoid d. pectoralis major
16. Which of these muscles would you expect to be especially well developed in a boxer known for his powerful jab (punching straight ahead)?
    a. biceps brachii c. trapezius e. supinator
    b. brachialis d. triceps brachii
17. Which of these muscles is an antagonist of the triceps brachii?
    a. biceps brachii c. latissimus dorsi e. supinator
    b. anconeus d. brachioradialis
18. The posterior group of forearm muscles is responsible for
    a. flexion of the wrist.
    b. flexion of the fingers.
    c. extension of the fingers.
    d. Both a and b are correct.
    e. All of these are correct.

19. Which of these muscles is an intrinsic hand muscle that moves the thumb?
    a. flexor pollicis brevis
    b. flexor digiti minimi brevis
    c. flexor pollicis longus
    d. extensor pollicis longus
    e. All of these are correct.
20. Given these muscles:
    (1) iliopsoas
    (2) rectus femoris
    (3) sartorius
    Which of the muscles flex the hip?
    a. 1 c. 1,3 e. 1,2,3
    b. 1,2 d. 2,3
21. Which of these muscles is found in the medial compartment of the thigh?
    a. rectus femoris c. gracilis e. semitendinosus
    b. sartorius d. vastus medialis
22. Which of these is *not* a muscle that can flex the knee?
    a. biceps femoris c. gastrocnemius e. sartorius
    b. vastus medialis d. gracilis
23. The ______________ muscles evert the foot, whereas the ______________ muscles invert the foot.
    a. fibularis (longus and brevis), gastrocnemius and soleus
    b. fibularis (longus and brevis), tibialis anterior and extensor hallucis longus
    c. tibialis anterior and extensor hallucis longus, fibularis longus and brevis
    d. tibialis anterior and extensor hallucis longus, flexor digitorum longus and flexor hallucis longus
    e. flexor digitorum longus and flexor hallucis longus, gastrocnemius and soleus
24. Which of these muscles causes plantar flexion of the foot?
    a. tibialis anterior d. soleus
    b. extensor digitorum longus e. sartorius
    c. fibularis (peroneus) tertius

Answers in Appendix E

CRITICAL THINKING

1. For each of the following muscles, (1) describe the movement the muscle produces, and (2) name the muscles that act as synergists and antagonists for them: longus capitis, erector spinae, coracobrachialis.
2. Consider only the effect of the brachioradialis muscle for these questions: If a weight is held in the hand and the forearm is flexed, what type of lever system is in action? If the weight is placed on the forearm? Which system can lift more weight, and how far?
3. A patient was involved in a rear-end auto collision, resulting in a whiplash injury to the head (hyperextension). What neck muscles might be injured in this type of accident? What is the easiest way to prevent such an injury in an automobile accident?
4. During surgery, a branch of a patient's facial nerve was accidentally cut on one side of the face. After the operation, the lower eyelid and the corner of the patient's mouth drooped on that side. What muscles were affected?
5. When a person becomes unconscious, the tongue muscles relax and the tongue tends to retract, or fall back, and obstruct the airway. Which tongue muscle is responsible? How can this be prevented or reversed?
6. The mechanical support of the head of the humerus in the glenoid fossa is weakest in the inferior direction. What muscles help prevent dislocation of the shoulder when a person carries a heavy weight, such as a suitcase?
7. Speedy Sprinter started a 200-meter dash and fell to the ground in pain. Examination of her right leg revealed the following symptoms: inability to plantar flex the foot against resistance, normal ability to evert the foot, abnormal dorsiflexion of the foot, and abnormal bulging of the calf muscles. Explain the nature of her injury.
8. What muscles are required to turn this page?

Answers in Appendix F

Visit this book's website at **www.mhhe.com/seeley10** for chapter quizzes, interactive learning exercises, and other study tools.

Mc Graw Hill
connect plus+
ANATOMY & PHYSIOLOGY

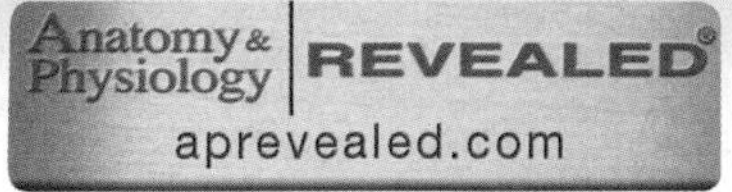
Anatomy & Physiology
REVEALED
aprevealed.com

11

Functional Organization of Nervous Tissue

From thinking and feeling to breathing, moving and eating, virtually everything our body does is controlled by the nervous system. The nervous system is one of the two major control systems in the body, with the endocrine system being the second. The nervous system is made up of the brain, spinal cord, nerves, and sensory receptors. We begin the study of the nervous system in this chapter by focusing on the physiology of nervous tissue. Chapters 12 and 13 discuss the anatomy of the spinal cord and spinal nerves and the brain and cranial nerves. Chapter 14 focuses on the integration of the nervous system components and discusses the effects of aging on the nervous system. Chapter 15 concentrates on the anatomy and physiology of the special senses. Finally, chapter 16 explores the autonomic nervous system anatomy and physiology. By the time you've reviewed the nervous system section, you will definitely understand why you jump out of your seat and your heart pounds when the scary movie music plays and the bad guy pops out of the shower!

Learn to Predict

Once she turned 21, Amanda expected good times ahead. So why could she barely manage to climb the two flights of steps to her chemistry class? When she started experiencing weakness in her left hand, Amanda consulted a physician. After conducting numerous tests, Amanda's physician told her she had multiple sclerosis (MS), a condition in which the myelin sheaths of motor and sensory neurons in the brain and spinal cord are gradually destroyed. By combining what you learned about the histology, physiology, and gross anatomy of the muscular system in chapters 9 and 10 with new information about nervous tissue organization in this chapter, explain why MS made it difficult for Amanda to walk up stairs and led to her hand weakness. Also, predict how Amanda's condition is likely to change over the next several years.

Photo: Light photomicrograph of pyramid-shaped neurons (green) growing on a fibrous network (yellow) in the central nervous system.

Module 7
Nervous System

Anatomy & Physiology
REVEALED
aprevealed.com

11.1 Functions of the Nervous System

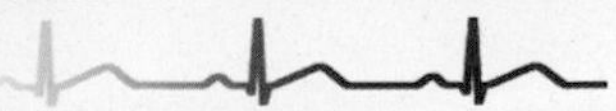

LEARNING OUTCOME

After reading this section, you should be able to

A. Explain the functions of the nervous system.

The regulatory and coordinating activities of the nervous system are necessary for the human body to function normally. The nervous system allows for this in the following ways:

1. *Maintaining homeostasis.* The trillions of cells in the human body do not function independently of each other but must work together to maintain homeostasis. For example, heart cells must contract at a rate that ensures adequate delivery of blood to all tissues of the body. The nervous system can stimulate or inhibit these activities to help maintain homeostasis.
2. *Receiving sensory input.* Sensory receptors monitor numerous external and internal stimuli. We are aware of sensations from some stimuli, such as sight, hearing, taste, smell, touch, pain, body position, and temperature. Other stimuli, such as blood pH, blood gases, and blood pressure, are processed at an unconscious level.
3. *Integrating information.* The brain and spinal cord are the major organs for processing sensory input and initiating responses. The input may produce an immediate response, be stored as memory, or be ignored.
4. *Controlling muscles and glands.* Skeletal muscles normally contract only when stimulated by the nervous system; thus, the nervous system controls the major movements of the body by controlling skeletal muscle. Some smooth muscle, such as that in the walls of blood vessels, contracts only when stimulated by the nervous system or by hormones (see chapter 18). Cardiac muscle and some smooth muscle, such as that in the wall of the stomach, contract autorhythmically—that is, no external stimulation is necessary for each contraction event. Although the nervous system does not initiate contraction in these muscles, it can cause the contractions to occur more rapidly or more slowly. Finally, the nervous system controls the secretions from many glands, including sweat glands, salivary glands, and the glands of the digestive system.
5. *Establishing and maintaining mental activity.* The brain is the center of mental activities, including consciousness, thinking, memory, and emotions.

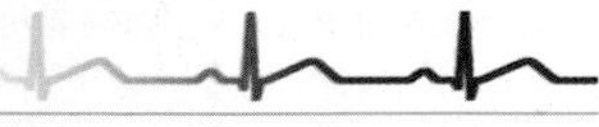

ASSESS YOUR PROGRESS

1. *List and give examples of the general functions of the nervous system.*

11.2 Divisions of the Nervous System

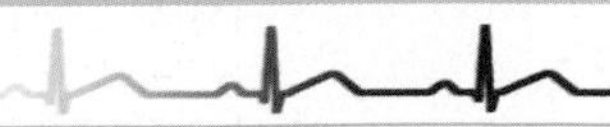

LEARNING OUTCOMES

After reading this section, you should be able to

A. List the divisions of the nervous system and describe the characteristics of each.

B. Differentiate between the somatic and the autonomic nervous systems.

C. Contrast the general functions of the CNS and the PNS.

Humans have only one nervous system, even though some of its subdivisions are referred to as separate systems (figure 11.1). The **central nervous system (CNS)** consists of the brain and the spinal cord. The brain is located within the skull, and the spinal cord is located within the vertebral canal formed by the vertebrae (see chapter 7). The brain and spinal cord are continuous with each other at the foramen magnum.

The **peripheral nervous system (PNS)** consists of all the nervous tissue outside the CNS. It includes sensory receptors, nerves, ganglia, and plexuses. **Sensory receptors** are the endings of neurons, or separate, specialized cells that detect temperature, pain, touch, pressure, light, sound, odor, and other stimuli. Sensory receptors are located in the skin, muscles, joints, internal organs, and specialized sensory organs, such as the eyes and ears. A **nerve** is a bundle of nerve fibers, called axons, and their sheaths; it connects the CNS to sensory

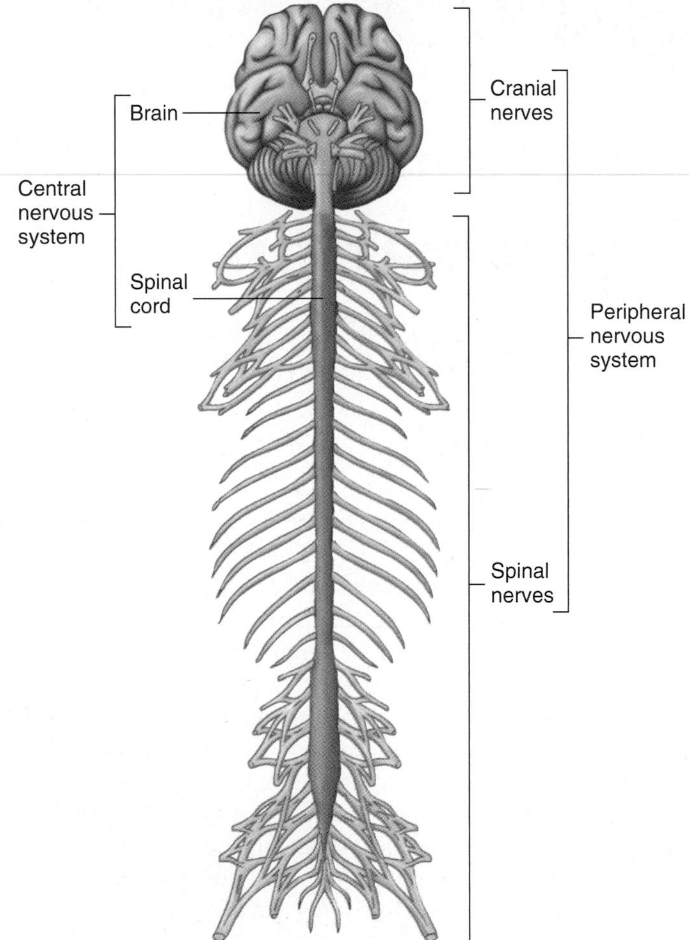
Brain
Central nervous system
Spinal cord
Cranial nerves
Peripheral nervous system
Spinal nerves

FIGURE 11.1 Nervous System
The central nervous system (CNS) consists of the brain and spinal cord. The peripheral nervous system (PNS) consists of cranial nerves, which arise from the brain, and spinal nerves, which arise from the spinal cord. The nerves, which are shown cut in the illustration, actually extend throughout the body.

receptors, muscles, and glands. There are 12 pairs of **cranial nerves** that originate from the brain and 31 pairs of **spinal nerves** that originate from the spinal cord (figure 11.1). A **ganglion** (gang′glē-on; pl. *ganglia*, gang′glē-ă; knot) is a collection of neuron cell bodies located outside the CNS. A **plexus** (plek′sus; braid) is an extensive network of axons and, in some cases, neuron cell bodies, located outside the CNS.

The PNS has two functional subdivisions: The **sensory division,** or *afferent* ("toward") *division*, transmits electrical signals, called **action potentials,** from the sensory receptors to the CNS (figure 11.2). The cell bodies of sensory neurons are located in dorsal root ganglia near the spinal cord (figure 11.3*a*) or in ganglia near the origin of certain cranial nerves. The **motor division,** or *efferent* (away) *division*, transmits action potentials from the CNS to effector organs, such as muscles and glands.

The sensory division of the PNS detects stimuli and transmits information in the form of action potentials to the CNS (see figure 11.2). The CNS is the major site for processing information, initiating responses, and integrating mental processes. It is much like a computer

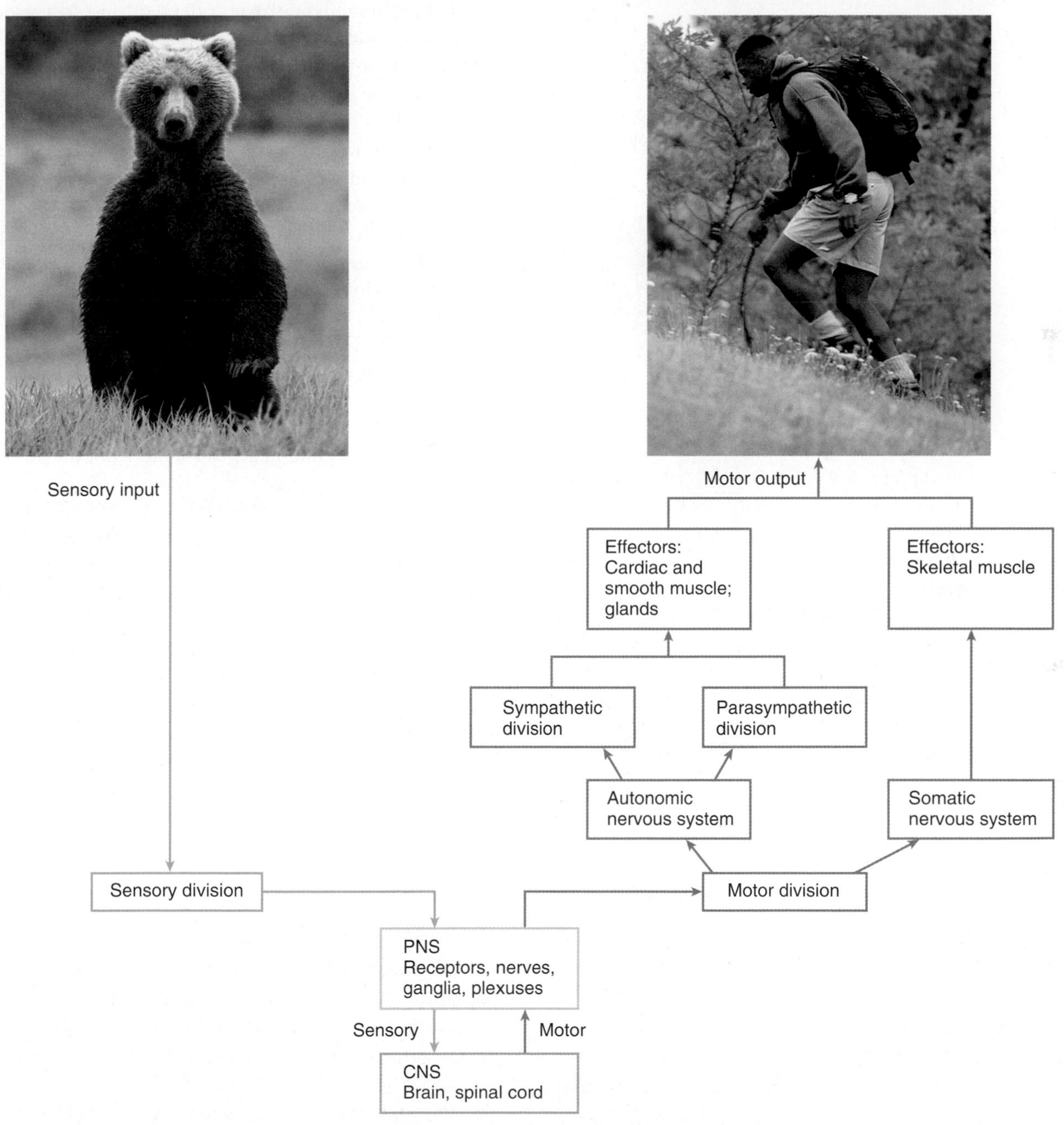
Sensory input
Motor output
Effectors: Cardiac and smooth muscle; glands
Effectors: Skeletal muscle
Sympathetic division
Parasympathetic division
Autonomic nervous system
Somatic nervous system
Sensory division
Motor division
PNS Receptors, nerves, ganglia, plexuses
Sensory
Motor
CNS Brain, spinal cord

FIGURE 11.2 Organization of the Nervous System

The sensory division of the peripheral nervous system (PNS) detects stimuli and conducts action potentials to the central nervous system (CNS). The CNS interprets incoming action potentials and initiates action potentials that are conducted through the motor division to produce a response. The motor division is divided into the somatic nervous system and the autonomic nervous system. The enteric nervous system is an independent branch of the PNS and is not illustrated in this figure.

FUNDAMENTAL **Figure**

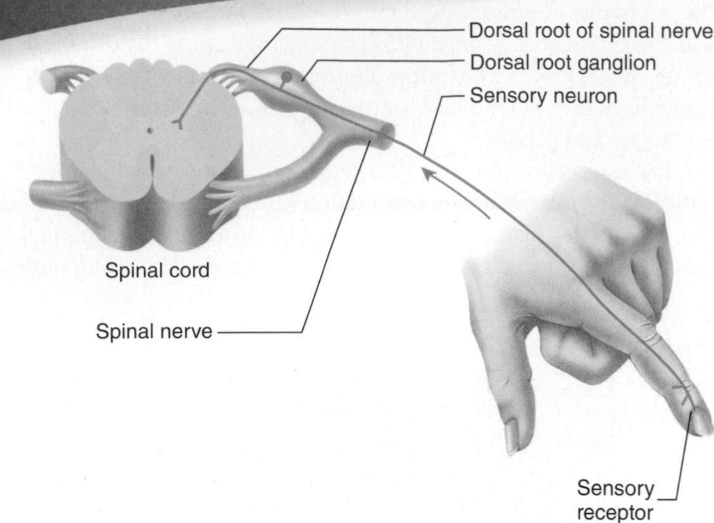
Dorsal root of spinal nerve
Dorsal root ganglion
Sensory neuron
Spinal cord
Spinal nerve
Sensory receptor

(a) **Sensory division**

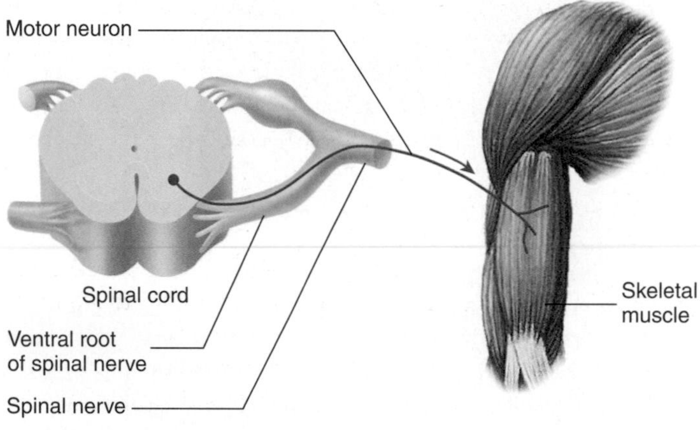
Motor neuron
Spinal cord
Ventral root of spinal nerve
Spinal nerve
Skeletal muscle

(b) **Somatic nervous system**

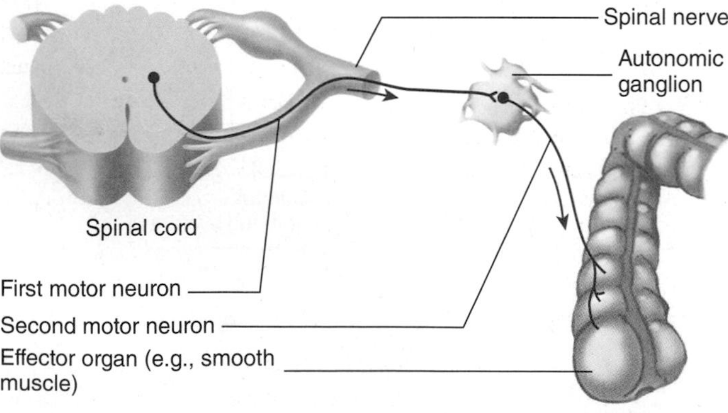
Spinal nerve
Autonomic ganglion
Spinal cord
First motor neuron
Second motor neuron
Effector organ (e.g., smooth muscle)
Large intestine

(c) **Autonomic nervous system**

FIGURE 11.3 AP|R **Divisions of the Peripheral Nervous System**
(*a*) Sensory division. A neuron with its cell body (green dot) in a dorsal root ganglion. (*b*) Somatic nervous system. The neuron (purple) extends from the spinal cord (CNS) to skeletal muscle. (*c*) Autonomic nervous system. Two neurons are in series between the CNS and the effector organ or cells (smooth muscle or glands). The first neuron has its cell body (red dot) in the CNS, and the second neuron has its cell body in an autonomic ganglion.

in its ability to receive input, process and store information, and generate responses. The motor division of the PNS conducts action potentials from the CNS to muscles and glands.

The motor division is divided into two subdivisions: a voluntary division called the **somatic** (sō-mat′ik; bodily) **nervous system** and an involuntary division called the **autonomic** (aw-tō-nom′ik; self-governing) **nervous system (ANS).** The somatic nervous system allows us to consciously control movements of our skeletal muscles through action potentials that originate in the CNS and are transmitted by the somatic nervous system to the same skeletal muscles (figure 11.3*b*). The cell bodies of somatic motor neurons are located within the CNS, and their axons extend through nerves to form connections with skeletal muscle cells. Each of these connections, called a **synapse** (sin′aps), is the junction of a neuron with another cell. The neuromuscular junction, the synapse between a neuron and a skeletal muscle fiber, is discussed in detail in chapter 9. Neurons can also form synapses with other neurons, smooth muscle cells, cardiac muscle cells, and gland cells.

The ANS controls our unconscious activities, such as contractions of smooth muscle, cardiac muscle, and secretion by certain glands. This contrasts with the somatic nervous system, which controls conscious thought and movement. The ANS has two sets of neurons in a series between the CNS and the effector organs (figure 11.3*c*). Cell bodies of the first neurons are within the CNS, and they send their axons to autonomic ganglia, where neuron cell bodies of the second neurons are located. Synapses exist between the first and second neurons within the autonomic ganglia, and the axons of the second neurons extend from the autonomic ganglia to the effector organs.

The ANS is subdivided into the sympathetic division and the parasympathetic division. In general, the **sympathetic division** is most active during physical activity, whereas the **parasympathetic division** regulates resting functions, such as digesting food or emptying the urinary bladder.

The **enteric nervous system (ENS)** consists of plexuses within the wall of the digestive tract (see figure 24.4). A unique feature of enteric neurons is that they monitor and control the digestive tract independently of the CNS through local reflexes. However, the CNS can override enteric functions via parasympathetic and sympathetic actions. Hence, the ENS is an independent subdivision of the PNS that is integrated with the ANS. See chapters 16 and 24 for details on the enteric nervous system.

ASSESS YOUR PROGRESS

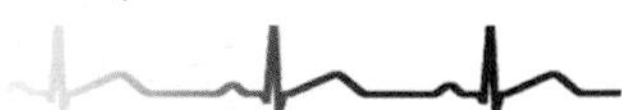

2. *Name the components of the CNS and the PNS.*
3. *What are the following: sensory receptor, nerve, ganglion, plexus?*
4. *Based on the direction they transmit action potentials, what are the two subcategories of the PNS?*
5. *Based on the structures they supply, what are the two subcategories of the motor division?*
6. *Where are the cell bodies of sensory, somatic motor, and autonomic neurons located? What is a synapse?*
7. *What are the subcategories of the ANS?*
8. *Compare the general functions of the CNS and the PNS.*

11.3 Cells of the Nervous System

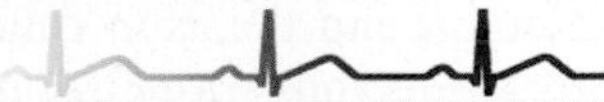

LEARNING OUTCOMES

After reading this section, you should be able to

A. Describe the structure of neurons and the functions of their components.

B. Classify neurons based on structure and function.

C. Describe the location, structure, and functions of neuroglia.

D. Discuss the function of the myelin sheath and describe its formation in the CNS and PNS.

The two types of cells that make up the nervous system are neurons and supportive cells. Neurons receive stimuli, conduct action potentials, and transmit signals to other neurons or effector organs. These supportive cells are called **neuroglia** (noo-rog′lē-ă; nerve glue), or *glial* (glī′ăl, glē′ăl) *cells,* and they support and protect neurons and perform other functions. Neuroglia account for over half of the brain's weight, and there can be 10 to 50 times more neuroglia than neurons in various parts of the brain.

Neuron Structure

Neurons, or *nerve cells,* receive stimuli and transmit action potentials to other neurons or to effector organs. They are organized to form complex networks that perform the functions of the nervous system. There are three parts to a neuron: a cell body and two types of cellular projections. The cell body is called the **neuron cell body,** or *soma* (sō′mă; body); as with any other type of cell, the cell body's nucleus is the source of information for protein synthesis. One type of cellular projection is called a **dendrite** (den′drīt; tree), referring to its branching organization. The other type of cellular projection is called the **axon** (ak′son; axis), referring to the straight alignment and uniform diameter of most axons. Axons are also called *nerve fibers.*

Neuron Cell Body

Each neuron cell body contains a single, relatively large, and centrally located nucleus with a prominent nucleolus. Extensive rough endoplasmic reticulum (ER) and the Golgi apparatuses surround the nucleus, and mitochondria and other organelles are present. Large numbers of intermediate filaments (neurofilaments) and microtubules form bundles that organize the cytoplasm into different regions. The neurofilaments separate abundant rough ER, called **Nissl** (nis′l) **bodies,** which are located primarily in the cell body and dendrites. Nissl bodies are the primary site of protein synthesis in neurons.

Predict 2

If an axon has been severed, so that it is no longer connected to its neuron cell body, what will be the effect on the distal and proximal portions of the axon? Explain your prediction.

Dendrites

Dendrites are short, often highly branched cytoplasmic extensions that are tapered from their bases at the neuron cell body to their tips (figure 11.4). Many dendrite surfaces have small extensions, called **dendritic spines,** where axons of other neurons form synapses with the dendrites. Dendrites receive input from other neurons' axons and from the environment. When stimulated, they generate small electric currents, which are conducted to the neuron cell body.

Axons

In most neurons, a single axon arises from a cone-shaped area of the neuron cell body called the **axon hillock.** The beginning of the axon is called the **initial segment.** Action potentials are generated at the **trigger zone,** which consists of the axon hillock and the part of the axon nearest the cell body. An axon can remain as a single structure or can branch to form collateral axons, or side branches (figure 11.4). Each axon has a constant diameter, but axons can vary in length from a few millimeters to more than 1 meter. The cytoplasm of an axon is sometimes called the **axoplasm,** and its plasma membrane is called the **axolemma** (*lemma,* husk). Axons terminate by branching to form small extensions with enlarged ends called **presynaptic terminals.** Within the presynaptic terminals are numerous small, membrane-bound secretory vesicles that contain chemicals called **neurotransmitters.** Action potentials conducted along the axon to the presynaptic terminal stimulate exocytosis of the neurotransmitters from their vesicles into the synapse. Then neurotransmitters cross the synaptic cleft to stimulate or inhibit the postsynaptic cell.

Axon transport mechanisms can move cytoskeletal proteins (see chapter 3), organelles (such as mitochondria), and vesicles containing neuropeptides to be secreted (see chapter 17) down the axon through the axoplasm to the presynaptic terminals (*anterograde*). In addition, damaged organelles, recycled plasma membrane, and substances taken in by endocytosis can be transported up the axon to the neuron cell body (*retrograde*). The movement of materials within the axon is necessary for its normal function, but it also provides a way for infectious agents and harmful substances to be transported from the periphery to the CNS. For example, rabies and herpes viruses can enter damaged axons in the skin and be transported within the axons to the CNS.

Types of Neurons

Neurons are classified according to either their function or their structure. The functional classification is based on the direction in which action potentials are conducted. **Sensory neurons** (*afferent neurons*) conduct action potentials toward the CNS; **motor neurons** (*efferent neurons*) conduct action potentials away from the CNS toward muscles or glands. **Interneurons** conduct action potentials from one neuron to another within the CNS.

The structural classification scheme is based on the arrangement of the processes that extend from the neuron cell body. The three major structural categories of neurons are multipolar, bipolar, and pseudo-unipolar.

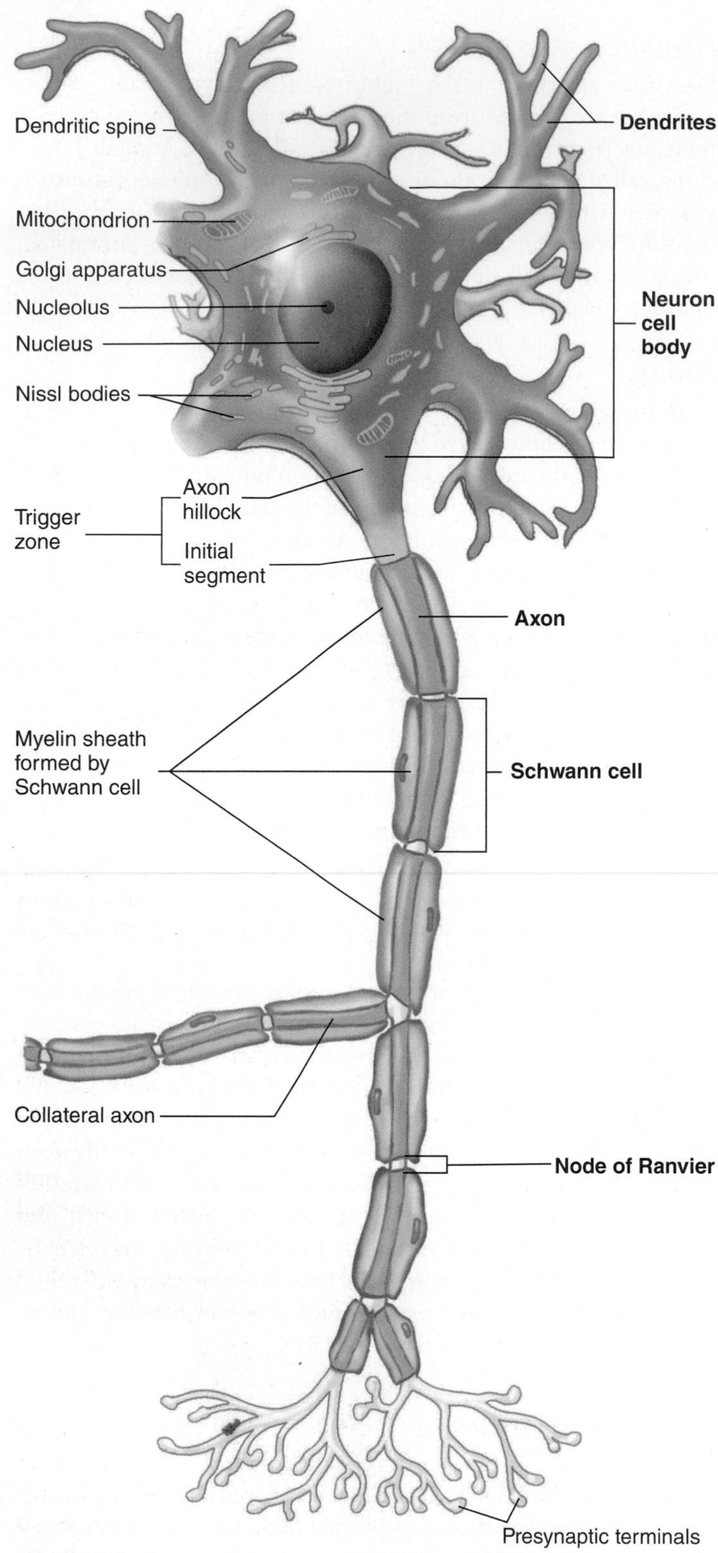

Dendritic spine
Dendrites
Mitochondrion
Golgi apparatus
Nucleolus
Nucleus
Neuron cell body
Nissl bodies
Trigger zone
Axon hillock
Initial segment
Axon
Myelin sheath formed by Schwann cell
Schwann cell
Collateral axon
Node of Ranvier
Presynaptic terminals

FIGURE 11.4 AP|R **Neuron**

The structural features of a neuron are a cell body and two types of cell projections: dendrites and an axon.

Multipolar neurons have many dendrites and a single axon. The dendrites vary in number and in their degree of branching (figure 11.5*a*). Most of the neurons within the CNS and motor neurons are multipolar.

Bipolar neurons have two processes: one dendrite and one axon (figure 11.5*b*). The dendrite is often specialized to receive the stimulus, and the axon conducts action potentials to the CNS. Bipolar neurons are located in some sensory organs, such as in the retina of the eye and in the nasal cavity.

Pseudo-unipolar neurons have a single process extending from the cell body (figure 11.5*c*). This process divides into two branches a short distance from the cell body. One branch extends to the CNS, and the other extends to the periphery and has dendrite-like sensory receptors. The two branches function as a single axon. The sensory receptors respond to stimuli, producing action potentials that are transmitted to the CNS. Most sensory neurons are pseudo-unipolar. According to a functional definition of a dendrite, the branch of a pseudo-unipolar neuron that extends from the periphery to the neuron cell body can be classified as a dendrite because it conducts action potentials toward the neuron cell body. However, this branch is usually referred to as an axon, for two reasons: It cannot be distinguished from an axon on the basis of its structure, and it conducts action potentials in the same fashion as an axon.

Neuroglia of the CNS

Neuroglia are the major supporting cells in the CNS; they participate in forming a permeability barrier between the blood and the neurons, phagocytize foreign substances, produce cerebrospinal fluid, and form myelin sheaths around axons. There are four types of CNS neuroglia, each with unique structural and functional characteristics. Refer to table 11.1 for a description and a figure of each neuroglial cell.

Astrocytes

Astrocytes (as′trō-sītz; *aster*, star) are neuroglia that are star-shaped because cytoplasmic processes extend from the cell body. These extensions widen and spread out to form foot processes, which cover the surfaces of blood vessels (table 11.1), neurons, and the pia mater. (The pia mater is a membrane covering the outside of the brain and spinal cord.) Astrocytes have an extensive cytoskeleton of microfilaments (see chapter 3), which enables them to form a supporting framework for blood vessels and neurons.

Astrocytes help regulate the extracellular composition of brain fluid. They do this by releasing chemicals that promote the formation of tight junctions (see chapter 4) between the endothelial cells of capillaries. The endothelial cells with their tight junctions form the **blood-brain barrier,** which determines what substances can pass from the blood into the nervous tissue of the brain and spinal cord. The blood-brain barrier protects neurons from toxic substances in the blood, allows the exchange of nutrients and waste products between neurons and the blood, and prevents fluctuations in blood composition from affecting brain functions.

Astrocytes aid both beneficial and detrimental responses to tissue damage in the CNS. Almost all injuries to CNS tissue induce **reactive astrocytosis,** in which astrocytes wall off the injury site and help limit the spread of inflammation to the surrounding healthy tissue. Reactive scar-forming astrocytes also limit the regeneration of the axons of injured neurons.

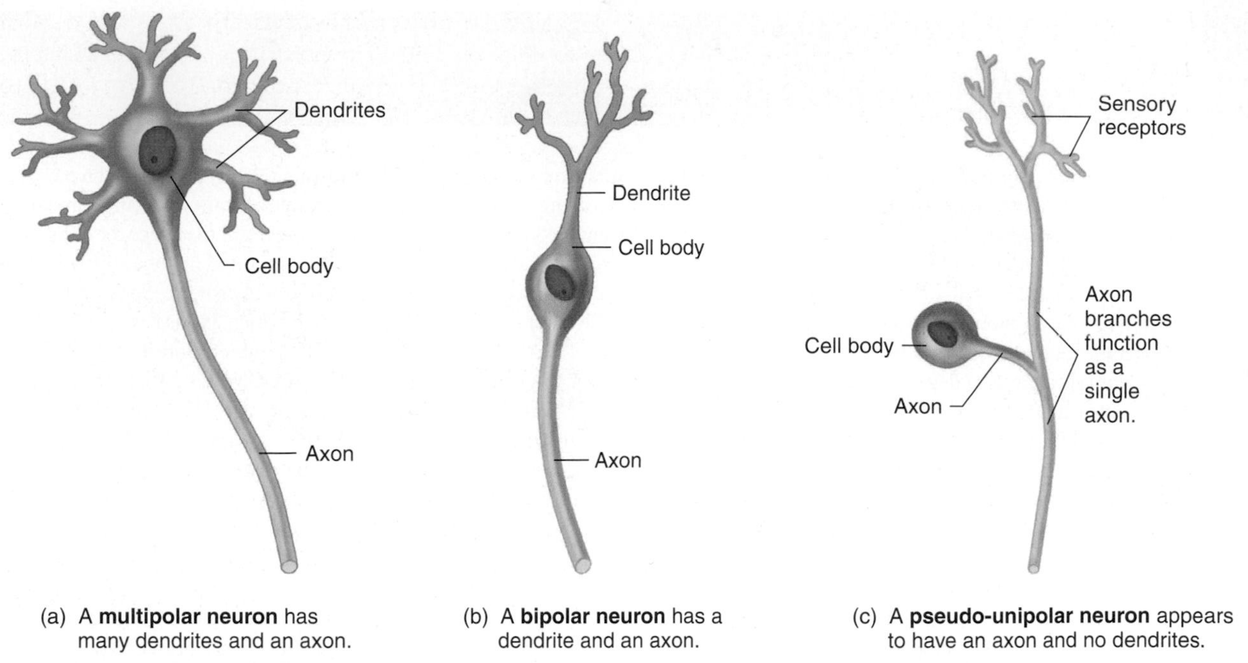

Dendrites
Cell body
Axon
Dendrite
Cell body
Axon
Sensory receptors
Cell body
Axon
Axon branches function as a single axon.
(a) A **multipolar neuron** has many dendrites and an axon.
(b) A **bipolar neuron** has a dendrite and an axon.
(c) A **pseudo-unipolar neuron** appears to have an axon and no dendrites.

FIGURE 11.5 Structural Classes of Neurons
Neurons are classified structurally by the number of cellular projections extending from their cell bodies. Dendrites and sensory receptors are specialized to receive stimuli, and axons are specialized to conduct action potentials.

Other astrocyte functions include releasing chemicals that promote the development of synapses and helping regulate synaptic activity through the synthesis, absorption, and recycling of neurotransmitters.

Ependymal Cells

Ependymal (ep-en′di-măl) **cells** line the ventricles (cavities) of the brain and the central canal of the spinal cord (table 11.1). Specialized ependymal cells and blood vessels form the **choroid plexuses** (ko′royd plek′sŭs-ez; table 11.1), which are located within certain regions of the ventricles. The choroid plexuses secrete the cerebrospinal fluid that flows through the ventricles of the brain (see chapter 13). The free surface of the ependymal cells frequently bears patches of cilia that help move cerebrospinal fluid through the brain cavities. Ependymal cells also have long processes at their basal surfaces that extend deep into the brain and the spinal cord and seem, in some cases, to have astrocyte-like functions.

Microglia

Microglia (mī-krog′lē-ă) are neuroglia in the CNS that become mobile and phagocytic in response to inflammation. They phagocytize necrotic tissue, microorganisms, and other foreign substances that invade the CNS (table 11.1). Numerous microglia migrate to areas damaged by infection, trauma, or stroke and perform phagocytosis. A pathologist can identify these damaged areas in the CNS during an autopsy because large numbers of microglia are found in them.

Oligodendrocytes

Oligodendrocytes (ol′i-gō-den′drō-sītz) have cytoplasmic extensions that can surround axons. If the cytoplasmic extensions wrap many times around the axons, they form an insulating material called a **myelin** (mī′ĕ-lin) **sheath.** A single oligodendrocyte can form myelin sheaths around portions of several axons (table 11.1).

Neuroglia of the PNS

Schwann cells are neuroglia in the PNS that wrap around axons. If a Schwann cell wraps many times around an axon, it forms a myelin sheath. However, unlike oligodendrocytes, each Schwann cell forms a myelin sheath around a portion of only one axon (table 11.1).

Satellite cells surround neuron cell bodies in sensory and autonomic ganglia (table 11.1). Besides providing support and nutrition to the neuron cell bodies, satellite cells protect neurons from heavy-metal poisons, such as lead and mercury, by absorbing them and reducing their access to the neuron cell bodies.

Myelinated and Unmyelinated Axons

Cytoplasmic extensions of the Schwann cells in the PNS and of the oligodendrocytes in the CNS surround axons to form either myelinated or unmyelinated axons. Myelin protects and electrically insulates axons from one another. In addition, action potentials travel along myelinated axons more rapidly than along unmyelinated axons (see "Propagation of Action Potentials" in section 11.5).

In **myelinated axons,** the extensions from Schwann cells or oligodendrocytes repeatedly wrap around a segment of an axon to form a series of tightly wrapped membranes rich in phospholipids, with little cytoplasm sandwiched between the membrane layers (figure 11.6*a*). The tightly wrapped membranes constitute the myelin sheath and give myelinated axons a white appearance because of the high lipid concentration. The myelin sheath is not

Clinical IMPACT — Nervous Tissue Response to Injury

When a nerve is cut, either it eventually heals or it is permanently interrupted. The final outcome depends on the severity of the injury and on its treatment.

Several degenerative changes result when a nerve is cut (figure 11A). Within about 3–5 days, the axons in the part of the nerve distal to the cut break into irregular segments and degenerate. This occurs because the neuron cell body produces the substances essential to maintain the axon, and these substances have no way of reaching parts of the axon distal to the point of damage. Eventually, the distal part of the axon completely degenerates. As the axons degenerate, the myelin part of the Schwann cells around them also degenerates, and macrophages invade the area to phagocytize the myelin. The Schwann cells then enlarge, undergo mitosis, and finally form a column of cells along the regions once occupied by the axons. The columns of Schwann cells are essential for the growth of new axons. If the ends of the regenerating axons encounter a Schwann cell column, they grow more rapidly, and reinnervation of their target is likely. If the ends of the axons do not encounter the columns, they fail to reinnervate their target.

The end of each regenerating axon forms several axonal sprouts. It normally takes about 2 weeks for the axonal sprouts to enter the Schwann cell columns. However, only one of the sprouts from each severed neuron forms an axon. The other branches degenerate. After the axons grow through the Schwann cell columns, new myelin sheaths form and the neurons reinnervate the structures they previously supplied.

Treatment strategies that increase the probability of reinnervation involve bringing the ends of the severed nerve close together surgically. When a section of nerve is destroyed as a result of trauma, a surgeon can perform a nerve transplant to replace the damaged segment. The transplanted nerve eventually degenerates, but it does provide Schwann cell columns through which axons can grow.

The regeneration of damaged nerve tracts within the CNS is very limited, especially when compared with the regeneration of nerves in the PNS. In part, the difference may result from the oligodendrocytes, which exist only in the CNS. An oligodendrocyte has several processes, each of which forms part of a myelin sheath. The cell bodies of the oligodendrocytes are a short distance from the axons they ensheathe, and fewer oligodendrocytes than Schwann cells are present. Consequently, when the myelin degenerates following damage, no column of cells remains in the CNS to act as a guide for the growing axons.

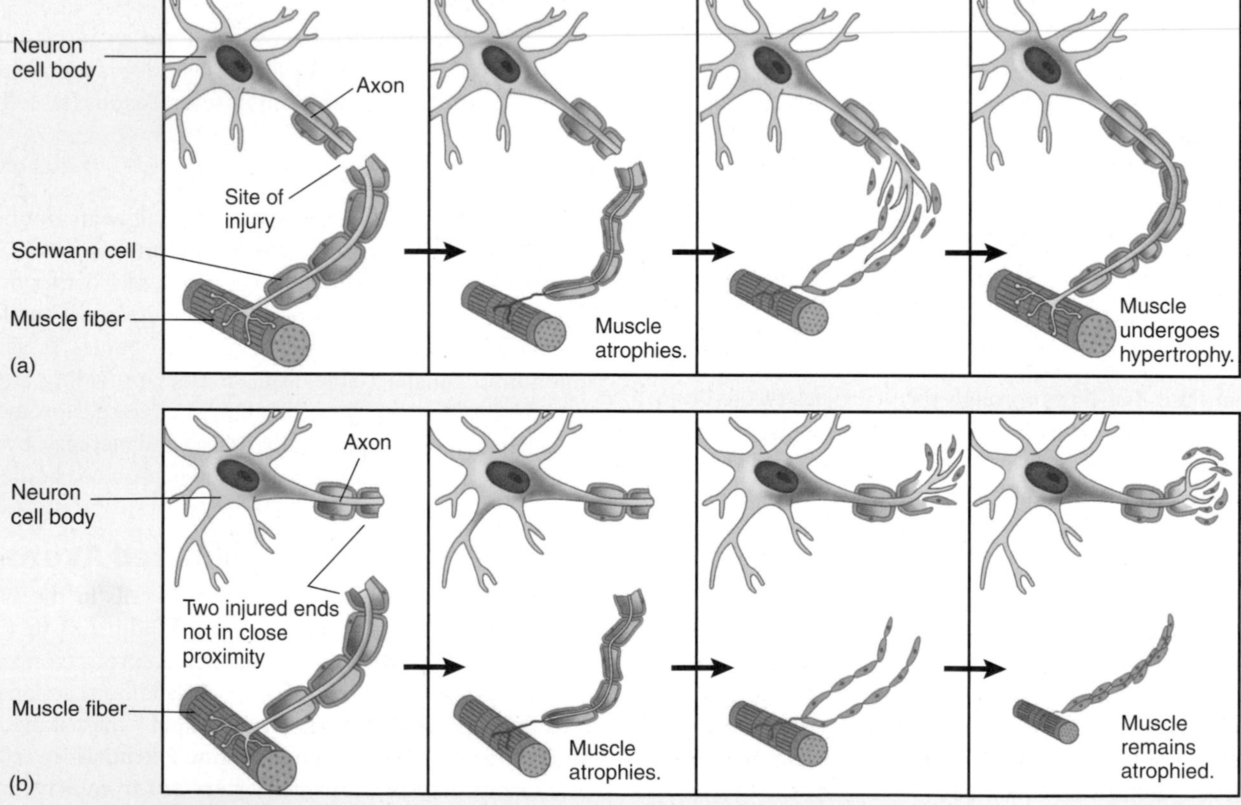
Neuron cell body
Axon
Site of injury
Schwann cell
Muscle fiber
(a)
Muscle atrophies.
Muscle undergoes hypertrophy.
Axon
Neuron cell body
Two injured ends not in close proximity
Muscle fiber
(b)
Muscle atrophies.
Muscle remains atrophied.

FIGURE 11A Responses to Injury in an Axon
(*a*) When the two ends of an injured axon are aligned in close proximity, healing and regeneration of the axon are likely to occur. Without stimulation from the nerve, the muscle is paralyzed and atrophies (shrinks in size). After reinnervation, the muscle can become functional and hypertrophy (increase in size). (*b*) When the two ends of an injured axon are not aligned in close proximity, regeneration is unlikely to occur. Without innervation from the nerve, muscle function is completely lost, and the muscle remains atrophied.

TABLE 11.1 Types of Neuroglial Cells APR

Neuroglial Cells	Function
CNS	
Astrocytes 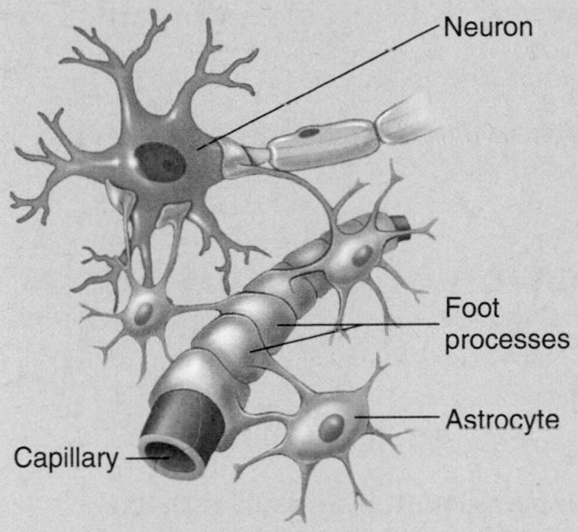 Neuron Foot processes Astrocyte Capillary	Astrocyte foot processes cover the surfaces of neurons, blood vessels, and the pia mater membrane of the brain and spinal cord. The astrocytes provide structural support and play a role in regulating what substances from the blood reach the neurons.
Ependymal cells 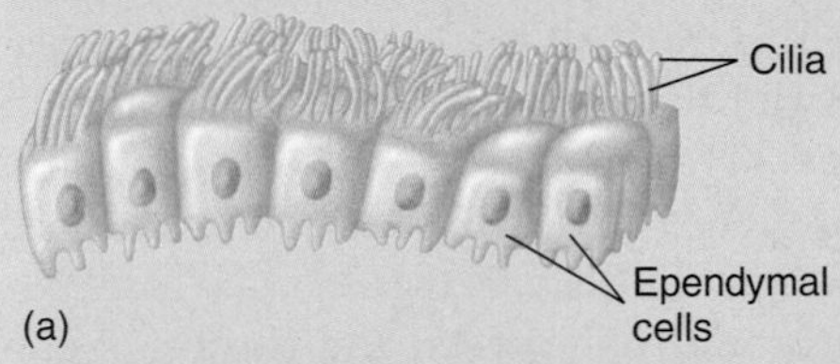 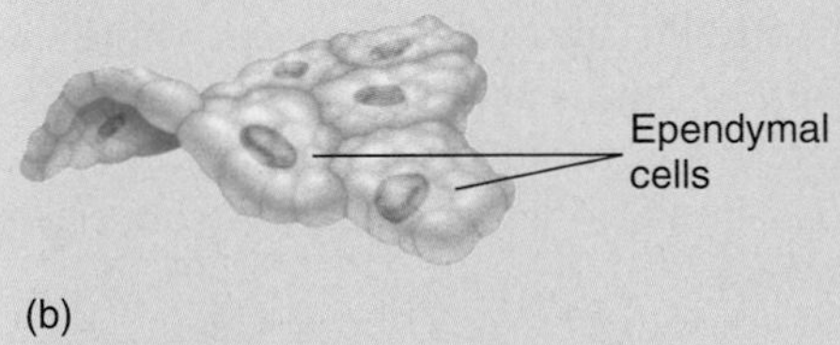 Cilia Ependymal cells (a) Ependymal cells (b)	(*a*) Ciliated ependymal cells lining the ventricles of the brain and the central canal of the spinal cord help move cerebrospinal fluid. (*b*) Ependymal cells on the surface of the choroid plexus secrete cerebrospinal fluid.
Microglia 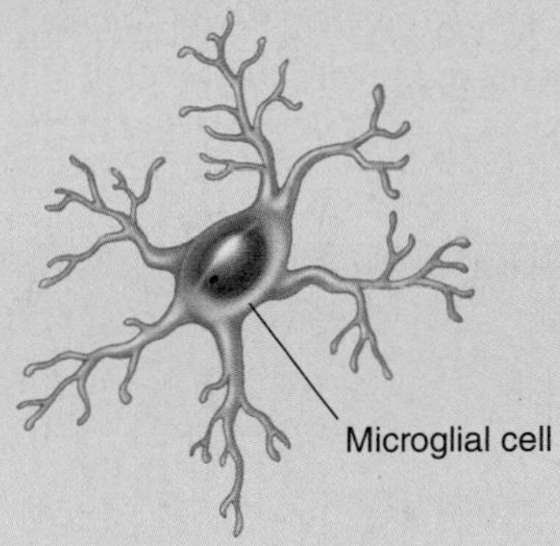 Microglial cell	Microglia are phagocytic cells within the CNS.
Oligodendrocytes 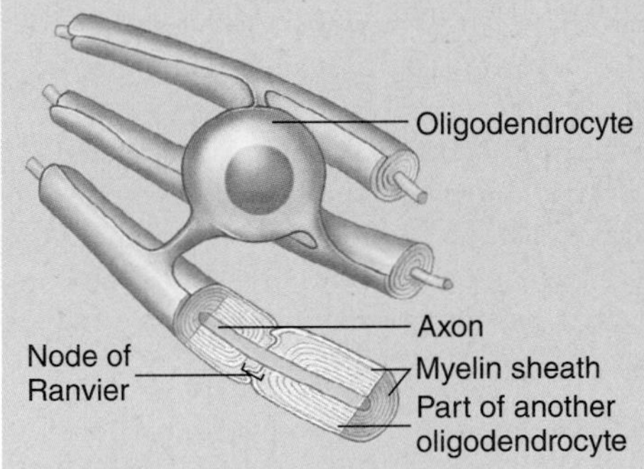 Oligodendrocyte Axon Myelin sheath Part of another oligodendrocyte Node of Ranvier	Extensions from oligodendrocytes form part of the myelin sheaths of several axons within the CNS.
PNS	
Schwann cells and satellite cells 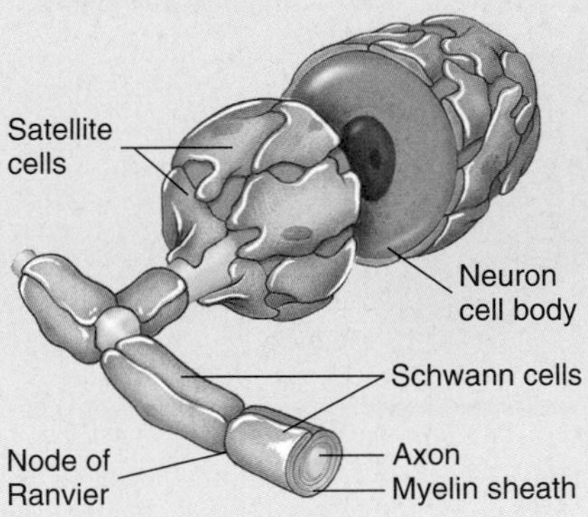 Satellite cells Neuron cell body Schwann cells Axon Myelin sheath Node of Ranvier	Neuron cell bodies within ganglia are surrounded by satellite cells. Schwann cells form the myelin sheath of an axon within the PNS.

continuous but is interrupted every 0.3–1.5 mm. At these locations are slight constrictions where the myelin sheaths of adjacent cells dip toward the axon but do not cover it, leaving a bare area 2–3 μm in length. These interruptions in the myelin sheath are the **nodes of Ranvier** (ron′vē-ā). Although the axon at a node of Ranvier is not covered with myelin, Schwann cells or oligodendrocytes extend across the node and connect to each other.

Unmyelinated axons rest in invaginations of the Schwann cells or oligodendrocytes (figure 11.6*b*). The cell's plasma membrane surrounds each axon but does not wrap around it many times. Thus, each axon is surrounded by a series of Schwann cells, and each Schwann cell can simultaneously surround more than one unmyelinated axon.

ASSESS YOUR PROGRESS

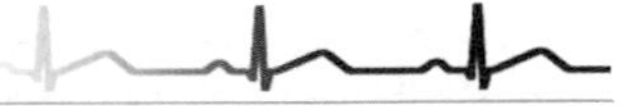

9. *Describe and give the function of a neuron cell body, a dendrite, and an axon.*

10. *What is the function of the trigger zone?*

11. *What is the role of a neurotransmitter? Where is it stored?*
12. *Describe the three types of neurons based on function.*
13. *Explain the three types of neurons based on structure, and give an example of where each type is found.*
14. *What characteristic makes neuroglia different from neurons?*
15. *Which neuroglia are found in the CNS? In the PNS?*
16. *Which type of neuroglia supports neurons and blood vessels and promotes formation of the blood-brain barrier? What is the blood-brain barrier, and what is its function?*
17. *Name the different kinds of neuroglia that are responsible for the following functions: production of cerebrospinal fluid, phagocytosis, production of myelin sheaths in the CNS, production of myelin sheaths in the PNS, support of neuron cell bodies in the PNS.*
18. *What is a myelin sheath? How is it formed in the CNS? In the PNS?*
19. *How do myelinated axons differ from unmyelinated axons?*

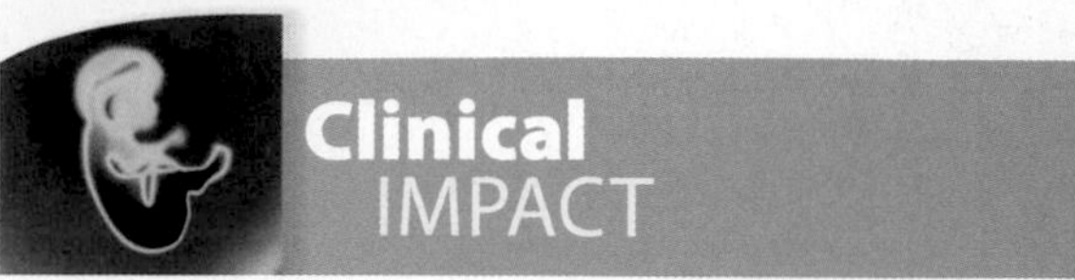
Clinical
IMPACT

Importance of Myelin Sheaths

Myelin sheaths begin to form late in fetal development. The process continues rapidly until the end of the first year after birth and continues more slowly thereafter. The development of myelin sheaths is associated with the infant's continuing development of more rapid and better coordinated responses.

The importance of myelinated fibers is dramatically illustrated in diseases that gradually destroy the myelin sheath, such as multiple sclerosis and some cases of diabetes mellitus. Action potential transmission is slowed, resulting in impaired control of skeletal and smooth muscles. In severe cases, action potential transmission can become completely blocked.

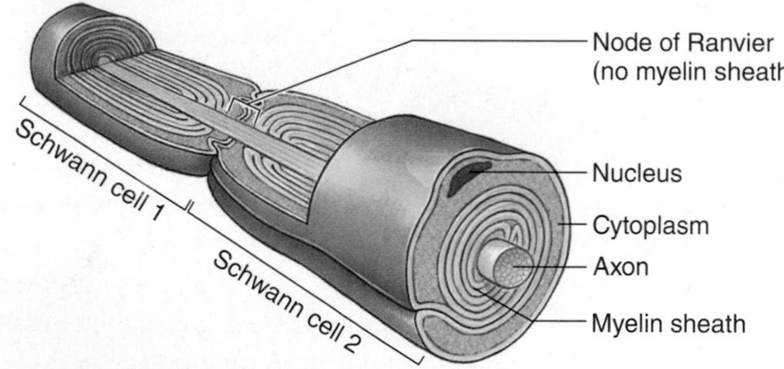
Node of Ranvier
(no myelin sheath)
Schwann cell 1
Schwann cell 2
Nucleus
Cytoplasm
Axon
Myelin sheath

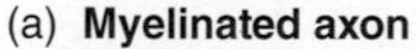
(a) **Myelinated axon**

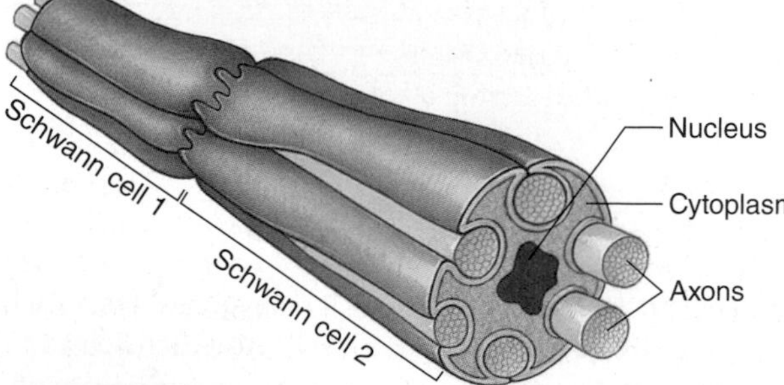
Schwann cell 1
Schwann cell 2
Nucleus
Cytoplasm
Axons

(b) **Unmyelinated axons**

FIGURE 11.6 Comparison of Myelinated and Unmyelinated Axons
(*a*) Myelinated axon with two Schwann cells forming part of the myelin sheath around a single axon. Each Schwann cell surrounds part of one axon. (*b*) Unmyelinated axons with two Schwann cells surrounding several axons in parallel formation. Each Schwann cell surrounds part of several axons.

11.4 Organization of Nervous Tissue

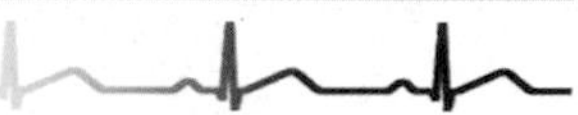

LEARNING OUTCOMES

After reading this section, you should be able to

A. Distinguish between gray matter and white matter.

B. Describe the components of gray matter and white matter in the CNS and PNS.

Nervous tissue is organized in both the CNS and the PNS so that axons form bundles, and neuron cell bodies and their relatively short dendrites are grouped together. Therefore, both the CNS and the PNS contain areas of gray matter and areas of white matter. **Gray matter** consists of groups of neuron cell bodies and their dendrites, where there is very little myelin. In the CNS, gray matter on the surface of the brain is called the **cortex,** and clusters of gray matter located deeper within the brain are called **nuclei.** In the PNS, a cluster of neuron cell bodies is called a **ganglion** (gang′glē-on; pl. *ganglia*; a swelling or knot). **White matter** consists of bundles of parallel axons with their myelin sheaths, which are whitish in color. White matter of the CNS forms **nerve tracts,** or *conduction pathways*, which propagate action potentials from one area of the CNS to another. In the PNS, bundles of axons and their connective tissue sheaths are called nerves.

ASSESS YOUR PROGRESS

20. *What makes up gray matter and white matter?*
21. *Describe and state the location of the following: nerve tracts, nerves, the brain cortex, nuclei, ganglia.*

Predict 3

A 75-year-old man was found unconscious in his bathroom after falling and hitting his head. He survived for several hours but died later in the hospital. An autopsy was performed to determine the exact cause of death. Evidence indicated that the man had suffered two strokes, both due to blocked blood vessels. One had occurred a few weeks earlier; the other had occurred very recently and may have led to the fall. Autopsy findings also indicated that, when the man hit his head, some damage to his brain occurred as well. Based on what you know about inflammation and the cellular structure of the brain, describe what the pathologist found in each of the damaged areas of the brain.

11.5 Electrical Signals

LEARNING OUTCOMES

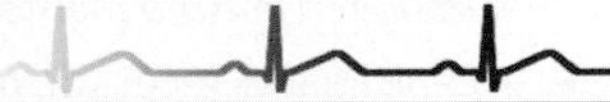

After reading this section, you should be able to

A. Describe a resting membrane potential and explain how it is created and maintained.

B. Explain the processes that can change the resting membrane potential.

C. Describe the characteristics of a graded potential.

D. Describe the creation of an action potential and explain how it is propagated.

E. Discuss the all-or-none principle as it applies to action potentials.

F. Explain the characteristics and purpose of the refractory period.

G. Explain the factors that determine action potential frequency and the five levels of stimulation.

H. Describe the effect of myelination on the speed of action potential propagation, as well as other factors that affect the speed of action potential conduction.

Like computers, humans depend on electrical signals to communicate and integrate information. The electrical signals produced by cells, called action potentials, are an important means by which cells transfer information from one part of the body to another. For example, stimuli—such as light, sound, and pressure—act on specialized sensory cells in the eye, ear, and skin to produce action potentials, which are conducted from these cells to the spinal cord and brain. Action potentials originating within the brain and spinal cord travel to muscles and certain glands to regulate their activities.

Our ability to perceive our environment, perform complex mental activities, and respond to stimuli depends on action potentials. For example, interpreting the action potentials received from sensory cells produces the sensations of sight, hearing, and touch. Complex mental activities, such as conscious thought, memory, and emotions, result from action potentials. The contraction of muscles and the secretion of certain glands occur in response to action potentials generated within them.

A basic knowledge of the electrical properties of cells is necessary for understanding many of the normal functions and pathologies of the body. Electrical properties are due to two major characteristics:

1. Ionic concentration differences across the plasma membrane
2. Permeability characteristics of the plasma membrane

Ionic Concentration Differences Across the Plasma Membrane

As you learned in chapter 9, excitable cells operate through ion movements across the plasma membrane. As you will see, many of the principles you studied for skeletal muscles also apply to neurons.

Table 11.2 lists the concentration differences for positively charged ions (cations) and negatively charged ions (anions) between the intracellular and extracellular fluids. The concentration of sodium ions (Na^+) and chloride ions (Cl^-) is much greater outside the cell than inside. The concentration of potassium ions (K^+) and negatively charged molecules, such as proteins, is much greater inside the cell than outside. Additionally, the concentration of other molecules that contain phosphate is higher inside the cell. Note that a steep concentration gradient (see chapter 3) exists for Na^+ from the outside to the inside of the cell. Also, a steep concentration gradient exists for K^+ from the inside to the outside of the cell.

TABLE 11.2 Representative Concentrations of the Principal Cations and Anions in Extracellular and Intracellular Fluids in the Human Body

Ions	Intracellular Fluid (mEq/L)*	Extracellular Fluid (mEq/L)
Cations (Positive)		
Potassium (K^+)	148	5
Sodium (Na^+)	10	142
Calcium (Ca^{2+})	<1	5
Others	41	3
TOTAL	200	155
Anions (Negative)		
Proteins	56	16
Chloride (Cl^-)	4	103
Others	140	36
TOTAL	200	155

* See appendix C for an explanation of milliequivalents (mEq).

Permeability Characteristics of the Plasma Membrane

Differences in intracellular and extracellular concentrations of ions result primarily from the sodium-potassium pump and the permeability characteristics of the plasma membrane.

Neurons expend energy to maintain an uneven distribution of ions across the plasma membrane. The **sodium-potassium pump** uses ATP to pump K^+ against its concentration gradient and keep it in high concentration inside the cell and to pump Na^+ against its concentration gradient and keep it in high concentration outside the cell (see figure 3.21). Three Na^+ are transported out of the cell and two K^+ are transported into the cell for each ATP molecule used.

As noted in chapter 3, the plasma membrane is selectively permeable, thus allowing some, but not all, substances to pass through it. Negatively charged proteins are synthesized inside the cell, and because of their large size and their solubility characteristic they cannot readily diffuse across the plasma membrane; therefore, they remain inside the cell (figure 11.7*a*). Negatively charged Cl^- is repelled by the negatively charged proteins and other negatively charged ions inside the cell. Chloride ions diffuse through the plasma membrane and accumulate outside it, resulting in a higher concentration of Cl^- outside the cell than inside.

Ions must pass through the plasma membrane through ion channels. The two major types of ion channels are leak ion channels and gated ion channels.

Leak Ion Channels

Leak ion channels, or *nongated ion channels,* are always open and are responsible for the permeability of the plasma membrane to ions when the plasma membrane is unstimulated, or at rest (figure 11.7*a*). Each ion channel is specific for one type of ion, although the specificity is not absolute. The number of each type of leak ion channel in the plasma membrane determines the permeability characteristics of the resting plasma membrane to different types of ions. The plasma membrane is more permeable to K^+ and Cl^- and much less permeable to Na^+ because the membrane has many more K^+ and Cl^- leak ion channels than Na^+ leak ion channels.

Gated Ion Channels

Gated ion channels are closed until opened by specific signals. By opening and closing, these channels can change the permeability of the plasma membrane. There are three major types of gated ion channels:

1. *Ligand-gated ion channels.* **Ligand-gated ion channels** are stimulated to open when a chemical such as a neurotransmitter or hormone (the ligand) binds to the receptor portion of the extracellular component of the ion channel. The membrane-spanning part forms an ion channel. When a ligand binds to the receptor site, the ion channel opens or closes. For example, the neurotransmitter acetylcholine released from the presynaptic terminal of a neuron is a chemical signal that can bind to a ligand-gated Na^+ channel in the membrane of a muscle fiber. As a result, the Na^+ channel opens, allowing Na^+ to enter the fiber (see figure 3.10). Ligand-gated ion channels exist for Na^+, K^+, Ca^{2+}, and Cl^-, and these channels are common in nervous and muscle tissues, as well as in glands.
2. *Voltage-gated ion channels.* **Voltage-gated ion channels** open and close in response to small voltage changes across the plasma membrane. In an unstimulated cell, the inside of the cell is negatively charged relative to the outside. This charge difference can be measured in units called **millivolts (mV; 1 mV = 1/1000 V).** For reference, a "double-A" battery generates 1.5 V of electricity. When a cell is stimulated, the permeability of the plasma membrane changes because gated ion channels open or close. The movement of ions into or out of the cell changes the charge difference across the plasma membrane, which, in turn, causes voltage-gated ion channels to open or close. Voltage-gated channels specific for Na^+ and K^+ are most numerous in electrically excitable tissues, but voltage-gated Ca^{2+} channels are also important, especially in smooth muscle and cardiac muscle fibers (see chapters 9 and 20).
3. *Other gated ion channels.* Gated ion channels that respond to stimuli other than ligands or voltage changes are present in specialized electrically excitable tissues. Examples include touch receptors, which respond to mechanical stimulation of the skin, and temperature receptors, which respond to temperature changes in the skin.

ASSESS **YOUR PROGRESS**

22. *Describe the concentration differences for Na^+ and K^+ that exist across the plasma membrane.*
23. *Explain how the sodium-potassium pump works to move ions.*
24. *Describe leak ion channels and gated ion channels. How are they responsible for the permeability of a resting versus a stimulated plasma membrane?*
25. *Define* ligand, receptor, *and* receptor site.
26. *What kinds of stimuli cause gated ion channels to open or close?*

Establishing the Resting Membrane Potential

Intracellular fluid is electrically neutral because the number of positively charged cations is equal to the number of negatively charged anions (table 11.2). Similarly, extracellular fluid is electrically neutral. However, there is a difference in *charge* across the plasma membrane because of the uneven distribution of positive and negative ions across it. For simplicity, we can say that the inside of the cell is negative compared with the outside of the cell (figure 11.7*b*). The plasma membrane, therefore, is said to be **polarized** because there are opposite charges, or poles, across the membrane.

The electrical charge difference across the plasma membrane is called a **potential difference.** In an unstimulated, or resting, cell, the potential difference is called the **resting membrane potential.** It can be measured using an oscilloscope or a voltmeter connected to microelectrodes positioned inside and outside the plasma membrane (figure 11.7). The resting membrane potential of neurons is approximately −70 mV, and that of skeletal muscle fibers is approximately −90 mV (see chapter 9). By convention, the potential difference is reported as a negative number because the inside of the plasma membrane is negative compared with the outside. The greater the charge difference across the plasma membrane, the greater the potential difference. A cell with a resting membrane potential of −90 mV has a greater charge difference between the inside of the cell membrane and the outside of the cell membrane than a cell with a resting membrane potential of −70 mV.

The resting membrane potential results from two characteristics of neurons:

1. The permeability characteristics of the resting plasma membrane
2. Differences in concentration of ions between the intracellular and the extracellular fluids

The plasma membrane is more permeable to K^+ because of a higher proportion of K^+ leak ion channels compared with leak channels for other ions. Positively charged K^+ can therefore diffuse down its concentration gradient from inside to outside the cell (figure 11.7). Negatively charged proteins and other molecules cannot diffuse through the plasma membrane with the K^+. As K^+ diffuses out of the cell, the loss of positive charges makes the inside the plasma membrane more negative. Because opposite charges attract, K^+ is attracted back toward the cell. The accumulation of K^+ outside the plasma membrane makes the outside of the plasma membrane positive relative to the inside. The resting membrane potential is an equilibrium that is established when the tendency for K^+ to diffuse out of the cell, because of the K^+ concentration gradient, is equal to the tendency for K^+ to move into the cell because of the attraction of the positively charged K^+ to negatively charged proteins and other molecules.

Predict 4

Given that tissue A has significantly more K^+ leak ion channels than tissue B, which tissue has the larger resting membrane potential?

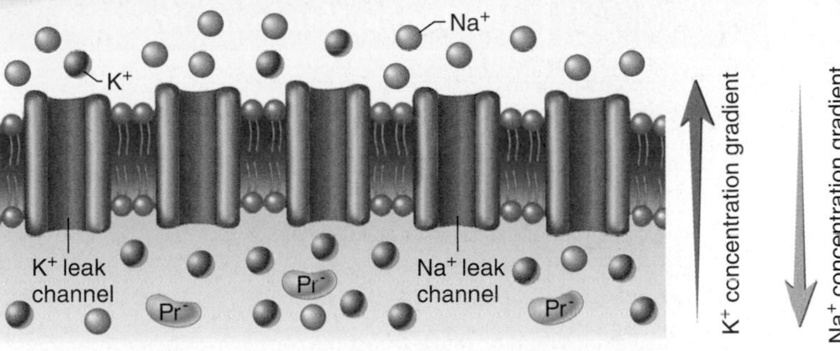
Na+
K+
K+ leak channel
Na+ leak channel
Pr-
K+ concentration gradient
Na+ concentration gradient

1 In a resting cell, there is a higher concentration of K^+ (*purple circles*) inside the cell membrane and a higher concentration of Na^+ (*pink circles*) outside the cell membrane. Because the membrane is not permeable to negatively charged proteins (*green*) they are isolated to inside of the cell membrane.

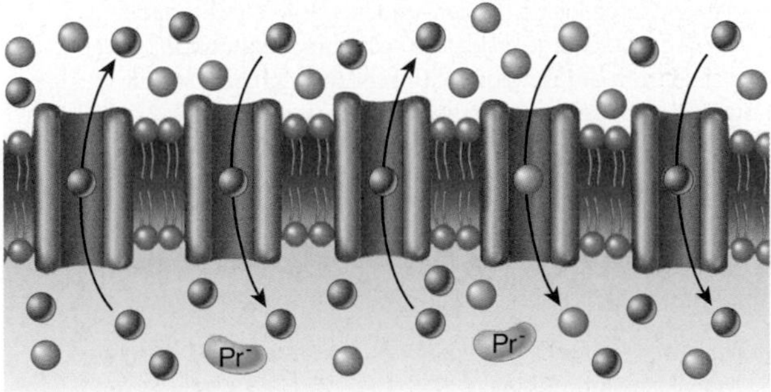
Pr-

2 There are more K^+ leak channels than Na^+ leak channels. In the resting cell, only the leak channels are opened; the gated channels (not shown) are closed. Because of the ion concentration differences across the membrane, K^+ diffuses out of the cell down its concentration gradient and Na^+ diffuses into the cell down its concentration gradient. The tendency for K^+ to diffuse out of the cell is opposed by the tendency of the positively charged K^+ to be attracted back into the cell by the negatively charged proteins.

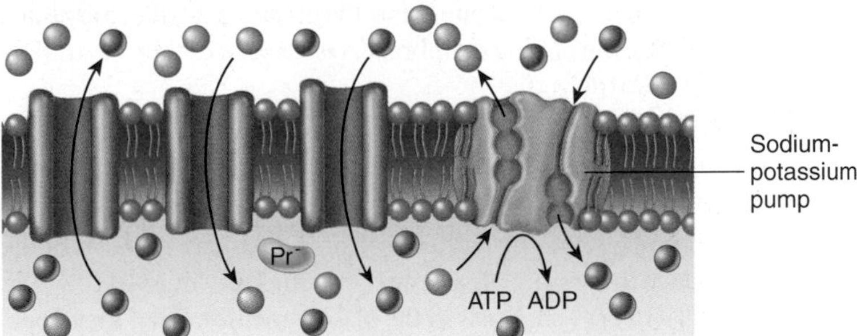
Pr-
ATP ADP
Sodium-potassium pump

3 The sodium-potassium pump helps maintain the differential levels of Na^+ and K^+ by pumping three Na^+ out of the cell in exchange for two K^+ into the cell. The pump is driven by ATP hydrolysis. The resting membrane potential is established when the movement of K^+ out of the cell is equal to the movement of K^+ into the cell.

(a)

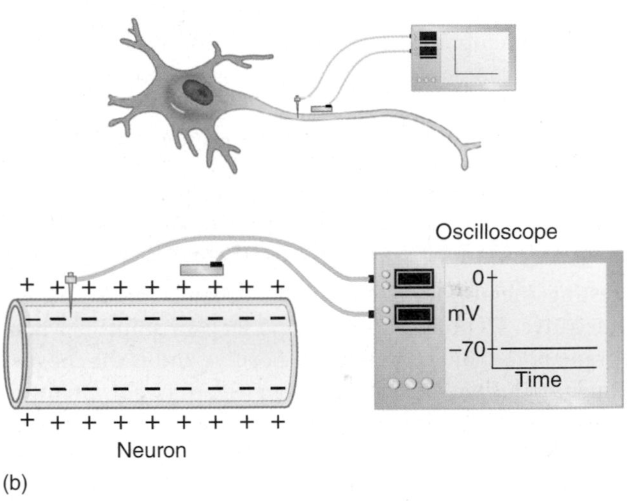
Oscilloscope
0
mV
–70
Time
Neuron

(b)

FIGURE 11.7 **AP|R** **Resting Membrane Potential**

(*a*) In a resting cell, there is a higher concentration of K^+ inside the cell and a higher concentration of Na^+ outside the cell. In a resting cell, only the leak ion channels are open; the gated ion channels are closed. There are many more K^+ leak ion channels than Na^+ leak ion channels. As a result, K^+ diffuses out of the cell down its concentration gradient. The membrane is not permeable to the negatively charged proteins inside the cell. The tendency for the K^+ to diffuse to the outside of the cell down its concentration gradient is opposed by the tendency for the positively charged K^+ to be attracted back into the cell by the negatively charged proteins. A small amount of Na^+ diffuses into the cell (not shown). The sodium-potassium pump helps maintain the differential levels of Na^+ and K^+ by pumping three Na^+ out of the cell in exchange for two K^+ into the cell. The pump is driven by ATP hydrolysis. (*b*) The recording electrode is inside the cell; the reference electrode is outside. A potential difference of about −70 mV is recorded, with the inside of the plasma membrance negative with respect to the outside of the membrane.

Other ions, such as Na^+, Cl^-, and Ca^{2+}, have a minor influence on the resting membrane potential, but the major influence is from K^+. Because the resting plasma membrane is 50–100 times less permeable to Na^+ than to K^+, very little Na^+ can diffuse from the outside to the inside of the resting cell. The resting plasma membrane is not very permeable to Ca^{2+}, either. The plasma membrane is relatively permeable to Cl^-, but these negatively charged ions are repelled by the negative charge inside the cell.

The resting membrane potential is proportional to the tendency for K^+ to diffuse out of the cell, not to the actual rate of flow for K^+. At equilibrium, very few of these ions pass through the plasma membrane because their movement out of the cell is opposed by the negative charge inside the cell. Still, some Na^+ and K^+ diffuse continuously across the plasma membrane, although at a low rate. The large concentration gradients for Na^+ and K^+ would eventually disappear without the continuous activity of the sodium-potassium pump.

As already noted, the sodium-potassium pump maintains the normal concentration gradients for Na^+ and K^+ across the plasma membrane. The pump is also responsible for a small portion of the resting membrane potential, usually less than 15 mV, because it transports approximately three Na^+ out of the cell and two K^+ into the cell for each ATP molecule used (see figure 3.21). The outside of the plasma membrane becomes more positively charged than the inside, because more positively charged ions are pumped out of the cell than are pumped into it.

The characteristics responsible for the resting membrane potential are summarized in table 11.3.

TABLE 11.3 Characteristics Responsible for the Resting Membrane Potential

1. The concentration of K^+ is higher inside the cell than outside, and the concentration of Na^+ is higher outside the cell than inside.
2. Due to the K^+ leak channels, the plasma membrane is 50–100 times more permeable to K^+ than to other positively charged ions, such as Na^+.
3. The plasma membrane is impermeable to large, intracellular, negatively charged molecules, such as proteins. In other words, these anions are "trapped" inside the cell.
4. Potassium ions tend to diffuse across the plasma membrane from the inside to the outside of the cell.
5. Because negatively charged molecules cannot follow the positively charged K^+, a small negative charge develops inside the plasma membrane.
6. The negative charge inside the cell attracts positively charged K^+. When the negative charge inside the cell is great enough to prevent additional K^+ from diffusing out of the cell through the plasma membrane, an equilibrium is established.
7. The charge difference across the plasma membrane at equilibrium is reflected as a difference in potential, which is measured in millivolts (mV).
8. The resting membrane potential is proportional to the potential for K^+ to diffuse out of the cell but not to the actual rate of flow for K^+.
9. At equilibrium, very little movement of K^+ or other ions takes place across the plasma membrane.

Changing the Resting Membrane Potential

The resting membrane potential can become more positive or more negative. **Depolarization** (dē-pō′lăr-i-zā′shŭn) is when the membrane potential becomes more positive and is the movement of the membrane potential closer to zero (figure 11.8*a*). On the other hand, **hyperpolarization** (hī′per-pō′lăr-i-zā′shŭn) is when the membrane potential becomes more negative and is the movement of the membrane potential further away from zero (figure 11.8*b*). The charge difference across a particular "stretch" of membrane is changed when ions move across the plasma membrane due to a change in the ion concentration gradients or ion permeability of the plasma membrane.

Potassium Ions

We will now consider some scenarios where the resting membrane potential is influenced by K^+. Under normal action potential conditions, K^+ diffuses out of the cell; however, changes in the K^+ concentration gradient and permeability could change the resting membrane potential. An increase in the extracellular concentration of K^+ decreases its concentration gradient because the concentration of K^+ is normally lower outside than inside a cell. As a consequence, the tendency for K^+ to diffuse out of the cell decreases, and a smaller negative charge inside the cell is required to oppose the diffusion of K^+ out of the cell. At this new equilibrium, the smaller charge difference across the plasma membrane changes the resting membrane potential to a more positive level than normal. Thus, we say the cell is depolarized.

In contrast, a decrease in the extracellular concentration of K^+ increases the K^+ concentration gradient. As a result, the tendency for K^+ to diffuse out of the cell increases, and a larger negative charge inside the cell is required to resist that diffusion. At this new equilibrium, the larger charge difference across the plasma membrane changes the resting membrane potential to a more negative level than normal. Thus, we say the cell is hyperpolarized.

Although K^+ leak ion channels allow some K^+ to diffuse across the plasma membrane, the resting membrane is not completely permeable to K^+. However, there are gated K^+ channels in the plasma membrane; if they open, membrane permeability to K^+ increases, and more K^+ diffuses out of the cell. The increased tendency for K^+ to diffuse out of the cell is opposed by the greater negative charge that develops inside the plasma membrane, resulting in hyperpolarization.

Sodium Ions

In a resting cell, the membrane is not very permeable to Na^+ because there are few Na^+ leak ion channels; therefore, changes in the concentration of Na^+ on either side of the plasma membrane do not influence the resting membrane potential very much. However, there are gated Na^+ channels in the plasma membrane; if they open, membrane permeability to Na^+ increases (see figure 3.10). Sodium then diffuses into the cell, because the concentration gradient for Na^+ is from the outside to the inside of the cell. As Na^+ diffuses into the cell, the inside of the membrane becomes more positive, resulting in depolarization. This is the typical situation when a neuron responds to a stimulus.

Calcium Ions

Calcium ions alter membrane potentials in two ways: (1) by affecting voltage-gated Na^+ channels and (2) by entering cells through gated Ca^{2+} channels. Tight regulation of voltage-gated Na^+ channels is important for the synchronization of membrane permeability to Na^+. It seems that closed voltage-gated Na^+ channels are stabilized

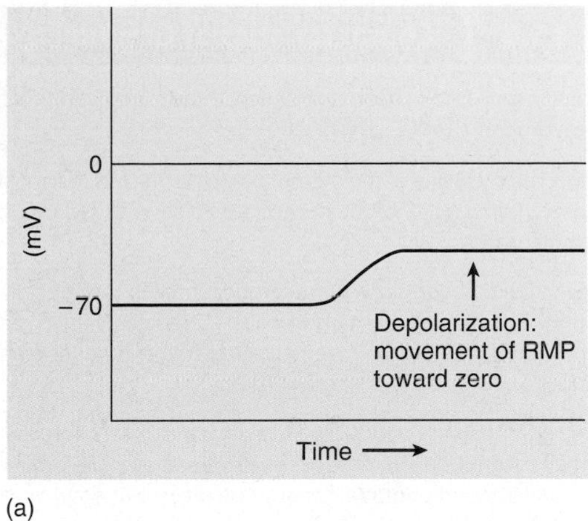
0
(mV)
−70
Depolarization: movement of RMP toward zero
Time

(a)

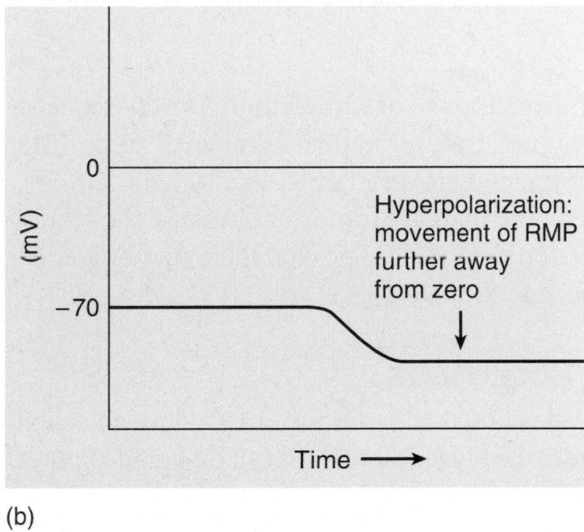
0
(mV)
−70
Hyperpolarization: movement of RMP further away from zero
Time

(b)

FIGURE 11.8 Depolarization and Hyperpolarization of the Resting Membrane Potential (RMP)
(*a*) In depolarization, the charge inside the plasma membrane becomes more positive. (*b*) In hyperpolarization, the charge inside the plasma membrane becomes more negative.

by Ca^{2+} and thus are sensitive to changes in the extracellular concentration of Ca^{2+}. Positively charged Ca^{2+} in the extracellular fluid is attracted to the negatively charged groups of proteins within the voltage-gated Na^+ channels. If the extracellular concentration of Ca^{2+} decreases, these ions diffuse away from the voltage-gated Na^+ channels, causing the channels to open. If the extracellular concentration of Ca^{2+} increases, it binds to voltage-gated Na^+ channels, causing them to close. Therefore, normal levels of Ca^{2+} in the extracellular fluid are crucial to keep voltage-gated Na^+ channels closed until the neuron fires an action potential.

› Predict 5

Predict the effect of a decrease in the extracellular concentration of Ca^{2+} on the resting membrane potential.

Changes in the permeability of Ca^{2+} can change the resting membrane potential. If the plasma membrane becomes permeable to Ca^{2+} by the opening of gated Ca^{2+} channels, Ca^{2+} diffuses into the cell, and the inside of the membrane becomes more positive, resulting in depolarization. This role of Ca^{2+} is important at the axon terminal (see section 11.6).

Chloride Ions

Changes in the permeability of Cl^- can change the resting membrane potential. If the plasma membrane becomes more permeable to Cl^- by the opening of gated Cl^- channels, then Cl^- diffuses into the cell, and the inside of the membrane becomes more negative, resulting in hyperpolarization.

ASSESS YOUR PROGRESS

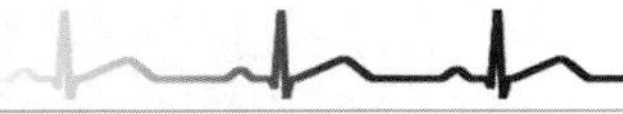

27. *What is the resting membrane potential? What does it result from? Is the outside of the plasma membrane positively or negatively charged relative to the inside?*

28. *What ion is the major influence on the resting membrane potential? Explain its role.*

29. *How does the resting membrane establish an equilibrium?*

30. *What happens to cause depolarization and hyperpolarization? How do alterations in the K^+ concentration gradient; changes in membrane permeability to K^+, Na^+, or Cl^-; and changes in extracellular Ca^{2+} concentration affect depolarization and hyperpolarization?*

Graded Potentials

A **graded potential** is a change in the membrane potential that is localized to one area of the plasma membrane. These local disturbances in the membrane potential are called graded potentials (or local potentials) because the potential change can vary from small to large. Graded potentials can result from (1) chemical signals binding to their receptors, (2) changes in the voltage across the plasma membrane, (3) mechanical stimulation, (4) temperature changes, or (5) spontaneous opening of ion channels. Graded potentials often occur in the dendrites or cell body of a neuron.

A graded potential can be either a depolarization or a hyperpolarization (figure 11.8). A change in membrane permeability to Na^+, K^+, or other ions can produce a graded potential. For example, if a stimulus causes gated Na^+ channels to open, the diffusion of Na^+ into the cell results in depolarization. If a stimulus causes gated K^+ channels to open, the diffusion of K^+ out of the cell results in hyperpolarization.

The magnitude of graded potentials can vary from small to large, depending on the stimulus strength or on summation. For example, a weak stimulus can cause only a few gated Na^+ channels to open. A small amount of Na^+ diffuses into the cell and causes a small depolarization. A stronger stimulus can cause a greater number of gated Na^+ channels to open. A greater amount of Na^+ diffusing into the cell causes a larger depolarization (figure 11.9*a*).

Summation of graded potentials occurs when the effects produced by one graded potential are added onto the effects produced by another graded potential, which could lead to an action potential (figure 11.9*b*). For example, if a second stimulus is applied before the graded potential produced by the first stimulus has returned to the resting membrane potential, a larger depolarization results than would result from a single stimulus. The first stimulus

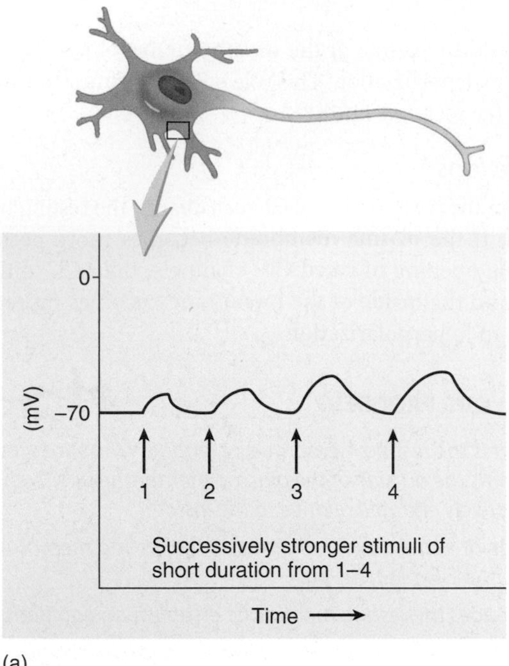
0
(mV)
−70
1
2
3
4
Successively stronger stimuli of short duration from 1–4
Time
(a)

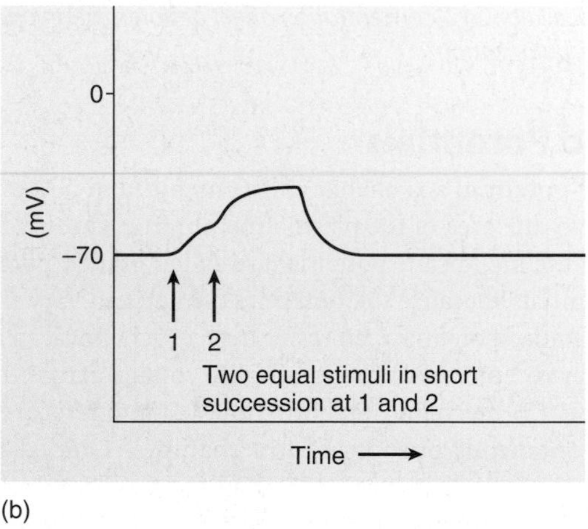
0
(mV)
−70
1
2
Two equal stimuli in short succession at 1 and 2
Time
(b)

FIGURE 11.9 Graded Potentials
(*a*) Graded potentials are proportional to the stimulus strength. A weak stimulus applied briefly causes a small depolarization, which quickly returns to the resting membrane potential (*1*). Progressively stronger stimuli result in larger depolarizations (*2–4*). (*b*) A stimulus applied to a cell causes a small depolarization. When a second stimulus is applied before the depolarization disappears, the depolarization caused by the second stimulus is added to the depolarization caused by the first to result in a larger depolarization.

causes gated Na^+ channels to open, and the second stimulus causes additional Na^+ channels to open. As a result, more Na^+ diffuses into the cell, producing a larger graded potential. Summation is discussed in greater detail in section 11.6.

Graded potentials spread, or are conducted, over the plasma membrane in a *decremental* fashion. That is, they rapidly decrease in magnitude as they spread over the surface of the plasma membrane, much as a teacher's voice spreads through a large lecture hall. At the front of the class, the teacher's voice can be heard easily but, the farther away a student sits, the harder it is to hear the teacher. Normally, a graded potential cannot be detected more than a few millimeters from the site of stimulation. As a consequence, a graded potential cannot transfer information over long distances from one part of the body to another.

Graded potentials are important because they can summate to generate action potentials. The characteristics of graded potentials are summarized in table 11.4.

TABLE 11.4 Characteristics of Graded Potentials
1. A stimulus causes ion channels to open, increasing the permeability of the membrane to Na^+, K^+, or Cl^-.
2. Increased permeability of the membrane to Na^+ results in depolarization. Increased permeability of the membrane to K^+ or Cl^- results in hyperpolarization.
3. The size of the graded potential is proportional to the strength of the stimulus. Graded potentials can also summate. Thus, a graded potential produced in response to several stimuli is larger than one produced in response to a single stimulus.
4. Graded potentials are conducted in a *decremental* fashion, meaning that their magnitude decreases as they spread over the plasma membrane. Graded potentials cannot be measured a few millimeters from the point of stimulation.
5. A depolarizing graded potential can cause an action potential.

Action Potentials

How do graded potentials summate to produce an action potential? Let's consider the bigger picture. Recall that graded potentials result from the detection of stimulus input to the neuron. For simplicity, let's follow a chemical stimulus to a neuron as it arrives at the dendrite. There, binding of a chemical stimulus quite often results in the opening of ligand-gated ion channels. If these channels are Na^+ channels, the receiving cell experiences a depolarizing graded potential. If enough Na^+ enters, the graded potentials can summate at the trigger zone in the axon hillock (figure 11.10). When the graded potentials summate to a level called **threshold,** an action potential results (figure 11.11). Threshold is the membrane potential at which voltage-gated Na^+ channels open. Because the trigger zone contains a much higher proportion of voltage-gated channels than other parts of the cell body, action potentials are initiated there.

An action potential has a **depolarization phase,** in which the membrane potential moves away from the resting state and becomes more positive, and a **repolarization phase,** in which the membrane potential returns toward the resting state and becomes more negative. After the repolarization phase, the plasma membrane may be slightly hyperpolarized for a short period called the **afterpotential.** An action potential is a large change in the membrane potential that propagates, without changing its magnitude, over long distances along the plasma membrane. Thus, action potentials can transfer information from one part of the body to another. It generally takes 1–2 milliseconds (ms; 1 ms = 0.001 s) for an action potential to occur. The characteristics of action potentials are summarized in table 11.5.

Depolarizing graded potentials that summate to threshold produce an action potential, but hyperpolarizing graded potentials can never reach threshold and do not produce action potentials. Thus, depolarizing graded potentials are stimulatory by triggering

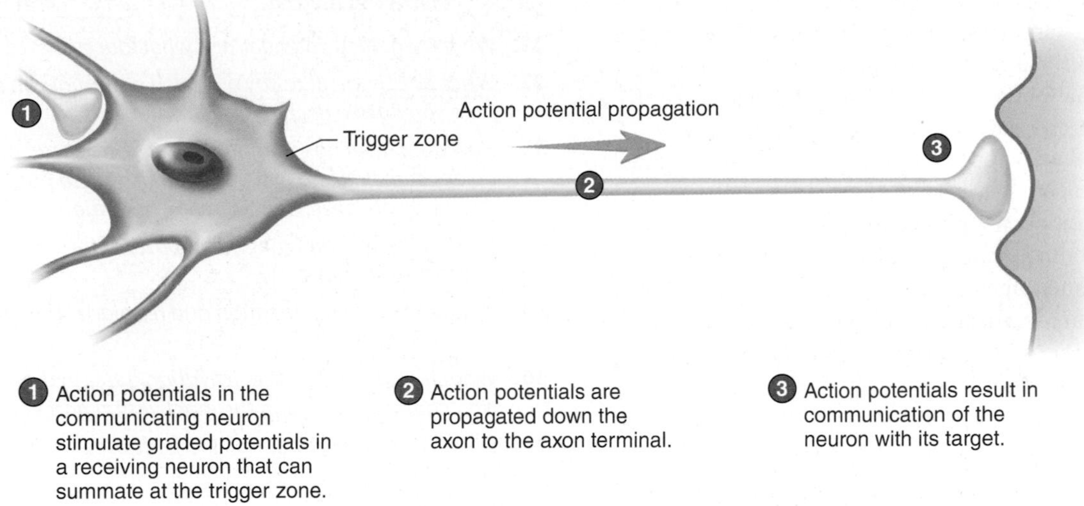
1
Trigger zone
Action potential propagation
2
3
1 Action potentials in the communicating neuron stimulate graded potentials in a receiving neuron that can summate at the trigger zone.
2 Action potentials are propagated down the axon to the axon terminal.
3 Action potentials result in communication of the neuron with its target.

PROCESS FIGURE 11.10 APR **Production of an Action Potential**

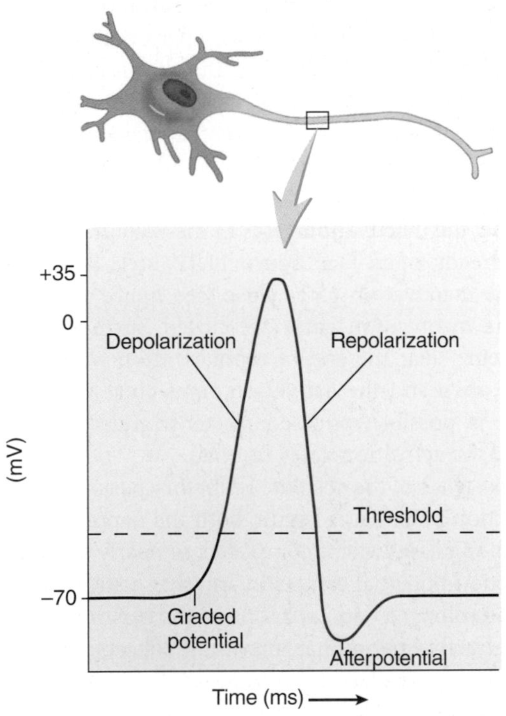
+35
0
–70
(mV)
Depolarization
Repolarization
Threshold
Graded potential
Afterpotential
Time (ms)

FIGURE 11.11 Action Potential

The action potential consists of a depolarization phase and a repolarization phase, often followed by a short period of hyperpolarization, called the afterpotential.

an action potential, whereas hyperpolarizing graded potentials are inhibitory by preventing an action potential.

The magnitude of a depolarizing graded potential affects the likelihood of generating an action potential. For example, a weak stimulus can produce a small depolarizing graded potential that does not reach threshold and therefore does not cause an action potential. A stronger stimulus, however, can produce a larger depolarizing graded potential that reaches threshold, resulting in an action potential.

All-or-None Principle

Action potentials occur according to the **all-or-none principle.** If a stimulus produces a depolarizing graded potential that is large enough to reach threshold, all the permeability changes responsible for an action potential proceed without stopping and are constant in magnitude (the "all" part). If a stimulus is so weak that the depolarizing graded potential does not reach threshold, few of the permeability changes occur. The membrane potential returns to its resting level after a brief period without producing an action potential (the "none" part). An action potential can be compared to the flash system of a camera. When the shutter is triggered (reaches threshold), the camera flashes (an action potential is produced), and each flash is the same brightness (magnitude; the "all" part) as previous flashes. If the shutter is depressed, but not triggered, no flash results (the "none" part).

Depolarization Phase

The increase in positive charge inside the plasma membrane caused by a depolarizing graded potential causes increasing numbers of voltage-gated Na^+ channels to open rapidly. As soon as a threshold depolarization is reached (usually at the axon hillock), many voltage-gated Na^+ channels begin to open. Sodium ions diffuse into the cell, and the resulting depolarization causes additional voltage-gated Na^+ channels to open. As a consequence, more Na^+ diffuses into the cell, causing a greater depolarization of the membrane, which in turn causes still more voltage-gated Na^+ channels to open. This is an example of a positive-feedback cycle, and it continues until most of the voltage-gated Na^+ channels in the plasma membrane are open.

Each voltage-gated Na^+ channel has two voltage-sensitive gates, called **activation gates** and **inactivation gates.** When the plasma membrane is at rest, the activation gates of the voltage-gated Na^+ channel are closed, and the inactivation gates are open (figure 11.12, *step 1*). Because the activation gates are closed, Na^+ cannot diffuse through the channels. When the graded potential reaches threshold, the change in the membrane potential causes many of the activation gates to open, and Na^+ can diffuse through the Na^+ channels into the cell.

When the plasma membrane is at rest, voltage-gated K^+ channels, which have one gate, are closed (figure 11.12, *step 1*). When the graded potential reaches threshold, the voltage-gated K^+ channels start to open at the same time as the voltage-gated Na^+ channels, but they open significantly more slowly than the voltage-gated Na^+ channels (figure 11.12, *step 2*). Because the voltage-gated K^+ channels open so slowly, only a small number of them are open, compared with the number of voltage-gated Na^+ channels. Depolarization occurs because more Na^+ diffuses into the cell than K^+ diffuses out of it.

Predict 6

Predict the effect of a reduced extracellular concentration of Na^+ on the magnitude of the action potential in an electrically excitable cell.

Repolarization Phase

As the membrane potential approaches its maximum depolarization, the inactivation gates of the voltage-gated Na^+ channels are triggered to close by the specific membrane potential. Thus, the permeability of the plasma membrane to Na^+ decreases. During the repolarization phase, the voltage-gated K^+ channels, which started to open slowly along with the voltage-gated Na^+ channels, continue to open (figure 11.12, *step 3*). Consequently, the permeability of the plasma membrane to Na^+ decreases, and the permeability to K^+ increases. The decreased diffusion of Na^+ into the cell and the increased diffusion of K^+ out of the cell cause repolarization.

At the end of repolarization, the return toward resting membrane potential causes the activation gates in the voltage-gated Na^+ channels to close and the inactivation gates to open. Although this change does not affect the diffusion of Na^+, it does return the voltage-gated Na^+ channels to their resting state (figure 11.12, *step 4*).

Afterpotential

In many cells, a period of hyperpolarization, or afterpotential, exists following each action potential. The afterpotential exists because the voltage-gated K^+ channels open and close more slowly than the voltage-gated Na^+ channels and remain open for a slightly longer time than it takes to bring the membrane potential back to its original resting level (figure 11.12, *step 4*). As the voltage-gated K^+ channels close, the original resting membrane potential is reestablished by the sodium-potassium pump (figure 11.12, *step 5*).

During an action potential, a small amount of Na^+ diffuses into the cell and a small amount of K^+ diffuses out of the cell. The sodium-potassium pump restores normal resting ion concentrations by transporting these ions in the opposite direction of their movement during the action potential. That is, Na^+ is pumped out of the cell and K^+ is pumped into the cell. The sodium-potassium pump is too slow to have an effect on either the depolarization or the repolarization phase of individual action potentials. As long as the Na^+ and K^+ concentrations remain unchanged across the plasma membrane, all the action potentials produced by a cell are identical. They all take the same amount of time, and they all exhibit the same magnitude.

ASSESS YOUR PROGRESS 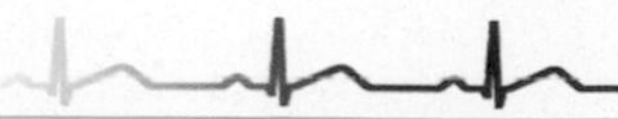

31. *What is a graded potential, and what four events can cause it?*

32. *What does it mean to say a graded potential can summate and then spread in a decremental fashion?*

33. *How does an action potential differ from a local potential? How do depolarizing and hyperpolarizing graded potentials affect the likelihood of generating an action potential?*

34. *Explain the "all" and the "none" parts of the all-or-none principle of action potentials.*

35. *What are the depolarization and repolarization phases of an action potential?*

36. *What happens when the activation gates in the voltage-gated Na^+ channels open and the inactivation gates close?*

37. *Describe the afterpotential and its cause.*

Refractory Period

Once an action potential is produced at a given point on the plasma membrane, the sensitivity of that area to further stimulation decreases for a time called the **refractory** (rē-frak′tōr-ē) **period.** The first part of the refractory period, during which complete insensitivity exists to another stimulus, is called the **absolute refractory period.** In many cells, it occurs from the beginning of the action potential until near the end of repolarization (figure 11.13). At the beginning of the action potential, depolarization occurs when the activation gates in the voltage-gated Na^+ channel open. At this time, the inactivation gates in the voltage-gated Na^+ channels are already open (see figure 11.12, *step 2*). Depolarization ends as the inactivation gates close (see figure 11.12, *step 3*). As long as the inactivation gates are closed, further depolarization cannot occur. Near the end of repolarization when the inactivation gates open and the activation gates close (see figure 11.12, *step 4*), it is possible, once again, to stimulate another action potential if the activation gates re-open.

The existence of the absolute refractory period guarantees that, once an action potential is begun, both the depolarization and the repolarization phases will be completed, or nearly completed, before another action potential can begin and that a strong stimulus cannot lead to prolonged depolarization of the plasma membrane. The absolute refractory period has important consequences for the rate at which action potentials can be generated and for the propagation of action potentials.

The second part of the refractory period, called the **relative refractory period,** follows the absolute refractory period. A stronger-than-threshold stimulus can initiate another action potential during the relative refractory period. Thus, after the absolute refractory period, but before the relative refractory period is completed, a sufficiently strong stimulus can produce another action potential. During the relative refractory period, the membrane is more permeable to K^+ because many voltage-gated K^+ channels are open (see figure 11.12, *step 4*). The relative refractory period ends when the voltage-gated K^+ channels close (see figure 11.12, *step 5*).

Predict 7

Which produces the most action potentials: a prolonged threshold stimulus or a prolonged, stronger-than-threshold stimulus of the same duration? Explain.

1 Resting membrane potential. Na^+ channels (*pink*) and most, but not all, K^+ channels (*purple*) are closed. The outside of the plasma membrane is positively charged compared to the inside.

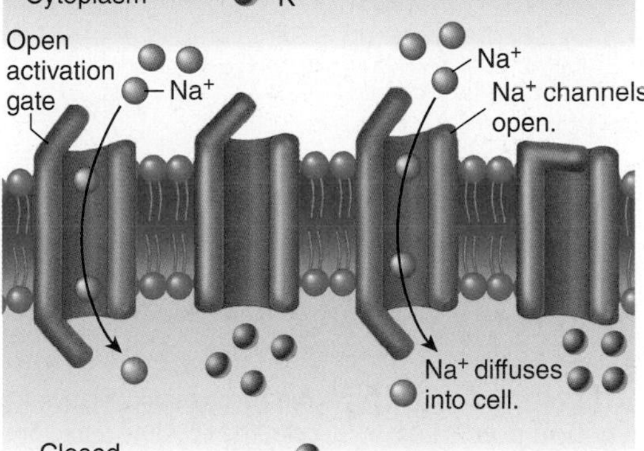

Closed activation gate
Na^+
Extracellular fluid
Voltage-gated Na^+ channel
Voltage-gated K^+ channel
Open inactivation gate
Cytoplasm
K^+

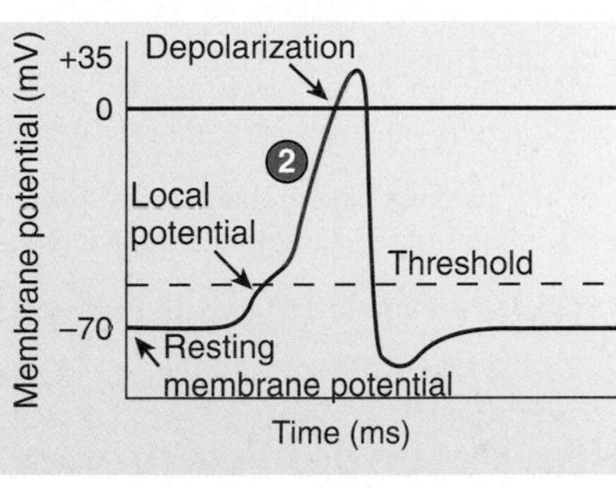

+35
0
–70
Membrane potential (mV)
Threshold
1
Resting membrane potential
Time (ms)

2 Depolarization. Na^+ channels open. K^+ channels begin to open. Depolarization results because the inward movement of Na^+ makes the inside of the membrane more positive.

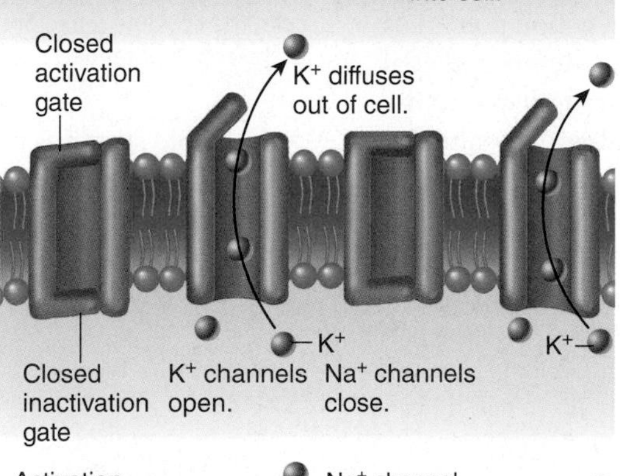

Open activation gate
Na^+
Na^+
Na^+ channels open.
Na^+ diffuses into cell.

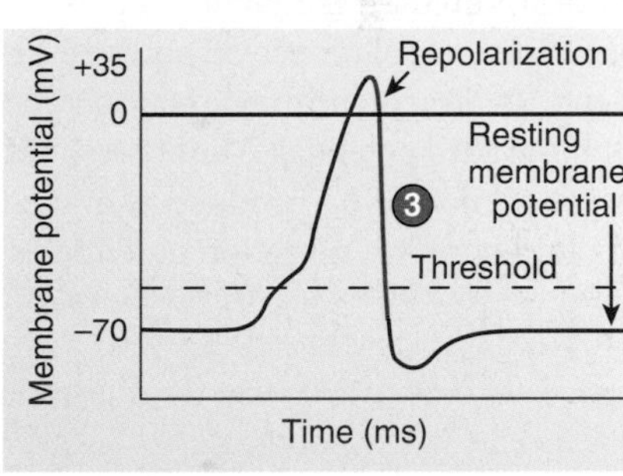

+35
0
–70
Membrane potential (mV)
Depolarization
2
Local potential
Threshold
Resting membrane potential
Time (ms)

3 Repolarization. Na^+ channels close and additional K^+ channels open. Na^+ movement into the cell stops, and K^+ movement out of the cell increases, causing repolarization.

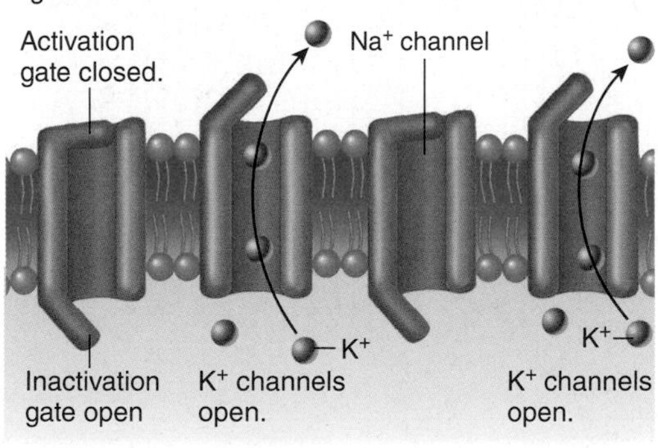

Closed activation gate
K^+ diffuses out of cell.
K^+
K^+
Closed inactivation gate
K^+ channels open.
Na^+ channels close.

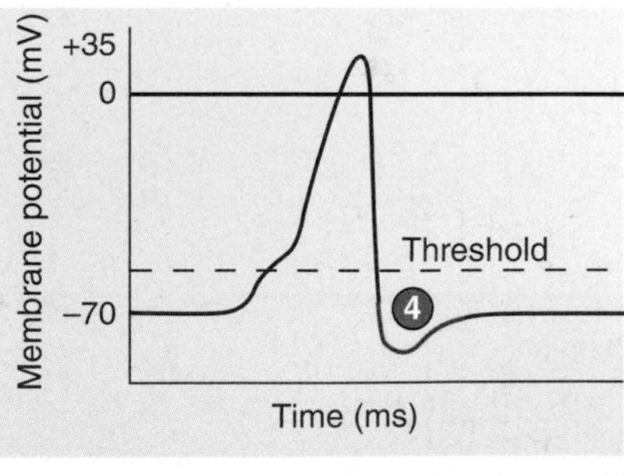

+35
0
–70
Membrane potential (mV)
Repolarization
Resting membrane potential
3
Threshold
Time (ms)

4 End of repolarization and afterpotential. Voltage-gated Na^+ channels are closed. Closure of the activation gates and opening of the inactivation gates reestablish the resting condition for Na^+ channels (see *step 1*). Diffusion of K^+ through voltage-gated channels produces the afterpotential.

Activation gate closed.
Na^+ channel
K^+
K^+
Inactivation gate open
K^+ channels open.
K^+ channels open.

+35
0
–70
Membrane potential (mV)
Threshold
4
Time (ms)

5 Resting membrane potential. The resting membrane potential is reestablished after the voltage-gated K^+ channels close.

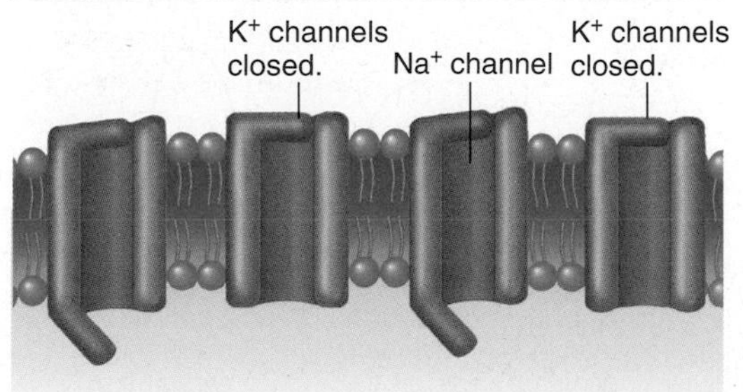

K^+ channels closed.
Na^+ channel
K^+ channels closed.

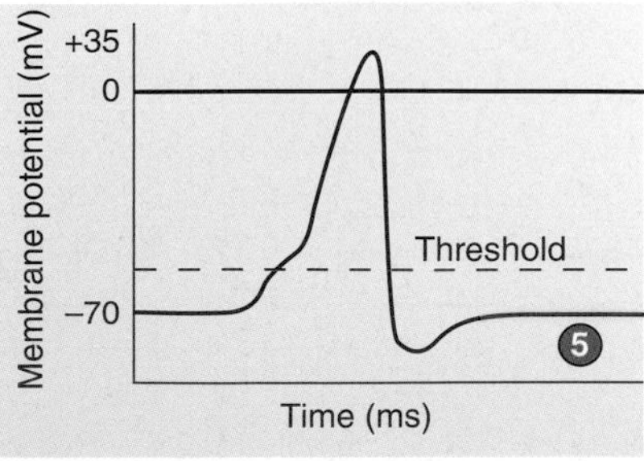

+35
0
–70
Membrane potential (mV)
Threshold
5
Time (ms)

PROCESS FIGURE 11.12 Voltage-Gated Ion Channels and the Action Potential

Step 1 illustrates the status of voltage-gated Na^+ and K^+ channels in a resting cell. Steps 2–5 show how the channels open and close to produce an action potential. Next to each step (far right), a graph shows in *red* the membrane potential resulting from the condition of the ion channels.

TABLE 11.5	Characteristics of Action Potentials
1.	Action potentials are produced when a graded potential reaches threshold.
2.	Action potentials are all-or-none.
3.	Depolarization is a result of increased membrane permeability to Na^+ and movement of Na^+ into the cell. Activation gates of the voltage-gated Na^+ channels open.
4.	Repolarization is a result of decreased membrane permeability to Na^+ and increased membrane permeability to K^+, which stops Na^+ movement into the cell and increases K^+ movement out of the cell. The inactivation gates of the voltage-gated Na^+ channels close, and the voltage-gated K^+ channels open.
5.	During the absolute refractory period, no action potential is produced by a stimulus, no matter how strong. During the relative refractory period, a stronger-than-threshold stimulus can produce an action potential.
6.	Action potentials are propagated and, for a given axon or muscle fiber, the magnitude of the action potential is constant.
7.	Stimulus strength determines the frequency of action potentials.

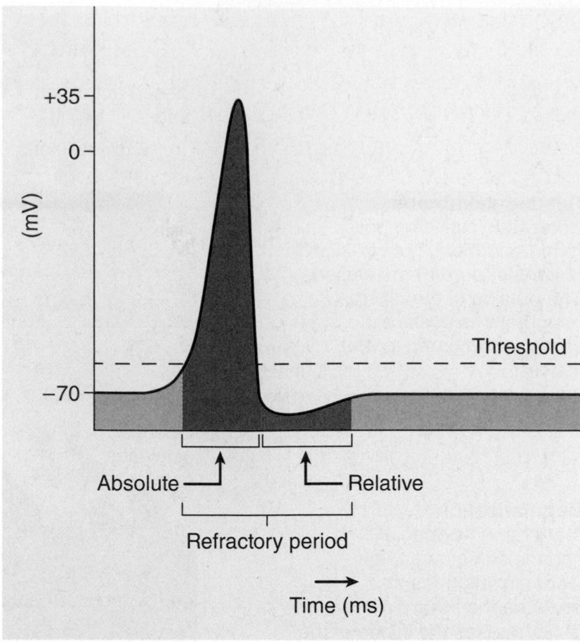
+35
0
(mV)
Threshold
–70
Absolute
Relative
Refractory period
Time (ms)

FIGURE 11.13 Refractory Period
The absolute and relative refractory periods of an action potential. In some cells, the absolute refractory period ends during the repolarization phase of the action potential.

Action Potential Frequency

The **action potential frequency** is the number of action potentials produced per unit of time in response to a stimulus. Action potential frequency is directly proportional to stimulus strength and to the size of the graded potential. A **subthreshold stimulus** is any stimulus not strong enough to produce a graded potential that reaches threshold. Therefore, no action potential is produced (figure 11.14). A **threshold stimulus** produces a graded potential that is just strong enough to reach threshold and cause the production of a single action potential. A **maximal stimulus** is just strong enough to produce a

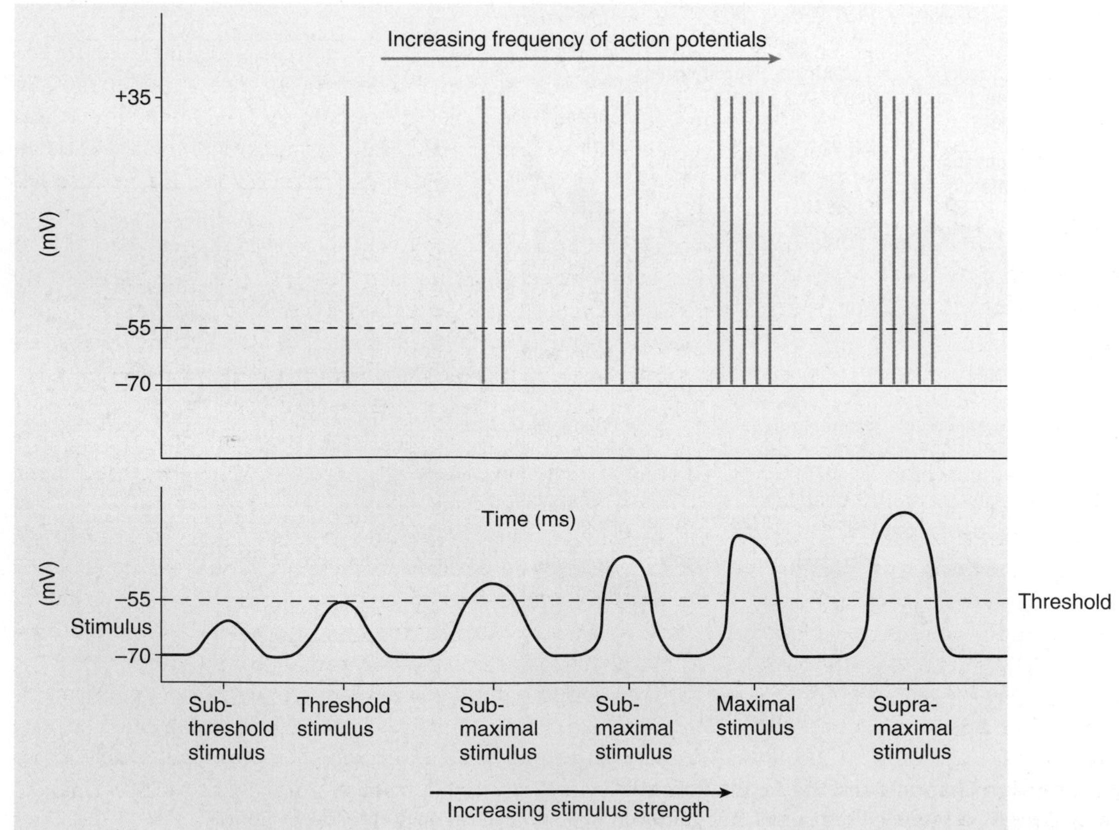
Increasing frequency of action potentials
+35
(mV)
–55
–70
Time (ms)
(mV)
–55
Stimulus
–70
Threshold
Sub-threshold stimulus
Threshold stimulus
Sub-maximal stimulus
Sub-maximal stimulus
Maximal stimulus
Supra-maximal stimulus
Increasing stimulus strength

FIGURE 11.14 Stimulus Strength and Action Potential Frequency
From left to right, each stimulus in the figure is stronger than the previous one. As stimulus strength increases, the frequency of action potentials increases until a maximal rate is produced. Thereafter, increasing stimulus strength does not increase action potential frequency due to the refractory period.

maximum frequency of action potentials. A **submaximal stimulus** includes all stimuli between threshold and the maximal stimulus strength. For submaximal stimuli, the action potential frequency increases in proportion to the strength of the stimulus because the size of the graded potential increases with stimulus strength. A **supramaximal stimulus** is any stimulus stronger than a maximal stimulus. Because an axon's ability to produce action potentials is limited, these stimuli cannot produce a greater frequency of action potentials than a maximal stimulus.

The duration of the absolute refractory period determines the maximum frequency of action potentials generated in an excitable cell. During the absolute refractory period, a second stimulus, no matter how strong, cannot stimulate an additional action potential. However, as soon as the absolute refractory period ends, it is possible for a second stimulus to cause the production of an action potential.

Predict 8

If the duration of the absolute refractory period of a neuron is 1 millisecond (ms), how many action potentials are generated by a maximal stimulus in 1 second?

Communication regarding the strength of stimuli cannot depend on the magnitudes of action potentials because, according to the all-or-none principle, the magnitudes of action potentials produced by weak and strong stimuli are always the same. Instead, the frequency of action potentials provides information about the strength of a stimulus. For example, a weak pain stimulus generates a low frequency of action potentials, whereas a stronger pain stimulus generates a higher frequency of action potentials. The ability to interpret a stimulus as mildly painful versus very painful depends, in part, on the frequency of action potentials generated by individual pain receptors.

The ability to stimulate muscle or gland cells also depends on action potential frequency. A low frequency of action potentials produces a weaker muscle contraction or less secretion than does a higher frequency. For example, a low frequency of action potentials in a muscle results in incomplete tetanus, and a high frequency results in complete tetanus (see chapter 9).

In addition to the frequency of action potentials, how long the action potentials are produced provides important information. For example, a pain stimulus of 1 second is interpreted differently than the same stimulus applied for 30 seconds.

Propagation of Action Potentials

A single action potential occurs in one very small area of the plasma membrane and does not occur over the entire membrane at one time. However, action potentials can **propagate,** or spread, across the plasma membrane because an action potential produced at one point on the plasma membrane stimulates the production of an action potential at the adjacent point of the same plasma membrane. Note that the same action potential does not travel down the entire length of an axon. Rather, a series of action potentials is generated down the length of the axon, like a long row of toppling dominos in which each domino knocks down the next one. Each domino falls, but no single domino actually travels the length of the row.

In a neuron, action potentials are normally produced at the trigger zone and propagate in one direction along the axon (figure 11.15, *step 1*). The location at which the next action potential is generated is different for unmyelinated and myelinated axons (see figure 11.6). In an unmyelinated axon, the next action potential is generated immediately adjacent to the previous action potential. When an action potential is produced, the inside of the membrane becomes more positive than the outside (figure 11.15, *step 2*). On the outside of the membrane, positively charged ions from the adjacent area are attracted to the negative charges at the site of the action potential. On the inside of the plasma membrane, positively charged ions at the site of the action potential are attracted to the adjacent negatively charged part of the membrane. The movement of positively charged ions is called a **local current,** or an *ionic current.* As a result of the local current, the part of the membrane immediately adjacent to the action potential depolarizes. That is, the outside of the membrane immediately adjacent to the action potential becomes more negative because of the loss of positive charges, and the inside becomes more positive because of the gain of positive charges. When the depolarization reaches threshold, an action potential is produced (figure 11.15, *step 3*). This type of action potential conduction in unmyelinated axons is called **continuous conduction.**

If an action potential is initiated at one end of an axon, it is propagated in one direction down the axon. The absolute refractory period ensures one-way propagation of an action potential because it prevents the local current from stimulating the production of an action potential in the reverse direction (figure 11.15, *step 4*).

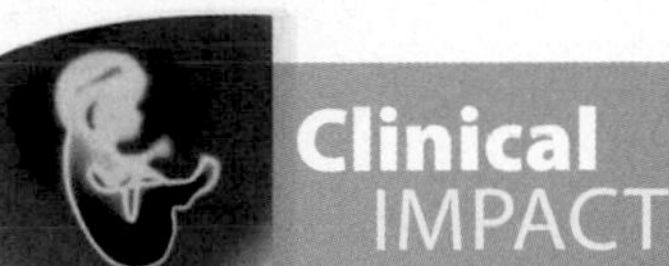

Clinical IMPACT

Effects of Abnormal Membrane Potentials

Several medical conditions can alter the physiology of membrane potentials. **Hypokalemia** (hī-pō-ka-lē′mē-ă) is a lower-than-normal concentration of K^+ in the blood or extracellular fluid. Reduced extracellular K^+ concentrations cause hyperpolarization of the resting membrane potential (see figure 11.8*b*). Thus, a greater-than-normal stimulus is required to depolarize the membrane to its threshold level and to initiate action potentials in neurons, skeletal muscle, and cardiac muscle. As the excitable tissues become less sensitive to stimulation, symptoms such as muscular weakness, an abnormal electrocardiogram, and sluggish reflexes result. The causes of hypokalemia include potassium depletion during starvation, alkalosis, and certain kidney diseases.

Hypocalcemia (hī-pō-kal-sē′mē-ă) is a lower-than-normal concentration of Ca^{2+} in the blood or extracellular fluid. Symptoms of hypocalcemia include nervousness and uncontrolled contraction of skeletal muscles, called **tetany** (tet′ă-nē). The symptoms are due to an increased membrane permeability to Na^+ that results because low blood levels of Ca^{2+} cause voltage-gated Na^+ channels in the membrane to open. Sodium ions diffuse into the cell, causing depolarization of the plasma membrane to threshold and initiating action potentials. The tendency for action potentials to occur spontaneously in nervous tissue and muscles accounts for the symptoms. Conditions that cause hypocalcemia include a lack of dietary calcium or vitamin D and a reduced secretion rate of a parathyroid gland hormone.

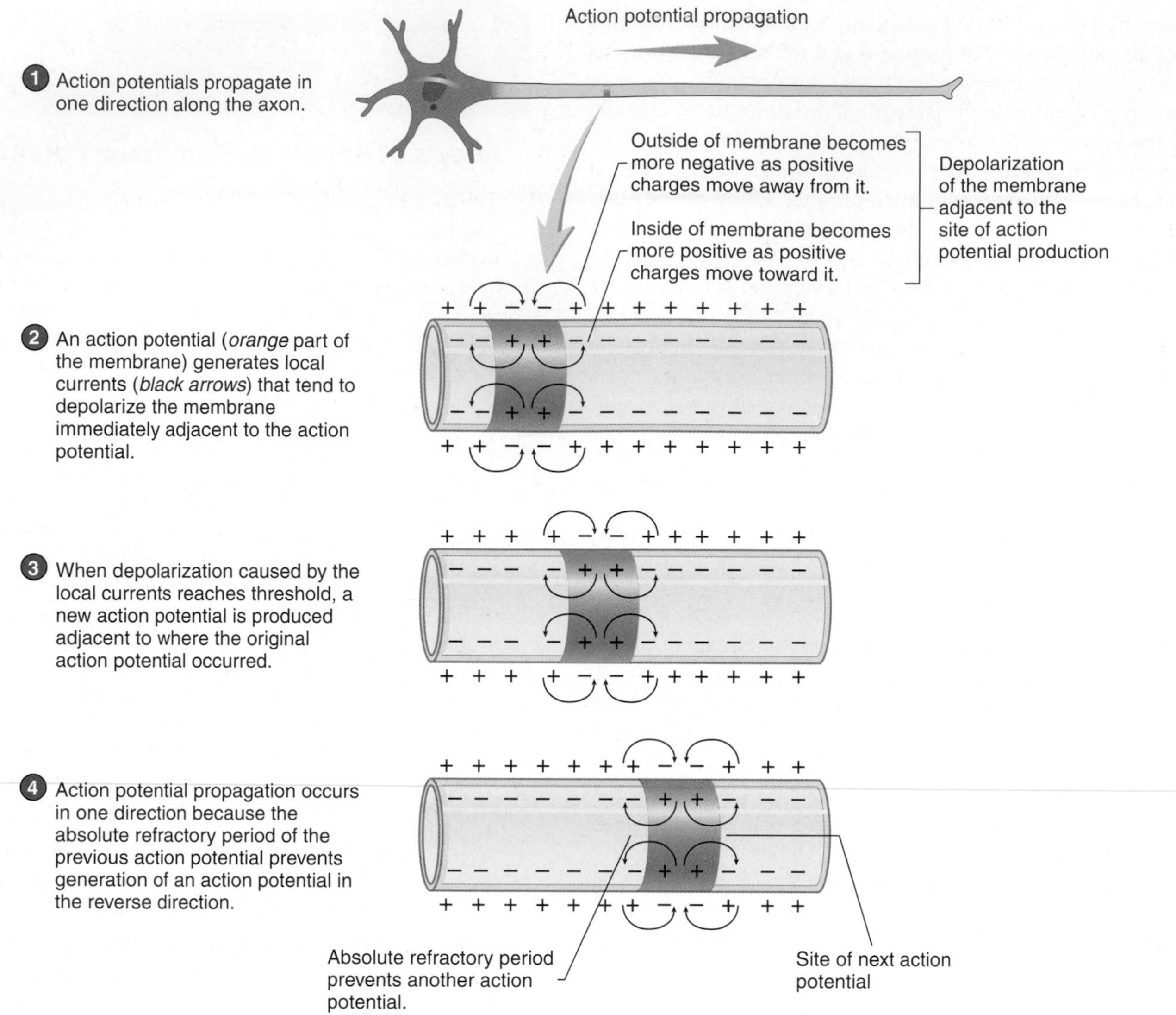
Action potential propagation
1 Action potentials propagate in one direction along the axon.
Outside of membrane becomes more negative as positive charges move away from it.
Inside of membrane becomes more positive as positive charges move toward it.
Depolarization of the membrane adjacent to the site of action potential production
2 An action potential (orange part of the membrane) generates local currents (black arrows) that tend to depolarize the membrane immediately adjacent to the action potential.
3 When depolarization caused by the local currents reaches threshold, a new action potential is produced adjacent to where the original action potential occurred.
4 Action potential propagation occurs in one direction because the absolute refractory period of the previous action potential prevents generation of an action potential in the reverse direction.
Absolute refractory period prevents another action potential.
Site of next action potential

PROCESS FIGURE 11.15 Action Potential Propagation in an Unmyelinated Axon

In a myelinated axon, an action potential is conducted from one node of Ranvier to another in a process called **saltatory conduction** (*saltare*, to leap). An action potential at one node of Ranvier generates local currents that flow toward the next node of Ranvier (figure 11.16, *step 1*). The lipids within the membranes of the myelin sheath act as a layer of insulation, forcing the local currents to flow from one node of Ranvier to the next. In addition, voltage-gated Na^+ channels are highly concentrated at the nodes of Ranvier. Therefore, the local current quickly flows to a node and stimulates the voltage-gated Na^+ channels to open, resulting in the production of an action potential (figure 11.16, *step 2*).

The speed of action potential conduction along an axon depends on the myelination of the axon. Action potentials are conducted more rapidly in myelinated than unmyelinated axons because they are formed quickly at each successive node of Ranvier (figure 11.16, *step 3*), instead of being propagated more slowly through every part of the axon's membrane, as in unmyelinated axons (see figure 11.15). Action potential conduction in a myelinated axon is like a grasshopper jumping; in an unmyelinated axon, it is like a grasshopper walking. The grasshopper (action potential) moves more rapidly by jumping. The generation of action potentials at nodes of Ranvier occurs so rapidly that as many as 30 successive nodes of Ranvier are simultaneously in some phase of an action potential.

The speed of action potential conduction is also affected by the thickness of the myelin sheath, which is determined by how many times oligodendrocytes or Schwann cells wrap around the axon. Heavily myelinated axons have a thicker myelin sheath and conduct action potentials more rapidly than lightly myelinated axons.

In addition to myelination, the diameter of an axon affects the speed of action potential conduction. Large-diameter axons conduct action potentials more rapidly than small-diameter axons because large-diameter axons have a greater surface area. Consequently, at a given site on an axon, more voltage-gated Na^+ channels open during depolarization, resulting in a greater local current flow, which more rapidly stimulates adjacent membrane areas.

Nerve fibers (axons) are classified according to their size and myelination. It is not surprising that the structure of nerve fibers reflects their functions. Type A fibers are large-diameter,

1 An action potential (*orange*) at a node of Ranvier generates local currents (*black arrows*). The local currents flow to the next node of Ranvier because the myelin sheath of the Schwann cell insulates the axon of the internode.

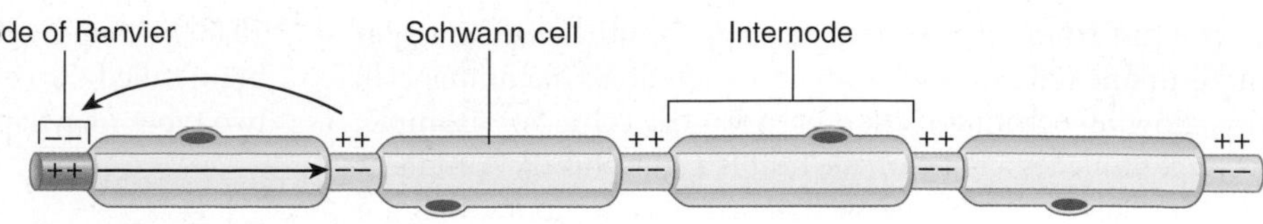

Node of Ranvier
Schwann cell
Internode

2 When the depolarization caused by the local currents reaches threshold at the next node of Ranvier, a new action potential is produced (*orange*).

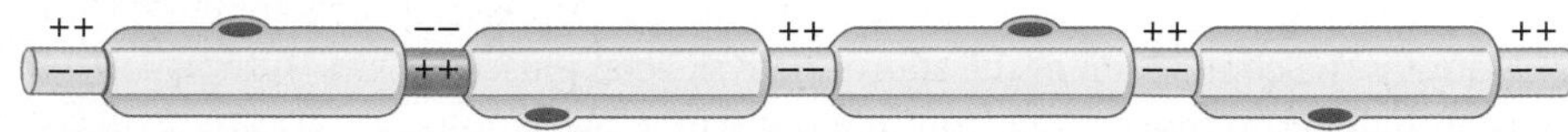

3 Action potential propagation is rapid in myelinated axons because the action potentials are produced at successive nodes of Ranvier (*1–5*) instead of at every part of the membrane along the axon.

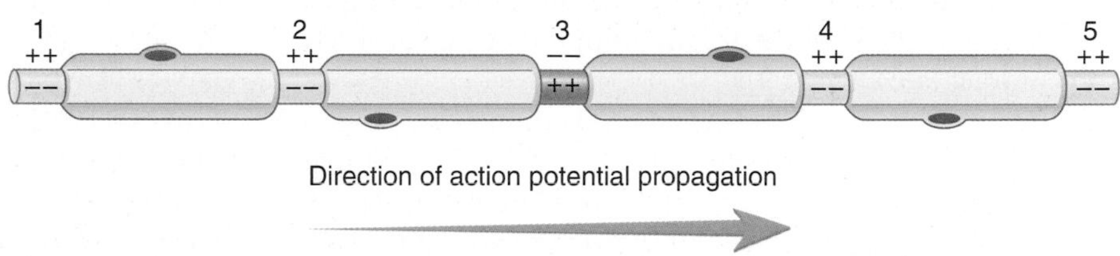

1
2
3
4
5
Direction of action potential propagation

PROCESS FIGURE 11.16 Saltatory Conduction: Action Potential Propagation in a Myelinated Axon
The gaps between the Schwann cells are exaggerated for clarity.

myelinated axons that conduct action potentials at 15–120 m/s. Motor neurons supplying skeletal muscles and most sensory neurons have type A fibers. Rapid response to the external environment is possible because of the rapid input of sensory information to the CNS and the rapid output of action potentials to skeletal muscles.

Type B fibers are medium-diameter, lightly myelinated axons that conduct action potentials at 3–15 m/s, and type C fibers are small-diameter, unmyelinated axons that conduct action potentials at 2 m/s or less. Types B and C fibers are primarily part of the ANS, which stimulates internal organs, such as the stomach, intestines, and heart. The responses necessary to maintain internal homeostasis, such as digestion, need not be as rapid as responses to the external environment.

ASSESS YOUR PROGRESS

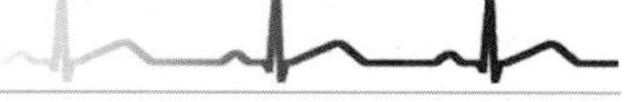

38. *Describe the absolute and relative refractory periods. Relate them to the depolarization and repolarization phases of the action potential.*

39. *What is action potential frequency? What two factors determine action potential frequency?*

40. *Describe subthreshold, threshold, maximal, submaximal, and supramaximal stimuli. What determines the maximum frequency of action potential generation?*

41. *What is a local current? How do local currents cause the propagation of action potentials in unmyelinated axons?*

42. *What prevents an action potential from reversing its direction of propagation?*

43. *Describe saltatory conduction of an action potential.*

44. *Compare the speed of action potential conduction in (a) heavily myelinated, lightly myelinated, and unmyelinated axons and (b) large-diameter and small-diameter axons.*

45. *Compare the function of type A nerve fibers with that of types B and C nerve fibers.*

11.6 The Synapse

LEARNING OUTCOMES

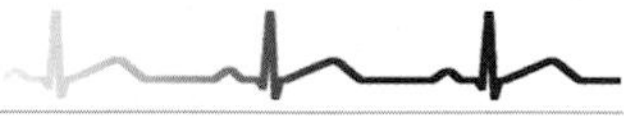

After reading this section, you should be able to

A. Describe the general structure and function of a synapse.

B. Distinguish between electrical and chemical synapses as to mode of operation and types of tissues where they are found.

C. Describe the release of a neurotransmitter in a chemical synapse, then its removal from the synapse.

D. Explain the effects of neurotransmitter binding to receptors in a chemical synapse.

E. Discuss the effects of neuromodulators in a chemical synapse.

F. Contrast excitatory and inhibitory postsynaptic potentials.

G. Explain the roles of presynaptic inhibition and of facilitation.

H. Describe the processes of spatial and temporal summation.

Just as the fire from one lit torch can light another torch, action potentials in one cell can stimulate action potentials in another cell, thereby allowing communication between the cells. For example, if your finger touches a hot pan, the heat is a stimulus that produces action potentials in sensory nerve fibers. The action potentials are propagated along the sensory fibers from the finger toward the CNS. For the CNS to get this information, the action potentials of the sensory neurons must produce action potentials in CNS neurons. After the CNS has received the information, it produces a response. One response is the contraction of the appropriate skeletal muscles that cause the finger to move away from the hot pan. CNS action potentials cause motor neurons to produce action potentials that are then transmitted by the motor neurons toward skeletal muscles. The action potentials of the motor neuron produce skeletal muscle action potentials, which are the stimuli that cause muscle fibers to contract (see chapter 9).

As stated in section 11.2, the synapse is the junction between two cells where they communicate with each other. The cell that transmits a signal toward the synapse is called the **presynaptic cell,** and the cell that receives the signal is called the **postsynaptic cell.**

The average presynaptic neuron synapses with about 1000 other neurons, but the average postsynaptic neuron has up to 10,000 synapses. Some postsynaptic neurons in the part of the brain called the cerebellum have up to 100,000 synapses. There are two types of synapses: electrical and chemical.

Electrical Synapses

Electrical synapses are gap junctions (see chapter 4) that allow a local current to flow between adjacent cells (figure 11.17). At these gap junctions, the membranes of adjacent cells are separated by a 2 nm gap spanned by structures called **connexons.** The connexons are groups of six tubular proteins, each called a connexin. The movements of ions through the connexons are local currents. Thus, an action potential in one cell produces a local current that generates an action potential in the adjacent cell almost as if the two cells had the same membrane. As a result, action potentials are conducted rapidly between cells, allowing the cells' activity to be synchronized. Electrical synapses are not common in the nervous system of vertebrates, but some do exist in humans, such as between adjacent cardiac muscle cells. Electrical synapses are also important in many types of smooth muscle. Coordinated contractions of these muscle cells occur when action potentials in one cell propagate to adjacent cells because of electrical synapses (see chapters 9 and 20).

1. Electrical synapses connect cardiac muscle cells.
2. An electrical synapse is a gap junction where the membranes of two cells are separated by a gap but connected by proteins called connexons.
3. An action potential (*orange arrow*) in the plasma membrane generates local currents (*black arrows*) that flow to adjacent parts of the plasma membrane and through the gap junction.
4. A local current stimulates the production of another action potential. Thus, the action potential propagates along the plasma membrane.
5. A local current flows through a gap junction and stimulates the production of an action potential in the adjacent cardiac muscle cell. Thus, the action potential propagates to the adjacent cell.

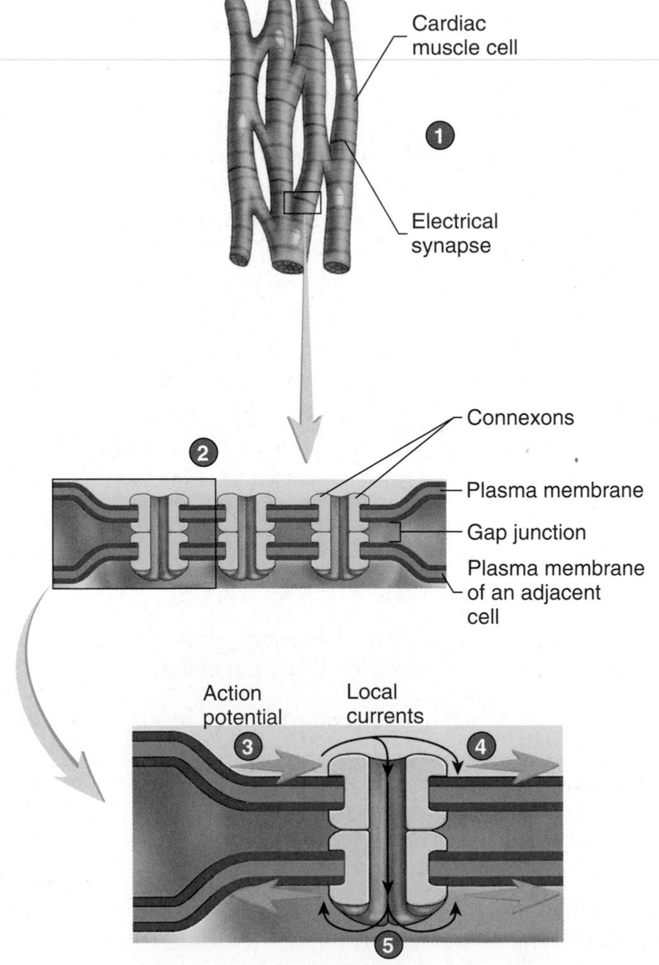

Cardiac muscle cell
1
Electrical synapse
Connexons
2
Plasma membrane
Gap junction
Plasma membrane of an adjacent cell
Action potential
Local currents
3
4
5

PROCESS FIGURE 11.17 Electrical Synapse

FUNDAMENTAL **Figure**

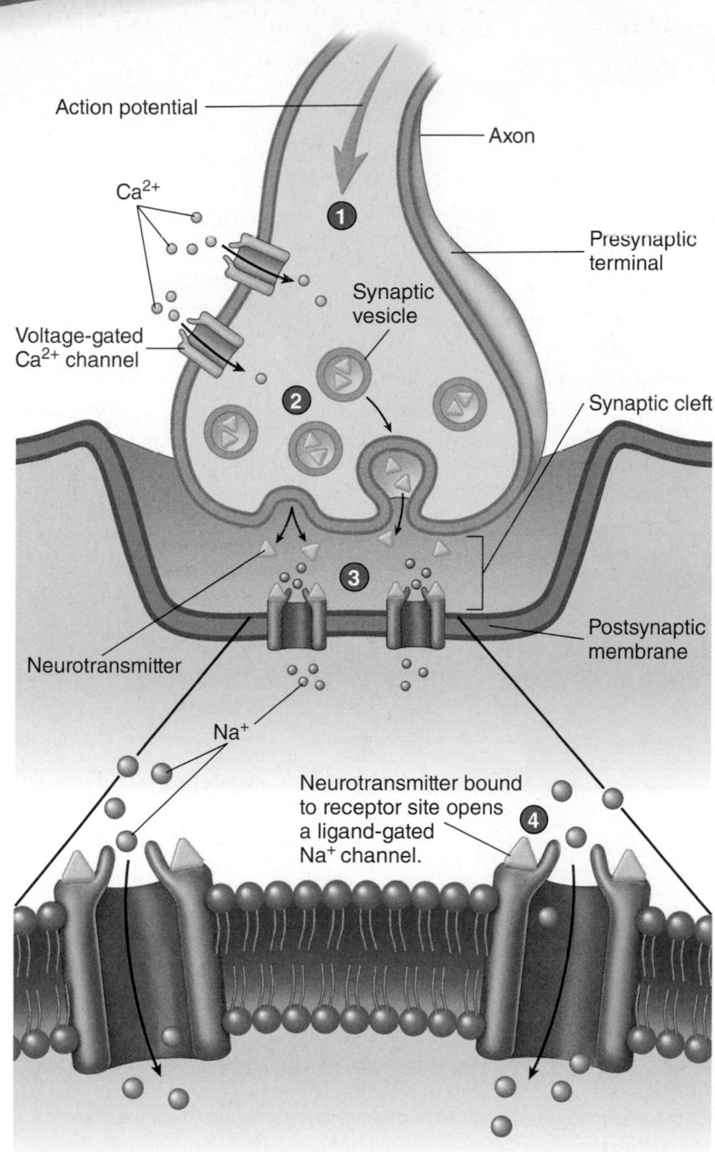
Action potential
Axon
Ca2+
Presynaptic terminal
Synaptic vesicle
Voltage-gated Ca2+ channel
Synaptic cleft
Postsynaptic membrane
Neurotransmitter
Na+
Neurotransmitter bound to receptor site opens a ligand-gated Na+ channel.
1
2
3
4

1 Action potentials arriving at the presynaptic terminal cause voltage-gated Ca^{2+} channels to open.

2 Ca^{2+} diffuses into the cell and causes synaptic vesicles to release neurotransmitter molecules.

3 Neurotransmitter molecules diffuse from the presynaptic terminal across the synaptic cleft.

4 Neurotransmitter molecules combine with their receptor sites and cause ligand-gated Na^+ channels to open. Na^+ diffuses into the cell (*shown in illustration*) or out of the cell (*not shown*) and causes a change in membrane potential.

PROCESS FIGURE 11.18 APIR **A Chemical Synapse**

A synapse consists of the end of a neuron (presynaptic terminal), a small space (synaptic cleft), and the postsynaptic membrane of another neuron or an effector cell, such as a muscle or gland cell.

Chemical Synapses

The essential components of a **chemical synapse** are the presynaptic terminal, the synaptic cleft, and the postsynaptic membrane (figure 11.18). The **presynaptic terminal** consists of the end of an axon. The space separating the axon ending and the cell with which it synapses is the **synaptic cleft.** The membrane of the postsynaptic cell opposed to the presynaptic terminal is the **postsynaptic membrane.** Postsynaptic cells are typically other neurons, muscle cells, or gland cells.

Neurotransmitter Release

In chemical synapses, action potentials do not pass directly from the presynaptic terminal to the postsynaptic membrane. Instead, the action potentials in the presynaptic terminal cause the release of neurotransmitters from its terminal.

Presynaptic terminals are specialized to produce and release neurotransmitters. The major cytoplasmic organelles within presynaptic terminals are mitochondria and numerous membrane-bound **synaptic vesicles,** which contain neurotransmitters, such as acetylcholine (figure 11.18). Each action potential arriving at the presynaptic terminal initiates a series of specific events, which result in the release of neurotransmitters. In response to an action potential, voltage-gated Ca^{2+} channels open, and Ca^{2+} diffuses into the presynaptic terminal. These ions cause synaptic vesicles to fuse with the presynaptic membrane and to release their neurotransmitters by exocytosis into the synaptic cleft.

Once neurotransmitters are released from the presynaptic terminal, they diffuse rapidly across the synaptic cleft, which is about 20 nm wide, and bind in a reversible fashion to specific receptors, such as ligand-gated ion channels, in the postsynaptic membrane (figure 11.18). Depending on the receptor type, this binding produces a depolarizing or hyperpolarizing graded potential in the postsynaptic membrane. For example, the binding of acetylcholine to ligand-gated Na^+ channels causes them to open, allowing Na^+ to diffuse into the postsynaptic cell. If the resulting depolarizing graded potential reaches threshold, an action potential is produced. On the other hand, the opening of K^+ or Cl^- channels results in a hyperpolarizing graded potential.

Predict 9

Is an action potential transmitted faster between cells connected by an electrical synapse or by a chemical synapse? Explain.

Neurotransmitter Removal

The interaction between a neurotransmitter and a receptor represents an equilibrium:

$$\text{Neurotransmitter} + \text{Receptor} \rightleftarrows \text{Neurotransmitter} - \text{Receptor complex}$$

When the neurotransmitter concentration in the synaptic cleft is high, many of the receptor molecules have neurotransmitter molecules bound to them; when the neurotransmitter concentration declines, the neurotransmitter molecules diffuse away from the receptor molecules.

Neurotransmitters have short-term effects on postsynaptic membranes because the neurotransmitter is rapidly destroyed or removed from the synaptic cleft. For example, in the neuromuscular junction (see chapter 9), the neurotransmitter acetylcholine is broken down by the enzyme **acetylcholinesterase** (as′e-til-kō-lin-es′ter-ās) to acetic acid and choline (figure 11.19*a*). Choline is then transported back into the presynaptic terminal and reacts with acetyl-CoA to

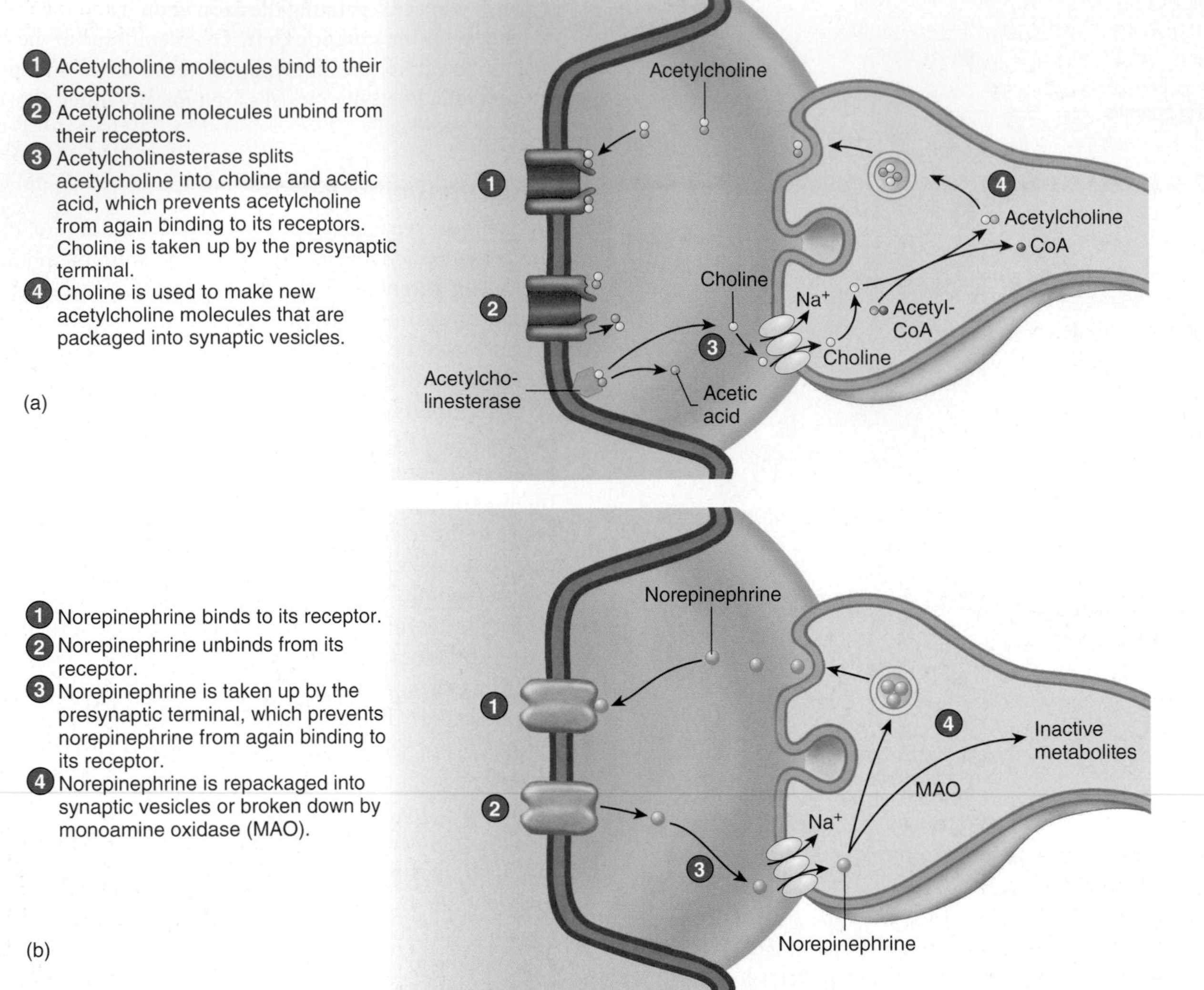

1 Acetylcholine molecules bind to their receptors.
2 Acetylcholine molecules unbind from their receptors.
3 Acetylcholinesterase splits acetylcholine into choline and acetic acid, which prevents acetylcholine from again binding to its receptors. Choline is taken up by the presynaptic terminal.
4 Choline is used to make new acetylcholine molecules that are packaged into synaptic vesicles.
(a)
Acetylcholine
Choline
Na+
Acetyl-CoA
Choline
Acetylcho-linesterase
Acetic acid
Acetylcholine
CoA
1 Norepinephrine binds to its receptor.
2 Norepinephrine unbinds from its receptor.
3 Norepinephrine is taken up by the presynaptic terminal, which prevents norepinephrine from again binding to its receptor.
4 Norepinephrine is repackaged into synaptic vesicles or broken down by monoamine oxidase (MAO).
(b)
Norepinephrine
Inactive metabolites
MAO
Na+
Norepinephrine

PROCESS FIGURE 11.19 Removal of Neurotransmitters from the Synaptic Cleft
(*a*) In some synapses, neurotransmitters are broken down by enzymes and recycled into the presynaptic terminal. (*b*) In other synapses, neurotransmitters are taken up whole into the presynaptic terminal.

form acetylcholine. Acetyl-CoA is synthesized by mitochondria as foods are metabolized to produce ATP (see chapter 25). Acetic acid can be absorbed from the synaptic cleft into the presynaptic terminal, or it can diffuse out of the synaptic cleft and be taken up by a variety of cells. Acetic acid can be used to synthesize acetyl-CoA.

When the neurotransmitter norepinephrine is released into the synaptic cleft, most of it is transported back into the presynaptic terminal, where it is repackaged into synaptic vesicles for reuse (figure 11.19*b*). The enzyme **monoamine oxidase (MAO;** mon-ō-am′īn ok′si-dās) inactivates some of the norepinephrine.

Diffusion of neurotransmitter molecules away from the synapse and into the extracellular fluid also limits the length of time the neurotransmitter molecules remain bound to their receptors. Norepinephrine in the circulation is taken up primarily by liver and kidney cells, where the enzymes monoamine oxidase and **catechol-O-methyltransferase** (kat′ĕ-kol-ō-meth-il-trans′fer-ās) convert it into inactive metabolites.

Receptor Molecules in Synapses

Receptor molecules in synapses are membrane-bound, ligand-activated receptors with highly specific receptor sites. Consequently, only neurotransmitter molecules or very closely related substances normally bind to their receptors. For example, acetylcholine binds to acetylcholine receptors but not to norepinephrine receptors, whereas norepinephrine binds to norepinephrine receptors but not to acetylcholine receptors. Any given cell does not have all possible receptors. Therefore, a neurotransmitter affects only the cells with receptors for that neurotransmitter.

A neurotransmitter can stimulate some cells but inhibit others. More than one type of receptor molecule exists for some neurotransmitters. Different cells respond differently to a neurotransmitter when these cells have different receptors. For example, norepinephrine can bind to one type of norepinephrine receptor to cause depolarization in one synapse and to another type of norepinephrine receptor to cause hyperpolarization in another synapse. Thus, norepinephrine is either stimulatory or inhibitory, depending on the type of norepinephrine receptor to which it binds and on the effect that receptor has on the permeability of the postsynaptic membrane.

Although neurotransmitter receptors are in greater concentrations on postsynaptic membranes, some receptors exist on presynaptic membranes. For example, norepinephrine released from the presynaptic membrane binds to receptors on both presynaptic and postsynaptic membranes. Its binding to the receptors of the presynaptic membrane decreases the release of additional synaptic vesicles. Norepinephrine can therefore modify its own release by binding to presynaptic receptors. A high frequency of presynaptic action potentials results in the release of fewer synaptic vesicles in response to later action potentials.

Neurotransmitters and Neuromodulators

Several substances have been identified as neurotransmitters, and others are suspected neurotransmitters. Scientists once thought that each neuron contains only one type of neurotransmitter; however, they now know that some neurons can secrete more than one type. If a neuron does produce more than one neurotransmitter, it secretes all of them from each of its presynaptic terminals. The physiological significance of presynaptic terminals that secrete more than one type of neurotransmitter has not been clearly established.

Neuromodulators are substances released from neurons that influence the likelihood of an action potential being produced in the postsynaptic cell. For example, a neuromodulator that decreases the release of an excitatory neurotransmitter from a presynaptic terminal decreases the likelihood of action potential production in the postsynaptic cell. Drugs can modulate the action of neurotransmitters at the synapse. For example, cocaine and amphetamines increase the release and block the reuptake of norepinephrine, resulting in overstimulation of postsynaptic neurons and deleterious effects on the body. Drugs that block serotonin reuptake are particularly effective at treating depression and behavioral disorders. A list of neurotransmitters and neuromodulators is presented in table 11.6.

ASSESS YOUR PROGRESS

46. *What are the components of a synapse? What is the purpose of a synapse?*

47. *What is an electrical synapse? Describe its operation.*

48. *Describe the release of neurotransmitter in a chemical synapse.*

49. *Name three ways to stop the effect of a neurotransmitter on the postsynaptic membrane. Give an example of each.*

50. *Why does a given type of neurotransmitter affect only certain types of cells? How can a neurotransmitter stimulate one type of cell but inhibit another type?*

51. *What is a neuromodulator? Give some examples of how drugs can modulate the action of neurotransmitters.*

Excitatory and Inhibitory Postsynaptic Potentials

The combination of neurotransmitters with their specific receptors causes either depolarization or hyperpolarization of the postsynaptic membrane. When depolarization occurs, the response is stimulatory, and the graded potential is called an **excitatory postsynaptic potential** (**EPSP;** figure 11.20*a*). EPSPs are important because the depolarization might reach threshold, thereby producing an action potential and a response from the cell. Neurons releasing neurotransmitter substances that cause EPSPs are **excitatory neurons.** In general, an EPSP occurs because the membrane has become more permeable to Na^+. For example, glutamate in the brain and acetylcholine in skeletal muscle can bind to their receptors, causing Na^+ channels to open. Because the concentration gradient is large for Na^+ and because the negative charge inside the cell attracts the positively charged Na^+, it diffuses into the cell and causes depolarization. If EPSPs cause a depolarizing graded potential that reaches threshold, an action potential is produced. For example, awareness of pain can occur only if action potentials generated by sensory neurons stimulate the production of action potentials in CNS neurons. Local anesthetics, such as procaine (novocaine), act at their site of application to prevent pain sensations. They do so by blocking voltage-gated Na^+ channels, which prevents action potentials from propagating along sensory neurons. Consequently, neurotransmitters are not released from the presynaptic terminals of the sensory neurons, and EPSPs are not produced in CNS neurons.

When the combination of a neurotransmitter with its receptor results in hyperpolarization of the postsynaptic membrane, the response is inhibitory, and the local hyperpolarization is called an **inhibitory postsynaptic potential** (**IPSP;** figure 11.20*b*). IPSPs are important because they decrease the likelihood of producing action potentials by moving the membrane potential farther from threshold. Neurons releasing neurotransmitter substances that cause IPSPs are called **inhibitory neurons.** The IPSP results from an increase in the permeability of the plasma membrane to Cl^- or K^+. For example, in the spinal cord, glycine binds to its receptors, directly causing Cl^- channels to open. Because Cl^- is more concentrated outside the cell than inside, when the membrane's permeability to Cl^- increases, it diffuses into the cell, causing the inside of the cell to become more negative and resulting in hyperpolarization. Acetylcholine can bind to its receptors in the heart, causing G protein–mediated opening of K^+ channels (see chapter 3). The concentration of K^+ is greater inside the cell than outside, and increased permeability of the membrane to K^+ allows K^+ to diffuse out of the cell. Consequently, the outside of the cell becomes more positive than the inside, resulting in hyperpolarization.

Presynaptic Inhibition and Facilitation

Many of the synapses of the CNS are **axoaxonic synapses,** meaning that the axon of one neuron synapses with the presynaptic terminal (axon) of another (figure 11.21*a*). Through axoaxonic synapses, one neuron can release a neuromodulator that influences the release of a neurotransmitter from the presynaptic terminal of another neuron.

In **presynaptic inhibition,** the amount of neurotransmitter released from the presynaptic terminal decreases. For example, sensory neurons for pain can release neurotransmitters from their presynaptic terminals and stimulate the postsynaptic membranes of neurons in the brain or spinal cord. Awareness of pain occurs only if action potentials are produced in the postsynaptic membranes of the CNS neurons. Enkephalins and endorphins released from inhibitory neurons of axoaxonic synapses can reduce or eliminate pain sensations by inhibiting the release of neurotransmitter from the presynaptic terminals of sensory neurons (figure 11.21*b*).

TABLE 11.6 Clinical Examples of Synaptic Function

Neurotransmitter/Neuromodulator	Clinical Examples
Acetylcholine Structure: Site of release: CNS synapses, ANS synapses, and neuromuscular junctions Effect: excitatory in the CNS and neuromuscular junctions; inhibitory or excitatory in ANS synapses	**Myasthenia Gravis** In the disease myasthenia gravis, the ability of skeletal muscle to respond to nervous system stimulation decreases, resulting in muscle weakness and even paralysis. Antibodies, proteins produced by the immune system, can attach to foreign substances, such as bacteria (see chapter 22). In myasthenia gravis, antibodies inappropriately attach to acetylcholine receptors. The antibodies link the receptors together, causing them to be removed from the plasma membrane faster than normal and decreasing the number of receptors. The antibodies also stimulate immune responses that lead to destruction of the postsynaptic membrane, which decreases the number of Na^+ channels in the synapse. Thus, the ability of ACh to stimulate action potential production decreases because there are fewer ligand-gated ACh receptors, and the ability to generate an action potential is reduced because there are fewer voltage-gated Na^+ channels.

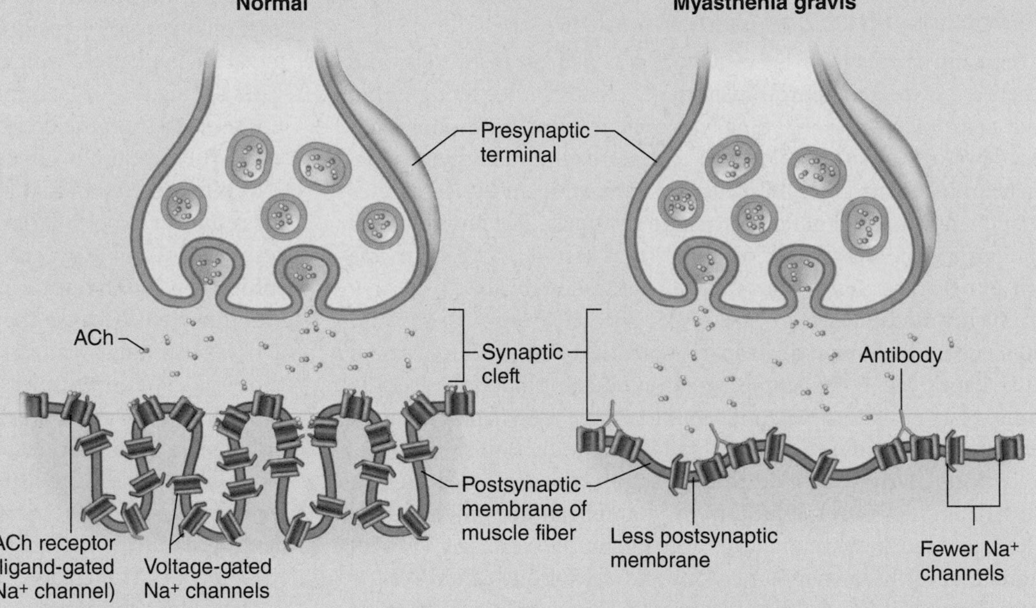

Normal
Myasthenia gravis
Presynaptic terminal
ACh
Synaptic cleft
Antibody
Postsynaptic membrane of muscle fiber
ACh receptor (ligand-gated Na+ channel)
Voltage-gated Na+ channels
Less postsynaptic membrane
Fewer Na+ channels

Neurotransmitter/Neuromodulator	Clinical Examples
Biogenic Amines *Serotonin* Structure: Site of release: CNS synapses Effect: both inhibitory and excitatory	**Antidepressant Therapy** Selective serotonin reuptake inhibitors (SSRIs), such as Prozac (fluoxetine hydrochloride) and Zoloft (sertraline HCl), are drugs commonly used to treat depression. They temporarily block serotonin transporters (symporters), which decreases serotonin transport back into presynaptic terminals, resulting in increased serotonin levels in synaptic clefts. In some people, the increased stimulation of the postsynaptic neuron by serotonin relieves depression.

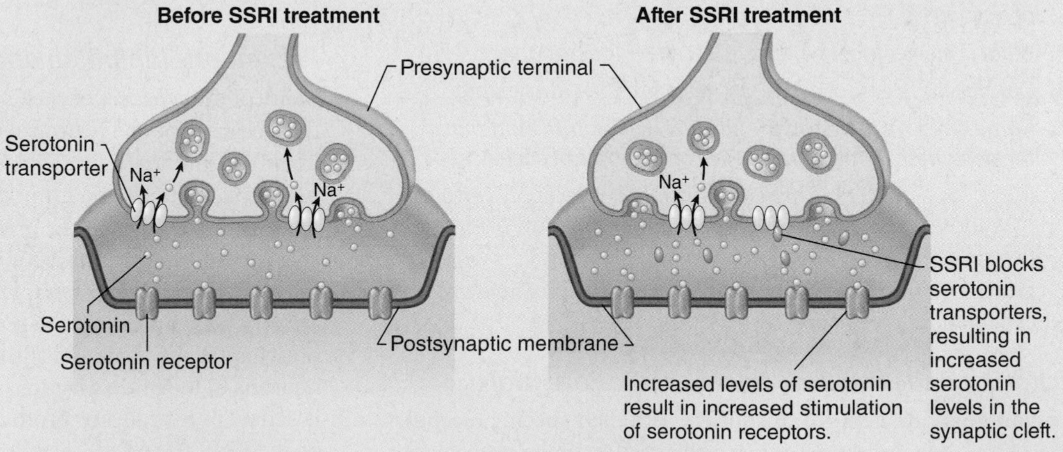

Before SSRI treatment
After SSRI treatment
Presynaptic terminal
Serotonin transporter
Na+
Na+
Na+
Serotonin
Serotonin receptor
Postsynaptic membrane
Increased levels of serotonin result in increased stimulation of serotonin receptors.
SSRI blocks serotonin transporters, resulting in increased serotonin levels in the synaptic cleft.

Anxiety Disorders

SSRIs are also used to treat panic disorders, such as obsessive-compulsive disorder (OCD), leading researchers to believe that OCD might be linked to abnormalities in serotonin function.

TABLE 11.6 Clinical Examples of Synaptic Function—Continued

Neurotransmitter/Neuromodulator	Clinical Examples
Serotonin (continued)	**Hallucinogens** The hallucinogenic drug D-lysergic acid diethylamide (LSD) blocks serotonin transporters in specific areas of the brain and produces hallucinogenic effects. Other drugs, such as ecstasy, are also hallucinogens that block serotonin transporters.
Dopamine Structure: HO, HO, NH_2 Site of release: selected CNS synapses; also found in some ANS synapses Effect: excitatory or inhibitory	**Drug Addiction** Cocaine blocks dopamine transporters (symporters), which increases dopamine levels in synaptic clefts, resulting in overstimulation of postsynaptic neurons. Although moderate levels of dopamine can cause euphoria, high levels of dopamine can produce psychotic effects. **Parkinson Disease** Parkinson disease results from the destruction of dopamine-producing neurons, and it is characterized by tremors and decreased voluntary motor control. Parkinson disease is treated with the drug L-Dopa, which increases the production of dopamine in the presynaptic terminals of remaining neurons. Another treatment option involves drugs that mimic the action of dopamine.
Norepinephrine Structure: OH, HO, CH, NH_2, CH_2, HO Site of release: selected CNS synapses and some ANS synapses Effect: excitatory	**Attention-Deficit Hyperactivity Disorder (ADHD)** ADHD is often treated with drugs that increase the level of excitatory neurotransmitters, such as norepinephrine, in the synaptic clefts. This is achieved by using selective norepinephrine reuptake inhibitors (SNRIs) that block norepinephrine transporters (symporters) and increase the levels of norepinephrine in the synaptic clefts. Additionally, simultaneously increasing the levels of dopamine in the synaptic cleft is a common treatment. The most familiar of the ADHD medications, Ritalin®, prevents reuptake of both norepinephrine and dopamine. The most recent information suggests that enhancing the nicotininc cholinergic pathways is also an effective treatment for ADHD. **Amphetamines** Amphetamines are drugs with excitatory effects on the CNS. They increase the levels of norepinephrine and dopamine in synaptic clefts by either blocking the reuptake of these neurotransmitters or promoting their release from synaptic vesicles. The CNS effects of amphetamines include decreased appetite, increased alertness, and enhanced ability to concentrate and perform physical tasks. Amphetamines are used to treat ADHD, clinical depression, narcolepsy, and chronic fatigue syndrome. An overdose of amphetamines can cause insomnia, tremors, anxiety, and panic.
Amino Acids *Gamma-Amino Butyric Acid (GABA)* Structure: H_3N^+, O, O^- Site of release: CNS synapses Effect: inhibitory effect on postsynaptic neurons; some presynaptic inhibition in the spinal cord	**Barbiturates** Certain GABA receptors are ligand-gated channels that permit the inflow of Cl^- when stimulated. GABA produces an inhibitory (or a hyperpolarizing) effect by binding to these receptors and promoting Cl^- inflow. Barbiturates enhance the binding of GABA to their receptors, resulting in the prolonged inhibition of postsynaptic neurons. These drugs are used as sedatives and anesthetics and as a treatment for epilepsy, which is characterized by excessive neuronal discharge.

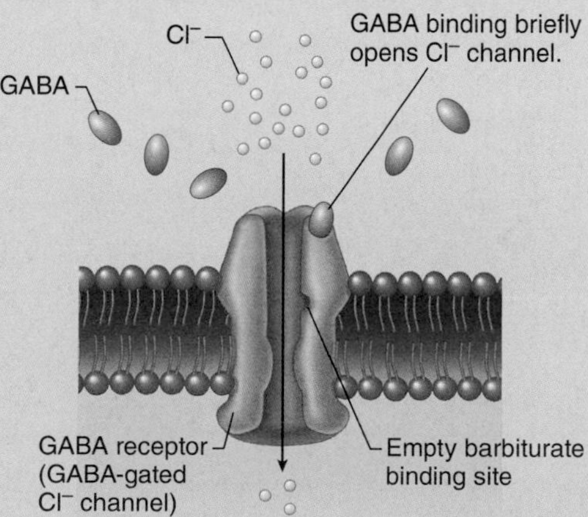
Without barbiturates
Cl^-
GABA
GABA binding briefly opens Cl^- channel.
GABA receptor (GABA-gated Cl^- channel)
Empty barbiturate binding site
The inflow of Cl^- causes inhibition of the postsynaptic neuron.

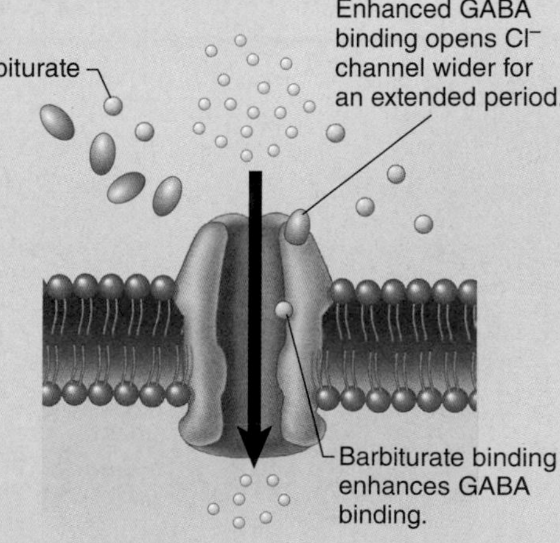
With barbiturates
Barbiturate
Enhanced GABA binding opens Cl^- channel wider for an extended period.
Barbiturate binding enhances GABA binding.
Greater inflow of Cl^- causes prolonged inhibition of the postsynaptic neuron.

TABLE 11.6 Clinical Examples of Synaptic Function—Continued

Neurotransmitter/Neuromodulator	Clinical Examples
Amino Acids *Gamma-Amino Butyric Acid (GABA) (continued)*	**Benzodiazepines** Benzodiazepines that are used in antianxiety drugs also have binding sites on certain GABA receptors. Their action is similar to that of barbiturates in that they enhance the binding of GABA to its receptor, producing an inhibitory effect. **Alcohol Dependence** Alcohol acts similarly to barbiturates to enhance the effect of GABA. As a result, the ligand-gated Cl^- channel becomes more permeable to Cl^-, producing an inhibitory effect. Chronic consumption of alcohol renders the GABA receptor less sensitive to both alcohol and GABA, resulting in increased alcohol dependence and alcohol withdrawal symptoms, such as anxiety, tremors, and insomnia. Alcohol withdrawal symptoms are often treated with benzodiazepines.
Glycine Structure: $H_2N{-}CH_2{-}COOH$ Site of release: CNS synapses Effect: inhibitory	**Strychnine Poisoning** Glycine receptors are similar to GABA receptors in that they act as ligand-gated channels permitting the inflow of Cl^-. The poison strychnine blocks glycine receptors, which increases the excitability of certain neurons by preventing their inhibition. Strychnine poisoning results in powerful muscle contractions and convulsions. Tetanus of respiratory muscles can cause death.
Glutamate Structure: $H_2N{-}CH(COOH){-}CH_2{-}CH_2{-}COOH$ Site of release: CNS synapses; in areas of the brain involved in learning and memory Effect: excitatory	**Stroke and Excitotoxicity** Glutamate is the major excitatory neurotransmitter of the CNS. Some glutamate receptors are ligand-gated Ca^{2+} channels. When stimulated, Ca^{2+} channels open, causing depolarization of postsynaptic membranes. Some glutamate is removed from the synapse by transporters in presynaptic terminals, whereas the bulk of it is removed by transporters (symporters) in neighboring astrocytes. When a person suffers a stroke, brain tissue is deprived of oxygen, and ATP levels decrease. This causes the secondary active transport of glutamate by the glutamate transporters to fail temporarily. As a result, glutamate accumulates in the synaptic clefts and causes excessive stimulation of postsynaptic neurons. Excessive movement of Ca^{2+} into neurons activates a variety of destructive processes, which can cause cell death. 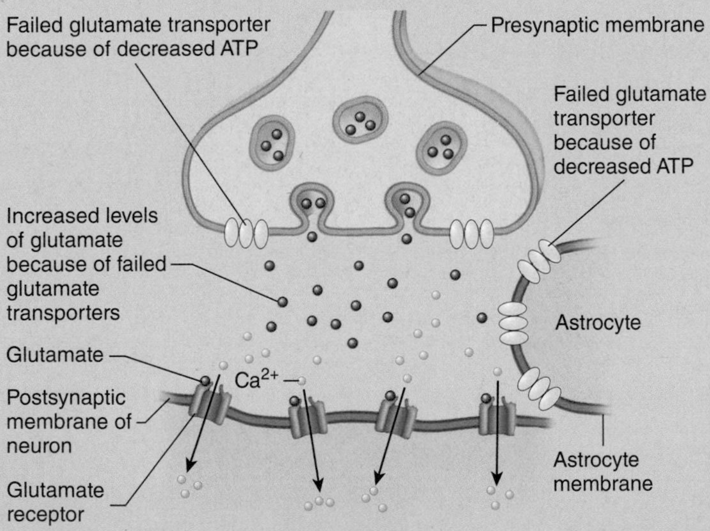 Failed glutamate transporter because of decreased ATP Presynaptic membrane Failed glutamate transporter because of decreased ATP Increased levels of glutamate because of failed glutamate transporters Astrocyte Glutamate Ca^{2+} Postsynaptic membrane of neuron Glutamate receptor Astrocyte membrane Excessive stimulation of postsynaptic neuron **Cognition** Glutamate is implicated in learning and memory. Drugs, such as Namenda (memantine), that target specific glutamate receptors are often used to treat Alzheimer disease.

TABLE 11.6 Clinical Examples of Synaptic Function—Continued

Neurotransmitter/Neuromodulator	Clinical Examples
Purines *Adenosine* Structure: [adenosine structural formula: NH_2, N, N, N, N, HO, O, H, H, OH, HO] Site of release: CNS synapses; in the areas of the brain involved in learning and memory Effect: inhibitory	**Neuroprotective Agent** Adenosine acts as both a neurotransmitter and a neuromodulator. Adenosine receptors are linked to G proteins (see figure 3.11). As a neurotransmitter, adenosine stimulates the opening of Cl^- and K^+ channels on postsynaptic membranes, thereby producing a hyperpolarizing effect. Acting as a neuromodulator, adenosine stimulates the closing of Ca^{2+} channels on presynaptic neurons, inhibiting neurotransmitter release. Adenosine production greatly increases during a stroke. It prevents the release of glutamate from presynaptic vesicles, which reduces the level of glutamate in synaptic clefts. It also hyperpolarizes the postsynaptic membranes of glutamate synapses, thereby countering the excitatory effects of glutamate. As a result, the damaging effects of glutamate during a stroke are diminished. The possibility of using adenosine as an antistroke agent is being investigated. **Caffeine** Adenosine produces drowsiness. Caffeine counters that effect by blocking adenosine receptors and promoting alertness. Caffeine also promotes cognition by blocking adenosine's inhibitory effect on glutamate function.
Neuropeptides *Substance P* Structure: polypeptide (10 amino acids) Site of release: descending pain pathways Effect: excitatory	**Pain Therapy** Substance P acts as a neurotransmitter and neuromodulator. The receptor for substance P is called a neurokinin receptor, which is linked to a G protein complex (see figure 3.11). Drugs such as morphine reduce pain by blocking the release of substance P.
Endorphins Structure: polypeptide (30 amino acids) Site of release: descending pain pathways Effect: inhibitory	**Opiates** Endorphins bind to endorphin receptors on presynaptic neurons and reduce pain by blocking the release of substance P. Endorphins also produce feelings of euphoria. Opiates such as morphine and heroin also bind to endorphin receptors, resulting in similar effects.
Gases *Nitric Oxide (NO)* Structure: N=O Site of release: CNS, nerves supplying the adrenal gland, penis Effect: excitatory	**Stroke Damage** During a stroke, rising glutamate levels act on postsynaptic neurons and cause the release of NO, which in high concentrations can be toxic to cells. Nitric oxide also diffuses out of postsynaptic neurons, enters neighboring cells, and damages them. **Treatment of Erectile Dysfunction** During sexual arousal, nerves release NO, causing the vasodilation of the blood vessels supplying the penis. Viagra (sildenafil citrate), which is used to treat erectile dysfunction, acts by prolonging the effect of NO on these blood vessels (see chapter 28).

Enkephalins and endorphins can block voltage-gated Ca^{2+} channels. Consequently, when action potentials reach the presynaptic terminal, the influx of Ca^{2+} that normally stimulates neurotransmitter release is blocked.

In **presynaptic facilitation,** the amount of neurotransmitter released from the presynaptic terminal increases. For example, serotonin, released in certain axoaxonic synapses, functions as a neuromodulator that increases the release of neurotransmitter from the presynaptic terminal by causing voltage-gated Ca^{2+} channels to open.

Spatial and Temporal Summation

Depolarizations produced in postsynaptic membranes are graded potentials. Within the CNS and in many PNS synapses, a single presynaptic action potential does not cause a graded potential in the postsynaptic membrane sufficient to reach threshold and produce an action potential. Instead, many presynaptic action potentials cause many graded potentials in the postsynaptic neuron. The graded potentials combine in summation at the trigger zone of the postsynaptic neuron, which is the normal site of action potential generation for

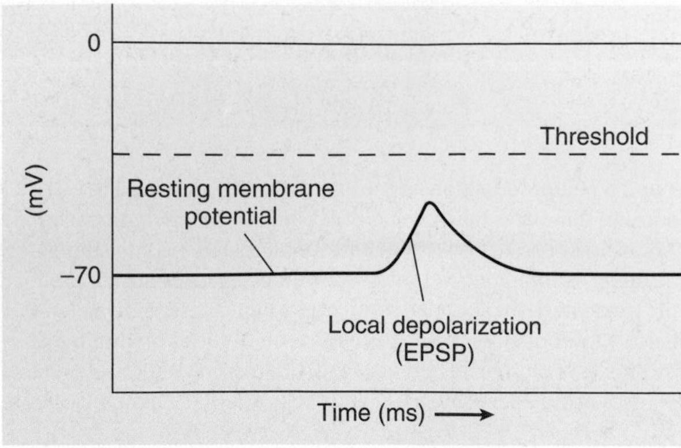
0
Threshold
(mV)
Resting membrane potential
–70
Local depolarization (EPSP)
Time (ms)

(a) **Excitatory postsynaptic potential (EPSP)**

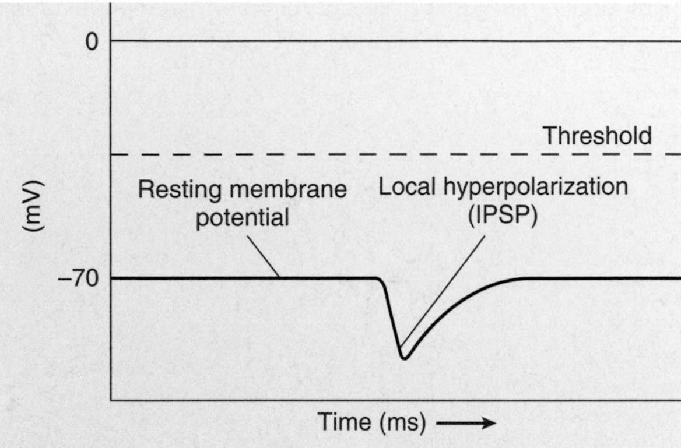
0
Threshold
(mV)
Resting membrane potential
Local hyperpolarization (IPSP)
–70
Time (ms)

(b) **Inhibitory postsynaptic potential (IPSP)**

FIGURE 11.20 Postsynaptic Potentials
(*a*) An excitatory postsynaptic potential (EPSP) is closer to threshold.
(*b*) An inhibitory postsynaptic potential (IPSP) is farther from threshold.

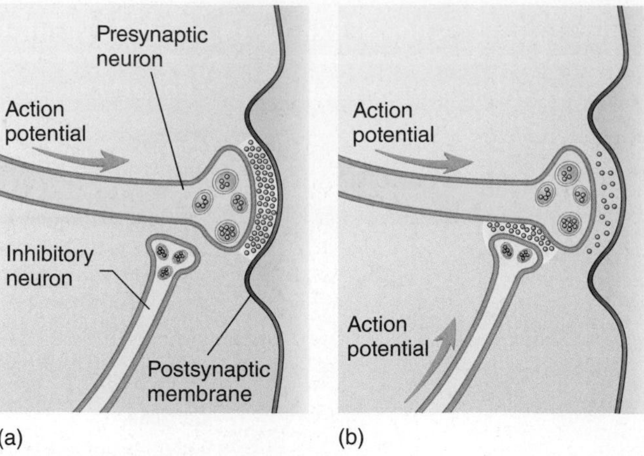
Presynaptic neuron
Action potential
Inhibitory neuron
Postsynaptic membrane
Action potential
Action potential
(a)
(b)

FIGURE 11.21 An Axoaxonic Synapse
(*a*) The inhibitory neuron of the axoaxonic synapse is inactive and has no effect on the release of neurotransmitter from the presynaptic terminal. (*b*) The inhibitory neuron of the axoaxonic synapse releases a neuromodulator, which reduces the amount of neurotransmitter released from the presynaptic terminal.

most neurons. If summation results in a graded potential that exceeds threshold at the trigger zone, an action potential is produced. Action potentials are readily produced at the trigger zone because the concentration of voltage-gated Na^+ channels is approximately seven times greater there than at the rest of the neuron cell body.

Two types of summation are possible: spatial summation and temporal summation. **Spatial summation** occurs when multiple action potentials arrive simultaneously at two different presynaptic terminals that synapse with the same postsynaptic neuron. In the postsynaptic neuron, each action potential causes a depolarizing graded potential that undergoes summation at the trigger zone. If the summated depolarization reaches threshold, an action potential is produced (figure 11.22*a*).

Temporal summation results when two or more action potentials arrive in very close succession at a single presynaptic terminal. The first action potential causes a depolarizing graded potential in the postsynaptic membrane that remains for a few milliseconds before it disappears, although its magnitude decreases through time. Temporal summation results when another action potential initiates another graded depolarization before the depolarization caused by the previous action potential returns to its resting value (see figure 11.9*b*). Subsequent action potentials cause depolarizations that summate with previous depolarizations. If the summated depolarizing graded potentials reach threshold at the trigger zone, an action potential is produced in the postsynaptic neuron (figure 11.22*b*).

Predict 10

Excitatory neurons A and B both synapse with neuron C. Neuron A releases a neurotransmitter, and neuron B releases the same type and amount of neurotransmitter plus a neuromodulator that produces EPSPs in neuron C. Action potentials produced in neuron A alone can result in action potential production in neuron C. Action potentials produced in neuron B alone also can cause action potential production in neuron C. Which results in more action potentials in neuron C, stimulation by only neuron A or stimulation by only neuron B? Explain.

Excitatory and inhibitory neurons can synapse with the same postsynaptic neuron. Spatial summation of EPSPs and IPSPs occurs in the postsynaptic neuron, and whether a postsynaptic action potential is initiated or not depends on which type of graded potential has the greatest influence on the postsynaptic membrane potential (figure 11.22*c*). If the EPSPs (local depolarizations) cancel the IPSPs (local hyperpolarizations) and summate to threshold, an action potential is produced. If the IPSPs prevent the EPSPs from summating to threshold, no action potential is produced.

The synapse is an essential structure for the integration carried out by the CNS. For example, action potentials propagated along axons from sensory organs to the CNS can allow the brain to perceive, or not perceive, a sensory stimulus. To produce the perception of a sensation, action potentials must be transmitted across synapses as they travel through the CNS to the cerebral cortex, where information is interpreted. Stimuli that do not result in action potential transmission across synapses are not perceived because the information never reaches the cerebral cortex. The brain doesn't perceive a large amount of sensory information as a result of complex integration.

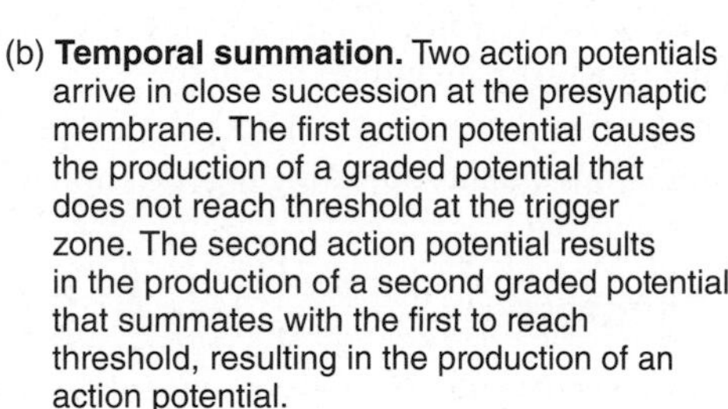
Action potential 1
Axon
Neuron cell body
Trigger zone
Axon
Axon
Action potential 2
+35
–55
–90
mV
Time
Action potentials
Trigger zone
Inhibitory
Excitatory
(with temporal summation)
Excitatory
Trigger zone
Inhibitory
Excitatory

(a) **Spatial summation.** Action potentials 1 and 2 cause the production of graded potentials at two different dendrites. These graded potentials summate at the trigger zone to produce a graded potential that exceeds threshold, resulting in an action potential.

(b) **Temporal summation.** Two action potentials arrive in close succession at the presynaptic membrane. The first action potential causes the production of a graded potential that does not reach threshold at the trigger zone. The second action potential results in the production of a second graded potential that summates with the first to reach threshold, resulting in the production of an action potential.

(c) **Combined spatial and temporal summation with both excitatory postsynaptic potentials (EPSPs) and inhibitory postsynaptic potentials (IPSPs).** An action potential is produced at the trigger zone when the graded potentials produced as a result of the EPSPs and IPSPs summate to reach threshold.

FIGURE 11.22 Summation

ASSESS YOUR PROGRESS

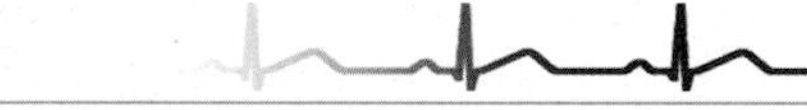

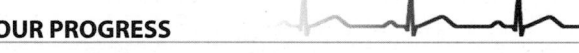

52. *Explain the production of EPSPs and IPSPs. Why are they important?*
53. *What are axoaxonic synapses?*
54. *Give an example of presynaptic inhibition. Describe presynaptic facilitation.*
55. *Distinguish between spatial summation and temporal summation. In what part of the neuron does summation take place?*
56. *How do EPSPs and IPSPs affect the likelihood that summation will result in an action potential?*

11.7 Neuronal Pathways and Circuits

LEARNING OUTCOMES

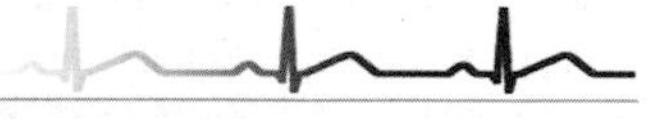

After reading this section, you should be able to

A. Contrast convergent and divergent neuron pathways.

B. Describe a reverberating circuit.

C. Explain a parallel after-discharge circuit.

The organizational patterns of neurons within the CNS vary from relatively simple to extremely complex. The simplest organization is a **serial** pathway, where the input travels along only one pathway. However, most pathways are more complex and are called **parallel** pathways, where the input travels along several pathways. The axon of a neuron can branch repeatedly to form synapses with many other neurons, and hundreds or even thousands of axons can synapse with the cell body and dendrites of a single neuron. Although their complexity varies, four basic patterns of parallel pathways can be recognized: convergent pathways, divergent pathways, reverberating circuits, and parallel after-discharge circuits.

In **convergent pathways,** many neurons converge and synapse with a smaller number of neurons (figure 11.23*a*). Convergence allows different parts of the nervous system to activate or inhibit the activity of neurons. For example, one part of the nervous system can stimulate the neurons responsible for making a muscle contract, whereas another part can inhibit those neurons. Through summation, muscle contraction can be activated if more converging neurons stimulate the production of EPSPs than converging neurons stimulate the production of IPSPs. Conversely, muscle contraction is inhibited if the converging neurons stimulate the production of more IPSPs than EPSPs.

In **divergent pathways,** a smaller number of presynaptic neurons synapse with a larger number of postsynaptic neurons to allow information transmitted in one neuronal pathway to diverge into two or more pathways (figure 11.23*b*). Diverging pathways allow one part of the nervous system to affect more than one other part of the nervous system. For example, sensory input to the central nervous system can go to both the spinal cord and the brain.

Reverberating circuits have a chain of neurons with synapses with previous neurons in the chain, making a positive-feedback loop. This allows action potentials entering the circuit to cause a neuron farther along in the circuit to produce an action potential more than once (figure 11.23*c*). This response, called **after-discharge,** prolongs the response to a stimulus. Once a reverberating circuit is stimulated, it continues to discharge until the synapses involved become fatigued or are inhibited by other neurons. Reverberating circuits play a role in neuronal circuits that control rhythmic activities. Respiration appears to be controlled by a reverberating circuit. In addition, neurons that spontaneously produce action potentials are common in the CNS and may activate reverberating circuits, which remain active awhile and include the sleep-wake cycle.

Parallel after-discharge circuits have neurons that stimulate several neurons in parallel organization, which stimulate a common output cell (figure 11.23*d*). These circuits are involved in complex neuronal processes, including mathematics and chemical conversions. The intricate functions carried out by the CNS are affected by the numerous circuits operating together and influencing the activity of one another.

ASSESS YOUR PROGRESS

57. *Diagram a convergent pathway, a divergent pathway, a reverberating circuit, and a parallel after-discharge circuit, and describe what is accomplished in each.*

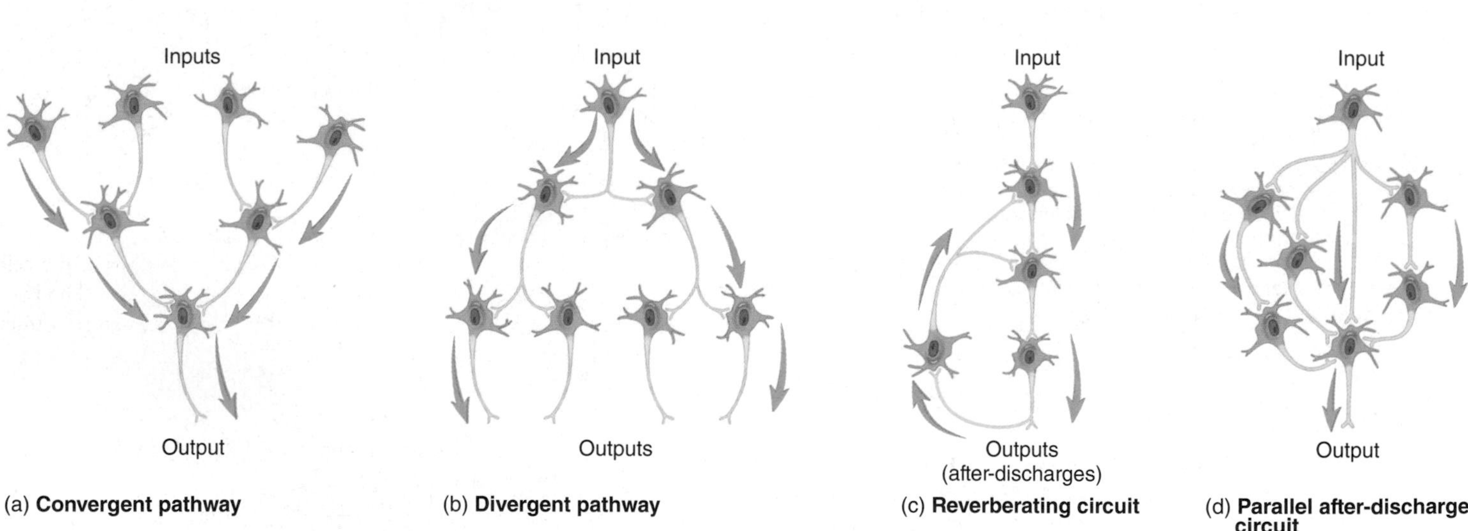
Inputs
Output
(a) Convergent pathway
Input
Outputs
(b) Divergent pathway
Input
Outputs
(after-discharges)
(c) Reverberating circuit
Input
Output
(d) Parallel after-discharge circuit

FIGURE 11.23 Neuronal Pathways and Circuits

The direction of action potential propagation is represented by the *orange arrows.* (*a*) General model of a convergent pathway; many neurons converge and synapse with a smaller number of neurons. (*b*) General model of a divergent pathway; a few neurons synapse with a larger number of neurons. (*c*) Simple model of a reverberating circuit; input action potentials result in the production of a larger number of output action potentials because neurons within the circuit are repeatedly stimulated to produce action potentials. (*d*) Simple model of a parallel after-discharge circuit; several neurons in parallel that integrate complex processes stimulate a common output cell.

Answer

Learn to Predict From page 361

Multiple sclerosis results from the destruction of myelin sheaths around axons of CNS neurons. As a result of her condition, Amanda is experiencing muscle weakness in her legs and left hand. In chapter 9 we learned that skeletal muscle contractions are stimulated by the nervous system and that the amount of muscle tension produced is determined by the frequency of the stimulation. This chapter explained that myelin sheaths function to increase the rate of action potential conduction in neurons and therefore increase the frequency of muscle stimulation. We can then conclude that the destruction of the myelin sheaths reduced the ability of Amanda's brain and spinal cord to communicate with the motor neurons supplying her leg and hand muscles. Because the brain and spinal cord neurons were conducting action potentials more slowly, the motor neurons were not being stimulated as quickly as normal. Therefore, Amanda's muscles could not produce as much tension as they could before the damage to the brain and spinal cord neurons occurred. Amanda's difficulty with climbing the steps suggests that the damaged spinal cord neurons include those that stimulate neurons that innervate muscles of her hip, thigh, and leg.

As the degeneration of myelin sheaths continues, Amanda is likely to experience increasing muscle weakness. Some muscle groups may become so weak that they cannot support Amanda's weight. She may have to use crutches, a cane, or eventually a wheelchair. Other muscle groups, including those involved in swallowing and breathing, will probably be compromised. In addition to motor functions, if Amanda's sensory neurons in her brain and spinal cord are affected, she may experience numbness, pain, and vision problems.

Answers to the rest of this chapter's Predict questions are in Appendix G.

Summary

11.1 Functions of the Nervous System (p. 362)

The nervous system detects external and internal stimuli (sensory input), processes and responds to sensory input (integration), controls body movements through skeletal muscles, maintains homeostasis by regulating other systems, and is the center for mental activities.

11.2 Divisions of the Nervous System (p. 362)

1. The nervous system has two anatomical divisions.
   - The central nervous system (CNS) consists of the brain and spinal cord and is encased in bone.
   - The peripheral nervous system (PNS), the nervous tissue outside of the CNS, consists of sensory receptors, nerves, ganglia, and plexuses.
2. The PNS has two divisions.
   - The sensory division transmits action potentials to the CNS and usually consists of single neurons that have their cell bodies in ganglia.
   - The motor division carries action potentials away from the CNS in cranial or spinal nerves.
3. The motor division has two subdivisions.
   - The somatic nervous system innervates skeletal muscle and is mostly under voluntary control. It consists of single neurons that have their cell bodies located within the CNS.
   - The autonomic nervous system (ANS) innervates cardiac muscle, smooth muscle, and glands. It has two sets of neurons between the CNS and effector organs. The first set has its cell bodies within the CNS, and the second set has its cell bodies within autonomic ganglia.
   - The ANS is subdivided into the sympathetic division, which is most active during physical activity, and the parasympathetic division, which regulates resting functions.
   - The enteric nervous system controls the digestive system.
4. The anatomical divisions perform different functions.
   - The PNS detects stimuli and transmits information to and receives information from the CNS.
   - The CNS processes, integrates, stores, and responds to information from the PNS.

11.3 Cells of the Nervous System (p. 365)

Neurons

1. Neurons receive stimuli and transmit action potentials.
2. Neurons have three components.
   - The cell body is the primary site of protein synthesis.
   - Dendrites are short, branched cytoplasmic extensions of the cell body that usually conduct electrical signals toward the cell body.
   - An axon is a cytoplasmic extension of the cell body that transmits action potentials to other cells.

Types of Neurons

1. Multipolar neurons have several dendrites and a single axon. Interneurons and motor neurons are multipolar.
2. Bipolar neurons have a single axon and dendrite and are components of sensory organs.
3. Pseudo-unipolar neurons have a single axon. Most sensory neurons are pseudo-unipolar.

Neuroglia of the CNS

1. Neuroglia are supportive cells that support and aid the neurons of the CNS and PNS.
2. Astrocytes provide structural support for neurons and blood vessels. Astrocytes influence the functioning of the blood-brain barrier and process substances that pass through it. Astrocytes isolate damaged tissue and limit the spread of inflammation. Astrocytes also help maintain synaptic function.

3. Ependymal cells line the ventricles and the central canal of the spinal cord. Some are specialized to produce cerebrospinal fluid.
4. Microglia phagocytize microorganisms, foreign substances, and necrotic tissue.
5. An oligodendrocyte forms myelin sheaths around the axons of several CNS neurons.

Neuroglia of the PNS

1. A Schwann cell forms a myelin sheath around part of the axon of a PNS neuron.
2. Satellite cells support and nourish neuron cell bodies within ganglia.

Myelinated and Unmyelinated Axons

1. Myelinated axons are wrapped by several layers of plasma membrane from Schwann cells (PNS) or oligodendrocytes (CNS). Spaces between the wrappings are the nodes of Ranvier. Myelinated axons conduct action potentials rapidly.
2. Unmyelinated axons rest in invaginations of Schwann cells (PNS) or oligodendrocytes (CNS). They conduct action potentials slowly.

11.4 Organization of Nervous Tissue (p. 370)

1. Nervous tissue can be grouped into white matter and gray matter.
   - White matter consists of myelinated axons; it propagates action potentials.
   - Gray matter consists of collections of neuron cell bodies or unmyelinated axons. Axons synapse with neuron cell bodies, which are functionally the site of integration in the nervous system.
2. White matter forms nerve tracts in the CNS and nerves in the PNS. Gray matter forms cortex and nuclei in the CNS and ganglia in the PNS.

11.5 Electrical Signals (p. 371)

Electrical properties of cells result from the ionic concentration differences across the plasma membrane and from the permeability characteristics of the plasma membrane.

Concentration Differences Across the Plasma Membrane

1. The sodium-potassium pump moves ions by active transport. Potassium is moved into the cell, and Na^+ is moved out of it.
2. The concentrations of K^+ and negatively charged proteins and other molecules are higher inside the cell, and the concentrations of Na^+ and Cl^- are higher outside the cell.
3. Negatively charged proteins and other negatively charged ions are synthesized inside the cell and cannot diffuse out of it; they repel negatively charged Cl^-.
4. The permeability of the plasma membrane to ions is determined by leak ion channels and gated ion channels.
   - Potassium ion leak channels are more numerous than Na^+ leak channels; thus, the plasma membrane is more permeable to K^+ than to Na^+ when at rest.
   - Gated ion channels in the plasma membrane include ligand-gated ion channels, voltage-gated ion channels, and other gated ion channels.

Establishing the Resting Membrane Potential

1. The resting membrane potential is a charge difference across the plasma membrane when the cell is in an unstimulated condition. The inside of the plasma membrane is negatively charged compared with the outside of the plasma membrane.
2. The resting membrane potential is due mainly to the tendency of positively charged K^+ to diffuse out of the cell, which is opposed by the negative charge that develops inside the plasma membrane.

Changing the Resting Membrane Potential

1. Depolarization, when the inside of the plasma membrane becomes more positive, can result from a decrease in the K^+ concentration gradient, a decrease in membrane permeability to K^+, an increase in membrane permeability to Na^+, an increase in membrane permeability to Ca^{2+}, or a decrease in extracellular Ca^{2+} concentration.
2. Hyperpolarization, when the inside of the plasma membrane becomes more negative, can result from an increase in the K^+ concentration gradient, an increase in membrane permeability to K^+, an increase in membrane permeability to Cl^-, a decrease in membrane permeability to Na^+, or an increase in extracellular Ca^{2+} concentration.

Graded Potentials

1. A graded potential is a small change in the resting membrane potential that is confined to a small area of the plasma membrane.
2. An increase in membrane permeability to Na^+ can cause graded depolarization, and an increase in membrane permeability to K^+ or Cl^- can result in graded hyperpolarization.
3. The term *graded potential* is used because a stronger stimulus produces a greater potential change than a weaker stimulus.
4. Graded potentials can summate, or add together.
5. A graded potential decreases in magnitude as the distance from the stimulation increases.

Action Potentials

1. An action potential is a larger change in the resting membrane potential that spreads over the entire surface of the cell.
2. Threshold is the membrane potential at which a graded potential depolarizes the plasma membrane sufficiently to produce an action potential.
3. Action potentials occur in an all-or-none fashion. If action potentials occur, they are of the same magnitude, no matter how strong the stimulus.
4. Depolarization occurs as the inside of the membrane becomes more positive because Na^+ diffuses into the cell through voltage-gated ion channels.
5. Repolarization is a return of the membrane potential toward the resting state. It occurs because voltage-gated Na^+ channels close and Na^+ diffusion into the cell slows to resting levels and because voltage-gated K^+ channels continue to open and K^+ diffuses out of the cell.
6. The afterpotential is a brief period of hyperpolarization following repolarization.

Refractory Period

1. The absolute refractory period is the time during an action potential when a second stimulus, no matter how strong, cannot initiate another action potential.
2. The relative refractory period follows the absolute refractory period and is the time during which a stronger-than-threshold stimulus can evoke another action potential.

Action Potential Frequency

1. The strength of stimuli affects the frequency of action potentials.
   - A subthreshold stimulus produces only a graded potential.
   - A threshold stimulus causes a graded potential that reaches threshold and results in a single action potential.

- A submaximal stimulus is greater than a threshold stimulus and weaker than a maximal stimulus. The action potential frequency increases as the strength of the submaximal stimulus increases.
- A maximal or a supramaximal stimulus produces a maximum frequency of action potentials.

2. A low frequency of action potentials represents a weaker stimulus than a high frequency.

Propagation of Action Potentials

1. An action potential generates local currents, which stimulate voltage-gated Na^+ channels in adjacent regions of the plasma membrane to open, producing a new action potential.
2. In an unmyelinated axon, action potentials are generated immediately adjacent to previous action potentials.
3. In a myelinated axon, action potentials are generated at successive nodes of Ranvier.
4. Reversal of the direction of action potential propagation is prevented by the absolute refractory period.
5. Action potentials propagate most rapidly in myelinated, large-diameter axons.

11.6 The Synapse (p. 383)

Electrical Synapses

1. Electrical synapses are gap junctions in which tubular proteins called connexons allow local currents to move between cells.
2. At an electrical synapse, an action potential in one cell generates a local current that causes an action potential in an adjacent cell.

Chemical Synapses

1. Anatomically, a chemical synapse has three components.
   - The enlarged ends of the axon are the presynaptic terminals containing synaptic vesicles.
   - The postsynaptic membranes contain receptors for the neurotransmitter.
   - The synaptic cleft is a space separating the presynaptic and postsynaptic membranes.
2. An action potential arriving at the presynaptic terminal causes the release of a neurotransmitter, which diffuses across the synaptic cleft and binds to the receptors of the postsynaptic membrane.
3. The effect of the neurotransmitter on the postsynaptic membrane is stopped in several ways.
   - The neurotransmitter is broken down by an enzyme.
   - The neurotransmitter is taken up by the presynaptic terminal.
   - The neurotransmitter diffuses out of the synaptic cleft.
4. Neurotransmitters are specific for their receptors. A neurotransmitter can be stimulatory in one synapse and inhibitory in another, depending on the type of receptor present.
5. Neuromodulators influence the likelihood that an action potential in a presynaptic terminal will result in an action potential in the membrane of a postsynaptic cell.
6. An excitatory postsynaptic potential (EPSP) is a depolarizing graded potential of the postsynaptic membrane. It can be caused by an increase in membrane permeability to Na^+.
7. An inhibitory postsynaptic potential (IPSP) is a hyperpolarizing graded potential of the postsynaptic membrane. It can be caused by an increase in membrane permeability to K^+ or Cl^-.
8. Presynaptic inhibition decreases neurotransmitter release. Presynaptic facilitation increases neurotransmitter release.

Spatial and Temporal Summation

1. Presynaptic action potentials through neurotransmitters produce graded potentials in postsynaptic neurons. The graded potential can summate to produce an action potential at the trigger zone.
2. Spatial summation occurs when two or more presynaptic terminals simultaneously stimulate a postsynaptic neuron.
3. Temporal summation occurs when two or more action potentials arrive in succession at a single presynaptic terminal.
4. Inhibitory and excitatory presynaptic neurons can converge on a postsynaptic neuron. The activity of the postsynaptic neuron is determined by the integration of the EPSPs and IPSPs produced in the postsynaptic neuron.

11.7 Neuronal Pathways and Circuits (p. 393)

1. Convergent pathways have many neurons synapsing with a few neurons.
2. Divergent pathways have a few neurons synapsing with many neurons.
3. Reverberating circuits have collateral branches of postsynaptic neurons synapsing with presynaptic neurons.
4. Parallel after-discharge circuits have neurons that stimulate several neurons arranged in parallel that stimulate a common output.

REVIEW AND COMPREHENSION

1. The part of the nervous system that controls smooth muscle, cardiac muscle, and glands is the
   a. somatic nervous system.
   b. autonomic nervous system.
   c. skeletal division.
   d. sensory division.
2. Motor neurons and interneurons are ________ neurons.
   a. pseudo-unipolar
   b. bipolar
   c. multipolar
   d. afferent
3. Cells found in the choroid plexuses that secrete cerebrospinal fluid are
   a. astrocytes.
   b. microglia.
   c. ependymal cells.
   d. oligodendrocytes.
   e. Schwann cells.
4. Neuroglia that are phagocytic within the central nervous system are
   a. oligodendrocytes.
   b. microglia.
   c. ependymal cells.
   d. astrocytes.
   e. Schwann cells.
5. Action potentials are conducted more rapidly
   a. in small-diameter axons than in large-diameter axons.
   b. in unmyelinated axons than in myelinated axons.
   c. along axons that have nodes of Ranvier.
   d. All of these are correct.
6. Clusters of neuron cell bodies within the peripheral nervous system are
   a. ganglia.
   b. fasciculi.
   c. nuclei.
   d. laminae.
7. Gray matter contains primarily
   a. myelinated fibers.
   b. neuron cell bodies.
   c. Schwann cells.
   d. oligodendrocytes.

8. Concerning concentration differences across the plasma membrane, there is
   a. more K^+ and Na^+ outside the cell than inside.
   b. more K^+ and Na^+ inside the cell than outside.
   c. more K^+ outside the cell than inside and more Na^+ inside the cell than outside.
   d. more K^+ inside the cell than outside and more Na^+ outside the cell than inside.
9. Compared with the inside of the resting plasma membrane, the outside surface of the membrane is
   a. positively charged.
   b. electrically neutral.
   c. negatively charged.
   d. continuously reversing, so that it is positive one second and negative the next.
   e. negatively charged whenever the sodium-potassium pump is operating.
10. Leak ion channels
    a. open in response to small voltage changes.
    b. open when a chemical signal binds to its receptor.
    c. are responsible for the ion permeability of the resting plasma membrane.
    d. allow substances to move into the cell but not out.
    e. All of these are correct.
11. The resting membrane potential results when the tendency for __________ to diffuse out of the cell is balanced by its attraction to opposite charges inside the cell.
    a. Na^+ c. Cl^-
    b. K^+ d. negatively charged protein
12. If the permeability of the plasma membrane to K^+ increases, the resting membrane potential difference ______________. This is called ______________.
    a. increases, hyperpolarization c. decreases, hyperpolarization
    b. increases, depolarization d. decreases, depolarization
13. Decreasing the extracellular concentration of K^+ affects the resting membrane potential by causing
    a. hyperpolarization.
    b. depolarization.
    c. no change.
14. Which of these terms is correctly matched with its definition or description?
    a. depolarization: membrane potential becomes more negative
    b. hyperpolarization: membrane potential becomes more negative
    c. hypopolarization: membrane potential becomes more negative
15. Which of these statements about ion movement through the plasma membrane is true?
    a. Movement of Na^+ out of the cell requires energy (ATP).
    b. When Ca^{2+} binds to proteins in ion channels, the diffusion of Na^+ into the cell is inhibited.
    c. Specific ion channels regulate the diffusion of Na^+ through the plasma membrane.
    d. All of these are true.
16. The *major* function of the sodium-potassium pump is to
    a. pump Na^+ into and K^+ out of the cell.
    b. generate the resting membrane potential.
    c. maintain the concentration gradients of Na^+ and K^+ across the plasma membrane.
    d. oppose any tendency of the cell to undergo hyperpolarization.
17. Graded potentials
    a. spread over the plasma membrane in decremental fashion.
    b. are not propagated for long distances.
    c. are confined to a small region of the plasma membrane.
    d. can summate.
    e. All of these are correct.
18. During the depolarization phase of an action potential, the permeability of the membrane
    a. to K^+ is greatly increased.
    b. to Na^+ is greatly increased.
    c. to Ca^{2+} is greatly increased.
    d. is unchanged.
19. During repolarization of the plasma membrane,
    a. Na^+ diffuses into the cell.
    b. Na^+ diffuses out of the cell.
    c. K^+ diffuses into the cell.
    d. K^+ diffuses out of the cell.
20. The absolute refractory period
    a. limits how many action potentials can be produced during a given period of time.
    b. prevents an action potential from starting another action potential at the same point on the plasma membrane.
    c. is the period of time when a strong stimulus can initiate a second action potential.
    d. Both a and b are correct.
    e. All of these are correct.
21. A subthreshold stimulus
    a. produces an afterpotential.
    b. produces a graded potential.
    c. causes an all-or-none response.
    d. produces more action potentials than a submaximal stimulus.
22. Neurotransmitter substances are stored in vesicles located in specialized portions of the
    a. neuron cell body.
    b. axon.
    c. dendrite.
    d. postsynaptic membrane.
23. In a chemical synapse,
    a. action potentials in the presynaptic terminal cause voltage-gated Ca^{2+} channels to open.
    b. neurotransmitters can cause ligand-gated Na^+ channels to open.
    c. neurotransmitters can be broken down by enzymes.
    d. neurotransmitters can be taken up by the presynaptic terminal.
    e. All of these are correct.
24. An inhibitory presynaptic neuron can affect a postsynaptic neuron by
    a. producing an IPSP in the postsynaptic neuron.
    b. hyperpolarizing the plasma membrane of the postsynaptic neuron.
    c. causing K^+ to diffuse out of the postsynaptic neuron.
    d. causing Cl^- to diffuse into the postsynaptic neuron.
    e. All of these are correct.
25. Summation
    a. is caused by combining two or more graded potentials.
    b. occurs at the trigger zone of the postsynaptic neuron.
    c. results in an action potential if it reaches the threshold potential.
    d. can occur when two action potentials arrive in close succession at a single presynaptic terminal.
    e. All of these are correct.

26. In convergent pathways,
    a. the response of the postsynaptic neuron depends on the summation of EPSPs and IPSPs.
    b. a smaller number of presynaptic neurons synapse with a larger number of postsynaptic neurons.
    c. information transmitted in one neuronal pathway can go into two or more pathways.
    d. All of these are correct.

Answers in Appendix E

CRITICAL THINKING

1. A child eats a whole bottle of salt (NaCl) tablets. What effect does this have on action potentials?
2. Some smooth muscle has the ability to contract spontaneously—that is, it contracts without any external stimulation. Propose an explanation for this ability based on what you know about membrane potentials. Assume that an action potential in a smooth muscle cell causes it to contract.
3. Assume that two nerve fibers have the same diameter, but one is myelinated and the other is unmyelinated. The conduction of an action potential is most energy-efficient along which type of fiber? (*Hint:* required ATP.)
4. Explain the consequences when an inhibitory neuromodulator is released from a presynaptic terminal and a stimulatory neurotransmitter is released from another presynaptic terminal, both of which synapse with the same neuron.
5. The speed of action potential propagation and synaptic transmission decreases with aging. List possible explanations.
6. Students in a veterinary school are given the following hypothetical problem: A dog ingests organophosphate poison, and the students are responsible for saving the animal's life. Organophosphate poisons bind to and inhibit acetylcholinesterase. Several substances the students could inject include the following: acetylcholine, curare (which blocks acetylcholine receptors), and potassium chloride. If you were a student in the class, what would you advise to save the animal?
7. Strychnine blocks receptor sites for inhibitory neurotransmitter substances in the CNS. Explain how strychnine can produce tetanus in skeletal muscles.
8. Alcohol affects the central nervous system by enhancing the effect of GABA at its receptor. GABA binds to GABA receptors and opens ligand-gated Cl^- channels. However, chronic consumption of alcohol makes the GABA receptor less sensitive to both alcohol and GABA, which increases alcohol dependence as well as alcohol withdrawal symptoms, such as anxiety, tremors, and insomnia. Benzodiazepines enhance the binding of GABA molecules to their receptors and thus are sometimes used to treat people with alcohol withdrawal symptoms. For synapses involving GABA, predict the effect of alcohol on the postsynaptic membranes; compare the effect of chronic alcohol consumption on the postsynaptic membranes in these synapses; and predict the effect of benzodiazepine treatment on the degree of polarization of postsynaptic membranes in people who are experiencing alcohol withdrawal symptoms.
9. The venom of many cobras contains a potent neurotoxin that binds to ligand-gated Na^+ channels, causing them to open. Unlike ACh, which binds to and then rapidly unbinds from ligand-gated Na^+ channels, the neurotoxin tends to remain bound to ligand-gated Na^+ channels. How does this neurotoxin affect the nervous system's ability to stimulate skeletal muscle contraction? How does it affect the ability of skeletal muscle fibers to respond to stimulation?
10. Epilepsy is a chronic disorder characterized by seizures that result when neurons in the brain produce excessive action potentials. Channelopathies are genetic disorders caused by mutations in ion channel genes, which result in ion channels that do not function normally. What kind of Na^+ or Ca^{2+} channelopathies might contribute to epilepsy?

Answers in Appendix F

Mc Graw Hill
connect plus+
ANATOMY & PHYSIOLOGY

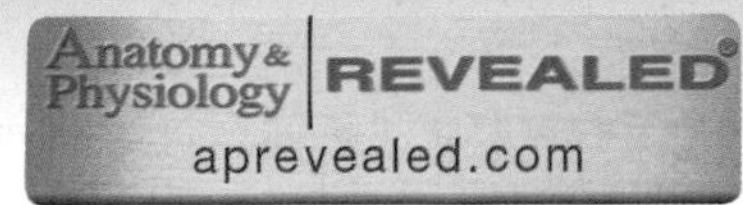
Anatomy & Physiology
REVEALED
aprevealed.com

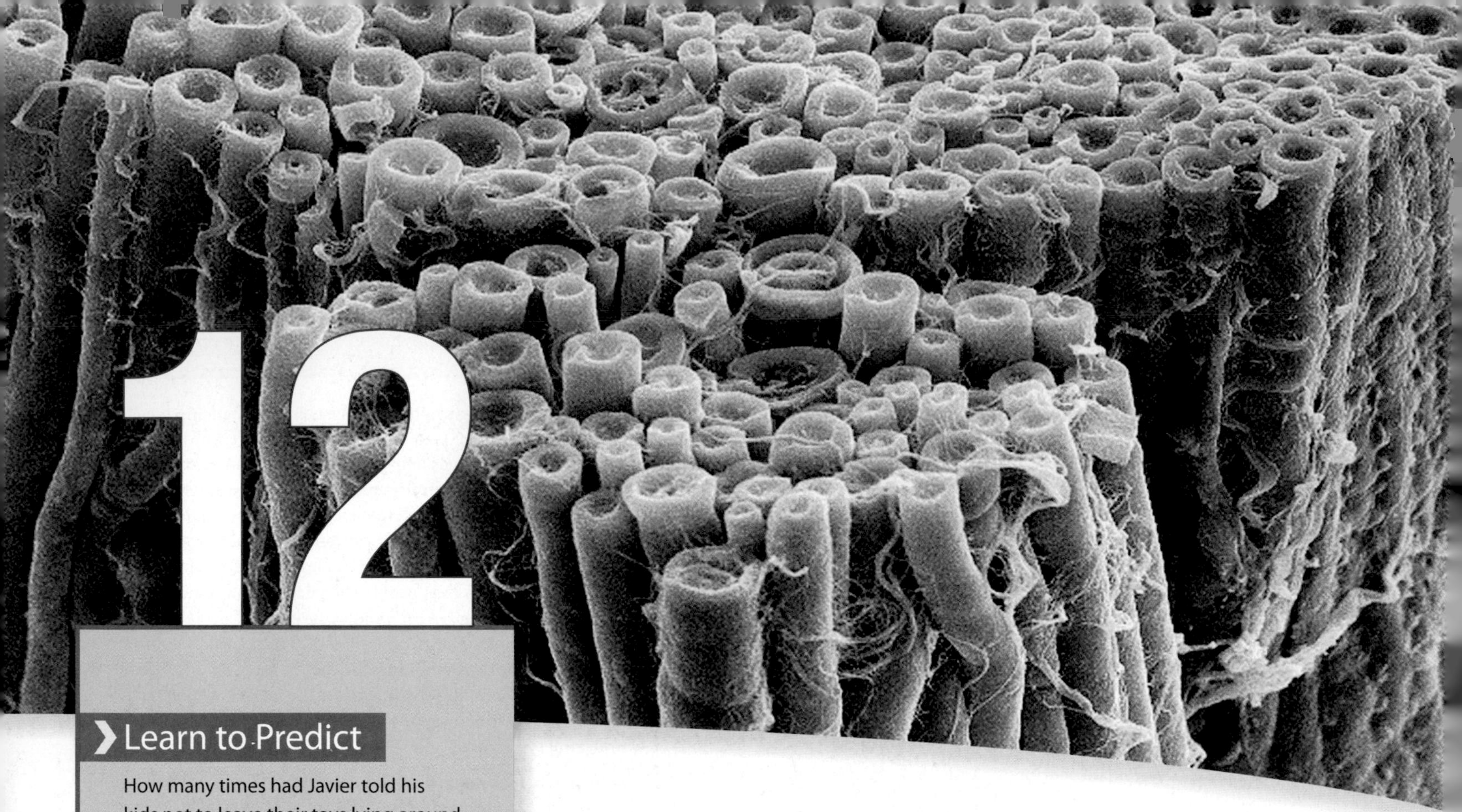

12

Learn to Predict

How many times had Javier told his kids not to leave their toys lying around on the floor? While walking down a dark hallway in the middle of the night, Javier stepped on a tiny race car with his right foot. He immediately withdrew his foot from the painful stimulus but then stepped on a second car with his left foot. Fortunately, he was able to shift his weight back to his other foot, therefore preventing a fall. After reading this chapter, explain the spinal reflexes that kept Javier on his feet. Describing those reflexes will require your knowledge of nervous system organization from chapter 11.

Photo: Colorized SEM of bundles of axons in a nerve.

Module 7
Nervous System

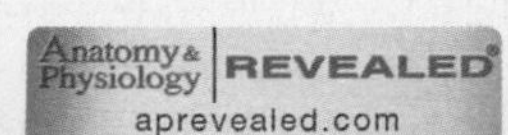
Anatomy & Physiology REVEALED
aprevealed.com

Spinal Cord and Spinal Nerves

Each year, over 10,000 people, usually males in their late teens and twenties, suffer spinal cord injuries. As we learn more about the spinal cord, there is growing hope for improved therapies for these injuries. The spinal cord and the associated spinal nerves play a central role in communication between the brain and the rest of the body.

Before focusing on the functions of the spinal cord and spinal nerves, it is important to understand their relationship to the nervous system as a whole. Recall from chapter 11 that the central nervous system (CNS) consists of the brain and spinal cord, whereas the peripheral nervous system (PNS) consists of the nerves and ganglia outside the CNS. The PNS includes 12 pairs of cranial nerves and 31 pairs of spinal nerves and their ganglia. The spinal cord is part of the CNS, and the spinal nerves are part of the PNS, so in this chapter we begin to see how the two systems work together. The PNS collects information from numerous sources both inside and outside the body and relays it through axons of sensory neurons to the CNS. The CNS receives this sensory information, integrates and evaluates the information, stores some information, and initiates reactions. Axons of motor neurons in the PNS relay information from the CNS to various parts of the body, primarily to muscles and glands, thereby regulating activity in those structures.

The spinal cord and spinal nerves are described in this chapter. The brain and cranial nerves are covered in chapter 13.

FUNDAMENTAL **Figure**

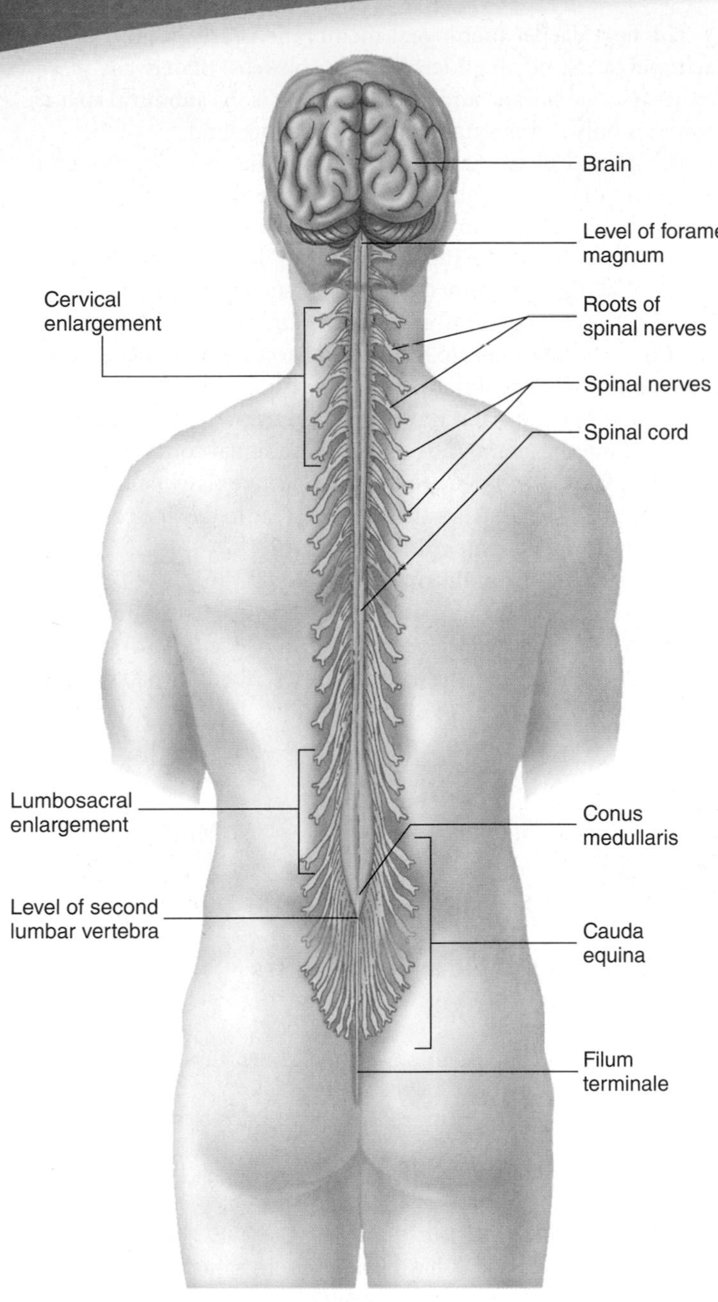
Brain
Level of foramen magnum
Cervical enlargement
Roots of spinal nerves
Spinal nerves
Spinal cord
Lumbosacral enlargement
Conus medullaris
Level of second lumbar vertebra
Cauda equina
Filum terminale
Posterior view

FIGURE 12.1 AP|R **Spinal Cord and Spinal Nerve Roots**

12.1 Spinal Cord

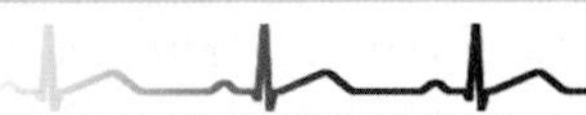

LEARNING OUTCOMES

After reading this section, you should be able to

A. Describe the general structure of the spinal cord.

B. Name the meninges (sing. meninx) and their related spaces surrounding the spinal cord.

C. Draw and label a cross section of the spinal cord with its dorsal and ventral nerve roots.

The **spinal cord** is the major communication link between the brain and the PNS inferior to the head. It integrates incoming information and produces responses through reflex mechanisms.

General Structure

The spinal cord extends from the foramen magnum to the level of the second lumbar vertebra (figure 12.1). It is considerably shorter than the vertebral column because it does not grow as rapidly during development. The spinal cord is composed of cervical, thoracic, lumbar, and sacral segments, named according to the portion of the vertebral column from which their nerves enter and exit. The spinal cord gives rise to 31 pairs of **spinal nerves,** which exit the vertebral column through intervertebral and sacral foramina (see figure 7.13). Each spinal nerve is a bundle of axons, Schwann cells, and connective tissue sheaths. The nerves from the lower segments descend some distance in the vertebral canal before they exit because the spinal cord is shorter than the vertebral column.

The spinal cord is larger in diameter at its superior end, and it gradually decreases in diameter toward its inferior end. Two enlargements occur where nerves supplying the upper and lower limbs enter and leave the spinal cord. The **cervical enlargement** in the inferior cervical region corresponds to the location where nerve fibers that supply the upper limbs enter and leave the spinal cord (figure 12.1). The **lumbosacral enlargement** in the inferior thoracic, lumbar, and superior sacral regions is the site where the nerve fibers supplying the lower limbs enter or leave the spinal cord.

Immediately inferior to the lumbosacral enlargement, the spinal cord tapers to form a conelike region called the **conus medullaris** (figure 12.1). Its tip is the inferior end of the spinal cord and extends to the level of the second lumbar vertebra. The nerves supplying the lower limbs and other inferior structures of the body arise from the lumbar and sacral regions. They exit the lumbosacral enlargement and conus medullaris, course inferiorly through the vertebral canal, and exit through the intervertebral and sacral foramina from the second lumbar to the fifth sacral vertebrae. The numerous roots (origins) of spinal nerves extending inferiorly from the lumbosacral enlargement and conus medullaris resemble a horse's tail and are therefore called the **cauda** (kaw′dă; tail) **equina** (ē-kwī′nă; horse; figure 12.1).

Meninges of the Spinal Cord

The spinal cord and brain are surrounded by connective tissue membranes called **meninges** (mĕ-nin′jēz; figure 12.2). The most superficial and thickest membrane is the **dura mater** (doo′ră mā′ter; tough mother). The dura mater forms a sac, often called the **thecal** (the′kal) **sac,** which surrounds the spinal cord. The thecal sac attaches to the rim of the foramen magnum and ends at the level of the second sacral vertebra. The spinal dura mater is continuous with the dura mater surrounding the brain and the connective tissue surrounding the spinal nerves. The dura mater around the spinal cord is separated from the periosteum of the vertebral canal by the **epidural space.** This is a true space between the walls of the vertebral canal and the dura mater of the spinal cord that contains spinal nerve roots, blood vessels, areolar connective tissue, and adipose tissue.

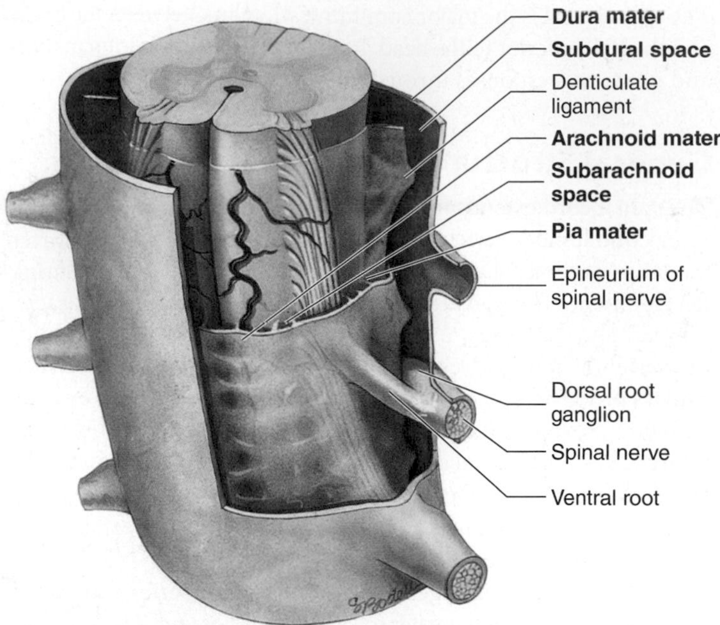
Dura mater
Subdural space
Denticulate ligament
Arachnoid mater
Subarachnoid space
Pia mater
Epineurium of spinal nerve
Dorsal root ganglion
Spinal nerve
Ventral root

(a) **Anterolateral view**

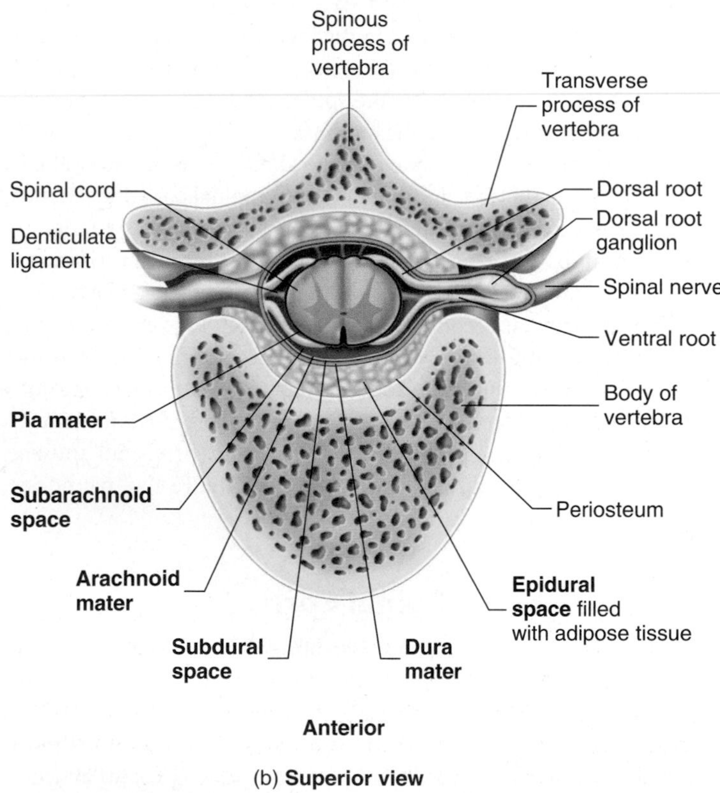
Posterior
Spinous process of vertebra
Transverse process of vertebra
Spinal cord
Denticulate ligament
Dorsal root
Dorsal root ganglion
Spinal nerve
Ventral root
Body of vertebra
Pia mater
Subarachnoid space
Periosteum
Arachnoid mater
Epidural space filled with adipose tissue
Subdural space
Dura mater
Anterior

(b) **Superior view**

FIGURE 12.2 AP|R **Meningeal Membranes Surrounding the Spinal Cord**

In contrast, the epidural space around the brain is only a potential space. **Epidural anesthesia** of the spinal nerves is often induced in women during childbirth by injecting anesthetics into the epidural space of the spinal cord.

The next deeper meningeal membrane is a very thin, wispy **arachnoid** (ă-rak′noyd; spiderlike—i.e., cobwebs) **mater.** The space between this membrane and the dura mater is the **subdural space;** it contains only a very small amount of serous fluid.

The third, deepest meningeal layer, the **pia** (pī′ă; affectionate) **mater** is bound very tightly to the surface of the spinal cord. Holding the spinal cord in place within the thecal sac are the denticulate ligaments and the filum terminale. The paired **denticulate** (den-tik′ū-lāt) **ligaments** are connective tissue septa extending from the lateral sides of the spinal cord to the dura mater (figure 12.2*b*). The term *denticulate* refers to having small teeth, and the denticulate ligaments attach to the dura mater by toothlike processes between the exits of the cervical and thoracic spinal nerves. The denticulate ligaments limit the lateral movement of the spinal cord. The **filum terminale** (fī′lŭm ter′mi-nal′ē) is a connective tissue strand that anchors the conus medullaris and the thecal sac to the first coccygeal vertebra, limiting their superior movement.

Between the arachnoid mater and the pia mater is the **subarachnoid space,** which contains weblike strands of the arachnoid mater, blood vessels, and **cerebrospinal** (ser′ĕ-brō-spī-năl, sĕ-rē′brō-spī-nal) **fluid (CSF),** which is described in chapter 13.

Clinical IMPACT

Introduction of Needles into the Subarachnoid Space

Several clinical procedures involve inserting a needle into the subarachnoid space at either the L3/L4 or L4/L5 level. The needle does not puncture the spinal cord because the cord extends only approximately to the second lumbar vertebra of the vertebral column, but the subarachnoid space extends to level S2 of the vertebral column. Nor does the needle damage the nerve roots of the cauda equina located in the subarachnoid space because the needle quite easily pushes them aside. In **spinal anesthesia,** or spinal block, drugs that block action potential transmission are introduced into the subarachnoid space to prevent pain sensations in the lower half of the body. There are advantages and disadvantages to spinal versus epidural anesthesia. In spinal anesthesia, the drugs are delivered directly to the CSF, so the anesthesia is generally stronger and takes effect faster than epidural anesthesia. With epidural anesthesia, the needle does not penetrate the dura mater, so the drugs must first diffuse into the CSF. However, an advantage is that the drugs can be readministered via a catheter (flexible tube) to maintain longer anesthesia. In some instances, a combination of spinal and epidural anesthesia is used. In a **lumbar puncture,** or *spinal tap,* CSF is removed from the subarachnoid space in order to examine it for infectious agents (meningitis) or for the presence of blood (hemorrhage) or to measure the CSF pressure. Sometimes clinicians inject a radiopaque substance into this area and take a **myelogram** (radiograph of the spinal cord) to visualize spinal cord defects or damage.

FUNDAMENTAL **Figure**

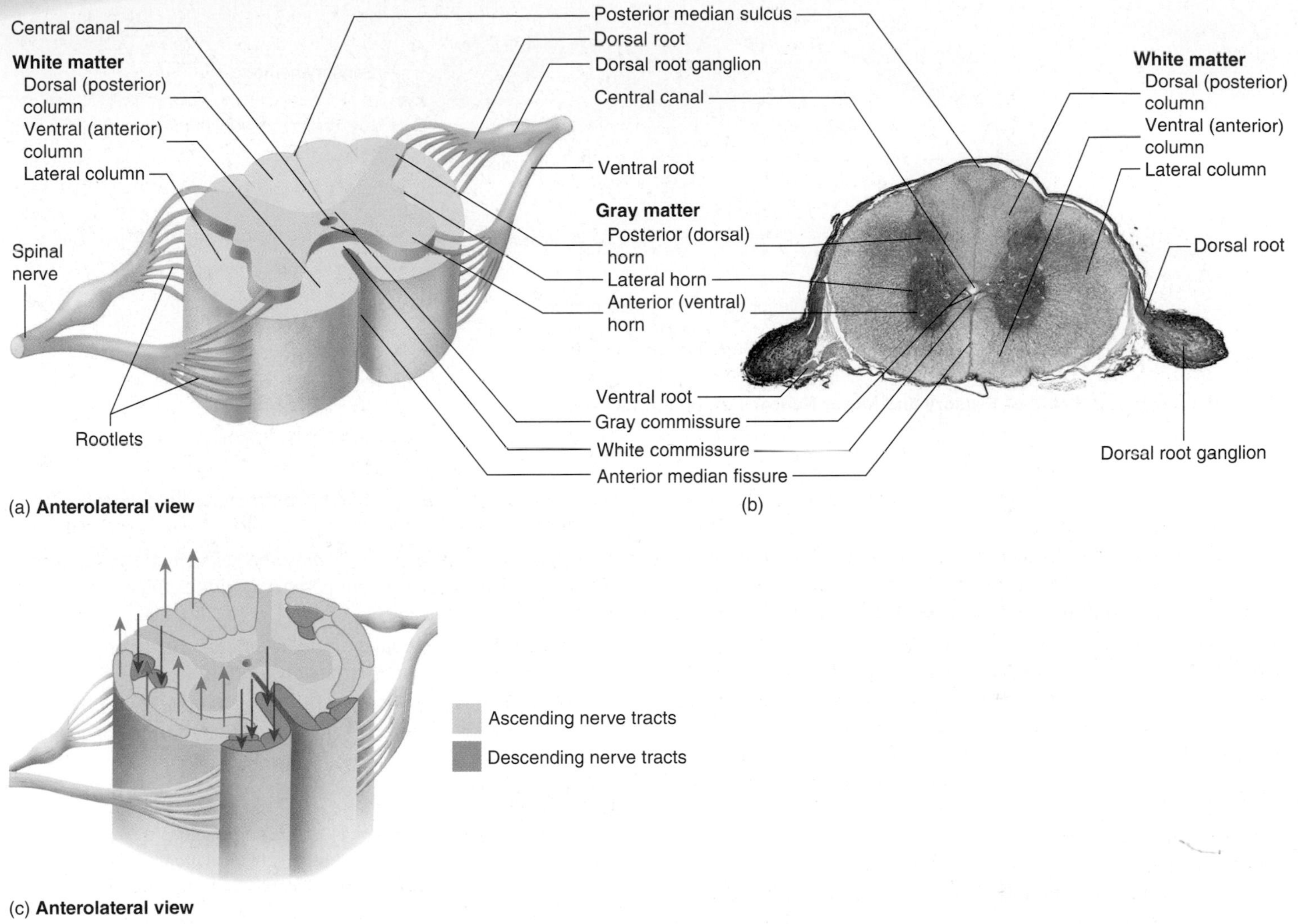

Central canal
White matter
Dorsal (posterior) column
Ventral (anterior) column
Lateral column
Spinal nerve
Rootlets
Posterior median sulcus
Dorsal root
Dorsal root ganglion
Central canal
Ventral root
Gray matter
Posterior (dorsal) horn
Lateral horn
Anterior (ventral) horn
Ventral root
Gray commissure
White commissure
Anterior median fissure
White matter
Dorsal (posterior) column
Ventral (anterior) column
Lateral column
Dorsal root
Dorsal root ganglion
(a) Anterolateral view
(b)
Ascending nerve tracts
Descending nerve tracts
(c) Anterolateral view

FIGURE 12.3 AP|R **Cross Section of the Spinal Cord**

(*a*) A segment of the spinal cord showing one dorsal and one ventral root on each side and the rootlets that form them. (*b*) Photograph of a cross section through the midlumbar region. The lighter tan *areas* are white matter, where myelinated tracts are located. The darker area is gray matter, where neuron cell bodies are located. (*c*) Ascending and descending tracts in the spinal cord. Ascending nerve tracts are green; descending nerve tracts are purple. The arrows indicate the direction of action potential propagation in each pathway.

Cross Section of the Spinal Cord

A cross section reveals that the spinal cord consists of a superficial white portion and a deep gray portion (figure 12.3*a*,*b*). The white matter consists of myelinated axons, which form nerve tracts, and the gray matter consists of neuron cell bodies, dendrites, and axons. An **anterior median fissure** and a **posterior median sulcus** are deep clefts partially separating the two halves of the cord. The white matter in each half of the spinal cord is organized into three **columns,** or *funiculi* (fū-nik′ū-lī), called the **ventral** (anterior), **dorsal** (posterior), and **lateral columns.** Each column of the spinal cord is subdivided into **tracts,** or *fasciculi* (fă-sik′ū-lī), also referred to as *pathways.* A collection of axons *inside* the CNS is called a tract, whereas *outside* the CNS it is called a *nerve.* Tracts have different myelination than nerves and lack the extensive connective tissue of nerves. Individual axons ascending to the brain or descending from the brain are usually grouped together within the tracts (figure 12.3*c*). Axons within a given tract carry basically the same type of information, although they may overlap to some extent. For example, one ascending tract carries action potentials related to pain and temperature sensations, whereas another carries action potentials related to light touch.

The central gray matter is organized into horns. Each half of the central gray matter of the spinal cord consists of a relatively thin **posterior** (dorsal) **horn** and a larger **anterior** (ventral) **horn.** Small **lateral horns** exist in the levels of the cord associated with the autonomic nervous system (see chapter 16). The two halves of the spinal cord are connected by **gray** and **white commissures** (figure 12.3*a*,*b*). The gray and white commissures contain axons that cross from one side of the spinal cord to the other. The **central canal,**

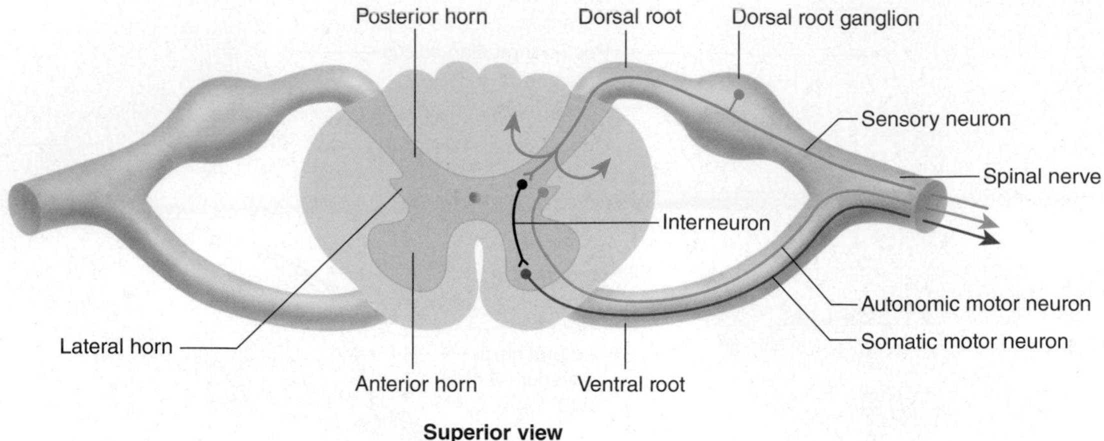
Posterior horn
Dorsal root
Dorsal root ganglion
Sensory neuron
Spinal nerve
Interneuron
Autonomic motor neuron
Somatic motor neuron
Lateral horn
Anterior horn
Ventral root
Superior view

FIGURE 12.4 Relationship of Sensory and Motor Neurons to the Spinal Cord

located in the center of the gray commissure, helps circulate CSF associated with the ventricular system (see chapter 13).

Spinal nerves arise from numerous rootlets along the dorsal and ventral surfaces of the spinal cord (figure 12.3*a*). Six to eight of these rootlets combine to form a **ventral root** on the ventral (anterior) side of the spinal cord, and another six to eight form a **dorsal root** on the dorsal (posterior) side of the cord at each segment. The ventral and dorsal roots extend laterally from the spinal cord, passing through the subarachnoid space, piercing the arachnoid mater and dura mater, and joining one another to form a spinal nerve. Each dorsal root contains a ganglion, called the **dorsal root ganglion,** or *spinal ganglion* (gang′glē-on; a swelling or knot).

The dorsal root ganglia are collections of cell bodies of the sensory neurons forming the dorsal roots of the spinal nerves (figure 12.4). The axons of these pseudo-unipolar neurons extend from various parts of the body and pass through spinal nerves to the dorsal root ganglia. The axons do not synapse in the dorsal root ganglion but pass through the dorsal root into the posterior horn of the spinal cord gray matter. The axons either synapse with interneurons in the posterior horn or pass into the white matter and ascend or descend in the spinal cord.

Motor neurons innervate muscles and glands. The cell bodies of the motor neurons are located in the anterior and lateral horns of the spinal cord gray matter (figure 12.4). The cell bodies of multipolar somatic motor neurons are in the anterior horn, also called the motor horn, and autonomic motor neuron cell bodies are in the lateral horn. Axons of the motor neurons form the ventral roots and pass into the spinal nerves. Thus, dorsal roots contain sensory axons, ventral roots contain motor axons, and spinal nerves have both sensory and motor axons.

ASSESS YOUR PROGRESS

1. *Where does the spinal cord begin and end? How many pairs of spinal nerves exit the spinal cord?*
2. *Name the meninges surrounding the spinal cord. What is found within each of these spaces: the epidural space, the subdural space, and the subarachnoid space?*
3. *What is the thecal sac? What two structures hold the thecal sac in the vertebral canal?*
4. *Describe the arrangement of gray and white matter in the spinal cord. What are the divisions of the gray matter and white matter? What are commissures?*
5. *Where are the cell bodies of somatic motor and autonomic motor neurons located in the gray matter?*
6. *What kinds of neurons are in the dorsal roots, in the ventral roots, and in the spinal nerves? What is found in the dorsal root ganglion?*

Predict 2

Explain why the dorsal root ganglia are larger in diameter than the dorsal roots, and describe the direction of action potential propagation in the spinal nerves, dorsal roots, and ventral roots.

12.2 Reflexes

LEARNING OUTCOMES

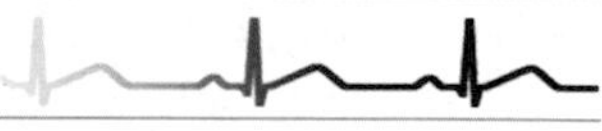

After reading this section, you should be able to

A. Describe the components of a monosynaptic and a polysynaptic reflex arc.

B. Compare and contrast the features of a stretch reflex, a Golgi tendon reflex, a withdrawal reflex, and a crossed extensor reflex.

The basic structural unit of the nervous system is the **neuron** (see section 11.3). The basic functional unit of the nervous system is the **reflex arc** because it is the smallest, simplest portion capable of receiving a stimulus and producing a response. Therefore, it is possible to learn much about nervous system functions by examining how reflex arcs receive stimuli and produce responses. The reflex arc generally has five basic components: (1) a sensory receptor, (2) a sensory neuron, (3) an interneuron, (4) a motor neuron, and (5) an effector organ (figure 12.5). The simplest reflex arcs do not involve interneurons.

FUNDAMENTAL **Figure**

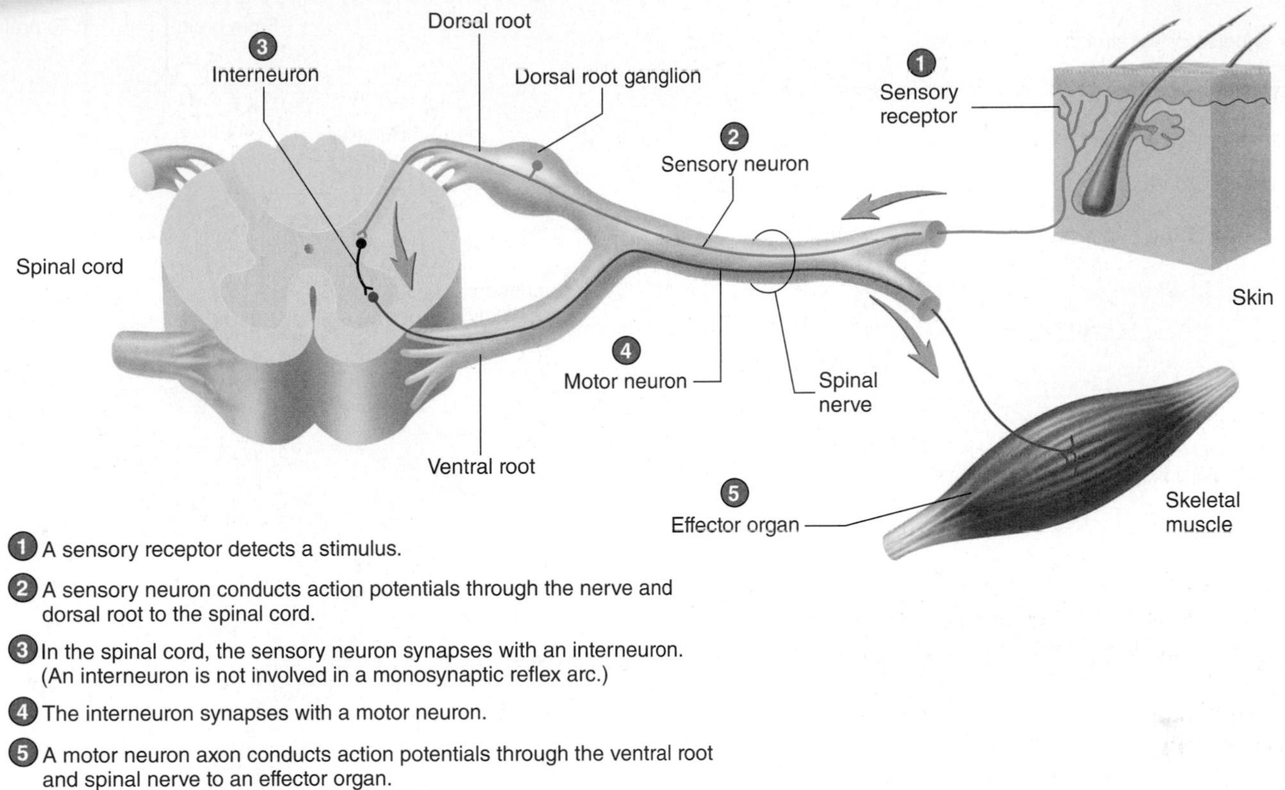

Dorsal root
3
Interneuron
Dorsal root ganglion
1
Sensory receptor
2
Sensory neuron
Spinal cord
Skin
4
Motor neuron
Spinal nerve
Ventral root
5
Effector organ
Skeletal muscle

1 A sensory receptor detects a stimulus.

2 A sensory neuron conducts action potentials through the nerve and dorsal root to the spinal cord.

3 In the spinal cord, the sensory neuron synapses with an interneuron. (An interneuron is not involved in a monosynaptic reflex arc.)

4 The interneuron synapses with a motor neuron.

5 A motor neuron axon conducts action potentials through the ventral root and spinal nerve to an effector organ.

PROCESS FIGURE 12.5 AP|R **Reflex Arc**
The parts of a reflex arc are labeled in the order in which action potentials pass through them. A polysynaptic reflex arc containing an interneuron is shown.

A **reflex** is an automatic response to a stimulus produced by a reflex arc. It occurs without conscious thought. Action potentials initiated in sensory receptors are transmitted along the axons of sensory neurons to the CNS, where the axons usually synapse with interneurons. Interneurons synapse with motor neurons, which send axons out of the spinal cord and through the PNS to muscles or glands, where the action potentials of the motor neurons cause these effector organs to respond.

Reflexes are homeostatic. For example, a number of reflexes, called autonomic reflexes, are responsible for maintaining relatively constant blood pressure, blood carbon dioxide levels, and water intake. Other reflexes, called somatic reflexes, remove the body from painful stimuli that would cause tissue damage or keep the body from suddenly falling or moving because of external forces.

Individual reflexes vary in complexity. **Monosynaptic reflexes** involve simple neuronal pathways in which sensory neurons synapse directly with motor neurons. **Polysynaptic reflexes** involve more complex pathways and integrative centers, with one or more interneurons between the sensory and motor neurons (figure 12.5).

Many reflexes are integrated within the spinal cord, and others are integrated within the brain. Some reflexes involve excitatory neurons and result in a response, as when a muscle contracts (see chapter 11). Other reflexes involve inhibitory neurons and result in the inhibition of a response, as when a muscle relaxes. In addition, higher brain centers influence reflexes by either suppressing or exaggerating them. The subsequent discussion describes three major spinal cord reflexes: the stretch reflex, the Golgi tendon reflex, and the withdrawal reflex.

Stretch Reflex

The simplest reflex is the **stretch reflex** (figure 12.6), in which muscles contract in response to a stretching force applied to them. The sensory receptor of this reflex is the **muscle spindle,** which consists of 3–10 small, specialized skeletal muscle fibers. The fibers are contractile only at their ends and are innervated by specific motor neurons called **gamma motor neurons** (the term *gamma* refers to motor neurons with small-diameter axons) originating from the spinal cord and controlling the sensitivity of the muscle spindle cells. Sensory neurons innervate the noncontractile centers of the muscle spindle cells. Axons of these sensory neurons extend to the spinal cord and synapse directly with motor neurons in the spinal cord called **alpha motor neurons,** which in turn innervate the muscle in which the muscle spindle is embedded. Neurons can be classified by the diameter of their axons, with alpha motor neurons having the largest diameter. The stretch reflex is a monosynaptic reflex because there is no interneuron between the sensory neuron and the alpha motor neuron.

Stretching a muscle also stretches muscle spindles located among the muscle fibers. The stretch stimulates the sensory neurons that innervate the center of each of the muscle spindles. The increased

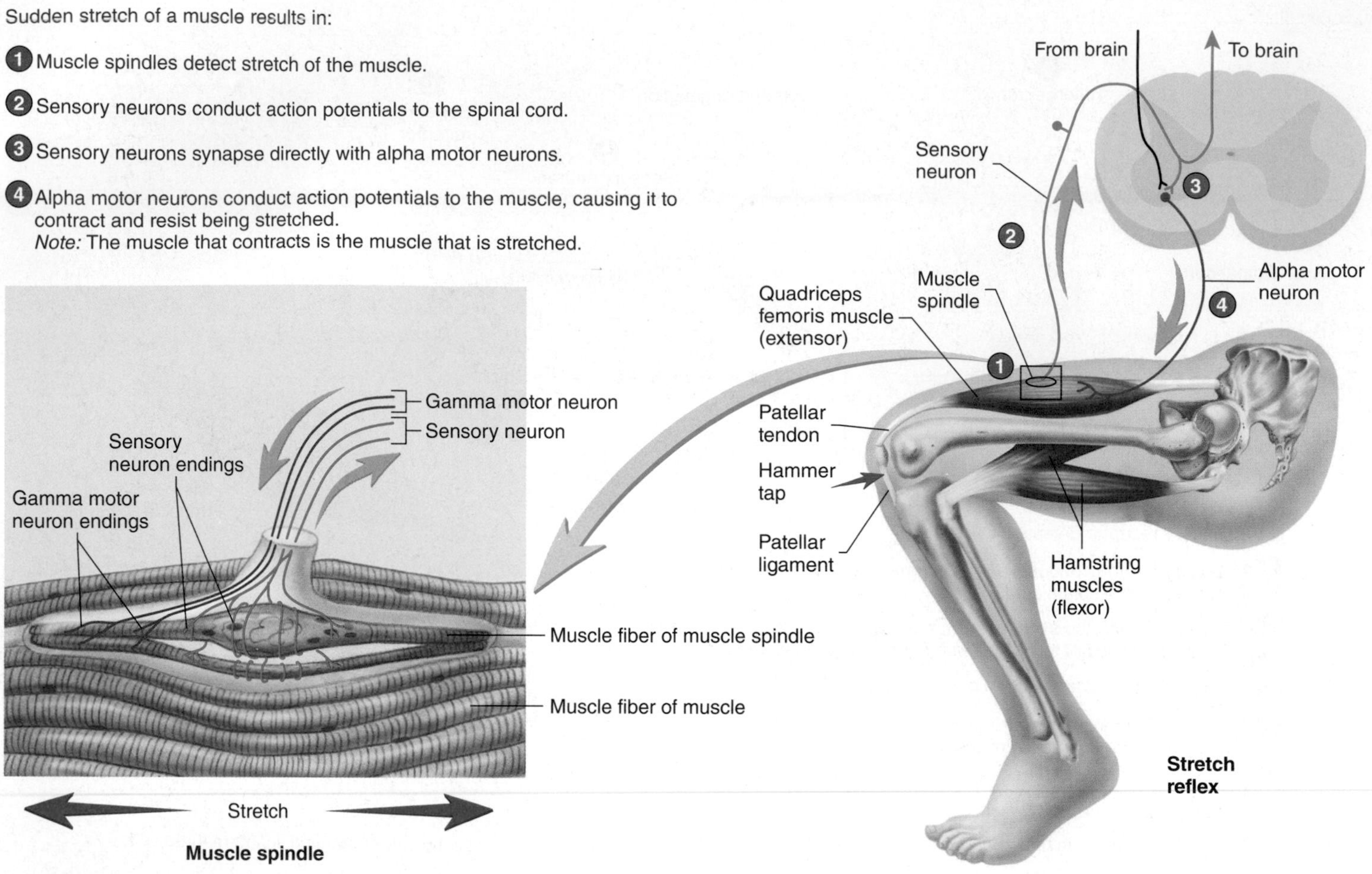

Sudden stretch of a muscle results in:

1. Muscle spindles detect stretch of the muscle.
2. Sensory neurons conduct action potentials to the spinal cord.
3. Sensory neurons synapse directly with alpha motor neurons.
4. Alpha motor neurons conduct action potentials to the muscle, causing it to contract and resist being stretched.
Note: The muscle that contracts is the muscle that is stretched.

From brain
To brain
Sensory neuron
Alpha motor neuron
Muscle spindle
Quadriceps femoris muscle (extensor)
Gamma motor neuron
Sensory neuron
Sensory neuron endings
Gamma motor neuron endings
Patellar tendon
Hammer tap
Patellar ligament
Hamstring muscles (flexor)
Muscle fiber of muscle spindle
Muscle fiber of muscle
Stretch
Muscle spindle
Stretch reflex

PROCESS FIGURE 12.6 Stretch Reflex

frequency of action potentials carried to the spinal cord by sensory neurons stimulates the alpha motor neurons in the spinal cord. The alpha motor neurons transmit action potentials to skeletal muscle, causing a rapid contraction of the stretched muscle, which opposes the stretch of the muscle. The postural muscles demonstrate the adaptive nature of this reflex. If a person is standing upright and then begins to tip slightly to one side, the postural muscles associated with the vertebral column on the other side are stretched. As a result, stretch reflexes are initiated in those muscles, which cause them to contract and reestablish normal posture.

Collateral axons from the sensory neurons of the muscle spindles also synapse with neurons whose axons contribute to ascending nerve tracts, which enable the brain to perceive that a muscle has been stretched (see figure 12.3*c*). Descending neurons within the spinal cord synapse with the neurons of the stretch reflex modifying their activity. This activity is important in maintaining posture and in coordinating muscle actions.

Gamma motor neurons are responsible for regulating the sensitivity of the muscle spindles. As a skeletal muscle contracts, the tension on the centers of muscle spindles within the muscle decreases because the muscle spindles passively shorten as the muscle shortens. The decrease in tension in the centers of the muscle

Clinical
IMPACT

Knee-Jerk Reflex

The **knee-jerk reflex,** or *patellar reflex,* is a classic example of the stretch reflex. Clinicians use this reflex to determine whether the higher CNS centers that normally influence this reflex are functional. When the patellar ligament is tapped, the tendons and muscles of the quadriceps femoris muscle group stretch. The muscle spindle fibers within these muscles also stretch, and the stretch reflex is activated. Consequently, contraction of the muscles extends the leg, producing the characteristic knee-jerk response. A greatly exaggerated stretch reflex indicates that the neurons within the brain that innervate the gamma motor neurons and enhance the stretch reflex are overly active. On the other hand, if the neurons that innervate the gamma motor neurons are depressed, the stretch reflex may be suppressed or absent. Absence of the stretch reflex may also indicate that the reflex pathway is not intact.

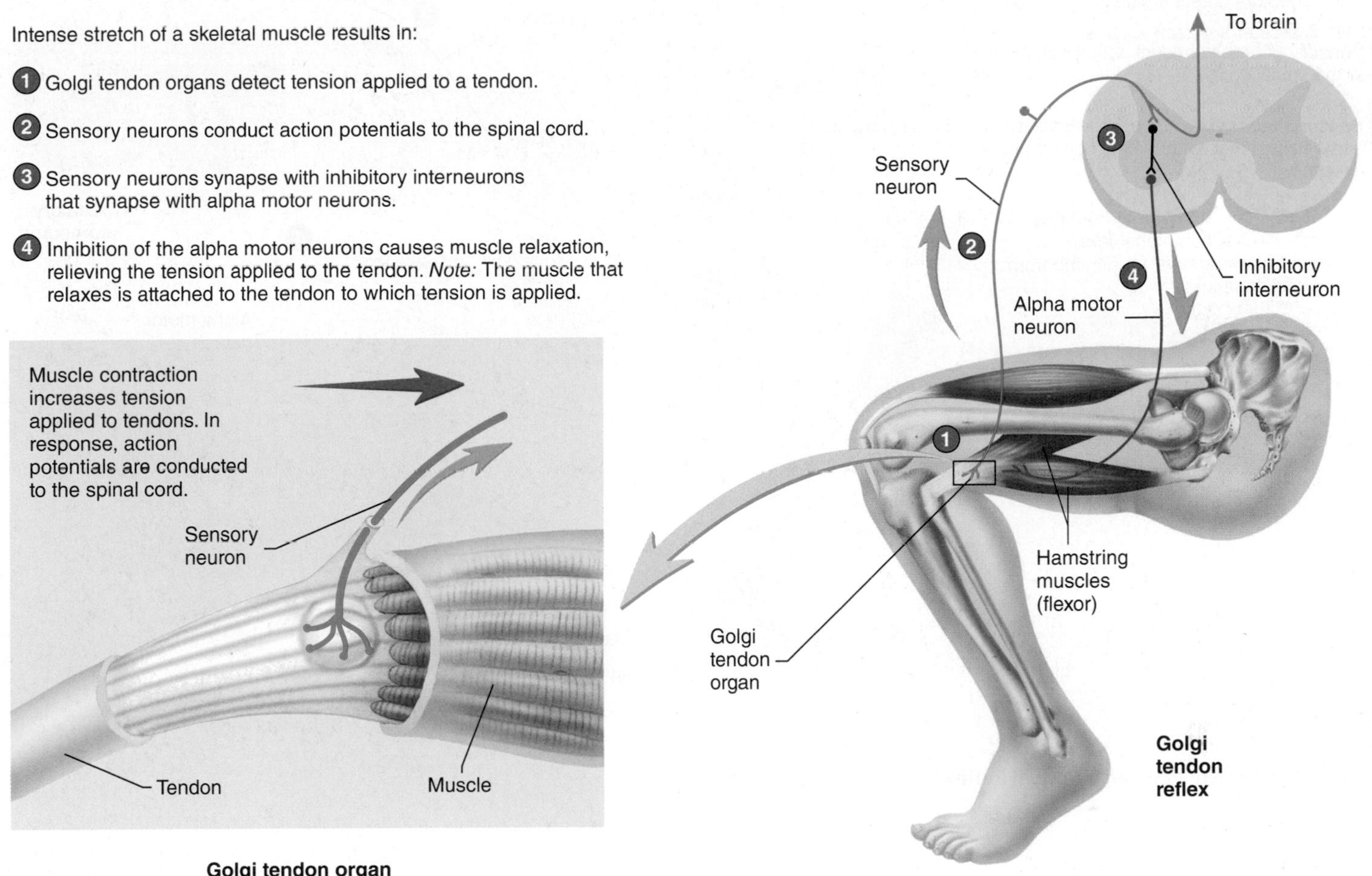
Intense stretch of a skeletal muscle results in:
1 Golgi tendon organs detect tension applied to a tendon.
2 Sensory neurons conduct action potentials to the spinal cord.
3 Sensory neurons synapse with inhibitory interneurons that synapse with alpha motor neurons.
4 Inhibition of the alpha motor neurons causes muscle relaxation, relieving the tension applied to the tendon. Note: The muscle that relaxes is attached to the tendon to which tension is applied.
Muscle contraction increases tension applied to tendons. In response, action potentials are conducted to the spinal cord.
Sensory neuron
Tendon
Muscle
Golgi tendon organ
To brain
Sensory neuron
Inhibitory interneuron
Alpha motor neuron
Hamstring muscles (flexor)
Golgi tendon organ
Golgi tendon reflex

PROCESS FIGURE 12.7 Golgi Tendon Reflex

spindles causes them to be less sensitive to stretch. Sensitivity is maintained because, while alpha motor neurons are stimulating the muscle to contract, gamma motor neurons are stimulating the muscle spindles to contract. The contraction of the muscle fibers at the ends of the muscle spindles pulls on the center part of the muscle spindles and maintains the proper tension. The activity of the muscle spindles helps control posture, muscle tension, and muscle length.

Golgi Tendon Reflex

The **Golgi tendon reflex** prevents contracting muscles from applying excessive tension to tendons. **Golgi tendon organs** are encapsulated nerve endings that have at their ends numerous branches with small swellings adjacent to bundles of collagen fibers in tendons. Golgi tendon organs are located near the muscle-tendon junction (figure 12.7). As a muscle contracts, the attached tendons stretch, resulting in increased tension in the tendon. The increased tension stimulates action potentials in the sensory neurons from the Golgi tendon organs. Golgi tendon organs have a high threshold and are sensitive only to intense stretch.

The sensory neurons of the Golgi tendon organs pass through the dorsal root to the spinal cord and enter the posterior gray matter, where they branch and synapse with inhibitory interneurons. The interneurons synapse with alpha motor neurons that innervate the muscle to which the Golgi tendon organ is attached. Applying a great amount of tension to the tendon stimulates the sensory neurons of the Golgi tendon organs. The sensory neurons stimulate the interneurons to release inhibitory neurotransmitters, which inhibit the alpha motor neurons of the associated muscle and cause it to relax. The sudden relaxation of the muscle reduces the tension applied to the muscle and tendons. This reflex protects muscles and tendons from damage caused by excessive tension. For example, a weight lifter who suddenly drops a heavy weight after straining to lift it does so, in part, because of the effect of the Golgi tendon reflex.

The muscles and tendons of the legs sustain tremendous amounts of tension, particularly in athletes. Frequently, an athlete's Golgi tendon reflex is inadequate to protect muscles and tendons from excessive tension. For example, the large muscles and sudden movements of football players and sprinters can make them vulnerable to relatively frequent hamstring pulls and calcaneal (Achilles) tendon injuries.

Withdrawal Reflex

The function of the **withdrawal reflex,** or *flexor reflex*, is to remove a limb or another body part from a painful stimulus. The sensory receptors are pain receptors (see chapter 15). Following painful stimuli, sensory neurons conduct action potentials through the dorsal root to the spinal cord, where the sensory neurons synapse with excitatory interneurons, which in turn synapse with alpha

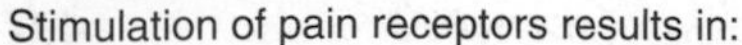

Stimulation of pain receptors results in:

1. Pain receptors detect a painful stimulus.
2. Sensory neurons conduct action potentials to the spinal cord.
3. Sensory neurons synapse with excitatory interneurons that synapse with alpha motor neurons.
4. Excitation of the alpha motor neurons results in contraction of the flexor muscles and withdrawal of the limb from the painful stimulus.

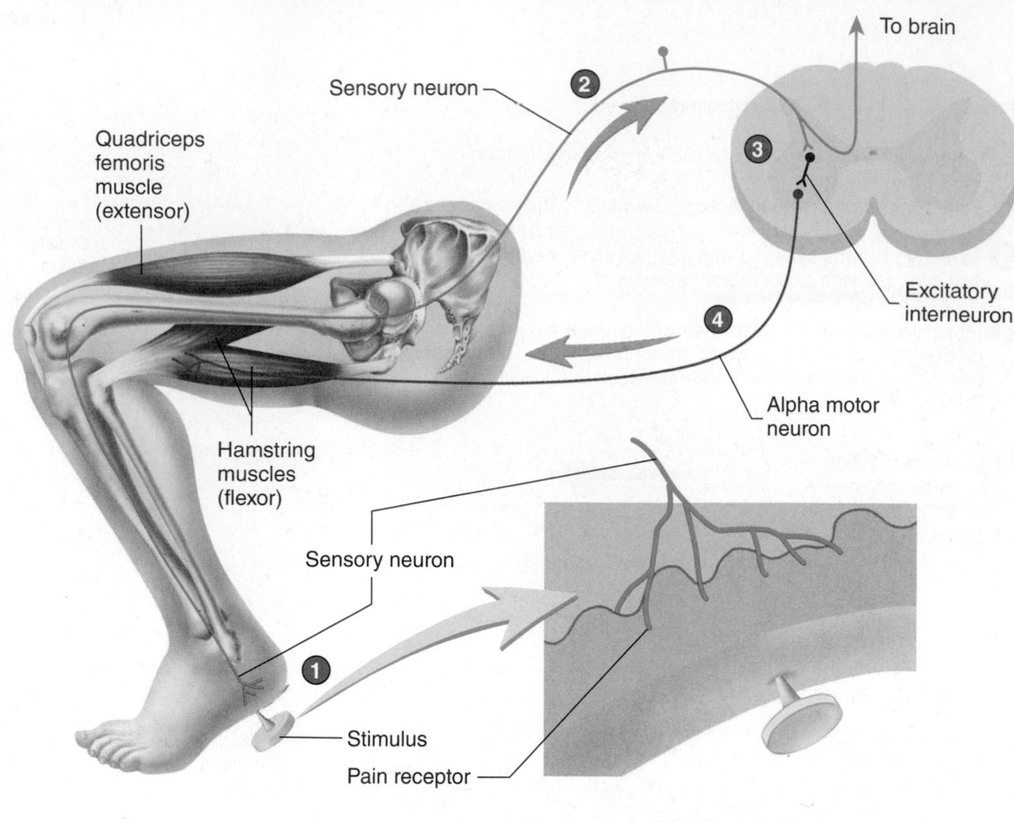

To brain
Sensory neuron
2
3
Quadriceps femoris muscle (extensor)
Excitatory interneuron
4
Alpha motor neuron
Hamstring muscles (flexor)
Sensory neuron
1
Stimulus
Pain receptor
Withdrawal reflex

PROCESS FIGURE 12.8 **Withdrawal Reflex**

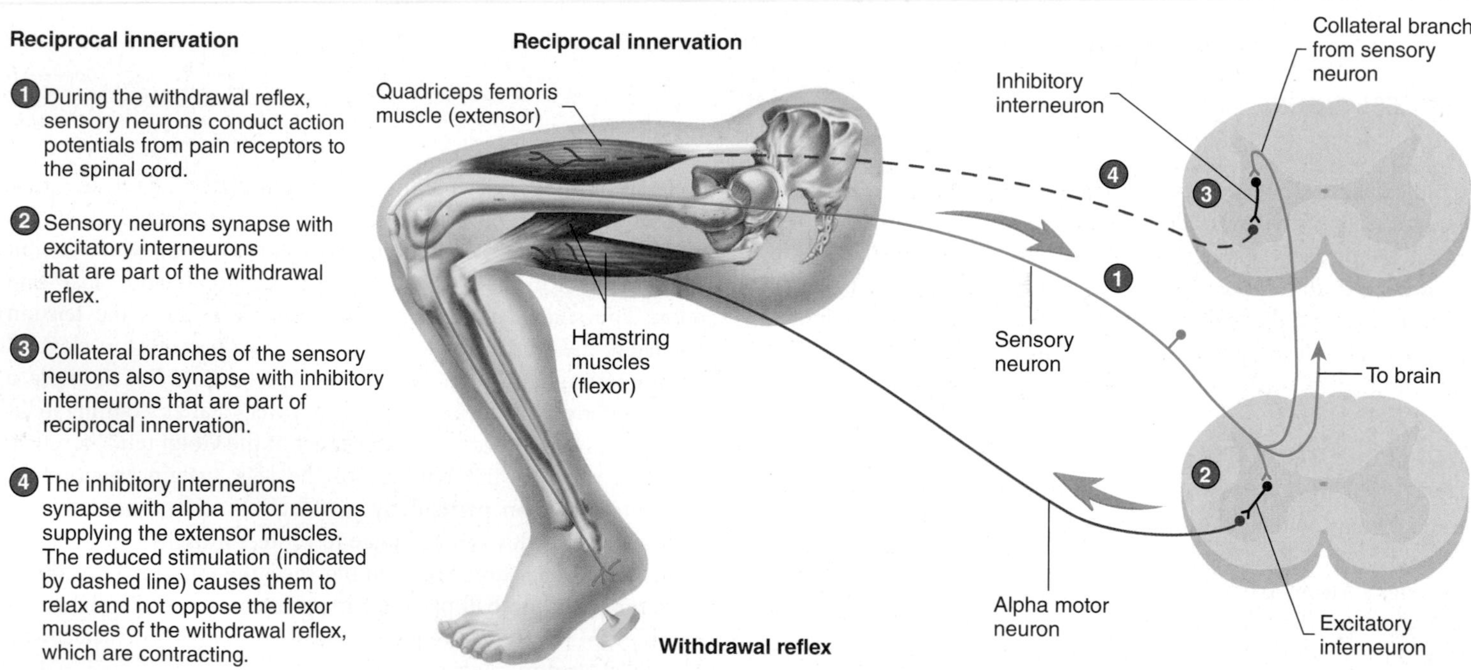

Reciprocal innervation
1 During the withdrawal reflex, sensory neurons conduct action potentials from pain receptors to the spinal cord.
2 Sensory neurons synapse with excitatory interneurons that are part of the withdrawal reflex.
3 Collateral branches of the sensory neurons also synapse with inhibitory interneurons that are part of reciprocal innervation.
4 The inhibitory interneurons synapse with alpha motor neurons supplying the extensor muscles. The reduced stimulation (indicated by dashed line) causes them to relax and not oppose the flexor muscles of the withdrawal reflex, which are contracting.
Reciprocal innervation
Quadriceps femoris muscle (extensor)
Inhibitory interneuron
Collateral branch from sensory neuron
4
3
1
Hamstring muscles (flexor)
Sensory neuron
To brain
2
Alpha motor neuron
Excitatory interneuron
Withdrawal reflex

PROCESS FIGURE 12.9 **Withdrawal Reflex with Reciprocal Innervation**

motor neurons (figure 12.8). The alpha motor neurons stimulate muscles, usually flexor muscles, that remove the limb from the source of the painful stimulus. Collateral branches of the sensory neurons synapse with ascending fibers to the brain, providing conscious awareness of the painful stimuli.

Reciprocal Innervation

Reciprocal innervation is a phenomenon that reinforces the efficiency of the withdrawal reflex (figure 12.9). Collateral axons of sensory neurons that carry action potentials from pain receptors synapse with inhibitory interneurons in the dorsal horn of the spinal cord.

The inhibitory interneurons synapse with and inhibit alpha motor neurons of extensor (antagonist) muscles. When the withdrawal reflex is initiated, flexor muscles contract and reciprocal innervation causes the extensor muscles to relax. This reduces the resistance to movement that the extensor muscles would otherwise generate.

Reciprocal innervation is also involved in the stretch reflex. When the stretch reflex causes a muscle to contract, reciprocal innervation causes opposing muscles to relax. In the patellar reflex, for example, the quadriceps femoris muscle contracts and the hamstring muscles relax.

Crossed Extensor Reflex

The **crossed extensor reflex** is another reflex associated with the withdrawal reflex (figure 12.10). Interneurons that stimulate alpha motor neurons, resulting in withdrawal of a limb, have collateral axons that extend through the white commissure to the opposite side of the spinal cord and synapse with alpha motor neurons that innervate extensor muscles in the opposite side of the body. When a withdrawal reflex is initiated in one lower limb, the crossed extensor reflex causes extension of the opposite lower limb.

The crossed extensor reflex is adaptive in that it helps prevent falls by shifting the weight of the body from the affected to the unaffected limb. For example, when you step on a sharp object with your right foot, you withdraw your right lower limb from the stimulus (withdrawal reflex) while extending your left lower limb (crossed extensor reflex). Therefore, your body weight shifts from the right to the left lower limb. Initiating a withdrawal reflex in both legs at the same time would cause you to fall.

Interactions with Spinal Cord Reflexes

Reflexes do not operate as isolated entities. Rather, because of divergent and convergent pathways (see chapter 11), their activities are integrated with the functions of the nervous system as a whole. Diverging branches of the sensory neurons or interneurons in a reflex arc send action potentials along ascending nerve tracts to the brain (figure 12.11). A pain stimulus, for example, not only initiates a withdrawal reflex, causing you to remove the affected body part from the painful stimulus, but also enables you to perceive the pain as a result of action potentials sent to your brain.

Crossed extensor reflex

1. During the withdrawal reflex, sensory neurons from pain receptors conduct action potentials to the spinal cord.
2. Sensory neurons synapse with excitatory interneurons that are part of the withdrawal reflex.
3. The excitatory interneurons that are part of the withdrawal reflex stimulate alpha motor neurons that innervate flexor muscles, causing withdrawal of the limb from the painful stimulus.
4. Collateral branches of the sensory neurons also synapse with excitatory interneurons that cross to the opposite side of the spinal cord as part of the crossed extensor reflex.
5. The excitatory interneurons that cross the spinal cord stimulate alpha motor neurons supplying extensor muscles in the opposite limb, causing them to contract and support body weight during the withdrawal reflex.

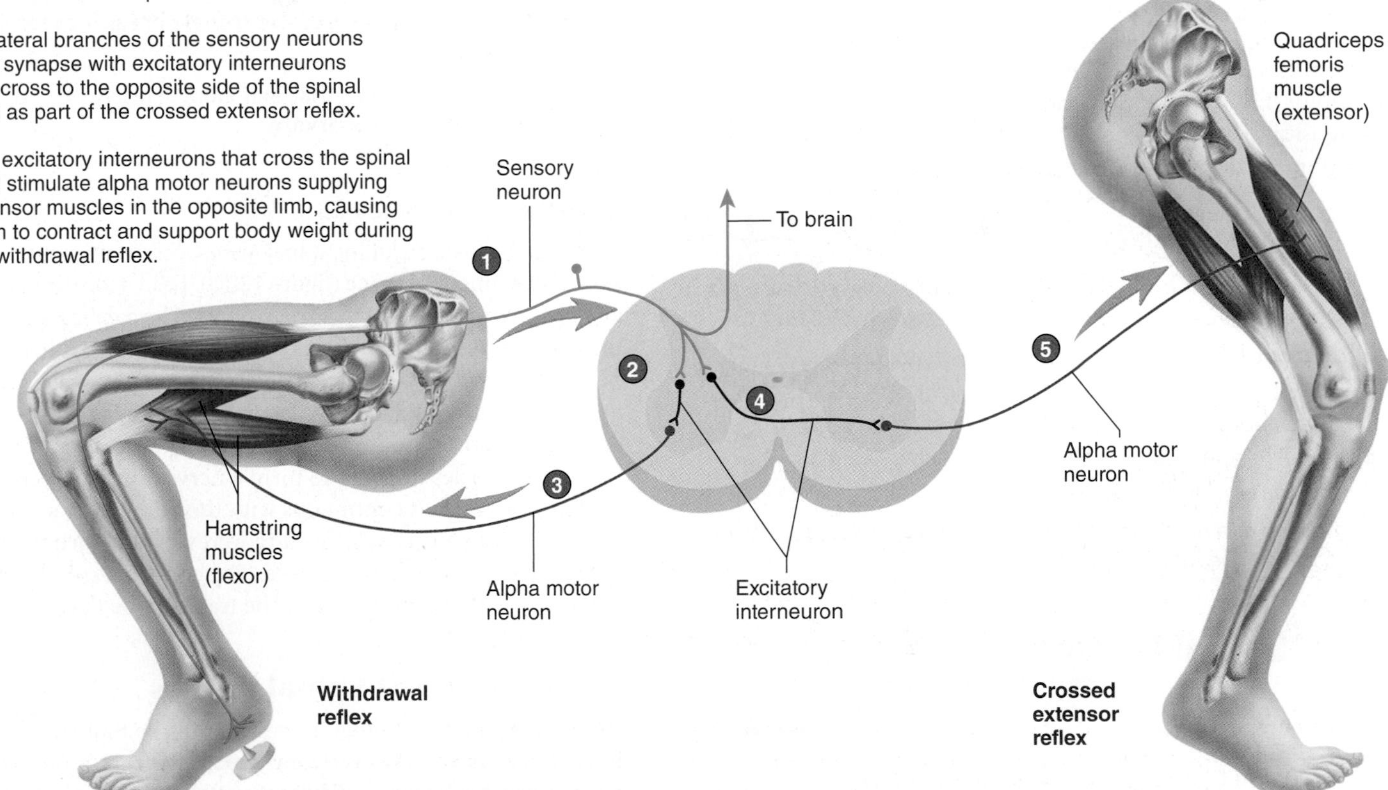
Sensory neuron
To brain
Quadriceps femoris muscle (extensor)
1
2
3
4
5
Alpha motor neuron
Hamstring muscles (flexor)
Alpha motor neuron
Excitatory interneuron
Withdrawal reflex
Crossed extensor reflex

PROCESS FIGURE 12.10 Withdrawal Reflex with Crossed Extensor Reflex

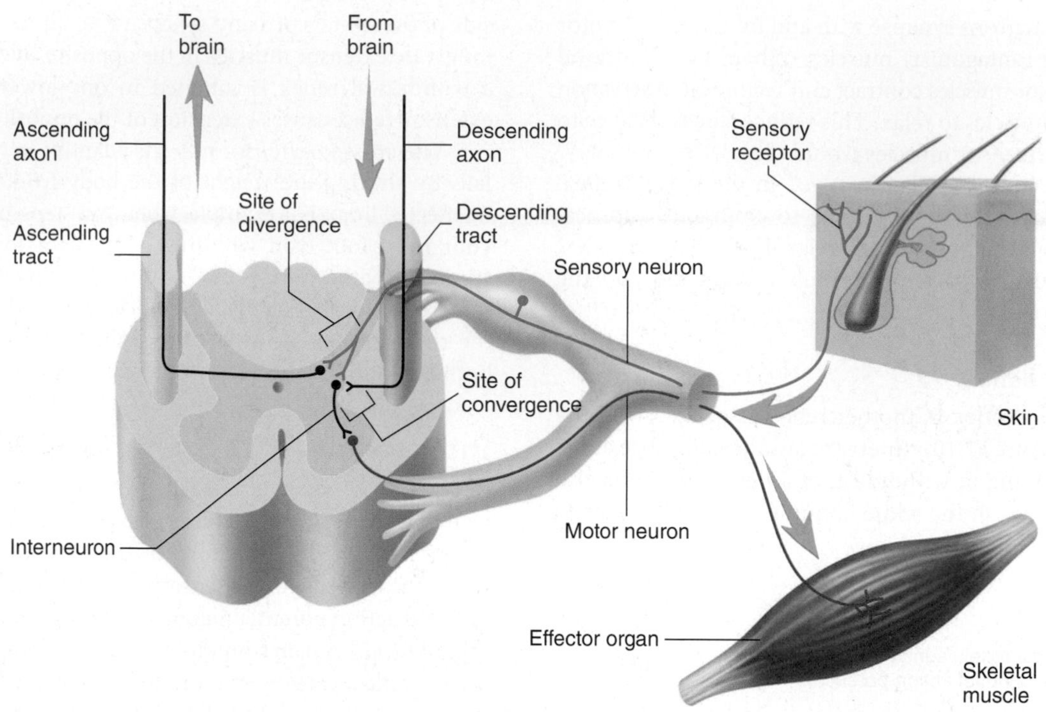
To brain
From brain
Ascending axon
Descending axon
Sensory receptor
Site of divergence
Ascending tract
Descending tract
Sensory neuron
Site of convergence
Skin
Interneuron
Motor neuron
Effector organ
Skeletal muscle

FIGURE 12.11 Spinal Reflex, with Ascending and Descending Axons
A diverging branch of a sensory neuron synapses with a neuron that extends to the brain in an ascending tract (*green*). An axon from the brain extends to the spinal cord in a descending tract (*purple*) and synapses with an interneuron, influencing its action on the motor neuron.

Axons within descending tracts from the brain carry action potentials to motor neurons in the anterior horn of the spinal cord, converging with neurons of reflex arcs. The neurotransmitters released from the axons of these tracts either stimulate or inhibit the motor neurons in the anterior horn. Neurotransmitters change the sensitivity of the reflex by stimulating (EPSP) or inhibiting (IPSP) the motor neurons (see chapter 11).

ASSESS YOUR PROGRESS

7. *Name the parts of a monosynaptic and of a polysynaptic reflex arc. What is a reflex? Explain how reflexes are homeostatic.*
8. *Describe the operation of a gamma motor neuron.*
9. *Contrast and give the functions of a stretch reflex and a Golgi tendon reflex. Describe the sensory receptors for each.*
10. *What is a withdrawal reflex? How do reciprocal innervation and the crossed extensor reflex assist the withdrawal reflex?*
11. *How do ascending and descending pathways relate to reflexes and other neuron functions?*

12.3 Spinal Nerves

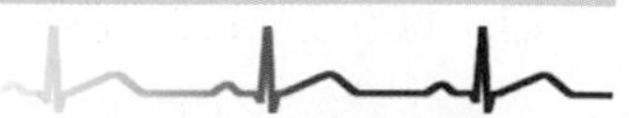

LEARNING OUTCOMES

After reading this section, you should be able to

A. Describe the connective tissue components of a nerve.
B. List the number and locations of the 31 pairs of spinal nerves.
C. Describe a dermatome and its clinical importance.
D. Explain the branching of the spinal nerves into rami and plexuses.
E. List the major nerves that exit each plexus and the body region they innervate.

Structure of Nerves

Nerves of the PNS, including spinal nerves, consist of axons, Schwann cells, and connective tissue (figure 12.12). Each axon, or nerve fiber, and its Schwann cell sheath are surrounded by a delicate connective tissue layer, the **endoneurium** (en-dō-noo′rē-ŭm). A heavier connective tissue layer, the **perineurium** (per-i-noo′rē-ŭm), surrounds groups of axons to form nerve **fascicles** (fas′i-klz). A third layer of dense connective tissue, the **epineurium** (ep-i-noo′rē-ŭm), binds the nerve fascicles together to form a nerve. The connective tissue of the epineurium is continuous with the dura mater surrounding the CNS. As an analogy, if the dura were the vest portion of a coat, the epineurium would be a sleeve. The connective tissue layers of nerves make them tougher than the tracts in the CNS.

Organization of Spinal Nerves

All of the 31 pairs of spinal nerves, except the first pair and those in the sacrum, exit the vertebral column through intervertebral foramina located between adjacent vertebrae (see figure 7.13). The first pair of spinal nerves exit between the skull and the first cervical vertebra. The nerves of the sacrum exit from the single bone of the

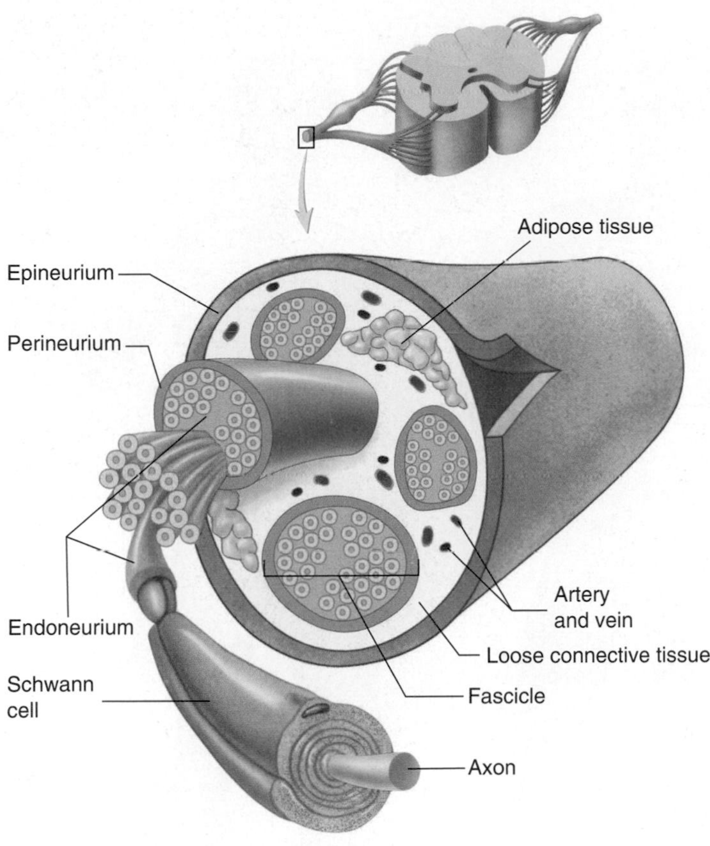
Adipose tissue
Epineurium
Perineurium
Endoneurium
Artery
and vein
Loose connective tissue
Schwann
cell
Fascicle
Axon

FIGURE 12.12 Structure of a Nerve

A nerve consists of axons surrounded by various layers of connective tissue: Epineurium surrounds the whole nerve, perineurium surrounds nerve fascicles, and endoneurium surrounds Schwann cells and axons. Loose connective tissue also surrounds the nerve fascicles.

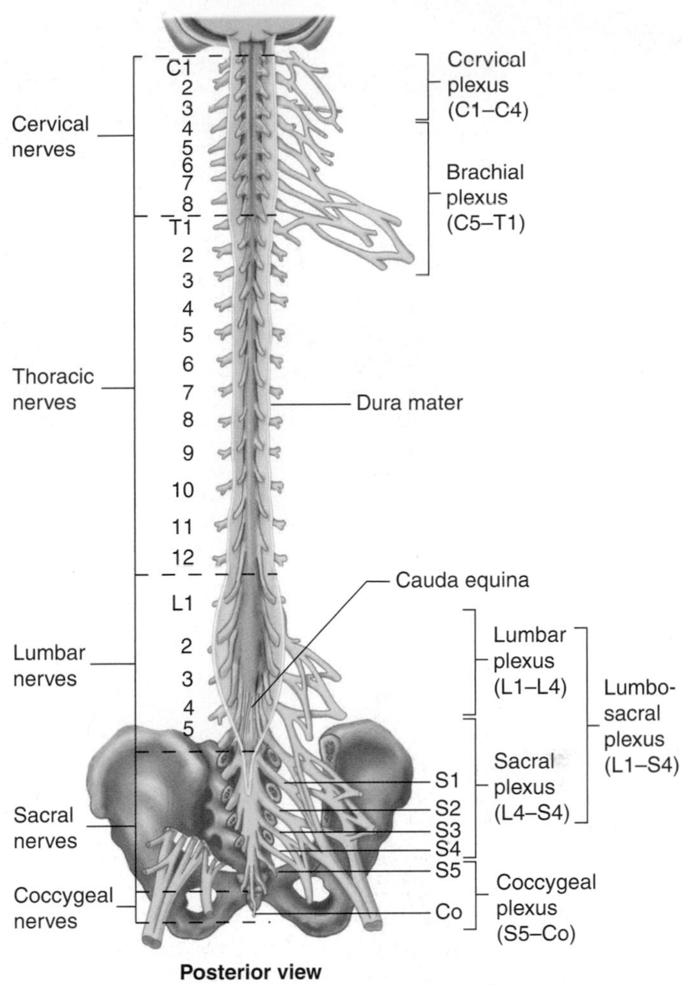
Cervical
nerves
C1
2
3
4
5
6
7
8
T1
2
3
4
5
6
7
8
9
10
11
12
L1
2
3
4
5
Cervical
plexus
(C1–C4)
Brachial
plexus
(C5–T1)
Thoracic
nerves
Dura mater
Cauda equina
Lumbar
nerves
Lumbar
plexus
(L1–L4)
Lumbo-
sacral
plexus
(L1–S4)
Sacral
plexus
(L4–S4)
Sacral
nerves
S1
S2
S3
S4
S5
Co
Coccygeal
nerves
Coccygeal
plexus
(S5–Co)
Posterior view

FIGURE 12.13 APR **Spinal Nerves and Plexuses**

The regional designations and the numbers of the spinal nerves are shown on the left. The plexuses formed by the spinal nerves are shown on the right.

sacrum through the sacral foramina (see chapter 7). Eight spinal nerve pairs exit the vertebral column in the cervical region, 12 in the thoracic region, 5 in the lumbar region, 5 in the sacral region, and 1 in the coccygeal region (figure 12.13). For convenience, each of the spinal nerves is designated by a letter and a number. The letter indicates the region of the vertebral column from which the nerve emerges: **C,** cervical; **T,** thoracic; **L,** lumbar; and **S,** sacral. The single coccygeal nerve is often not designated, but when it is the symbol **Co** is usually used. The number indicates the location in each region where the nerve emerges from the vertebral column, with the smallest number always representing the most superior origin. For example, the most superior nerve exiting the thoracic region of the vertebral column is designated T1. The cervical nerves are designated C1–C8, the thoracic nerves T1–T12, the lumbar nerves L1–L5, and the sacral nerves S1–S5.

The nerves arising from each region of the spinal cord and vertebral column supply specific regions of the body. A **dermatome** is the area of skin supplied with sensory innervation by a pair of spinal nerves. Each of the spinal nerves except C1 has a specific cutaneous sensory distribution. Figure 12.14 illustrates the **dermatomal** (der-mă-tō′măl) **map** for the sensory cutaneous distribution of the spinal nerves.

Predict 3

Loss of sensation in a dermatomal pattern can provide valuable information about the location of nerve damage. Using the dermatomal map in figure 12.14, predict the possible site of nerve damage for a patient who suffered whiplash in an automobile accident and subsequently developed anesthesia (no sensation) in the left arm, forearm, and hand.

Figure 12.15*a* depicts an idealized section through the trunk. Each spinal nerve has a dorsal and a ventral **ramus** (rā′mŭs; branch). Additional rami (rā′mī), called communicating rami, from the thoracic and upper lumbar spinal cord regions carry axons associated with the sympathetic division of the autonomic nervous system (see chapter 16). The **dorsal rami** (rā′mī) innervate most of the deep muscles of the dorsal trunk responsible for moving the vertebral column. They also innervate the connective tissue and skin near the midline of the back.

The **ventral rami** are distributed in two ways. In the thoracic region, the ventral rami form **intercostal** (between ribs) **nerves** (figure 12.15*a*), which extend along the inferior margin of each rib

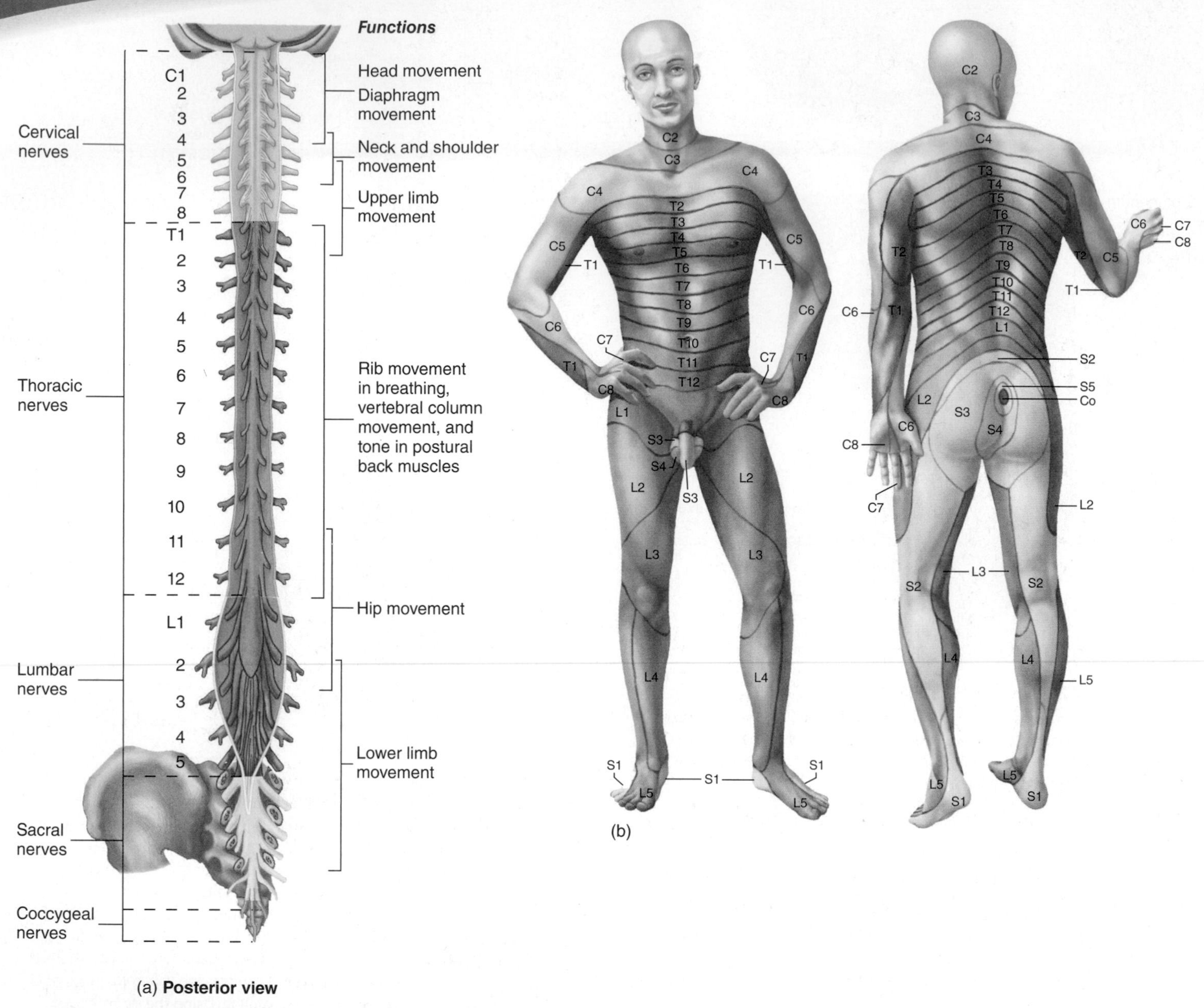
Functions
Head movement
Diaphragm movement
Neck and shoulder movement
Upper limb movement
Rib movement in breathing, vertebral column movement, and tone in postural back muscles
Hip movement
Lower limb movement
Cervical nerves
Thoracic nerves
Lumbar nerves
Sacral nerves
Coccygeal nerves
(a) Posterior view
(b)

FIGURE 12.14 AP|R **Spinal Cord and Dermatomal Map**
(*a*) Nerves of the spinal cord and their functions. The regions are color-coded. (*b*) Letters and numbers indicate the spinal nerves innervating a given region of skin (dermatome).

and innervate the intercostal muscles and the skin over the thorax. The ventral rami of the remaining spinal nerves form five major **plexuses** (plek′sŭs-ēz). The term *plexus* means braid and describes the organization produced by the intermingling of the nerves. The ventral rami of different spinal nerves, called the **roots,** join with each other to form a plexus. These roots should not be confused with the dorsal and ventral roots from the spinal cord, which are more medial. Nerves that arise from plexuses usually have axons from more than one spinal nerve and thus more than one level of the spinal cord. The ventral rami of spinal nerves C1–C4 form the cervical plexus, C5–T1 form the brachial plexus, L1–L4 form the lumbar plexus, L4–S4 form the sacral plexus, and S5 and the coccygeal nerve (Co) form the coccygeal plexus. An important functional aspect of these plexuses is that the intermingling of nerves from multiple spinal cord levels often minimizes the loss of control and feeling to a specific area of the body following spinal cord injury.

Several smaller somatic plexuses, such as the pudendal plexus in the pelvis, are derived from more distal branches of the spinal nerves. Some of the somatic plexuses are described later in this section. Autonomic plexuses (described in chapter 16) also exist in the thorax and abdomen. In the following discussion, we investigate the five major plexuses derived from the ventral rami of spinal nerves.

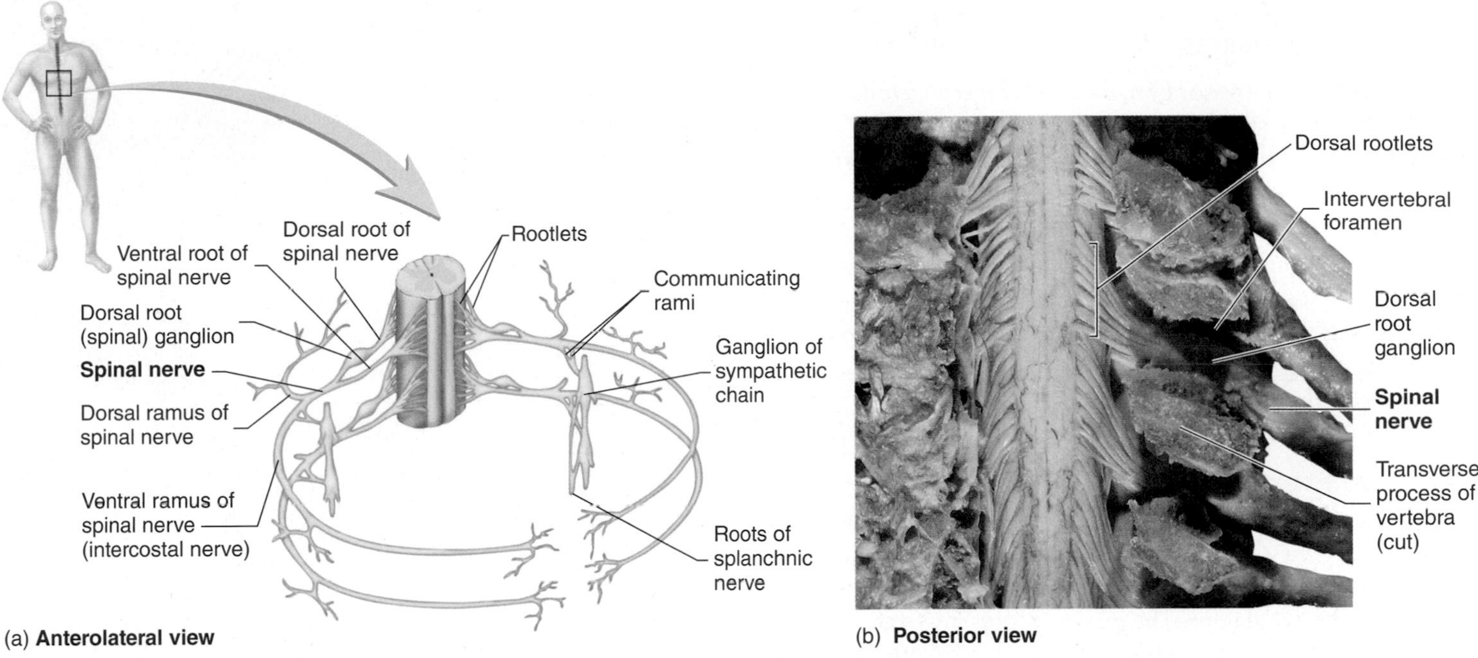

Ventral root of spinal nerve
Dorsal root of spinal nerve
Rootlets
Communicating rami
Dorsal root (spinal) ganglion
Spinal nerve
Ganglion of sympathetic chain
Dorsal ramus of spinal nerve
Ventral ramus of spinal nerve (intercostal nerve)
Roots of splanchnic nerve
(a) Anterolateral view
Dorsal rootlets
Intervertebral foramen
Dorsal root ganglion
Spinal nerve
Transverse process of vertebra (cut)
(b) Posterior view

FIGURE 12.15 AP|R **Spinal Nerves**

(*a*) Typical thoracic spinal nerves have dorsal and ventral roots, as well as dorsal, ventral, and communicating rami. Communicating rami connect to the sympathetic chain (see chapter 16). (*b*) Photograph of four dorsal roots in place along the vertebral column.

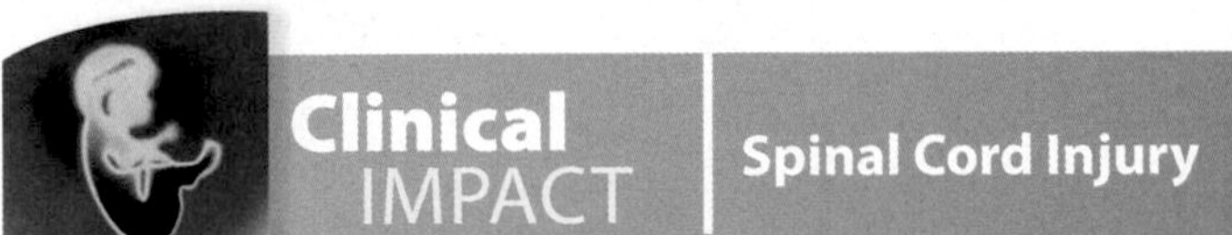

Clinical IMPACT — Spinal Cord Injury

Damage to the spinal cord can disrupt ascending tracts to the brain, resulting in the loss of sensation. Conversely, the disruption of descending tracts from the brain to motor neurons in the spinal cord can result in the loss of motor functions. About 10,000 new cases of **spinal cord injury** occur each year in the United States. Leading causes are automobile and motorcycle accidents, followed by gunshot wounds, falls, and swimming accidents. The primary mechanisms include concussion (an injury caused by a blow), contusion (an injury resulting in hemorrhage), and laceration (a tear or cut). Spinal cord injuries often involve excessive flexion, extension, rotation, or compression of the vertebral column. Most spinal cord injuries are acute contusions of the cord due to bone or disk displacement into the cord and involve a combination of excessive directional movements, such as simultaneous flexion and compression.

Spinal cord injury is classified according to the vertebral level at which the injury occurred, whether the entire cord or only a portion is damaged at that level, and the mechanism of injury. Most spinal cord injuries occur in the cervical region or at the thoracolumbar junction and are incomplete. Injuries in the cervical region above T1 are the most severe and can result in paralysis of all four limbs (quadriplegia or tetraplegia), with the abdominal and chest muscles also affected. Injuries at or below T1 can result in varying degrees of paralysis of the legs (paraplegia) and the abdomen, while retaining full function of the upper limbs.

At the time of spinal cord injury, two types of tissue damage occur: (1) primary, mechanical damage and (2) secondary, tissue damage extending into a much larger region of the cord than the primary damage. Secondary spinal cord damage begins within minutes of the primary damage and is caused by ischemia, edema, ion imbalances, the release of "excitotoxins" (such as glutamate), and inflammatory cell invasion. Secondary damage is the primary focus of current research.

Once an accident occurs, little can be done about the primary damage. On the other hand, much secondary damage can be prevented if promptly treated. Treatment with large doses of anti-inflammatory steroids, such as methylprednisolone, within 8 hours of the injury can dramatically lessen the secondary damage to the cord by reducing inflammation and edema. Additional treatments include structural realignment and stabilization of the vertebral column and decompression of the spinal cord. Rehabilitation is based on retraining the patient to use whatever residual connections exist across the site of damage.

For a long time, researchers thought the spinal cord was incapable of regeneration following severe damage. But they have now learned that most neurons of the adult spinal cord survive the injury and begin to regenerate, growing about 1 mm into the site of damage. Unfortunately, the neurons then regress to an inactive, atrophic state. A major block to adult spinal cord regeneration is the formation of a scar, consisting mainly of myelin and astrocytes, at the site of the injury. Myelin and other inhibitory factors, such as the protein Nogo, in the scar inhibit regeneration. Implantation of stem cells or other cell types, such as olfactory ensheathing glia and Schwann cells, can partially bridge the scar and stimulate some regeneration. Certain growth factors can also stimulate regeneration, and blocking inhibitory factors may be able to prevent the formation of the glial scar to allow axon regeneration. Current research continues to look for the right combination of cells and other factors to stimulate regeneration of the spinal cord following injury.

ASSESS YOUR PROGRESS

12. *List all the spinal nerves by name and number. Where do they exit the vertebral column?*
13. *What is a dermatome? Why are dermatomes clinically important?*
14. *Contrast the dorsal and ventral rami of the spinal nerves. What muscles do the dorsal rami innervate?*
15. *Describe the distribution of the ventral rami of the thoracic region.*
16. *What is a plexus? What happens to the axons of spinal nerves as they pass through a plexus? Why is this functionally important?*
17. *Name the five major spinal plexuses and the spinal nerves associated with each.*

Cervical Plexus

The **cervical plexus** is a relatively small plexus originating from spinal nerves C1–C4 (figure 12.16). Branches derived from this plexus innervate superficial neck structures, including several of the muscles attached to the hyoid bone. The cervical plexus innervates the skin of the neck and posterior portion of the head (see figure 12.14). An unusual part of the cervical plexus, the **ansa** (an′sah; bucket handle) **cervicalis,** is a loop between C1 and C3. Nerves to the infrahyoid muscles branch from the ansa cervicalis (see chapter 10).

One of the most important derivatives of the cervical plexus is the **phrenic** (fren′ik) **nerve,** which originates from spinal nerves C3–C5 and is derived from both the cervical and brachial plexuses. The phrenic nerves descend along each side of the neck to enter the thorax and then descend along the sides of the mediastinum to reach the diaphragm, which they innervate. Contraction of the diaphragm is largely responsible for a person's ability to breathe; therefore, damage to the phrenic nerve during surgery or compression of the nerve by a tumor at the base of the lung severely limits breathing. Cancer of the bronchus is the most common type of cancer in men, accounting for about 30% of all male cancers. It occurs most often in men who smoke cigarettes.

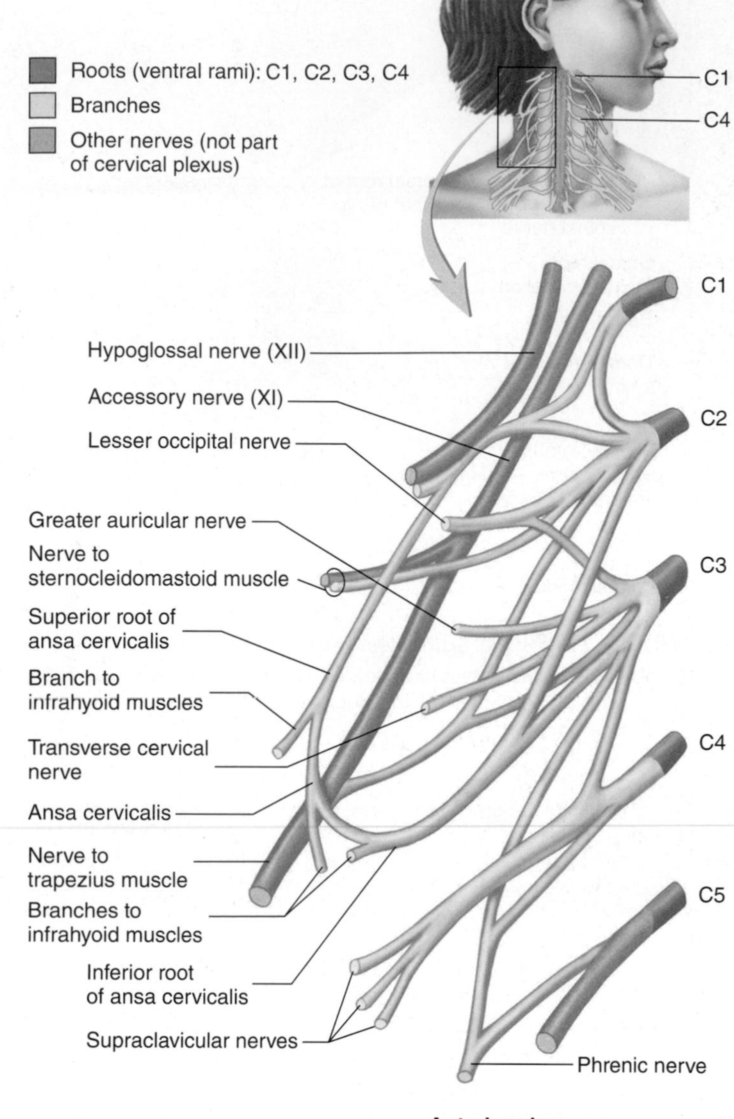
Roots (ventral rami): C1, C2, C3, C4
Branches
Other nerves (not part of cervical plexus)
C1
C4
C1
C2
C3
C4
C5
Hypoglossal nerve (XII)
Accessory nerve (XI)
Lesser occipital nerve
Greater auricular nerve
Nerve to sternocleidomastoid muscle
Superior root of ansa cervicalis
Branch to infrahyoid muscles
Transverse cervical nerve
Ansa cervicalis
Nerve to trapezius muscle
Branches to infrahyoid muscles
Inferior root of ansa cervicalis
Supraclavicular nerves
Phrenic nerve
Anterior view

FIGURE 12.16 AP|R **Cervical Plexus**
The roots of the plexus are formed by the ventral rami of spinal nerves C1–C4. Branches of the cervical plexus innervate the muscles (infrahyoid) and skin of the neck. The phrenic nerve (C3–C5) innervates the diaphragm.

> **Predict 4**
>
> Explain how damage to or compression of the right phrenic nerve affects the diaphragm. How would breathing be affected if the spinal cord were completely severed at the level of C2 versus at the level of C6?

Brachial Plexus

The **brachial plexus** originates from spinal nerves C5–T1 (figure 12.17). The five ventral rami that constitute the brachial plexus join to form three **trunks,** which separate into six **divisions** and then join again to create three **cords** (posterior, lateral, and medial) from which five **branches,** or nerves of the upper limb, emerge.

The five major nerves emerging from the brachial plexus to supply the upper limb are the axillary, radial, musculocutaneous, ulnar, and median nerves. The axillary nerve innervates part of the shoulder; the radial nerve innervates the posterior arm, forearm, and hand; the musculocutaneous nerve innervates the anterior arm; and the ulnar and median nerves innervate the anterior forearm and hand. Smaller nerves from the brachial plexus innervate the shoulder and pectoral muscles. Because of this anatomical organization, the entire upper limb can be anesthesized by injecting an anesthetic near the brachial plexus between the neck and the shoulder posterior to the clavicle. This is called **brachial anesthesia.**

Axillary Nerve

The **axillary** (ak′sil-ār-ē) **nerve** innervates the deltoid and teres minor muscles (figure 12.18). It also provides sensory innervation to the shoulder joint and to the skin over part of the shoulder.

Radial Nerve

The **radial nerve** emerges from the posterior cord of the brachial plexus and descends within the deep aspect of the posterior arm (figure 12.19). About midway down the shaft of the humerus, it

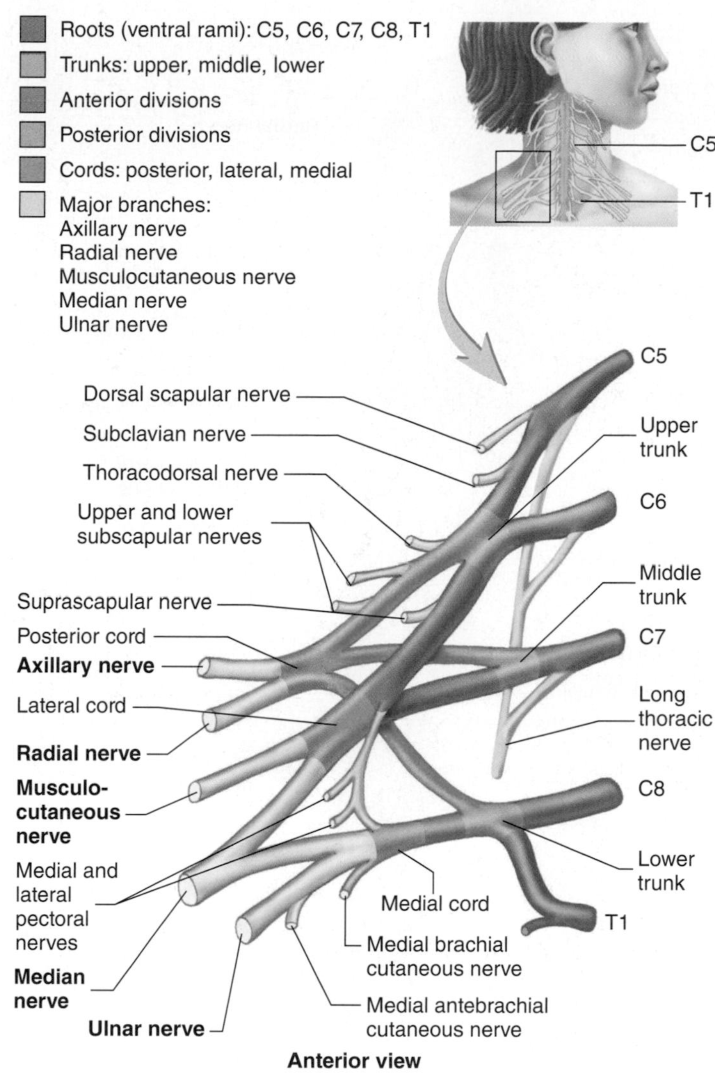

Roots (ventral rami): C5, C6, C7, C8, T1
Trunks: upper, middle, lower
Anterior divisions
Posterior divisions
Cords: posterior, lateral, medial
Major branches:
Axillary nerve
Radial nerve
Musculocutaneous nerve
Median nerve
Ulnar nerve
C5
T1
Dorsal scapular nerve
Subclavian nerve
Thoracodorsal nerve
Upper and lower subscapular nerves
Suprascapular nerve
Posterior cord
Axillary nerve
Lateral cord
Radial nerve
Musculo-cutaneous nerve
Medial and lateral pectoral nerves
Median nerve
Ulnar nerve
C5
Upper trunk
C6
Middle trunk
C7
Long thoracic nerve
C8
Lower trunk
T1
Medial cord
Medial brachial cutaneous nerve
Medial antebrachial cutaneous nerve
Anterior view

FIGURE 12.17 APR **Brachial Plexus**
The roots of the plexus are formed by the ventral rami of spinal nerves C5–T1 and join to form an upper, a middle, and a lower trunk. Each trunk divides into anterior and posterior divisions. The divisions join to form the posterior, lateral, and medial cords from which the major brachial plexus nerves arise. The major brachial plexus nerves include the axillary, radial, musculocutaneous, median, and ulnar nerves, which innervate the muscles and skin of the upper limb.

lies against the bone in the radial groove. The radial nerve innervates all of the extensor muscles of the upper limb, the supinator muscle, and the brachioradialis. Its cutaneous sensory distribution is to the posterior portion of the upper limb, including the posterior surface of the hand.

Musculocutaneous Nerve

The **musculocutaneous** (mŭs′kū-lō-kū-tā′nē-ŭs) **nerve** provides motor innervation to the anterior muscles of the arm, as well as cutaneous sensory innervation to part of the forearm (figure 12.20).

Ulnar Nerve

The **ulnar (ŭl′năr) nerve** innervates two forearm muscles plus most of the intrinsic hand muscles, except some associated with the thumb. Its sensory distribution is to the ulnar side of the hand (figure 12.21).

Axillary Nerve

Origin
Posterior cord of brachial plexus, C5–C6

Movements/Muscles Innervated

Laterally rotates arm
- *Teres minor*

Abducts arm
- *Deltoid*

Cutaneous (Sensory) Innervation
Inferior lateral shoulder

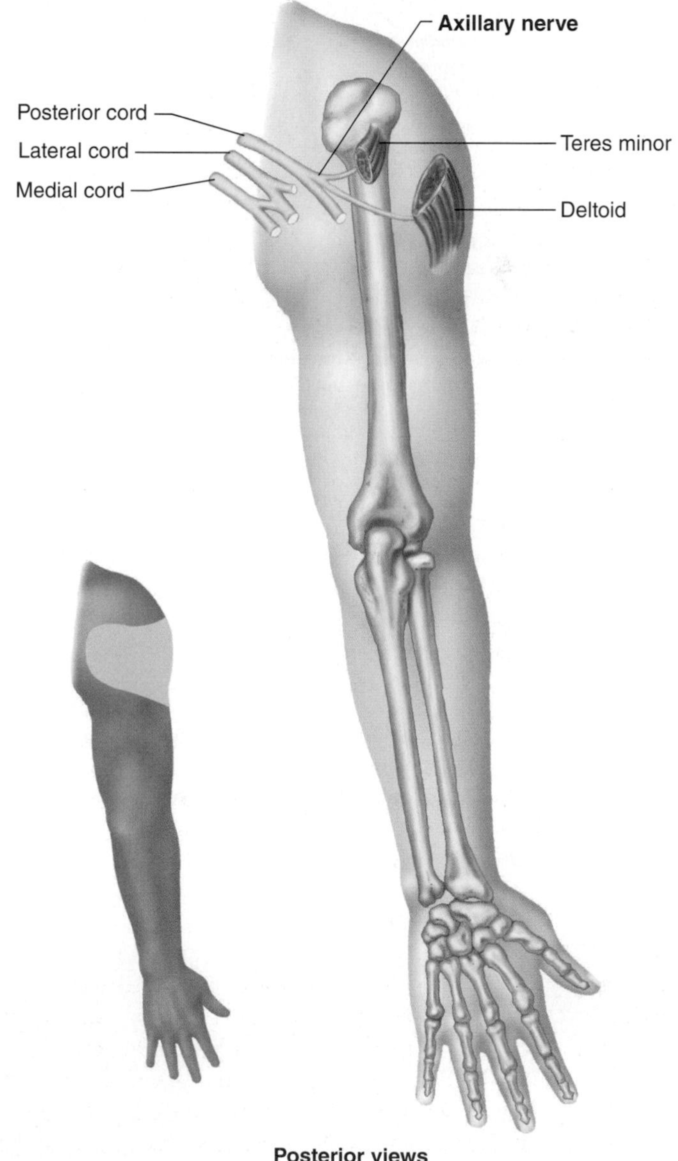

Axillary nerve
Posterior cord
Lateral cord
Medial cord
Teres minor
Deltoid
Posterior views

FIGURE 12.18 Axillary Nerve
The route of the axillary nerve and the muscles it innervates. The *inset* depicts the cutaneous (sensory) distribution of the nerve (blue area).

Radial Nerve

Origin

Posterior cord of brachial plexus, C5–T1

Movements/Muscles Innervated

Extends elbow
- *Triceps brachii*
- *Anconeus*

Flexes elbow
- *Brachialis (part; sensory only)*
- *Brachioradialis*

Extends and abducts wrist
- *Extensor carpi radialis longus*
- *Extensor carpi radialis brevis*

Supinates forearm and hand
- *Supinator*

Extends fingers
- *Extensor digitorum*
- *Extensor digiti minimi*
- *Extensor indicis*

Extends and adducts wrist
- *Extensor carpi ulnaris*

Abducts thumb
- *Abductor pollicis longus*

Extends thumb
- *Extensor pollicis longus*
- *Extensor pollicis brevis*

Cutaneous (Sensory) Innervation

Posterior surface of arm and forearm, lateral two-thirds of dorsum of hand

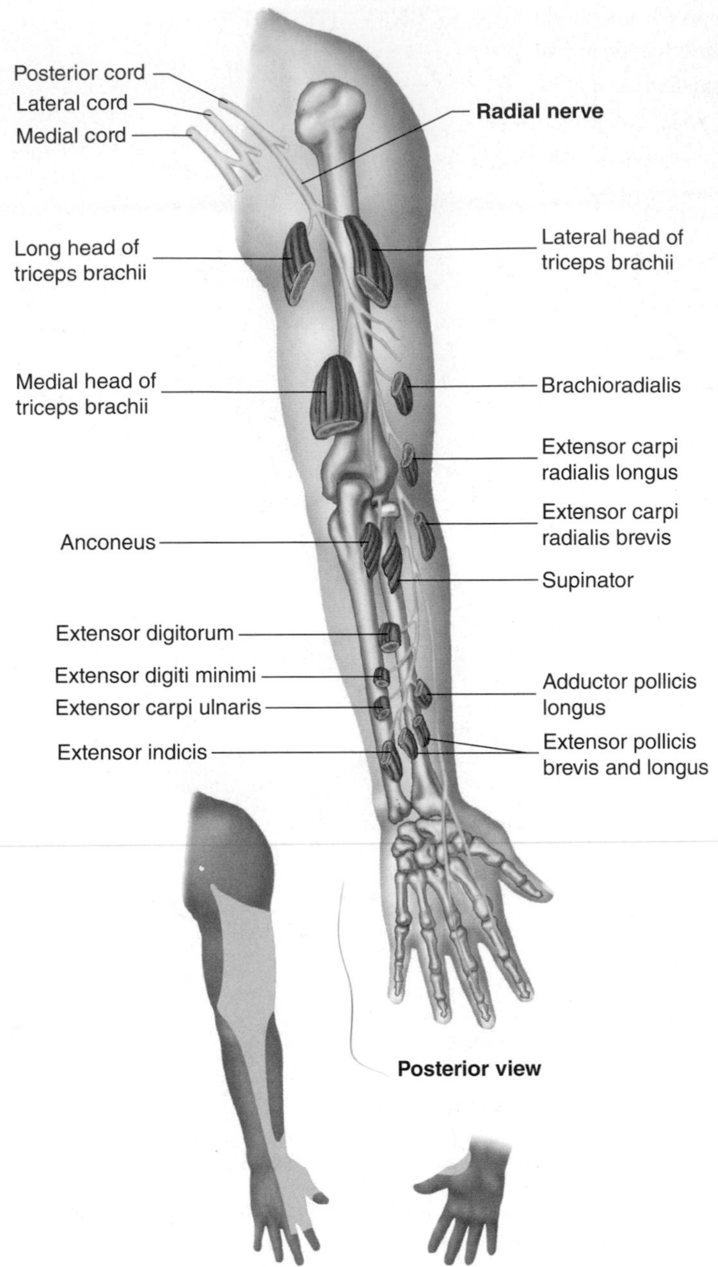
Posterior cord
Lateral cord
Medial cord
Radial nerve
Long head of triceps brachii
Lateral head of triceps brachii
Medial head of triceps brachii
Brachioradialis
Extensor carpi radialis longus
Extensor carpi radialis brevis
Anconeus
Supinator
Extensor digitorum
Extensor digiti minimi
Extensor carpi ulnaris
Extensor indicis
Adductor pollicis longus
Extensor pollicis brevis and longus
Posterior view
Posterior view
Anterior view

FIGURE 12.19 Radial Nerve
The route of the radial nerve and the muscles it innervates. The *insets* depict the cutaneous (sensory) distribution of the nerve (blue areas).

Clinical IMPACT — Radial Nerve Damage

Because the radial nerve lies near the humerus in the axilla, it can be damaged if it is compressed against the humerus. Improper use of crutches (pushing the crutch tightly into the axilla) can result in **crutch paralysis.** As the radial nerve is compressed between the top of the crutch and the humerus, the nerve is damaged and the muscles it innervates lose their function. The major symptom of crutch paralysis is **wrist drop,** in which the elbow, wrist, and fingers are constantly flexed because the extensor muscles of the wrist and fingers, which are innervated by the radial nerve, fail to function. There is also a loss of sensation over the back of the forearm and hand. Crutch paralysis is usually temporary as long as the patient begins to use the crutches correctly.

Musculocutaneous Nerve

Origin

Lateral cord of brachial plexus, C5–C7

Movements/Muscles Innervated

Flexes shoulder
- *Biceps brachii*
- *Coracobrachialis*

Flexes elbow and supinates forearm and hand
- *Biceps brachii*

Flexes elbow
- *Brachialis (also small amount of innervation from radial nerve)*

Cutaneous (Sensory) Innervation

Lateral surface of forearm

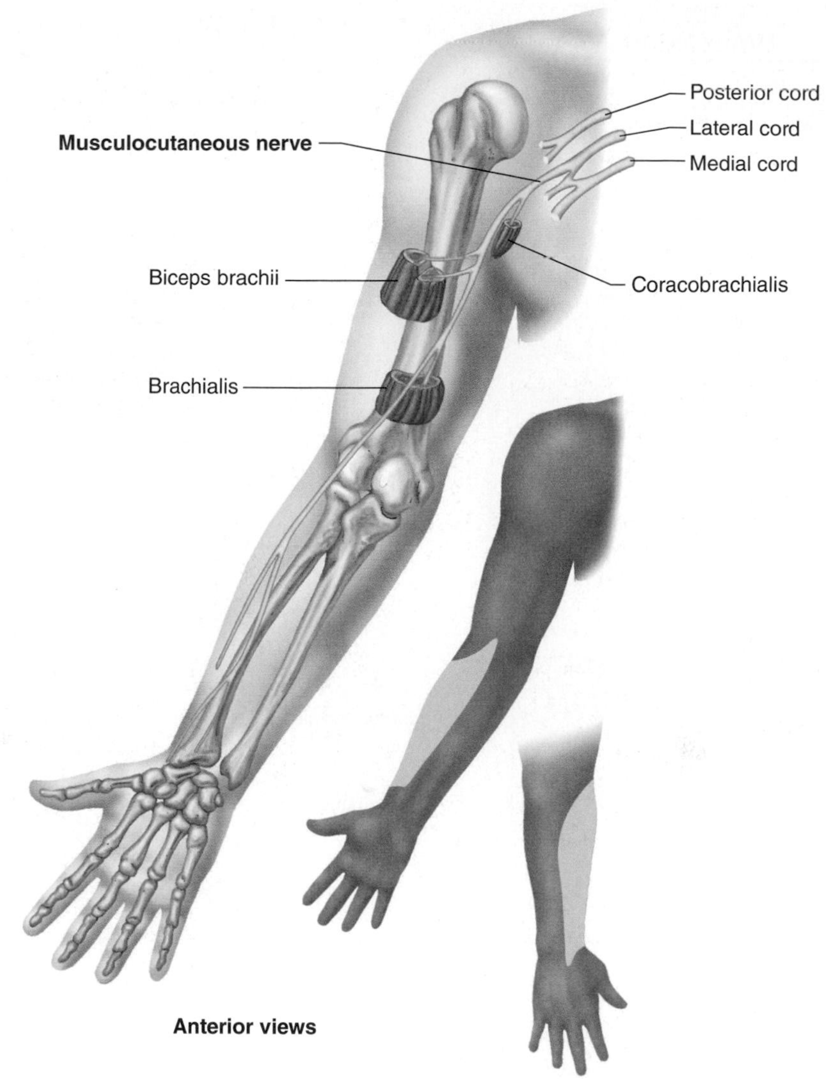
Posterior cord
Lateral cord
Medial cord
Musculocutaneous nerve
Biceps brachii
Coracobrachialis
Brachialis
Anterior views
Posterior view

FIGURE 12.20 Musculocutaneous Nerve
The route of the musculocutaneous nerve and the muscles it innervates. The *insets* depict the cutaneous (sensory) distribution of the nerve (blue areas).

Median Nerve

The **median nerve** innervates all but one of the flexor muscles of the forearm and most of the hand muscles at the base of the thumb, called the thenar area of the hand. The nerve's cutaneous sensory distribution is to the radial portion of the palm of the hand (figure 12.22).

Other Nerves of the Brachial Plexus

Several nerves, other than the five just described, arise from the brachial plexus (see figure 12.17). They supply most of the muscles acting on the scapula and arm and include the pectoral, long thoracic, thoracodorsal, subscapular, and suprascapular nerves. In addition, brachial plexus nerves innervate the skin of the medial arm and forearm.

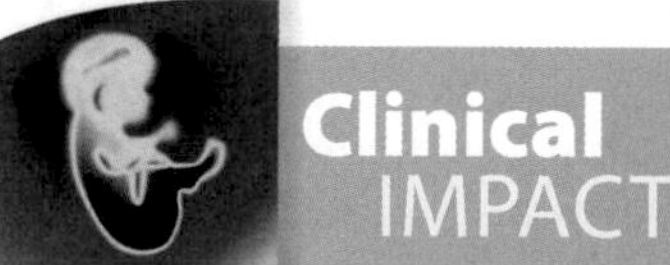

Clinical IMPACT

Ulnar Nerve Damage

The ulnar nerve is the most easily damaged of all the spinal nerves, but such damage is almost always temporary. The ulnar nerve passes posterior to the medial epicondyle of the humerus and can be felt just below the skin at this region. If the elbow is banged against a hard object, temporary ulnar nerve damage may occur, causing painful tingling to radiate down the ulnar side of the forearm and hand. Because of this sensation, we often call this area of the elbow the *funny bone* or *crazy bone*.

Ulnar Nerve

Origin

Medial cord of brachial plexus, C8–T1

Movements/Muscles Innervated

Flexes and adducts wrist
- *Flexor carpi ulnaris*

Flexes fingers
- *Part of the flexor digitorum profundus controlling the distal phalanges of little and ring fingers*

Adducts thumb
- *Adductor pollicis*

Controls hypothenar muscles
- *Flexor digiti minimi brevis*
- *Abductor digiti minimi*
- *Opponens digiti minimi*

Flexes metacarpophalangeal joints and extends interphalangeal joints
- *Two medial (ulnar) lumbricales*

Abducts and adducts fingers
- *Interossei*

Cutaneous (Sensory) Innervation

Medial third of hand, little finger, and medial half of ring finger

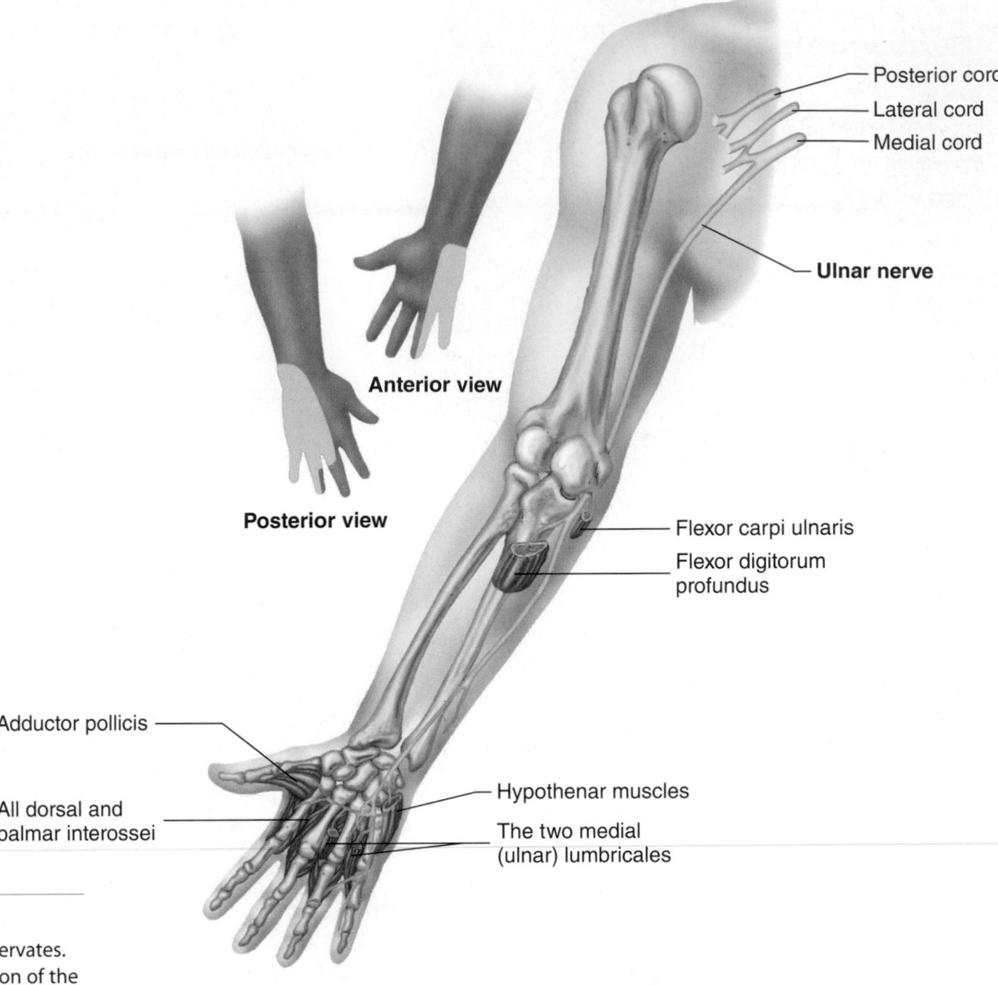
Posterior cord
Lateral cord
Medial cord
Ulnar nerve
Anterior view
Posterior view
Flexor carpi ulnaris
Flexor digitorum profundus
Adductor pollicis
All dorsal and palmar interossei
Hypothenar muscles
The two medial (ulnar) lumbricales
Anterior view

FIGURE 12.21 Ulnar Nerve
The route of the ulnar nerve and the muscles it innervates. The *insets* depict the cutaneous (sensory) distribution of the nerve (blue areas).

Case STUDY | Thoracic Outlet Syndrome

Sarah, age 26, noticed that over a period of time she experienced pain, tingling, and numbness in the ring finger and little finger of her right hand. She also felt pain in her elbow that radiated down the posteromedial portion of her forearm and hand, so she made an appointment with her physician. After careful examination of Sarah's upper limb, her physician ordered an x-ray of her neck. The x-ray disclosed a cervical rib on the right, which was attached to the C7 vertebra. Cervical ribs are extra ribs that occur in about 0.5% of the population, yet most people exhibit no symptoms. In Sarah's case, the extra rib was compressing the inferior roots of the brachial plexus (see figure 12.17), a condition called **thoracic outlet syndrome.** This syndrome is a group of related disorders that involve compression of the nerves and vessels in the "thoracic outlet" region of the lower neck and upper chest. The syndrome is generally caused by compression of the brachial plexus or the subclavian artery or vein. Compression can result from an anatomical abnormality, such as Sarah's extra rib, but to become symptomatic there is often an associated injury or overuse activity, such as repetitive overhead arm motions in sports and occupations, droopy shoulders, joint pressure from heavy backpacks or obesity, joint changes during pregnancy, and trauma, such as whiplash. Thoracic outlet syndrome can also result from these injuries and activities in the absence of a cervical rib. Treatment of thoracic outlet syndrome first involves physical therapy to improve shoulder alignment, range of motion, and strength. In more severe and persistent cases, surgery may be required to alleviate compression of the plexus or blood vessels.

Predict 5

Use figures 12.14, 12.17, and 12.19–12.21 to answer these questions:

a. Name the brachial plexus nerves supplying the skin of the hand.
b. Damage to which of these nerves could produce the symptoms in Sarah's hand?
c. What made Sarah's physician suspect thoracic outlet syndrome rather than damage to an individual nerve?
d. Thoracic outlet syndrome can also affect muscles, producing muscle weakness and paralysis. Muscles supplied by what nerves could be affected by a cervical rib?

Median Nerve

Origin

Medial and lateral cords of brachial plexus, C5–T1

Movements/Muscles Innervated

Pronates forearm and hand
- *Pronator teres*
- *Pronator quadratus*

Flexes and abducts wrist
- *Flexor carpi radialis*

Flexes wrist
- *Palmaris longus*

Flexes fingers
- *Part of flexor digitorum profundus controlling the distal phalanx of the middle and index fingers*
- *Flexor digitorum superficialis*

Controls thumb muscle
- *Flexor pollicis longus*

Controls thenar muscles
- *Abductor pollicis brevis*
- *Opponens pollicis*
- *Flexor pollicis brevis*

Flexes metacarpophalangeal joints and extends interphalangeal joints
- *Two lateral (radial) lumbricales*

Cutaneous (Sensory) Innervation

Lateral two-thirds of palm of hand, thumb, index and middle fingers, and the lateral half of ring finger and dorsal tips of the same fingers

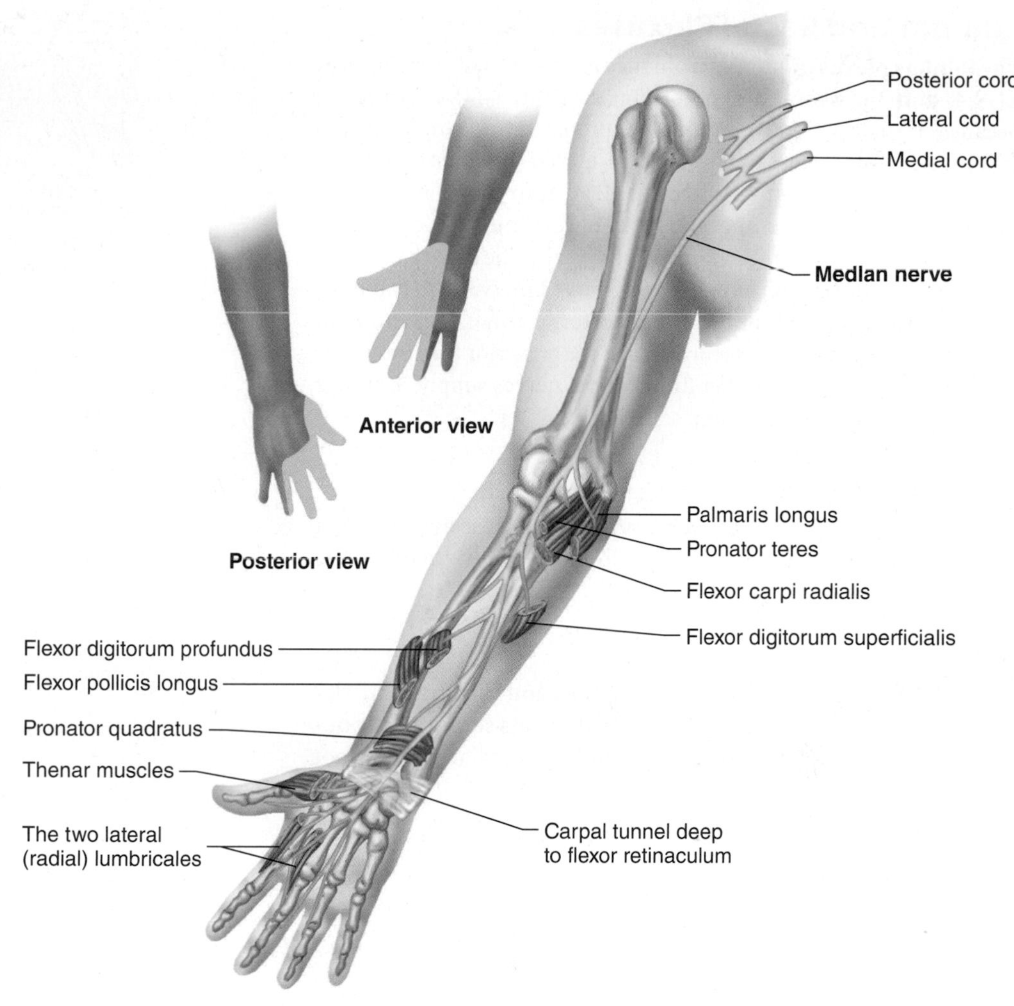

Posterior cord
Lateral cord
Medial cord
Median nerve
Anterior view
Posterior view
Palmaris longus
Pronator teres
Flexor carpi radialis
Flexor digitorum superficialis
Flexor digitorum profundus
Flexor pollicis longus
Pronator quadratus
Thenar muscles
The two lateral (radial) lumbricales
Carpal tunnel deep to flexor retinaculum
Anterior view

FIGURE 12.22 APR **Median Nerve**

The route of the median nerve and the muscles it innervates. The *insets* depict the cutaneous (sensory) distribution of the nerve (blue areas).

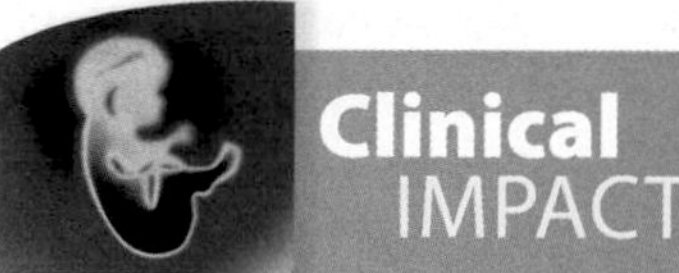

Clinical IMPACT — Median Nerve Damage

Damage to the median nerve occurs most commonly where it enters the wrist through the **carpal tunnel.** This tunnel is created by the concave organization of the carpal bones and the flexor retinaculum on the anterior surface of the wrist. Compression of the median nerve in this relatively narrow tunnel produces numbness, tingling, and pain in the fingers. In addition, the function of the thenar muscles, which are innervated by the median nerve, is reduced, resulting in weakness in thumb flexion and opposition. This condition is called **carpal tunnel syndrome.** In general, anything that compresses the median nerve in the carpal tunnel can cause the syndrome. Some people have smaller carpal tunnels and are thus more prone to nerve compression. This may partially explain why women are three times more likely to have carpal tunnel syndrome. The syndrome can also be caused by multiple additional factors, including wrist fractures, swelling during pregnancy, or diseases such as rheumatoid arthritis. Although repetitive movements—for example, on assembly lines or while playing a musical instrument—can contribute to carpal tunnel syndrome, it now appears that typing at a keyboard is not a major factor. Treatment usually begins with immobilization and reduction of inflammation; however, surgery is often needed to relieve the pressure.

Lumbar and Sacral Plexuses

The **lumbar plexus** originates from the ventral rami of spinal nerves L1–L4, and the **sacral plexus** originates from L4–S4. However, because of their close, overlapping relationship and similar distribution, the two plexuses are often considered together as a single **lumbosacral plexus** (L1–S4; figure 12.23). Four major nerves exit the lumbosacral plexus and enter the lower limb: the obturator, femoral, tibial, and common fibular (peroneal). The obturator nerve innervates the medial thigh; the femoral nerve innervates the anterior thigh; the tibial nerve innervates the posterior thigh, leg, and foot; and the common fibular nerve innervates the posterior thigh, anterior and lateral leg, and foot. Other lumbosacral nerves supply the lower back, hip, and lower abdomen.

Obturator Nerve

The **obturator** (ob′too-ră-tōr) **nerve** supplies the muscles that adduct the thigh. Its cutaneous sensory distribution is to the medial side of the thigh (figure 12.24).

Femoral Nerve

The **femoral nerve** innervates the iliopsoas and sartorius muscles and the quadriceps femoris group. Its cutaneous sensory distribution is the anterior and lateral thigh and the medial leg and foot (figure 12.25).

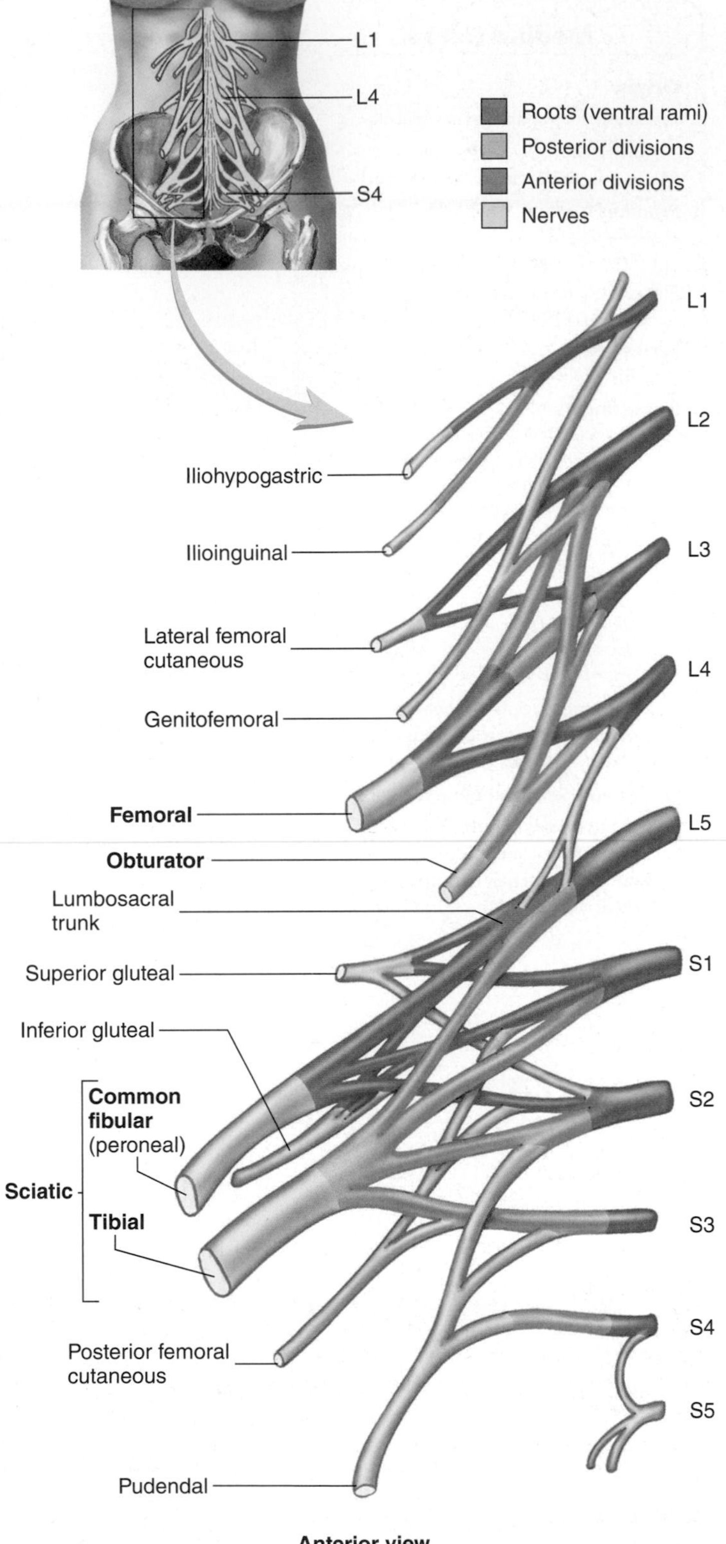
L1
L4
S4
Roots (ventral rami)
Posterior divisions
Anterior divisions
Nerves
L1
L2
L3
L4
L5
S1
S2
S3
S4
S5
Iliohypogastric
Ilioinguinal
Lateral femoral cutaneous
Genitofemoral
Femoral
Obturator
Lumbosacral trunk
Superior gluteal
Inferior gluteal
Common fibular (peroneal)
Tibial
Sciatic
Posterior femoral cutaneous
Pudendal
Anterior view

FIGURE 12.23 APR Lumbosacral Plexus

The roots of the plexus are formed by the ventral rami of the spinal nerves L1–S4 and form anterior and posterior divisions, which give rise to the lumbosacral nerves. The lumbosacral trunk joins the lumbar and sacral plexuses.

Predict 6

While moving a large box in his garage, Carl hurt his back. Within hours, the pain in his back was radiating down the medial side of his left thigh and knee. It was especially difficult for Carl to flex his left hip and extend his left knee. He could not stand the pain, so he went to the emergency room. Radiographs and an MRI revealed a bulging intervertebral disk. Which spinal nerve was most likely affected, and why were Carl's motor movements affected? Also, explain the referred pain.

Clinical
IMPACT

Sciatic Nerve Damage

If a person sits on a hard surface for a considerable time, the sciatic nerve may be compressed against the ischial portion of the coxal bone. When the person stands up, he or she feels a tingling sensation, described as "pins and needles," throughout the lower limb and often remarks that the limb has "gone to sleep." This condition is temporary, but the sciatic nerve can be seriously injured in a number of ways. A ruptured intervertebral disk or pressure from the uterus during pregnancy may compress the roots of the sciatic nerve. Other causes of sciatic nerve damage include hip injury, compression of the nerve by the piriformis muscle (piriformis syndrome), and an improperly administered injection in the hip region (see Clinical Impact, "Gluteal Injections," in section 7.3).

Obturator Nerve

Origin

Lumbosacral plexus, L2–L4

Movements/Muscles Innervated

Rotates thigh laterally

- *Obturator externus*

Adducts thigh

- *Adductor magnus (adductor part)*
- *Adductor longus*
- *Adductor brevis*

Adducts thigh and flexes knee

- *Gracilis*

Cutaneous (Sensory) Innervation

Superior medial side of thigh

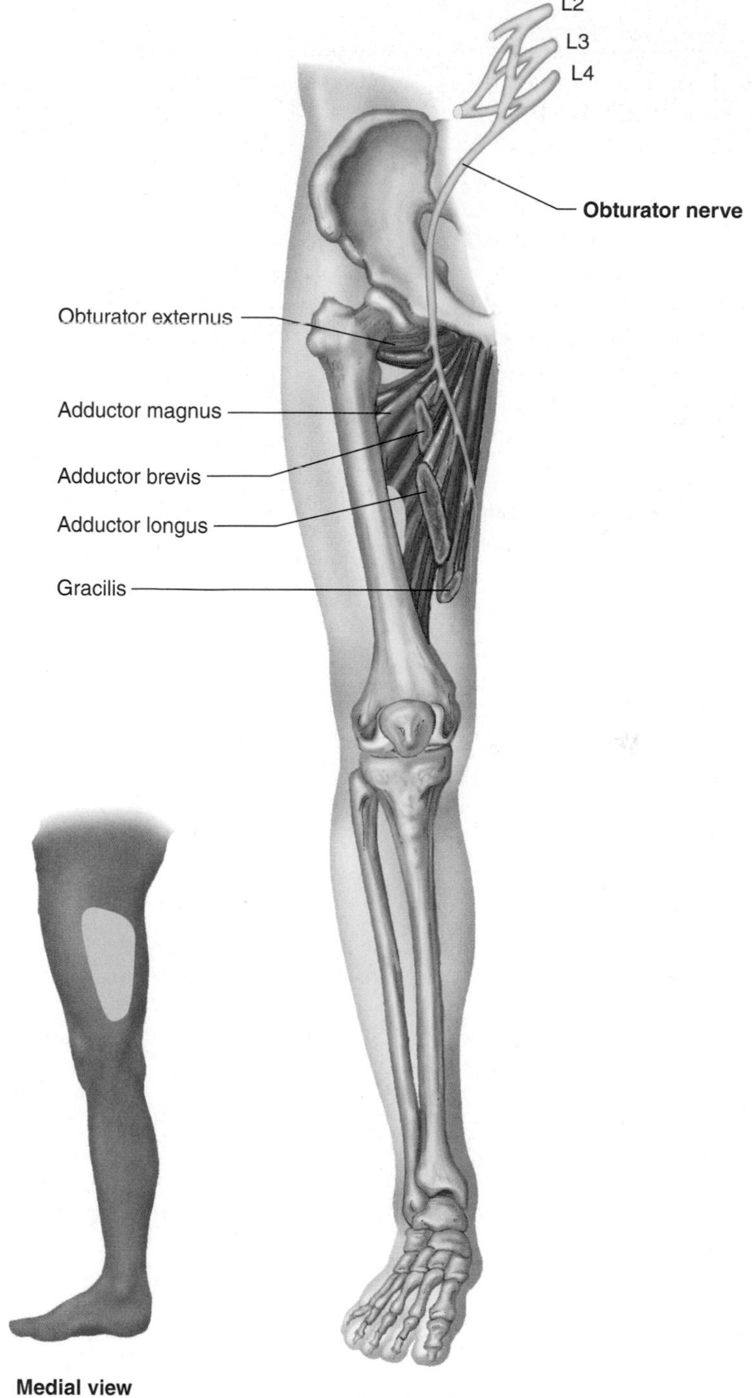
L2
L3
L4
Obturator nerve
Obturator externus
Adductor magnus
Adductor brevis
Adductor longus
Gracilis
Medial view
Anterior view

FIGURE 12.24 Obturator Nerve
The route of the obturator nerve and the muscles it innervates. The *inset* depicts the cutaneous (sensory) distribution of the nerve (blue area).

Tibial and Common Fibular Nerves

The **tibial** and **common fibular nerves** (*peroneal nerves*; per-ō-nē′ăl) originate from spinal segments L4–S3 and are bound together within a connective tissue sheath for the length of the thigh (figures 12.26 and 12.27; see figure 12.23). These two nerves, combined within the same sheath, are referred to jointly as the **sciatic** (sī-at′ik) **nerve,** or *ischiadic* (is-kē-ad′ik) *nerve* (see figure 12.23). The sciatic nerve, by far the largest peripheral nerve in the body, passes through the greater sciatic notch in the pelvis and descends in the posterior thigh to the popliteal fossa, where the two portions of the sciatic nerve separate.

The tibial nerve innervates most of the posterior thigh and leg muscle (see figure 12.26). It branches in the foot to form the **medial** and **lateral plantar** (plan′tăr) **nerves,** which innervate the plantar muscles of the foot and the skin over the sole of the foot. Another branch, the **sural** (soo′răl) **nerve,** supplies part of the cutaneous innervation over the calf of the leg and the plantar surface of the foot.

Femoral Nerve

Origin

Lumbosacral plexus, L2–L4

Movements/Muscles Innervated

Flexes hip
- *Psoas major*
- *Iliacus*
- *Pectineus*

Flexes hip and flexes knee
- *Sartorius*

Extends knee
- *Vastus lateralis*
- *Vastus intermedius*
- *Vastus medialis*

Extends knee and flexes hip
- *Rectus femoris*

Cutaneous (Sensory) Innervation

Anterior and lateral branches supply the anterior and lateral thigh; saphenous branch supplies the medial leg and foot

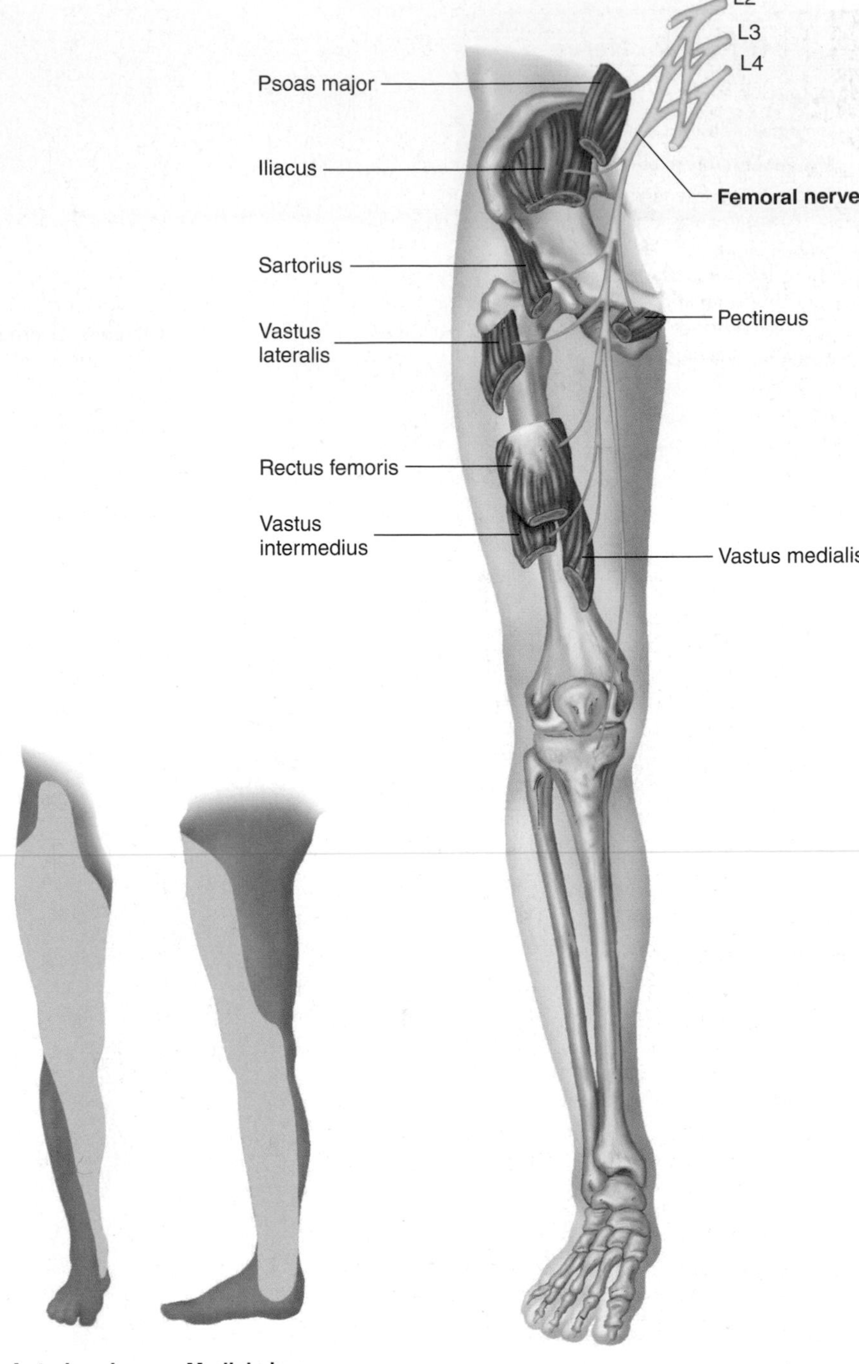
L2
L3
L4
Psoas major
Iliacus
Femoral nerve
Sartorius
Pectineus
Vastus lateralis
Rectus femoris
Vastus intermedius
Vastus medialis
Anterior view
Medial view
Anterior view

FIGURE 12.25 Femoral Nerve
The route of the femoral nerve and the muscles it innervates. The *insets* depict the cutaneous (sensory) distribution of the nerve (blue areas).

The common fibular nerve divides into the **deep** and **superficial fibular (peroneal) nerves.** These branches innervate the anterior and lateral muscles of the leg and foot. The cutaneous distribution of the common fibular nerve and its branches is the lateral and anterior leg and the dorsum of the foot (figure 12.27).

Other Lumbosacral Plexus Nerves

In addition to the nerves just described, the lumbosacral plexus gives rise to **gluteal nerves,** which supply the hip muscles that act on the femur, and the **pudendal** (pū-den′dăl) **nerve,** which supplies the muscles of the abdominal floor (see figure 12.23). The **iliohypogastric** (il′ē-ō-hī-pō-gas′trik), **ilioinguinal** (il′ē-ō-ing′gwi-năl), **genitofemoral** (jen′i-tō-fem′ŏ-răl), **cutaneous femoral,** and pudendal nerves innervate the skin of the suprapubic area, the external genitalia, the superior medial thigh, and the posterior thigh. The pudendal nerve plays a vital role in sexual stimulation and response. Branches of the pudendal nerve are anesthetized during childbirth before a doctor performs an **episiotomy** (e-piz-ē-ot′ō-mē, epis-ē-ot′ō-mē), a cut in the perineum that enlarges the opening of the birth canal.

Tibial Nerve

Origin

Lumbosacral plexus, L4–S3

Movements/Muscles Innervated

Extends hip and flexes knee
- *Biceps femoris (long head)*
- *Semitendinosus*
- *Semimembranosus*

Extends hip and adducts thigh
- *Adductor magnus (hamstring part)*

Plantar flexes foot
- *Plantaris*
- *Gastrocnemius*
- *Soleus*
- *Tibialis posterior*

Flexes knee
- *Popliteus*

Flexes toes
- *Flexor digitorum longus*
- *Flexor hallucis longus*

Cutaneous (Sensory) Innervation

None

Medial and Lateral Plantar Nerves

Origin

Tibial nerve

Movements/Muscles Innervated

Flex and adduct toes
- *Plantar muscles of foot*

Cutaneous (Sensory) Innervation

Sole of foot

Sural Nerve (Not Shown)

Origin

Tibial nerve

Movements/Muscles Innervated

None

Cutaneous (Sensory) Innervation

Lateral and posterior one-third of leg and lateral side of foot

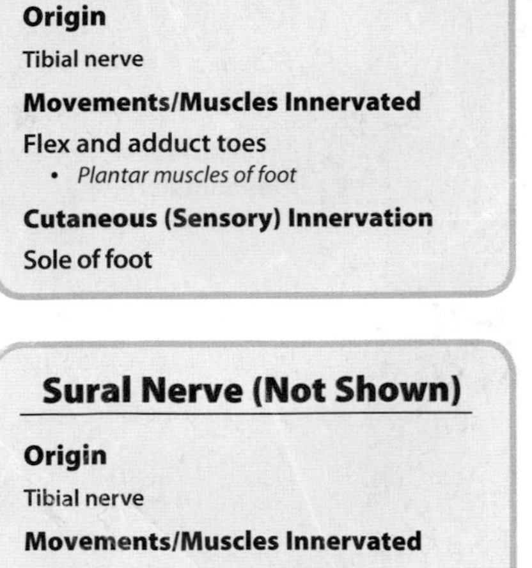
L4
L5
S1
S2
S3
Tibial nerve
Biceps femoris long head
Semimembranosus
Semitendinosus
Adduotor magnus
Gastrocnemius
Popliteus
Soleus
Flexor digitorum longus
Flexor hallucis longus
Tibialis posterior
Medial plantar nerve to plantar muscles
Lateral plantar nerve to plantar muscles
Anterior view
Posterior view
Posterior view

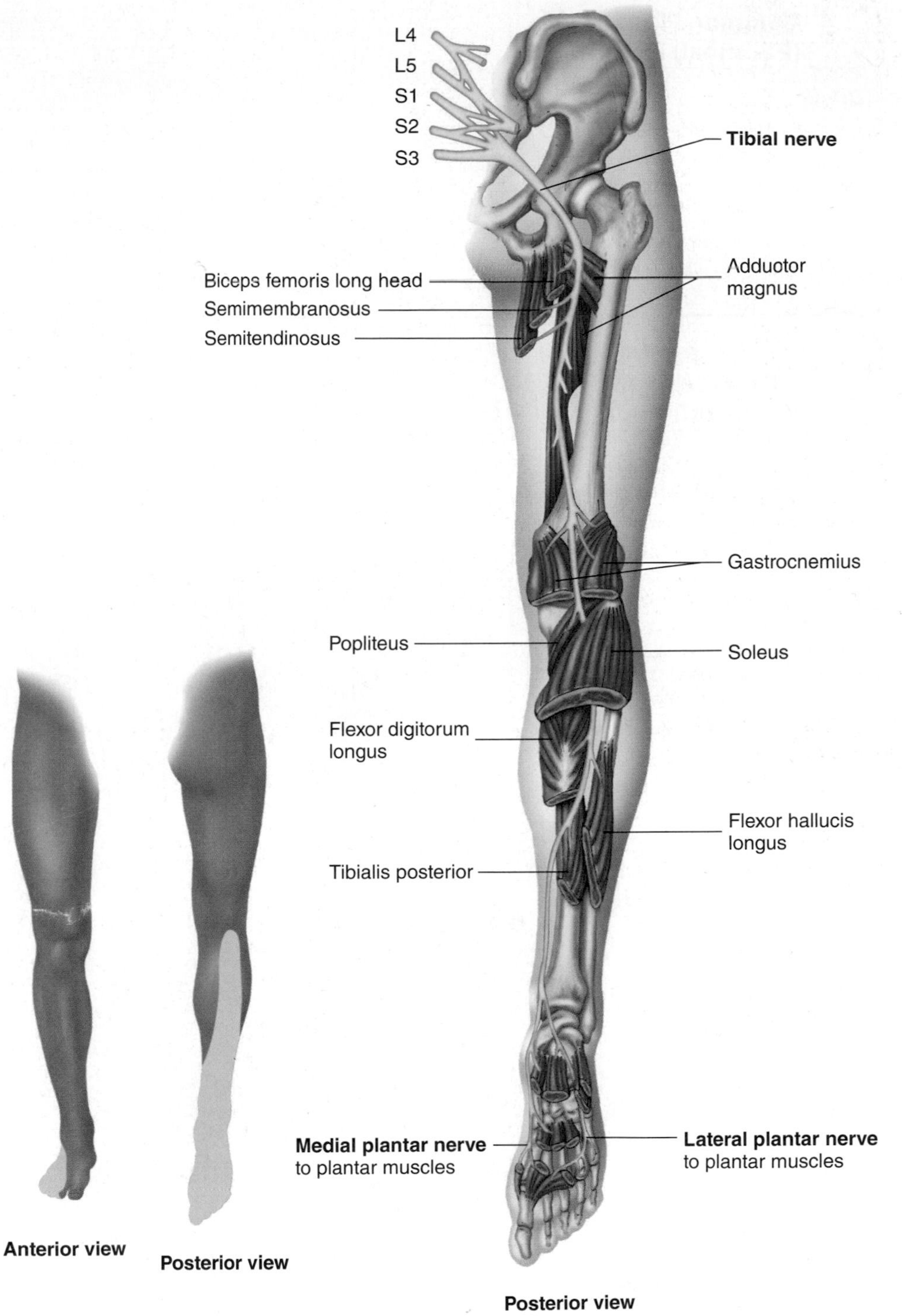

FIGURE 12.26 Tibial Nerve
The route of the tibial nerve and the muscles it innervates. The *insets* depict the cutaneous (sensory) distribution of the nerve (blue areas).

Coccygeal Plexus

The **coccygeal** (kok-sij′ē-ăl) **plexus** is a very small plexus formed from the ventral rami of spinal nerve S5 and the coccygeal nerve (Co). This small plexus supplies motor innervation to the muscles of the pelvic floor and sensory cutaneous innervation to the skin over the coccyx. The dorsal rami of the coccygeal nerves innervate some skin over the coccyx.

Common Fibular (Peroneal) Nerve

Origin
Lumbosacral plexus, L4–S2

Movements/Muscles Innervated
Extends hip and flexes knee
- *Biceps femoris (short head)*

Cutaneous (Sensory) Innervation
Lateral surface of knee

Deep Fibular (Peroneal) Nerve

Origin
Common fibular (peroneal) nerve

Movements/Muscles Innervated
Dorsiflexes foot
- *Tibialis anterior*
- *Fibularis tertius*

Extends toes
- *Extensor digitorum longus*
- *Extensor hallucis longus*
- *Extensor digitorum brevis*

Cutaneous (Sensory) Innervation
Great and second toe

Superficial Fibular (Peroneal) Nerve

Origin
Common fibular (peroneal) nerve

Movements/Muscles Innervated
Plantar flexes and everts foot
- *Fibularis longus*
- *Fibularis brevis*

Cutaneous (Sensory) Innervation
Dorsal anterior third of leg and dorsum of foot

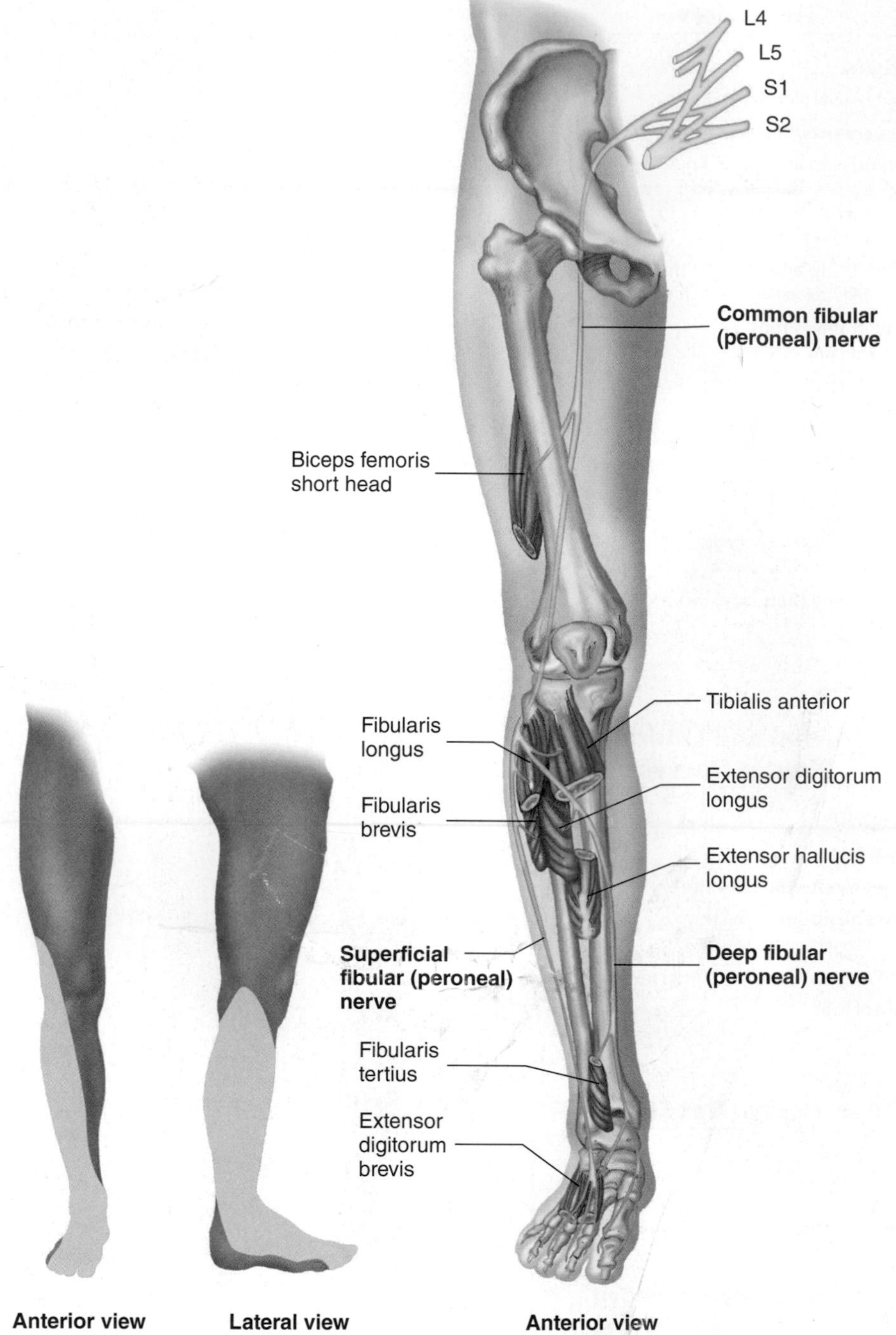
L4
L5
S1
S2
Common fibular (peroneal) nerve
Biceps femoris short head
Tibialis anterior
Fibularis longus
Extensor digitorum longus
Fibularis brevis
Extensor hallucis longus
Superficial fibular (peroneal) nerve
Deep fibular (peroneal) nerve
Fibularis tertius
Extensor digitorum brevis
Anterior view
Lateral view
Anterior view

FIGURE 12.27 Fibular Nerve
The route of the common fibular (peroneal) nerve and the muscles it innervates. The *insets* depict the cutaneous (sensory) distribution of the nerve (blue areas).

ASSESS YOUR PROGRESS

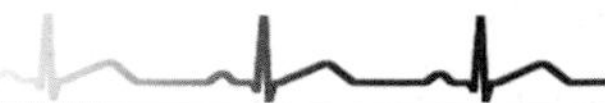

18. *Name the structures innervated by the cervical plexus. Describe the makeup of the phrenic nerve.*

19. *Name the five major nerves that emerge from the brachial plexus. List the muscles they innervate and the areas of skin they supply.*

20. *Name the four major nerves that arise from the lumbosacral plexus, and describe the muscles and skin areas they supply.*

21. *What is the name given to the tibial and common fibular nerves that are bound together?*

22. *Describe the structures innervated by the obturator and femoral nerves.*

23. *What structures are innervated by the coccygeal plexus?*

Clinical IMPACT — Bionic Sensors

Patients paralyzed by strokes or spinal cord lesions are able to regain some functions with the help of microcomputers that stimulate certain programmed activities, such as grasping and walking. The microcomputer initiates electrical impulses, which are conveyed through fine wire leads either to peripheral nerves or directly to the muscles responsible for the desired movement. The subtle movement of muscles not affected by the paralysis initiates the program. Sensors connected to the microcomputer are attached to the skin overlying functional muscles, where they are able to detect electrical activity associated with movement of the underlying muscles. For example, a person whose legs are paralyzed may have a sensor attached to the abdomen. The abdominal muscles normally involved in stabilizing and moving the pelvis during walking are stimulated by descending tracts when CNS centers initiate walking. The resultant abdominal muscle activity is detected by the sensor, which activates the program that stimulates the appropriate sequence of muscles in the lower limbs, and the paralyzed person walks. Similarly, a quadriplegic using subtle movements of the shoulder, neck, or face, where specific sensors can be placed, can initiate certain upper limb and grasping actions.

Diseases and Disorders

TABLE 12.1 Spinal Cord and Spinal Nerves

Condition	Description
SPINAL CORD DISORDERS	
Encephalitis	Inflammation of the brain caused by a virus and less often by bacteria or other agents; symptoms include fever, coma, and convulsions
Meningitis	Inflammation of meninges caused by viral or bacterial infection; symptoms include stiffness in the neck, headache, and fever; severe cases can cause paralysis, coma, or death
Rabies	Viral disease transmitted by an infected animal; brain infection results in abnormal excitability, aggression, paralysis, and death
Tetanus	Caused by bacterial neurotoxin; affects lower motor neurons in spinal cord and brainstem, leading to muscle contraction; prevents muscle relaxation; body becomes rigid, including "lockjaw"; death results from spasms in respiratory muscles
Multiple sclerosis	Autoimmune condition; may be initiated by viral infection; inflammation in brain and spinal cord with demyelination and sclerotic (hard) sheaths result in poor conduction of action potentials; symptoms include exaggerated reflexes, tremor, and speech defects
SPINAL NERVE DISORDERS	
Anesthesia	Loss of sensation; may be a pathological condition or may be induced temporarily to facilitate medical action
Neuritis	Inflammation of a nerve from a number of causes; in motor nerves, can result in loss of motor function; in sensory nerves, can result in anesthesia or neuralgia
Neuralgia	Nerve pain; involves severe spasms of throbbing or stabbing pain along the pathway of a nerve; can result from inflammation, nerve damage, or an unknown cause
Sciatica	Neuralgia of the sciatic nerve, with pain radiating down the back of the leg; most common cause is a herniated lumbar disk
Leprosy	Bacterial disease that kills skin and PNS cells; characterized by disfiguring nodules and tissue necrosis
Herpes	Family of diseases characterized by skin lesions due to herpes viruses in sensory ganglia; different viruses cause oral lesions (cold sores), sexually transmitted disease with lesions on genitalia, or chickenpox in children (shingles in adults)
Poliomyelitis	Viral infection of the CNS, but primarily damages somatic motor neurons, leaving muscles without innervation; leads to paralysis and atrophy
Diabetic neuropathy	Damage to nerves occurs in diabetic patients due to high blood sugar levels and decreased blood flow; occurs in over half of diabetics; symptoms develop slowly, often includes pain or numbness in extremities, but can be manifested in every organ system
Charcot-Marie-Tooth disease	One of the most common hereditary neurological disorders (named for three neurologists who described it); different forms caused by mutations that generally lead to damage of the myelin sheath; characterized primarily by muscle weakness, dysfunction, atrophy, and loss of sensation primarily in lower extremities
Neurofibromatosis	Genetic disorder; neurofibromas (benign tumors along peripheral nerve tract) occur in early childhood and result in skin growths
Myasthenia gravis	Autoimmune disorder affecting acetylcholine receptors; makes the neuromuscular junction less functional; muscle weakness and increased fatigue lead to paralysis

Go to www.mhhe.com/seeley10 for additional information on these pathologies.

Answer

Learn to Predict

From page 400

Javier reacted to the pain of stepping on the toy cars not once but twice. We know that each time he stepped on a toy he switched the leg that supported his weight, first from the right to the left and then from the left to the right. Chapter 12 described the withdrawal reflex that moves a limb away from a painful stimulus. Javier demonstrated this reflex when he stepped on the toys.

Javier displayed the withdrawal reflex and the crossed extensor reflex. Both of these reflexes are polysynaptic spinal reflexes, involving sensory neurons, interneurons, and motor neurons. Chapter 11 explained the functions of each of these neuron types. The sensory neurons delivered the painful stimulus to the central nervous system; the interneurons relayed the information between sensory and motor neurons; and the motor neurons regulated skeletal muscle activity.

When Javier stepped on the toy car with his right foot, he transferred nearly all of his weight to his right leg before the withdrawal reflex was activated. The painful stimulus of stepping on the toy car activated the withdrawal reflex, causing Javier to pick up his right leg. At the same time, Javier extended his left leg because of the crossed extensor reflex. However, as Javier extended his left leg, he stepped on another toy car, activating the withdrawal reflex in his left leg and the crossed extensor reflex in his right leg. The repeated sequence of withdrawing one leg and stepping down on the other foot prevented Javier from falling down.

Answers to the rest of this chapter's Predict questions are in Appendix G.

Summary

12.1 Spinal Cord (p. 401)

General Structure

1. The spinal cord gives rise to 31 pairs of spinal nerves. Nerves of the limbs enter and leave the spinal cord at the cervical and lumbosacral enlargements.
2. The spinal cord is shorter than the vertebral column. Nerves from the end of the spinal cord form the cauda equina.

Meninges of the Spinal Cord

Three meningeal layers surround the spinal cord: the dura mater, arachnoid mater, and pia mater.

Cross Section of the Spinal Cord

1. The spinal cord consists of peripheral white matter and central gray matter.
2. White matter is organized into columns, which are subdivided into nerve tracts, or fasciculi, which carry action potentials to and from the brain.
3. Gray matter is divided into horns.
   - The dorsal horns contain sensory axons that synapse with interneurons. The ventral horns contain the neuron cell bodies of somatic motor neurons, and the lateral horns contain the neuron cell bodies of autonomic motor neurons.
   - The gray and white commissures connect each half of the spinal cord.
4. The dorsal root conveys sensory input into the spinal cord, and the ventral root conveys motor output away from the spinal cord.

12.2 Reflexes (p. 404)

1. A reflex arc is the functional unit of the nervous system.
   - Sensory receptors respond to stimuli and produce action potentials in sensory neurons.
   - Sensory neurons propagate action potentials to the CNS.
   - Interneurons in the CNS synapse with sensory neurons and with motor neurons.
   - Motor neurons carry action potentials from the CNS to effector organs.
   - Effector organs, such as muscles or glands, respond to the action potentials.
2. Reflexes do not require conscious thought, and they produce a consistent and predictable result.
3. Reflexes are homeostatic.
4. Reflexes are integrated within the brain and spinal cord. Higher brain centers can suppress or exaggerate reflexes.

Stretch Reflex

Muscle spindles detect the stretch of skeletal muscles and cause the muscle to shorten reflexively.

Golgi Tendon Reflex

Golgi tendon organs respond to increased tension within tendons and cause skeletal muscles to relax.

Withdrawal Reflex

1. Activation of pain receptors causes muscles to contract and move some part of the body away from a painful stimulus.
2. Reciprocal innervation causes muscles that would oppose withdrawal to relax.
3. In the crossed extensor reflex, flexion of one limb caused by the withdrawal reflex stimulates the opposite limb to extend.

Interactions with Spinal Cord Reflexes

Convergent and divergent pathways interact with reflexes.

12.3 Spinal Nerves (p. 410)

Structure of Nerves

In the PNS, individual axons are surrounded by the endoneurium. Groups of axons, called fascicles, are bound together by the perineurium. The fascicles form the nerve and are held together by the epineurium.

Organization of Spinal Nerves

1. Eight cervical, 12 thoracic, 5 lumbar, 5 sacral pairs, and 1 coccygeal pair make up the spinal nerves.
2. Spinal nerves have specific cutaneous distributions called dermatomes.
3. Spinal nerves branch to form rami.
   - The dorsal rami supply the muscles and skin near the midline of the back.
   - The ventral rami in the thoracic region form intercostal nerves, which supply the thorax and upper abdomen. The remaining ventral rami join to form plexuses. Communicating rami supply sympathetic nerves.

Cervical Plexus

Spinal nerves C1–C4 form the cervical plexus, which supplies some muscles and the skin of the neck and shoulder. The phrenic nerves innervate the diaphragm.

Brachial Plexus

1. Spinal nerves C5–T1 form the brachial plexus, which supplies the upper limb.
2. The axillary nerve innervates the deltoid and teres minor muscles and the skin of the shoulder.
3. The radial nerve supplies the extensor muscles of the arm and forearm and the skin of the posterior surface of the arm, forearm, and hand.
4. The musculocutaneous nerve supplies the anterior arm muscles and the skin of the lateral surface of the forearm.
5. The ulnar nerve innervates most of the intrinsic hand muscles and the skin on the ulnar side of the hand.
6. The median nerve innervates the pronator and most of the flexor muscles of the forearm, most of the thenar muscles, and the skin of the radial side of the palm of the hand.
7. Other nerves supply most of the muscles that act on the arm, the scapula, and the skin of the medial arm and forearm.

Lumbar and Sacral Plexuses

1. Spinal nerves L1–S4 form the lumbosacral plexus.
2. The obturator nerve supplies the muscles that adduct the thigh and the skin of the medial thigh.
3. The femoral nerve supplies the muscles that flex the thigh and extend the leg and the skin of the anterior and lateral thigh and the medial leg and foot.
4. The tibial nerve innervates the muscles that extend the thigh and flex the leg and the foot. It also supplies the plantar muscles and the skin of the posterior leg and the sole of the foot.
5. The common fibular nerve and its branches supply the short head of the biceps femoris, the muscles that dorsiflex and plantar flex the foot, and the skin of the lateral and anterior leg and the dorsum of the foot.
6. In the thigh, the tibial nerve and the common fibular nerve are combined as the sciatic nerve.
7. Other lumbosacral nerves supply the lower abdominal muscles, the hip muscles, and the skin of the suprapubic area, external genitalia, and upper medial thigh.

Coccygeal Plexus

Spinal nerve S5 and the coccygeal nerve form the coccygeal plexus, which supplies the muscles of the pelvic floor and the skin over the coccyx.

REVIEW AND COMPREHENSION

1. The spinal cord extends from the
   a. medulla oblongata to the coccyx.
   b. level of the third cervical vertebra to the coccyx.
   c. level of the axis to the lowest lumbar vertebra.
   d. level of the foramen magnum to the second lumbar vertebra.
   e. axis to the sacral hiatus.
2. The structure that anchors the inferior end of the spinal cord to the coccyx is the
   a. conus medullaris.
   b. cauda equina.
   c. filum terminale.
   d. lumbar enlargement.
   e. posterior median sulcus.
3. Axons of sensory neurons synapse with the cell bodies of interneurons in the ________ of spinal cord gray matter.
   a. anterior horn
   b. lateral horn
   c. posterior horn
   d. gray commissure
   e. lateral columns
4. Given these components of a reflex arc:
   (1) effector organ
   (2) interneuron
   (3) motor neuron
   (4) sensory neuron
   (5) sensory receptor

   Choose the correct order an action potential follows after a sensory receptor is stimulated.
   a. 5,4,3,2,1
   b. 5,4,2,3,1
   c. 5,3,4,1,2
   d. 5,2,4,3,1
   e. 5,3,2,1,4
5. A reflex response accompanied by the conscious sensation of pain is possible because of
   a. convergent pathways.
   b. divergent pathways.
   c. a reflex arc that contains only one neuron.
   d. sensory perception in the spinal cord.
6. Several of the events that occur between the time a physician strikes a patient's patellar tendon with a rubber hammer and the time the quadriceps femoris contracts (knee-jerk reflex) are listed below:
   (1) increased frequency of action potentials in sensory neurons
   (2) stretch of the muscle spindles
   (3) increased frequency of action potentials in the alpha motor neurons
   (4) stretch of the quadriceps femoris
   (5) contraction of the quadriceps femoris

   Which list most closely describes the sequence of events as they normally occur?
   a. 4,1,2,3,5
   b. 4,1,3,2,5
   c. 1,4,3,2,5
   d. 4,2,1,3,5
   e. 4,2,3,1,5

7. ________ are responsible for regulating the sensitivity of the muscle spindle.
   a. Alpha motor neurons
   b. Sensory neurons
   c. Gamma motor neurons
   d. Golgi tendon organs
   e. Inhibitory interneurons
8. Which of these events occurs when a person steps on a tack with the right foot?
   a. The right foot is pulled away from the tack because of the Golgi tendon reflex.
   b. The left leg is extended to support the body because of the stretch reflex.
   c. The flexor muscles of the right thigh contract, and the extensor muscles of the right thigh relax because of reciprocal innervation.
   d. Extensor muscles contract in both thighs because of the crossed extensor reflex.
9. Damage to the dorsal ramus of a spinal nerve results in
   a. loss of sensation.
   b. loss of motor function.
   c. Both a and b are correct.
10. A collection of spinal nerves that join together after leaving the spinal cord is called a
    a. ganglion.
    b. nucleus.
    c. projection nerve.
    d. plexus.
11. A dermatome
    a. is the area of skin supplied by a pair of spinal nerves.
    b. exists for each spinal nerve except C1.
    c. can be used to locate the site of spinal cord or nerve root damage.
    d. All of these are correct.
12. Which of these nerves arises from the cervical plexus?
    a. median
    b. musculocutaneous
    c. phrenic
    d. obturator
    e. ulnar
13. The skin on the posterior surface of the hand is supplied by the
    a. median nerve.
    b. musculocutaneous nerve.
    c. ulnar nerve.
    d. axillary nerve.
    e. radial nerve.
14. The sciatic nerve is actually two nerves combined within the same sheath. The two nerves are the
    a. femoral and obturator.
    b. femoral and gluteal.
    c. common fibular (peroneal) and tibial.
    d. common fibular (peroneal) and obturator.
    e. tibial and gluteal.
15. The muscles of the anterior compartment of the thigh are supplied by the
    a. obturator nerve.
    b. gluteal nerve.
    c. sciatic nerve.
    d. femoral nerve.
    e. ilioinguinal nerve.

Answers in Appendix E

CRITICAL THINKING

1. Describe how stimulation of a neuron that has its cell body in the cerebrum can inhibit a reflex that is integrated within the spinal cord.
2. A cancer patient has his left lung removed. To reduce the space remaining where the lung is removed, the diaphragm on the left side is paralyzed to allow the abdominal viscera to push the diaphragm upward. What nerve is cut? Where is a good place to cut it, and when would the surgery be done?
3. During a difficult delivery, the baby's arm delivered first. The attending physician grasped the arm and forcefully pulled it. Later, a nurse observed that the baby could not abduct or adduct the medial four fingers, and flexion of the wrist was impaired. What nerve was damaged?
4. Two patients are admitted to the hospital. According to their charts, both have herniated disks that are placing pressure on the roots of the sciatic nerve. One patient has pain in the buttocks and the posterior aspect of the thigh. The other patient experiences pain in the posterior and lateral aspects of the leg and the lateral part of the ankle and foot. Explain how the same condition, a herniated disk, can produce such different symptoms.
5. In an automobile accident, a woman suffers a crushing hip injury. For each of the following conditions, state what nerve is damaged.
   a. unable to adduct the thigh
   b. unable to extend the knee
   c. unable to flex the knee
   d. loss of sensation from the skin of the anterior thigh
   e. loss of sensation from the skin of the medial thigh
6. A skier breaks his ankle. As part of his treatment, the ankle and leg are placed in a plaster cast. Unfortunately, the cast is too tight around the proximal portion of the leg and presses in against the neck of the fibula. Predict where the patient will experience tingling or numbness in the leg. Explain.
7. Cecil's motorcycle collided with a tree. When the ambulance arrived, he complained of loss of sensation and voluntary movement in his lower limbs, as well as impaired mobility in his upper limbs, especially his hands. Examination revealed that the accident had permanently damaged his spinal cord. Cecil was able to breathe on his own, and with exercise, movement of his upper limbs eventually improved, although the mobility of his hands was still impaired, and he never regained the use of his lower limbs. At which level was Cecil's spinal cord injured? Explain how you were able to determine your answer.

Answers in Appendix F

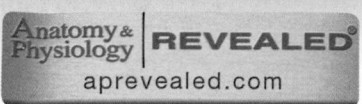

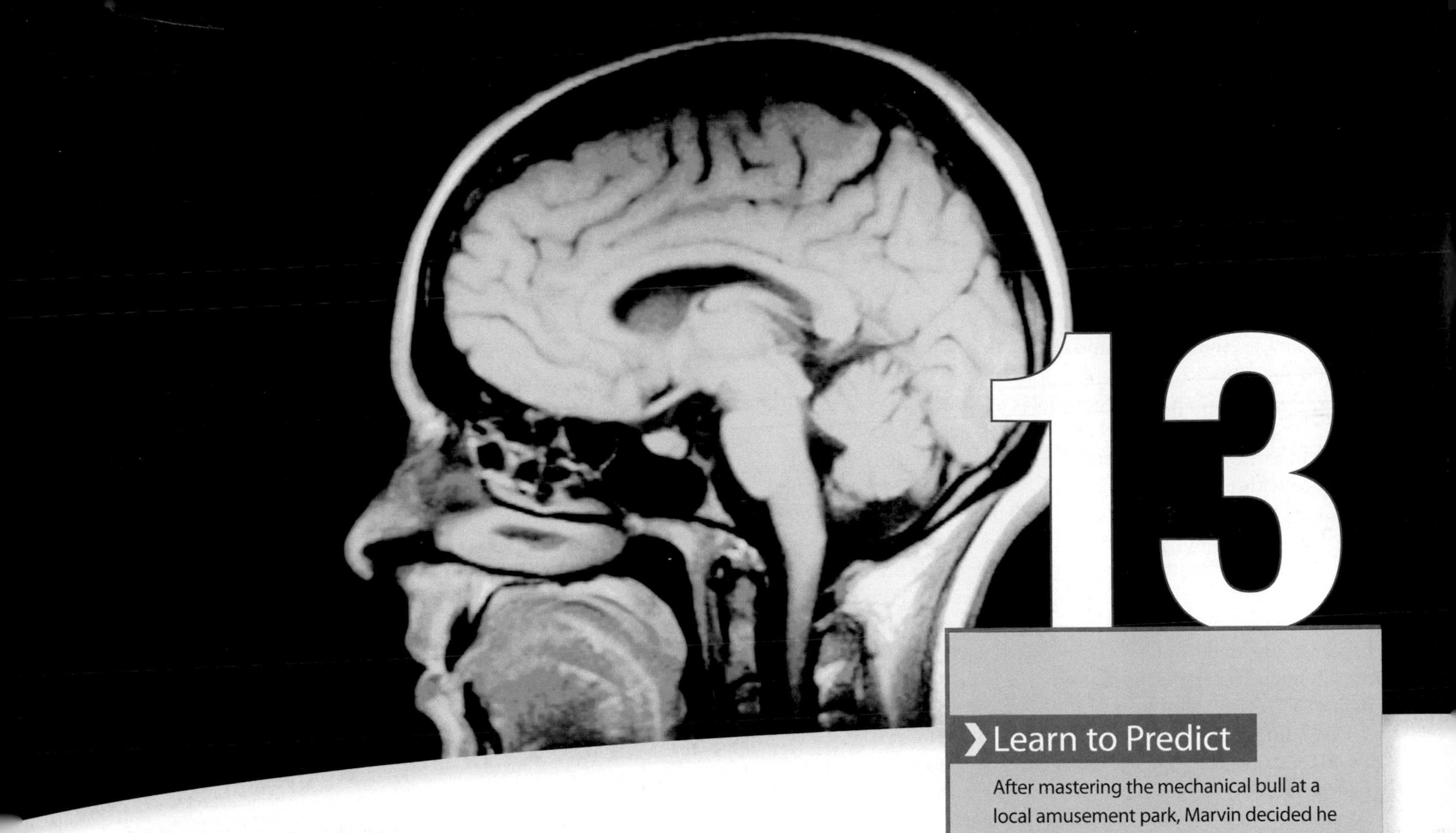

13

Brain and Cranial Nerves

The complexity of the human brain is mind-boggling. It contains around 100 billion neurons and another trillion neuroglial cells. Even more amazing is that each neuron has an average of 10,000 connections with other neurons. From that complexity arises the command and control of our bodies by the spinal cord and spinal nerves, described in chapter 12, and the cranial nerves, discussed in this chapter, that connect directly to the brain. Furthermore, these connections also, importantly, generate our perception of self. Current research is adding insight into the organization and functions of the brain, yet much remains a mystery. This chapter lays the foundation by describing the structure of the brain, its functional units, and its associated cranial nerves. The integration of brain functions is discussed in chapter 14.

Learn to Predict

After mastering the mechanical bull at a local amusement park, Marvin decided he was ready to compete in an amateur rodeo. But during the very first event, a live bull threw Marvin to the dirt and kicked him in the side of the face. Clinicians in the emergency room diagnosed a broken jaw, and they noted that Marvin's tongue deviated to the right when they asked him to stick it out. A magnetic resonance imaging (MRI) scan was performed to detect the presence of brain or cranial nerve damage. By recalling your knowledge of nervous system organization from chapter 11 and combining it with new information about the cranial nerves in this chapter, explain the results of Marvin's tongue protrusion test.

Photo: MRI scan showing the head and brain of a patient.

Module 7
Nervous System

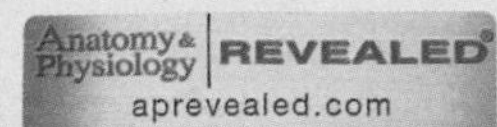
Anatomy & Physiology
REVEALED
aprevealed.com

13.1 Development of the CNS

LEARNING OUTCOMES

After reading this section, you should be able to

A. Describe the development of the neural tube and name the embryonic pouches and the adult brain structures that they become.

B. Explain the origin of the ventricles of the brain.

The brain is the part of the central nervous system (CNS) that is contained within the cranial cavity (figure 13.1). It consists of the brainstem, the cerebellum, the diencephalon, and the cerebrum (table 13.1). The brainstem includes the medulla oblongata, the pons, and the midbrain. We begin our study of the brain and the cranial nerves by describing how the CNS develops in the fetus.

The CNS forms from a flat plate of ectodermal tissue (see chapter 4), the **neural plate,** on the dorsal surface of the embryo, which is influenced in part by the underlying rod-shaped **notochord** (figure 13.2). The lateral sides of the neural plate become elevated as waves, forming **neural folds.** The crest of each fold is called a **neural crest,** and the center of the neural plate becomes the **neural groove.** The neural folds move toward each other in the midline, and the crests fuse to create a **neural tube** (figure 13.2). The cephalic portion of the neural tube becomes the brain, and the caudal portion becomes the spinal cord. **Neural crest cells** are cells that separate from the neural crests and give rise to sensory, autonomic, and enteric neurons of the peripheral nervous system. They also give rise to all the pigmented cells of the body, the adrenal medulla, the facial bones, and the dentin of the teeth.

A series of pouches develops in the anterior part of the neural tube, forming three brain regions in the early embryo (figure 13.3*a*): a **forebrain,** or *prosencephalon* (pros-en-sef′ă-lon); a **midbrain,** or *mesencephalon* (mez-en-sef′ă-lon); and a **hindbrain,** or *rhombencephalon* (rom-ben-sef′ă-lon). The pouch walls become the various portions of the adult brain (table 13.2). The forebrain divides into the **telencephalon** (tel-en-sef′ă-lon), which becomes the cerebrum, and the **diencephalon** (dī-en-sef′ă-lon). The midbrain remains a single structure as in the embryo, the **mesencephalon,** but the hindbrain divides into the **metencephalon** (met′en-sef′ă-lon), which becomes the pons and cerebellum, and the **myelencephalon** (mī′el-en-sef′ă-lon), which becomes the medulla oblongata (figure 13.3*b*,*c*).

The pouch cavities become fluid-filled **ventricles** (ven′tri-klz). The ventricles are continuous with each other and with the **central canal** of the spinal cord. The neural tube develops flexures that cause the brain to be oriented almost 90 degrees to the spinal cord.

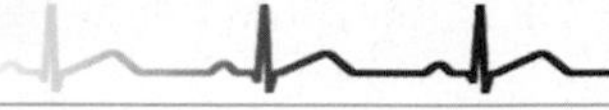

ASSESS YOUR PROGRESS

1. *Name the five pouches of the neural tube and the part of the adult brain that each division becomes.*
2. *What do the cavities of the neural tube become in the adult brain?*

FUNDAMENTAL **Figure**

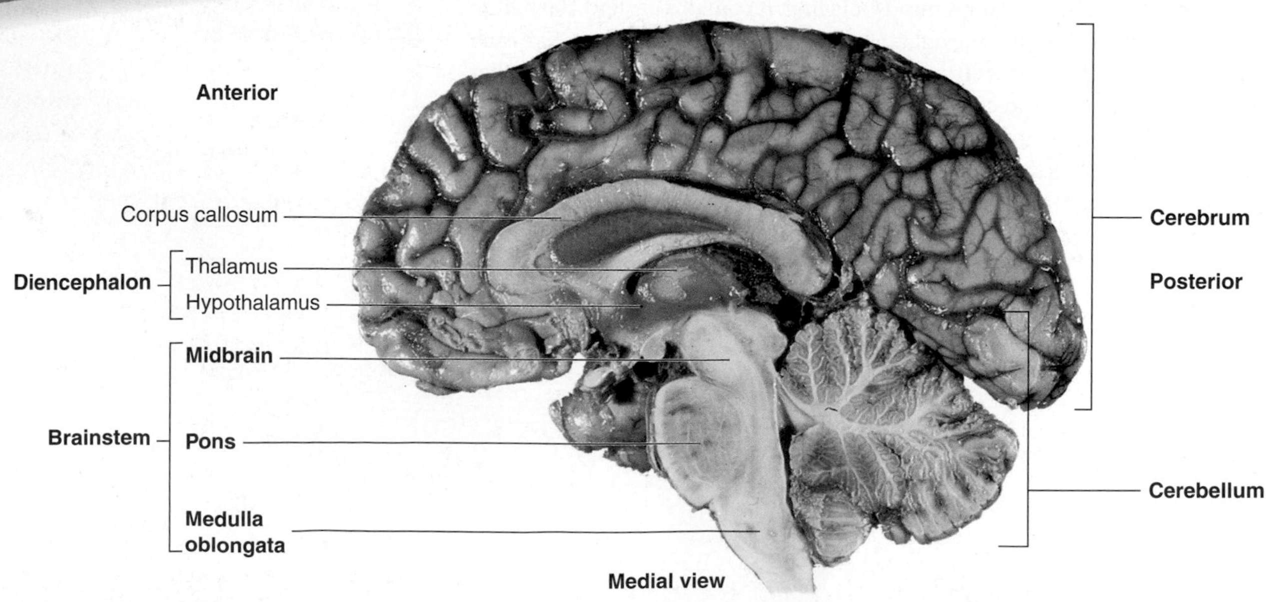

Anterior
Corpus callosum
Diencephalon
Thalamus
Hypothalamus
Midbrain
Brainstem
Pons
Medulla oblongata
Medial view
Cerebrum
Posterior
Cerebellum

FIGURE 13.1 AP|R **Regions of the Brain**

Medial view of a mid-saggital section of the right half of the brain.

TABLE 13.1 Divisions of the Brain and Their Functions

Division	Functions
Brainstem 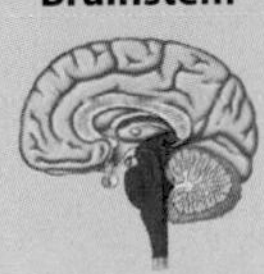	Connects the spinal cord to the cerebrum; consists of the medulla oblongata, pons, and midbrain, with the reticular formation scattered throughout the three regions; has many important functions, as listed under each subdivision; is the location of cranial nerve nuclei
Medulla oblongata 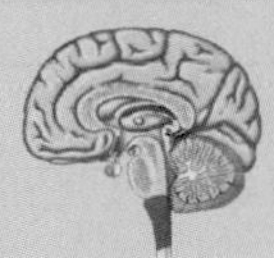	Pathway for ascending and descending nerve tracts; center for several important reflexes (e.g., heart rate, breathing, swallowing, vomiting)
Pons 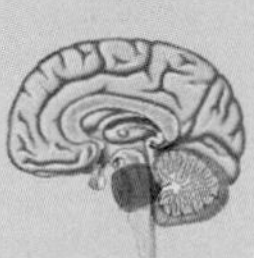	Contains ascending and descending nerve tracts; relays information between cerebrum and cerebellum; site of reflex centers
Midbrain 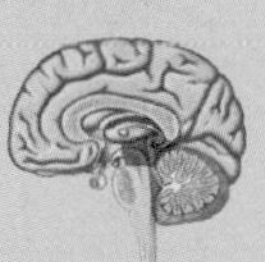	Contains ascending and descending nerve tracts; serves as visual reflex center; part of auditory pathway
Reticular formation	Scattered throughout brainstem; controls many brainstem activities, including motor control, pain perception, rhythmic contractions, and the sleep-wake cycle
Cerebellum 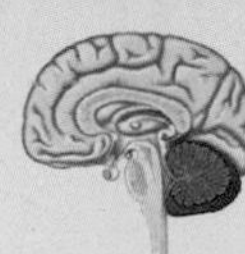	Controls muscle movement and tone; governs balance; regulates extent of intentional movement; involved in learning motor skills
Diencephalon 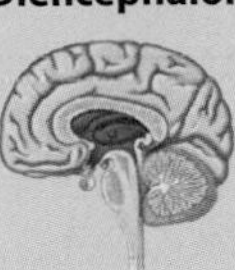	Connects the brainstem to the cerebrum; has many relay and homeostatic functions, as listed under each subdivision
Thalamus	Major sensory relay center; influences mood and movement
Subthalamus	Contains nerve tracts and nuclei
Epithalamus	Contains nuclei responding to olfactory stimulation and contains pineal gland
Hypothalamus	Major control center for maintaining homeostasis and regulating endocrine function
Cerebrum 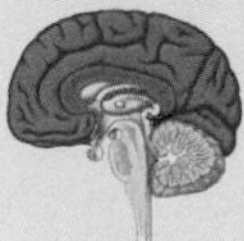	Controls conscious perception, thought, and conscious motor activity; can override most other systems
Basal nuclei	Controls muscle activity and posture; largely inhibits unintentional movement when at rest
Limbic system	Autonomic response to smell, emotion, mood, memory, and other such functions

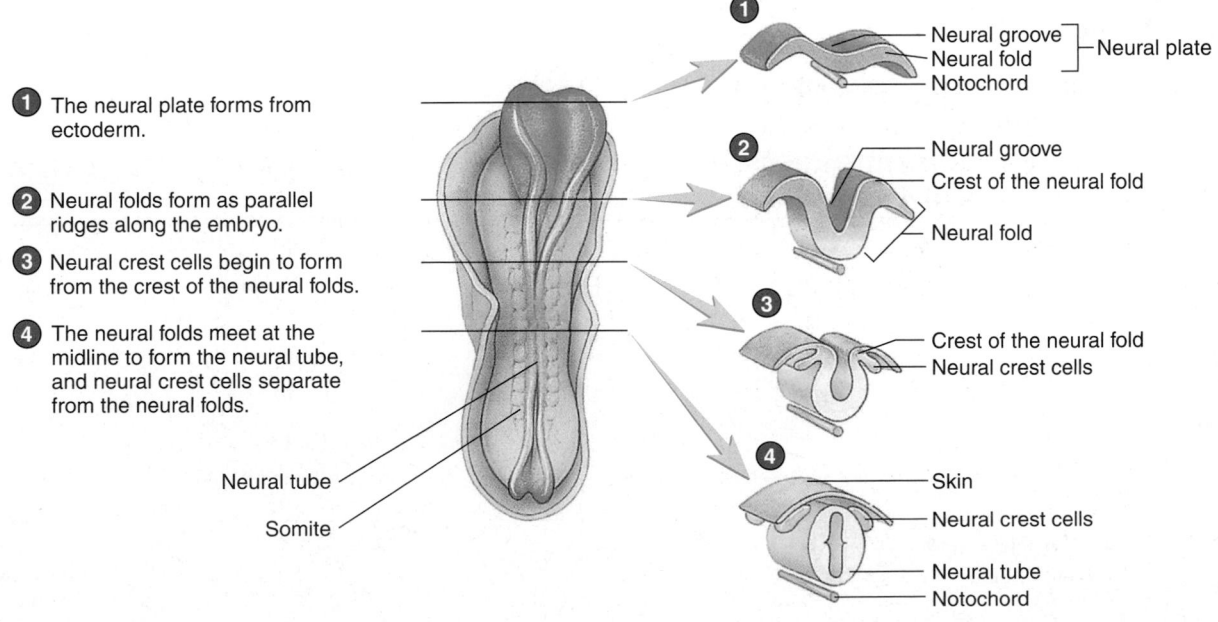

1. The neural plate forms from ectoderm.
2. Neural folds form as parallel ridges along the embryo.
3. Neural crest cells begin to form from the crest of the neural folds.
4. The neural folds meet at the midline to form the neural tube, and neural crest cells separate from the neural folds.

Neural tube
Somite
1
Neural groove
Neural fold
Notochord
Neural plate
2
Neural groove
Crest of the neural fold
Neural fold
3
Crest of the neural fold
Neural crest cells
4
Skin
Neural crest cells
Neural tube
Notochord

PROCESS FIGURE 13.2 Formation of the Neural Tube

A 21-day-old human embryo (superior view), with cross sections through the embryo shown to the right. The level of each transverse section is indicated by a line.

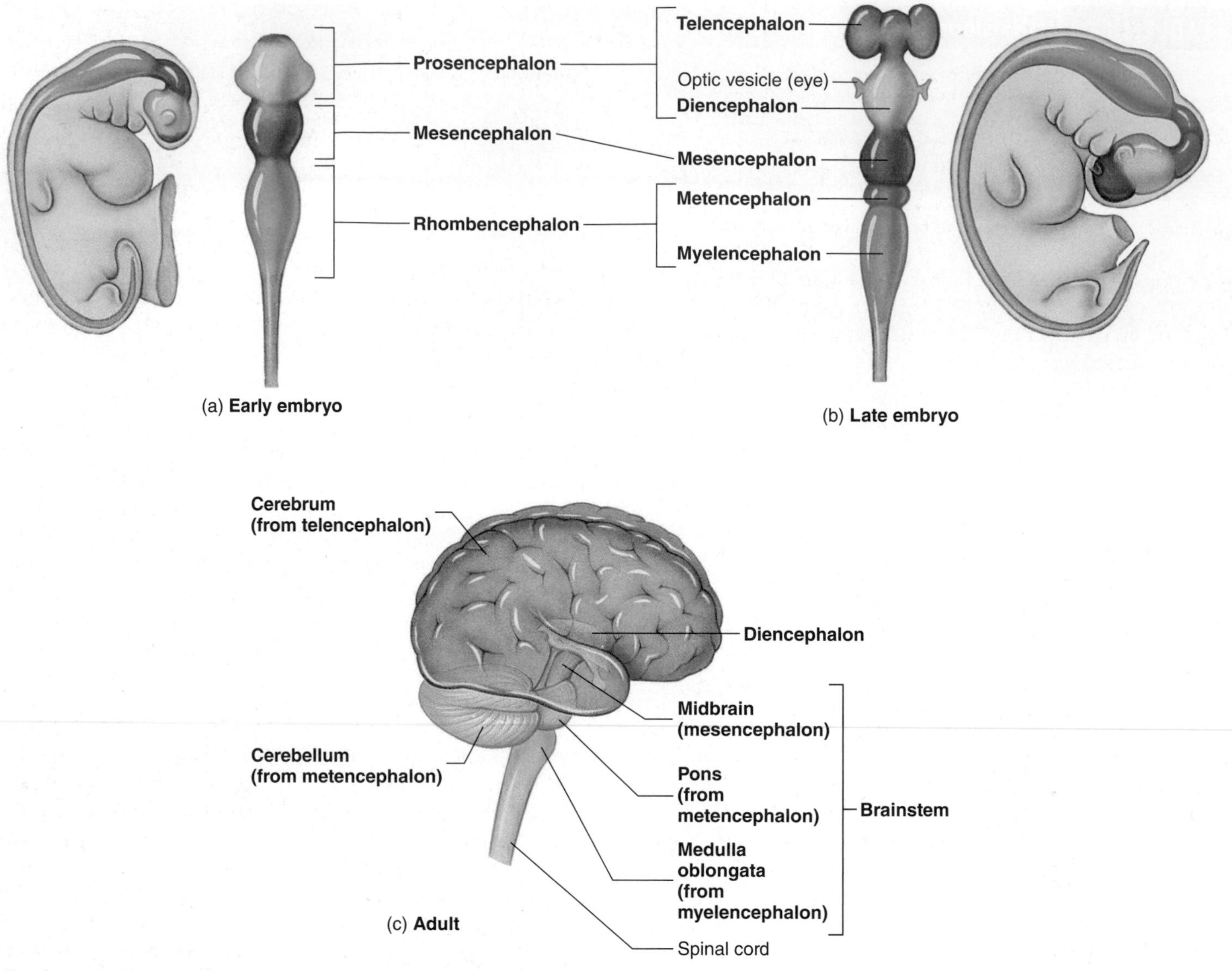
Prosencephalon
Mesencephalon
Rhombencephalon
(a) Early embryo
Telencephalon
Optic vesicle (eye)
Diencephalon
Mesencephalon
Metencephalon
Myelencephalon
(b) Late embryo
Cerebrum (from telencephalon)
Diencephalon
Midbrain (mesencephalon)
Cerebellum (from metencephalon)
Pons (from metencephalon)
Brainstem
Medulla oblongata (from myelencephalon)
(c) Adult
Spinal cord

FIGURE 13.3 Development of the Brain Segments and Ventricles

TABLE 13.2 Development of the Central Nervous System (see figure 13.3)

Early Embryo	Late Embryo	Adult	Cavity	Function
Prosencephalon (forebrain)	**Telencephalon**	**Cerebrum**	Lateral ventricles	Higher brain functions
	Diencephalon	**Diencephalon** (thalamus, subthalamus, epithalamus, hypothalamus)	Third ventricle	Relay center, autonomic nerve control, endocrine control
Mesencephalon (midbrain)	**Mesencephalon**	**Mesencephalon** (midbrain)	Cerebral aqueduct	Nerve pathways, reflex centers
Rhombencephalon (hindbrain)	**Metencephalon**	**Pons** and **cerebellum**	Fourth ventricle	Nerve pathways, reflex centers, muscle coordination, balance
	Myelencephalon	**Medulla oblongata**	Central canal	Nerve pathways, reflex centers

13.2 Brainstem

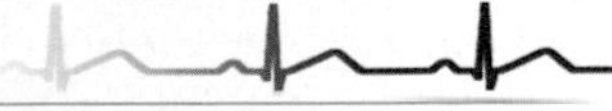

LEARNING OUTCOMES

After reading this section, you should be able to

A. List the parts of the brainstem and describe their structural characteristics.

B. Explain the functions of the parts of the brainstem.

The **brainstem** connects the spinal cord to the remainder of the brain. The brainstem consists of three parts: the medulla oblongata, pons, and midbrain (figure 13.4). In addition to these anatomical divisions, the reticular formation is a functional unit that spans all three divisions. The brainstem is responsible for many essential functions. Damage to small areas often causes death, because many reflexes essential for survival are integrated in the brainstem, whereas relatively large areas of the cerebrum or cerebellum may be damaged without life-threatening consequences.

Medulla Oblongata

The **medulla oblongata** (ob-long-gah′tă), often called the medulla, is about 3 cm long. It is the most inferior part of the brainstem and is continuous inferiorly with the spinal cord. The medulla oblongata contains sensory and motor tracts, cranial nerve nuclei, and related nuclei. Superficially, the spinal cord blends into the medulla oblongata, but internally several differences exist. In the medulla oblongata, the gray matter is organized into discrete **nuclei** (figure 13.4*b*), clusters of gray matter composed mostly of neuron cell bodies. This arrangement contrasts with that of the gray matter of the spinal cord, which extends as a continuous mass in the center of the cord. Several medullary nuclei function as centers for vital reflexes, such as those involved in regulating heart rate, blood vessel diameter, respiration, swallowing, vomiting, hiccuping, coughing, and sneezing.

Two prominent enlargements on the anterior surface of the medulla oblongata are called **pyramids** because they are broader near the pons and taper toward the spinal cord (figure 13.4*a*). The pyramids are formed by the large descending tracts involved in the conscious control of skeletal muscles. Near their inferior ends, most of the fibers of the descending tracts cross to the opposite side, or **decussate** (dē′kŭ-sāt, dē-kŭs′āt). This decussation accounts, in part, for the fact that each half of the brain controls the opposite half of the body. Its role as a conduction pathway is discussed in the description of ascending and descending tracts (see chapter 14).

Two rounded, oval structures, called **olives,** protrude from the anterior surface of the medulla oblongata just lateral to the superior ends of the pyramids (figure 13.4*a,b*). The olives are nuclei involved in functions such as balance, coordination, and modulation of sound from the inner ear (see chapter 15). Nuclei of cranial nerves V (trigeminal), VII (facial), IX (glossopharyngeal), X (vagus), XI (accessory), and XII (hypoglossal) are also located within the medulla oblongata (figure 13.4*c*). Note that some cranial nerves, such as V, VII, and X, have more than one nucleus in the brainstem and that some nuclei, such as the solitary nucleus and nucleus ambiguus, serve as nuclei for multiple cranial nerves.

Pons

The part of the brainstem just superior to the medulla oblongata is the **pons** (figure 13.4*a*). The pons contains ascending and descending tracts and several nuclei. The pontine nuclei, located in the anterior portion of the pons, relay information from the cerebrum to the cerebellum.

Nuclei for cranial nerves V (trigeminal), VI (abducens), VII (facial), and VIII (vestibulocochlear) are contained within the posterior pons. Other important pontine areas are the pontine sleep center, which initiates rapid eye movement sleep (see chapter 14), and the pontine respiratory center, which works with the respiratory centers in the medulla oblongata to help control respiratory movements (see chapter 23).

Midbrain

The **midbrain,** or *mesencephalon,* is the smallest region of the brainstem (figure 13.4*b*). It is located just superior to the pons and contains the nuclei of cranial nerves III (oculomotor), IV (trochlear), and V (trigeminal).

The **tectum** (tek′tŭm; roof; figure 13.5) of the midbrain consists of four nuclei that form mounds on the dorsal surface, collectively called **corpora** (kōr′pōr-ă; bodies) **quadrigemina** (kwah′dri-jem′i-nă; four twins). Each mound is called a **colliculus** (ko-lik′ū-lŭs; hill); the two superior mounds are **superior colliculi,** and the two inferior mounds are **inferior colliculi** (see figure 13.4*b*). The superior colliculi receive sensory input from visual, auditory, and tactile sensory systems and are involved in the reflex movements of the head, eyes, and body toward these stimuli, such as loud noises, flashing lights, or startling pain. For example, when a bright object suddenly appears in a person's field of vision, a reflex turns the eyes to focus on it; when a person hears a sudden, loud noise, a reflex turns the head and eyes toward it. The superior colliculi also receive input from the inferior colliculi and the cerebrum.

The inferior colliculi are involved in hearing and are an integral part of the auditory pathways in the CNS. Neurons conducting action potentials from the structures of the inner ear (see chapter 15) to the brain synapse in the inferior colliculi. Collateral fibers from the inferior colliculi to the superior colliculi provide auditory input that stimulates visual reflexes.

The **tegmentum** (teg-men′tŭm) of the midbrain largely consists of ascending tracts, such as the spinothalamic tract and the medial lemniscus, that carry sensory information from the spinal cord to the brain. The tegmentum also contains the red nuclei, the cerebral peduncles, and the substantia nigra. The paired **red nuclei** (figure 13.5) are so named because in fresh brain specimens they are pinkish in color as a result of an abundant blood supply. The red nuclei aid in the unconscious regulation and coordination of motor activities. **Cerebral peduncles** (pe-dŭng′klz, pē′dŭng-klz; the foot of a column) constitute the portion of the midbrain ventral to the tegmentum. They consist primarily of descending tracts, which carry motor information from the cerebrum to the brainstem and spinal cord. The **substantia nigra** (nī′gră; black substance) is a nuclear mass between the tegmentum and cerebral peduncles containing cytoplasmic melanin granules that give it a dark gray or black color (figure 13.5).

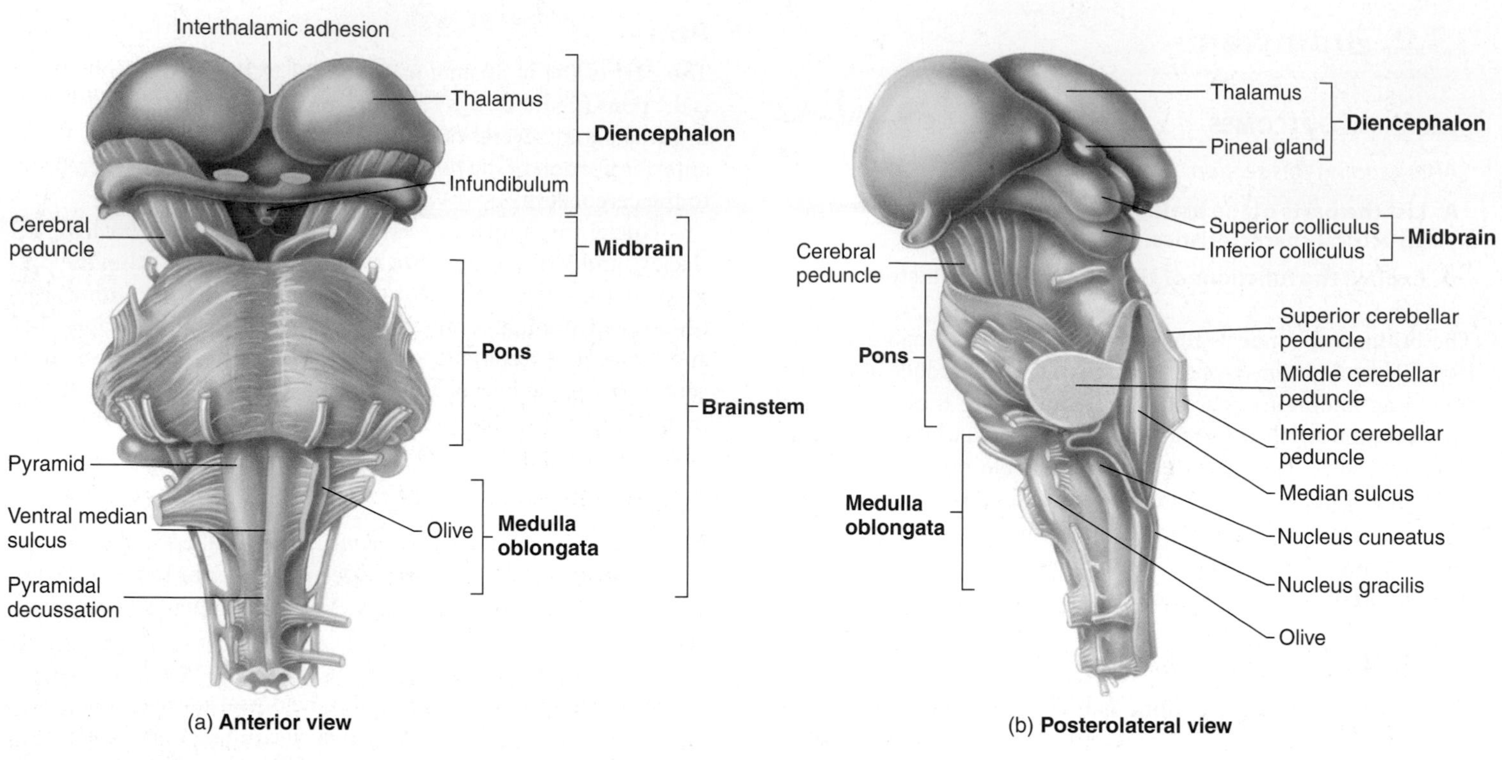
Interthalamic adhesion
Thalamus
Diencephalon
Infundibulum
Cerebral peduncle
Midbrain
Pons
Brainstem
Pyramid
Ventral median sulcus
Olive
Medulla oblongata
Pyramidal decussation
(a) Anterior view
Thalamus
Pineal gland
Diencephalon
Superior colliculus
Inferior colliculus
Midbrain
Cerebral peduncle
Superior cerebellar peduncle
Pons
Middle cerebellar peduncle
Inferior cerebellar peduncle
Median sulcus
Medulla oblongata
Nucleus cuneatus
Nucleus gracilis
Olive
(b) Posterolateral view

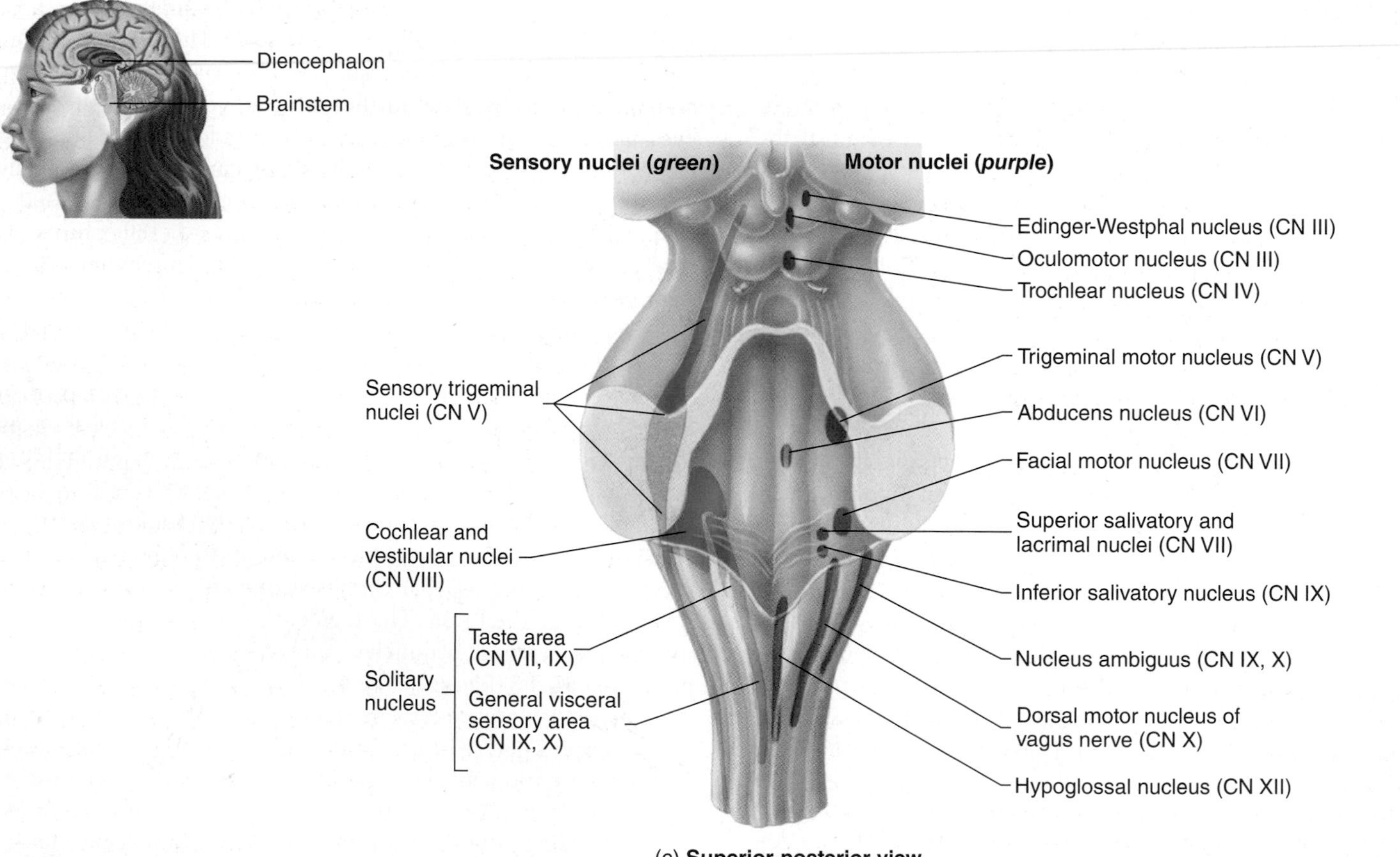
Diencephalon
Brainstem
Sensory nuclei (green)
Motor nuclei (purple)
Edinger-Westphal nucleus (CN III)
Oculomotor nucleus (CN III)
Trochlear nucleus (CN IV)
Trigeminal motor nucleus (CN V)
Sensory trigeminal nuclei (CN V)
Abducens nucleus (CN VI)
Facial motor nucleus (CN VII)
Cochlear and vestibular nuclei (CN VIII)
Superior salivatory and lacrimal nuclei (CN VII)
Inferior salivatory nucleus (CN IX)
Taste area (CN VII, IX)
Solitary nucleus
General visceral sensory area (CN IX, X)
Nucleus ambiguus (CN IX, X)
Dorsal motor nucleus of vagus nerve (CN X)
Hypoglossal nucleus (CN XII)
(c) Superior-posterior view

FIGURE 13.4 AP|R **Diencephalon and Brainstem**

(*a*) Anterior view. (*b*) Posterolateral view. The inset shows the location of the diencephalon (red) and the brainstem (blue). (*c*) Brainstem nuclei. The sensory nuclei are shown on the left (green). The motor nuclei are shown on the right (purple). Even though the nuclei are shown on only one side, each half of the brainstem has both sensory and motor nuclei. (CN = cranial nerve)

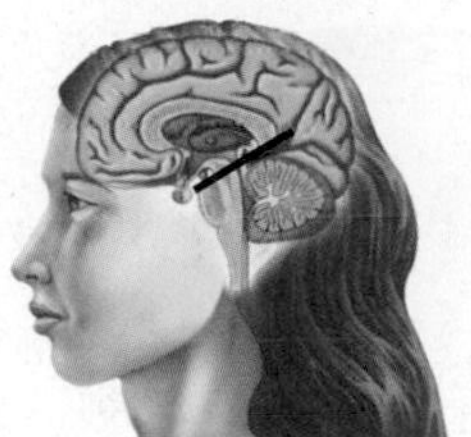

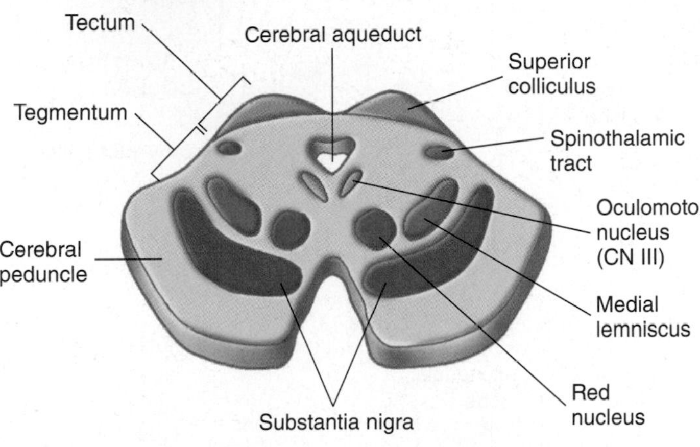
Tectum
Cerebral aqueduct
Superior colliculus
Tegmentum
Spinothalamic tract
Oculomotor nucleus (CN III)
Cerebral peduncle
Medial lemniscus
Red nucleus
Substantia nigra
Anterosuperior view of oblique section

FIGURE 13.5 Oblique Section Through the Midbrain
Inset shows the level of section.

The substantia nigra is interconnected with other basal nuclei of the cerebrum, described in section 13.5, and it is involved in maintaining muscle tone and coordinating movements.

Reticular Formation

The **reticular formation** is a diffuse system consisting of several loosely packed nuclei scattered throughout the length of the brainstem. The reticular formation receives axons from a large number of sources and especially from nerves that innervate the face. The reticular formation modulates and controls many functions mediated by the brainstem, only a few of which are listed here. Some reticular formation neurons send axons to the spinal cord in a motor tract that controls posture, whereas others send axons that reduce the transmission of pain signals from the spinal cord (see chapter 14). By modulating the activity of cranial nuclei within the brainstem, the reticular formation coordinates the rhythmic activities of swallowing, breathing, and the heart rate. Finally, the reticular formation controls the state of alertness and consciousness, including the sleep-wake cycle (see chapter 14).

› Predict 2

Willy was driving his car too fast on his way to school, when the car left the road and hit a guard rail. Elizabeth saw the accident and called 911 on her cell phone. Willy's injuries resulted in extensive blood loss and, while waiting for Life Flight, his heart rate became rapid, and his blood pressure fell to a very low level. Pallor (pale complexion) developed progressively, and his breathing was heavy. Which areas of Willy's brain were most important in integrating these responses?

ASSESS YOUR PROGRESS

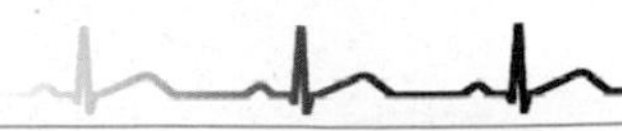

3. *What are the major functions of the medulla oblongata? What cranial nerves have nuclei in the medulla oblongata?*
4. *What activities do the pontine nuclei help control? What cranial nerves have nuclei in the pons?*
5. *What are the parts of the midbrain and their functions?*
6. *What is the function of the reticular formation?*

13.3 Cerebellum

LEARNING OUTCOME

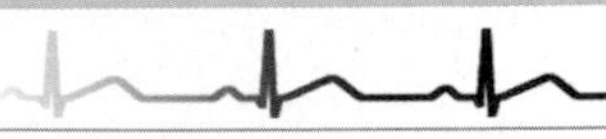

After reading this section, you should be able to

A. List the major regions of the cerebellum and describe the functions of each.

The **cerebellum** (ser-e-bel′ŭm; little brain) is attached to the brainstem posterior to the pons (figure 13.6). It communicates with other regions of the CNS through three large tracts: the **superior, middle,** and **inferior cerebellar peduncles** (see figure 13.4*b*), which connect the cerebellum to the midbrain, pons, and medulla oblongata, respectively. The cerebellum has a gray cortex and nuclei, with white medulla in between. (*Medulla* in this case is a general term meaning "the center of a structure.") The cerebellar cortex has ridges called **folia.** The white matter of the medulla resembles a branching tree and is called the **arbor vitae** (ar′bōr vī′te; tree of life). The **nuclei** of the cerebellum are located in the deep inferior center of the white matter.

The cerebellar cortex contains several cell types, including stellate, basket, granule, Golgi, and Purkinje cells. It also contains mossy fibers, which are afferent axons that branch extensively within the cerebellum. **Purkinje** (pŭr-kin′jē) **cells** are the largest and probably most interesting cells in the CNS. Purkinje cells receive 200,000 synapses, are inhibitory neurons, and are the only cerebellar cortex neurons that send axons to the cerebellar nuclei. The cerebellar cortex contains more neurons than the entire cerebral cortex.

The cerebellum consists of three parts: a small inferior part, the **flocculonodular** (flok′ū-lō-nod′ū-lăr) **lobe;** a narrow central **vermis** (worm-shaped); and two large **lateral hemispheres** (figure 13.6*b,c*). The flocculonodular lobe, the simplest part of the cerebellum, helps control balance and eye movements. The vermis and the medial portion of the lateral hemispheres are involved in controlling posture, locomotion, and fine motor coordination, thereby producing smooth, flowing movements. The major portions of the lateral hemispheres of the cerebellum function in concert with the frontal lobes of the cerebral cortex in planning, practicing, and learning complex movements.

Each lateral hemisphere is divided by a **primary fissure** into an **anterior lobe** and a **posterior lobe.** The lobes are subdivided into **lobules,** which contain the folia.

ASSESS YOUR PROGRESS

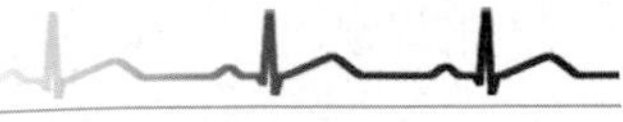

7. *How are the gray matter and white matter arranged in the cerebellum? What is the arbor vitae?*
8. *What are the major regions of the cerebellum? What are the primary functions of each?*

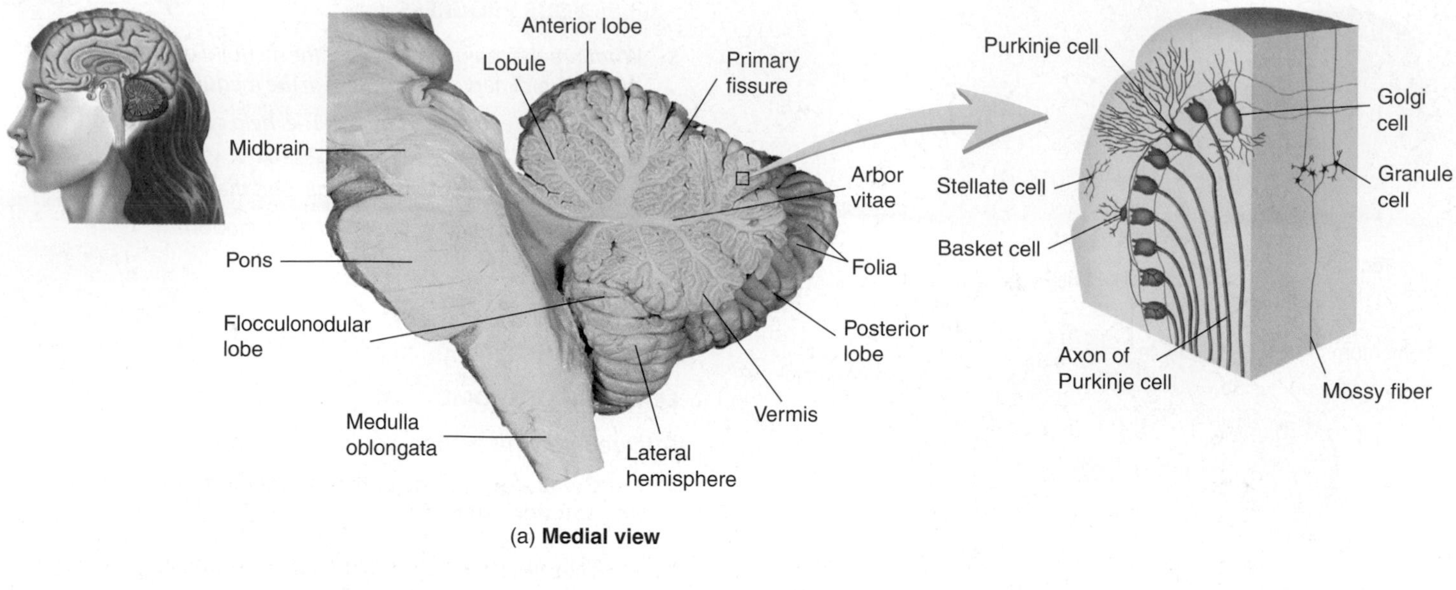
Anterior lobe
Lobule
Primary fissure
Midbrain
Arbor vitae
Folia
Pons
Flocculonodular lobe
Posterior lobe
Medulla oblongata
Lateral hemisphere
Vermis
Purkinje cell
Golgi cell
Stellate cell
Granule cell
Basket cell
Axon of Purkinje cell
Mossy fiber

(a) **Medial view**

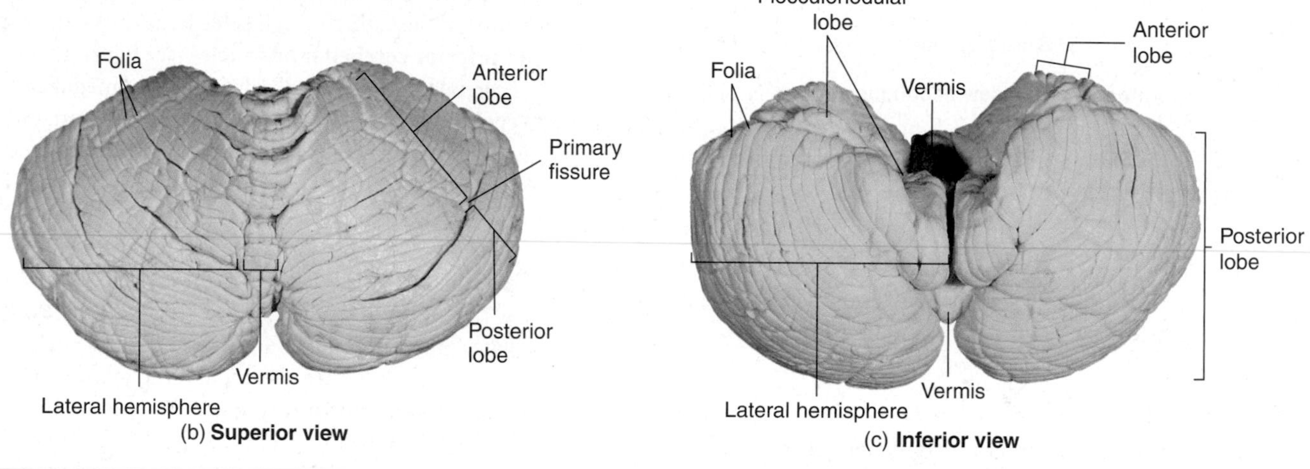
Folia
Anterior lobe
Primary fissure
Posterior lobe
Vermis
Lateral hemisphere
Flocculonodular lobe
Folia
Vermis
Anterior lobe
Posterior lobe
Vermis
Lateral hemisphere

(b) **Superior view**

(c) **Inferior view**

FIGURE 13.6 APR **Cerebellum**

(*a*) Right half of the cerebellum and brainstem as seen in a median section. Inset shows the histology of the cerebellum. (*b*) Superior view of the cerebellum. (*c*) Inferior view of the cerebellum.

13.4 Diencephalon

LEARNING OUTCOME

After reading this section, you should be able to

A. List the parts of the diencephalon and state their functions.

The **diencephalon** (dī-en-sef′ă-lon) is the part of the brain between the brainstem and the cerebrum (figure 13.7; see figures 13.1 and 13.4). Its main components are the thalamus, subthalamus, epithalamus, and hypothalamus.

Thalamus

The **thalamus** (thal′ă-mŭs; figure 13.7*a,b*) is by far the largest part of the diencephalon, constituting about four-fifths of its weight. It consists of a cluster of nuclei shaped somewhat like a yo-yo, with two large, lateral portions connected in the center by a small stalk called the **interthalamic adhesion,** or *intermediate mass*. The space surrounding the interthalamic adhesion and separating the two large portions of the thalamus is the third ventricle of the brain.

Except for the olfactory neurons, all sensory neurons that project to the cerebrum first synapse in the thalamus. Thalamic neurons then send projections to the appropriate areas of the cerebral cortex where sensory input is localized (see chapter 14). For this reason, the thalamus is considered the **sensory relay center** of the brain. Axons carrying auditory information synapse in the **medial geniculate** (je-nik′ū-lāt) **nucleus** of the thalamus. Axons carrying visual information synapse in the **lateral geniculate nucleus.** Most other sensory impulses synapse in the **ventral posterior nucleus.** Axons originating in the ventral posterior nucleus project to the **dorsal tier** of nuclei, which register pain (figure 13.7*b*). Other axons project to the cerebral cortex where sensory input is localized (see chapter 14). The **ventral anterior nucleus** and the **ventral lateral**

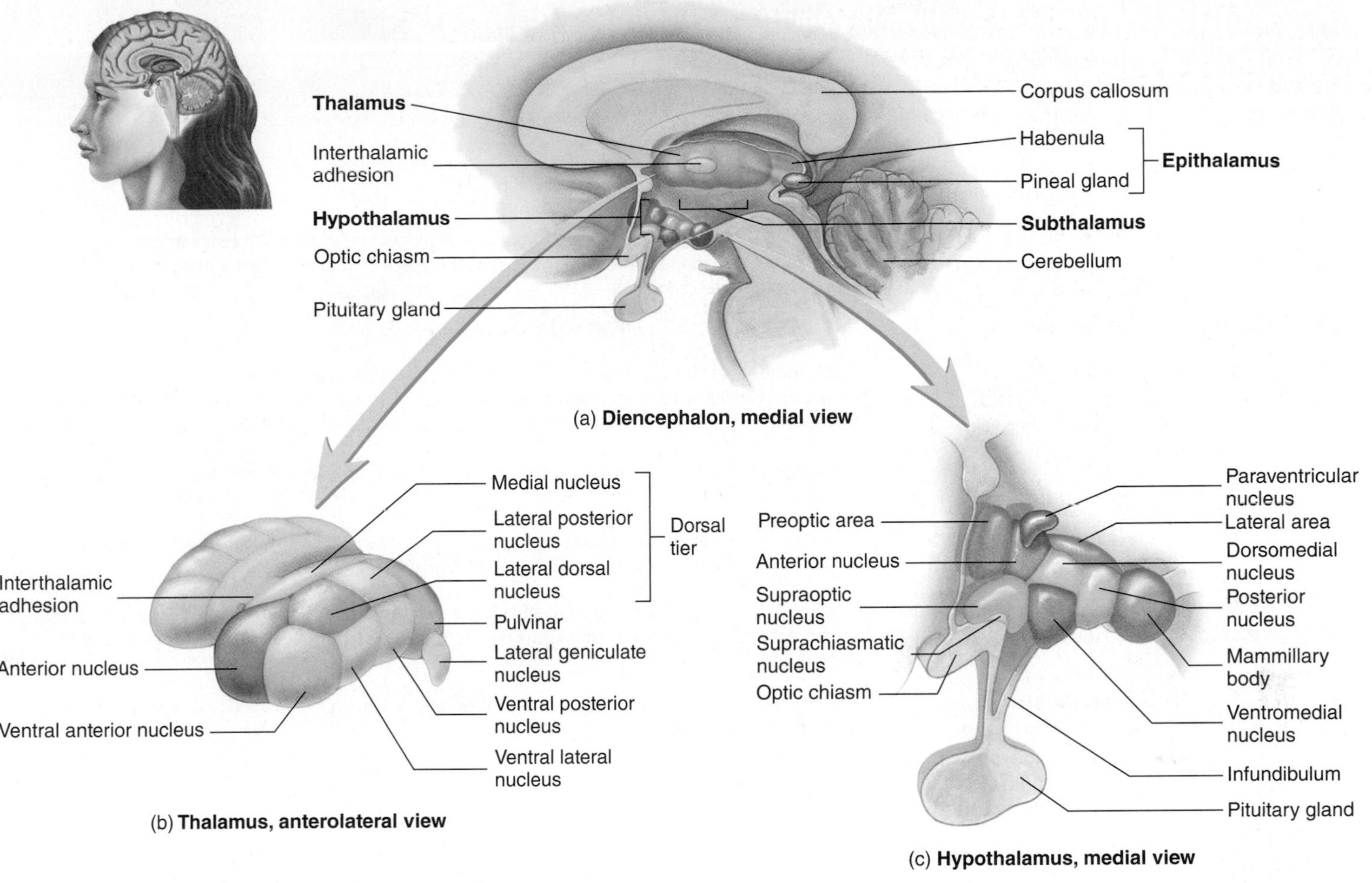
Thalamus
Interthalamic adhesion
Hypothalamus
Optic chiasm
Pituitary gland
Corpus callosum
Habenula
Pineal gland
Epithalamus
Subthalamus
Cerebellum
(a) Diencephalon, medial view
Medial nucleus
Lateral posterior nucleus
Lateral dorsal nucleus
Dorsal tier
Pulvinar
Lateral geniculate nucleus
Ventral posterior nucleus
Ventral lateral nucleus
Interthalamic adhesion
Anterior nucleus
Ventral anterior nucleus
(b) Thalamus, anterolateral view
Preoptic area
Anterior nucleus
Supraoptic nucleus
Suprachiasmatic nucleus
Optic chiasm
Paraventricular nucleus
Lateral area
Dorsomedial nucleus
Posterior nucleus
Mammillary body
Ventromedial nucleus
Infundibulum
Pituitary gland
(c) Hypothalamus, medial view

FIGURE 13.7 Diencephalon
(*a*) General overview of the right half of the diencephalon as seen in a median section. (*b*) Thalamus, showing the nuclei of the left half. (*c*) Hypothalamus, showing the nuclei and right half of the pituitary.

nucleus are involved with motor functions, communicating among the basal nuclei, the cerebellum, and the motor cortex (areas described later in this chapter).

The thalamus also influences mood and actions associated with strong emotions, such as fear and rage. The **anterior** and **medial nuclei** are connected to the limbic system and to the prefrontal cortex (described later in this chapter and in chapter 14). These nuclei are involved in mood modification. The **lateral dorsal nuclei,** which are connected to other thalamic nuclei and to the cerebral cortex, are involved in regulating emotions. The **lateral posterior nuclei** and the **pulvinar** (pŭl-vī′năr) also have connections to other thalamic nuclei and are involved in sensory integration.

Subthalamus

The **subthalamus** is a small area immediately inferior to the thalamus (figure 13.7*a*); it contains several ascending and descending tracts and the **subthalamic nuclei.** Small portions of both the red nucleus and the substantia nigra of the midbrain extend into this area.

The subthalamic nuclei are associated with the basal nuclei and are involved in controlling motor functions.

Epithalamus

The **epithalamus** is a small area superior and posterior to the thalamus (figure 13.7*a*). It consists of the habenula and the pineal gland. The **habenula** (hă-ben′ū-lă) is influenced by the sense of smell and is involved in emotional and visceral responses to odors. The **pineal** (pin′ē-ăl) **gland,** or *pineal body*, is shaped somewhat like a pinecone, from which the name *pineal* is derived. Pineal gland functions in humans are not fully understood, but they involve modulation of the sleep-wake cycle and other biorhythms.

Hypothalamus

The **hypothalamus,** the most inferior portion of the diencephalon (figure 13.7*a,c*), contains several small nuclei and tracts. The most conspicuous nuclei, called the **mammillary bodies,** appear as bulges on the ventral surface of the diencephalon. They are involved in olfactory reflexes and emotional responses to odors. They may also be involved in memory. A funnel-shaped stalk, the **infundibulum** (in-fŭn-dib′ū-lŭm), extends from the floor of the hypothalamus and connects it to the **pituitary gland.** The hypothalamus is a central controller of the endocrine system, because it regulates the

TABLE 13.3 Functions of the Hypothalamus

Function	Hypothalamic Nuclei	Description
Autonomic	Preoptic area and anterior nucleus (parasympathetic) Lateral area and posterior nucleus (sympathetic)	Helps control heart rate, urine release from the bladder, movement of food through the digestive tract, and blood vessel diameter
Endocrine	Paraventricular nucleus Supraoptic nucleus	Helps regulate pituitary gland secretions and influences metabolism, ion concentration, sexual development, and sexual functions; serves as the site of antidiuretic hormone and oxytocin production (see chapter 18)
Muscle control	Lateral area	Controls the muscles involved in swallowing; stimulates shivering
Regulation of body temperature	Preoptic area Anterior nucleus Posterior nucleus	Promotes heat loss when the hypothalamic temperature increases by increasing sweat production (anterior nucleus); promotes heat production when the hypothalamic temperature decreases by stimulating shivering (posterior nucleus); aspirin reduces fever by affecting the preoptic area
Regulation of food and water intake	Ventromedial nucleus Lateral area	Hunger center promotes eating; satiety center inhibits eating; thirst center promotes water intake
Emotions	Lateral area Medial area	Large range of emotional influences over body functions; directly involved in stress-related and psychosomatic illnesses and in feelings of fear and rage
Regulation of the sleep-wake cycle	Lateral area Suprachiasmatic nucleus	Coordinates responses to the sleep-wake cycle with the other areas of the brain (e.g., the reticular activating system); suprachiasmatic nucleus receives direct input from the eyes concerning light/dark cycles and is implicated in jet lag
Sexual development and behavior	Preoptic area Dorsomedial nucleus Ventromedial nucleus	Stimulates sexual development, sexual arousal, and sexual behavior; the preoptic area is larger in males than in females

pituitary gland's secretion of hormones, which influence functions as diverse as metabolism, reproduction, responses to stressful stimuli, and urine production (table 13.3; see chapter 18).

Sensory neurons that terminate in the hypothalamus provide input from (1) internal organs; (2) taste receptors of the tongue; (3) the limbic system, which is involved in many activities, including responses to smell; (4) specific cutaneous areas, such as the nipples and external genitalia; (5) the eyes; and (6) the prefrontal cortex of the cerebrum carrying information relative to "mood" through the thalamus. Efferent fibers from the hypothalamus extend into the brainstem and the spinal cord, where they synapse with neurons of the autonomic nervous system (see chapter 16). In this way, the hypothalamus serves as the major coordinating center of the autonomic nervous system. These connections regulate functions such as heart rate and digestive activities (table 13.3). Other fibers extend through the infundibulum to the posterior portion of the pituitary gland (see chapter 18); some extend to trigeminal and facial nerve nuclei to help control the muscles involved in swallowing; and some extend to motor neurons of the spinal cord to stimulate shivering.

Hypothalamic nuclei directly control body temperature by stimulating sweating or shivering. Other hypothalamic nuclei are involved in controlling thirst, hunger, and sex drive. The hypothalamus is also very important in a number of functions related to mood, motivation, and emotion. Sensations such as sexual pleasure, relaxation after a meal, rage, and fear are related to hypothalamic functions. This is one reason that strong emotional experiences may affect a person's desire or ability to eat, drink, or experience sexual pleasure, and vice versa. The hypothalamus also interacts with the reticular activating system in the brainstem to coordinate the sleep-wake cycle (table 13.3).

ASSESS YOUR PROGRESS

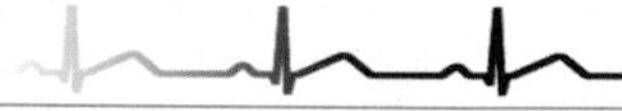

9. *What are the four main components of the diencephalon?*
10. *Which part serves as the sensory relay center of the brain? As the link between the nervous system and the endocrine system?*
11. *What is the role of the subthalamus? Name the parts of the epithalamus, and give their functions.*
12. *List and explain the functions of the hypothalamus.*

13.5 Cerebrum

LEARNING OUTCOMES

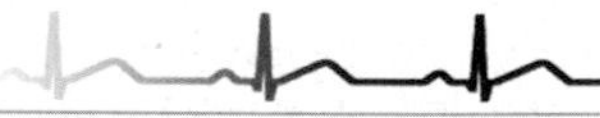

After reading this section, you should be able to

A. Describe the structure of the cerebrum, including the lobes, fissures, sulci, cerebral cortex, and cerebral medulla.
B. Relate the principle function of each lobe of the cerebrum.
C. List and describe the three types of tracts found in the cerebrum.
D. List the basal nuclei and explain their function.
E. State the parts and functions of the limbic system.

FUNDAMENTAL **Figure**

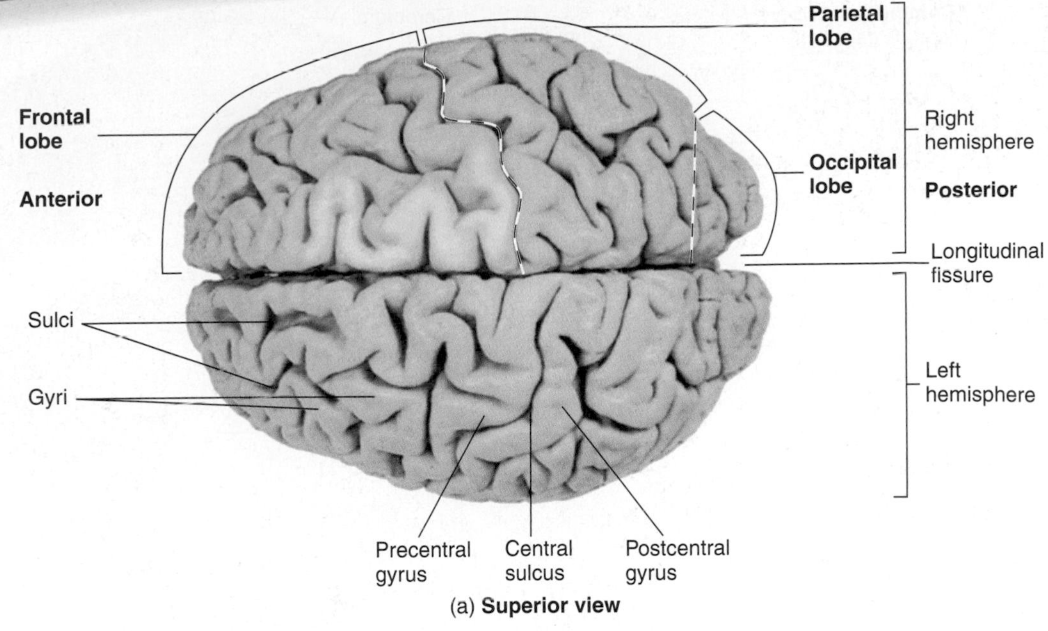
Parietal lobe
Frontal lobe
Anterior
Right hemisphere
Occipital lobe
Posterior
Longitudinal fissure
Sulci
Gyri
Left hemisphere
Precentral gyrus
Central sulcus
Postcentral gyrus

(a) **Superior view**

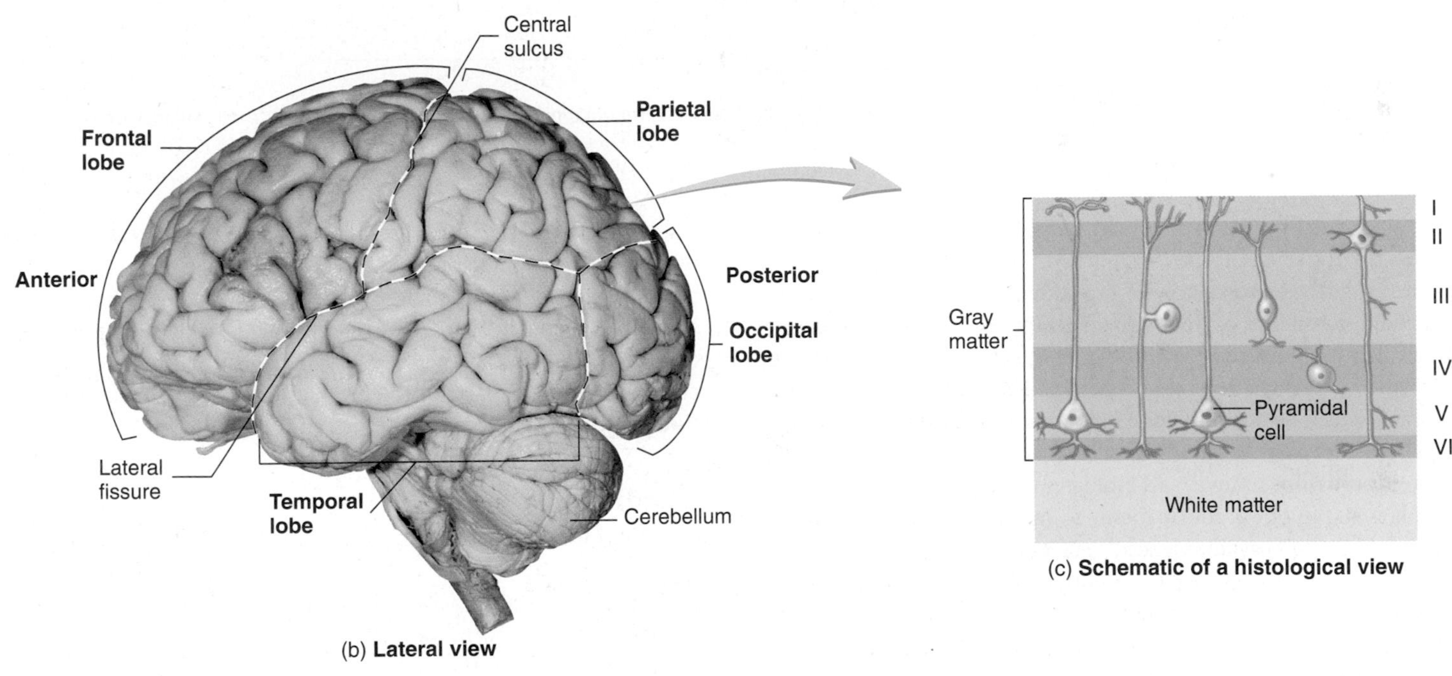
Central sulcus
Parietal lobe
Frontal lobe
Anterior
Posterior
Occipital lobe
Lateral fissure
Temporal lobe
Cerebellum
Gray matter
I
II
III
IV
V
VI
Pyramidal cell
White matter

(b) **Lateral view**

(c) **Schematic of a histological view**

FIGURE 13.8 AP|R **Cerebrum**

(*a*) Superior view of the two hemispheres. (*b*) Lateral view of the left cerebral hemisphere, showing the lobes. (*c*) Layers and cell types of the cerebral cortex.

The cerebrum (figure 13.8) is the part of the brain that most people think of when the term *brain* is mentioned. The cerebrum accounts for the largest portion of total brain weight, which is about 1200 g in females and 1400 g in males. Brain size is related to body size; larger brains are associated with larger bodies, not with greater intelligence.

The cerebrum is divided into left and right hemispheres by a **longitudinal fissure** (figure 13.8*a*). The most conspicuous features on the surface of each hemisphere are numerous folds called **gyri** (jī′rī; sing. *gyrus*), which greatly increase the surface area of the cortex. The grooves between the gyri are called **sulci** (sŭl′sī; sing. *sulcus*). The **central sulcus,** which extends across the lateral surface of the cerebrum from superior to inferior, is located about midway along the length of the brain. Anterior to the central sulcus is the **precentral gyrus,** which is the **primary motor cortex.** Posterior to the central sulcus is the **postcentral gyrus,** which is the **primary somatic sensory cortex** (see chapter 14). The general pattern of the gyri is similar in all normal human brains, but some variation exists between individuals and even between the two hemispheres of the same cerebrum.

Each cerebral hemisphere is divided into lobes, which are named for the skull bones overlying each of them (figure 13.8*b*). The

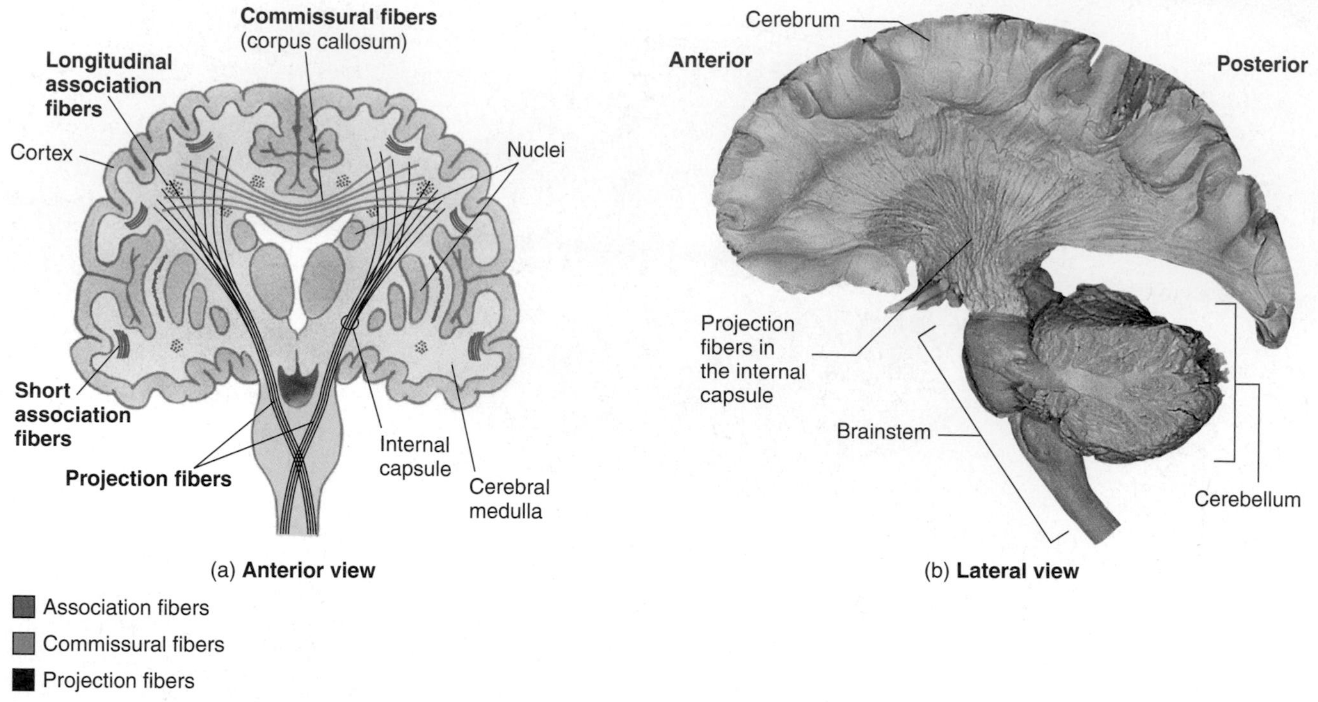
Commissural fibers
(corpus callosum)
Longitudinal
association
fibers
Cortex
Nuclei
Short
association
fibers
Projection fibers
Internal
capsule
Cerebral
medulla
(a) Anterior view
Association fibers
Commissural fibers
Projection fibers
Cerebrum
Anterior
Posterior
Projection
fibers in
the internal
capsule
Brainstem
Cerebellum
(b) Lateral view

FIGURE 13.9 Cerebral Medullary Tracts
(*a*) Frontal section of the brain showing commissural, association, and projection fibers. (*b*) Photograph of the left cerebral hemisphere from a lateral view with the cortex and association fibers removed to reveal the projection fibers of the internal capsule deep within the brain.

frontal lobe is important in voluntary motor function, motivation, aggression, the sense of smell, and mood. The anterior region, called the prefrontal cortex, is involved in personality and decision making (see chapter 14). The **parietal lobe** is the major center for receiving and evaluating most sensory information, except for smell, hearing, taste, and vision. The frontal and parietal lobes are separated by the central sulcus. The **occipital lobe** functions in receiving and integrating visual input and is not distinctly separate from the other lobes. The **temporal lobe** receives and evaluates input for smell and hearing and plays an important role in memory. Its anterior and inferior portions, called the "psychic cortex," are associated with such brain functions as abstract thought and judgment. The temporal lobe is separated from the rest of the cerebrum by a **lateral fissure,** and deep within the fissure is the **insula** (in′soo-lă; island). The insula receives and evaluates taste information and is often referred to as a fifth lobe.

The gray matter on the outer surface of the cerebrum is the **cerebral cortex,** and clusters of gray matter deep inside the brain are nuclei. The cerebral cortex contains a number of neuron types, named largely for their shape, such as fusiform cells, stellate cells, and pyramidal cells (figure 13.8*c*). These cells are distributed in layers within the cerebral cortex. The thickness of the cortex is not uniform throughout the cerebrum but ranges from two or three layers in the most "primitive" parts of the cortex to six layers in the more "advanced" regions.

The white matter of the brain between the cortex and the nuclei is the **cerebral medulla.** This term should not be confused with *medulla oblongata.* The cerebral medulla consists of tracts that connect areas of the cerebral cortex to each other or to other parts of the CNS (figure 13.9). The fibers in these tracts fall into three main categories: (1) **Association fibers** connect areas of the cerebral cortex within the same hemisphere. (2) **Commissural fibers** connect one cerebral hemisphere to the other. The largest bundle of commissural fibers connecting the two cerebral hemispheres is the **corpus callosum** (kōr′pŭs ka-lō′sŭm; see figure 13.1). (3) **Projection fibers** connect the cerebrum to other parts of the brain and spinal cord (figure 13.9). The projection fibers form the **internal capsule.**

Basal Nuclei

The **basal nuclei** are a group of functionally related nuclei located bilaterally in the inferior cerebrum, diencephalon, and midbrain (figure 13.10). These nuclei are involved in controlling motor functions (see chapter 14). The nuclei in the cerebrum are collectively called the **corpus striatum** (kōr′pŭs strī-ā′tŭm) and include the **caudate** (kaw′dāt; having a tail) **nucleus** and the **lentiform** (len′ti-fōrm; lens-shaped) **nucleus.** The lentiform nucleus, in turn, is divided into a lateral **putamen** (pū-tā′men) and the medial **globus pallidus** (glō′bŭs pal′li-dŭs; pale globe). The basal nuclei are the largest nuclei of the brain and occupy a large part of the cerebrum. The **subthalamic nucleus** and the **substantia nigra** function with the caudate and lentiform nuclei to control movement. The subthalamic nucleus is in the diencephalon, and the substantia nigra is in the midbrain.

Limbic System

Parts of the cerebrum and diencephalon are grouped together under the title **limbic system** (figure 13.11). The limbic system plays a central role in basic survival functions, such as memory, reproduction, and nutrition. It is also involved in interpreting sensory input and

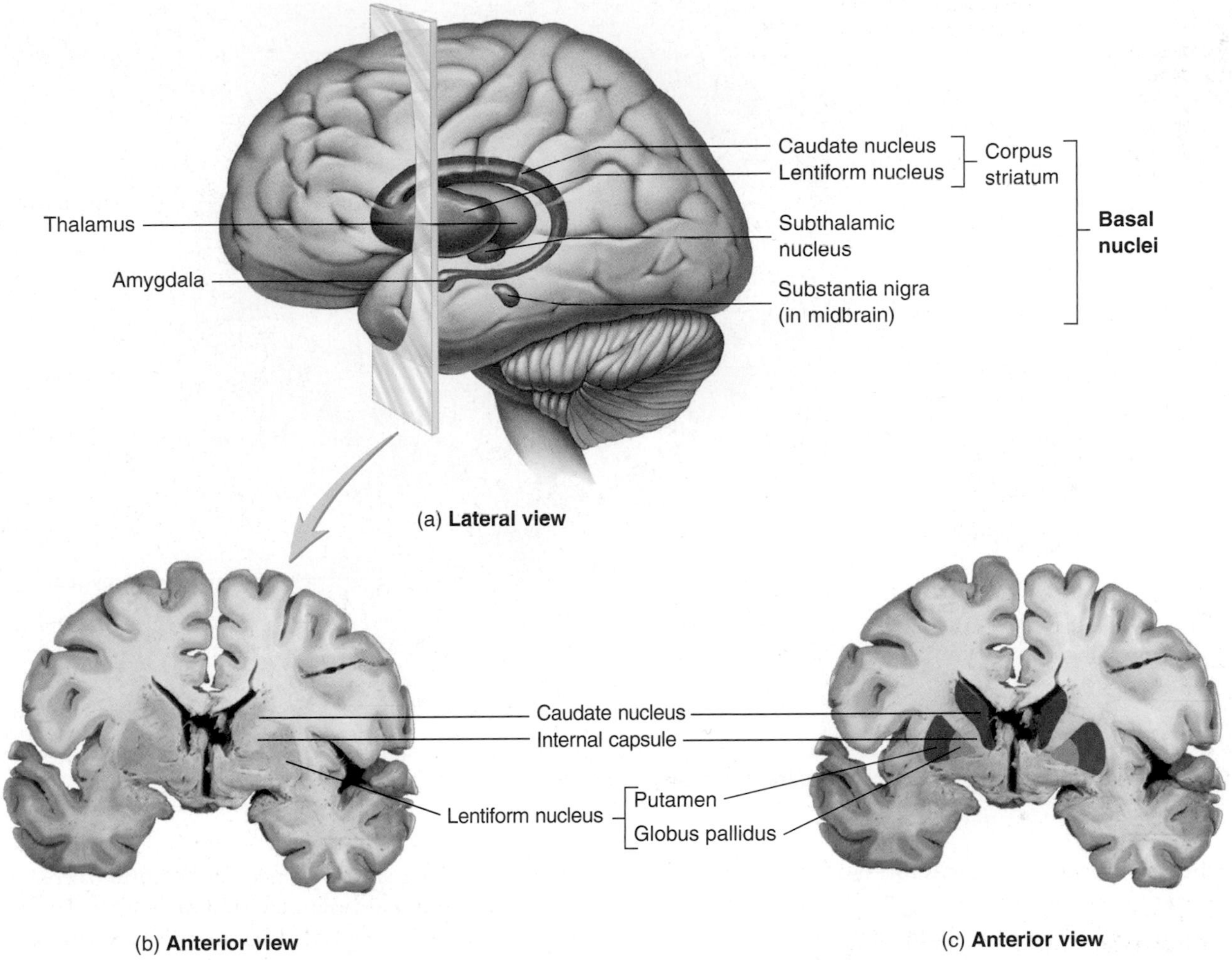

Caudate nucleus
Lentiform nucleus
Corpus striatum
Thalamus
Subthalamic nucleus
Basal nuclei
Amygdala
Substantia nigra (in midbrain)
(a) **Lateral view**

Caudate nucleus
Internal capsule
Lentiform nucleus
Putamen
Globus pallidus
(b) **Anterior view**
(c) **Anterior view**

FIGURE 13.10 Basal Nuclei of the Left Hemisphere
(*a*) "Transparent 3D" drawing of the basal nuclei inside the left hemisphere. (*b*) Photograph of a frontal section of the brain showing the basal nuclei and other structures. (*c*) The same photograph as in (*b*) but with nuclei highlighted.

emotions in general. The term *limbic* (lim′bik; border) refers to deep portions of the cerebrum that form a ring around the diencephalon. Structurally, the limbic system consists of (1) certain cerebral cortical areas, including the **cingulate** (sin′gū-lāt; to surround) **gyrus,** located along the inner surface of the longitudinal fissure just above the corpus callosum, and the **parahippocampal gyrus,** located on the medial side of the temporal lobe; (2) various nuclei, such as anterior nuclei of the thalamus, the habenula in the epithalamus, and the **dentate gyrus** of the **hippocampus;** (3) parts of the basal nuclei, such as the **amygdala;** (4) the hypothalamus, especially the mammillary bodies; (5) the **olfactory cortex;** and (6) tracts connecting the various cortical areas and nuclei, such as the **fornix,** which connects the hippocampus to the thalamus and mammillary bodies.

ASSESS YOUR PROGRESS

13. *Distinguish between gyri and sulci. What structures do the longitudinal fissure, central sulcus, and lateral fissure separate?*

14. *Describe the cerebral cortex and the cerebral medulla.*

15. *Name the five lobes of the cerebrum, and describe their locations and functions.*

16. *List the three categories of tracts in the cerebral medulla, and tell what each connects.*

17. *List the basal nuclei, and state their general function.*

18. *What are the parts of the limbic system? What are the functions of this system?*

13.6 Meninges, Ventricles, and Cerebrospinal Fluid

LEARNING OUTCOMES

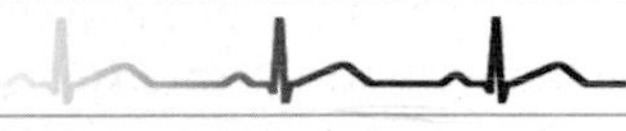

After reading this section, you should be able to

A. Describe the meninges and the spaces between them.

B. Identify the locations of the four ventricles and the structures that connect them.

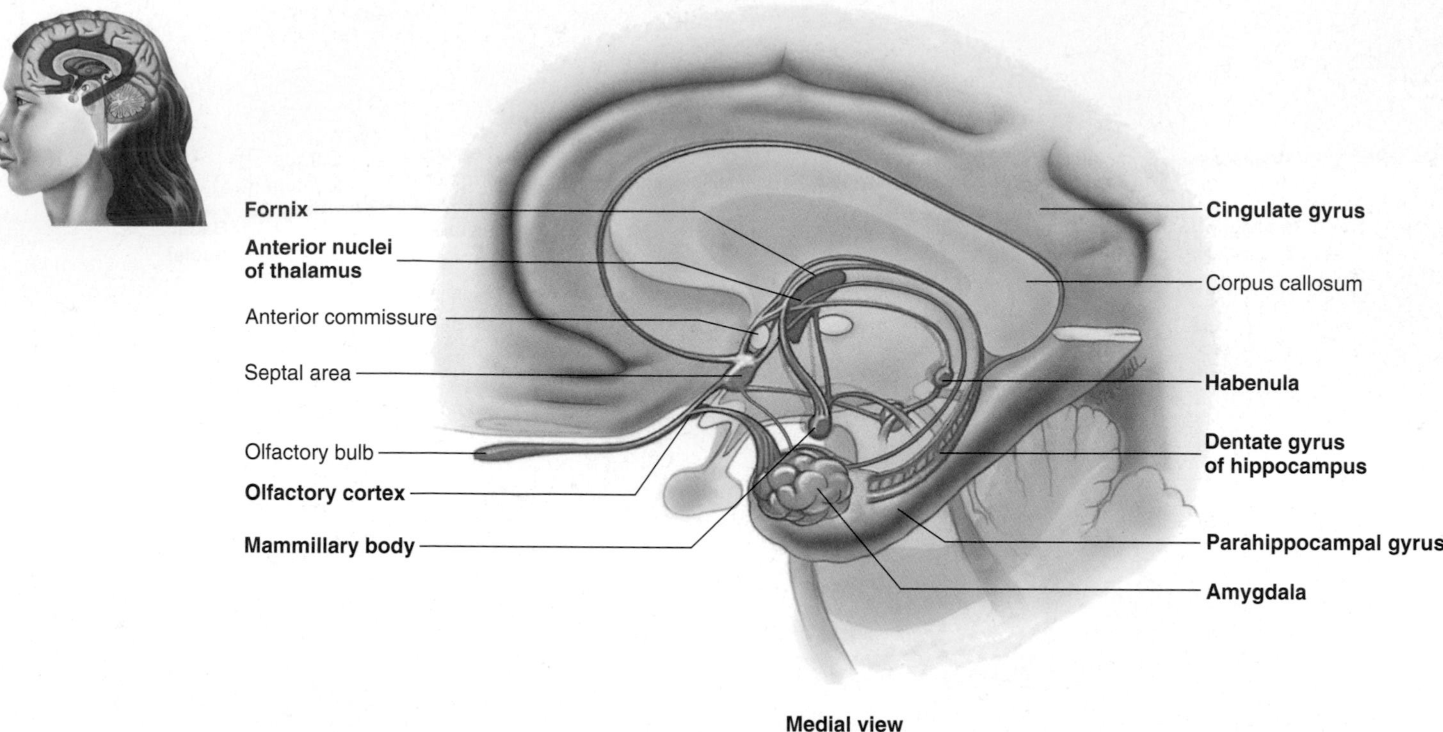
Fornix
Anterior nuclei
of thalamus
Anterior commissure
Septal area
Olfactory bulb
Olfactory cortex
Mammillary body
Cingulate gyrus
Corpus callosum
Habenula
Dentate gyrus
of hippocampus
Parahippocampal gyrus
Amygdala
Medial view

FIGURE 13.11 Limbic System and Associated Structures of the Right Hemisphere

C. Explain how cerebrospinal fluid is formed, circulated, and returned to the blood.

D. Describe the function of cerebrospinal fluid.

Meninges

Three connective tissue membranes, the **meninges** (mě-nin′jēz), surround and protect the brain and spinal cord (figure 13.12). The most superficial and thickest membrane is the **dura mater** (doo′ră mā′ter; tough mother), which is composed of dense irregular connective tissue. Within the vertebral canal, the dura mater is distinctly separate from the vertebrae, forming an **epidural space.** Within the cranial cavity, the dura mater tightly adheres to the cranial bones, so the epidural space of the cranial cavity is only a potential space. The dura mater within the cranial cavity consists of two layers. The outer layer, the **periosteal dura,** is the inner periosteum of the cranial bones. The inner layer, the **meningeal dura,** is continuous with the dura of the spinal cord. The meningeal dura is separated from the periosteal dura in several regions to form structures called dural folds and dural venous sinuses.

Dural folds are tough connective tissue partitions that extend into the major brain fissures. The dura mater and dural folds help hold the brain in place within the skull and keep it from moving around too freely. The largest of the dural folds is the **falx cerebri** (falks se-rē′brē; sickle-shaped), which lies in the longitudinal fissure and is anchored anteriorly to the crista galli of the ethmoid bone. The **tentorium cerebelli** (ten-tō′rē-ŭm, tent; ser′ĕ-bel′ē) is oriented horizontally between the cerebrum and the cerebellum. The **falx cerebelli** lies between the two cerebellar hemispheres.

Dural venous sinuses are spaces that form where the two layers of the dura mater are separated from each other. The largest of the sinuses, the **superior sagittal sinus,** forms between the falx cerebri and the periosteal dura and runs along the median plane (figure 13.12*b*). The dural venous sinuses are lined with endothelium and transport venous blood and cerebrospinal fluid (CSF; see "Cerebrospinal Fluid" later in this section) away from the brain. All veins draining blood from the brain empty into dural venous sinuses. The dural sinuses subsequently drain into the internal jugular veins, which are the major veins that exit the cranial cavity to carry blood back to the heart (see chapter 21).

The next meningeal membrane is the very thin, wispy **arachnoid** (ă-rak′noyd; spiderlike, as in cobwebs) **mater.** The space between this membrane and the dura mater is the **subdural space;** it contains only a very small amount of serous fluid. The third meningeal layer, the **pia** (pī′ă, pē′ă; affectionate) **mater** is bound very tightly to the surface of the brain. Between the arachnoid mater and the pia mater is the **subarachnoid space,** which contains weblike strands of arachnoid mater and the blood vessels supplying the brain. The subarachnoid space is filled with CSF.

Ventricles

The CNS forms as a hollow tube that may be quite reduced in some areas of the adult CNS and expanded in other areas (see section 13.1). The interior of the tube is lined with a single layer of epithelial cells called **ependymal** (ep-en′di-măl; see chapter 11) **cells.** Each cerebral hemisphere contains a relatively large cavity, the **lateral ventricle** (figure 13.13). The lateral ventricles are separated from each other by

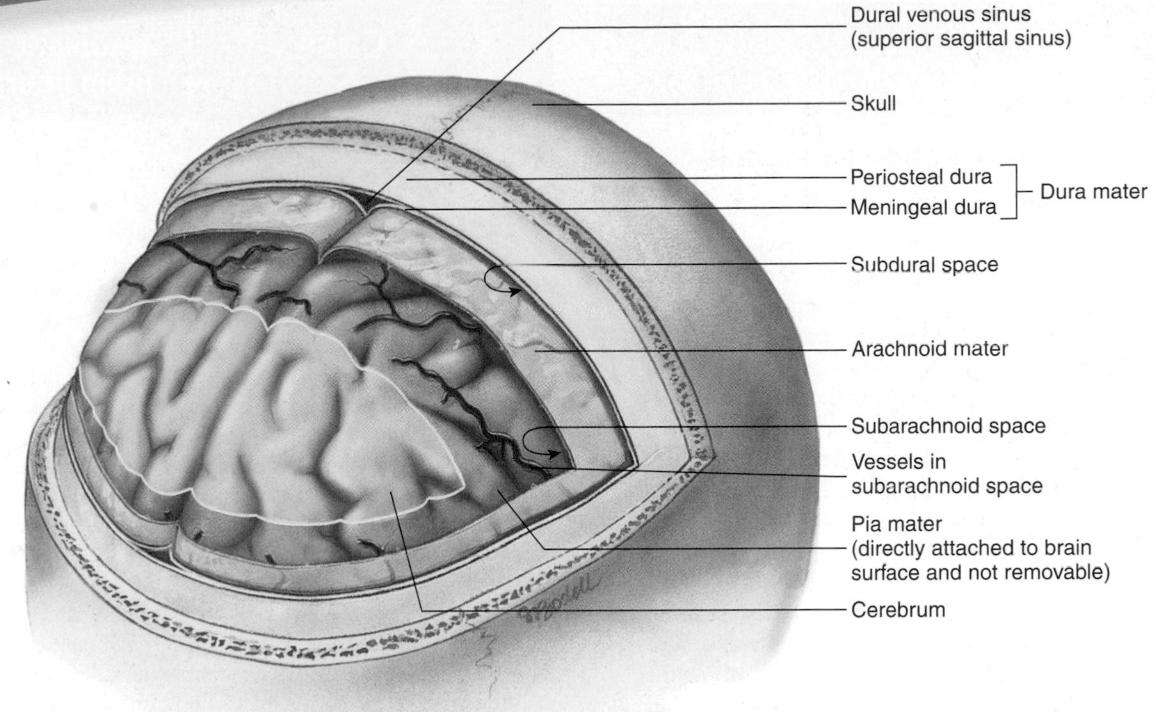
Dural venous sinus
(superior sagittal sinus)
Skull
Periosteal dura
Meningeal dura
Dura mater
Subdural space
Arachnoid mater
Subarachnoid space
Vessels in
subarachnoid space
Pia mater
(directly attached to brain
surface and not removable)
Cerebrum

(a) **Anterosuperior view**

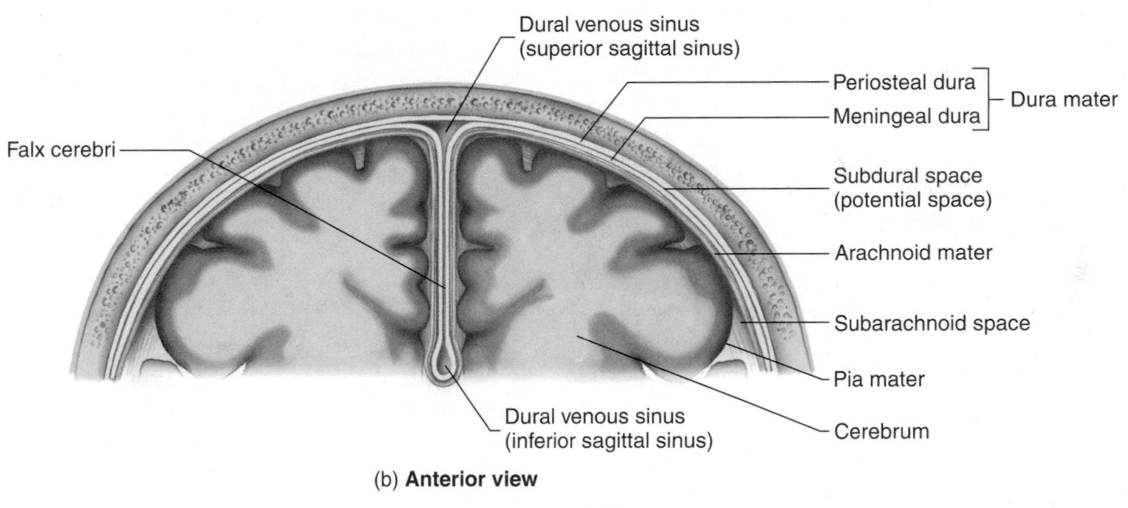
Dural venous sinus
(superior sagittal sinus)
Falx cerebri
Periosteal dura
Meningeal dura
Dura mater
Subdural space
(potential space)
Arachnoid mater
Subarachnoid space
Pia mater
Cerebrum
Dural venous sinus
(inferior sagittal sinus)

(b) **Anterior view**

FIGURE 13.12 APR **Meninges**

(*a*) Meningeal membranes surrounding the brain. (*b*) Frontal section of the head to show the meninges.

thin **septa pellucida** (sep′tă pe-loo′sid-ă; sing. *septum pellucidum*; translucent walls), which lie in the midline just inferior to the corpus callosum and are usually fused with each other. The lateral ventricles can be thought of as the first and second ventricles, but they are not designated as such. A smaller midline cavity, the **third ventricle,** is located in the center of the diencephalon between the two halves of the thalamus. The two lateral ventricles communicate with the third ventricle through two **interventricular foramina.** The **fourth ventricle** is in the inferior part of the pontine region and the superior region of the medulla oblongata at the base of the cerebellum. The third ventricle communicates with the fourth ventricle through a narrow canal, the **cerebral aqueduct,** which passes through the midbrain. The fourth ventricle is continuous with the central canal of the spinal cord, which extends nearly the full length of the cord. The fourth ventricle is also continuous with the subarachnoid space through two lateral apertures and one median aperture.

Clinical IMPACT

Traumatic Brain Injuries and Hematomas

Head injuries are classified as **open** (when some of the cranial cavity contents are exposed to the outside) or **closed** (when the cranial cavity remains intact; see Clinical Impact, "CSF and Skull Fractures"). Closed injuries are more common and usually result from the head striking a hard surface or an object striking the head. Such injuries may cause brain trauma—either **coup** (kū), occurring at the site of impact, or **contrecoup** (kon′tra-kū), occurring on the opposite side of the brain from the impact as a result of the brain moving within the skull. Traumatic brain injury is three times more common in males than in females. The most common traumatic brain injury (75–90%) is a **concussion,** characterized by immediate, but transient, impairment of neural function, such as loss of consciousness or blurred vision.

Traumatic brain injury may be diffuse or focal. **Diffuse brain injury** usually results from shaking, as when a child is shaken or when a person is thrown about in an automobile accident. As the name suggests, such injury is not localized to one focal point but involves damage to many small vessels and nerves, especially around the brainstem. **Focal traumatic brain injury** may involve cortical **contusions** (bruisings), caused by direct impact to the brain, or hemorrhages in or around the brain. Contusions are usually superficial and involve only the gyri.

Hemorrhagic brain injury is characterized by bleeding outside the dura (extradural or epidural), between the dura and the brain (subdural), or within the brain (intracerebral). A hemorrhage, which is bleeding, results in a hematoma, an accumulation of blood. **Extradural hemorrhages,** or *epidural hematomas,* occur in about 1–2% of major head injuries. They usually affect the middle cranial fossa and involve a tear in the middle meningeal artery (85%) or in the middle meningeal vein or dural sinus (15%). **Subdural hematomas** are much more common, occurring in 10–20% of major head injuries. They most commonly involve tears in the cortical veins or dural venous sinuses, occur in the superior portion of the cranial cavity, and appear within hours of the head injury. **Chronic subdural hematomas,** which involve slow bleeding over an extended period of time (weeks to months), are common in elderly people and in people who abuse alcohol. **Intracerebral hematomas** occur in about 2–3% of major head injuries and are often associated with contusions. They involve damage to small vessels within the brain itself and are most common in the frontal and temporal lobes. They occur within 3–10 days of the head trauma.

Traumatic brain injury has been called the "signature wound" of the Iraq and Afghanistan wars. A large number of soldiers have suffered from blasts, often from improvised explosive devices, which in previous wars would have been lethal without current armor and helmets. Unfortunately, effective treatment of traumatic brain injury remains difficult and is even more challenging for soldiers with physical and psychiatric symptoms that overlap with post-traumatic stress disorder.

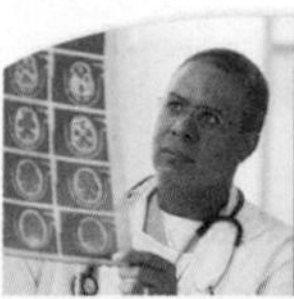

Case STUDY | Traumatic Brain Injury

The body of an 80-year-old woman was being examined by a hospital pathologist to determine the cause of death. Her medical records indicated no significant history of cardiovascular disease, stroke, Alzheimer disease, or cancer. She had been taken to the emergency room after being found not breathing, lying in her bathtub. Bleeding over the occipital region of her scalp led the pathologist to hypothesize that she had slipped and fallen while getting into the bath and had hit the back of her head on the edge of the tub. Because the pathologist suspected that a traumatic brain injury had caused her death, he focused most of his attention on the contents of the cranial cavity. The pathologist noted superficial bruising and bleeding of the scalp in the occipital region of the woman's head. He then opened the cranial cavity, where he noted a large subdural hematoma above the right frontal lobe of the woman's brain. In addition, he saw that the brain had shifted due to the extensive bleeding, so that the medulla oblongata had been pushed inferiorly (herniated) through the foramen magnum into the vertebral canal. After completing a thorough inspection of the remainder of the woman's body, the pathologist determined that a subdural hematoma resulting from traumatic brain injury had caused the woman's death.

Predict 3

a. Explain why the subdural hematoma was found in the frontal region of the brain, when the blow to the head occurred over the occipital region.

b. How did the herniation of the medulla oblongata into the vertebral canal affect brain function and contribute to the woman's death?

Cerebrospinal Fluid

Cerebrospinal (ser′ĕ-brō-spī-năl; sĕ-rē′brō-spī-năl) **fluid (CSF)** is a clear fluid similar to blood serum with most of the proteins removed. It bathes the brain and the spinal cord and provides a protective cushion around the CNS. CSF allows the brain to float within the cranial cavity, so that it does not rest directly on the surface of the skull or dura mater. In addition, it protects the brain against the shock of rapid head movements. It also provides some nutrients to CNS tissues.

About 80–90% of the CSF is produced by specialized ependymal cells within the lateral ventricles, with the remainder produced by similar cells in the third and fourth ventricles. These specialized ependymal cells, their support tissue, and the associated blood vessels are collectively called the **choroid** (kō′royd; lacy) **plexuses** (plek′sŭs-ez). The choroid plexuses are formed by invaginations of the vascular pia mater into the ventricles, thus producing a vascular connective tissue core covered by ependymal cells (figure 13.14 *inset, lower left*).

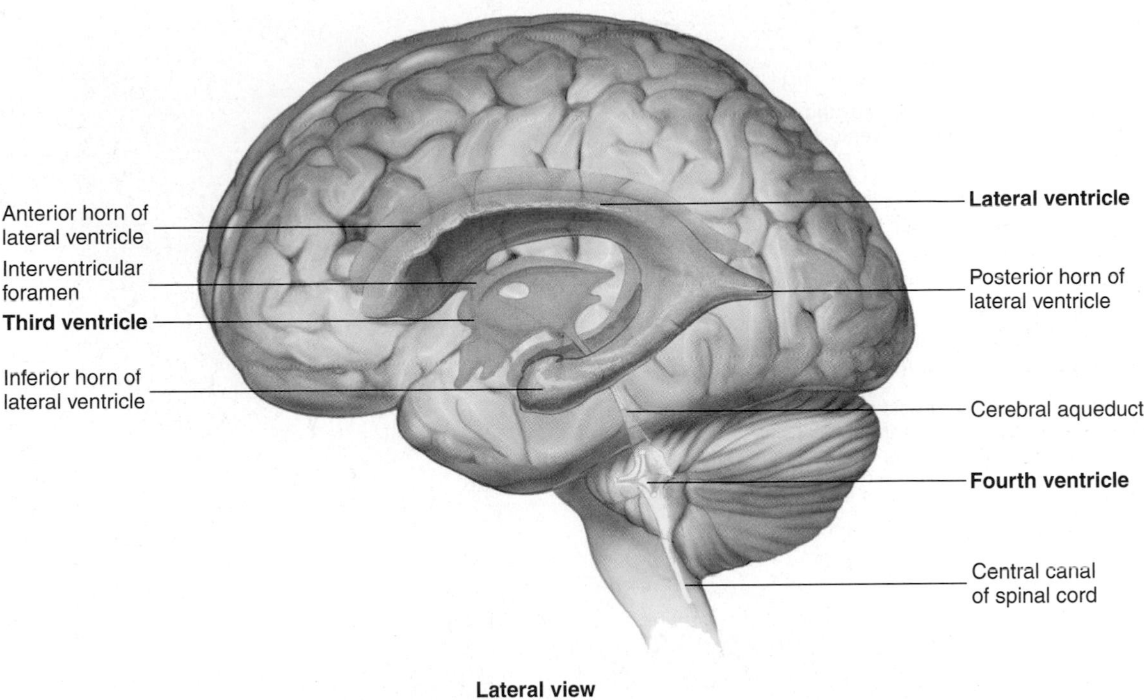
Anterior horn of
lateral ventricle
Interventricular
foramen
Third ventricle
Inferior horn of
lateral ventricle
Lateral ventricle
Posterior horn of
lateral ventricle
Cerebral aqueduct
Fourth ventricle
Central canal
of spinal cord
Lateral view

FIGURE 13.13 Ventricles of the Brain Viewed from the Left

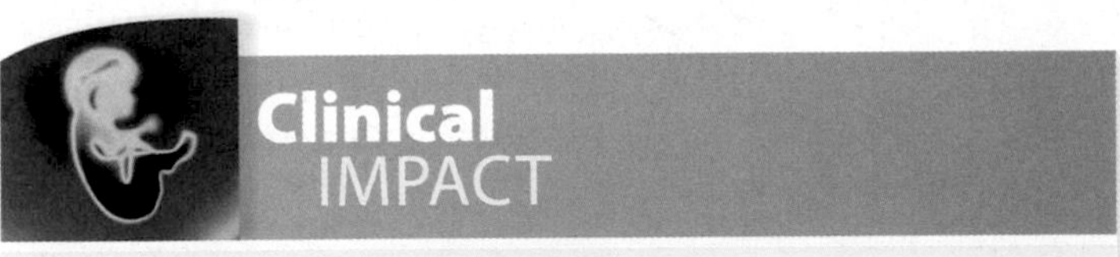
Clinical
IMPACT

CSF and Skull Fractures

Open head trauma involves a fracture or hole in the skull, which exposes the contents of the cranial cavity (brain, blood, and/or CSF) to the exterior. Head injuries involving scalp lacerations or damage to the eye, ear, or nose should be carefully evaluated for the possibility of open head trauma. In skull fractures in which the meninges are torn, CSF may leak from the nose if the fracture is in the frontal area or from the ear if the fracture is in the temporal area. Leakage of CSF indicates serious mechanical damage to the head and presents a risk for meningitis because bacteria may pass from the nose or ear through the tear and into the meninges. A skull fracture involving the base of the skull may result in cranial nerve damage where the nerves exit the skull. Open head trauma can result from falls, car accidents, and even sports. They are a good reason to use seat belts and wear helmets.

CSF is formed by the transport of water and solutes from the blood through a variety of mechanisms. The majority of the fluid enters the ventricles by following a Na^+ concentration gradient. Ependymal cells of the choroid plexus actively transport Na^+ into the ventricles, and water passively follows. Large molecules are transported by pinocytosis. The precise mechanisms for the transport of glucose and other substances into CSF remain unknown.

Endothelial cells of the blood vessels in the choroid plexuses, which are joined by tight junctions (see chapter 4), form the blood-brain barrier, or more correctly the blood-cerebrospinal fluid barrier (see section 13.7). Consequently, substances do not pass between the cells but must pass through the cells.

CSF fills the ventricles, the subarachnoid space of the brain and spinal cord, and the central canal of the spinal cord. Approximately 23 mL of fluid fills the ventricles, and 117 mL fills the subarachnoid space. The route taken by the CSF from its origin in the choroid plexuses to its return to the circulation is depicted in figure 13.14. The flow rate of CSF from its origin to the point at which it enters the bloodstream is about 0.4 mL/min. CSF passes from the lateral ventricles through the interventricular foramina into the third ventricle and then through the cerebral aqueduct into the fourth ventricle. It can exit the interior of the brain only from the fourth ventricle. One **median aperture** (foramen of Magendie), which opens through the roof of the fourth ventricle, and two **lateral apertures** (foramina of Luschka), which open through the walls, allow the CSF to pass from the fourth ventricle to the subarachnoid space. Some CSF continues to flow inferiorly into the central canal of the spinal cord. However, parts of the central canal are closed off in adults, so the amount circulating there is very small. Clusters of fingerlike protusions of arachnoid tissue called **arachnoid villi** form **arachnoid granulations,** which penetrate into the dural venous sinuses, especially the superior sagittal sinus. CSF passes into the blood-filled dural venous sinuses through these granulations. From the dural venous sinuses, the blood drains into the internal jugular veins to enter the veins of the general circulation; in this way, the CSF reenters the bloodstream.

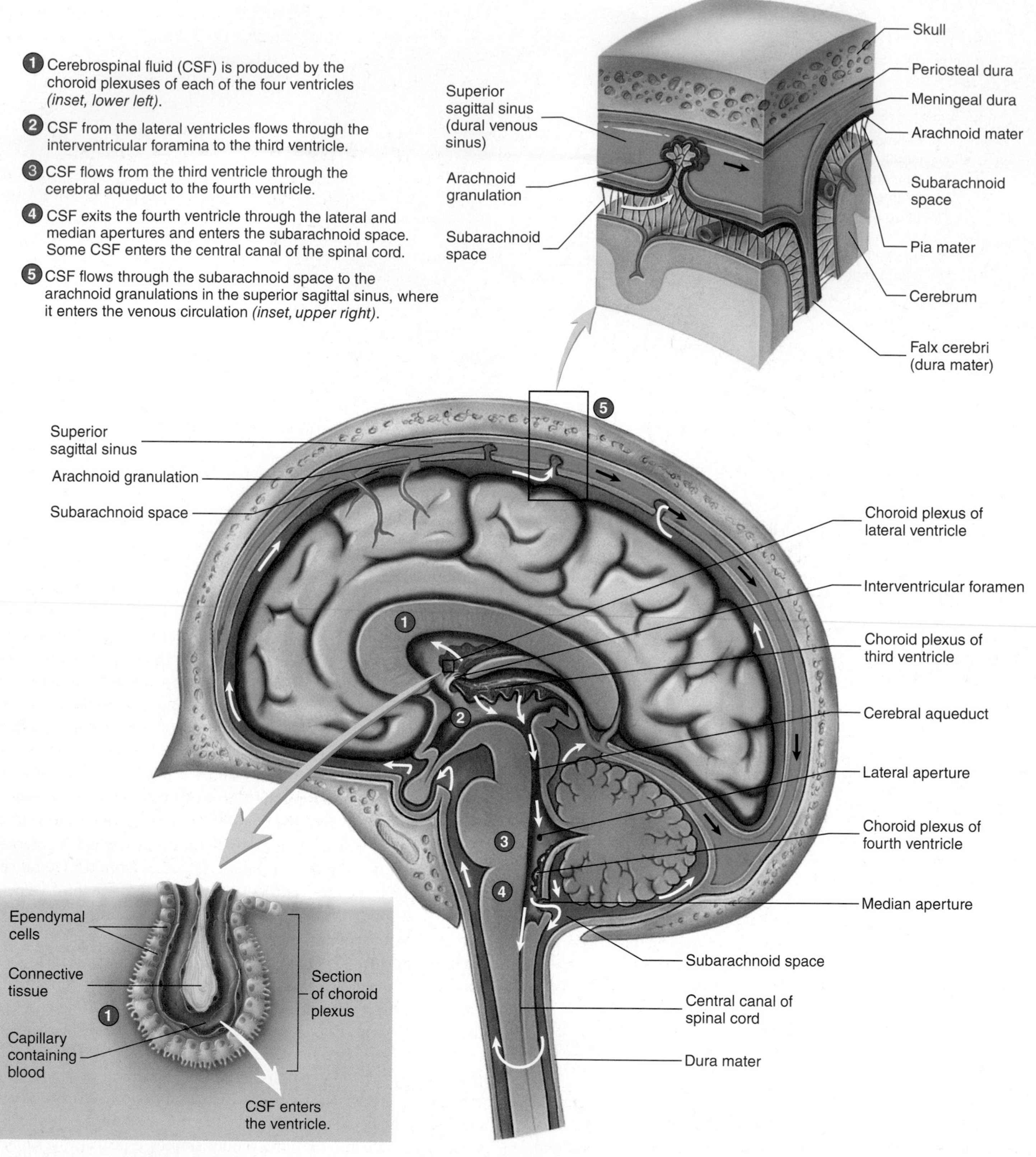

1 Cerebrospinal fluid (CSF) is produced by the choroid plexuses of each of the four ventricles *(inset, lower left)*.
2 CSF from the lateral ventricles flows through the interventricular foramina to the third ventricle.
3 CSF flows from the third ventricle through the cerebral aqueduct to the fourth ventricle.
4 CSF exits the fourth ventricle through the lateral and median apertures and enters the subarachnoid space. Some CSF enters the central canal of the spinal cord.
5 CSF flows through the subarachnoid space to the arachnoid granulations in the superior sagittal sinus, where it enters the venous circulation *(inset, upper right)*.
Superior sagittal sinus (dural venous sinus)
Arachnoid granulation
Subarachnoid space
Skull
Periosteal dura
Meningeal dura
Arachnoid mater
Subarachnoid space
Pia mater
Cerebrum
Falx cerebri (dura mater)
Superior sagittal sinus
Arachnoid granulation
Subarachnoid space
Choroid plexus of lateral ventricle
Interventricular foramen
Choroid plexus of third ventricle
Cerebral aqueduct
Lateral aperture
Choroid plexus of fourth ventricle
Median aperture
Subarachnoid space
Central canal of spinal cord
Dura mater
Ependymal cells
Connective tissue
Capillary containing blood
Section of choroid plexus
CSF enters the ventricle.
Sagittal section, medial view

PROCESS FIGURE 13.14 Choroid Plexuses and the Flow of CSF

CSF flows through the ventricles and subarachnoid space, as shown by the white arrows. Arrows going through the foramina in the wall and roof of the fourth ventricle represent the CSF entering the subarachnoid space. CSF passes back into the blood through the arachnoid granulations (white and black arrows), which penetrate the dural sinus. The black arrows show the direction of blood flow in the sinuses.

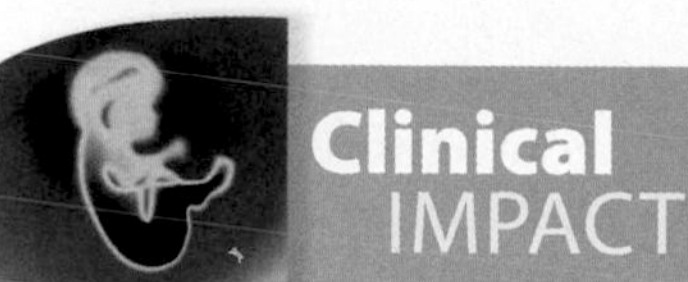

Clinical IMPACT

Hydrocephalus

The cerebral aqueduct may be blocked at the time of birth or may become blocked later in life because of a tumor growing in the brainstem. If the apertures of the fourth ventricle or the cerebral aqueduct are blocked, CSF can accumulate within the ventricles, resulting in a condition called **internal hydrocephalus** (*noncommunicating hydrocephalus*). The production of CSF continues, even when the passages that normally allow it to exit the brain are blocked. Consequently, fluid builds inside the brain, causing pressure, which compresses the nervous tissue and dilates the ventricles. Compression of the nervous tissue usually results in irreversible brain damage. If the skull bones are not completely ossified when the hydrocephalus occurs, the pressure may also severely enlarge the head.

Internal hydrocephalus can be treated successfully by placing a drainage tube (shunt) between the brain ventricles and the abdominal cavity to eliminate the high internal pressures. However, there is some risk of introducing infection into the brain through these shunts, and the shunts must be replaced as the person grows.

A subarachnoid hemorrhage may block the return of CSF to the circulation. If CSF accumulates in the subarachnoid space, the condition is called **external hydrocephalus** (*communicating hydrocephalus*). In this condition, pressure is applied to the brain externally, compressing neural tissues and causing brain damage. The condition usually resolves without treatment.

ASSESS YOUR PROGRESS

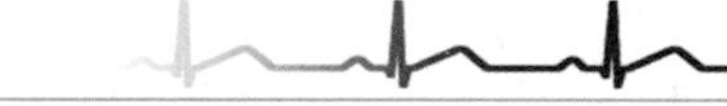

19. *Describe the three meninges that surround the brain. What are the falx cerebri, tentorium cerebelli, and falx cerebelli?*

20. *Describe the contents of the dural sinuses, subdural space, and subarachnoid space.*

21. *Name the four ventricles of the brain. Describe their locations and the connections between them. What are the septum pellucida?*

22. *Describe the production and circulation of CSF. Where does the CSF return to the blood?*

13.7 Blood Supply to the Brain

LEARNING OUTCOMES

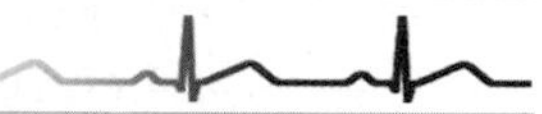

After reading this section, you should be able to

A. Describe how the brain is supplied with blood.

B. Explain the role of the blood-brain barrier.

The brain requires a tremendous amount of blood to maintain its normal functions. The brain has a very high metabolic rate, and brain cells are not capable of storing high-energy molecules for any length of time. In addition, brain cells depend almost entirely on glucose as their energy source (see chapter 25). Thus, the brain requires a constant blood supply to meet the demands of brain cells for both glucose and oxygen. Even though the brain accounts for only about 2% of total body weight, it receives approximately 15–20% of the blood pumped by the heart. Interruption of the brain's blood supply for only seconds can cause unconsciousness, and interruption for minutes can cause irreversible brain damage.

The brain's blood supply is illustrated in chapter 21 (see figures 21.10 and 21.11). Blood reaches the brain through the **internal carotid arteries,** which ascend to the head along the anterior-lateral part of the neck, and the **vertebral arteries,** which ascend along the posterior part of the neck, through the transverse foramina of the cervical vertebrae. The internal carotid arteries enter the cranial cavity through the carotid canals, and the vertebral arteries enter through the foramen magnum. The vertebral arteries join to form the **basilar artery,** which lies on the ventral surface of the pons. The basilar artery and the internal carotid arteries contribute to the **cerebral arterial circle** (circle of Willis). Branches from this circle and from the basilar artery supply blood to the brain.

The cerebral cortex on each side of the brain is supplied by three branches that arise from the cerebral arterial circle: the **anterior, middle,** and **posterior cerebral arteries.** The middle cerebral artery supplies most of the lateral surface of each cerebral hemisphere. The anterior cerebral artery supplies the medial portion of the parietal and frontal lobes. The posterior cerebral artery supplies the occipital lobe and the medial surface of the temporal lobe.

The arteries to the brain and their larger branches are located in the subarachnoid space. Small cortical arterial branches leave the subarachnoid space and enter the pia mater, where they branch extensively. Precapillary branches leave the pia mater and enter the tissue of the brain. Most of these branches are short and remain in the cortex. Fewer, longer branches extend into the medulla.

The arteries within the brain tissue quickly divide into capillaries. The epithelial cells of these capillaries are surrounded by the foot processes of astrocytes (see figure 11.6). The astrocytes promote the formation of tight junctions between the epithelial cells. The epithelial cells with their tight junctions form the **blood-brain barrier,** which

Clinical IMPACT

Drugs and the Blood-Brain Barrier

The permeability characteristics of the blood-brain barrier are an important consideration when developing drugs to affect the CNS. For example, Parkinson disease is caused by a lack of the neurotransmitter dopamine, which is normally produced by certain neurons of the brain. Lack of dopamine results in decreased muscle control and shaky movements. However, administering dopamine is not helpful because dopamine cannot cross the blood-brain barrier. Instead, physicians prescribe levodopa (L-dopa), a precursor to dopamine, because it can cross the blood-brain barrier. CNS neurons then convert levodopa to dopamine, which helps reduce the symptoms of Parkinson disease.

regulates the movement of materials from the blood into the brain. Materials that would enter many tissues by passing between the epithelial cells of capillaries cannot pass through the blood-brain barrier because of the tight junctions. Most materials that enter the brain pass through the epithelial cells. Water-soluble molecules, such as amino acids and glucose, require specific transporters to move across the plasma membranes of the epithelial cells by mediated transport (see chapter 3). However, lipid-soluble substances, such as nicotine and ethanol, can freely diffuse through the plasma membranes of the epithelial cells and enter the brain.

ASSESS **YOUR PROGRESS**

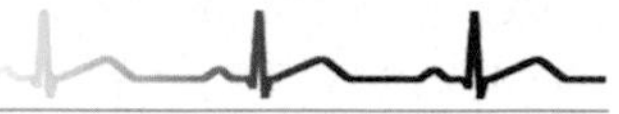

23. *Describe the blood supply to the brain. List the arteries supplying each part of the cerebral cortex.*
24. *Explain how the blood-brain barrier functions.*

13.8 Cranial Nerves

LEARNING OUTCOMES

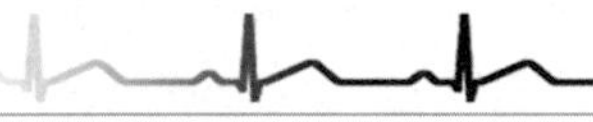

After reading this section, you should be able to

A. List the 12 cranial nerves and give the primary sensory, somatic motor, and/or parasympathetic functions of each.

B. Describe cranial reflexes.

Cranial nerves transmit and relay information to the brain analogous to the spinal nerves, except they do so by direct connections to the brain instead of the spinal cord.

A given cranial nerve may have one or more of three functions: (1) sensory, (2) somatic motor, and (3) parasympathetic (table 13.4). **Sensory** functions include the special senses, such as vision, and the more general senses, such as touch and pain. **Somatic** (sō-mat′ik) **motor** functions involve the control of skeletal muscles through motor neurons. The cranial nerves innervating skeletal muscles also contain proprioceptive sensory fibers, which convey action potentials to the CNS from those muscles. **Proprioception** (prō-prē-ō-sep′shun) informs the brain about the position of various body parts, including joints and muscles. However, because proprioception is the only sensory function of several otherwise somatic motor cranial nerves, that function is usually ignored, and the nerves are designated by convention as motor only. **Parasympathetic** function involves the regulation of glands, smooth muscles, and cardiac muscle. These functions are part of the autonomic nervous system and are discussed in chapter 16. Several of the cranial nerves have associated ganglia, and these ganglia are of two types: parasympathetic and sensory.

By convention, the 12 pairs of cranial nerves are indicated by Roman numerals (I–XII) from anterior to posterior (figure 13.15). The first 2 pairs of cranial nerves connect directly to the cerebrum (I) or the diencephalon (II). Nine pairs of cranial nerves connect to the brainstem. The remaining pair of cranial nerves (XI) is connected to the spinal cord and has no direct connection to brain structures.

TABLE 13.4 Functional Organization of the Cranial Nerves

Nerve Function	Cranial Nerve	
Sensory	I	Olfactory
	II	Optic
	VIII	Vestibulocochlear
Somatic motor	IV	Trochlear
	VI	Abducens
	XI	Accessory
	XII	Hypoglossal
Somatic motor and sensory	V	Trigeminal
Somatic motor and parasympathetic	III	Oculomotor
Somatic motor, sensory, and parasympathetic	VII	Facial
	IX	Glossopharyngeal
	X	Vagus

Characteristics of the Cranial Nerves

It may be helpful to refer to table 13.5 while reading the following descriptions of the cranial nerves. The **olfactory (I)** and **optic (II) nerves** are exclusively sensory and are involved in the special senses of smell and vision, respectively. The functions of these nerves are discussed in chapter 15.

The **oculomotor nerve (III)** innervates four of the six muscles that move the eyeball (the superior, inferior, and medial rectus muscles and the inferior oblique muscle; see chapter 10) and the levator palpebrae superioris muscle, which raises the superior eyelid. In addition, parasympathetic nerve fibers in the oculomotor nerve innervate smooth muscles in the eye and regulate the size of the pupil and the shape of the lens of the eye.

The **trochlear** (trōk′lē-ar) **nerve (IV)** is a somatic motor nerve that innervates one of the six eye muscles responsible for moving the eyeball (superior oblique).

The **trigeminal** (trī-jem′i-năl) **nerve (V)** has somatic motor, proprioceptive, and cutaneous sensory functions. It supplies motor innervation to the muscles of mastication, one middle ear muscle, one palatine muscle, and two throat muscles. In addition to the proprioception associated with its somatic motor functions, the trigeminal nerve also carries proprioceptive information from the temporomandibular joint, tongue, and cheek, which allows you to chew food without biting your tongue or cheek. Damage to the trigeminal nerve may impede chewing.

The trigeminal nerve has the greatest general sensory function of all the cranial nerves and is the only cranial nerve involved in **sensory cutaneous innervation** of the head. It also provides sensory innervation of blood vessels in the meninges that are associated with migraine (see chapter 14). Although the mechanisms are not fully understood, the trigeminal nerve is responsible for the pain of a headache. All other cutaneous innervation comes from spinal nerves (see figure 12.14). *Trigeminal* means "three twins," and the sensory distribution of the trigeminal nerve in the face is divided into three regions, each supplied by a branch of the nerve. The

FUNDAMENTAL **Figure**

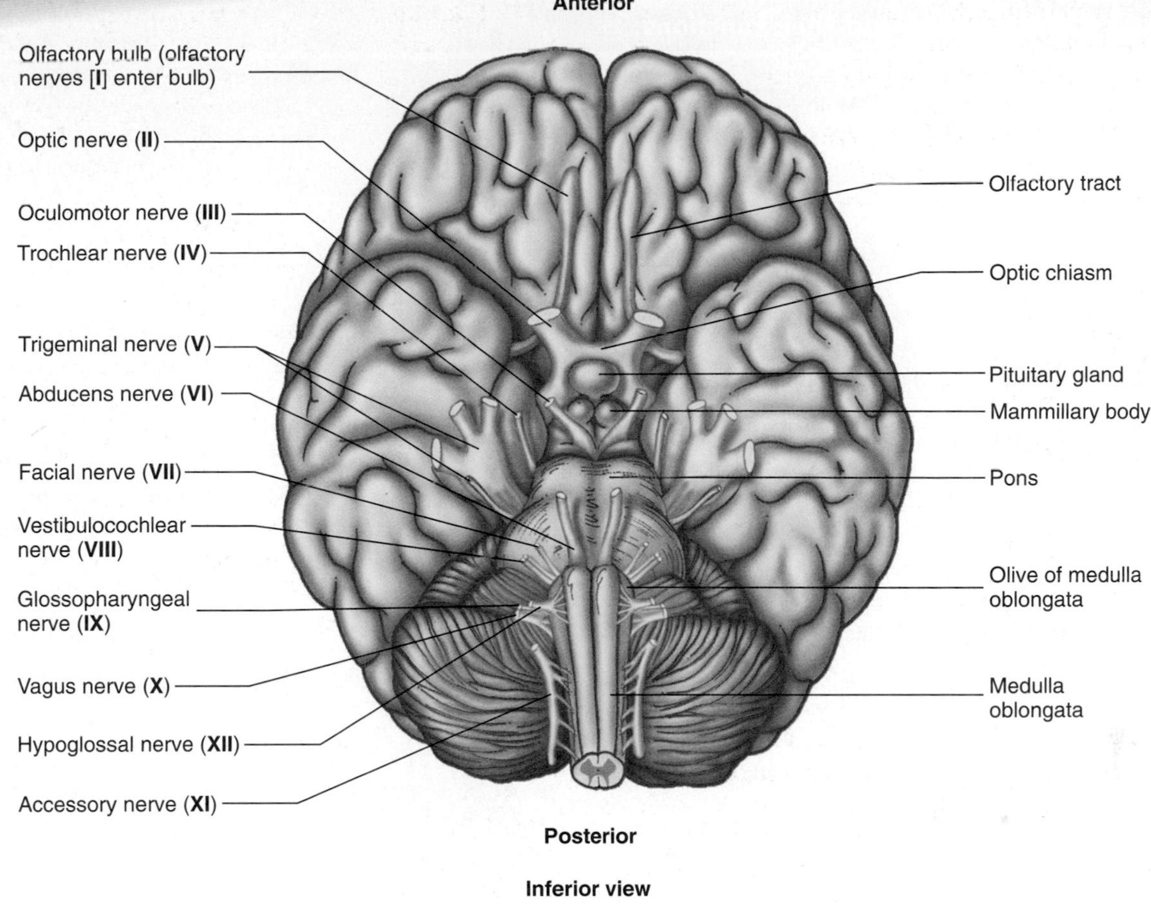
Anterior
Olfactory bulb (olfactory nerves [I] enter bulb)
Optic nerve (II)
Oculomotor nerve (III)
Trochlear nerve (IV)
Trigeminal nerve (V)
Abducens nerve (VI)
Facial nerve (VII)
Vestibulocochlear nerve (VIII)
Glossopharyngeal nerve (IX)
Vagus nerve (X)
Hypoglossal nerve (XII)
Accessory nerve (XI)
Olfactory tract
Optic chiasm
Pituitary gland
Mammillary body
Pons
Olive of medulla oblongata
Medulla oblongata
Posterior
Inferior view

FIGURE 13.15 AP|R **Inferior Surface of the Brain, Showing the Origins of the Cranial Nerves**

three branches—ophthalmic (V_1), maxillary (V_2), and mandibular (V_3)—arise directly from the trigeminal ganglion, which serves the same function as the dorsal root ganglia of the spinal nerves. Only the mandibular branch contains motor axons, which bypass the trigeminal ganglion, much as the ventral root of a spinal nerve bypasses a dorsal root ganglion.

In addition to these cutaneous functions, the maxillary and mandibular branches are important in dentistry. The maxillary nerve supplies sensory innervation to the maxillary teeth, palate, and gingiva (jin′ji-vă; gum). The mandibular branch supplies sensory innervation to the mandibular teeth, tongue, and gingiva. The various nerves innervating the teeth are referred to as **alveolar nerves** (al-vē′ō-lăr; socket).

Clinical IMPACT — Dental Anesthesia

Dentists inject anesthetic to block sensory transmission by the alveolar nerves. The superior alveolar nerves are not usually anesthetized directly because they are difficult to approach with a needle. For this reason, the maxillary teeth are usually anesthetized locally by inserting the needle beneath the oral mucosa surrounding the teeth. The inferior alveolar nerve is probably anesthetized more often than any other nerve in the body. To anesthetize this nerve, the dentist inserts the needle somewhat posterior to the patient's last molar and extends the needle near where the mandibular branch of the trigeminal nerve enters the mandibular foramen.

During an inferior alveolar block, several nondental nerves are usually anesthetized. The mental nerve, which supplies cutaneous innervation to the anterior lip and chin, is a distal branch of the inferior alveolar nerve. When the inferior alveolar nerve is blocked, the mental nerve is blocked also, resulting in a numb lip and chin. Nerves lying near the point where the inferior alveolar nerve enters the mandible are often also anesthetized during inferior alveolar anesthesia. For example, the lingual nerve can be anesthetized to produce a numb tongue. The facial nerve lies some distance from the inferior alveolar nerve, but in rare cases anesthetic can diffuse far enough posteriorly to anesthetize that nerve. The result is a temporary facial palsy (paralysis or paresis), with the injected side of the face drooping because of flaccid muscles; the condition disappears when the anesthesia wears off. If the facial nerve is cut by an improperly inserted needle, permanent facial palsy may occur.

TABLE 13.5 Cranial Nerves and Their Functions

Cranial Nerve	Foramen or Fissure*	Function	Consequences of Lesions to Nerve
I: Olfactory	Cribriform plate	Sensory Special sense of smell	Inability to smell
II: Optic	Optic foramen	Sensory Special sense of vision	Blindness on the affected side
III: Oculomotor	Superior orbital fissure	Motor[†] and parasympathetic Motor to eye muscles (superior, medial, and inferior rectus: inferior oblique) and upper eyelid (levator palpebrae superioris) Proprioceptive from those muscles Parasympathetic to the sphincter of the pupil (causing constriction) and the ciliary muscle of the lens (causing accommodation)	Pupil dilation; eye deviates inferiorly and laterally due to muscle paralysis, resulting in double vision; eyelid droops (ptosis); blurred vision due to loss of accommodation

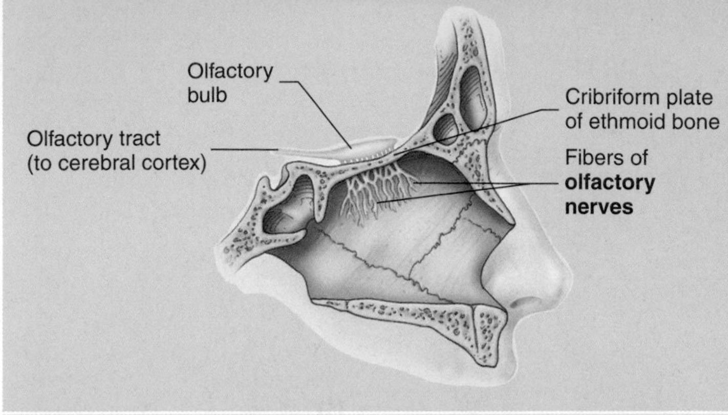

Olfactory bulb
Olfactory tract (to cerebral cortex)
Cribriform plate of ethmoid bone
Fibers of olfactory nerves

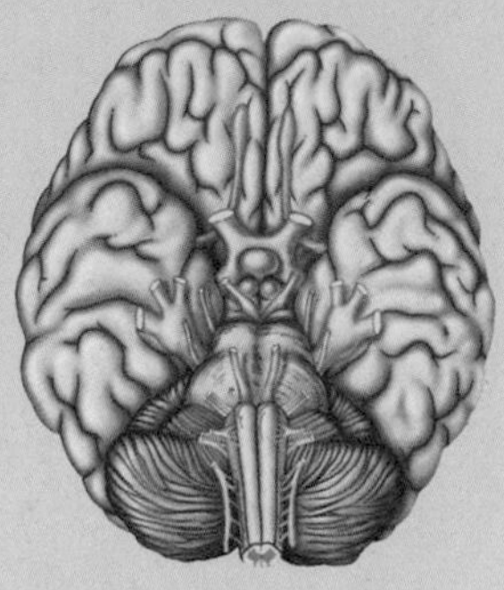

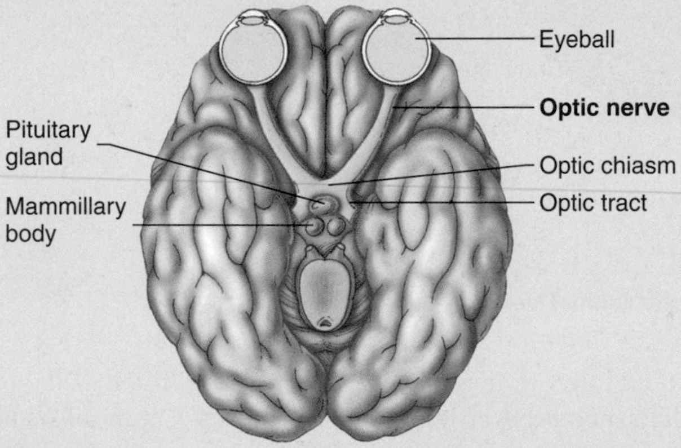

Eyeball
Optic nerve
Pituitary gland
Optic chiasm
Optic tract
Mammillary body

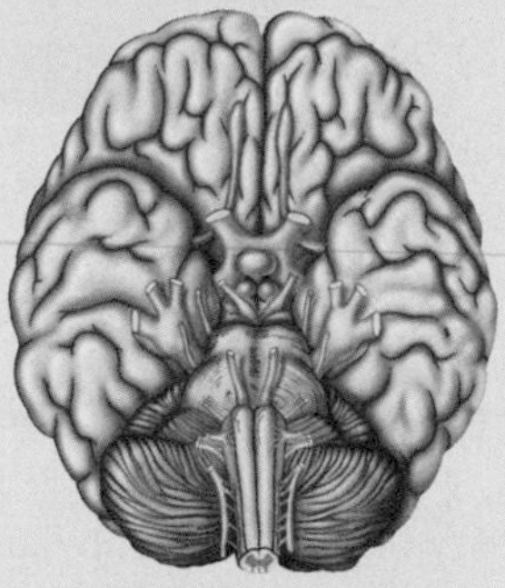

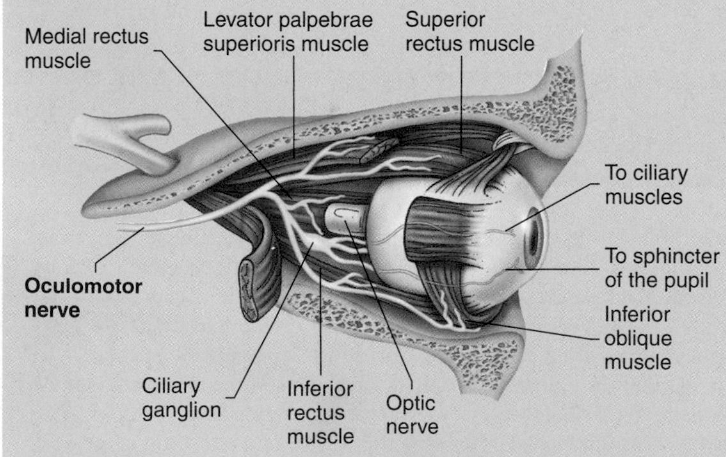

Medial rectus muscle
Levator palpebrae superioris muscle
Superior rectus muscle
To ciliary muscles
To sphincter of the pupil
Oculomotor nerve
Inferior oblique muscle
Ciliary ganglion
Inferior rectus muscle
Optic nerve

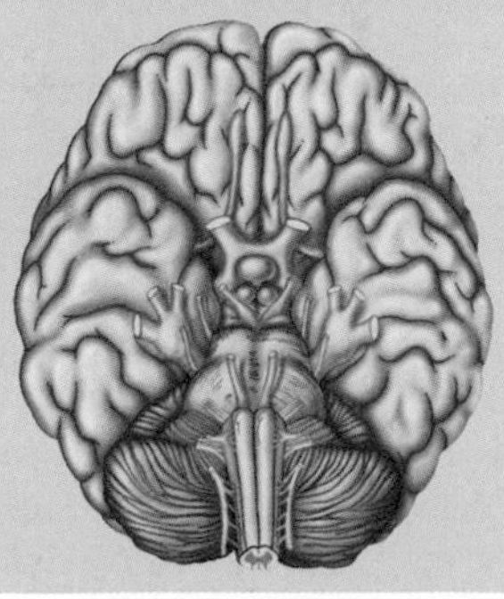

*Route of entry or exit from the skull.

[†]Proprioception is a sensory function, not a motor function; however, motor nerves to muscles also contain some proprioceptive afferent fibers from those muscles. Because proprioception is the only sensory information carried by some cranial nerves, these nerves are still considered "motor."

TABLE 13.5 Cranial Nerves and Their Functions—Continued

Cranial Nerve	Foramen or Fissure*	Function	Consequences of Lesions to Nerve
IV. Trochlear	Superior orbital fissure	Motor[†] Motor to one eye muscle (superior oblique) Proprioceptive from that muscle	Difficulty moving the eye inferiorly and laterally which leads to double vision
V. Trigeminal		Sensory	Trigeminal neuralgia; intense pain along the course of a branch of the nerve; loss of tactile sensation in the face; weakness in biting or clenching jaw
Ophthalmic branch (V_1)	Superior orbital fissure	Sensory from scalp, forehead, nose, upper eyelid, and cornea	
Maxillary branch (V_2)	Foramen rotundum	Sensory Sensory from palate, upper jaw, upper teeth and gums, nasopharynx, nasal cavity, skin and mucous membrane of cheek, lower eyelid, and upper lip	
Mandibular branch (V_3)	Foramen ovale	Sensory and motor[†] Sensory from lower jaw, lower teeth and gums, anterior two-thirds of tongue, mucous membrane of cheek, lower lip, skin of cheek and chin, auricle, and temporal region Motor to muscles of mastication (masseter, temporalis, medial and lateral pterygoids), soft palate (tensor veli palatini), throat (anterior belly of digastric, mylohyoid), and middle ear (tensor tympani) Proprioceptive from those muscles	

Superior oblique muscle
Trochlear nerve

Maxillary branch (V_2)
Ophthalmic branch (V_1)
Trigeminal ganglion
Trigeminal nerve
Sensory root
Motor root
Mandibular branch (V_3)
Chorda tympani (from facial nerve)
To muscles of mastication
Lingual nerve
Inferior alveolar nerve
Submandibular ganglion
To mylohyoid muscle
To skin of face
Superior alveolar nerves
Mental nerve

Trigeminal nerve
Ophthalmic branch (V_1)
Maxillary branch (V_2)
Mandibular branch (V_3)

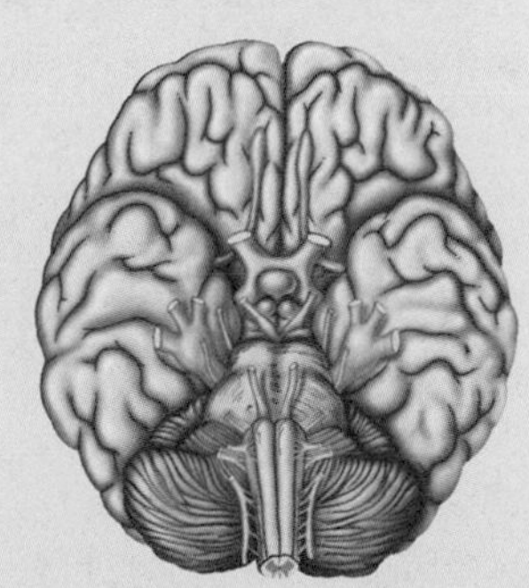

TABLE 13.5 Cranial Nerves and Their Functions—Continued

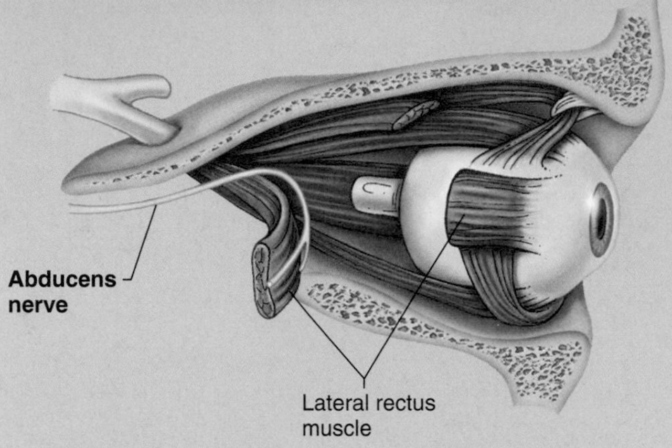

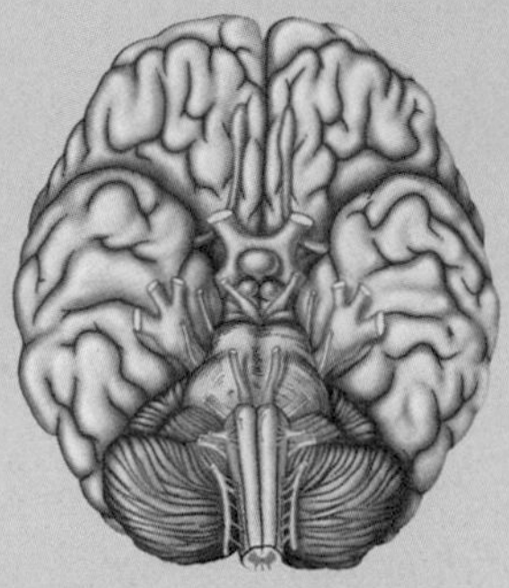

Cranial Nerve	Foramen or Fissure*	Function	Consequences of Lesions to Nerve
VI. Abducens	Superior orbital fissure	Motor† Motor to one eye muscle (lateral rectus) Proprioceptive from that muscle	Eye deviates medially (adducts) causing double vision
VII. Facial	Internal auditory canal Stylomastoid foramen	Sensory, motor,† and parasympathetic Sense of taste from anterior two-thirds of tongue, sensory from some of external ear and palate Motor to muscles of facial expression, throat (posterior belly of digastric, stylohyoid), and middle ear (stapedius) Proprioceptive from those muscles Parasympathetic to submandibular and sublingual salivary glands, lacrimal gland, and glands of nasal cavity and palate	Facial palsy; loss of taste sensation on the anterior two-thirds of tongue; decreased salivation

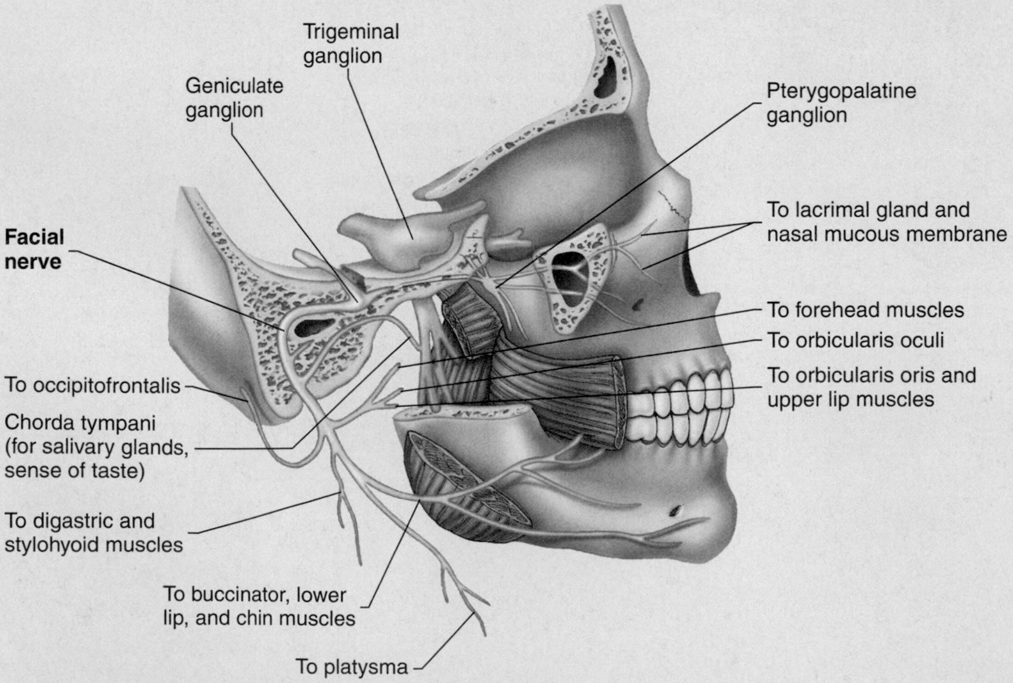

Abducens nerve
Lateral rectus muscle

Trigeminal ganglion
Geniculate ganglion
Pterygopalatine ganglion
Facial nerve
To lacrimal gland and nasal mucous membrane
To forehead muscles
To orbicularis oculi
To orbicularis oris and upper lip muscles
To occipitofrontalis
Chorda tympani (for salivary glands, sense of taste)
To digastric and stylohyoid muscles
To buccinator, lower lip, and chin muscles
To platysma

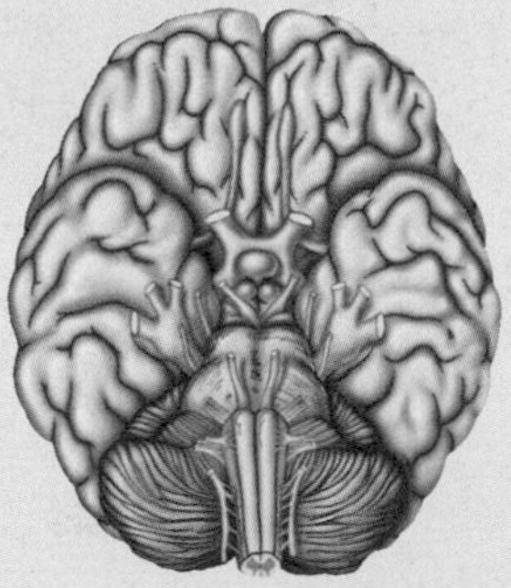

TABLE 13.5 Cranial Nerves and Their Functions—Continued

Cranial Nerve	Foramen or Fissure*	Function	Consequences of Lesions to Nerve
VIII. Vestibulocochlear	Internal auditory canal	Sensory Special senses of hearing (cochlear nerve) and balance (vestibular nerve)	Loss of hearing (cochlear nerve); loss of balance and equilibrium; nausea, vertigo, vomiting (vestibular nerve)
IX. Glossopharyngeal	Jugular foramen	Sensory, motor,[†] and parasympathetic Sense of taste from posterior one-third of tongue; sensory from pharynx, palatine tonsils, posterior one-third of tongue, middle ear, carotid sinus, and carotid body Motor to pharyngeal muscle (stylopharyngeus) Proprioceptive from pharyngeal muscle Parasympathetic to parotid salivary gland and glands of the posterior one-third of tongue	Difficulty swallowing; loss of taste sensation in posterior one-third of tongue; decreased salivation

Vestibular ganglion
Vestibular nerve
Vestibulocochlear nerve
Cochlear nerve
Spiral ganglion of cochlea

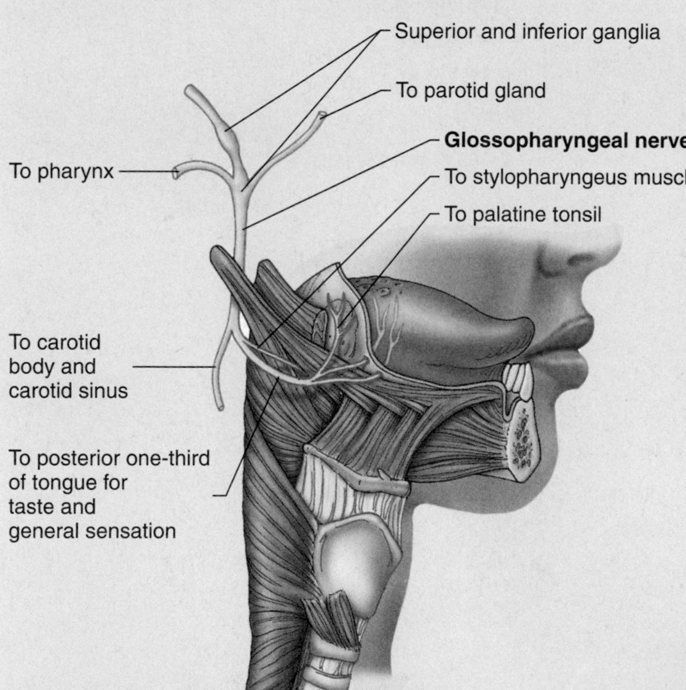
Superior and inferior ganglia
To parotid gland
Glossopharyngeal nerve
To stylopharyngeus muscle
To palatine tonsil
To pharynx
To carotid body and carotid sinus
To posterior one-third of tongue for taste and general sensation

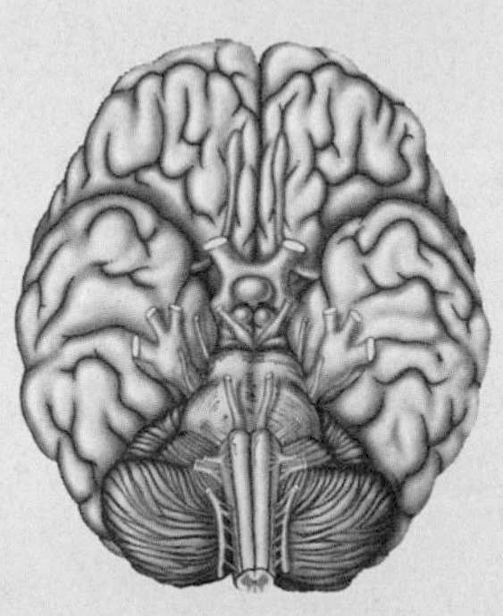

TABLE 13.5 Cranial Nerves and Their Functions—Continued

Cranial Nerve	Foramen or Fissure*	Function	Consequences of Lesions to Nerve
X. Vagus	Jugular foramen	Sensory, motor,[†] and parasympathetic Sensory from inferior pharynx, larynx, thoracic and abdominal organs; sense of taste from posterior tongue Motor to soft palate, pharynx, intrinsic laryngeal muscles (voice production), and an extrinsic tongue muscle (palatoglossus) Proprioceptive from those muscles Parasympathetic to thoracic and abdominal viscera	Difficulty swallowing and/or hoarseness; uvula deviates away from side of the dysfunction
XI. Accessory	Foramen magnum Jugular foramen	Motor[†] Motor to sternocleidomastoid and trapezius	Difficulty elevating the scapula or rotating the neck

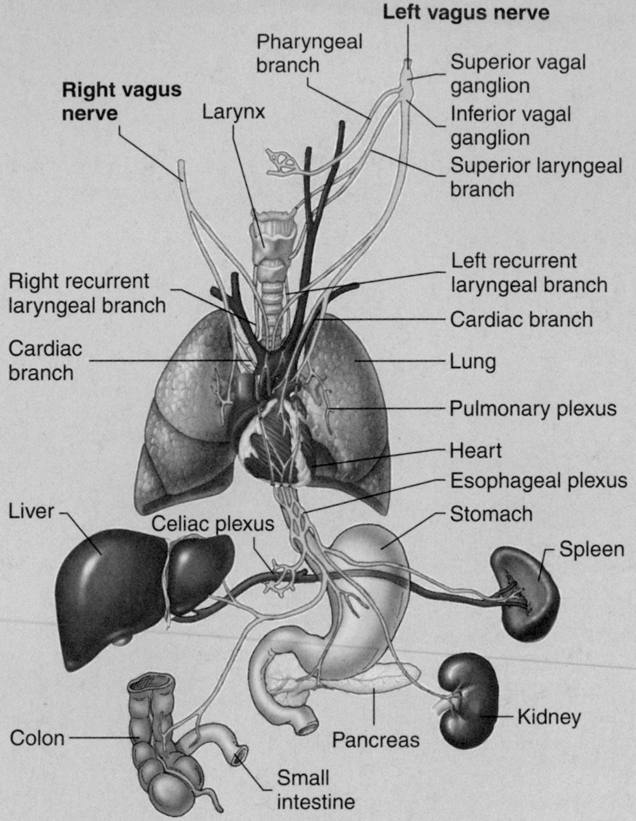

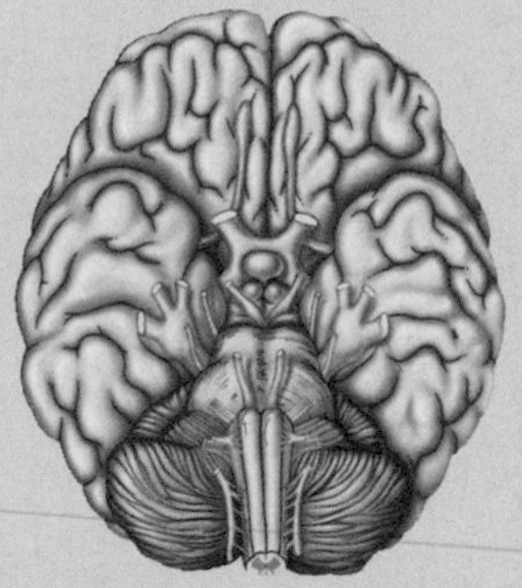

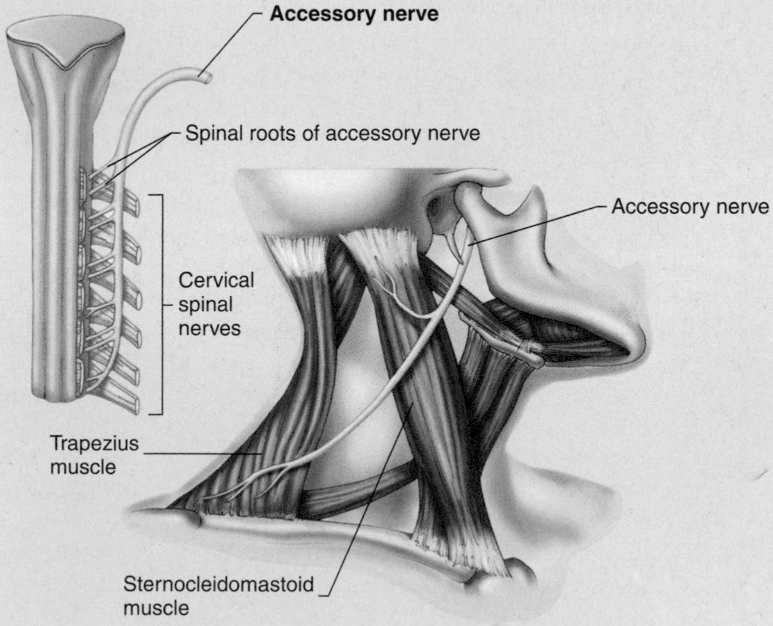

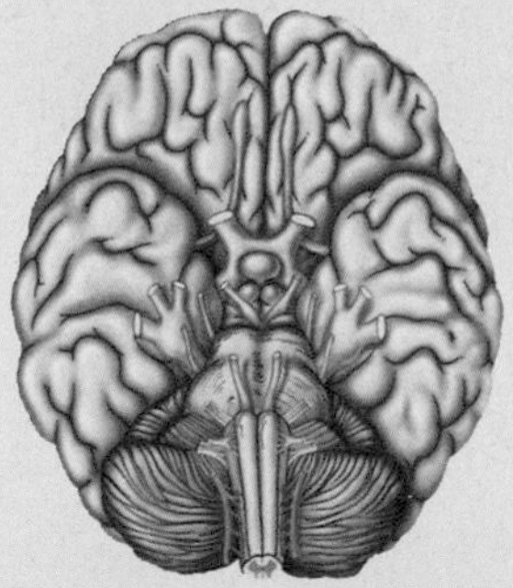

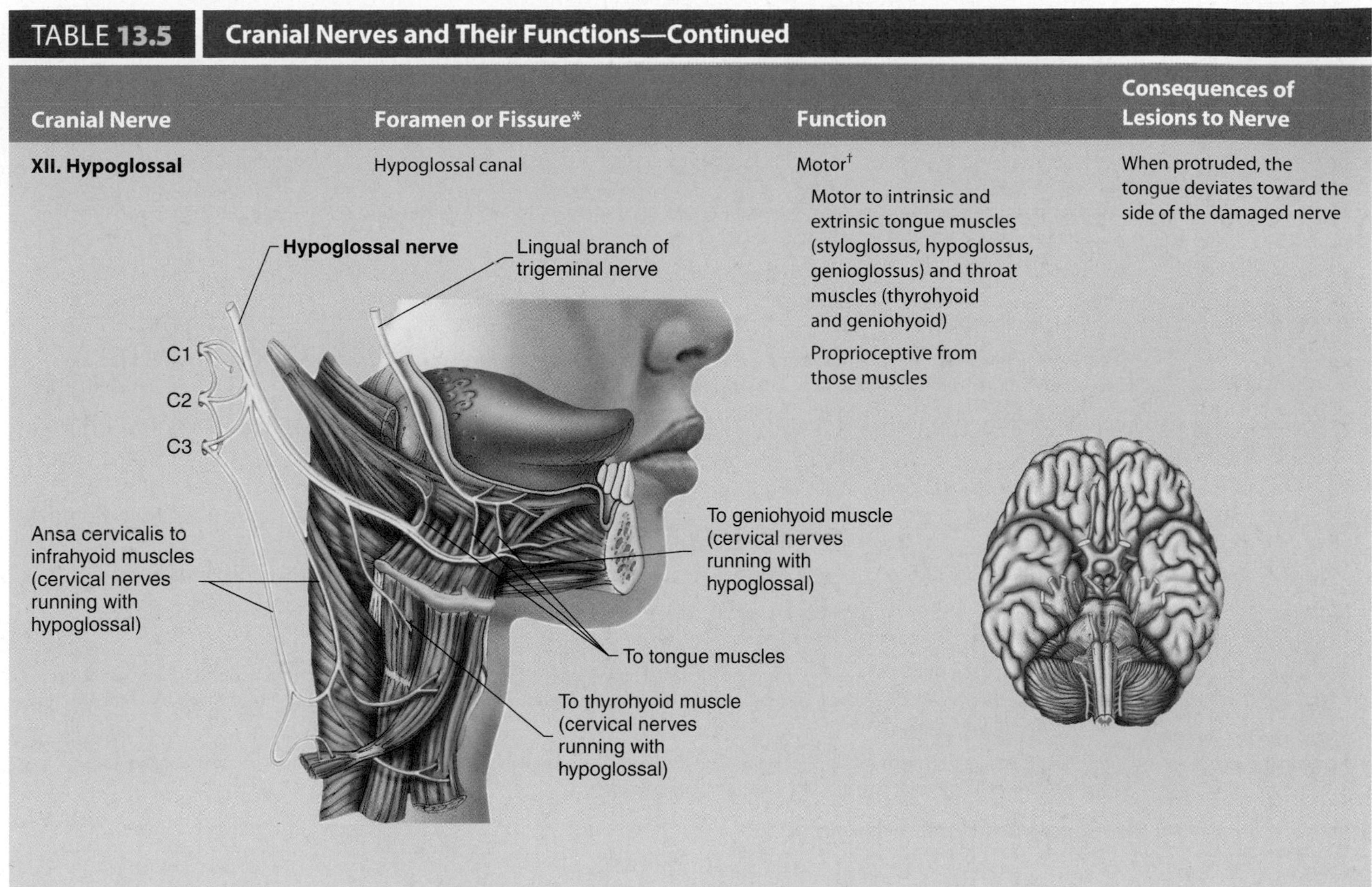

TABLE 13.5 Cranial Nerves and Their Functions—Continued

Cranial Nerve	Foramen or Fissure*	Function	Consequences of Lesions to Nerve
XII. Hypoglossal	Hypoglossal canal	Motor† Motor to intrinsic and extrinsic tongue muscles (styloglossus, hypoglossus, genioglossus) and throat muscles (thyrohyoid and geniohyoid) Proprioceptive from those muscles	When protruded, the tongue deviates toward the side of the damaged nerve

Hypoglossal nerve
Lingual branch of trigeminal nerve
C1
C2
C3
Ansa cervicalis to infrahyoid muscles (cervical nerves running with hypoglossal)
To geniohyoid muscle (cervical nerves running with hypoglossal)
To tongue muscles
To thyrohyoid muscle (cervical nerves running with hypoglossal)

The **superior alveolar nerves** to the maxillary teeth are derived from the maxillary branch of the trigeminal nerve, and the **inferior alveolar nerves** to the mandibular teeth are derived from the mandibular branch of the trigeminal nerve.

The **abducens** (ab-doo′senz) **nerve (VI),** like the trochlear nerve, is a somatic motor nerve that innervates one of the six muscles responsible for moving the eyeball (lateral rectus).

Predict 4

A drooping upper eyelid on one side of the face is a sign of possible oculomotor nerve damage. Describe how a patient could be tested for this type of damage by examining other oculomotor nerve functions. Describe the eye movements that distinguish among oculomotor, trochlear, and abducens nerve damage.

The **facial nerve (VII)** is somatic motor, sensory, and parasympathetic in function. It controls all the muscles of facial expression, a small muscle in the middle ear, and two hyoid muscles. It is sensory for the sense of taste in the anterior two-thirds of the tongue (see chapter 15). The facial nerve supplies parasympathetic innervation to the submandibular and sublingual salivary glands of the mouth and to the lacrimal glands of the eye.

The **vestibulocochlear** (ves-tib′ū-lō-kok′lē-ăr) **nerve (VIII),** like the olfactory and optic nerves, is exclusively sensory and transmits action potentials from the inner ear responsible for the special senses of hearing and balance (see chapter 15).

The **glossopharyngeal** (glos′ō-fă-rin′jē-ăl) **nerve (IX),** like the facial nerve, is somatic motor, sensory, and parasympathetic in function and has both sensory and parasympathetic ganglia. The glossopharyngeal nerve is somatic motor to one muscle of the pharynx and supplies parasympathetic innervation to the parotid salivary glands. The glossopharyngeal nerve is sensory for the sense of taste in the posterior one-third of the tongue. It also supplies tactile sensory innervation from the posterior tongue, middle ear, and pharynx, and it transmits sensory stimulation from receptors in the carotid arteries, which monitor blood pressure and blood carbon dioxide, oxygen, and pH levels (see chapter 21).

The **vagus** (vā′gŭs) **nerve (X),** like the facial and glossopharyngeal nerves, is somatic motor, sensory, and parasympathetic in function and has both sensory and parasympathetic ganglia. Most muscles of the soft palate, pharynx, and larynx are innervated by the vagus nerve. Damage to the laryngeal branches of the vagus nerve can interfere with normal speech. The vagus nerve is sensory for taste from the root of the tongue (see chapter 15). It is sensory for the inferior pharynx and the larynx and transmits sensory input from receptors in the aortic arch, which monitor blood pressure and the levels of carbon dioxide, oxygen, and pH in the blood (see chapter 21). In addition, the vagus nerve conveys sensory information from the thoracic and

Diseases and Disorders

TABLE 13.6 CNS and Cranial Nerves

Condition	Description
CNS DISORDERS*	
Cerebral aneurysm	Excessive dilation or ballooning of an artery; hemorrhaging (leaking) of the aneurysm leads to a hematoma within the brain or in the extradural, subdural, or subarachnoid spaces around the brain
Stroke	Loss of blood flow to the brain; caused by bleeding in the brain or meninges or a clot or spasm blocking cerebral blood vessels that results in a local area of cell death; symptoms include general intellectual deficiency, memory loss, short attention span, moodiness, disorientation, and irritability
Concussion	A blow to the head that produces momentary loss of consciousness; can lead to *postconscussion syndrome*, which can include headaches, fatigue, difficulties with mental tasks, depression, and personality changes
Cerebral compression	Increased intracranial pressure; can be caused by hematomas, hydrocephalus, tumors, edema from a severe head blow, or spinal cord injury

* For more CNS disorders, see Diseases and Disorders tables in chapters 12 and 14.

Condition	Description
CRANIAL NERVE DISORDERS	
Trigeminal neuralgia	Sharp pain in the face involving the trigeminal nerve; cause is unknown; can be triggered by touch near the mouth area
Migraine	Severe headache with associated sensory symptoms; cause unknown but involves the trigeminal nerve; can be triggered by environmental factors; tends to be hereditary
Facial palsy	Unilateral paralysis of the facial muscles involving the facial nerve; side of the face droops; can result from a stroke, infections, acute nerve damage, or a brain tumor; most common type is called *Bell palsy*, in which cause is unknown, but it is believed to be due to viral infection of the facial nerve
Herpes simplex I	Viral infection characterized by lesions (*cold sores*) on the lips or nasal region; virus can remain dormant in trigeminal ganglion; reactivated often at times of stress or reduced immune resistance

Go to: www.mhhe.com/seeley10 for additional information on these pathologies.

abdominal organs. The parasympathetic part of the vagus nerve is very important in regulating the functions of the thoracic and abdominal organs. It carries parasympathetic fibers to the heart and lungs in the thorax and to the digestive organs and kidneys in the abdomen.

The **accessory nerve (XI)** is a somatic motor nerve that has both cranial and spinal roots. The cranial root joins the vagus nerve (hence the name *accessory*) and participates in its function. However, the presence of the cranial root in the human brainstem is variable. The major component of the accessory nerve is the spinal root, which originates from the superior part of the cervical spinal cord, enters the cranial cavity through the foramen magnum, and then exits through the jugular foramen. The accessory nerve provides the major innervation to the sternocleidomastoid and trapezius muscles of the neck and shoulder.

The **hypoglossal nerve (XII)** is a somatic motor nerve that arises from the ventral surface of the medulla oblongata. It supplies the intrinsic tongue muscles, three of the four extrinsic tongue muscles, and the thyrohyoid and the geniohyoid muscles.

> **Predict 5**
>
> Injury to the accessory nerve may result in sternocleidomastoid muscle dysfunction, a condition called torticollis (sometimes called wry neck), in which the head is drawn to one side. If the head is turned to the left, does this position indicate injury to the left or the right spinal component of the accessory nerve?

Reflexes Involving Cranial Nerves

Reflexes integrated within the spinal cord were discussed in chapter 12. Many of the body's functions, especially those involved in maintaining homeostasis, involve reflexes that are integrated within the brain. Some of these reflexes, such as those involved in controlling heart rate (see chapter 20), blood pressure (see chapter 21), and respiration (see chapter 23), are integrated in the brainstem, and many involve cranial nerve X (vagus nerve).

Many of the brainstem reflexes are associated with cranial nerve function. The circuitry of most of these reflexes is too complex for our discussions. In general, these reflexes involve sensory input from the cranial nerves or spinal cord and the motor output of the cranial nerves.

Turning the eyes toward a flash of light, a sudden noise, or a touch on the skin is an example of a brainstem reflex. Moving the eyes to track a moving object is another complex brainstem reflex. Some of the sensory neurons from cranial nerve VIII form a reflex arc with neurons of cranial nerves V and VII, which send axons to muscles of the middle ear and dampen the effects of very loud, sustained noises on delicate inner ear structures (see chapter 15). Reflexes that occur during chewing allow the jaws to react to foods of various hardness and protect the teeth from breaking on very hard foods. Both the sensory and motor components of the reflex arc are carried by cranial nerve V. Reflexes involving input through cranial nerve V and output through cranial nerve XII move the tongue to position food between the teeth for chewing and then move the tongue out of the way, so that it is not bitten.

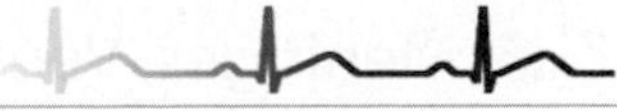

ASSESS **YOUR PROGRESS**

25. *What are the three major functions of the cranial nerves?*

26. *Which cranial nerves are sensory only? With what sense is each of these associated?*

27. *Name the cranial nerves that are somatic motor and proprioceptive only. What muscles or groups of muscles does each nerve supply? What is proprioception?*

28. *What cranial nerve provides the sensory cutaneous innervation of the face? How is this nerve important in dentistry? Name the muscles that would not function if this nerve were damaged.*

29. *Which four cranial nerves have a parasympathetic function? Describe the functions of each of these nerves.*

30. *Which cranial nerve leaves the head and neck region?*

31. *Give an example of reflex integration by cranial nerves.*

Answer

Learn to Predict ❮ From page 429

In addition to Marvin's broken jaw, the question suggested that he experienced damage to a cranial nerve, specifically the one associated with controlling the muscles of the tongue. Chapter 11 explained that somatic motor neurons innervate skeletal muscle, so the cranial nerve associated with motor control of the tongue is the affected nerve. Tables 13.4 and 13.5 summarize the functional organization of the cranial nerves. By reviewing these tables, we find that the hypoglossal nerve (XII) innervates intrinsic and extrinsic tongue muscles; therefore, this cranial nerve was damaged during Marvin's accident. More specifically, the damage was on the right side and caused paralysis of the tongue muscles on the right side. Thus, only the left side of the tongue can stick out. The hypoglossal nerve also innervates some throat muscles, such as the thyrohyoid and the geniohyoid muscles, which may also be affected.

Answers to the rest of this chapter's Predict questions are in Appendix G.

Summary

13.1 Development of the CNS (p. 430)

The brain and spinal cord develop from the neural tube. The ventricles and central canal develop from the lumen of the neural tube.

13.2 Brainstem (p. 433)

Medulla Oblongata

1. The medulla oblongata is continuous with the spinal cord and contains ascending and descending tracts.
2. The pyramids are tracts controlling voluntary muscle movement.
3. The olives are nuclei that function in equilibrium, coordination, and modulation of sound from the inner ear.
4. Medullary nuclei regulate the heart, blood vessels, respiration, swallowing, vomiting, coughing, sneezing, and hiccuping. The nuclei of cranial nerves V, VII, and IX–XII are in the medulla oblongata.

Pons

1. The pons is superior to the medulla oblongata.
2. Ascending and descending tracts pass through the pons.
3. Pontine nuclei regulate sleep and respiration. The nuclei of cranial nerves V–VIII are in the pons.

Midbrain

1. The midbrain is superior to the pons.
2. The midbrain contains the nuclei for cranial nerves III, IV, and V.
3. The tectum consists of four colliculi. The two inferior colliculi are involved in hearing, and the two superior colliculi in visual reflexes.
4. The tegmentum contains ascending tracts and the red nuclei, which are involved in motor activity.
5. The cerebral peduncles are the major descending motor pathway.
6. The substantia nigra connects to other basal nuclei and is involved with muscle tone and movement.

Reticular Formation

The reticular formation consists of nuclei scattered throughout the brainstem. The reticular system functions in many brainstem activities, including motor control, pain perception, rhythmic contractions, and the sleep-wake cycle.

13.3 Cerebellum (p. 435)

1. The cerebellar cortex contains more neurons than the cerebral cortex does. The Purkinje cells are the largest cells in the CNS.

2. The cerebellum has three parts, which control balance, gross motor coordination, and fine motor coordination.
3. The cerebellum corrects discrepancies between intended movements and actual movements.
4. The cerebellum can "learn" highly specific, complex motor activities.

13.4 Diencephalon (p. 436)

The diencephalon is located between the brainstem and the cerebrum.

Thalamus

1. The thalamus consists of two lobes connected by the interthalamic adhesion. The thalamus functions as an integration center.
2. Most sensory input synapses in the thalamus. Pain is registered in the thalamus.
3. The thalamus also has some motor functions.

Subthalamus

The subthalamus is inferior to the thalamus and is involved in motor function.

Epithalamus

The epithalamus is superior and posterior to the thalamus and contains the habenula, which influences emotions through the sense of smell. The pineal gland may play a role in the onset of puberty.

Hypothalamus

1. The hypothalamus, the most inferior portion of the diencephalon, contains several nuclei and tracts.
2. The mammillary bodies are reflex centers for olfaction.
3. The hypothalamus regulates many endocrine functions (e.g., metabolism, reproduction, response to stress, and urine production). The pituitary gland attaches to the hypothalamus.
4. The hypothalamus regulates body temperature, hunger, thirst, satiety, swallowing, and emotions.

13.5 Cerebrum (p. 438)

1. The cortex of the cerebrum is folded into ridges called gyri and grooves called sulci, or fissures.
2. The longitudinal fissure divides the cerebrum into left and right hemispheres. Each hemisphere has five lobes.
   - The frontal lobes are involved in smell, voluntary motor function, motivation, aggression, and mood.
   - The parietal lobes contain the major sensory areas receiving general sensory input, taste, and balance.
   - The occipital lobes contain the visual centers.
   - The temporal lobes receive olfactory and auditory input and are involved in memory, abstract thought, and judgment.
3. Tracts connect areas of the cortex within the same hemisphere (association fibers), between hemispheres (commissural fibers), and with other parts of the brain and the spinal cord (projection fibers).

Basal Nuclei

1. Basal nuclei include the corpus striatum, subthalamic nuclei, and substantia nigra.
2. The basal nuclei are important in controlling motor functions.

Limbic System

1. The limbic system includes parts of the cerebral cortex, basal nuclei, the thalamus, the hypothalamus, and the olfactory cortex.
2. The limbic system controls visceral functions through the autonomic nervous system and the endocrine system and is involved in emotions and memory.

13.6 Meninges, Ventricles, and Cerebrospinal Fluid (p. 441)

Meninges

1. The brain and spinal cord are covered by the dura, arachnoid, and pia mater.
2. The dura mater attaches to the skull and has two layers that can separate to form dural sinuses.
3. Beneath the arachnoid mater, the subarachnoid space contains CSF, which helps cushion the brain.
4. The pia mater attaches directly to the brain.

Ventricles

1. The lateral ventricles in the cerebrum are connected to the third ventricle in the diencephalon by the interventricular foramen.
2. The third ventricle is connected to the fourth ventricle in the pons by the cerebral aqueduct. The central canal of the spinal cord is connected to the fourth ventricle.

Cerebrospinal Fluid

1. CSF is produced from the blood in the choroid plexus of each ventricle. CSF moves from the lateral to the third and then to the fourth ventricle.
2. From the fourth ventricle, CSF enters the subarachnoid space through three apertures.
3. CSF leaves the subarachnoid space through arachnoid granulations and returns to the blood in the dural venous sinuses.

13.7 Blood Supply to the Brain (p. 447)

1. The brain receives blood from the internal carotid and vertebral arteries. The latter form the basilar artery. The basilar and internal carotid arteries contribute to the cerebral arterial circle. Branches from the circle and basilar artery supply the brain.
2. The blood-brain barrier is formed from the endothelial cells of the capillaries in the brain and the astrocytes in the brain tissue.

13.8 Cranial Nerves (p. 448)

Cranial nerves perform sensory, somatic motor, and parasympathetic functions.

Characteristics of the Cranial Nerves

1. The olfactory (I) and optic (II) nerves are involved in the senses of smell and vision, respectively.
2. The oculomotor nerve (III) innervates four of six extrinsic eye muscles and the upper eyelid. The oculomotor nerve also provides parasympathetic supply to the iris and lens of the eye.
3. The trochlear nerve (IV) controls one of the extrinsic eye muscles.
4. The trigeminal nerve (V) supplies the muscles of mastication, as well as a middle ear muscle, a palatine muscle, and two throat muscles. The trigeminal nerve has the greatest cutaneous sensory distribution of any cranial nerve. Two of the three trigeminal nerve branches innervate the teeth.
5. The abducens nerve (VI) controls one of the extrinsic eye muscles.
6. The facial nerve (VII) supplies the muscles of facial expression, an inner ear muscle, and two throat muscles. It is involved in the sense

of taste. It is parasympathetic to two sets of salivary glands and to the lacrimal glands.

7. The vestibulocochlear nerve (VIII) is involved in the senses of hearing and balance.
8. The glossopharyngeal nerve (IX) is involved in taste and supplies tactile sensory innervation from the posterior tongue, middle ear, and pharynx. It is also sensory for receptors that monitor blood pressure and gas levels in the blood. The glossopharyngeal nerve is parasympathetic to the parotid salivary glands.
9. The vagus nerve (X) innervates the muscles of the pharynx, palate, and larynx. It is also involved in the sense of taste. The vagus nerve is sensory for the pharynx and larynx and for receptors that monitor blood pressure and gas levels in the blood. The vagus nerve is sensory for thoracic and abdominal organs. The vagus nerve provides parasympathetic innervation to the thoracic and abdominal organs.
10. The accessory nerve (XI) has only a spinal component. It supplies the sternocleidomastoid and trapezius muscles.
11. The hypoglossal nerve (XII) supplies the intrinsic tongue muscles, three of four extrinsic tongue muscles, and two throat muscles.

Reflexes Involving Cranial Nerves

Many reflexes involved in homeostasis involve the cranial nerves and occur in the brainstem.

REVIEW AND COMPREHENSION

1. Which of these parts of the embryonic brain is correctly matched with the structure it becomes in the adult brain?
   a. mesencephalon—midbrain
   b. metencephalon—medulla oblongata
   c. myelencephalon—cerebrum
   d. telencephalon—pons and cerebellum
2. To separate the brainstem from the rest of the brain, a cut would have to be made between the
   a. medulla oblongata and pons.
   b. pons and midbrain.
   c. midbrain and diencephalon.
   d. thalamus and cerebrum.
   e. medulla oblongata and spinal cord.
3. Important centers for heart rate, blood pressure, respiration, swallowing, coughing, and vomiting are located in the
   a. cerebrum.
   b. medulla oblongata.
   c. midbrain.
   d. pons.
   e. cerebellum.
4. In which of these parts of the brain does decussation of descending tracts involved in the conscious control of skeletal muscles occur?
   a. cerebrum
   b. diencephalon
   c. midbrain
   d. pons
   e. medulla oblongata
5. The cerebral peduncles are a major descending motor pathway in the
   a. cerebrum.
   b. cerebellum.
   c. pons.
   d. midbrain.
   e. medulla oblongata.
6. The superior colliculi are involved in ____________, whereas the inferior colliculi are involved in ____________.
   a. hearing, visual reflexes
   b. visual reflexes, hearing
   c. balance, motor pathways
   d. motor pathways, balance
   e. respiration, sleep
7. The cerebellum communicates with other regions of the CNS through the
   a. flocculonodular lobe.
   b. cerebellar peduncles.
   c. vermis.
   d. lateral hemispheres.
   e. folia.
8. The major relay station for sensory input that projects to the cerebral cortex is the
   a. hypothalamus.
   b. thalamus.
   c. pons.
   d. cerebellum.
   e. midbrain.
9. The part of the diencephalon directly connected to the pituitary gland is the
   a. hypothalamus.
   b. epithalamus.
   c. subthalamus.
   d. thalamus.
10. Which of the following is a function of the hypothalamus?
    a. regulates autonomic nervous system functions
    b. regulates the release of hormones from the posterior pituitary
    c. regulates body temperature
    d. regulates food intake (hunger) and water intake (thirst)
    e. All of these are correct.
11. The grooves on the surface of the cerebrum are called the
    a. nuclei.
    b. commissures.
    c. tracts.
    d. sulci.
    e. gyri.
12. Which of these areas is located in the postcentral gyrus of the cerebral cortex?
    a. olfactory cortex
    b. visual cortex
    c. primary motor cortex
    d. primary somatic sensory cortex
    e. primary auditory cortex
13. Which of these cerebral lobes is important in voluntary motor function, motivation, aggression, sense of smell, and mood?
    a. frontal
    b. insula
    c. occipital
    d. parietal
    e. temporal
14. Fibers that connect areas of the cerebral cortex within the same hemisphere are
    a. projection fibers.
    b. commissural fibers.
    c. association fibers.
    d. All of these are correct.
15. The basal nuclei are located in the
    a. inferior cerebrum.
    b. diencephalon.
    c. midbrain.
    d. All of these are correct.
16. The most superficial of the meninges is a thick, tough membrane called the
    a. pia mater.
    b. dura mater.
    c. arachnoid mater.
    d. epidural mater.
17. The ventricles of the brain are interconnected. Which of these ventricles are *not* correctly matched with the structures that connect them?
    a. lateral ventricle to the third ventricle—interventricular foramina
    b. left lateral ventricle to right lateral ventricle—central canal
    c. third ventricle to fourth ventricle—cerebral aqueduct
    d. fourth ventricle to subarachnoid space—median and lateral apertures

18. Cerebrospinal fluid is produced by the ________, circulates through the ventricles, and enters the subarachnoid space. The cerebrospinal fluid leaves the subarachnoid space through the ________.
    a. choroid plexuses, arachnoid granulations
    b. arachnoid granulations, choroid plexuses
    c. dural venous sinuses, dura mater
    d. dura mater, dural venous sinuses
19. Water-soluble molecules of the blood plasma move across the blood-brain barrier by
    a. diffusion.
    b. endocytosis.
    c. exocytosis.
    d. symport.
    e. filtration.
20. The cranial nerve involved in chewing food is the
    a. trochlear (IV).
    b. trigeminal (V).
    c. abducens (VI).
    d. facial (VII).
    e. vestibulocochlear (VIII).
21. The cranial nerve responsible for focusing the eye (innervating the ciliary muscle of the eye) is the
    a. optic (II).
    b. oculomotor (III).
    c. trochlear (IV).
    d. abducens (VI).
    e. facial (VII).
22. The cranial nerve involved in moving the tongue is the
    a. trigeminal (V).
    b. facial (VII).
    c. glossopharyngeal (IX).
    d. accessory (XI).
    e. hypoglossal (XII).
23. The cranial nerve involved in feeling a toothache is the
    a. trochlear (IV).
    b. trigeminal (V).
    c. abducens (VI).
    d. facial (VII).
    e. vestibulocochlear (VIII).
24. From this list of cranial nerves:
    (1) olfactory (I)
    (2) optic (II)
    (3) oculomotor (III)
    (4) abducens (VI)
    (5) vestibulocochlear (VIII)

    Select the nerves that are sensory only.
    a. 1,2,3 b. 2,3,4 c. 1,2,5 d. 2,3,5 e. 3,4,5
25. From this list of cranial nerves:
    (1) trigeminal (V)
    (2) facial (VII)
    (3) glossopharyngeal (IX)
    (4) vagus (X)
    (5) hypoglossal (XII)

    Select the nerves involved in the sense of taste.
    a. 1,2,3 b. 1,4,5 c. 2,3,4 d. 2,3,5 e. 3,4,5
26. From this list of cranial nerves:
    (1) oculomotor (III)
    (2) trigeminal (V)
    (3) facial (VII)
    (4) vestibulocochlear (VIII)
    (5) glossopharyngeal (IX)
    (6) vagus (X)

    Select the nerves that are part of the parasympathetic division of the ANS.
    a. 1,2,4,5 b. 1,3,5,6 c. 1,4,5,6 d. 2,3,4,5 e. 2,3,5,6

Answers in Appendix E

CRITICAL THINKING

1. What happens to the developing brain if the CSF is not properly drained, resulting in early hydrocephalus?
2. A patient exhibits enlargement of the lateral and third ventricles, but no enlargement of the fourth ventricle. What do you conclude?
3. During a lumbar puncture (spinal tap), blood is discovered in the patient's CSF. What does this finding suggest?
4. Describe a clinical test to evaluate each of the 12 cranial nerves.
5. A patient presented to her physician with a loss of muscle tone and coordination. She was unable to perform coordinated tasks, such as touching the tip of her finger to her nose. Which part of her brain did the physician conclude was damaged?
6. A baseball player was accidentally struck on the bridge of his nose with a baseball, resulting in fractures of several facial bones. Soon after sustaining the injury, he noticed that he had lost his sense of smell. Explain how this probably happened.
7. Following a car accident in which he hit the left side of his head on the car door, Stanley developed diplopia (double vision) and was unable to move his left eye laterally (abduct the left eye). Explain how this injury caused Stanley's symptoms.
8. Over the past month, Andy has noticed that touching his upper lip, even during eating or drinking, produces intense pain in the area extending from the upper lip to just below his right eye. His ability to chew and swallow is not affected. Andy's dentist told him it was not a dental problem and he should see his physician. The physician explained that the condition might be temporary and that the cause is unknown. Explain what cranial nerve is responsible for these symptoms.
9. Afton, a teacher for the past 25 years, spent Saturday and Sunday skiing in very cold weather. The following Friday, she got up to get ready for school and was shocked to find that the right side of her face was drooping. She went to the emergency room, where tests could find no evidence of a stroke or a tumor. A physician diagnosed the condition and told Afton that it might be temporary. After about a week, her condition improved. What disorder did the physician diagnose, and what probably brought it on?

Answers in Appendix F

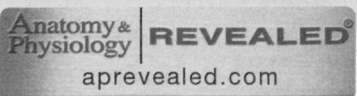

14

Integration of Nervous System Functions

There are thought to be more connections in the human brain than there are stars in the Milky Way. These connections allow you to read and understand this paragraph while disregarding the weight of the book in your hands, if you are holding it, or the weight of your forearms on the desk or on your lap. You are also probably not aware of the small noises around you or the clothes touching your body, until your attention is drawn to them. You certainly are not aware of changes in your blood pressure or your body fluid pH and blood glucose levels, yet your nervous system is actively processing all this sensory input and controlling the responses to that input.

In addition, the human brain is capable of other unique and complex functions—such as recording history, reasoning, and planning—to a degree unparalleled in the animal kingdom. Many of these functions can be studied only in humans. That is why much of human brain function remains elusive and is one of the most challenging frontiers of anatomy and physiology.

Learn to Predict

When told to put on blindfolds, the students in the school's new crime scene investigation lab became somewhat nervous—but it turned out to be a simple demonstration of how receptors in the skin detect hot and cold. Each blindfolded student was told to touch a hot object (55°C) with one hand and a cold object (5°C) with the other hand. After a few seconds, the instructor asked the students to identify which hand was touching each object. Were they able to answer correctly? After reading the chapter and recalling how spinal nerves and cranial nerves convey information to the brain, as described in chapters 12 and 13, determine whether the students could tell which hand was touching which object; then, keeping in mind the complexity of the brain, identify the parts of the CNS involved in making this distinction.

Photo: Distinct neurons can be observed by expression of different-colored fluorescent proteins in the brains of genetically engineered "brainbow" mice.

Module 7
Nervous System

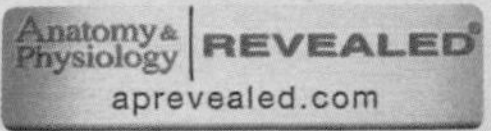
Anatomy & Physiology
REVEALED
aprevealed.com

14.1 Sensation

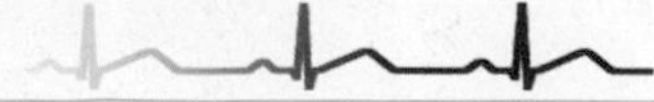

LEARNING OUTCOMES

After reading this section, you should be able to

A. List the types of somatic and visceral sensory receptors, tell where they are located, and describe how they function in sensation.

B. Describe the roles of receptor potentials and adaptation.

C. Differentiate between primary and secondary receptors, and between tonic and phasic receptors.

D. List the major ascending sensory tracts and state a function for each.

E. Describe the sensory and association areas of the cerebral cortex and discuss their interactions.

Sensation is the process initiated by stimuli acting on sensory receptors. **Perception** is the conscious awareness of those sensations. The brain constantly receives sensations as action potentials from a wide variety of sensory receptors that receive stimuli from both inside and outside the body. Sensory receptors respond to stimuli by generating action potentials that are propagated along nerves to the spinal cord and brain. Perception results when the brain interprets the sensation-generated action potentials in the cerebral cortex. Some other parts of the brain are involved in modulating sensations before they are perceived. For example, the thalamus and amygdala receive and integrate pain signals.

The **senses** are the means by which the brain receives information about the environment and the body. Historically, five senses were recognized: smell, taste, sight, hearing, and touch. Today, the senses are divided into two basic groups: general and special. The **general senses** have receptors distributed over a large part of the body. They are divided into two groups: somatic senses and visceral senses (table 14.1; figure 14.1). The **somatic senses,** including touch, pressure, itch, vibration, temperature, proprioception, and pain, provide sensory information about the body and the environment. **Proprioception** (prō-prē-ō-sep′-shun) is the perception of position and movement of parts of the body. The **visceral senses** provide information about various internal organs and consist primarily of pain and pressure.

The **special senses** are more specialized in structure, have specialized nerve endings, and are localized to specific organs (table 14.1). The special senses—smell, taste, sight, hearing, and balance—are considered in detail in chapter 15.

Not all of the sensory information detected by sensory receptors results in sensory perception. Some action potentials reach areas of the brain where they are not consciously perceived. Although we can be consciously aware of body position and movements, much of this sensory information is propagated to the cerebellum, where it is processed at an unconscious level. Sensory information from receptors that monitor blood pressure, blood oxygen, and pH levels are processed unconsciously in the medulla oblongata. For example, blood pressure must be regulated constantly to maintain homeostasis. If we had to regulate blood pressure consciously, we would not be able to think of much else. Homeostasis, therefore, is controlled largely without our conscious involvement.

TABLE 14.1 Classification of the Senses

Types of Sense	Nerve Endings	Receptor Type	Initiation of Response
GENERAL SENSES			
Somatic			
Touch		Mechanoreceptors	Compression of receptors
Stroking	Meissner corpuscle		
	Hair follicle receptor		
Texture	Merkel disk		
Vibration	Pacinian corpuscle		
Skin stretch	Ruffini end organ		
Itch, tickle	Free nerve endings		
Pressure	Merkel disk	Mechanoreceptors	Compression of receptors
Proprioception	Pacinian corpuscle	Mechanoreceptors	Compression of receptors
Temperature	Free nerve endings	Thermoreceptors	Temperature around nerve endings
	Cold receptors		
	Warm receptors		
Pain	Free nerve endings	Nociceptors	Irritation of nerve endings (e.g., mechanical, chemical, or thermal)
Visceral			
Pain	Free nerve endings	Nociceptors	Irritation of nerve endings
Pressure	Pacinian corpuscle	Mechanoreceptors	Compression of receptors
SPECIAL SENSES			
Smell	Specialized	Chemoreceptors	Binding of molecules to membrane receptors
Taste	Specialized	Chemoreceptors	Binding of molecules to membrane receptors
Sight	Specialized	Photoreceptors	Chemical change in receptors initiated by light
Hearing	Specialized	Mechanoreceptors	Bending of microvilli on receptor cells
Balance	Specialized	Mechanoreceptors	Bending of microvilli on receptor cells

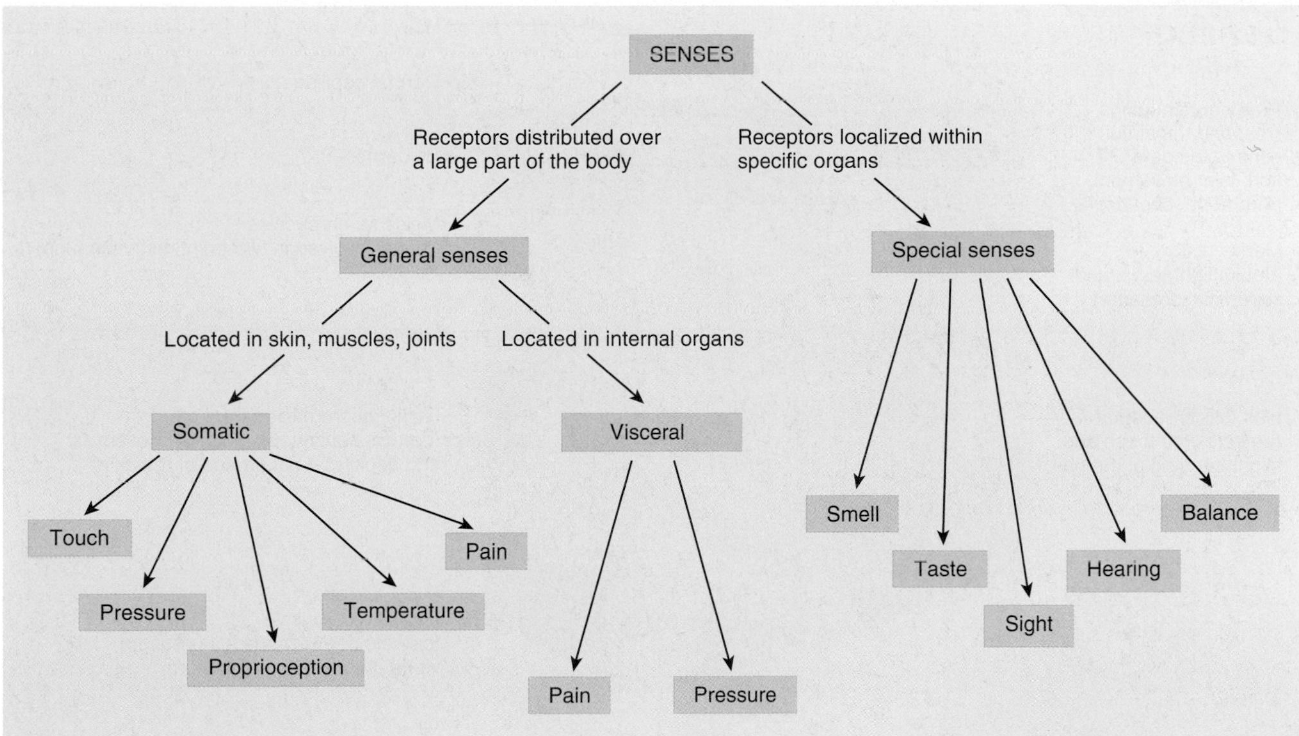

SENSES
Receptors distributed over a large part of the body
Receptors localized within specific organs
General senses
Special senses
Located in skin, muscles, joints
Located in internal organs
Somatic
Visceral
Touch
Pressure
Proprioception
Temperature
Pain
Pain
Pressure
Smell
Taste
Sight
Hearing
Balance

FIGURE 14.1 Classification of the Senses

Even the cerebral cortex screens much of what it receives and does not perceive many of the action potentials that reach it. In addition, humans exhibit selective awareness. We are more aware of sensations on which we focus our attention than on other sensations. If we were aware of all the sensory information that arrived at the cerebral cortex, we would probably not be able to function.

Sensory Receptors

Types of Sensory Receptors

The various senses depend on sensory receptors specialized to respond to specific types of stimuli (table 14.1). **Mechanoreceptors** respond to mechanical stimuli, such as compression, bending, or stretching of cells. The senses of touch, tickle, itch, vibration, pressure, proprioception, hearing, and balance all depend on a variety of mechanoreceptors. **Chemoreceptors** respond to chemicals. Smell and taste depend on chemoreceptors. **Thermoreceptors** respond to changes in temperature at the site of the receptor and are necessary for the sense of temperature. **Photoreceptors** respond to light striking the receptor cells and are necessary for vision. **Nociceptors** (nō-si-sep′ters; L., *noceo*, hurt), or *pain receptors*, respond to extreme mechanical, chemical, and thermal stimuli. Most sensory receptors typically respond to one type of stimulus, but some nociceptors respond to more than one.

At least eight major types of sensory receptors, differing in their structure and the types of stimuli to which they are most sensitive, are involved in general sensation (table 14.2). Many of these sensory receptors are associated with the skin (figure 14.2); others are associated with deeper structures, such as tendons, ligaments, and muscles; and some can be found in both the skin and deeper structures. In general, sensory receptors are classified into three groups based on their location: **Cutaneous receptors** are associated with the skin; **visceroreceptors** are associated with the viscera or organs: and **proprioceptors** are associated with joints, tendons, and other connective tissue. Cutaneous receptors provide information about the external environment, visceroreceptors provide information about the internal environment, and proprioceptors provide information about body position, movement, and the extent of stretch or the force of muscular contractions.

Structurally, the simplest and most common sensory receptors are the **free nerve endings** (figure 14.2), which are relatively unspecialized neuronal branches similar to dendrites. Free nerve endings are distributed throughout most parts of the body and are especially abundant in epithelial and connective tissues. These receptors are responsible for a number of sensations, including pain, temperature, itch, and movement.

The free nerve endings responsible for temperature detection respond to three types of sensations. One type, the **cold receptor,** increases its rate of action potential production as the skin is cooled. The cold receptor is also activated by menthol, which gives mint its cool taste. The second type, the **warm receptor,** increases its rate of action potential production as skin temperature increases. Both cold and warm receptors respond most strongly to changes in temperature. Cold receptors are 10 to 15 times more numerous than warm receptors in any given area of skin. The third type is a pain receptor that is stimulated by extreme cold or heat. At very cold temperatures (0°–12°C), only pain receptors are stimulated. As the temperature increases above 15°C, the pain sensation ends. Between 12°C and 35°C, cold fibers are stimulated, and, between

FUNDAMENTAL **Figure**

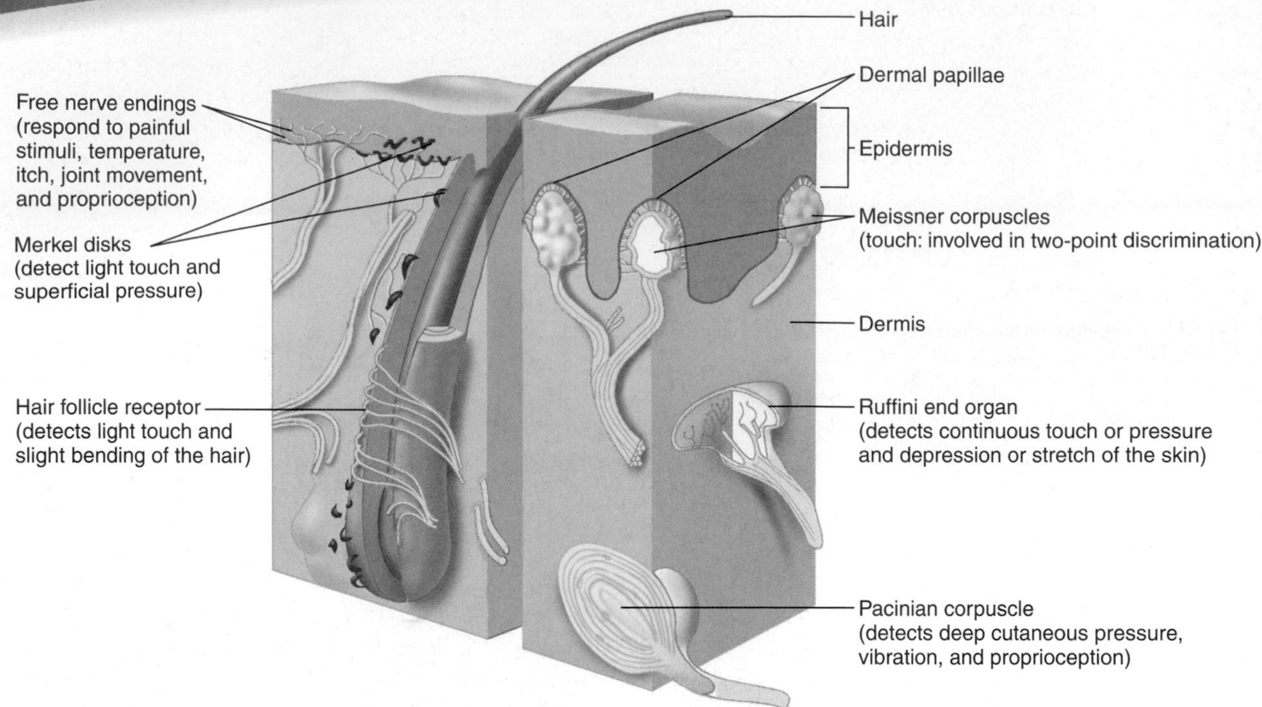

Hair
Dermal papillae
Epidermis
Meissner corpuscles
(touch: involved in two-point discrimination)
Dermis
Ruffini end organ
(detects continuous touch or pressure
and depression or stretch of the skin)
Pacinian corpuscle
(detects deep cutaneous pressure,
vibration, and proprioception)
Free nerve endings
(respond to painful
stimuli, temperature,
itch, joint movement,
and proprioception)
Merkel disks
(detect light touch and
superficial pressure)
Hair follicle receptor
(detects light touch and
slight bending of the hair)

FIGURE 14.2 **Sensory Nerve Endings in the Skin**

25°C and 47°C, nerve fibers from warm receptors are stimulated. Therefore, "comfortable" temperatures, between 25°C and 35°C, stimulate both warm and cold receptors. Temperatures above 47°C stimulate pain receptors but do not stimulate warm receptors. One pain receptor that is activated by high temperature is also activated by capsaicin, the chemical that gives chili peppers their hot taste.

Merkel (mer′kĕl) **disks,** or *tactile disks*, are more complex than free nerve endings (figure 14.2). A Merkel disk consists of axonal branches that end as flattened expansions, each associated with a specialized epithelial cell. These receptors are distributed throughout the basal layers of the epidermis just superficial to the basement membrane and are associated with dome-shaped mounds of thickened

TABLE 14.2 **Sensory Receptors**

Type of Receptor	Structure	Function
Free nerve ending	Branching, no capsule	Pain, itch, tickle, temperature, joint movement, and proprioception
Merkel disk	Consists of flattened expansions at the end of axons; each expansion is associated with a Merkel cell	Light touch and superficial pressure
Hair follicle receptor	Wrapped around hair follicles or extending along the hair axis; each axon supplies several hairs, and each hair receives branches from several neurons, resulting in considerable overlap	Light touch; responds to very slight bending of the hair
Pacinian corpuscle	Onion-shaped capsule composed of several cell layers with a single central nerve process	Deep cutaneous pressure, vibration, and proprioception
Meissner corpuscle	Several branches of a single axon associated with specialized Schwann cells and surrounded by a connective tissue capsule	Two-point discrimination
Ruffini end organ	Branching axon with numerous small, terminal knobs surrounded by a connective tissue capsule	Continuous touch or pressure; responds to depression or stretch of the skin
Muscle spindle	Three to 10 striated muscle fibers enclosed by a loose connective tissue capsule, striated only at the ends, with sensory nerve endings in the center	Proprioception associated with detection of muscle stretch; important for control of muscle tone
Golgi tendon organ	Surrounds a bundle of tendon fasciculi and is enclosed by a delicate connective tissue capsule; nerve terminations are branched, with small swellings applied to individual tendon fasciculi	Proprioception associated with the stretch of a tendon; important for control of muscle contraction

epidermis in hairy skin. Merkel disks are involved with the sensations of light touch and superficial pressure. These receptors can detect a skin displacement of less than 1 mm (1/25 inch).

Hair follicle receptors, or *hair end organs,* respond to very slight bending of the hair and are involved in light touch (figure 14.2). These receptors are extremely sensitive and require very little stimulation to elicit a response. The sensation, however, is not very well localized. The dendritic tree at the distal end of a sensory axon has several hair follicle receptors. The field of hairs innervated by these receptors overlaps with the fields of hair follicle receptors of adjacent axons. The considerable overlap in the endings of sensory neurons helps explain why light touch is not highly localized; however, because of converging signals within the CNS, it is very sensitive.

Pacinian (pa-sin′ē-an, pa-chin′ē-an) **corpuscles,** or *lamellated corpuscles,* are complex receptors that resemble an onion (figure 14.2). A single dendrite extends to the center of each Pacinian corpuscle. The corpuscles are located within the subcutaneous tissue, where they are responsible for deep cutaneous pressure and vibration. Pacinian corpuscles associated with the joints help relay proprioceptive information about joint positions.

Meissner (mīs′ner) **corpuscles,** or *tactile corpuscles,* are distributed throughout the dermal papillae (figure 14.2; see chapter 5) and are involved in two-point discrimination. **Two-point discrimination** (fine touch) is the ability to detect simultaneous stimulation of Meissner corpuscles in two distinct receptor fields by touching at two points on the skin (figure 14.3). The distance between two points that a person can detect as separate points of stimulation differs for various regions of the body. This sensation is important in evaluating the texture of objects. Meissner corpuscles are numerous and close together in the tongue and fingertips but are less numerous and more widely separated in other areas, such as the back.

Ruffini (rū-fē′nē) **end organs** are located in the dermis of the skin (see figure 14.2), primarily in the fingers. They respond to pressure on the skin directly superficial to the receptor and to stretch of adjacent skin. These receptors are important in responding to continuous touch or pressure.

Muscle spindles (figure 14.4) consist of 3–10 specialized muscle fibers that are located in skeletal muscles; they provide information about the length of the muscle (see "Stretch Reflex" in section 12.2). Muscle spindles are important to the control and tone of postural muscles. Brain centers act through descending tracts to either increase or decrease action potentials in gamma motor neurons. Stimulation of the gamma motor system, caused by stretch of the muscle, activates the stretch reflex, which in turn increases the tone of the muscles involved.

Golgi tendon organs are proprioceptive receptors associated with the fibers of a tendon near the junction between the muscle and the tendon (figure 14.5). They are activated by an increase in tendon tension, caused either by contraction of the muscle or by passive stretch of the tendon.

Responses of Sensory Receptors

Interaction of a stimulus with a sensory receptor produces a graded potential called a **receptor potential.** If the receptor potential reaches threshold, an action potential is produced and is propagated toward the CNS. Sensory receptor cells that conduct action potentials in response to the receptor potential are called **primary receptors** (figure 14.6*a*). Most sensory neurons, including all those in table 14.2, belong to this category. Other receptor cells, called **secondary receptors,** have no axons or have short, axonlike projections and generally produce receptor potentials (figure 14.6*b*). The receptor potentials cause the release of neurotransmitters from the receptor cell, which bind to receptor proteins on the membrane of a neuron. This causes a receptor potential in the neuron, which produces an action potential if threshold is reached. The receptor cells of the special senses of taste, hearing, and balance belong to this category.

Some sensations have the quality of **adaptation,** decreased sensitivity to a continued stimulus. After exposure to a certain

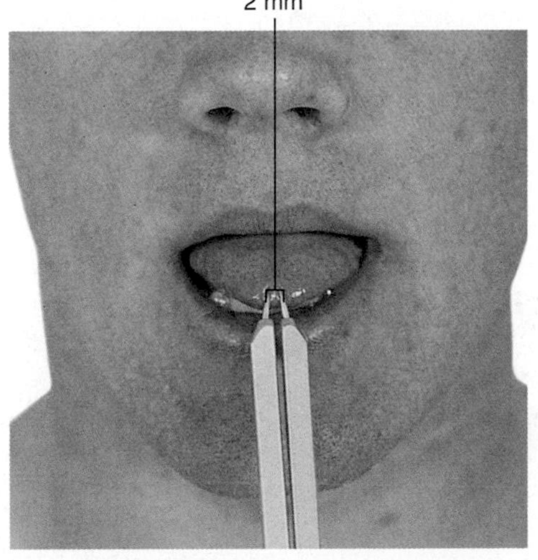
2 mm

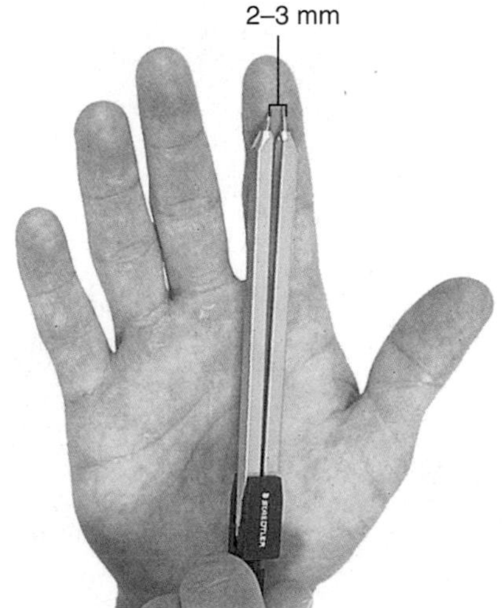
2–3 mm

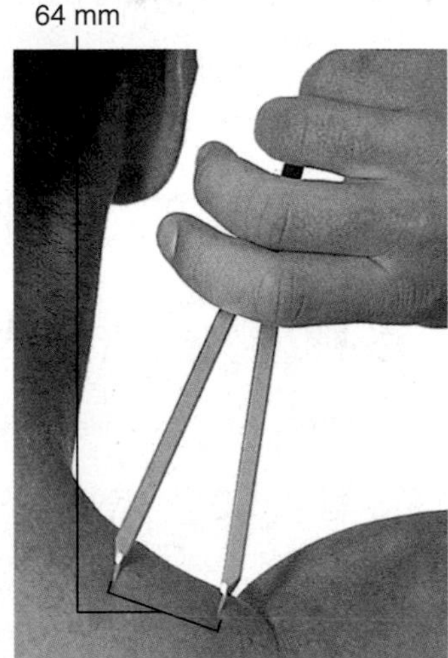
64 mm

FIGURE 14.3 Two-Point Discrimination

Two-point discrimination can be demonstrated by touching a person's skin with the two points of a compass. When the two points are closer together than the receptor field, the individual perceives only one point. When the two points of the compass are opened wider, the person becomes aware of two points. In each of these images, the person is detecting two points, but a greater distance between the compass points is required on the back.

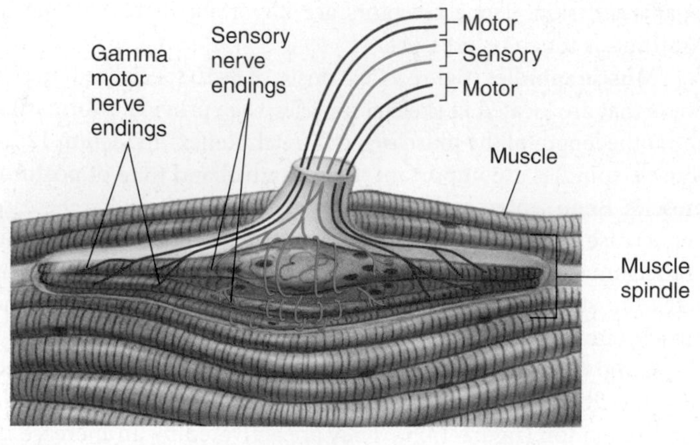
Gamma motor nerve endings
Sensory nerve endings
Motor
Sensory
Motor
Muscle
Muscle spindle

FIGURE 14.4 Muscle Spindle

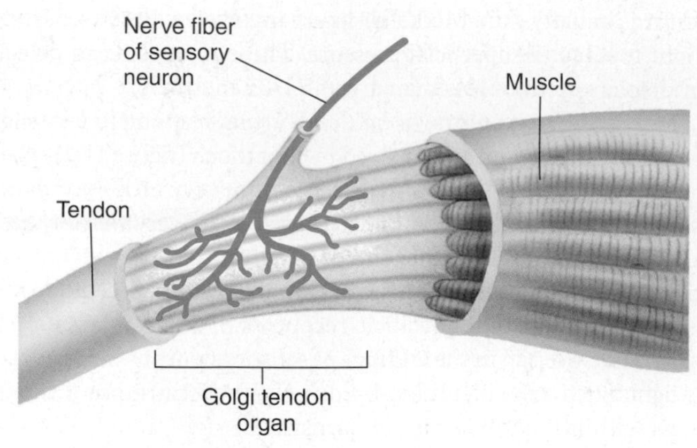
Nerve fiber of sensory neuron
Muscle
Tendon
Golgi tendon organ

FIGURE 14.5 Golgi Tendon Organ

stimulus strength for a time, the response of the receptors or the sensory pathways lessens from when the stimulus was first applied. The local graded depolarization that produces a receptor potential adapts, or returns to, its resting level, even though the stimulus is still applied. For example, when a person first gets dressed, tactile receptors and pathways relay information to the brain, creating an awareness that the clothes are touching the skin. After a time, the action potentials from the skin decrease, and the clothes are not perceived.

The rate of adaptation varies for different receptors, as occurs in proprioception. Proprioception provides information about the precise position and rate of movement of various body parts, the weight of an object being held in the hand, and the range of movement of a joint. This information is involved in activities such as walking, climbing stairs, shooting a basketball, driving a car, eating, and writing. Receptors for this system are located around joints and in muscles. Two types of proprioceptors provide positional

1. A mechanoreceptor (Pacinian corpuscle) is subjected to a pressure stimulus.
2. The sensory receptor responds by generating a graded potential, also known as the receptor potential.
3. When the graded potential reaches threshold, action potentials are generated in the axon. The action potentials are propagated toward the CNS.

(a)

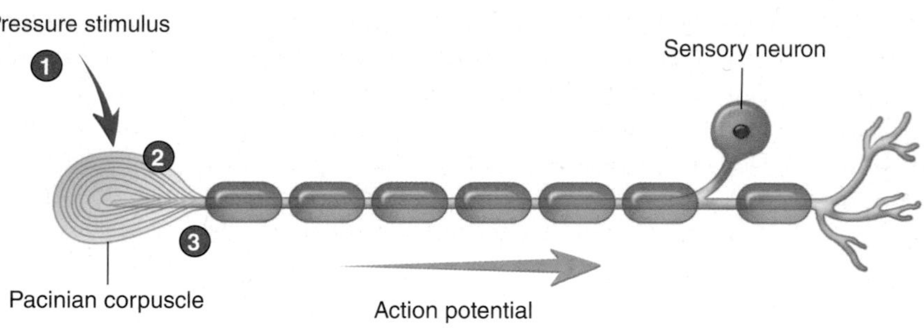
Pressure stimulus
1
2
3
Sensory neuron
Pacinian corpuscle
Action potential

1. A chemoreceptor (taste cell) is subjected to a chemical stimulus in the form of a salty substance.
2. The sensory receptor responds by generating a graded potential, also known as the receptor potential. In the case of taste cells, an action potential can also be generated.
3. The sensory receptor releases a neurotransmitter into the synaptic cleft, which in turn stimulates the sensory neuron.
4. When the stimulation reaches threshold, the neuron generates action potentials, which are propagated toward the CNS.

(b)

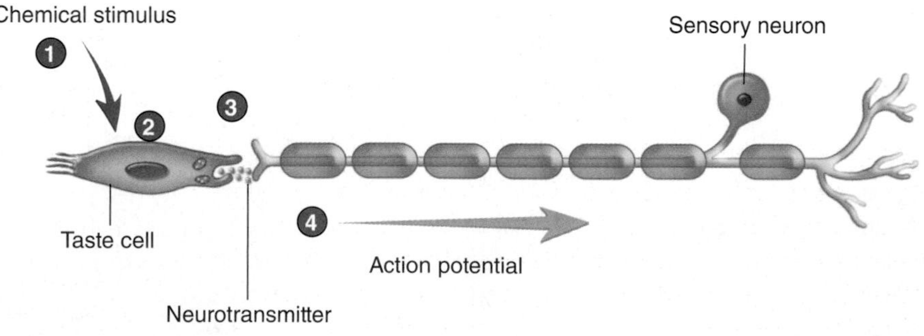
Chemical stimulus
1
2
3
Sensory neuron
Taste cell
4
Action potential
Neurotransmitter

PROCESS FIGURE 14.6 Comparison of Primary and Secondary Receptors

(*a*) A primary receptor has an axon that conducts action potentials in response to the receptor potential. (*b*) A secondary receptor has no axon, but the receptor potential results in the release of a neurotransmitter, which in turn stimulates a postsynaptic neuron.

information: tonic receptors and phasic receptors. **Tonic receptors,** or *slowly adapting receptors,* generate action potentials as long as a stimulus is applied and adapt very slowly. For example, information from slowly adapting receptors allows us to know where our little finger is at all times without having to look for it. **Phasic receptors,** or *rapidly adapting receptors,* adapt rapidly and are most sensitive to changes in stimuli. For example, information from rapidly adapting receptors allows us to know where our little finger is as it moves; thus, we can control its movement through space and predict where it will be in the next moment.

We are usually not conscious of tonic or phasic input because the higher brain centers ignore it most of the time. Through selective awareness, however, we can call up the information when we wish. For example, where is the thumb of your right hand at this moment? Were you aware of its position a few seconds ago?

ASSESS **YOUR PROGRESS**

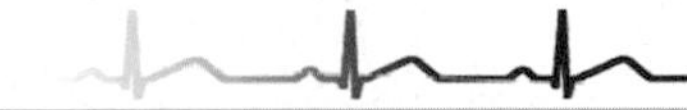

1. *In general, into what three groups can sensory receptors be classified? What does each provide information about?*
2. *List the eight major types of sensory receptors, indicate where they are located, and state the functions they perform.*
3. *What is the difference between primary and secondary receptors. What effect does a receptor potential have on them?*
4. *What is adaptation? Give an example. Describe tonic and phasic receptors.*

Sensory Pathways

The spinal cord and brainstem contain a number of sensory pathways, or tracts, that transmit information via action potentials from the periphery to various parts of the brain. These pathways are called *ascending* spinal pathways (table 14.3). Each pathway is involved with specific modalities (the type of information transmitted). The neurons that make up each pathway are associated with specific types of sensory receptors. For example, thermoreceptors in the skin generate action potentials that are propagated along the sensory pathway for pain and temperature, whereas Golgi tendon organs located in tendons generate action potentials that are propagated along the sensory pathway involved with proprioception.

The names of many ascending pathways in the CNS indicate their origin and termination. Each pathway is usually given a composite name in which the first half indicates its origin and the second half indicates its termination. Ascending pathways, therefore, usually begin with the prefix *spino-*, indicating that they originate in the spinal cord (figure 14.7). For example, a spinocerebellar (spī′nō-ser-e-bel′ar) tract originates in the spinal cord and terminates in the cerebellum. The names of other pathways indicate their location in the spinal cord. For example, the name dorsal-column/medial-lemniscal system is a combination of the pathway names in the spinal cord and brainstem.

The two major ascending pathways involved in the conscious perception of external stimuli are the spinothalamic tract of the anterolateral system and the dorsal-column/medial-lemniscal system (table 14.3). The pathways carrying sensory input that we are not consciously aware of include the spinocerebellar, spinomesencephalic, and spinoreticular tracts, the last two of which are part of the anterolateral system.

Anterolateral System

The **anterolateral system** is one of the two major systems that convey cutaneous sensory information to the brain. The anterolateral system includes the **spinothalamic tract,** the **spinoreticular tract,** and the **spinomesencephalic tract.** There is, however, considerable overlap among these three tracts within the anterolateral system. The spinothalamic tract carries pain and temperature information, as well as light touch and pressure, tickle, and itch sensations (figure 14.8).

From the peripheral receptor to the cerebral cortex, the signal passes through three neurons in sequence—the primary, secondary, and tertiary neurons. The **primary neuron** cell bodies of the spinothalamic tract are in the dorsal root ganglia. The primary neurons relay sensory input from the periphery to the posterior horn of the spinal cord, where they synapse with interneurons. The interneurons, which are not specifically named in the three-neuron sequence, synapse with secondary neurons. Axons from the **secondary neurons** cross to the *contralateral,* or opposite, side of the spinal cord through the anterior portion of the gray and white commissures and enter the spinothalamic tract, where they ascend to the thalamus. The secondary neurons synapse with cell bodies of tertiary neurons in the thalamus. **Tertiary neurons** from the thalamus project to the somatic sensory cortex.

The spinoreticular and spinomesencephalic tracts ascend with the spinothalamic tract through the spinal cord but then divert to brainstem and midbrain nuclei. Some neurons in the spinoreticular tracts do not cross over but ascend on the *ipsilateral,* or same, side of the spinal cord on which they enter. A portion of the spinomesencephalic tract, called the **spinotectal** (spī-nō-tek′tăl) **tract,** ends in

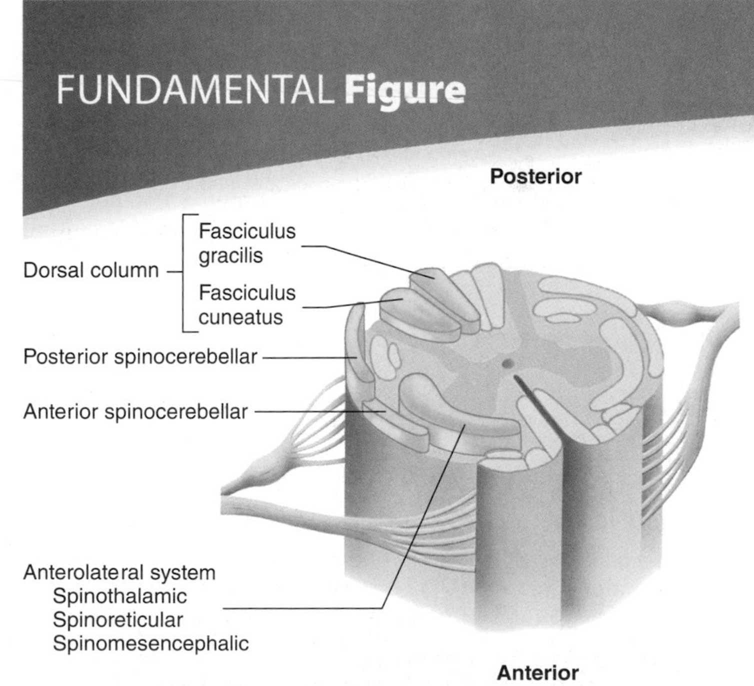
FUNDAMENTAL Figure
Posterior
Dorsal column
Fasciculus gracilis
Fasciculus cuneatus
Posterior spinocerebellar
Anterior spinocerebellar
Anterolateral system
Spinothalamic
Spinoreticular
Spinomesencephalic
Anterior

FIGURE 14.7 Ascending Pathways at the Cervical Level of the Spinal Cord

Ascending pathways (green) are labeled on the left side of the figure only, although they exist on both sides.

TABLE **14.3** **Ascending Spinal Pathways**

Pathway	Modality (Information Transmitted)	Origin	Primary Cell Body	Secondary Cell Body	Crossover
Anterolateral System					
Spinothalamic tract	Pain, temperature, light touch, pressure, tickle, and itch	Cutaneous receptors	Dorsal root ganglion	Posterior horn of spinal cord	Level at which primary neuron enters spinal cord for pain and temperature or 8–10 segments from where primary neuron enters spinal cord for light touch
Spinoreticular tract	Pain	Cutaneous receptors	Dorsal root ganglion	Posterior horn of spinal cord	Reticular formation
Spinomesencephalic tract (including spinotectal)	Pain and touch	Cutaneous receptors	Dorsal root ganglion	Posterior horn of spinal cord	At point of origin
Dorsal-column/ medial-lemniscal system					
Fasciculus gracilis	Proprioception, two-point discrimination, pressure, and vibration from inferior half of body	Joints, tendons, and muscles	Dorsal root ganglion	Medulla oblongata	Medulla oblongata
Fasciculus cuneatus	Proprioception, two-point discrimination, pressure, and vibration from superior half of body	Joints, tendons, and muscles	Dorsal root ganglion	Medulla oblongata	Medulla oblongata
Spinocerebellar Tract					
Posterior	Proprioception	Joints and tendons	Dorsal root ganglion	Posterior horn of spinal cord	Uncrossed
Anterior	Proprioception	Joints and tendons	Dorsal root ganglion	Posterior horn of spinal cord	Some uncrossed; some cross at point of entry, recross at cerebellum

the superior colliculi of the midbrain and transmits action potentials involved in reflexes that turn the head and eyes toward a point of cutaneous stimulation.

› Predict 2

Describe the clinical effect of a lesion on one side of the spinal cord that interrupts the spinothalamic tract of the anterolateral system.

Dorsal-Column/Medial-Lemniscal System

The **dorsal-column/medial-lemniscal** (lem-nis′kăl) **system** carries the sensations of two-point discrimination, proprioception, pressure, and vibration (figure 14.9). This system is named for the dorsal column of the spinal cord and the medial lemniscus, which is the continuation of the dorsal column in the brainstem. The term *lemniscus* means "ribbon" and refers to the thin, ribbonlike appearance of the pathway as it passes through the brainstem.

Primary neurons of the dorsal-column/medial-lemniscal system are located in the dorsal root ganglia. They are the largest cell bodies in the dorsal root ganglia, especially those for two-point discrimination. Many axons of the primary neurons of the dorsal-column/medial-lemniscal system enter the spinal cord, ascend its entire length without crossing to its opposite side, and synapse with secondary neurons located in the medulla oblongata. Others synapse in the thoracic portion of the spinal cord.

In the spinal cord, the dorsal-column/medial-lemniscal system is divided into two tracts (see figure 14.7) based on the source of the stimulus. The **fasciculus gracilis** (gras′i-lis; thin) conveys sensations from nerve endings below the midthoracic level, and the **fasciculus cuneatus** (kū′nē-ā′tŭs; wedge-shaped) conveys impulses from nerve endings above the midthorax. The fasciculus gracilis terminates by synapsing with secondary neurons in the **nucleus gracilis** or with neurons of the posterior spinocerebellar tracts. The fasciculus cuneatus primarily terminates by synapsing with secondary neurons in the **nucleus cuneatus.** Both the nucleus gracilis and the nucleus cuneatus are in the medulla oblongata. Secondary neurons exit the nucleus gracilis and the nucleus cuneatus, cross to the opposite side of the medulla through the decussations of the medial lemniscus, and ascend through the medial lemniscus to terminate in the thalamus. Tertiary neurons from the thalamus project to the primary somatic sensory cortex (see "Sensory Areas of the Cerebral Cortex," later in this chapter).

Tertiary Cell Body	Termination	Side of Body Where Fibers Terminate
Thalamus	Cerebral cortex	Contralateral
Reticular formation	Reticular formation, thalamus	Contralateral, ipsilateral
Superior colliculus	Mesencephalon (midbrain), superior colliculus	Contralateral
Thalamus	Cerebral cortex, cerebellum	Contralateral
Thalamus	Cerebral cortex, cerebellum	Contralateral
Cerebellum	Cerebellum	Ipsilateral
Cerebellum	Cerebellum	Ipsilateral

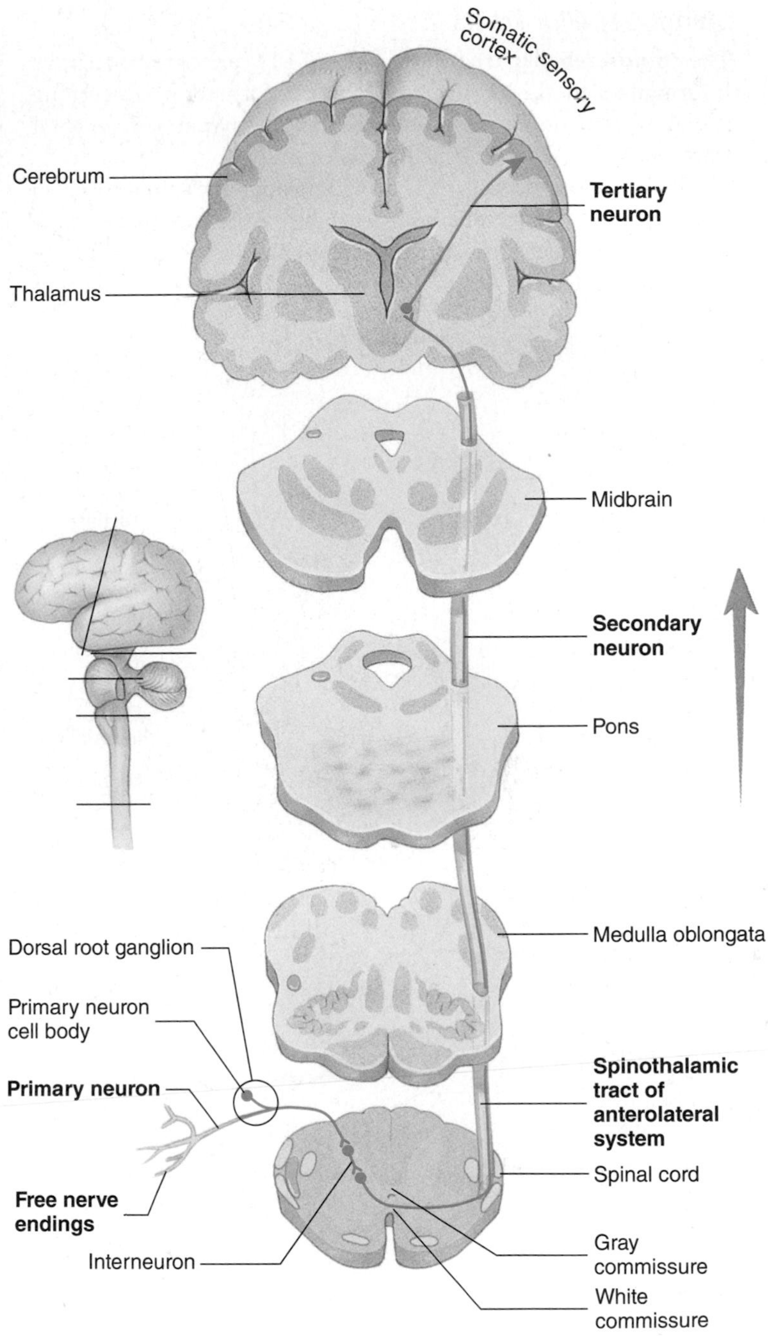
Somatic sensory cortex
Cerebrum
Tertiary neuron
Thalamus
Midbrain
Secondary neuron
Pons
Medulla oblongata
Dorsal root ganglion
Primary neuron cell body
Primary neuron
Spinothalamic tract of anterolateral system
Spinal cord
Free nerve endings
Interneuron
Gray commissure
White commissure

FIGURE 14.8 AP|R **Spinothalamic Tract of the Anterolateral System**

The spinothalamic tract of the anterolateral system transmits action potentials for pain and temperature. Primary neurons enter the spinal cord and synapse with interneurons that synapse with secondary neurons. The secondary neurons cross to the opposite side of the spinal cord, ascend to the thalamus, and synapse with tertiary neurons. The tertiary neurons connect to the somatic sensory cortex. Lines on the inset indicate levels of section. The orange arrow indicates the direction of action potentials.

> **Predict 3**
>
> Bill and Mary were involved in an accident, and each experienced loss of proprioception, fine touch, and vibration on the left side of the body below the waist. Physicians determined that Bill had damage to his spinal cord and that Mary had damage to her brainstem. Explain which side of the spinal cord was damaged in Bill and which side of the brainstem was damaged in Mary.

Trigeminothalamic Tract

As the fibers of the spinothalamic tracts pass through the brainstem, they are joined by fibers of the **trigeminothalamic tract.** The trigeminothalamic tract is made up primarily of afferent fibers from the trigeminal nerve, joined by a few tactile afferent fibers from the ear and tongue carried by cranial nerves VII, IX, and sometimes X. This tract carries the same sensory information as the spinothalamic tracts and dorsal-column/medial-lemniscal system but from the face, nasal cavity, and oral cavity, including the teeth. The trigeminothalamic tract is similar to the spinothalamic tracts and dorsal-column/medial-lemniscal system in that primary neurons from one side of the face synapse with secondary neurons, which cross to the opposite side of the brainstem. The secondary neurons synapse with tertiary neurons in the thalamus. Tertiary neurons from the thalamus project to the somatic sensory cortex.

Spinocerebellar Tracts

The **spinocerebellar tracts** (see figure 14.7) carry proprioceptive information to the cerebellum, where information concerning actual movements can be monitored and compared with cerebral information representing intended movements.

Two spinocerebellar tracts extend through the spinal cord: (1) The **posterior spinocerebellar tract** (figure 14.10) originates in the thoracic and upper lumbar regions and contains uncrossed nerve fibers that enter the cerebellum through the inferior cerebellar peduncles and (2) the **anterior spinocerebellar tract** carries information from the lower trunk and lower limbs and contains both crossed and uncrossed nerve fibers that enter the cerebellum through the superior cerebellar peduncle. The crossed fibers recross in the cerebellum. Both spinocerebellar tracts transmit proprioceptive information to the cerebellum from the same side of the body as the cerebellar hemisphere to which they project. Why the anterior spinocerebellar tract crosses twice to accomplish this feat is unknown. Much of the proprioceptive information carried from the lower limbs by the fasciculus gracilis of the dorsal-column/medial-lemniscal system is transferred by synapses in the inferior thorax to the spinocerebellar

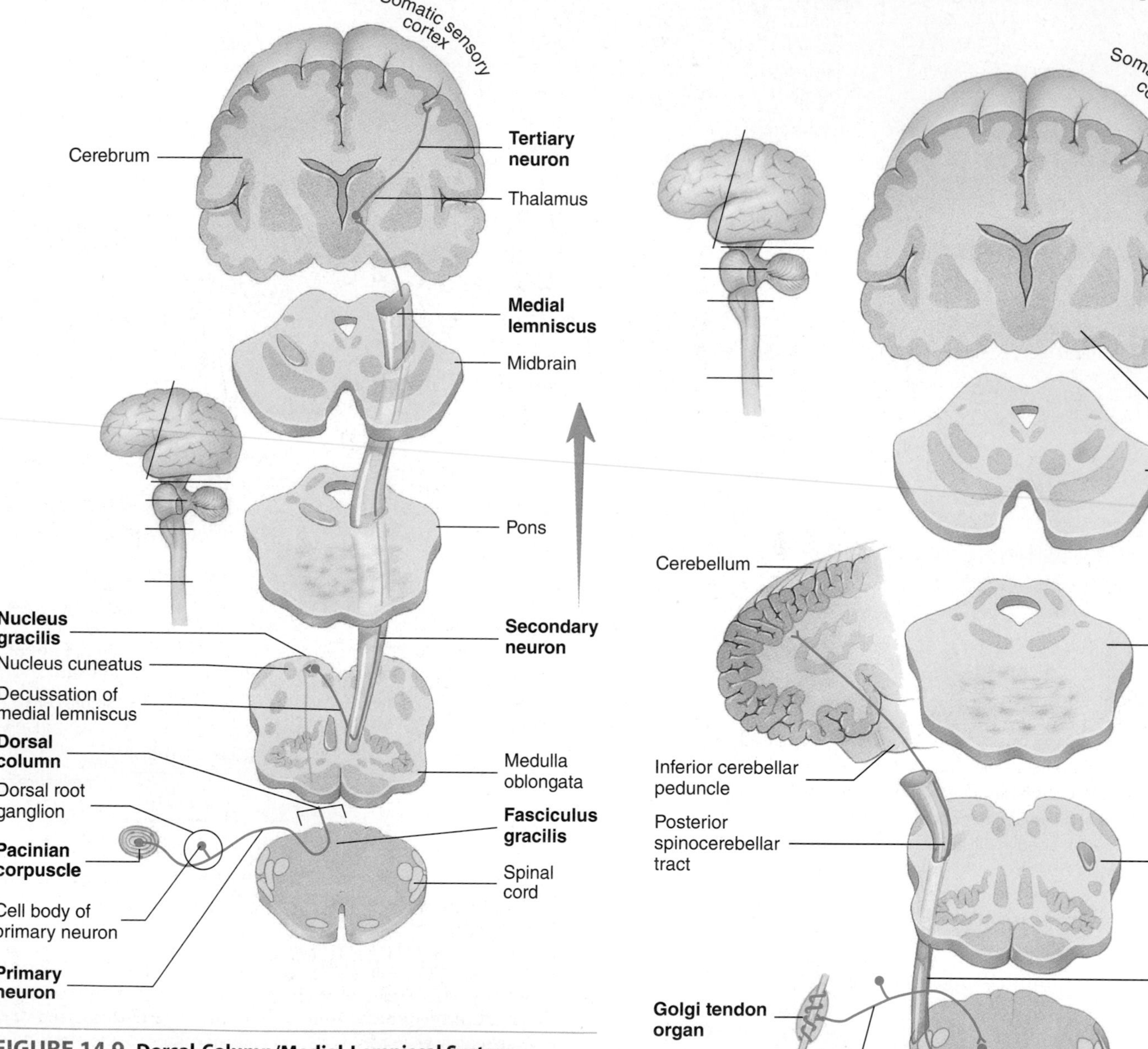
Somatic sensory cortex
Cerebrum
Tertiary neuron
Thalamus
Medial lemniscus
Midbrain
Pons
Nucleus gracilis
Nucleus cuneatus
Decussation of medial lemniscus
Secondary neuron
Dorsal column
Medulla oblongata
Dorsal root ganglion
Fasciculus gracilis
Pacinian corpuscle
Spinal cord
Cell body of primary neuron
Primary neuron
Somatic sensory cortex
Cerebrum
Midbrain
Cerebellum
Pons
Inferior cerebellar peduncle
Posterior spinocerebellar tract
Medulla oblongata
Secondary neuron
Golgi tendon organ
Primary neuron
Spinal cord

FIGURE 14.9 Dorsal-Column/Medial-Lemniscal System
The fasciculus gracilis and the fasciculus cuneatus (not illustrated) convey proprioception, pressure, vibration, and two-point discrimination. Primary neurons enter the spinal cord, ascend ipsilaterally, and synapse with secondary neurons in the medulla oblongata. The secondary neurons cross to the opposite side of the spinal cord, ascend to the thalamus, and synapse with tertiary neurons. The tertiary neurons connect to the somatic sensory cortex. Lines on the inset indicate levels of section. The orange arrow indicates direction of action potentials.

FIGURE 14.10 Posterior Spinocerebellar Tract
This tract transmits proprioceptive information from the thorax, upper limbs, and upper lumbar region to the cerebellum. Lines on the inset indicate levels of section. The orange arrow indicates direction of action potentials.

FUNDAMENTAL **Figure**

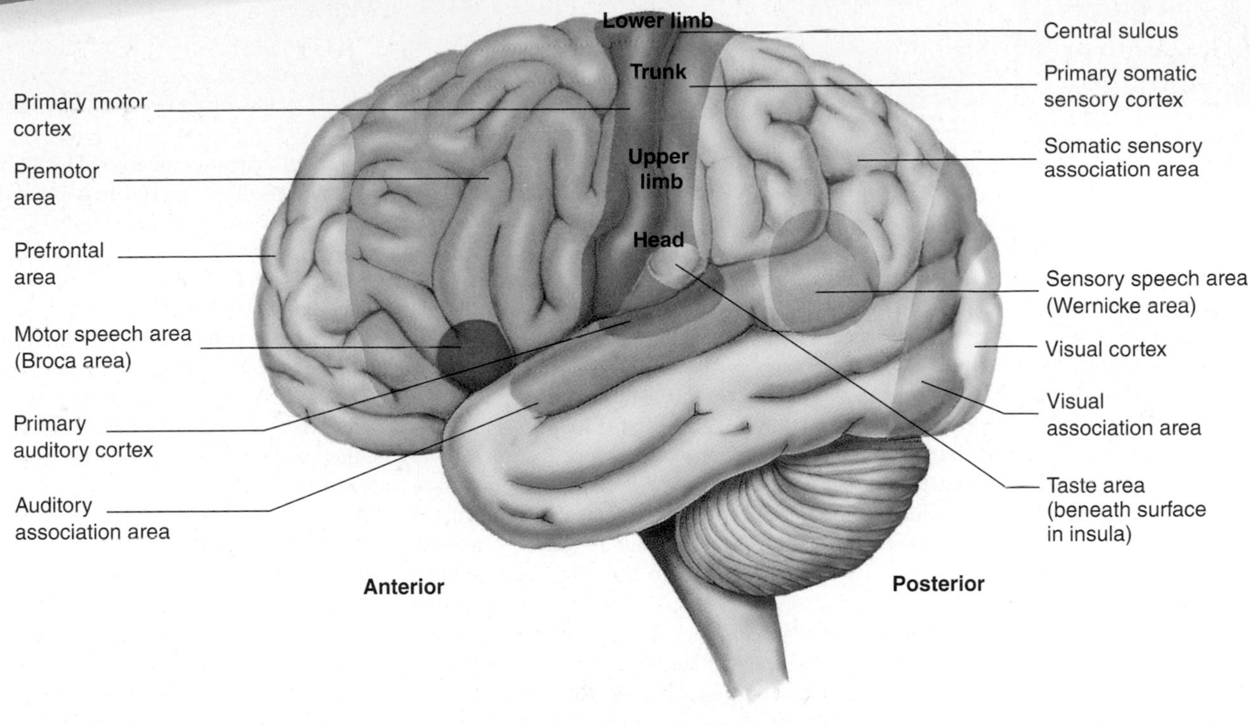

Lower limb
Trunk
Upper limb
Head
Central sulcus
Primary somatic sensory cortex
Somatic sensory association area
Primary motor cortex
Premotor area
Prefrontal area
Motor speech area (Broca area)
Primary auditory cortex
Auditory association area
Sensory speech area (Wernicke area)
Visual cortex
Visual association area
Taste area (beneath surface in insula)
Anterior
Posterior
Lateral view

FIGURE 14.11 APR **Functional Regions of the Lateral Side of the Left Cerebral Cortex**

system and enters the cerebellum as unconscious proprioceptive information. The spinocerebellar tracts convey very little information from the upper limbs to the cerebellum. Proprioception from the upper limbs is projected to the thalamus. This input enters the cerebellum through the inferior peduncle from the cuneate nucleus of the dorsal-column/medial-lemniscal system. Therefore, the dorsal-column/medial-lemniscal system is involved not only in conscious awareness of proprioception but also in unconscious neuromuscular functions.

Predict 4

Most of the neurons from the fasciculus gracilis synapse in the inferior thorax and enter the spinocerebellar system, whereas most of the neurons from the fasciculus cuneatus synapse in the nucleus cuneatus and then continue to the thalamus and cerebrum. Therefore, it can be deduced that most of the proprioception from the lower limbs is unconscious and most of the proprioception from the upper limbs is conscious. Explain why this difference in the two sets of limbs is of value.

Descending Pathways Modifying Sensation

Descending pathways "descend" from the brain to the spinal cord. The corticospinal tract (described in section 14.2) and other descending tracts send collateral branches to the thalamus, reticular formation, trigeminal nuclei, and spinal cord. Neuromodulators, such as endorphins and enkephalin, released from axons originating in these CNS regions, decrease the frequency of action potentials in sensory tracts (see the discussion of presynaptic inhibition in chapter 11). Through this route, the cerebral cortex or other brain regions may reduce the conscious perception of sensations, including pain.

ASSESS **YOUR PROGRESS**

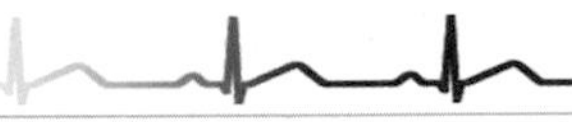

5. *What are the functions of the anterolateral and dorsal-column/medial-lemniscal systems? Describe where the neurons of these systems cross over and synapse.*
6. *How is the trigeminothalamic tract different from the spinothalamic tract?*
7. *What kind of information is carried by the spinocerebellar tracts? Where do the anterior and posterior spinocerebellar tracts originate? Do these tracts terminate on the same or opposite side of the body from where they originate?*
8. *How do descending pathways modulate sensation?*

Sensory Areas of the Cerebral Cortex

Figure 14.11 depicts a lateral view of the left cerebral cortex with some of its functional areas labeled. Sensory pathways project to specific regions of the cerebral cortex, called **primary sensory areas,** where these sensations are perceived. The primary sensory areas of the cerebral cortex must be intact in order for conscious perception, localization, and identification of a stimulus to occur.

The **primary somatic sensory cortex,** or *general sensory area,* occupies most of the postcentral gyrus (see chapter 13). The terms *area* and *cortex* are often used interchangeably to refer to the same functional region of the cerebral cortex. Fibers carrying general sensory input, such as pain, pressure, and temperature, synapse in the thalamus, and thalamic neurons relay the information to the primary somatic sensory cortex.

The primary somatic sensory cortex is organized topographically relative to the general plan of the body (figure 14.12). For example,

Clinical IMPACT | Pain

Pain is a sensation characterized by a group of unpleasant and complex perceptual and emotional experiences that trigger autonomic, psychological, and somatic motor responses. Pain sensation has two components: (1) rapidly conducted action potentials carried by large-diameter, myelinated axons, resulting in sharp, well-localized, pricking or cutting pain, followed by (2) more slowly propagated action potentials, carried by smaller, less heavily myelinated axons, resulting in diffuse burning or aching pain. Research indicates that pain receptors have very uniform sensitivity, which does not change dramatically from one instant to another. The variations in pain sensation that we experience result from the mechanisms by which pain receptors are stimulated, differences in the integration of action potentials from the pain receptors, and complex interactions in the cerebral cortex, cingulate gyrus, and thalamus, where the emotional component of pain is registered. Neurons in the cerebral cortex respond to pain stimuli selectively based on prior experience and context. Stress, for example, can reduce pain perception.

Although the dorsal-column/medial-lemniscal system contains no pain fibers, tactile and mechanoreceptors are often activated by the same stimuli that affect pain receptors. Action potentials from tactile receptors provide information that allows the pain sensation to be localized. Superficial pain is highly localized, in part, because of the simultaneous stimulation of pain receptors and mechanoreceptors in the skin. Deep, or visceral, pain is not highly localized (diffuse) because of fewer mechanoreceptors in the deeper structures, and it is normally perceived as a diffuse pain.

Dorsal-column/medial-lemniscal system neurons are involved in what is called the **gate-control theory** of pain control. Primary neurons of the dorsal-column/medial-lemniscal system send out collateral branches that synapse with interneurons in the posterior horn of the spinal cord. These interneurons have an inhibitory effect on secondary neurons of the spinothalamic tract. Thus, pain action potentials traveling through the spinothalamic tract can be suppressed by action potentials that originate in neurons of the dorsal-column/medial-lemniscal system. The arrangement may act as a "gate" for pain action potentials transmitted in the spinothalamic tract. Increased activity in the dorsal-column/medial-lemniscal system tends to close the gate, thereby reducing pain action potentials transmitted in the spinothalamic tract. Descending pathways from the cerebral cortex or other brain regions can also regulate this "gate."

The gate-control theory may explain the physiological basis for the following methods that have been used to reduce the intensity of chronic pain: electrical stimulation of the dorsal-column/medial-lemniscal neurons, transcutaneous electrical stimulation (applying a weak electrical stimulus to the skin), acupuncture, massage, and exercise. The frequency of action potentials that are transmitted in the dorsal-column/medial-lemniscal system is increased when the skin is rubbed vigorously and when the limbs are moved. This may explain why vigorously rubbing a large area around the source of pricking pain tends to reduce the intensity of the pain. Exercise normally decreases the sensation of pain, and exercise programs are important in managing chronic pain not associated with illness. Acupuncture may lessen pain through the action of a gating mechanism that inhibits pain transmission upward in the spinal cord.

Analgesics are pain-relieving medications that act in much the same way as gate control. Some analgesics act in the periphery to reduce inflammation and the activation of peripheral nerves; others block the transmission of pain sensations in the spinal cord from primary neurons to neurons of the ascending pathways. Other analgesics function at the level of the cerebral cortex to modulate pain.

Referred Pain

Referred pain is a painful sensation in a region of the body that is not the source of the pain stimulus. Most commonly, patients sense referred pain in the skin or other superficial structures when internal organs are damaged or inflamed. This sensation usually occurs because both the area of skin to which the pain is referred and the visceral area that is damaged are innervated by neurons that project to the same area of the cerebral cortex. The brain cannot distinguish between the two sources of painful stimuli, and the painful sensation is referred to the most superficial structures innervated by the converging neurons. This referral may occur because the number of receptors is much greater in superficial structures than in deep structures and the brain is more "accustomed" to dealing with superficial stimuli.

Referred pain is clinically useful in diagnosing the actual cause of a painful stimulus. Heart attack victims often feel cutaneous pain radiating from the left shoulder down the arm. Other examples of referred pain are shown in figure 14A.

Phantom Pain

Phantom pain occurs in people who have had appendages amputated or a structure, such as a tooth, removed. Many of these people perceive pain (which may be intense) or other sensations in the amputated structure as if it were still in place. If a neuron pathway that transmits action potentials is stimulated at any point along that pathway, action potentials are initiated and propagated toward the CNS. Integration results in the perception of pain that is projected to the site of the sensory receptors, even if those sensory receptors are no longer present. A similar phenomenon can be easily demonstrated by bumping the ulnar nerve where it crosses the elbow (the funny bone). Even though the neurons are stimulated at the elbow, we often feel a sensation of pain in the fourth and fifth digits.

A factor that may be important in phantom pain is the lack of touch, pressure, and proprioceptive impulses from the amputated limb. Those action potentials suppress the transmission of pain action potentials in the pain pathways, as explained by the gate-control theory of pain. When a limb is amputated, the inhibitory effect of sensory information is removed. As a consequence, the intensity of phantom pain may increase. Another factor in phantom pain may be that the cerebral cortex retains an image of the amputated body part.

Chronic Pain

Chronic pain is long-lasting. Some chronic pain has a known cause, such as tissue damage, as in the case of arthritis. Other chronic pain cannot be associated with tissue damage and has no known cause.

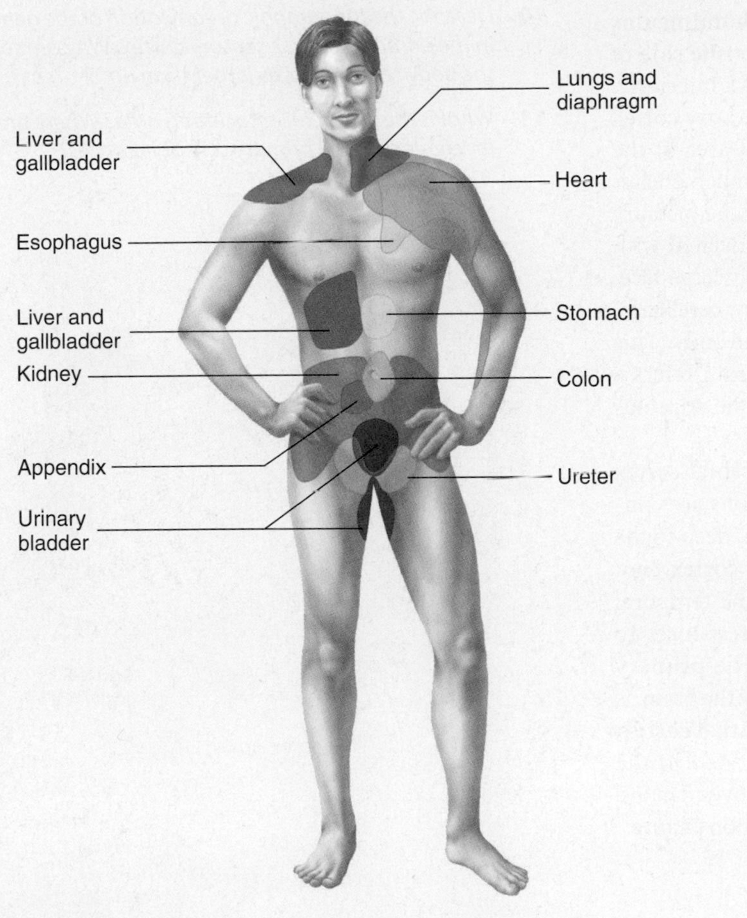
Lungs and diaphragm
Liver and gallbladder
Heart
Esophagus
Liver and gallbladder
Stomach
Kidney
Colon
Appendix
Ureter
Urinary bladder

FIGURE 14A Areas of Referred Pain on the Body Surface
Pain from the indicated internal organs is referred to the surface areas shown.

Pain is important in warning us of potentially injurious conditions because pain receptors are stimulated when tissues are injured. However, chronic pain, such as migraine headaches, localized facial pain, or back pain, can be very debilitating, and the pain loses its value of providing information about the condition of the body. People suffering from chronic pain often feel helpless and hopeless, and they may become dependent on drugs. The pain can interfere with vocational pursuits; therefore, many chronic pain sufferers are unemployed or even housebound and socially isolated. They may be easily frustrated or angered and exhibit symptoms of major depression, all qualities associated with **chronic pain syndrome.** Over 2 million people in the United States at any given time experience chronic pain sufficient to impair activity.

Chronic pain may originate with acute pain associated with an injury or may develop for no apparent reason. How sensory signals are processed in the thalamus and cerebrum may determine whether the input is evaluated as mere discomfort, minor pain, or severe pain and how much distress is associated with the sensation. The brain actively regulates the amount of pain information that gets through to the level of perception, thereby suppressing much of the input. If this dampening system becomes less functional, pain perception may increase. Other nervous system factors, such as the loss of some sensory modalities from an area or habituation of pain transmission, which may remain even after the stimulus is removed, may intensify otherwise normal pain sensations. Treatment often requires a multidisciplinary approach, including surgery or psychotherapy. Some sufferers respond well to drug therapy, but certain drugs, such as opiates, have a diminishing effect the longer they are used and may become addictive.

Sensitization In Chronic Pain

Tissue damage within an area of injury, such as the skin, can cause increased sensitivity in the nerve endings in the area of damage, a condition called **peripheral sensitization.** One class of pain receptors is not activated by traditional noxious stimuli but is recruited only when tissues become inflamed. These receptors, once activated, add to the total barrage of sensory signals to the brain and intensify the sensation of pain.

The CNS may also respond to tissue damage by decreasing its pain threshold, thereby increasing its sensitivity to pain. This condition, called **central sensitization,** apparently results from a specific subset of receptors that is only recruited during repetitive neuron firing, as occurs when a person experiences intense pain sensations. These receptors maintain a chronic, hyperexcitable state in the CNS that can result in chronic pain.

This information concerning peripheral and central sensitization and the knowledge that sensitization involves neuronal and chemical receptors not normally involved in sensation may lead to the discovery of new drugs for treating chronic pain. Rather than searching for new analgesics, which may decrease a broad range of sensations, researchers now have an opportunity to develop a new class of drugs that may block sensitization without diminishing other sensations, including normal pain.

› Predict 5

A man has constipation that is causing distension and painful cramping in his colon. What kind of pain does he experience (local or diffuse), and where does he perceive it? Explain.

sensory impulses conducting input from the feet project to the most superior portion of the primary somatic sensory cortex, and sensory impulses from the face project to the most inferior portion. The pattern of the primary somatic sensory cortex in each hemisphere is arranged in the form of an upside-down half **homunculus** (hō-mŭngk′ū-lŭs; little human) representing the opposite side of the body, with the feet located superiorly and the head inferiorly. The size of various regions of the primary somatic sensory cortex is related to the number of sensory receptors in that area of the body. For example, the density of sensory receptors is much greater in the face than in the legs; therefore, a greater area of the primary somatic sensory cortex contains sensory neurons associated with the face, and the homunculus has a disproportionately large face. Cutaneous sensations, although integrated within the cerebrum, are perceived as though they were on the surface of the body. This phenomenon, called **projection,** indicates that the brain refers a cutaneous sensation to the superficial site at which the stimulus interacts with the sensory receptors.

There are other primary sensory areas of the cerebral cortex (see figure 14.11). The **taste area,** where taste sensations are consciously perceived in the cortex, is located in the insula, deep to the inferior end of the postcentral gyrus. The **olfactory cortex** (not shown in figure 14.11) is on the inferior surface of the temporal lobe and is where both conscious and unconscious responses to odor are perceived and processed (see chapter 15). The **primary auditory cortex,** where auditory stimuli are processed by the brain, is located in the superior part of the temporal lobe. The **visual cortex,** where portions of visual images are processed, is located in the occipital lobe. In the visual cortex, color, shape, and movement are processed separately rather than as a complete "color motion picture." These sensory areas are discussed more fully in chapter 15.

Sensory Processing

Cortical areas immediately adjacent to the primary sensory areas are called **association areas.** The **somatic sensory association area** is posterior to the primary somatic sensory cortex, and the **visual association area** is anterior to the visual cortex (see figure 14.11). These areas function in the process of recognition. When sensory action potentials originating in the retina of the eye reach the visual cortex, the image is perceived. Action potentials then pass from the visual cortex to the visual association area, where the present visual information is compared with past visual experience ("Have I seen this before?"). On the basis of this comparison, the visual association area "decides" whether the visual input is recognized and passes judgment concerning the significance of the input. For example, when surveying a crowd, we generally pay less attention to a person we have never seen before than to someone we know.

The visual association area, like other association areas of the cortex, has reciprocal connections with other parts of the cortex that influence decisions. For example, the visual association area has input from the frontal lobe, where emotional value is placed on the visual input. Because of these numerous connections, visual information is judged several times as it passes beyond the visual association area. This may be one of the reasons that two individuals' perceptions of the same painting can be vastly different.

ASSESS YOUR PROGRESS

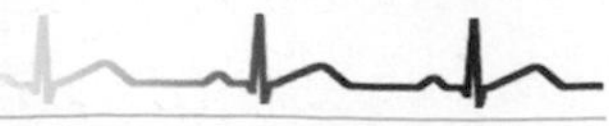

9. *Where are the locations of the primary sensory areas of the cerebral cortex, and what are the functions of each area?*
10. *Describe the topographic organization of the general body plan in the primary somatic sensory cortex. Why are some areas of the body represented as larger than other areas?*
11. *What is the role of an association area? Where are they located in reference to the primary sensory cortices?*

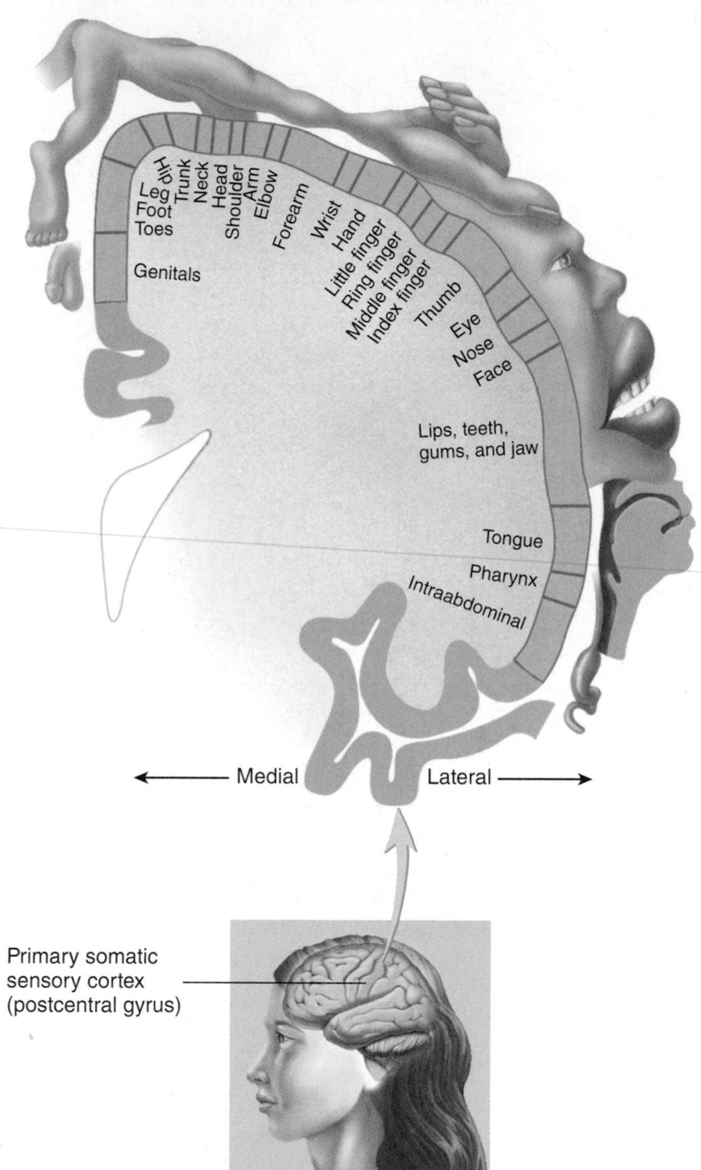
Hip
Leg
Foot
Toes
Trunk
Neck
Head
Shoulder
Arm
Elbow
Forearm
Wrist
Hand
Little finger
Ring finger
Middle finger
Index finger
Thumb
Genitals
Eye
Nose
Face
Lips, teeth, gums, and jaw
Tongue
Pharynx
Intraabdominal
Medial
Lateral
Primary somatic sensory cortex (postcentral gyrus)

FIGURE 14.12 AP|R **Topography of the Primary Somatic Sensory Cortex**

Cerebral cortex seen in frontal section on the left side of the brain. The figure of the body (homunculus) depicts the nerve distributions; the size of each body region shown indicates relative innervation. The cortex occurs on both sides of the brain but appears on only one side in this illustration. The inset shows the somatic sensory region of the left hemisphere (green).

14.2 Control of Skeletal Muscles

LEARNING OUTCOMES 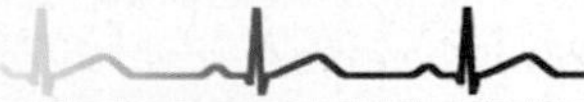

After reading this section, you should be able to

A. Describe the primary motor area of the cerebral cortex and discuss how it interacts with other parts of the frontal lobe.

B. Distinguish between upper and lower motor neurons, and between direct and indirect tracts.

C. Explain how the basal nuclei and the cerebellum regulate motor function.

The motor system of the brain and spinal cord is responsible for maintaining the body's posture and balance; for moving the trunk, head, limbs, and eyes; and for communicating through facial expressions and speech. Reflexes mediated through the spinal cord (see chapter 12) and the brainstem (see chapter 13) are responsible for some body movements. These are called **involuntary movements** because they occur without conscious thought. **Voluntary movements,** on the other hand, are consciously activated to achieve a specific goal, such as walking or typing. Although consciously activated, the details of most voluntary movements occur automatically once learned. Thus, a toddler who is just learning to walk must concentrate on every step. However, once the toddler starts walking, he or she does not have to think about the moment-to-moment control of every muscle because neural circuits in the reticular formation and spinal cord automatically control the limbs. After learning a complex task, such as typing, people can perform it relatively automatically.

Voluntary movements depend on upper and lower motor neurons. **Upper motor neurons** connect to lower motor neurons directly or through interneurons. The cell bodies of upper motor neurons are in the cerebral cortex. **Lower motor neurons** have axons that leave the central nervous system and extend through peripheral nerves to supply skeletal muscles. The cell bodies of lower motor neurons are located in the anterior horns of the spinal cord gray matter and in the cranial nerve nuclei of the brainstem.

Voluntary movements require the following steps:

1. The initiation of most voluntary movement begins in the premotor areas of the cerebral cortex and involves stimulation of upper motor neurons.
2. The axons of the upper motor neurons form the descending tracts. They stimulate lower motor neurons, which stimulate skeletal muscles to contract.
3. The cerebral cortex interacts with the basal nuclei and cerebellum in planning, coordinating, and executing movements.

Motor Areas of the Cerebral Cortex

Body movements are controlled by several motor areas of the brain. Motor pathways from the **primary motor cortex,** or *primary motor area* (see figure 14.11), control many voluntary movements, especially the fine motor movements of the hands. The primary motor cortex occupies the precentral gyrus (see chapter 13). Only about 30% of upper motor neurons are located in the primary motor cortex. Another 30% are in the premotor area, and the rest are in the primary somatic sensory cortex.

The cortical functions of the primary motor cortex are arranged topographically according to the general body plan—similar to the topographic arrangement of the primary somatic sensory cortex (figure 14.13). The neuron cell bodies controlling motor functions of the feet are in the most superior and medial portions of the precentral gyrus, whereas those for the face are in the inferior region. Muscle groups with many motor units are represented by relatively large areas of the primary motor cortex. For example,

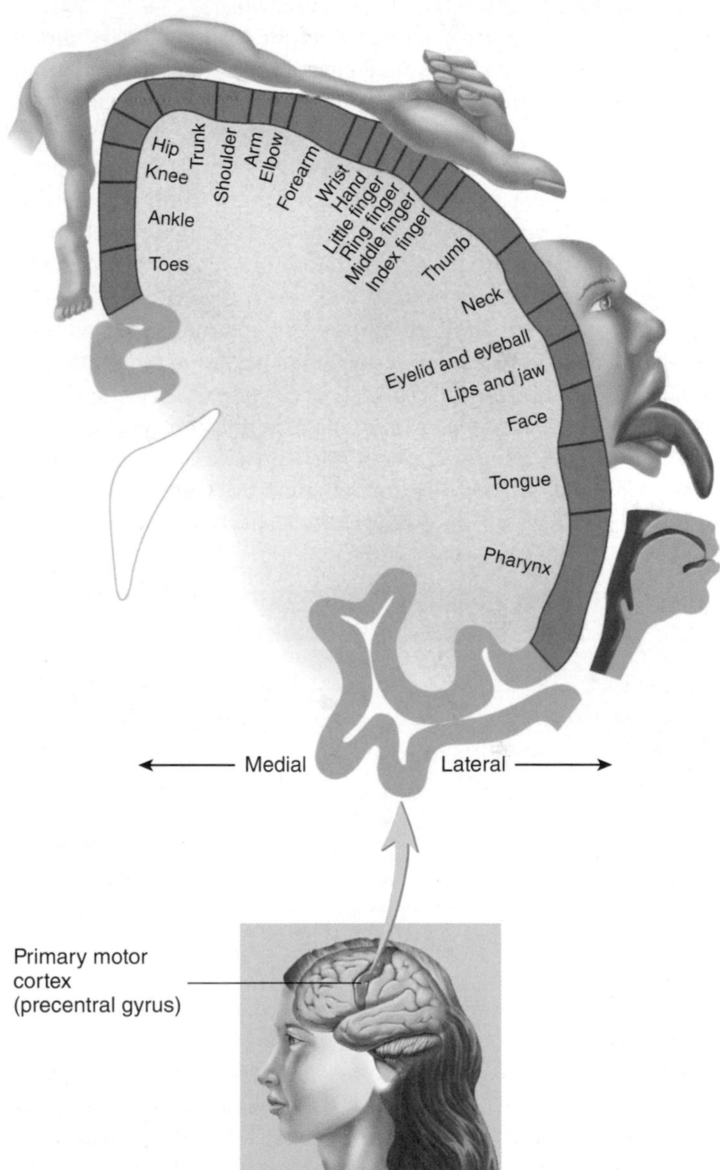
Hip
Knee
Trunk
Shoulder
Arm
Elbow
Forearm
Wrist
Hand
Little finger
Ring finger
Middle finger
Index finger
Thumb
Ankle
Toes
Neck
Eyelid and eyeball
Lips and jaw
Face
Tongue
Pharynx
Medial
Lateral
Primary motor cortex (precentral gyrus)

FIGURE 14.13 Topography of the Primary Motor Cortex

Cerebral cortex seen in frontal section on the left side of the brain. The figure of the body (homunculus) depicts the nerve distributions; the size of each body region shown indicates relative innervation. The cortex occurs on both sides of the brain but appears on only one side in this illustration. The inset shows the motor region of the left hemisphere (purple).

muscles performing precise movements, such as those controlling the hands and face, have many motor units, each of which has a small number of muscle fibers. Multiple-motor-unit summation (see chapter 9) can precisely control the force of contraction of these muscles, because only a few muscle fibers at a time are recruited. Muscle groups with few motor units are represented by relatively small areas of the primary motor cortex, even if the muscles innervated are quite large. For example, the muscles controlling movements of the thigh and leg have proportionately fewer motor units than the muscles of the hand, but they have many more and much larger muscle fibers per motor unit. Thigh and leg muscles are less precisely controlled because the activation of a motor unit stimulates the contraction of many large muscle fibers.

The **premotor area,** located anterior to the primary motor cortex (see figure 14.11), is the staging area where motor functions are organized before they are initiated in the primary motor cortex. For example, if a person decides to take a step, the neurons of the premotor area are stimulated first. The determination is made in the premotor area as to which muscles must contract, in what order, and to what degree. Action potentials are then passed to the upper motor neurons in the primary motor cortex, which actually initiate the planned movements.

The motivation and foresight to plan and initiate movements occur in the next most anterior portion of the brain, the **prefrontal area,** an association area that is well developed only in primates and especially in humans. It is involved in the motivation and regulation of emotional behavior and mood. The large size of this area of the brain in humans may account for our emotional complexity and our relatively well-developed capacity to think ahead and feel motivated.

ASSESS **YOUR PROGRESS**

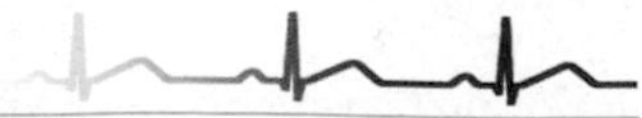

12. *Compare upper motor neurons with lower motor neurons.*

13. *Where are the primary motor, premotor, and prefrontal areas of the cerebral cortex located? Explain the sequential nature of their functions.*

14. *Why are some areas of the body represented as larger than other areas on the topographic map of the primary motor cortex?*

Motor Pathways

Motor pathways, or tracts, are *descending* pathways containing axons that carry action potentials from regions of the cerebrum or cerebellum to the brainstem or spinal cord. The names of descending pathways are based on their origin and termination. Much like the names of ascending pathways, the prefix indicates a pathway's origin, and the suffix indicates its destination. For example, the corticospinal tract is a motor pathway that originates in the cerebral cortex and terminates in the spinal cord (figure 14.14).

The descending motor fibers are divided into two groups: direct pathways and indirect pathways (table 14.4; figure 14.15). The **direct pathways,** also called the *pyramidal* (pi-ram′i-dal) *system,* are involved in maintaining muscle tone and controlling the speed and precision of skilled movements, primarily fine movements involved in dexterity. Most of the **indirect pathways,** sometimes called the *extrapyramidal system,* are involved in less precise control of motor functions, especially those associated with overall body coordination and cerebellar function, such as posture. Many of the indirect pathways are phylogenetically older and control more "primitive" movements of the trunk and proximal portions of the limbs. The direct pathways, which

TABLE 14.4 Descending Spinal Pathways

Pathway	Functions Controlled	Examples of Movements Controlled	Origin	Crossover
Direct	Conscious, skilled movements			
Corticospinal tract	Movements below the head, especially of the hands			
Lateral	Movements of the neck, trunk, upper and lower limbs, especially the fingers	Typing and push-ups	Cerebral cortex	Inferior end of the medulla oblongata
Anterior	Movements of the neck and trunk	Moving with a hula hoop	Cerebral cortex	Level of the lower motor neuron
Corticobulbar tract	Movements of the head and face	Facial expression and chewing	Cerebral cortex	Varies for the different cranial nerves
Indirect	Unconscious movements			
Rubrospinal	Movement coordination	Positioning of digits and the palm of the hand when reaching out to grasp	Red nucleus	Midbrain
Vestibulospinal	Maintenance of upright posture and balance	Extension of the upper limbs when falling	Vestibular nucleus	Uncrossed
Reticulospinal	Posture adjustment and walking	Maintenance of posture when standing on one foot	Reticular formation	Some uncrossed; some cross at termination
Tectospinal	Movements of the head and neck in response to visual and auditory reflexes	Movement of the head and neck away from a sudden flash of light	Superior colliculus	Midbrain

FUNDAMENTAL **Figure**

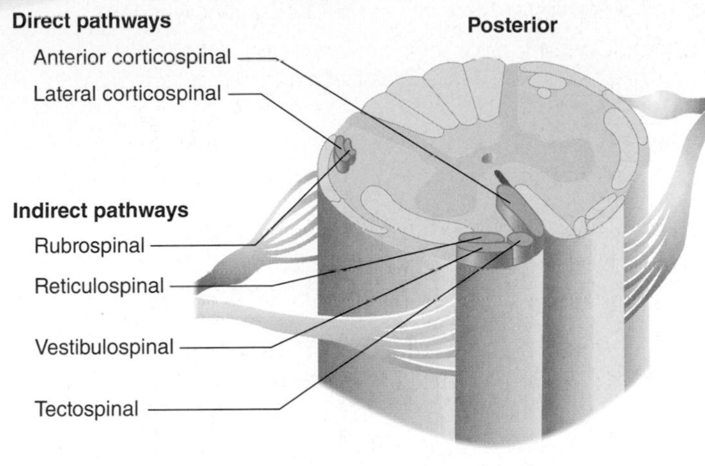
Direct pathways
Anterior corticospinal
Lateral corticospinal
Indirect pathways
Rubrospinal
Reticulospinal
Vestibulospinal
Tectospinal
Posterior
Anterior

FIGURE 14.14 APR **Descending Pathways at the Cervical Level of the Spinal Cord**

Descending pathways (purple) are labeled on the left side of the figure only, although they exist on both sides.

exist only in mammals, may be thought of as overlying the indirect pathways and are more involved in finely controlled movements of the face and distal portions of the limbs. Some indirect pathways, such as those from the basal nuclei and cerebellum, help in fine control of the direct pathways.

Termination	Side of Body Where Fibers Terminate
Anterior horn of the spinal cord	Contralateral
Anterior horn of the spinal cord	Contralateral
Cranial nerve nuclei in the brainstem (lower motor neuron)	Contralateral
Anterior horn of the spinal cord	Contralateral
Anterior horn of the spinal cord	Ipsilateral
Anterior horn of the spinal cord	Ipsilateral or contralateral
Cranial nerve nucleus in the medulla oblongata and anterior horn of the upper levels of the spinal cord (lower motor neurons that turn the head and neck)	Contralateral

Clinical
IMPACT

Amyotrophic Lateral Sclerosis

Amyotrophic (ă-mī-ō-trō′fik) **lateral sclerosis (ALS),** also called Lou Gehrig disease, usually affects people between the ages of 40 and 70. About 10% of ALS cases are inherited. The condition begins with weakness and clumsiness and progresses within 2–5 years to loss of muscle control as both upper and lower motor neurons are selectively destroyed. The causes of noninherited ALS remain unknown, but the inherited form results from a mutation in DNA coding for the enzyme **superoxide dismutase (SOD);** SOD is involved in eliminating the free radical superoxide from the body.

Free radicals are molecules that readily accept electrons, which makes them highly reactive. They can strip electrons from proteins, lipids, or nucleic acids, thereby destroying their functions and resulting in cell dysfunction or death. Free-radical damage has been implicated in ALS, arteriosclerosis, arthritis, cancer, and aging. Superoxide, one of the most important and toxic free radicals, forms as oxygen reacts with other free radicals. Although oxygen is critical for aerobic respiration, it is also dangerous to tissues. SOD catalyzes the conversion of superoxide to hydrogen peroxide, which is then converted by catalase to oxygen and water. Apparently, if SOD is defective, superoxide is not degraded and can destroy cells. Motor neurons appear to be particularly sensitive to superoxide attack.

Direct Pathways

Direct pathways (figure 14.16) are so named because upper motor neurons in the cerebral cortex, whose axons form these pathways, synapse directly with lower motor neurons in the brainstem or spinal cord. They are also called the pyramidal system because the fibers of these pathways form the medullary **pyramids.** The direct pathways include groups of nerve fibers arrayed into two tracts: the **corticospinal tract,** which is involved in direct cortical control of movements below the head, and the **corticobulbar tract,** which is involved in direct cortical control of movements in the head and neck.

The corticospinal tract consists of axons of upper motor neurons located in the primary motor and premotor areas of the frontal lobes and the somatic sensory parts of the parietal lobes. They descend through the **internal capsules** and the cerebral peduncles of the midbrain to the pyramids of the medulla oblongata. At the inferior end of the medulla, 75–85% of the corticospinal fibers cross to the opposite side of the CNS through the **pyramidal decussation,** which is visible on the anterior surface of the inferior medulla. The crossed fibers descend in the **lateral corticospinal tract** of the spinal cord (figure 14.16). The remaining 15–25% descend uncrossed in the **anterior corticospinal tract** and decussate near the level where they synapse with lower motor neurons. The anterior corticospinal tracts supply the neck and upper limbs, and the lateral corticospinal tracts supply all levels of the body (table 14.4).

Most of the corticospinal fibers synapse with interneurons in the lateral portions of the spinal cord central gray matter. The

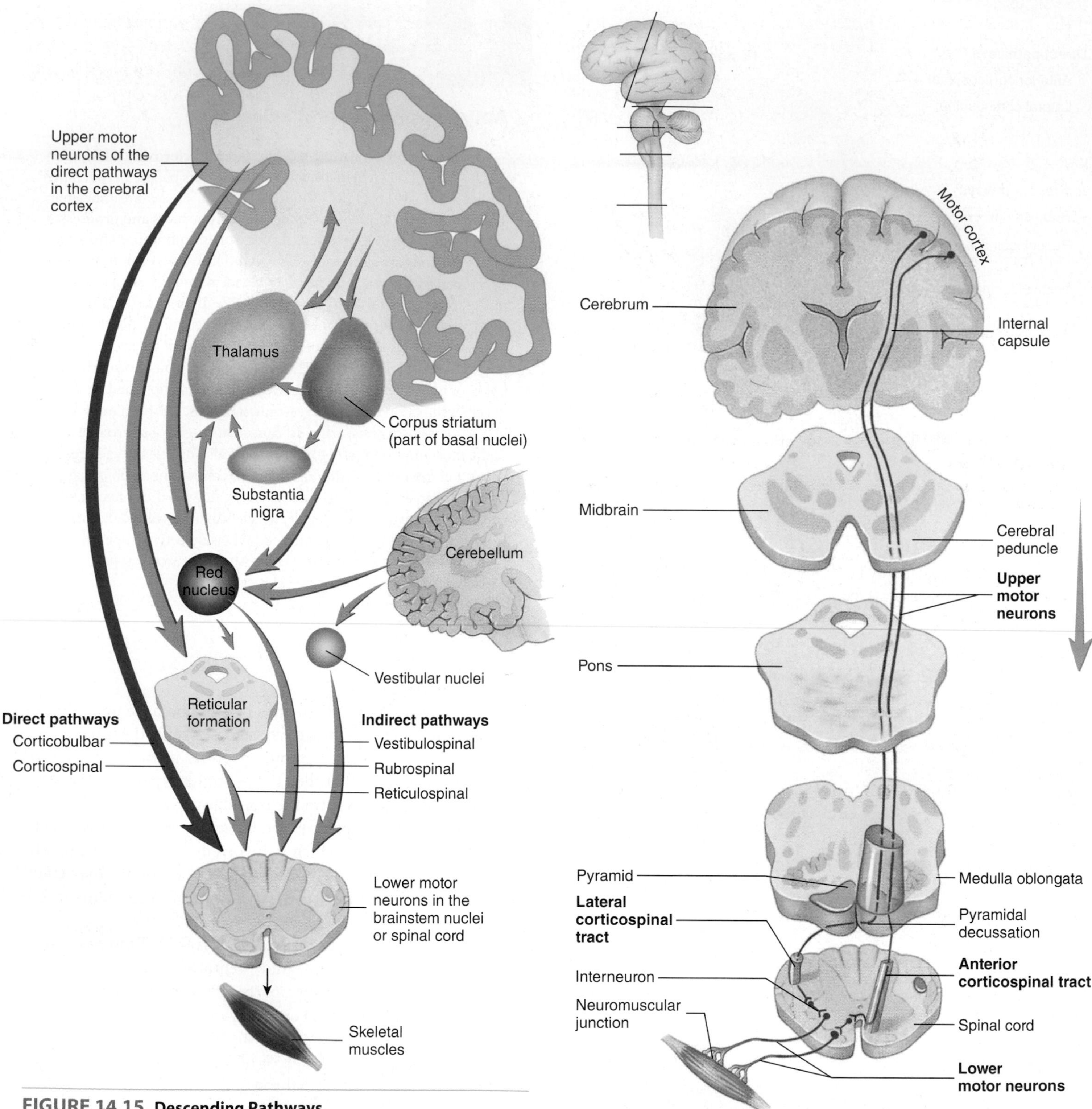

Upper motor neurons of the direct pathways in the cerebral cortex
Thalamus
Corpus striatum (part of basal nuclei)
Substantia nigra
Red nucleus
Cerebellum
Vestibular nuclei
Reticular formation
Direct pathways
Corticobulbar
Corticospinal
Indirect pathways
Vestibulospinal
Rubrospinal
Reticulospinal
Lower motor neurons in the brainstem nuclei or spinal cord
Skeletal muscles
Motor cortex
Cerebrum
Internal capsule
Midbrain
Cerebral peduncle
Upper motor neurons
Pons
Pyramid
Medulla oblongata
Lateral corticospinal tract
Pyramidal decussation
Interneuron
Anterior corticospinal tract
Neuromuscular junction
Spinal cord
Lower motor neurons

FIGURE 14.15 Descending Pathways
The direct pathways (corticobulbar and corticospinal) are indicated by the dark purple arrow. The indirect pathways and their interconnections are indicated by the light purple arrows.

FIGURE 14.16 Direct Pathways
Lateral and anterior corticospinal tracts are responsible for movement below the head. Upper motor neuron axons descend to the medulla oblongata. Most axons decussate in the medulla oblongata and descend in the lateral corticospinal tracts in the spinal cord. Some axons continue as the anterior corticospinal tracts and decussate in the spinal cord. Upper motor neurons synapse with interneurons that synapse with lower motor neurons. Lines on the inset indicate levels of section. The orange arrow indicates direction of action potentials.

interneurons, in turn, synapse with the lower motor neurons of the anterior horn that innervate primarily distal limb muscles.

Damage to the corticospinal tracts results in reduced muscle tone, clumsiness, and weakness but not in complete paralysis, even if the damage is bilateral. Experiments have demonstrated that bilateral sectioning of the medullary pyramids results in (1) loss of contact-related activities, such as tactile placing of the foot and grasping, (2) defective fine movements, and (3) hypotonia (reduced tone). These and other experimental data support the conclusion that the corticospinal system is superimposed over the older, indirect pathways and that it has many parallel functions. The main function of the direct pathways is to add speed and agility to conscious movements, especially of the hands, and to provide a high degree of fine motor control, as in movements of individual fingers. Spinal cord lesions that affect both the direct and the indirect pathways result in complete paralysis.

The corticobulbar tracts are analogous to the corticospinal tracts. The corticobulbar tracts extend to the brainstem (*bulbar*, brainstem) and innervate the head, whereas the corticospinal tracts extend to the spinal cord and innervate the rest of the body. Cells that contribute to the corticobulbar tracts are in regions of the cortex similar to those of the corticospinal tracts. Corticobulbar tracts follow the same basic route as the corticospinal system down to the level of the brainstem. At that point, most corticobulbar fibers terminate in the **cranial nerve nuclei,** where they synapse with interneurons and lower motor neurons. These nuclei give rise to the nerves that control tongue movements, mastication, facial expression, some eye movements, and palatine, pharyngeal, and laryngeal movements.

Indirect Pathways

The indirect pathways (figure 14.17) are so named because axons from motor neurons of the cerebrum and cerebellum first synapse in an intermediate nucleus in the brainstem rather than directly with lower motor neurons. The indirect pathways begin with the neurons in those brainstem nuclei. They do not pass through the pyramids or through the corticobulbar tracts and, therefore, are sometimes called extrapyramidal. The major tracts are the rubrospinal, vestibulospinal, reticulospinal, and tectospinal tracts. Many interconnections and feedback loops are present in this system.

Neurons of the **rubrospinal tract** begin in the red nucleus, which is located at the boundary between the diencephalon and the midbrain. The tract decussates in the midbrain and descends in the lateral column of the spinal cord. The red nucleus receives input from both the motor cortex and the cerebellum. Lesions in the red nucleus result in *intention tremors* (action tremors) similar to those

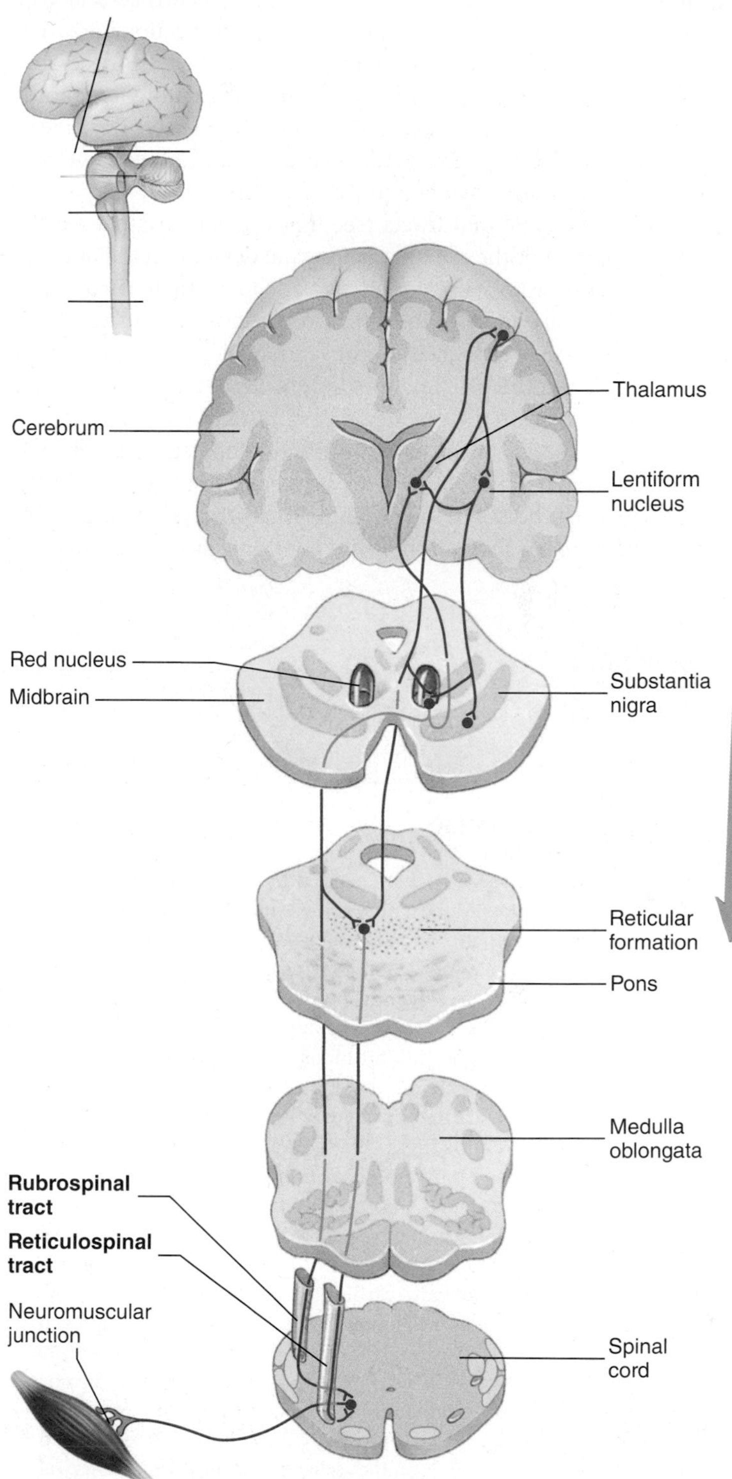
Cerebrum
Thalamus
Lentiform nucleus
Red nucleus
Midbrain
Substantia nigra
Reticular formation
Pons
Medulla oblongata
Rubrospinal tract
Reticulospinal tract
Neuromuscular junction
Spinal cord

FIGURE 14.17 Indirect Pathways
Examples of indirect pathways are rubrospinal and reticulospinal tracts. Neurons from the cerebrum project to neurons in brainstem nuclei. The axons of these neurons form the reticulospinal and rubrospinal tracts. Lines on the inset indicate levels of section. The orange arrow indicates direction of action potentials.

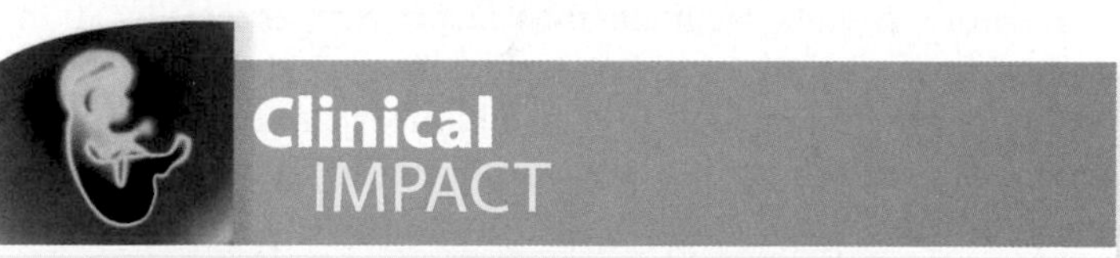
Clinical IMPACT

Brown-Séquard Syndrome

A lesion of the spinal cord that destroys half the cord at a specific level (hemisection) results in a very specific syndrome called the **Brown-Séquard syndrome.** Characteristics of the syndrome include contralateral loss of pain and temperature, because fibers entering the anterolateral system cross over where they enter the spinal cord, and ipsilateral loss of proprioception, two-point discrimination, and most upper motor control, because the dorsal-column/medial-lemniscal system ascends ipsilaterally and the lateral corticospinal tract descends ipsilaterally.

seen in cerebellar lesions (see Diseases and Disorders table, later in this chapter). The function of the red nucleus, therefore, is closely related to cerebellar function. The rubrospinal tract is the one indirect tract that is very closely related to the direct, corticospinal tract. It terminates in the lateral portion of the spinal cord central gray matter with the corticospinal tract, and it transmits action potentials involved in the comparator function of the cerebellum (described later in this chapter). It plays a major role in regulating fine motor control of muscles in the distal part of the upper limbs. Damage to the rubrospinal tract impairs forearm and hand movements but does not greatly affect general body movements.

The **vestibulospinal tracts** (see figure 14.14) originate in the vestibular nuclei of the medulla oblongata, descend in the anterior column, and synapse with interneurons and lower motor neurons in the ventromedial portion of the spinal cord central gray matter. Their fibers preferentially influence neurons innervating extensor muscles in the trunk and the proximal portion of the lower limbs and are involved primarily in the maintenance of upright posture. The vestibular nuclei receive major input from the vestibular nerve, which is involved in maintaining balance (see chapter 15), and the cerebellum.

Neuron cell bodies of the **reticulospinal tract** (see figure 14.14) are in the reticular formation of the pons and medulla oblongata. Their axons descend in the anterior portion of the lateral column and synapse with interneurons and lower motor neurons in the ventromedial portion of the spinal cord central gray matter. The reticulospinal tract maintains posture through the action of trunk and proximal upper and lower limb muscles during certain movements. For example, when a person who is standing lifts one foot off the ground, the weight of the body shifts to the other limb. During this type of movement, the reticulospinal tract apparently enhances the functions of the alpha motor neurons in the crossed extensor reflex, so that balance is maintained.

The **tectospinal tract** originates in the superior colliculus, which has been called the *tectum*, or roof, of the mesencephalon. This tract controls reflex movement of the head to bright lights, noises, and rapid movements.

Another major portion of the indirect pathways involves the basal nuclei (see figure 14.15). The basal nuclei have a number of connections within the brain, and they interact with indirect pathways, such as the rubrospinal tract, by which they modulate motor functions.

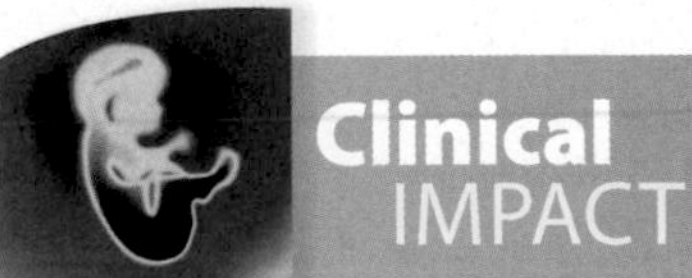
Clinical
IMPACT

Parkinson disease

Parkinson disease is characterized by muscular rigidity; loss of facial expression; tremor; a slow, shuffling gait; and general lack of movement. The disease usually occurs after age 55. A resting tremor, called "pill-rolling," is characteristic of Parkinson disease; it consists of circular movement of the opposed thumb and index fingertips. The increased muscular rigidity in Parkinson disease results from defective inhibition of some of the basal nuclei by the substantia nigra, one of the basal nuclei in the midbrain. Parkinson disease is caused by the death of dopamine-containing neurons in the substantia nigra. However, the cause of the neuronal death is not known. Dopamine is an inhibitory neurotransmitter that is required to modulate motor activity.

Parkinson disease can be treated with levodopa (lē-vō-dō′pă; L-dopa), a precursor to dopamine, or more effectively with Sinemet®, a combination of L-dopa and carbidopa (kar-bi-dō′pă). Carbidopa is a decarboxylase inhibitor that prevents L-dopa from breaking down before it can reach the brain. Because of the long-term side effects associated with levodopa, including dyskinesias, other dopamine agonists, such as ropinirole and pramipexole, are being examined. Researchers have discovered a protein, called **glial cell line–derived neurotrophic factor (GDNF),** that selectively promotes the survival of dopamine-secreting neurons. Alternatively, chronic stimulation of the globus pallidus (part of the lentiform nucleus) with an electrical pulse generator has shown some success. Treatment of the disorder by transplanting fetal tissues or stem cells from adult tissues that are capable of producing dopamine is also under investigation.

ASSESS YOUR PROGRESS

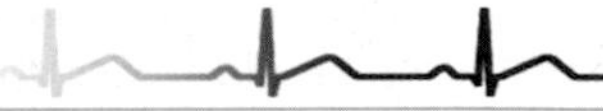

15. *What are the structural and functional differences between direct and indirect pathways?*
16. *What two tracts form the direct pathways? What area of the body is supplied by each tract?*
17. *Describe the location of the neurons in each tract, as well as where they synapse.*
18. *Name the structures and tracts that form the indirect pathways. What functions do they control?*

Modifying and Refining Motor Activities

Basal Nuclei

The **basal nuclei** (see figure 13.10) are important in planning, organizing, and coordinating motor movements and posture. Complex neural circuits link the basal nuclei with each other, with the thalamus, and with the cerebral cortex. These connections form several feedback loops, some of them stimulatory and others inhibitory. The stimulatory circuits facilitate muscle activity, especially at the beginning of a voluntary movement, such as rising from a sitting position or beginning to walk. The inhibitory circuits facilitate the actions of the stimulatory circuits by inhibiting muscle activity in antagonist muscles. Inhibitory circuits also decrease muscle tone when the body, limbs, and head are at rest (eliminating random and "unwanted" movements of the trunk and limbs). Disorders of the basal nuclei result in difficulty in rising from a sitting position and initiating walking. People with basal nuclei disorders exhibit increased muscle tone and exaggerated, uncontrolled movements when they are at rest. A specific feature of some basal nuclei disorders is a *resting tremor*, slight shaking of the hands when not performing a task. Parkinson disease and cerebral palsy are basal nuclei disorders.

Cerebellum

The **cerebellum** (ser-e-bel′ŭm; see figure 13.6) consists of three functional parts: the vestibulocerebellum, or flocculonodular lobe;

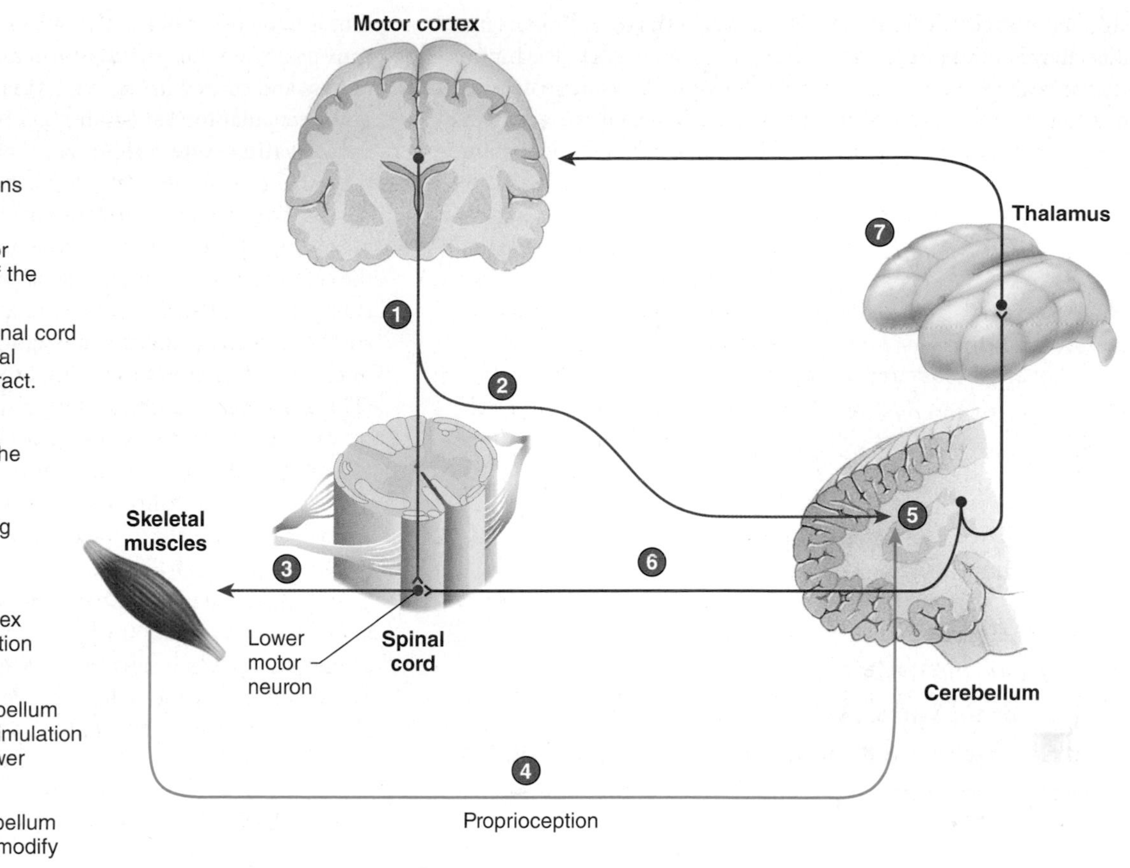

Motor cortex
Thalamus
Skeletal muscles
Lower motor neuron
Spinal cord
Cerebellum
Proprioception
1 The motor cortex sends action potentials to lower motor neurons in the spinal cord.
2 Action potentials from the motor cortex inform the cerebellum of the intended movement.
3 Lower motor neurons in the spinal cord send action potentials to skeletal muscles, causing them to contract.
4 Proprioceptive signals from the skeletal muscles and joints to the cerebellum convey information concerning the status of the muscles and the structure being moved during contraction.
5 The cerebellum compares the information from the motor cortex with the proprioceptive information from the skeletal muscle joints.
6 Action potentials from the cerebellum to the spinal cord modify the stimulation from the motor cortex to the lower motor neurons.
7 Action potentials from the cerebellum are sent to the motor cortex to modify its motor activity.

PROCESS FIGURE 14.18 Cerebellar Comparator Function

the spinocerebellum; and the cerebrocerebellum. The vestibulocerebellum receives direct input from the vestibular structures, especially the semicircular canals (see chapter 15), and sends axons to the vestibular nuclei of the brainstem. It helps maintain muscle tone in postural muscles. It also helps control balance, especially during movements, and coordinate eye movement.

The **vermis** and the medial portion of the **lateral hemisphere** are referred to jointly as the **spinocerebellum.** The spinocerebellum helps accomplish fine motor coordination of simple movements by means of its comparator function. A **comparator** is a sensing device that compares the data from two sources—in this case, the motor cortex and peripheral structures (figure 14.18). Action potentials from the motor cortex descend into the spinal cord to initiate voluntary movements. At the same time, action potentials are carried from the motor cortex to the cerebellum to give the cerebellar neurons information representing the intended movement. In addition, action potentials from proprioceptive neurons ascend through the spinocerebellar tracts and simultaneously arrive at the cerebellum. Proprioceptive neurons innervate the joints and tendons of the structure being moved, such as the elbow or knee, and provide information about the position of the body or body parts. These action potentials give the cerebellar neurons information from the periphery about the actual movements. The cerebellum compares the action potentials from the motor cortex with those from the moving structures. That is, it compares the intended movement with the actual movement. If a difference is detected, the cerebellum sends action potentials through the thalamus to the motor cortex and to the spinal cord to correct the discrepancy. The result is smooth, coordinated movements.

The comparator function coordinates simple movements, such as touching your nose. However, rapid, complex movements require much greater coordination and training. The **cerebrocerebellum,** which consists of the lateral two-thirds of the lateral hemispheres, communicates with the motor, premotor, and prefrontal portions of the cerebral cortex in planning and practicing rapid, complex motor actions. The connections from the cerebrum to the cerebellum constitute a large portion of the axons in the cerebral peduncles. Because of the cerebrocerebellum, with training, a person can perform highly skilled and rapid movements more quickly than would be possible with only the comparator function of the cerebellum. In these cases, the cerebellum participates with the cerebrum in learning highly specialized movements, such as playing the piano or swinging a baseball bat. The cerebrocerebellum is also involved in cognitive functions, such as rhythm, conceptualization of time intervals, some word associations, and solutions to pegboard puzzles—tasks once thought to occur only in the cerebrum.

Cerebellar dysfunction results in (1) decreased muscle tone, (2) balance impairment, (3) a tendency to overshoot when reaching for or touching an object, and (4) an intention tremor, which is shaking in the hands that occurs only while attempting to perform a task.

Although the cerebellum and basal nuclei both control motor functions, they have opposite effects, and they exhibit opposite symptoms when injured. For example, cerebellar dysfunction results in decreased muscle tone and an intention tremor, whereas basal nuclei dysfunction often results in increased muscle tone and a resting tremor.

ASSESS YOUR PROGRESS 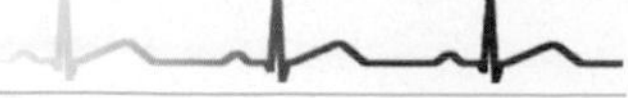

19. *What are the functions of the basal nuclei?*
20. *What are the general symptoms of basal nuclei disorders?*
21. *What are the three functional parts of the cerebellum, and what are the functions of each?*
22. *Explain the comparator activities of the spinocerebellum.*
23. *What are the general symptoms of cerebellar dysfunction?*

14.3 Brainstem Functions

LEARNING OUTCOMES 

After reading this section, you should be able to

A. Describe the sensory input from the brainstem.

B. Explain the role of the reticular activating system (RAS).

C. Discuss the motor output and reflexes of the brainstem.

The major ascending and descending pathways project through the brainstem. In addition, the brainstem contains the nuclei of cranial nerves III–X and XII (see table 13.4 and figure 13.4*c*) and the nuclei of the reticular formation. Only collateral branches of cranial nerve II (optic nerve) project to brainstem nuclei. Cranial nerves I (olfactory nerve) and XI (spinal accessory nerve) do not have projections to or nuclei in the brainstem.

Sensory Input Projecting Through the Brainstem

The brainstem receives sensory input from collateral branches of ascending spinal cord pathways and from the axons of cranial nerves II (vision), V (tactile sensation from the face, nasal cavity, and oral cavity), VII (taste), VIII (hearing and balance), IX (taste and tactile sensation in the throat), and X (taste, tactile sensation in the larynx, and visceral sensation in the thorax and abdomen). Cranial nerves V, VII, VIII, IX, and X have sensory nuclei in the brainstem. Many of these nuclei are involved in the special senses (see chapter 15). The brainstem nuclei associated with cranial nerve II are involved in visual reflexes.

As noted earlier, fibers of the spinothalamic tracts passing through the brainstem are joined by fibers of the trigeminothalamic tract. This tract carries tactile sensations, such as pain and temperature, two-point discrimination, and light touch, from the face, the nasal cavity, and the oral cavity, including the teeth.

RAS Functions of the Brainstem

Scattered throughout the brainstem is a group of nuclei collectively called the **reticular formation,** which is involved in regulating cyclical motor functions. Collateral branches of trigeminothalamic tract neurons project to the reticular formation, where they stimulate wakefulness and consciousness, including the sleep-wake cycle. This part of the reticular formation and its connections constitute the **reticular activating system (RAS).**

Also projecting to the RAS are collateral branches of cranial nerves II (optic), V (trigeminal), and VIII (vestibulocochlear); ascending tactile sensory pathways; and descending neurons from the cerebrum. Visual and acoustic stimuli, as well as mental activites, stimulate the RAS to help maintain alertness and attention. A stimulus such as a sudden flash of bright light, a ringing alarm clock, the smell of coffee, or a feather touching the face can arouse consciousness (figure 14.19). The removal of auditory, visual, and other stimuli may lead to drowsiness or sleep. For example, consider what happens to students during a monotonous lecture in a dark lecture hall. The RAS controls the brain's level of arousal or consciousness. Damage to RAS cells of the reticular formation can result in a lack of consciousness or coma.

Certain drugs can either depress or stimulate the RAS. General anesthetics and many tranquilizers depress it. On the other hand, ammonia (smelling salts) and other irritants stimulate trigeminal nerve endings in the nose. As a result, action potentials travel to the reticular formation and the cerebral cortex to arouse an unconscious patient.

Predict 6

Luke tried to go to sleep, but a dripping bathroom faucet kept him awake. Explain.

Motor Output and Reflexes Projecting Through the Brainstem

The brainstem is an important conduit and integration site for descending motor pathways and reflexes. As discussed earlier, the direct descending pathways originate in the cerebral cortex and pass directly through the brainstem (corticospinal tracts) or synapse in cranial nerve motor nuclei to initiate movements within the head, such as eye movements (corticobulbar tract). Others synapse with brainstem nuclei, which in turn send descending fibers into the spinal cord (indirect pathways). Descending fibers from the reticular formation constitute one of the body's most important motor pathways. Fibers from the reticular formation are critical in controlling many functions, such as respiratory movements and cardiac rhythms. Cranial nerves III, IV, V, VI, VII, IX, X, and XII all have motor nuclei in the brainstem. Functionally, the motor output projecting through the brainstem can be classified into two categories: somatic motor and parasympathetic.

Somatic Motor Output and Reflexes

Cranial nerves III (oculomotor), IV (trochlear), and VI (abducens) can be controlled by the cerebrum to direct eye movement. Action potentials reaching the superior colliculi from the cerebrum are involved in the visual tracking of moving objects. Visual tracking with both eyes to the right involves the lateral rectus muscle and abducens nerve of the right eye and the medial rectus muscle and oculomotor nerve of the left eye. Coordination of these two nerves and muscles

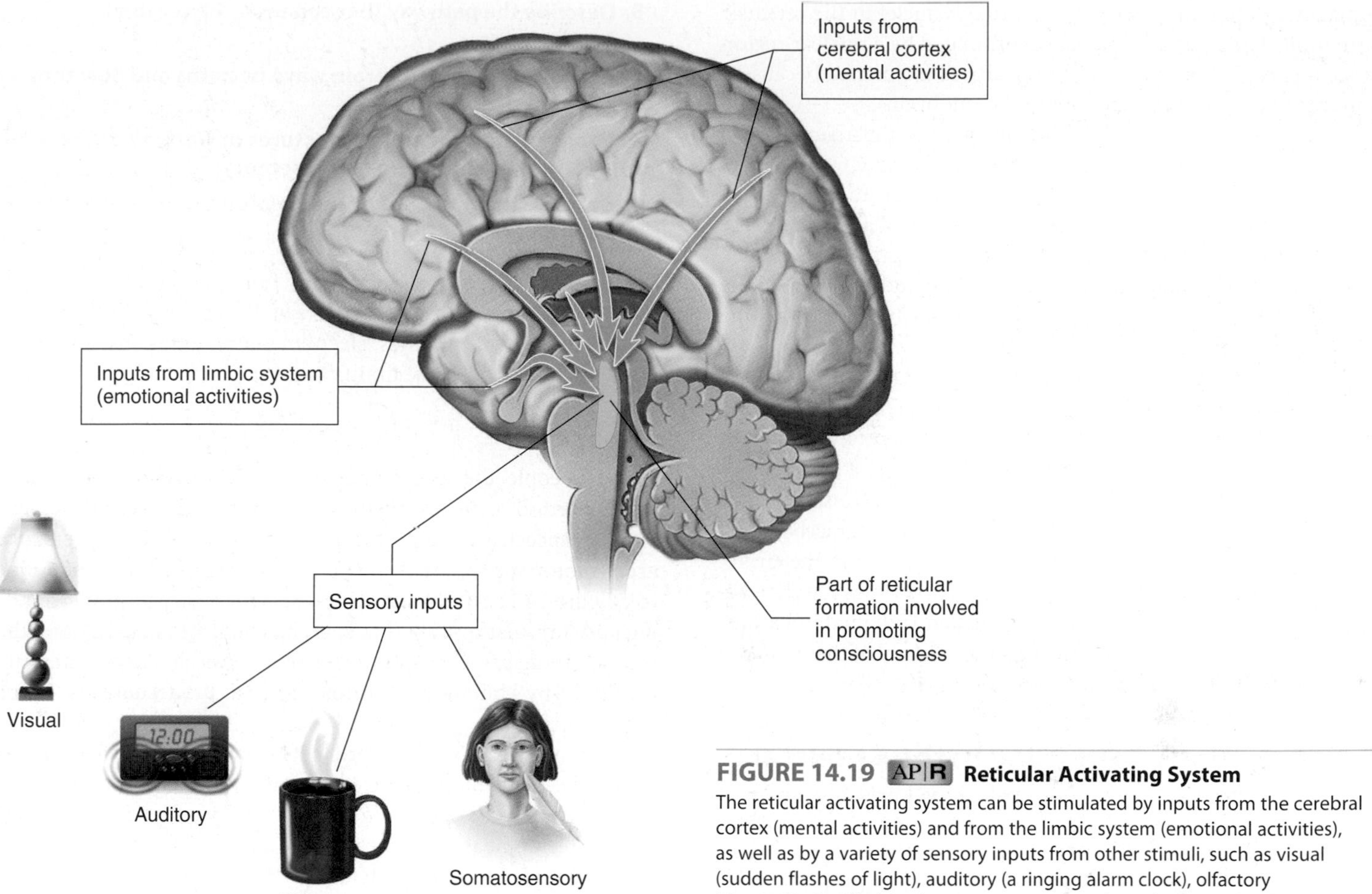
Inputs from
cerebral cortex
(mental activities)
Inputs from limbic system
(emotional activities)
Sensory inputs
Part of reticular
formation involved
in promoting
consciousness
Visual
12:00
Auditory
Olfactory
Somatosensory

FIGURE 14.19 APR **Reticular Activating System**
The reticular activating system can be stimulated by inputs from the cerebral cortex (mental activities) and from the limbic system (emotional activities), as well as by a variety of sensory inputs from other stimuli, such as visual (sudden flashes of light), auditory (a ringing alarm clock), olfactory (the smell of coffee), and somatosensory (touching the face) stimuli.

requires nuclei of the reticular formation. The eye muscles, along with the muscles of the neck (mainly the trapezius and sternocleidomastoid, innervated by cranial nerve IX), can also be involved in reflexes that are initiated in the superior colliculi in response to visual and auditory stimuli. Collateral branches from the optic tract (II) synapse in the superior colliculi of the midbrain. Axons from the superior colliculi project to cranial nerve nuclei III, IV, and VI and to the upper cervical part of the spinal cord (motor neurons of XI), where they stimulate the motor neurons involved in turning the eyes and head toward a visual stimulus. Similarly, the superior colliculi also receive input from auditory pathways, which can initiate a reflex that turns the eyes and head toward a sudden noise. Axons in the spinomesencephalic pathway of the spinal cord (see table 14.3) also project to the superior colliculi. Input from this pathway initiates a reflex that turns the eyes and head toward a tactile stimulus on the body.

Motor fibers from cranial nerve V (trigeminal) innervate the muscles of mastication and control chewing; however, once the chewing cycle is initiated, the reticular formation regulates the cycle. Even though the initiation of chewing can be under conscious control, the presence of food in the mouth initiates a reflex between the sensory nuclei and the motor nucleus of cranial nerve V, thereby starting the chewing cycle. Other reflexes in the trigeminal nerve system detect how hard or soft an item is in the mouth and adjust the bite accordingly. Reflexes between the trigeminal sensory nuclei and the motor nucleus of cranial nerve XII (hypoglossal) control the tongue to help place the food between the teeth for chewing, while keeping the tongue out of harm's way.

Cranial nerve VII (facial), which innervates the muscles of facial expression, is controlled by the cerebrum and is very important in communication.

Somatic motor fibers from cranial nerves IX (glossopharyngeal) and X (vagus) supply muscles of the pharynx and larynx associated with swallowing and speech. Swallowing, once initiated under conscious control, continues as a reflex. The pharyngeal muscles of swallowing are largely innervated by the vagus nerve and, to a smaller extent, by the glossopharyngeal nerve.

Unlike swallowing, which is largely a reflex, speech is highly controlled by the cerebrum. The vagus nerve innervates the muscles of the larynx responsible for voice production and controls the pharyngeal and most palatine muscles responsible for moving the soft palate during speech. The complex movements of the tongue during speech are controlled by the hypoglossal nerve (XII), which innervates nearly all the muscles of the tongue.

Parasympathetic Output and Reflexes

The constriction of the pupil involves parasympathetic stimulation through the oculomotor (III) nerve. The visual reflexes resulting in pupil constriction are coordinated through nuclei in the reticular formation. These reflexes are also coordinated by a nuclear region in the diencephalon called the pretectal area (so named because it is in front of the tectum, the roof of the midbrain).

Tactile sensory input from the nasal cavity via the trigeminal (V) nerve can initiate a sneeze reflex. Tactile sensory input from the oral cavity via the trigeminal nerve informs the cerebrum that food or some other object is in the mouth. The presence of an object in the mouth—even a nonfood item, such as a marble—stimulates a reflex between the trigeminal sensory nuclei and the motor nuclei of the facial (VII) and glossopharyngeal (IX) nerves, which innervate the salivary glands to stimulate salivation.

Sensory input from the glossopharyngeal nerve (IX) conveys tactile information from the back of the tongue, the soft palate, and the throat (pharynx) to the brainstem. Mechanical stimulation of these areas can initiate a gag reflex, whereas other stimulation of the throat can initiate a cough reflex. Sensory input from the vagus nerve (X) conveys tactile information from the larynx (voicebox) and the thoracic and abdominal viscera. Tactile input from the larynx can also initiate a cough reflex. In addition, the vagus nerve (X) is involved in many complex reflexes associated with vital functions, such as heart rate, respiration, and digestion. Many of these involve the reticular formation and are discussed in later chapters.

Vital Functions Controlled in the Brainstem

Many vital functions, such as heart rate, blood pressure, and respiration, are regulated by nuclei in the brainstem. When a person is involved in a serious accident or is extremely ill, these vital functions may be affected. Therefore, many emergency medical procedures are designed to evaluate brainstem function.

ASSESS YOUR PROGRESS

24. *Which cranial nerves provide sensory input to brainstem nuclei?*

25. *What is the reticular formation? What are the roles of the reticular activating system?*

26. *Discuss the somatic motor output and reflexes from the brainstem.*

27. *Describe the parasympathetic reflexes that involve the brainstem.*

28. *What are some vital functions that are regulated by the brainstem?*

Predict 7

Some types of epilepsy are treated by placing an electronic device inside the neck to stimulate the vagus nerve. Minor injury to the vagus nerve during implantation of the device can lead to hoarseness. Why is this so? Can you predict two other consequences of minor injury to the vagus nerve?

14.4 Higher Brain Functions

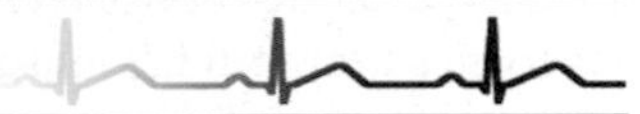

LEARNING OUTCOMES

After reading this section, you should be able to

A. Compare and contrast the two cortical areas required for speech.

B. Describe the pathway that connects the cerebral hemispheres.

C. Describe the types of brain wave patterns and how they relate to sleep.

D. Compare and contrast the features of working, short-term, and both types of long-term memory.

E. Describe the functions of the limbic system, including the impact of olfactory stimuli.

The human brain is capable of many functions besides awareness of sensory input and control of skeletal muscles. Speech, mathematical and artistic abilities, sleep, memory, emotions, and judgment are functions of the brain.

Speech

In most people, the speech area is in the left cerebral cortex. Two major cortical areas are involved in speech: the **Wernicke area** (sensory speech area), a portion of the parietal lobe, and the **Broca area** (motor speech area) in the inferior part of the frontal lobe (see figure 14.11). The Wernicke area is necessary for understanding and formulating coherent speech. The Broca area initiates the complex series of movements necessary for speech. The two areas are connected by a bundle of neurons known as the **arcuate fasciculus** (figure 14.20*a*).

For someone to speak a word that he or she sees, as when reading aloud, the following sequence of events must take place: Action potentials from the eyes reach the primary visual cortex, where the word is seen. The word is then recognized in the visual association area and understood in parts of the Wernicke area. Next, action potentials representing the word are conducted through association fibers that connect the Wernicke and Broca areas. In the Broca area, the word is formulated as it will be spoken. Action potentials are then propagated to the premotor area, where the movements are programmed, and finally to the primary motor cortex, where the proper movements are triggered (figure 14.20*b*).

The sequence of events required to repeat a word that has been heard is similar. The information passes from the ears to the primary auditory cortex and then to the auditory association area, where the word is recognized, and continues to the Wernicke area, where the word is understood. From the Wernicke area, it follows the same route as for speaking words that are seen.

Predict 8

Vern, age 75, is recovering from a stroke that caused some right-side paralysis. He understands verbal commands and instructions, but his speech is hesitant and distorted. In addition, Vern's facial expressions, especially on the right side, are limited, and he has some difficulty chewing and swallowing. Explain these manifestations.

Communication Between the Right and Left Hemispheres

The cortex of the right cerebral hemisphere controls muscular activity in and receives sensory input from the left half of the body. The left cerebral hemisphere controls muscles in and receives sensory input from the right half of the body. Sensory information received

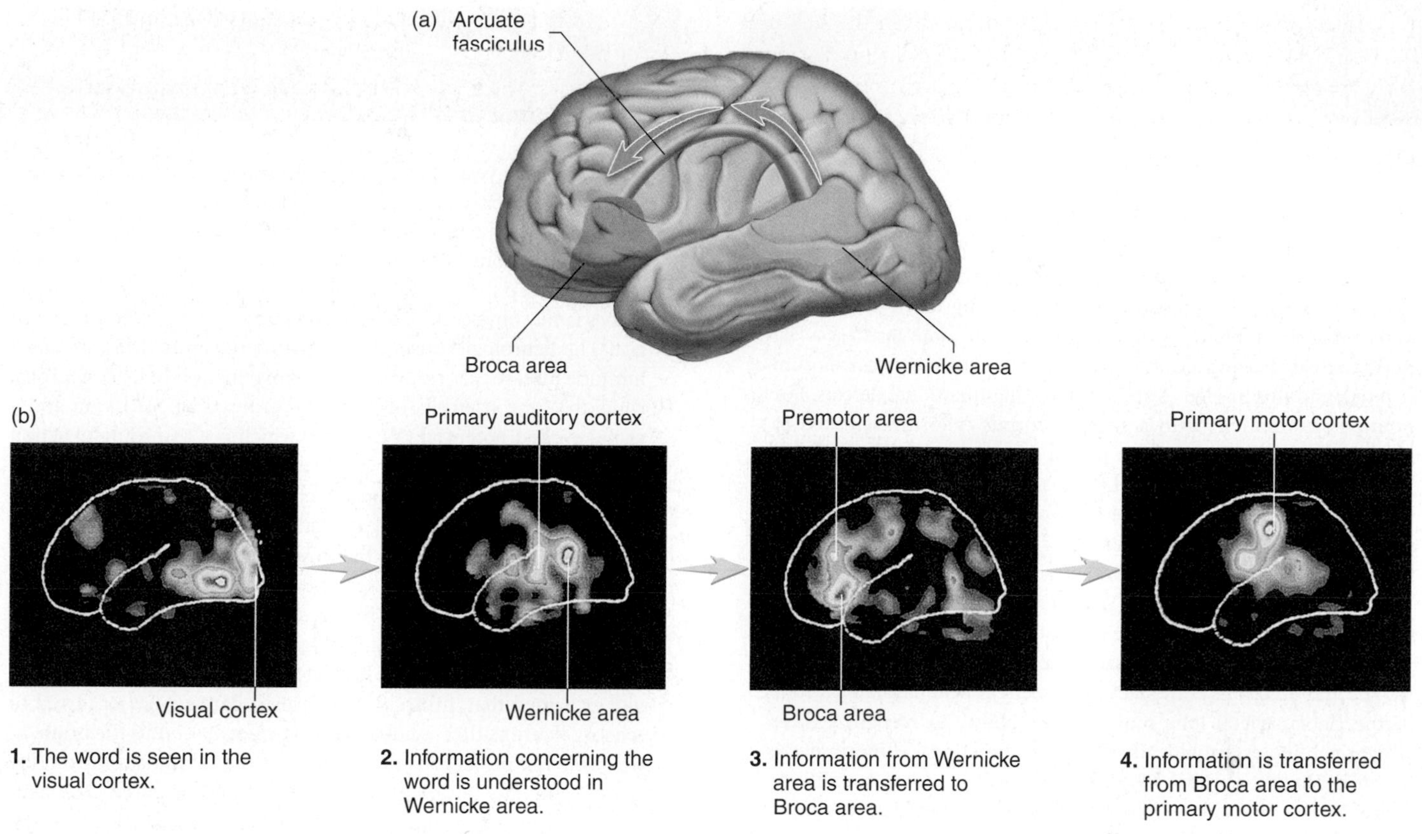
(a) Arcuate fasciculus
Broca area
Wernicke area
(b)
Primary auditory cortex
Premotor area
Primary motor cortex
Visual cortex
Wernicke area
Broca area
1. The word is seen in the visual cortex.
2. Information concerning the word is understood in Wernicke area.
3. Information from Wernicke area is transferred to Broca area.
4. Information is transferred from Broca area to the primary motor cortex.

FIGURE 14.20 Demonstration of Cortical Activities During Speech
(*a*) The arcuate fasciculus connects the two key areas involved in speech, the Broca area and the Wernicke area. The orange arrows indicate direction of action potentials. (*b*) Steps 1–4 show the pathway followed when reading words aloud. Positron emission tomography (PET) scans show the areas of the brain that are most active during various phases of speech. The highest level of brain activity is indicated in red, with successively lower levels represented by yellow, green, and blue.

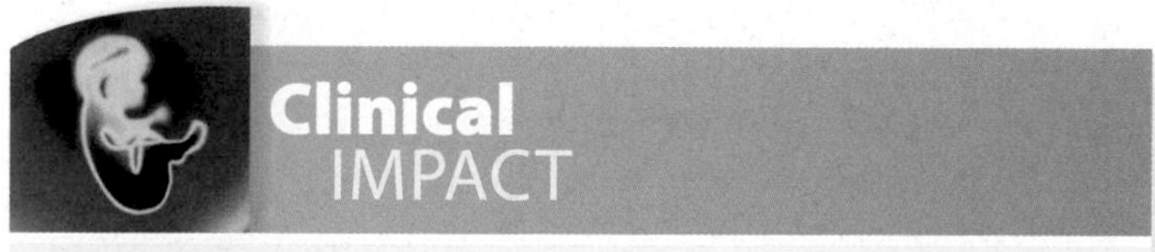
Clinical
IMPACT

Aphasia

Damage to the language areas of the cerebral cortex may result in **aphasia** (ă-fā′zē-ă), absent or defective speech or language comprehension. The most common cause is a stroke. It is estimated that 25–40% of stroke survivors exhibit aphasia.

There are several types of aphasia, depending on the site of the lesion. **Receptive aphasia** (Wernicke aphasia) is caused by a lesion in the Wernicke area. This condition is characterized by defective auditory and visual comprehension of language, defective naming of objects, and repetition of spoken sentences. Both **jargon aphasia,** in which a person may speak fluently but unintelligibly, and **conduction aphasia,** characterized by poor repetition but relatively good comprehension, can result from a lesion in the tracts between the Wernicke and Broca areas. **Anomic** (ă-nō′mik) **aphasia,** caused by the isolation of the Wernicke area from the parietal or temporal association areas, is characterized by fluent but circular speech resulting from poor word-finding ability. **Expressive aphasia** (Broca aphasia), caused by a lesion in the Broca area, is characterized by hesitant and distorted speech.

by the cortex of one hemisphere is shared with the other through connections between the two hemispheres called **commissures** (kom′i-shūrz; joining together). The largest of these commissures is the **corpus callosum** (kōr′pūs kă-lō′sŭm), which is a broad band of tracts at the base of the longitudinal fissure (see figure 13.1).

Language and perhaps other functions, such as artistic activities, are not shared equally between the left and right cerebral hemispheres. The left hemisphere is more involved in analytical skills, such as mathematics and speech. The right hemisphere is involved in activities such as spatial perception, the recognition of faces, and musical ability.

Brain Waves and Sleep

Different levels of consciousness can be revealed by different patterns of electrical activity in the brain. Electrodes placed on a person's scalp and attached to a recording device can record the brain's electrical activity, producing an **electroencephalogram** (ē-lek′trō-en-sef′ă-lō-gram; **EEG;** figure 14.21). These electrodes are not sensitive enough to detect individual action potentials, but they can detect the simultaneous action potentials in large numbers of neurons. As a result, the EEG displays wavelike patterns known as **brain waves.** Brain waves are produced continuously, but their intensity and frequency differ from time to time based on the state of brain activity. Most of the time, EEG patterns from a given individual are irregular, with no

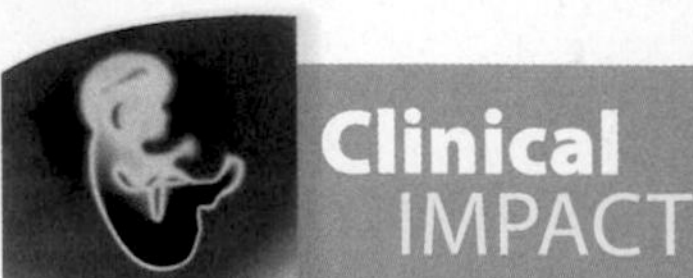

Clinical IMPACT

Hemisphere Dominance

For most functions and in most people, dominance of one cerebral hemisphere over the other is probably not very important, because the two hemispheres are in constant communication through the corpus callosum, literally allowing the right hand to know what the left hand is doing. However, in some cases, as a last resort to treat epileptic seizures, people have had their corpus callosum severed, and this has led to some interesting functional defects. For example, if a patient with a severed corpus callosum is asked to reach behind a screen to touch one of several items with one hand without being able to see it and then is asked to point out the same object with the other hand, the person cannot do it. Tactile information from the left hand enters the right somatic sensory cortex, but that information is not transferred to the left hemisphere, which controls the right hand. As a result, the left hemisphere cannot direct the right hand to the correct object.

Additional evidence that each brain hemisphere is responsible for specific functions is the fact that a person who has had a stroke in the right parietal lobe may lose the ability to recognize faces while retaining essentially all other brain functions. A more severe lesion in the left parietal lobe can take away a person's ability to recognize the right side of his or her body or even to identify simple objects, a defect called **amorphosynthesis** (ă-mōr′fō-sin′thĕ-sis). Some people with a similar lesion in the right cerebral hemisphere tend to ignore the left half of the world, including the left half of their own bodies. These people may completely ignore a person who is to their left but react normally when the person moves to their right. They may also fail to dress the left half of their bodies or to eat the food on the left half of their plates.

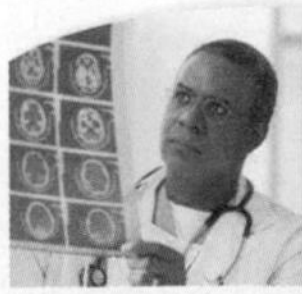

Case STUDY — Multiple Sclerosis

Betty, a 32-year-old woman, woke up one morning with weakness in her lower limbs. By that afternoon, she had become completely exhausted, with a familiar ache in her left eye and tingling sensations in her fingers. At the end of the day, she could barely stand up, and her left eye was blurry and increasingly painful. Betty's family physician made an appointment for her with a neurologist. The neurologist suspected **multiple sclerosis (MS),** an autoimmune disease that results in the demyelination of CNS neurons, which become sclerotic, or hard. He ordered an MRI scan and a **visual evoked potential (VEP) test,** which is a recording of action potentials in the optic nerve "evoked" in response to visual stimuli. During the VEP test, electrodes were placed on Betty's scalp, and she was asked to respond to visual stimuli, first with her normal eye covered by an eye patch and then with her normal eye uncovered.

The MRI scan showed lesions in the optic nerve and white matter of the brain, and the VEP test showed optic nerve dysfunction. These results confirmed a diagnosis of multiple sclerosis accompanied by **optic neuritis,** or inflammation of the optic nerve. Betty's neurologist explained that, in multiple sclerosis, damage has occurred to sensory neurons that conduct sensory information to the brain, as well as to motor neurons that conduct motor stimulation from the brain to the muscles. Betty asked if she could be cured. The neurologist replied that she would experience periods of remission, which might be interrupted by symptomatic periods. He also warned that, with each successive episode, neurons might become progressively more damaged.

Predict 9

a. What effect does demyelination have on action potential propagation along axons? How does this explain blurred vision?

b. How would you explain the weakness and tingling that Betty experienced?

particular pattern, because, although the normal brain is active, most of its electrical activity is not synchronous. At other times, however, specific patterns can be detected.

The different levels of consciousness in an awake and a sleeping person are marked by different types of brain wave patterns. These regular patterns are classified as alpha, beta, theta, or delta waves (figure 14.21). **Alpha waves** are observed in a normal person who is awake but in a quiet, resting state with the eyes closed. **Beta waves** have a higher frequency than alpha waves and occur during intense mental activity. **Theta waves** usually occur in children, but they can also occur in adults who are experiencing frustration or who have certain brain disorders. **Delta waves** occur in infants, in patients with severe brain disorders, and in people who are in deep sleep.

Brain wave patterns vary during the four stages of sleep (figure 14.21*c,d*). During the beginning of sleep, a rapid transition takes place from a beta rhythm to an alpha rhythm. As sleep deepens, progressively more delta waves occur. A sleeping person arouses several times during a period of sleep. Dreaming occurs during periods when eye movement can be observed in a sleeping person; this state is called **rapid eye movement (REM) sleep.**

Distinct types of EEG patterns can be detected in patients with specific brain disorders, such as epileptic seizures. Similar to EEG patterns, neurologists can also measure **evoked potentials,** which are electrical responses caused by light, sound, or somatosensory stimuli. EEGs and evoked potentials are useful tools to diagnose neurological disorders and determine the appropriate treatment.

ASSESS YOUR PROGRESS

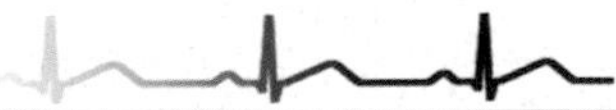

29. *Trace the sequence of events that must occur for a person to repeat a word that he or she hears.*

30. *Name the largest pathway that connects the right and left cerebral hemispheres.*

31. *What functions are localized in each cerebral hemisphere?*

32. *What does an EEG measure?*

33. *What conditions produce alpha, beta, theta, and delta waves, respectively?*

34. *Explain how brain waves change during sleep. What is REM sleep?*

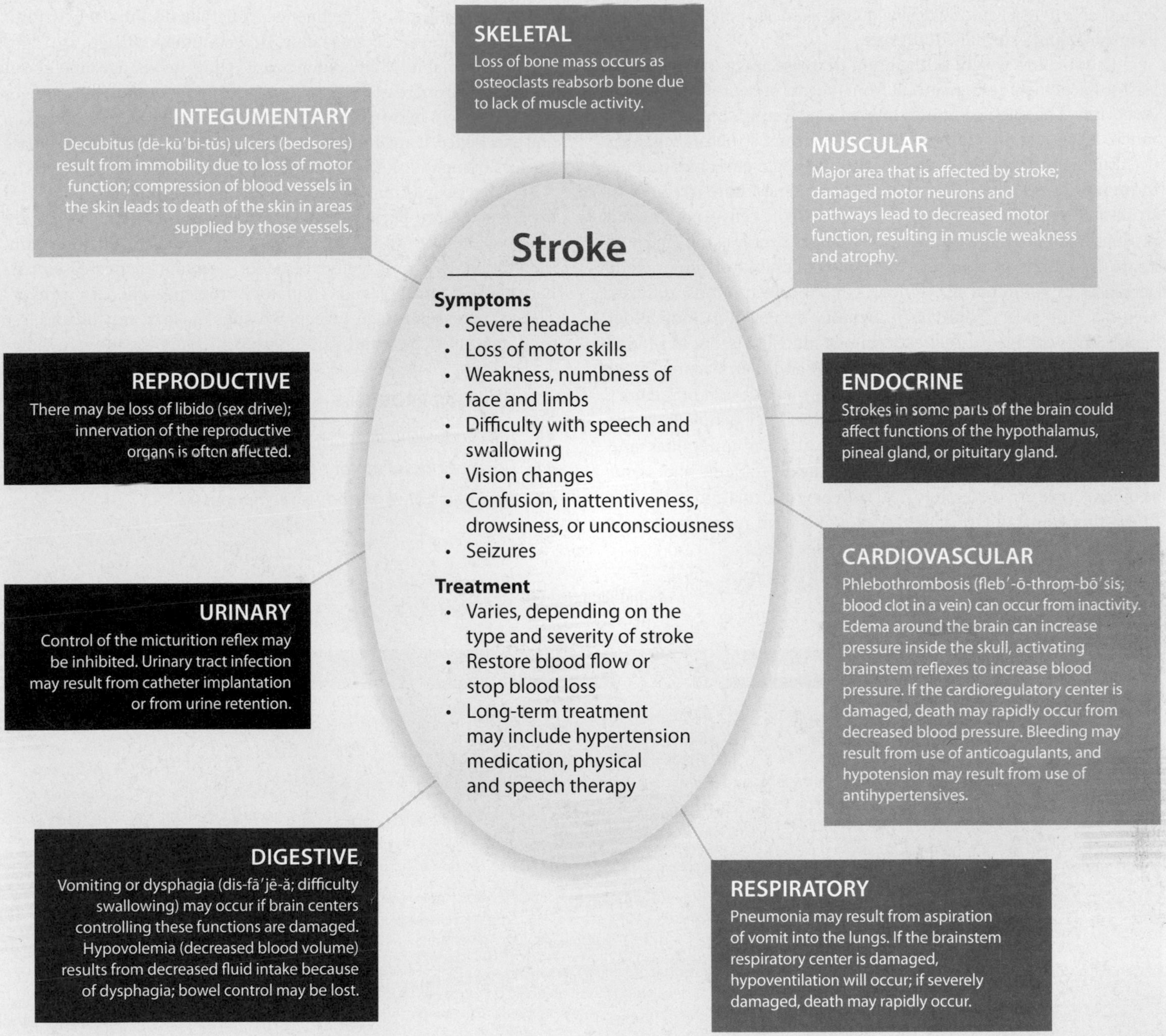

SKELETAL
Loss of bone mass occurs as osteoclasts reabsorb bone due to lack of muscle activity.

INTEGUMENTARY
Decubitus (dē-kū′bi-tŭs) ulcers (bedsores) result from immobility due to loss of motor function; compression of blood vessels in the skin leads to death of the skin in areas supplied by those vessels.

MUSCULAR
Major area that is affected by stroke; damaged motor neurons and pathways lead to decreased motor function, resulting in muscle weakness and atrophy.

Stroke

Symptoms
- Severe headache
- Loss of motor skills
- Weakness, numbness of face and limbs
- Difficulty with speech and swallowing
- Vision changes
- Confusion, inattentiveness, drowsiness, or unconsciousness
- Seizures

Treatment
- Varies, depending on the type and severity of stroke
- Restore blood flow or stop blood loss
- Long-term treatment may include hypertension medication, physical and speech therapy

REPRODUCTIVE
There may be loss of libido (sex drive); innervation of the reproductive organs is often affected.

ENDOCRINE
Strokes in some parts of the brain could affect functions of the hypothalamus, pineal gland, or pituitary gland.

URINARY
Control of the micturition reflex may be inhibited. Urinary tract infection may result from catheter implantation or from urine retention.

CARDIOVASCULAR
Phlebothrombosis (fleb′-ō-throm-bō′sis; blood clot in a vein) can occur from inactivity. Edema around the brain can increase pressure inside the skull, activating brainstem reflexes to increase blood pressure. If the cardioregulatory center is damaged, death may rapidly occur from decreased blood pressure. Bleeding may result from use of anticoagulants, and hypotension may result from use of antihypertensives.

DIGESTIVE
Vomiting or dysphagia (dis-fā′jē-ă; difficulty swallowing) may occur if brain centers controlling these functions are damaged. Hypovolemia (decreased blood volume) results from decreased fluid intake because of dysphagia; bowel control may be lost.

RESPIRATORY
Pneumonia may result from aspiration of vomit into the lungs. If the brainstem respiratory center is damaged, hypoventilation will occur; if severely damaged, death may rapidly occur.

General responses to neurological damage include drowsiness, disorientation, inattentiveness, loss of consciousness, and even seizures. Depression, due to either neurological damage or discouragement, is also common. Slight dilation of the pupils, short and shallow respiration, and increased pulse rate and blood pressure are all signs of Scott's anxiety about his current condition and his immediate future. Because he lost consciousness, Scott would not remember the last few minutes of what he was watching on television when he had his stroke. People in these circumstances are often worried about how they are going to deal with work tomorrow. They often have no idea that the motor and sensory losses may be permanent or that a long stretch of therapy and rehabilitation lies ahead.

Predict 10

Given that Scott exhibited weakness in his right limbs and loss of pain and temperature sensation in his right lower limb and left side of his face, which side of the brainstem was most severely affected by the stroke? Explain your answer.

reflex, the more it is affected by age. As reflexes slow, older people are less able to react automatically, quickly, and accurately to changes in internal and external conditions.

The size and weight of the brain decrease as a person ages. At least some of these changes result from the loss of neurons within the cerebrum. The remaining neurons can apparently compensate for much of this loss. In addition to loss of neurons, structural changes occur in the remaining neurons. Neuron plasma membranes become more rigid, the endoplasmic reticulum becomes more irregular in structure, neurofibrillary tangles develop in the cells, and amyloid plaques form in synapses. All these changes decrease the ability of neurons to function. Age-related changes in brain function include decreases in voluntary movement, conscious sensations, reflexes, memory, and sleep. Short-term memory decreases in most older people. This change varies greatly among individuals, but in general such changes are slow until about age 60 and then become more rapid, especially after age 70. However, the total amount of memory loss is normally not great for most people. Older people have the most difficulty assimilating information that is unfamiliar and presented verbally and rapidly. Some of these problems may occur as older people are required to deal with new information in the face of existing, contradictory memories. Long-term memory appears to be unaffected or even improved in older people.

As with short-term memory, thinking, which includes problem solving, planning, and intelligence, generally declines slowly until age 60 but more rapidly thereafter. However, these changes are slight and quite variable. Many older people show no change, and about 10% show an increase in thinking ability. Many of these changes are affected by a person's background, education, health, motivation, and experience. It appears that continued mental activity can decrease the loss of mental skills with aging.

Older people tend to require more time to fall asleep and experience more periods of waking during the night, which are also of greater duration. Factors that can affect sleep include pain, indigestion, rhythmic leg movements, sleep apnea, decreased urinary bladder capacity, and circulatory problems. There is, on average, an increase in stage 1 sleep, which is the least restful, and less time spent in stage 4 and REM sleep, which are the most restful.

ASSESS YOUR PROGRESS

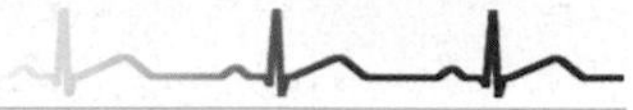

40. *How does aging affect sensory function?*

41. *How does loss of motor neurons affect muscle mass?*

42. *Does aging always produce memory loss?*

Diseases and Disorders

TABLE 14.5 Integrated CNS Disorders*

Condition	Description
CNS MOVEMENT DISORDERS	
Dyskinesias	General term for movement disorders; often involve basal nuclei
Parkinson disease	Caused by a lesion in basal nuclei; characterized by muscular rigidity, resting tremor, general lack of movement, and a slow, shuffling gait
Huntington disease	Dominant hereditary disorder; causes progressive degeneration of basal nuclei; characterized by involuntary movements
Cerebral palsy	General term for defects in motor functions or coordination due to brain damage caused by abnormal development or injury during pregnancy, birth, or first years of life; symptoms include increased muscle tone, resting tremors, difficulty speaking and swallowing, and slow, writhing, aimless movements
Cerebellar lesions	Often result in movements that are ataxic (jerky) and dysmetric (poor estimation of movement distance and force, such as overshooting when reaching for an object); can cause an intention tremor (i.e., the more one tries to control a movement, the greater the tremor becomes), in contrast to the resting tremor of Parkinson disease, which is reduced during purposeful movement
OTHER CNS DISORDERS	
Alzheimer disease	Mental deterioration, or dementia; usually affects older people; involves loss of neurons in the cerebral cortex; symptoms include general intellectual deficiency, memory loss, short attention span, moodiness, disorientation, and irritability
Tay-Sachs disease	Hereditary lipid-storage disorder of infants; primarily affects the CNS; symptoms include paralysis, blindness, and death
Epilepsy	Seizures involving a sudden, massive neuronal discharge, which may result in involuntary muscle contractions (convulsions)
Headaches	May be due to inflammation, dental irritations, eye disorders, tension in head and neck muscles, or unknown causes
Dyslexia	Reading deficiency; reading level is below that expected based on overall intelligence; symptoms vary but include confusion between similar letters, such as *b* and *d*; cause is not known
Stroke	Caused by bleeding in the brain or a clot or spasm blocking cerebral blood vessels, resulting in a local area of cell death; symptoms include loss of speech, numbness, and paralysis

* For more CNS disorders, see "Diseases and Disorders" tables in chapters 12 and 13.

Go to www.mhhe.com/seeley10 for additional information on these pathologies.

Answer

Learn to Predict ❮ From page 461

The students touched a hot (55°C) object with one hand and a cold (5°C) object with the other hand. We learned in this chapter that thermoreceptors allow us to distinguish between hot and cold stimuli. In section 14.1 we read that cold receptors respond to temperatures between 12°C and 35°C, whereas warm receptors respond to temperatures between 25°C and 47°C. However, temperatures both below 12°C and above 47°C stimulate pain receptors. Therefore, the students would most likely not be able to distinguish between the cold and hot objects. They would just feel pain.

The second part of the question asked us to identify the part of the CNS involved in making this distinction. In chapter 12 we learned that sensory neurons carry information from receptors in the skin to the spinal cord, which is then conveyed to the brain. This chapter explained more specifically that the sensation of pain and temperature are carried through the spinothalamic system to the thalamus, then to the primary sensory cortex. Chapter 13 taught us that the primary sensory cortex is located in the postcentral gyrus of the cerebrum. Stimulation of neurons in specific areas of the primary sensory cortex allows a person to perceive a particular sensation and locate the site of stimulation. The ability to recognize the stimulus also involves association areas of the cerebrum near the primary sensory cortex.

Answers to the rest of this chapter's Predict questions are in Appendix G.

Summary

14.1 Sensation (p. 462)

1. Sensation requires a stimulus, a sensory receptor, and conduction of an action potential to the CNS.
2. Perception is the conscious awareness of stimuli received by sensory receptors.
3. The senses include general senses (somatic and visceral) and special senses.
4. The somatic senses include touch, pressure, temperature, proprioception, and pain.
5. The visceral senses are primarily pain and pressure.
6. The special senses are smell, taste, sight, hearing, and balance.

Sensory Receptors

1. Receptors include mechanoreceptors, chemoreceptors, thermoreceptors, photoreceptors, and nociceptors.
2. Free nerve endings detect light touch, pain, itch, tickle, and temperature.
3. Merkel disks respond to light touch and superficial pressure.
4. Hair follicle receptors wrap around the hair follicle and are involved in the sensation of light touch when the hair is bent.
5. Pacinian corpuscles, located in the subcutaneous tissue, detect pressure. In joints, they serve a proprioceptive function.
6. Meissner corpuscles, located in the dermis, are responsible for two-point discriminative touch.
7. Ruffini end organs are involved in continuous touch or pressure.
8. Muscle spindles, located in skeletal muscle, are proprioceptors.
9. Golgi tendon organs, embedded in tendons, respond to changes in tension.
10. A stimulus produces a receptor potential in a sensory receptor. Primary receptors have axons that transmit action potentials toward the CNS. Secondary receptors have no axons but release neurotransmitters.
11. Adaptation is decreased sensitivity to a continued stimulus. Tonic receptors adapt slowly; phasic receptors adapt rapidly.

Sensory Pathways

1. Ascending pathways carry conscious and unconscious sensations. The two major ascending systems are the anterolateral and the dorsal-column/medial-lemniscal systems.
2. In the anterolateral system,
   - The spinothalamic tract carries pain, temperature, light touch, pressure, tickle, and itch sensations.
   - The spinothalamic tracts are formed by primary neurons that enter the spinal cord and synapse with secondary neurons. The secondary neurons conducting pain and temperature sensations cross the spinal cord and ascend to the thalamus, where they synapse with tertiary neurons that project to the somatic sensory cortex. Axons conducting other sensations, such as light touch or itch, may ascend several spinal cord levels before crossing.
   - The ascending neurons of the spinoreticular tract ascend both contralaterally and ipsilaterally, project to the reticular formation, and influence the level of consciousness.
   - The ascending neurons of the spinomesencephalic tract carry action potentials from cutaneous pain receptors. The spinomesencephalic tract also contributes to eye reflexes.
3. The dorsal-column/medial-lemniscal system carries the sensations of two-point discrimination, proprioception, pressure, and vibration. Primary neurons enter the spinal cord and ascend to the medulla, where they synapse with secondary neurons. Secondary neurons cross over and project to the thalamus. Tertiary neurons extend from there to the somatic sensory cortex.
4. The trigeminothalamic tract carries sensory information from the face, nose, and mouth.
5. In the spinocerebellar system and other tracts,
   - The spinocerebellar tracts carry unconscious proprioception to the cerebellum from the same side of the body.

- Neurons of the dorsal-column/medial-lemniscal system synapse with the neurons that carry proprioceptive information to the cerebellum.

6. Descending pathways can reduce conscious perception of sensations.

Sensory Areas of the Cerebral Cortex

1. Sensory pathways project to primary sensory areas in the cerebral cortex.
2. Sensory areas are organized topographically in the primary somatic sensory cortex in the postcentral gyrus.

Sensory Processing

Association areas of the cerebral cortex process sensory input from the primary sensory areas.

14.2 Control of Skeletal Muscles (p. 475)

1. Upper motor neurons are located in the cerebral cortex, cerebellum, and brainstem. Lower motor neurons are found in the cranial nuclei or the anterior horn of the spinal cord gray matter.
2. Upper motor neurons in the cerebral cortex and other brain areas project to lower motor neurons in the brainstem and spinal cord.

Motor Areas of the Cerebral Cortex

1. The primary motor cortex, premotor area, and prefrontal area are staging areas for motor function.
2. The motor cortex is organized topographically in the precentral gyrus.

Motor Pathways

1. The direct pathways maintain muscle tone and control fine, skilled movements in the face and distal limbs. The indirect pathways control conscious and unconscious muscle movements in the trunk and proximal limbs.
2. The corticospinal tracts control muscle movements below the head.
   - About 75–85% of the upper motor neurons of the corticospinal tracts cross over in the medulla to form the lateral corticospinal tracts in the spinal cord.
   - The remaining upper motor neurons pass through the medulla oblongata to form the anterior corticospinal tracts, which cross over in the spinal cord.
   - The upper motor neurons of both tracts synapse with interneurons, which then synapse with lower motor neurons in the spinal cord.
3. The corticobulbar tracts innervate the head muscles. Upper motor neurons synapse with interneurons in the reticular formation, which in turn synapse with lower motor neurons in the cranial nerve nuclei.
4. The indirect pathways include the rubrospinal, vestibulospinal, and reticulospinal tracts, as well as fibers from the basal nuclei.
5. The indirect pathways are involved in conscious and unconscious trunk and proximal limb muscle movements, posture, and balance.

Modifying and Refining Motor Activities

1. Basal nuclei are important in planning, organizing, and coordinating motor movements and posture.
2. The cerebellum has three parts.
   - The vestibulocerebellum controls balance and eye movement.
   - The spinocerebellum has a comparator function that corrects discrepancies between intended movements and actual movements.
   - The cerebrocerebellum can "learn" highly specific complex motor activities.

14.3 Brainstem Functions (p. 482)

The brainstem contains nuclei for cranial nerves III–X and XII and nuclei of the reticular formation.

Sensory Input Projecting Through the Brainstem

The brainstem receives sensory input from ascending spinal cord pathways and from the axons of cranial nerves.

RAS Functions of the Brainstem

Collateral branches of cranial nerves II, V, and VIII project to the reticular activating system (RAS) of the brainstem, where they stimulate wakefulness and consciousness.

Motor Output and Reflexes Projecting Through the Brainstem

1. Descending spinal pathways either pass directly through the brainstem or synapse with brainstem nuclei.
2. The brainstem controls several somatic motor and parasympathetic reflexes.

Vital Functions Controlled in the Brainstem

The brainstem controls many vital functions, including heart rate, blood pressure, and respiration.

14.4 Higher Brain Functions (p. 484)

Speech

1. The speech area is in the left cerebral cortex in most people.
2. The Wernicke area comprehends and formulates speech.
3. The Broca area receives input from the Wernicke area and sends impulses to the premotor and motor areas, which cause the muscle movements required for speech.

Communication Between the Right and Left Hemispheres

1. Each cerebral hemisphere controls and receives input from the opposite side of the body.
2. The right and left hemispheres are connected by commissures. The largest commissure is the corpus callosum, which allows the sharing of information between hemispheres.
3. In most people, the left hemisphere is dominant, controlling speech and analytical skills. The right hemisphere controls spatial and musical abilities.

Brain Waves and Sleep

1. Electroencephalograms (EEGs) record the electrical activity of the brain as alpha, beta, theta, and delta waves.
2. Some brain disorders can be detected with EEGs.
3. Sleep patterns are characterized by specific EEGs.

Memory

1. Three stages of memory exist: working, short-term, and long-term.
2. Short-term memory requires long-term potentiation.
3. Long-term memory is converted from short-term by consolidation.
4. The two types of memory are declarative and procedural.

Limbic System

1. The limbic system includes the olfactory cortex, deep cortical regions, and nuclei.
2. The limbic system is involved with emotions, motivation, mood, visceral functions, and memory. Olfactory stimulation is a major influence.

14.5 Effects of Aging on the Nervous System (p. 490)

1. A general decline in sensory and motor functions occurs as a person ages.
2. Short-term memory decreases in most older people.
3. Thinking ability does not decrease in most older people.

REVIEW AND COMPREHENSION

1. Nociceptors respond to
   a. changes in temperature at the site of the receptor.
   b. compression, bending, or stretching of cells.
   c. painful mechanical, chemical, or thermal stimuli.
   d. light striking a receptor cell.
2. Which of these types of sensory receptors respond to pain, itch, tickle, and temperature?
   a. Merkel disks
   b. Meissner corpuscles
   c. Ruffini end organs
   d. free nerve endings
   e. Pacinian corpuscles
3. Which of these types of sensory receptors are involved with proprioception?
   a. free nerve endings
   b. Golgi tendon organs
   c. muscle spindles
   d. Pacinian corpuscles
   e. All of these are correct.
4. The sensory receptors in the dermis and hypodermis responsible for sensing continuous touch or pressure are
   a. Merkel disks.
   b. Meissner corpuscles.
   c. Ruffini end organs.
   d. free nerve endings.
   e. Pacinian corpuscles.
5. Decreased sensitivity to a continued stimulus is called
   a. adaptation.
   b. projection.
   c. translation.
   d. conduction.
   e. phantom pain.
6. Secondary neurons in the spinothalamic tracts synapse with tertiary neurons in the
   a. medulla oblongata.
   b. gray matter of the spinal cord.
   c. cerebellum.
   d. thalamus.
   e. midbrain.
7. If the spinothalamic tract on the right side of the spinal cord is severed,
   a. pain sensations below the damaged area on the right side are eliminated.
   b. pain sensations below the damaged area on the left side are eliminated.
   c. temperature sensations are unaffected.
   d. neither pain sensations nor temperature sensations are affected.
8. Fibers of the dorsal-column/medial-lemniscal system
   a. carry the sensations of two-point discrimination, proprioception, pressure, and vibration.
   b. cross to the opposite side in the medulla oblongata.
   c. are divided into the fasciculus gracilis and the fasciculus cuneatus in the spinal cord.
   d. include secondary neurons that exit the medulla and synapse in the thalamus.
   e. All of these are correct.
9. Tertiary neurons in both the spinothalamic tract and the dorsal-column/medial-lemniscal system
   a. project to the somatic sensory cortex.
   b. cross to the opposite side in the medulla oblongata.
   c. are found in the spinal cord.
   d. connect to quaternary neurons in the thalamus.
   e. are part of a descending pathway.
10. Unlike the anterolateral and dorsal-column/medial-lemniscal systems, the spinocerebellar tracts
    a. are descending tracts.
    b. transmit information from the same side of the body as the side of the CNS to which they project.
    c. have four neurons in each pathway.
    d. carry only pain sensations.
    e. have primary neurons that synapse in the thalamus.
11. General sensory inputs (pain, pressure, temperature) to the cerebrum end in the
    a. precentral gyrus.
    b. postcentral gyrus.
    c. central sulcus.
    d. corpus callosum.
    e. arachnoid mater.
12. Neurons from which area of the body occupy the greatest area of the somatic sensory cortex?
    a. foot
    b. leg
    c. torso
    d. arm
    e. face
13. A cutaneous nerve to the hand is severed at the elbow. The distal end of the nerve at the elbow is then stimulated. The person reports
    a. no sensation because the receptors are gone.
    b. a sensation only in the region of the elbow.
    c. a sensation "projected" to the hand.
    d. a vague sensation on the side of the body containing the cut nerve.
14. Which of these areas of the cerebral cortex is involved in providing the motivation and foresight to plan and initiate movements?
    a. primary motor cortex
    b. primary somatic sensory cortex
    c. prefrontal area
    d. premotor area
    e. basal nuclei
15. Which of these pathways is *not* an ascending (sensory) pathway?
    a. spinothalamic tract
    b. corticospinal tract
    c. dorsal-column/medial-lemniscal system
    d. trigeminothalamic tract
    e. spinocerebellar tract
16. The ________ tracts innervate the head muscles.
    a. corticospinal
    b. rubrospinal
    c. vestibulospinal
    d. corticobulbar
    e. dorsal-column/medial-lemniscal

17. Most fibers of the corticospinal tract
   a. decussate in the medulla oblongata.
   b. synapse in the pons.
   c. descend in the rubrospinal tract.
   d. begin in the cerebellum.
18. A person with a spinal cord injury is suffering from paresis (partial paralysis) in the right lower limb. Which of these pathways is probably involved?
   a. left lateral corticospinal tract
   b. right lateral corticospinal tract
   c. left dorsal-column/medial-lemniscal system
   d. right dorsal-column/medial-lemniscal system
19. Which of these pathways is *not* an indirect (extrapyramidal) pathway?
   a. reticulospinal tract
   b. corticobulbar tract
   c. rubrospinal tract
   d. vestibulospinal tract
20. The indirect (extrapyramidal) system is concerned with
   a. posture.
   b. trunk movements.
   c. proximal limb movements.
   d. All of these are correct.
21. The major effect of the basal nuclei is
   a. to act as a comparator for motor coordination.
   b. to decrease muscle tone and inhibit unwanted muscular activity.
   c. to affect emotions and emotional responses to odors.
   d. to modulate pain sensations.
22. Which part of the cerebellum is correctly matched with its function?
   a. vestibulocerebellum—planning and learning rapid, complex movements
   b. spinocerebellum—comparator function
   c. cerebrocerebellum—balance
   d. None of these are correct.
23. Given the following events:
   1. Action potentials from the cerebellum go to the motor cortex and spinal cord.
   2. Action potentials from the motor cortex go to lower motor neurons and the cerebellum.
   3. Action potentials from proprioceptors go to the cerebellum.

   Arrange the events in the order they occur in the cerebellar comparator function.
   a. 1,2,3 b. 1,3,2 c. 2,1,3 d. 2,3,1 e. 3,2,1
24. The brainstem
   a. consists of ascending and descending pathways.
   b. contains cranial nerve nuclei III–X and XII.
   c. has nuclei and connections that form the reticular activating system.
   d. has many important reflexes, some of which are necessary for survival.
   e. has all of these features.
25. Given these areas of the cerebral cortex:
   1. Broca area
   2. premotor area
   3. primary motor cortex
   4. Wernicke area

   If a person hears and understands a word and then says the word out loud, in what order are the areas used?
   a. 1,4,2,3 b. 1,4,3,2 c. 3,1,4,2 d. 4,1,2,3 e. 4,1,3,2
26. The main connection between the right and left hemispheres of the cerebrum is the
   a. intermediate mass.
   b. corpus callosum.
   c. vermis.
   d. unmyelinated nuclei.
   e. thalamus.
27. Which of these activities is associated with the left cerebral hemisphere in most people?
   a. sensory input from the left side of the body
   b. mathematics and speech
   c. spatial perception
   d. recognition of faces
   e. musical ability
28. The limbic system is involved in the control of
   a. sleep and wakefulness.
   b. posture.
   c. higher intellectual processes.
   d. emotion, mood, and sensations of pain or pleasure.
   e. hearing.
29. Long-term memory involves
   a. a change in the cytoskeleton of neurons.
   b. an increased number of dendritic spines.
   c. cAMP signaling pathways that increase gene transcription.
   d. specific protein synthesis.
   e. All of these are correct.
30. Concerning long-term memory,
   a. declarative (explicit) memory involves the development of skills, such as riding a bicycle.
   b. procedural (implicit, or reflexive) memory involves the retention of facts, such as names, dates, or places.
   c. much of declarative (explicit) memory is lost through time.
   d. declarative (explicit) memory is stored primarily in the cerebellum and premotor area of the cerebrum.
   e. All of these are correct.

Answers in Appendix E

CRITICAL THINKING

1. Describe all the sensations and perceptions involved when a woman picks up an apple and bites into it. Explain which of these sensations are special and which are general. What types of receptors are involved?
2. Some student nurses are at a party. Because they love anatomy and physiology so much, they are discussing adaptation of the special senses. They make the following observations:
   a. When entering a room, people easily notice an odor, such as brewing coffee. A few minutes later, the odor might be barely detectable, if at all, no matter how hard they try to smell it.
   b. When entering a room, people can detect the sound of a ticking clock. Later the sound is not noticed until people make a conscious effort to hear it. Then the ticking is easy to hear.

   Explain the basis for each of these observations.
3. A patient suffered a loss of two-point discrimination and proprioception on the right side of the body. Voluntary movement of muscles was not affected, and pain and temperature sensations were normal. Is it possible to conclude that the right side of the spinal cord was damaged?
4. A patient has a lesion in the central core of the spinal cord. Physicians suspect that the fibers that decussate and are associated with the lateral spinothalamic tracts are affected in the area of the lesion. What observations are consistent with that diagnosis?
5. A person who was injured in a car accident exhibits the following symptoms: extreme paresis on the right side, including the arm and leg; reduction of pain sensation on the left side; and normal tactile sensation on both sides. Which tracts are damaged? Where in the spinal cord did the patient suffer tract damage?
6. A patient with a cerebral lesion exhibits loss of fine motor control of the left hand, arm, forearm, and shoulder. All other motor and sensory functions appear to be intact. Describe the location of the lesion as precisely as possible.
7. A patient suffers brain damage in an automobile accident. Physicians suspect that the cerebellum is the part of the brain affected. On the basis of what you know about cerebellar function, how can they confirm that the cerebellum is involved?
8. Woody Knothead was accidentally struck in the head with a baseball bat. He fell to the ground, unconscious. Later, when he regained consciousness, he could not remember any of the events that happened 10 minutes before the accident. Explain. What complications might develop at a later time?
9. Perry is a 93-year-old man who uses his computer to communicate with family and friends and to write poems and essays. One day last week, his daughter noticed that Perry was unable to use the computer keyboard normally with his right hand, and this ability deteriorated further over the next few hours. Perry was also experiencing muscle weakness on the right side, and soon he could not support himself with his right lower limb without using a cane; later in the day, he could hardly move his right lower limb at all. Concerned about a stroke, Perry's daughter took him to the emergency room, where radiographs and an MRI revealed a subdural hematoma. Explain how a subdural hematoma could be responsible for Perry's condition.

Answers in Appendix F

Visit this book's website at **www.mhhe.com/seeley10** for chapter quizzes, interactive learning exercises, and other study tools.

Mc Graw Hill
connect plus+
ANATOMY & PHYSIOLOGY

Anatomy & Physiology | REVEALED®
aprevealed.com

15

Learn to Predict

Freddy, a 67-year-old father and grandfather, was sitting on his couch with his hand over his left ear. His whole family was visiting and the noise level was pretty high in the living room. "What's wrong?" asked his wife. "I never realized how loud our family is," he laughed. For most of his life, Freddy suffered from complete hearing loss in his left ear. Recently, he underwent a surgery that replaced two auditory ossicles in his left ear and now his hearing has been restored. Freddy is hearing his family in a whole new way. After reading about the process of hearing in this chapter, explain the reason for Freddy's hearing loss, how this affected his ability to locate the direction of noises, and how his hearing was restored.

The Special Senses

Historically, physiologists thought humans had just five senses: smell, taste, sight, hearing, and touch. Today they recognize many more. The original sense of "touch" has been categorized into multiple types of *general senses*, including pressure, touch, pain, and others. The general senses are described in chapter 14. Smell, taste, sight, and hearing are now classified as *special senses*. The sense of balance has been added to that category. This chapter describes the five special senses—olfaction, taste, the visual system, hearing, and balance.

Photo: Photograph of an isolated cochlea from the inner ear.

Module 7
Nervous System

Anatomy & Physiology REVEALED
aprevealed.com

15.1 Olfaction

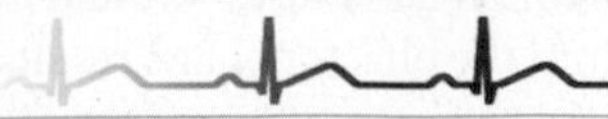

LEARNING OUTCOMES

After reading this section, you should be able to

A. Describe olfactory neurons and explain how airborne molecules can stimulate action potentials in olfactory nerves.

B. Locate the areas of the brain where olfaction is processed.

C. Explain the processes involved in olfactory adaptation.

Olfaction (ol-fak′shŭn), the sense of smell, occurs in response to odors that stimulate sensory receptors in the extreme superior region of the nasal cavity, called the **olfactory region** (figure 15.1*a*). The olfactory region is lined with a specialized epithelium called the **olfactory epithelium.** (The rest of the nasal cavity is involved in respiration, and its major anatomical features are described in chapter 23.)

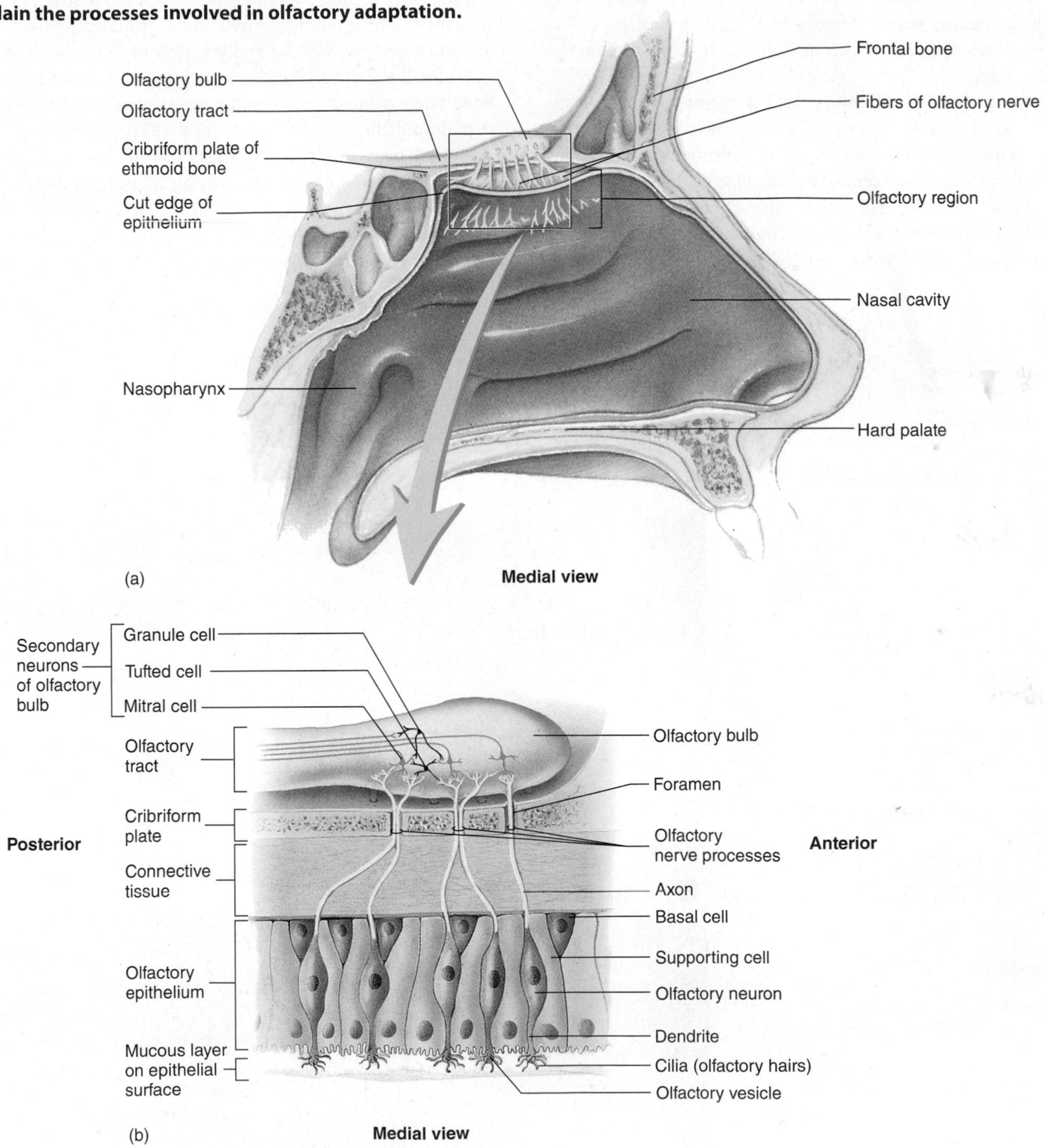

Olfactory bulb
Olfactory tract
Cribriform plate of ethmoid bone
Cut edge of epithelium
Nasopharynx
Frontal bone
Fibers of olfactory nerve
Olfactory region
Nasal cavity
Hard palate
(a)
Medial view
Secondary neurons of olfactory bulb
Granule cell
Tufted cell
Mitral cell
Olfactory tract
Cribriform plate
Connective tissue
Olfactory epithelium
Mucous layer on epithelial surface
Posterior
Olfactory bulb
Foramen
Olfactory nerve processes
Anterior
Axon
Basal cell
Supporting cell
Olfactory neuron
Dendrite
Cilia (olfactory hairs)
Olfactory vesicle
(b)
Medial view

FIGURE 15.1 APR **Olfactory Region, Epithelium, and Bulb**

(*a*) The lateral wall of the nasal cavity (cut in sagittal section), showing the olfactory region and olfactory bulb. (*b*) Close-up of the olfactory epithelium, showing the olfactory nerve processes passing through the cribriform plate and the fine structure of the olfactory bulb.

Olfactory Epithelium and Bulb

The olfactory epithelium contains approximately 10 million **olfactory neurons** (figure 15.1*b*). Olfactory nerves, formed by the axons of these bipolar neurons, project through numerous small foramina in the bony cribriform plate (see chapter 7) to the **olfactory bulbs.** The **axons** synapse with secondary neurons of the olfactory bulb. From the bulbs, **olfactory tracts** project to the cerebral cortex.

The dendrites of olfactory neurons extend to the epithelial surface of the nasal cavity, and their ends are modified into bulbous enlargements called **olfactory vesicles** (figure 15.1*b*). These vesicles possess cilia called **olfactory hairs,** which lie in a thin mucous film on the epithelial surface.

Airborne molecules enter the nasal cavity and are dissolved in the fluid covering the olfactory epithelium. Some of these molecules, referred to as **odorants** (ō′dŏr-ants), bind to transmembrane odorant receptor molecules (chemoreceptors) of the olfactory hair membranes (figure 15.2). A G protein, associated with each odorant receptor, is activated by the binding of the odorant. The α subunit of the activated G protein binds to and activates adenylate cyclase, which in turn catalyzes the formation of cyclic AMP (cAMP) from ATP. cAMP in these cells causes Na^+ and Ca^{2+} channels to open. The influx of ions into the olfactory hairs results in depolarization and the production of action potentials in the olfactory neurons.

Odorant receptors are composed of seven transmembrane polypeptide subunits produced by a large family of genes. Combinations of subunits from approximately 1000 different odorant receptors can be produced. These receptors can react to odorants of different sizes, shapes, and functional groups. These capabilities, together with multiple intracellular pathways involving G proteins, adenylate cyclase, and ion channels, allow for a wide variety of detectable smells—about 4000 for the average person. Some researchers have grouped this wide range of smells into seven primary classes: (1) camphoraceous (e.g., mothballs), (2) musky, (3) floral, (4) pepperminty, (5) ethereal (e.g., fresh pears), (6) pungent, and (7) putrid. However, other studies point to the possibility of as many as 50 primary odors.

In most people, the threshold for detecting odors is extremely low, so very few odorant molecules are required to trigger a response. Apparently, there is rather low specificity in the olfactory epithelium,

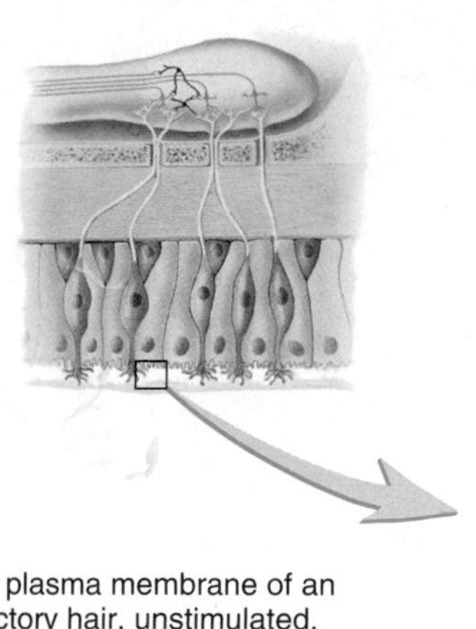

1 The plasma membrane of an olfactory hair, unstimulated. The gated ion channel is closed.

2 An odorant binds to a specific odorant receptor.

3 The associated G protein is activated. The α, β, and γ subunits dissociate. The α subunit of the G protein binds to and activates adenylate cyclase.

4 Adenylate cyclase catalyzes the conversion of ATP to cyclic AMP (cAMP).

5 cAMP opens ion channels, such as Na^+ and Ca^{2+} (not shown) channels.

6 Ions entering the olfactory hair cause depolarization of the neuron.

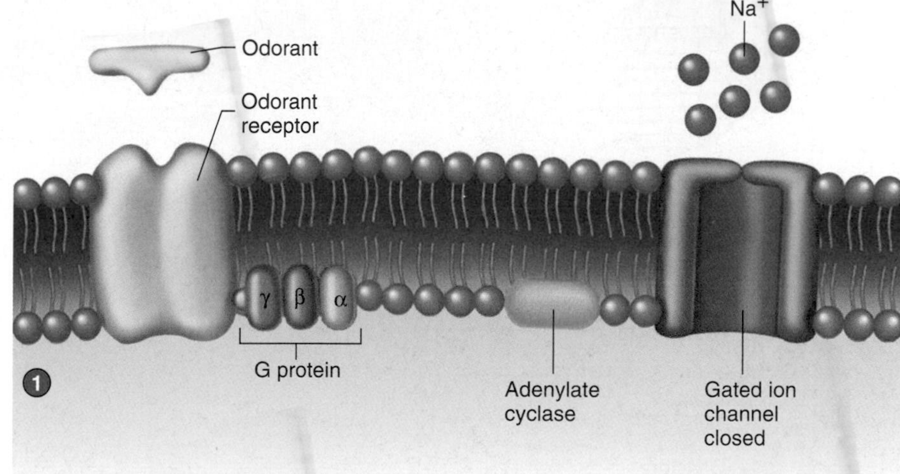
Odorant
Odorant receptor
Na+
γ β α
G protein
1
Adenylate cyclase
Gated ion channel closed

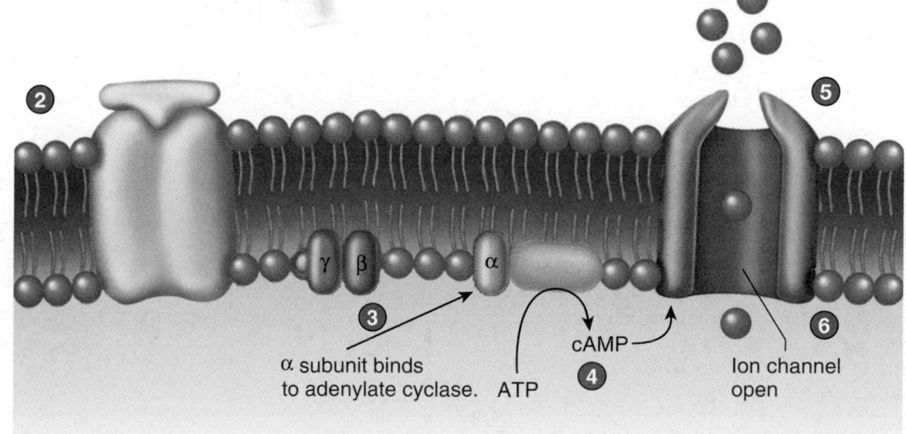
2
5
γ β
α
3
α subunit binds to adenylate cyclase.
ATP
4
cAMP
6
Ion channel open

PROCESS FIGURE 15.2 Action of Odorant Binding to Membrane Receptor of Olfactory Hair

so that a given receptor may react to more than one type of odorant. However, if odorant receptors become saturated with odorants, they no longer respond to odorant molecules. This adaptation makes a person less sensitive to an odorant after being exposed to it for a short time. For example, when you first enter a movie theater, the distinctive odor of popcorn is very noticeable, but it becomes almost unnoticeable after you have been in the theater for a short time.

The primary olfactory neurons have the most exposed nerve endings of any neurons, and they are constantly being replaced. The entire olfactory epithelium, including the olfactory neurons, degenerates and is lost from the surface about every 2 months. Lost olfactory cells are replaced by a proliferation of **basal cells** in the olfactory epithelium. This replacement of olfactory neurons is unique among neurons, most of which are permanent cells that have a very limited ability to replicate (see chapter 4).

Neuronal Pathways for Olfaction

Axons from the olfactory neurons form the olfactory nerves (cranial nerve I), which enter the olfactory bulbs (see figure 15.1*b*), where they synapse with either **mitral** (mī′trăl; triangular) **cells** or **tufted cells.** The mitral and tufted cells relay olfactory information to the brain through the olfactory tracts and synapse with **granule cells** in the olfactory bulb. Olfactory bulb neurons also receive input from nerve cell processes entering the olfactory bulb from the brain. As a result of input from both mitral cells and the brain, olfactory bulb neurons can modify olfactory information before it leaves the olfactory bulb. This modification enhances the adaptation occurring in the odorant receptors.

Olfaction is the only major sensation that is relayed directly to the cerebral cortex without first passing through the thalamus. Each olfactory tract terminates in either an area of the brain called the **olfactory cortex** (figure 15.3) or **secondary olfactory areas.** The olfactory cortex is in the temporal lobe and is involved in the conscious perception of smell. Secondary olfactory areas located in the frontal lobe include the medial and intermediate olfactory areas. The **medial olfactory area** is responsible for visceral and emotional reactions to odors and has connections to the limbic system, through which it connects to the hypothalamus. Axons extend from the **intermediate olfactory area** along the olfactory tract to the bulb and synapse with the olfactory bulb neurons, thus constituting a major mechanism by which sensory information is modulated within the olfactory bulb. The feedback from the intermediate olfactory area is mostly inhibitory, further enhancing the rapid adaptation of the olfactory system. Adaptation at the receptor level, the olfactory bulb level, and the CNS level makes the olfactory system insensitive to an odorant after a brief exposure.

ASSESS **YOUR PROGRESS**

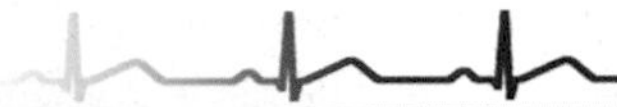

1. *Where are olfactory neurons located? Explain their structure.*
2. *Describe the initiation of an action potential in an olfactory neuron. Name all of the structures and cells that the action potential encounters on its way to the olfactory cortex.*
3. *What is unique about olfactory neurons with respect to replacement?*
4. *Where is the olfactory cortex located? Where are the secondary olfactory areas located, and what are their functions?*

1. Processes of the olfactory nerves, formed by the axons of the olfactory neurons, project through the foramina in the cribriform plate to the olfactory bulb.
2. Axons of neurons in the olfactory bulb project through the olfactory tract to the olfactory cortex or secondary olfactory areas.
3. The olfactory cortex is involved in the conscious perception of smell.
4. The medial olfactory area is involved in the visceral and emotional reaction to odors.
5. The intermediate olfactory area receives input from the medial olfactory area and olfactory cortex.
6. Axons from the intermediate olfactory area project along the olfactory tract to the olfactory bulb. Action potentials carried by those axons modulate the activity of the neurons in the olfactory bulb.

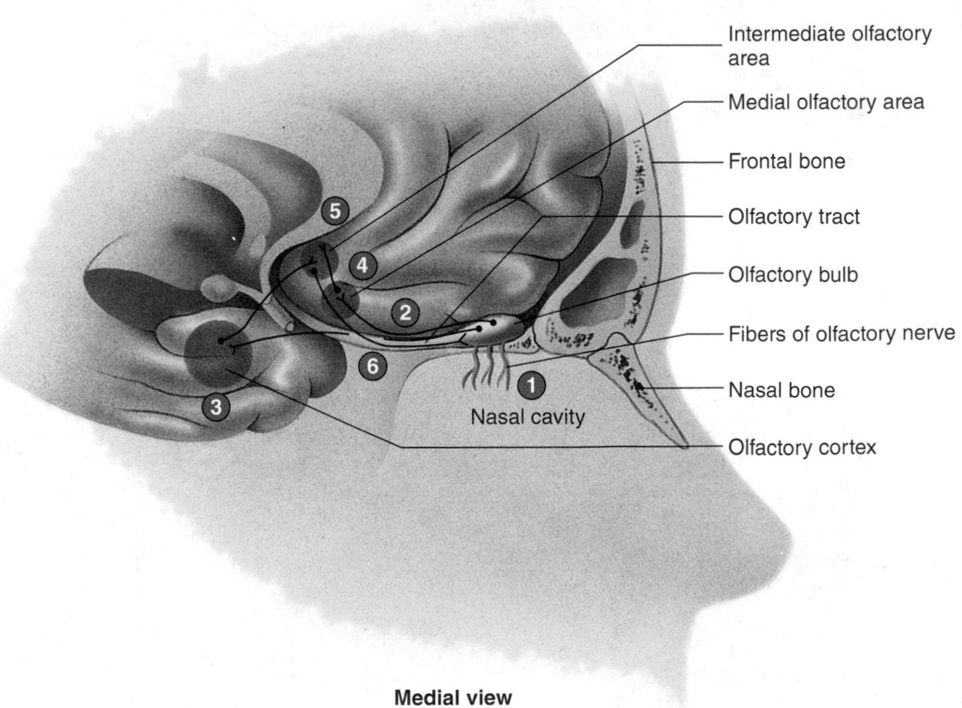
Intermediate olfactory area
Medial olfactory area
Frontal bone
Olfactory tract
Olfactory bulb
Fibers of olfactory nerve
Nasal bone
Olfactory cortex
Nasal cavity
1
2
3
4
5
6
Medial view

PROCESS FIGURE 15.3 Olfactory Cortex and Neuronal Pathways

15.2 Taste

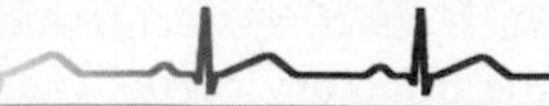

LEARNING OUTCOMES

After reading this section, you should be able to

A. List the types of papillae and outline the structure and function of a taste bud.

B. List the five major tastants. Relate the relationship of taste to smell.

C. Describe the formation of an action potential in a taste cell for each type of tastant.

D. Trace the pathway of the action potential to the taste area of the cerebral cortex.

The sensory structures that detect **taste,** or *gustatory,* stimuli are the **taste buds.** Most taste buds are in specialized portions of the tongue called **papillae** (pă-pil′ē). However, taste buds are also located on other areas of the tongue, the palate, and even the lips and throat, especially in children. The four major types of papillae are named according to their shape (figure 15.4): **filiform** (fil′i-form; filament-shaped), **vallate** (val′āt; surrounded by a wall), **foliate** (fō′lē-āt; leaf-shaped), and **fungiform** (fŭn′ji-fōrm; mushroomshaped). Taste buds (figure 15.4*c*–*e*) are associated with vallate, foliate, and fungiform papillae. Filiform papillae are the most numerous papillae on the surface of the tongue but have no taste buds. Rather, filiform papillae provide a rough surface on the tongue, allowing it to manipulate food more easily.

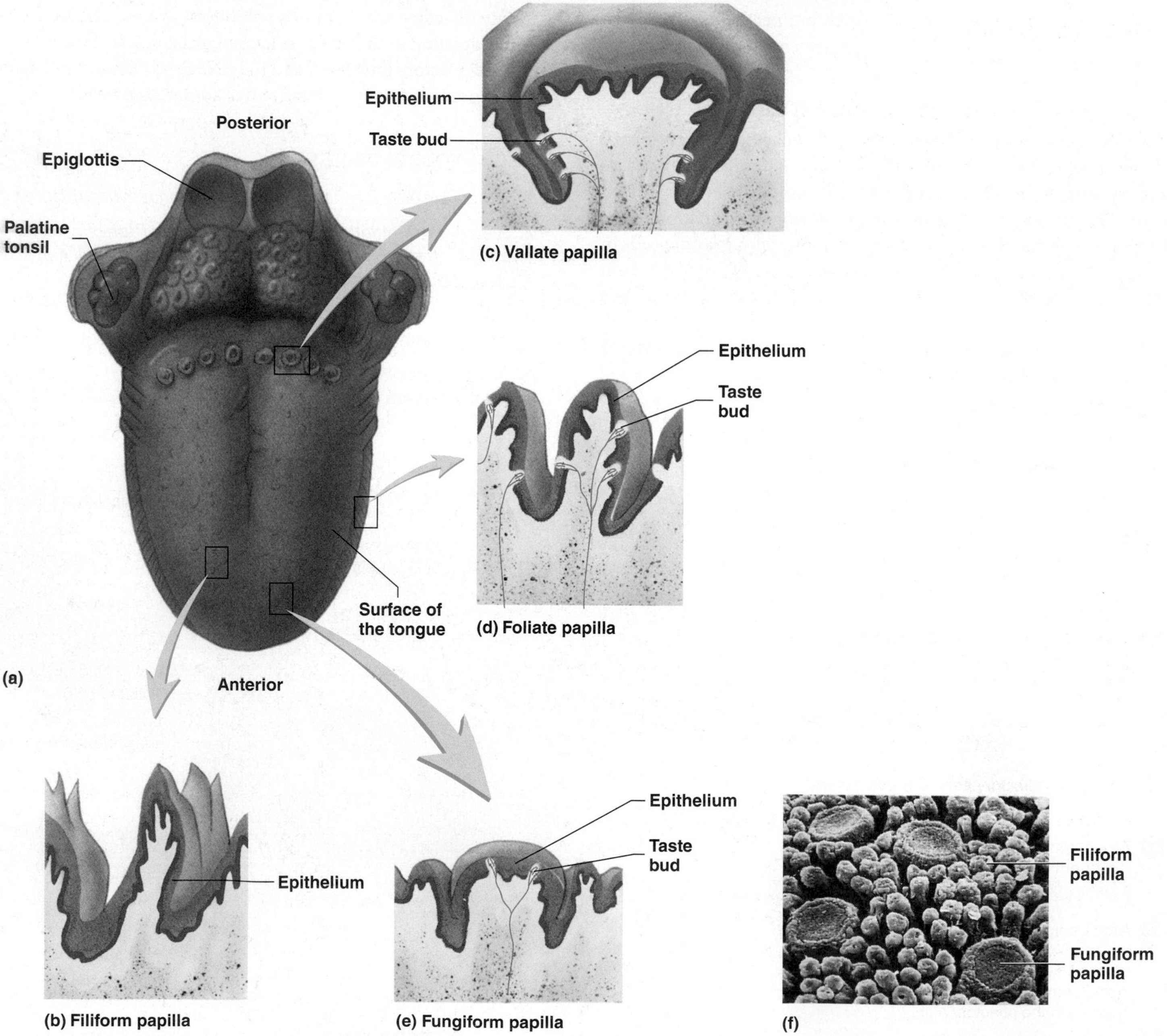

FIGURE 15.4 AP|R **Papillae**

(*a*) Surface of the tongue. (*b*) Filiform papilla. (*c*) Vallate papilla. (*d*) Foliate papilla. (*e*) Fungiform papilla. (*f*) Scanning electron micrograph of the surface of the tongue.

Vallate papillae are the largest but least numerous of the papillae. Eight to 12 of these papillae form a V-shaped row along the border between the anterior and posterior parts of the tongue (figure 15.4*a*). Foliate papillae are distributed in folds on the sides of the tongue and contain the most sensitive of the taste buds. They are most numerous in young children and decrease with age. In adults, they are located mostly posteriorly. Fungiform papillae are scattered irregularly over the entire superior surface of the tongue, appearing as small, red dots interspersed among the far more numerous filiform papillae.

Histology of Taste Buds

Taste buds are oval structures embedded in the epithelium of the tongue and mouth (figure 15.5). Each of the 10,000 taste buds on a person's tongue consists of three major types of specialized epithelial cells. The sensory cells of each taste bud consist of about 50 **taste cells,** or *gustatory cells.* The remaining two cell types, which are nonsensory cells, are **basal cells** and **supporting cells.** Like olfactory cells, the taste cells are replaced continuously, each having a normal life span of about 10 days. Each taste cell has several microvilli, called **taste hairs,** or *gustatory hairs,* extending from its apex into a tiny opening in the epithelium called the **taste pore,** or *gustatory pore.*

Function of Taste

Substances called **tastants** (tās′tants), dissolved in saliva, enter the taste pores and, by various mechanisms, cause the taste cells to depolarize. These cells do not have classic axons but have short connections with secondary sensory neurons (see chapter 14). Those connections have some characteristics of chemical synapses. Neurotransmitters (apparently, including ATP) are released from the taste cells and stimulate action potentials in the axons of sensory neurons associated with them.

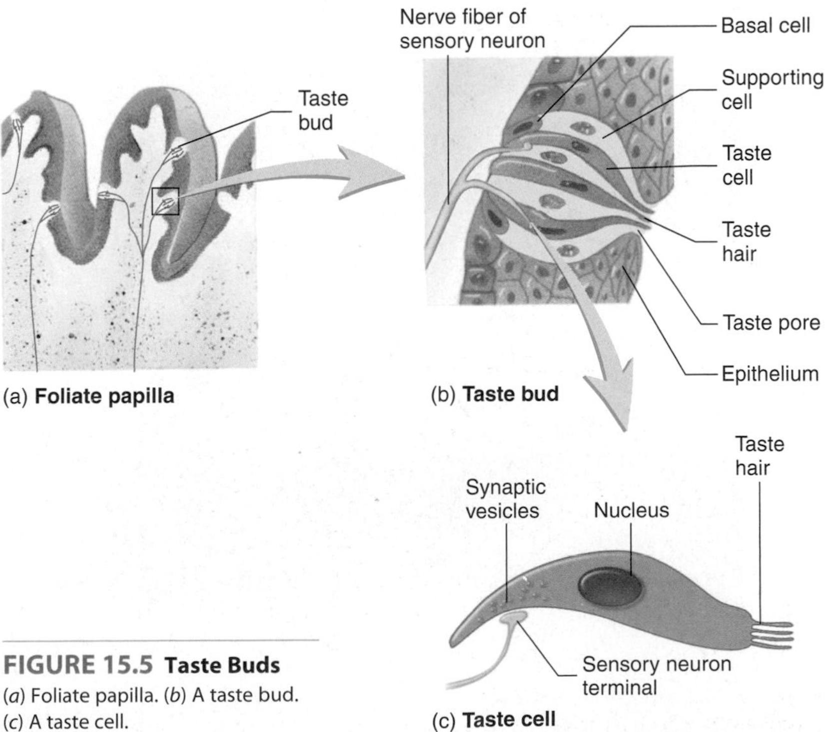
Taste bud
Nerve fiber of sensory neuron
Basal cell
Supporting cell
Taste cell
Taste hair
Taste pore
Epithelium
(a) Foliate papilla
(b) Taste bud
Taste hair
Synaptic vesicles
Nucleus
Sensory neuron terminal
(c) Taste cell

FIGURE 15.5 Taste Buds
(*a*) Foliate papilla. (*b*) A taste bud. (*c*) A taste cell.

Five major tastants are generally recognized: salt, sour, sweet, bitter, and umami. The taste of **salt** results when Na^+ diffuses through Na^+ channels (figure 15.6*a*) of the taste hairs or other cell surfaces of taste cells, resulting in depolarization of the cells. The **sour** taste results when hydrogen ions (H^+) of acids cause depolarization of taste cells by one of three mechanisms (figure 15.6*b*): (1) They can enter the cell directly through H^+ channels, (2) they can bind to ligand-gated K^+ channels and block the exit of K^+ from the cell, or (3) they can open ligand-gated channels for other positive ions and allow them to diffuse into the cell. **Sweet** and **bitter** tastants bind to receptors (figure 15.6*c,d*) on the taste hairs of taste cells and cause depolarization through a G protein mechanism. A taste called **umami** (ū-ma′mē; Japanese term loosely translated as savory) results when amino acids, such as glutamate, bind to receptors (figure 15.6*e*) on taste hairs of taste cells and cause depolarization through a G protein mechanism.

Other factors can influence the sense of taste. The texture of food in the oral cavity also affects the perception of taste. Hot or cold food may interfere with the taste buds' ability to function in tasting food. If a cold fluid is held in the mouth, the body warms the fluid and the taste becomes enhanced. On the other hand, adaptation is very rapid for taste. This adaptation apparently occurs both at the level of the taste bud and within the CNS. Adaptation may begin within 1 or 2 seconds after a taste sensation is perceived, and complete adaptation may occur within 5 minutes.

Although all taste buds are able to detect all five of the basic tastes, each taste cell is usually most sensitive to one. As with olfaction, however, the specificity of the receptor molecules is not perfect. For example, artificial sweeteners have different chemical structures than the sugars they are designed to replace, and some are many times more powerful than natural sugars in stimulating taste sensations. Presumably, our ability to perceive many different tastes is achieved through combinations of the five basic taste sensations.

Many of the sensations thought of as taste are strongly influenced by olfactory sensations. To demonstrate this phenomenon, pinch your nose while trying to taste something. With olfaction blocked, it is difficult to distinguish between the tastes of a piece of apple and a piece of potato. This is one reason that a person with a cold has a reduced sensation of taste.

Thresholds vary for the five primary tastes. Sensitivity for bitter substances is the highest; sensitivities for sweet and salty tastes are the lowest. Sugars, some other carbohydrates, and certain proteins produce sweet tastes; many proteins and amino acids produce umami tastes; acids produce sour tastes; metal ions tend to produce salty tastes; and alkaloids (bases) produce bitter tastes. Many alkaloids are poisonous; thus, the high sensitivity for bitter tastes may be protective. On the other hand, humans tend to crave sweet, salty, and umami tastes, perhaps in response to the body's need for sugars, carbohydrates, proteins, and minerals.

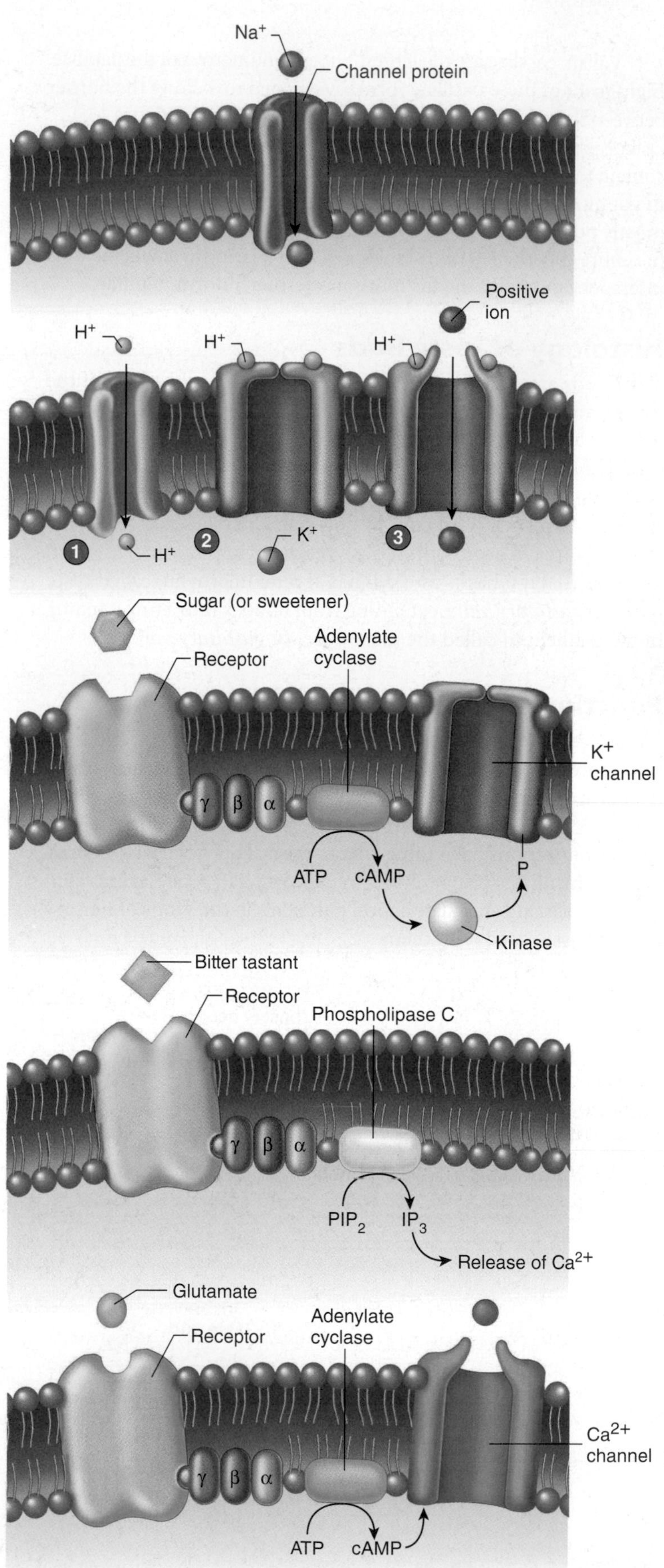
Na+
Channel protein
Positive ion
H+
H+
H+
1
H+
2
K+
3
Sugar (or sweetener)
Receptor
Adenylate cyclase
K+ channel
γ β α
ATP cAMP
P
Kinase
Bitter tastant
Receptor
Phospholipase C
γ β α
PIP2 IP3
Release of Ca2+
Glutamate
Receptor
Adenylate cyclase
Ca2+ channel
γ β α
ATP cAMP

(a) **Salt:** Sodium ions diffuse through Na^+ channels, resulting in depolarization.

(b) **Sour (acid):** Hydrogen ions (H^+) from acids can cause depolarization by one of three mechanisms: (1) They can enter the cell directly through H^+ channels, (2) they can bind to gated K^+ channels, closing the gate and preventing K^+ from leaving the cell, or (3) they can open ligand-gated channels for other positive ions.

(c) **Sweet:** Sugars, such as glucose, or artificial sweeteners bind to receptors and cause the cell to depolarize by means of a G protein mechanism. The α subunit of the G protein activates adenylate cyclase, which produces cAMP. cAMP activates a kinase that phosphorylates K^+ channels. The K^+ channels close, resulting in depolarization.

(d) **Bitter:** Bitter tastants, such as quinine, bind to receptors and cause depolarization of the cell through a G protein mechanism. The α subunit of the G protein activates phospholipase C, which converts phosphoinositol (PIP_2) to inositol triphosphate (IP_3). IP_3 causes Ca^{2+} release from intracellular stores and depolarization of the cell.

(e) **Glutamate (umami):** Amino acids, such as glutamate, bind to receptors and cause depolarization through a G protein mechanism. The α subunit of the G protein activates adenylate cyclase, which catalyzes the conversion of ATP to cAMP. cAMP opens Ca^{2+} channels. The influx of Ca^{2+} causes depolarization of the cell.

FIGURE 15.6 Actions of the Major Tastants

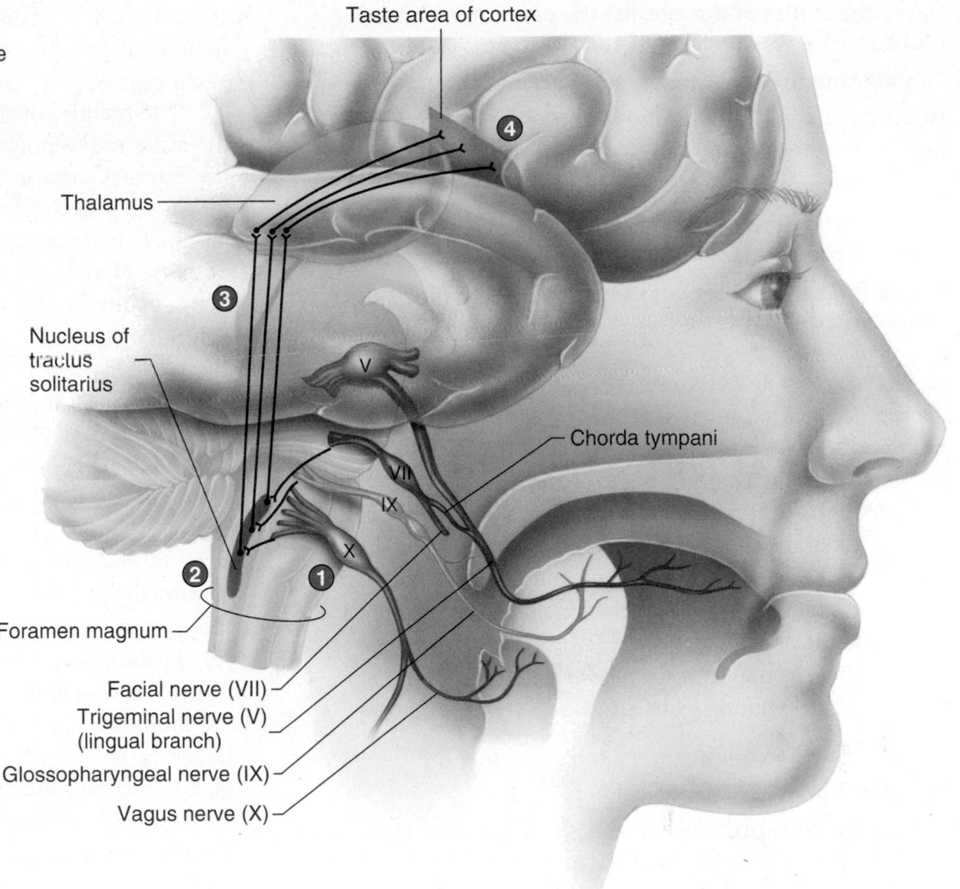
1 Axons of sensory neurons, which synapse with taste receptors, pass through cranial nerves VII, IX, and X and through the ganglion of each nerve (enlarged portion of each nerve).
2 The axons enter the brainstem and synapse in the nucleus of the tractus solitarius.
3 Axons from the nucleus of the tractus solitarius synapse in the thalamus.
4 Axons from the thalamus terminate in the taste area of the cerebral cortex.
Taste area of cortex
Thalamus
Nucleus of tractus solitarius
V
Chorda tympani
VII
IX
X
Foramen magnum
Facial nerve (VII)
Trigeminal nerve (V) (lingual branch)
Glossopharyngeal nerve (IX)
Vagus nerve (X)

PROCESS FIGURE 15.7 Pathways for the Sense of Taste
The facial nerve (anterior two-thirds of the tongue), glossopharyngeal nerve (posterior one-third of the tongue), and vagus nerve (root of the tongue) all carry taste sensations. The trigeminal nerve carries tactile sensations from the anterior two-thirds of the tongue. The chorda tympani from the facial nerve (carrying taste input) joins the trigeminal nerve.

Neuronal Pathways for Taste

Taste sensations are carried by three cranial nerves. The **chorda tympani** (kōr′dă tim′pă-ne; so named because it crosses over the surface of the tympanic membrane of the middle ear), a branch of the facial nerve (VII), transmits taste sensations from the anterior two-thirds of the tongue, except from the vallate papillae. The glossopharyngeal nerve (IX) carries taste sensations from the posterior one-third of the tongue, the vallate papillae, and the superior pharynx. In addition to these two major nerves, the vagus nerve (X) carries a few fibers for taste sensation from the epiglottis.

These nerves extend from the taste buds to the tractus solitarius of the medulla oblongata (figure 15.7). Fibers from this nucleus decussate and extend to the thalamus. Neurons from the thalamus project to the taste area of the cerebral cortex, which is located in the insula, deep to the inferior end of the postcentral gyrus of the parietal lobe.

ASSESS YOUR PROGRESS

5. *Name and describe the four kinds of papillae on the tongue. Which ones are associated with taste buds?*
6. *Describe the structure of a taste bud.*
7. *What are the five primary tastes? Describe how each type of tastant causes depolarization of a taste cell.*
8. *Starting with the taste hair, name the structures and cells that an action potential would encounter on the way to the taste area of the cerebral cortex.*
9. *How is the sense of taste related to the sense of smell?*

Predict 2

Ernie has difficulty swallowing, loss of taste sensation in the posterior one-third of the right half of his tongue, and decreased salivation, especially on the right side. Describe the sensory portion of the affected neuropathway in Ernie.

15.3 Visual System

LEARNING OUTCOMES

After reading this section, you should be able to

A. List the accessory structures of the eye and explain their functions.

B. Describe the chambers of the eye and the fluids they contain.

C. **Name the tunics of the eye, list the parts of each tunic, and describe the function of each part.**
D. **Explain the differences between rods and cones.**
E. **Describe the structure of the retina and explain how light stimulates action potentials in the optic nerves.**
F. **Relate how images are focused on the retina.**
G. **Describe the pathway an action potential will travel from the rods and cones to the visual cortex.**
H. **Relate the arrangement of the visual field to binocular vision and depth perception.**

The visual system includes the eyes, the accessory structures, and the optic nerves (II), tracts, and pathways. The **eye** includes the eyeball (the globe of the eye) and the lens. The eyes respond to light and initiate afferent action potentials, which are transmitted from the eyes to the brain by the optic nerves and tracts. We obtain much of our information about the world through the visual system. For example, education is largely based on visual input and depends on our ability to read words and numbers. Visual input includes information about light and dark, movement, and color. Visual stimuli can come from far greater distances than can stimuli for any other sense. The eye can detect light originating from stars billions of miles away.

Accessory Structures

Accessory structures protect, lubricate, move, and in other ways aid the function of the eye. These structures include the eyebrows, eyelids, eyelashes, conjunctiva, lacrimal apparatus, and extrinsic eye muscles (figure 15.8).

Eyebrows

The **eyebrows** prevent perspiration from running down the forehead and into the eyes and irritating them. Eyebrows also help shade the eyes from direct sunlight.

Eyelids

The **eyelids,** or *palpebrae* (pal-pē′brē), with their associated lashes, protect the eyes from foreign objects. The space between the two eyelids is called the **palpebral fissure,** and the angles where the eyelids join at the medial and lateral margins of the eye are called **canthi** (kan′thī; corners of the eye; figure 15.8). The medial canthus contains a small, reddish-pink mound called the **caruncle** (kar′ŭng-kl; mound of tissue), which houses some modified sebaceous and sweat glands.

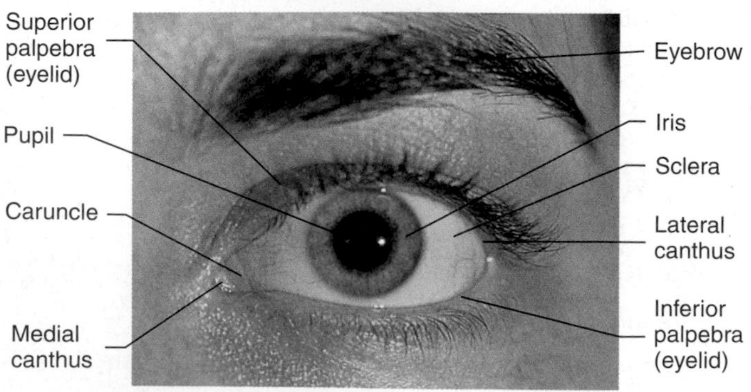
Superior palpebra (eyelid)
Pupil
Caruncle
Medial canthus
Eyebrow
Iris
Sclera
Lateral canthus
Inferior palpebra (eyelid)

FIGURE 15.8 Accessory Structures of the Left Eye

The eyelids consist of five layers of tissue (figure 15.9). From the outer to the inner surface, they are (1) a thin layer of skin on the external surface; (2) a thin layer of areolar connective tissue; (3) a layer of skeletal muscle consisting of the orbicularis oculi and levator palpebrae superioris muscles; (4) a crescent-shaped layer of dense connective tissue called the **tarsal** (tar′săl) **plate,** which helps maintain the shape of the eyelid; and (5) the palpebral conjunctiva (described in the next section), which lines the inner surface of the eyelid and the anterior surface of the eyeball.

If an object suddenly approaches the eye, the eyelids protect the eye by rapidly closing and then opening (blink reflex). Blinking, which normally occurs about 25 times per minute, also helps keep the eye lubricated by spreading tears over the surface. Movements of the eyelids are a function of skeletal muscles. The orbicularis oculi muscle closes the lids, and the levator palpebrae superioris elevates the upper lid (see figure 10.9). The eyelids also help regulate the amount of light entering the eye.

Eyelashes (see figures 15.8 and 15.9) are attached as a double or triple row of hairs to the free edges of the eyelids. **Ciliary glands** are modified sweat glands that open into the follicles of the eyelashes to keep them lubricated. A **sty** forms when one of these glands becomes inflamed. **Meibomian** (mī-bō′mē-an) **glands,** or *tarsal glands,* are sebaceous glands near the inner margins of the eyelids; they produce **sebum** (sē′bŭm), which lubricates the lids and restrains tears from flowing over the margin of the eyelids. A **chalazion** (ka-lā′zē-on), or *meibomian cyst,* is an infection or a blockage of a meibomian gland.

Conjunctiva

The **conjunctiva** (kon-jŭnk-tī′vă) is a thin, transparent mucous membrane. The **palpebral conjunctiva** covers the inner surface of the eyelids, and the **bulbar conjunctiva** covers the anterior white surface of the eye (figure 15.9). The points at which the palpebral and bulbar conjunctivae meet are the superior and inferior **conjunctival fornices** (sing. *fornix*). The secretions of the conjuctiva help lubricate the surface of the eye. **Conjunctivitis** is an inflammation of the conjunctiva caused by an infection or other irritation. One type of conjunctivitis caused by a bacterium is **acute contagious conjunctivitis,** also called *pinkeye.*

Lacrimal Apparatus

The **lacrimal** (lak′ri-măl) **apparatus** (figure 15.10) consists of a lacrimal gland situated in the superolateral corner of the orbit and a nasolacrimal duct beginning in the inferomedial corner of the orbit. The **lacrimal gland** is innervated by parasympathetic fibers from the facial nerve (VII). The gland produces tears, which leave the gland through several **lacrimal ducts** and pass over the anterior surface of the eyeball. The gland produces tears constantly at the rate of about 1 mL/day to moisten the surface of the eye, lubricate the eyelids, and wash away foreign objects. Tears are mostly water, along with some salts, mucus, and lysozyme, an enzyme that kills certain bacteria. Most of the fluid produced by the lacrimal glands evaporates from the surface of the eye, but excess tears are collected

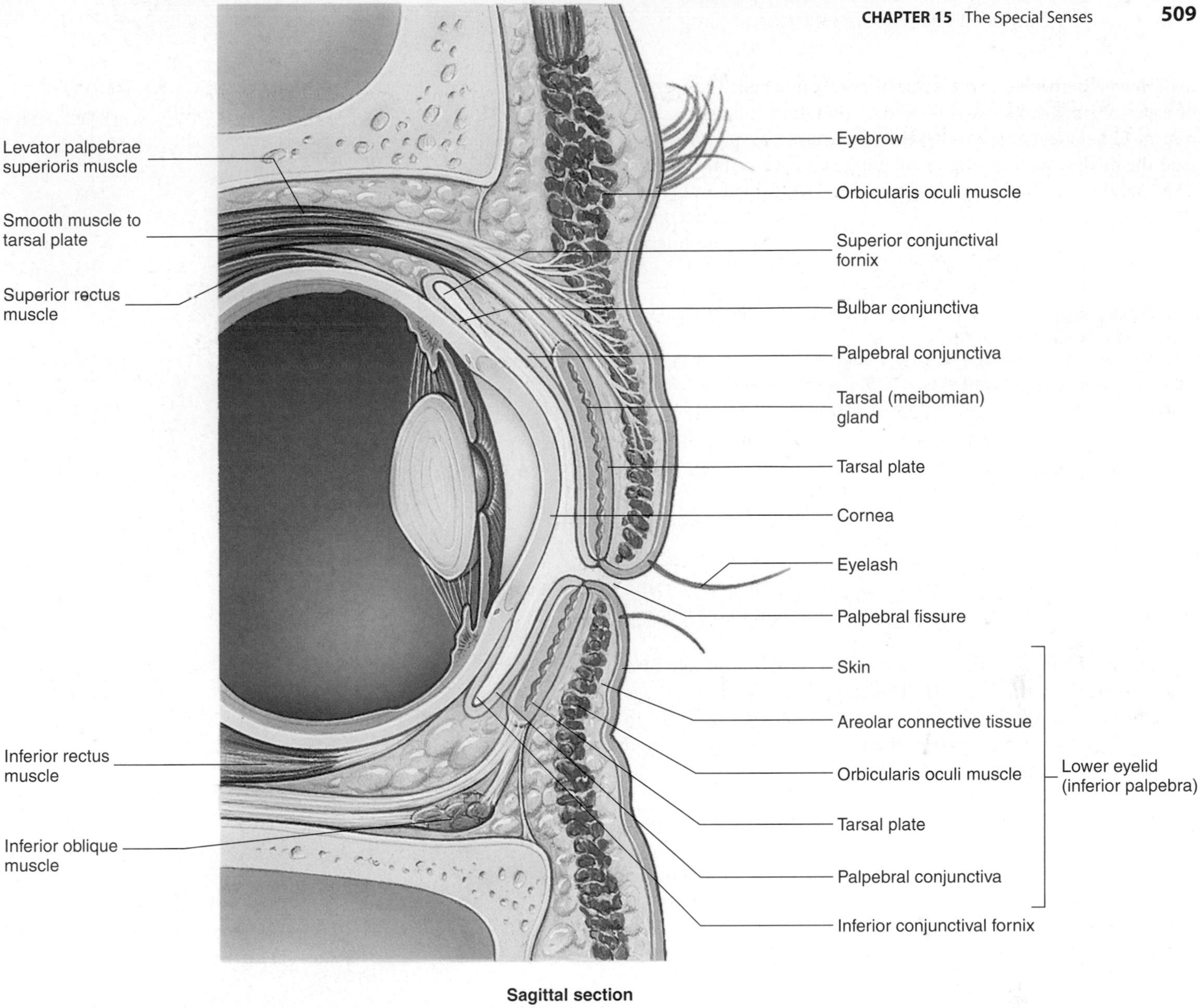

Levator palpebrae superioris muscle
Smooth muscle to tarsal plate
Superior rectus muscle
Inferior rectus muscle
Inferior oblique muscle
Eyebrow
Orbicularis oculi muscle
Superior conjunctival fornix
Bulbar conjunctiva
Palpebral conjunctiva
Tarsal (meibomian) gland
Tarsal plate
Cornea
Eyelash
Palpebral fissure
Skin
Areolar connective tissue
Orbicularis oculi muscle
Tarsal plate
Palpebral conjunctiva
Inferior conjunctival fornix
Lower eyelid (inferior palpebra)

Sagittal section

FIGURE 15.9 APIR **Sagittal Section Through the Eye, Showing Its Accessory Structures**

1. Tears are produced in the lacrimal gland and exit the gland through several lacrimal ducts.
2. The tears pass over the surface of the eye.
3. Tears enter the lacrimal canaliculi.
4. Tears are carried through the lacrimal sac to the nasolacrimal duct.
5. Tears enter the nasal cavity from the nasolacrimal duct.

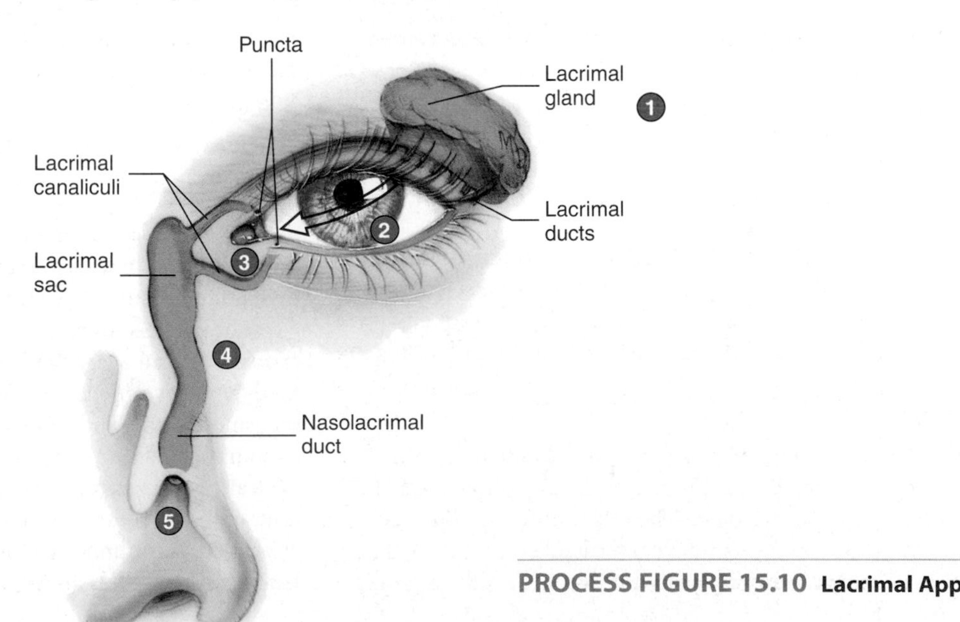
Puncta
Lacrimal gland
1
Lacrimal canaliculi
Lacrimal ducts
2
3
Lacrimal sac
4
Nasolacrimal duct
5

PROCESS FIGURE 15.10 Lacrimal Apparatus

in the medial corner of the eye by the **lacrimal canaliculi.** The opening of each lacrimal canaliculus is called a **punctum** (pŭngk′tŭm; pl. *puncta*). The upper and lower eyelids each have a punctum located near the medial canthus on a small lump called a **lacrimal papilla.** The lacrimal canaliculi open into a **lacrimal sac,** which drains into the **nasolacrimal duct** (figure 15.10). The nasolacrimal duct opens into the inferior meatus of the nasal cavity beneath the inferior nasal concha (see chapter 23).

> **Predict 3**
> Explain why it is often possible to "taste" medications, such as eyedrops, that have been placed into the eyes. Why does a person's nose "run" when he or she cries?

Extrinsic Eye Muscles

Six **extrinsic muscles** of the eye (figures 15.11 and 15.12; see chapter 10) cause the eyeball to move. Four of these muscles run more or less straight anteroposteriorly: the superior, inferior, medial, and lateral **rectus muscles.** Two other muscles, the superior and inferior **oblique muscles,** are positioned at an angle to the globe of the eye.

The movements of the eye can be described graphically by a figure resembling the letter *H*. The clinical test for normal eye movement is therefore called the **H test.** A person's inability to move the eye toward one part of the *H* may indicate dysfunction of either an extrinsic eye muscle or the cranial nerve to the muscle. (The actions of the eye muscles are listed in table 10.8.)

The superior oblique muscle is innervated by the trochlear nerve (IV). The nerve is so named because the superior oblique muscle goes around a little pulley, or trochlea, in the superomedial corner of the orbit. The lateral rectus muscle is innervated by the abducens nerve (VI), so named because the lateral rectus muscle abducts the eye. The other four extrinsic eye muscles are innervated by the oculomotor nerve (III).

ASSESS YOUR PROGRESS

10. *Describe the structures and state the functions of the eyebrows, eyelids, and eyelashes.*
11. *How do the conjunctiva, lacrimal apparatus, and extrinsic eye muscles aid in the function of the eye?*

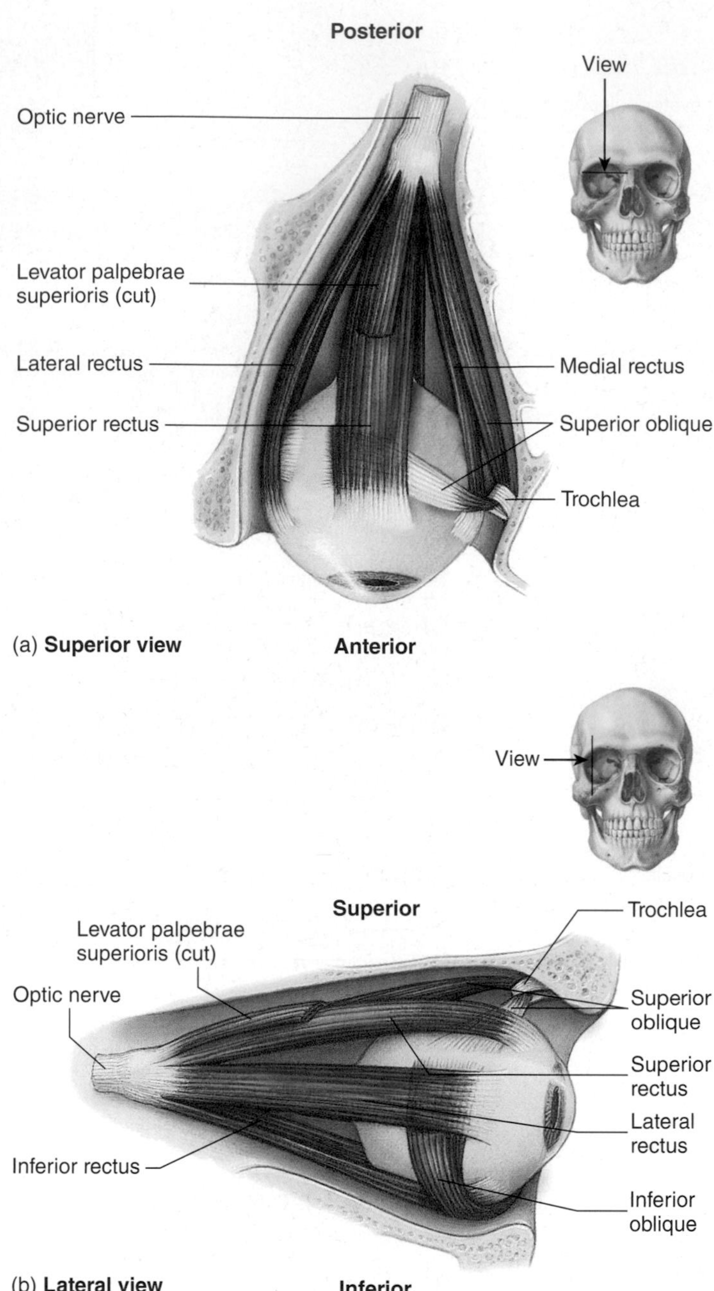
Posterior
View
Optic nerve
Levator palpebrae superioris (cut)
Lateral rectus
Superior rectus
Medial rectus
Superior oblique
Trochlea
(a) Superior view
Anterior
View
Superior
Trochlea
Levator palpebrae superioris (cut)
Optic nerve
Superior oblique
Superior rectus
Lateral rectus
Inferior rectus
Inferior oblique
(b) Lateral view
Inferior

FIGURE 15.11 Extrinsic Muscles of the Eye

Anatomy of the Eye

The wall of the eyeball is composed of three **tunics,** or layers (figure 15.13). The outer **fibrous tunic** consists of the sclera and cornea; the middle **vascular tunic** consists of the choroid, ciliary body, and iris; and the inner **nervous tunic** consists of the retina.

Fibrous Tunic

The **sclera** (sklēr′ă) is the firm, opaque, white, outer layer of the posterior five-sixths of the eyeball. It consists of dense collagenous connective tissue with elastic fibers. The sclera helps maintain the shape of the eyeball, protects its internal structures, and provides an attachment point for the muscles that move it. Usually, a small portion of the sclera can be seen as the "white of the eye" when the eye and its surrounding structures are intact (see figure 15.8). The bulbar conjunctiva is loosely attached to the sclera.

The sclera is continuous anteriorly with the cornea. The **cornea** (kōr′nē-ă) is an avascular, transparent structure that permits light to enter the eye and bends, or refracts, that light as part of the eye's focusing system. The cornea consists of a connective tissue matrix containing collagen, elastic fibers, and proteoglycans, with a layer of stratified squamous epithelium covering the outer surface and a layer of simple squamous epithelium on the inner surface. The outer epithelium is continuous with the bulbar conjunctiva over the sclera. Large collagen fibers are white, whereas smaller collagen fibers and

Levator palpebrae superioris muscle
Smooth muscle to tarsal plate
Superior rectus muscle
Inferior rectus muscle
Inferior oblique muscle
Eyebrow
Orbicularis oculi muscle
Superior conjunctival fornix
Bulbar conjunctiva
Palpebral conjunctiva
Tarsal (meibomian) gland
Tarsal plate
Cornea
Eyelash
Palpebral fissure
Skin
Areolar connective tissue
Orbicularis oculi muscle
Tarsal plate
Palpebral conjunctiva
Inferior conjunctival fornix
Lower eyelid (inferior palpebra)

Sagittal section

FIGURE 15.9 APR **Sagittal Section Through the Eye, Showing Its Accessory Structures**

1. Tears are produced in the lacrimal gland and exit the gland through several lacrimal ducts.
2. The tears pass over the surface of the eye.
3. Tears enter the lacrimal canaliculi.
4. Tears are carried through the lacrimal sac to the nasolacrimal duct.
5. Tears enter the nasal cavity from the nasolacrimal duct.

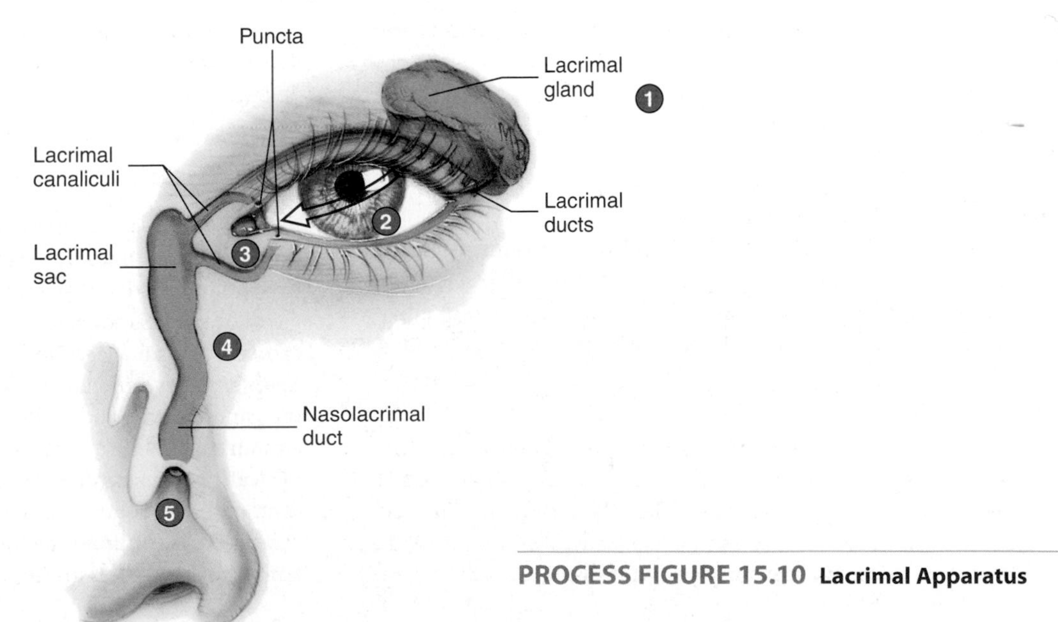
Puncta
Lacrimal gland
1
Lacrimal canaliculi
Lacrimal ducts
2
3
Lacrimal sac
4
Nasolacrimal duct
5

PROCESS FIGURE 15.10 Lacrimal Apparatus

in the medial corner of the eye by the **lacrimal canaliculi.** The opening of each lacrimal canaliculus is called a **punctum** (pŭngk′tŭm; pl. *puncta*). The upper and lower eyelids each have a punctum located near the medial canthus on a small lump called a **lacrimal papilla.** The lacrimal canaliculi open into a **lacrimal sac,** which drains into the **nasolacrimal duct** (figure 15.10). The nasolacrimal duct opens into the inferior meatus of the nasal cavity beneath the inferior nasal concha (see chapter 23).

Predict 3

Explain why it is often possible to "taste" medications, such as eyedrops, that have been placed into the eyes. Why does a person's nose "run" when he or she cries?

Extrinsic Eye Muscles

Six **extrinsic muscles** of the eye (figures 15.11 and 15.12; see chapter 10) cause the eyeball to move. Four of these muscles run more or less straight anteroposteriorly: the superior, inferior, medial, and lateral **rectus muscles.** Two other muscles, the superior and inferior **oblique muscles,** are positioned at an angle to the globe of the eye.

The movements of the eye can be described graphically by a figure resembling the letter *H.* The clinical test for normal eye movement is therefore called the **H test.** A person's inability to move the eye toward one part of the *H* may indicate dysfunction of either an extrinsic eye muscle or the cranial nerve to the muscle. (The actions of the eye muscles are listed in table 10.8.)

The superior oblique muscle is innervated by the trochlear nerve (IV). The nerve is so named because the superior oblique muscle goes around a little pulley, or trochlea, in the superomedial corner of the orbit. The lateral rectus muscle is innervated by the abducens nerve (VI), so named because the lateral rectus muscle abducts the eye. The other four extrinsic eye muscles are innervated by the oculomotor nerve (III).

ASSESS YOUR PROGRESS

10. *Describe the structures and state the functions of the eyebrows, eyelids, and eyelashes.*
11. *How do the conjunctiva, lacrimal apparatus, and extrinsic eye muscles aid in the function of the eye?*

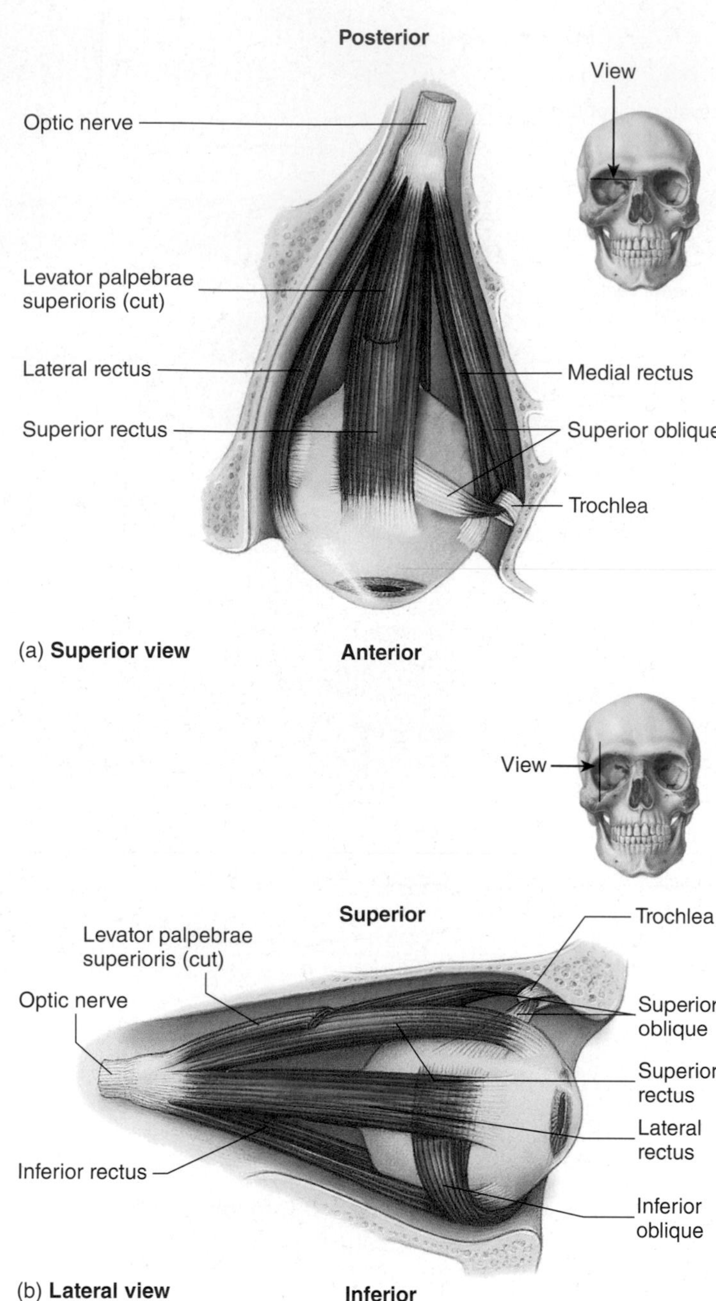
Posterior
View
Optic nerve
Levator palpebrae superioris (cut)
Lateral rectus
Superior rectus
Medial rectus
Superior oblique
Trochlea
(a) Superior view
Anterior
View
Superior
Trochlea
Levator palpebrae superioris (cut)
Optic nerve
Superior oblique
Superior rectus
Lateral rectus
Inferior rectus
Inferior oblique
(b) Lateral view
Inferior

FIGURE 15.11 Extrinsic Muscles of the Eye

Anatomy of the Eye

The wall of the eyeball is composed of three **tunics,** or layers (figure 15.13). The outer **fibrous tunic** consists of the sclera and cornea; the middle **vascular tunic** consists of the choroid, ciliary body, and iris; and the inner **nervous tunic** consists of the retina.

Fibrous Tunic

The **sclera** (sklēr′ă) is the firm, opaque, white, outer layer of the posterior five-sixths of the eyeball. It consists of dense collagenous connective tissue with elastic fibers. The sclera helps maintain the shape of the eyeball, protects its internal structures, and provides an attachment point for the muscles that move it. Usually, a small portion of the sclera can be seen as the "white of the eye" when the eye and its surrounding structures are intact (see figure 15.8). The bulbar conjunctiva is loosely attached to the sclera.

The sclera is continuous anteriorly with the cornea. The **cornea** (kōr′nē-ă) is an avascular, transparent structure that permits light to enter the eye and bends, or refracts, that light as part of the eye's focusing system. The cornea consists of a connective tissue matrix containing collagen, elastic fibers, and proteoglycans, with a layer of stratified squamous epithelium covering the outer surface and a layer of simple squamous epithelium on the inner surface. The outer epithelium is continuous with the bulbar conjunctiva over the sclera. Large collagen fibers are white, whereas smaller collagen fibers and

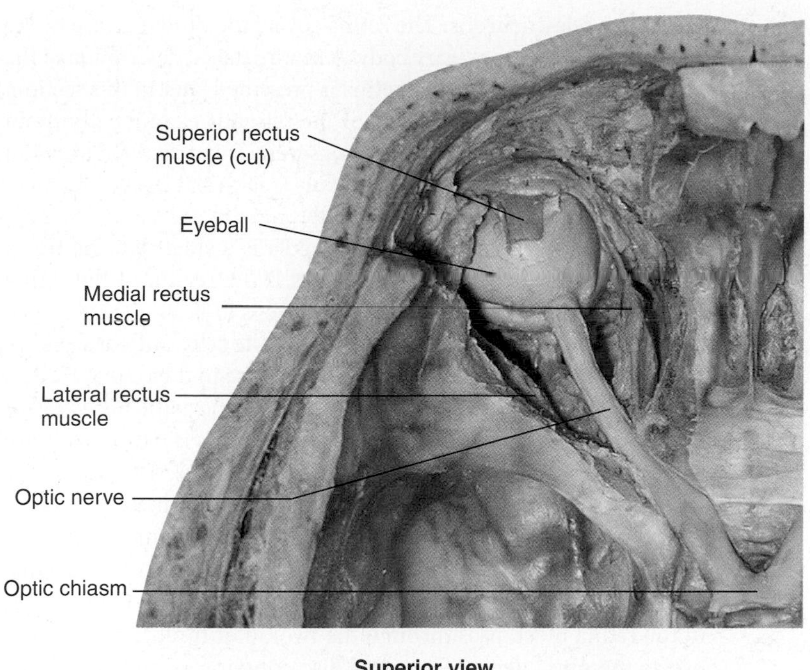
Superior rectus muscle (cut)
Eyeball
Medial rectus muscle
Lateral rectus muscle
Optic nerve
Optic chiasm
Superior view

FIGURE 15.12 Photograph of the Eye and Its Associated Structures

proteoglycans are transparent. The cornea is transparent, rather than white, like the sclera, in part because fewer large collagen fibers and more proteoglycans are present in the cornea than in the sclera. The transparency of the cornea also results from its low water content. In the presence of water, proteoglycans trap water and expand, which scatters light. In the absence of water, the proteoglycans decrease in size and do not interfere with the passage of light through the matrix.

The central part of the cornea receives oxygen from the outside air. Contact lenses worn for long periods must therefore be permeable, so that air can reach the cornea. The most common eye injuries are cuts or tears of the cornea caused by a stick, a stone, or some other foreign object hitting the cornea. Extensive injury to the cornea may cause connective tissue deposition, thereby making the cornea opaque. The cornea was one of the first organs to be transplanted. Several characteristics make it relatively easy to transplant: It is easily accessible and relatively easily removed; it is avascular, so it does not require extensive circulation, as other tissues do; and it is less immunologically active and therefore less likely to be rejected than other tissues are.

FUNDAMENTAL **Figure**

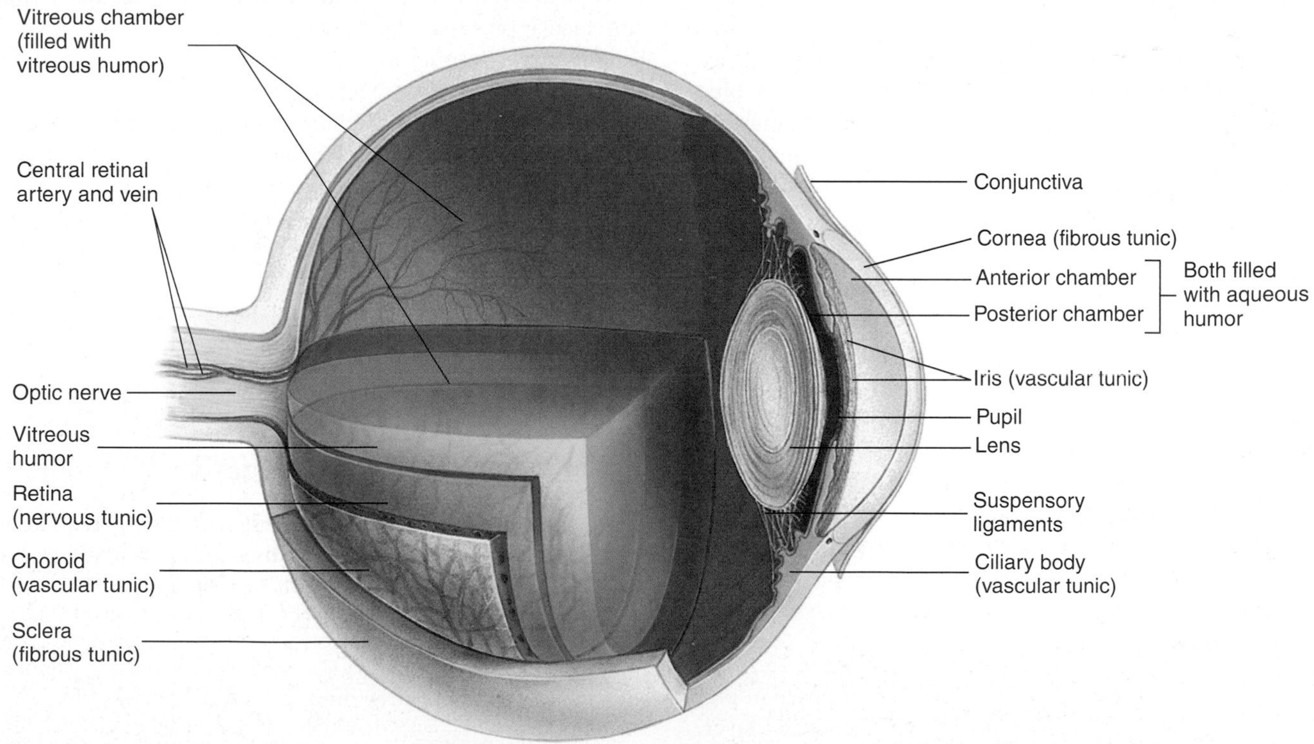
Vitreous chamber (filled with vitreous humor)
Central retinal artery and vein
Optic nerve
Vitreous humor
Retina (nervous tunic)
Choroid (vascular tunic)
Sclera (fibrous tunic)
Conjunctiva
Cornea (fibrous tunic)
Anterior chamber
Posterior chamber
Both filled with aqueous humor
Iris (vascular tunic)
Pupil
Lens
Suspensory ligaments
Ciliary body (vascular tunic)

FIGURE 15.13 Layers of the Eyeball, Sagittal Section

Vascular Tunic

The middle tunic of the eyeball is called the vascular tunic because it contains most of the blood vessels of the eyeball (figure 15.13). The arteries of the vascular tunic are derived from a number of arteries called **short ciliary arteries,** which pierce the sclera in a circle around the optic nerve. These arteries are branches of the **ophthalmic** (of-thal′mik) **artery,** which is a branch of the internal carotid artery. The vascular tunic contains a large number of melanin-containing pigment cells and appears black in color. The portion of the vascular tunic associated with the sclera of the eye is the **choroid** (ko′royd). The term *choroid* means "membrane" and suggests that this layer is relatively thin (0.1–0.2 mm thick). Anteriorly, the vascular tunic consists of the ciliary body and the iris.

The **ciliary** (sil′ē-ar-ē) **body** is continuous with the choroid, and the **iris** is attached at its lateral margins to the ciliary body (figure 15.14*a,b*). The ciliary body consists of an outer **ciliary ring** and an inner group of **ciliary processes,** which are attached to the lens by **suspensory ligaments.** The ciliary body contains smooth muscles called the **ciliary muscles,** which are arranged with the outer muscle fibers oriented radially and the central fibers oriented circularly. The ciliary muscles function as a sphincter, and contraction of these muscles can change the shape of the lens. (This function is described in more detail later in this section.) The ciliary processes are a complex of capillaries and cuboidal epithelium that produces aqueous humor.

The iris is the "colored part" of the eye, and its color differs from person to person. A large amount of melanin in the iris causes it to appear brown or even black. Less melanin results in light brown, green, or gray irises. Even less melanin causes the eyes to appear blue. If there is no pigment in the iris, as occurs in albinism, the iris appears pink because blood vessels in the eye reflect light back to the iris.

The genetics of eye color are quite complex. Many genes affect eye color, which explains the complexity of eye colors and inheritance patterns. Interestingly, although many newborn babies have blue eyes, their eye color changes over the first year of life. As melanin production increases during the first year, it accumulates in the iris, resulting in the more permanent eye color.

The iris is a contractile structure, consisting mainly of smooth muscle, surrounding an opening called the **pupil.** Light enters the eye through the pupil, and the iris regulates the amount of light by controlling the size of the pupil. The iris contains two groups of smooth muscles: a circular group called the **sphincter pupillae** (pū-pil′ē) and a radial group called the **dilator pupillae** (figure 15.14*c,d*). The sphincter pupillae are innervated by parasympathetic fibers from the oculomotor nerve (III). When they contract, the iris decreases, or constricts, the size of the pupil. The dilator pupillae are innervated by sympathetic fibers. When they contract, the pupil is dilated. The ciliary muscles, sphincter pupillae, and dilator pupillae are sometimes referred to as the *intrinsic eye muscles.*

Retina

The **retina** is the nervous tunic of the eyeball (see figure 15.13). It consists of the outer **pigmented layer,** which is composed of pigmented simple cuboidal epithelium, and the inner **neural layer,** which responds to light. The neural layer contains numerous photoreceptor cells: 120 million **rods** and 6 or 7 million **cones,** as well as numerous relay neurons. The retina covers the inner surface of the eyeball posterior to the ciliary body. A more detailed description of the histology and function of the retina is presented later in this section.

When the posterior region of the retina is examined with an **ophthalmoscope** (of-thal′mō-skōp), several important features can be observed (figure 15.15*a*). The **macula** (mak′ū-lă) is a small yellow spot, approximately 4 mm in diameter, near the center of the posterior retina. In the center of the macula is a small pit, the **fovea** (fō′vē-ă) **centralis.** The fovea centralis is the region of the retina where light is most focused when the eye is looking directly at an object. The fovea centralis contains only cone cells, and the cells are more tightly packed there than any where else in the retina. Hence, the fovea centralis is the portion of the retina with the greatest visual acuity (the ability to see fine images). This is why objects are best seen when viewed directly in front of the eye. The **optic disc** is a white spot just medial to the macula through which the central retinal artery enters and the central retinal vein exits the eyeball. Branches from these vessels spread over the surface of the retina. This is also the spot where nerve processes from the neural layer of the retina meet, pass through the two outer tunics, and exit the eye as the optic nerve. The optic disc contains no photoreceptor cells and does not respond to light; therefore, it is called the **blind spot** of the eye (figure 15.15*b*).

Chambers of the Eye

The interior of the eye is divided into three chambers: the **anterior chamber,** the **posterior chamber,** and the **vitreous** (vit′rē-ŭs) **chamber** (*postremal chamber*; see figure 15.13). The anterior chamber lies between the cornea and the iris, and the smaller posterior chamber lies between the iris and the lens (see figure 15.14*a*). These two chambers are filled with **aqueous humor,** which helps maintain intraocular pressure. The pressure within the eyeball keeps it inflated and is largely responsible for maintaining the eyeball's shape. The aqueous humor also refracts light and provides nutrition for the structures of the anterior chamber, such as the avascular cornea. Aqueous humor is produced by the ciliary processes as a blood filtrate and is returned to the circulation through a venous ring at the base of the cornea called the **scleral venous sinus** (canal of Schlemm;

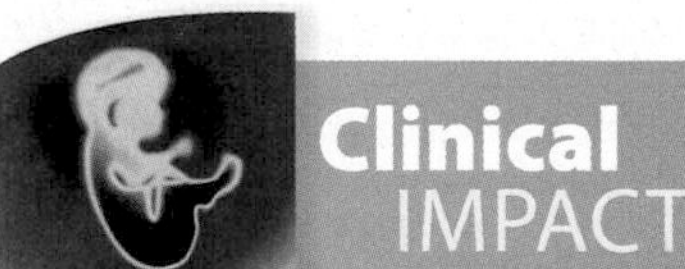
Clinical
IMPACT

Ophthalmoscopic Examination of the Retina

Using an ophthalmoscope to examine the posterior retina can reveal some general body disorders. **Hypertension,** or high blood pressure, results in "nicking" (compression) of the retinal veins where the abnormally pressurized arteries cross them. **Increased cerebrospinal fluid (CSF) pressure** associated with hydrocephalus may cause the optic disc to swell, a condition referred to as **papilledema** (pă-pil-e-dē′mă). Furthermore, **cataracts** (opacity of the lens; see the Diseases and Disorders table, "Vision," later in this chapter) are usually discovered or confirmed by ophthalmoscopic examination.

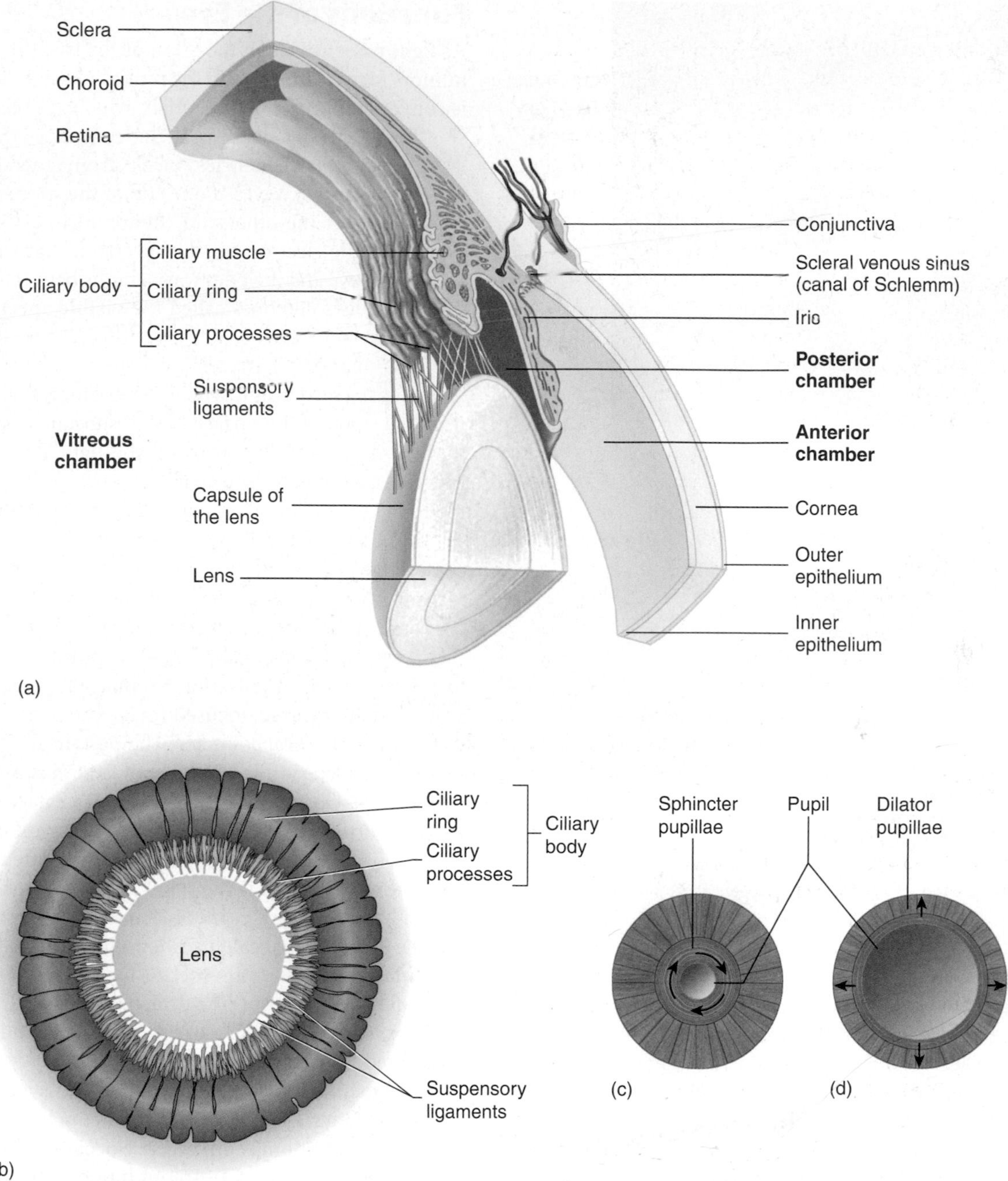
Sclera
Choroid
Retina
Ciliary body
Ciliary muscle
Ciliary ring
Ciliary processes
Suspensory ligaments
Vitreous chamber
Capsule of the lens
Lens
Conjunctiva
Scleral venous sinus (canal of Schlemm)
Iris
Posterior chamber
Anterior chamber
Cornea
Outer epithelium
Inner epithelium
(a)
Ciliary ring
Ciliary processes
Ciliary body
Lens
Suspensory ligaments
(b)
Sphincter pupillae
Pupil
Dilator pupillae
(c)
(d)

FIGURE 15.14 Lens, Cornea, Iris, and Ciliary Body
(*a*) The orientation is the same as in figure 15.13. (*b*) The lens and ciliary body. (*c*) The sphincter pupillae muscles of the iris constrict the pupil. (*d*) The dilator pupillae muscles of the iris dilate the pupil.

see figure 15.14*a*). The production and removal of aqueous humor result in the "circulation" of aqueous humor and maintenance of a constant intraocular pressure. If circulation of the aqueous humor is inhibited, a defect called **glaucoma** (glaw-kō′mā), characterized by an abnormal increase in intraocular pressure, can result.

The vitreous chamber of the eye is much larger than the anterior and posterior chambers. It is almost completely surrounded by the retina and is filled with a transparent, jellylike substance called **vitreous humor.** Vitreous humor is not produced as rapidly as the aqueous humor is, and its turnover is extremely slow. Like the aqueous humor, the vitreous humor helps maintain intraocular pressure and therefore the shape of the eyeball, and it holds the lens and retina in place. It also functions in the refraction of light in the eye.

Lens

The **lens** is an unusual biological structure. It is transparent and biconvex, with the greatest convexity on its posterior side. The lens's anterior surface consists of a layer of cuboidal epithelial cells, and

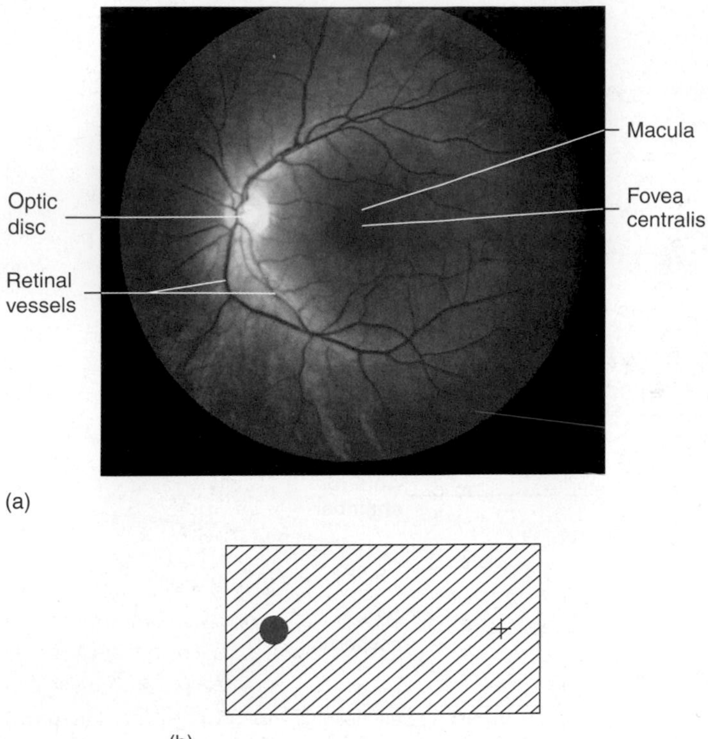
Macula
Optic disc
Fovea centralis
Retinal vessels

(a)

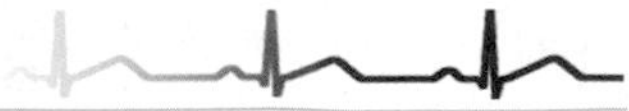

(b)

FIGURE 15.15 Ophthalmoscopic View of the Left Retina
(*a*) The posterior wall of the retina as seen when looking through the pupil. Notice the vessels entering the eye through the optic disc (the optic nerve) and the macula with the fovea centralis (the part of the retina with the greatest visual acuity). (*b*) Demonstration of the blind spot. Close your right eye. Hold the figure in front of your left eye and stare at the +. Move the figure toward your eye. At a certain point, when the image of the spot is over the optic disc, the red spot seems to disappear.

its posterior region contains very long, columnar epithelial cells called **lens fibers.** Cells from the anterior epithelium proliferate and give rise to the lens fibers at the equator of the lens. The lens fibers lose their nuclei and other cellular organelles and accumulate a set of proteins called **crystallines.** This crystalline lens is covered by a highly elastic, transparent **capsule.**

The lens is suspended between the posterior chamber and the vitreous chamber by the suspensory ligaments of the lens, which are connected from the ciliary body to the lens capsule.

ASSESS YOUR PROGRESS

12. *Name the three tunics of the eye, describe the parts of the tunics, and explain their functions.*
13. *How does the pupil constrict? How does it dilate?*
14. *What is the blind spot? What are the fovea centralis and macula?*
15. *Name the three chambers of the eye and the substances that fill each chamber.*
16. *What is the function of the ciliary process and the scleral venous sinus?*
17. *Describe the structure of the lens, and explain how it is held in place.*

Functions of the Eye

As light passes through the pupil of the iris, the cornea, lens, and humors focus the light onto the retina. The light striking the retina is converted into action potentials, which are relayed to the brain.

The electromagnetic spectrum comprises the entire range of wavelengths, or frequencies, of electromagnetic radiation, from very short gamma waves at one end of the spectrum to the longest radio waves at the other end (figure 15.16). **Visible light** is the portion of the electromagnetic spectrum that can be detected by the human eye and includes wavelengths between 380 and 750 nm. This range is sometimes called the **visible spectrum.** Within the visible spectrum, each color has a different wavelength.

An important characteristic of light rays is that they can bend. As light passes from air to a denser substance, such as glass or water, its speed slows. If the surface of that substance is at an angle other than 90 degrees to the direction the light rays are traveling, the rays bend, because the speed of light varies as it encounters the new medium. This bending of light is called **refraction.**

If the surface of a lens is concave, with the lens thinnest in the center, the light rays diverge as a result of refraction. If the surface is convex, with the lens thickest in the center, the light rays tend to converge. As light rays converge, they finally reach a point at which they cross. This point is called the **focal point (FP),** and causing light to converge is called **focusing.** No image forms exactly at the focal point, but an inverted, focused image can form on a surface some distance past the focal point. How far past the focal point the focused image forms depends on a number of factors. A biconvex lens causes light to focus closer to the lens than does a lens with a single convex surface. Furthermore, the more nearly spherical the lens, the closer to the lens the light is focused; the more flattened the biconcave lens, the more distant is the point where the light is focused.

If light rays strike an object that is not transparent, they bounce off the surface. This phenomenon is called **reflection.** If the surface is very smooth, such as the surface of a mirror, the light rays bounce off in a specific direction. If the surface is rough, the light rays are reflected in several directions and produce a more diffuse reflection.

The focusing system of the eye projects a clear image on the retina. Light rays converge as they pass from the air through the convex cornea. Additional convergence occurs as light encounters the aqueous humor, lens, and vitreous humor. The greatest contrast in media density is between the air and the cornea; therefore, the greatest amount of convergence occurs at that point. However, the shape of the cornea and its distance from the retina are fixed, so that the cornea cannot make any adjustment in the location of the focal point. Fine adjustment in focal point location is accomplished by changing the shape of the lens. In general, focusing can be accomplished in two ways: (1) by keeping the shape of the lens constant and moving it nearer or farther from the point at which the image will be focused, as occurs in a camera, microscope, or telescope, or (2) by keeping the distance constant and changing the shape of the lens, which is the technique used in the eye.

As light rays enter the eye and are focused, the image formed just past the focal point is inverted. Action potentials that represent the inverted image are passed to the visual cortex of the cerebrum, where the brain interprets them as being right side up.

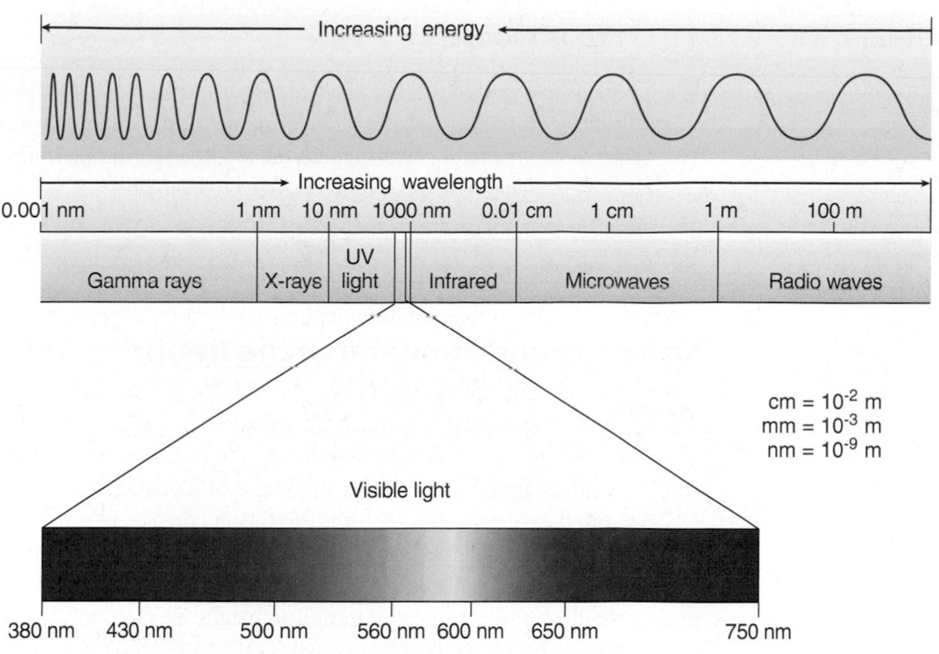
Increasing energy
Increasing wavelength
0.001 nm
1 nm
10 nm
1000 nm
0.01 cm
1 cm
1 m
100 m
Gamma rays
X-rays
UV light
Infrared
Microwaves
Radio waves
cm = 10-2 m
mm = 10-3 m
nm = 10-9 m
Visible light
380 nm
430 nm
500 nm
560 nm
600 nm
650 nm
750 nm

FIGURE 15.16 Electromagnetic Spectrum
The visible light spectrum is enlarged to show the wavelengths of the various colors.

Because the visual image is inverted when it reaches the retina, the image of the world focused on the retina is upside down. However, the brain processes information from the retina so that the world is perceived the way "it really is." If, as an experiment, a person wears glasses that invert the image entering the eye, he or she sees the world upside down for a few days, after which the brain adjusts to the new input to set the world right side up again. If the glasses are then removed, another adjustment period is required before the brain makes the world "right" again.

When the ciliary muscles are relaxed, the suspensory ligaments of the ciliary body maintain elastic pressure on the lens, thereby keeping it relatively flat and allowing for distant vision (figure 15.17*a*). The condition in which the lens is flattened so that nearly parallel rays from a distant object are focused on the retina is referred to as **emmetropia** (em-ĕ-trō′pē-ă) and is the normal resting condition of the lens. The point at which the lens does not have to thicken for focusing to occur is called the **far point of vision** and is normally 20 feet or more from the eye.

When an object is brought closer than 20 feet to the eye, three events bring the image into focus on the retina: accommodation by the lens, constriction of the pupil, and convergence of the eyes.

1. *Accommodation.* The shape of the lens changes, depending on whether the eye is focusing on an object that is near or farther away. The process of changing the shape of the lens is referred to as **accommodation.** When the eye focuses on a nearby object, the ciliary muscles contract as a result of parasympathetic stimulation from the oculomotor nerve (III). This sphincter-like contraction pulls the choroid toward the lens to reduce the tension on the suspensory ligaments, allowing the lens to assume a more spherical form because of its own elastic nature (figure 15.17*b*). The more spherical lens then has a more convex surface, causing greater refraction of light.

   As light strikes a solid object, the rays are reflected in every direction from the object's surface. However, only a small portion of the light rays reflected from a solid object pass through the pupil and enter the eye. An object far away from the eye appears small, compared with a nearby object, because only nearly parallel light rays enter the eye from a distant object (figure 15.17*a*). Converging rays leaving an object closer to the eye can also enter the eye (figure 15.17*b*), and the object appears larger.

   When rays from a distant object reach the lens, they do not have to be refracted to any great extent to be focused on the retina, and the lens can remain fairly flat. When an object is closer to the eye, the more obliquely directed rays must be refracted to a greater extent to be focused on the retina.

   As an object is brought closer and closer to the eye, accommodation becomes more and more difficult, because the lens cannot become any more convex. At some point, the eye can no longer focus the object, and it is seen as a blur. The point at which this blurring occurs, called the **near point of vision,** is usually 2–3 inches from the eye for a child, 4–6 inches for a young adult, 20 inches for a 45-year-old adult, and 60 inches for an 80-year-old adult. The increase in the near point of vision that develops with age is called **presbyopia.** It occurs because the lens becomes less flexible with increasing age, and it is the reason some older people say they could read with no problem if only they had longer arms.
2. *Pupil constriction.* **Depth of focus** is the greatest distance through which an object (for example, a newspaper that you are reading) can be moved and still remain in focus on the retina. The main factor affecting depth of focus is the size of the pupil. If the pupil diameter is small, the depth of focus is greater than if the pupil diameter is large. Therefore, with a smaller pupil opening, an object may be moved slightly nearer or farther from the eye without disturbing its focus. This is particularly important when viewing an object at close range, because the interest in detail is much greater, and thus the acceptable margin for error is smaller. When the pupil is constricted, the light entering the eye tends to pass more nearly through the center of the lens and is more accurately focused than light passing through the edges of the lens. Pupil diameter also regulates the amount of light entering the eye. The dimmer the light, the greater the pupil diameter must be. Therefore, as the pupil constricts during close vision, more light is required on the object being observed.
3. *Convergence.* Because the light rays entering the eyes from a distant object are nearly parallel, both pupils can pick up the light rays when the eyes are directed more or less straight ahead. As an object moves closer, however, the eyes must be rotated medially, so that the object is kept focused on corresponding areas of each retina. Otherwise, the object appears blurry. This medial rotation of the eyes, called **convergence,** is accomplished

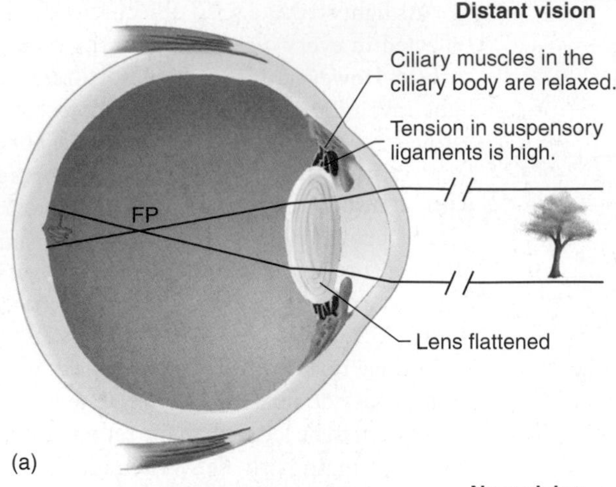
Distant vision
Ciliary muscles in the ciliary body are relaxed.
Tension in suspensory ligaments is high.
FP
Lens flattened
(a)

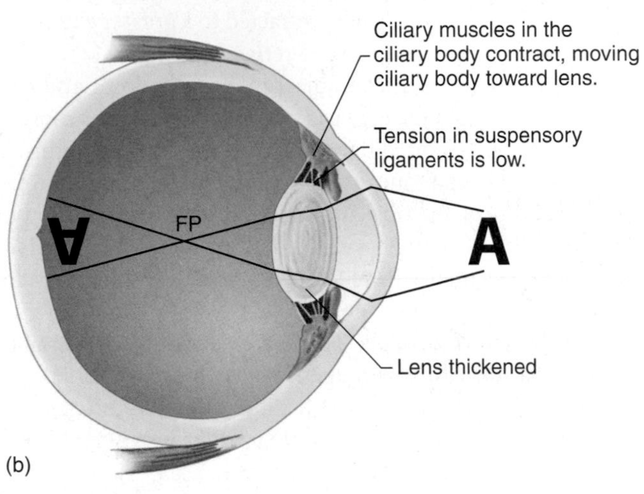
Near vision
Ciliary muscles in the ciliary body contract, moving ciliary body toward lens.
Tension in suspensory ligaments is low.
FP
A
Lens thickened
(b)

FIGURE 15.17 Focus and Accommodation
The focal point (FP) is where light rays cross. (*a*) Distant vision. The lens is flattened, and the image is focused on the retina. (*b*) Accommodation for near vision. The lens is more rounded, and the image is focused on the retina.

by a reflex that stimulates the medial rectus muscle of each eye. To observe convergence, have someone stand facing you. Ask the person to reach out one hand and extend an index finger as far in front of his or her face as possible. Then have the person keep the gaze fixed on the finger while slowly bringing the finger in toward his or her nose until finally touching it. Notice the movement of the person's pupils. What happens?

Predict 4

During a recent Alaskan cruise, Max and his grandfather were staring at a glacier some distance from their tour boat. Suddenly, Max's mother interrupted them to show them a piece of ice one of the deck hands had scooped from the water in the bay. Max was able to focus on the piece of ice and see its clear blue features, but his grandfather had to reach for his glasses. What changes occurred in their eyes between the time they were staring at the glacier and the time they were looking at the piece of ice? Explain why Max's grandfather reached for his glasses.

ASSESS YOUR PROGRESS

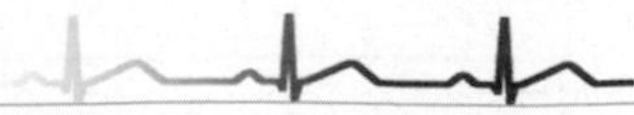

18. *What causes light to refract? What is a focal point? What is emmetropia?*
19. *What three processes occur to focus an object on the retina that is closer than 20 feet? Explain what occurs in each process.*
20. *How are far point of vision and near point of vision determined?*
21. *Describe the image of an object as it is focused on the retina.*

Structure and Function of the Retina

Leonardo da Vinci, speaking of the eye, said, "Who would believe that so small a space could contain the images of all the universe?" The retina of each eyeball, which gives us the potential to see the whole world, is about the size and thickness of a postage stamp.

As stated earlier in this section, the retina consists of a neural layer and a pigmented layer (figure 15.18). The neural layer, in turn, has several sublayers: three neuron layers (composed of photoreceptor cells, bipolar cells, and ganglionic cells, respectively) and two plexiform (networklike) layers, where the neurons of adjacent layers synapse with each other. The outer plexiform layer is between the photoreceptor and bipolar cell layers. The inner plexiform layer is between the bipolar and ganglionic cell layers (figure 15.18*a*).

The pigmented layer, or pigmented epithelium, consists of a single layer of cells filled with the pigment melanin. Together with the pigment in the choroid, the pigmented layer provides a black matrix that enhances visual acuity by isolating individual photoreceptors and reducing light scattering. However, this pigmentation is not strictly necessary for vision. People with albinism (lack of pigment) can see, although their visual acuity is reduced because of some light scattering.

The portion of the neural layer nearest the pigmented layer consists of photoreceptor cells called rods and cones (figure 15.18*b*).

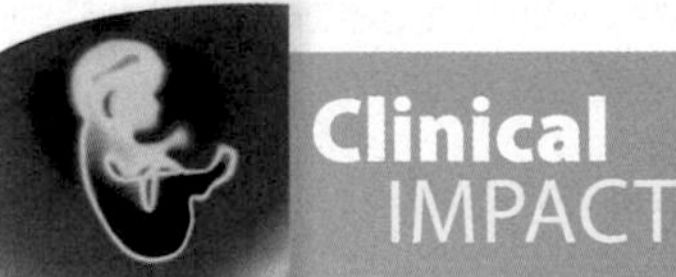

Clinical IMPACT

Eye Pigment

The eyeball is a closed chamber that allows light to enter only through the pupil. The light is absorbed by the pigmented inner lining of the eyeball; thus, looking into it is like looking into a dark room. The pupil appears black because of the pigment in the choroid and the pigmented portion of the retina. However, if a bright light is directed into the pupil, the reflected light is red because of the blood vessels on the surface of the retina. This is why the pupils of a person looking directly at a flash camera often appear red in a photograph.

People with albinism lack the pigment melanin, and the pupil always appears red because the eyeball's inner lining lacks melanin to absorb light and prevent it from being reflected from the back of the eyeball. The diffusely lighted blood vessels in the interior of the eyeball contribute to the red color of the pupil.

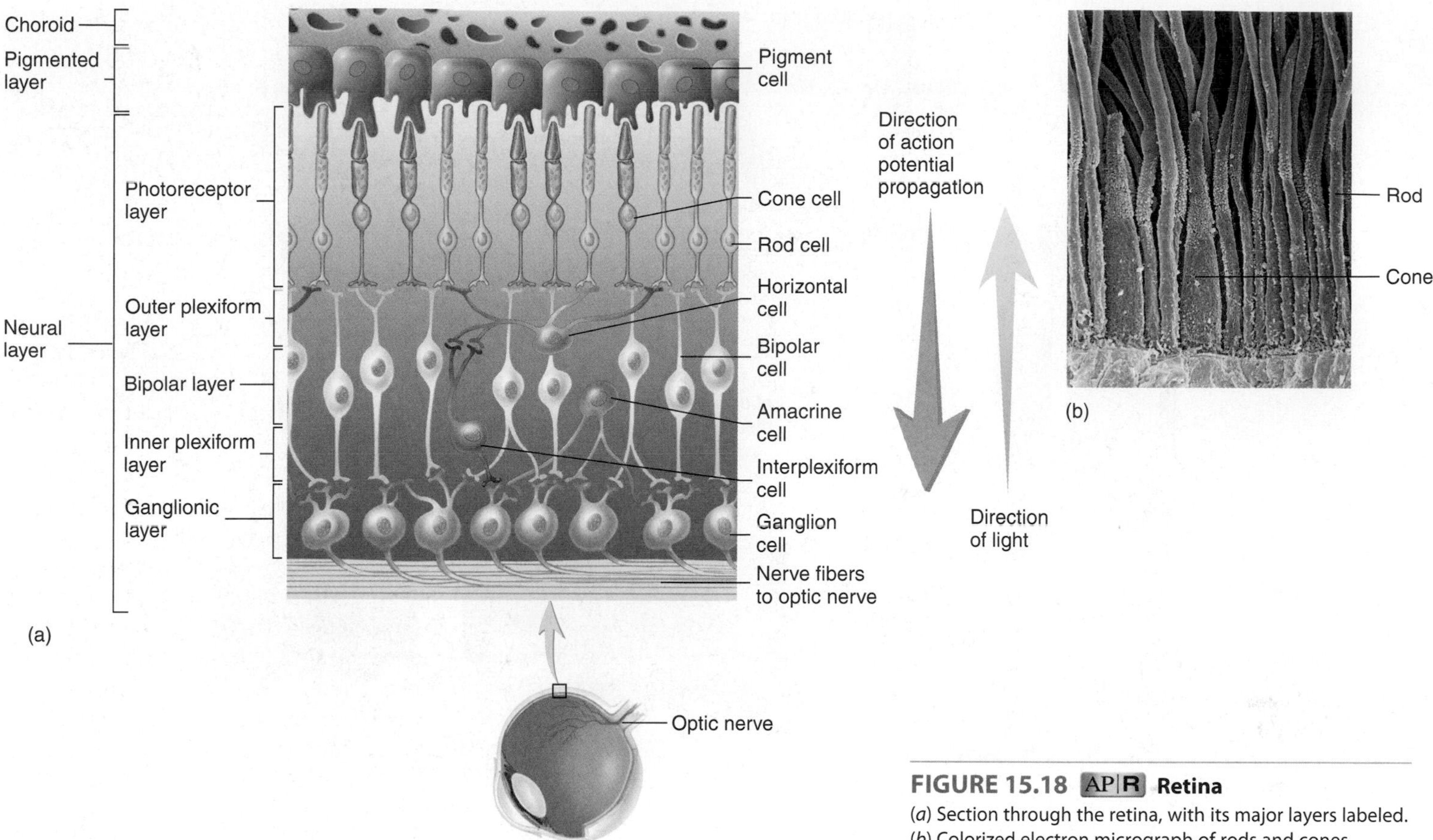
Choroid
Pigmented layer
Neural layer
Photoreceptor layer
Outer plexiform layer
Bipolar layer
Inner plexiform layer
Ganglionic layer
(a)
Pigment cell
Cone cell
Rod cell
Horizontal cell
Bipolar cell
Amacrine cell
Interplexiform cell
Ganglion cell
Nerve fibers to optic nerve
Optic nerve
Direction of action potential propagation
Direction of light
Rod
Cone
(b)

FIGURE 15.18 AP|R **Retina**
(*a*) Section through the retina, with its major layers labeled. (*b*) Colorized electron micrograph of rods and cones.

The rods and cones are sensitive to stimulation from visible light. The light-sensitive portion of each photoreceptor cell is adjacent to the pigmented layer.

Rods

Rods are bipolar photoreceptor cells that are involved in noncolor vision; they are responsible for vision under conditions of reduced light (table 15.1). The modified, dendritic, light-sensitive part of a rod cell is cylindrical, with no taper from base to apex (figure 15.19*a*). This rod-shaped photoreceptive part contains about 700 double-layered membranous discs. The discs contain **rhodopsin** (rō-dop′sin), a purple pigment that consists of the protein **opsin** covalently bound to a yellow photosensitive pigment called **retinal** (derived from vitamin A). Retinal alternates between two conformations (shapes): 11-*cis*- and all-*trans*-, depending on exposure to light. Opsin is a protein much like a channel protein, consisting of seven transmembrane helical regions. An extracellular plug closes the external opening of the opsin "channel." G proteins and cyclic GMP (cGMP) phosphodiesterase enzymes are also associated with the disc membranes. Sodium ion channels are located in the outer membrane of the rod cell outer segment (figure 15.19*d*).

Figure 15.20 depicts the changes that rhodopsin undergoes in response to light. In the resting (dark) state, 11-*cis*-retinal is tightly bound to the internal surface of opsin. The extracellular plug helps keep retinal from being lost from the cell out of the opsin "channel." Cyclic GMP is attached to the Na^+ channel, keeping it open. As

TABLE 15.1 Rods and Cones

Photoreceptive End	Photoreceptive Molecule	Function	Location
Rod			
Cylindrical	Rhodopsin	Noncolor vision; vision under conditions of low light	Over most of retina; none in fovea centralis
Cone			
Conical	Iodopsin	Color vision; visual acuity	Numerous in fovea centralis and macula; sparse over rest of retina

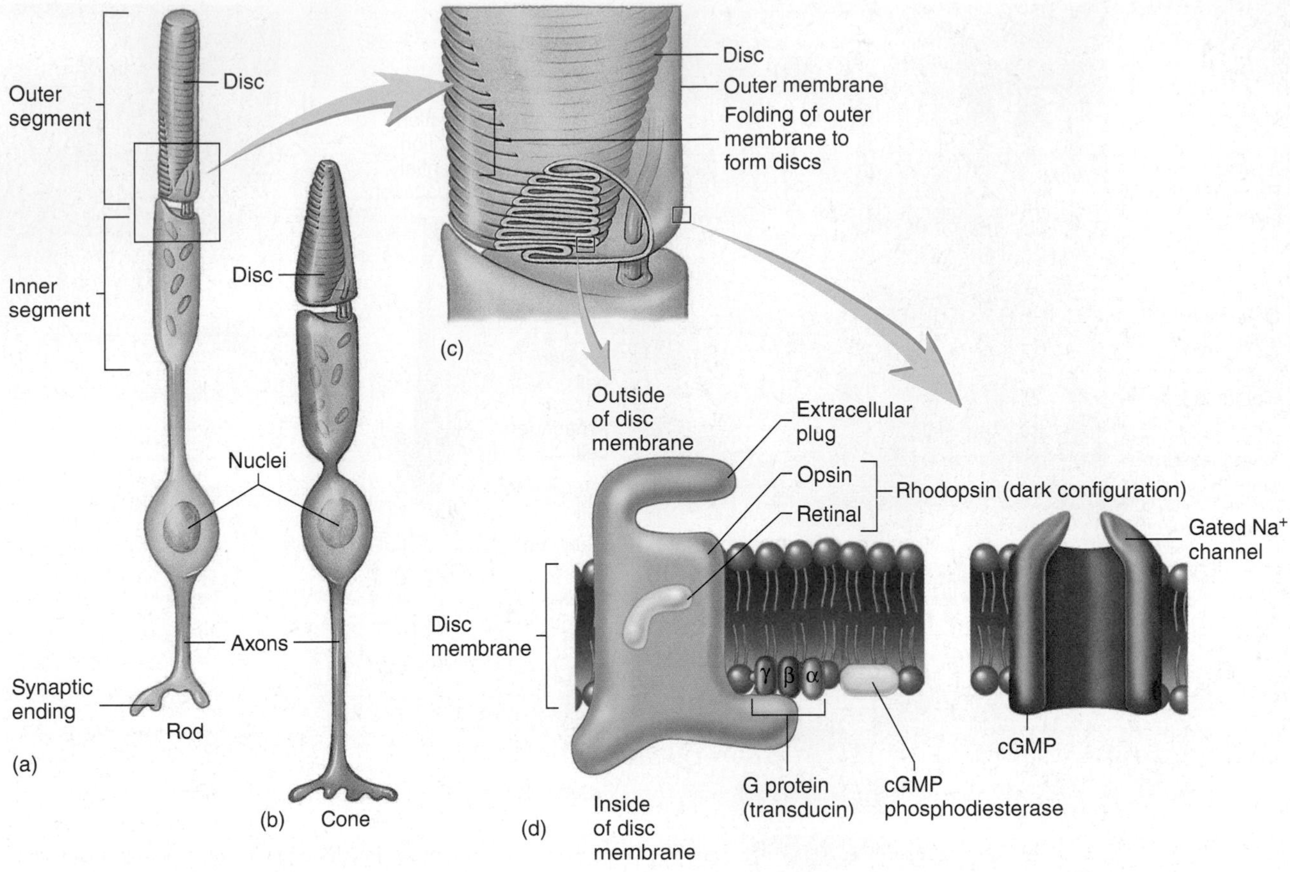
Outer segment
Disc
Inner segment
Disc
Nuclei
Axons
Synaptic ending
Rod
(a)
(b) Cone
Disc
Outer membrane
Folding of outer membrane to form discs
(c)
Outside of disc membrane
Extracellular plug
Opsin
Retinal
Rhodopsin (dark configuration)
Gated Na+ channel
Disc membrane
γ β α
cGMP
G protein (transducin)
cGMP phosphodiesterase
(d)
Inside of disc membrane

FIGURE 15.19 Sensory Receptor Cells of the Retina
(*a*) Rod cell. (*b*) Cone cell. (*c*) Enlargement of the discs in the outer segment. (*d*) Enlargement of one of the discs, showing the relation of rhodopsin and a cGMP gated Na^+ channel in the membrane.

rod cells absorb light, retinal changes shape from 11-*cis*-retinal to all-*trans*-retinal. This change causes opsin to also change shape (dark to light state). The changes in opsin activate a G protein called **transducin** (trans-doo′sin), which activates a cyclic GMP phosphodiesterase. The cGMP phosphodiesterase catalyzes the conversion of cGMP to GMP. This reaction causes cGMP to diffuse away from Na^+ channels, resulting in closure of the Na^+ channels and hyperpolarization of the cell (figure 15.21).

This hyperpolarization in the photoreceptor cells is somewhat remarkable because most neurons respond to stimuli by depolarizing. When photoreceptor cells are not exposed to light and are in a resting, nonactivated state, gated Na^+ channels in their membranes are open, and Na^+ flows into the cell. This influx of Na^+, referred to as the dark current, causes the photoreceptor cells to release the neurotransmitter glutamate from their presynaptic terminals (figure 15.21). Glutamate binds to receptors on the postsynaptic membranes of the bipolar cells of the retina, causing them to hyperpolarize. Thus, glutamate causes an inhibitory postsynaptic potential (IPSP) in the bipolar cells. The influx of Na^+ is offset by the efflux of K^+ through nongated K^+ channels. Equilibrium of Na^+ and K^+ in the cell is maintained by a sodium-potassium pump.

To summarize, when photoreceptor cells are exposed to light, the Na^+ channels close, less Na^+ enters the cell, and the amount of glutamate released from the presynaptic terminals decreases. As a result, the hyperpolarization in the bipolar cells decreases, and they depolarize sufficiently to release neurotransmitters, which stimulate ganglionic cells to generate action potentials. The number of Na^+ channels that close and the degree to which they close are proportional to the amount of light exposure.

At the final stage of this light-initiated reaction, retinal is completely released from the opsin. This free retinal may then be converted back to vitamin A, from which it was originally derived. The total vitamin A/retinal pool is in equilibrium, so that under normal conditions the amount of free retinal is relatively constant. To create more rhodopsin, the altered retinal must be converted back to its original shape, a reaction that requires energy. Once the retinal resumes its original shape, its recombination with opsin is spontaneous, and the newly formed rhodopsin can again respond to light.

The adjustment of the eyes to changes in light, as occurs when going from a darkened building into the sunlight or vice versa, is called **light** and **dark adaptation.** This adjustment is accomplished by changes in the amount of available rhodopsin. In bright light, excess rhodopsin is broken down, so that not as much is available to initiate action potentials, and the eyes become adapted to bright light. Conversely, in a dark room more rhodopsin is produced, making the retina more light-sensitive.

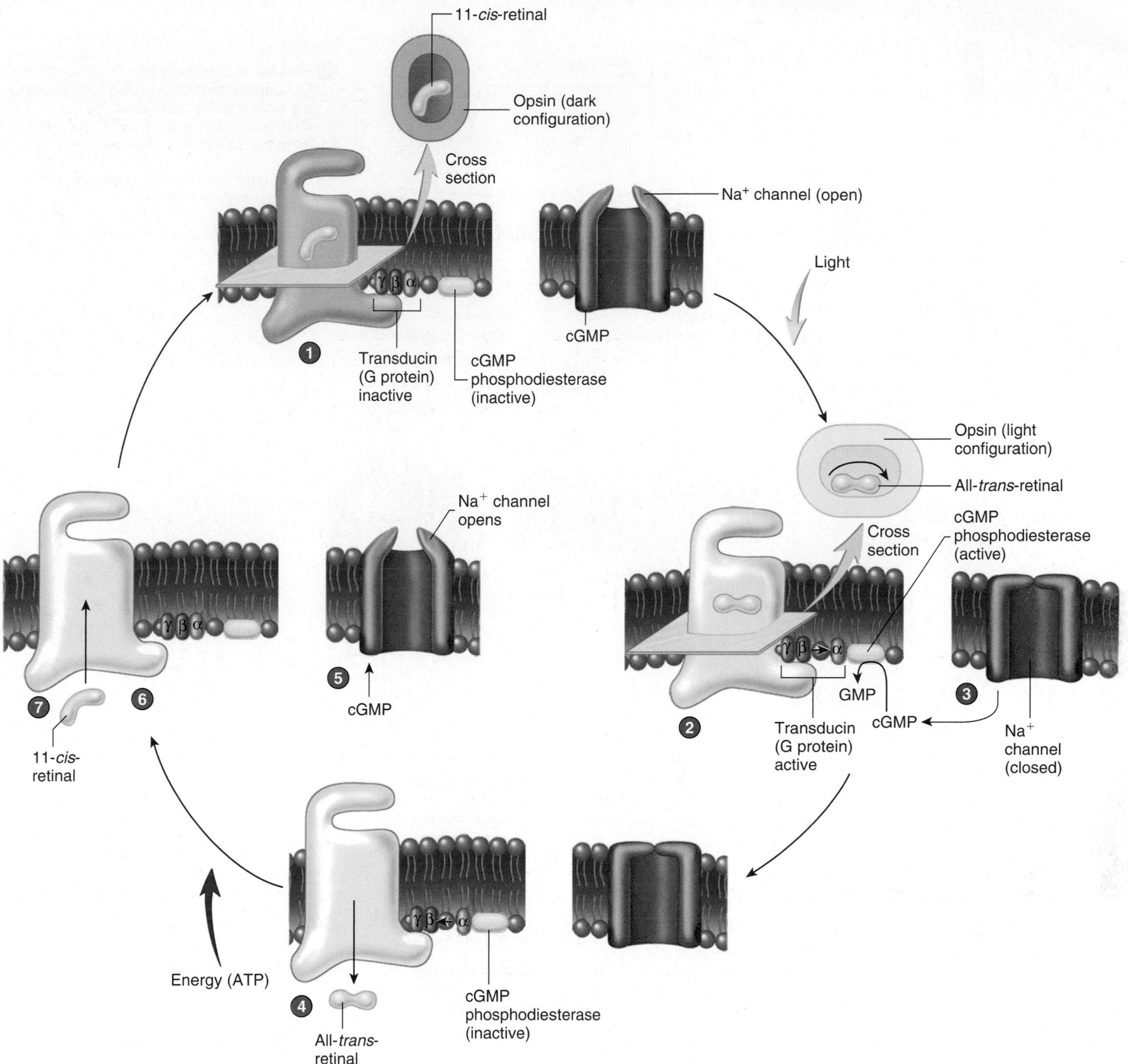
11-*cis*-retinal
Opsin (dark configuration)
Cross section
Na⁺ channel (open)
Transducin (G protein) inactive
cGMP phosphodiesterase (inactive)
cGMP
Light
Opsin (light configuration)
All-*trans*-retinal
cGMP phosphodiesterase (active)
Cross section
GMP
cGMP
Transducin (G protein) active
Na⁺ channel (closed)
Na⁺ channel opens
cGMP
11-*cis*-retinal
Energy (ATP)
All-*trans*-retinal
cGMP phosphodiesterase (inactive)

1. Retinal (in an inactive configuration called 11-*cis*-) is attached inside opsin to make rhodopsin. Cyclic GMP attached to a Na^+ channel keeps it open.
2. Light causes retinal to change shape from 11-*cis*-retinal to all-*trans*-retinal. This change causes opsin also to change shape (from its dark to light configuration). This activated rhodopsin activates a G protein (called transducin), which activates a cGMP phosphodiesterase.
3. The cGMP phosphodiesterase catalyzes the conversion of cGMP to GMP, which decreases cGMP concentration and causes the cGMP to diffuse away from the Na^+ channels, closing these channels.
4. All-*trans*-retinal detaches from opsin. Transducin is reassociated, and cGMP phosphodiesterase is inactivated.
5. Cyclic GMP concentration increases, and cGMP attaches to Na^+ channels, causing them to open.
6. All-*trans*-retinal is converted to 11-*cis*-retinal, a process that requires energy.
7. 11-*cis*-retinal attaches to opsin, which returns to its original (*dark*) configuration.

PROCESS FIGURE 15.20 Rhodopsin Cycle

FUNDAMENTAL **Figure**

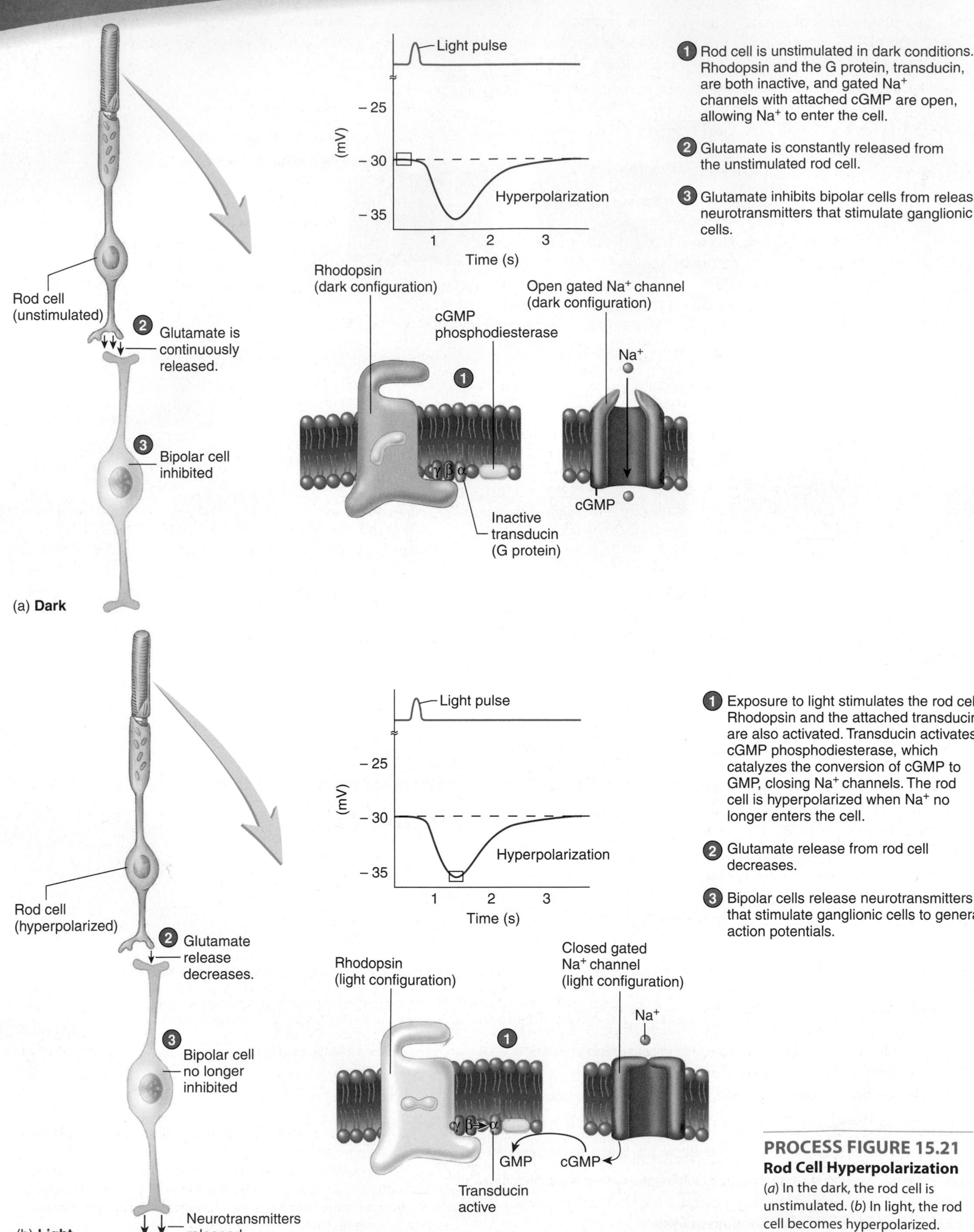

Light pulse
(mV)
– 25
– 30
– 35
Hyperpolarization
1
2
3
Time (s)
1 Rod cell is unstimulated in dark conditions. Rhodopsin and the G protein, transducin, are both inactive, and gated Na⁺ channels with attached cGMP are open, allowing Na⁺ to enter the cell.
2 Glutamate is constantly released from the unstimulated rod cell.
3 Glutamate inhibits bipolar cells from releasing neurotransmitters that stimulate ganglionic cells.
Rod cell (unstimulated)
2 Glutamate is continuously released.
3 Bipolar cell inhibited
Rhodopsin (dark configuration)
cGMP phosphodiesterase
Open gated Na⁺ channel (dark configuration)
Na⁺
1
γ β α
cGMP
Inactive transducin (G protein)
(a) Dark
Light pulse
(mV)
– 25
– 30
– 35
Hyperpolarization
1
2
3
Time (s)
1 Exposure to light stimulates the rod cell. Rhodopsin and the attached transducin are also activated. Transducin activates cGMP phosphodiesterase, which catalyzes the conversion of cGMP to GMP, closing Na⁺ channels. The rod cell is hyperpolarized when Na⁺ no longer enters the cell.
2 Glutamate release from rod cell decreases.
3 Bipolar cells release neurotransmitters that stimulate ganglionic cells to generate action potentials.
Rod cell (hyperpolarized)
2 Glutamate release decreases.
3 Bipolar cell no longer inhibited
Rhodopsin (light configuration)
Closed gated Na⁺ channel (light configuration)
Na⁺
1
γ β α
GMP
cGMP
Transducin active
Neurotransmitters released
(b) Light

PROCESS FIGURE 15.21
Rod Cell Hyperpolarization
(*a*) In the dark, the rod cell is unstimulated. (*b*) In light, the rod cell becomes hyperpolarized.

Light and dark adaptation also involves pupil reflexes. In dim light, the pupil enlarges to allow more light into the eye; in bright light, the pupil constricts to allow less light into the eye. In addition, during light conditions, rod function decreases and cone function increases, whereas the opposite happens during dark conditions. This occurs because rod cells are more sensitive to light than cone cells and because the rhodopsin in rods is depleted more rapidly than the visual pigment in cones.

Cones

Cones are bipolar photoreceptor cells with a conical, light-sensitive part that tapers slightly from base to apex (see figure 15.19*b*). The outer segments of the cone cells, like those of the rods, consist of double-layered discs. The discs are slightly more numerous and more closely stacked in the cones than in the rods.

Color vision and visual acuity are the functions of cone cells (table 15.1). Color is a function of the wavelength of light, and each color results from a certain wavelength within the visible spectrum. Even though rods are very sensitive to light, they cannot detect color, and sensory input that ultimately reaches the brain from these cells is interpreted by the brain as shades of gray. Cones require relatively bright light to function. As a result, as the light decreases, so does the color of objects that can be seen until, under conditions of very low illumination, the objects appear gray. This occurs because, as the light decreases, fewer cone cells respond to the dim light.

Cone cells contain a visual pigment, **iodopsin** (ī-ō-dop′sin), which consists of retinal combined with a photopigment opsin protein. Three major types of color-sensitive opsin exist—blue, red, and green; each closely resembles the opsin proteins of rod cells but with somewhat different amino acid sequences. These color photopigments function in much the same manner as rhodopsin; however, whereas rhodopsin responds to the entire spectrum of visible light, each iodopsin is sensitive to a much narrower spectrum.

As can be seen in figure 15.22, considerable overlap occurs in the wavelength of light to which the blue, green, and red pigments are sensitive; however, each pigment absorbs light of a certain range of wavelengths. As light of a given wavelength, representing a certain color, strikes the retina, all cone cells containing photopigments capable of responding to that wavelength generate action potentials. Because of the overlap among the three types of cones, especially between the green and red pigments, different proportions of cone cells respond to each wavelength, thus allowing color perception over a wide range. Color is interpreted in the visual cortex of the occipital lobe as combinations of sensory input originating from cone cells. For example, when orange light strikes the retina, 99% of the red-sensitive cones respond, 42% of the green-sensitive cones respond, and no blue cones respond. When yellow light strikes the retina, the response shifts, so that a greater number of green-sensitive cones respond. The variety of combinations created allows humans to distinguish among several million gradations of light and shades of color.

Not everyone sees the same red. Two forms of the red photopigment are common in humans. Approximately 60% of people have the amino acid serine in position 180 of the red opsin protein, whereas 40% have alanine in that position. That subtle difference in the protein results in slightly different absorption characteristics (figure 15.22). Even though we were each taught to recognize red when we see a certain color, apparently we do not all see that color in quite the same way.

Distribution of Rods and Cones in the Retina

In addition to their role in color vision, cones are involved in visual acuity. The macula and especially the fovea centralis function when visual acuity is required, such as for focusing on the words of this page. The fovea centralis has about 35,000 cones and no rods. The rest of the macula has more cones than rods. However, the 120 million rods are 10–20 times more plentiful than cones over most of the remaining retina. They are more highly concentrated away from the macula and are more important in low-light conditions.

Predict 5

Explain why at night a person may notice a movement "out of the corner of the eye" but, when the person tries to focus on the area of movement, nothing seems to be there.

Inner Layers of the Retina

The middle and inner cell layers of the neural layer of the retina consist of two major neuron types: bipolar cells and ganglion cells (see figure 15.18*a*). The rod and cone photoreceptor cells synapse with **bipolar cells,** which in turn synapse with **ganglion cells.** Except in the area of the fovea centralis, axons from the ganglion cells pass over the inner surface of the retina, converge at the **optic disc;** and exit the eye as the **optic nerve (II).** The fovea centralis is devoid of

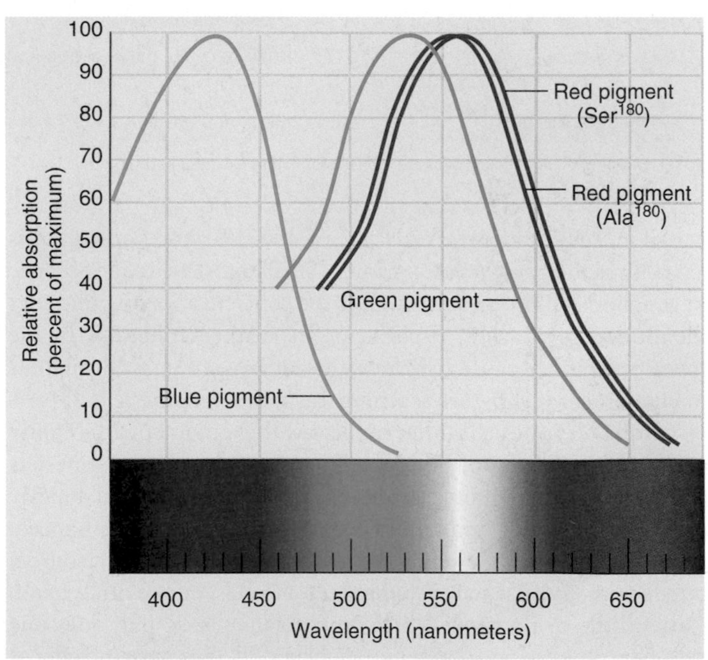
Relative absorption (percent of maximum)
100
90
80
70
60
50
40
30
20
10
0
Red pigment (Ser180)
Red pigment (Ala180)
Green pigment
Blue pigment
400
450
500
550
600
650
Wavelength (nanometers)

FIGURE 15.22 Wavelengths to Which Blue, Green, and Red Pigments Are Sensitive
The red pigment exists in two chemical forms. One, found in 60% of the population, has the amino acid serine at position 180; the other, found in 40% of the population, has an alanine at position 180. Each red pigment has a slightly different wavelength sensitivity.

Diseases and Disorders

TABLE 15.2 Vision

Condition	Description
Infections	
Conjunctivitis (kon-jŭnk-ti-vi′tis)	Inflammation of conjunctiva, usually from bacterial infection; one form, pinkeye, occurs primarily in children
Trachoma (tră-kō′mă)	Type of conjunctivitis caused by *Chlamydia*; leading cause of infectious blindness in the world; transmitted by contact or flies
Neonatal gonorrheal ophthalmia	*Neisseria gonorrhoeae* infection in eyes of newborn infant whose mother has gonorrhea; can cause blindness if the infant is not treated with silver nitrate, tetracycline, or erythromycin drops soon after birth
Stye (stī)	Infection of eyelash hair follicle
Defects of Focus, Alignment, or Color Vision	
Myopia (mī-ō′pē-ă)	Nearsightedness—ability to see close but not distant objects; caused when refractive power of cornea and lens is too great relative to length of eye
Hyperopia (hī-per-ō′pē-ă)	Farsightedness—ability to see distant but not close objects; caused when cornea is too flat or lens has too little refractive power relative to length of eye
Presbyopia (prez-bē-ō′pē-ă)	Decrease in near vision, a normal part of aging
Astigmatism (ă-stig′mă-tizm)	Cornea or lens is not uniformly curved, so image is not sharply focused
Strabismus (stra-biz′mŭs)	One or both eyes are misdirected; can result from weak eye muscles
Diplopia (di-plō′pē-ă)	Double vision
Color blindness	Complete or partial absence of perception of one or more colors (see figure 15B); most forms are more frequent in males
Blindness	
Cataract (kat′ă-rakt)	Clouding of lens as a result of advancing age, infection, or trauma; most common cause of blindness in the world
Macular degeneration	Loss of sharp central vision, peripheral vision maintained; leading cause of legal blindness in older Americans; most causes not known
Glaucoma (glaw-kō′mă)	Excessive pressure buildup in aqueous humor; may destroy retina or optic nerve, resulting in blindness
Diabetic retinopathy	Involves optic nerve degeneration, cataracts, retinal detachment; often caused by blood vessel degeneration and hemorrhage
Retinal detachment	Separation of sensory retina from pigmented retina; relatively common problem; may result in vision loss

Go to: www.mhhe.com/seeley10 for additional information on these pathologies.

ganglion cell processes, resulting in a small depression in this area—thus the name *fovea*, meaning "small pit." Due to the absence of ganglion cell processes as well as the concentration of cone cells mentioned previously, visual acuity is further enhanced in the fovea centralis because light rays do not have to pass through as many tissue layers before reaching the photoreceptor cells.

Rod and cone cells differ in the way they interact with bipolar and ganglion cells. One bipolar cell receives input from numerous rods, and one ganglion cell receives input from several bipolar cells, so that spatial summation occurs and the signal is enhanced. This system allows awareness of stimuli from very dim light sources but decreases visual acuity in these cells. Cones, on the other hand, exhibit little or no convergence on bipolar cells, so that one cone cell may synapse with only one bipolar cell. This system reduces light sensitivity but enhances visual acuity.

The area from which a ganglion cell receives input, called the **receptive field,** is roughly circular, with a smaller portion in the center, called the receptive field center, and a larger surrounding area. The receptive fields in the fovea centralis are very small, compared with the receptive fields of more peripheral parts of the retina. The small size of these receptive fields is consistent with the visual acuity of the fovea centralis.

Two types of receptive fields exist: on-center ganglion cells and off-center ganglion cells. On-center cells generate more action potentials when light is directed onto the receptive field center. Off-center neurons generate more action potentials when light is turned off in the receptive field center or when light shines on the surrounding area. These receptive fields respond primarily to contrasts in light (edges) rather than to the absolute intensity of light.

Within the inner layers of the retina, interneurons modify the signals from the photoreceptor cells before the signal ever leaves the retina (see figure 15.18). **Horizontal cells** form the outer plexiform layer and synapse with photoreceptor cells and bipolar cells. **Amacrine** (am′ă-krin) **cells** form the inner plexiform layer and synapse with bipolar and ganglion cells. **Interplexiform cells** form the bipolar layer and synapse with amacrine, bipolar, and horizontal cells to form a feedback loop. The interneurons are either excitatory or inhibitory on the cells with which they synapse. These interneurons enhance borders and contours, thereby increasing the intensity at boundaries, such as the edge of a dark object against a light background.

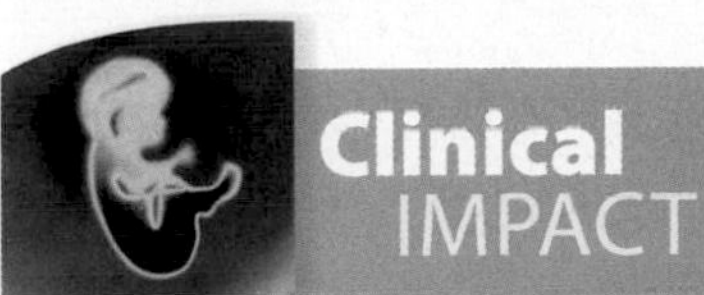

Clinical IMPACT | Visual Acuity

Visual acuity is the eye's ability to focus an image on the retina so that a clear image is perceived. Many factors affect visual acuity, including the shape of the eyeball and the flexibility of the lens. When a person's visual acuity is tested, a chart with rows of letters of decreasing size is placed 20 feet from the eye, and the person is asked to read the row of the smallest letters he or she can see clearly. The rows of letters are standardized for normal vision at 20 feet. If the person can clearly see the row of letters identified as 20 feet, his or her vision is considered 20/20, meaning the person can see at 20 feet what a person with normal vision can see at 20 feet. On the other hand, if the person can clearly see the row of letters identified as 40 feet, his or her vision is 20/40, meaning the person can see at 20 feet what a person with normal vision can see at 40 feet. In this case, a defect in visual acuity is diagnosed, and corrective lenses are usually prescribed. Some common visual acuity defects are myopia, hyperopia, presbyopia, and astigmatism.

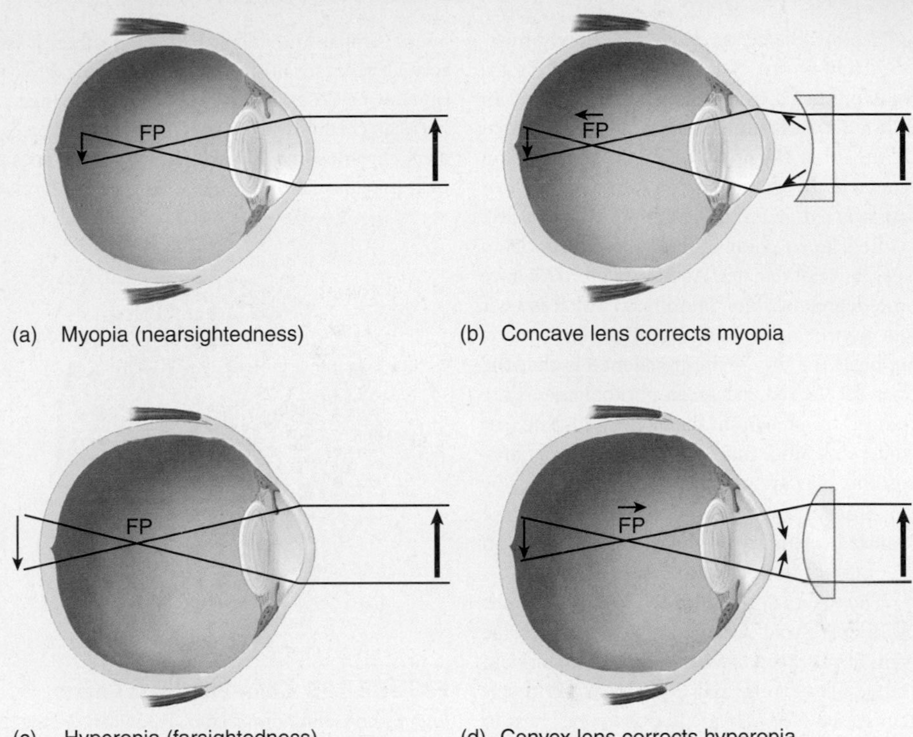
FP
FP
FP
FP
(a) Myopia (nearsightedness)
(b) Concave lens corrects myopia
(c) Hyperopia (farsightedness)
(d) Convex lens corrects hyperopia

FIGURE 15A Visual Disorders and Their Correction

Myopia

People who have **myopia** (mī-ō′pē-ă), or nearsightedness, can see close objects clearly, but distant objects appear blurry. Myopia is a defect of the eye in which the focusing system, the cornea and lens, is optically too powerful, or the eyeball is too long (axial myopia). As a result, the focal point is too near the lens, and the image is focused in front of the retina (figure 15A*a*).

Myopia is corrected by a concave lens that counters the refractive power of the eye. Concave lenses cause the light rays coming to the eye to diverge and are therefore called "minus" lenses (figure 15A*b*).

Another technique for correcting mild myopia is **radial keratotomy** (ker′ă-tot′ō-mē), which consists of making a series of four to eight radiating cuts in the cornea. The cuts are intended to weaken the dome of the cornea slightly, so that it becomes more flattened and eliminates the myopia. One problem with the technique is that it is difficult to predict how much flattening will occur. In one study of 400 patients 5 years after the surgery, 55% had normal vision, 28% were still somewhat myopic, and 17% had become hyperopic. Another problem is that some patients are bothered by glare following radial keratotomy, because the slits apparently do not heal evenly.

Over the past decade an alternative procedure, **lasix,** or *laser corneal sculpturing,* has become more common for correcting myopia. Lasix is a laser surgery procedure in which a thin portion of the cornea is etched away to make the cornea less convex. The advantage of this procedure is that the results can be predicted more accurately than can those of radial keratotomy.

Hyperopia

People who have **hyperopia** (hī-per-ō′pē-ă), or farsightedness, can see distant objects clearly, but close objects appear blurry. In hyperopia, the cornea and lens system is optically too weak or the eyeball is too short. The image is focused behind the retina (figure 15A*c*).

Hyperopia can be corrected by convex lenses that cause light rays to converge as they approach the eye (figure 15A*d*). Such lenses are called "plus" lenses.

Presbyopia

Presbyopia (prez-bē-ō′pē-ă) is the normal, presently unavoidable degeneration of the accommodation power of the eye associated with aging. It occurs because the lens becomes sclerotic and less flexible. The eye is presbyopic when the near point of vision has increased beyond 9 inches. The average age for the onset of presbyopia is the mid-40s. Avid readers and people who engage in fine, close work may develop the symptoms earlier.

Presbyopia can be corrected by wearing "reading glasses" for close work and removing them to see at a distance. Because constantly removing and replacing glasses is annoying and because reading glasses hamper vision of only a few feet away, many people wear **bifocals,** which have different lenses in the top and the bottom, or **progressive lenses,** which have a graded lens.

Astigmatism

Astigmatism (ă-stig′mă-tizm) is a type of refractive error that affects the quality of focus. If the cornea or lens is not uniformly curved, the light rays do not focus at a single point, but fall as a blurred circle. Regular astigmatism can be corrected by glasses formed with the opposite curvature gradation. In irregular astigmatism, the abnormal form of the cornea fits no specific pattern and is very difficult to correct with glasses.

Clinical GENETICS | Color Blindness

Color blindness results from the dysfunction of one or more of the three photopigments (red, green, blue) involved in color vision. If one pigment is dysfunctional and the other two are functional, the condition is called **dichromatism.** An example of dichromatism is red-green color blindness (figure 15B).

Red-green color blindness is common in males, but not females. About 7% of males have some degree of color blindness, which is over eight times more common than in females. The basis for this male prevalence is that the genes for the red and green photopigments are arranged in tandem on the X chromosome (see chapter 29). Since males have only one X chromosome, they are more likely to be affected by an X-linked mutation than females, who have a higher probability of having a good gene on one of their two X chromosomes.

The vast majority, over 95%, of color blindness involves red-green color vision, not blue vision. The reason can be traced to the tandem arrangement of the red and green photopigment genes. Not only are the two genes next to each other, but they are also nearly identical; differences in only 3 of the 360 amino acids determine the red versus green wavelength absorption characteristics. Because the red and green genes are so similar and adjacent to each other, it is relatively easy for mistakes to occur during development as DNA is replicated and exchanged between chromosomes (see chapter 29). Hence, an X chromosome may lack one or both genes, or it may have a hybrid gene containing exons from both red and green genes, which may or may not alter their degree of functionality. The blue photopigment gene is rarely associated with color blindess, since it is not adjacent to another photopigment gene. It is also not X-linked, so it is equally rare in males and females.

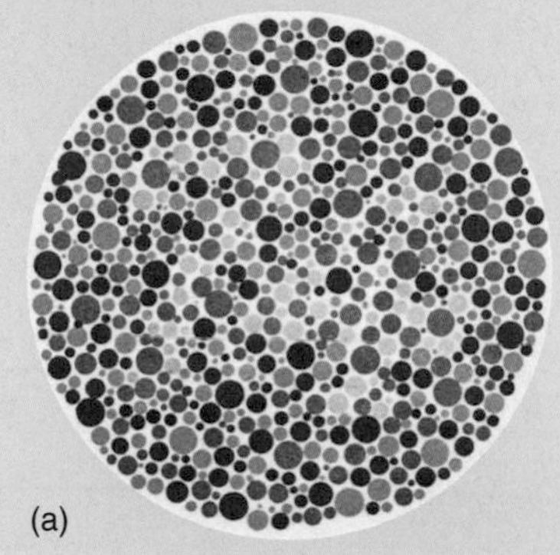

(a)

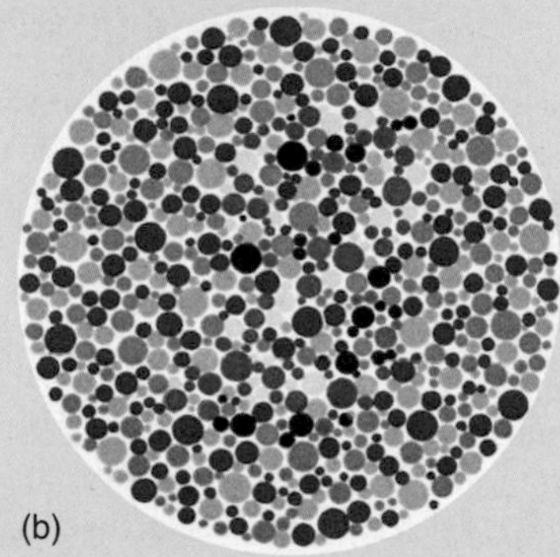

(b)

FIGURE 15B Color Blindness Charts

(*a*) A person with normal color vision can see the number 74, whereas a person with red-green color blindness sees the number 21. (*b*) A person with normal color vision can see the number 5, whereas a person with red-green color blindness sees the number 2.

ASSESS YOUR PROGRESS

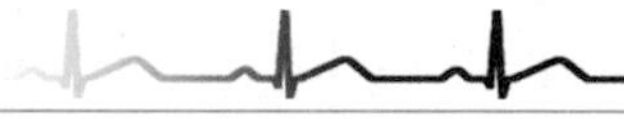

22. *What function do the pigmented layer of the retina and the pigment of the choroid perform?*

23. *Describe the changes that occur in a rod cell after light strikes rhodopsin. How does rhodopsin reform? Why is the response of a rod cell to a stimulus unusual?*

24. *How do dark and light adaptation occur?*

25. *What are the three types of cone cells? How do they produce the color we see?*

26. *Describe the arrangement of rods and cones in the fovea centralis, the macula, and the peripheral parts of the retina.*

27. *Starting with a rod or cone cell, name the cells or structures that an action potential encounters while traveling to the optic nerve.*

Neuronal Pathways for Vision

Figure 15.23 shows the neuronal pathways that transmit action potentials generated by light from the time light enters the eye until it reaches the area of the cerebrum where vision is perceived. The optic nerve leaves the eye and exits the orbit through the optic foramen to enter the cranial cavity. Just inside the cranial cavity and just anterior to the pituitary gland, the optic nerves are connected to each other at the **optic chiasm** (kī′azm). Ganglion cell axons from the nasal retina (the medial portion of the retina) cross through the optic chiasm and project to the opposite side of the brain. Ganglion cell axons from the temporal retina (the lateral portion of the retina) pass through the optic nerves and project to the brain on the same side of the body without crossing.

Beyond the optic chiasm, the route of the ganglionic axons is called the **optic tract** (figure 15.23*a*,*b*,*d*). Most of the optic tract axons terminate in the **lateral geniculate nucleus** of the thalamus. However, some axons do not terminate in the thalamus but separate from the optic tract to terminate in the **superior colliculi,** the center for visual reflexes (see chapter 13). Neurons of the lateral geniculate nucleus form the fibers of the **optic radiations,** which project to the **visual cortex** in the **occipital lobe.** Neurons of the visual cortex integrate the messages coming from the retina into a single message, translate that message into a mental image, and then transfer the image to other parts of the brain, where it is evaluated and either acted on or ignored.

1. Each visual field is divided into a temporal and a nasal half.
2. After passing through the lens, light from each half of a visual field projects to the opposite side of the retina.
3. An optic nerve consists of axons extending from the retina to the optic chiasm.
4. In the optic chiasm, axons from the nasal part of the retina cross and project to the opposite side of the brain. Axons from the temporal part of the retina do not cross.
5. An optic tract consists of axons that have passed through the optic chiasm (with or without crossing) to the thalamus.
6. The axons synapse in the lateral geniculate nuclei of the thalamus. Collateral branches of the axons in the optic tracts synapse in the superior colliculi.
7. An optic radiation consists of axons from thalamic neurons that project to the visual cortex.
8. The right part of each visual field (*dark green* and *light blue*) projects to the left side of the brain, and the left part of each visual field (*light green* and *dark blue*) projects to the right side of the brain.

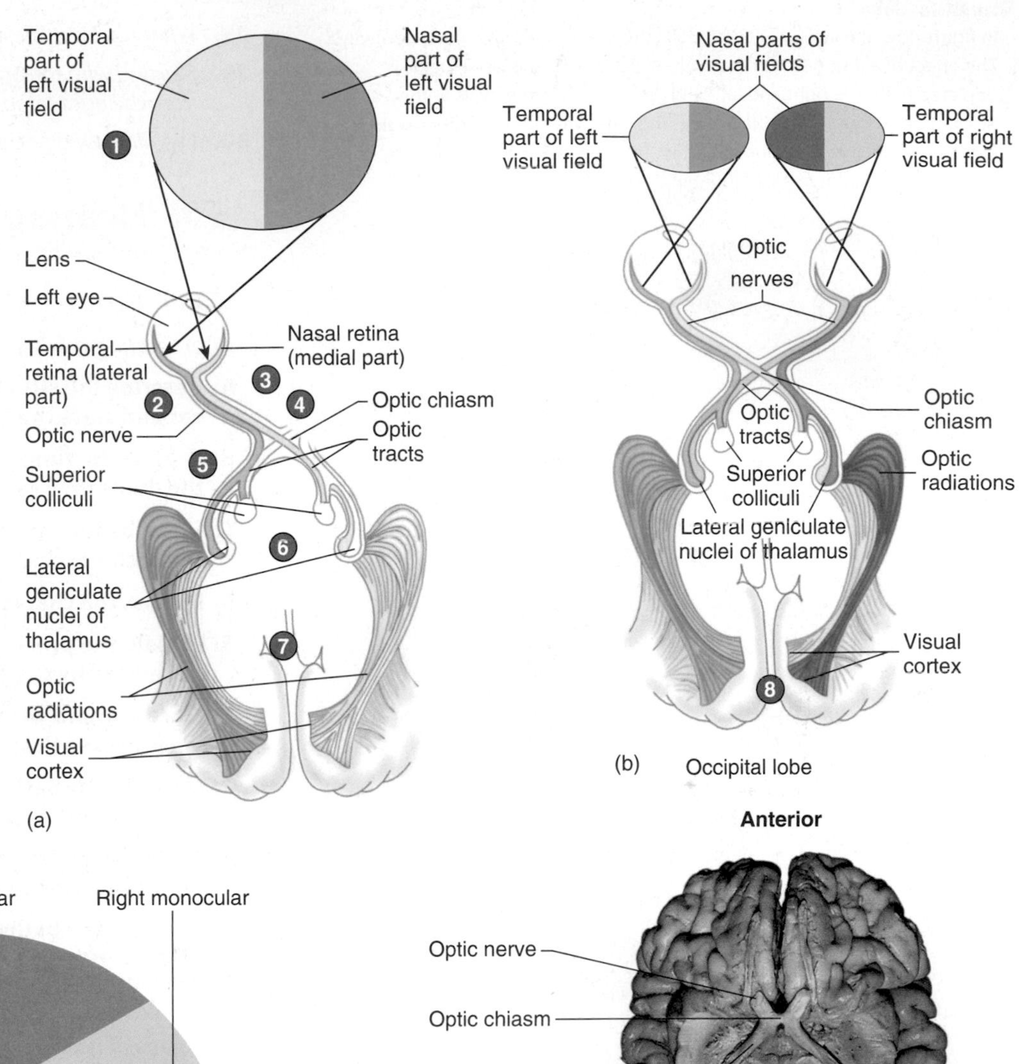
Left visual field
Temporal part of left visual field
Nasal part of left visual field
Lens
Left eye
Temporal retina (lateral part)
Nasal retina (medial part)
Optic nerve
Optic chiasm
Optic tracts
Superior colliculi
Lateral geniculate nuclei of thalamus
Optic radiations
Visual cortex
(a)
Nasal parts of visual fields
Temporal part of left visual field
Temporal part of right visual field
Optic nerves
Optic chiasm
Optic tracts
Optic radiations
Superior colliculi
Lateral geniculate nuclei of thalamus
Visual cortex
(b)
Occipital lobe

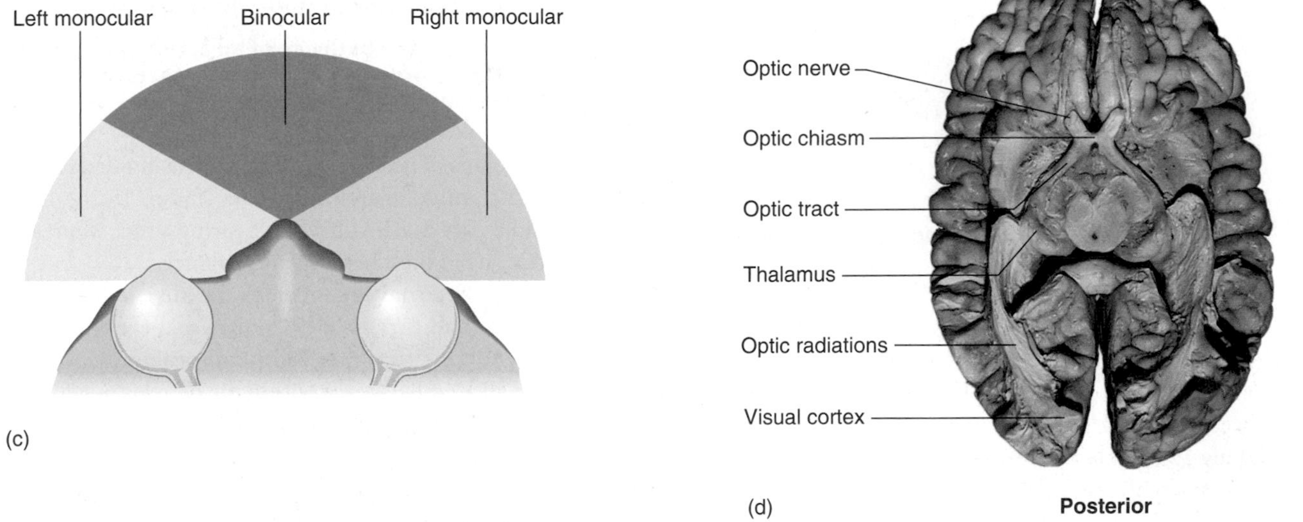
Left monocular
Binocular
Right monocular
(c)
Anterior
Optic nerve
Optic chiasm
Optic tract
Thalamus
Optic radiations
Visual cortex
(d)
Posterior

PROCESS FIGURE 15.23 APR **Visual Pathways**

(*a*) Pathways for the left eye (superior view). (*b*) Pathways for both eyes (superior view). (*c*) Overlap of the fields of vision (superior view). (*d*) Inferior view of the brain, showing the visual nerves, tracts, and pathways.

The projections of ganglion cells from the retina can be related to the **visual fields** (figure 15.23*c*). You can observe the visual field by closing one eye. Everything that you can see with the one open eye is the visual field of that eye. The visual field of each eye can be divided into temporal (lateral) and nasal (medial) parts. The temporal part of the visual field projects onto the nasal retina, whereas the nasal part projects to the temporal retina. The projections and nerve pathways are arranged such that images entering the eye from the right part of each visual field project to the left side of the brain. Conversely, the left part of each visual field projects to the right side of the brain.

Predict 6

In figure 15C, the lines at *A* and *B* depict two lesions on the visual pathways. The effect of a lesion at *A* in the optic radiations on the visual fields is depicted (with the right and left fields separated) in the ovals at the bottom of the figure. The black areas indicate what parts of the visual fields are defective. Describe the effect that the lesion at *B* has on the visual fields (see figure 15.23 for help).

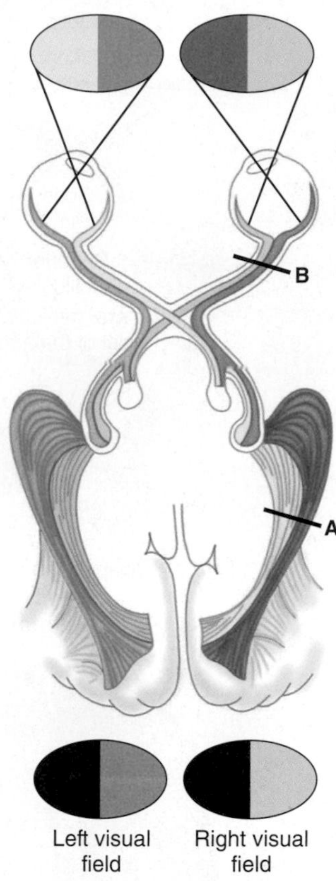
B
A
Left visual field
Right visual field

FIGURE 15C Effects of Lesions on Visual Pathways

Because the optic chiasm lies just anterior to the pituitary gland, a pituitary tumor can put pressure on the optic chiasm and cause visual defects. Because the nerve fibers crossing in the optic chiasm are carrying information from the temporal halves of the visual fields, a person with optic chiasm damage cannot see objects in the temporal halves of the visual fields. This condition, called **tunnel vision,** is often an early sign of a pituitary tumor.

The visual fields of the eyes partially overlap (see figure 15.23*c*). The region of overlap—the area seen with both eyes at the same time—is the area of **binocular vision.** Because humans see the same object with both eyes, the image of the object reaches the retina of one eye at a slightly different angle from that of the other. With experience, the brain can interpret these differences in angle, so that distance can be judged quite accurately. Thus, binocular vision gives us **depth perception,** the ability to distinguish between near and far objects and to judge their distance.

ASSESS **YOUR PROGRESS** 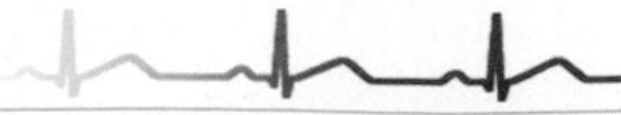

28. *What is a visual field?*

29. *Starting with the optic nerve, trace the action potential going from the right temporal visual field to the visual cortex.*

30. *Explain how binocular vision allows for depth perception.*

15.4 Hearing and Balance

LEARNING OUTCOMES 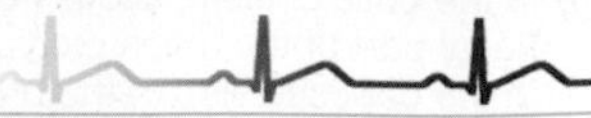

After reading this section, you should be able to

A. Describe the structures of the external ear and middle ear and state the function of each.

B. Explain the tunnels and chambers of the inner ear and the fluids they contain.

C. Describe the anatomy of the cochlea and, in particular, the cochlear duct.

D. Discuss the characteristics of sound.

E. Explain the process by which an action potential is generated by the hair cells of the spiral organ.

F. Trace a sound wave through the external, middle, and inner ears to the spiral organ.

G. Describe the pathway of an action potential from the hair cells of the spiral organ to the auditory cortex.

H. Explain how the structures of the vestibule and semicircular canals function in static and dynamic equilibrium.

I. Describe the pathway of an action potential from the balance organs to the various parts of the central nervous system and eyes.

The organs of hearing and balance are divided into three parts: the external, middle, and inner ears (figure 15.24). The external and middle ears are involved in hearing only, whereas the inner ear functions in both hearing and balance.

The **external ear** includes the **auricle** (aw′ri-kl; ear) and the **external auditory canal.** The external ear terminates medially at the **tympanic** (tim-pan′ik) **membrane,** or *eardrum.* The **middle ear** is an air-filled space within the petrous portion of the temporal bone that contains the **auditory ossicles.** The **inner ear** houses the sensory organs for both hearing and balance. It consists of interconnecting, fluid-filled tunnels and chambers within the petrous portion of the temporal bone.

Auditory Structures and Their Functions

External Ear

The auricle, or *pinna* (pin′ă), is the fleshy part of the external ear on the outside of the head; it consists primarily of elastic cartilage covered with skin (figure 15.25). Its shape helps collect sound waves and direct them toward the external auditory canal. The external auditory canal is lined with hairs and **ceruminous** (sě-roo′mi-nŭs) **glands,** which produce **cerumen,** a modified sebum commonly called

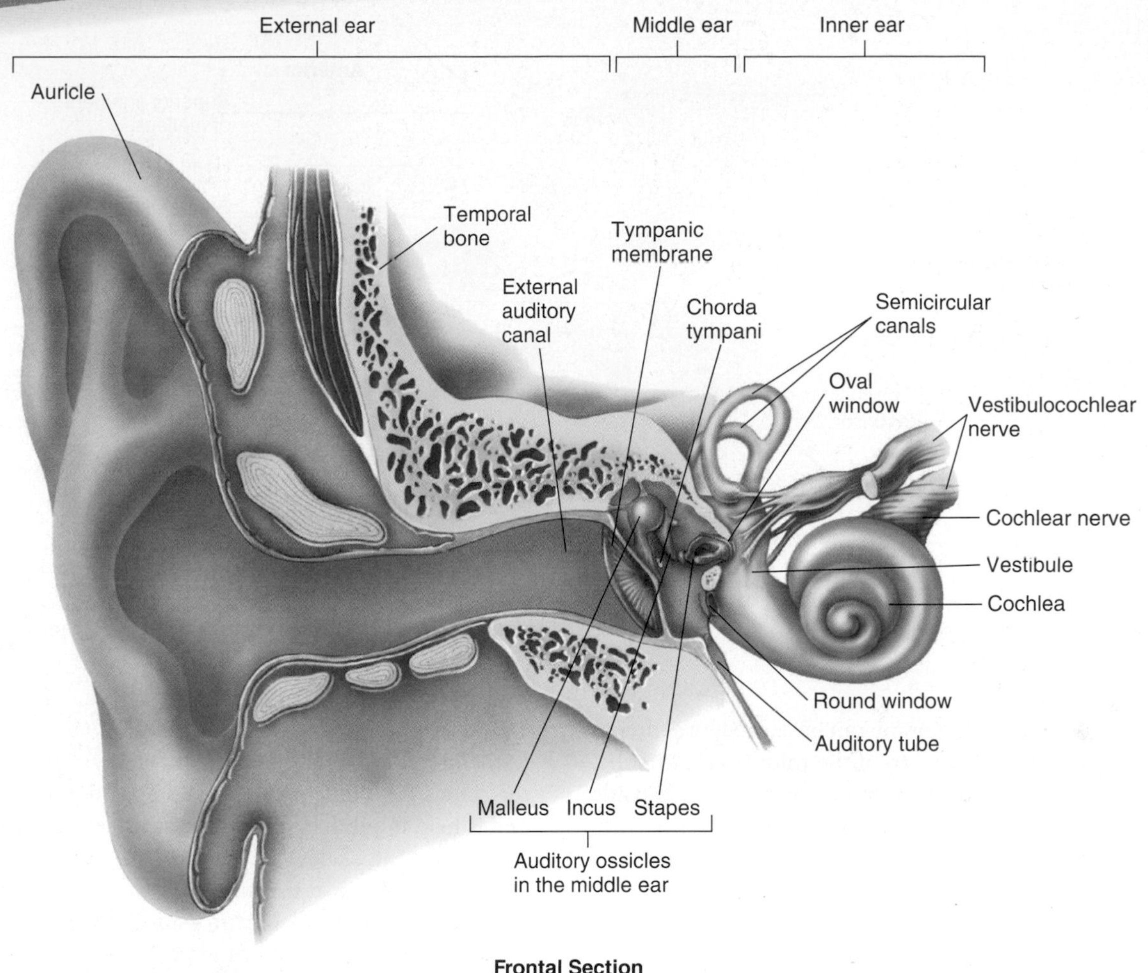
External ear
Middle ear
Inner ear
Auricle
Temporal bone
Tympanic membrane
External auditory canal
Chorda tympani
Semicircular canals
Oval window
Vestibulocochlear nerve
Cochlear nerve
Vestibule
Cochlea
Round window
Auditory tube
Malleus
Incus
Stapes
Auditory ossicles in the middle ear
Frontal Section

FIGURE 15.24 APR **External, Middle, and Inner Ears**

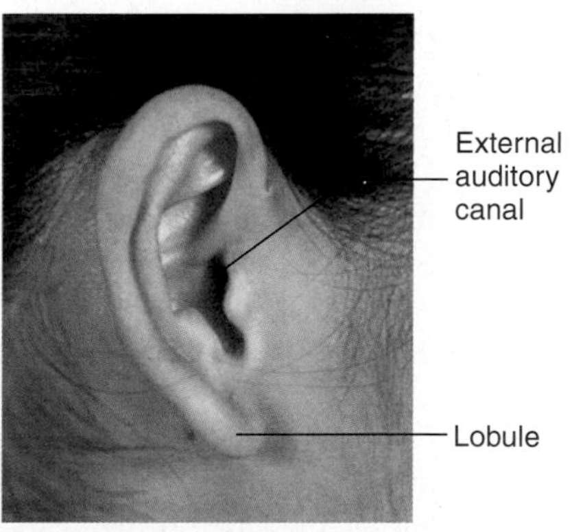
External auditory canal
Lobule

FIGURE 15.25 Structures of the Auricle of the Right Ear

earwax. The hairs and cerumen help prevent foreign objects from reaching the delicate tympanic membrane. However, overproduction of cerumen may block the canal.

The tympanic membrane is a thin, semitransparent, nearly oval membrane that separates the external ear from the middle ear. It consists of three layers: a simple cuboidal epithelium on the inner surface and a thin stratified squamous epithelium on the outer surface, with a layer of connective tissue between. Sound waves reaching the tympanic membrane through the external auditory canal cause it to vibrate.

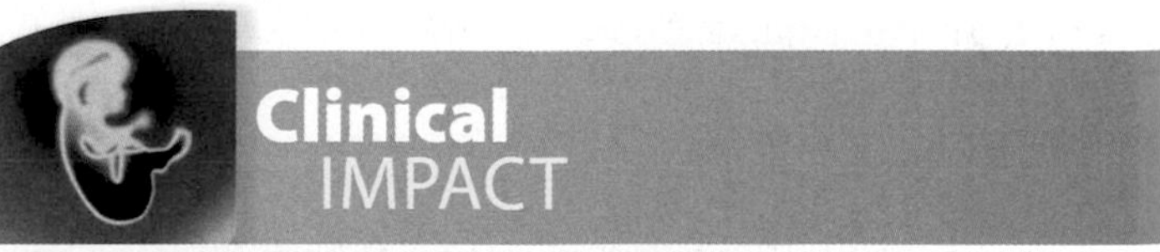
Clinical IMPACT

Function of the Chorda Tympani

A structure you might be somewhat surprised to find in the middle ear is the **chorda tympani,** a branch of the facial nerve carrying taste impulses from the anterior two-thirds of the tongue. It crosses over the inner surface of the tympanic membrane (see figure 15.24). The chorda tympani has nothing to do with hearing but is just passing through. However, this nerve can be damaged during ear surgery or by a middle ear infection, resulting in loss of taste sensation from the anterior two-thirds of the tongue on the side innervated by that nerve.

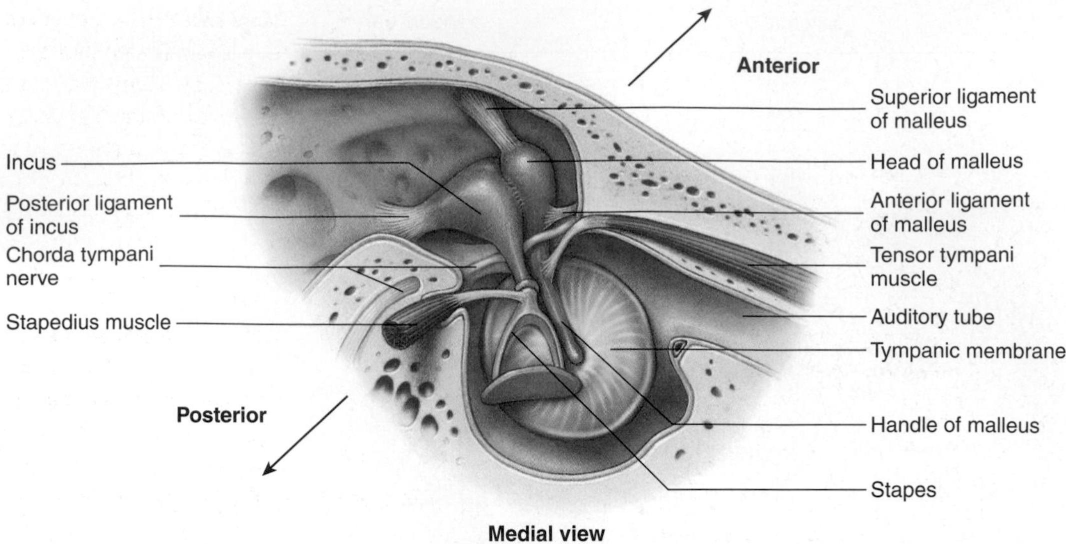
Anterior
Superior ligament of malleus
Head of malleus
Anterior ligament of malleus
Tensor tympani muscle
Auditory tube
Tympanic membrane
Handle of malleus
Stapes
Incus
Posterior ligament of incus
Chorda tympani nerve
Stapedius muscle
Posterior
Medial view

FIGURE 15.26 Auditory Ossicles and Muscles of the Middle Ear
Medial view of the middle ear (as though viewed from the inner ear), showing the three auditory ossicles with their ligaments and the two muscles of the middle ear: the tensor tympani and the stapedius.

Rupture of the tympanic membrane can be caused by a foreign object thrust into the ear, an infection of the middle ear, or sufficient differential pressure between the middle ear and the outside air, as occurs when changing altitude in an airplane or diving into deep water. Rupture of the tympanic membrane may result in hearing impairment.

Middle Ear

Medial to the tympanic membrane is the air-filled cavity of the middle ear (see figure 15.24). Two covered openings, the round and oval windows, on the medial side of the middle ear separate it from the inner ear. Two additional openings provide air passages from the middle ear. One passage opens into the **mastoid air cells** in the mastoid process of the temporal bone. The other passageway, the **auditory tube,** or *pharyngotympanic tube* (also called the eustachian [ū-stā′shŭn] tube), opens into the pharynx and equalizes air pressure between the outside air and the middle ear cavity. Unequal pressure between the middle ear and the outside environment can distort the tympanic membrane, dampen its vibrations, and make hearing difficult. Distortion of the tympanic membrane, which occurs under these conditions, also stimulates pain fibers associated with it. Because of this distortion, when a person changes altitude, sounds seem muffled and the eardrum may become painful. Swallowing, yawning, chewing, and holding the nose and mouth shut while gently forcing air out of the lungs can relieve distortion of the tympanic membrane. These actions open the auditory tube, which allows air to pass through the auditory tube and equalizes air pressure on each side of the eardrum.

The middle ear contains three auditory ossicles (see figure 15.24)—the **malleus** (mal′ē-ŭs; hammer), the **incus** (ing′kŭs; anvil), and the **stapes** (stā′pēz; stirrup)—which transmit vibrations from the tympanic membrane to the **oval window.** The handle of the malleus is attached to the inner surface of the tympanic membrane, and vibration of the membrane causes the malleus to vibrate as well. The head of the malleus is attached by a very small synovial joint to the incus, which in turn is attached by a small synovial joint to the stapes. The foot plate of the stapes fits into the oval window and is held in place by a flexible **annular ligament.**

Two small skeletal muscles originate from bone around the middle ear and insert onto auditory ossicles (figure 15.26). The **tensor tympani** (ten′sōr tim′păn-ē) muscle is attached to the malleus and is innervated by the trigeminal nerve (V). The **stapedius** (stā-pē′dē-ŭs) muscle is attached to the stapes and is innervated by the facial nerve (VII).

Inner Ear

The tunnels and chambers inside the temporal bone are called the **bony labyrinth** (lab′i-rinth; maze; figure 15.27). Because the bony labyrinth consists of tunnels within the bone, it cannot easily be removed and examined separately. The bony labyrinth is lined with endosteum; when the inner ear is shown separately (figure 15.28*a*), the endosteum is what is depicted. Inside the bony labyrinth is a similarly shaped but smaller set of membranous tunnels and chambers called the **membranous labyrinth.** The inner surface of the endosteum and the outer surface of the membranous labyrinth are covered with a very thin layer of cells called the **perilymphatic cells** (figure 15.28*b*). The membranous labyrinth is filled with a clear fluid called **endolymph,** and the space between the membranous labyrinth and bony labyrinth is filled with a fluid called **perilymph.** Perilymph is very similar to cerebrospinal fluid, but endolymph has a high concentration of K^+ and a low concentration of Na^+, which is opposite from the composition of perilymph and cerebrospinal fluid.

The bony labyrinth is divided into three regions: cochlea, vestibule, and semicircular canals. The **vestibule** (ves′ti-bool) and **semicircular canals** are primarily involved in balance, and the **cochlea** (kok′lē-ă) functions in hearing. The membranous labyrinth of the cochlea is divided into three parts: the scala vestibuli, the scala tympani, and the cochlear duct.

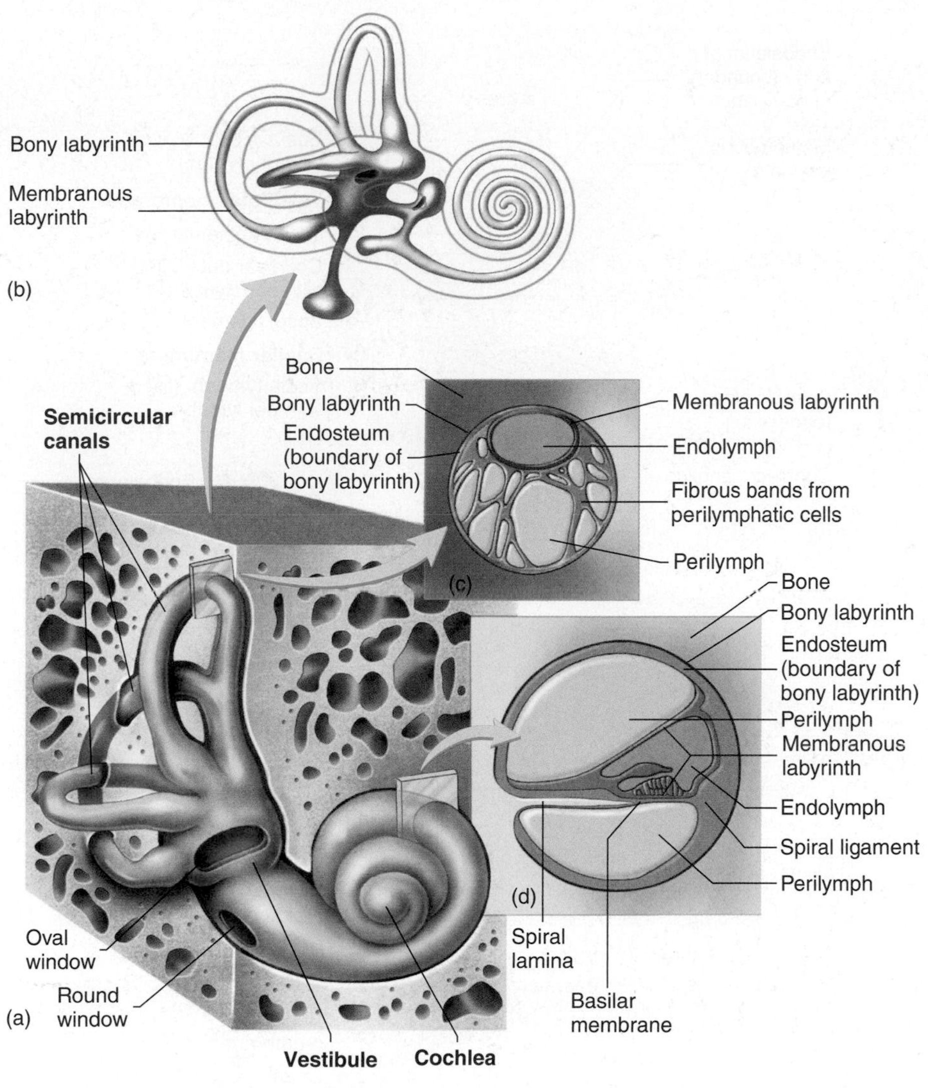
Bony labyrinth
Membranous labyrinth
(b)
Semicircular canals
Bone
Bony labyrinth
Endosteum (boundary of bony labyrinth)
Membranous labyrinth
Endolymph
Fibrous bands from perilymphatic cells
Perilymph
(c)
Bone
Bony labyrinth
Endosteum (boundary of bony labyrinth)
Perilymph
Membranous labyrinth
Endolymph
Spiral ligament
Perilymph
(d)
Spiral lamina
Basilar membrane
Oval window
Round window
(a)
Vestibule
Cochlea

FIGURE 15.27 Inner Ear: Bony and Membranous Labyrinths

(*a*) The structures of the inner ear are embedded in the temporal bone. (*b*) The membranous labyrinth seen within the outline of the bony labyrinth. (*c*) A cross section through a semicircular canal and (*d*) a cross section through the cochlea show the relationship between the bony and membranous labyrinths.

The **scala vestibuli** (skā′lă ves-tib′ū-lē; figure 15.28*b*) extends from the oval window to the **helicotrema** (hel′ikō-trē′mă) at the apex of the cochlea. The **scala tympani** (tim′pănē) extends from the helicotrema, back from the apex, parallel to the scala vestibuli, to the membrane of the **round window.**

The scala vestibuli and the scala tympani are the perilymph-filled spaces between the walls of the bony and membranous labyrinths. The wall of the membranous labyrinth that borders the scala vestibuli is called the **vestibular membrane** (*Reissner membrane*); the wall of the membranous labyrinth bordering the scala tympani is the **basilar membrane** (figure 15.28*b,c*). The **cochlear duct,** or *scala media,* is formed by the space between the vestibular membrane and the basilar membrane and is filled with endolymph.

The vestibular membrane consists of a double layer of squamous epithelium and is the simplest region of the membranous labyrinth. The vestibular membrane is so thin that it has little or no mechanical effect on the transmission of sound waves through the inner ear; therefore, the perilymph and endolymph on the two sides of the vestibular membrane can be thought of mechanically as one fluid, even though they are chemically different. The basilar membrane is somewhat more complex and is of much greater physiological interest in relation to the mechanics of hearing. It has an acellular portion, consisting of collagen fibers, ground substance, and sparsely dispersed elastic fibers, and a cellular portion, composed of a thin layer of vascular connective tissue overlaid with simple squamous epithelium.

The basilar membrane is attached at one side to the bony **spiral lamina,** which projects from the sides of the **modiolus** (mō-dī′ō-lus), the bony core of the cochlea, like the threads of a screw. At the other side, the basilar membrane is attached to the lateral wall of the bony labyrinth by the **spiral ligament,** a local thickening of the endosteum. The distance between the spiral lamina and the spiral ligament (i.e., the width of the basilar membrane) increases from 0.04 mm near the oval window to 0.5 mm near the helicotrema. The collagen fibers of the basilar membrane are oriented across the membrane between the spiral lamina and the spiral ligament, somewhat like the strings of a piano. The collagen fibers near the oval window are both shorter and thicker than those near the helicotrema. The diameter of the collagen fibers in the membrane decreases as the basilar membrane widens. As a result, the basilar membrane near the oval window is short and stiff, and it responds to high-frequency vibrations, whereas the part near the helicotrema is wide and limber and responds to low-frequency vibrations.

The cells inside the cochlear duct are highly modified to form a structure called the **spiral organ,** or *organ of Corti* (figure 15.28*b,c*). The spiral organ contains supporting epithelial cells and specialized sensory cells called **hair cells** (figure 15.28*d*), which have hairlike projections at their apical ends. In children, these projections consist of one cilium (kinocilium) and about 80 very long microvilli, often referred to as **stereocilia,** but in adults the cilium is absent from most hair cells. The hair cells are arranged in four long rows extending the length of the cochlear duct. Each row contains 3500–4000 hair cells. The inner row consists of **inner hair cells,** which are the hair cells primarily responsible for hearing. The outer three rows contain **outer hair cells,** which are involved in regulating the tension of the basilar membrane. The outer hair cells are separated from the inner hair cells by a gap in the basilar membrane (figure 15.28*c*). The stereocilia of one inner hair cell form a conical group called a **hair bundle** (figure 15.29*a*). The length of each stereocilium within a hair bundle increases gradually from one side of the hair cell to the other. The stereocilia of an outer hair cell are arranged in a curved line (figure 15.29*b*). The tips of the longest stereocilia of the outer hair cells are embedded within an acellular, gelatinous shelf called the **tectorial** (tek-tōr′ē-ăl) **membrane,** which is attached to the spiral lamina.

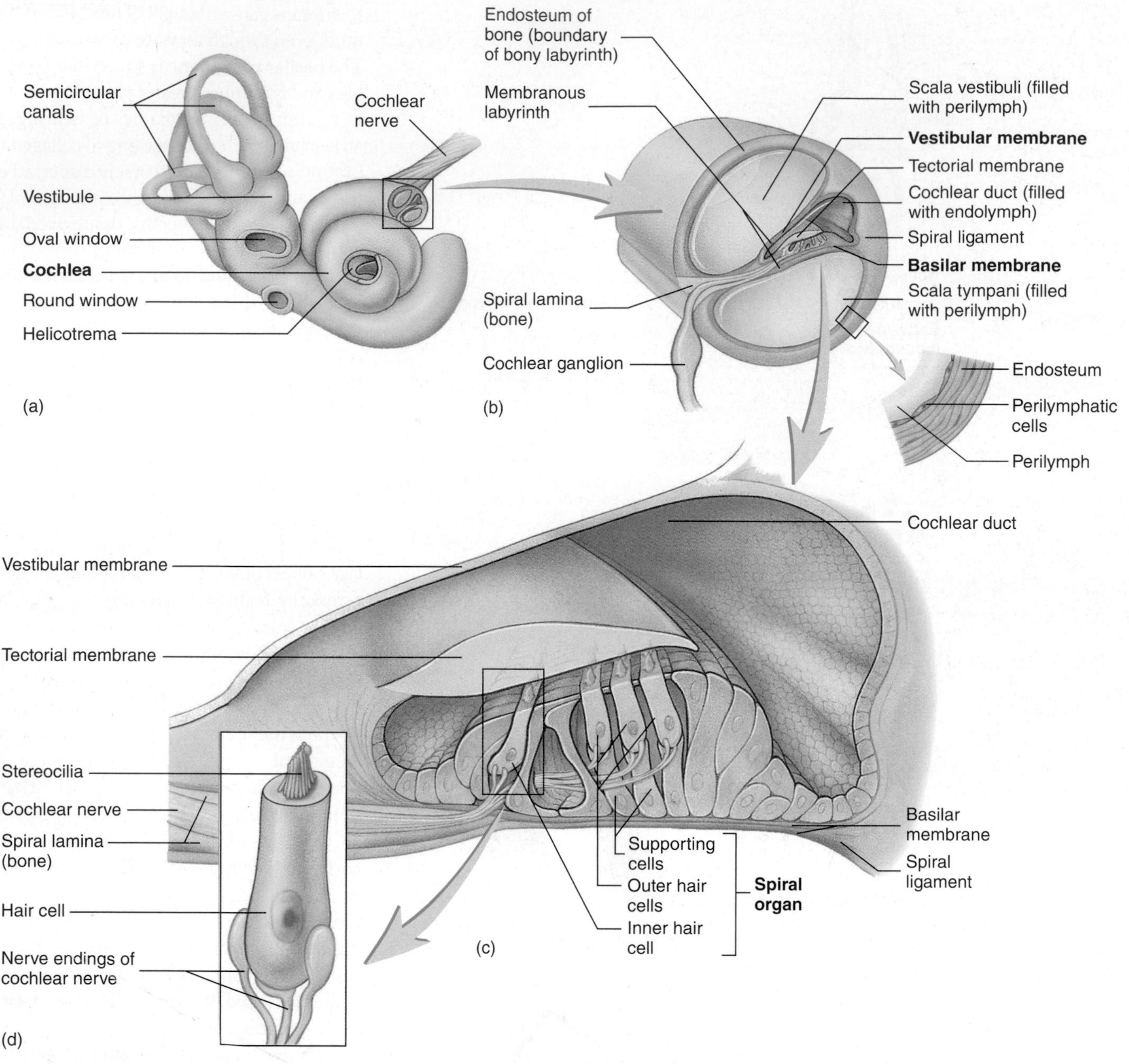
Semicircular canals
Cochlear nerve
Vestibule
Oval window
Cochlea
Round window
Helicotrema
(a)
Endosteum of bone (boundary of bony labyrinth)
Membranous labyrinth
Scala vestibuli (filled with perilymph)
Vestibular membrane
Tectorial membrane
Cochlear duct (filled with endolymph)
Spiral ligament
Basilar membrane
Scala tympani (filled with perilymph)
Spiral lamina (bone)
Cochlear ganglion
(b)
Endosteum
Perilymphatic cells
Perilymph
Cochlear duct
Vestibular membrane
Tectorial membrane
Stereocilia
Cochlear nerve
Spiral lamina (bone)
Hair cell
Nerve endings of cochlear nerve
(d)
Supporting cells
Outer hair cells
Inner hair cell
Spiral organ
Basilar membrane
Spiral ligament
(c)

FIGURE 15.28 AP|R **Cochlea**

(*a*) Inner ear structures. The outer surface (gray) is the endosteum lining the inner surface of the bony labyrinth. (*b*) Cross section of the cochlea. The outer layer is the endosteum lining the inner surface of the bony labyrinth. The membranous labyrinth (pink) is very small in the cochlea and consists of the vestibular and basilar membranes. The space between the membranous and bony labyrinths consists of two parallel tunnels: the scala vestibuli and the scala tympani. (*c*) Enlarged section of the cochlear duct (membranous labyrinth). (*d*) Individual sensory hair cell, greatly enlarged.

A **tip link** connects the tip of each stereocilium in a hair bundle to the side of the next longer stereocilium (figure 15.30). Each tip link is a *gating spring*, a pair of microtubule strands that attaches to the gate of a gated K^+ channel. The gated K^+ channels of hair cells open mechanically. As the stereocilia bend, the tip link pulls the K^+ gate open (figure 15.31). The response time for such a mechanism is very brief and much faster than for a gating mechanism involving the synthesis of intracellular chemical signals, such as cAMP.

Hair cells have no axons, but the basilar regions of each hair cell are covered by synaptic terminals of sensory neurons. The cell bodies of afferent neurons are located within the cochlear modiolus and are grouped into a **cochlear ganglion,** or *spiral ganglion* (see figure 15.28*b*). Afferent fibers of these neurons join to form the **cochlear nerve.** This nerve then joins the vestibular nerve to become the **vestibulocochlear nerve (VIII),** which traverses the internal auditory canal and enters the cranial cavity.

(a)

(b)

FIGURE 15.29 **Scanning Electron Micrograph of Cochlear Hair Cell Stereocilia**
(*a*) Hair bundle of one inner hair cell. (*b*) Stereocilia of three outer hair cells.

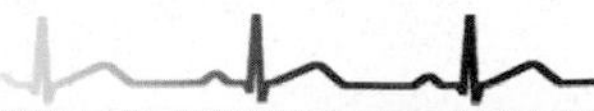

ASSESS YOUR PROGRESS

31. *Name the three regions of the ear, and list each region's parts.*
32. *Describe the relationship among the tympanic membrane, the ear ossicles, and the oval window.*
33. *What are the functions of the external auditory canal and the auditory tube?*
34. *Explain how the membranous labyrinth of the cochlea is divided into three compartments. What is found in each compartment?*
35. *Describe the structure of the spiral organ.*
36. *Explain the differences between inner and outer hair cells.*
37. *Relate how tip links function.*
38. *How is the cochlear nerve formed?*

Auditory Function

Vibration of matter, such as air, water, or a solid material, creates sound. No sound occurs in a vacuum. When a person speaks, the vocal cords vibrate, causing the air passing out of the lungs to vibrate. The vibrations consist of bands of compressed air followed by bands of less compressed air (figure 15.32*a*). These vibrations are propagated through the air as sound waves, somewhat as ripples are propagated over the surface of water. **Volume,** or loudness, is a function of wave *amplitude,* or height, measured in decibels (db; figure 15.32*b*). The greater the amplitude, the louder the sound. **Pitch** is a function of the wave *frequency* (i.e., the number of waves or cycles per second) measured in hertz (Hz; figure 15.32*c*). The higher the frequency, the higher the pitch. The normal range of human hearing is 20–20,000 Hz and 0 or more db. Sounds louder than 125 db are painful to the ear. Normal human speech ranges in volume between 250 and 8000 Hz. This is the range that is used when testing for hearing impairment, because it is the most important for communication.

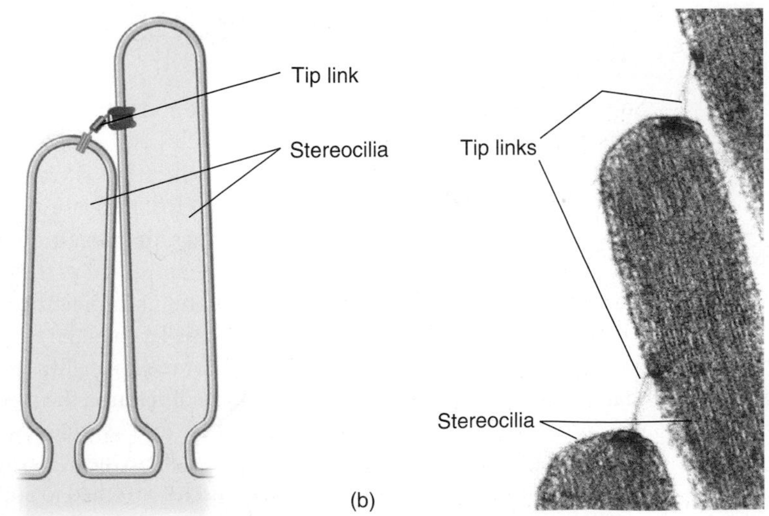
Tip link
Stereocilia
Tip links
Stereocilia
(a)
(b)

Tip links
Stereocilia
(c)

FIGURE 15.30 **Tip Links on Inner Hair Cell Stereocilia**
(*a*) Diagram of two stereocilia connected by one tip link. (*b*) Transmission electron micrograph of three stereocilia and two tip links. (*c*) Scanning electron micrograph of the tops of stereocilia, showing tip links.

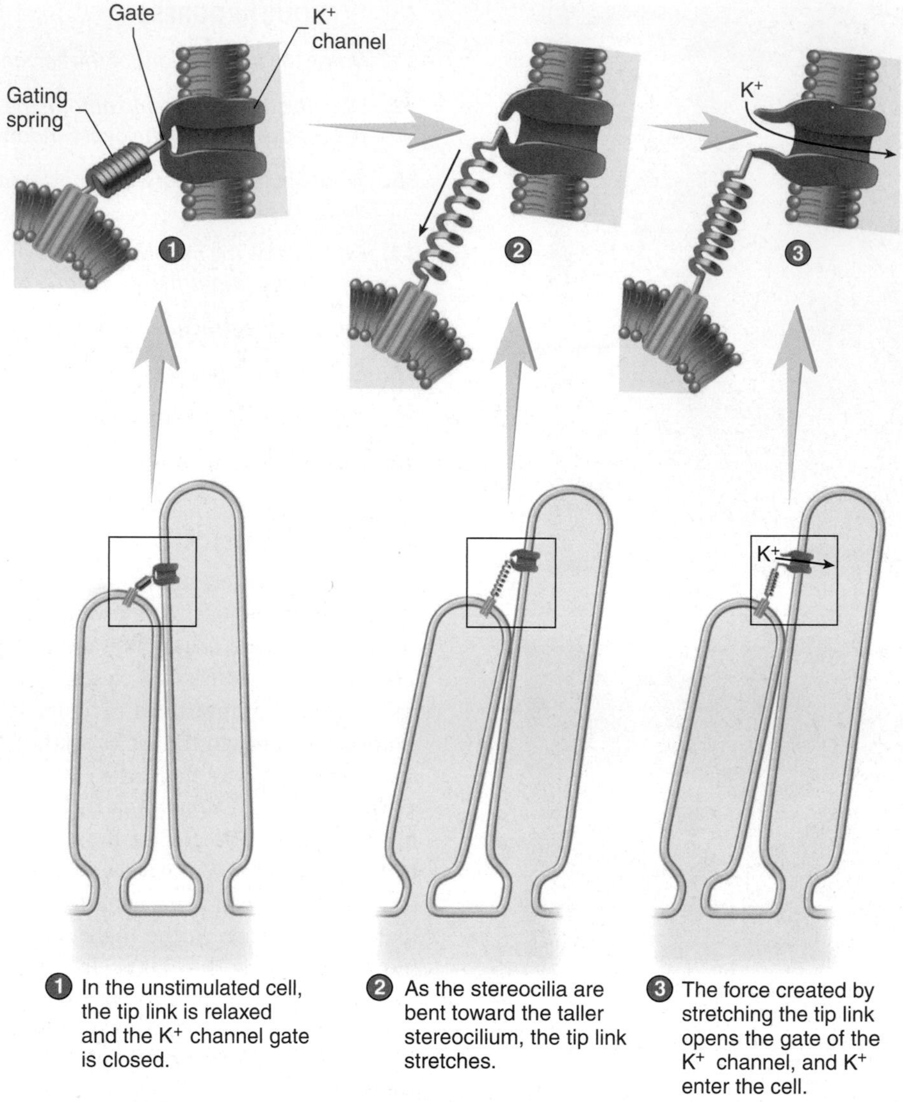
Gate
K+ channel
Gating spring
K+
1
2
3
K+
1 In the unstimulated cell, the tip link is relaxed and the K+ channel gate is closed.
2 As the stereocilia are bent toward the taller stereocilium, the tip link stretches.
3 The force created by stretching the tip link opens the gate of the K+ channel, and K+ enter the cell.

PROCESS FIGURE 15.31 Action of the Tip Link to Open a K^+ Channel When Two Stereocilia Bend

Timbre (tam′br, tim′br) is the resonance quality or overtones of a sound. A "pure" sound wave would be represented by a smooth sigmoid curve, but such a wave is extremely rare in nature. The sounds made by musical instruments and the human voice are not smooth sigmoid curves, but rough, jagged curves formed by numerous, superimposed curves of various amplitudes and frequencies. The roughness of the curve accounts for the timbre. Timbre is the difference in quality between, for example, an oboe and a French horn playing a note at the same pitch and volume. The steps involved in the mechanical part of hearing are illustrated in figure 15.33.

External Ear

The auricle collects sound waves, which are then conducted through the external auditory canal toward the tympanic membrane. Sound waves travel relatively slowly in air (332 m/s), and a significant time interval may elapse between the time a sound wave reaches one ear and the time it reaches the other. The brain can interpret this interval to determine the direction from which a sound is coming.

Middle Ear

Sound waves strike the tympanic membrane and cause it to vibrate. This vibration in turn causes the three ossicles of the middle ear to vibrate and, by this mechanical linkage, vibration is transferred to the oval window. More force is required to cause vibration in a liquid, such as the perilymph of the inner ear, than in air; thus, the vibrations reaching the perilymph must be amplified as they cross the middle ear. The footplate of the stapes and its annular ligament, which occupy the oval window, are much smaller than the tympanic membrane. The area of the tympanic membrane is roughly 20 times that of the oval window. Because of this size difference, the mechanical force of vibration is amplified about 20-fold as it passes from the tympanic membrane through the ossicles and to the oval window.

The tensor tympani and stapedius muscles, attached to auditory ossicles, reflexively dampen excessively loud sounds (see figure 15.26). This so-called **sound attenuation reflex** protects the delicate ear structures from damage by loud noises. The sound attenuation reflex responds most effectively to low-frequency sounds and can

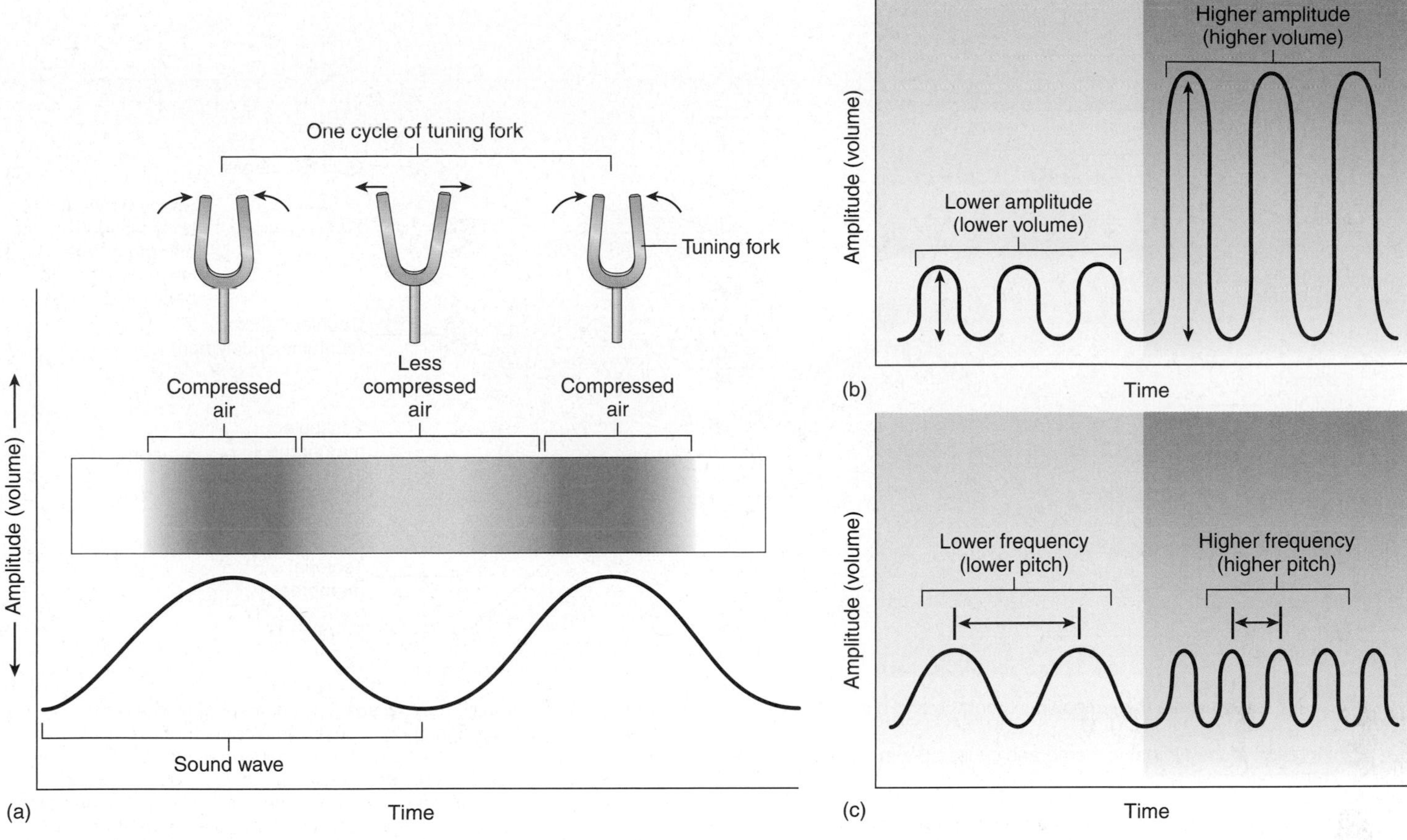
One cycle of tuning fork
Tuning fork
Less compressed air
Compressed air
Compressed air
Amplitude (volume)
Sound wave
(a)
Time
Higher amplitude (higher volume)
Lower amplitude (lower volume)
Amplitude (volume)
(b)
Time
Lower frequency (lower pitch)
Higher frequency (higher pitch)
Amplitude (volume)
(c)
Time

FIGURE 15.32 Sound Waves

(*a*) When an object, such as a tuning fork, vibrates, its movements alternate between compressing the air and decompressing the air, or making the air less compressed, thus producing sound. The human vocal cords function in the same way. Each sound wave consists of a region of compressed air between two regions of less compressed air (blue bars). The sigmoid (S-shaped) waves correspond to the regions of more compressed air (peaks) and less compressed air (troughs). The green shadowed area represents the width of one cycle (distance between peaks). (*b*) Low- and high-volume sound waves. Compare the relative lengths of the arrows indicating the wave height (amplitude). (*c*) Lower- and higher-pitch sound. Compare the relative number of peaks (frequency) within a given time interval (between arrows).

reduce by a factor of 100 the energy reaching the oval window. The facial nerve and stapedius are primarily involved in the sound attenuation reflex. The trigeminal nerve and tensor tympani are only stimulated by extremely loud noise. The reflex is too slow to prevent damage from a sudden noise, such as a gunshot, and it cannot function effectively for longer than about 10 minutes in response to prolonged noise.

Inner Ear

As the stapes vibrates, it produces waves in the perilymph of the scala vestibuli (figure 15.33). Vibrations of the perilymph are transmitted through the thin vestibular membrane and cause simultaneous vibrations of the endolymph. The mechanical effect is as though the perilymph and endolymph were a single fluid. Vibration of the endolymph causes distortion of the basilar membrane. Waves in the perilymph of the scala vestibuli are transmitted also through the helicotrema and into the scala tympani. However, because the helicotrema is very small, this transmitted vibration is probably of little consequence. Distortions of the basilar membrane, together with weaker waves coming through the helicotrema, cause waves in the scala tympani perilymph and ultimately result in vibration of the membrane of the round window. Vibration of the round window membrane is important to hearing because it acts as a mechanical release for waves from within the cochlea. If this window were solid, it would reflect the waves, which would interfere with and dampen later waves. The round window also allows the relief of pressure in the perilymph, because fluid is not compressible, thereby preventing compression damage to the spiral organ.

The distortion of the basilar membrane is most important to hearing. As this membrane distorts, the hair cells resting on the basilar membrane move relative to the tectorial membrane, which remains stationary. The inner hair cell microvilli bend as they move against the tectorial membrane.

The apical portion of each hair cell is surrounded by endolymph, and the basal portion of the cell is surrounded by perilymph. Endolymph has a high K^+ concentration, similar to the intracellular K^+ concentration of hair cells. Perilymph has a low concentration of K^+, similar to that of other extracellular fluid. The intracellular charge of hair cells compared with perilymph is −60 mV. The charge of the endolymph is +80 mV compared with the perilymph. This charge difference is called the **endocochlear potential.** Consequently, the intracellular charge of hair cells compared with endolymph is

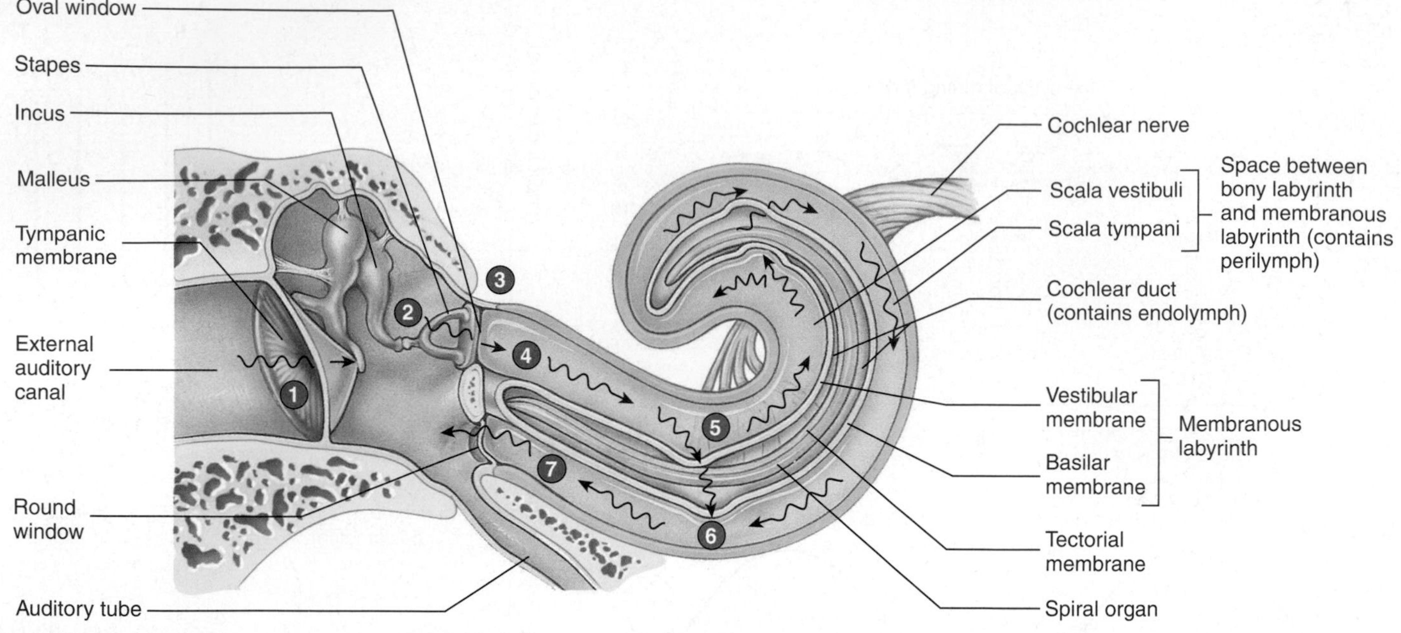
Oval window
Stapes
Incus
Malleus
Tympanic membrane
External auditory canal
Round window
Auditory tube
Cochlear nerve
Scala vestibuli
Scala tympani
Space between bony labyrinth and membranous labyrinth (contains perilymph)
Cochlear duct (contains endolymph)
Vestibular membrane
Basilar membrane
Membranous labyrinth
Tectorial membrane
Spiral organ
1
2
3
4
5
6
7

1. Sound waves strike the tympanic membrane and cause it to vibrate.
2. Vibration of the tympanic membrane causes the malleus, the incus, and the stapes to vibrate.
3. The foot plate of the stapes vibrates in the oval window.
4. Vibration of the foot plate causes the perilymph in the scala vestibuli to vibrate.
5. Vibration of the perilymph causes the vestibular membrane to vibrate, which causes vibrations in the endolymph.
6. Vibration of the endolymph causes displacement of the basilar membrane. Short waves (high pitch), cause displacement of the basilar membrane near the oval window, and longer waves (low pitch) cause displacement of the basilar membrane some distance from the oval window. Movement of the basilar membrane is detected in the hair cells of the spiral organ, which are attached to the basilar membrane. Vibrations of the perilymph in the scala vestibuli and of the basilar membrane are transferred to the perilymph of the scala tympani.
7. Vibrations in the perilymph of the scala tympani are transferred to the round window, where they are dampened.

PROCESS FIGURE 15.33 AP|R **Effect of Sound Waves on Cochlear Structures**

-140 mV, which is a large charge difference. Therefore, when K^+ channels open, K^+ flows into the hair cells because it is attracted to the negative charge inside the hair cell, even though the intracellular concentration of K^+ is about the same as in the endolymph. The movement of the K^+ causes depolarization of the hair cells. This is a rare instance in which an increase in K^+ permeability of the plasma membrane of a cell results in depolarization.

In the unstimulated hair cell, approximately 15% of the gated K^+ channels are open, and the resting membrane potential of the cell is approximately -60 mV. If the hair bundle is bent toward the shortest stereocilium (negative stimulus), the tip link attached to the K^+ channel gates slackens, allowing the open K^+ channels to close, and the cell hyperpolarizes. If the hair bundle is bent toward the longest stereocilium (positive stimulus), the tip link pulls additional K^+ channel gates open, and K^+ rushes into the cell. The influx of K^+ into the hair cell causes a slight depolarization of the cell. This depolarization causes voltage-gated Ca^{2+} channels to open. Calcium ions rush into the cell, causing a further depolarization. The cell depolarizes by a total of about 10 mV. Depolarization of hair cells results in the increased release of neurotransmitters, which increases the action potential frequency in the afferent neurons. Hyperpolarization decreases neurotransmitter release and decreases action potential frequency in afferent neurons. Depolarization also opens voltage-gated K^+ channels in the basal portion of the hair cell. Potassium ions tend to leave the cell, causing the cell to repolarize.

The neurotransmitter released by the inner hair cells is apparently glutamate, but other neurotransmitters may also be involved. The release of neurotransmitters from the inner hair cells induces action potentials in the cochlear neurons that synapse on the hair cells. The cell bodies of those neurons are located in the cochlear ganglion.

The part of the basilar membrane that distorts as a result of endolymph vibration depends on the pitch of the sound that created the vibration and, as a result, on the vibration frequency within the endolymph. The location of the optimal amount of basilar membrane vibration produced by a given pitch is determined by two factors: the width of the basilar membrane and the length and diameter of the collagen fibers stretching across the membrane at each level along the cochlear duct (figure 15.34). Higher-pitched tones cause optimal vibration near the base, and lower-pitched tones cause optimal vibration near the apex of the basilar membrane. As the basilar

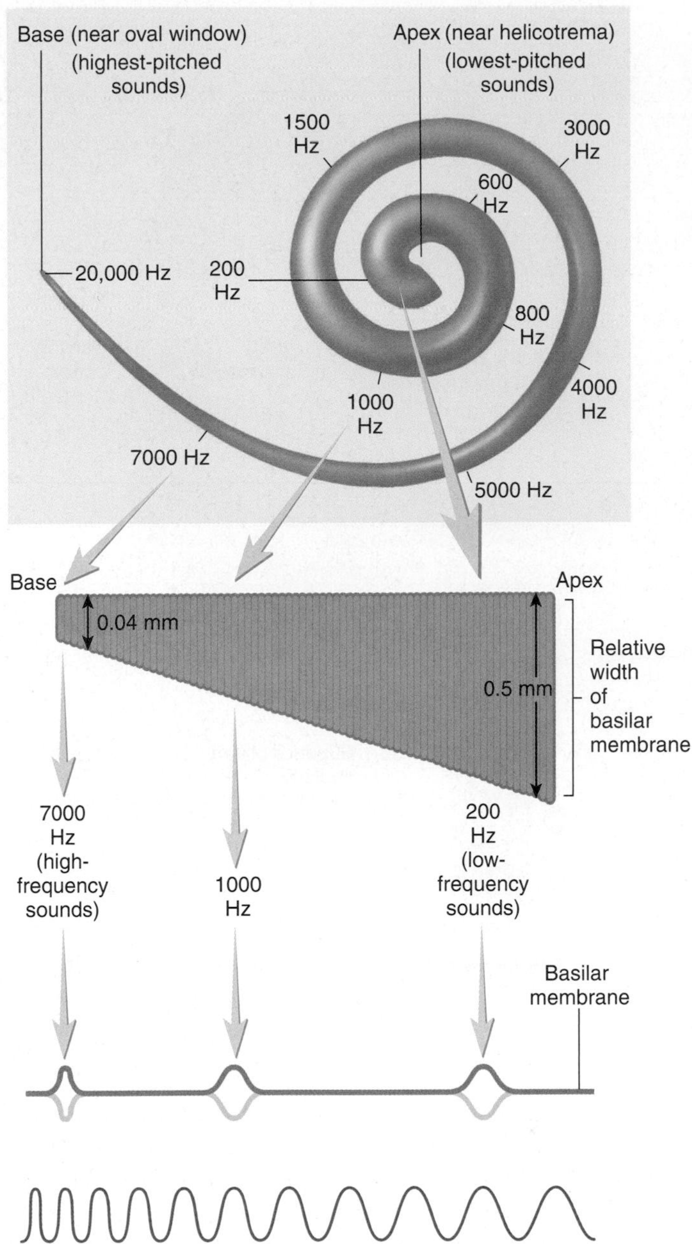
Base (near oval window)
(highest-pitched sounds)
Apex (near helicotrema)
(lowest-pitched sounds)
1500 Hz
3000 Hz
600 Hz
20,000 Hz
200 Hz
800 Hz
4000 Hz
1000 Hz
7000 Hz
5000 Hz
Base
Apex
0.04 mm
0.5 mm
Relative width of basilar membrane
7000 Hz (high-frequency sounds)
1000 Hz
200 Hz (low-frequency sounds)
Basilar membrane

FIGURE 15.34 Effect of Sound Waves on Points Along the Basilar Membrane

Points of maximum vibration along the basilar membrane result from stimulation by sounds of various frequencies (in hertz).

membrane vibrates, hair cells along a large part of the basilar membrane are stimulated. In areas of minimum vibration, the amount of stimulation may not reach threshold. In other areas, a low frequency of afferent action potentials may be transmitted, whereas in the optimally vibrating regions of the basilar membrane a high frequency of action potentials is initiated.

There are approximately twice as many nerve cells in the cochlear ganglion as there are hair cells. Over 90% of the afferent axons synapse with inner hair cells—about 10–30 axons per hair cell. Only a few, small-diameter afferent axons synapse with the three rows of outer hair cells. However, the outer cells receive input from efferent axons.

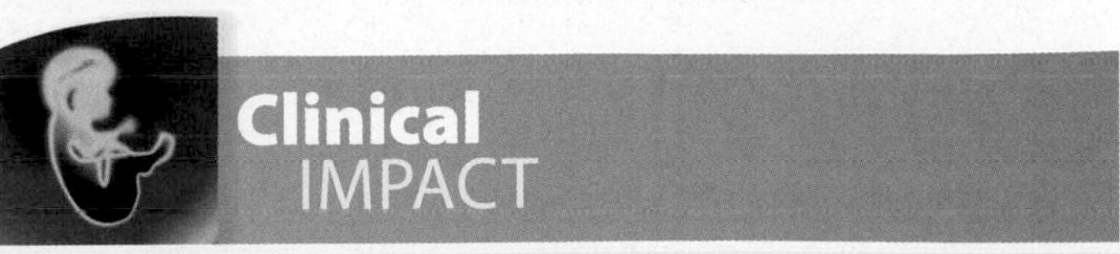
Clinical IMPACT

Loud Noises and Hearing Loss

Prolonged or frequent exposure to excessively loud noises can cause degeneration of the spiral organ at the base of the cochlea, resulting in high-frequency deafness. For example, loud music (amplified to 120 db) can impair hearing, although the actual amount of damage can vary from person to person. The defects may not be detectable on routine diagnosis, but they include decreased sensitivity to sound in specific narrow frequency ranges and decreased ability to discriminate between two pitches.

Loud music, however, is not as harmful as the sudden sound of a nearby gunshot at 140 db. This sound is too sudden for the attenuation reflex to protect the inner ear structures, and the intensity is great enough to cause auditory damage. In fact, gunshot noise is the most common recreational cause of serious hearing loss.

Action potentials from those efferent axons stimulate the contraction of actin filaments within the hair cells, causing them to shorten. This adjustment in the height of the outer hair cells, attached to both the basilar membrane and the tectorial membrane, fine-tunes the tension of the basilar membrane and the distance between the basilar membrane and the tectorial membrane. Additional sensitivity can be adjusted within the inner hair cells. The mechanically gated K^+ channels are attached to actin filaments inside the cell. Those filaments can move the K^+ channels along the plasma membrane, tightening or loosening the tip links. By this means, hair cells are tuned to very specific frequencies. Likewise, the response of an inner hair cell to stimulation is graded and can increase up to a point of saturation (when all the K^+ channels are open maximally). Inner hair cells are also very sensitive. A 100 nm (1-degree) deflection of the stereocilia results in a response that is 90% of maximum.

Afferent action potentials conducted by cochlear nerve fibers from all along the spiral organ terminate in the **superior olivary nucleus** in the medulla oblongata (figure 15.35; see chapter 13). These action potentials are compared with one another, and the strongest action potential, corresponding to the area of maximum basilar membrane vibration, is taken as standard. Efferent action potentials then are sent from the superior olivary nucleus back to the spiral organ to all regions where the maximum vibration did not occur. These action potentials inhibit the hair cells from initiating additional action potentials in the sensory neurons. Thus, only action potentials from regions of maximum vibration are sent to the cortex, where they become consciously perceived.

By this process, tones are localized along the cochlea. As a result of this localization, neurons along a given portion of the cochlea send action potentials to the cerebral cortex only in response to specific pitches. Action potentials near the base of the basilar membrane stimulate neurons in a certain part of the auditory cortex, which interpret the stimulus as a high-pitched sound, whereas action potentials from the apex stimulate a different part of the cortex, which interprets the stimulus as a low-pitched sound.

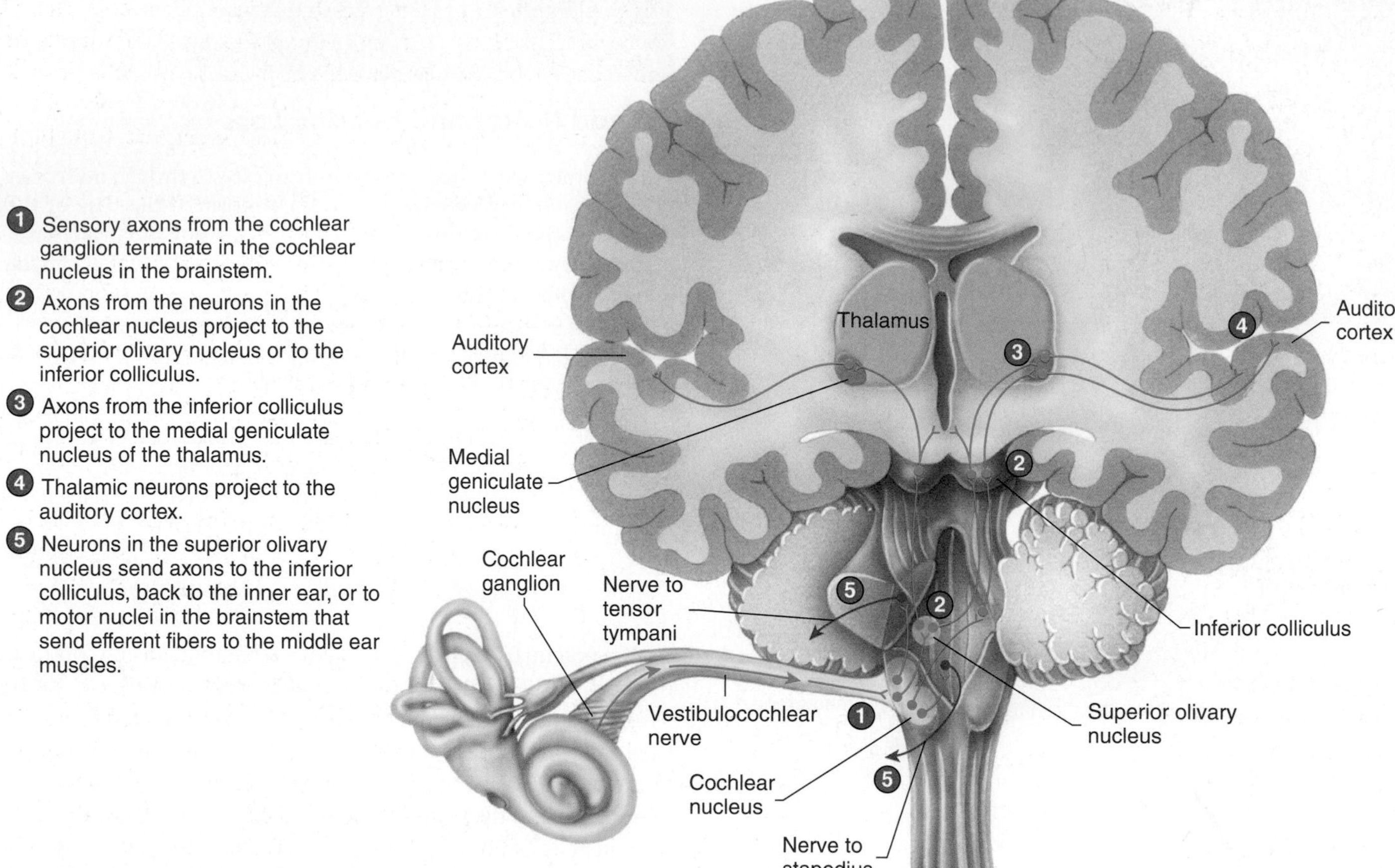

1 Sensory axons from the cochlear ganglion terminate in the cochlear nucleus in the brainstem.
2 Axons from the neurons in the cochlear nucleus project to the superior olivary nucleus or to the inferior colliculus.
3 Axons from the inferior colliculus project to the medial geniculate nucleus of the thalamus.
4 Thalamic neurons project to the auditory cortex.
5 Neurons in the superior olivary nucleus send axons to the inferior colliculus, back to the inner ear, or to motor nuclei in the brainstem that send efferent fibers to the middle ear muscles.
Thalamus
Auditory cortex
Auditory cortex
Medial geniculate nucleus
Cochlear ganglion
Nerve to tensor tympani
Inferior colliculus
Vestibulocochlear nerve
Superior olivary nucleus
Cochlear nucleus
Nerve to stapedius
Frontal section

PROCESS FIGURE 15.35 Central Nervous System Pathways for Hearing

> **Predict 7**
>
> Suggest some possible sites and mechanisms within the auditory structures to explain why certain people have "perfect pitch" and other people are "tone deaf."

Sound volume, or loudness, is a function of sound wave amplitude. As high-amplitude sound waves reach the ear, the perilymph, endolymph, and basilar membrane vibrate more intensely, and the hair cells are stimulated more intensely. As a result of the increased stimulation, more hair cells send action potentials at a higher frequency to the cerebral cortex, where this information is perceived as a greater sound volume.

> **Predict 8**
>
> Explain why it is much easier to perceive subtle musical tones when music is played somewhat softly, as opposed to very loudly.

Neuronal Pathways for Hearing

The vestibulocochlear (VIII) nerve transmits the senses of both hearing and balance. The vestibulocochlear nerve functions as two separate nerves carrying information from two separate but closely related structures. (*Vestibular* refers to the vestibule of the inner ear, which is involved in balance, and *cochlear* refers to the cochlea, the portion of the inner ear involved in hearing.)

The auditory pathways within the CNS are very complex, with both crossed and uncrossed tracts (figure 15.35). Therefore, unilateral CNS damage usually has little effect on hearing. The neurons from the cochlear ganglion synapse with CNS neurons in the dorsal or ventral **cochlear nucleus** in the superior medulla oblongata near the inferior cerebellar peduncle. These neurons in turn either synapse in or pass through the superior olivary nucleus. Neurons terminating in this nucleus may synapse with efferent neurons returning to the cochlea to modulate pitch perception. Nerve fibers from the superior olivary nucleus also project to the trigeminal nerve (V) nucleus, which controls the tensor tympani, and the facial nerve (VII) nucleus, which controls the stapedius muscle. This reflex pathway dampens loud sounds by initiating contractions of these muscles (the sound attenuation reflex). Neurons synapsing in the superior olivary nucleus may also join other ascending neurons to the cerebral cortex.

Ascending neurons from the superior olivary nucleus travel in the **lateral lemniscus.** All ascending fibers synapse in the **inferior colliculi,** and neurons from there project to the **medial geniculate**

nucleus of the **thalamus,** where they synapse with neurons that project to the cortex. These neurons terminate in the **auditory cortex** in the dorsal portion of the temporal lobe within the lateral fissure and, to a lesser extent, on the superolateral surface of the temporal lobe (see chapter 13). Neurons from the inferior colliculus also project to the **superior colliculus,** where reflexes that turn the head and eyes in response to loud sounds are initiated.

ASSESS YOUR PROGRESS

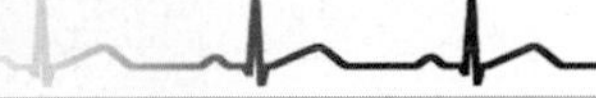

39. *Contrast volume, pitch, and timbre.*
40. *Starting with the auricle, trace a sound wave into the inner ear to the point at which action potentials are generated in the cochlear nerve.*
41. *What is the importance of the sound attenuation reflex?*
42. *Where do higher-pitched tones cause vibration of the basilar membrane? Lower-pitched tones?*
43. *Describe the neuronal pathways for hearing, from the cochlear nerve to the cerebral cortex.*

Balance

The organs of balance are divided structurally and functionally into two parts. The first, the **static labyrinth,** consists of the **utricle** (oo′tri-kl) and the **saccule** (sak′ūl) of the vestibule and is primarily involved in evaluating the position of the head relative to gravity, although the system also responds to linear acceleration or deceleration, as when a person is in a car that is increasing or decreasing speed. The second, the **dynamic labyrinth,** is associated with the semicircular canals and is involved in evaluating movements of the head.

Most of the utricular and saccular walls consist of simple cuboidal epithelium. However, the utricle and saccule each contain a specialized patch of epithelium about 2–3 mm in diameter called the **utricular macula** and the **saccular macula** (mak′ū-lă; figure 15.36*a,b*). The utricular macula is oriented parallel to the base of the skull, and the saccular macula is perpendicular to the base of the skull.

The maculae resemble the spiral organ and consist of columnar supporting cells and hair cells. The "hairs" of these cells, which consist of numerous microvilli, called **stereocilia,** and one cilium, called a **kinocilium** (kī-nō-sil′ē-ŭm), are embedded in the otolithic membrane, a gelatinous mass weighted by the presence of **otoliths** (ō′tō-liths) composed of protein and calcium carbonate (figure 15.36*b,d*). The gelatinous mass moves in response to gravity, bending the hair cells and initiating action potentials in the associated neurons. The stereocilia function much as the stereocilia of cochlear hair cells do, with tip links connected to gated K^+ channels. Deflection of the hairs toward the kinocilium results in depolarization of the hair cell, whereas deflection of the hairs away from the kinocilium results in hyperpolarization of the hair cell. If the head is tipped, otoliths move in response to gravity and stimulate certain hair cells (figure 15.37).

The hair cells have no axons but synapse directly with neurons of the vestibulocochlear nerve (VIII). The hair cells release several neurotransmitters, one of which is glutamate. Research indicates that other neurotransmitters may also be involved.

The hair cells are constantly being stimulated at a low level by the otolith-weighted covering of the maculae; however, as this covering moves in response to gravity, the intensity of hair cell stimulation changes. This pattern of stimulation—and the subsequent pattern

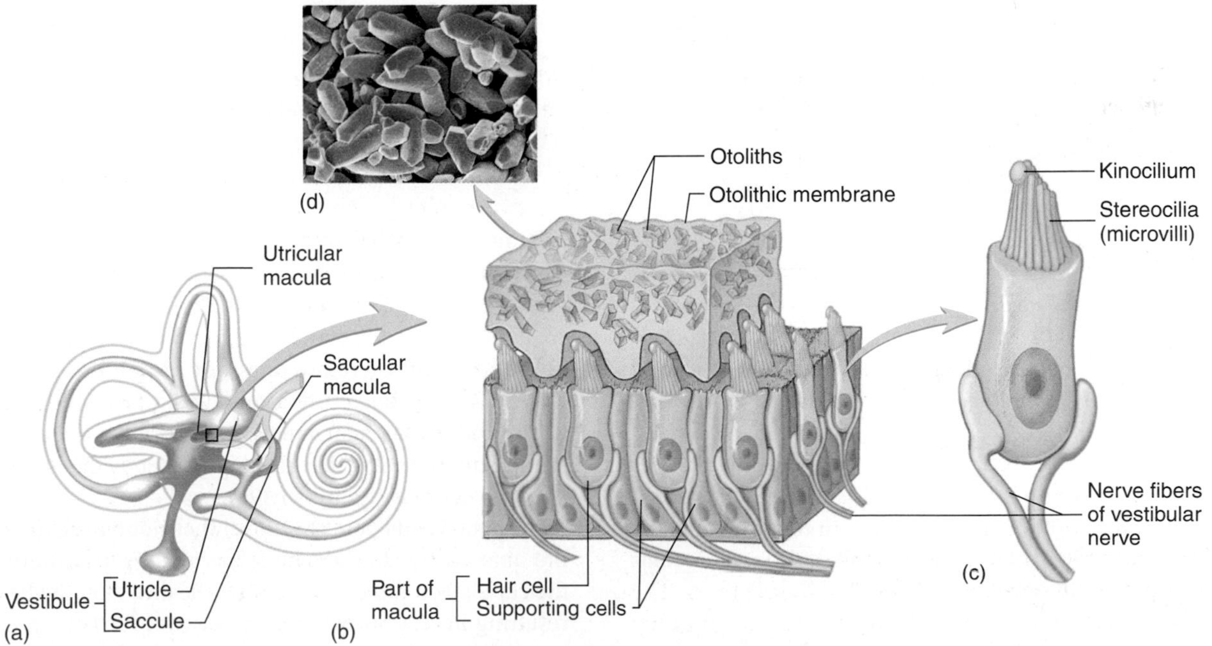

(d)
Utricular macula
Saccular macula
Vestibule
Utricle
Saccule
(a)
Otoliths
Otolithic membrane
Part of macula
Hair cell
Supporting cells
(b)
Kinocilium
Stereocilia (microvilli)
Nerve fibers of vestibular nerve
(c)

FIGURE 15.36 Structure of the Utricular and Saccular Maculae
(*a*) Vestibule showing the location of the utricular and saccular maculae. (*b*) Enlargement of the utricular macula, showing hair cells and otoliths within it. (*c*) Enlarged hair cell, showing the kinocilium and stereocilia. (*d*) Colorized scanning electron micrograph of otoliths.

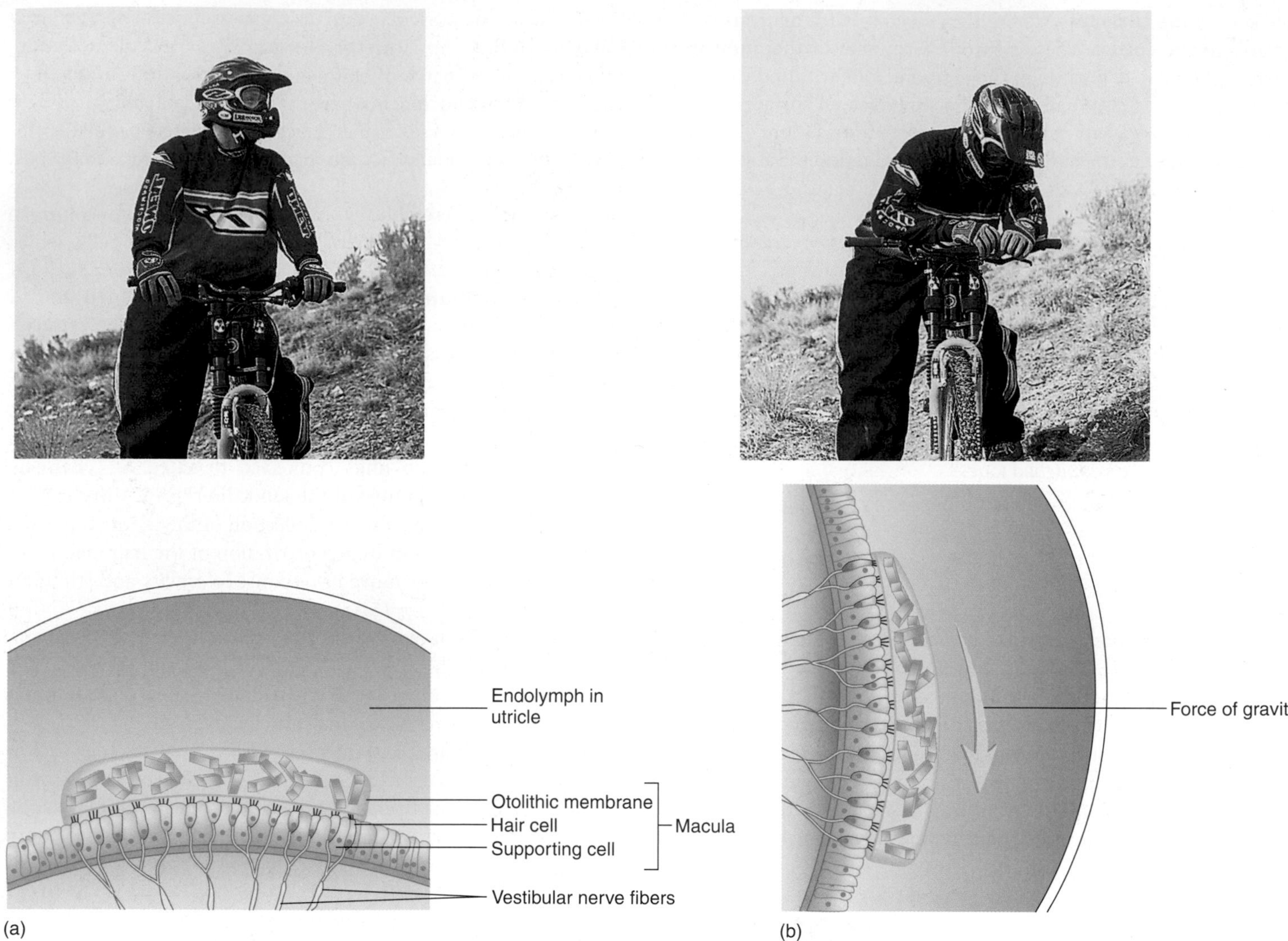
Endolymph in utricle
Otolithic membrane
Hair cell
Supporting cell
Macula
Vestibular nerve fibers
(a)
Force of gravity
(b)

FIGURE 15.37 Function of the Vestibule in Maintaining Balance
(*a*) In an upright position, the otolithic membrane does not move. (*b*) As the position of the head changes, as when a person bends over, gravity causes the otolithic membrane to move.

of action potentials from the numerous hair cells of the maculae—can be translated by the brain into specific information about head position or acceleration. Much of this information is not perceived consciously but is dealt with subconsciously. The body responds by making subtle tone adjustments in the muscles of the back and neck, which are intended to restore the head to its proper neutral, balanced position.

The dynamic labyrinth consists of three **semicircular canals** at nearly right angles to one another, one lying nearly in the transverse plane, one in the coronal plane, and one in the sagittal plane (figure 15.38*a;* see chapter 1). The arrangement of the semicircular canals enables a person to detect movement in all directions. The base of each semicircular canal is expanded into an **ampulla.** Within each ampulla, the epithelium is specialized to form a **crista ampullaris** (kris′tă am-pū-lar′ŭs), which is structurally and functionally very similar to the sensory epithelium of the maculae. Each crista consists of a ridge or crest of epithelium with a curved, gelatinous mass, the **cupula** (koo′poo-lă), suspended over the crest. The hairlike processes of the crista hair cells (which are stereocilia similar to those of the hair cells in the maculae and the spiral organ) are embedded in the cupula (figure 15.38*b–d*). The cupula contains no otoliths and therefore does not respond to gravitational pull. Instead, the cupula is a float that is displaced by fluid movements within the semicircular canals. Endolymph movement within each semicircular canal moves the cupula, bends the hairs, and initiates action potentials (figure 15.39*b*).

As the head begins to move, the endolymph does not move at the same rate as the semicircular canals. This difference displaces the cupula in a direction opposite the direction the head is moving, resulting in relative movement between the cupula and the endolymph (figure 15.39). As movement continues, the fluid of the semicircular canals begins to move and catches up with the cupula, and stimulation stops. As the head stops moving, the endolymph continues to move because of its momentum, displacing the cupula

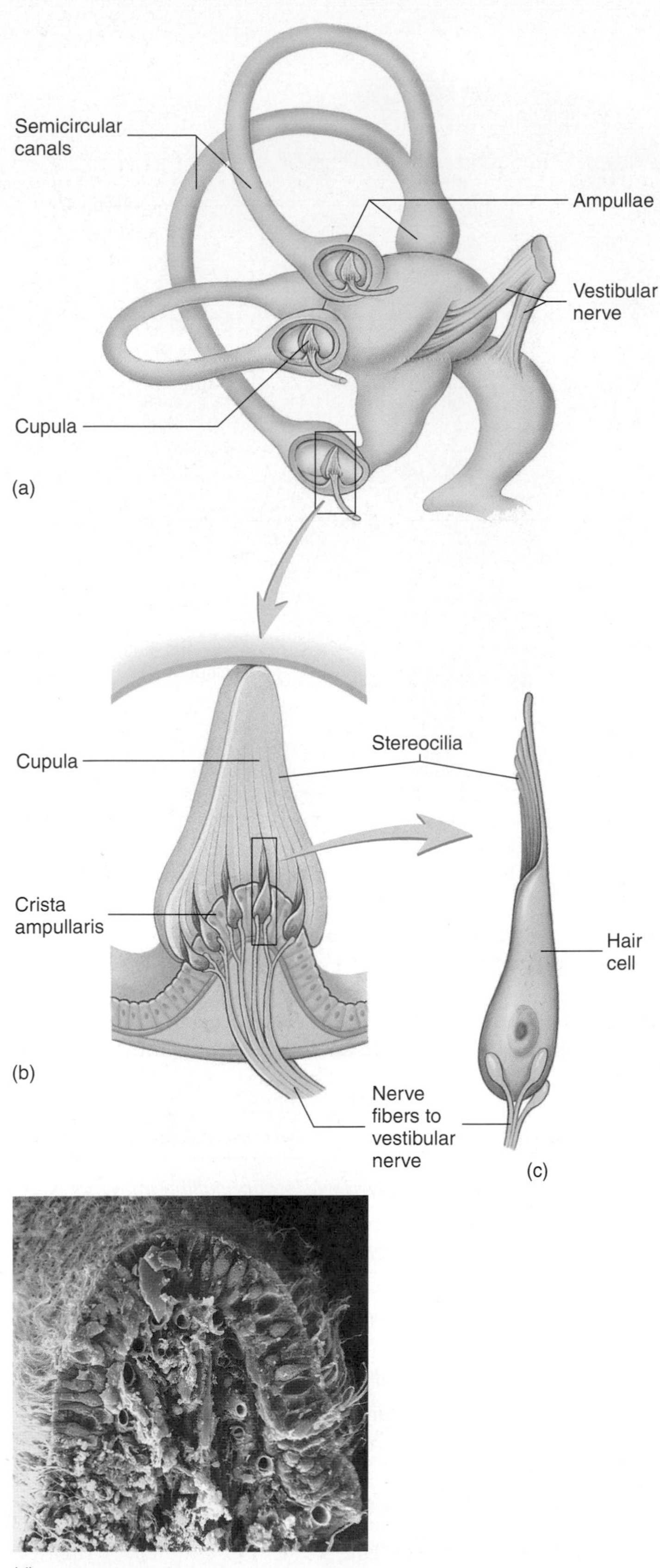
Semicircular canals
Ampullae
Vestibular nerve
Cupula
(a)
Cupula
Stereocilia
Crista ampullaris
Hair cell
(b)
Nerve fibers to vestibular nerve
(c)
(d)

FIGURE 15.38 AP|R **Semicircular Canals**

(*a*) Semicircular canals, showing the location of the crista ampullaris in the ampullae. (*b*) Enlargement of the crista ampullaris, showing the cupula and hair cells. (*c*) Enlargement of a hair cell. (*d*) SEM of a crista ampullaris with hair cells.

(a)

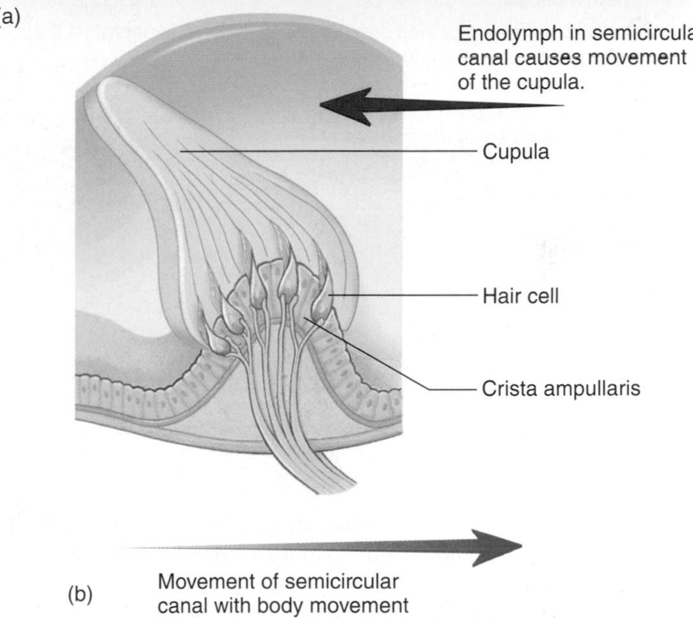
Endolymph in semicircular canal causes movement of the cupula.
Cupula
Hair cell
Crista ampullaris
(b)
Movement of semicircular canal with body movement

FIGURE 15.39 Function of the Semicircular Canals

The crista ampullaris responds to fluid movements within the semicircular canals. (*a*) When a person is still, the cupula is stationary, as when a gymnast is standing (see figure 15.38*b*). (*b*) As a person moves (for example when the gymnast moves into a new position), the semicircular canals begin to move with the body (blue arrow), but the endolymph tends to remain stationary relative to the movement (red arrow pointing in opposite direction of body and semicircular canal movement), and the cupula is displaced by the endolymph in a direction opposite the direction of movement.

in the same direction as the head was moving. Because displacement of the cupula is most intense when the rate of head movement changes, this system detects changes in the rate of movement rather than movement alone. As with the static labyrinth, the information the brain obtains from the dynamic labyrinth is largely subconscious.

Neuronal Pathways for Balance

Neurons synapsing on the hair cells of the maculae and cristae ampullares converge into the **vestibular ganglion,** where their cell bodies are located (figure 15.40). Sensory fibers from these neurons

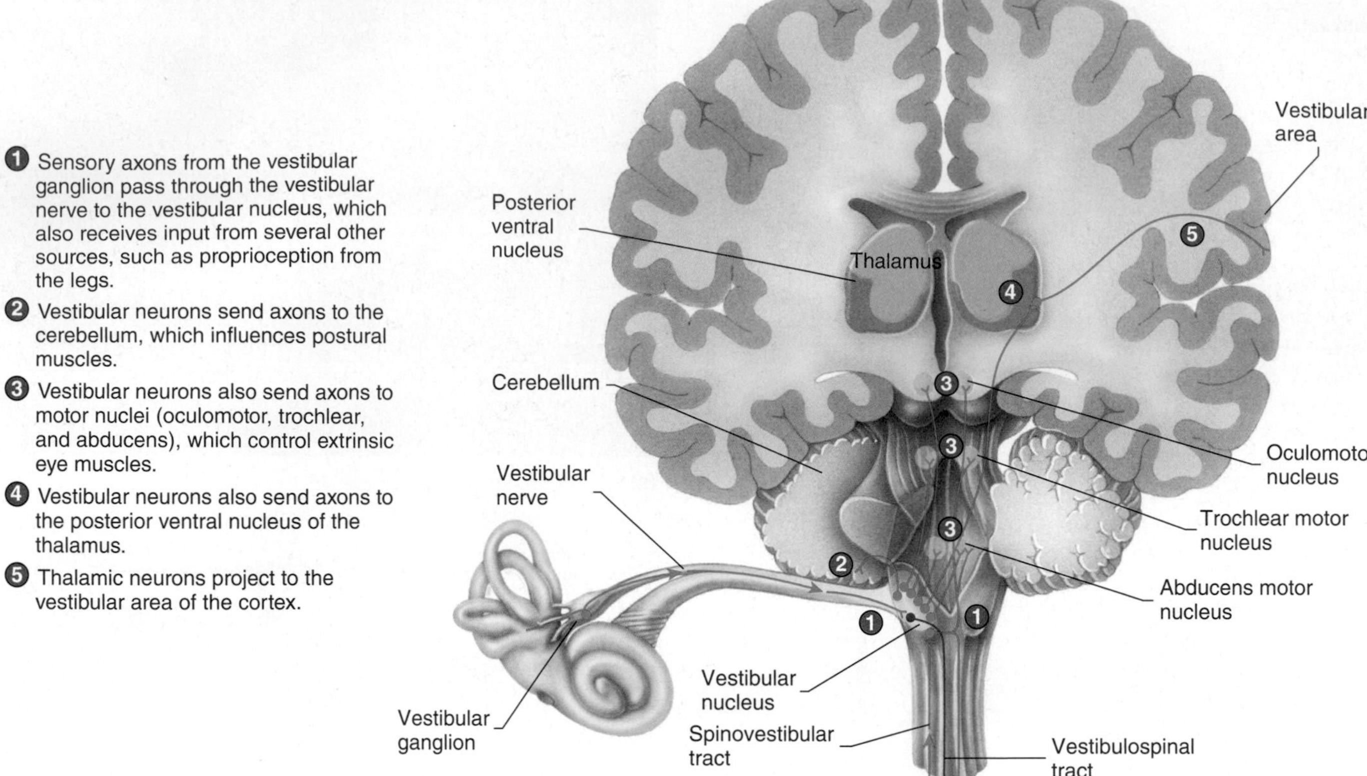

1 Sensory axons from the vestibular ganglion pass through the vestibular nerve to the vestibular nucleus, which also receives input from several other sources, such as proprioception from the legs.
2 Vestibular neurons send axons to the cerebellum, which influences postural muscles.
3 Vestibular neurons also send axons to motor nuclei (oculomotor, trochlear, and abducens), which control extrinsic eye muscles.
4 Vestibular neurons also send axons to the posterior ventral nucleus of the thalamus.
5 Thalamic neurons project to the vestibular area of the cortex.
Posterior ventral nucleus
Thalamus
Vestibular area
Cerebellum
Oculomotor nucleus
Vestibular nerve
Trochlear motor nucleus
Abducens motor nucleus
Vestibular nucleus
Vestibular ganglion
Spinovestibular tract
Vestibulospinal tract
Frontal section

PROCESS FIGURE 15.40 Central Nervous System Pathways for Balance

join sensory fibers from the cochlear ganglion to form the vestibulocochlear nerve (VIII) and terminate in the **vestibular nucleus** within the medulla oblongata. Axons run from this nucleus to numerous areas of the CNS, such as the spinal cord, the cerebellum, the cerebral cortex, and the nuclei controlling extrinsic eye muscles.

Balance is a complex process not simply confined to one type of input. In addition to vestibular sensory input, the vestibular nucleus receives input from proprioceptive neurons throughout the body, as well as from the visual system. In sobriety tests, people are asked to close their eyes while their balance is evaluated, because alcohol affects the proprioceptive and vestibular components of balance (cerebellar function) to a greater extent than it does the visual portion.

Reflex pathways exist between the dynamic part of the vestibular system and the nuclei controlling the extrinsic eye muscles (oculomotor, trochlear, and abducens). A reflex pathway allows a person to maintain visual fixation on an object while the head is in motion. To demonstrate this function, try spinning a person around about 10 times in 20 seconds, then stopping him or her and observing eye movements. The reaction is most pronounced if the individual's head is tilted forward about 30 degrees while spinning, thus bringing the lateral semicircular canals into the horizontal plane. A slight oscillatory movement of the eyes occurs. The eyes track in the direction of motion and return with a rapid recovery movement before repeating the tracking motion. This oscillation of the eyes is called **nystagmus** (nis-tag′mŭs). If then asked to walk in a straight line, the person deviates in the direction of rotation; if asked to point to an object, his or her finger deviates in the direction of rotation.

ASSESS YOUR PROGRESS

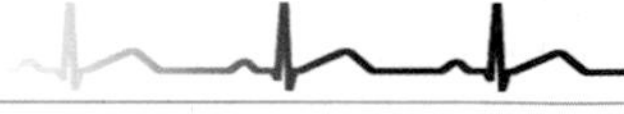

44. *What structures are involved with static equilibrium? What is static equilibrium?*

45. *Describe how the utricular macula and saccular macula function in static equilibrium.*

46. *What structures are involved with dynamic equilibrium? What is dynamic equilibrium?*

47. *Describe the crista ampullaris and its mode of operation.*

48. *Describe the neuronal pathways for balance.*

15.5 Effects of Aging on the Special Senses

LEARNING OUTCOME

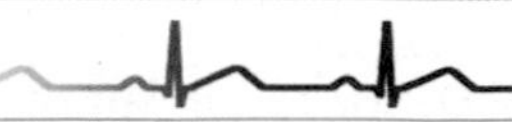

After reading this section, you should be able to

A. Describe changes that occur with the special senses because of aging.

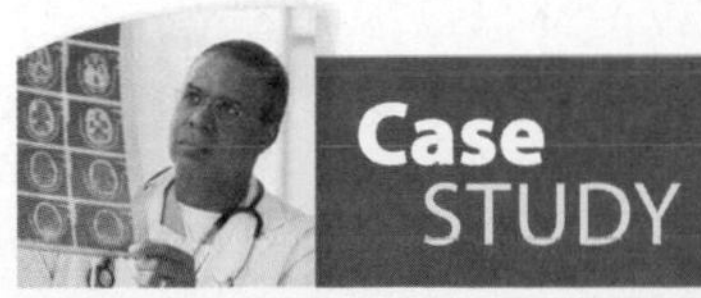

Case STUDY — Motion Sickness

Earl booked his first trip on a charter fishing boat. Crossing the bar was invigorating, and Earl was surprised that all those warnings about seasickness did not seem to apply to him. At last, the boat arrived at the fishing site, the engine was cut, and the sea anchor was set. As the boat drifted, it began to roll and pitch. For the first time, Earl noticed the unpleasant mixture of smelly bait and diesel fumes. Earl felt a little light-headed and a bit drowsy; then he began to feel nauseated, and his face became pale. "I'm seasick," he realized. Trying to fish seemed to worsen his condition. Eventually, his nausea intensified and he leaned over the boat rail and vomited into the ocean. "Improves the fishing," the ship owner shouted to him cheerily. After that, Earl felt distinctly better, and his condition continued to improve. He found that looking at the horizon rather than at the water helped his condition. He enjoyed the rest of the trip and even caught a couple of fish.

Seasickness is a form of **motion sickness,** which consists of nausea, weakness, and other dysfunctions resulting from stimulation of the semicircular canals during motion, as may occur while riding in a boat, an automobile, an airplane, a swing, or an amusement park ride. Motion sickness can progress to vomiting and incapacitation. It occurs because the brain simultaneously perceives differing sensory input from the semicircular canals, eyes, and proprioceptors in the lower limbs. Motion sickness can be decreased by closing the eyes or looking at a distant object, such as the horizon.

Various drugs are used to treat motion sickness. Antiemetics, such as anticholinergic or antihistamine medications, can counter the nausea and vomiting. Scopolamine is an anticholinergic drug that blocks acetylcholine-mediated transmission in the parasympathetic nervous system. Scopolamine can be administered transdermally in the form of a patch placed on the skin behind the ear (Transdermal-Scop). A patch lasts about 3 days. It depresses parasympathetic activity within areas of the CNS, such as the hypothalamus, in response to vestibular stimulation. Unfortunately, scopolamine also depresses other CNS functions, which may cause side effects such as restlessness, agitation, psychosis, mania, and Parkinson-like tremors. Cyclizine (Marezine), dimenhydrinate (Dramamine), and diphenhydramine (Benadryl) are antihistamines that affect the neural pathways from the vestibule.

Predict 9

Explain why closing your eyes can help decrease motion sickness. Why might looking at the horizon help?

People experience only a slight loss in the ability to detect odors as they age. However, their ability to identify specific odors correctly decreases, especially in men over age 70.

In general, the sense of taste decreases as people age because the number of sensory receptors decreases and the brain's ability to interpret taste sensations declines. Responses to taste also change in some elderly people who are fighting cancer. One side effect of radiation treatment and chemotherapy is gastrointestinal discomfort, which causes patients to lose their appetite because of conditioned taste aversions.

The lenses of the eyes lose flexibility as a person ages, because the connective tissue of the lenses becomes more rigid. Consequently, the lenses' ability to change shape is at first reduced and eventually completely lost. Recall that this condition, called presbyopia, is the most common age-related change in the eyes. In addition, the number of cones decreases, especially in the fovea centralis. These changes cause a gradual decline in visual acuity and color perception. The most common visual problem in older people requiring medical treatment, such as surgery, is the development of cataracts. Macular degeneration, which affects visual acuity in the center of the visual field, is the leading

Diseases and Disorders

TABLE 15.3 Hearing and Balance

Condition	Description
Conductive hearing loss	Mechanical deficiency in transmission of sound waves from external ear to spiral organ
Sensorineural hearing loss	Deficiencies of spiral organ or nerve pathways
Otosclerosis	Type of conductive hearing loss resulting from spongy bone growing over oval window and immobilizing stapes; can be surgically corrected
Tinnitus (ti-nī′tus)	Phantom sound sensations, such as ringing in ears; a common problem
Otitis media	Low-grade fever, lethargy, irritability, and pulling at ear; in extreme cases, can damage or rupture tympanic membrane; common in young children
Inner ear infection	Can decrease detection of sound and maintenance of balance; may be caused by chronic middle ear infections
Motion sickness	Nausea and weakness caused when information to brain from semicircular canals conflicts with information from eyes and position sensors in back and lower limbs
Meniere disease	Vertigo, hearing loss, tinnitus, and a feeling of fullness in the affected ear; most common disease involving dizziness from inner ear; cause unknown but may involve a fluid abnormality in ears

Go to www.mhhe.com/seeley10 for additional information on these pathologies.

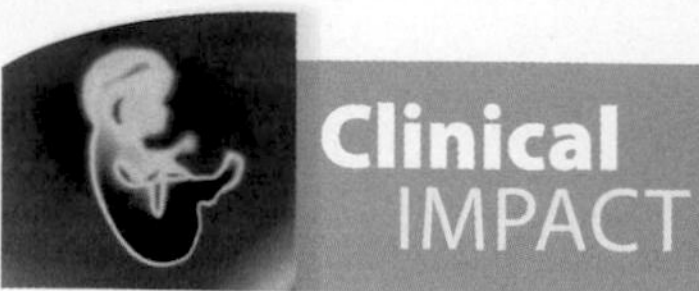

Clinical IMPACT — Hearing Impairment and Functional Replacement of the Ear

The term **hearing-impaired** refers to any type or degree of hearing loss. Two types of hearing impairment have been identified: conductive and sensorineural. In **conductive hearing loss,** the spiral organ and neuronal pathways for hearing function normally, but there is a mechanical deficiency in the transmission of sound waves from the external ear to the spiral organ. Conductive hearing loss often can be treated—for example, by removing earwax blocking the external auditory canal or by replacing or repairing the auditory ossicles. If the degree of conductive hearing loss does not justify surgical treatment, or if treatment does not resolve the hearing loss, a hearing aid may be worn to help transmit the amplified (louder) sound waves through the conductive blockage and provide normal stimulation to the spiral organ.

Sensorineural hearing loss involves the spiral organ or neuronal pathways. Sound waves are transmitted normally to the spiral organ, but the nervous system's ability to respond to the sound waves is impaired. People with sensorineural hearing loss commonly use hearing aids, which produce amplified sound waves that stimulate the spiral organ more than normal, helping overcome the perception of reduced sound volume. Sound clarity also improves with sound amplification but may never be perceived as normal.

The term **deaf** refers to sensorineural hearing loss so profound that the sense of hearing is nonfunctional, with or without amplification, for ordinary purposes of life. Stimulation of the spiral organ or hearing nerve pathways can help deaf people hear.

Research is being conducted on ways to replace the hearing pathways with electrical circuits. One approach involves directly stimulating the cochlear nerve using electrical impulses. Types of sensorineural deafness in which the hair cells of the spiral organ are impaired can now be partially corrected by implanting a prosthesis consisting of a microphone for picking up the initial sound waves, a microelectronic processor for converting the sound into electrical signals, a transmission system for relaying the signals to the inner ear, and a long, slender electrode threaded into the cochlea. This electrode delivers electrical signals directly to the endings of the cochlear nerve (figure 15D). High-frequency sounds are picked up by the microphone and transmitted through specific circuits to terminate near the oval window, whereas low-frequency sounds are transmitted farther up the cochlea to cochlear nerve endings near the helicotrema.

For patients with vestibulocochlear nerve damage, research is underway to develop a technique for implanting electrodes directly into the cochlear nucleus of the brainstem. These implanted electrodes would be of various lengths, so that they could stimulate parts of the cochlear nucleus, at various depths from the surface, that respond to sounds of different frequencies.

1 A receiver, a transmitter, and an antenna are implanted under the skin near the auricle.

2 A small lead from the transmitter is fed through the external auditory canal, tympanic membrane, and middle ear into the cochlea.

3 In the cochlea, the cochlear nerve can be directly stimulated by electrical impulses from the receiver.

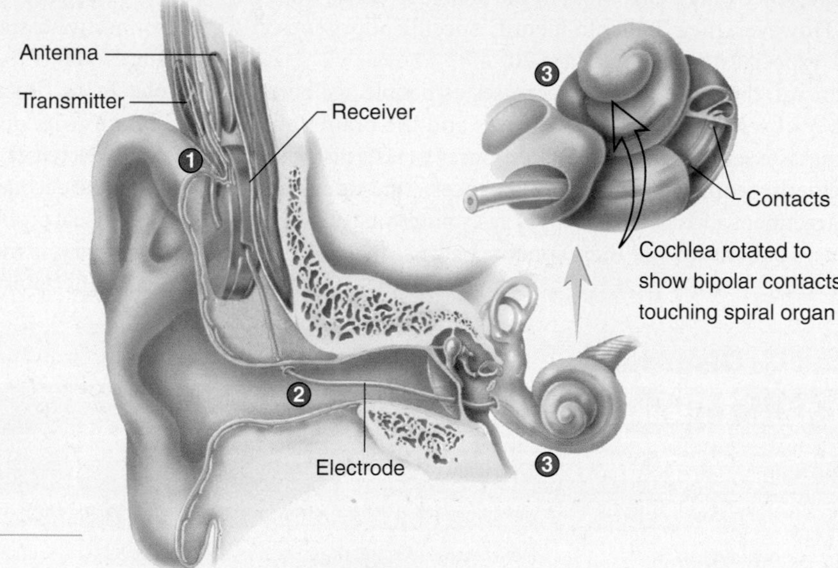
Antenna
Transmitter
Receiver
1
2
Electrode
3
3
Contacts
Cochlea rotated to show bipolar contacts touching spiral organ

FIGURE 15D Cochlear Implant

cause of vision loss in people over the age of 60. Other age-related defects affecting vision include glaucoma and diabetic retinopathy.

As people age, the number of hair cells in the cochlea decreases, leading to age-related hearing loss, called **presbyacusis** (prez′bē-ă-koo′sis). This decline does not occur equally in both ears, however. As a result, because direction is determined by comparing sounds coming into each ear, elderly people may experience a decreased ability to localize the origin of certain sounds. In some people, this leads to a general sense of disorientation. In addition, CNS defects in the auditory pathways can result in difficulty understanding sounds with echoes or background noise. Such a deficit makes it difficult for elderly people to understand rapid or broken speech.

With age, the number of hair cells in the saccule, utricle, and ampullae decreases. The number of otoliths also declines. As a result, elderly people experience decreased sensitivity to gravity, acceleration, and rotation, leading to dizziness (instability) and vertigo (a feeling of spinning). Many elderly people feel they cannot maintain posture and are prone to fall.

ASSESS YOUR PROGRESS

49. *Explain the changes in smell, taste, vision, hearing, and balance that occur with aging.*

Answer

Learn to Predict **From page 500**

We learn that Freddy's surgery replaced two of the three auditory ossicles. Recall that the auditory ossicles distribute sound waves from the tympanic membrane to the oval window of the inner ear, where the sound waves can then stimulate hair cells of the cochlear duct. If Freddy's auditory ossicles were not functional, we can predict that his loss of hearing was conductive hearing loss due to the absence of functional auditory ossicles. To address the question of sound location, recall that the brain can determine the direction from which a sound is coming because of the time interval between when the sound reaches one ear and the time it reaches the other. Since Freddy had complete hearing loss in one ear, his ability to determine the direction from which a sound came was probably diminished. Finally, by replacing the damaged auditory ossicles, Freddy's hearing was restored, because sound waves can now be conducted from Freddy's tympanic membrane to his oval window.

Answers to the rest of this chapter's Predict questions are in Appendix G.

Summary

15.1 Olfaction (p. 501)

Olfaction is the sense of smell.

Olfactory Epithelium and Bulb

1. Olfactory neurons in the olfactory epithelium are bipolar neurons. Their distal ends are enlarged as olfactory vesicles, which have long cilia.
2. The cilia have receptors that respond to dissolved substances. There are approximately 1000 different odorant receptors.
3. The receptors activate a G protein complex, which opens ion channels.
4. At least 7 (but perhaps as many as 50) primary odors exist. The olfactory neurons have a very low threshold and adapt rapidly.

Neuronal Pathways for Olfaction

1. Axons from the olfactory neurons extend as olfactory nerves to the olfactory bulb, where they synapse with mitral and tufted cells. Axons from these cells form the olfactory tracts and synapse with granule cells. The olfactory bulb neurons can modulate output to the olfactory tracts.
2. The olfactory tracts terminate in the olfactory cortex of the temporal lobe or secondary olfactory areas of the frontal lobe. The olfactory cortex is involved in the conscious perception of smell, the intermediate olfactory area with modulating smell, and the medial olfactory area with visceral and emotional responses to smell.

15.2 Taste (p. 504)

Taste buds are usually associated with vallate, foliate, and fungiform papillae. Filiform papillae do not have taste buds.

Histology of Taste Buds

1. Taste buds consist of basal cells, supporting cells, and taste cells.
2. The taste cells have taste hairs that extend into taste pores.

Function of Taste

1. Receptors on the taste hairs detect dissolved substances.
2. Five basic types of taste exist: salty, sour, sweet, bitter, and umami.

Neuronal Pathways for Taste

1. The facial nerve carries taste sensations from the anterior two-thirds of the tongue, the glossopharyngeal nerve from the posterior one-third of the tongue, and the vagus nerve from the epiglottis.
2. The neural pathways for taste extend from the medulla oblongata to the thalamus and to the cerebral cortex.

15.3 Visual System (p. 507)

Accessory Structures

1. The eyebrows prevent perspiration from entering the eyes and help shade the eyes.
2. The eyelids consist of five tissue layers. They protect the eyes from foreign objects and help lubricate the eyes by spreading tears over their surface.
3. The conjunctiva covers the inner eyelid and the anterior part of the eye.
4. Lacrimal glands produce tears, which flow across the surface of the eye. Excess tears enter the lacrimal canaliculi and reach the nasal cavity through the nasolacrimal canal. Tears lubricate and protect the eye.
5. The extrinsic eye muscles move the eyeball.

Anatomy of the Eye

1. The fibrous tunic is the outer layer of the eyeball. It consists of the sclera and then cornea.

- The sclera is the posterior four-fifths of the eyeball. It is white connective tissue that maintains the shape of the eyeball and provides a site for muscle attachment.
- The cornea is the anterior one-fifth of the eye. It is transparent and refracts light that enters the eye.

2. The vascular tunic is the middle layer of the eyeball.
   - The iris is smooth muscle regulated by the autonomic nervous system. It controls the amount of light entering the pupil.
   - The ciliary muscles control the shape of the lens. They are smooth muscles regulated by the autonomic nervous system. The ciliary process produces aqueous humor.
3. The retina is the nervous tunic of the eyeball and contains neurons sensitive to light.
   - The fovea centralis is the area of greatest visual acuity.
   - The optic disc is the location through which nerves exit and blood vessels enter the eye. It has no photosensory cells and is therefore the blind spot of the eye.
4. The eyeball has three chambers: anterior, posterior, and vitreous.
   - The anterior and posterior chambers are filled with aqueous humor, which circulates and leaves by way of the scleral venous sinus.
   - The vitreous chamber is filled with vitreous humor.
5. The lens is held in place by the suspensory ligaments, which are attached to the ciliary muscles.

Functions of the Eye

1. Light is the portion of the electromagnetic spectrum that humans can see.
2. When light travels from one medium to another, it can bend, or refract. Light striking a concave surface refracts outward (divergence). Light striking a convex surface refracts inward (convergence).
3. Converging light rays meet at the focal point and are said to be focused.
4. The cornea, aqueous humor, lens, and vitreous humor all refract light. The cornea is responsible for most of the convergence, whereas the lens can adjust the focal point by changing shape.
   - Relaxation of the ciliary muscles causes the lens to flatten into its normal resting condition, called emmetropia.
   - Contraction of the ciliary muscles causes the lens to become more spherical. This change in lens shape enables the eye to focus on objects that are nearby, a process called accommodation.
5. The far point of vision is the distance at which the eye no longer has to change shape to focus on an object. The near point of vision is the closest an object can come to the eye and still be focused.
6. The pupil becomes smaller during accommodation, increasing the depth of focus.

Structure and Function of the Retina

1. The pigmented layer of the retina provides a black backdrop for increasing visual acuity.
2. Rods are responsible for vision in low illumination (night vision).
   - A pigment called rhodopsin is split by light into retinal and opsin, producing hyperpolarization in the rod.
   - Light adaptation is caused by the reduction of rhodopsin; dark adaptation is caused by the production of rhodopsin.
3. Cones are responsible for color vision and visual acuity.
   - Cones are of three types, each with a different type of iodopsin photopigment. The pigments are most sensitive to blue, red, and green lights.
   - Perception of many colors results from mixing the ratio of the different types of cones that are active at a given moment.
4. Most visual images are focused on the fovea centralis, which has a very high concentration of cones. Moving away from the fovea centralis, fewer cones are present; the periphery of the retina contains mostly rods.
5. The rods and the cones synapse with bipolar cells, which in turn synapse with ganglion cells, which form the optic nerves.
6. Bipolar and ganglion cells in the retina can modify information sent to the brain.
7. Ganglion cells have receptive fields with on-centers or off-centers. This arrangement enhances contrast.

Neuronal Pathways for Vision

1. Ganglion cell axons extend to the lateral geniculate ganglion of the thalamus, where they synapse. From there, neurons form the optic radiations that project to the visual cortex.
2. Neurons from the nasal visual field (temporal retina) of one eye and the temporal visual field (nasal retina) of the opposite eye project to the same cerebral hemisphere. Axons from the nasal retina cross in the optic chiasm, and axons from the temporal retina remain uncrossed.
3. Depth perception is the ability to judge relative distances of an object from the eyes and is a property of binocular vision. Binocular vision results because a slightly different image is seen by each eye.

15.4 Hearing and Balance (p. 526)

The external and middle ears are involved in hearing only, whereas the inner ear functions in both hearing and balance.

Auditory Structures and Their Functions

1. The external ear consists of the auricle and the external auditory canal.
2. The middle ear connects the external and inner ears.
   - The tympanic membrane is stretched across the external auditory canal.
   - The malleus, incus, and stapes connect the tympanic membrane to the oval window of the inner ear.
   - The auditory tube connects the middle ear to the pharynx and functions to equalize pressure.
   - The middle ear is connected to the mastoid air cells.
3. The bony labyrinth of the inner ear is a canal system within the temporal bone that contains perilymph and the membranous labyrinth.
   - Endolymph is inside the membranous labyrinth.
   - The bony labyrinth has three parts: the semicircular canals, the vestibule (containing the utricle and the saccule), and the cochlea.
4. The cochlea is a spiral-shaped canal within the temporal bone.
   - The cochlea is divided into three compartments by the vestibular and basilar membranes. The scala vestibuli and scala tympani contain perilymph. The cochlear duct contains endolymph and the spiral organ.
   - The spiral organ consists of inner and outer hair cells that attach to the tectorial membrane.

Auditory Function

1. Sound waves are funneled by the auricle down the external auditory canal, causing the tympanic membrane to vibrate.
2. The tympanic membrane vibrations are passed along the auditory ossicles to the oval window of the inner ear.
3. Movement of the stapes in the oval window causes the perilymph, vestibular membrane, and endolymph to vibrate, producing movement of the basilar membrane.
4. Movement of the basilar membrane causes bending of the stereocilia of inner hair cells in the spiral organ.
5. Bending of the stereocilia pulls on tip links, which open K^+ channels.
6. Potassium ions entering the hair cell depolarize the cell, which opens Ca^{2+} channels. Calcium ions entering the cell cause further depolarization.
7. Depolarization causes the release of glutamate, generating action potentials in the vestibulocochlear nerve.

8. Some vestibulocochlear nerve axons synapse in the superior olivary nucleus. Efferent neurons from this nucleus project back to the cochlea, where they regulate the perception of pitch.
9. The round window protects the inner ear from pressure buildup and dissipates sound waves.

Neuronal Pathways for Hearing

1. Axons from the vestibulocochlear nerve synapse in the medulla oblongata. Neurons from the medulla oblongata project axons to the inferior colliculi, where they synapse. Neurons from this point project to the thalamus and synapse. Thalamic neurons extend to the auditory cortex.
2. Efferent neurons project to the cranial nerve nuclei responsible for controlling the muscles that dampen sound in the middle ear.

Balance

1. The static labyrinth is involved in evaluating the position of the head relative to gravity and detecting linear acceleration and deceleration.
   - The utricle and saccule in the inner ear contain maculae. The maculae consist of hair cells with the hairs embedded in a membrane that contains otoliths.
   - The ololithic membrane moves in response to gravity.
2. Dynamic balance evaluates movements of the head.
   - Three semicircular canals at right angles to one another are present in the inner ear. The ampulla of each semicircular canal contains the crista ampullaris, which has hair cells with hairs embedded in a gelatinous mass, the cupula.
   - When the head moves, endolymph within the semicircular canal moves the cupula.

Neuronal Pathways for Balance

1. Axons from the maculae and the cristae ampullares extend to the vestibular nucleus of the medulla oblongata. Fibers from the medulla run to the spinal cord, cerebellum, cortex, and nuclei that control the extrinsic eye muscles.
2. Balance also depends on proprioception and visual input.

15.5 Effects of Aging on the Special Senses (p. 540)

Elderly people experience a functional decline in all the special senses: olfaction, taste, vision, hearing, and balance. These declines can result in loss of appetite, visual impairment, disorientation, and risk of falling.

REVIEW AND COMPREHENSION

1. Which of these statements is *not* true with respect to olfaction?
   a. Olfactory sensation is relayed directly to the cerebral cortex without passing through the thalamus.
   b. Olfactory neurons are replaced about every 2 months.
   c. The olfactory cortex is involved in the conscious perception of smell.
   d. The medial olfactory area of the cortex is responsible for visceral and emotional reactions to odors.
   e. The olfactory cortex is in the occipital lobe of the cerebrum.
2. Taste cells
   a. are found only on the tongue.
   b. extend through tiny openings called taste buds.
   c. have no axons but release neurotransmitters when stimulated.
   d. have axons that extend directly to the taste area of the cerebral cortex.
3. Which of these is *not* one of the basic tastes?
   a. spicy b. salt c. bitter d. umami e. sour
4. Which of these types of papillae have no taste buds associated with them?
   a. vallate b. filiform c. foliate d. fungiform
5. The fibrous tunic of the eye includes the
   a. conjunctiva. b. sclera. c. choroid. d. iris. e. retina.
6. The ciliary body
   a. contains smooth muscles that attach to the lens by suspensory ligaments.
   b. produces the vitreous humor.
   c. is part of the iris of the eye.
   d. is part of the sclera.
   e. All of these are correct.
7. The lens normally focuses light onto the
   a. optic disc. b. iris. c. macula. d. cornea. e. ciliary body.
8. Given these structures:
   (1) lens
   (2) aqueous humor
   (3) vitreous humor
   (4) cornea

   Which of the following arrangements lists the structures in the order that light entering the eye encounters them?
   a. 1,2,3,4 b. 1,4,2,3 c. 4,1,2,3 d. 4,2,1,3 e. 4,3,2,1
9. Contraction of the smooth muscle in the ciliary body causes the
   a. lens to flatten. b. pupil to constrict. c. lens to become more spherical. d. pupil to dilate.
10. Given these events:
    (1) bipolar cells depolarize
    (2) glutamate release from presynaptic terminals of photoreceptor cells decreases
    (3) light strikes photoreceptor cells
    (4) photoreceptor cells are depolarized
    (5) photoreceptor cells are hyperpolarized

    Choose the arrangement that lists the correct order of events, starting with the photoreceptor cells in the resting, nonactivated state.
    a. 1,2,3,4,5 b. 2,4,3,5,1 c. 3,4,2,5,1 d. 4,3,5,2,1 e. 5,3,4,1,2
11. Given these neurons in the retina:
    (1) bipolar cells (2) ganglionic cells (3) photoreceptor cells

    Choose the arrangement that lists the correct order of the cells encountered by light as it enters the eye and travels toward the pigmented layer of the retina.
    a. 1,2,3 b. 1,3,2 c. 2,1,3 d. 2,3,1 e. 3,1,2
12. Which of these photoreceptor cells is *not* correctly matched with its function?
    a. rods—vision in low light
    b. cones—color vision
    c. rods—visual acuity
13. Concerning dark adaptation,
    a. the amount of rhodopsin increases.
    b. the pupils constrict.
    c. it occurs more rapidly than light adaptation.
    d. All of these are correct.

14. In the retina are cones that are most sensitive to a particular color. Given this list of colors:
    (1) red (2) yellow (3) green (4) blue

    Indicate which colors correspond to specific types of cones.
    a. 2,3 b. 3,4 c. 1,2,3 d. 1,3,4 e. 1,2,3,4
15. Given these areas of the retina:
    (1) macula (3) optic disc
    (2) fovea centralis (4) periphery of the retina

    Choose the arrangement that lists the areas according to the density of cones, starting with the area that has the highest density of cones.
    a. 1,2,3,4 b. 1,3,2,4 c. 2,1,4,3 d. 2,4,1,3 e. 3,4,1,2
16. Axons in the optic nerve from the right eye
    a. all go to the right occipital lobe.
    b. all go to the left occipital lobe.
    c. all go to the thalamus.
    d. go mostly to the thalamus, but some go to the superior colliculus.
    e. go partly to the right occipital lobe and partly to the left occipital lobe.
17. A person with an abnormally powerful focusing system is ________ and uses a ________ to correct his or her vision.
    a. nearsighted, concave lens c. farsighted, concave lens
    b. nearsighted, convex lens d. farsighted, convex lens
18. Which of these structures is found within or is a part of the external ear?
    a. oval window c. ossicles e. cochlear duct
    b. auditory tube d. external auditory canal
19. Given these ear bones:
    (1) incus (2) malleus (3) stapes

    Choose the arrangement that lists the ear bones in order from the tympanic membrane to the inner ear.
    a. 1,2,3 b. 1,3,2 c. 2,1,3 d. 2,3,1 e. 3,2,1
20. Given these structures:
    (1) perilymph (3) vestibular membrane
    (2) endolymph (4) basilar membrane

    Which of the following arrangements lists the structures in the order sound waves coming from the outside encounter them in producing sound?
    a. 1,3,2,4 b. 1,4,2,3 c. 2,3,1,4 d. 2,4,1,3 e. 3,4,2,1
21. The spiral organ is within the
    a. cochlear duct. c. scala tympani. e. semicircular canals.
    b. scala vestibuli. d. vestibule.
22. An increase in the loudness of sound occurs as a result of an increase in the ________ of the sound wave.
    a. frequency c. resonance
    b. amplitude d. both a and b
23. Interpretation of different sounds is possible because of the ability of the ________ to vibrate at different frequencies and stimulate the ________.
    a. vestibular membrane, vestibular nerve
    b. vestibular membrane, spiral organ
    c. basilar membrane, vestibular nerve
    d. basilar membrane, spiral organ
24. Which structure is a specialized receptor within the utricle?
    a. macula c. spiral organ
    b. crista ampullaris d. cupula
25. Damage to the semicircular canals affects the ability to detect
    a. sound.
    b. the position of the head relative to the ground.
    c. the movement of the head in all directions.
    d. All of these are correct.

Answers in Appendix E

CRITICAL THINKING

1. An elderly man with normal vision develops cataracts. He is surgically treated by removing the lenses of his eyes. What kind of glasses should he wear to compensate for the removal of his lenses?
2. Perhaps you have heard that eating carrots is good for the eyes. What is the basis for this claim?
3. A man stares at a black clock on a white wall for several minutes. Then he shifts his view and looks at only the blank white wall. Although he is no longer looking at the clock, he sees a light clock against a dark background. Explain what is happening.
4. Describe the results of a lesion of the optic chiasm.
5. Explain how several hours of reading can cause eyestrain, or eye fatigue. Describe what structures are involved.
6. Persistent exposure to loud noise can cause loss of hearing, especially for high-frequency sounds. What part of the ear is probably damaged? Be as specific as possible.
7. Professional divers are subject to increased pressure as they descend to the bottom of the ocean. Sometimes this pressure can lead to damage to the ear and loss of hearing. Describe the normal mechanisms that adjust for changes in pressure, suggest some conditions that might interfere with pressure adjustment, and explain how the increased pressure might cause loss of hearing.
8. If a vibrating tuning fork is placed against the mastoid process of the temporal bone, the vibrations are perceived as sound, even if the external auditory canal is plugged. Explain how this happens.

Answers in Appendix F

Visit this book's website at **www.mhhe.com/seeley10** for chapter quizzes, interactive learning exercises, and other study tools.

Mc Graw Hill
connect plus+
ANATOMY & PHYSIOLOGY

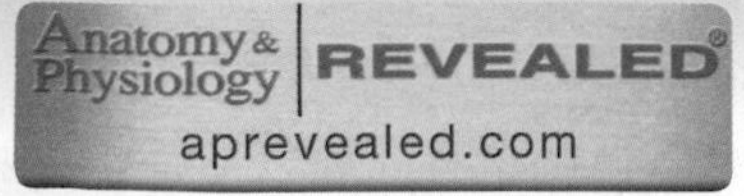
Anatomy & Physiology
REVEALED
aprevealed.com

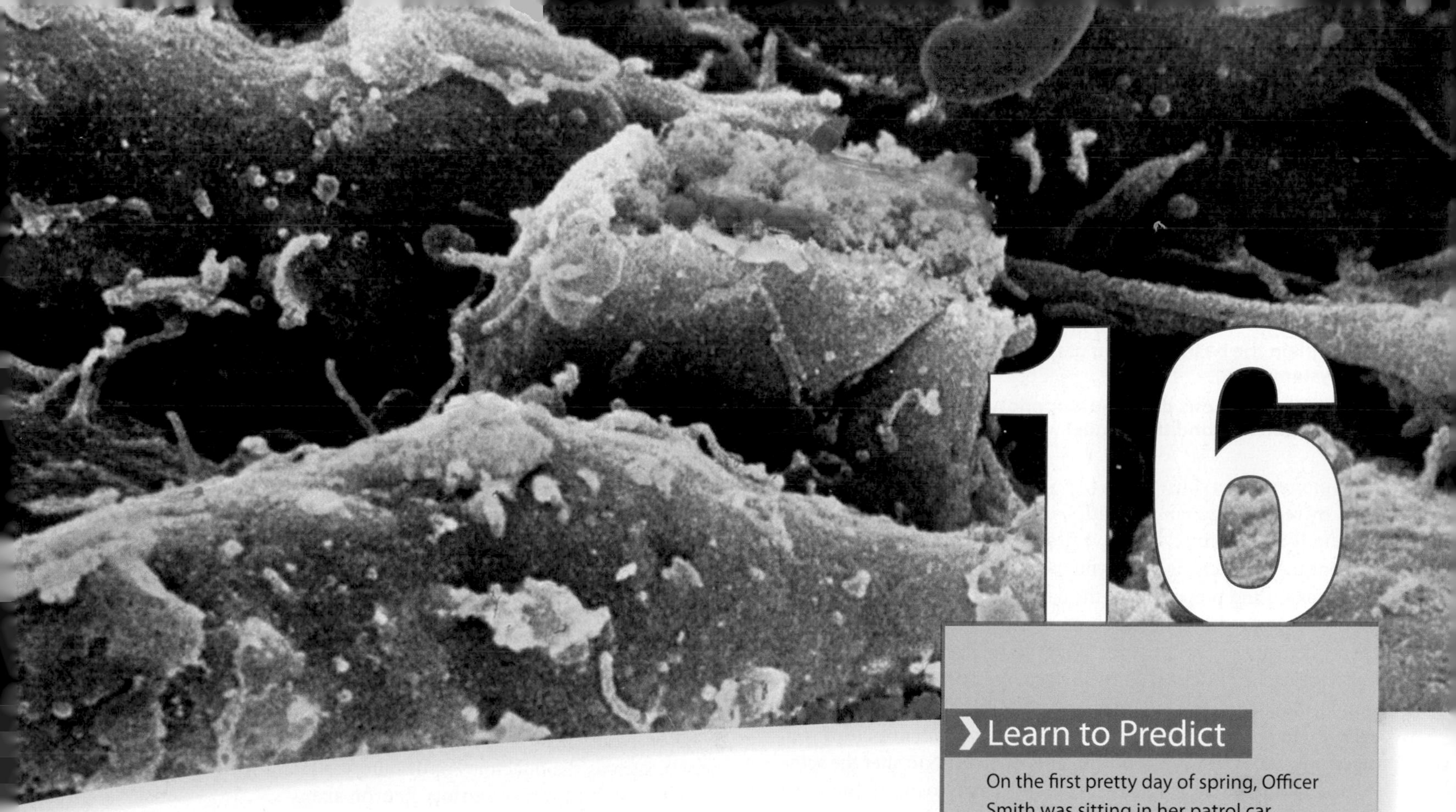

16

Autonomic Nervous System

Learn to Predict

On the first pretty day of spring, Officer Smith was sitting in her patrol car, observing traffic near the town's busy park, when a sports car suddenly sped past. Realizing that families with young children may be crossing the same street on their way to the park, Officer Smith quickly pursued the speeding car. By recalling information about the integration of nervous system functions from previous chapters and reading chapter 16, describe in general the neural pathways involved in Officer Smith's quick reactions—the movement of her skeletal muscles, her increased heart rate, and decreased movements and secretions in her digestive system.

During a picnic on a sunny spring day, it is easy to concentrate on the delicious food and the pleasant surroundings. Maintaining homeostasis requires no conscious thought. The autonomic nervous system (ANS) helps keep body temperature at a constant level by controlling the activity of sweat glands and the amount of blood flowing through the skin. The ANS also helps regulate the complex activities necessary for digesting food. Absorbed nutrients travel to the tissues through the bloodstream because the ANS controls the heart rate, which helps maintain the necessary blood pressure. Without the ANS, all of the activities necessary to maintain homeostasis would be overwhelming.

This chapter examines the autonomic nervous system. A functional knowledge of the ANS enables you to predict general responses to a variety of stimuli, explain responses to changes in environmental conditions, comprehend symptoms that result from abnormal autonomic functions, and understand how drugs affect the ANS.

Photo: Colorized scanning electromicrograph (SEM) of cells of the adrenal medulla.

Module 7
Nervous System

Anatomy & Physiology
REVEALED
aprevealed.com

16.1 Overview of the Autonomic Nervous System

LEARNING OUTCOMES 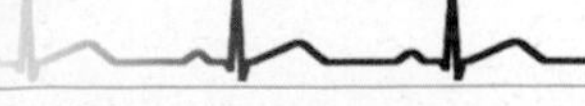

After reading this section, you should be able to

A. Explain the basic function of the autonomic nervous system (ANS).

B. List the divisions of the autonomic nervous system and describe the conditions under which each is more influential.

The autonomic nervous system (ANS) maintains homeostasis of the body by regulating many activities, including heart rate, breathing rate, body temperature, digestive processes, and urinary functions. Imagine the many changes that your body experiences during the day, from waking and preparing for the day to exercising and completing other daily tasks. All of these changes in activity involve differences in energy demands by your tissues. When you are very active, your contracting skeletal muscles require more energy reserves. When you are resting, the energy demands of your skeletal muscles decrease, but the energy demands of other tissues, such as the smooth muscle of your digestive tract, increase. It is the function of the ANS to alter the activity of smooth muscle, cardiac muscle, and glands to match the needs of the different tissues of the body during varying levels of activity.

Recall from chapter 11 that the motor division of the nervous system controls the effectors (muscles and glands) of the body. The motor division consists of two parts: the somatic motor system, which regulates the activities of skeletal muscle (see chapter 14) and the ANS, which regulates the activity of all the other effectors (smooth muscle, cardiac muscle, and glands). The ANS is further subdivided into the enteric nervous system, the sympathetic division, and the parasympathetic division. The enteric nervous system consists of nervous tissue of the digestive tract. The sympathetic and parasympathetic divisions regulate the activity of effectors throughout the body, but each division influences the tissues under different conditions. The sympathetic division, often referred to as the fight-or-flight division, has more influence on effectors under conditions of increased physical activity or stress, whereas the parasympathetic division has more influence under conditions of rest and is often referred to as the rest-and-digest division.

ASSESS **YOUR PROGRESS** 

1. *Describe the function of the ANS.*
2. *List the divisions of the ANS. Under what conditions would each division be more influential?*

16.2 Contrasting the Somatic and Autonomic Nervous Systems

LEARNING OUTCOMES 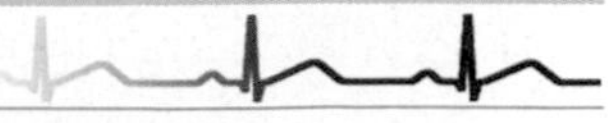

After reading this section, you should be able to

A. Describe the structural and functional differences between the somatic nervous system and the ANS.

B. Describe the relationship between preganglionic and postganglionic neurons. Contrast somatic and autonomic motor neurons with sensory neurons.

The peripheral nervous system (PNS) is composed of sensory and motor neurons. **Sensory neurons** carry action potentials from the periphery to the central nervous system (CNS), and **motor neurons** carry action potentials from the CNS to the periphery. Motor neurons that innervate skeletal muscle are called somatic motor neurons, and they are part of the somatic nervous system. Motor neurons that innervate smooth muscle, cardiac muscle, and glands are called autonomic motor neurons, and they are part of the autonomic nervous system (ANS). In this section, we compare these two systems. The rest of the chapter focuses on the structure and function of the autonomic nervous system.

The axons of autonomic, somatic, and sensory neurons are in the same nerves, but their proportions vary from nerve to nerve. For example, nerves innervating smooth muscle, cardiac muscle, and glands, such as the vagus nerves, consist primarily of axons of sensory neurons and autonomic motor neurons. Nerves innervating skeletal muscles, such as the sciatic nerves, consist primarily of axons of sensory neurons and somatic motor neurons. Some cranial nerves, such as the olfactory, optic, and vestibulocochlear nerves, are composed entirely of sensory neuron axons.

In the somatic nervous system, the cell bodies of somatic motor neurons are in the CNS, and their axons extend from there to skeletal muscle (figure 16.1*a*). The ANS, on the other hand, has two neurons in a series extending between the CNS and the innervated organs (figure 16.1*b*). The first neuron of the series is called the **preganglionic neuron.** Its cell body is located in the CNS within either the brainstem or the lateral horn of the spinal cord gray matter, and its axon extends to an autonomic ganglion located outside the CNS. The **autonomic ganglion** contains the cell body of the second neuron of the series, which is called the **postganglionic neuron.** Preganglionic neurons synapse with postganglionic neurons in the autonomic ganglia. The axons of postganglionic neurons extend from autonomic ganglia to effector organs, where they synapse with their target tissues.

Many movements controlled by the somatic nervous system are conscious, whereas ANS functions are unconsciously controlled. The effect of somatic motor neurons on skeletal muscle is always excitatory, but the effect of autonomic motor neurons on target tissues can be either excitatory or inhibitory. For example, after a meal, the ANS can stimulate stomach activities but, during exercise, the ANS can inhibit those activities. Table 16.1 summarizes the differences between the somatic nervous system and the ANS.

Sensory neurons are not classified as somatic or autonomic. These neurons propagate action potentials from sensory receptors to the CNS and can provide information for reflexes mediated through the somatic nervous system or the ANS. For example, stimulation of pain receptors can initiate somatic reflexes, such as the withdrawal reflex, and autonomic reflexes, such as an increase in heart rate. Although some sensory neurons primarily affect somatic functions and others primarily influence autonomic functions, functional overlap makes attempts to classify sensory neurons as either somatic or autonomic meaningless.

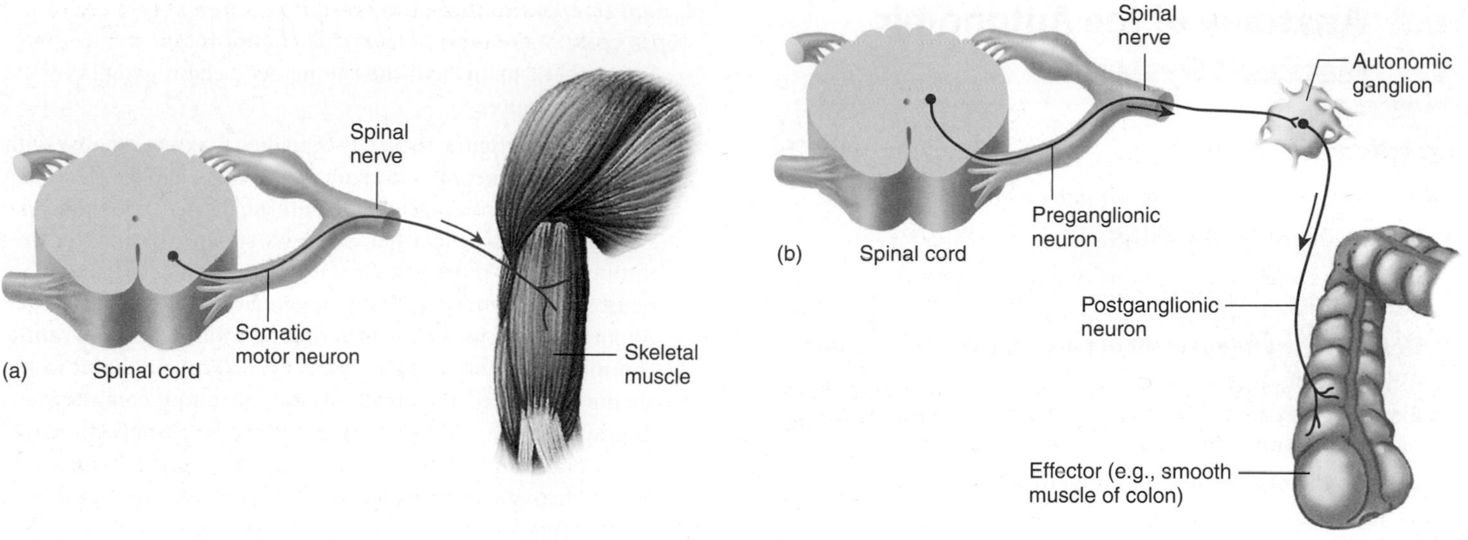
Spinal nerve
Somatic motor neuron
(a) Spinal cord
Skeletal muscle
Spinal nerve
Autonomic ganglion
(b) Spinal cord
Preganglionic neuron
Postganglionic neuron
Effector (e.g., smooth muscle of colon)

FIGURE 16.1 Organization of Somatic and Autonomic Nervous System Neurons
(*a*) The cell body of the somatic motor neuron is in the CNS, and its axon extends to a skeletal muscle. (*b*) The cell body of the preganglionic neuron is in the CNS, and its axon extends to an autonomic ganglion and synapses with a postganglionic neuron. The postganglionic neuron extends to and synapses with its effector.

ASSESS YOUR PROGRESS

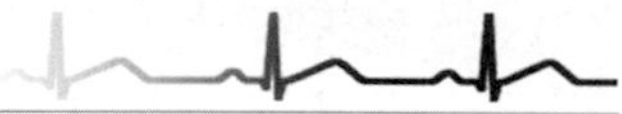

3. *Contrast the somatic nervous system with the ANS for each of the following:*
   a. number of neurons between the CNS and the effector organ
   b. location of neuron cell bodies
   c. structures each innervates
   d. inhibitory or excitatory effects
   e. conscious or unconscious control
   f. neurotransmitter(s) used
4. *Differentiate between preganglionic neurons and postganglionic neurons.*
5. *Why are sensory neurons not classified as somatic or autonomic?*

TABLE 16.1 Comparison of the Somatic and Autonomic Nervous Systems

	Somatic Nervous System	Autonomic Nervous System
Effector	Skeletal muscle	Smooth muscle, cardiac muscle, and glands
Regulation	Controls all conscious and unconscious movements of skeletal muscle	Unconscious regulation, although influenced by conscious mental functions
Response to Stimulation	Skeletal muscle contracts.	Target tissues are stimulated or inhibited.
Neuron Arrangement	One neuron extends from the central nervous system (CNS) to skeletal muscle.	There are two neurons in series; the preganglionic neuron extends from the CNS to an autonomic ganglion, and the postganglionic neuron extends from the autonomic ganglion to the target tissue.
Neuron Cell Body Location	Neuron cell bodies are in motor nuclei of the cranial nerves and in the ventral horn of the spinal cord.	Preganglionic neuron cell bodies are in autonomic nuclei of the cranial nerves and in the lateral part of the spinal cord; postganglionic neuron cell bodies are in autonomic ganglia.
Number of Synapses	One synapse between the somatic motor neuron and the skeletal muscle	Two synapses; first is in the autonomic ganglia, and second is at the target tissue
Axon Sheaths	Myelinated	Preganglionic axons are myelinated; postganglionic axons are unmyelinated.
Neurotransmitter Substance	Acetylcholine	Acetylcholine is released by preganglionic neurons; either acetylcholine or norepinephrine is released by postganglionic neurons.
Receptor Molecules	Receptor molecules for acetylcholine are nicotinic.	In autonomic ganglia, receptor molecules for acetylcholine are nicotinic; in target tissues, receptor molecules for acetylcholine are muscarinic, whereas receptor molecules for norepinephrine are either α- or β-adrenergic.

16.3 Anatomy of the Autonomic Nervous System

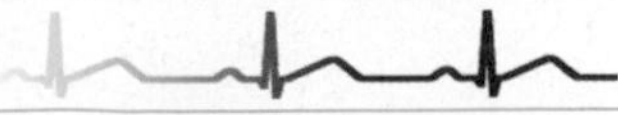

LEARNING OUTCOMES

After reading this section, you should be able to

A. **List the divisions of the ANS.**
B. **Describe the arrangement of sympathetic neurons and ganglia.**
C. **Describe the arrangement of parasympathetic neurons and ganglia.**
D. **Explain what an autonomic nerve plexus is and list the major autonomic nerve plexuses in the body.**
E. **Discuss the organization of the ENS.**

As described earlier,the ANS is subdivided into the **sympathetic division,** the **parasympathetic division,** and the **enteric** (en-ter′ik; bowels) **nervous system (ENS).** The sympathetic and parasympathetic divisions differ structurally in (1) the location of their preganglionic neuron cell bodies within the CNS and (2) the location of their autonomic ganglia. The ENS is a complex network of neuron cell bodies and axons within the wall of the digestive tract. The ENS is considered part of the ANS because sympathetic and parasympathetic neurons are an important part of it. The anatomy of the sympathetic and parasympathetic divisions of the ANS will be described first, followed by a description of the ENS.

Sympathetic Division

The cell bodies of sympathetic preganglionic neurons are in the lateral horns of the spinal cord gray matter between the first thoracic (T1) segment and the second lumbar (L2) segment (figure 16.2). Because of the location of the preganglionic cell bodies, the sympathetic division is sometimes called the *thoracolumbar division.* The axons of the preganglionic neurons exit through the ventral roots of spinal nerves T1–L2, course through the spinal nerves for a short distance, leave these nerves, and project to sympathetic ganglia.

There are two types of sympathetic ganglia: sympathetic chain ganglia and collateral ganglia. **Sympathetic chain ganglia** are connected to each other and are so named because they form a chain along the left and right sides of the vertebral column. They are also called *paravertebral* (alongside the vertebral column) *ganglia* because of their location. Although the sympathetic division originates in the thoracic and lumbar vertebral regions, the sympathetic chain ganglia extend into the cervical and sacral regions. As a result of ganglia fusion during fetal development, there are typically 3 pairs of cervical ganglia, 11 pairs of thoracic ganglia, 4 pairs of lumbar ganglia, and 4 pairs of sacral ganglia. **Collateral** (meaning "accessory") **ganglia** are unpaired ganglia located in the abdominopelvic cavity. They are also called *prevertebral ganglia* because they are anterior to the vertebral column.

The axons of preganglionic neurons are small in diameter and myelinated. The short connection between a spinal nerve and a sympathetic chain ganglion through which the preganglionic axons pass is called a **white ramus communicans** (rā′mŭs kŏ-mū′ni-kans; pl. *rami communicantes,* rā′mī kŏ-mū-ni-kan′tēz) because of the whitish color of the myelinated axons (figure 16.3).

Sympathetic axons exit the sympathetic chain ganglia by the following four routes:

1. *Spinal nerves* (figure 16.3*a*). Preganglionic axons synapse with postganglionic neurons in sympathetic chain ganglia. They can synapse at the same level that the preganglionic axons enter the sympathetic chain, or they can pass superiorly or inferiorly through one or more ganglia and synapse with postganglionic neurons in a sympathetic chain ganglion at a different level. Axons of the postganglionic neurons pass through a **gray ramus communicans** and reenter a spinal nerve. Postganglionic axons are not myelinated, thereby giving the gray ramus communicans its grayish color. All spinal nerves receive postganglionic axons from a gray ramus communicans. The postganglionic axons then project through the spinal nerve to the skin and blood vessels of skeletal muscles.

FUNDAMENTAL **Figure**

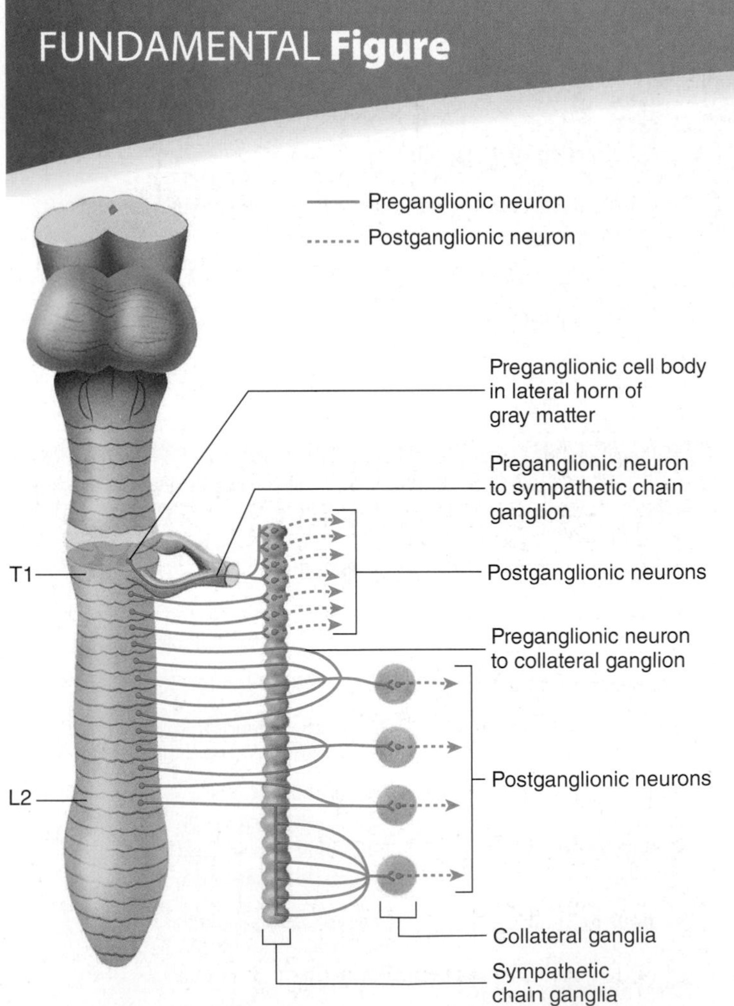
Preganglionic neuron
Postganglionic neuron
Preganglionic cell body in lateral horn of gray matter
Preganglionic neuron to sympathetic chain ganglion
Postganglionic neurons
Preganglionic neuron to collateral ganglion
Postganglionic neurons
Collateral ganglia
Sympathetic chain ganglia
T1
L2

FIGURE 16.2 APR **Sympathetic Division**

The location of sympathetic preganglionic (solid blue) and postganglionic (dotted blue) neurons. The preganglionic cell bodies are in the lateral gray matter of the thoracic and lumbar parts of the spinal cord. The cell bodies of the postganglionic neurons are within the sympathetic chain ganglia or within collateral ganglia.

2. *Sympathetic nerves* (figure 16.3*b*). Preganglionic axons enter the sympathetic chain and synapse in a sympathetic chain ganglion at the same or a different level with postganglionic neurons. The postganglionic axons leaving the sympathetic chain ganglion form **sympathetic nerves,** which supply organs in the thoracic cavity.
3. *Splanchnic* (splangk′nik) *nerves* (figure 16.3*c*). Some preganglionic axons enter sympathetic chain ganglia and, without synapsing, exit at the same or a different level to form **splanchnic nerves.** Those preganglionic axons extend to collateral ganglia, where they synapse with postganglionic neurons. Axons of the postganglionic neurons leave the collateral ganglia through small nerves that extend to effectors in the abdominopelvic cavity.
4. *Innervation to the adrenal gland* (figure 16.3*d*). The splanchnic nerve innervation to the adrenal glands is different from other ANS nerves because axons of the preganglionic neurons do not synapse in sympathetic chain ganglia or in collateral ganglia. Instead, the axons pass through those ganglia and synapse with cells in the medulla of the adrenal gland. The **adrenal medulla** (me-dool′ă) is the inner portion of the adrenal gland and consists of specialized cells derived during embryonic development from neural crest cells (see figure 13.2), which are the same cells that give rise to the postganglionic cells of the ANS. Adrenal medullary cells are round, have no axons or dendrites, and are divided into two groups. About 80% of the cells secrete **epinephrine** (ep′i-nef′rin), also called *adrenaline* (ă-dren′ă-lin), and about 20% secrete **norepinephrine** (nōr′ep-i-nef′rin), also called *noradrenaline* (nōr-ă-dren′ă-lin). Stimulation of these cells by preganglionic axons causes the release of epinephrine and norepinephrine. These substances circulate in the blood and affect all tissues having receptors to which they can bind. The general response to epinephrine and norepinephrine released from the adrenal medulla is to prepare the individual for physical activity. Secretions of the adrenal medulla are considered hormones because they are released into the general circulation and travel some distance to the effectors (see chapters 17 and 18).

ASSESS YOUR PROGRESS

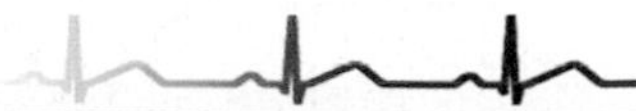

6. *Where are the cell bodies of sympathetic preganglionic neurons located?*
7. *What types of axons (preganglionic or postganglionic, myelinated or unmyelinated) are found in the white and gray rami communicantes?*
8. *Where do preganglionic axons synapse with the postganglionic neurons in spinal and sympathetic nerves?*
9. *Where do preganglionic axons that form splanchnic nerves (except those to the adrenal gland) synapse with postganglionic neurons?*
10. *What is unusual about the splanchnic nerve innervation to the adrenal gland? What do the specialized cells of the adrenal medulla secrete, and what is the effect of these substances?*
11. *Describe the lengths of the preganglionic and postganglionic neurons.*

Parasympathetic Division

The cell bodies of parasympathetic preganglionic neurons are located either within cranial nerve nuclei in the brainstem or within the lateral parts of the gray matter in the sacral region of the spinal cord from S2 to S4 (figure 16.4). For that reason, the parasympathetic division is sometimes called the *craniosacral* (krā′nē-ō-sā′krăl) *division.*

Axons of the parasympathetic preganglionic neurons from the brain are in cranial nerves III, VII, IX, and X and from the spinal cord in **pelvic splanchnic nerves.** The preganglionic axons course through these nerves to **terminal ganglia,** where they synapse with postganglionic neurons. The axons of the postganglionic neurons extend relatively short distances from the terminal ganglia to the effectors. The terminal ganglia are either near or embedded within the walls of the organs innervated by the parasympathetic neurons. Many of the parasympathetic ganglia are small, but some, such as those in the wall of the digestive tract, are large.

Table 16.2 summarizes the structural differences between the sympathetic and parasympathetic divisions.

TABLE 16.2 Comparison of the Sympathetic and Parasympathetic Divisions

	Sympathetic Division	Parasympathetic Division
Location of Preganglionic Cell Body	Lateral horns of spinal cord gray matter (T1–L2)	Brainstem and lateral parts of spinal cord gray matter (S2–S4)
Outflow from the CNS	Spinal nerves Sympathetic nerves Splanchnic nerves	Cranial nerves Pelvic splanchnic nerves
Ganglia	Sympathetic chain ganglia along spinal cord for spinal and sympathetic nerves; collateral ganglia for splanchnic nerves	Terminal ganglia near or on effector organ
Number of Postganglionic Neurons for Each Preganglionic Neuron	Many (much divergence)	Few (less divergence)
Relative Length of Neurons	Short preganglionic Long postganglionic	Long preganglionic Short postganglionic

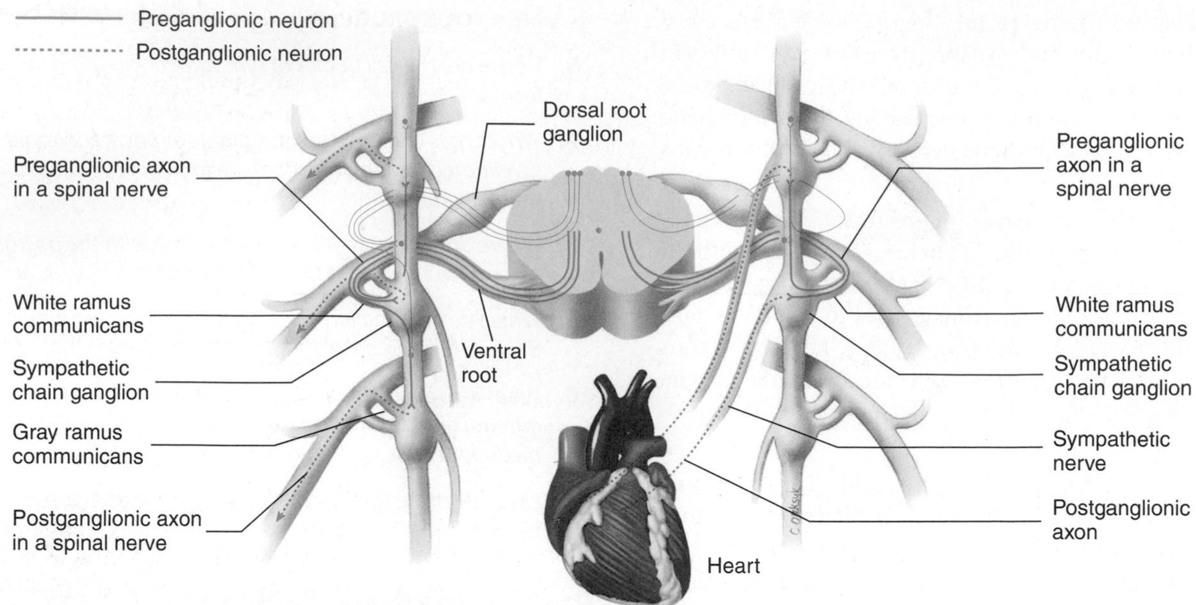
Preganglionic neuron
Postganglionic neuron
Dorsal root ganglion
Preganglionic axon in a spinal nerve
White ramus communicans
Sympathetic chain ganglion
Gray ramus communicans
Postganglionic axon in a spinal nerve
Ventral root
Preganglionic axon in a spinal nerve
White ramus communicans
Sympathetic chain ganglion
Sympathetic nerve
Postganglionic axon
Heart

(a) Preganglionic axons from a spinal nerve pass through a white ramus communicans into a sympathetic chain ganglion. Some axons synapse with a postganglionic neuron at the level of entry; others ascend or descend to other levels before synapsing. Each postganglionic axon exits the sympathetic chain through a gray ramus communicans and enters a spinal nerve.

(b) Part (*b*) is like part (*a*), except that each postganglionic neuron exits a sympathetic chain ganglion through a sympathetic nerve.

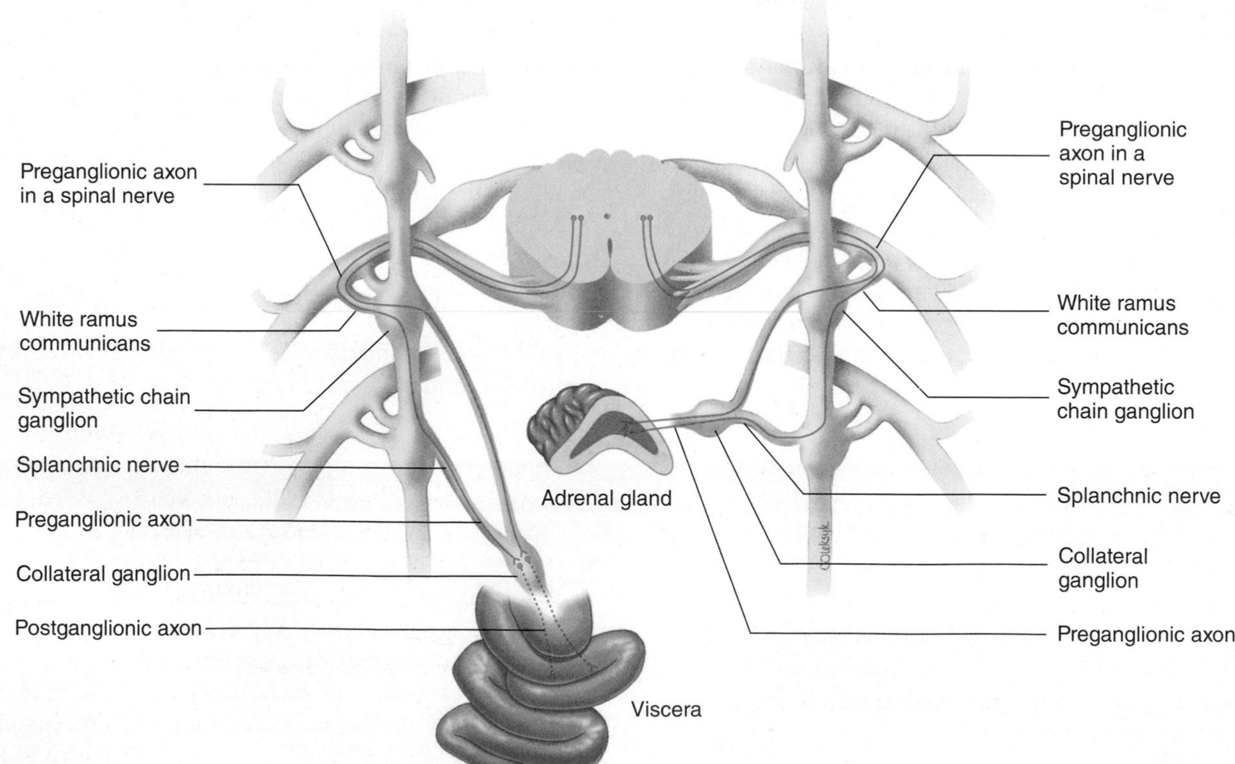
Preganglionic axon in a spinal nerve
White ramus communicans
Sympathetic chain ganglion
Splanchnic nerve
Preganglionic axon
Collateral ganglion
Postganglionic axon
Adrenal gland
Viscera
Preganglionic axon in a spinal nerve
White ramus communicans
Sympathetic chain ganglion
Splanchnic nerve
Collateral ganglion
Preganglionic axon

(c) Preganglionic neurons do not synapse in the sympathetic chain ganglia, but exit in splanchnic nerves and extend to a collateral ganglion, where they synapse with postganglionic neurons.

(d) Part (*d*) is like part (*c*), except that the preganglionic axons extend to the adrenal medulla, where they synapse with specialized adrenal medullary cells.

FIGURE 16.3 APIR Routes Taken by Sympathetic Axons

Preganglionic axons are illustrated as solid lines and postganglionic axons as dashed lines.

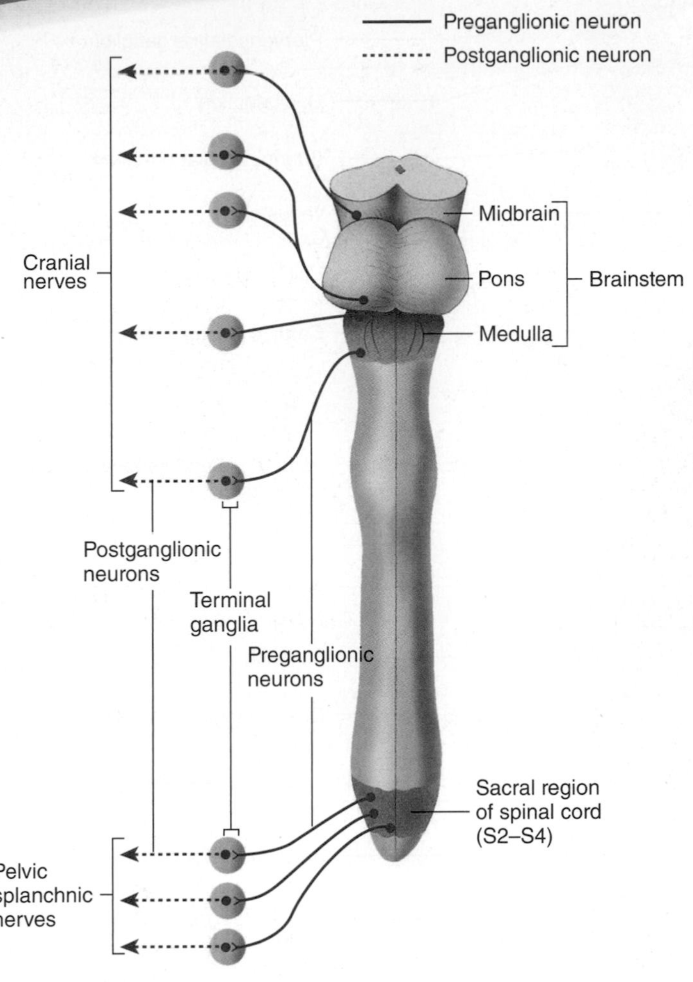
Preganglionic neuron
Postganglionic neuron
Cranial nerves
Midbrain
Pons
Medulla
Brainstem
Postganglionic neurons
Terminal ganglia
Preganglionic neurons
Sacral region of spinal cord (S2–S4)
Pelvic splanchnic nerves

FIGURE 16.4 APR **Parasympathetic Division**

The location of parasympathetic preganglionic (solid red) and postganglionic (dotted red) neurons. The preganglionic neuron cell bodies are in the brainstem and the lateral gray matter of the sacral part of the spinal cord, and the postganglionic neuron cell bodies are within terminal ganglia.

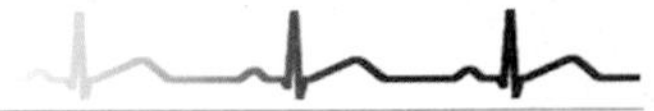

ASSESS **YOUR PROGRESS**

12. *Where are the cell bodies of parasympathetic preganglionic neurons located?*

13. *In what structures do parasympathetic preganglionic neurons synapse with postganglionic neurons? Where are these structures located?*

14. *What nerves are formed by the axons of parasympathetic preganglionic neurons?*

15. *Describe the lengths of the preganglionic and postganglionic neurons of each division.*

Autonomic Nerve Plexuses and Distribution of Autonomic Nerve Fibers

In the previous sections, the major pathways for preganglionic and postganglionic neurons of the sympathetic and parasympathetic divisions were described. This section will detail the distribution of the autonomic nerve fibers to the target organs. In some cases, the postganglionic axons extend directly through nerves to the target organ. In other cases, the postganglionic axons become part of autonomic nerve plexuses. **Autonomic nerve plexuses** are complex, interconnected neural networks formed by neurons of the sympathetic and parasympathetic divisions. Axons of sensory neurons also contribute to these plexuses. The autonomic nerve plexuses are typically named according to the organs they supply or the blood vessels along which they are found. For example, the cardiac plexus supplies the heart, and the thoracic aortic plexus is found along the thoracic aorta. Plexuses following the route of blood vessels are a major means by which autonomic axons are distributed throughout the body. Autonomic nerve plexuses are associated with both the sympathetic and parasympathetic divisions.

Sympathetic Division Distribution

Sympathetic outflow is through spinal nerves, sympathetic nerves, and splanchnic nerves. Branches of these nerves either extend to effectors or join autonomic nerve plexuses to be distributed to effectors. The major means by which sympathetic postganglionic axons reach organs include the following:

1. *Spinal nerves.* From all levels of the sympathetic chain, some postganglionic axons project through gray rami communicates to spinal nerves. The axons extend to the same structures innervated by the spinal nerves and supply sweat glands in the skin, the smooth muscle in the blood vessels of the skin, and the smooth muscle of the arrector pili. See figure 12.14 for the distribution of spinal nerves to the skin.
2. *Head and neck nerve plexuses.* Most of the sympathetic nerve supply to the head and neck is derived from the superior cervical sympathetic chain ganglion (figure 16.5). Postganglionic axons of sympathetic nerves form plexuses that extend superiorly to the head and inferiorly to the neck. The plexuses give off branches to supply sweat glands in the skin, the smooth muscle in the blood vessels of the skin, and the smooth muscle of the arrector pili. Axons from the plexuses also join branches of the trigeminal nerves (cranial nerve V) to supply the skin of the face, the salivary glands, the iris, and the ciliary muscles of the eye.
3. *Thoracic nerve plexuses.* The sympathetic nerve supply for organs of the thorax is mainly derived from the cervical and upper five thoracic sympathetic chain ganglia. Postganglionic axons in sympathetic nerves contribute to the **cardiac plexus,** supplying the heart; the **pulmonary plexus,** supplying the lungs; and other thoracic plexuses (figure 16.5).
4. *Abdominopelvic nerve plexuses.* Sympathetic chain ganglia from T5 and below mainly supply the abdominopelvic organs. The preganglionic axons of splanchnic nerves synapse with postganglionic neurons in the collateral ganglia of abdominopelvic nerve plexuses. Postganglionic axons from the collateral ganglia innervate smooth muscle and glands in the abdominopelvic organs. There are several abdominopelvic nerve plexuses (figure 16.5). The **celiac** (sē′lē-ak) **plexus** has two large celiac ganglia and other, smaller ganglia. It supplies the diaphragm, stomach, spleen, liver, gallbladder, adrenal glands, kidneys, testes, and ovaries. The **superior mesenteric** (mez-en-ter′ik) **plexus** includes the superior mesenteric ganglion and supplies the pancreas,

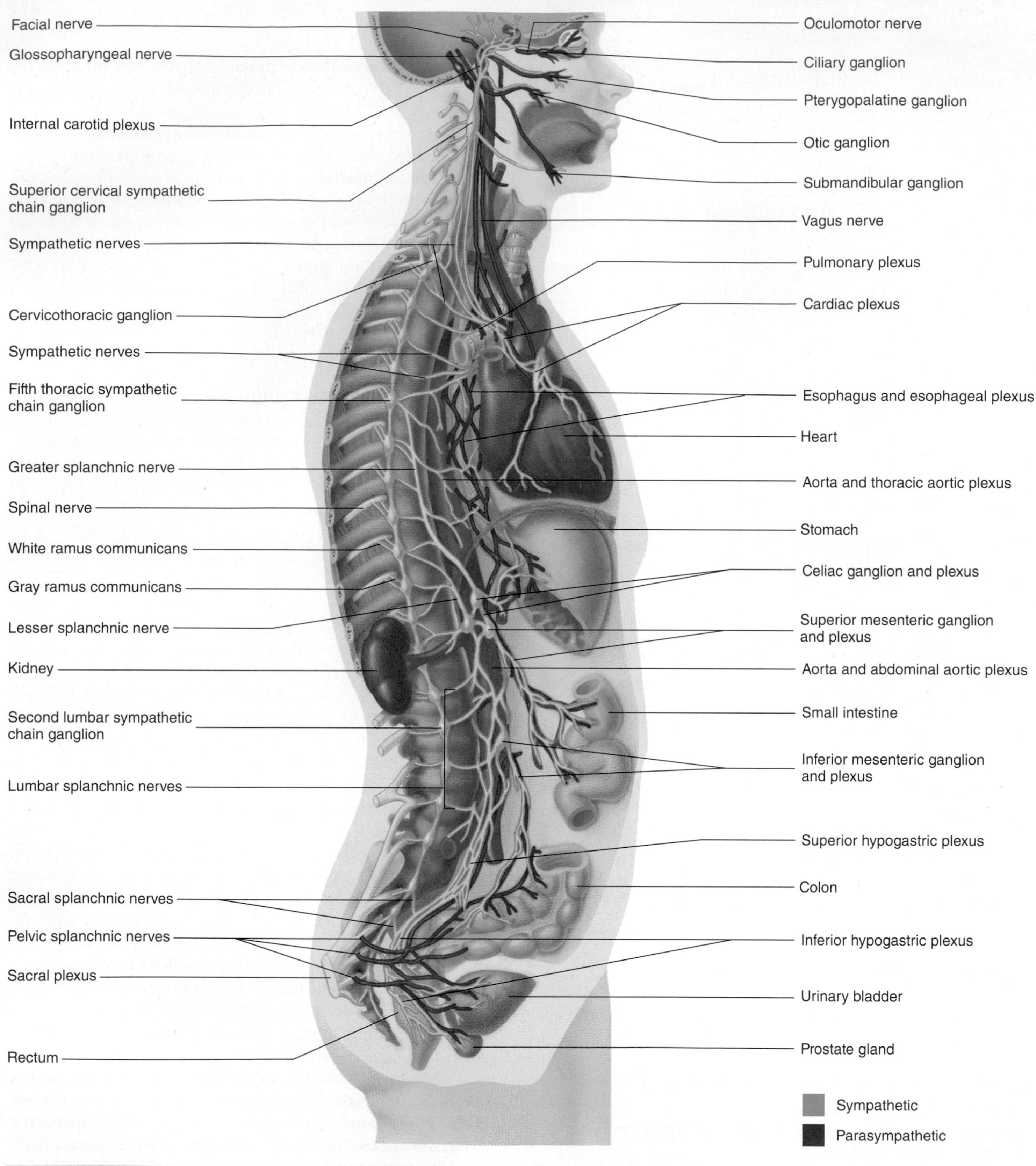

Facial nerve
Glossopharyngeal nerve
Internal carotid plexus
Superior cervical sympathetic chain ganglion
Sympathetic nerves
Cervicothoracic ganglion
Sympathetic nerves
Fifth thoracic sympathetic chain ganglion
Greater splanchnic nerve
Spinal nerve
White ramus communicans
Gray ramus communicans
Lesser splanchnic nerve
Kidney
Second lumbar sympathetic chain ganglion
Lumbar splanchnic nerves
Sacral splanchnic nerves
Pelvic splanchnic nerves
Sacral plexus
Rectum
Oculomotor nerve
Ciliary ganglion
Pterygopalatine ganglion
Otic ganglion
Submandibular ganglion
Vagus nerve
Pulmonary plexus
Cardiac plexus
Esophagus and esophageal plexus
Heart
Aorta and thoracic aortic plexus
Stomach
Celiac ganglion and plexus
Superior mesenteric ganglion and plexus
Aorta and abdominal aortic plexus
Small intestine
Inferior mesenteric ganglion and plexus
Superior hypogastric plexus
Colon
Inferior hypogastric plexus
Urinary bladder
Prostate gland
Sympathetic
Parasympathetic

FIGURE 16.5 Distribution of Autonomic Nerve Fibers

small intestine, ascending colon, and transverse colon. The **inferior mesenteric plexus** includes the inferior mesenteric ganglion and supplies the transverse colon to the rectum. The **superior** and **inferior hypogastric plexuses** supply the descending colon to the rectum, urinary bladder, and reproductive organs in the pelvis.

Parasympathetic Division Distribution

Parasympathetic outflow is through cranial and pelvic splanchnic nerves. Branches of these nerves either supply organs or join nerve plexuses to be distributed to organs. The major means by which parasympathetic postganglionic axons reach effectors include the following:

1. *Cranial nerves supplying the head and neck.* Three pairs of cranial nerves have parasympathetic preganglionic axons that extend to terminal ganglia in the head. Postganglionic neurons from the terminal ganglia supply nearby structures. The following are the parasympathetic cranial nerves, their terminal ganglia, and the structures innervated (figure 16.5; see table 13.5):
   a. The **oculomotor (III) nerve,** through the **ciliary** (sil′ē-ar-ē) **ganglion,** supplies the ciliary muscles and the iris of the eye.
   b. The **facial (VII) nerve,** through the **pterygopalatine** (ter′i-gō-pal′ă-tīn) **ganglion,** supplies the lacrimal gland and the mucosal glands of the nasal cavity and palate. The facial nerve, through the **submandibular ganglion,** also supplies the submandibular and sublingual salivary glands.
   c. The **glossopharyngeal (IX) nerve,** through the **otic** (ō′tik) **ganglion,** supplies the parotid salivary gland.
2. *The vagus nerve and thoracic nerve plexuses.* Although cranial nerve X, the **vagus nerve,** has somatic motor and sensory functions in the head and neck, its parasympathetic distribution is to the thorax and abdomen. Preganglionic axons extend through the vagus nerves to the thorax. Within the thorax, the axons pass through branches of the vagus nerves to contribute to the cardiac plexus, which supplies the heart, and the pulmonary plexus, which supplies the lungs. The vagus nerves continue down the esophagus and give off branches to form the **esophageal plexus.**
3. *Abdominal nerve plexuses.* After the esophageal plexus passes through the diaphragm, some of the vagal preganglionic axons supply terminal ganglia in the wall of the stomach, whereas others contribute to the celiac and superior mesenteric plexuses. Through these plexuses, the preganglionic axons supply terminal ganglia in the walls of the gallbladder, biliary ducts, pancreas, small intestine, ascending colon, and transverse colon.
4. *Pelvic splanchnic nerves and pelvic nerve plexuses.* The parasympathetic preganglionic axons whose cell bodies are in the S2–S4 region of the spinal cord pass to the ventral rami of spinal nerves and enter the pelvic splanchnic nerves. The pelvic splanchnic nerves supply terminal ganglia in the transverse colon to the rectum, and they contribute to the hypogastric plexus. The hypogastric plexus and its derivatives supply the lower colon, rectum, urinary bladder, and organs of the reproductive system in the pelvis.

Sensory Neurons in Autonomic Nerve Plexuses

Although not strictly part of the ANS, the axons of sensory neurons run alongside ANS axons within ANS nerves and plexuses. Some of these sensory neurons are part of reflex arcs regulating organ activities. Sensory neurons also transmit pain and pressure sensations from organs to the CNS. The cell bodies of these sensory neurons are found in the dorsal root ganglia and in the sensory ganglia of certain cranial nerves, which are swellings on the nerves close to their attachment to the brain.

ASSESS **YOUR PROGRESS**

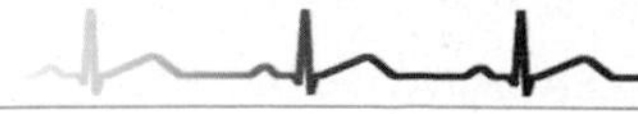

16. *From what are sympathetic autonomic nerve plexuses formed? How are they normally named?*
17. *Describe the four major ways by which sympathetic axons pass from sympathetic chain ganglia to effectors. Name four thoracic and four abdominopelvic autonomic nerve plexuses.*
18. *List the four major means by which parasympathetic axons reach effectors. List the cranial nerves and ganglia that supply the head and neck.*
19. *What are the roles of sensory neurons in the ANS?*

Enteric Nervous System

The enteric nervous system consists of nerve plexuses within the wall of the digestive tract (see chapter 24). The plexuses have contributions from three sources: (1) sensory neurons that connect the digestive tract to the CNS, (2) ANS motor neurons that connect the CNS to the digestive tract, and (3) enteric neurons, which are confined to the enteric plexuses. The CNS is capable of monitoring the digestive tract and controlling its smooth muscle and glands through autonomic reflexes (see section 16.5). For example, sensory neurons detect stretch of the digestive tract, and action potentials are transmitted to the CNS. In response, the CNS sends action potentials to glands in the digestive tract, causing them to secrete.

There are three major types of enteric neurons:

1. *Enteric sensory neurons* detect changes in the chemical composition of the contents of the digestive tract or detect stretch of the digestive tract wall.
2. *Enteric motor neurons* stimulate or inhibit smooth muscle contraction and gland secretion.
3. *Enteric interneurons* connect enteric sensory and motor neurons to each other.

A unique feature of enteric neurons is that they are capable of monitoring and controlling the digestive tract independently of the CNS through local reflexes (see section 16.5). For example, stretch of the digestive tract is detected by enteric sensory neurons, which stimulate enteric interneurons. The enteric interneurons stimulate enteric motor neurons, which stimulate glands to secrete. Although the enteric nervous system is capable of controlling the activities of the digestive tract completely independently of the CNS, the two systems normally work together.

ASSESS **YOUR PROGRESS**

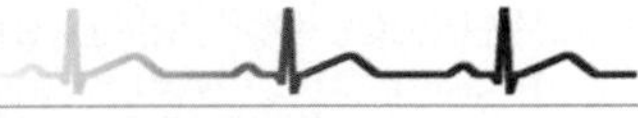

20. *What is the enteric nervous system (ENS), and where is it located?*
21. *What three sources contribute to ENS plexuses?*
22. *Name three major types of enteric neurons. How do enteric neurons monitor and control the digestive tract?*

16.4 Physiology of the Autonomic Nervous System

LEARNING OUTCOMES

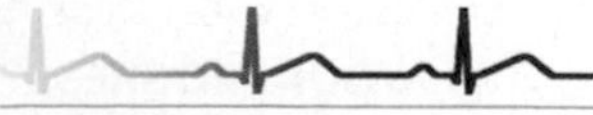

After reading this section, you should be able to

A. Explain dual innervation of the ANS.

B. Describe the role of the sympathetic division during activity or stress and the general effects on the body.

C. Describe the role of the parasympathetic division during rest and the general effects on the body.

D. Differentiate between cholinergic and adrenergic neurons as to the neurotransmitter secreted and the type of neuron that secretes the neurotransmitter.

E. Contrast the two types of cholinergic receptors.

F. Describe the types of adrenergic receptors and their subtypes.

G. List effectors and the type(s) of receptors they have.

Sympathetic Versus Parasympathetic Activity

The sympathetic and parasympathetic divisions of the ANS maintain homeostasis by adjusting body functions to match the level of physical activity. The ANS innervates most organs via sympathetic and parasympathetic fibers (figure 16.6). Examples of dually innervated organs are the gastrointestincal tract, heart, urinary bladder, and reproductive tract. However, dual innervation of organs by both divisions of the ANS is not universal. For example, sweat glands and blood vessels are innervated by sympathetic neurons almost exclusively. In addition, where dual innervation exists, one division may be more predominant than the other division. For example, parasympathetic innervation of the gastrointestinal tract is more extensive and exerts a greater influence than does sympathetic innervation.

In cases in which both sympathetic and parasympathetic neurons innervate a single organ, the sympathetic division has a major influence under conditions of physical activity or stress, whereas the parasympathetic division has a greater influence under resting conditions. However, the sympathetic division is not inactive during resting conditions; rather, it plays a major role during rest by maintaining blood pressure and body temperature.

In general, during physical exercise the sympathetic division shunts blood and nutrients to structures that are active and decreases the activity of the nonessential organs. This is sometimes referred to as the **fight-or-flight response** (see Clinical Impact, "Biofeedback, Meditation, and the Fight-or-Flight Response"). Typical responses produced by the sympathetic division during exercise include the following:

1. Increased heart rate and force of contraction increase blood pressure and the movement of blood.
2. Vasodilation of blood vessels of muscle occurs during exercise. As skeletal or cardiac muscle contracts, oxygen and nutrients are used and waste products are produced. The decrease in oxygen and nutrients and the accumulation of waste products stimulate vasodilation (see chapter 21). Vasodilation is beneficial because it increases blood flow, bringing needed oxygen and nutrients and removing waste products. Too much vasodilation, however, can lower blood pressure, thereby decreasing blood flow. Conversely, increased stimulation of blood vessels of skeletal muscle by sympathetic nerves during exercise causes vasoconstriction, which prevents a drop in blood pressure (see chapter 21).
3. Vasoconstriction occurs in blood vessels of tissues not involved in exercise. For example, vasoconstriction in the abdominopelvic organs reduces blood flow through them, making more blood available for the exercising tissues.
4. Dilation of air passageways increases airflow into and out of the lungs.
5. Increased breakdown of stored energy sources occurs during exercise. Skeletal muscle cells and liver cells (hepatocytes) are stimulated to break down glycogen to glucose. Skeletal muscle cells use the glucose, and liver cells release it into the blood for use by other tissues. Adipocytes break down triglycerides and release fatty acids into the blood, which are used as an energy source by skeletal muscle and cardiac muscle.

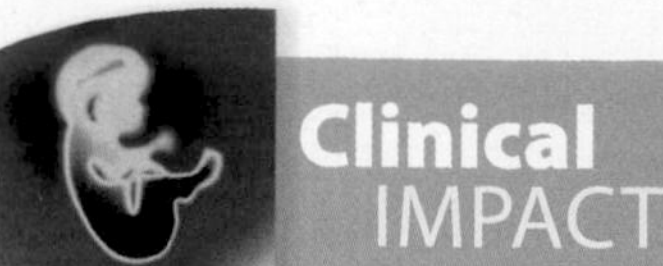

Clinical IMPACT

Biofeedback, Meditation, and the Fight-or-Flight Response

Biofeedback is a technique that uses electronic instruments or other methods to monitor and change subconscious activities, many of which are regulated by the ANS. For example, skin temperature, heart rate, and brain waves can be monitored electronically. Then, by watching the monitor and using biofeedback techniques, a person can learn to consciously reduce his or her heart rate and blood pressure and regulate blood flow in the limbs. For example, some people claim they can prevent the onset of migraine headaches or reduce their intensity by learning to dilate the blood vessels in the skin of their forearms and hands. Increased blood vessel dilation increases skin temperature, which is correlated with a decrease in the severity of the migraine. Some people use biofeedback methods to relax by learning to reduce their heart rate or change the pattern of their brain waves. Biofeedback techniques can also be used to reduce the severity of some stomach ulcers, high blood pressure, anxiety, and depression.

Meditation is another technique that influences autonomic functions. Some people use meditation techniques to reduce heart rate, blood pressure, the severity of ulcers, and other symptoms associated with stress.

The fight-or-flight response occurs when an individual is subjected to stress, as in a threatening, frightening, embarrassing, or exciting situation. Whether a person confronts or avoids the stressful situation, the nervous system and the endocrine system are involved, either consciously or unconsciously. The autonomic part of the fight-or-flight response results in a general increase in sympathetic activity, including heart rate, blood pressure, sweating, and other responses that prepare the individual for physical activity. The fight-or-flight response is adaptive because it also enables a person to resist or move away from a threatening situation.

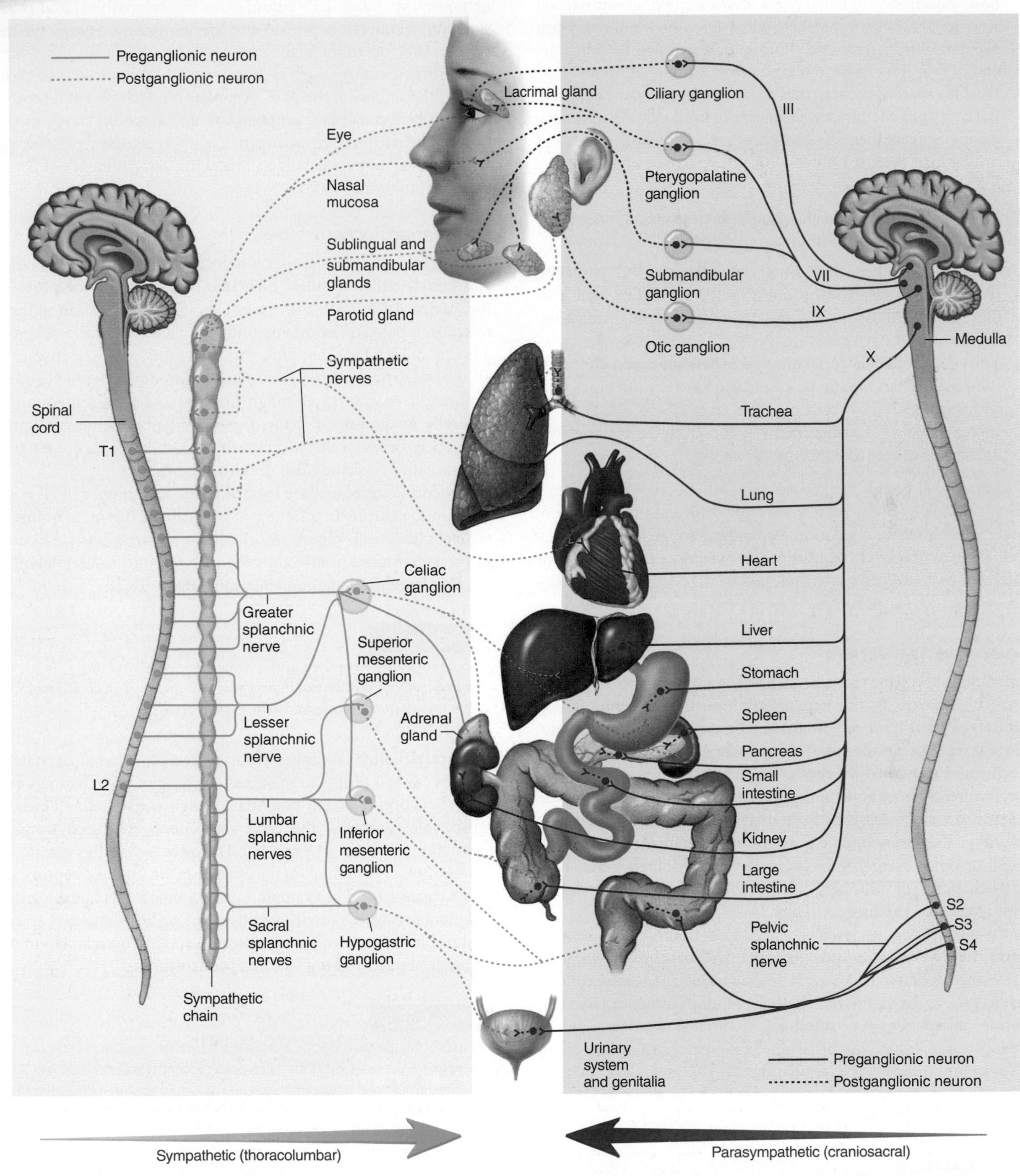
Preganglionic neuron
Postganglionic neuron
Lacrimal gland
Ciliary ganglion
III
Eye
Nasal
mucosa
Pterygopalatine
ganglion
Sublingual and
submandibular
glands
Submandibular
ganglion
VII
Parotid gland
IX
Otic ganglion
Medulla
X
Sympathetic
nerves
Spinal
cord
T1
Trachea
Lung
Heart
Celiac
ganglion
Greater
splanchnic
nerve
Liver
Superior
mesenteric
ganglion
Stomach
Adrenal
gland
Spleen
Lesser
splanchnic
nerve
Pancreas
Small
intestine
L2
Lumbar
splanchnic
nerves
Inferior
mesenteric
ganglion
Kidney
Large
intestine
S2
S3
S4
Sacral
splanchnic
nerves
Hypogastric
ganglion
Pelvic
splanchnic
nerve
Sympathetic
chain
Urinary
system
and genitalia
Preganglionic neuron
Postganglionic neuron
Sympathetic (thoracolumbar)
Parasympathetic (craniosacral)

FIGURE 16.6 AP|R **Innervation of Organs by the ANS**

Preganglionic fibers are indicated by solid lines, and postganglionic fibers are indicated by dashed lines.

6. Body temperature increases as exercising muscles generate heat. Vasodilation of blood vessels in the skin brings warm blood close to the surface, where heat is lost to the environment. Sweat gland activity increases, resulting in increased sweat production, and evaporation of the sweat removes additional heat.
7. During exercise, the activities of nonessential organs decrease. For example, the process of digesting food slows as digestive glands decrease their secretions and the contractions of smooth muscle that mix and move food through the gastrointestinal tract decrease.

Increased activity of the parasympathetic division is generally consistent with resting conditions, as in the following examples:

1. The parasympathetic division regulates digestion by stimulating the secretions of glands, promoting the mixing of food with digestive enzymes and bile, and moving materials through the digestive tract.
2. The parasympathetic division controls defecation and urination.
3. Increased parasympathetic stimulation lowers the heart rate, which lowers blood pressure.
4. Increased parasympathetic activity constricts air passageways, decreasing air movement through them.

In summary, although neither the sympathetic nor the parasympathetic divisions are chronically active, or chronically inactive, there are different levels of regulation, depending on the body's activity level. These differences in regulation of the same organs are the result of the unique neurotransmitters released by the postganglionic fibers and the specific receptors on the target cells.

Neurotransmitters

Sympathetic and parasympathetic nerve endings secrete one of two neurotransmitters. If the neuron secretes acetylcholine, it is a **cholinergic** (kol-in-er′jik) **neuron;** if it secretes norepinephrine (epinephrine), it is an **adrenergic** (ad-rĕ-ner′jik) **neuron.** Adrenergic neurons are so named because at one time they were believed to secrete adrenaline, or epinephrine. All preganglionic neurons of the sympathetic and parasympathetic divisions and all postganglionic neurons of the parasympathetic division are cholinergic. Almost all postganglionic neurons of the sympathetic division are adrenergic, but a few postganglionic neurons that innervate thermoregulatory sweat glands are cholinergic (figure 16.7).

In recent years, substances in addition to the regular neurotransmitters have been extracted from ANS neurons. These substances include nitric oxide; fatty acids, such as eicosanoids; peptides, such as gastrin, somatostatin, cholecystokinin, vasoactive intestinal peptide, enkephalins, and substance P; and monoamines, such as dopamine, serotonin, and histamine. The specific roles that many of these compounds play in regulating the ANS is unclear, but they appear to function as either neurotransmitters or neuromodulator substances (see chapter 11).

Receptors

Receptors for acetylcholine and norepinephrine are located in the plasma membranes of certain cells. The combination of neurotransmitter and receptor functions as a signal to cells, causing them to respond. Depending on the type of cell, the response is excitatory or inhibitory. Table 16.3 summarizes the sympathetic and parasympathetic effects, as well as the specific receptor types, for the various effectors of the body.

Other substances, such as drugs, can also interact with receptors to alter the activity of the ANS. **Agonists** bind to specific receptors and activate them, whereas **antagonists** bind to specific receptors and prevent them from being activated. Examples of these types of drugs are described in more detail in Clinical Impact, "Influence of Drugs on the Autonomic Nervous System."

Cholinergic Receptors

Cholinergic receptors are receptors to which acetylcholine binds and are classified as either **nicotinic** (nik-ō-tin′ik) **receptors** or **muscarinic** (mŭs-kă-rin′ik) **receptors.** The classification of these receptors is based on laboratory findings that nicotine, an alkaloid in tobacco, can bind to some cholinergic receptors, whereas muscarine, an alkaloid extracted from poisonous mushrooms, can bind to other cholinergic receptors. Although nicotine and muscarine are not naturally in the human body, these substances demonstrate the differences between the two classes of cholinergic receptors and, thus, are used to distinguish the two.

Nicotinic receptors are located in the membranes of all postganglionic neurons in autonomic ganglia and the membranes of skeletal muscle cells (figure 16.7*a*). Muscarinic receptors are located in the membranes of effector cells that respond to acetylcholine released from postganglionic neurons (figure 16.7*b*).

Predict 2

Would structures innervated by the sympathetic division or the parasympathetic division be affected after the consumption of nicotine? After the consumption of muscarine? Explain.

Acetylcholine binding to nicotinic receptors has an excitatory effect because it results in the direct opening of Na^+ channels and the production of action potentials. When acetylcholine binds to muscarinic receptors, the cell's response is mediated through G proteins (see chapters 3 and 17). The response is either excitatory or inhibitory, depending on the effector in which the receptors are found (table 16.3). For example, acetylcholine binds to muscarinic receptors in cardiac muscle, thereby reducing heart rate, and acetylcholine binds to muscarinic receptors in smooth muscle cells of the stomach, thus increasing its rate of contraction.

Predict 3

Bethanechol (be-than′ĕ-kol) chloride is a drug that binds to muscarinic receptors. Explain why this drug can promote emptying of the urinary bladder. Which of the following side effects would you predict: abdominal cramps, asthmatic attack, decreased tear production, decreased salivation, dilation of the pupils, or sweating?

Adrenergic Receptors

Adrenergic receptors are receptors to which norepinephrine or epinephrine binds. They are located in the plasma membranes of

(a) **Nicotinic receptors,** which respond to acetylcholine (ACh), are located on the cell bodies of both sympathetic and parasympathetic postganglionic neurons in autonomic ganglia. Binding of ACh to nicotinic receptors has an excitatory effect.

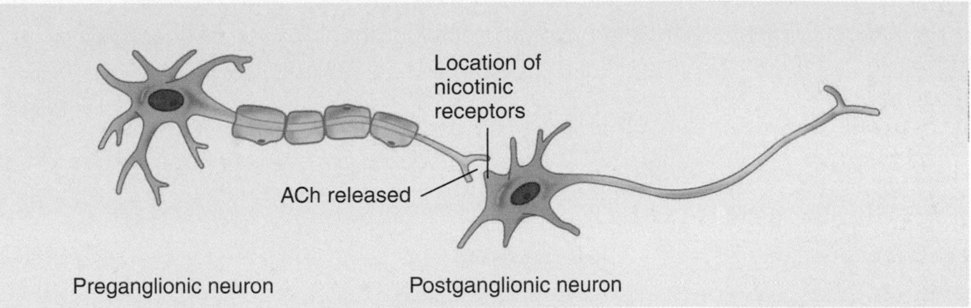
Location of
nicotinic
receptors
ACh released
Preganglionic neuron
Postganglionic neuron

(b) **Muscarinic receptors,** which respond to ACh, are located on the cells of all parasympathetic effectors and some sympathetic effectors, such as sweat glands. Binding of ACh to muscarinic receptors may be excitatory or inhibitory, depending on the specific effectors.

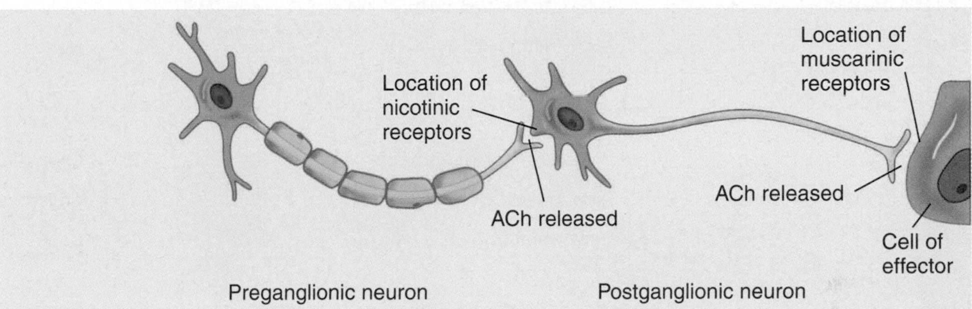
Location of
nicotinic
receptors
Location of
muscarinic
receptors
ACh released
ACh released
Cell of
effector
Preganglionic neuron
Postganglionic neuron

(c) **Adrenergic receptors,** which respond to norepinephrine (NE), are located on most sympathetic effectors. Binding of NE to adrenergic receptors may be excitatory or inhibitory, depending on the specific effector.

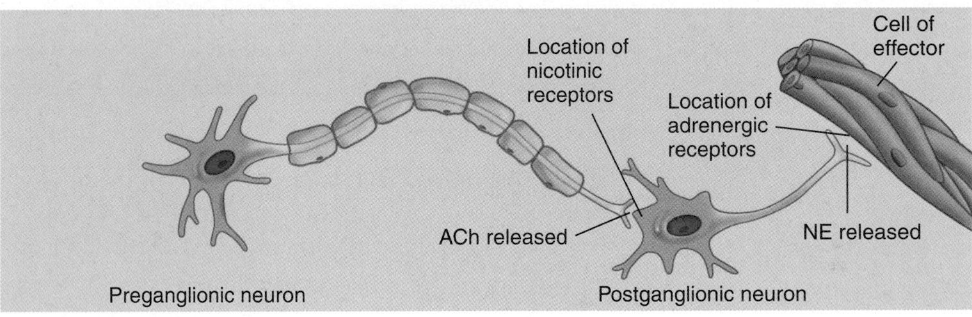
Location of
nicotinic
receptors
Location of
adrenergic
receptors
Cell of
effector
ACh released
NE released
Preganglionic neuron
Postganglionic neuron

FIGURE 16.7 Locations of ANS Receptors

effectors innervated by the sympathetic division (figure 16.7*c*). The response of cells to norepinephrine or epinephrine binding to adrenergic receptors is mediated through G proteins (see chapters 3 and 17). Depending on the effector, the activation of G proteins can result in excitatory or inhibitory responses (table 16.3).

Adrenergic receptors are subdivided into two major categories: **alpha (α) receptors** and **beta (β) receptors.** Epinephrine has a greater effect than norepinephrine on most α and β receptors. The main subtypes of alpha receptors are α_1- and α_2-adrenergic receptors; the subtypes of beta receptors are β_1- and β_2-adrenergic receptors.

Adrenergic receptors can be stimulated in two ways: by the nervous system or by epinephrine and norepinephrine released from the adrenal gland. Sympathetic postganglionic neurons release norepinephrine, which stimulates adrenergic receptors within synapses (figure 16.7*c*). For example, blood vessels are continually stimulated to contract through the release of norepinephrine. Increased stimulation causes further constriction and reduces blood flow, whereas decreased stimulation results in dilation and increases blood flow. The control of blood vessel diameter plays an important role in regulating blood flow and blood pressure (see chapter 20).

Epinephrine and norepinephrine released from the adrenal glands and carried to effectors by the blood can bind to adrenergic receptors located in the plasma membranes of cells that are not involved in synapses. For example, during exercise epinephrine and norepinephrine bind to β_2 receptors and cause blood vessels in skeletal muscles to dilate.

TABLE 16.3 Effects of the Sympathetic and Parasympathetic Divisions on Various Effectors

Effector	Sympathetic Effects and Receptor Type*	Parasympathetic Effects and Receptor Type*
Adipose tissue	Lipid breakdown and release of fatty acids (α_2, β_1)	None
Arrector pili muscle	Contraction (α_1)	None
Blood (platelets)	Increased coagulation (α_2)	None
Blood vessels		
Arterioles (carry blood to tissues)		
Digestive organs	Constriction (α_1)	None
Heart	Constriction (α_1), dilation (β_2)[†]	None
Kidneys	Constriction (α_1, α_2), dilation (β_1, β_2)	None
Lungs	Constriction (α_1), dilation (β_2)	None
Skeletal muscle	Constriction (α_1), dilation (β_2)	None
Skin	Constriction (α_1, α_2)	None
Veins (carry blood away from tissues)	Constriction (α_1, α_2), dilation (β_2)	
Eye		
Ciliary muscle	Relaxation for far vision (β_2)	Contraction for near vision (m)
Pupil	Dilation (α_1)[‡]	Constriction (m)[‡]
Gallbladder	Relaxation (β_2)	Contraction (m)
Glands		
Adrenal	Release of epinephrine and norepinephrine (n)	None
Gastric	Decreased gastric secretion (α_2)	Increased gastric secretion (m)
Lacrimal	Slight tear production (α)	Increased tear secretion (m)
Pancreas	Decreased insulin secretion (α_2)	Increased insulin secretion (m)
	Decreased exocrine secretion (α)	Increased exocrine secretion (m)
Salivary	Constriction of blood vessels and slight production of thick, viscous saliva (α_1)	Dilation of blood vessels and thin, copious saliva (m)
Sweat		
Apocrine	Thick, organic secretion (m)	None
Merocrine	Watery sweat from most of the skin (m), sweat from palms and soles (α_1)	None
Heart	Increased rate and force of contraction (β_1, β_2)	Decreased rate of contraction (m)
Liver	Glucose released into blood (α_1, β_2)	None
Lungs	Dilated air passageways (β_2)	Constricted air passageways (m)
Metabolism	Increased up to 100% (α, β)	None
Sex organs	Ejaculation (α_1), erection[§]	Erection (m)
Skeletal muscles	Breakdown of glycogen to glucose (β_2)	None
Stomach and intestines		
Wall	Decreased tone (α_1, α_2, β_2)	Increased motility (m)
Sphincter	Increased tone (α_1)	Decreased tone (m)
Urinary bladder		
Wall (detrusor)	None	Contraction (m)
Neck of bladder	Contraction (α_1)	Relaxation (m)
Internal urinary sphincter	Contraction (α_1)	Relaxation (m)

*Receptor subtypes are indicated. The receptors are α_1- and α_2-adrenergic, β_1- and β_2-adrenergic, nicotinic cholinergic (n), and muscarinic cholinergic (m).

[†]Normally, blood flow increases through coronary arteries because of increased demand by cardiac tissue for oxygen (local control of blood flow is discussed in chapter 21). In experiments that isolate the coronary arteries, sympathetic nerve stimulation, acting through α-adrenergic receptors, causes vasoconstriction. The β-adrenergic receptors are relatively insensitive to sympathetic nerve stimulation but can be activated by epinephrine released from the adrenal gland and by drugs. As a result, coronary arteries vasodilate.

[‡]Contraction of the radial muscles of the iris causes the pupil to dilate. Contraction of the circular muscles causes the pupil to constrict (see chapter 15).

[§]Decreased stimulation of alpha receptors by the sympathetic division can cause vasodilation of penile blood vessels, resulting in an erection.

Clinical IMPACT — Influence of Drugs on the Autonomic Nervous System

Some drugs that affect the ANS have important value in treating certain diseases because they can increase or decrease activities normally controlled by the ANS. Chemicals that affect the ANS are also found in medically hazardous substances, such as tobacco and insecticides.

Both direct- and indirect-acting drugs influence the ANS. Direct-acting drugs bind to ANS receptors to produce their effects. For example, agonists, or *stimulating agents*, bind to specific receptors and activate them, and antagonists, or *blocking agents*, bind to specific receptors and prevent them from being activated. Here we focus on direct-acting drugs, but it should be noted that some indirect-acting drugs produce a stimulatory effect by causing the release of neurotransmitters or by preventing neurotransmitters from breaking down metabolically. Other indirect-acting drugs produce an inhibitory effect by preventing the biosynthesis or release of neurotransmitters.

Drugs That Bind to Nicotinic Receptors

Drugs that bind to nicotinic receptors and activate them are **nicotinic agents.** Nicotinic agents bind to the nicotinic receptors on all postganglionic neurons within autonomic ganglia and produce stimulation. Although these agents have little therapeutic value and are mainly of interest to researchers, nicotine is medically important because of its presence in tobacco.

Responses to nicotine are variable and depend on the amount taken into the body. Because nicotine stimulates the postganglionic neurons of both the sympathetic and parasympathetic divisions, the variability of its effects largely results from the opposing actions of these divisions. For example, in response to the nicotine in a cigarette, the heart rate may either increase or decrease. Heart rate rhythm tends to become less regular as a result of the simultaneous actions on the sympathetic division, which increase the heart rate, and the parasympathetic division, which decrease the heart rate. Blood pressure tends to increase because of the constriction of blood vessels, which are almost exclusively innervated by sympathetic neurons. In addition to its influence on the ANS, nicotine also affects the CNS; therefore, not all of its effects can be explained on the basis of action on the ANS. Nicotine is extremely toxic, and even small amounts can be lethal.

Drugs that bind to and block nicotinic receptors are called **ganglionic blocking agents** because they block the effect of acetylcholine on both parasympathetic and sympathetic postganglionic neurons. However, the effect of these substances on the sympathetic division overshadows their effect on the parasympathetic division. For example, trimethaphan camsylate (trī-meth′ă-fan kam′sil-āt), used to treat high blood pressure, blocks the sympathetic stimulation of blood vessels, causing the blood vessels to dilate, which decreases blood pressure. Ganglionic blocking agents have limited uses because they affect both sympathetic and parasympathetic ganglia. Whenever possible, more selective drugs, which affect receptors of effectors, are now used.

Drugs That Bind to Muscarinic Receptors

Drugs that bind to and activate muscarinic receptors are **muscarinic agents,** or *parasympathomimetic* (par-ă-sim′pă-thō-mi-met′ik) *agents.* These drugs activate the muscarinic receptors of effectors of the parasympathetic division and the muscarinic receptors of sweat glands, which are innervated by the sympathetic division. One of these drugs, muscarine, causes increased sweating, increased secretion of glands in the digestive system, decreased heart rate, constriction of the pupils, and contraction of respiratory, digestive, and urinary system smooth muscles. Bethanechol (be-than′ē-kol) chloride is a muscarinic agent used to stimulate the urinary bladder following surgery because the general anesthetics used for surgery can temporarily inhibit a person's ability to urinate.

Drugs that bind to and block the action of muscarinic receptors are **muscarinic blocking agents,** or *parasympathetic blocking agents.* For example, the activation of muscarinic receptors causes the constriction of air passageways. Ipratropium (i-pră-trō′pē-ŭm) is used to treat chronic obstructive pulmonary disease because it blocks muscarinic receptors, which promotes the relaxation of air passageways. Atropine (at′rō-pēn) is used to block parasympathetic reflexes associated with the surgical manipulation of organs.

Drugs That Bind to Alpha and Beta Receptors

Drugs that activate adrenergic receptors are **adrenergic agents,** or *sympathomimetic* (sim′pă-thō-mi-met′ik) *agents.* Drugs such as phenylephrine (fen-il-ef′rin) stimulate alpha receptors, which are numerous in the smooth muscle cells of certain blood vessels, especially in the digestive tract and the skin. These drugs increase blood pressure by causing vasoconstriction. On the other hand, albuterol (al-bū′ter-ol) is a drug that selectively activates beta receptors in bronchiolar smooth muscle. β-adrenergic-stimulating agents are sometimes used to dilate bronchioles in respiratory disorders, such as asthma.

Drugs that bind to and block the action of alpha receptors are **α-adrenergic-blocking agents.** For example, prazosin (prā′zō-sin) hydrochloride is used to treat hypertension. By binding to alpha receptors in the smooth muscle of blood vessel walls, prazosin hydrochloride blocks the normal effects of norepinephrine released from sympathetic postganglionic neurons. Thus, the blood vessels relax, and blood pressure decreases.

Propranolol (prō-pran′ō-lōl) is an example of a **β-adrenergic-blocking agent.** These drugs are sometimes used to treat high blood pressure, some types of cardiac arrhythmias, and heart attacks. Blockage of the beta receptors within the heart prevents sudden increases in heart rate and thus decreases the probability of arrhythmic contractions.

Future Research

Present knowledge of the ANS is more complicated than the broad outline presented here. In fact, each of the major receptor types has subtype receptors. For example, α-adrenergic receptors are subdivided into the following subgroups: α_{1A}-, α_{1B}-, α_{2A}-, and α_{2B}-adrenergic receptors. The exact number of subtypes in humans is not yet known; however, their existence suggests the possibility of designing drugs that affect only one subtype, such as a drug that affects the blood vessels of the heart but not other blood vessels. Such drugs could produce specific effects but avoid undesirable side effects because they would act only on specific effectors.

Case STUDY | Eyedrops

Sally is a 50-year-old woman with diabetes. Every year, she has her eyes examined by an ophthalmologist because damage to the retina can be associated with diabetes. In addition to checking her vision using an eye chart, her doctor examines the insides of her eyes using an ophthalmoscope (see figure 15.15*a*). To see inside Sally's eyes better, the doctor applies eyedrops that cause the pupils to enlarge. The diameter of a typical pupil in normal room light is approximately 3–4 mm, whereas a dilated pupil measures 7–8 mm. Because Sally has complained in the past about light sensitivity as a result of the pupil-dilating eyedrops, the doctor applies pupil-constricting eyedrops after the eye exam.

Predict 4

a. Review the anatomy of the iris of the eye in chapter 15 (see section 15.3). Do the radial muscles or the circular muscles of the iris cause the pupil to dilate?
b. Which division of the ANS controls the radial muscles? The circular muscles?
c. Four types of drugs act on the receptors of target organs of the ANS (see Clinical Impact, "Influence of Drugs on the Autonomic Nervous System"). Which of these drugs could explain dilation of Sally's pupils? (*Hint:* See table 16.3.)
d. A side effect of the pupil-dilating eyedrops is blurred vision—that is, an inability to see close objects clearly. Based on this observation, what causes the blurred vision and which type of drug is in the pupil-dilating eyedrops?
e. Which type of drug could be in the eyedrops that reversed the dilation of Sally's pupils?
f. A side effect of the pupil-constricting eyedrops is bloodshot eyes. Based on this observation, which type of drug is in the pupil-constricting eyedrops?

ASSESS YOUR PROGRESS

23. *Give two exceptions to the generalization that organs are innervated by both divisions of the ANS.*
24. *Use the fight-or-flight response to describe the effects of sympathetic stimulation.*
25. *What are the major roles of the parasympathetic division? Give several examples.*
26. *Describe how a neuron would be classified as cholinergic or adrenergic. Which ANS neurons are cholinergic, and which are adrenergic?*
27. *Name the two major types of cholinergic receptors. Where are they located?*
28. *When acetylcholine binds to each receptor subtype, does it result in an excitatory or an inhibitory cell response?*
29. *Where are adrenergic receptors located? Name the two major types of adrenergic receptors. What are the subtypes of each?*
30. *In what two ways are adrenergic receptors stimulated?*

16.5 Regulation of the Autonomic Nervous System

LEARNING OUTCOMES

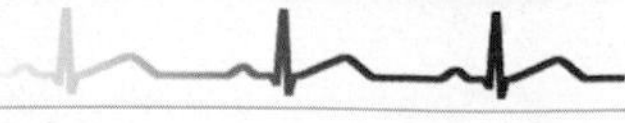

After reading this section, you should be able to

A. Explain the importance of autonomic reflexes to maintaining homeostasis.
B. Describe several autonomic reflexes.
C. Explain how a local reflex differs from other types of reflexes.

An individual's ability to maintain homeostasis is dependent on the regulatory activity of the ANS. Much ANS regulation occurs through autonomic reflexes, but input from the cerebrum, the hypothalamus, and other areas of the brain allows conscious thoughts and actions, emotions, and other CNS activities to influence autonomic functions.

Autonomic reflexes, like other reflexes, involve sensory receptors, sensory neurons, interneurons, motor neurons, and effector cells (figure 16.8; see chapter 12). For example, **baroreceptors** (stretch receptors) in the walls of large arteries near the heart detect changes in blood pressure, and sensory neurons transmit information from the baroreceptors through the glossopharyngeal and vagus nerves to the medulla oblongata. Interneurons in the medulla oblongata integrate the information, and action potentials are produced in autonomic neurons that extend to the heart. If baroreceptors detect a change in blood pressure, autonomic reflexes change the heart rate, which returns the blood pressure to normal. A sudden increase in blood pressure initiates a parasympathetic reflex, which inhibits cardiac muscle cells and reduces the heart rate, thereby bringing blood pressure down toward its normal value. Conversely, a sudden decrease in blood pressure initiates a sympathetic reflex, which stimulates the heart to increase its rate and force of contraction, thereby increasing blood pressure.

Predict 5

Brad ran a 6.2 km race for his cross-country team. The weather was cool, but he started to sweat during the race and continued to sweat for a short time afterward. Evaporation of the sweat from his body then caused him to feel cold. Explain how the autonomic nervous system functioned to control Brad's body temperature during and after the race.

Other autonomic reflexes help regulate blood pressure (see chapter 21). For example, numerous sympathetic neurons transmit a low but relatively constant frequency of action potentials that stimulate blood vessels throughout the body, keeping them partially constricted. If the vessels constrict further, blood pressure increases; if they dilate, blood pressure decreases. Thus, altering the frequency of action potentials delivered to blood vessels along sympathetic neurons can either raise or lower blood pressure.

The brainstem and the spinal cord contain important autonomic reflex centers responsible for maintaining homeostasis (table 16.4). However, the hypothalamus is in overall control of the ANS. Almost any type of autonomic response can be evoked by stimulating a part of the hypothalamus, which in turn stimulates ANS centers in the brainstem or spinal cord. Although there is overlap, stimulation of

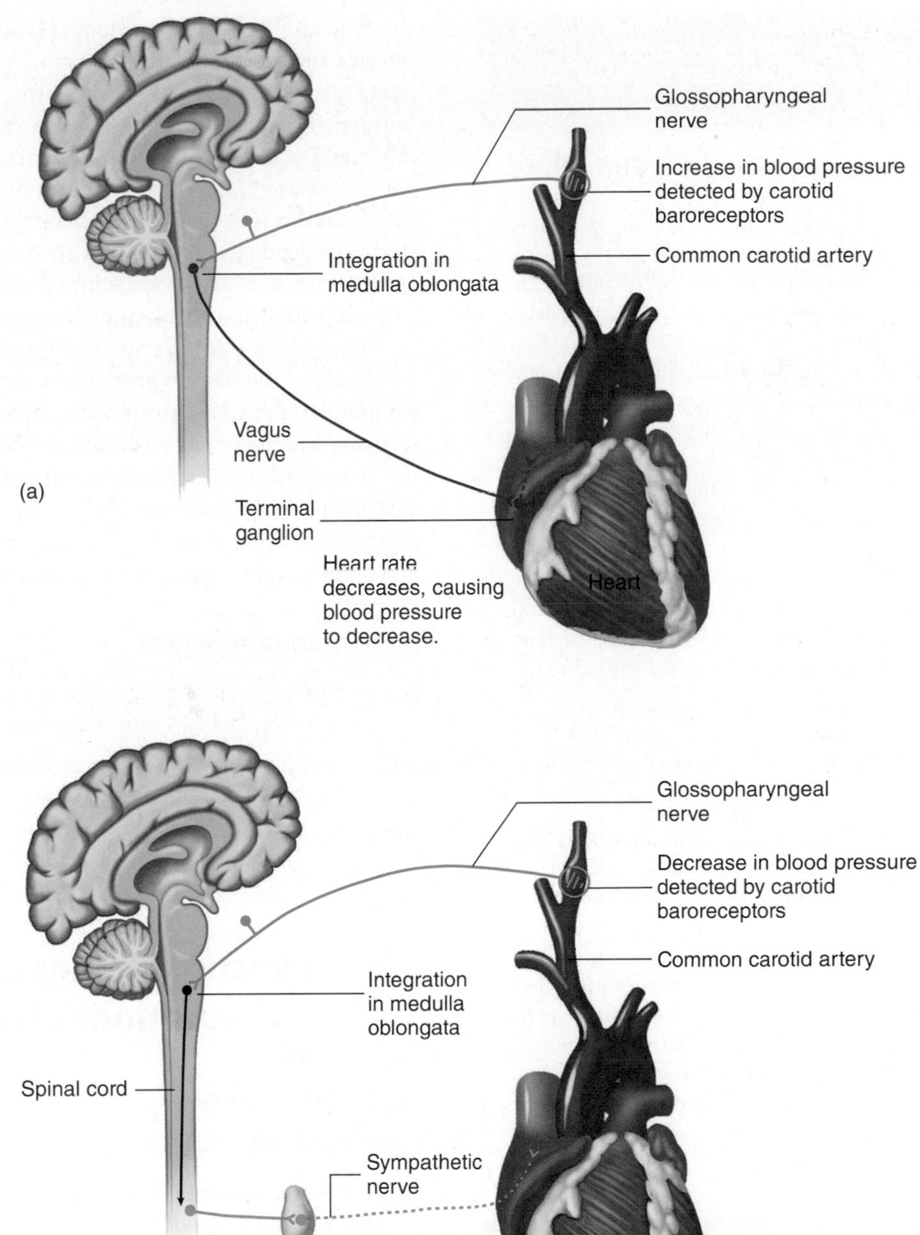
Glossopharyngeal nerve
Increase in blood pressure detected by carotid baroreceptors
Common carotid artery
Integration in medulla oblongata
Vagus nerve
(a)
Terminal ganglion
Heart rate decreases, causing blood pressure to decrease.
Heart
Glossopharyngeal nerve
Decrease in blood pressure detected by carotid baroreceptors
Common carotid artery
Integration in medulla oblongata
Spinal cord
Sympathetic nerve
Heart rate increases, causing blood pressure to increase.
Sympathetic chain ganglia
(b)

FIGURE 16.8 APR **Autonomic Reflexes**

Sensory input from the carotid baroreceptors is sent along the glossopharyngeal nerves to the medulla oblongata. The input is integrated in the medulla oblongata, and motor output is sent to the heart. (*a*) Parasympathetic reflex. Increased blood pressure results in increased stimulation of the heart by the vagus nerves, which increases inhibition of the heart and lowers the heart rate. (*b*) Sympathetic reflex. Decreased blood pressure results in increased stimulation of the heart by sympathetic nerves, which in turn increases stimulation of the heart and increases the heart rate and the force of contraction.

the posterior hypothalamus produces sympathetic responses, whereas stimulation of the anterior hypothalamus produces parasympathetic responses. In addition, the hypothalamus monitors and controls body temperature.

The hypothalamus has connections with the cerebrum and is an important part of the limbic system, which plays an important role in emotions. The hypothalamus integrates thoughts and emotions to produce ANS responses. For example, pleasant thoughts of a delicious banquet initiate increased secretion by salivary glands and by glands within the stomach, as well as increased smooth muscle contractions within the digestive system. These responses are controlled by parasympathetic neurons. Emotions such as anger increase blood pressure by increasing the heart rate and constricting the blood vessels through sympathetic stimulation.

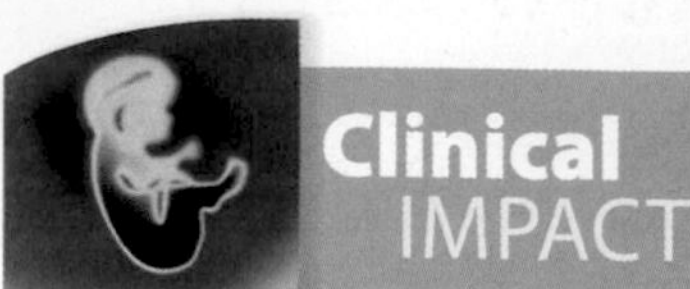

Clinical IMPACT

Effects of Spinal Cord Injury on ANS Functions

Spinal cord injury can damage nerve tracts, resulting in the loss of sensation and motor control below the level of the injury. Spinal cord injury also interrupts the control of autonomic neurons by ANS centers in the brain. For the parasympathetic division, effector organs innervated through the sacral region of the spinal cord are affected, but most effectors still have normal parasympathetic function because they are innervated by the vagus nerve. For the sympathetic division, brain control of sympathetic neurons is lost below the site of the injury. The higher the level of injury, the greater the number of body parts affected.

Immediately after spinal cord injury, spinal cord reflexes, including ANS reflexes, below the level of the injury are lost. With time, the reflex centers in the spinal cord become functional again. This recovery is particularly important for reflexes involving urination and defecation. Autonomic reflexes mediated through the vagus nerves or the enteric nervous system are not affected by spinal cord injury.

Predict 6

Important sensory receptors that monitor blood pressure are located in arteries in the chest above the heart and in arteries in the head. When Sarah does a headstand, gravity causes the blood in her arteries to flow toward her head, and the blood pressure in the arteries of her chest and head increases. When Sarah crouches for a short time and then stands up rapidly, gravity causes the blood in her arteries to flow away from her neck and head and toward her feet, and the blood pressure in the arteries of her chest and head decreases. How do the sympathetic reflexes that control blood vessels respond when Sarah does a headstand, and how do the sympathetic reflexes that control blood vessels respond if Sarah suddenly stands up after crouching for a short time?

TABLE 16.4 Higher Parts of the CNS Influencing Autonomic Functions

CNS Component	Effects on ANS
Cerebrum and limbic system	Thoughts and emotions can influence ANS functions through the hypothalamus
Hypothalamus	ANS integrating center that interacts with the cerebrum, limbic system, brainstem, and spinal cord; also regulates body temperature
Brainstem	ANS reflex centers for controlling pupil size, accommodation, tear production, salivation, coughing, swallowing, digestive activities, heart rate and force of contraction, blood vessel diameter, and respiration
Spinal cord	ANS reflex centers for regulating defecation, urination, penile and clitoral erection, and ejection

The enteric nervous system is involved with both autonomic reflexes and local reflexes that regulate the activity of the digestive tract. Autonomic reflexes help control the digestive tract because sensory neurons of the enteric plexuses supply the CNS with information about intestinal contents, and ANS neurons to the enteric plexuses affect the responses of smooth muscle and glands within the digestive tract wall. For example, sensory neurons detecting stretch of the digestive tract wall send action potentials to the CNS. In response, the CNS sends action potentials out the ANS, causing smooth muscle in the digestive tract wall to contract.

The neurons of the enteric nervous system also operate independently of the CNS to produce local reflexes. A **local reflex** does not involve the CNS, but produces an involuntary, unconscious, stereotypical response to a stimulus. For example, sensory neurons not connected to the CNS detect stretch of the digestive tract wall. These sensory neurons send action potentials through the enteric plexuses to motor neurons, causing smooth muscle to contract or relax. See chapter 24 for more information on local reflexes.

ASSESS YOUR PROGRESS

31. *Describe the autonomic reflex that maintains blood pressure by altering the heart rate or the diameter of blood vessels.*
32. *What part of the CNS stimulates ANS reflexes and integrates thoughts and emotions to produce ANS responses?*
33. *What is a local reflex? How is it different from other autonomic reflexes? How does the ENS produce local reflexes?*

16.6 Functional Generalizations About the Autonomic Nervous System

LEARNING OUTCOME

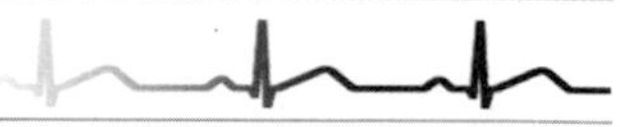

After reading this section, you should be able to

A. Explain opposite effects, cooperative effects, and general versus localized effects for the ANS.

Several generalizations can be made about the influence of the ANS on effector organs. However, bear in mind that most of the following generalizations have exceptions.

Stimulatory Versus Inhibitory Effects

Both divisions of the ANS produce stimulatory and inhibitory effects. For example, the parasympathetic division stimulates contraction of the urinary bladder, causing urination, but inhibits the heart, causing a decrease in heart rate. The sympathetic division stimulates smooth muscle contraction in blood vessel walls, causing vasoconstriction, but inhibits smooth muscle contractions in the lungs, causing dilation of lung air passageways. Thus, it is *not* true that one division of the ANS is always stimulatory and the other is always inhibitory.

Diseases and Disorders

TABLE 16.5 Autonomic Nervous System

Condition	Description
Raynaud disease	Spasmodic contraction of blood vessels and poor circulation, particularly in the fingers, resulting in cold hands prone to ulcerations and gangrene; results from blood vessels that show exaggerated sensitivity to sympathetic innervations; treatments include preganglionic denervation (cutting the preganglionic neurons)
Hyperhidrosis (hī′per-hī-drō′sis)	Excessive sweating caused by exaggerated sympathetic innervation of the sweat glands
Achalasia (ak-ă-lā′zē-ă)	Difficulty in swallowing and in controlling contraction of the esophagus where it enters the stomach; caused by abnormal parasympathetic regulation of the swallowing reflex
Dysautonomia (dis′aw-tō-nō′mē-ă)	Autosomal recessive genetic disorder characterized by reduced tear gland secretion, poor vasomotor control, trouble swallowing, and other symptoms
Herschsprund disease	Also known as *megacolon*, a functional obstruction in the lower colon and rectum that inhibits peristaltic contractions, leading to accumulation of feces in the affected area; caused by ineffective parasympathetic stimulation and a predominance of sympathetic stimulation of the colon

Go to www.mhhe.com/seeley10 for additional information on these pathologies.

Opposite Effects

When a *single* structure is innervated by both autonomic divisions, the two divisions usually produce opposite effects on the structure. As a consequence, the ANS is capable of both increasing and decreasing the activity of the structure. In the gastrointestinal tract, for example, parasympathetic stimulation increases secretion from glands, whereas sympathetic stimulation decreases secretion. However, in a few instances the effect of the two divisions is not clearly opposite. For example, both divisions of the ANS increase salivary secretion: The parasympathetic division initiates the production of a large volume of thin, watery saliva, and the sympathetic division causes the secretion of a small volume of viscous saliva.

Cooperative Effects

One autonomic division can coordinate the activities of *different* structures. For example, the parasympathetic division stimulates the pancreas to release digestive enzymes into the small intestine and stimulates contractions of the small intestine to mix the digestive enzymes with the food. These responses enhance the digestion and absorption of the food.

Alternatively, both divisions of the ANS can act together to coordinate the activity of *different* structures. In the male, the parasympathetic division initiates erection of the penis, whereas the sympathetic division stimulates the release of secretions from male reproductive glands and helps initiate ejaculation in the male reproductive tract.

General Versus Localized Effects

The sympathetic division has a more general effect than the parasympathetic division because activation of the sympathetic division often causes secretion of both epinephrine and norepinephrine from the adrenal medulla. These hormones circulate in the blood and stimulate effectors throughout the body. Because circulating epinephrine and norepinephrine can persist for a few minutes before being broken down, they can also produce an effect for a longer time than the direct stimulation of effectors by postganglionic sympathetic axons.

The sympathetic division diverges more than the parasympathetic division. Each sympathetic preganglionic neuron synapses with many postganglionic neurons, whereas each parasympathetic preganglionic neuron synapses with about two postganglionic neurons. Consequently, stimulation of sympathetic preganglionic neurons can result in greater stimulation of an effector.

Sympathetic stimulation often activates many different kinds of effectors at the same time as a result of CNS stimulation or epinephrine and norepinephrine release from the adrenal medulla. It is possible, however, for the CNS to selectively activate effectors. For example, vasoconstriction of cutaneous blood vessels in a cold hand is not always associated with an increased heart rate or other responses controlled by the sympathetic division.

ASSESS YOUR PROGRESS

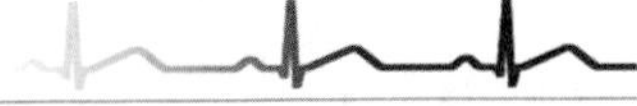

34. *What kinds of effects, excitatory or inhibitory, do the sympathetic and parasympathetic divisions produce?*

35. *When a single organ is innervated by both divisions of the ANS, how would you describe their effects?*

36. *Explain how the ANS coordinates the activities of different organs.*

37. *Which ANS division produces the most generalized effects? How does this happen?*

Answer

Learn to Predict ❮ From page 547

Describing the neural pathways involved in Officer Smith's muscular movements, increased heart rate and changes in digestive functions, requires an understanding of the somatic motor pathways described in chapter 14, the visual pathway described in chapter 15, and the autonomic pathways described in this chapter. The changes in Officer Smith's activity were initiated when she saw the speeding car. Action potentials were conducted from her eyes via the optic nerve (cranial nerve II) to the occipital lobe of the cerebral cortex, where conscious awareness of the situation occurred. The changes in her muscle movements occurred when action potentials were then conducted to the frontal lobe of the cerebrum, where voluntary motor activity is controlled. Action potentials initiated in the motor area of the frontal lobe traveled through descending pathways of the spinal cord. Somatic motor neurons carried action potentials from the spinal cord to the involved skeletal muscles, allowing the movements necessary for Officer Smith to drive her car.

In addition, action potentials also traveled from the cerebral cortex to the limbic system, including the hypothalamus, and from the hypothalamus to motor pathways of the autonomic nervous system. In the autonomic nervous system, action potentials were conducted to the sympathetic chain ganglia. Sympathetic neurons that regulate heart rate carried the action potentials to the heart, causing Officer Smith's heart rate to increase. Sympathetic neurons that regulate digestive function passed through the sympathetic chain ganglia to activate the splanchnic nerves and synapsed in the collateral ganglia. These neurons then conducted the action potentials to the digestive tract, where they decreased the activity of enteric neurons and decreased digestive secretion and motility.

Answers to the rest of this chapter's Predict questions are in Appendix G.

Summary

16.1 Overview of the Autonomic Nervous System (p. 548)

1. The autonomic nervous system (ANS) maintains homeostasis of the body by regulating many activities including heart rate, breathing rate, body temperature, digestive and urinary function.
2. The sympathetic division has more influence on effectors under conditions of increased physical activity or stress, whereas the parasympathetic division has more influence under conditions of rest.

16.2 Contrasting the Somatic and Autonomic Nervous Systems (p. 548)

1. The cell bodies of somatic motor neurons are located in the CNS, and their axons extend to skeletal muscles, where they have an excitatory effect that is usually consciously controlled.
2. The cell bodies of the preganglionic neurons of the ANS are located in the CNS and extend to ganglia, where they synapse with postganglionic neurons. The postganglionic axons extend to smooth muscle, cardiac muscle, or glands and have an excitatory or inhibitory effect, which is usually unconsciously controlled.

16.3 Anatomy of the Autonomic Nervous System (p. 550)

Sympathetic Division

1. Preganglionic cell bodies are in the lateral horns of the spinal cord gray matter from T1 to L2.
2. Preganglionic axons pass through the ventral roots to the white rami communicantes to the sympathetic chain ganglia. From there, four courses are possible:
   - Preganglionic axons synapse (at the same or a different level) with postganglionic neurons, which exit the ganglia through the gray rami communicantes and enter spinal nerves.
   - Preganglionic axons synapse (at the same or a different level) with postganglionic neurons, which exit the ganglia through sympathetic nerves.
   - Preganglionic axons pass through the chain ganglia without synapsing to form splanchnic nerves. Preganglionic axons then synapse with postganglionic neurons in collateral ganglia.
   - Preganglionic axons synapse with the cells of the adrenal medulla.

Parasympathetic Division

1. Preganglionic cell bodies are in nuclei in the brainstem or the lateral parts of the spinal cord gray matter from S2 to S4.
   - Preganglionic axons from the brain pass to ganglia through cranial nerves.
   - Preganglionic axons from the sacral region pass through the pelvic splanchnic nerves to the ganglia.
2. Preganglionic axons pass to terminal ganglia within the wall of or near the organ that is innervated.

Autonomic Nerve Plexuses and Distribution of Autonomic Nerve Fibers

1. Sympathetic axons reach organs through spinal nerves, head and neck nerve plexuses, thoracic nerve plexuses, and abdominopelvic nerve plexuses.
2. Parasympathetic axons reach organs through cranial nerves, thoracic nerve plexuses, abdominopelvic nerve plexuses, and pelvic splanchnic nerves.
3. Sensory neurons run alongside sympathetic and parasympathetic neurons within nerves and nerve plexuses.

Enteric Nervous System

1. The enteric nerve plexus is within the wall of the digestive tract.
2. The enteric plexus consists of sensory neurons, ANS motor neurons, and enteric neurons.

16.4 Physiology of the Autonomic Nervous System (p. 556)

Sympathetic Versus Parasympathetic Activity

1. Most organs are innervated by both the sympathetic and parasympathetic divisions.
2. Sympathetic activity generally prepares the body for physical activity, whereas parasympathetic activity is more important for resting functions.

Neurotransmitters

1. Acetylcholine is released by cholinergic neurons (all preganglionic neurons, all parasympathetic postganglionic neurons, and some sympathetic postganglionic neurons).
2. Norepinephrine is released by adrenergic neurons (most sympathetic postganglionic neurons).

Receptors

1. Acetylcholine binds to nicotinic receptors (found in all postganglionic neurons) and muscarinic receptors (found in all parasympathetic and some sympathetic effectors).
2. Norepinephrine and epinephrine bind to alpha and beta receptors (found in most sympathetic effectors).
3. Activation of nicotinic receptors is excitatory, whereas activation of muscarinic, alpha, or beta receptors is either excitatory or inhibitory.

16.5 Regulation of the Autonomic Nervous System (p. 562)

1. Autonomic reflexes control most of the activity of visceral organs, glands, and blood vessels.
2. Autonomic reflex activity can be influenced by the hypothalamus and higher brain centers.
3. The sympathetic and parasympathetic divisions can influence the activities of the enteric nervous system through autonomic reflexes. The enteric nervous system can function independently of the CNS through local reflexes.

16.6 Functional Generalizations About the Autonomic Nervous System (p. 564)

1. Both divisions of the ANS produce stimulatory and inhibitory effects.
2. Usually, each division produces an opposite effect on a given organ.
3. Either division alone or both working together can coordinate the activities of different structures.
4. The sympathetic division produces more generalized effects than the parasympathetic division.

REVIEW AND COMPREHENSION

1. Given these phrases:
   (1) neuron cell bodies in the nuclei of cranial nerves
   (2) neuron cell bodies in the lateral gray matter of the spinal cord (S2–S4)
   (3) two synapses between the CNS and effectors
   (4) regulates smooth muscle

   Which of the phrases are true for the autonomic nervous system?
   a. 1,3 b. 2,4 c. 1,2,3 d. 2,3,4 e. 1,2,3,4
2. Given these structures:
   (1) collateral ganglion
   (2) sympathetic chain ganglion
   (3) white ramus communicans
   (4) splanchnic nerve

   Choose the arrangement that lists the structures in the order an action potential travels through them on the way from a spinal nerve to an effector.
   a. 1,3,2,4 b. 1,4,2,3 c. 3,1,4,2 d. 3,2,4,1 e. 4,3,1,2
3. The white ramus communicans contains
   a. preganglionic sympathetic fibers.
   b. postganglionic sympathetic fibers.
   c. preganglionic parasympathetic fibers.
   d. postganglionic parasympathetic fibers.
4. The cell bodies of the postganglionic neurons of the sympathetic division are located in the
   a. sympathetic chain ganglia.
   b. collateral ganglia.
   c. terminal ganglia.
   d. dorsal root ganglia.
   e. both a and b.
5. Splanchnic nerves
   a. are part of the parasympathetic division.
   b. have preganglionic neurons that synapse in the collateral ganglia.
   c. exit from the cervical region of the spinal cord.
   d. travel from the spinal cord to the sympathetic chain ganglia.
   e. All of these are correct.
6. Which of the following statements regarding the adrenal gland is true?
   a. The parasympathetic division stimulates the adrenal gland to release acetylcholine.
   b. The parasympathetic division stimulates the adrenal gland to release epinephrine.
   c. The sympathetic division stimulates the adrenal gland to release acetylcholine.
   d. The sympathetic division stimulates the adrenal gland to release epinephrine.
7. The parasympathetic division
   a. is also called the craniosacral division.
   b. has preganglionic axons in cranial nerves.
   c. has preganglionic axons in pelvic splanchnic nerves.
   d. has ganglia near or in the wall of effectors.
   e. All of these are correct.
8. Which of these is *not* a part of the enteric nervous system?
   a. ANS motor neurons
   b. neurons located only in the digestive tract
   c. sensory neurons
   d. somatic motor neurons
9. Sympathetic axons reach organs through all of the following *except*
   a. abdominopelvic nerve plexuses.
   b. head and neck nerve plexuses.
   c. thoracic nerve plexuses.
   d. pelvic splanchnic nerves.
   e. spinal nerves.
10. Which of these cranial nerves does *not* contain parasympathetic fibers?
    a. oculomotor (III)
    b. facial (VII)
    c. glossopharyngeal (IX)
    d. trigeminal (V)
    e. vagus (X)

11. Which of these events is expected if the sympathetic division is activated?
    a. Secretion of watery saliva increases.
    b. Tear production increases.
    c. Air passageways dilate.
    d. Glucose release from the liver decreases.
    e. All of these are correct.
12. Which of the following statements concerning the preganglionic neurons of the ANS is true?
    a. All parasympathetic preganglionic neurons secrete acetylcholine.
    b. Only parasympathetic preganglionic neurons secrete acetylcholine.
    c. All sympathetic preganglionic neurons secrete norepinephrine.
    d. Only sympathetic preganglionic neurons secrete norepinephrine.
13. When acetylcholine binds to nicotinic receptors,
    a. the cell's response is mediated by G proteins.
    b. the response can be excitatory or inhibitory.
    c. Na^+ channels open.
    d. the binding occurs at the effectors.
    e. All of these are correct.
14. Nicotinic receptors are located in
    a. postganglionic neurons of the parasympathetic division.
    b. postganglionic neurons of the sympathetic division.
    c. membranes of skeletal muscle cells.
    d. Both a and b are correct.
    e. all of these sites.
15. The sympathetic division
    a. is always stimulatory.
    b. is always inhibitory.
    c. is usually under conscious control.
    d. generally opposes the actions of the parasympathetic division.
    e. Both a and c are correct.
16. A sudden increase in blood pressure
    a. initiates a sympathetic reflex that decreases heart rate.
    b. initiates a local reflex that decreases heart rate.
    c. initiates a parasympathetic reflex that decreases heart rate.
    d. Both a and b are correct.
    e. Both b and c are correct.
17. Which of these structures is innervated almost exclusively by the sympathetic division?
    a. gastrointestinal tract
    b. heart
    c. urinary bladder
    d. reproductive tract
    e. blood vessels

Answers in Appendix E

CRITICAL THINKING

1. When a person is startled or sees a "pleasurable" object, the pupils of the eyes may dilate. What division of the ANS is involved in this reaction? Describe the nerve pathway involved.
2. Reduced secretion from salivary and lacrimal glands can indicate damage to what nerve?
3. Patients with diabetes mellitus can develop autonomic neuropathy, which is damage to parts of the autonomic nerves. Given the following parts of the ANS—vagus nerve, oculomotor nerve, splanchnic nerve, pelvic splanchnic nerve, outflow of gray ramus—match the part with the symptom it would produce if the part were damaged:
    a. impotence
    b. subnormal sweat production
    c. relaxed stomach muscles and delayed emptying of the stomach
    d. diminished pupil reaction (constriction) to light
    e. bladder paralysis with urinary retention
4. A patient has been exposed to the organophosphate pesticide malathion, which inactivates acetylcholinesterase. Which of the following symptoms would you predict: blurring of vision, excess tear formation, frequent or involuntary urination, pallor (pale skin), muscle twitching, or cramps? Would atropine be an effective drug to treat the symptoms (see Clinical Impact, "Influence of Drugs on the Autonomic Nervous System," for the action of atropine)? Explain.
5. Epinephrine is routinely mixed with local anesthetic solutions. Why?
6. A drug blocks the sympathetic division's effect on the heart. Careful investigation reveals that, after administration of the drug, normal action potentials are produced in the sympathetic preganglionic and postganglionic neurons. Also, injection of norepinephrine produces a normal response in the heart. Explain, in as many ways as you can, the mode of action of the unknown drug.
7. A drug is known to decrease heart rate. After cutting the white rami of T1–T4, the drug still causes the heart rate to decline. After cutting the vagus nerves, the drug no longer affects heart rate. Which division of the ANS does the drug affect? Does the drug affect the synapse between preganglionic and postganglionic neurons, the synapse between postganglionic neurons and effectors, or the CNS? Is the drug's effect excitatory or inhibitory?
8. Make a list of the responses controlled by the ANS in (a) a person who is extremely angry and (b) a person who has just finished eating and is relaxing.

Answers in Appendix F

Mc Graw Hill
connect plus+
ANATOMY & PHYSIOLOGY

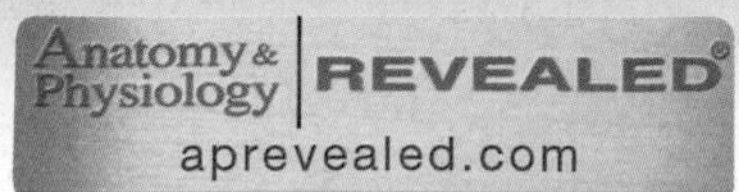
Anatomy & Physiology
REVEALED
aprevealed.com

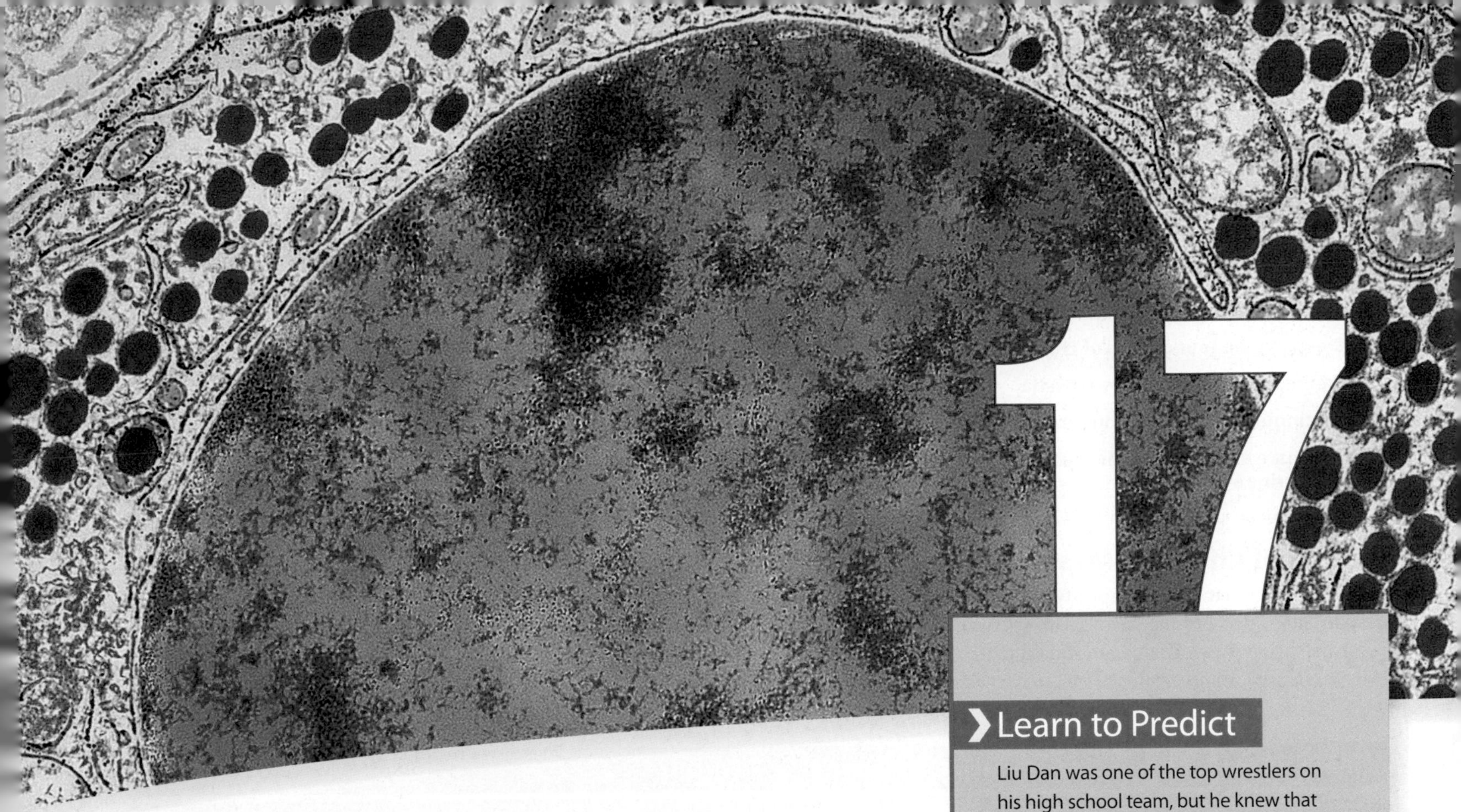

17

Functional Organization of the Endocrine System

"Time's up. Please hand in your exams." These words may cause anxiety in some students' minds. Their hearts start pounding, and they might even feel shaky and short of breath. Why? The culprit for these sensations is an "adrenaline rush." Adrenaline, or epinephrine, is one of hundreds of chemical messengers, called hormones, that circulate in the body. These chemical messengers have differences, but they all share the fundamental property of transmitting signals to target cells. This chapter focuses on one of the two major control systems in the body, the endocrine system. The other major control system is the nervous system. Here, we present the general principles of hormones; chapter 18 then discusses specific hormones and their functions.

Learn to Predict

Liu Dan was one of the top wrestlers on his high school team, but he knew that his small size would make it difficult to compete at the college level. He had tried lifting weights, but that did not seem to be working. So Liu decided to try something he had never thought he would do—he began taking anabolic steroids. At first he was excited, because his muscles were larger and he felt stronger. After a few more weeks, though, he started noticing some troublesome changes: His pectoral muscles were getting larger, but they looked more feminine than masculine; his testes had shrunk; and he'd had some frightening episodes that could only be described as temper tantrums. After reading this chapter, explain why anabolic steroids were able to alter muscle tissue growth and cause unintended changes in other tissues of the body.

Photo: Colorized transmission electron micrograph of an endocrine cell from the anterior pituitary gland. The secretory vesicles (*brown*) contain hormones.

Module 8
Endocrine System

Anatomy & Physiology
REVEALED
aprevealed.com

17.1 Principles of Chemical Communication

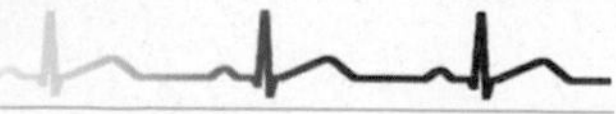

LEARNING OUTCOMES

After reading this section, you should be able to

A. Describe the four classes of chemical messengers.

B. Define hormone and target tissue.

C. Distinguish between endocrine and exocrine glands.

D. Compare and contrast the nervous system with the endocrine system.

Classes of Chemical Messengers

The body has a remarkable capacity for maintaining homeostasis despite having to coordinate the activities of over 75 trillion cells. The principal means by which this coordination occurs is through chemical messengers, some produced by the nervous system and others produced by the endocrine system. **Chemical messengers** allow cells to communicate with each other to regulate body activities.

Most chemical messengers are produced by a specific collection of cells or by a gland. Recall from chapter 4 that a gland is an organ consisting of epithelial cells that specialize in **secretion,** which is the controlled release of chemicals from a cell. This text identifies four classes of chemical messengers based on the source of the chemical messenger and its mode of transport in the body (table 17.1). In this section, we describe chemical messengers in terms of how they function. But it is important to note that some chemical messengers fall into more than one functional category. For example, prostaglandins are listed in multiple categories because they have diverse functions and cannot be categorized in just one class. Therefore, the study of the endocrine system includes several of the following categories:

1. *Autocrine chemical messengers.* An autocrine chemical messenger stimulates the cell that originally secreted it. Good examples of autocrine chemical messengers are those secreted by white blood cells during an infection. Several types of white blood cells can stimulate their own replication, so that the total number of white blood cells increases rapidly (see chapter 22).

TABLE 17.1 Classes of Chemical Messengers

Chemical Messenger	Description	Example	
Autocrine	Secreted by cells in a local area; influences the activity of the same cell from which it was secreted	Eicosanoids (prostaglandins, thromboxanes, prostacyclins, leukotrienes)	Chemical messenger Autocrine
Paracrine	Produced by a wide variety of tissues and secreted into extracellular fluid; has a localized effect on other tissues	Somatostatin, histamine, eicosanoids	Chemical messenger Paracrine
Neurotransmitter	Produced by neurons; secreted into a synaptic cleft by presynaptic nerve terminals; travels short distances; influences postsynaptic cells	Acetylcholine, epinephrine	Neuron Neurotransmitter
Endocrine	Secreted into the blood by specialized cells; travels some distance to target tissues; results in coordinated regulation of cell function	Thyroid hormones, growth hormone, insulin, epinephrine, estrogen, progesterone, testosterone, prostaglandins	Hormone Endocrine

2. *Paracrine chemical messengers.* Paracrine chemical messengers act locally on nearby cells. These chemical messengers are secreted by one cell type into the extracellular fluid and affect surrounding cells. An example of a paracrine chemical messenger is histamine, released by certain white blood cells during allergic reactions. Histamine stimulates vasodilation in nearby blood vessels.
3. *Neurotransmitters.* Neurotransmitters are chemical messengers secreted by neurons that activate an adjacent cell, whether it is another neuron, a muscle cell, or a glandular cell. Neurotransmitters are secreted into a synaptic cleft, rather than into the bloodstream (see chapter 11). Therefore, in the strictest sense neurotransmitters are paracrine agents, but for our purposes it is most appropriate to consider them as a separate category.
4. *Endocrine chemical messengers.* Endocrine chemical messengers are secreted into the bloodstream by certain glands and cells, which together constitute the endocrine system. These chemical messengers travel through the general circulation to their target cells.

Characteristics of the Endocrine System

The **endocrine system** is composed of **endocrine glands** and specialized endocrine cells located throughout the body (figure 17.1). Endocrine glands secrete minute amounts of chemical messengers called **hormones** (hor′mōnz) into the bloodstream, rather than into a duct. Hormones then travel a distance from their source through the bloodstream to specific sites called **target tissues,** or *effectors,* where they produce a coordinated response of the target tissues. Thus, the term **endocrine** (en′dō-krin), derived from the Greek words *endo,* meaning "within," and *krino,* "to secrete," appropriately describes this system.

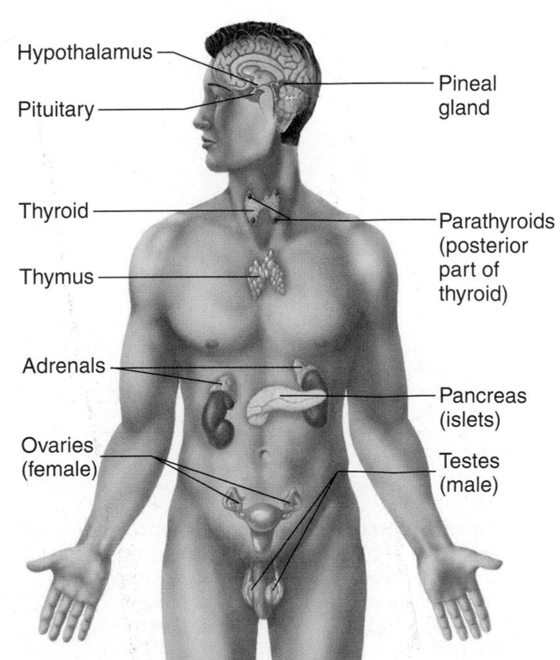
Hypothalamus
Pituitary
Pineal gland
Thyroid
Parathyroids (posterior part of thyroid)
Thymus
Adrenals
Pancreas (islets)
Ovaries (female)
Testes (male)

FIGURE 17.1 AP|R **Major Endocrine Glands and Tissues**

Endocrine glands are not to be confused with **exocrine glands.** Exocrine glands have ducts that carry their secretions to the outside of the body, or into a hollow organ, such as the stomach or intestines. Examples of exocrine secretions are saliva, sweat, breast milk, and digestive enzymes.

The study of the endocrine system, known as **endocrinology,** is the topic of this chapter as well as chapter 18. In this chapter, we present the general principles of hormones and, in chapter 18, we discuss specific hormones and their functions. Before focusing on the endocrine system, it is important to understand its relationship to the body's other major regulatory system, the nervous system.

Comparison of the Nervous and Endocrine Systems

Together, the nervous system and the endocrine system regulate and coordinate the activities of essentially all body structures to achieve and maintain homeostasis. As you learned in chapter 11, the nervous system functions as a communication system, transmitting messages in the form of action potentials along the axons of neurons and in the form of neurotransmitters at synapses between neurons and the cells they control. The endocrine system is more like the postal service delivering a mass mailing. Every residence receives the mail, but only those residents who understand the language of the letters can interpret them. The endocrine system sends information to the cells it controls in the form of hormones, which are carried by the bloodstream to all parts of the body. Cells with receptors for those hormones respond to them, whereas cells lacking receptors do not.

Given that both the nervous system and the endocrine system control their targets with chemical messengers, what is the difference between the two systems? In fact, it is difficult to completely separate the two systems because they have many similarities:

1. Both systems have structures associated with the brain. In chapter 13, the hypothalamus is discussed as a critical area of the brain responsible for many functions, including neural functions and hormone production. For example, the hypothalamus detects body temperature and sends impulses to either the sweat glands or skeletal muscle, depending on whether the body is too hot or too cold. On the other hand, the hypothalamus sends hormones to the pituitary gland that regulate the secretion of hormones from the pituitary. In addition, hypothalamic neurons synthesize and secrete two hormones, antidiuretic hormone and oxytocin, directly into the blood. Thus, anatomically, the hypothalamus plays a role in both the nervous and endocrine systems.
2. In many cases, the nervous system uses certain molecules as neurotransmitters, and the endocrine system may use the same molecules as hormones. For example, when a neuron secretes epinephrine into a synaptic cleft, it is a neurotransmitter. In contrast, when cells of the adrenal gland secrete epinephrine into the bloodstream, it is a hormone.
3. The two systems work together to regulate critical body processes. For example, the introduction to this chapter pointed out that epinephrine, the hormone, is important in stressful situations. However, the initial, immediate release of epinephrine, the neurotransmitter, in times of crisis is from the nervous system. Thus, the two systems work almost simultaneously.

4. Some neurons secrete hormones. In this case, rather than the neuron communicating directly with another cell, the neuron releases a chemical messenger that enters the bloodstream and functions as a hormone. To help distinguish these chemical messengers from neurotransmitters and other hormones, they are often called **neuropeptides,** or *neurohormones*. An example of a neuropeptide is the labor-inducing hormone oxytocin.
5. Both neurotransmitters and hormones can affect their targets through receptors linked to G proteins.

In addition to the similarities between the nervous and endocrine systems, there are some important differences:

1. *Mode of transport.* The endocrine system secretes hormones, which are transported in the bloodstream, whereas the nervous system secretes neurotransmitters, which are released directly onto their target cells.
2. *Speed of response.* In general, the nervous system responds faster than the endocrine system. However, it is not accurate to say the endocrine system responds slowly; rather, it responds *more slowly* than the nervous system. Neurotransmitters, such as acetylcholine, are delivered to their target cells in milliseconds, whereas some hormones are delivered to their target cells in seconds.
3. *Duration of response.* The nervous system typically activates its targets quickly and only for as long as action potentials are sent to the target. The target cells' response is terminated shortly after action potentials cease. In contrast, the endocrine system tends to have longer-lasting effects. Hormones remain in the bloodstream for minutes, days, or even weeks and activate their target tissues as long as they are present in the circulation. The target tissue products may remain active for a substantial length of time.

In summary, the hormones secreted by most endocrine glands can be described as **amplitude-modulated signals,** which consist mainly of fluctuations in the concentration of hormones in the bloodstream (figure 17.2), over a period of time ranging from minutes to hours. On the other hand, the all-or-none action potentials carried along axons can be described as **frequency-modulated signals** (figure 17.2), which vary in frequency but not amplitude. A low frequency of action potentials is a weak stimulus, whereas a high frequency of action potentials is a strong stimulus. Thus, both the nervous system and the endocrine system work to maintain homeostasis using chemical messengers, but they differ from each other in the way their chemical messengers work to activate their target cells.

ASSESS YOUR PROGRESS

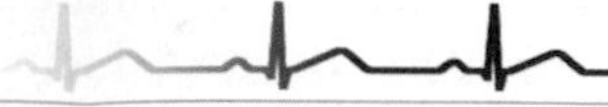

1. *Name and describe the four classes of chemical messengers.*
2. *How does an endocrine gland differ from an exocrine gland?*
3. *Describe the similarities between the nervous system and the endocrine system.*
4. *In what ways does the nervous system differ from the endocrine system?*

17.2 Hormones

LEARNING OUTCOMES

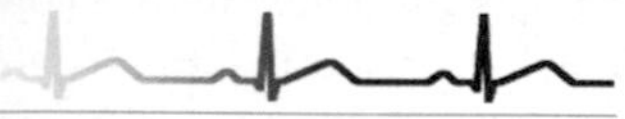

After reading this section, you should be able to

A. Describe the common characteristics of all hormones.

B. Define *binding protein, bound hormone,* and *free hormone* and discuss the effect of binding proteins on circulating hormone levels.

C. List and describe the two chemical categories of hormones.

D. Explain the influence of the chemical nature of a hormone on its transport in the blood, its removal from circulation, and its life span.

E. Describe the three main patterns of hormone secretion.

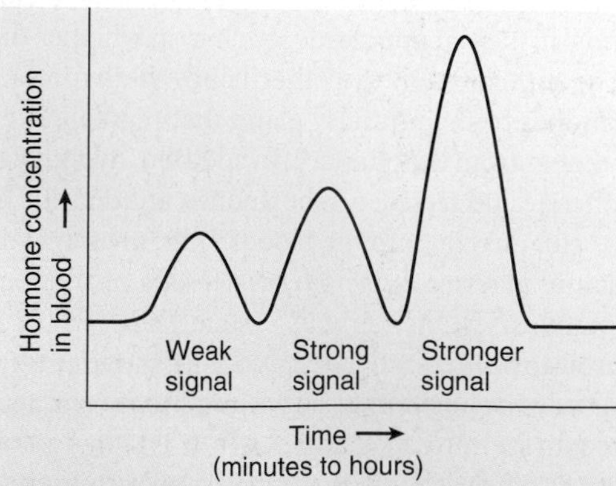

Hormone concentration in blood →
Weak signal
Strong signal
Stronger signal
Time →
(minutes to hours)

Amplitude-modulated system. The concentration of the hormone determines the strength of the signal and the magnitude of the response. For most hormones, a small concentration of a hormone is a weak signal and produces a small response, whereas a larger concentration is a stronger signal and results in a greater response.

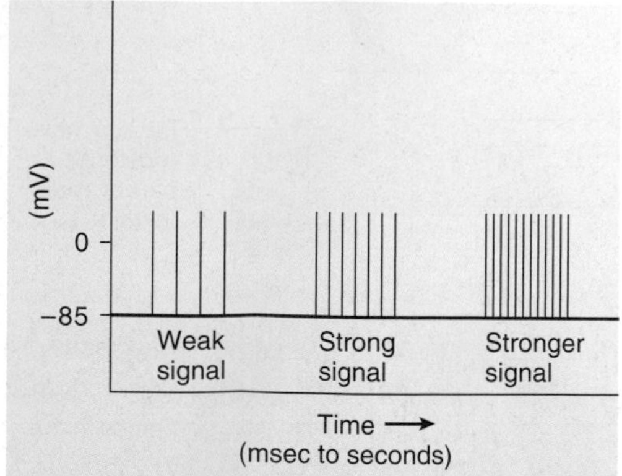

(mV)
0
−85
Weak signal
Strong signal
Stronger signal
Time →
(msec to seconds)

Frequency-modulated system. The strength of the signal depends on the frequency, not the size, of the action potentials. All action potentials are the same size in a given tissue. A low frequency of action potentials is a weak stimulus, and a higher frequency is a stronger stimulus.

FIGURE 17.2 Regulatory Systems

The word *hormone* is derived from the Greek word *hormon*, which means "set into motion."

General Characteristics of Hormones

Hormones share several characteristics:

1. *Stability.* For hormones to activate their targets continuously, they must remain active in the circulation long enough to arrive at their target cells. This means that hormone levels remain stable in the bloodstream; however, some hormones are more stable than others. The life span of a given hormone varies with its chemical nature. A hormone's life span can be expressed as its **half-life,** which is the amount of time it takes for 50% of the circulating hormone to be removed from the circulation and excreted. Some hormones have a short half-life, whereas others have a much longer half-life. For example, thyrotropin-releasing hormone (TRH) is a three-amino-acid hormone with a short half-life. Because of TRH's composition, it is quickly degraded in the circulation and can activate only the target cells it can reach within 2 minutes of being secreted. On the other hand, cortisol is a steroid hormone with a longer half-life, 90 minutes. Due to its lipid-soluble nature, it is not easily degraded and can activate target cells for more than an hour.
2. *Communication.* Hormones must be able to interact with their target tissue in a specific manner in order to activate a coordinated set of events. For example, the formation of reproductive organs in the fetus is activated by reproductive steroid hormones. Without proper functioning of the male reproductive steroid testosterone, a newborn will have the outward appearance of a female despite being genetically male. Hormones must therefore be able to regulate specific cellular pathways once they arrive at their targets.
3. *Distribution.* Hormones are transported by the blood to many locations and therefore have the potential to activate any cell in the body, including those far away from where they were produced. However, the blood contains many hydrolytic enzymes and is an aqueous solution. These factors can present a challenge when transporting hormones to their targets. Small, water-soluble hormones are quickly digested by enzymes in the blood, because, with their small size, they become inactive with very little alteration in their structure. In addition, they are easily filtered from the blood in the kidneys because they are so small. Still other hormones, such as lipid-soluble hormones, have low solubility in the blood plasma. Their chemical nature does not allow them to dissolve in the blood. Thus, some hormones require a chaperone to arrive safely at their target. Hormones requiring a transport chaperone bind to blood proteins called **binding proteins** and are then called **bound hormones.** For small hormones, the binding protein protects them from degradation by hydrolytic enzymes and from being filtered from the blood in the kidney. For lipid-soluble hormones that are insoluble in plasma, being bound to a binding protein causes them to become more water-soluble. Hormones bind only to selective binding proteins. For example, thyroid hormones bind to a specific binding protein, transthyretin; testosterone binds to a different type of binding protein, called testosterone-binding globulin; and progesterone binds to yet another type of binding protein, called progesterone-binding globulin.

The binding of hormones to binding proteins is reversible. Hormones dissociate from their binding proteins at their target tissues and are then called **free hormones.** It is important to note that some hormones always exist as free hormones because they do not have specific binding proteins to which they attach. Thus, some hormones are "always free," whereas other hormones are "sometimes free." The binding protein's affinity for its hormone determines the concentration of free hormones.

The reversible binding of hormones to their binding proteins is important because only free hormones are able to diffuse through capillary walls and bind to target tissues. The bound hormone thus serves as a reservoir for the hormone. If blood levels of the hormone begin to decline, some of the bound hormone is released from the binding proteins. Because of this reservoir function of the bound hormones, the circulating concentration of free hormones remains more stable than that of hormones that do not use binding proteins (figure 17.3). Consequently, hormones that bind to binding proteins tend to have longer half-lives than hormones that do not require binding proteins (figure 17.3).

Chemical Nature of Hormones

Hormones fit into two chemical categories: lipid-soluble hormones and water-soluble hormones, a distinction based on their chemical behavior. For example, recall from chapter 3 that the plasma membrane is a selectively permeable phospholipid bilayer that excludes water-soluble molecules but allows lipid-soluble molecules to pass through. Therefore, the entire basis of a hormone's interaction with its target is dependent on the hormone's chemical nature (table 17.2).

Within the two chemical categories, hormones can be subdivided into groups based on their chemical structures. Steroid hormones are derived from cholesterol and thyroid hormones are derived from the amino acid tyrosine, whereas other hormones are categorized as amino acid derivatives, peptides, or proteins, including glycoproteins.

Lipid-Soluble Hormones

Lipid-soluble hormones are nonpolar and include steroid hormones, thyroid hormones, and fatty acid derivative hormones, such as certain eicosanoids.

Transport of Lipid-Soluble Hormones

Because of their small size and low solubility in aqueous fluids, lipid-soluble hormones travel in the bloodstream bound to binding proteins, proteins that chaperone the hormone. As a result, the rate at which lipid-soluble hormones are degraded or eliminated from the circulation is greatly reduced and their life spans range from a few days to as long as weeks.

Without the binding proteins, the lipid-soluble hormones would quickly diffuse out of capillaries and be degraded by enzymes of the liver and lungs or be removed from the body by the kidneys. Circulating hydrolytic enzymes can also metabolize free lipid-soluble hormones. The breakdown products are then excreted in the urine or the bile.

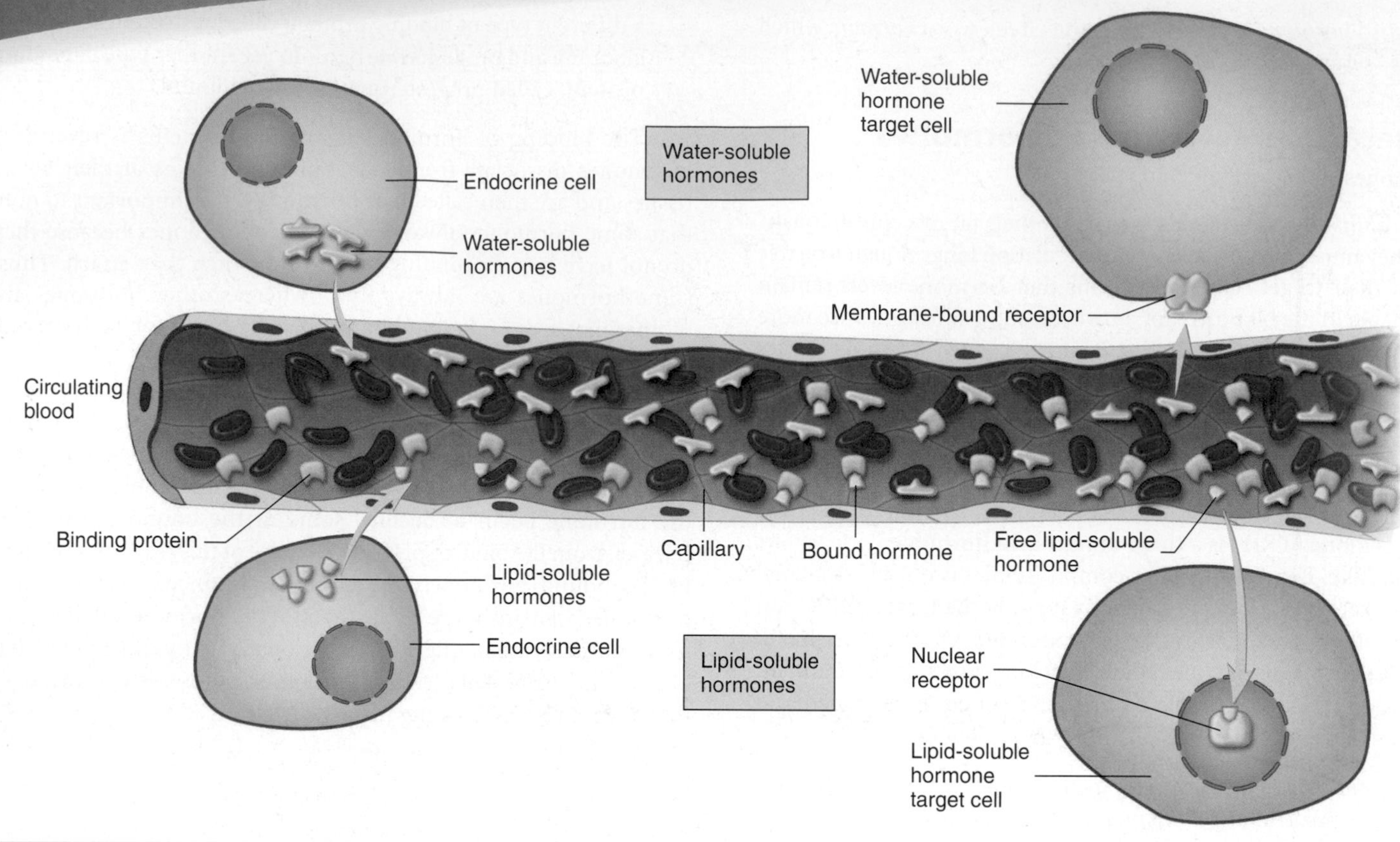
Water-soluble hormones
Endocrine cell
Water-soluble hormones
Water-soluble hormone target cell
Membrane-bound receptor
Circulating blood
Binding protein
Capillary
Bound hormone
Free lipid-soluble hormone
Lipid-soluble hormones
Endocrine cell
Lipid-soluble hormones
Nuclear receptor
Lipid-soluble hormone target cell

FIGURE 17.3 Effect of Binding Proteins

(*a*) Free hormones (those that circulate freely in the blood) immediately activate target cells once they are delivered from the blood. Thus, the blood levels of these hormones tend to fluctuate to a greater degree than the levels of hormones that attach to binding proteins; water-soluble hormones bind their receptors, which are membrane bound. (*b*) Hormones that are transported in the blood attached to binding proteins circulate in the blood as bound or free hormones. As the concentration of free hormones decreases, bound hormones are released from the binding proteins. This provides a chronic, stable supply of hormone and, thus, more consistent control of target cells. This is especially important for hormones that regulate basal metabolism. Lipid-soluble hormones bind their receptors in either the cytoplasm or the nucleus.

Additionally, lipid-soluble hormones are removed from the circulation when certain enzymes in the liver attach water-soluble molecules to the hormones, a process called **conjugation** (kon-jŭ-gā′shŭn). These conjugation molecules are usually sulfate or glucuronic acid. Once the hormones are conjugated, the kidneys and liver excrete them into the urine and bile at a greater rate.

Water-Soluble Hormones

Water-soluble hormones are polar molecules; they include protein hormones, peptide hormones, and most amino acid derivative hormones.

Transport of Water-Soluble Hormones

Because water-soluble hormones can dissolve in blood, many circulate as free hormones, meaning that most of them dissolve directly into the blood and are delivered to their target tissue without binding to a binding protein. Because many water-soluble hormones are quite large, they do not readily diffuse through the walls of all capillaries; therefore, they tend to diffuse from the blood into tissue spaces more slowly. The capillaries of organs that are regulated by protein hormones are usually very porous, or *fenestrated* (see chapter 21). On the other hand, other water-soluble hormones are quite small and require attachment to a larger protein, a binding protein, to avoid being filtered out of the blood.

All hormones are destroyed either in the circulation or at their target cells. The destruction and elimination of hormones limit the length of time they are active, and body processes change quickly when hormones are secreted and remain functional for only short periods.

Water-soluble hormones, such as proteins, peptides, and amino acid derivatives, have relatively short half-lives because they are rapidly degraded by enzymes, called proteases, within the bloodstream. The kidneys then remove the hormone breakdown products from the blood. Target cells also destroy water-soluble hormones when the hormones are internalized via endocytosis. Once the hormones are inside the target cell, lysosomal enzymes degrade them. Often, the target cell recycles the amino acids of peptide and protein hormones and uses them to synthesize new proteins. Hormones with short half-lives normally have concentrations that change rapidly within the blood and tend to regulate activities that have a rapid onset and short duration.

TABLE 17.2 Chemical Nature of Hormones

Chemical Nature	Examples	Structures
Lipid-Soluble Hormones	Steroids (all cholesterol-based)	Testosterone, aldosterone Aldosterone 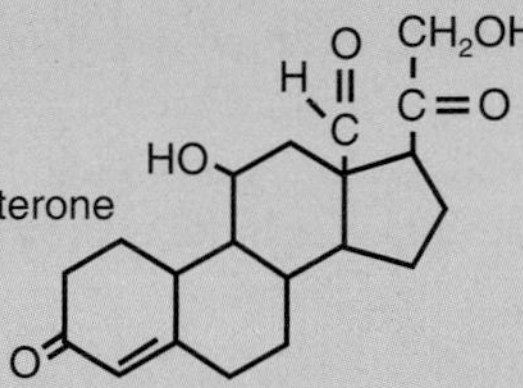 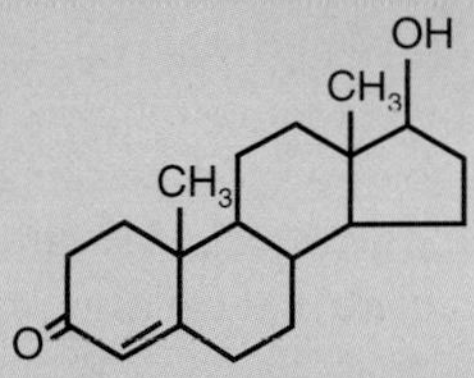 Testosterone
	Amino acid derivative (only one example of lipid-soluble)	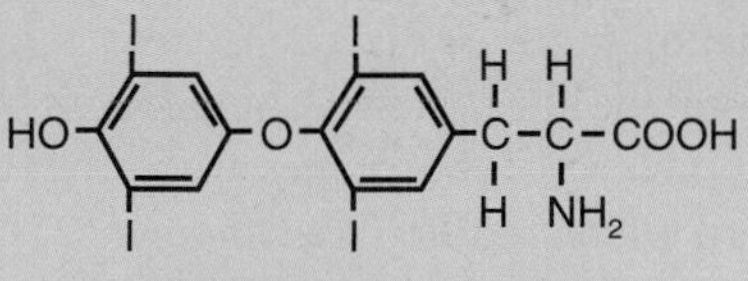 Thyroid hormone (thyroxine) Tetraiodothyronine or thyroxine (T_4)
	Fatty acid derivatives	Prostaglandins Fatty acid derivative (formed from a fatty acid) 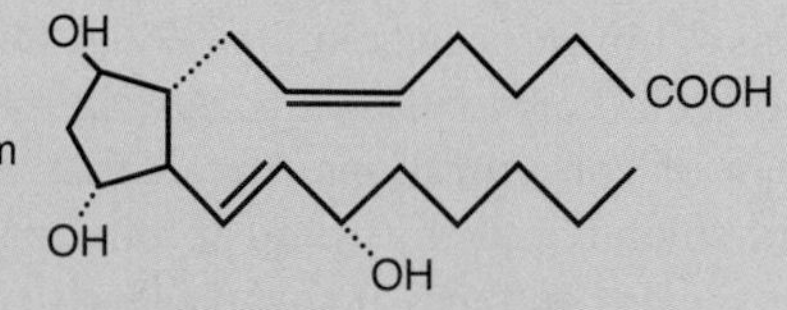 Prostaglandin $F_{2\alpha}$ ($PGF_{2\alpha}$)
Water-Soluble Hormones	Proteins	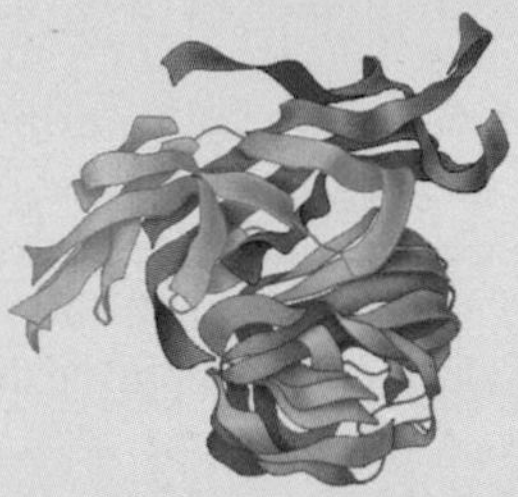 Thyroid-stimulating hormone, growth hormone Thyroid-stimulating hormone (a glycoprotein) 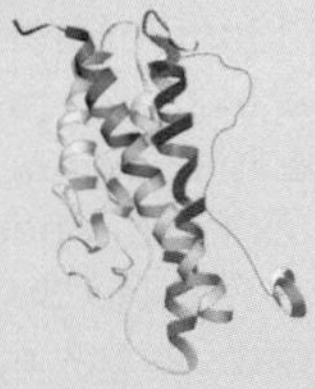 Growth hormone
	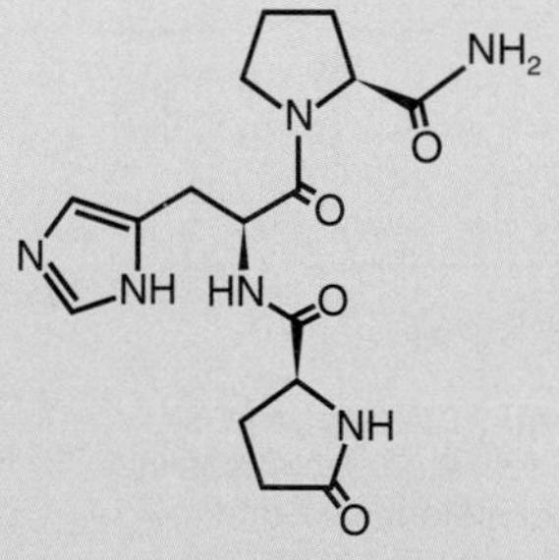 Peptides	Insulin, thyrotropin-releasing hormone Thyrotropin-releasing hormone Insulin: A chain: Gly-Ile-Val-Glu-Gln-Cys-Cys-Thr-Ser-Ile-Cys-Ser-Leu-Tyr-Gln-Leu-Glu-Asn-Tyr Cys-Asn B chain: Phe-Val-Asn-Gln-His-Leu-Cys-Gly-Ser-His-Leu-Val-Glu-Ala-Leu-Tyr-Leu-Val-Cys-Gly-Glu-Arg-Gly-Phe-Phe-Tyr-Thr-Pro-Lys-Thr 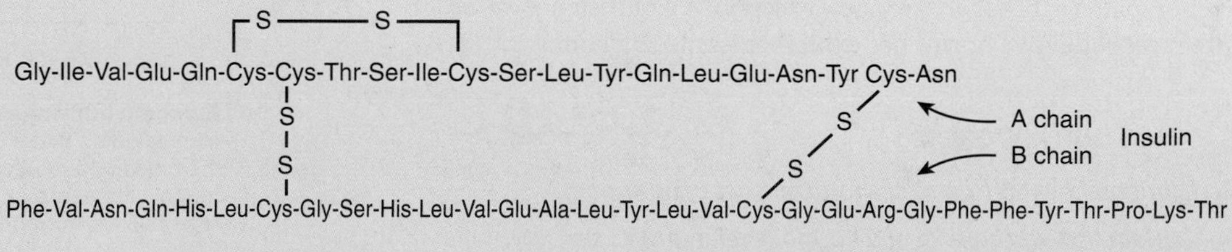
Amino Acid Derivatives	Epinephrine	Epinephrine 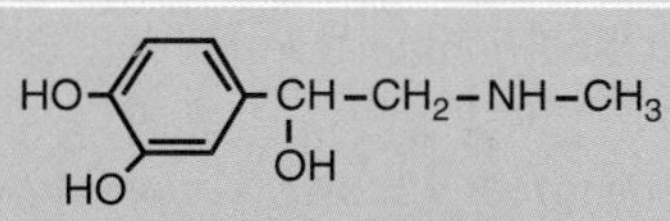

However, some water-soluble hormones are more stable in the circulation than others. In many instances, protein and peptide hormones are made more stable by having a carbohydrate attached to them and are then called glycoproteins. Other hormones are made more stable by having a modified terminal end. These modifications protect them from protease activity to a greater extent than water-soluble hormones lacking such modifications (table 17.2). In addition, some water-soluble hormones attach to binding proteins and therefore circulate in the plasma longer than free water-soluble hormones do.

Patterns of Hormone Secretion

As a result of the variation in transport and removal of lipid-soluble hormones and water-soluble hormones, the blood levels of hormones differ. In addition, blood levels of hormones are further determined by the overall pattern of secretion. The three main patterns of hormone secretion are chronic, acute, and episodic (figure 17.4):

- **Chronic hormone secretion** results in relatively constant blood levels of hormone over long periods of time and is exemplified by thyroid hormones, which vary in the blood within a small range of concentrations. Recall that thyroid hormones are lipid-soluble and thus bind to binding proteins, which also helps maintain them at chronic levels.
- **Acute hormone secretion** occurs when the hormone's concentration changes dramatically and irregularly, and its circulating levels differ with each stimulus. This secretion pattern is represented by the amino acid derivative epinephrine, which is released in large amounts in response to stress or physical exercise. In addition, because epinephrine is small and usually circulates as a free hormone, it has a short half-life, which contributes to the fact that blood levels of epinephrine drop significantly within a few minutes of its secretion.
- **Episodic hormone secretion** occurs when hormones are secreted at fairly regular intervals and concentrations. This pattern is often observed in steroid reproductive hormones, which fluctuate over a month in cyclic fashion during the human reproductive years. Additionally, because steroid hormones also often have binding proteins, they have longer half-lives than other hormones, which contributes to their relative stablity in the circulation.

In general, lipid-soluble hormones exhibit the two regular secretion patterns (chronic and episodic), whereas because of their short half-life water-soluble hormones tend to exhibit the irregular (acute) secretion pattern, but there are a few exceptions. For instance, some protein reproductive hormones exhibit episodic secretion.

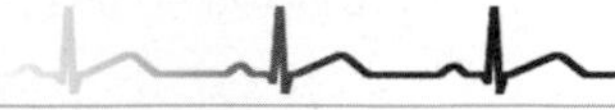

ASSESS **YOUR PROGRESS**

5. *What are the three general characteristics of hormones?*
6. *Explain how the half-life of a hormone relates to its stability.*
7. *Why do some hormones require a binding protein during transport in the blood?*
8. *What effect does a bound hormone have on the concentration of a free hormone in the blood?*

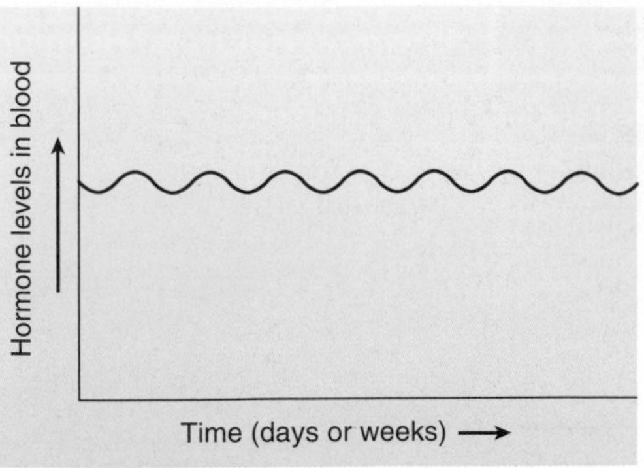
Hormone levels in blood
Time (days or weeks) ⟶

(a) **Chronic hormone secretion.** A relatively stable concentration of hormone is maintained in the circulating blood over a relatively long period. This pattern is exemplified by the thyroid hormones.

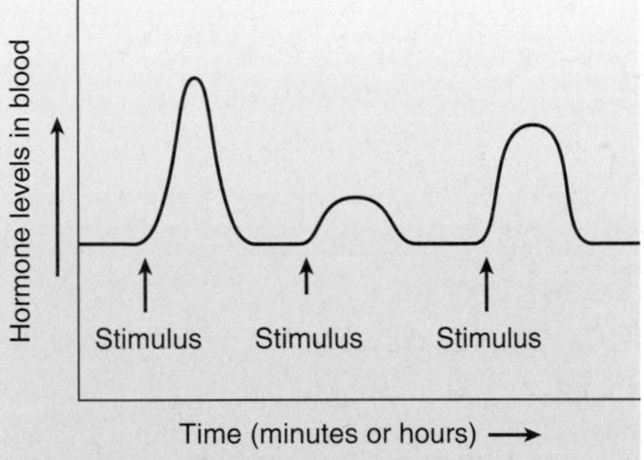
Hormone levels in blood
Stimulus
Stimulus
Stimulus
Time (minutes or hours) ⟶

(b) **Acute hormone secretion.** A hormone rapidly increases in the blood for a short time in response to a specific stimulus—for example, insulin (the blood sugar–regulating hormone) secretion following a meal. Note that the size of the stimulus arrow represents its strength. A smaller stimulus does not activate as much hormone secretion as a larger stimulus.

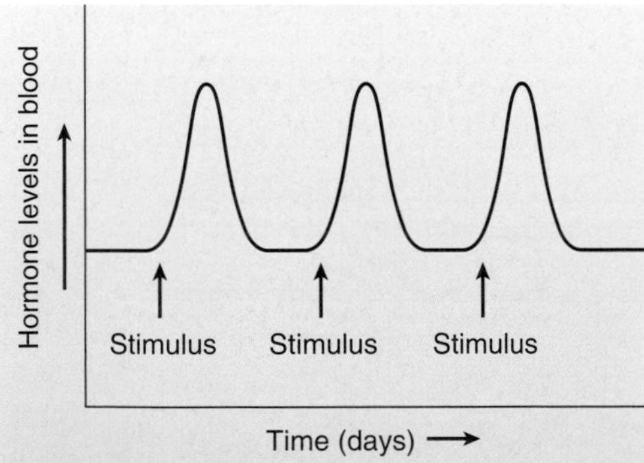
Hormone levels in blood
Stimulus
Stimulus
Stimulus
Time (days) ⟶

(c) **Episodic hormone secretion.** A hormone is stimulated so that it increases and decreases in the blood at a relatively consistent time and to roughly the same amount. Examples are the reproductive hormones regulating menstruation.

FIGURE 17.4 Patterns of Hormone Secretion

The overall pattern of hormone secretion is a result of the secretion, transport, and removal of the hormones, which is further dependent on the chemical nature of the hormones.

9. *What are the two chemical categories of hormones? Give examples of both types.*
10. *Describe how the chemical nature of a hormone affects its transport in the blood, its removal from circulation, and its half-life.*
11. *What happens to the half-life when a hormone binds to a binding protein? What kinds of hormones bind to binding proteins?*
12. *Why do organs regulated by protein hormones have fenestrated capillaries?*
13. *What kinds of activities are regulated by hormones with a short half-life? With a long half-life?*
14. *Describe chronic, acute, and episodic patterns of hormone secretion.*

17.3 Control of Hormone Secretion

LEARNING OUTCOMES

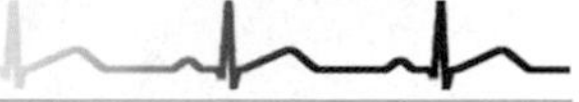

After reading this section, you should be able to

A. List and describe the three stimulatory influences on hormone secretion and give examples of each.

B. List and describe the three inhibitory influences on hormone secretion and give examples of each.

C. Describe the major mechanisms that maintain blood hormone levels.

Three types of stimuli regulate hormone release: humoral, neural, and hormonal. No matter what stimulus releases the hormone, however, the blood level of most hormones fluctuates within a homeostatic range through negative-feedback mechanisms (see chapter 1). In a few instances, positive-feedback systems also regulate blood hormone levels.

Stimulation of Hormone Release

Control by Humoral Stimuli

Blood-borne molecules can directly stimulate the release of some hormones. These molecules are referred to as **humoral stimuli** because they circulate in the blood, and the word *humoral* refers to body fluids, including blood. These hormones are sensitive to the blood levels of a particular substance, such as glucose, calcium, or sodium. When the blood level of the particular molecule changes, the hormone is released in response to the molecule's concentration (figure 17.5). For instance, if a runner has just finished a long race during hot weather, he may not produce urine for up to 12 hours after the race because his elevated concentration of blood solutes stimulates the release of a water-conservation hormone called antidiuretic hormone (ADH). Similarly, elevated blood glucose levels directly stimulate insulin secretion by the pancreas, and elevated blood potassium levels directly stimulate aldosterone release by the adrenal cortex.

Control by Neural Stimuli

The second type of hormone regulation involves **neural stimuli** of endocrine glands. Following action potentials, neurons release a neurotransmitter into the synapse with hormone-producing cells. In some cases, the neurotransmitter stimulates the cells to increase hormone secretion. Figure 17.6 illustrates the neural control of hormone secretion from cells of an endocrine gland. For example, in response to stimuli, such as stress or exercise, the sympathetic division of the autonomic nervous system (see chapter 16) stimulates the adrenal gland to secrete epinephrine and norepinephrine, which help the body respond to the stimulus. Responses include an elevated heart rate and increased blood flow through the exercising muscles. When the stimulus is no longer present, the neural

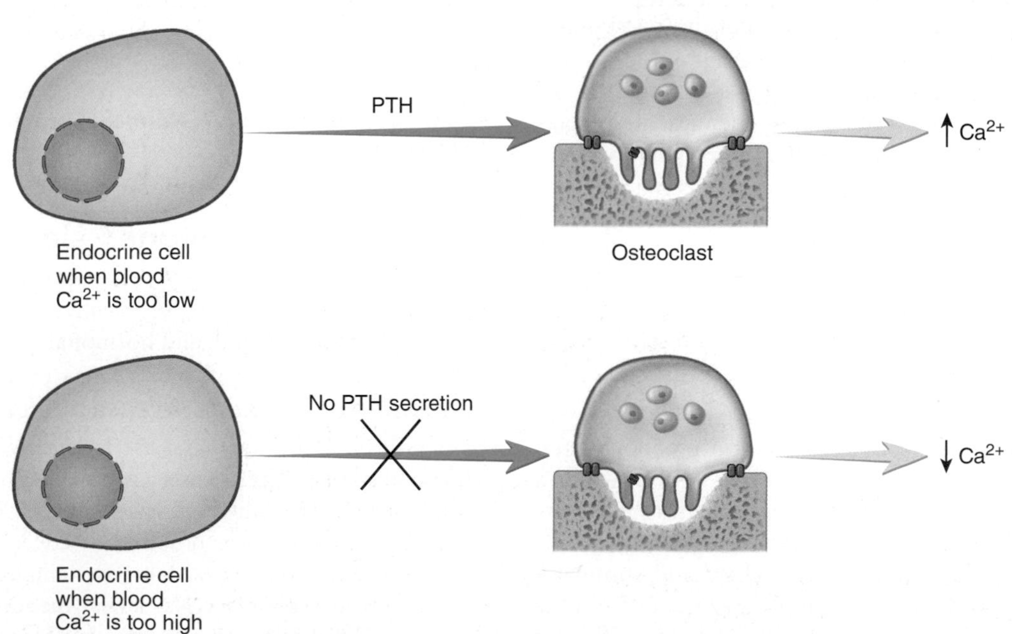
PTH
↑ Ca^{2+}
Endocrine cell when blood Ca^{2+} is too low
Osteoclast
No PTH secretion
↓ Ca^{2+}
Endocrine cell when blood Ca^{2+} is too high

FIGURE 17.5 Humoral Regulation of Hormone Secretion

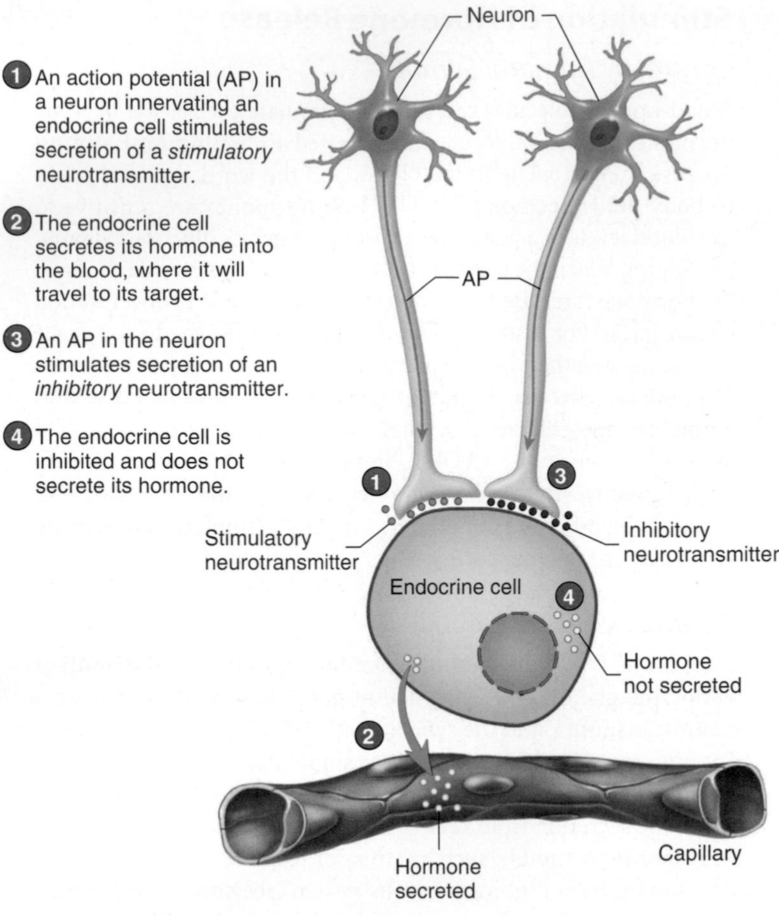
Neuron
1 An action potential (AP) in a neuron innervating an endocrine cell stimulates secretion of a *stimulatory* neurotransmitter.
2 The endocrine cell secretes its hormone into the blood, where it will travel to its target.
3 An AP in the neuron stimulates secretion of an *inhibitory* neurotransmitter.
4 The endocrine cell is inhibited and does not secrete its hormone.
AP
Stimulatory neurotransmitter
Inhibitory neurotransmitter
Endocrine cell
Hormone not secreted
Hormone secreted
Capillary

PROCESS FIGURE 17.6 Control of Hormone Secretion by Direct Neural Innervation

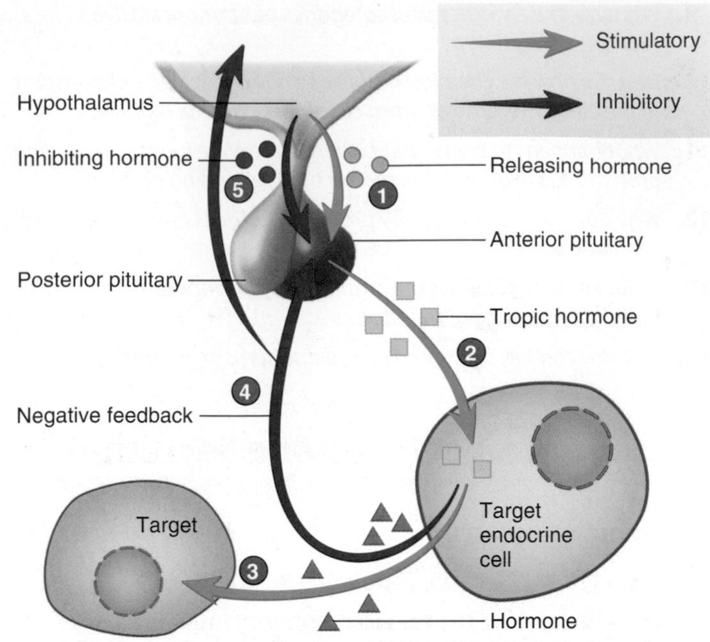
Stimulatory
Inhibitory
Hypothalamus
Inhibiting hormone
Releasing hormone
Anterior pituitary
Posterior pituitary
Tropic hormone
Negative feedback
Target
Target endocrine cell
Hormone

1 Neurons in the hypothalamus release stimulatory hormones, called releasing hormones. Releasing hormones travel in the blood to the anterior pituitary gland.

2 Releasing hormones stimulate the release of tropic hormones from the anterior pituitary, which travel in the blood to their target endocrine cell.

3 The target endocrine cell secretes its hormone into the blood, where it travels to its target and produces a response.

4 The hormone from the target endocrine cell also inhibits the hypothalamus and anterior pituitary from secreting the releasing hormone and the tropic hormone. This is negative feedback.

5 In some instances, the hypothalamus can also secrete inhibiting hormones, which prevent the secretion of anterior pituitary tropic hormones.

PROCESS FIGURE 17.7 Hormonal Regulation of Hormone Secretion

stimulation declines, and the secretion of epinephrine and norepinephrine decreases.

Some neurons secrete chemical messengers directly into the blood when they are stimulated, making these chemical messengers hormones. These hormones are neuropeptides. Some neuropeptides stimulate hormone secretion from other endocrine cells and are called **releasing hormones,** a term usually reserved for hormones from the hypothalamus.

Control by Hormonal Stimuli

The third type of regulation uses **hormonal stimuli.** It occurs when a hormone is secreted that, in turn, stimulates the secretion of other hormones (figure 17.7). The most common examples are hormones from the anterior pituitary gland, called **tropic** (trō′pik) **hormones.** Many tropic hormones are part of a complex process in which a releasing hormone from the hypothalamus stimulates the release of a tropic hormone from the pituitary gland. The pituitary tropic hormone then travels to a third endocrine gland and stimulates the release of a third hormone. For example, hormones from the hypothalamus and anterior pituitary regulate the secretion of thyroid hormones from the thyroid gland.

Inhibition of Hormone Release

Stimulating hormone secretion is important, but so is inhibiting hormone release. This process involves the same three types of stimuli: humoral, neural, and hormonal.

Inhibition of Hormone Release by Humoral Stimuli

Often when a hormone's release is sensitive to the presence of a humoral stimulus, there exists a companion hormone whose release is inhibited by the same humoral stimulus. Usually, the companion hormone's effects oppose those of the secreted hormone and counteract the secreted hormone's action. For example, to raise blood pressure (as influenced by low blood Na^+ levels), the adrenal cortex secretes the hormone aldosterone in response to low blood pressure. However, if blood pressure goes up (as influenced by high blood Na^+ levels), the

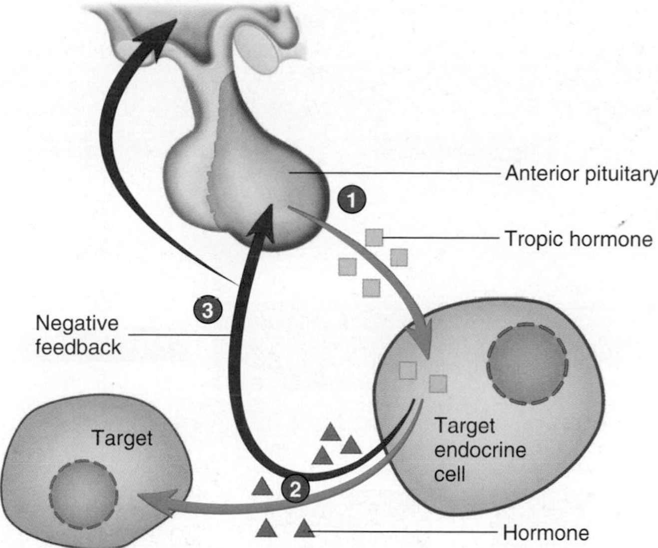
Anterior pituitary
Tropic hormone
Negative feedback
Target
Target endocrine cell
Hormone
1
2
3

(a) **Negative feedback by hormones**

1 The anterior pituitary gland secretes a tropic hormone, which travels in the blood to the target endocrine cell.

2 The hormone from the target endocrine cell travels to its target.

3 The hormone from the target endocrine cell also has a negative-feedback effect on the anterior pituitary and hypothalamus and decreases secretion of the tropic hormone.

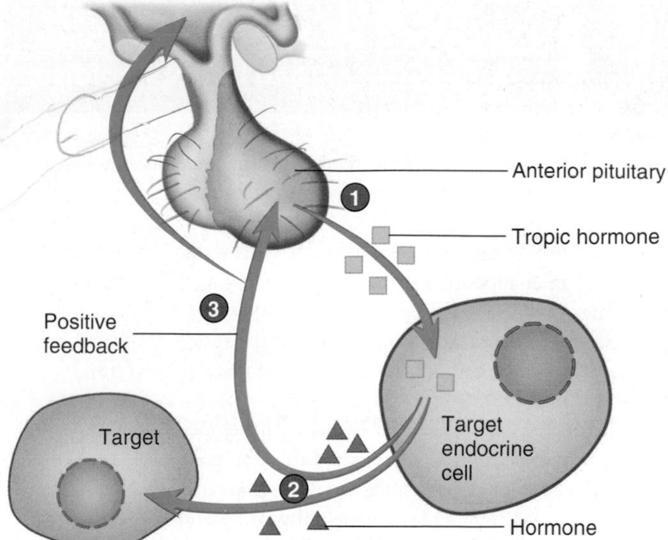
Anterior pituitary
Tropic hormone
Positive feedback
Target
Target endocrine cell
Hormone
1
2
3

(b) **Positive feedback by hormones**

1 The anterior pituitary gland secretes a tropic hormone, which travels in the blood to the target endocrine cell.

2 The hormone from the target endocrine cell travels to its target.

3 The hormone from the target endocrine cell also has a positive-feedback effect on the anterior pituitary and increases secretion of the tropic hormone.

PROCESS FIGURE 17.8 Negative and Positive Feedback

atria of the heart secrete the hormone atrial natriuretic peptide (ANP), which lowers blood pressure. Therefore, aldosterone and ANP work together to maintain homeostasis of blood pressure.

Inhibition of Hormone Release by Neural Stimuli

Neurons inhibit targets just as often as they stimulate targets. If the neurotransmitter is inhibitory, the target endocrine gland does not secrete its hormone. Hormones from the hypothalamus that prevent the secretion of tropic hormones from the pituitary gland are called **inhibiting hormones.**

Inhibition of Hormone Release by Hormonal Stimuli

Some hormones prevent the secretion of other hormones, which is a common mode of hormone regulation. For example, thyroid hormones can control their own blood levels by inhibiting their pituitary tropic hormone. Without the original stimulus, less thyroid hormone is released.

Regulation of Hormone Levels in the Blood

Two major mechanisms maintain hormone levels in the blood within a homeostatic range: negative feedback and positive feedback (see chapter 1).

1. *Negative feedback.* Most hormones are regulated by a negative-feedback mechanism, whereby the hormone's secretion is inhibited by the hormone itself once blood levels have reached a certain point and there is adequate hormone to activate the target cell. The hormone may inhibit the action of other, stimulatory hormones to prevent the secretion of the hormone in question. Thus, it is a self-limiting system (figure 17.8*a*). For example, thyroid hormones inhibit the secretion of TRH from the hypothalamus and TSH from the anterior pituitary.
2. *Positive feedback.* Some hormones, when stimulated by a tropic hormone, promote the synthesis and secretion of the tropic hormone in addition to stimulating their target cell. In turn, this stimulates further secretion of the original hormone. Thus, it is a self-propagating system (figure 17.8*b*). For example, prolonged estrogen stimulation promotes a release of luteinizing hormone.

ASSESS **YOUR PROGRESS**

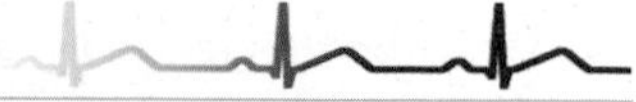

15. *Describe and give examples of the three major ways hormone secretion is stimulated and inhibited.*

16. *Is hormone secretion generally regulated by negative-feedback or positive-feedback mechanisms?*

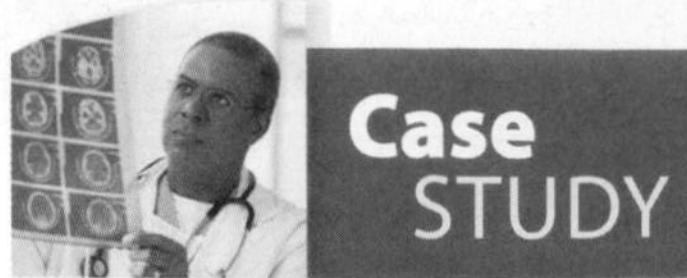

Case STUDY | Negative Feedback and Hypothyroidism

Josie owns a business and works hard to manage her employees and make time for her family. Over several months, she frequently felt weak, was often unable to concentrate, and felt cold when others did not. In addition, she began to gain weight, even though she had a small appetite. Finally, after noticing a large lump in her neck inferior and lateral to her larynx, Josie decided to see her physician. A blood sample was taken, and the results indicated low levels of thyroid hormones (hypothyroidism; see chapter 18), high levels of TSH, and low levels of iodine.

The doctor concluded that Josie had developed a goiter, or an enlarged thyroid gland. He explained that the goiter had probably formed because her dietary intake of iodine was too low over a prolonged time. Without iodine, Josie's thyroid gland was unable to synthesize thyroid hormones. Thus, in response to low thyroid hormone levels, the anterior pituitary gland continued to secrete the tropic hormone TSH, which caused the thyroid gland to keep getting bigger and bigger. In addition, the hypothalamus continued to stimulate the anterior pituitary in the absence of thyroid hormones.

Josie was treated with radioactive iodine (^{131}I) atoms, which were actively transported into her thyroid cells, where the radiation helped shrink her thyroid gland back to normal size. Subsequently, Josie had to take dietary iodine supplements and thyroid hormone supplements until her thyroid gland was able to produce thyroid hormones on its own again.

Predict 2

a. Name and explain the mechanism controlling TSH in Josie's blood. Why were TSH levels high and the levels of thyroid hormones low prior to treatment?
b. Explain why the doctor could tell that Josie's condition was not the result of a tumor in the thyroid gland.
c. What role did a lack of iodine play in Josie's condition?
d. After treatment with iodine, predict how Josie's blood levels of thyroid hormones, as well as TSH, changed.
e. Explain why Josie had to take thyroid hormone supplements. Will Josie have to do this for the rest of her life?

17.4 Hormone Receptors and Mechanisms of Action

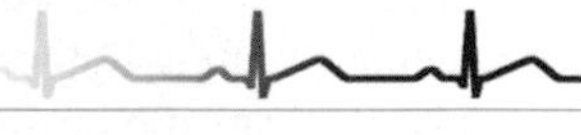

LEARNING OUTCOMES

After reading this section, you should be able to

A. **Describe the general properties of a receptor and how a target cell may decrease or increase its sensitivity to a hormone.**

B. **Explain the mechanisms of action for the two types of receptor classes.**

C. **Define *amplification* and explain how, despite small hormone concentrations, water-soluble hormones can cause rapid responses.**

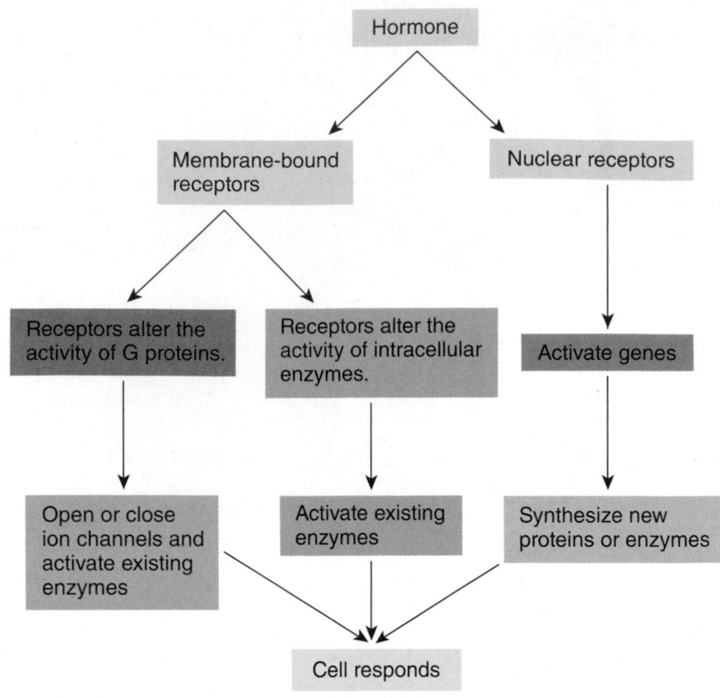
Hormone
Membrane-bound receptors
Nuclear receptors
Receptors alter the activity of G proteins.
Receptors alter the activity of intracellular enzymes.
Activate genes
Open or close ion channels and activate existing enzymes
Activate existing enzymes
Synthesize new proteins or enzymes
Cell responds

FIGURE 17.9 Overview of Responses to Hormones Binding to Their Receptors

Hormones exert their actions by binding to proteins called **receptors** (figure 17.9). A hormone can stimulate only the cells that have the receptor for that hormone. The portion of each receptor molecule where a hormone binds is called a **receptor site,** and the shape and chemical characteristics of each receptor site allow only a specific type of hormone to bind to it. The tendency for each type of hormone to bind to one type of receptor, and not to others, is called **specificity** (figure 17.10). For example, insulin binds to insulin receptors, but not to receptors for thyroid hormones. However, some hormones, such

Clinical IMPACT

Agonists and Antagonists

Drugs with structures similar to those of specific hormones may compete with those hormones for their receptors (see chapter 3). A drug that binds to a hormone receptor and activates it is called an **agonist.** A drug that binds to a hormone receptor and inhibits its action is called an **antagonist.**

For example, certain drugs mimic epinephrine and can bind to its receptors. Some of these drugs, called epinephrine agonists, activate epinephrine receptors. In fact, the drugs in asthma inhalers often work by mimicking epinephrine and causing the smooth muscle in lung bronchioles to relax. In contrast, some antistroke medications are epinephrine antagonists that prevent epinephrine-stimulated platelet aggregation and thus prevent the blockage of blood vessels.

FUNDAMENTAL **Figure**

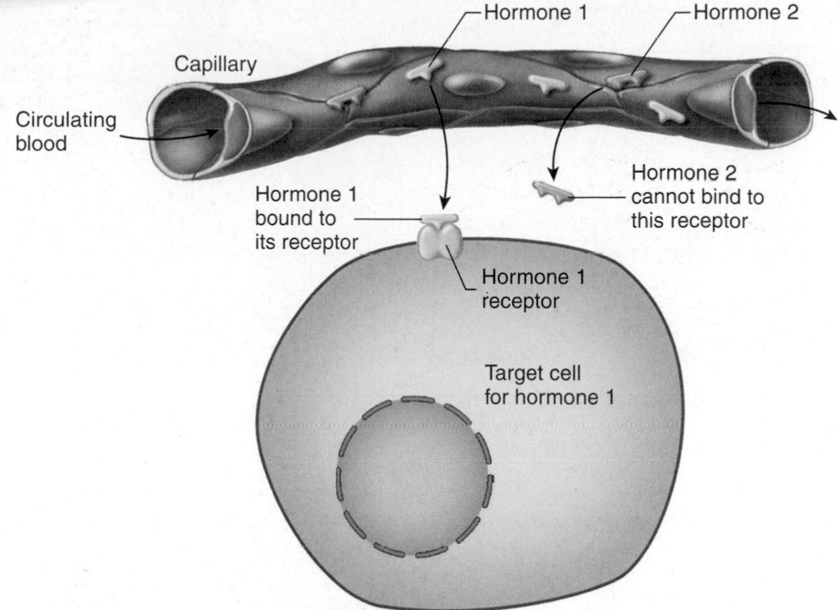
Hormone 1
Hormone 2
Capillary
Circulating blood
Hormone 2 cannot bind to this receptor
Hormone 1 bound to its receptor
Hormone 1 receptor
Target cell for hormone 1

FIGURE 17.10 Target Tissue Specificity and Response

Hormones bind (physically attach) to receptor proteins. The shape and chemical nature of each receptor site allow only certain hormones to bind. This relationship is called specificity. Additionally, in order for a target cell to respond to its hormone, the hormone must bind to its receptor.

as epinephrine, can bind to a "family" of receptors that are structurally similar. Because hormone receptors have a high affinity for the hormones that bind to them, only a small concentration of a given hormone is needed to activate a significant number of its receptors.

Decrease in Receptor Number

Target tissues' sensitivity to hormone levels can change, for various reasons. Changing the receptor number at a target ensures an optimal target tissue response to a hormone. For example, the response of some target tissues rapidly decreases over time through desensitization. This happens because the cells' nutrient and energy supplies become depleted, causing the cells to lose the ability to respond to the hormone. Densensitization occurs when the number of receptors rapidly decreases after exposure to certain hormones, a phenomenon called **down-regulation** (figure 17.11*a*). Because most receptor molecules are degraded over time, a decrease in their synthesis rate reduces the total number of receptor molecules in a cell. Often, the target cells internalize the receptors and destroy them. For example, experimental exposure of anterior pituitary cells to the reproductive hormone gonadotropin-releasing hormone (GnRH) causes the number of receptor molecules for GnRH in the pituitary gland cells to decrease dramatically after several hours after exposure to the hormone. This down-regulation causes the pituitary gland to become less sensitive to additional GnRH. Therefore, to ensure that the GnRH receptors are not down-regulated and that the pituitary gland remains responsive, the hypothalamus releases brief pulses of GnRH approximately once each hour. In this way, the reproductive system stays active and is less likely to stop working. From an organismal standpoint, reproduction is one of the body's most protected functions.

Increase in Receptor Number

In addition to down-regulation, a target tissue can also periodically increase sensitivity through up-regulation. **Up-regulation** results from an increase in the rate of receptor molecule synthesis in the target cells (figure 17.11*b*). An example of up-regulation is the increased number of receptors for luteinizing hormone (LH) in ovary cells during each menstrual cycle. Follicle-stimulating hormone (FSH) secreted by the pituitary gland increases the rate of LH receptor synthesis in ovary cells. This is important because a surge in LH will cause release of the egg. Thus, a tissue's exposure to one hormone can increase its sensitivity to a second hormone by causing the up-regulation of hormone receptors. This type of multilevel manipulation allows hormone levels to be very precisely controlled and enables the timing of certain processes to be tightly regulated.

Predict 3

Ovaries secrete the hormone estrogen in greater amounts after menstruation and a few days before ovulation. Among its many effects, estrogen causes the up-regulation of receptors in the uterus for another hormone secreted by the ovaries, called progesterone. Progesterone, which is secreted after ovulation, prepares the uterine wall for possible implantation of an embryo. Pregnancy cannot occur unless the embryo implants in the uterine wall. Predict the consequence of the ovaries' secreting too little estrogen.

Classes of Receptors

Lipid-soluble and water-soluble hormones bind to their own classes of receptors. Table 17.3 provides an overview of receptor type and mechanism of action.

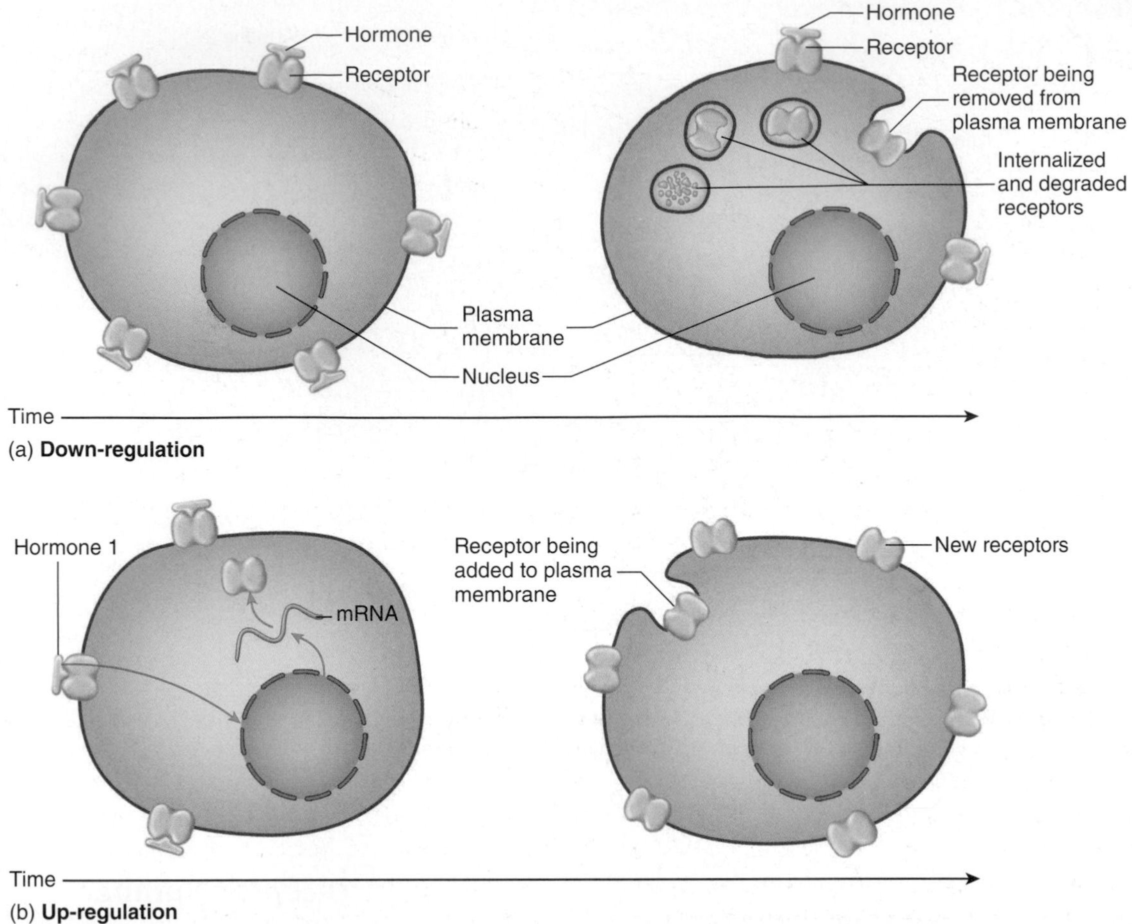
Hormone
Receptor
Plasma membrane
Nucleus
Hormone
Receptor
Receptor being removed from plasma membrane
Internalized and degraded receptors
Time
(a) Down-regulation
Hormone 1
mRNA
Receptor being added to plasma membrane
New receptors
Time
(b) Up-regulation

FIGURE 17.11 Down-Regulation and Up-Regulation of Target Cell Receptors
(*a*) Down-regulation occurs when the number of hormone receptors in a target cell decreases. Often, the target cells internalize the receptors and destroy them.
(*b*) Up-regulation occurs when the number of receptors for a hormone in a target cell increases. Often, the hormone stimulates the synthesis of receptors in the target cells.

TABLE 17.3 Hormone Receptor Types and Mechanisms of Action

Hormone Category	Receptor Type	Hormone Examples	Mechanism of Action*
Lipid-soluble	Nuclear	Steroid hormones Testosterone Estrogen Progesterone Aldosterone Cortisol Thyroid hormone Vitamin D	Bind hormone to receptor within cell followed by hormone-receptor complex attachment to hormone response element on DNA; results in synthesis of mRNA specific to the particular hormone
Water-soluble	Membrane-bound	Luteinizing hormone Follicle-stimulating hormone Thyroid-stimulating hormone Adrenocorticotropic hormone Glucagon Oxytocin Antidiuretic hormone Calcitonin Parathyroid hormone	Activate G proteins Stimulate synthesis of cAMP
		Epinephrine	Open ion channels
		Insulin Growth hormone Prolactin	Phosphorylate intracellular proteins

*Hormones often exhibit more than one mechanism of action. For simplicity, hormones are listed in only one category.

FUNDAMENTAL **Figure**

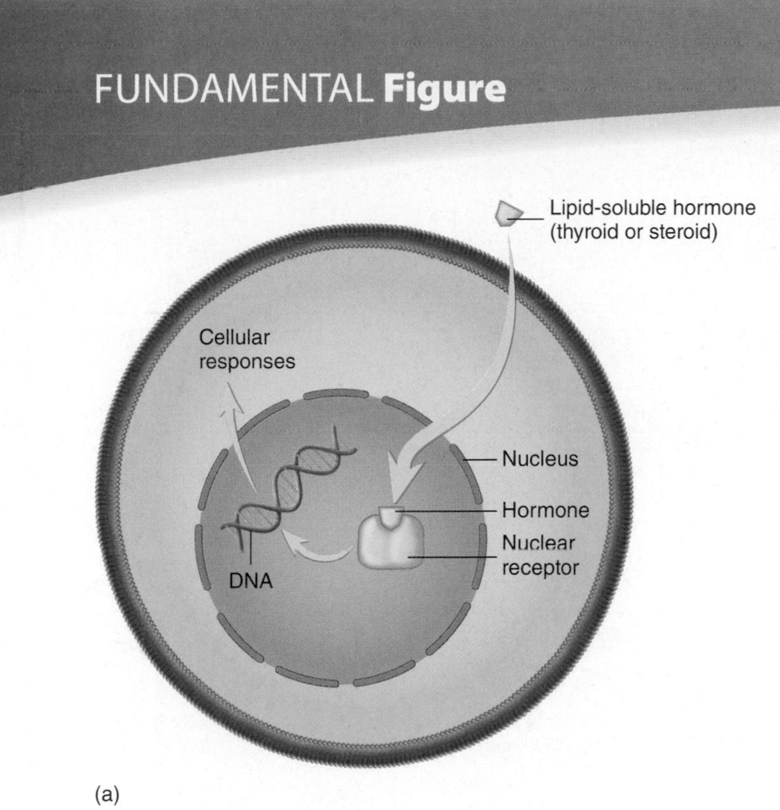
Lipid-soluble hormone
(thyroid or steroid)
Cellular
responses
Nucleus
Hormone
Nuclear
receptor
DNA

(a)

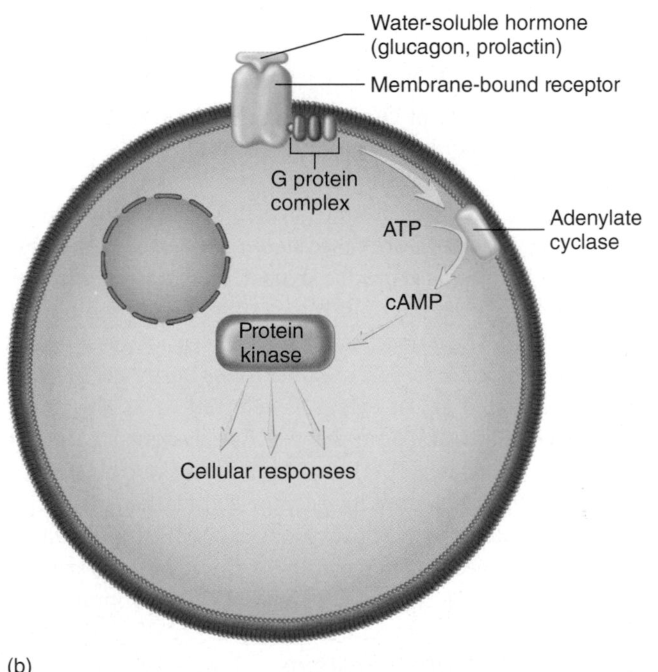
Water-soluble hormone
(glucagon, prolactin)
Membrane-bound receptor
G protein
complex
ATP
Adenylate
cyclase
cAMP
Protein
kinase
Cellular responses

(b)

FIGURE 17.12 General Comparison of Nuclear and Membrane-Bound Receptors

(*a*) *Lipid-soluble* hormones diffuse through the plasma membrane of their target cell and bind to a cytoplasmic receptor or a nuclear receptor. In the nucleus, the combination of the hormone and the receptor initiates protein synthesis, described later in this chapter. (*b*) *Water-soluble* hormones bind to the external portion of membrane-bound receptors, which are integral membrane proteins on their target cell.

1. *Lipid-soluble hormones bind to nuclear receptors.* Lipid-soluble hormones are relatively small. They diffuse through the plasma membrane and bind to **nuclear receptors,** which are most often found in the cell nucleus (figure 17.12*a*). Nuclear receptors can also be located in the cytoplasm, but then they move to the nucleus when activated. When hormones bind to nuclear receptors, the hormone-receptor complex interacts with DNA in the nucleus or with cellular enzymes to regulate the transcription of particular genes in the target tissue. Thyroid hormones and steroid hormones (testosterone, estrogen, progesterone, aldosterone, and cortisol) bind to nuclear receptors.
2. *Water-soluble hormones bind to membrane-bound receptors.* Water-soluble hormones are large molecules and cannot pass through the plasma membrane. Instead, they interact with **membrane-bound receptors,** which are proteins that extend across the plasma membrane, with their hormone-binding sites exposed on the plasma membrane's outer surface (figure 17.12*b*). When a hormone binds to a receptor on the outside of the plasma membrane, the hormone-receptor complex initiates a response inside the cell. Hormones that bind to membrane-bound receptors include proteins, peptides, and some amino acid derivatives, such as epinephrine and norepinephrine.

ASSESS **YOUR PROGRESS**

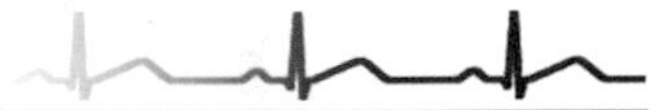

17. *What characteristics of a hormone receptor make it specific for one type of hormone?*
18. *What is down-regulation, and what may cause it to occur? Give an example of down-regulation in the body.*
19. *What is up-regulation, and what may cause it to occur? Give an example of up-regulation in the body.*
20. *What are the two classes of hormone receptors? How do they differ in the chemical category of hormones that will bind to them? Give examples of the types of hormones that bind to each type of receptor.*

Action of Nuclear Receptors

After lipid-soluble hormones diffuse across the plasma membrane and bind to their receptors, the hormone-receptor complex binds to DNA to produce a response (figure 17.13). The receptors that bind to DNA have fingerlike projections that recognize and bind to specific nucleotide sequences in the DNA called **hormone-response elements.** The combination of the hormone and its receptor forms a **transcription factor,** because, when the hormone-receptor complex binds to the hormone-response element, it activates the transcription of specific **messenger ribonucleic acid (mRNA)** molecules. The mRNA molecules then move to the cytoplasm to be translated into specific proteins at the ribosomes. The newly synthesized proteins produce the cell's response to the hormone. For example, testosterone stimulates the synthesis of the proteins that are responsible for male secondary sex characteristics, such as the formation of muscle mass and the typical male body structure. The steroid hormone aldosterone affects its target cells in the kidneys by stimulating the synthesis of proteins that increase the rate of Na^+ and K^+ transport. The result is a reduction in the amount of Na^+ and an increase in the amount of K^+ lost in the urine. Other hormones

1. Lipid-soluble hormones diffuse through the plasma membrane.
2. Lipid-soluble hormones bind to cytoplasmic receptors and travel to the nucleus or bind to nuclear receptors. Some lipid-soluble hormones bind receptors in the cytoplasm and then move into the nucleus.
3. The hormone-receptor complex binds to a hormone response element on the DNA, acting as a transcription factor.
4. The binding of the hormone-receptor complex to DNA stimulates the synthesis of messenger RNA (mRNA), which codes for specific proteins.
5. The mRNA leaves the nucleus, passes into the cytoplasm of the cell, and binds to ribosomes, where it directs the synthesis of specific proteins.
6. The newly synthesized proteins produce the cell's response to the lipid-soluble hormones—for example, the secretion of a new protein.

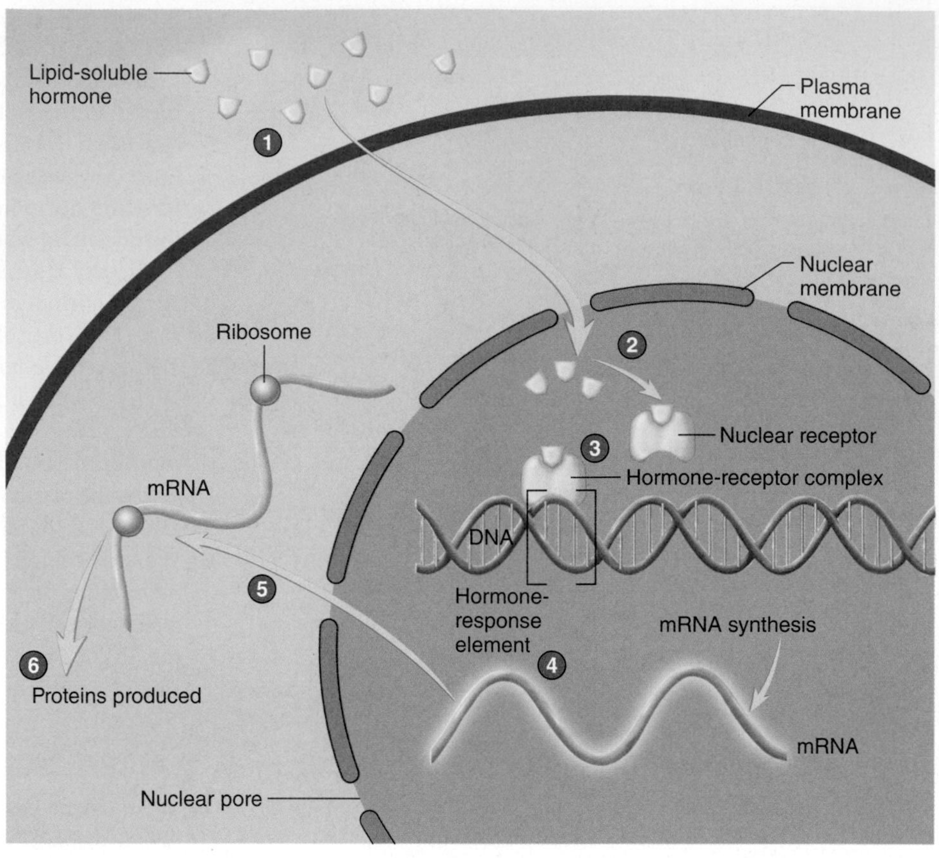

Lipid-soluble hormone
Plasma membrane
1
Nuclear membrane
Ribosome
2
Nuclear receptor
3
Hormone-receptor complex
mRNA
DNA
5
Hormone-response element
mRNA synthesis
4
6
Proteins produced
mRNA
Nuclear pore

PROCESS FIGURE 17.13 AP|R **Nuclear Receptor Model**

that produce responses through nuclear receptor mechanisms include thyroid hormones and vitamin D.

Target cells that synthesize new protein molecules in response to hormonal stimuli normally have a latent period of several hours between the time the hormones bind to their receptors and the time responses are observed. During this latent period, mRNA and new proteins are synthesized. Hormone-receptor complexes are eventually degraded within the cell, limiting the length of time hormones influence the cells' activities, and the cells slowly return to their previous functional states.

ASSESS YOUR PROGRESS

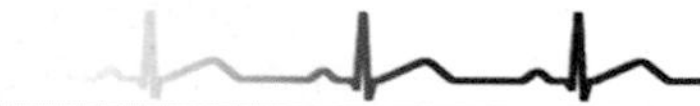

21. *Describe how a hormone that crosses the plasma membrane interacts with its receptor and how it affects protein synthesis.*
22. *Why is there normally a latent period between the time a hormone binds to its receptor and the time responses are observed?*
23. *What eventually limits the processes activated by the nuclear receptor mechanism?*

Membrane-Bound Receptors and Signal Amplification

Membrane-bound receptors have peptide chains that are anchored in the phospholipid bilayer of the plasma membrane (see chapter 3). Membrane-bound receptors activate responses in two ways: (1) Some receptors alter the activity of G proteins at the inner surface of the plasma membrane. (2) Other receptors directly alter the activity of intracellular enzymes. These intracellular pathways elicit specific responses in cells, including the production of **intracellular mediators.** Some intracellular mediators are called *second messengers*. An intracellular mediator is a chemical produced inside a cell once a hormone or another chemical messenger binds to certain membrane-bound receptors (table 17.4). The intracellular mediator then activates specific cellular processes inside the cell in response to the hormone. In some cases, this coordinated set of events is referred to as a **second-messenger system.** For example, in chapter 11 we discussed that cyclic adenosine monophosphate (cAMP) is a common second messenger produced when a ligand binds to its receptor. Rather than the ligand entering the cell to activate a cellular process, cAMP stimulates the cellular process. This mechanism is usually employed by water-soluble hormones that are unable to cross the target cell's membrane.

Receptors That Activate G Proteins

Many membrane-bound receptors produce responses through the action of G proteins (see figures 3.11 and 17.12). G proteins consist of three subunits; from largest to smallest, they are called alpha (α), beta (β), and gamma (γ); figure 17.14*a*, *step 1*). The G proteins are so named because one of the subunits binds to guanine nucleotides. In the inactive state, a guanine diphosphate (GDP) molecule is bound to the α subunit of each G protein. In the active state, guanine triphosphate (GTP) is bound to the α subunit.

TABLE 17.4 Common Intracellular Mediators

Intracellular Mediator	Example of Cell Type	Example of Response
Cyclic guanine monophosphate (cGMP)	Kidney cells	Increased Na^+ and water excretion by the kidneys
Cyclic adenosine monophosphate (cAMP)	Liver cells	Increased breakdown of glycogen and release of glucose into the circulatory system
Calcium ions (Ca^{2+})	Smooth muscle cells	Contraction of smooth muscle cells
Inositol triphosphate (IP_3)	Smooth muscle cells	Contraction of certain smooth muscle cells in response to epinephrine
Diacylglycerol (DAG)	Smooth muscle cells	Contraction of certain smooth muscle cells in response to epinephrine
Nitric oxide (NO)	Smooth muscle cells	Relaxation of smooth muscle cells of blood vessels, resulting in vasodilation

After a hormone binds to the receptor on the outside of a cell, the receptor changes shape (figure 17.14*a*, *step 2*). As a result, the receptor binds to a G protein on the inner surface of the plasma membrane, and GDP is released from the α subunit. Guanine triphosphate (GTP) binds to the α subunit, thereby activating it (figure 17.14*a*, *step 3*). As a result, G proteins separate from the receptor, and the activated α subunit separates from the β and γ subunits. The activated α subunit can alter the activity of molecules within the plasma membrane

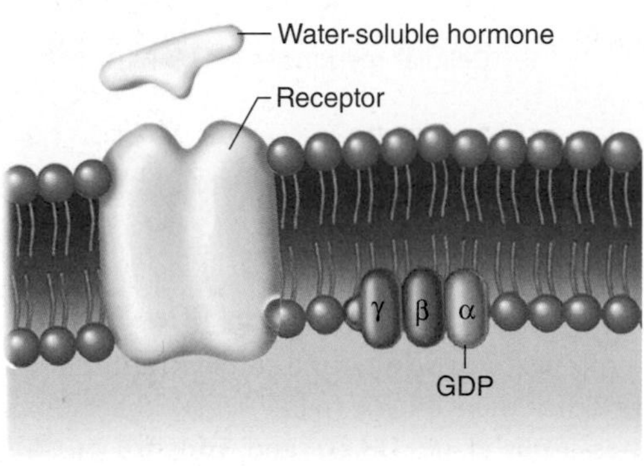
Water-soluble hormone
Receptor
γ
β
α
GDP

1 Before the hormone binds to its receptor, the G protein consists of three subunits, with GDP attached to the α subunit, and freely floats in the plasma membrane.

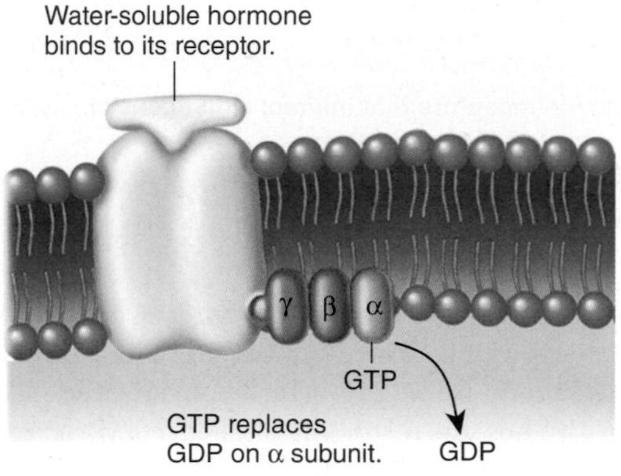
Water-soluble hormone binds to its receptor.
γ
β
α
GTP
GTP replaces GDP on α subunit.
GDP

2 After the hormone binds to its membrane-bound receptor, the receptor changes shape, and the G protein binds to it. GTP replaces GDP on the α subunit of the G protein.

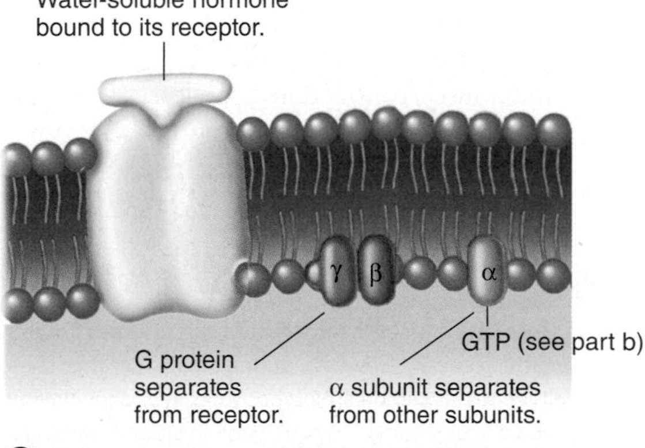
Water-soluble hormone bound to its receptor.
γ
β
α
G protein separates from receptor.
GTP (see part b)
α subunit separates from other subunits.

3 The G protein separates from the receptor. The GTP-linked α subunit activates cellular responses, which vary among target cells.

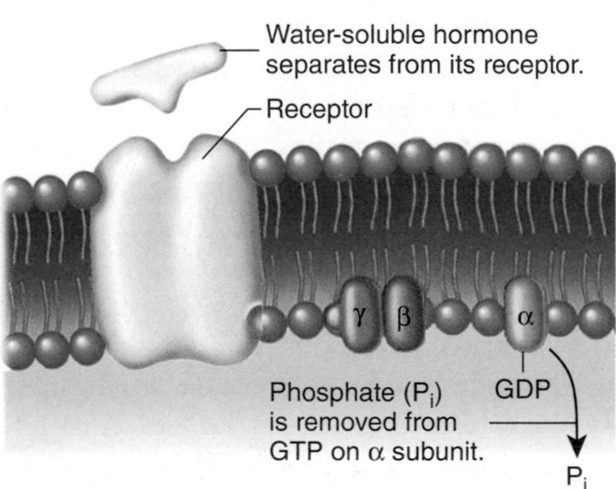
Water-soluble hormone separates from its receptor.
Receptor
γ
β
α
Phosphate (P_i) is removed from GTP on α subunit.
GDP
P_i

4 When the hormone separates from the receptor, additional G proteins are no longer activated. Inactivation of the α subunit occurs when phosphate (P_i) is removed from the GTP, leaving GDP bound to the α subunit.

(a) **Membrane-bound receptors activating G proteins**

PROCESS FIGURE 17.14 APR **Membrane-Bound Receptors Activating G Proteins and Interacting with Adenylate Cyclase**

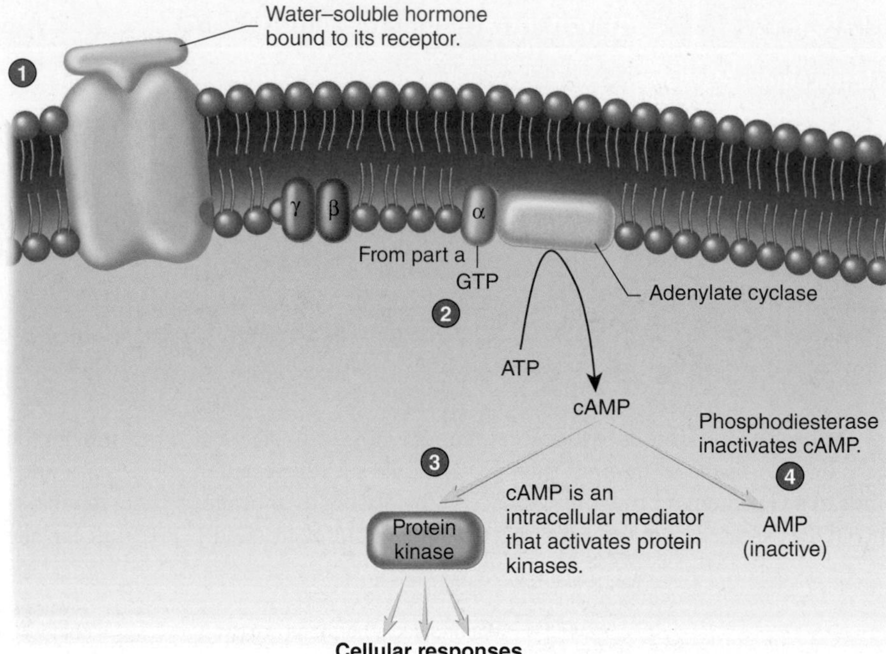
1 After a water-soluble hormone binds to its receptor, the G protein is activated.
2 The activated α subunit, with GTP bound to it, binds to and activates an adenylate cyclase enzyme, so that it converts ATP to cAMP.
3 The cAMP can activate protein kinase enzymes, which phosphorylate specific enzymes activating them. The chemical reactions catalyzed by the activated enzymes produce the cell's response.
4 Phosphodiesterase enzymes inactivate cAMP by converting cAMP to AMP.
Water–soluble hormone bound to its receptor.
γ
β
α
From part a
GTP
Adenylate cyclase
ATP
cAMP
Phosphodiesterase inactivates cAMP.
Protein kinase
cAMP is an intracellular mediator that activates protein kinases.
AMP (inactive)
Cellular responses

(b) **Membrane-bound receptors that interact with adenylate cyclase**

PROCESS FIGURE 17.14 (continued)

or inside the cell, thus producing cellular responses. After a short time, the activated α subunit is turned off, because the G protein removes a phosphate group from GTP, which converts it to GDP (figure 17.14*a*, *step 4*). Thus, the α subunit is called a GTPase. The α subunit then recombines with the β and γ subunits.

G Proteins That Interact with Adenylate Cyclase

Activated α subunits of G proteins can alter the activity of enzymes inside the cell membrane. For example, activated α subunits can influence the rate of cAMP formation by activating or inhibiting **adenylate cyclase** (a-den′i-lāt sī′ klās), an enzyme that converts ATP to cAMP (figure 17.14*b*). Cyclic AMP functions as a second messenger, an intracellular mediator that carries out cellular metabolic processes in response to hormonal activation. For example, cAMP binds to protein kinases and activates them. **Protein kinases** are enzymes that regulate the activity of other enzymes by attaching phosphates to them, a process called **phosphorylation.** Depending on the enzyme, phosphorylation increases or decreases the enzyme's activity. The amount of time cAMP is present to produce a response in a cell is limited. An enzyme in the cytoplasm, called **phosphodiesterase** (fos′fō-dī-es′ter-ās), breaks down cAMP to AMP. Once cAMP levels drop, the enzymes in the cell are no longer stimulated.

Cyclic AMP can elicit many different responses in the body, because each cell type possesses a unique set of enzymes. For example, the hormone glucagon increases blood glucose levels when it binds to receptors on the surface of liver cells. Binding to the receptor activates G proteins and causes an increase in cAMP synthesis. Cyclic AMP then stimulates the activity of enzymes that break down glycogen into glucose for release from liver cells (figure 17.14*b*).

G Proteins That Activate Other Intracellular Mediators

G proteins can also alter the concentration of intracellular mediators other than cAMP (table 17.4). For example, **diacylglycerol** (dī-as-il-glis′er-ol; **DAG**) and **inositol** (in-ō′si-tōl, in-ō′si-tol) **triphosphate** (**IP_3**) are intracellular mediators that are influenced by G proteins (figure 17.15). Epinephrine binds to certain membrane-bound receptors in some types of smooth muscle. The combination activates a G protein mechanism, which in turn increases the activity of the enzyme phospholipase C. Phospholipase C converts phosphoinositol diphosphate (PIP_2), a constituent of the plasma membrane, to DAG and IP_3, which are released into the cytosol. DAG activates enzymes that synthesize prostaglandins, which increase smooth muscle contraction. IP_3 releases Ca^{2+} from the endoplasmic reticulum or opens Ca^{2+} channels in the plasma membrane, allowing the ions to enter the cytoplasm and increase the contraction of the smooth muscle cells.

Predict 4

As long as the smooth muscle cells in the airways of the lungs are relaxed, breathing is easy. However, during asthma attacks, these smooth muscle cells contract, and breathing becomes very difficult. Some of the drugs used to treat asthma increase cAMP in smooth muscle cells. Explain some of the different ways in which these drugs work.

G Proteins That Open Ion Channels

Some activated α subunits of G proteins can combine with ion channels, causing them to open or close (figure 17.16). For example, epinephrine activates α subunits to open Ca^{2+} channels in smooth

1. Epinephrine binds to its receptor in the smooth muscle plasma membrane.
2. The G protein is activated. The activated α subunit with GTP bound to it separates from the γ and β subunits.
3. The activated subunit then binds with phospholipase C, which acts on phosphoinositol diphosphate (PIP_2) and produces inositol triphosphate (IP_3) and diacylglycerol (DAG).
4. IP_3 releases Ca^{2+} from the endoplasmic reticulum or opens Ca^{2+} channels in the plasma membrane. Calcium ions then regulate enzyme activity.
5. DAG regulates enzymes such as protein kinases and those that synthesize prostaglandin. These responses increase smooth muscle contraction.

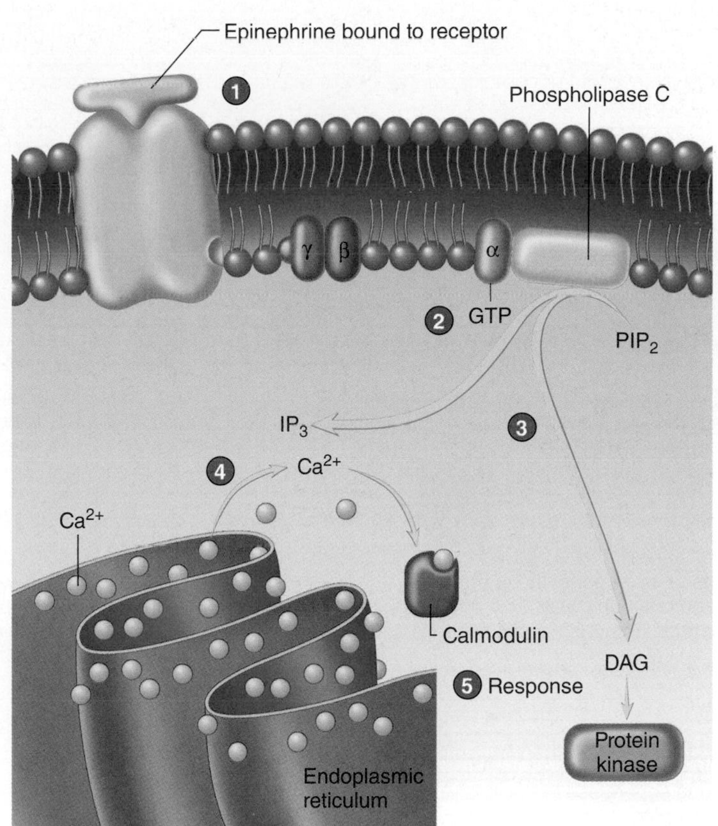
Epinephrine bound to receptor
1
Phospholipase C
γ
β
α
2 GTP
PIP2
IP3
3
4
Ca2+
Ca2+
Calmodulin
DAG
5 Response
Protein kinase
Endoplasmic reticulum

PROCESS FIGURE 17.15 AP|R **Membrane-Bound Receptors Activating G Proteins to Increase the Synthesis of DAG and IP_3**
Epinephrine receptors in some smooth muscle cells are associated with G proteins.

muscle cells, allowing Ca^{2+} to move into those cells. The Ca^{2+} ions combine with calmodulin (kal-mod′ū-lin) proteins, and the calcium-calmodulin complexes activate enzymes that cause the smooth muscle cells to contract (figure 17.16, *steps 1 and 2*; see figure 9.24). After a short time, the activated α subunit is inactivated, because GTP is converted to GDP, and muscle contraction ceases. The α subunit then recombines with the β and γ subunits (figure 17.16, *steps 3 and 4*).

Receptors That Directly Activate Intracellular Mediators

Cyclic guanine (gwahn′ēn) **monophosphate (cGMP),** an intracellular mediator, is synthesized in response to a hormone binding to a membrane-bound receptor (figure 17.17). The hormone binds to its receptor, and the combination activates an enzyme called **guanylate cyclase** (gwahn′i-lāt sī′klās) located at the inner surface of the plasma membrane. The guanylate cyclase converts guanine triphosphate (GTP) to cGMP and two inorganic phosphate groups. The cGMP molecules then combine with specific enzymes in the cytoplasm of the cell and activate them. In turn, the activated enzymes produce the cell's response to the hormone. For example, atrial natriuretic hormone, secreted by the heart atria, binds with its receptor in the plasma membrane of kidney cells. The result is an increase in the rate of cGMP synthesis at the inner surface of the plasma membrane (figure 17.17). Cyclic GMP influences the action of enzymes in the kidney cells, which increases the kidneys' rate of Na^+ and water excretion (see chapter 26). The cGMP is present in the cell for only a limited amount of time, because phosphodiesterase breaks down cGMP to GMP. Consequently, the length of time a hormone increases cGMP synthesis and has an effect on a cell is brief, once the hormone is no longer present.

Receptors That Phosphorylate Intracellular Proteins

Some hormones bind to membrane-bound receptors, and the portion of the receptor on the inner surface of the plasma membrane acts as a kinase enzyme that phosphorylates several specific proteins (figure 17.18). Some of the phosphorylated proteins are part of the membrane-bound receptor; others are in the cytoplasm of the cell. The phosphorylated proteins influence the activity of other enzymes in the cytoplasm. For example, insulin binds to its membrane-bound receptor, resulting in the phosphorylation of parts of the receptor on the inner surface of the plasma membrane and the phosphorylation of certain other intracellular proteins. The phosphorylated proteins enable the very powerful effect of insulin to allow glucose entry into cells.

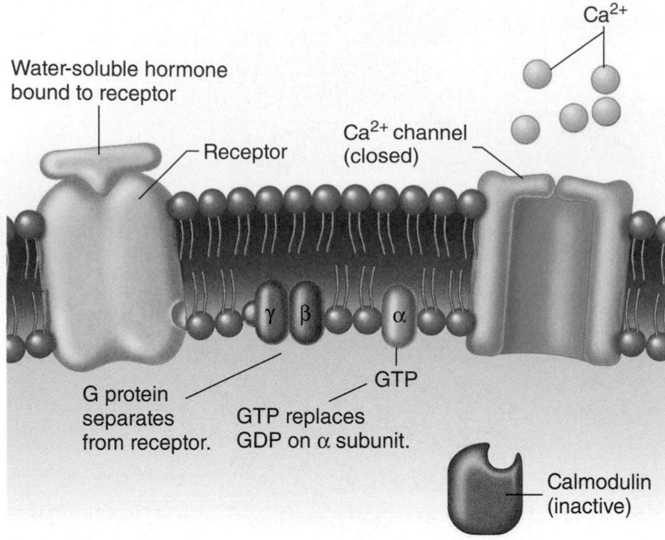
Ca2+
Water-soluble hormone bound to receptor
Receptor
Ca2+ channel (closed)
γ
β
α
G protein separates from receptor.
GTP
GTP replaces GDP on α subunit.
Calmodulin (inactive)

1 After a hormone binds to its receptor, the G protein is activated. The activated α subunit with GTP bound to it separates from the γ and β subunits.

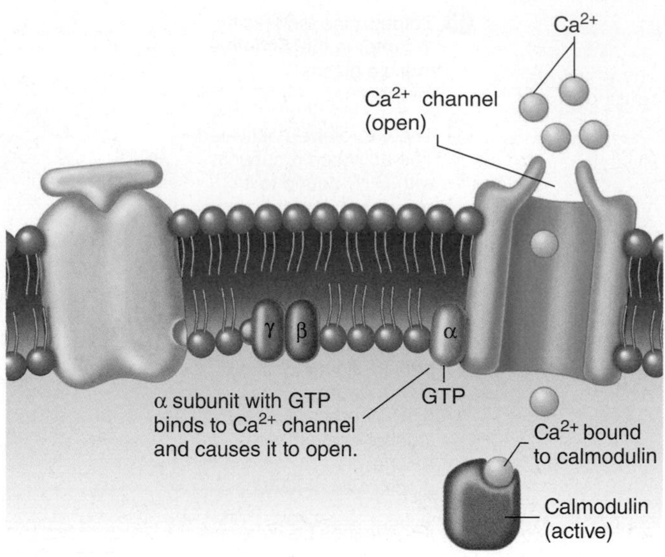
Ca2+
Ca2+ channel (open)
γ
β
α
α subunit with GTP binds to Ca2+ channel and causes it to open.
GTP
Ca2+ bound to calmodulin
Calmodulin (active)

2 The α subunit, with GTP bound to it, binds to the Ca^{2+} channel, and the combination causes the Ca^{2+} channel to open. The Ca^{2+} diffuses into the cell and combines with calmodulin. The combination of Ca^{2+} with calmodulin produces the cell's response to the hormone.

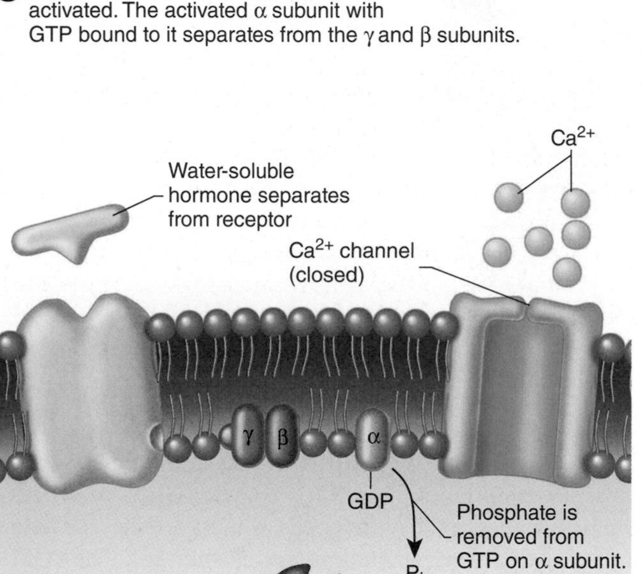
Ca2+
Water-soluble hormone separates from receptor
Ca2+ channel (closed)
γ
β
α
GDP
Phosphate is removed from GTP on α subunit.
Pi
Calmodulin (inactive)

3 Phosphate is removed from the GTP bound to the α subunit, leaving GDP bound to the α subunit. The α subunit can no longer stimulate a cellular response; it separates from the Ca^{2+} channel, and the channel closes.

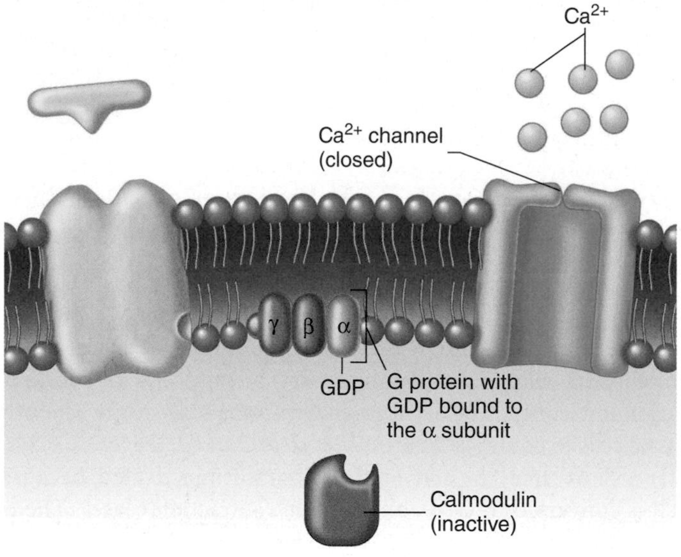
Ca2+
Ca2+ channel (closed)
γ
β
α
GDP
G protein with GDP bound to the α subunit
Calmodulin (inactive)

4 The α subunit recombines with the γ and β subunits.

FIGURE 17.16 G Proteins Opening Ion Channels

Receptors That Directly Alter the Activity of Intracellular Enzymes

Some hormones bind to membrane-bound receptors and directly change the activity of an intracellular enzyme. The altered enzyme activity regulates the synthesis of intracellular mediators or results in the phosphorylation of intracellular proteins. The intracellular mediators or phosphorylated proteins activate processes that produce the cells' response to the chemical signals. Intracellular enzymes controlled by membrane-bound receptors can be part of the receptor, or they can be separate molecules. The intracellular mediators act as chemical signals that move from the enzymes that produce them into the cytoplasm of the cell, where they activate the processes that produce the cell's response. For example, the hormone glucagon activates enzymes that release glucose into the circulation from cells within the liver.

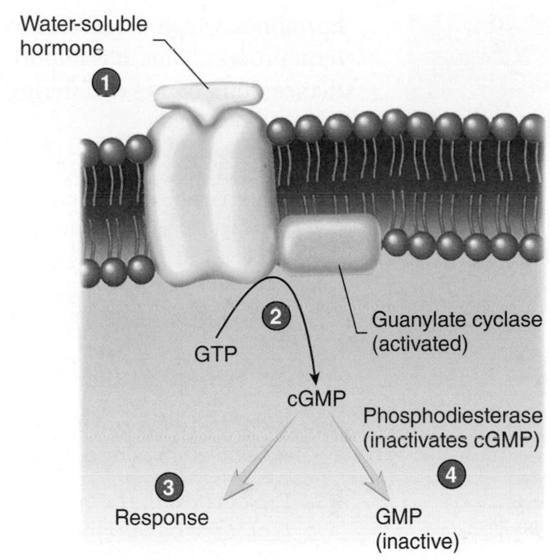
1 The water-soluble hormone binds to its receptor.
2 At the inner surface of the plasma membrane, guanylate cyclase is activated to produce cGMP from GTP.
3 Cyclic GMP is an intracellular mediator that alters the activity of other intracellular enzymes to produce the response of the cell.
4 Phosphodiesterase is an enzyme that converts cGMP to inactive GMP.
Water-soluble hormone
Guanylate cyclase (activated)
GTP
cGMP
Phosphodiesterase (inactivates cGMP)
Response
GMP (inactive)

PROCESS FIGURE 17.17 Membrane-Bound Receptor Directly Activating an Intracellular Mediator

Some second messengers are produced when the hormone binds to its receptor, which activates the second messenger–producing enzyme.

Signal Amplification

The rate and magnitude at which a hormone's response is elicited are determined by its mechanism of action at the receptor. Nuclear receptors work by activating protein synthesis, which for some hormones can take several hours (see "Action of Nuclear Receptors" earlier in this section). However, hormones that stimulate the synthesis of second messengers can produce an almost instantaneous response, because the second messenger influences existing enzymes. In other words, the response proteins are already present. Additionally, each receptor produces thousands of second messengers, leading to a cascade effect and ultimately **amplification** of the hormonal signal. With amplification, a single hormone activates many second messengers, each of which activates enzymes that produce an enormous amount of final product (figure 17.19). The efficiency of this second-messenger amplification is virtually unparalleled in the body and can be thought of as an "army of molecules" launching an offensive. In a war, the general gives the signal to attack, and thousands of soldiers carry out the order. The general alone could not kill thousands of enemies. Likewise, one hormone could not single-handedly produce millions of final products within a few seconds. However, with amplification, one hormone has an army of molecules working simultaneously to produce the final products.

Both nuclear receptor and membrane-bound receptor hormone systems are effective, but each is more suited to one type of response than another. For example, the reason epinephrine is effective in a fight-or-flight situation is that it can turn on the target cell responses within a few seconds. If running away from an immediate threat depended on producing new proteins, a process that can take several hours, many of us would have already perished. On the other hand, pregnancy maintenance is mediated by steroids, long-acting

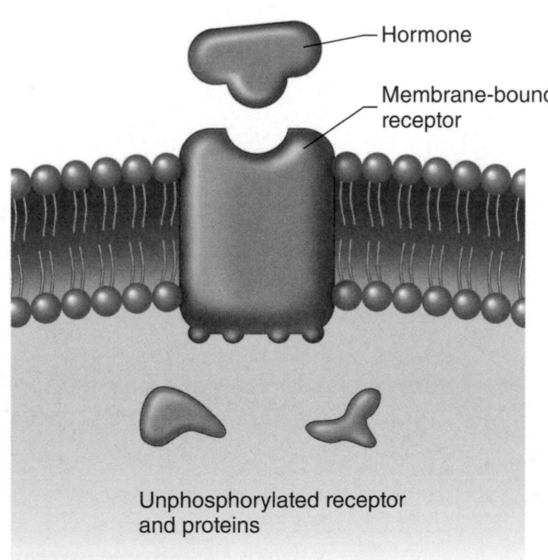
Hormone
Membrane-bound receptor
Unphosphorylated receptor and proteins
1 The membrane-bound receptor and other proteins have sites that can be phosphorylated at the inner surface of the plasma membrane. When the receptor is not bound to a hormone, these sites remain unphosphorylated.

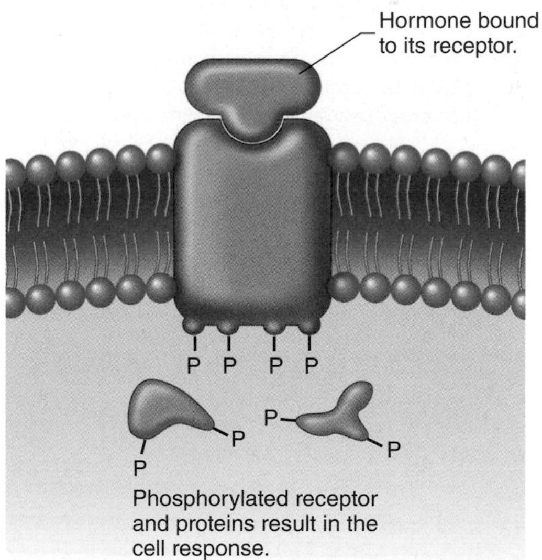
Hormone bound to its receptor.
P P P P
P
P
P
P
Phosphorylated receptor and proteins result in the cell response.
2 The hormone binds to its receptor site on the outside of the plasma membrane. The receptor acts as an enzyme that phosphorylates the sites on the receptor and associated proteins. These phosphorylated proteins produce a response inside the cell.

PROCESS FIGURE 17.18 Membrane-Bound Receptors That Phosphorylate Intracellular Proteins

FUNDAMENTAL **Figure**

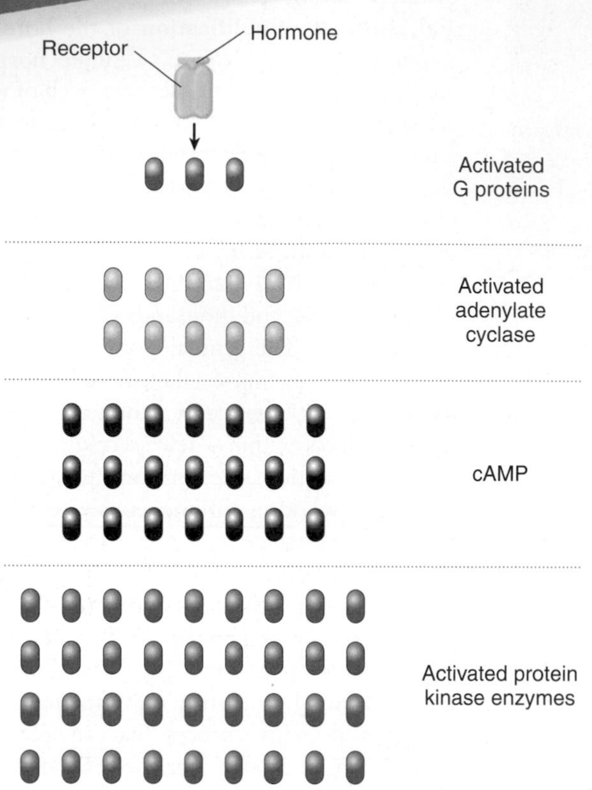
Receptor
Hormone
Activated G proteins
Activated adenylate cyclase
cAMP
Activated protein kinase enzymes

FIGURE 17.19 Cascade Effect
The combination of a hormone with a membrane-bound receptor activates several G proteins. The G proteins, in turn, activate many inactive adenylate cyclase enzymes, which cause the synthesis of a large number of cAMP molecules. The large number of cAMP molecules, in turn, activate many inactive protein kinase enzymes, which produce a rapid and amplified response.

hormones, which is reflected by the fact that pregnancy is a long-term process. Thus, it is important for our body to have hormones that can function over differing time scales.

Predict 5

Of membrane-bound receptors and nuclear receptors, which is better adapted for mediating a response that lasts a considerable length of time, and which is better for mediating a response with a rapid onset and short duration? Explain why.

ASSESS YOUR PROGRESS

24. *What two ways can a membrane-bound receptor use to activate cellular response?*
25. *Explain how the hormone-receptor complex can alter the G proteins on the inner surface of the plasma membrane. Which subunit of the G protein alters the activity of molecules inside the plasma membrane or inside the cell?*
26. *List four intracellular mediators affected by G proteins.*
27. *Describe how G proteins can alter the permeability of the plasma membrane and how they can alter the synthesis of an intracellular mediator, such as cAMP. Give examples.*
28. *Describe how a hormone can combine with a membrane-bound receptor, directly change enzyme activity inside the cell, and increase phosphorylation of intracellular proteins. Give examples.*
29. *What limits the activity of intracellular mediators, such as cGMP, and phosphorylated proteins?*
30. *Explain the cascade effect for the second-messenger model of hormone action. Does the second-messenger amplification produce a slow or rapid response?*

Answer

Learn to Predict **From page 569**

After using anabolic steroids, Liu Dan's muscles increased in size, but he also experienced some unintended changes, including slight breast development, reduced testes size, and mood swings. We learned in this chapter that steroid hormones are all derived from cholesterol, a type of lipid, and are thus lipid-soluble. We also know that hormones, chemical messengers produced by the endocrine system, travel through the body via the bloodstream until they arrive at their target tissues. The cells of target tissues possess specific receptors that the hormones bind to and initiate changes in cellular metabolism or cell growth. Steroid hormones, because they are lipid-soluble, bind to intracellular receptors in the cell's cytoplasm or nucleus. Once the steroid hormone has bound to its receptor, the hormone-receptor complex stimulates increased gene expression and therefore increased protein production. In muscle cells, this leads to the increased muscle mass that Liu Dan experienced.

However, other tissues of the body contain receptors that can also bind the anabolic steroids, so using anabolic steroids can cause some unintended effects. The anabolic steroids Liu Dan used also led to the abnormal breast tissue growth, decreased testes size, and behavioral changes. Thus, although it may be tempting to take anabolic steroids to increase muscle mass, the risk for side effects is simply too great.

Answers to the rest of this chapter's Predict questions are in Appendix G.

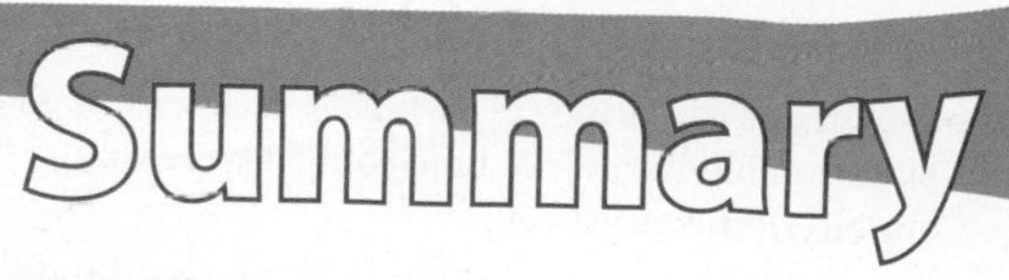
Summary

17.1 Principles of Chemical Communication (p. 570)

Classes of Chemical Messengers

1. The four classes of chemical messengers are autocrine, paracrine, neurotransmitter, and endocrine.
2. Endocrine chemical messengers are called hormones.

Characteristics of the Endocrine System

1. The endocrine system includes glands and specialized endocrine cells that secrete hormones into the bloodstream.
2. A hormone is a chemical messenger that is secreted into the blood, travels to a distant target tissue, and binds to specific receptors to produce a coordinated set of events in that target tissue.

Comparison of the Nervous and Endocrine Systems

1. The endocrine system and the nervous system are closely related.
   - They share anatomical structures in the brain.
   - They share molecules that are both neurotransmitters and hormones.
   - They cooperate to regulate important processes.
   - They both have chemical messengers that bind to the same receptor type.
2. The endocrine system and the nervous system have important differences.
   - Neurotransmitters deliver their chemical messengers directly to their target, whereas hormones travel in the bloodstream.
   - The endocrine system is slower than the nervous system.
   - The endocrine system has longer-lasting effects than the nervous system.

17.2 Hormones (p. 572)

General Characteristics of Hormones

1. Hormones have several characteristics in common: stability, communication, and distribution.
   - The length of time a hormone is active in the circulation is termed its half-life.
   - The half-life of some hormones is prolonged because they circulate in the blood bound to binding proteins.
2. Hormones not bound to binding proteins are called free hormones, and they can interact with their receptor.

Chemical Nature of Hormones

1. There are two chemical categories of hormones: lipid-soluble and water-soluble.
2. Lipid-soluble hormones include steroids, thyroid hormones, and some fatty acid derivatives.
   - Most lipid-soluble hormones are transported bound to binding proteins. Thus, their half-life extends from minutes to weeks.
   - Lipid-soluble hormones are removed from the circulation by conjugation to sulfate or glucuronic acid, which then allows them to be excreted in the bile.
3. Water-soluble hormones include proteins, peptides, and amino acid derivatives.
   - Water-soluble hormones circulate freely in the blood.
   - Proteases degrade protein and peptide hormones in the circulation; the breakdown products are then excreted in the urine. However, some water-soluble hormones have chemical modifications, such as the addition of a carbohydrate group, which prolongs their half-life.

Patterns of Hormone Secretion

The three main patterns of hormone secretions are chronic, acute, and episodic.

1. Chronic hormone secretion results in hormones whose circulating levels are relatively constant.
2. Acute hormone secretion results in hormone levels that can vary dramatically.
3. Episodic hormone secretion results in a cyclic pattern of hormone release.

17.3 Control of Hormone Secretion (p. 577)

Stimulation of Hormone Release

Three types of stimuli result in hormone secretion: humoral, neural, and hormonal.

1. Humoral stimulation is exhibited by hormones that are sensitive to circulating blood levels of certain molecules, such as glucose or calcium.
2. Neural stimuli cause hormone secretion in direct response to action potentials in neurons, as occurs during stress or exercise. Hormones from the hypothalamus that cause the release of other hormones are called releasing hormones.
3. Hormonal stimulation of other hormone secretion is common in the endocrine system. Hormones from the anterior pituitary that stimulate hormones from other endocrine glands are called tropic hormones.

Inhibition of Hormone Release

Although the stimulus of hormone secretion is important, inhibition is equally important.

1. Humoral substances can inhibit the secretion of hormones.
2. Neural stimuli can prevent hormone secretion.
3. Inhibiting hormones prevent hormone release.

Regulation of Hormone Levels in the Blood

Two processes regulate the overall blood levels of hormones: negative feedback and positive feedback.

1. Negative feedback prevents further hormone secretion once a set point is achieved.
2. Positive feedback is a self-promoting system whereby the stimulation of hormone secretion increases over time.

17.4 Hormone Receptors and Mechanisms of Action (p. 580)

Decrease in Receptor Number

1. Hormones stimulate their targets by binding to proteins in the target cell called receptors.
2. A target cell may decrease its sensitivity to a hormone through desensitization, which can occur through a decrease in receptor number, a process called down-regulation.

Increase in Receptor Number

1. A target cell may increase its sensitivity to a hormone through sensitization, which can occur through an increase in receptor number, a process called up-regulation.

Classes of Receptors

The two groups of hormones each have their own class of receptors.

1. Lipid-soluble hormones bind to nuclear receptors located inside the nucleus of the target cell.
2. Water-soluble hormones bind to membrane-bound receptors, which are integral membrane proteins.

Action of Nuclear Receptors

1. Nuclear receptors have portions that allow them to bind to the DNA in the nucleus once the hormone is bound.
   - The hormone-receptor complex activates genes, which in turn activate the DNA to produce mRNA.
   - The mRNA increases the synthesis of certain proteins that produce the target cell's response.
2. Nuclear receptors cannot respond immediately, because it takes time to produce the mRNA and the protein.

Membrane-Bound Receptors and Signal Amplification

1. Membrane-bound receptors activate a cascade of events once the hormone binds.
2. Some membrane-bound receptors are associated with membrane proteins called G proteins.
   - Hormone binds to a membrane-bound receptor, and G proteins are activated.
   - The α subunit of the G protein binds to ion channels and causes them to open or change the rate of synthesis of intracellular mediators, such as cAMP, cGMP, IP_3, and DAG.
3. Intracellular enzymes can be activated directly, which in turn causes the synthesis of intracellular mediators, such as cGMP, or adds a phosphate group to intracellular enzymes, which alters their activity.
4. Second-messenger systems act rapidly, because they act on already existing enzymes to amplify the signal.

REVIEW AND COMPREHENSION

1. When comparing the endocrine system and the nervous system, the endocrine system generally
   a. is faster-acting than the nervous system.
   b. produces effects that are of shorter duration.
   c. uses blood-borne chemical messengers.
   d. produces more localized effects.
   e. relies less on chemical messengers.
2. Given this list of molecule types:
   (1) nucleic acid derivatives
   (2) fatty acid derivatives
   (3) peptides
   (4) proteins
   (5) phospholipids

   Which could be hormone molecules?
   a. 1,2,3
   b. 2,3,4
   c. 1,2,3,4
   d. 2,3,4,5
   e. 1,2,3,4,5
3. Which of these can regulate the secretion of a hormone from an endocrine tissue?
   a. other hormones
   b. negative-feedback mechanisms
   c. humoral substances in the blood
   d. the nervous system
   e. All of these are correct.
4. Hormones are released into the blood
   a. at relatively constant levels.
   b. in large amounts in response to a stimulus.
   c. in an episodic fashion.
   d. All of these are correct.
5. Lipid-soluble hormones readily diffuse through capillary walls, whereas water-soluble hormones, such as proteins, must
   a. pass through capillary cells.
   b. pass through pores in the capillary endothelium.
   c. be moved out of the capillary by active transport.
   d. remain in the blood.
   e. be broken down to amino acids before leaving the blood.
6. Concerning the half-life of hormones,
   a. lipid-soluble hormones generally have a longer half-life.
   b. hormones with a shorter half-life regulate activities with a slow onset and long duration.
   c. hormones with a shorter half-life are maintained at more constant levels in the blood.
   d. lipid-soluble hormones are degraded rapidly by enzymes in the circulatory system.
   e. water-soluble hormones usually combine with plasma proteins.
7. Given these observations:
   (1) A hormone affects only a specific tissue (not all tissues).
   (2) A tissue can respond to more than one hormone.
   (3) Some tissues respond rapidly to a hormone, whereas others take many hours to respond.

   Which of these observations can be explained by the characteristics of hormone receptors?
   a. 1
   b. 1,2
   c. 2,3
   d. 1,3
   e. 1,2,3
8. Which of these is *not* a means by which hormones are eliminated from the circulatory system?
   a. excreted into urine or bile
   b. bound to binding proteins
   c. enzymatically degraded in the blood (metabolism)
   d. actively transported into cells
   e. conjugated with sulfate or glucuronic acid
9. Down-regulation
   a. produces a decrease in the number of receptors in the target cells.
   b. produces an increase in target cells' sensitivity to a hormone.
   c. is found in target cells that respond to hormones that are maintained at constant levels.
   d. occurs partly because of an increase in receptor synthesis by the target cell.
   e. All of these are correct.
10. Activated G proteins can
    a. cause ion channels to open or close.
    b. activate adenylate cyclase.
    c. inhibit the synthesis of cAMP.
    d. alter the activity of IP_3.
    e. All of these are correct.

11. Given these events:
    (1) GTP is converted to GDP.
    (2) The subunit separates from the β and γ units.
    (3) GDP is released from the α subunit.

    List the order in which the events occur after a hormone binds to a membrane-bound receptor.
    a. 1,2,3
    b. 1,3,2
    c. 2,3,1
    d. 3,2,1
    e. 3,1,2
12. Which of these can limit a cell's response to a hormone?
    a. phosphodiesterase
    b. converting GTP to GDP
    c. decreasing the number of receptors
    d. blocking binding sites
    e. All of these are correct.
13. Given these events:
    (1) The α subunit of a G protein interacts with Ca^{2+} channels.
    (2) Calcium ions diffuse into the cell.
    (3) The α subunit of a G protein is activated.

    Choose the arrangement that lists the events in the order they occur after a hormone binds to a receptor on a smooth muscle cell.
    a. 1,2,3
    b. 1,3,2
    c. 2,1,3
    d. 3,1,2
    e. 3,2,1
14. Given these events:
    (1) cAMP is synthesized.
    (2) The α subunit of G protein is activated.
    (3) Phosphodiesterase breaks down cAMP.

    Choose the arrangement that lists the events in the order they occur after a hormone binds to a receptor.
    a. 1,2,3
    b. 1,3,2
    c. 2,1,3
    d. 2,3,1
    e. 3,2,1
15. When a hormone binds to a nuclear receptor,
    a. DNA produces mRNA.
    b. G proteins are activated.
    c. the hormone-receptor complex causes ion channels to open or close.
    d. the cell's response is faster than when a hormone binds to a membrane-bound receptor.
    e. the hormone is usually a large, water-soluble molecule.
16. Given these events:
    (1) activation of cAMP
    (2) activation of genes
    (3) alteration of enzyme activity

    Which of these events can occur when a hormone binds to a nuclear hormone receptor?
    a. 1
    b. 1,2
    c. 2,3
    d. 1,2,3

Answers in Appendix E

CRITICAL THINKING

1. Consider a hormone that is secreted in large amounts at a given interval, modified chemically by the liver, and excreted by the kidneys at a rapid rate, thus making the half-life of the hormone in the circulatory system very short. The hormone therefore rapidly increases in the blood and then decreases rapidly. Suppose that a patient has both liver and kidney disease, and predict the consequences on the blood levels of that hormone.
2. Consider a hormone that increases the concentration of a substance in the circulatory system. If a tumor begins to produce that substance in large amounts in an uncontrolled fashion, predict the effect on the hormone's secretion rate.
3. How could you determine whether a hormone-mediated response has resulted from the intracellular mediator mechanism or the nuclear receptor mechanism?
4. If a hormone affects a target tissue through a membrane-bound receptor that has a G protein associated with it, predict the consequences if a genetic disease causes the α subunit of the G protein to have a structure that prevents it from binding to GTP.
5. For a hormone that binds to a membrane-bound receptor and has cAMP as the intracellular mediator, predict and explain the consequences if a patient takes a drug that strongly inhibits phosphodiesterase.
6. When an individual is confronted with a potentially harmful situation, the adrenal glands release epinephrine (adrenaline). Epinephrine prepares the body for action by increasing the heart rate and blood glucose levels. Explain the advantages or disadvantages associated with a shorter half-life for epinephrine, such as when it is released as a neurotransmitter, and those associated with a longer half-life, such as when it is secreted as a hormone.
7. Thyroid hormones are important in regulating the body's basal metabolic rate. Thyroid hormones are lipid-soluble and have a long half-life. What are the advantages and disadvantages of a long half-life for thyroid hormones, compared with a short half-life?
8. Predict the effect on LH and FSH secretion if a small tumor in the hypothalamus continuously secretes large concentrations of GnRH. Given that LH and FSH regulate the function of the male and female reproductive systems, state whether the condition will increase or decrease the activity of these systems.
9. Predict some consequences of trying to use a skin patch to administer insulin, a protein hormone, to a person who has diabetes mellitus.

Answers in Appendix F

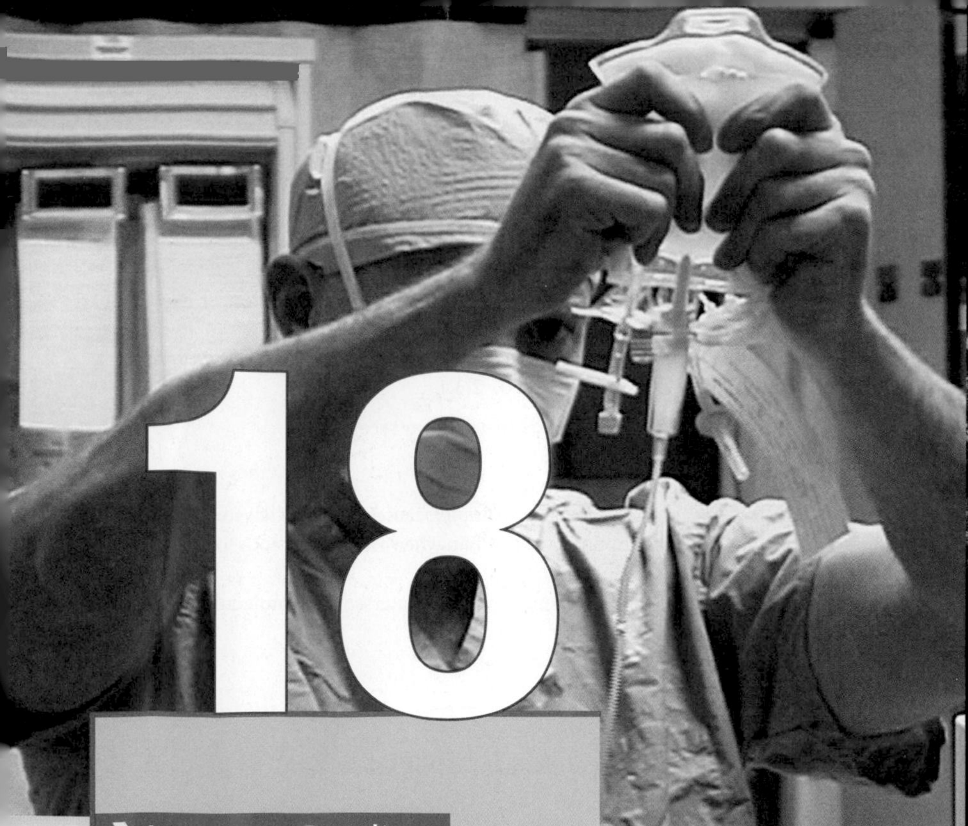

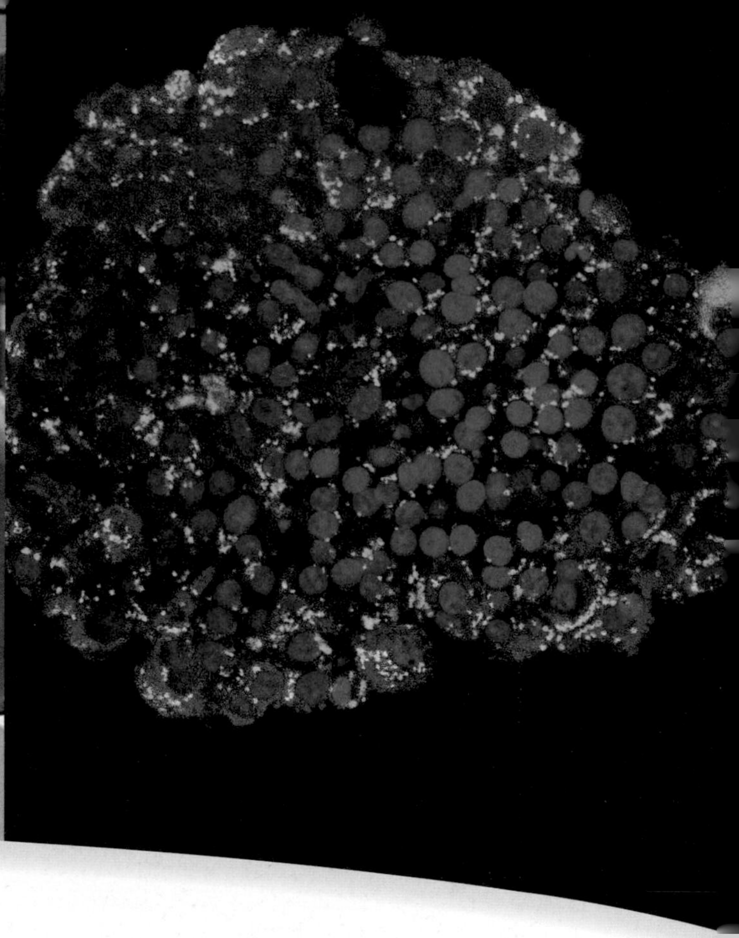

18

Learn to Predict

Dylan, a 10-year-old boy, was constantly hungry and was losing weight rapidly in spite of his unusually large food intake. Dylan was also constantly thirsty and urinated frequently. In addition, he felt weak and lethargic, and his breath occasionally had a distinctive sweet, or acetone, odor. Dylan's parents tried to make sure he ate a healthy diet, but Dylan was sneaking candy and soft drinks when his parents were not around. After reading this chapter and learning about the body's hormones, what type of hormonal imbalance do you think is responsible for Dylan's symptoms? What effect would eating candy and drinking sugary soda have on Dylan's health?

Endocrine Glands

As described in chapter 17, the nervous and endocrine systems work together to regulate and coordinate the activities of nearly all the other body structures. When either system fails to function properly, conditions can rapidly deviate from homeostasis, and disease may result. One of the most common endocrine system disorders is insulin-dependent diabetes mellitus. You may know someone who has this condition, but, as recently as the early 1900s, people who developed the disease died because no effective treatment was available. As physiologists learned more about the function of endocrine glands and the nature of their hormones, successful treatments were developed for diabetes mellitus, as well as for many other endocrine disorders. This chapter examines each of the organs of the endocrine system, describes the hormones they secrete, and explains how hormone secretion is regulated so that homeostasis is maintained.

Photo: The surgeon in this photo is transfusing donor islet cells into a diabetic patient. The islet cells may take residence in the pancreas and secrete insulin for the patient. Note the new islet cells in the right-hand photo. They are now functioning normally. This patient will never again need to inject insulin.

Module 8
Endocrine System

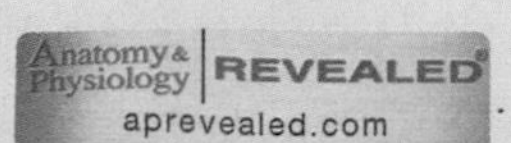
Anatomy & Physiology
REVEALED
aprevealed.com

18.1 Overview of the Endocrine System

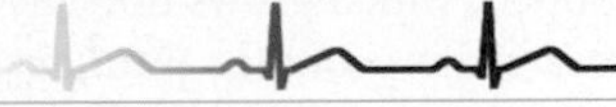

LEARNING OUTCOMES

After reading this section, you should be able to

A. Explain the types of information that are necessary to understand endocrine function.

B. Describe the 10 regulatory functions of the endocrine system.

Recall from chapter 17 that the endocrine system is composed of glands and cells that secrete hormones into the bloodstream (see figure 17.1). The hormones are secreted in response to neural, hormonal, or humoral stimuli and then travel to target cells and regulate homeostasis. In order to understand completely how the endocrine system regulates body functions, you need to know the various endocrine glands, their hormones and mechanisms of action. In addition, many disorders of the body are caused by either hypersecretion or hyposecretion of hormones.

The main regulatory functions of the endocrine system are the following:

1. *Metabolism.* The endocrine system regulates the rate of nutrient utilization and energy production.
2. *Control of food intake and digestion.* The endocrine system regulates the level of satiation (fullness) and the breakdown of food into individual nutrients.
3. *Tissue development.* The endocrine system influences the development of tissues, such as those of the nervous system.
4. *Ion levels.* The endocrine system helps regulate blood pH, as well as Na^+, K^+, and Ca^{2+} concentrations in the blood.
5. *Water balance.* The endocrine system regulates water balance by controlling the solute concentration of the blood.
6. *Heart rate and blood pressure changes.* The endocrine system helps regulate the heart rate and blood pressure and prepare the body for physical activity.
7. *Control of blood glucose and other nutrients.* The endocrine system regulates the levels of glucose and other nutrients in the blood.
8. *Control of reproductive functions.* The endocrine system controls the development and functions of the reproductive systems in males and females.
9. *Uterine contractions and milk release.* The endocrine system regulates uterine contractions during delivery and stimulates milk release from the breasts in lactating females.
10. *Immune system function.* The endocrine system helps control the production of immune cells.

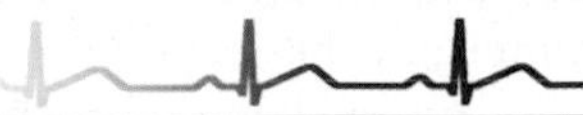

ASSESS YOUR PROGRESS

1. *What pieces of information are needed to understand how the endocrine system regulates body functions?*
2. *List 10 regulatory functions of the endocrine system.*

18.2 Pituitary Gland and Hypothalamus

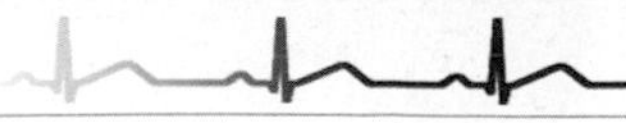

LEARNING OUTCOMES

After reading this section, you should be able to

A. Describe the location and structure of the pituitary gland.

B. Explain the physical, neural, and vascular connections between the hypothalamus and the pituitary gland.

C. Describe how the hypothalamus regulates hormone secretion from the pituitary gland.

D. List the hormones produced by the hypothalamus and state their structural type, target tissues, and actions.

E. List the hormones produced by the anterior pituitary gland and state their structural type, target tissues, and actions.

F. Explain the nature of a tropic hormone.

G. Describe the conditions that result from over- and undersecretion of pituitary hormones.

The **pituitary** (pi-too′i-tār-rē) **gland,** or *hypophysis* (hī-pof′i-sis; an undergrowth), secretes nine major hormones that regulate numerous body functions and the secretory activity of several other endocrine glands.

The pituitary gland and the **hypothalamus** (hī′pō-thal′ă-mŭs) of the brain are major sites of nervous and endocrine system interaction (figure 18.1). The hypothalamus regulates the secretory activity of the pituitary gland; indeed, the posterior pituitary is an extension of the hypothalamus. The hypothalamus, in turn, is influenced by hormones, sensory information entering the central nervous system, and emotions.

Structure of the Pituitary Gland

The pituitary gland is roughly 1 cm in diameter, weighs 0.5–1.0 g, and rests in the sella turcica of the sphenoid bone (figure 18.1*a*). It is located inferior to the hypothalamus and is connected to it by a stalk of tissue called the **infundibulum** (in-fŭn-dib′u-lŭm).

The pituitary gland is divided functionally into two parts: the **posterior pituitary gland,** or *neurohypophysis* (noor′ō-hī-pof′i-sis), and the **anterior pituitary gland,** or *adenohypophysis* (ad′ĕ-nō-hī-pof′i-sis; *adeno*, gland).

Posterior Pituitary

The posterior pituitary is called the neurohypophysis because it is continuous with the brain (*neuro-* refers to the nervous system). During embryonic development, the posterior pituitary forms from an outgrowth of the inferior part of the brain in the area of the hypothalamus (see chapter 29). The outgrowth of the brain forms the infundibulum, and the distal end of the infundibulum enlarges to form the posterior pituitary (figure 18.1*b*). Because the posterior pituitary is an extension of the nervous system, its secretions are called **neuropeptides,** or **neurohormones** (noor-ō-hōr′mōnz).

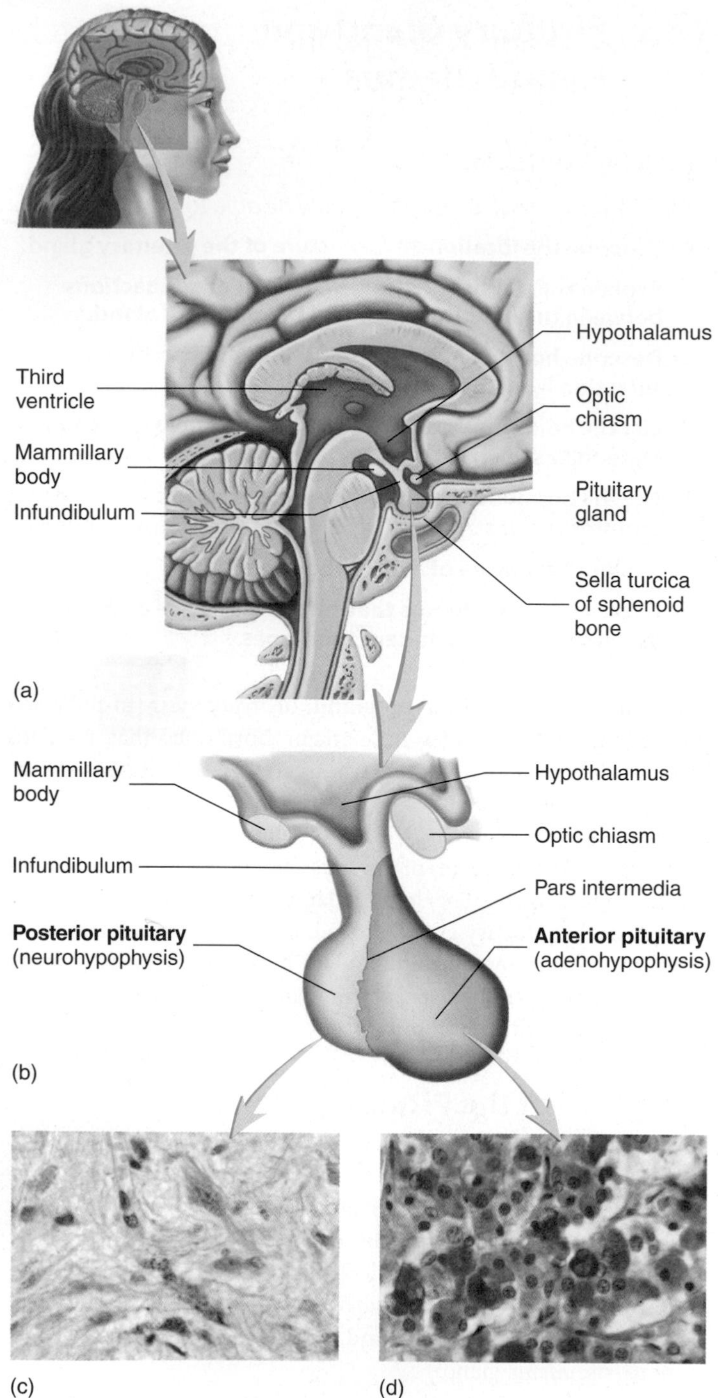
Hypothalamus
Third ventricle
Optic chiasm
Mammillary body
Pituitary gland
Infundibulum
Sella turcica of sphenoid bone
(a)
Mammillary body
Hypothalamus
Optic chiasm
Infundibulum
Pars intermedia
Posterior pituitary (neurohypophysis)
Anterior pituitary (adenohypophysis)
(b)
(c)
(d)

FIGURE 18.1 AP|R **Subdivisions of the Pituitary Gland**

(*a*) A midsagittal section of the head through the pituitary gland, showing the location of the hypothalamus of the brain and the pituitary gland. The pituitary gland is in a depression called the sella turcica in the floor of the skull. It is connected to the hypothalamus by the infundibulum. (*b*) The pituitary gland is divided into the anterior pituitary gland and the posterior pituitary gland. The posterior pituitary consists of the enlarged distal end of the infundibulum, which connects the posterior pituitary to the hypothalamus. (*c,d*) Histology of the pituitary gland: (*c*) The posterior pituitary consists of axon terminals, whereas (*d*) the anterior pituitary consists of groups of secretory cells.

Anterior Pituitary

The anterior pituitary arises as an outpocketing of the roof of the embryonic oral cavity called the pituitary diverticulum, or Rathke pouch, which grows dorsally toward the posterior pituitary. As it nears the posterior pituitary, the pituitary diverticulum loses its connection with the oral cavity and becomes the anterior pituitary, which includes an area called the pars intermedia that is not functional in adult humans (figure 18.2). Because the anterior pituitary is derived from epithelial tissue of the embryonic oral cavity, not from neural tissue, the hormones secreted from the anterior pituitary are not neurohormones.

Relationship of the Pituitary Gland to the Brain: The Hypothalamus

The pituitary is regulated, in part, by hormones produced and secreted by neurons in the hypothalamus. Some of these hypothalamic hormones are delivered to the anterior pituitary via a circulatory system called a portal system. Unlike most blood vessels that follow a prescribed pattern of artery → arteriole → capillary network → venule → vein, portal vessels directly connect a primary capillary network to a secondary capillary network. The **hypothalamohypophysial** (hī′pō-thal′ă-mō-hī′pō-fiz′ē-ăl) **portal system** is one of the major portal systems in the body. The others include the hepatic portal system and the renal nephron portal systems (see chapters 21 and 26). The hypothalamohypophysial portal system extends from the floor of the hypothalamus to the anterior pituitary (figure 18.3). The primary capillary network in the hypothalamus is supplied with blood from arteries that deliver blood to the hypothalamus. From the primary capillary network, the hypothalamohypophysial portal vessels carry blood to a secondary capillary network in the anterior pituitary. Veins from the secondary capillary network eventually merge with the general circulation.

Neurohormones, produced and secreted by neurons of the hypothalamus, enter the primary capillary network and are carried to the secondary capillary network. There the neurohormones leave the blood and act on cells of the anterior pituitary. Each acts as either a **releasing hormone,** increasing the secretion of a specific anterior pituitary hormone, or an **inhibiting hormone,** decreasing the secretion of a specific anterior pituitary hormone. In response to the releasing hormones, anterior pituitary cells secrete hormones that enter the secondary capillary network and are carried by the general circulation to their target tissues. Thus, the hypothalamohypophysial portal system provides a means by which the hypothalamus, using neurohormones as chemical messengers, regulates the secretory activity of the anterior pituitary (figure 18.3).

Several major releasing and inhibiting hormones are produced and released from hypothalamic neurons (table 18.1). **Growth hormone–releasing hormone (GHRH)** is a small peptide that stimulates the secretion of growth hormone from the anterior pituitary gland, and **growth hormone–inhibiting hormone (GHIH),** also called *somatostatin* (sō′mă-tō-stat′in), is a small peptide that inhibits growth hormone secretion. **Thyrotropin-releasing hormone (TRH)** is a small peptide that stimulates the secretion of thyroid-stimulating hormone from the anterior pituitary gland. **Corticotropin-releasing hormone (CRH)** is a peptide that stimulates

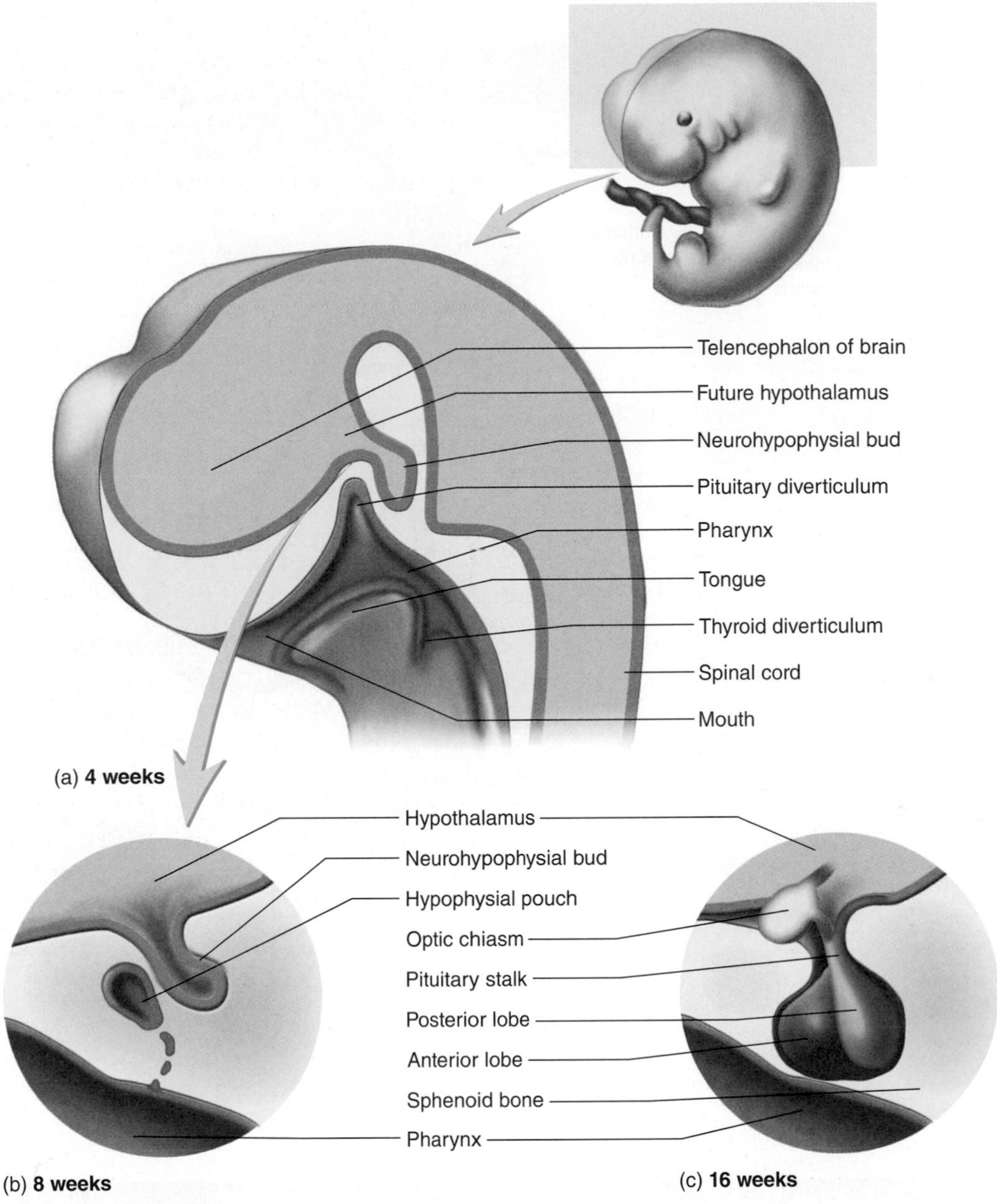
Telencephalon of brain
Future hypothalamus
Neurohypophysial bud
Pituitary diverticulum
Pharynx
Tongue
Thyroid diverticulum
Spinal cord
Mouth
(a) 4 weeks
Hypothalamus
Neurohypophysial bud
Hypophysial pouch
Optic chiasm
Pituitary stalk
Posterior lobe
Anterior lobe
Sphenoid bone
Pharynx
(b) 8 weeks
(c) 16 weeks

FIGURE 18.2 Development of the Pituitary Gland
The posterior pituitary develops as a downward growth of the hypothalamus. The anterior pituitary develops as an outpocketing of the embryonic gut called the pituitary diverticulum; the pars intermedia is not shown in this figure.

adrenocorticotropic hormone from the anterior pituitary gland. **Gonadotropin-releasing hormone (GnRH)** is a small peptide that stimulates luteinizing hormone and follicle-stimulating hormone from the anterior pituitary gland. **Prolactin-releasing hormone (PRH)** and **prolactin-inhibiting hormone (PIH)** regulate the secretion of prolactin from the anterior pituitary gland. Secretions of the anterior pituitary gland are described later in this section.

There is no portal system to carry hypothalamic neuropeptides to the posterior pituitary. Neurohormones released from the posterior pituitary are produced by neurosecretory neurons that have cell bodies located in the hypothalamus. The axons of these neurons extend from the hypothalamus through the infundibulum into the posterior pituitary and form a tract called the **hypothalamohypophysial tract** (figure 18.4). Neurohormones produced in the hypothalamus pass down these axons in tiny vesicles and are stored in secretory vesicles in the enlarged ends of the axons. Action potentials originating in the neuron cell bodies in the hypothalamus are propagated along the axons to the axon terminals in the posterior pituitary. The action potentials cause the release of neurohormones from the axon terminals, and they enter the circulatory system. Secretions of the posterior pituitary gland are described later in this section.

FUNDAMENTAL Figure

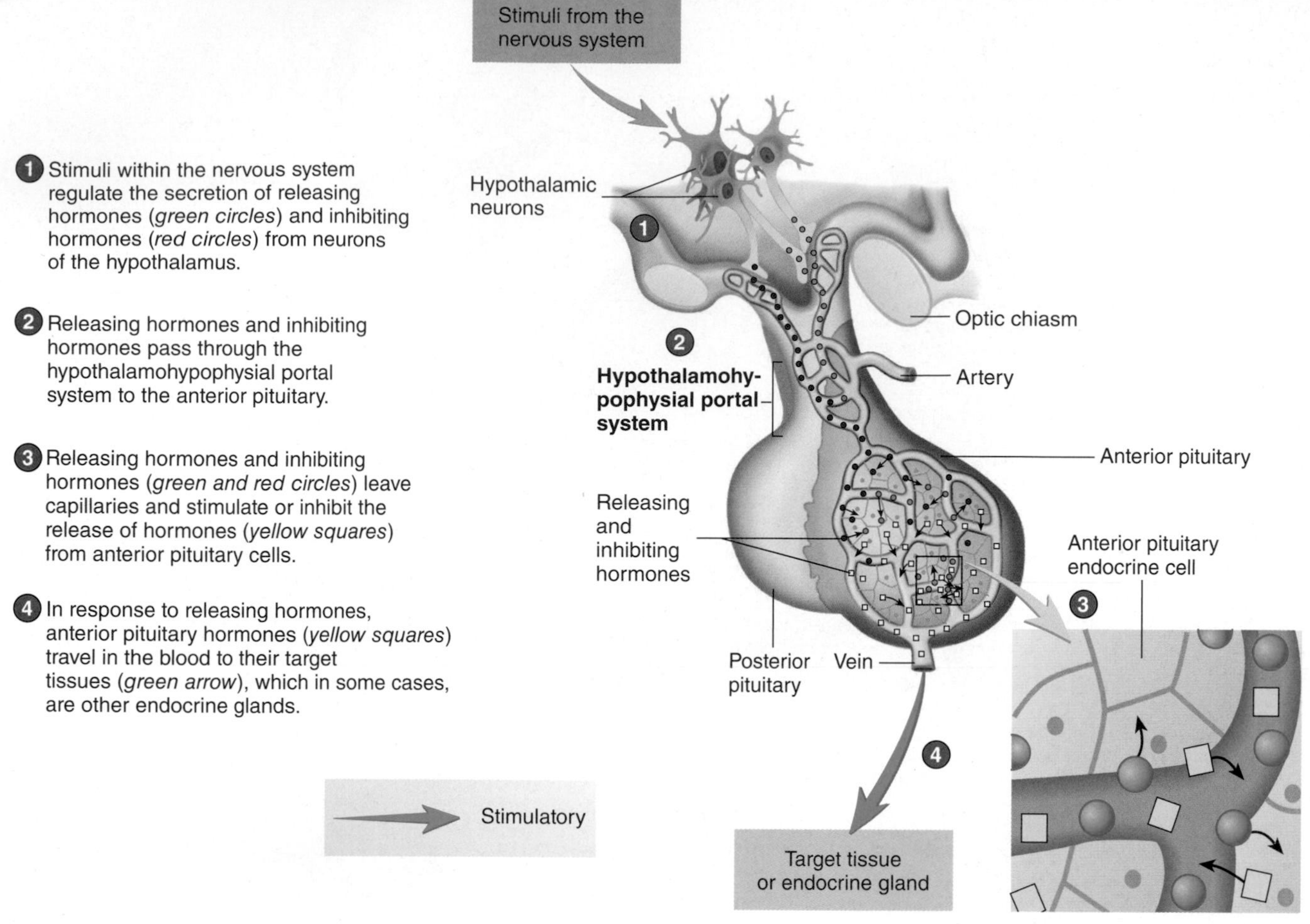

Stimuli from the nervous system
1 Stimuli within the nervous system regulate the secretion of releasing hormones (*green circles*) and inhibiting hormones (*red circles*) from neurons of the hypothalamus.
2 Releasing hormones and inhibiting hormones pass through the hypothalamohypophysial portal system to the anterior pituitary.
3 Releasing hormones and inhibiting hormones (*green and red circles*) leave capillaries and stimulate or inhibit the release of hormones (*yellow squares*) from anterior pituitary cells.
4 In response to releasing hormones, anterior pituitary hormones (*yellow squares*) travel in the blood to their target tissues (*green arrow*), which in some cases, are other endocrine glands.
Hypothalamic neurons
Optic chiasm
Hypothalamohypophysial portal system
Artery
Anterior pituitary
Releasing and inhibiting hormones
Anterior pituitary endocrine cell
Posterior pituitary
Vein
Stimulatory
Target tissue or endocrine gland

PROCESS FIGURE 18.3 Hypothalamic Control of the Anterior Pituitary

TABLE 18.1 Hormones of the Hypothalamus

Hormones	Structure	Target Tissue	Response
Growth hormone–releasing hormone (GHRH)	Peptide	Anterior pituitary cells that secrete growth hormone	Increased growth hormone secretion
Growth hormone–inhibiting hormone (GHIH), or somatostatin	Small peptide	Anterior pituitary cells that secrete growth hormone	Decreased growth hormone secretion
Thyrotropin-releasing hormone (TRH)	Small peptide	Anterior pituitary cells that secrete thyroid-stimulating hormone	Increased thyroid-stimulating hormone secretion
Corticotropin-releasing hormone (CRH)	Peptide	Anterior pituitary cells that secrete adrenocorticotropic hormone	Increased adrenocorticotropic hormone secretion
Gonadotropin-releasing hormone (GnRH)	Small peptide	Anterior pituitary cells that secrete luteinizing hormone and follicle-stimulating hormone	Increased secretion of luteinizing hormone and follicle-stimulating hormone
Prolactin-releasing hormone (PRH)	Unknown	Anterior pituitary cells that secrete prolactin	Increased prolactin secretion
Prolactin-inhibiting hormone (PIH)	Dopamine (amino acid derivative)	Anterior pituitary cells that secrete prolactin	Decreased prolactin secretion

FUNDAMENTAL **Figure**

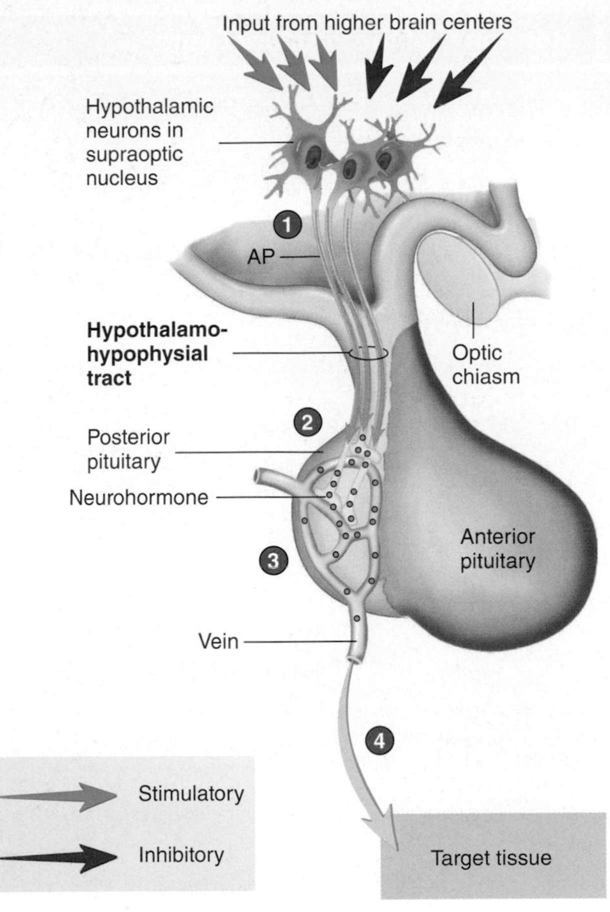

Input from higher brain centers
Hypothalamic neurons in supraoptic nucleus
AP
Hypothalamo-hypophysial tract
Optic chiasm
Posterior pituitary
Neurohormone
Anterior pituitary
Vein
Stimulatory
Inhibitory
Target tissue
1 Stimuli within the nervous system cause hypothalamic neurons to either increase or decrease their action potential frequency.
2 Action potentials are conducted by axons of the hypothalamic neurons through the hypothalamohypophysial tract to the posterior pituitary. The axon endings of neurons store neurohormones in the posterior pituitary.
3 In the posterior pituitary gland, action potentials cause the release of neurohormones (blue circles) from axon terminals into the circulatory system.
4 The neurohormones pass through the circulatory system and influence the activity of their target tissues.

PROCESS FIGURE 18.4 Secretion of Posterior Pituitary Hormones

ASSESS **YOUR PROGRESS**

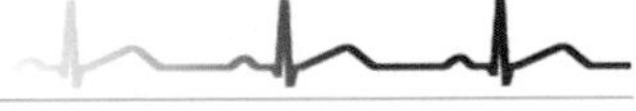

3. *Where is the pituitary gland located? Contrast the embryonic origins of the anterior pituitary and the posterior pituitary.*
4. *What is a portal system? Describe the hypothalamohypophysial portal system.*
5. *How does the hypothalamus regulate the secretion of anterior pituitary hormones?*
6. *List the releasing and inhibiting hormones that are produced and released from hypothalamic neurons.*
7. *Describe the hypothalamohypophysial tract, including the production of neurohormones in the hypothalamus and their subsequent release from the posterior pituitary gland.*

Predict 2

Surgical removal of the posterior pituitary in experimental animals results in clear symptoms of a hormone shortage, but they can be temporary. Surgical removal of the anterior pituitary, in contrast, results in many manifestations and a permanent shortage of several hormones. Explain these results.

Hormones of the Pituitary Gland

This section describes the hormones secreted from the pituitary gland (table 18.2), their effects on the body, and the mechanisms that regulate their secretion rate. In addition, some major consequences of abnormal hormone secretion are stressed.

Posterior Pituitary Hormones

The posterior pituitary gland stores and secretes two neurohormones, antidiuretic hormone and oxytocin. A separate population of neurons secretes each neurohormone.

Antidiuretic Hormone

Antidiuretic (an′tē-d-ī-ū-ret′ik) **hormone (ADH)** is so named because it prevents (*anti-*) the output of large amounts of urine (*diuresis*). The technical name for ADH is **vasopressin** (vā-sō-pres′in, vas-ō-pres′in) because it constricts blood vessels and raises blood pressure when large amounts are released. ADH molecules are synthesized predominantly by neurosecretory neuron cell bodies in the supraoptic nuclei of the hypothalamus and are transported within the axons of the hypothalamohypophysial tract from the supraoptic nuclei to the posterior pituitary, where they are stored in axon terminals. Action potentials in these neurons stimulate the release of ADH into the blood, where it is carried to its primary target tissue, the kidney tubules. Kidney tubules are the sites of urine production in the kidneys. ADH promotes the reabsorption of water from kidney tubules, which reduces urine volume (see chapter 26).

TABLE 18.2 Hormones of the Pituitary Gland

Hormones	Structure	Target Tissue	Response
Posterior Pituitary (Neurohypophysis)			
Antidiuretic hormone (ADH)	Small peptide	Kidneys	Increased water reabsorption (less water is lost in the form of urine)
Oxytocin	Small peptide	Uterus; mammary glands	Increased uterine contractions; increased milk expulsion from mammary glands; unclear function in males
Anterior Pituitary (Adenohypophysis)			
Growth hormone (GH), or somatotropin	Protein	Most tissues	Increased growth in tissues; increased amino acid uptake and protein synthesis; increased breakdown of lipids and release of fatty acids from cells; increased glycogen synthesis and increased blood glucose levels; increased somatomedin production
Thyroid-stimulating hormone (TSH)	Glycoprotein	Thyroid gland	Increased thyroid hormone secretion
Adrenocorticotropic hormone (ACTH)	Peptide	Adrenal cortex	Increased glucocorticoid hormone secretion
Lipotropins	Peptides	Adipose tissues	Increased lipid breakdown
β endorphins	Peptides	Brain, but not all target tissues are known	Analgesia in the brain; inhibition of gonadotropin-releasing hormone secretion
Melanocyte-stimulating hormone (MSH)	Peptide	Melanocytes in the skin	Increased melanin production in melanocytes to make the skin darker in color
Luteinizing hormone (LH)	Glycoprotein	Ovaries in females; testes in males	Ovulation and progesterone production in ovaries; testosterone synthesis and support for sperm cell production in testes
Follicle-stimulating hormone (FSH)	Glycoprotein	Follicles in ovaries in females; seminiferous tubules in males	Follicle maturation and estrogen secretion in ovaries; sperm cell production in testes
Prolactin	Protein	Ovaries and mammary glands in females	Milk production in lactating women; increased response of follicle to LH and FSH; unclear function in males

The secretion rate for ADH changes in response to alterations in blood osmolality and blood volume (figure 18.5). The **osmolality** of a solution increases as the concentration of solutes in the solution increases. Specialized neurons, called **osmoreceptors** (os′mō-rē-sep′terz), synapse with the ADH neurosecretory neurons in the hypothalamus. Osmoreceptors are sensitive to changes in blood osmolality. When blood osmolality increases, the frequency of action potentials in the osmoreceptors increases, resulting in a greater frequency of action potentials in the axons of ADH neurosecretory neurons. As a consequence, ADH secretion increases. ADH stimulates the kidney tubules to retain water, which reduces blood osmolality and resists any further increase in the osmolality of body fluids.

As the osmolality of the blood decreases, the action potential frequency in the osmoreceptors and the neurosecretory neurons decreases. Thus, less ADH is secreted from the posterior pituitary gland, and the decreased ADH secretion causes an increased volume of water to be eliminated in the form of urine.

Urine volume increases within minutes to a few hours in response to the consumption of a large volume of water. In contrast, urine volume decreases and urine concentration increases within hours if little water is consumed. ADH regulates these changes in urine formation by controlling the permeability of kidney tubules to water. Its effect is to maintain the osmolality and the volume of the extracellular fluid within a normal range of values.

Because ADH regulates blood volume, its secretion is also controlled by blood pressure changes. Sensory receptors that detect changes in blood pressure send action potentials through sensory nerve fibers of the vagus nerve that eventually communicate these changes to the ADH neurosecretory neurons. A decrease in blood pressure, which normally accompanies a decrease in blood volume, causes an increased action potential frequency in the neurosecretory neurons and increased ADH secretion, which stimulates the kidneys to retain water. Because the water in urine is derived from blood as it passes through the kidneys, ADH slows any further reduction in blood volume.

An increase in blood pressure decreases the action potential frequency in the ADH neurosecretory neurons. This leads to the secretion of less ADH from the posterior pituitary. As a result, the volume of urine produced by the kidneys increases (figure 18.5). Even small changes in blood osmolality influence ADH secretion. Larger changes in blood pressure are required to influence ADH secretion. The effect of ADH on the kidney and its role in regulating extracellular osmolality and volume are described in greater detail in chapters 26 and 27.

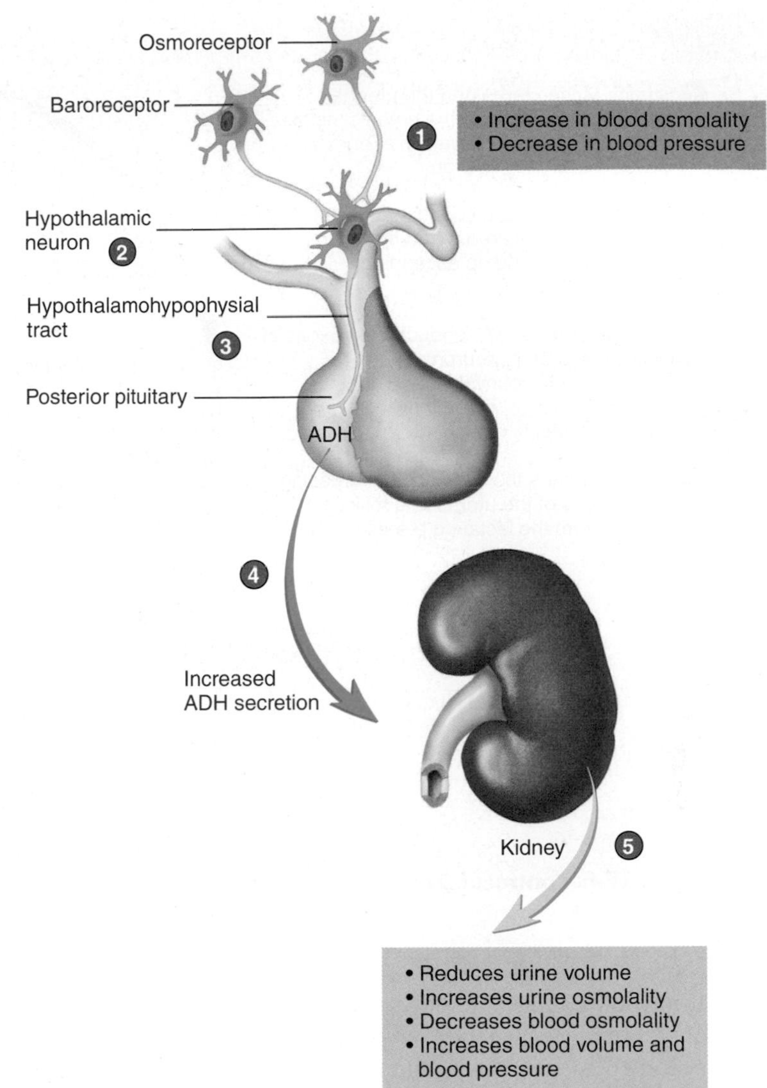

Osmoreceptor
Baroreceptor
1
• Increase in blood osmolality
• Decrease in blood pressure
Hypothalamic neuron 2
Hypothalamohypophysial tract 3
Posterior pituitary
ADH
4
Increased ADH secretion
Kidney 5
• Reduces urine volume
• Increases urine osmolality
• Decreases blood osmolality
• Increases blood volume and blood pressure
1 Osmoreceptors in the hypothalamus detect changes in blood osmolality, and baroreceptors detect changes in blood pressure and change the firing rate of action potentials in axons of the vagus nerve to the hypothalamus.
2 An increase in osmolality and a decrease in blood pressure increase action potentials in ADH-secreting neurons.
3 Action potentials are carried by axons of ADH-secreting neurons through the hypothalamohypophysial tract to the posterior pituitary.
4 In the posterior pituitary, action potentials cause the release of ADH from the axon terminals into the circulatory system.
5 Increasing ADH acts on the kidney tubules to increase water reabsorption, resulting in reduced urine volume, increased urine osmolality, and decreased blood osmolality. This helps maintain blood osmolality and volume.

PROCESS FIGURE 18.5 Control of Antidiuretic Hormone (ADH) Secretion
Small changes in blood osmolality are important in regulating ADH secretion. Larger changes in blood pressure are required to influence ADH secretion.

Predict 3

After his school's football team won the division championship, Luke went to a local bar with some friends and drank too much beer. Fortunately, one of his friends served as a designated driver. The next morning, Luke wondered why he was thirsty and felt somewhat dehydrated. His roommate, an anatomy and physiology student, pointed out that alcohol inhibits ADH secretion from the posterior pituitary. The roommate then explained why Luke was thirsty and dehydrated. What was the explanation?

Oxytocin

Oxytocin (ok-sē-tō′sin) is synthesized by neurosecretory neuron cell bodies in the paraventricular nuclei of the hypothalamus and then is transported through axons to the posterior pituitary, where it is stored in the axon terminals.

Oxytocin stimulates smooth muscle cells of the uterus. This neurohormone plays an important role in the expulsion of the fetus during delivery by stimulating uterine smooth muscle contraction. It also causes contraction of uterine smooth muscle in nonpregnant women, primarily during menses and sexual intercourse. The uterine contractions help expel the uterine epithelium and small amounts of blood during menses and can facilitate the movement of sperm cells through the uterus after sexual intercourse. Oxytocin is also responsible for milk ejection in lactating females by promoting the contraction of smooth-muscle-like cells surrounding the alveoli of the mammary glands. In addition, oxytocin is associated with maternal nurturing and bonding (see chapter 29). Although little is known about the specific effects of oxytocin in males, evidence suggests that it promotes sperm movement during ejaculation and pair bonding.

Stretch of the uterus, mechanical stimulation of the cervix, and stimulation of the nipples of the breast when a baby nurses activate nervous reflexes that stimulate oxytocin release. Action potentials are carried by sensory neurons from the uterus and from the nipples to the spinal cord. Action potentials then travel up the

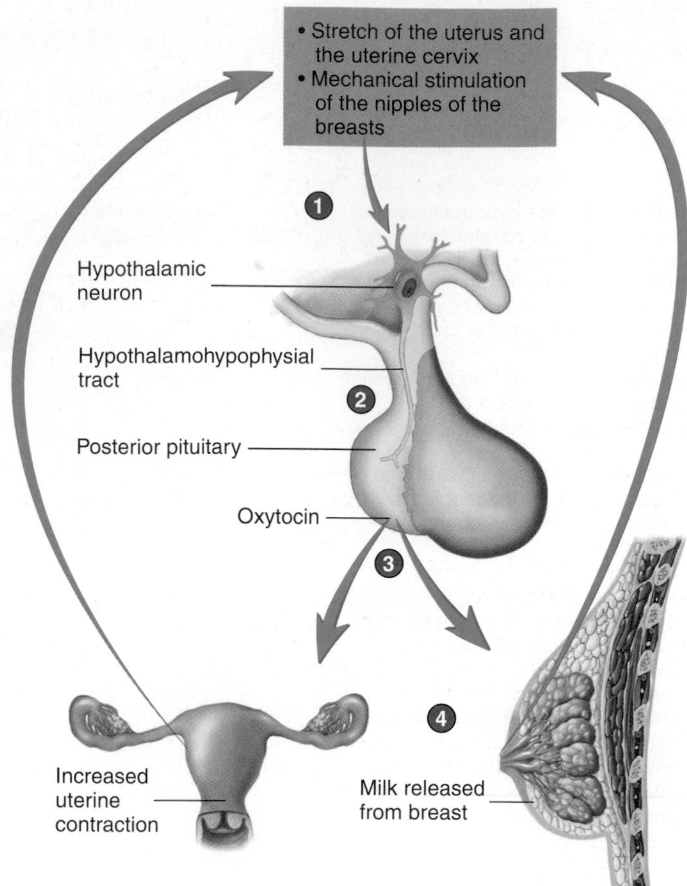

1 Stretch of the uterus and the uterine cervix or stimulation of the breasts' nipples increases action potentials in axons of oxytocin-secreting neurons.
2 Action potentials are conducted by sensory neurons from the uterus and breast to the spinal cord and up ascending tracts to the hypothalamus.
3 Action potentials are conducted by axons of oxytocin-secreting neurons in the hypothalamohypophysial tract to the posterior pituitary, where they increase oxytocin secretion.
4 Oxytocin enters the circulation, increasing contractions of the uterus and milk ejection from the lactating breast.
• Stretch of the uterus and the uterine cervix
• Mechanical stimulation of the nipples of the breasts
Hypothalamic neuron
Hypothalamohypophysial tract
Posterior pituitary
Oxytocin
Increased uterine contraction
Milk released from breast

PROCESS FIGURE 18.6 Control of Oxytocin Secretion

spinal cord to the hypothalamus, where they stimulate an increase in the action potential frequency in the axons of oxytocin-secreting neurons. The action potentials pass along the axons in the hypothalamohypophysial tract to the posterior pituitary, where they cause the axon terminals to secrete oxytocin (figure 18.6). The role of oxytocin in the reproductive system is described in greater detail in chapter 29.

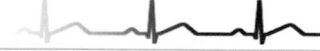

ASSESS YOUR PROGRESS

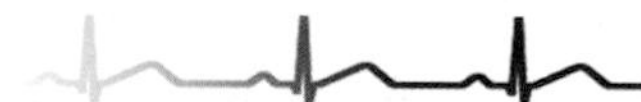

8. *Where is ADH produced, from where is it secreted, and what is its target tissue?*
9. *When ADH levels increase, how are urine volume, blood osmolarity, and blood volume affected?*
10. *What two factors will cause changes in ADH secretion rates? Name the types of sensory cells that respond to alterations in those factors.*
11. *Where is oxytocin produced, from where is it secreted, and what are its target tissues?*
12. *What effects does oxytocin have on its target tissues? What factors stimulate the secretion of oxytocin?*

Anterior Pituitary Hormones

Releasing and inhibiting hormones that pass from the hypothalamus through the hypothalamohypophysial portal system to the anterior pituitary influence anterior pituitary secretions. For some anterior pituitary hormones, the hypothalamus produces both releasing hormones and inhibiting hormones. For others, regulation is primarily by releasing hormones (see table 18.1).

The hormones secreted from the anterior pituitary are proteins, glycoproteins, or polypeptides. They are transported in the circulatory system, have a half-life measured in minutes, and bind to membrane-bound receptor molecules on their target cells. For the most part, each hormone is secreted by a separate cell type. Adrenocorticotropic hormone and lipotropin are exceptions because these hormones are derived from the same precursor protein.

Many hormones from the anterior pituitary gland are **tropic** (trō′pik) **hormones,** which stimulate the secretion of other hormones from the target tissues. Tropic hormones also control the growth of target tissues. The anterior pituitary hormones include the protein hormones, growth hormone, and prolactin; the peptide hormone adrenocorticotropic hormone and related substances; and the glycoprotein hormones, luteinizing hormone, follicle-stimulating hormone, and thyroid-stimulating hormone.

Growth Hormone

Growth hormone (GH), or *somatotropin*, stimulates growth in most tissues and plays an important role in determining how tall a person becomes. It also regulates metabolism. GH plays an important role in regulating blood nutrient levels after a meal and during periods of fasting. GH increases the movement of amino acids into cells, favors their incorporation into proteins, and slows protein breakdown. GH increases lipolysis (lipid breakdown) and the release of fatty acids from adipocytes into the blood. Fatty acids then can be used as energy sources to drive chemical reactions, including anabolic reactions, by other cells. GH also increases glucose synthesis by the liver, which releases glucose into the blood. The increased use of lipids as an energy source accompanies a decrease in glucose usage. Overall, GH activates the use of lipids to promote growth and protein synthesis.

GH binds directly to membrane-bound receptors on target cells (see chapter 17), such as adipocytes, to produce responses. These responses are called the direct effects of GH and include the increased breakdown of lipids and the decreased use of glucose as an energy source.

GH also has indirect effects on some tissues. It increases the production of a number of polypeptides, primarily by the liver but also by skeletal muscle and other tissues. These polypeptides are called **somatomedins** (sō′mă-tō-mē′dinz). The best-known somatomedins are **insulin-like growth factors (IGFs).** They are so named because of their structural similarity to insulin and because the receptor molecules function through a mechanism similar to that of the insulin receptors. IGFs have paracrine effects and circulate in the blood until they bind to receptors on target tissues. Somatomedins stimulate growth in cartilage and bone and increase the synthesis of protein in skeletal muscles. Like somatomedins, growth hormone and growth factors bind to membrane-bound receptors that phosphorylate intracellular proteins (see chapter 17).

Two neurohormones released from the hypothalamus regulate the secretion of GH (figure 18.7). One factor, growth hormone–releasing hormone (GHRH), stimulates the secretion of GH, and the other, growth hormone–inhibiting hormone (GHIH), inhibits the secretion of GH. Stimuli that influence GH secretion act on the hypothalamus to increase or decrease the secretion of the releasing and inhibiting hormones. Low blood glucose levels and other stressors stimulate the secretion of GH, and high blood glucose levels inhibit the secretion of GH. Rising blood levels of certain amino acids also increase GH secretion.

In most people, a rhythm of GH secretion occurs. Daily peak levels of GH are correlated with deep sleep. A chronically elevated blood GH level during periods of rapid growth does not occur, although children tend to have somewhat higher blood levels of GH than adults. In addition to GH, factors such as genetics, nutrition, and sex hormones influence growth.

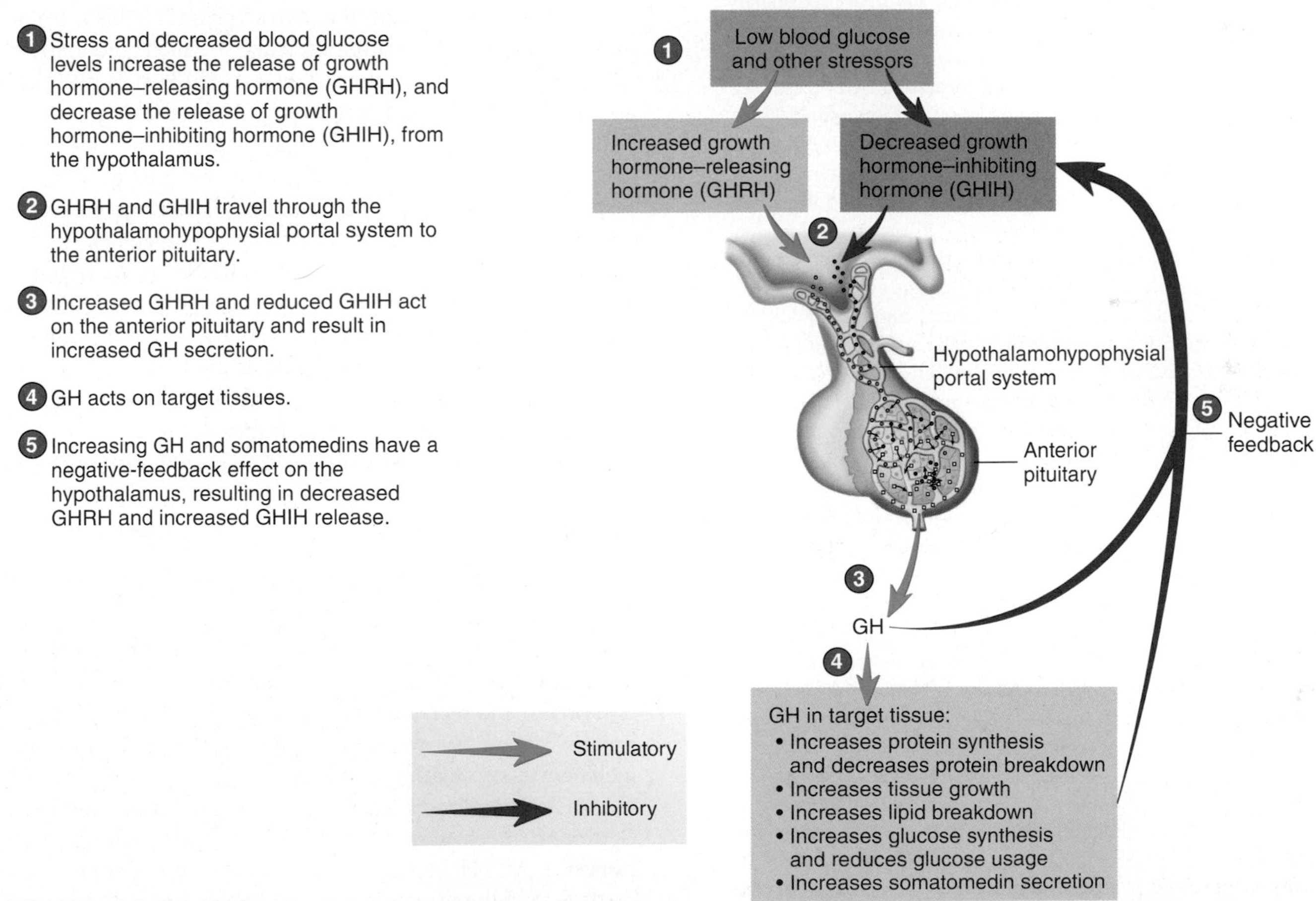
1 Stress and decreased blood glucose levels increase the release of growth hormone–releasing hormone (GHRH), and decrease the release of growth hormone–inhibiting hormone (GHIH), from the hypothalamus.
2 GHRH and GHIH travel through the hypothalamohypophysial portal system to the anterior pituitary.
3 Increased GHRH and reduced GHIH act on the anterior pituitary and result in increased GH secretion.
4 GH acts on target tissues.
5 Increasing GH and somatomedins have a negative-feedback effect on the hypothalamus, resulting in decreased GHRH and increased GHIH release.
1
Low blood glucose and other stressors
Increased growth hormone–releasing hormone (GHRH)
Decreased growth hormone–inhibiting hormone (GHIH)
2
Hypothalamohypophysial portal system
Anterior pituitary
5 Negative feedback
3
GH
4
GH in target tissue:
• Increases protein synthesis and decreases protein breakdown
• Increases tissue growth
• Increases lipid breakdown
• Increases glucose synthesis and reduces glucose usage
• Increases somatomedin secretion
Stimulatory
Inhibitory

PROCESS FIGURE 18.7 Control of Growth Hormone (GH) Secretion
Secretion of GH is controlled by two neurohormones released from the hypothalamus: growth hormone–releasing hormone (GHRH), which stimulates GH secretion, and growth hormone–inhibiting hormone (GHIH), which inhibits GH secretion. Stress increases GHRH secretion and inhibits GHIH secretion. High levels of GH have a negative-feedback effect on the production of GHRH by the hypothalamus.

Several pathological conditions are associated with abnormal GH secretion. In general, hypersecretion or hyposecretion of GH is caused by tumors in the hypothalamus or pituitary, the synthesis of structurally abnormal GH, the liver's inability to produce somatomedins, or the lack of functional receptors in target tissues. The consequences of hypersecretion and hyposecretion of GH are described in the Clinical Impact, "Growth Hormone and Growth Disorders."

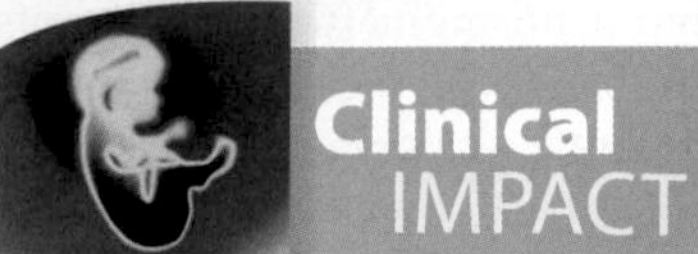

Clinical IMPACT

Growth Hormone and Growth Disorders

Chronic hyposecretion, or insufficient secretion, of GH in infants and children leads to a condition called **pituitary dwarfism** (dwōrf′izm). Insufficient amounts of GH delay bone growth, resulting in short stature. However, the bones usually have a normal shape, and people with this condition exhibit normal intelligence, in contrast to those who have dwarfism caused by hyposecretion of thyroid hormones. Other symptoms resulting from the lack of GH include mild obesity and retarded development of adult reproductive functions.

There are two types of pituitary dwarfism: (1) In approximately two-thirds of the cases, GH and other anterior pituitary hormones are secreted in reduced amounts. The decrease in other anterior pituitary hormones can result in additional disorders caused by reduced secretion of thyroid hormones, adrenal cortex hormones, and reproductive hormones. (2) In the remaining approximately one-third of cases, GH secretion is reduced, and the secretion of other anterior pituitary hormones is closer to normal. Therefore, these individuals do not experience additional hormone-related disorders, and normal reproduction is possible for them. In adults, no obvious pathology is associated with hyposecretion of GH, although some evidence suggests that lack of GH can lead to reduced bone mineral content in adults.

The gene responsible for determining the structure of GH has been transferred successfully from human cells to bacterial cells, which can produce GH that is identical to human GH. GH produced in this fashion is available to treat patients who suffer from reduced GH secretion, especially children.

Chronic hypersecretion, or excessive secretion, of GH also leads to disorder, depending on whether the hypersecretion occurs before or after complete ossification of the epiphyseal plates in the skeletal system. Chronic hypersecretion of GH before the epiphyseal plates have ossified causes exaggerated and prolonged growth in long bones, a condition called **giantism** (jī′an-tizm). Some individuals thus affected have grown to be 8 feet tall or more.

In adults, chronically elevated GH levels result in **acromegaly** (ak-rō-meg′ă-lē). No height increase occurs because the epiphyseal plates have ossified. But the condition does result in an increased diameter of the fingers, toes, hands, and feet; the deposition of heavy bony ridges above the eyes; and a prominent jaw. The influence of GH on soft tissues results in a bulbous or broad nose, an enlarged tongue, thickened skin, and sparse subcutaneous adipose tissue. Nerves are frequently compressed as a result of the proliferation of connective tissue. Because GH spares glucose usage, chronic hyperglycemia results, frequently leading to diabetes mellitus and severe atherosclerosis. Treatment for chronic hypersecretion of GH often involves the surgical removal or irradiation of a GH-producing tumor.

Predict 4

Mr. Hoops has a son who wants to be a basketball player almost as much as Mr. Hoops wants him to be one. Mr. Hoops knows a little bit about growth hormone and asks his son's doctor if she would prescribe some for his son, so that he can grow tall. What do you think the doctor tells Mr. Hoops?

Prolactin

Prolactin (prō-lak′tin) plays an important role in milk production by the mammary glands of lactating females. It binds to a membrane-bound receptor, which is linked to a kinase that phosphorylates intracellular proteins. The phosphorylated proteins produce the response in the cell. Prolactin also can enhance progesterone secretion by the ovaries after ovulation. No role for this hormone has been clearly established in males. Several hypothalamic neuropeptides can be involved in the complex regulation of prolactin secretion. One neurohormone is prolactin-releasing hormone (PRH), and another is prolactin-inhibiting hormone (PIH). The regulation of gonadotropin and prolactin secretion and their specific effects are explained more fully in chapter 28.

Thyroid-Stimulating Hormone

Thyroid-stimulating hormone (TSH), also called *thyrotropin* (thī-rot′rō-pin, thī-rō-trō′pin), stimulates the synthesis and secretion of thyroid hormones from the thyroid gland. TSH is a glycoprotein dimer consisting of two subunits, α and β, which bind to membrane-bound receptors of the thyroid gland. The α subunit is common among the glyoprotein hormones, TSH, luteinizing hormone, and follicle-stimulating hormone. It is the β subunit that dictates the specificity of each of the glycoprotein hormones. The receptors respond through a G protein mechanism that increases the intracellular chemical messenger cAMP (see chapter 17).

TSH secretion is controlled by TRH from the hypothalamus and by thyroid hormones from the thyroid gland. TRH binds to membrane-bound receptors in cells of the anterior pituitary gland and activates G proteins, which results in increased TSH secretion. In contrast, thyroid hormones inhibit both TRH and TSH secretion. TSH is secreted in an episodic fashion and its blood levels are highest at night, but the blood levels of thyroid hormones are maintained within a narrow range of values (see "Thyroid Hormones," in section 18.3).

Adrenocorticotropic Hormone and Related Substances

Adrenocorticotropic (ă-drē′nō-kōr′ti-kō-trō′pik) **hormone (ACTH)** is one of several peptide hormones derived from a precursor protein called **proopiomelanocortin** (prō-ō′pē-ō-mel′ă-nō-kōr′tin) and secreted from the anterior pituitary. This precursor protein gives rise to ACTH, lipotropins, β endorphins, and melanocyte-stimulating hormone.

ACTH binds to membrane-bound receptors and activates a G protein mechanism that increases cAMP, which produces a response. ACTH increases the secretion of hormones, primarily cortisol, from the adrenal cortex. ACTH and melanocyte-stimulating hormone also bind to melanocytes in the skin and increase skin pigmentation (see chapter 5). In pathological conditions such as chronic adrenocortical insufficiency (Addison disease), the adrenal cortex degenerates, usually due to an autoimmune condition (see chapter 22). Blood levels of ACTH and related hormones are chronically elevated, and the skin becomes markedly darker. Regulation of ACTH secretion and the effects of the hypersecretion and hyposecretion of ACTH are described in section 18.5.

The **lipotropins** (li-pō-trō′pinz) secreted from the anterior pituitary bind to membrane-bound receptor molecules on adipocytes. They cause lipid breakdown and the release of fatty acids into the circulatory system.

The **β endorphins** (en′dōr-finz) have the same effects as opiate drugs, such as morphine, and they can play a role in analgesia (pain relief) in response to stress and exercise. Other functions have been proposed for the β endorphins, including the regulation of body temperature, food intake, and water balance. Both ACTH and β endorphin secretions increase in response to stress and exercise.

Melanocyte-stimulating hormone (MSH) binds to membrane-bound receptors on skin melanocytes and stimulates increased melanin deposition in the skin. The regulation of MSH secretion and its function in humans are not well understood, although studies have shown that MSH is also important in regulating appetite and sexual behavior.

Luteinizing Hormone and Follicle-Stimulating Hormone

Gonadotropins (gō′nad-ō-trō′pinz) are glycoprotein hormones capable of promoting the growth and function of the **gonads,** the ovaries and testes. The two major gonadotropins secreted from the anterior pituitary are **luteinizing** (loo′tē-i-nīz-ing) **hormone (LH)** and **follicle-stimulating hormone (FSH).** LH and FSH, along with prolactin, play important roles in regulating reproduction.

LH and FSH secreted into the blood bind to membrane-bound receptors, increase the intracellular synthesis of cAMP through G protein mechanisms, and stimulate the production of **gametes** (gam′ēts)—sperm cells in the testes and oocytes in the ovaries. LH and FSH also control the production of reproductive hormones—estrogens and progesterone in the ovaries and testosterone in the testes.

LH and FSH are released from anterior pituitary cells under the influence of the hypothalamic-releasing hormone gonadotropin-releasing hormone (GnRH). Gonadal steroid hormones are also critical regulators of the gonadotropins and exhibit a complex cycle of hormone interactions, which are further described in chapter 29.

ASSESS YOUR PROGRESS

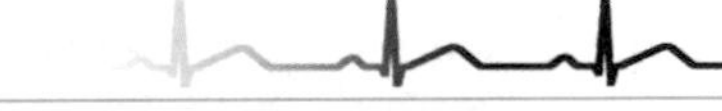

13. *Structurally, what kinds of hormones are released from the anterior pituitary gland? Do these hormones bind to plasma proteins? How long is their half-life, and how do they activate their target tissues?*

14. *What effects do stress, blood amino acid levels, and blood glucose levels have on GH secretion?*

15. *Describe the effects of GH on its target tissues.*

16. *What stimulates the release of somatomedins? Where are they produced, and what are their effects?*

17. *What pathological conditions are the result of hypersecretion of GH? Describe their symptoms.*

18. *What pathological conditions are the result of hyposecretion of GH? Describe their symptoms.*

19. *For each of the following hormones secreted by the anterior pituitary gland, name the target tissue and the hormone's effect on its target tissue: GH, prolactin, ACTH, LH, and FSH.*

20. *How are ACTH, lipotropins, β endorphins, and MSH related? What are the functions of these hormones?*

21. *What is a gonadotropin? Name two gonadotropins produced by the anterior pituitary gland, and explain their functions.*

18.3 Thyroid Gland

LEARNING OUTCOMES

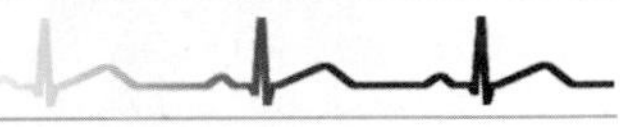

After reading this section, you should be able to

A. Describe the structure of the thyroid gland.

B. Explain the processes for the synthesis, secretion, and blood transport of T_3 and T_4.

C. Describe the mechanism of action and the effects of T_3 and T_4 in the body.

D. Relate how thyroid hormone secretion is regulated.

E. Describe the effects of hyposecretion and hypersecretion of thyroid hormones and the pathological conditions that cause the abnormalities.

F. Describe the role of calcitonin in the maintenance of blood calcium levels and in bone health.

The **thyroid gland** is composed of two lobes connected by a narrow band of thyroid tissue called the **isthmus.** The lobes are lateral to the upper portion of the trachea just inferior to the larynx, and the isthmus extends across the anterior aspect of the trachea (figure 18.8*a*). The thyroid gland is one of the largest endocrine glands, with a weight of approximately 20 g. Because it is highly vascular, it is a darker red than surrounding tissues.

The thyroid gland contains numerous **follicles,** which are small spheres whose walls are composed of a single layer of cuboidal epithelial cells (figure 18.8*b,c*). The center of each thyroid follicle is filled with a gelatinous material called colloid. Colloid is composed of a highly concentrated protein called **thyroglobulin** (thī-rō-glob′ū-lin), which is synthesized and secreted by cells of the thyroid follicle. Large amounts of the thyroid hormones are stored in the thyroglobulin molecules.

Between the follicles, a delicate network of loose connective tissue contains numerous capillaries. Scattered **parafollicular** (par-ă-fo-lik′ū-lăr) **cells** lie between the follicles and among the cells that make up the walls of the follicle. The parafollicular cells secrete **calcitonin** (kal-si-tō′nin), which plays a role in reducing the concentration of calcium in the body fluids when calcium levels become elevated.

Thyroid Hormones

The thyroid hormones include **triiodothyronine** (trī-ī′ō-dō-thī′rō-nēn), commonly called $\mathbf{T_3}$, and **tetraiodothyronine** (tet′ră-ī′ō-dō-thī′rō-nēn). A more common name for tetraiodothyronine is **thyroxine** (thī-rok′sēn; thī-rok′sin), or even more commonly $\mathbf{T_4}$. T_4 is the precursor for T_3, and both are major secretory products of the thyroid gland, consisting of 10% T_3 and 80% T_4, respectively

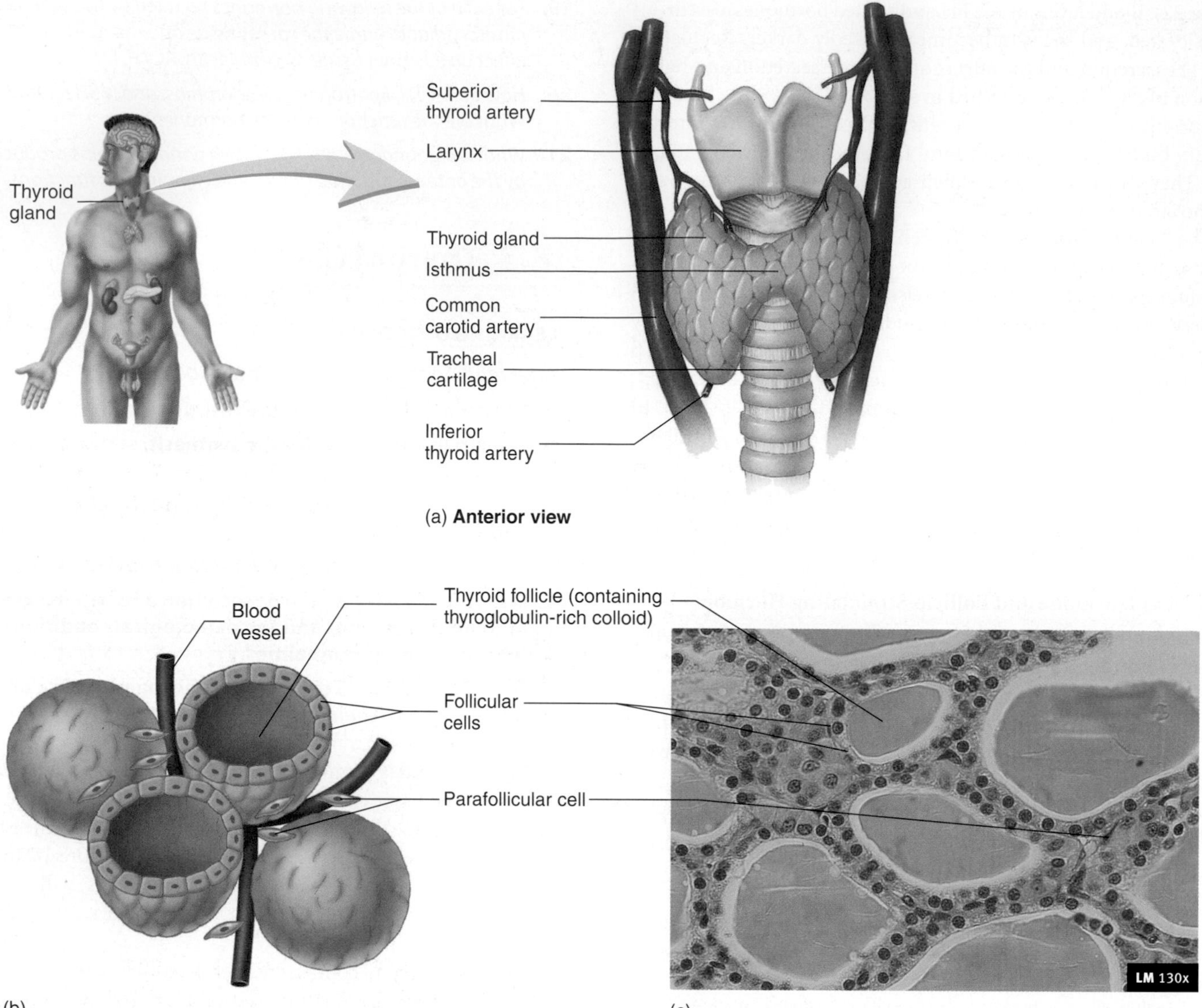
Thyroid gland
Superior thyroid artery
Larynx
Thyroid gland
Isthmus
Common carotid artery
Tracheal cartilage
Inferior thyroid artery
(a) Anterior view
Blood vessel
Thyroid follicle (containing thyroglobulin-rich colloid)
Follicular cells
Parafollicular cell
LM 130x
(b)
(c)

FIGURE 18.8 AP|R **Anatomy and Histology of the Thyroid Gland**

(*a*) Anterior view of the thyroid gland. (*b*) Histology of the thyroid gland. The gland is made up of many spheric thyroid follicles containing thyroglobulin-rich colloid. Parafollicular cells are in the tissue between the thyroid follicles. (*c*) Low-power photomicrograph of thyroid follicles.

(table 18.3). Although calcitonin (10%) is secreted by the parafollicular cells of the thyroid gland, T_3 and T_4 are considered the thyroid hormones because they are more clinically important and because they are secreted from the thyroid follicles.

T_3 and T_4 Synthesis

Thyroid-stimulating hormone (TSH) from the anterior pituitary stimulates thyroid hormone synthesis and secretion. TSH causes an increase in the synthesis of T_3 and T_4, which are then stored inside the thyroid follicles as part of thyroglobulin. TSH also causes T_3 and T_4 to be released from thyroglobulin and enter the circulatory system. Because iodine is a component of T_3 and T_4, an adequate amount of iodine in the diet is required for thyroid hormone synthesis. The following events in the thyroid follicles result in T_3 and T_4 synthesis and secretion (figure 18.9):

1. Iodide ions (I^-) are taken up by thyroid follicle cells via secondary active transport by a sodium-iodide symporter. The active transport of the I^- is against a concentration gradient of approximately 30-fold in healthy individuals.
2. Thyroglobulins, which contain numerous tyrosine molecules, are synthesized within the cells of the follicles.
3. Nearly simultaneously, the I^- are oxidized to form iodine (I), and either one or two iodine atoms are bound to some of the tyrosine molecules of thyroglobulin by the enzyme, thyroid peroxidase.
4. This occurs close to the time the thyroglobulin molecules are secreted by exocytosis into the lumen of the follicle. As a result, the secreted thyroglobulin contains iodinated tyrosine amino acids with either one iodine atom (monoiodotyrosine) or two iodine atoms (diiodotyrosine).

TABLE 18.3 Hormones of the Thyroid and Parathyroid Glands

Hormones	Structure	Target Tissue	Response
Thyroid Gland			
Secreted by *Thyroid Follicles*			
Thyroid hormones (triiodothyronine and tetraiodothyronine)	Amino acid derivative	Most cells of the body	Increased metabolic rate; essential for normal growth and maturation
Secreted by *Parafollicular Cells*			
Calcitonin	Polypeptide	Bone	Decreased rate of breakdown of bone by osteoclasts; prevention of a large increase in blood Ca^{2+} levels
Parathyroid Gland			
Parathyroid hormone (PTH)	Polypeptide	Bone, kidneys, small intestine	Increased rate of breakdown of bone by osteoclasts; increased reabsorption of Ca^{2+} in kidneys; increased absorption of Ca^{2+} from the small intestine; increased vitamin D synthesis; increased blood Ca^{2+} levels

5. Within the colloid in the lumen of the follicle, two diiodotyrosine molecules of thyroglobulin combine to form tetraiodothyronine (T_4), or one monoiodotyrosine and one diiodotyrosine molecule combine to form triiodothyronine (T_3). Large amounts of T_3 and T_4 are stored within the thyroid follicles as part of thyroglobulin. A reserve sufficient to supply thyroid hormones for approximately 2–4 months is stored in this form.
6. Thyroglobulin is taken into the thyroid follicle cells by endocytosis.

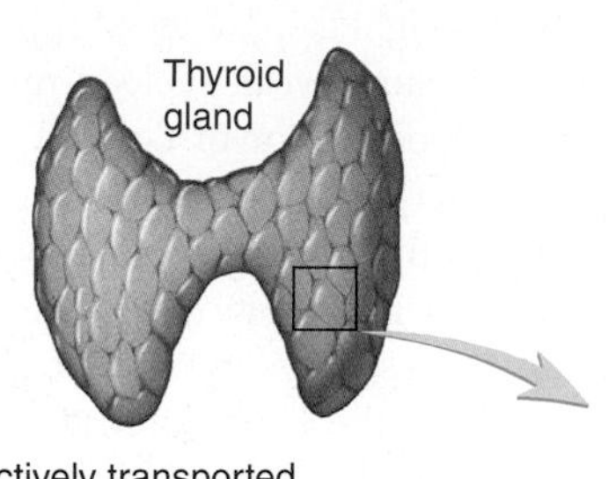
Thyroid gland

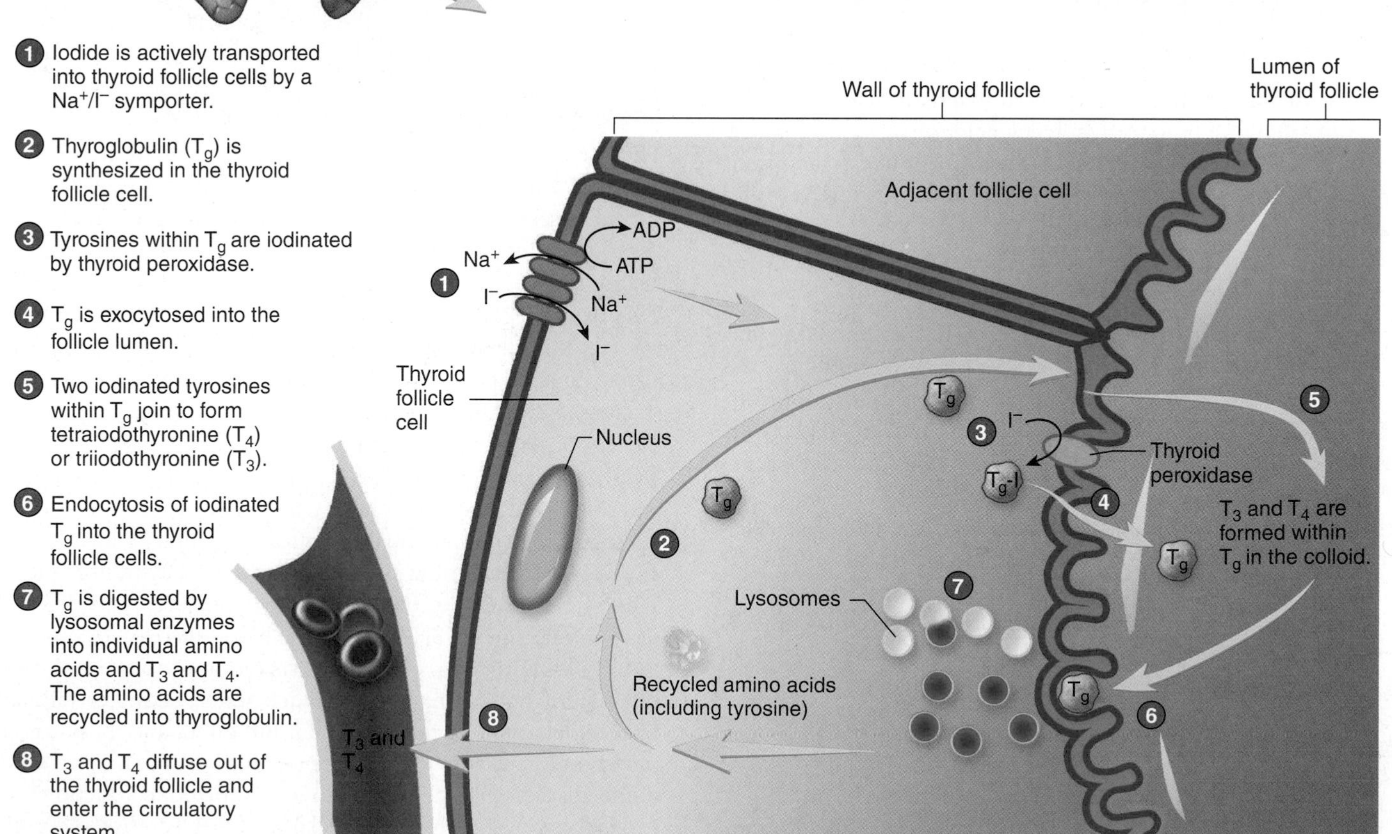
1 Iodide is actively transported into thyroid follicle cells by a Na+/I− symporter.
2 Thyroglobulin (Tg) is synthesized in the thyroid follicle cell.
3 Tyrosines within Tg are iodinated by thyroid peroxidase.
4 Tg is exocytosed into the follicle lumen.
5 Two iodinated tyrosines within Tg join to form tetraiodothyronine (T4) or triiodothyronine (T3).
6 Endocytosis of iodinated Tg into the thyroid follicle cells.
7 Tg is digested by lysosomal enzymes into individual amino acids and T3 and T4. The amino acids are recycled into thyroglobulin.
8 T3 and T4 diffuse out of the thyroid follicle and enter the circulatory system.
Wall of thyroid follicle
Lumen of thyroid follicle
Adjacent follicle cell
Na+
I−
ADP
ATP
Na+
I−
Thyroid follicle cell
Nucleus
Tg
Tg-I
Thyroid peroxidase
T3 and T4 are formed within Tg in the colloid.
Lysosomes
Recycled amino acids (including tyrosine)
T3 and T4

FIGURE 18.9 Biosynthesis of Thyroid Hormones
The numbered steps describe the synthesis and secretion of T_3 and T_4 from the thyroid gland.

7. Lysosomes fuse with the endocytotic vesicles. Proteolytic enzymes break down thyroglobulin to release T_3 and T_4. The remaining amino acids of thyroglobulin are recycled to synthesize more thyroglobulin.
8. T_3 and T_4 either diffuse through the plasma membranes or are carried by specific transporters of the follicular cells into the interstitial spaces and finally into the capillaries of the thyroid gland.

Transport in the Blood

T_3 and T_4 are transported in combination with plasma proteins in the circulatory system. Approximately 75% of the circulating T_3 and T_4 are bound to **thyroxine-binding globulin (TBG),** which is synthesized by the liver, and 20–30% are bound to other plasma proteins, including albumin. T_3 and T_4, bound to these plasma proteins, form a large reservoir of circulating thyroid hormones, and the half-life of these hormones is greatly increased because of this binding. After thyroid gland removal in experimental animals, it takes approximately 1 week for T_3 and T_4 levels in the blood to decrease by 50%. As free T_3 and T_4 levels decrease in the interstitial spaces, additional T_3 and T_4 dissociate from the plasma proteins to maintain the levels in the tissue spaces. When sudden secretion of T_3 and T_4 occurs, the excess binds to the plasma proteins. As a consequence, the concentration of thyroid hormones in the tissue spaces fluctuates very little.

Approximately 40% of the T_4 is converted to T_3 in the body tissues. This conversion can be important in the action of thyroid hormones on their target tissues because T_3 is the major hormone that interacts with target cells. T_3 is several times more potent than T_4 because of its higher affinity for the thyroid hormone receptor.

Much of the circulating T_4 is eliminated from the body by being converted to tetraiodothyroacetic acid, or by being sulfated or glucuronated, and then excreted in the urine or bile. In addition, a large amount is converted to an inactive form of T_3, rapidly metabolized, and excreted.

Mechanism of Action of T_3 and T_4

T_3 and T_4 interact with their target tissues in a fashion similar to that of the steroid hormones. They are transported through the plasma membrane into the cytoplasm of cells. Within cells, they bind to receptor molecules in the nuclei. Thyroid hormones combined with their receptor molecules interact with DNA in the nucleus to influence regulatory genes and regulate protein synthesis. The newly synthesized proteins within the target cells mediate the cells' response to thyroid hormones. It takes up to a week after the administration of thyroid hormones for a maximal response to develop, and new protein synthesis occupies much of that time.

Effects of T_3 and T_4

T_3 and T_4 affect nearly every tissue in the body, but not all tissues respond identically. Metabolism is primarily affected in some tissues, and growth and maturation are influenced in others.

The normal rate of metabolism for an individual depends on an adequate supply of thyroid hormone, which increases the rate at which glucose, lipids, and protein are metabolized. The metabolic rate can increase 60–100% when blood T_3 and T_4 are elevated, whereas low levels of T_3 and T_4 lead to the opposite effect. Because the increased rate of metabolism produces heat, normal body temperature is partly due to adequate thyroid hormones. Thyroid hormones increase the activity of Na^+–K^+ pumps, which give off heat as a "by-product." T_3 and T_4 also alter the number and activity of mitochondria, resulting in greater ATP synthesis and thus heat production.

In addition to metabolism, T_3 and T_4 regulate the normal growth and maturation of organs. For example, the growth of bone, hair, teeth, connective tissue, and nervous tissue requires thyroid hormone. One reason tissues require thyroid hormones for normal growth is that T_3 and T_4 play a permissive role for GH, which means that GH does not have its normal effect on target tissues if T_3 and T_4 are not present.

Failure to maintain homeostatic amounts of thyroid hormone dramatically affects the body's functions. Hypersecretion of T_3 and T_4 increases the rate of metabolism. High body temperature, weight loss, increased appetite, rapid heart rate, and an enlarged thyroid gland are major symptoms.

Hyposecretion of T_3 and T_4 decreases the rate of metabolism. Low body temperature, weight gain, reduced appetite, reduced heart rate, reduced blood pressure, weak skeletal muscles, and apathy are major symptoms. Hyposecretion of T_3 and T_4 that occurs during development causes a decreased metabolic rate, abnormal nervous system development, abnormal growth, and abnormal maturation of tissues. The consequence is neonatal hypothyroidism characterized by mental retardation, short stature, and specific physical deformities.

The specific effects of the hyposecretion and hypersecretion of thyroid hormones are outlined in table 18.4.

Regulation of Thyroid Hormone Secretion

Thyrotropin-releasing hormone (TRH) from the hypothalamus and TSH from the anterior pituitary function together to increase T_3 and T_4 secretion from the thyroid gland (figure 18.10). Stress and exposure to cold cause increased TRH secretion, and prolonged fasting decreases TRH secretion. TRH stimulates the secretion of TSH from the anterior pituitary. TSH travels to the thyroid gland, where it stimulates the synthesis and secretion of T_3 and T_4. TSH also causes hypertrophy (increased cell size) and hyperplasia (increased cell number) of the thyroid gland, which helps explain the abnormal thyroid gland growth called goiter (table 18.5). Decreased blood levels of TSH lead to decreased T_3 and T_4 secretion and to thyroid gland atrophy. T_3 and T_4 have a negative-feedback effect on the hypothalamus and anterior pituitary gland. As T_3 and T_4 levels increase in the circulatory system, they inhibit TRH and TSH secretion. If the thyroid gland is removed or if T_3 and T_4 secretion declines, TSH levels in the blood increase dramatically.

Hypothyroidism, or reduced secretion of thyroid hormones, can result from having an iodine deficiency, taking certain drugs, or being exposed to chemicals that inhibit T_3 and T_4 synthesis. It can also be caused by inadequate secretion of TSH, by an autoimmune disease that depresses thyroid hormone function, or by surgical removal of the thyroid gland. Hypersecretion of T_3 and T_4 can result from the synthesis of an immunoglobulin that stimulates TSH receptors and acts like TSH, from TSH-secreting tumors of the pituitary gland, and from thyroid tumors.

TABLE 18.4 Effects of Hyposecretion and Hypersecretion of Thyroid Hormones

Hypothyroidism	Hyperthyroidism
Decreased metabolic rate, low body temperature, cold intolerance	Increased metabolic rate, high body temperature, heat intolerance
Weight gain, reduced appetite	Weight loss, increased appetite
Reduced activity of sweat and sebaceous glands; dry, cold skin	Copious sweating; warm, flushed skin
Reduced heart rate, reduced blood pressure, dilated and enlarged heart	Rapid heart rate, elevated blood pressure, abnormal electrocardiogram
Weak, untoned skeletal muscles; sluggish movements	Weak skeletal muscles that exhibit tremors, quick movements with exaggerated reflexes
Constipation	Bouts of diarrhea
Myxedema (swelling of the face and body) as a result of subcutaneous mucoprotein deposits	Exophthalmos (protruding eyes) as a result of connective tissue proliferation and other deposits behind the eye
Apathy, somnolence	Hyperactivity, insomnia, restlessness, irritability, short attention span
Coarse hair; rough, dry skin	Soft, smooth hair and skin
Decreased iodide uptake	Increased iodide uptake
Possible goiter (enlargement of the thyroid gland) due to loss of negative feedback	Almost always a goiter

1. Stress and hypothermia cause TRH to be released from neurons within the hypothalamus. It passes through the hypothalamohypophysial portal system to the anterior pituitary.
2. TRH causes cells of the anterior pituitary to secrete TSH, which passes through the general circulation to the thyroid gland.
3. TSH causes increased synthesis and release of T_3 and T_4 into the general circulation.
4. T_3 and T_4 act on target tissues to produce a response.
5. T_3 and T_4 also have an inhibitory effect on the secretion of TRH from the hypothalamus and TSH from the anterior pituitary.

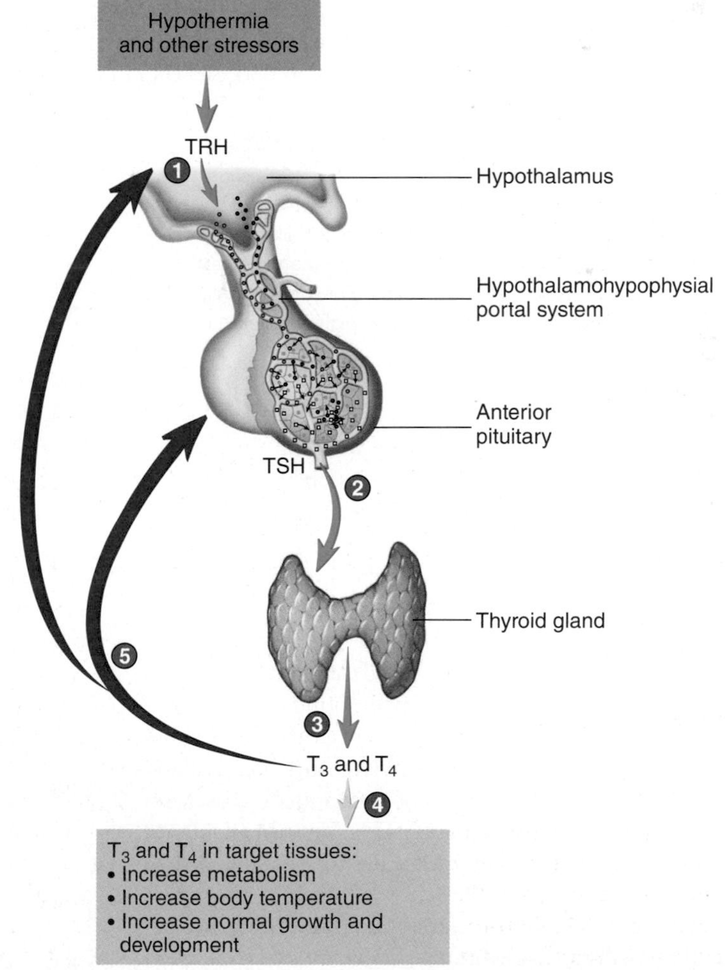
Hypothermia and other stressors
TRH
1
Hypothalamus
Hypothalamohypophysial portal system
Anterior pituitary
TSH
2
Thyroid gland
5
3
T_3 and T_4
4
T_3 and T_4 in target tissues:
• Increase metabolism
• Increase body temperature
• Increase normal growth and development

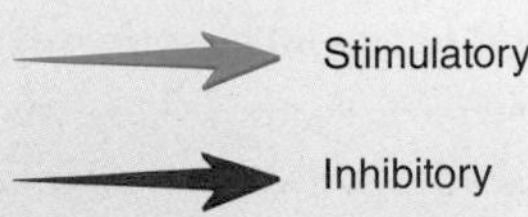
Stimulatory
Inhibitory

FIGURE 18.10 Regulation of Thyroid Hormone (T_3 and T_4) Secretion

TABLE 18.5 Abnormal Thyroid Conditions

Cause	Description
Hypothyroidism	
Iodine deficiency	Causes inadequate T_3 and T_4 synthesis, which results in elevated thyroid-stimulating hormone (TSH) secretion; thyroid gland enlarges (goiter) as a result of TSH stimulation; T_3 and T_4 frequently remain in the low to normal range
Goitrogenic (goiter-causing) substances	Inhibit T_3 and T_4 synthesis; found in certain drugs and in small amounts in certain plants, such as cabbage
Neonatal hypothyroidism	Caused by maternal iodine deficiency or congenital errors in thyroid hormone synthesis; results in mental retardation and a short, malformed appearance
Pituitary insufficiency	Results from lack of TSH secretion; often associated with inadequate secretion of other anterior pituitary hormones
Hashimoto disease	Autoimmune disease in which thyroid hormone secretion can be normal or depressed
Lack of thyroid gland	Partial or complete surgical removal or drug-induced destruction of the thyroid gland as a treatment for Graves disease (hyperthyroidism)
Hyperthyroidism	
Graves disease	Characterized by goiter and exophthalmos; apparently an autoimmune disease; most patients have a TSH-like immunoglobulin, called thyroid-stimulating immunoglobulin (TSI), in their plasma
Tumors—benign adenoma or cancer	Result in either normal secretion or hypersecretion of thyroid hormones (rarely hyposecretion)
Thyroiditis—a viral infection	Produces painful swelling of the thyroid gland with normal or slightly increased T_3 and T_4 production
Elevated TSH levels	Result from a pituitary tumor
Thyroid storm	Sudden release of large amounts of T_3 and T_4; caused by surgery, stress, infections, or other, unknown factors

Predict 5

Becky has lost 30 pounds over the past several months, even though her appetite has been good and she has been eating more than usual. She complains to her physician that she is nervous and restless, has a short attention span, becomes fatigued easily but cannot sleep well, moves compulsively, and sweats excessively. Her physician notes that she also exhibits tachycadia. Suspecting hyperthyroidism, he orders a blood test, which indicates elevated levels of T_3 and T_4 and low levels of TSH. Becky also has a TSH-like immunoglobulin in her plasma. Explain these results.

Calcitonin

Parafollicular cells, or C cells, are dispersed between the thyroid follicles throughout the thyroid gland. These cells secrete the hormone calcitonin in response to increased calcium levels in the blood.

The primary target tissue for calcitonin is bone (see chapter 6). Calcitonin binds to membrane-bound receptors, decreases osteoclast activity, and lengthens the life span of osteoblasts. The resulting bone deposition leads to decreases in blood calcium and phosphate levels.

Calcitonin's importance in regulating blood Ca^{2+} levels is unclear; it may be most important in juveniles in promoting bone growth. The rate of calcitonin secretion increases in response to elevated blood Ca^{2+} levels, and calcitonin may prevent large increases in blood Ca^{2+} levels following a meal. Blood levels of calcitonin decrease with age to a greater extent in females than in males. Also, the incidence of osteoporosis increases with age and occurs to a greater degree in females than in males.

Interestingly, complete thyroidectomy does not result in high blood Ca^{2+} levels, possibly because the regulation of blood Ca^{2+} levels by vitamin D and other hormones, such as parathyroid hormone, compensates for the loss of calcitonin in individuals who have undergone a thyroidectomy. No pathological condition is directly associated with a lack of calcitonin secretion. Some evidence suggests that calcitonin may play a role in regulating food intake by decreasing appetite. Clinically, calcitonin nasal sprays have been effective in the management of postmenopausal osteoporosis.

ASSESS YOUR PROGRESS

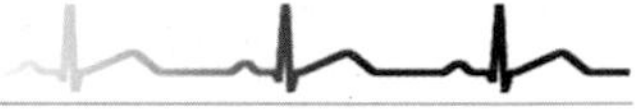

22. *Where is the thyroid gland located? Describe the follicles and the parafollicular cells within the thyroid. What hormones do they produce?*
23. *Starting with the uptake of iodide by the follicles, describe the production and secretion of thyroid hormones (T_3 and T_4).*
24. *How are the thyroid hormones transported in the blood? What effect does this transport have on their half-life?*
25. *What are the target tissues of thyroid hormones? By what mechanism do thyroid hormones alter the activities of their target tissues? What effects are produced?*
26. *Starting in the hypothalamus, explain how chronic exposure to cold, food deprivation, or stress can affect thyroid hormone production.*
27. *Diagram two negative-feedback mechanisms involving hormones that regulate the production of thyroid hormones.*
28. *What is a goiter? What can cause one to develop?*
29. *What conditions cause hypothyroidism? Describe the effects of hyposecretion.*
30. *What conditions cause hyperthyroidism? Describe the effects of hypersecretion.*
31. *What effect does calcitonin have on osteoclasts, osteoblasts, and blood calcium levels? What stimulus can cause an increase in calcitonin secretion?*

Clinical
IMPACT

Goiter and Exophthalmos

An abnormal enlargement of the thyroid gland is called a **goiter**. Goiters can result from either hypothyroidism or hyperthyroidism. An **iodine-deficiency goiter** results when dietary iodine intake is so low that there is not enough iodine to synthesize T_3 and T_4 (table 18.5). As a result, blood levels of T_3 and T_4 decrease, and the person may exhibit symptoms of hypothyroidism. At the same time, the reduced negative feedback of T_3 and T_4 on the anterior pituitary and hypothalamus results in elevated TSH secretion. TSH causes hypertrophy and hyperplasia of the thyroid gland and increased thyroglobulin synthesis, even though there is not enough iodine to synthesize T_3 and T_4. Consequently, the thyroid gland enlarges.

Historically, iodine-deficiency goiters were common in people inhabiting areas where the soil was depleted of iodine, called "goiter belts." Consequently, plants grown in these areas had little iodine in them and caused iodine-deficient diets. In the United States, iodized salt has nearly eliminated iodine-deficiency goiters. However, iodine-deficiency diseases are still common throughout the world. The World Health Organization has called them the most common preventable cause of mental defects, and hypothyroidism may be the most common endemic disease on the planet.

A **toxic goiter** secretes excess T_3 and T_4, and it can result from elevated TSH secretion or elevated TSH-like immunoglobulin (see Graves disease in table 18.5). Toxic goiter results in symptoms of hyperthyroidism. Protruding eyes, a condition called **exophthalmos,** often accompany hyperthyroidism. The deposition of excess connective tissue proteins behind the eyes makes the eyes move anteriorly; consequently, they appear to be larger than normal.

Graves disease is the most common cause of hyperthyroidism (see Systems Pathology, later in this chapter). Elevated T_3 and T_4 levels resulting from this condition suppress TSH and TRH, but the T_3 and T_4 levels remain elevated. Exophthalmos is common. Treatment often involves antithyroid drugs that reduce T_3 and T_4 secretion, partial destruction of the thyroid gland using radioactive iodine, and partial or complete surgical removal of the thyroid gland. These treatments are followed by the oral administration of the appropriate amount of T_3 and T_4. Unfortunately, removal of the thyroid gland normally does not reverse exophthalmos.

18.4 Parathyroid Glands

LEARNING OUTCOMES

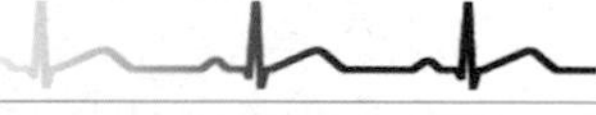

After reading this section, you should be able to

A. Describe the location and structure of the parathyroid glands.

B. Explain the mechanism of parathyroid hormone action and its effects.

C. Describe the causes and symptoms of hypoparathyroidism and hyperparathyroidism.

The **parathyroid** (par-ă-thī′royd) **glands** are usually embedded in the posterior part of each lobe of the thyroid gland and are made up of two cell types: chief cells and oxyphils. The chief cells secrete parathyroid hormone, but the function of the oxyphils is unknown. Usually, four parathyroid glands are present, with their cells organized in densely packed masses, or cords, rather than in follicles (figure 18.11). In some cases, one or more of the parathyroid glands do not become embedded in the thyroid gland and remain in the nearby connective tissue.

Parathyroid hormone (PTH), also called *parathormone*, is a polypeptide hormone that is important in regulating calcium levels

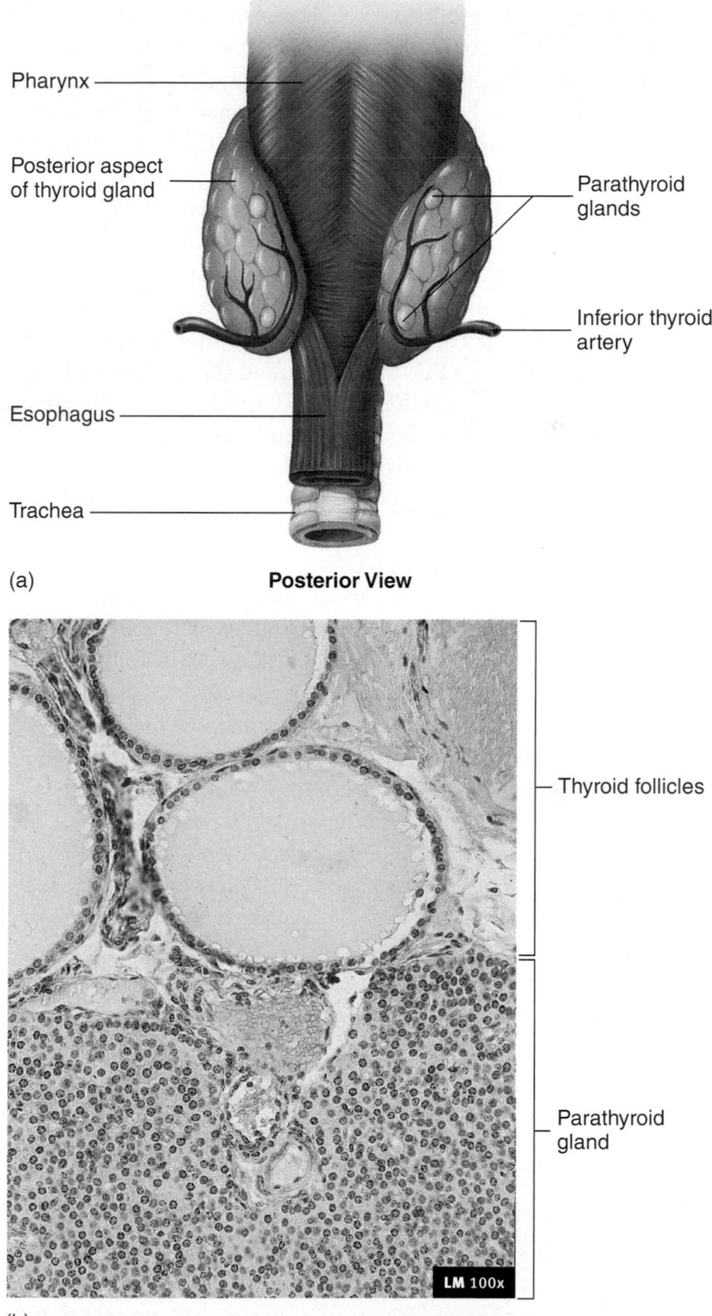
Pharynx
Posterior aspect of thyroid gland
Parathyroid glands
Inferior thyroid artery
Esophagus
Trachea
(a)
Posterior View
Thyroid follicles
Parathyroid gland
LM 100x
(b)

FIGURE 18.11 APR **Anatomy and Histology of the Parathyroid Glands**
(*a*) The parathyroid glands are embedded in the posterior part of the thyroid gland. (*b*) The parathyroid glands are composed of densely packed cords of cells.

in body fluids (see table 18.3). Its major target tissues are bone, the kidneys, and the small intestine. PTH binds to membrane-bound receptors and activates a G protein mechanism that increases intracellular cAMP levels in target tissues. Without functional parathyroid glands, the ability to adequately regulate blood calcium levels is lost.

PTH stimulates osteoclast activity in bone and can cause the number of osteoclasts to increase. The increased osteoclast activity results in bone reabsorption and the release of calcium and phosphate, causing an increase in blood calcium levels. Osteoclasts have no PTH receptors, but osteoblasts and red bone marrow stem cells do. PTH binds to receptors on osteoblasts, which then promote an increase in osteoclast activity (see chapter 6).

PTH induces calcium reabsorption within the kidneys, so that less calcium leaves the body in urine. It also increases the enzymatic formation of active vitamin D in the kidneys. Calcium is actively absorbed by the epithelial cells of the small intestine, and the synthesis of transport proteins in the intestinal cells requires active vitamin D. PTH increases the rate of active vitamin D synthesis, which in turn increases the rate of calcium and phosphate absorption in the intestine, thereby elevating blood levels of calcium.

Although PTH increases the release of phosphate ions (PO_4^{3-}) from bone and increases PO_4^{3-} absorption in the small intestine, it increases PO_4^{3-} excretion in the kidney. The overall effect of PTH is to decrease blood phosphate levels. A simultaneous increase in both Ca^{2+} and PO_4^{3-} is undesirable because it allows calcium phosphate to precipitate in the body's soft tissues, where they cause irritation and inflammation.

The regulation of PTH secretion and the role of PTH and calcitonin in regulating blood Ca^{2+} levels are outlined in figure 18.12. The primary stimulus for the secretion of PTH is a decrease in blood Ca^{2+} levels, whereas elevated blood Ca^{2+} levels inhibit PTH secretion. This regulation keeps blood Ca^{2+} levels fluctuating within a normal range of values. Both hypersecretion and hyposecretion of PTH cause serious symptoms (table 18.6). The regulation of blood Ca^{2+} levels is discussed more thoroughly in chapter 27.

Inactive parathyroid glands result in **hypocalcemia**, abnormally low levels of calcium in the blood. Reduced extracellular calcium levels cause voltage-gated Na^+ channels in plasma membranes to open, which increases the permeability of plasma membranes to Na^+. As a consequence, Na^+ diffuses into cells and causes depolarization (see chapter 11). Symptoms of hypocalcemia are nervousness, muscle spasms, cardiac arrhythmia, and convulsions. Extreme cases may lead to tetany of skeletal muscles, including the respiratory muscles, which can cause death.

Predict 6

A patient with a malignant tumor had his thyroid gland removed. What effect does this removal have on blood levels of Ca^{2+}? If the parathyroid glands are inadvertently removed along with the thyroid gland, death can result because the muscles of respiration undergo sustained contractions. Explain.

ASSESS YOUR PROGRESS

32. *Where are the parathyroid glands located, and what hormone do they produce?*
33. *What are the major target tissues for parathyroid hormone?*
34. *What effect does PTH have on osteoclasts, osteoblasts, the kidneys, the small intestine, and bone?*
35. *How are blood calcium and phosphate levels regulated by PTH?*
36. *What can cause hypoparathyroidism? Describe the symptoms.*
37. *What can cause hyperparathyroidism? Describe the symptoms.*

18.5 Adrenal Glands

LEARNING OUTCOMES

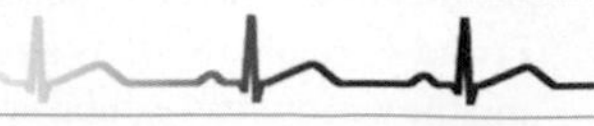

After reading this section, you should be able to

A. **Relate the location of the adrenal glands and explain the embryological origins of the two parts of the glands.**
B. **Describe the mechanisms and actions of the hormones secreted by the adrenal medulla.**
C. **Name the layers of the adrenal cortex, the type of product secreted by each layer, and the predominant hormone of each layer.**
D. **Describe the individual target tissues and their responses to the hormones of the adrenal cortex.**
E. **Explain the role of ACTH in the regulation of the adrenal cortex hormones.**
F. **Discuss the causes and symptoms of hyposecretion and hypersecretion of adrenal cortex hormones.**

The **adrenal** (ă-drē′năl) **glands,** also called the *suprarenal* (soo′pră-rē′năl) *glands*, are near the superior poles of the kidneys. Like the kidneys, they are retroperitoneal, and they are surrounded by abundant adipose tissue. The adrenal glands are enclosed by a connective tissue capsule and have a well-developed blood supply (figure 18.13*a*).

The adrenal glands are composed of an inner **medulla** and an outer **cortex**, which are derived from two separate embryonic tissues. The adrenal medulla arises from neural crest cells, which also give rise to postganglionic neurons of the sympathetic division of the autonomic nervous system (see chapters 16 and 29). Unlike most glands of the body, which develop from invaginations of epithelial tissue, the adrenal cortex is derived from mesoderm.

Trabeculae of the connective tissue capsule penetrate the adrenal gland in several locations, and numerous small blood vessels course within the trabeculae to supply the gland. The adrenal medulla consists of closely packed polyhedral cells centrally located in the gland (figure 18.13*b*). The adrenal cortex is composed of smaller cells and forms three indistinct layers: the **zona glomerulosa** (glō-mār′ū-lōs-ă), the **zona fasciculata** (fa-sik′ū-lă-tă), and the **zona reticularis** (re-tik′ū-lăr′is). These three layers are functionally and structurally specialized. The zona glomerulosa, located immediately beneath the capsule, is composed of small clusters of cells and secretes aldosterone. Beneath the zona glomerulosa is the thickest part of the adrenal cortex, the zona fasciculata, which secretes cortisol. In this layer, the cells form long columns, or fascicles, that extend from the surface toward the medulla of the gland. The deepest layer of the adrenal cortex, the zona reticularis, secretes androgens and is a thin layer of irregularly arranged cords of cells.

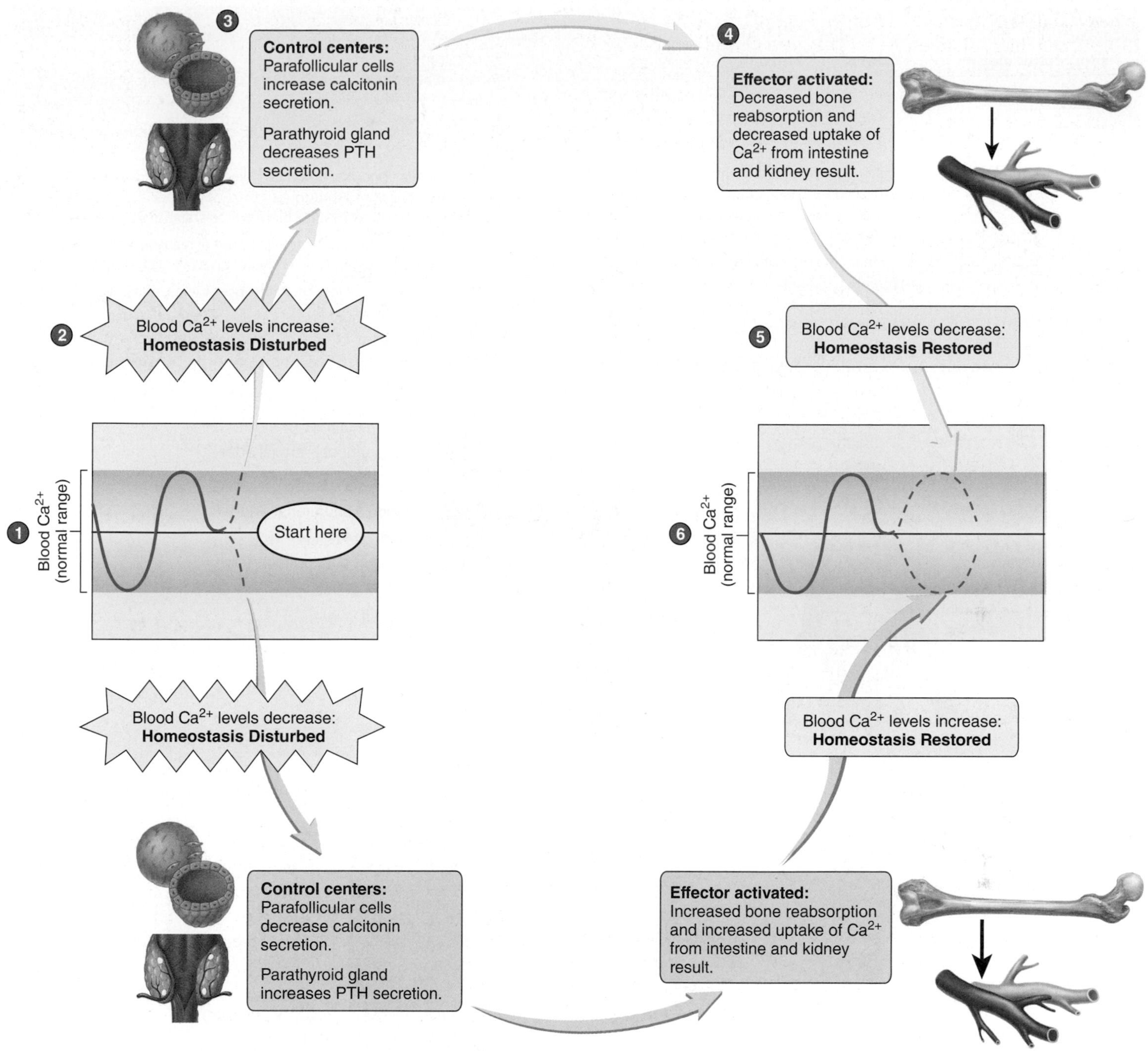

3
Control centers:
Parafollicular cells increase calcitonin secretion.
Parathyroid gland decreases PTH secretion.
4
Effector activated:
Decreased bone reabsorption and decreased uptake of Ca2+ from intestine and kidney result.
2
Blood Ca2+ levels increase:
Homeostasis Disturbed
5
Blood Ca2+ levels decrease:
Homeostasis Restored
1
Blood Ca2+ (normal range)
Start here
6
Blood Ca2+ (normal range)
Blood Ca2+ levels decrease:
Homeostasis Disturbed
Blood Ca2+ levels increase:
Homeostasis Restored
Control centers:
Parafollicular cells decrease calcitonin secretion.
Parathyroid gland increases PTH secretion.
Effector activated:
Increased bone reabsorption and increased uptake of Ca2+ from intestine and kidney result.

HOMEOSTASIS FIGURE 18.12 Regulation of Blood Ca^{2+} Concentration by Parathyroid Hormone (PTH)
(1) Blood Ca^{2+} is within its normal range. (2) Blood Ca^{2+} level increases outside the normal range. (3) The parafollicular cells and the parathyroid gland cells detect elevated blood Ca^{2+}. The parafollicular cells secrete calcitonin; the parathyroid gland cells decrease PTH secretion. (4) There is less bone reabsorption and less uptake of Ca^{2+} from both the kidney and the small intestine. (5) Blood Ca^{2+} level drops back to its normal range. (6) Homeostasis is restored. Observe the response to a drop in blood Ca^{2+} by following the red arrows.

Hormones of the Adrenal Medulla

The adrenal medulla secretes two major hormones: **Epinephrine** (ep′i-nef′rin; *adrenaline*; ă-dren′ă-lin) accounts for 80% of adrenal medulla hormones, and **norepinephrine** (nōr′ep-i-nef′rin; *noradrenaline*; nor-ă-dren′ă-lin) accounts for 20% (table 18.7). Epinephrine and norepinephrine are closely related. In fact, norepinephrine is a precursor to the formation of epinephrine. Because the adrenal medulla consists of cells derived from the same cells that give rise to postganglionic sympathetic neurons, its secretory products are neurohormones.

Epinephrine and norepinephrine combine with adrenergic receptors, which are membrane-bound receptors in target cells.

TABLE 18.6 Causes and Symptoms of Hyposecretion and Hypersecretion of Parathyroid Hormone

	Cause	Symptoms
Hypoparathyroidism	Accidental removal during thyroidectomy	Hypocalcemia Increased neuromuscular excitability; possible tetany, laryngospasm, and death from asphyxiation Flaccid heart muscle; possible cardiac arrhythmia Diarrhea
Hyperparathyroidism	Primary hyperparathyroidism: a result of abnormal parathyroid function—adenomas of the parathyroid gland (90%), hyperplasia of parathyroid idiopathic (unknown cause) cells (9%), or carcinomas (1%) Secondary hyperparathyroidism: caused by conditions that reduce blood Ca^{2+} levels, such as inadequate Ca^{2+} in the diet, inadequate levels of vitamin D, pregnancy, or lactation	Hypercalcemia or normal blood Ca^{2+} levels; calcium carbonate salts may be deposited throughout the body, especially in the renal tubules (kidney stones), lungs, blood vessels, and gastric mucosa Bones weakened and eaten away as a result of reabsorption; some cases are first diagnosed when a radiograph is taken of a broken bone Neuromuscular system less excitable; possible muscular weakness Increased force of contraction of cardiac muscle; at very high blood Ca^{2+} levels, possible cardiac arrest during contraction Constipation

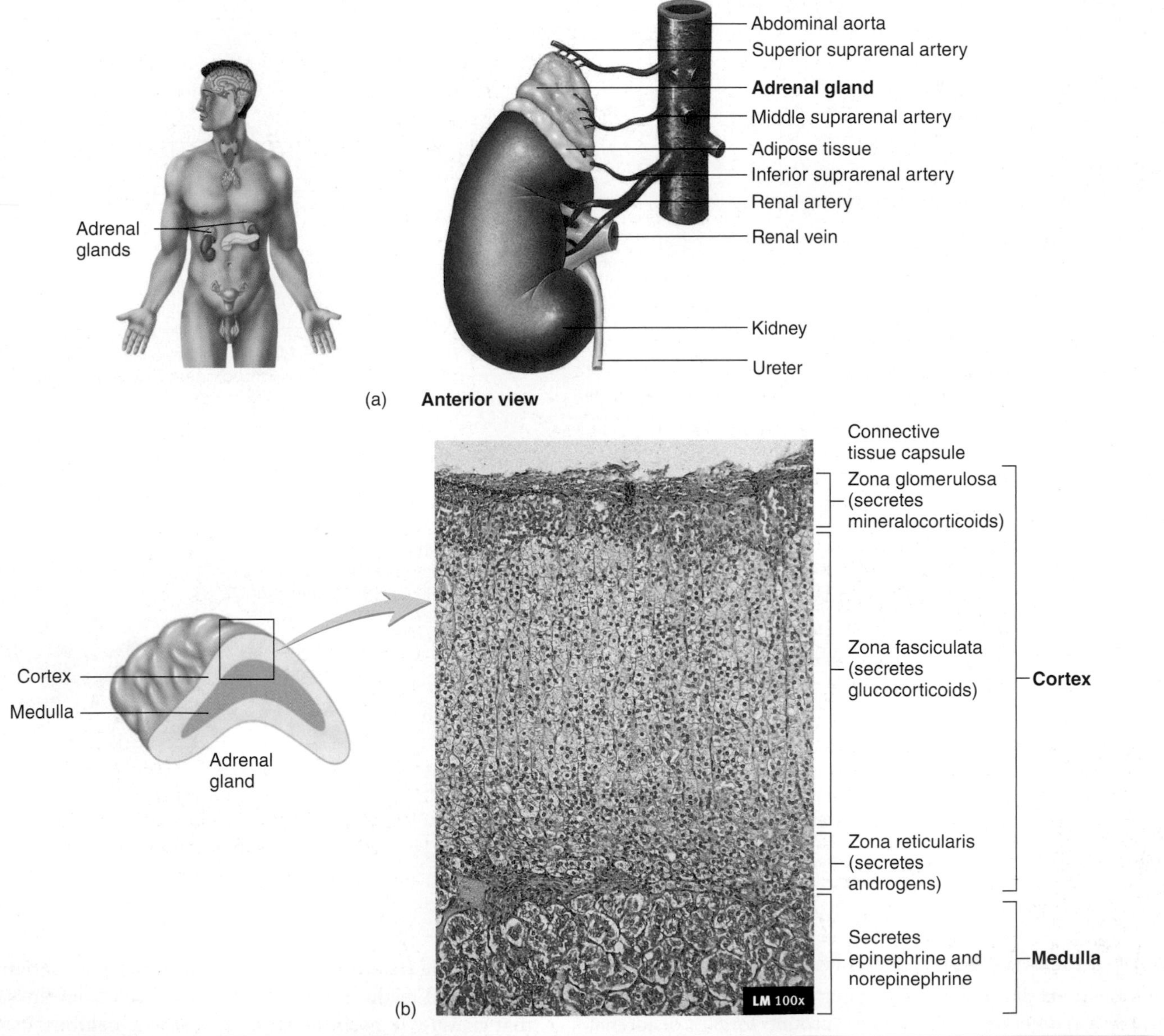

Adrenal glands
Abdominal aorta
Superior suprarenal artery
Adrenal gland
Middle suprarenal artery
Adipose tissue
Inferior suprarenal artery
Renal artery
Renal vein
Kidney
Ureter
(a) Anterior view
Cortex
Medulla
Adrenal gland
Connective tissue capsule
Zona glomerulosa (secretes mineralocorticoids)
Zona fasciculata (secretes glucocorticoids)
Zona reticularis (secretes androgens)
Secretes epinephrine and norepinephrine
Cortex
Medulla
LM 100x
(b)

FIGURE 18.13 APR Anatomy and Histology of Adrenal Glands
(*a*) An adrenal gland is at the superior pole of each kidney. (*b*) The adrenal glands have an outer cortex and an inner medulla. The cortex is surrounded by a connective tissue capsule and consists of three layers: the zona glomerulosa, the zona fasciculata, and the zona reticularis.

TABLE 18.7 Hormones of the Adrenal Gland

Hormones	Structure	Target Tissue	Response
Adrenal Medulla			
Epinephrine primarily; norepinephrine	Amino acid derivatives	Heart, blood vessels, liver, adipose cells	Increased cardiac output; increased blood flow to skeletal muscles and to the heart (see chapter 20); vasoconstriction of blood vessels, especially in the viscera and skin; increased release of glucose and fatty acids into the blood; in general, preparation for physical activity
Adrenal Cortex			
Mineralocorticoids (aldosterone)	Steroids	Kidney	Increased Na^+ reabsorption and K^+ and H^+ excretion; enhanced water reabsorption
Glucocorticoids (cortisol)	Steroids	Most tissues	Increased protein and lipid breakdown; increased glucose production; inhibition of immune response and decreased inflammation
Androgens	Steroids	Many tissues	Of minor importance in males; in females, development of some secondary sex characteristics, such as axillary and pubic hair

They are classified as either α-adrenergic or β-adrenergic receptors, and each of these categories has subcategories that affect target tissues differently. All of the adrenergic receptors function through G protein mechanisms. In general, the α-adrenergic receptors cause Ca^{2+} channels to open, cause the release of Ca^{2+} from the endoplasmic reticulum by activating phospholipase enzymes, open K^+ channels, decrease cAMP synthesis, or stimulate the synthesis of eicosanoid molecules, such as prostaglandins. The β-adrenergic receptors all increase cAMP synthesis. A complete description of epinephrine and norepinephrine is not included in this chapter; rather, the effects of these hormones are described in the context of the body systems (see chapters 16, 20, 21, 24, and 26).

Epinephrine increases blood glucose levels. It binds to membrane-bound receptors in liver cells. Cyclic AMP, in turn, activates enzymes that catalyze the breakdown of glycogen to glucose and the release of glucose from the liver cells into the blood. Epinephrine also increases the breakdown of glycogen in muscle cells, but muscle cells do not release glucose into the blood because glucose is metabolized in muscle cells. Epinephrine also increases lipid breakdown in adipose tissue, and fatty acids are released into the blood. The fatty acids can be taken up and metabolized by tissues such as skeletal and cardiac muscle. Epinephrine and norepinephrine increase the heart's rate and force of contraction and cause blood vessels to constrict in the skin, kidneys, gastrointestinal tract, and other viscera. Also, epinephrine causes dilation of blood vessels in skeletal muscles and cardiac muscle.

Secretion of adrenal medullary hormones prepares the individual for physical activity and is a major component of the fight-or-flight response (see chapter 16). This response results in reduced activity in organs not essential for physical activity, as well as increased blood flow and metabolic activity in organs that participate in physical activity. In addition, it mobilizes nutrients that can be used to sustain physical exercise.

The effects of epinephrine and norepinephrine are short-lived because they are rapidly metabolized, excreted, or taken up by tissues. Their half-life in the circulatory system is measured in minutes.

The release of adrenal medullary hormones primarily occurs in response to stimulation by sympathetic neurons because the adrenal medulla is a specialized part of the autonomic nervous system. Several conditions, including emotional excitement, injury, stress, exercise, and low blood glucose levels, lead to the release of adrenal medullary neuropeptides (figure 18.14).

ASSESS **YOUR PROGRESS** 

38. *Where are the adrenal glands located? Describe the embryonic origins of the adrenal medulla and the adrenal cortex.*
39. *Name two hormones secreted by the adrenal medulla, and list the effects of these hormones.*
40. *List several conditions that can stimulate the production of adrenal medullary hormones. What role does the nervous system play in the release of these hormones? How does this role relate to the embryonic origin of the adrenal medulla?*

Hormones of the Adrenal Cortex

The adrenal cortex secretes three hormone types: **mineralocorticoids** (min′er-al-ō-kōr′ti-koydz), **glucocorticoids** (gloo-kō-kōr′ti-koydz), and **androgens** (an′drō-jenz; table 18.7). All have a similar structure in that they are steroids, highly specialized lipids derived from cholesterol. Because these hormones are lipid-soluble, they are not stored in the adrenal gland cells but diffuse from the cells as they are synthesized. Adrenal cortical hormones are transported in the blood in combination with specific plasma proteins; they are metabolized in the liver and excreted in the bile and urine. The hormones of the adrenal cortex bind to nuclear receptors and stimulate the synthesis of specific proteins responsible for producing the cell's responses.

Mineralocorticoids

The major secretory products of the zona glomerulosa are the mineralocorticoids. **Aldosterone** (al-dos′ter-ōn) is produced in the greatest amounts, although other, closely related mineralocorticoids are also secreted. Aldosterone is secreted under low blood pressure conditions; it increases the rate of sodium reabsorption by the kidneys, thereby increasing blood levels of sodium. Sodium

1. Stress, physical activity, and low blood glucose levels act as stimuli to the hypothalamus, resulting in increased sympathetic nervous system activity.
2. An increased frequency of action potentials conducted through the sympathetic division of the autonomic nervous system stimulates the adrenal medulla to secrete epinephrine and some norepinephrine into the circulatory system.
3. Epinephrine and norepinephrine act on their target tissues to produce responses.

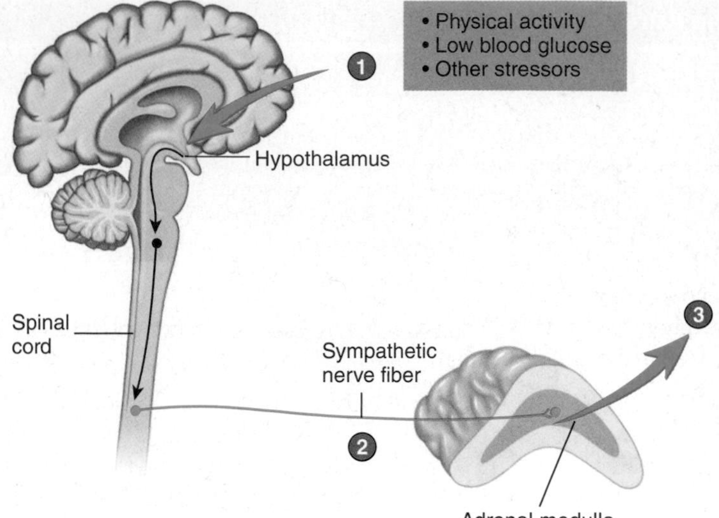
1
• Physical activity
• Low blood glucose
• Other stressors
Hypothalamus
Spinal cord
Sympathetic nerve fiber
2
Adrenal medulla

3

Epinephrine and norepinephrine in the target tissues:
- Increase the release of glucose from the liver into the blood
- Increase the release of fatty acids from adipose tissue into the blood
- Increase heart rate
- Decrease blood flow through blood vessels of most internal organs
- Increase blood flow through blood vessels of skeletal muscle and the heart
- Increase blood pressure
- Decrease the function of visceral organs
- Increase the metabolic rate of skeletal muscles

PROCESS FIGURE 18.14 Regulation of Adrenal Medulla Secretions
Stress, physical exercise, and low blood glucose levels cause increased activity of the sympathetic nervous system, which increases epinephrine and norepinephrine secretion from the adrenal medulla.

reabsorption can result in increased water reabsorption by the kidneys and an increase in blood volume. Aldosterone increases K^+ excretion into the urine by the kidneys, thereby decreasing blood levels of K^+. It also increases the rate of H^+ excretion into the urine. When aldosterone is secreted in high concentrations, reduced blood levels of K^+ and alkalosis (elevated pH of body fluids) may result. The specific effects of aldosterone and the mechanisms controlling aldosterone secretion are discussed along with kidney functions in chapters 26 and 27 and with cardiovascular system functions in chapter 21.

Glucocorticoids

The zona fasciculata of the adrenal cortex secretes primarily **glucocorticoid hormones,** the major one being **cortisol** (kōr′ti-sol). The numerous target tissues and responses to the glucocorticoids are listed in table 18.8. The responses are classified as metabolic, developmental, or anti-inflammatory. Glucocorticoids increase lipid catabolism, decrease glucose and amino acid uptake in skeletal muscle, increase **gluconeogenesis** (gloo′kō-nē-ō-jen′ĕ-sis; the synthesis of glucose from precursor molecules, such as amino acids in the liver), and increase protein degradation. Thus, a major effect of glucocorticoids is an increased use of lipids and proteins to provide energy for cells.

Another effect of glucocorticoids is an increase in blood glucose levels and glycogen deposits in cells. The glucose and glycogen are a reservoir of molecules that can be metabolized rapidly. Glucocorticoids are also required for the maturation of tissues, such as fetal lungs, and for the development of receptor molecules in target tissues for epinephrine and norepinephrine. In addition, glucocorticoids decrease the intensity of the inflammatory and immune responses by decreasing both the number of white blood cells and the secretion of inflammatory chemicals from tissues. This anti-inflammatory effect is important under conditions of stress, when the rate of glucocorticoid secretion is relatively high. Synthetic glucocorticoids are often used to suppress the immune response in people suffering from autoimmune conditions and in transplant recipients (see chapter 22).

ACTH is necessary to maintain the secretory activity of the adrenal cortex, which rapidly atrophies without this hormone. Corticotropin-releasing hormone (CRH) released from the hypothalamus stimulates the anterior pituitary to secrete ACTH. The zona fasciculata is very sensitive to ACTH, and it responds by

TABLE 18.8 Target Tissues and Their Responses to Glucocorticoid Hormones

Target Tissues	Responses
Peripheral tissues, such as skeletal muscle, liver, and adipose tissue	Inhibit glucose use; stimulate the formation of glucose from amino acids and, to some degree, from lipids (gluconeogenesis) in the liver, which results in elevated blood glucose levels; stimulates glycogen synthesis in cells; mobilize lipids by increasing lipolysis, which results in the release of fatty acids into the blood and an increased rate of fatty acid metabolism; increase protein breakdown and decrease protein synthesis
Immune tissues	Anti-inflammatory; depress antibody production, white blood cell production, and the release of inflammatory components in response to injury; suppress the immune system
Target cells for epinephrine	Receptor molecules for epinephrine and norepinephrine decrease without adequate amounts of glucocorticoid hormone

increasing cortisol secretion. The regulation of ACTH and cortisol secretion is outlined in figure 18.15. Both ACTH and cortisol inhibit CRH secretion from the hypothalamus by negative feedback. In addition, high concentrations of cortisol in the blood inhibit ACTH secretion from the anterior pituitary, and low concentrations stimulate it. This negative-feedback loop is important in maintaining blood cortisol levels within a controlled range of concentration. Stress and hypoglycemia (low levels of glucose in the blood) trigger a large increase in CRH release from the hypothalamus by causing a rapid increase in blood levels of cortisol. In addition, CRH levels vary significantly throughout the day. Table 18.9 outlines several abnormalities associated with the hyposecretion or hypersecretion of adrenal hormones.

Predict 7

Cortisone, a drug similar to cortisol, is sometimes given to people who have severe allergies or extensive inflammation or to those who suffer from autoimmune diseases. Taking this substance chronically can damage the adrenal cortex. Explain how this damage can occur.

Adrenal Androgens

Some adrenal steroids, including **androstenedione** (an-drō-stēn′dī-ōn), are weak androgens. *Androgen* is a generic term for steroid hormones that cause the development of male secondary sex characteristics (see chapter 28). Most androgens are secreted by the reproductive system (see chapter 28). Adrenal androgens are secreted by the zona reticularis and converted by peripheral tissues to the more potent androgen testosterone. Adrenal androgens stimulate pubic and axillary hair growth and sex drive in females. However, the effects of adrenal androgens in males are negligible, in comparison with testosterone secreted by the testes.

ASSESS YOUR PROGRESS

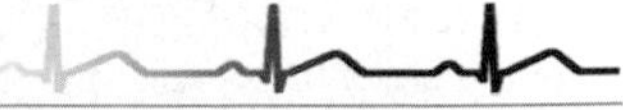

41. *Describe the three layers of the adrenal cortex, and name the hormones produced by each layer.*
42. *Name the target tissue of aldosterone, and list the effects of an increase in aldosterone secretion on the concentration of ions in the blood.*
43. *Describe the effects produced by an increase in cortisol secretion. Starting with the hypothalamus, describe how stress or low blood glucose levels can stimulate cortisol release.*
44. *List the possible causes of hyposecretion of adrenal cortex hormones, and describe the symptoms.*
45. *List the possible causes of hypersecretion of adrenal cortex hormones, and describe the symptoms.*
46. *What effects do adrenal androgens have on males and on females?*

1. Corticotropin-releasing hormone (CRH) is released from hypothalamic neurons in response to stress or low blood glucose and passes, by way of the hypothalamohypophysial portal system, to the anterior pituitary.
2. In the anterior pituitary, CRH binds to and stimulates cells that secrete adrenocorticotropic hormone (ACTH).
3. ACTH binds to membrane-bound receptors on cells of the adrenal cortex and stimulates the secretion of glucocorticoids, primarily cortisol.
4. Cortisol acts on target tissues, resulting in increased lipid and protein breakdown, increased glucose levels, and anti-inflammatory effects.
5. Cortisol has a negative-feedback effect because it inhibits CRH release from the hypothalamus and ACTH secretion from the anterior pituitary.

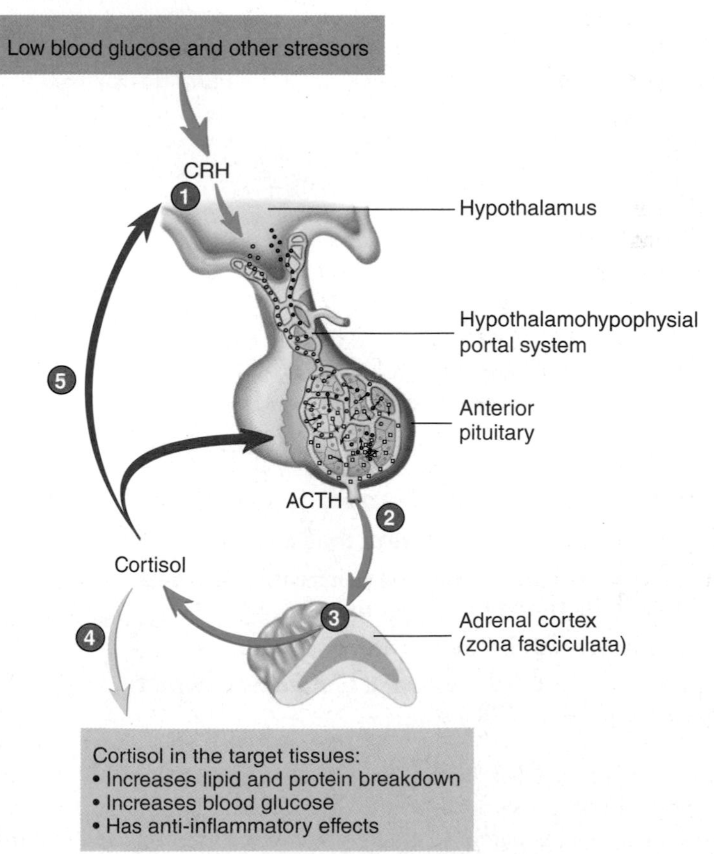
Low blood glucose and other stressors
CRH
1
Hypothalamus
Hypothalamohypophysial portal system
5
Anterior pituitary
ACTH
2
Cortisol
3
Adrenal cortex (zona fasciculata)
4
Cortisol in the target tissues:
• Increases lipid and protein breakdown
• Increases blood glucose
• Has anti-inflammatory effects

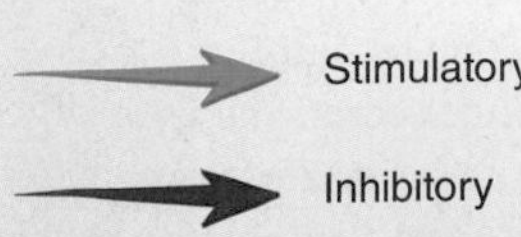
Stimulatory
Inhibitory

PROCESS FIGURE 18.15 **Regulation of Cortisol Secretion**

TABLE 18.9 Symptoms of Hyposecretion and Hypersecretion of Adrenal Cortex Hormones

	Cause	Symptoms
Hyposecretion		
Mineralocorticoids (aldosterone)	Removal of gland or loss of function or Addison disease (low levels of aldosterone *and* cortisol)	Hyponatremia (low blood levels of sodium) Hypokalemia (high blood levels of potassium) Acidosis Low blood pressure Tremors and tetany of skeletal muscles Polyuria
Glucocorticoids (cortisol)	Removal of gland or loss of function	Hypoglycemia (low blood glucose levels) Depressed immune system Proteins and lipids from diet unused, resulting in weight loss Loss of appetite, nausea, vomiting Bronzing of skin due to increased pigmentation (if ACTH levels are elevated)
Androgens		In women, reduction of pubic and axillary hair
Hypersecretion		
Mineralocorticoids (aldosterone)	Tumor in gland or aldosteronism	Slight hypernatremia (high blood levels of sodium) Hypokalemia (low blood levels of potassium) Alkalosis High blood pressure Weakness of skeletal muscles Acidic urine
Glucocorticoids (cortisol)	Tumor in gland or Cushing syndrome (high cortisol *and* androgens)	Hyperglycemia (high blood glucose levels; adrenal diabetes; leads to diabetes mellitus) Depressed immune system Destruction of tissue proteins, causing muscle atrophy and weakness, osteoporosis, weak capillaries (easy bruising), thin skin, and impaired wound healing; mobilization and redistribution of lipids, causing depletion of adipose tissue from limbs and deposition in face (moon face), neck (buffalo hump), and abdomen (Cushing syndrome) Emotional effects, including euphoria and depression
Androgens	Tumor in gland or adrenogenital syndrome	In women, hirsutism (excessive facial and body hair), acne, increased sex drive, regression of breast tissue, and loss of regular menses

18.6 Pancreas

LEARNING OUTCOMES

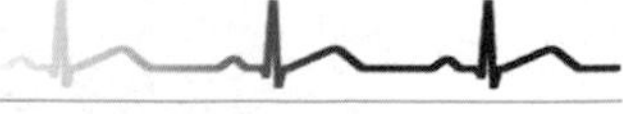

After reading this section, you should be able to

A. Describe the location and structure of the pancreas.

B. Name the hormones secreted by the pancreatic islets and describe their effects on their target tissues.

C. Explain how pancreatic hormones are regulated.

D. Compare and contrast the causes and symptoms of type 1 and type 2 diabetes mellitus.

The **pancreas** (pan′krē-us) lies behind the peritoneum between the greater curvature of the stomach and the duodenum. It is an elongated structure approximately 15 cm long, weighing approximately 85–100 g. The head of the pancreas lies near the duodenum, and its body and tail extend toward the spleen.

The pancreas is both an exocrine gland and an endocrine gland. The exocrine portion consists of **acini** (as′i-nī), which produce pancreatic juice, and a duct system, which carries the pancreatic juice to the small intestine (see chapter 24). The endocrine part, consisting of **pancreatic islets** (islets of Langerhans; figure 18.16), secretes hormones that enter the circulatory system.

Between 500,000 and 1 million pancreatic islets are dispersed among the ducts and acini of the pancreas. Each islet is composed of **alpha (α) cells** (20%), which secrete **glucagon,** a small polypeptide hormone, and **beta (β) cells** (75%), which secrete **insulin,** a small protein hormone consisting of two polypeptide chains bound together. The remaining 5% of cell types are either immature cells that produce secretions with digestive functions or **delta (δ) cells,** which secrete somatostatin, a small polypeptide hormone. Nerves from both divisions of the autonomic nervous system innervate the pancreatic islets, and a well-developed capillary network surrounds each islet.

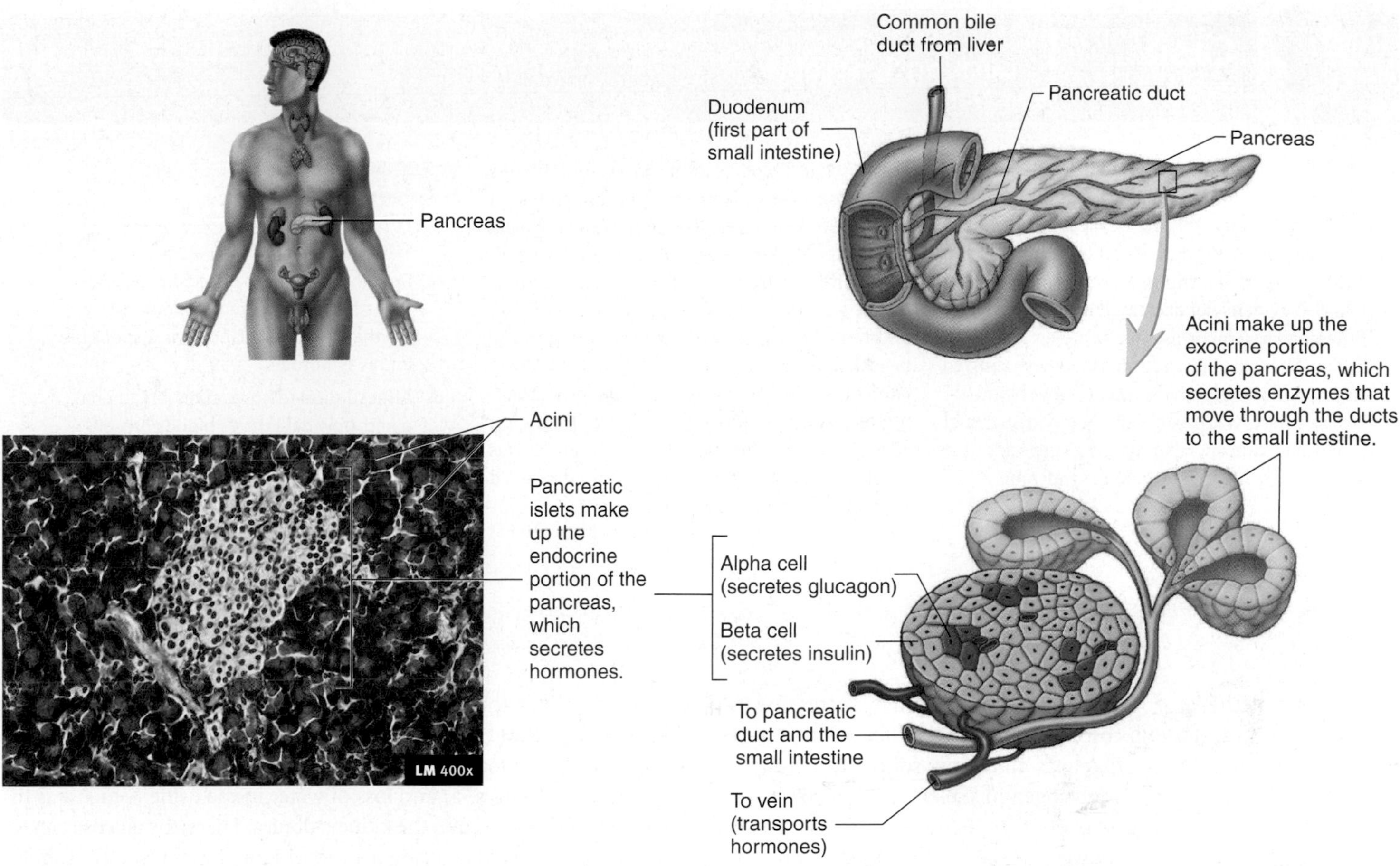
Pancreas
Common bile duct from liver
Duodenum (first part of small intestine)
Pancreatic duct
Pancreas
Acini make up the exocrine portion of the pancreas, which secretes enzymes that move through the ducts to the small intestine.
Acini
Pancreatic islets make up the endocrine portion of the pancreas, which secretes hormones.
LM 400x
Alpha cell (secretes glucagon)
Beta cell (secretes insulin)
To pancreatic duct and the small intestine
To vein (transports hormones)

FIGURE 18.16 APR **Histology of the Pancreatic Islets**
A pancreatic islet consists of clusters of specialized cells among the acini of the exocrine portion of the pancreas. The stain used for this slide does not distinguish between alpha and beta cells.

TABLE 18.10 Pancreatic Hormones

Cells in Islets	Hormone	Structure	Target Tissue	Response
Beta (β)	Insulin	Protein	Especially liver, skeletal muscle, adipose tissue	Increased uptake and use of glucose and amino acids
Alpha (α)	Glucagon	Polypeptide	Primarily liver	Increased breakdown of glycogen; release of glucose into the circulatory system
Delta (δ)	Somatostatin	Peptide	Alpha and beta cells (some somatostatin is produced in the hypothalamus)	Inhibition of insulin and glucagon secretion

Effect of Insulin and Glucagon on Their Target Tissues

The pancreatic hormones play an important role in regulating the concentration of critical nutrients in the circulatory system, especially glucose and amino acids (table 18.10). The major target tissues of insulin are the liver, adipose tissue, the skeletal muscles, and the satiety center within the hypothalamus of the brain. The **satiety** (sa′-tī-ĕ-tē) **center** is a collection of neurons in the hypothalamus that controls appetite, but insulin does not directly affect most areas of the nervous system. The specific effects of insulin on these target tissues are listed in table 18.11.

Insulin

Insulin molecules bind to membrane-bound receptors on target cells. Once insulin molecules bind to their receptors, the receptors cause specific proteins in the membrane to become phosphorylated. Part of the cells' response to insulin is to increase the number of transport proteins in the plasma membrane for glucose and amino acids. Finally, insulin and its receptors enter the cell by endocytosis. The insulin molecules are released from the insulin receptors and broken down within the cell, and the insulin receptor returns to the plasma membrane.

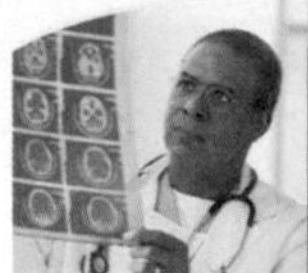

Case STUDY — Cushing Syndrome

Ethan noticed that he had gained a substantial amount of weight over the past few months and that he was feeling weak. His physician observed that the body fat distribution was mainly in his trunk, face, and neck. There was also evidence of decreased muscle mass, and Ethan had several bruises on his upper and lower limbs. Results of a routine blood test showed elevated blood glucose levels and low blood K^+ levels. There was no observable evidence that Ethan had cancer. His physician suspected that he was suffering from Cushing syndrome.

A second blood sample was taken. Ethan's blood cortisol levels were very high, and his blood ACTH levels were very low. Based on these data, Ethan's physician explained that he was probably suffering from an adrenal gland tumor, which was secreting large amounts of cortisol, and that the tumor was not responding to the negative-feedback mechanisms that normally control cortisol secretion. Subsequently, imaging techniques revealed a tumor in Ethan's left adrenal gland. After his left adrenal gland was surgically removed, Ethan's symptoms decreased dramatically over the next few weeks.

Predict 8

a. Why did the physician suspect Cushing syndrome?
b. Explain why Ethan's physician concluded that a hormone-secreting tumor in one of Ethan's adrenal glands was responsible for his symptoms.
c. After surgical removal of his left adrenal gland, how did Ethan's blood cortisol and ACTH levels change?
d. How would the data from the second blood sample have been different if Ethan's condition had been due to a hormone-secreting tumor in his anterior pituitary gland?

In general, the target tissue responds to insulin by increasing its ability to take up and use glucose and amino acids. Glucose molecules that are not needed immediately as an energy source to maintain cell metabolism are stored as glycogen in skeletal muscle, the liver, and other tissues and are converted to lipids in adipose tissue. Amino acids can be broken down and used as an energy source, or they can be converted to protein. Without insulin, the ability of these tissues to take up and use glucose and amino acids is minimal.

The normal regulation of blood glucose levels requires insulin. Blood glucose levels can increase dramatically when too little insulin is secreted or when insulin receptors do not respond to it (see Clinical Impact, "Diabetes Mellitus"). In the absence of insulin, the movement of glucose and amino acids into cells declines dramatically, even though blood levels of these molecules may increase to very high levels. The hypothalamic satiety center requires insulin in order to take up glucose. In the absence of insulin, the satiety center cannot detect the presence of glucose in the extracellular fluid, even when glucose levels are high. The result is an intense sensation of hunger in spite of high blood glucose levels, a condition called *polyphagia* (pol-ē-fā′jē-ă). High blood glucose levels also cause increased urine volume (*polyuria*; pol-ē-ū′rē-ă) and loss of water in the urine. Glucose is filtered from the blood into the kidney tubules. There, the glucose creates an osmotic gradient favoring the movement of water into the tubules and its subsequent loss in urine (see chapter 26). Elevated blood glucose levels also increase blood osmolality, resulting in an increased sensation of thirst (*polydipsia*; pol-ē-dip′sē-ă; see chapter 27).

When too much insulin is secreted, blood glucose levels can fall very low because too much insulin causes target tissues to rapidly take up glucose from the blood. Although the nervous system, except for cells of the satiety center, is not a target tissue for insulin, the nervous system depends primarily on blood glucose as an energy source. Consequently, low blood glucose levels cause changes in the function of the CNS, including dizziness, loss of cognitive function, and in extreme cases, loss of consciousness.

TABLE 18.11 Effect of Insulin and Glucagon on Target Tissues

Target Tissue	Response to Insulin	Response to Glucagon
Skeletal muscle, cardiac muscle, cartilage, bone, fibroblasts, leukocytes, and mammary glands	Increased glucose uptake and glycogen synthesis; increased uptake of certain amino acids	Little effect
Liver	Increased glycogen synthesis; increased use of glucose for energy (glycolysis)	Rapid increase in the breakdown of glycogen to glucose (glycogenolysis) and release of glucose into the blood; increased formation of glucose (gluconeogenesis) from amino acids and, to some degree, from lipids; increased metabolism of fatty acids, resulting in more ketones in the blood
Adipose cells	Increased glucose uptake, glycogen synthesis, lipid synthesis, and fatty acid uptake; increased glycolysis	High concentrations cause breakdown of lipids (lipolysis); probably unimportant under most conditions
Nervous system	Little effect, except increased glucose uptake in the satiety center	No effect

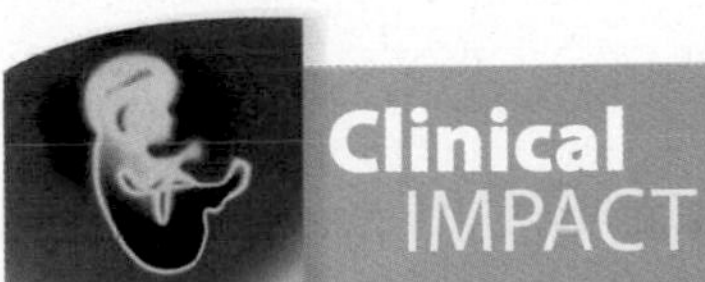

Clinical IMPACT — Diabetes Mellitus

Diabetes mellitus results primarily from the inadequate secretion of insulin or the inability of tissues to respond to insulin. **Type 1 diabetes mellitus,** also called *insulin-dependent diabetes mellitus (IDDM),* affects approximately 5–10% of people with diabetes mellitus and results from diminished insulin secretion. It develops as a result of autoimmune destruction of the pancreatic islets, and symptoms appear after approximately 90% of the islets have been destroyed. Type 1 diabetes mellitus most commonly develops in young people. Heredity may play a role in the condition, although the initiation of pancreatic islet destruction may involve a viral infection of the pancreas.

Type 2 diabetes mellitus, also called *noninsulin-dependent diabetes mellitus (NIDDM),* results from the inability of tissues to respond to insulin. Type 2 diabetes mellitus usually develops in people older than 40–45 years of age, although it is being observed more frequently in much younger patients.

Type 2 diabetes mellitus is more common than type 1 diabetes mellitus. Approximately 90–95% of people who have diabetes mellitus have type 2. People with type 2 diabetes mellitus have a reduced number of functional receptors for insulin, or one or more of the enzymes activated by the insulin receptor are defective. Thus, glucose uptake by cells is very slow, which results in elevated blood glucose after a meal. Obesity is common, although not universal, in patients with type 2 diabetes mellitus. Elevated blood glucose levels cause adipose cells to convert glucose to lipid, even though the rate at which adipose cells take up glucose is impaired. Increased blood glucose and increased urine production lead to hyperosmolality of blood and dehydration of cells. The poor use of nutrients and dehydration of cells lead to lethargy, fatigue, and periods of irritability. Elevated blood glucose levels affect the endothelial tissue of blood vessels, as well as the nervous system's ability to respond to tactile sensation. The combination of these effects results in recurrent injury and infection, especially in distal tissues, such as the feet.

Patients with type 2 diabetes mellitus do not experience sudden, large increases in blood glucose and severe tissue wasting, as occurs with type 1 diabetes mellitus, because a slow rate of glucose uptake does occur, even though the insulin receptors are defective. In some people with type 2 diabetes mellitus, insulin production eventually decreases because pancreatic islet cells atrophy; then type 1 diabetes mellitus develops. Approximately 25–30% of patients with type 2 diabetes mellitus take insulin, 50% take oral medication to increase insulin secretion and improve the efficiency of glucose utilization, and the remainder control blood glucose levels with exercise and diet alone.

Glucose tolerance tests are used to diagnose diabetes mellitus. In general, the test involves feeding the patient a large amount of glucose after a period of fasting and then collecting blood samples for a few hours afterward. A sustained increase in blood glucose levels strongly indicates that the person has diabetes mellitus.

Too much insulin relative to the amount of glucose ingested leads to insulin shock. The high levels of insulin cause target tissues to take up glucose at a very high rate. As a result, the concentration of blood glucose rapidly falls to a low level. Because the nervous system depends on glucose as its major source of energy, neurons malfunction, leading to nervous system responses, such as disorientation, confusion, and convulsions. Too much insulin, too little food intake after an insulin injection, or increased metabolism of glucose due to excess physical exercise can cause insulin shock in a diabetic patient.

Keeping blood glucose within normal levels at all times can prevent damage to blood vessels and reduced nerve function in patients with either type of diabetes mellitus. However, doing so requires increased attention to diet and frequent blood glucose testing to ensure that blood glucose levels do not fall too low and lead to insulin shock.

Clinical GENETICS — Type 2 Diabetes Mellitus

Most people with diabetes mellitus have type 2. Type 2 diabetes mellitus has a genetic basis, and it appears that several genes can make people more susceptible to developing the condition. For example, individuals whose close relatives have type 2 diabetes mellitus have an increased risk of developing it. In addition, type 2 diabetes mellitus is more prevalent among certain populations. For example, it is more common in Native Americans than in Caucasians, African-Americans, and Hispanics.

The pathways activated by the insulin receptor are complex, and genes on 10 different chromosomes that code for proteins in those pathways have been associated with the development of type 2 diabetes mellitus. Antibodies that bind to insulin receptors and make them nonfunctional or reduced numbers of functional insulin receptors can cause type 2 diabetes mellitus. In most cases, however, the insulin receptor is normal, but mutations in the genes that code for enzymes activated by the combination of insulin and its receptors result in a reduced response to insulin.

Type 2 diabetes mellitus involves a gradual failure of cells to respond to insulin and to take up glucose. Therefore, people who inherit genes that make them susceptible to type 2 diabetes mellitus are likely to develop the condition later in life. Also, symptoms are more likely to develop in people with an unhealthy lifestyle, which includes a diet high in calories and simple sugars and a sedentary tendency. A high percentage of people who have type 2 diabetes mellitus are obese. The severity of the condition may decrease in response to weight loss. An unhealthy lifestyle is also associated with a recent trend toward the development of diabetes mellitus in younger people.

The "thrifty genotype" hypothesis suggests that type 2 diabetes mellitus may be more common today because the genes that make people susceptible to the condition were once beneficial. For example, during periods of famine, the ability to store body fat and to have altered glucose metabolism may have been advantageous, but today, when food is abundant, having these genes increases the likelihood of developing type 2 diabetes mellitus.

Glucagon

Glucagon primarily influences the liver, although it has some effect on skeletal muscle and adipose tissue (table 18.11). Glucagon binds to membrane-bound receptors, activates G proteins, and increases cAMP synthesis. In general, glucagon causes the breakdown of glycogen and increases glucose synthesis in the liver. It also increases the breakdown of lipids. The amount of glucose released from the liver into the blood increases dramatically after glucagon secretion increases. Because glucagon is secreted into the hepatic portal circulation, which carries blood from the intestines and pancreas to the liver, glucagon is delivered in a relatively high concentration to the liver, where it has its major effect. The liver also rapidly metabolizes it. Thus, glucagon has less effect on skeletal muscles and adipose tissue than on the liver.

Regulation of Pancreatic Hormone Secretion

Blood levels of nutrients, neural stimulation, and hormones control the secretion of insulin. *Hyperglycemia*, or elevated blood levels of glucose, directly stimulates insulin secretion from β cells. *Hypoglycemia*, or low blood levels of glucose, is directly related to low insulin secretion. In addition, the sympathetic nervous system can directly reduce insulin secretion. Thus, blood glucose levels play a major role in regulating insulin secretion. Certain amino acids also stimulate insulin secretion by acting directly on the β cells. After a meal, when glucose and amino acid levels in the circulatory system are elevated, insulin secretion increases. During periods of fasting, when blood glucose levels are low, the rate of insulin secretion declines (figure 18.17).

The autonomic nervous system also controls insulin secretion. Parasympathetic stimulation associated with food intake acts with the elevated blood glucose levels to increase insulin secretion. Sympathetic innervation inhibits insulin secretion and helps prevent a rapid fall in blood glucose levels. Because most tissues, except nervous tissue, require insulin to take up glucose, sympathetic stimulation maintains blood glucose levels in a normal range during periods of physical activity or excitement. This response is important for maintaining normal nervous system function.

Gastrointestinal hormones involved with regulating digestion, such as gastrin, secretin, and cholecystokinin (see chapter 24), increase insulin secretion. Somatostatin inhibits insulin and glucagon secretion, but the factors that regulate somatostatin secretion are not clear. It can be released in response to food intake, in which case somatostatin may prevent the oversecretion of insulin.

Predict 9

Explain why the increase in insulin secretion in response to parasympathetic stimulation and gastrointestinal hormones is consistent with the maintenance of blood glucose levels in the circulatory system.

Low blood glucose levels stimulate glucagon secretion, and high blood glucose levels inhibit it. Certain amino acids and sympathetic stimulation also increase glucagon secretion. After a high-protein meal, amino acids increase both insulin and glucagon secretion. Insulin causes target tissue uptake of amino acids for protein synthesis, and glucagon increases the process of glucose synthesis from amino acids in the liver (gluconeogenesis). Both protein synthesis and the use of amino acids to maintain blood glucose levels result from the low, but simultaneous, secretion of insulin and glucagon induced by meals high in protein.

Predict 10

Compare the regulation of glucagon and insulin secretion after a meal high in carbohydrates, after a meal low in carbohydrates but high in proteins, and during physical exercise.

ASSESS YOUR PROGRESS

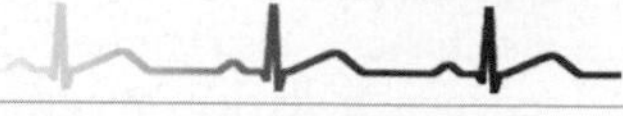

47. *Where is the pancreas located? Describe the exocrine and endocrine parts of this gland and the secretions produced by each portion.*
48. *Name the target tissues for insulin and glucagon, and list their effects on the target tissues.*
49. *How does insulin affect the satiety center of the hypothalamus?*
50. *What effect do blood glucose levels, blood amino acid levels, the autonomic nervous system, and somatostatin have on insulin and glucagon secretion?*
51. *Describe the causes and symptoms of type 1 diabetes mellitus and type 2 diabetes mellitus.*

18.7 Hormonal Regulation of Nutrient Utilization

LEARNING OUTCOMES

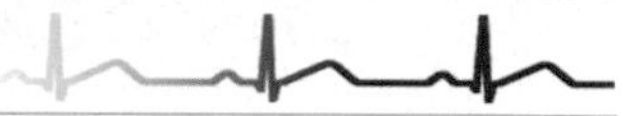

After reading this section, you should be able to

A. **Explain the interactions of insulin, glucagon, cortisol, GH, and epinephrine immediately after a meal and 1–2 hours after a meal.**
B. **Describe the nervous and hormonal interactions during exercise that will provide enough energy to cells.**

Two situations—after a meal and during exercise—illustrate how several hormones function together to regulate blood nutrient levels. After a meal and under resting conditions, secretion of glucagon, cortisol, GH, and epinephrine is reduced (figure 18.18). Both increasing blood glucose levels and parasympathetic stimulation elevate insulin secretion to increase the uptake of glucose, amino acids, and lipids by target tissues. Substances not immediately used for cell metabolism are stored. Glucose is converted to glycogen in skeletal muscle and the liver, and it is used for lipid synthesis in adipose tissue and the liver. The rapid uptake and storage of glucose prevent too large an increase in blood glucose levels. Amino acids are incorporated into proteins, and lipids that were ingested as part of the meal are stored in adipose tissue and the liver. If the meal is high in protein, a small amount of glucagon is secreted, thereby increasing the rate at which the liver uses amino acids to form glucose.

Within 1–2 hours after the meal, absorption of digested materials from the digestive tract decreases, and blood glucose levels decline.

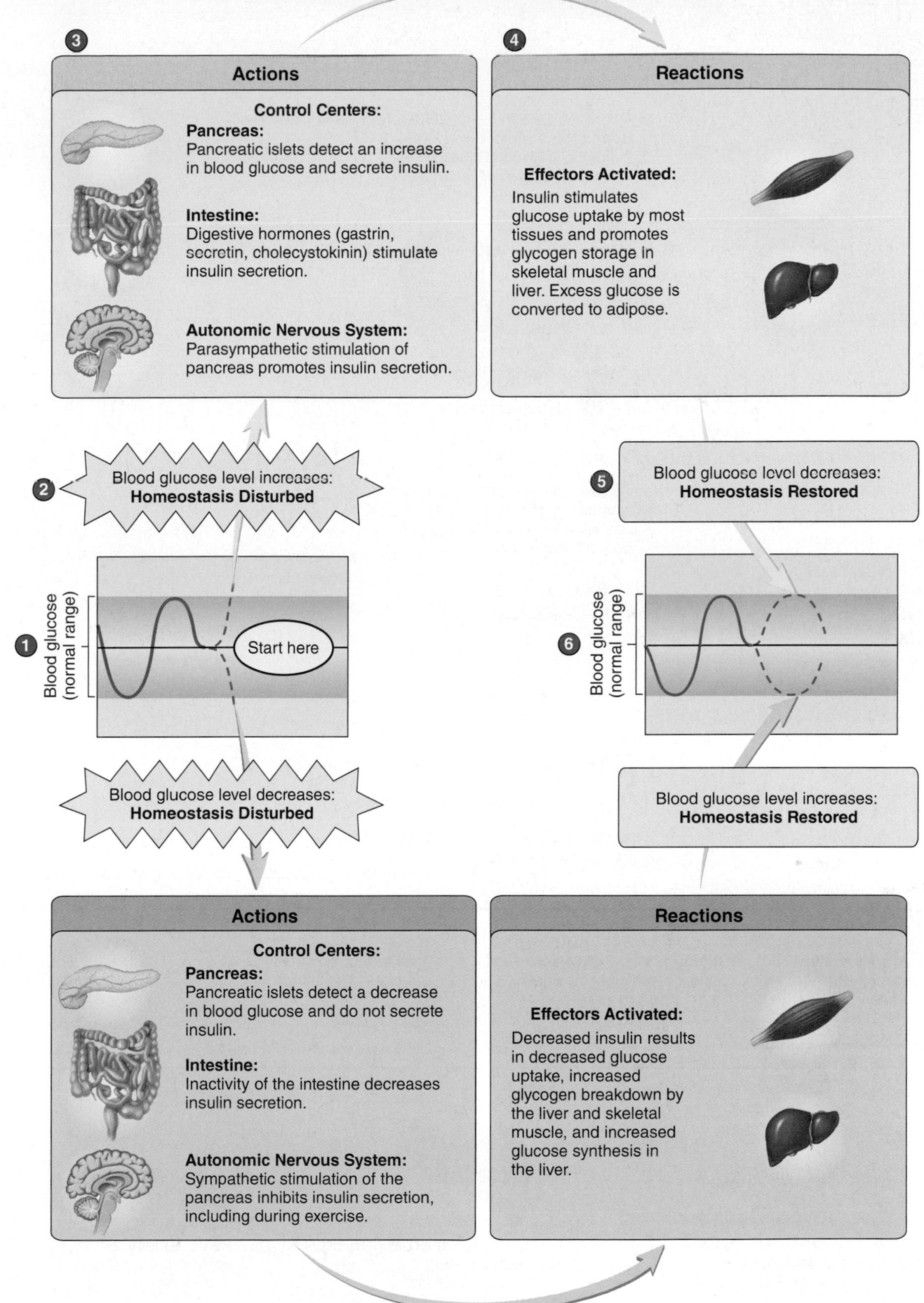

3
Actions
Control Centers:
Pancreas:
Pancreatic islets detect an increase in blood glucose and secrete insulin.
Intestine:
Digestive hormones (gastrin, secretin, cholecystokinin) stimulate insulin secretion.
Autonomic Nervous System:
Parasympathetic stimulation of pancreas promotes insulin secretion.
4
Reactions
Effectors Activated:
Insulin stimulates glucose uptake by most tissues and promotes glycogen storage in skeletal muscle and liver. Excess glucose is converted to adipose.
2
Blood glucose level increases:
Homeostasis Disturbed
5
Blood glucose level decreases:
Homeostasis Restored
1
Blood glucose (normal range)
Start here
6
Blood glucose (normal range)
Blood glucose level decreases:
Homeostasis Disturbed
Blood glucose level increases:
Homeostasis Restored
Actions
Control Centers:
Pancreas:
Pancreatic islets detect a decrease in blood glucose and do not secrete insulin.
Intestine:
Inactivity of the intestine decreases insulin secretion.
Autonomic Nervous System:
Sympathetic stimulation of the pancreas inhibits insulin secretion, including during exercise.
Reactions
Effectors Activated:
Decreased insulin results in decreased glucose uptake, increased glycogen breakdown by the liver and skeletal muscle, and increased glucose synthesis in the liver.

HOMEOSTASIS FIGURE 18.17 Regulation of Insulin Secretion
(1) Blood glucose is within its normal range. (2) Blood glucose level increases outside the normal range. (3) Pancreatic islets secrete insulin in direct response to elevated blood glucose. Digestive hormones and parasympathetic activity also stimulate insulin secretion. (4) Most tissues take up glucose when insulin binds to its receptors on the tissue. Liver and skeletal muscle cells convert glucose to glycogen. (5) Blood glucose level drops back to the normal range. (6) Homeostasis is restored. Observe the response to a drop in blood glucose by following the red arrows.

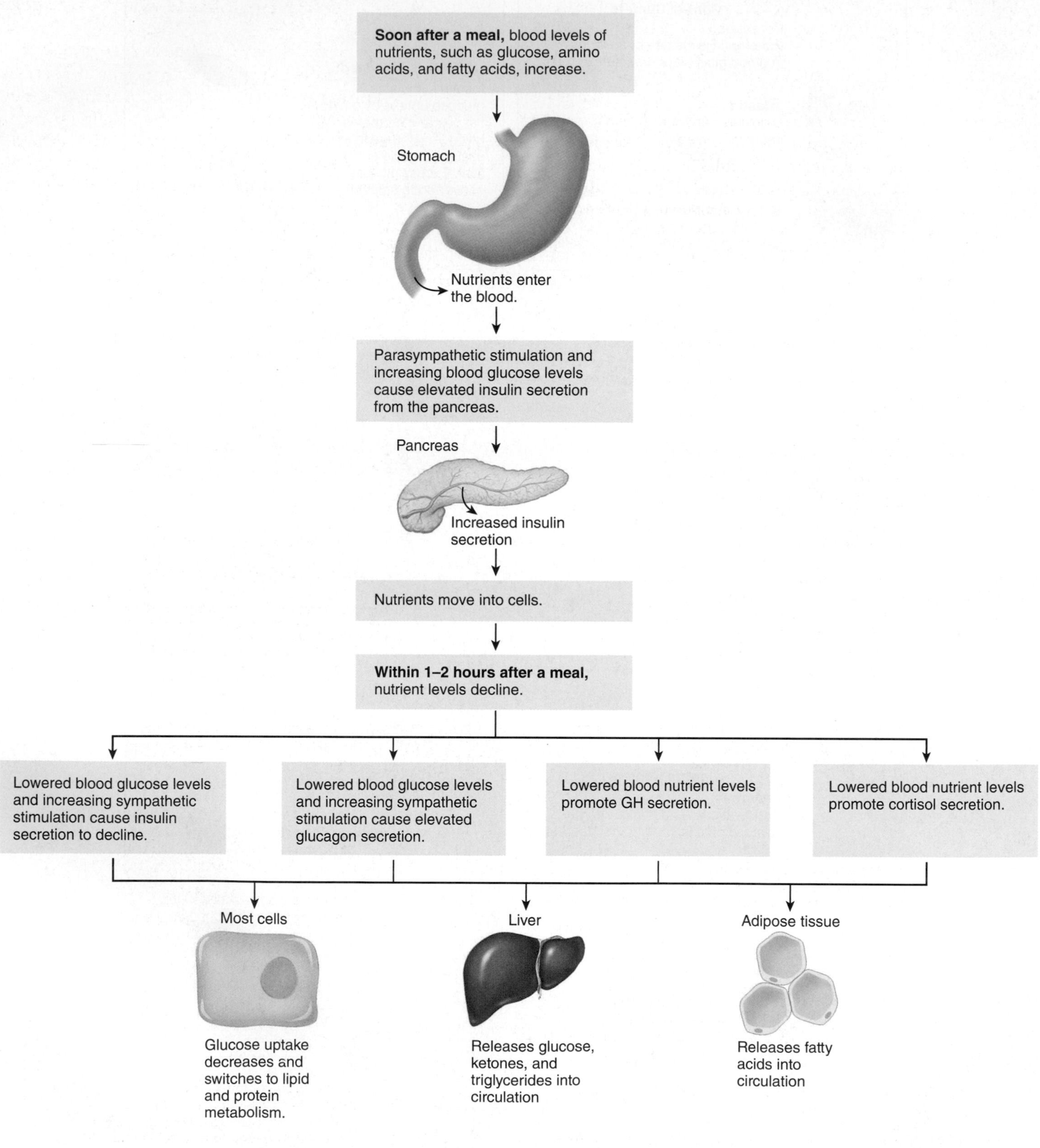
Soon after a meal, blood levels of nutrients, such as glucose, amino acids, and fatty acids, increase.
Stomach
Nutrients enter the blood.
Parasympathetic stimulation and increasing blood glucose levels cause elevated insulin secretion from the pancreas.
Pancreas
Increased insulin secretion
Nutrients move into cells.
Within 1–2 hours after a meal, nutrient levels decline.
Lowered blood glucose levels and increasing sympathetic stimulation cause insulin secretion to decline.
Lowered blood glucose levels and increasing sympathetic stimulation cause elevated glucagon secretion.
Lowered blood nutrient levels promote GH secretion.
Lowered blood nutrient levels promote cortisol secretion.
Most cells
Liver
Adipose tissue
Glucose uptake decreases and switches to lipid and protein metabolism.
Releases glucose, ketones, and triglycerides into circulation
Releases fatty acids into circulation

FIGURE 18.18 Regulation of Blood Nutrient Levels After a Meal
Blood nutrient levels are maintained immediately after a meal and for several hours afterward.

As a result, secretion of glucagon, GH, cortisol, and epinephrine increases. As blood glucose declines, insulin secretion decreases, and the rate of glucose entry into the target tissues for insulin decreases. Glycogen, stored in cells, is converted back to glucose and used as an energy source. The liver releases glucose into the blood. The decreased uptake of glucose by most tissues, combined with its release from the liver, helps maintain blood glucose at levels necessary for normal brain function. Cells that use less glucose start using more lipids and proteins. Adipose tissue releases fatty acids, and the liver releases triglycerides (in lipoproteins) and ketones into the blood. Tissues take up these substances from the blood and use them for energy. Lipid molecules are a major source of energy for most tissues when blood glucose levels are low.

The interactions of insulin, GH, glucagon, epinephrine, and cortisol are excellent examples of negative-feedback mechanisms. When blood glucose levels are high, these hormones cause the rapid uptake and storage of glucose, amino acids, and lipids. When blood glucose levels are low, they cause the release of glucose and a switch to lipid and protein metabolism as a source of energy for most tissues.

During exercise, skeletal muscles require energy to support the contraction process (see chapter 9). Although metabolism of intracellular nutrients can sustain muscle contraction for a short time, additional energy sources are required during prolonged activity. Sympathetic nervous system activity, which increases during exercise, stimulates the release of epinephrine from the adrenal medulla and the release of glucagon from the pancreas (figure 18.19). These hormones induce the conversion of glycogen to glucose in the liver and the release of glucose into the blood, thus providing skeletal muscles with a source of energy. Because epinephrine and glucagon have short half-lives, they can rapidly adjust blood glucose levels for varying conditions of activity.

Exercise

Muscle

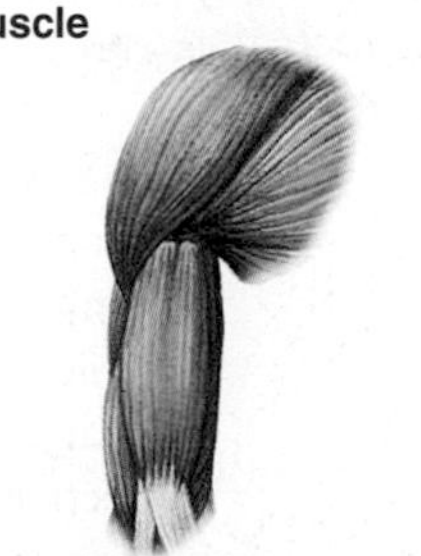

Liver

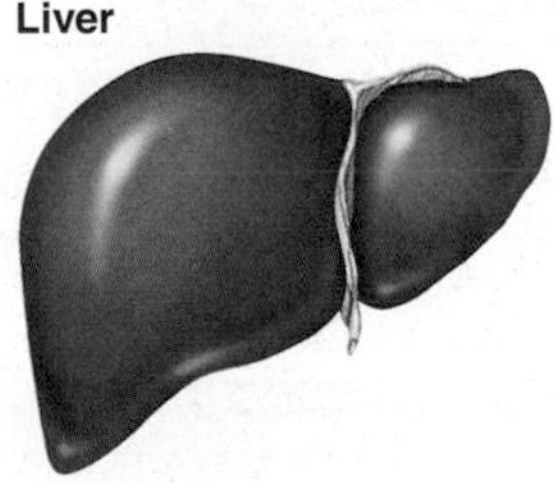

Adipose tissue

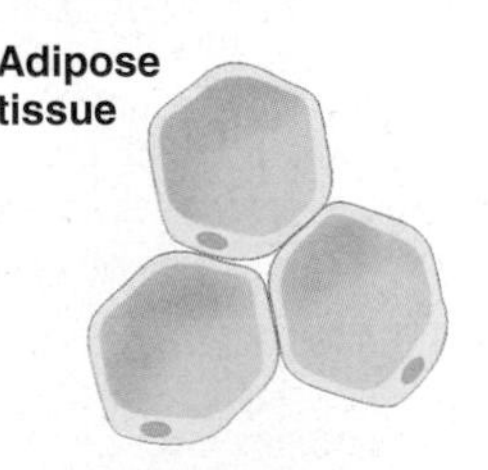

Blood nutrients

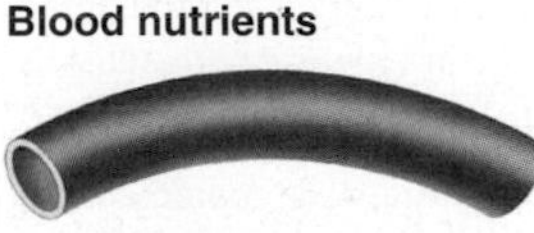

Short-term exercise

Sympathetic stimulation increases epinephrine secretion from the adrenal medulla and glucagon secretion from the pancreas. It also inhibits insulin secretion from the pancreas.

Epinephrine increases the rate at which glycogen in **muscle cells** is broken down to glucose. Because muscle cells use glucose as an energy source, the muscle cells take up less glucose from the blood and increase their rate of fatty acid metabolism.

Epinephrine and glucagon increase glycogen breakdown to glucose molecules in the **liver.** The glucose molecules are released into the circulatory system.

Epinephrine and sympathetic stimulation also increase the breakdown of **lipid** and the release of fatty acids from adipose tissue.

Blood glucose levels are maintained for normal nervous system function.

Prolonged exercise

Increased ACTH and GH are released from the anterior pituitary. ACTH stimulates increased cortisol secretion from the adrenal cortex.

Cortisol increases protein breakdown in **muscle** and the **liver** to amino acids and increases glucose synthesis from amino acids and from some components of fat, such as glycerol.

Cortisol increases the breakdown of **lipids** and the use of fatty acids by **muscle cells** as an energy source.

GH slows the breakdown of proteins and helps conserve them, thus increasing the dependence on lipids as an energy source.

Blood glucose levels are maintained for normal nervous system function and the body is increasingly dependent on lipids and proteins for energy sources.

FIGURE 18.19 Regulation of Blood Nutrient Levels During Exercise

During sustained activity, glucose released from the liver and other tissues is not adequate to support muscle activity, and the danger exists that blood glucose levels will become too low to support brain function. A decrease in insulin prevents the uptake of glucose by most tissues, thus conserving glucose for the brain. In addition to other functions, epinephrine, glucagon, cortisol, and GH cause an increase in fatty acids, triglycerides, and ketones in the blood. Because GH increases protein synthesis and slows the breakdown of proteins, muscle proteins are not used as an energy source. Consequently, glucose metabolism decreases, and lipid metabolism in skeletal muscles increases. At the end of a long race, for example, muscles rely to a large extent on lipid metabolism for energy.

Predict 11

Explain why long-distance runners may not have much of a "kick" left when they try to sprint to the finish line.

ASSESS YOUR PROGRESS

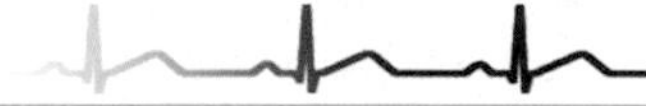

52. *Describe the hormonal effects that occur immediately after a meal to cause nutrients to move into cells and be stored.*

53. *What occurs hormonally 1–2 hours after a meal that causes stored materials to be released and used for energy?*

54. *During exercise, how does sympathetic nervous system activity regulate blood glucose levels? Name five hormones that interact to ensure that the brain and the muscles have adequate energy sources during exercise, and explain the role of each.*

18.8 Hormones of the Reproductive System

LEARNING OUTCOMES

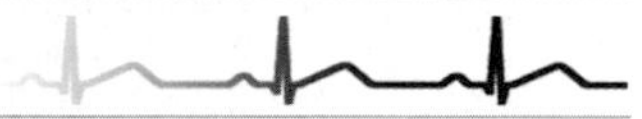

After reading this section, you should be able to

A. List and describe the functions of the hormones secreted by the testes and ovaries.

B. Explain how the anterior pituitary regulates secretion by the testes and ovaries.

C. Explain how the placenta acts as a temporary endocrine gland.

Reproductive hormones are secreted primarily from the ovaries, testes, placenta, and pituitary gland (table 18.12). These hormones are discussed in chapter 28. The main endocrine glands of the male reproductive system are the testes. The functions of the testes depend on the secretion of FSH and LH from the anterior pituitary gland. The main hormone secreted by the testes is testosterone, an androgen. **Testosterone** regulates the production of sperm cells by the testes and the development and maintenance of male reproductive organs and secondary sex characteristics. The testes secrete another hormone, called **inhibin**, which inhibits the secretion of FSH from the anterior pituitary gland.

The main endocrine glands of the female reproductive system are the ovaries. Like the testes, the functions of the ovaries depend on the secretion of FSH and LH from the anterior pituitary gland. The main hormones secreted by the ovaries are **estrogen** and **progesterone**. These hormones, along with FSH and LH, control the female reproductive cycle, prepare the mammary glands for lactation, and maintain pregnancy. Estrogen and progesterone are also responsible for the development of the female reproductive organs and female secondary sex characteristics. Like the testes, the ovaries secrete inhibin, which inhibits FSH secretion.

During the first one-third of pregnancy, the placenta secretes an LH-like substance that is necessary to maintain pregnancy (see chapter 28). Throughout most of pregnancy, the ovaries and placenta secrete increasing amounts of estrogen and progesterone, which are also necessary to maintain pregnancy. In addition, the ovaries secrete **relaxin**, which increases the flexibility of the connective tissue of the symphysis pubis and helps dilate the cervix of the uterus. This facilitates delivery by making the birth canal larger.

ASSESS YOUR PROGRESS

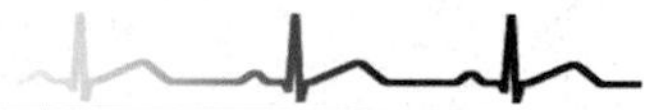

55. *List the hormones secreted by the testes, and describe their functions.*

TABLE 18.12 Hormones of the Reproductive Organs

Hormones	Structure	Target Tissue	Response
Testes			
Testosterone	Steroid	Most cells	Aids in spermatogenesis, development of genitalia, maintenance of functional reproductive organs, secondary sex characteristics, and sexual behavior
Inhibin	Polypeptide	Anterior pituitary gland	Inhibits FSH secretion
Ovaries			
Estrogen	Steroid	Most cells	Aids in uterine and mammary gland development and function, maturation of genitalia, secondary sex characteristics, sexual behavior, and menstrual cycle
Progesterone	Steroid	Most cells	Aids in uterine and mammary gland development and function, maturation of genitalia, secondary sex characteristics, and menstrual cycle
Inhibin	Polypeptide	Anterior pituitary gland	Inhibits FSH secretion
Relaxin	Polypeptide	Connective tissue cells	Increases the flexibility of connective tissue in the pelvic area, especially the symphysis pubis

56. *List the hormones secreted by the ovaries, and describe their functions.*
57. *What hormones from the anterior pituitary gland regulate secretion by the testes and ovaries? Explain how these hormones function.*
58. *What hormones are secreted by the placenta to help maintain pregnancy and facilitate birth?*

18.9 Hormones of the Pineal Gland

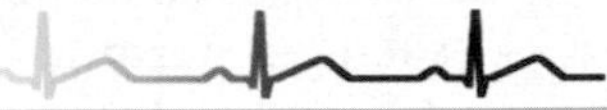

LEARNING OUTCOMES

After reading this section, you should be able to

A. List the hormones secreted by the pineal gland, and describe their possible functions.

B. Explain the photoperiod and its relationship to pineal gland secretion.

The **pineal** (pin′ē-ăl) **gland** in the epithalamus of the brain secretes hormones that act on the hypothalamus and the gonads to inhibit reproductive functions, such as by inhibiting the secretion of certain reproductive hormones. Two substances have been proposed as secretory products: **melatonin** (mel-ă-tōn′in) and **arginine vasotocin** (ar′ji-nēn vā-sō-tō′sin; table 18.13). Melatonin can decrease GnRH secretion from the hypothalamus and may inhibit reproductive functions through this mechanism. It may also help regulate sleep cycles by increasing the tendency to sleep.

Arginine vasotocin works with melatonin to regulate the function of the reproductive system in some animals. Evidence for the role of melatonin is more extensive, however.

In some animals, pineal secretions are regulated by the **photoperiod,** the amount of daylight and darkness that occurs each day and changes with the seasons of the year (figure 18.20). For example, increased daylight initiates action potentials in the retina of the eye, which are propagated to the brain and cause a decrease in the action potentials sent first to the spinal cord and then through sympathetic neurons to the pineal gland, resulting in decreased secretions. In the dark, action potentials delivered by sympathetic neurons to the pineal gland increase, stimulating the secretion of pineal hormones. Humans secrete larger amounts of melatonin at night than during the day. In animals that breed in the spring, the increased length of a day decreases pineal secretions. Because pineal secretions inhibit reproductive functions in these species, the increased length of a day results in hypertrophy of the reproductive structures.

Melatonin's role in regulating reproductive functions in humans is not clear, but some researchers recommend its use to enhance sleep.

TABLE 18.13 Other Hormones and Hormonelike Substances

Chemical Messenger	Structure	Target Tissue	Response
Pineal Gland			
Melatonin	Amino acid derivative	At least the hypothalamus	Inhibition of gonadotropin-releasing hormone secretion, thereby inhibiting reproduction; significance is not clear in humans; may help regulate sleep-wake cycles
Arginine vasotocin	Peptide	Possibly the hypothalamus	Possible inhibition of gonadotropin-releasing hormone secretion
Thymus			
Thymosin	Peptide	Immune tissues	Development and function of the immune system
Several Tissues (autocrine and paracrine chemical messengers)			
Eicosanoids			
Prostaglandins	Modified fatty acid	Most tissues	Mediation of the inflammatory response; increased uterine contractions; involved in ovulation, possible inhibition of progesterone synthesis; blood coagulation; other functions
Thromboxanes	Modified fatty acid	Most tissues	Mediation of the inflammatory response; other functions, including blood clotting
Prostacyclins	Modified fatty acid	Most tissues	Mediation of the inflammatory response; other functions, including blood clotting
Leukotrienes	Modified fatty acid	Most tissues	Mediation of the inflammatory response; other functions, including blood clotting
Enkephalins and endorphins	Peptides	Nervous system	Reduction of pain sensation; other functions
Epidermal growth factor	Protein	Many tissues	Stimulation of division in many cell types; embryonic development
Fibroblast growth factor	Protein	Many tissues	Stimulation of cell division in many cell types; embryonic development
Interleukin-2	Protein	Certain immune-competent cells	Stimulation of cell division of T lymphocytes

1. Light entering the eye stimulates neurons in the retina to fire action potentials.
2. Action potentials are transmitted to the hypothalamus.
3. Action potentials from the hypothalamus are transmitted through the sympathetic division to the pineal gland.
4. A *decrease* in light (darkness) results in increased sympathetic stimulation of the pineal gland and increased melatonin secretion. An *increase* in light results in decreased sympathetic stimulation of the pineal gland and decreased melatonin secretion.
5. Melatonin inhibits GnRH secretion from the hypothalamus and may help regulate sleep cycles.

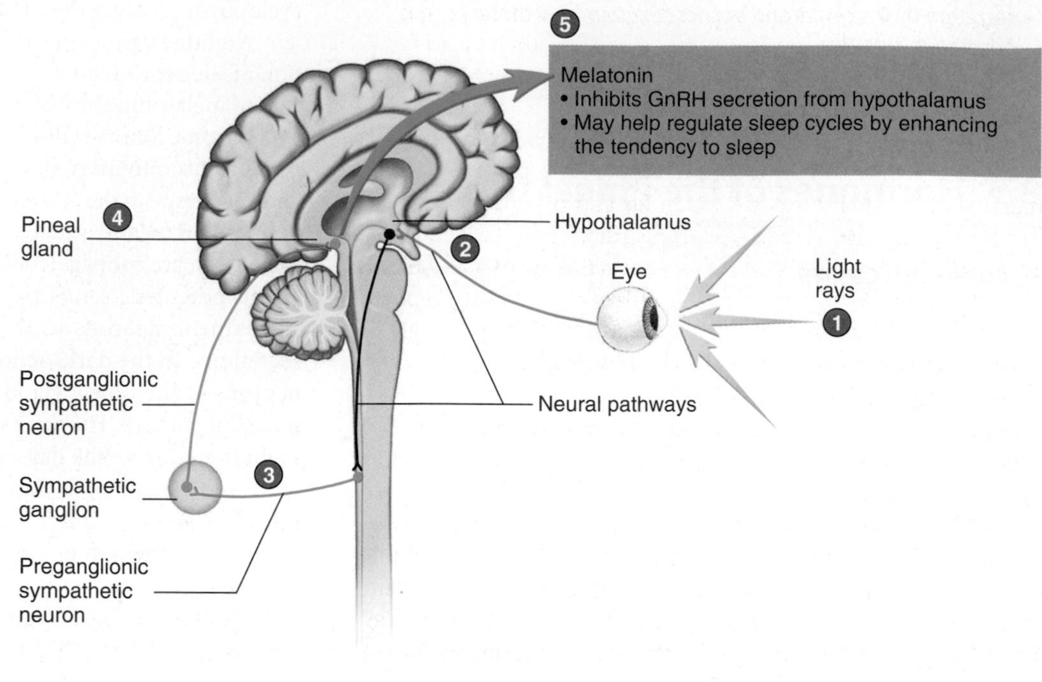
5
Melatonin
• Inhibits GnRH secretion from hypothalamus
• May help regulate sleep cycles by enhancing the tendency to sleep
Pineal gland
4
Hypothalamus
2
Eye
Light rays
1
Postganglionic sympathetic neuron
Neural pathways
Sympathetic ganglion
3
Preganglionic sympathetic neuron

PROCESS FIGURE 18.20 Regulation of Melatonin Secretion from the Pineal Gland
Light entering the eye inhibits the release of melatonin from the pineal gland, and dark stimulates the release of melatonin.

However, because melatonin causes atrophy of reproductive structures in some species, undesirable side effects on the reproductive system may be possible for people who take supplemental melatonin.

The function of the pineal gland in humans is not clear, but tumors that destroy the pineal gland correlate with early sexual development, and tumors that result in pineal hormone secretion correlate with retarded development of the reproductive system. It is not clear, however, if the pineal gland controls the onset of puberty.

ASSESS **YOUR PROGRESS**

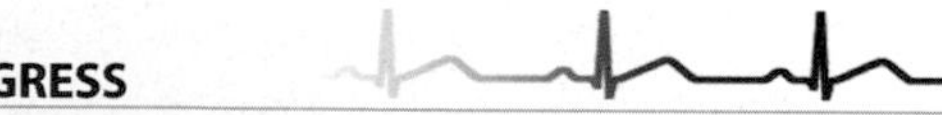

59. *Where is the pineal gland located? Name the hormones it produces and their possible effects.*

60. *Explain the relationship between the photoperiod and pineal gland secretion.*

18.10 Other Hormones and Chemical Messengers

LEARNING OUTCOMES

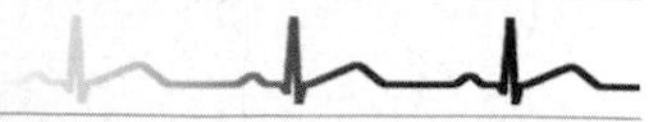

After reading this section, you should be able to

A. Describe the functions of the hormones secreted by the thymus, digestive system, heart, and kidneys.

B. Differentiate between autocrine and paracrine chemical messengers.

C. Give examples of both autocrine and paracrine chemical messengers and describe their functions.

Hormones of the Thymus

The **thymus** (thī′mŭs) is in the neck and superior to the heart in the thorax; it secretes a hormone called **thymosin** (thī′mō-sin; table 18.13). Both the thymus and thymosin play an important role in the development and maturation of the immune system, as discussed in chapter 22.

Hormones of the Digestive Tract

Several hormones are released from the digestive tract. They regulate digestive functions by influencing the activity of the stomach, intestines, liver, and pancreas (see chapter 24).

Hormonelike Chemicals

Some chemical messengers differ from hormones in that they are not secreted from discrete endocrine glands, their effects are local rather than systemic, or their functions are not understood adequately to explain their role in the body. **Autocrine chemical messengers** are released from cells that influence the same cell from which they are released. **Paracrine chemical messengers** are released from one cell type, diffuse short distances, and influence the activity of another cell type, which is the target tissue. Certain molecules sometimes function in an autocrine fashion and sometimes in a paracrine fashion. This distinction is similar to that for differentiating when a particular molecule is acting as a neurotransmitter and when it is acting as a hormone. The difference lies in the mode of transport (see chapter 17).

Examples of autocrine chemical messengers include chemical mediators of inflammation derived from the fatty acid arachidonic (ă-rak-i-don′ik) acid, such as eicosanoids and modified phospholipids. The eicosanoids include prostaglandins (pros′tă-glan-dinz), thromboxanes (throm′bok-zānz), prostacyclins (pros-tă-sī′klinz), and leukotrienes (loo′kō-trī′enz). An example of a modified phospholipid is platelet-activating factor (see chapter 19). Paracrine chemical messengers include substances that play a role in modulating the sensation of pain, such as **endorphins** (en′dōr-finz), **enkephalins** (en-kef′ă-linz), prostaglandins, and several peptide growth factors, such as **epidermal growth factor**, **fibroblast growth factor,** and **interleukin-2** (in-ter-loo′kin; table 18.13).

Prostaglandins, thromboxanes, prostacyclins, and leukotrienes are released from injured cells and are responsible for initiating some of the symptoms of inflammation (see chapter 22), in addition to being released from certain healthy cells. For example, prostaglandins are involved in the regulation of uterine contractions during menstruation and childbirth, the process of ovulation, the inhibition of progesterone synthesis by the corpus luteum, the regulation of coagulation, kidney function, and the modification of the effect of other hormones on their target tissues. Pain receptors are stimulated directly by prostaglandins and other inflammatory compounds, or prostaglandins cause vasodilation of blood vessels, which is associated with headaches. Anti-inflammatory drugs, such as aspirin, inhibit prostaglandin synthesis and, as a result, reduce inflammation and pain. These examples are paracrine chemical messengers because they are synthesized and secreted by the cells near their target cells. Once prostaglandins enter the circulatory system, they are metabolized rapidly.

Three classes of peptide molecules, which are endogenously produced analgesics, bind to the same receptors as morphine. They include enkephalins, endorphins, and **dynorphins** (dī′nōr-finz). They are produced in several body sites, such as in parts of the brain, pituitary gland, spinal cord, and intestines. They act as neurotransmitters in some neurons of both the central and the peripheral nervous systems and as hormones or paracrine regulatory substances. In general, they moderate the sensation of pain (see chapter 14). Decreased sensitivity to painful stimuli during exercise and stress may result from the increased secretion of these substances.

Several proteins can be classified as growth factors. They generally function as paracrine chemical messengers because they are secreted near their target tissues. Epidermal growth factor stimulates cell divisions in a number of tissues and plays an important role in embryonic development. Interleukin-2 stimulates the proliferation of T lymphocytes and plays a very important role in immune responses (see chapter 22).

The number of hormonelike substances in the body is large, and only a few of them have been mentioned here. Chemical communication among body cells is complex, well-developed, and necessary to maintain homeostasis. Investigations into chemical regulation increase our knowledge of body functions—knowledge that can be used to develop new treatments for pathological conditions.

ASSESS YOUR PROGRESS

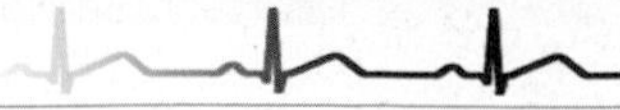

61. *What hormone is secreted by the thymus? What is its function?*
62. *What function do the hormones secreted by the stomach and small intestine perform?*
63. *What is the difference between an autocrine and a paracrine chemical messenger?*
64. *List eicosanoids and modified phospholipids that function as autocrine chemical messengers, and explain how they work.*
65. *List examples of paracrine chemical messengers that play a role in modulating pain or are peptide growth factors.*
66. *Describe the paracrine functions of prostaglandins and how anti-inflammatory drugs can reduce pain and inflammation.*

18.11 Effects of Aging on the Endocrine System

LEARNING OUTCOME

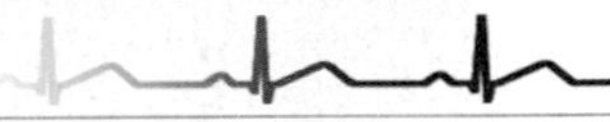

After reading this section, you should be able to

A. Describe major age-related changes that occur in the endocrine system.

Age-related changes in the endocrine system are not the same for all of the endocrine glands. Some, but not all, undergo a gradual decrease in secretory activity. In addition, some decreases in the secretory activity of endocrine glands appear to be secondary to a decrease in physical activity as people age.

GH secretion decreases as people grow older. The decrease is greater in people who do not exercise, and it may not occur at all in those who exercise regularly. Decreasing GH secretion may explain the gradual decrease in lean body mass. For example, bone mass and muscle mass decrease as GH levels decline. At the same time, the proportion of adipose tissue increases.

Melatonin secretion decreases in aging people. This decrease may influence age-related changes in sleep patterns and the secretory patterns of other hormones, such as GH and testosterone.

The secretion of thyroid hormones decreases slightly with increasing age, and the T_3/T_4 ratio decreases. However, this may be less of a decrease in the secretory activity of the thyroid gland than a compensation for the decreased lean body mass in aging people. Age-related damage to the thyroid gland by the immune system can occur and is more common in women than in men. As a result, approximately 10% of elderly women's thyroid glands do not produce enough T_3 and T_4.

Parathyroid hormone secretion does not appear to decrease with age. Blood levels of Ca^{2+} may decline slightly because of reduced dietary calcium intake and vitamin D levels. The greatest risk is loss of bone matrix as parathyroid hormone increases to maintain blood levels of Ca^{2+} within their normal range.

Systems PATHOLOGY | Graves Disease (Hyperthyroidism)

3:17 PM
69%
Name: Grace
Gender: Female
Age: 36
Comments
Grace manages a business, has several employees, and works hard to make time for her husband and two children. Over several months, she slowly noticed that she was sweating excessively and appeared flushed. In addition, she often felt her heart pounding, was much more nervous than usual, and found it difficult to concentrate. Then Grace began to feel weak and lose weight, even though her appetite was greater than normal. Her family recognized some of these changes and noticed that her eyes seemed larger than usual. They encouraged her to see her physician. After an examination and some blood tests, Grace was diagnosed with Graves disease, a type of hyperthyroidism.

Background Information

Graves disease is caused by altered regulation of hormone secretion—specifically, the elevated secretion of thyroid hormones from the thyroid gland. In approximately 95% of Graves disease cases, the immune system produces an unusual antibody type, which binds to receptors on the cells of the thyroid follicle and stimulates them to secrete increased amounts of thyroid hormone. The secretion of the releasing hormone and thyroid-stimulating hormone is inhibited by elevated thyroid hormones. However, the antibody is produced in large amounts and is not inhibited by thyroid hormones. A very elevated rate of thyroid hormone secretion is therefore maintained. In addition, the size of the thyroid gland increases, and connective tissue components are deposited behind the eyes, causing them to bulge (figure 18A). Enlargement of the thyroid gland is called a goiter.

Grace was treated with radioactive iodine (^{131}I) atoms that were actively transported into thyroid cells, where they destroyed a substantial portion of the thyroid gland. Data indicate that this treatment has few side effects and is effective in treating most cases of Graves disease. Other options include (1) drugs that inhibit the synthesis and secretion of thyroid hormones and (2) surgery to remove part of the thyroid gland.

Predict 12

Explain why removal of part of the thyroid gland is an effective treatment for Graves disease.

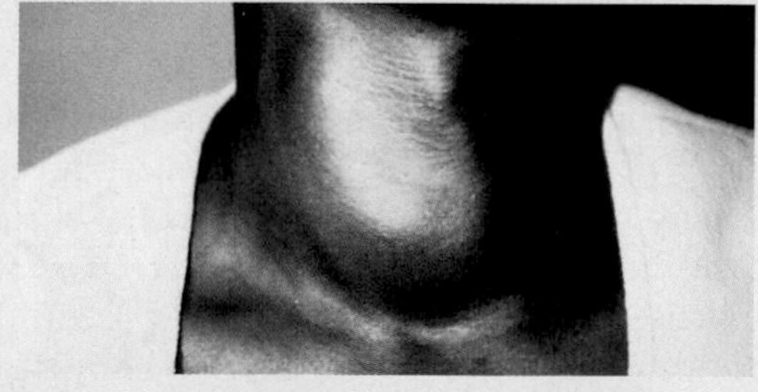

(a)

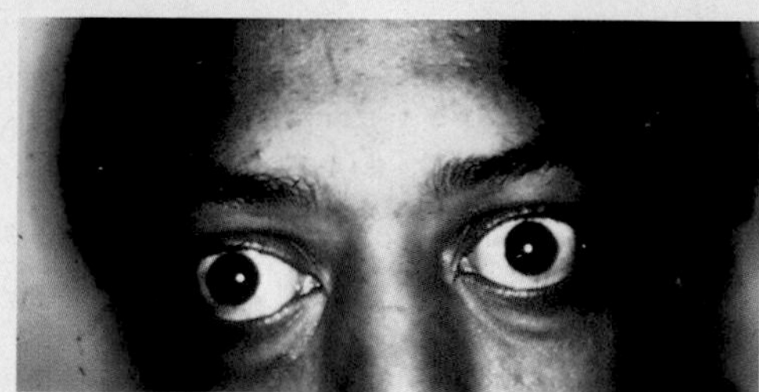

(b)

Figure 18A

(*a*) A goiter and (*b*) protruding eyes are symptoms of hyperthyroidism.

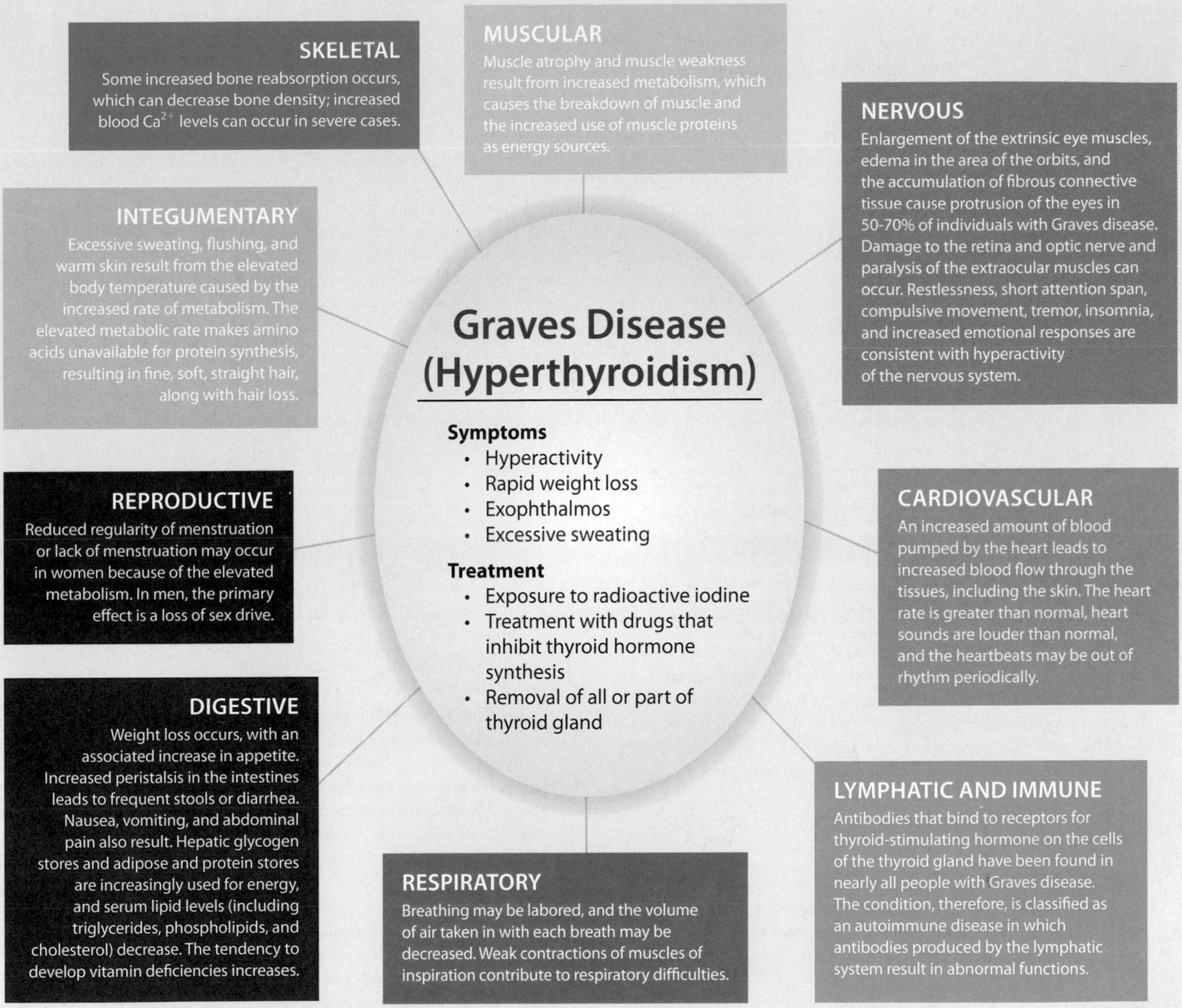

Graves Disease (Hyperthyroidism)

Symptoms
- Hyperactivity
- Rapid weight loss
- Exophthalmos
- Excessive sweating

Treatment
- Exposure to radioactive iodine
- Treatment with drugs that inhibit thyroid hormone synthesis
- Removal of all or part of thyroid gland

SKELETAL
Some increased bone reabsorption occurs, which can decrease bone density; increased blood Ca^{2+} levels can occur in severe cases.

MUSCULAR
Muscle atrophy and muscle weakness result from increased metabolism, which causes the breakdown of muscle and the increased use of muscle proteins as energy sources.

NERVOUS
Enlargement of the extrinsic eye muscles, edema in the area of the orbits, and the accumulation of fibrous connective tissue cause protrusion of the eyes in 50-70% of individuals with Graves disease. Damage to the retina and optic nerve and paralysis of the extraocular muscles can occur. Restlessness, short attention span, compulsive movement, tremor, insomnia, and increased emotional responses are consistent with hyperactivity of the nervous system.

INTEGUMENTARY
Excessive sweating, flushing, and warm skin result from the elevated body temperature caused by the increased rate of metabolism. The elevated metabolic rate makes amino acids unavailable for protein synthesis, resulting in fine, soft, straight hair, along with hair loss.

REPRODUCTIVE
Reduced regularity of menstruation or lack of menstruation may occur in women because of the elevated metabolism. In men, the primary effect is a loss of sex drive.

CARDIOVASCULAR
An increased amount of blood pumped by the heart leads to increased blood flow through the tissues, including the skin. The heart rate is greater than normal, heart sounds are louder than normal, and the heartbeats may be out of rhythm periodically.

DIGESTIVE
Weight loss occurs, with an associated increase in appetite. Increased peristalsis in the intestines leads to frequent stools or diarrhea. Nausea, vomiting, and abdominal pain also result. Hepatic glycogen stores and adipose and protein stores are increasingly used for energy, and serum lipid levels (including triglycerides, phospholipids, and cholesterol) decrease. The tendency to develop vitamin deficiencies increases.

RESPIRATORY
Breathing may be labored, and the volume of air taken in with each breath may be decreased. Weak contractions of muscles of inspiration contribute to respiratory difficulties.

LYMPHATIC AND IMMUNE
Antibodies that bind to receptors for thyroid-stimulating hormone on the cells of the thyroid gland have been found in nearly all people with Graves disease. The condition, therefore, is classified as an autoimmune disease in which antibodies produced by the lymphatic system result in abnormal functions.

Reproductive hormone secretion gradually declines in elderly men, and women experience menopause, as described in chapter 28.

The ability to regulate blood glucose levels does not decline with age. However, there is an age-related probability of developing type 2 diabetes mellitus for those who have the familial tendency, and the incidence of the condition is correlated with age-related increases in body weight.

Secretion of thymosin from the thymus decreases with age. Fewer immature lymphocytes are able to mature and become functional, and the immune system becomes less effective in protecting the body. Thus, people's susceptibility to infection and to cancer increases.

ASSESS YOUR PROGRESS

67. *Describe age-related changes in the secretion of the following hormones and the consequences of each: GH, melatonin, thyroid hormones, reproductive hormones, and thymosin.*

68. *Which hormone does not appear to decrease with age?*

Answer

Learn to Predict ❮ From page 594

The first piece of information is that, although Dylan is always hungry and eating, he is losing weight. You learned in this chapter that two of the most important hormones for metabolism and blood glucose regulation are insulin and glucagon. The disease most often affiliated with disruptions in insulin regulation is diabetes mellitus. Normally, insulin allows for glucose to enter the body's cells for energy production, but in cases of diabetes mellitus, insufficient insulin causes excessive circulating blood glucose. When coupled with Dylan's other symptoms, chronic thirst and urination, there is a clear link to diabetes mellitus. Recall from chapter 3 that membrane transport proteins can be saturated by their transport molecule. The reason Dylan is always thirsty is that too much glucose is filtered out of his blood in his kidneys to be reabsorbed; the excess filtered glucose saturates its transport molecule. In the filtrate, the glucose has an osmotic effect and prevents the kidneys from conserving water. The sweet, or acetone, breath is derived from the fact that, when starved from glucose, cells begin to catalyze lipids, and the by-products of this metabolism are acetone and other molecules chemically related to acetone. Dylan was treated with injectable insulin and began to feel much better. However, if Dylan keeps eating candy and drinking sugary soda, elevated blood glucose levels will continue to dehydrate Dylan and his neurons can become dehydrated. This would cause him to feel irritable and unwell. Dylan may also experience a sudden weight gain because of sugar intake while injecting insulin. His cells will be able to use the extra glucose and may convert it to body fat. Dylan should eat a healthy diet and avoid sugary foods.

Answers to the rest of this chapter's Predict questions are in Appendix G.

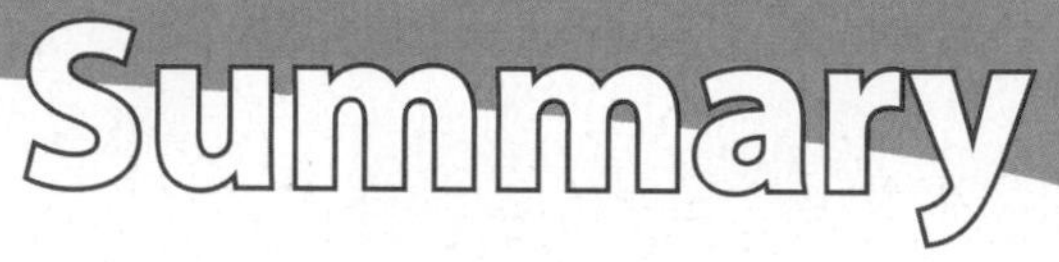
Summary

18.1 Overview of the Endocrine System (p. 595)

The main regulatory functions of the endocrine system are metabolism, control of food intake and digestion, tissue maturation, ion regulation, water balance, heart rate and blood pressure regulation, control of blood glucose and other nutrients, control of reproductive functions, uterine contractions and milk release, and immune system regulation.

18.2 Pituitary Gland and Hypothalamus (p. 595)

1. The pituitary gland secretes at least nine hormones that regulate numerous body functions as well as other endocrine glands.
2. The hypothalamus regulates pituitary gland activity through neurohormones and action potentials.

Structure of the Pituitary Gland

1. The posterior pituitary develops from the floor of the brain and consists of the infundibulum and the neurohypophysis.
2. The anterior pituitary develops from the roof of the mouth.

Relationship of the Pituitary Gland to the Brain: The Hypothalamus

1. The hypothalamohypophysial portal system connects the hypothalamus and the anterior pituitary.
   - Neurohormones are produced in hypothalamic neurons.
   - Through the portal system, the neurohormones inhibit or stimulate hormone production in the anterior pituitary.
2. The hypothalamohypophysial tract connects the hypothalamus and the posterior pituitary.
   - Neurohormones are produced in hypothalamic neurons.
   - The neurohormones move down the axons of the tract and are secreted from the posterior pituitary.

Hormones of the Pituitary Gland

1. ADH promotes water retention by the kidneys.
2. Oxytocin promotes uterine contractions during delivery and causes milk ejection in lactating women.
3. GH is sometimes called somatotropin.
   - GH stimulates growth in most tissues and regulates metabolism.
   - GH stimulates the uptake of amino acids and their conversion into proteins and stimulates the breakdown of lipids and the synthesis of glucose.
   - GH stimulates the production of somatomedins; together, they promote bone and cartilage growth.
   - GH secretion increases in response to an increase in blood amino acids, low blood glucose, or stress.
   - GH is regulated by GHRH and GHIH or by somatostatin.
4. TSH, or thyrotropin, causes the release of thyroid hormones.
5. ACTH is derived from proopiomelanocortin; it stimulates cortisol secretion from the adrenal cortex and increases skin pigmentation.
6. Several hormones in addition to ACTH are derived from proopiomelanocortin.
   - Lipotropins cause lipid breakdown.
   - β endorphins play a role in analgesia.
   - MSH increases skin pigmentation.
7. LH and FSH are major gonadotropins.
   - Both hormones regulate the production of gametes and reproductive hormones (testosterone in males, estrogen and progesterone in females).
   - GnRH from the hypothalamus stimulates LH and FSH secretion.
8. Prolactin stimulates milk production in lactating females. Prolactin-releasing hormone and prolactin-inhibiting hormone from the hypothalamus affect prolactin secretion.

18.3 Thyroid Gland (p. 605)

1. The thyroid gland is just inferior to the larynx.
2. The thyroid gland is composed of small, hollow balls of cells called follicles, which contain thyroglobulin.
3. Parafollicular cells are scattered throughout the thyroid gland.

Thyroid Hormones

1. T_3 and T_4 synthesis occurs in thyroid follicles.
   - Iodide ions are taken into the follicles by active transport, oxidized, and bound to tyrosine molecules in thyroglobulin.
   - Thyroglobulin is secreted into the follicle lumen. Tyrosine molecules with iodine combine to form T_3 and T_4 thyroid hormones.
   - Thyroglobulin is taken into the follicular cells and broken down; T_3 and T_4 diffuse from the follicles to the blood.
2. T_3 and T_4 are transported in the blood.
   - T_3 and T_4 bind to thyroxine-binding globulin and other plasma proteins.
   - The plasma proteins prolong the half-life of T_3 and T_4 and regulate the levels of T_3 and T_4 in the blood.
   - Approximately one-third of the T_4 is converted into functional T_3.
3. T_3 and T_4 bind with nuclear receptor molecules and initiate new protein synthesis.
4. T_3 and T_4 affect nearly every tissue in the body.
   - T_3 and T_4 increase the rate of glucose, lipid, and protein metabolism in many tissues, thus increasing body temperature.
   - Normal growth of many tissues is dependent on T_3 and T_4.
5. TRH and TSH regulate T_3 and T_4 secretion.
   - Increased TSH from the anterior pituitary increases T_3 and T_4 secretion.
   - TRH from the hypothalamus increases TSH secretion. TRH increases as a result of chronic exposure to cold, food deprivation, and stress.
   - T_3 and T_4 inhibit TSH and TRH secretion.

Calcitonin

1. An increase in blood calcium levels stimulates calcitonin secretion by the parafollicular cells.
2. Calcitonin decreases blood calcium and phosphate levels by inhibiting osteoclasts.

18.4 Parathyroid Glands (p. 611)

1. The parathyroid glands are embedded in the thyroid gland.
2. PTH increases blood calcium levels.
   - PTH stimulates osteoclasts.
   - PTH promotes calcium reabsorption by the kidneys and the formation of active vitamin D by the kidneys.
   - Active vitamin D increases calcium absorption by the intestine.
3. A decrease in blood calcium stimulates PTH secretion.

18.5 Adrenal Glands (p. 612)

1. The adrenal glands are near the superior poles of the kidneys.
2. The adrenal medulla arises from neural crest cells and functions as part of the sympathetic nervous system. The adrenal cortex is derived from mesoderm.
3. The adrenal medulla is composed of closely packed cells.
4. The adrenal cortex is divided into three layers: the zona glomerulosa, the zona fasciculata, and the zona reticularis.

Hormones of the Adrenal Medulla

1. Epinephrine accounts for 80% and norepinephrine for 20% of the adrenal medulla hormones.
   - Epinephrine increases blood glucose levels, the use of glycogen and glucose by skeletal muscle, and heart rate and force of contraction. It also causes vasoconstriction in the skin and viscera and vasodilation in skeletal and cardiac muscle.
   - Norepinephrine and epinephrine stimulate cardiac muscle and cause the constriction of most peripheral blood vessels.
2. The adrenal medulla hormones prepare the body for physical activity.
3. Release of adrenal medulla hormones is mediated by the sympathetic nervous system in response to emotions, injury, stress, exercise, and low blood glucose.

Hormones of the Adrenal Cortex

1. The zona glomerulosa secretes the mineralocorticoids, especially aldosterone. Aldosterone acts on the kidneys to increase sodium and to decrease potassium and hydrogen levels in the blood.
2. The zona fasciculata secretes glucocorticoids, especially cortisol.
   - Cortisol increases lipid and protein breakdown, increases glucose synthesis from amino acids, decreases the inflammatory response, and is necessary for the development of some tissues.
   - ACTH from the anterior pituitary stimulates cortisol secretion. CRH from the hypothalamus stimulates ACTH release. Low blood glucose levels and stress stimulate CRH secretion.
3. The zona reticularis secretes androgens. In females, androgens stimulate axillary and pubic hair growth and sex drive.

18.6 Pancreas (p. 618)

1. The pancreas, located along the small intestine and the stomach, is both an exocrine and an endocrine gland.
2. The exocrine portion of the pancreas consists of a complex duct system, which ends in small sacs, called acini, that produce pancreatic digestive juices.
3. The endocrine portion consists of the pancreatic islets. Each islet is composed of alpha cells, which secrete glucagon; beta cells, which secrete insulin; and delta cells, which secrete somatostatin.

Effect of Insulin and Glucagon on Their Target Tissues

1. Insulin's target tissues are the liver, adipose tissue, muscle, and the satiety center in the hypothalamus. The nervous system is not a target tissue, but it does rely on blood glucose levels maintained by insulin.
2. Insulin increases the uptake of glucose and amino acids by cells. Glucose is used for energy or is stored as glycogen. Amino acids are used for energy or are converted to glucose or proteins.
3. Glucagon's target tissue is mainly the liver.
4. Glucagon causes the breakdown of glycogen and lipids for use as an energy source.

Regulation of Pancreatic Hormone Secretion

1. Insulin secretion increases because of elevated blood glucose levels, an increase in some amino acids, parasympathetic stimulation, and gastrointestinal hormones. Sympathetic stimulation decreases insulin secretion.
2. Glucagon secretion is stimulated by low blood glucose levels, certain amino acids, and sympathetic stimulation.
3. Somatostatin inhibits insulin and glucagon secretion.

18.7 Hormonal Regulation of Nutrient Utilization (p. 622)

1. After a meal, the following events take place:
   - High glucose levels inhibit glucagon, cortisol, GH, and epinephrine, which reduces the release of glucose from tissues.
   - Insulin secretion increases as a result of the high blood glucose levels, thereby increasing the uptake of glucose, amino acids, and lipids, which are used for energy or stored.
   - Sometime after the meal, blood glucose levels drop. Glucagon, GH, cortisol, and epinephrine levels increase, insulin levels decrease, and glucose is released from tissues.
   - Adipose tissue releases fatty acids, triglycerides, and ketones, which most tissues use for energy.
2. During exercise, the following events occur:
   - Sympathetic activity increases epinephrine and glucagon secretion, causing a release of glucose into the blood.
   - Low blood sugar levels, caused by the uptake of glucose by skeletal muscles, stimulate epinephrine, glucagon, GH, and cortisol secretion, causing an increase in fatty acids, triglycerides, and ketones in the blood, all of which are used for energy.

18.8 Hormones of the Reproductive System (p. 626)

The ovaries, testes, placenta, and pituitary gland secrete reproductive hormones.

18.9 Hormones of the Pineal Gland (p. 627)

The pineal gland produces melatonin and arginine vasotocin, which can inhibit reproductive maturation and may regulate sleep-wake cycles.

18.10 Other Hormones and Chemical Messengers (p. 628)

1. The thymus produces thymosin, which is involved in the development of the immune system.
2. The digestive tract produces several hormones that regulate digestive functions.
3. Autocrine and paracrine chemical messengers are produced by many cells of the body and usually have a local effect on body functions.
4. Eicosanoids, such as prostaglandins, prostacyclins, thromboxanes, and leukotrienes, are derived from fatty acids and mediate inflammation and other functions. Endorphins, enkephalins, and dynorphins are analgesic substances. Growth factors influence cell division and growth in many tissues, and interleukin-2 influences cell division in the T cells of the immune system.

18.11 Effects of Aging on the Endocrine System (p. 629)

A gradual decrease in the secretion rate occurs for most, but not all, hormones. Some of these decreases are related to gradual decreases in physical activity.

REVIEW AND COMPREHENSION

1. The pituitary gland
   a. develops from the floor of the brain.
   b. develops from the roof of the mouth.
   c. is stimulated by neurohormones produced in the midbrain.
   d. secretes only three major hormones.
   e. Both a and b are correct.
2. The hypothalamohypophysial portal system
   a. contains one capillary bed.
   b. carries hormones from the anterior pituitary to the body.
   c. carries hormones from the posterior pituitary to the body.
   d. carries hormones from the hypothalamus to the anterior pituitary.
   e. carries hormones from the hypothalamus to the posterior pituitary.
3. Which of these hormones is *not* secreted into the hypothalamohypophysial portal system?
   a. GHRH
   b. TRH
   c. PIH
   d. GnRH
   e. ACTH
4. Which of these stimulates the secretion of ADH?
   a. elevated blood osmolality
   b. decreased blood osmolality
   c. release of hormones from the hypothalamus
   d. ACTH
   e. increased blood pressure
5. Oxytocin is responsible for
   a. preventing milk release from the mammary glands.
   b. preventing goiter.
   c. causing contraction of the uterus.
   d. maintaining normal calcium levels.
   e. increasing the metabolic rate.
6. Growth hormone
   a. increases the usage of glucose.
   b. increases the breakdown of lipids.
   c. decreases the synthesis of proteins.
   d. decreases the synthesis of glycogen.
   e. All of these are correct.
7. Which of these hormones stimulates somatomedin secretion?
   a. FSH
   b. GH
   c. LH
   d. prolactin
   e. TSH
8. Hypersecretion of growth hormone
   a. results in giantism if it occurs in children.
   b. causes acromegaly in adults.
   c. increases the probability that a person will develop diabetes.
   d. can lead to severe atherosclerosis.
   e. All of these are correct.
9. LH and FSH
   a. are produced in the hypothalamus.
   b. production is increased by TSH.
   c. promote the production of gametes and reproductive hormones.
   d. inhibit the production of prolactin.
   e. All of these are correct.
10. T_3 and T_4
    a. require iodine for their production.
    b. are made from the amino acid tyrosine.
    c. are transported in the blood bound to thyroxine-binding globulin.
    d. All of these are correct.

11. Which of these symptoms is associated with hyposecretion of thyroid hormones?
    a. hypertension
    b. nervousness
    c. diarrhea
    d. weight loss with either normal or increased food intake
    e. decreased metabolic rate
12. Choose the statement that most accurately predicts the long-term effect of exposure to a substance that prevents the active transport of iodide by the thyroid gland.
    a. Large amounts of T_3 and T_4 accumulate within the thyroid follicles, but little is released.
    b. The person exhibits hypothyroidism.
    c. The anterior pituitary secretes smaller amounts of TSH.
    d. The circulating levels of T_3 and T_4 increase.
13. Calcitonin
    a. is secreted by the parathyroid glands.
    b. levels increase when blood calcium levels decrease.
    c. causes blood calcium levels to decrease.
    d. insufficiency results in weak bones and tetany.
14. Parathyroid hormone secretion increases in response to
    a. a decrease in blood calcium levels.
    b. increased production of parathyroid-stimulating hormone from the anterior pituitary.
    c. increased secretion of parathyroid-releasing hormone from the hypothalamus.
    d. increased secretion of calcitonin.
    e. decreased secretion of ACTH.
15. If parathyroid hormone levels increase, which of these conditions is expected?
    a. Osteoclast activity increases.
    b. Calcium absorption from the small intestine is inhibited.
    c. Calcium reabsorption from the urine is inhibited.
    d. Less active vitamin D forms in the kidneys.
    e. All of these are correct.
16. The adrenal medulla
    a. produces steroids.
    b. secretes cortisol as its major product.
    c. decreases its secretions during exercise.
    d. forms from a modified portion of the sympathetic division of the ANS.
    e. All of these are correct.
17. In the condition in which a benign tumor results in hypersecretion of hormones from the adrenal medulla, expected symptoms include
    a. hypotension.
    b. bradycardia (slow heart rate).
    c. pallor (decreased blood flow to the skin).
    d. lethargy.
    e. hypoglycemia.
18. Which of these is *not* a hormone secreted by the adrenal cortex?
    a. aldosterone
    b. androgens
    c. cortisol
    d. epinephrine
19. If aldosterone secretions increase,
    a. blood potassium levels increase.
    b. blood hydrogen levels increase.
    c. acidosis results.
    d. blood sodium levels decrease.
    e. blood volume increases.
20. Glucocorticoids (cortisol)
    a. increase the breakdown of lipids.
    b. increase the breakdown of proteins.
    c. increase blood glucose levels.
    d. decrease inflammation.
    e. All of these are correct.
21. Which of these is (are) expected in Cushing syndrome (hypersecretion of adrenal cortex hormones)?
    a. loss of hair in women
    b. deposition of adipose tissue in the face, neck, and abdomen
    c. low blood glucose
    d. low blood pressure
    e. All of these are correct.
22. Within the pancreas, the pancreatic islets produce
    a. insulin.
    b. glucagon.
    c. digestive enzymes.
    d. Both a and b are correct.
    e. All of these are correct.
23. Insulin increases
    a. the uptake of glucose by its target tissues.
    b. the breakdown of protein.
    c. the breakdown of lipids.
    d. glycogen breakdown in the liver.
    e. All of these are correct.
24. Glucagon
    a. primarily affects the liver.
    b. causes glycogen to be stored.
    c. causes blood glucose levels to decrease.
    d. decreases lipid metabolism.
    e. performs all of these functions.
25. When blood glucose levels increase, the secretion of which of these hormones increases?
    a. glucagon
    b. insulin
    c. GH
    d. cortisol
    e. epinephrine
26. If a person who has diabetes mellitus has forgotten to take an insulin injection, the symptoms that may soon appear include
    a. acidosis.
    b. hyperglycemia.
    c. increased urine production.
    d. lethargy and fatigue.
    e. All of these are correct.
27. Which of the following is *not* a hormone produced by the ovaries?
    a. estrogen
    b. progesterone
    c. prolactin
    d. inhibin
    e. relaxin
28. Melatonin
    a. is produced by the posterior pituitary.
    b. production increases as day length increases.
    c. inhibits the development of the reproductive system.
    d. increases GnRH secretion from the hypothalamus.
    e. decreases the tendency to sleep.
29. Which of these substances, produced by many body tissues, can promote inflammation, pain, and vasodilation of blood vessels?
    a. endorphins
    b. enkephalins
    c. thymosin
    d. epidermal growth factor
    e. prostaglandins
30. Which of these secretions does *not* decrease with aging of the endocrine system?
    a. GH secretion
    b. melatonin secretion
    c. thyroid hormone secretion
    d. parathyroid hormone secretion

Answers in Appendix E

CRITICAL THINKING

1. The hypothalamohypophysial portal system connects the hypothalamus with the anterior pituitary. Why is such a special circulatory system advantageous?
2. A patient exhibits polydipsia (thirst), polyuria (excess urine production), and urine with a low specific gravity (contains few ions and no glucose). If you wanted to reverse the symptoms, would you administer insulin, glucagon, ADH, or aldosterone? Explain.
3. A patient complains of headaches and visual disturbances. A casual glance reveals enlarged finger bones, a heavy deposition of bone over the eyes, and a prominent jaw. The doctor determines that the headaches and visual disturbances result from increased pressure within the skull and that the presence of a pituitary tumor is affecting hormone secretion. Name the hormone causing the problem, and explain why increased pressure exists within the skull.
4. Most laboratories are able to determine blood levels of TSH, T_3, and T_4. Given that ability, design a method of determining whether hyperthyroidism in a patient results from a pituitary abnormality or from the production of a nonpituitary thyroid stimulatory substance.
5. Over the past year, Julie has gradually gained weight. The increase in adipose tissue is distributed over her trunk, face, and neck, and her muscle mass appears to be decreased. Julie also feels weak and bruises easily. Her physician suspects Cushing syndrome and orders a series of blood tests. The results reveal elevated blood levels of cortisol and ACTH. There is no evidence of an extrapituitary source of ACTH. Predict the cause of Julie's condition and the treatments that are likely to be recommended.
6. An anatomy and physiology instructor asks two students to predict a patient's response to chronic vitamin D deficiency. One student claims the person would suffer from hypocalcemia. The other student claims the calcium levels would remain within their normal range, although at the low end, and that bone reabsorption would occur to the point that advanced osteomalacia might occur. With whom do you agree, and why?
7. A patient arrives at the emergency room in an unconscious condition. A medical emergency bracelet reveals that he has diabetes. The patient is in either diabetic coma or insulin shock. How can you tell which, and what treatment do you recommend for each condition?
8. Predict some of the consequences of exposure to intense and prolonged stress.
9. Katie was getting nervous. At 16, she was the only one in her group of friends who had not started menstruating. Katie had always dreamed of having three beautiful children someday and she was worried. Her mother took her to see Dr. Josephine, who ordered several blood tests. When the results came back, Dr. Josephine gently explained to Katie and her mother that Katie would never be able to have children and would never menstruate. Dr. Josephine then asked Katie to wait in the outer room while she spoke privately to her mother. She explained to Katie's mom that Katie has androgen insensitivity syndrome. Though Katie is genetically male and her gonads produce more of the male reproductive hormone, testosterone, than the female reproductive hormone, estrogen, Katie did not reflect the tissue changes expected. What malfunction in Katie's body would cause this? Why does Katie's body look feminine if she is genetically male?

Answers in Appendix F

Visit this book's website at **www.mhhe.com/seeley10** for chapter quizzes, interactive learning exercises, and other study tools.

Mc Graw Hill
connect plus+
ANATOMY & PHYSIOLOGY

Anatomy & Physiology
REVEALED
aprevealed.com

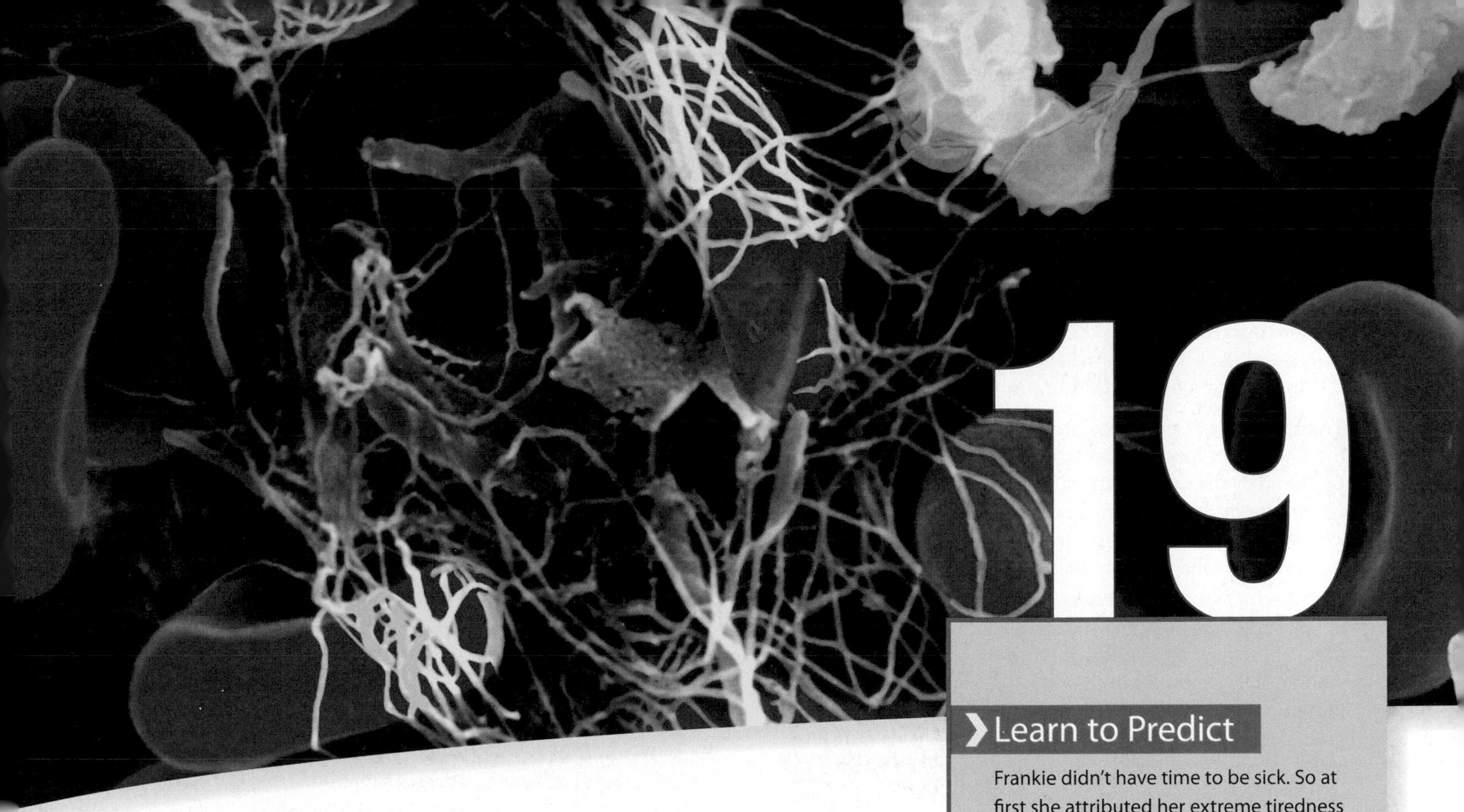

19

Cardiovascular System

BLOOD

Historically, many cultures around the world, both ancient and modern, have believed in the magical qualities of blood. Some societies consider blood the "essence of life" because the uncontrolled loss of it can result in death. Blood has also been thought to define character and emotions. For example, people from prominent families are sometimes described as "bluebloods," whereas criminals are said to have "bad blood." Common expressions allege that anger causes the blood to "boil" and that fear results in blood "curdling." The scientific study of blood reveals characteristics as fascinating as any of these fantasies. Blood performs many functions essential to life and can often reveal much about our health.

Blood is one component of the **cardiovascular system,** which also consists of the heart and the blood vessels. The cardiovascular system connects the various tissues of the body. The heart pumps blood through vessels extending throughout the body (often referred to as the *circulatory system*), and the blood delivers nutrients and picks up waste products at the body tissues. This chapter focuses on the blood, whereas chapters 20 and 21 discuss the heart and the blood vessels, respectively.

Learn to Predict

Frankie didn't have time to be sick. So at first she attributed her extreme tiredness to the stress of being a 40-year-old single mother of two teenagers while working full-time and attending school part-time. However, when she started experiencing significant abdominal pain, she consulted her doctor, who ordered several tests. The results indicated a low red blood cell (RBC) count with microcytic RBCs, a high reticulocyte count, low hemoglobin and hematocrit levels, and evidence of hemoglobin in her feces. After reading this chapter and recalling what you learned about the endocrine system in chapters 17 and 18, explain Frankie's symptoms and test results.

Photo: Colorized scanning electron micrograph of a blood clot. The red discs are red blood cells, the blue particles are platelets, and the yellow strands are fibrin.

Module 9
Cardiovascular System

Anatomy & Physiology
REVEALED
aprevealed.com

19.1 Functions of Blood

LEARNING OUTCOME

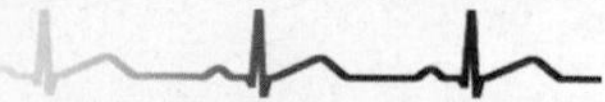

After reading this section, you should be able to

A. List and explain the ways blood helps maintain homeostasis in the body.

Blood helps maintain homeostasis in several ways:

1. *Transport of gases, nutrients, and waste products.* Oxygen enters the blood in the lungs and is carried to the cells. Carbon dioxide, produced by the cells, is carried in the blood to the lungs, from which it is expelled. The blood transports ingested nutrients, ions, and water from the digestive tract to the cells, and the blood transports the cells' waste products to the kidneys for elimination.
2. *Transport of processed molecules.* Many substances are produced in one part of the body and transported in the blood to another part, where they are modified. For example, the precursor to vitamin D is produced in the skin (see chapter 5) and transported by the blood to the liver and then to the kidneys for processing into active vitamin D. The blood then transports active vitamin D to the small intestine, where it promotes the uptake of calcium. Another example involves lactic acid produced by skeletal muscles during anaerobic respiration (see chapter 9). The blood carries lactic acid to the liver, where it is converted into glucose.
3. *Transport of regulatory molecules.* The blood carries the hormones and many of the enzymes that regulate body processes from one part of the body to another.
4. *Regulation of pH and osmosis.* Buffers (see chapter 2), which help keep the blood's pH within its normal range of 7.35–7.45, are in the blood. The osmotic composition of blood is also critical for maintaining normal fluid and ion balance.
5. *Maintenance of body temperature.* Movement of warm blood from the interior of the body, to its surface where heat is released, is one of several mechanisms that help regulate body temperature.
6. *Protection against foreign substances.* Certain cells and chemicals in the blood make up an important part of the immune system, protecting against foreign substances, such as microorganisms and toxins.
7. *Clot formation.* When blood vessels are damaged, blood clotting protects against excessive blood loss. When tissues are damaged, the blood clot that forms is also the first step in tissue repair and the restoration of normal function (see chapter 4).

ASSESS YOUR PROGRESS

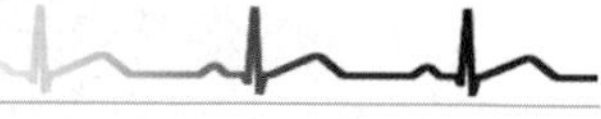

1. *List the ways that blood helps maintain homeostasis in the body.*
2. *What substances are transported by the blood?*
3. *What is the normal pH range of the blood?*
4. *How does the blood provide protection?*

19.2 Composition of Blood

LEARNING OUTCOMES

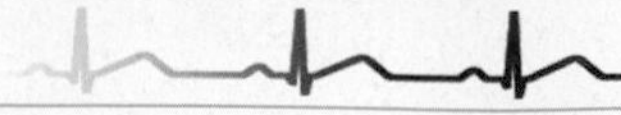

After reading this section, you should be able to

A. List the components of blood.

B. Relate the average total blood volume for males and for females.

Blood is a type of connective tissue consisting of a liquid matrix containing cells and cell fragments. The liquid matrix is the plasma, and the cells and cell fragments are the formed elements. The plasma makes up 55% of the total blood volume, and the formed elements make up 45% (figure 19.1). The total blood volume in the average adult is about 4–5 L in females and 5–6 L in males. Blood makes up about 8% of the total weight of the body.

ASSESS YOUR PROGRESS

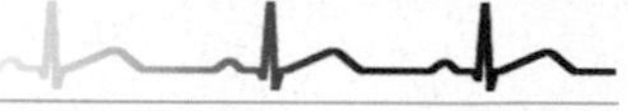

5. *What are the two major components of blood? What portion of the total blood volume does each compose?*
6. *What is the average total blood volume for males and for females?*

19.3 Plasma

LEARNING OUTCOMES

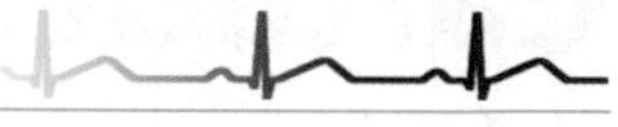

After reading this section, you should be able to

A. Name the components of blood plasma.

B. List the three major plasma proteins and describe their functions.

Plasma (plaz′mă) is the liquid part of blood. It is a pale yellow fluid that consists of about 91% water and 9% other substances, such as proteins, ions, nutrients, gases, waste products, and regulatory substances (table 19.1). Plasma is a **colloid** (kol′oyd), which is a liquid containing suspended substances that do not settle out of solution. Most of the suspended substances are plasma proteins. Based on molecular size and charge, the plasma proteins can be classified as albumin, globulins, and fibrinogen. Almost all of the plasma proteins are produced by the liver or blood cells, a notable exception being protein hormones. **Albumin** (al-bū′min) makes up 58% of the plasma proteins and is important in regulating the movement of water between the tissues and the blood. Because albumin does not pass easily from the blood into tissues, it plays an important role in maintaining blood colloid osmotic pressure (see chapters 21 and 26). Albumins also bind and transport other molecules in the blood, such as fatty acids, bilirubin, and thyroid hormones. **Globulins** (glob′ū-linz) account for 38% of the plasma proteins. The globulins are subdivided into α, β, and γ globulins. Many substances in the blood are transported by globulins; as part of immunity, globulins provide protection against microorganisms

FUNDAMENTAL **Figure**

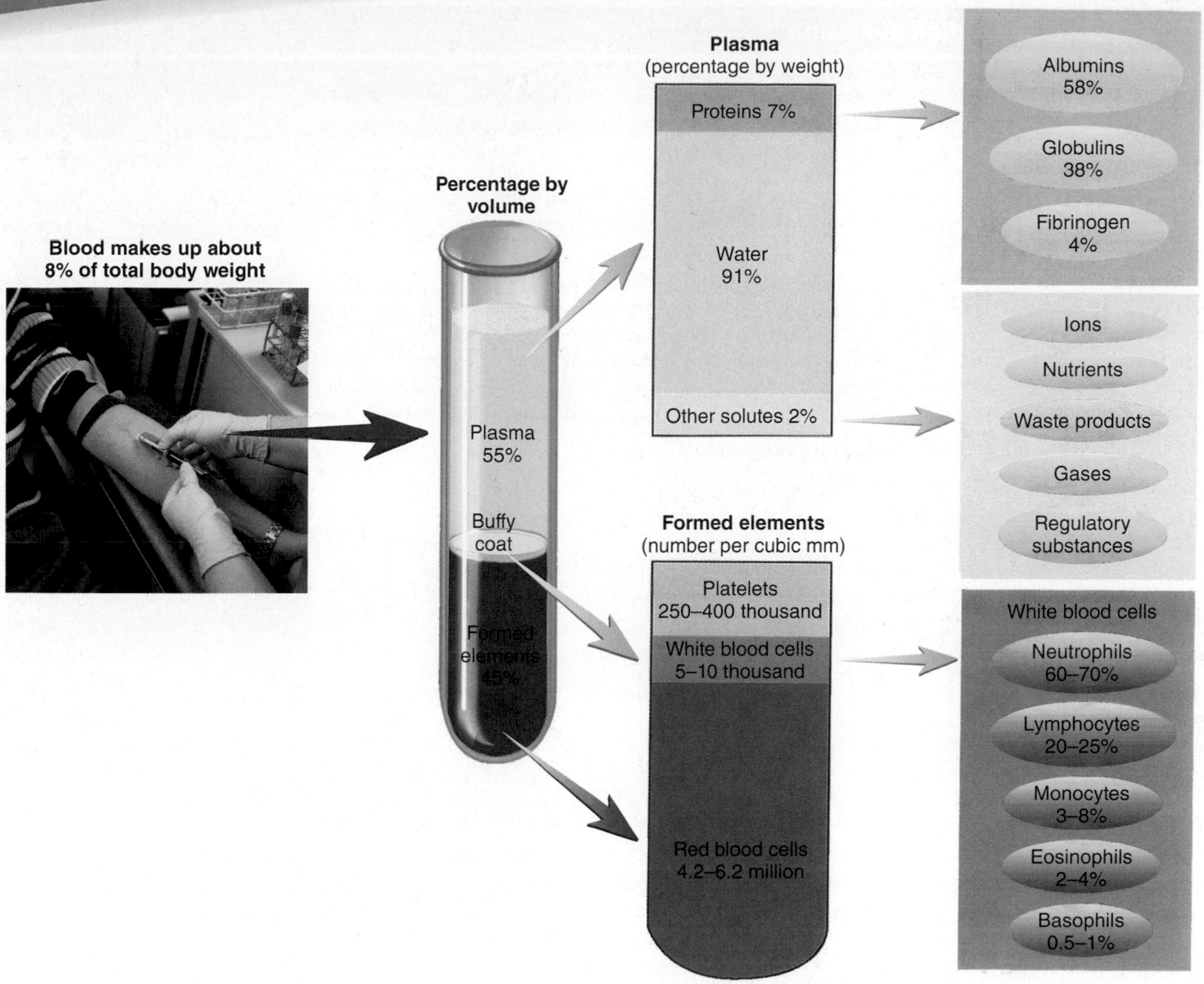

Blood makes up about 8% of total body weight
Percentage by volume
Plasma 55%
Buffy coat
Formed elements 45%
Plasma (percentage by weight)
Proteins 7%
Water 91%
Other solutes 2%
Albumins 58%
Globulins 38%
Fibrinogen 4%
Ions
Nutrients
Waste products
Gases
Regulatory substances
Formed elements (number per cubic mm)
Platelets 250–400 thousand
White blood cells 5–10 thousand
Red blood cells 4.2–6.2 million
White blood cells
Neutrophils 60–70%
Lymphocytes 20–25%
Monocytes 3–8%
Eosinophils 2–4%
Basophils 0.5–1%

FIGURE 19.1 APR **Composition of Blood**
Approximate values for the components of blood in a normal adult.

(table 19.1; see chapter 22). **Fibrinogen** (fī-brin′ō-jen) constitutes 4% of the plasma proteins and is responsible for the formation of blood clots (see "Coagulation" in section 19.5). **Serum** (ser′um; whey) is plasma without the clotting factors.

The levels of water, proteins, and other substances in the blood, such as ions, nutrients, waste products, gases, and regulatory substances, are maintained within narrow limits. Normally, the amount of water taken in through the digestive tract closely matches the amount of water lost through the kidneys, lungs, digestive tract, and skin. Therefore, plasma volume remains relatively constant. Suspended or dissolved substances in the blood come from the liver, kidneys, intestines, endocrine glands, and immune tissues, such as the lymph nodes and spleen. Oxygen enters the blood in the lungs and leaves the blood as it flows through tissues. Carbon dioxide enters the blood from the tissues and leaves the blood as it flows through the lungs.

ASSESS YOUR PROGRESS

7. *What is plasma, and what does it consist of? Why is plasma a colloid?*
8. *What are the three major plasma proteins, and what roles do they play in the blood?*
9. *Explain how plasma volume remains relatively constant.*

19.4 Formed Elements

LEARNING OUTCOMES

After reading this section, you should be able to

A. List the three kinds of formed elements using both their common and technical names.

TABLE 19.1 Composition of Plasma

Components	Function
Water	Acts as a solvent and suspending medium for blood components
Plasma Proteins	
Albumin	Partly responsible for blood viscosity and osmotic pressure; acts as a buffer; transports fatty acids, free bilirubin, and thyroid hormones
Globulins	
α	Protect tissues from damage by inflammation (alpha-1 antitrypsin); transport thyroid hormones (thyroid-binding globulin), cortisol (transcortin), and testosterone and estrogen (sex hormone–binding globulin); transport lipids (e.g., cholesterol in high-density lipoproteins); convert ferrous iron (Fe^{2+}) to ferric iron (Fe^{3+}), which promotes iron transport by transferrin (ceruloplasmin); transport hemoglobin released from damaged red blood cells (haptoglobin)
β	Transport iron (transferrin); transport lipids (beta-lipoproteins), especially cholesterol in low-density lipoproteins; involved with immunity (complement); prevent blood loss (coagulation proteins)
γ	Involved in immunity (most antibodies are γ globulins, but some are β or α globulins)
Fibrinogen	Functions in blood clotting
Ions	
Sodium, potassium, calcium, magnesium, chloride, iron, phosphate, hydrogen, hydroxide, bicarbonate	Involved in osmosis, membrane potentials, and acid-base balance
Nutrients	
Glucose, amino acids, triglycerides, cholesterol	Source of energy and basic "building blocks" of more complex molecules
Vitamins	Promote enzyme activity
Waste Products	
Urea, uric acid, creatinine, ammonia salts	Breakdown products of protein metabolism; excreted by the kidneys
Bilirubin	Breakdown product of red blood cells; excreted as part of the bile from the liver into the small intestine
Lactic acid	End product of anaerobic respiration; converted to glucose by the liver
Gases	
Oxygen	Necessary for aerobic respiration; terminal electron acceptor in electron-transport chain
Carbon dioxide	Waste product of aerobic respiration; as bicarbonate, helps buffer blood
Nitrogen	Inert
Regulatory Substances	Enzymes catalyze chemical reactions; hormones stimulate or inhibit many body functions

B. Describe the origin and production of the formed elements.

C. Describe the structure and function of hemoglobin and relate which gases associate with hemoglobin and how.

D. Compare fetal and adult hemoglobin as to structure and affinity for oxygen.

E. Discuss the life history of red blood cells.

F. Compare the structures and functions of the five types of white blood cells.

G. Describe the origin and structure of platelets.

H. Relate the functions of platelets in preventing blood loss.

About 95% of the volume of the **formed elements** consists of red blood cells, or *erythrocytes* (ĕ-rith′rō-sītz). The remaining 5% consists of white blood cells, or *leukocytes* (loo′kō-sītz), and cell fragments called platelets, or *thrombocytes* (throm′bō-sītz). Table 19.2 illustrates the formed elements of the blood. In healthy adults, white blood cells are the only formed elements possessing nuclei; red blood cells and platelets lack nuclei.

Production of Formed Elements

The process of blood cell production is called **hematopoiesis** (hē′mă-tō-poy-ē′sis, hem′ă-to-poy-ē′sis), or *hemopoiesis* (hē′mō-poy-ē′sis). In the embryo and fetus, this process occurs in tissues such as the yolk sac, liver, thymus, spleen, lymph nodes, and red bone marrow. After birth, hematopoiesis is confined primarily to red bone marrow, with some lymphatic tissue helping produce lymphocytes (see chapter 22). In young children, nearly all the marrow is red bone marrow. In adults, however, red marrow is confined to the ribs, sternum, vertebrae, pelvis, proximal femur, and proximal humerus. Yellow marrow replaces red marrow in other body locations (see chapter 6).

TABLE 19.2 Formed Elements of the Blood

Cell Type	Illustration	Description	Function
Red Blood Cell		Biconcave disc; no nucleus; contains hemoglobin, which colors the cell red; 7.5 μm in diameter	Transports oxygen and carbon dioxide
White Blood Cells		Spherical cells with a nucleus	Five types of white blood cells, each with specific functions
Granulocytes			
Neutrophil		Nucleus with two to five lobes connected by thin filaments; cytoplasmic granules stain a light pink or reddish-purple; 10–12 μm in diameter	Phagocytizes microorganisms and other substances
Eosinophil		Nucleus often bilobed; cytoplasmic granules stain orange-red or bright red; 11–14 μm in diameter	Attacks certain worm parasites; releases chemicals that modulate inflammation; negatively impacts airways during asthma attacks
Basophil		Nucleus with two indistinct lobes; cytoplasmic granules stain blue-purple; 10–12 μm in diameter	Releases histamine, which promotes inflammation, and heparin, which prevents clot formation
Agranulocytes			
Lymphocyte		Round nucleus; cytoplasm forms a thin ring around the nucleus; 6–14 μm in diameter	Produces antibodies and other chemicals responsible for destroying microorganisms; contributes to allergic reactions, graft rejection, tumor control, and regulation of the immune system
Monocyte		Nucleus round, kidney-shaped, or horseshoe-shaped; contains more cytoplasm than lymphocyte does; 12–20 μm in diameter	Phagocytic cell in the blood; leaves the blood and becomes a macrophage, which phagocytizes bacteria, dead cells, cell fragments, and other debris within tissues
Platelet		Cell fragment surrounded by plasma membrane and containing granules; 2–4 μm in diameter	Forms platelet plugs; releases chemicals necessary for blood clotting

All the formed elements of the blood are derived from a single population of stem cells called **hemocytoblasts,** located in the red bone marrow. Hemocytoblasts are precursor cells capable of dividing to produce daughter cells that can differentiate into various types of blood cells (figure 19.2). When a hemocytoblast divides, one daughter cell remains a hemocytoblast while the other daughter cell differentiates to form one of two types of intermediate stem cells: a **myeloid stem cell** or a **lymphoid stem cell.** Red blood cells, platelets and most of the white blood cells develop from myeloid stem cells. Specifically, myeloid stem cells can differentiate into **proerythroblasts** (prō-ē-rith′rō-blastz), from which red blood cells develop; **myeloblasts** (mī′ĕ-lō-blastz), from which basophils, eosinophils, and neutrophils develop; **monoblasts** (mon′ō-blastz), from which monocytes develop; and **megakaryoblasts** (meg-ă-kar′ē-ō-blastz), from which platelets develop. Lymphocytes develop from lymphoid stem cells.

Chemical signals regulate the development of the different types of formed elements. These chemical signals include **colony-stimulating factors (CSFs)** and hormones transported to the bone marrow through the blood or substances released by bone marrow cells. **Erythropoietin (EPO)** is an example of a hormone, secreted by endocrine cells of the kidneys, that stimulates myeloid stem cells to develop into red blood cells.

ASSESS **YOUR PROGRESS** 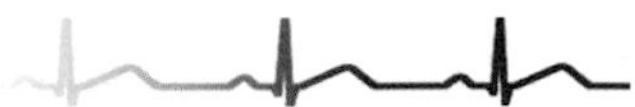

10. *Name the three general types of formed elements in the blood, using both their common and technical names.*
11. *What is hematopoiesis? Where does the process occur before birth? After birth? What type of stem cell are all formed elements derived from? Distinguish between myeloid stem cells and lymphoid stem cells.*
12. *What types of formed elements develop from each of the following cells: proerythroblasts, myeloblasts, lymphoblasts, monoblasts, and megakaryocytes?*

Red Blood Cells

Red blood cells (RBCs), or *erythrocytes,* are about 700 times more numerous than white blood cells and 17 times more numerous than platelets in the blood (figure 19.3*a*). Males have about 5.4 million red blood cells per microliter (μL; 1 mm^3, or 10^{-6} L) of blood (range:

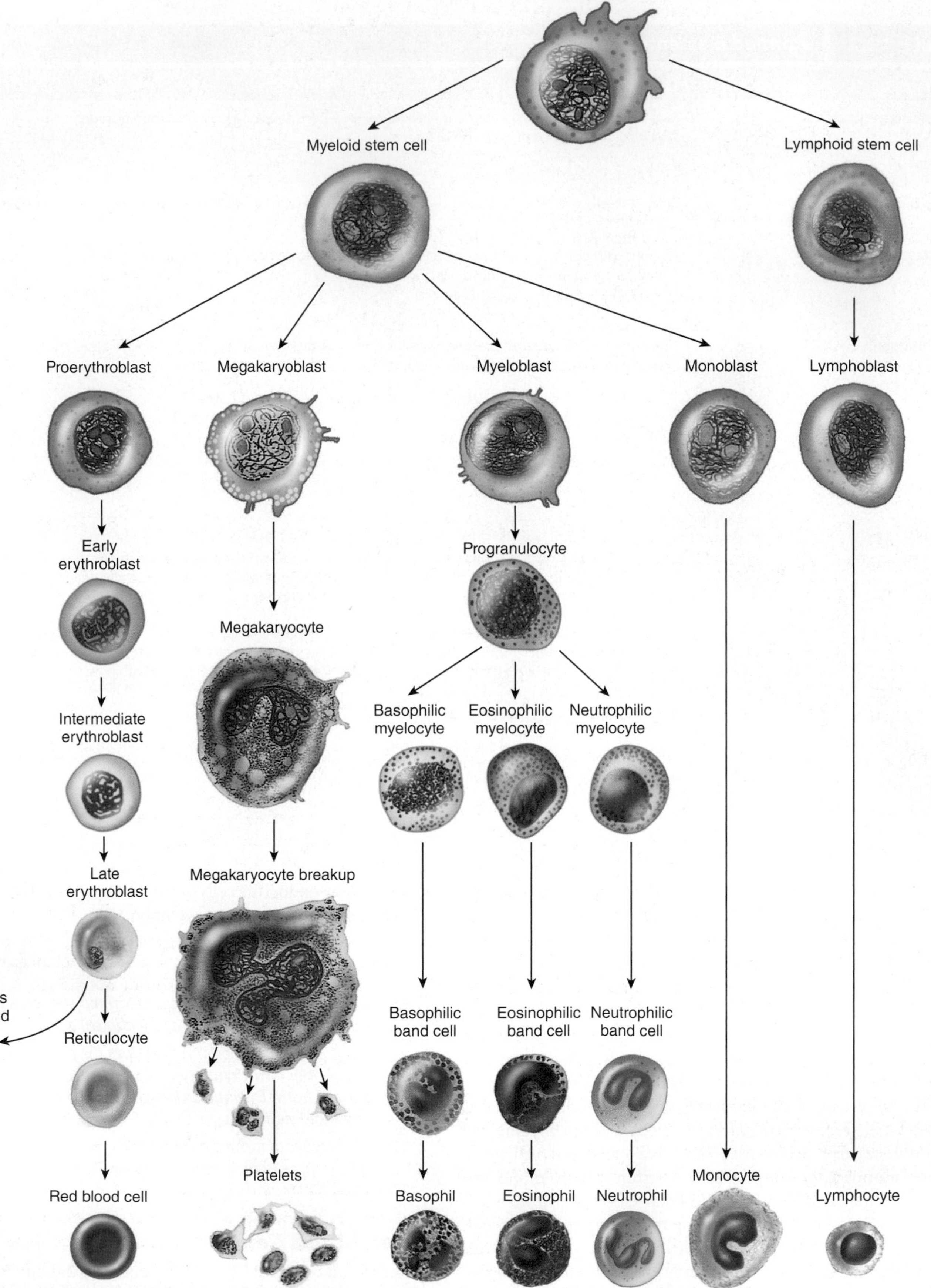
Stem cell (hemocytoblast)
Myeloid stem cell
Lymphoid stem cell
Proerythroblast
Megakaryoblast
Myeloblast
Monoblast
Lymphoblast
Early erythroblast
Progranulocyte
Megakaryocyte
Intermediate erythroblast
Basophilic myelocyte
Eosinophilic myelocyte
Neutrophilic myelocyte
Late erythroblast
Megakaryocyte breakup
Nucleus extruded
Reticulocyte
Basophilic band cell
Eosinophilic band cell
Neutrophilic band cell
Platelets
Monocyte
Red blood cell
Basophil
Eosinophil
Neutrophil
Lymphocyte

FIGURE 19.2 APR **Hematopoiesis**
Stem cells give rise to the cell lines that produce the formed elements. The production of red blood cells (far left column) is called erythropoiesis.

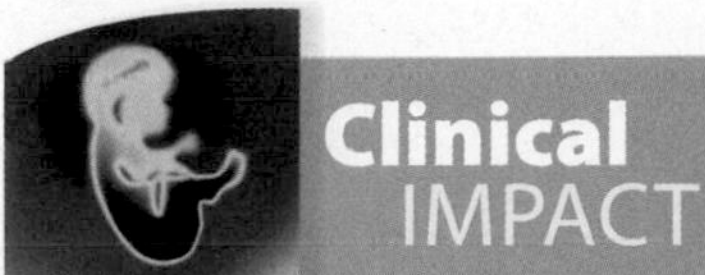
Clinical
IMPACT

Stem Cells and Cancer Therapy

Many cancer therapies affect the type of rapidly dividing cells found in tumors. However, an undesirable side effect of such therapies can be the destruction of nontumor cells that are dividing, such as the stem cells and their derivatives in red bone marrow. After being treated for cancer, some patients are prescribed growth factors to stimulate the rapid regeneration of the red bone marrow. Although not a cure for cancer, the growth factors can speed recovery from the side effects of cancer therapy.

Some types of leukemia and genetic immune deficiency diseases can be treated with a bone marrow transplant that provides stem cells to the patient. To avoid tissue rejection, families with a history of these disorders can freeze the umbilical cord blood of their newborn children. The cord blood, which contains many stem cells, can be used instead of bone marrow.

4.6–6.2 million), whereas females have about 4.8 million/μL (range: 4.2–5.4 million). Red blood cells cannot move of their own accord; they are passively moved by forces that cause the blood to circulate.

Structure

Normal red blood cells are discs about 7.5 μm in diameter, and they are biconcave, meaning that their edges are thicker than their center (figure 19.3*b*). The biconcave shape increases the cell's surface area compared with a flat disc of the same size. The greater surface area allows gases to move into and out of the red blood cell more rapidly. In addition, the red blood cell can bend or fold around its thin center, thereby decreasing its size and enabling it to pass more easily through smaller blood vessels.

Red blood cells are derived from specialized cells that lose their nuclei and nearly all their cellular organelles during maturation. The main component of the red blood cell is the pigmented protein **hemoglobin** (hē-mō-glō′bin), which occupies about one-third of the total cell volume and accounts for its red color. Other red blood cell contents include lipids, adenosine triphosphate (ATP), and the enzyme carbonic anhydrase.

Functions

The primary functions of red blood cells are to transport oxygen from the lungs to the various body tissues and to transport carbon dioxide from the tissues to the lungs. Approximately 98.5% of the oxygen in the blood is transported in combination with the hemoglobin in the red blood cells, and the remaining 1.5% is dissolved in the water part of the plasma. If red blood cells rupture, the hemoglobin leaks out into the plasma and becomes nonfunctional because the shape of the molecule changes as a result of denaturation (see chapter 2). Red blood cell rupture followed by hemoglobin release is called **hemolysis** (hē-mol′i-sis). Hemolysis occurs in hemolytic anemia (see Diseases and Disorders table, later in this chapter), transfusion reactions, hemolytic disease of the newborn, and malaria.

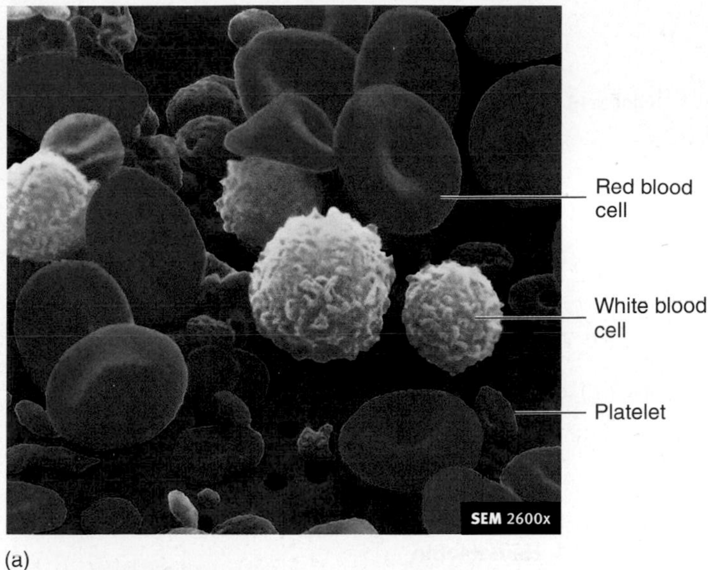
Red blood cell
White blood cell
Platelet
SEM 2600x

(a)

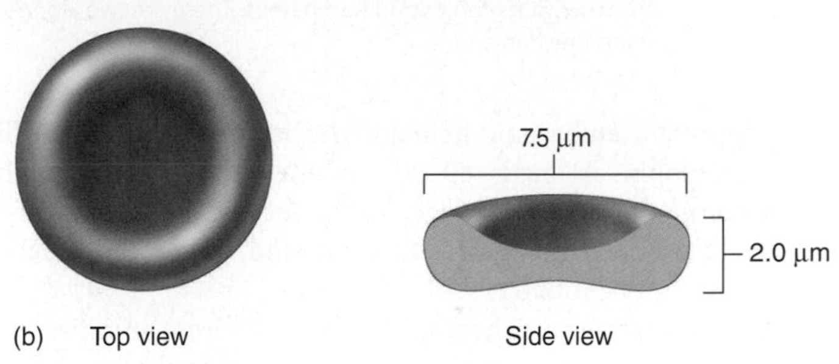
7.5 μm
2.0 μm

(b) Top view Side view

FIGURE 19.3 AP|R **Formed Elements**

(*a*) Color-enhanced scanning electron micrograph of formed elements: red blood cells (red doughnut shapes), white blood cells (yellow), and platelets (red, irregular shapes). (*b*) Shape and dimensions of a red blood cell.

Carbon dioxide is transported in the blood in three major ways: Approximately 7% is dissolved in the plasma; approximately 23% is combined with hemoglobin; and 70% is in the form of bicarbonate ions. The bicarbonate ions (HCO_3^-) are produced when carbon dioxide (CO_2) and water (H_2O) combine to form carbonic acid (H_2CO_3), which dissociates to form hydrogen (H^+) and bicarbonate ions. The combination of carbon dioxide and water is catalyzed by an enzyme, **carbonic anhydrase,** which is located primarily within red blood cells.

$$\underset{\text{Carbon dioxide}}{CO_2} + \underset{\text{Water}}{H_2O} \overset{\text{Carbonic anhydrase}}{\rightleftarrows} \underset{\text{Carbonic acid}}{H_2CO_3} \rightleftarrows \underset{\text{Hydrogen ion}}{H^+} + \underset{\text{Bicarbonate ion}}{HCO_3^-}$$

Hemoglobin

Hemoglobin consists of four polypeptide chains and four heme groups. Each polypeptide chain, called a **globin** (glō′bin), is bound to one **heme** (hēm). Each heme is a red-pigment molecule containing one iron atom (figure 19.4). There are three forms of hemoglobin: embryonic, fetal, and adult. Embryonic hemoglobin is the first type of hemoglobin produced during development. By the third month

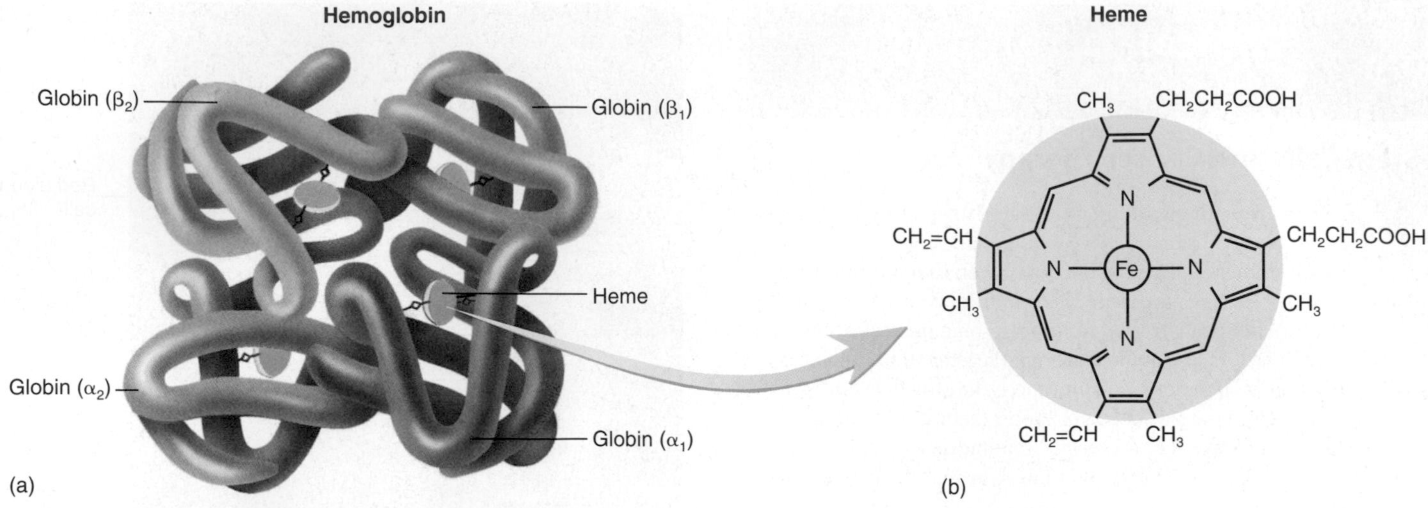
Hemoglobin
Globin (β_2)
Globin (β_1)
Heme
Globin (α_2)
Globin (α_1)
(a)
Heme
CH_3
CH_2CH_2COOH
CH_2=CH
N
Fe
CH_2CH_2COOH
CH_3
CH_3
CH_2=CH
CH_3
(b)

FIGURE 19.4 Hemoglobin

(*a*) Hemoglobin consists of four globins and four hemes. There are two alpha (α) globins and two beta (β) globins. A heme is associated with each globin. (*b*) Each heme contains one iron atom.

of development, embryonic hemoglobin has been replaced with fetal hemoglobin. At birth, 60–90% of the hemoglobin is adult hemoglobin. At 2 to 4 years of age, fetal hemoglobin makes up less than 2% of the hemoglobin and, in adulthood, only traces of fetal hemoglobin can be found.

The different forms of hemoglobin have different affinities for, or abilities to bind with, oxygen. Embryonic and fetal hemoglobin have a higher affinity for oxygen than adult hemoglobin does. In the embryo and fetus, hemoglobin picks up oxygen from the mother's blood at the placenta. Even though placental blood contains less oxygen than air in the mother's lungs, adequate amounts of oxygen are picked up because of the higher affinity of embryonic and fetal hemoglobin for oxygen. After birth, hemoglobin picks up oxygen from the air in the baby's lungs.

Although embryonic, fetal, and adult hemoglobin each have four globins, the types of globins are different. There are nine types of globins, each with a slightly different amino acid composition. For example, there are two types of alpha globins, which differ from each other by one amino acid. Because they are so similar, they are usually referred to simply as alpha globins. There are also a beta globin, two kinds of gamma globins, a delta globin, and three kinds of embryonic globins. Most adult hemoglobin has two alpha globins (one of each type) and two beta globins (figure 19.4). Fetal hemoglobin has two alpha globins (one of each type) and two gamma globins (one of each type).

There are nine globin genes, each of which codes for one of the globins. The globin genes are active during different stages of development. The embryonic globin genes are active first, but they become inactive as fetal globin genes become active. In a similar fashion, the fetal globin genes become inactive as the adult globin genes become active.

Predict 2

What would happen to a fetus if hemoglobin of the maternal blood had an affinity for oxygen that was equal to or greater than the hemoglobin of fetal blood?

Each oxygen molecule that is transported by hemoglobin is associated with an iron atom; therefore, iron is necessary for normal hemoglobin function. The adult human body normally contains about 4 g of iron, two-thirds of which is associated with hemoglobin. Small amounts of iron are regularly lost from the body in waste products, such as urine and feces. Females lose additional iron as a result of menstrual bleeding and, therefore, require more dietary iron than males do. Dietary iron is absorbed into the circulation from the upper part of the intestinal tract. Stomach acid and vitamin C in food increase iron absorption by converting ferric iron (Fe^{3+}) to ferrous iron (Fe^{2+}), which is more readily absorbed.

When hemoglobin is exposed to oxygen, one oxygen molecule can become associated with each heme group. This oxygenated form of hemoglobin is called **oxyhemoglobin** (ok′sē-hē-mō-glō′bin). The oxyhemoglobin in one red blood cell transports about 1 billion molecules of oxygen. Each heme molecule binds to one oxygen molecule; there are four heme molecules per hemoglobin and 280 million hemoglobin molecules per red blood cell. Hemoglobin containing no oxygen is called **deoxyhemoglobin.** Oxyhemoglobin is bright red, whereas deoxyhemoglobin has a darker red color.

Hemoglobin transports carbon dioxide, which does not combine with the iron atoms but is attached to the globin molecule. This hemoglobin form is **carbaminohemoglobin** (kar-bam′i-nō-hē-mō-glō′bin). The transport of oxygen and carbon dioxide by the blood is discussed more fully in chapter 23.

Additionally, hemoglobin transports nitric oxide, which is produced by the endothelial cells lining the blood vessels. In the lungs, at the same time that heme picks up oxygen, in each β-globin a sulfur-containing amino acid, cysteine, binds with a nitric oxide molecule to form *S*-nitrosothiol (nī-trōs′ō-thī-ol; SNO). When oxygen is released in tissues, so is the nitric oxide, which functions as a chemical messenger that induces the relaxation of the smooth muscle of blood vessels. By affecting the amount of nitric oxide in tissues, hemoglobin may play a role in regulating blood pressure because the relaxation of blood vessels results in decreased blood pressure (see chapter 21).

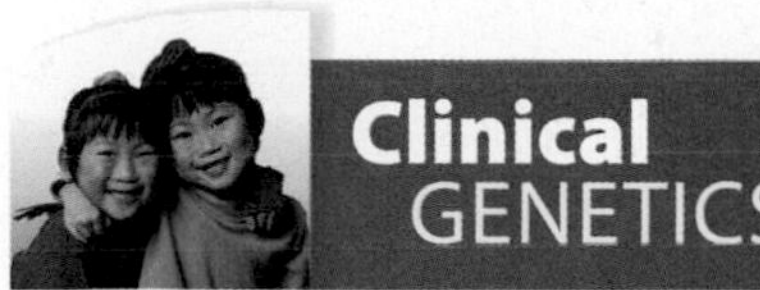

Clinical GENETICS — Sickle-Cell Disease

Sickle-cell disease is a disorder in which red blood cells become sickle-shaped. It results from a mutation in the gene that codes for the beta globin chain of hemoglobin. The mutation is a change in one nucleotide in the DNA that leads to a change in one amino acid in beta globin. The single amino acid change in beta globin has a dramatic effect on hemoglobin. When blood oxygen levels decrease, as when oxygen diffuses away from hemoglobin in tissue capillaries, the abnormal hemoglobin molecules join together, causing a change in red blood cell shape (figure 19A). When blood oxygen levels increase, as in the lungs, the abnormal hemoglobin molecules separate, and red blood cells can resume their normal shape.

Sickle-shaped red blood cells are less able to squeeze through small capillaries. Consequently, they become lodged in capillaries, blocking blood flow through them. This causes a further decrease in oxygen levels, which promotes more sickling. As oxygen levels decrease further, more capillary blockage is promoted, and so on. After repeated cycles of sickling, red blood cells lose their ability to resume their normal shape. This increases the number of sickled cells.

The major consequence of sickle-cell disease is tissue damage resulting from reduced blood flow through tissues. As tissues are deprived of blood, the most common symptom is pain, which is often severe. In addition, spleen and liver enlargement, kidney and lung damage, and stroke can occur. Priapism (prī′ă-pizm), a prolonged, painful erection due to venous blockage, can develop in men. Sickle-shaped red blood cells are also likely to rupture, which can result in hemolytic anemia (see the Diseases and Disorders table, later in this chapter).

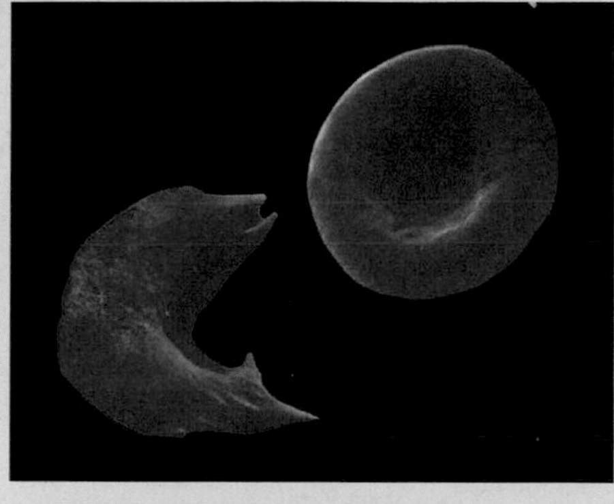

FIGURE 19A Sickle-Cell Disease
Red blood cells in a person with sickle-cell disease appear normal in oxygenated blood. In deoxygenated blood, hemoglobin changes shape and causes the cells to become sickle-shaped and rigid.

Sickle-cell disease is an autosomal recessive disorder. Only individuals who have two mutated beta globin alleles express the disease. Individuals who are heterozygous have a normal beta globin allele and produce sufficient amounts of normal beta globin, so their red blood cells do not usually become sickle-shaped. Heterozygotes are carriers (see chapter 29) and are said to have **sickle-cell trait.**

Sickle-cell disease is an example of a genetic disorder in which the heterozygote has a better ability to survive under certain circumstances than homozygous individuals. Carriers (heterozygotes) with sickle-cell trait have increased resistance to malaria. Malaria is a disease caused by a parasitic protozoan that reproduces inside red blood cells. The parasite is usually transmitted from one person to another through the bite of a mosquito. The red blood cells of people with sickle-cell trait tend to rupture before the parasite successfully reproduces. Therefore, those people are less likely to contract malaria, and the disease is much milder if they do.

The highest percentage of people with sickle-cell trait occurs in populations exposed to malaria or whose ancestors were exposed to malaria. In certain parts of Africa where malaria is rampant, the percentage of sickle-cell carriers can be as high as 50%. In the United States, 8% of African-Americans are sickle-cell carriers, and 0.8% have sickle-cell disease. The mutant gene can also be found in other groups, but at lower frequencies.

Treatment for sickle-cell disease attempts to reduce the blockage of blood vessels, alleviate pain, and prevent infections. Hydroxyurea (hī-drok′sē-ū-rē′ă) stimulates the production of gamma (fetal) globins. When the gamma globins combine with defective beta globins, the formation of sickle-shaped cells slows. Bone marrow transplants can cure sickle-cell disease, but such transplants can be dangerous and even fatal. Gene therapy is under investigation.

Various types of poisons affect the hemoglobin molecule. Carbon monoxide (CO), which is produced by the incomplete combustion of gasoline, binds very strongly to the iron of hemoglobin to form the relatively stable compound **carboxyhemoglobin** (kar-bok′sē-hē-mō-glō′bin). As a result of the stable binding of carbon monoxide, hemoglobin cannot transport oxygen. Nausea, headache, unconsciousness, and death are possible consequences of prolonged exposure to carbon monoxide. Interestingly, carbon monoxide is found in cigarette smoke, and the blood of smokers can contain 5–15% carboxyhemoglobin.

Life History of Red Blood Cells

Under normal conditions, about 2.5 million red blood cells are destroyed every second. This loss seems staggering, but it represents only 0.00001% of the total 25 trillion red blood cells contained in the normal adult circulation. Homeostasis is maintained by replacing the 2.5 million cells lost every second with an equal number of new red blood cells. Thus, approximately 1% of the total number of red blood cells is replaced each day.

The process by which new red blood cells are produced is called **erythropoiesis** (ĕ-rith′rō-poy-ē′sis; see figure 19.2). The time required to produce a single red blood cell is about 4 days. Myeloid stem cells, derived from hemocytoblasts, give rise to **proerythroblasts.** After several mitotic divisions, proerythroblasts become **early erythroblasts.** Proerythroblasts are also called *basophilic erythroblasts* because they stain with a basic dye. The dye binds to the large numbers of ribosomes necessary for the production of hemoglobin, giving the cytoplasm a purplish color. Early erythroblasts give rise to **intermediate erythroblasts.** These cells are also called *polychromatic erythroblasts* because they stain different colors with basic and acidic dyes. For example, when an acidic dye is used, intermediate erythroblasts stain a reddish color when it interacts with the hemoglobin accumulating in the cytoplasm. Intermediate erythroblasts continue to produce hemoglobin, and then most of their ribosomes and other organelles degenerate. The resulting **late erythroblasts** have a reddish color because about one-third of the cytoplasm is hemoglobin.

The late erythroblasts lose their nuclei by a process of extrusion to become immature red blood cells, called **reticulocytes** (re-tik′ū-lō-sītz). *Reticulocyte* refers to a reticulum, or network, that can be observed in the cytoplasm when a special staining technique is used. The reticulum is artificially produced by the reaction of the dye with the few remaining ribosomes in the reticulocyte. Reticulocytes are released from the bone marrow into the circulating blood, which normally consists of mature red blood cells and 1–3% reticulocytes. Within 2 days, the ribosomes in the cells degenerate, and the reticulocytes become mature red blood cells.

Predict 3

During a local Red Cross blood drive, Juan donated one unit of blood (about 500 mL). Predict how his reticulocyte count changed during the week after he donated blood, and explain why the change occurred.

Cell division requires the B vitamins folate and B_{12}, which are necessary for the synthesis of DNA (see chapter 3). Hemoglobin production requires iron. Consequently, adequate amounts of folate, vitamin B_{12}, and iron are necessary for normal red blood cell production.

Red blood cell production is stimulated by low blood oxygen levels, which result from several conditions: decreased numbers of red blood cells, decreased or defective hemoglobin, diseases of the lungs, high altitude, inability of the cardiovascular system to deliver blood to tissues, and increased tissue demands for oxygen—for example, during endurance exercises.

Low blood oxygen levels stimulate red blood cell production by increasing the formation of the glycoprotein **erythropoietin** (ĕ-rith-rō-poy′ĕ-tin), a hormone produced mostly by the kidneys (figure 19.5). Erythropoietin stimulates red bone marrow to produce more red blood cells by increasing the number of proerythroblasts formed and by decreasing the time required for red blood cells to mature. Thus, when blood oxygen levels decrease, erythropoietin production increases, which increases red blood cell production. The greater number of red blood cells increases the blood's ability to transport oxygen. This negative-feedback mechanism returns blood oxygen levels to normal and maintains homeostasis by increasing the delivery of oxygen to tissues. Conversely, if blood oxygen levels rise, less erythropoietin is released, and red blood cell production decreases.

Predict 4

Cigarette smoke produces carbon monoxide. If a nonsmoker smoked a pack of cigarettes a day for a few weeks, what would happen to the number of red blood cells in the person's blood? Explain.

Red blood cells normally stay in the circulation for about 120 days in males and 110 days in females. These cells have no nuclei and therefore cannot produce new proteins or divide. As their existing proteins, enzymes, plasma membrane components, and other structures degenerate, the red blood cells are less able to transport oxygen, and their plasma membranes become more fragile. Eventually, the red blood cells rupture as they squeeze through a tight spot in the circulation.

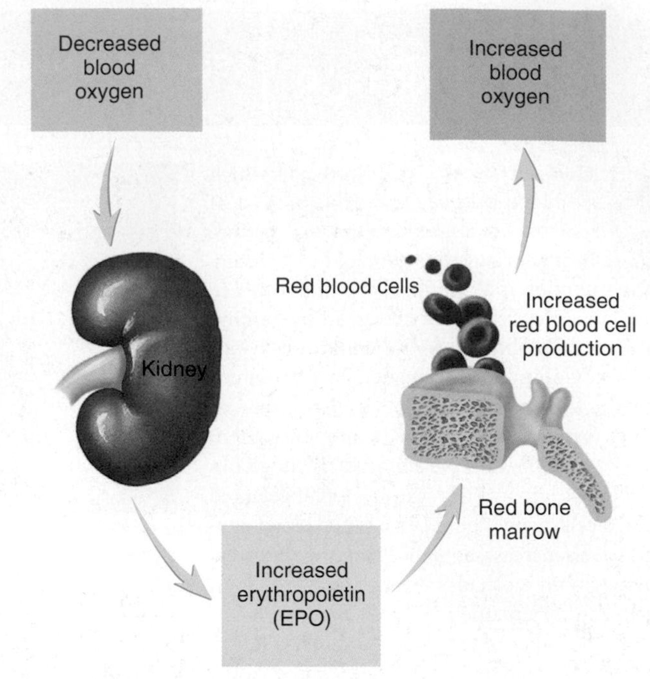
Decreased blood oxygen
Increased blood oxygen
Red blood cells
Increased red blood cell production
Kidney
Red bone marrow
Increased erythropoietin (EPO)

FIGURE 19.5 Red Blood Cell Production
In response to decreased blood oxygen, the kidneys release erythropoietin into the bloodstream. The increased erythropoietin stimulates red blood cell production in the red bone marrow. This process increases blood oxygen levels, restoring homeostasis.

Macrophages located in the spleen, liver, and other lymphatic tissue take up the hemoglobin released from ruptured red blood cells (figure 19.6). Within a macrophage, lysosomal enzymes digest the hemoglobin to yield amino acids, iron, and bilirubin. The globin part of hemoglobin is broken down into its component amino acids, most of which are reused in the production of other proteins. Iron atoms released from heme can be carried by the blood to red bone marrow, where they are incorporated into new hemoglobin molecules. The non-iron part of the heme groups is converted to **biliverdin** (bil-i-ver′din) and then to **bilirubin** (bil-i-roo′bin), which is released into the plasma. Bilirubin binds to albumin and is transported to liver cells. This bilirubin, called **free bilirubin,** is taken up by the liver cells and conjugated, or joined, to glucuronic acid to form **conjugated bilirubin,** which is more water-soluble than free bilirubin. The conjugated bilirubin becomes part of the **bile,** which is the fluid secreted from the liver into the small intestine. In the intestine, bacteria convert bilirubin into the pigments that give feces its characteristic brownish color. Some of these pigments are absorbed from the intestine, modified in the kidneys, and excreted in the urine, thus contributing to the characteristic yellowish color of urine. **Jaundice** (jawn′dis) is a yellowish staining of the skin and the sclerae of the eyes caused by a buildup of bile pigments in the circulation and interstitial spaces. Any process that causes increased destruction of red blood cells can cause jaundice, such as damage by toxins, genetic defects in red blood cell plasma membranes,

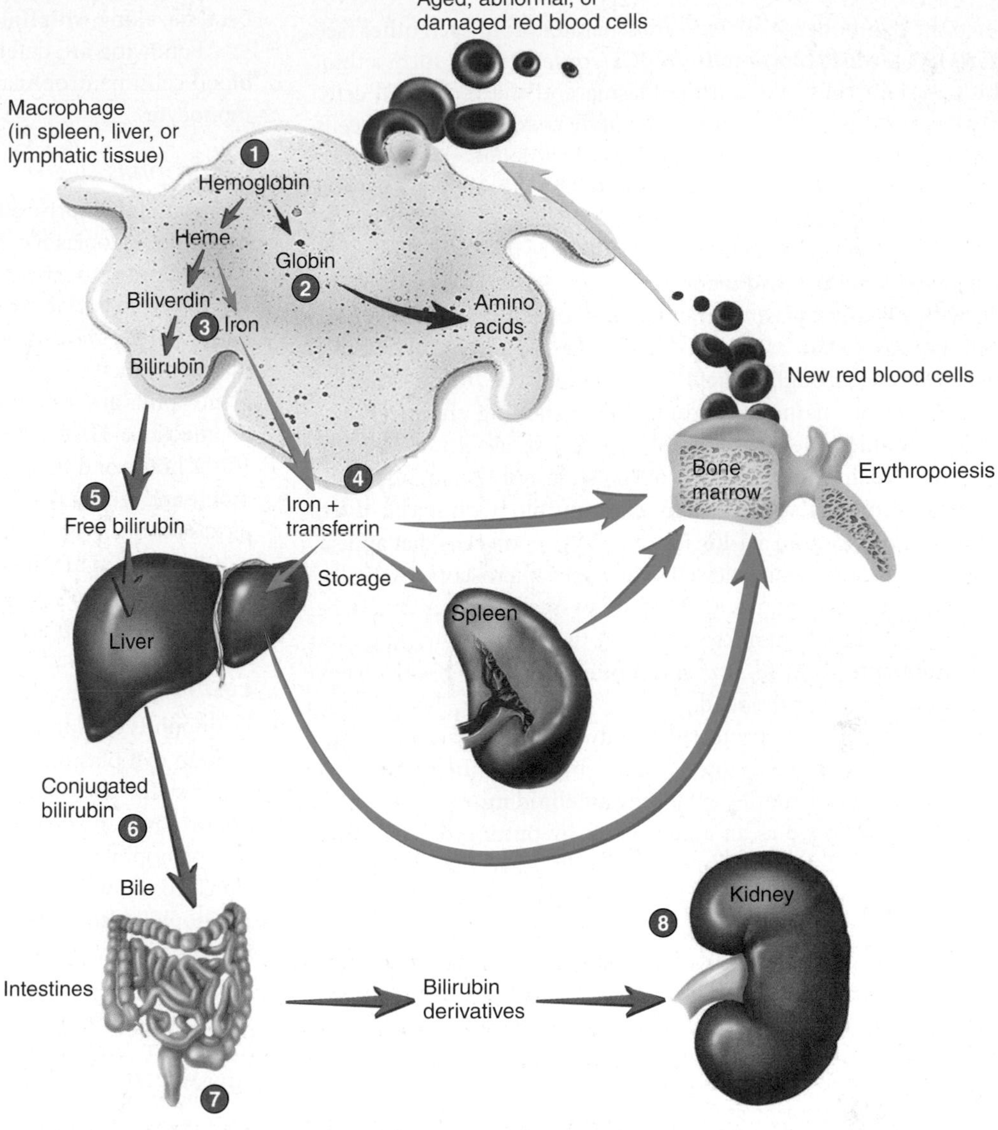
Aged, abnormal, or damaged red blood cells
Macrophage (in spleen, liver, or lymphatic tissue)
1
Hemoglobin
Heme
Globin
2
Biliverdin
3 Iron
Amino acids
Bilirubin
New red blood cells
Bone marrow
Erythropoiesis
4
5
Free bilirubin
Iron + transferrin
Storage
Liver
Spleen
Conjugated bilirubin
6
Bile
Kidney
8
Intestines
Bilirubin derivatives
7

1. Hemoglobin is broken down by macrophages into heme and globin chains.
2. The globin chains of hemoglobin are broken down to individual amino acids (*red arrows*) and are metabolized or used to build new proteins.
3. The heme of hemoglobin releases the iron. The heme is then converted into biliverdin, which is converted into bilirubin.
4. Blood transports iron in combination with transferrin to various tissues for storage or to the red bone marrow where it is used in the production of new hemoglobin (*green arrows*).
5. Blood transports free bilirubin to the liver (*blue arrows*).
6. Conjugated bilirubin is excreted as part of the bile into the small intestine. Bacteria of the intestine break down the bilirubin.
7. Some bilirubin derivatives contribute to the color of feces.
8. Other bilirubin derivatives are reabsorbed from the intestine into the blood. These derivatives are excreted from the kidneys and contribute to the color of urine.

PROCESS FIGURE 19.6 APR **Hemoglobin Breakdown**
Macrophages break down hemoglobin, and the breakdown products are used or excreted.

infections, and immune reactions. Other causes of jaundice are dysfunction or destruction of liver tissue and blockage of the duct system that drains bile from the liver (see chapter 24).

ASSESS YOUR PROGRESS

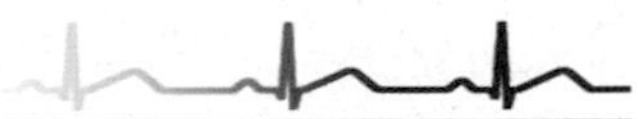

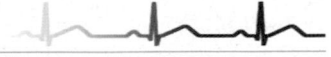

13. *What is the normal amount of red blood cells in a male? In a female?*
14. *How does the shape of red blood cells enable them to exchange gases and move through blood vessels more easily?*
15. *What is the main component of a red blood cell? What is the primary function of red blood cells?*
16. *Give the percentage for each of the ways that oxygen and carbon dioxide are transported in the blood. What is the function of carbonic anhydrase?*
17. *Describe the two basic parts of a hemoglobin molecule. Which part is associated with iron? What gases are transported by each part?*
18. *What is the significance of fetal hemoglobin's difference from adult hemoglobin?*
19. *Describe the process of erythropoiesis, beginning with hemocytoblasts in the red bone marrow.*
20. *What is erythropoietin, where is it produced, what causes it to be produced, and what effect does it have on red blood cell production?*
21. *How long do red blood cells normally stay in circulation? Where are red blood cells removed from the blood? List the three breakdown products of hemoglobin, and explain what happens to them.*

White Blood Cells

When the components of blood are separated from each other (see figure 19.1), **white blood cells (WBCs),** or *leukocytes*, form a thin, white layer of cells between the plasma and the red blood cells. This layer is often referred to as the *buffy coat.* White blood cells lack hemoglobin but have a nucleus. In stained preparations, white blood cells attract stain, whereas red blood cells remain relatively unstained (figure 19.7; table 19.2).

White blood cells are named according to their appearance in stained preparations. **Granulocytes** (gran′yū-lō-sitz) are white blood cells with large cytoplasmic granules and lobed nuclei (table 19.2). Their granules stain with dyes that make the cells more visible when viewed through a light microscope. The three types of granulocytes are named according to the staining characteristics of their granules: **Neutrophils** (nu′trō-filz) stain with acidic and basic dyes, **eosinophils** (ē-ō-sin′ō-filz) stain red with acidic dyes, and **basophils** (bā′sō-filz) stain dark purple with basic dyes. **Agranulocytes** (ă-gran′yū-lō-sītz) are white blood cells that appear to have no granules when viewed with a light microscope. Actually, agranulocytes have granules, but they are so small that they cannot be seen easily with the light microscope. The two types of agranulocytes are **lymphocytes** (lim′fō-sītz) and **monocytes** (mon′ō-sītz). They have nuclei that are not lobed.

White blood cells protect the body against invading microorganisms and remove dead cells and debris from the body. Most white blood cells are motile, exhibiting **ameboid movement,** which is the ability to move as an ameba does, by putting out irregular cytoplasmic projections. White blood cells leave the circulation and enter tissues by the process of **diapedesis** (dī′ă-pĕ-dē′sis), in which they become thin and elongated and slip between or through the cells of blood vessel walls. The white blood cells can then be attracted to foreign materials or dead cells within the tissue by **chemotaxis** (kē-mō-tak′sis; see chapter 22). At the site of an infection, white blood cells accumulate and phagocytize bacteria, dirt, and dead cells; then they die. The accumulation of dead white blood cells and bacteria, along with fluid and cell debris, is called **pus.**

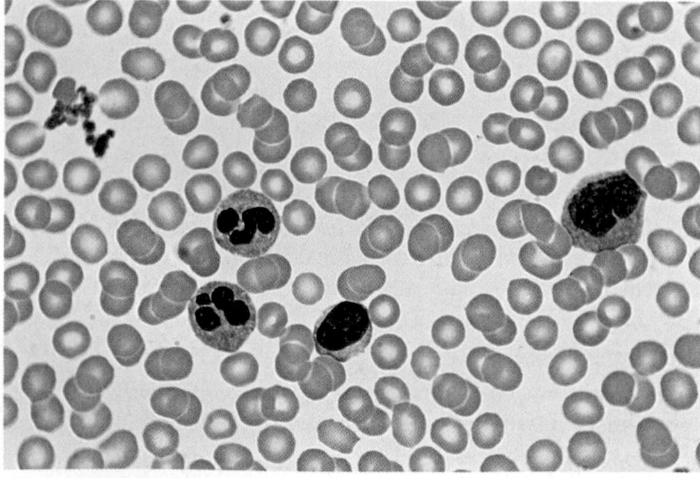

FIGURE 19.7 Standard Blood Smear
A thin film of blood is spread on a microscope slide and stained. The white blood cells have pink-colored cytoplasm and purple-colored nuclei. The red blood cells do not have nuclei. The center of a red blood cell appears whitish because light more readily shines through the thin center of the disc than through the thicker edges. The platelets are purple cell fragments.

Following are detailed descriptions of the five types of white blood cells: neutrophils, eosinophils, basophils, lymphocytes, and monocytes.

Neutrophils

Neutrophils comprise 60–70% of white blood cells (table 19.2). They have small cytoplasmic granules that stain with both acidic and basic dyes. Commonly, their nuclei are lobed, with the number of lobes varying from two to five. Neutrophils are often called *polymorphonuclear* (pol′ē-mōr-fō-noo′klē-ăr) *neutrophils*, or *PMNs*, to indicate that their nuclei can occur in more than one (*poly*) form (*morph*). Neutrophils are usually the first of the white blood cells to respond to infection. They normally remain in the circulation for about 10–12 hours and then move into other tissues, where they become motile and seek out and phagocytize bacteria, antigen-antibody complexes (antigens and antibodies bound together), and other foreign matter. Neutrophils also secrete a class of enzymes called **lysozymes** (lī′sō-zīmz), which are capable of destroying certain bacteria. Neutrophils usually survive 1–2 days after leaving the circulation.

Eosinophils

Eosinophils comprise 2–4% of white blood cells (table 19.2). They contain cytoplasmic granules that stain bright red with eosin, an acidic stain. They often have a two-lobed nucleus. Eosinophils are important in the defense against certain worm parasites. Although the eosinophils are not able to phagocytize the large parasites, they attach to the worms and release substances that kill the parasites. Eosinophils also increase in number in tissues experiencing inflammation, such as during allergic reactions. Eosinophils apparently modulate the inflammatory response by producing enzymes that destroy inflammatory chemicals, such as histamine. However, research has shown that eosinophils have harmful effects on respiratory airways in certain forms of asthma.

Basophils

Basophils comprise 0.5–1% of white blood cells (table 19.2). They contain large cytoplasmic granules that stain blue or purple with basic dyes. Basophils, like eosinophils and neutrophils, leave the circulation and migrate through the tissues. They increase in number in both allergic and inflammatory reactions. Basophils contain large amounts of **histamine** (see chapter 22), which they release within tissues to increase inflammation. They also release **heparin,** which inhibits blood clotting.

Lymphocytes

Lymphocytes comprise 20–25% of white blood cells (table 19.2). They are the smallest white blood cells, usually slightly larger in diameter than red blood cells. The lymphocytic cytoplasm consists of only a thin, sometimes imperceptible, ring around the nucleus. Although lymphocytes originate in red bone marrow, they migrate through the blood to lymphatic tissues, where they can proliferate and produce more lymphocytes. The majority of the body's total lymphocyte population is in the lymphatic tissues: the lymph nodes, spleen, tonsils, lymphatic nodules, and thymus.

Although they cannot be identified by standard microscopic examination, a number of different kinds of lymphocytes play important roles in immunity (see chapter 22). For example, **B cells** can be stimulated by bacteria or toxins to divide and form cells that produce proteins called **antibodies.** Antibodies can attach to bacteria and activate mechanisms that destroy the bacteria. **T cells** protect against viruses and other intracellular microorganisms by attacking and destroying the cells in which they are found. In addition, T cells are involved in the destruction of tumor cells and in tissue graft rejections.

Monocytes

Monocytes comprise 3–8% of white blood cells (table 19.2). They are typically the largest of the white blood cells. Monocytes normally remain in the circulation for about 3 days. Then they leave the circulation, become transformed into macrophages, and migrate through various tissues, where they phagocytize bacteria, dead cells, cell fragments, and other debris. An increase in the number of monocytes is often associated with chronic infection. Macrophages also stimulate responses from other cells in two ways: (1) by releasing chemical messengers and (2) by phagocytizing and processing foreign substances, which are presented to lymphocytes. The responses of these other cells help protect against microorganisms and other foreign substances (see chapter 22).

ASSESS YOUR PROGRESS

22. *What are the two major functions of white blood cells? Define ameboid movement, diapedesis, and chemotaxis.*
23. *Describe the morphology of the five types of white blood cells.*
24. *Name the two white blood cells that function primarily as phagocytic cells. What are lysozymes?*
25. *Which white blood cell defends against parasitic worms?*
26. *Which white blood cell releases histamine and promotes inflammation?*
27. *B cells and T cells are examples of which type of white blood cell? How do these cells protect against bacteria and viruses?*

Predict 5

Based on their morphology, identify each of the white blood cells shown in figure 19.8.

Platelets

Platelets, or *thrombocytes* (table 19.2; see figure 19.7), are minute fragments of cells consisting of a small amount of cytoplasm surrounded by a plasma membrane. Platelets are roughly disc-shaped and average about 3 μm in diameter. Glycoproteins and proteins on their surface allow platelets to attach to other molecules, such as collagen in connective tissue. Some of these surface molecules, as well as molecules released from granules in the platelet cytoplasm, play important roles in controlling blood loss. The platelet cytoplasm also contains actin and myosin, which can cause contraction of the platelet (see "Clot Retraction and Dissolution" in section 19.5).

The life expectancy of platelets is about 5–9 days. They are produced within the red marrow and are derived from **megakaryocytes** (meg-ă-kar′ē-ō-sitz), which are extremely large cells—up to 100 μm in diameter. Small fragments of these cells break off and enter the circulation as platelets.

Platelets play an important role in preventing blood loss by (1) forming platelet plugs that seal holes in small vessels and (2) promoting the formation and contraction of clots that help seal off larger wounds in the vessels.

ASSESS YOUR PROGRESS

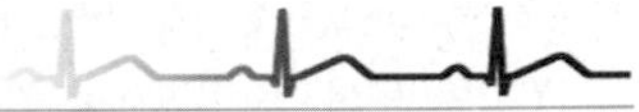

28. *What is a platelet? How do platelets form?*
29. *What are the two major roles of platelets in preventing blood loss?*

19.5 Hemostasis

LEARNING OUTCOMES

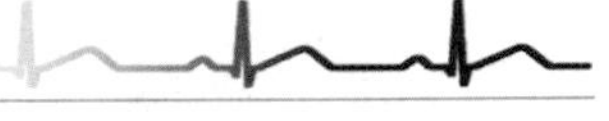

After reading this section, you should be able to

A. Explain the three processes that can lead to hemostasis: vascular spasm, platelet plug formation, and coagulation.
B. Describe how aspirin affects the action of platelets.
C. Describe the regulation of clot formation and how clots are removed.

Hemostasis (hē′mō-stā-sis, hē-mos′tă-sis), the stoppage of bleeding, is very important to the maintenance of homeostasis. If not stopped, excessive bleeding from a cut or torn blood vessel can result in a positive-feedback cycle, consisting of ever-decreasing blood volume

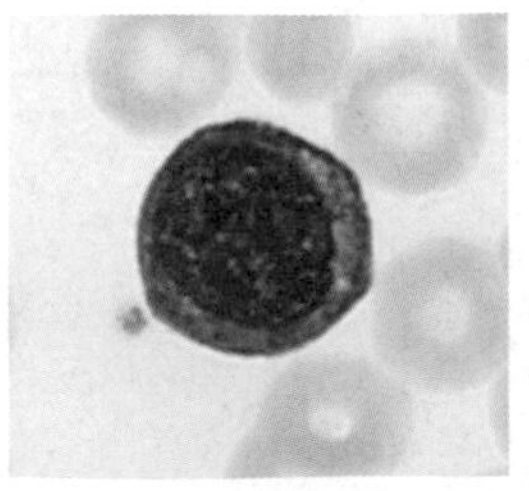

(a)

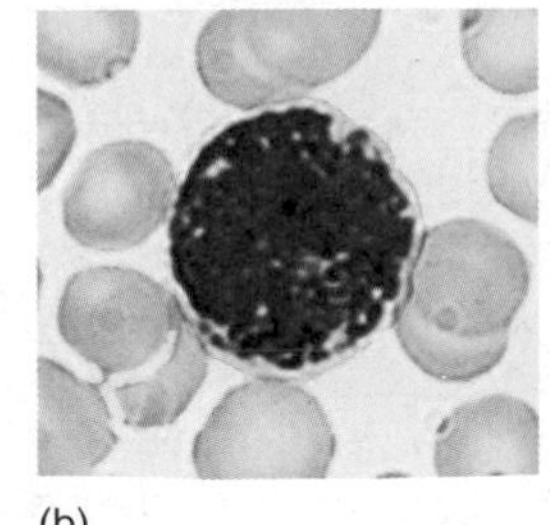

(b)

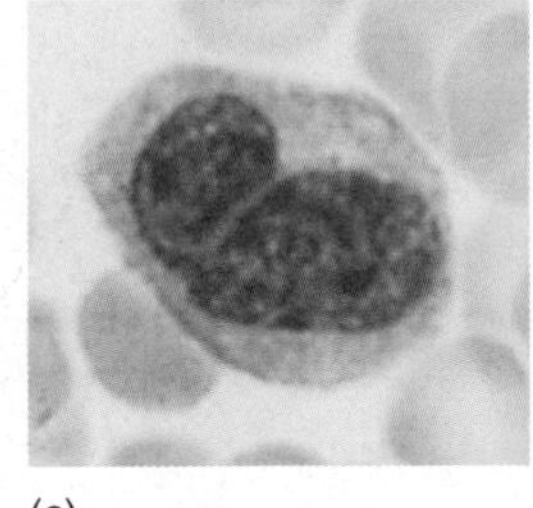

(c)

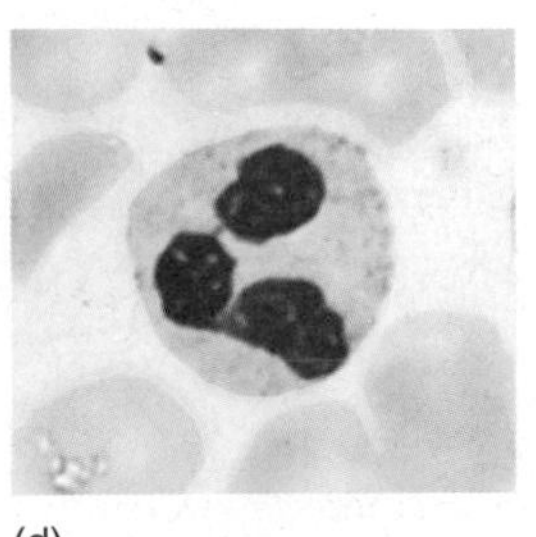

(d)

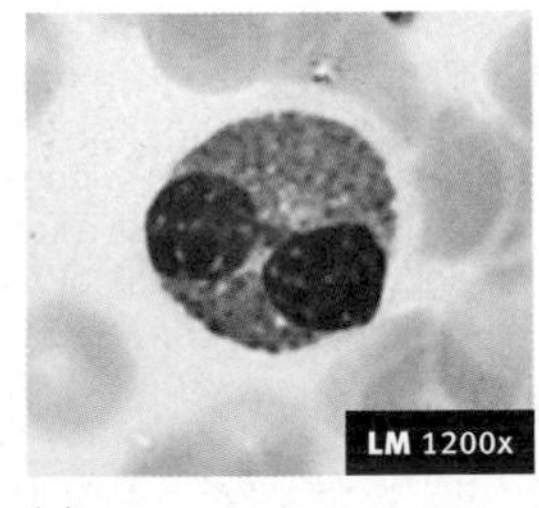
LM 1200x

(e)

FIGURE 19.8 APR **Identification of White Blood Cells**
See Predict question 5.

and blood pressure that disrupts homeostasis and results in death. Fortunately, when a blood vessel is damaged, a series of events helps prevent excessive blood loss. Hemostasis involves vascular spasm, platelet plug formation, and coagulation.

Vascular Spasm

Vascular spasm is the immediate but temporary constriction of a blood vessel that results when smooth muscle within the wall of the vessel contracts. This constriction can close small vessels completely and stop the flow of blood through them. Damage to blood vessels can activate nervous system reflexes that cause vascular spasms. Chemicals released by cells of the damaged vessel as well as platelets also produce vascular spasms. For example, during the formation of a platelet plug, platelets release **thromboxanes** (throm′bok-zānz), which are derived from certain prostaglandins, and endothelial cells release the peptide **endothelin** (en-dō′thē-lin), both of which lead to constriction of the vessel.

Platelet Plug Formation

A **platelet plug** is an accumulation of platelets that can seal small breaks in blood vessels. Platelet plug formation is very important in maintaining the integrity of the circulatory system because small tears occur in the smaller vessels and capillaries many times each day, and platelet plug formation quickly closes them. People who lack the normal number of platelets tend to develop numerous small hemorrhages in their skin and internal organs.

The formation of a platelet plug can be described as a series of steps, but in actuality many of the steps take place simultaneously (figure 19.9):

1. **Platelet adhesion** occurs when platelets bind to collagen exposed by blood vessel damage. Most platelet adhesion is mediated through **von Willebrand factor (vWF),** which is a protein produced and secreted by blood vessel endothelial cells. Von Willebrand factor forms a bridge between exposed collagen of the blood vessel wall and platelets by binding to platelet surface receptors and collagen. In addition, other platelet surface receptors can bind directly to collagen.
2. After platelets adhere to collagen, they become activated; in the **platelet release reaction,** adenosine diphosphate (ADP), thromboxanes, and other chemicals are extruded from the platelets by exocytosis. The ADP and thromboxane bind to their respective receptors on the surfaces of other platelets, activating them. These activated platelets release additional chemicals, thereby producing a cascade of chemical release by the platelets. Thus, more and more platelets become activated. This is an example of positive feedback.
3. As platelets become activated, they change shape and express fibrinogen receptors that can bind to fibrinogen, a plasma protein. In **platelet aggregation,** fibrinogen forms a bridge between the fibrinogen receptors of different platelets, resulting in a platelet plug.

In addition to forming a platelet plug, activated platelets also release phospholipids (platelet factor III) and coagulation factor V, which are important in clot formation.

Coagulation

Vascular spasms and platelet plugs alone are not sufficient to close large tears or cuts. When a blood vessel is severely damaged, **coagulation** (kō-ag-ū-lā′shŭn), or blood clotting, results in the formation of a clot. A **blood clot** is a network of threadlike protein fibers, called **fibrin,** that traps blood cells, platelets, and fluid (figure 19.10).

1 Platelet adhesion occurs when von Willebrand factor connects exposed collagen to platelets.

2 During the platelet release reaction, ADP, thromboxanes, and other chemicals are released and activate other platelets.

3 Platelet aggregation occurs when fibrinogen receptors on activated platelets bind to fibrinogen, connecting the platelets to one another. The accumulating mass of platelets forms a platelet plug.

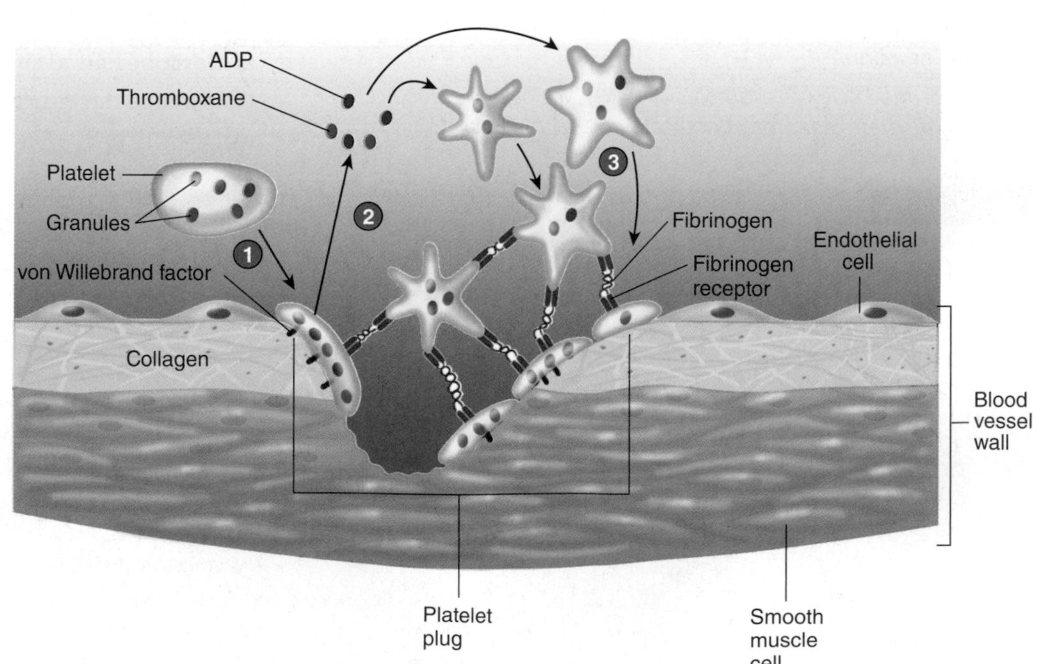
ADP
Thromboxane
Platelet
Granules
von Willebrand factor
1
2
3
Fibrinogen
Fibrinogen receptor
Endothelial cell
Collagen
Blood vessel wall
Platelet plug
Smooth muscle cell

PROCESS FIGURE 19.9 Platelet Plug Formation

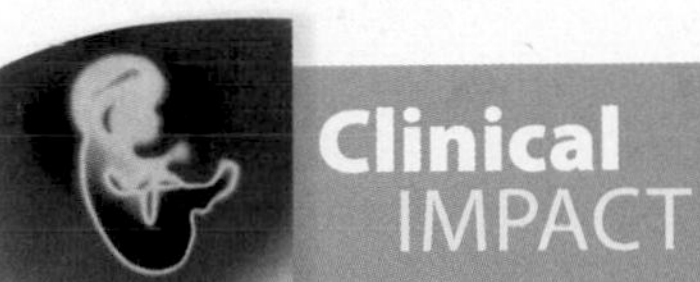

Clinical IMPACT — Clinical Importance of Taking Aspirin

Platelet activation results in platelet plug formation and the production of chemicals, such as phospholipids, that are important for blood clotting. Alternatively, the inhibition of platelet activation reduces the formation of blood clots. Understanding how this occurs requires knowledge of the chemical behavior of the **eicosanoids**, a group that includes prostaglandins, thromboxanes, and leukotrienes, the compounds involved in platelet activation. In humans, arachidonic acid is the most common precursor molecule for the eicosanoids. The enzyme cyclooxygenase (COX) converts arachidonic acid into a prostaglandin that can be converted into thromboxane. However, the actions of COX are inhibited by aspirin, which inhibits prostaglandin and thromboxane synthesis. As a result, aspirin reduces platelet activation.

Taking aspirin can have harmful or beneficial effects, depending on the circumstances. If an expectant mother ingests aspirin near the end of pregnancy, thromboxane synthesis is inhibited and several effects are possible. The mother can experience excessive bleeding after delivery because of decreased platelet function, and the baby can exhibit numerous localized hemorrhages called **petechiae** (pe-tē′kē-ē) over the surface of its body as a result of decreased platelet function. If the quantity of ingested aspirin is large, the infant, the mother, or both may die as a result of hemorrhage.

On the other hand, platelet plugs and blood clots can block blood vessels, producing heart attacks and strokes. Therefore, suspected heart attack victims are routinely given aspirin en route to the emergency room to reduce further clotting. The United States Preventive Services Task Force (USPSTF) and the American Heart Association (AHA) recommend low-dose aspirin therapy (75–160 mg/day) for all men and women at high risk for cardiovascular disease. Determining risk involves analyzing many factors and should be done in consultation with a physician. The decreased risk for cardiovascular disease from aspirin therapy must be weighed against the increased risk for hemorrhagic stroke and gastrointestinal bleeding.

The drug Plavix (clopidogrel bisulfate) reduces the activation of platelets by blocking the ADP receptors on the surface of platelets. It is used to prevent clotting and, with other anticlotting drugs, to treat heart attacks.

The formation of a blood clot depends on a number of proteins, called **clotting factors,** or *coagulation factors,* found within plasma (table 19.3). Normally, the clotting factors are in an inactive state and do not cause clotting. After injury, the clotting factors are activated. The activation of clotting factors is a complex process involving many chemical reactions, some of which require calcium ions (Ca^{2+}) and molecules on the surface of activated platelets, such as phospholipids and factor V.

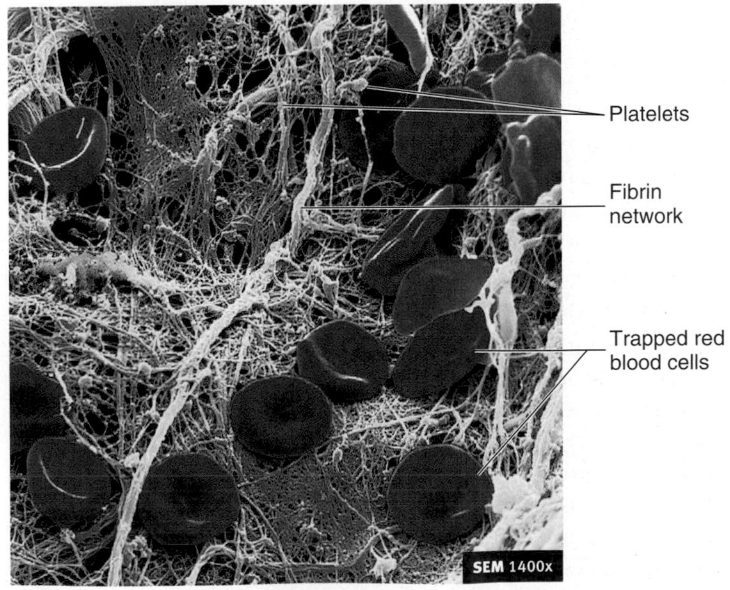
Platelets
Fibrin network
Trapped red blood cells
SEM 1400x

FIGURE 19.10 Blood Clot
A blood clot consists of fibrin, which traps red blood cells, platelets, and fluid.

Predict 6

Why is it advantageous for clot formation to involve molecules on the surface of activated platelets?

The activation of clotting factors begins with the extrinsic and intrinsic pathways (figure 19.11). These pathways converge to form the common pathway, which results in the formation of a fibrin clot.

Extrinsic Pathway

The extrinsic pathway is so named because it begins with chemicals that are outside of, or extrinsic to, the blood (figure 19.11). Damaged tissues release a mixture of lipoproteins and phospholipids called **thromboplastin** (throm-bō-plas′tin), also known as *tissue factor (TF)* or factor III. Thromboplastin, in the presence of Ca^{2+}, forms a complex with factor VII that activates factor X, which is the beginning of the common pathway.

Intrinsic Pathway

The intrinsic pathway is so named because it begins with chemicals that are inside, or intrinsic to, the blood (figure 19.11). Damage to blood vessels can expose collagen in the connective tissue beneath the epithelium lining the blood vessel. When plasma factor XII comes into contact with collagen, factor XII is activated, and it stimulates factor XI, which in turn activates factor IX. Activated factor IX joins with factor VIII, platelet phospholipids, and Ca^{2+} to activate factor X, which is the beginning of the common pathway.

Although the extrinsic and intrinsic pathways were once considered distinct, we now know that the extrinsic pathway can activate the clotting factors in the intrinsic pathway. The thromboplastin/factor VII complex from the extrinsic pathway can stimulate the formation of activated factor IX in the intrinsic pathway.

TABLE 19.3 Clotting Factors

Factor Number	Name (Synonym)	Description and Function
I	Fibrinogen	Plasma protein synthesized in the liver; converted to fibrin in the common pathway
II	Prothrombin	Plasma protein synthesized in the liver (requires vitamin K); converted to thrombin in the common pathway
III	Thromboplastin (tissue factor)	Mixture of lipoproteins released from damaged tissue; required in the extrinsic pathway
IV	Calcium ion	Required throughout the clotting sequence
V	Proaccelerin (labile factor)	Plasma protein synthesized in the liver; activated form functions in the intrinsic and extrinsic pathways
VI		Once thought to be involved but no longer accepted as playing a role in clotting; apparently the same as activated factor V
VII	Serum prothrombin conversion accelerator (stable factor, proconvertin)	Plasma protein synthesized in the liver (requires vitamin K); functions in the extrinsic pathway
VIII	Antihemophilic factor (antihemophilic globulin)	Plasma protein synthesized in megakaryocytes and endothelial cells; required in the intrinsic pathway
IX	Plasma thromboplastin component (Christmas factor)	Plasma protein synthesized in the liver (requires vitamin K); required in the intrinsic pathway
X	Stuart factor (Stuart-Prower factor)	Plasma protein synthesized in the liver (requires vitamin K); required in the common pathway
XI	Plasma thromboplastin antecedent	Plasma protein synthesized in the liver; required in the intrinsic pathway
XII	Hageman factor	Plasma protein required in the intrinsic pathway
XIII	Fibrin-stabilizing factor	Protein found in plasma and platelets; required in the common pathway
Platelet Factors		
I	Platelet accelerator	Same as plasma factor V
II	Thrombin accelerator	Accelerates thrombin and fibrin production
III		Phospholipids necessary for the intrinsic and extrinsic pathways
IV		Binds heparin, which prevents clot formation

Common Pathway

On the surface of platelets, activated factor X, factor V, platelet phospholipids, and Ca^{2+} combine to form **prothrombinase,** or *prothrombin activator.* Prothrombinase converts the soluble plasma protein **prothrombin** to the enzyme **thrombin.** Thrombin converts the soluble plasma protein fibrinogen to the insoluble protein fibrin. Fibrin forms the fibrous network of the clot. Thrombin stimulates factor XIII activation, which is necessary to stabilize the clot.

Thrombin can activate many of the clotting proteins, such as factor XI and prothrombinase. Thus, a positive-feedback system operates whereby thrombin production stimulates the production of additional thrombin. Thrombin also has a positive-feedback effect on coagulation by stimulating platelet activation.

Many of the factors involved in clot formation require vitamin K for their production (table 19.3). Humans rely on two sources for vitamin K. About half comes from the diet, and half comes from bacteria within the large intestine. Antibiotics taken to fight bacterial infections sometimes kill these intestinal bacteria, thereby reducing vitamin K levels and causing bleeding. Vitamin K supplements may be necessary for patients on prolonged antibiotic therapy. Newborns lack these intestinal bacteria; thus, they routinely receive a vitamin K injection at birth. Infants can also obtain vitamin K from food, such as milk.

The absorption of vitamin K from the large intestine requires the presence of bile because vitamin K is fat-soluble. Therefore, disorders involving an obstruction of bile flow to the intestine can interfere with vitamin K absorption and lead to insufficient clotting. Liver diseases that result in the decreased synthesis of clotting factors can also cause insufficient clot formation.

Control of Clot Formation

Without control, clot formation would spread from the point of initiation through the entire circulatory system. Furthermore, blood vessels in a healthy person contain rough areas that can stimulate clot formation, and small amounts of prothrombin are constantly being converted into thrombin. To prevent unwanted clotting, the blood contains several **anticoagulants** (an′tē-kō-ag′ū-lantz), which prevent clotting factors from initiating clot formation under normal circumstances. Only when clotting factor concentrations exceed a given threshold in a local area does clot formation occur. At the site of injury, so many clotting factors are activated that the anti-coagulants are unable to prevent clot formation. However, away

FUNDAMENTAL Figure

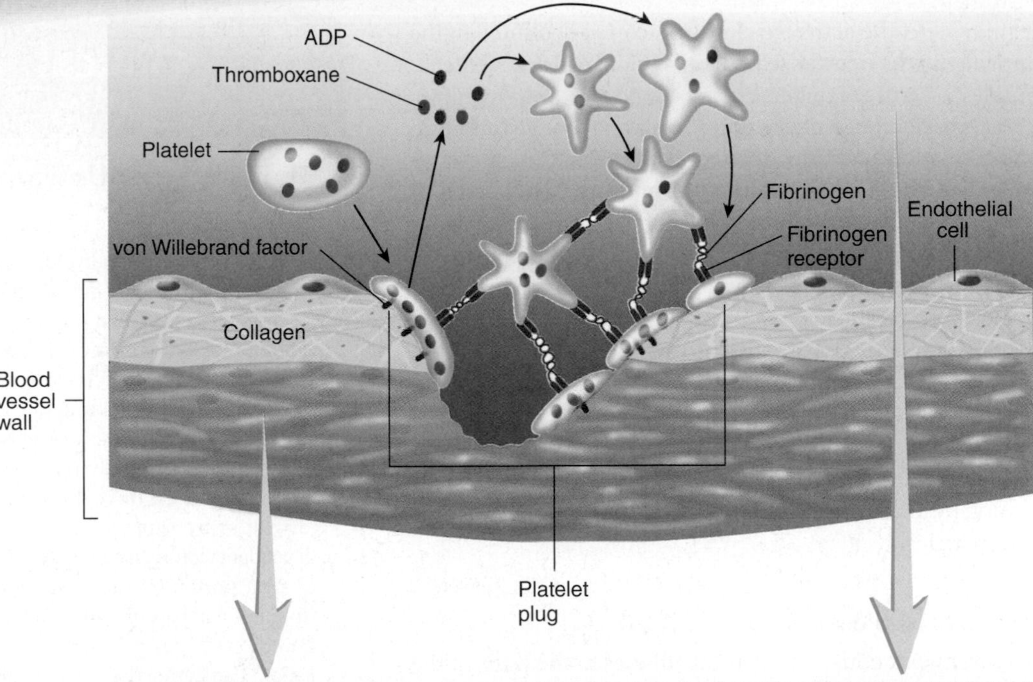

ADP
Thromboxane
Platelet
von Willebrand factor
Collagen
Blood vessel wall
Fibrinogen
Fibrinogen receptor
Endothelial cell
Platelet plug

1. The extrinsic pathway of clotting is stimulated by thromboplastin, released by damaged tissue.
2. The intrinsic pathway of clotting starts when inactive factor XII, which is in the blood, is activated by coming into contact with a damaged blood vessel.
3. Activation of the extrinsic or intrinsic pathway results in the production of activated factor X.
4. Activated factor X, factor V, phospholipids, and Ca^{2+} form prothrombinase.
5. Prothrombinase converts prothrombin to thrombin.
6. Thrombin converts fibrinogen to fibrin (the clot).
7. Thrombin activates clotting factors, promoting clot formation and stabilizing the fibrin clot.

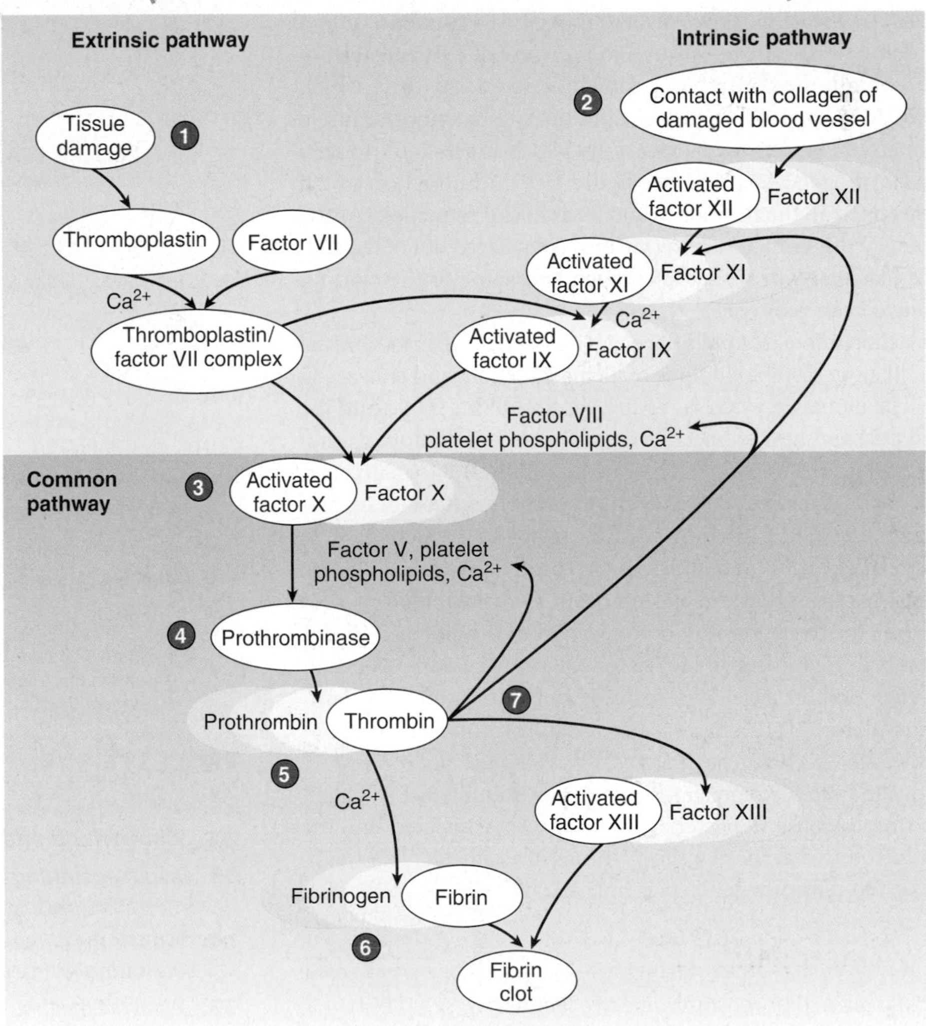

Extrinsic pathway
Intrinsic pathway
Tissue damage
1
Contact with collagen of damaged blood vessel
2
Activated factor XII
Factor XII
Thromboplastin
Factor VII
Activated factor XI
Factor XI
Ca2+
Thromboplastin/ factor VII complex
Activated factor IX
Factor IX
Factor VIII platelet phospholipids, Ca2+
Common pathway
3
Activated factor X
Factor X
Factor V, platelet phospholipids, Ca2+
4
Prothrombinase
Prothrombin
Thrombin
7
5
Activated factor XIII
Factor XIII
Fibrinogen
Fibrin
6
Fibrin clot

PROCESS FIGURE 19.11 Clot Formation

In a sequence of chemical reactions, activated clotting factors (white ovals) activate inactive clotting factors (yellow ovals). Clot formation begins through either the extrinsic or the intrinsic pathway. The common pathway starts with factor X and results in a fibrin clot.

from the injury site, the activated clotting factors are diluted in the blood, anticoagulants neutralize them, and clotting is prevented.

Examples of anticoagulants in the blood are antithrombin, heparin, and prostacyclin. **Antithrombin,** a plasma protein produced by the liver, slowly inactivates thrombin. Heparin, produced by basophils and endothelial cells, works with antithrombin to rapidly inactivate thrombin. **Prostacyclin** (pros-tă-sī′klin) is a prostaglandin derivative produced by endothelial cells. It counteracts the effects of thrombin by causing vasodilation and inhibiting the release of clotting factors from platelets.

Anticoagulants are also important when blood is outside the body. They prevent the clotting of blood used in transfusions and laboratory blood tests. Besides heparin, examples include **ethylenediaminetetraacetic** (eth′il-ēn-dī′ă-mēn-tet-ră-ă-sē′tik) **acid (EDTA)** and sodium citrate. EDTA and sodium citrate prevent clot formation by binding to Ca^{2+}, thus making the ions inaccessible for clotting reactions.

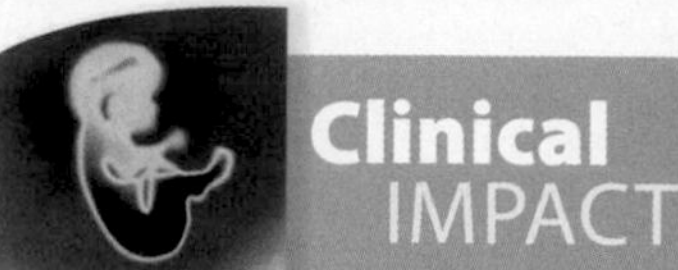

Clinical IMPACT

The Danger of Unwanted Clots

When platelets encounter damaged or diseased areas on the walls of blood vessels or the heart, an attached clot called a **thrombus** (throm′bŭs) may form. A thrombus that breaks loose and begins to float through the circulation is called an **embolus** (em′bō-lŭs). Both thrombi and emboli can cause death if they block vessels that supply blood to essential organs, such as the heart, brain, or lungs. Abnormal clotting can be prevented or hindered by administering an anticoagulant, such as heparin, which acts rapidly. Warfarin (war′fă-rin), commonly referred to by the brand name Coumadin® (koo′mă-din), acts more slowly than heparin. Coumadin prevents clot formation by suppressing the liver's production of vitamin K–dependent clotting factors (II, VII, IX, and X). Warfarin was first used as a rat poison by causing rats to bleed to death. In small doses, Coumadin is a proven, effective anticoagulant in humans. However, caution is necessary with anticoagulant treatment because the patient can hemorrhage internally or bleed excessively when cut.

Clot Retraction and Dissolution

The fibrin meshwork constituting a clot adheres to the walls of the blood vessel. Once a clot has formed, it begins to condense into a denser, compact structure through the process of **clot retraction.** Platelets contain the contractile proteins actin and myosin, which operate in a similar fashion to actin and myosin in smooth muscle (see chapter 9). Platelets form extensions, which attach to fibrinogen through fibrinogen receptors (see figure 19.9). Contraction of the extensions pulls on the fibrinogen and leads to clot retraction. As the clot retracts, a fluid called **serum** (sēr′ŭm) is squeezed out of the clot. Serum is plasma from which fibrinogen and some of the clotting factors have been removed.

Retraction of the clot pulls the edges of the damaged blood vessel together, helping stop blood flow, reducing infection, and enhancing healing. The damaged vessel is repaired as fibroblasts move into the damaged area and new connective tissue forms. In addition, epithelial cells around the wound proliferate and fill in the torn area.

The clot is usually dissolved within a few days after clot formation by a process called **fibrinolysis** (fī-bri-nol′i-sis), which involves the activity of **plasmin** (plaz′min), an enzyme that hydrolyzes fibrin. Plasmin forms from inactive plasminogen, a normal blood protein produced by the liver. Plasmin becomes part of the clot as it forms. Plasmin is activated by thrombin, factor XII, tissue plasminogen activator (t-PA), urokinase, and lysosomal enzymes released from damaged tissues (figure 19.12). In disorders resulting from the blockage of a vessel by a clot, such as a heart attack, dissolving the clot can restore blood flow and reduce damage to tissues. For example, t-PA, urokinase, or streptokinase (a bacterial enzyme) can be injected into the blood or introduced at the clot site by means of a catheter. These substances convert plasminogen to plasmin, which breaks down the clot.

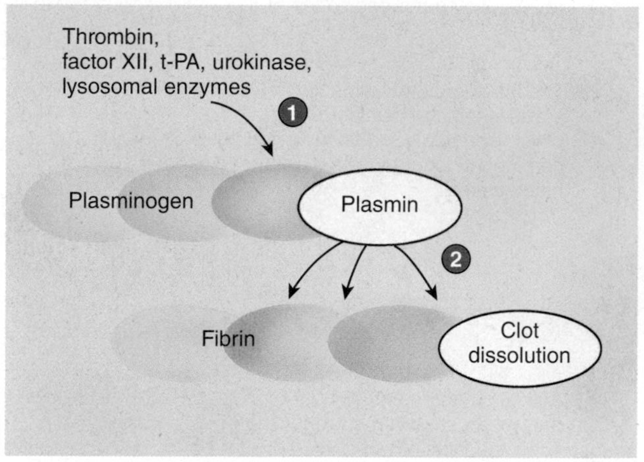
Thrombin,
factor XII, t-PA, urokinase,
lysosomal enzymes
1
Plasminogen
Plasmin
2
Fibrin
Clot dissolution

1. Inactive plasminogen is converted to the active enzyme plasmin.
2. Plasmin breaks the fibrin molecules, and therefore the clot, into smaller pieces, which are washed away in the blood or phagocytized.

PROCESS FIGURE 19.12 Fibrinolysis

ASSESS YOUR PROGRESS

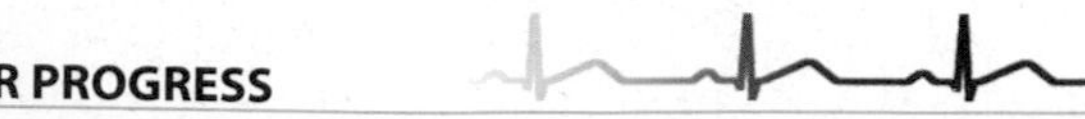

30. *What is a vascular spasm? Name two factors that produce it. What is the source of thromboxanes and endothelin?*
31. *What is the function of a platelet plug? Describe the process of platelet plug formation. How are platelets important to clot formation?*
32. *What is a clot, and what is its function?*
33. *What are clotting factors? What vitamin is required to produce many clotting factors?*
34. *What is the difference between extrinsic and intrinsic activation of clotting? What factor is activated by both pathways?*
35. *What are the three reactions that occur in the common pathway of clotting? What ion is a necessary part of the clotting process?*
36. *What is the function of anticoagulants in blood? Name three anticoagulants in blood, and explain how they prevent clot formation.*

37. *Describe the process of clot retraction. What is serum?*

38. *What is fibrinolysis? How does it occur?*

Predict 7

Cedric's doctor recommended taking a small amount of aspirin each morning because Cedric has substantial atherosclerotic plaques in his coronary arteries. One morning, Cedric took his aspirin as usual, but that afternoon he was transported to the emergency room because of a coronary thrombosis. The ER team administered t-PA, and Cedric recovered quickly. What contributed to the rapid improvement in his condition?

19.6 Blood Grouping

LEARNING OUTCOMES

After reading this section, you should be able to

A. Explain the basis of the ABO blood group system and how incompatibilities occur.

B. Describe the Rh blood group and its connection to hemolytic disease of the newborn (HDN).

If large quantities of blood are lost during surgery or due to injury, the patient can go into shock and die unless red blood cells are replaced to restore the blood's oxygen-carrying capacity. In this event, a transfusion or an infusion is required. A **transfusion** is the transfer of blood or blood components from one individual to another. An **infusion** is the introduction of a fluid other than blood, such as a saline or glucose solution, into the blood. In many cases, the return of blood volume to normal levels is all that is necessary to prevent shock. Eventually, the body produces enough red blood cells to replace those that were lost.

Early attempts to transfuse blood from one person to another were often unsuccessful because they resulted in transfusion reactions, characterized by clotting within blood vessels, kidney damage, and death. We now know that transfusion reactions are caused by interactions between antigens and antibodies (see chapter 22). In brief, the surfaces of red blood cells have molecules called **antigens** (an′ti-jenz), and the plasma includes proteins called **antibodies.** Antibodies are very specific, meaning that each antibody can bind only to a certain antigen. When the antibodies in the plasma bind to the antigens on the surfaces of the red blood cells, they form molecular bridges that connect the red blood cells. As a result, **agglutination** (ă-gloo-ti-nā′shŭn), or clumping, of the cells occurs. The combination of the antibodies with the antigens can also initiate reactions that cause hemolysis. Because the antigen-antibody combinations can cause agglutination, the antigens are often called **agglutinogens** (ă-gloo-tin′ō-jenz), and the antibodies are called **agglutinins** (ă-gloo′ti-ninz).

The antigens on the surface of red blood cells have been categorized into **blood groups,** and more than 35 blood groups, most of them rare, have been identified. For transfusions, the ABO and Rh blood groups are among the most important. Other well-known groups are the Lewis, Duffy, MNSs, Kidd, Kell, and Lutheran groups.

ABO Blood Group

The **ABO blood group** system is used to categorize human blood based on the presence or absence of ABO antigens on the surface of red blood cells. In the ABO blood group, type A blood has type A antigens, type B blood has type B antigens, type AB blood has both A and B antigens, and type O blood has neither A nor B antigens on the surface of red blood cells (figure 19.13). The ABO blood group is an example of codominance in that the A and B antigens can be expressed at the same time (see chapter 29).

Plasma from type A blood contains anti-B antibodies, which act against type B antigens; plasma from type B blood contains anti-A antibodies, which act against type A antigens. Type AB blood has neither type of antibody, and type O blood has both anti-A and anti-B antibodies.

The ABO blood types do not exist in equal numbers. In Caucasians in the United States, the distribution is type O, 47%; type A, 41%; type B, 9%; and type AB, 3%. Among African-Americans, the distribution is type O, 46%; type A, 27%; type B, 20%; and type AB, 7%.

Normally, antibodies do not develop against an antigen unless the body is exposed to that antigen. Scientists are unsure exactly how this exposure occurs. One possible explanation for the production of anti-A and/or anti-B antibodies is that type A or B antigens on bacteria or food in the digestive tract stimulate the formation of antibodies against antigens that are different from the body's own antigens. In support of this explanation, anti-A and anti-B antibodies are not found in the blood until about 2 months after birth. It is possible that an infant with type A blood produces anti-B antibodies against the B antigens on bacteria or food. Meanwhile, an infant with A antigens does not produce antibodies against the A antigens on bacteria or food because mechanisms exist in the body to prevent the production of antibodies that react with the body's own antigens (see chapter 22).

When a blood transfusion is performed, the **donor** is the person who gives blood, and the **recipient** is the person who receives it. Usually, a recipient can successfully receive blood from a donor as long as they both have the same blood type. For example, a person with type A blood can receive blood from a person with type A blood. No ABO transfusion reaction occurs because the recipient has no anti-A antibodies against the type A antigen. On the other hand, if type A blood were donated to a person with type B blood, a transfusion reaction would occur because the person with type B blood has anti-A antibodies against the type A antigen, causing agglutination (figure 19.14).

Historically, people with type O blood have been called *universal donors* because they can usually give blood to the other ABO blood types without causing an ABO transfusion reaction. Their red blood cells have no ABO surface antigens and therefore do not react with a recipient's anti-A or anti-B antibodies. For example, if a person with type A blood receives type O blood, the type O red blood cells do not react with the anti-B antibodies in the recipient's blood.

The term *universal donor* is misleading, however. Transfusion of type O blood can still produce a transfusion reaction in one of two ways: First, other blood groups can cause a transfusion reaction. Second, antibodies in the donor's blood can react with antigens in the recipient's blood. For example, type O blood has anti-A and anti-B antibodies. If type O blood is transfused into a person with type A blood, the anti-A antibodies (in the type O blood) react

	Antigen A	Antigen B	Antigens A and B	Neither antigen A nor B
Red blood cells				
Plasma	Anti-B antibody	Anti-A antibody	Neither anti-A nor anti-B antibodies	Anti-A and anti-B antibodies

Type A
Red blood cells with type A surface antigens and plasma with anti-B antibodies

Type B
Red blood cells with type B surface antigens and plasma with anti-A antibodies

Type AB
Red blood cells with both type A and type B surface antigens and neither anti-A nor anti-B plasma antibodies

Type O
Red blood cells with neither type A nor type B surface antigens but both anti-A and anti-B plasma antibodies

FIGURE 19.13 ABO Blood Groups
For simplicity, only parts of the anti-A and anti-B antibodies are illustrated. Each antibody has five identical, Y-shaped arms (see chapter 22).

against the A antigens (in the type A blood). Usually, such reactions are not serious because the antibodies in the donor's blood are diluted in the larger volume of the recipient's blood, and few reactions take place. Blood banks separate donated blood into several products, such as packed red blood cells, plasma, platelets, and cryoprecipitate, which contains von Willebrand factor, clotting factors, and fibrinogen. This process allows the donated blood to be used by multiple recipients, each of whom may need only one of the blood components. Type O packed red blood cells are unlikely to cause an ABO transfusion reaction when given to a person with a different blood type because the transfusion fluid contains concentrated red blood cells with very little plasma containing anti-A and anti-B antibodies.

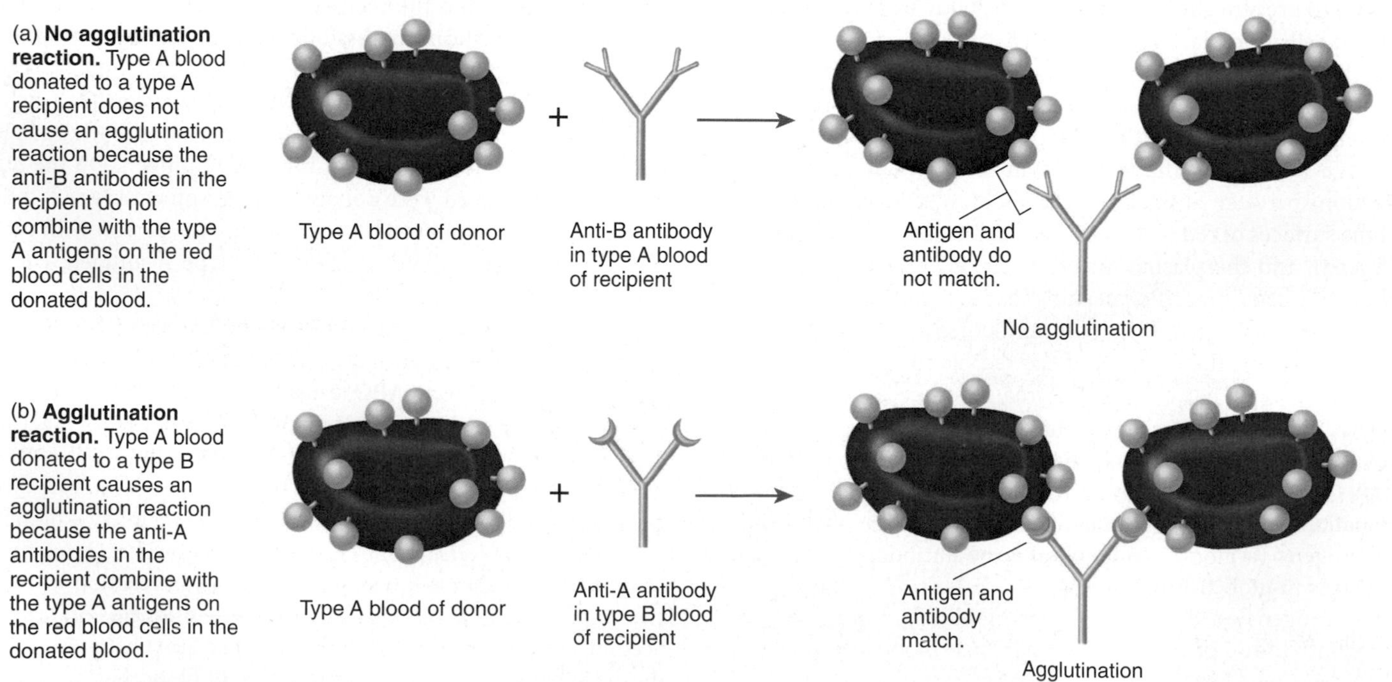
(a) **No agglutination reaction.** Type A blood donated to a type A recipient does not cause an agglutination reaction because the anti-B antibodies in the recipient do not combine with the type A antigens on the red blood cells in the donated blood.
+
Type A blood of donor
Anti-B antibody in type A blood of recipient
Antigen and antibody do not match.
No agglutination
(b) **Agglutination reaction.** Type A blood donated to a type B recipient causes an agglutination reaction because the anti-A antibodies in the recipient combine with the type A antigens on the red blood cells in the donated blood.
+
Type A blood of donor
Anti-A antibody in type B blood of recipient
Antigen and antibody match.
Agglutination

FIGURE 19.14 Agglutination Reaction
For simplicity, only parts of the anti-A and anti-B antibodies are illustrated. Each antibody has five identical, Y-shaped arms (see chapter 22).

Rh Blood Group

The **Rh blood group** is so named because it was first studied in rhesus monkeys. People are Rh-positive if they have a certain Rh antigen (the D antigen) on the surface of their red blood cells, and people are Rh-negative if they do not have this Rh antigen. About 85% of Caucasians in the United States and 88% of African-Americans are Rh-positive. The ABO blood type and the Rh blood type are usually expressed together. For example, a person designated type A in the ABO blood group and Rh-positive is said to be A-positive. The rarest combination in the United States is AB-negative, which occurs in less than 1% of the population.

Antibodies against the Rh antigen do not develop unless an Rh-negative person is exposed to Rh-positive blood. This can occur either through a transfusion or when blood crosses the placenta to a mother from her fetus.

Rh incompatibility can pose a major problem in a pregnancy when the mother is Rh-negative and the fetus is Rh-positive. If fetal blood leaks through the placenta and mixes with the mother's blood, the mother becomes sensitized to the Rh antigen. The mother then produces anti-Rh antibodies that cross the placenta and cause agglutination and hemolysis of fetal red blood cells. This disorder is called **hemolytic** (hē-mō-lit′ik) **disease of the newborn (HDN),** or *erythroblastosis fetalis* (ĕ-rith′rō-blas-tō′sis fē-ta′lis; figure 19.15). In the mother's first pregnancy, there is often no problem. The leakage of fetal blood is usually the result of a tear in the placenta that takes place either late in the pregnancy or during delivery. Thus, there is not sufficient time for the mother to produce enough anti-Rh antibodies to harm the fetus. However, if sensitization occurs, it can cause problems in a subsequent pregnancy. First, once a woman is sensitized and produces anti-Rh antibodies, she may continue to produce the antibodies throughout her life. Thus, in a subsequent pregnancy, anti-Rh antibodies may already be present. Second, and especially dangerous in a subsequent pregnancy with an Rh-positive fetus, if any fetal blood leaks into the mother's blood, she rapidly produces large amounts of anti-Rh antibodies, resulting in HDN. Because HDN can be fatal to the fetus, the levels of anti-Rh antibodies in the mother should be monitored. If they increase to unacceptable levels, the fetus should be tested to determine the severity of the HDN. In severe cases, a transfusion to replace lost red blood cells can be performed through the umbilical cord, or the baby can be delivered if mature enough.

Prevention of HDN is often possible if the Rh-negative mother is injected with a specific type of antibody preparation, called Rh_o(D) immune globulin (RhoGAM), which contains antibodies against Rh antigens. The injection can be given during the pregnancy, before delivery, or immediately after each delivery, miscarriage, or abortion. The injected antibodies bind to the Rh antigens of any fetal red blood cells that may have entered the mother's blood. This treatment inactivates the fetal Rh antigens and prevents sensitization of the mother. However, if sensitization has already occurred, the treatment is ineffective.

ASSESS YOUR PROGRESS

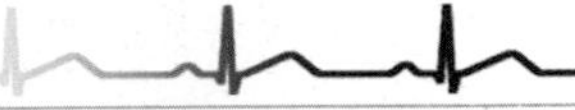

39. *What are blood groups, and how do they cause transfusion reactions? What is agglutination?*
40. *What kinds of antigens and antibodies are found in each of the four ABO blood types?*
41. *Why is a person with type O blood considered a universal donor?*
42. *What does it mean to be Rh-positive?*
43. *What Rh blood types must the mother and the fetus have before HDN can occur?*
44. *Why does HDN usually not develop in the first pregnancy?*

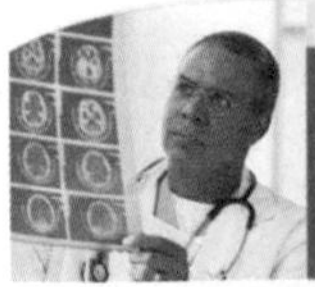

Case STUDY — Treatment of Hemolytic Disease of the Newborn

Billy was born with hemolytic disease of the newborn (HDN). He was treated with exchange transfusion, erythropoietin, and phototherapy. An exchange transfusion replaced Billy's blood with donor blood. In this procedure, as the donor's blood was transfused into Billy, his blood was withdrawn. During fetal development, the increased rate of red blood cell destruction caused by the mother's anti-Rh antibodies results in lower than normal numbers of red blood cells, a condition called **anemia** (ă-nē′mē-ă). It also results in increased levels of bilirubin. Although high levels of bilirubin can damage the brain by killing nerve cells, this is not usually a problem in the fetus because the bilirubin is removed by the placenta. Following birth, bilirubin levels can increase because red blood cells continue to lyse, and the newborn's liver is unable to handle the large bilirubin load. However, in phototherapy, blood that passes through the skin is exposed to blue or white lights, which break down bilirubin to less toxic compounds that the newborn's liver can remove.

Predict 8

Answer the following questions about Billy's treatment for HDN.

a. What is the purpose of giving Billy a transfusion?
b. What is the benefit of the exchange transfusion?
c. Explain the reason for giving Billy erythropoietin.
d. Just before birth, would Billy's erythropoietin levels have been higher or lower than those of a fetus without HDN?
e. After birth, but before treatment, did Billy's erythropoietin levels increase or decrease?
f. When treating HDN with an exchange transfusion, should the donor's blood be Rh-positive or Rh-negative? Explain.
g. Does giving an Rh-positive newborn a transfusion of Rh-negative blood change the newborn's blood type? Explain.

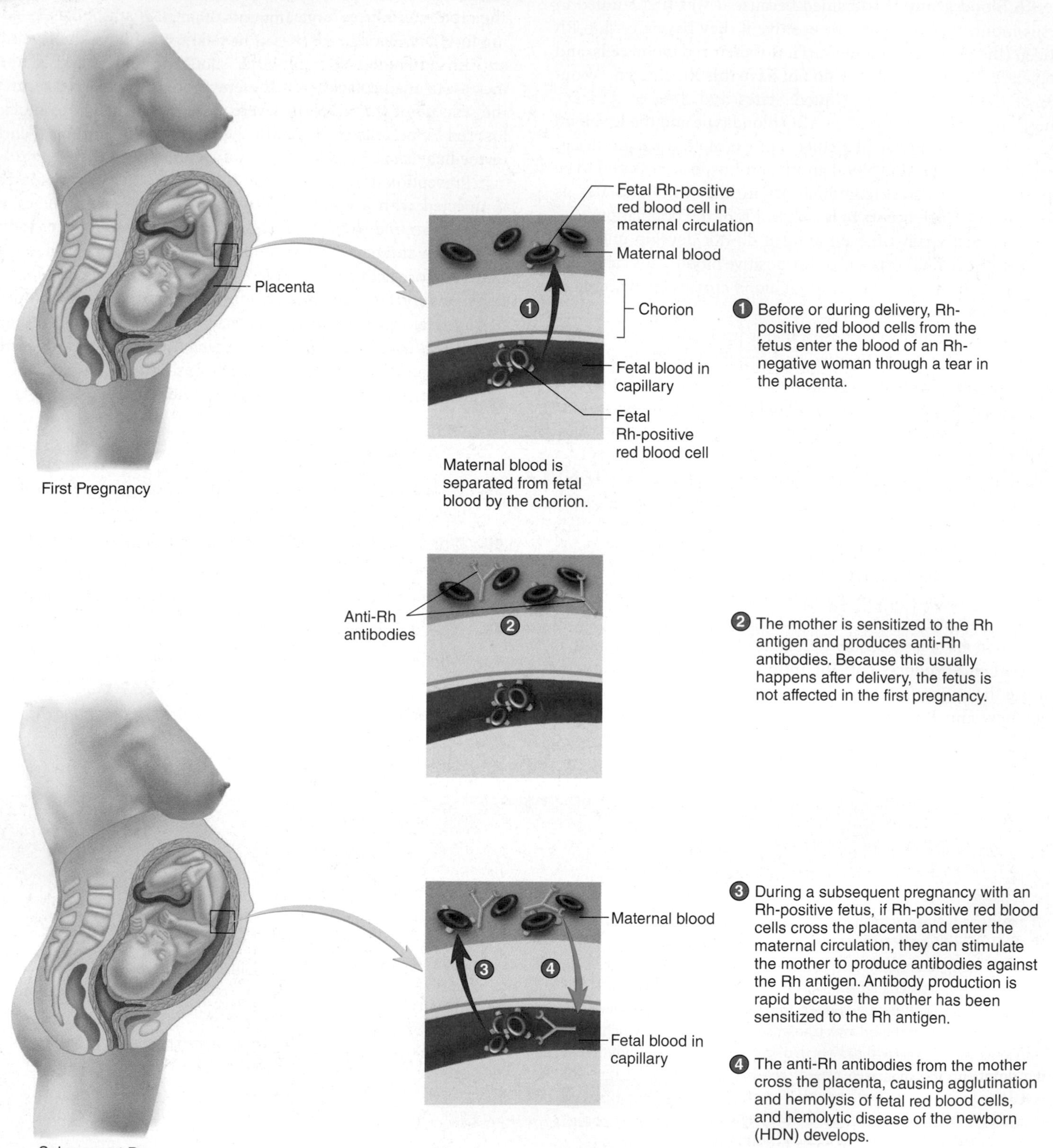
Placenta
Fetal Rh-positive red blood cell in maternal circulation
Maternal blood
Chorion
Fetal blood in capillary
Fetal Rh-positive red blood cell
Maternal blood is separated from fetal blood by the chorion.
First Pregnancy
1 Before or during delivery, Rh-positive red blood cells from the fetus enter the blood of an Rh-negative woman through a tear in the placenta.
Anti-Rh antibodies
2 The mother is sensitized to the Rh antigen and produces anti-Rh antibodies. Because this usually happens after delivery, the fetus is not affected in the first pregnancy.
Maternal blood
Fetal blood in capillary
Subsequent Pregnancy
3 During a subsequent pregnancy with an Rh-positive fetus, if Rh-positive red blood cells cross the placenta and enter the maternal circulation, they can stimulate the mother to produce antibodies against the Rh antigen. Antibody production is rapid because the mother has been sensitized to the Rh antigen.
4 The anti-Rh antibodies from the mother cross the placenta, causing agglutination and hemolysis of fetal red blood cells, and hemolytic disease of the newborn (HDN) develops.

PROCESS FIGURE 19.15 Hemolytic Disease of the Newborn (HDN)

19.7 Diagnostic Blood Tests

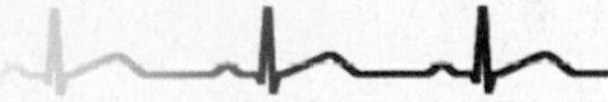

LEARNING OUTCOMES

After reading this section, you should be able to

A. Describe diagnostic blood tests and the normal values for the tests.

B. Give examples of disorders that produce abnormal test values.

Type and Crossmatch

To prevent transfusion reactions, blood is typed. **Blood typing** determines the ABO and Rh blood groups of the blood sample. Typically, the cells are separated from the serum and then tested with known antibodies to determine the type of antigen on the cell surface. For example, if a patient's blood cells agglutinate when mixed with anti-A antibodies but do not agglutinate when mixed with anti-B antibodies, the cells have type A antigen. In a similar fashion, the serum is mixed with known cell types (antigens) to determine the type of antibodies in the serum. Normally, donor blood must match the ABO and Rh type of the recipient.

The International Society of Blood Transfusion recognizes 29 important blood groups, including the ABO and Rh groups. Because any of these blood groups can cause a transfusion reaction, a crossmatch is performed. In a **crossmatch,** the donor's blood cells are mixed with the recipient's serum, and the donor's serum is mixed with the recipient's cells. The donor's blood is considered safe for transfusion only if no agglutination occurs in either match.

Complete Blood Count

A **complete blood count (CBC)** is an analysis of blood that provides much useful information. A CBC consists of a red blood count, hemoglobin and hematocrit measurements, a white blood count, and a differential white blood count.

Red Blood Count

Blood cell counts are usually performed with an electronic instrument, but they can also be done manually with a microscope. A **red blood count (RBC)** is the number (expressed in millions) of red blood cells per microliter of blood. A normal RBC for a male is 4.6–6.2 million/μL of blood; for a female, a normal RBC is 4.2–5.4 million/μL of blood. The condition called **erythrocytosis** (ĕ-rith′rō-sī-tō′sis) is an overabundance of red blood cells (see the Diseases and Disorders table, later in this chapter).

Hemoglobin Measurement

A **hemoglobin measurement** determines the amount of hemoglobin in a given volume of blood, usually expressed as grams of hemoglobin per 100 mL of blood. The normal hemoglobin count for a male is 14–18 g/100 mL of blood, and for a female it is 12–16 g/100 mL of blood. Abnormally low hemoglobin is an indication of anemia (see the Diseases and Disorders table).

Hematocrit Measurement

The percentage of the total blood volume that is composed of red blood cells is the **hematocrit** (hē′mă-tō-krit, hem′ă-tō-krit). One way to determine hematocrit is to place blood in a tube and spin it in a centrifuge. The formed elements, which are heavier than the plasma, are forced to one end of the tube (figure 19.16). Of these, the white blood cells and platelets form a thin, whitish layer, called the buffy coat, between the plasma and the red blood cells. The red blood cells account for 40–54% of the total blood volume in males and 38–47% in females.

The number and size of red blood cells affect the hematocrit measurement. **Normocytes** (nōr′mō-sītz) are normal-sized red blood cells with a diameter of 7.5 mm. **Microcytes** (mī′krō-sītz) are smaller than normal, with a diameter of 6 μm or less, and **macrocytes** (mak′krō-sītz) are larger than normal, with a diameter of 9 μm or greater. Blood disorders can result in an abnormal hematocrit measurement because they cause red blood cell numbers to be abnormally high or low or cause the red blood cells themselves to be abnormally small or large (see the Diseases and Disorders table). A decreased hematocrit indicates that the volume of red blood cells is less than normal. This can result from a decreased number of normocytes or a normal number of microcytes. For example, inadequate iron in the diet can impair hemoglobin production. Consequently, during their formation, red blood cells do not fill with hemoglobin, and they remain smaller than normal.

White Blood Count

A **white blood count (WBC)** measures the total number of white blood cells in the blood. Normally, 5000–10,000 white blood cells are present in each microliter of blood. **Leukopenia** (loo-kō-pē′nē-ă)

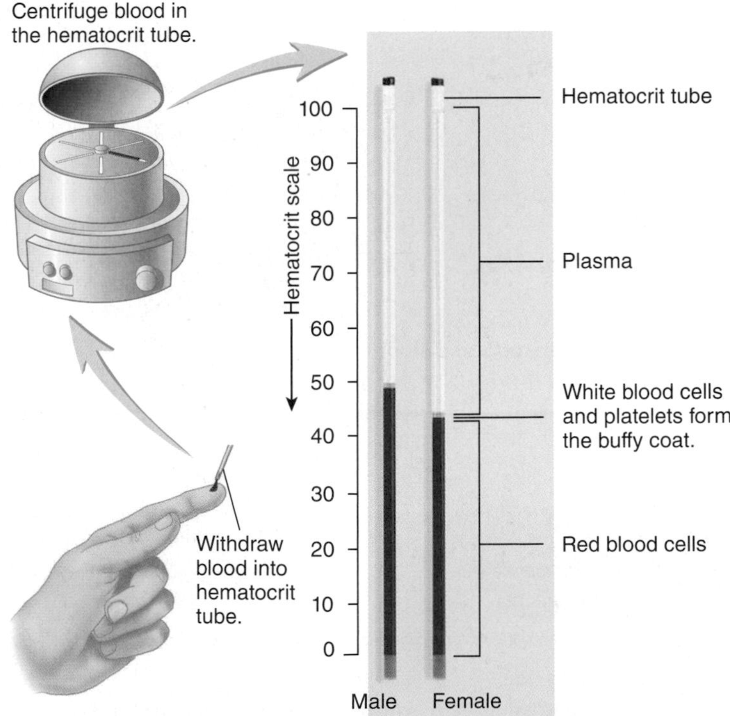
Centrifuge blood in the hematocrit tube.
Withdraw blood into hematocrit tube.
Hematocrit scale
100
90
80
70
60
50
40
30
20
10
0
Hematocrit tube
Plasma
White blood cells and platelets form the buffy coat.
Red blood cells
Male
Female

FIGURE 19.16 Hematocrit

Blood is withdrawn into a capillary tube and spun in a centrifuge. The blood is separated into plasma and red blood cells, with a narrow layer of white blood cells and platelets forming in between. The hematocrit is the percentage of the total blood volume that is composed of red blood cells. It does not include the white blood cells and platelets. Normal hematocrits for a male and a female are shown.

Diseases and Disorders

TABLE 19.4 Blood

Condition	Description
Erythrocytosis	
Relative erythrocytosis	Overabundance of red blood cells due to decreased blood volume, as may result from dehydration, diuretics, or burns
Primary erythrocytosis (polycythemia vera)	Stem cell defect of unknown cause; results in overproduction of red blood cells, granulocytes, and platelets; signs include low erythropoietin levels and enlarged spleen; increased blood viscosity and blood volume can cause clogging of the capillaries and hypertension
Secondary erythrocytosis	Overabundance of red blood cells resulting from decreased oxygen supply, as occurs at high altitudes, in chronic obstructive pulmonary disease, and in congestive heart failure; decreased oxygen delivery to the kidney stimulates the secretion of erythropoietin, resulting in increased blood viscosity and blood volume that can cause clogging of the capillaries and hypertension
Anemia	Deficiency of hemoglobin in the blood
Iron-deficiency anemia	Caused by insufficient intake or absorption of iron or by excessive iron loss; leads to reduced hemoglobin production
Folate-deficiency anemia	Folate is important in DNA synthesis; inadequate folate in the diet results in a reduction in cell division and therefore a reduced number of red blood cells
Pernicious anemia	Secondary folate-deficiency anemia caused by inadequate amounts of vitamin B_{12}, which is important for folate synthesis
Hemorrhagic anemia	Results from blood loss due to trauma, ulcers, or excessive menstrual bleeding
Hemolytic anemia	Occurs when red blood cells rupture or are destroyed at an excessive rate; causes include inherited defects, exposure to certain drugs or snake venom, response to artificial heart valves, autoimmune disease, and hemolytic disease of the newborn
Aplastic anemia	Caused by an inability of the red bone marrow to produce red blood cells, usually as a result of damage to stem cells after exposure to certain drugs, chemicals, or radiation
Thalassemia	Autosomal recessive disease that results in insufficient production of globin part of hemoglobin
Leukemia	Cancers of the red bone marrow in which one or more white blood cell types is produced; cells are usually immature or abnormal and lack normal immunological functions
Thrombocytopenia	Reduction in the number of platelets that leads to chronic bleeding through small vessels and capillaries; causes include genetics, autoimmune disease, infections, and decreased platelet production resulting from pernicious anemia, drug therapy, radiation therapy, or leukemias
Clotting Disorders	
Disseminated intravascular coagulation (DIC)	Clotting throughout the vascular system, followed by bleeding; may develop when normal regulation of clotting by anticoagulants is overwhelmed, as occurs due to massive tissue damage; also caused by alteration of the lining of the blood vessels resulting from infections or snakebites
Von Willebrand disease	Most common inherited bleeding disorder; platelet plug formation and the contribution of activated platelets to blood clotting are impaired; treatments are injection of von Willebrand factor or administration of drugs that increase von Willebrand factor levels in blood, which helps platelets adhere to collagen and become activated
Hemophilia	Genetic disorder in which clotting is abnormal or absent; each of the several types results from deficiency or dysfunction of a clotting factor; most often a sex-linked trait that occurs almost exclusively in males
Infectious Diseases of the Blood	
Septicemia (blood poisoning)	Spread of microorganisms and their toxins by the blood; often the result of a medical procedure, such as insertion of an intravenous tube; release of toxins by bacteria can cause septic shock, producing decreased blood pressure and possibly death
Malaria	Caused by a protozoan introduced into blood by *Anopheles* mosquito; symptoms include chills and fever produced by toxins released when the protozoan causes red blood cells to rupture
Infectious mononucleosis	Caused by Epstein-Barr virus, which infects salivary glands and lymphocytes; symptoms include fever, sore throat, and swollen lymph nodes, all probably produced by the immune system response to infected lymphocytes
Acquired immunodeficiency syndrome (AIDS)	Caused by human immunodeficiency virus (HIV), which infects lymphocytes and suppresses immune system

Go to: www.mhhe.com/seeley10 for additional information on these pathologies.

is a lower than normal WBC resulting from depression or destruction of the red marrow. Viral infections, radiation, drugs, tumors, and vitamin deficiencies (B_{12} or folate) can cause leukopenia. **Leukocytosis** (loo′kō-sī-tō′sis) is an abnormally high WBC. **Leukemia** (loo-kē′mē-ă; a cancer of the red marrow) often results in leukocytosis, but the white blood cells have an abnormal structure and function as well. Bacterial infections can also cause leukocytosis by stimulating neutrophils to increase in number.

Differential White Blood Count

A **differential white blood count** determines the percentage of each of the five kinds of white blood cells. Normally, neutrophils account for 60–70%; lymphocytes, 20–30%; monocytes, 2–8%; eosinophils, 1–4%; and basophils, 0.5–1%. A differential WBC can provide insight into a patient's condition. For example, in patients with bacterial infections the neutrophil count is often greatly increased, whereas in patients with allergic reactions the eosinophil and basophil counts are elevated.

Clotting

The blood's ability to clot can be assessed by the platelet count and the prothrombin time measurement.

Platelet Count

A normal **platelet count** is 250,000–400,000 platelets per microliter of blood. In the condition called **thrombocytopenia** (throm′bō-sī-tō-pē′nē-ă), the platelet count is greatly reduced, resulting in chronic bleeding through small vessels and capillaries. It can be caused by decreased platelet production as a result of hereditary disorders, lack of vitamin B_{12}, drug therapy, or radiation therapy.

Prothrombin Time Measurement

Prothrombin time measurement expresses how long it takes for the blood to start clotting, which is normally 9–12 seconds. Prothrombin time is determined by adding thromboplastin to whole plasma. Thromboplastin is a chemical released from injured tissues that starts the process of clotting (see figure 19.11). Prothrombin time is officially reported as the International Normalized Ratio (INR), which standardizes the time blood takes to clot based on the slightly different thromboplastins used by different labs. Because many clotting factors must be activated to form fibrin, a deficiency of any one of them can cause the prothrombin time to be abnormal. Vitamin K deficiency, certain liver diseases, and drug therapy can increase prothrombin time.

Blood Chemistry

The composition of materials dissolved or suspended in the plasma can be used to assess the functioning of many of the body's systems. For example, high blood glucose levels can indicate that the pancreas is not producing enough insulin; high blood urea nitrogen (BUN) can be a sign of reduced kidney function; increased bilirubin can indicate liver dysfunction or hemolysis; and high cholesterol levels can signify an increased risk for cardiovascular disease. A number of blood chemistry tests are routinely done when a blood sample is taken, and additional tests are available.

ASSESS YOUR PROGRESS

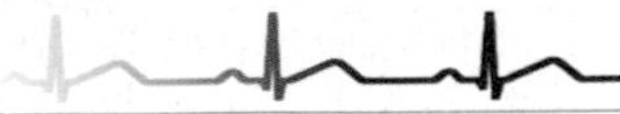

45. *What occurs in a type and crossmatch?*

46. *What tests are included in a CBC? Give the normal value, and name a disorder that would cause an abnormal test result for each.*

47. *What are the normal values for a platelet count and a prothrombin time measurement? Name a disorder that would cause an abnormal result for each test.*

48. *What are some examples of blood chemistry tests?*

Predict 9

When a patient complains of acute pain in the abdomen, the physician suspects appendicitis, which is often caused by a bacterial infection of the appendix. What blood test should be done to support the diagnosis?

Answer

Learn to Predict

From page 637

Frankie's feeling of fatigue and her blood test results are consistent with anemia. A low red blood cell count with microcytic cells, low hemoglobin and a low hematocrit are all indicators of iron deficiency anemia.

The increased reticulocyte count indicated an increased rate of red blood cell production. But, if red blood cell production was increased, why was Frankie's red blood cell count still low? We learned in this chapter that red blood cell production is regulated by the hormone erythropoietin. Specifically, reduced red blood cell numbers, as indicated by Frankie's blood test, caused less oxygen to be transported to her kidneys. Consequently, her kidneys secreted more erythropoietin, which resulted in increased red blood cell production in the red bone marrow. Because of Frankie's iron deficiency, which caused hemoglobin synthesis to slow, the newly synthesized red blood cells were smaller than normal, or microcytic. Remember, Frankie also complained of intense abdominal pain. The evidence of hemoglobin in her feces suggested that Frankie is losing blood into her digestive system, which, considering her abdominal pain, would be consistent with having an ulcer. Frankie's doctor would need to order additional tests to confirm the presence of ulcers before determining treatment.

Answers to the rest of this chapter's Predict questions are in Appendix G.

Summary

19.1 Functions of Blood (p. 638)

1. Blood transports gases, nutrients, waste products, processed molecules, and regulatory molecules.
2. Blood is involved in the regulation of pH, osmosis, and body temperature.
3. Blood protects against disease and initiates tissue repair.

19.2 Composition of Blood (p. 638)

Blood is a type of connective tissue that consists of plasma and formed elements.

19.3 Plasma (p. 638)

1. Plasma is mostly water (91%) and contains proteins, such as albumin (maintains osmotic pressure), globulins (function in transport and immunity), fibrinogen (involved in clot formation), and hormones and enzymes (involved in regulation).
2. Plasma contains ions, nutrients, waste products, and gases.

19.4 Formed Elements (p. 639)

The formed elements are red blood cells (erythrocytes), white blood cells (leukocytes), and platelets (cell fragments).

Production of Formed Elements

1. In the embryo and fetus, the formed elements are produced in a number of locations.
2. After birth, red bone marrow becomes the source of the formed elements.
3. All formed elements are derived from hemocytoblast, which gives rise to two intermediate stem cells: myeloid stem cells and lymphoid stem cells. Myeloid stem cells give rise to red blood cells, platelets, and most of the white blood cells. Lymphoid stem cells give rise to lymphocytes.

Red Blood Cells

1. Red blood cells are biconcave discs containing hemoglobin and carbonic anhydrase.
   - A hemoglobin molecule consists of four heme and four globin molecules. The heme molecules transport oxygen, and the globin molecules transport carbon dioxide and nitric oxide. Iron is required for oxygen transport.
   - Carbonic anhydrase is involved with the transport of carbon dioxide.
2. Erythropoiesis is the production of red blood cells.
   - Stem cells in red bone marrow eventually give rise to late erythroblasts, which lose their nuclei and are released into the blood as reticulocytes. Loss of the endoplasmic reticulum by a reticulocyte produces a red blood cell.
   - In response to low blood oxygen, the kidneys produce erythropoietin, which stimulates erythropoiesis.
3. Hemoglobin from ruptured red blood cells is phagocytized by macrophages. The hemoglobin is broken down, and heme becomes bilirubin, which is secreted in bile.

White Blood Cells

1. White blood cells protect the body against microorganisms and remove dead cells and debris.
2. Five types of white blood cells exist.
   - Neutrophils are small, phagocytic cells.
   - Eosinophils attack certain worm parasites and modulate inflammation.
   - Basophils release histamine and are involved with increasing the inflammatory response.
   - Lymphocytes are important in immunity, including the production of antibodies.
   - Monocytes leave the blood, enter tissues, and become large, phagocytic cells called macrophages.

Platelets

Platelets, or thrombocytes, are cell fragments pinched off from megakaryocytes in the red bone marrow.

19.5 Hemostasis (p. 649)

Hemostasis, the stoppage of bleeding, is very important to the maintenance of homeostasis.

Vascular Spasm

Vasoconstriction of damaged blood vessels reduces blood loss.

Platelet Plug Formation

1. Platelets repair minor damage to blood vessels by forming platelet plugs.
   - In platelet adhesion, platelets bind to collagen in damaged tissues.
   - In the platelet release reaction, platelets release chemicals that activate additional platelets.
   - In platelet aggregation, platelets bind to one another to form a platelet plug.
2. Platelets also release chemicals involved with coagulation.

Coagulation

1. Coagulation is the formation of a blood clot.
2. The first stage of coagulation occurs through the extrinsic or intrinsic pathway. Both pathways end with the production of activated factor X.
   - The extrinsic pathway begins with the release of thromboplastin from damaged tissues.
   - The intrinsic pathway begins with the activation of factor XII.
3. Activated factor X, factor V, phospholipids, and Ca^{2+} form prothrombinase.
4. Prothrombin is converted to thrombin by prothrombinase.
5. Fibrinogen is converted to fibrin by thrombin. The insoluble fibrin forms the clot.

Control of Clot Formation

1. Heparin and antithrombin inhibit thrombin activity. Therefore, fibrinogen is not converted to fibrin, and clot formation is inhibited.
2. Prostacyclin counteracts the effects of thrombin.

Clot Retraction and Dissolution

1. Clot retraction results from the contraction of platelets, which pull the edges of damaged tissue closer together.
2. Serum, which is plasma minus fibrinogen and some clotting factors, is squeezed out of the clot.
3. Factor XII, thrombin, tissue plasminogen activator, and urokinase activate plasmin, which dissolves fibrin (the clot).

19.6 Blood Grouping (p. 655)

1. Blood groups are determined by antigens on the surface of red blood cells.
2. Antibodies can bind to red blood cell antigens, resulting in agglutination or hemolysis of red blood cells.

ABO Blood Group

1. Type A blood has A antigens, type B blood has B antigens, type AB blood has A and B antigens, and type O blood has neither A nor B antigens.
2. Type A blood has anti-B antibodies, type B blood has anti-A antibodies, type AB blood has neither anti-A nor anti-B antibodies, and type O blood has both anti-A and anti-B antibodies.
3. Mismatching the ABO blood group results in a transfusion reaction.

Rh Blood Group

1. Rh-positive blood has certain Rh antigens (the D antigen), whereas Rh-negative blood does not.
2. Antibodies against the Rh antigen are produced by an Rh-negative person when the person is exposed to Rh-positive blood.
3. The Rh blood group is responsible for hemolytic disease of the newborn.

19.7 Diagnostic Blood Tests (p. 659)

Type and Crossmatch

Blood typing determines the ABO and Rh blood groups of a blood sample. A crossmatch tests for agglutination reactions between donor and recipient blood.

Complete Blood Count

A complete blood count consists of the following: red blood count, hemoglobin measurement (grams of hemoglobin per 100 mL of blood), hematocrit measurement (percent volume of red blood cells), white blood count, and differential white blood count (the percentage of each type of white blood cell).

Clotting

Platelet count and prothrombin time measurement assess the blood's ability to clot.

Blood Chemistry

The composition of materials dissolved or suspended in plasma (e.g., glucose, urea nitrogen, bilirubin, and cholesterol) can be used to assess the functioning and status of the body's systems.

REVIEW AND COMPREHENSION

1. Which of these is a function of blood?
   a. clot formation
   b. protection against foreign substances
   c. maintenance of body temperature
   d. regulation of pH and osmosis
   e. All of these are correct.
2. Which of these is *not* a component of plasma?
   a. nitrogen
   b. sodium ions
   c. platelets
   d. water
   e. urea
3. Which of these plasma proteins plays an important role in maintaining the osmotic concentration of the blood?
   a. albumin
   b. fibrinogen
   c. platelets
   d. hemoglobin
   e. globulins
4. Red blood cells
   a. are the least numerous formed element in the blood.
   b. are phagocytic cells.
   c. are produced in the yellow marrow.
   d. do not have a nucleus.
   e. All of these are correct.
5. Given these ways of transporting carbon dioxide in the blood:
   (1) bicarbonate ions
   (2) combined with blood proteins
   (3) dissolved in plasma

   Choose the arrangement that lists them in the correct order from largest to smallest percentage of carbon dioxide transported.
   a. 1,2,3
   b. 1,3,2
   c. 2,3,1
   d. 2,1,3
   e. 3,1,2
6. Each hemoglobin molecule can become associated with ________ oxygen molecule(s).
   a. one
   b. two
   c. three
   d. four
   e. an unlimited number of
7. Erythropoietin
   a. is produced mainly by the heart.
   b. inhibits the production of red blood cells.
   c. production increases when blood oxygen decreases.
   d. production is inhibited by testosterone.
   e. All of these are correct.
8. Which of these changes occur(s) in the blood in response to the initiation of a vigorous exercise program?
   a. increased erythropoietin production
   b. increased concentration of reticulocytes
   c. decreased bilirubin formation
   d. both a and b
   e. All of these are correct.
9. Which of the components of hemoglobin is correctly matched with its fate following the destruction of a red blood cell?
   a. heme—reused to form a new hemoglobin molecule
   b. globin—broken down into amino acids
   c. iron—mostly secreted in bile
   d. All of these are correct.
10. The blood cells that protect against worm parasites are
    a. eosinophils.
    b. basophils.
    c. neutrophils.
    d. monocytes.
    e. lymphocytes.
11. The most numerous type of white blood cell, whose primary function is phagocytosis, is
    a. eosinophils.
    b. basophils.
    c. neutrophils.
    d. monocytes.
    e. lymphocytes.
12. Monocytes
    a. are the smallest white blood cells.
    b. increase in number during chronic infections.
    c. give rise to neutrophils.
    d. produce antibodies.
13. The smallest white blood cells, which include B cells and T cells, are
    a. eosinophils.
    b. basophils.
    c. neutrophils.
    d. monocytes.
    e. lymphocytes.
14. Platelets
    a. are derived from megakaryocytes.
    b. are cell fragments.
    c. have surface molecules that attach to collagen.
    d. play an important role in clot formation.
    e. All of these are correct.
15. Given these processes in platelet plug formation:
    (1) platelet adhesion
    (2) platelet aggregation
    (3) platelet release reaction

    Choose the arrangement that lists the processes in the correct order after a blood vessel is damaged.
    a. 1,2,3
    b. 1,3,2
    c. 3,1,2
    d. 3,2,1
    e. 2,3,1
16. A constituent of blood plasma that forms the network of fibers in a clot is
    a. fibrinogen.
    b. tissue factor.
    c. platelets.
    d. thrombin.
    e. prothrombinase.
17. Given these chemicals:
    (1) activated factor XII
    (2) fibrinogen
    (3) prothrombinase
    (4) thrombin

    Choose the arrangement that lists the chemicals in the order they are used during clot formation.
    a. 1,3,4,2
    b. 2,3,4,1
    c. 3,2,1,4
    d. 3,1,2,4
    e. 3,4,2,1

18. The extrinsic pathway
    a. begins with the release of thromboplastin (tissue factor).
    b. leads to the production of activated factor X.
    c. requires Ca^{2+}.
    d. All of these are correct.
19. The chemical involved in the breakdown of a clot (fibrinolysis) is
    a. antithrombin.
    b. fibrinogen.
    c. heparin
    d. plasmin.
    e. sodium citrate.
20. A person with type A blood
    a. has anti-A antibodies.
    b. has type B antigens.
    c. will have a transfusion reaction if given type B blood.
    d. All of these are correct.
21. In the United States, the most common blood type is
    a. A positive.
    b. B positive.
    c. O positive.
    d. O negative.
    e. AB negative.

Answers in Appendix E

CRITICAL THINKING

1. In hereditary hemolytic anemia, massive destruction of red blood cells occurs. Would you expect the reticulocyte count to be above or below normal? Explain why one of the symptoms of the disease is jaundice. In 1910, physicians discovered that hereditary hemolytic anemia can be treated successfully by removing the spleen. Explain why this treatment is effective.
2. Red Packer, a physical education major, wanted to improve his performance in an upcoming marathon race. About 6 weeks before the race, 500 mL of blood was removed from his body, and the formed elements were separated from the plasma. The formed elements were frozen, and the plasma was reinfused into his body. Just before the competition, the formed elements were thawed and injected into his body. Explain why this procedure, called blood doping or blood boosting, would help Red's performance. Suggest any possible bad effects.
3. Ben has an infected prostate. His physician prescribed several antibiotics before finding one that was effective. Results of the most recent blood tests indicate that Ben is anemic. After analyzing the following measurements of Ben's blood, identify the type of anemia he has.

	Ben's Values	Normal Values
Red blood count	3.1 million RBCs/mm^3	4.6–6.2 million RBCs/mm^3
Reticulocyte count	0.4%	1–3%
Red blood cells	Normocytic, 7.5 mm	Normocytic, 7.5 mm
White blood count	2800 WBCs/mm^3	5000–10,000 WBCs/mm^3
Hemoglobin	9.0 g/100 mL	14–16.5 g/100 mL
Hematocrit	27%	40–54%
Prothrombin time	20 seconds	11–15 seconds
Platelets	200,000/mm^3	250,000–400,000 platelets/mm^3

Which of the following conclusions is most consistent with the results? Explain.
    a. pernicious anemia
    b. iron-deficiency anemia
    c. aplastic anemia
    d. hemorrhagic anemia
    e. vitamin B_{12} deficiency
4. Some people habitually use barbiturates to depress feelings of anxiety. Barbiturates cause hypoventilation, a slower than normal rate of breathing, because they suppress the respiratory centers in the brain. What happens to the red blood count of a habitual user of barbiturates? Explain.
5. According to an old saying, "good food makes good blood." Name three substances in the diet that are essential for "good blood." What blood disorders develop if these substances are absent from the diet?
6. Grace has a plasma membrane defect in her red blood cells that makes them more susceptible to rupturing. Her red blood cells are destroyed faster than they can be replaced. Are her RBC, hemoglobin, hematocrit, and bilirubin levels below normal, normal, or above normal? Explain.
7. Pam lives in Los Angeles, not far from the beach. She traveled by plane with her fiance, Alex, to Jackson Hole, which is approximately 6000 feet above sea level. They took hikes of increasing length for each of 4 days and rested on the fifth day. On the sixth day, she and Alex hiked to the top of Table Mountain, which is approximately 11,000 feet above sea level. Which of the following was (were) apparent in Pam on the day they climbed Table Mountain? Explain.
    (1) Pam's rate of erythropoietin secretion was increasing.
    (2) Pam's erythrocyte count was increasing.
    (3) Pam's reticulocyte count was increasing.
    (4) Pam's platelet count was increasing.
    a. 1,2,3,4
    b. 1,2,3
    c. 1,2
    d. 2,3
    e. 1

Answers in Appendix F

Visit this book's website at **www.mhhe.com/seeley10** for chapter quizzes, interactive learning exercises, and other study tools.

Mc Graw Hill
connect plus+
ANATOMY & PHYSIOLOGY

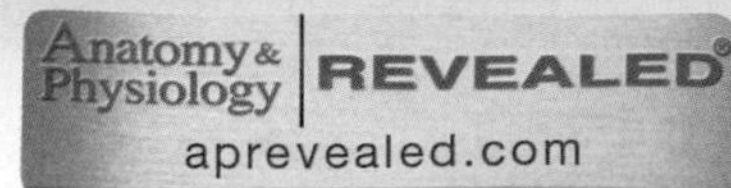
Anatomy & Physiology
REVEALED
aprevealed.com

20

Cardiovascular System

THE HEART

Almost 400 years ago, scientists established that the heart's pumping action maintains the continuous circulation of blood throughout the body. Our current understanding of the detailed function of this amazing pump, its regulation, and modern treatments for heart disease is, in comparison, very recent.

The heart of a healthy 70 kg person pumps approximately 7200 L (approximately 1900 gallons) of blood each day at a rate of 5 L/min. For most people, the heart continues to pump for more than 75 years. During periods of vigorous exercise, the amount of blood pumped per minute increases dramatically, but the person's life is in danger if the heart loses its ability to pump blood for even a few minutes. **Cardiology** (kar-dē-ol′ō-jē) is the medical specialty concerned with diagnosing and treating heart disease.

Learn to Predict

Grandpa Stan never missed watching his grandson pitch. One day, while climbing into the bleachers, he had difficulty breathing and knew he'd better see a doctor. Using a stethoscope, Stan's regular physician could hear an irregular swooshing after the first heart sound, so he referred Stan to a cardiologist. The cardiologist conducted a series of exams and determined that Stan had an incompetent bicuspid valve. By recalling some information from chapter 19 and reading chapter 20, identify the major functional changes in the heart that result from an incompetent valve, and explain how these changes led to Stan's symptoms.

Photo: Photograph of the chordae tendineae attached to the papillary muscles of a ventricle.

Module 9
Cardiovascular System

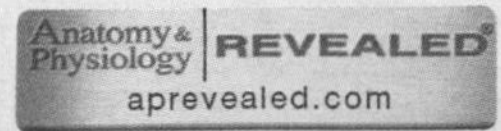
Anatomy & Physiology
REVEALED
aprevealed.com

20.1 Functions of the Heart

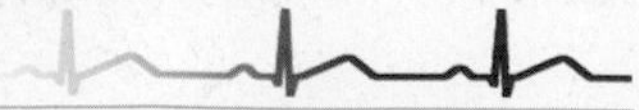

LEARNING OUTCOME

After reading this section, you should be able to

A. List the major functions of the heart.

The heart is actually two pumps in one. The right side of the heart receives blood from the body and pumps blood through the **pulmonary** (pŭl′mō-nār-ē) **circulation,** which carries blood to the lungs and returns it to the left side of the heart. In the lungs, carbon dioxide diffuses from the blood into the lungs, and oxygen diffuses from the lungs into the blood. The left side of the heart pumps blood through the **systemic circulation,** which delivers oxygen and nutrients to all the remaining tissues of the body. From those tissues, carbon dioxide and other waste products are carried back to the right side of the heart (figure 20.1).

The following are the functions of the heart:

1. *Generating blood pressure.* Contractions of the heart generate blood pressure, which is responsible for moving blood through the blood vessels.
2. *Routing blood.* The heart separates the pulmonary and systemic circulations and ensures better oxygenation of the blood flowing to the tissues.
3. *Ensuring one-way blood flow.* The valves of the heart ensure a one-way flow of blood through the heart and blood vessels.
4. *Regulating blood supply.* The rate and force of heart contractions change to meet the metabolic needs of the tissues, which vary depending on such conditions as rest, exercise, and changes in body position.

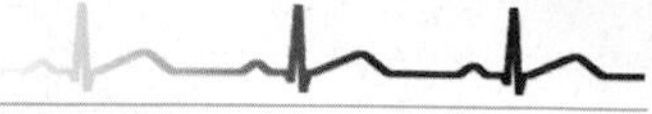

ASSESS **YOUR PROGRESS**

1. *State the four functions of the heart.*

FUNDAMENTAL **Figure**

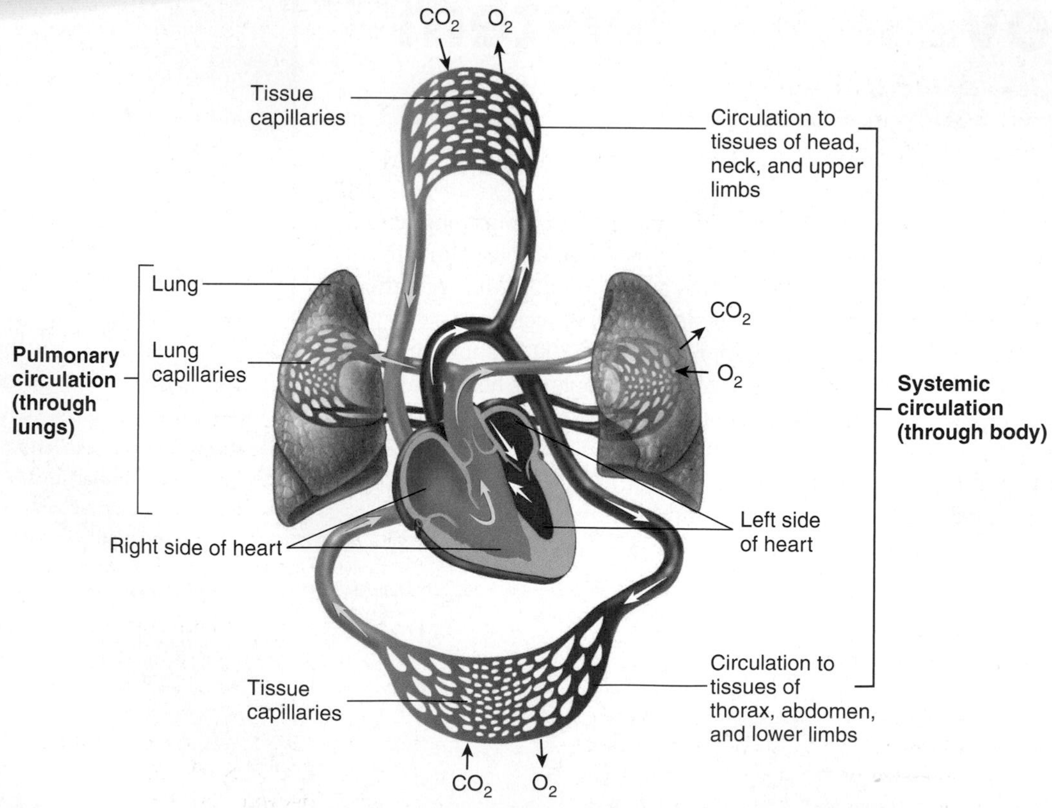
CO2
O2
Tissue capillaries
Circulation to tissues of head, neck, and upper limbs
Lung
Pulmonary circulation (through lungs)
Lung capillaries
CO2
O2
Systemic circulation (through body)
Left side of heart
Right side of heart
Tissue capillaries
Circulation to tissues of thorax, abdomen, and lower limbs
CO2
O2

FIGURE 20.1 AP|R **Systemic and Pulmonary Circulation**

The circulatory system consists of the pulmonary and systemic circulations. The right side of the heart pumps blood through vessels to the lungs and back to the left side of the heart through the pulmonary circulation. The left side of the heart pumps blood through vessels to the tissues of the body and back to the right side of the heart through the systemic circulation.

20.2 Size, Shape, and Location of the Heart

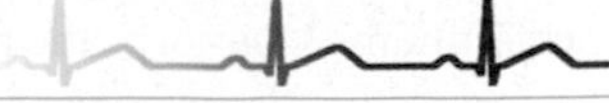

LEARNING OUTCOMES

After reading this section, you should be able to

A. Cite the size, shape, and location of the heart.

B. Explain why knowing the heart's location is important.

The adult heart is shaped like a blunt cone and is approximately the size of a closed fist, with an average mass of 250 g in females and 300 g in males. It is larger in physically active adults than in other healthy adults. The heart generally decreases in size after approximately age 65, especially in people who are not physically active. The blunt, rounded point of the heart is the **apex;** the larger, flat part at the opposite end of the heart is the **base.**

The heart is located in the **mediastinum** (me′dē-as-tī′nŭm; see figure 1.14), a midline partition of the thoracic cavity that also contains the trachea, the esophagus, the thymus, and associated structures.

It is important for health professionals to know the location of the heart in the thoracic cavity. Positioning a stethoscope to hear the heart sounds and positioning electrodes to record an electrocardiogram from chest leads depend on this knowledge. Effective **cardiopulmonary resuscitation** (kar′dē-ō-pŭl′mo-nār-ē rē-sŭs′i-tā-shŭn; **CPR**) also depends on a reasonable knowledge of the position of the heart.

The heart lies obliquely in the mediastinum, with its base directed posteriorly and slightly superiorly and its apex directed anteriorly and slightly inferiorly. The apex is also directed to the left, so that approximately two-thirds of the heart's mass lies to the left of the midline of the sternum (figure 20.2). The base of the heart is located deep to the sternum and extends to the second intercostal space. The apex is located deep to the fifth intercostal space, approximately 7–9 centimeters (cm) to the left of the sternum and medial to the midclavicular line, a perpendicular line that extends down from the middle of the clavicle.

Clinical IMPACT

Cardiopulmonary Resuscitation

When the heart suddenly stops beating, cardiopulmonary resuscitation (CPR) can save lives. CPR consists of firm and rhythmic compression of the chest combined with artificial ventilation of the lungs. Applying pressure to the sternum compresses the chest wall, which also compresses the heart and causes it to pump blood. CPR can provide an adequate blood supply to the heart wall and brain until emergency medical assistance arrives.

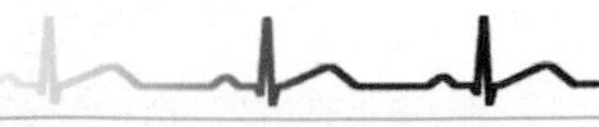

ASSESS YOUR PROGRESS

2. *What is the approximate size and shape of the heart?*
3. *Where is the heart located? How does this knowledge assist in performing CPR?*

20.3 Anatomy of the Heart

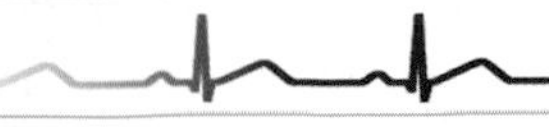

LEARNING OUTCOMES

After reading this section, you should be able to

A. Describe the structure of the pericardium.

B. List the layers of the heart wall and describe the structure and function of each.

C. Relate the large veins and arteries that enter and exit the heart.

D. Describe the location and blood flow through the coronary arteries and cardiac veins.

E. Review the structure and functions of the chambers of the heart.

F. Name the valves of the heart and state their locations and functions.

Pericardium

The **pericardium** (per-i-kar′dē-ŭm), or *pericardial sac,* is a double-layered, closed sac that surrounds the heart (figure 20.3). It consists

Clinical IMPACT

Pericarditis and Cardiac Tamponade

Pericarditis (per′i-kar-dī′tis) is an inflammation of the serous pericardium. The cause is frequently unknown, but it can result from infection, diseases of connective tissue, or damage due to radiation treatment for cancer. The condition can cause extremely painful sensations that are referred to the back and chest and can be confused with a myocardial infarction (heart attack). Pericarditis can lead to fluid accumulation within the pericardial sac.

Cardiac tamponade (tam-pŏ-nād′) is a potentially fatal condition in which a large volume of fluid or blood accumulates in the pericardial cavity and compresses the heart from the outside. Although the heart is a powerful muscle, it relaxes passively. When it is compressed by fluid within the pericardial cavity, it cannot expand when the cardiac muscle relaxes. Consequently, it cannot fill with blood during relaxation; therefore, it cannot pump blood. Cardiac tamponade can cause a person to die quickly unless the fluid is removed. Causes of cardiac tamponade include rupture of the heart wall following a myocardial infarction, rupture of blood vessels in the pericardium after a malignant tumor has invaded the area, damage to the pericardium due to radiation therapy, and trauma, such as that resulting from a traffic accident.

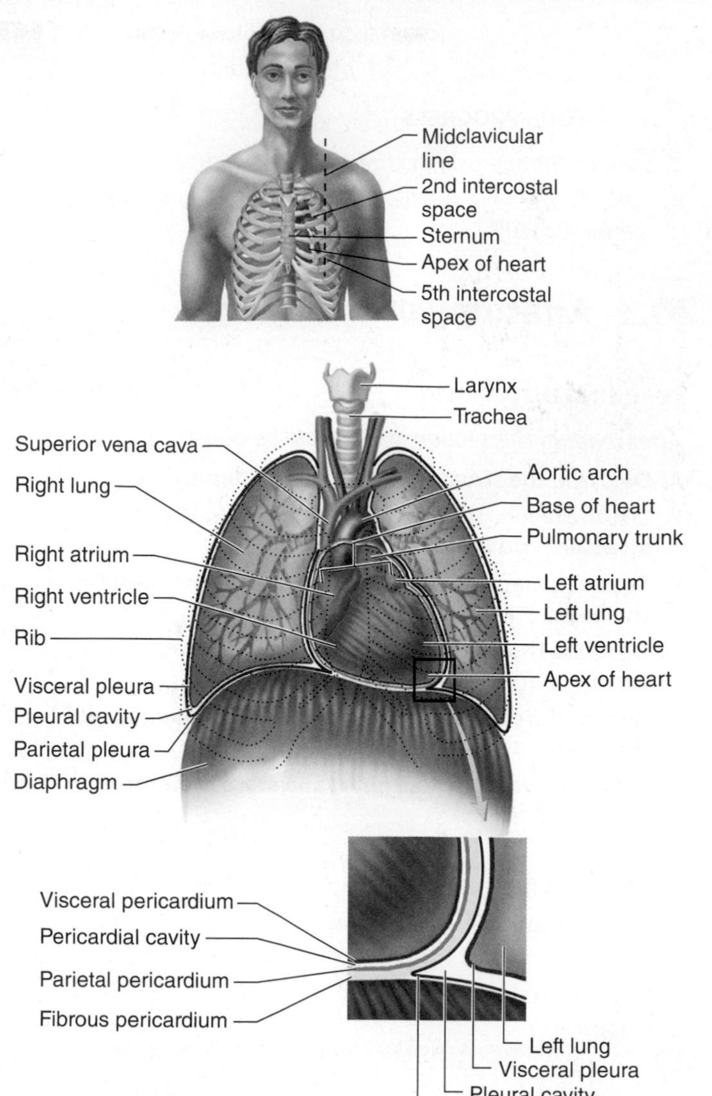
Midclavicular line
2nd intercostal space
Sternum
Apex of heart
5th intercostal space
Larynx
Trachea
Superior vena cava
Right lung
Aortic arch
Base of heart
Pulmonary trunk
Right atrium
Right ventricle
Left atrium
Left lung
Rib
Left ventricle
Apex of heart
Visceral pleura
Pleural cavity
Parietal pleura
Diaphragm
Visceral pericardium
Pericardial cavity
Parietal pericardium
Fibrous pericardium
Left lung
Visceral pleura
Pleural cavity
Parietal pleura
(a)
Anterior view

of a tough, fibrous connective tissue outer layer called the **fibrous pericardium** and a thin, transparent, inner layer of simple squamous epithelium called the **serous pericardium.** The fibrous pericardium prevents overdistension of the heart and anchors it within the mediastinum. Superiorly, the fibrous pericardium is continuous with the connective tissue coverings of the great vessels, and inferiorly it is attached to the surface of the diaphragm (see figure 20.2*a*).

The part of the serous pericardium lining the fibrous pericardium is the **parietal pericardium,** and the part covering the heart surface is the **visceral pericardium,** or *epicardium* (figure 20.3). The parietal and visceral portions of the serous pericardium are continuous with each other where the great vessels enter or leave the heart. The **pericardial cavity,** the space between the visceral and parietal pericardia, is filled with a thin layer of serous **pericardial fluid,** which helps reduce friction as the heart moves within the pericardial sac.

Even though the pericardium contains fibrous connective tissue, it can accommodate changes in heart size by gradually enlarging. The pericardial cavity can also increase in volume to hold a significant volume of pericardial fluid.

Predict 2

Over the weekend, Tony, 22 years old, developed severe chest pains that became worse with deep inhalations and when lying down. As his condition worsened over the next day, Tony became anxious and feared that he might be having a heart attack, so he had a friend drive him to the emergency room. The ER physician identified a low-grade fever, tachycardia (increased heart rate), and a weak and rapid pulse. A chest x-ray showed distension of the jugular veins and pericardial space. The ER physician diagnosed pericarditis, which was probably due to a viral infection, and performed pericardiocentesis to drain the excess fluid from the pericardium. Explain the manifestations that the physician observed, describe how she drained the excess fluid with a needle, and name the body layers the needle penetrated.

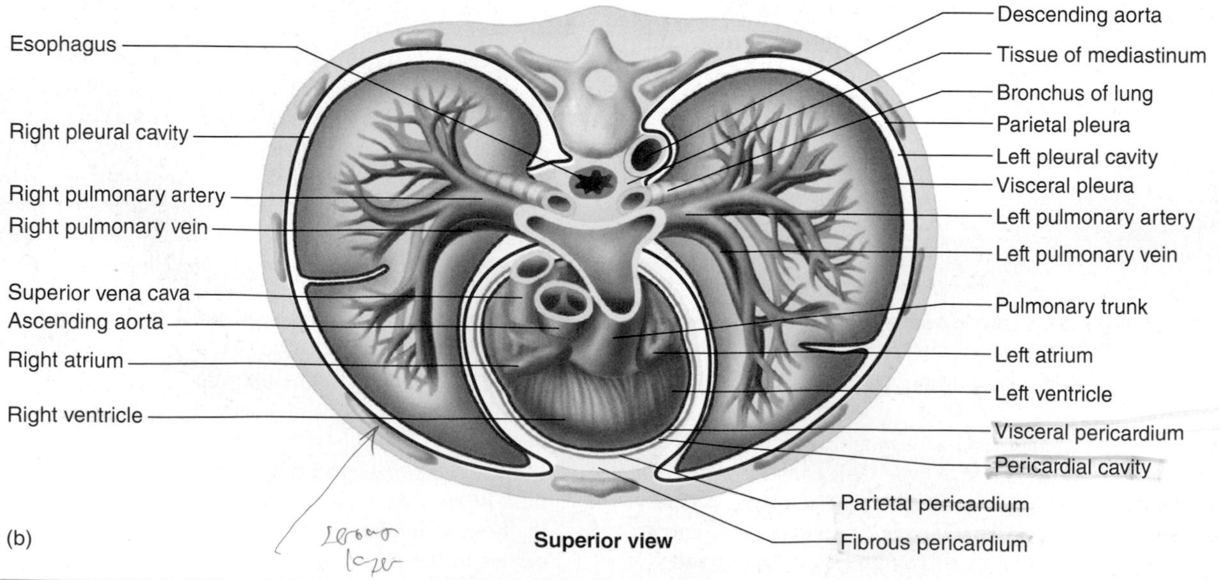
Esophagus
Right pleural cavity
Right pulmonary artery
Right pulmonary vein
Superior vena cava
Ascending aorta
Right atrium
Right ventricle
Descending aorta
Tissue of mediastinum
Bronchus of lung
Parietal pleura
Left pleural cavity
Visceral pleura
Left pulmonary artery
Left pulmonary vein
Pulmonary trunk
Left atrium
Left ventricle
Visceral pericardium
Pericardial cavity
Parietal pericardium
Fibrous pericardium
(b)
Superior view

FIGURE 20.2 AP|R **Location of the Heart in the Thorax**

(*a*) The heart lies in the thoracic cavity between the lungs, deep to and slightly to the left of the sternum. The base of the heart, located deep to the sternum, extends to the second intercostal space, and the apex of the heart is deep to the fifth intercostal space, approximately 7–9 cm to the left of the sternum, or where the midclavicular line intersects with the fifth intercostal space (see inset). (*b*) Cross section of the thorax, showing the position of the heart in the mediastinum and its relationship to other structures.

20.4 Route of Blood Flow Through the Heart

LEARNING OUTCOME

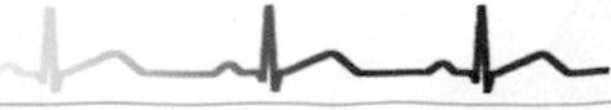

After reading this section, you should be able to

A. Relate the flow of blood through the heart, naming the chambers, valves, and vessels in the correct order.

Blood flow through the heart is depicted in figure 20.10. Even though it is more convenient to discuss blood flow through the heart one side at a time, it is important to understand that both atria contract at about the same time and both ventricles contract at about the same time. This concept is particularly important when considering electrical activity, pressure changes, and heart sounds.

Blood enters the right atrium from the systemic circulation, which returns blood from all the tissues of the body. Blood flows from an area of higher pressure in the systemic circulation to the right atrium, which has a lower pressure. Most of the blood in the right atrium then passes into the right ventricle as the ventricle relaxes following the previous contraction. The right atrium then contracts, and most of the blood remaining in the atrium is pushed into the ventricle to complete right ventricular filling.

Contraction of the right ventricle pushes blood against the tricuspid valve, forcing it closed, and against the pulmonary semilunar valve, forcing it open, thus allowing blood to enter the pulmonary trunk.

The pulmonary trunk branches to form the **pulmonary arteries** (see figure 20.5), which carry blood to the lungs, where carbon dioxide is released and oxygen is picked up (see chapters 21 and 23). Blood returning from the lungs enters the left atrium through the four pulmonary veins. The blood passing from the left atrium to the left ventricle opens the bicuspid valve, and contraction of the left atrium completes left ventricular filling.

Contraction of the left ventricle pushes blood against the bicuspid valve, closing it, and against the aortic semilunar valve, opening it and allowing blood to enter the aorta. Blood flowing through the aorta is distributed to all parts of the body, except to the parts of the lungs supplied by the pulmonary blood vessels (see chapter 23).

ASSESS YOUR PROGRESS

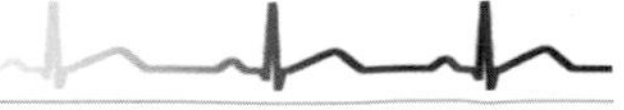

15. *Starting at the venae cavae and ending at the aorta, trace the flow of blood through the heart.*

20.5 Histology

LEARNING OUTCOMES

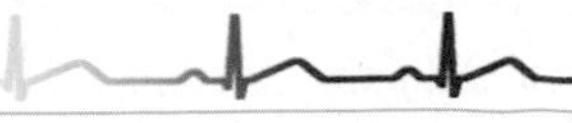

After reading this section, you should be able to

A. Describe the structure and functions of the heart skeleton.

B. Relate the structural and functional characteristics of cardiac muscle cells.

C. Compare and contrast cardiac muscle and skeletal muscle.

D. Explain the structure and function of the conducting system of the heart.

Heart Skeleton

The **heart skeleton** consists of a plate of fibrous connective tissue between the atria and the ventricles. This connective tissue plate forms **fibrous rings** around the atrioventricular and semilunar valves and provides solid support for them, reinforcing the valve openings (figure 20.11). The fibrous connective tissue plate also serves as electrical insulation between the atria and the ventricles and provides a rigid site for attachment of the cardiac muscles.

Cardiac Muscle

Cardiac muscle cells are elongated, branching cells that have one, or occasionally two, centrally located nuclei. Cardiac muscle cells contain actin and myosin myofilaments organized to form sarcomeres, which join end-to-end to form myofibrils (see chapter 9). The actin and myosin myofilaments are responsible for muscle contraction, and their organization gives cardiac muscle a striated (banded) appearance. The striations are less regularly arranged and less numerous than in skeletal muscle (figure 20.12).

Cardiac muscle has a **smooth sarcoplasmic reticulum,** but it is not as regularly arranged as in skeletal muscle fibers, and there are no dilated cisternae, as in skeletal muscle. The sarcoplasmic reticulum comes into close association at various points with membranes of **transverse (T) tubules.** The T tubules of cardiac muscle are found near the Z disks of the sarcomeres, instead of where the actin and myosin overlap, as in skeletal muscle. The T tubules in cardiac muscle are larger in diameter than in skeletal muscle, and extensions of T tubules are parallel with the sarcoplasmic reticulum. The loose association between the sarcoplasmic reticulum and the T tubules is partly responsible for the slow onset of contraction and the prolonged contraction phase in cardiac muscle. Depolarizations of the cardiac muscle plasma membrane are not carried from the surface of the cell to the sarcoplasmic reticulum as efficiently as they are in skeletal muscle, and Ca^{2+} must diffuse a greater distance from the sarcoplasmic reticulum to the actin myofilaments. In addition, some Ca^{2+} enters the cardiac muscle cells from the extracellular fluid and from the T tubules.

Adenosine triphosphate (ATP) provides the energy for cardiac muscle contraction; as in other tissues, ATP production depends on oxygen availability. Cardiac muscle, however, cannot develop a large oxygen deficit, which would interfere with heart function by causing muscular fatigue and cessation of cardiac muscle contraction. Cardiac muscle cells are rich in mitochondria, which perform oxidative metabolism at a rate rapid enough to sustain normal myocardial energy requirements. The extensive capillary network provides an adequate oxygen supply to the cardiac muscle cells.

Cardiac muscle cells are organized in spiral bundles or sheets. The cells are bound end-to-end and laterally to adjacent cells by specialized cell-to-cell contacts called **intercalated** (in-ter′kă-lā-ted) **disks** (figure 20.12). The membranes of the intercalated disks have folds, and the adjacent cells fit together, thus greatly increasing contact between them. Specialized plasma membrane structures

FUNDAMENTAL **Figure**

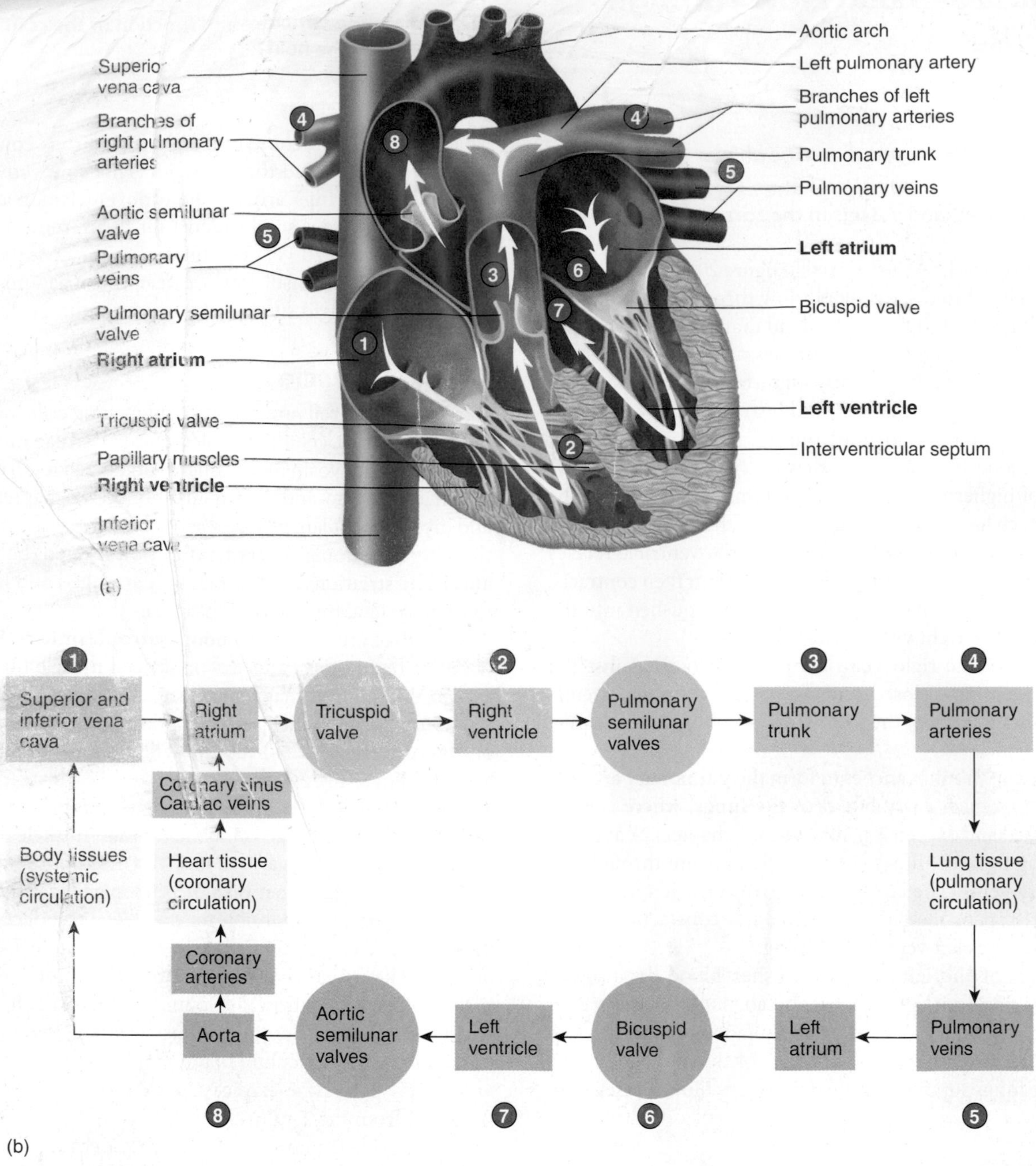

Superior vena cava
Branches of right pulmonary arteries
Aortic semilunar valve
Pulmonary veins
Pulmonary semilunar valve
Right atrium
Tricuspid valve
Papillary muscles
Right ventricle
Inferior vena cava
(a)
Aortic arch
Left pulmonary artery
Branches of left pulmonary arteries
Pulmonary trunk
Pulmonary veins
Left atrium
Bicuspid valve
Left ventricle
Interventricular septum
1
Superior and inferior vena cava
Right atrium
Tricuspid valve
2
Right ventricle
Pulmonary semilunar valves
3
Pulmonary trunk
4
Pulmonary arteries
Coronary sinus
Cardiac veins
Body tissues (systemic circulation)
Heart tissue (coronary circulation)
Lung tissue (pulmonary circulation)
Coronary arteries
Aorta
Aortic semilunar valves
Left ventricle
Bicuspid valve
Left atrium
Pulmonary veins
8
7
6
5
(b)

PROCESS FIGURE 20.10 Blood Flow Through the Heart

(*a*) Frontal section of the heart revealing the four chambers and the direction of blood flow (purple numbers). (*b*) Diagram listing, in order, the structures through which blood flows in the systemic, pulmonary, and coronary circulations. The heart valves are indicated by circles; deoxygenated blood appears blue, and oxygenated blood appears red.

called **desmosomes** (dez′mō-sōmz) hold the cells together, and **gap junctions** allow cytoplasm to flow freely between cells, resulting in areas of low electrical resistance between the cells. This enables action potentials to pass easily from one cell to the next (see figure 4.2). Electrically, the cardiac muscle cells behave as a single unit, and the heart's highly coordinated contractions depend on this functional characteristic.

Conducting System

A conducting system relays action potentials through the heart. This system consists of modified cardiac muscle cells that form two nodes (knots or lumps) and a conducting bundle (figure 20.13). The two nodes are contained within the walls of the right atrium and are named according to their position in the atrium. The **sinoatrial (SA) node** is medial to the opening of the superior vena

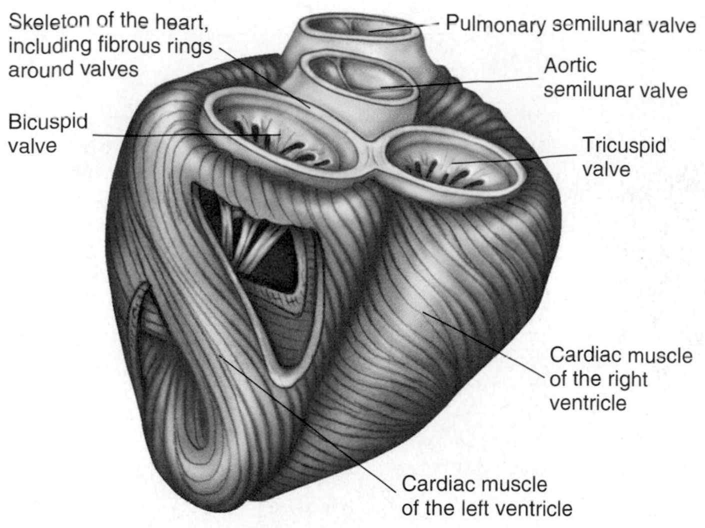
Skeleton of the heart, including fibrous rings around valves
Pulmonary semilunar valve
Aortic semilunar valve
Bicuspid valve
Tricuspid valve
Cardiac muscle of the right ventricle
Cardiac muscle of the left ventricle

FIGURE 20.11 Heart Skeleton
The skeleton of the heart consists of fibrous connective tissue rings, which surround the heart valves and separate the atria from the ventricles. Cardiac muscle attaches to the fibrous connective tissue. The muscle fibers are arranged so that contraction of the ventricles produces a wringing motion and the distance between the apex and the base of the heart shortens.

cava, and the **atrioventricular (AV) node** is medial to the right atrioventricular valve. The AV node gives rise to a conducting bundle of the heart, the **atrioventricular (AV) bundle** (bundle of His). This bundle passes through a small opening in the fibrous skeleton to reach the interventricular septum, where it divides to form the **right** and **left bundle branches,** which extend beneath the endocardium on each side of the interventricular septum to the apex of both the right and the left ventricles.

The inferior terminal branches of the bundle called **Purkinje** (per-kin′jē) **fibers,** are large-diameter cardiac muscle fibers. They have fewer myofibrils than most cardiac muscle cells and do not contract as forcefully. Intercalated disks are well developed between the Purkinje fibers and contain numerous gap junctions. As a result of these structural modifications, action potentials travel along the Purkinje fibers much more rapidly than through other cardiac muscle tissue.

Unlike skeletal muscle cells that require neural stimulation for contraction, cardiac muscle cells have the intrinsic capacity to spontaneously generate action potentials for contraction. Because cells of the SA node spontaneously generate action potentials at a greater frequency than other cardiac muscle cells, these cells are called the **pacemaker** of the heart. The SA node is made up of specialized, small-diameter cardiac muscle cells that merge with the other cardiac muscle cells of the right atrium. Once action potentials are produced, they spread from the SA node to adjacent cardiac muscle fibers of the atrium. Preferential pathways conduct action potentials from the SA node to the AV node at a greater velocity than they are transmitted in the remainder of the atrial muscle fibers. However, such pathways cannot be distinguished structurally from the remainder of the atrium. Thus, it is the activity of the SA node that causes the heart to contract spontaneously and rhythmically.

When the heart beats under resting conditions, approximately 0.04 second is required for action potentials to travel from the SA node to the AV node. Within the AV node, action potentials are propagated slowly, compared with the remainder of the conducting system. The slow rate of action potential conduction in the AV node is due, in part, to the smaller-diameter muscle fibers and fewer gap junctions in their intercalated disks. Like the other specialized

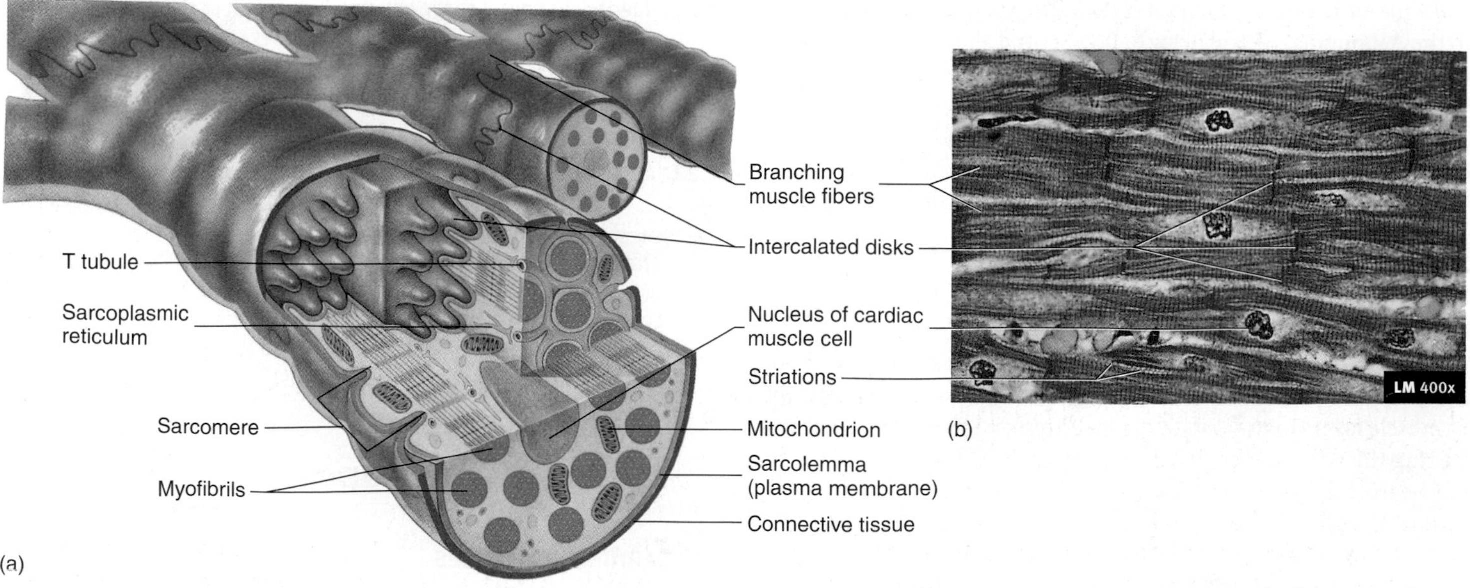
T tubule
Sarcoplasmic reticulum
Sarcomere
Myofibrils
(a)
Branching muscle fibers
Intercalated disks
Nucleus of cardiac muscle cell
Striations
Mitochondrion
Sarcolemma (plasma membrane)
Connective tissue
LM 400x
(b)

FIGURE 20.12 AP|R Histology of the Heart
(*a*) Cardiac muscle cells are branching cells with centrally located nuclei. As in skeletal muscle, sarcomeres join end-to-end to form myofibrils, and mitochondria provide ATP for contraction. The cells are joined to one another by intercalated disks. Gap junctions in the intercalated disks allow action potentials to pass from one cardiac muscle cell to the next. Sarcoplasmic reticulum and T tubules are visible but are not as numerous as they are in skeletal muscle. (*b*) A light micrograph of cardiac muscle tissue. The cardiac muscle fibers appear striated because of the arrangement of the individual myofilaments.

1 Action potentials originate in the sinoatrial (SA) node (the pacemaker) and travel across the wall of the atrium (*arrows*) from the SA node to the atrioventricular (AV) node.

2 Action potentials pass through the AV node and along the atrioventricular (AV) bundle, which extends from the AV node, through the fibrous skeleton, into the interventricular septum.

3 The AV bundle divides into right and left bundle branches, and action potentials descend to the apex of each ventricle along the bundle branches.

4 Action potentials are carried by the Purkinje fibers from the bundle branches to the ventricular walls and papillary muscles.

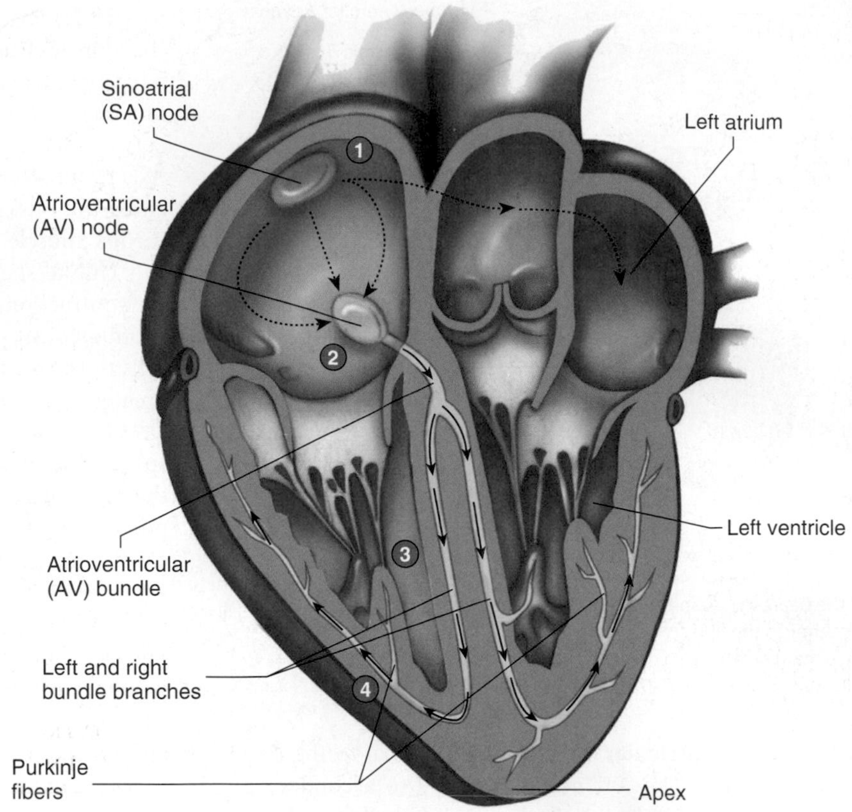
Sinoatrial
(SA) node
Left atrium
Atrioventricular
(AV) node
1
2
3
4
Left ventricle
Atrioventricular
(AV) bundle
Left and right
bundle branches
Purkinje
fibers
Apex

PROCESS FIGURE 20.13 APR **Conducting System of the Heart**

conducting fibers in the heart, they have fewer myofibrils than most cardiac muscle cells. As a consequence, a delay of 0.11 second occurs from the time action potentials reach the AV node until they pass to the AV bundle. A total delay of 0.15 second allows for completion of the atrial contraction before ventricular contraction begins.

After action potentials pass from the AV node to the highly specialized conducting bundles, the velocity of conduction increases dramatically. The action potentials pass through the left and right bundle branches and through the individual Purkinje fibers that penetrate the myocardium of the ventricles (figure 20.13).

Because of the arrangement of the conducting system in the ventricles, the first part of the ventricular myocardium that is stimulated is the inner wall of the ventricles near the apex. Thus, ventricular contraction begins at the apex and progresses throughout the ventricles toward the base of the heart. The spiral arrangement of muscle layers in the wall of the heart results in a wringing action. During the process, the distance between the apex and the base of the heart decreases and blood is forced upwards from the apex toward the great vessels at the base of the heart (see figure 20.10).

ASSESS YOUR PROGRESS

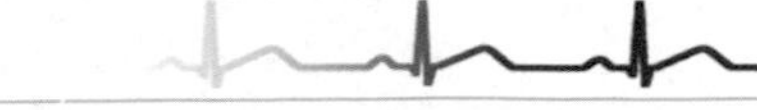

16. *What is the heart skeleton composed of? What are its functions?*

17. *Compare and contrast cardiac muscle and skeletal muscle.*

18. *Why does cardiac muscle have slow onset of contraction and prolonged contraction?*

19. *What anatomical features are responsible for the ability of cardiac muscle cells to contract as a unit?*

20. *Identify the parts of the conducting system of the heart. Explain how the conducting system coordinates contraction of the atria and ventricles.*

21. *Explain why Purkinje fibers conduct action potentials more rapidly than other cardiac muscle cells.*

22. *Relate why the SA node is the pacemaker of the heart.*

Predict 3

Because of a reduced blood supply to the AV node, the delay in the conduction of action potentials from the SA node to the AV node is increased slightly. All other areas of the conducting system of the heart are functioning normally. Predict how this affects the normal rhythm of the heart.

20.6 Electrical Properties

LEARNING OUTCOMES

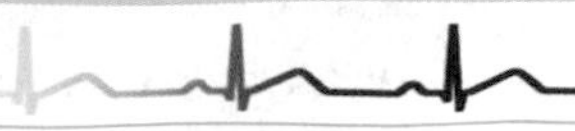

After reading this section, you should be able to

A. Summarize the characteristics of action potentials in cardiac muscle.

B. Explain what is meant by the autorhythmicity of cardiac muscle and relate it to the pacemaker potential.

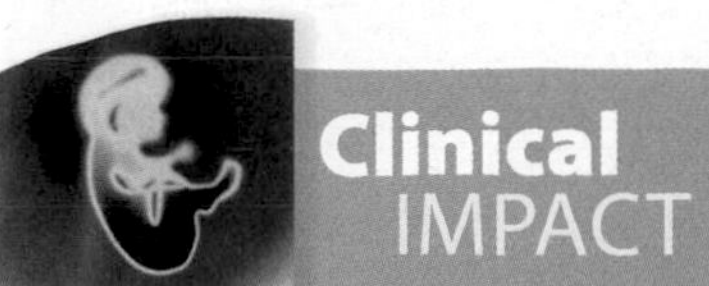

Clinical IMPACT — Angina, Infarctions, and the Treatment of Blocked Coronary Arteries

Angina pectoris (an′ji-nă, an-jī′nă pek′tō-ris) is chest pain that results from a reduced blood supply to cardiac muscle. The pain is temporary and, if blood flow is restored, little permanent change or damage results. Angina pectoris is characterized by chest discomfort deep to the sternum, often described as heaviness, pressure, or moderately severe pain. It is often mistaken for indigestion. The pain can also be referred to the neck, lower jaw, left arm, and left shoulder (see chapter 14).

Most often, angina pectoris results from narrowed and hardened coronary arterial walls. The reduced blood flow results in a reduced supply of oxygen to cardiac muscle cells. As a consequence, the limited anaerobic respiration of cardiac muscle results in a reduced pH in affected areas of the heart, which stimulates pain receptors. The pain is predictably associated with exercise because the increased pumping of the heart requires more oxygen, and the narrowed blood vessels cannot supply it.

Rest and drugs, such as nitroglycerin, frequently relieve angina pectoris. Nitroglycerin dilates the blood vessels, including the coronary arteries. Consequently, the drug increases the oxygen supply to cardiac muscle and reduces the heart's workload. Because peripheral arteries are dilated, the heart has to pump blood against a lower pressure, and the need for oxygen decreases. The heart also pumps less blood because blood tends to remain in the dilated blood vessels and less blood is returned to the heart.

Myocardial infarction (mī-ō-kar′dē-ăl in-fark′shŭn) results when a prolonged lack of blood flow to a part of the cardiac muscle leads to a lack of oxygen and ultimately cellular death. Symptoms of myocardial infarction include chest pain that radiates into the left shoulder and arm, shortness of breath, nausea, vomiting, and sweating. Interestingly, these symptoms are common in men, but women may experience very different symptoms. Over 40% of women who have suffered a myocardial infarction did not experience chest pain. Symptoms for women include sudden fatigue, dizziness, and abdominal pain.

Myocardial infarctions vary in severity, depending on the amount of cardiac muscle and the part of the heart that is affected. If blood supply to cardiac muscle is reestablished within 20 minutes, no permanent damage occurs. If the oxygen deficiency lasts longer, cell death results. However, within 30–60 seconds after blockage of a coronary blood vessel, functional changes are obvious. The electrical properties of the cardiac muscle are altered, and the heart's ability to function properly is lost.

The most common cause of myocardial infarction is thrombus formation that blocks a coronary artery. Coronary arteries narrowed by **atherosclerotic** (ath′er-ō-skler-ot′ik) **lesions** increase the risk for myocardial infarction. Atherosclerotic lesions partially block blood vessels, resulting in turbulent blood flow, and the surfaces of the lesions are rough. These changes increase the probability of thrombus formation.

Blocked blood vessels can be treated using various medical techniques. **Angioplasty** (an′jē-ō-plas-tē) is a process whereby a surgeon threads a small balloon, usually into the femoral artery (see chapter 21), through the aorta and into a coronary artery. After entering the partially occluded coronary artery, the balloon is inflated, flattening the atherosclerotic deposits against the vessel walls and opening the occluded blood vessel. This technique improves the function of cardiac muscle in patients experiencing inadequate blood flow to the cardiac muscle through the coronary arteries. However, some controversy exists about its effectiveness. At least in some patients, dilation of the coronary arteries can reverse within a few weeks or months, and blood clots can form in coronary arteries following angioplasty. To help prevent future blockage, a metal-mesh tube called a **stent** is inserted into the vessel. Although the stent is better able to hold the vessel open, it, too, can eventually become blocked. Small, rotating blades and lasers are also used to remove lesions from coronary vessels.

Coronary bypass is a surgical procedure that relieves the effects of obstructed coronary arteries. This technique involves taking healthy segments of blood vessels from other parts of the patient's body and using them to bypass obstructions in the coronary arteries. The technique is common in cases of severe occlusion in specific parts of coronary arteries.

Enzymes are used to break down blood clots that form in the coronary arteries and cause myocardial infarctions. The major enzymes used are **tissue plasminogen** (plaz-min′ō-jen) **activator (t-PA),** and **urokinase** (ūr-ō-kī′nās). These enzymes activate plasminogen, an inactive form of an enzyme in the body that breaks down the fibrin of clots. The strategy calls for administering these drugs to people suffering from myocardial infarctions as soon as possible following the onset of symptoms. Removal of the occlusions produced by clots reestablishes blood flow to the cardiac muscle and reduces the amount of cardiac muscle permanently damaged by the occlusions.

C. Explain the importance of a long refractory period in cardiac muscle.

D. Describe the waves and intervals of an electrocardiogram.

Cardiac muscle cells—like other electrically excitable cells, such as neurons and skeletal muscle fibers—have a **resting membrane potential (RMP).** The resting membrane potential depends on a low permeability of the plasma membrane to Na^+ and Ca^{2+} and a higher permeability to K^+. When neurons, skeletal muscle fibers, and cardiac muscle cells are depolarized to their threshold level, action potentials result (see chapter 11).

Action Potentials

Like action potentials in skeletal muscle, those in cardiac muscle exhibit depolarization followed by repolarization of the resting membrane potential. Alterations in membrane channels are responsible for the changes in the permeability of the plasma membrane that produce the action potentials. Action potentials in cardiac muscle last longer than those in skeletal muscle, and the membrane channels differ somewhat from those in skeletal muscle. In contrast to action potentials in skeletal muscle, which take less than 2 milliseconds (ms) to complete, action potentials in cardiac muscle take approximately 200–500 ms to complete (figure 20.14).

FUNDAMENTAL **Figure**

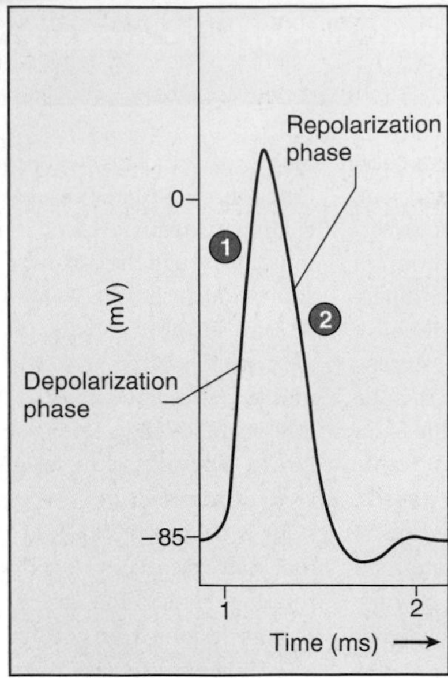
Repolarization phase
0
(mV)
1
2
Depolarization phase
−85
1
2
Time (ms)

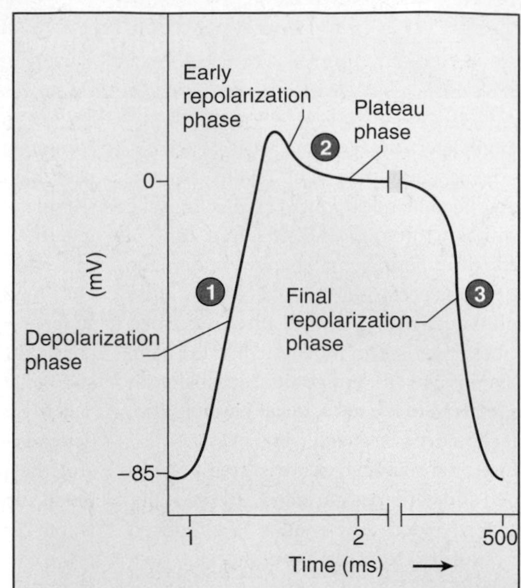
Early repolarization phase
Plateau phase
2
0
(mV)
1
Final repolarization phase
3
Depolarization phase
−85
1
2
500
Time (ms)

Permeability changes during an action potential in skeletal muscle:

1. **Depolarization phase**
   - Voltage-gated Na^+ channels open.
   - Voltage-gated K^+ channels begin to open.
2. **Repolarization phase**
   - Voltage-gated Na^+ channels close.
   - Voltage-gated K^+ channels continue to open.
   - Voltage-gated K^+ channels close at the end of repolarization and return the membrane potential to its resting value.

Permeability changes during an action potential in cardiac muscle:

1. **Depolarization phase**
   - Voltage-gated Na^+ channels open.
   - Voltage-gated K^+ channels close.
   - Voltage-gated Ca^{2+} channels begin to open.
2. **Early repolarization and plateau phases**
   - Voltage-gated Na^+ channels close.
   - Some voltage-gated K^+ channels open, causing early repolarization.
   - Voltage-gated Ca^{2+} channels are open, producing the plateau by slowing further repolarization.
3. **Final repolarization phase**
   - Voltage-gated Ca^{2+} channels close.
   - Many voltage-gated K^+ channels open.

PROCESS FIGURE 20.14 Comparison of Action Potentials in Skeletal and Cardiac Muscle
(*a*) An action potential in skeletal muscle consists of depolarization and repolarization phases. (*b*) An action potential in cardiac muscle consists of depolarization, early repolarization, plateau, and final repolarization phases. Cardiac muscle does not repolarize as rapidly as skeletal muscle (indicated by the break in the curve) because of the plateau phase.

In cardiac muscle, the action potential consists of a rapid **depolarization phase,** followed by a rapid but partial **early repolarization phase.** Then a prolonged period of slow repolarization occurs, called the **plateau phase.** At the end of the plateau phase, a more rapid **final repolarization phase** takes place, during which the membrane potential returns to its resting level (figure 20.14).

Depolarization is the result of changes in membrane permeability to Na^+, K^+, and Ca^{2+}. Membrane channels, called **voltage-gated Na^+ channels,** open, bringing about the depolarization phase of the action potential. As the voltage-gated Na^+ channels open, Na^+ diffuses into the cell, causing rapid depolarization until the cell is depolarized to approximately +20 millivolts (mV).

The voltage change occurring during depolarization affects other ion channels in the plasma membrane. Several types of **voltage-gated K^+ channels** exist, each of which opens and closes at different membrane potentials, causing changes in membrane permeability to K^+. For example, at rest, the movement of K^+ through open voltage-gated K^+ channels is primarily responsible for establishing the resting membrane potential in cardiac muscle cells. Depolarization causes these voltage-gated K^+ channels to close, thereby decreasing membrane permeability to K^+. Depolarization also causes **voltage-gated Ca^{2+}** channels to begin to open. These changes contribute to depolarization. Compared with Na^+ channels, the Ca^{2+} channels open and close slowly.

Repolarization is also the result of changes in membrane permeability to Na^+, K^+, and Ca^{2+}. Early repolarization occurs when the voltage-gated Na^+ channels and some voltage-gated Ca^{2+} channels close, and a small number of voltage-gated K^+ channels open. Sodium ion movement into the cell slows, and some K^+ moves out of the cell. The plateau phase occurs as voltage-gated Ca^{2+} channels remain open, and the movement of Ca^{2+} and some Na^+ through the voltage-gated Ca^{2+} channels into the cell counteracts the potential change produced by the movement of K^+ out of the cell. The plateau phase ends, and final repolarization begins as the voltage-gated Ca^{2+} channels close and many more voltage-gated K^+ channels open. Thus, Ca^{2+} and Na^+ stop diffusing into the cell, and the tendency for K^+ to diffuse out of the cell increases. These permeability changes cause the membrane potential to return to its resting level.

Action potential propagation in cardiac muscle differs from that in skeletal muscle. First, action potentials in cardiac muscle are conducted from cell to cell, whereas action potentials in skeletal muscle fibers are conducted along the length of a single muscle fiber, but not from fiber to fiber. Gap junctions of intercalated disks in cardiac muscle allow for the cell-to-cell conduction. Second, action potential propagation is slower in cardiac muscle than in skeletal muscle because cardiac muscle cells are smaller in diameter and much shorter than skeletal muscle fibers. Although the gap junctions allow the transfer of action potentials between cardiac muscle cells, they slow the rate of action potential conduction between the cardiac muscle cells.

The movement of Ca^{2+} through the plasma membrane, including the membranes of the T tubules, into cardiac muscle cells stimulates the release of Ca^{2+} from the sarcoplasmic reticulum, a process called **calcium-induced calcium release (CICR).** When an action potential occurs in a cardiac muscle cell, Ca^{2+} enters the cell and binds to receptors in the membranes of the sarcoplasmic reticulum, resulting in the opening of Ca^{2+} channels. Calcium ions then move out of the sarcoplasmic reticulum and activate the interaction between actin and myosin to produce contraction of the cardiac muscle cells.

Autorhythmicity of Cardiac Muscle

The heart is said to be **autorhythmic** (aw′tō-rith′mik) because it stimulates itself (*auto*) to contract at regular intervals (*rhythmic*). If the heart is removed from the body and maintained under physiological conditions with the proper nutrients and temperature, it will continue to beat autorhythmically for a long time.

In the SA node, pacemaker cells generate action potentials spontaneously and at regular intervals. These action potentials spread through the conducting system of the heart to other cardiac muscle cells, causing voltage-gated Na^+ channels to open. As a result, action potentials are produced and the cardiac muscle cells contract.

When a spontaneously developing local potential, called the **pacemaker potential,** reaches threshold, then action potentials are generated in the SA node (figure 20.15). Changes in ion movement into and out of the pacemaker cells cause the pacemaker potential. Sodium ions cause depolarization by moving into the cells through specialized nongated Na^+ channels. A decreasing permeability to K^+ also causes depolarization as less K^+ moves out of the cells. The decreasing K^+ permeability occurs due to the voltage changes at the end of the previous action potential. As a result of the depolarization, voltage-gated Ca^{2+} channels open, and the movement of Ca^{2+} into the pacemaker cells causes further depolarization. When the pacemaker potential reaches threshold, many voltage-gated Ca^{2+} channels open. In pacemaker cells, the movement of Ca^{2+} into the cells is primarily responsible for the depolarization phase of the action potential. This is different from other cardiac muscle cells, where the movement of Na^+ into the cells is primarily responsible for depolarization. Repolarization occurs, as in other cardiac muscle cells, when the voltage-gated Ca^{2+} channels close and the voltage-gated K^+ channels open. After the resting membrane potential is reestablished, production of another pacemaker potential starts the generation of the next action potential.

Although most cardiac muscle cells respond to action potentials produced by the SA node, some cardiac muscle cells in the conducting system can also generate spontaneous action potentials. Normally, the SA node controls the rhythm of the heart because

Permeability changes in pacemaker cells

1 Pacemaker potential

- A small number of Na^+ channels are open.
- Voltage-gated K^+ channels that opened in the repolarization phase of the previous action potential are closing.
- Voltage-gated Ca^{2+} channels begin to open.

2 Depolarization phase

- Voltage-gated Ca^{2+} channels are open.
- Voltage-gated K^+ channels are closed.

3 Repolarization phase

- Voltage-gated Ca^{2+} channels close.
- Voltage-gated K^+ channels open.

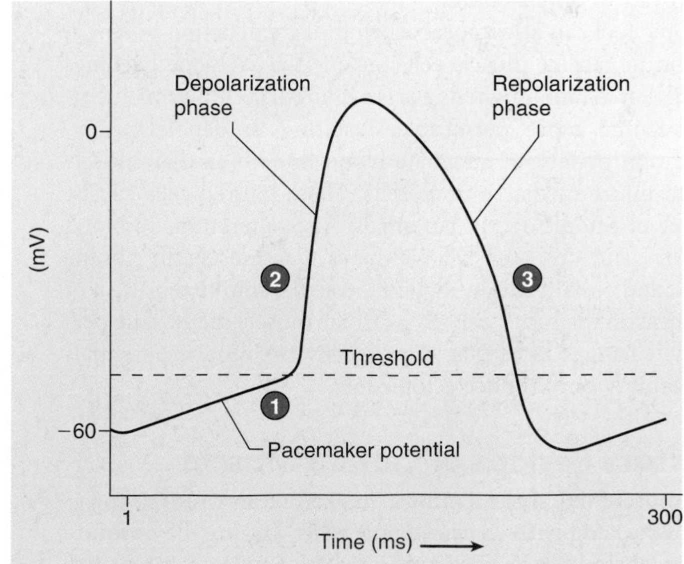

Depolarization phase
Repolarization phase
0
(mV)
2
3
Threshold
1
−60
Pacemaker potential
1
300
Time (ms)

PROCESS FIGURE 20.15 SA Node Action Potential

The production of action potentials by the SA node is responsible for the autorhythmicity of the heart.

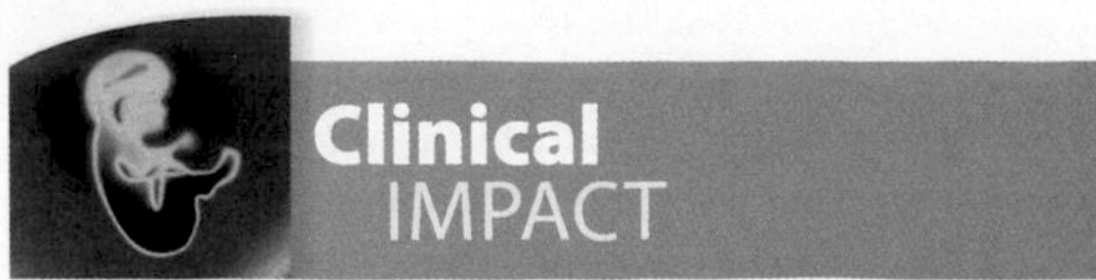
Clinical
IMPACT

Drugs That Block Calcium Channels

Various chemical agents, such as nifedipine and verapamil (ver-ap′ă-mil), block voltage-gated Ca^{2+} channels. Voltage-gated Ca^{2+} channel–blocking agents prevent the movement of Ca^{2+} through voltage-gated Ca^{2+} channels into the cell; for that reason, they are called **calcium channel blockers.** Some calcium channel blockers are widely used to treat various cardiac disorders, including tachycardia and certain arrhythmias (table 20.1). Calcium channel blockers slow the development of the pacemaker potential and thus reduce the heart rate. If action potentials arise prematurely within the SA node or other areas of the heart, calcium channel blockers reduce that tendency. Calcium channel blockers also reduce the amount of work performed by the heart because less Ca^{2+} enters cardiac muscle cells to activate the contractile mechanism. On the other hand, epinephrine and norepinephrine increase the heart rate and its force of contraction by opening voltage-gated Ca^{2+} channels.

its pacemaker cells generate action potentials at a faster rate than other potential pacemaker cells, producing a heart rate of 70–80 beats per minute (bpm). An **ectopic focus** (ek-top′ik fō′kŭs; pl. *foci*, fō′sī) is any part of the heart other than the SA node that generates a heartbeat. For example, if the SA node does not function properly, the part of the heart that can produce action potentials at the next highest frequency is the AV node, which produces a heart rate of 40–60 bpm. Another cause of an ectopic focus is blockage of the conducting pathways between the SA node and other parts of the heart. For example, if action potentials do not pass through the AV node, an ectopic focus can develop in an AV bundle, resulting in a heart rate of only 30 bpm.

Ectopic foci can also appear when the rate of action potential generation in cardiac muscle cells outside the SA node becomes enhanced. For example, when cells are injured, their plasma membranes become more permeable, resulting in depolarization. Inflammation or lack of adequate blood flow to cardiac muscle tissue can injure cardiac muscle cells. These injured cells can be the source of ectopic action potentials. Also, alterations in blood levels of K^+ and Ca^{2+} can change the cardiac muscle membrane potential, and certain drugs, such as those that mimic the effect of epinephrine on the heart, can alter cardiac muscle membrane permeability. Changes in cardiac muscle cells' membrane potentials or permeability can produce ectopic foci.

Refractory Periods of Cardiac Muscle

Cardiac muscle, like skeletal muscle, has **refractory** (rē-frak′tōr-ē) **periods** associated with its action potentials. During the **absolute refractory period,** the cardiac muscle cell is completely insensitive to further stimulation. During the **relative refractory period,** the cell is sensitive to stimulation, but a greater stimulation than normal is required to cause an action potential. Because the plateau phase of the action potential in cardiac muscle delays repolarization to the resting membrane potential, the refractory period is prolonged. The long refractory period ensures that contraction and most of relaxation are complete before another action potential can be initiated. This prevents tetanic contractions in cardiac muscle and is responsible for rhythmic contractions.

> **Predict 4**
>
> Predict the consequences if cardiac muscle could undergo tetanic contraction.

Electrocardiogram

Action potentials conducted through the myocardium during the cardiac cycle produce electrical currents that can be measured at the body surface. Electrodes placed on the body surface and attached to an appropriate recording device can detect small voltage changes resulting from action potentials in the cardiac muscle. The electrodes do not detect individual action potentials; rather, they detect a summation of all the action potentials transmitted by the cardiac muscle cells through the heart at a given time. The summated record of the cardiac action potentials is an **electrocardiogram** (**ECG** or **EKG;** figure 20.16).

The ECG is not a direct measurement of mechanical events in the heart, and neither the force of contraction nor blood pressure can be determined from it. However, each deflection in the ECG record indicates an electrical event within the heart that is correlated with a subsequent mechanical event. Consequently, electrocardiography

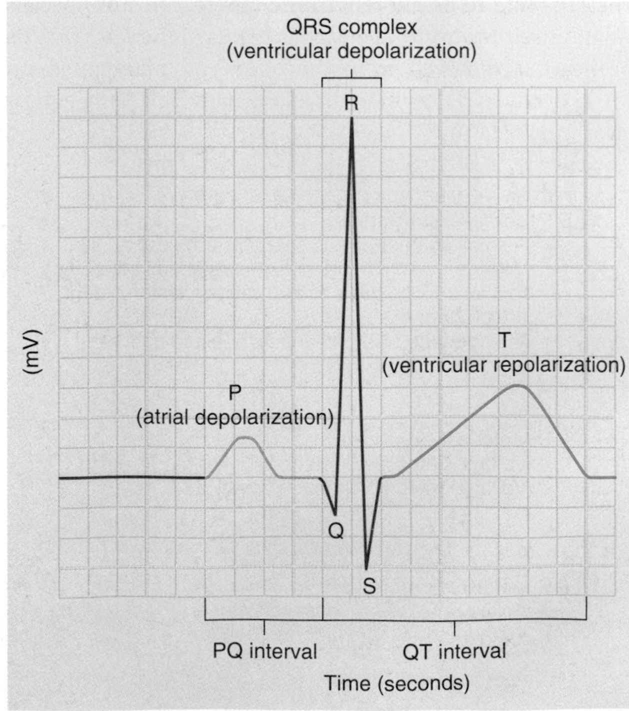
QRS complex
(ventricular depolarization)
R
(mV)
P
(atrial depolarization)
T
(ventricular repolarization)
Q
S
PQ interval
QT interval
Time (seconds)

FIGURE 20.16 Electrocardiogram

The major waves and intervals of an electrocardiogram. Each thin, horizontal line on the ECG recording represents 1 mV, and each thin, vertical line represents 0.04 second.

Clinical
IMPACT

Alterations in the Electrocardiogram

Elongation of the PR interval can result from three events: (1) a delay in action potential conduction through the atrial muscle because of damage, such as that caused by **ischemia** (is-kē′mē-ă), which is obstruction of the blood supply to the walls of the heart; (2) a delay in action potential conduction through atrial muscle because of a dilated atrium; or (3) a delay in action potential conduction through the AV node and bundle because of ischemia, compression, or necrosis of the AV node or bundle. These conditions result in slow conduction of action potentials through the bundle branches. An unusually long QT interval reflects the abnormal conduction of action potentials through the ventricles, which can result from myocardial infarctions or from an abnormally enlarged left or right ventricle.

Altered forms of the electrocardiogram due to cardiac abnormalities include complete heart block, premature ventricular contraction, bundle branch block, atrial fibrillation, and ventricular fibrillation (figure 20.17).

is extremely valuable in diagnosing a number of abnormal cardiac rhythms (arrhythmias; table 20.1) and other abnormalities, particularly because it is painless, easy to record, and noninvasive (does not require surgery). In addition to abnormal heart rates and rhythms, ECG analysis can reveal abnormal conduction pathways, hypertrophy or atrophy of portions of the heart, and the approximate location of damaged cardiac muscle (table 20.1).

The normal ECG consists of a P wave, a QRS complex, and a T wave (figure 20.16). The **P wave,** which is the result of action potentials that cause depolarization of the atrial myocardium, signals the onset of atrial contraction. The **QRS complex** is composed of three individual waves: the Q, R, and S waves. The QRS complex results from ventricular depolarization and signals the onset of ventricular contraction. The **T wave** represents repolarization of the ventricles and precedes ventricular relaxation. A wave representing repolarization of the atria cannot be seen because it occurs during the QRS complex.

The time between the beginning of the P wave and the beginning of the QRS complex is the PQ interval, commonly called the PR interval because the Q wave is often very small. During the PR interval, which lasts approximately 0.16 second, the atria contract and begin to relax. The ventricles begin to depolarize at the end of the PR interval. The QT interval extends from the beginning of the QRS complex to the end of the T wave, lasting approximately 0.36 second. The QT interval represents the approximate length of time required for the ventricles to contract and begin to relax.

ASSESS YOUR PROGRESS

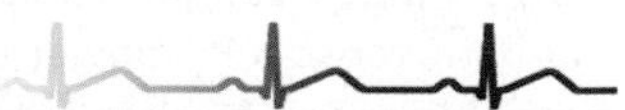

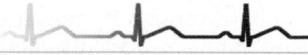

23. *For cardiac muscle action potentials, describe ion movement during the depolarization, early repolarization, plateau, and final repolarization phases.*
24. *Why is cardiac muscle referred to as autorhythmic? What are ectopic foci?*
25. *How does the depolarization of pacemaker cells differ from the depolarization of other cardiac cells? What is the pacemaker potential?*
26. *How is the prolonged refractory period generated in cardiac muscle? What is the advantage of a prolonged refractory period?*
27. *What does an ECG measure? Name the waves and intervals produced by an ECG, and state what events occur during each wave and interval.*

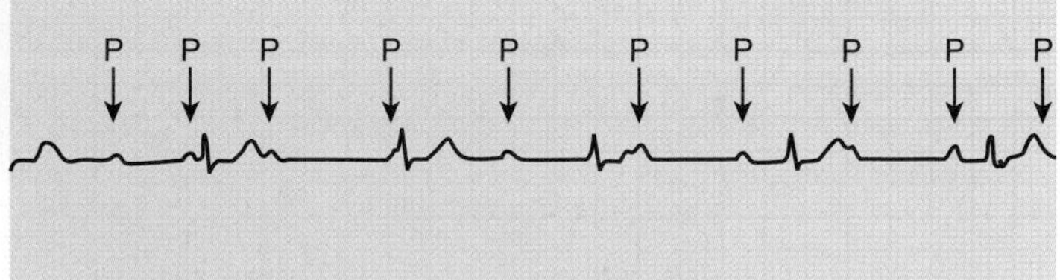
P
P
P
P
P
P
P
P
P
P

Complete heart block (P waves and QRS complexes are not coordinated)

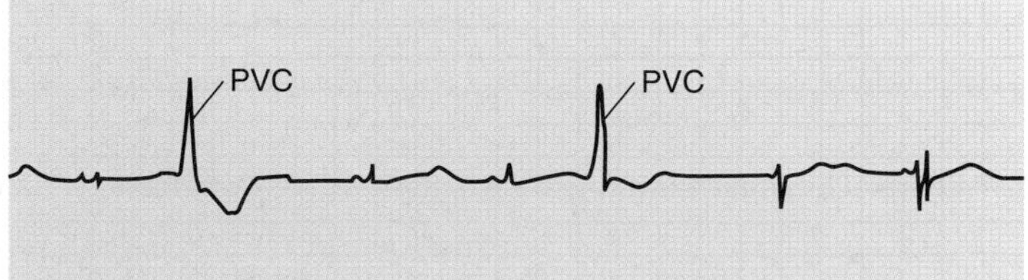
PVC
PVC

Premature ventricular contraction (PVC) (no P waves precede PVCs)

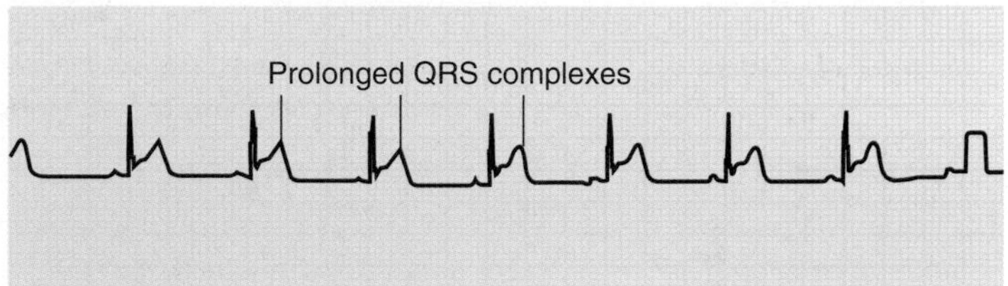
Prolonged QRS complexes

Bundle branch block

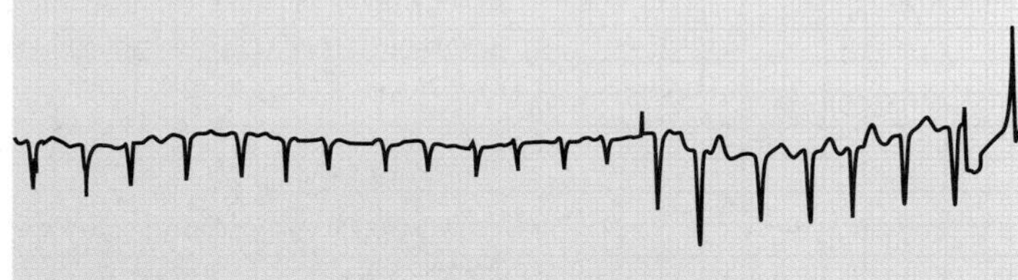

Atrial fibrillation (no clear P waves and rapid QRS complexes)

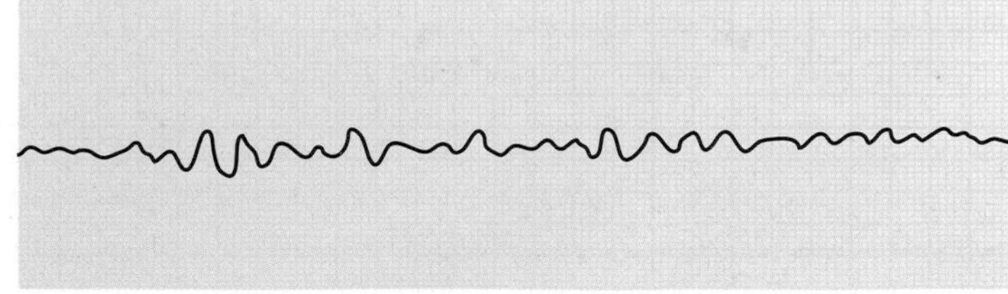

Ventricular fibrillation (no P, QRS, or T waves)

FIGURE 20.17 Alterations in an Electrocardiogram

TABLE 20.1 Major Cardiac Arrhythmias

Conditions	Symptoms	Possible Causes
Abnormal Heart Rhythms		
Tachycardia	Heart rate in excess of 100 beats per minute (bpm)	Elevated body temperature; excessive sympathetic stimulation; toxic conditions
Paroxysmal atrial tachycardia	Sudden increase in heart rate to 95–150 bpm for a few seconds or even for several hours; P wave precedes every QRS complex; P wave inverted and superimposed on T wave	Excessive sympathetic stimulation; abnormally elevated permeability of slow channels
Ventricular tachycardia	Frequently causes fibrillation	Often associated with damage to AV node or ventricular muscle
Abnormal Rhythms Resulting from Ectopic Action Potentials		
Atrial flutter	300 P waves/min; 125 QRS complexes/min, resulting in two or three P waves (atrial contraction) for every QRS complex (ventricular contraction)	Ectopic action potentials in the atria
Atrial fibrillation	No P waves; normal QRS complexes; irregular timing; ventricles constantly stimulated by atria; reduced pumping effectiveness and filling time	Ectopic action potentials in the atria
Ventricular fibrillation	No QRS complexes; no rhythmic contraction of the myocardium; many patches of asynchronously contracting ventricular muscle	Ectopic action potentials in the ventricles
Bradycardia	Heart rate less than 60 bpm	Elevated stroke volume in athletes; excessive vagal stimulation; carotid sinus syndrome
Sinus arrhythmia	Heart rate varies 5% during respiratory cycle and up to 30% during deep respiration.	Cause not always known; occasionally caused by ischemia or inflammation or associated with cardiac failure
SA node block	Cessation of P wave; new low heart rate due to AV node acting as pacemaker; normal QRS complex and T wave	Ischemia; tissue damage due to infarction; causes unknown
AV Node Block		
First-degree	PR interval greater than 0.2 second	Inflammation of AV bundle
Second-degree	PR interval 0.25–0.45 second; some P waves trigger QRS complexes and others do not; 2:1, 3:1, and 3:2 P wave/QRS complex ratios may occur	Excessive vagal stimulation
Third-degree (complete heart block)	P wave dissociated from QRS complex; atrial rhythm approximately 100 bpm; ventricular rhythm less than 40 bpm	Ischemia of AV nodal fibers or compression of AV bundle
Premature atrial contractions	Occasional shortened intervals between contractions; frequently occurs in healthy people P wave superimposed on QRS complex	Excessive smoking; lack of sleep; too much caffeine; alcoholism
Premature ventricular contractions (PVCs)	Prolonged QRS complex; exaggerated voltage because only one ventricle may depolarize; inverted T wave; increased probability of fibrillation	Ectopic foci in ventricles; lack of sleep; too much caffeine, irritability; occasionally occurs with coronary thrombosis

Abbreviations: SA = sinoatrial; AV = atrioventricular.

20.7 Cardiac Cycle

LEARNING OUTCOMES

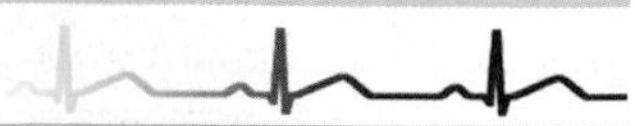

After reading this section, you should be able to

A. Describe the cardiac cycle and the relationship among the contraction of each of the chambers, the opening and closing of valves, the pressure in each of the chambers, the phases of the electrocardiogram, and the heart sounds.

B. Discuss the heart sounds and their significance.

The right and left halves of the heart can be viewed as two separate pumps that work together. Each pump consists of a "primer pump" (the atrium) and a "power pump" (the ventricle). Both atrial primer pumps complete the filling of the ventricles with blood, and both ventricular power pumps produce the major force that causes blood to flow through the pulmonary and systemic arteries. The term **cardiac cycle** refers to the repetitive pumping process that begins with the onset of cardiac muscle contraction and ends with the beginning of the next contraction (figure 20.18; table 20.2).

Pressure changes produced within the heart chambers as a result of cardiac muscle contraction move blood from areas of higher pressure to areas of lower pressure.

The duration of the cardiac cycle varies considerably among humans and during an individual's lifetime. It can be as short as 0.25–0.3 second in a newborn or as long as 1 or more seconds in a well-trained athlete. The normal cardiac cycle of 0.7–0.8 second depends on the capability of cardiac muscle to contract and on the functional integrity of the conducting system.

As you study the events of the cardiac cycle listed in table 20.2, keep in mind that the term **systole** (sis′tō-lē) means to contract, and **diastole** (dī-as′tō-lē) means to dilate. Therefore, **atrial systole** is contraction of the atrial myocardium, and **atrial diastole** is relaxation of the atrial myocardium. Similarly, **ventricular systole** is contraction of the ventricular myocardium, and **ventricular diastole** is relaxation of the ventricular myocardium. When the terms *systole* and *diastole* are used alone, they refer to ventricular systole and diastole.

During the cardiac cycles, changes in chamber pressure and the opening and closing of the heart valves determine the direction of blood movement. As the cardiac cycle is described, it is important to focus on these pressure changes and heart valve movements. Before we start, it is also important to have a clear image of the state of the heart. At the beginning of the cardiac cycle, the atria and ventricles are relaxed, the AV valves are open, and the semilunar valves are closed.

At rest, most of the blood movement into the chambers is a passive process resulting from the greater blood pressure in the veins than in the heart chambers. As the blood moves into the atria, much of it flows into the ventricles because the AV valves are open and atrial pressure is slightly greater than ventricular pressure. This time period when blood is passively moving into the ventricles is called *passive ventricular filling*.

The SA node generates an AP that stimulates atrial contraction and begins the cardiac cycle. As the atria contract, they carry out the primer pump function by forcing more blood into the ventricles. This period is referred to as *active ventricular filling* (figure 20.18, *step 1*).

The AP passes to the AV node, down the AV bundle, bundle branches, and Purkinje fibers, stimulating ventricular systole. As the ventricles contract, ventricular pressures increase, causing blood to flow toward the atria and close the AV valves. Ventricular contraction continues and ventricular pressures rise; however, because all the valves are closed, no blood flows from the ventricles. This brief interval is called the *period of isovolumetric contraction* because the volume of blood in the ventricles does not change, even though the ventricles are contracting (figure 20.18, *step 2*). Ventricular contraction continues, and ventricular pressures become greater than the pressures in the pulmonary trunk and aorta. As a result, the semilunar valves are pushed open, and blood flows from the ventricles into those arteries. This time period, when blood moves from the ventricles into the arteries, is called the *period of ejection* (figure 20.18, *step 3*).

As ventricular diastole begins, the ventricles relax, and ventricular pressures decrease below the pressures in the pulmonary trunk and aorta. Consequently, blood begins to flow back toward the ventricles, causing the semilunar valves to close (figure 20.18, *step 4*). With closure of the semilunar valves, all the heart valves are closed, and no blood flows into the relaxing ventricles during the *period of isovolumetric relaxation*.

Atrial diastole began during ventricular systole, and as the atria relaxed blood flowed into them from the veins. As the ventricles continue to relax, ventricular pressures drop below atrial pressures, and the AV valves open. Passive ventricular filling begins again (figure 20.18, *step 5*). Note that, once the ventricles have fully relaxed, the state of the heart is the same as when the cardiac cycle began, all chambers are relaxed, the AV valves are open and the semilunar valves are closed. With the next stimulus from the SA node, another cardiac cycle will begin (figure 20.18, *step 1*).

Events Occurring During the Cardiac Cycle

Figure 20.19 graphs the main events of the cardiac cycle and should be examined from top to bottom:

- An ECG indicates the electrical events that cause contraction and relaxation of the atria and ventricles.
- The pressure graph shows the pressure changes within the left atrium, left ventricle, and aorta resulting from atrial and ventricular contraction and relaxation. Although pressure changes in the right side of the heart are not shown, they are similar to those in the left side, only lower.
- The volume graph presents the changes in left ventricular volume as blood flows into and out of the left ventricle as a result of the pressure changes.
- The sound graph records the closing of valves caused by blood flow.

See also figure 20.18 for illustrations of the valves and blood flow and table 20.2 for a summary of the events of each period.

Atrial Systole and Active Ventricular Filling

Before the cardiac cycle begins, all chambers are relaxed and blood is flowing from the veins into the atria and passively into the ventricles. Depolarization of the SA node generates action potentials, which spread over the atria, producing the P wave and stimulating both atria to contract (atrial systole). The atria contract during the last one-third of ventricular diastole and complete ventricular filling.

Under most conditions, the atria function primarily as reservoirs, and the ventricles can pump sufficient blood to maintain homeostasis even if the atria do not contract at all. During exercise, however, the heart pumps 300–400% more blood than during rest. As heart rate increases during exercise, atrial contraction is important for ventricular filling because less time is available for passive ventricular filling. Therefore, it is during exercise that the pumping action of the atria becomes important for maintaining the pumping efficiency of the heart.

Ventricular Systole: Period of Isovolumetric Contraction

Completion of the QRS complex initiates contraction of the ventricles. Ventricular pressure rapidly increases, resulting in closure of the AV valves. During the previous ventricular diastole, the

TABLE 20.2 Summary of the Events of the Cardiac Cycle

	Atrial Systole Ventricular Diastole Active Ventricular Filling	Atrial Diastole Ventricular Systole Period of Isovolumetric Contraction
Time Period	Contraction of the atria pumps blood into the ventricles.	The ventricles begin to contract, but ventricular volume does not change.
Condition of Valves	The semilunar valves are closed; the AV valves are opened (figure 20.18, *step 1*).	The semilunar valves are closed; the AV valves are closed (figure 20.18, *step 2*).
ECG	The P wave is completed and the atria are stimulated to contract. Action potentials are delayed in the AV node for 0.11 second, allowing time for the atria to contract. The **QRS complex** begins as action potentials are propagated from the AV node to the ventricles.	The QRS complex is completed, and the ventricles are depolarized. As a result, the ventricles begin to contract. Atrial repolarization is masked by the QRS complex. The atria are relaxed (atrial diastole).
Atrial Pressure Graph	Atrial contraction (systole) causes an increase in atrial pressure, and blood is forced to flow from the atria into the ventricles.	Atrial pressure decreases in the relaxed atria. When atrial pressure is less than venous pressure, blood flows into the atria. Atrial pressure increases briefly as the contracting ventricles push blood back toward the atria.
Ventricular Pressure Graph	Atrial contraction (systole) and the movement of blood into the ventricles cause a slight increase in ventricular pressure.	Ventricular contraction causes an increase in ventricular pressure, which causes blood to flow toward the atria, closing the AV valves. Ventricular pressure increases rapidly.
Aortic Pressure Graph	Aortic pressure gradually decreases as blood runs out of the aorta into other systemic blood vessels.	Just before the semilunar valves open, pressure in the aorta decreases to its lowest value, called the **diastolic pressure** (approximately 80 mm Hg).
Volume Graph	Atrial contraction (systole) completes ventricular filling during the last one-third of diastole. The amount of blood in a ventricle at the end of ventricular diastole is called the **end-diastolic volume.**	During the **period of isovolumetric contraction,** ventricular volume does not change because the semilunar and AV valves are closed.
Sound Graph		Blood flowing from the ventricles toward the atria closes the AV valves. Vibrations of the valves and the turbulent flow of blood produce the **first heart sound,** which marks the beginning of ventricular systole.

Abbreviation: AV = atrioventricular.

ventricles were filled with 120–130 mL of blood, which is called the **end-diastolic volume.** Ventricular volume does not change during the period of isovolumetric contraction because all the heart valves are closed.

Ventricular Systole: Period of Ejection

As soon as ventricular pressures exceed the pressures in the aorta and pulmonary trunk, the semilunar valves open. The aortic semilunar valve opens at approximately 80 mm Hg ventricular pressure, whereas the pulmonary semilunar valve opens at approximately 8 mm Hg. Although the pressures are different, both valves open at nearly the same time.

As blood flows from the ventricles during the period of ejection, the left ventricular pressure continues to climb to approximately 120 mm Hg, and the right ventricular pressure increases to approximately 25 mm Hg. The larger left ventricular pressure causes blood to flow throughout the body (systemic circulation), whereas the lower right ventricular pressure causes blood to flow through the lungs

Atrial Diastole		
Ventricular Systole	**Ventricular Diastole**	
Period of Ejection	**Period of Isovolumetric Relaxation**	**Passive Ventricular Filling**
The ventricles continue to contract, and blood is pumped out of the ventricles.	The ventricles relax, but ventricular volume does not change.	Blood flows into the ventricles because blood pressure is higher in the veins and atria than in the relaxed ventricles.
The semilunar valves are opened; the AV valves are closed (figure 20.18, *step 3*).	The semilunar valves are closed; the AV valves are closed (figure 20.18, *step 4*).	The semilunar valves are closed; the AV valves are opened (see figure 20.18, *step 5*).
The **T wave** results from ventricular repolarization.	The T wave is completed, and the ventricles are repolarized. The ventricles relax.	The **P wave** is produced when the SA node generates action potentials and a wave of depolarization begins to propagate across the atria.
Atrial pressure increases gradually as blood flows from the veins into the relaxed atria.	Atrial pressure continues to increase gradually as blood flows from the veins into the relaxed atria.	After the AV valves open, atrial pressure decreases as blood flows out of the atria into the relaxed ventricles.
Ventricular pressure becomes greater than pressure in the aorta as the ventricles continue to contract. The semilunar valves are pushed open as blood flows out of the ventricles. Ventricular pressure peaks as the ventricles contract maximally; then pressure decreases as blood flow out of the ventricles decreases.	Elastic recoil of the aorta pushes blood back toward the heart, causing the semilunar valves to close. After closure of the semilunar valves, the pressure in the relaxing ventricles rapidly decreases.	No significant change occurs in ventricular pressure during this time period.
As ventricular contraction forces blood into the aorta, pressure in the aorta increases to its highest value, called the **systolic pressure** (approximately 120 mm Hg).	After the semilunar valves close, elastic recoil of the aorta causes a slight increase in aortic pressure, producing the **dicrotic notch,** or *incisura*.	Aortic pressure gradually decreases as blood runs out of the aorta into other systemic blood vessels.
After the semilunar valves open, blood volume decreases as blood flows out of the ventricles during the **period of ejection.** The amount of blood left in a ventricle at the end of the period of ejection is called the **end-systolic volume.**	During the **period of isovolumetric relaxation,** ventricular volume does not change because the semilunar and AV valves are closed.	After the AV valves open, blood flows from the atria and veins into the ventricles because of pressure differences. Most ventricular filling occurs during the first one-third of diastole. Little ventricular filling occurs during the middle one-third of diastole.
	Blood flowing from the ventricles toward the aorta and pulmonary trunk closes the semilunar valves. Vibrations of the valves and the turbulent flow of blood produce the **second heart sound,** which marks the beginning of ventricular diastole.	Sometimes the turbulent flow of blood into the ventricles produces a **third heart sound.**

(pulmonary circulation). Even though the pressure generated by the left ventricle is much higher than that of the right ventricle, the amount of blood pumped by each is almost the same.

Predict 5

Which ventricle has the thickest wall? Why is it important for each ventricle to pump approximately the same volume of blood?

During the first part of ejection, blood flows rapidly out of the ventricles. Toward the end of ejection, ventricular pressure decreases due to reduced blood flow, despite continued ventricular contraction. By the end of ejection, the volume has decreased to 50–60 mL, which is called the **end-systolic volume.**

Ventricular Diastole: Period of Isovolumetric Relaxation

Completion of the T wave results in ventricular repolarization and relaxation. The already decreasing ventricular pressure falls very

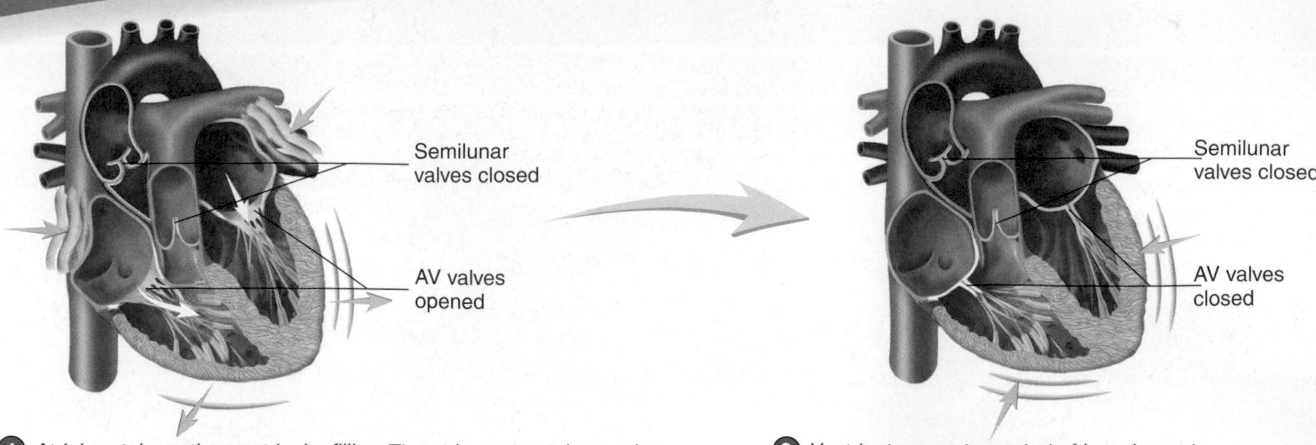
Semilunar valves closed
AV valves opened
Semilunar valves closed
AV valves closed

(1) *Atrial systole: active ventricular filling*. The atria contract, increasing atrial pressure and completing ventricular filling while the ventricles are relaxed.

(2) *Ventricular systole: period of isovolumetric contraction.* The atria are relaxed, and blood flows into them from the veins. Ventricular contraction causes ventricular pressure to increase and causes the AV valves to close, which is the beginning of ventricular systole. The semilunar valves were closed in the previous diastole and remain closed during this period.

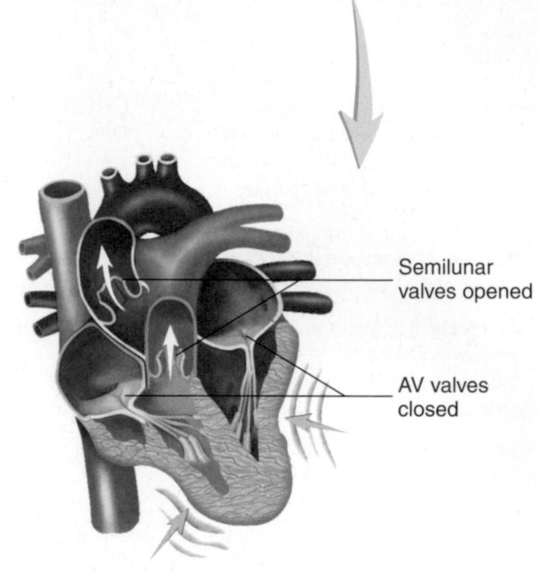
Semilunar valves opened
AV valves closed

(3) *Ventricular systole: period of ejection.* Continued ventricular contraction causes a greater increase in ventricular pressure, which pushes blood out of the ventricles, causing the semilunar valves to open.

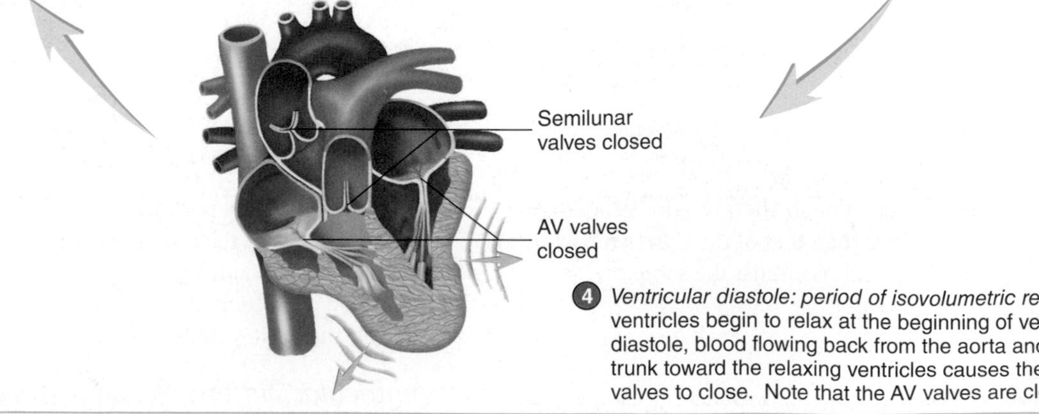
Semilunar valves closed
AV valves closed

(4) *Ventricular diastole: period of isovolumetric relaxation.* As the ventricles begin to relax at the beginning of ventricular diastole, blood flowing back from the aorta and pulmonary trunk toward the relaxing ventricles causes the semilunar valves to close. Note that the AV valves are closed also.

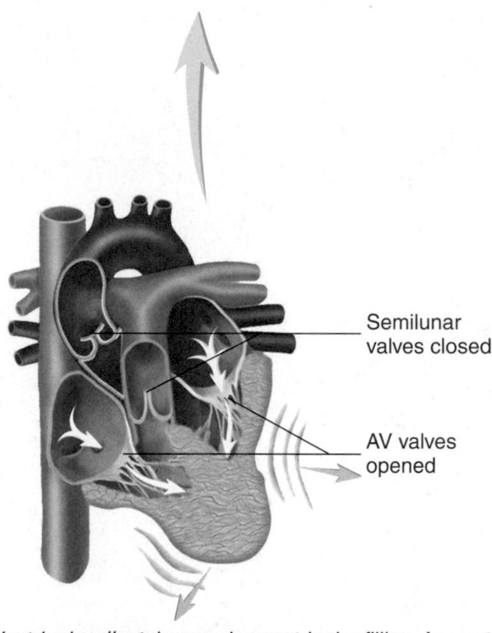
Semilunar valves closed
AV valves opened

(5) *Ventricular diastole: passive ventricular filling*. As ventricular relaxation continues, the AV valves open, and blood flows from the atria into the relaxing ventricles, accounting for most of the ventricular filling.

PROCESS FIGURE 20.18 Cardiac Cycle

The cardiac cycle is a repeating series of contraction and relaxation that moves blood through the heart (*AV* = atrioventricular).

FUNDAMENTAL **Figure**

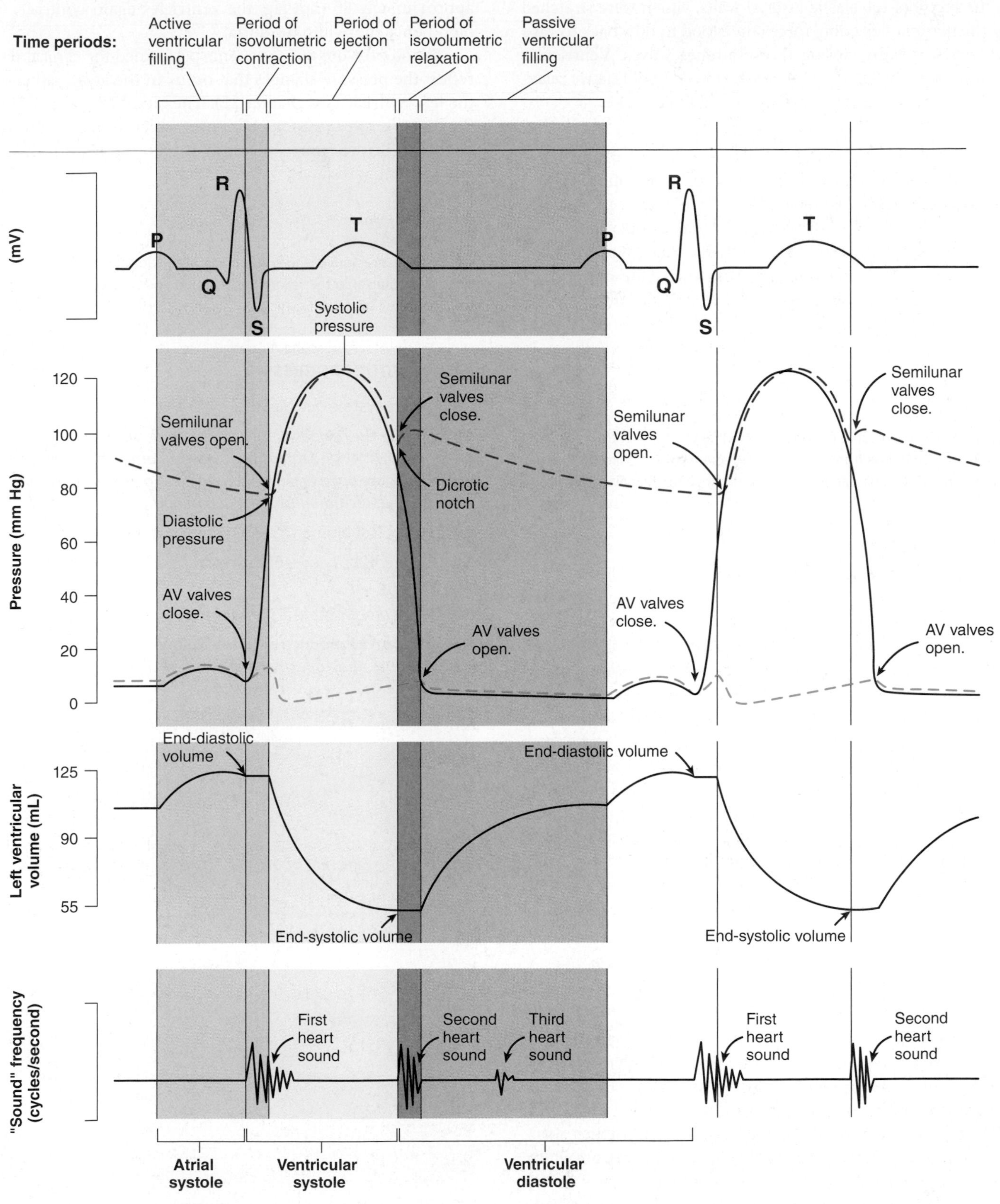

Time periods:
Active ventricular filling
Period of isovolumetric contraction
Period of ejection
Period of isovolumetric relaxation
Passive ventricular filling
(mV)
P
R
Q
S
T
Systolic pressure
Pressure (mm Hg)
120
100
80
60
40
20
0
Semilunar valves open.
Diastolic pressure
AV valves close.
Semilunar valves close.
Dicrotic notch
AV valves open.
Left ventricular volume (mL)
125
90
55
End-diastolic volume
End-systolic volume
"Sound" frequency (cycles/second)
First heart sound
Second heart sound
Third heart sound
Atrial systole
Ventricular systole
Ventricular diastole

FIGURE 20.19 AP|R **Events Occurring During the Cardiac Cycle**

The cardiac cycle is divided into five time periods (top). From top to bottom: the electrocardiogram; pressure changes for the left atrium (blue line), left ventricle (black line), and aorta (red line); left ventricular volume curve; and heart sounds.

rapidly as the ventricles suddenly relax. When the ventricular pressures fall below the pressures in the aorta and pulmonary trunk, the recoil of the elastic arterial walls, which were stretched during the period of ejection, forces the blood to flow back toward the ventricles, thereby closing the semilunar valves. Ventricular volume does not change during the period of isovolumetric relaxation because all the heart valves are closed at this time.

Ventricular Diastole: Passive Ventricular Filling

During ventricular systole and the period of isovolumetric relaxation, the relaxed atria fill with blood. As ventricular pressure drops below atrial pressure, the atrioventricular valves open and allow blood to flow from the atria into the ventricles. Blood flows from the area of higher pressure in the veins and atria toward the area of lower pressure in the relaxed ventricles. Most ventricular filling occurs during the first one-third of ventricular diastole. At the end of passive ventricular filling, the ventricles are approximately 70% filled.

Predict 6

Fibrillation is abnormal, rapid contractions of different parts of the heart that prevent the heart muscle from contracting as a single unit. Explain why atrial fibrillation does not immediately cause death but ventricular fibrillation does.

Heart Sounds

The pumping heart produces distinct sounds, as revealed by using a stethoscope (figure 20.19, *bottom*). These sounds are best heard by applying the stethoscope at particular sites in relation to the heart valves (figure 20.20). The **first heart sound** is a low-pitched sound, often described as "lubb." It is caused by vibration of the atrioventricular valves and surrounding fluid as the valves close at the beginning of ventricular systole. The **second heart sound** is a higher-pitched sound often described as "dupp." It results from closure of the aortic and pulmonary semilunar valves, at the beginning of ventricular diastole. Systole is therefore approximately the time between the first and second heart sounds. Diastole, which lasts somewhat longer, is approximately the time between the second heart sound and the next first heart sound.

A faint **third heart sound** can be heard in some normal people, particularly those who are thin and young. It is caused by blood flowing in a turbulent fashion into the ventricles, and it can be detected near the end of the first one-third of diastole.

Aortic Pressure Curve

The elastic walls of the aorta are stretched as blood is ejected into the aorta from the left ventricle. Aortic pressure remains slightly below ventricular pressure during this period of ejection. As ventricular pressure drops below that in the aorta, blood flows back toward the ventricle because of the elastic recoil of the aorta. Consequently, the aortic semilunar valve closes, and pressure within the aorta increases slightly, producing a **dicrotic** (dī-krot′ik) **notch** in the aortic pressure curve (see figure 20.19). The term *dicrotic* means double-beating; when increased pressure caused by recoil is large, a double pulse can be felt. The dicrotic notch is also called an *incisura* (in′sī-soo′ră; a cutting into). Aortic pressure then gradually falls throughout the rest of ventricular diastole as blood flows through the peripheral vessels. When aortic pressure has fallen to approximately 80 mm Hg, the ventricles again contract, forcing blood once more into the aorta.

Blood pressure measurements performed for clinical purposes reflect the pressure changes that occur in the aorta rather than in the left ventricle (see chapter 21). The blood pressure in the aorta fluctuates between systolic pressure, which is about 120 mm Hg, and diastolic pressure, which is about 80 mm Hg, for the average young adult at rest.

Predict 7

Predict the pressure changes that occur in the aorta, the left ventricle, and the left atrium after the second heart sound and before the first heart sound of the next cardiac cycle.

ASSESS YOUR PROGRESS

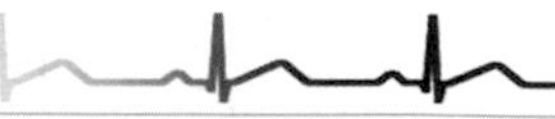

28. *Define* systole *and* diastole.
29. *List the five periods of the cardiac cycle (see figure 20.18, figure 20.19, and table 20.2). State whether the AV and semilunar valves are open or closed during each period and the phase of the electrocardiogram for each period.*
30. *Define* isovolumetric. *When does most ventricular filling occur?*
31. *Differentiate between end-diastolic volume and end-systolic volume.*
32. *What produces the first and the second heart sounds?*
33. *Explain the production of the following in the aorta: systolic pressure, diastolic pressure, and the dicrotic notch (incisura).*

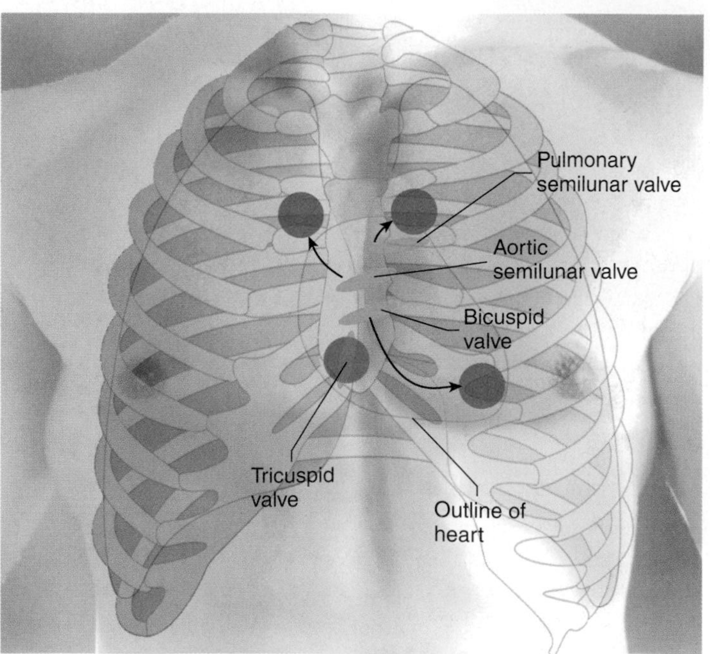
Pulmonary semilunar valve
Aortic semilunar valve
Bicuspid valve
Tricuspid valve
Outline of heart

FIGURE 20.20 Location of the Heart Valves in the Thorax
Surface markings of the heart in the male. The positions of the four heart valves are indicated by blue ellipses, and the sites where the sounds of the valves are best heard with the stethoscope are indicated by pink circles.

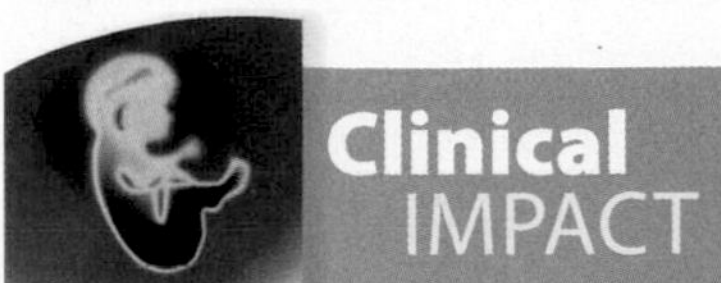

Clinical IMPACT Abnormal Heart Sounds

Heart sounds provide important information about the normal function of the heart and help clinicians diagnose cardiac abnormalities. Abnormal heart sounds are called **murmurs** (mer′merz), and certain murmurs are important indicators of specific cardiac abnormalities. For example, an **incompetent valve** (also called a valvular insufficiency) leaks significantly. After an incompetent valve closes, blood flows through it but in a reverse direction, called regurgitation. Regurgitation results in turbulence, which causes a gurgling or swishing sound immediately after the valve closes. An incompetent tricuspid valve or bicuspid valve makes a swish sound immediately after the first heart sound, and the first heart sound may be muffled. An incompetent aortic or pulmonary semilunar valve results in a swish sound immediately after the second heart sound.

Stenosed (sten′ōzd) **valves** have an abnormally narrow opening and produce abnormal heart sounds. Blood flows through stenosed valves very turbulently and produces a rushing sound before the valve closes. For example, a stenosed atrioventricular valve produces a rushing sound immediately before the first heart sound, and a stenosed semilunar valve produces a rushing sound immediately before the second heart sound.

Inflammation of the heart valves, resulting from a condition such as rheumatic fever, can cause valves to become either incompetent or stenosed. In addition, myocardial infarctions that make papillary muscles nonfunctional can cause bicuspid or tricuspid valves to be incompetent. Heart murmurs also result from congenital abnormalities in the hearts of infants. Two examples are septal defects in the heart and patent ductus arteriosus (see chapter 29).

Either incompetent or stenosed valves increase the amount of work the cardiac muscle must perform. Consequently, these conditions can lead to heart failure.

20.8 Mean Arterial Blood Pressure

LEARNING OUTCOMES

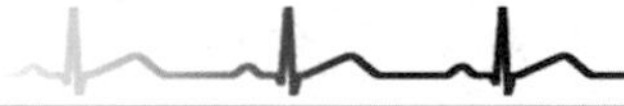

After reading this section, you should be able to

A. Define *mean arterial pressure, cardiac output,* and *peripheral resistance.*

B. Explain the role of MAP in causing blood flow.

Blood pressure is necessary to move the blood and therefore is critical to the maintenance of homeostasis. Blood flows from areas of higher to lower pressure. For example, during one cardiac cycle, blood flows from the higher pressure in the aorta, resulting from contraction of the left ventricle, through the systemic circulation, toward the lower pressure in the relaxed right atrium.

Mean arterial pressure (MAP) is slightly less than the average of the systolic and diastolic pressures in the aorta. It is proportional to cardiac output times peripheral resistance. **Cardiac output (CO),** or *minute volume*, is the amount of blood pumped by the heart per minute, and **peripheral resistance (PR)** is the total resistance against which blood must be pumped:

$$\text{MAP} = \text{CO} \times \text{PR}$$

Changes in cardiac output and peripheral resistance can alter mean arterial pressure (figure 20.21). Cardiac output is discussed in this chapter, and peripheral resistance is explained in chapter 21.

Cardiac output is equal to heart rate times stroke volume. **Heart rate (HR)** is the number of times the heart beats (contracts) per minute. **Stroke volume (SV)** is the volume of blood pumped during each heartbeat (cardiac cycle), and it is equal to end-diastolic volume minus end-systolic volume. During diastole, blood flows from the atria into the ventricles, and end-diastolic volume normally increases to approximately 125 mL. After the ventricles partially empty during systole, end-systolic volume decreases to approximately 55 mL. Thus, the stroke volume is equal to 70 mL (125 – 55).

To better understand stroke volume, imagine that you are squeezing a sponge under a running water faucet. As you relax your fingers, the sponge fills with water; as you contract your fingers, the sponge releases water. Even after you have squeezed the sponge, some water remains in it. In this analogy, the amount of water you squeeze out of the sponge (stroke volume) is the difference between the amount of water in the sponge when your hand is relaxed (end-diastolic volume) and the amount left in the sponge after you squeeze it (end-systolic volume).

Stroke volume can be increased by increasing end-diastolic volume or by decreasing end-systolic volume (figure 20.21). During exercise, end-diastolic volume increases because of an increase in **venous return,** which is the amount of blood returning to the heart from the systemic circulation. End-systolic volume decreases because the heart contracts more forcefully. For example, stroke volume can increase from a resting value of 70 mL to an exercising value of 115 mL by increasing end-diastolic volume to 145 mL and decreasing end-systolic volume to 30 mL.

Under resting conditions, the heart rate is approximately 72 bpm, and the stroke volume is approximately 70 mL/beat, although these values can vary considerably from person to person. Therefore, the cardiac output is

$$\begin{aligned}\text{CO} &= \text{HR} \times \text{SV} \\ &= 72 \text{ bpm} \times 70 \text{ mL/beat} \\ &= 5040 \text{ mL/min (approximately 5 L/min)}\end{aligned}$$

During exercise, the heart rate can increase to 190 bpm, and the stroke volume can increase to 115 mL. Consequently, cardiac output is

$$\begin{aligned}\text{CO} &= 190 \text{ bpm} \times 115 \text{ mL/beat} \\ &= 21{,}850 \text{ mL/min (approximately 22 L/min)}\end{aligned}$$

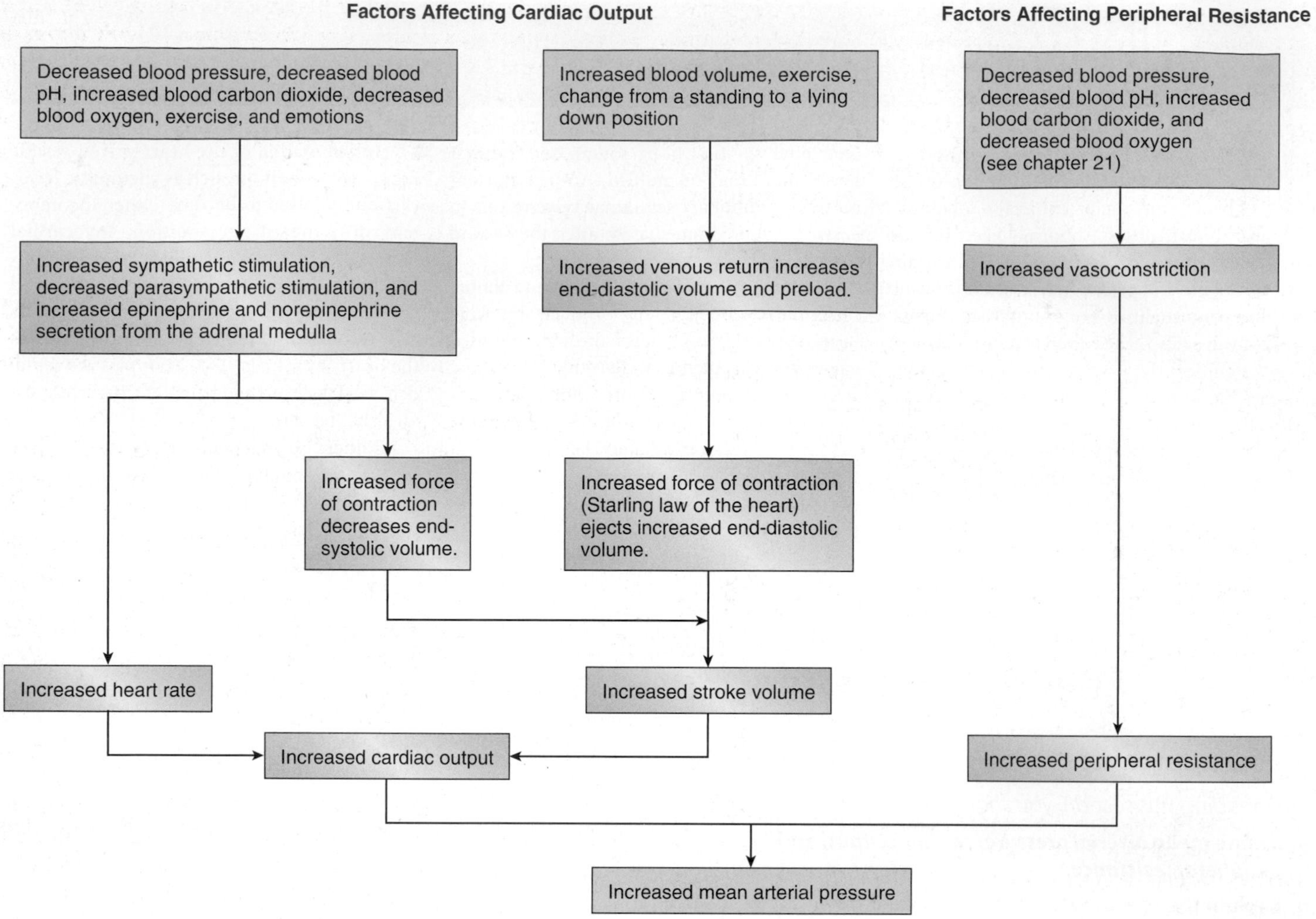

Factors Affecting Cardiac Output

Factors Affecting Peripheral Resistance

Decreased blood pressure, decreased blood pH, increased blood carbon dioxide, decreased blood oxygen, exercise, and emotions

Increased blood volume, exercise, change from a standing to a lying down position

Decreased blood pressure, decreased blood pH, increased blood carbon dioxide, and decreased blood oxygen (see chapter 21)

Increased sympathetic stimulation, decreased parasympathetic stimulation, and increased epinephrine and norepinephrine secretion from the adrenal medulla

Increased venous return increases end-diastolic volume and preload.

Increased vasoconstriction

Increased force of contraction decreases end-systolic volume.

Increased force of contraction (Starling law of the heart) ejects increased end-diastolic volume.

Increased heart rate

Increased stroke volume

Increased cardiac output

Increased peripheral resistance

Increased mean arterial pressure

FIGURE 20.21 Factors Affecting Mean Arterial Pressure
Mean arterial pressure is regulated by controlling cardiac output and peripheral resistance.

The difference between cardiac output when a person is at rest and maximum cardiac output is called the **cardiac reserve.** The greater a person's cardiac reserve, the greater his or her capacity for doing exercise. Cardiovascular disease and lack of exercise can reduce cardiac reserve and affect a person's quality of life. Exercise can greatly increase cardiac reserve by increasing cardiac output. In well-trained athletes, stroke volume during exercise can increase to over 200 mL/beat, resulting in cardiac outputs of 40 L/min or more.

ASSESS YOUR PROGRESS

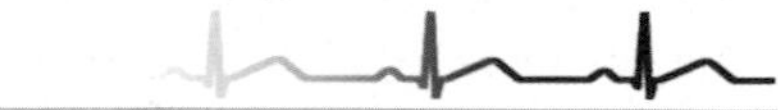

34. *Explain the relationship among mean arterial pressure, cardiac output, and peripheral resistance. Define each.*
35. *Explain the role of MAP in causing blood flow.*
36. *What is stroke volume, and what are two ways to increase it?*
37. *What is cardiac reserve? How can exercise influence cardiac reserve?*

20.9 Regulation of the Heart

LEARNING OUTCOMES

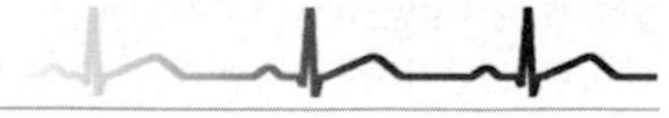

After reading this section, you should be able to

A. Describe intrinsic regulation of the heart.

B. Relate the three types of extrinsic regulation of the heart and their effects.

To maintain homeostasis, the amount of blood pumped by the heart must vary dramatically. For example, during exercise, cardiac output can increase several times over resting values. Intrinsic and extrinsic regulatory mechanisms control cardiac output. **Intrinsic regulation** results from the heart's normal functional characteristics and does not depend on either neural or hormonal regulation. It functions whether the heart is in place in the body or is removed and maintained outside the body under proper conditions. On the other hand, **extrinsic regulation** involves neural and hormonal control. Neural regulation of the heart results from sympathetic and parasympathetic reflexes, and the major hormonal regulation comes from epinephrine and norepinephrine secreted by the adrenal medulla.

Intrinsic Regulation

The force of contraction produced by cardiac muscle is related to the degree of stretch of the cardiac muscle fibers. As venous return increases, end-diastolic volume increases (figure 20.21). A greater end-diastolic volume increases the stretch of the ventricular walls. The extent to which the ventricular walls are stretched is sometimes called the **preload.** An increased preload increases cardiac output, and a decreased preload decreases cardiac output.

The length-versus-tension relationship in cardiac muscle is similar to that in skeletal muscle. Skeletal muscle, however, is normally stretched to nearly its optimal length before contraction, whereas cardiac muscle cells are not stretched to the point at which they contract with a maximal force (see chapter 9). Thus, an increased preload causes the cardiac muscle cells to contract with a greater force and produce a greater stroke volume. This relationship between preload and stroke volume is commonly referred to as the **Starling law of the heart**, and it describes the relationship between changes in the pumping effectiveness of the heart and changes in preload (figure 20.21). Venous return can decrease to a value as low as 2 L/min or increase to as much as 24 L/min, which has a major effect on the preload.

Afterload is the pressure the contracting left ventricle must produce to overcome the pressure in the aorta and move blood into the aorta. Although the heart's pumping effectiveness is greatly influenced by relatively small changes in the preload, it is very insensitive to large changes in afterload. Aortic blood pressure must increase to more than 170 mm Hg before it hampers the ventricles' ability to pump blood.

During exercise, blood vessels in exercising skeletal muscles dilate and allow more blood to flow through the vessels. The increased blood flow increases oxygen and nutrient delivery to the exercising muscles. In addition, skeletal muscle contractions repeatedly compress veins and cause blood to flow more rapidly from the skeletal muscles toward the heart. As blood flows rapidly through skeletal muscles and back to the heart, venous return to the heart increases, increasing the preload. The increased preload causes an increased force of cardiac muscle contraction, which increases stroke volume. The increase in stroke volume results in increased cardiac output, and the volume of blood flowing to the exercising muscles increases. When a person rests, venous return to the heart decreases because arteries in the skeletal muscles constrict and because muscular contractions no longer repeatedly compress the veins. As a result, blood flow through skeletal muscles decreases, and preload and cardiac output decrease.

Extrinsic Regulation

The heart is innervated by both **parasympathetic** and **sympathetic** nerve fibers (see figure 16.5). They influence the pumping action of the heart by affecting both heart rate and stroke volume. However, the influence of parasympathetic stimulation on the heart is much less than that of sympathetic stimulation. Sympathetic stimulation can increase cardiac output by 50–100% over resting values, whereas parasympathetic stimulation can cause only a 10–20% decrease.

Extrinsic regulation of the heart keeps blood pressure, blood oxygen levels, blood carbon dioxide levels, and blood pH within their normal ranges of values. For example, if blood pressure suddenly decreases, extrinsic mechanisms detect the decrease and initiate responses that increase cardiac output to bring blood pressure back into its normal range.

Parasympathetic Control

Parasympathetic nerve fibers are in the **vagus nerves.** Preganglionic fibers of the vagus nerve extend from the brainstem to terminal ganglia within the wall of the heart, and postganglionic fibers extend from the ganglia to the SA node, AV node, coronary blood vessels, and atrial myocardium.

Parasympathetic stimulation has an inhibitory influence on the heart, primarily by decreasing the heart rate. When a person is at rest, continuous parasympathetic stimulation inhibits the heart to some degree. An increase in heart rate during exercise results, in part, from decreased parasympathetic stimulation. Strong parasympathetic stimulation can decrease the heart rate below resting levels by at least 20–30 bpm, but it has little effect on stroke volume. In fact, if venous return remains constant while the heart is inhibited by parasympathetic stimulation, stroke volume can actually increase. The longer time between heartbeats allows the heart to fill to a greater capacity, resulting in an increased preload, which in turn increases stroke volume.

Acetylcholine, the neurotransmitter produced by postganglionic parasympathetic neurons, binds to ligand-gated channels that cause cardiac plasma membranes to become more permeable to K^+. As a consequence, the membrane hyperpolarizes. Heart rate decreases because the hyperpolarized membrane takes longer to depolarize and cause an action potential.

Sympathetic Control

Sympathetic nerve fibers originate in the thoracic region of the spinal cord as preganglionic neurons. These neurons synapse with postganglionic neurons of the inferior **cervical** and upper **thoracic sympathetic chain ganglia,** which project to the heart as **cardiac nerves** (figure 20.22; see chapter 16). The postganglionic sympathetic nerve fibers innervate the SA and AV nodes, the coronary blood vessels, and the atrial and ventricular myocardia.

Sympathetic stimulation increases both the heart rate and the force of muscular contraction. In response to strong sympathetic stimulation, the heart rate can increase to 250 or, occasionally, 300 bpm. Stronger contractions can also increase stroke volume. The increased force of contraction resulting from sympathetic stimulation causes a lower end-systolic volume in the heart; therefore, the heart empties to a greater extent (see figure 20.21).

Predict 8

What effect does sympathetic stimulation have on stroke volume if the venous return remains constant? Dilation of the coronary blood vessels occurs in response to an increased heart rate and stroke volume. Explain the functional advantage of that effect.

1. Sensory neurons (green) carry action potentials from baroreceptors and carotid body chemoreceptors to the cardioregulatory center. Chemoreceptors in the medulla oblongata also influence the cardioregulatory center.
2. The cardioregulatory center controls the frequency of action potentials in the parasympathetic neurons (red) extending to the heart through the vagus nerves. The parasympathetic neurons decrease the heart rate.
3. The cardioregulatory center controls the frequency of action potentials in the sympathetic neurons (blue). The sympathetic neurons extend through the cardiac nerves and increase the heart rate and the stroke volume.
4. The cardioregulatory center influences the frequency of action potentials in the sympathetic neurons (blue) extending to the adrenal medulla. The sympathetic neurons increase the secretion of epinephrine and some norepinephrine into the systemic circulation. Epinephrine and norepinephrine increase the heart rate and stroke volume.

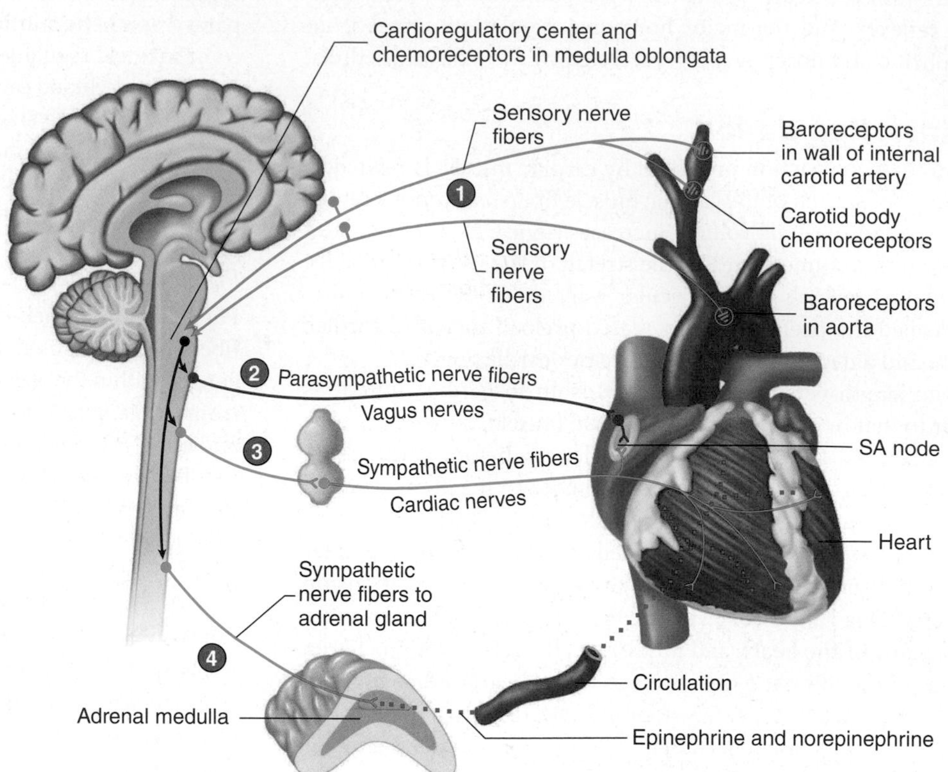
Cardioregulatory center and chemoreceptors in medulla oblongata
Sensory nerve fibers
1
Baroreceptors in wall of internal carotid artery
Carotid body chemoreceptors
Sensory nerve fibers
Baroreceptors in aorta
2
Parasympathetic nerve fibers
Vagus nerves
SA node
3
Sympathetic nerve fibers
Cardiac nerves
Heart
Sympathetic nerve fibers to adrenal gland
4
Circulation
Adrenal medulla
Epinephrine and norepinephrine

PROCESS FIGURE 20.22 AP|R **Baroreceptor and Chemoreceptor Reflexes**
Sensory neurons (green) carry action potentials from sensory receptors to the medulla oblongata. Sympathetic (blue) and parasympathetic (red) neurons exit the spinal cord or medulla oblongata and extend to the heart to regulate its function. Epinephrine and norepinephrine from the adrenal gland also help regulate the heart's action (*SA* = sinoatrial).

Limitations exist, however, to the relationship between increased heart rate and cardiac output. If the heart rate becomes too fast, ventricular diastole does not last long enough to allow complete ventricular filling, end-diastolic volume decreases, and stroke volume actually decreases. In addition, if the heart rate increases beyond a critical level, the strength of contraction decreases, probably because metabolites accumulate in cardiac muscle cells. The limit of the heart's ability to increase the volume of blood pumped is 170–250 bpm in response to intense sympathetic stimulation.

Sympathetic stimulation of the ventricular myocardium plays a significant role in regulating its contraction force when a person is at rest. Sympathetic stimulation maintains the strength of ventricular contraction at a level approximately 20% greater than it would be without sympathetic stimulation.

Norepinephrine, the postganglionic sympathetic neurotransmitter, increases the rate and degree of cardiac muscle depolarization, so that both the frequency and the amplitude of the action potentials increase. The effect of norepinephrine on the heart involves its association with cell surface β-adrenergic receptors. This combination causes a G protein–mediated synthesis and accumulation of cAMP in the cytoplasm of cardiac muscle cells. Cyclic-AMP increases the permeability of the plasma membrane to Ca^{2+}, primarily by opening calcium channels in the plasma membrane.

Increased sympathetic stimulation causes coronary arteries to constrict to some degree. However, increased metabolism of cardiac muscle, in response to sympathetic stimulation, allows metabolic by-products to accumulate in cardiac muscle, which causes coronary blood vessels to dilate. The dilation effect of these metabolites predominates (see chapter 21).

Hormonal Control

Epinephrine and norepinephrine released from the adrenal medulla can markedly influence the heart's pumping effectiveness. Epinephrine has essentially the same effect on cardiac muscle as norepinephrine, increasing the rate and force of heart contractions (see figure 20.21).

The secretion of epinephrine and norepinephrine is controlled by sympathetic stimulation of the adrenal medulla; it occurs in response to increased physical activity, emotional excitement, or other stressful conditions. Many stimuli that increase sympathetic stimulation of the heart also increase the release of epinephrine and norepinephrine from the adrenal medulla (see chapter 18). Epinephrine and norepinephrine travel in the blood through the vessels of the heart to the cardiac muscle cells, where they bind to β-adrenergic receptors and stimulate cAMP synthesis. Epinephrine takes a longer time to act on the heart than sympathetic stimulation does, but the effect lasts longer.

ASSESS YOUR PROGRESS

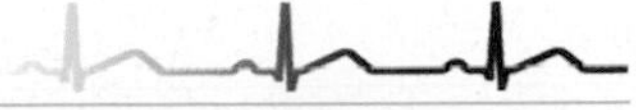

38. *What is venous return? Explain how it affects preload. How does preload affect cardiac output? State the Starling law of the heart.*

39. *Define* afterload, *and describe its effect on the pumping effectiveness of the heart.*

40. *What part of the brain regulates the heart? Describe the autonomic nerve supply to the heart.*

41. *What affects do parasympathetic stimulation and sympathetic stimulation have on heart rate, force of contraction, and stroke volume?*

42. *What neurotransmitters are released by the parasympathetic and sympathetic postganglionic neurons of the heart? What effects do they have on membrane permeability and excitability?*

43. *Name the two main hormones that affect the heart. Where are they produced, what causes their release, and what effects do they have on the heart?*

20.10 The Heart and Homeostasis

LEARNING OUTCOMES

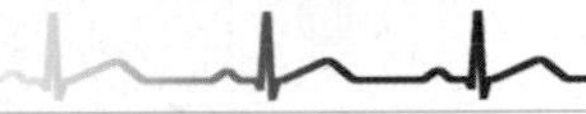

After reading this section, you should be able to

A. Describe how changes in blood pressure, pH, carbon dioxide, and oxygen affect the function of the heart.

B. Explain how extracellular ion concentration and body temperature affect the function of the heart.

The pumping efficiency of the heart plays an important role in maintaining homeostasis. Blood pressure in the systemic vessels must be high enough to allow nutrient and waste product exchange across the walls of the capillaries and to meet metabolic demands. In addition, the heart's activity must be regulated because the metabolic activities of the tissues change under such conditions as exercise and rest. Baroreceptor reflexes regulate blood pressure, and chemoreceptor reflexes help regulate the heart's activity.

Effect of Blood Pressure

Baroreceptor (bar′ō-rē-sep′ter, bar′ō-rē-sep′tōr) **reflexes** detect changes in blood pressure and lead to changes in heart rate and force of contraction. Stretch receptors, the sensory receptors of the baroreceptor reflexes, are in the walls of certain large arteries, such as the internal carotid arteries and the aorta. These stretch receptors measure blood pressure (figure 20.22). The anatomy of these sensory structures and their afferent pathways are described in chapter 21.

Afferent neurons project primarily through the glossopharyngeal (cranial nerve IX) and vagus (cranial nerve X) nerves from the baroreceptors to an area in the medulla oblongata called the **cardioregulatory center,** where sensory action potentials are integrated (figure 20.22). The part of the cardioregulatory center that increases heart rate is called the **cardioacceleratory center,** and the part that decreases heart rate is called the **cardioinhibitory center.** Efferent action potentials then travel from the cardioregulatory center to the heart through both the sympathetic and the parasympathetic divisions of the autonomic nervous system.

Elevated blood pressure within the internal carotid arteries and aorta causes their walls to stretch, thereby stimulating an increase in action potential frequency in the baroreceptors (figure 20.23). At normal blood pressures (80–120 mm Hg), afferent action potentials are sent from the baroreceptors to the medulla oblongata at a relatively constant frequency. When blood pressure rises, the arterial walls are stretched farther, and the afferent action potential frequency increases. When blood pressure decreases, the arterial walls are stretched to a lesser extent, and the afferent action potential frequency decreases. In response to elevated blood pressure, the baroreceptor reflexes reduce sympathetic stimulation and increase parasympathetic stimulation of the heart, causing the heart rate to slow. Decreased blood pressure causes decreased parasympathetic and increased sympathetic stimulation of the heart, resulting in an increased heart rate and force of contraction. Withdrawal of parasympathetic stimulation is primarily responsible for increases in heart rate up to approximately 100 bpm. Larger increases in heart rate, especially during exercise, result from sympathetic stimulation. The baroreceptor reflexes are homeostatic because they keep the blood pressure within a narrow range of values that is adequate to maintain blood flow to the tissues.

Effect of pH, Carbon Dioxide, and Oxygen

Chemoreceptor (kē′mō-rē-sep′tor) **reflexes** help regulate the heart's activity. Chemoreceptors sensitive to changes in pH and carbon dioxide levels exist within the medulla oblongata. A drop in pH and a rise in carbon dioxide decrease parasympathetic and increase sympathetic stimulation of the heart, resulting in increased heart rate and force of contraction (figure 20.24).

The increased cardiac output causes greater blood flow through the lungs, where carbon dioxide is eliminated from the body. This helps lower the blood carbon dioxide level to within its normal range and helps increase blood pH.

Chemoreceptors primarily sensitive to blood oxygen levels are found in the carotid and aortic bodies. These small structures are located near large arteries close to the brain and heart, and they monitor blood flowing to the brain and the rest of the body. A dramatic decrease in blood oxygen levels, as occurs during asphyxiation, activates the carotid and aortic body chemoreceptor reflexes. In carefully controlled experiments, it is possible to isolate the effects of the carotid and aortic body chemoreceptor reflexes from other reflexes, such as the medullary chemoreceptor reflexes. These experiments indicate that a reduction in blood oxygen results in decreased heart rate and increased vasoconstriction. The vasoconstriction causes blood pressure to rise, which promotes blood delivery despite the decrease in heart rate. The carotid and aortic body chemoreceptor reflexes may protect the heart for a short time by slowing the heart rate, thereby reducing its need for oxygen. The carotid and aortic body chemoreceptor reflexes normally do not function independently of other regulatory mechanisms. When all the regulatory mechanisms function together, large, prolonged decreases in blood oxygen levels increase the heart rate. Low blood oxygen levels also increase stimulation of respiratory movements (see chapter 23).

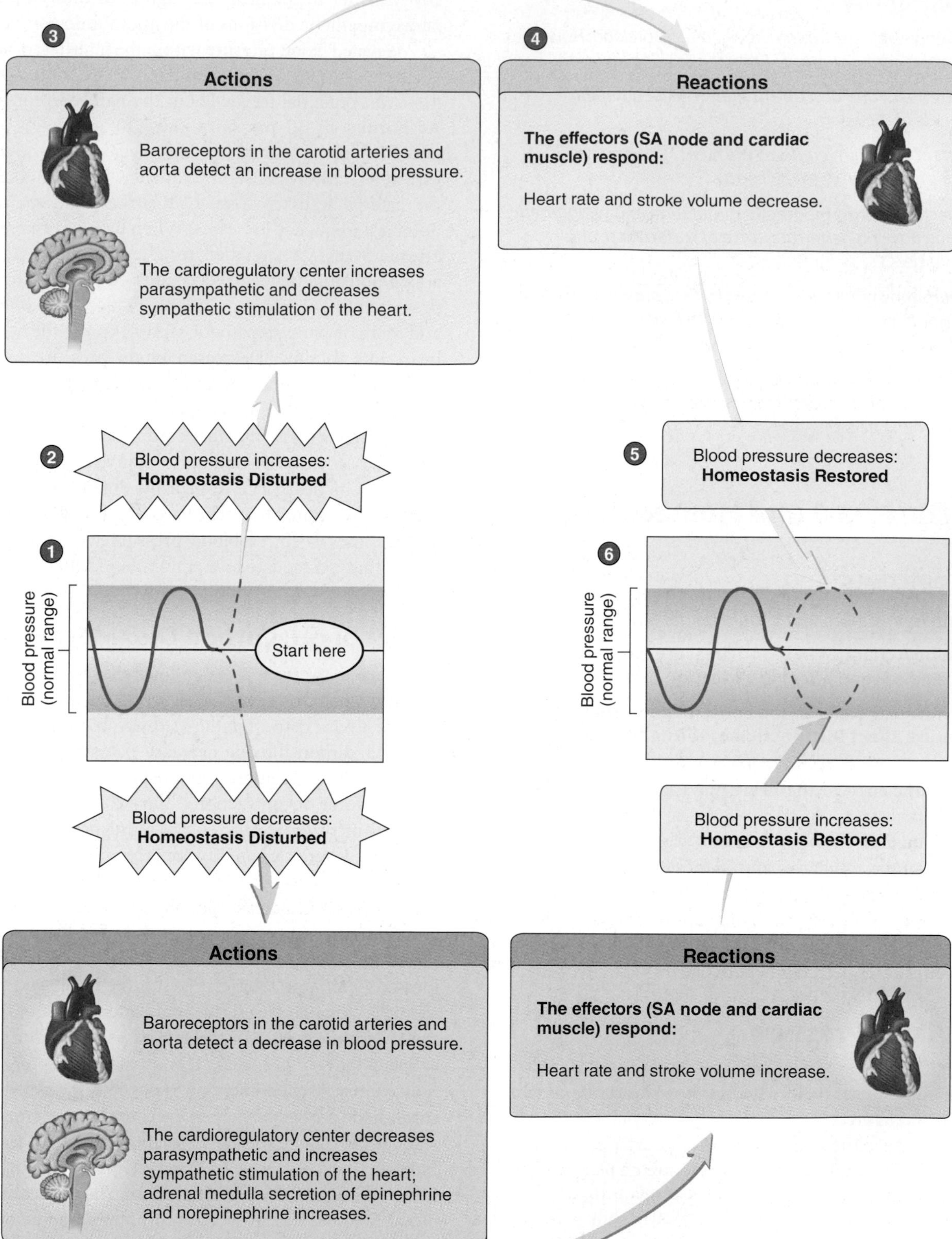
3
Actions
Baroreceptors in the carotid arteries and aorta detect an increase in blood pressure.
The cardioregulatory center increases parasympathetic and decreases sympathetic stimulation of the heart.
4
Reactions
The effectors (SA node and cardiac muscle) respond:
Heart rate and stroke volume decrease.
2
Blood pressure increases:
Homeostasis Disturbed
5
Blood pressure decreases:
Homeostasis Restored
1
Blood pressure (normal range)
Start here
6
Blood pressure (normal range)
Blood pressure decreases:
Homeostasis Disturbed
Blood pressure increases:
Homeostasis Restored
Actions
Baroreceptors in the carotid arteries and aorta detect a decrease in blood pressure.
The cardioregulatory center decreases parasympathetic and increases sympathetic stimulation of the heart; adrenal medulla secretion of epinephrine and norepinephrine increases.
Reactions
The effectors (SA node and cardiac muscle) respond:
Heart rate and stroke volume increase.

HOMEOSTASIS FIGURE 20.23 Summary of the Baroreceptor Reflex

The baroreceptor reflex maintains homeostasis in response to changes in blood pressure. (1) Blood pressure is within its normal range. (2) Blood pressure increases outside the normal range, which causes homeostasis to be disturbed. (3) Baroreceptors in the carotid arteries and aorta detect the increase in blood pressure, and the cardioregulatory center in the brain alters autonomic stimulation of the heart. (4) Heart rate and stroke volume decrease. (5) These changes cause blood pressure to decrease. (6) Blood pressure returns to its normal range; homeostasis is restored. Observe the responses to a decrease in blood pressure outside its normal range by following the *red arrows*.

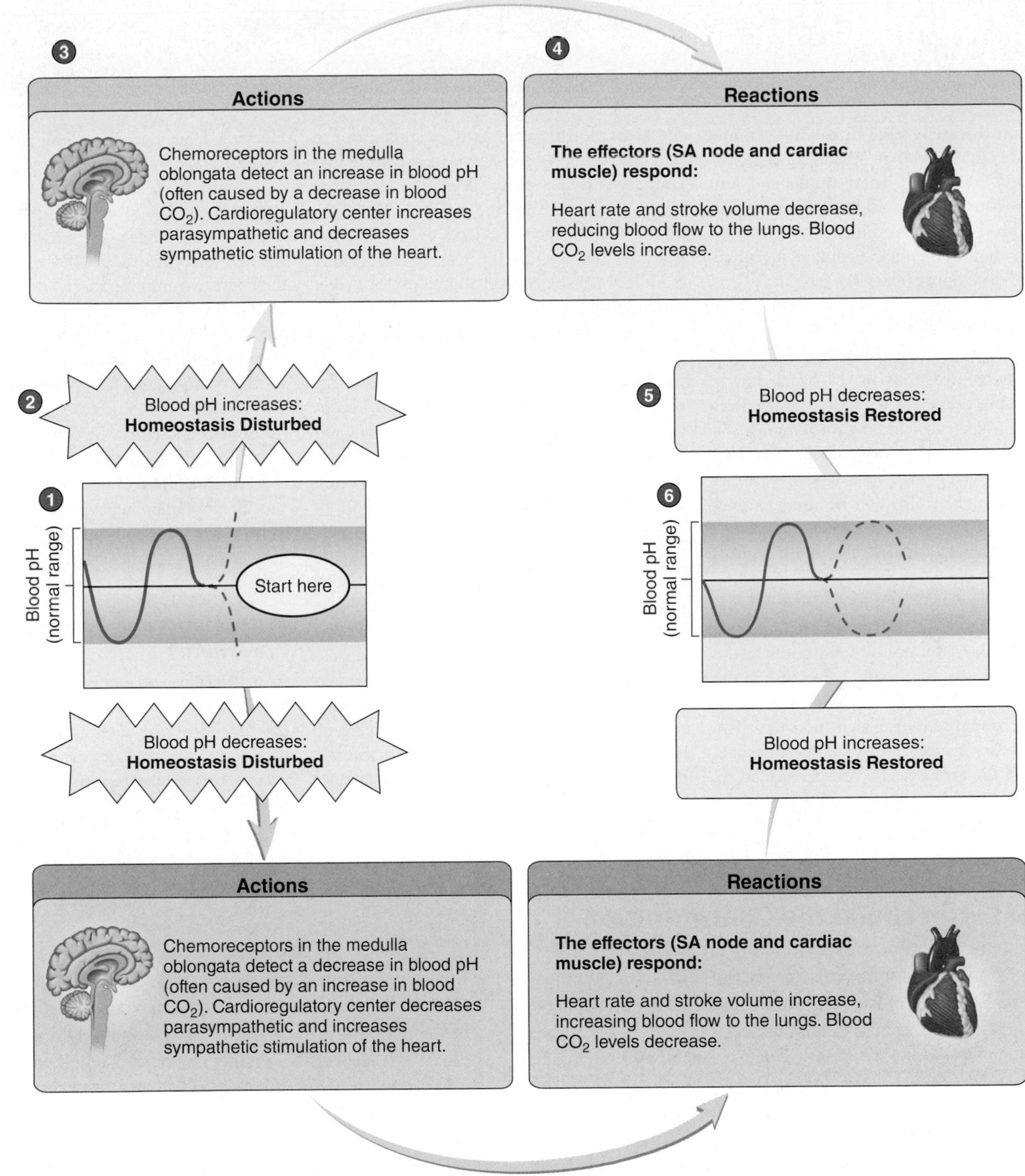
3
Actions
Chemoreceptors in the medulla oblongata detect an increase in blood pH (often caused by a decrease in blood CO_2). Cardioregulatory center increases parasympathetic and decreases sympathetic stimulation of the heart.
4
Reactions
The effectors (SA node and cardiac muscle) respond:
Heart rate and stroke volume decrease, reducing blood flow to the lungs. Blood CO_2 levels increase.
2
Blood pH increases:
Homeostasis Disturbed
5
Blood pH decreases:
Homeostasis Restored
1
Blood pH (normal range)
Start here
6
Blood pH (normal range)
Blood pH decreases:
Homeostasis Disturbed
Blood pH increases:
Homeostasis Restored
Actions
Chemoreceptors in the medulla oblongata detect a decrease in blood pH (often caused by an increase in blood CO_2). Cardioregulatory center decreases parasympathetic and increases sympathetic stimulation of the heart.
Reactions
The effectors (SA node and cardiac muscle) respond:
Heart rate and stroke volume increase, increasing blood flow to the lungs. Blood CO_2 levels decrease.

HOMEOSTASIS FIGURE 20.24 Summary of the Chemoreceptor Reflex

The chemoreceptor reflex maintains homeostasis in response to changes in blood CO_2 and H^+ concentrations (pH). (1) Blood pH is within its normal range. (2) Blood pH increases outside the normal range. (3) Chemoreceptors in the medulla oblongata detect increased blood pH. Control centers in the brain decrease sympathetic stimulation of the heart and adrenal medulla. (4) Heart rate and stroke volume decrease, reducing blood flow to the lungs. (5) These changes cause blood pH to decrease (as a result of increase in blood CO_2). (6) Blood pH returns to its normal range; homeostasis is restored. Observe the responses to a decrease in blood pH outside its normal range by following the *red arrows*.

Increased inflation of the lungs stimulates stretch receptors in the lungs. Afferent action potentials from these stretch receptors influence the cardioregulatory center, which causes the heart rate to increase. The reduced oxygen levels that exist at high altitudes can cause an increase in heart rate even when blood carbon dioxide levels remain low. However, the carotid and aortic body chemoreceptor reflexes are more important in regulating respiration (see chapter 23) and blood vessel constriction (see chapter 21) than heart rate.

Clinical IMPACT — Treatment and Prevention of Heart Disease

Heart disease is a life-threatening condition that affects people of all walks of life. Fortunately, medications and surgical procedures are available to treat many forms of heart disease. Preventive measures have also been identified to help people reduce their chance of developing heart disease.

Heart Medications

Digitalis (dij-i-tal′is, dij-i-ta′lis) slows and strengthens contractions of the heart muscle by increasing the amount of Ca^{2+} that enters cardiac muscle cells and by prolonging the action potentials' refractory period. This drug is frequently given to people who have heart failure, although it also can be used to treat atrial tachycardia.

Nitroglycerin (nī-trō-glis′er-in) causes dilation of all the veins and arteries, including coronary arteries, without an increase in heart rate or stroke volume. When all blood vessels dilate, a greater volume of blood pools in the dilated blood vessels, causing a decrease in the venous return to the heart. The flow of blood through coronary arteries also increases. The reduced preload causes cardiac output to decline, decreasing the amount of work performed by the heart. Nitroglycerin is frequently given to patients who have coronary artery disease that restricts coronary blood flow. The decreased work performed by the heart reduces the amount of oxygen required by the cardiac muscle. Consequently, the heart does not suffer from lack of oxygen, and angina pectoris does not develop.

Beta-adrenergic blocking agents reduce the rate and strength of cardiac muscle contractions, reducing the heart's demand for oxygen. These blocking agents bind to receptors for norepinephrine and epinephrine and prevent these substances from having their normal effects. These blocking agents are often used to treat rapid heart rate, certain types of arrhythmia, and hypertension.

Calcium channel blockers reduce the rate at which Ca^{2+} diffuses into cardiac and smooth muscle cells. Because the action potentials that produce cardiac muscle contractions depend in part on the flow of Ca^{2+} into cardiac muscle cells, calcium channel blockers can be used to control the force of heart contractions and to reduce arrhythmia, tachycardia, and hypertension. Because the entry of Ca^{2+} into smooth muscle cells causes contraction, calcium channel blockers dilate coronary blood vessels and can be used to treat angina pectoris.

Antihypertensive (an′tē-hī-per-ten′siv) **agents** comprise several drugs used to treat hypertension. These drugs lower blood pressure and therefore reduce the work required by the heart to pump blood. In addition, reduced blood pressure decreases the risk for heart attack and stroke. Drugs used to treat hypertension include those that reduce the activity of the sympathetic nervous system, dilate arteries and veins, increase urine production (diuretics), and block the conversion of angiotensinogen to angiotensin I.

Anticoagulants (an′tē-kō-ag′ū-lantz) prevent clot formation in people who have damaged heart valves or blood vessels or in those who have had a myocardial infarction. Aspirin functions as a weak anticoagulant.

Instruments and Selected Procedures

An **artificial pacemaker** is an instrument, placed beneath the skin, equipped with an electrode that extends to the heart and provides an electrical stimulus at a set frequency. Artificial pacemakers are used in patients whose natural

Effect of Extracellular Ion Concentration

The ions that affect cardiac muscle function are the same ions (K^+, Ca^{2+}, and Na^+) that influence membrane potentials in other electrically excitable tissues. However, cardiac muscle responds to these ions differently than nerve or skeletal muscle tissue does. For example, the extracellular levels of Na^+ rarely deviate enough from normal to significantly affect cardiac muscle function.

Excess K^+ in cardiac tissue causes the heart rate and stroke volume to decrease. A twofold increase in extracellular K^+ results in **heart block,** which is the loss of action potential conduction through the heart. The excess K^+ in the extracellular fluid causes partial depolarization of the resting membrane potential, resulting in a reduced amplitude of action potentials and, because of the reduced amplitude, a decreased rate at which action potentials are conducted along cardiac muscle cells. As the conduction rates decrease, ectopic action potentials can occur. In many cases, partially depolarized cardiac muscle cells spontaneously produce action potentials because the membrane potential reaches threshold. Elevated blood levels of K^+ can produce enough ectopic action potentials to cause fibrillation. The reduced action potential amplitude also results in less Ca^{2+} entering the sarcoplasm of the cell; thus, the strength of cardiac muscle contraction lessens.

Although the extracellular concentration of K^+ is normally small, a reduction in extracellular K^+ causes the resting membrane potential to become hyperpolarized; as a consequence, it takes longer for the membrane to depolarize to threshold. Ultimately, the reduction in extracellular K^+ results in a decrease in heart rate. The force of contraction is not affected, however.

A rise in the extracellular concentration of Ca^{2+} produces a greater force of cardiac contraction because of a higher influx of Ca^{2+} into the sarcoplasm during action potential generation. Elevated plasma Ca^{2+} levels have an indirect effect on heart rate because they reduce the frequency of action potentials in nerve fibers, thus reducing sympathetic and parasympathetic stimulation of the heart (see chapter 11). Generally, elevated blood Ca^{2+} levels lower the heart rate.

A low blood Ca^{2+} level increases the heart rate, although the effect is imperceptible until blood Ca^{2+} levels are reduced to approximately one-tenth of their normal value. The reduced extracellular Ca^{2+} levels cause Na^+ channels to open, which allows Na^+ to diffuse more readily into the cell, resulting in depolarization and action potential generation. However, reduced Ca^{2+} levels usually cause death due to tetany of skeletal muscles before they decrease enough to markedly influence the heart's function.

heart pacemakers do not produce a heart rate high enough to sustain normal physical activity. Artificial pacemakers can increase the heart rate as physical activity increases. Pacemakers can also detect cardiac arrest, extreme arrythmias, or fibrillation. In response, strong stimulation of the heart by the pacemaker may restore heart function.

A **heart-lung machine** temporarily substitutes for a patient's heart and lungs. It oxygenates the blood, removes carbon dioxide, and pumps blood throughout the body. Heart-lung machines have made possible many surgeries on the heart and lungs.

Heart valve replacement or repair is a surgical procedure performed on valves that are so deformed and scarred from conditions such as endocarditis that they have become severely incompetent or stenosed. Substitute valves made of synthetic materials, such as plastic or Dacron, are effective; valves transplanted from pigs are also used.

A **heart transplant** is a surgical procedure in which the heart of a recently deceased donor is transplanted to the recipient, and the recipient's diseased heart is removed. A heart transplant is possible only if the characteristics of the donor closely match those of the recipient (see chapter 22). People who have received heart transplants must continue to take drugs that suppress their immune responses for the rest of their lives. If they do not, the immune system will reject the transplanted heart.

An **artificial heart** is a mechanical pump that replaces the heart. Although still experimental and not able to substitute permanently for the heart, artificial hearts have been used to keep patients alive until a donor heart can be found.

Cardiac assistance involves temporarily implanting a mechanical device that assists the heart in pumping blood. In some cases, the decreased workload on the heart provided by the device appears to promote recovery of failing hearts, whereupon the device has been successfully removed. In **cardiomyoplasty**, a piece of a back muscle (latissimus dorsi) is wrapped around the heart and stimulated to contract in synchrony with the heart.

Prevention of Heart Disease

Heart disease is a major cause of death. Several precautions can help prevent heart disease. Proper nutrition is important in reducing the risk for heart disease (see chapter 25).

The recommended diet is low in fats, especially saturated fats and cholesterol, and low in refined sugar. The diet should be high in fiber, whole grains, fruits, and vegetables. Total food intake should be limited to avoid obesity, and sodium chloride intake should be reduced.

Smoking and excessive alcohol consumption should be avoided. Smoking increases the risk for heart disease at least ten-fold, and excessive alcohol consumption substantially increases the risk for heart disease.

Chronic stress, frequent emotional upsets, and lack of physical exercise can increase the risk for cardiovascular disease. Remedies include relaxation techniques and aerobic exercise programs involving gradual increases in the duration and difficulty of activities such as walking, swimming, jogging, and aerobic dancing.

Hypertension (hī′per-ten′shŭn), abnormally high systemic blood pressure, affects about one-fifth of the U.S. population. People are advised to have their blood pressure measured regularly because hypertension does not produce obvious symptoms. If hypertension cannot be controlled by diet and exercise, blood pressure–lowering drugs are prescribed. The cause of hypertension is unknown in most cases.

Some data suggest that taking an aspirin daily reduces the chance of having a heart attack. Aspirin inhibits the synthesis of prostaglandins in platelets, thereby helping prevent clot formation (see chapter 19).

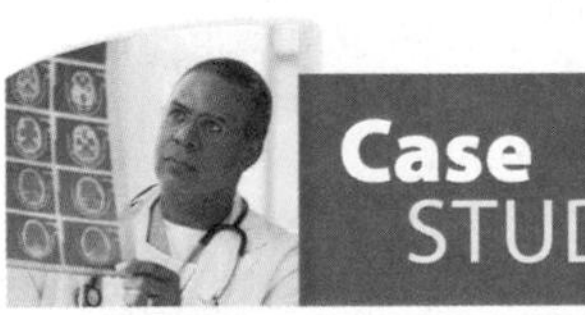

Case STUDY Aortic Valve Stenosis

Norma is a 62-year-old woman who had rheumatic fever when she was 12 years old. She has had a heart murmur ever since. Norma went to her doctor, complaining of fatigue; dizziness, especially on rising from a sitting or lying position; and pain in her chest when she exercises. Her doctor listened to Norma's heart and determined she has a systolic murmur (see Clinical Impact, "Abnormal Heart Sounds"). Norma's blood pressure (90/65 mm Hg) and heart rate (55 beats/min) were lower than normal. Norma's doctor referred her to a cardiologist, who did additional tests. An electrocardiogram indicated she has left ventricular hypertrophy. Imaging techniques confirmed the left ventricular hypertrophy and indicated a stenosed aortic semilunar valve. The cardiologist explained to Norma that the rheumatic fever she had as a child damaged her aortic semilunar valve and that the valve's condition had gradually become worse. The cardiologist recommended surgical replacement of Norma's aortic semilunar valve. Otherwise, she is likely to develop heart failure.

Predict 9

a. What effect does Norma's stenosed valve have on stroke volume?

b. Norma has left ventricular hypertrophy, which means the left ventricle is enlarged and has thicker walls than normal. Explain how that condition developed.

c. Explain Norma's low blood pressure.

d. Explain why Norma becomes dizzy on rising from a sitting or lying position (*Hint:* venous return).

e. Predict how Norma's heart rate changes on standing (see figure 20.23).

f. Norma experiences chest pain when she exercises, a condition called angina pectoris (see Clinical Impact, "Angina, Infarctions, and the Treatment of Blocked Coronary Arteries"). Why doesn't she feel this pain at rest?

Systems PATHOLOGY | Myocardial Infarction

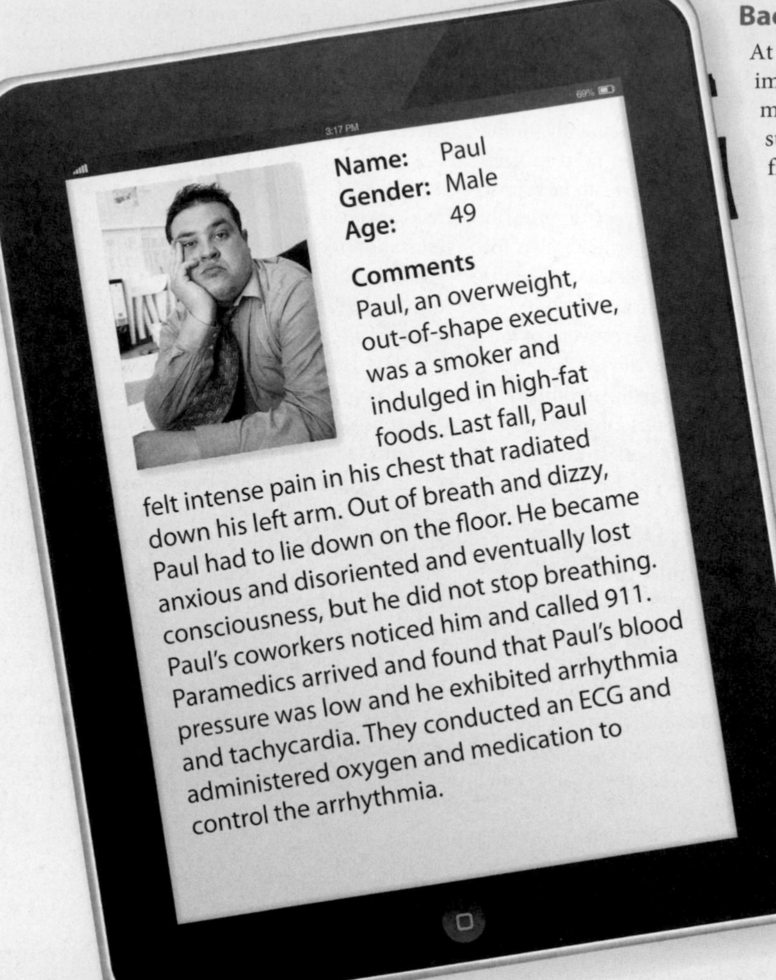
Name: Paul
Gender: Male
Age: 49
Comments
Paul, an overweight, out-of-shape executive, was a smoker and indulged in high-fat foods. Last fall, Paul felt intense pain in his chest that radiated down his left arm. Out of breath and dizzy, Paul had to lie down on the floor. He became anxious and disoriented and eventually lost consciousness, but he did not stop breathing. Paul's coworkers noticed him and called 911. Paramedics arrived and found that Paul's blood pressure was low and he exhibited arrhythmia and tachycardia. They conducted an ECG and administered oxygen and medication to control the arrhythmia.

Background Information

At the hospital, Paul was given tissue plasminogen activator (t-PA) to improve blood flow to the damaged area of the heart by activating plasminogen, which dissolves blood clots. Over the next few days, enzymes, such as creatine phosphokinase, increased in Paul's blood, which confirmed that cardiac muscle had been damaged by an infarction.

In the hospital, Paul began to experience shortness of breath because of pulmonary edema, and after a few days he developed pneumonia. He was treated for the pneumonia and gradually improved over the next few weeks. An angiogram (an′jē-ō-gram) performed several days after Paul's infarction indicated that he had suffered damage to a significant part of the lateral wall of his left ventricle. Although blood flow in some of Paul's coronary arteries was seriously restricted, his physicians did not believe that angioplasty or bypass surgery was necessary.

Paul experienced a myocardial infarction, in which a thrombosis in one of the branches of the left coronary artery reduced the blood supply to the lateral wall of the left ventricle, resulting in ischemia of the left ventricle wall. The fact that t-PA was an effective treatment supports the conclusion that the infarction was caused by a thrombosis. An ischemic area of the heart wall is not able to contract normally; therefore, Paul's heart's pumping effectiveness was dramatically reduced. The reduced pumping capacity was responsible for the low blood pressure, which decreased the blood flow to Paul's brain, resulting in confusion, disorientation, and unconsciousness.

Low blood pressure, increasing blood carbon dioxide levels, pain, and anxiousness increased sympathetic stimulation of the heart and adrenal glands. Increased sympathetic stimulation of the adrenal medulla caused the release of epinephrine. Increased parasympathetic stimulation of the heart resulted from pain sensations. In such cases, the heart is periodically arrhythmic due to the combined effects of parasympathetic stimulation, sympathetic stimulation, and the release of epinephrine and norepinephrine from the adrenal glands. In addition, the ischemic areas of the left ventricle produces ectopic beats.

Pulmonary edema resulted from the increased pressure in Paul's pulmonary veins because of the left ventricle's inability to pump blood. The edema allowed bacteria to infect the lungs and cause pneumonia.

Paul's heart began to beat rhythmically in response to medication because the infarction had not damaged the conducting system of the heart, which is an indication that no permanent arrhythmias had developed. (Permanent arrhythmias indicate damage to cardiac muscle specialized to conduct action potentials in the heart.)

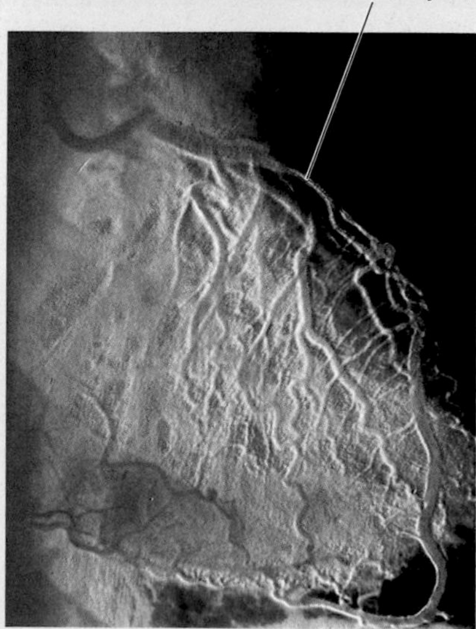
Occluded coronary artery

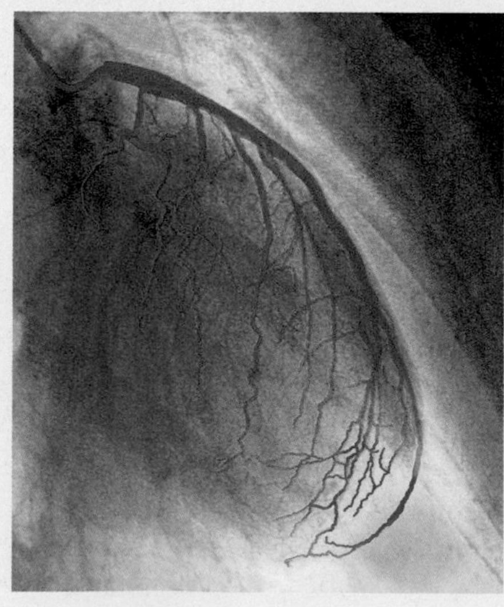

Figure 20A Angiograms

An angiogram is usually obtained by placing a catheter into a blood vessel and injecting a dye that can be detected with x-rays. Note the occluded (blocked) coronary blood vessel in this angiogram, which has been computer-enhanced to show colors. The angiogram on the right is of a normal heart.

INTEGUMENTARY

Pallor of the skin results from intense constriction of systemic blood vessels, including those in the skin.

MUSCULAR

Skeletal muscle activity is reduced because of lack of blood flow to the brain and because blood is shunted from blood vessels that supply skeletal muscles to blood vessels that supply the heart and brain.

NERVOUS

Decreased blood flow to the brain, decreased blood pressure, and pain due to ischemia of heart muscle result in increased sympathetic and decreased parasympathetic stimulation of the heart. Loss of consciousness occurs when the blood flow to the brain decreases so that not enough oxygen is available to maintain normal brain function, especially in the reticular activating system.

URINARY

Blood flow to the kidneys decreases dramatically in response to sympathetic stimulation. If the kidneys become ischemic, kidney tubules can be damaged, resulting in acute renal failure and reduced urine production. Increased blood urea nitrogen, increased blood levels of K^+, and edema are indications that the kidneys cannot eliminate waste products and excess water. If damage is not too great, the period of reduced urine production may last up to 3 weeks; then the rate of urine production slowly returns to normal as the kidney tubules heal.

Myocardial Infarction

Symptoms

- Chest pain that radiates down left arm
- Tightness and pressure in chest
- Difficulty breathing
- Nausea and vomiting
- Dizziness and fatigue

Treatment

- Restore blood flow to cardiac muscle
- Medication to reduce blood clotting (aspirin) and increase blood flow (t-PA)
- Supplemental oxygen to restore normal levels to heart tissue
- Prevention and control of hypertension
- Angioplasty or bypass surgery

ENDOCRINE

When blood pressure becomes low, antidiuretic hormone (ADH) is released from the posterior pituitary gland, and renin, released from the kidneys, activates the renin-angiotensinogen-aldosterone mechanism. ADH, secreted in large amounts, and angiotensin II cause vasoconstriction of systemic blood vessels. ADH and aldosterone act on the kidneys to retain water and electrolytes. Increased blood volume increases venous return, which results in increased stroke volume and blood pressure unless damage to the heart is very severe.

DIGESTIVE

Blood flow to the digestive system that decreases to very low levels often results in increased nausea and vomiting.

RESPIRATORY

Decreased blood pressure results in decreased blood flow to the lungs. The decrease in gas exchange leads to increased blood CO_2 levels, acidosis, and decreased blood O_2 levels. Initially, respiration becomes deep and labored because of the elevated CO_2 levels, decreased blood pH, and depressed O_2 levels. If the blood O_2 levels decrease too much, the person loses consciousness. Pulmonary edema can result when the pumping effectiveness of the left ventricle is substantially reduced.

LYMPHATIC AND IMMUNE

White blood cells, including macrophages, move to the damaged area of cardiac muscle and phagocytize any dead cardiac muscle cells.

Analysis of the electrocardiogram, blood pressure measurements, and an angiogram (figure 20A) indicated that the infarction was located on the left side of Paul's heart. Paul's lifestyle correlated with an increased probability of myocardial infarction in several ways: lack of physical exercise, overweight, smoking, and stress.

Paul's physician made it very clear that he was lucky to have survived a myocardial infarction and recommended a weight-loss program and a low-sodium and low-fat diet. The physician also suggested that Paul stop smoking. She explained that Paul would have to take medication for high blood pressure if his blood pressure did not decrease in response to the recommended changes. The physician recommended an aerobic exercise program for Paul after a period of recovery. She also advised Paul to seek ways to reduce the stress associated with his job and recommended that he take a small amount of aspirin regularly to reduce the probability of thrombosis. Because aspirin inhibits prostaglandin synthesis, it reduces the tendency for blood to clot. Paul followed the doctor's recommendations; after several months, his blood pressure was normal and he began to feel better than he had in years.

Predict 10

Severe ischemia in the wall of a ventricle can cause the death of cardiac muscle cells. Inflammation develops around the necrotic (dead) tissue, and macrophages invade the necrotic tissue and phagocytize dead cells. At the same time, blood vessels and connective tissue grow into the area and begin to deposit connective tissue to replace the necrotic tissue. Suppose that Paul had been recovering from his myocardial infarction for about a week when suddenly his blood pressure decreased to very low levels and he died within a very short time. At autopsy, a large amount of blood was found in the pericardial sac, and the wall of the left ventricle was ruptured. Explain.

Effect of Body Temperature

Under resting conditions, the temperature of cardiac muscle normally does not change dramatically, although alterations in temperature influence the heart rate. Small increases in cardiac muscle temperature cause the heart rate to speed up, and decreases in temperature cause the heart rate to slow. For example, during exercise or fever, increased heart rate and force of contraction accompany temperature elevations, but the heart rate drops under conditions of hypothermia. During heart surgery, body temperature is sometimes reduced dramatically on purpose to slow the heart rate and other metabolic functions.

ASSESS YOUR PROGRESS

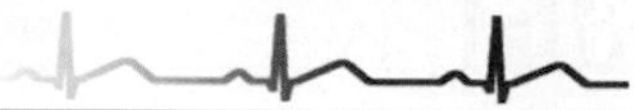

44. *Explain how the nervous system detects and responds to each of the following:*
a. a decrease in blood pressure
b. an increase in blood carbon dioxide level
c. a decrease in blood pH
d. a decrease in blood oxygen level

45. *Describe the baroreceptor reflex and the heart's response to an increase in venous return.*

46. *What effect does an increase or a decrease in extracellular potassium, calcium, and sodium ions have on the heart's rate and force of contraction?*

47. *What effect does temperature have on heart rate?*

20.11 Effects of Aging on the Heart

LEARNING OUTCOME

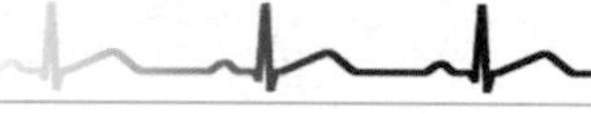

After reading this section, you should be able to

A. List the major age-related changes that affect the heart.

Aging results in gradual changes in heart function, which are minor under resting conditions but become more significant in response to exercise or age-related diseases. Under resting conditions, the mechanisms that regulate the heart compensate effectively for most of the age-related changes.

Hypertrophy of the left ventricle is a common age-related change. This appears to result from a gradual increase in the pressure in the aorta, against which the left ventricle must pump blood, and a gradual increase in the stiffness of cardiac muscle tissue. The elevated aortic pressure results from a gradual reduction in arterial elasticity, leading to increased stiffness of the aorta and other large arteries. Myocardial cells accumulate lipids, and the number of collagen fibers increases in cardiac tissue. These changes make the cardiac muscle tissue stiffer and less compliant. The increased volume of the left ventricle can sometimes result in higher left atrial pressure and increased pulmonary capillary pressure. This can cause pulmonary edema and a tendency for older people to feel out of breath when they exercise strenuously.

The maximum heart rate gradually declines, as can be roughly predicted by the following formula:

$$\text{Maximum heart rate} = 220 - \text{Age of individual}$$

The rate at which cardiac muscle breaks down ATP increases, and the rate of Ca^{2+} transport decreases. The maximum rate at which cardiac muscle can carry out aerobic respiration also decreases. In addition, the degree to which epinephrine and norepinephrine can increase the heart rate declines. These changes lead to longer contraction and relaxation times for cardiac muscle and a decrease in the maximum heart rate. Both the resting and maximum cardiac outputs slowly decline as people age; by 85 years of age, the cardiac output may have decreased by 30–60%.

Age-related changes also occur in the connective tissue of the heart valves. The connective tissue becomes less flexible, and Ca^{2+} deposits increase. The result is an increased tendency for the heart valves to function abnormally. The aortic semilunar valve is especially likely to become stenosed, but other heart valves, such as the bicuspid valve, may become either stenosed or incompetent.

The atrophy and replacement of cells of the left bundle branch and a decrease in the number of SA node cells alter the electrical conducting system of the heart and lead to a higher rate of cardiac arrhythmias in elderly people.

The enlarged and thickened cardiac muscle, especially in the left ventricle, requires more oxygen to pump the same amount of blood pumped by a younger heart. This change is not significant unless the coronary circulation is diminished by coronary artery disease. However, the development of coronary artery disease is age-related, as is congestive heart disease. Approximately 10% of elderly people over 80 have congestive heart failure, and a major contributing factor is coronary artery disease. Because of age-related changes in the heart, many elderly people are limited in their ability to respond to emergencies, infections, blood loss, and stress.

Exercise has many beneficial effects on the heart. Regular aerobic exercise improves the heart's functional capacity at all ages, provided the person has no other conditions that cause the extra workload on the heart to be harmful.

ASSESS YOUR PROGRESS

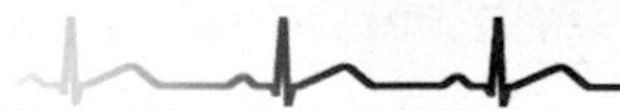

48. *Explain how age-related changes affect the function of the left ventricle.*

49. *Describe age-related changes in the heart rate.*

50. *Describe how increasing age affects the function of the conducting system and the heart valves.*

51. *Discuss the effect of two age-related heart diseases on the function of the aging heart.*

Diseases and Disorders

TABLE 20.3 Heart

Condition	Description
Inflammation of Heart Tissue	
Endocarditis	Inflammation of the endocardium; affects the valves more severely than other areas of the endocardium; may lead to scarring, causing stenosed or incompetent valves
Pericarditis	Inflammation of the pericardium; see Clinical Impact, "Pericarditis and Cardiac Tamponade"
Cardiomyopathy	Disease of the myocardium of unknown cause or occurring secondarily to other disease; results in weakened cardiac muscle, causing all chambers of the heart to enlarge; may eventually lead to congestive heart failure
Rheumatic heart disease	Results from a streptococcal infection in young people; toxin produced by the bacteria can cause rheumatic fever several weeks after the infection that can result in rheumatic endocarditis
Reduced Blood Flow to Cardiac Muscle	
Coronary heart disease	Reduces the amount of blood the coronary arteries can deliver to the myocardium
Coronary thrombosis	Formation of blood clot in a coronary artery
Myocardial infarction	Damaged cardiac muscle tissue resulting from lack of blood flow to the myocardium; often referred to as a heart attack; see Clinical Impact, "Angina, Infarctions, and the Treament of Blocked Coronary Arteries"
Congenital Heart Diseases (occur at birth)	
Septal defect	Hole in the septum between the left and right sides of the heart, allowing blood to flow from one side of the heart to the other and greatly reducing the heart's pumping effectiveness
Patent ductus arteriosus	Ductus arteriosus fails to close after birth, allowing blood to flow from the aorta to the pulmonary trunk under a higher pressure, which damages the lungs; also, the left ventricle must work harder to maintain adequate systemic pressure
Stenosis of the heart valve	Narrowed opening through one or more of the heart valves; aortic or pulmonary semilunar stenosis increases the heart's workload; bicuspid valve stenosis causes blood to back up in the left atria and lungs, resulting in edema of the lungs; tricuspid valve stenosis results in similar blood flow problems and edema in the peripheral tissues
Incompetent heart valve	Heart valves do not close correctly, and blood flows through in the reverse direction; see Clinical Impact, "Abnormal Heart Sounds"
Cyanosis (sī-ā-nō′sis; *cyan,* blue + *osis,* condition of)	Symptom of inadequate heart function in babies with congenital heart disease; the infant's skin appears blue because of low oxygen levels in the blood in peripheral blood vessels
Heart Failure	Progressive weakening of the heart muscle, reducing the heart's pumping action; hypertension leading to heart failure due to increased afterload; advanced age, malnutrition, chronic infections, toxins, severe anemias, hyperthyroidism, and hereditary factors can lead to heart failure

Go to: www.mhhe.com/seeley10 for additional information on these pathologies.

Learn to Predict ❮ From page 665

We learned in this chapter that the heart valves maintain a one-way flow of blood through the heart—from the atria to the ventricles. We also learned that an incompetent valve is one that leaks, or allows some blood to flow in the opposite direction—from the ventricles to the atria. An irregular swooshing noise following the first heart sound, as noted by Stan's regular physician, is a typical sign of an incompetent valve. The first heart sound is produced when the bicuspid and tricuspid valves close. The swooshing sound is the regurgitation of blood into the atria. The cardiologist determined that the bicuspid valve was incompetent, resulting in abnormal blood flow on the left side of the heart.

Stan's difficulty breathing resulted from the abnormal blood flow caused by his incompetent valve. After reviewing the blood flow through the heart in this chapter, we are aware that blood entering the left atrium is returning from the lungs through the pulmonary veins. As a result of the incompetent valve, the pressure in the left atrium, which is normally low, increases substantially during ventricular systole. The increased left atrial pressure causes the pressure in the pulmonary veins and pulmonary capillaries to increase. As a result, fluid leaks from the pulmonary capillaries into the lungs, causing pulmonary edema, or fluid accumulation in the lungs, making it difficult for Stan to breathe.

Answers to the rest of this chapter's Predict questions are in Appendix G.

Summary

20.1 Functions of the Heart (p. 666)

The heart produces the force that causes blood to circulate.

20.2 Size, Shape, and Location of the Heart (p. 667)

1. The heart is approximately the size of a closed fist and is shaped like a blunt cone.
2. The heart lies obliquely in the mediastinum, with its base directed posteriorly and slightly superiorly and its apex directed anteriorly, inferiorly, and to the left.
3. The base is deep to the second intercostal space, and the apex extends to the fifth intercostal space.

20.3 Anatomy of the Heart (p. 667)

The heart consists of two atria and two ventricles.

Pericardium

1. The pericardium is a sac that surrounds the heart and consists of the fibrous pericardium and the serous pericardium.
2. The fibrous pericardium helps hold the heart in place.
3. The serous pericardium reduces friction as the heart beats. It consists of the following parts:
   - The parietal pericardium lines the fibrous pericardium.
   - The visceral pericardium lines the exterior surface of the heart.
   - The pericardial cavity lies between the parietal and visceral pericardia and is filled with pericardial fluid, which reduces friction as the heart beats.

Heart Wall

1. The heart wall has three layers:
   - The outer epicardium (visceral pericardium) provides protection against the friction of rubbing organs.
   - The middle myocardium is responsible for contraction.
   - The inner endocardium reduces the friction resulting from blood passing through the heart.
2. The inner surfaces of the atria are mainly smooth. The auricles have muscular ridges called pectinate muscles.
3. The ventricles have ridges called trabeculae carneae.

External Anatomy and Coronary Circulation

1. Each atrium has a flap called an auricle.
2. The coronary sulcus separates the atria from the ventricles. The interventricular grooves separate the right and left ventricles.
3. The inferior and superior venae cavae and the coronary sinus enter the right atrium. The four pulmonary veins enter the left atrium.
4. The pulmonary trunk exits the right ventricle, and the aorta exits the left ventricle.
5. Coronary arteries branch off the aorta to supply the heart. Blood returns from the heart tissues to the right atrium through the coronary sinus and cardiac veins.

Heart Chambers and Valves

1. The interatrial septum separates the atria from each other, and the interventricular septum separates the ventricles.
2. The tricuspid valve separates the right atrium and ventricle. The bicuspid valve separates the left atrium and ventricle. The chordae tendineae attach the papillary muscles to the atrioventricular valves.
3. The semilunar valves separate the aorta and pulmonary trunk from the ventricles.

20.4 Route of Blood Flow Through the Heart (p. 675)

1. Blood from the body flows through the right atrium into the right ventricle and then to the lungs.
2. Blood returns from the lungs to the left atrium, enters the left ventricle, and is pumped back to the body.

20.5 Histology (p. 675)

Heart Skeleton

The fibrous heart skeleton supports the openings of the heart, electrically insulates the atria from the ventricles, and provides a point of attachment for heart muscle.

Cardiac Muscle

1. Cardiac muscle cells are branched and have a centrally located nucleus. Actin and myosin are organized to form sarcomeres. The sarcoplasmic reticulum and T tubules are not as organized as in skeletal muscle.
2. Cardiac muscle cells are joined by intercalated disks, which allow action potentials to move from one cell to the next. Thus, cardiac muscle cells function as a unit.
3. Cardiac muscle cells have a slow onset of contraction and a prolonged contraction time caused by the length of time required for Ca^{2+} to move to and from the myofibrils.
4. Cardiac muscle is well supplied with blood vessels that support aerobic respiration.
5. Cardiac muscle aerobically uses glucose, fatty acids, and lactic acid to produce ATP for energy. Cardiac muscle does not develop a significant oxygen deficit.

Conducting System

1. The SA node and the AV node are in the right atrium.
2. The AV node is connected to the bundle branches in the interventricular septum by the AV bundle.
3. The bundle branches give rise to Purkinje fibers, which supply the ventricles.
4. The SA node is made up of small-diameter cardiac muscle cells that initiate action potentials, which spread across the atria and cause them to contract.
5. Action potentials are slowed in the AV node, allowing the atria to contract and blood to move into the ventricles. Then the action potentials travel through the AV bundles and bundle branches to the Purkinje fibers, causing the ventricles to contract, starting at the apex. The AV node is also made up of small-diameter cardiac muscle fibers.

20.6 Electrical Properties (p. 678)

Action Potentials

1. After depolarization and partial repolarization, a plateau is reached, during which the membrane potential only slowly repolarizes.

2. The movement of Na^+ through the voltage-gated Na^+ channels causes depolarization.
3. During depolarization, voltage-gated K^+ channels close, and voltage-gated Ca^{2+} channels begin to open.
4. Early repolarization results from closure of the voltage-gated Na^+ channels and the opening of some voltage-gated K^+ channels.
5. The plateau exists because voltage-gated Ca^{2+} channels remain open.
6. The rapid phase of repolarization results from closure of the voltage-gated Ca^+ channels and the opening of many voltage-gated K^+ channels.
7. The entry of Ca^{2+} into cardiac muscle cells causes Ca^{2+} to be released from the sarcoplasmic reticulum to trigger contractions.

Autorhythmicity of Cardiac Muscle

1. Cardiac pacemaker muscle cells are autorhythmic because of the spontaneous development of a pacemaker potential.
2. The pacemaker potential results from the movement of Na^+ and Ca^{2+} into the pacemaker cells.
3. Ectopic foci are areas of the heart that regulate heart rate under abnormal conditions.

Refractory Periods of Cardiac Muscle

Cardiac muscle has a prolonged depolarization and thus a prolonged refractory period, which allows time for the cardiac muscle to relax before the next action potential causes a contraction.

Electrocardiogram

1. An ECG records only the electrical activities of the heart.
   - Depolarization of the atria produces the P wave.
   - Depolarization of the ventricles produces the QRS complex. Repolarization of the atria occurs during the QRS complex.
   - Repolarization of the ventricles produces the T wave.
2. Based on the magnitude of the ECG waves and the time between waves, ECGs can be used to diagnose heart abnormalities.

20.7 Cardiac Cycle (p. 684)

1. The cardiac cycle involves repetitive contraction and relaxation of the heart chambers.
2. Blood moves through the circulatory system from areas of higher pressure to areas of lower pressure. Contraction of the heart produces the pressure.
3. The cardiac cycle is divided into five periods:
   - Active ventricular filling results when the atria contract and pump blood into the ventricles.
   - Although the ventricles are contracting, during the period of isovolumetric contraction, ventricular volume does not change because all the heart valves are closed.
   - During the period of ejection, the semilunar valves open, and blood is ejected from the heart.
   - Although the heart is relaxing, during the period of isovolumetric relaxation, ventricular volume does not change because all the heart valves are closed.
   - Passive ventricular filling results when blood flows from the higher pressure in the veins and atria to the lower pressure in the relaxed ventricles.

Events Occurring During the Cardiac Cycle

1. Most ventricular filling occurs when blood flows from the higher pressure in the veins and atria to the lower pressure in the relaxed ventricles.
2. Contraction of the atria completes ventricular filling.
3. Contraction of the ventricles closes the AV valves, opens the semilunar valves, and ejects blood from the heart.
4. The volume of blood in a ventricle just before it contracts is the end-diastolic volume. The volume of blood after contraction is the end-systolic volume.
5. Relaxation of the ventricles results in the closing of the semilunar valves, the opening of the AV valves, and the movement of blood into the ventricles.

Heart Sounds

1. Closure of the atrioventricular valves produces the first heart sound.
2. Closure of the semilunar valves produces the second heart sound.
3. Turbulent flow of blood into the ventricles can be heard in some people, producing a third heart sound.

Aortic Pressure Curve

1. Contraction of the ventricles forces blood into the aorta, producing the peak systolic pressure.
2. Blood pressure in the aorta falls to the diastolic level as blood flows out of the aorta.
3. Elastic recoil of the aorta maintains pressure in the aorta and produces the dicrotic notch.

20.8 Mean Arterial Blood Pressure (p. 691)

1. Mean arterial pressure is the average blood pressure in the aorta. Adequate blood pressure is necessary to ensure delivery of blood to the tissues.
2. Mean arterial pressure is proportional to cardiac output (amount of blood pumped by the heart per minute) times peripheral resistance (total resistance to blood flow through blood vessels).
3. Cardiac output is equal to stroke volume times heart rate.
4. Stroke volume, the amount of blood pumped by the heart per beat, is equal to end-diastolic volume minus end-systolic volume.
   - Venous return is the amount of blood returning to the heart. Increased venous return increases stroke volume by increasing end-diastolic volume.
   - Increased force of contraction increases stroke volume by decreasing end-systolic volume.
5. Cardiac reserve is the difference between resting and exercising cardiac output.

20.9 Regulation of the Heart (p. 692)

Intrinsic Regulation

1. Venous return is the amount of blood that returns to the heart during each cardiac cycle.
2. The Starling law of the heart describes the relationship between preload and the stroke volume of the heart. An increased preload causes the cardiac muscle cells to contract with a greater force and produce a greater stroke volume.

Extrinsic Regulation

1. The cardioregulatory center in the medulla oblongata regulates parasympathetic and sympathetic nervous control of the heart.
2. Parasympathetic stimulation is supplied by the vagus nerve.
   - Parasympathetic stimulation decreases heart rate.
   - Postganglionic neurons secrete acetylcholine, which increases membrane permeability to K^+, producing hyperpolarization of the membrane.

3. Sympathetic stimulation is supplied by the cardiac nerves.
   - Sympathetic stimulation increases heart rate and force of contraction (stroke volume).
   - Postganglionic neurons secrete norepinephrine, which increases membrane permeability to Na^+ and Ca^{2+} and produces depolarization of the membrane.
4. Epinephrine and norepinephrine are released into the blood from the adrenal medulla as a result of sympathetic stimulation.
   - The effects of epinephrine and norepinephrine on the heart are long-lasting, compared with those of neural stimulation.
   - Epinephrine and norepinephrine increase the rate and force of heart contraction.

20.10 The Heart and Homeostasis (p. 695)

Effect of Blood Pressure

1. Baroreceptors monitor blood pressure.
2. In response to a decrease in blood pressure, the baroreceptor reflexes increase sympathetic stimulation and decrease parasympathetic stimulation of the heart, resulting in increased heart rate and force of contraction.

Effect of pH, Carbon Dioxide, and Oxygen

1. Chemoreceptors monitor blood carbon dioxide, pH, and oxygen levels.
2. In response to increased carbon dioxide and decreased pH, medullary chemoreceptor reflexes increase sympathetic stimulation and decrease parasympathetic stimulation of the heart.
3. Carotid body chemoreceptor receptors stimulated by low oxygen levels result in decreased heart rate and vasoconstriction.
4. All regulatory mechanisms functioning together in response to low blood pH, high blood carbon dioxide, and low blood oxygen levels usually produce increased heart rate and vasoconstriction. Decreased oxygen levels stimulate an increase in heart rate indirectly by stimulating respiration, and the stretch of the lungs activates a reflex that increases sympathetic stimulation of the heart.

Effect of Extracellular Ion Concentration

1. An increase or a decrease in extracellular K^+ decreases heart rate.
2. Increased extracellular Ca^{2+} increases force of contraction of the heart and decreases heart rate. Decreased Ca^{2+} levels produce the opposite effect.

Effect of Body Temperature

Heart rate increases when body temperature increases, and it decreases when body temperature decreases.

20.11 Effects of Aging on the Heart (p. 702)

1. Aging results in gradual changes in heart function, which are minor under resting conditions but more significant during exercise.
2. Hypertrophy of the left ventricle is a common age-related condition.
3. The maximum heart rate declines so that, by age 85, the cardiac output may be decreased by 30–60%.
4. There is an increased tendency for valves to function abnormally and for arrhythmias to occur.
5. Because increased oxygen consumption is required to pump the same amount of blood, age-related coronary artery disease is more severe.
6. Exercise improves the functional capacity of the heart at all ages.

REVIEW AND COMPREHENSION

1. Which of these structures returns blood to the right atrium?
   a. coronary sinus
   b. inferior vena cava
   c. superior vena cava
   d. Both b and c are correct.
   e. All of these are correct.
2. The valve located between the right atrium and the right ventricle is the
   a. aortic semilunar valve.
   b. pulmonary semilunar valve.
   c. tricuspid valve.
   d. bicuspid (mitral) valve.
3. The papillary muscles
   a. are attached to chordae tendineae.
   b. are found in the atria.
   c. contract to close the foramen ovale.
   d. are attached to the semilunar valves.
   e. surround the openings of the coronary arteries.
4. Given these blood vessels:
   (1) aorta
   (2) inferior vena cava
   (3) pulmonary trunk
   (4) pulmonary vein

   Choose the arrangement that lists the vessels in the order a red blood cell would encounter them going from the systemic veins to the systemic arteries.
   a. 1,3,4,2
   b. 2,3,4,1
   c. 2,4,3,1
   d. 3,2,1,4
   e. 3,4,2,1
5. The bulk of the heart wall is
   a. epicardium.
   b. pericardium.
   c. myocardium.
   d. endocardium.
   e. exocardium.
6. Cardiac muscle has
   a. sarcomeres.
   b. a sarcoplasmic reticulum.
   c. transverse tubules.
   d. many mitochondria.
   e. All of these are correct.
7. Action potentials pass from one cardiac muscle cell to another
   a. through gap junctions.
   b. by a special cardiac nervous system.
   c. because of the large voltage of the action potentials.
   d. because of the plateau phase of the action potentials.
   e. by neurotransmitters.
8. During the transmission of action potentials through the conducting system of the heart, there is a temporary delay at the
   a. bundle branches.
   b. Purkinje fibers.
   c. AV node.
   d. SA node.
   e. AV bundle.

9. Given these structures of the conducting system of the heart:
   (1) atrioventricular bundle
   (2) AV node
   (3) bundle branches
   (4) Purkinje fibers
   (5) SA node

   Choose the arrangement that lists the structures in the order an action potential passes through them.
   a. 2,5,1,3,4
   b. 2,5,3,1,4
   c. 2,5,4,1,3
   d. 5,2,1,3,4
   e. 5,2,4,3,1
10. Purkinje fibers
    a. are specialized cardiac muscle cells.
    b. conduct impulses much more slowly than ordinary cardiac muscle.
    c. conduct action potentials through the atria.
    d. connect the SA node and the AV node.
    e. ensure that ventricular contraction starts at the base of the heart.
11. T waves on an ECG represent
    a. depolarization of the ventricles.
    b. repolarization of the ventricles.
    c. depolarization of the atria.
    d. repolarization of the atria.
12. The greatest amount of ventricular filling occurs during
    a. the first one-third of diastole.
    b. the middle one-third of diastole.
    c. the last one-third of diastole.
    d. ventricular systole.
13. While the semilunar valves are open during a normal cardiac cycle, the pressure in the left ventricle is
    a. higher than the pressure in the aorta.
    b. lower than the pressure in the aorta.
    c. the same as the pressure in the left atrium.
    d. lower than the pressure in the left atrium.
14. Blood flows neither into nor out of the ventricles during
    a. the period of isovolumetric contraction.
    b. the period of isovolumetric relaxation.
    c. diastole.
    d. systole.
    e. Both a and b are correct.
15. Stroke volume is the
    a. amount of blood pumped by the heart per minute.
    b. difference between end-diastolic and end-systolic volume.
    c. difference between the amount of blood pumped at rest and that pumped at maximum output.
    d. amount of blood pumped from the atria into the ventricles.
16. Cardiac output is defined as
    a. blood pressure times peripheral resistance.
    b. peripheral resistance times heart rate.
    c. heart rate times stroke volume.
    d. stroke volume times blood pressure.
    e. blood pressure minus peripheral resistance.
17. Pressure in the aorta is at its lowest
    a. at the time of the first heart sound.
    b. at the time of the second heart sound.
    c. just before the AV valves open.
    d. just before the semilunar valves open.
18. Just after the dicrotic notch on the aortic pressure curve,
    a. the pressure in the aorta is greater than the pressure in the left ventricle.
    b. the pressure in the left ventricle is greater than the pressure in the aorta.
    c. the pressure in the left atrium is greater than the pressure in the left ventricle.
    d. the pressure in the left atrium is greater than the pressure in the aorta.
    e. blood pressure in the aorta is 0 mm Hg.
19. The “lubb” sound (first heart sound) is caused by the
    a. closing of the AV valves.
    b. closing of the semilunar valves.
    c. blood rushing out of the ventricles.
    d. filling of the ventricles.
    e. ventricular contraction.
20. Increased venous return results in
    a. increased stroke volume.
    b. increased cardiac output.
    c. decreased heart rate.
    d. Both a and b are correct.
21. Parasympathetic nerve fibers are found in the ________ nerves and release ________ at the heart.
    a. cardiac, acetylcholine
    b. cardiac, norepinephrine
    c. vagus, acetylcholine
    d. vagus, norepinephrine
22. Increased parasympathetic stimulation of the heart
    a. increases the force of ventricular contraction.
    b. increases the rate of depolarization in the SA node.
    c. decreases heart rate.
    d. increases cardiac output.
23. Because of the baroreceptor reflex, when normal arterial blood pressure decreases, the
    a. heart rate decreases.
    b. stroke volume decreases.
    c. frequency of afferent action potentials from baroreceptors decreases.
    d. cardioregulatory center stimulates parasympathetic neurons.
24. A decrease in blood pH and an increase in blood carbon dioxide levels result in
    a. increased heart rate.
    b. increased stroke volume.
    c. increased sympathetic stimulation of the heart.
    d. increased cardiac output.
    e. All of these are correct.
25. An increase in extracellular potassium levels can cause
    a. an increase in stroke volume.
    b. an increase in force of contraction.
    c. a decrease in heart rate.
    d. Both a and b are correct.

Answers in Appendix E

CRITICAL THINKING

1. Explain why the walls of the ventricles are thicker than the walls of the atria.
2. In most tissues, peak blood flow occurs during systole and decreases during diastole. In heart tissue, however, the opposite is true, and peak blood flow occurs during diastole. Explain this difference.
3. Explain why it is more efficient for contraction of the ventricles to begin at the apex of the heart than at the base.
4. Predict the consequences for the heart's pumping effectiveness if numerous ectopic foci in the ventricles produce action potentials.
5. A patient has tachycardia. Would you recommend a drug that prolongs or shortens the plateau of cardiac muscle cell action potentials?
6. Many endurance-trained athletes have a decreased resting heart rate, compared with that of nonathletes. Explain why an endurance-trained athlete's resting heart rate decreases rather than increases.
7. A doctor lets you listen to a patient's heart with a stethoscope at the same time that you feel the patient's pulse. Once in a while, you hear two heartbeats very close together, but you feel only one pulse beat. Later, the doctor tells you that the patient has an ectopic focus in the right atrium. Explain why you hear two heartbeats very close together. The doctor also tells you that the patient exhibits a pulse deficit (the number of pulse beats felt is fewer than the number of heartbeats heard). Explain why a pulse deficit occurs.
8. Explain why it is sufficient to replace the ventricles, but not the atria, in artificial heart transplantation.
9. A friend tells you an ECG revealed that her son has a slight heart murmur. Should you be convinced that he has a heart murmur? Explain.
10. An experiment on a dog was performed in which the mean arterial blood pressure was monitored before and after the common carotid arteries were partially clamped (at time A). The results are graphed here:

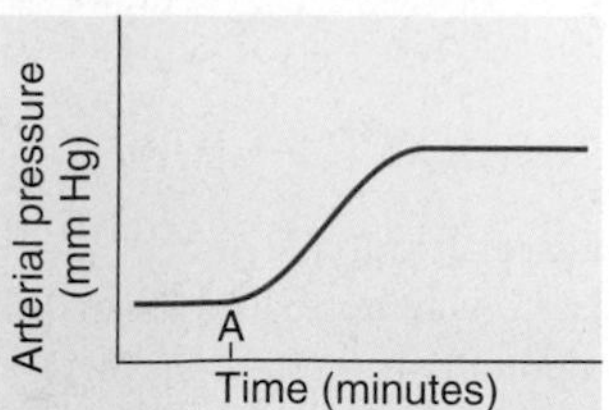
Arterial pressure (mm Hg)
A
Time (minutes)

Explain the change in mean arterial blood pressure. (*Hint:* Baroreceptors are located in the internal carotid arteries, which are superior to the site of clamping of the common carotid arteries.)

11. During hemorrhagic shock (caused by loss of blood), blood pressure may fall dramatically, although the heart rate is elevated. Explain why blood pressure falls despite the increase in heart rate.

Answers in Appendix F

Visit this book's website at **www.mhhe.com/seeley10** for chapter quizzes, interactive learning exercises, and other study tools.

Mc Graw Hill
connect plus+
ANATOMY & PHYSIOLOGY

Anatomy & Physiology
REVEALED
aprevealed.com

21

Cardiovascular System

BLOOD VESSELS AND CIRCULATION

Complex urban water systems seem rather simple when compared with the intricacy and coordinated functions of blood vessels. The heart is the pump that provides the major force causing blood to circulate, and the blood vessels are the pipes that carry blood to the body tissues and back to the heart. In addition, the blood vessels participate in regulating blood pressure and directing blood flow to the body's most active tissues. Both the blood vessels and the heart are regulated to ensure that the blood pressure is high enough to cause blood flow sufficient to meet the tissues' metabolic needs.

Learn to Predict

T. J. and Tyler were building a treehouse. While searching for a board in a pile of used lumber, T. J. stepped on a rusty nail, which penetrated deeply into his foot, causing it to bleed. Neither T. J. nor Tyler wanted to tell their parents about the accident, but after 3 days, T. J. developed septic shock. His foot had become infected, and the infection had spread into his bloodstream. After reading this chapter and recalling information about the structure and function of the heart, described in chapter 20, explain how T. J.'s blood volume, blood pressure, heart rate, and stroke volume changed due to septic shock. Also, explain how blood flow in the periphery changed and how it affected T. J.'s appearance. Finally, explain the consequences if T. J.'s blood pressure remained abnormally low for a prolonged period of time.

Photo: Photograph of medical professional measuring a patient's blood pressure. Blood pressure is the result of the heart forcing blood through the blood vessels.

Module 9
Cardiovascular System

Anatomy & Physiology
REVEALED
aprevealed.com

21.1 Functions of the Circulatory System

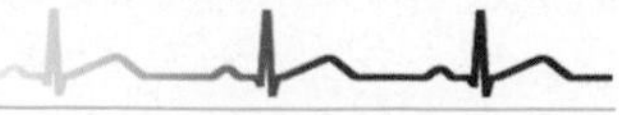

LEARNING OUTCOMES

After reading this section, you should be able to

A. Distinguish between pulmonary and systemic vessels.

B. Recite the functions of the circulatory system.

The circulatory system comprises two sets of blood vessels: pulmonary vessels and systemic vessels. **Pulmonary vessels** transport blood from the right ventricle, through the lungs, and back to the left atrium. **Systemic vessels** transport blood through all parts of the body from the left ventricle and back to the right atrium (see figure 20.1). Whereas the heart provides the major force that causes blood to circulate, the circulatory system has five unique functions:

1. *Carries blood.* Blood vessels carry blood from the heart to almost all the body tissues and back to the heart.
2. *Exchanges nutrients, waste products, and gases with tissues.* Nutrients and oxygen diffuse from blood vessels to cells in all areas of the body. Waste products and carbon dioxide diffuse from the cells, where they are produced, to blood vessels.
3. *Transports substances.* Hormones, components of the immune system, molecules required for coagulation, enzymes, nutrients, gases, waste products, and other substances are transported in the blood to all areas of the body.
4. *Helps regulate blood pressure.* The circulatory system and the heart work together to maintain blood pressure within a normal range of values.
5. *Directs blood flow to tissues.* The circulatory system directs blood to tissues when increased blood flow is required to maintain homeostasis.

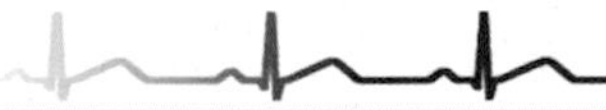

ASSESS YOUR PROGRESS

1. *What is the difference between pulmonary and systemic vessels?*

2. *Describe the five functions of the circulatory system.*

21.2 Structural Features of Blood Vessels

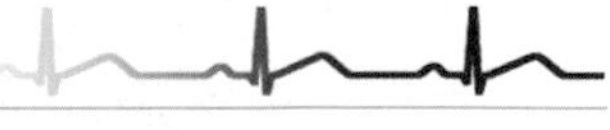

LEARNING OUTCOMES

After reading this section, you should be able to

A. List the types of capillaries, arteries, and veins.

B. Describe the structure and function of capillaries, arteries, and veins.

C. Describe the innervation of the blood vessel walls.

D. Discuss age-related changes to blood vessels.

The three main types of blood vessels are arteries, capillaries, and veins. These vessels form a continuous passageway for blood flow from the heart, through the body tissues, and back to the heart.

Arteries carry blood away from the heart. The ventricles pump blood from the heart into large, elastic arteries that branch repeatedly to form many progressively smaller arteries. As they become smaller, the artery walls undergo a gradual transition from having a large amount of elastic tissue and a smaller amount of smooth muscle to having a smaller amount of elastic tissue and a relatively large amount of smooth muscle. Although the arteries form a continuum from the largest to the smallest branches, they are normally classified as (1) elastic arteries, (2) muscular arteries, or (3) arterioles.

Blood flows from arterioles into **capillaries,** the most common blood vessel type. Most of the exchange that occurs between the blood and interstitial spaces occurs across the walls of capillaries. Their walls are the thinnest of all the blood vessels. Blood flows through capillaries slowly.

From the capillaries, blood flows into **veins,** vessels that carry blood toward the heart. When compared with arteries, the walls of the veins are thinner and contain less elastic tissue and fewer smooth muscle cells (see figure 21.5). As the blood returns to the heart, it flows through veins with thicker walls and greater diameters. Veins are classified as (1) venules, (2) small veins, or (3) medium or large veins.

Capillaries

All blood vessels have an internal lining of simple squamous epithelial cells called the **endothelium** (en-dō-thē′lē-ŭm), which is continuous with the endocardium of the heart.

The capillary wall consists primarily of endothelial cells (figure 21.1), which rest on a basement membrane. Outside the basement membrane is a delicate layer of loose connective tissue that merges with the connective tissue surrounding the capillary.

Scattered along the length of the capillary are **pericapillary cells** closely associated with the endothelial cells. These scattered cells lie between the basement membrane and the endothelial cells and are fibroblasts, macrophages, or undifferentiated smooth muscle cells.

Most capillaries range from 7 μm to 9 μm in diameter, and they branch without changing in diameter. Capillaries are variable in length, but in general they are approximately 1 mm long. Red blood cells flow through most capillaries single file and are frequently folded as they pass through the smaller-diameter capillaries.

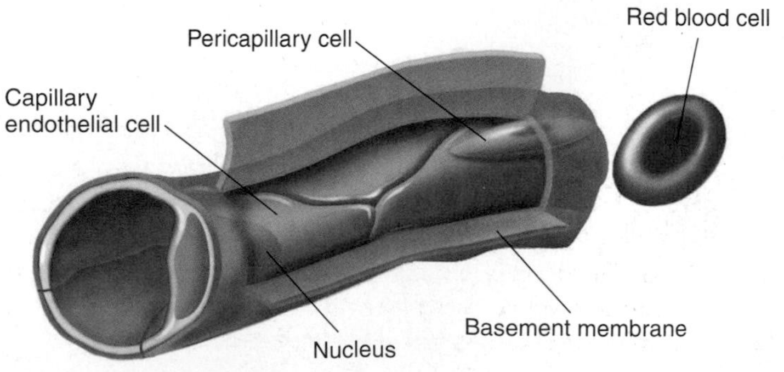
Pericapillary cell
Red blood cell
Capillary endothelial cell
Basement membrane
Nucleus

FIGURE 21.1 Capillary
Section of a capillary, showing that it is composed primarily of flattened endothelial cells.

Types of Capillaries

Capillaries are classified as continuous, fenestrated, or sinusoidal, depending on their diameter and permeability characteristics. **Continuous capillaries** are approximately 7–9 μm in diameter, and their walls exhibit no gaps between the endothelial cells (figure 21.2*a*). Continuous capillaries are less permeable to large molecules than are other capillary types; they exist in muscle, nervous tissue, and many other locations.

In **fenestrated** (fen′es-trā′ted) **capillaries,** endothelial cells have numerous fenestrae (figure 21.2*b*). The **fenestrae** (fe-nes′trē; windows) are areas approximately 70–100 nm in diameter in which the cytoplasm is absent and the plasma membrane consists of a porous diaphragm that is thinner than the normal plasma membrane. In some capillaries, the diaphragm is not present. Fenestrated capillaries are in tissues where capillaries are highly permeable, such as the intestinal villi, ciliary processes of the eyes, choroid plexuses of the central nervous system, and glomeruli of the kidneys.

Sinusoidal (sī-nŭ-soy′dăl) **capillaries** are larger in diameter than either continuous or fenestrated capillaries, and their basement membrane is less prominent (figure 21.2*c*) or completely absent. Their fenestrae are larger than those in fenestrated capillaries, and gaps can exist between endothelial cells. The sinusoidal capillaries occur in such places as endocrine glands, where large molecules cross their walls.

Sinusoids are large-diameter sinusoidal capillaries. Their basement membrane is sparse and often missing, and their structure suggests that large molecules and sometimes cells can move readily across their walls between the endothelial cells (figure 21.2*c*). Sinusoids are common in the liver and bone marrow. Macrophages are closely associated with the endothelial cells of the liver sinusoids. **Venous sinuses** are similar in structure to the sinusoidal capillaries but even larger in diameter. They exist primarily in the spleen, and there are large gaps between the endothelial cells that make up their walls.

Substances cross capillary walls by diffusing through or between the endothelial cells or through fenestrae. Lipid-soluble substances, such as oxygen and carbon dioxide, and small, water-soluble molecules readily diffuse through the plasma membrane. Larger water-soluble substances must pass through the fenestrae or gaps between the endothelial cells. In addition, transport by pinocytosis occurs, but little is known about its role in the capillaries. The walls of the capillaries are effective permeability barriers because red blood cells and large, water-soluble molecules, such as proteins, cannot readily pass through them.

Capillary Network

Arterioles supply blood to each capillary network (figure 21.3). Blood flows from arterioles through **metarterioles** (met′ar-ter′e-olz), vessels with isolated smooth muscle cells along their walls. Blood then flows from a metarteriole into a **thoroughfare channel,** a vessel that extends in a relatively direct fashion from a metarteriole to a venule. Blood flow through thoroughfare channels is relatively continuous. Several capillaries branch from the thoroughfare channels, and in these branches blood flow is regulated by **precapillary sphincters,** smooth muscle cells located at the origin of the branches (figure 21.3). Blood then flows through the capillary network into the venules. The ends of capillaries closest to the arterioles are **arterial capillaries,** and the ends closest to venules are **venous capillaries.**

Capillary networks are more numerous and more extensive in highly metabolic tissues, such as in the lungs, liver, kidneys, skeletal

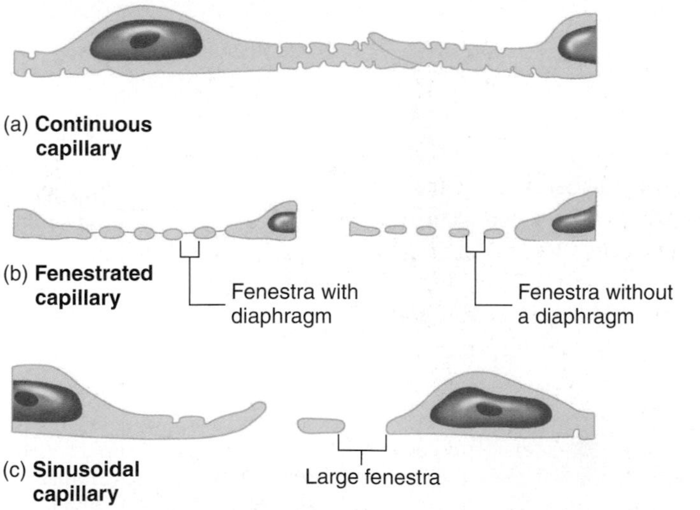

(a) **Continuous capillary**

(b) **Fenestrated capillary**

Fenestra with diaphragm

Fenestra without a diaphragm

(c) **Sinusoidal capillary**

Large fenestra

FIGURE 21.2 Structure of Capillary Walls

(*a*) Continuous capillaries have no gaps between endothelial cells and no fenestrae. They are common in muscle, nervous, and connective tissue.
(*b*) Fenestrated capillaries have fenestrae 7–100 nm in diameter, covered by thin, porous diaphragms, which are not present in some capillaries. They are found in intestinal villi, ciliary processes of the eyes, choroid plexuses of the central nervous system, and glomeruli of the kidneys.
(*c*) Sinusoidal capillaries have larger fenestrae without diaphragms and can have gaps between endothelial cells. They are found in bone marrow, the liver, the spleen, and the lymphatic organs.

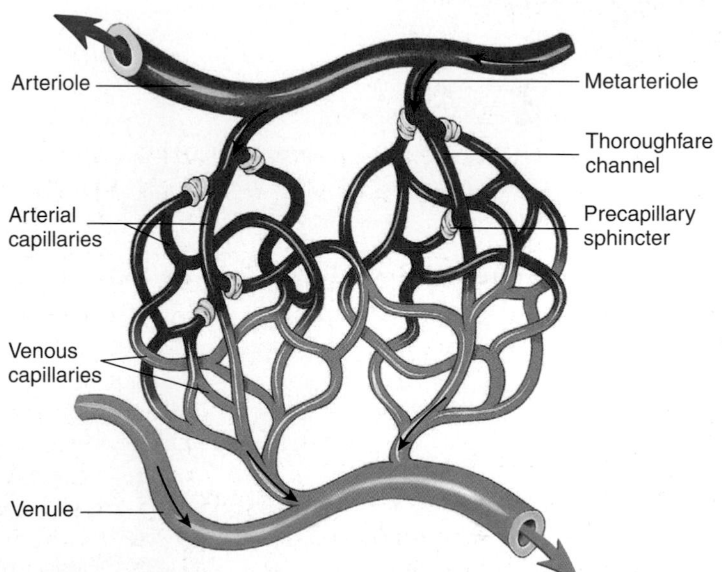
Arteriole
Metarteriole
Thoroughfare channel
Arterial capillaries
Precapillary sphincter
Venous capillaries
Venule

FIGURE 21.3 Capillary Network

A capillary network stems from an arteriole. Blood flows from the arteriole, through metarterioles, through the capillary network, to venules. Smooth muscle cells, called precapillary sphincters, regulate blood flow through the capillaries. Blood flow decreases when the precapillary sphincters constrict and increases when they dilate.

muscle, and cardiac muscle. Capillaries in the skin function in thermoregulation, and heat loss results from the flow of a large volume of blood through them. Capillary networks in the dermis of skin have many more thoroughfare channels than capillary networks in cardiac or skeletal muscle. The major function of the capillaries in these muscle tissues is nutrient and waste product exchange.

Arteriovenous Anastomoses

Arteriovenous anastomoses (ă-nas′tō-mō′sēz) allow blood to flow from arterioles to small veins without passing through capillaries. A **glomus** (glō′mŭs; pl. *glomera*, glom′er-ă) is an arteriovenous anastomosis that consists of arterioles with abundant smooth muscle in their walls. The vessels are branched and coiled and are surrounded by connective tissue sheaths.

Glomera are present in large numbers in the sole of the foot, the palm of the hand, the terminal phalanges, and the nail beds. The glomera help regulate body temperature. As body temperature decreases, glomera constrict and less blood flows through them, reducing the rate of heat loss from the body. As body temperature increases, glomera dilate and more blood flows through them, increasing the rate of heat loss from the body. **Pathologic arteriovenous anastomoses** can result from injury or tumors. They cause the direct flow of blood from arteries to veins and, if they are sufficiently large, can lead to heart failure because of the tremendous increase in venous return to the heart.

ASSESS **YOUR PROGRESS**

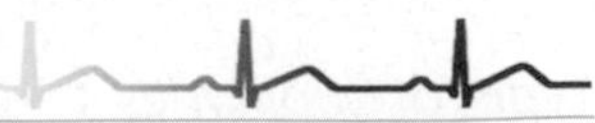

3. *In what direction, relative to the heart, is blood carried by arteries and by veins?*
4. *Name, in order, all the types of blood vessels, starting at the heart, going into the tissues, and returning to the heart.*
5. *Describe the general structure of a capillary.*
6. *Compare the structure of the three types of capillaries. Explain the various ways that materials pass through capillary walls.*
7. *Describe a capillary network. Where is the smooth muscle that regulates blood flow into and through the capillary network located? What is the function of a thoroughfare channel?*
8. *Contrast the function of capillaries in the skin with the function of capillaries in muscle tissue.*
9. *Define* arteriovenous anastomosis *and* glomus, *and explain their functions.*

Structure of Arteries and Veins

General Features

Except for the capillaries and the venules, the blood vessel walls consist of three relatively distinct layers, which are most evident in the muscular arteries and least evident in the veins. From the lumen to the outer wall of the blood vessels, the layers, or **tunics** (too′niks), are (1) the tunica intima, (2) the tunica media, and (3) the tunica adventitia, or tunica externa (figures 21.4 and 21.5).

FUNDAMENTAL **Figure**

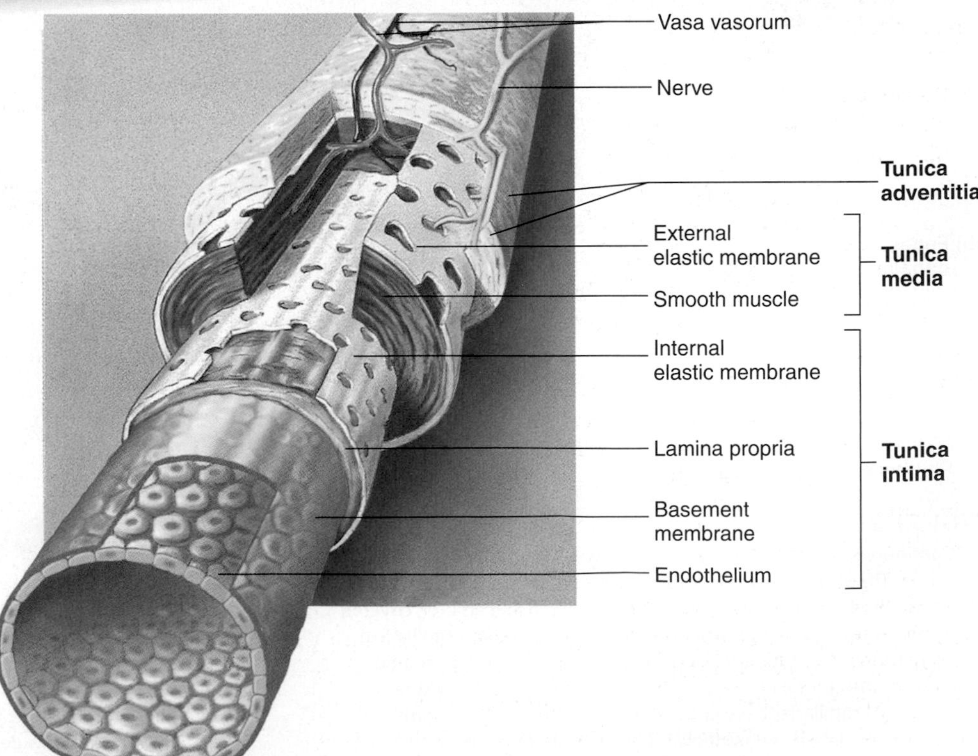
Vasa vasorum
Nerve
Tunica adventitia
External elastic membrane
Smooth muscle
Tunica media
Internal elastic membrane
Lamina propria
Basement membrane
Endothelium
Tunica intima

FIGURE 21.4 AP|R **Histology of a Blood Vessel**
The layers, or tunics, of the blood vessel wall are the tunica intima, media, and adventitia. Vasa vasorum are blood vessels that supply blood to the wall of the blood vessel.

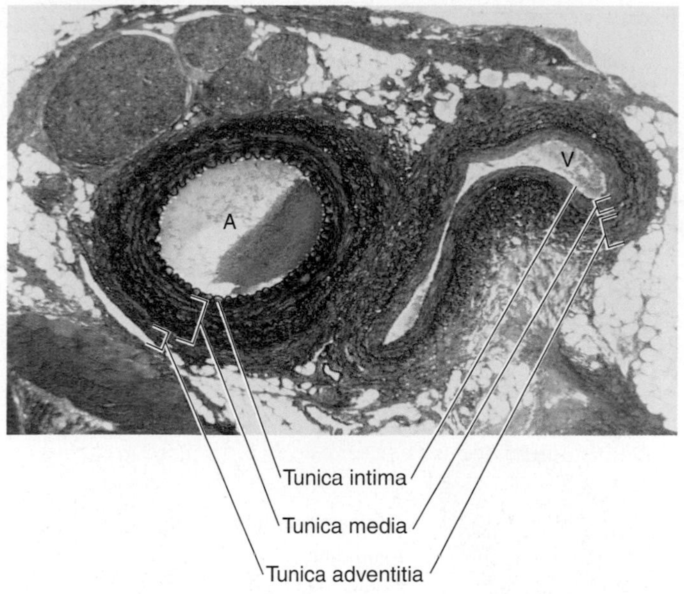
V
A
Tunica intima
Tunica media
Tunica adventitia

FIGURE 21.5 Comparison of an Artery and a Vein
The typical structure of a medium-sized artery (A) and a vein (V). Note that the artery has a thicker wall than the vein. The predominant layer in the wall of the artery is the tunica media, with its circular layers of smooth muscle. The predominant layer in the wall of the vein is the tunica adventitia, and the tunica media is thinner than in the artery.

The **tunica intima** consists of endothelium, a delicate connective tissue basement membrane, a thin layer of connective tissue called the lamina propria, and a fenestrated layer of elastic fibers called the **internal elastic membrane.** The internal elastic membrane separates the tunica intima from the next layer, the tunica media.

The **tunica media,** or middle layer, consists of smooth muscle cells arranged circularly around the blood vessel. The amount of blood flowing through a blood vessel can be regulated by contraction or relaxation of the smooth muscle in the tunica media. A decrease in blood flow results from **vasoconstriction** (vā′sō-kon-strik′shŭn, vas′ō-kon-strik′shŭn), a decrease in blood vessel diameter caused by smooth muscle contraction. An increase in blood flow is produced by **vasodilation** (vā′sō-dī-lā′shŭn, vas-ō-dī-lā′shŭn), an increase in blood vessel diameter resulting from smooth muscle relaxation. The tunica media also contains variable amounts of elastic and collagen fibers, depending on the size of the vessel. An **external elastic membrane,** which separates the tunica media from the tunica adventitia, can be identified at the outer border of the tunica media in some arteries. In addition, a few longitudinally oriented smooth muscle cells occur in some arteries near the tunica intima.

The **tunica adventitia** (too′ni-kă ad-ven-tish′ă) is composed of connective tissue, which varies from dense connective tissue near the tunica media to loose connective tissue that merges with the connective tissue surrounding the blood vessels.

The relative thickness and composition of each layer vary with the diameter of the blood vessel and its type. The transition from one vessel type to another is gradual, as are the structural changes.

Types of Arteries

Elastic Arteries

Elastic arteries have the largest diameters (figure 21.6*a*) and are often called *conducting arteries.* Blood pressure is relatively high in these vessels, and it fluctuates between systolic and diastolic values. Elastic arteries have a greater amount of elastic tissue and a smaller amount of smooth muscle in their walls, compared with other arteries. The elastic fibers are responsible for the elastic characteristics of the blood vessel wall, but collagenous connective tissue determines the degree to which the arterial wall can stretch.

The tunica intima is relatively thick. The elastic fibers of the internal and external elastic membranes merge and are not recognizable as distinct layers. The tunica media consists of a meshwork of elastic fibers with interspersed, circular smooth muscle cells and some collagen fibers. The tunica adventitia is relatively thin.

Muscular Arteries

Muscular arteries include medium-sized and small arteries. The walls of medium-sized arteries are relatively thick, compared with their diameter, mainly because the tunica media contains 25–40 layers of smooth muscle (figure 21.6*b*). The tunica intima of the muscular arteries has a well-developed internal elastic membrane. The tunica adventitia is composed of a relatively thick layer of collagenous connective tissue that blends with the surrounding connective tissue. Muscular arteries are frequently called *distributing arteries* because the smooth muscle cells allow them to partially regulate blood supply to different body regions by either constricting or dilating.

Smaller muscular arteries range from 40 μm to 300 μm in diameter. Those that are 40 μm in diameter have approximately three or four layers of smooth muscle in their tunica media, whereas arteries that are 300 μm across have essentially the same structure as the larger muscular arteries. The small muscular arteries are adapted for vasodilation and vasoconstriction.

Arterioles

Arterioles (ar-tēr′ē-ōlz) transport blood from small arteries to capillaries and are the smallest arteries in which the three tunics can be identified (see figure 21.3). They range in diameter from approximately 40 μm to as small as 9 μm. The tunica intima has no observable internal elastic membrane, and the tunica media consists of one or two layers of circular smooth muscle cells. Arterioles, like the small arteries, are capable of vasodilation and vasoconstriction.

Types of Veins

Venules and Small Veins

Venules (ven′oolz, vē′noolz), with a diameter of up to 50 μm, are tubes composed of endothelium resting on a delicate basement membrane. Their structure, except for their diameter, is very similar to that of capillaries (see figure 21.1). A few isolated smooth muscle cells exist outside the endothelial cells, especially in the larger venules. As the vessels increase to 0.2–0.3 mm in diameter, the smooth muscle cells form a continuous layer; the vessels are then called **small veins.** The small veins also have a tunica adventitia composed of collagenous connective tissue.

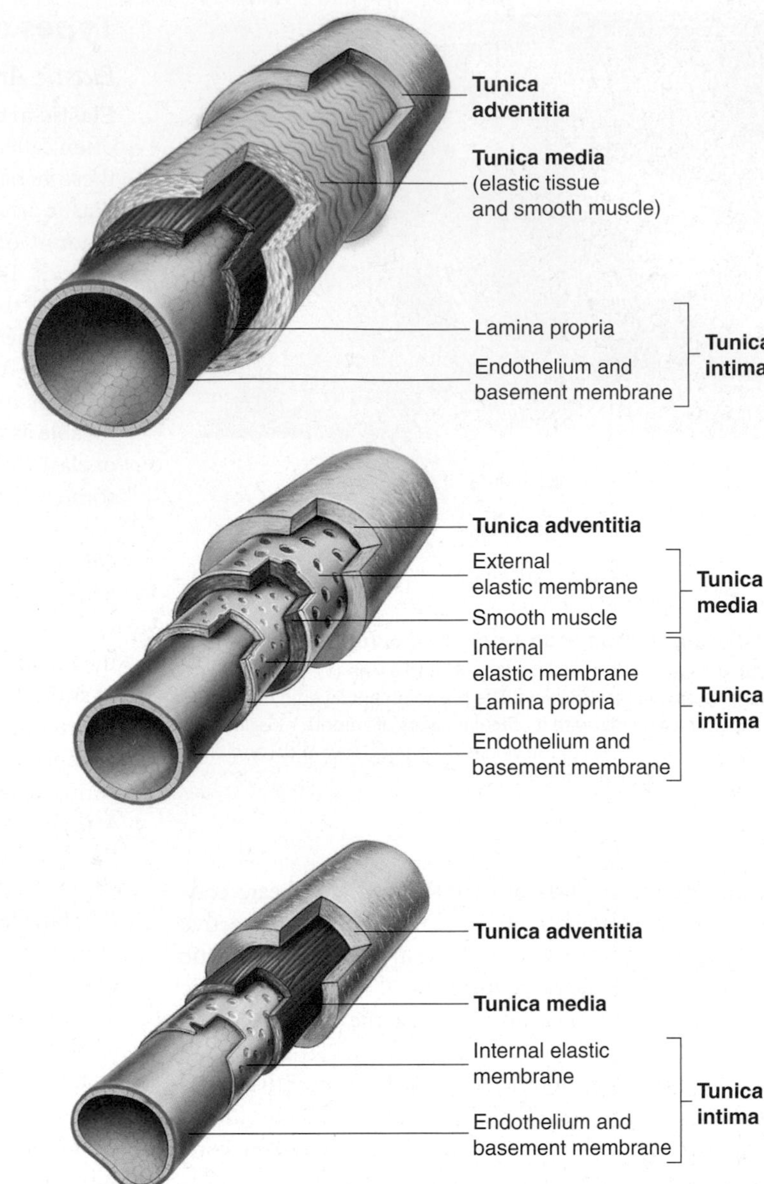
(a) **Elastic arteries.** The tunica media is mostly elastic connective tissue. Elastic arteries recoil when stretched, which prevents blood pressure from falling rapidly.
Tunica adventitia
Tunica media (elastic tissue and smooth muscle)
Lamina propria
Endothelium and basement membrane
Tunica intima
(b) **Muscular arteries.** The tunica media is a thick layer of smooth muscle. Muscular arteries regulate blood flow to different regions of the body.
Tunica adventitia
External elastic membrane
Smooth muscle
Tunica media
Internal elastic membrane
Lamina propria
Endothelium and basement membrane
Tunica intima
(c) **Medium and large veins.** All three tunics are present. The tunica media is thin but can regulate vessel diameter because blood pressure in the venous system is low. The predominant layer is the tunica adventitia.
Tunica adventitia
Tunica media
Internal elastic membrane
Endothelium and basement membrane
Tunica intima

FIGURE 21.6 Structural Comparison of Blood Vessel Types

The venules collect blood from the capillaries (see figure 21.3) and transport it to small veins, which in turn transport it to medium veins. Nutrient exchange occurs across the venule walls but, as the walls of the small veins increase in thickness, the degree of nutrient exchange decreases.

Medium and Large Veins

Most of the veins observed in gross anatomical dissections, except for the large veins, are **medium veins.** They collect blood from small veins and deliver it to large veins. The **large veins** transport blood from the medium veins to the heart. Their tunica intima is thin and consists of endothelial cells, a relatively thin layer of collagenous connective tissue, and a few scattered elastic fibers. The tunica media is also thin and is composed of a thin layer of circularly arranged smooth muscle cells containing some collagen fibers and a few sparsely distributed elastic fibers. The tunica adventitia, which is composed of collagenous connective tissue, is the predominant layer (figure 21.6*c*).

Portal Veins

Portal (pōr′tăl; door) **veins** begin in a primary capillary network, extend some distance, and end in a secondary capillary network. There is no pumping mechanism like the heart between the two capillary networks. There are two systems of portal veins in humans: (1) The hepatic portal veins carry blood from the capillaries in the gastrointestinal tract and spleen to dilated capillaries, called sinusoids, in the liver (see figure 21.27). (2) The hypothalamohypophysial portal veins carry blood from the hypothalamus of the brain to the anterior pituitary gland (see figure 18.3).

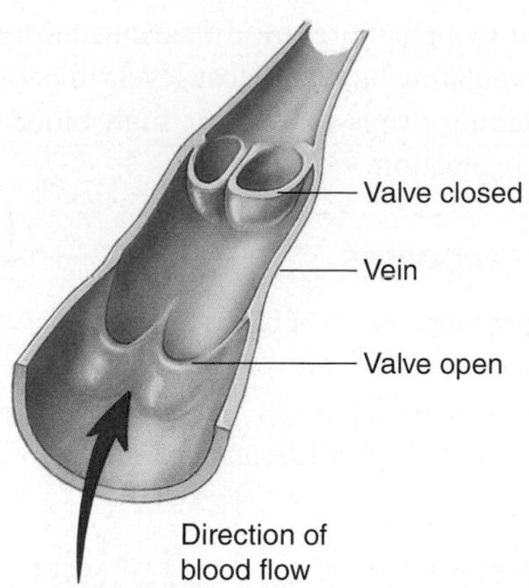
Valve closed
Vein
Valve open
Direction of blood flow

FIGURE 21.7 Valves
Folds in the tunica intima form the valves of veins, which allow blood to flow toward the heart but not in the opposite direction.

Valves

Veins that have diameters greater than 2 mm contain **valves,** which allow blood to flow toward the heart, but not in the opposite direction (figure 21.7). The valves consist of folds in the tunica intima that form two flaps shaped like the semilunar valves of the heart. The two folds overlap in the middle of the vein, so that, when blood attempts to flow in a reverse direction, the valves occlude the vessel. The medium veins contain many valves, and the number of valves is greater in veins of the lower limbs than in veins of the upper limbs.

Vasa Vasorum

For arteries and veins greater than 1 mm in diameter, nutrients cannot diffuse from the lumen of the vessel to all the layers of the wall. Therefore, nutrients are supplied to their walls by way of small blood vessels called **vasa vasorum** (vā′să vā′sor-ŭm), which penetrate from the exterior of the vessel to form a capillary network in the tunica adventitia and the tunica media (see figure 21.4).

ASSESS YOUR PROGRESS

10. *Name the three layers of a blood vessel. What kinds of tissue are in each layer?*
11. *List the types of arteries and veins. Compare the amount of elastic fibers and smooth muscle in each type of artery and vein.*
12. *Describe portal veins. Name two examples.*
13. *In which type of blood vessels are valves found? What is their function?*
14. *What is the vasa vasorum? What is its function?*

Neural Innervation of Blood Vessels

The walls of most blood vessels are richly innervated by unmyelinated sympathetic nerve fibers (see figure 21.4). Some blood vessels, such as those in the penis and clitoris, are innervated by parasympathetic fibers. The nerve fibers branch to form plexuses in the tunica adventitia, and nerve terminals containing neurotransmitter vesicles project among the smooth muscle cells of the tunica media. Synapses consist of several enlargements of each of the nerve fibers among the smooth muscle cells. Small arteries and arterioles are innervated to a greater extent than other blood vessel types. Sympathetic stimulation causes blood vessels to constrict; parasympathetic stimulation causes blood vessels in the penis and clitoris to dilate.

The smooth muscle cells of blood vessels act to some extent in unison. Gap junctions exist between adjacent smooth muscle cells; as a consequence, stimulation of a few smooth muscle cells in the vessel wall results in constriction of a relatively large segment of the blood vessel.

A few myelinated sensory neurons innervate some blood vessels and function as baroreceptors. They monitor stretch in the blood vessel wall and detect changes in blood pressure.

Clinical IMPACT

Varicose Veins, Phlebitis, and Gangrene

The veins of the lower limbs are subject to certain disorders. **Varicose veins** result when the veins of the lower limbs are stretched to the point that the valves become incompetent. Because of the stretching of the vein walls, the flaps of the valves no longer overlap to prevent the backflow of blood. As a consequence, the venous pressure is greater than normal in the veins of the lower limbs, resulting in edema. Blood flow in the veins can become sufficiently stagnant that the blood clots. This condition can result in **phlebitis** (fle-bī′tis), which is inflammation of the veins. If the inflammation is severe and blood flow becomes stagnant in a large area, it can lead to **gangrene** (gang′grēn), tissue death caused by a reduction in or loss of blood supply. Some people have a genetic propensity to develop varicose veins. The condition is further encouraged by activities that increase the pressure in the veins. One such condition is pregnancy, in which the venous pressure in the lower limbs increases because of compression by the expanded uterus. Also, standing in place for prolonged periods can lead to varicose veins.

Aging of the Arteries

The walls of all arteries undergo changes as people age, although some arteries change more rapidly than others and some people are more susceptible to change. The most significant change occurs in the large elastic arteries, such as the aorta, in the large arteries that carry blood to the brain, and in the coronary arteries; the age-related changes described here refer to these blood vessel types. Muscular arteries exhibit age-related changes, but these are less dramatic and seldom disrupt normal vessel function.

Arteriosclerosis (ar-tēr′ē-ō-skler-o′sis; hardening of the arteries) consists of degenerative changes in arteries that make them less elastic. These changes occur in many individuals, and they become more severe with advancing age. Arteriosclerosis greatly increases resistance to blood flow. Therefore, advanced arteriosclerosis reduces the normal circulation of blood and greatly increases the work performed by the heart.

Arteriosclerosis involves general hypertrophy of the tunica intima, including the internal elastic membrane, and hypertrophy of the tunica media. For example, when arteriosclerosis is associated with hypertension, the amount of smooth muscle and elastic tissue in the arterial walls increases. The elastic tissue can form concentric layers in the tunica intima, which becomes less elastic. Also some of the smooth muscle cells of the tunica media can ultimately be replaced by collagen fibers. Arteriosclerosis in some older people can involve the formation of calcium deposits in the tunica media of the arteries, primarily those of the lower limbs, with little or no encroachment on their lumens. The calcium deposits reduce the vessels' elasticity.

Atherosclerosis (ath′er-ō-skler-ō′sis) is the deposition of material in the walls of arteries to form distinct plaques. It is a common type of arteriosclerosis. Like the other types, atherosclerosis is related to age and certain risk factors. Atherosclerosis affects primarily medium and larger arteries, including the coronary arteries. The plaques form when macrophages containing cholesterol accumulate in the tunica intima, and smooth muscle cells of the tunica media proliferate (figure 21.8). After the plaques enlarge, they consist of smooth muscle cells, white blood cells, lipids (including cholesterol), and, in the largest plaques, fibrous connective tissue and calcium deposits. The plaques narrow the lumens of blood vessels and make their walls less elastic. Atherosclerotic plaques can become so large that they severely restrict or block blood flow through arteries. In addition, the plaques are sites of thrombosis and embolism formation.

Some investigators propose that arteriosclerosis is not a pathological process but an aging or wearing-out process. Evidence also suggests that arteriosclerosis may be caused by inflammation, possibly a result of autoimmune disease. Atherosclerosis has been studied extensively, and many risk factors have been associated with the development of atherosclerotic plaques. These risk factors include being elderly, being a male, being a postmenopausal woman, having a family history of atherosclerosis, smoking cigarettes, having hypertension, having diabetes mellitus, having increased blood LDL and cholesterol levels, being overweight, leading a sedentary lifestyle, and having high blood triglyceride levels. Avoiding the environmental factors that influence atherosclerosis slows the development of atherosclerotic plaques. In some cases, the severity of the plaques can be reduced by behavioral modifications and/or drug therapy. For example, regulating blood glucose levels in people with diabetes mellitus and taking drugs that lower high blood cholesterol can provide some protection.

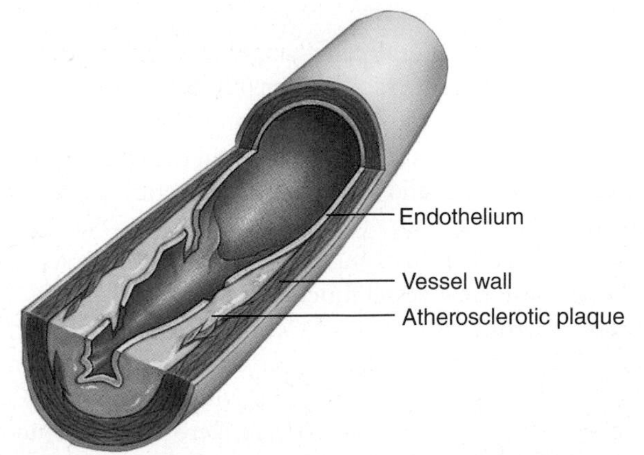
Endothelium
Vessel wall
Atherosclerotic plaque

FIGURE 21.8 Atherosclerotic Plaque in an Artery
Atherosclerotic plaques develop within the tissue of the artery wall.

ASSESS YOUR PROGRESS

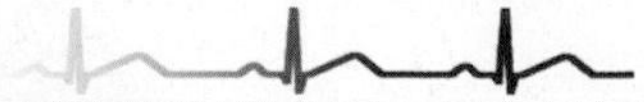

15. *Describe the innervation of blood vessel walls. Which types of vessels have the most innervation?*
16. *Describe the changes that occur in arteries due to aging. In which vessels do the most significant changes occur?*

21.3 Pulmonary Circulation

LEARNING OUTCOME

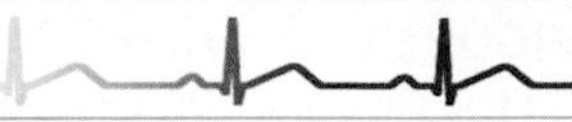

After reading this section, you should be able to

A. Trace the path of blood flow in pulmonary circulation.

The **pulmonary** (pŭl′mō-nār-ē; relating to the lungs) **circulation** is the system of blood vessels that carries blood from the right ventricle of the heart to the lungs and back to the left atrium of the heart. The heart pumps blood from the right ventricle into a short vessel (about 5 cm long) called the **pulmonary trunk** (figure 21.9). The pulmonary trunk then branches into the right and left **pulmonary arteries,** one transporting blood to each lung. Within the lungs, gas exchange occurs between the air in the lungs and the blood. Two **pulmonary veins** exit each lung and enter the left atrium (see figure 20.10).

ASSESS YOUR PROGRESS

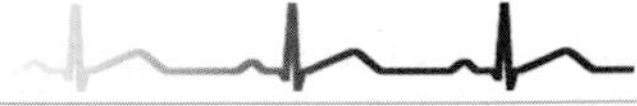

17. *Name, in order, the vessels of pulmonary circulation, beginning with the right ventricle.*

21.4 Systemic Circulation: Arteries

LEARNING OUTCOME

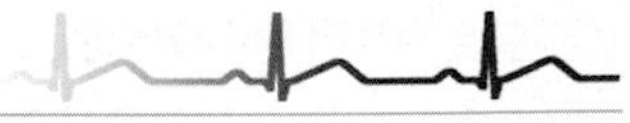

After reading this section, you should be able to

A. List the major arteries that supply each of the body areas.

The **systemic circulation** is the system of vessels that carries blood from the left ventricle of the heart to the tissues of the body and back to the right atrium. Oxygenated blood entering the heart from the pulmonary veins passes through the left atrium into the left ventricle and from the left ventricle into the aorta. Blood flows from the aorta to all parts of the body (figure 21.9).

Aorta

All arteries of the systemic circulation are derived either directly or indirectly from the **aorta** (ā-ōr′tă), which is usually divided into three general parts: the ascending aorta, the aortic arch, and the descending aorta. The descending aorta is divided further into a thoracic aorta and an abdominal aorta (see figure 21.15).

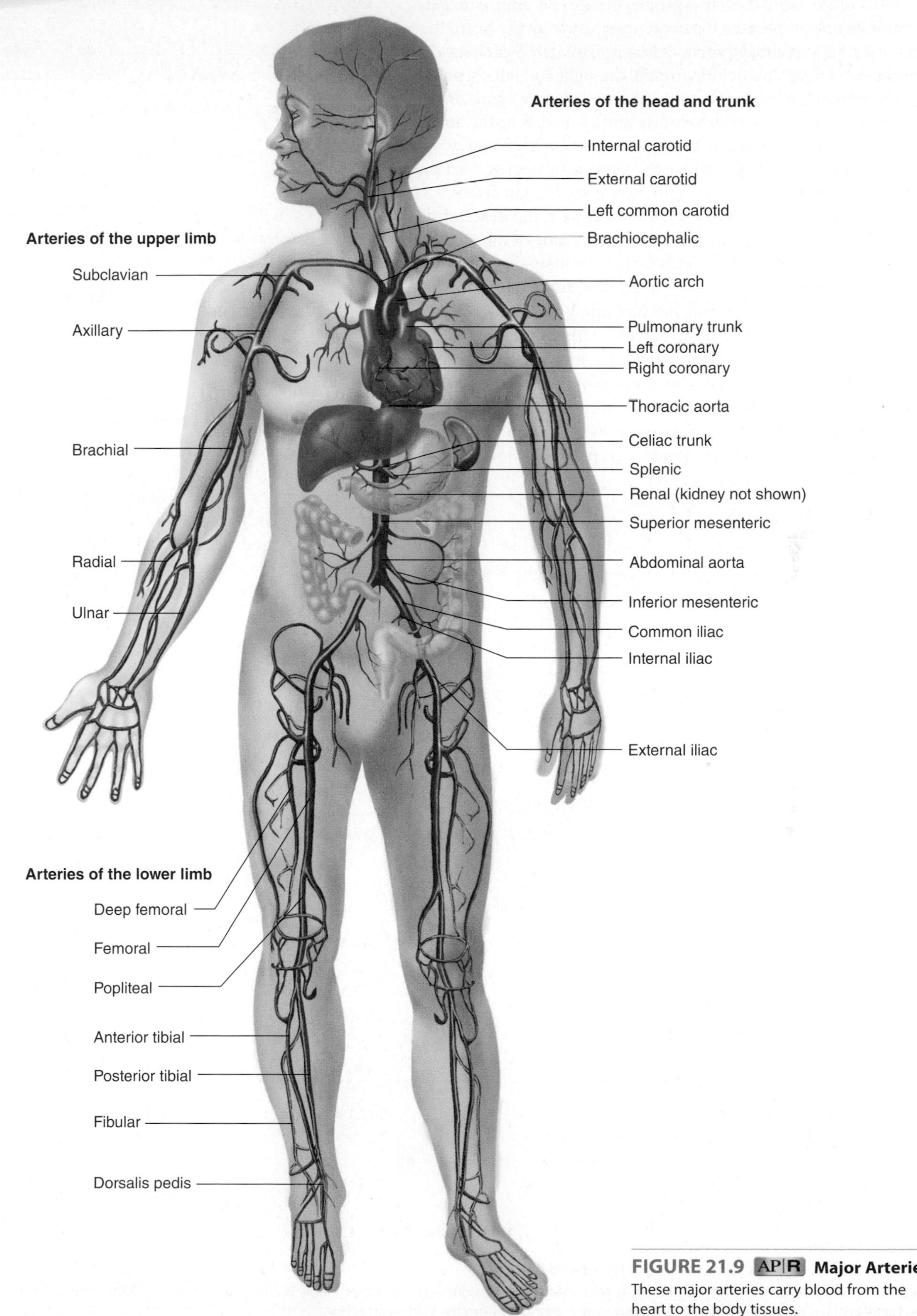
Arteries of the head and trunk
Internal carotid
External carotid
Left common carotid
Brachiocephalic
Aortic arch
Pulmonary trunk
Left coronary
Right coronary
Thoracic aorta
Celiac trunk
Splenic
Renal (kidney not shown)
Superior mesenteric
Abdominal aorta
Inferior mesenteric
Common iliac
Internal iliac
External iliac
Arteries of the upper limb
Subclavian
Axillary
Brachial
Radial
Ulnar
Arteries of the lower limb
Deep femoral
Femoral
Popliteal
Anterior tibial
Posterior tibial
Fibular
Dorsalis pedis

FIGURE 21.9 APR **Major Arteries**
These major arteries carry blood from the heart to the body tissues.

At its origin from the left ventricle, the aorta is approximately 2.8 cm in diameter. Because it passes superiorly from the heart, this part is called the **ascending aorta.** It is approximately 5 cm long and has only two arteries branching from it: the right and left **coronary arteries,** which supply blood to the cardiac muscle (see figure 20.6*a*).

The aorta then arches posteriorly and to the left as the **aortic arch.** Three major arteries that carry blood to the head and upper limbs originate from the aortic arch: the brachiocephalic artery, the left common carotid artery, and the left subclavian artery.

The next part of the aorta is the longest part, the **descending aorta.** It extends through the thorax in the left side of the mediastinum and through the abdomen to the superior margin of the pelvis. The **thoracic aorta** is the portion of the descending aorta located in the thorax. It has several branches that supply various structures between the aortic arch and the diaphragm. The **abdominal aorta** is the part of the descending aorta between the diaphragm and the point at which the aorta divides into the two **common iliac** (il′ē-ak; relating to the flank area) **arteries.** The abdominal aorta has several branches that supply the abdominal wall and organs. Its terminal branches, the common iliac arteries, supply blood to the pelvis and lower limbs.

Coronary Arteries

The **coronary** (kōr′o-năr-ē; encircling the heart like a crown) **arteries,** which are the only branches of the ascending aorta, are described in chapter 20.

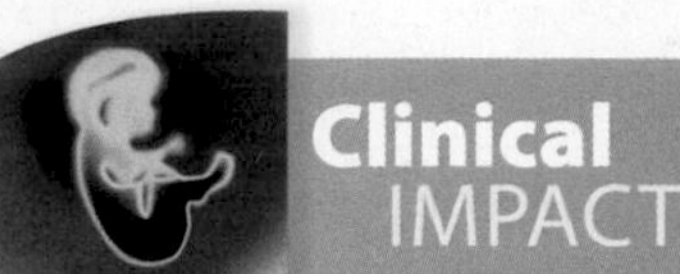
Clinical
IMPACT

Trauma and the Aorta

Trauma that ruptures the aorta is almost immediately fatal. Trauma can also lead to an **aneurysm** (an′ū-rizm), a bulge caused by a weakened spot in the aortic wall. Once the aneurysm forms, it is likely to enlarge and may rupture. The weakened aortic wall may leak blood slowly into the thorax and must be corrected surgically. The majority of traumatic aortic arch ruptures occur during automobile accidents when the body is thrown with great force into the steering wheel, the dashboard, or some other object. This type of injury is effectively prevented by shoulder-type safety belts and air bags.

Arteries of the Head and Neck

The first vessel to branch from the aortic arch is the **brachiocephalic** (brā′kē-ō-se-fal′ik; arm and head) **artery** (figure 21.10). This short artery branches at the level of the clavicle to form the **right common carotid** (ka-rot′id) **artery,** which transports blood to the right side of the head and neck, and the **right subclavian** (sŭb-klā′vē-an; below the clavicle) **artery,** which transports blood to the right upper limb (see figures 21.9, 21.10, and 21.12).

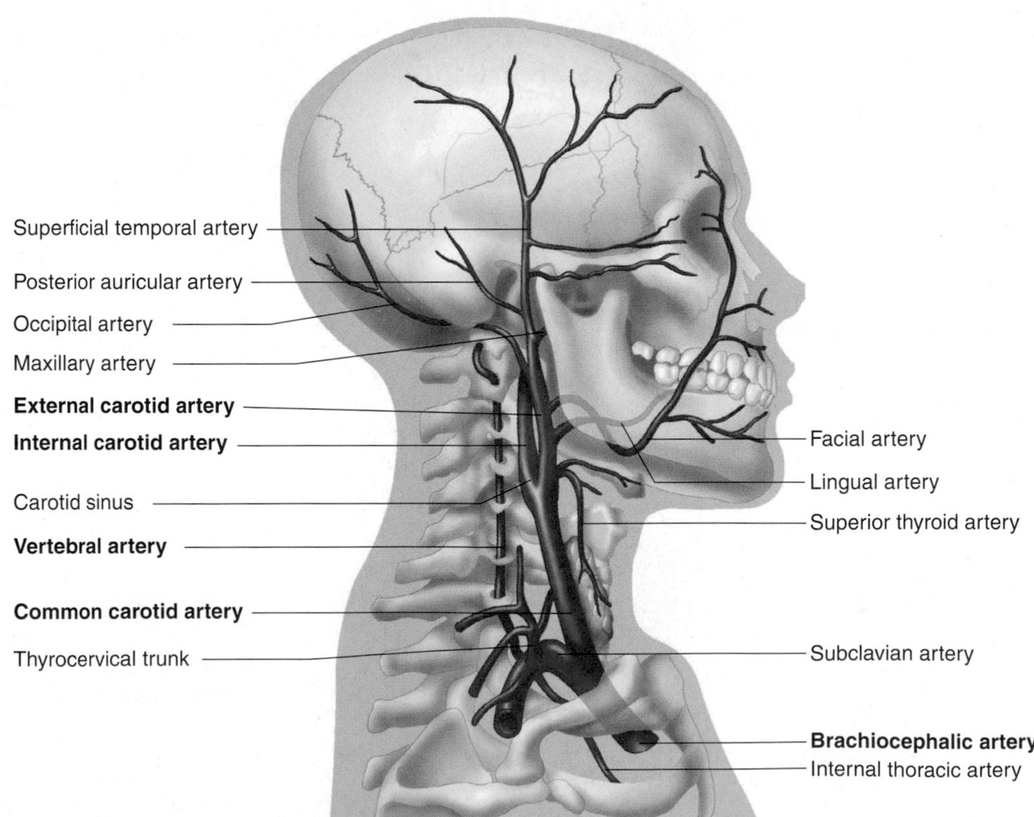
Superficial temporal artery
Posterior auricular artery
Occipital artery
Maxillary artery
External carotid artery
Internal carotid artery
Carotid sinus
Vertebral artery
Common carotid artery
Thyrocervical trunk
Facial artery
Lingual artery
Superior thyroid artery
Subclavian artery
Brachiocephalic artery
Internal thoracic artery

FIGURE 21.10 APR **Arteries of the Head and Neck**
The brachiocephalic artery, the right common carotid artery, and the right vertebral artery supply the head and neck. The right common carotid artery branches from the brachiocephalic artery, and the vertebral artery branches from the subclavian artery.

The second and third branches of the aortic arch are the **left common carotid artery,** which transports blood to the left side of the head and neck, and the **left subclavian artery,** which transports blood to the left upper limb.

The common carotid arteries extend superiorly, without branching, along each side of the neck, from their base to the inferior angle of the mandible. There each common carotid artery branches into **internal** and **external carotid arteries** (figure 21.10; see figure 21.12). At the point of bifurcation on each side of the neck, the common carotid artery and the base of the internal carotid artery are dilated slightly to form the **carotid sinus,** which is important in monitoring blood pressure (baroreceptor reflex). The external carotid arteries have several branches that supply the structures of the neck and face (table 21.1; see figures 21.10 and 21.12). The internal carotid arteries, together with the vertebral arteries, which are branches of the subclavian arteries, supply the brain (see figures 21.10, 21.11, and 21.12; table 21.1).

Predict 2

The term *carotid* means to put to sleep, implying that, if the carotid arteries are occluded for even a short time, the patient can lose consciousness (go to sleep). The blood supply to the brain is extremely important to brain function. Elimination of this supply for even a relatively short time can result in permanent brain damage because the brain is dependent on oxidative metabolism and quickly malfunctions in the absence of oxygen. What is the physiological significance of arteriosclerosis, which slowly reduces blood flow through the carotid arteries?

Branches of the subclavian arteries, the **left** and **right vertebral arteries,** pass through the transverse foramina of the cervical vertebrae and enter the cranial cavity through the foramen magnum. They give off arteries to the cerebellum and then unite to form a single, midline **basilar** (bas′i-lăr) **artery** (figures 21.11 and 21.12; table 21.1). The basilar artery gives off branches to the pons and the cerebellum and then branches to form the **posterior cerebral arteries,** which supply the posterior part of the cerebrum (see figure 21.11).

The internal carotid arteries enter the cranial vault through the carotid canals and terminate by forming the **middle cerebral arteries,** which supply large parts of the lateral cerebral cortex (see figure 21.11). Posterior branches of these arteries, called the **posterior communicating arteries,** unite with the posterior cerebral arteries. Anterior branches, called the **anterior cerebral arteries,** supply blood to the frontal lobes of the brain. The anterior cerebral arteries are in turn connected by an **anterior communicating artery,** which completes a circle around the pituitary gland and the base of the brain called the **cerebral arterial circle** (circle of Willis; see figures 21.11 and 21.12).

Arteries of the Upper Limb

The three major arteries of the upper limb, the **subclavian, axillary,** and **brachial arteries,** are a continuum rather than a branching system. The subclavian artery is located deep to the clavicle. The axillary artery is the continuation of the subclavian artery in the axilla. The brachial artery is the continuation of the axillary artery as it passes into the arm (figures 21.13 and 21.14; table 21.2).

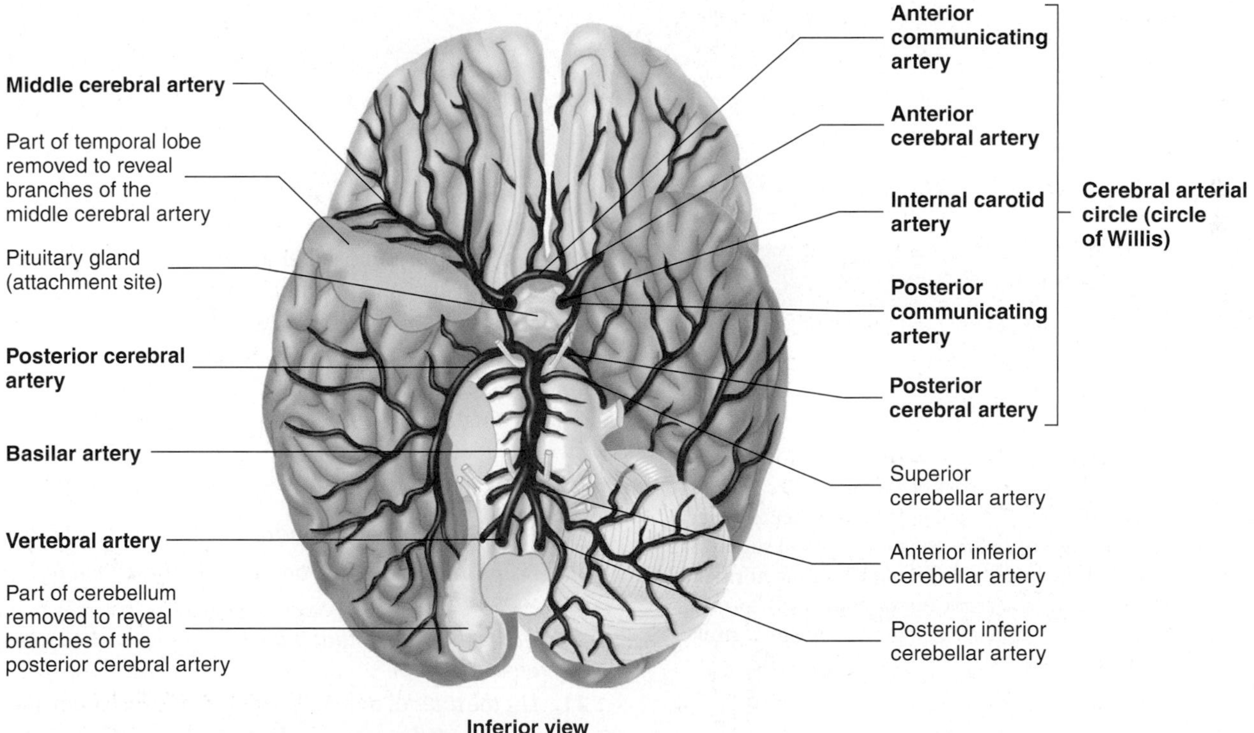
Middle cerebral artery
Part of temporal lobe removed to reveal branches of the middle cerebral artery
Pituitary gland (attachment site)
Posterior cerebral artery
Basilar artery
Vertebral artery
Part of cerebellum removed to reveal branches of the posterior cerebral artery
Anterior communicating artery
Anterior cerebral artery
Internal carotid artery
Posterior communicating artery
Posterior cerebral artery
Cerebral arterial circle (circle of Willis)
Superior cerebellar artery
Anterior inferior cerebellar artery
Posterior inferior cerebellar artery
Inferior view

FIGURE 21.11 APR **Cerebral Arterial Circle (Circle of Willis)**
The internal carotid and vertebral arteries carry blood to the brain. The vertebral arteries join to form the basilar artery. Branches of the internal carotid arteries and the basilar artery supply blood to the brain and complete a circle of arteries around the pituitary gland and the base of the brain called the cerebral arterial circle (circle of Willis).

TABLE 21.1 Arteries of the Head and Neck (figures 21.10, 21.11, and 21.12)

Arteries	Tissues Supplied
Common Carotid Arteries	Head and neck by branches listed below
External Carotid	
Superior thyroid	Neck, larynx, and thyroid gland
Lingual	Tongue, mouth, and submandibular and sublingual glands
Facial	Mouth, pharynx, and face
Occipital	Posterior head and neck and meninges around posterior brain
Posterior auricular	Middle and inner ear, head, and neck
Ascending pharyngeal	Deep neck muscles, middle ear, pharynx, soft palate, and meninges around posterior brain
Superficial temporal	Temple, face, and anterior ear
Maxillary	Middle and inner ears, meninges, lower jaw and teeth, upper jaw and teeth, temple, external eye structures, face, palate, and nose
Internal Carotid	
Posterior communicating	Joins the posterior cerebral artery
Anterior cerebral	Anterior portions of the cerebrum; forms the anterior communicating arteries
Middle cerebral	Most of the lateral surface of the cerebrum
Vertebral Arteries (branches of the subclavian arteries)	
Anterior spinal	Anterior spinal cord
Posterior inferior cerebellar	Cerebellum and fourth ventricle
Basilar Artery (formed by junction of vertebral arteries)	
Anterior interior cerebellar	Cerebellum
Superior cerebellar	Cerebellum and midbrain
Posterior cerebral	Posterior portions of the cerebrum

Clinical IMPACT

Stroke

A **stroke** is a sudden neurological disorder often caused by decreased blood supply to a part of the brain. It can occur as a result of a thrombosis, an embolism, or a hemorrhage. Any one of these conditions can reduce the brain's blood supply or cause trauma to a part of the brain. As a result, the tissue normally supplied by the arteries becomes **necrotic** (nĕ-krot′ik; dead), forming an infarct in the affected areas. The neurological results of a stroke are described in chapter 14.

The brachial artery divides at the elbow into **ulnar** and **radial arteries,** which form two arches within the palm of the hand. The **superficial palmar arch** is formed by the ulnar artery and is completed by anastomosing with the radial artery. The **deep palmar arch** is formed by the radial artery and is completed by anastomosing with the ulnar artery. This arch is not only deep to the superficial arch but proximal as well.

Digital (dij′i-tăl; relating to the digits—the fingers and the thumb) **arteries** branch from each of the two palmar arches and unite to form single arteries on the medial and lateral sides of each digit.

ASSESS **YOUR PROGRESS**

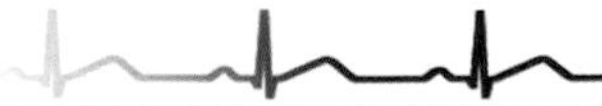

18. *Name the parts of the aorta.*
19. *Name the arteries that branch from the ascending aorta to supply the heart.*
20. *Name the arteries that branch from the aorta to supply the head and neck.*
21. *List the arteries that are part of, and branch from, the cerebral arterial circle.*
22. *Name the arteries that branch from the aorta to supply the upper limbs.*
23. *List, in order, the arteries that travel through the upper limb to the digits.*

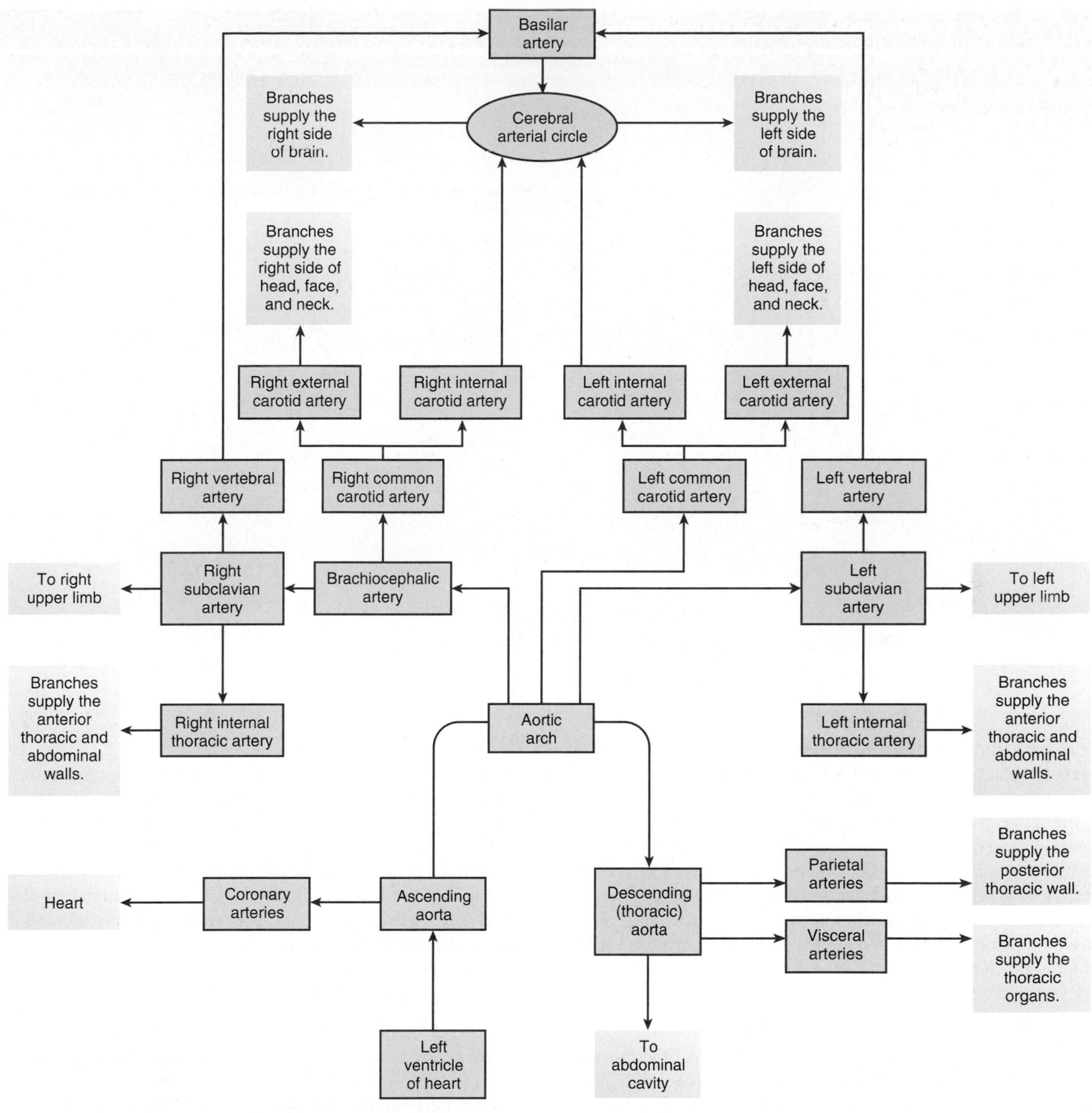
Basilar artery
Branches supply the right side of brain.
Cerebral arterial circle
Branches supply the left side of brain.
Branches supply the right side of head, face, and neck.
Branches supply the left side of head, face, and neck.
Right external carotid artery
Right internal carotid artery
Left internal carotid artery
Left external carotid artery
Right vertebral artery
Right common carotid artery
Left common carotid artery
Left vertebral artery
To right upper limb
Right subclavian artery
Brachiocephalic artery
Left subclavian artery
To left upper limb
Branches supply the anterior thoracic and abdominal walls.
Right internal thoracic artery
Aortic arch
Left internal thoracic artery
Branches supply the anterior thoracic and abdominal walls.
Branches supply the posterior thoracic wall.
Parietal arteries
Heart
Coronary arteries
Ascending aorta
Descending (thoracic) aorta
Visceral arteries
Branches supply the thoracic organs.
Left ventricle of heart
To abdominal cavity

FIGURE 21.12 Major Arteries of the Head and Thorax

Thoracic Aorta and Its Branches

The branches of the thoracic aorta are divided into two groups: the **visceral branches** supplying the thoracic organs and the **parietal branches** supplying the thoracic wall (figure 21.15*a,b*; table 21.3). The visceral branches supply a portion of the lungs, including the bronchi and bronchioles (see chapter 23), the esophagus, and the pericardium. Even though a large quantity of blood flows to the lungs through the pulmonary arteries, the bronchi and bronchioles require a separate oxygenated blood supply through small bronchial branches from the thoracic aorta.

The walls of the thorax are supplied with blood by the **intercostal** (in-ter-kos′tăl; between the ribs) **arteries,** which consist of two sets:

TABLE 21.2 **Arteries of the Upper Limb (figures 21.13 and 21.14)**

Arteries	Tissues Supplied
Subclavian Arteries (right subclavian originates from the brachiocephalic artery, and left subclavian originates directly from the aorta)	
Vertebral	Spinal cord and cerebellum form the basilar artery (see table 21.1)
Internal thoracic	Diaphragm, mediastinum, pericardium, anterior thoracic wall, and anterior abdominal wall
Thyrocervical trunk	Inferior neck and shoulder
Axillary Arteries (continuation of subclavian)	
Thoracoacromial	Pectoral region and shoulder
Lateral thoracic	Pectoral muscles, mammary gland, and axilla
Subscapular	Scapular muscles
Brachial Arteries (continuation of axillary arteries)	
Deep brachial	Arm and humerus
Radial	Forearm
Deep palmar arch	Hand and fingers
Digital arteries	Fingers
Ulnar	Forearm
Superficial palmar arch	Hand and fingers
Digital arteries	Fingers

the anterior intercostals and the posterior intercostals. The **anterior intercostals** are derived from the **internal thoracic arteries,** which are branches of the subclavian arteries and lie on the inner surface of the anterior thoracic wall (figure 21.15*a,b*; table 21.3). The **posterior intercostals** are derived as bilateral branches directly from the descending aorta. The anterior and posterior intercostal arteries lie along the inferior margin of each rib and anastomose with each other approximately midway between the ends of the ribs. **Superior phrenic** (fren′ik; to the diaphragm) **arteries** supply blood to the diaphragm.

Abdominal Aorta and Its Branches

The branches of the abdominal aorta, like those of the thoracic aorta, are divided into visceral and parietal parts (figures 21.15*a,c* and 21.16; table 21.3). The visceral arteries are in turn divided into paired and unpaired branches. Three major unpaired branches exist: the **celiac** (sē′lē-ak; belly) **trunk,** the **superior mesenteric** (mez-en-ter′ik; relating to the mesenteries) **artery,** and the **inferior mesenteric artery** (see figure 21.15*a,c*). Each has several major branches supplying the abdominal organs.

The paired visceral branches of the abdominal aorta supply the kidneys, adrenal glands, and gonads (testes and ovaries). The parietal arteries of the abdominal aorta supply the diaphragm and the abdominal wall (figure 21.16).

Arteries of the Pelvis

The abdominal aorta divides at the level of the fifth lumbar vertebra into two **common iliac arteries.** They divide to form the **external iliac arteries,** which enter the lower limbs, and the **internal iliac arteries,** which supply the pelvic area. Visceral branches supply the pelvic organs, such as the urinary bladder, rectum, uterus, and vagina. Parietal branches supply blood to the walls and floor of the pelvis; the lumbar, gluteal, and proximal thigh muscles; and the external genitalia (figures 21.16 and 21.17; table 21.4).

Arteries of the Lower Limb

The arteries of the lower limb form a continuum similar to that of the arteries of the upper limb. The **external iliac artery** becomes the **femoral** (fem′ŏ-răl; relating to the thigh) **artery** in the thigh, which becomes the **popliteal** (pop-lit′ē-ăl, pop-li-tē′ăl; ham, the hamstring area posterior to the knee) **artery** in the popliteal space. The popliteal artery gives off the **anterior tibial artery** just inferior to the knee and then continues as the **posterior tibial artery.** The anterior tibial artery becomes the **dorsalis pedis artery** at the foot. The posterior tibial artery gives off the **fibular artery,** or *peroneal artery,* and then gives rise to **medial** and **lateral plantar** (plan′tăr; the sole of the foot) **arteries,** which in turn give off **digital branches** to the toes. The arteries of the lower limb are illustrated in figures 21.17 and 21.18 and are listed in table 21.5.

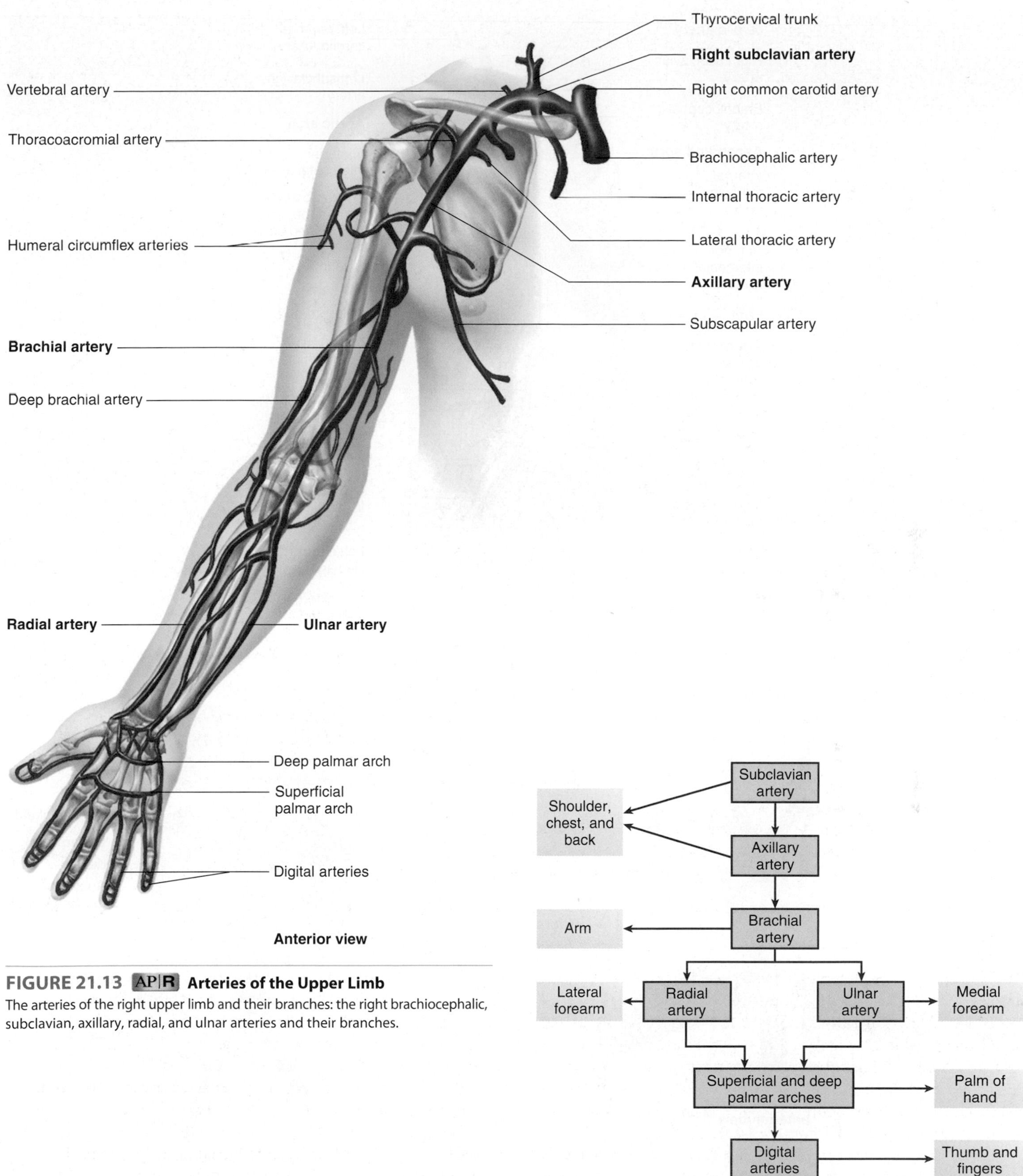
Thyrocervical trunk
Right subclavian artery
Right common carotid artery
Vertebral artery
Thoracoacromial artery
Brachiocephalic artery
Internal thoracic artery
Humeral circumflex arteries
Lateral thoracic artery
Axillary artery
Subscapular artery
Brachial artery
Deep brachial artery
Radial artery
Ulnar artery
Deep palmar arch
Superficial palmar arch
Digital arteries
Anterior view
Subclavian artery
Shoulder, chest, and back
Axillary artery
Arm
Brachial artery
Lateral forearm
Radial artery
Ulnar artery
Medial forearm
Superficial and deep palmar arches
Palm of hand
Digital arteries
Thumb and fingers

FIGURE 21.13 AP|R **Arteries of the Upper Limb**
The arteries of the right upper limb and their branches: the right brachiocephalic, subclavian, axillary, radial, and ulnar arteries and their branches.

FIGURE 21.14 Major Arteries of the Shoulder and Upper Limb

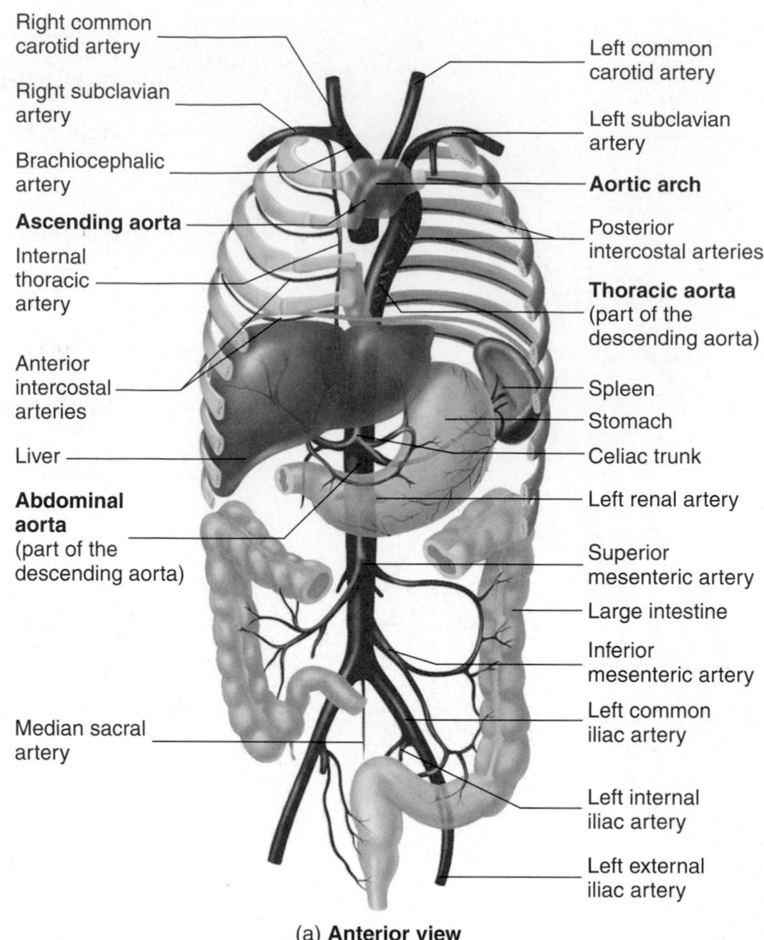
Right common carotid artery
Right subclavian artery
Brachiocephalic artery
Ascending aorta
Internal thoracic artery
Anterior intercostal arteries
Liver
Abdominal aorta (part of the descending aorta)
Median sacral artery
Left common carotid artery
Left subclavian artery
Aortic arch
Posterior intercostal arteries
Thoracic aorta (part of the descending aorta)
Spleen
Stomach
Celiac trunk
Left renal artery
Superior mesenteric artery
Large intestine
Inferior mesenteric artery
Left common iliac artery
Left internal iliac artery
Left external iliac artery

(a) **Anterior view**

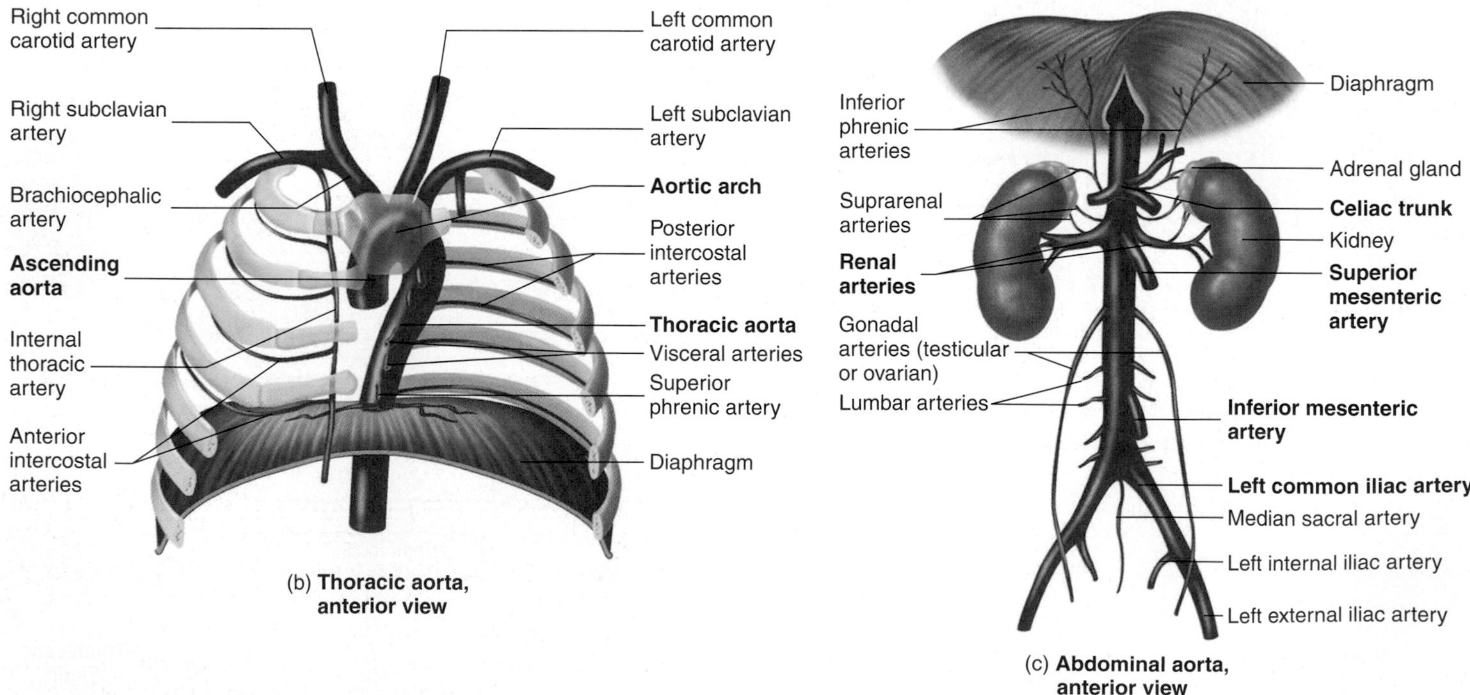
Right common carotid artery
Right subclavian artery
Brachiocephalic artery
Ascending aorta
Internal thoracic artery
Anterior intercostal arteries
Left common carotid artery
Left subclavian artery
Aortic arch
Posterior intercostal arteries
Thoracic aorta
Visceral arteries
Superior phrenic artery
Diaphragm
Inferior phrenic arteries
Suprarenal arteries
Renal arteries
Gonadal arteries (testicular or ovarian)
Lumbar arteries
Diaphragm
Adrenal gland
Celiac trunk
Kidney
Superior mesenteric artery
Inferior mesenteric artery
Left common iliac artery
Median sacral artery
Left internal iliac artery
Left external iliac artery

(b) **Thoracic aorta, anterior view**

(c) **Abdominal aorta, anterior view**

FIGURE 21.15 Branches of the Aorta

(*a*) The aorta is considered in three portions: the ascending aorta, the aortic arch, and the descending aorta. The descending aorta consists of the thoracic aorta and the abdominal aorta. (*b*) The thoracic aorta. (*c*) The abdominal aorta.

TABLE 21.3 Thoracic and Abdominal Aorta (see figures 21.15 and 21.16)

Arteries	Tissues Supplied
Thoracic Aorta	
Visceral Branches	
Bronchial	Lung tissue
Esophageal	Esophagus
Parietal Branches	
Intercostal	Thoracic wall
Superior phrenic	Superior surface of diaphragm
Abdominal Aorta	
Visceral Branches	
Unpaired	
Celiac trunk	
Left gastric	Stomach and esophagus
Common hepatic	
Gastroduodenal	Stomach and duodenum
Right gastric	Stomach
Hepatic	Liver
Splenic	Spleen and pancreas
Left gastroepiploic	Stomach
Superior mesenteric	Pancreas, small intestine, and colon
Inferior mesenteric	Descending colon and rectum
Paired	
Suprarenal	Adrenal gland
Renal	Kidney
Gonadal	
Testicular (male)	Testis and ureter
Ovarian (female)	Ovary, ureter, and uterine tube
Parietal Branches	
Inferior phrenic	Adrenal gland and inferior surface of diaphragm
Lumbar	Lumbar vertebrae and back muscles
Median sacral	Inferior vertebrae
Common iliac	
External iliac	Lower limb (see table 21.5)
Internal iliac	Lower back, hip, pelvis, urinary bladder, vagina, uterus, rectum, and external genitalia (see table 21.4)

TABLE 21.4 Arteries of the Pelvis (figures 21.16 and 21.17)

Arteries	Tissues Supplied
Internal Iliac	Pelvis through the branches listed below
Visceral Branches	
Middle rectal	Rectum
Vaginal	Vagina and uterus
Uterine	Uterus, vagina, uterine tube, and ovary
Parietal Branches	
Lateral sacral	Sacrum
Superior gluteal	Muscles of the gluteal region
Obturator	Pubic region, deep groin muscles, and hip joint
Internal pudendal	Rectum, external genitalia, and floor of pelvis
Inferior gluteal	Inferior gluteal region, coccyx, and proximal thigh

ASSESS YOUR PROGRESS

24. *Name the two types of branches arising from the thoracic aorta. What structures are supplied from each group?*
25. *What areas of the body are supplied by the paired arteries that branch from the abdominal aorta? The unpaired arteries? Name the three major unpaired arteries.*
26. *Name the arteries that branch from the aorta to supply the pelvic area. List the organs of the pelvis that are supplied by branches of these arteries.*
27. *List, in order, the arteries that travel from the aorta to the digits of the lower limbs.*

21.5 Systemic Circulation: Veins

LEARNING OUTCOME

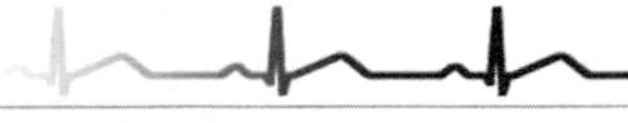

After reading this section, you should be able to

A. List the major veins that carry blood away from each of the body areas.

Three major veins return blood from the body to the right atrium: the **coronary sinus,** returning blood from the walls of the heart (see figures 20.6*b* and 20.7); the **superior vena cava** (vē′nă kă′vă, kā′vă; venous cave), returning blood from the head, neck, thorax, and upper limbs; and the **inferior vena cava,** returning blood from the abdomen, pelvis, and lower limbs (figure 21.19).

In a very general way, the smaller veins follow the same course as the arteries, and many are given the same names. The veins, however, are more numerous and more variable. The larger veins often follow a very different course and have names different from the arteries.

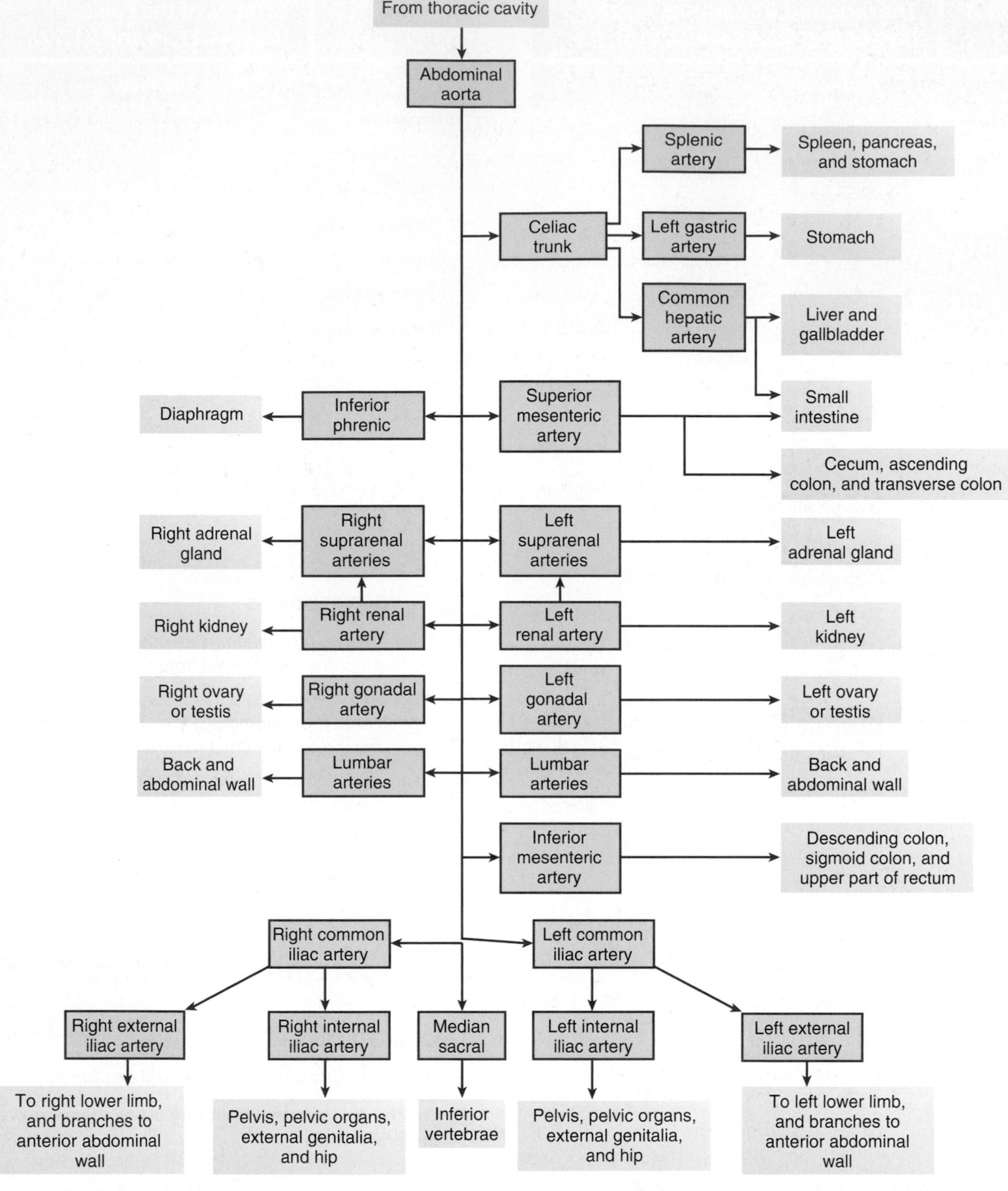
From thoracic cavity
Abdominal aorta
Splenic artery
Spleen, pancreas, and stomach
Celiac trunk
Left gastric artery
Stomach
Common hepatic artery
Liver and gallbladder
Diaphragm
Inferior phrenic
Superior mesenteric artery
Small intestine
Cecum, ascending colon, and transverse colon
Right adrenal gland
Right suprarenal arteries
Left suprarenal arteries
Left adrenal gland
Right kidney
Right renal artery
Left renal artery
Left kidney
Right ovary or testis
Right gonadal artery
Left gonadal artery
Left ovary or testis
Back and abdominal wall
Lumbar arteries
Lumbar arteries
Back and abdominal wall
Inferior mesenteric artery
Descending colon, sigmoid colon, and upper part of rectum
Right common iliac artery
Left common iliac artery
Right external iliac artery
Right internal iliac artery
Median sacral
Left internal iliac artery
Left external iliac artery
To right lower limb, and branches to anterior abdominal wall
Pelvis, pelvic organs, external genitalia, and hip
Inferior vertebrae
Pelvis, pelvic organs, external genitalia, and hip
To left lower limb, and branches to anterior abdominal wall

FIGURE 21.16 Major Arteries of the Abdomen and Pelvis
Visceral branches include those that are unpaired (celiac trunk, superior mesenteric, and inferior mesenteric) and those that are paired (renal, suprarenal, testicular, and ovarian). Parietal branches include inferior phrenic, lumbar, and median sacral.

The three major types of veins are superficial veins, deep veins, and sinuses. In general, the superficial veins of the limbs are larger than the deep veins, whereas in the head and trunk the opposite is the case. Venous sinuses occur primarily in the cranial cavity and the heart.

Veins Draining the Heart

The **cardiac veins,** which transport blood from the walls of the heart and return it through the coronary sinus to the right atrium, are described in chapter 20.

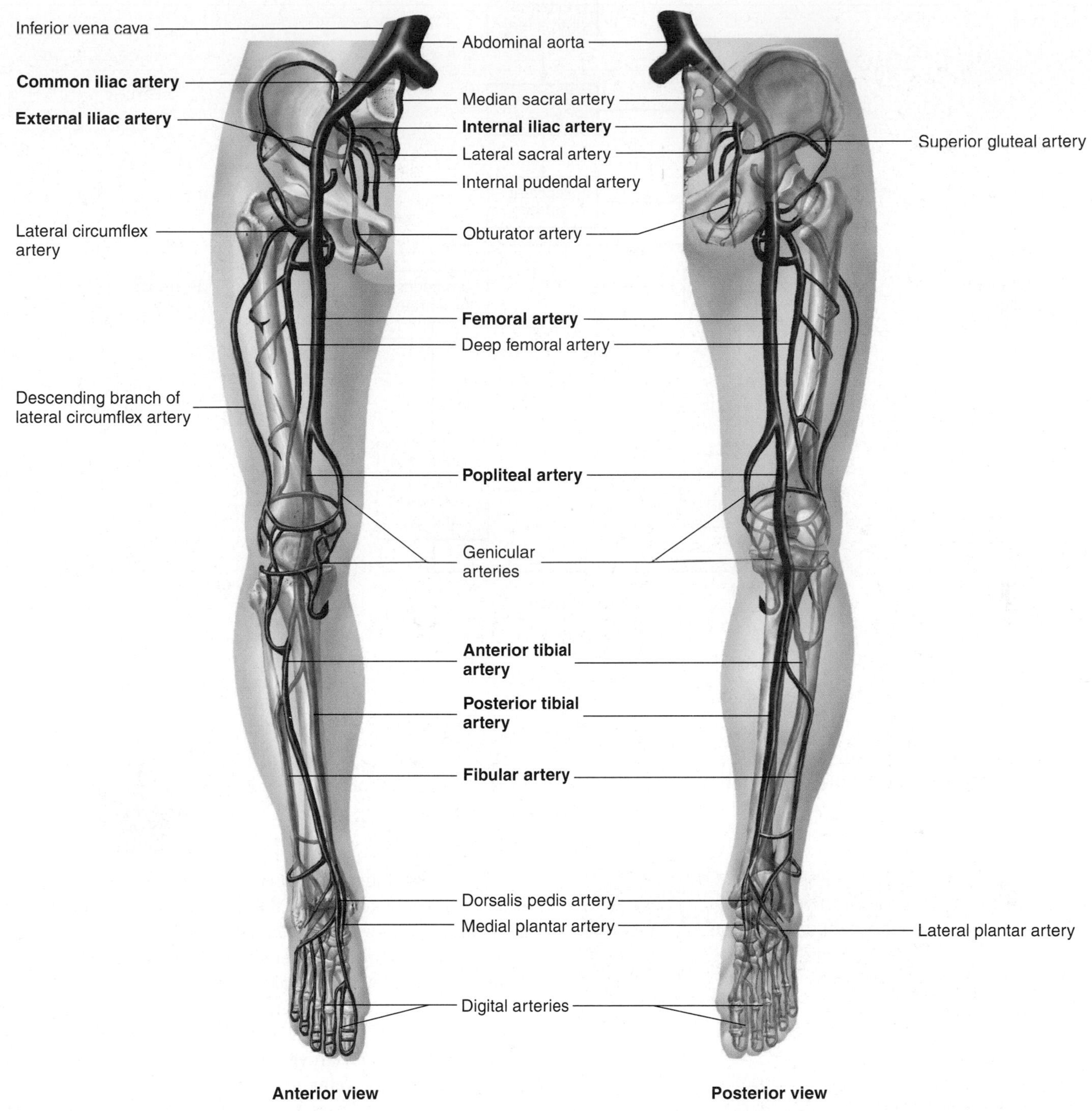

Inferior vena cava
Common iliac artery
External iliac artery
Lateral circumflex artery
Descending branch of lateral circumflex artery
Abdominal aorta
Median sacral artery
Internal iliac artery
Lateral sacral artery
Internal pudendal artery
Obturator artery
Femoral artery
Deep femoral artery
Popliteal artery
Genicular arteries
Anterior tibial artery
Posterior tibial artery
Fibular artery
Dorsalis pedis artery
Medial plantar artery
Digital arteries
Superior gluteal artery
Lateral plantar artery
Anterior view
Posterior view

FIGURE 21.17 Arteries of the Pelvis and Lower Limb
The internal and external iliac arteries and their branches. The internal iliac artery supplies the pelvis and hip, and the external iliac artery supplies the lower limb through the femoral artery.

Veins of the Head and Neck

The two pairs of major veins that drain blood from the head and neck are the **external** and **internal jugular** (jŭg′ū-lar; neck) **veins**. The external jugular veins are the more superficial of the two sets, and they drain blood primarily from the posterior head and neck. The external jugular vein drains into the subclavian vein. The internal jugular veins are much larger and deeper than the external jugular veins. They drain blood from the cranial cavity and the anterior head, face, and neck.

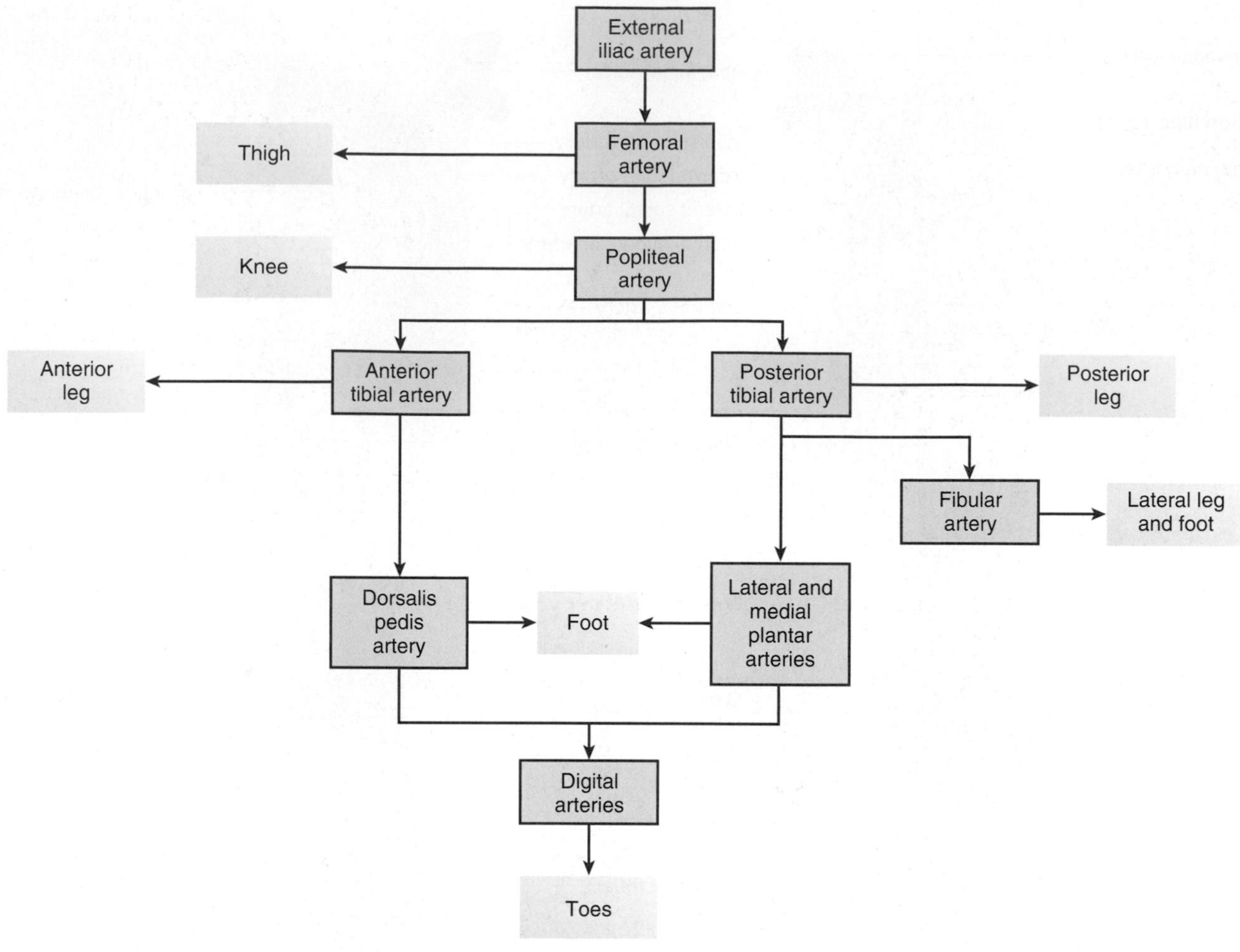
External iliac artery
Femoral artery
Thigh
Popliteal artery
Knee
Anterior tibial artery
Anterior leg
Posterior tibial artery
Posterior leg
Fibular artery
Lateral leg and foot
Dorsalis pedis artery
Foot
Lateral and medial plantar arteries
Digital arteries
Toes

FIGURE 21.18 **Major Arteries of the Lower Limb**

TABLE 21.5 Arteries of the Lower Limb (figures 21.17 and 21.18)

Arteries	Tissues Supplied
Femoral	Thigh, external genitalia, and anterior abdominal wall
Deep femoral	Thigh, knee, and femur
Popliteal (continuation of the femoral artery)	
Posterior tibial	Knee and leg
Fibular (peroneal)	Calf and peroneal muscles and ankle
Medial plantar	Plantar region of foot
Digital	Digits of foot
Lateral plantar	Plantar region of foot
Digital	Digits of foot
Anterior tibial	Knee and leg
Dorsalis pedis	Dorsum of foot
Digital	Digits of foot

The internal jugular vein is formed primarily as the continuation of the **venous sinuses** of the cranial cavity. The venous sinuses are actually spaces within the dura mater surrounding the brain (see chapter 13). They are depicted in figure 21.20 and listed in table 21.6.

Once the internal jugular veins exit the cranial cavity, they receive several venous tributaries that drain the external head and face (figures 21.21 and 21.22; table 21.7). The **internal jugular veins** join the **subclavian veins** on each side of the body to form the **brachiocephalic veins.**

Veins of the Upper Limb

The **cephalic** (se-fal′ik; toward the head), **basilic** (ba-sil′ik), and **brachial veins** are responsible for draining most of the blood from the upper limbs (figures 21.23 and 21.24; table 21.8). Many of the tributaries of the cephalic and basilic veins in the forearm and hand can be seen through the skin. Because of the considerable variation in the tributary veins of the forearm and hand, they often are left unnamed. The basilic vein of the arm becomes the **axillary vein** as it courses through the axillary region. The axillary vein then becomes the **subclavian vein** at the margin of the first rib. The cephalic vein enters the axillary vein.

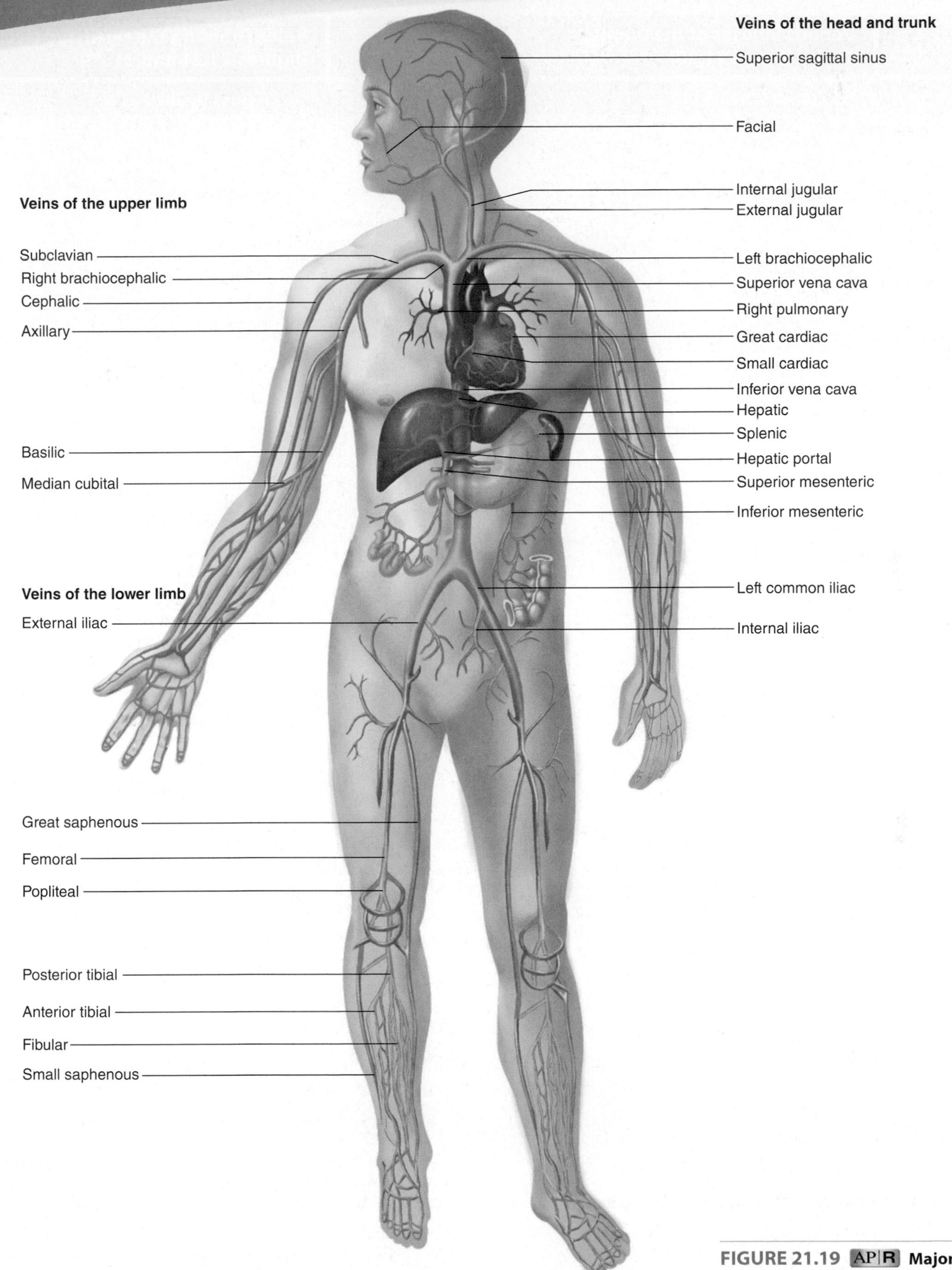
Veins of the head and trunk
Superior sagittal sinus
Facial
Internal jugular
External jugular
Left brachiocephalic
Superior vena cava
Right pulmonary
Great cardiac
Small cardiac
Inferior vena cava
Hepatic
Splenic
Hepatic portal
Superior mesenteric
Inferior mesenteric
Left common iliac
Internal iliac
Veins of the upper limb
Subclavian
Right brachiocephalic
Cephalic
Axillary
Basilic
Median cubital
Veins of the lower limb
External iliac
Great saphenous
Femoral
Popliteal
Posterior tibial
Anterior tibial
Fibular
Small saphenous
Anterior view

FIGURE 21.19 APR **Major Veins**
These veins carry blood from the body tissues to the heart.

TABLE 21.6 Venous Sinuses of the Cranial Cavity (figure 21.20)

Veins	Tissues Drained
Internal Jugular Vein	
Sigmoid sinus	
Superior and inferior petrosal sinuses	Anterior portion of cranial cavity
Cavernous sinus	
Ophthalmic veins	Orbit
Transverse sinus	
Occipital sinus	Central floor of posterior fossa of skull
Superior sagittal sinus	Superior portion of cranial cavity and brain
Straight sinus	
Inferior sagittal sinus	Deep portion of longitudinal fissure

TABLE 21.7 Veins Draining the Head and Neck (figures 21.21 and 21.22)

Veins	Tissues Drained
Brachiocephalic	
Internal jugular	Brain
Lingual	Tongue and mouth
Superior thyroid	Thyroid and deep posterior facial structures (also empties into external jugular)
Facial	Superficial and anterior facial structures
External jugular	Superficial surface of posterior head and neck

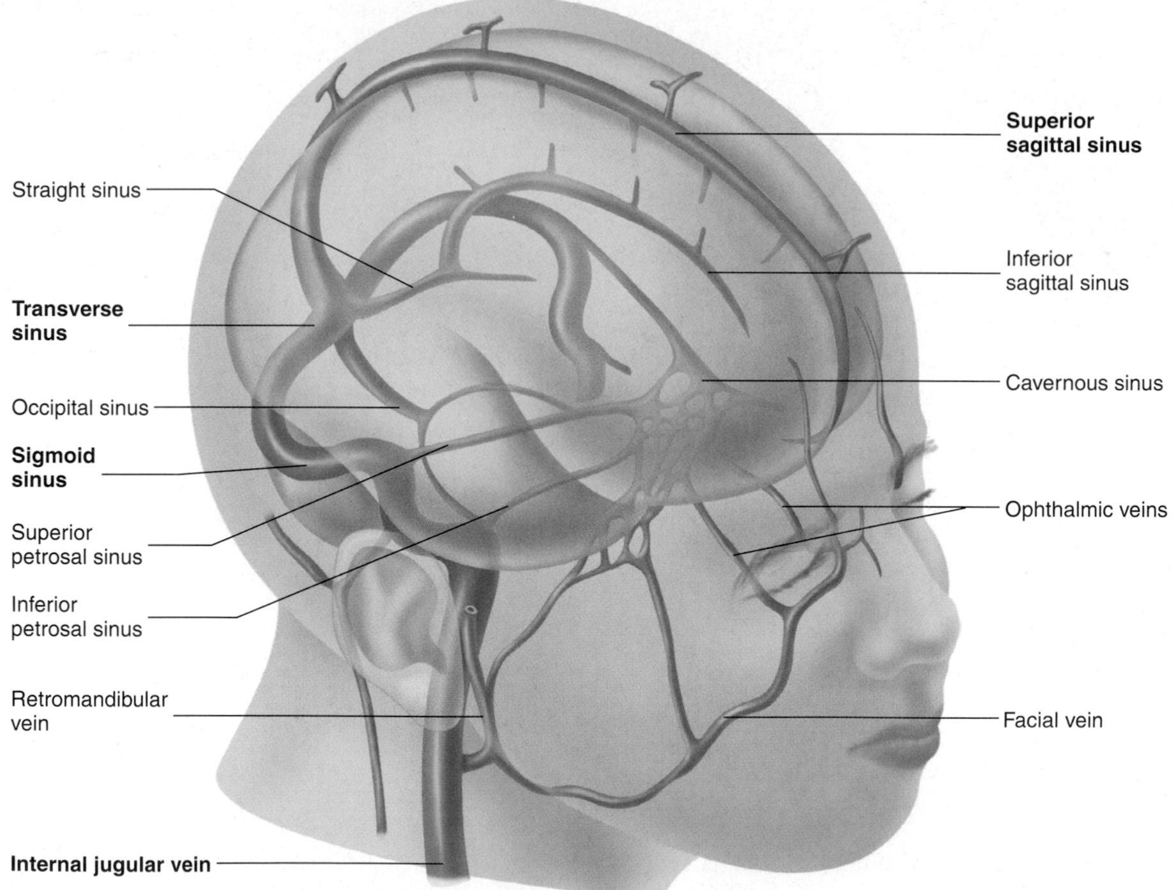

Straight sinus
Transverse sinus
Occipital sinus
Sigmoid sinus
Superior petrosal sinus
Inferior petrosal sinus
Retromandibular vein
Internal jugular vein
Superior sagittal sinus
Inferior sagittal sinus
Cavernous sinus
Ophthalmic veins
Facial vein

FIGURE 21.20 Venous Sinuses Associated with the Brain

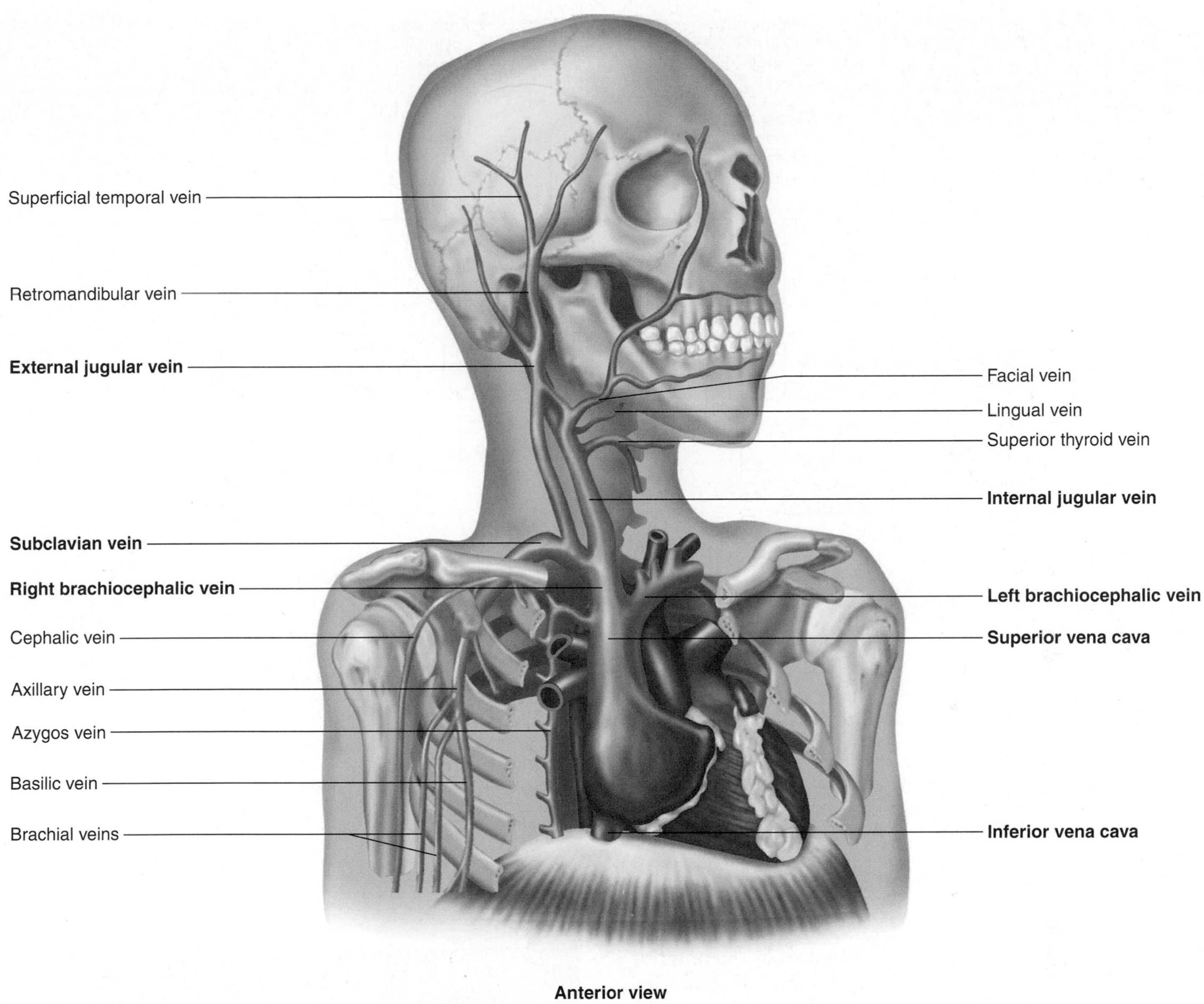
Superficial temporal vein
Retromandibular vein
External jugular vein
Subclavian vein
Right brachiocephalic vein
Cephalic vein
Axillary vein
Azygos vein
Basilic vein
Brachial veins
Facial vein
Lingual vein
Superior thyroid vein
Internal jugular vein
Left brachiocephalic vein
Superior vena cava
Inferior vena cava
Anterior view

FIGURE 21.21 APR **Veins of the Head and Neck**
The right brachiocephalic vein and its tributaries. The major veins draining the head and neck are the internal and external jugular veins.

The **median cubital** (kū′bi-tăl; pertaining to the elbow) **vein** is a variable vein that usually connects the cephalic vein or its tributaries with the basilic vein. In many people, this vein is quite prominent on the anterior surface of the upper limb at the level of the elbow (cubital fossa) therefore, it is often used as a site for drawing blood from a patient.

The deep veins draining the upper limb follow the same course as the arteries. Thus, the **radial** and **ulnar veins** are named for the arteries they attend. They are usually paired, with one small vein lying on each side of the artery, and they have numerous connections with one another and with the superficial veins. The radial and ulnar veins empty into the **brachial veins,** which accompany the brachial artery and empty into the axillary vein (figures 21.23 and 21.24).

Veins of the Thorax

Three major veins return blood from the thorax to the superior vena cava: the right and left brachiocephalic veins and the **azygos** (az′ī-gos; unpaired) **vein.** The thoracic drainage to the brachiocephalic veins is through the anterior thoracic wall by way of the

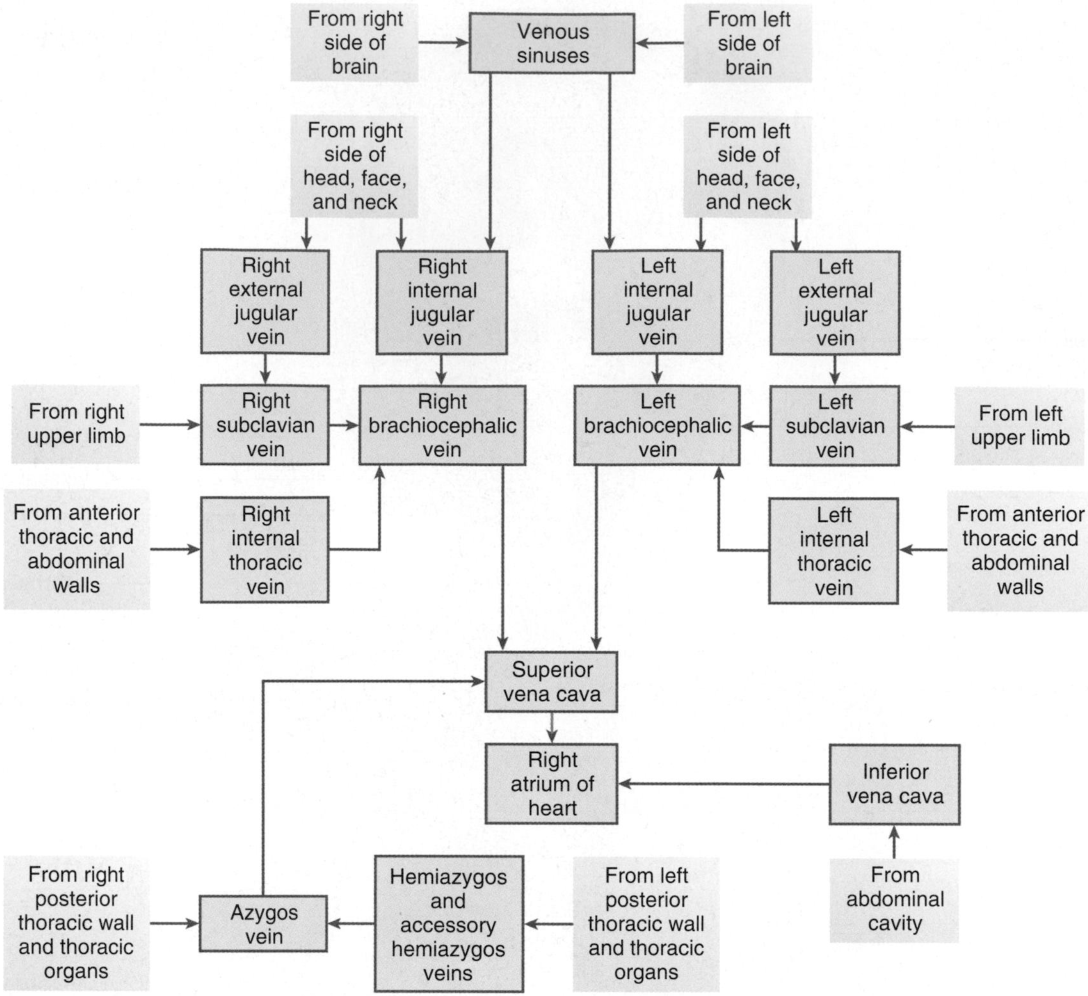
From right side of brain
Venous sinuses
From left side of brain
From right side of head, face, and neck
From left side of head, face, and neck
Right external jugular vein
Right internal jugular vein
Left internal jugular vein
Left external jugular vein
From right upper limb
Right subclavian vein
Right brachiocephalic vein
Left brachiocephalic vein
Left subclavian vein
From left upper limb
From anterior thoracic and abdominal walls
Right internal thoracic vein
Left internal thoracic vein
From anterior thoracic and abdominal walls
Superior vena cava
Right atrium of heart
Inferior vena cava
From abdominal cavity
From right posterior thoracic wall and thoracic organs
Azygos vein
Hemiazygos and accessory hemiazygos veins
From left posterior thoracic wall and thoracic organs

FIGURE 21.22 Major Veins of the Head and Thorax

internal thoracic veins. They receive blood from the **anterior intercostal veins.** Blood from the posterior thoracic wall is collected by **posterior intercostal veins** that drain into the azygos vein on the right and the **hemiazygos vein** (hem′ē-az′ī-gos) or the **accessory hemiazygos vein** on the left. The hemiazygos and accessory hemiazygos veins empty into the azygos vein, which drains into the superior vena cava. The thoracic veins are listed in table 21.9 and illustrated in figure 21.25 (also see figure 21.22).

ASSESS YOUR PROGRESS

28. *What are the three major veins that return blood to the right atrium?*

29. *What veins collect blood from the heart muscle?*

30. *List the two pairs of major veins that drain blood from the head and neck. Describe venous sinuses. To what large vein do the venous sinuses connect?*

31. *List the major deep and superficial veins of the upper limb.*

32. *List the three major veins that return blood from the thorax to the superior vena cava.*

TABLE 21.8 Veins of the Upper Limb (figures 21.23 and 21.24)

Veins	Tissues Drained
Subclavian (continuation of the axillary vein)	
Axillary (continuation of the basilic vein)	
Cephalic	Lateral arm, forearm, and hand (superficial veins of the forearm and hand are variable)
Brachial (paired, deep veins)	Deep structures of the arm
Radial	Deep forearm
Ulnar	Deep forearm
Basilic	Medial arm, forearm, and hand (superficial veins of the forearm and hand are variable)
Median cubital	Connects basilic and cephalic veins
Deep and superficial palmar venous arches	Drain into superficial and deep veins of the forearm
Digital	Fingers

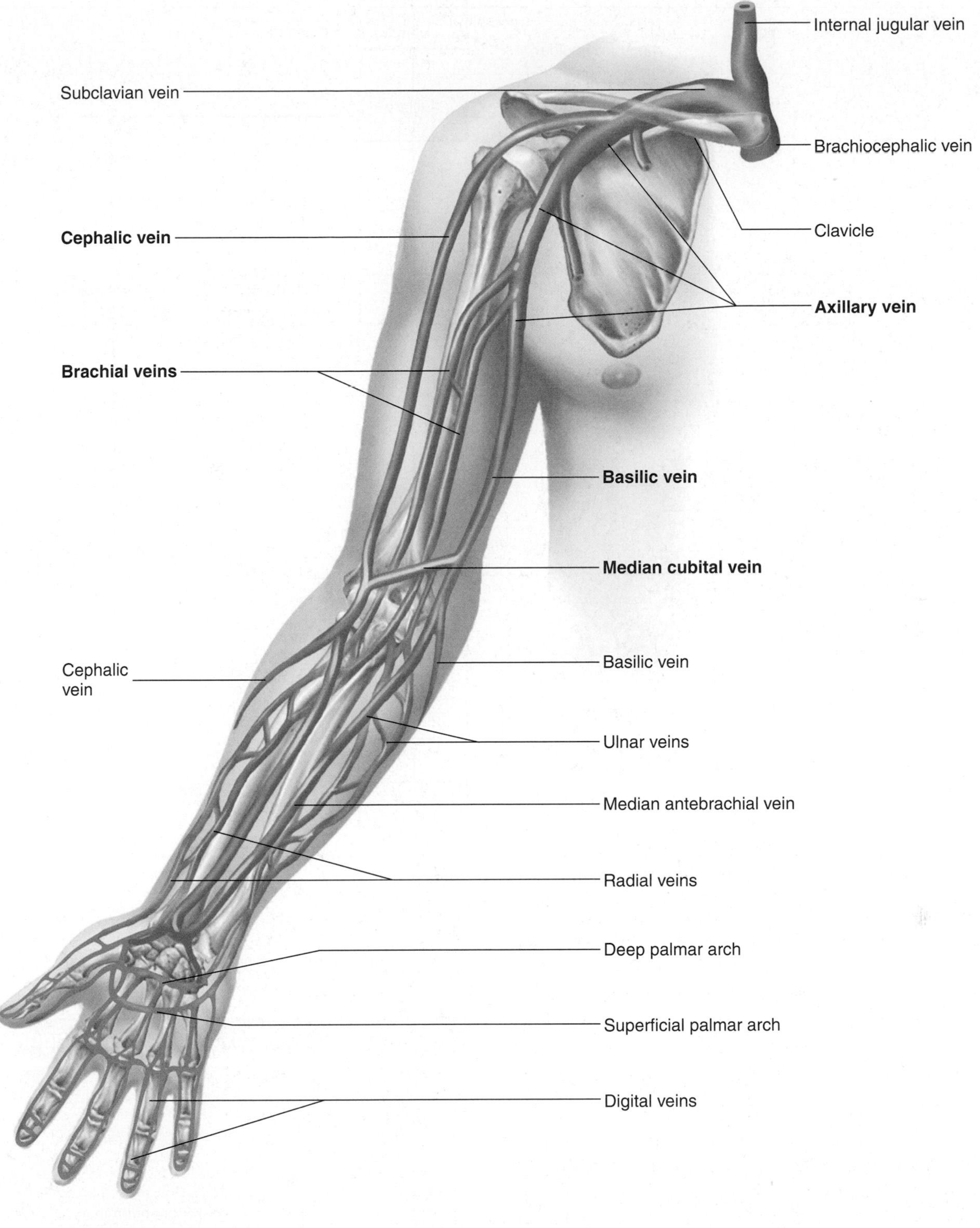
Internal jugular vein
Subclavian vein
Brachiocephalic vein
Clavicle
Cephalic vein
Axillary vein
Brachial veins
Basilic vein
Median cubital vein
Cephalic vein
Basilic vein
Ulnar veins
Median antebrachial vein
Radial veins
Deep palmar arch
Superficial palmar arch
Digital veins
Anterior view

FIGURE 21.23 Veins of the Upper Limb
The subclavian vein and its tributaries. The major veins draining the superficial structures of the limb are the cephalic and basilic veins. The brachial veins drain the deep structures.

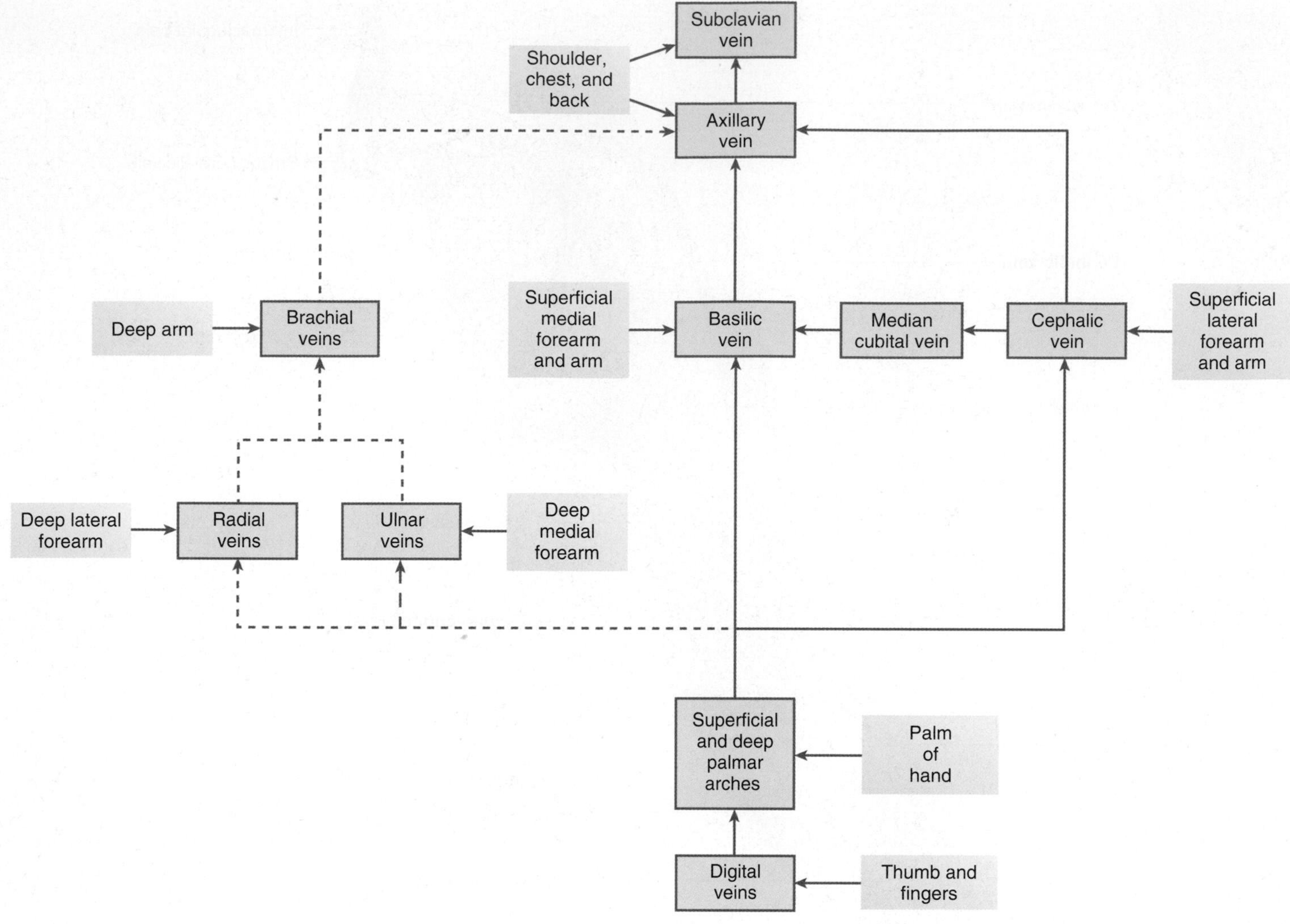
Subclavian vein
Shoulder, chest, and back
Axillary vein
Deep arm
Brachial veins
Superficial medial forearm and arm
Basilic vein
Median cubital vein
Cephalic vein
Superficial lateral forearm and arm
Deep lateral forearm
Radial veins
Ulnar veins
Deep medial forearm
Superficial and deep palmar arches
Palm of hand
Digital veins
Thumb and fingers

FIGURE 21.24 Major Veins of the Shoulder and Upper Limb
The deep veins, which carry far less blood than the superficial veins, are indicated by *dashed lines*.

Veins of the Abdomen and Pelvis

Blood from the posterior abdominal wall drains into the **ascending lumbar veins.** These veins are continuous superiorly with the hemiazygos on the left and the azygos on the right. Blood from the rest of the abdomen, pelvis, and lower limbs returns to the heart through the inferior vena cava. The gonads (testes and ovaries), kidneys, and adrenal glands are the only abdominal organs outside the pelvis that drain directly into the inferior vena cava. The **internal iliac veins** drain the pelvis and join the **external iliac veins** from the lower limbs to form the **common iliac veins,** which unite to form the inferior vena cava. The major abdominal and pelvic veins are listed in table 21.10 and illustrated in figure 21.26; also see figure 21.28.

TABLE 21.9 Veins of the Thorax (figure 21.25)

Veins	Tissues Drained
Superior Vena Cava	
Brachiocephalic	
Azygos	Right side, posterior thoracic wall and posterior abdominal wall; esophagus, bronchi, pericardium, and mediastinum
Hemiazygos	Left side, inferior posterior thoracic wall and posterior abdominal wall; esophagus and mediastinum
Accessory hemiazygos	Left side, superior posterior thoracic wall

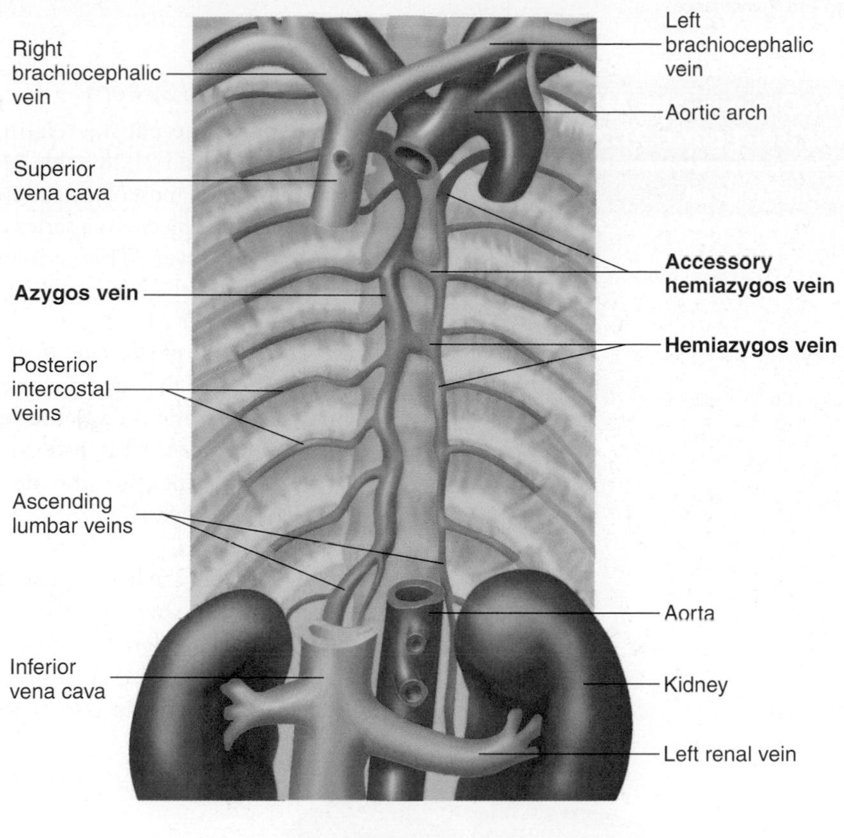
Right brachiocephalic vein
Left brachiocephalic vein
Aortic arch
Superior vena cava
Accessory hemiazygos vein
Azygos vein
Hemiazygos vein
Posterior intercostal veins
Ascending lumbar veins
Aorta
Kidney
Inferior vena cava
Left renal vein
Anterior view

FIGURE 21.25 APR **Veins of the Thorax**

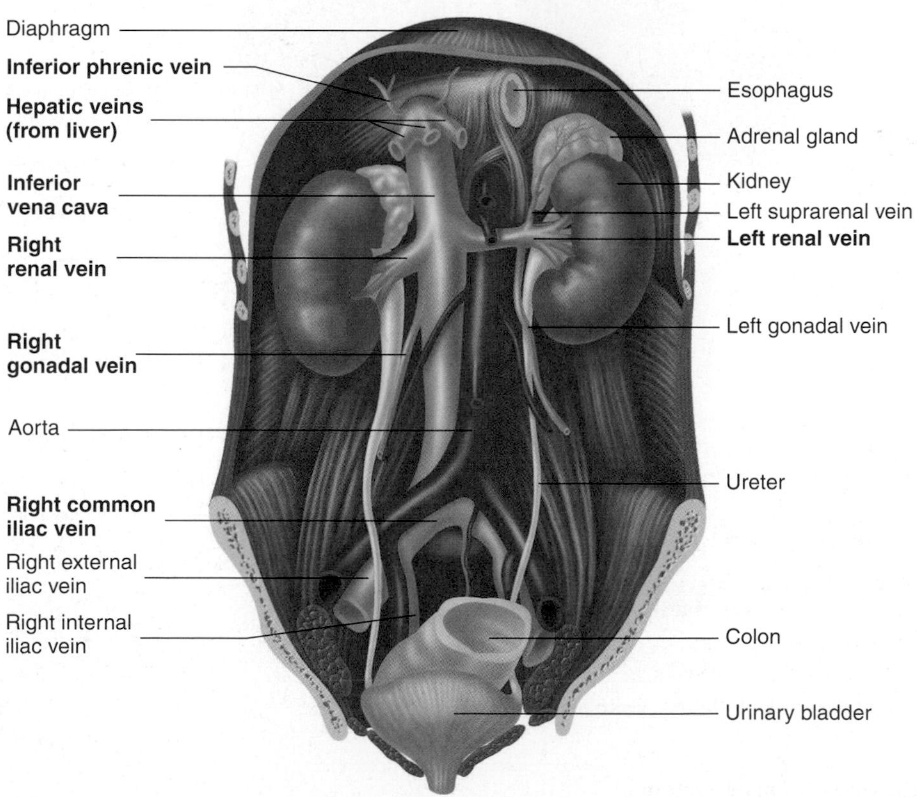
Diaphragm
Inferior phrenic vein
Esophagus
Hepatic veins (from liver)
Adrenal gland
Kidney
Inferior vena cava
Left suprarenal vein
Left renal vein
Right renal vein
Left gonadal vein
Right gonadal vein
Aorta
Ureter
Right common iliac vein
Right external iliac vein
Right internal iliac vein
Colon
Urinary bladder

FIGURE 21.26 Inferior Vena Cava and Its Tributaries
The hepatic veins transport blood to the inferior vena cava from the hepatic portal system, which ends as a series of blood sinusoids in the liver (also see figure 21.27).

TABLE 21.10 Veins Draining the Abdomen and Pelvis (see figures 21.26 and 21.28)

Veins	Tissues Drained
Inferior Vena Cava	
Hepatic	Liver (see the section "Hepatic Portal System")
Common iliac	
External iliac	Lower limb (see table 21.12)
Internal iliac	Pelvis and its viscera
Ascending lumbar	Posterior abdominal wall (empties into common iliac, azygos, and hemiazygos veins)
Renal	Kidney
Suprarenal	Adrenal gland
Gonadal	
Testicular (male)	Testis
Ovarian (female)	Ovary
Phrenic	Diaphragm

Hepatic Portal System

The **hepatic** (he-pat′ik; relating to the liver) **portal system** (figures 21.27 and 21.28; table 21.11) carries blood drained from capillaries within most of the abdominal viscera, such as the stomach, intestines, and spleen, to a series of dilated capillaries, called sinusoids, in the liver. This system delivers nutrients and other substances absorbed from the stomach or small intestine to the liver (see chapter 24).

The **hepatic portal vein,** the largest vein of the system, is formed by the union of the **superior mesenteric vein,** which drains the small intestine, and the **splenic vein,** which drains the spleen. The splenic vein receives the **inferior mesenteric vein,** which drains part of the large intestine, and the **pancreatic veins,** which drain the pancreas. The hepatic portal vein also receives gastric veins before entering the liver.

Blood from the liver sinusoids is collected into **central veins,** which empty into **hepatic veins.** Blood from the cystic veins, which drain the gallbladder, also enters the hepatic veins. The hepatic veins join the inferior vena cava. Blood entering the liver through the hepatic portal vein is rich with nutrients collected from the

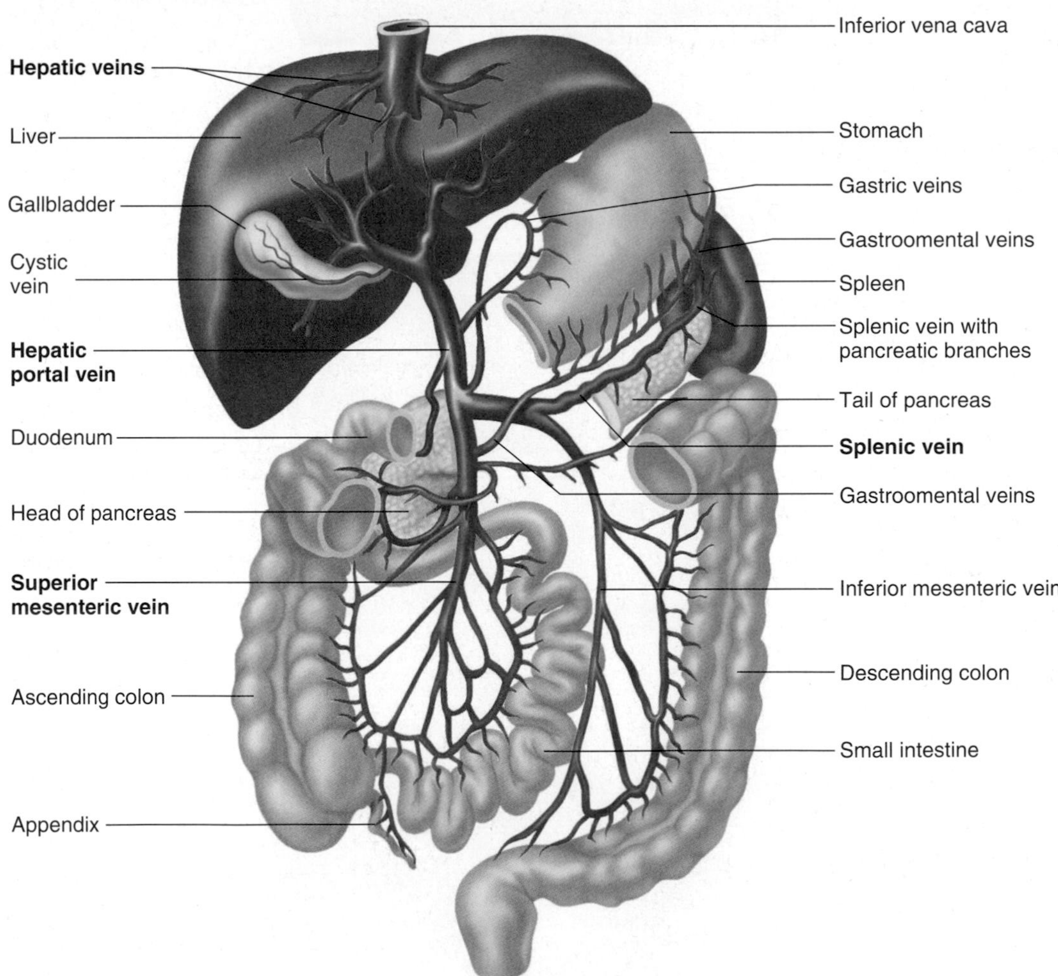
Inferior vena cava
Hepatic veins
Liver
Stomach
Gastric veins
Gallbladder
Gastroomental veins
Cystic vein
Spleen
Splenic vein with pancreatic branches
Hepatic portal vein
Tail of pancreas
Duodenum
Splenic vein
Gastroomental veins
Head of pancreas
Superior mesenteric vein
Inferior mesenteric vein
Descending colon
Ascending colon
Small intestine
Appendix

FIGURE 21.27 Veins of the Hepatic Portal System

The hepatic portal system begins as capillary beds in the stomach, pancreas, spleen, small intestine, and large intestine. The veins of the hepatic portal system converge on the hepatic portal vein, which carries blood to a series of capillaries (sinusoids) in the liver. Hepatic veins carry blood from capillaries in the liver to the inferior vena cava (also see figure 21.26).

small intestine, but it can also contain a number of toxic substances harmful to the body tissues. Within the liver, the nutrients are either taken up and stored or modified chemically so they can be used by other cells of the body (see chapter 24). The liver cells also help remove toxic substances by altering their structure or making them water-soluble, a process called **biotransformation.** The water-soluble substances can then be transported in the blood to the kidneys, which excrete them in the urine (see chapter 26).

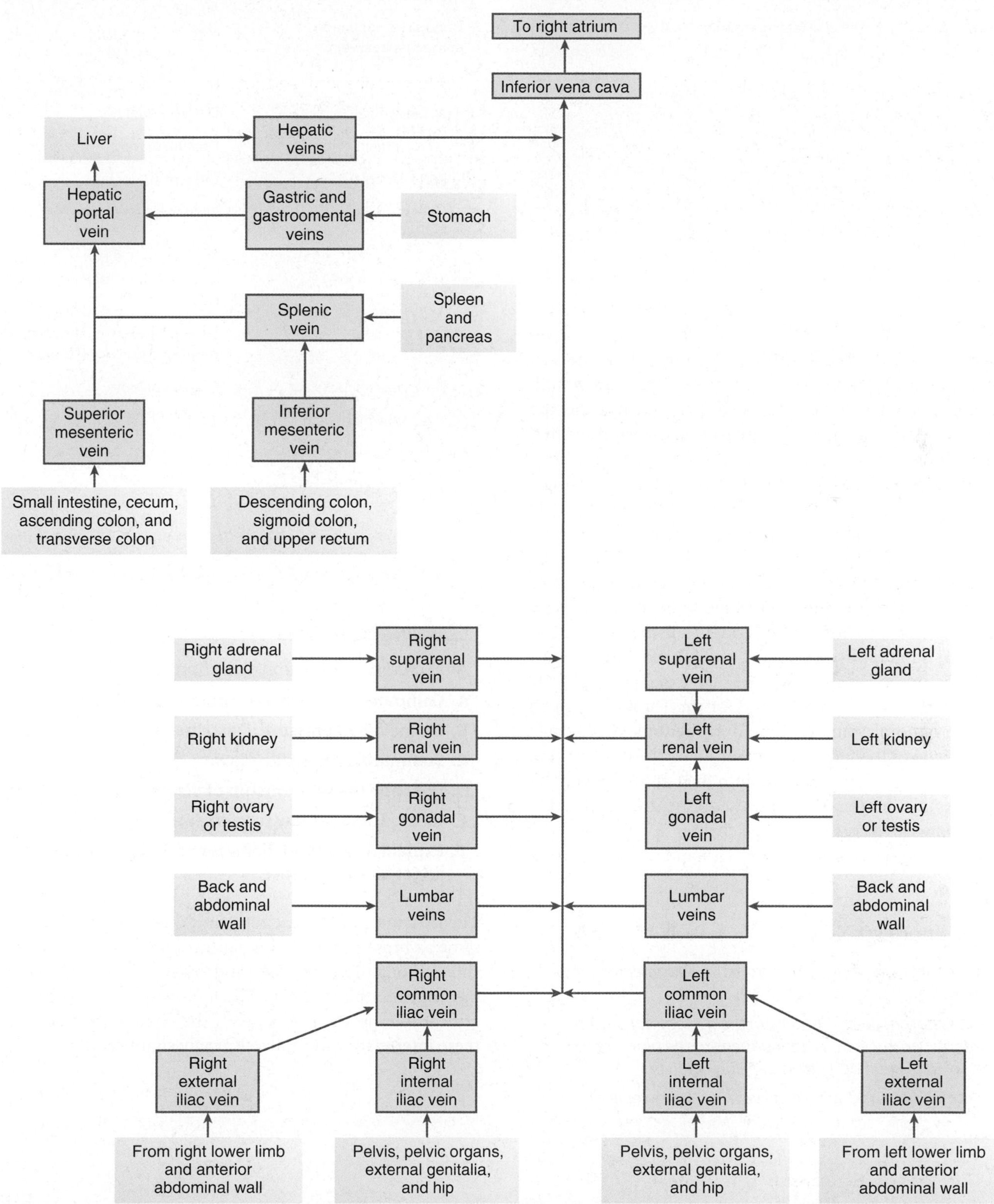
To right atrium
Inferior vena cava
Liver
Hepatic veins
Hepatic portal vein
Gastric and gastroomental veins
Stomach
Splenic vein
Spleen and pancreas
Superior mesenteric vein
Inferior mesenteric vein
Small intestine, cecum, ascending colon, and transverse colon
Descending colon, sigmoid colon, and upper rectum
Right adrenal gland
Right suprarenal vein
Left suprarenal vein
Left adrenal gland
Right kidney
Right renal vein
Left renal vein
Left kidney
Right ovary or testis
Right gonadal vein
Left gonadal vein
Left ovary or testis
Back and abdominal wall
Lumbar veins
Lumbar veins
Back and abdominal wall
Right common iliac vein
Left common iliac vein
Right external iliac vein
Right internal iliac vein
Left internal iliac vein
Left external iliac vein
From right lower limb and anterior abdominal wall
Pelvis, pelvic organs, external genitalia, and hip
Pelvis, pelvic organs, external genitalia, and hip
From left lower limb and anterior abdominal wall

FIGURE 21.28 Major Veins of the Abdomen and Pelvis

TABLE 21.11 Hepatic Portal System (figures 21.27 and 21.28)

Veins	Tissues Drained
Hepatic Portal	
Superior mesenteric	Small intestine and most of the colon
Splenic	Spleen
Inferior mesenteric	Descending colon and rectum
Pancreatic	Pancreas
Gastroomental	Stomach
Gastric	Stomach
Cystic	Gallbladder

Veins of the Lower Limb

The veins of the lower limb, like those of the upper limb, consist of superficial and deep groups. The distal deep veins of each limb are paired and follow the same path as the arteries, whereas the proximal deep veins are unpaired. The **anterior** and **posterior tibial veins** are paired and accompany the anterior and posterior tibial arteries. They unite just inferior to the knee to form the single **popliteal vein,** which ascends through the thigh and becomes the **femoral vein.** The femoral vein becomes the external iliac vein. **Fibular veins,** or *peroneal* (per-ō-nē′ăl) *veins,* are also paired in each leg and accompany the fibular arteries. They empty into the posterior tibial veins just before those veins contribute to the popliteal vein.

The superficial veins consist of the great and small saphenous veins. The **great saphenous** (să-fē′nŭs; visible) **vein,** the longest vein of the body, originates over the dorsal and medial side of the foot and ascends along the medial side of the leg and thigh to empty into the femoral vein. The **small saphenous vein** begins over the lateral side of the foot and ascends along the posterior leg to the popliteal space, where it empties into the popliteal vein. The saphenous veins can be removed and used as a source of blood vessels for coronary bypass surgery (see chapter 20). The veins of the lower limb are illustrated in figures 21.29 and 21.30 and listed in table 21.12.

ASSESS YOUR PROGRESS

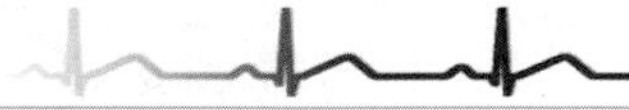

33. *Explain the three ways that blood from the abdomen returns to the heart.*
34. *List the vessels that carry blood from the abdominal organs to the hepatic portal vein. What happens to the blood of the hepatic portal system as it filters through the liver?*
35. *List the major deep and superficial veins of the lower limbs.*

TABLE 21.12 Veins of the Lower Limb (figures 21.29 and 21.30)

Veins	Tissues Drained
External Iliac Vein (continuation of the femoral vein)	
Femoral (continuation of the popliteal vein)	Thigh
Popliteal	
Anterior tibial	Deep anterior leg
Dorsal vein of foot	Dorsum of foot
Posterior tibial	Deep posterior leg
Plantar veins	Plantar region of foot
Fibular (peroneal)	Deep lateral leg and foot
Small saphenous	Superficial posterior leg and lateral side of foot
Great saphenous	Superficial anterior and medial leg, thigh, and dorsum of foot
Dorsal vein of foot	Dorsum of foot
Dorsal venous arch	Foot
Digital veins	Toes

21.6 Dynamics of Blood Circulation

LEARNING OUTCOMES

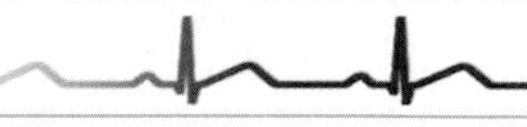

After reading this section, you should be able to

- **A. Compare laminar and turbulent blood flow.**
- **B. Define *blood pressure*. Describe how it is measured.**
- **C. Summarize Poiseuille's law.**
- **D. Describe the relationship of viscosity to blood flow.**
- **E. Relate Laplace's law to critical closing pressure.**
- **F. Explain how vessel diameter and vascular compliance affect blood pressure.**

The dynamics of blood circulating through blood vessels are the same as those of water flowing through pipes. The interrelationships among pressure, flow, and resistance, as well as the control mechanisms that regulate blood pressure and blood flow, play critical roles in the functions of the circulatory system. Many of these interrelationships are clinically significant.

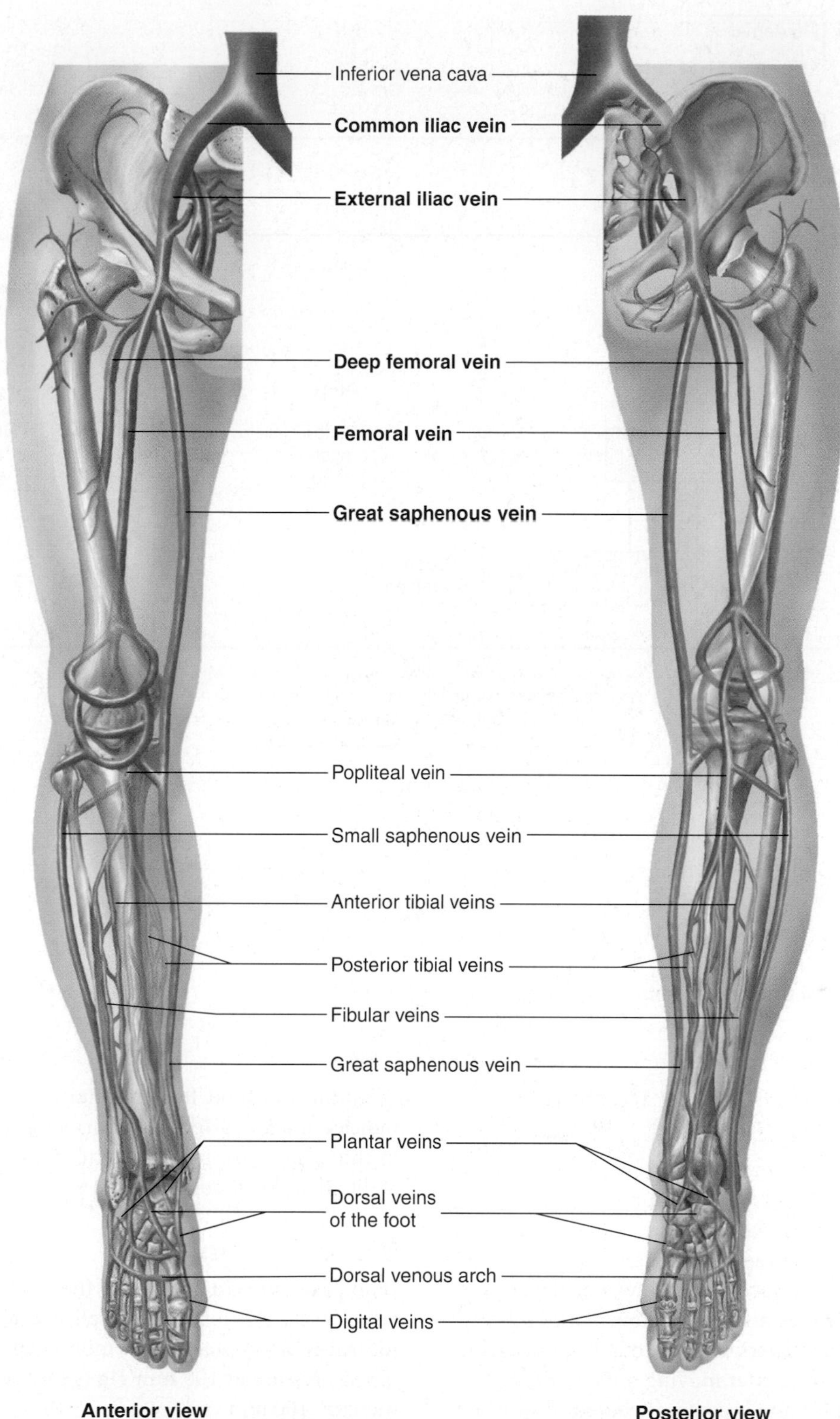
Inferior vena cava
Common iliac vein
External iliac vein
Deep femoral vein
Femoral vein
Great saphenous vein
Popliteal vein
Small saphenous vein
Anterior tibial veins
Posterior tibial veins
Fibular veins
Great saphenous vein
Plantar veins
Dorsal veins of the foot
Dorsal venous arch
Digital veins
Anterior view
Posterior view

FIGURE 21.29 APR **Veins of the Pelvis and Lower Limb**
The right common iliac vein and its tributaries.

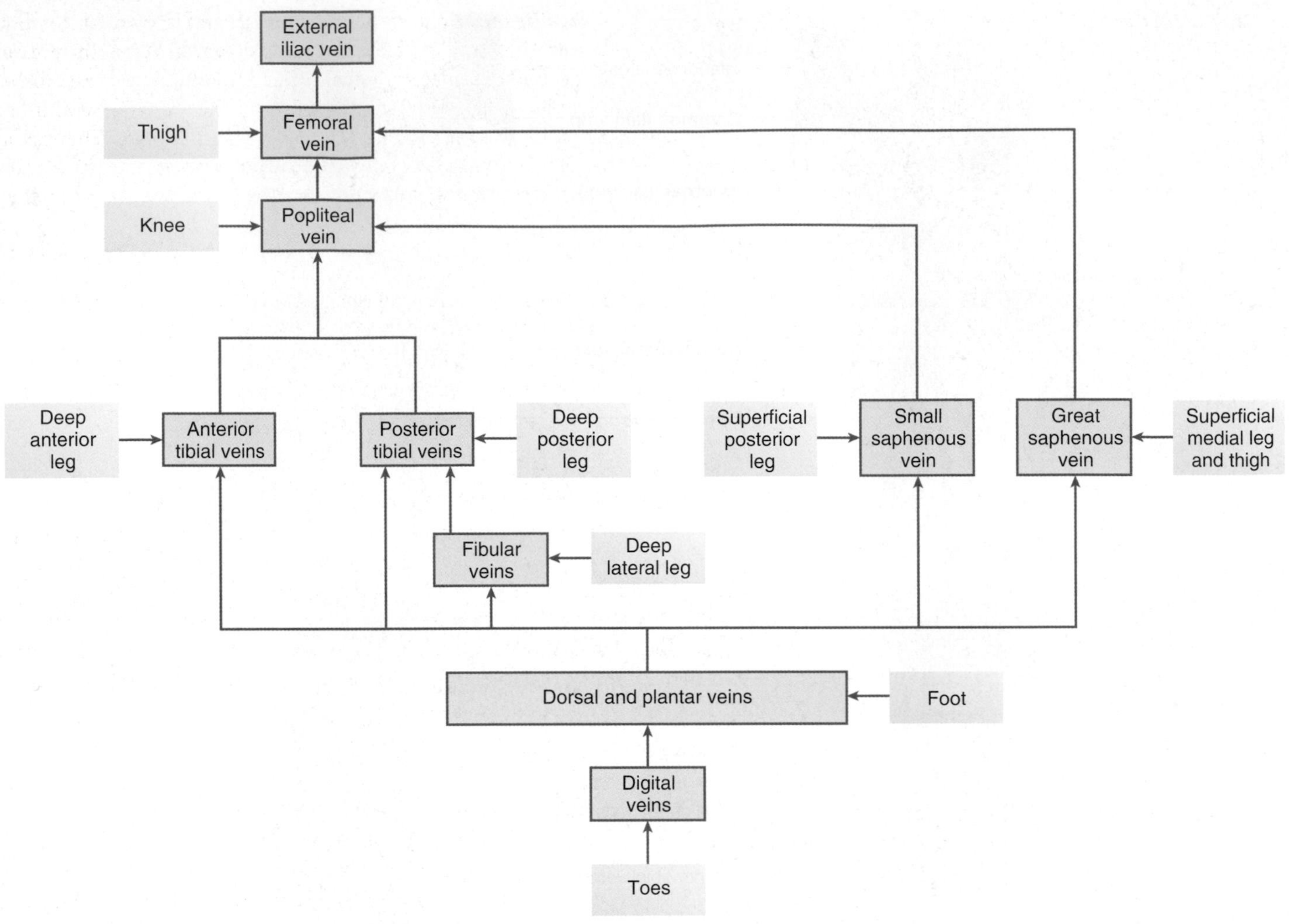
External iliac vein
Thigh
Femoral vein
Knee
Popliteal vein
Deep anterior leg
Anterior tibial veins
Posterior tibial veins
Deep posterior leg
Superficial posterior leg
Small saphenous vein
Great saphenous vein
Superficial medial leg and thigh
Fibular veins
Deep lateral leg
Dorsal and plantar veins
Foot
Digital veins
Toes

FIGURE 21.30 Major Veins of the Lower Limb

Laminar and Turbulent Flow in Vessels

Fluid, including blood, tends to flow through long, smooth-walled tubes in a streamlined fashion called **laminar flow** (figure 21.31*a*). Fluid behaves as if it were composed of a large number of concentric layers. The layer nearest the wall of the tube experiences the greatest resistance to flow because it moves against the stationary wall. The innermost layers slip over the surface of the outermost layers and experience less resistance to movement. Thus, flow in a vessel consists of movement of concentric layers, with the outer layer moving most slowly and the layer at the center moving most rapidly.

Laminar flow is interrupted and becomes **turbulent flow** when the rate of flow exceeds a critical velocity or when the fluid passes a constriction, a sharp turn, or a rough surface (figure 21.31*b*). Vibrations of the liquid and blood vessel walls during turbulent flow cause the sounds heard when blood pressure is measured using a blood pressure cuff. Turbulent flow is also common as blood flows past the valves in the heart and is partially responsible for the heart sounds (see chapter 20).

Turbulent flow of blood through vessels occurs primarily in the heart and to a lesser extent where arteries branch. Sounds caused by turbulent blood flow in arteries are not normal and usually indicate that the artery is abnormally constricted. Turbulent flow in abnormally constricted arteries may indicate an increased probability that thromboses will develop.

Blood Pressure

Blood pressure is a measure of the force blood exerts against blood vessel walls. An instrument called a **mercury (Hg) manometer** measures blood pressure in millimeters of mercury (mm Hg). A blood pressure of 100 mm Hg is great enough to lift a column of mercury 100 mm.

Blood pressure can be measured directly by inserting a **cannula** (tube) into a blood vessel and connecting a manometer or an electronic pressure transducer to it. Electronic transducers are very sensitive and can precisely detect rapid fluctuations in pressure.

Placing a catheter into a blood vessel or into a chamber of the heart to monitor pressure changes is possible but not appropriate for routine clinical examinations. Health professionals most often use the **auscultatory** (aws-kŭl′tă-tō′rē) **method** to measure blood pressure. They wrap a blood pressure cuff connected to a **sphygmomanometer**

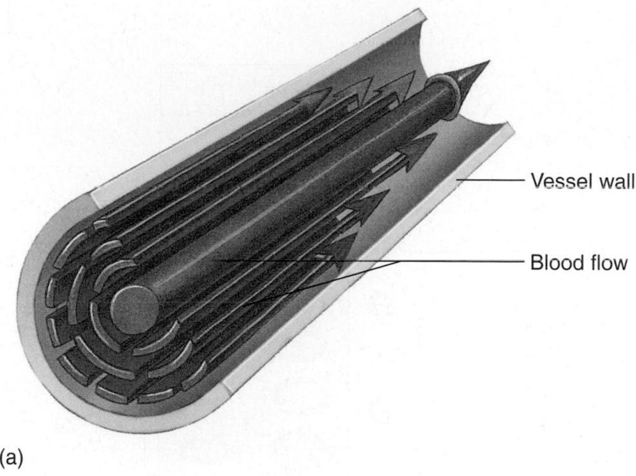
Vessel wall
Blood flow
(a)

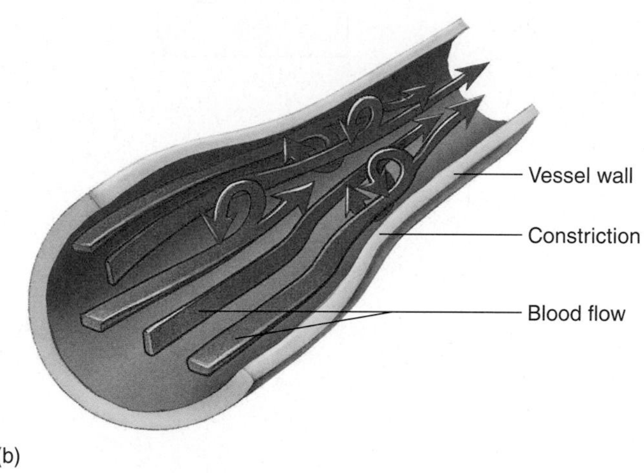
Vessel wall
Constriction
Blood flow
(b)

FIGURE 21.31 Laminar and Turbulent Flow
(*a*) In laminar flow, fluid flows in long, smooth-walled tubes as if it were composed of a large number of concentric layers. (*b*) Turbulent flow is caused by numerous small currents flowing crosswise or obliquely to the long axis of the vessel, resulting in flowing whorls and eddy currents.

(sfig′mō-mă-nom′ĕ-ter) around a patient's arm just above the elbow and place a stethoscope over the brachial artery (figure 21.32). Some sphygmomanometers have mercury manometers, and others have digital manometers, but they all measure pressure in terms of millimeters of mercury. The blood pressure cuff is inflated until the brachial artery is completely collapsed. Because no blood flows through the constricted area at this point, no sounds can be heard through the stethoscope. Then the pressure in the cuff is gradually lowered. As soon as it declines below the systolic pressure, blood flows through the constricted area during systole. The blood flow produces vibrations in the blood and surrounding tissues that can be heard through the stethoscope. These sounds are called **Korotkoff** (kō-rot′kof) **sounds,** and the pressure at which a Korotkoff sound is first heard represents the **systolic pressure.**

As the pressure in the blood pressure cuff is lowered still more, the Korotkoff sounds change tone and loudness. When the pressure has dropped until continuous laminar blood flow is reestablished, the sound disappears completely. The pressure at which continuous laminar flow is reestablished is the **diastolic pressure.** This method for determining systolic and diastolic pressures is not entirely accurate, but its results are within 10% of methods that are more direct.

ASSESS **YOUR PROGRESS**

36. *Describe laminar flow and turbulent flow through a tube. What conditions cause turbulent flow of blood?*

37. *What creates blood pressure? Describe the auscultatory method of measuring blood pressure.*

38. *What are Korotkoff sounds?*

Blood Flow and Poiseuille's Law

The **rate** at which blood or any other liquid flows through a tube can be expressed as the volume that passes a specific point per unit of time. Blood flow is usually reported in either milliliters (mL) or liters (L) per minute. For example, when a person is resting, the **cardiac output** of the heart is approximately 5 L/min; thus, the rate of blood flow through the aorta is approximately 5 L/min. The rate of blood flow in a vessel can be described by the following equation:

$$\text{Flow} = \frac{P_1 - P_2}{R}$$

where P_1 and P_2 are the pressures in the vessel at points one and two, respectively, and R is the resistance to flow. Blood always flows from an area of higher pressure to an area of lower pressure; the greater the pressure difference, the greater the rate of flow. For example, the average blood pressure in the aorta (P_1) is greater than the blood pressure in the vessels of the relaxed right atrium (P_2). Therefore, blood flows from the aorta to tissues and from tissues to the right atrium. If the heart should stop contracting, the pressure in the aorta would become equal to that in the right atrium, and blood would no longer flow.

The flow of blood, resulting from a pressure difference between the two ends of a blood vessel, is opposed by a resistance to flow. As the resistance increases, blood flow decreases; as the resistance decreases, blood flow increases. The factors that affect resistance include blood viscosity, vessel length, and vessel diameter and can be represented mathematically as follows:

$$\text{Resistance} = \frac{128vl}{\pi D^4}$$

where v is the viscosity of blood, l is the length of the vessel, and D is the diameter of the vessel. Both 128 and π are constants and for practical purposes the length of the blood vessel is constant. Thus, the diameter of the blood vessel and the viscosity of the blood determine resistance. The viscosity of blood changes slowly.

When the equation for resistance is combined with the equation for flow, the following relationship, called **Poiseuille's** (pwah-zuh′yes) **law,** results:

$$\text{Flow} = \frac{P_1 - P_2}{R} = \frac{\pi(P_1 - P_2)\,D^4}{128vl}$$

FUNDAMENTAL **Figure**

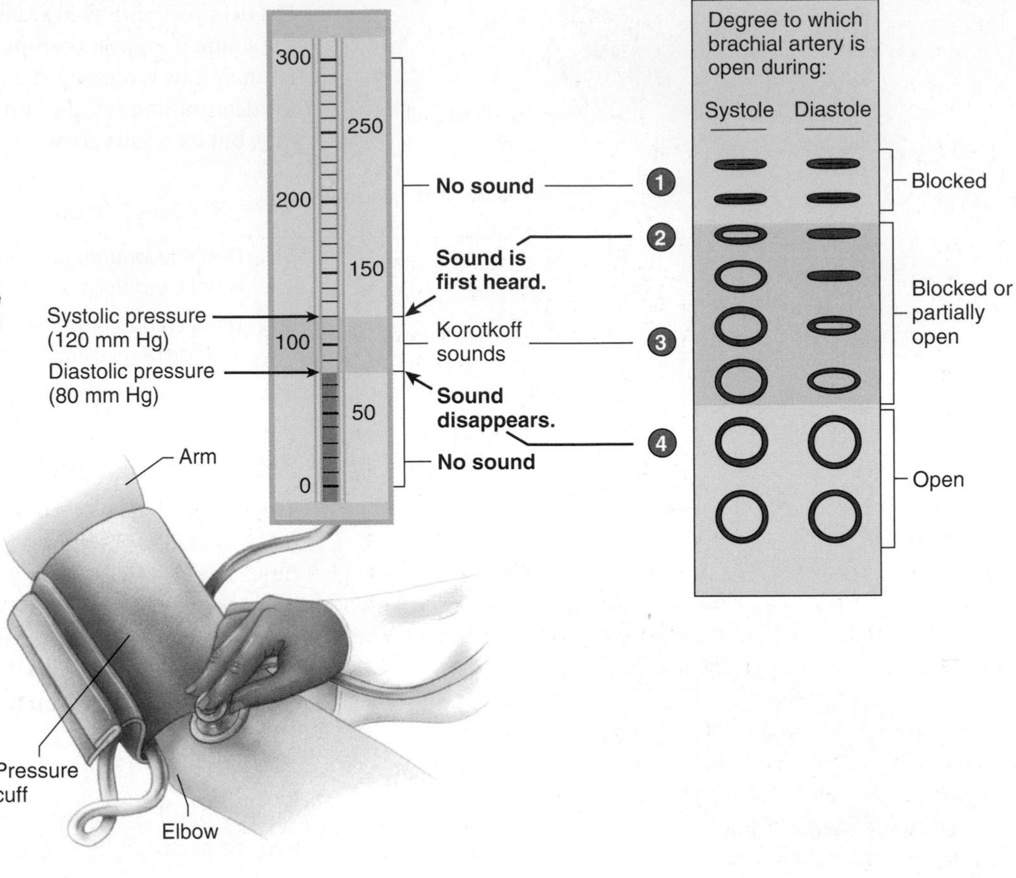

1 When the cuff pressure is high enough to keep the brachial artery closed, no blood flows through it, and no sound is heard.
2 When cuff pressure decreases and is no longer able to keep the brachial artery closed, blood is pushed through the partially opened brachial artery, producing turbulent blood flow and a sound. **Systolic pressure** is the pressure at which a sound is first heard.
3 As cuff pressure continues to decrease, the brachial artery opens even more during systole. At first, the artery is closed during diastole, but as cuff pressure continues to decrease, the brachial artery partially opens during diastole. Turbulent blood flow during systole produces Korotkoff sounds, although the pitch of the sounds changes as the artery becomes more open.
4 Eventually, cuff pressure decreases below the pressure in the brachial artery, and it remains open during systole and diastole. Nonturbulent flow is reestablished, and no sounds are heard. **Diastolic pressure** is the pressure at which the sound disappears.
300
250
200
150
100
50
0
Systolic pressure (120 mm Hg)
Diastolic pressure (80 mm Hg)
Arm
Pressure cuff
Elbow
No sound
Sound is first heard.
Korotkoff sounds
Sound disappears.
No sound
Degree to which brachial artery is open during:
Systole
Diastole
Blocked
Blocked or partially open
Open

PROCESS FIGURE 21.32 **Blood Pressure Measurement**

According to Poiseuille's law, a small change in the diameter of a vessel dramatically changes the resistance to flow, and therefore the amount of blood that flows through it. This is apparent when reviewing the Poiseuille's law formula. The diameter (D) is raised to the fourth power and therefore has a great impact on the overall calculation of flow. Vasoconstriction decreases the diameter of a vessel, increases resistance to flow, and decreases blood flow through the vessel. For example, decreasing the diameter of a vessel by half increases the resistance to flow 16-fold and decreases flow 16-fold. Vasodilation increases the diameter of a vessel, decreases resistance to flow, and increases blood flow through the vessel.

Changes in blood pressure and blood vessel diameter produce the major changes in blood flow through blood vessels. During exercise, heart rate and stroke volume increase, causing blood pressure in the aorta to increase. In addition, blood vessels in skeletal muscles vasodilate, and resistance to flow decreases. As a consequence, a dramatic increase in blood flow through blood vessels in exercising skeletal muscles occurs.

Viscosity (vis-kos′i-tē) is a measure of a liquid's resistance to flow. As the viscosity of a liquid increases, the pressure required to force it to flow also increases. The viscosity of liquids is commonly determined by considering the viscosity of distilled water as 1 and then comparing the viscosity of other liquids with that. Using this procedure, whole blood has a viscosity of 3.0–4.5, which means that about three times as much pressure is required to force whole blood through a given tube at the same rate as forcing water through the same tube.

The viscosity of blood is influenced largely by **hematocrit** (hē′mă-tō-krit, hem′ă-tō-krit), which is the percentage of the total blood volume composed of red blood cells (see chapter 19). As the hematocrit increases, the viscosity of blood increases logarithmically. Blood with a hematocrit of 45% has a viscosity about three times that of water, whereas blood with a very high hematocrit of 65% has a viscosity about seven to eight times that of water. The plasma proteins have only a minor effect on the viscosity of blood, but dehydration or uncontrolled production of red blood cells can increase the hematocrit and the viscosity of blood substantially. Viscosity above the normal range increases the workload on the heart. If this workload is great enough, heart failure can result.

Predict 3

Predict the effect of each of the following conditions on blood flow: (a) vasoconstriction of blood vessels in the skin in response to cold exposure, (b) vasodilation of blood vessels in the skin in response to elevated body temperature, and (c) erythrocytosis, which results in a greatly increased hematocrit.

Critical Closing Pressure and Laplace's Law

Each blood vessel exhibits a **critical closing pressure,** the pressure below which the vessel collapses and blood flow through the vessel stops. When a person is in shock, blood pressure can decrease below the critical closing pressure in vessels (see Clinical Impact, "Circulatory Shock," later in this chapter). As a consequence, the blood vessels collapse, and flow ceases. The tissues supplied by those vessels can become necrotic because of the lack of blood supply. **Laplace's** (la-plas′ez) **law,** which helps explain the critical closing pressure, states that the force that stretches the vascular wall is proportional to the diameter of the vessel times the blood pressure. As the pressure in a vessel decreases, the force that stretches the vessel wall also decreases. Some minimum force is required to keep the vessel open. If the pressure decreases, so that the force is below that minimum requirement, the vessel closes. As the pressure in a vessel increases, the force that stretches the vessel wall also increases.

Laplace's law is expressed by the following formula:

$$F = D \times P$$

where F is force, D is vessel diameter, and P is pressure.

According to Laplace's law, as the diameter of a vessel increases, the force applied to the vessel wall increases, even if the pressure remains constant. Sometimes a part of an arterial wall becomes weakened and a bulge, called an **aneurysm,** forms in it. The force applied to the weakened part is greater than at other points along the blood vessel because its diameter is greater. The greater force causes the weakened vessel wall to bulge even more, further increasing the pressure on it. This series of events can proceed until the vessel finally ruptures. Ruptured aneurysms in the blood vessels of the brain or in the aorta are often fatal.

Predict 4

Richard does not know it, but he has an aneurysm at the base of his left middle cerebral artery. One of Richard's favorite activities is to take a hot sauna bath and then jump into cold water. Richard does not realize that this causes rapid vasoconstriction of his cutaneous blood vessels. How will this activity affect Richard's aneurysm?

Vascular Compliance

Compliance (kom-plī′ans) is the tendency for blood vessel volume to increase as blood pressure increases. The more easily the vessel wall stretches, the greater is its compliance. The less easily the vessel wall stretches, the smaller is its compliance.

Compliance is expressed by the following formula:

$$\text{Compliance} = \frac{\text{Increase in volume (mL)}}{\text{Increase in pressure (mm Hg)}}$$

TABLE 21.13 Distribution of Blood Volume in Blood Vessels

Vessels		Total Blood Volume (%)
Systemic		
Veins		64
Large veins	(39%)	
Small veins	(25%)	
Arteries		15
Large arteries	(8%)	
Small arteries	(5%)	
Arterioles	(2%)	
Capillaries		5
	TOTAL IN SYSTEMIC VESSELS	84
Pulmonary vessels		9
Heart		7
	TOTAL BLOOD VOLUME	100

Vessels with a large compliance exhibit a large increase in volume when the pressure increases a small amount. Vessels with a small compliance do not show a large increase in volume when the pressure increases.

Venous compliance is approximately 24 times greater than arterial compliance. As venous pressure increases, the volume of the veins increases greatly. Consequently, veins act as storage areas, or reservoirs, for blood because their large compliance allows them to hold much more blood than other areas of the vascular system (table 21.13).

ASSESS YOUR PROGRESS

39. *Describe the relationship among blood flow, blood pressure, and resistance.*
40. *According to Poiseuille's law, what effects do viscosity, blood vessel diameter, and blood vessel length have on resistance? On blood flow?*
41. *Define* viscosity, *and state the effect of hematocrit on viscosity.*
42. *State Laplace's law. How does it relate to critical closing pressure and aneurysm?*
43. *What is vascular compliance? Do veins or arteries have greater compliance? Explain why.*

21.7 Physiology of the Systemic Circulation

LEARNING OUTCOMES

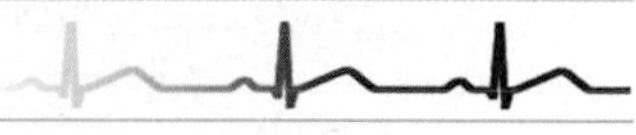

After reading this section, you should be able to

A. List the percent distribution of blood in each of the systemic vessel types.

B. Explain the relationship between cross-sectional area of blood vessels and the rate of blood flow.

C. Explain how blood pressure and resistance to flow change as blood flows through the blood vessels.

D. Define *pulse pressure* and list locations on the body surface where the pulse can be detected.

E. Describe the exchange of materials across a capillary wall.

F. Explain how preload, venous tone, and gravity affect cardiac output.

The anatomy of the circulatory system, the dynamics of blood flow, and the regulatory mechanisms that control the heart and blood vessels determine the physiological characteristics of the circulatory system. The entire circulatory system maintains adequate blood flow to all tissues.

Approximately 84% of the total blood volume is contained in the systemic blood vessels. Most of that blood is in the veins (64%), which are the vessels with the largest compliance. Smaller volumes of blood are in the arteries (15%) and the capillaries (5%; table 21.13).

Cross-Sectional Area of Blood Vessels

If the cross-sectional area of each blood vessel type is determined and multiplied by the number of each type of blood vessel, the result is the total cross-sectional area for each blood vessel type. For example, only one aorta exists, and it has a cross-sectional area of 5 square centimeters (cm^2). On the other hand, millions of capillaries exist, and each has a very small cross-sectional area. However, the total cross-sectional area of all capillaries is 2500 cm^2, which is much greater than the cross-sectional area of the aorta (figure 21.33).

The velocity of blood flow is greatest in the aorta, but the total cross-sectional area is small. In contrast, the total cross-sectional area of the capillaries is large, but the velocity of blood flow is low. As the veins become larger in diameter, their total cross-sectional area decreases, and the velocity of blood flow increases. The relationship between blood vessel diameter and velocity of blood flow is much like a stream that flows rapidly through a narrow gorge but more slowly through a broad plane (figure 21.33).

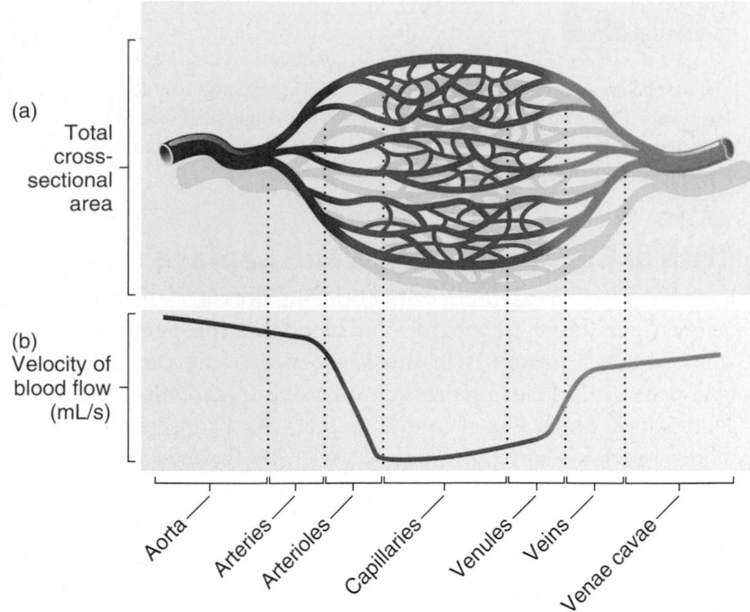
(a)
Total cross-sectional area
(b)
Velocity of blood flow (mL/s)
Aorta
Arteries
Arterioles
Capillaries
Venules
Veins
Venae cavae

FIGURE 21.33 Blood Vessel Area and Velocity of Blood Flow
(*a*) A schematic representing the total cross-sectional area for each of the major blood vessel types. The total cross-sectional area of all the capillaries is much greater (2500 cm^2) than that of the aorta (5 cm^2), although the cross-sectional area of each capillary is much smaller than that of the aorta. (*b*) Blood velocity decreases dramatically in arterioles, capillaries, and venules and is greater in the aorta and the large veins. As the total cross-sectional area increases, the velocity of blood flow decreases.

Pressure and Resistance

The left ventricle forcefully ejects blood from the heart into the aorta. Because the heart's pumping action is pulsatile, the aortic pressure fluctuates between a systolic pressure of 120 mm Hg and a diastolic pressure of 80 mm Hg (table 21.14; figure 21.34). As blood flows from arteries through the capillaries and the veins, the pressure falls progressively to a minimum of approximately 0 mm Hg or even slightly lower by the time it returns to the right atrium.

The decrease in arterial pressure in each part of the systemic circulation is directly proportional to the resistance to blood flow. In other words, the greater the resistance in a blood vessel, the more rapidly the pressure decreases as blood flows through it. Resistance is small in the aorta, so the average pressure at the end of the aorta is nearly the same as at the beginning of the aorta, about 100 mm Hg. The resistance in medium arteries, which are as small as 3 mm in diameter, is also small, so their average pressure is only decreased to 95 mm Hg. In the smaller arteries, however, the resistance to blood flow is greater; by the time blood reaches the arterioles, the average pressure is approximately 85 mm Hg. Within the arterioles, the resistance to flow is greater than in any other part of the systemic

TABLE 21.14 Blood Pressure Classification in Adults

	Systolic Blood Pressure (mm Hg)	Diastolic Blood Pressure (mm Hg)
Normal blood pressure	<120	<80
Prehypertension	120–139	80–89
Stage 1 hypertension	140–159	90–99
Stage 2 hypertension	⩾160	⩾100

Source: The Seventh Report of the Joint National Committee on Prevention, Detection, Evaluation, and Treatment of High Blood Pressure, U.S. Dept. of Health and Human Services, NIH Publication No. 03–5233, May 2003.

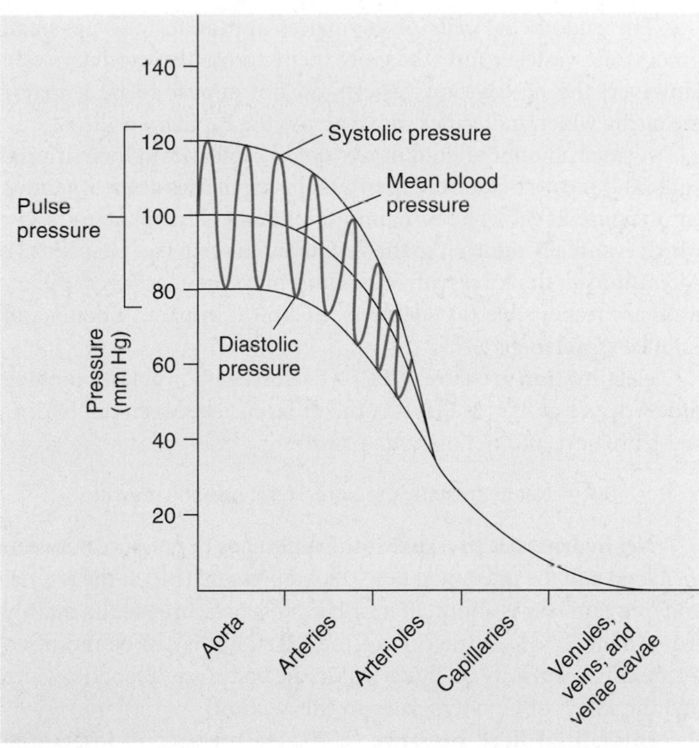
140
120
Pulse
pressure
100
80
60
40
20
Pressure
(mm Hg)
Systolic pressure
Mean blood
pressure
Diastolic
pressure
Aorta
Arteries
Arterioles
Capillaries
Venules,
veins, and
venae cavae

FIGURE 21.34 Blood Pressure in the Major Blood Vessel Types
In small arteries and arterioles, blood pressure fluctuations between systole and diastole are reduced. No fluctuations in blood pressure occur in capillaries and veins.

circulation; at their ends, the average pressure is only approximately 30 mm Hg. The resistance is also fairly high in the capillaries. The blood pressure at the arterial end of the capillaries is approximately 30 mm Hg, and it decreases to approximately 10 mm Hg at the venous end. Resistance to blood flow in the veins is small because of their relatively large diameter; by the time the blood reaches the right atrium in the venous system, the average pressure has decreased from 10 mm Hg to approximately 0 mm Hg.

The muscular arteries and arterioles are capable of constricting or dilating in response to autonomic and hormonal stimulation, altering resistance and blood flow. If vessels constrict, resistance to blood flow increases, less blood flows through the constricted blood vessels, and blood is shunted to other, nonconstricted areas of the body. Muscular arteries help control the amount of blood flowing to each body region, and arterioles regulate blood flow through specific tissues. Constriction of an arteriole decreases blood flow through the local area it supplies, and vasodilation increases blood flow.

ASSESS **YOUR PROGRESS**

44. *List the percent distribution of blood in the large arteries, small arteries, arterioles, capillaries, small veins, and large veins.*

45. *Describe the total cross-sectional areas of the aorta, arteries, arterioles, capillaries, venules, veins, and vena cavae.*

46. *Describe how the rate changes as blood flows through the aorta to the venae cavae.*

47. *Describe the changes in resistance and blood pressure as blood flows through the aorta to the venae cavae.*

48. *Explain how constriction and dilation of muscular arteries shunt blood from one area of the body to another and how constriction and dilation of arterioles change blood flow through local areas.*

Pulse Pressure

The difference between systolic and diastolic pressures is called **pulse pressure** (figure 21.34). For example, in a healthy, young adult at rest, systolic pressure is approximately 120 mm Hg, and diastolic pressure is approximately 80 mm Hg; thus, the pulse pressure is approximately 40 mm Hg. Two major factors influence pulse pressure: stroke volume of the heart and vascular compliance. When stroke volume decreases, pulse pressure also decreases; when stroke volume increases, pulse pressure increases. For example, during exercise, such as running, the stroke volume increases; as a consequence, the pulse pressure also increases. After running, the pulse pressure gradually returns to its resting value as the stroke volume of the heart decreases.

The compliance of blood vessels decreases as a person ages. Arteries in older people become less elastic, or arteriosclerotic, causing the pressure in the aorta to rise more rapidly and to a greater degree during systole and to fall more rapidly to its diastolic value. Thus, for a given stroke volume, systolic pressure and pulse pressure are higher as vascular compliance decreases.

The pulse pressure caused by the ejection of blood from the left ventricle into the aorta produces a pressure wave, or **pulse,** that travels rapidly along the arteries. Its rate of transmission is approximately 15 times greater in the aorta (7–10 m/s) and 100 times greater in the distal arteries (15–35 m/s) than the velocity of blood flow.

The pulse is important clinically because health professionals can determine heart rate, rhythmicity, and other characteristics by feeling it. The pulse can be felt at 10 major locations on each side of the body where large arteries are close to the surface (figure 21.35).

On the head and neck, a pulse can be felt in three arteries: the common carotid artery in the neck, the superficial temporal artery immediately anterior to the ear, and the facial artery at the point where it crosses the inferior border of the mandible approximately midway between the angle and the genu.

On the upper limb, a pulse can also be felt in three arteries: the axillary artery in the axilla, the brachial artery on the medial side of the arm slightly proximal to the elbow, and the radial artery on the lateral side of the anterior forearm just proximal to the wrist. The **radial pulse,** taken at the radial artery, is traditionally used because it is the most easily accessible artery in the body.

In the lower part of the body, a pulse can be felt in four locations: the femoral artery in the groin, the popliteal artery just proximal to the knee, and the dorsalis pedis artery and posterior tibial artery at the ankle.

As the pulse passes through the smallest arteries and arterioles, it is gradually damped, so that the fluctuation between the systolic and diastolic pressures becomes smaller until the difference is almost absent at the end of the arterioles (see figure 21.34). At the beginning of the capillary, there is a steady pressure of close to 30 mm Hg, which is adequate to force blood through the capillaries if the precapillary sphincters dilate.

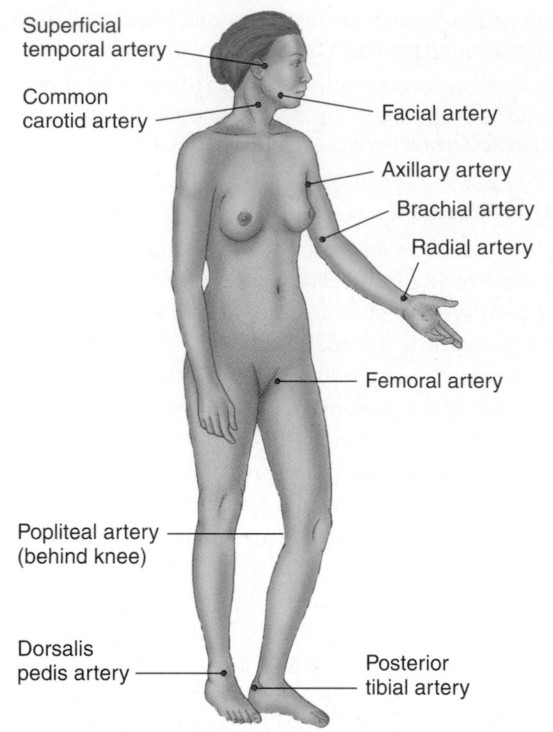
Superficial temporal artery
Common carotid artery
Facial artery
Axillary artery
Brachial artery
Radial artery
Femoral artery
Popliteal artery (behind knee)
Dorsalis pedis artery
Posterior tibial artery

FIGURE 21.35 Major Points at Which the Pulse Can Be Monitored
Each pulse point is named after the artery on which it occurs.

Capillary Exchange and Regulation of Interstitial Fluid Volume

Approximately 10 billion capillaries exist in the body. The heart and blood vessels maintain blood flow through those capillaries and support **capillary exchange,** which is the movement of substances into and out of capillaries. Capillary exchange is the process by which cells receive everything they need to survive and to eliminate metabolic waste products. If blood flow through capillaries is not maintained, cells cannot survive.

By far, the most important means by which capillary exchange occurs is **diffusion.** Oxygen, hormones, and nutrients, such as glucose and amino acids, diffuse from a higher concentration in capillaries to a lower concentration in the interstitial fluid. Waste products, including carbon dioxide, diffuse from a higher concentration in the interstitial fluid to a lower concentration in the capillaries. Lipid-soluble molecules, such as oxygen, carbon dioxide, steroid-hormones, and fatty acids, diffuse through the plasma membranes of the endothelial cells of the capillaries. Water-soluble substances, such as glucose and amino acids, diffuse through intercellular spaces or through fenestrations of capillaries. In a few areas of the body, such as the spleen and liver, the spaces between the endothelial cells are large enough to allow proteins to pass through them. In other areas, the connections between endothelial cells are extensive, and few molecules pass between the endothelial cells; such is the case in the capillaries of the brain that form the blood-brain barrier. In these capillaries, mediated transport moves water-soluble substances across the capillary walls (see chapter 13 for a description of the blood-brain barrier).

The endothelial cells of capillaries appear to take up small pinocytotic vesicles and transport them across the capillary wall. However, the pinocytotic vesicles do not appear to be a major means by which molecules move across the capillary wall.

A small amount of fluid moves out of capillaries at their arterial ends, and most of that fluid reenters the capillaries at their venous ends (figure 21.36). The remaining fluid enters lymphatic vessels, which eventually return it to the venous circulation (see chapter 22). Alterations in the forces affecting fluid movement across capillary walls are responsible for edema (see Clinical Impact, "Edema and Capillary Exchange").

Net filtration pressure (NFP) is the force responsible for moving fluid across capillary walls. It is the difference between net hydrostatic pressure and net osmotic pressure:

$$\text{NFP} = \text{Net hydrostatic pressure} - \text{Net osmotic pressure}$$

Net hydrostatic pressure is the difference in pressure between the blood and the interstitial fluid. Blood pressure (BP) at the arterial end of a capillary is about 30 mm Hg. This pressure results mainly from the force of contraction of the heart, but it can be modified by the effect of gravity on fluids within the body (see "Blood Pressure and the Effect of Gravity," later in this section).

Interstitial fluid pressure (IFP), the pressure of interstitial fluid within the tissue spaces, is –3 mm Hg. IFP is a negative number because of the suction effect produced by the lymphatic vessels as they pump excess fluid from the tissue spaces. The lymphatic system is described in chapter 22. Here, it is only necessary to understand that excess interstitial fluid enters lymphatic capillaries and is eventually returned to the blood.

At the arterial end of capillaries, the net hydrostatic pressure that moves fluid across the capillary walls into the tissue spaces is the difference between BP and IFP:

$$\begin{aligned}\text{Net hydrostatic pressure} &= \text{BP} - \text{IFP}\\ &= 30 - (-3)\\ &= 33 \text{ mm Hg}\end{aligned}$$

Net osmotic pressure is the difference in osmotic pressure between the blood and the interstitial fluid. The osmotic pressure caused by the plasma proteins is called the **blood colloid osmotic pressure (BCOP),** and the osmotic pressure caused by proteins in the interstitial fluid is called the **interstitial colloid osmotic pressure (ICOP).** Large proteins do not pass freely through the capillary walls, and the difference in protein concentrations between the blood and the interstitial fluid is responsible for osmosis. Ions and small molecules do not make a significant contribution to osmosis across the capillary wall because they pass freely through it and their concentrations are approximately the same in the blood as in the interstitial fluid.

The BCOP (28 mm Hg) is several times larger than the ICOP (8 mm Hg) because of the presence of albumin and other proteins in the plasma (see chapter 19). Therefore, the net osmotic pressure is equal to BCOP – ICOP:

$$\begin{aligned}\text{Net osmotic pressure} &= \text{BCOP} - \text{ICOP}\\ &= 28 - 8\\ &= 20 \text{ mm Hg}\end{aligned}$$

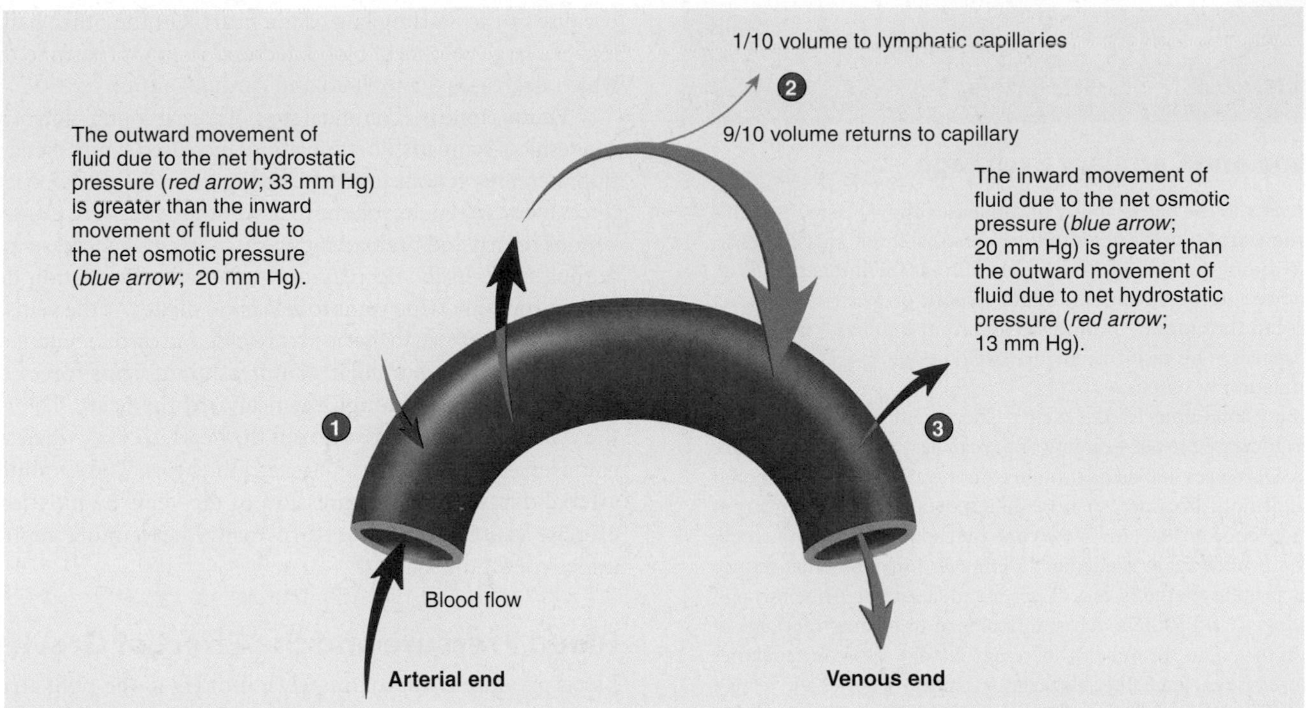
1/10 volume to lymphatic capillaries
2
9/10 volume returns to capillary
The outward movement of fluid due to the net hydrostatic pressure (*red arrow*; 33 mm Hg) is greater than the inward movement of fluid due to the net osmotic pressure (*blue arrow*; 20 mm Hg).
The inward movement of fluid due to the net osmotic pressure (*blue arrow*; 20 mm Hg) is greater than the outward movement of fluid due to net hydrostatic pressure (*red arrow*; 13 mm Hg).
1
3
Blood flow
Arterial end
Venous end

1 At the arterial end of the capillary, the **net hydrostatic pressure** is greater than the **net osmotic pressure.** When the net osmotic pressure is subtracted from the net hydrostatic pressure, the result is a positive **net filtration pressure** that causes fluid to move out of the capillary.

33 mm Hg (Net hydrostatic pressure)
−20 mm Hg (Net osmotic pressure)
13 mm Hg (Net filtration pressure)

2 Approximately nine-tenths of the fluid that leaves the capillary at its arterial end reenters the capillary at its venous end. About one-tenth of the fluid passes into the lymphatic capillaries.

3 At the venous end of the capillary, the **net hydrostatic pressure** is less than the **net osmotic pressure.** When the net osmotic pressure is subtracted from the net hydrostatic pressure, the result is a negative **net filtration pressure** that causes fluid to move into the capillary.

13 mm Hg (Net hydrostatic pressure)
−20 mm Hg (Net osmotic pressure)
− 7 mm Hg (Net filtration pressure)

PROCESS FIGURE 21.36 Fluid Exchange Across the Walls of Capillaries
Pressure differences exist between the inside and the outside of capillaries at their arterial and venous ends.

The greater the osmotic pressure of a fluid, the greater the tendency for water to move into that fluid (see chapter 3). The net osmotic pressure results in the osmosis of water into the capillary because water has a greater tendency to move into the blood than into the interstitial fluid.

The net filtration pressure at the arterial end of the capillary is equal to the net hydrostatic pressure, which moves fluid out of the capillary, minus the net osmotic pressure, which moves fluid into the capillary. Since the net hydrostatic pressure is greater than the net osmotic pressure at the arterial end of the capillary, there is a net movement of fluid out of the capillary:

$$\begin{aligned} \text{NFP} &= \text{Net hydrostatic pressure} - \text{Net osmotic pressure} \\ &- 33 - 20 \\ &= 13 \text{ mm Hg} \end{aligned}$$

Between the arterial and venous ends of capillaries, the blood pressure decreases from about 30 mm Hg to 10 mm Hg. This causes a reduction in the net hydrostatic pressure moving fluid out of the venous end of the capillary:

$$\begin{aligned} \text{Net hydrostatic pressure} &= \text{BP} - \text{IFP} \\ &= 10 - (-3) \\ &= 13 \text{ mm Hg} \end{aligned}$$

The concentration of proteins within capillaries and the concentration of proteins within interstitial fluid do not change significantly because only a small amount of fluid passes from the capillaries into the tissue spaces. Therefore, the net osmotic pressure moving fluid into capillaries by osmosis is still approximately 20 mm Hg. At the venous end of capillaries, the NFP now causes fluid to reenter the capillary:

$$\begin{aligned} \text{NFP} &= \text{Net hydrostatic pressure} - \text{Net osmotic pressure} \\ &= 13 - 20 \\ &= -7 \text{ mm Hg} \end{aligned}$$

Exchange of fluid across the capillary wall and movement of fluid into lymphatic capillaries keep the volume of the interstitial fluid within a narrow range of values. Disruptions in the movement of fluid across the wall of the capillary can result in edema, or swelling, as a result of increased interstitial fluid volume.

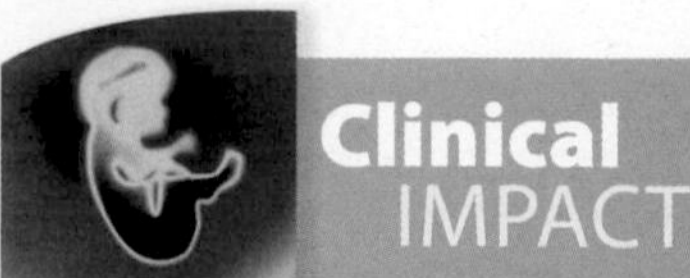
Clinical
IMPACT

Edema and Capillary Exchange

Increases in the permeability of capillaries allow plasma proteins to move from capillaries into the interstitial fluid. This causes an increase in the colloid osmotic pressure of the interstitial fluid. An increase in the interstitial colloid osmotic pressure causes a net increase in the amount of fluid moving from capillaries into interstitial spaces. The result is edema, swelling due to excessive fluid accumulation in tissues.

Many conditions lead to edema. Chemical mediators of inflammation increase the permeability of the capillary walls and can cause edema. Decreases in plasma protein concentration reduce the blood colloid osmotic pressure, so more fluid moves out of the capillary at its arterial end and less fluid moves into the capillary at its venous end. The result once again is edema. Severe liver infections that reduce plasma protein synthesis, loss of protein molecules in urine through the kidneys, and protein starvation all lead to edema. Blockage of veins, as in venous thrombosis, increases blood pressure in capillaries and can cause edema. Either blockage or removal of lymphatic vessels, as occurs when lymph nodes are suspected of being cancerous, allows fluid to accumulate in the interstitial spaces and results in edema.

ASSESS YOUR PROGRESS

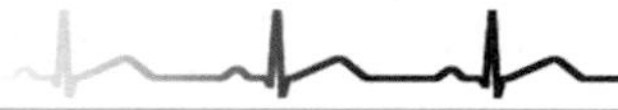

49. *What is pulse pressure? How do stroke volume and vascular compliance affect pulse pressure?*

50. *List the locations on the body where the pulse can easily be detected.*

51. *What is the most important means by which capillary exchange occurs?*

52. *Describe the factors that influence the movement of fluid from capillaries into the tissues.*

53. *What is the main force for the return of fluids at the venous end of capillaries?*

54. *What happens to the fluid in the tissues? What is edema?*

Predict 5

Edema often results from a disruption in the normal inwardly and outwardly directed pressures across the capillary wall. On the basis of what you know about fluid movement across the wall of the capillary and the regulation of capillary blood pressure, explain why large fluctuations in arterial blood pressure do not cause significant edema, whereas small increases in venous pressure can lead to edema.

Functional Characteristics of Veins

Cardiac output depends on the preload, which is determined by the volume of blood that enters the heart from the veins (see chapter 20). Therefore, the factors that affect flow in the veins are of great importance to the overall function of the cardiovascular system. If the volume of blood is increased because of a rapid transfusion, the amount of blood flow to the heart through the veins increases. This increases the preload, which causes the cardiac output to increase because of the Starling law of the heart. On the other hand, rapid loss of a large volume of blood decreases venous return to the heart, which decreases the preload and cardiac output.

Venous tone is a continual state of partial contraction of the veins as a result of sympathetic stimulation. Increased sympathetic stimulation increases venous tone by causing the veins to constrict, which forces the large venous volume to flow toward the heart. Consequently, venous return and preload increase, causing an increase in cardiac output. Conversely, decreased sympathetic stimulation decreases venous tone, allowing veins to relax and dilate. As the veins fill with blood, venous return to the heart, preload, and cardiac output decrease.

The periodic muscular compression of veins forces blood to flow more rapidly through them toward the heart. The valves in the veins prevent flow away from the heart, so that, when veins are compressed, blood is forced toward the heart. The combination of arterial dilation and compression of the veins by muscles during exercise causes blood to return to the heart more rapidly than under conditions of rest.

Blood Pressure and the Effect of Gravity

Blood pressure is approximately 0 mm Hg in the right atrium and approximately 100 mm Hg in the aorta. However, the pressure in the vessels above and below the heart is affected by gravity. While a person is standing, the pressure in the venules of the feet can be as much as 90 mm Hg, instead of the usual 10 mm Hg. Arterial pressure is influenced by gravity to the same degree; thus, the arterial ends of the capillaries can have a pressure of 110 mm Hg rather than 30 mm Hg. The normal pressure difference between the arterial and the venous ends of capillaries remains the same, so that blood continues to flow through the capillaries. The major effect of the high pressure in the feet and legs when a person stands for a prolonged time without moving is edema. Without muscular movement, the pressure at the venous end of the capillaries increases. Up to 15–20% of the total blood volume can pass through the walls of the capillaries into the interstitial spaces of the lower limbs during 15 minutes of standing still.

When a person changes position from lying down to standing, the blood pressure in the veins of the lower limbs increases. Because of the structure of their walls, the compliance of veins is approximately 24 times greater than the compliance of arteries. The increased blood pressure causes the distensible (compliant) veins to expand but has little effect on the arteries. Venous return decreases because less blood is returning to the heart as the veins are filling with blood. As venous return decreases, cardiac output and blood pressure decrease (see chapter 20). If negative-feedback mechanisms do not compensate and cause blood pressure to increase, the delivery of blood to the brain is not adequate to maintain homeostasis, and the person may feel dizzy or even faint.

ASSESS YOUR PROGRESS

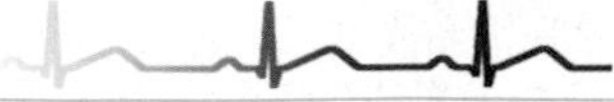

55. *How do blood volume and venous tone affect cardiac output?*

56. *What effect does standing still for a prolonged time have on the blood pressure in the feet and in the head? Explain why this effect occurs.*

57. *Why does a person feel dizzy if he or she stands up too quickly from sitting or lying down?*

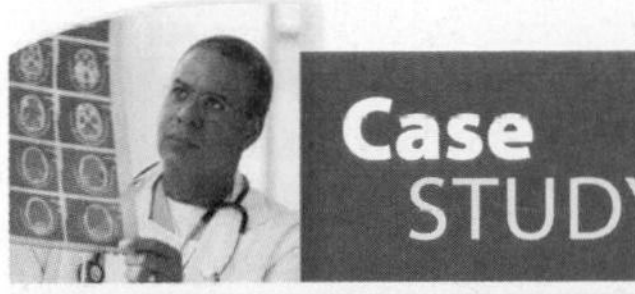

Case STUDY Venous Thrombosis

Harry is a 55-year-old college professor who teaches a night class in a small town about 50 miles from his home. One night, as he walked to his car after class, Harry noticed that his right leg was uncomfortable. When he arrived home, about 90 minutes later, Harry realized that the calf of his right leg had become very swollen. When he extended his knee and plantar flexed his foot, the pain in his right leg increased. Harry thought this might be a serious condition, so he drove to the hospital.

In the emergency room, technicians performed a Doppler test, which monitors the flow of blood through blood vessels. The test confirmed that a thrombus had formed in one of the deep veins of Harry's right leg. His pain and edema were consistent with the presence of a venous thrombosis.

Harry was admitted to the hospital, and his physician prescribed intravenous (IV) heparin. About 4 a.m., Harry experienced an increase in his respiratory rate, his breathing became labored, he felt pain in his chest and back, and his arterial oxygen levels decreased. In response to these changes, Harry's physician increased the amount of heparin. The chest pain subsided, and Harry's respiratory movements improved over the next 24 hours. The next day, a CT scan revealed pulmonary emboli, but no infarctions of the lung. The edema in Harry's leg also slowly improved.

Harry remained in the hospital for several days, during which heparin was continued and then oral coumadin was prescribed. Frequent blood samples were taken to determine Harry's prothrombin time (see chapter 19). After about a week, Harry was released from the hospital. His physician, however, prescribed oral coumadin for at least several months. In addition, Harry was required to have his prothrombine time checked periodically.

Predict 6

a. Explain why edema and pain developed in response to a thrombus in a deep vein of Harry's right leg.

b. If a thrombus in the posterior tibial vein gave rise to an embolus, name in order the parts of the circulatory system the embolus would pass through before lodging in a blood vessel in the lungs. Explain why the lungs are the most likely places the embolus will lodge.

c. Predict the effect of pulmonary emboli on the right ventricle's ability to pump blood.

d. Predict the effect of pulmonary emboli on blood oxygen levels, on the left ventricle's ability to pump blood, and on systemic blood pressure. What responses would be activated by this change in blood pressure? (*Hint:* See figure 21.40.)

e. Explain why Harry's physician prescribed heparin and coumadin and why coumadin was continued long after the venous thrombosis and lung emboli had dissolved.

21.8 Control of Blood Flow in Tissues

LEARNING OUTCOMES

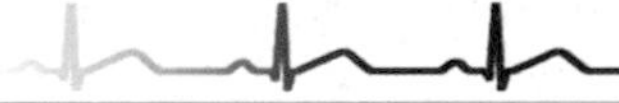

After reading this section, you should be able to

A. Explain how local mechanisms regulate blood flow.

B. Explain how nervous and hormonal mechanisms control blood flow.

Blood flow provided to the tissues by the circulatory system is highly controlled and matched closely to the metabolic needs of tissues. Mechanisms that control blood flow through tissues are classified as (1) local control and (2) nervous and hormonal control (table 21.15).

Local Control of Blood Flow in Tissues

Blood flow is much greater in some organs than in others. For example, blood flow through the brain, kidneys, and liver is relatively high. By contrast, blood flow through resting skeletal muscles is not high, but it is greater than that through other tissue types because skeletal muscle constitutes 35–40% of the total body mass. However, blood flow through exercising skeletal muscles can increase up to 20-fold, and the flow through the viscera, including the kidneys and liver, either remains the same or decreases. Local control of blood flow is achieved by the periodic relaxation and contraction of precapillary sphincters regulating blood flow through the tissues. In most tissues, blood flow is proportional to the metabolic needs of the tissue; therefore, as the activity of skeletal muscle increases, blood flow increases to supply the greater need for oxygen and other nutrients. Blood flow also increases in response to a buildup of metabolic end products.

In some tissues, blood flow serves purposes other than delivering nutrients and removing waste products. In the skin, blood flow dissipates heat from the body. In the kidneys, it eliminates metabolic waste products, regulates water balance, and controls the pH of body fluids. Among other functions, blood flow delivers nutrients that enter the blood from the small intestine to the liver for processing.

Functional Characteristics of the Capillary Bed

The innervation of the metarterioles and the precapillary sphincters in capillary beds is sparse. Local factors primarily regulate these structures. As the rate of metabolism increases in a tissue, blood flow through its capillaries increases. The precapillary sphincters relax, allowing blood to flow into the local capillary bed. Blood flow can increase sevenfold to eightfold as a result of vasodilation of the metarterioles and the precapillary sphincters in response to an increased rate of metabolism.

Vasodilator substances are produced in the extracellular fluid as the rate of metabolism increases. These chemicals include carbon

TABLE 21.15 Homeostasis: Control of Blood Flow

Stimulus	Response
Local Control	
Regulation by Metabolic Need of Tissues	
Increased vasodilator substances (e.g., CO_2, lactic acid, adenosine, adenosine monophosphate, adenosine diphosphate, endothelium-derived relaxation factor, K^+, decreased pH) or decreased O_2 and nutrients (e.g., glucose, amino acids, fatty acids, and other nutrients) as a result of increased metabolism	Relaxation of precapillary sphincters and subsequent increase in blood flow through capillaries
Decreased vasodilator substances and a reduced need for O_2 and nutrients	Contraction of precapillary sphincters and subsequent decrease in blood flow through capillaries
Autoregulation	
Increased blood pressure	Contraction of precapillary sphincters to maintain constant capillary blood flow
Decreased blood pressure	Relaxation of precapillary sphincters to maintain constant capillary blood flow
Long-Term Local Blood Flow	
Increased metabolic activity of tissues over a long period	Increased diameter and number of capillaries
Decreased metabolic activity of tissues over a long period	Decreased diameter and number of capillaries
Nervous Control	
Increased physical activity or increased sympathetic activity	Constriction of blood vessels in skin and viscera
Increased body temperature detected by neurons of the hypothalamus	Dilation of blood vessels in skin (see chapter 5)
Decreased body temperature detected by neurons of the hypothalamus	Constriction of blood vessels in skin (see chapter 5)
Decreased skin temperature below a critical value	Dilation of blood vessels in skin (protects skin from extreme cold)
Anger or embarrassment	Dilation of blood vessels in skin of face and upper thorax
Hormonal Control (reinforces increased activity of the sympathetic nervous system)	
Increased physical activity and increased sympathetic activity, causing release of epinephrine and small amounts of norepinephrine from adrenal medulla	Constriction of blood vessels in skin and viscera; dilation of blood vessels in skeletal and cardiac muscle

dioxide, lactic acid, adenosine, adenosine monophosphate, adenosine diphosphate, endothelium-derived relaxation factor (EDRF), K^+, and H^+. Once produced, the vasodilator substances diffuse from the tissues supplied by the capillary to the area of the precapillary sphincter, the metarterioles, and the arterioles to cause vasodilation (figure 21.37).

Lack of oxygen and nutrients can also be important in regulating blood flow in tissues. For example, oxygen and nutrients are required to maintain vascular smooth muscle contraction. An increased rate of metabolism decreases the amount of oxygen and nutrients in the tissues. Smooth muscle cells of the precapillary sphincter relax in response to a lack of oxygen and nutrients, resulting in vasodilation (figure 21.37).

Blood flow through capillaries is not continuous, but cyclic. The cyclic fluctuation is the result of periodic contraction and relaxation of the precapillary sphincters, called **vasomotion** (vā-sō-mō′shŭn, vas-ō-mō′shŭn). Blood flows through the capillaries until the by-products of metabolism are reduced in concentration and until nutrient supplies to precapillary smooth muscles are replenished. Then the precapillary sphincters constrict and remain constricted until the by-products of metabolism increase and nutrients decrease (figure 21.37).

Autoregulation of Blood Flow

Arterial pressure can change over a wide range, whereas blood flow through tissues remains relatively constant. The maintenance of blood flow by tissues is called **autoregulation** (aw′tō-reg-ū-lā′shŭn). Between arterial pressures of approximately 75 mm Hg and 175 mm Hg, blood flow through tissues remains within 10–15% of its normal value. The mechanisms responsible for autoregulation are the same as those for vasomotion. The need for oxygen and nutrients and the buildup of metabolic by-products cause precapillary sphincters to dilate, and blood flow through tissues increases if a minimum blood pressure exists. On the other hand, once the supply of oxygen and nutrients to tissues is adequate, the precapillary sphincters constrict, and blood flow through the tissues decreases, even if blood pressure is very high.

Clinical
IMPACT

Hypertension

Hypertension, or high blood pressure, affects nearly 30% of the population at sometime in their lives. Generally, a person is considered hypertensive if the systolic blood pressure is greater than 140 mm Hg and the diastolic pressure is greater than 90 mm Hg. However, current methods of evaluation take into consideration diastolic and systolic pressures in determining whether a person is suffering from hypertension (see table 21.14). In addition, normal blood pressure is age-dependent, so classification of an individual as hypertensive also depends on the person's age.

Chronic hypertension has an adverse effect on the function of both the heart and the blood vessels. Hypertension requires the heart to work harder than normal. This extra work leads to hypertrophy of the cardiac muscle, especially in the left ventricle, and can result in heart failure. Hypertension also increases the rate at which arteriosclerosis develops. Arteriosclerosis, in turn, increases the probability that blood clots, or thromboemboli (throm'bō-em'bō-lī), will form and that blood vessels will rupture. Common medical problems associated with hypertension are cerebral hemorrhage, coronary infarction, hemorrhage of renal blood vessels, and poor vision caused by burst blood vessels in the retina.

Some conditions leading to hypertension include a decrease in functional kidney mass, excess aldosterone or angiotensin production, and increased resistance to blood flow in the renal arteries. All of these conditions lead to an increase in total blood volume, which causes cardiac output to increase. Increased cardiac output forces blood to flow through tissue capillaries, causing the precapillary sphincters to constrict. Thus, increased blood volume increases cardiac output and peripheral resistance, both of which result in greater blood pressure.

Although many known conditions result in hypertension, roughly 90% of the diagnosed cases are called **idiopathic hypertension,** or *essential hypertension*, which means that the cause of the condition is unknown. Drugs that dilate blood vessels (called vasodilators), drugs that increase the rate of urine production (called diuretics), and drugs that decrease cardiac output are normally used to treat idiopathic hypertension. The vasodilator drugs increase the rate of blood flow through the kidneys and thus increase urine production; the diuretics increase urine production as well. Increased urine production reduces blood volume, which reduces blood pressure. Substances that decrease cardiac output, such as β-adrenergic blocking agents, decrease the heart rate and force of contraction. In addition to these treatments, low-salt diets are normally recommended to reduce the amount of sodium chloride (NaCl) and water absorbed from the intestine into the bloodstream.

Long-Term Local Blood Flow

The long-term regulation of blood flow through tissues is matched closely to the tissues' metabolic requirements. If the metabolic activity of a tissue increases and remains elevated for an extended period, the diameter and the number of capillaries in the tissue increase, and local blood flow increases. An example is the increased density of capillaries in the well-trained skeletal muscles of athletes, compared with that in poorly trained skeletal muscles.

The availability of oxygen to a tissue can be a major factor in determining the adjustment of the tissue's vascularity to its long-term metabolic needs. If oxygen is scarce, capillaries increase in diameter and in number but, if oxygen levels remain elevated in a tissue, the vascularity decreases.

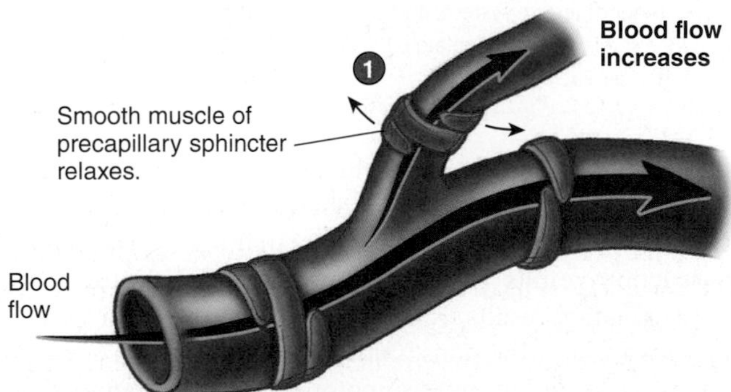
1
Blood flow increases
Smooth muscle of precapillary sphincter relaxes.
Blood flow

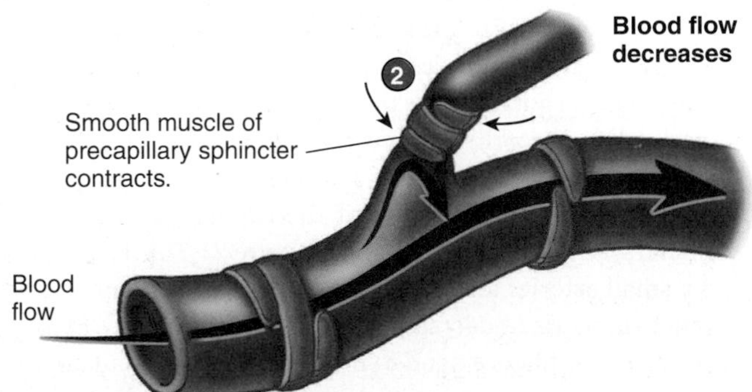
2
Blood flow decreases
Smooth muscle of precapillary sphincter contracts.
Blood flow

① **Vasodilation of precapillary sphincters**
Precapillary sphincters relax as the tissue concentration of O_2 and nutrients, such as glucose, amino acids, and fatty acids, decreases. The sphincters also relax as the concentration of vasodilator substances, such as CO_2, lactic acid, adenosine, adenosine monophosphate, adenosine diphosphate, nitric oxide, and K^+, increase, and as the pH decreases.

② **Constriction of precapillary sphincters**
Precapillary sphincters contract as the tissue concentration of O_2 and nutrients, such as glucose, amino acids, and fatty acids, increases. The sphincters also contract as the tissue concentration of metabolic by-products, such as CO_2, lactic acid, adenosine, adenosine monophosphate, adenosine diphosphate, nitric oxide, and K^+, decrease, and as the pH increases.

PROCESS FIGURE 21.37 APR **Local Control of Blood Flow Through Capillary Beds**

Clinical IMPACT

Occlusion of Blood Vessels and Collateral Circulation

Occlusion, or blockage, of a blood vessel leads to an increase in the diameter of smaller blood vessels that bypass the occluded vessel. In many cases, the development of these collateral vessels is marked. For example, if the femoral artery is occluded, the small vessels that bypass the occluded vessel become greatly enlarged, and an adequate blood supply to the lower limb is often reestablished over a period of weeks. If the occlusion is sudden and so complete that tissues supplied by a blood vessel suffer from ischemia (lack of blood flow), cell necrosis (death) can occur. In this instance, collateral circulation does not have a chance to develop before necrosis sets in.

Nervous and Hormonal Control of Blood Flow in Tissues

Nervous control of arterial blood pressure is important in the minute-to-minute regulation of blood flow in tissues. The blood pressure must be adequate to cause blood to flow through capillaries while at rest, during exercise, or in response to circulatory shock, during which blood pressure becomes very low. For example, during exercise, increased arterial blood pressure is needed to sustain increased blood flow through the capillaries of skeletal muscles, in which precapillary sphincters have dilated. The increased blood flow is needed to supply oxygen and nutrients to the exercising skeletal muscles.

Nervous regulation also provides a means by which blood can be shunted from one large area of the circulatory system to another. For example, in response to blood loss, blood flow to the viscera and the skin is reduced dramatically. This helps maintain the arterial blood pressure within a range sufficient to allow adequate blood flow through the capillaries of the brain and cardiac muscle.

Nervous regulation by the autonomic nervous system, particularly the sympathetic division, can function rapidly (within 1–30 seconds). Sympathetic vasomotor fibers innervate all the blood vessels of the body except the capillaries, the precapillary sphincters, and most metarterioles (figure 21.38). The innervation of the small arteries and arterioles allows the sympathetic nervous system to increase or decrease resistance to blood flow. Sympathetic vasoconstrictor fibers extend to most parts of the circulatory system, but they are less prominent in skeletal muscle, cardiac muscle, and the brain and more prominent in the kidneys, the digestive tract, the spleen, and the skin.

An area of the lower pons and upper medulla oblongata, called the **vasomotor** (vā-sō-mō′ter, vas-ō-mō′ter) **center** (figure 21.38), is tonically active. A low frequency of action potentials is transmitted continually through the sympathetic vasoconstrictor fibers. As a consequence, the peripheral blood vessels are partially constricted, a condition called **vasomotor tone.**

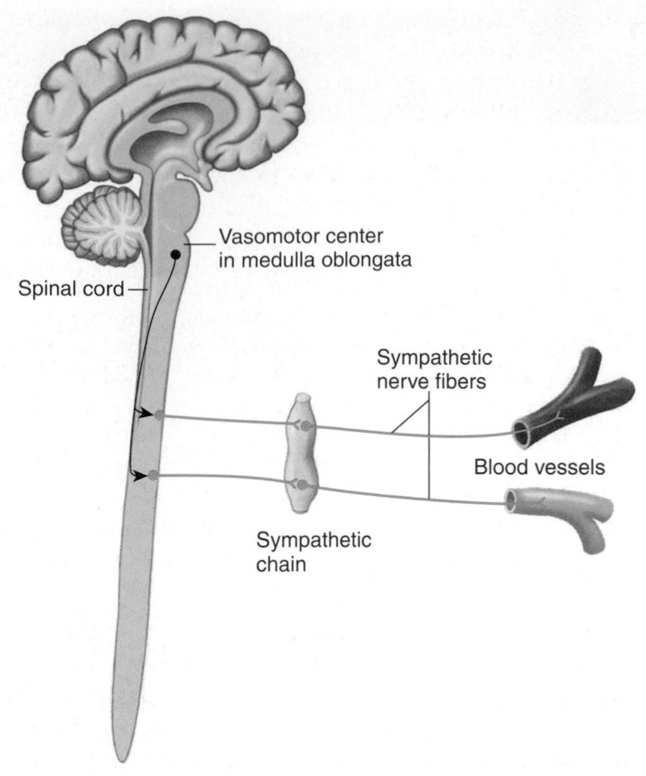
Vasomotor center
in medulla oblongata
Spinal cord
Sympathetic
nerve fibers
Blood vessels
Sympathetic
chain

FIGURE 21.38 Nervous Regulation of Blood Vessels
Most blood vessels are innervated by sympathetic nerve fibers. The vasomotor center within the medulla oblongata plays a major role in regulating the frequency of action potentials in nerve fibers that innervate blood vessels.

Predict 7

A strong athlete just finished a 1-mile run and sat down to have a drink with her friends. During the run, her blood pressure was not dramatically elevated, but her cardiac output increased greatly. After the run, her cardiac output decreased dramatically, but her blood pressure decreased only to its resting level. Predict how the sympathetic stimulation of her large veins, the arteries in her digestive system, and the arteries in her skeletal muscles changes while she is relaxing. Explain why this is consistent with the decrease in her cardiac output.

Part of the vasomotor center inhibits vasomotor tone. Thus, the vasomotor center consists of an excitatory part, which is tonically active, and an inhibitory part, which can induce vasodilation. Vasoconstriction results from an increase in vasomotor tone, whereas vasodilation results from a decrease in vasomotor tone.

Areas throughout the pons, midbrain, and diencephalon can either stimulate or inhibit the vasomotor center. For example, the hypothalamus can exert either strong excitatory or strong inhibitory effects on the vasomotor center. Increased body temperature detected by temperature receptors in the hypothalamus causes vasodilation of blood vessels in the skin (see chapter 5). The cerebral cortex can also either excite or inhibit the vasomotor center. For example, action potentials that originate in the cerebral cortex during periods of emotional excitement activate hypothalamic centers, which in turn increase vasomotor tone (table 21.15).

The neurotransmitter for the vasoconstrictor fibers is norepinephrine, which binds to α-adrenergic receptors on vascular smooth muscle cells to cause vasoconstriction. Sympathetic action potentials also cause the release of epinephrine and norepinephrine into the blood from the adrenal medulla. These hormones are transported in the blood to all parts of the body. In most vessels, they cause vasoconstriction, but in some vessels, especially those in skeletal muscle, epinephrine binds to β-adrenergic receptors, which are present in larger numbers, and can cause the skeletal muscle blood vessels to dilate.

ASSESS **YOUR PROGRESS**

58. *How is local control of blood flow in tissues achieved?*

59. *Explain how vasodilator substances, oxygen, and nutrients are involved with local control of blood flow.*

60. *What is vasomotion? What is autoregulation of local blood flow?*

61. *How is long-term regulation of blood flow through tissues accomplished?*

62. *Describe nervous and hormonal control of blood flow. Under what conditions is nervous control of blood flow important? What is vasomotor tone?*

21.9 Regulation of Mean Arterial Pressure

LEARNING OUTCOMES

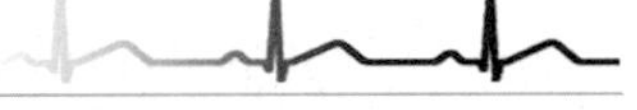

After reading this section, you should be able to

A. Recall the definitions of *mean arterial pressure, cardiac output,* and *peripheral resistance.*

B. Relate the factors that determine mean arterial pressure.

C. Describe the short-term and the long-term mechanisms that regulate arterial blood pressure.

Blood flow to all areas of the body depends on the maintenance of adequate pressure in the arteries. As long as arterial blood pressure is adequate, local control of blood flow is appropriately matched to tissues' metabolic needs. Blood flow through tissues cannot be adequate if arterial blood pressure is too low. Inadequate blood flow throughout the body due to the failure of mechanisms to maintain normal blood pressure is **circulatory shock.** If normal blood pressure is not maintained, damage to the body tissues can lead to death (see Clinical Impact, "Circulatory Shock," later in this chapter). On the other hand, if arterial blood pressure is too high, the heart and blood vessels may be damaged.

Mean arterial pressure (MAP) is slightly less than the average of systolic and diastolic pressures because diastole lasts longer than systole. MAP is approximately 70 mm Hg at birth, slightly less than 100 mm Hg from adolescence to middle age, and 110–130 mm Hg in healthy older persons (see table 21.14).

Cardiac output (CO) is the volume of blood pumped by the heart each minute. It is equal to the **heart rate (HR)** times the **stroke volume (SV). Peripheral resistance (PR)** is the resistance to blood flow in all the blood vessels. The body's MAP is proportional to the cardiac output times the peripheral resistance:

$$\text{MAP} = \text{CO} \times \text{PR or MAP} = \text{HR} \times \text{SV} \times \text{PR}$$

As indicated by the equations, blood pressure is influenced by three factors: heart rate, stroke volume, and peripheral resistance. An increase in any one of these elevates blood pressure. Conversely, a decrease in any one of them reduces blood pressure.

Because stroke volume depends on the amount of blood entering the heart, regulatory mechanisms that control blood volume also affect blood pressure. For example, an increase in blood volume increases venous return, which increases preload, and the increased preload increases stroke volume.

When blood pressure suddenly drops because of hemorrhage or some other cause, the control systems respond by increasing blood pressure to a value consistent with life and by increasing blood volume to its normal value. Two major types of control systems operate to achieve these responses: (1) those that respond in the short term and (2) those that respond in the long term. The short-term regulatory mechanisms respond quickly but begin to lose their capacity to regulate blood pressure a few hours to a few days after blood pressure is maintained at homeostatic values. This occurs because sensory receptors adapt to the altered pressures. Long-term regulation of blood pressure is controlled primarily by mechanisms that influence kidney function. Those mechanisms do not adapt rapidly to altered blood pressures.

Short-Term Regulation of Blood Pressure

The short-term, rapidly acting mechanisms controlling blood pressure include the baroreceptor reflexes, the adrenal medullary mechanism, chemoreceptor reflexes, and the central nervous system's ischemic response. Some of these reflex mechanisms operate on a minute-to-minute basis and help regulate blood pressure within a narrow range of values. Other mechanisms respond primarily to emergency situations (see figures 21.41 and 21.43).

Baroreceptor Reflexes

Baroreceptor reflexes are very important in regulating blood pressure on a minute-to-minute basis. They detect even small changes in blood pressure and respond quickly. However, they are not as important as other mechanisms in regulating blood pressure over long periods of time.

Baroreceptors, or *pressoreceptors*, are sensory receptors sensitive to stretch. They are scattered along the walls of most of the large arteries of the neck and thorax and are most numerous in the area of the carotid sinus at the base of the internal carotid artery and in the walls of the aortic arch. Action potentials travel from the carotid sinus baroreceptors through the glossopharyngeal (IX) nerves to the cardioregulatory and vasomotor centers in the medulla oblongata and from the aortic arch through the vagus (X) nerves to the medulla oblongata (figure 21.39). Stimulation of baroreceptors in the carotid sinus activates the **carotid sinus reflex,** and stimulation of baroreceptors in the aortic arch activates the **aortic arch reflex.** Both of these reflexes are baroreceptor reflexes, and they help keep blood pressure within homeostatic values.

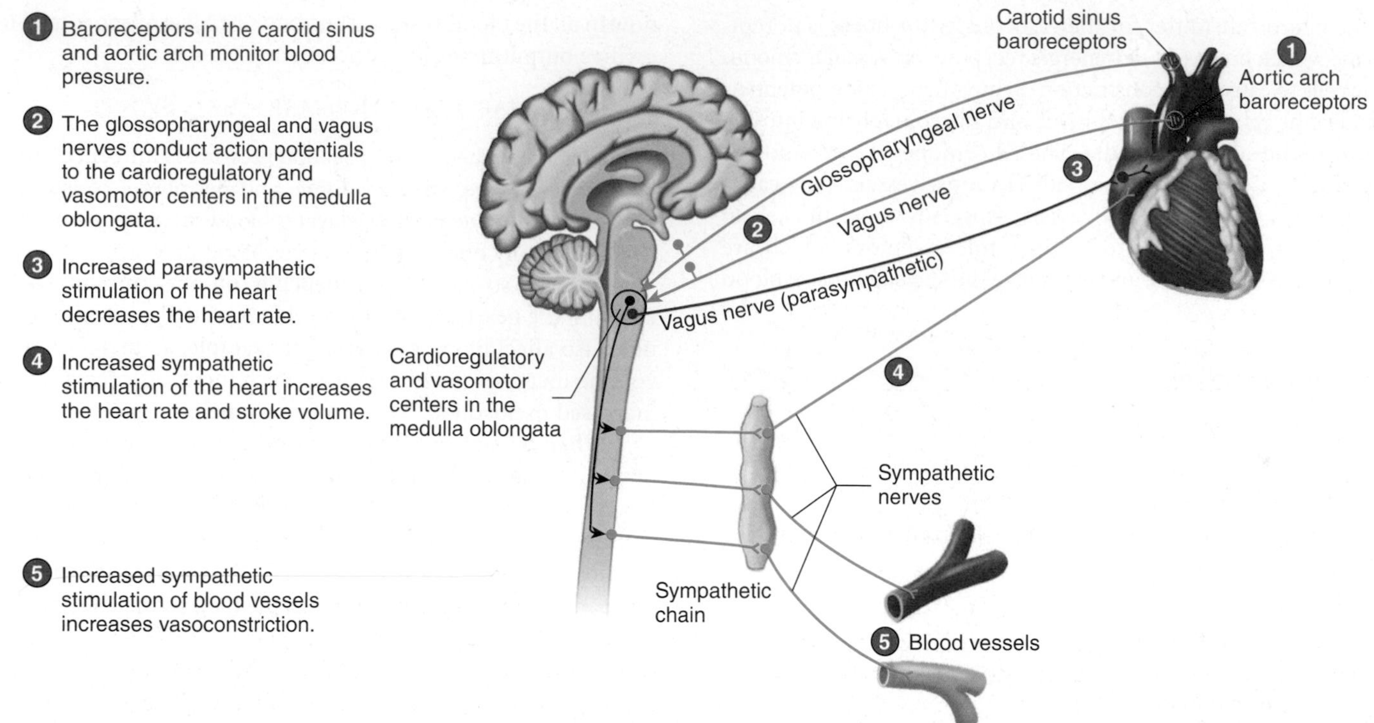
1 Baroreceptors in the carotid sinus and aortic arch monitor blood pressure.
2 The glossopharyngeal and vagus nerves conduct action potentials to the cardioregulatory and vasomotor centers in the medulla oblongata.
3 Increased parasympathetic stimulation of the heart decreases the heart rate.
4 Increased sympathetic stimulation of the heart increases the heart rate and stroke volume.
5 Increased sympathetic stimulation of blood vessels increases vasoconstriction.
Carotid sinus baroreceptors
Aortic arch baroreceptors
Glossopharyngeal nerve
Vagus nerve
Vagus nerve (parasympathetic)
Cardioregulatory and vasomotor centers in the medulla oblongata
Sympathetic nerves
Sympathetic chain
Blood vessels

PROCESS FIGURE 21.39 Baroreceptor Reflex Control of Blood Pressure

An increase in blood pressure increases parasympathetic stimulation of the heart and decreases sympathetic stimulation of the heart and blood vessels, resulting in a decrease in blood pressure. A decrease in blood pressure decreases parasympathetic stimulation of the heart and increases sympathetic stimulation of the heart and blood vessels, resulting in an increase in blood pressure.

In the carotid sinus and the aortic arch, normal blood pressure partially stretches the arterial wall, so that the baroreceptors produce a constant but low frequency of action potentials. Increased pressure in the blood vessels stretches the vessel walls, increasing the frequency of action potentials produced by the baroreceptors. Conversely, a decrease in blood pressure reduces the stretch of the arterial wall, causing a decrease in the frequency of action potentials produced by the baroreceptors.

A sudden increase in blood pressure causes the frequency of action potentials produced in the baroreceptors to also increase. In response to these changes, the vasomotor center decreases sympathetic stimulation of blood vessels, and the cardioregulatory center increases parasympathetic stimulation of the heart. As a result, peripheral blood vessels dilate, heart rate decreases, and blood pressure decreases (figure 21.40).

Similarly, a sudden decrease in blood pressure causes the frequency of action potentials produced by the baroreceptors to also decrease. In response, the vasomotor center increases sympathetic stimulation of the blood vessels, and the cardioregulatory center increases sympathetic stimulation and decreases parasympathetic stimulation of the heart. As a result, peripheral blood vessels constrict, heart rate and stroke volume increase, and blood pressure increases (see figures 21.39 and 21.40).

The carotid sinus and aortic arch baroreceptor reflexes are important in regulating blood pressure moment to moment. When a person rises rapidly from sitting or lying to a standing position, blood pressure in the neck and thoracic regions drops dramatically because of the pull of gravity on the blood. This reduction can cause blood flow to the brain to become so sluggish that dizziness or loss of consciousness results. The falling blood pressure activates the baroreceptor reflexes, which reestablish normal blood pressure within a few seconds. A healthy person may experience only a temporary sensation of dizziness.

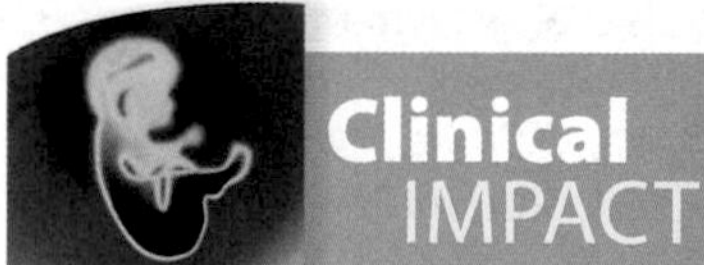
Clinical
IMPACT

Carotid Sinus Syndrome

Occasionally, applying pressure to the carotid arteries in the upper neck results in a dramatic decrease in blood pressure. This condition, called **carotid sinus syndrome,** is most common in patients with advanced arteriosclerosis of the carotid artery. In such patients, even a tight collar can apply enough pressure to the region of the carotid sinuses to stimulate the baroreceptors. The increased action potentials from the baroreceptors initiate reflexes that decrease vasomotor tone and increase parasympathetic action potentials to the heart. The decreased peripheral resistance and heart rate cause blood pressure to decline dramatically. As a consequence, blood flow to the brain decreases to such a low level that the person becomes dizzy or may even faint. People suffering from this condition must avoid applying external pressure to the neck region. If the carotid sinus becomes too sensitive, surgical destruction of the innervation to the carotid sinuses may be necessary.

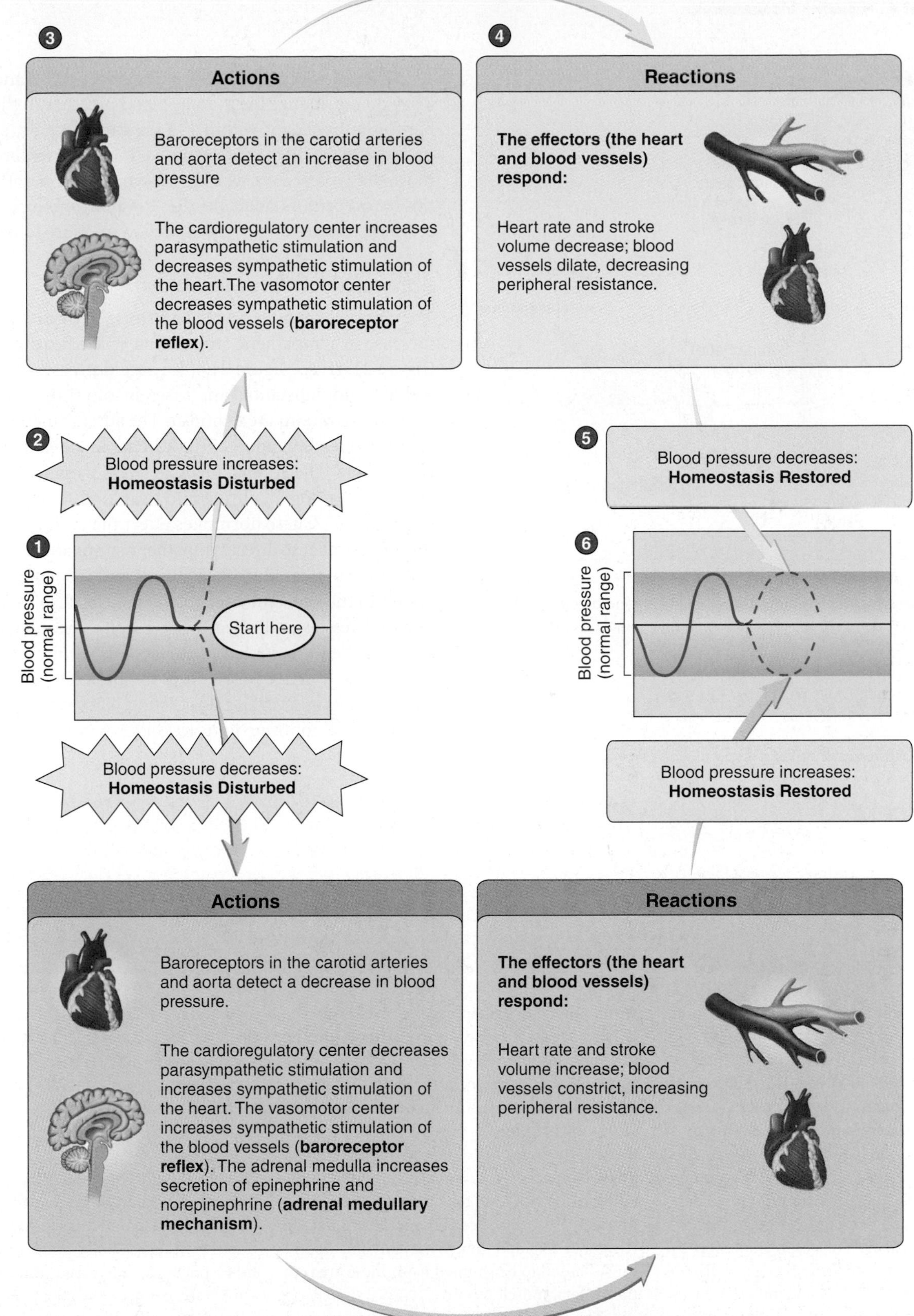
3
Actions
Baroreceptors in the carotid arteries and aorta detect an increase in blood pressure
The cardioregulatory center increases parasympathetic stimulation and decreases sympathetic stimulation of the heart.The vasomotor center decreases sympathetic stimulation of the blood vessels (baroreceptor reflex).
4
Reactions
The effectors (the heart and blood vessels) respond:
Heart rate and stroke volume decrease; blood vessels dilate, decreasing peripheral resistance.
2
Blood pressure increases: Homeostasis Disturbed
5
Blood pressure decreases: Homeostasis Restored
1
Blood pressure (normal range)
Start here
6
Blood pressure (normal range)
Blood pressure decreases: Homeostasis Disturbed
Blood pressure increases: Homeostasis Restored
Actions
Baroreceptors in the carotid arteries and aorta detect a decrease in blood pressure.
The cardioregulatory center decreases parasympathetic stimulation and increases sympathetic stimulation of the heart. The vasomotor center increases sympathetic stimulation of the blood vessels (baroreceptor reflex). The adrenal medulla increases secretion of epinephrine and norepinephrine (adrenal medullary mechanism).
Reactions
The effectors (the heart and blood vessels) respond:
Heart rate and stroke volume increase; blood vessels constrict, increasing peripheral resistance.

HOMEOSTASIS FIGURE 21.40 Summary of the Baroreceptor Effects on Blood Pressure

(1) Blood pressure is within its normal range. (2) Blood pressure increases outside the normal range, which causes homeostasis to be disturbed. (3) Baroreceptors detect the increase in blood pressure. The cardioregulatory and vasomotor centers in the brain respond to changes in blood pressure. (4) Nervous and hormonal changes alter the activity of cardiac muscle of the heart and smooth muscle of the blood vessels (effectors), causing heart rate and stroke volume to decrease and blood vessels to dilate. (5) These changes cause blood pressure to decrease. (6) Blood pressure returns to its normal range, and homeostasis is restored. Observe the responses to a decrease in blood pressure outside its normal range by following the *red arrows*. For more information on the baroreceptor reflex, see figure 21.39; for the adrenal medullary mechanism, see figure 21.41.

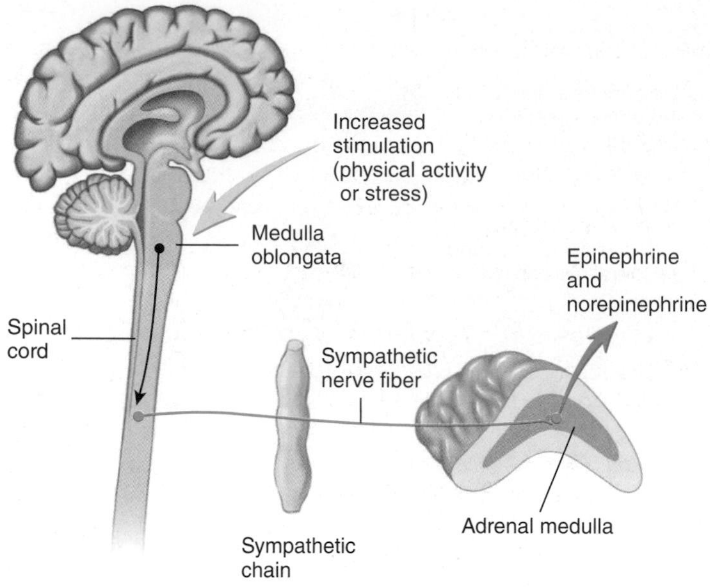
Increased
stimulation
(physical activity
or stress)
Medulla
oblongata
Epinephrine
and
norepinephrine
Spinal
cord
Sympathetic
nerve fiber
Adrenal medulla
Sympathetic
chain

FIGURE 21.41 Adrenal Medullary Mechanism
Stimuli that increase sympathetic stimulation of the heart and blood vessels also result in increased sympathetic stimulation of the adrenal medulla and result in epinephrine and some norepinephrine secretion.

Predict 8

Explain how the baroreceptor reflexes respond when a person does a headstand.

The baroreceptor reflexes are short-term and rapid-acting. They do not change the average blood pressure in the long run. The baroreceptors adapt within 1–3 days to any new, sustained blood pressure to which they are exposed. If blood pressure is elevated for more than a few days, as is the case in a person with hypertension, the baroreceptors adapt to the elevated pressure, and the baroreceptor reflexes do not reduce blood pressure to its original value.

Adrenal Medullary Mechanism

The adrenal medullary mechanism is activated by a substantial increase in sympathetic stimulation of the heart and blood vessels (figure 21.41; see figure 21.40). Large decreases in blood pressure, sudden and substantial increases in physical activity, and other stressful conditions are examples. The adrenal medullary mechanism results from stimulation of the adrenal medulla by the sympathetic nerve fibers. The adrenal medulla releases epinephrine and smaller amounts of norepinephrine into the bloodstream (figure 21.41; see figure 21.40). These hormones affect the cardiovascular system in a fashion similar to direct sympathetic stimulation, causing increased heart rate, increased stroke volume, and vasoconstriction in blood vessels to the skin and viscera. Epinephrine also causes vasodilation of blood vessels of the heart. The adrenal medullary mechanism is short-term and rapid-acting. It responds within seconds to minutes and is usually active for minutes to hours. Other hormonal mechanisms are long-term and slow-acting. They respond within minutes to hours and continue to function for many hours to days.

Clinical IMPACT — Blood Flow Through Tissues During Exercise

Exercise greatly increases blood flow through muscles and keeps blood flow through other organs at a value just adequate to supply their metabolic needs. During exercise, blood flow through skeletal muscles can be 15–20 times greater than through resting muscles. The increase in blood flow ensures that the skeletal muscles receive adequate oxygen and nutrients to sustain activity and remove metabolic waste. Local, nervous, and hormonal regulatory mechanisms are responsible for the increased blood flow. When skeletal muscle is resting, only 20–25% of the capillaries in the skeletal muscle are open, whereas during exercise 100% of the capillaries are open.

Low blood oxygen levels resulting from greatly increased muscular activity and the release of vasodilator substances, such as lactic acid, CO_2, and K^+, cause the dilation of precapillary sphincters. Increased sympathetic stimulation and epinephrine released from the adrenal medulla cause vasoconstriction in the blood vessels of the skin and viscera, but only some vasoconstriction in the blood vessels of skeletal muscles. Even though some vasoconstriction in skeletal muscle blood vessels occurs, the total resistance to blood flow in skeletal muscle decreases because all the capillaries are open. Blood flow through the skeletal muscles is also enhanced because the increased resistance to blood flow in the skin and viscera causes blood to be shunted from these areas to the skeletal muscles.

The movement of skeletal muscles compresses veins in a cyclic fashion and greatly increases venous return to the heart. In addition, veins undergo some constriction, which reduces the total volume of blood in the veins without dramatically increasing resistance to blood flow. The resulting increase in the preload and increased sympathetic stimulation of the heart lead to elevated heart rate and stroke volume, which increases cardiac output. As a consequence, blood pressure usually increases by 20–60 mm Hg, which also helps sustain the increased blood flow through skeletal muscle blood vessels.

As previously mentioned, blood flow through the skin decreases at the beginning of exercise in response to sympathetic stimulation. However, as body temperature increases in response to increased muscular activity, temperature receptors in the hypothalamus are stimulated. As a result, action potentials in sympathetic nerve fibers causing vasoconstriction decrease, allowing vasodilation of blood vessels in the skin. As a consequence, the skin turns a red or pinkish color, and a great deal of excess heat is lost as blood flows through the dilated blood vessels.

Chemoreceptor Reflexes

The **chemoreceptor** (kē′mō-rē-sep′tor) **reflexes** help maintain homeostasis when blood oxygen levels decrease or when blood carbon dioxide levels increase and pH decreases (figures 21.42 and 21.43). Chemoreceptors are located in **carotid bodies,** small organs approximately 1–2 mm in diameter, which lie near the carotid sinuses, and in several **aortic bodies** lying adjacent to the aorta. Afferent nerve fibers pass to the medulla oblongata through the glossopharyngeal nerve (IX) from the carotid bodies and through the vagus nerve (X) from the aortic bodies.

The chemoreceptors receive an abundant blood supply. However, when oxygen availability decreases in the chemoreceptor cells, the frequency of action potentials increases and stimulates the vasomotor center, resulting in increased vasomotor tone. The chemoreceptors act under emergency conditions and do not regulate the cardiovascular system under resting conditions. They normally do not respond strongly unless blood oxygen decreases markedly. The chemoreceptor cells are also stimulated by increased carbon dioxide and decreased blood pH to increase vasomotor tone, which causes the mean arterial pressure to rise. The elevated mean arterial pressure increases blood flow through tissues in which blood vessels do not constrict, such as the brain and cardiac muscle. Thus, the reflex helps provide adequate oxygen to the brain and the heart when blood oxygen levels in the blood decrease.

Central Nervous System's Ischemic Response

Elevated blood pressure in response to lack of blood flow to the medulla oblongata of the brain is called the **central nervous system (CNS) ischemic response.** The CNS ischemic response does not play an important role in regulating blood pressure under normal conditions. It functions primarily in response to emergency situations, such as when blood flow to the brain is severely restricted or when blood pressure falls below approximately 50 mm Hg.

Reduced blood flow results in decreased oxygen, increased carbon dioxide, and reduced pH within the medulla oblongata. Neurons of the vasomotor center are strongly stimulated. As a result, the vasomotor center stimulates vasoconstriction, and blood pressure rises dramatically.

The increase in blood pressure that occurs in response to CNS ischemia increases blood flow to the CNS, provided the blood vessels are intact. However, if severe ischemia lasts longer than a few minutes, metabolism in the brain fails because of the lack of oxygen. The vasomotor center becomes inactive, and extensive vasodilation occurs in the periphery as vasomotor tone decreases. Prolonged ischemia of the medulla oblongata leads to a massive decline in blood pressure and ultimately death.

Summary of Short-Term Regulation of Blood Pressure

In most circumstances throughout the day, the baroreceptor reflex is the most important short-term regulatory mechanism for maintaining blood pressure. The adrenal medullary mechanism plays a role during exercise and emergencies. The chemoreceptor mechanism is more important when blood oxygen levels are reduced, such as at high altitudes or when carbon dioxide is elevated or pH is reduced. Thus, it is more important in emergency situations. The CNS ischemic response is activated only in rare, emergency conditions when the brain receives too little oxygen.

1. Chemoreceptors in the carotid and aortic bodies monitor blood O_2, CO_2, and pH.
2. Chemoreceptors in the medulla oblongata monitor blood CO_2 and pH.
3. Decreased blood O_2, increased CO_2, and decreased pH decrease parasympathetic stimulation of the heart, which increases the heart rate.
4. Decreased blood O_2, increased CO_2, and decreased pH increase sympathetic stimulation of the heart, which increases the heart rate and stroke volume.
5. Decreased blood O_2, increased CO_2, and decreased pH increase sympathetic stimulation of blood vessels, which increases vasoconstriction.

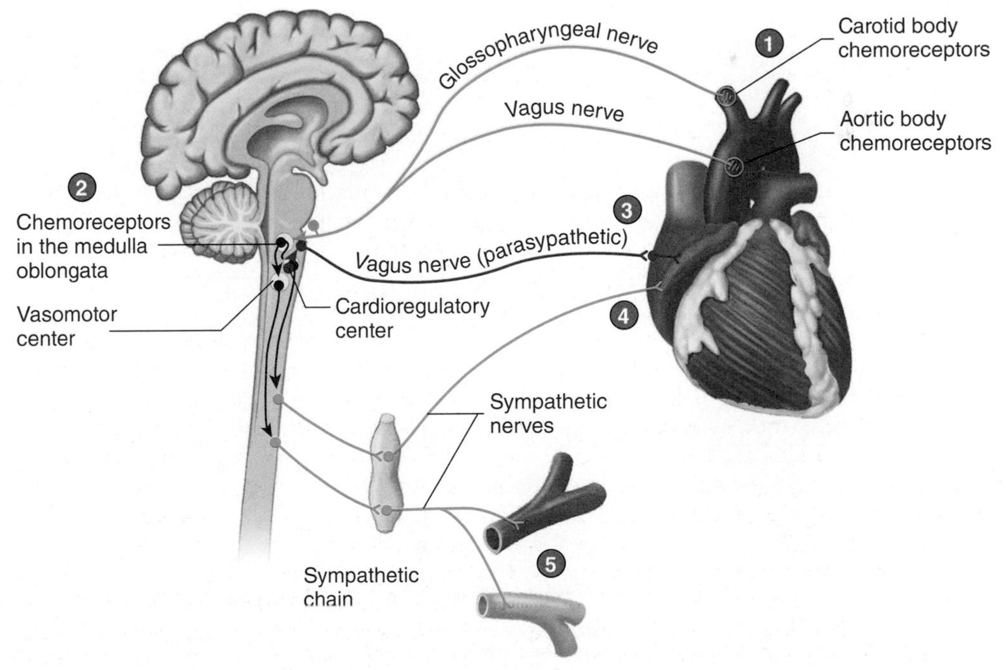

Glossopharyngeal nerve
1
Carotid body chemoreceptors
Vagus nerve
Aortic body chemoreceptors
2
Chemoreceptors in the medulla oblongata
3
Vagus nerve (parasypathetic)
Vasomotor center
Cardioregulatory center
4
Sympathetic nerves
Sympathetic chain
5

PROCESS FIGURE 21.42 Chemoreceptor Reflex

An increase in blood CO_2 and a decrease in pH and O_2 result in an increased heart rate and vasoconstriction. A decrease in blood CO_2 and an increase in blood pH result in a decreased heart rate and vasodilation.

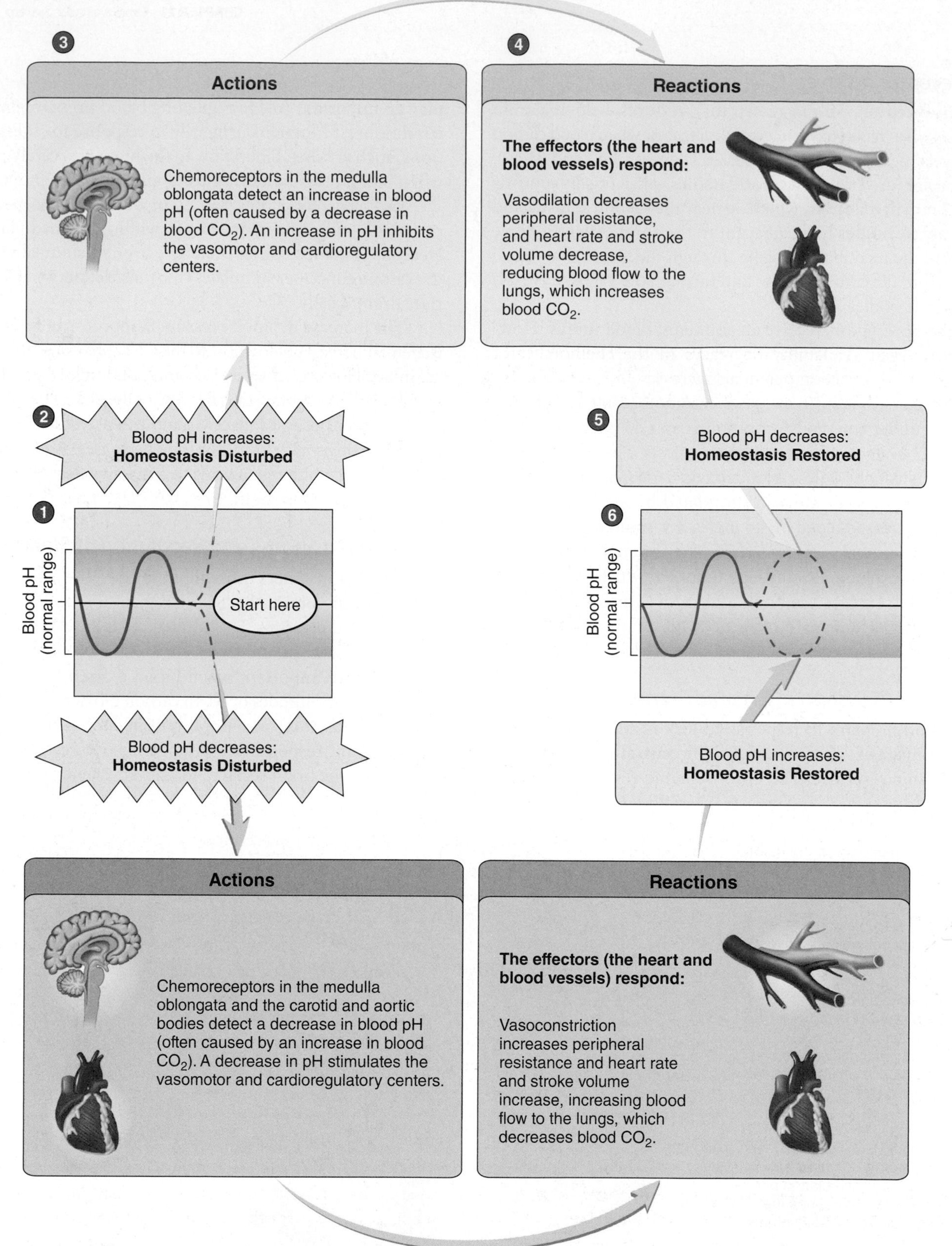

3
Actions
Chemoreceptors in the medulla oblongata detect an increase in blood pH (often caused by a decrease in blood CO2). An increase in pH inhibits the vasomotor and cardioregulatory centers.
4
Reactions
The effectors (the heart and blood vessels) respond:
Vasodilation decreases peripheral resistance, and heart rate and stroke volume decrease, reducing blood flow to the lungs, which increases blood CO2.
2
Blood pH increases:
Homeostasis Disturbed
5
Blood pH decreases:
Homeostasis Restored
1
Blood pH (normal range)
Start here
6
Blood pH (normal range)
Blood pH decreases:
Homeostasis Disturbed
Blood pH increases:
Homeostasis Restored
Actions
Chemoreceptors in the medulla oblongata and the carotid and aortic bodies detect a decrease in blood pH (often caused by an increase in blood CO2). A decrease in pH stimulates the vasomotor and cardioregulatory centers.
Reactions
The effectors (the heart and blood vessels) respond:
Vasoconstriction increases peripheral resistance and heart rate and stroke volume increase, increasing blood flow to the lungs, which decreases blood CO2.

HOMEOSTASIS FIGURE 21.43 Summary of the Effects of pH and Gases on Blood Pressure

(1) Blood pH is within its normal range. (2) Blood pH increases outside the normal range, which causes homeostasis to be disturbed. (3) Chemoreceptors detect the increase in blood pH. The cardioregulatory and vasomotor centers in the brain are inhibited. (4) Nervous and hormonal changes alter the activity of cardiac muscle of the heart and smooth muscle of the blood vessels (effectors), causing heart rate and stroke volume to decrease and blood vessels to dilate, reducing blood flow to the lungs, which increases blood CO_2. (5) These changes cause blood pH to decrease. (6) Blood pH returns to its normal range, and homeostasis is restored. Observe the responses to a decrease in blood pH outside its normal range by following the *red arrows*. For more information on the chemoreceptor reflex, see figure 21.42; for the central nervous system ischemic response, see the text.

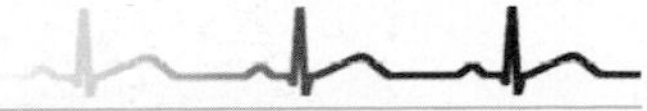

ASSESS **YOUR PROGRESS**

63. *Explain the relationship among mean arterial pressure, cardiac output, and peripheral resistance.*
64. *What are the two major control systems that provide homeostasis of blood pressure? Give a definition of each.*
65. *Where are baroreceptors located? Describe the response of the baroreceptor reflexes when blood pressure increases and decreases.*
66. *Elaborate on the adrenal medullary control mechanism.*
67. *Where are the chemoreceptors for oxygen, CO_2, and, pH levels located? Discuss what happens when oxygen levels in the blood decrease markedly.*
68. *Describe the CNS ischemic response. Under what conditions does this mechanism operate?*
69. *What mechanism is most important for short-term regulation of blood pressure under resting conditions?*

Long-Term Regulation of Blood Pressure

Long-term (slow-acting) regulation of blood pressure involves the regulation of blood concentration and volume by the kidneys, the movement of fluid across the wall of blood vessels, and alterations in the volume of the blood vessels. Some of the long-term regulatory mechanisms begin to respond in minutes, but they continue to function for hours, days, or longer. They adjust blood pressure precisely and keep it within a narrow range of values for years. Major regulatory mechanisms include the renin-angiotensin-aldosterone mechanism, the antidiuretic hormone (vasopressin) mechanism, the atrial natriuretic mechanism, the fluid shift mechanism, and the stress-relaxation response.

Renin-Angiotensin-Aldosterone Mechanism

The kidneys increase urine output as the blood volume and arterial pressure increase, and they decrease urine output as the blood volume and arterial pressure decrease. Increased urine output reduces blood volume and blood pressure, and decreased urine output resists a further decrease in blood volume and blood pressure. Controlling urine output is an important means by which blood pressure is regulated, and it continues to operate until blood pressure is precisely within its normal range of values. The **renin-angiotensin-aldosterone mechanism** helps regulate blood pressure by altering blood volume.

The kidneys release an enzyme, called **renin** (rē′nin), into the circulatory system (figure 21.44; see chapter 26) from specialized structures called the **juxtaglomerular** (jŭks′tă-glŏ-mer′ū-lăr) **apparatuses.** Renin acts on a plasma protein, synthesized by the liver, called **angiotensinogen** (an′jē-ō-ten-sin′ō-jen) to split a fragment off one end. The fragment, called **angiotensin** (an-jē-ō-ten′sin) **I,** contains 10 amino acids. Another enzyme, called **angiotensin-converting enzyme,** found primarily in small blood vessels of the lungs, cleaves 2 additional amino acids from angiotensin I to produce a fragment consisting of 8 amino acids, called **angiotensin II,** or *active angiotensin.*

Angiotensin II causes vasoconstriction in arterioles and, to some degree, in veins. As a result, it increases peripheral resistance and venous return to the heart, both of which raise blood pressure. Angiotensin II also stimulates aldosterone secretion from the adrenal cortex. **Aldosterone** (al-dos′ter-ōn) acts on the kidneys to increase the reabsorption of Na^+ and Cl^- from the filtrate into the extracellular fluid. If antidiuretic hormone (ADH; see chapter 18) is present, water moves by osmosis with the Na^+ and Cl^-. Consequently, aldosterone causes the kidneys to retain solutes, such as Na^+ and Cl^-, and water. The result is increased blood volume by decreasing the production of urine and conserving water (see chapter 26). Angiotensin II also increases salt appetite, thirst, and ADH secretion.

Decreased blood pressure stimulates renin secretion, and increased blood pressure decreases renin secretion. The renin-angiotensin-aldosterone mechanism is important in maintaining blood pressure on a daily basis. It also reacts strongly under conditions of circulatory shock, but it requires many hours to become maximally effective. Its onset is not as fast as that of nervous reflexes or the adrenal medullary response, but its duration is longer. Once renin is secreted, it remains active for approximately 1 hour, and the effect of aldosterone lasts much longer (many hours).

Angiotensin-converting enzyme (ACE) inhibitors are a class of drugs that inhibit angiotensin-converting enzyme, which converts angiotensin I to angiotensin II. These drugs were first identified as components of the venom of pit vipers. Subsequently, several ACE inhibitors were synthesized. ACE inhibitors are commonly administered to combat hypertension.

Some stimuli can directly stimulate aldosterone secretion. For example, an increased plasma ion concentration of K^+ and a reduced plasma concentration of Na^+ directly stimulate aldosterone secretion from the adrenal cortex (see chapters 18 and 27). Aldosterone regulates the concentration of these ions in the plasma. A decreased blood pressure and an elevated K^+ concentration occur during plasma loss, during dehydration, and in response to tissue damage, such as burns and crushing injuries.

Antidiuretic Hormone (Vasopressin) Mechanism

The **antidiuretic hormone (vasopressin) mechanism** works in harmony with the renin-angiotensin-aldosterone mechanism in response to changes in blood pressure (figure 21.45). Baroreceptors are sensitive to changes in blood pressure. Decreases in blood pressure detected by the baroreceptors result in the release of antidiuretic hormone (ADH) from the posterior pituitary, although the blood pressure must decrease substantially before the mechanism is activated.

ADH acts directly on blood vessels to cause vasoconstriction, although it is not as potent as other vasoconstrictors. Within minutes after a rapid and substantial decline in blood pressure, ADH is released in sufficient quantities to help reestablish normal blood pressure. ADH also decreases the rate of urine production by the kidneys, thereby helping maintain blood volume and blood pressure.

Neurons of the hypothalamus are sensitive to changes in the solute concentration of the plasma. Even small increases directly stimulate hypothalamic neurons that increase ADH secretion (figure 21.45; see chapter 26). Increases in the concentration of the plasma, as occur during dehydration, and decreases in blood pressure, as happens after plasma loss, such as in extensive burns or crushing injuries, stimulate ADH secretion.

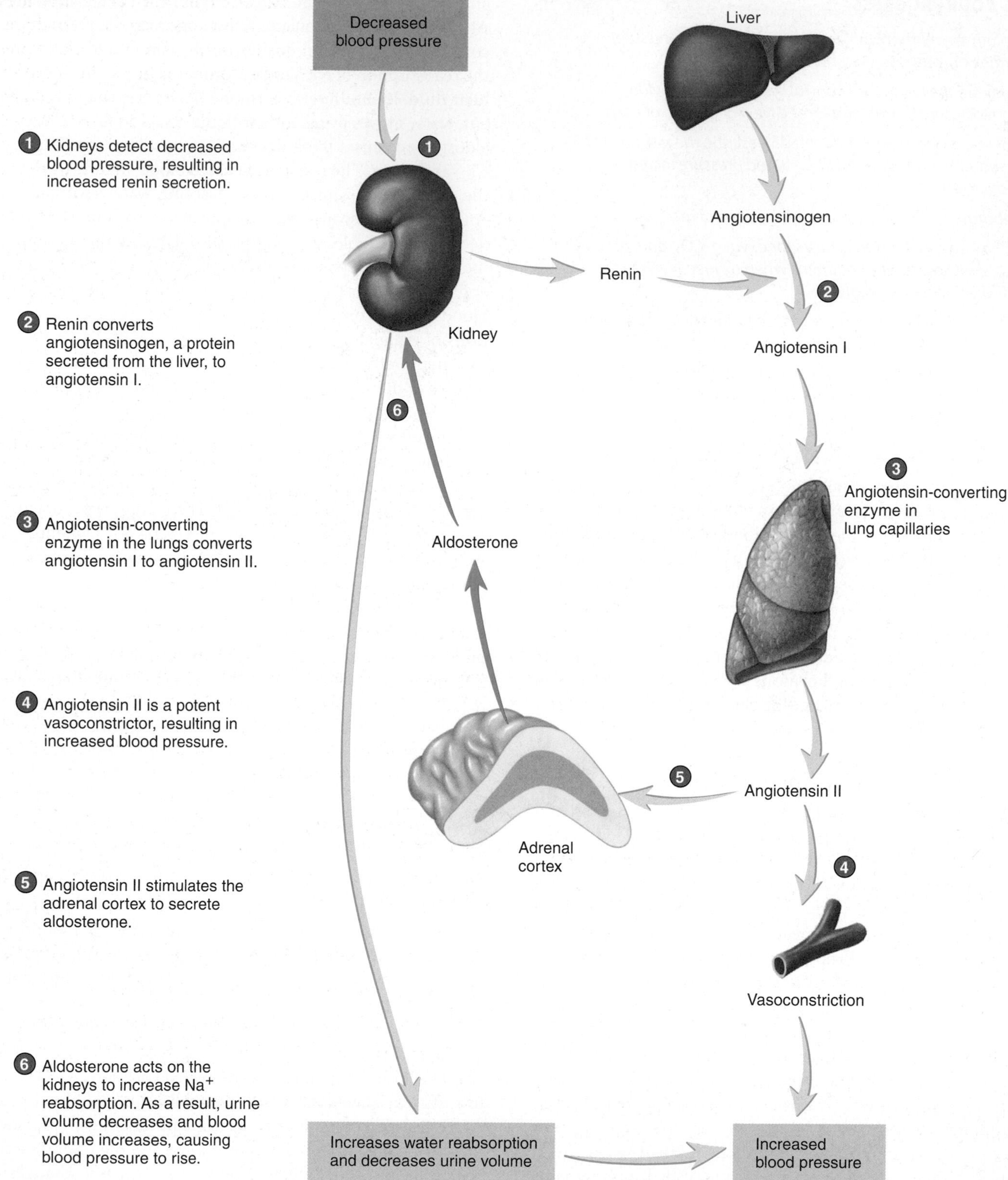
Decreased blood pressure
Liver
1 Kidneys detect decreased blood pressure, resulting in increased renin secretion.
1
Angiotensinogen
Renin
2
Kidney
2 Renin converts angiotensinogen, a protein secreted from the liver, to angiotensin I.
Angiotensin I
6
3
Angiotensin-converting enzyme in lung capillaries
3 Angiotensin-converting enzyme in the lungs converts angiotensin I to angiotensin II.
Aldosterone
4 Angiotensin II is a potent vasoconstrictor, resulting in increased blood pressure.
5
Angiotensin II
Adrenal cortex
4
5 Angiotensin II stimulates the adrenal cortex to secrete aldosterone.
Vasoconstriction
6 Aldosterone acts on the kidneys to increase Na+ reabsorption. As a result, urine volume decreases and blood volume increases, causing blood pressure to rise.
Increases water reabsorption and decreases urine volume
Increased blood pressure

PROCESS FIGURE 21.44 Renin-Angiotensin-Aldosterone Mechanism
The kidneys detect decreased blood pressure and increase renin secretion. The result is vasoconstriction, increased water reabsorption, and decreased urine volume, changes that maintain blood pressure.

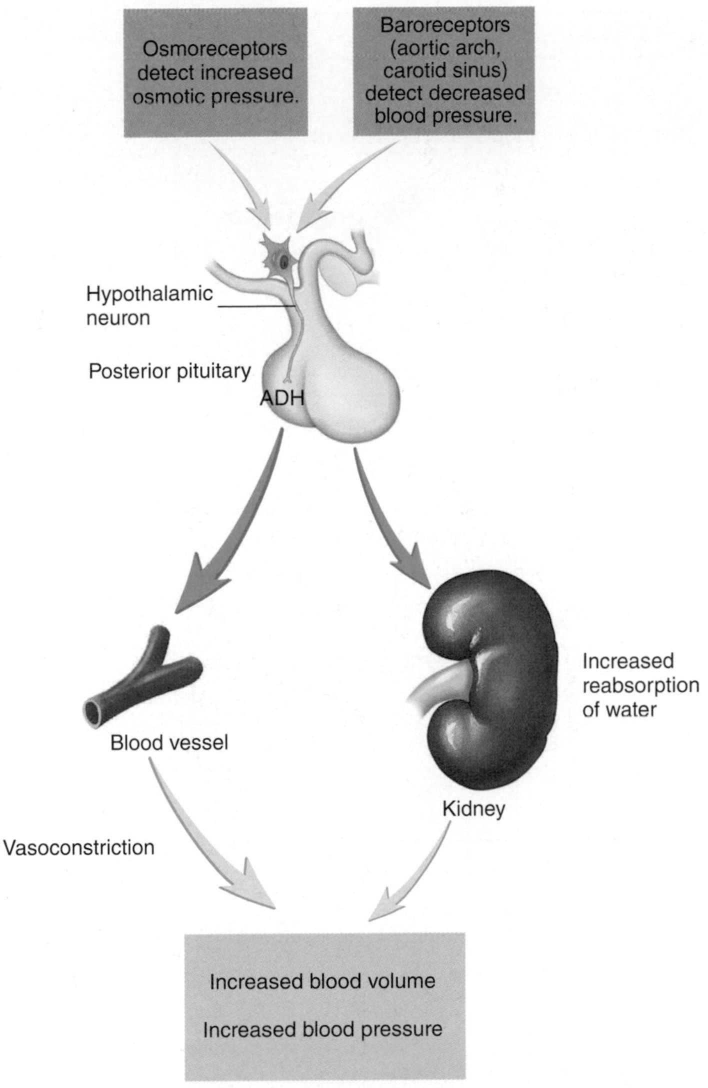
Osmoreceptors detect increased osmotic pressure.
Baroreceptors (aortic arch, carotid sinus) detect decreased blood pressure.
Hypothalamic neuron
Posterior pituitary
ADH
Increased reabsorption of water
Blood vessel
Kidney
Vasoconstriction
Increased blood volume
Increased blood pressure

FIGURE 21.45 Antidiuretic Hormone (Vasopressin) Mechanism
Increases in the osmolality of blood or decreases in blood pressure result in antidiuretic hormone (ADH) secretion. ADH increases water reabsorption by the kidneys, and large amounts of ADH result in vasoconstriction. These changes maintain blood pressure.

Atrial Natriuretic Mechanism

A polypeptide called **atrial natriuretic** (ā′trē-ăl nā′trē-ū-ret′ik) **hormone (ANH)** is released from cells in the atria of the heart. A major stimulus for its release is increased venous return, which stretches atrial cardiac muscle cells. Atrial natriuretic hormone acts on the kidneys to increase the rate of urine production and Na^+ loss in the urine. It also dilates arteries and veins. Loss of water and Na^+ in the urine causes the blood volume to decrease, which decreases venous return, and vasodilation results in a decrease in peripheral resistance. These effects cause a decrease in blood pressure.

The renin-angiotensin-aldosterone, ADH, and atrial natriuretic mechanisms work simultaneously to help regulate blood pressure by controlling urine production by the kidneys. If blood pressure drops below 50 mm Hg, the volume of urine produced by the kidneys is reduced to nearly zero. If blood pressure is increased to 200 mm Hg, the urine volume produced is approximately six to eight times greater than normal. The mechanisms that regulate blood pressure in the long term are summarized in figure 21.46.

Fluid Shift Mechanism

The **fluid shift mechanism** begins to act within a few minutes but requires hours to achieve its full functional capacity. It plays a very important role when dehydration develops over several hours, or when a large volume of saline is administered over several hours. The fluid shift mechanism occurs in response to small changes in pressures across capillary walls. As blood pressure increases, some fluid is forced from the capillaries into the interstitial spaces. This movement of fluid helps prevent the development of high blood pressure. As blood pressure falls, interstitial fluid moves into capillaries, and this fluid movement resists a further decline in blood pressure. Fluid shift is a powerful mechanism by which blood pressure is maintained, because the interstitial volume acts as a reservoir and is in equilibrium with the large volume of intercellular fluid.

Stress-Relaxation Response

A **stress-relaxation response** is characteristic of smooth muscle cells (see chapter 9). When blood volume suddenly declines, blood pressure also decreases, reducing the force applied to smooth muscle cells in blood vessel walls. As a result, during the next few minutes to an hour, the smooth muscle cells contract, reducing the volume of the blood vessels and thus resisting a further decline in blood pressure. Conversely, when blood volume increases rapidly, as occurs during a transfusion, blood pressure increases, and smooth muscle cells of the blood vessel walls relax, resulting in a more gradual increase in blood pressure. The stress-relaxation mechanism is most effective when changes in blood pressure occur over a period of many minutes.

ASSESS YOUR PROGRESS

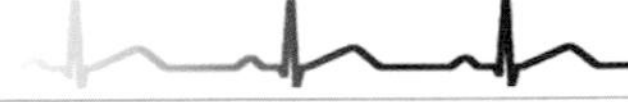

70. *What stimulates renin secretion in the kidneys?*

71. *For each of these chemicals—angiotensin II, aldosterone, antidiuretic hormone, and atrial natriuretic hormone—state where each is produced and how each affects the circulatory system.*

72. *What is fluid shift, and what does it accomplish?*

73. *Describe the stress-relaxation response of a blood vessel.*

Predict 9

Explain the various mechanisms that regulate blood pressure in response to the rapid loss of a large volume of blood, compared with the loss of the same volume of blood over a period of several hours.

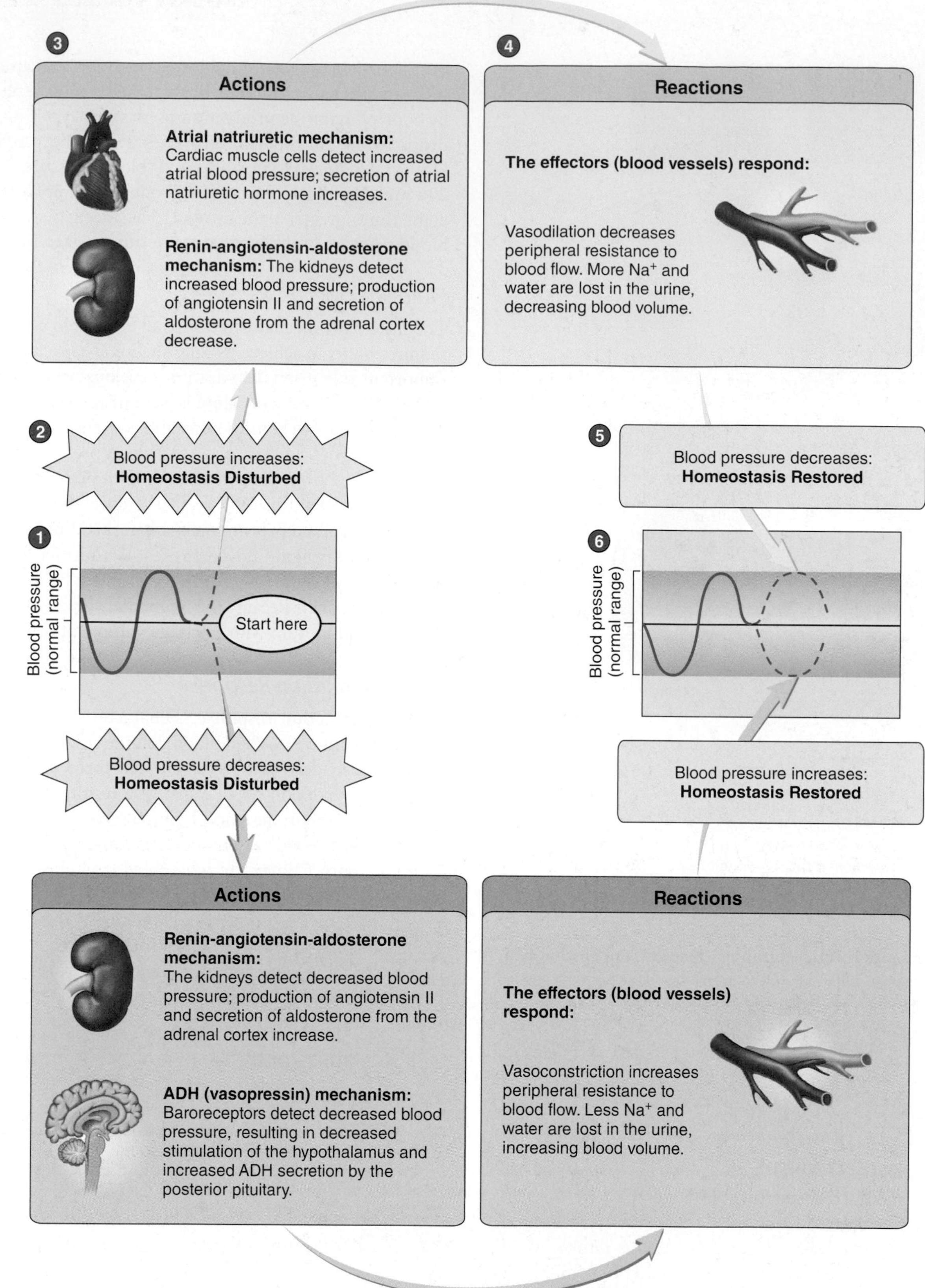

3
Actions
Atrial natriuretic mechanism:
Cardiac muscle cells detect increased atrial blood pressure; secretion of atrial natriuretic hormone increases.
Renin-angiotensin-aldosterone mechanism: The kidneys detect increased blood pressure; production of angiotensin II and secretion of aldosterone from the adrenal cortex decrease.
4
Reactions
The effectors (blood vessels) respond:
Vasodilation decreases peripheral resistance to blood flow. More Na+ and water are lost in the urine, decreasing blood volume.
2
Blood pressure increases:
Homeostasis Disturbed
5
Blood pressure decreases:
Homeostasis Restored
1
Blood pressure (normal range)
Start here
6
Blood pressure (normal range)
Blood pressure decreases:
Homeostasis Disturbed
Blood pressure increases:
Homeostasis Restored
Actions
Renin-angiotensin-aldosterone mechanism:
The kidneys detect decreased blood pressure; production of angiotensin II and secretion of aldosterone from the adrenal cortex increase.
ADH (vasopressin) mechanism:
Baroreceptors detect decreased blood pressure, resulting in decreased stimulation of the hypothalamus and increased ADH secretion by the posterior pituitary.
Reactions
The effectors (blood vessels) respond:
Vasoconstriction increases peripheral resistance to blood flow. Less Na+ and water are lost in the urine, increasing blood volume.

HOMEOSTASIS FIGURE 21.46 Summary of Long-Term (Slow-Acting) Blood Pressure Control Mechanisms

(1) Blood pressure is within its normal range. (2) Blood pressure increases outside the normal range, which causes homeostasis to be disturbed. (3) Increased blood pressure is detected by cardiac muscle cells and the kidneys (receptors). The heart and kidneys (control center) respond to increased blood pressure by secretion of hormones. (4) Blood vessels of the body and the kidneys (effectors) respond to the hormones by dilating or adjusting blood volume through urine formation. (5) These changes cause blood pressure to decrease. (6) Blood pressure returns to its normal range, and homeostasis is restored. Observe the responses to a decrease in blood pressure outside its normal range by following the *red arrows*. For more information on the renin-angiotensin-aldosterone mechanism, see figure 21.44; for the antidiuretic hormone mechanism, see figure 21.45; for the atrial natriuretic mechanism, see figure 27.6.

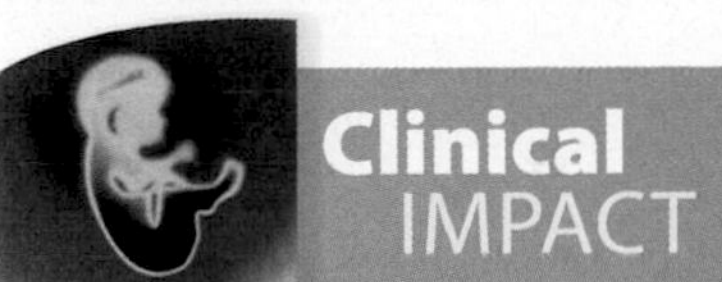
Clinical
IMPACT

Circulatory Shock

Circulatory shock is inadequate blood flow throughout the body due to the failure of the mechanisms that maintain normal blood pressure. As a consequence, tissues suffer damage due to lack of oxygen. Severe shock can damage vital body tissues to the extent that the patient dies.

Depending on its severity, circulatory shock can be divided into three stages: (1) compensated shock, (2) progressive shock, and (3) irreversible shock. All types of circulatory shock exhibit one or more of these stages, regardless of their cause. Several causes of circulatory shock exist. Hemorrhagic, or hypovolemic, shock, caused by excessive blood loss, is used here to illustrate the characteristics of each stage.

In **compensated shock,** blood pressure decreases only a moderate amount, and the mechanisms that regulate blood pressure successfully reestablish normal blood pressure and blood flow. The baroreceptor reflexes, chemoreceptor reflexes, and ischemia within the medulla oblongata initiate strong sympathetic responses that result in intense vasoconstriction and increased heart rate. As blood volume decreases, the stress-relaxation response of blood vessels causes them to contract and helps sustain blood pressure. In response to reduced blood flow through the kidneys, increased amounts of renin are released. The elevated renin release causes a greater rate of angiotensin II formation, resulting in vasoconstriction and increased aldosterone release from the adrenal cortex. Aldosterone, in turn, promotes water and salt retention by the kidneys, thereby conserving water. In addition, ADH is released from the posterior pituitary gland and enhances the kidneys' retention of water. Because of the fluid shift mechanism, water also moves from the interstitial spaces and the intestinal lumen to restore normal blood volume. An intense sensation of thirst increases water intake, also helping elevate normal blood volume. In mild cases of compensated shock, baroreceptor reflexes can compensate for blood loss until blood volume is restored but, in more severe cases, all the mechanisms described are required to compensate for the blood loss.

In **progressive shock,** the regulatory mechanisms are inadequate to compensate for the reduction in blood volume. As a consequence, a positive-feedback cycle develops. As circulatory shock worsens, regulatory mechanisms become even less able to compensate for the increasing severity. The cycle proceeds until the next stage of shock is reached or until medical treatment assists the regulatory mechanisms in reestablishing adequate blood flow to the tissues.

During progressive shock, blood pressure declines to a level that is inadequate for maintaining blood flow to cardiac muscle; thus, the heart begins to deteriorate. Tissues subject to severe ischemia release substances that are toxic to the heart. When blood pressure declines to a very low level, blood begins to clot in the small vessels. Eventually, blood vessels begin to dilate due to decreased sympathetic activity and a lack of oxygen in capillary beds. Capillary permeability increases under ischemic conditions, allowing fluid to leave the blood vessels and enter the interstitial spaces. Finally, intense tissue deterioration begins in response to inadequate blood flow.

Without medical intervention, progressive shock leads to **irreversible shock,** which results in death, regardless of the medical treatment applied. In this stage, the damage to tissues, including cardiac muscle, is so extensive that the patient is destined to die, even if adequate blood volume is reestablished and blood pressure is elevated to its normal value. Irreversible shock is characterized by decreasing heart function and progressive dilation of and increased permeability of peripheral blood vessels.

Patients suffering from circulatory shock are normally placed in a horizontal position, usually with the head slightly lower than the feet. Oxygen is often supplied. **Replacement therapy,** which includes transfusions of whole blood, plasma, artificial solutions called plasma substitutes, and physiological saline solutions, is administered to increase blood volume. In some circumstances, drugs that enhance vasoconstriction are also administered. Patients in anaphylactic (an′ă-fī-lak′tik; allergic) shock are given anti-inflammatory substances, such as glucocorticoids and antihistamines. The basic objective in treating shock is to reverse the condition and prevent it from progressing to the irreversible stage, as well as to reestablish normal blood flow through tissues.

Several types of shock can be classified by cause:

- **Hemorrhagic shock** is caused by external or internal bleeding that reduces blood volume.
- **Plasma loss shock** is reduced blood volume due to loss of plasma into the interstitial spaces and greatly increased blood viscosity. Causes of plasma loss shock include intestinal obstruction, in which a large amount of plasma moves from the blood into the intestines, and severe burns, which cause large amounts of plasma to be lost from the burned surface.
- **Dehydration** results from a severe and prolonged shortage of fluid intake.
- **Severe diarrhea** or **vomiting** causes a loss of plasma through the intestinal wall.
- **Neurogenic shock** is a rapid loss of vasomotor tone, leading to vasodilation so extensive that blood pressure declines severely.
- **Anesthesia** includes deep general anesthesia and spinal anesthesia that decrease the activity of the medullary vasomotor center or the sympathetic nerve fibers.
- **Brain damage** leads to an ineffective medullary vasomotor function.
- **Emotional shock** (*vasovagal syncope*) stems from emotions that cause strong parasympathetic stimulation of the heart and results in vasodilation in skeletal muscles and in the viscera.
- **Anaphylactic shock** results from an allergic response in which the release of inflammatory substances increases vasodilation and capillary permeability.
- **Septic shock,** or *blood poisoning*, results from peritoneal, systemic, and gangrenous infections that cause the release of toxic substances into the circulatory system, depressing the activity of the heart and leading to vasodilation and increased capillary permeability.
- **Cardiogenic shock** occurs when the heart stops pumping in response to various conditions, such as heart attack or electrocution.

Learn to Predict ❮ From page 709

After reading the Clinical Impact, "Circulatory Shock," in this chapter, we learned that circulatory shock is inadequate blood flow throughout the body. More specifically, septic shock results from infections that cause the release of toxic substances into the circulatory system that depress heart activity, cause vasodilation, and increase capillary permeability. After T. J. developed septic shock, we would expect his blood volume to decrease as fluid moved from the more permeable capillaries to the interstitial spaces. The reduction in blood volume would lead to a drop in his blood pressure, stimulating the baroreceptor reflex mechanism and subsequently an increase in heart rate. We would also expect T. J.'s stroke volume to decrease with a drop in blood volume. Increased sympathetic stimulation would cause vasoconstriction of blood vessels as T. J.'s body tried to maintain normal blood pressure. With the reduction in blood flow through the skin, T. J. would appear very pale. If T. J.'s blood pressure is not maintained, he could progress to irreversible shock, which is lethal.

Answers to the rest of this chapter's Predict questions are in Appendix G.

21.1 Functions of the Circulatory System (p. 710)

1. The circulatory system carries blood from the heart to the tissues of the body and returns the blood to the heart.
2. The circulatory system allows for nutrient, waste, and gas exchange with the tissues.
3. The circulatory system transports other substances (hormones, enzymes, etc.) through the body.
4. The circulatory system regulates blood pressure and blood flow to the tissues.

21.2 Structural Features of Blood Vessels (p. 710)

1. Blood flows from the heart through elastic arteries, muscular arteries, and arterioles to the capillaries.
2. Blood returns to the heart from the capillaries through venules, small veins, and large veins.

Capillaries

1. The entire circulatory system is lined with simple squamous epithelium called endothelium. Capillaries consist only of endothelium.
2. Capillaries are surrounded by loose connective tissue, the adventitia, that contains pericapillary cells.
3. Three types of capillaries exist.
   - The walls of continuous capillaries have no gaps between the endothelial cells.
   - Fenestrated capillaries have pores, called fenestrae, that extend completely through the cell.
   - Sinusoidal capillaries are large-diameter capillaries with large fenestrae.
4. Materials pass through the capillaries in several ways: between the endothelial cells, through the fenestrae, and through the plasma membrane.
5. Blood flows from arterioles through metarterioles and then through the capillary network. Venules drain the capillary network.
   - Smooth muscle in the arterioles, metarterioles, and precapillary sphincters regulates blood flow into the capillaries.
   - Blood can pass rapidly through the thoroughfare channel.
6. Arteriovenous anastomoses allow blood to flow from arteries to veins without passing through the capillaries. They function in temperature regulation.

Structure of Arteries and Veins

Except for capillaries and venules, blood vessels have three layers.

1. The inner tunica intima consists of endothelium, a basement membrane, and an internal elastic lamina.
2. The tunica media, the middle layer, contains circular smooth muscle and elastic fibers.
3. The outer tunica adventitia is connective tissue.

Types of Arteries

1. Large elastic arteries are thin-walled with large diameters. The tunica media has many elastic fibers and little smooth muscle.
2. Muscular arteries are thick-walled with small diameters. The tunica media has abundant smooth muscle and some elastic fibers.
3. Arterioles are the smallest arteries. The tunica media consists of smooth muscle cells and a few elastic fibers.

Types of Veins

1. Venules are composed of endothelium surrounded by a few smooth muscle cells.
2. Small veins are venules covered with a layer of smooth muscle.
3. Medium-sized veins and large veins contain less smooth muscle and fewer elastic fibers than arteries of the same size.
4. Valves prevent the backflow of blood in the veins.
5. Vasa vasorum are blood vessels that supply the tunica adventitia and tunica media.

Neural Innervation of Blood Vessels

Sympathetic nerve fibers supply the smooth muscle of the tunica media.

Aging of the Arteries

Arteriosclerosis results from a loss of elasticity in the aorta, large arteries, and coronary arteries.

21.3 Pulmonary Circulation (p. 716)

The pulmonary circulation moves blood to and from the lungs. The pulmonary trunk arises from the right ventricle and divides to form the pulmonary arteries, which project to the lungs. From the lungs, the pulmonary veins return to the left atrium.

21.4 Systemic Circulation: Arteries (p. 716)

Arteries carry blood from the left ventricle of the heart to all parts of the body.

Aorta

The aorta leaves the left ventricle to form the ascending aorta, aortic arch, and descending aorta (consisting of the thoracic and abdominal aortae).

Coronary Arteries

Coronary arteries supply the heart.

Arteries of the Head and Neck

1. The brachiocephalic, left common carotid, and left subclavian arteries branch from the aortic arch to supply the head and the upper limbs. The brachiocephalic artery divides to form the right common carotid and the right subclavian arteries. The vertebral arteries branch from the subclavian arteries.
2. The common carotid arteries and the vertebral arteries supply the head.
   - The common carotid arteries divide to form the external carotids, which supply the face and mouth, and the internal carotids, which supply the brain.
   - The vertebral arteries join within the cranial cavity to form the basilar artery, which supplies the brain.

Arteries of the Upper Limb

1. The subclavian artery continues (without branching) as the axillary artery and then as the brachial artery. The brachial artery divides into the radial and ulnar arteries.
2. The radial artery supplies the deep palmar arch, and the ulnar artery supplies the superficial palmar arch. Both arches give rise to the digital arteries.

Thoracic Aorta and Its Branches

The thoracic aorta has visceral branches that supply the thoracic organs and parietal branches that supply the thoracic wall.

Abdominal Aorta and Its Branches

1. The abdominal aorta has visceral branches that supply the abdominal organs and parietal branches that supply the abdominal wall.
2. The visceral branches are paired and unpaired. The paired arteries supply the kidneys, adrenal glands, and gonads. The unpaired arteries supply the stomach, spleen, and liver (celiac trunk); the small intestine and upper part of the large intestine (superior mesenteric); and the lower part of the large intestine (inferior mesenteric).

Arteries of the Pelvis

1. The common iliac arteries arise from the abdominal aorta, and the internal iliac arteries branch from the common iliac arteries.
2. The visceral branches of the internal iliac arteries supply the pelvic organs, and the parietal branches supply the pelvic wall and floor and the external genitalia.

Arteries of the Lower Limb

1. The external iliac arteries branch from the common iliac arteries.
2. The external iliac artery continues (without branching) as the femoral artery and then as the popliteal artery. The popliteal artery divides to form the anterior and posterior tibial arteries.
3. The posterior tibial artery gives rise to the fibular (peroneal) and plantar arteries. The plantar arteries form the plantar arch from which the digital arteries arise.

21.5 Systemic Circulation: Veins (p. 725)

1. The three major veins returning blood to the heart are the superior vena cava (head, neck, thorax, and upper limbs), the inferior vena cava (abdomen, pelvis, and lower limbs), and the coronary sinus (heart).
2. Veins are of three types: superficial, deep, and sinuses.

Veins Draining the Heart

Coronary veins enter the coronary sinus or the right atrium.

Veins of the Head and Neck

1. The internal jugular veins drain the venous sinuses of the anterior head and neck.
2. The external jugular veins and the vertebral veins drain the posterior head and neck.

Veins of the Upper Limb

1. The deep veins are the small ulnar and radial veins of the forearm, which join the brachial veins of the arm. The brachial veins drain into the axillary vein.
2. The superficial veins are the basilic, cephalic, and median cubital. The basilic vein becomes the axillary vein, which then becomes the subclavian vein. The cephalic vein drains into the axillary vein.

Veins of the Thorax

The left and right brachiocephalic veins and the azygos veins return blood to the superior vena cava.

Veins of the Abdomen and Pelvis

1. Ascending lumbar veins from the abdomen join the azygos and hemiazygos veins.
2. Vessels from the kidneys, adrenal gland, and gonads directly enter the inferior vena cava.
3. Vessels from the stomach, intestines, spleen, and pancreas connect with the hepatic portal vein. The hepatic portal vein transports blood to the liver for processing. Hepatic veins from the liver join the inferior vena cava.

Veins of the Lower Limb

1. The deep veins are the fibular (peroneal), anterior and posterior tibials, popliteal, femoral, and external iliac.
2. The superficial veins are the great and small saphenous veins.

21.6 Dynamics of Blood Circulation (p. 738)

The interrelationships among pressure, flow, resistance, and the control mechanisms that regulate blood pressure and blood flow play a critical role in the function of the circulatory system.

Laminar and Turbulent Flow in Vessels

Blood flow through vessels is normally streamlined, or laminar. Turbulent flow is disruption of laminar flow.

Blood Pressure

1. Blood pressure is a measure of the force exerted by blood against the blood vessel wall. Blood moves through vessels because of blood pressure.
2. Blood pressure can be measured by listening for Korotkoff sounds produced by turbulent flow in arteries as pressure is released from a blood pressure cuff.

Blood Flow and Poiseuille's Law

1. Blood flow is the amount of blood that moves through a vessel in a given period. Blood flow is directly proportional to pressure differences and inversely proportional to resistance.
2. Resistance is the sum of all the factors that inhibit blood flow. Resistance increases when viscosity increases and when blood vessels become smaller in diameter or increase in length.
3. Viscosity is the resistance of a liquid to flow. Most of the viscosity of blood results from red blood cells. The viscosity of blood increases when the hematocrit increases.

Critical Closing Pressure and Laplace's Law

1. As pressure in a vessel decreases, the force holding it open decreases, and the vessel tends to collapse. The critical closing pressure is the pressure at which a blood vessel closes.
2. Laplace's law states that the force acting on the wall of a blood vessel is proportional to the diameter of the vessel times blood pressure.

Vascular Compliance

Vascular compliance is a measure of the change in volume of blood vessels produced by a change in pressure. The venous system has a large compliance and acts as a blood reservoir.

21.7 Physiology of the Systemic Circulation (p. 743)

The greatest volume of blood is contained in the veins. The smallest volume is in the arterioles.

Cross-Sectional Area of Blood Vessels

As the diameter of vessels decreases, their total cross-sectional area increases, and the velocity of blood flow through them decreases.

Pressure and Resistance

Blood pressure averages 100 mm Hg in the aorta and drops to 0 mm Hg in the right atrium. The greatest drop occurs in the arterioles, which regulate blood flow through tissues.

Pulse Pressure

1. Pulse pressure is the difference between systolic and diastolic pressures. Pulse pressure increases when stroke volume increases or vascular compliance decreases.
2. Pulse pressure waves travel through the vascular system faster than the blood flows.
3. Pulse pressure can be used to take the pulse, which can serve as an indicator of heart rate and rhythm.

Capillary Exchange and Regulation of Interstitial Fluid Volume

1. Blood pressure, capillary permeability, and osmosis affect the movement of fluid from the capillaries.
2. A net movement of fluid occurs from the blood into the tissues. The fluid gained by the tissues is removed by the lymphatic system.

Functional Characteristics of Veins

Venous return to the heart increases because of an increase in blood volume, venous tone, and arteriole dilation.

Blood Pressure and the Effect of Gravity

In a standing person, hydrostatic pressure caused by gravity increases blood pressure below the heart and decreases pressure above the heart.

21.8 Control of Blood Flow in Tissues (p. 749)

Blood flow through tissues is highly controlled and matched closely to the metabolic needs of tissues.

Local Control of Blood Flow in Tissues

1. Blood flow through a tissue is usually proportional to the tissue's metabolic needs. Exceptions are tissues that perform functions that require additional blood.
2. Control of blood flow by the metarterioles and precapillary sphincters can be regulated by vasodilator substances or by lack of O_2 and nutrients.
3. Only large changes in blood pressure have an effect on blood flow through tissues.
4. If the metabolic activity of a tissue increases, the number and the diameter of capillaries in the tissue increase over time.

Nervous and Hormonal Control of Blood Flow in Tissues

1. The sympathetic nervous system (vasomotor center in the medulla) controls blood vessel diameter. Other brain areas can excite or inhibit the vasomotor center.
2. Vasomotor tone is a state of partial contraction of blood vessels.
3. The nervous system is responsible for routing the flow of blood and maintaining blood pressure.
4. Sympathetic action potentials stimulate epinephrine and norepinephrine release from the adrenal medulla, and these hormones cause vasoconstriction in most blood vessels.

21.9 Regulation of Mean Arterial Pressure (p. 753)

Mean arterial pressure (MAP) is proportional to cardiac output times peripheral resistance.

Short-Term Regulation of Blood Pressure

1. Baroreceptors are sensory receptors sensitive to stretch.
   - Baroreceptors are located in the carotid sinuses and the aortic arch.
   - The baroreceptor reflex changes peripheral resistance, heart rate, and stroke volume in response to changes in blood pressure.
2. Chemoreceptors are sensory receptors sensitive to O_2, CO_2, and pH levels in the blood.
3. Epinephrine and norepinephrine are released from the adrenal medulla as a result of sympathetic stimulation. They increase heart rate, stroke volume, and vasoconstriction.
4. The CNS ischemic response, which results from high CO_2 or low pH levels in the medulla, increases peripheral resistance.

Long-Term Regulation of Blood Pressure

1. In the renin-angiotensin-aldosterone mechanism, renin is released by the kidneys in response to low blood pressure. Renin promotes the production of angiotensin II, which causes vasoconstriction and an increase in aldosterone secretion.
2. The antidiuretic hormone (vasopressin) mechanism causes ADH release from the posterior pituitary in response to a substantial decrease in blood pressure. ADH acts directly on blood vessels to cause vasoconstriction.
3. The atrial natriuretic mechanism causes atrial natriuretic hormone release from the cardiac muscle cells when atrial blood pressure increases. It stimulates an increase in urine production, causing a decrease in blood volume and blood pressure.
4. The fluid shift mechanism causes fluid shift, which is the movement of fluid between the interstitial spaces and capillaries in response to changes in blood pressure to maintain blood volume.
5. The stress-relaxation response is an adjustment of the smooth muscles of blood vessels in response to a change in blood volume.

REVIEW AND COMPREHENSION

1. Given these blood vessels:
   (1) arteriole (3) elastic artery (5) vein
   (2) capillary (4) muscular artery (6) venule
   Choose the arrangement that lists the blood vessels in the order a red blood cell passes through them as it leaves the heart, travels to a tissue, and returns to the heart.
   a. 3,4,2,1,5,6 c. 4,3,1,2,5,6 e. 4,2,3,5,1,6
   b. 3,4,1,2,6,5 d. 4,3,2,1,6,5
2. Given these structures:
   (1) metarteriole (3) thoroughfare channel
   (2) precapillary sphincter
   Choose the arrangement that lists the structures in the order a red blood cell encounters them as it passes through a tissue.
   a. 1,3,2 b. 2,1,3 c. 2,3,1 d. 3,1,2 e. 3,2,1
3. In which of these blood vessels are elastic fibers present in the largest amounts?
   a. large arteries c. arterioles e. large veins
   b. medium arteries d. venules
4. Comparing and contrasting arteries and veins, veins have
   a. thicker walls.
   b. a greater amount of smooth muscle than arteries.
   c. a tunica media, but arteries do not.
   d. valves, but arteries do not.
   e. All of these are correct.
5. The structures that supply the walls of blood vessels with blood are
   a. venous shunts. d. vasa vasorum.
   b. tunic channels. e. coronary arteries.
   c. arteriovenous anastomoses.
6. Given these arteries:
   (1) basilar (3) internal carotid
   (2) common carotid (4) vertebral
   Which of these arteries have *direct* connections with the cerebral arterial circle (circle of Willis)?
   a. 1,2 b. 2,4 c. 1,3 d. 3,4 e. 2,3
7. Given these blood vessels:
   (1) axillary artery (4) radial artery
   (2) brachial artery (5) subclavian artery
   (3) brachiocephalic artery
   Choose the arrangement that lists the vessels in order, from the aorta to the right hand.
   a. 2,5,4,1 b. 5,2,1,4 c. 5,3,1,4,2 d. 3,5,1,2,4 e. 4,5,1,2,3
8. A branch of the aorta that supplies the liver, stomach, and spleen is the
   a. celiac trunk. d. superior mesenteric.
   b. common iliac. e. renal.
   c. inferior mesenteric.
9. Given these arteries:
   (1) common iliac (3) femoral
   (2) external iliac (4) popliteal
   Choose the arrangement that lists the arteries in order, from the aorta to the knee.
   a. 1,2,3,4 b. 1,2,4,3 c. 2,1,3,4 d. 2,1,4,3 e. 3,1,2,4
10. Given these veins:
    (1) brachiocephalic (3) superior vena cava
    (2) internal jugular (4) venous sinus
    Choose the arrangement that lists the veins in order, from the brain to the heart.
    a. 1,2,4,3 b. 2,4,1,3 c. 2,4,3,1 d. 4,2,1,3 e. 4,2,3,1
11. Blood returning from the arm to the subclavian vein passes through which of these veins?
    a. cephalic c. brachial e. All of these are correct.
    b. basilic d. Both a and b are correct.
12. Given these blood vessels:
    (1) inferior mesenteric vein (3) hepatic portal vein
    (2) superior mesenteric vein (4) hepatic vein
    Choose the arrangement that lists the vessels in order, from the small intestine to the inferior vena cava.
    a. 1,3,4 b. 1,4,3 c. 2,3,4 d. 2,4,3 e. 3,1,4
13. Given these veins:
    (1) small saphenous (3) fibular (peroneal)
    (2) great saphenous (4) posterior tibial
    Which are superficial veins?
    a. 1,2 b. 1,3 c. 2,3 d. 2,4 e. 3,4
14. Vascular compliance is
    a. greater in arteries than in veins.
    b. the increase in vessel volume divided by the increase in vessel pressure.
    c. the pressure at which blood vessels collapse.
    d. proportional to the diameter of the blood vessel times pressure.
    e. All of these are correct.
15. The resistance to blood flow is greatest in the
    a. aorta. c. capillaries. e. veins.
    b. arterioles. d. venules.
16. Veins
    a. increase their volume because of their large compliance.
    b. increase venous return to the heart when they vasodilate.
    c. vasodilate because of increased sympathetic stimulation.
    d. All of these are correct.
17. Local direct control of blood flow through a tissue
    a. maintains an adequate rate of flow despite large changes in arterial blood pressure.
    b. results from relaxation and contraction of precapillary sphincters.
    c. occurs in response to a buildup in CO_2 in the tissues.
    d. occurs in response to a decrease in oxygen in the tissues.
    e. All of these are correct.
18. An increase in mean arterial pressure can result from
    a. an increase in peripheral resistance.
    b. an increase in heart rate.
    c. an increase in stroke volume.
    d. All of these are correct.
19. When blood O_2 levels markedly decrease, the chemoreceptor reflex causes
    a. peripheral resistance to decrease. d. vasodilation.
    b. mean arterial blood pressure to increase. e. All of these are correct.
    c. vasomotor tone to decrease.
20. When blood pressure is suddenly decreased a small amount (10 mm Hg), which of these mechanisms are activated to restore blood pressure to normal levels?
    a. chemoreceptor reflexes c. CNS ischemic response
    b. baroreceptor reflexes d. All of these are correct.
21. A sudden release of epinephrine from the adrenal medulla
    a. increases heart rate.
    b. increases stroke volume.
    c. causes vasoconstriction in visceral blood vessels.
    d. All of these are correct.
22. In response to a decrease in blood pressure,
    a. ADH secretion increases.
    b. the kidneys decrease urine production.
    c. blood volume increases.
    d. All of these are correct.

23. In response to a decrease in blood pressure,
    a. more fluid than normal enters the tissues (fluid shift mechanism).
    b. smooth muscles in blood vessels relax (stress-relaxation response).
    c. the kidneys retain more salts and water than normal.
    d. All of these are correct.
24. A patient is found to have severe arteriosclerosis of the renal arteries, which has reduced renal blood pressure. Which of these is consistent with that condition?
    a. hypotension
    b. hypertension
    c. decreased vasomotor tone
    d. exaggerated sympathetic stimulation of the heart
    e. Both a and c are correct.
25. During exercise, the blood flow through skeletal muscle may increase up to 20-fold. However, the cardiac output does not increase that much. This occurs because of
    a. vasoconstriction in the viscera.
    b. vasoconstriction in the skin (at least temporarily).
    c. vasodilation of skeletal muscle blood vessels.
    d. Both a and b are correct.
    e. All of these are correct.

Answers in Appendix E

CRITICAL THINKING

1. For each of the following destinations, name all the arteries that a red blood cell would encounter if it started its journey in the left ventricle.
    a. posterior interventricular groove of the heart
    b. anterior neck to the brain (give two ways)
    c. posterior neck to the brain (give two ways)
    d. external skull
    e. tip of the fingers of the left hand (what other blood vessel would be encountered if the trip were through the right upper limb?)
    f. anterior compartment of the leg
    g. liver
    h. small intestine
    i. urinary bladder
2. For each of the following starting places, name all the veins that a red blood cell would encounter on its way back to the right atrium.
    a. anterior interventricular groove of the heart (give two ways)
    b. venous sinus near the brain
    c. external posterior of skull
    d. hand (return deep and superficial)
    e. foot (return deep and superficial)
    f. stomach
    g. kidney
    h. left inferior wall of the thorax
3. In a study of heart valve functions, it is necessary to inject a dye into the right atrium of the heart by inserting a catheter into a blood vessel and moving the catheter into the right atrium. What route would you suggest? If you wanted to do this procedure into the left atrium, what would you do differently?
4. All the blood that passes through the aorta, except the blood that flows into the coronary vessels, returns to the heart through the venae cavae. (*Hint:* The diameter of the aorta is 26 mm, and the diameter of a vena cava is 32 mm.) Explain why the resistance to blood flow in the aorta is greater than the resistance to blood flow in the venae cavae. Because the resistances are different, explain why blood flow can be the same.
5. As blood vessels increase in diameter, the amount of smooth muscle decreases and the amount of connective tissue increases. Explain why. (*Hint:* Remember Laplace's law.)
6. A very short nursing student is asked to measure the blood pressure of a very tall person. She decides to measure the blood pressure at the level of the tall person's foot while he is standing. What artery does she use? After taking the blood pressure, she decides that the tall person is suffering from hypertension because the systolic pressure is 200 mm Hg. Is her diagnosis correct? Why or why not?
7. David was suffering from severe cirrhosis of the liver and hepatitis. Over a period of time, he developed severe edema. Explain how decreased liver function can result in edema.
8. During hyperventilation, CO_2 is "blown off," and CO_2 levels in the blood decrease. What effect does this decrease have on blood pressure? Explain. What symptoms do you expect to see as a result?
9. Epinephrine causes vasodilation of blood vessels in cardiac muscle but vasoconstriction of blood vessels in the skin. Explain why this is a beneficial arrangement.
10. While Dick and Jane were backpacking on a trail in Yellowstone Park, they encountered a grizzly bear cub that seemed amazingly tame. However, while Dick tried to feed the cub, its mother appeared and attacked him. Jane escaped by climbing a tree, but Dick received several deep lacerations (cuts) and lost a lot of blood over the next several hours. Jane helped him reach medical aid, and he survived. Which of the following mechanisms was (were) activated to help keep Dick alive? Explain your choice.
    (1) baroreceptor mechanism
    (2) CNS ischemic response
    (3) renin-angiotensin-aldosterone mechanism
    (4) fluid shift mechanism
    (5) antidiuretic hormone mechanism
    (6) adrenal medullary response
    a. 1,2,3,4,5,6 b. 1,3,4,5,6 c. 1,6 d. 1,4,6 e. 1
11. Mr. Wilson, age 85, lives in a care facility, where he is not very mobile and is often lethargic. One day an aide noticed that he was sitting in his usual chair but appeared to be unconscious. She took his pulse, which was 140 beats/minute and then took his blood pressure, which was 190/130. Which of the following conditions is (are) consistent with these observations?
    a. stroke
    b. activation of the CNS ischemic response
    c. heart attack
    d. shock
    e. a and b

Answers in Appendix F

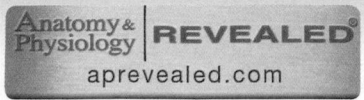

22

Lymphatic System and Immunity

One of the basic themes of life is that many organisms consume other organisms—or use them to survive. For example, a parasite lives on or in another organism called the host. The host provides the parasite with the conditions and food necessary for survival. Humans are host to many kinds of organisms, including microorganisms (bacteria, viruses, fungi, and protozoans), insects, and worms. Hookworms, for example, can live in the sheltered environment of the human intestines, where they feed on blood. Often, parasites harm humans, causing disease and sometimes death. However, our bodies have ways to resist or destroy harmful microorganisms. The lymphatic system and immunity are the body's defense systems against threats arising both inside and outside the body.

Learn to Predict

The next time autumn leaves drift by Maddie's window, she may just let them lie. During the last week of September, city workers sprayed chemicals on the trees along the street in front of her house. In October, when the leaves from those trees began to fall, Maddie raked them up and stuffed them into several large garbage bags. The next day, her hands were red and itchy, and by the day after that, the itching had become intense. In addition, Maddie's hands were covered with blisterlike lesions and so swollen that she could hardly move her fingers. Maddie suspected she was having an allergic reaction. Was she right? After reading about the lymphatic system and immunity in this chapter and connecting that information with what you learned about the peripheral circulation in chapter 21, explain the probable cause of Maddie's symptoms.

Photo: Photograph of a young woman raking leaves.

Module 10
Lymphatic System

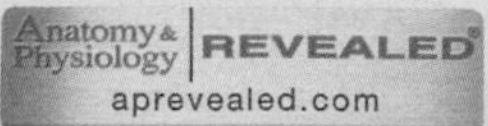
Anatomy & Physiology
REVEALED
aprevealed.com

22.1 Functions of the Lymphatic System

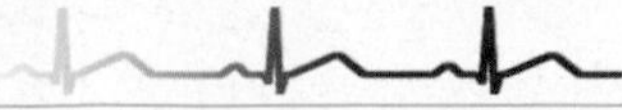

LEARNING OUTCOME

After reading this section, you should be able to

A. Describe the functions of the lymphatic system.

The lymphatic system has three main functions:

1. *Fluid balance.* Approximately 30 L of fluid pass from the blood capillaries into the interstitial fluid each day, whereas only 27 L pass from the interstitial fluid back into the capillaries. If the extra 3 L of fluid were to remain in the interstitial fluid, edema would result, causing tissue damage and eventual death. Instead, the 3 L of fluid enter the lymphatic capillaries, where it is called **lymph** (limf), and passes through the lymphatic vessels back to the blood. In addition to water, lymph contains solutes derived from two sources: (a) substances in plasma, such as ions, nutrients, gases, and some proteins, which pass from blood capillaries into the interstitial fluid, and (b) substances derived from cells, such as hormones, enzymes, and waste products.
2. *Lipid absorption.* The lymphatic system absorbs lipids and other substances from the digestive tract (see chapter 24) through lymphatic vessels called **lacteals** (lak′tē-ălz) located in the lining of the small intestine. Lipids enter the lacteals and pass through the lymphatic vessels to the venous circulation. The lymph passing through these lymphatic vessels, called **chyle** (kīl), appears white because of its lipid content.
3. *Defense.* Microorganisms and other foreign substances are filtered from lymph by lymph nodes and from blood by the spleen. In addition, lymphocytes and other cells are capable of destroying microorganisms and other foreign substances. Because the lymphatic system fights infections, and filters blood and lymph to remove microorganisms, many infectious diseases produce symptoms associated with the lymphatic system (see the Diseases and Disorders table, later in this chapter).

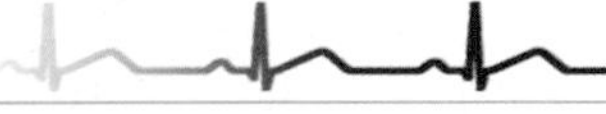

ASSESS YOUR PROGRESS

1. *Describe the three functions of the lymphatic system.*

22.2 Anatomy of the Lymphatic System

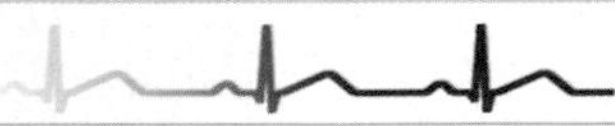

LEARNING OUTCOMES

After reading this section, you should be able to

A. List the parts of the lymphatic system.

B. Describe the structure of lymphatic vessels.

C. Explain how lymph is formed and transported through lymphatic vessels.

D. Distinguish between lymphatic tissue and a lymphatic organ.

E. Describe the structure and function of tonsils, lymph nodes, the spleen, and the thymus.

The **lymphatic** (lim-fat′ik) **system** includes lymph, lymphatic vessels, lymphatic tissue, lymphatic nodules, lymph nodes, the tonsils, the spleen, and the thymus (figure 22.1).

Lymphatic Vessels

The lymphatic vessels are essential for the maintenance of fluid balance. They begin as small, dead-end tubes called **lymphatic capillaries** (figure 22.2*a*). Fluids tend to move out of blood capillaries into tissue spaces (see "Capillary Exchange and Regulation of Interstitial Fluid Volume," section 21.7). Excess fluid passes through the tissue spaces and enters lymphatic capillaries to become lymph. Lymphatic capillaries are in most tissues of the body. Exceptions are the central nervous system, the bone marrow, and tissues without blood vessels, such as cartilage, epidermis, and the cornea. A superficial group of lymphatic capillaries is in the dermis of the skin and the subcutaneous tissue. A deep group of lymphatic capillaries drains the muscles, joints, viscera, and other deep structures.

Lymphatic capillaries differ from blood capillaries in that they lack a basement membrane, and the cells of the simple squamous epithelium slightly overlap and are loosely attached to one another (figure 22.2*b*). This structure facilitates their function in two ways: First, the lymphatic capillaries are far more permeable than blood capillaries, so nothing in the interstitial fluid is excluded from entering the lymphatic capillaries. Second, the lymphatic capillary epithelium functions as a series of one-way valves that allow fluid to enter the capillary but prevent it from passing back into the interstitial spaces.

The lymphatic capillaries join to form larger **lymphatic vessels,** which resemble small veins. The inner layer of the lymphatic vessel consists of endothelium surrounded by an elastic membrane; the middle layer consists of smooth muscle cells and elastic fibers; and the outer layer is a thin layer of fibrous connective tissue.

Small lymphatic vessels have a beaded appearance because they have one-way valves along their lengths that are similar to the valves of veins (see figure 22.2*b*). When a lymphatic vessel is compressed, the valves close and prevent backward movement of lymph; as a consequence, the lymph moves forward through the lymphatic vessel.

Three major mechanisms are responsible for moving lymph through lymphatic vessels:

1. *Contraction of lymphatic vessels.* In many parts of the body, lymphatic vessels pump lymph. The unidirectional valves divide lymphatic vessels into a series of chambers, which function as "primitive hearts." Lymph moves into a chamber, smooth muscle in the chamber wall contracts, and lymph moves into the next chamber. Some of the smooth muscle cells in the walls of lymphatic vessels are pacemaker cells (see "Electrical Properties of Smooth Muscle," chapter 9). The pacemaker cells spontaneously depolarize, resulting in periodic contraction of the lymphatic vessels.
2. *Contraction of skeletal muscles.* When surrounding muscle cells contract, lymphatic vessels are compressed, causing lymph to move.
3. *Thoracic pressure changes.* During inspiration, pressure in the thoracic cavity decreases, lymphatic vessels expand, and lymph flows into them. During expiration, pressure in the thoracic cavity increases, and lymphatic vessels are compressed, causing lymph to move.

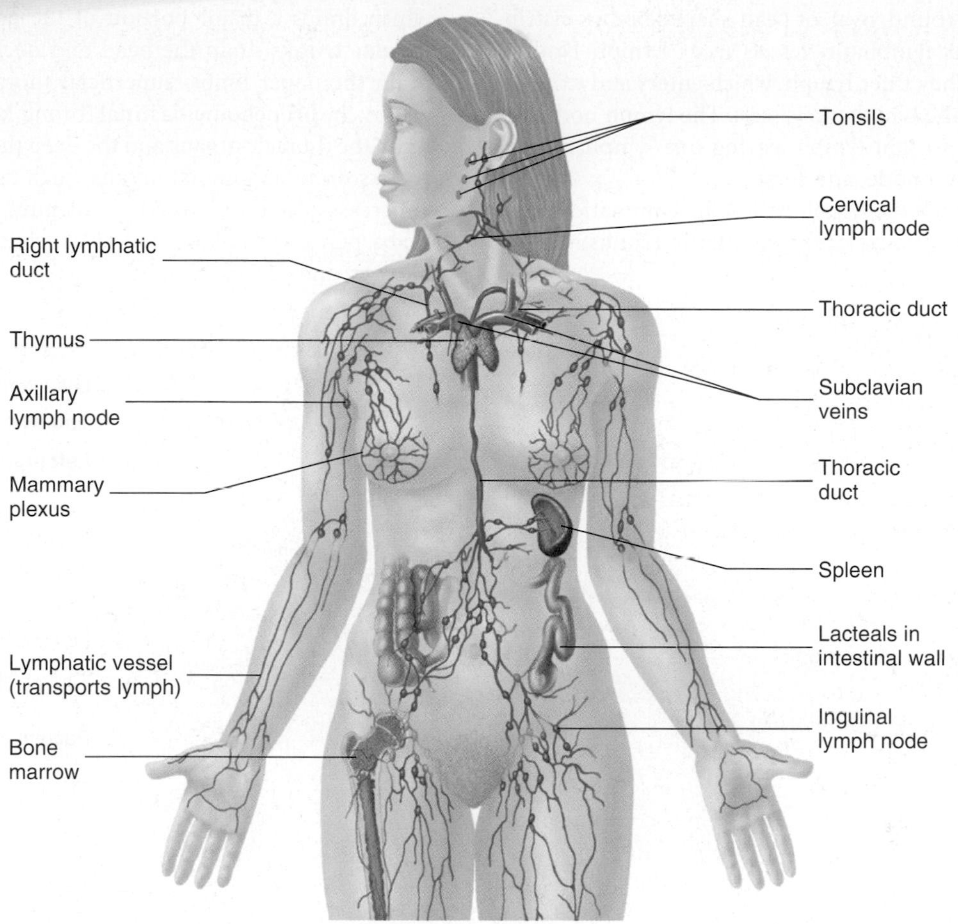
Tonsils
Cervical lymph node
Right lymphatic duct
Thoracic duct
Thymus
Axillary lymph node
Subclavian veins
Thoracic duct
Mammary plexus
Spleen
Lacteals in intestinal wall
Lymphatic vessel (transports lymph)
Inguinal lymph node
Bone marrow

FIGURE 22.1 AP|R **Lymphatic System**

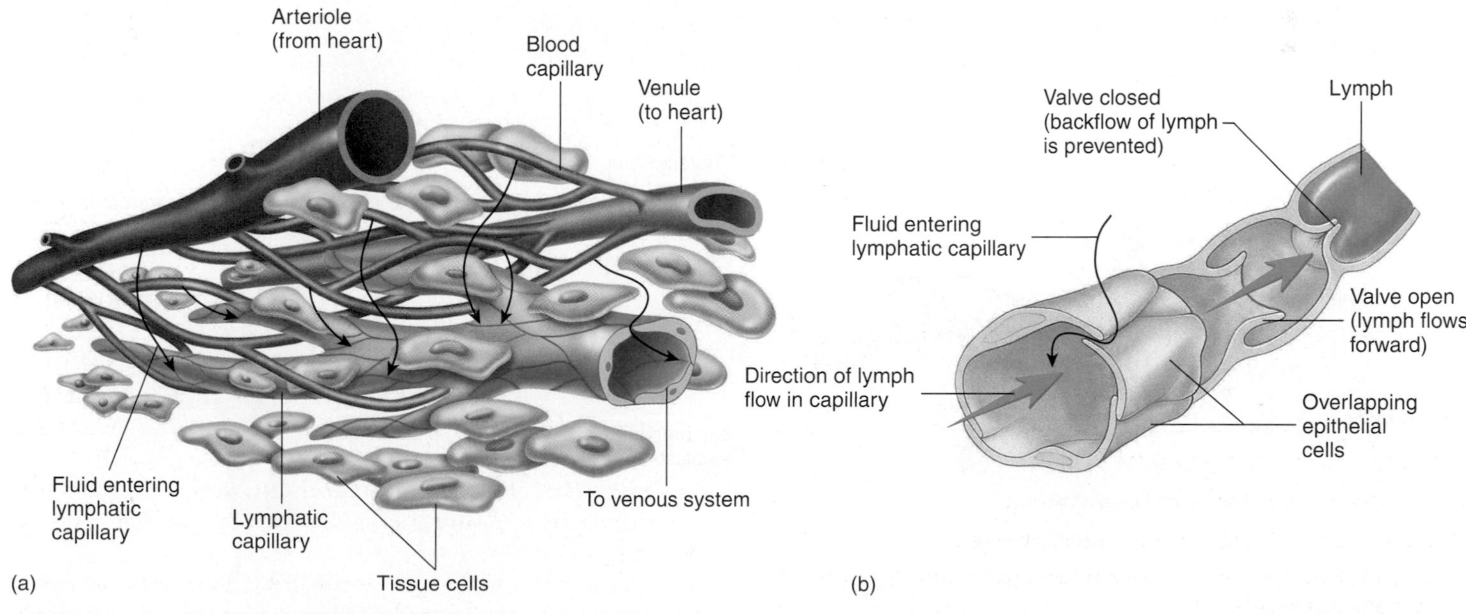
Arteriole (from heart)
Blood capillary
Venule (to heart)
Fluid entering lymphatic capillary
Lymphatic capillary
Tissue cells
To venous system
(a)
Valve closed (backflow of lymph is prevented)
Lymph
Fluid entering lymphatic capillary
Valve open (lymph flows forward)
Direction of lymph flow in capillary
Overlapping epithelial cells
(b)

FIGURE 22.2 Lymph Formation and Movement

(*a*) Fluid from blood capillaries moves into tissues and from tissues into lymphatic capillaries to form lymph. (*b*) The overlap of epithelial cells of the lymphatic capillary allows fluid to enter but prevents it from moving back into the tissue. Valves, located farther along in lymphatic vessels, also ensure one-way flow of lymph.

Lymph nodes are round, oval, or bean-shaped bodies distributed along the various lymphatic vessels (see "Lymph Nodes," later in this section). They filter lymph, which enters and exits the lymph nodes through the lymphatic vessels. The lymph nodes are connected in a series, so that lymph leaving one lymph node is carried to another lymph node, and so on.

After passing through the lymph nodes, the lymphatic vessels converge to form larger vessels called **lymphatic trunks,** each of which drains a major portion of the body (figure 22.3*a,b*). The **jugular trunks** drain the head and neck; the **subclavian trunks** drain the upper limbs, superficial thoracic wall, and mammary glands; the **bronchomediastinal** (brong′kō-mē′dē-as-tī′năl) **trunks** drain the thoracic organs and the deep thoracic wall; the **intestinal trunks** drain abdominal organs, such as the intestines, stomach, pancreas, spleen, and liver; and the **lumbar trunks** drain the lower limbs, pelvic and abdominal walls, pelvic organs, ovaries or testes, kidneys, and adrenal glands.

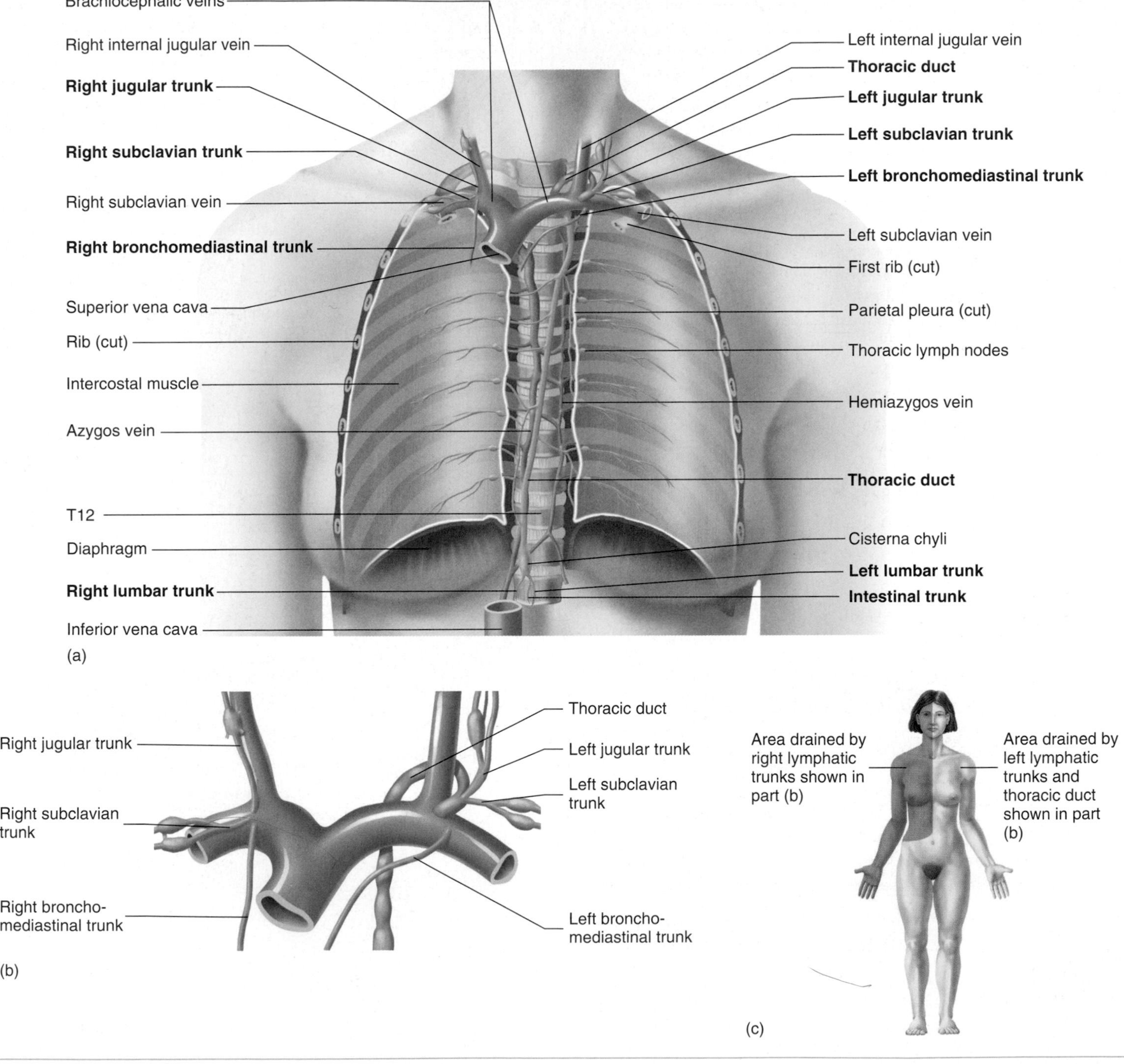
Brachiocephalic veins
Right internal jugular vein
Right jugular trunk
Right subclavian trunk
Right subclavian vein
Right bronchomediastinal trunk
Superior vena cava
Rib (cut)
Intercostal muscle
Azygos vein
T12
Diaphragm
Right lumbar trunk
Inferior vena cava
(a)
Left internal jugular vein
Thoracic duct
Left jugular trunk
Left subclavian trunk
Left bronchomediastinal trunk
Left subclavian vein
First rib (cut)
Parietal pleura (cut)
Thoracic lymph nodes
Hemiazygos vein
Thoracic duct
Cisterna chyli
Left lumbar trunk
Intestinal trunk
Right jugular trunk
Right subclavian trunk
Right broncho-mediastinal trunk
(b)
Thoracic duct
Left jugular trunk
Left subclavian trunk
Left broncho-mediastinal trunk
Area drained by right lymphatic trunks shown in part (b)
Area drained by left lymphatic trunks and thoracic duct shown in part (b)
(c)

FIGURE 22.3 AP|R **Lymph Drainage into Veins**

(*a*) Anterior view of the major lymphatic vessels in the thorax and abdomen. (*b*) Close-up view of the lymphatic vessels from which lymph enters the blood. (*c*) Regions of the body drained by the right and left lymphatic vessels.

The lymphatic trunks either connect to large veins in the thorax or join to yet larger vessels called **lymphatic ducts,** which then connect to the large veins. The connections of the lymphatic trunks and ducts to veins are quite variable. Many connect at the junction of the internal jugular and subclavian veins, but connections on the subclavian, the jugular, and even the brachiocephalic veins exist.

On the right side, the jugular, subclavian, and bronchomediastinal trunks typically join a right thoracic vein separately (figure 22.3*b*). However, in about 20% of people, the three trunks join to form a short duct 1 cm in length called the **right lymphatic duct** (see figure 22.1), which joins a right thoracic vein. These trunks drain the right side of the head, upper-right limb, and right thorax (figure 22.3*c*).

The **thoracic duct** (figure 22.3*a*,*b*) drains lymph from the right side of the body inferior to the thorax and the entire left side of the body (figure 22.3*c*). The thoracic duct is the largest lymphatic vessel. It is approximately 38–45 cm in length, extending from the twelfth thoracic vertebra to the base of the neck (figure 22.3*a*). The jugular and subclavian trunks join the thoracic duct. The bronchomediastinal trunk sometimes connects to the thoracic duct but typically joins a vein. The intestinal and lumbar trunks, which drain lymph inferior to the diaphragm, supply the inferior end of the thoracic duct. They can directly join the thoracic duct or merge to form a network that connects to the thoracic duct. In a small proportion of cases, the lymphatic trunks form a sac called the **cisterna chyli** (sis-ter′nă kī′lī).

In summary, lymph enters lymphatic capillaries, which converge to form larger lymphatic vessels. Lymph passes through the lymphatic vessels and through associated lymph nodes, where it is filtered. Lymphatic vessels converge to form larger lymphatic trunks, which drain lymph from major regions of the body. Lymphatic trunks empty directly into thoracic veins or combine to form larger lymphatic ducts, which empty into thoracic veins.

ASSESS **YOUR PROGRESS**

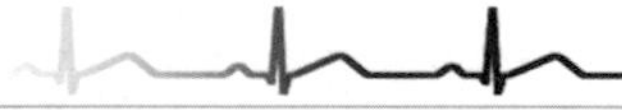

2. *Name the parts of the lymphatic system.*
3. *How is lymph formed?*
4. *Describe the structure of a lymphatic capillary. Why is it easy for fluid and other substances to enter a lymphatic capillary?*
5. *What is the function of valves in lymphatic vessels? Name three mechanisms responsible for moving lymph through the lymphatic vessels.*
6. *What are lymphatic trunks and ducts? Name the largest lymphatic vessel. What is the cisterna chyli?*
7. *What areas of the body are drained by the right lymphatic trunks, left lymphatic trunks, and thoracic duct?*

Lymphatic Tissue and Organs

Lymphatic organs contain **lymphatic tissue,** which consists primarily of lymphocytes but also includes macrophages, dendritic cells, reticular cells, and other cell types. **Lymphocytes** are white blood cells (see chapter 19). The two types of lymphocytes, **B cells** and **T cells,** originate from red bone marrow and are carried by the blood to lymphatic organs and other tissues. When the body is exposed to microorganisms or other foreign substances, the lymphocytes divide, increase in number, and become part of the immune response that destroys microorganisms and foreign substances. In addition to cells, lymphatic tissue is composed of very fine collagen fibers, called **reticular fibers** produced by **reticular cells.** Lymphocytes and other cells attach to these fibers. When lymph or blood passes through lymphatic organs, the fiber network traps microorganisms and other particles, filtering the fluid.

Lymphatic tissue that is not surrounded by a connective tissue capsule is called nonencapsulated, whereas lymphatic tissue surrounded by a capsule is said to be encapsulated. Aggregations of nonencapsulated lymphatic tissue, called **mucosa-associated lymphoid tissue (MALT),** are found in and beneath the mucous membranes lining the digestive, respiratory, urinary, and reproductive tracts. In these locations, the lymphatic tissue is well positioned to intercept microorganisms as they enter the body. Examples of MALT include diffuse lymphatic tissue, lymphatic nodules, and the tonsils. Lymphatic organs with a capsule include lymph nodes, the spleen, and the thymus.

Diffuse Lymphatic Tissue and Lymphatic Nodules

Diffuse lymphatic tissue contains dispersed lymphocytes, macrophages, and other cells; has no clear boundary; and blends with surrounding tissues (figure 22.4). It is located deep to mucous membranes, around lymphatic nodules, and within the lymph nodes and spleen.

Lymphatic nodules are denser arrangements of lymphatic tissue organized into compact, somewhat spherical structures, ranging in size from a few hundred microns to a few millimeters or more in diameter (figure 22.4). Lymphatic nodules are numerous in the loose connective tissue of the digestive, respiratory, urinary, and reproductive systems. **Peyer patches** are aggregations of lymphatic nodules in the distal half of the small intestine and the appendix. In addition to MALT, lymphatic nodules are found within lymph nodes and the spleen, where they are usually referred to as **lymphatic follicles.**

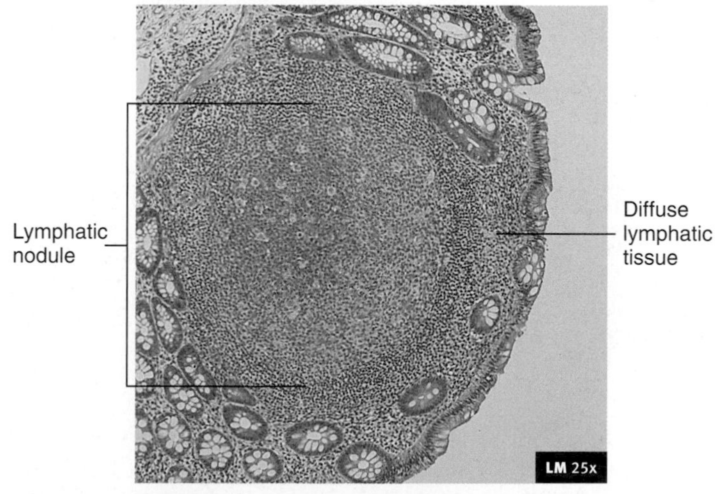
Lymphatic nodule
Diffuse lymphatic tissue
LM 25x

FIGURE 22.4 APIR **Diffuse Lymphatic Tissue and a Lymphatic Nodule**
Diffuse lymphatic tissue surrounding a lymphatic nodule in the small intestine.

Tonsils

Tonsils are large groups of lymphatic nodules and diffuse lymphatic tissue located deep to the mucous membranes within the pharynx (throat; figure 22.5). The tonsils protect against bacteria and other potentially harmful material entering the pharynx from the nasal or oral cavity. In adults, the tonsils decrease in size and eventually may disappear.

There are three groups of tonsils, but the **palatine tonsils** are usually the ones referred to. They are relatively large, oval lymphatic masses on each side of the junction between the oral cavity and the pharynx. The **pharyngeal** (fă-rin′jē-ăl) **tonsil** is a collection of somewhat closely aggregated lymphatic nodules near the junction between the nasal cavity and the pharynx. When the pharyngeal tonsil is enlarged, it is commonly called the **adenoid** (ad′ĕ-noyd; glandlike) or **adenoids** (ad′ĕ-noydz). An enlarged pharyngeal tonsil can interfere with normal breathing. The **lingual tonsil** is a loosely associated collection of lymphatic nodules on the posterior surface of the tongue.

Sometimes the palatine or pharyngeal tonsils become chronically infected and must be removed. The lingual tonsil becomes infected less often than the other tonsils and is more difficult to remove.

ASSESS YOUR PROGRESS

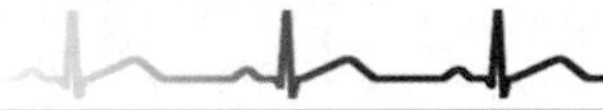

8. *What are the functions of lymphocytes and reticular fibers in lymphatic tissue?*
9. *What is mucosa-associated lymphoid tissue (MALT)? In what way is the location of MALT beneficial?*
10. *Distinguish among lymphatic tissue, lymphatic nodules, Peyer patches, and lymphatic follicles.*
11. *Describe the structure, function, and location of the tonsils.*

Lymph Nodes

Lymph nodes are small, round or bean-shaped structures ranging from 1 mm to 25 mm long. They are distributed along the course of the lymphatic vessels (figure 22.6; see figure 22.1). Lymph nodes filter the lymph, removing bacteria and other materials. In addition, lymphocytes congregate, function, and proliferate within lymph nodes.

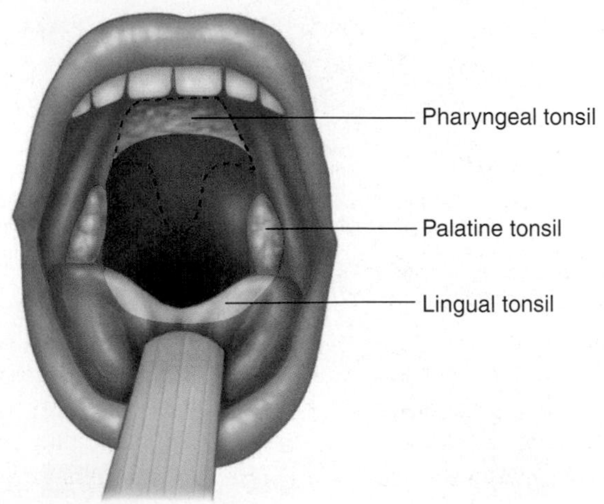
Pharyngeal tonsil
Palatine tonsil
Lingual tonsil

FIGURE 22.5 APR **Tonsils**
Anterior view of the oral cavity, showing the tonsils. Part of the palate is removed (dotted line) to show the pharyngeal tonsil.

Lymph nodes are categorized as superficial or deep. **Superficial lymph nodes** are in the subcutaneous tissue beneath the skin, and **deep lymph nodes** are everywhere else. Most superficial and deep lymph nodes are located near or on blood vessels. Approximately 450 lymph nodes are distributed throughout the body. Cervical and head nodes (about 70) filter lymph from the head and neck, axillary nodes (about 30) filter lymph from the upper limbs and superficial thorax, thoracic nodes (about 100) filter lymph from the thoracic wall and organs, abdominopelvic nodes (about 230) filter lymph from the abdomen and pelvis, and inguinal and popliteal nodes (about 20) filter lymph from the lower limbs and the superficial pelvis.

A dense connective tissue **capsule** surrounds each lymph node (figure 22.6). Extensions of the capsule, called **trabeculae** (tră-bek′ū-lē), form a delicate internal skeleton in the lymph node. Reticular fibers extend from the capsule and trabeculae to form a fibrous network throughout the node. In some areas of the lymph node, lymphocytes and macrophages are packed around the reticular fibers to form lymphatic tissue; in other areas, the reticular fibers extend across open spaces to form **lymphatic sinuses.** The lymphatic tissue and sinuses within the node are arranged into two somewhat indistinct layers, an outer cortex and an inner medulla. The **cortex** consists of a subcapsular sinus, beneath the capsule, and cortical sinuses, which are separated by diffuse lymphatic tissue, trabeculae, and lymphatic nodules. The inner **medulla** is organized into branching, irregular strands of diffuse lymphatic tissue, the **medullary cords,** separated by medullary sinuses.

Lymph nodes are the only structures that filter lymph. **Afferent lymphatic vessels** carry lymph to the lymph nodes, where it is filtered, and **efferent lymphatic vessels** carry lymph away from the nodes. Lymph from afferent lymphatic vessels enters the subcapsular sinus, filters through the cortex and medulla, and exits the lymph node through efferent lymphatic vessels. The efferent vessels of one lymph node may become the afferent vessels of another node or may converge to form lymphatic trunks, which carry lymph to the blood at thoracic veins.

Macrophages lining the lymphatic sinuses remove bacteria and other foreign substances from the lymph as it slowly filters through the sinuses. Microorganisms and other foreign substances in the lymph can stimulate lymphocytes throughout the lymph node to undergo cell division, with proliferation especially evident in the lymphatic nodules of the cortex. These areas of rapid lymphocyte division are called **germinal centers.** The newly produced lymphocytes are released into the lymph and eventually reach the bloodstream, where they circulate. Subsequently, the lymphocytes can leave the blood and enter other lymphatic tissues.

Spleen

The **spleen** is roughly the size of a clenched fist and is located on the left, superior part of the abdominal cavity (figure 22.7). The average weight of the adult spleen is 180 g in males and 140 g in females. The size and weight of the spleen tend to decrease in older people, but in certain diseases the spleen can achieve a weight of 2000 g or more. For example, about 50% of individuals who have

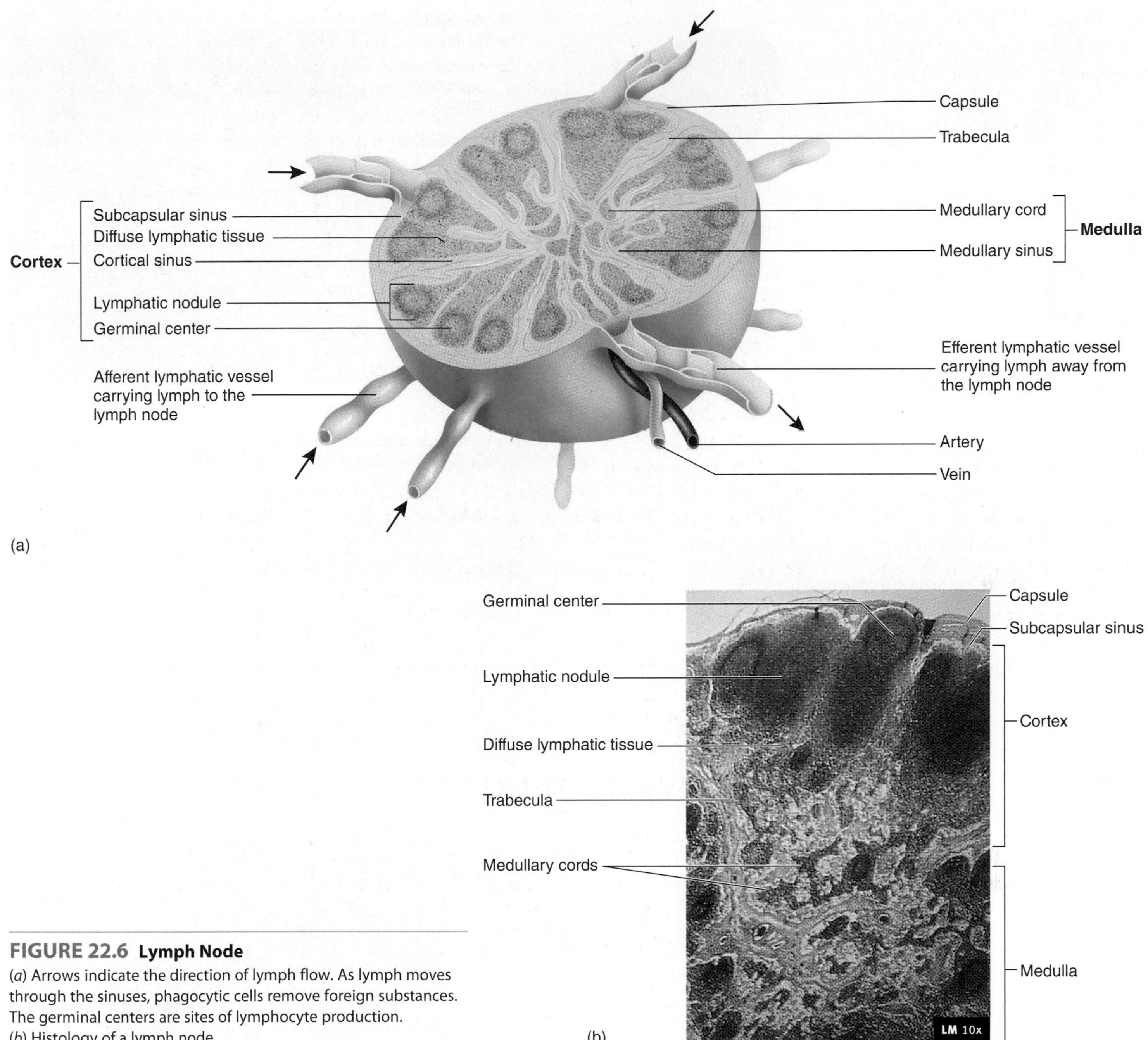
Capsule
Trabecula
Medullary cord
Medullary sinus
Medulla
Subcapsular sinus
Diffuse lymphatic tissue
Cortical sinus
Cortex
Lymphatic nodule
Germinal center
Afferent lymphatic vessel carrying lymph to the lymph node
Efferent lymphatic vessel carrying lymph away from the lymph node
Artery
Vein
(a)
Germinal center
Lymphatic nodule
Diffuse lymphatic tissue
Trabecula
Medullary cords
Capsule
Subcapsular sinus
Cortex
Medulla
LM 10x
(b)

FIGURE 22.6 Lymph Node
(*a*) Arrows indicate the direction of lymph flow. As lymph moves through the sinuses, phagocytic cells remove foreign substances. The germinal centers are sites of lymphocyte production.
(*b*) Histology of a lymph node.

infectious mononucleosis develop an enlarged spleen as a result of increased numbers of defense cells. Usually, the spleen will reduce in size following the infection.

The spleen has an outer capsule of dense irregular connective tissue and a small amount of smooth muscle. Bundles of connective tissue fibers from the capsule form trabeculae, which extend into the organ, subdividing it into small, interconnected compartments. Arteries, veins, and lymphatic vessels extend through the trabeculae to supply the compartments, which are filled with white and red pulp. **White pulp** is lymphatic tissue surrounding the arteries within the spleen. **Red pulp** is associated with the veins. It consists of a fibrous network, filled with macrophages and red blood cells, and enlarged capillaries that connect to the veins. Approximately one-fourth of the volume of the spleen is white pulp, and three-fourths is red pulp.

Branches of the **splenic** (splen′ik) **artery** enter the spleen at the **hilum,** and their branches follow the various trabeculae into the spleen (figure 22.7*a*,*b*). From the trabeculae, arterial branches extend into the white pulp, which consists of the periarterial lymphatic sheath and lymphatic nodules (figure 22.7*c*). The **periarterial lymphatic sheath** is composed of diffuse lymphatic tissue surrounding arteries and arterioles extending to lymphatic nodules. Arterioles enter lymphatic nodules and give rise to capillaries supplying the red pulp, which consists of the splenic cords and the venous sinuses. The **splenic cords** are a network of reticular cells that produce reticular fibers (see chapter 4).

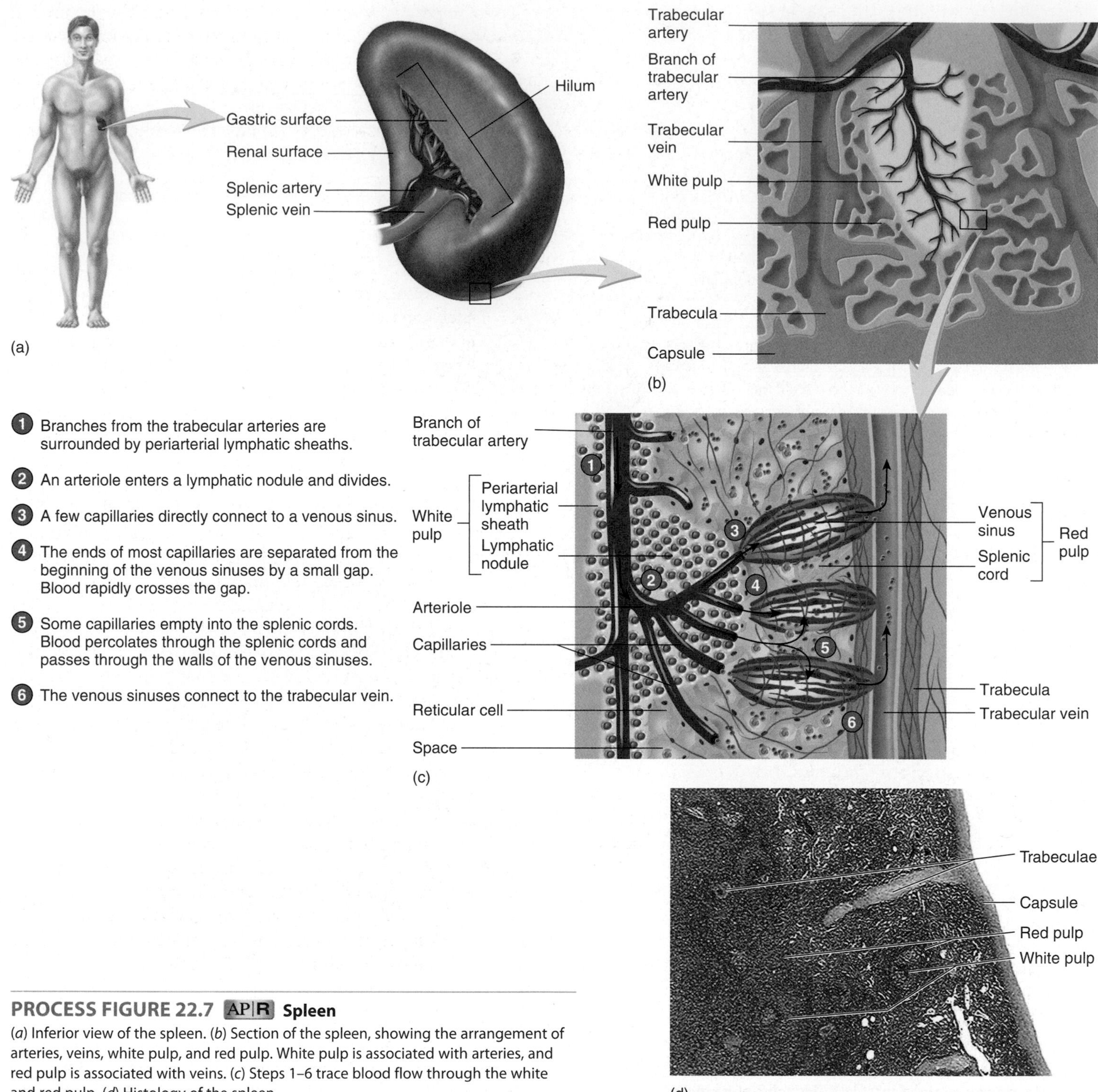
Gastric surface
Renal surface
Splenic artery
Splenic vein
Hilum
(a)
Trabecular artery
Branch of trabecular artery
Trabecular vein
White pulp
Red pulp
Trabecula
Capsule
(b)
1 Branches from the trabecular arteries are surrounded by periarterial lymphatic sheaths.
2 An arteriole enters a lymphatic nodule and divides.
3 A few capillaries directly connect to a venous sinus.
4 The ends of most capillaries are separated from the beginning of the venous sinuses by a small gap. Blood rapidly crosses the gap.
5 Some capillaries empty into the splenic cords. Blood percolates through the splenic cords and passes through the walls of the venous sinuses.
6 The venous sinuses connect to the trabecular vein.
Branch of trabecular artery
White pulp
Periarterial lymphatic sheath
Lymphatic nodule
Arteriole
Capillaries
Reticular cell
Space
Venous sinus
Splenic cord
Red pulp
Trabecula
Trabecular vein
(c)
Trabeculae
Capsule
Red pulp
White pulp
(d)

PROCESS FIGURE 22.7 AP|R **Spleen**

(*a*) Inferior view of the spleen. (*b*) Section of the spleen, showing the arrangement of arteries, veins, white pulp, and red pulp. White pulp is associated with arteries, and red pulp is associated with veins. (*c*) Steps 1–6 trace blood flow through the white and red pulp. (*d*) Histology of the spleen.

The spaces between the reticular cells are occupied by splenic macrophages and blood cells that have come from the capillaries. The **venous sinuses** are enlarged capillaries between the splenic cords. The venous sinuses typically connect to trabecular veins, which unite to form vessels that leave the spleen to form the **splenic vein.**

Although blood flow through the spleen can take from a few seconds to an hour or more, most blood flows through the spleen rapidly. The rapid flow results from the movement of blood from the ends of capillaries into the beginning of the venous sinuses. Although a few capillaries connect directly to venous sinuses, most capillaries are separated by a small gap (figure 22.7*c*). Slower blood flow occurs when blood leaves the ends of the capillaries, enters the splenic cords, percolates through them, and passes through the walls of the venous sinuses.

The spleen destroys defective red blood cells, detects and responds to foreign substances in the blood, and acts as a blood reservoir. As

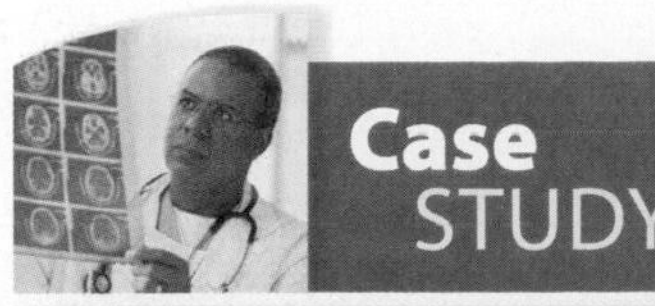

Case STUDY Lymphedema

Cindy is a 40-year-old woman who has been diagnosed with breast cancer. Before removing the cancerous tumor from her left breast, her surgeon injected a dye and a radioactive tracer, technetium-99, at the tumor site. The dye and tracer enabled the surgeon to find **sentinel lymph nodes,** those nodes closest to the tumor. The sentinel lymph nodes were sampled for cancer. When cancer cells were found in all of them, the surgeon removed the axillary lymph nodes from under Cindy's left arm.

A few days after the surgery, Cindy noticed that the skin on her left arm felt tight and the arm felt heavy. In addition, her wedding ring was tighter than usual. These symptoms were due to an abnormal accumulation of fluid in Cindy's upper limb, called **lymphedema** (limf′e-dē′mă), caused by the removal of her axillary lymph nodes. In the United States, the most common cause of lymphedema is removal of or damage to lymph nodes and vessels as a result of cancer surgery or radiation treatment. Approximately 10–20% of women whose axillary lymph nodes have been removed develop lymphedema.

Predict 2

a. What is the rationale for testing sentinel lymph nodes for cancer?

b. What was the rationale for removing Cindy's axillary lymph nodes?

c. Why does removing the axillary lymph nodes result in lymphedema?

d. Exercise can help reduce lymphedema. Explain.

e. A compression bandage or garment can help reduce lymphedema. Explain.

f. In intermittent pneumatic pump compression therapy, the pressure of a garment enclosing a limb increases and decreases periodically. In addition, the pressure increases sequentially from the distal part to the proximal part of a limb. How does this therapy help reduce lymphedema?

red blood cells age, they lose their ability to bend and fold. Consequently, the cells can rupture as they pass slowly through the meshwork of the splenic cords or the intercellular slits of the venous sinus walls. Splenic macrophages then phagocytize the cellular debris.

Foreign substances in the blood passing through the spleen can stimulate an immune response because of the presence in the white pulp of specialized lymphocytes (see section 22.5). There are high concentrations of T cells in the periarterial lymphatic sheath and B cells in the lymphatic nodules.

The splenic cords of the spleen are a limited reservoir for blood. For example, during exercise splenic volume can be reduced by 40–50%. The resulting small increase in circulating red blood cells can promote better oxygen delivery to muscles during exercise or emergency situations. Physiologists do not presently know if this reduction results from the contraction of smooth muscle within the capsule, from the contraction of smooth muscle (myofibroblast) within the trabeculae, or from the reduced blood flow through the spleen caused by the constriction of blood vessels.

Although the ribs protect the spleen, it is often ruptured in traumatic abdominal injuries. A ruptured spleen can cause severe bleeding, shock, and death. Surgical intervention may stop the bleeding. Cracks in the spleen are repaired using sutures and blood clotting agents. Mesh wrapped around the spleen can hold it together. If these techniques do not stop the bleeding, a **splenectomy** (splē-nek′tō-mē), removal of the spleen, may be necessary. After removal of the spleen, other lymphatic organs and the liver compensate for the loss of its functions.

Thymus

The **thymus** (thī′mŭs) is a bilobed gland (figure 22.8) located in the superior mediastinum, the partition dividing the thoracic cavity into the left and right parts. The thymus increases in size until the first year of life, after which it remains approximately the same size until 60 years of age, when it decreases in size (see section 22.9).

Each lobe of the thymus is surrounded by a thin connective tissue capsule. Trabeculae extend from the capsule into the substance of the gland, dividing it into **lobules.** Unlike other lymphatic tissue, which has a fibrous network of reticular fibers, the framework of thymic tissue consists of epithelial cells. The processes of the epithelial cells are joined by desmosomes, and the cells form small, irregularly shaped compartments filled with lymphocytes. Near the capsule and trabeculae, the lymphocytes are numerous and form dark-staining areas of the lobules called the cortex. A lighter-staining, central portion of the lobules, called the medulla, has fewer lymphocytes. The medulla also contains rounded epithelial structures, called **thymic corpuscles** (Hassall corpuscles). For more than 150 years, the function of these structures was unknown, but current research has shown that thymic corpuscles function in the development of **regulatory T cells.** Regulatory T cells suppress the body's immune response and protect against autoimmune diseases (see section 22.5).

The thymus is the site for the maturation of T cells. **Thymosin,** a hormone secreted by the thymus, is important in the T-cell maturation process. Large numbers of lymphocytes are produced in the thymus, but most degenerate. The lymphocytes that survive the maturation process are capable of reacting to foreign substances, but they normally do not react to and destroy healthy body cells (see "Origin and Development of Lymphocytes," in section 22.5). These surviving thymic lymphocytes migrate to the medulla, enter the blood, and travel to other lymphatic tissues.

ASSESS YOUR PROGRESS

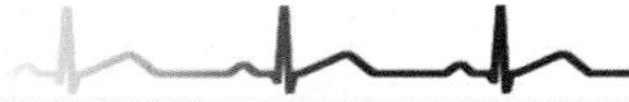

12. *Where are lymph nodes found? Describe the parts of a lymph node, and explain how lymph flows through a lymph node.*
13. *Describe the functions of lymph nodes. What is a germinal center?*
14. *Where is the spleen located? Describe the structure of the spleen.*
15. *Explain how the spleen performs its three main functions.*
16. *Where is the thymus located? Describe its structure and function.*

Overview of the Lymphatic System

Figure 22.9 summarizes the processes performed by the lymphatic system. Lymphatic capillaries and vessels remove fluid from tissues and absorb lipids from the small intestine. Lymph nodes filter lymph, and the spleen filters blood.

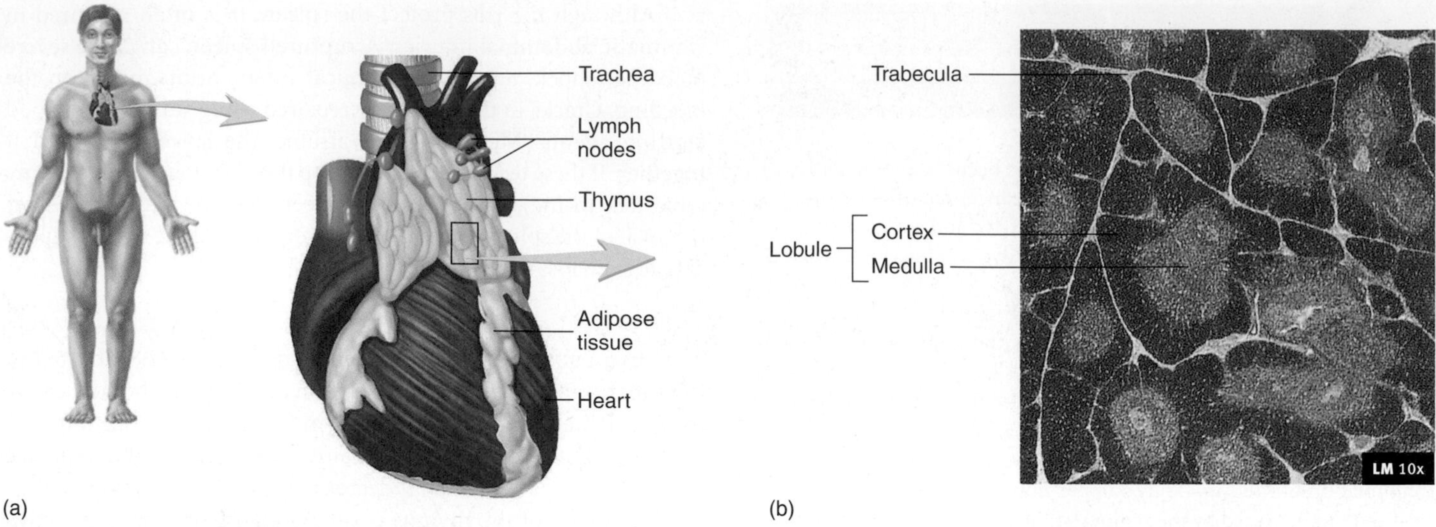
Trachea
Lymph nodes
Thymus
Adipose tissue
Heart
(a)
Trabecula
Lobule
Cortex
Medulla
LM 10x
(b)

FIGURE 22.8 APR **Thymus**
(*a*) Location and shape of the thymus in relation to the heart. (*b*) Histology of thymic lobules, showing the outer cortex and the inner medulla.

Figure 22.9 also illustrates two types of lymphocytes, B cells and T cells. B cells originate and mature in red bone marrow. Pre-T cells are produced in red bone marrow and migrate to the thymus, where they mature to become T cells. B cells from red bone marrow and T cells from the thymus circulate to, and populate, other lymphatic tissues.

B cells and T cells are responsible for much of immunity. In response to infections, B cells and T cells increase in number and circulate to lymphatic and infected tissues. How B cells and T cells protect the body is discussed in section 22.5.

22.3 Immunity

LEARNING OUTCOMES

After reading this section, you should be able to

A. Define the concepts of specificity and memory as they apply to immunity.

B. Distinguish between the general characteristics of innate immunity and adaptive immunity.

Immunity is the ability to resist damage from foreign substances, such as microorganisms; harmful chemicals, such as toxins released by microorganisms; and internal threats, such as cancer cells. Immunity is categorized as **innate immunity** (also called *nonspecific resistance*) and **adaptive immunity** (also called *specific immunity*), although the two systems are fully integrated in the body. In innate immunity, the body recognizes and destroys certain foreign substances, but the response to them is the same each time the body is exposed. In adaptive immunity, the body recognizes and destroys foreign substances, but the response to them is faster and stronger each time the foreign substance is encountered.

Specificity and memory are characteristics of adaptive immunity, but not of innate immunity. **Specificity** is the ability of adaptive immunity to recognize a particular substance. For example, innate immunity can act against bacteria in general, whereas adaptive immunity can distinguish among various kinds of bacteria. **Memory** is the ability of adaptive immunity to "remember" previous encounters with a particular substance. As a result, the response is faster, stronger, and longer-lasting.

Innate immunity includes body defenses that are present at birth and genetically determined. In innate immunity, each time the body is exposed to a substance, the response is the same because specificity and memory of previous encounters do not apply. For example, each time a bacterial cell is introduced into the body, it is phagocytized with the same speed and efficiency. Innate immunity is present to some degree in all multicellular organisms.

Adaptive immunity includes body defenses that are acquired through a person's lifetime, depending on exposure to different microorganisms. In adaptive immunity, the response during the second exposure is faster and stronger than the response to the first exposure because the immune system "remembers" the bacteria from the first exposure. For example, following the first exposure to the bacteria, the body can take many days to destroy them. During this time, the bacteria damage tissues and produce the symptoms of disease. Following the second exposure to the same bacteria, the response is rapid and effective. Bacteria are destroyed before any symptoms develop, and the person is said to be **immune.** Adaptive immunity is unique to vertebrates.

Innate and adaptive immunity are intimately linked. Most importantly, mediators of innate immunity are required to initiate and regulate adaptive immunity.

ASSESS YOUR PROGRESS

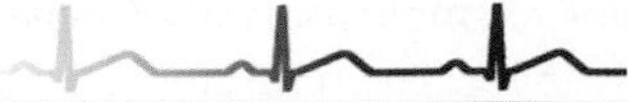

17. *What is immunity?*

18. *Why do specificity and memory relate to adaptive immunity but not to innate immunity?*

19. *What are the differences between innate immunity and adaptive immunity?*

1. Lymphatic capillaries remove fluid from tissues. The fluid becomes lymph (see figure 22.2*a*).

2. Lymph flows through lymphatic vessels, which have valves that prevent the backflow of lymph (see figure 22.2*b*).

3. Lymph nodes filter lymph (see figure 22.6) and are sites where lymphocytes respond to infections.

4. Lymph enters the thoracic duct or the right lymphatic duct.

5. Lymph enters the blood.

6. Lacteals in the small intestine (see figure 24.16*c*) absorb lipids, which enter the thoracic duct (see figure 22.3*a*).

7. Chyle, which is lymph containing lipids, enters the blood.

8. The spleen (see figure 22.7) filters blood and is a site where lymphocytes respond to infections.

9. Lymphocytes (pre-B and pre-T cells) originate from stem cells in the red bone marrow (see figure 22.12). The pre-B cells become mature B cells in the red bone marrow and are released into the blood. The pre-T cells enter the blood and migrate to the thymus.

10. The thymus (see figure 22.8) is where pre-T cells derived from red bone marrow increase in number and become mature T cells that are released into the blood (see figure 22.12).

11. B cells and T cells from the blood enter and populate all lymphatic tissues. These lymphocytes can remain in tissues or pass through them and return to the blood. B cells and T cells can also respond to infections by dividing and increasing in number (see figures 22.18 and 22.22). Some of the newly formed cells enter the blood and circulate to other tissues.

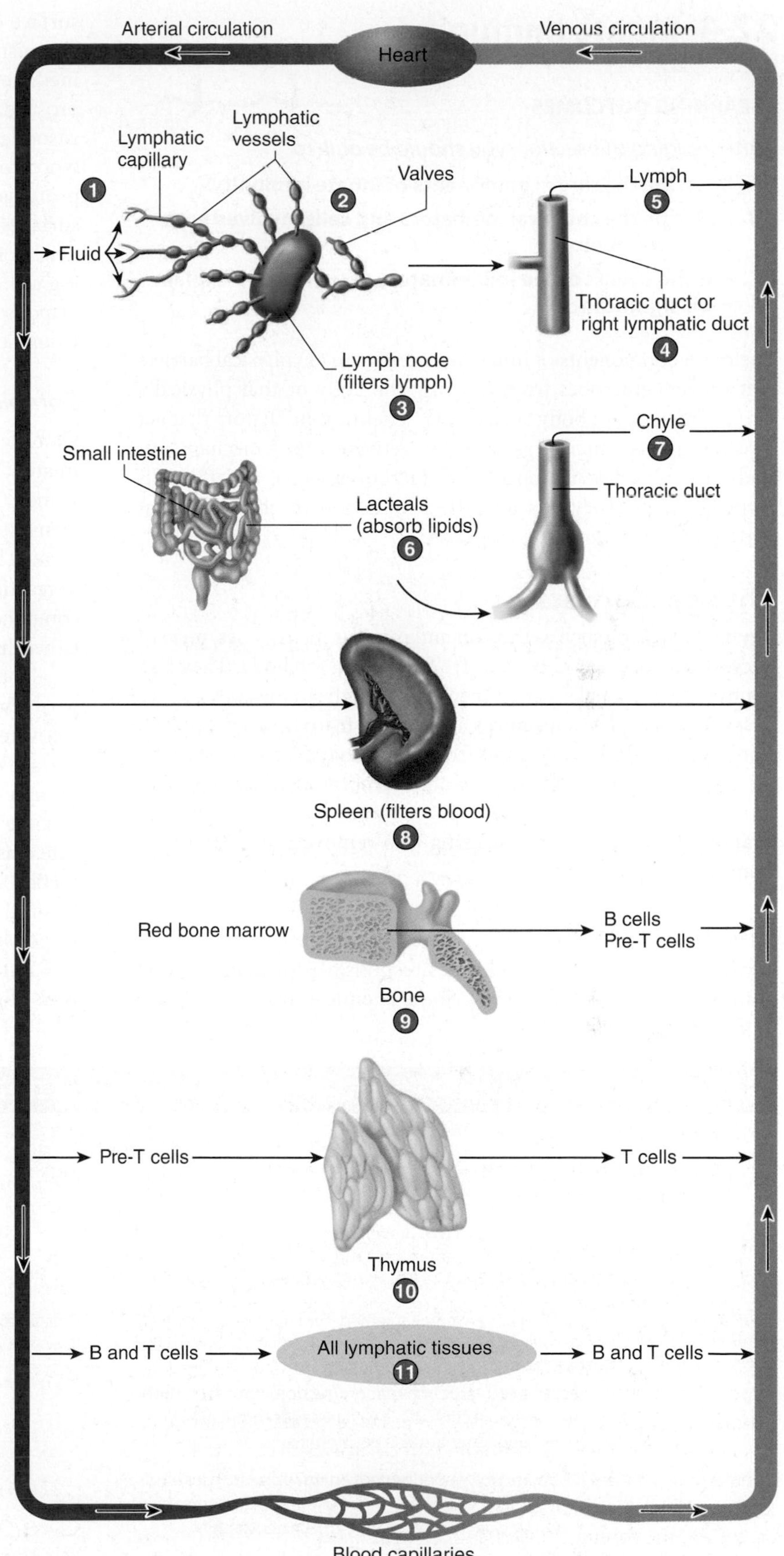

Arterial circulation
Heart
Venous circulation
Lymphatic capillary
Lymphatic vessels
1
Fluid
2
Valves
Lymph
5
Thoracic duct or right lymphatic duct
4
Lymph node (filters lymph)
3
Small intestine
Chyle
7
Thoracic duct
Lacteals (absorb lipids)
6
Spleen (filters blood)
8
Red bone marrow
B cells
Pre-T cells
Bone
9
Pre-T cells
T cells
Thymus
10
B and T cells
All lymphatic tissues
11
B and T cells
Blood capillaries

PROCESS FIGURE 22.9 APR **Overview of the Lymphatic System**

22.4 Innate Immunity

LEARNING OUTCOMES

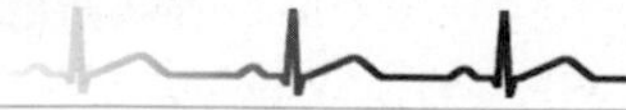

After reading this section, you should be able to

A. Describe the three components of innate immunity.

B. Describe the chemical mediators and cells involved with innate immunity.

C. List the events of the inflammatory response and explain their significance.

The main components of innate immunity are (1) physical barriers that prevent microbes from entering the body or that physically remove them from body surfaces; (2) chemical mediators that act directly against microorganisms or activate other mechanisms, leading to the destruction of the microorganisms; and (3) cells involved in phagocytosis and the production of chemicals that participate in the immune response.

Physical Barriers

Physical barriers, such as the skin and mucous membranes, prevent microorganisms and chemicals from entering the body. They also remove microorganisms and other substances from the body surface in several ways. The substances are washed from the eyes by tears, from the mouth by saliva, and from the urinary tract by urine. In the respiratory tract, ciliated mucous membranes sweep microbes trapped in the mucus to the back of the throat, where they are swallowed. Coughing and sneezing also remove microorganisms from the respiratory tract.

Chemical Mediators

Chemical mediators are molecules responsible for many aspects of innate immunity (table 22.1). Some chemical mediators on the surface of cells, such as lysozyme, sebum, and mucus, kill microorganisms or prevent them from entering the cells. Other chemical mediators, such as histamine, complement, and eicosanoids (e.g., prostaglandins and leukotrienes), promote inflammation by causing vasodilation, increasing vascular permeability, attracting white blood cells, and stimulating phagocytosis. **Cytokines** (si′tō-kīnz) are proteins or peptides secreted by cells that bind to receptors on cell surfaces, stimulating a response. They usually bind to receptors on neighboring cells, but sometimes they bind to receptors on the secreting cell. Cytokines regulate the intensity and duration of immune responses and stimulate the proliferation and differentiation of cells. Examples of cytokines are interferons, interleukins, and lymphokines.

Complement

Complement is a group of about 20 proteins that make up approximately 10% of the globulin part of plasma. They include proteins named C1–C9 and factors B, D, and P (properdin). Normally, complement proteins circulate in the blood in an inactive, nonfunctional form. They become activated in the **complement cascade,** a series of reactions in which each component of the series activates the next component (figure 22.10). The complement cascade begins through either the alternative pathway or the classical pathway.

The **alternative pathway,** part of innate immunity, is initiated when the complement protein C3 becomes spontaneously active. Activated C3 normally is quickly inactivated by proteins on the surface of the body's cells. Activated C3 can combine with some foreign substances, such as part of a bacterial cell or virus. It can become stabilized and cause activation of the complement cascade. The **classical pathway** is part of adaptive immunity, discussed in section 22.5. The activation of complement by the alternative and classical pathways is one of several ways in which innate and adaptive responses are integrated.

Activated complement proteins provide protection in several ways (figure 22.10). They can form a **membrane attack complex**

TABLE 22.1 Chemical Mediators of Innate Immunity and Their Functions

Chemical	Description
Surface chemicals	Lysozymes (in tears, saliva, nasal secretions, and sweat) lyse cells; acid secretions (sebum in the skin and hydrochloric acid in the stomach) prevent microbial growth or kill microorganisms; mucus on the mucous membranes traps microorganisms until they can be destroyed.
Histamine	Histamine is an amine released from mast cells, basophils, and platelets; histamine causes vasodilation, increases vascular permeability, stimulates gland secretions (especially mucus and tear production), causes smooth muscle contraction of airway passages (bronchioles) in the lungs, and attracts eosinophils.
Kinins	Kinins are polypeptides derived from plasma proteins; kinins cause vasodilation, increase vascular permeability, stimulate pain receptors, and attract neutrophils.
Interferons	Interferons are proteins, produced by most cells, that interfere with virus production and infection.
Complement	Complement is a group of plasma proteins that increase vascular permeability, stimulate the release of histamine, activate kinins, lyse cells, promote phagocytosis, and attract neutrophils, monocytes, macrophages, and eosinophils.
Prostaglandins	Prostaglandins are a group of lipids (PGEs, PGFs, thromboxanes, and prostacyclins), some of which cause smooth muscle relaxation and vasodilation, increase vascular permeability, and stimulate pain receptors.
Leukotrienes	Leukotrienes are a group of lipids, produced primarily by mast cells and basophils, that cause prolonged smooth muscle contraction (especially in the lung bronchioles), increase vascular permeability, and attract neutrophils and eosinophils.
Pyrogens	Pyrogens are chemicals, released by neutrophils, monocytes, and other cells, that stimulate fever production.

Abbreviations: PGE = prostaglandin E; PGF = prostaglandin F.

(MAC), which produces a channel through the plasma membrane. MAC formation begins when activated C3 attaches to a plasma membrane, stimulating a series of reactions activating C5–C9. The main component of a MAC is activated C9 molecules, which change shape, attach to each other, and form a channel through the membrane. When MACs form in the plasma membrane of a nucleated cell, Na^+ and water enter the cell through the channel and cause the cell to lyse. When MACs form in the outer membrane of certain bacteria (Gram negative), an enzyme called lysozyme passes through the channel and digests the bacterial cell wall. When the wall breaks apart, the bacterial cell undergoes lysis.

Complement proteins can also attach to the surface of bacterial cells and stimulate macrophages to phagocytize the bacteria. This process is called **opsonization.** In addition, complement proteins attract immune system cells to sites of infection and promote inflammation.

Interferons

Interferons (in-ter-fēr′onz) are proteins that protect the body against viral infection and perhaps some forms of cancer. After a virus infects a cell, viral replication can occur. Viral nucleic acids and proteins, which are produced using the infected cell's organelles, are assembled into new viruses. The new viruses are released from the infected cell to infect other cells. Because infected cells usually stop their normal functions or die during viral replication, viral infections are clearly harmful to the body. Fortunately, viruses and other substances can stimulate infected cells to produce interferons, which neither protect the cell that produces them nor act directly against viruses. Instead, interferons bind to the surface of neighboring cells and stimulate them to produce antiviral proteins. These antiviral proteins stop viral reproduction in the neighboring cells by preventing the production of viral nucleic acids and proteins.

1. The classical pathway begins when an antigen–antibody complex binds to C1. The C1–antigen–antibody complex activates C4.
2. Activated C4 forms a complex with C2 that activates C3.
3. The alternative pathway begins when C3 is spontaneously activated.
4. Foreign substances and factors B, D, and P stabilize activated C3.
5. Once C3 is activated, the classical and alternative pathways are the same. C3 activates C5, C5 activates C6, C6 activates C7, C7 activates C8, and C8 activates C9.
6. Activated C3–C7 promote phagocytosis, inflammation, and chemotaxis (attracts cells).
7. Activated C5–C9 combine to form a membrane attack complex (MAC), which forms a channel through the plasma membrane (only C9 of MAC is shown).

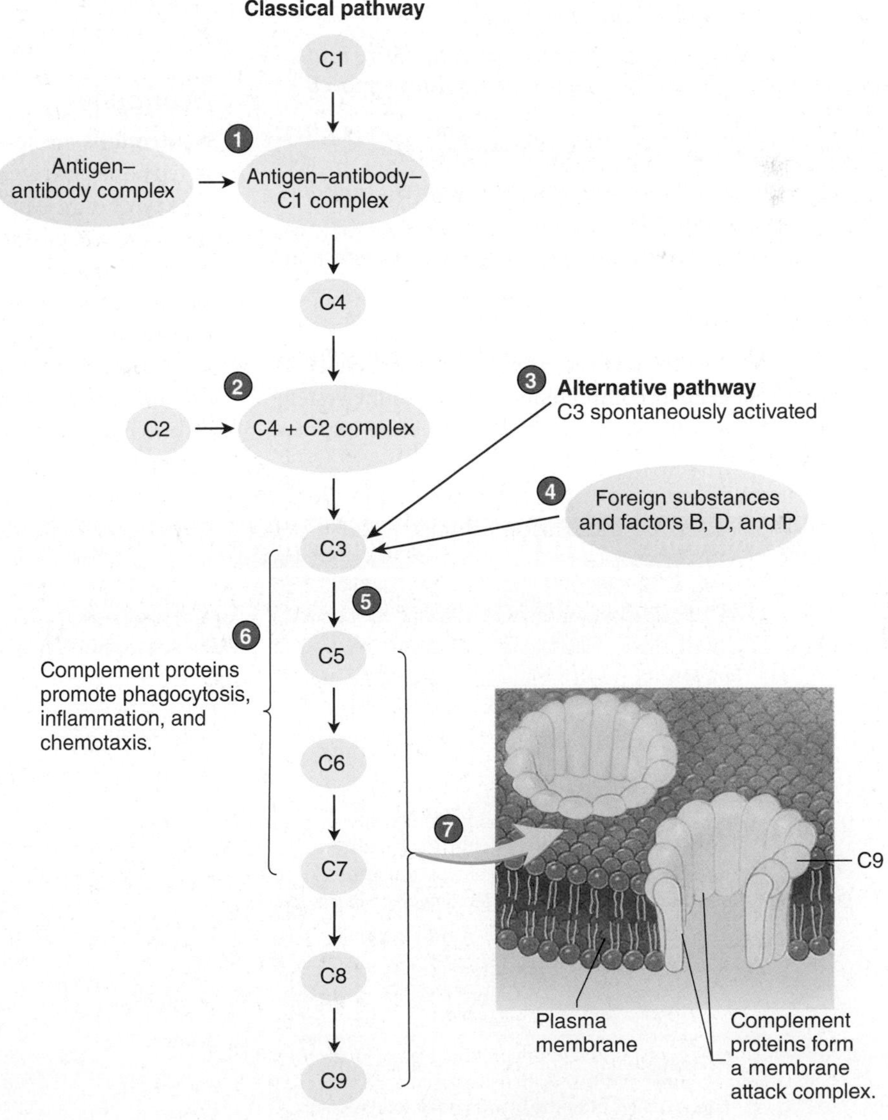
Classical pathway
C1
1
Antigen–antibody complex
Antigen–antibody–C1 complex
C4
2
C2
C4 + C2 complex
3
Alternative pathway
C3 spontaneously activated
4
Foreign substances and factors B, D, and P
C3
5
6
Complement proteins promote phagocytosis, inflammation, and chemotaxis.
C5
C6
C7
C8
C9
7
C9
Plasma membrane
Complement proteins form a membrane attack complex.

PROCESS FIGURE 22.10 Complement Cascade
Complement proteins circulate in the blood in an inactive form. Only the activated proteins are shown. The complement cascade begins with either the alternative pathway or the classical pathway.

Interferon viral resistance is innate rather than adaptive, and the same interferons act against many different viruses. Infection by one kind of virus can actually protect against infection by other kinds of viruses. Some interferons also play a role in activating immune cells, such as macrophages and natural killer cells.

Because viruses induce some cancers, interferons may play a role in controlling cancers. Interferons activate macrophages and natural killer cells (a type of lymphocyte) that attack tumor cells. Through genetic engineering, interferons are produced in sufficient quantities for clinical use and, along with other therapies, have been effective in treating certain viral infections and cancers. For example, interferons are used to treat hepatitis C, a viral disorder that can cause cirrhosis and cancer of the liver, and to treat genital warts caused by the herpes virus. Interferons are also approved to treat Kaposi sarcoma, a cancer that can develop in AIDS patients.

ASSESS YOUR PROGRESS

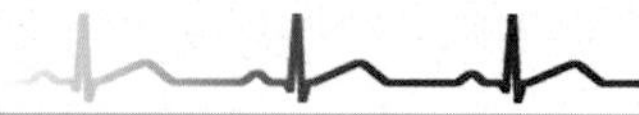

20. *List the three components of innate immunity.*
21. *Name two physical barriers that prevent microorganisms from entering the body. In what ways are microorganisms removed from body surfaces?*
22. *What roles do cytokines play as chemical mediators?*
23. *What is complement? In what two ways is it activated? How does complement provide protection?*
24. *What are interferons? How do they protect against viral infection?*

White Blood Cells

White blood cells and the cells derived from them (see table 19.2) are the most important cellular components of the immune system (table 22.2). White blood cells are produced in red bone marrow and lymphatic tissue and are released into the blood, where they are transported throughout the body. To be effective, white blood cells must move into the tissues where they are needed. **Chemotactic** (kē-mō-tak′tik) **factors** are parts of microbes or chemicals released by tissue cells that act as chemical signals to attract white blood cells. Important chemotactic factors include complement, leukotrienes, kinins, and histamine. They diffuse from the area where they are released. White blood cells can detect small differences in chemotactic factor concentration and move from areas of lower chemotactic factor concentration to areas of higher concentration. Thus, they move toward the source of these substances, an ability called **chemotaxis.** White blood cells can move by ameboid motion over the surface of cells, can squeeze between cells, and can sometimes pass directly through other cells.

Phagocytosis (fag-ō-sī-tō′sis) is the endocytosis and destruction of particles by cells called **phagocytes** (see figure 3.17). The particles can be microorganisms or their parts, foreign substances, or dead cells from the body. The most important phagocytic cells are neutrophils and macrophages.

Neutrophils

Neutrophils are small, phagocytic cells produced in large numbers in red bone marrow and released into the blood, where they circulate for a few hours. Approximately 126 billion neutrophils per day leave the blood and pass through the wall of the digestive tract, where they provide phagocytic protection. The neutrophils are then eliminated as part of the feces. Neutrophils are usually the first cells to enter infected tissues in large numbers. They release chemical signals, such as cytokines and chemotactic factors, that increase the inflammatory response by recruiting and activating other immune cells. Neutrophils often die after a single phagocytic event.

TABLE 22.2 Immune System Cells and Their Primary Functions

Cell	Primary Function	Cell	Primary Function
Innate Immunity		**Adaptive Immunity**	
Neutrophil	Phagocytosis and inflammation; usually the first cell to leave the blood and enter infected tissues	B cell	After activation, differentiates to become plasma cell or memory B cell
Monocyte	Leaves the blood and enters tissues to become a macrophage	Plasma cell	Produces antibodies that are directly or indirectly responsible for destroying the antigen
Macrophage	Most effective phagocyte; important in later stages of infection and in tissue repair; located throughout the body to "intercept" foreign substances; processes antigens; involved in the activation of B cells and T cells	Memory B cell	Quick and effective response to an antigen against which the immune system has previously reacted; responsible for adaptive immunity
Basophil	Motile cell that leaves the blood, enters tissues, and releases chemicals that promote inflammation	Cytotoxic T cell	Responsible for destroying cells by lysis or by producing cytokines
Mast cell	Nonmotile cell in connective tissues that promotes inflammation through the release of chemicals	Helper T cell	Activates B cells and cytotoxic T cells
Eosinophil	Enters tissues from the blood and defends against parasitic infections; participates in inflammation associated with asthma and allergies	Regulatory T cell	Inhibits B cells, helper T cells, and cytotoxic T cells
Natural killer cell	Lyses tumor and virus-infected cells	Memory T cell	Quick and effective response to an antigen against which the immune system has previously reacted; responsible for adaptive immunity
		Dendritic cell	Processes antigen and is involved in the activation of B cells and T cells

Neutrophils also release lysosomal enzymes that kill microorganisms and cause tissue damage and inflammation. **Pus** is an accumulation of dead neutrophils, dead microorganisms, debris from dead tissue, and fluid.

Macrophages

Macrophages are large phagocytic cells. Macrophages are derived from monocytes that leave the blood, enter tissues, enlarge about fivefold, and increase their number of lysosomes and mitochondria. Macrophages have longer life spans than neutrophils. In addition, they can ingest more and larger phagocytic particles than neutrophils. Macrophages usually accumulate in tissues after neutrophils do, and they are responsible for most of the phagocytic activity in the late stages of an infection, including cleaning up dead neutrophils and other cellular debris. In addition to their phagocytic role, macrophages produce a variety of chemicals, such as interferons, prostaglandins, and complement, that enhance the immune response (see table 22.1).

Macrophages are beneath the free surfaces of the body, such as in the skin (dermis), subcutaneous tissue, mucous membranes, and serous membranes, and around blood and lymphatic vessels. Macrophages provide protection in these areas by trapping and destroying microorganisms entering the tissues.

If microbes do gain entry to the blood or lymphatic system, macrophages are waiting within enlarged spaces, called **sinuses,** to phagocytize them. Blood vessels in the spleen, bone marrow, and liver have sinuses, as do lymph nodes. Within the sinuses, reticular cells produce a fine network of reticular fibers that slows the flow of blood or lymph and provides a large surface area for the attachment of macrophages. In addition, macrophages are on the endothelial lining of the sinuses.

Because macrophages on the reticular fibers and endothelial lining of the sinuses were among the first macrophages studied, these cells are referred to as the **reticuloendothelial system.** However, researchers now recognize that macrophages are derived from monocytes and are in locations other than the sinuses. Because monocytes and macrophages have a single, unlobed nucleus, they are also called the **mononuclear phagocytic system.** Sometimes macrophages are given specific names—for instance, *dust cells* in the lungs, *Kupffer cells* in the liver, and *microglia* in the central nervous system.

Basophils and Mast Cells

Basophils and mast cells play important roles in stimulating inflammation. **Basophils,** which are derived from red bone marrow, are motile white blood cells that can leave the blood and enter infected tissues. **Mast cells,** which are also derived from red bone marrow, are nonmotile cells in connective tissue, especially near capillaries. Like macrophages, mast cells are located at points where microorganisms may enter the body, such as the skin, lungs, digestive tract, and urogenital tract.

Basophils and mast cells can be activated through innate immunity (e.g., by complement) or through adaptive immunity (see "Effects of Antibodies" in section 22.5). When activated, they release chemicals—for example, histamine and leukotrienes—that produce an inflammatory response or activate other mechanisms, such as smooth muscle contraction in the lungs.

Eosinophils

Eosinophils are produced in red bone marrow, enter the blood, and within a few minutes enter tissues. Eosinophil numbers increase in response to parasitic infections. Eosinophils secrete enzymes that effectively kill some parasites. Also, eosinophil numbers greatly increase in the case of an allergic reaction with much inflammation.

Natural Killer (NK) Cells

Natural killer (NK) cells, a type of lymphocyte produced in red bone marrow, account for up to 15% of lymphocytes. NK cells recognize classes of cells, such as tumor cells or virus-infected cells in general, rather than specific tumor cells or cells infected by a specific virus. For this reason, and because NK cells do not exhibit a memory response, they are classified as part of innate immunity. NK cells use a variety of methods to kill their target cells, including releasing chemicals that damage plasma membranes and cause the cells to lyse.

ASSESS **YOUR PROGRESS**

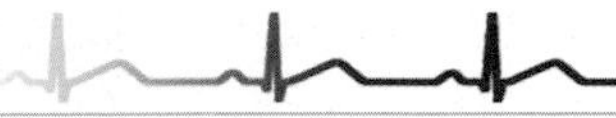

25. *Define* chemotactic factor, chemotaxis, *and* phagocytosis.

26. *What are the functions of neutrophils and macrophages? What is pus?*

27. *What effects are produced by the chemicals released from basophils, mast cells, and eosinophils?*

28. *How do NK cells function?*

Inflammatory Response

The **inflammatory response** is a complex sequence of events involving many of the chemical mediators and cells of innate immunity. Trauma, burns, chemicals, and infections can damage tissues, resulting in inflammation. Here we use a bacterial infection to illustrate an inflammatory response (figure 22.11). Bacteria enter the tissue, causing damage that stimulates the release or activation of chemical mediators, such as histamine, complement, kinins, and eicosanoids (e.g., prostaglandins and leukotrienes). The chemical mediators produce several effects: (1) Vasodilation increases blood flow and takes phagocytes and other white blood cells to the area; (2) phagocytes leave the blood and enter the tissue; and (3) increased vascular permeability allows fibrinogen and complement to enter the tissue from the blood. Fibrinogen is converted to fibrin, which prevents the spread of infection by walling off the infected area. Complement further enhances the inflammatory response and attracts additional phagocytes. The process of releasing chemical mediators and attracting phagocytes and other white blood cells continues until the bacteria are destroyed. Phagocytes, such as neutrophils and macrophages, remove microorganisms and dead tissue, and the damaged tissues are repaired.

Inflammation can be local or systemic. **Local inflammation** is an inflammatory response confined to a specific area of the body (see chapter 4). Symptoms of local inflammation include redness, heat, and swelling due to increased blood flow and increased vascular permeability, as well as pain caused by swelling and by chemical mediators acting on pain receptors. The tissue destruction, swelling, and pain lead to loss of function.

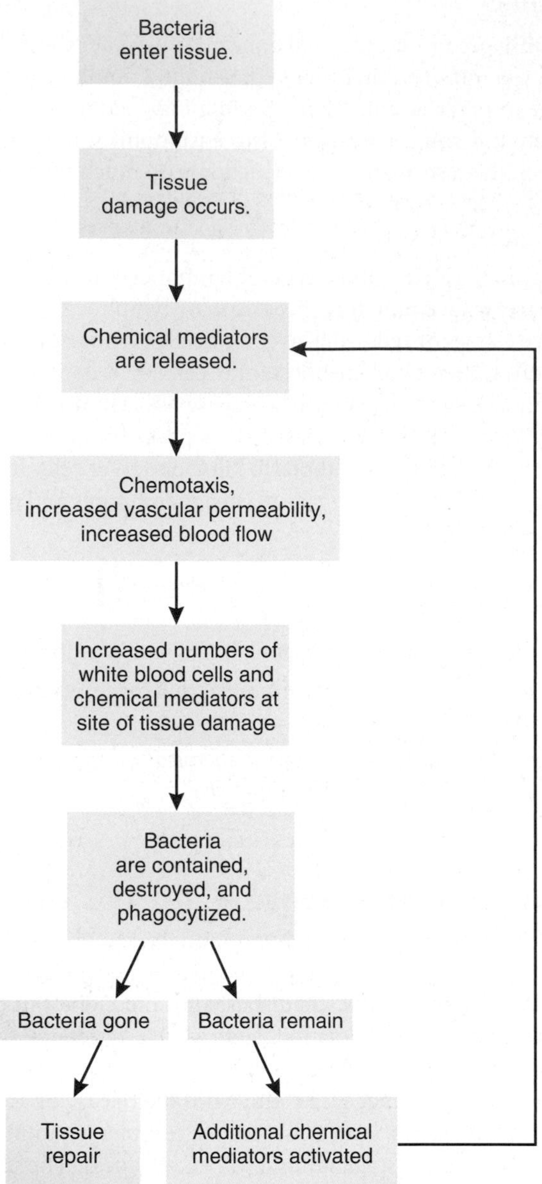
Bacteria enter tissue.
Tissue damage occurs.
Chemical mediators are released.
Chemotaxis, increased vascular permeability, increased blood flow
Increased numbers of white blood cells and chemical mediators at site of tissue damage
Bacteria are contained, destroyed, and phagocytized.
Bacteria gone
Bacteria remain
Tissue repair
Additional chemical mediators activated

FIGURE 22.11 APIR **Inflammatory Response**
Bacteria cause tissue damage and the release of chemical mediators, which initiate inflammation and phagocytosis, resulting in the destruction of the bacteria. If any bacteria remain, additional chemical mediators are activated. After all the bacteria have been destroyed, the tissue is repaired.

Systemic inflammation is an inflammatory response that occurs in many parts of the body. In addition to the local symptoms at the sites of inflammation, three additional features can be present:

1. Red bone marrow produces and releases large numbers of neutrophils, which promote phagocytosis.
2. **Pyrogens** (pī′rō-jenz; fire-producing), chemicals released by microorganisms, macrophages, neutrophils, and other cells, stimulate fever production. Pyrogens affect the body's temperature-regulating mechanism in the hypothalamus. As a consequence, heat production and heat conservation increase, raising body temperature. Fever promotes the activities of the immune system, such as phagocytosis, and inhibits the growth of some microorganisms.
3. In severe cases of systemic inflammation, increased vascular permeability is so widespread that large amounts of fluid are lost from the blood into the tissues. The decreased blood volume can cause shock and death.

ASSESS **YOUR PROGRESS**

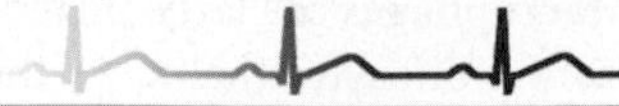

29. *What kinds of tissue damage can result in inflammation?*
30. *Describe the events that take place during an inflammatory response.*
31. *What are the symptoms of local inflammation and of systemic inflammation?*

22.5 Adaptive Immunity

LEARNING OUTCOMES

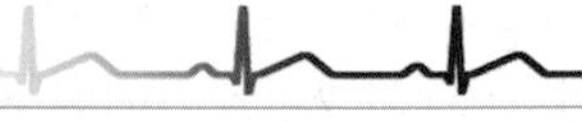

After reading this section, you should be able to

A. Define *antigen* and describe the two groups of antigens.
B. Explain the role of haptens in allergic reactions.
C. Describe the origin, development, activation, proliferation, and inhibition of lymphocytes.
D. Describe the function of major histocompatibility complex (MHC) molecules in immunity.
E. Distinguish between MHC class I molecules and MHC class II molecules.
F. Define *antibody-mediated immunity* and *cell-mediated immunity* and name the cells responsible for each.
G. Diagram the structure of an antibody and describe the effects produced by antibodies.
H. Discuss the primary and secondary responses to an antigen and explain the basis for long-lasting immunity.
I. Describe the types and functions of T cells.

Adaptive immunity can recognize, respond to, and remember a particular substance. Substances that stimulate adaptive immunity are called **antigens** (an′ti-jenz). They are usually large molecules with a molecular weight of 10,000 or more.

Antigens are divided into two groups: foreign antigens and self-antigens. **Foreign antigens** are not produced by the body but are introduced from outside it. Components of bacteria, viruses, and other microorganisms are examples of foreign antigens. Other foreign antigens include pollen, animal dander (scaly, dried skin), feces of house dust mites, foods, and drugs. Many of these trigger an **allergic reaction,** an overreaction of the immune system in some people. Transplanted tissues and organs that contain foreign antigens cause rejection of the transplant. **Self-antigens** are molecules the body produces to stimulate an adaptive immune system response. The response to self-antigens can be beneficial or harmful. For example, the recognition of tumor antigens can result in tumor destruction, whereas **autoimmune disease** can develop when self-antigens stimulate unwanted tissue destruction. An example is rheumatoid arthritis, which destroys the tissues within joints.

Adaptive immunity can be divided into antibody-mediated immunity and cell-mediated immunity. **Antibody-mediated immunity** involves proteins called **antibodies,** which are found in fluids outside cells, such as the plasma of blood, interstitial fluid, and lymph. B cells give rise to cells that produce antibodies. **Cell-mediated immunity** involves the actions of a second type of lymphocyte, called T cells. Several subpopulations of T cells exist, each responsible for a particular aspect of cell-mediated immunity. For example, **cytotoxic T cells** are responsible for producing the effects of cell-mediated immunity. **Helper T cells** and **regulatory T cells** can promote or inhibit the activities of both antibody-mediated immunity and cell-mediated immunity.

Table 22.3 summarizes and contrasts the main features of innate immunity and the two categories of adaptive immunity.

ASSESS YOUR PROGRESS

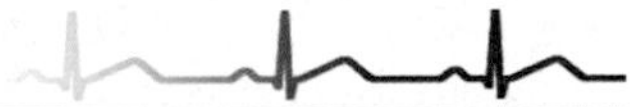

32. *Define* antigen. *Distinguish between a foreign antigen and a self-antigen.*

33. *What are allergic reactions and autoimmune diseases?*

34. *What is a hapten? How can a hapten cause an allergic reaction?*

35. *What are the two types of adaptive immunity?*

Clinical
IMPACT

Haptens and Allergic Reactions

Haptens (hap′tenz), often referred to as incomplete antigens, are small molecules (of low molecular weight) that can combine with large molecules, such as blood proteins, to stimulate an adaptive immune response. In many cases, however, haptens lead to allergic reactions (see the Diseases and Disorders table, later in this chapter). For example, penicillin, a common antibiotic prescribed to combat bacterial infections, is a hapten that can break down and bind to other molecules in the blood. The combined molecule can then stimulate an allergic reaction that ranges from a rash and fever to severe symptoms that can lead to death. It is estimated that 20% of patients have allergic reactions when administered penicillin. Research indicates that the likelihood of a reaction increases with subsequent prescriptions. Skin tests are available to determine a patient's susceptibility to an allergic reaction to penicillin.

Origin and Development of Lymphocytes

All blood cells, including lymphocytes, are derived from stem cells in the red bone marrow (see chapter 19). The process of blood cell formation begins during embryonic development and continues throughout life. Some stem cells give rise to pre-T cells, which migrate through the blood to the thymus, where they divide and are processed into T cells (figure 22.12; see figure 22.9). The thymus produces hormones, such as **thymosin,** which stimulate T-cell maturation. Other stem cells produce pre-B cells, which are processed in the red bone marrow into B cells. A **positive selection** process results in the survival of pre-B and pre-T cells that are capable of an immune response. Cells that are incapable of an immune response die.

TABLE 22.3 Comparison of Innate and Adaptive Immunity

		ADAPTIVE IMMUNITY	
Characteristics	**INNATE IMMUNITY**	**Antibody-Mediated Immunity**	**Cell-Mediated Immunity**
Primary cells	Neutrophils, eosinophils, basophils, mast cells, monocytes, and macrophages	B cells	T cells
Origin of cells	Red bone marrow	Red bone marrow	Red bone marrow
Site of maturation	Red bone marrow (neutrophils, eosinophils, basophils, and monocytes) and tissues (mast cells and macrophages)	Red bone marrow	Thymus
Location of mature cells	Blood, connective tissue, and lymphatic tissue	Blood and lymphatic tissue	Blood and lymphatic tissue
Primary secretory products	Histamine, kinins, complement, prostaglandins, leukotrienes, and interferons	Antibodies	Cytokines
Primary actions	Inflammatory response and phagocytosis	Protection against extracellular antigens (bacteria, toxins, parasites, and viruses outside cells)	Protection against intracellular antigens (viruses, intracellular bacteria, and intracellular fungi) and tumors; regulates antibody-mediated immunity and cell-mediated immunity responses (helper T and regulatory T cells)
Hypersensitivity reactions	None	Immediate hypersensitivity (atopy, anaphylaxis, cytotoxic reactions, and immune complex disease)	Delayed hypersensitivity (allergic reaction to infection or contact hypersensitivity)

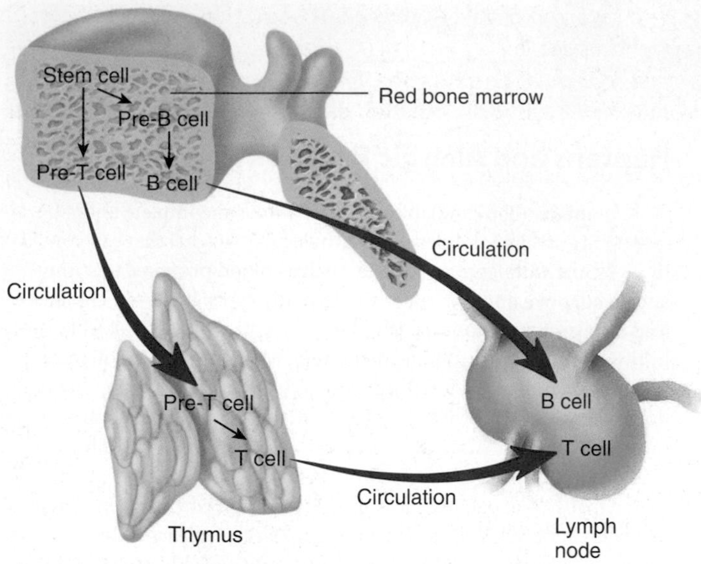
Stem cell
Pre-B cell
Pre-T cell
B cell
Red bone marrow
Circulation
Circulation
Pre-T cell
T cell
B cell
T cell
Circulation
Thymus
Lymph node

FIGURE 22.12 APR **Origin and Processing of B Cells and T Cells**
Pre-B cells and pre-T cells originate from stem cells in red bone marrow. The pre-B cells remain in the red bone marrow and become B cells. The pre-T cells circulate to the thymus, where they become T cells. Both B cells and T cells circulate to other lymphatic tissues, such as lymph nodes, where they can divide and increase in number in response to antigens.

The B cells and T cells that can respond to antigens are composed of small groups of identical lymphocytes called **clones.** Although each clone can respond only to a particular antigen, such a large number of clones exist that the immune system can react to most molecules. Some of the clones can also respond to self-antigens. A **negative selection** process eliminates or suppresses clones acting against self-antigens, thereby preventing the destruction of a person's own cells. Although the negative selection process occurs mostly during prenatal development, it continues throughout life (see "Inhibition of Lymphocytes," later in this section).

B cells are released from red bone marrow, T cells are released from the thymus, and both types of cells move through the blood to lymphatic tissue. There are approximately five T cells for every B cell in the blood. These lymphocytes live for a few months to many years and continually circulate between the blood and the lymphatic tissues. Antigens can come into contact with and activate lymphocytes, resulting in cell divisions that increase the number of lymphocytes able to recognize the antigen. These lymphocytes can circulate in blood and lymph to reach antigens in tissues throughout the body.

Lymphocytes mature into functional cells in the **primary lymphatic organs,** which are the red bone marrow and thymus. In the **secondary lymphatic organs and tissues,** lymphocytes interact with each other, antigen-presenting cells, and antigens to produce an immune response. The secondary lymphatic organs and tissues include the diffuse lymphatic tissue, lymphatic nodules, tonsils, lymph nodes, and spleen.

Activation of Lymphocytes

Antigens activate lymphocytes in different ways, depending on the type of lymphocyte and the type of antigen involved. Despite these differences, however, two general principles of lymphocyte activation exist: (1) Lymphocytes must be able to recognize the antigen; (2) after recognition, the lymphocytes must increase in number to destroy the antigen.

Antigenic Determinants and Antigen Receptors

If an adaptive immune system response is to occur, lymphocytes must recognize an antigen. However, lymphocytes do not interact with an entire antigen. Instead, **antigenic determinants,** or *epitopes* (ep′i-tōps), are specific regions of a given antigen recognized by a lymphocyte, and each antigen has many different antigenic determinants (figure 22.13). All the lymphocytes of a given clone have, on their surfaces, identical proteins called **antigen receptors,** which combine with a specific antigenic determinant. The immune system response to an antigen with a particular antigenic determinant is similar to the lock-and-key model for enzymes (see chapter 2), and any given antigenic determinant can combine only with a specific antigen receptor. The **T-cell receptor** consists of two polypeptide chains, which are subdivided into a variable and a constant region (figure 22.14). The variable region can bind to an antigen. The many different types of T-cell receptors respond to different antigens because they have different variable regions. The **B-cell receptor,** consisting of four polypeptide chains with two identical variable regions, is a type of antibody.

Major Histocompatibility Complex Molecules

Although some antigens bind to their receptors and directly activate B cells and some T cells, most lymphocyte activation involves glycoproteins on the surfaces of cells, called **major histocompatibility complex (MHC) molecules.** MHC molecules are glycoproteins found on the plasma membranes of most of the body's cells. Each MHC molecule has a variable region that can bind to foreign and self-antigens. The immune system cannot directly respond to an antigen inside a cell. Instead, MHC molecules display antigens produced or processed inside the cell on the cell's surface. Two classes of MHC molecules are present in the body: MHC class I and MHC class II.

MHC class I molecules are found on nucleated cells; they display antigens produced inside the cell on the cells surface (figure 22.15*a*).

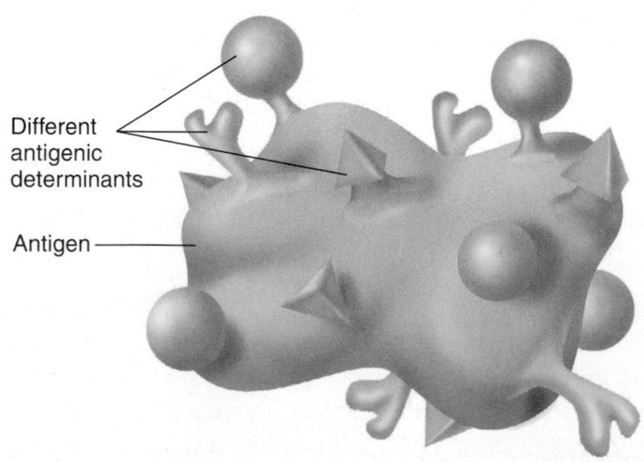
Different
antigenic
determinants
Antigen

FIGURE 22.13 Antigenic Determinants
An antigen has many antigenic determinants to which lymphocytes can respond.

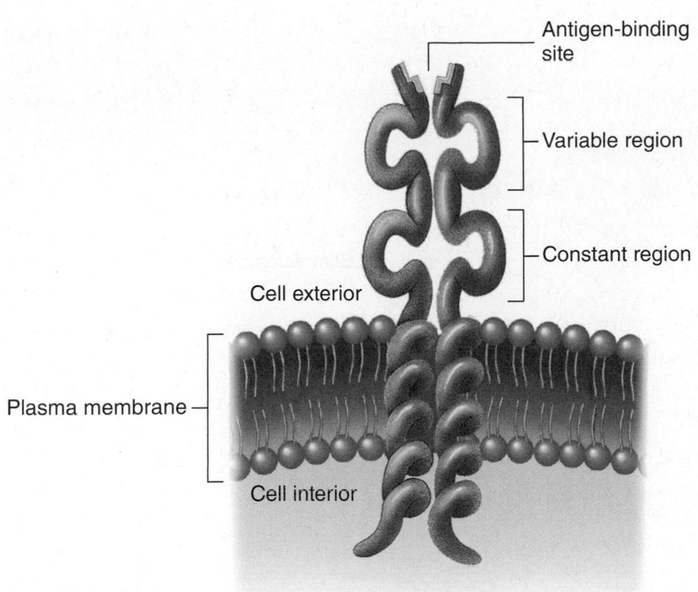

Antigen-binding site
Variable region
Constant region
Cell exterior
Plasma membrane
Cell interior

FIGURE 22.14 T-Cell Receptor
A T-cell receptor consists of two polypeptide chains. The variable region of each type of T-cell receptor is specific for a given antigen. The constant region attaches the T-cell receptor to the plasma membrane.

For example, viruses reproduce inside a cell, forming viral proteins that are foreign antigens. Some of these viral proteins are broken down in the cytoplasm. The protein fragments enter the rough endoplasmic reticulum and combine with MHC class I molecules to form complexes that move through the Golgi apparatus to be distributed on the cell's surface (see chapter 3).

MHC class I/antigen complexes on the surface of cells can bind to T-cell receptors on the surface of T cells. This combination is a signal that activates T cells. Activated T cells can destroy infected cells, which effectively stops viral replication (see "Cell-Mediated Immunity," later in this section). Thus, the MHC class I/antigen complex functions as a signal, or "red flag," that prompts the immune system to destroy the displaying cell. In essence, the cell is displaying a sign that says, "Kill me!" This process is said to be **MHC-restricted** because both the antigen and the individual organism's own MHC molecule are required.

Predict 3

In mouse A, T cells can respond to virus X. If these T cells are transferred to mouse B, which is infected with virus X, will the T cells respond to the virus? Explain.

The same process that moves foreign protein fragments to the cell's surface can also inadvertently transport self-protein fragments (figure 22.15*a*). As part of normal protein metabolism, cells continually break down old proteins and synthesize new ones. Some self-protein fragments that result from protein breakdown can combine with MHC class I molecules and be displayed on the surface of the cell, thus becoming self-antigens. Normally, the immune system does not respond to self-antigens in combination with MHC molecules because the lymphocytes that could respond have been eliminated or inactivated (see "Inhibition of Lymphocytes," later in this section).

MHC class II molecules are found on **antigen-presenting cells,** which include B cells, macrophages, monocytes, and dendritic cells. **Dendritic** (den-drit′ik) **cells** are large, motile cells with long cytoplasmic extensions. These cells are scattered throughout most tissues (except the brain), with their highest concentrations in lymphatic tissues and the skin. Dendritic cells in the skin are often called **Langerhans cells.**

Antigen-presenting cells can take in foreign antigens by endocytosis (figure 22.15*b*). Within the endocytotic vesicle, the antigen is broken down into fragments to form processed antigens. Vesicles from the Golgi apparatus containing MHC class II molecules combine with the endocytotic vesicles. The MHC class II molecules and processed antigens combine, and the MHC class II/antigen complexes are transported to the cell's surface, where they are displayed to other immune cells.

MHC class II/antigen complexes on the cell's surface can bind to T-cell receptors on the surfaces of T cells. The presentation of antigen using MHC class II molecules is MHC-restricted because both the antigen and the individual's own MHC class II molecule are required. Unlike MHC class I molecules, however, this display does not result in the destruction of the antigen-presenting cell. Instead, the MHC class II/antigen complex is a "rally around the flag" signal that stimulates other immune system cells to respond to the antigen. The displaying cell is like Paul Revere, who spread the alarm for the militia to arm and organize. The militia then went out and killed the enemy. For example, when the lymphocytes of the B-cell clone that can recognize the antigen come into contact with the MHC class II/antigen complex, they are stimulated to divide. The activities of these lymphocytes, such as the production of antibodies, then destroy the antigen.

MHC molecules are genetically determined. Due to the complexity of the genes involved in MHC protein production, the variation in MHC molecules is great in the population. As a result, it is rare for two people to have exactly the same MHC molecules, except in the case of identical twins. MHC molecule differences are of major concern when tissue is transplanted from one individual to another. Genetically similar individuals (siblings or parent and offspring) are more likely to have similar MHC molecules and, therefore, a better "match" for tissue transplants (see Clinical Impact, "Transplant Rejection").

ASSESS YOUR PROGRESS

36. *Describe the origin and development of B cells and T cells.*
37. *What are lymphocyte clones? What is the difference between positive and negative lymphocyte selection?*
38. *What are the primary lymphatic organs? What are the secondary lymphatic organs and tissues?*
39. *Define* antigenic determinant *and* antigen receptor. *How are they related to each other?*
40. *What types of cells display MHC class I and class II antigen complexes, and what happens as a result?*
41. *What types of antigens are displayed by MHC class I molecules? By MHC class II molecules?*
42. *What does MHC-restricted mean?*

FUNDAMENTAL **Figure**

1. Foreign proteins or self-proteins within the cytosol are broken down into fragments that are antigens.
2. Antigens are transported into the rough endoplasmic reticulum.
3. Antigens combine with MHC class I molecules.
4. The MHC class I/antigen complex is transported to the Golgi apparatus, packaged into a vesicle, and transported to the plasma membrane.
5. Foreign antigens combined with MHC class I molecules stimulate cell destruction.
6. Self-antigens combined with MHC class I molecules do not stimulate cell destruction.

1. A foreign antigen is ingested by endocytosis and is within a vesicle.
2. The antigen is broken down into fragments to form processed foreign antigens.
3. The vesicle containing the processed foreign antigens fuses with vesicles produced by the Golgi apparatus that contain MHC class II molecules. Processed foreign antigens and MHC class II molecules combine.
4. The MHC class II/antigen complex is transported to the plasma membrane.
5. The displayed MHC class II/antigen complex can stimulate immune cells.

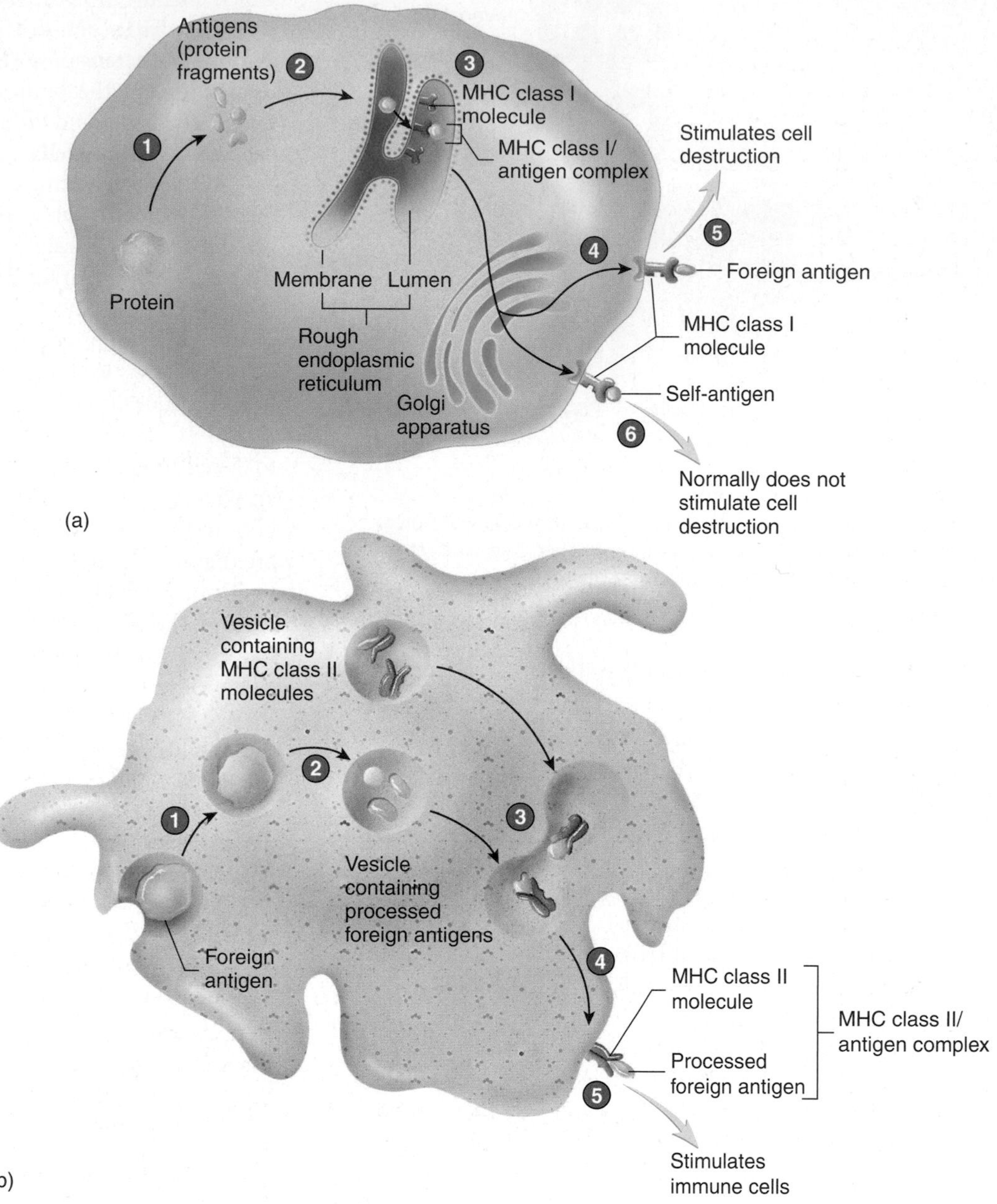
Antigens (protein fragments)
MHC class I molecule
MHC class I/ antigen complex
Stimulates cell destruction
Foreign antigen
MHC class I molecule
Self-antigen
Protein
Membrane
Lumen
Rough endoplasmic reticulum
Golgi apparatus
Normally does not stimulate cell destruction
(a)
Vesicle containing MHC class II molecules
Vesicle containing processed foreign antigens
Foreign antigen
MHC class II molecule
Processed foreign antigen
MHC class II/ antigen complex
Stimulates immune cells
(b)

PROCESS FIGURE 22.15 AP|R **Antigen Processing**

(*a*) Foreign proteins, such as viral proteins, or self-proteins in the cytosol are processed and presented at the cell surface by MHC class I molecules.
(*b*) Foreign antigens are taken into an antigen-presenting cell, processed, and presented at the cell surface by MHC class II molecules.

> **Predict 4**
>
> Antibodies bind to a foreign antigen, resulting in removal of that foreign antigen from the body. Explain what happens to antibody production as the foreign antigens decrease.

Costimulation

The combination of an MHC class II/antigen complex with an antigen receptor is usually only the first signal necessary to produce a response from a B cell or a T cell. In many cases, **costimulation** by additional signals is also required. Costimulation is accomplished by cytokines released from cells and by molecules attached to the surfaces of cells (figure 22.16*a*). Cytokines produced by lymphocytes are often called **lymphokines** (lim′fō-kīnz). Table 22.4 lists important cytokines and their functions.

Certain pairs of surface molecules can also be involved in costimulation (figure 22.16*b*). When the surface molecule on one cell combines with the surface molecule on another, the combination

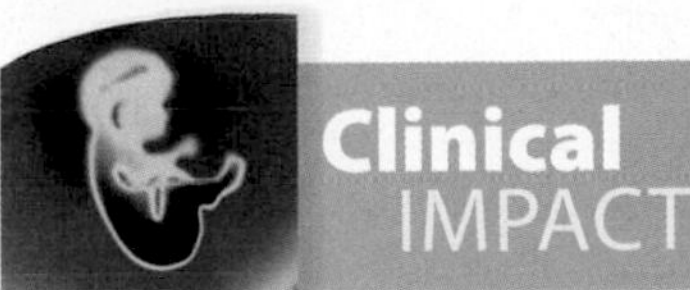

Clinical IMPACT

Transplant Rejection

Genes that code for the production of MHC molecules are generally called major histocompatibility complex genes. Histocompatibility is the tissues' ability (Gr. *histo*) to get along (compatibility) when tissues are transplanted from one individual to another. In humans, the major histocompatibility complex genes are often referred to as **human leukocyte antigen (HLA) genes** because they were first identified in leukocytes. The HLA genes control the production of MHC antigens, which are found on the plasma membrane of cells. Millions of possible combinations of the HLA genes exist, and it is very rare for two individuals (except identical twins) to have the same set of HLA genes. The closer the relationship between two people, the greater the likelihood they share the same HLA genes.

The immune system can distinguish between self cells and foreign cells because they are both marked with MHCs. Rejection of a transplanted tissue is caused by a normal immune system response to the foreign MHCs. **Acute rejection** occurs several weeks after transplantation and results from a delayed hypersensitivity reaction and cell lysis. Lymphocytes and macrophages infiltrate the area, a strong inflammatory response occurs, and the foreign tissue is destroyed. If acute rejection does not develop, **chronic rejection** may occur at a later time. In chronic rejection, immune complexes form in the arteries supplying the graft, the blood supply fails, and the transplanted tissue is rejected.

Graft rejection can occur in two different directions. In **host-versus-graft rejection,** the recipient's immune system recognizes the donor's tissue as foreign and rejects the transplant. In **graft-versus-host rejection,** the donor tissue recognizes the recipient's tissue as foreign, and the transplant rejects the recipient, causing destruction of the recipient's tissues and death.

To reduce graft rejection, a tissue match is performed. Only tissues with MHCs similar to the recipient's have a chance of acceptance. Even when the match is close, immunosuppressive drugs must be administered throughout the person's life to prevent rejection. Unfortunately, the person then has a drug-produced immunodeficiency and is more susceptible to infections. An exact match is possible only for a graft from one part to another part of a person's body or between identical twins.

can act as a signal that stimulates one of the cells to respond, or the combination can hold the cells together. Typically, several kinds of surface molecules are necessary to produce a response. For example, a molecule called B7 on macrophages must bind with a molecule called CD28 on helper T cells before the helper T cells can respond to the antigen presented by the macrophage. In addition, helper T cells have a glycoprotein called CD4, which helps connect helper T cells to the macrophage by binding to MHC class II molecules. For this reason, helper T cells are sometimes referred to as *CD4 cells* or *T4 cells.* In a similar fashion, cytotoxic T cells are sometimes called *CD8 cells* or *T8 cells* because they have a glycoprotein called CD8, which helps connect cytotoxic T cells to cells displaying MHC class I molecules. The CD designation stands for "cluster of differentiation," which is a system used to classify many surface molecules.

Lymphocyte Proliferation

Before exposure to an antigen, the number of lymphocytes in a clone is too small to produce an effective response against the antigen. Exposure to an antigen results in an increase in lymphocyte number. First, the number of helper T cells increases. This is important because the increased number of helper T cells responding to the antigen can find and stimulate B cells or cytotoxic T cells. Second, the number of B cells or cytotoxic T cells increases. This is important because these cells are responsible for the immune response that destroys the antigen.

1. *Proliferation of helper T cells* (figure 22.17). Antigen-presenting cells use MHC class II molecules to present processed antigens to helper T cells. Only the helper T cells with the T-cell receptors that can bind to the antigen respond. These helper T cells respond to the MHC class II/antigen complex and costimulation by dividing. As a result, the number of helper T cells that recognize the antigen increases.
2. *Proliferation and activation of B cells or cytotoxic T cells.* Typically, the proliferation and activation of B cells or cytotoxic T cells involve helper T cells. This process is illustrated in figure 22.18 for B cells (see "Cell-Mediated Immunity," later in this section, for activation of cytotoxic T cells). The clone of B cells that can recognize a particular antigen has B-cell receptors that can bind to that antigen. The antigens and receptors enter the B cell by receptor-mediated endocytosis. The antigens break down into fragments to form processed antigens that combine with MHC class II molecules. B cells use MHC class II/antigen complexes to present antigens to the helper T cells that increased in number in response to the same antigen. These helper T cells stimulate the B cells to divide. Many of the resulting daughter cells differentiate to become specialized cells, called **plasma cells,** which produce antibodies. The increased number of cells, each producing antibodies, can produce an immune response that destroys the antigens (see "Effects of Antibodies," later in this section).

Inhibition of Lymphocytes

Tolerance is a state of unresponsiveness of lymphocytes to a specific antigen. The most important function of tolerance is to prevent the immune system from responding to self-antigens, although foreign antigens can also induce tolerance. The need to maintain tolerance and to avoid the development of autoimmune disease is obvious. Tolerance can be induced in three primary ways:

1. *Deletion of self-reactive lymphocytes.* During prenatal development and after birth, stem cells in red bone marrow and the thymus give rise to immature lymphocytes that develop into mature lymphocytes capable of an immune response. When immature lymphocytes are exposed to antigens, instead of responding in ways that cause elimination of the antigen, they respond by dying. Because immature lymphocytes are exposed to self-antigens, this process eliminates self-reactive lymphocytes. In addition, immature lymphocytes that escape deletion during their development and become mature, self-reacting lymphocytes can still be deleted most likely by the activities of regulatory T cells.

(a) Helper T cells are activated by a first signal and by costimulation. The first signal is the binding of the MHC class II/ antigen complex to the T-cell receptor. Costimulation is an additional signal, such as molecules released from another cell. For example, macrophages release cytokines that bind to receptors on helper T cells, resulting in costimulation.

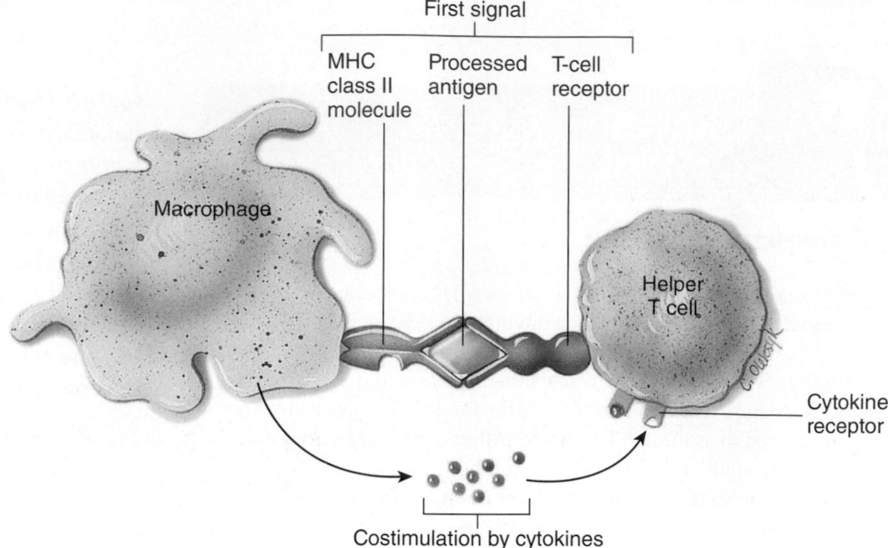
First signal
MHC class II molecule
Processed antigen
T-cell receptor
Macrophage
Helper T cell
Cytokine receptor
Costimulation by cytokines

(b) Other costimulatory signals are the combining of surface molecules between cells, such as the binding of a B7 molecule of the macrophage with a CD28 molecule of the helper T cell. The CD4 molecule of the helper T cell binds to the macrophage's MHC class II molecule and helps hold the cells together.

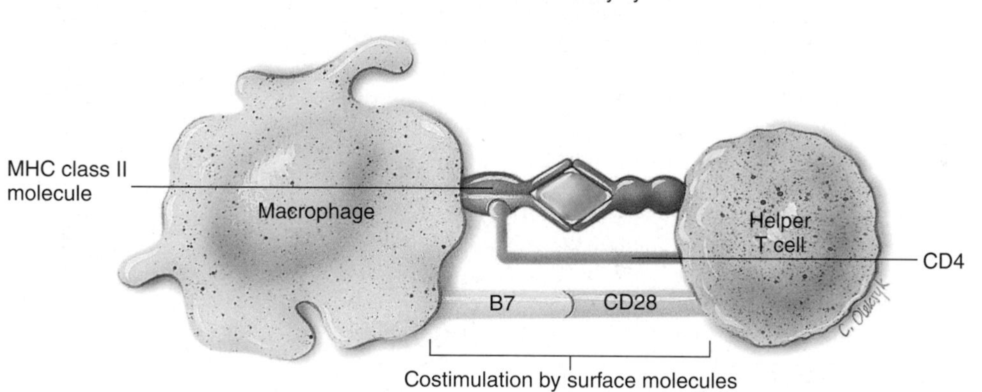
MHC class II molecule
Macrophage
Helper T cell
CD4
B7
CD28
Costimulation by surface molecules

FIGURE 22.16 Costimulation

TABLE 22.4 Cytokines and Their Functions

Cytokine*	Description
Interferon alpha (IFNα)	Prevents viral replication and inhibits cell growth; secreted by virus-infected cells
Interferon beta (IFNβ)	Prevents viral replication, inhibits cell growth, and decreases the expression of major histocompatibility complex (MHC) class I and II molecules; secreted by virus-infected fibroblasts
Interferon gamma (IFNγ)	About 20 different proteins that activate macrophages and natural killer (NK) cells, stimulate adaptive immunity by increasing the expression of MHC class I and II molecules, and prevent viral replication; secreted by helper T, cytotoxic T, and NK cells
Interleukin-1 (IL-1)	Costimulation of B cells and T cells; promotes inflammation through prostaglandin production and induces fever acting through the hypothalamus (pyrogen); secreted by macrophages, B cells, and fibroblasts
Interleukin-2 (IL-2)	Costimulation of B cells and T cells and activation of macrophages and NK cells; secreted by helper T cells
Interleukin-4 (IL-4)	Plays a role in allergic reactions by activation of B cells, resulting in the production of immunoglobulin E (IgE); secreted by helper T cells
Interleukin-5 (IL-5)	Part of the response against parasites by stimulating eosinophil production; secreted by helper T cells
Interleukin-8 (IL-8)	Chemotactic factor that promotes inflammation by attracting neutrophils and basophils; secreted by macrophages
Interleukin-10 (IL-10)	Inhibits the secretion of interferon gamma and interleukins; secreted by regulatory T cells
Interleukin-15 (IL-15)	Promotes inflammation and activates memory T cells and natural killer cells
Lymphotoxin	Kills target cells; secreted by cytotoxic T cells
Perforin	Makes a hole in the membrane of target cells, resulting in lysis of the cell; secreted by cytotoxic T cells
Tumor necrosis factor α (TNFα)	Activates macrophages and promotes fever (pyrogen); secreted by macrophages

*Some cytokines were named according to the laboratory test first used to identify them; however, these names are rarely an accurate description of the actual function of the cytokine.

1. Antigen-presenting cells, such as macrophages, take in, process, and display antigens on the cell's surface.
2. The antigen is bound to an MHC class II molecule, which functions to present the processed antigen to a T-cell receptor of a helper T cell for recognition.
3. Costimulation occurs by a CD4 glycoprotein of the helper T cell or by cytokines. The macrophage secretes a cytokine called interleukin-1.
4. Interleukin-1 attaches to interleukin receptors and stimulates the helper T cell to secrete the cytokine interleukin-2 and to produce interleukin-2 receptors.
5. The helper T cell stimulates itself to divide when interleukin-2 binds to interleukin-2 receptors.
6. The "daughter" helper T cells resulting from this division can be stimulated to divide again if they are exposed to the same antigen that stimulated the "parent" helper T cell. This greatly increases the number of helper T cells.
7. The increased number of helper T cells can facilitate the activation of B cells or effector T cells.

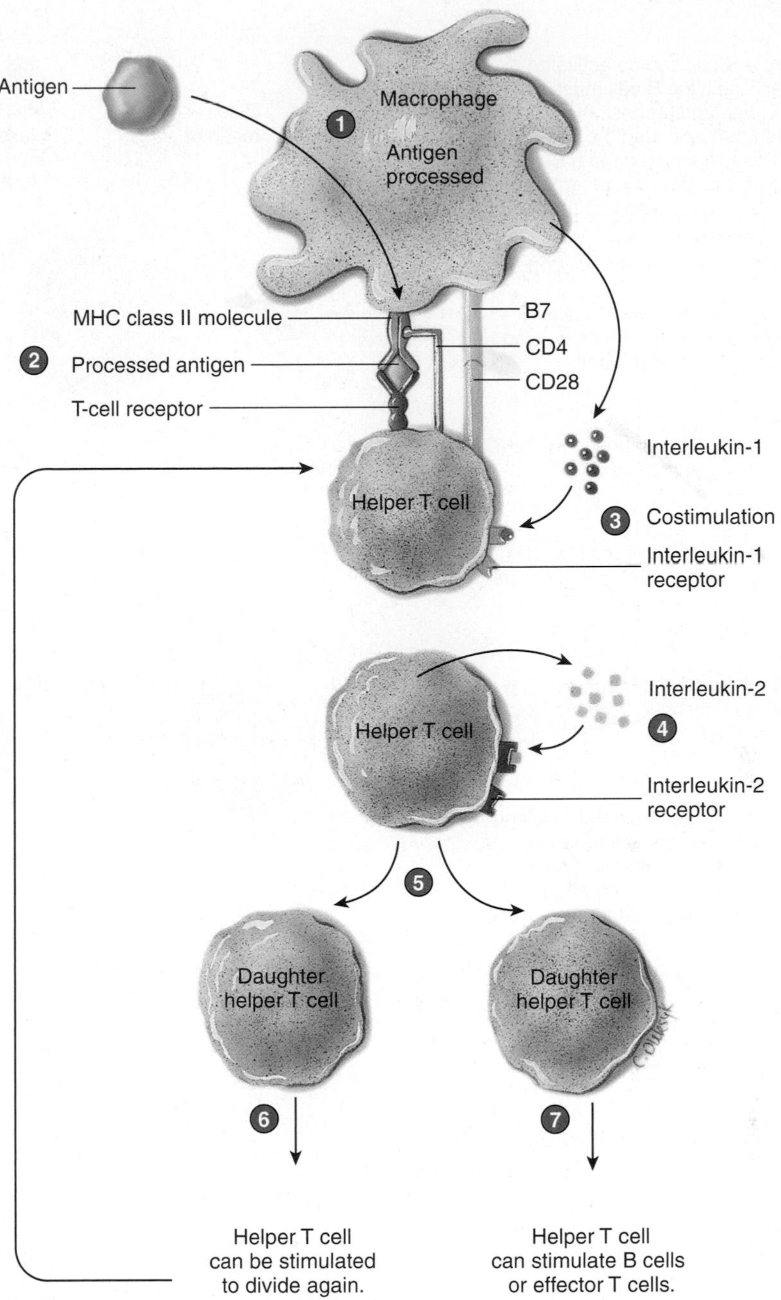
Antigen
1
Macrophage
Antigen processed
MHC class II molecule
2
Processed antigen
T-cell receptor
B7
CD4
CD28
Interleukin-1
Helper T cell
3
Costimulation
Interleukin-1 receptor
Interleukin-2
Helper T cell
4
Interleukin-2 receptor
5
Daughter helper T cell
Daughter helper T cell
6
7
Helper T cell can be stimulated to divide again.
Helper T cell can stimulate B cells or effector T cells.

PROCESS FIGURE 22.17 Proliferation of Helper T Cells
An antigen-presenting cell (macrophage) stimulates helper T cells to divide and produce cytokines.

2. *Prevention of the activation of lymphocytes.* For lymphocytes to be activated, two signals are usually required: (1) the MHC/antigen complex binding with an antigen receptor and (2) costimulation. Preventing either of these events stops lymphocyte activation. For example, blocking, altering, or deleting an antigen receptor prevents activation. **Anergy** (an′er-jē; without working) is a condition of inactivity in which a B cell or T cell does not respond to an antigen. Anergy develops when an MHC-antigen complex binds to an antigen receptor and no costimulation occurs. For example, if a helper T cell encounters a self-antigen on a cell that cannot provide costimulation, the T cell is turned off. It is likely that only antigen-presenting cells can provide costimulation.
3. *Activation of regulatory T cells.* **Regulatory T cells,** also called *suppressor T cells*, are a poorly understood group of T cells that are defined by their ability to suppress immune responses. Regulatory T cells develop in the thymic corpuscles and then enter the tissues of the body. Regulatory T cells reduce adaptive immune response by releasing cytokines that inhibit helper T cells and cytotoxic T cells, thereby regulating the activities of both antibody-mediated immunity and cell-mediated immunity. Decreasing the production or activity of cytokines can suppress

1 Before a B cell can be activated by a helper T cell, the B cell must endocytize and process the same antigen that activated the helper T cell. The antigen binds to a B-cell receptor, and both the receptor and the antigen are taken into the cell by endocytosis.

2 The B cell uses an MHC class II molecule to present the processed antigen to the helper T cell.

3 The T-cell receptor binds to the MHC class II/antigen complex.

4 Costimulation of the B cell by CD4 and other surface molecules occurs.

5 Costimulation by interleukins (cytokines) released from the helper T cell occurs.

6 The B cell divides, and the resulting daughter cells divide, and so on, eventually producing many cells (only two are shown here) that recognize the same antigen.

7 Many of the daughter cells differentiate to become plasma cells, which produce antibodies. Antibodies are part of the immune response that eliminates the antigen.

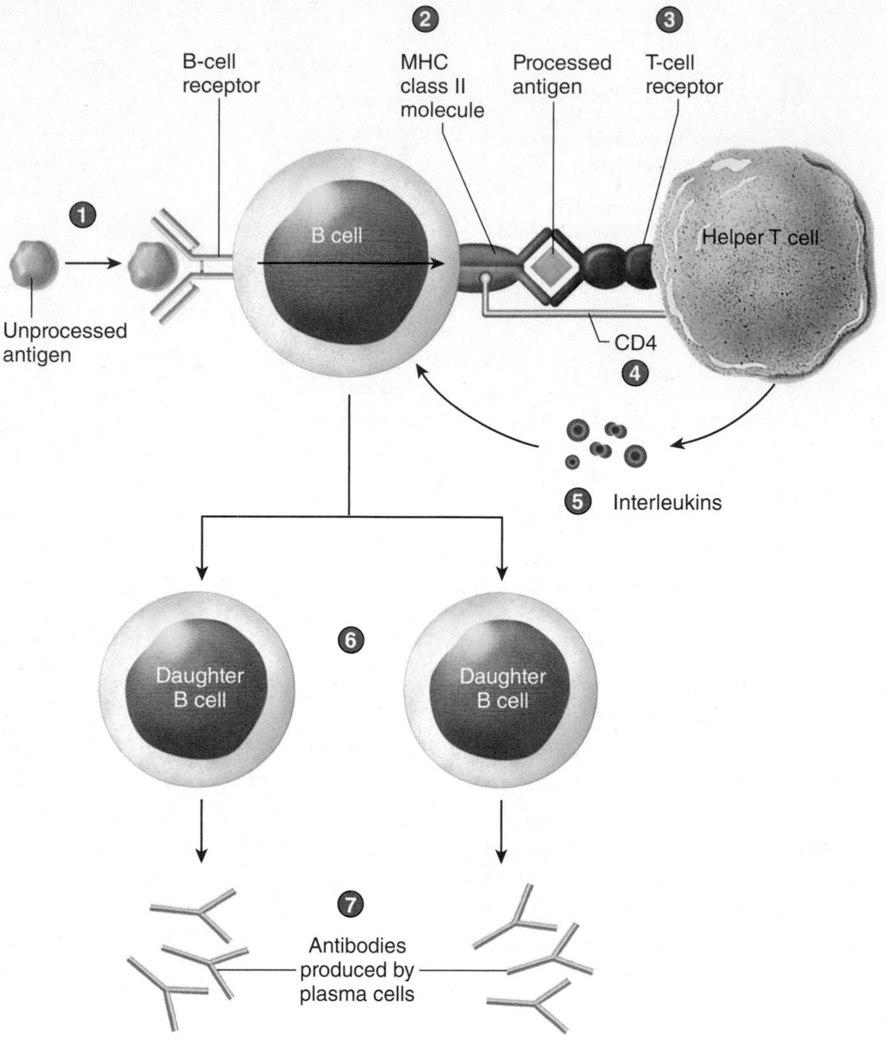
B-cell receptor
2
MHC class II molecule
Processed antigen
3
T-cell receptor
1
B cell
Helper T cell
Unprocessed antigen
CD4
4
5 Interleukins
Daughter B cell
6
Daughter B cell
7
Antibodies produced by plasma cells

PROCESS FIGURE 22.18 Proliferation of B Cells
A helper T cell stimulates a B cell to divide and produce antibodies.

the immune system. For example, cyclosporine, a drug used to prevent the rejection of transplanted organs, inhibits the production of interleukin-2. Conversely, genetically engineered interleukins can be used to stimulate the immune system. Administering interleukin-2 has promoted the destruction of cancer cells in some cases by increasing the activities of cytotoxic T cells.

ASSESS YOUR PROGRESS

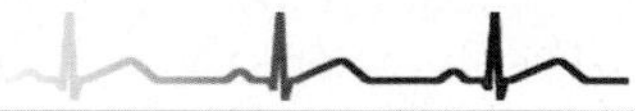

43. *What is costimulation? State two ways it can happen.*

44. *Why are helper T cells sometimes called CD4 or T4 cells? Why are cytotoxic T cells sometimes called CD8 or T8 cells?*

45. *Describe how antigen-presenting cells stimulate an increase in the number of helper T cells. Why is this important?*

46. *Describe how helper T cells stimulate an increase in the number of B cells or T cells. Why is this important?*

47. *What is tolerance? Explain three ways it is accomplished.*

Antibody-Mediated Immunity

Exposure of the body to an antigen can lead to the activation of B cells and to the production of antibodies, which are responsible for destroying the antigen. Because antibodies occur in body fluids, antibody-mediated immunity is effective against extracellular antigens, such as bacteria, viruses, protozoans, fungi, parasites, and toxins, when they are outside cells. Antibody-mediated immunity can also cause immediate hypersensitivity reactions (see the Diseases and Disorders table, later in this chapter).

Structure of Antibodies

Antibodies are proteins produced in response to an antigen. Large numbers of antibodies exist in plasma, although plasma also contains other proteins. On the basis of protein type and associated lipids, plasma proteins are separated into albumin and alpha-(α), beta-(β), and gamma-(γ)globulin parts. As a group, antibodies are sometimes called *gamma globulins* because they are found mostly in the γ-globulin part of plasma, or **immunoglobulins (Ig),** because they are globulin proteins involved in immunity.

The five general classes of antibodies are denoted IgG, IgM, IgA, IgE, and IgD (table 22.5). All classes of antibodies have a similar structure, consisting of four polypeptide chains: two identical heavy chains and two identical light chains (figure 22.19). Each light chain is attached to a heavy chain, and the ends of the combined heavy and light chains form the **variable region** of the antibody, which is the part that combines with the antigenic determinant of the antigen. Different antibodies have different variable regions, and they are specific for different antigens. The rest of the antibody is the **constant region,** which is responsible for the activities of antibodies, such as the ability to activate complement or to attach the antibody to cells such as macrophages, basophils, mast cells, and eosinophils. All the antibodies of a particular class have nearly the same constant regions.

Effects of Antibodies

Antibodies can directly affect antigens in two ways. The antibody can bind to the antigenic determinant and interfere with the antigen's ability to function (figure 22.20*a*). Alternatively, the antibody can combine with an antigenic determinant on two different antigens, rendering the antigens ineffective (figure 22.20*b*). The ability of antibodies to join antigens together is the basis for many clinical tests, such as blood typing, because, when enough antigens are bound together, they become visible as a clump or a precipitate.

Although antibodies can directly affect antigens, most of their effectiveness results from other mechanisms. When an antibody (IgG or IgM) combines with an antigen through the variable region, the constant region can activate the complement cascade through the classical pathway (figure 22.20*c*; see figure 22.10). Activated complement stimulates inflammation; attracts neutrophils, monocytes, macrophages, and eosinophils to sites of infection; and kills bacteria by lysis.

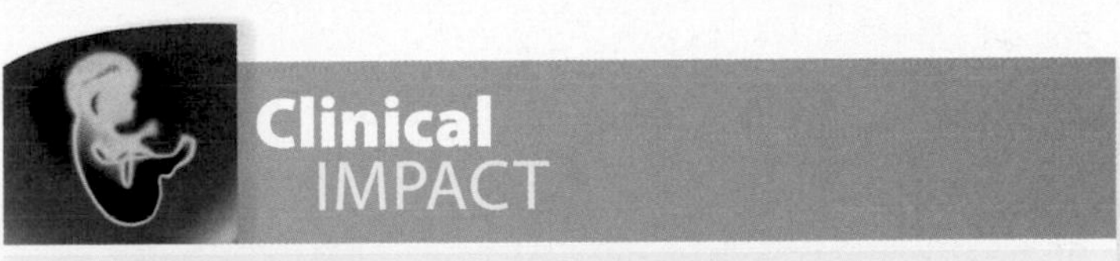
Clinical
IMPACT

Uses of Monoclonal Antibodies

A **monoclonal antibody** is a pure antibody preparation that is specific for only one antigen. A monoclonal antibody preparation can be produced by injecting a laboratory animal with a specific antigen. The antigen activates a B-cell clone against the antigen. The B cells are removed from the animal and fused with tumor cells, which divide to form large numbers of cells. The tumor cells of a given clone produce only one kind of antibody.

Monoclonal antibodies are used for determining pregnancy and for diagnosing diseases, such as gonorrhea, syphilis, hepatitis, rabies, and cancer. These tests are specific and rapid because the monoclonal antibodies bind only to the antigen being tested. Monoclonal antibodies have been used to treat some autoimmune diseases and reduce the chances of tissue rejection after transplant. They may also be used as treatments for certain cancers, such as breast cancer (see section 22.8).

Antibodies (IgE) can initiate an inflammatory response (figure 22.20*d*). The antibodies attach to mast cells or basophils through their constant region. When antigens combine with the variable region of the antibodies, the mast cells or basophils release chemicals through exocytosis, and inflammation results. For example, people who have hay fever inhale antigens (usually, plant pollen), which are then absorbed through the respiratory mucous membrane. The combination of the antigens with antibodies stimulates mast cells to release inflammatory chemicals, such as histamine. The resulting localized inflammatory response produces swelling and increased mucus production in the respiratory tract.

Opsonins (op′sŏ-ninz) are substances that make an antigen more susceptible to phagocytosis. An antibody (IgG) acts as an

TABLE 22.5 Classes of Antibodies and Their Functions

Antibody	Total Serum Antibody (%)	Description	Structure
IgG	80–85	Activates complement and promotes phagocytosis; can cross the placenta and provide immune protection to the fetus and newborn; responsible for Rh reactions, such as hemolytic disease of the newborn	IgG
IgM	5–10	Activates complement and acts as an antigen-binding receptor on the surface of B cells; responsible for transfusion reactions in the ABO blood system; often the first antibody produced in response to an antigen	IgM
IgA	15	Secreted into saliva, into tears, and onto mucous membranes to provide protection on body surfaces; found in colostrum and milk to provide immune protection to newborns	IgA
IgE	0.002	Binds to mast cells and basophils and stimulates the inflammatory response	IgE
IgD	0.2	Functions as antigen-binding receptors on B cells	IgD

Heavy chain
Light chain

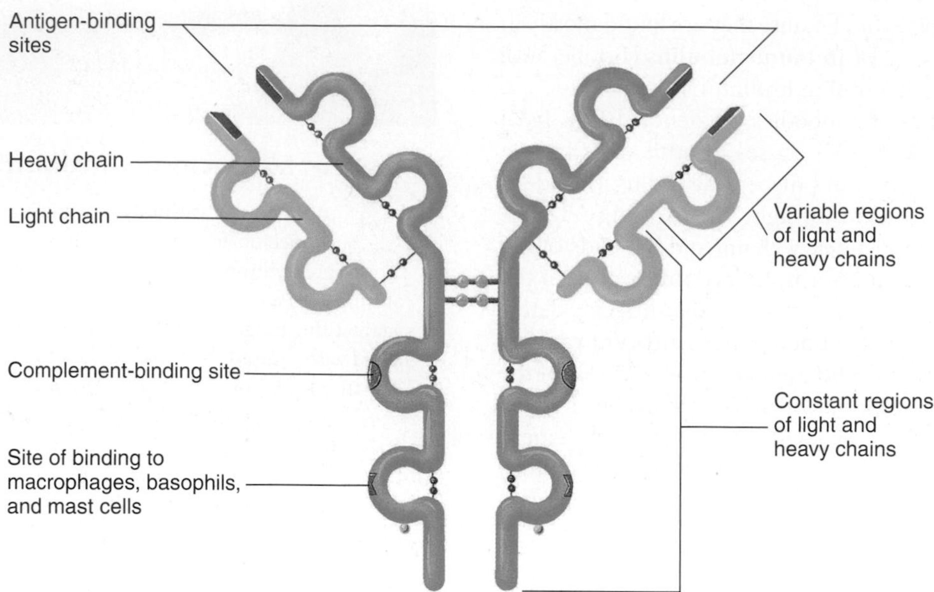
Antigen-binding sites
Heavy chain
Light chain
Variable regions of light and heavy chains
Complement-binding site
Constant regions of light and heavy chains
Site of binding to macrophages, basophils, and mast cells

FIGURE 22.19 Structure of an Antibody
Antibodies consist of two heavy and two light polypeptide chains. The variable region of the antibody binds to the antigen. The constant region of the antibody can activate the classical pathway of the complement cascade. The constant region can also attach the antibody to the plasma membrane of cells such as macrophages, basophils, and mast cells.

opsonin by connecting to an antigen through the variable region of the antibody and to a macrophage through the constant region of the antibody. The macrophage then phagocytizes the antigen and the antibody (figure 22.20*e*).

Antibody Production

The production of antibodies after the first exposure to an antigen is different from that after a second or subsequent exposure. The first exposure of a B cell to an antigen for which it is specific produces the **primary response,** which includes a series of cell divisions, cell differentiation, and antibody production. The B-cell receptors on the surface of B cells are antibodies, usually IgM and IgD. The receptors have the same variable region as the antibodies that are eventually produced by the B cell. Before stimulation by an antigen, B cells are small lymphocytes. After activation, the B cells undergo a series of divisions to produce large lymphocytes. Some of these enlarged cells become **plasma cells,** which produce antibodies, and others revert to small lymphocytes and become **memory B cells** (figure 22.21). Usually, IgM is the first antibody produced in response to an antigen, but later other classes of antibodies are produced as well. The primary response normally takes 3–14 days to produce enough antibodies to be effective against the antigen. In the meantime, disease symptoms usually develop because the antigen has had time to cause tissue damage.

The **secondary response,** or *memory response,* occurs when the immune system is exposed to an antigen against which it has already produced a primary response. When exposed to the antigen, memory B cells rapidly divide to produce plasma cells, which produce large amounts of antibody. The secondary response provides better protection than the primary response for two reasons: (1) The time required to start producing antibodies is less (hours to a few days), and (2) the amount of antibody produced is much larger. As a consequence, the antigen is quickly destroyed, no disease symptoms develop, and the person is immune.

The secondary response also includes the formation of new memory B cells, which protect against additional exposures to the antigen. Memory B cells are the basis for adaptive immunity. After destruction of the antigen, plasma cells die, the antibodies they released are degraded, and antibody levels decline to the point at which they can no longer provide adequate protection. Memory B cells persist for many years—for life, in some cases. However, if memory cell production is not stimulated or if the memory B cells produced are short-lived, repeated infections of the same disease are possible. For example, the same cold virus can cause the common cold more than once in the same person.

ASSESS YOUR PROGRESS

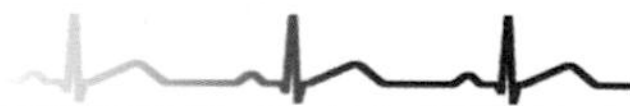

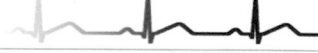

48. *What type of lymphocyte is responsible for antibody-mediated immunity? What are the functions of antibody-mediated immunity?*

49. *What are the functions of the variable and constant regions of an antibody?*

50. *List the five classes of antibodies, and state their functions.*

51. *Describe the different ways that antibodies participate in destroying antigens.*

52. *What are plasma cells and memory B cells, and how do they function?*

53. *What are the primary and secondary antibody responses? Why doesn't the primary response prevent illness, whereas the secondary response does?*

FUNDAMENTAL **Figure**

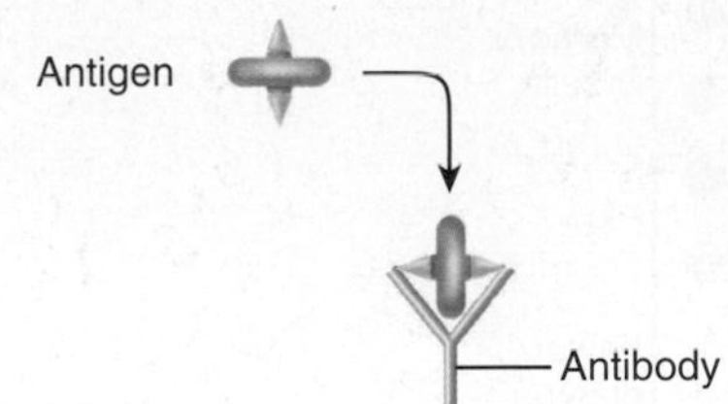
Antigen
Antibody

(a) **Inactivate the antigen.** An antibody binds to an antigen and inactivates it.

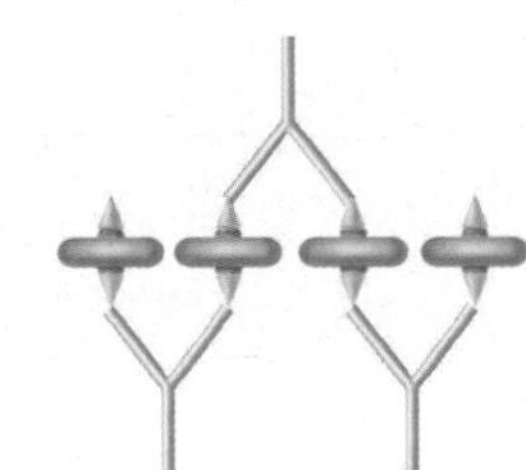

(b) **Bind antigens together.** Antibodies bind several antigens together.

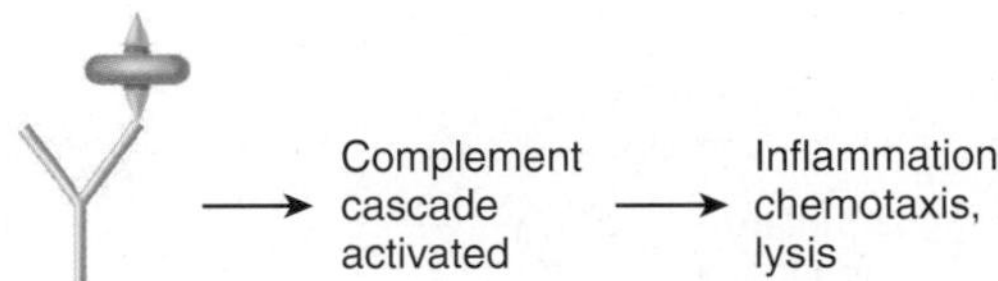
Complement cascade activated
Inflammation, chemotaxis, lysis

(c) **Activate the complement cascade.** An antigen binds to an antibody. As a result, the antibody can activate complement proteins, which can produce inflammation, chemotaxis, and lysis.

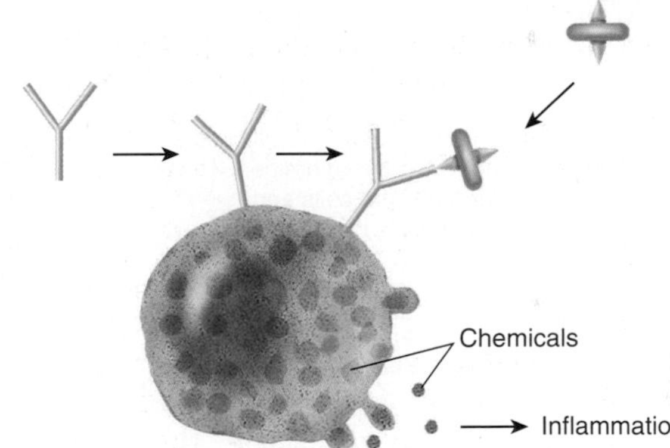
Chemicals
Inflammation
Mast cell or basophil

(d) **Initiate the release of inflammatory chemicals.** An antibody binds to a mast cell or a basophil. When an antigen binds to the antibody, it triggers the release of chemicals that cause inflammation.

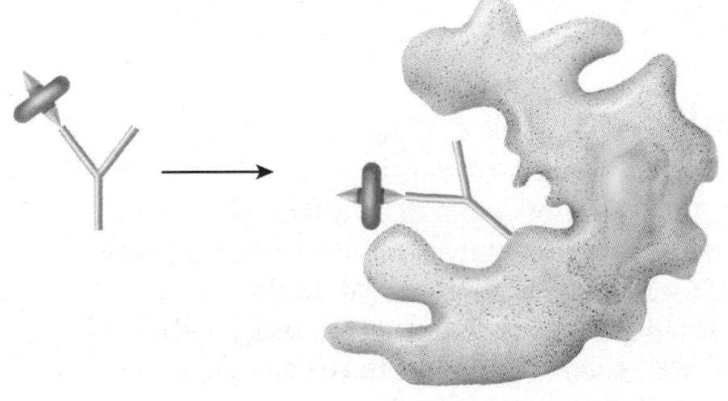
Macrophage

(e) **Facilitate phagocytosis.** An antibody binds to an antigen and then to a macrophage, which phagocytizes the antibody and antigen.

FIGURE 22.20 Effects of Antibodies

Antibodies directly affect antigens by inactivating the antigens or binding the antigens together. Antibodies indirectly affect antigens by activating other mechanisms through the constant region of the antibody. Indirect mechanisms include activation of complement, increased inflammation resulting from the release of inflammatory chemicals from mast cells or basophils, and increased phagocytosis resulting from antibody attachment to macrophages.

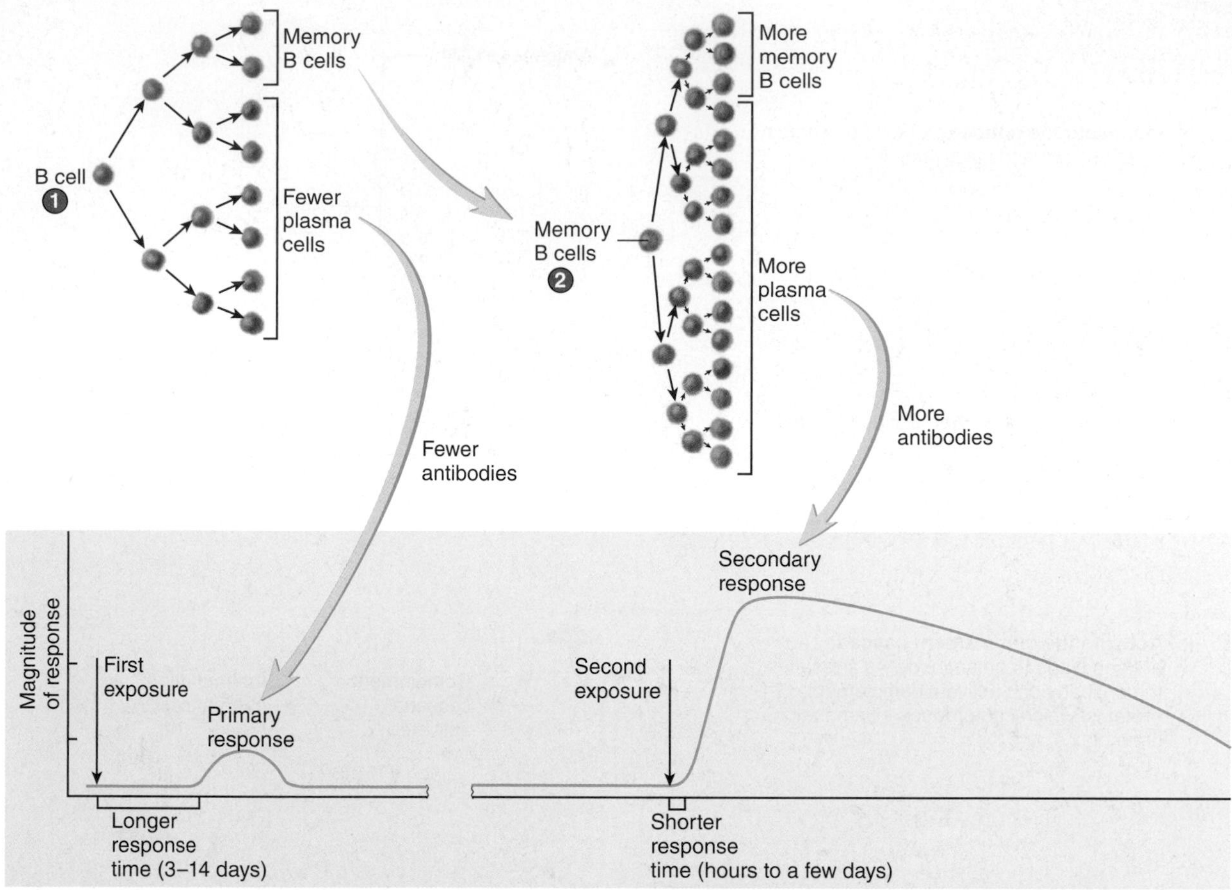

Memory B cells
B cell
1
Fewer plasma cells
Memory B cells
2
More memory B cells
More plasma cells
More antibodies
Fewer antibodies
Secondary response
Magnitude of response
First exposure
Primary response
Second exposure
Longer response time (3–14 days)
Shorter response time (hours to a few days)

1 **Primary response.** The primary response occurs when a B cell is first activated by an antigen. The B cell proliferates to form plasma cells and memory cells. The plasma cells produce antibodies.

2 **Secondary response.** The secondary response occurs when another exposure to the same antigen causes the memory cells to rapidly form plasma cells and additional memory cells. The secondary response is faster and produces more antibodies than the primary response.

PROCESS FIGURE 22.21 Antibody Production

Predict 5

One theory for long-lasting immunity assumes that humans are continually exposed to the disease-causing agent. Explain how this exposure can produce lifelong immunity.

Cell-Mediated Immunity

Cell-mediated immunity is most effective against intracellular microorganisms through the action of cytotoxic T cells. Cell-mediated immunity involves delayed hypersensitivity reactions and the control of tumors (see the Diseases and Disorders table in this chapter).

Antibody-mediated immunity is not effective against intracellular microorganisms, such as viruses, fungi, intracellular bacteria, and parasites, because antibodies cannot cross the plasma membrane. However, cell-mediated immunity is effective against these intracellular microorganisms because it destroys the cells in which the microorganisms are located. For example, viruses enter cells and direct the cells to make new viruses, which are then released to infect other cells. Thus, cells are turned into virus-manufacturing cells. Cell-mediated immunity fights viral infections by destroying virally infected cells. When viruses infect cells, some viral proteins are broken down and become processed foreign antigens that are combined with MHC class I molecules and displayed on the surface of the infected cells (figure 22.22). T cells can distinguish between virally infected cells and noninfected cells because MHC class I/antigen complexes are on the surface of infected cells, but not on the surface of uninfected cells. Binding of the T-cell receptor to the MHC class I/antigen complex is a signal for activating cytotoxic T cells. Costimulation by other surface molecules, such as CD8, also occurs. Helper T cells provide costimulation by releasing cytokines, such as interleukin-2, which stimulates activation and cell division of T cells. However, unlike their interactions with macrophages and B cells, helper T cells do not connect to T cells through MHC class II/antigen complexes or other surface molecules.

1. An MHC class I molecule displays an antigen, such as a viral protein, on the surface of a target cell.

2. The activation of a cytotoxic T cell begins when the T-cell receptor binds to the MHC class I/antigen complex.

3. There is costimulation of the cytotoxic T cell by CD8 and other surface molecules.

4. There is costimulation by cytokines, such as interleukin-2, released from helper T cells.

5. The activated cytotoxic T cell divides, the resulting daughter cells divide, and so on, eventually producing many cytotoxic T cells (only two are shown here).

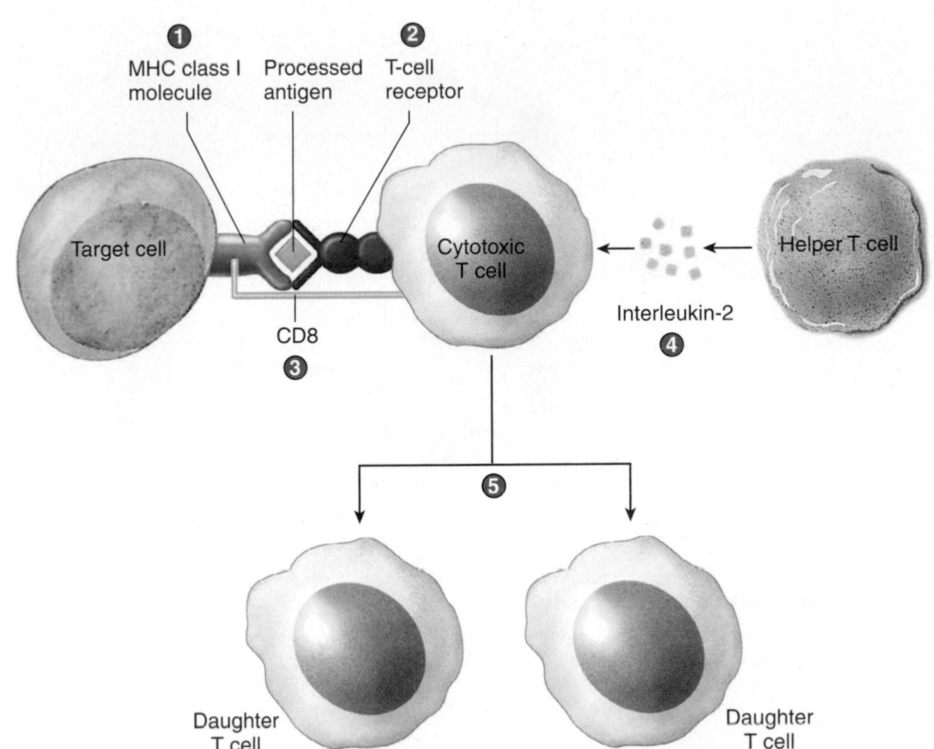
1
MHC class I molecule
Processed antigen
2
T-cell receptor
Target cell
Cytotoxic T cell
Helper T cell
Interleukin-2
4
CD8
3
5
Daughter T cell
Daughter T cell

PROCESS FIGURE 22.22 APIR **Proliferation of Cytotoxic T Cells**

An increased number of helper T cells results in greater stimulation of cytotoxic T cells. In cell-mediated responses, helper T cells are activated and stimulated to divide in the same fashion as in antibody-mediated responses (see figure 22.17).

After T cells are activated by an antigen on the surface of a target cell, they undergo a series of divisions to produce cytotoxic T cells and memory T cells (figure 22.23). The cytotoxic T cells are responsible for the cell-mediated immune response. Memory T cells can provide a secondary response and long-lasting immunity in the same fashion as memory B cells.

Cytotoxic T Cells

Cytotoxic T cells have two main effects: They lyse cells and produce cytokines (figure 22.23). Cytotoxic T cells can come into contact with other cells and cause them to lyse. Virus-infected cells have viral antigens, tumor cells have tumor antigens, and tissue transplants have foreign antigens on their surfaces that can stimulate cytotoxic T-cell activity. A cytotoxic T cell binds to a target cell and releases chemicals that cause the target cell to lyse. The major method of lysis involves a protein called **perforin,** which is similar to the complement protein C9 (see figure 22.10). Perforin forms a channel in the plasma membrane of the target cell through which water enters the cell, causing lysis. The cytotoxic T cell then moves on to destroy additional target cells.

In addition to lysing cells, cytotoxic T cells release cytokines that activate additional components of the immune system. For example, one important function of cytokines is the recruitment of cells, such as macrophages. These cells are then responsible for phagocytosis and inflammation.

Predict 6

In patients with acquired immunodeficiency syndrome (AIDS), the HIV virus infects and destroys helper T cells. Patients often die of pneumonia caused by an intracellular fungus (*Pneumocystis carinii*) or of Kaposi sarcoma, which is characterized by tumorous growths in the skin and lymph nodes. Explain what is happening.

ASSESS YOUR PROGRESS

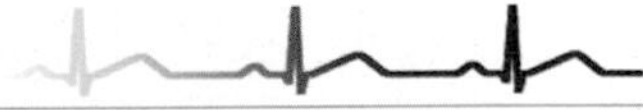

54. *What type of lymphocyte is responsible for cell-mediated immunity? What are the functions of cell-mediated immunity?*

55. *How do intracellular microorganisms stimulate cytotoxic T cells? What role do helper T cells play in this process?*

56. *State the two main responses of cytotoxic T cells.*

57. *How is long-lasting immunity achieved in cell-mediated immunity?*

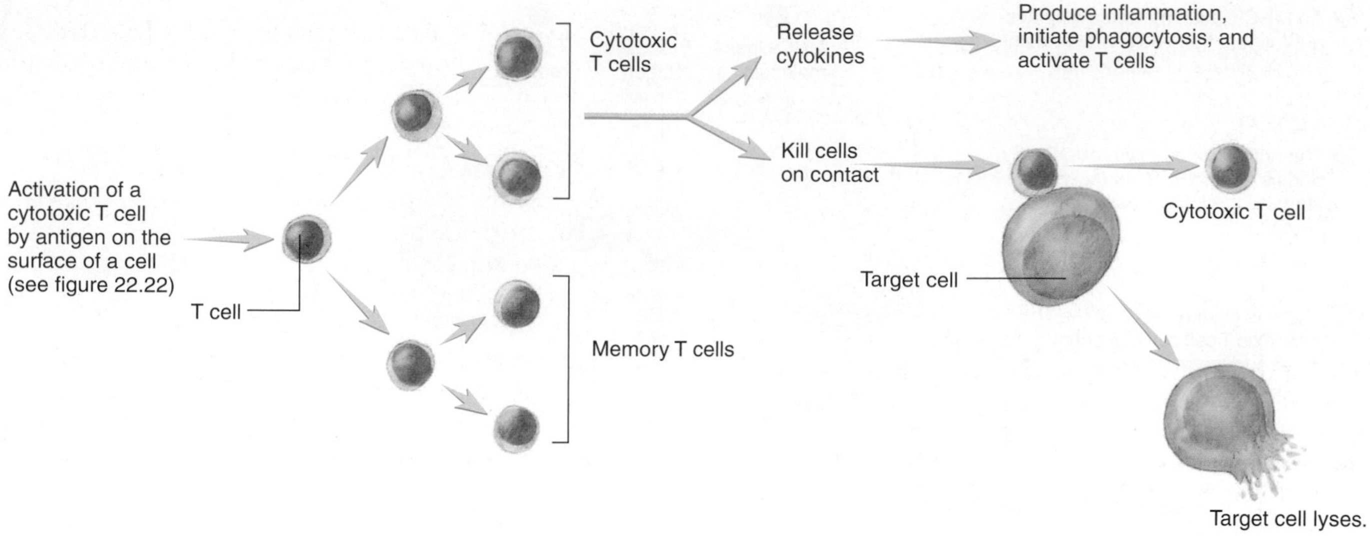
Activation of a cytotoxic T cell by antigen on the surface of a cell (see figure 22.22)
T cell
Cytotoxic T cells
Memory T cells
Release cytokines
Kill cells on contact
Produce inflammation, initiate phagocytosis, and activate T cells
Cytotoxic T cell
Target cell
Target cell lyses.

FIGURE 22.23 Stimulation and Effects of T Cells
When activated, cytotoxic T cells form many additional cytotoxic T cells, as well as memory T cells. The cytotoxic T cells release cytokines that promote the destruction of the antigen or cause the lysis of target cells, such as virus-infected cells, tumor cells, or transplanted cells. The memory T cells are responsible for the secondary response.

Clinical GENETICS — Gluten-Sensitive Enteropathy

Gluten-sensitive enteropathy, also called *celiac* (sē′lē-ăk) *disease*, is a malabsorption disorder, meaning that nutrients are poorly absorbed. Gluten-sensitive enteropathy results from damage to the lining of the small intestine, specifically the fingerlike projections, called villi, that increase the surface area for nutrient absorption (see figure 24.16). In a healthy intestine, the lining resembles a shag carpet. In gluten-sensitive enteropathy, the epithelium has become damaged, and the intestinal villi are flattened and inflamed.

Gluten-sensitive enteropathy is usually characterized by gastrointestinal symptoms, such as diarrhea, painful abdominal cramping, bloating, and intestinal gas. Prolonged gluten-sensitive enteropathy leads to additional complications, including anemia, osteoporosis, and neurological problems, in part due to nutritional deficiencies.

Gluten-sensitive enteropathy occurs in about 1 in 133 people, but the frequency may actually be even higher because the widely varying symptoms and severity of the disease make diagnosis difficult.

Gluten-sensitive enteropathy is an autoimmune disease and is often associated with other autoimmune diseases, such as systemic lupus erythematosus (see Systems Pathology, later in this chapter). The damage to the intestinal lining is caused by an inappropriate immune response, which is triggered by the gluten proteins in wheat, barley, and rye. Although neither rice nor corn contains gluten proteins, gluten is often hidden as an additive in prepared foods and sauces.

Gluten-sensitive enteropathy has a genetic component. Most patients have mutated variants of some of the MHC class II genes. As a result, abnormal MHC class II molecules are produced and can bind to digested fragments of gluten. However, the genetics of gluten-sensitive enteropathy are complex and not fully understood. For example, the variant MHC alleles alone are not sufficient to cause the disease, since many people who have the variant alleles do not develop gluten-sensitive enteropathy. Furthermore, the genetic expression of the disease is influenced by variable environmental factors because the onset and severity of the disease can be triggered by unknown factors at any time in life.

Gluten-sensitive enteropathy results from both adaptive and innate immune responses. On the surface of antigen-presenting cells, the MHC class II/gluten complex is presented to helper T cells to initiate an adaptive immune response (see figure 22.15*b*). The adaptive immune response includes antibody production and the activation of cytotoxic T cells. The innate immune response promotes inflammation through the activation of the alternative complement pathway and the release of cytokines, such as interleukin-15 (IL-15), from macrophages, dendritic cells, and other cells. In addition, exposure to gluten can activate natural killer cells and dendritic cells. The result of these immune responses is a deleterious attack on the epithelial lining of the small intestine, leading to the damaged villi common to gluten-sensitive enteropathy.

The only treatment for gluten-sensitive enteropathy is a strict, gluten-free diet. Before embarking on a life-long gluten-free diet, however, it is important to have a definitive diagnosis. Tests for higher than normal levels of antibodies produced in gluten-sensitive enteropathy, such as anti-tissue transglutaminase, and a biopsy of the small intestine to examine the villi are recommended. In the future, early genetic diagnosis and manipulation of the immune response may be able to reduce the sensitivity to gluten.

22.6 Acquired Adaptive Immunity

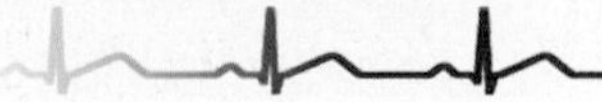

LEARNING OUTCOME

After reading this section, you should be able to

A. Explain the four ways that adaptive immunity can be acquired.

Unlike innate immunity, adaptive immunity is not necessarily present at birth. Instead, adaptive immunity must be acquired. Adaptive immunity can be broken down into four types, based on the way it is acquired: active natural, active artificial, passive natural, and passive artificial (figure 22.24). **Active immunity** results when an individual is exposed to an antigen (either naturally or artificially) and the response of the individual's immune system is the cause of the immunity. **Passive immunity** occurs when another person or an animal develops immunity and the immunity is transferred to a nonimmune individual. *Natural* and *artificial* refer to the method of exposure. *Natural exposure* implies that contact with an antigen or antibody occurs as part of everyday living and is not deliberate. Artificial exposure, also called **immunization,** is the deliberate introduction of an antigen or antibody into the body.

How long the immunity lasts differs for active and passive immunity. Active immunity can persist for a few weeks (e.g., the common cold) to a lifetime (e.g., whooping cough, polio). Immunity can be long-lasting if enough memory B cells or memory T cells are produced and persist to respond to later antigen exposure. Passive immunity is not long-lasting because the individual does not produce his or her own memory cells. Because active immunity can last longer than passive immunity, it is the preferred method. However, passive immunity is preferred when immediate protection is needed.

Active Natural Immunity

Natural exposure to an antigen, such as a disease-causing microorganism, can cause the immune system to mount an adaptive immune response against the antigen and achieve **active natural immunity.** Because the individual is not immune during the first exposure, he or she usually develops the symptoms of the disease. Interestingly, exposure to an antigen does not always produce symptoms. For example, many people exposed to the poliomyelitis virus at an early age have an immune system response, produce poliomyelitis antibodies, but do not exhibit any disease symptoms.

Active Artificial Immunity

In **active artificial immunity,** an antigen is deliberately introduced into a person's body to stimulate the immune system. This process is called immunization, or **vaccination,** and the introduced antigen is a **vaccine.** A vaccine is usually administered by injection. Examples of vaccinations are the DPT injection against diphtheria, pertussis (whooping cough), and tetanus and the MMR injection against mumps, measles, and rubella (German measles).

A vaccine usually consists of a part of a microorganism, a dead microorganism, or a live, altered microorganism. The antigen has been changed so that it will stimulate an immune response but will not cause the disease symptoms. Because active artificial immunity produces long-lasting immunity without disease symptoms, it is the preferred method of acquiring adaptive immunity.

FUNDAMENTAL **Figure**

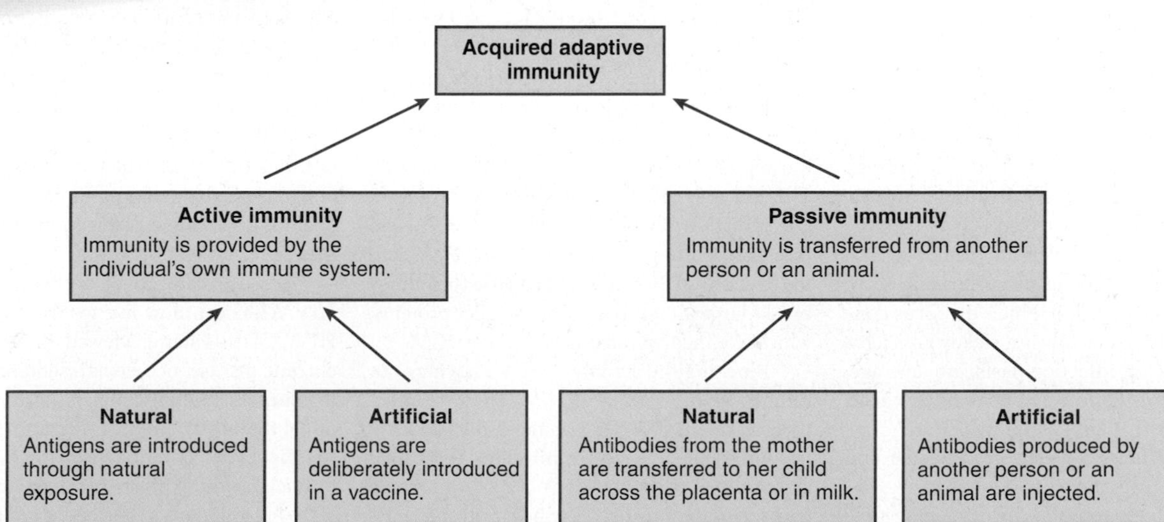
Acquired adaptive immunity
Active immunity
Immunity is provided by the individual's own immune system.
Passive immunity
Immunity is transferred from another person or an animal.
Natural
Antigens are introduced through natural exposure.
Artificial
Antigens are deliberately introduced in a vaccine.
Natural
Antibodies from the mother are transferred to her child across the placenta or in milk.
Artificial
Antibodies produced by another person or an animal are injected.

FIGURE 22.24 Ways to Acquire Adaptive Immunity

Clinical IMPACT — Acquired Immunodeficiency Syndrome

Acquired immunodeficiency syndrome (AIDS) is a life-threatening disease caused by the **human immunodeficiency virus (HIV).** HIV is transmitted from an infected person to a noninfected person in body fluids, such as blood, semen, or vaginal secretions. The major methods of transmission are through unprotected sexual contact, through contaminated needles used by intravenous drug users, through tainted blood products, and from a pregnant woman to her fetus. Evidence indicates that household, school, and work contacts do not result in transmission. Reduced exposure to HIV is the best prevention for its transmission. Practices such as abstinence, the use of latex condoms, monogamy, and avoiding sharing needles are effective ways to reduce exposure to HIV. Medical professionals should also use care when handling body fluids, such as wearing latex gloves.

HIV infection begins when a protein on the surface of the virus, called gp120, binds to a CD4 molecule on the surface of a cell. The CD4 molecule is found primarily on helper T cells, and it normally enables helper T cells to adhere to other lymphocytes—for example, during antigen presentation. Certain monocytes, macrophages, neurons, and neuroglia also have CD4 molecules. Once attached to the CD4 molecules, the virus injects its genetic material (RNA) and enzymes into the cell and begins to replicate. Copies of the virus are manufactured using the organelles and materials within the cell. Replicated viruses escape from the cell and infect other cells.

Following infection by HIV, within 3 weeks to 3 months, many patients develop mononucleosis-like symptoms, such as fever, sweats, fatigue, muscle and joint aches, headache, sore throat, diarrhea, rash, and swollen lymph nodes. Within 1–3 weeks, these symptoms disappear as the immune system responds to the virus by producing antibodies and activating cytotoxic T cells that kill HIV-infected cells. However, the immune system is not able to eliminate HIV completely, and by about 6 months a kind of "set point" is achieved in which the virus continues to replicate at a low but steady rate. This chronic stage of infection lasts, on average, 8–10 years, and the infected person feels good and exhibits few, if any, symptoms.

Although helper T cells are infected and destroyed during the chronic stage of HIV infection, the body responds by producing large numbers of helper T cells. Nonetheless, over a period of years the HIV numbers gradually increase, and helper T cell numbers decrease. Normally, approximately 1200 helper T cells are present per cubic millimeter of blood. An HIV-infected person is diagnosed with AIDS when one or more of the following conditions appear: The helper T cell count falls below 200 cells/mm^3, an opportunistic infection occurs, or Kaposi sarcoma develops.

Opportunistic infections involve organisms that normally do not cause disease but do so when the immune system is depressed. Without helper T cells, cytotoxic T and B cell activation is impaired, and adaptive resistance is suppressed. Examples of opportunistic infections include pneumocystis (noo-mō-sis′tis) pneumonia (caused by an intracellular fungus, *Pneumocystis carinii*), tuberculosis (caused by an intracellular bacterium, *Mycobacterium tuberculosis*), syphilis (caused by a sexually transmitted bacterium, *Treponema pallidum*), candidiasis (kan-di-dī′ă-sis; a yeast infection of the mouth or vagina caused by *Candida albicans*), and protozoans that cause severe, persistent diarrhea. Kaposi sarcoma is a type of cancer that produces lesions in the skin, lymph nodes, and visceral organs. AIDS symptoms resulting from the effects of HIV on the nervous system include motor retardation, behavioral changes, progressive dementia, and possibly psychosis.

A cure for AIDS has yet to be discovered. Management of AIDS can be divided into two categories: (1) management of secondary infections or malignancies associated with AIDS and (2) control of HIV replication. In order for HIV to replicate, the viral RNA is used to make viral DNA, which is inserted into the host cell's DNA. The inserted viral DNA directs the production of new viral RNA and proteins, which are assembled to form new HIV. Key steps in the replication of HIV require viral enzymes. The enzyme **reverse transcriptase** promotes the formation of viral DNA from viral RNA, and **integrase** (in′te-grās) inserts the viral DNA into the host cell's DNA. A viral **protease** (prō′tē-ās) breaks large viral proteins into smaller proteins, which are incorporated into the new HIV.

Blocking the activity of HIV enzymes can inhibit the replication of HIV. The first effective treatment of AIDS was the drug azidothymidine (AZT; az′i-dō-thī′mi-dēn), also called zidovudine (zī-dō′voo-dēn). AZT is a **reverse transcriptase inhibitor,** which prevents HIV RNA from producing viral DNA. AZT can delay the onset of AIDS but does not appear to increase the survival time of AIDS patients. However, the number of babies who contract AIDS from their HIV-infected mothers can be dramatically reduced by giving AZT to the mothers during pregnancy and to the babies following birth.

Protease inhibitors are drugs that interfere with viral proteases. The current treatment for suppressing HIV replication is **highly active antiretroviral therapy (HAART).** This therapy uses drugs from at least two classes of antivirals. Treatment may involve combining three drugs, such as two reverse transcriptase inhibitors and one protease inhibitor, because HIV is unlikely to develop resistance to all three drugs. This strategy has proven very effective in reducing the death rate from AIDS and partially restoring health in some individuals.

Still in the research stage are **integrase inhibitors,** which prevent the insertion of viral DNA into the host cell's DNA. Another advance in AIDS treatment is a test for measuring **viral load,** which measures the number of viral RNA molecules in a milliliter of blood. The actual level of HIV is one-half the RNA count because each HIV has two RNA strands. Viral load is a good predictor of how soon a person will develop AIDS. If viral load is high, the onset of AIDS is likely to occur sooner than if the viral load is low. It is also possible to detect developing viral resistance by an increase in viral load. In response, a change in drug dose or type may slow viral replication. Current treatment goals are to keep viral load below 500 RNA molecules per milliliter of blood.

Effective treatment for AIDS is not the same as a cure. Even if viral load decreases to the point that the virus is undetected in the blood, the virus still remains in cells throughout the body. The virus may eventually mutate and escape drug suppression. The long-term goal for deterring AIDS is to develop a vaccine that prevents HIV infection.

Because of improved treatment, people with HIV/AIDS can now live for many years. Thus, HIV/AIDS is being viewed increasingly as a chronic disease, not a death sentence. Working together, a multidisciplinary team of occupational therapists, physical therapists, nutritionists/dieticians, psychologists, infectious disease physicians, and others can help patients with HIV/AIDS have a better quality of life.

Predict 7

Some vaccination procedures require a booster shot, another dose of the original vaccine given sometime after the original dose was administered. Why are booster shots administered?

Passive Natural Immunity

Passive natural immunity results when antibodies are transferred from a mother to her child across the placenta before birth. During her life, the mother has been exposed to many antigens, either naturally or artificially, and she has antibodies against many of these antigens that protect her and the developing fetus against disease. Some of the antibodies (IgG) can cross the placenta and enter the fetal blood. Following birth, the antibodies protect the baby for the first few months. Eventually, the antibodies break down, and the baby must rely on his or her own immune system. If the mother nurses her baby, antibodies (IgA) in the mother's milk may also provide some protection for the baby.

Passive Artificial Immunity

Achieving **passive artificial immunity** usually begins with vaccinating an animal, such as a horse. After the animal's immune system responds to the antigen, antibodies (and sometimes T cells) are removed from the animal and injected into the human requiring immunity. In some cases, a human who has developed immunity through natural exposure or vaccination can serve as a source of antibodies. Passive artificial immunity provides immediate protection for the individual receiving the antibodies and is therefore preferred when time might not be available for the individual to develop his or her own immunity. However, this technique provides only temporary immunity because the antibodies are used or eliminated by the recipient.

Antiserum is the general term for serum, which is plasma minus the clotting factors, that contains antibodies responsible for passive artificial immunity. Antisera are available against microorganisms that cause diseases, such as rabies, hepatitis, and measles; bacterial toxins, such as those that cause tetanus, diphtheria, and botulism; and venoms from poisonous snakes and black widow spiders.

ASSESS YOUR PROGRESS

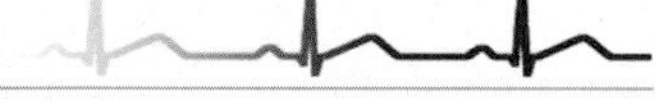

58. *Distinguish between active and passive immunity.*
59. *State four general ways of acquiring adaptive immunity. Which two provide the longest-lasting immunity?*
60. *What type of immunity occurs when a child is given the chickenpox vaccine? What type of immunity does a nursing baby obtain?*
61. *What type of immunity occurs when a person is given the rabies antisera? What type of immunity occurs if a person had the chickenpox as a child?*

22.7 Overview of Immune Interactions

LEARNING OUTCOME

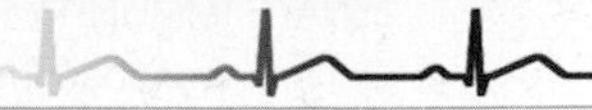

After reading this section, you should be able to

A. Explain how innate, antibody-mediated, and cell-mediated immunity can function together to eliminate an antigen.

Although the immune system can be described in terms of innate, antibody-mediated, and cell-mediated immunity, these categories are artificial divisions used to emphasize particular aspects of immunity. Actually, there is only one immune system, but its responses often involve components of more than one type of immunity (figure 22.25). For example, although adaptive immunity can recognize and remember specific antigens, once recognition has occurred the antigen is destroyed with the help of many innate immunity activities, including inflammation and phagocytosis.

ASSESS YOUR PROGRESS

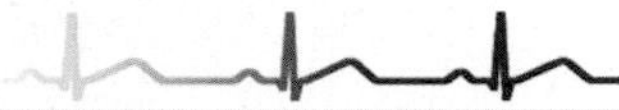

62. *Describe how interactions among innate, antibody-mediated, and cell-mediated immunity can protect the body from an antigen.*

22.8 Immunotherapy

LEARNING OUTCOME

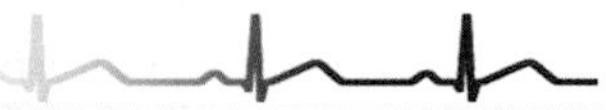

After reading this section, you should be able to

A. Define and give examples of *immunotherapy*.

Knowledge of how the immune system operates has produced two fundamental benefits: (1) an understanding of the cause and progression of many diseases and (2) the development or proposed development of methods to prevent, stop, or even reverse diseases. In this section, we discuss the second benefit, **immunotherapy,** which treats disease by altering immune system function or by directly attacking harmful cells. Some types of immunotherapy attempt to boost immune system function in general. For example, administering cytokines or other agents can promote inflammation and activate immune cells, which can help destroy tumor cells. On the other hand, sometimes inhibiting the immune system is helpful. For example, multiple sclerosis is an autoimmune disease in which the immune system treats self-antigens as foreign antigens, thereby destroying the myelin that covers axons. The cytokine interferon beta (IFNβ) blocks the expression of MHC molecules that display self-antigens and is used to treat multiple sclerosis.

Some immunotherapy methods take a more specific approach. For example, vaccination can prevent many diseases (see section 22.6).

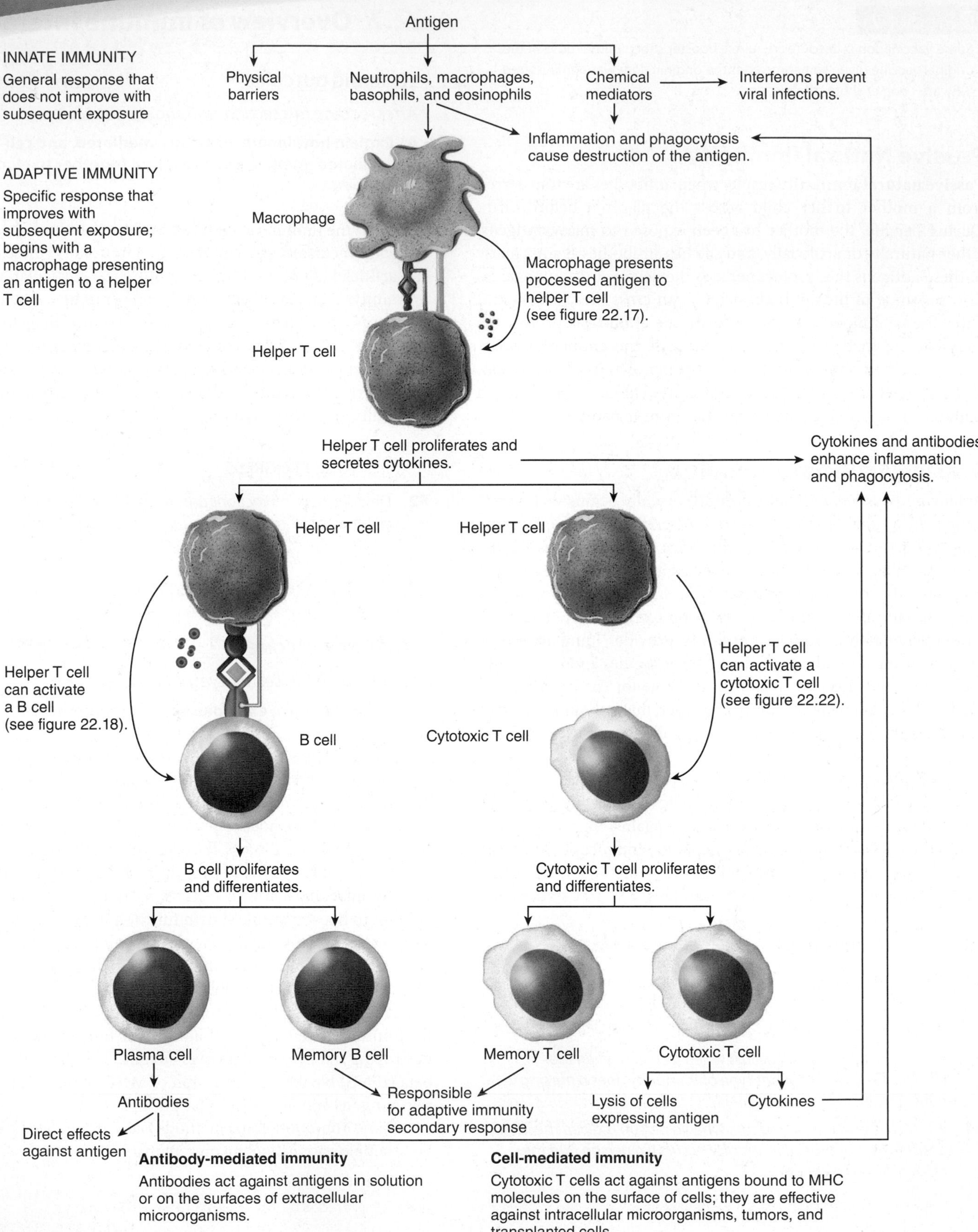
Antigen
INNATE IMMUNITY
General response that does not improve with subsequent exposure
Physical barriers
Neutrophils, macrophages, basophils, and eosinophils
Chemical mediators
Interferons prevent viral infections.
Inflammation and phagocytosis cause destruction of the antigen.
ADAPTIVE IMMUNITY
Specific response that improves with subsequent exposure; begins with a macrophage presenting an antigen to a helper T cell
Macrophage
Macrophage presents processed antigen to helper T cell (see figure 22.17).
Helper T cell
Helper T cell proliferates and secretes cytokines.
Cytokines and antibodies enhance inflammation and phagocytosis.
Helper T cell
Helper T cell
Helper T cell can activate a B cell (see figure 22.18).
Helper T cell can activate a cytotoxic T cell (see figure 22.22).
B cell
Cytotoxic T cell
B cell proliferates and differentiates.
Cytotoxic T cell proliferates and differentiates.
Plasma cell
Memory B cell
Memory T cell
Cytotoxic T cell
Antibodies
Responsible for adaptive immunity secondary response
Lysis of cells expressing antigen
Cytokines
Direct effects against antigen
Antibody-mediated immunity
Antibodies act against antigens in solution or on the surfaces of extracellular microorganisms.
Cell-mediated immunity
Cytotoxic T cells act against antigens bound to MHC molecules on the surface of cells; they are effective against intracellular microorganisms, tumors, and transplanted cells.

FIGURE 22.25 Immune Interactions

The major interactions and responses of innate and adaptive immunity to an antigen.

Diseases and Disorders

TABLE 22.6 Lymphatic System and Immunity

Condition	Description
	LYMPHATIC SYSTEM
Infections	
Lymphadenitis (lim-fad′ĕ-nī′tis)	Inflammation of the lymph nodes; nodes become enlarged and tender as microorganisms are trapped and destroyed
Lymphangitis (lim-fan-jī′tis)	Inflammation of the lymphatic vessels; often results in visible red streaks in the skin that extend from the site of infection
Bubonic (bū-bon′ik) plague	Enlarged lymph nodes caused by bacterial infection (transferred by flea bites from rats); without treatment, bacteria enter the blood, and death occurs rapidly due to septicemia; known as the Black Death in the Middle Ages
Lymphedema (limf′e-dē′ma)	
Abnormal accumulation of lymph in tissues, often the limbs; 70–90% of cases in women; can be inherited or caused by developmental defects (primary lymphedema), disease, or damage to the lymphatic system (secondary lymphedema)	
Elephantiasis (el-ĕ-fan-tī′ă-sis)	Caused by long, slender roundworms transferred to humans by mosquito bites; adult worms lodge in lymphatic vessels and block lymph flow, so that a limb can become permanently swollen and enlarged; major cause of lymphedema worldwide
Lymphedema following cancer treatment	Caused by removal of lymph nodes near a tumor by surgery or radiation therapy; sentinel lymph nodes (those nearest the tumor) are first examined for cancer cells
Lymphoma (lim-fō′mă)	Cancer of lymphocytes that often begins in lymph nodes; immune system becomes depressed, with increased susceptibility to infections
	IMMUNITY
Immediate Hypersensitivities	
Symptoms occur within a few minutes of exposure to an antigen because antibodies are already present from prior exposure.	
Hay fever	Often caused by inhalation of plant pollen antigens
Asthma (az′mă)	Antigen combines with antibodies on mast cells or basophils in the lungs, which then release inflammatory chemicals that cause constriction of the air tubes, so that the patient has trouble breathing
Immune complex disease	Caused by excessive formation of immune complexes (combinations of antigens and IgG or IgM), which activate too much complement; results in acute inflammatory response and tissue damage; examples include serum sickness, some autoimmune diseases, chronic graft rejection, and Arthus reactions (localized reactions)
Urticaria (er′ti-kar′i-ă)	Skin rash or localized swelling; can be caused by an ingested antigen; also called hives
Anaphylaxis (an′ă-fī-lak′sis)	Systemic allergic reaction, often resulting from insect sting or drugs, such as penicillin; chemicals released from mast cells and basophils cause systemic vasodilation, increased vascular permeability, drop in blood pressure, and possibly death
Delayed Allergic Reactions	
Symptoms occur in hours to days following exposure to the antigen because these types of reactions involve migration of T cells to the antigen, followed by release of cytokines.	
Poison ivy and poison oak	Antigen absorbed by epithelial cells, which are then destroyed by T cells, causing inflammation and tissue destruction; itching can be intense
Autoimmune Diseases	
Similar to allergic reactions, except that the immune system incorrectly treats self-antigens as foreign antigens. Many types of autoimmune diseases exist, including type 1 diabetes, gluten-sensitive enteropathy, rheumatoid arthritis, multiple sclerosis, systemic lupus erythematosus, and Graves disease.	
Congenital Immunodeficiencies	
They usually involve failure of the fetus to form adequate numbers of B cells, T cells, or both.	
Severe combined immunodeficiency (SCID)	Both B cells and T cells fail to form; unless patient is kept in a sterile environment or provided with a compatible bone marrow transplant, death from infection results
Acquired Immunodeficiencies	
They have many causes—for example, diseases, stress, and drugs.	
Acquired immunodeficiency syndrome (AIDS)	Life-threatening disease caused by the human immunodeficiency virus (HIV); HIV is transmitted in body fluids; infection begins when the virus binds to the CD4 protein found primarily on helper T cells; without helper T cells, cytotoxic T-cell and B-cell activation is impaired, and adaptive immunity is suppressed; course of HIV infection varies, with most people surviving 10 or more years
Transplanted tissue rejection	Caused by a normal immune response to foreign antigens encoded by the major histocompatilbility complex genes, also called human leukocyte antigen (HLA) genes; drugs that suppress the immune system must be administered for life to prevent graft rejection

Go to: www.mhhe.com/seeley10 for additional information on these pathologies.

Systems PATHOLOGY | Systemic Lupus Erythematosus

3:17 PM
Name: Lucy
Gender: Female
Age: 30
Comments
Lucy, a divorced mother of two, has been working full-time the past few years but has decided to complete her nursing degree. Lucy was diagnosed with lupus when she was 25 and knew that the added stress of college could cause her condition to worsen. Sure enough, by midterm her attendance and performance on assignments was erratic as her energy level and emotional state alternated between highs and lows. Near the end of the semester she developed a rash on her face and a large red lesion on her arm. Knowing Lucy's situation, her instructor suggested she receive an incomplete and finish the coursework later that summer.

Background Information

Systemic lupus erythematosus (SLE) is an autoimmune disease, meaning that tissues and cells are damaged by the body's own immune system. The name describes the skin rash that is characteristic of the disease (figure 22A). The term *lupus* means "wolf" and originally referred to eroded (as if gnawed by a wolf) lesions of the skin. *Erythematosus* refers to redness of the skin resulting from inflammation.

In SLE, a large variety of antibodies are produced that recognize self-antigens, such as nucleic acids, phospholipids, coagulation factors, red blood cells, and platelets. The combination of the antibodies with self-antigens forms immune complexes that circulate throughout the body and are deposited in various tissues, where they stimulate inflammation and tissue destruction. Thus, SLE can affect many body systems, as the term *systemic* implies. For example, the most common antibodies act against DNA released from damaged cells. Normally, the liver removes the DNA, but sometimes DNA and antibodies form immune complexes that tend to be deposited in the kidneys and other tissues. Approximately 40–50% of individuals with SLE develop renal disease. In some cases, the antibodies can bind to antigens on cells, causing the cells to lyse. For example, antibodies binding to red blood cells cause hemolysis and anemia.

The cause of SLE is unknown. The most popular hypothesis suggests that a viral infection disrupts the function of regulatory T cells, resulting in loss of tolerance to self-antigens. The picture is probably more complicated, however, because not all SLE patients have reduced numbers of regulatory T cells. In addition, some patients have decreased numbers of the helper T cells that normally stimulate regulatory T-cell activity.

Genetic factors probably contribute to the development of the disease. The likelihood of developing SLE is much higher if a family member also has it. In addition, family members of SLE patients who do not have SLE are much more likely to have DNA antibodies than the general population does.

Approximately 1 of every 2000 individuals in the United States has SLE. The first symptoms usually appear between 15 and 25 years of age and affect women approximately nine times as often as men. A low-grade fever is present in most cases of active SLE. The progress of the disease is unpredictable, with flare-ups followed by periods of remission. The survival after diagnosis is greater than 90% after 10 years. The most frequent causes of death are kidney failure, central nervous system dysfunction, infections, and cardiovascular disease.

No cure for SLE exists, nor is there one standard of treatment, because the course of the disease is highly variable and patient histories differ widely. Treatment usually begins with mild medications and proceeds to increasingly

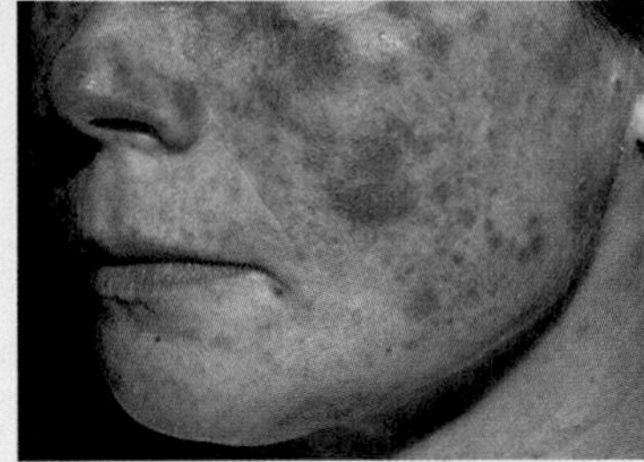

Figure 22A Systemic Lupus Erythematosus
The butterfly rash results from inflammation in the skin.

The ability to produce monoclonal antibodies may result in effective treatments for tumors. If an antigen unique to tumor cells can be found, monoclonal antibodies can deliver radioactive isotopes, drugs, toxins, enzymes, or cytokines that kill the tumor cell directly or activate the immune system to kill the cell. Unfortunately, so far researchers have found no antigen on tumor cells that is not also present on normal cells. Nonetheless, this approach may be useful if damage to normal cells is minimal. For example, tumor cells may have more surface antigens of a particular type than normal cells, resulting in greater treatment delivery. Tumor cells may also be more susceptible to damage, or normal cells may be better able to recover from the treatment.

One problem with monoclonal antibody delivery systems is that the immune system recognizes the monoclonal antibody as a foreign antigen. After the first exposure, a memory response quickly destroys the monoclonal antibodies, rendering the treatment ineffective. In a process called **humanization,** the monoclonal antibodies

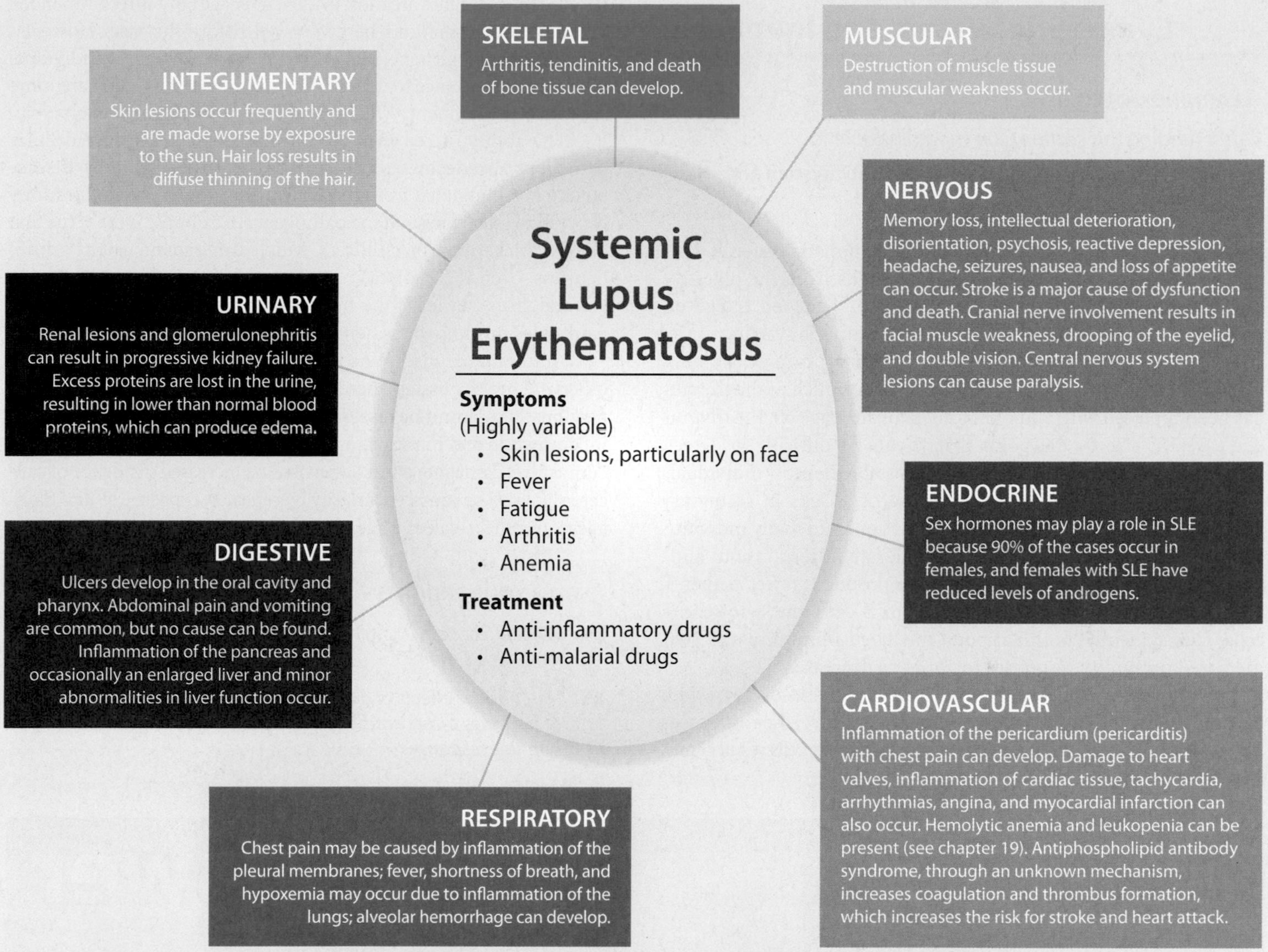

Systemic Lupus Erythematosus

Symptoms
(Highly variable)

- Skin lesions, particularly on face
- Fever
- Fatigue
- Arthritis
- Anemia

Treatment

- Anti-inflammatory drugs
- Anti-malarial drugs

SKELETAL
Arthritis, tendinitis, and death of bone tissue can develop.

MUSCULAR
Destruction of muscle tissue and muscular weakness occur.

INTEGUMENTARY
Skin lesions occur frequently and are made worse by exposure to the sun. Hair loss results in diffuse thinning of the hair.

NERVOUS
Memory loss, intellectual deterioration, disorientation, psychosis, reactive depression, headache, seizures, nausea, and loss of appetite can occur. Stroke is a major cause of dysfunction and death. Cranial nerve involvement results in facial muscle weakness, drooping of the eyelid, and double vision. Central nervous system lesions can cause paralysis.

URINARY
Renal lesions and glomerulonephritis can result in progressive kidney failure. Excess proteins are lost in the urine, resulting in lower than normal blood proteins, which can produce edema.

ENDOCRINE
Sex hormones may play a role in SLE because 90% of the cases occur in females, and females with SLE have reduced levels of androgens.

DIGESTIVE
Ulcers develop in the oral cavity and pharynx. Abdominal pain and vomiting are common, but no cause can be found. Inflammation of the pancreas and occasionally an enlarged liver and minor abnormalities in liver function occur.

CARDIOVASCULAR
Inflammation of the pericardium (pericarditis) with chest pain can develop. Damage to heart valves, inflammation of cardiac tissue, tachycardia, arrhythmias, angina, and myocardial infarction can also occur. Hemolytic anemia and leukopenia can be present (see chapter 19). Antiphospholipid antibody syndrome, through an unknown mechanism, increases coagulation and thrombus formation, which increases the risk for stroke and heart attack.

RESPIRATORY
Chest pain may be caused by inflammation of the pleural membranes; fever, shortness of breath, and hypoxemia may occur due to inflammation of the lungs; alveolar hemorrhage can develop.

potent therapies as conditions warrant. Aspirin and nonsteroidal anti-inflammatory drugs are used to suppress inflammation. Antimalarial drugs are prescribed to treat skin rash and arthritis in SLE, but the mechanism of action is unknown. Patients who do not respond to these drugs and those who have severe SLE are helped by glucocorticoids. Although glucocorticoids effectively treat inflammation, they can produce undesirable side effects, including suppression of normal adrenal gland functions. In patients with life-threatening SLE, very high doses of glucocorticoids are used.

Predict 8

The red lesion Lucy developed on her arm is called purpura (pŭr′poo-ră), and it is caused by bleeding into the skin. The lesions gradually change color and disappear in 2–3 weeks. Explain how SLE produces purpura.

are modified to resemble human antibodies. This approach has allowed monoclonal antibodies to sneak past the immune system.

Some uses of monoclonal antibodies to treat tumors are yielding promising results. For example, monoclonal antibodies with radioactive iodine (^{131}I) have caused the regression of B-cell lymphomas with few side effects. Herceptin, a monoclonal antibody, binds to a growth factor that is overexpressed in 25–30% of primary breast cancers. The antibodies "tag" cancer cells, which are then lysed by natural killer cells. Herceptin slows disease progression and increases survival time, but it is not a cure for breast cancer.

Many other immunotherapy approaches are being studied, and more treatments that use the immune system are sure to be developed.

ASSESS YOUR PROGRESS

63. *What is immunotherapy? Give some examples.*

22.9 Effects of Aging on the Lymphatic System and Immunity

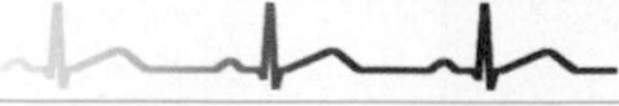

LEARNING OUTCOME

After reading this section, you should be able to

A. Describe how aging affects the lymphatic system and immunity.

Aging appears to have little effect on the lymphatic system's ability to remove fluid from tissues, absorb lipids from the digestive tract, or remove defective red blood cells from the blood. However, aging has a severe impact on the adaptive immune system.

With age, people eventually lose the ability to produce new, mature T cells in the thymus. By the age of 40, much of the thymus has been replaced with adipose tissue and, after age 60, the thymus decreases in size to the point that it can be difficult to detect. Although the number of T cells remains stable in most individuals due to the replication (not maturation) of T cells in secondary lymphatic tissues, the T cells are less functional. In many individuals, the ability of helper T cells to proliferate in response to antigens decreases. Thus, antigen exposure produces fewer helper T cells, which results in less stimulation of B cells and cytotoxic T cells. Consequently, both antibody-mediated immunity and cell-mediated immunity responses to antigens decrease.

Both primary and secondary antibody responses decrease with age. More antigen is required to produce a response, the response is slower, less antibody is produced, and fewer memory cells result. Thus, a person's ability to resist infections and develop immunity decreases. Because these declines are most evident after age 60, it is recommended that regular vaccinations be given well before that age. However, vaccinations can be beneficial at any age, especially if the individual has reduced resistance to infection. For example, the elderly are more susceptible to influenza (flu) and should be vaccinated every year.

The ability of cell-mediated immunity to resist intracellular pathogens also decreases with age. Some pathogens cause disease but are not eliminated from the body; with age, decreased immunity can lead to reactivation of the pathogen. An example is the virus that causes chickenpox in children, which can remain latent within neurons, even if the disease seems to have disappeared. Later in life, the virus can leave the neurons and infect skin cells, causing painful lesions known as herpes zoster, or shingles.

Autoimmune disease occurs when immune responses destroy otherwise healthy tissue. There is very little increase in the number of new-onset autoimmune diseases in the elderly. However, the chronic inflammation and immune responses that begin earlier in life have a cumulative, damaging effect. Likewise, the increased incidence of cancer is likely to be caused primarily by repeated exposure to and damage from cancer-causing agents rather than by decreased immunity.

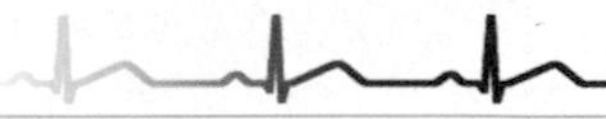

ASSESS YOUR PROGRESS

64. *What effect does aging have on the major functions of the lymphatic system?*

65. *Describe the effects of aging on B cells and T cells. Give examples of how they affect antibody-mediated and cell-mediated immune responses.*

Answer

Learn to Predict **From page 769**

Maddie was correct; she was having an allergic, or hypersensitivity, reaction to one of the chemicals sprayed on the trees. After reviewing the Diseases and Disorders table in this chapter, we can conclude that her symptoms are most consistent with a delayed hypersensitivity response since the itching and swelling did not occur until a day after she had been exposed to the chemicals. The chemicals on the leaves acted as haptens and combined with proteins, which were processed by the MHC class I processing system in the skin cells of her hands. In this chapter we learned that cytotoxic T cells can combine with MHC class I molecules on the surface of skin cells, destroying the cells. In addition, cytokines are released that attract macrophages to the area, thereby enhancing phagocytosis and inflammation.

Answers to the rest of this chapter's Predict questions appear in Appendix G.

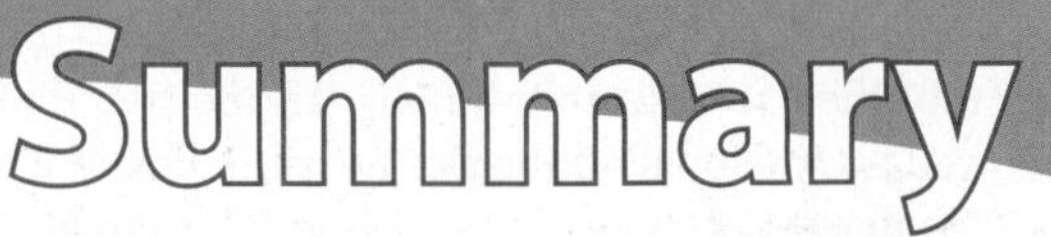
Summary

22.1 Functions of the Lymphatic System (p. 770)

The lymphatic system maintains fluid balance in tissues, absorbs lipids from the small intestine, and defends against microorganisms and foreign substances.

22.2 Anatomy of the Lymphatic System (p. 770)

The lymphatic system consists of lymph, lymphatic vessels, lymphatic tissue, lymphatic nodules, lymph nodes, the tonsils, the spleen, and the thymus.

Lymphatic Vessels

1. Lymphatic vessels carry lymph away from tissues.
2. Lymphatic capillaries lack a basement membrane and have loosely overlapping epithelial cells. Fluids and other substances easily enter lymphatic capillaries.
3. Lymphatic capillaries join to form lymphatic vessels.
   - Lymphatic vessels have valves that ensure a one-way flow of lymph.
   - Contraction of lymphatic vessel smooth muscle, contraction of skeletal muscle, and thoracic pressure changes move the lymph.
4. Lymph nodes are found along the lymphatic vessels. After passing through lymph nodes, lymphatic vessels form lymphatic trunks and lymphatic ducts.
5. Lymphatic trunks and ducts empty into the blood at thoracic veins (junctions of the internal jugular and subclavian veins).
   - Lymph from the right thorax, the right-upper limb, and the right side of the head and the neck enters the right thoracic veins.
   - Lymph from the lower limbs, pelvis, and abdomen; the left thorax; the left-upper limb; and the left side of the head and the neck enters the left thoracic veins.
6. The jugular, subclavian, and bronchomediastinal trunks may unite to form the right lymphatic duct.
7. The thoracic duct is the largest lymphatic vessel.
8. The intestinal and lumbar trunks may converge on the cisterna chyli, a sac that joins the inferior end of the thoracic duct.

Lymphatic Tissue and Organs

1. Lymphatic tissue is reticular connective tissue that contains lymphocytes and other cells.
2. Lymphatic tissue can be surrounded by a capsule (lymph nodes, spleen, thymus).
3. Lymphatic tissue can be nonencapsulated (diffuse lymphatic tissue, lymphatic nodules, tonsils). Mucosa-associated lymphoid tissue (MALT) is nonencapsulated lymphatic tissue located in and below the mucous membranes of the digestive, respiratory, urinary, and reproductive tracts.
4. Diffuse lymphatic tissue consists of dispersed lymphocytes and has no clear boundaries.
5. Lymphatic nodules are small aggregates of lymphatic tissue (e.g., Peyer patches in the small intestine).
6. The tonsils
   - The tonsils are large groups of lymphatic nodules in the oral cavity and nasopharynx.
   - The three groups of tonsils are the palatine, pharyngeal, and lingual tonsils.
7. Lymph nodes
   - Lymphatic tissue in the node is organized into the cortex and the medulla. Lymphatic sinuses extend through the lymphatic tissue.
   - Substances in lymph are removed by phagocytosis, or they stimulate lymphocytes (or both).
   - Lymphocytes leave the lymph nodes and circulate to other tissues.
8. The spleen
   - The spleen is in the left superior side of the abdomen.
   - Foreign substances stimulate lymphocytes in the white pulp of the spleen (periarterial lymphatic sheath and lymphatic nodules).
   - Foreign substances and defective red blood cells are removed from the blood by phagocytes in the red pulp of the spleen (splenic cords and venous sinuses).
   - The spleen is a limited reservoir for blood.
9. The thymus
   - The thymus is a gland in the superior mediastinum and is divided into a cortex and a medulla.
   - Lymphocytes in the cortex are separated from the blood by reticular cells.
   - Lymphocytes produced in the cortex migrate through the medulla, enter the blood, and travel to other lymphatic tissues, where they can proliferate.

Overview of the Lymphatic System

See figure 22.9.

22.3 Immunity (p. 778)

Immunity is the ability to resist the harmful effects of microorganisms and other foreign substances.

22.4 Innate Immunity (p. 780)

Physical Barriers

Physical barriers prevent the entry of microbes (skin and mucous membranes) or remove them (tears, saliva, and mucus).

Chemical Mediators

1. Chemical mediators promote phagocytosis and inflammation.
2. Complement can be activated by either the alternative or the classical pathway. Complement lyses cells, increases phagocytosis, attracts immune system cells, and promotes inflammation.
3. Interferons prevent viral replication. Interferons are produced by virally infected cells and move to other cells, which are then protected.

White Blood Cells

1. Chemotactic factors are parts of microorganisms or chemicals that are released by damaged tissues. Chemotaxis is the ability of white blood cells to move to tissues that release chemotactic factors.
2. Phagocytosis is the ingestion and destruction of materials.
3. Neutrophils are small phagocytic white blood cells.
4. Macrophages are large phagocytic white blood cells.
   - Macrophages can engulf more than neutrophils can.
   - Macrophages in connective tissue protect the body at locations where microbes are likely to enter, and macrophages clean blood and lymph.
5. Basophils and mast cells release chemicals that promote inflammation.
6. Eosinophils defend against parasitic worms.
7. Natural killer cells lyse tumor cells and virus-infected cells.

Inflammatory Response

1. The inflammatory response can be initiated in many ways.
   - Chemical mediators cause vasodilation and increase vascular permeability, which allows the entry of other chemical mediators.
   - Chemical mediators attract phagocytes.
   - The numbers of chemical mediators and phagocytes increase until the cause of the inflammation is destroyed. Then the tissue undergoes repair.
2. Local inflammation produces redness, heat, swelling, pain, and loss of function. Symptoms of systemic inflammation include an increase in neutrophil numbers, fever, and shock.

22.5 Adaptive Immunity (p. 784)

1. Antigens are large molecules that stimulate an adaptive immune response.
2. B cells are responsible for antibody-mediated immunity. T cells are involved with cell-mediated immunity.

Origin and Development of Lymphocytes

1. B cells and T cells originate in red bone marrow. T cells are processed in the thymus, and B cells are processed in bone marrow.
2. Positive selection ensures the survival of lymphocytes that can react against antigens, and negative selection eliminates lymphocytes that react against self-antigens.
3. A clone is a group of identical lymphocytes that can respond to a specific antigen.
4. B cells and T cells move to lymphatic tissue from their processing sites. They continually circulate from one lymphatic tissue to another.
5. The primary lymphatic organs (red bone marrow and the thymus) are where lymphocytes mature into functional cells. Secondary lymphatic organs and tissues are where lymphocytes produce an immune response.

Activation of Lymphocytes

1. The antigenic determinant is the specific part of the antigen to which the lymphocyte responds. The antigen receptor (T-cell receptor or B-cell receptor) on the surface of lymphocytes combines with the antigenic determinant.
2. MHC class I molecules display antigens on the surface of nucleated cells, resulting in the destruction of the cells.
3. MHC class II molecules display antigens on the surface of antigen-presenting cells, resulting in the activation of immune cells.
4. MHC-antigen complex and costimulation are usually necessary to activate lymphocytes. Costimulation involves cytokines and certain surface molecules.
5. Antigen-presenting cells stimulate the proliferation of helper T cells, which stimulate the proliferation of B cells or cytotoxic T cells.

Inhibition of Lymphocytes

1. Tolerance is suppression of the immune system's response to an antigen.
2. Tolerance is produced by the deletion of self-reactive cells, by the prevention of lymphocyte activation, and by the activation of regulatory T cells.

Antibody-Mediated Immunity

1. Antibodies are proteins.
   - The variable region of an antibody combines with the antigen. The constant region activates complement or binds to cells.
   - Five classes of antibodies exist: IgG, IgM, IgA, IgE, and IgD.
2. Antibodies affect the antigen in many ways.
   - Antibodies bind to the antigen and interfere with antigen activity or bind the antigens together.
   - Antibodies act as opsonins (substances that increase phagocytosis) by binding to the antigen and to macrophages.
   - Antibodies can activate complement through the classical pathway.
   - Antibodies attach to mast cells or basophils and cause the release of inflammatory chemicals when the antibody combines with the antigen.
3. The primary response results from the first exposure to an antigen. B cells form plasma cells, which produce antibodies and memory B cells.
4. The secondary response results from exposure to an antigen after a primary response, and memory B cells quickly form plasma cells and additional memory B cells.

Cell-Mediated Immunity

1. Cells infected with intracellular microorganisms process antigens that combine with MHC class I molecules.
2. Cytotoxic T cells are stimulated to divide, producing more cytotoxic T cells and memory T cells, when MHC class I/antigen complexes are presented to T-cell receptors. Cytokines released from helper T cells also stimulate cytotoxic T cells.
3. Cytotoxic T cells lyse virus-infected cells, tumor cells, and tissue transplants.
4. Cytotoxic T cells produce cytokines, which promote phagocytosis and inflammation.

22.6 Acquired Adaptive Immunity (p. 799)

1. Active natural immunity results from natural exposure to an antigen.
2. Active artificial immunity results from deliberate exposure to an antigen.
3. Passive natural immunity results from the transfer of antibodies from a mother to her fetus or baby.
4. Passive artificial immunity results from the transfer of antibodies (or cells) from an immune animal to a nonimmune animal.

22.7 Overview of Immune Interactions (p. 801)

Innate immunity, antibody-mediated immunity, and cell-mediated immunity can function together to eliminate an antigen.

22.8 Immunotherapy (p. 801)

Immunotherapy treats diseases by stimulating or inhibiting the immune system.

22.9 Effects of Aging on the Lymphatic System and Immunity (p. 806)

1. Aging has little effect on the lymphatic system's ability to remove fluid from tissues, absorb lipids from the digestive tract, or remove defective red blood cells from the blood.
2. Decreased helper T-cell proliferation results in decreased antibody-mediated and cell-mediated immune responses to antigens.
3. Primary and secondary antibody responses decrease with age.
4. The ability to resist intracellular pathogens decreases with age.

REVIEW AND COMPREHENSION

1. The lymphatic system
   a. removes excess fluid from tissues.
   b. absorbs lipids from the digestive tract.
   c. defends the body against microorganisms and other foreign substances.
   d. All of these are correct.
2. Which of the following statements is correct?
   a. Lymphatic vessels do not have valves.
   b. Lymphatic vessels empty into lymph nodes.
   c. Lymph from the right-lower limb passes into the right jugular or subclavian vein.
   d. Lymph from the jugular and subclavian trunks empties into the cisterna chyli.
   e. All of these are correct.

3. The tonsils
   a. consist of large groups of lymphatic nodules.
   b. protect against bacteria.
   c. can become chronically infected.
   d. decrease in size in adults.
   e. All of these are correct.
4. Lymph nodes
   a. filter lymph.
   b. are where lymphocytes divide and increase in number.
   c. contain a network of reticular fibers.
   d. contain lymphatic sinuses.
   e. All of these are correct.
5. Which of these statements about the spleen is *not* correct?
   a. The spleen has white pulp associated with the arteries.
   b. The spleen has red pulp associated with the veins.
   c. The spleen destroys defective red blood cells.
   d. The spleen is surrounded by trabeculae located outside the capsule.
   e. The spleen is a limited reservoir for blood.
6. The thymus
   a. increases in size in adults.
   b. produces lymphocytes that move to other lymphatic tissue.
   c. is located in the abdominal cavity.
   d. All of these are correct.
7. Which of these is an example of innate immunity?
   a. Tears and saliva wash away microorganisms.
   b. Basophils release histamine and leukotrienes.
   c. Neutrophils phagocytize a microorganism.
   d. The complement cascade is activated.
   e. All of these are correct.
8. Neutrophils
   a. enlarge to become macrophages.
   b. account for most of the dead cells in pus.
   c. are usually the last cell type to enter infected tissues.
   d. are usually located in lymphatic and blood sinuses.
9. Macrophages
   a. are large, phagocytic cells that outlive neutrophils.
   b. develop from mast cells.
   c. often die after a single phagocytic event.
   d. have the same function as eosinophils.
   e. All of these are correct.
10. Which of these cells is the most important in the release of histamine, which promotes inflammation?
    a. monocyte
    b. macrophage
    c. eosinophil
    d. mast cell
    e. natural killer cell
11. Which of these conditions does *not* occur during the inflammatory response?
    a. release of histamine and other chemical mediators
    b. chemotaxis of phagocytes
    c. entry of fibrinogen into tissues from the blood
    d. vasoconstriction of blood vessels
    e. increased permeability of blood vessels
12. Antigens
    a. are foreign substances introduced into the body.
    b. are molecules produced by the body.
    c. stimulate an adaptive immune system response.
    d. All of these are correct.
13. B cells
    a. are processed in the thymus.
    b. originate in red bone marrow.
    c. once released into the blood remain in the blood.
    d. are responsible for cell-mediated immunity.
    e. All of these are correct.
14. MHC molecules
    a. are glycoproteins.
    b. attach to the plasma membrane.
    c. have a variable region that can bind to foreign antigens and self-antigens.
    d. may form an MHC-antigen complex that activates T cells.
    e. All of these are correct.
15. Antigen-presenting cells can
    a. take in foreign antigens.
    b. process antigens.
    c. use MHC class II molecules to display the antigens.
    d. stimulate other immune system cells.
    e. All of these are correct.
16. Which of these participates in costimulation?
    a. cytokines
    b. complement
    c. antibodies
    d. histamine
    e. natural killer cells
17. Helper T cells
    a. respond to antigens from macrophages.
    b. respond to cytokines from macrophages.
    c. stimulate B cells with cytokines.
    d. All of these are correct.
18. The most important function of tolerance is to
    a. increase lymphocyte activity.
    b. increase complement activation.
    c. prevent the immune system from responding to self-antigens.
    d. prevent excessive immune response to foreign antigens.
    e. process antigens.
19. Variable amino acid sequences on the arms of the antibody molecule
    a. make the antibody specific for a given antigen.
    b. enable the antibody to activate complement.
    c. enable the antibody to attach to basophils and mast cells.
    d. are part of the constant region.
    e. All of these are correct.
20. Antibodies
    a. prevent antigens from binding together.
    b. promote phagocytosis.
    c. inhibit inflammation.
    d. block complement activation.
    e. block the function of opsonins.
21. The secondary antibody response
    a. is slower than the primary response.
    b. produces fewer antibodies than the primary response.
    c. prevents disease symptoms from occurring.
    d. occurs because of cytotoxic T cells.
22. The type of lymphocyte responsible for the secondary antibody response is the
    a. memory B cell.
    b. B cell.
    c. T cell.
    d. helper T cell.
23. The largest percentage of antibodies in the blood are
    a. IgA.
    b. IgD.
    c. IgE.
    d. IgG.
    e. IgM.

24. Antibody-mediated immunity
    a. works best against intracellular antigens.
    b. regulates the activity of T cells.
    c. cannot be transferred from one person to another.
    d. is responsible for immediate hypersensitivity reactions.

25. The activation of cytotoxic T cells can result in
    a. lysis of virus-infected cells.
    b. production of cytokines.
    c. production of memory T cells.
    d. All of these are correct.

Answers in Appendix E

CRITICAL THINKING

1. A patient is suffering from edema in the right-lower limb. Explain why elevating and massaging the limb help remove the excess fluid.

2. If the thymus of an experimental animal is removed immediately after its birth, the animal exhibits the following characteristics: (a) an increased susceptibility to infections, (b) decreased numbers of lymphocytes in lymphatic tissue, and (c) a greatly decreased ability to reject grafts. Explain these observations.

3. If the thymus of an adult experimental animal is removed, the following observations can be made: (a) No immediate effect occurs, and (b) after 1 year, decreases occur in the number of lymphocytes in the blood, the ability to reject grafts, and the ability to produce antibodies. Explain these observations.

4. Adjuvants are substances that slow but do not stop the release of an antigen from an injection site into the blood. Suppose injection A is given without an adjuvant and injection B of the same amount of antigen is given with an adjuvant that causes antigen to be released over a period of 2–3 weeks. Does injection A or injection B result in the greater amount of antibody production? Explain.

5. Tetanus is caused by bacteria that enter the body through wounds in the skin. The bacteria produce a toxin that causes spastic muscle contractions. Death often results from failure of the respiratory muscles. A patient goes to the emergency room after stepping on a nail. If the patient has been vaccinated against tetanus, he or she is given a tetanus booster shot, which consists of the toxin altered so that it is harmless. A patient who has never been vaccinated against tetanus is given an antiserum shot against tetanus. Explain the rationale for this treatment strategy. Sometimes both a booster and an antiserum shot are given, but at different locations on the body. Explain why both vaccinations are given and why they are injected in different locations.

6. An infant appears healthy until about 9 months of age, when he develops severe bacterial infections, one after another. Fortunately, the infections are treated successfully with antibiotics. When infected with the measles and other viral diseases, the infant recovers without unusual difficulty. Explain the different immune responses to these infections. Why did it take so long for this disorder to become apparent? (*Hint:* Consider IgG.)

7. A patient has many allergic reactions. As part of the treatment scheme, doctors try to identify the allergen that stimulates the immune system's response. A series of solutions, each containing an allergen that commonly causes a reaction, is composed. Each solution is injected into the skin at different locations on the patient's back. The following results are obtained: (a) At one location, the injection site becomes red and swollen within a few minutes; (b) at another injection site, swelling and redness appear 2 days later; and (c) no redness or swelling develops at the other sites. Explain what happened for each observation by describing what part of the immune system was involved and what caused the redness and swelling.

8. Ivy Hurtt developed a poison ivy rash after a camping trip. Her doctor prescribed a cortisone ointment to relieve the inflammation. A few weeks later, Ivy scraped her elbow, which became inflamed. Because she had some of the cortisone ointment left over, she applied it to the scrape. Was the ointment an effective treatment for the poison ivy? Was the ointment an appropriate treatment for the scrape?

9. Billie Boards was riding her skateboard down a newly paved road from the top of a hill. Part of the way down, she fell and skinned her knees on the asphalt. From the following list, choose the immune response(s) that occurred in the next several hours.
   (1) increased capillary permeability
   (2) chemotaxis of neutrophils
   (3) coagulation
   (4) release of mediators of inflammation
   (5) increased mitosis of B lymphocytes
   a. 1,2,3,4,5 b. 1,2,3,4 c. 1,2,3 d. 1,2

10. Upon first exposure to an antigen, a sequence of events results in antigen processing and an increase in the number of helper T cells. Given the following list of events, select the sequence that results in an increased number of helper T cells.
    (1) Unprocessed extracellular antigen is ingested by a macrophage.
    (2) The MHC II complex is presented at the cell surface.
    (3) Costimulation occurs.
    (4) Interleukin-1 is released from macrophages, and interleukin-2 is released from T lymphocytes.
    (5) A specific helper T cell recognizes and binds to the MHC II complex.
    (6) Mitosis of helper T cells takes place.
    (7) The ingested antigen is broken down to fragments.
    (8) The processed antigen and MHC II molecules are joined and transported to the cell surface.
    a. 1,7,8,2,5,3,4,6
    b. 1,7,8,5,2,3,4,6
    c. 7,8,1,2,5,3,4,6
    d. 1,2,3,4,5,6,7,8
    e. 1,8,7,2,3,5,6,4

Answers in Appendix F

Visit this book's website at **www.mhhe.com/seeley10** for chapter quizzes, interactive learning exercises, and other study tools.

Mc Graw Hill
connect plus+
ANATOMY & PHYSIOLOGY

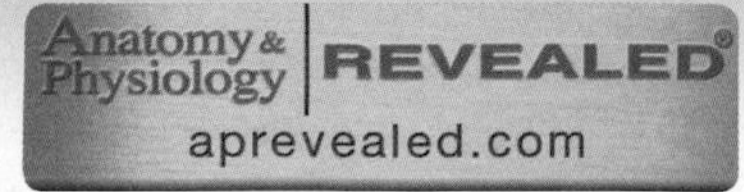
Anatomy & Physiology
REVEALED
aprevealed.com

23

Respiratory System

Learn to Predict

Flashing lights at 2 a.m. alerted the neighbors that something was wrong at the Theron home. Mr. Theron, who has moderate emphysema, could not stop coughing, so his wife called 911. In the emergency room, a physician listened to Mr. Theron's respiratory sounds and concluded that his left lung had collapsed. By using some information about the peripheral circulation from chapter 21 and reading chapter 23, explain how emphysema affected Mr. Theron's breathing, what caused his lung to collapse, and how the physician was able to detect the collapsed lung.

If you've ever had a severe chest cold, bronchitis, or pneumonia, you might be able to relate to what Mr. Theron of this chapter's "Learn to Predict" experienced when his left lung collapsed. If we stop breathing, or breathing becomes difficult, within seconds we feel a strong need for air. From our first breath at birth, the rate and depth of our breathing are unconsciously matched to our activities, whether studying, sleeping, talking, eating, or exercising. Breathing is so characteristic of life that, along with the pulse, it is one of the first vital signs checked to determine whether an unconscious person is alive.

Breathing is necessary because all living cells of the body require oxygen and produce carbon dioxide. The respiratory system exchanges these gases between the air and the blood, and the cardiovascular system transports them between the lungs and the body cells. Without healthy respiratory and cardiovascular systems, the capacity to carry out normal activity is reduced.

Photo: Colorized scanning electron micrograph of the lung, showing alveoli, which are small air chambers where gas exchange takes place between the air and the blood.

Module 11
Respiratory System

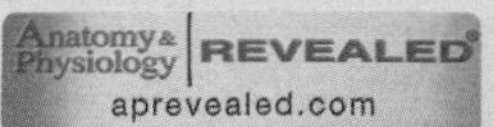
Anatomy & Physiology
REVEALED
aprevealed.com

23.1 Functions of the Respiratory System

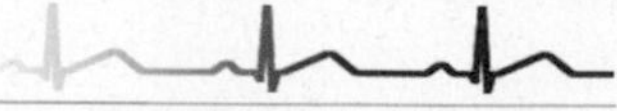

LEARNING OUTCOME

After reading this section, you should be able to

A. Describe the functions of the respiratory system.

The respiratory system helps us in what we commonly call breathing but is more appropriately termed **respiration.** Respiration includes four processes: (1) ventilation, the movement of air into and out of the lungs; (2) gas exchange between the air in the lungs and the blood, sometimes called *external respiration*; (3) the transport of oxygen and carbon dioxide in the blood; and (4) gas exchange between the blood and the tissues, sometimes called *internal respiration*.

It can be confusing to hear the term *respiration* alone because sometimes it also refers to cellular metabolism, or **cellular respiration** (discussed in chapter 25); in fact, the two processes are related. Breathing provides the oxygen needed in cellular respiration to make ATP from glucose. Breathing also rids the body of potentially toxic carbon dioxide, the waste produced during cellular respiration.

In addition to respiration, the respiratory system performs the following functions:

1. *Regulation of blood pH.* The respiratory system can alter blood pH by changing blood carbon dioxide levels.
2. *Production of chemical mediators.* The lungs produce an enzyme called angiotensin-converting enzyme (ACE), which is an important component of blood pressure regulation (discussed in chapter 26).
3. *Voice production.* Air moving past the vocal folds makes sound and speech possible.
4. *Olfaction.* The sensation of smell occurs when airborne molecules are drawn into the nasal cavity (discussed in chapter 15).
5. *Protection.* The respiratory system provides protection against some microorganisms by preventing them from entering the body and removing them from respiratory surfaces.

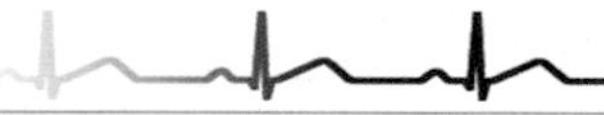

ASSESS YOUR PROGRESS

1. *What are the four processes of respiration?*
2. *Explain the functions of the respiratory system.*

23.2 Anatomy and Histology of the Respiratory System

LEARNING OUTCOMES

After reading this section, you should be able to

A. Describe the anatomy of the respiratory passages, beginning at the nose and ending with the alveoli.

B. Describe the structural and functional anatomy of the respiratory system.

C. State the parts of the respiratory membrane.

D. Describe the structure of the lungs, including the blood and lymphatic supply.

E. Explain the roles of the thoracic wall and the pleura in respiration.

F. Explain how contraction of the muscles of respiration causes changes in thoracic volume during quiet breathing and during labored breathing.

The respiratory system consists of the external nose, the nasal cavity, the pharynx, the larynx, the trachea, the bronchi, and the lungs (figure 23.1). The diaphragm and the muscles of the thoracic and abdominal walls are responsible for respiratory movements. (Although air frequently passes through it, the oral cavity is considered part of the digestive system rather than the respiratory system.) There are two ways of classifying the parts of the respiratory system: structurally and functionally. Structurally, the respiratory system is divided into the upper respiratory tract and the lower respiratory tract. The **upper respiratory tract** includes the external nose, the nasal cavity, the pharynx with its associated structures, and the larynx; the **lower respiratory tract** includes the trachea, the bronchi and smaller bronchioles, and the lungs. However, these are not official anatomical terms, and they have several alternate definitions. For example, one classification system places the larynx in the lower respiratory tract.

Functionally, the respiratory system is divided into two regions. The **conducting zone** is exclusively for air movement and extends from the nose to the bronchioles. The **respiratory zone** is within the lungs and is where gas exchange between air and blood takes place.

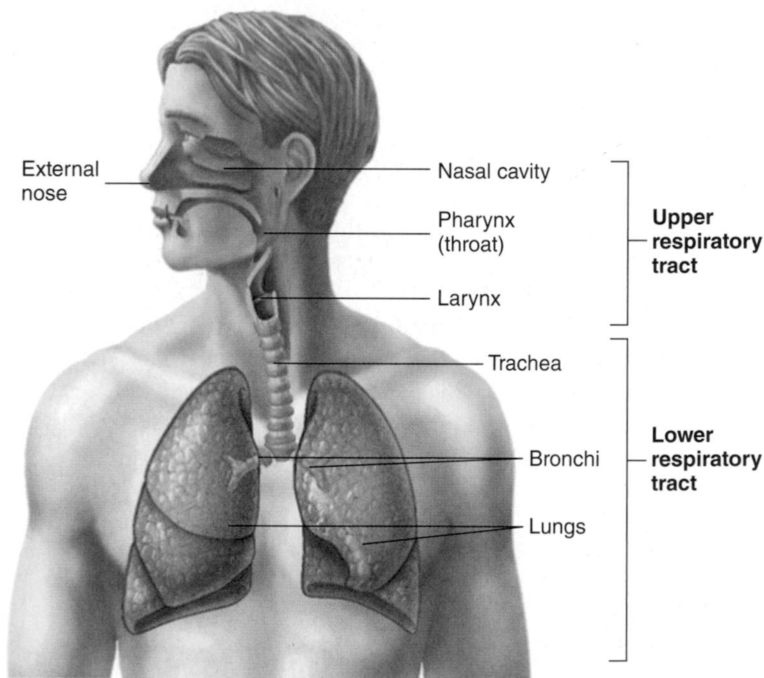
External nose
Nasal cavity
Pharynx (throat)
Larynx
Upper respiratory tract
Trachea
Bronchi
Lungs
Lower respiratory tract

FIGURE 23.1 APIR **Respiratory System**

The upper respiratory tract consists of the external nose, the nasal cavity, the pharynx (throat) and its associated structures, and the larynx. The lower respiratory tract consists of the trachea, the bronchi and smaller bronchioles, and the lungs.

Conducting Zone

The conducting zone consists of respiratory system structures adapted for air movement, cleaning, warming, and humidification. Gases are simply moved from the external environment to the area where they interact with blood.

Nose

The **nose,** or *nasus* (nā′sŭs), consists of the external nose and the nasal cavity. The **external nose** is the visible structure that forms a prominent feature of the face. The largest part of the external nose is composed of hyaline cartilage plates (see figure 7.9*b*). The bridge of the nose consists of the nasal bones plus extensions of the frontal and maxillary bones.

The **nasal cavity** extends from the nares to the choanae (figure 23.2). The **nares** (nā′res; sing. *naris*), or *nostrils*, are the external openings of the nasal cavity, and the **choanae** (kō′an-ē) are the openings into the pharynx. The anterior part of the nasal cavity, just inside each naris, is the **vestibule** (ves′ti-bool; entry room). The vestibule is lined with stratified squamous epithelium, which is continuous with the stratified squamous epithelium of the skin. The **hard palate** (pal′ăt) is formed by the palatine process of the maxillae and the palatine bone. It is covered by a highly vascular mucous membrane that forms the floor of the nasal cavity. It separates the nasal cavity from the oral cavity. The **nasal septum** is a partition dividing the nasal cavity into right and left parts (see figure 7.9*a*). The anterior part of the nasal septum is cartilage, and the posterior part consists of the vomer bone and the perpendicular plate of the ethmoid bone. A deviated nasal septum occurs when the septum bulges to one side.

Three bony ridges, called **conchae** (kon′kē; resembling a conch shell), modify the lateral walls of the nasal cavity. Beneath each concha is a passageway called a **meatus** (mē-ā′tŭs; tunnel or passageway). Within the superior and middle meatuses are openings from the various **paranasal sinuses** (see figure 7.10), and the opening of a **nasolacrimal** (nā-zō-lak′ri-măl) **duct** is within each inferior meatus (see figure 15.10).

The nasal cavity has five functions:

1. *Serves as a passageway for air.* The nasal cavity remains open even when the mouth is full of food.
2. *Cleans the air.* The vestibule is lined with hairs, which trap some of the large particles of dust in the air. The nasal septum and nasal conchae increase the surface area of the nasal cavity and make airflow within the cavity more turbulent, thereby increasing the likelihood that air will come into contact with the mucous membrane lining the nasal cavity. This mucous membrane consists of pseudostratified ciliated columnar epithelium with goblet cells, which secrete a layer of mucus. The mucus traps debris in the air, and the cilia on the surface of the mucous membrane sweep the mucus posteriorly to the pharynx, where it is swallowed and eliminated by the digestive system.
3. *Humidifies and warms the air.* Moisture from the mucous epithelium and from excess tears that drain into the nasal cavity through the nasolacrimal duct is added to the air as it passes through the nasal cavity. Warm blood flowing through the mucous membrane warms the air within the nasal cavity before it passes into the pharynx, thus preventing damage to the rest of the respiratory passages due to cold air.
4. *Contains the olfactory epithelium.* The olfactory epithelium, the sensory organ for smell, is located in the most superior part of the nasal cavity (see figure 15.1).
5. *Helps determine voice sound.* The nasal cavity and paranasal sinuses are resonating chambers for speech. For example, most people know immediately when you have a cold because your voice sounds different.

Pharynx

The **pharynx** (far′ingks; throat) is the common opening of both the digestive and the respiratory systems. It receives air from the nasal cavity and receives air, food, and drink from the oral cavity. Inferiorly, the pharynx is connected to the respiratory system at the larynx and to the digestive system at the esophagus. The pharynx is divided into three regions: the nasopharynx, the oropharynx, and the laryngopharynx (figure 23.2*a*).

The **nasopharynx** (nā′zō-far′ingks) is located posterior to the choanae and superior to the **soft palate,** which is an incomplete muscle and connective tissue partition separating the nasopharynx from the oropharynx. The **uvula** (ū′vū-lă; grape) is the posterior extension of the soft palate. The soft palate prevents swallowed materials from entering the nasopharynx and nasal cavity. The nasopharynx is lined with a mucous membrane containing pseudostratified ciliated columnar epithelium with goblet cells. Debris-laden mucus from the nasal cavity is moved through the nasopharynx and swallowed; digestive juices help kill any swallowed pathogens. Two auditory tubes from the middle ears open into the nasopharynx (figure 23.2*a*; see figure 15.24). Air passes through them to equalize air pressure between the atmosphere and the middle ears. The posterior surface of the nasopharynx contains the pharyngeal tonsil, or *adenoid* (ad′ĕ-noyd), which helps defend the body against infection (see chapter 22). An enlarged pharyngeal tonsil can interfere with normal breathing and the passage of air through the auditory tubes.

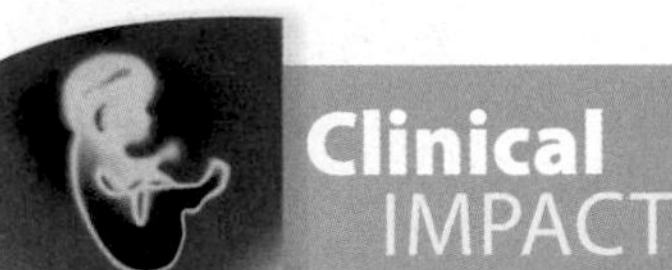
Clinical
IMPACT

Sinusitis

Sinusitis (sī-nŭ-sī′tis) is an inflammation of the mucous membrane of any sinus, especially one or more of the paranasal sinuses. Viral infections, such as the common cold, can cause mucous membranes to become inflamed and swollen and to produce excess mucus. As a result, the sinus opening into the nasal cavity is partially or completely blocked, allowing mucus to accumulate within the sinus, which can promote a bacterial infection. Treatments include taking antibiotics and promoting sinus drainage with decongestants, hydration, and steam inhalation. Sinusitis can also result from swelling caused by allergies or by polyps that obstruct the sinus opening into the nasal cavity.

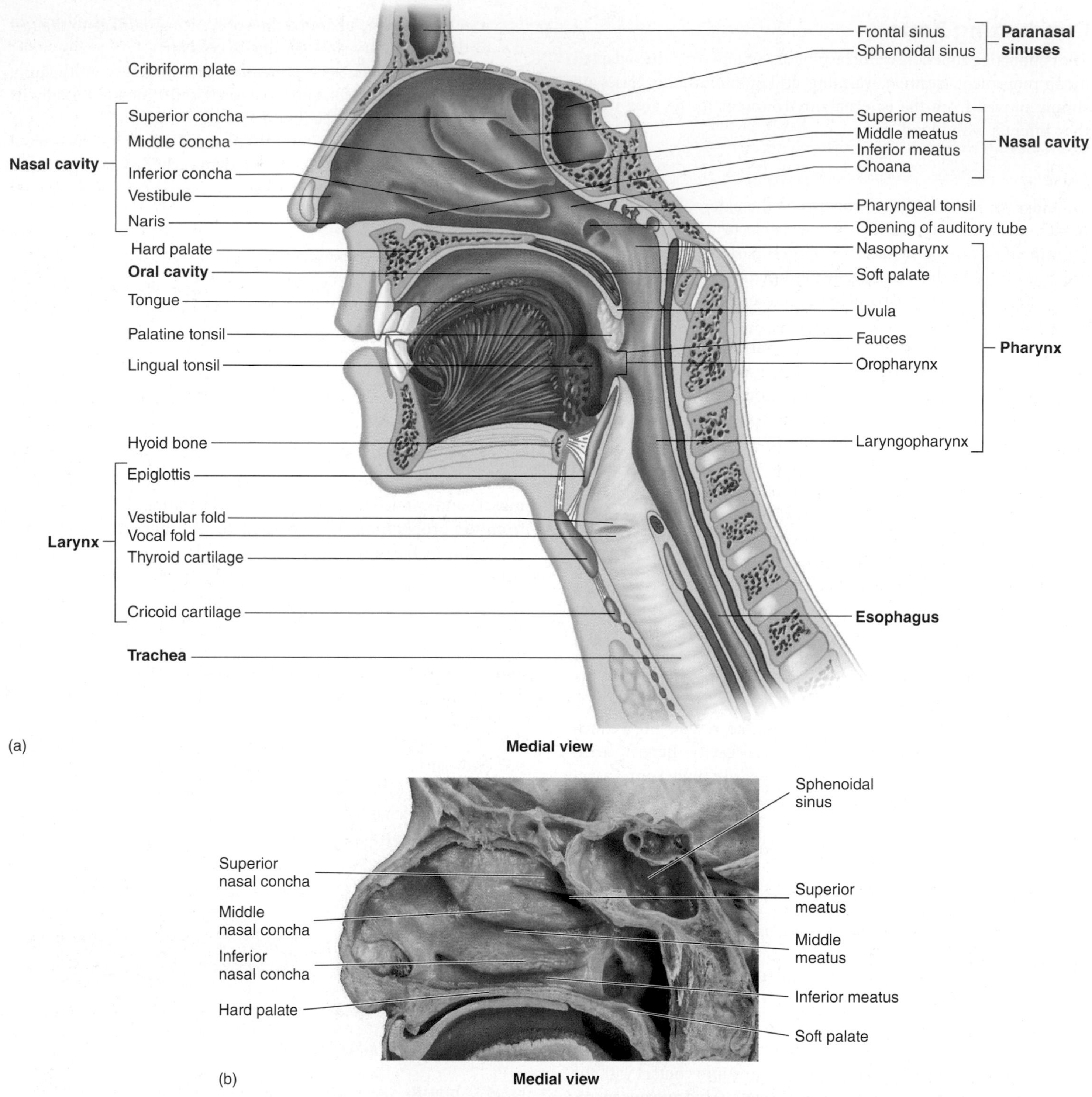
Frontal sinus
Sphenoidal sinus
Paranasal sinuses
Cribriform plate
Nasal cavity
Superior concha
Middle concha
Inferior concha
Vestibule
Naris
Superior meatus
Middle meatus
Inferior meatus
Choana
Nasal cavity
Pharyngeal tonsil
Opening of auditory tube
Hard palate
Oral cavity
Tongue
Palatine tonsil
Lingual tonsil
Nasopharynx
Soft palate
Uvula
Fauces
Oropharynx
Pharynx
Hyoid bone
Laryngopharynx
Larynx
Epiglottis
Vestibular fold
Vocal fold
Thyroid cartilage
Cricoid cartilage
Esophagus
Trachea
(a)
Medial view
Sphenoidal sinus
Superior nasal concha
Middle nasal concha
Inferior nasal concha
Hard palate
Superior meatus
Middle meatus
Inferior meatus
Soft palate
(b)
Medial view

FIGURE 23.2 APIR **Nasal Cavity and Pharynx**
(*a*) Sagittal section through the nasal cavity and pharynx. (*b*) Photograph of sagittal section of the head.

The **oropharynx** (ōr′ō-far′ingks) extends from the soft palate to the epiglottis. The oral cavity opens into the oropharynx through the **fauces** (faw′sēz). Thus, air, food, and drink all pass through the oropharynx. Moist stratified squamous epithelium lines the oropharynx and protects it against abrasion. Two sets of tonsils, called the palatine tonsils and the lingual tonsils, are located near the fauces.

The **laryngopharynx** (lă-ring′gō-far-ingks) extends from the tip of the epiglottis to the esophagus and passes posterior to the larynx. Food and drink pass through the laryngopharynx to the esophagus. A small amount of air is usually swallowed with the food and drink. The laryngopharynx is lined with moist stratified squamous epithelium.

ASSESS YOUR PROGRESS

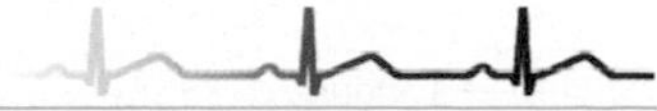

3. *Name the parts of the upper and lower respiratory tracts.*
4. *Explain how the conducting zone differs from the respiratory zone.*
5. *Describe the structures of the nasal cavity.*
6. *What are the five functions of the nasal cavity?*
7. *Name the three regions of the pharynx. With what other structures does each part communicate?*

Larynx

The **larynx** (lar′ingks) is located in the anterior part of the throat and extends from the base of the tongue to the trachea (figure 23.2*a*). It is a passageway for air between the pharynx and the trachea. The larynx is connected by membranes and/or muscles superiorly to the hyoid bone and consists of an outer casing of nine cartilages connected to one another by muscles and ligaments (figure 23.3). Six of the nine cartilages are paired, and three are unpaired. The largest of the cartilages is the unpaired **thyroid** (shield; refers to the shape of the cartilage) **cartilage,** or *Adam's apple.*

The most inferior cartilage of the larynx is the unpaired **cricoid** (krī′koyd; ring-shaped) **cartilage,** which forms the base of the larynx on which the other cartilages rest.

The third unpaired cartilage is the **epiglottis** (ep-i-glot′is; on the glottis). It is attached to the thyroid cartilage and projects superiorly as a free flap toward the tongue. The epiglottis differs from the other cartilages in that it consists of elastic rather than hyaline cartilage.

The paired **arytenoid** (ar-i-tē′noyd; ladle-shaped) **cartilages** articulate with the posterior, superior border of the cricoid cartilage, and the paired **corniculate** (kōr-nik′ū-lāt; horn-shaped) **cartilages** are attached to the superior tips of the arytenoid cartilages. The paired **cuneiform** (kū′nē-i-fōrm; wedge-shaped) **cartilages** are contained in a mucous membrane anterior to the corniculate cartilages (figure 23.3*b*).

Two pairs of ligaments extend from the anterior surface of the arytenoid cartilages to the posterior surface of the thyroid cartilage. The superior ligaments are covered by a mucous membrane called the **vestibular folds,** or *false vocal cords* (figures 23.3*c* and 23.4*a,b*).

The inferior ligaments are covered by a mucous membrane called the **vocal folds,** or *true vocal cords* (figure 23.4). The vocal

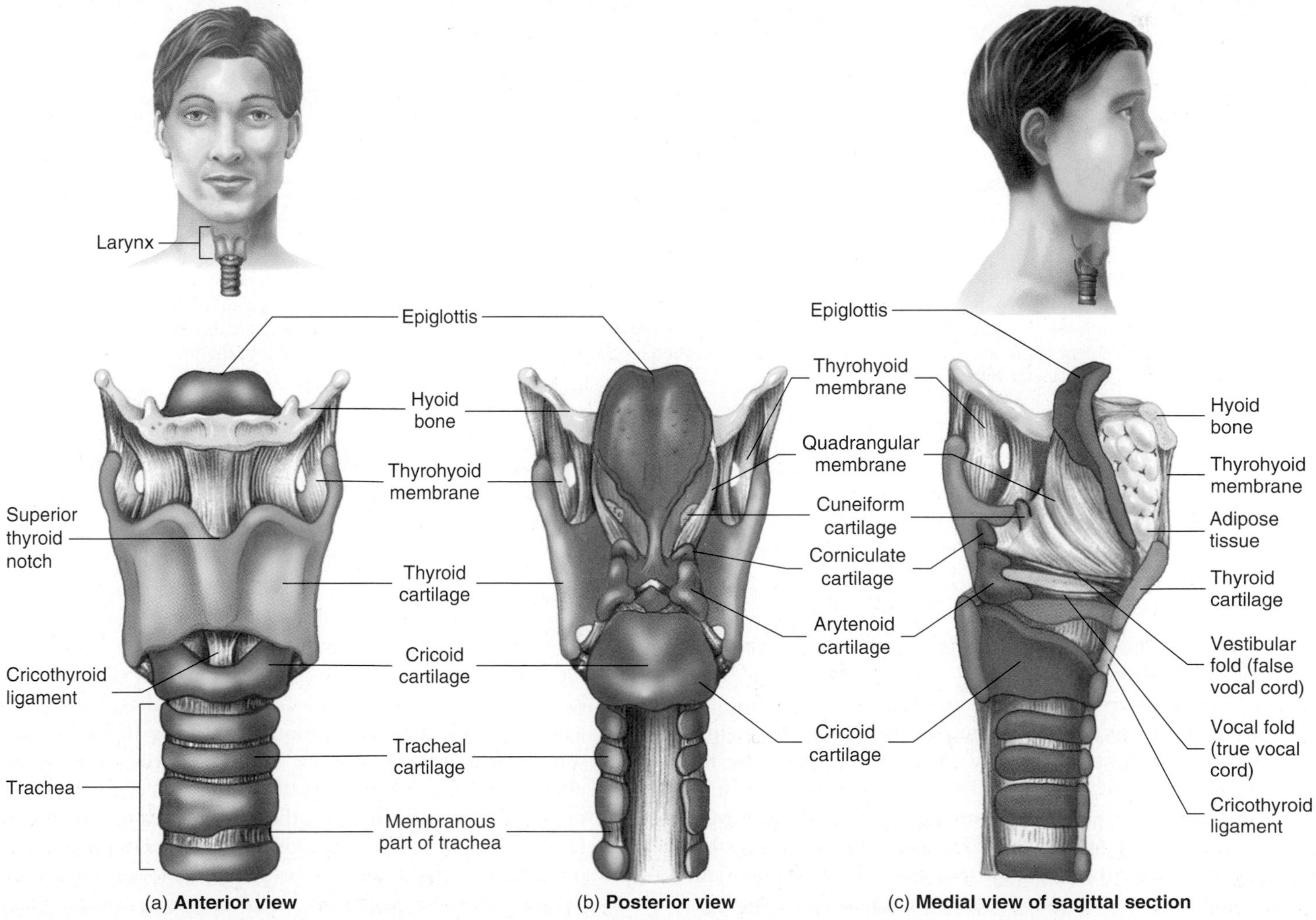

Larynx
Epiglottis
Hyoid bone
Thyrohyoid membrane
Superior thyroid notch
Thyroid cartilage
Cricothyroid ligament
Cricoid cartilage
Tracheal cartilage
Trachea
Membranous part of trachea
Quadrangular membrane
Cuneiform cartilage
Corniculate cartilage
Arytenoid cartilage
Adipose tissue
Vestibular fold (false vocal cord)
Vocal fold (true vocal cord)

(a) **Anterior view** (b) **Posterior view** (c) **Medial view of sagittal section**

FIGURE 23.3 APR **Anatomy of the Larynx**

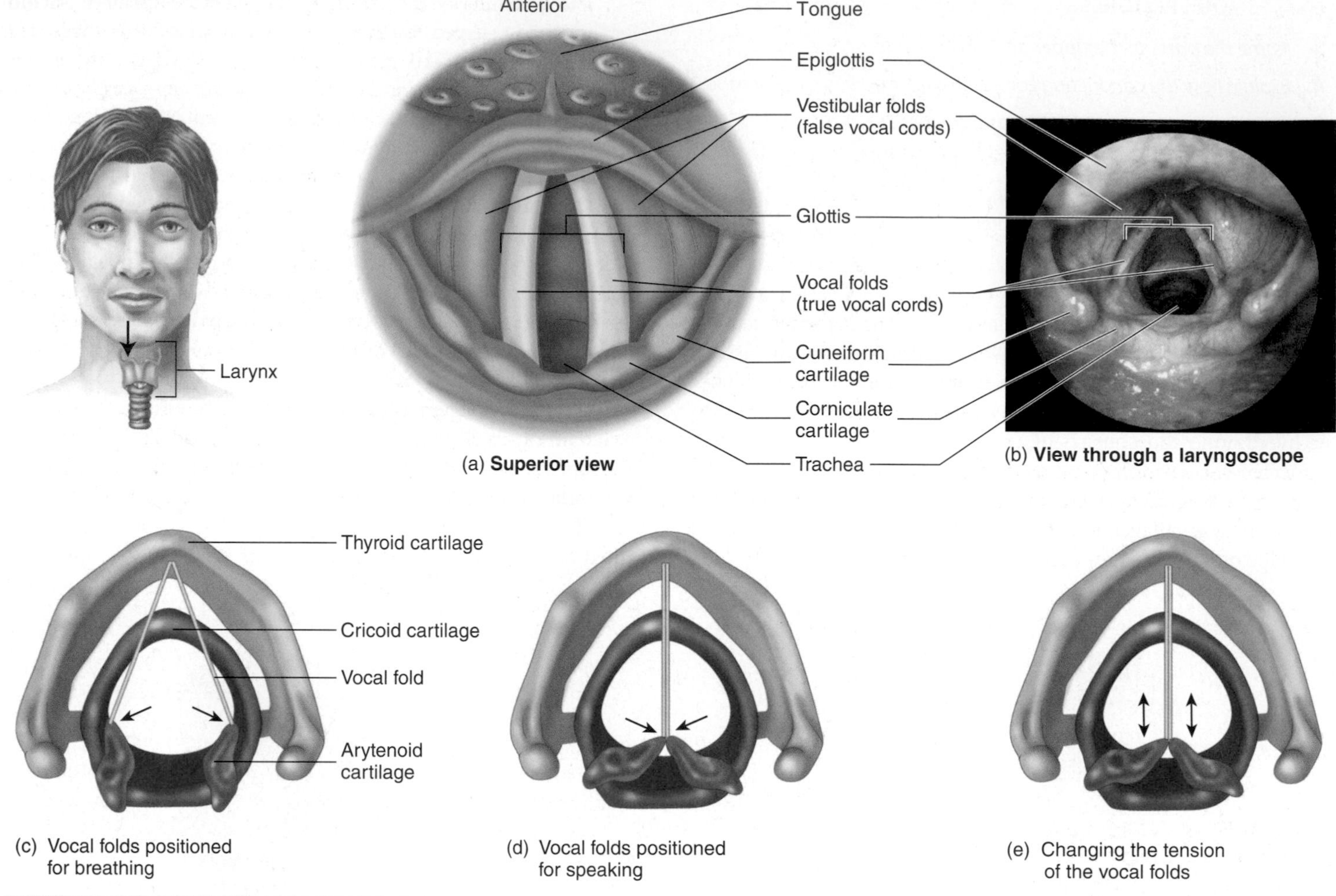

Anterior
Tongue
Epiglottis
Vestibular folds
(false vocal cords)
Glottis
Vocal folds
(true vocal cords)
Cuneiform
cartilage
Corniculate
cartilage
Trachea
Larynx
(a) Superior view
(b) View through a laryngoscope
Thyroid cartilage
Cricoid cartilage
Vocal fold
Arytenoid
cartilage
(c) Vocal folds positioned for breathing
(d) Vocal folds positioned for speaking
(e) Changing the tension of the vocal folds

FIGURE 23.4 Vestibular Vocal Folds

(*a*) Relationship of the vestibular folds to the vocal folds and the laryngeal cartilages. (*b*) Laryngoscopic view of the vestibular and vocal folds. (*c*) Lateral rotation of the arytenoid cartilages moves the vocal folds laterally for breathing. (*d*) Medial rotation of the arytenoid cartilages moves the vocal folds medially for speaking. (*e*) Anterior/posterior movement of the arytenoid cartilages changes the length and tension of the vocal folds, altering the pitch of sounds. Arrows show the direction of viewing the vestibular and vocal folds.

folds and the opening between them are called the **glottis** (glot′is; figure 23.4). The vestibular folds and the vocal folds are lined with stratified squamous epithelium. The remainder of the larynx is lined with pseudostratified ciliated columnar epithelium. An inflammation of the vocal folds is called **laryngitis** (lar-in-jī′tis).

The larynx performs four important functions:

1. The thyroid and cricoid cartilages maintain an open passageway for air movement.
2. The larynx prevents swallowed materials from entering the lower respiratory tract and regulates the passage of air into and out of the lower respiratory tract. During swallowing, the epiglottis tips posteriorly until it lies below the horizontal plane and covers the opening into the larynx (see "Swallowing Phases," section 24.7). Thus, food and liquid slide over the epiglottis toward the esophagus. However, the most important method for preventing materials from entering the larynx is closure of the vestibular and vocal folds, which move medially and come together. The closure of the vestibular and vocal folds can also prevent the passage of air, as when a person holds his or her breath or increases air pressure within the lungs prior to coughing or sneezing.
3. The vocal folds are the primary source of sound production. Air moving past the vocal folds causes them to vibrate and produce sound. The force of air moving past the vocal folds determines the amplitude of the vibration and the loudness of the sound. The greater the amplitude of the vibration, the louder the sound. The frequency of vibrations determines pitch, with higher-frequency vibrations producing higher-pitched sounds and lower-frequency fibrations producing lower-pitched sounds. Variations in the length of the vibrating segments of the vocal folds affect the frequency of the vibrations. Higher-pitched tones are produced when only the anterior parts of the folds vibrate, and progressively lower tones result when longer sections of the folds vibrate. Because males usually have longer vocal folds than females, most males have lower-pitched voices. The sound produced by the vibrating vocal folds

is modified by the tongue, lips, teeth, and other structures to form words. Interestingly, a person whose larynx has been removed because of carcinoma of the larynx can produce sound by swallowing air and causing the esophagus to vibrate.

Movement of the arytenoid and other cartilages is controlled by skeletal muscles, thereby changing the position and length of the vocal folds. When a person is simply breathing, lateral rotation of the arytenoid cartilages abducts the vocal folds, which allows greater movement of air (figure 23.4*c*). Medial rotation of the arytenoid cartilages adducts the vocal folds, places them in position for producing sounds, and changes the tension on them (figure 23.4*d*). Anterior movement of the arytenoid cartilages decreases the length and tension of the vocal folds, lowering pitch. Posterior movement of the arytenoid cartilages increases the length and tension of the vocal folds, increasing pitch (figure 23.4*e*).

4. The pseudostratified ciliated columnar epithelium lining the larynx produces mucus, which traps debris in air. The cilia move the mucus and debris into the pharynx.

> **Predict 2**
>
> Jake told his girlfriend that the rollercoaster did not bother him, but during the ride he let loose with a long, high-pitched scream. Explain how the muscles of respiration, the muscles that control the vocal folds, and the muscles that move the epiglottis produced Jake's scream.

ASSESS **YOUR PROGRESS**

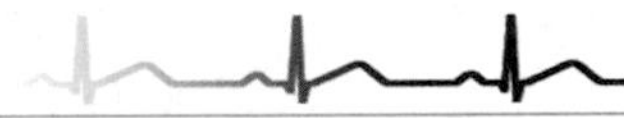

8. *Name and describe the three unpaired cartilages of the larynx. What are their functions?*
9. *Distinguish between the vestibular and vocal folds. How are sounds of different loudness and pitch produced by the vocal folds?*
10. *How does the position of the arytenoid cartilages change when a person is simply breathing versus making low-pitched and high-pitched sounds?*
11. *What are the four functions of the larynx?*

Trachea

The **trachea** (trā′kē-ă), or windpipe, is a membranous tube attached to the larynx (see figure 23.2). It consists of dense regular connective tissue and smooth muscle reinforced with 15–20 C-shaped pieces of hyaline cartilage. The cartilages support the anterior and lateral sides of the trachea (figure 23.5*a*). They protect the trachea and maintain an open passageway for air. The posterior wall of the trachea is devoid of cartilage; it contains an elastic ligamentous membrane and bundles of smooth muscle called the **trachealis** (trā′kē-ā-lis) **muscle.** Contraction of the smooth muscle can narrow the diameter of the trachea. During coughing, this action causes air to move more rapidly through the trachea, which helps expel mucus and foreign objects. The esophagus lies immediately posterior to the cartilage-free posterior wall of the trachea.

> **Predict 3**
>
> Explain what happens to the shape of the trachea when a person swallows a large mouthful of food. Why is this change of shape advantageous?

The mucous membrane lining the trachea consists of pseudostratified ciliated columnar epithelium with numerous goblet cells (figure 23.5*b*). The goblet cells produce mucus, which traps inhaled foreign particles. The cilia move the mucus and foreign particles into the larynx, from which they enter the pharynx and are swallowed (figure 23.5*b,c*). Constant, long-term irritation to the trachea, as occurs in smokers, can cause the tracheal epithelium to become moist stratified squamous epithelium that lacks cilia and goblet cells. Consequently, the normal function of the tracheal epithelium is lost.

The trachea has an inside diameter of 12 mm and a length of 10–12 cm, descending from the larynx to the level of the fifth thoracic vertebra (see figure 23.6). The trachea divides to form two smaller tubes called **main bronchi,** or *primary bronchi* (brong′kī; sing. *bronchus*, brong′kŭs; windpipe), each of which extends to a lung. The most inferior tracheal cartilage forms a ridge called the **carina** (kă-rī′nă), which separates the openings into the main bronchi. The carina is an important radiological landmark. In addition, the mucous membrane of the carina is very sensitive to mechanical

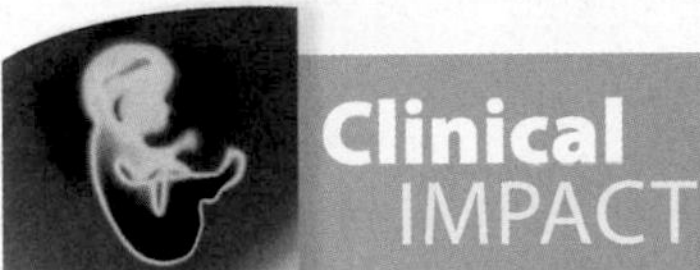

Clinical IMPACT

Establishing Airflow

In cases of extreme emergency when the upper air passageway is blocked by a foreign object so that the person cannot breathe, quick reaction is required to save the person's life. **Abdominal thrusts,** or the *Heimlich maneuver*, are designed to force an object out of the air passage by the sudden application of pressure to the abdomen. The person who performs the technique stands behind the patient, with his or her arms under the patient's arms and hands over the patient's abdomen between the navel and the rib cage. With one hand formed into a fist and the other hand over it, the rescuer suddenly pulls both hands toward the abdomen with an upward motion. This technique, if done properly, forces air up the trachea and dislodges most foreign objects.

There are other ways to establish airflow, but they should be performed only by trained medical personnel. In **intubation,** a tube is passed through the mouth or nose into the pharynx and then through the larynx to the trachea. Sometimes it is necessary to make an opening through which to pass the tube. The preferred point of entry in emergency cases is through the membrane between the cricoid and thyroid cartilages, a procedure called a **cricothyrotomy** (krī′kō-thī-rot′ō-mē).

A **tracheostomy** (trā′-kē-os′tō-mē; *tracheo-* + *stoma*, mouth) is an operation performed to make an opening into the trachea, commonly between the second and third cartilage rings. Usually, the opening is intended to be permanent, and a tube is inserted into the trachea to allow airflow and provide a way to remove secretions. The term **tracheotomy** (trā-kē-ot′ō-mē; *tracheo-* + *tome*, incision) refers to the actual cutting into the trachea, but sometimes the terms *tracheostomy* and *tracheotomy* are used interchangeably. In emergencies, opening the air passageway through the trachea is not advisable because it may disturb the arteries, nerves, and thyroid gland overlying the anterior surface of the trachea.

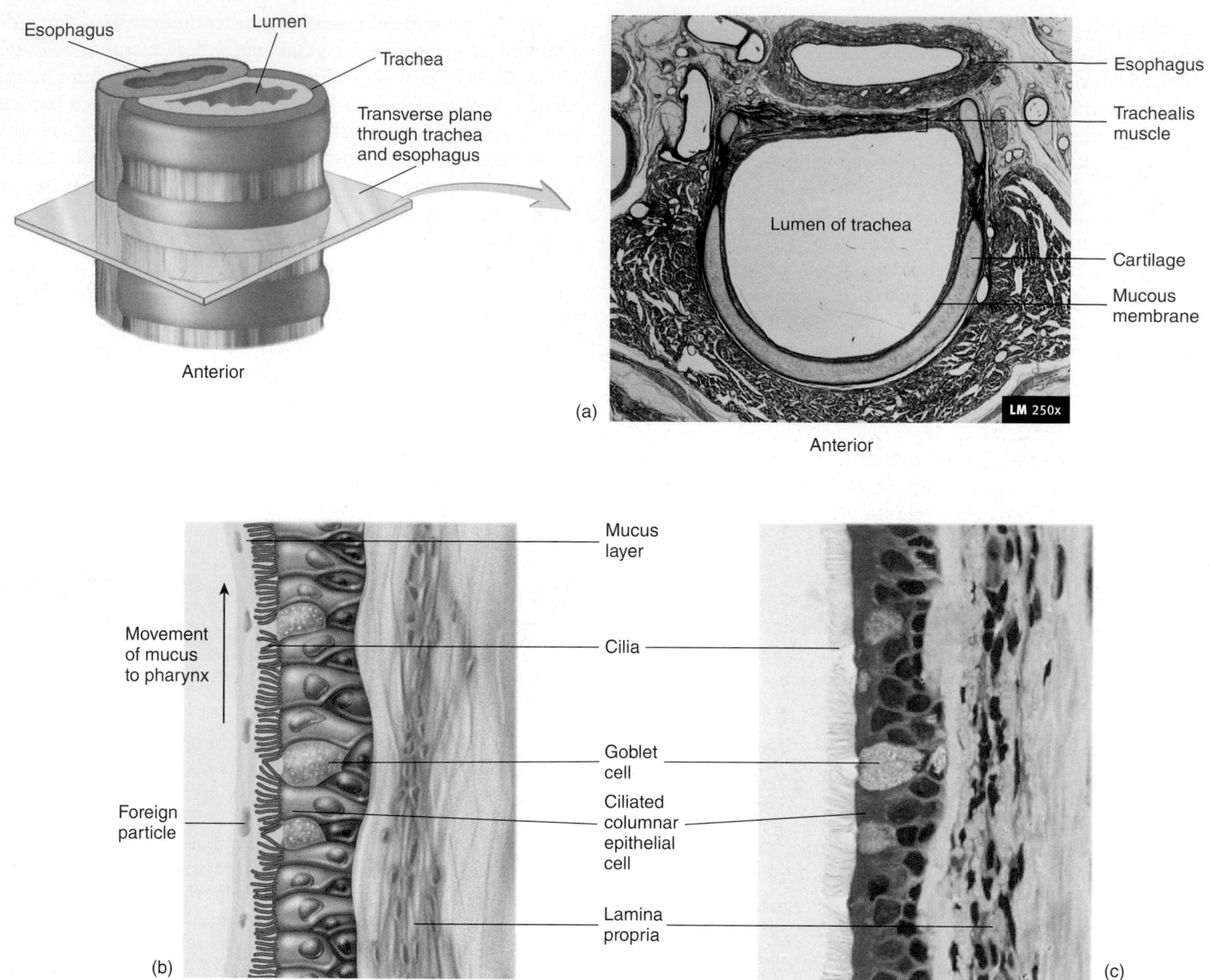
Esophagus
Lumen
Trachea
Transverse plane through trachea and esophagus
Anterior
Esophagus
Trachealis muscle
Lumen of trachea
Cartilage
Mucous membrane
LM 250x
Anterior
(a)
Mucus layer
Movement of mucus to pharynx
Cilia
Goblet cell
Foreign particle
Ciliated columnar epithelial cell
Lamina propria
(b)
(c)

FIGURE 23.5 AP|R **Trachea**

(*a*) Light micrograph of a transverse section of the trachea. The esophagus is posterior to the trachea, next to the smooth muscle connecting the ends of the C-shaped cartilages of the trachea. (*b*) Mucus, produced by the goblet cells, traps particles in the air. Movement of the cilia moves the mucus and particles to the laryngopharynx. (*c*) Light micrograph of the surface of the mucous membrane lining the trachea. Goblet cells are interspersed between ciliated cells.

stimulation, and materials reaching the carina stimulate a powerful cough reflex. Materials in the air passageways inferior to the carina do not usually stimulate a cough reflex.

Tracheobronchial Tree

Beginning with the trachea, all the respiratory passageways are called the **tracheobronchial** (trā′kē-ō-brong′kē-ăl) **tree** (figure 23.6). The trachea divides to form main bronchi, which in turn divide to form smaller and smaller bronchi, leading eventually to many microscopically small tubes and sacs. The right main bronchus is larger in diameter and more directly in line with the trachea than the left main bronchus. As a result, swallowed objects that accidentally enter the lower respiratory tract are most likely to become lodged in the right main bronchus.

The main bronchi divide into **lobar bronchi,** or *secondary bronchi*, within each lung. Two lobar bronchi exist in the left lung, and three exist in the right lung. The lobar bronchi, in turn, give rise to **segmental bronchi,** or *tertiary bronchi*. The bronchi continue to branch, finally giving rise to **bronchioles** (brong′kē-ōlz), which are less than 1 mm in diameter. The bronchioles also subdivide several times to become even smaller **terminal bronchioles.** Approximately 16 generations of branching occur from the trachea to the terminal bronchioles.

As the air passageways of the lungs become smaller, the structure of their walls changes. Like the trachea, the main bronchi are supported by C-shaped cartilage connected by smooth muscle. In the lobar bronchi, the C-shaped cartilages are replaced with cartilage plates, and smooth muscle forms a layer between the cartilage and

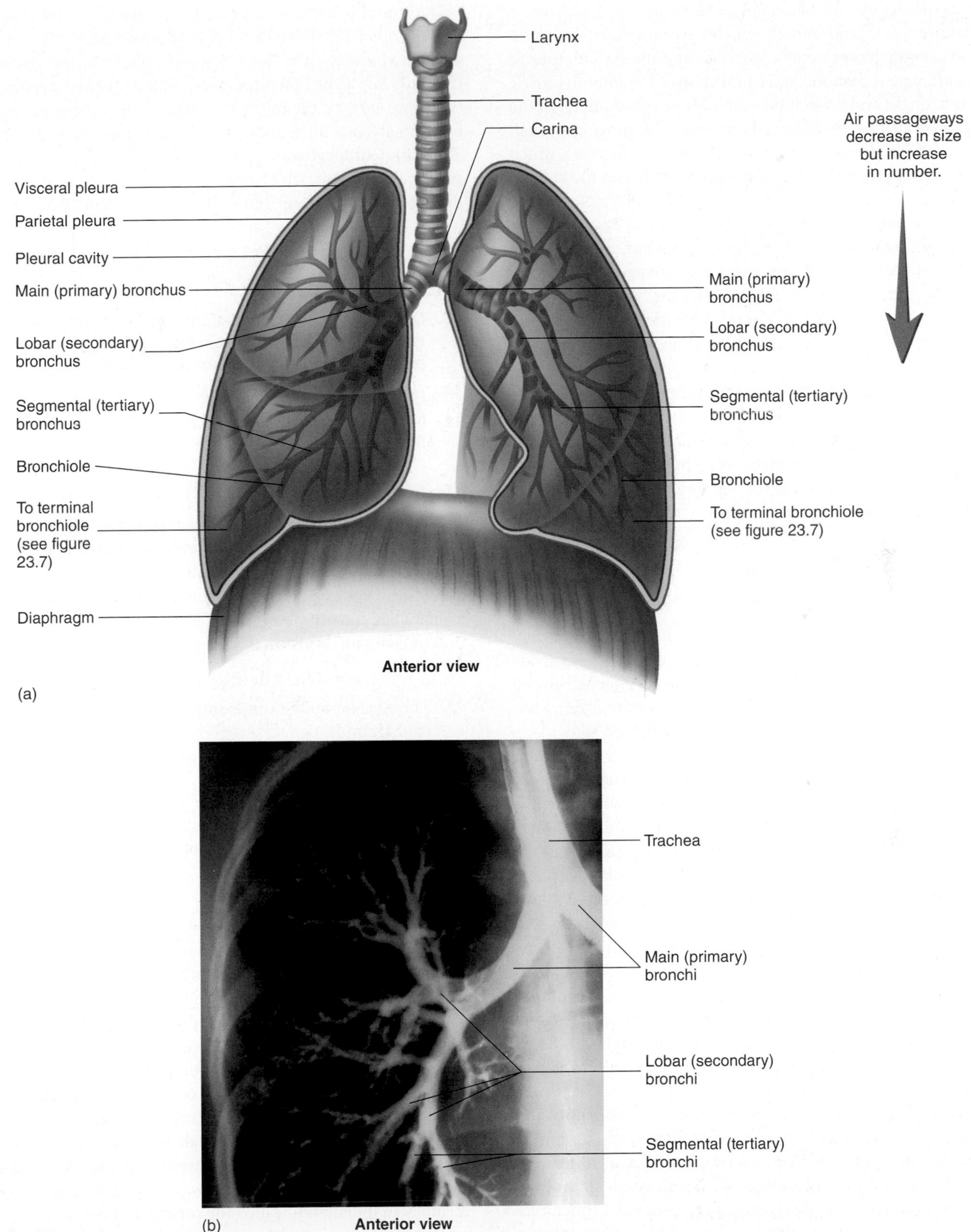
Larynx
Trachea
Carina
Air passageways decrease in size but increase in number.
Visceral pleura
Parietal pleura
Pleural cavity
Main (primary) bronchus
Lobar (secondary) bronchus
Segmental (tertiary) bronchus
Bronchiole
To terminal bronchiole (see figure 23.7)
Diaphragm
Main (primary) bronchus
Lobar (secondary) bronchus
Segmental (tertiary) bronchus
Bronchiole
To terminal bronchiole (see figure 23.7)
Anterior view
(a)
Trachea
Main (primary) bronchi
Lobar (secondary) bronchi
Segmental (tertiary) bronchi
(b)
Anterior view

FIGURE 23.6 AP|R **Tracheobronchial Tree**

(*a*) The conducting zone of the tracheobronchial tree begins at the trachea and ends at the terminal bronchioles. (*b*) A bronchogram is a radiograph of the tracheobronchial tree. A contrast medium, which makes the passageways visible, is injected through a catheter after a topical anesthetic is applied to the mucous membranes of the nose, pharynx, larynx, and trachea.

the mucous membrane. As the bronchi become smaller, the cartilage becomes more sparse, and smooth muscle becomes more abundant. The terminal bronchioles have no cartilage, and the smooth muscle layer is prominent. Relaxation and contraction of the smooth muscle within the bronchi and bronchioles can change the diameter of the air passageways and thereby change the volume of air moving through them. For example, during exercise the diameter can increase, which reduces the resistance to airflow and thereby increases the volume of air moved. However, during an **asthma attack,** contraction of the smooth muscle in the bronchi and bronchioles can decrease their diameter, resulting in increased resistance to airflow, which greatly reduces air movement. In severe cases, air movement can be so restricted that the patient dies. Fortunately, medications, such as albuterol (al-bū′ter-ol), help counteract the effects of an asthma attack by promoting smooth muscle relaxation in the walls of terminal bronchioles, so that air can flow more freely.

The bronchi are lined with pseudostratified ciliated columnar epithelium. The larger bronchioles are lined with ciliated simple columnar epithelium, which changes to ciliated simple cuboidal epithelium in the terminal bronchioles. The ciliated epithelium functions as a mucus-cilia escalator, trapping debris in the air and moving it to the larynx.

Respiratory Zone

Alveoli

Once gas exchange between inspired air and blood is anatomically possible, the respiratory zone begins. The terminal bronchioles divide to form **respiratory bronchioles,** which have a few attached alveoli (figure 23.7*a*). **Alveoli** (al-vē′ō-lī; hollow cavity) are small, air-filled chambers where gas exchange between the air and blood takes place. As the respiratory bronchioles divide to form smaller respiratory bronchioles, the number of attached alveoli increases. The respiratory bronchioles give rise to **alveolar** (al-vē′ō-lăr) **ducts,** which are like long, branching hallways with many open doorways. The "doorways" open into alveoli, which become so numerous that the alveolar duct wall is little more than a succession of alveoli. The alveolar ducts end as two or three **alveolar sacs,** which are chambers connected to two or more alveoli (figure 23.7*b*). Approximately seven generations of branching occur from the terminal bronchioles to the alveolar ducts.

The tissue surrounding the alveoli contains elastic fibers, which allow the alveoli to expand during inspiration and recoil during expiration. The lungs are very elastic and, when inflated, are capable of expelling air and returning to their original, uninflated state. Even when not inflated, however, the lungs retain some air, which gives them a spongy quality.

The walls of respiratory bronchioles consist of collagenous and elastic connective tissue with bundles of smooth muscle. The epithelium in the respiratory bronchioles is a simple cuboidal epithelium. The alveolar ducts and alveoli consist of simple squamous epithelium. Although the epithelium of the alveoli and respiratory bronchioles is not ciliated, debris from the air can be removed by macrophages that move over the surfaces of the cells. The macrophages do not accumulate in the respiratory zone because they either move into nearby lymphatic vessels or enter terminal bronchioles, thereby becoming entrapped in mucus that is swept to the pharynx.

Approximately 300 million alveoli are in the two lungs. The average diameter of an alveolus is approximately 250 μm, and its wall is extremely thin. Two types of cells form the alveolar wall (figure 23.8*a*). **Type I pneumocytes** are thin squamous epithelial cells that form 90% of the alveolar surface. Most of the gas exchange between alveolar air and the blood takes place through these cells. **Type II pneumocytes** are round or cube-shaped secretory cells that produce surfactant, which makes it easier for the alveoli to expand during inspiration (see "Lung Recoil" in section 23.3).

ASSESS **YOUR PROGRESS** 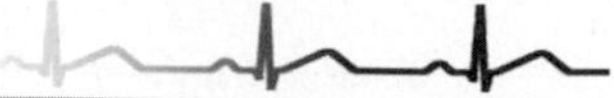

12. *Explain the branching of the tracheobronchial tree.*
13. *Describe the arrangement of cartilage, smooth muscle, and epithelium in the tracheobronchial tree. Explain why breathing becomes more difficult during an asthma attack.*
14. *How is debris removed from the tracheobronchial tree?*
15. *Name the two types of cells in the alveolar wall, and state their functions.*

The Respiratory Membrane

The **respiratory membrane of the respiratory zone** in the lungs is where gas exchange between the air and blood takes place. It is formed mainly by the alveolar walls and surrounding pulmonary capillaries (figure 23.8*b*), with some contribution by respiratory bronchioles and alveolar ducts. The respiratory membrane is very thin to facilitate diffusion of gases. It consists of several layers:

1. A thin layer of fluid lining the alveolus
2. The alveolar epithelium composed of simple squamous epithelium
3. The basement membrane of the alveolar epithelium
4. A thin interstitial space
5. The basement membrane of the capillary endothelium
6. The capillary endothelium, composed of simple squamous epithelium

Lungs

The **lungs** are the principal organs of respiration, and on a volume basis they are among the largest organs of the body. Each lung is conical in shape, with its base resting on the diaphragm and its apex extending to a point approximately 2.5 cm superior to the clavicle. The right lung is larger than the left and weighs an average of 620 g, whereas the left lung weighs an average of 560 g.

The **hilum** (hī′lŭm) is a region on the medial surface of the lung where structures, such as the main bronchus, blood vessels, nerves, and lymphatic vessels, enter or exit the lung. All the structures passing through the hilum are referred to as the **root of the lung.**

The right lung has three **lobes,** and the left lung has two and includes an indentation called the **cardiac notch** (figure 23.9). This arrangement allows the heart to lie between the lungs. The lobes are separated by deep, prominent **fissures** on the surface of the lung, and each lobe is supplied by a lobar bronchus. The lobes are subdivided into **bronchopulmonary segments,** which are supplied by the segmental bronchi. There are 9 bronchopulmonary segments in the

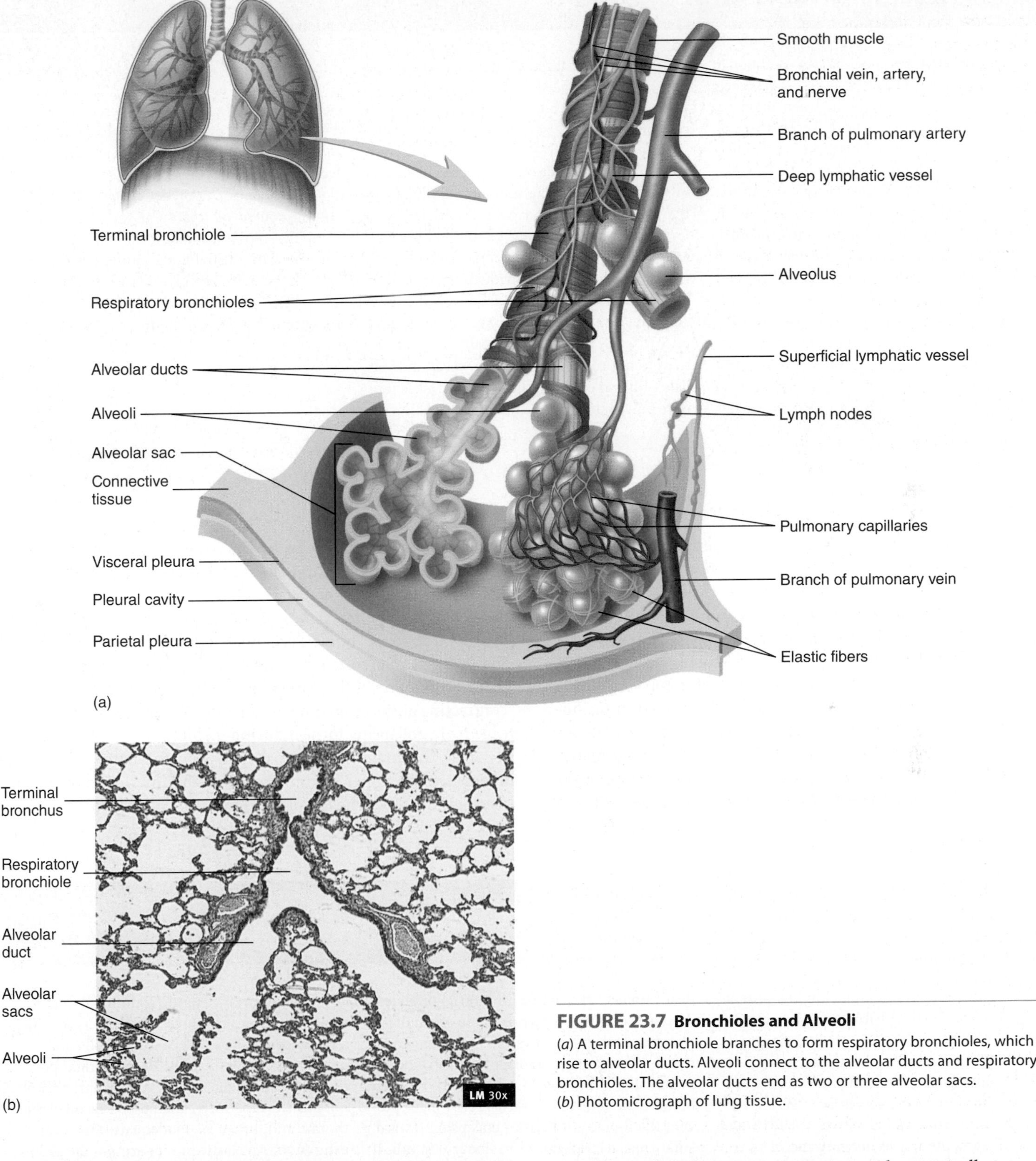
Smooth muscle
Bronchial vein, artery, and nerve
Branch of pulmonary artery
Deep lymphatic vessel
Terminal bronchiole
Alveolus
Respiratory bronchioles
Superficial lymphatic vessel
Alveolar ducts
Lymph nodes
Alveoli
Alveolar sac
Connective tissue
Pulmonary capillaries
Visceral pleura
Branch of pulmonary vein
Pleural cavity
Parietal pleura
Elastic fibers
(a)
Terminal bronchus
Respiratory bronchiole
Alveolar duct
Alveolar sacs
Alveoli
LM 30x
(b)

FIGURE 23.7 Bronchioles and Alveoli

(*a*) A terminal bronchiole branches to form respiratory bronchioles, which give rise to alveolar ducts. Alveoli connect to the alveolar ducts and respiratory bronchioles. The alveolar ducts end as two or three alveolar sacs.
(*b*) Photomicrograph of lung tissue.

left lung and 10 in the right lung. The bronchopulmonary segments are separated from each other by connective tissue partitions, which are not visible as surface fissures. Because major blood vessels and bronchi do not cross the connective tissue partitions, individual diseased bronchopulmonary segments can be surgically removed, leaving the rest of the lung relatively intact. The bronchopulmonary segments are subdivided into **lobules** by incomplete connective tissue walls. The lobules are supplied by the bronchioles.

Clinical GENETICS — Alpha-1 Antitrypsin Deficiency

Emphysema (em-fi-zē′mă) is a condition in which lung alveoli become progressively enlarged as the walls between them are destroyed. Individuals who have emphysema experience shortness of breath and coughing.

Cigarette smoking is the major risk factor for emphysema. Chemicals in cigarette smoke damage lung tissues and stimulate inflammation. As part of the inflammatory response, neutrophils and macrophages release **proteases,** which are enzymes that break down proteins. Proteases in the lungs protect against some bacteria and foreign substances, but too much protease activity can result in the breakdown of lung tissue proteins, especially elastin in elastic fibers. **Alpha-1 anti-trypsin (AAT),** which is synthesized in the liver, is a **protease inhibitor (Pi).** Normally, AAT inhibits protease activity, preventing the destruction of lung tissue. However, excess protease production stimulated by cigarette smoke can cause lung damage, leading to emphysema.

Approximately 1–2% of emphysema cases are due to a deficiency of AAT caused by defects in the AAT gene. The mutated gene reduces the amount of secreted AAT. Multiple alleles for AAT have been identified. Individuals who are homozygous for the normal allele produce normal levels of AAT. Individuals with one copy of the normal allele and one copy of the most common abnormal allele have about 60% of normal levels of AAT. This is sufficient activity to prevent protease damage. However, individuals with two copies of the *abnormal* allele produce only about 15–20% of normal AAT levels. If these individuals smoke, the development of emphysema is accelerated by 10–15 years. Other variant alleles cause different levels of AAT. The most severe form results in no AAT and the development of emphysema by age 30, even in nonsmokers.

Treatment of AAT deficiency follows the normal course of treatment for emphysema. Stopping smoking reduces the destruction of lung tissue by removing the stimulus for excess protease activity. Drugs, such as danazol and tamoxifen, can stimulate increased AAT production in the liver. In addition, patients may receive intravenous infusions of AAT, a process called **alpha-1 antitrypsin augmentation.**

Thoracic Wall and Muscles of Respiration

The **thoracic wall** consists of the thoracic vertebrae, ribs, costal cartilages, sternum, and associated muscles (see chapters 7 and 10). The **thoracic cavity** is the space enclosed by the thoracic wall and the **diaphragm** (dī′ă-fram; partition). Recall from chapter 10 that the diaphragm is a sheet of skeletal muscle separating the thoracic cavity from the abdominal cavity. The diaphragm and other skeletal muscles associated with the thoracic wall are responsible for **ventilation** (air movement; figure 23.10). The **muscles of inspiration** include the diaphragm, external intercostals, pectoralis minor, and scalenes. Contraction of the diaphragm is responsible for approximately two-thirds of the increase in thoracic volume during inspiration. The external intercostals, pectoralis minor, and scalene muscles also increase thoracic volume by elevating the ribs. The **muscles of expiration** consist of the muscles that depress the ribs and sternum, such as the internal intercostals and transverse thoracis, which are assisted by the abdominal muscles. Although the internal intercostals and the transverse thoracis are most active during expiration, and the external intercostals are most active during inspiration, the primary function of these muscles is to stiffen the thoracic wall by contracting at the same time. In this way, they prevent the thoracic cage from collapsing inward during inspiration.

The diaphragm is dome-shaped, and the base of the dome attaches to the inner circumference of the inferior thoracic cage (see figure 10.18). The top of the dome is a flat sheet of connective

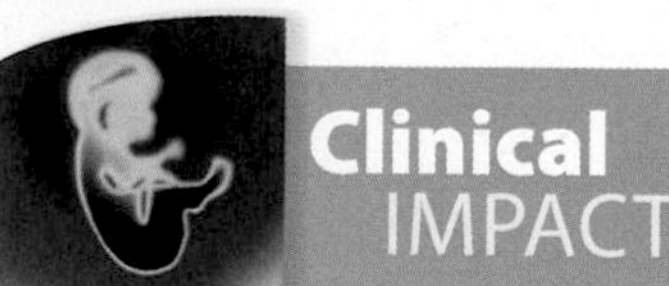

Clinical IMPACT — Effect of Spinal Cord Injury on Ventilation

The diaphragm is supplied by the phrenic nerves, which arise from spinal nerves C3–C5 (see figure 12.16), descend along each side of the neck to enter the thorax, and pass to the diaphragm. The intercostal muscles are supplied by the intercostal nerves (see figure 12.15), which arise from spinal nerves T1–T11 and extend along the spaces between the ribs.

Spinal cord injury superior to the origin of the phrenic nerves causes paralysis of the diaphragm and intercostal muscles and results in death unless artificial respiration is provided. A spinal cord injury inferior to the origin of the phrenic nerves causes paralysis of the intercostal muscles. Even though the diaphragm can function maximally, ventilation is drastically reduced because the intercostal muscles no longer prevent the thoracic wall from collapsing inward. Vital capacity is reduced to about 300 mL. If the spinal cord is injured inferior to the origin of the intercostal nerves, both the diaphragm and the intercostal muscles function normally.

The importance of the abdominal muscles in breathing can be observed in a person with a spinal cord injury that causes flaccid paralysis of the abdominal muscles. In the upright position, the abdominal organs and diaphragm are not pushed superiorly, and passive recoil of the thorax and lungs is inadequate for normal expiration. An elastic binder around the abdomen can help such patients. When a person is lying down, the weight of the abdominal organs can assist in expiration.

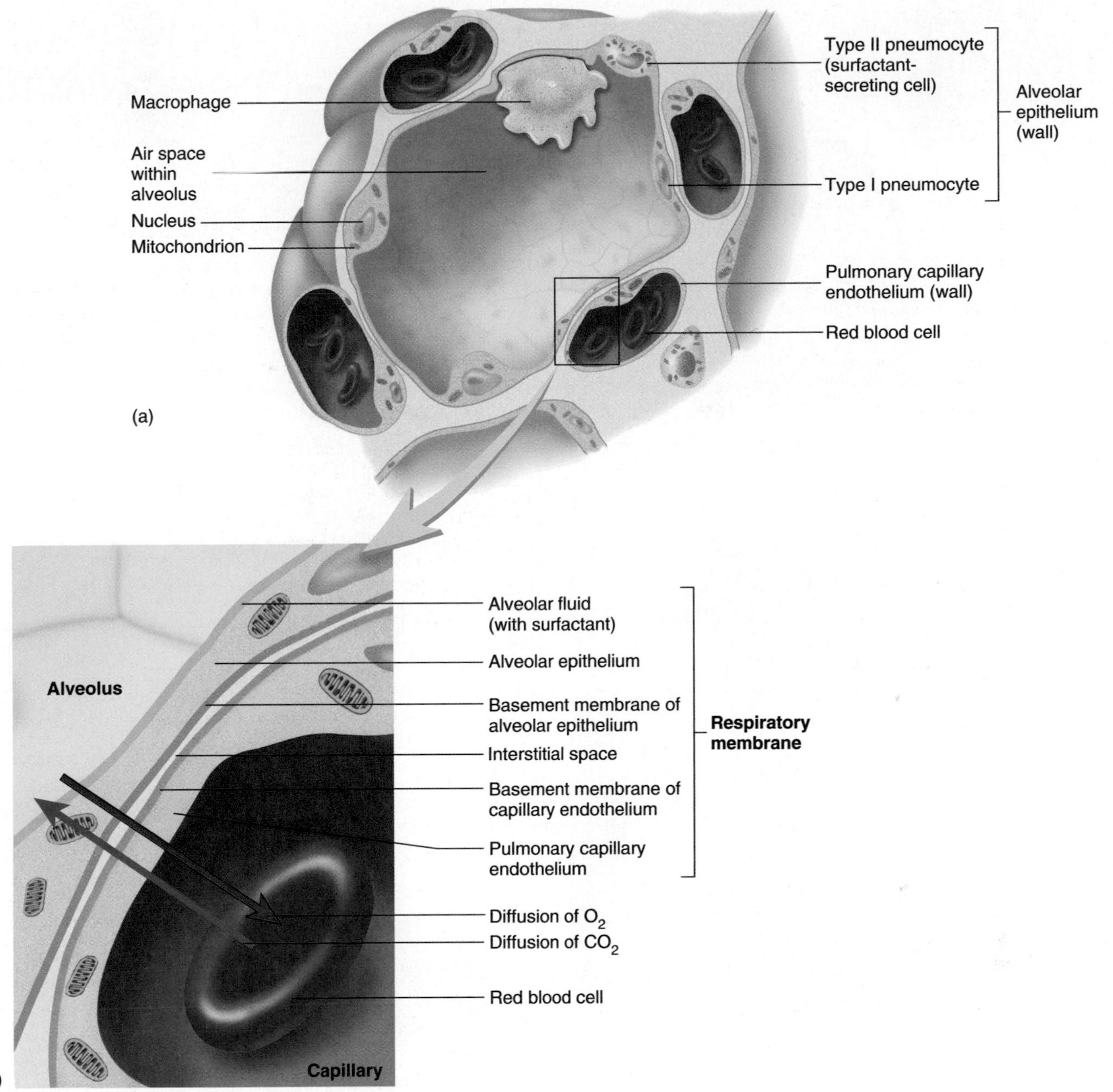

Macrophage
Air space within alveolus
Nucleus
Mitochondrion
Type II pneumocyte (surfactant-secreting cell)
Type I pneumocyte
Alveolar epithelium (wall)
Pulmonary capillary endothelium (wall)
Red blood cell
(a)
Alveolus
Capillary
Alveolar fluid (with surfactant)
Alveolar epithelium
Basement membrane of alveolar epithelium
Interstitial space
Basement membrane of capillary endothelium
Pulmonary capillary endothelium
Respiratory membrane
Diffusion of O_2
Diffusion of CO_2
Red blood cell
(b)

FIGURE 23.8 APR **Alveolus and the Respiratory Membrane**

(*a*) Section of an alveolus, showing the air-filled interior and thin walls composed of simple squamous epithelium. The alveolus is surrounded by elastic connective tissue and pulmonary capillaries. (*b*) Diffusion of oxygen and carbon dioxide across the six thin layers of the respiratory membrane.

tissue called the **central tendon.** In normal, quiet inspiration, contraction of the diaphragm results in inferior movement of the central tendon with very little change in the overall shape of the dome. Inferior movement of the central tendon can occur because the abdominal muscles relax, allowing the abdominal organs to move out of the way of the diaphragm. However, as the depth of inspiration increases, the abdominal organs prevent the central tendon from moving inferiorly. Continued contraction of the diaphragm causes it to flatten as the lower ribs are elevated. In addition, other muscles of inspiration can elevate the ribs. As the ribs are elevated, the costal cartilages allow lateral rib movement and lateral expansion of the thoracic cavity (figure 23.11). The ribs slope inferiorly from the vertebrae to the sternum, and elevation of the ribs also increases the anterior-posterior dimension of the thoracic cavity.

Expiration during quiet breathing occurs when the diaphragm and external intercostals relax and the elastic properties of the thorax and lungs cause a passive decrease in thoracic volume. In addition, contraction of the abdominal muscles helps push the abdominal organs and the diaphragm in a superior direction.

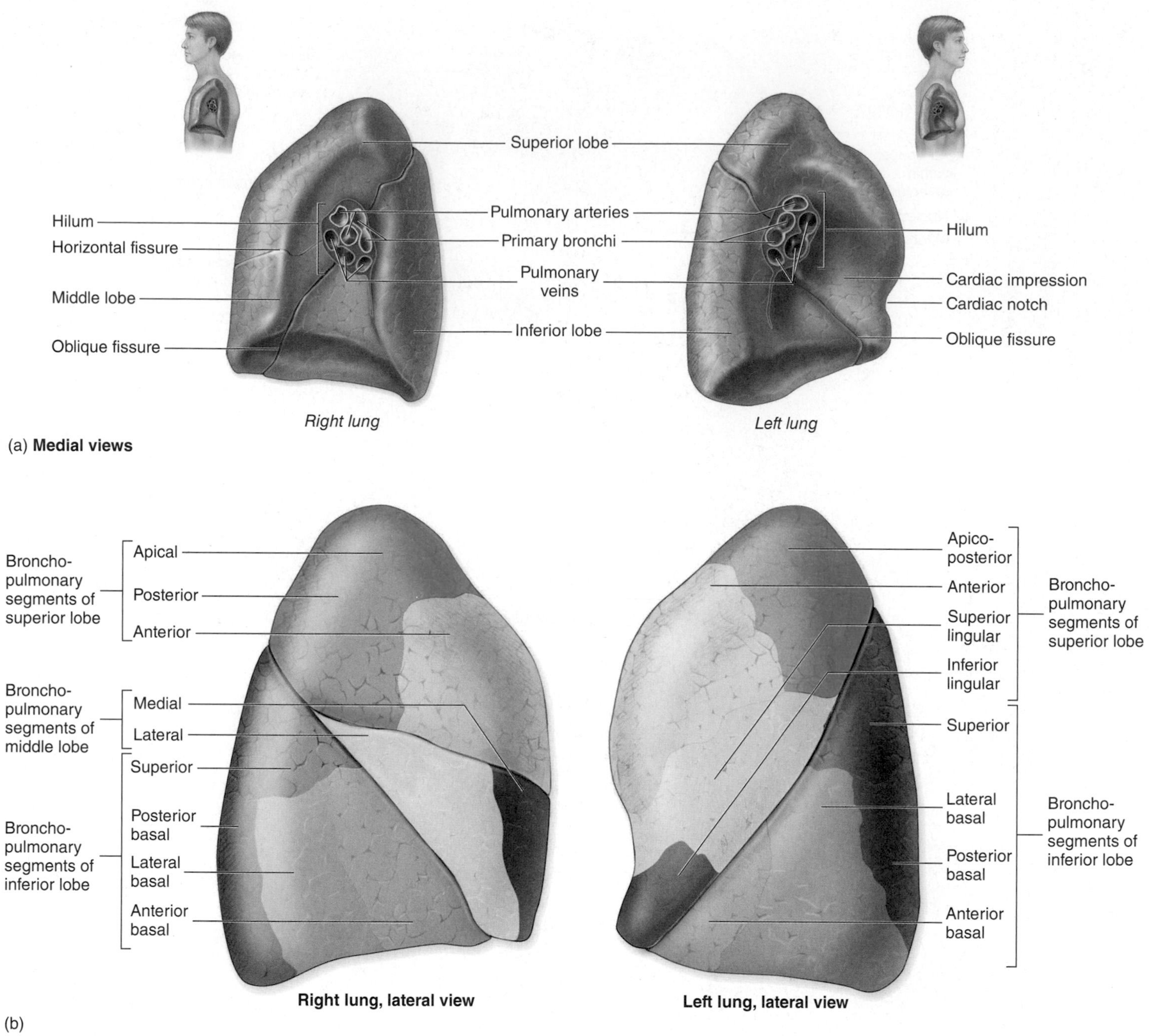
Superior lobe
Hilum
Horizontal fissure
Middle lobe
Oblique fissure
Pulmonary arteries
Primary bronchi
Pulmonary veins
Inferior lobe
Hilum
Cardiac impression
Cardiac notch
Oblique fissure
Right lung
Left lung
(a) Medial views
Broncho-pulmonary segments of superior lobe
Apical
Posterior
Anterior
Broncho-pulmonary segments of middle lobe
Medial
Lateral
Broncho-pulmonary segments of inferior lobe
Superior
Posterior basal
Lateral basal
Anterior basal
Apico-posterior
Anterior
Superior lingular
Inferior lingular
Broncho-pulmonary segments of superior lobe
Superior
Lateral basal
Posterior basal
Anterior basal
Broncho-pulmonary segments of inferior lobe
Right lung, lateral view
Left lung, lateral view
(b)

FIGURE 23.9 Lobes and Bronchopulmonary Segments of the Lungs

(*a*) Gross anatomy of the lungs, showing the lung lobes and bronchi. The right lung is divided into three lobes by the horizontal and oblique fissures. The left lung is divided into two lobes by the oblique fissure. A main bronchus supplies each lung. A lobar bronchus supplies each lung lobe, and segmental bronchi supply the bronchopulmonary segments (not visible). (*b*) Bronchopulmonary segments are supplied by segmental bronchi.

Several differences can be recognized between normal, quiet breathing and labored breathing. During labored breathing, all of the inspiratory muscles are active, and they contract more forcefully than during quiet breathing, causing a greater increase in thoracic volume (see figure 23.10*b*). During labored breathing, forceful contraction of the internal intercostals and the abdominal muscles produces a more rapid and greater decrease in thoracic volume than would be produced by the passive recoil of the thorax and lungs.

Pleura

The lungs are contained within the thoracic cavity, but each lung is surrounded by a separate **pleural** (ploor′ăl; relating to the ribs) **cavity** formed by the pleural serous membranes (figure 23.12). The **mediastinum** (mē′dē-as-tī′nŭm), a midline partition formed by the heart, trachea, esophagus, and associated structures, separates the pleural cavities. The **parietal pleura** covers the inner thoracic wall, the superior surface of the diaphragm, and the mediastinum.

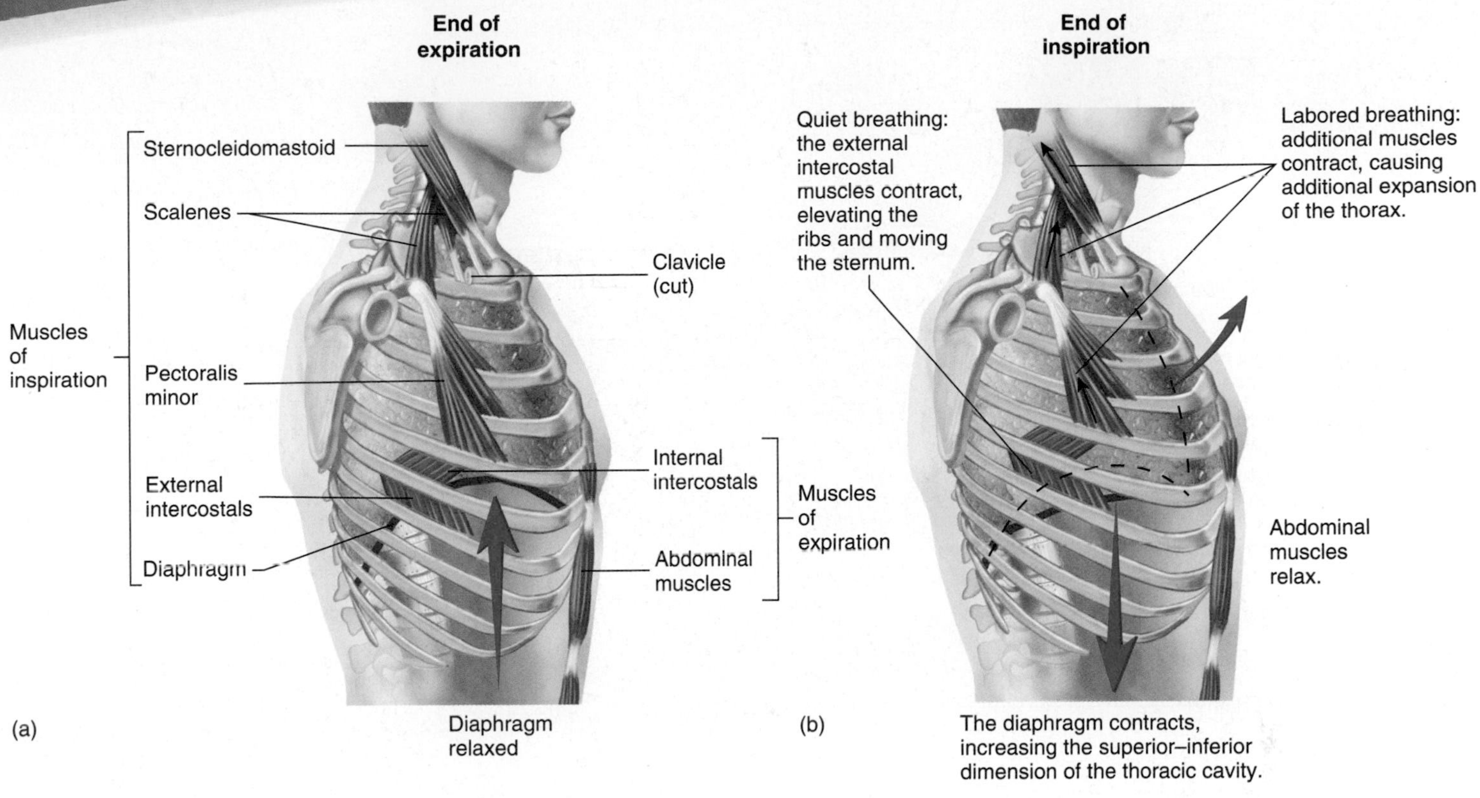
End of expiration
End of inspiration
Sternocleidomastoid
Scalenes
Muscles of inspiration
Pectoralis minor
External intercostals
Diaphragm
Clavicle (cut)
Internal intercostals
Abdominal muscles
Muscles of expiration
Quiet breathing: the external intercostal muscles contract, elevating the ribs and moving the sternum.
Labored breathing: additional muscles contract, causing additional expansion of the thorax.
Abdominal muscles relax.
(a)
Diaphragm relaxed
(b)
The diaphragm contracts, increasing the superior–inferior dimension of the thoracic cavity.

FIGURE 23.10 Effect of the Respiratory Muscles on Thoracic Volume
(*a*) Muscles of respiration at the end of expiration. (*b*) Muscles of respiration at the end of inspiration.

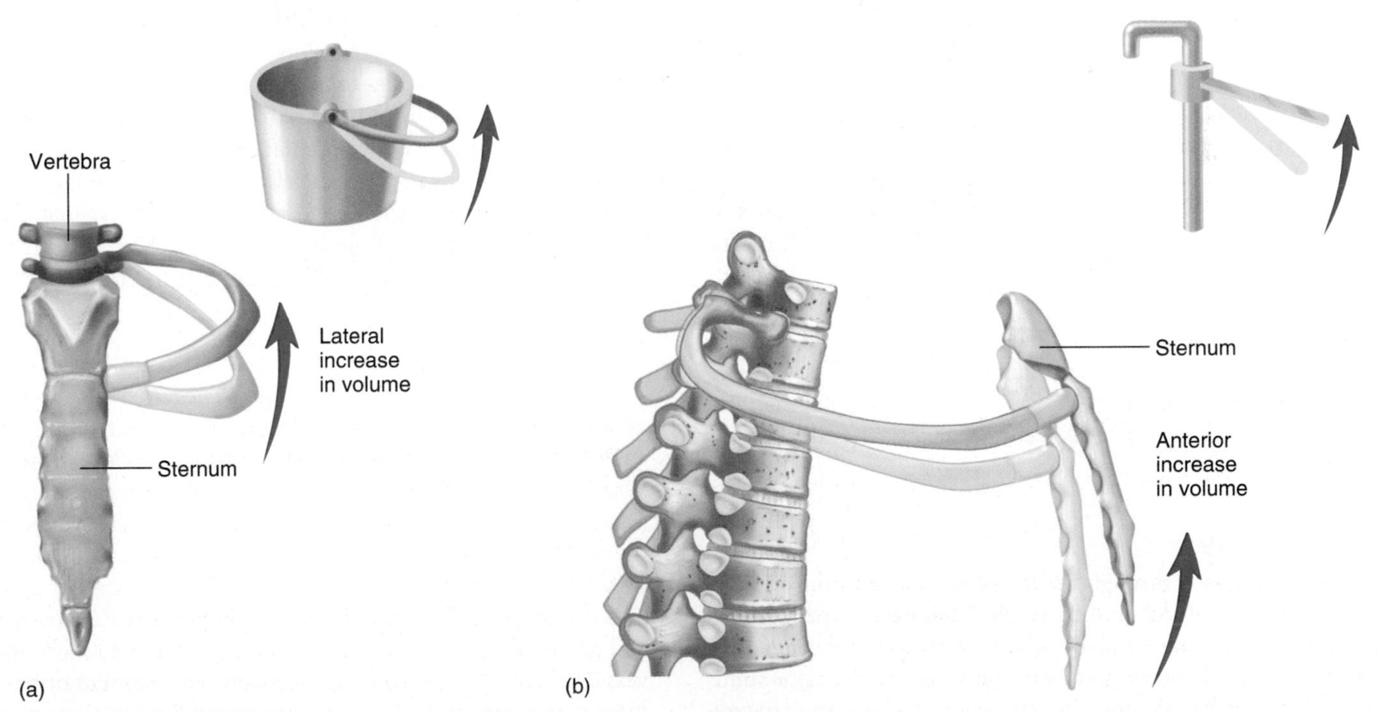
Vertebra
Lateral increase in volume
Sternum
Sternum
Anterior increase in volume
(a)
(b)

FIGURE 23.11 AP|R Effect of Rib and Sternum Movement on Thoracic Volume
(*a*) Elevation of the rib in the "bucket-handle" movement increases thoracic volume laterally. (*b*) As the rib is elevated, rotation of the rib in the "pump-handle" movement increases thoracic volume anteriorly.

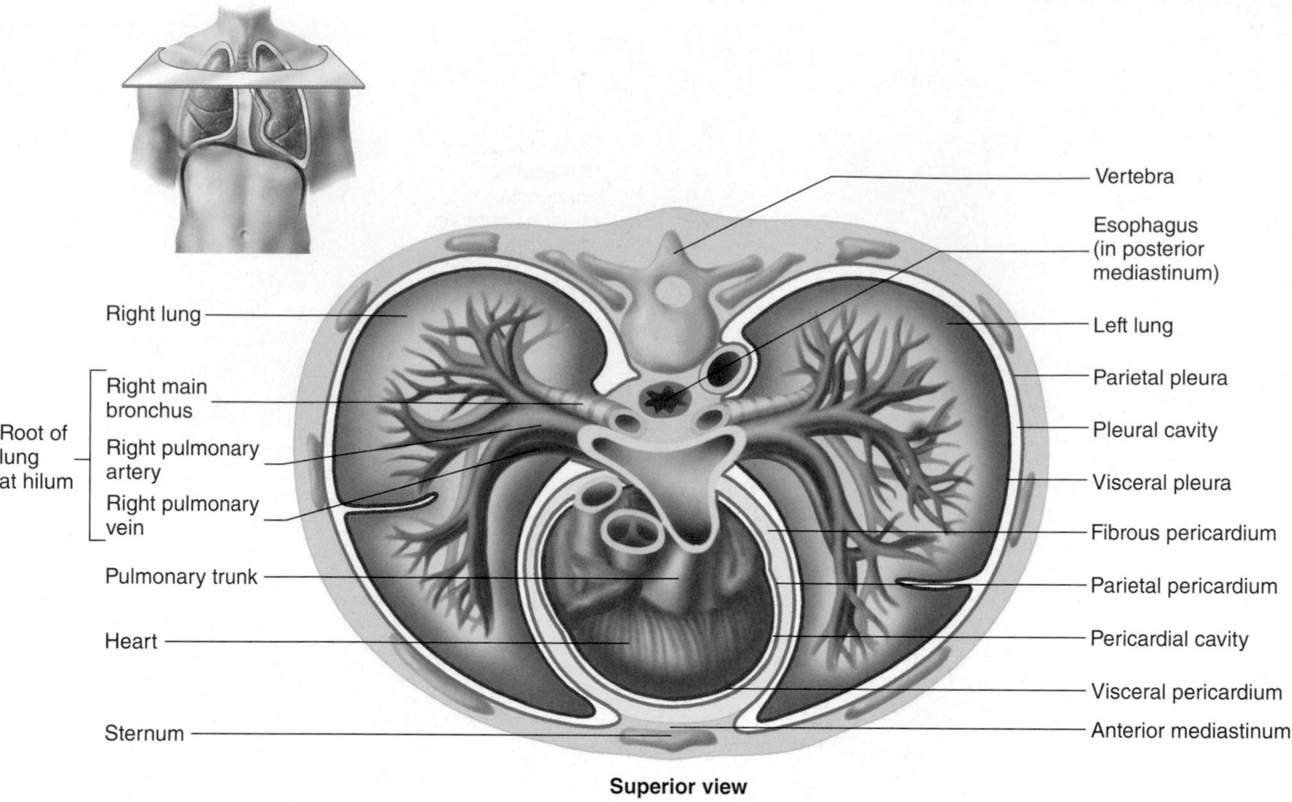
Vertebra
Esophagus (in posterior mediastinum)
Left lung
Parietal pleura
Pleural cavity
Visceral pleura
Fibrous pericardium
Parietal pericardium
Pericardial cavity
Visceral pericardium
Anterior mediastinum
Right lung
Root of lung at hilum
Right main bronchus
Right pulmonary artery
Right pulmonary vein
Pulmonary trunk
Heart
Sternum
Superior view

FIGURE 23.12 Pleural Cavities and Membranes
Transverse section of the thorax, showing the relationship of the pleural cavities to the thoracic organs. Each lung is surrounded by a pleural cavity. The parietal pleura lines the wall of each pleural cavity, and the visceral pleura covers the surface of the lungs. The space between the parietal and visceral pleurae is small and filled with pleural fluid.

At the hilum, the parietal pleura is continuous with the **visceral pleura,** which covers the surface of the lung.

The pleural cavity is filled with pleural fluid, which is produced by the pleural membranes. The pleural fluid has two functions: (1) It acts as a lubricant, allowing the parietal and visceral pleural membranes to slide past each other as the lungs and the thorax change shape during ventilation, and (2) it helps hold the parietal and visceral pleural membranes together, thus adhering the lungs to the thoracic wall. When thoracic volume changes during ventilation, lung volume changes because the parietal pleura is attached to the diaphragm and inner thoracic wall, and the visceral pleura is attached to the lungs. The pleural fluid is analogous to a thin film of water between two sheets of glass (the visceral and parietal pleurae); the glass sheets can easily slide over each other, but it is difficult to separate them.

Blood Supply

Blood that has passed through the lungs and picked up oxygen is called **oxygenated blood,** and blood that has passed through the tissues and released some of its oxygen is called **deoxygenated blood.** Two blood flow routes to the lungs exist. The major route takes deoxygenated blood to the lungs, where it is oxygenated (see chapter 21). The deoxygenated blood flows through pulmonary arteries to pulmonary capillaries, becomes oxygenated, and returns to the heart through pulmonary veins. The other route takes oxygenated blood to the tissues of the bronchi down to the respiratory bronchioles. The oxygenated blood flows from the thoracic aorta through bronchial arteries to capillaries, where oxygen is released. Deoxygenated blood from the proximal part of the bronchi returns to the heart through the bronchial veins and the azygos venous system (see chapter 21). More distally, the venous drainage from the bronchi enters the pulmonary veins. Thus, the oxygenated blood returning from the alveoli in the pulmonary veins is mixed with a small amount of deoxygenated blood returning from the bronchi.

Lymphatic Supply

The lungs have two lymphatic supplies. The **superficial lymphatic vessels** are deep to the visceral pleura; they drain lymph from the superficial lung tissue and the visceral pleura. The **deep lymphatic vessels** follow the bronchi; they drain lymph from the bronchi and associated connective tissues. No lymphatic vessels are located in the walls of the alveoli. Both the superficial and deep lymphatic vessels exit the lung at the hilum.

Phagocytic cells within the lungs phagocytize carbon particles and other debris from inspired air and move them to the lymphatic vessels. In an older person, especially one who smokes or has lived most of his or her life in a city with air pollution, these particles accumulate and cause the surface of the lungs to become gray or black. In addition, cancer cells from the lungs can sometimes spread to other parts of the body through the lymphatic vessels.

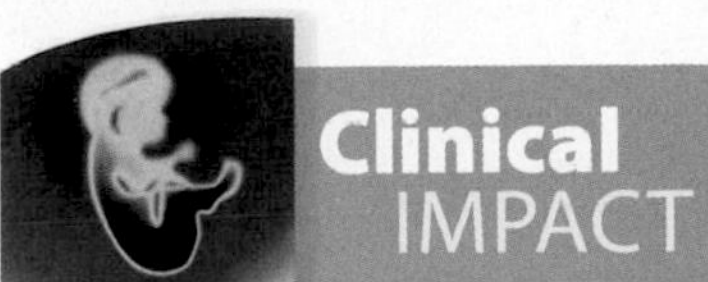

Clinical IMPACT — Cough and Sneeze Reflexes

Both the cough reflex and the sneeze reflex dislodge foreign matter or irritating material from the respiratory passages. The bronchi and trachea contain sensory receptors that detect such substances and initiate action potentials that pass along the vagus nerves to the medulla oblongata, where the cough reflex is triggered.

The movements resulting in a cough occur as follows: Approximately 2.5 L of air are inspired; the vestibular and vocal folds close tightly to trap the inspired air in the lungs; the abdominal muscles contract to force the abdominal contents up against the diaphragm; and the muscles of expiration contract forcefully. As a consequence, the pressure in the lungs increases to 100 mm Hg or more. Then the vestibular and vocal folds open suddenly, the soft palate is elevated, and air rushes from the lungs and out the oral cavity at a high velocity, carrying foreign particles with it.

The sneeze reflex is similar to the cough reflex, but it differs in several ways. The irritation that initiates the sneeze reflex occurs in the nasal passages instead of in the trachea and bronchi, and the action potentials travel along the trigeminal nerves to the medulla oblongata, where the reflex is triggered. During the sneeze reflex, the soft palate is depressed, so that air is directed primarily through the nasal passages, although a considerable amount passes through the oral cavity. The rapidly flowing air dislodges particulate matter from the nasal passages and can propel it a considerable distance from the nose.

About 17–25% of people have a photic sneeze reflex, which is stimulated by exposure to bright light, such as the sun. The sneeze reflex may be connected to the pupillary reflex, which causes the pupils to constrict in response to bright light. Researchers speculate that the complicated "wiring" of the pupillary and sneeze reflexes are intermixed in some people, so that, when bright light activates a pupillary reflex, it also activates a sneeze reflex. Sometimes the photic sneeze reflex is fancifully called ACHOO, which stands for *a*utosomal dominant *c*ompelling *h*elio-*o*phthalmic *o*utburst. As the name suggests, the reflex is inherited as an autosomal dominant trait. A person needs to inherit only one copy of the gene to have a photic sneeze reflex.

ASSESS YOUR PROGRESS

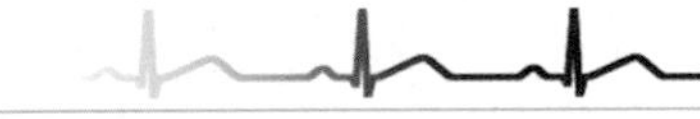

16. *List the layers of the respiratory membrane.*
17. *Distinguish among a lung, a lung lobe, a bronchopulmonary segment, and a lobule. How are they related to the tracheobronchial tree?*
18. *How many lobes are in the right lung and in the left lung? Why is there a difference in the number of lobes?*
19. *List the muscles of inspiration, and describe their role in quiet inspiration. List the muscles of expiration, and describe their role in quiet expiration. How does this change during labored breathing?*
20. *Name the pleurae of the lungs. What is their function?*
21. *What are the two major routes of blood flow to and from the lungs? What is the function of each route?*
22. *Describe the lymphatic supply of the lungs.*

23.3 Ventilation

LEARNING OUTCOMES

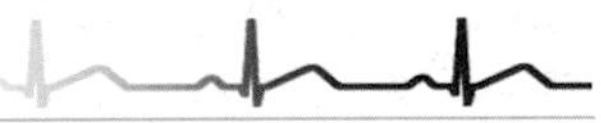

After reading this section, you should be able to

A. Describe the changes in alveolar pressure that are responsible for the movement of air into and out of the lungs.

B. Explain how surfactant and pleural pressure prevent the collapse of the lungs and how changes in pleural pressure cause changes in alveolar volume.

Pressure Differences and Airflow

Ventilation is the process of moving air into and out of the lungs. The flow of air into the lungs requires a pressure gradient from the outside of the body to the alveoli. This pressure gradient is provided by atmospheric pressure—the combined force of all the gases that make up the air we breathe. Airflow from the lungs requires a pressure gradient in the opposite direction, which is created by the components of the respiratory system. The physics of airflow in tubes, such as the ones that make up the respiratory passages, is the same as that of the flow of blood in blood vessels (see chapter 21). Thus, the following relationships hold:

$$F = \frac{P_1 - P_2}{R}$$

where F is airflow (milliliters per minute) in a tube, P_1 is pressure at point 1, P_2 is pressure at point 2, and R is resistance to airflow.

Air moves through tubes because of a pressure difference. When P_1 is greater than P_2, gas flows from P_1 to P_2 at a rate that is proportional to the pressure difference. For example, during inspiration, air pressure outside the body is greater than air pressure in the alveoli, and air flows through the trachea and bronchi to the alveoli.

Pressure and Volume

The pressure of a gas, such as air, in a container at a constant temperature is described according to **Boyle's law:**

$$P = k/V$$

where P is gas pressure, k is a constant for a given temperature, and V is the volume of the container. Body temperature in humans can be considered a constant. Thus, Boyle's law states that, in a container, such as the thoracic cavity or an alveolus, pressure is inversely proportional to volume. As volume increases, pressure decreases; as volume decreases, pressure increases (table 23.1).

TABLE 23.1 Gas Laws

Description	Importance
Boyle's Law The pressure of a gas is inversely proportional to its volume at a given temperature.	Air flows from areas of higher to lower pressure. When alveolar volume increases, causing pleural pressure to decrease below atmospheric pressure, air moves into the lungs. When alveolar volume decreases, causing pleural pressure to increase above atmospheric pressure, air moves out of the lungs.
Dalton's Law The partial pressure of a gas in a mixture of gases is the percentage of the gas in the mixture.	Gases move from areas of higher to lower partial pressure. The greater the difference in partial pressure between two points, the greater the rate of gas movement. Maintaining partial pressure differences ensures gas movement.
Henry's Law The concentration of a gas dissolved in a liquid is equal to the partial pressure of the gas over the liquid times the solubility coefficient of the gas.	Only a small amount of the gases in air dissolves in the fluid lining the alveoli. Carbon dioxide, however, is 24 times more soluble than oxygen; therefore, carbon dioxide exits through the respiratory membrane more readily than oxygen enters.

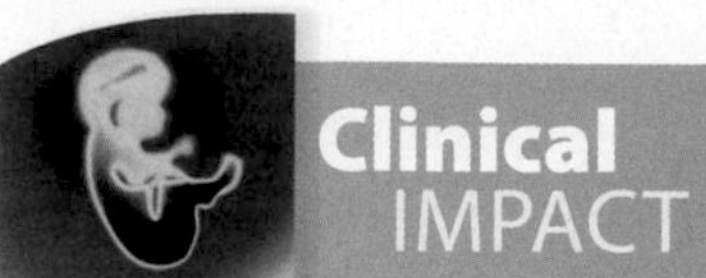

Clinical IMPACT

Disorders That Decrease the Diameter of Air Passageways

The flow of air decreases when the resistance to airflow is increased by conditions that reduce the diameter of the respiratory passageways. According to Poiseuille's law (see chapter 21), the resistance to airflow is proportional to the diameter (*D*) of a tube. Thus, a small change in diameter results in a large change in resistance, which greatly decreases airflow. For example, asthma results in the release of inflammatory chemicals, such as leukotrienes, that cause severe constriction of the bronchioles. Emphysema produces increased airway resistance because the bronchioles are obstructed as a result of inflammation and because damaged bronchioles collapse during expiration, thus trapping air within the alveoli. Cancer can also occlude respiratory passages as the tumor replaces lung tissue.

When there is increased resistance, increasing the pressure difference between alveoli and the atmosphere can help maintain airflow. Within limits, this can be accomplished by increased contraction of the muscles of respiration.

Airflow Into and Out of Alveoli

Respiratory physiologists express the pressure within the respiratory system in the following ways:

1. **Barometric air pressure (P_B),** which is atmospheric air pressure outside the body, is assigned a value of zero. Thus, whether at sea level with a pressure of 760 mm Hg or at 10,000 feet above sea level on a mountaintop with a pressure of 523 mm Hg, P_B is always assigned a value of zero.
2. Respiratory pressures are measured in reference to barometric air pressure. For example, **intra-alveolar pressure (P_{alv})** is the pressure inside an alveolus. Because the pressure inside the alveolus decreases to approximately 759 mm Hg upon inhalation, P_{alv} is usually expressed as -1 mm Hg. When you exhale, the P_{alv} increases to about 761 mm Hg, so it is usually expressed as $+1$ mm Hg. Thus, for simplicity, P_{alv} is always given as its difference from P_B—either 1 mm Hg above P_B or 1 mm Hg below P_B.

The movement of air into and out of the lungs results from changes in thoracic volume, which cause changes in alveolar volume. The changes in alveolar volume produce changes in intra-alveolar pressure. The pressure difference between barometric air pressure and intra-alveolar pressure ($P_B - P_{alv}$) results in air movement. The details of this process during quiet breathing are as follows:

1. *End of expiration* (figure 23.13, *step 1*). At the end of expiration, barometric air pressure and intra-alveolar pressure are equal. Therefore, no air moves into or out of the lungs.
2. *During inspiration* (figure 23.13, *step 2*). As inspiration begins, contraction of inspiratory muscles increases thoracic volume, which results in expansion of the lungs and an increase in alveolar volume (see the section "Changing Alveolar Volume"). The increased alveolar volume causes a decrease in intra-alveolar pressure below barometric air pressure to approximately -1 mm Hg. Air flows into the lungs because barometric air pressure is greater than intra-alveolar pressure.
3. *End of inspiration* (figure 23.13, *step 3*). At the end of inspiration, the thorax stops expanding, the alveoli stop expanding, and intra-alveolar pressure becomes equal to barometric air pressure because of airflow into the lungs. No movement of air occurs after intra-alveolar pressure becomes equal to barometric pressure, but the volume of the lungs is larger than at the end of expiration.
4. *During expiration* (figure 23.13, *step 4*). During expiration, the volume of the thorax decreases as the diaphragm relaxes, and the thorax and lungs recoil. The decreased thoracic volume results in a decrease in alveolar volume and an increase in intra-alveolar pressure over barometric air pressure to approximately $+1$ mm Hg. Air flows out of the lungs because intra-alveolar pressure is greater than barometric air pressure. As expiration ends, the decrease in thoracic volume stops, and the alveoli stop changing size. The process repeats, beginning at step 1.

ASSESS YOUR PROGRESS

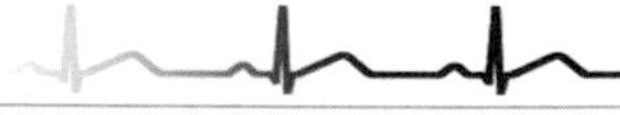

23. *What is ventilation?*

24. *How do pressure differences and resistance affect airflow through a tube?*

25. *What happens to the pressure within a container when the volume of the container increases? Whose law describes this relationship?*

26. *What are the assigned values for barometric air pressure and for intra-alveolar pressure?*

27. *Describe the process of making intra-alveolar pressure changes that occurs during quiet breathing.*

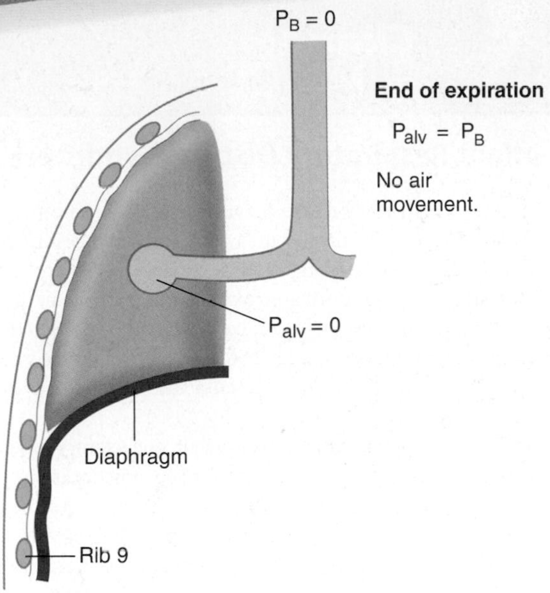

$P_B = 0$
End of expiration
$P_{alv} = P_B$
No air movement.
$P_{alv} = 0$
Diaphragm
Rib 9

1 At the end of expiration, intra-alveolar pressure (P_{alv}) is equal to barometric air pressure (P_B) and there is no air movement.

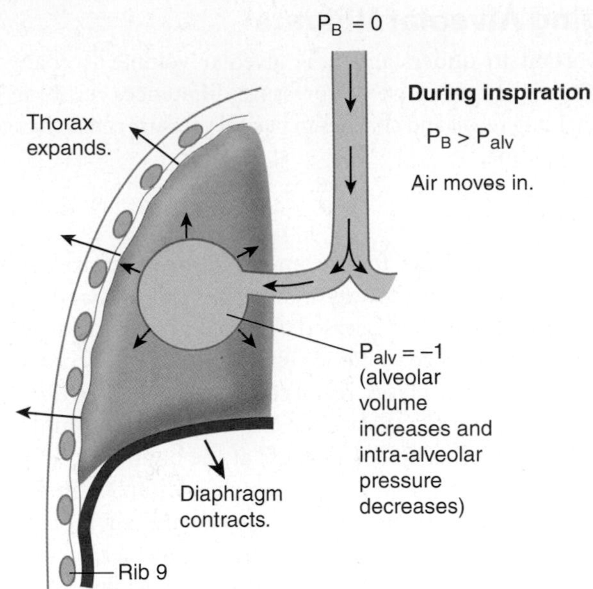

$P_B = 0$
Thorax expands.
During inspiration
$P_B > P_{alv}$
Air moves in.
$P_{alv} = -1$ (alveolar volume increases and intra-alveolar pressure decreases)
Diaphragm contracts.
Rib 9

2 During inspiration, increased thoracic volume results in increased alveolar volume and decreased intra-alveolar pressure. Barometric air pressure is greater than intra-alveolar pressure, and air moves into the lungs.

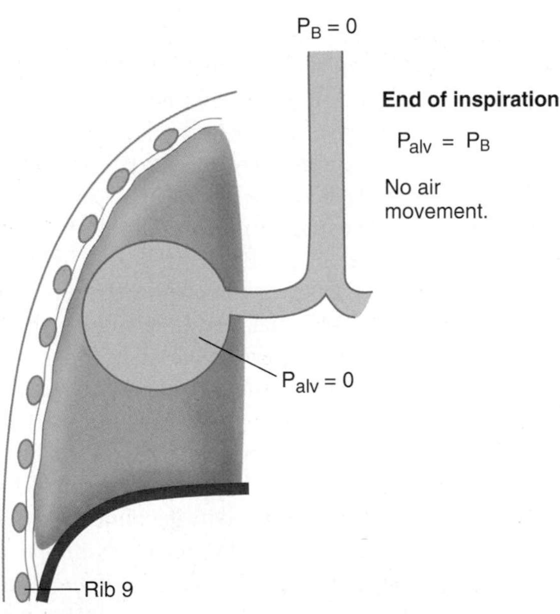

$P_B = 0$
End of inspiration
$P_{alv} = P_B$
No air movement.
$P_{alv} = 0$
Rib 9

3 At the end of inspiration, intra-alveolar pressure is equal to barometric air pressure and there is no air movement.

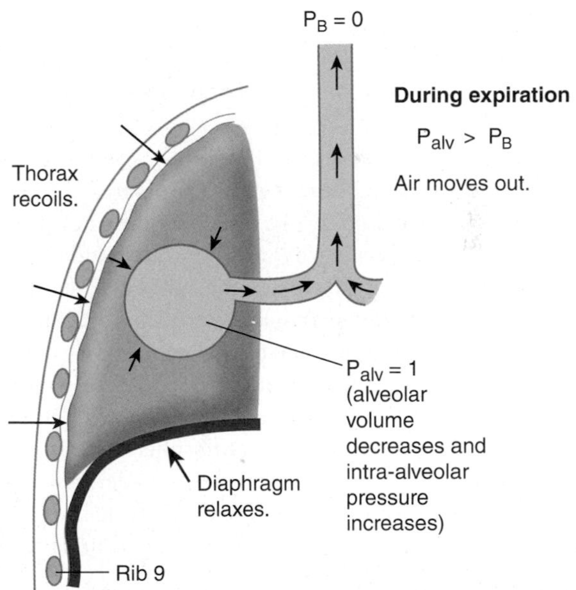

$P_B = 0$
Thorax recoils.
During expiration
$P_{alv} > P_B$
Air moves out.
$P_{alv} = 1$ (alveolar volume decreases and intra-alveolar pressure increases)
Diaphragm relaxes.
Rib 9

4 During expiration, decreased thoracic volume results in decreased alveolar volume and increased intra-alveolar pressure. Intra-alveolar pressure is greater than barometric air pressure, and air moves out of the lungs.

PROCESS FIGURE 23.3 AP|R **Intra-alveolar Pressure Changes During Inspiration and Expiration**

The combined space of all the alveoli is represented by a large "bubble." The alveoli are actually microscopic and would not be visible at the scale of this illustration.

Changing Alveolar Volume

It is important to understand how alveolar volume is changed because these changes cause the pressure differences resulting in ventilation. Lung recoil and changes in pleural pressure cause changes in alveolar volume.

Lung Recoil

Lung recoil is the tendency for the lungs to decrease in size as they are stretched, similar to the way a stretched rubber band snaps back to its original size when released. A lung decreases in size when its alveoli decrease in size (volume). Expanded alveoli decrease in size for two reasons: (1) elastic recoil caused by the elastic fibers in the alveolar walls and (2) surface tension of the film of fluid that lines the alveoli. Surface tension occurs at the boundary between water and air because the polar water molecules are attracted to one another more than they are attracted to the air molecules. Consequently, the water molecules are drawn together, forming a droplet. Because the water molecules of the alveolar fluid are also attracted to the surface of the alveoli, formation of a droplet causes the alveoli to collapse, thus producing fluid-filled alveoli with smaller volumes than air-filled alveoli.

Surfactant (ser-fak′tănt; *surf*ace *act*ing *agent*) is a mixture of lipoprotein molecules produced by the type II pneumocytes of the alveolar epithelium. The surfactant molecules form a monomolecular layer over the surface of the fluid within the alveoli to reduce the surface tension. With surfactant, the force produced by surface tension is approximately 3 mm Hg; without surfactant, the force can be as high as 30 mm Hg. Thus, surfactant greatly reduces the tendency of the lungs to collapse.

Pleural Pressure

Pleural pressure (P_{pl}) is the pressure in the pleural cavity. When pleural pressure is less than intra-alveolar pressure, the alveoli tend to expand. This principle can be understood by considering a balloon. The balloon expands when the pressure inside the balloon is greater than the pressure outside the balloon. The pressure inside the balloon is normally increased by blowing forcefully into it. However, a pressure difference can also be achieved by decreasing the pressure outside the balloon. For example, if the balloon is placed in a chamber from which air is removed, the pressure around the balloon becomes lower than atmospheric pressure, and the balloon expands. The lower the pressure outside the balloon, the greater the tendency for the higher pressure inside the balloon to cause it to expand. In a similar fashion, decreasing pleural pressure can result in expansion of the alveoli.

In a normal individual, the alveoli are always expanded. This is because there is a negative pleural pressure that is lower than intra-alveolar pressure. At the end of a normal expiration, pleural pressure is –4 mm Hg, and intra-alveolar pressure is 0 mm Hg. Pleural pressure is lower than intra-alveolar pressure because of a suction effect caused by fluid removal by the lymphatic system and by lung recoil. As the lungs recoil, the visceral pleura is pulled against the parietal pleura. Normally, the visceral pleura will not pull away from the parietal pleura because pleural fluid holds them together. Nonetheless, this pull decreases pressure in the pleural cavity, an effect that can be appreciated by putting water on the palms of your hands and then placing them together. As you gently pull your hands apart, you will feel a sensation of negative pressure.

The difference between intra-alveolar pressure and pleural pressure is called **transpulmonary pressure.** When pleural pressure is lower than intra-alveolar pressure, the alveoli tend to expand. However, they do not expand to the point of bursting because the lungs' tendency to recoil places an opposite force on alveolar expansion. Therefore, the alveoli expand when the pull of the visceral pleura against the parietal pleura is stronger than the opposite pull of lung recoil. In other words, when the transpulmonary pressure is around 4 mm Hg, the alveoli are expanded.

However, if the transpulmonary pressure decreases, the alveoli collapse. Specifically, pleural pressure increases and becomes equal to barometric air pressure. This happens when the thoracic wall or lung is pierced, and the pleural cavity is connected to the outside. Therefore, the alveoli collapse, lung recoil is unopposed, and the entire lung collapses and pulls away from the thoracic wall. Because the two pleural cavities are separated by the mediastinum, one lung can collapse while the other remains inflated.

This pleural pressure increase, called a **pneumothorax,** occurs when air is introduced into the pleural cavity through an opening in the thoracic wall or lung. A pneumothorax has several possible causes: penetrating trauma by a knife, a bullet, a broken rib, or another object; nonpenetrating trauma, such as a blow to the chest; a medical procedure, such as insertion of a catheter to withdraw pleural fluid; disease, such as an infection or emphysema; or some other, unknown cause.

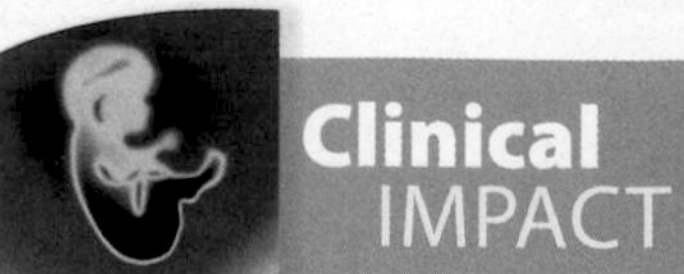

Clinical IMPACT

Infant Respiratory Distress Syndrome

Infant respiratory distress syndrome is common in premature infants, especially those with a gestation age of less than 7 months. This condition occurs because surfactant is not produced in adequate quantities until approximately 7 months of development. Thereafter, the amount produced increases as the fetus matures. Pregnant women who are likely to deliver prematurely can be given cortisol, which crosses the placenta into the fetus and stimulates surfactant synthesis.

If a newborn produces insufficient surfactant, the lungs tend to collapse. Thus, the muscles of respiration must exert a great deal of energy to keep the lungs inflated, and even then ventilation is inadequate. Without specialized treatment, most babies with this disease die soon after birth as a result of inadequate ventilation of the lungs and fatigued respiratory muscles. Infants with infant respiratory distress syndrome are treated with pressurized air, which delivers oxygen-rich air to the lungs. The pressure helps keep the alveoli inflated. In addition, surfactant administered with the pressurized air can reduce surface tension in the alveoli. Surfactant can be obtained from cow, pig, and human lungs; from human amniotic fluid; or from genetically modified bacteria. Synthetic surfactant is also available.

The most common symptoms of a pneumothorax are chest pain and shortness of breath. Treatment depends on the cause and severity. In patients with mild symptoms, the pneumothorax may resolve on its own. In other cases, a chest tube that aspirates the pleural cavity and restores a negative pressure can cause reexpansion of the lung. Surgery may also be necessary to close the opening into the pleural cavity.

In a **tension pneumothorax,** the pressure within the thoracic cavity is always higher than barometric air pressure. A tissue flap or an air passageway forms a flutter valve, which allows air to enter the pleural cavity during inspiration but does not allow it to exit during expiration. The result is an increase in air and pressure within the pleural cavity, which can compress blood vessels returning blood to the heart, causing decreased venous return, low blood pressure, and inadequate delivery of oxygen to tissues. The insertion of a large-bore needle into the pleural cavity allows air to escape and releases the pressure.

Pressure Changes During Inspiration and Expiration

At the end of a normal expiration, pleural pressure is −4 mm Hg, and intra-alveolar pressure is equal to barometric pressure (0 mm Hg). During normal, quiet inspiration, pleural pressure decreases to −7 mm Hg (figure 23.14, *step 1*). Consequently, the alveolar volume increases, intra-alveolar pressure decreases below barometric air pressure, and air flows into the lungs. As air flows into the lungs, intra-alveolar pressure increases and becomes equal to barometric pressure at the end of inspiration (figure 23.14, *steps 2* and *3*).

The decrease in pleural pressure during inspiration occurs for two reasons. First, because changing volume affects pressure (Boyle's law), the increased volume of the thoracic cavity causes decreased pleural pressure. Second, as the thoracic cavity expands, the lungs expand because they adhere to the inner thoracic wall through the pleurae. As the lungs expand, their tendency to recoil increases, resulting in an increased suction effect and a lowering of pleural pressure. The tendency for the lungs to recoil increases as the

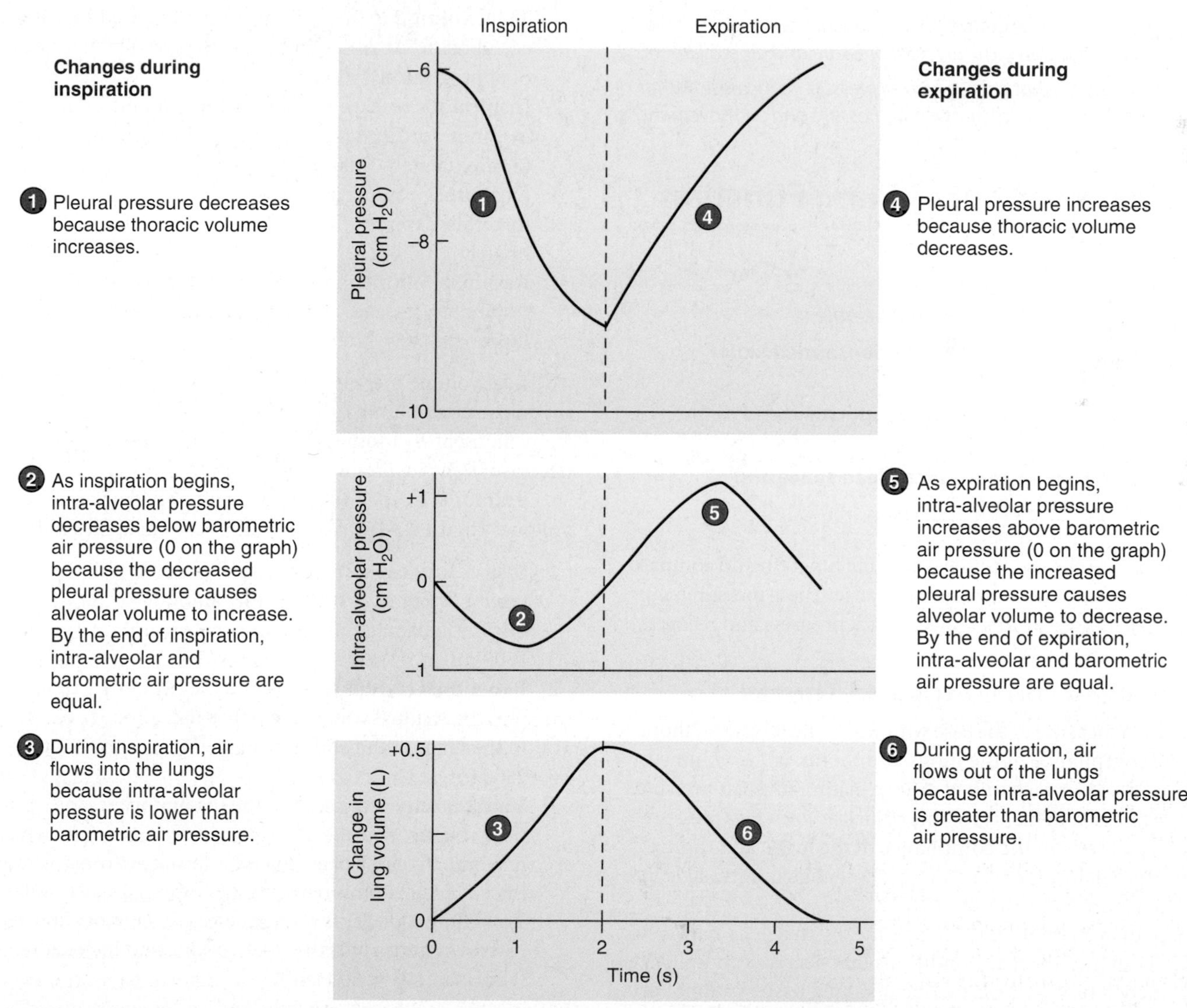
Inspiration
Expiration
Changes during inspiration
Changes during expiration
1 Pleural pressure decreases because thoracic volume increases.
4 Pleural pressure increases because thoracic volume decreases.
Pleural pressure (cm H_2O)
−6
−8
−10
2 As inspiration begins, intra-alveolar pressure decreases below barometric air pressure (0 on the graph) because the decreased pleural pressure causes alveolar volume to increase. By the end of inspiration, intra-alveolar and barometric air pressure are equal.
5 As expiration begins, intra-alveolar pressure increases above barometric air pressure (0 on the graph) because the increased pleural pressure causes alveolar volume to decrease. By the end of expiration, intra-alveolar and barometric air pressure are equal.
Intra-alveolar pressure (cm H_2O)
+1
0
−1
3 During inspiration, air flows into the lungs because intra-alveolar pressure is lower than barometric air pressure.
6 During expiration, air flows out of the lungs because intra-alveolar pressure is greater than barometric air pressure.
Change in lung volume (L)
+0.5
0
0
1
2
3
4
5
Time (s)

PROCESS FIGURE 23.14 Dynamics of a Normal Breathing Cycle

lungs are stretched, similar to the increased force generated in a stretched rubber band.

During expiration, pleural pressure increases because of decreased thoracic volume and decreased lung recoil (figure 23.14, *step 4*). As pleural pressure increases, alveolar volume decreases, intra-alveolar pressure increases above barometric air pressure, and air flows out of the lungs. As air flows out of the lungs, intra-alveolar pressure decreases and becomes equal to barometric pressure at the end of expiration (figure 23.14, *steps 5* and *6*).

ASSESS YOUR PROGRESS

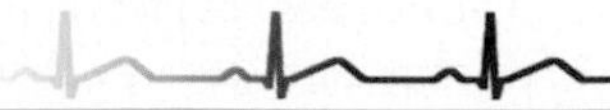

28. *What is lung recoil, and what two factors cause it?*

29. *How does surfactant reduce lung recoil? What happens if the alveoli have insufficient surfactant?*

30. *What is pleural pressure? What happens to alveolar volume when pleural pressure decreases? What causes pleural pressure to be lower than intra-alveolar pressure?*

31. *How does a pneumothorax cause a lung to collapse? How does a pneumothorax affect the chest cavity?*

32. *During inspiration, what causes pleural pressure to decrease? What effect does this have on intra-alveolar pressure and air movement?*

33. *During expiration, what causes pleural pressure to increase? What effect does this have on intra-alveolar pressure and air movement?*

23.4 Measurement of Lung Function

LEARNING OUTCOMES

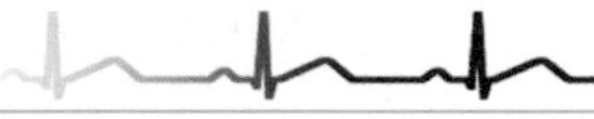

After reading this section, you should be able to

A. Define *compliance, minute ventilation,* and *alveolar ventilation.*

B. List the pulmonary volumes and capacities, and define each of them.

C. Distinguish between anatomical dead space and physiological dead space.

A variety of tests can be used to measure lung function and compare the patient's measurements with a normal range. These measurements can be used to diagnose diseases and to track progress and recovery.

Compliance of the Lungs and Thorax

Compliance is a measure of the ease with which the lungs and thorax expand. The compliance of the lungs and thorax is the volume by which they increase for each unit of change in intra-alveolar pressure. It is usually expressed in liters (volume of air) per mm Hg (pressure), and for a normal person the compliance of the lungs and thorax is 0.18 L/mm Hg. That is, for every 1 mm Hg change in intra-alveolar pressure, the volume changes by 0.18 L.

A lower than normal compliance means that it is harder to expand the lungs and thorax. Conditions that decrease compliance include the deposition of inelastic fibers in lung tissue (pulmonary fibrosis), the collapse of the alveoli (infant respiratory distress syndrome and pulmonary edema), increased resistance to airflow caused by airway obstruction (asthma, bronchitis, and lung cancer), and deformities of the thoracic wall that reduce its ability to expand and allow the thoracic volume to increase (kyphosis and scoliosis).

The greater the compliance, the easier it is for a change in pressure to cause expansion of the lungs and thorax. For example, emphysema sometimes causes the destruction of elastic lung tissue. This reduces the elastic recoil force of the lungs, thereby making expansion of the lungs easier and resulting in a higher than normal compliance. Pulmonary diseases that result in decreased compliance can markedly affect the total amount of energy required for ventilation and increase the total amount of energy expended by the body up to 30%.

Pulmonary Volumes and Capacities

Spirometry (spī-rom′ĕ-trē) is the process of measuring volumes of air that move into and out of the respiratory system, and a **spirometer** (spī-rom′ĕ-ter) is the device used to measure these pulmonary volumes. The four pulmonary volumes and representative values for a young adult male are graphed in figure 23.15 and listed here:

1. **Tidal volume** is the volume of air inspired or expired with each breath. At rest, quiet breathing results in a tidal volume of approximately 500 mL.
2. **Inspiratory reserve volume** is the amount of air that can be inspired forcefully after inspiration of the tidal volume (approximately 3000 mL at rest).
3. **Expiratory reserve volume** is the amount of air that can be forcefully expired after expiration of the tidal volume (approximately 1100 mL at rest).
4. **Residual volume** is the volume of air still remaining in the respiratory passages and lungs after the most forceful expiration (approximately 1200 mL).

The tidal volume increases when a person is more active. Because the maximum volume of the respiratory system does not change from moment to moment, an increase in tidal volume causes a decrease in the inspiratory and expiratory reserve volumes.

Pulmonary capacities are the sum of two or more pulmonary volumes (figure 23.15). Some pulmonary capacities follow:

1. **Inspiratory capacity** is the tidal volume plus the inspiratory reserve volume, which is the amount of air a person can inspire maximally after a normal expiration (approximately 3500 mL at rest).
2. **Functional residual capacity** is the expiratory reserve volume plus the residual volume, which is the amount of air remaining in the lungs at the end of a normal expiration (approximately 2300 mL at rest).
3. **Vital capacity** is the sum of the inspiratory reserve volume, the tidal volume, and the expiratory reserve volume, which is the maximum volume of air a person can expel from the respiratory tract after a maximum inspiration (approximately 4600 mL).
4. **Total lung capacity** is the sum of the inspiratory and expiratory reserve volumes plus the tidal volume and the residual volume (approximately 5800 mL).

Factors such as gender, age, body size, and physical conditioning cause variations in respiratory volumes and capacities from one

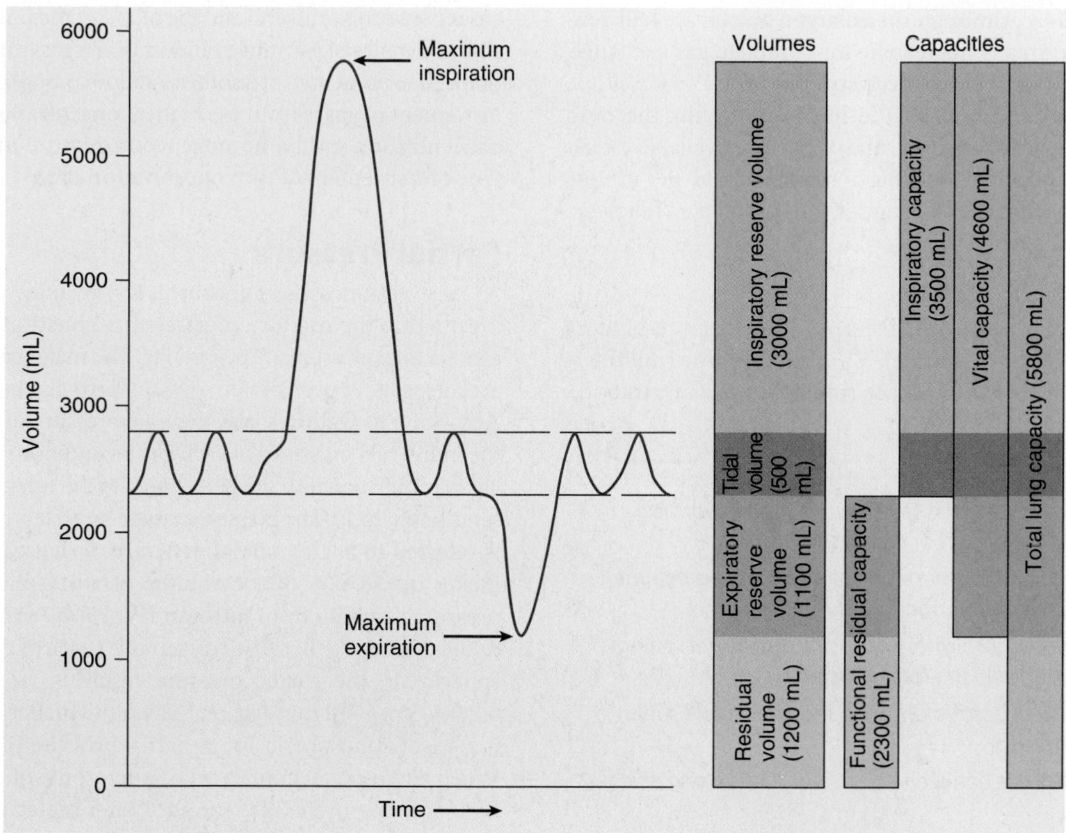
6000
5000
4000
3000
2000
1000
0
Volume (mL)
Time
Maximum inspiration
Maximum expiration
Volumes
Capacities
Inspiratory reserve volume (3000 mL)
Tidal volume (500 mL)
Expiratory reserve volume (1100 mL)
Residual volume (1200 mL)
Functional residual capacity (2300 mL)
Inspiratory capacity (3500 mL)
Vital capacity (4600 mL)
Total lung capacity (5800 mL)

FIGURE 23.15 Lung Volumes and Lung Capacities
The tidal volume during resting conditions is represented.

individual to another. For example, the vital capacity of adult females is usually 20–25% less than that of adult males. The vital capacity reaches its maximum amount in young adults and gradually decreases in the elderly. Tall people usually have a greater vital capacity than short people, and thin people have a greater vital capacity than obese people. Well-trained athletes can have a vital capacity 30–40% above that of untrained people. In patients whose respiratory muscles are paralyzed by spinal cord injury or diseases such as poliomyelitis or muscular dystrophy, vital capacity can be reduced to values not consistent with survival (less than 500–1000 mL). Factors that reduce compliance also reduce vital capacity.

The **forced expiratory vital capacity** is a simple and clinically important pulmonary test. The patient inspires maximally and then exhales maximally into a spirometer as rapidly as possible. The volume of air expired at the end of the test is the vital capacity. The spirometer also records the volume of air that enters it per second. The **forced expiratory volume in 1 second (FEV_1)** is the amount of air expired during the first second of the test. In some conditions, the vital capacity may not be dramatically affected, but how rapidly air is expired can be greatly decreased. Airway obstruction, caused by asthma, the collapse of bronchi in emphysema, or a tumor, as well as disorders that reduce the ability of the lungs or chest wall to deflate, such as pulmonary fibrosis, silicosis, kyphosis, and scoliosis, can cause a decreased FEV_1.

Minute Ventilation and Alveolar Ventilation

Minute ventilation is the total amount of air moved into and out of the respiratory system each minute; it is equal to the tidal volume times the respiratory rate. **Respiratory rate,** or *respiratory frequency*, is the number of breaths taken per minute. Because the resting tidal volume is approximately 500 mL and the respiratory rate is approximately 12 breaths per minute, minute ventilation averages approximately 6 L/min.

Although minute ventilation measures the amount of air moving into and out of the respiratory system per minute, it is not a measure of the amount of air available for gas exchange because gas exchange takes place mainly in the alveoli and to a lesser extent in the alveolar ducts and respiratory bronchioles. The part of the respiratory system where gas exchange does not take place is called the dead space. A distinction can be made between anatomical and physiological dead space. **Anatomical dead space,** which measures 150 mL, is formed by the nasal cavity, pharynx, larynx, trachea, bronchi, bronchioles, and terminal bronchioles. **Physiological dead space** is the anatomical dead space plus the volume of any alveoli in which gas exchange is less than normal. In a healthy person, anatomical and physiological dead spaces are nearly the same, meaning that few nonfunctional alveoli exist. However, in patients with emphysema, alveolar walls degenerate, and small alveoli combine to form larger alveoli. The result is fewer alveoli, but alveoli with an increased volume and decreased surface area. Although the enlarged alveoli are still ventilated, their surface area is inadequate for complete gas exchange, and the physiological dead space increases.

During inspiration, much of the inspired air fills the dead space first before reaching the alveoli and, thus, is unavailable for gas exchange. The volume of air available for gas exchange per minute is called **alveolar ventilation ($\dot{V}_A$)**, and it is calculated as follows:

$$\dot{V}_A = f(V_T - V_D)$$

where $\dot{V}_A$ is alveolar ventilation (milliliters per minute), f is respiratory rate (frequency; breaths per minute), V_T is tidal volume (milliliters per respiration), and V_D is dead space (milliliters per respiration).

ASSESS YOUR PROGRESS

34. *What is compliance? What is the effect on lung expansion when compliance increases or decreases?*
35. *Distinguish among tidal volume, inspiratory reserve volume, expiratory reserve volume, and residual volume.*
36. *Differentiate among inspiratory capacity, functional residual capacity, vital capacity, and total lung capacity.*
37. *What is forced expiratory volume in 1 second, and why it is clinically important?*
38. *What is the difference between minute ventilation and alveolar ventilation?*
39. *What is dead space? Contrast anatomical dead space with physiological dead space.*

Predict 4

What is the alveolar ventilation of a resting person with a tidal volume of 500 mL, a dead space of 150 mL, and a respiratory rate of 12 breaths per minute? Suppose the person exercises, so that tidal volume increases to 4000 mL, dead space increases to 300 mL due to dilation of the respiratory passageways, and respiratory rate increases to 24 breaths per minute. What is the alveolar ventilation then? How is the change in alveolar ventilation beneficial for doing exercise?

23.5 Physical Principles of Gas Exchange

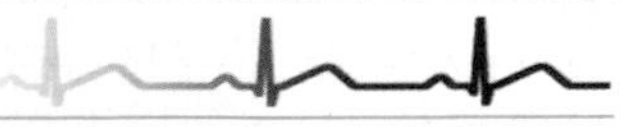

LEARNING OUTCOMES

After reading this section, you should be able to

A. **Explain partial pressure and its relationship to the concentration of gases in the body. Describe the partial pressure gradients for oxygen and carbon dioxide.**
B. **Explain the factors that affect gas movement through the respiratory membrane.**
C. **Describe the relationship between alveolar ventilation and pulmonary capillary perfusion.**

Ventilation supplies atmospheric air to the alveoli. The next step in the process of respiration (external respiration) is the diffusion of gases between the alveoli and the blood in the pulmonary capillaries. The molecules of gas move randomly; if a gas is in higher concentration at one point than at another, random motion ensures that the net movement of gas is from the higher concentration toward the lower concentration until a homogeneous mixture of gases is achieved. One measurement of the concentration of gases is partial pressure.

Partial Pressure

At sea level, atmospheric pressure is approximately 760 mm Hg, which means that the mixture of gases that constitutes atmospheric air exerts a total pressure of 760 mm Hg. The major components of dry air are nitrogen (approximately 79%) and oxygen (approximately 21%). According to **Dalton's law,** the total pressure of a gas is the sum of the individual pressures of each gas. In addition, the pressure of each gas is proportional to the percentage of the total that each gas represents (table 23.1). The pressure exerted by each type of gas in a mixture is referred to as the **partial pressure** of that gas. Because nitrogen makes up 78.62% of the volume of atmospheric air, the partial pressure resulting from nitrogen (P_{N_2}) is 0.7862 times 760 mm Hg, or 597.5 mm Hg. Because oxygen is 20.84% of the volume of atmospheric air, the partial pressure resulting from oxygen (P_{O_2}) is 0.2084 times 760 mm Hg, or 158.4 mm Hg. Because carbon dioxide is .04% of atmospheric air, its partial pressure (P_{CO_2}) is 0.3 mm Hg. When air comes in contact with water, some of the water turns into a gas and evaporates into the air. Water molecules in gaseous form also exert a partial pressure. This partial pressure (P_{H_2O}) is sometimes referred to as **water vapor pressure** (table 23.2). In other words, atmospheric air can be represented by the following equation:

$$P_{O_2} + P_{CO_2} + P_{N_2} + P_{H_2O} = 760 \text{ mm Hg}$$

The compositions of alveolar air and expired air are not identical to the composition of dry atmospheric air for three reasons. First, air entering the respiratory system during inspiration is humidified; second, oxygen diffuses from the alveoli into the blood, and carbon dioxide diffuses from the pulmonary capillaries into the alveoli; third, the air within the alveoli is only partially replaced with atmospheric air during each inspiration.

Diffusion of Gases Into and Out of Liquids

Gas molecules move from the air into a liquid, or from a liquid into the air, because of partial pressure gradients. If the partial pressure of a gas in the air is greater than in the liquid, gas molecules diffuse into the liquid. On the other hand, if the partial pressure of a gas in the air is less than in the liquid, gas molecules diffuse out of the liquid. If the partial pressure of a gas in the air is equal to that in the liquid, the gas in the air is in equilibrium with the gas in the liquid, and no gas diffuses into or out of the liquid. At a given temperature, **Henry's law** describes the concentration of a gas at equilibrium in a liquid (see table 23.1):

$$\text{Concentration of dissolved gas} = \text{Pressure of gas} \times \text{Solubility coefficient}$$

The solubility coefficient is a measure of how easily the gas dissolves in the liquid. In water, the solubility coefficient for oxygen is 0.024; for carbon dioxide, it is 0.57. Thus, carbon dioxide is approximately 24 times more soluble in water than oxygen is.

TABLE 23.2 Partial Pressures of Gases at Sea Level

	Dry Air		Humidified Air		Alveolar Air		Expired Air	
Gases	**mm Hg**	**%**	**mm Hg**	**%**	**mm Hg**	**%**	**mm Hg**	**%**
Nitrogen	597.5	78.62	563.4	74.09	569.0	74.9	566.0	74.5
Oxygen	158.4	20.84	149.3	19.67	104.0	13.6	120.0	15.7
Carbon dioxide	0.3	0.04	0.3	0.04	40.0	5.3	27.0	3.6
Water vapor	0.0	0.00	47.0	6.20	47.0	6.2	47.0	6.2

ASSESS **YOUR PROGRESS**

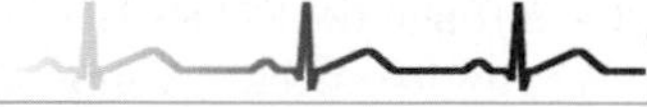

40. *According to Dalton's law, what is the partial pressure of a mixture of gases? What is water vapor pressure?*

41. *Why are the compositions of inspired, alveolar, and expired air different?*

42. *According to Henry's law, how do partial pressure and solubility of a gas affect its concentration in a liquid?*

Predict 5

As a scuba diver descends, the pressure of the water on the body prevents normal expansion of the lungs. To compensate, the diver breathes pressurized air, which has a greater pressure than air at sea level. What effect does the increased pressure have on the amount of gas dissolved in the diver's body fluids? A scuba diver who suddenly ascends to the surface from a great depth can develop decompression sickness (the bends), in which bubbles of nitrogen gas form. The expanding bubbles damage tissues or block blood flow through small blood vessels. Explain why the bubbles develop.

Diffusion of Gases Through the Respiratory Membrane

The factors that influence the rate of gas diffusion through the respiratory membrane are (1) the thickness of the membrane; (2) the diffusion coefficient of the gas in the substance of the membrane, which is approximately the same as the diffusion coefficient for the gas through water; (3) the surface area of the membrane; and (4) the partial pressure gradient of the gas across the membrane.

Respiratory Membrane Thickness

Increasing the thickness of the respiratory membrane decreases the rate of diffusion. The thickness of the respiratory membrane normally averages 0.6 μm, but diseases can cause an increase in the thickness. If the thickness of the respiratory membrane increases two or three times, the rate of gas exchange markedly decreases. The most common cause of increased respiratory membrane thickness is an accumulation of fluid in the alveoli, known as *pulmonary edema*. Alveolar fluid accumulation is usually caused by failure of the left side of the heart. Left side heart failure increases venous pressure in the pulmonary capillaries and causes fluid to accumulate in the alveoli. Conditions that result in inflammation of the lung tissues, such as tuberculosis, pneumonia, or advanced silicosis, can also cause fluid accumulation within the alveoli.

Diffusion Coefficient

The **diffusion coefficient** is a measure of how easily a gas diffuses into and out of a liquid or tissue, taking into account the solubility of the gas in the liquid and the size of the gas molecule (molecular weight). For example, if the diffusion coefficient of oxygen is assigned a value of 1, the relative diffusion coefficient of carbon dioxide is 20, which means carbon dioxide diffuses through the respiratory membrane about 20 times more readily than oxygen does. In other words, CO_2 is 24 times more soluble in water than oxygen is; when the chemical characteristics of the two molecules are also considered, the diffusion *rate* is calculated to be 20:1 of CO_2 to O_2.

When the respiratory membrane becomes progressively damaged as a result of disease, its capacity for allowing oxygen to move into the blood is often so impaired that death from oxygen deprivation results before the diffusion of carbon dioxide can be reduced. If life is being maintained by extensive oxygen therapy, which increases the concentration of oxygen in the lung alveoli, the reduced capacity for the diffusion of carbon dioxide across the respiratory membrane can result in substantial increases in carbon dioxide in the blood.

Surface Area

In a healthy adult, the total surface area of the respiratory membrane is approximately 70 m^2 (approximately the floor area of a 25- by 30-foot room). Several respiratory diseases, including emphysema and lung cancer, cause a decrease in the surface area of the respiratory membrane. Even small decreases in this surface area adversely affect the respiratory exchange of gases during strenuous exercise. When the total surface area of the respiratory membrane is decreased to one-third or one-fourth of normal, the exchange of gases is significantly restricted, even under resting conditions.

A decreased surface area for gas exchange can also result from the surgical removal of lung tissue, the destruction of lung tissue by cancer, the degeneration of the alveolar walls by emphysema, or the replacement of lung tissue by connective tissue due to tuberculosis. More acute conditions that cause the alveoli to fill with fluid also reduce the surface area for gas exchange because the increased thickness of the respiratory membrane caused by the fluid accumulation makes the alveoli nonfunctional. This may occur in pneumonia or in pulmonary edema resulting from failure of the left ventricle.

Partial Pressure Gradient

The partial pressure gradient of a gas across the respiratory membrane is the difference between the partial pressure of the gas in the alveoli and the partial pressure of the gas in the blood of the pulmonary capillaries. When the partial pressure of a gas is greater on one side of the respiratory membrane than on the other side, net diffusion occurs from the higher to the lower partial pressure (see figure 23.8*b*). Normally, the partial pressure of oxygen (Po_2) is greater in the alveoli than in the blood of the pulmonary capillaries, and the partial pressure of carbon dioxide (Pco_2) is greater in the blood than in the alveolar air.

By increasing alveolar ventilation, the partial pressure gradient for oxygen and carbon dioxide can be increased. The greater volume of atmospheric air exchanged with the residual volume raises alveolar Po_2, lowers alveolar Pco_2, and thus promotes gas exchange. Conversely, inadequate ventilation causes a lower than normal partial pressure gradient for oxygen and carbon dioxide, resulting in inadequate gas exchange.

Relationship Between Alveolar Ventilation and Pulmonary Capillary Perfusion

Under normal conditions, alveolar ventilation and **pulmonary capillary perfusion,** which is blood flow through pulmonary capillaries, is such that effective gas exchange occurs between the air and the blood. During exercise, effective gas exchange is maintained because both ventilation and cardiac output increase.

The normal relationship between alveolar ventilation and pulmonary capillary perfusion can be disrupted in two ways. First, alveolar ventilation may exceed the blood's ability to pick up oxygen, which can happen because of inadequate cardiac output after a heart attack. Second, alveolar ventilation may not be great enough to provide the oxygen needed to oxygenate the blood flowing through the pulmonary capillaries. For example, constriction of the bronchioles in asthma can decrease air delivery to the alveoli.

Blood that is not completely oxygenated is called shunted blood. Two sources of shunted blood exist in the lungs. An **anatomical shunt** results when deoxygenated blood from the bronchi and bronchioles mixes with blood in the pulmonary veins (see "Blood Supply" in section 23.2). The other source of shunted blood is blood that passes through pulmonary capillaries but does not become fully oxygenated. The **physiological shunt** is the combination of deoxygenated blood from the anatomical shunt and deoxygenated blood that passes through the pulmonary capillaries. Normally, 1–2% of cardiac output makes up the physiological shunt.

Any condition that decreases gas exchange between the alveoli and the blood can increase the amount of shunted blood. For example, obstruction of the bronchioles in conditions such as asthma can decrease ventilation beyond the obstructed areas. The result is a large increase in shunted blood because the blood flowing through the pulmonary capillaries in the obstructed area remains deoxygenated. In pneumonia or pulmonary edema, a buildup of fluid in the alveoli results in poor gas diffusion and less oxygenated blood.

When a person is standing, greater blood flow and ventilation occur in the base of the lung than in the top of the lung because of the effects of gravity. Arterial pressure at the base of the lung is 22 mm Hg greater than at the top of the lung because of hydrostatic pressure caused by gravity (see chapter 21). This greater pressure increases blood flow and distends blood vessels. The decreased pressure at the top of the lung results in less blood flow and less distension in blood vessels, some of which are even collapsed during diastole.

During exercise, cardiac output and ventilation increase. The increased cardiac output raises pulmonary blood pressure throughout the lung, which increases blood flow. However, blood flow increases most at the top of the lung, because the increased pressure expands the less distended vessels and opens the collapsed vessels. Thus, gas exchange at the top of the lung is more effective because of greater blood flow.

Although gravity is the major factor affecting regional blood flow in the lung, under certain circumstances alveolar Po_2 can also have an effect. In most tissues, low Po_2 results in increased blood flow through the tissues (see chapter 21). In the lung, low Po_2 has the opposite effect, causing arterioles to constrict and reducing blood flow. This response reroutes blood away from areas of low oxygen toward parts of the lung that are better oxygenated. For example, if a bronchus becomes partially blocked, ventilation of alveoli past the blockage site decreases, which in turn decreases gas exchange between the air and blood. The effect of this decreased gas exchange on overall gas exchange in the lungs is reduced by rerouting the blood to better-ventilated alveoli.

ASSESS YOUR PROGRESS

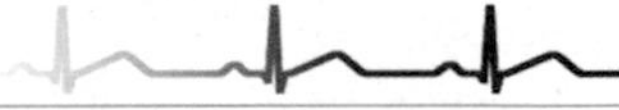

43. *Describe the four factors that affect the diffusion of gases through the respiratory membrane. Give examples of diseases that decrease diffusion by altering these factors.*
44. *Does oxygen or carbon dioxide diffuse more easily through the respiratory membrane?*
45. *What effect do alveolar ventilation and pulmonary capillary perfusion have on gas exchange?*
46. *What are the anatomical shunt and the physiological shunt?*
47. *What are the effects of gravity and alveolar Po_2 on blood flow in the lung?*

Predict 6

Even people in "good shape" may have trouble breathing at high altitudes. Explain how this can happen, even when ventilation of the lungs increases.

23.6 Oxygen and Carbon Dioxide Transport in the Blood

LEARNING OUTCOMES

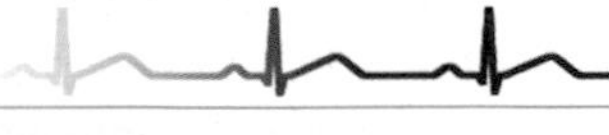

After reading this section, you should be able to

A. Describe the partial pressure gradients of oxygen and carbon dioxide.

B. Explain how oxygen and carbon dioxide are transported in the blood.

C. Discuss the factors that affect oxygen and carbon dioxide transport in the blood.

D. Explain the carbon dioxide exchange in the lungs and at the tissues.

E. Contrast fetal hemoglobin with maternal hemoglobin.

Once oxygen diffuses through the respiratory membrane into the blood, most of it combines reversibly with hemoglobin, and a smaller amount dissolves in the plasma. Hemoglobin transports oxygen from the pulmonary capillaries through the blood vessels to the tissue capillaries, where some of the oxygen is released. The oxygen diffuses from the blood to tissue cells, where it is used in aerobic respiration (see chapter 25).

Cells produce carbon dioxide during aerobic respiration. The carbon dioxide diffuses from the cells into the tissue capillaries. Once carbon dioxide enters the blood, it is transported in three ways: dissolved in the plasma, in combination with hemoglobin, or in the form of bicarbonate ions (HCO_3^-).

Oxygen Partial Pressure Gradients

The Po_2 within the alveoli averages approximately 104 mm Hg, whereas the Po_2 in blood flowing into the pulmonary capillaries is approximately 40 mm Hg (figure 23.16, *step 1*). Consequently, oxygen diffuses down its partial pressure gradient from the alveoli into the pulmonary capillary blood. By the time blood flows through the first third of the pulmonary capillary beds, an equilibrium has been achieved, and the Po_2 in the blood is 104 mm Hg, which is equivalent to the Po_2 in the alveoli. Even with the greater velocity of blood flow associated with exercise, by the time blood reaches the venous ends of the pulmonary capillaries, the Po_2 in the capillaries has achieved the same value as that in the alveoli (figure 23.16, *step 2*).

Blood leaving the pulmonary capillaries has a Po_2 of 104 mm Hg, but blood leaving the lungs in the pulmonary veins has a Po_2 of approximately 95 mm Hg. This decrease in the Po_2 occurs because the blood from the pulmonary capillaries mixes with deoxygenated (shunted) blood from the bronchial veins (figure 23.16, *step 3*).

The blood that enters the arterial end of the tissue capillaries has a Po_2 of approximately 95 mm Hg. The Po_2 of the interstitial fluid, in contrast, is close to 40 mm Hg and is probably near 20 mm Hg in the individual cells. Oxygen diffuses from the tissue capillaries to the interstitial fluid and from the interstitial fluid into the body's cells, where it is used in aerobic respiration. Because the cells use oxygen continuously, a constant partial pressure gradient exists for oxygen from the tissue capillaries to the cells (figure 23.16, *steps 4* and *5*).

Carbon Dioxide Partial Pressure Gradients

Carbon dioxide is continually produced as a by-product of cellular respiration, and a partial pressure gradient is established from tissue cells to the blood within the tissue capillaries. The intracellular Pco_2 is approximately 46 mm Hg, and the interstitial fluid Pco_2 is approximately 45 mm Hg. At the arterial end of the tissue capillaries, the Pco_2 is close to 40 mm Hg. As blood flows through the tissue capillaries, carbon dioxide diffuses from a higher Pco_2 to a lower Pco_2 until an equilibrium in Pco_2 is established. At the venous end of the capillaries, blood has a Pco_2 of 45 mm Hg (figure 23.16, *step 5*).

After blood leaves the venous end of the capillaries, it is transported through the cardiovascular system to the lungs. At the arterial end of the pulmonary capillaries, the Pco_2 is 45 mm Hg. Because the Pco_2 is approximately 40 mm Hg in the alveoli, carbon dioxide diffuses from the pulmonary capillaries into the alveoli. At the venous end of the pulmonary capillaries, the Pco_2 has again decreased to 40 mm Hg (figure 23.16, *steps 1* and *2*).

Hemoglobin and Oxygen Transport

Approximately 98.5% of the oxygen transported in the blood from the lungs to the tissues is transported in combination with hemoglobin in red blood cells, and the remaining 1.5% is dissolved in the water part of the plasma. The combination of oxygen with hemoglobin is reversible. In the pulmonary capillaries, oxygen binds to hemoglobin; in the tissue spaces, oxygen diffuses away from hemoglobin and enters the tissues.

Effect of Po_2

The **oxygen-hemoglobin dissociation curve** describes the percent saturation of hemoglobin in the blood at different blood Po_2 values. Hemoglobin is 100% saturated with oxygen when four oxygen molecules are bound to each hemoglobin molecule in the blood. There are four heme groups in a hemoglobin molecule (see chapter 19), and an oxygen molecule is bound to each heme group. Hemoglobin is 50% saturated with oxygen when there is an average of two oxygen molecules bound to each hemoglobin molecule.

The Po_2 in the blood leaving the pulmonary capillaries is normally 104 mm Hg. At that partial pressure, hemoglobin is 98% saturated (figure 23.17*a*). Decreases in the Po_2 in the pulmonary capillaries have a relatively small effect on hemoglobin saturation, as shown by the fairly flat shape of the upper part of the oxygen-hemoglobin dissociation curve. Even if the blood Po_2 decreases from 104 mm Hg to 60 mm Hg, hemoglobin is still 90% saturated. Hemoglobin is very effective at picking up oxygen in the lungs, even if the Po_2 in the pulmonary capillaries decreases significantly.

In a resting person, the normal blood Po_2 leaving the tissue capillaries is 40 mm Hg, and hemoglobin is 75% saturated. Thus, 23% (98% – 75%) of the oxygen picked up in the lungs is released from hemoglobin and diffuses into the tissues (figure 23.17*b*). The 75% of oxygen still bound to the hemoglobin is an oxygen reserve, which can be released if blood Po_2 decreases further. In the tissues, a relatively small change in blood Po_2 results in a relatively large change in hemoglobin saturation, as shown by the steep slope of the oxygen-hemoglobin dissociation curve. For example, during vigorous exercise, the Po_2 in skeletal muscle capillaries can decline to levels as low as 15 mm Hg because of the increased use of oxygen during aerobic respiration in skeletal muscle cells (see chapter 9). At a Po_2 of 15 mm Hg, hemoglobin is only 25% saturated, resulting in the release of 73% (98% – 25%) of the oxygen picked up in the lungs (figure 23.17*c*). Thus, as tissues use more oxygen, hemoglobin releases more oxygen to those tissues.

ASSESS YOUR PROGRESS

48. *Describe the partial pressure of oxygen and carbon dioxide in the alveoli, lung capillaries, tissue capillaries, and tissues.*

49. *How do these pressures account for the movement of oxygen and carbon dioxide between air and blood, and between blood and tissues?*

50. *Name the two ways oxygen is transported in the blood, and state the percentage of total oxygen transport for which each method is responsible.*

51. *How does the oxygen-hemoglobin dissociation curve explain the uptake of oxygen in the lungs and the release of oxygen in tissues?*

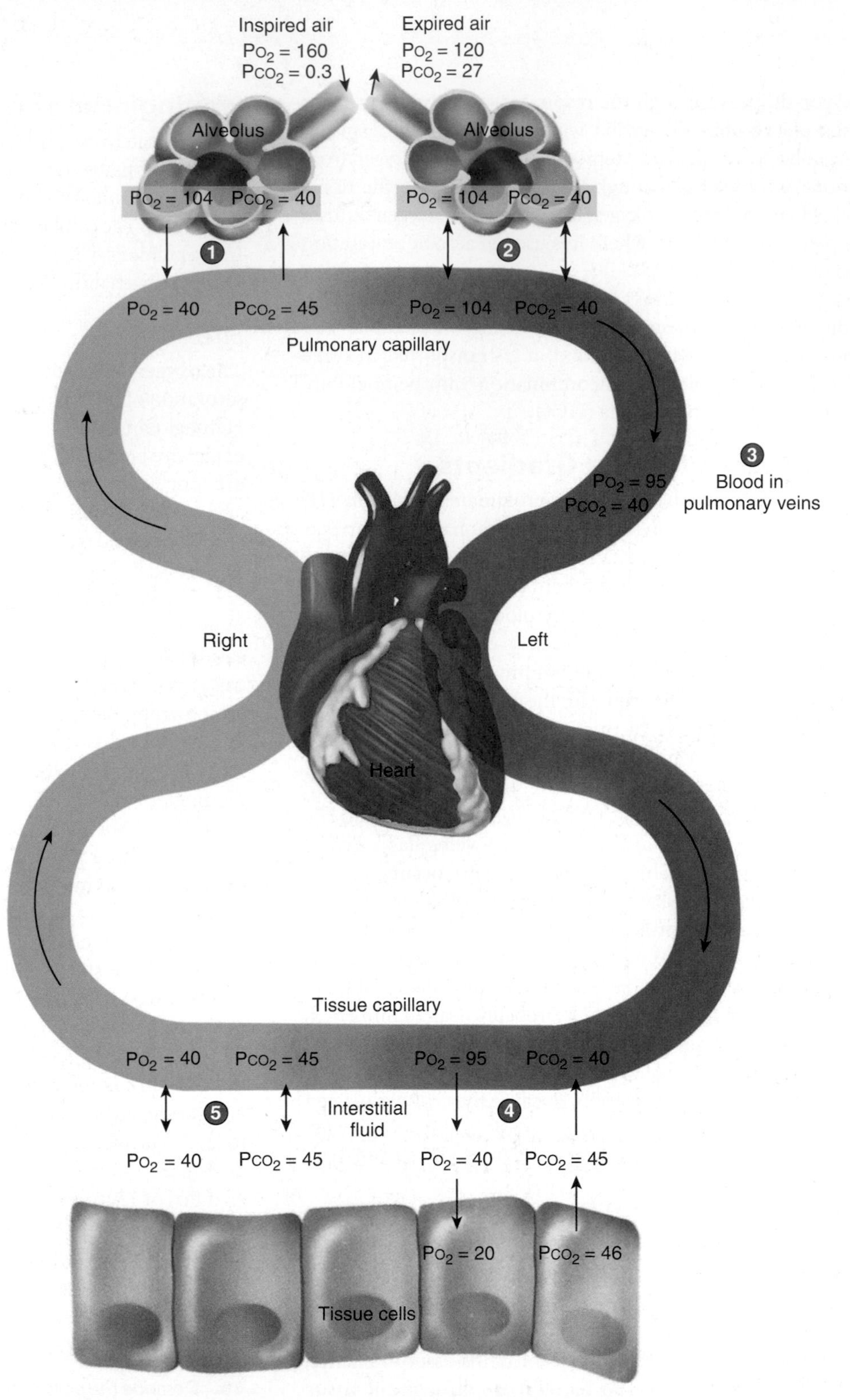

Inspired air
P_{O_2} = 160
P_{CO_2} = 0.3
Expired air
P_{O_2} = 120
P_{CO_2} = 27
Alveolus
Alveolus
P_{O_2} = 104 P_{CO_2} = 40
P_{O_2} = 104 P_{CO_2} = 40
1
2
P_{O_2} = 40 P_{CO_2} = 45
P_{O_2} = 104 P_{CO_2} = 40
Pulmonary capillary
1 Oxygen diffuses into the arterial ends of pulmonary capillaries, and CO_2 diffuses into the alveoli because of differences in partial pressures.
2 As a result of diffusion at the venous ends of pulmonary capillaries, the P_{O_2} in the blood is equal to the P_{O_2} in the alveoli, and the P_{CO_2} in the blood is equal to the P_{CO_2} in the alveoli.
3 The P_{O_2} of blood in the pulmonary veins is less than in the pulmonary capillaries because of mixing with deoxygenated blood from veins draining the bronchi and bronchioles.
P_{O_2} = 95
P_{CO_2} = 40
3 Blood in pulmonary veins
Right
Left
Heart
Tissue capillary
4 Oxygen diffuses out of the arterial ends of tissue capillaries, and CO_2 diffuses out of the tissue because of differences in partial pressures.
5 As a result of diffusion at the venous ends of tissue capillaries, the P_{O_2} in the blood is equal to the P_{O_2} in the tissue, and the P_{CO_2} in the blood is equal to the P_{CO_2} in the tissue. Go back to *step 1*.
P_{O_2} = 40 P_{CO_2} = 45
P_{O_2} = 95 P_{CO_2} = 40
5
Interstitial fluid
4
P_{O_2} = 40 P_{CO_2} = 45
P_{O_2} = 40 P_{CO_2} = 45
P_{O_2} = 20 P_{CO_2} = 46
Tissue cells

PROCESS FIGURE 23.16 AP|R **Gas Exchange**

Partial pressure gradients of oxygen and carbon dioxide between the alveoli and the pulmonary capillaries and between the tissues and the tissue capillaries are responsible for gas exchange. All partial pressures shown are expressed in mm Hg.

Effect of pH, P_{CO_2}, and Temperature

In addition to P_{O_2}, three other factors influence the degree to which oxygen binds to hemoglobin: blood pH, P_{CO_2}, and temperature (figure 23.18). As the pH of the blood declines, the amount of oxygen bound to hemoglobin at any given P_{O_2} also declines. This occurs because decreased pH results from an increase in H^+, and the H^+ combines with the protein part of the hemoglobin molecule and changes its three-dimensional structure, causing a decrease in the hemoglobin's ability to bind oxygen. Conversely, an increase in blood pH results in an increase in hemoglobin's ability to bind

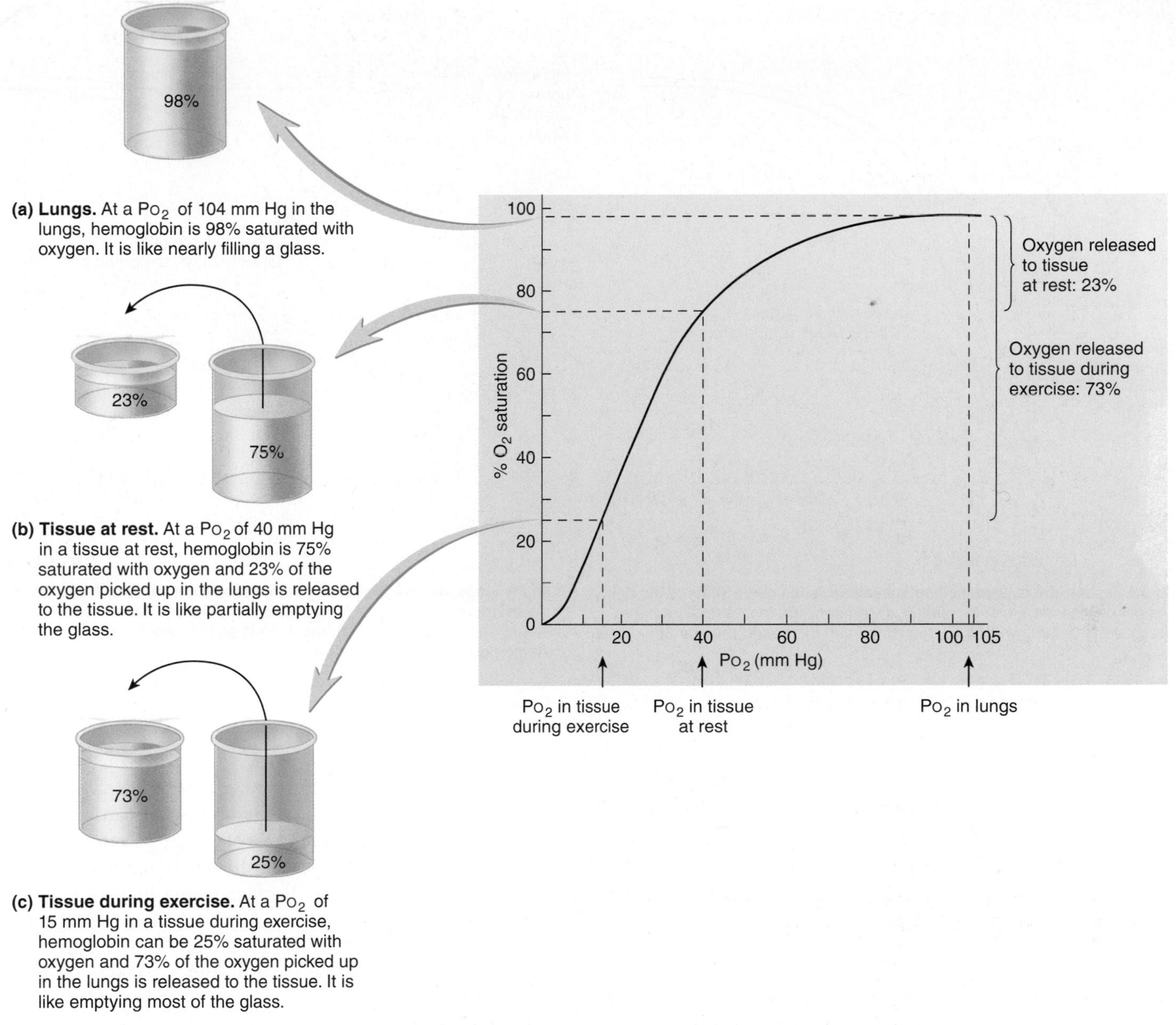

98%
(a) Lungs. At a Po_2 of 104 mm Hg in the lungs, hemoglobin is 98% saturated with oxygen. It is like nearly filling a glass.
23%
75%
(b) Tissue at rest. At a Po_2 of 40 mm Hg in a tissue at rest, hemoglobin is 75% saturated with oxygen and 23% of the oxygen picked up in the lungs is released to the tissue. It is like partially emptying the glass.
73%
25%
(c) Tissue during exercise. At a Po_2 of 15 mm Hg in a tissue during exercise, hemoglobin can be 25% saturated with oxygen and 73% of the oxygen picked up in the lungs is released to the tissue. It is like emptying most of the glass.
% O_2 saturation
100
80
60
40
20
0
20
40
60
80
100 105
Po_2 (mm Hg)
Oxygen released to tissue at rest: 23%
Oxygen released to tissue during exercise: 73%
Po_2 in tissue during exercise
Po_2 in tissue at rest
Po_2 in lungs

FIGURE 23.17 Oxygen-Hemoglobin Dissociation Curve

The oxygen-hemoglobin dissociation curve shows the percent saturation of hemoglobin as a function on Po_2. The ability of hemoglobin to pick up oxygen in the lungs and release it in the tissues is like a glass filling and emptying.

oxygen. The effect of pH on the oxygen-hemoglobin dissociation curve is called the **Bohr effect,** after its discoverer, Christian Bohr.

An increase in Pco_2 also decreases hemoglobin's ability to bind oxygen because of the effect of carbon dioxide on pH. Within red blood cells, an enzyme called **carbonic anhydrase** catalyzes this reversible reaction:

$$\underset{\text{Carbon dioxide}}{CO_2} + \underset{\text{Water}}{H_2O} \overset{\text{Carbonic anhydrase}}{\rightleftarrows} \underset{\text{Carbonic acid}}{H_2CO_3} \rightleftarrows \underset{\text{Hydrogen ion}}{H^+} + \underset{\text{Bicarbonate ion}}{HCO_3^-}$$

As carbon dioxide levels increase, more H^+ is produced, and the pH declines. As carbon dioxide levels decline, the reaction proceeds in the opposite direction, resulting in a decrease in H^+ concentration and an increase in pH. Thus, changes in carbon dioxide levels indirectly produce a Bohr effect by altering pH. In addition, carbon dioxide can directly affect hemoglobin's ability to bind oxygen, to a small extent. When carbon dioxide binds to the α- and β-globin chains of hemoglobin (see chapter 19), hemoglobin's ability to bind oxygen decreases.

As blood passes through tissue capillaries, carbon dioxide enters the blood from the tissues. As a consequence, blood carbon dioxide levels increase, pH decreases, and hemoglobin has less affinity for

FUNDAMENTAL **Figure**

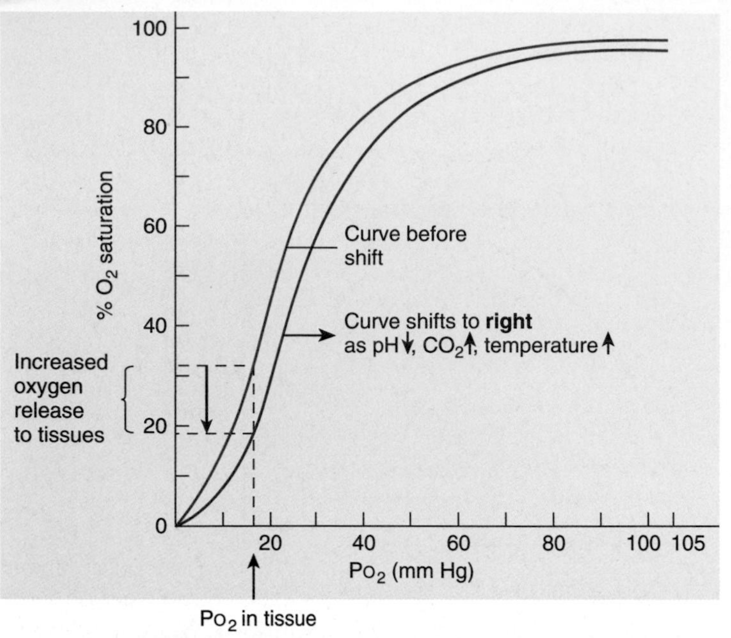

100
80
60
40
20
0
% O_2 saturation
Increased oxygen release to tissues
Curve before shift
Curve shifts to **right** as pH↓, CO_2↑, temperature↑
20
40
60
80
100 105
Po_2 (mm Hg)
Po_2 in tissue

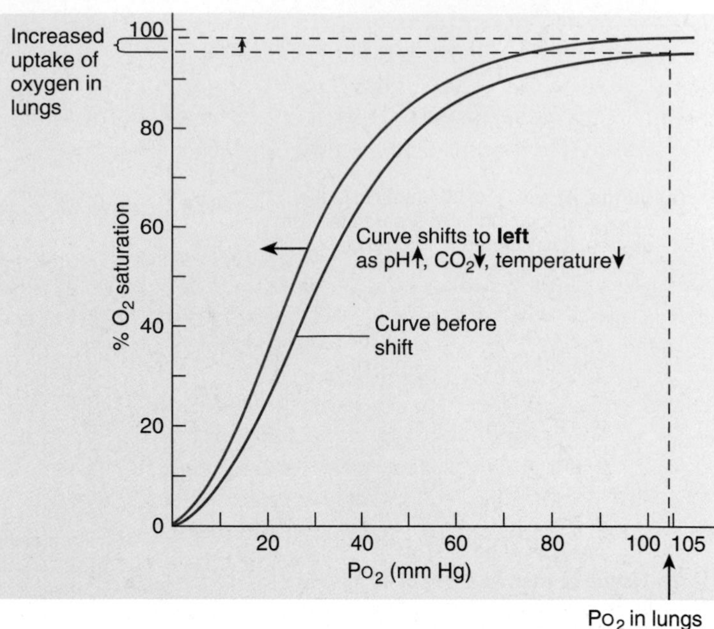

Increased uptake of oxygen in lungs
100
80
60
40
20
0
% O_2 saturation
Curve shifts to **left** as pH↑, CO_2↓, temperature↓
Curve before shift
20
40
60
80
100 105
Po_2 (mm Hg)
Po_2 in lungs

(a) **In the tissues, the oxygen–hemoglobin dissociation curve shifts to the right.** As pH decreases, Pco_2 increases, or temperature increases, the curve (*red*) shifts to the right (*blue*), resulting in an increased release of oxygen.

(b) **In the lungs, the oxygen–hemoglobin dissociation curve shifts to the left.** As pH increases, Pco_2 decreases, or temperature decreases, the curve (*blue*) shifts to the left (*red*), resulting in an increased ability of hemoglobin to pick up oxygen.

FIGURE 23.18 Effects of Shifting the Oxygen-Hemoglobin Dissociation Curve

oxygen in the tissue capillaries. Therefore, a greater amount of oxygen is released in the tissue capillaries than would be released if carbon dioxide were not present. When blood is returned to the lungs and passes through the pulmonary capillaries, carbon dioxide leaves the capillaries and enters the alveoli. As a result, carbon dioxide levels in the pulmonary capillaries are reduced, pH increases, and hemoglobin's affinity for oxygen increases.

An increase in temperature also decreases oxygen's tendency to remain bound to hemoglobin. Therefore, elevated temperatures resulting from increased metabolism increase the amount of oxygen released into the tissues by hemoglobin. In less metabolically active tissues in which the temperature is lower, less oxygen is released from hemoglobin.

When hemoglobin's affinity for oxygen decreases, the oxygen-hemoglobin dissociation curve is shifted to the right, and hemoglobin releases more oxygen (figure 23.18*a*). During exercise, when carbon dioxide and acidic substances, such as lactic acid, accumulate and the temperature increases in the tissue spaces, the oxygen-hemoglobin curve shifts to the right. Under these conditions, as much as 75–85% of the oxygen is released from the hemoglobin. In the lungs, however, the curve shifts to the left because of the lower carbon dioxide levels, lower temperature, and lower lactic acid levels. Therefore, hemoglobin's affinity for oxygen increases, and it becomes easily saturated (figure 23.18*b*).

During resting conditions, approximately 5 mL of oxygen are transported to the tissues in each 100 mL of blood, and cardiac output is approximately 5000 mL/min. Consequently, 250 mL of oxygen are delivered to the tissues each minute. During exercise, this value can increase up to 15 times. Oxygen transport can be increased threefold because of a greater degree of oxygen release from hemoglobin in the tissue capillaries, and the rate of oxygen transport is increased another five times because of the increase in cardiac output. Consequently, the volume of oxygen delivered to the tissues can be as high as 3750 mL/min (15 × 250 mL/min). Highly trained athletes can increase this volume to as high as 5000 mL/min.

Predict 7

In carbon monoxide (CO) poisoning, CO binds to hemoglobin, thereby decreasing the uptake of oxygen by hemoglobin. In addition, when CO binds to hemoglobin, the oxygen-hemoglobin dissociation curve shifts to the left. How does this shift affect the ability of tissues to get oxygen? Explain.

Effect of BPG

As red blood cells metabolize glucose for energy, they produce a by-product called **2,3-bisphosphoglycerate** (**BPG;** formerly called diphosphoglycerate). BPG binds to hemoglobin, reducing its affinity for oxygen, which increases its ability to release oxygen. A potent trigger for increased BPG production is low blood oxygen. For example, barometric pressure is lower at high altitudes than at sea level, causing both the partial pressure of oxygen in the alveoli and the percent saturation of blood with oxygen in the pulmonary capillaries to be lower. Consequently, the blood holds less oxygen for delivery to tissues. BPG helps increase oxygen delivery to tissues because higher levels of BPG increase the release of oxygen in tissues (the oxygen-hemoglobin dissociation curve shifts to the right). On the other hand, when blood is removed from the body and stored in a blood bank, the BPG levels in the stored blood decrease. As BPG levels decrease, the blood becomes unsuitable for transfusion after approximately 6 weeks because the hemoglobin releases less oxygen to the tissues. Banked blood is, therefore, discarded after 6 weeks of storage.

❯ Predict 8

If a person lacks the enzyme necessary for BPG synthesis, does he or she exhibit anemia (a lower than normal number of red blood cells) or erythrocytosis (a higher than normal number of red blood cells)? Explain.

Fetal Hemoglobin

Fetal hemoglobin is a unique type of hemoglobin found in fetuses only (see chapter 19). As fetal blood circulates through the placenta, oxygen is released from the mother's blood into the fetal blood, and carbon dioxide is released from fetal blood into the mother's blood. Fetal blood is very efficient at picking up oxygen for several reasons:

1. The concentration of fetal hemoglobin is approximately 50% greater than the concentration of maternal hemoglobin.
2. Fetal hemoglobin is different from maternal hemoglobin. Its oxygen-hemoglobin dissociation curve is to the left of the maternal oxygen-hemoglobin dissociation curve. Thus, for a given Po_2, fetal hemoglobin can hold on more tightly to oxygen than maternal hemoglobin can.
3. BPG has little effect on fetal hemoglobin. That is, BPG does not cause fetal hemoglobin to release oxygen.

ASSESS **YOUR PROGRESS**

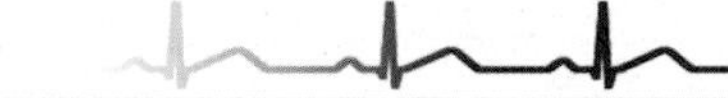

52. *What is the Bohr effect? How is it related to blood carbon dioxide?*

53. *Why is it advantageous for the oxygen-hemoglobin dissociation curve to shift to the left in the lungs and to the right in tissues?*

54. *How does temperature affect oxygen's tendency to bind to hemoglobin?*

55. *How does BPG affect the release of oxygen from hemoglobin?*

56. *Why is fetal hemoglobin's affinity for oxygen greater than that of maternal hemoglobin?*

❯ Predict 9

How does the movement of CO_2 from fetal blood into maternal blood increase the movement of oxygen from maternal blood into fetal blood? (*Hint:* Consider the shift of the oxygen-hemoglobin dissociation curve.)

Transport of Carbon Dioxide

Carbon dioxide is transported in the blood in three forms: as HCO_3^- dissolved in the plasma or the red blood cells, as CO_2 dissolved in the plasma, or as CO_2 bound to hemoglobin.

Carbon Dioxide Exchange in Tissues

Carbon dioxide diffuses from tissues into the plasma of blood (figure 23.19*a*). Most of the carbon dioxide diffuses from tissues into red blood cells. Inside the red blood cells, the carbon dioxide reacts with water to form carbonic acid, a reaction catalyzed by carbonic anhydrase. Carbonic acid then dissociates to form bicarbonate ions (HCO_3^-). About 70% of blood carbon dioxide is transported in the form of HCO_3^- dissolved in either the red blood cells or the plasma. Another 7% is transported as CO_2 dissolved in the plasma, and approximately 23% is transported bound to hemoglobin.

Removing HCO_3^- from inside the red blood cells promotes carbon dioxide transport because, as the HCO_3^- concentration decreases, more carbon dioxide combines with water to form additional HCO_3^- and H^+ (see "Reversible Reactions," chapter 2). In a process called **chloride shift** (figure 23.19*a*, *step 4*), antiporters exchange Cl^- for HCO_3^-. This exchange maintains electrical balance in the red blood cells and plasma as HCO_3^- diffuses out of, and Cl^- diffuses into, red blood cells.

Hydrogen ions bind to hemoglobin (figure 23.19*a*, *step 6*), resulting in three effects: (1) The transport of carbon dioxide increases because, as H^+ concentration decreases, more carbon dioxide combines with water to form additional HCO_3^- and H^+; (2) the pH inside the red blood cells does not decrease because hemoglobin is a buffer, preventing an increase in H^+ concentration; and (3) the affinity of hemoglobin for oxygen decreases. Hemoglobin releases oxygen in tissue capillaries because of decreased Po_2 (see figure 23.17). Hemoglobin's decreased affinity for oxygen shifts the oxygen-hemoglobin dissociation curve to the right (the Bohr effect; see figure 23.18*a*) and results in an increase in the release of oxygen from hemoglobin.

Approximately 23% of blood carbon dioxide is transported bound to hemoglobin. Many carbon dioxide molecules bind in a reversible fashion to the α- and β-globin chains of hemoglobin molecules (figure 23.19*a*, *step 7*). Carbon dioxide's ability to bind to hemoglobin is affected by the amount of oxygen bound to hemoglobin. The smaller the amount of oxygen bound to hemoglobin, the greater the amount of carbon dioxide able to bind to it, and vice versa. This relationship is called the **Haldane effect.** In tissues, as hemoglobin releases oxygen, the hemoglobin gains an increased ability to pick up carbon dioxide.

Carbon Dioxide Exchange in the Lungs

Carbon dioxide diffuses from red blood cells and plasma into the alveoli (figure 23.19*b*). As carbon dioxide levels in the red blood cells decrease, carbonic acid is converted to carbon dioxide and water. In response, HCO_3^- join with H^+ to form carbonic acid. As HCO_3^- and H^+ concentrations decrease because of this reaction, HCO_3^- enters red blood cells in exchange for Cl^-, and H^+ is released from hemoglobin. Hemoglobin picks up oxygen in pulmonary capillaries because of increased Po_2 (see figure 23.17). The release of H^+ from hemoglobin increases hemoglobin's affinity for oxygen, shifting the oxygen-hemoglobin curve to the left (Bohr effect; see figure 23.18*b*). Oxygen from the alveoli diffuses into the pulmonary capillaries and into the red blood cells, where it binds with hemoglobin. Carbon dioxide is released from hemoglobin and diffuses out of the red blood cells into the alveoli. As hemoglobin binds to oxygen, it more readily releases carbon dioxide (Haldane effect).

FUNDAMENTAL **Figure**

1 In the tissues, carbon dioxide (CO_2) diffuses into the plasma and into red blood cells. Some of the carbon dioxide remains in the plasma.

2 In red blood cells, carbon dioxide reacts with water (H_2O) to form carbonic acid (H_2CO_3) in a reaction catalyzed by the enzyme carbonic anhydrase (CA).

3 Carbonic acid dissociates to form bicarbonate ions (HCO_3^-) and hydrogen ions (H^+).

4 In the chloride shift, as HCO_3^- diffuse out of the red blood cells, electrical neutrality is maintained by the diffusion of chloride ions (Cl^-) into them.

5 Oxygen (O_2) is released from hemoglobin (Hb). Oxygen diffuses out of red blood cells and plasma into the tissue.

6 Hydrogen ions combine with hemoglobin, which promotes the release of oxygen from hemoglobin (Bohr effect).

7 Carbon dioxide combines with hemoglobin. Hemoglobin that has released oxygen readily combines with carbon dioxide (Haldane effect).

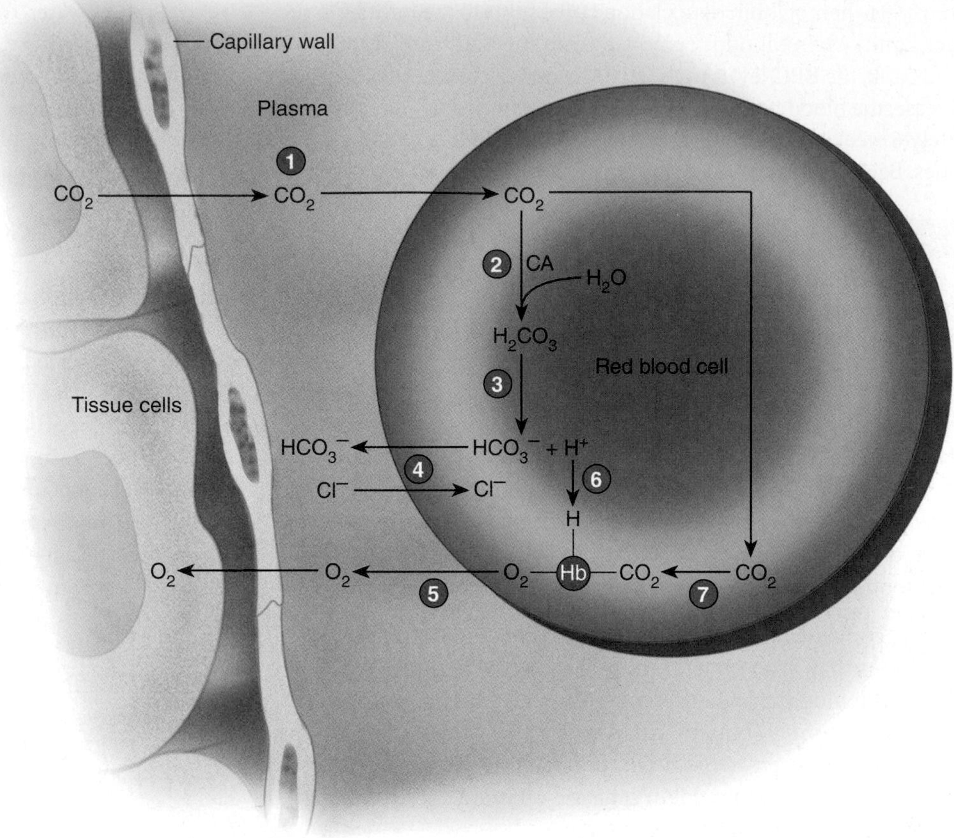

Capillary wall
Plasma
1
CO2
CO2
CO2
2
CA
H2O
H2CO3
3
Red blood cell
Tissue cells
HCO3−
HCO3− + H+
4
Cl−
Cl−
6
H
O2
O2
O2
Hb
CO2
CO2
5
7

(a) **Gas exchange in the tissues**

1 In the lungs, carbon dioxide (CO_2) diffuses from red blood cells and plasma into the alveoli.

2 Carbonic anhydrase catalyzes the formation of CO_2 and H_2O from H_2CO_3.

3 Bicarbonate ions and H^+ combine to replace H_2CO_3.

4 In the chloride shift, as HCO_3^- diffuse into red blood cells, electrical neutrality is maintained by the diffusion of chloride ions (Cl^-) out of them.

5 Oxygen diffuses into the plasma and into red blood cells. Some of the oxygen remains in the plasma. Oxygen binds to hemoglobin.

6 Hydrogen ions are released from hemoglobin, which promotes the uptake of oxygen by hemoglobin (Bohr effect).

7 Carbon dioxide is released from hemoglobin. Hemoglobin that is bound to oxygen readily releases carbon dioxide (Haldane effect).

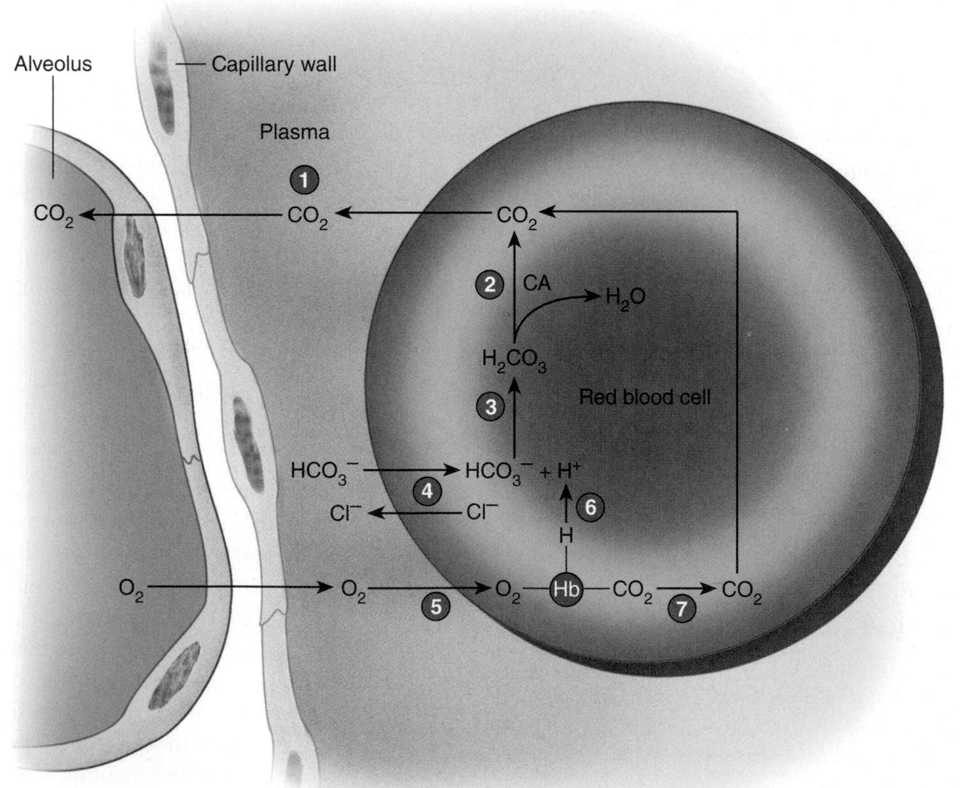

Alveolus
Capillary wall
Plasma
1
CO2
CO2
CO2
2
CA
H2O
H2CO3
3
Red blood cell
HCO3−
HCO3− + H+
4
Cl−
Cl−
6
H
O2
O2
O2
Hb
CO2
CO2
5
7

(b) **Gas exchange in the lungs**

PROCESS FIGURE 23.19 Gas Exchange

Carbon Dioxide and Blood pH

Blood pH refers to the pH in plasma, not inside red blood cells. As plasma carbon dioxide levels increase, H^+ levels increase, and blood pH decreases. An important function of the respiratory system is to regulate blood pH by changing plasma carbon dioxide levels (see chapter 27). Hyperventilation decreases plasma carbon dioxide, and hypoventilation increases it.

ASSESS YOUR PROGRESS

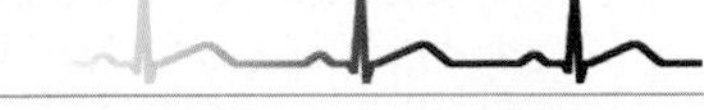

57. *How does the lowering HCO_3^- concentrations inside red blood cells affect carbon dioxide transport?*

58. *What is the chloride shift, and what does it accomplish?*

59. *Name three effects produced by H^+ binding to hemoglobin.*

60. *What is the Haldane effect?*

61. *What effect does blood carbon dioxide level have on blood pH?*

Predict 10

Explain the effect of (1) hyperventilation and (2) holding one's breath on blood pH.

23.7 Regulation of Ventilation

LEARNING OUTCOMES

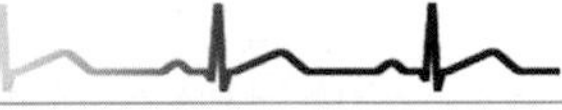

After reading this section, you should be able to

A. Describe the respiratory areas of the brainstem and how they produce a rhythmic pattern of ventilation.

B. Explain how the cerebral cortex and limbic system can affect ventilation.

C. Explain how blood pH, carbon dioxide, and oxygen levels affect ventilation.

D. Discuss the Hering-Breuer reflex and its importance.

E. Describe the effect of exercise on ventilation.

The basic rhythm of ventilation is controlled by neurons within the medulla oblongata that stimulate the muscles of respiration. The recruitment of muscle fibers and the more frequent stimulation of muscle fibers result in stronger contractions of the muscles and increased depth of respiration. The rate of respiration is determined by how frequently the respiratory muscles are stimulated.

Respiratory Areas in the Brainstem

Neurons involved with respiration are aggregated in certain parts of the brainstem. Scientists have learned that neurons that are active during inspiration are intermingled with those that are active during expiration.

The **medullary respiratory center** in the brainstem consists of two **dorsal respiratory groups,** each forming a longitudinal column of cells located bilaterally in the dorsal part of the medulla oblongata, and two **ventral respiratory groups,** each forming a longitudinal column of cells located bilaterally in the ventral part of the medulla oblongata (figure 23.20). Although the dorsal and ventral respiratory groups are bilaterally paired, cross-communication takes place between the pairs, so respiratory movements are symmetrical. In addition, communication occurs between the dorsal and ventral respiratory groups.

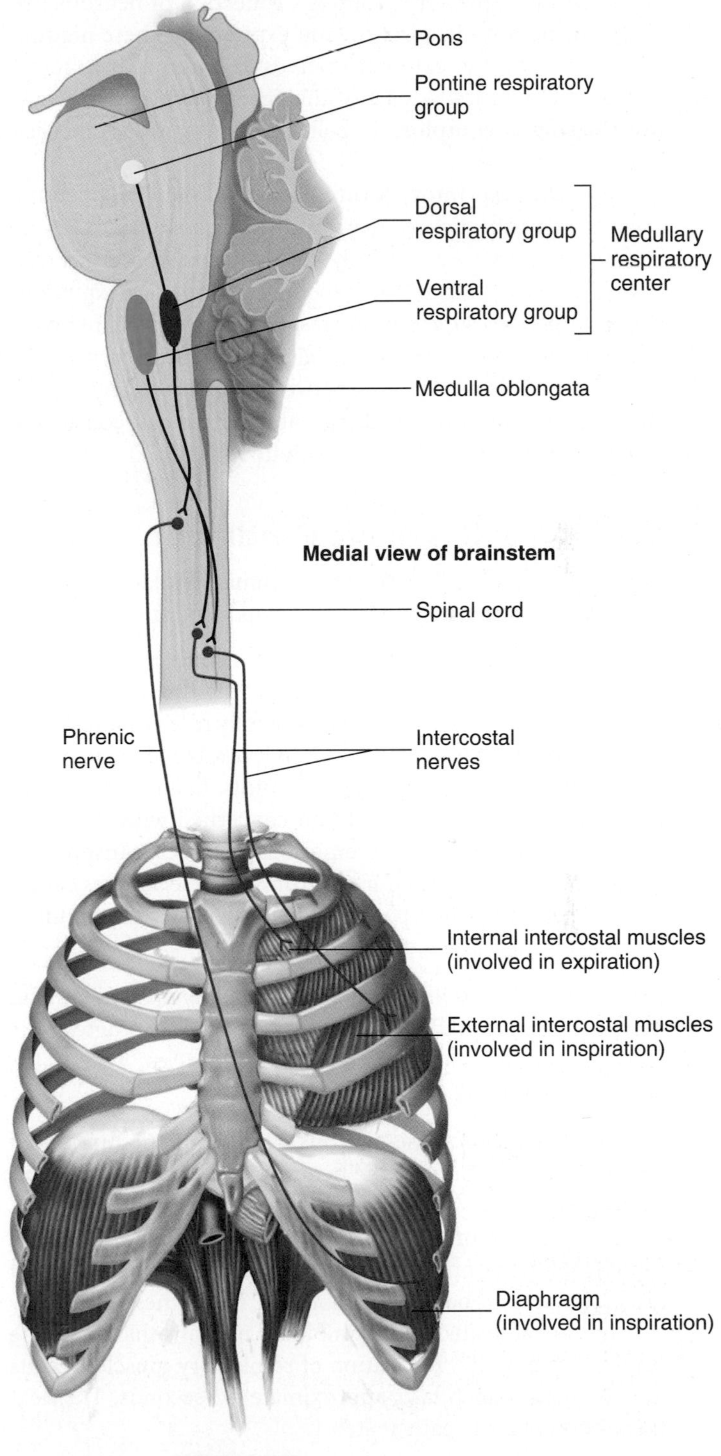
Pons
Pontine respiratory group
Dorsal respiratory group
Ventral respiratory group
Medullary respiratory center
Medulla oblongata
Medial view of brainstem
Spinal cord
Phrenic nerve
Intercostal nerves
Internal intercostal muscles (involved in expiration)
External intercostal muscles (involved in inspiration)
Diaphragm (involved in inspiration)
Anterior view

FIGURE 23.20 Respiratory Structures in the Brainstem
This figure shows the relationship of respiratory structures to each other and to the nerves innervating the muscles of respiration on the left side of the body.

Each dorsal respiratory group is a collection of neurons that are most active during inspiration, but some are active during expiration. The dorsal respiratory groups are primarily responsible for stimulating contraction of the diaphragm. They receive input from other parts of the brain and peripheral receptors, which allows the modification of respiration.

Each ventral respiratory group is a collection of neurons that are active during both inspiration and expiration. These neurons primarily stimulate the external intercostal, internal intercostal, and abdominal muscles. A part of the ventral respiratory group, the **pre-Bötzinger complex,** is believed to establish the basic rhythm of respiration.

The **pontine respiratory group,** formerly called the pneumotaxic center, is a collection of neurons in the pons (figure 23.20). Some of the neurons are active only during inspiration, some only during expiration, and others during both inspiration and expiration. The precise function of the pontine respiratory group is unknown, but it has connections with the medullary respiratory center and appears to play a role in switching between inspiration and expiration, thus fine-tuning the breathing pattern. It is not considered essential for the generation of the respiratory rhythm.

Generation of Rhythmic Ventilation

One explanation for the generation of rhythmic ventilation involves the integration of stimuli that start and stop inspiration:

1. *Starting inspiration.* Neurons in the medullary respiratory center spontaneously establish the basic rhythm of ventilation. The medullary respiratory center constantly receives stimulation from receptors that monitor blood gas levels, blood temperature, and the movements of muscles and joints. In addition, stimulation from the parts of the brain concerned with voluntary respiratory movements and emotions can occur. Inspiration starts when the combined input from all these sources causes the production of action potentials in the neurons that stimulate respiratory muscles.
2. *Increasing inspiration.* Once inspiration begins, more and more neurons are gradually activated. The result is progressively stronger stimulation of the respiratory muscles, which lasts for approximately 2 seconds.
3. *Stopping inspiration.* The neurons stimulating the muscles of respiration also stimulate the neurons in the medullary respiratory center that are responsible for stopping inspiration. The neurons responsible for stopping inspiration also receive input from the pontine respiratory group, stretch receptors in the lungs, and probably other sources. When these inhibitory neurons are activated, they inhibit the neurons that stimulate respiratory muscles. Relaxation of respiratory muscles results in expiration, which lasts approximately 3 seconds. The next inspiration begins again at step 1.

Although the medullary neurons establish the basic rate and depth of ventilation, their activities can be influenced by input from other parts of the brain and by input from peripherally located receptors.

Cerebral and Limbic System Control

Through the cerebral cortex, it is possible to consciously or unconsciously increase or decrease the rate and depth of the respiratory movements (figure 23.21*a*). For example, during talking or singing, air movement is controlled to produce sounds, as well as to facilitate gas exchange.

Apnea (ap′nē-ă) is the absence of breathing. A person may stop breathing voluntarily. As the period of voluntary apnea increases, a greater and greater urge to breathe develops. That urge is primarily associated with increasing P_{CO_2} levels in the arterial blood. Finally, the P_{CO_2} reaches levels that cause the respiratory center to override the conscious influence from the cerebrum. Occasionally, people are able to hold their breath until the blood P_{O_2} declines to a level low enough that they lose consciousness. After consciousness is lost, the respiratory center resumes its normal automatic control of respiration.

On the other hand, voluntary hyperventilation can decrease blood P_{CO_2} levels sufficiently to cause vasodilation of the peripheral blood vessels and a decrease in blood pressure (see chapter 21). Dizziness or a giddy feeling can result because the decreased blood pressure results in a decreased rate of blood flow to the brain, and therefore less oxygen is delivered to the brain.

Emotions acting through the limbic system of the brain can also affect the respiratory center (figure 23.21*a*). For example, strong emotions can cause hyperventilation or produce the sobs and gasps of crying.

ASSESS YOUR PROGRESS 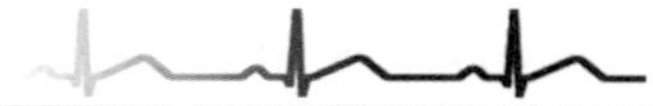

62. *Name the three respiratory groups, and describe their main functions.*

63. *How is rhythmic ventilation generated?*

64. *Explain how the cerebral cortex and limbic system can exert control over ventilation. What is apnea?*

Chemical Control of Ventilation

The respiratory system maintains blood oxygen and carbon dioxide concentrations and blood pH within normal values. Any deviation of these parameters from their normal range has a marked influence on respiratory movements. Changes in oxygen and carbon dioxide concentrations and in pH are superimposed on the neural mechanisms that establish rhythmic ventilation.

Chemoreceptors

Chemoreceptors are specialized neurons that respond to changes in chemicals in solution. The chemoreceptors involved in regulating respiration respond to changes in hydrogen ion concentrations, changes in P_{O_2}, or both (figures 23.21*b,c* and 23.22). **Central chemoreceptors** are located bilaterally and ventrally in the **chemosensitive area** of the medulla oblongata, and they are connected to the respiratory center. **Peripheral chemoreceptors** are found in the carotid and aortic bodies. These structures are small vascular sensory organs encapsulated in connective tissue and located near the carotid sinuses and the aortic arch (see chapter 21). The respiratory center is connected to the carotid body chemoreceptors through the glossopharyngeal nerve (IX) and to the aortic body chemoreceptors by the vagus nerve (X).

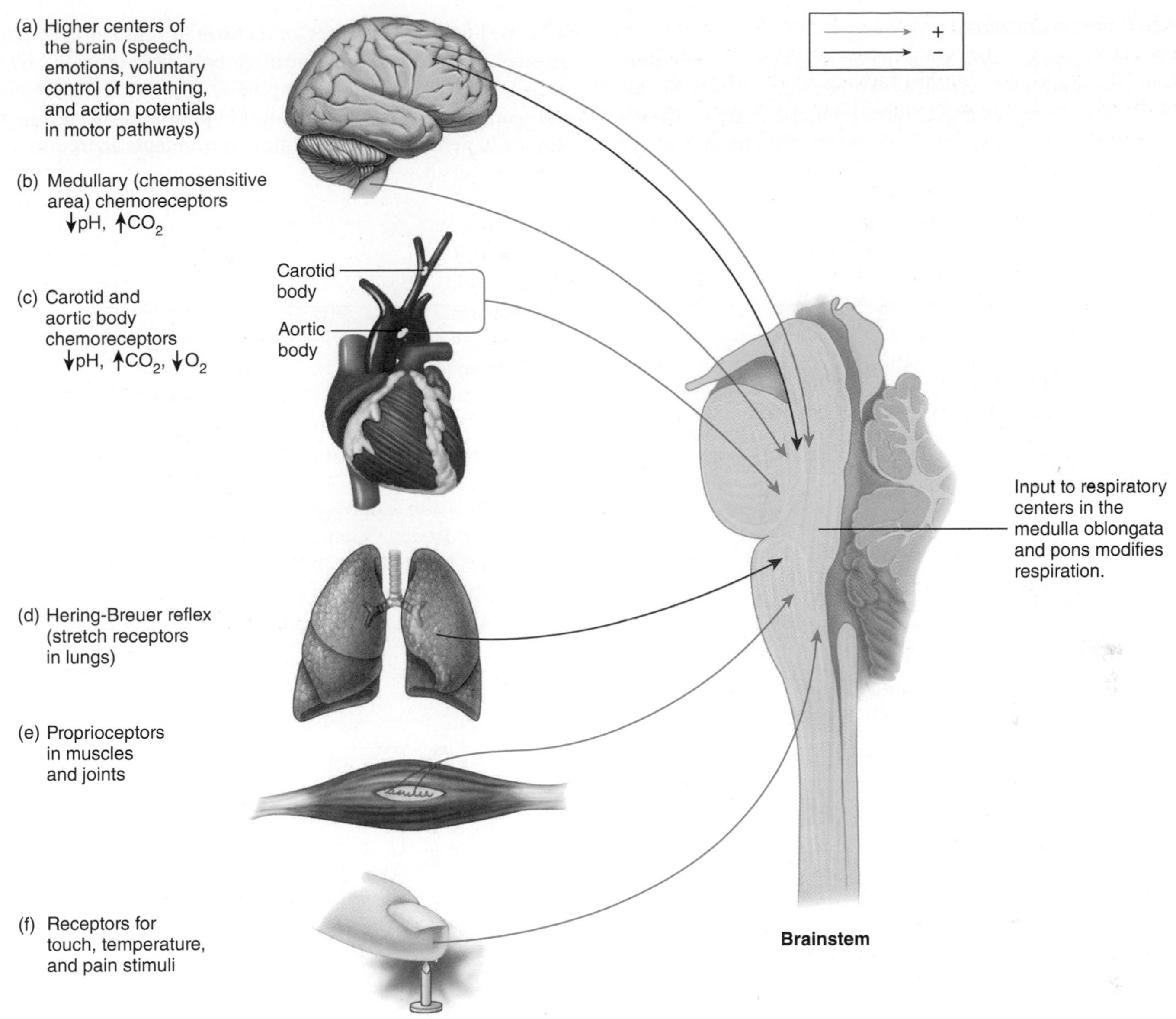
+
–
(a) Higher centers of the brain (speech, emotions, voluntary control of breathing, and action potentials in motor pathways)
(b) Medullary (chemosensitive area) chemoreceptors
↓pH, ↑CO_2
(c) Carotid and aortic body chemoreceptors
↓pH, ↑CO_2, ↓O_2
Carotid body
Aortic body
(d) Hering-Breuer reflex (stretch receptors in lungs)
(e) Proprioceptors in muscles and joints
(f) Receptors for touch, temperature, and pain stimuli
Input to respiratory centers in the medulla oblongata and pons modifies respiration.
Brainstem

FIGURE 23.21 Major Regulatory Mechanisms of Ventilation
A plus sign indicates an increase in ventilation, and a minus sign indicates a decrease in ventilation.

Effect of pH

The chemosensitive area of the medulla oblongata responds indirectly to changes in blood pH, whereas the carotid and aortic bodies respond directly. Hydrogen ions do not easily cross the blood-brain barrier or the blood-cerebrospinal fluid barrier (see chapter 11) to affect the chemosensitive area, but they do easily cross from the blood to the carotid and aortic bodies.

The chemosensitive area detects changes in blood pH through changes in blood carbon dioxide, which easily diffuses across the blood-brain barrier and the blood-cerebrospinal fluid barrier. For example, if blood carbon dioxide levels increase, carbon dioxide diffuses across the blood-brain barrier into the cerebrospinal fluid. The carbon dioxide combines with water to form carbonic acid, which dissociates into H^+ and HCO_3^-. The increased concentration of H^+ lowers the pH and stimulates the chemosensitive area, which then stimulates the respiratory center, resulting in a greater rate and depth of breathing. Consequently, carbon dioxide levels decrease as carbon dioxide is eliminated from the body, and blood pH increases to normal levels.

Maintaining body pH levels within normal parameters is necessary for the proper functioning of cells. Because changes in carbon dioxide levels can change pH, the respiratory system plays an important role in acid-base balance. For example, if blood pH decreases, the respiratory center is stimulated, so that carbon dioxide is eliminated, and blood pH increases back to normal levels. Conversely, if blood pH increases, the respiratory rate decreases, and carbon dioxide levels increase, causing blood pH to decrease back to normal levels. The respiratory system's role in maintaining pH is considered in greater detail in chapter 27.

Effect of Carbon Dioxide

Blood carbon dioxide levels are a major regulator of respiration during resting conditions, as well as at times when carbon dioxide levels are elevated—for example, during intense exercise. Even a small increase in carbon dioxide in the bloodstream triggers a large increase in the rate and depth of ventilation. For example, an increase in P_{CO_2} of 5 mm Hg causes an increase in ventilation of 100%. A greater than normal amount of carbon dioxide in the blood is called **hypercapnia** (hī-per-kap′nē-ă). Conversely, a lower than normal carbon dioxide level, called **hypocapnia** (hī-pō-kap′nē-ă), results in periods when respiratory movements are reduced or do not occur at all.

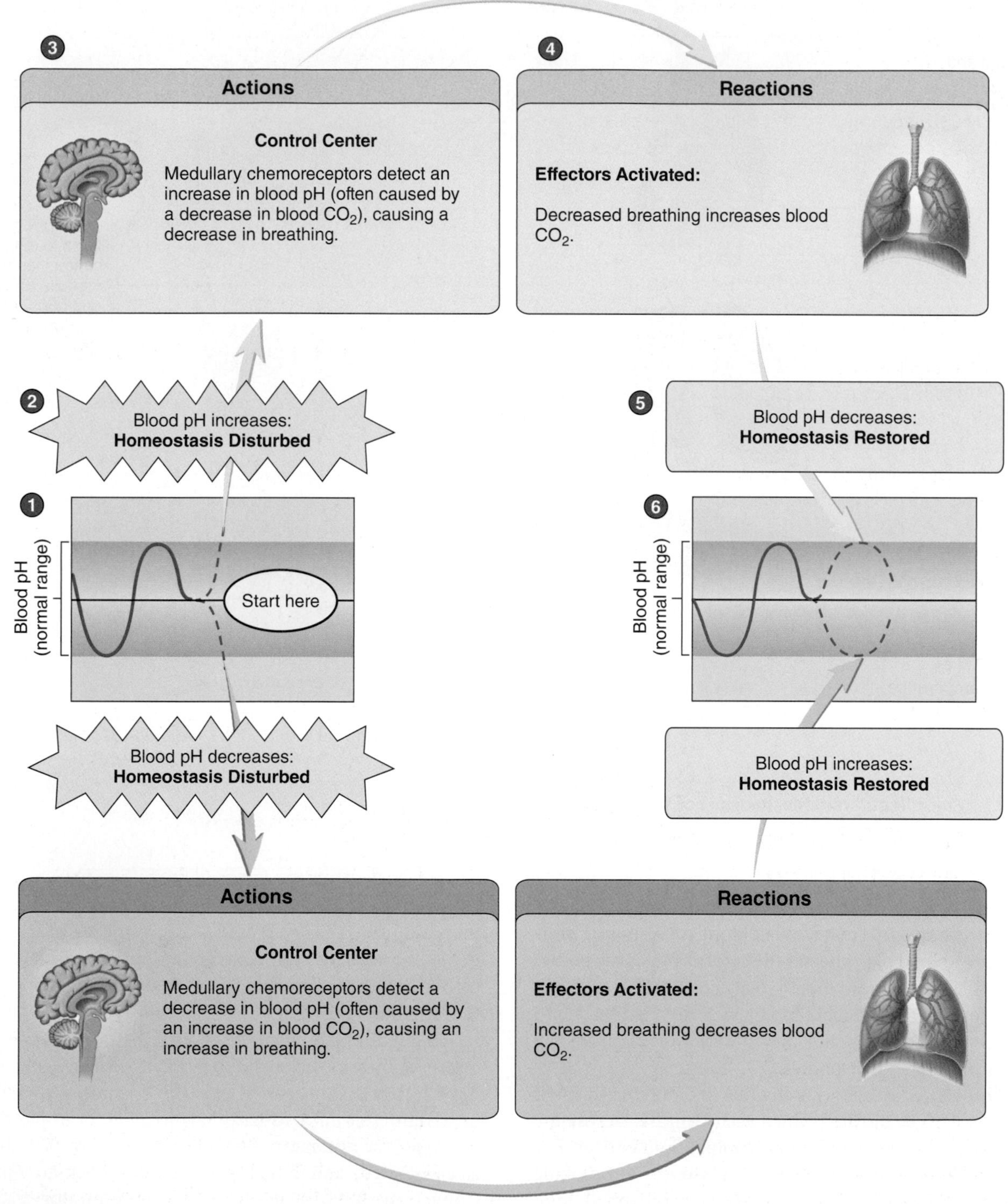
3
Actions
Control Center
Medullary chemoreceptors detect an increase in blood pH (often caused by a decrease in blood CO2), causing a decrease in breathing.
4
Reactions
Effectors Activated:
Decreased breathing increases blood CO2.
2
Blood pH increases:
Homeostasis Disturbed
5
Blood pH decreases:
Homeostasis Restored
1
Blood pH (normal range)
Start here
6
Blood pH (normal range)
Blood pH decreases:
Homeostasis Disturbed
Blood pH increases:
Homeostasis Restored
Actions
Control Center
Medullary chemoreceptors detect a decrease in blood pH (often caused by an increase in blood CO2), causing an increase in breathing.
Reactions
Effectors Activated:
Increased breathing decreases blood CO2.

HOMEOSTASIS FIGURE 23.22 Regulation of Blood pH and CO_2

(1) Blood pH is in its normal range. (2) Blood pH increases outside the normal range. (3) The medullary chemoreceptors detect an increase in blood pH. (4) The breathing rate decreases. (5) Blood pH returns to the normal range. (6) Homeostasis is restored. Follow the red arrows when blood pH decreases.

The chemoreceptors in the chemosensitive area of the medulla oblongata and in the carotid and aortic bodies respond to changes in carbon dioxide because of the effects of carbon dioxide on blood pH (figure 23.22). The chemosensitive area in the medulla oblongata is far more important in regulating P_{CO_2} and pH than are the carotid and aortic bodies. The carotid and aortic bodies are responsible for, at most, 15–20% of the total response to changes in P_{CO_2} or pH. During intense exercise, however, the carotid bodies respond more rapidly to changes in blood pH than does the chemosensitive area of the medulla.

Effect of Oxygen

Changes in P_{O_2} can affect respiration, although P_{CO_2} levels detected by the chemosensitive area are responsible for most of the changes in respiration (figure 23.22). A decrease in oxygen below its normal values is called **hypoxia** (hī-pok′sē-ă). If P_{O_2} levels in the arterial blood are markedly reduced while the pH and P_{CO_2} are held constant, an increase in ventilation rate occurs. However, within a normal range of P_{O_2} levels, the effect of oxygen on the regulation of respiration is small. Only after arterial P_{O_2} decreases to approximately 50% of its normal value does it begin to have a large stimulatory effect on respiratory movements.

At first, it is somewhat surprising that small changes in P_{O_2} do not cause changes in respiratory rate. But the reason becomes clear if we consider the oxygen-hemoglobin dissociation curve. Because of the S shape of the curve, at any P_{O_2} above 80 mm Hg nearly all of the hemoglobin is saturated with oxygen. Consequently, until P_{O_2} levels change significantly, the oxygen-carrying capacity of the blood is unaffected.

The carotid and aortic body chemoreceptors respond to decreased P_{O_2} by increasing stimulation of the respiratory center, thus keeping it active despite decreasing oxygen levels. However, if P_{O_2} decreases sufficiently, the respiratory center can fail to function, resulting in death.

ASSESS YOUR PROGRESS

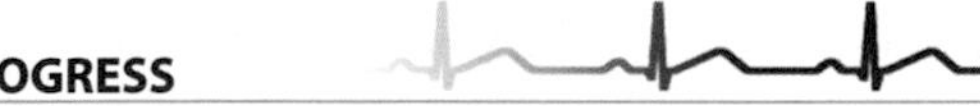

65. *Where are central chemoreceptors and peripheral chemoreceptors? Which are most important for regulating blood pH and carbon dioxide level? How does this change during intense exercise?*
66. *Define* hypercapnia *and* hypocapnia.
67. *How does a decrease in blood pH affect respiratory rate? How does a decrease in carbon dioxide affect respiratory rate?*
68. *What is hypoxia? Why must arterial P_{O_2} change significantly before it affects respiratory rate?*

Hering-Breuer Reflex

The **Hering-Breuer** (her′ing-broy′er) **reflex** limits the degree to which inspiration proceeds and prevents overinflation of the lungs (see figure 23.21*d*). This reflex depends on stretch receptors in the walls of the bronchi and bronchioles of the lungs. Action potentials are initiated in these stretch receptors when the lungs are inflated and are passed along sensory neurons within the vagus nerves to the medulla oblongata. The action potentials have an inhibitory influence on the respiratory center and result in expiration.

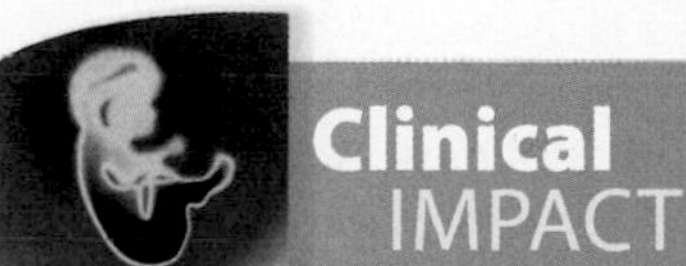

Clinical IMPACT

Importance of Reduced P_{O_2}

Carbon dioxide is much more important than oxygen as a regulator of normal alveolar ventilation, but under certain circumstances a reduced P_{O_2} in the arterial blood plays an important stimulatory role. During conditions of shock when blood pressure is very low, the P_{O_2} in arterial blood can drop low enough to strongly stimulate carotid and aortic body sensory receptors. At high altitudes, where barometric air pressure is low, the P_{O_2} in arterial blood can also drop to levels low enough to stimulate the carotid and aortic bodies. Although P_{O_2} levels in the blood are reduced, the respiratory system's ability to eliminate carbon dioxide is not greatly affected by low barometric air pressure. Thus, blood carbon dioxide levels become lower than normal because of the increased alveolar ventilation initiated in response to low P_{O_2}.

In people with emphysema, the destruction of the respiratory membrane allows less oxygen to move into the blood. The resulting low arterial P_{O_2} levels stimulate an increased rate and depth of respiration. At first, arterial P_{CO_2} levels may be unaffected by the reduced surface area of the respiratory membrane because carbon dioxide diffuses across the respiratory membrane 20 times more readily than does oxygen. However, if alveolar ventilation increases to the point that carbon dioxide exchange increases above normal, arterial carbon dioxide becomes lower than normal. More severe emphysema, in which the surface area of the respiratory membrane is reduced to a minimum, can decrease carbon dioxide exchange to the point that arterial carbon dioxide becomes elevated.

As expiration proceeds, the stretch receptors are no longer stimulated, and the decreased inhibitory effect on the respiratory center allows inspiration to begin again.

In infants, the Hering-Breuer reflex plays a role in regulating the basic rhythm of breathing and in preventing overinflation of the lungs. In adults, however, the reflex is important only when the tidal volume is large, such as during exercise.

Effect of Exercise on Ventilation

The mechanisms by which ventilation is regulated during exercise are controversial, and no single factor can account for all of the observed responses. Ventilation during exercise is divided into two phases:

1. *Ventilation increases abruptly.* At the onset of exercise, ventilation immediately increases. This initial increase can be as much as 50% of the total increase that occurs during exercise. The immediate increase in ventilation occurs too quickly to be explained by changes in metabolism or blood gases. As axons pass from the motor cortex of the cerebrum through the motor pathways, numerous collateral fibers project into the reticular formation of the brain. During exercise, action potentials in the motor pathways stimulate skeletal muscle contractions, and action potentials in the collateral fibers stimulate the respiratory center (see figure 23.21*e*).

Furthermore, during exercise, body movements stimulate proprioceptors in the joints of the limbs. Action potentials from the proprioceptors pass along sensory nerve fibers to the spinal cord and along ascending nerve tracts (the dorsal-column/medial-lemniscal system) of the spinal cord to the brain. Collateral fibers project from these ascending pathways to the respiratory center in the medulla oblongata. Movement of the limbs has a strong stimulatory influence on the respiratory center (see figure 23.21*e*).

A learned component may also exist in the ventilation response during exercise. After a period of training, the brain "learns" to match ventilation with the intensity of the exercise. Well-trained athletes match their respiratory movements more efficiently with their level of physical activity than do untrained individuals. Thus, the centers of the brain involved in learning have an indirect influence on the respiratory center, but the exact mechanism for this kind of regulation is unclear.

2. *Ventilation increases gradually.* After the immediate increase in ventilation, a gradual increase occurs and levels off within 4–6 minutes after the onset of exercise. Factors responsible for the immediate increase in ventilation may play a role in the gradual increase as well.

Despite large changes in oxygen consumption and carbon dioxide production during exercise, the *average* arterial Po_2, Pco_2, and pH remain constant and close to resting levels as long as the exercise is aerobic (see chapter 9). This suggests that changes in blood gases and pH do not play an important role in regulating ventilation during aerobic exercise. However, during exercise, the values of arterial Po_2, Pco_2, and pH rise and fall more than at rest. Thus, even though the average values do not change, their oscillations may be a signal for helping control ventilation.

The highest level of exercise that can be performed without causing a significant change in blood pH is called the **anaerobic threshold.** If the exercise intensity is high enough to exceed the anaerobic threshold, skeletal muscles produce and release lactic acid into the blood. The resulting change in blood pH stimulates the carotid bodies, resulting in increased ventilation. In fact, ventilation can increase so much that arterial Pco_2 decreases below resting levels and arterial Po_2 increases above resting levels.

Other Modifications of Ventilation

The activation of touch, thermal, and pain receptors can also affect the respiratory center because higher brain centers take over (see figure 23.21*f*). For example, irritants in the nasal cavity can initiate a sneeze reflex, and irritants in the lungs can stimulate a cough reflex. An increase in body temperature can stimulate increased ventilation because metabolism is elevated and more CO_2 is produced, which then needs to be expelled from the body.

ASSESS YOUR PROGRESS

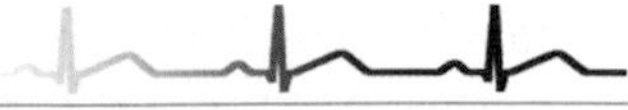

69. *Describe the Hering-Breuer reflex and its function.*

70. *What mechanisms regulate ventilation at the onset of exercise and then during exercise? What is the anaerobic threshold?*

Predict 11

Suppose that cold water is suddenly splashed on you. Describe your respiratory response. In the past, newborn babies were sometimes swatted on the buttocks. Explain the rationale for this procedure.

23.8 Respiratory Adaptations to Exercise

LEARNING OUTCOME

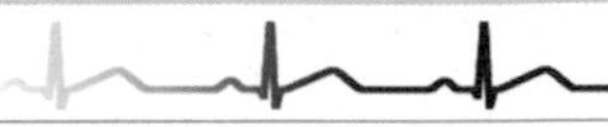

After reading this section, you should be able to

A. Describe the changes in the respiratory system that result from exercise training.

In response to training, athletic performance increases because the cardiovascular and respiratory systems become more efficient at delivering oxygen and picking up carbon dioxide. Ventilation does not limit performance in most individuals because ventilation can increase to a greater extent than does cardiovascular function.

After training, vital capacity increases slightly, and residual volume decreases slightly. Tidal volume at rest and during submaximal exercise does not change. At maximal exercise, however, tidal volume increases. After training, the respiratory rate at rest or during submaximal exercise is slightly lower than in an untrained person but, at maximal exercise, the respiratory rate is generally increased.

Minute ventilation is affected by the changes in tidal volume and respiratory rate. After training, minute ventilation is essentially unchanged or slightly reduced at rest and is slightly reduced during submaximal exercise. Minute ventilation is greatly increased at maximal exercise. For example, an untrained person's minute ventilation of 120 L/min can increase to 150 L/min after training. Increases to 180 L/min are typical of highly trained athletes.

Gas exchange between the alveoli and the blood increases at maximal exercise following training. The increased minute ventilation results in increased alveolar ventilation. In addition, increased cardiovascular efficiency allows greater blood flow through the lungs, especially the superior parts of the lungs.

ASSESS YOUR PROGRESS

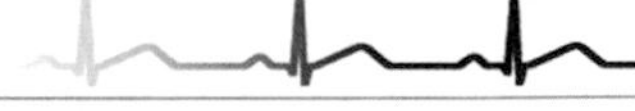

71. *What effect does training have on resting, submaximal, and maximal tidal volumes and on minute ventilation?*

23.9 Effects of Aging on the Respiratory System

LEARNING OUTCOME

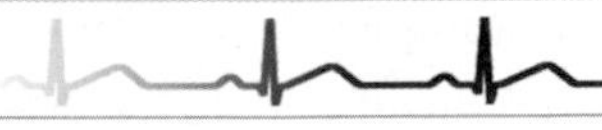

After reading this section, you should be able to

A. Describe the effects of aging on the respiratory system.

Most aspects of the respiratory system are affected by aging. However, even though vital capacity, maximum ventilation rates, and gas exchange decrease with age, the elderly can engage in light to moderate exercise because the respiratory system has a large reserve capacity.

Diseases and Disorders

TABLE 23.3 Respiratory System

Condition	Description
	RESPIRATORY DISORDERS
Bronchi and Lungs	
Bronchitis (brong-kī′tis)	Inflammation of the bronchi caused by irritants, such as cigarette smoke or infection; swelling impairs breathing; bronchitis can progress to emphysema
Emphysema (em-fi-sē′mă)	Destruction of alveolar walls; increased coughing increases pressure on the alveoli, causing rupture and destruction; loss of alveoli decreases surface area for gas exchange and decreases the lungs' ability to expel air; progression can be slowed, but there is no cure; in combination with bronchitis, the condition is known as chronic obstructive pulmonary disease (COPD)
Adult respiratory distress syndrome (ARDS)	Caused by damage to the respiratory membrane, which promotes inflammation; amount of surfactant is reduced, and fluid fills the alveoli, lessening gas exchange; ARDS usually develops after an injurious event, such as inhaling smoke from a fire or breathing toxic fumes
Cystic fibrosis (fi-brō′sis)	Genetic disorder that affects mucus secretions throughout the body due to an abnormal transport protein; mucus is much more viscous and accumulates in ducts and tubes, such as the bronchioles; airflow is restricted, and infections are more likely
Pulmonary fibrosis	Replacement of lung tissue with fibrous connective tissue, making the lungs less elastic; exposure to asbestos or coal dust is a common cause
Lung cancer	Occurs in the epithelium of the respiratory tract; can easily spread to other parts of the body because of the rich blood and lymphatic supply to the lungs;
Asthma	See Systems Pathology, later in this chapter
Circulatory System	
Thrombosis of the pulmonary arteries	Blood clot in lung blood vessels, causing inadequate blood flow through the pulmonary capillaries, which affects respiratory function
Anemia	Reduced hemoglobin lowers oxygen-carrying capacity of blood
Carbon monoxide poisoning	Carbon monoxide binds more strongly to hemoglobin than oxygen does and prevents already-bound oxygen from entering tissues
Nervous System	
Sudden infant death syndrome (SIDS)	Most frequent cause of death of infants between 2 weeks and 1 year of age; cause is still unknown, but at-risk babies can be placed on monitors that warn if breathing stops
Paralysis of the respiratory muscles	Damage to the spinal cord in the cervical or thoracic region interrupts nervous signals to the muscles of respiration
Thoracic Wall	Decreased elasticity of the thoracic wall prevents it from expanding to full capacity and reduces air movement; two spinal curvature conditions that reduce elasticity of the thoracic wall are scoliosis (skō-lē-ō′sis) and kyphosis (kī-fō′sis)
	INFECTIOUS DISEASES OF THE RESPIRATORY SYSTEM
Upper Respiratory Tract	
Strep throat	Caused by streptococcal bacteria (*Streptococcus pyogenes*); characterized by inflammation of the pharynx and fever
Diphtheria (dif-thēr′ē-ă)	Caused by the bacterium *Corynebacterium diphtheriae*; a grayish membrane forms in the throat and can completely block respiratory passages; DPT immunization for children partially targets diphtheria
Common cold	Results from a viral infection
Lower Respiratory Tract	
Whooping cough (pertussis; per-tŭs′is)	Caused by the bacterium *Bordetella pertussis*, which destroys cilia lining the respiratory epithelium, allowing mucus to accumulate; leads to a very severe cough; DPT immunization for children partially targets pertussis
Tuberculosis (tū-ber′kyū-lō′sis)	Caused by the bacterium *Clostridium tuberculosis*, which forms small, lumplike lesions called tubercles; immune system targets tubercles and causes larger lesions; certain strains of tuberculosis are resistant to antibiotics
Pneumonia (noo-mō′nē-ă)	Can be caused by a number of bacterial or viral infections of the lungs that cause fever, difficulty in breathing, and chest pain; edema in the lungs decreases their inflation ability and reduces gas exchange
Flu (influenza; in-flū-en′ză)	Viral infection of the respiratory system; does not affect the digestive system, as is commonly misunderstood; causes chills, fever, headache, and muscle aches
Fungal diseases	Fungal spores enter the respiratory tract attached to dust particles, usually resulting in minor respiratory infections that in some cases can spread to other parts of the body; examples are histoplasmosis and coccidioidomycosis

Go to: www.mhhe.com/seeley10 for additional information on these pathologies.

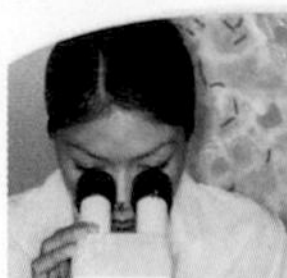

Systems PATHOLOGY | Asthma

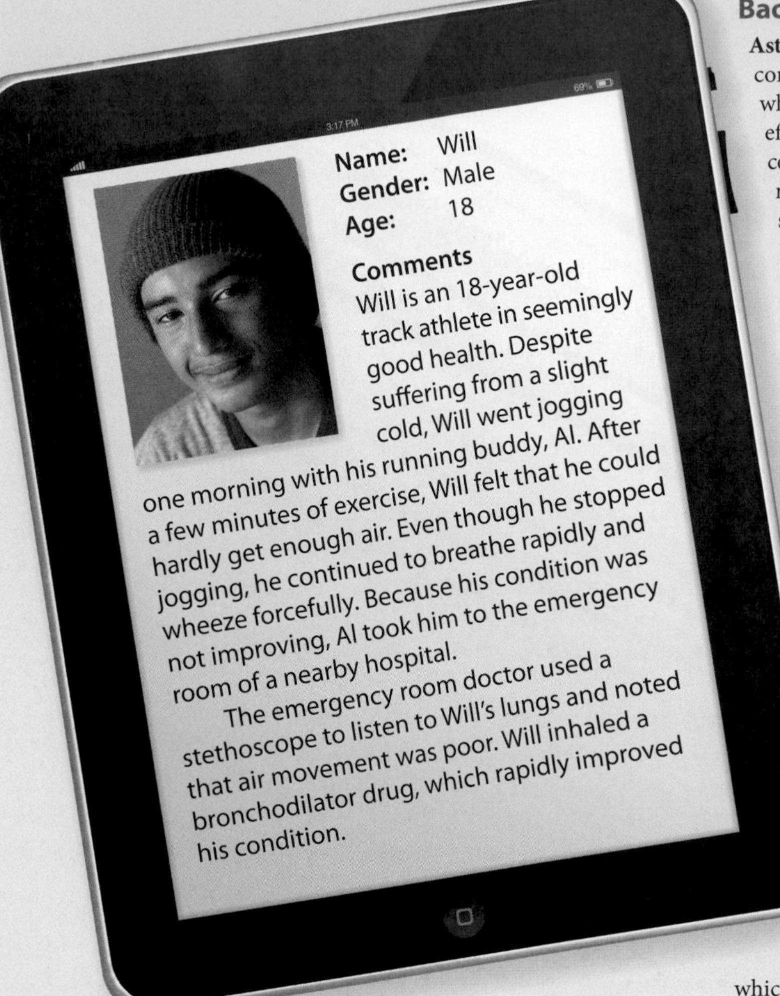
3:17 PM
69%
Name: Will
Gender: Male
Age: 18
Comments
Will is an 18-year-old track athlete in seemingly good health. Despite suffering from a slight cold, Will went jogging one morning with his running buddy, Al. After a few minutes of exercise, Will felt that he could hardly get enough air. Even though he stopped jogging, he continued to breathe rapidly and wheeze forcefully. Because his condition was not improving, Al took him to the emergency room of a nearby hospital.
The emergency room doctor used a stethoscope to listen to Will's lungs and noted that air movement was poor. Will inhaled a bronchodilator drug, which rapidly improved his condition.

Background Information

Asthma (az′mă; difficult breathing) is characterized by abnormally increased constriction of the trachea and bronchi in response to various stimuli, which results in narrowed air passageways and decreased ventilation efficiency. Symptoms include rapid and shallow breathing, wheezing, coughing, and shortness of breath (figure 23A). In contrast to many other respiratory disorders, the symptoms of asthma typically reverse either spontaneously or with therapy.

There is no definitive pathological feature or diagnostic test for asthma, but three important characteristics of the disease are chronic airway inflammation, airway hyperreactivity, and airflow obstruction. The inflammation results in tissue damage, edema, and mucus buildup, which can block airflow through the bronchi. Airway hyperreactivity means that the smooth muscle in the trachea and bronchi contracts greatly in response to a stimulus, thus decreasing the diameter of the airway and increasing resistance to airflow. The effects of inflammation and airway hyperreactivity combine to cause airflow obstruction (figure 23B).

Many cases of asthma appear to be associated with a chronic inflammatory response by the immune system. The number of immune cells in the bronchi, including mast cells, eosinophils, neutrophils, macrophages, and lymphocytes, increases. Inflammation appears to be linked to airway hyperreactivity by some chemical mediators released by immune cells (e.g., leukotrienes, prostaglandins, and interleukins), which increase the airway's sensitivity to stimulation and cause smooth muscle contraction.

The stimuli that prompt airflow obstruction in asthma vary from one individual to another. Some asthmatics react to particular allergens, which are foreign substances that evoke an inappropriate immune system response (see chapter 22). Examples include inhaled pollen, animal dander, and dust mites. Many cases of asthma are caused by an allergic reaction to substances in the droppings and carcasses of cockroaches, which may explain the higher rate of asthma in poor, urban areas. However, other inhaled substances, such as chemicals in the workplace or cigarette smoke, can provoke an asthma attack without stimulating an allergic reaction.

Figure 23A

Strenuous exercise is one of the many factors that can bring on an asthma attack.

Asthmatic bronchiole:
Note how constricted it is.

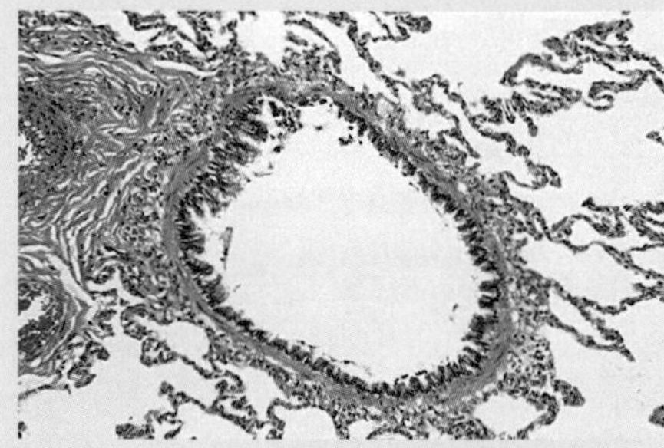

Normal bronchiole:
Note how clear it is.

Figure 23B

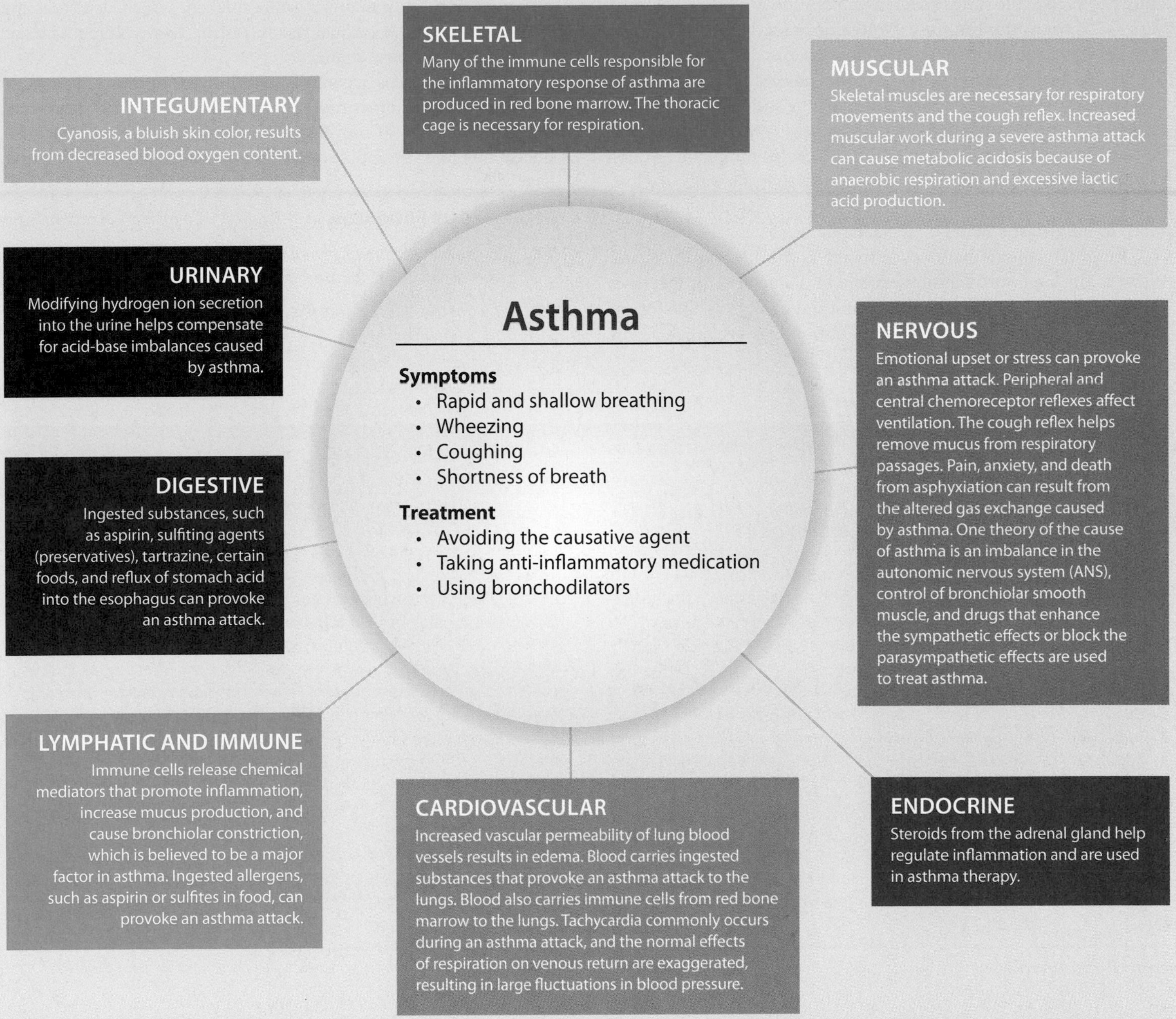

SKELETAL
Many of the immune cells responsible for the inflammatory response of asthma are produced in red bone marrow. The thoracic cage is necessary for respiration.

INTEGUMENTARY
Cyanosis, a bluish skin color, results from decreased blood oxygen content.

MUSCULAR
Skeletal muscles are necessary for respiratory movements and the cough reflex. Increased muscular work during a severe asthma attack can cause metabolic acidosis because of anaerobic respiration and excessive lactic acid production.

URINARY
Modifying hydrogen ion secretion into the urine helps compensate for acid-base imbalances caused by asthma.

Asthma

Symptoms
- Rapid and shallow breathing
- Wheezing
- Coughing
- Shortness of breath

Treatment
- Avoiding the causative agent
- Taking anti-inflammatory medication
- Using bronchodilators

NERVOUS
Emotional upset or stress can provoke an asthma attack. Peripheral and central chemoreceptor reflexes affect ventilation. The cough reflex helps remove mucus from respiratory passages. Pain, anxiety, and death from asphyxiation can result from the altered gas exchange caused by asthma. One theory of the cause of asthma is an imbalance in the autonomic nervous system (ANS), control of bronchiolar smooth muscle, and drugs that enhance the sympathetic effects or block the parasympathetic effects are used to treat asthma.

DIGESTIVE
Ingested substances, such as aspirin, sulfiting agents (preservatives), tartrazine, certain foods, and reflux of stomach acid into the esophagus can provoke an asthma attack.

LYMPHATIC AND IMMUNE
Immune cells release chemical mediators that promote inflammation, increase mucus production, and cause bronchiolar constriction, which is believed to be a major factor in asthma. Ingested allergens, such as aspirin or sulfites in food, can provoke an asthma attack.

CARDIOVASCULAR
Increased vascular permeability of lung blood vessels results in edema. Blood carries ingested substances that provoke an asthma attack to the lungs. Blood also carries immune cells from red bone marrow to the lungs. Tachycardia commonly occurs during an asthma attack, and the normal effects of respiration on venous return are exaggerated, resulting in large fluctuations in blood pressure.

ENDOCRINE
Steroids from the adrenal gland help regulate inflammation and are used in asthma therapy.

Over 200 substances have been associated with occupational asthma. An asthma attack can also be stimulated by ingested substances, such as aspirin; nonsteroidal anti-inflammatory compounds, such as ibuprofen (ī-bū′prō-fen); sulfites in food preservatives; and tartrazine (tar′tră-zēn) in food colorings. Asthmatics can substitute acetaminophen (as-et-ă-mē′nō-fen, a-set-ă-min′ō-fen; e.g., Tylenol) for aspirin.

Other stimuli, such as strenuous exercise (especially in cold weather) can precipitate an asthma attack. Such episodes can often be avoided by using a bronchodilator prior to exercise. Viral infections, emotional upset, stress, air pollution, and even reflux of stomach acid into the esophagus are known to elicit an asthma attack.

Treatment of asthma involves avoiding the causative stimulus and taking medications. Steroids and mast cell–stabilizing agents, which prevent the release of chemical mediators from mast cells, can reduce airway inflammation. Bronchodilators are used to increase airflow.

Predict 12
It is not usually necessary to assess arterial blood gases when diagnosing and treating asthma. However, this information can sometimes be useful in severe asthma attacks. Suppose that Will had a Po_2 of 60 mm Hg and a Pco_2 of 30 mm Hg when he first went to the emergency room. Explain how that could happen.

Vital capacity decreases with age because of a decreased ability to fill the lungs (decreased inspiratory reserve volume) and a decreased ability to empty the lungs (decreased expiratory reserve volume). As a result, maximum minute ventilation rates decrease, which in turn decreases the ability to perform intense exercise. These changes are related to weakening of respiratory muscles and to decreased compliance of the thoracic cage caused by the stiffening of cartilage and ribs. Lung compliance actually increases with age, but this effect is offset by the decreased thoracic cage compliance. Lung compliance increases because parts of the alveolar walls are lost, which reduces lung recoil. No significant age-related changes take place in lung elastic fibers or surfactant.

Residual volume increases with age as the alveolar ducts and many of the larger bronchioles expand in diameter. This increases the dead space, which decreases the amount of air available for gas exchange (alveolar ventilation). In addition, gas exchange across the respiratory membrane is reduced because parts of the alveolar walls are lost, which decreases the surface area available for gas exchange, and the remaining walls thicken, which decreases the diffusion of gases. A gradual rise in resting tidal volume with age compensates for these changes.

With age, mucus accumulates within the respiratory passageways because it becomes more viscous and because the number of cilia and their rate of movement decrease. As a consequence, the elderly are more susceptible to respiratory infections and bronchitis.

ASSESS YOUR PROGRESS

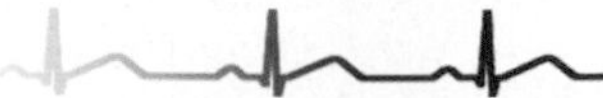

72. *Why do vital capacity, alveolar ventilation, and the diffusion of gases across the respiratory membrane decrease with age?*

73. *Why are the elderly more likely to develop respiratory infections and bronchitis?*

Answer

Learn to Predict

From page 811

Mr. Theron suffers from emphysema, a respiratory disorder that results in the destruction of alveoli. This chapter explained that the alveoli form the respiratory membrane, the site of gas exchange between the atmosphere and the blood. Alveolar destruction would directly reduce the respiratory membrane surface and therefore gas exchange. As a consequence, Mr. Theron has exaggerated respiratory movements to compensate for the reduction in surface area. Blood Po_2 is an important stimulus for the respiratory center, and the increased respiratory movements keep the ventilation just adequate to maintain blood Po_2 in the low normal range. Because CO_2 diffuses across the respiratory membrane at a faster rate than O_2, the elevated respiration required to maintain blood Po_2 causes too much CO_2 to be expired, resulting in his blood Pco_2 level dropping below normal.

This chapter explained that a pneumothorax, or the introduction of air into the pleural cavity through an opening in the thoracic wall or lung, can cause a lung to collapse. Due to his emphysema, Mr. Theron's lung collapsed when alveoli near the surface of the lung ruptured, allowing air to enter the pleural space. The physician was able to diagnose Mr. Theron's collapsed lung by listening for respiratory sounds with a stethoscope. He detected respiratory sounds in the right lung but not in the left lung. We would expect Mr. Theron's respiratory movements to be even more exaggerated, since the respiratory membrane was reduced by half when his left lung collapsed. The reduction in the respiratory membrane would cause a drop in Po_2 and an increase in Pco_2, both of which would stimulate respiratory centers to increase ventilation.

Answers to the rest of this chapter's Predict questions are in Appendix G.

Summary

23.1 Functions of the Respiratory System (p. 812)

1. Respiration includes the movement of air into and out of the lungs, the exchange of gases between the air and the blood, the transport of gases in the blood, and the exchange of gases between the blood and tissues.
2. Other functions of the respiratory system are regulation of blood pH, production of chemical mediators, voice production, olfaction, and protection against some microorganisms.

23.2 Anatomy and Histology of the Respiratory System (p. 812)

Conducting Zone

Nose

1. The nose consists of the external nose and the nasal cavity.
2. The bridge of the nose is bone, and most of the external nose is cartilage.

3. Openings of the nasal cavity
   - The nares open to the outside, and the choanae lead to the pharynx.
   - The paranasal sinuses and the nasolacrimal duct open into the nasal cavity.
4. Parts of the nasal cavity
   - The nasal cavity is divided by the nasal septum.
   - The anterior vestibule contains hairs that trap debris.
   - The nasal cavity is lined with pseudostratified ciliated columnar epithelium that traps debris and moves it to the pharynx.
   - The superior part of the nasal cavity contains the olfactory epithelium.
5. The nasal cavity serves as a passageway for air; cleans and humidifies air; is the location for the sense of smell; and, with the paranasal sinuses, functions as a resonating chamber for speech.

Pharynx

1. The nasopharynx joins the nasal cavity through the internal choanae and contains the openings to the auditory tube and the pharyngeal tonsils.
2. The oropharynx joins the oral cavity and contains the palatine and lingual tonsils.
3. The laryngopharynx opens into the larynx and the esophagus.

Larynx

1. Cartilage
   Three of the cartilages are unpaired. The thyroid cartilage and cricoid cartilage form most of the larynx. The epiglottis covers the opening of the larynx during swallowing.
   Six of the cartilages are paired. The vocal folds attach to the arytenoid cartilages.
2. The larynx maintains an open air passageway, regulates the passage of swallowed materials and air, produces sounds, and removes debris from the air.
3. Sounds are produced as the vocal folds vibrate when air passes through the larynx. Tightening the folds produces sounds of different pitches by controlling the length of the fold, which is allowed to vibrate.

Trachea

1. The trachea connects the larynx to the main bronchi.
2. The trachealis muscle regulates the diameter of the trachea.

Tracheobronchial Tree

1. The trachea divides to form two main bronchi, which extend to the lungs. The main bronchi divide to form lobar bronchi, which divide to form segmental bronchi, which divide to form bronchioles, which divide to form terminal bronchioles.
2. The trachea to the terminal bronchioles is a passageway for air movement.
   - The area from the trachea to the terminal bronchioles is ciliated to facilitate the removal of inhaled debris.
   - Cartilage helps hold the tube system open (from the trachea to the bronchioles).
   - Smooth muscle controls the diameter of the tubes (terminal bronchioles).

Respiratory Zone

Alveoli

1. Terminal bronchioles divide to form respiratory bronchioles, which give rise to alveolar ducts. Air-filled chambers called alveoli open into the respiratory bronchioles and alveolar ducts. The alveolar ducts end as alveolar sacs, which are chambers that connect to two or more alveoli.
2. Gas exchange occurs between the respiratory bronchioles and the alveoli.
3. The components of the respiratory membrane are a film of water, the walls of the alveolus and the capillary, and an interstitial space.

Lungs

1. The thoracic cavity contains two lungs.
2. The lungs are divided into lobes, bronchopulmonary segments, and lobules.

Thoracic Wall and Muscles of Respiration

1. The thoracic wall consists of vertebrae, ribs, the sternum, and muscles that allow expansion of the thoracic cavity.
2. Contraction of the diaphragm increases thoracic volume.
3. Muscles can elevate the ribs and increase thoracic volume or depress the ribs and decrease thoracic volume.

Pleura

The pleural membranes surround the lungs and protect against friction.

Blood Supply

1. Deoxygenated blood is transported to the lungs through the pulmonary arteries, and oxygenated blood leaves through the pulmonary veins.
2. Oxygenated blood is mixed with a small amount of deoxygenated blood from the bronchi.

Lymphatic Supply

The superficial and deep lymphatic vessels drain lymph from the lungs.

23.3 Ventilation (p. 827)

Pressure Differences and Airflow

1. Ventilation is the movement of air into and out of the lungs.
2. Air moves from an area of higher pressure to an area of lower pressure.

Pressure and Volume

Pressure is inversely related to volume.

Airflow Into and Out of Alveoli

1. Inspiration results when barometric air pressure is greater than intra-alveolar pressure.
2. Expiration results when barometric air pressure is less than intra-alveolar pressure.

Changing Alveolar Volume

1. Lung recoil causes alveoli to collapse.
   - Lung recoil results from elastic fibers and water surface tension.
   - Surfactant reduces water surface tension.
2. Pleural pressure is the pressure in the pleural cavity.
   - A negative pleural pressure can cause the alveoli to expand.
   - Pneumothorax is an opening between the pleural cavity and the air that causes a loss of pleural pressure.
3. Changes in thoracic volume cause changes in pleural pressure, resulting in changes in alveolar volume, intra-alveolar pressure, and airflow.

23.4 Measurement of Lung Function (p. 832)

Compliance of the Lungs and Thorax

1. Compliance is a measure of lung expansion caused by intra-alveolar pressure.
2. Reduced compliance means that it is more difficult than normal to expand the lungs.

Pulmonary Volumes and Capacities

1. Four pulmonary volumes exist: tidal volume, inspiratory reserve volume, expiratory reserve volume, and residual volume.
2. Pulmonary capacities are the sum of two or more pulmonary volumes and include inspiratory capacity, functional residual capacity, vital capacity, and total lung capacity.
3. The forced expiratory vital capacity measures vital capacity while the individual exhales as rapidly as possible.

Minute Ventilation and Alveolar Ventilation

1. Minute ventilation is the total amount of air moved in and out of the respiratory system per minute.
2. Dead space is the part of the respiratory system where gas exchange does not take place.
3. Alveolar ventilation is how much air per minute enters the parts of the respiratory system where gas exchange takes place.

23.5 Physical Principles of Gas Exchange (p. 834)

Partial Pressure

1. Partial pressure is the contribution of a gas to the total pressure of a mixture of gases (Dalton's law).
2. Water vapor pressure is the partial pressure produced by water.
3. Atmospheric air, alveolar air, and expired air have different compositions.

Diffusion of Gases Into and Out of Liquids

The concentration of a dissolved gas in a liquid is determined by its pressure and by its solubility coefficient (Henry's law).

Diffusion of Gases Through the Respiratory Membrane

1. The respiratory membrane is thin and has a large surface area that facilitates gas exchange.
2. The rate of diffusion of gases through the respiratory membrane depends on its thickness, the diffusion coefficient of the gas, the surface area of the membrane, and the partial pressure of the gases in the alveoli and the blood.

Relationship Between Alveolar Ventilation and Pulmonary Capillary Perfusion

1. Increased alveolar ventilation or increased pulmonary capillary perfusion increases gas exchange.
2. The physiological shunt is the deoxygenated blood returning from the lungs.

23.6 Oxygen and Carbon Dioxide Transport in the Blood (p. 836)

Oxygen Partial Pressure Gradients

1. Oxygen moves from the alveoli (P_{O_2} = 104 mm Hg) into the blood (P_{O_2} = 40 mm Hg). Blood is almost completely saturated with oxygen when it leaves the capillary.
2. The P_{O_2} in the blood decreases (P_{O_2} = 95 mm Hg) when it mixes with deoxygenated blood.
3. Oxygen moves from the tissue capillaries (P_{O_2} = 95 mm Hg) into the tissues (P_{O_2} = 40 mm Hg).

Carbon Dioxide Partial Pressure Gradients

1. Carbon dioxide moves from the tissues (P_{CO_2} = 45 mm Hg) into tissue capillaries (P_{CO_2} = 40 mm Hg).
2. Carbon dioxide moves from the pulmonary capillaries (P_{CO_2} = 45 mm Hg) into the alveoli (P_{CO_2} = 40 mm Hg).

Hemoglobin and Oxygen Transport

1. Oxygen is transported by hemoglobin (98.5%) and is dissolved in plasma (1.5%).
2. The oxygen-hemoglobin dissociation curve shows that hemoglobin is almost completely saturated when P_{O_2} is 80 mm Hg or above. At lower partial pressures, the hemoglobin releases oxygen.
3. Hemoglobin's ability to hold oxygen decreases because of a shift of the oxygen-hemoglobin dissociation curve to the right due to decreased pH (Bohr effect), increased carbon dioxide, or increased temperature.
4. Hemoglobin's ability to hold oxygen increases because of a shift of the oxygen-hemoglobin dissociation curve to the left due to increased pH (Bohr effect), decreased carbon dioxide, or decreased temperature.
5. The substance 2,3-bisphosphoglycerate increases hemoglobin's ability to release oxygen.
6. Fetal hemoglobin has a higher affinity for oxygen than maternal hemoglobin does.

Transport of Carbon Dioxide

1. Carbon dioxide is transported dissolved in plasma (7%), as HCO_3^- dissolved in plasma and in red blood cells (70%), and bound to hemoglobin (23%).
2. In tissue capillaries, the following events occur:
   - Carbon dioxide combines with water inside red blood cells to form carbonic acid, which dissociates to form HCO_3^-. Decreasing HCO_3^- concentrations promote carbon dioxide transport.
   - Exchange of Cl^- for HCO_3^- occurs between plasma and red blood cells; this is called the chloride shift.
   - Hydrogen ions binding to hemoglobin promote carbon dioxide transport, prevent a pH change in red blood cells, and produce a Bohr effect.
   - In the Haldane effect, the smaller the amount of oxygen bound to hemoglobin, the greater the amount of carbon dioxide bound to it, and vice versa.
3. In pulmonary capillaries, the events occurring in the tissue capillaries are reversed.

23.7 Regulation of Ventilation (p. 843)

Respiratory Areas in the Brainstem

1. The medullary respiratory center consists of the dorsal and ventral respiratory groups.
   - The dorsal respiratory groups stimulate the diaphragm.
   - The ventral respiratory groups stimulate the intercostal and abdominal muscles.
2. The pontine respiratory group is involved with switching between inspiration and expiration.

Generation of Rhythmic Ventilation

1. Neurons in the medullary respiratory center establish the basic rhythm of ventilation.
2. When stimuli from receptors or other parts of the brain exceed a threshold level, inspiration begins.
3. As respiratory muscles are stimulated, neurons that stop inspiration are stimulated. When the stimulation of these neurons exceeds a threshold level, inspiration is inhibited.

Cerebral and Limbic System Control

Ventilation can be voluntarily controlled and can be modified by emotions.

Chemical Control of Ventilation

1. Carbon dioxide is the major regulator of ventilation. An increase in carbon dioxide or a decrease in pH can stimulate the chemosensitive area, causing a greater rate and depth of ventilation.
2. Oxygen levels in the blood affect ventilation when a 50% or greater decrease from normal exists. Decreased oxygen is detected by receptors in the carotid and aortic bodies, which then stimulate the respiratory center.

Hering-Breuer Reflex

Stretch of the lungs during inspiration can inhibit the respiratory center and contribute to a cessation of inspiration.

Effect of Exercise on Ventilation

1. Collateral fibers from motor neurons and from proprioceptors stimulate the respiratory centers.
2. Chemosensitive mechanisms and learning fine-tune the effects produced through the motor neurons and proprioceptors.

Other Modifications of Ventilation

Touch, thermal, and pain sensations can modify ventilation.

23.8 Respiratory Adaptations to Exercise (p. 848)

Tidal volume, respiratory rate, minute ventilation, and gas exchange between the alveoli and blood remain unchanged or slightly lower at rest or during submaximal exercise but increase at maximal exercise.

23.9 Effects of Aging on the Respiratory System (p. 848)

1. Vital capacity and maximum minute ventilation decrease with age because of weakened respiratory muscles and decreased thoracic cage compliance.
2. Residual volume and dead space increase because of the enlarged diameter of respiratory passageways. As a result, alveolar ventilation decreases.
3. An increase in resting tidal volume compensates for decreased alveolar ventilation, loss of alveolar walls (surface area), and thickening of alveolar walls.
4. The ability to remove mucus from the respiratory passageways decreases with age.

REVIEW AND COMPREHENSION

1. The nasal cavity
   a. has openings for the paranasal sinuses.
   b. has a vestibule, which contains the olfactory epithelium.
   c. is connected to the pharynx by the nares.
   d. has passageways called conchae.
   e. is lined with squamous epithelium, except for the vestibule.
2. The larynx
   a. connects the oropharynx to the trachea.
   b. has three unpaired and six paired cartilages.
   c. contains the vocal folds.
   d. contains the vestibular folds.
   e. All of these are correct.
3. Terminal bronchioles branch to form
   a. the alveolar duct.
   b. alveoli.
   c. bronchioles.
   d. respiratory bronchioles.
4. During an asthma attack, a person has difficulty breathing because of constriction of the
   a. trachea.
   b. bronchi.
   c. terminal bronchioles.
   d. alveoli.
   e. respiratory membrane.
5. During quiet expiration, the
   a. abdominal muscles relax.
   b. diaphragm moves inferiorly.
   c. external intercostal muscles contract.
   d. thorax and lungs passively recoil.
   e. All of these are correct.
6. The parietal pleura
   a. covers the surface of the lung.
   b. covers the inner surface of the thoracic cavity.
   c. is the connective tissue partition that divides the thoracic cavity into right and left pleural cavities.
   d. covers the inner surface of the alveoli.
   e. is the membrane across which gas exchange occurs.
7. Contraction of the bronchiolar smooth muscle has which of these effects?
   a. A smaller pressure gradient is required to get the same rate of airflow, compared with normal bronchioles.
   b. It increases airflow through the bronchioles.
   c. It increases resistance to airflow.
   d. It increases alveolar ventilation.
8. During expiration, the intra-alveolar pressure is
   a. lower than the pleural pressure.
   b. greater than the barometric pressure.
   c. lower than the barometric pressure.
   d. unchanged.
9. Normally, which of the following keeps the lungs from collapsing?
   a. surfactant
   b. pleural pressure
   c. elastic recoil
   d. Both a and b are correct.
10. Immediately after the creation of an opening through the thorax into the pleural cavity,
   a. air flows through the hole and into the pleural cavity.
   b. air flows through the hole and out of the pleural cavity.
   c. air flows neither out nor in.
   d. the lung protrudes through the hole.

11. Compliance of the lungs and thorax
    a. is the volume by which the lungs and thorax change for each unit change of intra-alveolar pressure.
    b. increases in emphysema.
    c. decreases because of lack of surfactant.
    d. All of these are correct.
12. Given these lung volumes:
    (1) tidal volume = 500 mL
    (2) residual volume = 1000 mL
    (3) inspiratory reserve volume = 2500 mL
    (4) expiratory reserve volume = 1000 mL
    (5) dead space = 1000 mL

    The vital capacity is
    a. 3000 mL.
    b. 3500 mL.
    c. 4000 mL.
    d. 5000 mL.
    e. 6000 mL.
13. Alveolar ventilation is the
    a. tidal volume times the respiratory rate.
    b. minute ventilation plus the dead space.
    c. amount of air available for gas exchange in the lungs.
    d. vital capacity divided by the respiratory rate.
    e. inspiratory reserve volume times minute ventilation.
14. The rate of diffusion of a gas across the respiratory membrane increases as the
    a. respiratory membrane becomes thicker.
    b. surface area of the respiratory membrane decreases.
    c. partial pressure gradient of the gas across the respiratory membrane increases.
    d. diffusion coefficient of the gas decreases.
    e. All of these are correct.
15. Oxygen is mostly transported in the blood
    a. dissolved in plasma.
    b. bound to blood proteins.
    c. within HCO_3^-.
    d. bound to the heme portion of hemoglobin.
16. The oxygen-hemoglobin dissociation curve is adaptive because it
    a. shifts to the right in the pulmonary capillaries and to the left in the tissue capillaries.
    b. shifts to the left in the pulmonary capillaries and to the right in the tissue capillaries.
    c. does not shift.
17. Carbon dioxide is mostly transported in the blood
    a. dissolved in plasma.
    b. bound to blood proteins.
    c. within HCO_3^-.
    d. bound to the heme portion of hemoglobin.
    e. bound to the globin portion of hemoglobin.
18. The chloride shift
    a. promotes the transport of carbon dioxide in the blood.
    b. occurs when Cl^- replaces HCO_3^- within red blood cells.
    c. maintains electrical neutrality in red blood cells and the plasma.
    d. All of these are correct.
19. Which of these parts of the brainstem is correctly matched with its main function?
    a. ventral respiratory groups—stimulate the diaphragm
    b. dorsal respiratory groups—limit inflation of the lungs
    c. pontine respiratory group—is involved in the switch between inspiration and expiration
    d. All of these are correct.
20. The chemosensitive area
    a. stimulates the respiratory center when blood carbon dioxide levels increase.
    b. stimulates the respiratory center when blood pH increases.
    c. is located in the pons.
    d. stimulates the respiratory center when blood oxygen levels increase.
    e. All of these are correct.
21. Blood oxygen levels
    a. are more important than carbon dioxide levels in the regulation of respiration.
    b. need to change only slightly to cause a change in respiration.
    c. are detected by sensory receptors in the carotid and aortic bodies.
    d. All of these are correct.

Answers in Appendix E

CRITICAL THINKING

1. A person's vital capacity is measured while standing and while lying down. What difference, if any, in the measurement do you predict and why?
2. Ima Diver wanted to do some underwater exploration. Instead of buying expensive SCUBA equipment, she obtained a long hose and an innertube. She attached one end of the hose to the innertube so that the end was always out of the water, and she inserted the other end of the hose in her mouth and went diving. What happened to her alveolar ventilation and why? How can she compensate for this change? How does diving affect lung compliance and the work of ventilation?
3. The bacteria that cause gangrene (*Clostridium perfringens*) are anaerobic microorganisms that do not thrive in the presence of oxygen. Hyperbaric oxygenation (HBO) treatment places a person in a chamber containing oxygen at three to four times normal atmospheric pressure. Explain how HBO helps treat gangrene.
4. One technique for artificial respiration is mouth-to-mouth resuscitation. The rescuer takes a deep breath, blows air into the patient's mouth, and then lets air flow out. The process is repeated. Explain the following: (1) Why do the patient's lungs expand? (2) Why does air move out of the patient's lungs? (3) What effect do the P_{O_2} and the P_{CO_2} of the rescuer's air have on the victim?
5. The left phrenic nerve supplies the left side of the diaphragm, and the right phrenic nerve supplies the right side. Damage to the left phrenic nerve results in paralysis of the left side of the diaphragm. During inspiration, does the left side of the diaphragm move superiorly, move inferiorly, or stay in place?
6. Suppose that the thoracic wall is punctured at the end of a normal expiration, producing a pneumothorax. Does the thoracic wall move inward, move outward, or not move at all?

7. During normal, quiet respiration, when does the maximum rate of diffusion of oxygen in the pulmonary capillaries occur? When does the maximum rate of diffusion of carbon dioxide occur?
8. Experimental evidence suggests that the overuse of erythropoietin (EPO; see chapter 19) reduces athletic performance. What side effects of EPO abuse reduce exercise stamina?
9. Predict what would happen to tidal volume if (a) the vagus nerves were cut, (b) the phrenic nerves were cut, or (c) the intercostal nerves were cut.
10. You and your physiology instructor are trapped in an overturned ship. To escape, you must swim under water a long distance. You tell your instructor it would be a good idea to hyperventilate before making the escape attempt. Your instructor calmly replies, "What good would that do, since your pulmonary capillaries are already 100% saturated with oxygen?" What should you do and why?
11. Ima Anxious was hysterical and hyperventilating, so a doctor made her breathe into a paper bag. An especially astute student said to the doctor, "When Ima was hyperventilating, she was reducing blood carbon dioxide levels; when she breathed into the paper bag, carbon dioxide was trapped in the bag, and she was rebreathing it, thus causing blood carbon dioxide levels to increase. As Ima's blood carbon dioxide levels increased, her urge to breathe should have increased. Instead, she began to breathe more slowly. Please explain." How do you think the doctor responded? (*Hint:* Recall that the effect of decreased blood carbon dioxide on the vasomotor center results in vasodilation and a sudden decrease in blood pressure.)

Answers in Appendix F

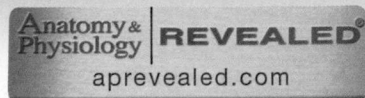

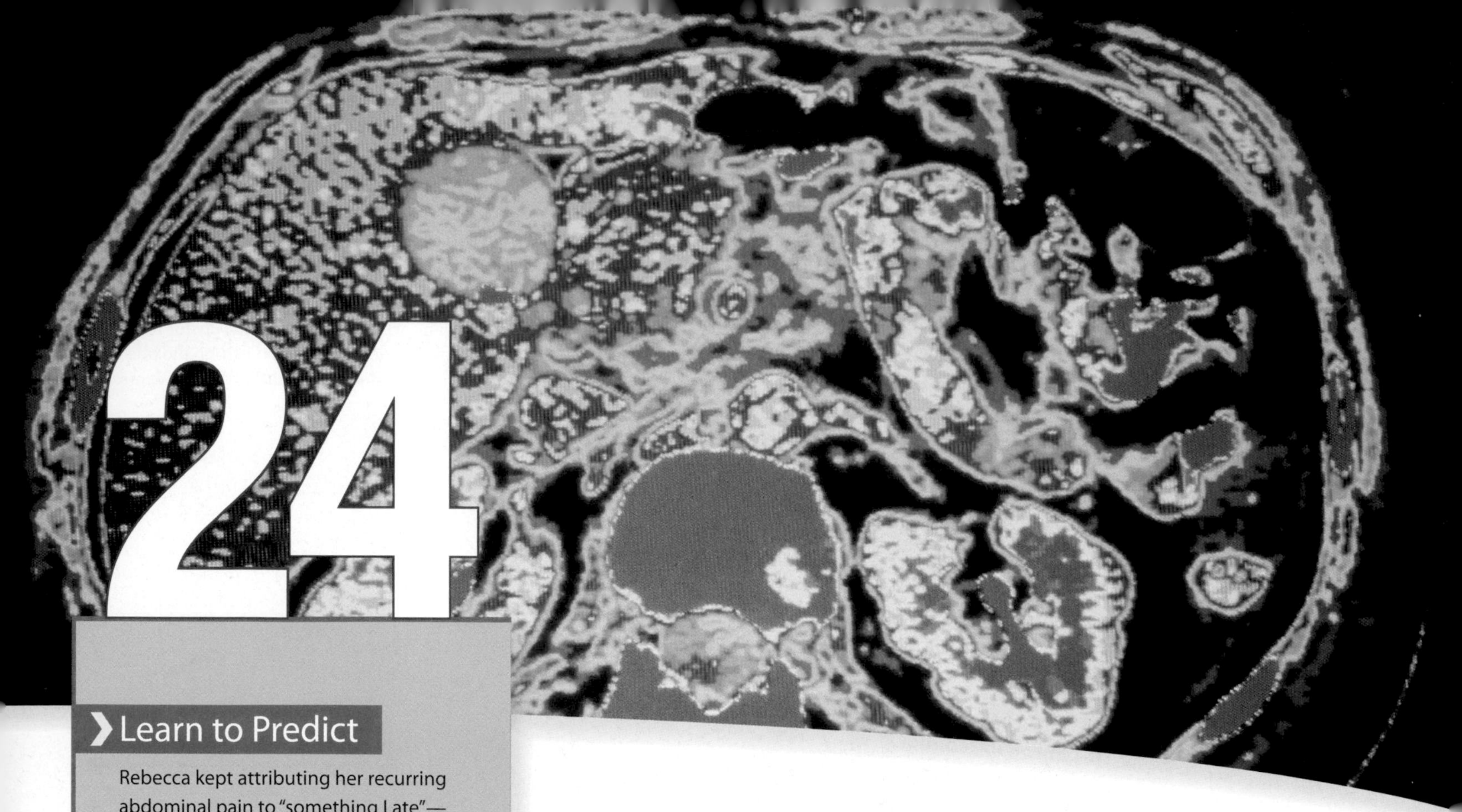

24

Learn to Predict

Rebecca kept attributing her recurring abdominal pain to "something I ate"—and she was partly right about that. Several times during the past year, eating high-fat meals had led to episodes of serious abdominal pain. During the most recent attack, the discomfort become so intense that Rebecca went to the emergency room, where she was given medication to relieve the pain. Still, over the next few hours, her skin took on a yellowish tint, and the next morning she had diarrhea and clay-colored feces. Following lab tests and ultrasonography, a physician diagnosed gallstones and recommended the removal of Rebecca's gallbladder. By combining your background knowledge from chapters 21 and 22 with new information about the digestive system in this chapter, explain how gallstones led to Rebecca's pain and other symptoms.

Digestive System

Almost everyone loves to eat, and we all must eat to stay alive. Throughout history, food and drink have provided not only nourishment but also the foundation for many social gatherings. Although it's not something we often think about while enjoying our pizza and favorite beverage, the body has an amazing digestive system that includes its own quality control and waste disposal methods.

Every cell of the body needs nourishment, yet most cells cannot leave their position in the body and travel to a food source. Therefore, the food must be converted to a usable form and delivered. To do this, the digestive system is specialized to ingest food, propel it through the digestive tract, digest it, and absorb water, electrolytes, and nutrients. The digestive process involves a choreographed mixing of food with digestive juices that include strong acids, detergent-like bile salts, and activated enzymes. The body then maximizes absorption of digested nutrients. Once these useful substances are absorbed, they are transported through the circulatory system to cells, which use them for energy or as new molecules for building and maintaining tissues and organs. Indeed, the digestive system is the body's "meals on wheels."

Photo: Colorized computed tomographic (CT) scan of an axial section through the abdomen. The image shows a gallstone (pink) obstructing the bile duct where it leaves the gallbladder (green circle). The large yellow region around the gallbladder is the liver. A spine vertebra (pink) is visible in bottom center, with C-shaped kidneys (yellow) on either side.

Module 12
Digestive System

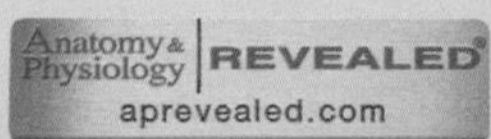
Anatomy & Physiology
REVEALED
aprevealed.com

24.1 Anatomy of the Digestive System

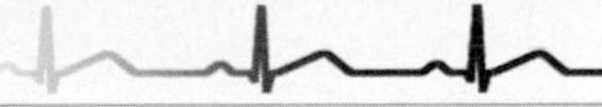

LEARNING OUTCOME

After reading this section, you should be able to

A. List the regions of the digestive tract.

The **digestive system** (figure 24.1) consists of the **digestive tract,** a tube extending from the mouth to the anus, and its associated **accessory organs**—primarily glands located outside the digestive tract that secrete fluids into it. The digestive tract is also called the *alimentary tract,* or *alimentary canal.* Technically, the term **gastrointestinal** (gas′trō-in-tes′tin-ăl; **GI**) **tract** refers only to the stomach and intestines, but it is often used as a synonym for *digestive tract.*

The regions of the digestive tract include the following:

1. *Oral cavity,* or mouth, with the salivary glands and tonsils as accessory organs
2. *Pharynx,* or throat
3. *Esophagus*
4. *Stomach*
5. *Small intestine,* consisting of the duodenum, jejunum, and ileum, with the liver, gallbladder, and pancreas as major accessory organs
6. *Large intestine,* including the cecum, colon, rectum, and anal canal
7. *Anus*

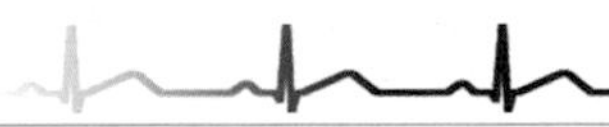

ASSESS YOUR PROGRESS

1. *List the regions of the digestive tract, from beginning to end.*

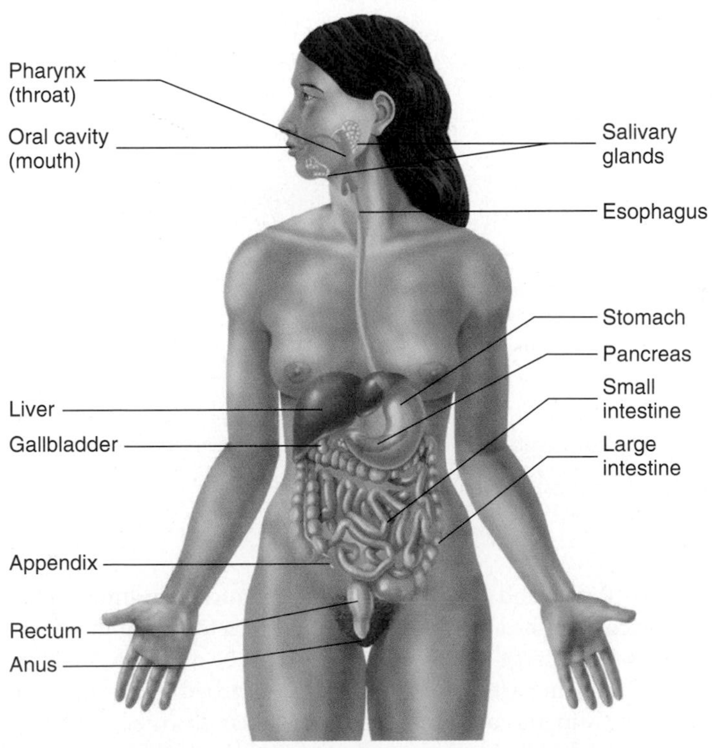
Pharynx (throat)
Oral cavity (mouth)
Salivary glands
Esophagus
Stomach
Pancreas
Small intestine
Large intestine
Liver
Gallbladder
Appendix
Rectum
Anus

FIGURE 24.1 Digestive System

24.2 Functions of the Digestive System

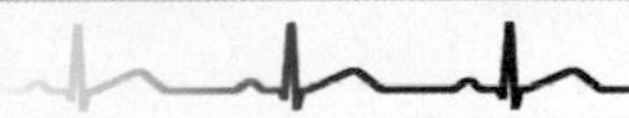

LEARNING OUTCOMES

After reading this section, you should be able to

A. Describe the major functions of the digestive system.

B. State which digestive functions occur in the different regions of the digestive tract.

The major functions of the digestive system are listed in table 24.1 and described here:

1. *Ingestion* is the intake of solid or liquid food into the stomach. The normal route of ingestion is through the oral cavity.
2. *Mastication* is the process by which the teeth chew food in the mouth. Digestive enzymes cannot easily penetrate solid food particles and can work effectively only on particle surfaces. It is vital, therefore, that solid foods be mechanically broken down by mastication into smaller particles to increase the total surface area of food for digestion.
3. *Propulsion* is the movement of food from one end of the digestive tract to the other. The total time it takes food to travel the length of the digestive tract is usually about 24–36 hours. Each segment of the digestive tract is specialized to assist in moving its contents from the oral end to the anal end:
   a. **Swallowing,** or *deglutition* (dē′gloo-tish′ŭn), moves liquids or a soft mass of food and liquid, called a **bolus** (bō′lŭs), from the oral cavity into the esophagus.
   b. **Peristalsis** (per-i-stal′sis; figure 24.2) propels material through most of the digestive tract. **Peristaltic** (per-i-stal′tik) **waves** are muscular contractions consisting of a wave of relaxation of the circular muscles in front of the bolus, followed by a wave of strong contraction of the circular muscles behind the bolus, which force the bolus along the digestive tube. Each peristaltic wave travels the length of the esophagus in about 10 seconds. Peristaltic waves in the small intestine usually travel only short distances.
   c. **Mass movements** are contractions that move material in some parts of the large intestine. Mass movements extend over much larger parts of the digestive tract than peristaltic movements.
4. *Mixing.* Some contractions do not propel food from one end of the digestive tract to the other but, rather, move it back and forth within the digestive tract to *mix* it with digestive secretions and help break it into smaller pieces. **Segmental contractions** are mixing contractions that occur in the small intestine (figure 24.3).
5. *Secretion.* As food moves through the digestive tract, secretions are added to lubricate, liquefy, buffer, and digest the food. **Mucus,** secreted along the entire digestive tract, lubricates the food and the lining of the tract. The mucus coats and protects the epithelial cells of the digestive tract from mechanical abrasion, stomach acid, and digestive enzymes. The secretions also contain large amounts of **water,** which liquefies the food, making it easier to digest and absorb. Water also moves into the intestine by osmosis. Liver secretions break large lipid droplets into much smaller droplets, which makes the digestion and absorption of

TABLE 24.1 Functions of the Digestive System

Organ	Functions
Oral cavity	*Ingestion.* Solid food and fluids are taken into the digestive tract through the oral cavity. *Taste.* Tastants dissolved in saliva stimulate taste buds in the tongue. *Mastication.* Movement of the mandible by the muscles of mastication causes the teeth to break food into smaller pieces. The tongue and cheeks help place the food between the teeth. *Digestion.* Amylase in saliva begins carbohydrate (starch) digestion. *Swallowing.* The tongue forms food into a bolus and pushes the bolus into the pharynx. *Communication.* The lips, cheeks, teeth, and tongue are involved in speech. The lips change shape as part of facial expressions. *Protection.* Mucin and water in saliva provide lubrication, and lysozyme (an enzyme that lyses cells) kills microorganisms. Nonkeratinized stratified squamous epithelium prevents abrasion.
Pharynx	*Swallowing.* The involuntary phase of swallowing moves the bolus from the oral cavity to the esophagus. Materials are prevented from entering the nasal cavity by the soft palate and kept out of the lower respiratory tract by the epiglottis and vestibular folds. *Breathing.* Air passes from the nasal or oral cavity through the pharynx to the lower respiratory tract. *Protection.* Mucus provides lubrication. Nonkeratinized stratified squamous epithelium prevents abrasion.
Esophagus	*Propulsion.* Peristaltic contractions move the bolus from the pharynx to the stomach. The lower esophageal sphincter limits reflux of the stomach contents into the esophagus. *Protection.* Glands produce mucus, which provides lubrication and protects the inferior esophagus from stomach acid.
Stomach	*Storage.* Rugae allow the stomach to expand and hold food until it can be digested. *Digestion.* Protein digestion begins as a result of the actions of hydrochloric acid and pepsin. *Absorption.* Absorption of a few substances (e.g., water, alcohol, aspirin) takes place in the stomach. *Mixing and propulsion.* Mixing waves churn ingested materials and stomach secretions into chyme. Peristaltic waves move the chyme into the small intestine. *Protection.* Mucus provides lubrication and prevents digestion of the stomach wall. Stomach acid kills most microorganisms.
Small intestine	*Neutralization.* Bicarbonate ions from the pancreas and bile from the liver neutralize stomach acid to form a pH environment suitable for pancreatic and intestinal enzymes. *Digestion.* Enzymes from the pancreas and the lining of the small intestine complete the breakdown of food molecules. Bile salts from the liver emulsify lipids. *Absorption.* The circular folds, villi, and microvilli increase surface area. Most nutrients are actively or passively absorbed. Most of the ingested water or the water in digestive tract secretions is absorbed. *Mixing and propulsion.* Segmental contractions mix the chyme, and peristaltic contractions move the chyme into the large intestine. *Excretion.* Bile from the liver contains bilirubin and excess cholesterol. *Protection.* Mucus provides lubrication, prevents digestion of the intestinal wall, and protects the small intestine from stomach acid. Peyer patches protect against microorganisms.
Large intestine	*Absorption.* The proximal half of the colon absorbs salts (e.g., sodium chloride), water, and vitamins (e.g., K) produced by bacteria. *Storage.* The distal half of the colon holds feces until they are eliminated. *Mixing and propulsion.* Slight segmental mixing occurs. Mass movements propel feces toward the anus, and defecation eliminates the feces. *Protection.* Mucus provides lubrication; mucus and bicarbonate ions protect against acids produced by bacteria.

lipids possible. **Enzymes** secreted by the oral cavity, stomach, small intestine, and pancreas break down large food molecules into smaller molecules that can be absorbed by the intestinal wall.

6. *Digestion* is the breakdown of large organic molecules into their component parts: carbohydrates into monosaccharides, proteins into amino acids, and triglycerides into fatty acids and glycerol. Digestion consists of **mechanical digestion,** which involves the mastication and mixing of food, and **chemical digestion,** which is accomplished by digestive enzymes secreted along the digestive tract. Large organic molecules must be digested into their component parts before they can be absorbed by the digestive tract. Minerals and water are not broken down before being absorbed. Vitamins are also absorbed without digestion; in fact, they lose their function if their structure is altered by digestion.

1. A wave of smooth muscle relaxation moves ahead of the bolus, allowing the digestive tract to expand.
2. A wave of contraction of the smooth muscle behind the bolus propels it through the digestive tract.

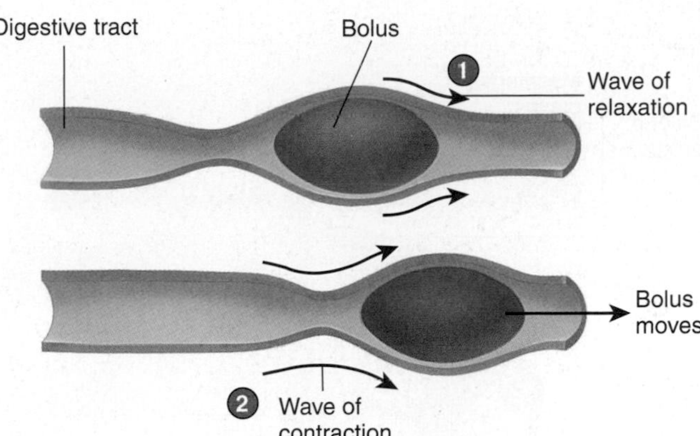

Digestive tract
Bolus
1
Wave of relaxation
Bolus moves
2
Wave of contraction

PROCESS FIGURE 24.2 Peristalsis

1. A secretion introduced into the digestive tract or food within the tract begins in one location.
2. Segments of the digestive tract alternate between contraction and relaxation.
3. Material (*brown*) in the intestine is spread out in both directions from the site of introduction.
4. The secretion or food is spread out in the digestive tract and becomes more diffuse (*lighter color*) through time.

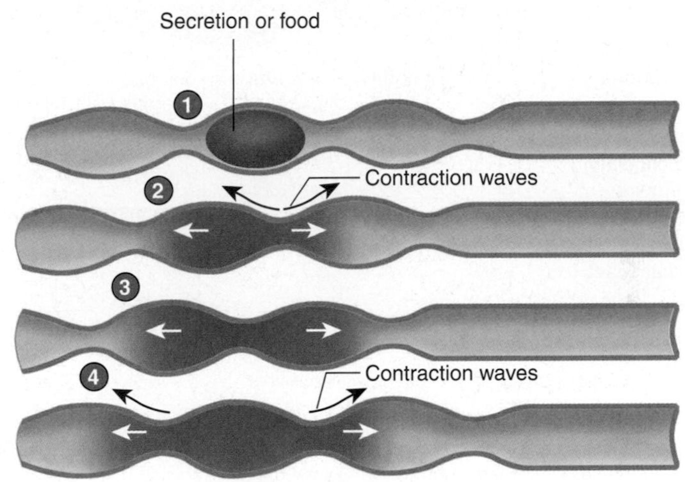

Secretion or food
1
2
Contraction waves
3
4
Contraction waves

PROCESS FIGURE 24.3 Segmental Contractions

7. *Absorption* is the movement of molecules out of the digestive tract and into the circulation or into the lymphatic system. The mechanism by which absorption occurs depends on the type of molecule involved. Molecules pass out of the digestive tract by diffusion, facilitated diffusion, active transport, symport, or endocytosis (see chapter 3).
8. *Elimination* is the process by which the waste products of digestion are removed from the body. During this process, which occurs primarily in the large intestine, water and salts are absorbed, changing the material in the digestive tract from liquefied to semi-solid. These semi-solid waste products, called **feces,** are then eliminated from the digestive tract by the process of **defecation.**

ASSESS YOUR PROGRESS

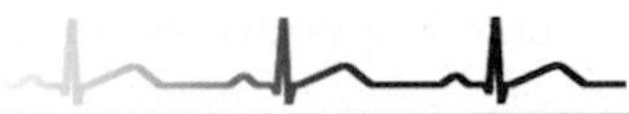

2. *Describe each of the functions involved in the normal functions of the digestive system.*
3. *Explain the three types of propulsion through the digestive tract.*
4. *What is the difference between mechanical digestion and chemical digestion?*
5. *What digestive functions occur in the stomach? In the small intestine?*

24.3 Histology of the Digestive Tract

LEARNING OUTCOMES

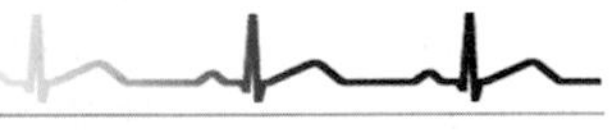

After reading this section, you should be able to

A. Describe the histology of the digestive tract.

B. List the types of glands associated with the digestive tract.

Figure 24.4 depicts a generalized view of digestive tract histology. The digestive tube consists of four major tunics, or layers: an internal mucosa and an external serosa, with a submucosa and a muscularis in between. These four tunics are present in all areas of the digestive tract, from the esophagus to the anus. Three major types of glands are associated with the intestinal tract: (1) unicellular mucous glands in the mucosa, (2) multicellular glands in the mucosa and submucosa, and (3) multicellular glands (accessory glands) outside the digestive tract.

Mucosa

The innermost tunic, the **mucosa** (mū-kō′să), or *mucous membrane,* consists of three layers: (1) the inner **mucous epithelium,** which is moist stratified squamous epithelium in the mouth, oropharynx,

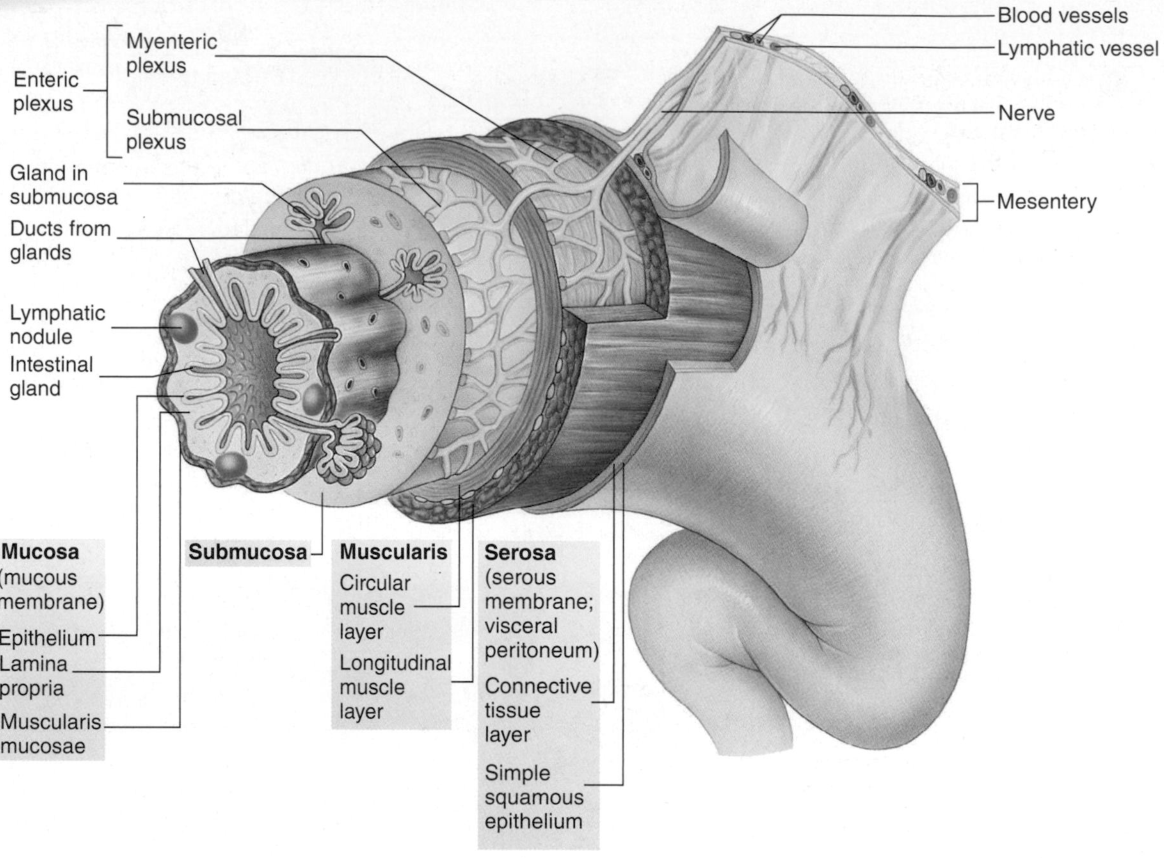
Enteric plexus
Myenteric plexus
Submucosal plexus
Gland in submucosa
Ducts from glands
Lymphatic nodule
Intestinal gland
Blood vessels
Lymphatic vessel
Nerve
Mesentery
Mucosa (mucous membrane)
Epithelium
Lamina propria
Muscularis mucosae
Submucosa
Muscularis
Circular muscle layer
Longitudinal muscle layer
Serosa (serous membrane; visceral peritoneum)
Connective tissue layer
Simple squamous epithelium

FIGURE 24.4 Digestive Tract Histology

The four tunics are the mucosa, the submucosa, the muscularis, and a serosa or an adventitia. Glands may exist along the digestive tract as part of the epithelium, as glands within the submucosa, or as large glands outside the digestive tract.

esophagus, and anal canal and simple columnar epithelium in the remainder of the digestive tract; (2) a loose connective tissue called the **lamina propria** (lam′i-nă prō′prē-ă); and (3) a thin outer layer of smooth muscle, the **muscularis mucosae.** The epithelium extends deep into the lamina propria in many places to form **intestinal glands** and **crypts.** Two types of specialized cells in the mucosa are mechanoreceptors involved in peristaltic reflexes and chemoreceptors that detect the chemical composition of food.

Submucosa

Beneath the mucosa lies the **submucosa,** a thick connective tissue layer containing nerves, blood vessels, lymphatic vessels, and small glands. A network of nerve cells in the submucosa forms the **submucosal plexus** (plek′sŭs), or *Meissner plexus,* consisting of axons, many scattered neuron cell bodies, and neuroglial cells. Axons from the submucosal plexus extend to cells in epithelial intestinal glands, stimulating their secretion. The esophagus and stomach lack a submucosal plexus, but the plexus is extensive throughout the rest of the digestive tract.

Muscularis

The next tunic is the **muscularis,** which consists of an inner layer of circular smooth muscle and an outer layer of longitudinal smooth muscle. Two exceptions are the upper esophagus, where the muscles are striated, and the stomach, which has three layers of smooth muscle. Another nerve plexus, the **myenteric** (mī-en-ter′ik) **plexus,** or *Auerbach plexus,* also consists of axons, many scattered neuron cell bodies, and neuroglial cells and is between these two muscle layers (figure 24.4). The myenteric plexus, which is much more extensive than the submucosal plexus, controls the motility of the intestinal tract.

Within the myenteric plexus, specialized **interstitial cells** form a network of pacemakers, which promote rhythmic contractions of smooth muscle along the digestive tract. These cells also help transmit signals from neurons to muscles to regulate movement. Dysfunction of these pacemakers decreases motility in the digestive tract.

Together, the submucosal and myenteric plexuses constitute the **enteric nervous system (ENS),** or the *enteric* (en-tĕr′ik) *plexus,* which is extremely important in controlling secretion and movement (see section 24.4).

Serosa or Adventitia

The fourth layer of the digestive tract is either a **serosa** or an **adventitia** (ad-ven-tish′ă; foreign or coming from outside), depending on the structure of the layer. Parts of the digestive tract that protrude into the peritoneal cavity have a serosa as the outermost layer. This serosa, or serous membrane, is called the visceral peritoneum. It consists of a thin layer of connective tissue and a simple squamous epithelium. When the outer layer of the digestive tract is derived from adjacent connective tissue, the tunic is called the adventitia and consists of a connective tissue covering that blends with the surrounding connective tissue. These areas include the esophagus and the retroperitoneal organs (see section 24.5).

ASSESS YOUR PROGRESS

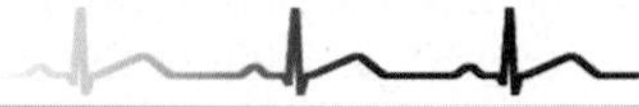

6. *What are the major tunics of the digestive tract wall, listed from inside to outside?*
7. *What types of tissue are found in each tunic?*
8. *In what tunics of the digestive tract are the submucosa and myenteric plexuses found? What are their functions?*
9. *How do the serosa and adventitia differ?*

24.4 Regulation of the Digestive System

LEARNING OUTCOME

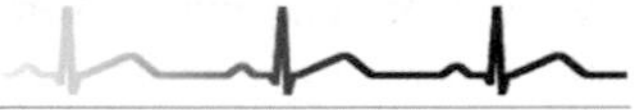

After reading this section, you should be able to

A. Describe the overall neural and chemical regulation of the digestive system.

Elaborate nervous and chemical mechanisms regulate the movement, secretion, absorption, and elimination processes.

Nervous Regulation of the Digestive System

The **enteric nervous system (ENS)** is an extensive network of the submucosal and myenteric plexuses within the walls of the digestive tract (figure 24.4). This network of neurons and associated neuroglial cells is a division of the autonomic nervous system (see chapter 16). The ENS contains more neurons than the spinal cord. Most of the nervous control of the digestive tract is local, occurring as a result of local reflexes within the ENS, and some is more general, mediated largely by the parasympathetic division of the ANS through the vagus nerves and by sympathetic nerves (see chapter 16).

Local neuronal control of the digestive tract occurs within the ENS. There are three major types of enteric neurons: (1) Enteric sensory neurons detect changes in the chemical composition of digestive tract contents or detect mechanical changes, such as stretch of the digestive tract wall; (2) enteric motor neurons stimulate or inhibit smooth muscle contraction and glandular secretion in the digestive system; and (3) enteric interneurons connect enteric sensory and motor neurons. The ENS functions through **local reflexes** to control activities within specific, short regions of the digestive tract. The ENS is capable of controlling the complex peristaltic and mixing movements, as well as blood flow to the digestive tract, without any outside influences. The importance of the ENS is highlighted by the poor intestinal motility observed in patients with **Hirschprung disease,** or *megacolon,* who lack a subset of enteric neurons.

Although the ENS can control the activities of the digestive tract independent of the CNS, the two systems normally work together. Autonomic innervation from the CNS can increase or decrease ENS activity.

Control of the digestive system by the CNS occurs when reflexes are activated by stimuli originating either in the digestive tract or in the CNS. From within the digestive system, action potentials are carried by sensory neurons in the vagus and sympathetic nerves to the CNS, where the reflexes are integrated. Reflexes within the CNS can be activated by the sight, smell, or taste of food, which stimulate hunger. An example is increased salivation and pancreatic secretions when food is seen or smelled. All of these reflexes influence activity in parasympathetic neurons of the CNS. Parasympathetic neurons extend to the digestive tract through the vagus nerves to control responses or alter the activity of the ENS and local reflexes. Some sympathetic neurons inhibit muscle contraction and secretion in the digestive system and decrease blood flow to the digestive system.

Chemical Regulation of the Digestive System

Over 30 neurotransmitters are associated with the ENS. Two major ENS neurotransmitters are acetylcholine and norepinephrine. In general, acetylcholine stimulates and norepinephrine inhibits digestive tract motility and secretions. Another major ENS neurotransmitter is serotonin, which stimulates digestive tract motility.

In addition to neural release, serotonin is also produced by endocrine cells within the digestive tract wall. Over 95% of the serotonin in the body is found in the digestive tract, so drugs that increase serotonin levels and function, such as antidepressants (see chapter 11) and chemotherapeutics used for cancer treatment, can also affect digestive tract activity. An unintended consequence of many cancer therapies is nausea, due to increased serotonin release

Clinical IMPACT

Enteric Neurons

Hirschprung disease, also called *megacolon,* is a painful developmental disorder caused by the absence of enteric neurons in the distal large intestine. Mutations in the *RET* gene have been identified in patients with Hirschprung disease. The *RET* gene encodes a receptor that is normally activated by the growth factors required for the survival and differentiation of a subset of enteric neurons. The mutations in *RET* that lead to loss of receptor function result in loss of enteric neurons, which results in poor intestinal motility and severe constipation. Conversely, a different set of mutations in the *RET* gene is linked to an inherited cancer called **multiple endocrine neoplasia type 2 (MEN2).** In contrast to the loss of function due to Hirschprung mutations, the MEN2 mutations cause a gain of *RET* receptor function, so that it is active even in the absence of growth factors. Hence, two types of mutations in the same gene result in two very different syndromes. Rapid DNA tests are used to screen patients and family members for suspected Hirschprung and MEN2 mutations.

from endocrine cells in the digestive tract. Serotonin binds to a subset of serotonin receptors, called 5HT3 receptors, on sensory terminals of the vagus nerve, which stimulates the vomiting center in the brain. This results in the nausea and vomiting associated with chemotherapy and radiotherapy. 5HT3 receptor blockers, such as ondansetron (ōn-dan′sē-tron), are commonly used to alleviate nausea.

A number of hormones, such as gastrin and secretin, are secreted by endocrine cells in the digestive system and are carried through the circulation to target organs of the digestive system or to target tissues in other systems. These hormones help regulate many digestive tract functions, as well as the secretions of associated glands, such as the liver and pancreas.

In addition to the hormones produced by the digestive system that enter the circulation, other paracrine chemicals, such as histamine, are released locally within the digestive tract, where they influence the activity of nearby cells. These localized chemical regulators help local reflexes within the ENS control local digestive tract environments, such as pH levels.

ASSESS YOUR PROGRESS

10. *Describe the roles of the ENS, CNS, and ANS in controlling the digestive system.*
11. *What chemical mechanisms regulate the digestive system?*

24.5 Peritoneum

LEARNING OUTCOME

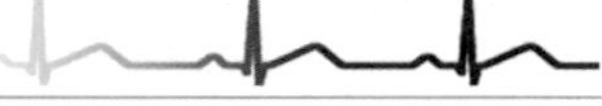

After reading this section, you should be able to

A. Describe the peritoneum and its function.

The walls and organs of the abdominal cavity are lined with **serous membranes.** These membranes are very smooth and secrete a serous fluid, which provides a lubricating film between the layers of membranes. The membranes and fluid reduce friction as organs move within the abdomen. The serous membrane that covers the organs is the **visceral peritoneum** (per′i-tō-nē′ŭm; to stretch over), and the one that covers the interior surface of the wall of the abdominal cavity is the **parietal peritoneum** (figure 24.5). Serous membranes also line other organs of the body. **Peritonitis** is a potentially life-threatening inflammation of the peritoneal membranes. The inflammation can result from chemical irritation by substances, such as bile, that have escaped from a damaged digestive tract or from infection originating in the digestive tract, as when the appendix ruptures. The main symptoms of peritonitis are acute abdominal pain and tenderness that are worsened by movement. An accumulation of excess serous fluid in the peritoneal cavity, called **ascites** (ă-sī′tēz), can occur in peritonitis. Ascites can also accompany starvation, alcoholism, or liver cancer.

Connective tissue sheets called **mesenteries** (mes′en-ter′ēz; middle intestine) hold many of the organs in place within the abdominal cavity. The mesenteries consist of two layers of serous membranes with a thin layer of loose connective tissue between them. They provide a route by which vessels and nerves can pass from the abdominal wall to the organs. Other abdominal organs lie against the abdominal wall, have no mesenteries, and are referred to as **retroperitoneal** (re′trō-per′i-tō-nē′ăl; behind the peritoneum; see chapter 1). The retroperitoneal organs include the duodenum, pancreas, ascending colon, descending colon, rectum, kidneys, adrenal glands, and urinary bladder.

Although *mesentery* is a general term referring to the serous membranes attached to the abdominal organs, it is also applied specifically to the mesentery associated with the small intestine, sometimes called the **mesentery proper.** The mesenteries of parts of the colon are the **transverse mesocolon,** which extends from the transverse colon to the posterior body wall, and the **sigmoid mesocolon.** The vermiform appendix has its own little mesentery, called the **mesoappendix.**

The mesentery connecting the lesser curvature of the stomach and the proximal end of the duodenum to the liver and diaphragm is called the **lesser omentum** (ō-men′tŭm; membrane of the bowels), and the mesentery extending as a fold from the greater curvature and then to the transverse colon is called the **greater omentum** (figure 24.5). The greater omentum forms a long, double fold of mesentery that extends inferiorly from the stomach over the surface of the small intestine. Because of this folding, a cavity called the **omental bursa** (ber′să; pocket) forms between the two layers of mesentery. A large amount of adipose tissue accumulates in the greater omentum, and it is sometimes referred to as the "fatty apron." The greater omentum has considerable mobility in the abdomen.

Predict 2

If you placed a pin through the greater omentum, through how many layers of simple squamous epithelium would the pin pass?

The **coronary ligament** attaches the liver to the diaphragm. Unlike other mesenteries, the coronary ligament has a wide space in the center, the bare area of the liver, where no peritoneum exists. The **falciform ligament** attaches the liver to the anterior abdominal wall (figure 24.5*b*).

ASSESS YOUR PROGRESS

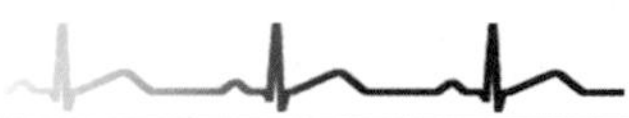

12. *Where are the visceral peritoneum and parietal peritoneum found? Define and give examples of* retroperitoneal organs.
13. *What is the function of the peritoneum?*
14. *What are the mesenteries? Name and describe the location of the mesenteries in the abdominal cavity.*

24.6 Oral Cavity

LEARNING OUTCOMES

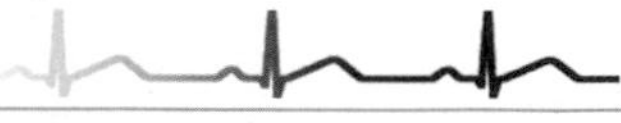

After reading this section, you should be able to

A. Describe the oral cavity and the structure and function of the lips, cheeks, palate, and tongue.

B. Outline the structure and types of adult teeth and describe the process of mastication.

C. Compare the structures and locations of the major salivary glands and describe the composition and functions of saliva and the control of its release.

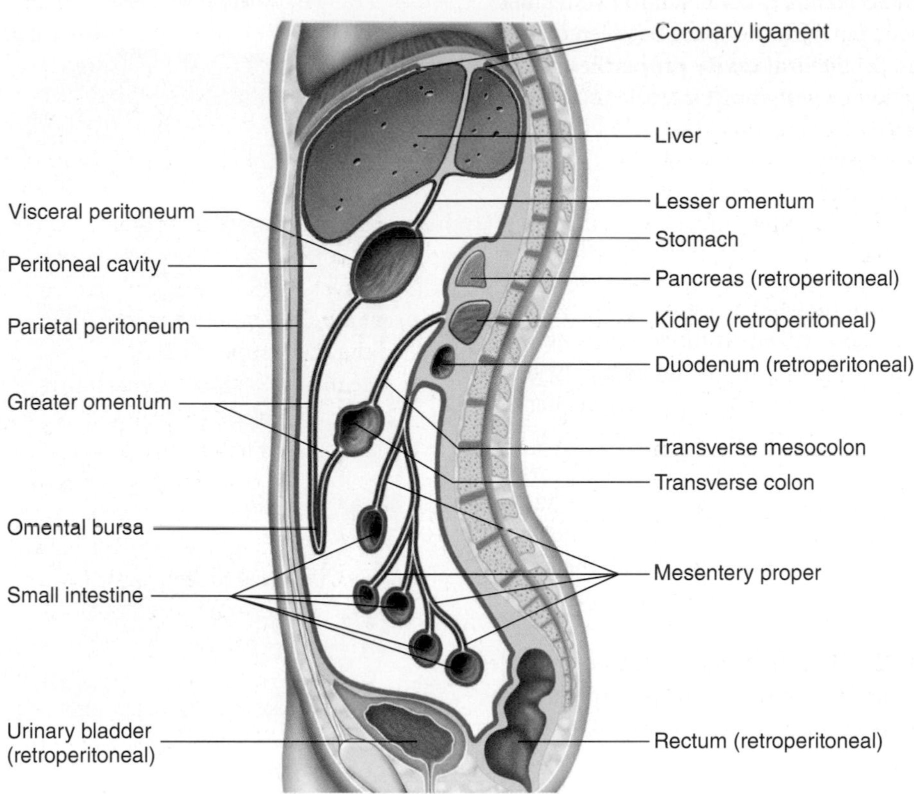

Coronary ligament
Liver
Visceral peritoneum
Lesser omentum
Stomach
Peritoneal cavity
Pancreas (retroperitoneal)
Parietal peritoneum
Kidney (retroperitoneal)
Duodenum (retroperitoneal)
Greater omentum
Transverse mesocolon
Transverse colon
Omental bursa
Mesentery proper
Small intestine
Urinary bladder (retroperitoneal)
Rectum (retroperitoneal)
(a) Medial view

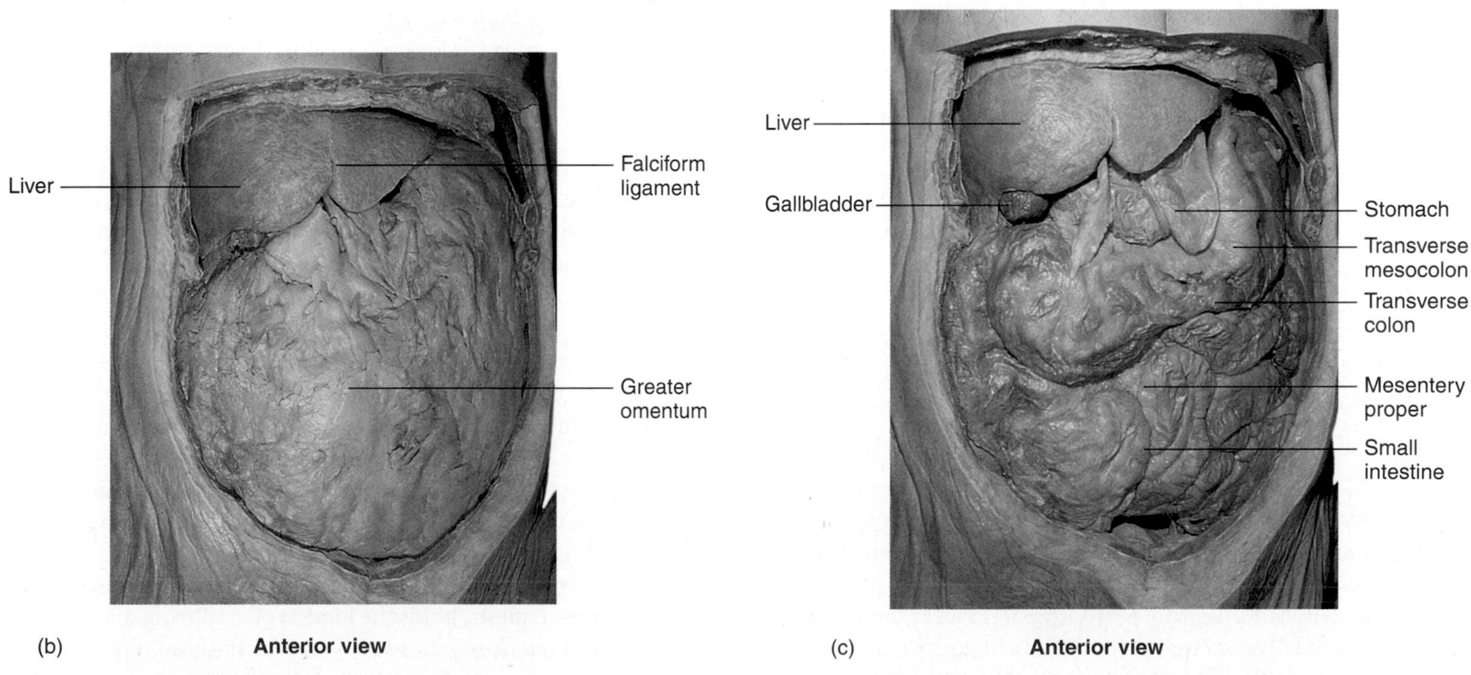

Liver
Falciform ligament
Greater omentum
(b) Anterior view
Liver
Gallbladder
Stomach
Transverse mesocolon
Transverse colon
Mesentery proper
Small intestine
(c) Anterior view

FIGURE 24.5 APIR **Peritoneum and Mesenteries**

(*a*) Sagittal section through the trunk, showing the peritoneum and mesenteries associated with some abdominal organs. (*b*) Photograph of the abdomen of a cadaver, with the greater omentum in place. (*c*) Photograph of the abdomen of a cadaver, with the greater omentum removed to reveal the underlying viscera.

The **oral cavity** (figure 24.6), or *mouth,* is divided into two regions: (1) The **vestibule** (ves′ti-bool; entry) is the space between the lips or cheeks and the teeth, and (2) the **oral cavity proper** lies medial to the teeth. The oral cavity is lined with moist stratified squamous epithelium, which protects against abrasion.

Lips, Cheeks, and Palate

The **lips,** or *labia* (lā′bē-ă; figure 24.6), are muscular structures formed mostly by the **orbicularis oris** (ōr-bik′ū-lā′ris ōr′is) **muscle** (see figure 10.9) and connective tissue. The outer surfaces of the lips are covered by skin. The keratinized stratified epithelium of the skin is thin at the margin of the lips and is not as highly keratinized as the epithelium of the surrounding skin (see chapter 5); consequently, it is more transparent than the epithelium over the rest of the body. The color from the underlying blood vessels shows through the relatively transparent epithelium, giving the lips a reddish-pink to dark red appearance, depending on the overlying pigment. At the internal margin of the lips, the epithelium is continuous with the moist stratified squamous epithelium of the mucosa in the oral cavity.

One or more **labial frenula** (fren′ū-lă; sing. *frenulum*), which are mucosal folds, extend from the alveolar process of the maxilla to the upper lip and from the alveolar process of the mandible to the lower lip.

The **cheeks** form the lateral walls of the oral cavity. They consist of an interior lining of moist stratified squamous epithelium and an exterior covering of skin. The substance of the cheek includes the **buccinator muscle** (see chapter 10), which flattens the cheek against the teeth, and the **buccal fat pad,** which rounds out the profile on the side of the face.

The lips and cheeks are important in mastication and speech. They help manipulate food within the oral cavity and hold it in place while the teeth crush or tear it. They also help form words when we speak. A large number of the muscles of facial expression are involved in moving the cheeks and lips (see chapter 10).

The roof of the oral cavity is called the **palate.** The palate separates the oral and nasal cavities and prevents food from passing into the nasal cavity during chewing and swallowing. The palate consists of two parts (figure 24.6). The anterior, bony part is the **hard palate** (see chapter 7). The posterior, nonbony part is the **soft palate,** which consists of skeletal muscle and connective tissue. The **uvula** (ū′vū-lă; a grape) is a posterior projection from the soft palate. The posterior boundary of the oral cavity is the **fauces** (faw′sēz), or *throat,* which is the opening into the pharynx. The **palatine tonsils** are in the lateral wall of the fauces (see chapter 22).

Tongue

The **tongue** is a large, muscular organ that occupies most of the oral cavity proper when the mouth is closed. Its major attachment in the oral cavity is through its posterior part. The anterior part of the tongue is relatively free, except for attachment to the floor of the mouth by a thin fold of tissue called the **lingual** (tongue) **frenulum.** The muscles associated with the tongue are divided into two categories: **Intrinsic muscles** are within the tongue itself, and **extrinsic muscles** are outside the tongue but attached to it.

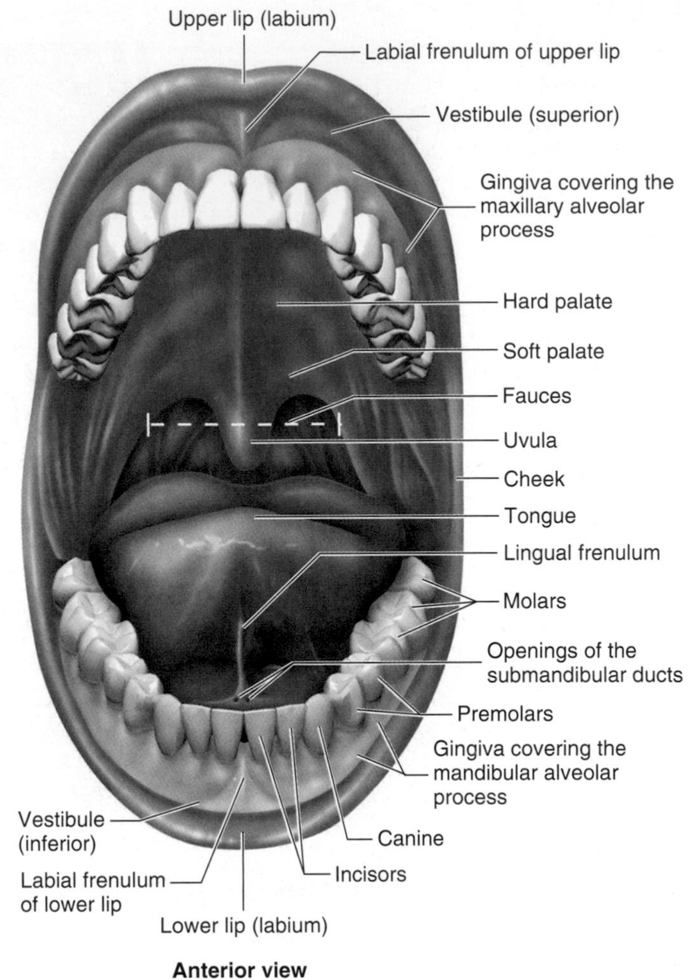
Upper lip (labium)
Labial frenulum of upper lip
Vestibule (superior)
Gingiva covering the maxillary alveolar process
Hard palate
Soft palate
Fauces
Uvula
Cheek
Tongue
Lingual frenulum
Molars
Openings of the submandibular ducts
Premolars
Gingiva covering the mandibular alveolar process
Canine
Incisors
Vestibule (inferior)
Labial frenulum of lower lip
Lower lip (labium)
Anterior view

FIGURE 24.6 APR **Oral Cavity**

The intrinsic muscles are largely responsible for changing the shape of the tongue, such as flattening and elevating it during drinking and swallowing. The extrinsic tongue muscles protrude and retract the tongue, move it from side to side, and change its shape (see chapter 10).

A groove called the **terminal sulcus** divides the tongue into two parts. The part anterior to the terminal sulcus accounts for about two-thirds of the surface area and is covered by papillae, some of which contain taste buds (see chapter 15). The posterior one-third of the tongue is devoid of papillae and has only a few scattered taste buds. Instead, it has a few small glands and a large amount of lymphatic tissue, which form the **lingual tonsil** (see chapter 22). Moist stratified squamous epithelium covers the tongue.

The tongue moves food in the mouth and, in cooperation with the lips and gums, holds the food in place during mastication. It also plays a major role in swallowing. In addition, the tongue is a major sensory organ for taste (see chapter 15) and one of the primary organs of speech. Patients with cancer of the tongue often have part or all of their tongue removed. These patients can learn to speak fairly well but have difficulty chewing and swallowing.

Teeth

Adults normally have 32 **teeth,** which are distributed in two **dental arches:** the maxillary arch and the mandibular arch. The teeth in the right and left halves of each dental arch are roughly mirror images of each other. As a result, the teeth are apportioned into four quadrants: right-upper, left-upper, right-lower, and left-lower. The teeth in each quadrant include one central and one lateral **incisor,** one **canine,** first and second **premolars,** and first, second, and third **molars** (figure 24.7*a*). The third molars are often called *wisdom teeth* because they usually appear in the late teens or early twenties, when a person is old enough to have acquired some wisdom. In people with small dental arches, the third molars may not have room to erupt into the oral cavity and remain embedded within the jaw. Embedded wisdom teeth are referred to as impacted and may cause pain or irritation. Usually, the impacted wisdom teeth are surgically removed.

The teeth of the adult mouth are called **permanent teeth,** or *secondary teeth.* Most of them are replacements for **deciduous** (dē-sid′ū-ŭs) **teeth,** or *primary teeth,* also called *milk teeth,* which are lost during childhood (figure 24.7*b*). The deciduous teeth erupt (the crowns appear within the oral cavity) between about 6 months and 24 months of age (figure 24.7*b*). The permanent teeth begin replacing the deciduous teeth at about 5 years, and the process is completed by about 11 years.

Each tooth consists of a **crown** with one or more **cusps** (points), a **neck,** and a **root** (figure 24.8). The **clinical crown** is the part of the tooth exposed in the oral cavity. The **anatomical crown** is the entire enamel-covered part of the tooth. The center of the tooth is a **pulp cavity,** which is filled with blood vessels, nerves, and connective tissue called **pulp.** The pulp cavity within the root is called the **root canal.** The nerves and blood vessels of the tooth enter and exit the pulp through a hole at the point of each root called the **apical foramen.** The pulp cavity is surrounded by living, cellular, calcified tissue called **dentin.** The dentin of the tooth crown is covered by an extremely hard, nonliving, acellular substance called **enamel,** which protects the tooth against abrasion and acids produced by bacteria in the mouth. The surface of the dentin in the root is covered with a bonelike substance called **cementum,** which helps anchor the tooth in the jaw.

The teeth are set in **alveoli** (al-vē′ō-lī; sockets) along the alveolar processes of the mandible and maxilla. Dense fibrous connective tissue and stratified squamous epithelium, referred to as the **gingiva** (jin′ji-vă; gums), cover the alveolar processes (see figure 24.6). **Periodontal** (per′ē-ō-don′tăl; around a tooth) **ligaments** secure the teeth in the alveoli.

The teeth play an important role in mastication and assist in speech.

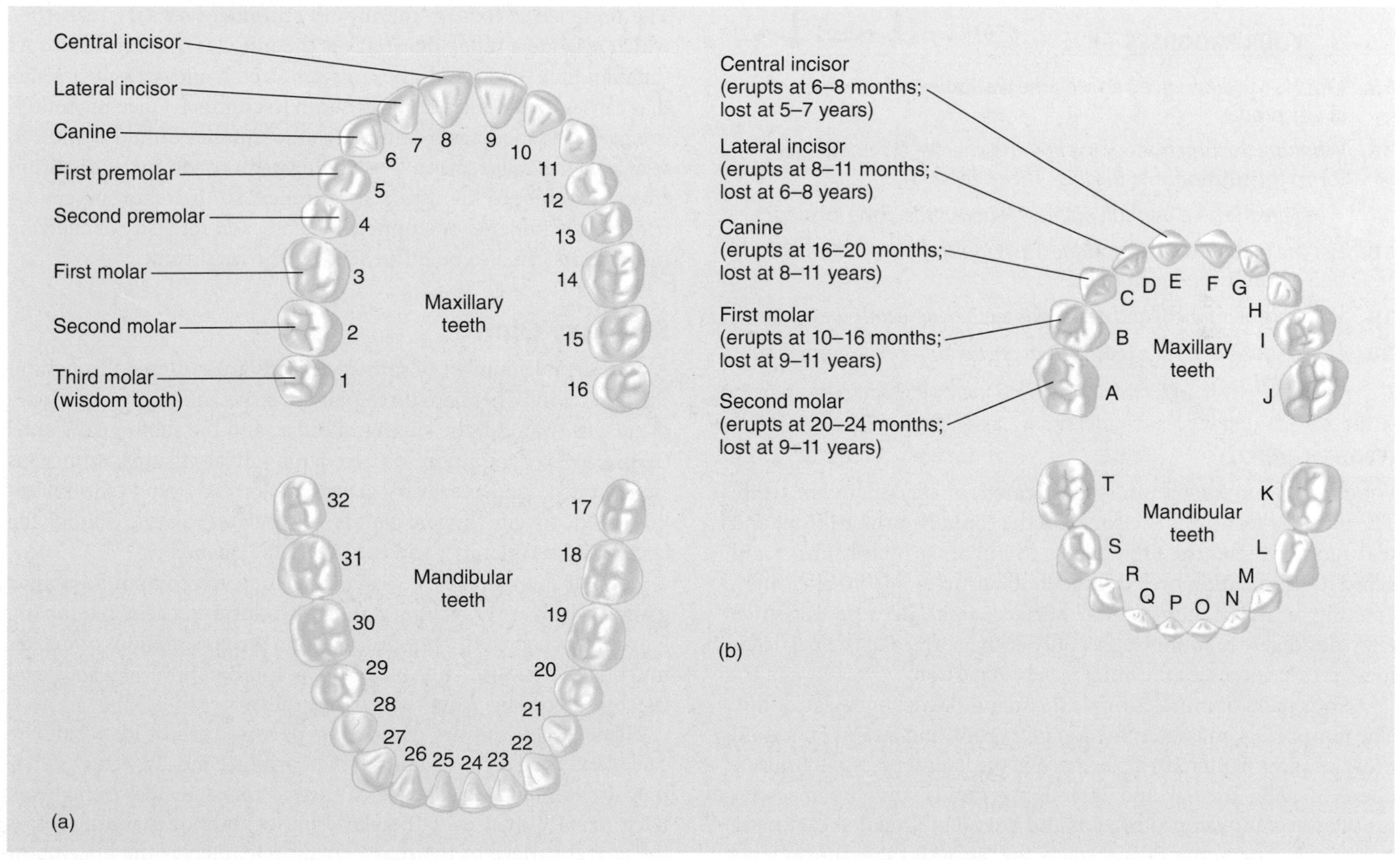
Central incisor
Lateral incisor
Canine
First premolar
Second premolar
First molar
Second molar
Third molar
(wisdom tooth)
1 2 3 4 5 6 7 8 9 10 11 12 13 14 15 16
Maxillary
teeth
17 18 19 20 21 22 23 24 25 26 27 28 29 30 31 32
Mandibular
teeth
(a)
Central incisor
(erupts at 6–8 months;
lost at 5–7 years)
Lateral incisor
(erupts at 8–11 months;
lost at 6–8 years)
Canine
(erupts at 16–20 months;
lost at 8–11 years)
First molar
(erupts at 10–16 months;
lost at 9–11 years)
Second molar
(erupts at 20–24 months;
lost at 9–11 years)
A B C D E F G H I J
Maxillary
teeth
K L M N O P Q R S T
Mandibular
teeth
(b)

FIGURE 24.7 Teeth
(*a*) Permanent teeth. (*b*) Deciduous teeth. Dental professionals have developed a "universal" numbering and lettering system for convenience in identifying individual teeth.

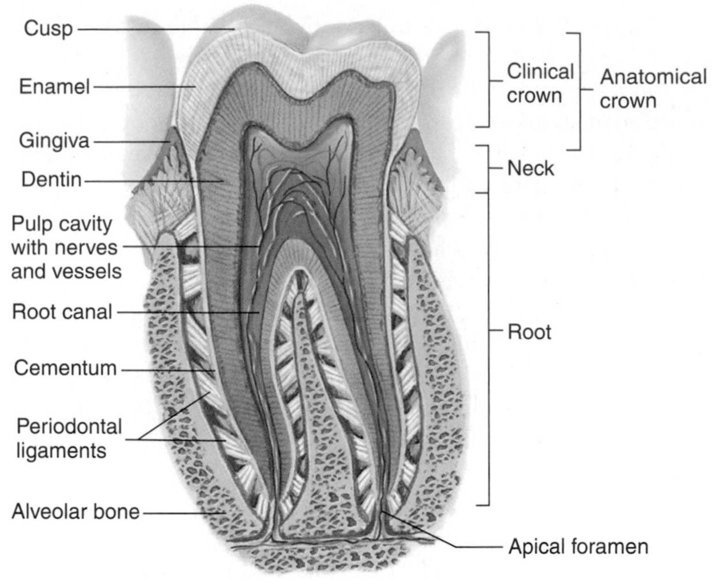
Cusp
Enamel
Gingiva
Dentin
Pulp cavity with nerves and vessels
Root canal
Cementum
Periodontal ligaments
Alveolar bone
Clinical crown
Anatomical crown
Neck
Root
Apical foramen

FIGURE 24.8 Molar Tooth in Place in the Alveolar Bone
A tooth consists of a crown (anatomical and clinical), a neck, and a root. The root is covered with cementum, and the tooth is held in the socket by periodontal ligaments. Nerves and vessels enter and exit the tooth through the apical foramen.

Clinical IMPACT

Dental Diseases

Dental caries, or tooth decay, is a breakdown of enamel by bacterial acids on the tooth surface. Because the enamel is nonliving and cannot repair itself, a dental filling is necessary to prevent further damage. If the decay reaches the pulp cavity, with its rich supply of nerves, a toothache may result. Sometimes, when decay has reached the pulp cavity, a dentist must perform a procedure called a root canal, which consists of removing the pulp from the tooth.

Periodontal disease is the inflammation and degradation of the periodontal ligaments, gingiva, and alveolar bone. This disease is the most common cause of tooth loss in adults. **Gingivitis** (jin-ji-vī′tis) is an inflammation of the gingiva, often caused by food deposited in gingival crevices and not promptly removed by brushing and flossing. Gingivitis may eventually lead to periodontal disease. **Pyorrhea** (pī-ō-rē′ă) is a periodontal disease accompanied by pus. **Halitosis** (hal-i-tō′sis), or bad breath, often occurs with periodontal disease and pyorrhea.

ASSESS **YOUR PROGRESS**

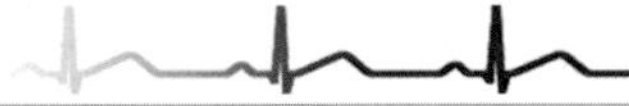

15. *What is the difference between the vestibule and the oral cavity proper?*

16. *What are the functions of the lips and cheeks? What muscle forms the substance of the lips? The cheeks?*

17. *What are the hard and soft palates? Where is the uvula located?*

18. *List the functions of the tongue. Distinguish between intrinsic and extrinsic tongue muscles.*

19. *What are permanent and deciduous teeth? Name the types of teeth.*

20. *List the three parts of a tooth. What are dentin, enamel, cementum, and pulp?*

Mastication

Food taken into the mouth is **masticated,** or *chewed,* by the teeth. The anterior teeth (the incisors and the canines) primarily cut and tear food, whereas the premolars and molars primarily crush and grind it. Mastication breaks large food particles into smaller ones, creating a much larger total surface area. Because digestive enzymes digest food molecules only at the surface of the particles, mastication increases the efficiency of digestion.

Four pairs of muscles move the mandible during mastication. The **temporalis, masseter, medial pterygoid,** and **lateral pterygoid** muscles (see chapter 10) close the jaw; the lateral pterygoid muscle opens it. The medial and lateral pterygoids and the masseter muscles accomplish protraction and lateral and medial excursion of the jaw. The temporalis retracts the jaw. All these movements are involved in tearing, crushing, and grinding food.

The **mastication reflex,** or *chewing reflex,* is integrated in the medulla oblongata and controls the basic movements of chewing. The presence of food in the mouth stimulates sensory receptors, which activate a reflex that relaxes the muscles of mastication. As the mandible is lowered, the muscles stretch and activate a reflex that causes the muscles of mastication to contract. Once the mouth is closed, the food again stimulates the muscles of mastication to relax, and the cycle repeats. Descending pathways from the cerebrum strongly influence the mastication reflex, so that chewing can be consciously initiated or stopped. The rate and intensity of chewing movements can also be influenced by the cerebrum.

Salivary Glands

A considerable number of **salivary glands** are scattered throughout the oral cavity. There are three pairs of large, multicellular salivary glands: the parotid, the submandibular, and the sublingual glands (figure 24.9). In addition to these large salivary glands, numerous small, coiled, tubular salivary glands are located deep to the epithelium of the tongue (lingual glands), the palate (palatine glands), the cheeks (buccal glands), and the lips (labial glands).

All of the major large salivary glands are compound **acinar glands,** which are branching glands with clusters of acini resembling grapes (see chapter 4). They produce thin serous secretions or thicker mucous secretions. Thus, saliva is a combination of serous and mucous secretions from the various salivary glands.

The largest salivary glands, the **parotid** (pă-rot′id; beside the ear) **glands,** are serous glands, which produce mostly watery saliva; they are located just anterior to the ear on each side of the head. Each **parotid duct** exits the gland on its anterior margin, crosses the lateral surface of the masseter muscle, pierces the buccinator muscle, and enters the oral cavity adjacent to the second upper molar (figure 24.9). A viral infection can cause the parotid glands to become inflamed and swollen, making the cheeks quite large.

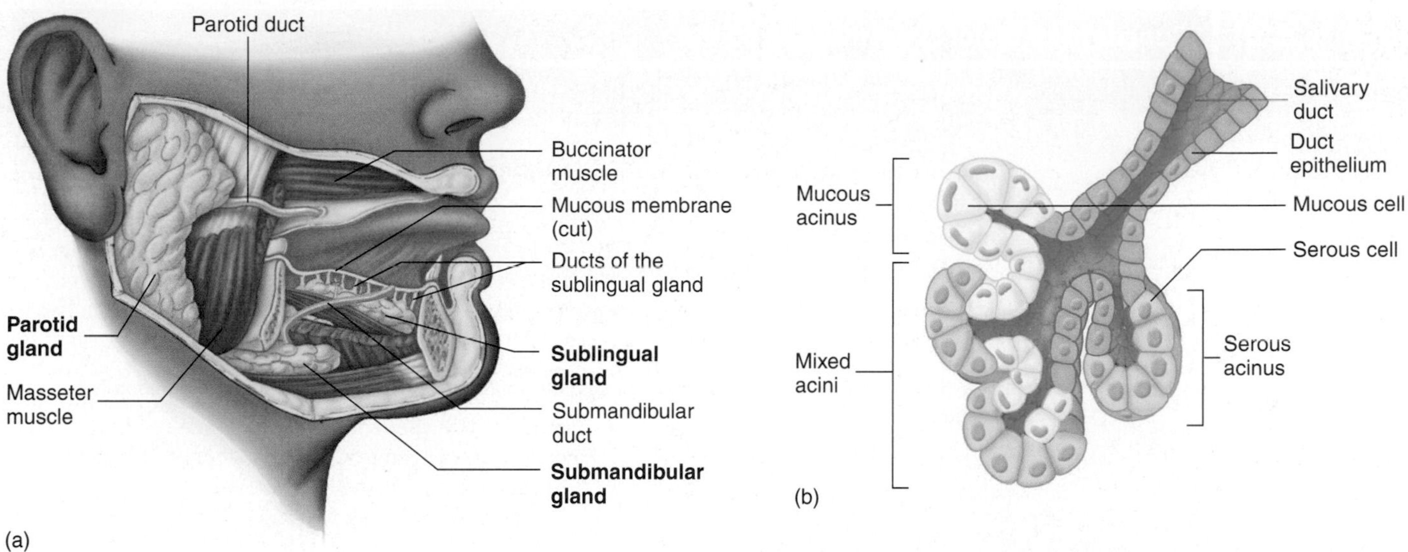
Parotid duct
Buccinator muscle
Mucous membrane (cut)
Ducts of the sublingual gland
Parotid gland
Sublingual gland
Masseter muscle
Submandibular duct
Submandibular gland
(a)
Salivary duct
Duct epithelium
Mucous acinus
Mucous cell
Serous cell
Mixed acini
Serous acinus
(b)

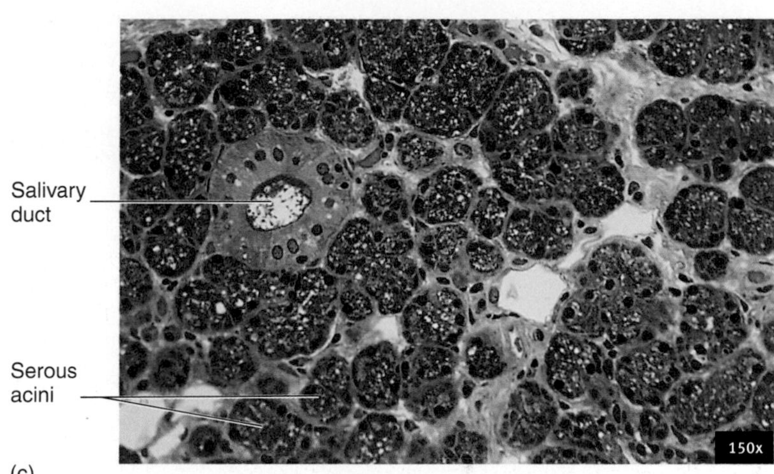
Salivary duct
Serous acini
(c)
150x

FIGURE 24.9 Salivary Glands

(*a*) The large salivary glands are the parotid glands, the submandibular glands, and the sublingual glands. The parotid duct extends anteriorly from the parotid gland. (*b*) An idealized schematic illustrates the histology of the large salivary glands. The figure is representative of all the glands and does not depict any specific salivary gland. (*c*) Photomicrograph of the parotid gland.

The most common type of viral infection results in **mumps.** The virus causing mumps can also infect other tissues, including the testes, which can result in sterility in an adult male.

The **submandibular** (below the mandible) **glands** are mixed glands with more serous than mucous acini. Each gland can be felt as a soft lump along the inferior border of the posterior half of the mandible. A submandibular duct exits each gland, passes anteriorly deep to the mucous membrane on the floor of the oral cavity, and opens into the oral cavity beside the frenulum of the tongue (see figure 24.6).

The **sublingual** (below the tongue) **glands,** the smallest of the three large, paired salivary glands, are mixed glands containing some serous acini but consisting primarily of mucous acini. They lie immediately below the mucous membrane in the floor of the oral cavity. These glands do not have single, well-defined ducts like those of the submandibular and parotid glands. Instead, each sublingual gland opens into the floor of the oral cavity through 10–12 small ducts.

The salivary glands secrete a large amount of **saliva** at the rate of about 1–1.5 L/day. Saliva has multiple roles: It helps keep the oral cavity moist, which is needed for normal speech and for the suspension of food molecules in solution so they can be tasted; it begins the process of digestion; and it has protective functions.

The serous part of saliva, produced mainly by the parotid and submandibular glands, contains a digestive enzyme called **salivary amylase** (am′il-ās; starch-splitting enzyme), which breaks the covalent bonds between glucose molecules in starch and other polysaccharides to produce the disaccharides maltose and isomaltose (table 24.2). The release of maltose and isomaltose gives starches a sweet taste. However, food spends very little time in the mouth, so only about 3–5% of the total carbohydrates are digested there. Most starchy foods come from plants and are therefore covered by cellulose, making them inaccessible to salivary amylase. Cooking and thoroughly chewing food destroy the cellulose covering and increase the efficiency of the digestive process.

Saliva prevents bacterial infection in the mouth by washing it. In addition, the bicarbonate ions in saliva act as a buffer to neutralize the acids produced by oral bacteria. This reduces the harmful effects of bacterial acids on tooth enamel. Saliva also contains **lysozyme,** an enzyme that has a weak antibacterial action, and immunoglobulin A,

TABLE 24.2 Functions of Major Digestive Secretions

Fluid or Enzyme	Function
Oral Cavity Secretions	
Serous saliva (mostly water, bicarbonate ions)	Moistens food and mucous membrane; neutralizes bacterial acids; flushes bacteria from oral cavity; has weak antibacterial activity
Salivary amylase	Digests carbohydrates
Mucus	Lubricates food; protects digestive tract from digestion by enzymes
Lingual lipase	Digests a minor amount of lipids
Esophagus Secretions	
Mucus	Lubricates esophagus; protects lining of esophagus from abrasion and allows food to move more smoothly through esophagus
Gastric Secretions	
Hydrochloric acid	Antibacterial; decreases stomach pH to activate pepsinogen to pepsin
Pepsin*	Digests protein into smaller peptide chains; activates pepsinogen
Mucus	Protects stomach lining from digestion
Intrinsic factor	Binds to vitamin B_{12} and aids in its absorption
Gastric lipase	Digests a minor amount of lipids
Liver Secretions	
Bile Bile salts Bile pigments (bilirubin) Cholesterol Mucus Lecithin	Bile salts emulsify lipids, making them available to lipases, and help make end products soluble and available for absorption by the intestinal mucosa; many of the other bile contents are waste products transported to the intestines for disposal
Pancreas Secretions	
Trypsin*	Digests proteins (cleaves at arginine or lysine amino acids); activates trypsinogen and other digestive enzymes
Chymotrypsin*	Digests proteins (cleaves at hydrophobic amino acids)
Carboxypeptidase*	Digests proteins (removes amino acids from the carboxyl end of proteins)
Pancreatic amylase	Digests carbohydrates (hydrolyzes starches and glycogen to form maltose and isomaltose)
Pancreatic lipase	Digests lipids (breaks down triglycerides into monoglycerides and free fatty acids)
Ribonuclease	Digests ribonucleic acid (hydrolyzes phosphodiester bonds)
Deoxyribonuclease	Digests deoxyribonucleic acid (hydrolyzes phosphodiester bonds)
Bicarbonate ions	Neutralize acid from stomach; provide appropriate pH for pancreatic enzymes
Small Intestine Secretions	
Mucus	Protects duodenum from stomach acid, gastric enzymes, and intestinal enzymes; provides adhesion for fecal matter; protects intestinal wall from bacterial action and acid produced in the feces
Peptidases†	Split amino acids from polypeptides
Enterokinase†	Activates trypsin from trypsinogen
Sucrase†	Splits sucrose into glucose and fructose
Maltase†	Splits maltose into two glucose molecules
Isomaltase†	Splits isomaltose into two glucose molecules
Lactase†	Splits lactose into glucose and galactose

*These enzymes are secreted as inactive forms and then activated.
†These enzymes remain in the microvilli.

which helps prevent bacterial infection. Any lack of salivary gland secretion increases the risk for ulceration and infection of the oral mucosa and for caries (cavities) in the teeth.

The mucous secretions of the submandibular and sublingual glands contain a large amount of **mucin** (mū′sin), a proteoglycan that gives a lubricating quality to the secretions of the salivary glands.

Salivary gland secretion is stimulated by both the parasympathetic and the sympathetic nervous systems, but the parasympathetic system is the more important. Salivary nuclei in the brainstem increase salivary secretions by sending action potentials through parasympathetic fibers of the facial (VII) and glossopharyngeal (IX) cranial nerves in response to a variety of stimuli, such as tactile stimulation in the oral cavity or certain tastes, especially sour. Higher centers of the brain also affect salivary gland activity. Odors that trigger thoughts of food or the sensation of hunger can increase saliva secretion.

ASSESS YOUR PROGRESS

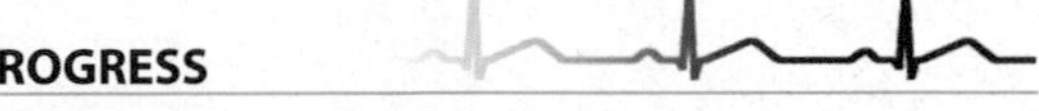

21. *List the muscle of mastication and the actions they produce. Describe the mastication reflex.*
22. *Name and give the location of the three largest salivary glands. What are the other types of salivary glands called?*
23. *What are the functions of saliva? What substances are contained in saliva?*
24. *What is the difference between serous and mucous saliva?*
25. *Describe the stimuli that stimulate the release of saliva. What nerves are involved?*

24.7 Swallowing

LEARNING OUTCOMES

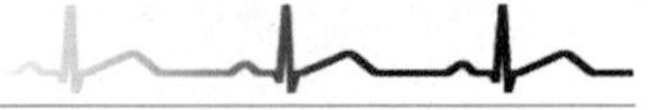

After reading this section, you should be able to

A. List the parts of the pharynx involved with digestion.

B. Describe the structure of the esophagus.

C. Explain the three phases of swallowing.

Swallowing involves distinct phases in the pharynx and esophagus (figure 24.10).

Pharynx

The **pharynx,** described in detail in chapter 23, consists of three parts: the nasopharynx, oropharynx, and laryngopharynx. Normally, only the oropharynx and laryngopharynx transmit food. The **oropharynx** communicates with the nasopharynx superiorly, with the larynx and **laryngopharynx** inferiorly, and with the mouth anteriorly. The laryngopharynx extends from the oropharynx to the esophagus and is posterior to the larynx. The epiglottis covers the opening of the larynx and keeps food and drink from entering the larynx. The posterior walls of the oropharynx and laryngopharynx consist of three muscles: the superior, middle, and inferior **pharyngeal constrictors,** which are arranged like three stacked flowerpots, one inside the other. The oropharynx and the laryngopharynx are lined with moist stratified squamous epithelium, and the nasopharynx is lined with ciliated pseudostratified columnar epithelium.

Esophagus

The **esophagus** is the part of the digestive tract that extends between the pharynx and the stomach. It is about 25 cm long and lies in the mediastinum, anterior to the vertebrae and posterior to the trachea. It passes through the esophageal hiatus (opening) of the diaphragm and ends at the stomach. The esophagus transports food from the pharynx to the stomach.

The esophagus has thick walls consisting of the four tunics common to the digestive tract: mucosa, submucosa, muscularis, and adventitia. The muscular tunic has an outer longitudinal layer and an inner circular layer, as is true of most parts of the digestive tract, but it differs by having skeletal muscle in the superior part of the esophagus and smooth muscle in the inferior part. An **upper esophageal sphincter** and a **lower esophageal sphincter,** at the upper and lower ends of the esophagus, respectively, regulate the movement of materials into and out of the esophagus. The mucosal lining of the esophagus is moist stratified squamous epithelium. Numerous mucous glands in the submucosal layer produce a thick, lubricating mucus, which passes through ducts to the surface of the esophageal mucosa.

Swallowing Phases

Swallowing, or *deglutition,* is divided into three phases: voluntary, pharyngeal, and esophageal. During the **voluntary phase** (figure 24.10, *step 1*), a bolus of food is formed in the mouth and pushed by the tongue against the hard palate, until it is forced toward the posterior part of the mouth and into the oropharynx.

The **pharyngeal phase** of swallowing (figure 24.10, *steps 2–4*) is a reflex initiated by the stimulation of tactile receptors in the area of the oropharynx. Afferent action potentials travel through the trigeminal (V) and glossopharyngeal (IX) nerves to the **swallowing center** in the medulla oblongata. There, they initiate action potentials in motor neurons, which pass through the trigeminal (V), glossopharyngeal (IX), vagus (X), and accessory (XI) nerves to the soft palate and pharynx. This phase of swallowing begins with the elevation of the soft palate, which closes the passage between the nasopharynx and oropharynx. The pharynx elevates to receive the bolus of food from the mouth and moves the bolus down the pharynx into the esophagus. The superior, middle, and inferior pharyngeal constrictor muscles contract in succession, forcing the food through the pharynx. At the same time, the upper esophageal sphincter relaxes, the elevated pharynx opens the esophagus, and food is pushed into the esophagus. This phase of swallowing is unconscious and is controlled automatically, even though the muscles involved are skeletal. The pharyngeal phase of swallowing lasts about 1–2 seconds.

During the pharyngeal phase, the vestibular folds and vocal cords close, and the **epiglottis** (ep-i-glot′is; on the glottis) is tipped posteriorly, so that the epiglottic cartilage covers the opening into the larynx, and the larynx is elevated. These movements prevent food from passing into the larynx.

Predict 3

Why is it important to close the opening between the nasopharynx and the oropharynx during swallowing? What may happen if a person emits an explosive burst of laughter while trying to swallow a liquid? Predict the consequences of trying to swallow and speak at the same time.

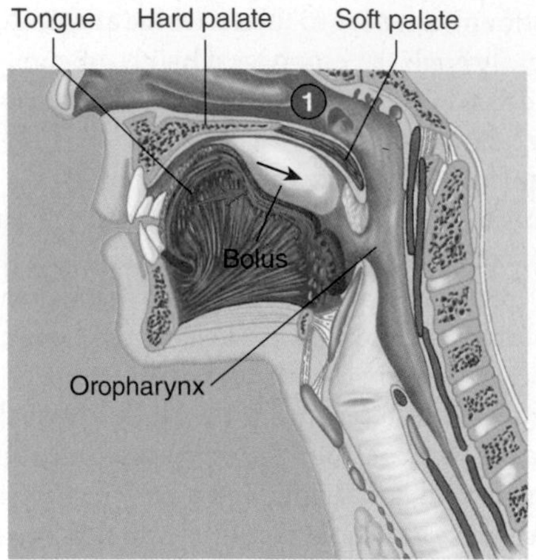
Tongue
Hard palate
Soft palate
1
Bolus
Oropharynx

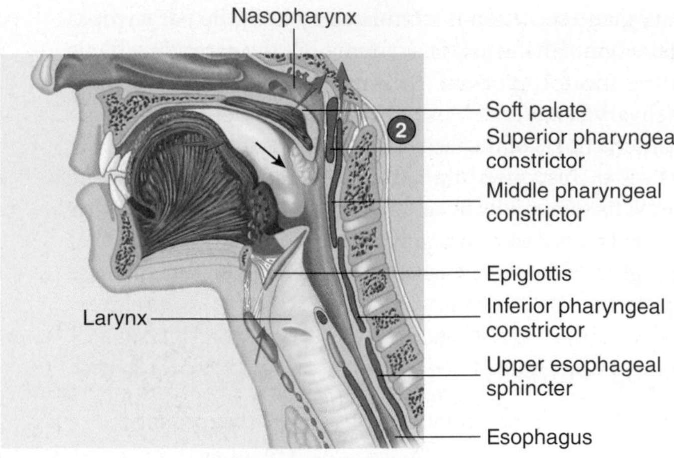
Nasopharynx
2
Soft palate
Superior pharyngeal constrictor
Middle pharyngeal constrictor
Epiglottis
Larynx
Inferior pharyngeal constrictor
Upper esophageal sphincter
Esophagus

1 During the **voluntary phase**, a bolus of food (*yellow*) is pushed by the tongue against the hard and soft palates and posteriorly toward the oropharynx (*blue arrow* indicates tongue movement; *black arrow* indicates movement of the bolus). *Tan:* bone; *purple:* cartilage; *red:* muscle.

2 During the **pharyngeal phase**, the soft palate is elevated, closing off the nasopharynx. The pharynx and larynx are elevated (*blue arrows* indicate muscle movement; *green arrow* indicates elevation of the larynx).

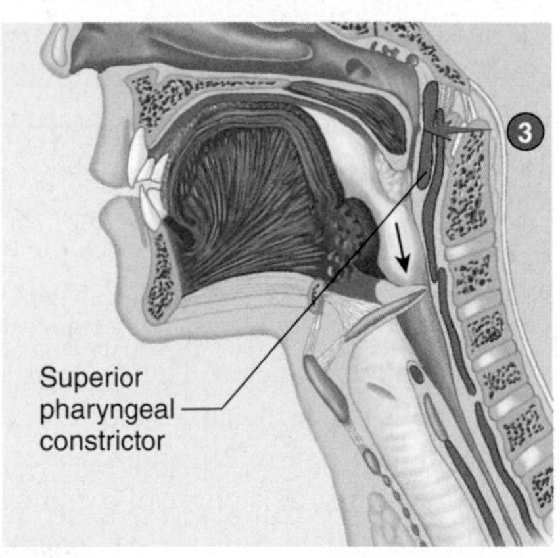
3
Superior pharyngeal constrictor

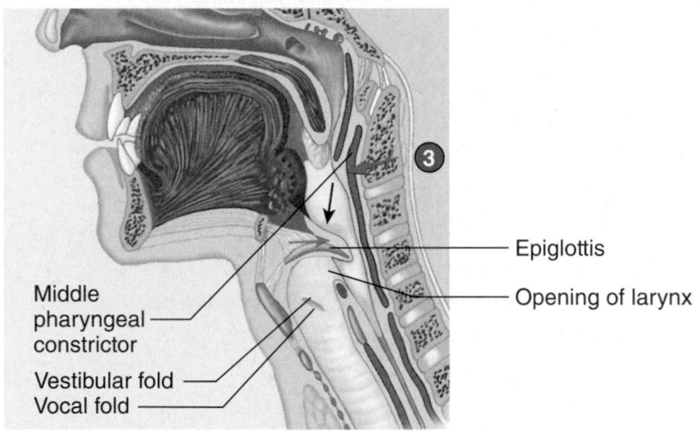
3
Epiglottis
Opening of larynx
Middle pharyngeal constrictor
Vestibular fold
Vocal fold

3 Successive constriction of the pharyngeal constrictors from superior to inferior (*blue arrows*) forces the bolus through the pharynx and into the esophagus. As this occurs, the vestibular and vocal folds expand medially to close the passage of the larynx. The epiglottis (*green arrow*) is bent down over the opening of the larynx largely by the force of the bolus pressing against it.

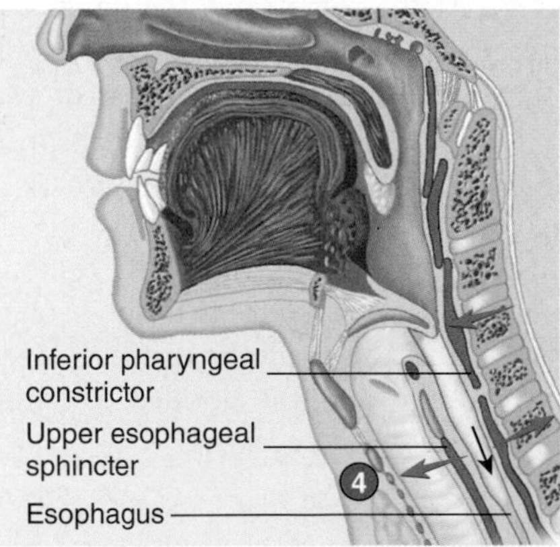
Inferior pharyngeal constrictor
Upper esophageal sphincter
4
Esophagus

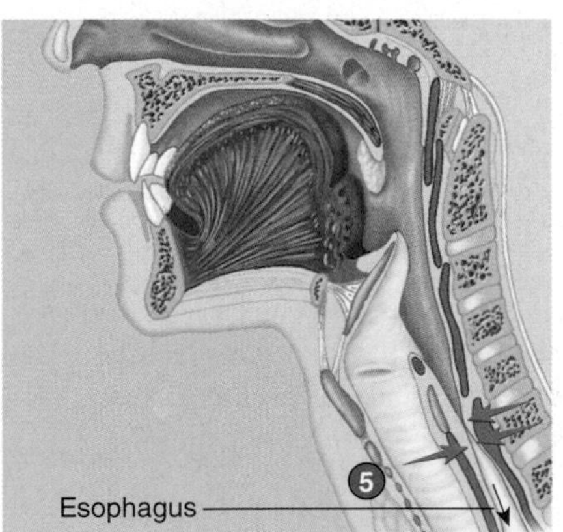
Esophagus
5

4 As the inferior pharyngeal constrictor contracts, the upper esophageal sphincter relaxes (outwardly directed *blue arrows*), allowing the bolus to enter the esophagus.

5 During the **esophageal phase**, the bolus is moved by peristaltic contractions of the esophagus toward the stomach (inwardly directed *blue arrows*).

PROCESS FIGURE 24.10 Three Phases of Swallowing (Deglutition)

The **esophageal phase** of swallowing (figure 24.10, *step 5*), which takes about 5–8 seconds, is responsible for moving food from the pharynx to the stomach. Muscular contractions in the wall of the esophagus occur in peristaltic waves. Gravity helps move food, especially liquids, through the esophagus. However, the peristaltic contractions in the esophagus are forceful enough to allow a person to swallow even while doing a headstand or floating in the zero-gravity environment of space.

As the peristaltic waves and the food bolus approach the stomach, the lower esophageal sphincter in the esophagus relaxes. This sphincter is not anatomically distinct from the rest of the esophagus, but it can be identified physiologically because it remains tonically constricted to prevent the reflux of stomach contents into the lower part of the esophagus.

The presence of food in the esophagus stimulates the myenteric plexus, which controls the peristaltic waves. Food in the esophagus also stimulates tactile receptors, which send afferent impulses to the medulla oblongata through the vagus nerves. Motor impulses, in turn, pass along the vagal efferent fibers to the striated and smooth muscles within the esophagus, thereby stimulating their contractions and reinforcing the peristaltic contractions.

ASSESS **YOUR PROGRESS**

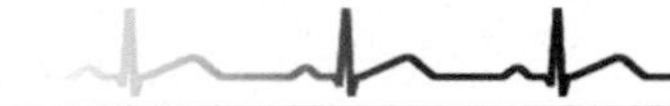

26. *Name the parts of the pharynx involved with digestion. What are pharyngeal constrictors?*

27. *Where is the esophagus located? Describe the tunics of the esophageal wall and the esophageal sphincters.*

28. *What are the three phases of swallowing? Sequentially list the processes involved in the last two phases and describe how they are regulated.*

24.8 Stomach

LEARNING OUTCOMES

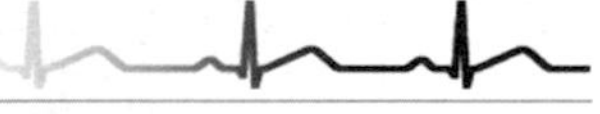

After reading this section, you should be able to

A. Outline the anatomical and histological characteristics of the stomach.

B. Describe stomach secretions, their function, and their regulation.

C. Describe gastric movements and their regulation.

The **stomach** is an enlarged segment of the digestive tract that primarily functions as a storage and mixing chamber. It is located in the left superior part of the abdomen (see figure 24.1). Its shape and size vary from person to person, even within the same individual from time to time, depending on food content and body posture. Nonetheless, several general anatomical features can be described.

Anatomy of the Stomach

The opening from the esophagus into the stomach is the **gastroesophageal opening,** or *cardiac* (located near the heart) *opening,* and the region of the stomach around the cardiac opening is the *cardiac part* (figure 24.11). The **lower esophageal sphincter,** also called the *cardiac sphincter,* surrounds the cardiac opening. Recall that, although this is an important structure in the normal function of the stomach, it is a physiological constrictor only and cannot be seen anatomically. A part of the stomach to the left of the cardiac part, the **fundus** (fŭn′dŭs), is actually superior to the cardiac opening. The largest part of the stomach is the **body,** which turns to the right, creating a **greater curvature** and a **lesser curvature.** The body narrows to form the funnel-shaped **pyloric** (pī-lōr′ik; gatekeeper) **part** of the stomach. The wider part of the funnel, toward the body of the stomach, is the **pyloric antrum.** The narrow part of the funnel is the **pyloric canal.** The pyloric canal opens through the **pyloric orifice** into the small intestine. The pyloric orifice is surrounded by the **pyloric sphincter,** or *pylorus,* a relatively thick ring of smooth muscle, which helps regulate the movement of gastric contents into the small intestine. **Hypertrophic pyloric stenosis** is a common defect of the stomach in infants, in which the pyloric sphincter is greatly thickened and thus interferes with normal stomach emptying.

Histology of the Stomach

The **serosa,** or *visceral peritoneum,* is the outermost tunic of the stomach. It consists of an outer layer of simple squamous epithelium and an inner layer of connective tissue. The **muscularis** of the stomach consists of three layers: an outer longitudinal layer, a middle circular layer, and an inner oblique layer (figure 24.11*a*). In some areas of the stomach, such as the fundus, the three layers blend with one another and cannot be separated. Deep to the muscular layer are the submucosa and the mucosa, which are thrown into large folds called **rugae** (roo′gē; wrinkles) when the stomach is empty. These folds allow the mucosa and submucosa to stretch, and the folds disappear as the stomach volume increases as it is filled.

The stomach is lined with simple columnar epithelium. The epithelium forms numerous, tubelike **gastric pits,** which are the openings for the **gastric glands** (figure 24.11*b*). The epithelial cells of the stomach are of five types. The first type, **surface mucous cells,** are found on the surface and lining the gastric pit.

Surface mucous cells protect the stomach wall from being damaged by acid and digestive enzymes. These cells produce an alkaline mucus on their surface that neutralizes the acid and is a barrier to the digestive enzymes. The surface mucous cells are connected by tight junctions, which provide an additional barrier that prevents acids and enzymes from reaching deeper tissues. In addition, when surface mucous cells are damaged, they are rapidly replaced.

The remaining four cell types are in the gastric glands. They are **mucous neck cells,** which produce mucus; **parietal cells,** which produce hydrochloric acid and intrinsic factor; **chief cells,** which produce pepsinogen; and **endocrine cells,** which produce regulatory hormones and paracrine factors. The mucous neck cells are located near the openings of the glands, whereas the parietal, chief, and endocrine cells are interspersed in the deeper parts of the glands. There are several types of endocrine cells. Enterochromaffin-like cells produce histamine, which stimulates acid secretion by parietal cells. Gastrin-containing cells secrete gastrin, and somatostatin-containing cells secrete somatostatin, which inhibits gastrin and insulin secretion.

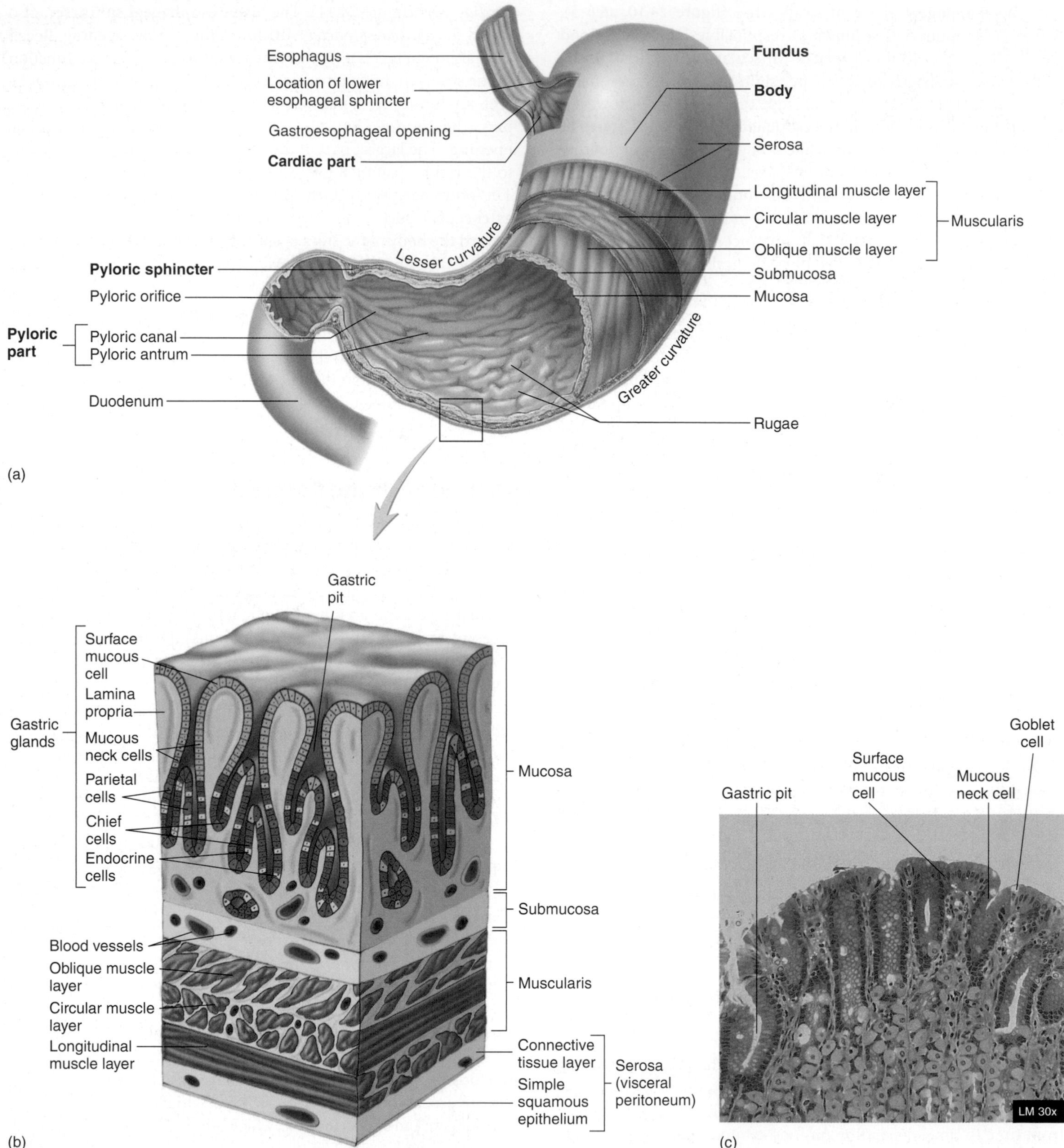
Esophagus
Location of lower esophageal sphincter
Gastroesophageal opening
Cardiac part
Fundus
Body
Serosa
Longitudinal muscle layer
Circular muscle layer
Oblique muscle layer
Muscularis
Submucosa
Mucosa
Lesser curvature
Greater curvature
Pyloric sphincter
Pyloric orifice
Pyloric part
Pyloric canal
Pyloric antrum
Duodenum
Rugae
(a)
Gastric pit
Gastric glands
Surface mucous cell
Lamina propria
Mucous neck cells
Parietal cells
Chief cells
Endocrine cells
Mucosa
Submucosa
Blood vessels
Oblique muscle layer
Circular muscle layer
Longitudinal muscle layer
Muscularis
Connective tissue layer
Simple squamous epithelium
Serosa (visceral peritoneum)
(b)
Goblet cell
Surface mucous cell
Mucous neck cell
Gastric pit
LM 30x
(c)

FIGURE 24.11 AP|R **Anatomy and Histology of the Stomach**
(*a*) Cutaway section reveals the muscular layers and internal anatomy of the stomach. (*b*) A section of the stomach wall illustrates its histology, including several gastric pits and glands. (*c*) Photomicrograph of gastric glands.

Secretions of the Stomach

Once food enters the stomach, it is mixed with stomach secretions to form a semifluid material called **chyme** (kīm; juice). The primary function of the stomach is to store and mix the chyme. Although some digestion and absorption occur in the stomach, they are not its major functions.

Stomach secretions include mucus, hydrochloric acid, gastrin, histamine, intrinsic factor, and pepsinogen. Pepsinogen is the inactive form of the protein-digesting enzyme pepsin.

The surface mucous cells and mucous neck cells secrete a viscous, alkaline mucus that covers the surface of the epithelial cells, forming a layer 1–1.5 mm thick. The thick layer of mucus lubricates and protects the epithelial cells of the stomach wall from the damaging effect of the acidic chyme and pepsin. Irritation of the stomach mucosa stimulates the secretion of a greater volume of mucus.

Parietal cells in the gastric glands of the pyloric region secrete intrinsic factor and a concentrated solution of hydrochloric acid. **Intrinsic factor** is a glycoprotein that binds with vitamin B_{12}, making the vitamin more readily absorbed in the ileum. Vitamin B_{12} is important in deoxyribonucleic acid (DNA) synthesis, which is especially important for continual red blood cell production. A lack of vitamin B_{12} absorption leads to pernicious anemia (see chapter 19).

Hydrochloric acid produces the low pH of the stomach's contents, which is normally between 1 and 3. Although the hydrochloric acid secreted into the stomach has a minor digestive effect on ingested food, one of its main functions is to kill bacteria that are ingested with essentially everything humans put into their mouths. However, some pathogenic bacteria may avoid digestion in the stomach, because they have an outer coat that resists stomach acids.

The low pH of the stomach's contents has additional functions. It stops carbohydrate digestion by inactivating salivary amylase. Stomach acid also denatures many proteins, so that proteolytic enzymes can reach internal peptide bonds. The acid environment provides the proper pH for the activation and function of pepsin.

Hydrogen ions are derived from carbon dioxide and water, which enter the parietal cell from its serosal surface, the side opposite the lumen of the gastric pit (figure 24.12). Once inside the cell, carbonic anhydrase catalyzes the reaction between carbon dioxide and water to form carbonic acid. Some of the carbonic acid molecules then dissociate to form H^+ and HCO_3^- (bicarbonate ion). Hydrogen ions are then actively transported across the mucosal surface of the parietal cell into the lumen of the stomach by a H^+–K^+ exchange pump, often called a **proton pump.** Drugs that block the proton pump are used to lower gastric acid levels. The pump moves H^+ by active transport against a steep concentration gradient, and Cl^- diffuses from the cell through ion channels in the plasma membrane. Diffusion of Cl^- into the gastric gland duct balances the positively charged H^+ to reduce the amount of energy needed to transport the H^+ against both a concentration gradient and an electrical gradient. Bicarbonate ions move down their concentration gradient from the parietal cell into the extracellular fluid. During this process, HCO_3^- is exchanged for Cl^- through an antiporter, which is located in the plasma membrane, and the Cl^- subsequently moves into the cell. This results in an elevated blood pH in the veins that carry blood away from the stomach, called the **alkaline tide.**

Chief cells within the gastric glands secrete **pepsinogen** (pep-sin'ō-jen). Pepsinogen is packaged in **zymogen** (zī-mō-jen) **granules,** which are released by exocytosis when pepsinogen secretion is stimulated. *Zymogen* is the term for an inactive enzyme. Once pepsinogen enters the lumen of the stomach, hydrochloric acid and previously formed pepsin molecules convert it to **pepsin.** Pepsin exhibits optimal enzymatic activity at a pH of 3 or less. Pepsin catalyzes the cleavage of some covalent bonds in proteins, thus breaking them into smaller peptide chains.

Regulation of Stomach Secretion

Approximately 2–3 L of gastric secretions (gastric juice) are produced each day. The amount and type of food entering the stomach and small intestine dramatically affect the quantity of gastric secretions, but up to 700 mL are secreted as a result of a typical meal. Both nervous and hormonal mechanisms regulate gastric secretions. The neural mechanisms involve reflexes integrated within the medulla oblongata and local reflexes integrated within the ENS. In addition, higher brain centers influence the reflexes. The chemical messengers that regulate stomach secretions include the hormones gastrin, secretin, and cholecystokinin (table 24.3), as well as the paracrine chemical messenger histamine.

The regulation of stomach secretion is divided into three phases: cephalic, gastric, and intestinal. The cephalic phase can be viewed as the "get started" phase, when stomach secretions are increased in anticipation of incoming food. This is followed by the "go for it" gastric phase, when most of the stimulation of secretion occurs. Finally, the intestinal phase is the "slow down" phase, during which stomach secretion decreases.

1. **Cephalic phase.** In the cephalic phase of gastric regulation, several types of stimuli act on the centers within the medulla oblongata to influence gastric secretions (figure 24.13*a*). These stimuli include the taste and smell of food, the stimulation of tactile receptors during the process of chewing and swallowing, and pleasant thoughts of food. Action potentials are sent from the medulla oblongata along parasympathetic neurons within the vagus (X) nerves to the stomach. Within the stomach wall, the preganglionic neurons stimulate the postganglionic neurons in the ENS. The postganglionic neurons, which are primarily cholinergic, stimulate secretory activity in the cells of the stomach mucosa.

   Parasympathetic stimulation of the stomach mucosa results in the release of the neurotransmitter acetylcholine, which increases the secretory activity of both the parietal and the chief cells and stimulates the secretion of **gastrin** (gas'trin) and **histamine** from endocrine cells. The gastrin released into the circulation travels to the parietal cells, where it stimulates additional hydrochloric acid and pepsinogen secretion. In addition, gastrin stimulates enterochromaffin-like cells to release histamine, which stimulates parietal cells to secrete hydrochloric acid. Histamine acts as both a local paracrine chemical messenger and a hormone in the blood to stimulate gastric gland secretory activity. Acetylcholine, histamine, and gastrin working together cause a greater secretion of hydrochloric acid than any of them does separately. Of the three, histamine has the greatest stimulatory effect. Drugs that block the actions of histamine are used to lower acid levels.

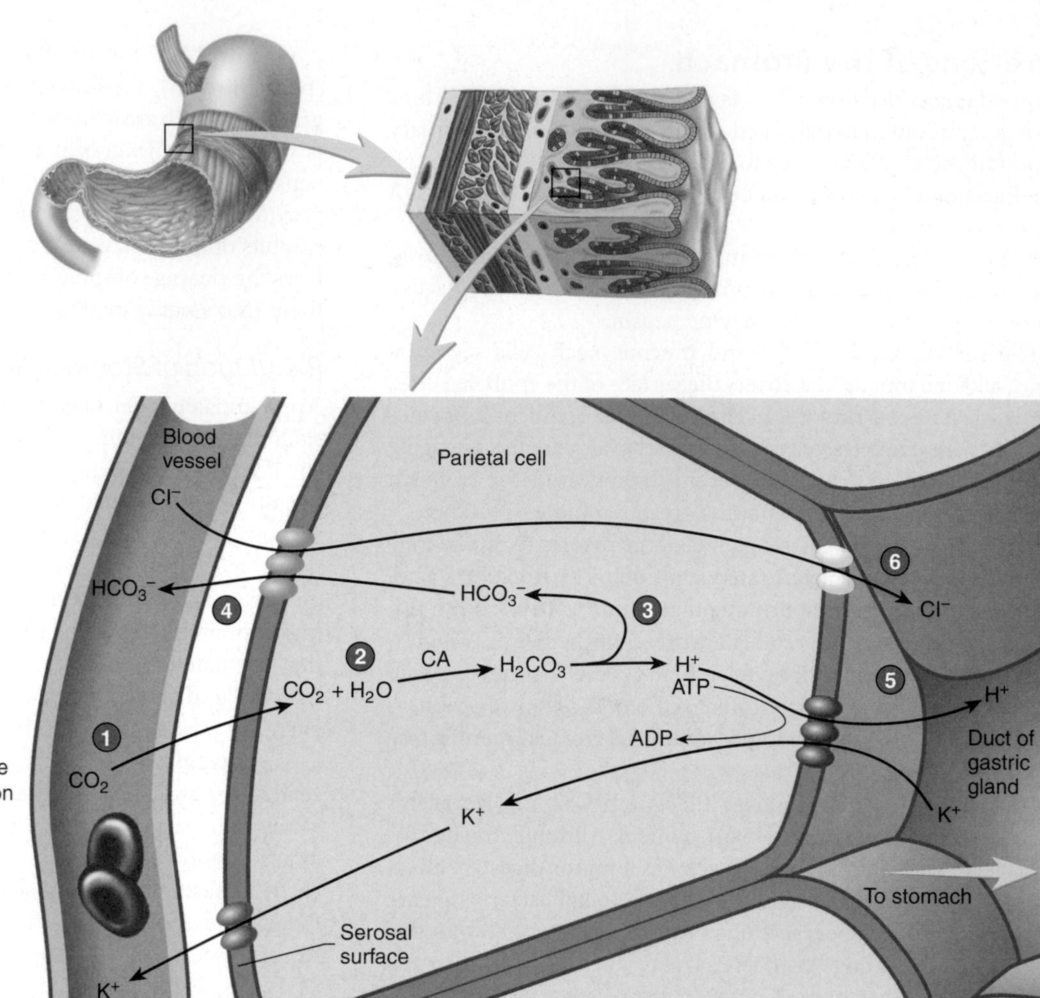

1 Carbon dioxide (CO_2) diffuses into the parietal cell.
2 Carbon dioxide combines with water (H_2O) in an enzymatic reaction that is catalyzed by carbonic anhydrase (CA) to form carbonic acid (H_2CO_3).
3 Carbonic acid dissociates into a bicarbonate ion (HCO_3^-) and a hydrogen ion (H^+).
4 Bicarbonate ions are transported back into the bloodstream. An antiporter in the plasma membrane exchanges HCO_3^- for a chloride ion (Cl^-).
5 A H^+–K^+ pump moves H^+ into the duct of the gastric gland and K^+ into the parietal cell.
6 Chloride ions diffuse into the gastric gland duct.
Blood vessel
Parietal cell
Cl^-
HCO_3^-
HCO_3^-
CO_2
$CO_2 + H_2O$
CA
H_2CO_3
H^+
ATP
ADP
Cl^-
H^+
Duct of gastric gland
K^+
K^+
K^+
To stomach
Serosal surface

PROCESS FIGURE 24.12 APR **Hydrochloric Acid Production by Parietal Cells in the Gastric Glands of the Stomach**

TABLE 24.3 Functions of the Gastrointestinal Hormones

Site of Production	Method of Stimulation	Secretory Effects	Motility Effects
Gastrin			
Stomach	Distension; partially digested proteins, autonomic stimulation, ingestion of alcohol or caffeine	Increases gastric secretion	Causes a minor increase in gastric motility
Secretin			
Duodenum	Acidity of chyme	Decreases gastric secretion; stimulates pancreatic and bile secretions high in bicarbonate ions	Decreases gastric motility
Cholecystokinin			
Duodenum	Fatty acids and peptides	Slightly decreases gastric secretion; stimulates pancreatic secretions high in digestive enzymes; causes contraction of the gallbladder and relaxation of the hepatopancreatic ampullar sphincter	Strongly decreases gastric motility

2. **Gastric phase.** The greatest volume of gastric secretions is produced during the gastric phase of gastric regulation. The presence of food in the stomach initiates the gastric phase (figure 24.13*b*). The primary stimuli are distension of the stomach and the presence of amino acids and peptides in the stomach.

FUNDAMENTAL **Figure**

Cephalic Phase

1. The taste, smell, or thought of food or tactile sensations of food in the mouth stimulate the medulla oblongata (*light green arrows*).
2. Vagus nerves carry parasympathetic action potentials to the stomach (*pink arrow*), where enteric plexus neurons are activated.
3. Postganglionic neurons stimulate secretion by parietal and chief cells and stimulate gastrin and histamine secretion by endocrine cells.
4. Gastrin is carried through the circulation back to the stomach (*dark green arrow*), where, along with histamine, it stimulates secretion.

Taste, smell, or thought of food (chemoreceptors)
Tactile sensation in mouth
1
Hypothalamus
Medulla oblongata
Secretions stimulated
4
Vagus nerves
2
3
Histamine
Gastrin
Circulation
Stomach

Gastric Phase

1. Distention of the stomach stimulates mechanoreceptors (stretch receptors) and activates a parasympathetic reflex. Action potentials generated by the mechanoreceptors are carried by the vagus nerves to the medulla oblongata (*light green arrow*).
2. The medulla oblongata increases action potentials in the vagus nerves that stimulate secretions by parietal and chief cells and stimulate gastrin and histamine secretion by endocrine cells (*pink arrow*).
3. Distention of the stomach also activates local reflexes that increase stomach secretions (*orange arrow*).
4. Gastrin is carried through the circulation back to the stomach (*dark green arrow*), where, along with histamine, it stimulates secretion.

Vagus nerves
Medulla oblongata
1
Secretions stimulated
2
Secretions stimulated
4
Distention
3
Histamine
Gastrin
Circulation
Stomach

Intestinal Phase

1. Chyme in the duodenum with a pH less than 2 or containing fat digestion products (lipids) inhibits gastric secretions by three mechanisms (2–4).
2. Chemoreceptors in the duodenum are stimulated by H^+ (low pH) or lipids. Action potentials generated by the chemoreceptors are carried by the vagus nerves to the medulla oblongata (*light green arrow*), where they inhibit parasympathetic action potentials (*pink arrow*), thereby decreasing gastric secretions.
3. Local reflexes activated by H^+ or lipids also inhibit gastric secretion (*orange arrows*).
4. Secretin and cholecystokinin produced by the duodenum (*dark red arrows*) decrease gastric secretions in the stomach.

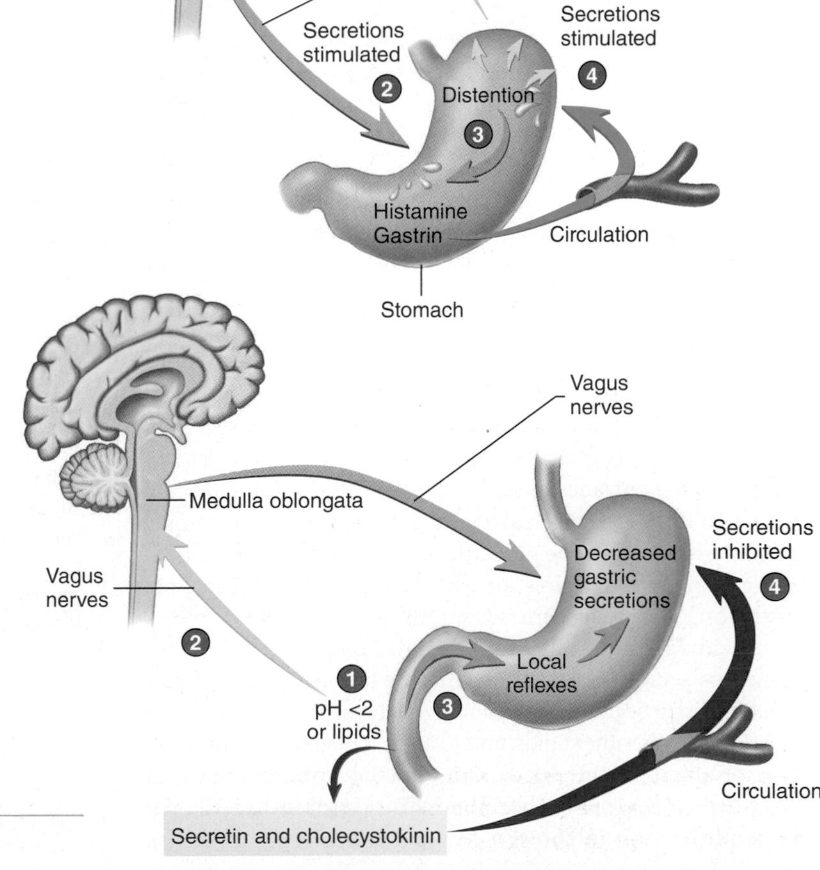
Vagus nerves
Medulla oblongata
Vagus nerves
2
Decreased gastric secretions
Secretions inhibited
4
1
pH <2 or lipids
3
Local reflexes
Circulation
Secretin and cholecystokinin

PROCESS FIGURE 24.13 **Phases of Stomach Secretion**

Clinical IMPACT — Peptic Ulcer

Approximately 10% of people in the United States will develop a **peptic ulcer** during their lifetime. Peptic ulcers are caused when the gastric juices (acid and pepsin) digest the mucosal lining of the digestive tract. Approximately 80% of peptic ulcers occur on the duodenal side of the pyloric sphincter, but peptic ulcers can also occur in the stomach (gastric ulcers) or esophagus (esophageal ulcers).

Nearly all peptic ulcers are due to infection by a specific bacterium, *Helicobacter pylori,* which is also linked to gastritis and gastric cancer. Because stress, diet, smoking, and alcohol cause excess acid secretion in the stomach, these lifestyle patterns were deemed responsible for ulcers for many years. Although these factors can contribute to ulcers, it is now clear that the root cause is *H. pylori.*

The presence of bacteria in the stomach mucosa was first discovered in 1892, but the finding was met with severe skepticism. In 1982, an Australian doctor, Barry Marshall, was finally able to culture an unusual bacterium, *H. pylori,* from stomach biopsies. To prove his belief that this bacterium can cause gastritis and ulcers, Marshall did something that no one should do at home. He drank a solution of *H. pylori* and subsequently developed gastric inflammation. Luckily, antibiotic treatment was able to cure him. In 2005, along with his colleague, Dr. Robin Warren, he received the Nobel Prize in Physiology or Medicine for his discovery.

Antibiotic treatment to eradicate *H. pylori* is the best therapy for ulcers. A combination of antibiotics and antacids cures 95% of gastric and 74% of duodenal ulcers within 2 months, with less than a 10% recurrence rate. By contrast, the previous conventional treatment using antacids yields only temporary relief, with about 90% recurrence within a year. Other treatments involve drugs that prevent histamine-stimulated acid secretion or that directly inhibit the proton pumps that secrete the acid. Such treatments are effective only for short-term relief, not for long-term treatment.

Most bacteria cannot survive in the stomach. Hence, *H. pylori* is one of the most pervasive of human pathogens because it inhabits a niche without competition. Estimates suggest that well over half of the world's population is infected with *H. pylori.* The infection rate in the United States is about 1% per year of age—for example, 30% of all 30-year-olds are infected. In developing countries, nearly all people over age 25 are infected. This may contribute to the high rates of stomach cancer in some of those countries.

Analyses of the *H. pylori* DNA sequences from various ethnic and geographic populations suggest that *H. pylori* infection has been present in humans for over 150,000 years, yet only about 15–20% exhibit gastric problems attributed to *H. pylori.* What triggers the development of ulcers is a major unanswered question. It seems likely that both *H. pylori* infection and conditions that elevate acid secretion or damage the stomach wall, such as stress or the excessive ingestion of alcohol or aspirin, contribute to the development of an ulcer. For example, if a person is highly stressed, elevated sympathetic activity may inhibit duodenal gland secretion and increase the person's susceptibility to ulcers in the duodenum by reducing the protective coating of mucus on the duodenal wall.

Distension of the stomach wall, especially in the body or fundus, stimulates mechanoreceptors. Action potentials generated by these receptors initiate reflexes that involve both the CNS and the ENS. These reflexes result in acetylcholine release and the cascade of events that increase secretion, as in the cephalic phase. The presence of partially digested proteins or moderate amounts of alcohol or caffeine in the stomach also stimulates gastrin secretion.

When the pH of the stomach contents falls below 2, increased gastric secretion produced by distension of the stomach is blocked. This negative-feedback mechanism limits the secretion of gastric juice.

3. **Intestinal phase.** The intestinal phase of gastric secretion primarily inhibits gastric secretions (figure 24.13*c*). It is controlled by the entrance of acidic chyme into the duodenum of the small intestine, which activates both neural and hormonal mechanisms. When the pH of the chyme entering the duodenum drops to 2 or below, or when the chyme contains lipid digestion products, gastric secretions are inhibited.

   Acidic solutions in the duodenum cause the release of the hormone **secretin** (se-krē′tin) into the bloodstream. Secretin inhibits gastric secretion by inhibiting both parietal and chief cells.

   Fatty acids, other lipids, and to a lesser degree protein digestion products in the duodenum and the proximal jejunum initiate the release of the hormone **cholecystokinin** (kō′lē-sis-tō-kī′nin), which inhibits gastric secretion.

   The inhibition of gastric secretion is also under nervous control. The **enterogastric reflex** consists of a local reflex and a reflex integrated within the medulla oblongata that reduce gastric secretion. Distension of the duodenal wall, the presence of irritating substances in the duodenum, reduced pH, and hypertonic or hypotonic solutions in the duodenum activate the enterogastric reflex.

To summarize, gastric acid secretion is controlled by negative-feedback loops involving nerves and hormones. During the gastric phase, high acid levels in the stomach trigger a decrease in additional acid secretion. Then, during the intestinal phase, acidic chyme entering the duodenum triggers a decrease in stomach acid secretion. These negative-feedback loops ensure that the acidic chyme entering the duodenum is neutralized, which is required for the digestion of food by pancreatic enzymes and for the prevention of peptic ulcer formation.

Predict 4

Alice, age 85, reported periods of laryngitis that began when she awakened in the morning and lasted a few days. Alice's physician used a laryngoscope to examine her vocal folds and upper trachea, which appeared inflamed. He prescribed an antacid and a drug to decrease H^+ secretion and told her to take the medications prior to going to bed at night. He also advised Alice to avoid eating just before bedtime. Explain the cause of her laryngitis and why the medications should relieve the symptoms.

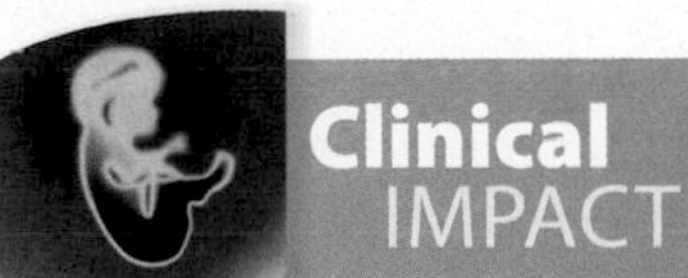
Clinical
IMPACT

Gastroesophageal Reflux

Gastroesophageal reflux is the reflux of acidic chyme from the stomach into the esophagus. Gastroesophageal reflux is commonly called *heartburn* because the refluxed acid causes a painful, burning sensation in the chest. The pain is usually short-lived but may be confused with the pain of an ulcer or a heart attack. The lower esophageal sphincter normally prevents acid reflux. Overeating (especially fatty and fried foods), lying down immediately after a meal, consuming too much alcohol or caffeine, smoking, and wearing extremely tight clothing can all cause gastroesophageal reflux. Gastroesophageal reflux commonly occurs in infants, but they usually outgrow it by their first birthday.

Chronic reflux more than twice a week in infants or adults is more serious and is called **gastroesophageal reflux disease (GERD).** GERD in young infants can be difficult to diagnose. Women commonly experience GERD during pregnancy because of increased abdominal pressure from the fetus and higher levels of the hormone progesterone, which relaxes the lower esophageal sphincter.

For most adults, lifestyle changes and medications that decrease gastric acid secretion are sufficient to relieve the symptoms of GERD. Antacids that buffer gastric acid can also alleviate minor discomfort. One class of drugs acts by blocking the H_2 histamine receptors on parietal cells. H_2 receptors are different from the H_1 receptors involved in allergic reactions. Drugs that block allergic reactions do not affect histamine-mediated stomach acid secretion, and vice versa. Cimetidine (Tagamet®) and ranitidine (Zantac®) are histamine receptor antagonists that prevent histamine from binding to receptors on parietal cells. These chemicals are extremely effective inhibitors of gastric acid secretion. Cimetidine, one of the most commonly prescribed drugs, is also used to treat gastric acid hypersecretion associated with gastritis and gastric ulcers. The most effective inhibitors of gastric acid secretion are the proton pump inhibitors, such as omeprazole (Prilosec®) and lansoprazole (Prevacid®). These drugs inhibit the proton pumps on parietal cells, thus preventing acid secretion into the stomach. If not treated, GERD can lead to serious complications, including esophageal ulcers, scarring that constricts the esophagus, and esophageal cancer.

Movements of the Stomach

Stomach Filling

As food enters the stomach, the rugae flatten and the stomach volume increases up to 20-fold. This expansion allows the stomach to accommodate a large amount of food with very little increased pressure, until the stomach nears maximum capacity. Relaxation of the rugae is mediated by a reflex integrated within the medulla oblongata that inhibits muscle tone and pressure is further minimized by the ability of smooth muscle to stretch without an increase in tension (see chapter 9).

Mixing of Stomach Contents

Ingested food is thoroughly mixed with stomach gland secretions to form chyme. This mixing is accomplished by gentle **mixing waves,** which are peristaltic-like contractions that occur about every 20 seconds, proceeding from the body of the stomach toward the pyloric sphincter. **Peristaltic waves** occur less frequently, are significantly more powerful than mixing waves, and force the chyme near the periphery of the stomach toward the pyloric sphincter. The more solid material near the center of the stomach is pushed superiorly toward the cardiac part for further digestion (figure 24.14). Roughly 80% of the contractions are mixing waves, and 20% are peristaltic waves. The back-and-forth movement of the chyme effectively mixes the ingested food with gastric juice.

Stomach Emptying

The amount of time food remains in the stomach depends on a number of factors, including the type and volume of food. Liquids exit the stomach within 1½–2½ hours after ingestion. After a typical meal, the stomach is usually empty within 3–4 hours. The pyloric sphincter normally remains partially closed because of mild tonic contraction. Each peristaltic contraction is strong enough to force a small amount of chyme through the pyloric opening and into the duodenum. The peristaltic contractions responsible for moving chyme through the partially closed pyloric opening are called the **pyloric pump.** In general, increased motility leads to increased emptying. In an empty stomach, peristaltic contractions that approach tetanic contractions can occur for about 2–3 minutes. The contractions are increased by low blood glucose levels and are strong enough to create uncomfortable sensations called **hunger pangs.** Hunger pangs usually begin 12–24 hours after a meal, in less time for some people. If nothing is ingested, hunger pangs reach their maximum intensity within 3–4 days and then become progressively weaker.

Regulation of Stomach Emptying

If the stomach empties too fast, the efficiency of digestion and absorption is reduced, and acidic gastric contents dumped into the

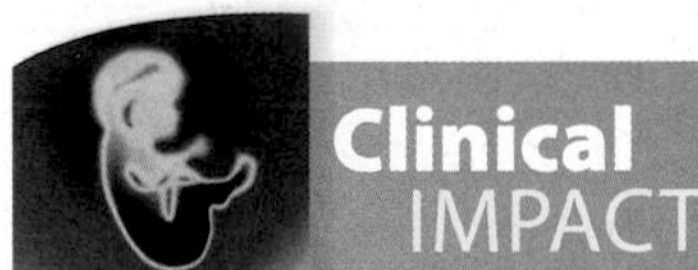
Clinical
IMPACT

Vomiting

Vomiting is usually a protective mechanism against the ingestion of toxic or harmful substances. Vomiting can result from irritation (e.g., overdistension or overexcitation) anywhere along the digestive tract. Action potentials travel through the vagus nerve and spinal visceral afferent nerves to the vomiting center in the medulla oblongata. Once the vomiting center is stimulated and the reflex is initiated, the following events occur: (1) A deep breath is taken; (2) the hyoid bone and larynx are elevated, opening the upper esophageal sphincter; (3) the opening of the larynx is closed; (4) the soft palate is elevated, closing the connection between the oropharynx and the nasopharynx; (5) the diaphragm and abdominal muscles are forcefully contracted, strongly compressing the stomach and increasing the intragastric pressure; (6) the lower esophageal sphincter is relaxed; and (7) the gastric contents are forced out of the stomach, through the esophagus and oral cavity, to the outside.

1 A mixing wave initiated in the body of the stomach progresses toward the pyloric sphincter (*pink arrows directed inward*).

2 The more fluid part of the chyme is pushed toward the pyloric sphincter (*blue arrows*), whereas the more solid center of the chyme squeezes past the peristaltic constriction back toward the body of the stomach (*orange arrow*).

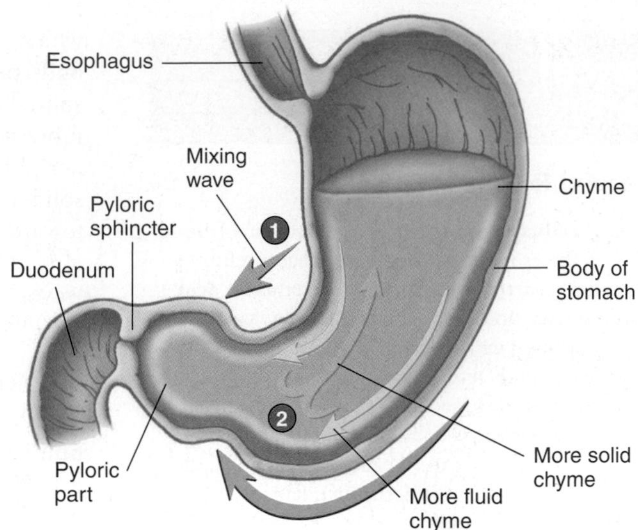
Esophagus
Mixing wave
Pyloric sphincter
Duodenum
Chyme
Body of stomach
Pyloric part
More fluid chyme
More solid chyme
1
2

3 Peristaltic waves (*purple arrows*) move in the same direction and in the same way as the mixing waves but are stronger.

4 Again, the more fluid part of the chyme is pushed toward the pyloric region (*blue arrows*), whereas the more solid center of the chyme squeezes past the peristaltic constriction back toward the body of the stomach (*orange arrow*).

5 Peristaltic contractions force a few milliliters of the mostly fluid chyme through the pyloric opening into the duodenum (*small red arrows*). Most of the chyme, including the more solid portion, is forced back toward the body of the stomach for further mixing (*yellow arrow*).

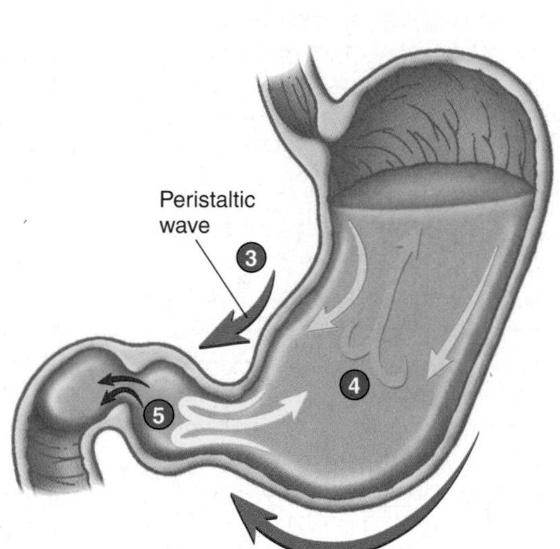
Peristaltic wave
3
4
5

PROCESS FIGURE 24.14 Movements in the Stomach

duodenum may damage its lining. However, if the rate of emptying is too slow, the highly acidic contents of the stomach may damage the stomach wall and reduce the rate at which nutrients are digested and absorbed.

To prevent these two extremes, stomach emptying is regulated. The neural mechanisms that stimulate stomach secretions are also involved with increasing stomach motility. The major stimulus for both motility and secretion is distension of the stomach wall. Increased stomach motility increases stomach emptying. Conversely, the hormonal and neural mechanisms associated with the duodenum that decrease gastric secretions also decrease gastric motility and increase constriction of the pyloric sphincter. The enterogastric reflex and the hormone cholecystokinin are major inhibitors of gastric motility. The result is a decrease in the rate of stomach emptying.

A meal of polysaccharide carbohydrates (starch and glycogen) has the fastest clearance time from the stomach, typically 1 hour. For comparison, a meal heavy with dietary fats and proteins takes up to 6 hours to clear from the stomach. A major reason for the fast clearance of carbohydrates is that they do not increase the release of cholecystokinin, which is a major inhibitor of stomach emptying.

ASSESS YOUR PROGRESS

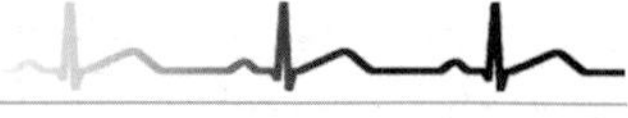

29. *Describe the parts of the stomach. List the tunics of the stomach wall. How is the stomach wall different from the esophagus wall?*

30. *What are gastric pits and gastric glands?*

31. *Name the types of cells in the stomach and the secretions they produce. What are the functions of the secretions?*

32. *Describe the three phases of regulation of stomach secretion.*

33. *How are gastric secretions inhibited? Why is this inhibition necessary?*

34. *As the stomach fills, why does the pressure not greatly increase until maximum volume is reached?*

35. *Name the two kinds of stomach movements. How are stomach movements regulated by hormones and nervous control?*

24.9 Small Intestine

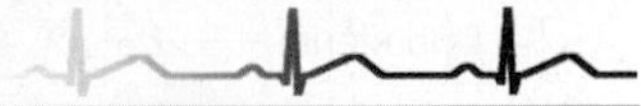

LEARNING OUTCOMES

After reading this section, you should be able to

A. List the sections of the small intestine and describe the characteristics that account for its large surface area.

B. Name the four major cell types of the duodenal mucosa and describe their functions.

C. Describe the secretions and movements of the small intestine.

The **small intestine** consists of three parts: the duodenum, the jejunum, and the ileum (figure 24.15). The entire small intestine is about 6 m long (range: 4.6–9 m). The duodenum is about 25 cm long (*duodenum* means 12, suggesting that it is 12 inches long). The jejunum, constituting about two-fifths of the total length of the small intestine, is about 2.5 m long. The ileum, constituting three-fifths of the small intestine, is about 3.5 m long. Two major accessory glands, the liver and the pancreas, are associated with the duodenum.

The small intestine is where the greatest amount of digestion and absorption occurs. Each day, about 9 L of water enter the digestive system. This includes water that is ingested and fluid secretions produced by glands along the length of the digestive tract. Most of the water (8–8.5 L) moves by osmosis, along with the absorbed solutes, out of the small intestine. A small part (0.5–1 L) enters the colon.

Anatomy and Histology of the Small Intestine

Duodenum

The **duodenum** (doo-ō-dē′nŭm, doo-od′ĕ-nŭm) nearly completes a 180-degree arc as it curves within the abdominal cavity (figure 24.16*a*), and the head of the pancreas lies within this arc. The duodenum begins with a short, superior part, where it exits the pylorus of the stomach, and ends in a sharp bend, where it joins the jejunum. Within the duodenum, about two-thirds of the way down the descending part, are two small mounds: the **major duodenal papilla** and the **minor duodenal papilla.** Ducts from the liver and/or pancreas open at these papillae.

The surface of the duodenum has several modifications that increase its area about 600-fold to allow for more efficient digestion and absorption of food. The mucosa and submucosa form a series of folds called the **circular folds,** or *plicae* (plī′sē) *circulares* (figure 24.16*a*), which run perpendicular to the long axis of the digestive tract. Tiny, fingerlike projections of the mucosa form numerous **villi** (vil′ī), which are 0.5–1.5 mm in length (figure 24.16*c*). Each villus is covered by simple columnar epithelium and contains a blood capillary network and a lymphatic capillary called a **lacteal** (lak′tē-ăl; figure 24.16*d*). Most of the cells that make up the surface of the villi have numerous cytoplasmic extensions (about 1 μm long) called **microvilli,** which further increase the surface area (figure 24.16*e*). The combined microvilli on the entire epithelial surface form the **brush border.** These various modifications greatly increase the surface area of the small intestine, thereby greatly enhancing absorption.

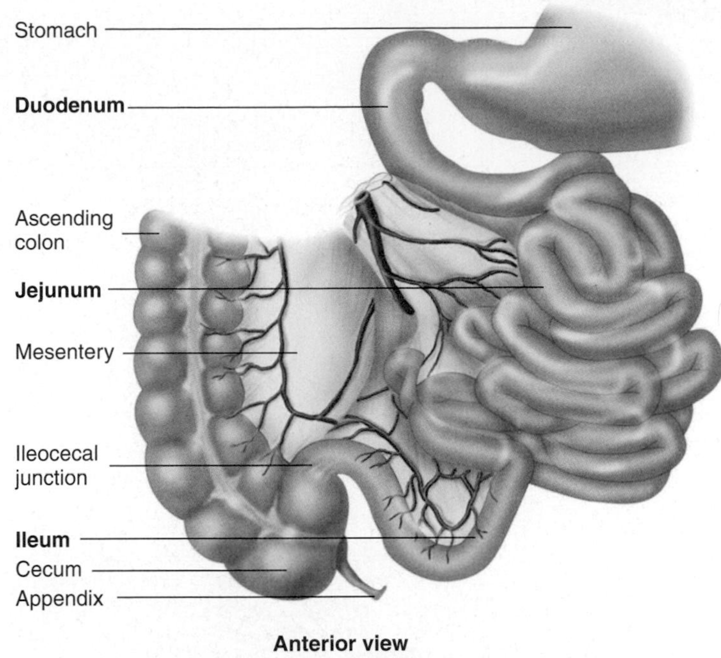
Stomach
Duodenum
Ascending colon
Jejunum
Mesentery
Ileocecal junction
Ileum
Cecum
Appendix
Anterior view

FIGURE 24.15 APR **Small Intestine**

The mucosa of the duodenum is simple columnar epithelium with four major cell types: (1) **Absorptive cells** are cells with microvilli that produce digestive enzymes and absorb digested food; (2) **goblet cells** produce a protective mucus; (3) **granular cells,** or *Paneth cells,* may help protect the intestinal epithelium from bacteria; and (4) **endocrine cells** produce regulatory hormones. The epithelial cells are produced within tubular invaginations of the mucosa, called **intestinal glands,** or *crypts of Lieberkühn,* at the base of the villi. The absorptive and goblet cells migrate from the intestinal glands to cover the surface of the villi and are eventually shed from its tip. The granular and endocrine cells remain in the bottom of the glands. The submucosa of the duodenum contains coiled, tubular mucous glands called **duodenal glands,** or *Brunner glands,* which open into the base of the intestinal glands.

Jejunum and Ileum

The **jejunum** (jĕ-joo′nŭm) and **ileum** (il′ē-ŭm) are similar in structure to the duodenum (see figure 24.15). However, progressing from the duodenum through the ileum, there are gradual decreases in the diameter of the small intestine, the thickness of the intestinal wall, the number of circular folds, and the number of villi. The duodenum and jejunum are the major sites of nutrient absorption, although some absorption occurs in the ileum. Lymphatic nodules called **Peyer patches** are numerous in the mucosa and submucosa of the ileum. Peyer patches and other mucosa-associated lymphoid tissue in the digestive tract initiate immune responses against microorganisms that enter the mucosa from ingested food (see chapter 22).

The site where the ileum connects to the large intestine is the **ileocecal junction.** It has a ring of smooth muscle, the **ileocecal sphincter,** and a one-way **ileocecal valve,** which allow intestinal contents to move from the ileum to the large intestine, but not in the opposite direction (see figures 24.15 and 24.25).

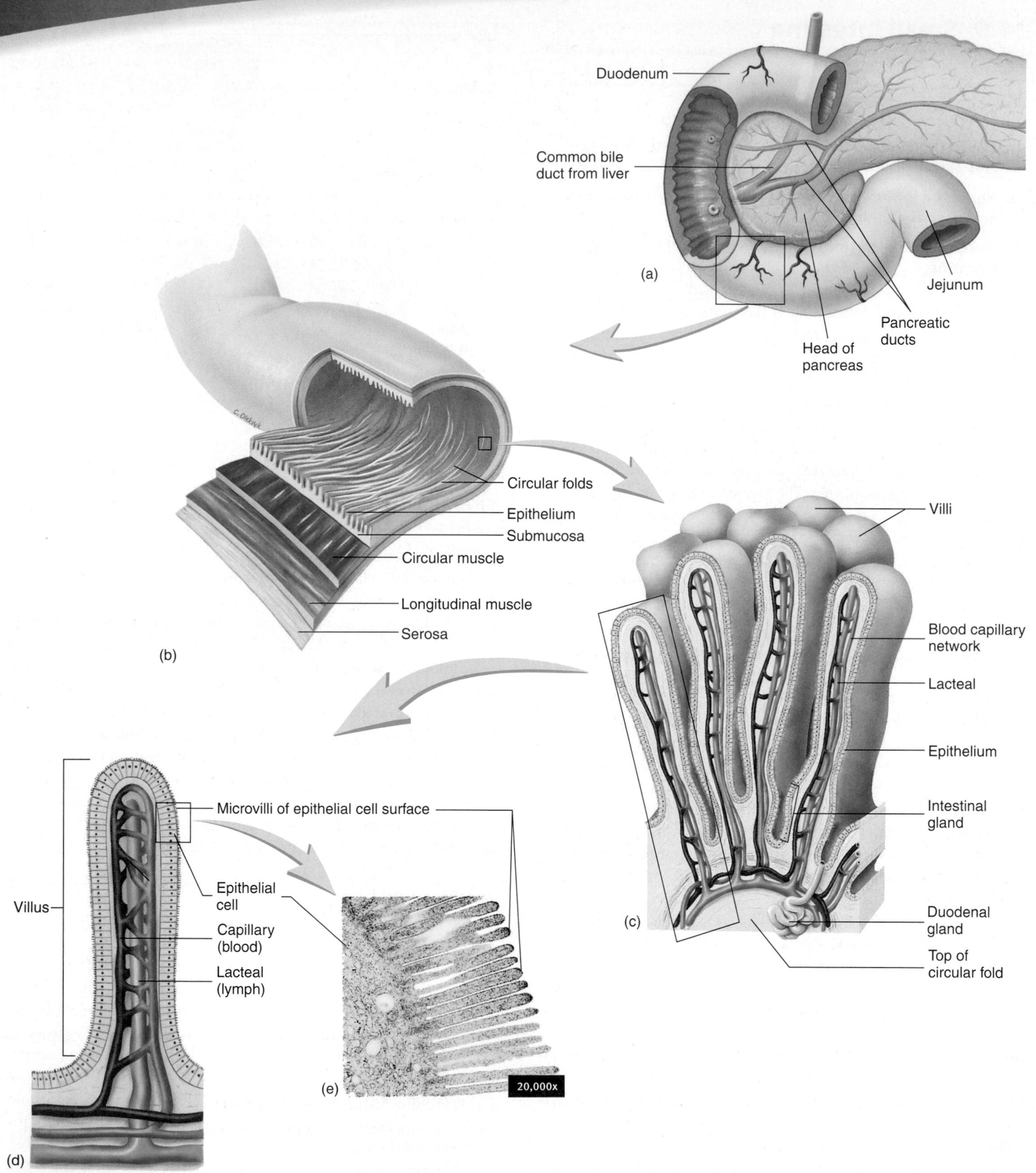

Duodenum
Common bile duct from liver
(a)
Jejunum
Pancreatic ducts
Head of pancreas
C. Ordway
Circular folds
Epithelium
Submucosa
Circular muscle
Longitudinal muscle
Serosa
(b)
Villi
Blood capillary network
Lacteal
Epithelium
Intestinal gland
Duodenal gland
Top of circular fold
(c)
Villus
Microvilli of epithelial cell surface
Epithelial cell
Capillary (blood)
Lacteal (lymph)
(d)
(e)
20,000x

FIGURE 24.16 Anatomy and Histology of the Duodenum
(*a*) Ducts from the liver and pancreas empty into the duodenum. (*b*) Wall of the duodenum, showing the circular folds. (*c*) The villi on a circular fold. (*d*) A single villus, showing the lacteal and capillary network. (*e*) Transmission electron micrograph of microvilli on the surface of a villus.

Secretions of the Small Intestine

The mucosa of the small intestine produces secretions that contain primarily mucus, electrolytes, and water. Intestinal secretions lubricate and protect the intestinal wall from the acidic chyme and the action of digestive enzymes. They also keep the chyme in the small intestine in a liquid form to facilitate the digestive process (see table 24.2). The intestinal mucosa produces most of the secretions that enter the small intestine, but the secretions of the liver and the pancreas also enter the small intestine and play essential roles in digestion. The pancreas secretes most of the digestive enzymes entering the small intestine. The intestinal mucosa also produces enzymes, but these remain associated with the intestinal epithelial surface.

The duodenal glands, intestinal glands, and goblet cells secrete large amounts of mucus. This mucus protects the wall of the intestine from the irritating effects of acidic chyme and from the digestive enzymes that enter the duodenum from the pancreas. The intestinal mucosa releases secretin and cholecystokinin, which stimulate hepatic and pancreatic secretions (see figures 24.21 and 24.24).

The vagus nerve, secretin, and chemical or tactile irritation of the duodenal mucosa stimulate secretion from the duodenal glands. Goblet cells produce mucus in response to the tactile and chemical stimulation of the mucosa.

Enzymes of the intestinal mucosa are bound to the membranes of the absorptive cell microvilli. These surface-bound enzymes include **disaccharidases,** which break down disaccharides to monosaccharides, and **peptidases,** which hydrolyze the peptide bonds between small amino acid chains (see table 24.2). Although these enzymes are not secreted into the intestine, they influence the digestive process significantly, and the large surface area of the intestinal epithelium brings these enzymes into contact with the intestinal contents. Small molecules, which are breakdown products of digestion, are absorbed through the microvilli and enter the circulatory or lymphatic system.

Movement in the Small Intestine

The mixing and propulsion of chyme are the primary mechanical functions of the small intestine. These functions are performed with the aid of segmental and peristalic contractions accomplished by the smooth muscle in the wall of the small intestine and propagated for short distances. Segmental contractions (see figure 24.3) mix the intestinal contents, and peristaltic contractions propel the intestinal contents along the digestive tract. A few peristaltic contractions may proceed the entire length of the intestine. Frequently, intestinal peristaltic contractions are continuations of peristaltic contractions that begin in the stomach. These contractions both mix and propel substances through the small intestine as the wave of contraction proceeds. The contractions move at a rate of about 1 cm/min. The movements are slightly faster at the proximal end of the small intestine and slightly slower at the distal end. It usually takes 3–5 hours for chyme to move from the pyloric region to the ileocecal junction.

Local mechanical and chemical stimuli are especially important in regulating the motility of the small intestine. Smooth muscle contraction increases in response to distension of the intestinal wall. Solutions that are either hypertonic or hypotonic, solutions with a low pH, and certain products of digestion, such as amino acids and peptides, also stimulate contractions of the small intestine. Local reflexes, which are integrated within the ENS of the small intestine, mediate the intestine's response to these mechanical and chemical stimuli. Stimulation through parasympathetic nerve fibers may also increase the intestine's motility, but the parasympathetic influences in the intestine are not as important as those in the stomach.

The ileocecal sphincter at the juncture between the ileum and the large intestine remains mildly contracted most of the time, but peristaltic waves reaching it from the small intestine cause it to relax and allow the chyme to move from the small intestine into the cecum. Cecal distension, however, initiates a local reflex that causes more intense constriction of the ileocecal sphincter. Closure of the sphincter facilitates digestion and absorption in the small intestine by slowing the rate of chyme movement from the small intestine into the large intestine and prevents material from returning to the ileum from the cecum.

Predict 5

Amos suffers from intermittent pain in the epigastric area that begins about 2 or 3 hours after eating. The pain is relieved by taking an antacid. An endoscopic exam identified duodenal ulcers and Amos's physician recommended antacids and an antibiotic. Amos wondered why he could not control the condition with antacids alone, but his physician was worried about perforation of the duodenum. Explain why Amos's physician prescribed both antacids and antibiotics. How could the lack of antibiotics lead to perforation of the duodenum?

ASSESS YOUR PROGRESS

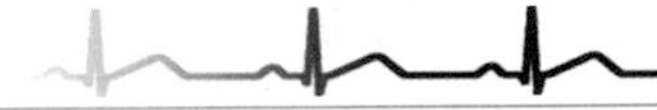

36. *Name and describe the three parts of the small intestine.*
37. *What are the circular folds, villi, and microvilli in the small intestine? What are their functions?*
38. *Name the four types of cells found in the duodenal mucosa, and state their functions.*
39. *What are the functions of the intestinal glands and duodenal glands? State the factors that stimulate secretion from the duodenal glands and from goblet cells.*
40. *List the enzymes of the small intestine wall, and give their functions.*
41. *What are the two kinds of movement of the small intestine? How are they regulated?*
42. *What is the function of the ileocecal sphincter and valve?*

24.10 Liver

LEARNING OUTCOMES

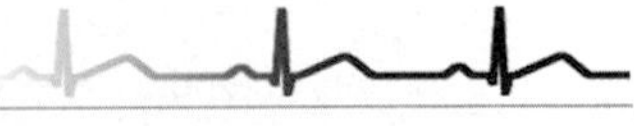

After reading this section, you should be able to

A. Describe the anatomy, histology, and ducts of the liver.

B. Describe the major functions of the liver and explain how they are regulated.

Anatomy of the Liver

The **liver** is the largest internal organ of the body, weighing about 1.36 kg (3 pounds). It is in the right-upper quadrant of the abdomen, tucked against the inferior surface of the diaphragm (figure 24.17; see figure 24.1). The liver consists of two major lobes, the **right lobe** and **left lobe,** which are separated by a connective tissue septum, the falciform ligament. Two minor lobes, the **caudate lobe** and the **quadrate lobe,** can be seen from an inferior view, along with the porta.

The **porta** (gate) is on the inferior surface of the liver, where the various vessels, ducts, and nerves enter and exit the liver. The **hepatic** (he-pat′ik) **portal vein,** the **hepatic artery,** and a small hepatic nerve plexus enter the liver through the porta (figure 24.17*b*). Lymphatic vessels and two hepatic ducts, one each from the right and left lobes, exit the liver at the porta. The hepatic ducts transport bile out of the liver. The right and left hepatic ducts unite to form a single **common hepatic duct** (figure 24.18). The **cystic duct** from the gallbladder joins the common hepatic duct to form the **common bile duct,** which joins the pancreatic duct at the **hepatopancreatic ampulla** (hĕ-pat′ō-pan-crē-at′ik am-pul′lă), an enlargement where the hepatic and pancreatic ducts come together. The hepatopancreatic ampulla empties into the duodenum at the major duodenal papilla (figure 24.18). A smooth muscle sphincter surrounds the common bile duct where it enters the hepatopancreatic ampulla. The gallbladder is a small sac on the inferior surface of the liver that stores bile. Bile can flow from the gallbladder through the cystic duct into the common bile duct, or it can flow back up the cystic duct into the gallbladder.

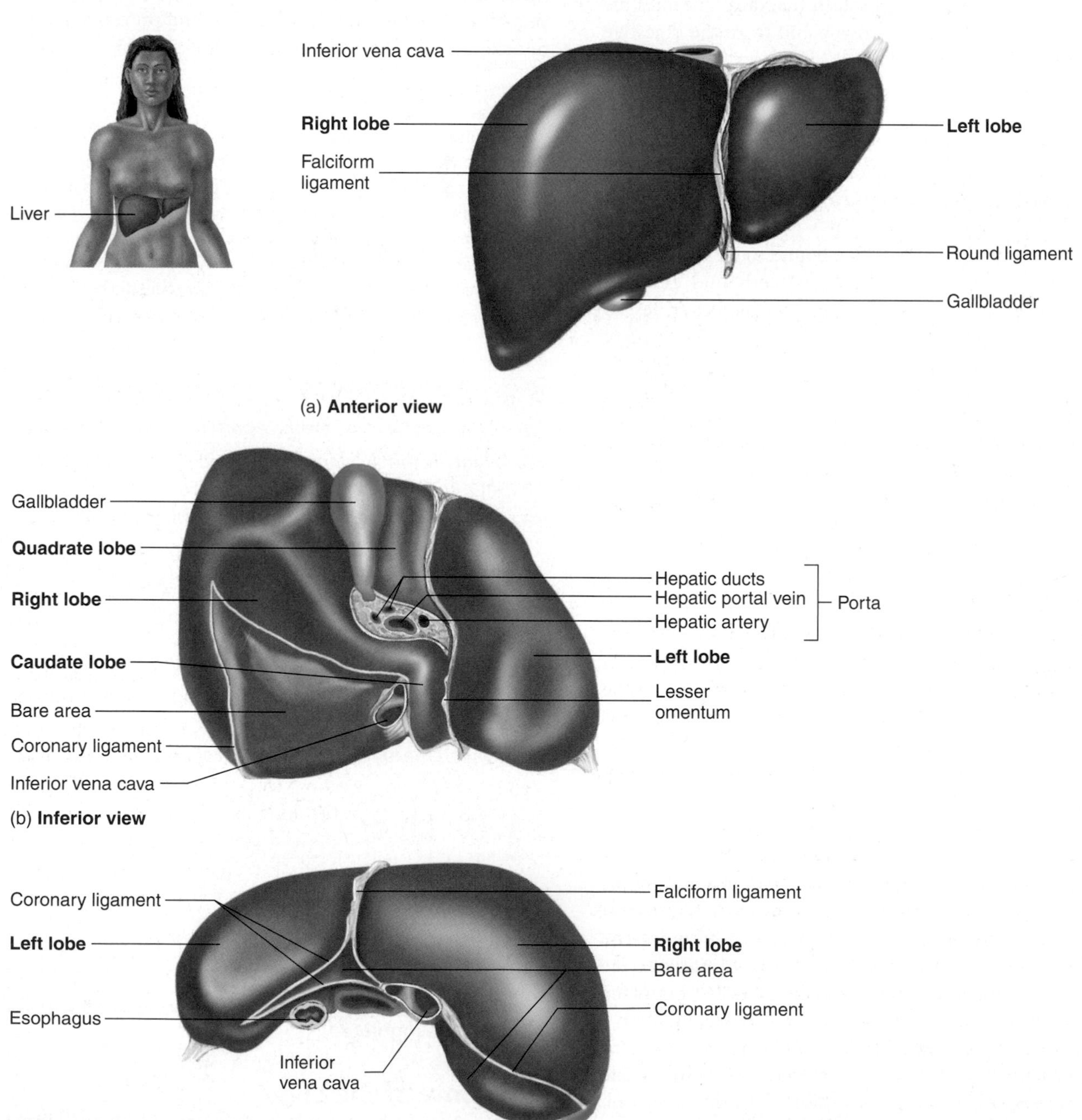
Liver
Inferior vena cava
Right lobe
Falciform ligament
Left lobe
Round ligament
Gallbladder
(a) Anterior view
Gallbladder
Quadrate lobe
Right lobe
Caudate lobe
Bare area
Coronary ligament
Inferior vena cava
Hepatic ducts
Hepatic portal vein
Hepatic artery
Porta
Left lobe
Lesser omentum
(b) Inferior view
Coronary ligament
Left lobe
Esophagus
Inferior vena cava
Falciform ligament
Right lobe
Bare area
Coronary ligament
(c) Superior view

FIGURE 24.17 APR **Liver**

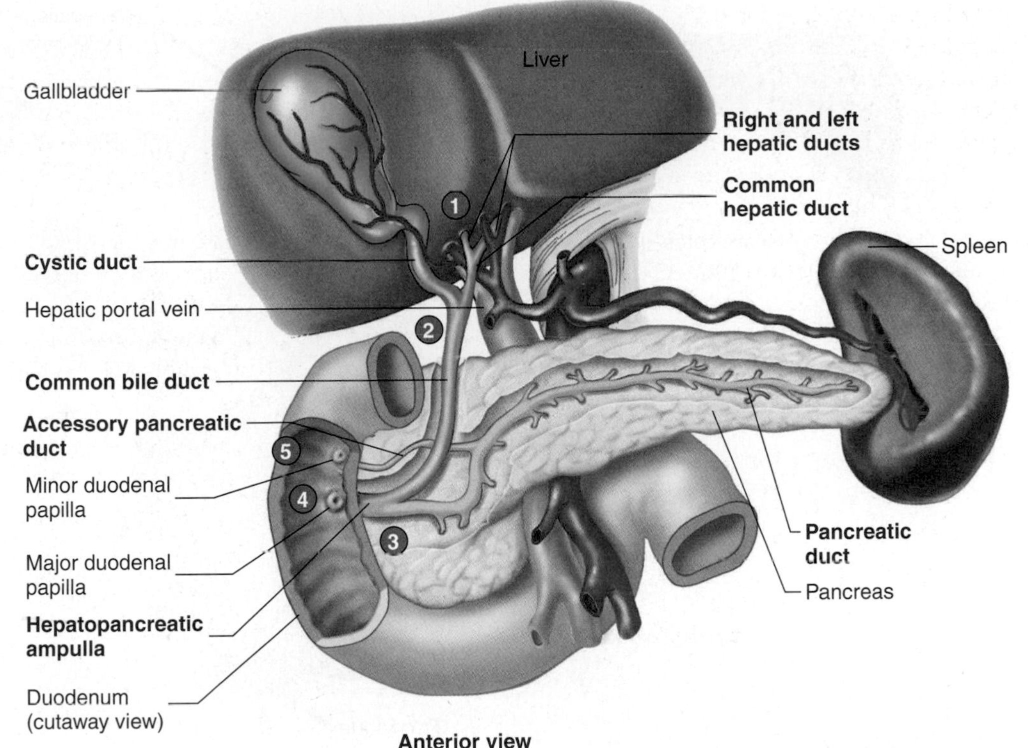

1 The hepatic ducts, which carry bile from the liver lobes, combine to form the common hepatic duct.
2 The common hepatic duct combines with the cystic duct from the gallbladder to form the common bile duct.
3 The common bile duct and the pancreatic duct combine to form the hepatopancreatic ampulla.
4 The hepatopancreatic ampulla empties bile and pancreatic secretions into the duodenum at the major duodenal papilla.
5 The accessory pancreatic duct empties pancreatic secretions into the duodenum at the minor duodenal papilla.
Gallbladder
Liver
Right and left hepatic ducts
Common hepatic duct
Spleen
Cystic duct
Hepatic portal vein
Common bile duct
Accessory pancreatic duct
Minor duodenal papilla
Major duodenal papilla
Hepatopancreatic ampulla
Duodenum (cutaway view)
Pancreatic duct
Pancreas
Anterior view

PROCESS FIGURE 24.18 Flow of Bile and Pancreatic Secretions Through the Duct System of the Liver, Gallbladder, and Pancreas

Histology of the Liver

A connective tissue capsule and visceral peritoneum cover the liver, except for the **bare area,** a small area on the diaphragmatic surface that lacks a visceral peritoneum and is surrounded by the coronary ligament (see figure 24.17*c*). At the porta, the connective tissue capsule sends a branching network of septa (walls) into the substance of the liver to provide its main support. Vessels, nerves, and ducts follow the connective tissue branches throughout the liver.

The connective tissue septa divide the liver into hexagon-shaped **lobules** with a **portal triad** at each corner (figure 24.19). The triads are so named because three vessels—the hepatic portal vein, hepatic artery, and hepatic duct—are located in them. Hepatic nerves and lymphatic vessels, often too small to be seen easily in light micrographs, are also located in these areas. A **central vein** is in the center of each lobule. Central veins of the lobules unite to form **hepatic veins,** which exit the liver on its posterior and superior surfaces and empty into the inferior vena cava (figure 24.19).

Hepatic cords radiate out from the central vein of each lobule like the spokes of a wheel. The hepatic cords are composed of **hepatocytes,** the functional cells of the liver. The spaces between the hepatic cords are blood channels called **hepatic sinusoids.** The sinusoids are lined with a very thin, irregular squamous endothelium consisting of two cell populations: (1) extremely thin, sparse **endothelial cells** and (2) **hepatic phagocytic cells,** Kupffer cells. A cleftlike lumen, the **bile canaliculus** (kan-ă-lik′ū-lŭs; little canal), lies between the cells within each cord (figure 24.19).

Hepatocytes have six major functions: (1) bile production, (2) storage, (3) interconversion of nutrients, (4) detoxification, (5) phagocytosis, and (6) synthesis of blood components. Nutrient-rich, deoxygenated blood from the viscera enters the hepatic sinusoids from branches of the hepatic portal vein and mixes with nutrient-depleted, oxygenated blood from the hepatic arteries (figure 24.20). From the blood, the hepatocytes can take up the oxygen and nutrients, which are stored, detoxified, used for energy, or used to synthesize new molecules. Molecules produced by or modified in the hepatocytes are released into the hepatic sinusoids or into the bile canaliculi.

Mixed blood in the hepatic sinusoids flows to the central vein, where it exits the lobule and then exits the liver through the hepatic veins. **Bile,** produced by the hepatocytes and consisting primarily of metabolic by-products, flows through the bile canaliculi toward the hepatic triad and exits the liver through the hepatic ducts. Blood, therefore, flows from the triad toward the center of each lobule, whereas bile flows away from the center of the lobule toward the triad.

In the fetus, special blood vessels bypass the liver sinusoids. The remnants of fetal blood vessels can be seen in the adult as the round ligament (ligamentum teres) and the ligamentum venosum (see chapter 29).

Functions of the Liver

The liver performs important digestive and excretory functions, stores and processes nutrients, detoxifies harmful chemicals, and synthesizes new molecules.

Bile Production

The liver produces and secretes about 600–1000 mL of bile each day (see table 24.2). Bile contains no digestive enzymes, but it plays a

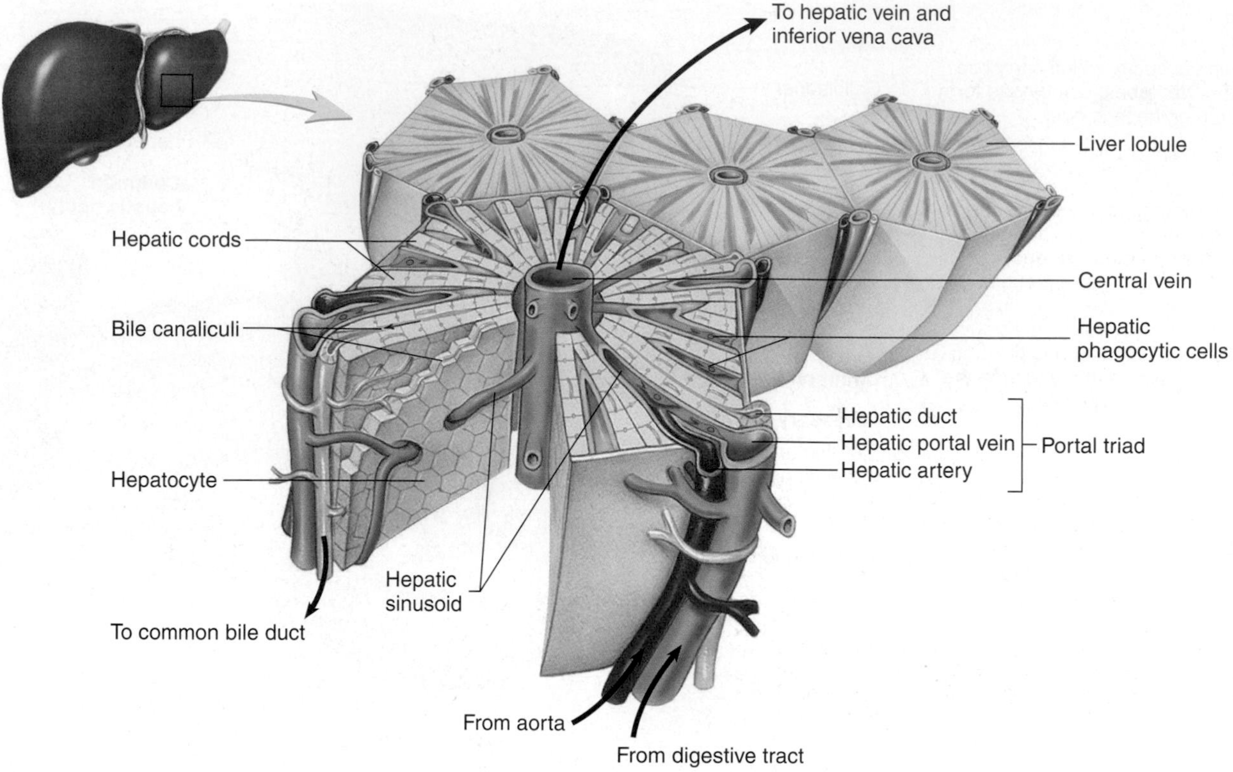
To hepatic vein and
inferior vena cava
Liver lobule
Hepatic cords
Central vein
Bile canaliculi
Hepatic
phagocytic cells
Hepatic duct
Hepatic portal vein
Hepatic artery
Portal triad
Hepatocyte
Hepatic
sinusoid
To common bile duct
From aorta
From digestive tract

FIGURE 24.19 APIR **Histology of the Liver**
The liver consists of hexagon-shaped lobules with a portal triad at each corner. A central vein is located in the center of each lobule.

role in digestion because it neutralizes and dilutes stomach acid and emulsifies lipids. The pH of chyme as it leaves the stomach is too low for the normal function of pancreatic enzymes. Bile helps neutralize the acidic chyme and bring the pH up to a level at which pancreatic enzymes can function. **Bile salts** emulsify lipids. Bile also contains excretory products, such as bile pigments; one bile pigment is bilirubin, which results from the breakdown of hemoglobin. Bile also contains cholesterol, lipids, lipid-soluble hormones, and lecithin.

Neural and hormonal stimuli regulate the secretion and release of bile (figure 24.21). Bile secretion by the liver is increased by parasympathetic stimulation through the vagus nerve. Secretin, which is released from the duodenum, also stimulates bile secretion, primarily by increasing the water and bicarbonate ion content of bile. Cholecystokinin stimulates gallbladder contractions to release bile into the duodenum.

Bile salts also increase bile secretion through a positive-feedback system. Over 90% of bile salts are reabsorbed in the ileum and carried in the blood by the hepatic portal circulation. Upon their return to the liver, the bile salts stimulate further bile secretion and are once again secreted into the bile. This recycling process reduces the loss of bile salts in the feces. Bile secretion into the duodenum continues until the duodenum empties.

Storage

Hepatocytes can remove sugar from the blood and store it in the form of **glycogen.** They can also store lipids, vitamins (A, B_{12}, D, E, and K), copper, and iron. This storage function is usually short-term, and the amount of stored material in the hepatocytes—hence their size—fluctuates during the day.

Hepatocytes help maintain blood sugar levels within very narrow limits. If a large amount of sugar enters the general circulation after a meal, the blood osmolality will increase, resulting in hyperglycemia. Under normal conditions, this is prevented because the blood from the small intestine passes through the hepatic portal vein to the liver, where hepatocytes remove glucose and other substances from the blood, store them, and then secrete them back into the circulation when needed.

Nutrient Interconversion

The interconversion of nutrients is another important function of the liver. Ingested nutrients are not always present in the proportion needed by the tissues. In this case, the liver can convert some nutrients into others. For example, if a person is on a diet that is excessively high in protein, an oversupply of amino acids and an undersupply of lipids and carbohydrates may be delivered to the liver. The hepatocytes break down the amino acids and cycle many of them through metabolic pathways, so that they can be used to produce adenosine triphosphate, lipids, and glucose (see chapter 25).

Hepatocytes also transform substances that cannot be used by most cells into more readily usable substances. For example, they combine ingested dietary fats with choline and phosphorus in the liver to produce phospholipids, which are essential components of

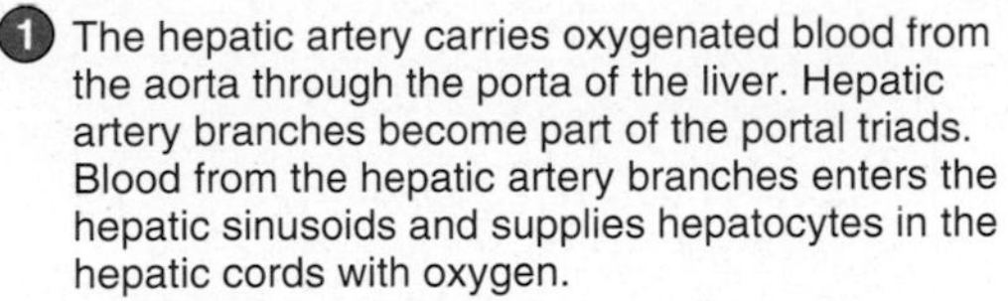
Heart
Inferior vena cava
Aorta
Hepatic veins
Central vein
3
Hepatic cords composed of hepatocytes
Hepatic duct
Hepatic portal vein
Hepatic artery
Portal triad
Liver lobule
4
Bile canaliculi
Hepatic sinusoid
Portal triad
Hepatic duct branch
Hepatic portal vein branch
Hepatic artery branch
Porta of liver
Hepatic ducts
Hepatic portal vein
Hepatic artery
5
2
1
Bile
Nutrient-rich, deoxygenated blood
Oxygenated blood
Oxygenated blood
Small intestine
Hepatic ducts
Hepatic portal vein
Hepatic artery
Inferior vena cava
Inferior view

1. The hepatic artery carries oxygenated blood from the aorta through the porta of the liver. Hepatic artery branches become part of the portal triads. Blood from the hepatic artery branches enters the hepatic sinusoids and supplies hepatocytes in the hepatic cords with oxygen.

2. The hepatic portal vein carries nutrient-rich, deoxygenated blood from the intestines through the porta of the liver. Hepatic portal vein branches become part of the portal triads. Blood from the hepatic portal vein branches enters the hepatic sinusoids and supplies hepatocytes in the hepatic cords with nutrients.

3. Blood in the hepatic sinusoids that comes from the hepatic artery and hepatic portal vein picks up plasma proteins, processed molecules, and waste products produced by the hepatocytes of the hepatic cords. The hepatic sinusoids empty into central veins. The central veins connect to hepatic veins, which connect to the inferior vena cava.

4. Bile produced by hepatocytes in the hepatic cords enters bile canaliculi, which connect to hepatic duct branches that are part of the portal triads.

5. The hepatic duct branches converge to form the left and right hepatic ducts, which carry bile out the porta of the liver.

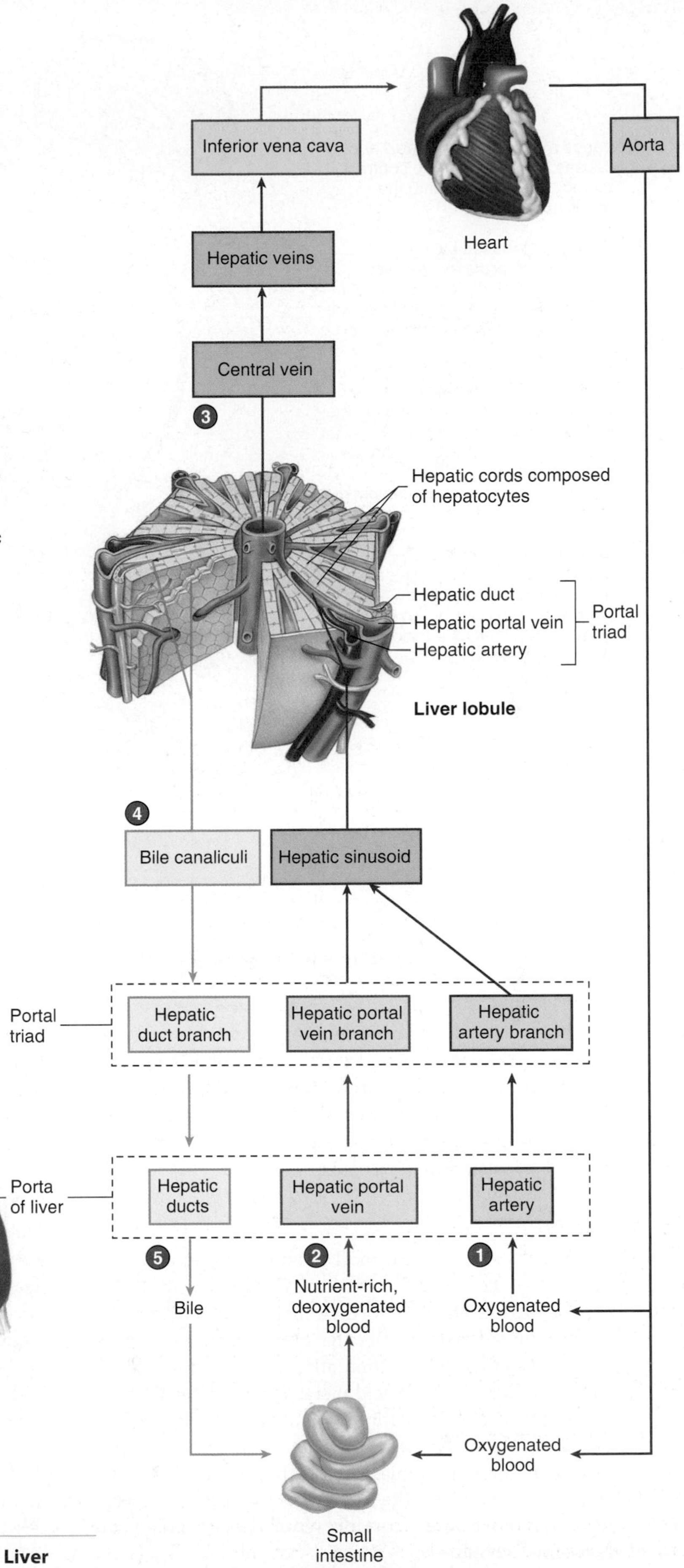

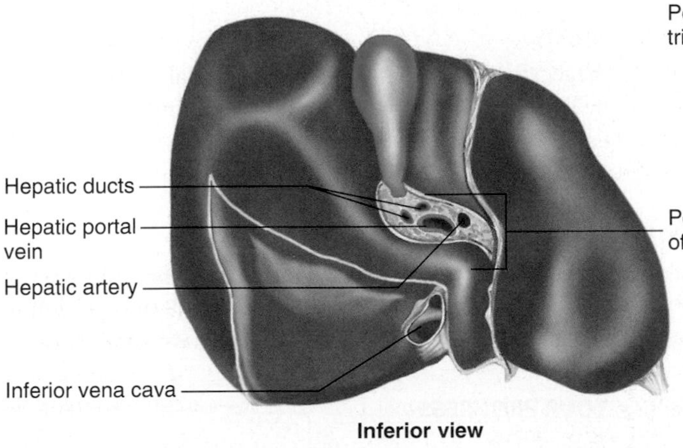

PROCESS FIGURE 24.20 Blood and Bile Flow Through the Liver

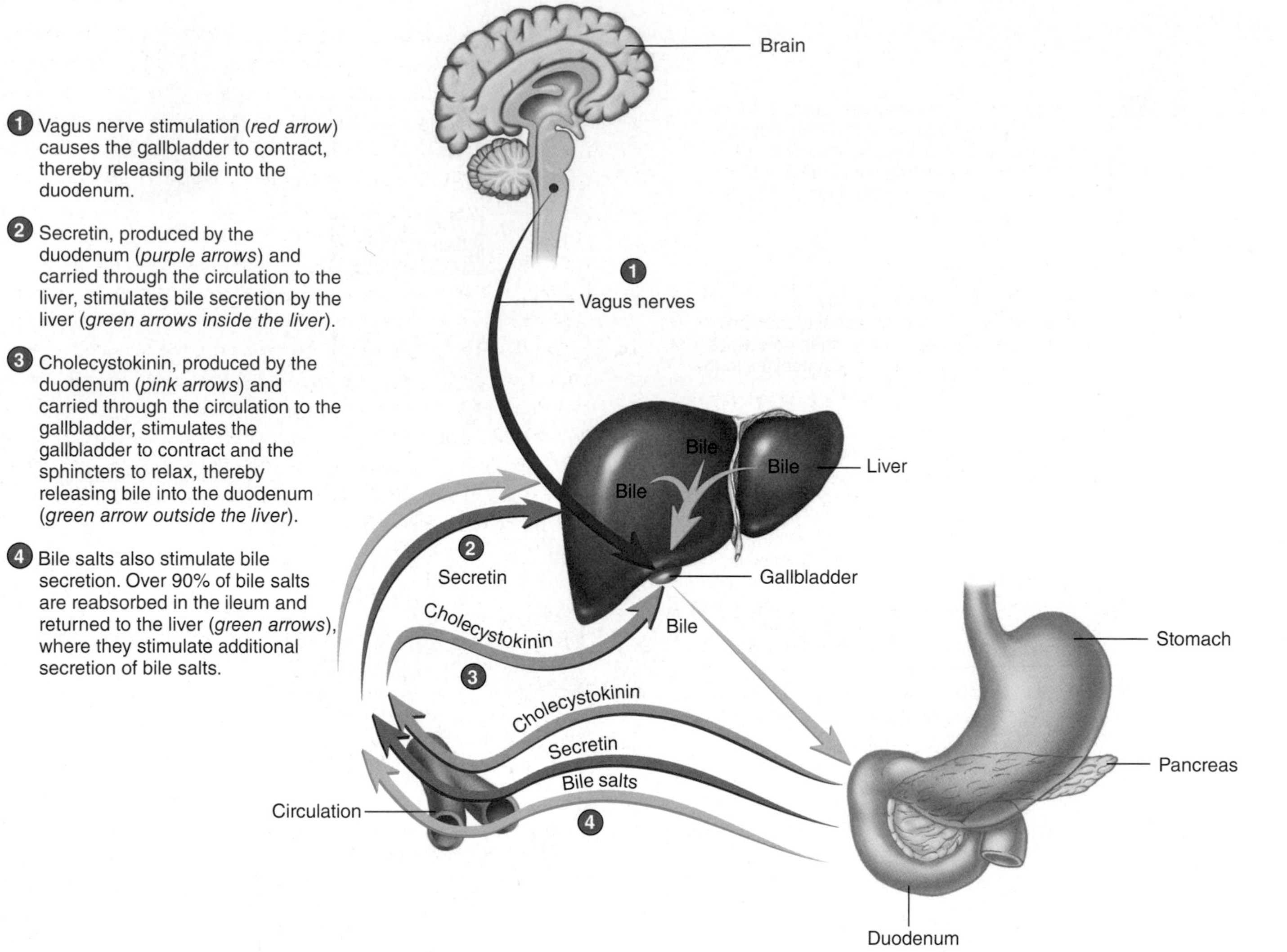

1. Vagus nerve stimulation (*red arrow*) causes the gallbladder to contract, thereby releasing bile into the duodenum.
2. Secretin, produced by the duodenum (*purple arrows*) and carried through the circulation to the liver, stimulates bile secretion by the liver (*green arrows inside the liver*).
3. Cholecystokinin, produced by the duodenum (*pink arrows*) and carried through the circulation to the gallbladder, stimulates the gallbladder to contract and the sphincters to relax, thereby releasing bile into the duodenum (*green arrow outside the liver*).
4. Bile salts also stimulate bile secretion. Over 90% of bile salts are reabsorbed in the ileum and returned to the liver (*green arrows*), where they stimulate additional secretion of bile salts.

PROCESS FIGURE 24.21 Control of Bile Secretion and Release

plasma membranes. In addition, vitamin D is hydroxylated in the liver hepatocytes. The hydroxylated form of vitamin D, which is the major circulating form of vitamin D, is transported through the circulation to the kidneys, where it is again hydroxylated. The double-hydroxylated vitamin D is the active form of the vitamin, which functions in calcium maintenance.

Detoxification

Many ingested substances are harmful to body cells. In addition, the body itself produces many by-products of metabolism that, if accumulated, are toxic. The liver forms a major line of defense by altering the structure of many of these harmful substances to make them less toxic or to make their elimination easier. Ammonia, for example, a by-product of amino acid metabolism, is toxic and not readily removed from the circulation by the kidneys. Hepatocytes remove ammonia from the circulation and convert it to urea, which is less toxic than ammonia. Urea is then secreted into the circulation and eliminated by the kidneys in the urine. The liver hepatocytes also remove other substances from the circulation and excrete them into the bile.

Phagocytosis

Hepatic phagocytic cells (Kupffer cells), which lie along the sinusoid walls of the liver, phagocytize "worn-out" and dying red and white blood cells, some bacteria, and other debris that enters the liver through the circulation.

Synthesis

The liver can produce its own new compounds, including blood proteins such as albumins, fibrinogen, globulins, heparin, and clotting factors, which are released into the circulation (see chapter 19).

ASSESS YOUR PROGRESS

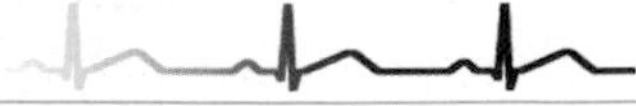

43. *Describe the lobes of the liver. What is the porta?*

44. *Diagram the duct system from the liver, gallbladder, and pancreas that empties into the major duodenal papilla.*

45. *Describe the flow of blood to and through the liver. Describe the flow of bile away from the liver.*

46. *Explain and give examples of the major functions of the liver.*

47. *What stimulates bile secretion from the liver?*

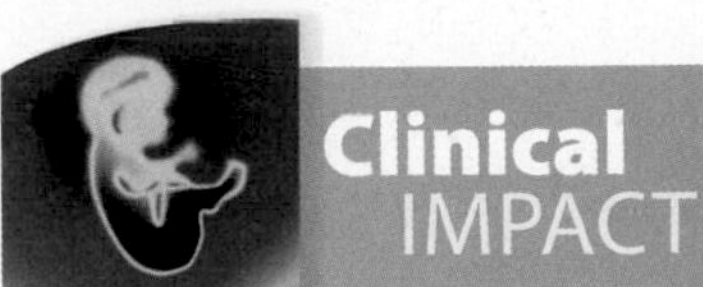
Clinical
IMPACT

Hepatitis, Cirrhosis, and Liver Damage

Hepatitis is an inflammation of the liver that can be caused by alcohol consumption or a viral infection. Hepatitis A, also called infectious hepatitis, is responsible for about 30% of all hepatitis cases in the United States. Hepatitis B, also called serum hepatitis, is a more chronic infection and is responsible for half the hepatitis cases in the United States. The remainder of U.S. cases (about 20%) are hepatitis C, also called non-A and non-B hepatitis. Hepatitis C is caused by one or more virus types that cannot be identified in blood tests, and it is spread by blood transfusions or sexual intercourse. If hepatitis is not treated, liver cells die and are replaced by scar tissue, resulting in loss of liver function. Liver failure and death can occur.

Cirrhosis (sir-rō′sis) of the liver involves the death of hepatocytes and their replacement by fibrous connective tissue. The liver becomes pale in color (the term *cirrhosis* means a tawny or orange condition) because of the presence of excess white connective tissue. It also becomes firmer, and the surface becomes nodular. The buildup of connective tissue can impede blood flow through the liver. The loss of hepatocytes eliminates the function of the liver, often resulting in jaundice, in which the skin and eyes appear yellowish due to buildup of bile pigments in the blood and interstitial fluid. Cirrhosis develops in many alcoholics and may result from biliary obstruction, hepatitis, or nutritional deficiencies.

Under most conditions, mature hepatocytes can proliferate and replace lost parts of the liver. However, if the liver is severely damaged, the hepatocytes may not have enough regenerative power to replace the lost parts. In this case, a liver transplant may be necessary. The liver also maintains an undifferentiated stem cell population, called "oval" cells, which gives rise to two cell lines, one forming bile duct epithelium and one producing hepatocytes. Researchers hope that these stem cells can one day be used to reconstitute a severely damaged liver. It may even become possible to remove stem cells from a person with hemophilia, genetically engineer the cells to produce the missing clotting factors, and then reintroduce the altered stem cells into the person's liver.

24.11 Gallbladder

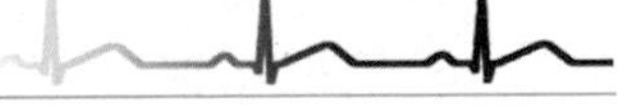

LEARNING OUTCOMES

After reading this section, you should be able to

A. Discuss the structure and function of the gallbladder.

B. State what stimulates the release of bile.

The **gallbladder** is a saclike structure on the inferior surface of the liver; it is about 8 cm long and 4 cm wide (see figure 24.18). Three tunics form the gallbladder wall: (1) an inner mucosa folded into rugae that allow the gallbladder to expand; (2) a muscularis, which is a layer of smooth muscle that allows the gallbladder to contract; and (3) an outer covering of serosa. The cystic duct connects the gallbladder to the common bile duct.

The liver continually secretes bile, which flows to the gallbladder, where 40–70 mL of bile can be stored. While the bile is in the gallbladder, water and electrolytes are absorbed, and bile salts and pigments become as much as 5–10 times more concentrated than they were when secreted by the liver. Shortly after a meal, the gallbladder contracts in response to stimulation by cholecystokinin and, to a lesser degree, in response to vagal stimulation, thereby dumping large amounts of concentrated bile into the small intestine (figure 24.21).

Cholesterol, secreted by the liver, may precipitate in the gallbladder to produce **gallstones.** Cholesterol is not soluble in water and is ordinarily kept in solution by bile salts. Gallstones can form when the bile contains excess cholesterol due to a high-cholesterol diet or due to cholesterol in the gallbladder. Occasionally, a gallstone passes out of the gallbladder and enters the cystic duct, blocking the release of bile. This condition interferes with normal digestion, and often the gallstone must be surgically removed. If the gallstone moves far enough down the duct, it can also block the pancreatic duct, resulting in pancreatitis.

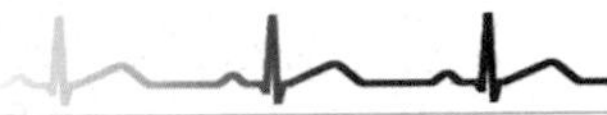

ASSESS YOUR PROGRESS

48. *Describe the three tunics of the gallbladder wall.*
49. *What is the function of the gallbladder? What stimulates the release of bile from the gallbladder?*

24.12 Pancreas

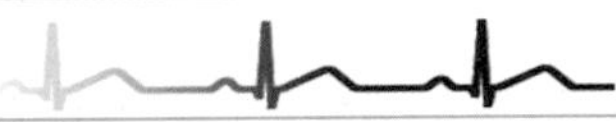

LEARNING OUTCOMES

After reading this section, you should be able to

A. Describe the anatomy, histology, and ducts of the pancreas.

B. List the secretions of the pancreas, their functions, and their regulation.

Anatomy of the Pancreas

The **pancreas** is a complex organ composed of both endocrine and exocrine tissues that perform several functions. The pancreas consists of a **head,** located within the curvature of the duodenum (figure 24.22*a*), a **body,** and a **tail,** which extends to the spleen.

The endocrine part of the pancreas consists of **pancreatic islets,** or *islets of Langerhans* (figure 24.22*b*). The islet cells produce insulin and glucagon, which are very important in controlling the blood levels of nutrients, such as glucose and amino acids, and somatostatin, which regulates insulin and glucagon secretion and may inhibit growth hormone secretion (see chapter 18).

The exocrine part of the pancreas is a compound acinar gland (see chapter 4). The **acini** (as′i-nī; figure 24.22*b*) produce digestive enzymes. Clusters of acini form lobules that are separated by thin septa. Lobules are connected by small **intercalated ducts** to **intra-lobular ducts,** which leave the lobules to join **interlobular ducts** between the lobules. The interlobular ducts attach to the main **pancreatic duct,** which joins the common bile duct at the hepatopancreatic ampulla, or *Vater's ampulla* (figure 24.22*a*; see figure 24.18). The hepatopancreatic ampulla empties into the duodenum at the major

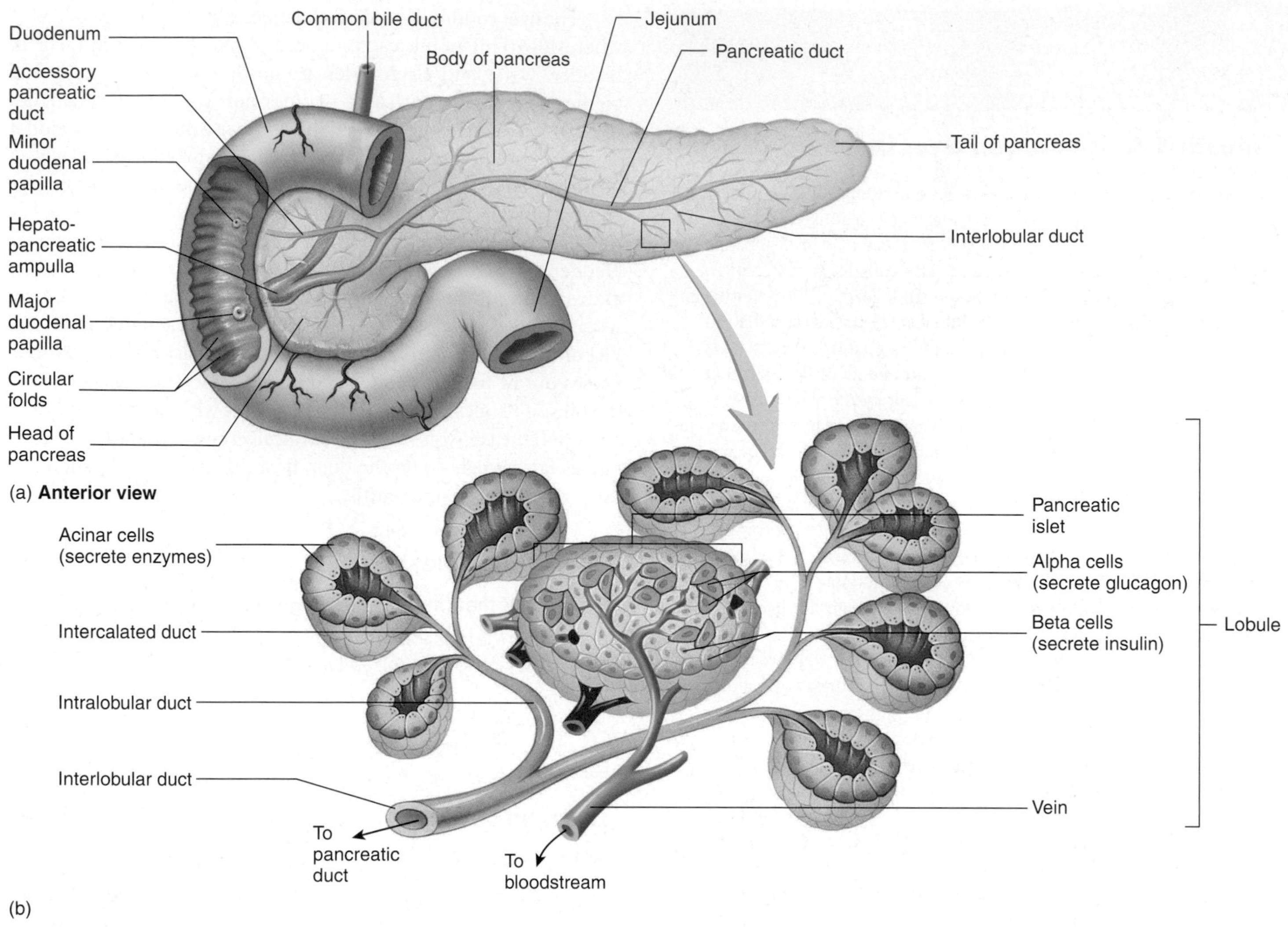
Duodenum
Accessory pancreatic duct
Minor duodenal papilla
Hepato-pancreatic ampulla
Major duodenal papilla
Circular folds
Head of pancreas
Common bile duct
Body of pancreas
Jejunum
Pancreatic duct
Tail of pancreas
Interlobular duct
(a) Anterior view
Acinar cells (secrete enzymes)
Intercalated duct
Intralobular duct
Interlobular duct
To pancreatic duct
To bloodstream
Pancreatic islet
Alpha cells (secrete glucagon)
Beta cells (secrete insulin)
Lobule
Vein
(b)

FIGURE 24.22 APR **Anatomy and Histology of the Duodenum and Pancreas**
(*a*) The head of the pancreas lies within the duodenal curvature, with the pancreatic duct emptying into the duodenum. (*b*) Histology of the pancreas, showing both the acinar cells and the pancreatic duct system.

duodenal papilla. A smooth muscle sphincter, the **hepatopancreatic ampullar sphincter,** or *sphincter of Oddi,* regulates the opening of the ampulla. In most people, an accessory pancreatic duct opens at the minor duodenal papilla. The ducts are lined with simple cuboidal epithelium, and the epithelial cells of the acini are pyramid-shaped. A smooth muscle sphincter surrounds the pancreatic duct where it enters the hepatopancreatic ampulla.

Pancreatic Secretions

The exocrine secretions of the pancreas, called **pancreatic juice,** have an aqueous component and an enzymatic component. Pancreatic juice is delivered to the small intestine through the pancreatic ducts, where it functions in digestion.

The **aqueous component** is produced principally by columnar epithelial cells that line the smaller ducts of the pancreas. It contains Na^+ and K^+ in about the same concentration found in extracellular fluid. Bicarbonate ions (HCO_3^-) are a major part of the aqueous component, and they neutralize the acidic chyme that enters the small intestine from the stomach. The increased pH caused by pancreatic secretions in the duodenum stops pepsin digestion but provides the proper environment for the function of pancreatic enzymes. Bicarbonate ions are actively secreted by the duct epithelium, and water follows passively to make the pancreatic juice isotonic.

The cellular mechanism responsible for the secretion of HCO_3^- is diagrammed in figure 24.23. The production of HCO_3^- in the pancreas is similar to the production of H^+ in the gastric gland (see figure 24.12). Both processes require the enzyme carbonic anhydrase and the exchange of HCO_3^- and Cl^-, but the final products are different. An alkaline solution is produced in the pancreas, whereas an acidic solution is produced in the stomach.

The acinar cells of the pancreas produce the **enzyme-rich** pancreatic juice. The enzymes are important in digesting all major classes of food. Without the enzymes produced by the pancreas, lipids, proteins, and carbohydrates cannot be adequately digested (see tables 24.1 and 24.2).

1 Water (H_2O) and carbon dioxide (CO_2) combine under the influence of carbonic anhydrase (CA) to form carbonic acid (H_2CO_3).

2 Carbonic acid dissociates to form hydrogen ions (H^+) and bicarbonate ions (HCO_3^-).

3 The H^+ is exchanged for sodium ions (Na^+) by an antiporter. Sodium ions are removed by the Na^+–K^+ pump.

4 The HCO_3^- is transported into the intercalated ducts in exchange for Cl^-, which returns to the lumen by a channel. Sodium ions and H_2O follow the HCO_3^- into the ducts.

5 The ions and H_2O move through the intercalated duct toward the interlobular duct.

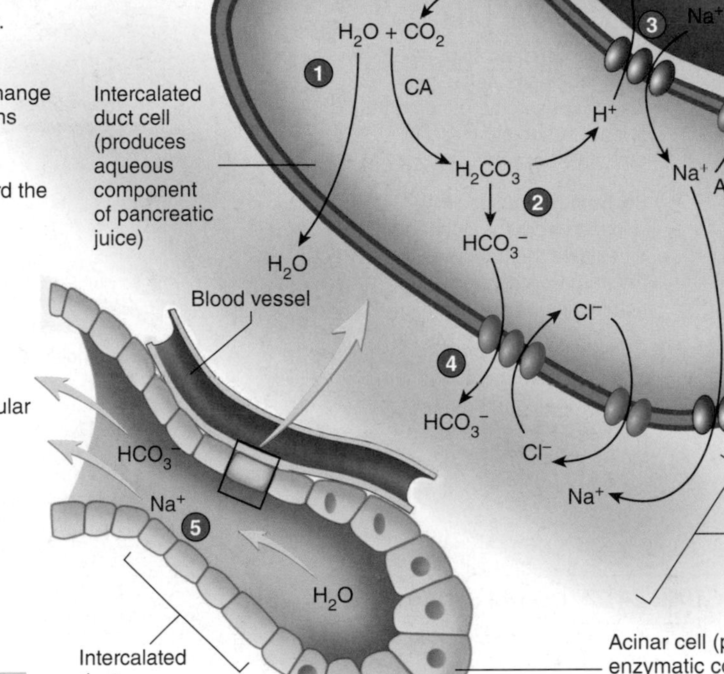

Blood vessel
H+
CO2
H2O + CO2
CA
H2CO3
H+
HCO3−
Na+
ATP
K+
ADP
Cl−
HCO3−
Cl−
Na+
Intercalated duct cell (produces aqueous component of pancreatic juice)
H2O
Blood vessel
To interlobular duct
HCO3−
Na+
H2O
Intercalated duct
Intercalated duct lumen
Acinar cell (produces enzymatic component of pancreatic juice)

PROCESS FIGURE 24.23 Bicarbonate Ion Production in the Pancreas

The proteolytic pancreatic enzymes, which digest proteins, are secreted in inactive forms, whereas many of the other enzymes are secreted in active form. The major proteolytic enzymes are **trypsin, chymotrypsin,** and **carboxypeptidase.** They are secreted in their inactive forms as trypsinogen, chymotrypsinogen, and procarboxypeptidase and are activated by the removal of certain peptides from the larger precursor proteins. If these were produced in their active forms, they would digest the tissues producing them. The proteolytic enzyme **enterokinase** (en′tĕr-ō-kī′nās), which is attached to the brush border of the small intestine, activates trypsinogen. Trypsin then activates more trypsinogen, as well as chymotrypsinogen and procarboxypeptidase.

Pancreatic juice also contains **pancreatic amylase,** which continues the polysaccharide digestion initiated in the oral cavity. In addition, pancreatic juice contains a lipid-digesting enzyme called **pancreatic lipase,** which breaks down lipids into monoglycerides and free fatty acids. Also present in pancreatic juice are enzymes that degrade DNA and RNA to their component nucleotides, **deoxyribonucleases** and **ribonucleases,** respectively.

Pancreatitis is an inflammation of the pancreas involving the release of pancreatic enzymes within the pancreas itself, which digest pancreatic tissue. Pancreatitis can result from alcoholism, the use of certain drugs, pancreatic duct blockage, cystic fibrosis, viral infection, or pancreatic cancer. Symptoms can range from mild abdominal pain to systemic shock and coma.

Regulation of Pancreatic Secretion

Both hormonal and neural mechanisms control the exocrine secretions of the pancreas (figure 24.24). An acidic chyme in the duodenum stimulates the release of secretin. Secretin stimulates the secretion of a watery, bicarbonate ion–rich solution from the pancreas into the duodenum. Bicarbonate ions increase the pH of chyme in the duodenum, so that the duodenum is not damaged by acidic pH. In addition, pancreatic and brush border enzymes do not function at a low pH.

Predict 6

Explain why secretin production in response to acidic chyme and its stimulation of bicarbonate ion secretion constitute a negative-feedback mechanism.

Cholecystokinin stimulates the release of bile from the gallbladder and the secretion of pancreatic juice rich in digestive enzymes. As mentioned earlier, the major stimulus for the release of cholecystokinin is the presence of fatty acids and other lipids in the duodenum.

Parasympathetic stimulation through the vagus (X) nerves also stimulates the secretion of pancreatic juices rich in pancreatic enzymes, and sympathetic impulses inhibit secretion. The effect of vagal stimulation on pancreatic juice secretion is greatest during the cephalic and gastric phases of stomach secretion.

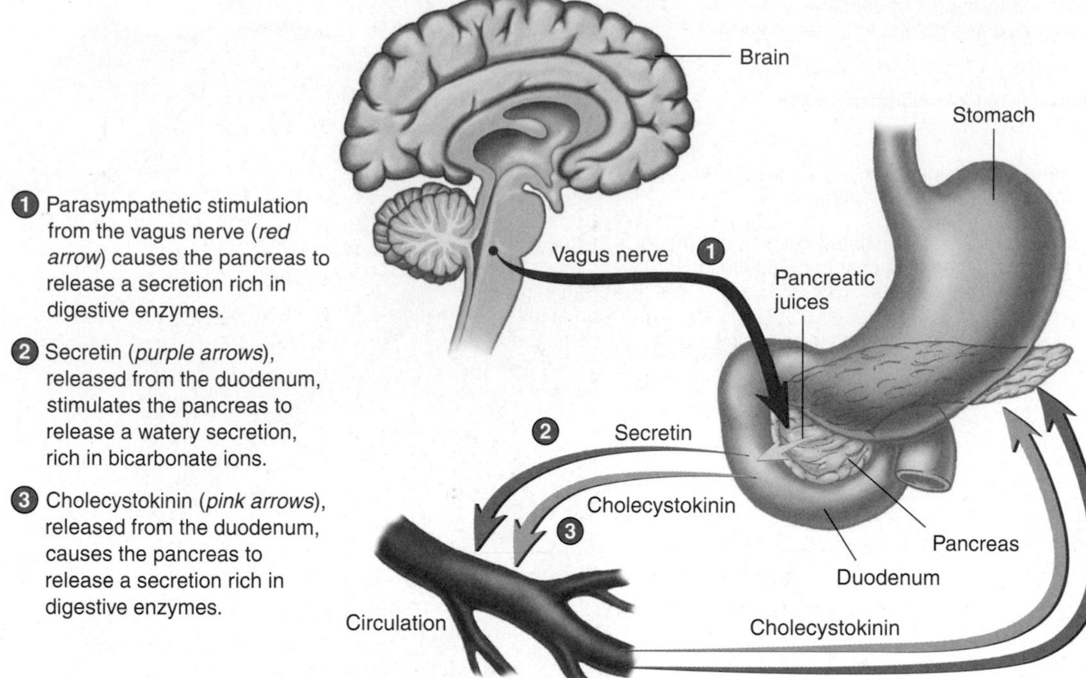

1 Parasympathetic stimulation from the vagus nerve (red arrow) causes the pancreas to release a secretion rich in digestive enzymes.
2 Secretin (purple arrows), released from the duodenum, stimulates the pancreas to release a watery secretion, rich in bicarbonate ions.
3 Cholecystokinin (pink arrows), released from the duodenum, causes the pancreas to release a secretion rich in digestive enzymes.
Brain
Stomach
Vagus nerve 1
Pancreatic juices
2 Secretin
Cholecystokinin
3
Pancreas
Duodenum
Circulation
Cholecystokinin
Secretin

PROCESS FIGURE 24.24 **Control of Pancreatic Secretion**

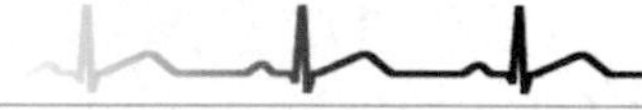

ASSESS YOUR PROGRESS

50. *Describe the parts of the pancreas responsible for endocrine and exocrine secretions. Diagram the duct system of the pancreas.*

51. *Name the two kinds of exocrine secretions produced by the pancreas. What stimulates their production, and what is their function?*

52. *What enzymes are present in pancreatic juice? Explain the function of each.*

24.13 Large Intestine

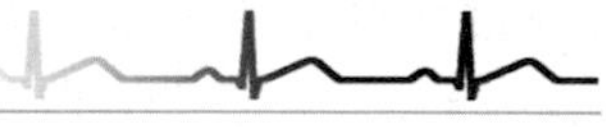

LEARNING OUTCOMES

After reading this section, you should be able to

A. List the parts of the large intestine and describe its anatomy and histology.

B. Describe the major functions of the large intestine and explain how defecation is regulated.

The **large intestine** is the portion of the digestive tract extending from the ileocecal junction to the anus. It consists of the cecum, colon, rectum, and anal canal. Normally, 18–24 hours are required for material to pass through the large intestine, in contrast to the 3–5 hours required for chyme to move through the small intestine. Thus, the movements of the colon are more sluggish than those of the small intestine. While in the colon, chyme is converted to feces. The formation of feces involves the absorption of water and salts, secretion of mucus, and extensive action of microorganisms. The colon stores the feces until they are eliminated by defecation. About 1500 mL of chyme enter the cecum each day, but more than 90% of the volume is reabsorbed, so that only 80–150 mL of feces are normally eliminated by defecation.

Anatomy of the Large Intestine

Cecum

The **cecum** (sē′kŭm) is the proximal end of the large intestine, where it meets the small intestine at the ileocecal junction. The cecum extends inferiorly about 6 cm past the ileocecal junction in the form of a blind sac (figure 24.25). Attached to the cecum is a smaller, blind tube about 9 cm long called the **vermiform** (ver′mi-fōrm; worm-shaped) **appendix.** The walls of the appendix contain many lymphatic nodules, which contribute to immune functions.

Colon

The **colon** (kō′lon), about 1.5–1.8 m long, consists of four parts: the ascending colon, transverse colon, descending colon, and sigmoid colon (figure 24.25). The **ascending colon** extends superiorly from the cecum and ends at the right colic flexure (hepatic flexure) near the right inferior margin of the liver. The **transverse colon** extends from the right colic flexure to the left colic flexure (splenic flexure), and the **descending colon** extends from the left colic flexure to the superior opening of the true pelvis, where it becomes the sigmoid colon. The **sigmoid colon** forms an S-shaped tube that extends into the pelvis and ends at the rectum.

The circular muscle layer of the colon is complete, but the longitudinal muscle layer is incomplete. Rather than completely enveloping the intestinal wall, the longitudinal layer forms three bands, called the **teniae coli** (tē′nē-ē kō′lī), that run the length of the colon (figure 24.26*a*; see figure 24.25). Contractions of the teniae coli cause pouches called **haustra** (haw′stră; to draw up) to form along the length of the colon, giving it a puckered appearance. Small, lipid-filled connective tissue pouches called **omental appendages** are attached to the outer surface of the colon along its length.

The mucosal lining of the large intestine consists of simple columnar epithelium. This epithelium is not formed into folds or villi like that of the small intestine but has numerous, straight, tubular glands called **crypts** (figure 24.26*b–d*). The crypts, which are somewhat similar to the intestinal glands of the small intestine, are composed of three cell types—absorptive, goblet, and granular cells. The major difference is that, in the large intestine, goblet cells predominate, and the other two cell types are greatly reduced in number.

Rectum

The **rectum** is a straight, muscular tube that begins at the termination of the sigmoid colon and ends at the anal canal (see figure 24.25). The mucosal lining of the rectum is simple columnar epithelium, and the muscular tunic is relatively thick, compared with the rest of the digestive tract.

Anal Canal

The last 2–3 cm of the digestive tract is the **anal canal** (see figure 24.25). It begins at the inferior end of the rectum and ends at the **anus** (external digestive tract opening). The smooth muscle layer of the anal canal is even thicker than that of the rectum and forms the **internal anal sphincter** at its superior end. Skeletal muscle forms the **external anal sphincter** at the inferior end of the canal. The epithelium of the superior part of the anal canal is simple

Clinical
IMPACT

Appendicitis

Appendicitis is an inflammation of the vermiform appendix that usually occurs because of an obstruction of the appendix. Secretions from the appendix cannot pass the obstruction and accumulate, resulting in enlargement and pain. Bacteria in the area can cause infection of the appendix. Symptoms include sudden abdominal pain, particularly in the right-lower portion of the abdomen; slight fever; loss of appetite; constipation or diarrhea; nausea; and vomiting. In the right-lower quadrant of the abdomen, about one-third the distance along a line from the right anterior superior iliac spine to the umbilicus, is an area called the McBurney point. This area of the body surface becomes very tender in patients with acute appendicitis because of pain referred from the inflamed appendix. Each year, 500,000 people in the United States experience appendicitis. The usual treatment is surgical removal of the appendix, called an appendectomy. If the appendix bursts, the infection can spread throughout the peritoneal cavity, causing peritonitis, with life-threatening results.

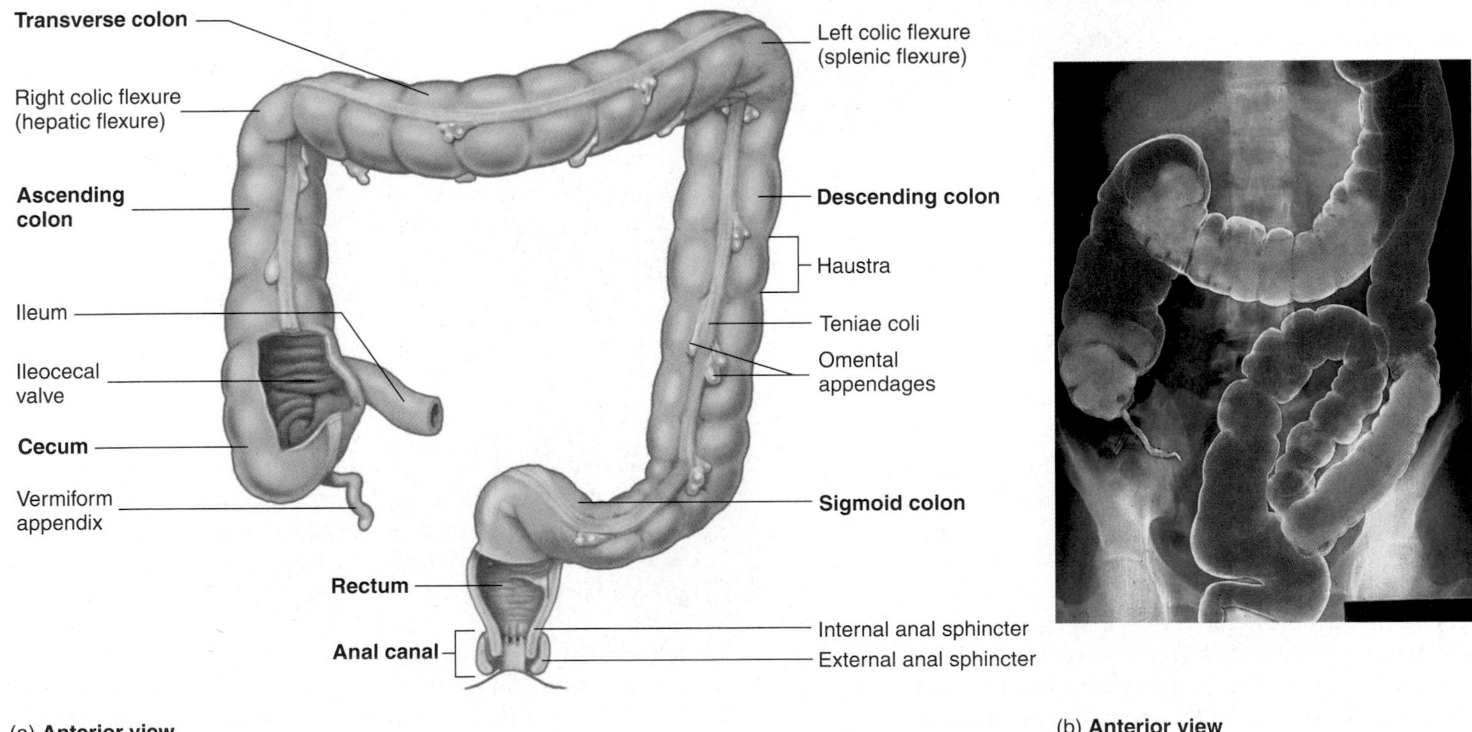
Transverse colon
Right colic flexure (hepatic flexure)
Ascending colon
Ileum
Ileocecal valve
Cecum
Vermiform appendix
Rectum
Anal canal
Left colic flexure (splenic flexure)
Descending colon
Haustra
Teniae coli
Omental appendages
Sigmoid colon
Internal anal sphincter
External anal sphincter
(a) Anterior view
(b) Anterior view

FIGURE 24.25 APR **Large Intestine**

(*a*) The large intestine consists of the cecum, colon, rectum, and anal canal. The teniae coli and omental appendages are along the length of the colon.
(*b*) Radiograph of the large intestine following a barium enema.

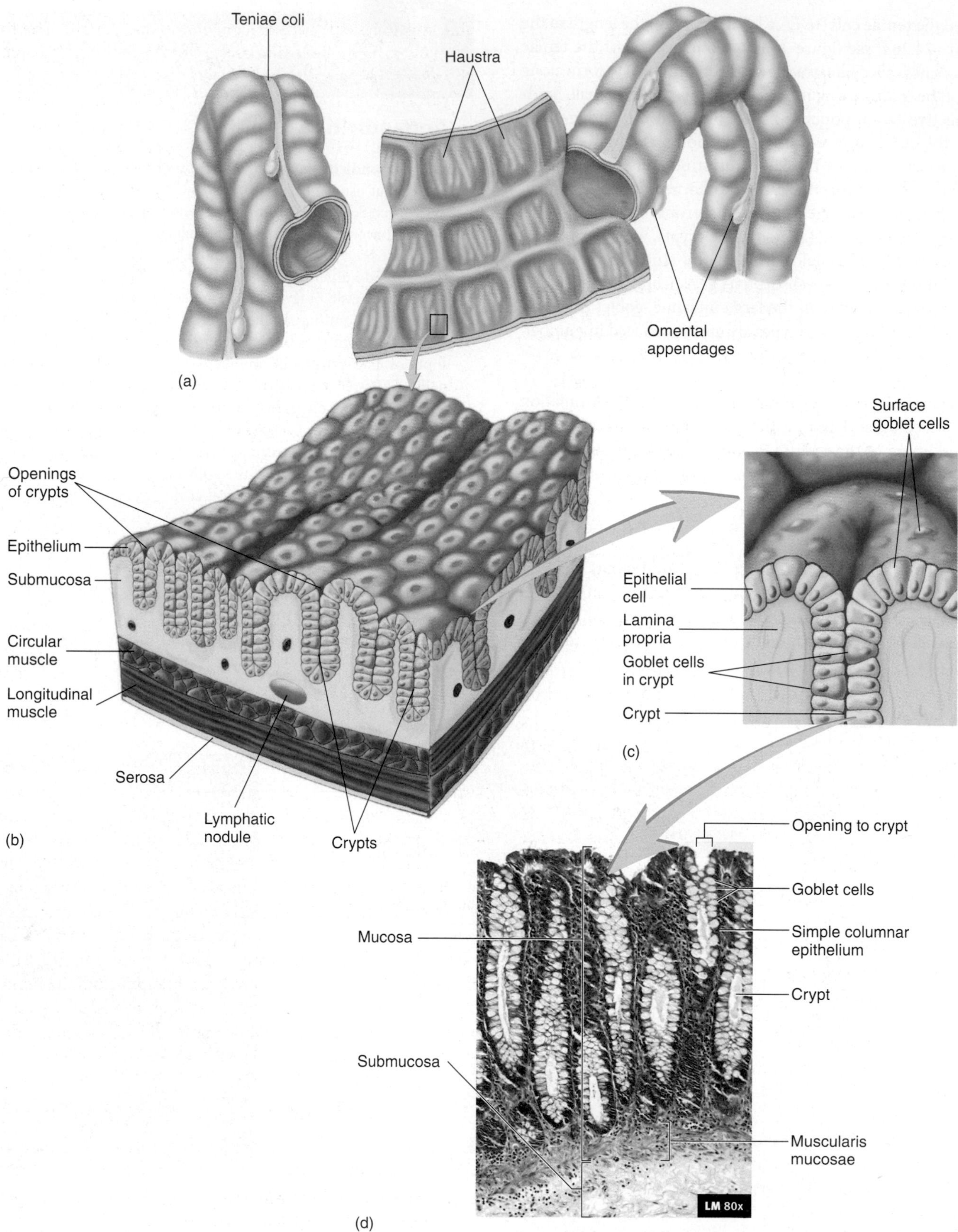
Teniae coli
Haustra
Omental appendages
(a)
Openings of crypts
Epithelium
Submucosa
Circular muscle
Longitudinal muscle
Serosa
Lymphatic nodule
Crypts
(b)
Surface goblet cells
Epithelial cell
Lamina propria
Goblet cells in crypt
Crypt
(c)
Opening to crypt
Goblet cells
Simple columnar epithelium
Crypt
Mucosa
Submucosa
Muscularis mucosae
LM 80x
(d)

FIGURE 24.26 Histology of the Large Intestine

(*a*) Section of the transverse colon cut open to show the inner surface. (*b*) Enlargement of the inner surface, showing openings of the crypts. (*c*) Higher magnification of a single crypt. (*d*) Photomicrograph showing the histology of the large intestine wall.

columnar, and that of the inferior part is stratified squamous. Rectal veins that supply the anal canal can become enlarged or inflamed, a condition known as **hemorrhoids.** Hemorrhoids cause pain, itching, and bleeding around the anus. They can usually be treated by changes in diet or medications.

Secretions of the Large Intestine

The mucosa of the colon has numerous goblet cells scattered along its length and numerous crypts lined almost entirely with goblet cells. Little enzymatic activity is associated with secretions of the colon because the major secretory product is mucus (see tables 24.1 and 24.2). Mucus lubricates the wall of the colon and helps the fecal matter stick together. Tactile stimuli and irritation of the colon wall trigger local enteric reflexes that increase mucous secretion. Parasympathetic stimulation also increases the secretory rate of the goblet cells.

A molecular pump exchanges HCO_3^- for Cl^- in epithelial cells of the colon in response to acid produced by colic bacteria. Another pump exchanges Na^+ for H^+. Water leaves the lumen of the colon through osmosis as Na^+ and Cl^- move into the epithelial cells.

The feces that leave the digestive tract consist of water, solid substances (e.g., undigested food), microorganisms, and sloughed-off epithelial cells. An abnormally frequent discharge of watery feces is called **diarrhea** (see Systems Pathology, later in this chapter).

Numerous microorganisms inhabit the colon. They reproduce rapidly and ultimately constitute about 30% of the dry weight of the feces. Some bacteria in the intestine synthesize vitamin K, which is passively absorbed in the colon, and break down a small amount of cellulose to glucose. However, the glucose cannot be absorbed in the large intestine.

Bacterial actions in the colon produce gases called **flatus** (flā′tŭs; blowing). The amount of flatus depends partly on the bacterial population in the colon and partly on the type of food consumed. For example, beans, which contain certain complex carbohydrates, are well known for their flatus-producing effect.

Movement in the Large Intestine

Segmental mixing movements occur in the colon much less often than in the small intestine. Peristaltic waves are largely responsible for moving chyme along the ascending colon. At widely spaced intervals (normally three or four times each day), large parts of the transverse and descending colon undergo several strong contractions, called **mass movements.** Each mass movement contraction extends over a much longer part of the digestive tract ($\geq$20 cm) than does a peristaltic contraction and propels the colon contents a considerable distance toward the anus (figure 24.27). Mass movements are very common after meals because they are initiated by the presence of food in the stomach or duodenum. Mass movements are most common about 15 minutes after breakfast. They usually persist for 10–30 minutes and then stop for perhaps half a day. Local reflexes in the ENS, called **gastrocolic reflexes** if initiated by the stomach, or **duodenocolic reflexes** if initiated by the duodenum, integrate mass movements.

The gastrocolic and duodenocolic reflexes promote peristalsis of the small and large intestines, including mass movements. These reflexes are mediated by parasympathetic reflexes, local reflexes, and hormones, such as cholecystokinin and gastrin. The thought or smell of food, distension of the stomach, and the movement of chyme into the duodenum can stimulate them.

During defecation, the contractions that move feces toward the anus must be coordinated with the relaxation of the internal and external anal sphincters. The movement of feces from the colon into the rectum distends the rectal wall, which stimulates the defecation reflex. The **defecation reflex** consists of local and parasympathetic reflexes (figure 24.27). Local reflexes cause weak contractions of the distal colon and rectum and relaxation of the internal anal sphincter. Parasympathetic reflexes are responsible for most of the defecation reflex. Action potentials produced in response to distension of the rectal wall travel along afferent nerve fibers to the defecation reflex center (S2–S4) in the conus medullaris of the spinal cord. Then efferent action potentials are initiated that return through nerves to the colon and rectum, reinforcing peristaltic contractions and relaxation of the internal anal sphincter.

Action potentials from the sacral spinal cord ascend to the brain, where parts of the brainstem and hypothalamus inhibit or facilitate reflex activity in the spinal cord. In addition, action potentials ascend to the cerebrum, where awareness of the need to defecate is realized. The external anal sphincter is composed of skeletal muscle and is under conscious cerebral control. If this sphincter is relaxed voluntarily, feces are expelled. On the other hand, increased contraction of the external anal sphincter prevents defecation. The defecation reflex persists for only a few minutes and quickly declines. Generally, the reflex is reinitiated after a period that may be as long as several hours. Mass movements in the colon are usually the reason for reinitiation of the defecation reflex.

Contraction of the internal and external anal sphincters prevents defecation. Resting sphincter pressure results from tonic muscle contractions, mostly of the internal anal sphincter. In response to increased abdominal pressure, reflexes mediated through the spinal cord cause contractions of the external anal sphincter. Thus, the untimely expulsion of feces during coughing or exertion is avoided.

Defecation can be initiated by voluntary actions that stimulate a defecation reflex. This "straining" includes a large inspiration of air, followed by closure of the larynx and forceful contraction of the abdominal muscles. As a consequence, the pressure in the abdominal cavity increases and forces feces into the rectum. Stretch of the rectum initiates a defecation reflex, and input from the brain overrides the reflexive contraction of the external anal sphincter stimulated by increased abdominal pressure. The increased abdominal pressure also helps push feces through the rectum.

ASSESS **YOUR PROGRESS**

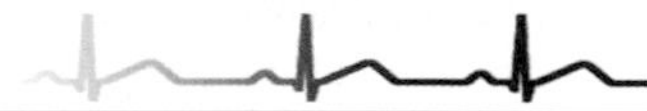

53. *Describe the parts of the large intestine. What are teniae coli, haustra, and crypts?*
54. *Explain the difference in structure between the internal anal sphincter and the external anal sphincter.*
55. *Name the substances secreted and absorbed in the large intestine.*
56. *What is the role of microorganisms in the colon?*
57. *What kinds of movements occur in the colon? Describe the defecation reflex.*

1. The thought or smell of food, distention of the stomach, and the movement of chyme into the duodenum can stimulate the gastrocolic and duodenocolic reflexes (*green arrows*).
2. The gastrocolic and duodenocolic reflexes stimulate mass movements in the colon, which propel the contents of the colon toward the rectum (*orange arrow*).
3. Distention of the rectum by feces stimulates local defecation reflexes. These reflexes cause contractions of the colon and rectum (*brown arrow*), which move feces toward the anus.
4. Local reflexes cause relaxation of the internal anal sphincter (*brown arrow*).
5. Distention of the rectum by feces stimulates parasympathetic reflexes. Action potentials are propagated to the defecation reflex center located in the spinal cord (*yellow arrow*).
6. Action potentials stimulate contraction of the colon and rectum and relaxation of the internal anal sphincter (*purple arrows*).
7. Action potentials are propagated through ascending nerve tracts to the brain (*blue arrow*).
8. Descending nerve tracts from the brain regulate the defecation reflex center (*pink arrow*).
9. Action potentials from the brain control the external anal sphincter (*purple arrow*).

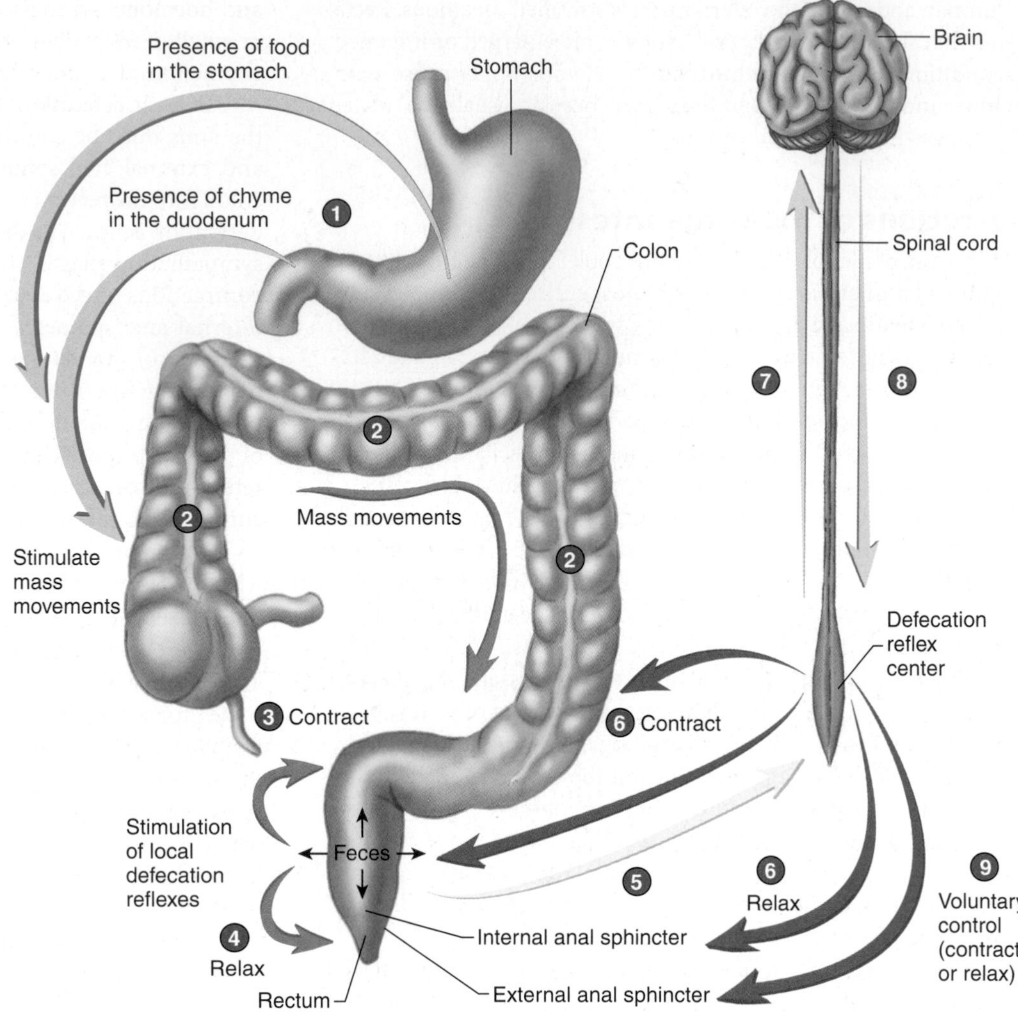
Presence of food in the stomach
Stomach
Brain
Presence of chyme in the duodenum
1
Colon
Spinal cord
7
8
2
2
Mass movements
2
Stimulate mass movements
Defecation reflex center
3 Contract
6 Contract
Stimulation of local defecation reflexes
Feces
5
6 Relax
9 Voluntary control (contract or relax)
4 Relax
Internal anal sphincter
Rectum
External anal sphincter

PROCESS FIGURE 24.27 Control of Defecation

24.14 Digestion and Absorption

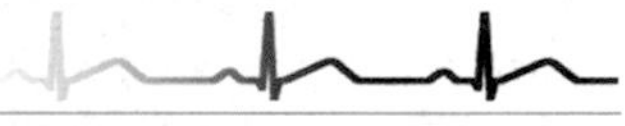

LEARNING OUTCOMES

After reading this section, you should be able to

A. Describe the chemical digestion and absorption of carbohydrates, lipids, and proteins.

B. Explain the transport of water and ions through the intestinal wall.

Digestion is the breakdown of food to molecules small enough to be absorbed into the circulation. **Mechanical digestion** breaks large food particles into smaller ones. **Chemical digestion** is the breaking of covalent chemical bonds in organic molecules by digestive enzymes. Carbohydrates break down into monosaccharides, lipids break down into fatty acids and monoglycerides, and proteins break down into amino acids. However, some molecules (e.g., vitamins, minerals, and water) are not broken down. Digestion begins in the oral cavity and continues in the stomach, but most digestion occurs in the proximal end of the small intestine, especially in the duodenum.

Absorption is the means by which molecules are moved out of the digestive tract into the circulation for distribution throughout the body. The absorption of certain molecules can occur all along the digestive tract. A few chemicals, such as nitroglycerin, can be absorbed through the thin mucosa of the oral cavity below the tongue. Some small molecules (e.g., alcohol and aspirin) can diffuse through the stomach epithelium into the circulation. Although some absorption occurs in the ileum, most takes place in the duodenum and jejunum.

Some molecules can be absorbed by diffusion, whereas others must be transported across the intestinal wall. Transport requires transport proteins, which work by facilitated diffusion, active-transport, or secondary-transport mechanisms, such as symport and antiport. The epithelial cells that form the intestinal wall have two distinct sides with different transport proteins on each side. The side that faces the digestive tract lumen is called the **apical membrane,** and the side that faces the circulation is called the **basolateral membrane.** The transport proteins in these membranes are responsible for the one-way movement of molecules from the digestive tract to the rest of the body.

Case STUDY — Spinal Cord Injury and Defecation

Dan, a 17-year-old male, was driving home late at night after a ski trip when he missed a sharp curve and crashed. He suffered traumatic injury at the T11 level of the spinal cord, with complete paralysis of both lower limbs. As a result, Dan became incontinent and unable to control his bowel movements.

Approximately 10,000 new spinal cord injuries occur per year in the United States. About 80% of those injuries involve men, usually in their late teens or twenties. The most common cause is motor vehicle accidents, followed by violence, falls, and sports.

Loss of the ability to control defecation affects the quality of life of most spinal cord injury patients, at least temporarily. The spinal cord is required for a normal defecation reflex and for voluntary control of the external anal sphincter (figure 24.27). In terms of their effect on defecation, spinal cord injuries can be divided into two groups: injuries that occur above the conus medullaris and those that damage the conus medullaris where the defecation reflex center is located. Immediately following a spinal cord injury, loss of reflexes below the level of the injury, called **spinal shock,** occurs. However, the reflexes usually become functional again, and the defecation reflex may be depressed for a few weeks but eventually returns.

Predict 7

Explain how an enema stimulates defecation.

Once the digestive products have been absorbed, they are transported to other parts of the body by two routes. Water, ions, and water-soluble digestion products, such as glucose and amino acids, enter the hepatic portal system and travel to the liver. The products of lipid metabolism are coated with proteins and transported into lymphatic capillaries called lacteals (see figure 24.16*c,d*). The lacteals are connected by lymphatic vessels to the thoracic duct (see chapter 22), which empties into the left subclavian vein. The protein-coated lipid products then travel in the circulation to adipose tissue or to the liver.

Carbohydrates

Ingested **carbohydrates** consist primarily of polysaccharides, such as starches; disaccharides, such as sucrose (table sugar) and lactose (milk sugar); and monosaccharides, such as glucose and fructose (the sugar found in many fruits). During digestion, polysaccharides break down into smaller chains and finally into disaccharides and monosaccharides. Disaccharides break down into monosaccharides. A minor amount of carbohydrate digestion begins in the oral cavity with the partial digestion of starches by **salivary amylase** (am′il-ās). Digestion continues in the stomach until the food is well mixed with acid, which inactivates salivary amylase. Carbohydrate digestion is resumed in the small intestine by **pancreatic amylase** (figure 24.28). A series of **disaccharidases** that are bound to the microvilli of the intestinal epithelium digest disaccharides into monosaccharides. The major monosaccharide is glucose.

The monosaccharides glucose and galactose are taken up into intestinal epithelial cells by symport, powered by a Na^+ gradient (figure 24.29). The Na^+ gradient is generated by the **sodium-potassium pump** located on the basolateral membrane. Diffusion of Na^+ down its concentration gradient provides the energy to transport glucose or galactose across the plasma membrane. In contrast to glucose and galactose, the monosaccharide fructose is taken up by facilitated diffusion. Once inside the intestinal epithelial cell, monosaccharides are transported by facilitated diffusion to the capillaries of the intestinal villi and carried by the hepatic portal system to the liver, where the nonglucose sugars are converted to glucose. Glucose enters the cells through facilitated diffusion. The rate of glucose transport into most types of cells is greatly influenced by **insulin** and may increase 10-fold in its presence (see chapter 18).

Clinical IMPACT — Lactose Intolerance

Lactose intolerance is the inability to digest the lactose in milk and other dairy products. Adults in most of the world are lactose intolerant, although infants are not. Why can infants digest milk, whereas their parents cannot? The reason is that many adults lack the enzyme lactase. Lactase, present on the surface of absorptive cells in the intestinal mucosa, digests the disaccharide lactose down to two monosaccharides. Lactase is made at birth but is no longer synthesized after about age 6 in 5–15% of Europeans and 80–90% of Africans and Asians. Therefore, these people can no longer digest lactose. The major exceptions are people of northern European ancestry and some pastoral nomadic tribes in Africa and the Middle East. In these populations, a mutation in the promoter (see chapter 3) of the lactase gene permits the continued expression of lactase into adulthood. Normally, lactase production stops because the promoter is "turned off" as the infant ages, but the mutation allows the promoter to ignore this developmental switch.

Researchers believe that the dietary reliance on milk and milk products in some societies provided a selective advantage for lactase persistence. In the United States, most people are lactose tolerant, but intolerance is still one of the most common digestive tract disorders seen by primary care physicians. The main symptom of lactose intolerance is diarrhea due to fluid loss as water follows lactose through the digestive tract. In addition, a considerable amount of gas is generated from lactose metabolism by bacteria in the large intestine. Even though these colonic bacteria metabolize lactose to monosaccharides, it is too late for the monosaccharides to be absorbed. Gene therapy has proven successful in animal models of lactose intolerance, although at present the best treatment is simply to avoid foods containing lactose.

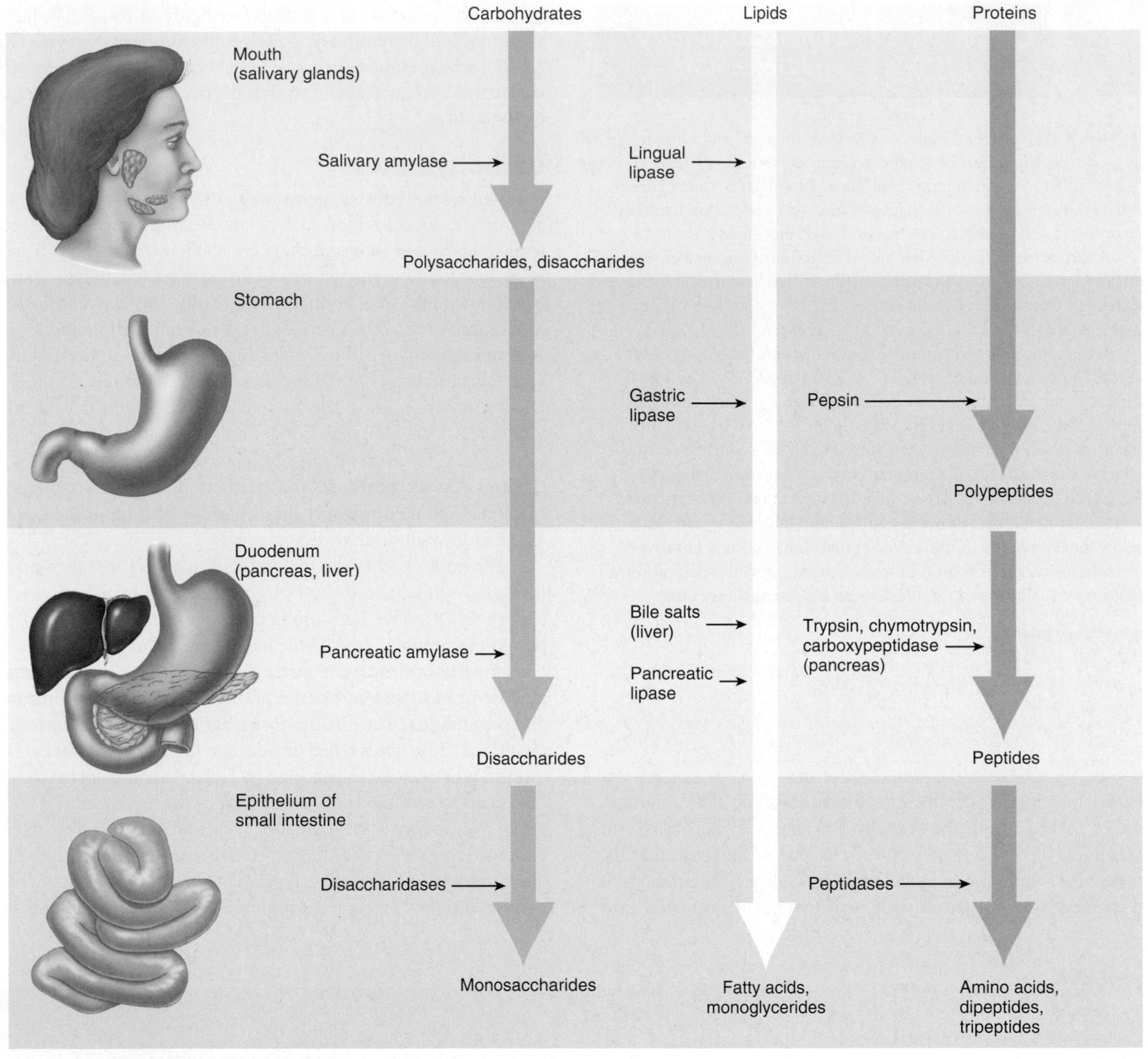
Carbohydrates
Lipids
Proteins
Mouth
(salivary glands)
Salivary amylase
Lingual
lipase
Polysaccharides, disaccharides
Stomach
Gastric
lipase
Pepsin
Polypeptides
Duodenum
(pancreas, liver)
Bile salts
(liver)
Trypsin, chymotrypsin,
carboxypeptidase
(pancreas)
Pancreatic amylase
Pancreatic
lipase
Disaccharides
Peptides
Epithelium of
small intestine
Disaccharidases
Peptidases
Monosaccharides
Fatty acids,
monoglycerides
Amino acids,
dipeptides,
tripeptides

FIGURE 24.28 AP|R **Digestion of the Three Major Food Types**

The enzymes involved in digesting carbohydrates, lipids, and proteins are depicted in relation to the region of the digestive tract where each functions.

Lipids

Lipids are molecules that are insoluble or only slightly soluble in water. They include triglycerides, phospholipids, cholesterol, steroids, and fat-soluble vitamins. **Triglycerides** (trī-glis′er-īdz) are the most common type of lipid and are often referred to as fats. They consist of three fatty acids bound to glycerol.

Lipase (lip′ās) digests lipid molecules (see figure 24.28). The vast majority of lipase is secreted by the pancreas and is referred to as **pancreatic lipase.** A minor amount of **lingual lipase,** which is secreted in the oral cavity and swallowed with food, digests a small amount (<10%) of lipid in the stomach. The stomach also produces very small amounts of **gastric lipase.** The primary products of lipase digestion are free fatty acids and monoglycerides. However, the lipases alone cannot efficiently digest lipids until the lipid droplets mix with bile salts in the small intestine. The bile salts are secreted by the liver and stored in the gallbladder until needed in the duodenum.

A key step in lipid digestion is **emulsification** (ē-mŭl′si-fi-kā′shŭn), by which bile salts transform large lipid droplets into much smaller droplets. Bile salts act as detergents to disrupt lipid droplets. Emulsification is necessary because the lipases are water-soluble and can digest lipids only by acting at the surface of the droplets.

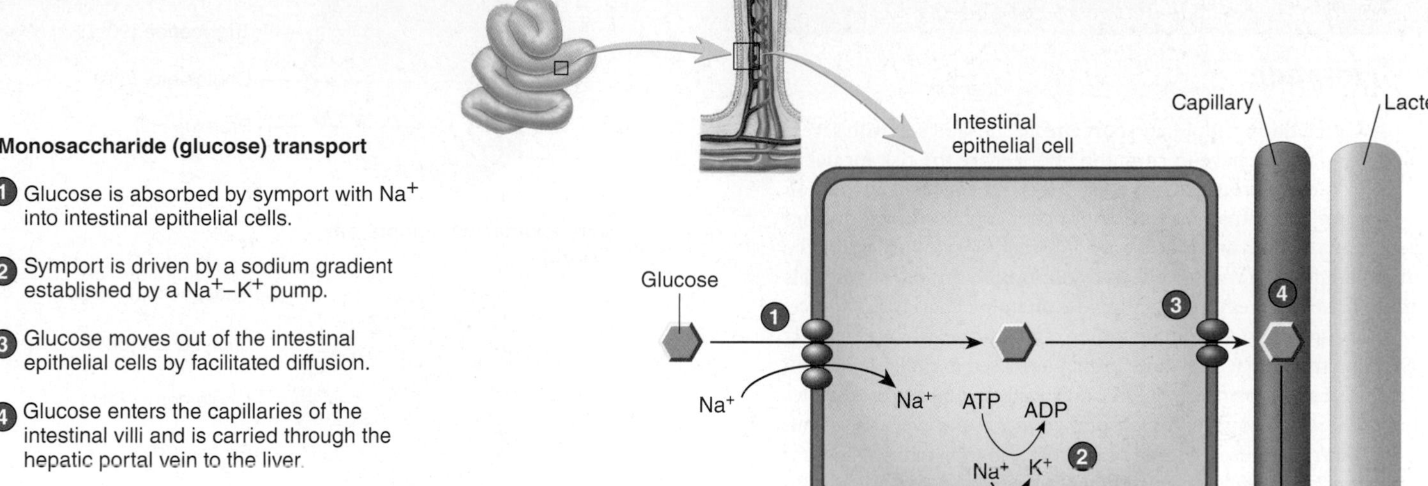
Monosaccharide (glucose) transport
1 Glucose is absorbed by symport with Na+ into intestinal epithelial cells.
2 Symport is driven by a sodium gradient established by a Na+–K+ pump.
3 Glucose moves out of the intestinal epithelial cells by facilitated diffusion.
4 Glucose enters the capillaries of the intestinal villi and is carried through the hepatic portal vein to the liver.
Villus
Intestinal epithelial cell
Capillary
Lacteal
Glucose
Na+
Na+
ATP
ADP
Na+
K+
To liver

PROCESS FIGURE 24.29 **Transport of Glucose Across the Intestinal Epithelium**

By decreasing the droplet size, emulsification increases the surface area of the lipid exposed to lipases and other digestive enzymes.

Once lipids are digested in the intestine, bile salts aggregate around the small droplets to form **micelles** (mī-selz′; small morsels; figure 24.30). The hydrophobic ends of the bile salts are directed toward the free fatty acids, cholesterol, and monoglycerides at the center of the micelle; the hydrophilic ends are directed outward toward the water environment. When a micelle comes in contact with the epithelial cells of the small intestine, the lipid contents of the micelle pass by simple diffusion through the plasma membrane of the epithelial cells. The bile salts are not absorbed until they reach the epithelium of the distal ileum.

Lipid transport

1. Bile salts surround fatty acids and monoglycerides to form micelles.
2. Micelles attach to the plasma membranes of intestinal epithelial cells, and the fatty acids and monoglycerides pass by simple diffusion into the intestinal epithelial cells.
3. Within the intestinal epithelial cell, the fatty acids and monoglycerides are converted to triglycerides; proteins coat the triglycerides to form chylomicrons, which move out of the intestinal epithelial cells by exocytosis.
4. The chylomicrons enter the lacteals of the intestinal villi and are carried through the lymphatic system to the general circulation.

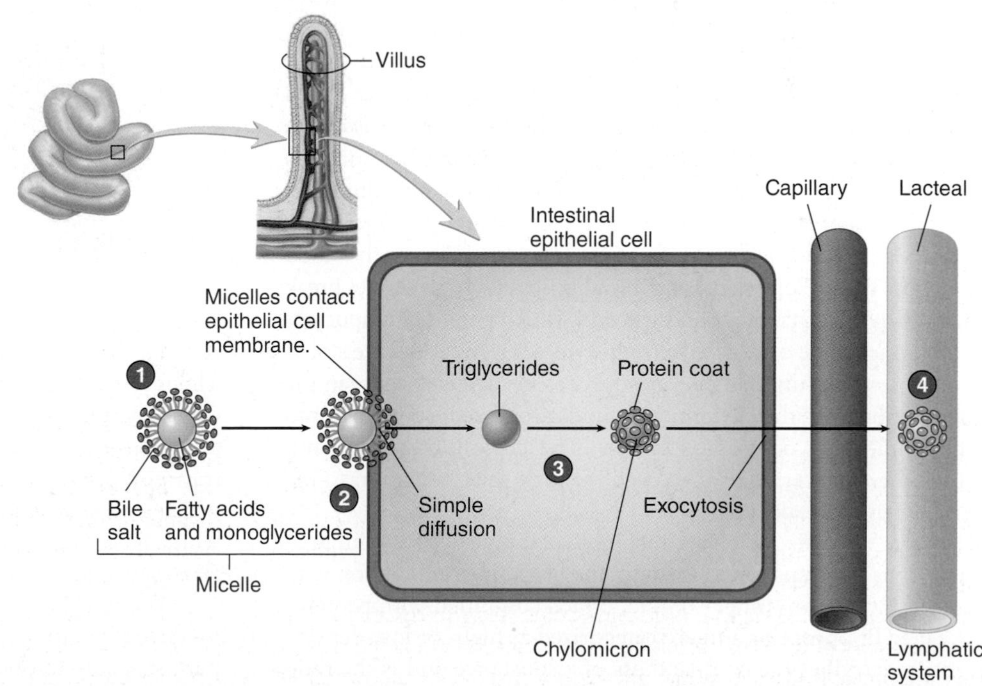
Villus
Intestinal epithelial cell
Capillary
Lacteal
Micelles contact epithelial cell membrane.
Triglycerides
Protein coat
Bile salt
Fatty acids and monoglycerides
Micelle
Simple diffusion
Exocytosis
Chylomicron
Lymphatic system

PROCESS FIGURE 24.30 **Transport of Lipids Across the Intestinal Epithelium**

Clinical
IMPACT

Rehydration

What do an endurance sport athlete and a person with severe diarrhea have in common? Both share the risk for dehydration from excessive water loss. An effective rehydration strategy for both situations is to drink water containing sodium and glucose. As sodium and glucose are absorbed by symport across the intestinal epithelium, water follows by osmosis. As an added value, this strategy also replaces ions and provides an immediate energy source.

Most sports drinks contain sodium and glucose, which efficiently rehydrate the athlete. The same principle is used in oral rehydration therapy for severe diarrhea. The World Health Organization estimates that millions of people in third world countries die every year from severe diarrhea caused by intestinal infections. Drinking a sodium and glucose solution is often sufficient to prevent dehydration until the infection clears. Since oral rehydration therapy was adopted as a main strategy for treating diarrhea, the annual number of deaths of children under 5 dropped worldwide from over 4 million in 1980 to about 1.5 million in 2000. Unfortunately, this simple, cheap treatment and the clean water it requires are unavailable in many areas of exploding populations and poor sanitary conditions.

Lipid Transport

Within the smooth endoplasmic reticulum of the intestinal epithelial cells, free fatty acids are combined with monoglyceride molecules to form triglycerides. Proteins synthesized in the epithelial cells attach to droplets of triglycerides, phospholipids, and cholesterol to form **chylomicrons** (kī-lō-mi′kronz). The chylomicrons leave the epithelial cells and enter the lacteals of the lymphatic system within the villi. Chylomicrons enter the lymphatic capillaries rather than the blood capillaries because the lymphatic capillaries lack a basement membrane and are more permeable to large particles, such as chylomicrons, which are about 0.3 mm in diameter. Chylomicrons contain about 90% triglyceride, 5% cholesterol, 4% phospholipid, and 1% protein (figure 24.31). They travel through the lymphatic system via the thoracic duct to the bloodstream and then by the blood to adipose tissue. Before entering the adipose cells, triglycerides break back down into fatty acids and glycerol, which enter the adipocytes and are once more converted to triglycerides. Triglycerides are stored in adipose tissue until an energy source is needed elsewhere in the body. In the liver, the chylomicron lipids are stored, converted into other molecules, or used as energy. The chylomicron remnant, minus the triglyceride, is conveyed through the circulation to the liver, where it breaks up.

Because lipids are either insoluble or only slightly soluble in water, they are transported through the blood in combination with proteins, which are water-soluble. Lipids combined with proteins are called **lipoproteins** and are categorized as high- or low-density. Density describes the compactness of a substance and is the ratio of mass to volume. Lipids are less dense than water and tend to float in water. Proteins, which are denser than water, tend to sink in water.

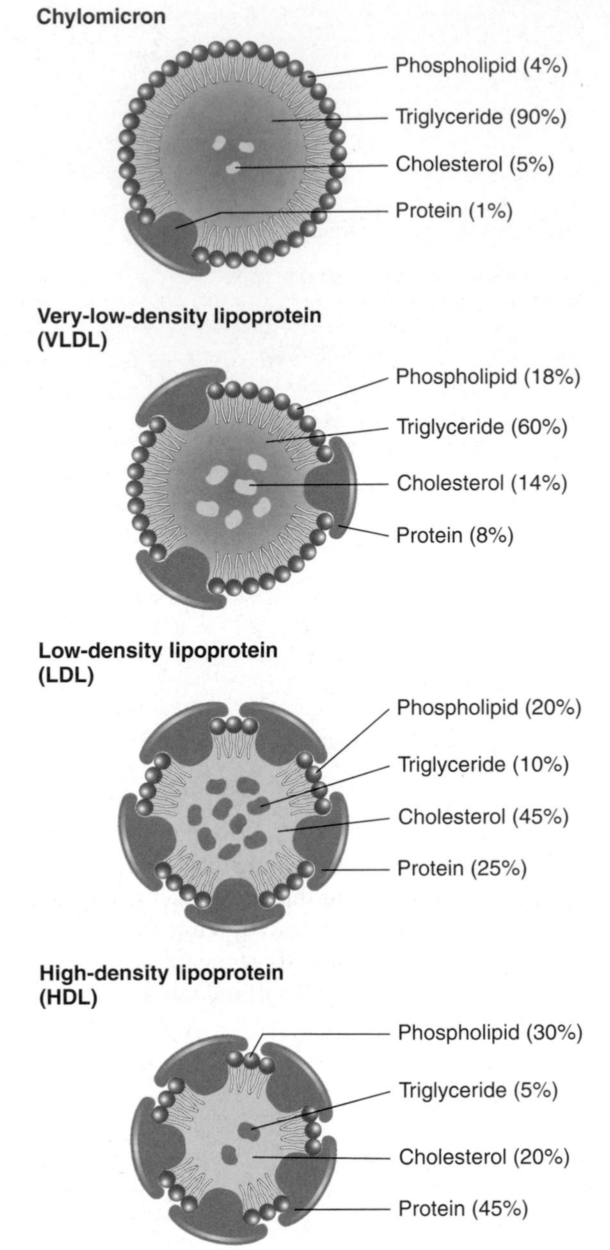
Chylomicron
Phospholipid (4%)
Triglyceride (90%)
Cholesterol (5%)
Protein (1%)
Very-low-density lipoprotein (VLDL)
Phospholipid (18%)
Triglyceride (60%)
Cholesterol (14%)
Protein (8%)
Low-density lipoprotein (LDL)
Phospholipid (20%)
Triglyceride (10%)
Cholesterol (45%)
Protein (25%)
High-density lipoprotein (HDL)
Phospholipid (30%)
Triglyceride (5%)
Cholesterol (20%)
Protein (45%)

FIGURE 24.31 Lipoproteins

A lipoprotein with a high lipid content has a very low density, whereas a lipoprotein with a high protein content has a relatively high density. Chylomicrons, which are made up of 99% lipid and only 1% protein, are lipoproteins with an extremely low density. The other major transport lipoproteins are **very low-density lipoprotein (VLDL),** which is 92% lipid and 8% protein; **low-density lipoprotein (LDL),** which is 75% lipid and 25% protein; and **high-density lipoprotein (HDL),** which is 55% lipid and 45% protein (figure 24.31).

About 15% of the cholesterol in the body is ingested in the food we eat. Eating foods containing saturated fatty acids can raise plasma cholesterol levels by stimulating LDL production and inhibiting LDL receptor production. Conversely, ingesting unsaturated fatty acids lowers plasma cholesterol. Replacing fats with

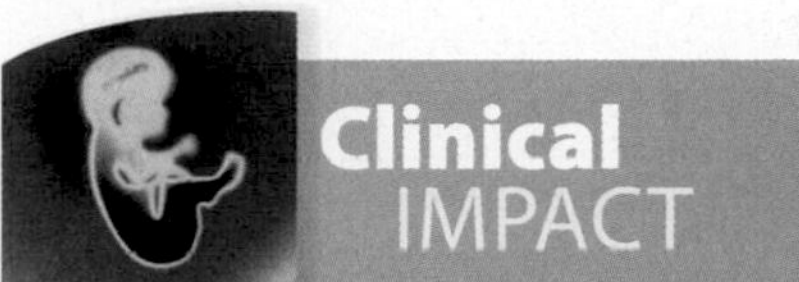
Clinical
IMPACT

Cystic Fibrosis

Cystic fibrosis is a hereditary disorder that occurs in 1 of every 2000 births and affects 33,000 people in the United States; it is the most common lethal genetic disorder among Caucasians. The most critical effects of the disease, accounting for 90% of the deaths, are on the respiratory system. Several other problems occur, however, in affected people. Because the disease is a disorder in a Cl^- transport channel protein—which affects chloride transport and, as a result, the movement of water—all exocrine glands are affected. The buildup of thick mucus in the pancreatic and hepatic ducts causes blockage of the ducts, so that bile salts and pancreatic digestive enzymes are prevented from reaching the duodenum. As a result, digestion is reduced, and fat-soluble vitamins are poorly absorbed due to the lack of bile to form micelles. The person suffers from vitamin A, D, E, and K deficiencies, which result in conditions such as night blindness, skin disorders, rickets, and excessive bleeding. Therapy includes administering the missing vitamins to the person and reducing dietary fat intake.

carbohydrates in the diet can also reduce blood cholesterol. The remaining 85% is manufactured in body cells, mostly in the liver and intestinal mucosa. Most of the cholesterol and other lipids taken into or manufactured in the liver leave the liver in the form of VLDL. Most of the triglycerides are removed from the VLDL to be stored in adipose tissue; as a result, VLDL becomes LDL.

The cholesterol in LDL is critical for the production of steroid hormones and bile salts in the liver. It is also an important component of plasma membranes. Abnormally low cholesterol levels may lead to weakened blood vessel walls and an increased risk for cerebral hemorrhage.

LDL is delivered to cells of various tissues through the circulation. Cells have **LDL receptors** in "pits" on their surfaces, which bind the LDL. Once LDL is bound to the receptors, the pits on the cell surface become endocytotic vesicles, and the cell takes in LDL by receptor-mediated endocytosis (figure 24.32). For example, each fibroblast has 20,000–50,000 LDL receptors on the surface. However, those receptors are confined to cell surface pits, which occupy only 2% of the cell surface. Once inside the cell, the endocytotic vesicle combines with a lysosome, and the LDL components are separated for use in the cell.

Cells not only take in cholesterol and other lipids from LDLs but also make their own cholesterol. When the combined intake and manufacture of cholesterol exceeds a cell's needs, a negative-feedback system reduces the amount of LDL receptors and cholesterol manufactured by the cell. Excess lipids are also packaged into HDLs by the cells. These are transported back to the liver for recycling or excretion in bile.

LDL is commonly considered "bad" because, when in excess, it deposits cholesterol in arterial walls. On the other hand, HDL is considered "good" because it transports cholesterol from the tissues via blood to the liver for removal from the body in the bile. A high HDL/LDL ratio in the bloodstream is related to a lower risk for heart disease. Low HDL levels are linked to obesity, and weight reduction increases HDL levels. Aerobic exercise can decrease LDL levels and increase HDL levels.

Proteins

Proteins are taken into the body from a number of dietary sources. **Pepsin** secreted by the stomach catalyzes the cleavage of covalent bonds in proteins to produce smaller polypeptide chains. Gastric pepsin digests as much as 10–20% of the total ingested protein. Once the proteins and polypeptide chains leave the stomach, proteolytic enzymes produced in the pancreas continue the digestive process and produce small peptide chains (see figure 24.28). These are broken down into tripeptides, dipeptides, and amino acids by **peptidases** bound to the microvilli of the small intestine. Each peptidase is specific for a certain peptide chain length or for a certain amino acid sequence.

Separate carrier molecules transport basic, acidic, and neutral amino acids into the epithelial cells. Acidic and most neutral amino acids enter by symport with a Na^+ gradient (figure 24.33), similar to the mechanism used for glucose transport. Basic amino acids enter the epithelial cells by facilitated diffusion. Dipeptides and tripeptides

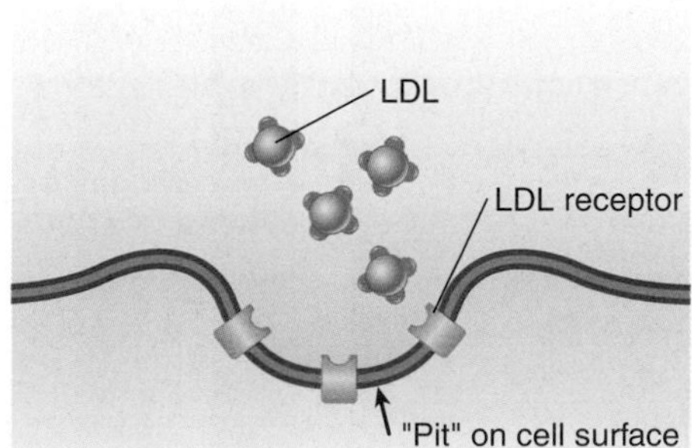
LDL
LDL receptor
"Pit" on cell surface

1 Cells have pits on the surface, which contain LDL receptors.

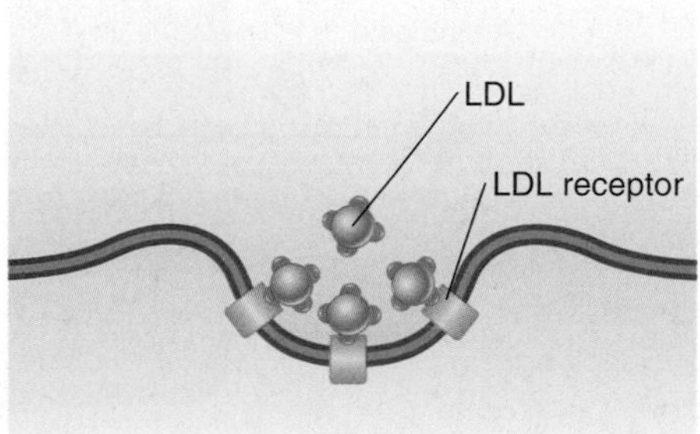
LDL
LDL receptor

2 LDL binds to the LDL receptors in the pits.

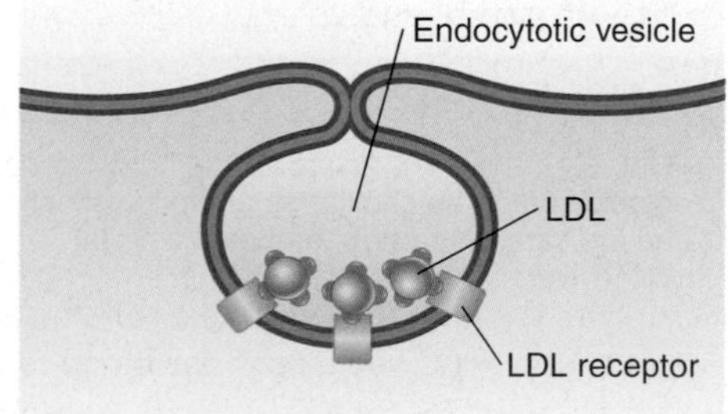
Endocytotic vesicle
LDL
LDL receptor

3 The LDL, bound to LDL receptors, is taken into the cell by endocytosis.

PROCESS FIGURE 24.32 Transport of LDL into Cells

Clinical GENETICS — Familial Hypercholesterolemia

Familial hypercholesterolemia (FH) is a common genetic disorder in Europe and North America that affects 1 out of 500 people. The clinical sign of the disease is increased blood levels of LDL cholesterol. The elevated cholesterol levels accelerate the development of atherosclerosis, which often leads to coronary artery disease and heart attacks among people in their forties and fifties. Another common feature of FH is the presence of **xanthomas** (zan-thō′măs), which are nodules of cholesterol and other lipids just under the skin, especially at joints.

FH is caused by mutations in the LDL receptor gene that result in defective LDL receptors. The LDL receptor normally removes cholesterol from the blood by transporting LDL cholesterol into cells. Once inside the cell, LDL cholesterol is metabolized, and cholesterol synthesis is inhibited by a negative-feedback mechanism. When less LDL cholesterol is transported into cells, blood LDL cholesterol rises for two reasons: (1) The normal removal of LDL cholesterol does not occur, and (2) there is less inhibition of cholesterol synthesis. The usual treatment for FH is statin drugs, which lower blood LDL levels by inhibiting the synthesis of cholesterol.

FH is an example of an incomplete dominance disorder, in which the dominant allele does not completely mask the effects of the recessive allele in the heterozygote (see chapter 29). The severity of FH depends on whether a person is homozygous dominant, heterozygous, or homozygous recessive. Homozygous dominant individuals have two mutant alleles and completely lack LDL receptors. These patients have the most severe form of FH. Their blood LDL cholesterol levels are very high, and they often have heart attacks in their teens. Most FH patients are heterozygous, with one normal and one mutant allele. These patients have half the number of normal LDL receptors and are at increased risk for heart attacks at midlife. Homozygous recessive individuals have two normal LDL receptor alleles and a normal number of receptors. Although genetic testing is not yet standard, automated assays are available for the more common mutations in the LDL receptor, and they can allow early diagnosis and treatment before life-threatening symptoms develop. In the future, FH will be an excellent candidate for gene therapy because it is a severe but well-studied single-gene defect.

enter intestinal epithelial cells by a H^+ symport mechanism analogous to Na^+ symport. The total amount of each amino acid that enters the intestinal epithelial cells as dipeptides or tripeptides is considerably more than the amount that enters as single amino acids. Once inside the cells, dipeptidases and tripeptidases split the dipeptides and tripeptides into their component amino acids. Individual amino acids then leave the epithelial cells and enter the hepatic portal system, which transports them to the liver (figure 24.33). The amino acids may be modified in the liver or released into the bloodstream and distributed throughout the body.

Amino acids are actively transported into the various cells of the body. This transport is stimulated by growth hormone and insulin. Most amino acids serve as building blocks to form new proteins (see chapter 2), but some amino acids may be used for energy.

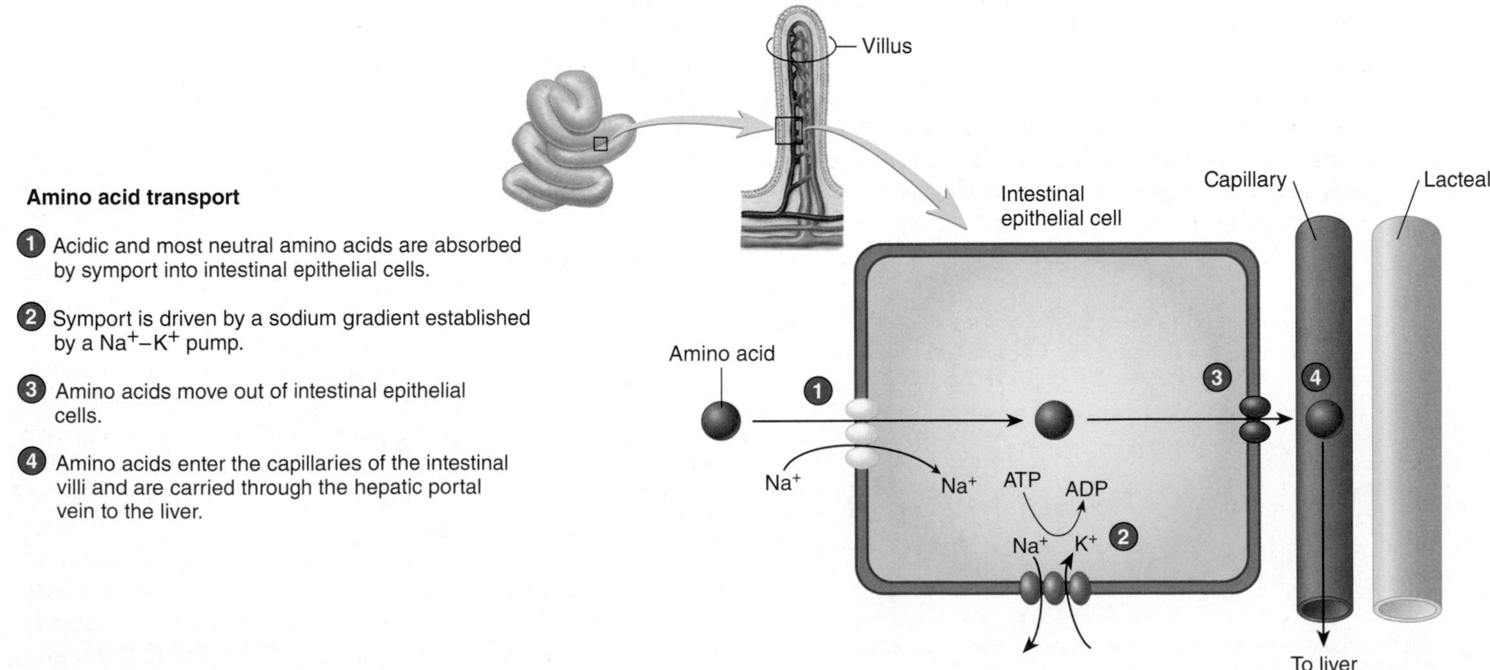
Villus
Amino acid transport
1 Acidic and most neutral amino acids are absorbed by symport into intestinal epithelial cells.
2 Symport is driven by a sodium gradient established by a Na+–K+ pump.
3 Amino acids move out of intestinal epithelial cells.
4 Amino acids enter the capillaries of the intestinal villi and are carried through the hepatic portal vein to the liver.
Intestinal epithelial cell
Capillary
Lacteal
Amino acid
Na+
Na+
ATP
ADP
Na+
K+
To liver

PROCESS FIGURE 24.33 Amino Acid Transport Across the Intestinal Epithelium

Water

About 9 L of **water** enter the digestive tract each day as a combination of ingested and secreted fluids. Of this 9L, about 92% is absorbed in the small intestine, and another 6–7% is absorbed in the large intestine (figure 24.34). Water can move in either direction across the wall of the small intestine. Osmotic gradients across the epithelium determine the direction of its diffusion. When the chyme is dilute, water is absorbed by osmosis across the intestinal wall into the blood. When the chyme is very concentrated and contains very little water, water moves by osmosis into the lumen of the small intestine. As nutrients are absorbed in the small intestine, its osmotic pressure decreases; as a consequence, water moves from the small intestine into the surrounding extracellular fluid. Water in the extracellular fluid can then enter the circulation. Because of the osmotic gradient produced as nutrients are absorbed in the small intestine, nearly all the water that enters the small intestine by way of the oral cavity, stomach, or intestinal secretions is reabsorbed.

Ions

Active transport mechanisms for **sodium** ions are present within the epithelial cells of the small intestine. **Potassium, calcium, magnesium,** and **phosphate** are also actively transported. **Chloride** ions move passively through the intestinal wall of the duodenum and the jejunum following the positively charged sodium ions, but chloride ions are actively transported from the ileum. Although calcium ions are actively transported along the entire length of the small intestine, vitamin D is required for that transport process. The absorption of calcium is under hormonal control, as are its excretion and storage. Parathyroid hormones, calcitonin, and vitamin D all play a role in regulating blood levels of calcium (see chapters 6, 18, and 27).

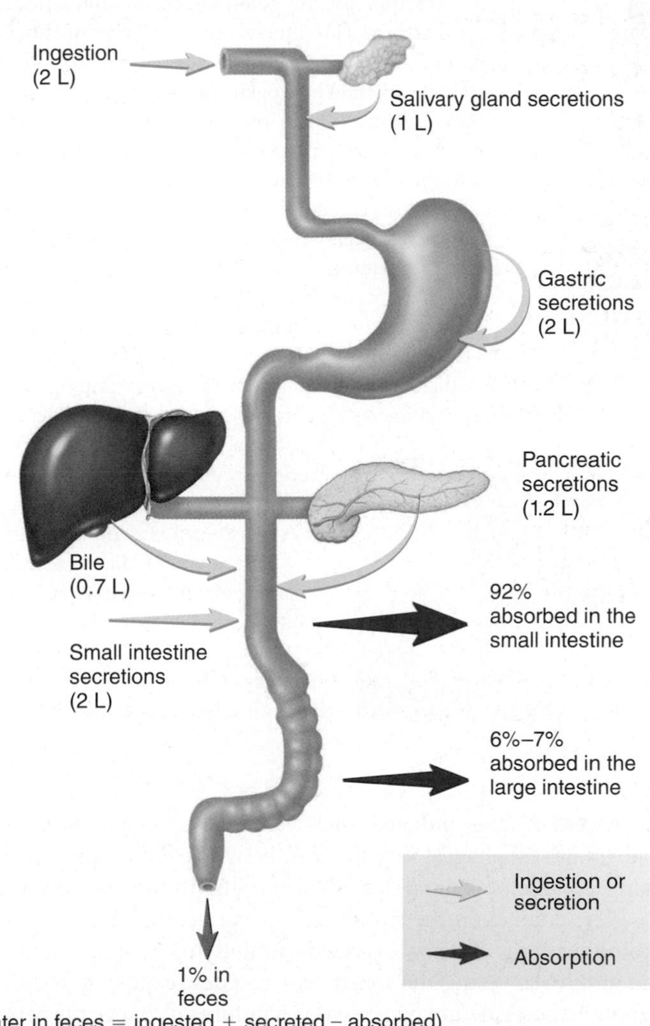
Ingestion
(2 L)
Salivary gland secretions
(1 L)
Gastric
secretions
(2 L)
Pancreatic
secretions
(1.2 L)
Bile
(0.7 L)
92%
absorbed in the
small intestine
Small intestine
secretions
(2 L)
6%–7%
absorbed in the
large intestine
Ingestion or
secretion
Absorption
1% in
feces
(Water in feces = ingested + secreted – absorbed)

FIGURE 24.34 Fluid Volumes in the Digestive Tract

ASSESS **YOUR PROGRESS**

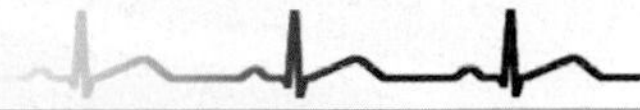

58. *Describe the mechanism of absorption and the route of transport for water-soluble and lipid-soluble molecules.*
59. *Describe the enzymatic digestion of carbohydrates, lipids, and proteins. List where each step of digestion occurs and the breakdown products of each step.*
60. *Explain how lipids are emulsified. Describe the role of micelles, chylomicrons, VLDLs, LDLs, and HDLs in the absorption and transport of lipids in the body.*
61. *Explain how tripeptides, dipeptides, and amino acids enter intestinal epithelial cells.*
62. *Describe the movement of water through the intestinal wall.*
63. *When and where are various ions absorbed?*

24.15 Effects of Aging on the Digestive System

LEARNING OUTCOME

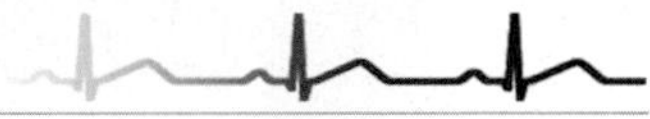

After reading this section, you should be able to

A. Discuss the effects of aging on the digestive system.

As a person ages, gradual changes occur throughout the digestive tract. The connective tissue layers of the digestive tract—the submucosa and serosa—tend to thin. The blood supply to the digestive tract decreases. The number of smooth muscle cells in the muscularis also decreases, resulting in reduced motility in the digestive tract. In addition, goblet cells within the mucosa secrete less mucus. Glands along the digestive tract, such as the gastric glands, the liver, and the pancreas, also tend to secrete less with age. However, these changes by themselves do not appreciably decrease the function of the digestive system.

Through the years, the digestive tract, like the skin and lungs, is directly exposed to materials from the outside environment. Some of those substances can cause mechanical damage to the digestive tract, and others are toxic to the tissues. Because the connective tissue of the digestive tract becomes thin with age and because the protective mucous covering is reduced, an elderly person's digestive tract becomes less and less protected from these outside influences. In addition, the mucosa of elderly people tends to heal more slowly following injury. Declines also occur in the liver's ability to detoxify certain chemicals, the hepatic phagocytic cells' ability to remove particulate contaminants, and the liver's ability to store glycogen. These problems worsen in people who smoke.

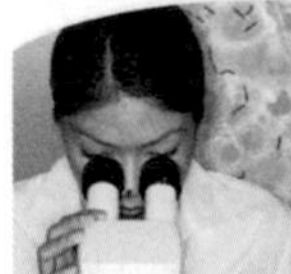

Systems PATHOLOGY | Diarrhea

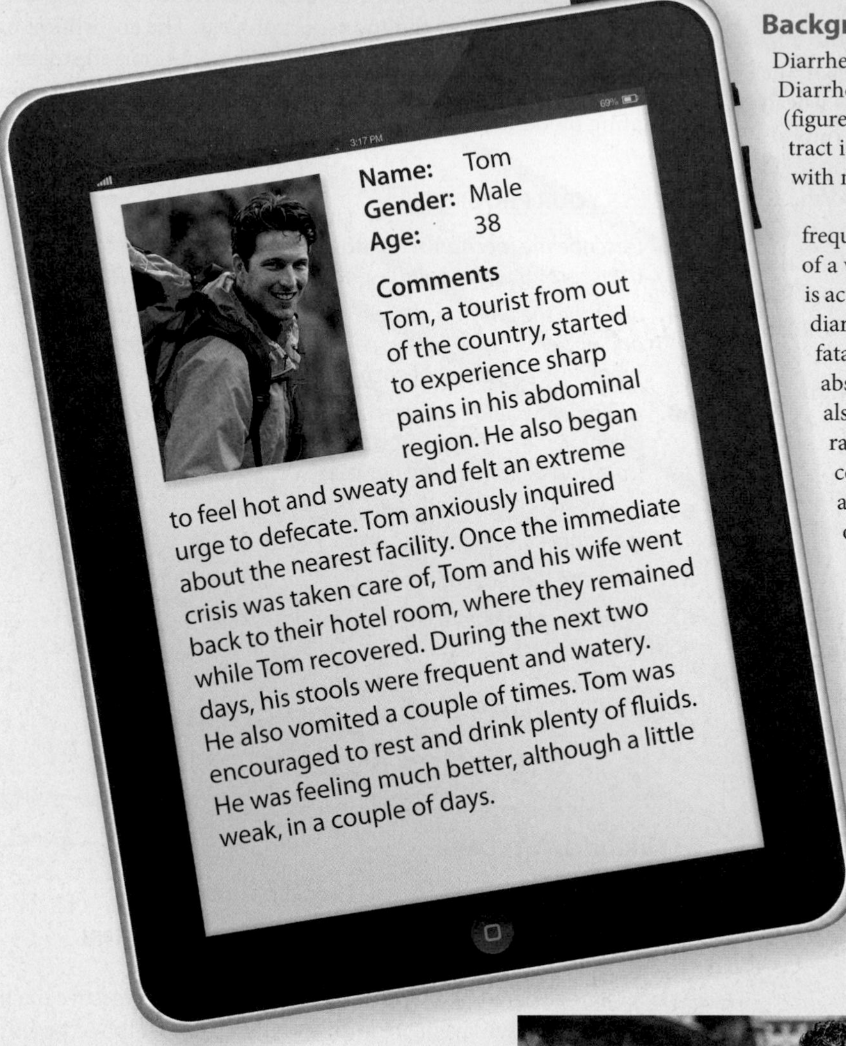
3:17 PM
69%
Name: Tom
Gender: Male
Age: 38
Comments
Tom, a tourist from out of the country, started to experience sharp pains in his abdominal region. He also began to feel hot and sweaty and felt an extreme urge to defecate. Tom anxiously inquired about the nearest facility. Once the immediate crisis was taken care of, Tom and his wife went back to their hotel room, where they remained while Tom recovered. During the next two days, his stools were frequent and watery. He also vomited a couple of times. Tom was encouraged to rest and drink plenty of fluids. He was feeling much better, although a little weak, in a couple of days.

Background Information

Diarrhea is one of the most common complaints in clinical medicine. Diarrhea affects more than half the tourists in developing countries (figure 24A), where it may result from eating food to which the digestive tract is not accustomed or from ingesting food or water contaminated with microorganisms.

Diarrhea is any change in bowel habits involving increased stool frequency or fluidity. It is not a disease in itself, but it can be a symptom of a wide variety of disorders. Diarrhea that lasts less than 2–3 weeks is acute diarrhea; diarrhea lasting longer is considered chronic. Acute diarrhea is usually self-limiting, but some forms of diarrhea can be fatal if not treated. Diarrhea results from either a decrease in fluid absorption in the intestine or an increase in fluid secretion. It can also be caused by increased bowel motor activity that moves chyme rapidly through the small intestine, so that more water enters the colon. Normally, about 600 mL of fluid enter the colon each day, and all but 150 mL are reabsorbed. The loss of more than 200 mL of fluid per day in the stool is considered abnormal.

Mucus secretion by the colon increases dramatically in response to diarrhea. This mucus contains large quantities of bicarbonate ions, which come from dissociation of carbonic acid into bicarbonate ions and hydrogen ions within the blood supply to the colon. The bicarbonate ions enter the mucus secreted by the colon, whereas the hydrogen ions remain in the circulation; as a result, the blood pH decreases. Thus, a condition called metabolic acidosis can develop (see chapter 27).

Diarrhea is usually caused by bacteria, viruses, amoebic parasites, or chemical toxins. Symptoms can begin from as little as 1–2 hours after bacterial toxins are ingested to as long as 24 hours or more for some strains of bacteria. Nearly any bacterial species is capable of causing diarrhea. Some types of bacterial diarrhea are associated with severe vomiting, whereas others are not. Some bacterial toxins also induce fever. Identifying the causal organism usually requires laboratory analysis of the food or stool but, in cases of acute diarrhea, the infectious agent is seldom identified.

Figure 24A

Treatment of diarrhea involves replacing lost fluids and ions. The diet should be limited to clear fluids during at least the first day or so. Medicines that may help combat diarrhea include bismuth subsalicylate (sŭb-să-lis′i-lāt),

The overall decline in the defenses of the digestive tract leaves elderly people more susceptible to infections and toxic agents. Elderly people are therefore more likely to develop ulcerations and cancers of the digestive tract. Colorectal cancers, for example, are the second leading cause of cancer deaths in the United States, with an estimated 135,000 new cases and 57,000 deaths each year.

Gastroesophageal reflux disorder increases with advancing age. It is probably the main reason that elderly people take antacids, H_2 antagonists, and proton pump inhibitors. Disorders that are not necessarily age-induced, such as hiatal hernia and irregular or inadequate esophageal motility, can be worsened by the effects of aging because of general decreased motility in the digestive tract.

The enamel on the surface of elderly people's teeth becomes thinner with age and may expose the underlying dentin. In addition, the gingiva covering the tooth root recedes, exposing additional dentin. Exposed dentin may become painful and change the person's eating habits. Many elderly people also lose teeth, which can have a marked effect on eating habits unless they are fitted with artificial

INTEGUMENTARY
Pallor is due to vasoconstriction of blood vessels in the skin, resulting from a decrease in blood volume. Pallor and sweating increase in response to abdominal pain and anxiety.
MUSCULAR
Muscular weakness may result due to ion loss, metabolic acidosis, fever, and general malaise. The stimulus to defecate may become so strong that it overcomes the voluntary control mechanisms.
URINARY
In the event of fluid loss from the intestine, the kidney is activated to compensate for metabolic acidosis by increasing hydrogen ion secretion and bicarbonate ion reabsorption.
Diarrhea
Symptoms
• Increased stool frequency
• Increased stool volume
• Increased stool fluidity
Treatments
• Replacing lost fluid and ions
• Eating a bland diet without dairy
NERVOUS
Local reflexes in the colon respond to increased colon fluid volume by stimulating mass movements and the defecation reflex. Abdominal pain, much of which is felt as referred pain, can occur as the result of inflammation and distention of the colon. Nervous system function decreases due to ion loss. Reduced blood volume stimulates a sensation of thirst in the CNS.
RESPIRATORY
Increased HCO_3^- secretion and H^+ absorption reduce blood pH. As the result of reduced blood pH, the rate of respiration increases to eliminate carbon dioxide, which helps eliminate excess H^+.
ENDOCRINE
A decrease in extracellular fluid volume, due to the loss of fluid in the feces, stimulates the release of hormones (antidiuretic hormone from the posterior pituitary and aldosterone from the adrenal cortex) that increase water retention and sodium reabsorption in the kidney. In addition, decreased extracellular fluid volume and anxiety result in increased release of epinephrine and norepinephrine from the adrenal medulla.
LYMPHATIC AND IMMUNE
White blood cells migrate to the colon in response to infection and inflammation. In the case of bacterial diarrhea, the immune response is initiated to begin production of antibodies against bacteria and bacterial toxins.
CARDIOVASCULAR
Movement of extracellular fluid into the colon results in decreased blood volume. The reduced blood volume activates the baroreceptor reflex, which maintains blood volume and blood pressure.

which increases mucus and HCO_3^- secretion and decreases pepsin activity, and loperamide (lō-per′ă-mīd), which slows intestinal motility. Patients should avoid milk and milk products. Breads, rice, and baked fish or chicken can be added to the diet as the person's condition improves. A normal diet can be resumed after 2–3 days.

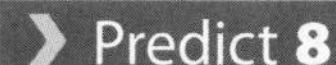

Predict 8

Predict the effects of prolonged diarrhea.

teeth. The muscles of mastication tend to become weaker; as a result, older people tend to chew their food less before swallowing.

Another age-related complication in the digestive system involves the way medications and other chemicals are absorbed from the digestive tract. The decreased mucous covering and the thinned connective tissue layers allow chemicals to pass more readily from the digestive tract into the circulatory system. However, a decline in the blood supply to the digestive tract hinders the absorption of such chemicals. Drugs administered to treat cancer, which occurs in many elderly people, may irritate the mucosa of the digestive tract, resulting in nausea and loss of appetite.

ASSESS YOUR PROGRESS

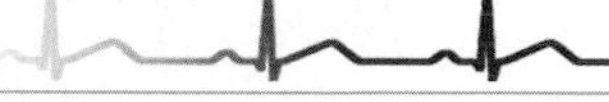

64. *List the effects of aging on the digestive system.*

65. *What are the effects of the overall decline in the defenses of the digestive tract with advancing age?*

66. *Explain why absorption of medication is a problem with aging.*

Diseases and Disorders

TABLE 24.4 Digestive System

Condition	Description
Stomach	
Vomiting	Contraction of the diaphragm and abdominal muscles and relaxation of the esophageal sphincters to forcefully expel gastric contents; vomiting reflex is initiated by irritation of the stomach or small intestine
Peptic ulcer	Lesions in the lining of the stomach or duodenum, usually due to infection by the bacterium *Helicobacter pylori;* stress, diet, smoking, or alcohol may be a predisposing factor; antibiotic therapy is the accepted treatment
Liver	
Cirrhosis (sir-ō′sis)	Characterized by damage and death of hepatic cells and replacement by connective tissue; results in loss of normal liver function and interference with blood flow through the liver; a common consequence of alcoholism
Hepatitis (hep-ă-tī′tis)	Inflammation of the liver that causes liver cell death and replacement by scar tissue; if not corrected, results in loss of liver function and eventually death; symptoms include nausea, abdominal pain, fever, chills, malaise, and jaundice; caused by any of seven distinct viruses
Hepatitis A	Infectious hepatitis; usually transmitted by poor sanitation practices or from mollusks living in contaminated waters
Hepatitis B	Serum hepatitis; usually transmitted through blood or other body fluids through either sexual contact or contaminated hypodermic needles
Hepatitis C	Often a chronic disease leading to cirrhosis and possibly cancer of the liver
Gallstones	Most often due to excess cholesterol in the bile; gallstones can enter the cystic duct, where they block the release of bile and/or pancreatic enzymes, which interferes with digestion
Intestine	
Inflammatory bowel disease (IBD)	Localized inflammatory degeneration that may occur anywhere along the digestive tract but most commonly involves the distal ileum and proximal colon; the intestinal wall often becomes thickened, constricting the lumen, with ulcers and fissures in the damaged areas; symptoms include diarrhea, abdominal pain, fever, fatigue, and weight loss; cause is unknown; treatments involve anti-inflammatory drugs, avoidance of foods that produce symptoms, and surgery in some cases; also called Crohn disease or ulcerative colitis
Irritable bowel syndrome (IBS)	Disorder of unknown cause marked by alternating bouts of constipation and diarrhea; may be linked to stress or depression; high familial incidence
Gluten enteropathy (celiac disease)	Malabsorption in the small intestine due to the effects of gluten, a protein in certain grains, especially wheat; the reaction can destroy newly formed epithelial cells, causing the intestinal villi to become blunted and decreasing the intestinal surface, which reduces absorption of nutrients
Constipation (kon-sti-pā′shŭn)	Slow movement of feces through the large intestine, causing the feces to become dry and hard because of increased fluid absorption while being retained; often results from inhibiting normal defecation reflexes; spasms of the sigmoid colon resulting from irritation can also result in slow feces movement and constipation; high-fiber diet can be preventative
Infections of the Digestive Tract	
Food poisoning	Caused by ingesting bacteria or toxins, such as *Staphylococcus aureus, Salmonella,* or *Escherichia coli;* symptoms include nausea, abdominal pain, vomiting, and diarrhea; in severe cases, death can occur
Typhoid (tī′foyd) fever	Caused by a virulent strain of the bacterium *Salmonella typhi,* which can cross the intestinal wall and invade other tissues; symptoms include severe fever, headaches, and diarrhea; usually transmitted through poor sanitation practices; leading cause of death in many developing countries
Cholera (kol′er-a)	Caused by a bacterium, *Vibrio cholerae,* in contaminated water; bacteria produce a toxin that stimulates the secretion of chloride, bicarbonate, and water into the large intestine, resulting in severe diarrhea; the loss of as much as 12–20 L of fluid per day causes shock, circulatory collapse, and even death; still a major health problem in parts of Asia
Giardiasis (jē-ar-dī′a-sis)	Caused by a protozoan, *Giardia lamblia,* invading the large intestine; symptoms include nausea, abdominal cramps, weakness, weight loss, and malaise; the protozoans are transmitted in the feces of humans and other animals, often by drinking from contaminated wilderness streams
Intestinal parasites	Common under conditions of poor sanitation; parasites include tapeworms, pinworms, hookworms, and roundworms
Diarrhea (dī-ă-rē′ă)	Intestinal mucosa secretes large amounts of water and ions due to irritation, inflammation, or infection; diarrhea moves feces out of the large intestine more rapidly and speeds recovery
Dysentery (dis′en-tār-ē)	Severe form of diarrhea with blood or mucus in the feces; can be caused by bacteria, protozoa, or amoebae

Go to www.mhhe.com/seeley10 for additional information on these pathologies.

Learn to Predict ❮ From page 858

Answer

In this chapter we learned that the gallbladder stores bile, a secretion of the liver that neutralizes stomach acids and emulsifies lipids. We also learned that gallstones may form when there is an abundance of cholesterol in the bile, such as from a high-cholesterol diet. Gallstones can block the cystic duct, blocking the flow of bile from the gallbladder to the duodenum. Eating food high in fat causes gallbladder contractions. Specifically, gallbladder contraction is stimulated hormonally by cholecystokinin from the duodenum and parasympathetic stimulation. If Rebecca's cystic duct were blocked by gallstones, the increased pressure in the contracting gallbladder would result in pain and inflammation.

But why did Rebecca's skin turn yellow? Recall that bile contains bilirubin, a yellow pigment produced from the breakdown of hemoglobin that when further processed in the intestine turns brown and contributes to the normal color of feces. The gallstones blocked Rebecca's common bile duct, preventing bile from passing from the liver to the duodenum and reducing the amount of bilirubin removed from the blood. Rebecca's skin turned yellow due to the accumulation of bile pigments in the blood. Also, the reduced volume of bile entering the duodenum resulted in poor emulsification of lipids. The lipids remained in the small intestine, causing distension of the intestine, and the undigested lipids passing through the small and large intestine were responsible for the diarrhea. The lack of bilirubin resulted in the clay-colored feces.

Answers to the rest of this chapter's Predict questions are in Appendix G.

Summary

24.1 Anatomy of the Digestive System (p. 859)

1. The digestive system consists of a digestive tube and its associated accessory organs.
2. The digestive system includes the oral cavity, pharynx, esophagus, stomach, small intestine, large intestine, and anus.
3. Accessory organs, such as the salivary glands, liver, gallbladder, and pancreas, are located along the digestive tract.

24.2 Functions of the Digestive System (p. 859)

The functions of the digestive system are ingestion, mastication, propulsion, mixing, secretion, digestion, absorption, and elimination.

24.3 Histology of the Digestive Tract (p. 861)

The digestive tract is composed of four tunics: mucosa, submucosa, muscularis, and serosa or adventitia.

Mucosa

1. The mucosa consists of a mucous epithelium, a lamina propria, and a muscularis mucosae.
2. The epithelium extends into the lamina propria to form intestinal glands.

Submucosa

The submucosa is a connective tissue layer containing the submucosal plexus, blood vessels, and small glands.

Muscularis

1. The muscularis consists of an inner layer of circular smooth muscle and an outer layer of longitudinal smooth muscle.
2. The myenteric plexus is between the two muscle layers.
3. Interstitial pacemaker cells are located throughout the myenteric plexus.

Serosa or Adventitia

The serosa or adventitia forms the outermost layer of the digestive tract.

24.4 Regulation of the Digestive System (p. 863)

Nervous, hormonal, and local chemical mechanisms regulate digestion.

Nervous Regulation of the Digestive System

Nervous regulation involves the ENS and CNS reflexes.

Chemical Regulation of the Digestive System

1. Over 30 neurotransmitters are associated with the ENS.
2. The digestive tract produces hormones that regulate digestion.
3. Other chemicals produced by the digestive tract exercise local control of digestion.

24.5 Peritoneum (p. 864)

1. The peritoneum is a serous membrane that lines the abdominal cavity and organs.
2. Mesenteries are peritoneum that extends from the body wall to many of the abdominal organs.
3. Retroperitoneal organs are located behind the peritoneum.

24.6 Oral Cavity (p. 864)

The oral cavity includes the vestibule and the oral cavity proper.

Lips, Cheeks, and Palate

1. The lips and cheeks are involved in facial expression, mastication, and speech.
2. The roof of the oral cavity is divided into the hard and soft palates.
3. The palatine tonsils are located in the lateral wall of the fauces.

Tongue

1. The tongue is involved in speech, taste, mastication, and swallowing.
2. The intrinsic tongue muscles change the shape of the tongue, and the extrinsic tongue muscles move the tongue.

3. The anterior two-thirds of the tongue is covered with papillae; the posterior one-third is devoid of papillae.

Teeth

1. Twenty deciduous teeth are replaced by 32 permanent teeth.
2. The types of teeth are incisors, canines, premolars, and molars.
3. A tooth consists of a crown, a neck, and a root.
4. The root is composed of dentin. Within the dentin of the root is the pulp cavity, which is filled with pulp, blood vessels, and nerves. The crown is dentin covered by enamel.
5. Periodontal ligaments hold the teeth in the alveoli.

Mastication

The muscles of mastication are the temporalis, masseter, medial pterygoid, and lateral pterygoid.

Salivary Glands

1. Salivary glands produce serous and mucous secretions.
2. The three pairs of large salivary glands are the parotid, submandibular, and sublingual.

24.7 Swallowing (p. 871)

Swallowing involves the pharynx and esophagus. It is divided into three phases.

Pharynx

The pharynx consists of the nasopharynx, oropharynx, and laryngopharynx.

Esophagus

1. The esophagus connects the pharynx to the stomach. The upper and lower esophageal sphincters regulate movement.
2. The esophagus consists of an outer adventitia, a muscular layer (longitudinal and circular), a submucosal layer (with mucous glands), and a stratified squamous epithelium.

Swallowing Phases

1. During the voluntary phase of swallowing, a bolus of food is moved by the tongue from the oral cavity to the pharynx.
2. The pharyngeal phase is a reflex caused by the stimulation of stretch receptors in the pharynx.
   - The soft palate closes the nasopharynx, and the epiglottis and vestibular folds close the opening into the larynx.
   - Pharyngeal muscles move the bolus to the esophagus.
3. The esophageal phase is a reflex initiated by the stimulation of stretch receptors in the esophagus. A wave of contraction (peristalsis) moves the food to the stomach.

24.8 Stomach (p. 873)

Anatomy of the Stomach

The openings of the stomach are the gastroesophageal (to the esophagus) and the pyloric (to the duodenum).

Histology of the Stomach

1. The wall of the stomach consists of an external serosa, a muscle layer (longitudinal, circular, and oblique), a submucosa, and simple columnar epithelium (surface mucous cells).
2. Rugae are the folds in the stomach when it is empty.
3. Gastric pits are the openings to the gastric glands, which contain mucous neck cells, parietal cells, chief cells, and endocrine cells.

Secretions of the Stomach

1. Mucus protects the stomach lining.
2. Pepsinogen is converted to pepsin, which digests proteins.
3. Hydrochloric acid promotes pepsin activity and kills microorganisms.
4. Intrinsic factor is necessary for vitamin B_{12} absorption.
5. The sight, smell, taste, or thought of food initiates the cephalic phase. Nerve impulses from the medulla stimulate hydrochloric acid, pepsinogen, gastrin, and histamine secretion.
6. Distension of the stomach, which stimulates gastrin secretion and activates CNS and local reflexes that promote secretion, initiates the gastric phase.
7. Acidic chyme, which enters the duodenum and stimulates neuronal reflexes and the secretion of hormones that inhibit gastric secretions, initiates the intestinal phase.

Movements of the Stomach

1. The stomach stretches and relaxes to increase volume.
2. Mixing waves mix the stomach contents with stomach secretions to form chyme.
3. Peristaltic waves move the chyme into the duodenum.
4. Gastrin and stretching of the stomach stimulate stomach emptying.
5. Chyme entering the duodenum inhibits movement through neuronal reflexes and the release of hormones.

24.9 Small Intestine (p. 881)

The small intestine is divided into the duodenum, jejunum, and ileum.

Anatomy and Histology of the Small Intestine

1. Circular folds, villi, and microvilli greatly increase the surface area of the intestinal lining.
2. Absorptive, goblet, and endocrine cells are in intestinal glands. Duodenal glands produce mucus.

Secretions of the Small Intestine

1. Mucus protects against digestive enzymes and stomach acids.
2. Digestive enzymes (disaccharidases and peptidases) are bound to the intestinal wall.
3. The vagus nerve, secretin, and chemical or tactile irritation stimulate intestinal secretion.

Movement in the Small Intestine

1. Segmental contractions mix intestinal contents. Peristaltic contractions move materials distally.
2. Stretch of smooth muscles, local reflexes, and the parasympathetic nervous system stimulate contractions. Distension of the cecum initiates a reflex that inhibits peristalsis.

24.10 Liver (p. 883)

Anatomy of the Liver

1. The liver has four lobes: right, left, caudate, and quadrate.
2. The liver is divided into lobules.
   - The hepatic cords are composed of columns of hepatocytes separated by the bile canaliculi.
   - The sinusoids are enlarged spaces filled with blood and lined with endothelium and hepatic phagocytic cells.

Histology of the Liver

1. The portal triads supply the lobules.
   - The hepatic arteries and the hepatic portal veins take blood to the lobules and empty into the sinusoids.
   - The sinusoids empty into central veins, which join to form the hepatic veins, which leave the liver.
   - Bile canaliculi converge to form hepatic ducts, which leave the liver.
2. Bile leaves the liver through the hepatic duct system.
   - The hepatic ducts receive bile from the lobules.
   - The cystic duct from the gallbladder joins the hepatic duct to form the common bile duct.
   - The common bile duct joins the pancreatic duct at the point at which it empties into the duodenum.

Functions of the Liver

1. The liver produces bile, which contains bile salts that emulsify lipids.
2. The liver stores and processes nutrients, detoxifies harmful chemicals, and synthesizes new molecules.
3. Hepatic phagocytic cells phagocytize red blood cells, bacteria, and other debris.
4. The liver produces blood proteins.

24.11 Gallbladder (p. 889)

1. The gallbladder is a small sac on the inferior surface of the liver.
2. The gallbladder stores and concentrates bile.
3. Cholecystokinin stimulates gallbladder contraction.

24.12 Pancreas (p. 889)

Anatomy of the Pancreas

1. The pancreas is both an endocrine and an exocrine gland. Its exocrine function is the production of digestive enzymes.
2. The pancreas is divided into lobules that contain acini. The acini connect to a duct system that eventually forms the pancreatic duct, which empties into the duodenum.

Pancreatic Secretions

Digestive enzymes and a watery bicarbonate solution that neutralizes acidic chyme.

Regulation of Pancreatic Secretion

Cholecystokinin and the vagus nerve stimulate the release of digestive enzymes. Secretin stimulates release of bicarbonate ions and water.

24.13 Large Intestine (p. 892)

Anatomy of the Large Intestine

1. The cecum forms a blind sac at the junction of the small and large intestines. The vermiform appendix is a blind tube off the cecum.
2. The ascending colon extends from the cecum superiorly to the right colic flexure. The transverse colon extends from the right to the left colic flexure. The descending colon extends inferiorly to join the sigmoid colon.
3. The sigmoid colon is an S-shaped tube that ends at the rectum.
4. Longitudinal smooth muscles of the large intestine wall are arranged into bands, called teniae coli, that contract to produce pouches called haustra.
5. The mucosal lining of the large intestine is simple columnar epithelium with mucus-producing crypts.
6. The rectum is a straight tube that ends at the anus.
7. An internal anal sphincter (smooth muscle) and an external anal sphincter (skeletal muscle) surround the anal canal.

Secretions of the Large Intestine

1. Mucus protects the intestinal lining.
2. Epithelial cells secrete HCO_3^-. Sodium is absorbed by active transport, and water is absorbed by osmosis.
3. Microorganisms are responsible for vitamin K production, gas production, and much of the bulk of feces.

Movement in the Large Intestine

1. Segmental movements mix the colon's contents.
2. Mass movements are strong peristaltic contractions that occur three or four times a day.
3. Defecation is the elimination of feces. Reflex activity moves feces through the internal anal sphincter. Voluntary activity regulates movement through the external anal sphincter.

24.14 Digestion and Absorption (p. 896)

1. Digestion is the breakdown of organic molecules into their component.
2. Absorption is the means by which molecules are moved out of the digestive tract and distributed throughout the body.
3. Transport from the intestinal epithelium occurs by two routes.
   - Water, ions, and water-soluble products of digestion are transported to the liver through the hepatic portal system.
   - The products of lipid digestion are transported through the lymphatic system to the circulatory system.

Carbohydrates

1. Carbohydrates consist of starches, glycogen, sucrose, lactose, glucose, and fructose.
2. Polysaccharides are broken down into monosaccharides by a number of different enzymes.
3. Monosaccharides are taken up by intestinal epithelial cells by symport that is powered by a Na^+ gradient or by facilitated diffusion.
4. The monosaccharides are carried to the liver, where the nonglucose sugars are converted to glucose.
5. Glucose is transported to the cells that require energy.
6. Glucose enters the cells through facilitated diffusion.
7. Insulin influences the rate of glucose transport.

Lipids

1. Lipids include triglycerides, phospholipids, steroids, and fat-soluble vitamins.
2. Lipase digests lipid molecules to form free fatty acids and monoglycerides.
3. Emulsification, the transformation of large lipid droplets into smaller droplets, is accomplished by bile salts.
4. Micelles form around lipid digestion products and move to epithelial cells of the small intestine, where the products pass into the cells by simple diffusion.
5. Within the epithelial cells, free fatty acids are combined with a monoglyceride to form triglycerides.
6. Proteins coat triglycerides, phospholipids, and cholesterol to form chylomicrons.
7. Chylomicrons enter lacteals within intestinal villi and are carried through the lymphatic system to the bloodstream.
8. Triglycerides are stored in adipose tissue, converted into other molecules, or used as energy.
9. Lipoproteins include chylomicrons, VLDL, LDL, and HDL.

10. LDL transports cholesterol to cells, and HDL transports it from cells to the liver.
11. LDLs are taken into cells by receptor-mediated endocytosis, which is controlled by a negative-feedback mechanism.

Proteins

1. Pepsin in the stomach breaks proteins into polypeptide chains.
2. Trypsin and other proteolytic enzymes from the pancreas produce smaller peptides.
3. Peptidases, bound to the microvilli of the small intestine, break down peptides.
4. Amino acids, dipeptides, and tripeptides are absorbed by symport that is powered by Na^+ or H^+ gradients or by facilitated diffusion.
5. Amino acids are transported to the liver, where the amino acids can be modified or released into the bloodstream.
6. Amino acids are actively transported into cells under the stimulation of growth hormone and insulin.
7. Amino acids are used as building blocks or for energy.

Water

Water can move in either direction across the wall of the small intestine, depending on the osmotic gradients across the epithelium.

Ions

1. Sodium, potassium, calcium, magnesium, and phosphate are actively transported.
2. Chloride ions move passively through the wall of the duodenum and jejunum but are actively transported from the ileum.
3. Calcium ions are actively transported, but vitamin D is required for transport, and the transport is under hormonal control.

24.15 Effects of Aging on the Digestive System (p. 903)

The mucous layer, the connective tissue, the muscles, and the secretions of the digestive tract all tend to decrease as a person ages. These changes make an older person more open to infections and toxic agents.

REVIEW AND COMPREHENSION

1. Which layer of the digestive tract is in direct contact with the food that is consumed?
   a. mucosa
   b. muscularis
   c. serosa
   d. submucosa
2. The ENS is found in
   a. the submucosa layer.
   b. the muscularis layer.
   c. the serosa layer.
   d. Both a and b are correct.
   e. All of these are correct.
3. Dentin
   a. forms the surface of the crown of the teeth.
   b. holds the teeth to the periodontal ligaments.
   c. is found in the pulp cavity.
   d. makes up most of the structure of the teeth.
   e. is harder than enamel.
4. The number of premolar deciduous teeth is
   a. 0. c. 4. e. 12.
   b. 2. d. 8.
5. Which of these glands does *not* secrete saliva into the oral cavity?
   a. submandibular gland c. sublingual gland
   b. pancreas d. parotid gland
6. The portion of the digestive tract in which digestion begins is the
   a. oral cavity.
   b. esophagus.
   c. stomach.
   d. duodenum.
   e. jejunum.
7. During swallowing,
   a. the movement of food results primarily from gravity.
   b. the swallowing center in the medulla oblongata is activated.
   c. food is pushed into the oropharynx during the pharyngeal phase.
   d. the soft palate closes off the opening into the larynx.
8. The stomach
   a. has large folds in the submucosa and mucosa called rugae.
   b. has two layers of smooth muscle in the muscularis tunic.
   c. opening from the esophagus is the pyloric opening.
   d. has an area closest to the duodenum called the fundus.
   e. All of these are correct.
9. Which of these stomach cell types is *not* correctly matched with its function?
   a. surface mucous cells—produce mucus
   b. parietal cells—produce hydrochloric acid
   c. chief cells—produce intrinsic factor
   d. endocrine cells—produce regulatory hormones
10. Why doesn't the stomach digest itself?
    a. The stomach wall is not composed of protein, so it is not affected by proteolytic enzymes.
    b. The digestive enzymes of the stomach are not strong enough to digest the stomach wall.
    c. The lining of the stomach wall has a protective layer of epithelial cells.
    d. The stomach wall is protected by large amounts of mucus.
11. Which of these hormones stimulates stomach secretions?
    a. cholecystokinin
    b. insulin
    c. gastrin
    d. secretin
12. Which of these structures increase the mucosal surface of the small intestine?
    a. circular folds
    b. villi
    c. microvilli
    d. length of the small intestine
    e. All of these are correct.
13. Which cells in the small intestine have digestive enzymes attached to their surfaces?
    a. mucous cells c. endocrine cells
    b. goblet cells d. absorptive cells

14. The hepatic sinusoids
    a. receive blood from the hepatic artery.
    b. receive blood from the hepatic portal vein.
    c. empty into the central veins.
    d. All of these are correct.
15. Which of the following might occur if a person suffers from a severe case of hepatitis that impairs liver function?
    a. Lipid digestion is difficult.
    b. By-products of hemoglobin breakdown accumulate in the blood.
    c. Plasma proteins decrease in concentration.
    d. Toxins in the blood increase.
    e. All of these occur.
16. The gallbladder
    a. produces bile.
    b. stores bile.
    c. contracts and releases bile in response to secretin.
    d. contracts and releases bile in response to sympathetic stimulation.
    e. Both b and c are correct.
17. The aqueous component of pancreatic secretions
    a. is secreted by the pancreatic islets.
    b. contains HCO_3^-.
    c. is released primarily in response to cholecystokinin.
    d. passes directly into the blood.
    e. All of these are correct.
18. Which of these is *not* a function of the large intestine?
    a. absorption of glucose
    b. absorption of certain vitamins
    c. absorption of water and salts
    d. production of mucus
19. Defecation
    a. can be initiated by stretch of the rectum.
    b. can occur as a result of mass movements.
    c. involves local reflexes.
    d. involves parasympathetic reflexes mediated by the spinal cord.
    e. All of these characteristics are true of defecation.
20. Which of these structures produces enzymes that digest carbohydrates?
    a. salivary glands
    b. pancreas
    c. lining of the small intestine
    d. Both a and b are correct.
    e. All of these are correct.
21. Bile
    a. is an important enzyme for the digestion of lipids.
    b. is made by the gallbladder.
    c. contains breakdown products from hemoglobin.
    d. emulsifies lipids.
    e. Both c and d are correct.
22. Micelles are
    a. lipids surrounded by bile salts.
    b. produced by the pancreas.
    c. released into lacteals.
    d. stored in the gallbladder.
    e. reabsorbed in the colon.
23. If the thoracic duct were tied off, which of these classes of nutrients would *not* enter the circulatory system at their normal rate?
    a. amino acids
    b. glucose
    c. lipids
    d. fructose
    e. nucleotides
24. Which of these lipoprotein molecules transports excess lipids from cells back to the liver?
    a. high-density lipoprotein (HDL)
    b. low-density lipoprotein (LDL)
    c. very low-density lipoprotein (VLDL)

Answers in Appendix E

CRITICAL THINKING

1. While anesthetized, patients sometimes vomit. Given that the anesthetic eliminates the swallowing reflex, explain why it is dangerous for an anesthetized patient to vomit.
2. Achlorhydria is a condition in which the stomach stops producing hydrochloric acid and other secretions. What effect would achlorhydria have on the digestive process? On red blood cell count?
3. Victor Worrystudent experienced the pain of a duodenal ulcer during final examination week. Explain what habits caused the ulcer, and recommend possible remedies.
4. Gallstones sometimes obstruct the common bile duct. What are the consequences of such a blockage?
5. A patient has a spinal cord injury at level L2. How does this injury affect the patient's ability to defecate? What components of the defecation response are still present, and which are lost?
6. The bacterium *Vibrio cholerae* produces cholera toxin, which activates a chloride channel in the intestinal epithelium. In contrast, mutations that inactivate the same channel cause cystic fibrosis. Explain how increased chloride channel activity causes severe diarrhea, whereas decreased activity causes the intestinal symptoms of cystic fibrosis.
7. Discuss why the most effective oral rehydration therapy is water containing sodium and glucose instead of water alone or water with fructose.
8. Would a patient with familial hypercholesterolemia (FH) benefit from dietary changes?

Answers in Appendix F

Visit this book's website at **www.mhhe.com/seeley10** for chapter quizzes, interactive learning exercises, and other study tools.

connect plus+
ANATOMY & PHYSIOLOGY

Anatomy & Physiology REVEALED
aprevealed.com

25

Learn to Predict

Sadie and David loved it when their mothers planned a picnic at the park. They enjoyed running and playing on the jungle gym. Today, Sadie begged her mom to pack a snack of chocolate chip cookies and grape soda. This snack was packed with calories that would give the children lots of energy, but otherwise, it had very little nutritional value. Sadie's mom explained that a snack of fruits, whole-wheat crackers, and water would be much better for everyone. After reading this chapter and recalling what you learned about nutrient digestion and absorption in chapter 24, predict the outcome if Sadie's mom allowed them to bring Sadie's suggested snack, rather than the snack her mother suggested they take to the park.

Photo: Photograph of two young children enjoying a healthy snack.

Nutrition, Metabolism, and Temperature Regulation

"You are what you eat," is a common phrase we all hear. Health claims about foods and food supplements bombard us every day. Meanwhile, obesity rates in U.S. children and adults have soared. Nutrition and weight maintenance are subjects of constant discussion on television, in newspapers, and around the water cooler. When choosing food, many of us are more concerned about its taste than its nutritional value. What happens if we do not obtain enough vitamins, or if we eat too much sugar and fat? Which new diets or diet supplements are ridiculous, and which ones have merit? A basic understanding of nutrition can answer these and other questions, so that we can develop a healthy diet. It also allows us to know which nutrition questions currently do not have good answers.

Module 12
Digestive System

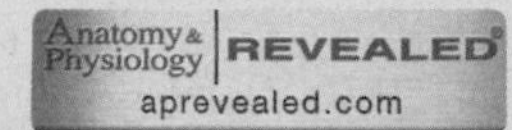

Anatomy & Physiology
REVEALED
aprevealed.com

25.1 Nutrition

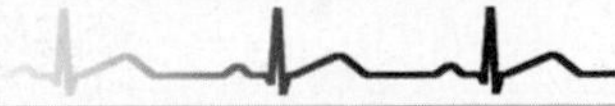

LEARNING OUTCOMES

After reading this section, you should be able to

A. Define *nutrition, nutrients, essential nutrients,* and *kilocalorie*.

B. Describe MyPlate and its use.

C. For carbohydrates, lipids, and proteins, describe their dietary sources, their uses in the body, and the daily recommended amounts of each in the diet.

D. List the important vitamins and minerals for body health. Give the function and symptoms of deficiency for each.

E. Discuss the Reference Daily Intake and the Daily Reference Value of food.

Nutrition is the process by which the body obtains and uses certain components of food. The process includes digestion, absorption, transportation, and cell metabolism. *Nutrition* also refers to the evaluation of food and drink requirements for normal body function.

Nutrients

Nutrients are the chemicals taken into the body that are used to produce energy, to provide building blocks for new molecules, and to function in other chemical reactions. Some important substances in food, such as nondigestible plant fibers, are not nutrients. Nutrients are divided into six major classes: carbohydrates, lipids, proteins, vitamins, minerals, and water. Carbohydrates, lipids, and proteins are the major organic nutrients, and they are broken down by enzymes into their components during digestion. Many of these subunits are broken down further to supply energy, whereas others are used as building blocks for new carbohydrates, lipids, and proteins. Carbohydrates, lipids, proteins, and water are required in fairly substantial quantities, whereas vitamins and minerals are required in only small amounts. Vitamins, minerals, and water are taken into the body without being digested.

Essential nutrients are chemicals that must be ingested because the body cannot manufacture them at all or cannot manufacture adequate amounts of them. The essential nutrients include certain amino acids and fatty acids, most vitamins, minerals, water, and a few carbohydrates. However, the term *essential* does not mean that the body requires only the essential nutrients. Other nutrients are necessary; if they are not part of the diet, they can be synthesized from the essential nutrients. Most of this synthesis takes place in the liver, which has a remarkable ability to transform and manufacture molecules.

Kilocalories

The body uses the energy stored within the chemical bonds of certain nutrients. A **calorie** (kal′ō-rē; **cal**) is the amount of energy (heat) necessary to raise the temperature of 1 g of water 1°C. A **kilocalorie** (kil′ō-kal-ō-rē; **kcal**) is 1000 calories and is used to express the larger amounts of energy supplied by foods and released through metabolism.

A kilocalorie is often called a *Calorie* (with a capital *C*). Unfortunately, this usage has been confused with the term *calorie* (with a lowercase *c*). Food labels and nutrition books commonly use *calorie* when *Calorie* (*kilocalorie*) is the proper term.

Most of the kilocalories supplied by food come from carbohydrates, proteins, or lipids (fats). For each gram of carbohydrate or protein the body metabolizes, about 4 kcal of energy are released. Fats contain more energy per unit of weight than carbohydrates and proteins and yield about 9 kcal/g. Table 25.1 lists the kilocalories supplied by some common foods. A typical diet in the United States consists of 50–60% carbohydrates, 35–45% fats, and 10–15% proteins. Table 25.1 also lists the carbohydrate, fat, and protein composition of some foods.

MyPlate

Every 5 years, the U.S. Department of Health and Human Services (HHS) and the Department of Agriculture (USDA) jointly recommend the types and amounts of food Americans should eat to be healthy. The latest recommendations, *Dietary Guidelines for Americans 2010,* were published in January 2011. In light of the increasing problem of obesity in the United States, the latest recommendations focus on two concepts: (1) balancing calorie intake to obtain and maintain a healthy weight and (2) increasing consumption of healthy, nutrient-rich foods. In June 2011, the USDA also introduced MyPlate, a new food icon to replace the former food guide icon, called MyPyramid. MyPlate (figure 25.1) is a simple visual reminder of how to build a healthy meal. The MyPlate icon shows a plate and glass with portions representing foods from the fruits, vegetables, grains, proteins, and dairy food groups.

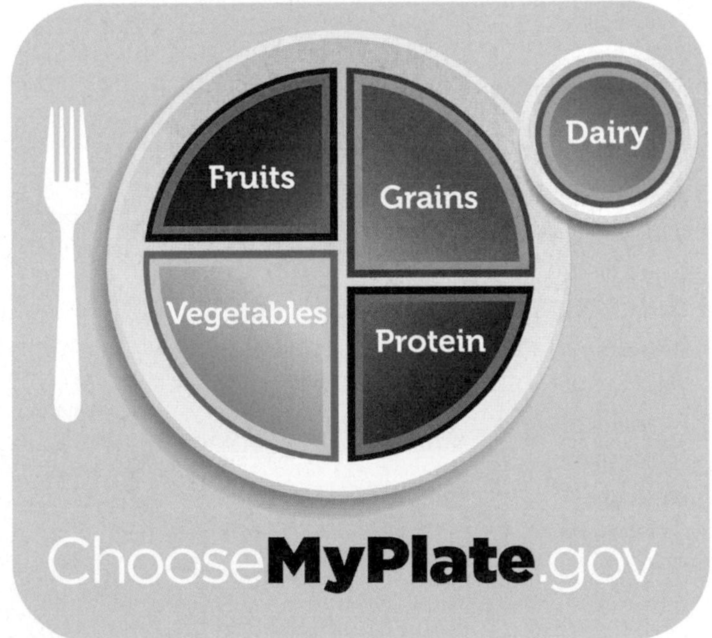
Fruits
Grains
Dairy
Vegetables
Protein
ChooseMyPlate.gov

FIGURE 25.1 MyPlate
The MyPlate icon provides a visual reminder for making choices at mealtime, by selecting healthy foods from five food groups. Half the meal should be fruits and vegetables. *Source: U.S. Department of Agriculture.*

TABLE 25.1 Nutrient Content of Some Typical Foods

Food	Quantity	Food Energy (kcal)	Carbohydrate (g)	Fat (g)	Protein (g)
Dairy Products					
Whole milk (3.3% fat)	1 cup	150	11	8	8
Low-fat milk (2% fat)	1 cup	120	12	5	8
Butter	1 tablespoon	100	——	12	——
Grains					
Bread, white enriched	1 slice	75	24	1	2
Bread, whole-wheat	1 slice	65	14	1	3
Fruits					
Apple	1	80	20	1	——
Banana	1	100	26	——	1
Orange	1	65	16	——	1
Vegetables					
Corn, canned	1 cup	140	33	1	4
Peas, canned	1 cup	150	29	1	8
Lettuce	1 cup	5	2	——	——
Celery	1 cup	20	5	——	1
Potato, baked	1 large	145	33	——	4
Meat, Fish, and Poultry					
Lean ground beef (10% fat)	3 ounces	185	——	10	23
Shrimp, french fried	3 ounces	190	9	9	17
Tuna, canned	3 ounces	170	——	7	24
Chicken breast, fried	3 ounces	160	1	5	26
Bacon	2 slices	85	——	8	4
Hot dog	1	170	1	15	7
Fast Foods					
McDonald's Egg McMuffin	1	300	30	12	18
McDonald's Big Mac	1	540	45	29	25
Taco Bell beef burrito	1	420	53	15	17
Arby's roast beef	1	360	37	14	22
Pizza Hut Super Supreme	1 slice	245	29	17	14
Long John Silver fish	2 pieces	520	32	34	24
Dairy Queen Oreo Cookie Blizzard, medium	1	680	100	25	14
Desserts					
Chocolate chip cookie	1	50	7	2	1
Apple pie	1 piece	135	49	14	3
Soft ice cream	1 cup	377	38	23	7
Beverages					
Cola soft drink	12 ounces	145	37	——	——
Beer	12 ounces	144	13	——	1
Wine	3-1/2 ounces	73	2	——	——
Hard liquor (86 proof)	1-1/2 ounces	105	——	——	——
Miscellaneous					
Egg	1	80	1	6	6
Mayonnaise	1 tablespoon	100	——	11	——
Sugar	1 tablespoon	45	12	——	——

To emphasize the importance of making healthy food choices, half the plate is fruits and vegetables. In addition to the MyPlate icons, the USDA also launched ChooseMyPlate.gov, a website that includes information on how to make healthy dietary choices.

ASSESS YOUR PROGRESS 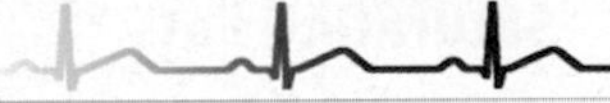

1. *What is nutrition, and what processes does it include?*
2. *Distinguish between a nutrient and an essential nutrient.*
3. *List the six major classes of nutrients.*
4. *Define* kilocalorie. *State the number of kilocalories supplied by a gram of carbohydrate, a gram of lipid, and a gram of protein.*
5. *List the five food groups shown in MyPlate. How is the importance of eating fruits and vegetables indicated in MyPlate?*

Carbohydrates

Sources in the Diet

Carbohydrates include monosaccharides, disaccharides, and polysaccharides (see chapter 2). Most of the carbohydrates humans ingest come from plants. An exception is lactose, which is found in milk and other dairy products.

The most common monosaccharides in the diet are glucose and fructose. Plants capture the energy in sunlight and use it to produce glucose, which can be found in vegetables. Fructose (fruit sugar) and galactose are isomers of glucose (see figure 2.14). Fructose is in fruits, berries, honey, and high-fructose corn syrup, which is used to sweeten soft drinks and desserts. Galactose is found in milk.

The disaccharide sucrose (table sugar) is what most people think of when they use the term *sugar.* Sucrose is composed of a glucose and a fructose molecule joined together, and its principal sources are sugarcane, sugar beets, maple sugar, and honey. Maltose (malt sugar), derived from germinating cereals, is a combination of two glucose molecules, and lactose (milk sugar) consists of a glucose and a galactose molecule (see figure 2.14).

Complex carbohydrates are polysaccharides: starch, glycogen, and cellulose. These polysaccharides consist of many glucose molecules bound together to form long chains. Starch is an energy-storage molecule found primarily in plants (vegetables, fruits, and grains). Glycogen is an energy-storage molecule in animals and is located primarily in muscle and in the liver. By the time meats have been processed, they contain little, if any, glycogen because it is used up by the dying muscle cells (see "Lactic Acid Fermentation" in section 25.3). Cellulose forms plant cell walls.

Uses of Carbohydrates in the Body

During digestion, polysaccharides and disaccharides are split into monosaccharides, which are absorbed into the blood (see chapter 24). Humans can digest starch and glycogen because our bodies produce enzymes that break the bonds between the glucose molecules of starch and glycogen. Humans are unable to digest cellulose because our bodies do not produce the enzymes that break the bonds between its glucose molecules. Instead, cellulose provides fiber, or "roughage," thereby increasing the bulk of feces and making it easier to defecate.

The liver converts fructose, galactose, and other monosaccharides absorbed by the blood into glucose. Glucose, whether absorbed from the digestive tract or synthesized in the liver, provides energy to produce **adenosine triphosphate (ATP)** molecules (see section 25.3). Because the brain relies almost entirely on glucose for its energy, the body carefully regulates blood glucose levels (see chapter 18).

Muscle and liver cells convert excess glucose into glycogen for storage. Because cells can store only a limited amount of glycogen, any additional glucose is converted into lipids and stored in adipose tissue. Glycogen can be rapidly converted back to glucose when energy is needed. For example, during exercise, muscles convert glycogen to glucose, and between meals the liver helps maintain blood sugar levels by converting glycogen to glucose, which is released into the blood.

In addition to serving as a source of energy, sugars have other functions. They form part of deoxyribonucleic acid (DNA), ribonucleic acid (RNA), and ATP molecules (see chapter 2). They also combine with proteins to form glycoproteins, such as the glycoprotein receptor molecules on the outer surface of the plasma membrane (see chapter 3).

Recommended Consumption of Carbohydrates

According to the Dietary Guidelines Advisory Committee, the **Acceptable Macronutrient Distribution Range (AMDR)** for carbohydrates is 45–65% of total kilocalories. Although a minimum level of carbohydrates has not been established, researchers believe that amounts of 100 g or less per day result in overuse of the body's proteins and lipids for energy. Because muscles are primarily protein, the use of proteins for energy can result in the breakdown of muscle tissue. The extensive use of lipids for energy can lead to acidosis (see chapter 18).

Complex carbohydrates are recommended in the diet because many starchy foods contain other valuable nutrients, such as vitamins and minerals, and because the slower rate of digestion and absorption of complex carbohydrates does not cause large increases and decreases in blood glucose levels, as the consumption of large amounts of simple sugars does. Foods primarily composed of simple sugars, such as soft drinks and candy, are rich in carbohydrates but have few other nutrients. For example, a typical soft drink is mostly sucrose, containing 9 teaspoons of sugar per 12 oz can. Consuming these kinds of foods in excess usually results in obesity and tooth decay.

ASSESS YOUR PROGRESS 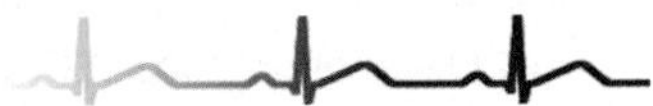

6. *What are the most common monosaccharides in the diet? What are the sources of the three common disaccharides: sucrose, maltose, and lactose?*
7. *Give three examples of complex carbohydrates. How does the body use them?*
8. *How does the body use glucose and other monosaccharides?*
9. *What is the recommended daily consumption of carbohydrates?*

Lipids

Sources in the Diet

About 95% of the lipids in the human diet are **triglycerides** (trī-glis′er-īdz). Triglycerides, which are sometimes called *triacylglycerols* (trī-as′il-glis′er-olz), consist of three fatty acids attached to a glycerol molecule (see chapter 2). Triglycerides are often referred to as fats, which are solid at room temperature, or oils, which are liquid at room temperature. Fats and oils can be categorized as saturated or unsaturated. **Saturated fats and oils** have only single covalent bonds between the carbon atoms of their fatty acids (see figure 2.17). They are found in the fats of meat (e.g., beef, pork), dairy products (e.g., whole milk, cheese, butter), eggs, coconut oil, and palm oil. **Unsaturated fats and oils** have one or more double covalent bonds between the carbon atoms of their fatty acids (see figure 2.17). **Monounsaturated fats** have one double bond, and **polyunsaturated fats** have two or more double bonds. Monounsaturated fats include olive and peanut oils; polyunsaturated fats are in fish, safflower, sunflower, and corn oils.

Unsaturated fatty acids can be classified according to the location of their first double bond at the omega (methyl) end of the fatty acid. The first double bond of an omega-3 fatty acid starts three carbon atoms after the omega end; an omega-6 fatty acid starts after six carbons; and an omega-9 fatty acid starts after nine carbons (see figure 2.17).

The remaining 5% of lipids include cholesterol and phospholipids, such as **lecithin** (les′i-thin). Cholesterol is a steroid (see chapter 2) found in high concentrations in liver and egg yolks, but it is also present in whole milk, cheese, butter, and meats. Cholesterol is not in plants. Phospholipids, major components of plasma membranes, are found in a variety of foods, including egg yolks.

Uses of Lipids in the Body

Triglycerides are important sources of energy that can be used to produce ATP. A gram of triglyceride delivers more than twice as many kilocalories as a gram of carbohydrate. Some cells, such as skeletal muscle cells, derive most of their energy from triglycerides.

After a meal, excess triglycerides that are not immediately used are stored in adipose tissue or the liver. Later, when energy is required, the triglycerides are broken down, and their fatty acids are released into the blood to be taken up and used by various tissues. In addition to storing energy, adipose tissue surrounds and pads organs. Adipose tissue located under the skin is an insulator, which prevents heat loss.

Cholesterol is an important molecule that has many functions in the body. It can either be obtained in food or manufactured by the liver and most other tissues. Cholesterol is a component of the plasma membrane, and it can be modified to form other useful molecules, such as bile salts and steroid hormones. Bile salts are necessary for lipid digestion and absorption. Steroid hormones include the reproductive hormones estrogen, progesterone, and testosterone.

The eicosanoids (ī′kō-să-noydz), which include prostaglandins and leukotrienes, are derived from fatty acids. The molecules are involved in activities such as inflammation, blood clotting, tissue repair, and smooth muscle contraction. Phospholipids, such as lecithin, are part of the plasma membrane and are used to construct the myelin sheath around the axons of neurons. Lecithin is also found in bile and helps emulsify fats.

Recommended Consumption of Lipids

The AMDR for lipids is 20–35% for adults, 25–35% for children and adolescents 4 to 18 years of age, and 30–35% for children 2 to 3 years of age. Saturated fats should amount to no more than 10% of total kilocalories or be as low as possible. Most dietary fat should come from sources of polyunsaturated and monounsaturated fats. Nutritionists recommend that people limit their cholesterol intake to 300 mg (the amount in one egg yolk) or less per day and keep their *trans* fat consumption as low as possible. These guidelines reflect the belief that excess fats, especially saturated fats, *trans* fats, and cholesterol, contribute to cardiovascular disease. The typical American diet derives 35–45% of its kilocalories from lipids, indicating that most Americans need to reduce their dietary fat consumption.

Most of the lecithin consumed in the diet is broken down in the digestive tract. The liver can manufacture all the lecithin necessary to meet the body's needs, so taking lecithin supplements is not necessary.

Alpha-linolenic (lin-ō-len′ik) **acid** is an omega-3 fatty acid, and **linoleic** (lin-ō-lē′ik) **acid** is an omega-6 fatty acid. They are **essential fatty acids,** which must be ingested because humans lack the enzymes necessary to synthesize them. Other fatty acids, such as omega-9 fatty acids, can be synthesized from essential fatty acids. Seeds, nuts, and legumes are good sources of alpha-linolenic and linoleic acids. Alpha-linolenic acid is in the green leaves of plants, and linoleic acid is in grains.

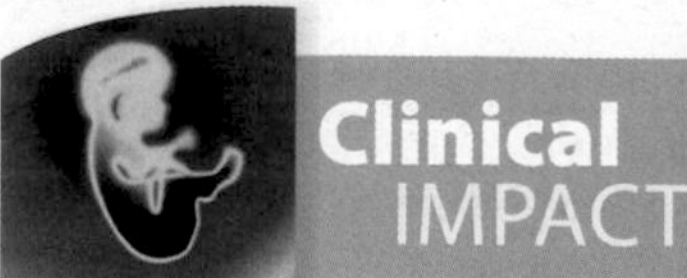

Clinical IMPACT

Saturating Fats

Solid fats, mainly shortening and margarine, work better than liquid oils for preparing some foods, such as pastries. Polyunsaturated vegetable oils can be changed from a liquid to a solid by making them more saturated—that is, by decreasing the number of double covalent bonds in their polyunsaturated fatty acids. To saturate an unsaturated oil, the oil can be *hydrogenated,* which means that hydrogen gas is bubbled through the oil to produce a change in molecular shape that solidifies the oil. The more saturated the product, the harder it becomes at room temperature. These processed fats are usually referred to as ***trans* fats.**

Unprocessed polyunsaturated fats are found mostly in the ***cis* form,** which means the hydrogen atoms are on the same side of the carbon–carbon double bond in their fatty acids (see figure 2.17). During hydrogenation, some of the hydrogen atoms are transferred to the opposite side of the double bond to make the ***trans* form,** characterized by one hydrogen atom on one side of the double bond and another on the opposite side. Processed foods and oils account for most of the *trans* fats in the American diet, although some *trans* fats occur naturally in food from animal sources. *Trans* fatty acids raise the concentration of low-density lipoproteins and lower the concentration of high-density lipoproteins in the blood (see chapter 24). These changes are associated with a greater risk for cardiovascular disease. Beginning in 2006, the Food and Drug Administration (FDA) required that food labels include a detailed list of the amounts of saturated and *trans* fats, allowing the consumer to make better dietary choices.

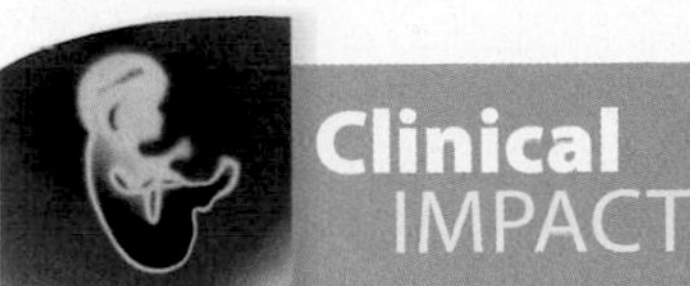

Clinical IMPACT

Fatty Acids and Blood Clotting

Essential fatty acids are used to synthesize prostaglandins that affect blood clotting. Linoleic acid can be converted to **arachidonic** (ă-rak-i-don′ik) **acid,** an omega-6 fatty acid used to produce thromboxanes, which increase blood clotting. Alpha-linolenic acid can be converted to **eicosapentaenoic** (ī′kō-să-pen-tă-nō′ik) **acid (EPA)** and **docasahexaenoic** (dō′kō-să-heks-ă-nō′ik) **acid (DHA),** omega-3 fatty acids that can be used to synthesize prostaglandins, which decrease blood clotting. People who eat foods rich in EPA and DHA, such as herring, salmon, tuna, and sardines, increase the synthesis of prostaglandins from EPA and DHA. Those who eat these fish two or more times per week have a lower risk for heart attack than those who do not, possibly because of reduced blood clotting. EPA and DHA are also known to reduce blood triglyceride levels. People who do not like to eat fish can take fish oil supplements as a source of EPA and DHA. Flaxseed is a source of alpha-linolenic acid, from which EPA and DHA can be synthesized. Whether or not flaxseed can provide the same benefits as EPA and DHA is under investigation. People who have bleeding disorders, take anticoagulants, or anticipate surgery should follow their physicians' advice regarding the use of these supplements because they can increase the risk for bleeding and hemorrhagic stroke.

ASSESS YOUR PROGRESS

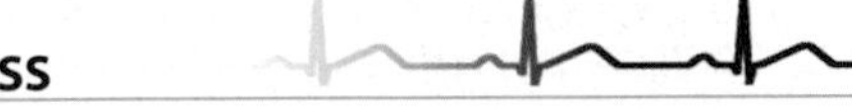

10. *What is the major source of lipids in the diet? What are some other sources?*
11. *Distinguish between saturated and unsaturated fats. What are trans fats?*
12. *How does the body use triglycerides, cholesterol, prostaglandins, and lecithin?*
13. *Describe the recommended dietary intake of lipids. List the essential fatty acids, and state food sources that contain them.*

Proteins

Sources in the Diet

Proteins are chains of amino acids (see chapter 2). Proteins in the body are constructed of 20 kinds of amino acids, which are divided into two groups: essential and nonessential. The body cannot synthesize **essential amino acids,** so they must be obtained in the diet. The nine essential amino acids are histidine, isoleucine, leucine, lysine, methionine, phenylalanine, threonine, tryptophan, and valine. **Nonessential amino acids,** which are necessary to construct our proteins, do not necessarily need to be ingested because they can be synthesized from essential amino acids.

A **complete protein** is a food that contains adequate amounts of all nine essential amino acids, whereas an **incomplete protein** does not. Examples of complete protein foods are meat, fish, poultry, milk, cheese, and eggs; incomplete proteins include leafy green vegetables, grains, and legumes (peas and beans). If two incomplete proteins, such as rice and beans, are ingested together, the amino acid composition of each complements the other, and a complete protein is created. Thus, a vegetarian diet, if balanced correctly, provides all the essential amino acids.

Uses of Proteins in the Body

The body uses essential and nonessential amino acids to synthesize proteins. Proteins perform numerous functions. For example, collagen provides structural strength in connective tissue, as does keratin in the skin, and the combination of actin and myosin makes muscle contraction possible. Enzymes regulate the rate of chemical reactions, and protein hormones regulate many physiological processes (see chapter 18). Proteins in the blood prevent changes in pH (buffers), promote blood clotting (coagulation factors), and transport oxygen and carbon dioxide (hemoglobin). Transport proteins (see chapter 3) move materials across plasma membranes, and other proteins in the plasma membrane function as receptor molecules. Antibodies, lymphokines, and complement are part of the immune system response that protects against microorganisms and other foreign substances.

The body also uses proteins for energy. As an energy source, proteins yield the same amount as carbohydrates. If excess proteins are ingested, the energy in the proteins can be stored by converting their amino acids into glycogen or lipids.

Recommended Consumption of Proteins

The AMDR for protein is 10–35% of total kilocalories. When protein intake is adequate, the synthesis and breakdown of proteins in a healthy adult occur at the same rate. The amino acids of proteins contain nitrogen, so saying that a person is in **nitrogen balance** means that the nitrogen content of ingested protein is equal to the nitrogen excreted in urine and feces. A starving person is in negative nitrogen balance because the nitrogen gained in the diet is less than that lost by excretion. In other words, when proteins are broken down for energy, more nitrogen is lost than is replaced in the diet. On the other hand, a growing child or a healthy pregnant woman is in positive nitrogen balance because more nitrogen is going into the body to produce new tissues than is lost by excretion.

ASSESS YOUR PROGRESS

14. *Distinguish between essential and nonessential amino acids, and between complete and incomplete protein foods.*
15. *Describe the functions of proteins in the body.*
16. *What is the AMDR of proteins? What is nitrogen balance? What would place a person in negative nitrogen balance?*

Vitamins

Vitamins (vīt′ă-minz; life-giving chemicals) are organic molecules that exist in minute quantities in food and are essential to normal metabolism (table 25.2). **Essential vitamins** cannot be produced by the body and must be obtained through the diet. Because no single food or nutrient provides all the essential vitamins, people should maintain a balanced diet by eating a variety of foods. The absence of an essential vitamin in the diet can result in a deficiency disease.

A few vitamins, such as vitamin K, are produced by intestinal bacteria, and a few others can be formed by the body from substances called provitamins. A **provitamin** is a part of a vitamin that the body can assemble or modify into a functional vitamin. For example, beta carotene is a provitamin that the body can form into vitamin A. The other provitamins are **7-dehydrocholesterol** (dē-hī′dro-kō-les′ter-ol), which can be converted to vitamin D, and **tryptophan** (trip′tō-fan), which can be converted to niacin.

Rather than breaking down vitamins by catabolism, the body uses them in their original or slightly modified forms. If the chemical structure of a vitamin is destroyed, its function is usually lost. For example, the chemical structure of many vitamins is destroyed by heat, as when food is overcooked.

Many vitamins function as **coenzymes,** which combine with enzymes to make the enzymes functional (see chapter 2). Without enzymes and their coenzymes, many chemical reactions would occur too slowly to support good health and life. For example, vitamins B_2 and B_3, biotin (bī′ō-tin), and pantothenic (pan-tō-then′ik) acid are critical for some of the chemical reactions involved in producing ATP. Folate (fō′lāt) and vitamin B_{12} are required for nucleic acid synthesis. Vitamins A, B_1, B_6, B_{12}, C, and D are necessary for growth. Vitamin K is necessary for the synthesis of proteins involved in blood clotting (table 25.2).

Vitamins are either fat-soluble or water-soluble. **Fat-soluble vitamins,** such as vitamins A, D, E, and K, dissolve in lipids and are absorbed from the intestine along with lipids. Some of them can be stored in the body for a long time. Because they can be stored, these vitamins can accumulate in the body to the point of toxicity. **Water-soluble vitamins,** such as the B vitamins and vitamin C, dissolve in water. They are absorbed from the water in the intestinal tract and typically remain in the body only a short time before being excreted in the urine.

Vitamins were discovered at the beginning of the twentieth century. They were found to be associated with certain foods known to protect people from diseases such as rickets and beriberi. In 1941, the first Food and Nutrition Board established the **Recommended Dietary Allowances (RDAs),** which are the nutrient intakes sufficient to meet the needs of nearly all people in certain age and gender groups. RDAs were established for different-aged males and females, starting with infants and continuing on to adults. RDAs were also set for pregnant and lactating women. The RDAs have been reevaluated every 4–5 years and updated when necessary on the basis of new information.

The RDAs establish a minimum intake of vitamins and minerals that should protect almost everyone (97%) in a given group from diseases caused by vitamin or mineral deficiencies. Although personal requirements can vary, the RDAs are a good benchmark. The further below the RDAs an individual's dietary intake is, the more likely that person is to develop a nutritional deficiency. On the other hand, consuming too much of some vitamins and minerals can be harmful. For example, the long-term ingestion of 3–10 times the RDA for vitamin A can cause bone and muscle pain, skin disorders, hair loss, and an enlarged liver. Consuming 5–10 times the RDA of vitamin D over the long term can result in calcium deposits in the kidneys, heart, and blood vessels, and consuming more than 2 g of vitamin C daily can cause stomach inflammation and diarrhea.

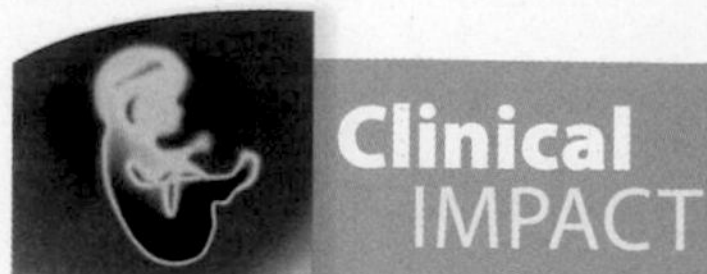

Clinical IMPACT

Free Radicals and Antioxidants

Free radicals are molecules, produced as part of normal metabolism, that are missing an electron. The loss of an electron from a molecule is called oxidation. Free radicals can replace the missing electron by taking an electron from cell molecules, such as lipids, proteins, or DNA, resulting in damage to the cell. Damage from free radicals may contribute to aging and certain diseases, such as atherosclerosis and cancer. However, substances called **antioxidants** may counteract these effects by donating an electron to free radicals and thus preventing the oxidation of cell components. Examples of antioxidants are beta carotene (provitamin A), vitamin C, and vitamin E.

Many studies have attempted to determine whether taking large doses of antioxidants is beneficial. The best evidence presently available does not support the claim that large doses of antioxidants can prevent chronic disease or otherwise improve health. On the other hand, the amount of antioxidants normally found in a balanced diet (including fruits and vegetables rich in antioxidants), combined with the complex mix of other chemicals found in food, can be beneficial.

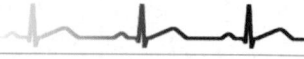

ASSESS YOUR PROGRESS

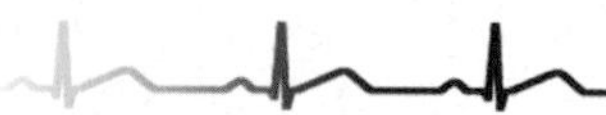

17. *What are vitamins, essential vitamins, and provitamins?*

18. *Name the fat-soluble vitamins and the water-soluble vitamins.*

19. *List some of the functions of vitamins in the body.*

20. *What are Recommended Dietary Allowances (RDAs)? Why are they useful?*

21. *What are some symptoms of specific vitamin deficiencies?*

Predict 2

What would happen if vitamins were broken down during digestion rather than being absorbed intact into the circulation?

Minerals

Minerals (min′er-ălz) are inorganic nutrients that are necessary for normal metabolic functions. The minerals are divided into two groups based on the amount required in the diet for good health. The daily requirement for **major minerals** is 100 mg or more daily; for **trace minerals,** less than 100 mg daily can suffice. The requirement for some trace minerals is unknown. Minerals constitute about 4–5% of total body weight and are components of coenzymes, a few vitamins, hemoglobin, and other organic molecules. Minerals are involved in a number of important functions, including establishing resting membrane potentials and generating action potentials, adding mechanical strength to bones and teeth, combining with organic molecules, and acting as coenzymes, buffers, and regulators of osmotic pressure. Table 25.3 lists important minerals and their functions.

TABLE 25.2 **Principal Vitamins**

Vitamin	Fat-Soluble (F) or Water-Soluble (W)	Source	Function	Symptoms of Deficiency	Reference Daily Intake (RDI)*
A (retinol)	F	From provitamin beta carotene found in yellow and green vegetables: preformed in liver, egg yolk, butter, and milk	Necessary for rhodopsin synthesis, normal health of epithelial cells, and bone and tooth growth	Rhodopsin deficiency, night blindness, retarded growth, skin disorders, and increased infection risk	5000 IU
B_1 (thiamine)	W	Yeast, grains, and milk	Involved in carbohydrate and amino acid metabolism, necessary for growth	Beriberi—muscle weakness (including cardiac muscle), neuritis, and paralysis	1.5 mg
B_2 (riboflavin)	W	Green vegetables, liver, wheat germ, milk, and eggs	Component of flavin adenine dinucleotide, involved in citric acid cycle	Eye disorders and skin cracking, especially at corners of the mouth	1.7 mg
B_3 (niacin)	W	Fish, liver, red meat, yeast, grains, peas, beans, and nuts	Component of nicotinamide adenine dinucleotide, involved in glycolysis and citric acid cycle	Pellagra—diarrhea, dermatitis, and nervous system disorders	20 mg
Pantothenic acid	W	Liver, yeast, green vegetables, grains, and intestinal bacteria	Constituent of coenzyme-A, glucose production from lipids and amino acids, and steroid hormone synthesis	Neuromuscular dysfunction and fatigue	10 mg
Biotin	W	Liver, yeast, eggs, and intestinal bacteria	Fatty acid and nucleic acid synthesis, movement of pyruvic acid into citric acid cycle	Mental and muscle dysfunction, fatigue, and nausea	30 µg
B_6 (pyridoxine)	W	Fish, liver, yeast, tomatoes, and intestinal bacteria	Involved in amino acid metabolism	Dermatitis, retarded growth, and nausea	2.0 mg
Folate	W	Liver, leafy green vegetables, and intestinal bacteria	Nucleic acid synthesis, hemopoiesis, and prevention of birth defects	Macrocytic anemia (enlarged red blood cells) and neural tube defects	0.4 mg
B_{12} (cobalamins)	W	Liver, red meat, milk, and eggs	Necessary for red blood cell production, some nucleic acid and amino acid metabolism	Pernicious anemia and nervous system disorders	6.0 µg
C (ascorbic acid)	W	Citrus fruit, tomatoes, and green vegetables	Collagen synthesis, general protein metabolism	Scurvy—defective bone formation and poor wound healing	60 mg
D (cholecalciferol, ergosterol)	F	Fish liver oil, enriched milk, and eggs; provitamin D converted by sunlight to cholecalciferol in the skin	Promotes calcium and phosphorus use, normal growth and bone and tooth formation	Rickets—poorly developed, weak bones; osteomalacia; and bone reabsorption	400 IU
E (tocopherol, tocotrienols)	F	Wheat germ; cottonseed, palm, and rice oils; grain; liver; and lettuce	Prevents oxidation of plasma membranes and DNA	Hemolysis of red blood cells	30 IU
K (phylloquinone)	F	Alfalfa, liver, spinach, vegetable oils, cabbage, and intestinal bacteria	Required for synthesis of a number of clotting factors	Excessive bleeding due to retarded blood clotting	80 µg

*Reference Daily Intakes for people over 4 years of age; IU = international units.

People ingest minerals alone or in combination with organic molecules, as well as obtain them from both animal and plant sources. However, mineral absorption from plants can be limited because the minerals tend to bind to plant fibers. Refined breads and cereals contain hardly any minerals or vitamins because the seeds used to make them are crushed, and the outer parts of the seeds housing

TABLE 25.3 Important Minerals

Mineral	Function	Symptoms of Deficiency	Reference Daily Intake (RDI)*
Calcium	Bone and teeth formation, blood clotting, muscle activity, nerve function	Spontaneous action potential generation in neurons, tetany	1000 mg
Chlorine	Blood acid-base balance; hydrochloric acid production in stomach	Acid-base imbalance	2.3 g[†]
Chromium	Associated with enzymes in glucose metabolism	Unknown	35 μg
Cobalt	Component of vitamin B_{12}, red blood cell production	Anemia	Unknown
Copper	Hemoglobin and melanin production, electron-transport system	Anemia and loss of energy	0.9 mg
Fluorine	Provides extra strength in teeth, prevents dental caries	No real pathology	4 mg
Iodine	Thyroid hormone production; maintenance of normal metabolic rate	Goiter and decrease in normal metabolism	150 μg
Iron	Component of hemoglobin; ATP production in electron-transport system	Anemia, decreased O_2 transport, energy loss	18 mg
Magnesium	Coenzyme constituent, bone formation, muscle and nerve function	Increased nervous system irritability, vasodilation, arrhythmias	420 mg
Manganese	Hemoglobin synthesis, growth, activation of several enzymes	Tremors, convulsions	2.3 mg
Molybdenum	Enzyme component	Unknown	45 μg
Phosphorus	Bone and teeth formation; energy transfer (ATP); component of nucleic acids	Loss of energy and cellular function	1250 mg
Potassium	Muscle and nerve function	Muscle weakness, abnormal electrocardiogram, alkaline urine	3.5 g
Selenium	Component of many enzymes	Unknown	55 μg
Sodium	Osmotic pressure regulation; nerve and muscle function	Nausea, vomiting, exhaustion, dizziness	1.5 g[†]
Sulfur	Component of hormones, several vitamins, proteins	Unknown	Unknown
Zinc	Component of several enzymes; CO_2 transport and metabolism; protein metabolism	Deficient CO_2 transport, deficient protein metabolism	11 mg

*Reference Daily Intakes for people over 4 years of age, except for sodium.
[†]3.8 g sodium chloride (table salt).

most of the minerals and vitamins, are discarded. Minerals and vitamins are often added to refined breads and cereals to compensate for their loss during the refinement process.

A balanced diet can provide all the vitamins and minerals required for good health for most people. Some nutritionists, however, recommend taking a once-a-day multiple vitamin and mineral supplement as insurance because many people's diets are not balanced.

ASSESS **YOUR PROGRESS**

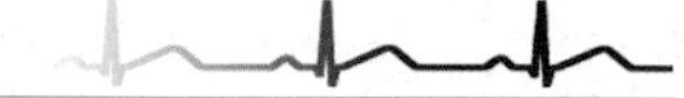

22. *What are minerals? What is the daily requirement for major minerals?*

23. *List the several minerals the body needs and a function of each.*

Daily Values for Nutrients

Daily Values are dietary values that appear on food labels to help consumers plan a healthy diet. Daily Values are based on two other sets of reference values: Reference Daily Intakes and Daily Reference Values:

- **Reference Daily Intakes (RDIs)** are based on the 1968 RDAs for certain vitamins and minerals. RDIs have been set for four categories of people: infants, toddlers, people over 4 years of age, and pregnant or lactating women. Generally, the RDIs are set to the highest 1968 RDA value of an age category. For example, the highest RDA for iron in males over 4 years of age is 10 mg/day and, for females over 4 years of age, 18 mg/day. Thus, the RDI for iron is set at 18 mg/day.

Nutrition Facts

Serving Size 1 oz. (28g/About 32 chips)
Servings Per Container 2.5

Amount Per Serving

Calories 160	Calories from Fat 90
	% Daily Value*
Total Fat 10g	**16%**
Saturated Fat 1.5g	**7%**
Cholesterol 0mg	**0%**
Sodium 170mg	**7%**
Total Carbohydrate 15g	**5%**
Dietary Fiber 1g	**4%**
Sugars less than 1g	
Protein 2g	

Vitamin A 0% • Vitamin C 0%
Calcium 2% • Iron 0%

* Percent Daily Values are based on a 2,000 calorie diet. Your daily values may be higher or lower depending on your calorie needs:

	Calories:	2,000	2,500
Total Fat	Less than	65g	80g
Sat Fat	Less than	20g	25g
Cholesterol	Less than	300mg	300mg
Sodium	Less than	2,400mg	2,400mg
Total Carbohydrate		300g	375g
Dietary Fiber		25g	30g

Calories per gram:
Fat 9 • Carbohydrate 4 • Protein 4

FIGURE 25.2 A Typical Food Label

- **Daily Reference Values (DRVs)** are set for total fat, saturated fat, cholesterol, total carbohydrate, dietary fiber, sodium, potassium, and protein.

Having two standards on food labels, RDIs for vitamins and minerals and DRVs for other nutrients, was thought to be more confusing for consumers than having one standard. Therefore, the RDIs and DRVs were combined to form the Daily Values.

The Daily Values appearing on food labels are based on a 2000 kcal reference diet, which approximates the weight maintenance requirements of postmenopausal women, women who exercise moderately, teenage girls, and sedentary men (figure 25.2). On large food labels, additional information is listed based on a daily intake of 2500 kcal, which is adequate for young men. However, government standards do not require that all possible Daily Values be listed on food labels.

The Daily Values for energy-producing nutrients are determined as a percentage of daily kilocalorie intake: 60% for carbohydrates, 30% for total fats, 10% for saturated fats, and 10% for proteins. The Daily Value for fiber is 14 g for each 1000 kcal of intake. The Daily Values for a nutrient in a 2000 kcal/day diet can be calculated on the basis of the recommended daily percentage of the nutrient and the kilocalories in a gram of the nutrient. For example, carbohydrates should compose 60% of a 2000 kcal/day diet, or 1200 kcal/day (0.60 × 2000). Since there are 4 kcal in a gram of carbohydrate, the Daily Value for carbohydrate is 300 g/day (1200/4).

The Daily Values for some nutrients are limited due to their link to certain diseases. Thus, the Daily Values for total fats are less than 65 g; for saturated fats, less than 20 g; and for cholesterol, less than 300 mg because these nutrients are associated with increased risk for heart disease. The Daily Value for sodium is less than 2400 mg because of its association with high blood pressure in some people.

For a particular food, the Daily Value is used to calculate the **Percent Daily Value (% Daily Value)** for some of the nutrients in one serving of the food (figure 25.2). For example, if a serving of food has 3 g of fat and the Daily Value for total fat is 65 g, the % Daily Value is 5% (3/65 = 0.05, or 5%). The Food and Drug Administration (FDA) requires % Daily Values to be on food labels, so that the public has useful and accurate dietary information.

Predict 3

One serving of a food contains 30 g of carbohydrate. What % Daily Value for carbohydrate would appear on the food label?

The % Daily Values for nutrients related to energy consumption are based on a 2000 kcal/day diet. For people who maintain their weight on a 2000 kcal/day diet, the total of the % Daily Values for each of these nutrients should be no more than 100%. However, for individuals consuming more or fewer than 2000 kcal/day, the total % Daily Values can be more or fewer than 100%. For example, for a person consuming 2200 kcal/day, the total of the % Daily Values for each of these nutrients should be no more than 110% because 2200/2000 = 1.10, or 110%.

When using the % Daily Values of a food to determine how the amounts of its nutrients fit into the overall diet, the number of servings in a container or package needs to be considered. For example, suppose a small (2.25-ounce) bag of corn chips has a % Daily Value of 16% for total fat. A person might suppose that eating the bag of chips accounts for 16% of total fat for the day. The bag, however, contains 2.5 servings. Therefore, if all the chips in the bag are consumed, they account for 40% (16% × 2.5) of the maximum recommended total fat.

ASSESS YOUR PROGRESS

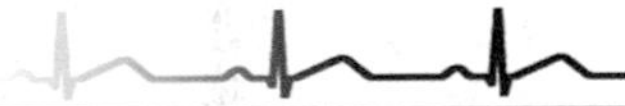

24. *What are the Reference Daily Intakes and the Daily Reference Values? When combined, what reference set of values is established?*
25. *Define* % Daily Values. *The % Daily Values appearing on food labels is based on how many kilocalories per day?*

25.2 Metabolism

LEARNING OUTCOMES

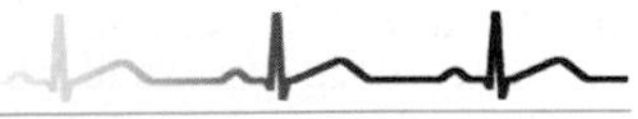

After reading this section, you should be able to

A. Define *metabolism, anabolism,* and *catabolism.*

B. Relate hydrogen atoms to energy.

Metabolism (mĕ-tab′ō-lizm; change) is the total of all the chemical reactions that occur in the body. It consists of **catabolism** (kă-tab′ō-lizm), the energy-releasing process by which large molecules are broken down into smaller molecules, and **anabolism** (ă-nab′ō-lizm), the energy-requiring process by which small molecules are joined to form larger molecules.

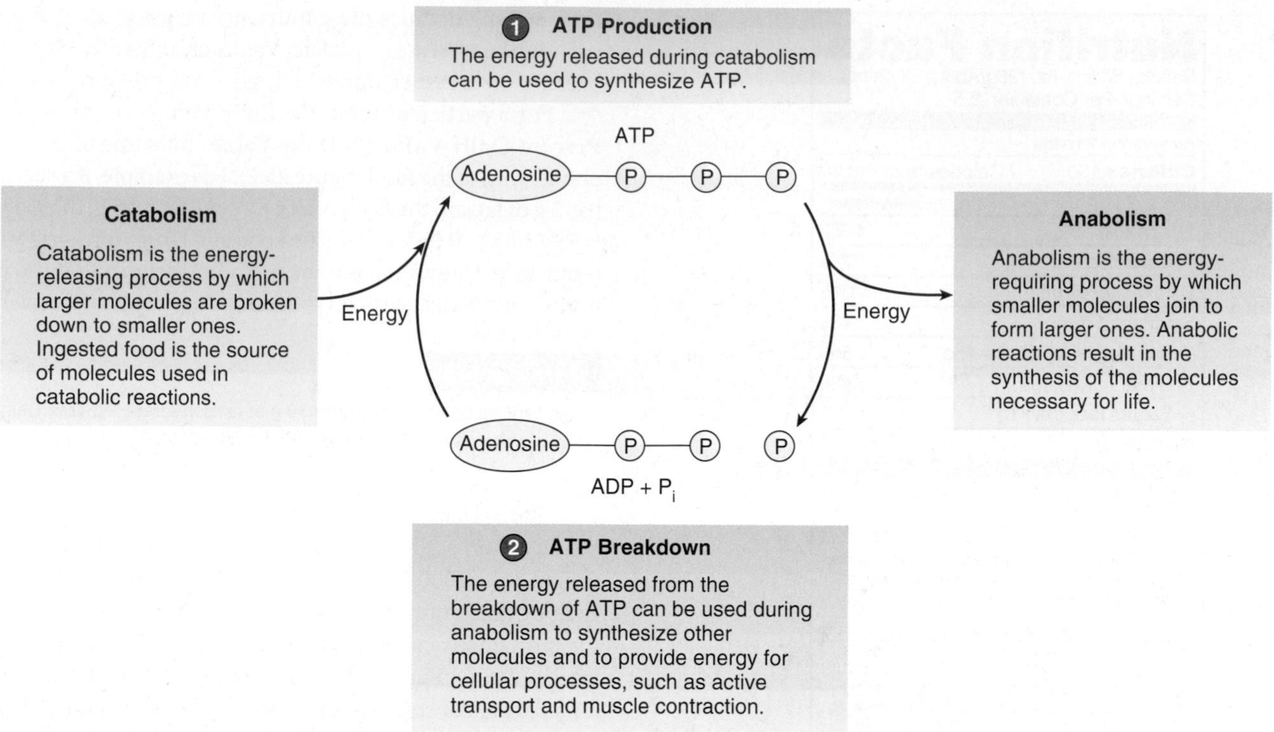
1 ATP Production
The energy released during catabolism can be used to synthesize ATP.
ATP
Adenosine
P
P
P
Catabolism
Catabolism is the energy-releasing process by which larger molecules are broken down to smaller ones. Ingested food is the source of molecules used in catabolic reactions.
Energy
Energy
Anabolism
Anabolism is the energy-requiring process by which smaller molecules join to form larger ones. Anabolic reactions result in the synthesis of the molecules necessary for life.
Adenosine
P
P
P
ADP + Pi
2 ATP Breakdown
The energy released from the breakdown of ATP can be used during anabolism to synthesize other molecules and to provide energy for cellular processes, such as active transport and muscle contraction.

PROCESS FIGURE 25.3 ATP Derived from Catabolic Reactions Drives Anabolic Reactions
Energy released by catabolism is required to form ATP from ADP and phosphate (P_i). Energy and a phosphate are given off when ATP is converted back to ADP during anabolism.

Catabolism begins during digestion and is concluded within individual cells. The energy derived from catabolism is used to drive anabolic reactions and processes such as active transport and muscle contraction. Anabolism occurs in all the body cells as they divide to form new cells, maintain their own intracellular structure, and produce molecules, such as hormones, neurotransmitters, and extracellular matrix molecules, for export.

Large nutrient molecules, such as carbohydrates, lipids, and proteins, are broken down by digestion into smaller molecules, such as glucose, amino acids, and fatty acids, which are absorbed from the digestive tract into the blood (see chapter 24). These smaller molecules are taken into cells, where they are catabolized, and the energy from them is used to combine adenosine diphosphate (ADP) and an inorganic phosphate group (P_i) to form ATP (figure 25.3):

$$\text{ADP} + \text{P}_i + \text{Energy} \rightarrow \text{ATP}$$

The energy in small nutrient molecules is used to produce many ATP molecules, each of which stores a small amount of energy. These smaller amounts of energy are more readily available for use in cells than is the larger amount of energy stored in nutrient molecules. ATP is often called the energy currency of the cell because, when it is spent, or broken down to ADP, energy becomes available for use by the cell. As a comparison, if a quarter represents an ATP molecule, then a $20 bill is analogous to a small nutrient molecule. The quarter (ATP) can be used in various vending machines (chemical reactions), but the $20 bill (nutrient molecule) cannot.

The chemical reactions responsible for transferring energy from the chemical bonds of nutrient molecules to ATP molecules involve oxidation-reduction reactions (see chapter 2). A molecule is reduced when it gains electrons and oxidized when it loses electrons. A nutrient molecule has many hydrogen atoms covalently bonded to the carbon atoms that form the "backbone" of the molecule. Because a hydrogen atom is composed of a H^+ (proton) and an electron, the nutrient molecule has many electrons and is, therefore, highly reduced. When a H^+ and an associated electron are lost from the nutrient molecule, the molecule loses energy and becomes oxidized. The energy in the electron is used to synthesize ATP. The major events of ATP synthesis are summarized in figure 25.4.

ASSESS YOUR PROGRESS

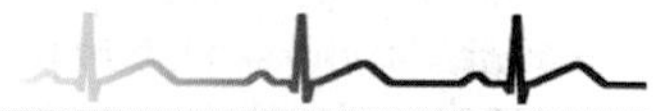

26. *How are metabolism, anabolism, and catabolism related? How is the energy derived from catabolism used to drive anabolic reactions?*
27. *How does the removal of hydrogen atoms from nutrient molecules result in a loss of energy from the nutrient molecule?*

25.3 Carbohydrate Metabolism

LEARNING OUTCOMES

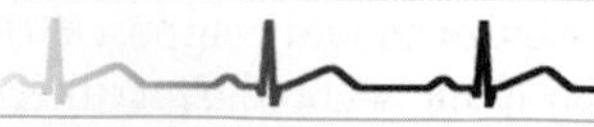

After reading this section, you should be able to

A. Describe glycolysis and name its products.

B. Describe the citric acid cycle and name its products.

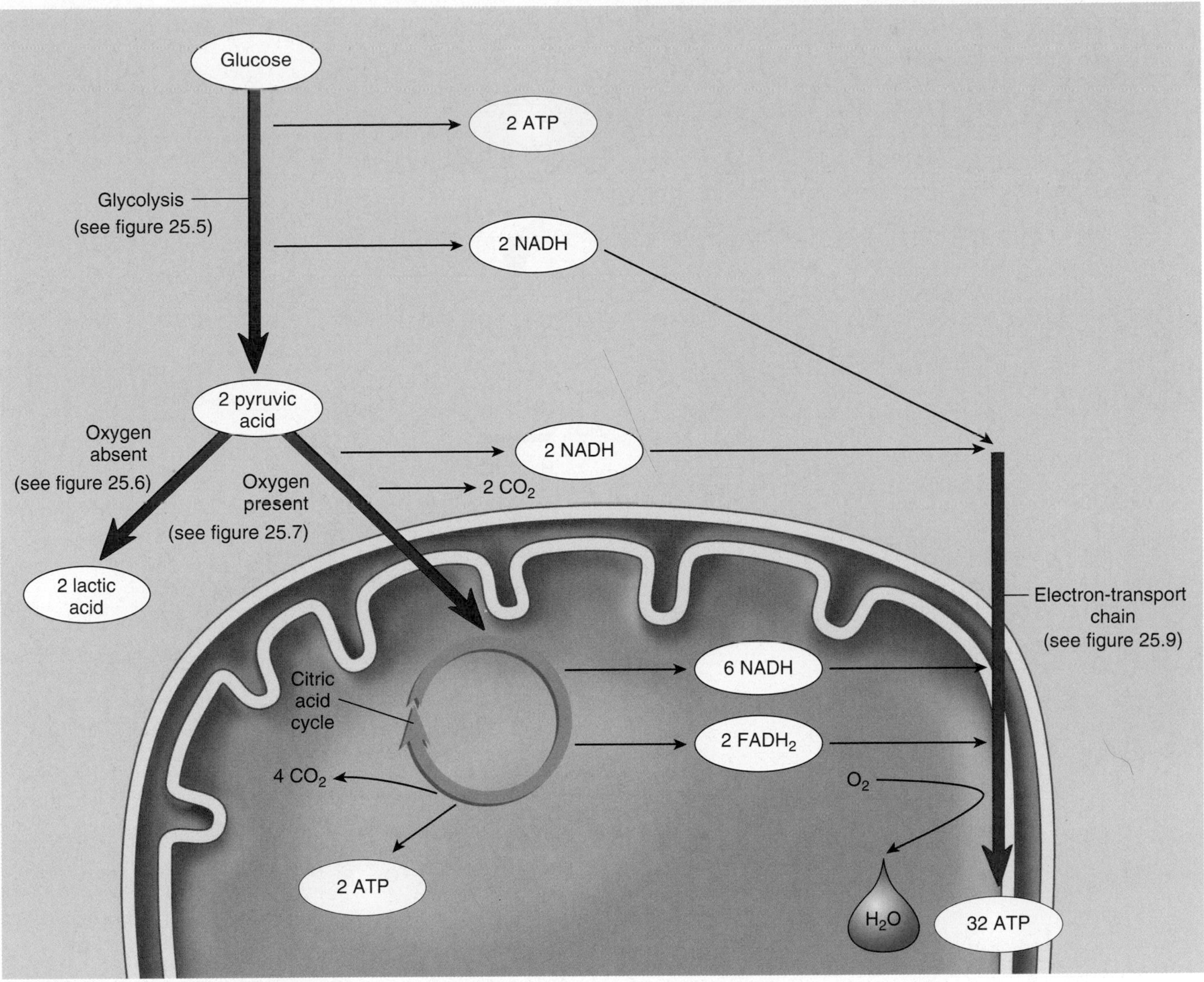
Glucose
2 ATP
Glycolysis
(see figure 25.5)
2 NADH
2 pyruvic acid
Oxygen absent
(see figure 25.6)
Oxygen present
(see figure 25.7)
2 NADH
2 CO_2
2 lactic acid
Electron-transport chain
(see figure 25.9)
Citric acid cycle
6 NADH
2 $FADH_2$
4 CO_2
O_2
2 ATP
H_2O
32 ATP

FIGURE 25.4 APR **ATP Synthesis**
The major events of ATP synthesis are glycolysis, the citric acid cycle, and the electron-transport chain.

C. Explain the electron-transport chain and how ATP is produced in the process.

D. Explain the difference between the number of ATP molecules produced by aerobic respiration and the number produced by lactic acid fermentation.

The breakdown products of carbohydrate digestion are monosaccharides. Of these, glucose is the most important in terms of cellular metabolism. Glucose is transported in the circulation to all the tissues of the body, where it serves as a source of energy. Any excess glucose in the blood following a meal can be used to form **glycogen** (glī′kō-jen; *glyks,* sweet), or it can be partially broken down and the components used to form lipids. Glycogen is a short-term energy-storage molecule that the body can store in limited amounts, whereas lipids are long-term energy-storage molecules that the body can store in large amounts. Most of the body's glycogen is in skeletal muscle and in the liver.

Glycolysis

Carbohydrate metabolism begins with **glycolysis** (glī-kol′i-sis), a series of chemical reactions in the cytosol that results in the breakdown of glucose into two **pyruvic** (pī-roo′vik) **acid** molecules (figure 25.5).

Glycolysis is divided into four phases:

1. *Input of ATP.* The first steps in glycolysis require the input of energy in the form of two ATP molecules (figure 25.5, *step 1*). The energy is necessary to make the glucose molecule, a relatively stable molecule, more reactive. A phosphate group is transferred from ATP to the glucose molecule, a process called **phosphorylation** (fos′fōr-i-lā′shŭn), to form glucose-6-phosphate. The glucose-6-phosphate atoms are rearranged to form fructose-6-phosphate, which is then converted to fructose-1,6-bisphosphate by the addition of another phosphate group from a second ATP molecule.

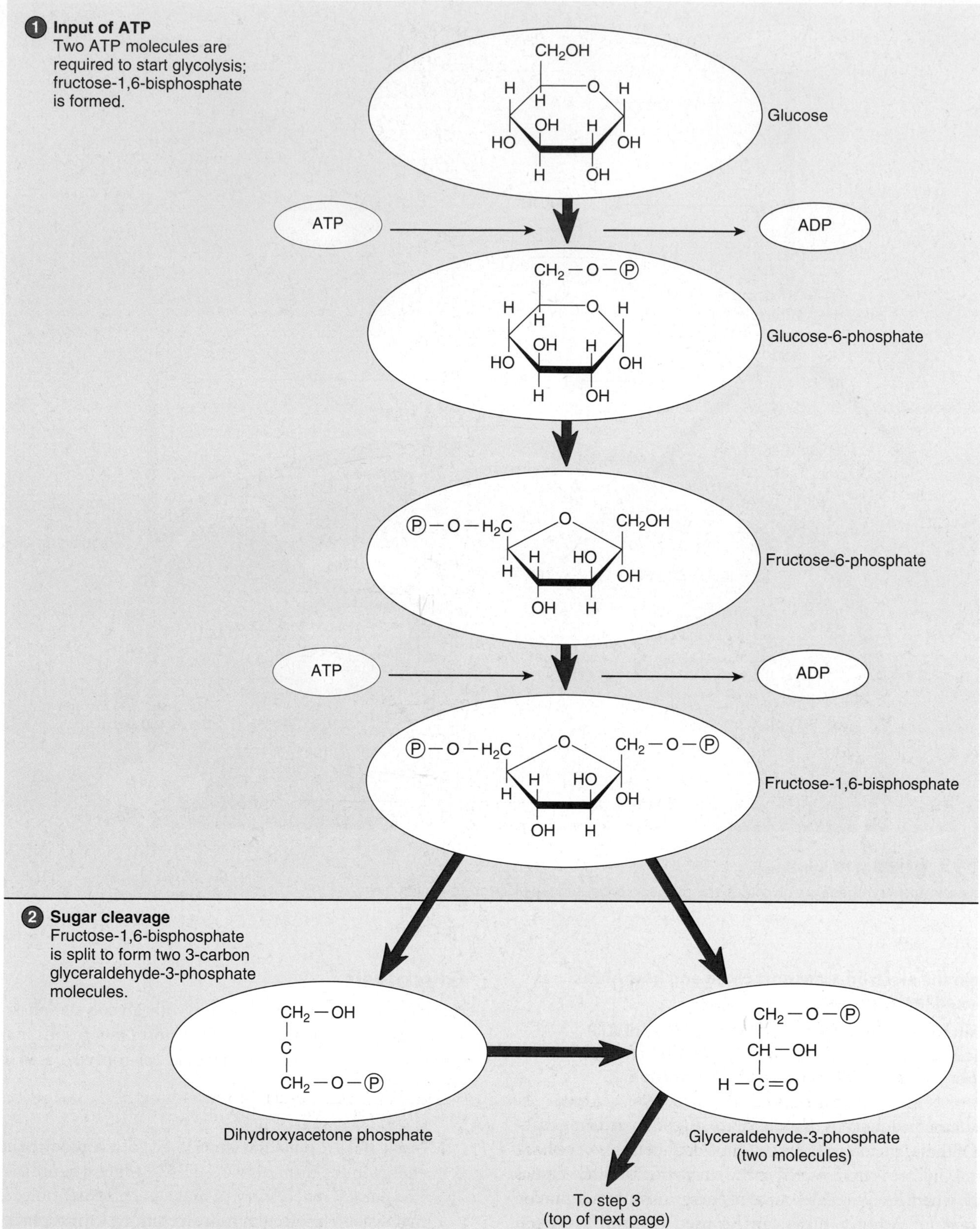

1 Input of ATP
Two ATP molecules are required to start glycolysis; fructose-1,6-bisphosphate is formed.
Glucose
ATP
ADP
Glucose-6-phosphate
Fructose-6-phosphate
ATP
ADP
Fructose-1,6-bisphosphate
2 Sugar cleavage
Fructose-1,6-bisphosphate is split to form two 3-carbon glyceraldehyde-3-phosphate molecules.
Dihydroxyacetone phosphate
Glyceraldehyde-3-phosphate (two molecules)
To step 3 (top of next page)

PROCESS FIGURE 25.5 Glycolysis

The chemical reactions of glycolysis take place in the cytosol.

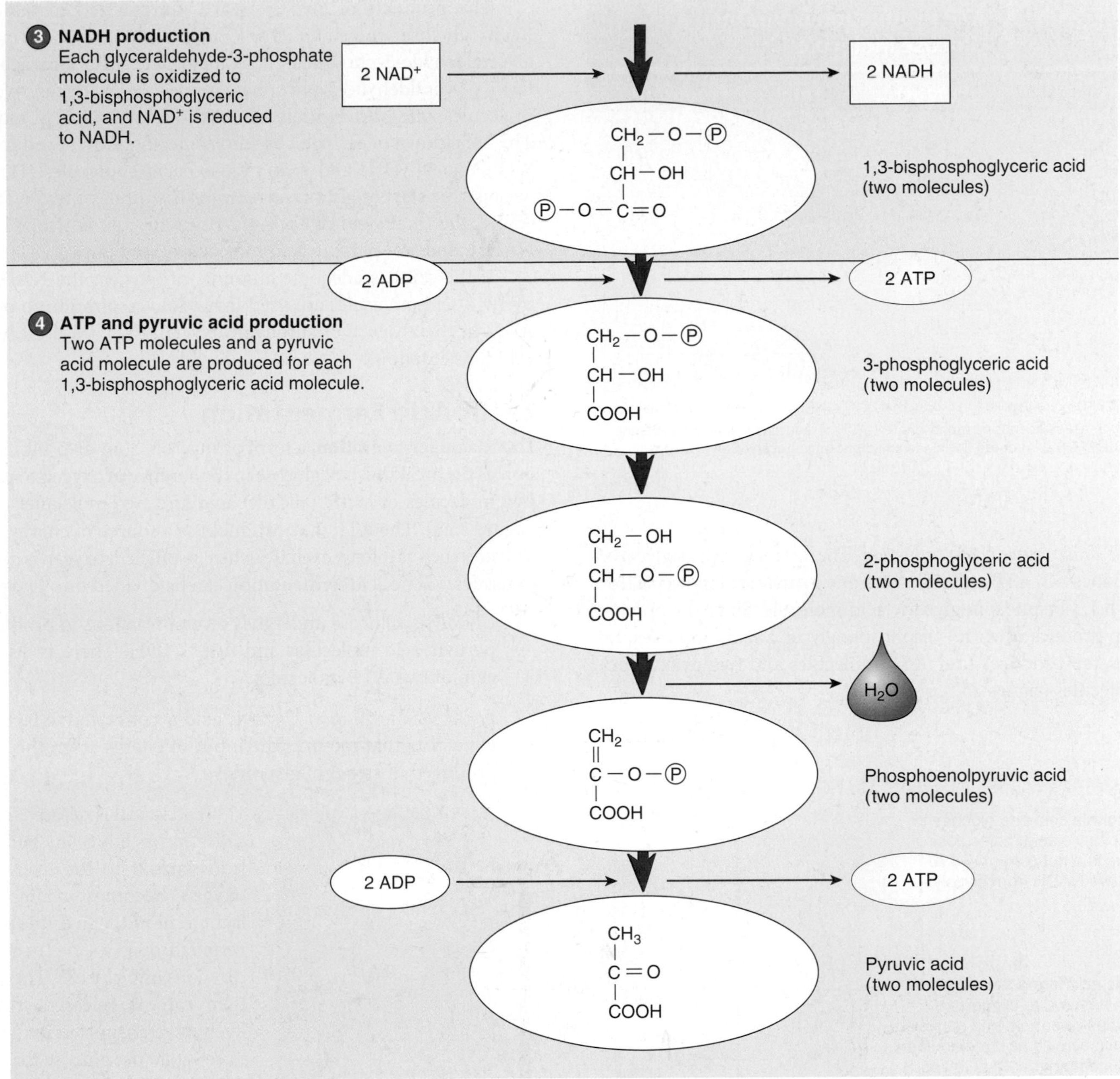

3 NADH production
Each glyceraldehyde-3-phosphate molecule is oxidized to 1,3-bisphosphoglyceric acid, and NAD+ is reduced to NADH.
2 NAD+
2 NADH
1,3-bisphosphoglyceric acid (two molecules)
2 ADP
2 ATP
4 ATP and pyruvic acid production
Two ATP molecules and a pyruvic acid molecule are produced for each 1,3-bisphosphoglyceric acid molecule.
3-phosphoglyceric acid (two molecules)
2-phosphoglyceric acid (two molecules)
H2O
Phosphoenolpyruvic acid (two molecules)
2 ADP
2 ATP
Pyruvic acid (two molecules)

PROCESS FIGURE 25.5 (continued)

2. *Sugar cleavage.* Fructose-1,6-bisphosphate is cleaved into two 3-carbon molecules, glyceraldehyde (glis-er-al′dĕ-hīd)-3-phosphate and dihydroxyacetone (dī′hī-drok-sē-as′e-tōn) phosphate. Dihydroxyacetone phosphate is rearranged to form a second glyceraldehyde-3-phosphate; consequently, two molecules of glyceraldehyde-3-phosphate result (figure 25.5, *step 2*).
3. *NADH production.* Each glyceraldehyde-3-phosphate molecule is oxidized (loses two electrons) to form 1,3- bisphosphoglyceric (biz′phos-fo-gli′sĕr′ik) acid, and **nicotinamide adenine** (nik-ō-tin′ă-mīd ad′ĕ-nēn) **dinucleotide** (NAD^+) is reduced (gains two electrons) to **NADH** (figure 25.5, *step 3*). Glyceraldehyde-3-phosphate also loses two H^+, one of which binds to NAD^+:

$$NAD^+ + 2e^- + 2H^+ \rightarrow NADH + H^+$$

NADH is referred to as an electron carrier molecule. The two high-energy electrons (e^-) gained by NADH can be used to produce ATP molecules through the electron-transport chain (see "Electron-Transport Chain," later in this section).

TABLE 25.4 ATP Production from One Glucose Molecule

Process	Product	Total ATP Produced*
Glycolysis	4 ATP	2 ATP (4 ATP produced minus 2 ATP to start)
	2 NADH	4 ATP
Acetyl-CoA production	2 NADH	6 ATP
Citric acid cycle	2 ATP	2 ATP
	6 NADH	18 ATP
	2 $FADH_2$	4 ATP
Total		36 ATP

*NADH and $FADH_2$ are used in the production of ATP in the electron-transport chain.
Abbreviations: ATP = adenosine triphosphate; NADH = reduced nicotinamide adenine dinucleotide; $FADH_2$ = reduced flavin adenine diphosphate; acetyl-CoA = acetyl coenzyme A.

4. *ATP and pyruvic acid production.* The last four steps of glycolysis produce two ATP molecules and one pyruvic acid molecule from each 1,3-bisphosphoglyceric acid molecule. Since the previous step produced two 1,3-bisphosphoglyceric acid molecules, this last step produces four ATP molecules and two pyruvic acid molecules (figure 25.5, *step 4*).

The products of glycolysis are summarized in table 25.4. Each glucose molecule that enters glycolysis forms two glyceraldehyde-3-phosphate molecules at the sugar cleavage phase. Each glyceraldehyde-3-phosphate molecule produces two ATP molecules, one NADH molecule, and one pyruvic acid molecule. The breakdown of each glucose molecule, therefore, produces four ATP, two NADH, and two pyruvic acid molecules. However, because the start of glycolysis requires the input of two ATP molecules, the final yield of each glucose molecule is two ATP, two NADH, and two pyruvic acid molecules (see figure 25.4).

If the cell has adequate amounts of oxygen, the NADH and pyruvic acid molecules are used in aerobic respiration to produce ATP. In the absence of sufficient oxygen, they are used in lactic acid fermentation.

Lactic Acid Fermentation

Lactic acid fermentation, a form of anaerobic (an-ār-ō′bik) respiration, is the breakdown of glucose in the absence of oxygen to produce two molecules of **lactic** (lak′tik) **acid** and two molecules of ATP (figure 25.6). The ATP thus produced is a source of energy during activities such as intense exercise, when insufficient oxygen is delivered to tissues. Lactic acid fermentation can be divided into two phases:

1. *Glycolysis.* Glucose undergoes several reactions to produce two pyruvic acid molecules and two NADH. There is also a net gain of two ATP molecules.
2. *Lactic acid formation.* Pyruvic acid is converted to lactic acid, a reaction that requires the input of energy from the NADH produced in step 3 of glycolysis.

Lactic acid is released from the cells that produce it, and blood transports it to the liver. When oxygen becomes available, the lactic acid in the liver can be converted through a series of chemical reactions into glucose. The glucose then can be released from the liver and transported in the blood to cells that use glucose for energy. This process of converting lactic acid to glucose is called the **Cori cycle.** Some of the reactions involved in the Cori cycle require energy derived from ATP that is produced by aerobic respiration. The oxygen necessary for synthesizing the ATP is part of the **oxygen deficit** (see chapter 9).

1. Glycolysis converts glucose to two pyruvic acid molecules. There is a net gain of two ATP and two NADH from glycolysis.

2. Lactic acid fermentation, which does not require oxygen, includes glycolysis and converts the two pyruvic acid molecules produced by glycolysis to two lactic acid molecules. This conversion requires energy, which is derived from the NADH generated in glycolysis.

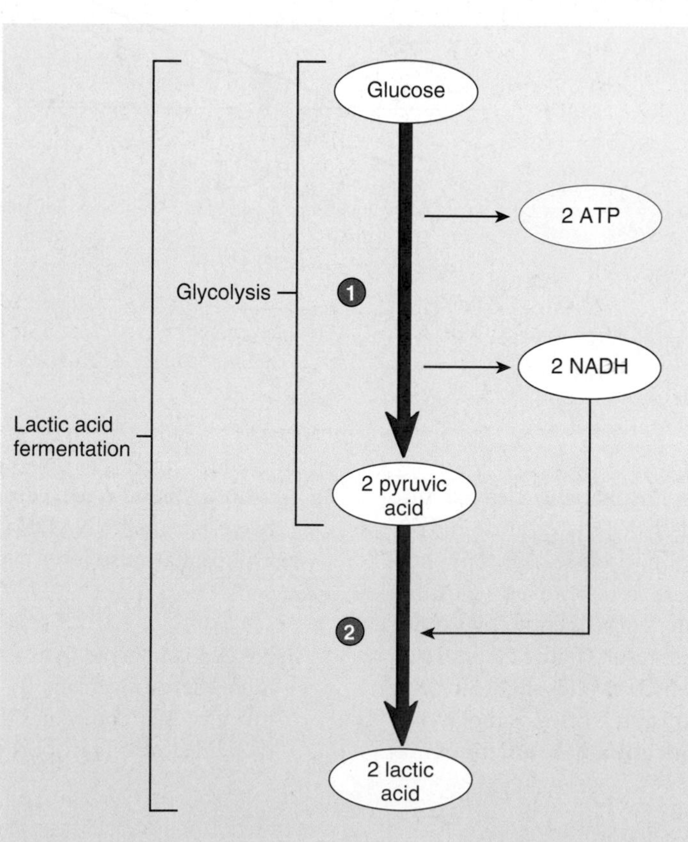
Glucose
2 ATP
Glycolysis
1
2 NADH
Lactic acid fermentation
2 pyruvic acid
2
2 lactic acid

PROCESS FIGURE 25.6
Lactic Acid Fermentation

FUNDAMENTAL **Figure**

1. Glycolysis in the cytosol converts glucose to two pyruvic acid molecules and produces two ATP and two NADH. The NADH can go to the electron-transport chain in the inner mitochondrial membrane.

2. The two pyruvic acid molecules produced in glycolysis are converted to two acetyl-CoA molecules, producing two CO_2 and two NADH. The NADH can go to the electron-transport chain.

3. The two acetyl-CoA molecules enter the citric acid cycle, which produces four CO_2, six NADH, two $FADH_2$, and two ATP. The NADH and $FADH_2$ can go to the electron-transport chain.

4. The electron-transport chain uses NADH and $FADH_2$ to produce 32 ATP. This process requires O_2, which combines with H^+ to form H_2O.

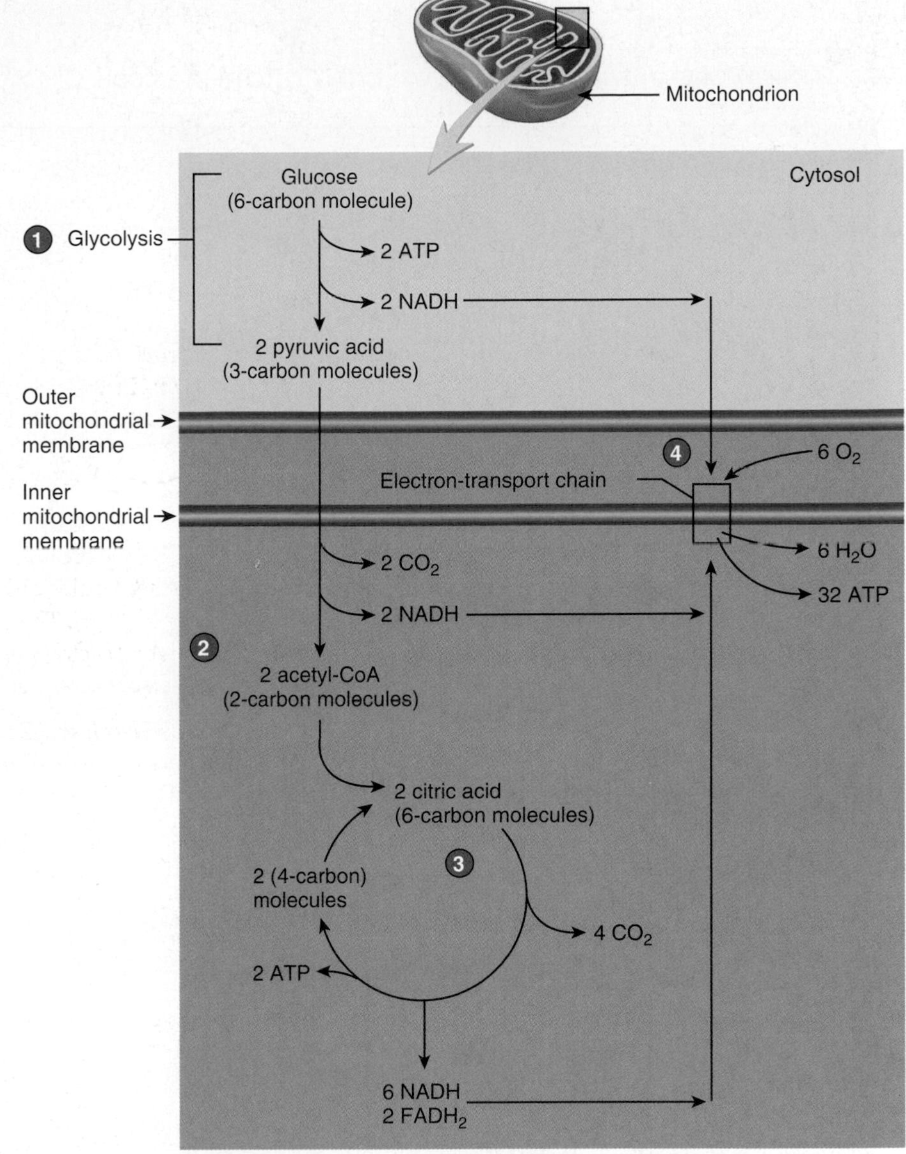

Mitochondrion
Cytosol
Glucose
(6-carbon molecule)
1 Glycolysis
2 ATP
2 NADH
2 pyruvic acid
(3-carbon molecules)
Outer mitochondrial membrane
Inner mitochondrial membrane
4
Electron-transport chain
6 O_2
6 H_2O
32 ATP
2 CO_2
2 NADH
2
2 acetyl-CoA
(2-carbon molecules)
2 citric acid
(6-carbon molecules)
3
2 (4-carbon) molecules
4 CO_2
2 ATP
6 NADH
2 $FADH_2$

PROCESS FIGURE 25.7 Aerobic Respiration
Aerobic respiration involves four phases: (1) glycolysis, (2) acetyl-CoA formation, (3) the citric acid cycle, and (4) the electron-transport chain. The number of carbon atoms in a molecule is indicated after the molecule's name. As glucose is broken down, the carbon atoms from glucose are incorporated into carbon dioxide.

ASSESS **YOUR PROGRESS**

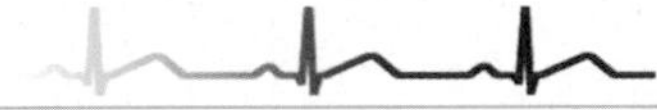

28. *Describe the four phases of glycolysis. What are the products of glycolysis?*

29. *What determines whether the pyruvic acid produced in glycolysis is used in aerobic respiration or lactic acid fermentation?*

30. *Describe the two phases of lactic acid fermentation. How many ATP molecules result from lactic acid fermentation?*

31. *What happens to the lactic acid produced in lactic acid fermentation once oxygen becomes available?*

Aerobic Respiration

Aerobic (ār-ō′bik) **respiration** is the breakdown of glucose in the presence of oxygen to produce carbon dioxide, water, and 36 ATP molecules. Most of the ATP molecules required to sustain life are produced through aerobic respiration, which can be considered in four phases. The first phase of aerobic respiration, as in anaerobic respiration, is glycolysis. The remaining phases are acetyl-CoA formation, the citric acid cycle, and the electron-transport chain (figure 25.7).

Acetyl-CoA Formation

In the second phase of aerobic respiration, pyruvic acid moves from the cytosol into a mitochondrion. Recall that a mitochondrion is separated into the intermembrane space and the matrix by the inner mitochondrial membrane (see figure 3.29). Within the matrix, enzymes remove a carbon and two oxygen atoms from the 3-carbon pyruvic acid molecule to form carbon dioxide and a 2-carbon acetyl

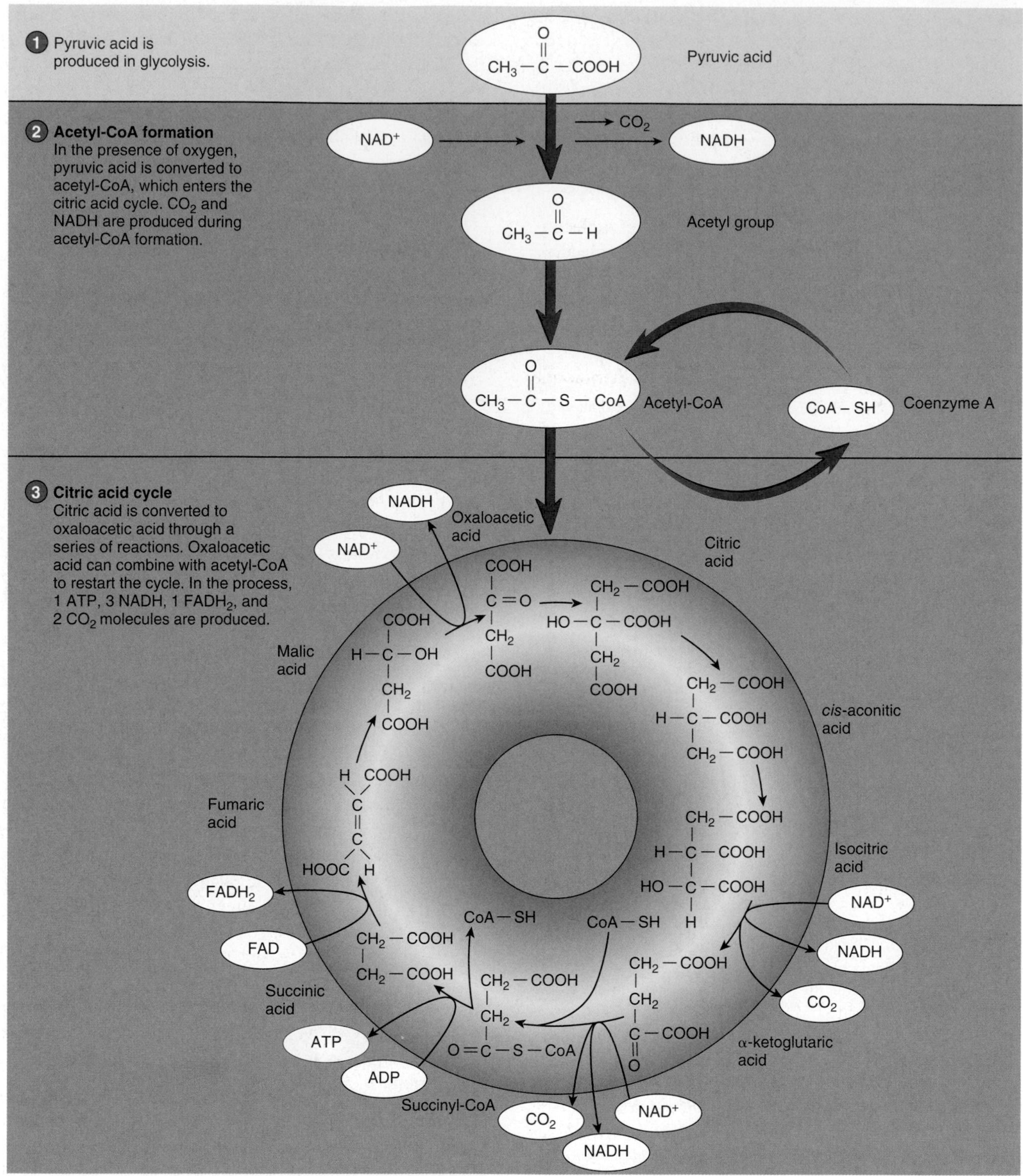

1 Pyruvic acid is produced in glycolysis.
Pyruvic acid
2 Acetyl-CoA formation
In the presence of oxygen, pyruvic acid is converted to acetyl-CoA, which enters the citric acid cycle. CO_2 and NADH are produced during acetyl-CoA formation.
NAD^+
CO_2
NADH
Acetyl group
Acetyl-CoA
CoA – SH
Coenzyme A
3 Citric acid cycle
Citric acid is converted to oxaloacetic acid through a series of reactions. Oxaloacetic acid can combine with acetyl-CoA to restart the cycle. In the process, 1 ATP, 3 NADH, 1 $FADH_2$, and 2 CO_2 molecules are produced.
Oxaloacetic acid
Citric acid
cis-aconitic acid
Isocitric acid
α-ketoglutaric acid
Succinyl-CoA
Succinic acid
Fumaric acid
Malic acid
$FADH_2$
FAD
ATP
ADP
NADH
CO_2

PROCESS FIGURE 25.8 Acetyl-CoA and the Citric Acid Cycle
Pyruvic acid from the cytosol is converted to acetyl-CoA in mitochondria. The acetyl-CoA enters the citric acid cycle.

(as′e-til) group (figure 25.8, *step 2*). Energy is released in the reaction and used to reduce NAD^+ to NADH. The acetyl group combines with coenzyme A (CoA) to form acetyl-CoA. For each two pyruvic acid molecules from glycolysis, two acetyl-CoA molecules, two carbon dioxide molecules, and two NADH are formed (see figure 25.4).

Citric Acid Cycle

The third phase of aerobic respiration is the **citric acid cycle,** which is named after the 6-carbon citric acid molecule formed in the first step of the cycle (see figure 25.8, *step 3*). It is also called the *Krebs cycle* after its discoverer, British biochemist Sir Hans Krebs. The citric acid cycle begins with the production of citric acid from the combination of acetyl-CoA and a 4-carbon molecule called **oxaloacetic** (ok′să-lō-ă-sē′tik) **acid.** A series of reactions occurs, resulting in the formation of another oxaloacetic acid molecule, which can start the cycle again by combining with another acetyl-CoA. During the reactions of the citric acid cycle, three important events occur:

1. *ATP production.* For each citric acid molecule, one ATP is formed.
2. *NADH and $FADH_2$ production.* For each citric acid molecule, three NAD^+ molecules are converted to NADH molecules, and one flavin (flā′vin) adenine dinucleotide (FAD) molecule is converted to $FADH_2$. The NADH and $FADH_2$ molecules are electron carriers that enter the electron-transport chain and are used to produce ATP.
3. *Carbon dioxide production.* Each 6-carbon citric acid molecule at the start of the cycle becomes a 4-carbon oxaloacetic acid molecule at the end of the cycle. Two carbon and four oxygen atoms from the citric acid molecule are used to form two carbon dioxide molecules. Thus, some of the carbon and oxygen atoms that make up food molecules, such as glucose, are eventually eliminated from the body as carbon dioxide. Humans literally breathe out part of the food they eat.

For each glucose molecule that begins aerobic respiration, two pyruvic acid molecules are produced in glycolysis, and they are converted into two acetyl-CoA molecules, each of which then enters the citric acid cycle. To determine the number of molecules produced from glucose by the citric acid cycle, two "turns" of the cycle must be counted; the results are two ATP, six NADH, two $FADH_2$, and four carbon dioxide molecules (see figure 25.4).

Electron-Transport Chain

The fourth phase of aerobic respiration involves the **electron-transport chain** (figure 25.9), which is a series of electron carriers in the inner mitochondrial membrane. Electrons are transferred from NADH and $FADH_2$ to the electron-transport carriers, and H^+ is released from NADH and $FADH_2$ (figure 25.9, *step 1*). After the loss of the electrons and the H^+, the oxidized NAD^+ and FAD are reused to transport additional electrons from the citric acid cycle to the electron-transport chain.

The electrons released from NADH and $FADH_2$ pass from one electron carrier to the next through a series of oxidation-reduction reactions. Three of the electron carriers also function as proton pumps, which move the H^+ from the mitochondrial matrix into the intermembrane space (figure 25.9, *step 2*). Each proton pump accepts an electron, uses some of the electron's energy to export a H^+, and passes the electron to the next electron carrier. The last electron carrier in the series collects the electrons and combines them with oxygen and H^+ to form water (figure 25.9, *step 3*):

$$1/2\ O_2 + 2\ H^+ + 2\ e^- \rightarrow H_2O$$

Without oxygen to accept the electrons, the reactions of the electron-transport chain cease, effectively stopping aerobic respiration.

The H^+ released from NADH and $FADH_2$ is moved from the intermembrane space to the matrix by active transport. As a result, the concentration of H^+ in the intermembrane space exceeds that of the matrix, and H^+ diffuses back into the matrix. The H^+ passes through certain channels formed by an enzyme called **ATP synthase.** As the H^+ diffuses down its concentration gradient, energy is released that is used to produce ATP (figure 25.9, *step 3*). This process is called **chemiosmosis** (kem-ē-os-mō′sis) because the chemical formation of ATP is coupled to a diffusion force similar to osmosis.

Predict 4

Many poisons function by blocking certain steps in metabolic pathways. For example, cyanide blocks the last step in the electron-transport chain. Explain why this blockage causes death.

Summary of ATP Production

For each glucose molecule, aerobic respiration produces a net gain of 36 ATP molecules: 2 from glycolysis, 2 from the citric acid cycle, and 32 from the NADH molecules and $FADH_2$ molecules that pass through the electron-transport chain (table 25.4). For each NADH molecule formed, 2 or 3 ATP molecules are produced by the electron-transport chain; for each $FADH_2$ molecule, 2 ATP molecules are produced.

The number of ATP molecules produced from each glucose molecule is a theoretical number. The calculations that predict 36 ATP molecules per glucose molecule assume that 2 H^+ are necessary for the formation of each ATP. If more than 2 are required, the efficiency of aerobic respiration decreases. In addition, the 2 NADH molecules produced by glycolysis in the cytosol cannot cross the inner mitochondrial membrane; thus, their electrons are donated to a shuttle molecule, which carries the electrons to the electron-transport chain. Depending on the shuttle molecule, each glycolytic NADH molecule can produce 2 or 3 ATP molecules. In skeletal muscle and the brain, 2 ATP molecules are produced for each NADH molecule, resulting in a total number of 36 ATP molecules; however, in the liver, kidneys, and heart, 3 ATP molecules may be produced for each NADH molecule, and the total number of ATP molecules formed may be higher. Also, it costs energy to get ADP and phosphates into the mitochondria and to get ATP out. Considering all these factors, each glucose molecule yields about 25 ATP molecules instead of 36.

Aerobic respiration produces 6 carbon dioxide molecules. Water molecules are reactants in some of the chemical reactions of aerobic respiration and products in others. Six water molecules are used, but 12 are formed, for a net gain of 6 water molecules. Thus, aerobic respiration can be summarized as follows:

$$C_6H_{12}O_6 + 6\ O_2 + 6\ H_2O + 36\ ADP + 36\ P_i \rightarrow$$
$$6\ CO_2 + 12\ H_2O + 36\ ATP$$

ASSESS YOUR PROGRESS

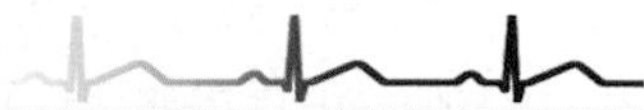

32. *Define* aerobic respiration, *and list its products. Describe the four phases of aerobic respiration.*
33. *Why is the citric acid cycle a cycle? What molecules are produced as a result of the citric acid cycle?*
34. *What is the function of the electron-transport chain? Describe the process of chemiosmosis that occurs during ATP production.*

FUNDAMENTAL **Figure**

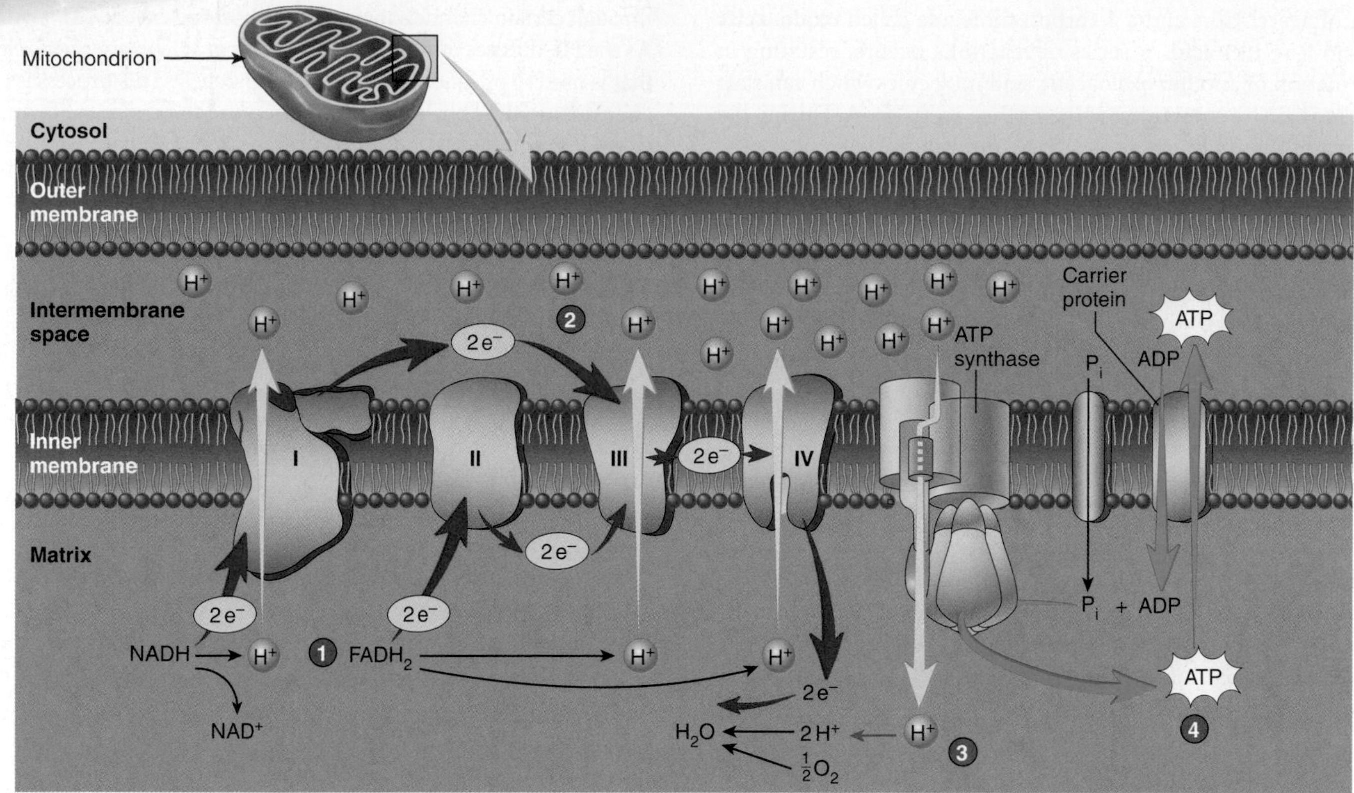

Mitochondrion
Cytosol
Outer membrane
Intermembrane space
Inner membrane
Matrix
H^+
$2e^-$
I
II
III
IV
NADH
NAD^+
$FADH_2$
H_2O
$2H^+$
$\frac{1}{2}O_2$
ATP synthase
Carrier protein
ATP
ADP
P_i
P_i + ADP
1
2
3
4

1. NADH or $FADH_2$ transfer their electrons to the electron-transport chain.

2. As the electrons move through the electron-transport chain, some of their energy is used to pump H^+ into the intermembrane space, resulting in a higher concentration of H^+ in the intermembrane space than in the matrix.

3. The H^+ diffuse back into the matrix through special channels (ATP synthase) that couple the H^+ movement with the production of ATP. The electrons, H^+, and O_2 combine to form H_2O.

4. ATP is transported out of the matrix by a carrier protein that exchanges ATP for ADP. A different carrier protein moves phosphate into the matrix.

PROCESS FIGURE 25.9 Electron-Transport Chain
The electron-transport chain in the mitochondrial inner membrane consists of four protein complexes (purple; numbered I to IV) with carrier proteins.

35. *In aerobic respiration, how many ATP molecules are produced from one molecule of glucose through glycolysis, the citric acid cycle, and the electron-transport chain?*

36. *Why is the total number of ATP produced in aerobic respiration listed as 36? Why could that number be different?*

37. *Write the summary equation for the aerobic breakdown of one glucose molecule.*

25.4 Lipid Metabolism

LEARNING OUTCOME

After reading this section, you should be able to

A. Describe the basic steps involved in using lipids as an energy source.

Lipids are the body's main energy-storage molecules. In a healthy person, lipids are responsible for about 99% of the body's energy storage, and glycogen accounts for about 1%. Although proteins serve as an energy source, they are not considered storage molecules because their breakdown normally involves the loss of molecules that perform other functions.

Lipids are stored primarily as triglycerides in adipose tissue. Synthesis and breakdown of triglycerides occur constantly; thus, the lipids present in adipose tissue today are not the same lipids that were there a few weeks ago. Between meals, when triglycerides are broken down in adipose tissue, some of the fatty acids produced are released into the blood, where they are called **free fatty acids.** Other tissues, especially skeletal muscle and the liver, use the free fatty acids as a source of energy.

The metabolism of fatty acids occurs by **beta-oxidation,** a series of reactions in which two carbon atoms are removed from the end of a fatty acid chain to form acetyl-CoA. The process of beta-oxidation continues to remove two carbon atoms at a time until the entire fatty acid chain is converted into acetyl-CoA molecules. Acetyl-CoA can enter the citric acid cycle and be used to generate ATP (figure 25.10).

Acetyl-CoA is also used in **ketogenesis** (kē-tō-jen′ĕ-sis), the formation of ketone bodies. In the liver, when large amounts of acetyl-CoA are produced, not all of the acetyl-CoA enters the citric acid cycle. Instead, two acetyl-CoA molecules combine to form a molecule of acetoacetic (as′e-tō-a-sē′tik) acid, which is converted mainly into β-hydroxybutyric (hī-drōk′sē-bu-tir′ik) acid and a smaller amount of acetone (as′e-tōn). Acetoacetic acid, β-hydroxybutyric acid, and acetone are called **ketone** (kē′tōn) **bodies;** they are released into the blood, where they travel to other tissues, especially skeletal muscle. In these tissues, the ketone bodies are converted back into acetyl-CoA, which enters the citric acid cycle to produce ATP.

The presence of small amounts of ketone bodies in the blood is normal and beneficial, but excessive production of ketone bodies is called **ketosis** (kē-tō′sis). If the increased number of acidic ketone bodies exceeds the capacity of the body's buffering systems, acidosis, a decrease in blood pH, can occur (see chapter 27). Conditions that increase lipid metabolism can speed the rate of ketone body formation. Examples are starvation (see Clinical Impact, "Starvation"), diets consisting mainly of proteins and lipids with few carbohydrates, and untreated diabetes mellitus. Ketone bodies are excreted by the kidneys and diffuse into the alveoli of the lungs. Because ketone bodies are excreted by the kidneys and lungs, the characteristics of untreated diabetes mellitus include ketone bodies in the urine and "acetone breath."

ASSESS YOUR PROGRESS

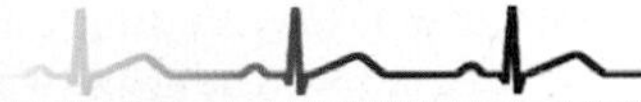

38. *What is beta-oxidation? Explain how it results in ATP production.*

39. *What are ketone bodies, how are they produced, and for what are they used? What occurs when there is an excess?*

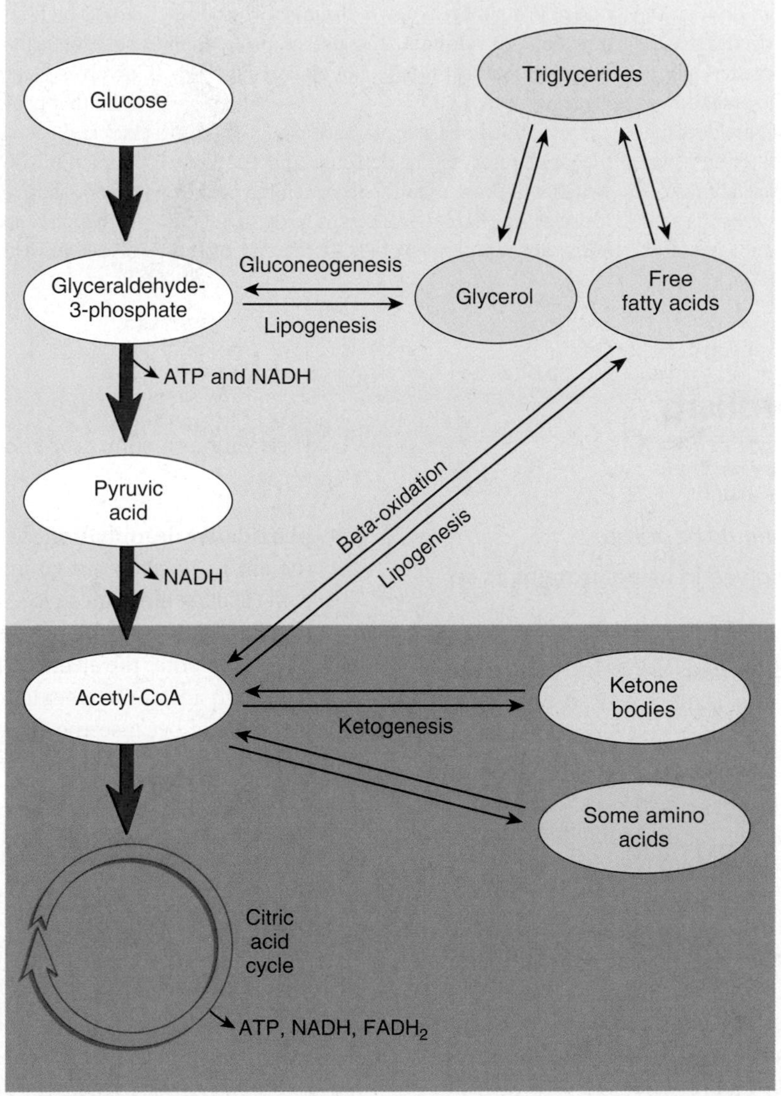

Glucose
Triglycerides
Glyceraldehyde-3-phosphate
Gluconeogenesis
Lipogenesis
Glycerol
Free fatty acids
ATP and NADH
Pyruvic acid
Beta-oxidation
Lipogenesis
NADH
Acetyl-CoA
Ketone bodies
Ketogenesis
Some amino acids
Citric acid cycle
ATP, NADH, $FADH_2$

FIGURE 25.10 Lipid Metabolism

Triglycerides are broken down into glycerol and fatty acids. Glycerol enters glycolysis to produce ATP. The fatty acids are broken down by beta-oxidation into acetyl-CoA, which enters the citric acid cycle to produce ATP and electron carriers. Acetyl-CoA can also be used to produce ketone bodies (ketogenesis). Lipogenesis is the production of lipids. Glucose is converted to glycerol, and amino acids are converted to acetyl-CoA molecules. Acetyl-CoA molecules can combine to form fatty acids. Glycerol and fatty acids join to form triglycerides.

Clinical
IMPACT

Starvation

Starvation is the inadequate intake of nutrients or the inability to metabolize or absorb nutrients. It has a number of causes, including prolonged fasting, anorexia, deprivation, and disease. But no matter what the cause, starvation follows the same course and consists of three phases. The first two phases take place even during relatively short periods of fasting or dieting, but the third phase occurs only in prolonged starvation and can end in death.

During the first phase of starvation, blood glucose levels are maintained through the production of glucose from glycogen, lipids, and proteins. At first, glycogen is broken down into glucose; however, only enough glycogen is stored in the liver to last a few hours. Thereafter, blood glucose levels are maintained by the breakdown of lipids and proteins. Triglycerides are decomposed into fatty acids and glycerol. Fatty acids can provide energy, especially for skeletal muscle, thus decreasing the use of glucose by tissues other than the brain. The brain cannot use fatty acids as an energy source, so the conservation of glucose is critical to normal brain function. Glycerol can be used to make a small amount of glucose, but most of the glucose is formed from the amino acids of proteins. In addition, some amino acids can be used directly for energy.

In the second phase, which can last for several weeks, lipids are the primary energy source. The liver metabolizes fatty acids into ketone bodies, which serve as a source of energy. After about a week of fasting, the brain begins to use ketone bodies, as well as glucose, for energy. This usage decreases the demand for glucose, and the rate of protein breakdown diminishes but does not stop. In addition, the use of proteins is selective; that is, proteins not essential for survival are used first.

The third phase of starvation begins when the lipid reserves are depleted and the body switches to proteins as the major energy source. Muscles, the largest source of protein in the body, are rapidly depleted. At the end of this phase, proteins essential for cell functions are broken down, and cell function degenerates.

In addition to weight loss, the symptoms of starvation include apathy, listlessness, withdrawal, and increased susceptibility to infection. Few people die directly from starvation because they usually die of an infectious disease first. Other signs of starvation include changes in hair color, flaky skin, and massive edema in the abdomen and lower limbs, causing the abdomen to appear bloated.

During starvation, the body's ability to consume normal volumes of food also decreases. Foods high in bulk but low in protein often cannot reverse the process of starvation. Intervention involves feeding the starving person low-bulk food that provides ample proteins and kilocalories and is fortified with vitamins and minerals. Starvation also results in dehydration; thus, rehydration is an important part of intervention. Even with intervention, a person may be so affected by disease or weakness that he or she cannot recover.

25.5 Protein Metabolism

LEARNING OUTCOME

After reading this section, you should be able to

A. Describe the basic steps involved in using proteins as an energy source.

Once absorbed into the body, amino acids are quickly taken up by cells, especially in the liver. Amino acids are used primarily to synthesize needed proteins (see chapter 3), and only secondarily as a source of energy (figure 25.11). Unlike glycogen and triglycerides, amino acids are not stored in the body.

As stated earlier in the chapter, the body can manufacture some amino acids. The synthesis of nonessential amino acids usually begins with keto acids. A keto acid can be converted into an amino acid by replacing its oxygen atom with an amine group (figure 25.12). Usually, this conversion is accomplished by transferring an amine group from an amino acid to the keto acid, a reaction called **transamination** (trans-am′i-nā′shŭn). For example, α-ketoglutaric acid (a keto acid) reacts with an amino acid to form glutamic acid (an amino acid; figure 25.13*a*). Most amino acids can undergo transamination to produce glutamic acid. The glutamic acid provides an amine group that is used to synthesize most of the nonessential amino acids. A few nonessential amino acids are formed from the essential amino acids by other chemical reactions.

If serving as a source of energy, amino acids can be used in two ways:

- In **oxidative deamination** (dē-am-i-nā′shŭn), or *deaminization* (dē-am′i-nizā′shŭn), an amine group is removed from an amino acid (usually glutamic acid), leaving ammonia and a keto acid (figure 25.13*b*). In the process, NAD^+ is reduced to NADH, which can enter the electron-transport chain to produce ATP. Although ammonia is toxic to cells, it does not accumulate to toxic levels because the liver converts it to urea, which the blood carries to the kidneys for elimination (figure 25.13*c;* see chapter 26).
- Amino acids can also be converted into the intermediate molecules of carbohydrate metabolism (see figure 25.11). These molecules are then metabolized to yield ATP. The conversion of an amino acid often begins with a transamination or oxidative deamination reaction, in which the amino acid is converted into a keto acid (see figure 25.12). The keto acid enters the citric acid cycle or is converted into pyruvic acid or acetyl-CoA.

ASSESS YOUR PROGRESS

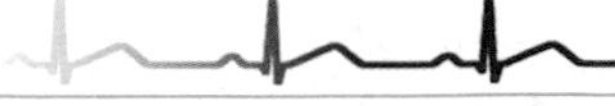

40. *What is accomplished by transamination and oxidative deamination?*

41. *How are proteins (amino acids) used to produce energy?*

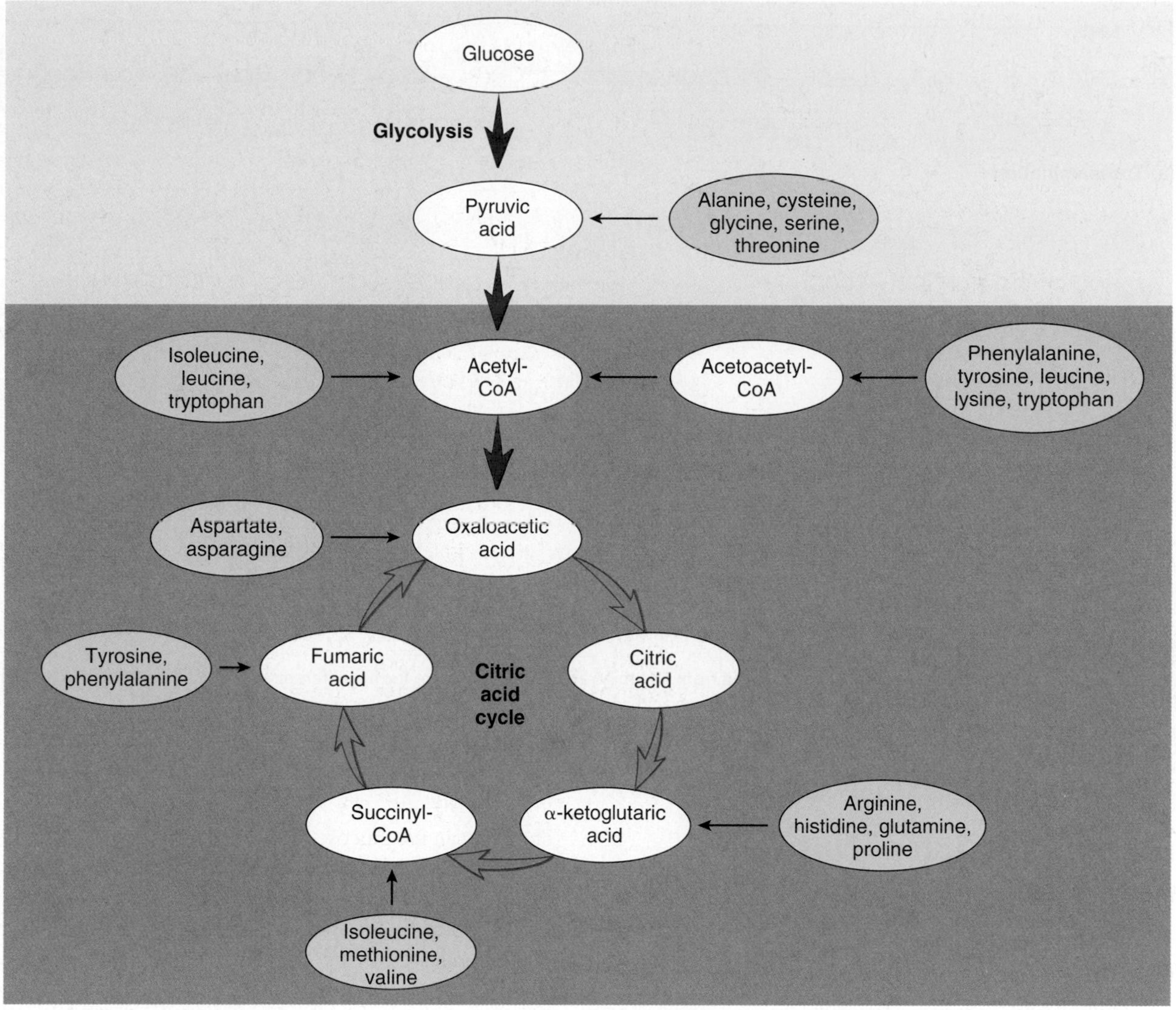

Glucose
Glycolysis
Pyruvic acid
Alanine, cysteine, glycine, serine, threonine
Isoleucine, leucine, tryptophan
Acetyl-CoA
Acetoacetyl-CoA
Phenylalanine, tyrosine, leucine, lysine, tryptophan
Aspartate, asparagine
Oxaloacetic acid
Tyrosine, phenylalanine
Fumaric acid
Citric acid cycle
Citric acid
Succinyl-CoA
α-ketoglutaric acid
Arginine, histidine, glutamine, proline
Isoleucine, methionine, valine

FIGURE 25.11 Amino Acid Metabolism
Amino acids (pink ovals) can enter carbohydrate metabolism at various points.

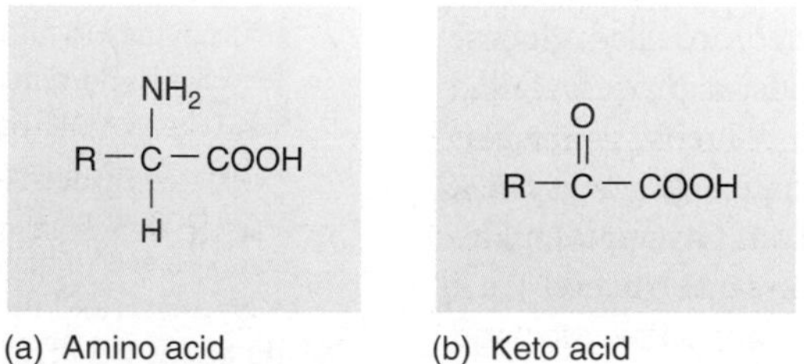
NH2
R—C—COOH
H
(a) Amino acid
O
R—C—COOH
(b) Keto acid

FIGURE 25.12 General Formulas of an Amino Acid and a Keto Acid
(*a*) Amino acid with a carboxyl group (—COOH), an amine group (NH_2), a hydrogen atom (H), and a group called "R," which represents the rest of the molecule. (*b*) Keto acid with a double-bonded oxygen replacing the amine group and the hydrogen atom of the amino acid.

$$R_1{-}CH(NH_2){-}COOH + HOOC{-}CH_2{-}CH_2{-}C(=O){-}COOH \xrightleftharpoons{\text{Enzymes}} R_1{-}C(=O){-}COOH + HOOC{-}CH_2{-}CH_2{-}CH(NH_2){-}COOH$$

Amino acid, α-ketoglutaric acid, α-keto acid, Glutamic acid

(a) **Transamination**

$$HOOC{-}CH_2{-}CH_2{-}CH(NH_2){-}COOH + H_2O \xrightarrow[\text{NAD}^+ \to \text{NADH}]{\text{Enzymes}} HOOC{-}CH_2{-}CH_2{-}C(=OH){-}COOH + NH_3$$

Glutamic acid, NAD^+, NADH, α-ketoglutaric acid, Ammonia

(b) **Oxidative deamination**

$$2\,NH_3 + CO_2 \xrightarrow{\text{Enzymes}} (NH_2)_2C{=}O + H_2O$$

Ammonia, Carbon dioxide, Urea, Water

(c) **Conversion of ammonia to urea**

FIGURE 25.13 Amino Acid Reactions

(*a*) Transamination reaction in which an amine group is transferred from an amino acid to a keto acid to form a different amino acid. (*b*) Oxidative deamination reaction in which an amino acid loses an amine group to become a keto acid and to form ammonia. In the process, NADH, which can be used to generate ATP, is formed. (*c*) Ammonia is converted to urea in the liver. (The actual conversion of ammonia to urea is more complex, involving a number of intermediate reactions that constitute the urea cycle.)

25.6 Interconversion of Nutrient Molecules

LEARNING OUTCOME

After reading this section, you should be able to

A. Define *glycogenesis, lipogenesis, glycogenolysis,* and *gluconeogenesis.*

Blood glucose enters most cells by facilitated diffusion and is immediately converted to glucose-6-phosphate, which cannot recross the plasma membrane (figure 25.14*a*). Glucose-6-phosphate then continues through glycolysis to produce ATP. However, if excess glucose is present (e.g., after a meal), it is used to form glycogen through a process called **glycogenesis** (glī-kō-jen′ĕ-sis). Most of the body's glycogen is contained in skeletal muscle and the liver.

Once glycogen stores, which are quite limited, are filled, glucose and amino acids are used to synthesize lipids, a process called **lipogenesis** (lip-ō-jen′ĕ-sis; see figure 25.10). Glucose molecules can be used to form glyceraldehyde-3-phosphate and acetyl-CoA. Amino acids can also be converted to acetyl-CoA. Glyceraldehyde-3-phosphate is converted to glycerol, and the 2-carbon acetyl-CoA molecules join together to form fatty acid chains. Glycerol and three fatty acids then combine to form triglycerides.

When glucose is needed, glycogen can be broken down into glucose-6-phosphate through a set of reactions called **glycogenolysis** (glī′kō-jĕ-nol′i-sis; figure 25.14*b*). In skeletal muscle, glucose-6-phosphate continues through glycolysis to produce ATP. The liver can use glucose-6-phosphate for energy or can convert it to glucose, which diffuses into the blood. Although the liver can release glucose into the blood, skeletal muscle cannot because it lacks the necessary enzymes to convert glucose-6-phosphate into glucose.

The release of glucose from the liver is necessary to maintain blood glucose levels between meals. Maintaining these levels is especially important to the brain, which normally uses only glucose for an energy source and consumes about two-thirds of the total

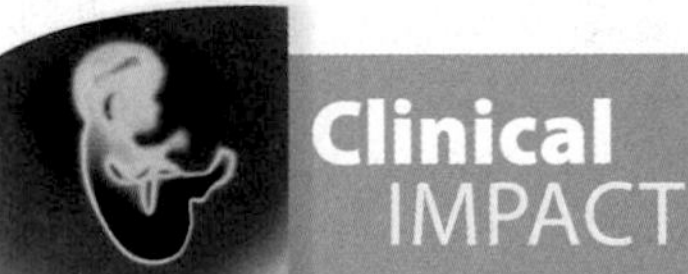

Clinical IMPACT

Alcoholism and Cirrhosis of the Liver

Enzymes in the liver convert ethanol (beverage alcohol) into acetyl-CoA; in the process, two NADH molecules are produced. The NADH molecules enter the electron-transport chain and are used to produce ATP molecules. Each gram of ethanol provides 7 kcal of energy. A high level of NADH in the cell results from the metabolism of ethanol, thereby inhibiting the production of NADH by glycolysis and the citric acid cycle. Consequently, carbohydrates and amino acids do not break down but are converted into lipids, which accumulate in the liver. Therefore, chronic alcohol abuse can result in **cirrhosis** (sir-rō′sis) **of the liver,** which involves lipid deposition, cell death, inflammation, and scar tissue formation. Because the liver is unable to carry out its normal functions, death can result.

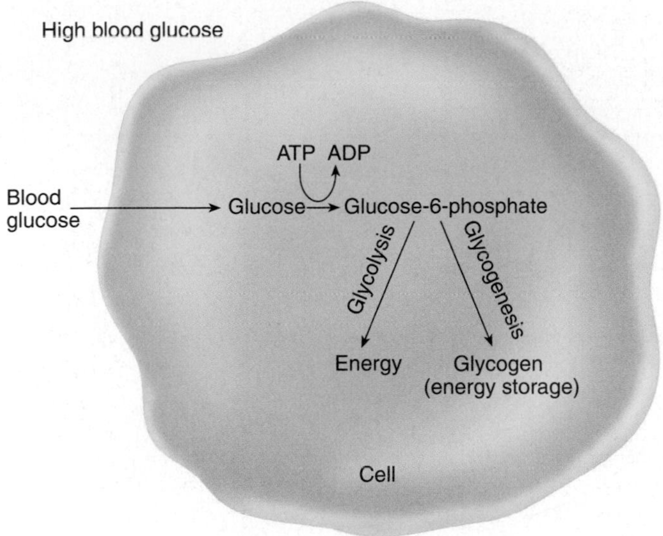
High blood glucose
Blood glucose
ATP ADP
Glucose
Glucose-6-phosphate
Glycolysis
Glycogenesis
Energy
Glycogen (energy storage)
Cell

(a) When blood glucose levels are high, glucose enters the cell and is phosphorylated to form glucose-6-phosphate, which can enter glycolysis or glycogenesis.

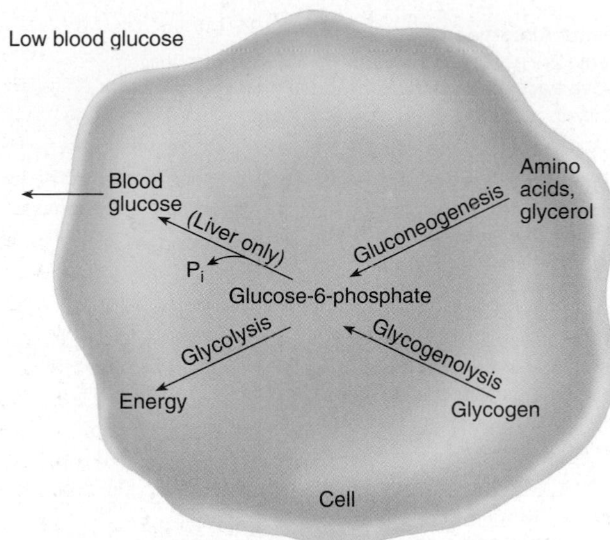
Low blood glucose
Blood glucose
(Liver only)
P_i
Amino acids, glycerol
Gluconeogenesis
Glucose-6-phosphate
Glycolysis
Glycogenolysis
Energy
Glycogen
Cell

(b) When blood glucose levels drop, glucose-6-phosphate can be produced through glycogenolysis or gluconeogenesis. Glucose-6-phosphate can enter glycolysis, or in the liver, the phosphate group can be removed and glucose released into the blood.

FIGURE 25.14 Interconversion of Nutrient Molecules

glucose used each day. When liver glycogen levels are inadequate to supply glucose, it is synthesized from molecules other than carbohydrates, such as amino acids from proteins and glycerol from triglycerides, in a process called **gluconeogenesis** (gloo′kō-nē-ō-jen′ĕ-sis). Most amino acids can be converted into citric acid cycle molecules, acetyl-CoA, or pyruvic acid (see figure 25.11). Through a series of chemical reactions, these molecules are converted into glucose. Glycerol can be converted to glyceraldehyde-3-phosphate, which in turn can be converted to glucose.

ASSESS **YOUR PROGRESS**

42. *Distinguish among the processes of glycogenesis, lipogenesis, glycogenolysis, and gluconeogenesis.*

25.7 Metabolic States

LEARNING OUTCOME

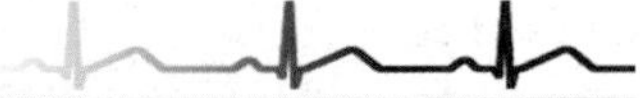

After reading this section, you should be able to

A. Differentiate between the absorptive and postabsorptive metabolic states.

The body experiences two major metabolic states: the absorptive state and the postabsorptive state. The regulation of these states is discussed in chapter 18 (see section 18.7). The **absorptive state** is the period immediately after a meal, when nutrients are being absorbed through the intestinal wall into the circulatory and lymphatic systems (figure 25.15). The absorptive state usually lasts about 4 hours after each meal, though the rate of absorption declines after 1-2 hours. During this time the cells use most of the glucose that enters the circulation for the energy they require. The remainder of the glucose is converted into glycogen or lipids. Most of the absorbed lipids are deposited in adipose tissue. Many of the absorbed amino acids are used by cells in protein synthesis, some are used for energy, and still others enter the liver and are converted into lipids or carbohydrates.

The **postabsorptive state** occurs late in the morning, late in the afternoon, or during the night after each absorptive state is concluded (figure 25.16). The maintenance of normal blood glucose levels is vital to the body's homeostasis, especially for normal functioning of the brain. Therefore, during the postabsorptive state, blood glucose levels are maintained by the conversion of other molecules to glucose. The first source of blood glucose during the postabsorptive state is the glycogen stored in the liver, but this glycogen supply can provide glucose for only about 4 hours. The glycogen stored in skeletal muscles can also be used during times of vigorous exercise. As glycogen stores are depleted, the body uses lipids as an energy source. The glycerol from triglycerides can be converted to glucose. The fatty acids from lipids can be converted to acetyl-CoA, moved into the citric acid cycle, and used as energy to produce ATP. In the liver, acetyl-CoA is used to produce ketone bodies, which other tissues use for energy. The use of fatty acids as an energy source partly eliminates the need to use glucose for energy so less glucose is removed from the blood and homeostasis is maintained. Amino acids can again be converted to glucose or used to produce energy, conserving blood glucose.

ASSESS **YOUR PROGRESS**

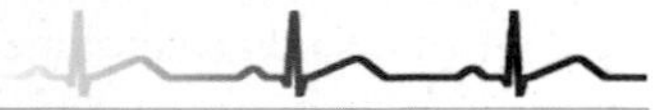

43. *When does the absorptive state occur?*

44. *What happens to glucose, lipids, and amino acids during the absorptive state?*

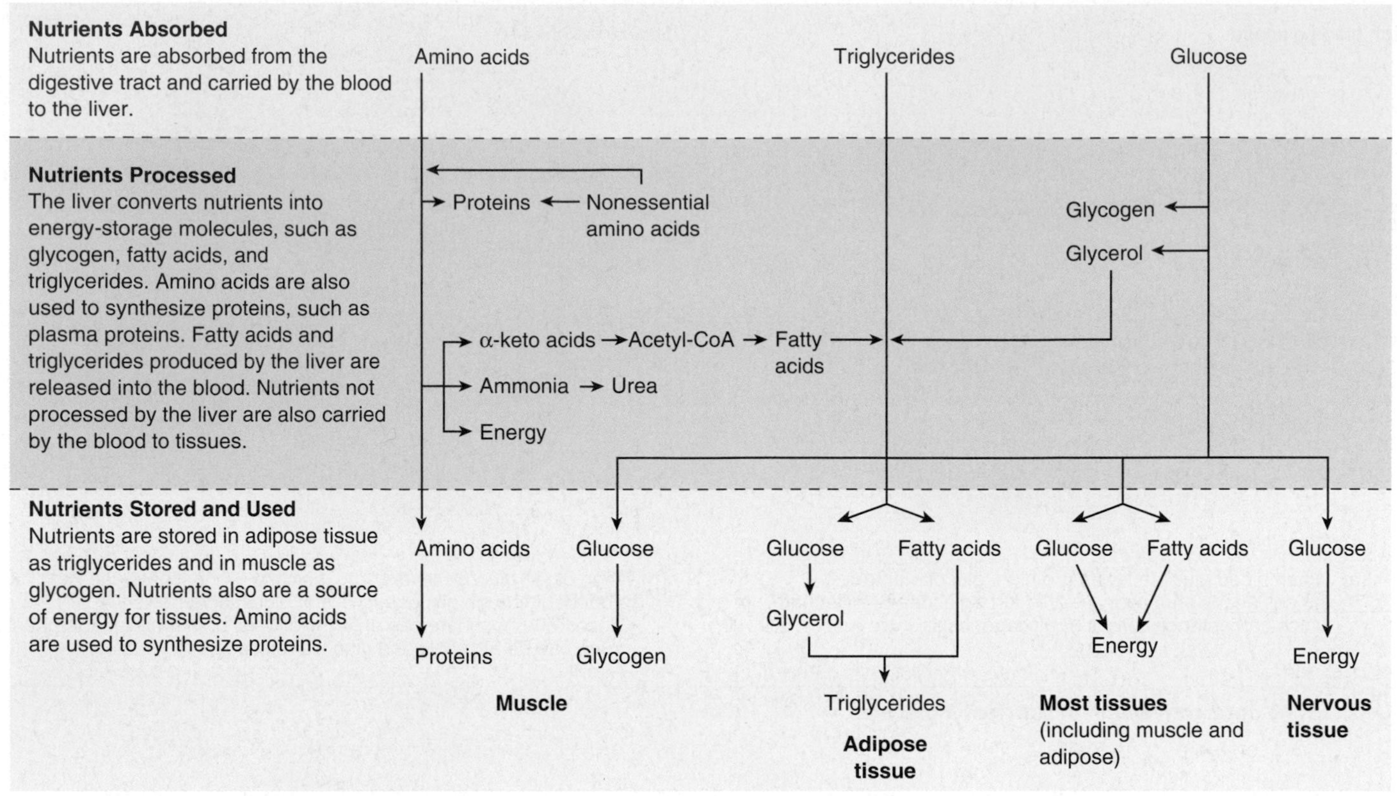

Nutrients Absorbed
Nutrients are absorbed from the digestive tract and carried by the blood to the liver.
Amino acids
Triglycerides
Glucose
Nutrients Processed
The liver converts nutrients into energy-storage molecules, such as glycogen, fatty acids, and triglycerides. Amino acids are also used to synthesize proteins, such as plasma proteins. Fatty acids and triglycerides produced by the liver are released into the blood. Nutrients not processed by the liver are also carried by the blood to tissues.
Proteins
Nonessential amino acids
Glycogen
Glycerol
α-keto acids
Acetyl-CoA
Fatty acids
Ammonia
Urea
Energy
Nutrients Stored and Used
Nutrients are stored in adipose tissue as triglycerides and in muscle as glycogen. Nutrients also are a source of energy for tissues. Amino acids are used to synthesize proteins.
Amino acids
Glucose
Glucose
Fatty acids
Glucose
Fatty acids
Glucose
Glycerol
Energy
Proteins
Glycogen
Energy
Muscle
Triglycerides
Adipose tissue
Most tissues (including muscle and adipose)
Nervous tissue

FIGURE 25.15 Events of the Absorptive State

Absorbed molecules, especially glucose, are used as sources of energy. Molecules not immediately needed for energy are stored: Glucose is converted to glycogen or triglycerides, triglycerides are deposited in adipose tissue, and amino acids are converted to triglycerides or carbohydrates.

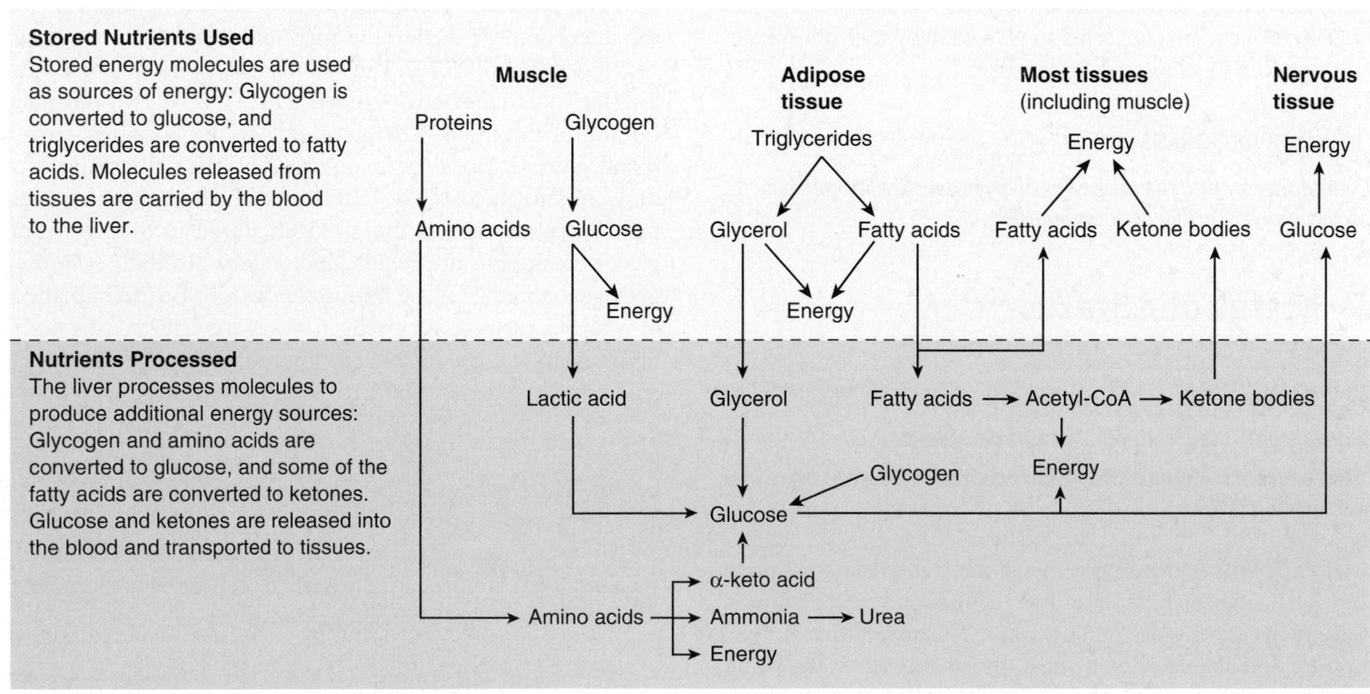

Stored Nutrients Used
Stored energy molecules are used as sources of energy: Glycogen is converted to glucose, and triglycerides are converted to fatty acids. Molecules released from tissues are carried by the blood to the liver.
Muscle
Adipose tissue
Most tissues (including muscle)
Nervous tissue
Proteins
Glycogen
Triglycerides
Energy
Energy
Amino acids
Glucose
Glycerol
Fatty acids
Fatty acids
Ketone bodies
Glucose
Energy
Energy
Nutrients Processed
The liver processes molecules to produce additional energy sources: Glycogen and amino acids are converted to glucose, and some of the fatty acids are converted to ketones. Glucose and ketones are released into the blood and transported to tissues.
Lactic acid
Glycerol
Fatty acids
Acetyl-CoA
Ketone bodies
Glycogen
Energy
Glucose
α-keto acid
Amino acids
Ammonia
Urea
Energy

FIGURE 25.16 Events of the Postabsorptive State

Stored energy molecules are used as sources of energy: Glycogen is converted to glucose; triglycerides are broken down to fatty acids, some of which are converted to ketones; and proteins are converted to glucose.

Clinical GENETICS — Newborn Screening of Metabolic Disorders

Metabolic disorders, sometimes called inborn errors of metabolism, are a large class of genetic disorders that result in biochemical defects. Metabolic disorders affect the body's ability to break down or use nutrients needed for energy, growth, and repair. Too little synthesis of certain substances or a buildup of toxic compounds can cause significant health problems. Although the frequency of any given individual disorder is rare, the overall incidence of metabolic disorders is estimated to be up to 1 in 1000 births.

Early detection through newborn screening is vital. Metabolic disorders can hinder early mental and physical development. Depending on the disorder, specific treatment can prevent or limit harm if it is started early. In the United States, most states require the screening of newborns. However, there is no national standard for newborn screening, so the specific disorders for which tests are performed vary from state to state. Although over several hundred genetic disorders are known, most are so rare that it is not cost-effective to test for them.

Table 25A lists the most common blood tests performed for metabolic disorders. All of the disorders listed are autosomal recessive.

TABLE 25A Metabolic Disorders

Disorder	Description	Effect	Treatment
Phenylketonuria (PKU)	Inability to metabolize the amino acid phenylalanine (see chapter 29)	Mental retardation	Restrict dietary phenylalanine.
Galactosemia	Inability to convert the sugar galactose to glucose, resulting in a buildup of galactose	Mental retardation, growth deficiency, cataracts, severe infections, death	Eliminate milk and other dairy products from the diet. Galactose is one of two sugars in lactose (milk sugar).
Biotinidase deficiency	Inability to separate the vitamin biotin from other chemicals, resulting in a biotin deficiency	Seizures, hearing loss, optic atrophy, mental retardation, poor muscle control	Take oral biotin supplements.
Maple syrup urine disease	Deficiency in an enzyme complex, resulting in an inability to metabolize the amino acids leucine, isoleucine, and valine	Mental retardation in those surviving past 3 months of age	Restrict dietary intake of the affected amino acids.
Homocystinuria	Defect in methionine metabolism, leading to an accumulation of homocysteine	Dislocated lenses of the eyes, mental retardation, skeletal abnormalities, abnormal blood clotting	Take high doses of vitamin B_6; eat methionine-restricted diet supplemented with cysteine.
Tyrosinemia	Deficiency in a series of enzymes that break down the amino acid tyrosine	Mild retardation, language skill difficulties, liver and kidney failure	Restrict dietary tyrosine and phenylalanine.

45. *When does the postabsorptive state occur?*

46. *Why is it important to maintain blood glucose levels during the postabsorptive state? Name three sources of this glucose.*

25.8 Metabolic Rate

LEARNING OUTCOMES

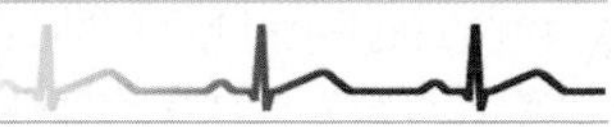

After reading this section, you should be able to

A. Define *metabolic rate* and describe the three major uses of metabolic energy in the body.

B. Explain how to maintain body weight.

Metabolic rate is the total amount of energy produced and used by the body per unit of time. A molecule of ATP exists for less than 1 minute before it is degraded back to ADP and inorganic phosphate. For this reason, ATP is produced in cells at about the same rate as it is used. Thus, in examining metabolic rate, ATP production and use can be roughly equated. Because most ATP production involves the use of oxygen, metabolic rate is usually estimated by measuring the amount of oxygen used per minute. One liter of oxygen consumed by the body is estimated to produce 4.825 kcal of energy.

The daily input of energy should equal the metabolic expenditure of energy; otherwise, a person will gain or lose weight. For a typical 23-year-old, 70 kg (154-pound) male to maintain his weight, the daily input should be 2700 kcal/day; for a typical 58 kg (128-pound) female of the same age, 2000 kcal/day is sufficient. A pound of body fat (adipose tissue) provides about 3500 kcal. Reducing kilocaloric intake by 500 kcal/day can result in the loss of 1 pound of body fat per week. Clearly, adjusting kilocaloric input is an important way to control body weight.

Not only the number of kilocalories ingested but also the proportion of dietary fat can affect body weight. To convert dietary fat into body fat, 3% of the energy in the dietary fat is used, leaving 97% for storage as body fat deposits. On the other hand, converting dietary carbohydrate to lipids requires 23% of the energy in the carbohydrate, leaving just 77% as body fat. If two people have the same kilocaloric intake, the one consuming the higher proportion of lipids is more likely to gain weight because fewer kilocalories are used to convert the dietary fat into body fat.

Metabolic energy is used in three ways: for basal metabolism, for the thermic effect of food, and for muscular activity.

Basal Metabolic Rate

The **basal metabolic rate (BMR)** is the energy needed to keep the resting body functional. It is expressed in expended kilocalories per square meter of body surface area per hour (kcal/m^2/hr). BMR is determined by measuring the oxygen consumption of a person who is awake but restful and has not eaten for 12 hours. The liters of oxygen consumed are then multiplied by 4.825 because each liter of oxygen used results in 4.825 kcal of energy. A typical BMR for a 70 kg (154-pound) male is 38 kcal/m^2/h.

In the average person, basal metabolism accounts for about 60% of energy expenditure. Basal metabolism supports active-transport mechanisms, muscle tone, maintenance of body temperature, beating of the heart, and other activities. A number of factors can affect the BMR. Muscle tissue is metabolically more active than adipose tissue, even at rest, so people with well-developed muscles have a higher BMR than other people. Younger people have a higher BMR than older people because of increased cell activity, especially during growth. Fever can increase BMR 7% for each degree Fahrenheit elevation in body temperature. During dieting or fasting, greatly reduced kilocaloric input can depress BMR, which apparently is a protective mechanism to prevent weight loss. Thyroid hormones can increase BMR on a long-term basis, and epinephrine can increase BMR on a short-term basis (see chapter 18). Males have a greater BMR than females because men have proportionately more muscle tissue and less adipose tissue than women do. During pregnancy, a woman's BMR can increase 20% because of the metabolic activity of the fetus.

Thermic Effect of Food

The second component of metabolic energy is the assimilation of food. When food is ingested, the accessory digestive organs and the intestinal lining produce secretions, the motility of the digestive tract increases, active transport increases, and the liver is involved in synthesizing new molecules. The energy cost of these events, called the **thermic effect of food,** accounts for about 10% of the body's energy expenditure.

Muscular Activity

Muscular activity consumes about 30% of the body's energy. Therefore, physical activity resulting from skeletal muscle movement requires the expenditure of energy. In addition, energy is needed for the increased contraction of the heart and muscles of respiration. The number of kilocalories expended in an activity depends almost entirely on the amount and duration of muscular work performed. Despite the fact that studying can make a person feel tired, intense mental concentration produces little change in BMR.

Energy loss through muscular activity is the only component of energy expenditure that a person can reasonably control. Comparing the number of kilocalories gained from food with the number of kilocalories lost in exercise reveals why losing weight can be difficult. For example, if brisk walking uses 225 kcal/h, it takes 20 minutes of brisk walking to burn off the 75 kcal in one slice of bread (75/225 = 0.33 h). Research suggests that a combination of appropriate physical activity and appropriate kilocaloric intake is the best way to maintain a healthy body composition and weight.

Predict 5

If watching TV uses 95 kcal/h, how long does it take to burn off the kilocalories in one cola or beer (see table 25.1)? If jogging at a pace of 6 mph uses 580 kcal/h, how long does it take to use the kilocalories in one cola or beer?

ASSESS YOUR PROGRESS

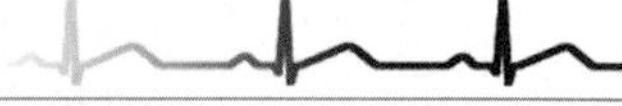

47. *What is metabolic rate? How is it measured?*
48. *What is BMR? What factors can alter BMR?*
49. *What is the thermic effect of food?*
50. *BMR, the thermic effect of food, and muscular activity each account for what percentage of total energy expenditure?*
51. *How are kilocaloric input and output adjusted to maintain body weight?*

25.9 Body Temperature Regulation

LEARNING OUTCOME

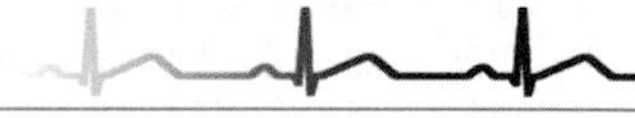

After reading this section, you should be able to

A. Describe heat production and regulation in the body.

Humans can maintain a relatively constant internal body temperature despite changes in the temperature of the surrounding environment. A constant body temperature is very important for homeostasis. For example, environmental temperatures are too low for normal enzyme function, so the heat produced by metabolism helps maintain body temperature at a steady level that is high enough for normal enzyme function. Excessively high body temperatures, on the other hand, can alter enzyme structure, resulting in loss of the enzyme's function.

Free energy is the total amount of energy liberated by the complete catabolism of food. It is usually expressed in terms of kilocalories (kcal) per mole of food consumed. For example, the complete catabolism of 1 mole of glucose (168 g; see chapter 2) releases 686 kcal of free energy. About 43% of the total energy released by catabolism is used to produce ATP and to accomplish biological work, such as anabolism, muscular contraction, and other cellular activities. The remaining energy is lost as **heat.**

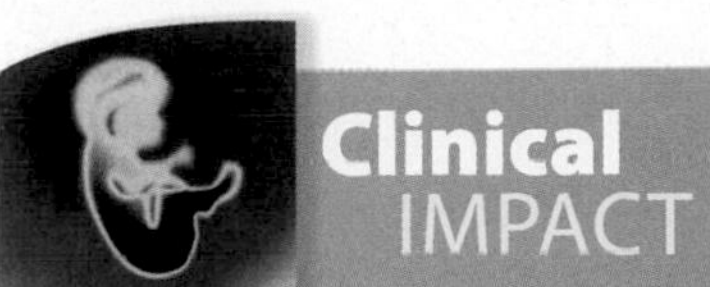

Clinical IMPACT — Obesity

Consuming more food than is necessary for the body's energy needs can cause a person to be overweight or even **obese.** These conditions can be defined on the basis of three parameters: body weight, body mass index, and percent body fat.

"Desirable body weight" is listed in the Metropolitan Life Insurance Table (1999) and indicates, for any height, the weight associated with a maximum life span. By these standards, being overweight is defined as weighing 10–20% more than the "desirable weight," and being obese is defined as weighing 20% or more than the "desirable weight."

Another standard for assessing body weight is body mass index (BMI), which can be calculated by dividing a person's weight (Wt) in kilograms by the square of his or her height (Ht) in meters:

$$BMI = Wt/Ht^2$$

A person whose BMI is greater than 25–27 is overweight, and one whose BMI value is greater than 30 is considered obese. About 26% of people in the United States have a BMI of 30 or greater.

In terms of the percent of total body weight contributed by body fat, 15% body fat or less in men and 25% body fat or less in women is associated with reduced health risks. Obesity is defined as more than 25% body fat in men and 30–35% in women.

Obesity can be further classified according to the number and size of adipocytes. The greater the amount of lipids stored in the adipocytes, the larger their size. In **hyperplastic obesity,** adipocytes are both more numerous and larger than normal. This type of obesity, which is associated with massive obesity, begins at an early age. In nonobese children, the number of adipocytes triples or quadruples between birth and 2 years of age and then remains relatively stable until puberty, when the number increases further. However, in obese children between 2 years of age and puberty, the number of adipocytes continues to increase.

Hypertrophic obesity results from a normal number of adipocytes that have increased in size. This more common type of obesity is associated with moderate obesity and typically develops in adults. People who were thin or of average weight and quite active when they were young become less active as they grow older. Although they no longer use as many kilocalories, they still take in the same amount of food as when they were younger, and they begin to gain weight. The unused kilocalories are turned into lipids, causing adipocytes to increase in size. At one time, scientists believed that the number of adipocytes did not increase after adulthood, but now we know that the number of adipocytes can increase in adults. Apparently, if all the existing adipocytes are filled to capacity with lipids, new adipocytes form to store the excess lipids. Once adipocytes form, however, dieting and weight loss do not decrease their number; rather, the adipocytes become smaller in size as their lipid content decreases.

The distribution of adipose tissue in obese individuals varies. Adipose tissue can be found mainly in the upper body, such as in the abdominal region, or it can be associated with the hips and buttocks. These distribution differences are clinically significant because upper body obesity is associated with an increased likelihood of diabetes mellitus, cardiovascular disease, stroke, and death.

People commonly believe that the main cause of obesity is overeating and, certainly, for obesity to occur, energy intake must exceed energy expenditure. However, comparing the kilocaloric intakes of obese and lean individuals at their usual weights reveals that, on a per-kilogram basis, obese people consume fewer kilocalories than lean people. Obesity occurs for many other reasons. In some cases, a specific physiological mechanism can be identified. For example, a tumor in the hypothalamus can stimulate overeating. In most cases, however, no specific cause is apparent. In fact, obesity can have more than one cause in the same person. Obesity seems to have a genetic component; if one or both parents are obese, their children are more likely to be obese also. In addition, environmental factors, such as eating habits, can play an important role, as in the case of adopted children who exhibit the obesity of their adoptive parents. Furthermore, psychological factors, such as overeating as a means for dealing with stress, can contribute to obesity.

Regulating body weight is actually a matter of regulating adipose tissue because most changes in body weight reflect changes in the amount of adipose tissue. According to the "set point" theory of weight control, the body maintains a certain amount of adipose tissue. If the amount varies below or above this level, mechanisms are activated to return the amount of body fat to its normal value.

When people lose a large amount of weight, their eating behavior changes. They become hyperresponsive to external food cues, think of food often, and cannot get enough to eat without gaining weight. This behavior is typical of both lean and obese individuals who are below their relative set point for weight. Other changes, such as a decrease in basal metabolic rate (BMR), take place in a person who has lost a large amount of weight. Most of this decline in BMR probably results from a decrease in muscle mass associated with weight loss. In addition, some evidence indicates that the amount of energy lost through exercise and the thermic effect of food are also reduced.

Because a person who has lost a large amount of weight has an increased appetite and a decreased ability to expend energy, it is no surprise that only a small percentage of obese people maintain weight loss over the long term. Instead, repeated cycles of weight loss followed by rapid regain are typical.

Current research indicates that body weight is determined by complicated genetic and metabolic factors that can go awry in many ways. Obesity is now regarded as a chronic condition that may respond to medication in much the same way that diabetes does. However, even though drugs can help, eating less and exercising more are still necessary for optimal health.

Predict 6

Why do we become warm during exercise? Why is shivering useful when it is cold?

The average normal body temperature is considered to be 37°C (98.6°F) when measured orally and 37.6°C (99.7°F) when measured rectally. Rectal temperature comes closer to the true core body temperature, but an oral temperature is more easily obtained in older children and adults and therefore is the preferred measure.

Our bodies exchange heat with the environment in a number of ways (figure 25.17). **Radiation** is the gain or loss of heat as infrared energy between two objects that are not in physical contact. For example, the body can gain heat by radiation from the sun, a hot coal, or the hot sand of a beach. On the other hand, the body can lose heat as radiation to cool vegetation or snow on the ground. **Conduction** is the exchange of heat between objects in direct contact with each other, such as the bottoms of the feet and the floor. **Convection** is a transfer of heat between the body and the air or water. A cool breeze causes air to move over the body, allowing body heat to be lost. **Evaporation** is the conversion of water from a liquid to a gas, a process that requires heat. The evaporation of 1 g of water from the body surface results in the loss of 580 cal of heat.

Body temperature is maintained by balancing heat gain with heat loss. If heat gain exceeds heat loss, body temperature increases; if heat loss exceeds heat gain, body temperature decreases. The body generates heat through normal metabolism as well as through the muscle contractions of shivering. The body also exchanges heat with the environment by radiation, conduction, or convection, depending on skin temperature and the environmental temperature. In addition, the body can lose heat to the environment through evaporation of perspiration from the skin.

The difference in temperature between the body and the environment determines the amount of heat exchanged between the two. The greater the temperature difference, the greater the rate of heat exchange. Control of the temperature difference is used to regulate body temperature. For example, if the environmental temperature is very cold, as on a cold winter day, a large temperature difference exists between the body and the environment, and therefore a large loss of heat occurs. Behaviorally, we can reduce the heat loss by seeking a warmer environment—for example, by going inside a heated house or putting on extra clothes. Physiologically, the body controls temperature difference through the dilation and constriction of blood vessels in the skin. When the blood vessels dilate, they bring warm blood to the surface of the body, raising skin temperature; conversely, vasoconstriction decreases blood flow, and skin temperature decreases (see figure 5.9).

When the environmental temperature is greater than body temperature, vasodilation brings warm blood to the skin, causing an increase in skin temperature, which decreases heat gain from the environment. At the same time, evaporation carries away excess heat to prevent heat gain and overheating.

Body temperature regulation is an example of a negative-feedback system controlled by a set point. A small area in the anterior part of the hypothalamus detects slight increases in body temperature through changes in blood temperature (figure 25.18). As a result, mechanisms that cause heat loss, such as vasodilation and sweating, are activated, and body temperature decreases. Alternatively, a small area in the posterior hypothalamus can detect slight decreases in body temperature and can initiate heat gain by increasing muscular activity (shivering) and vasoconstriction.

Radiation from sun and water
Evaporation
Convection from cool breeze
Radiation from sand
Conduction from hot sand

FIGURE 25.17 Heat Exchange
Heat exchange between a person and the environment occurs by radiation, conduction, convection, and evaporation. Arrows show the direction of net heat gain or loss in this environment.

Under some conditions, the set point of the hypothalamus changes. For example, during a fever, the set point is raised, heat-conserving and heat-producing mechanisms are stimulated, and body temperature increases. To recover from a fever, the set point is lowered to normal, heat-loss mechanisms are initiated, and body temperature decreases.

ASSESS YOUR PROGRESS

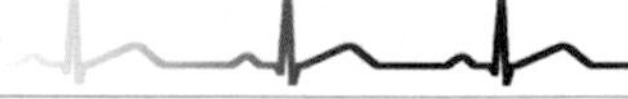

52. *What is free energy? How much free energy is lost as heat from the body?*
53. *What are four ways that heat is exchanged between the body and the environment?*
54. *How is body temperature maintained behaviorally in a cold environment? How is body temperature maintained physiologically in a hot environment?*
55. *How does the hypothalamus regulate body temperature?*

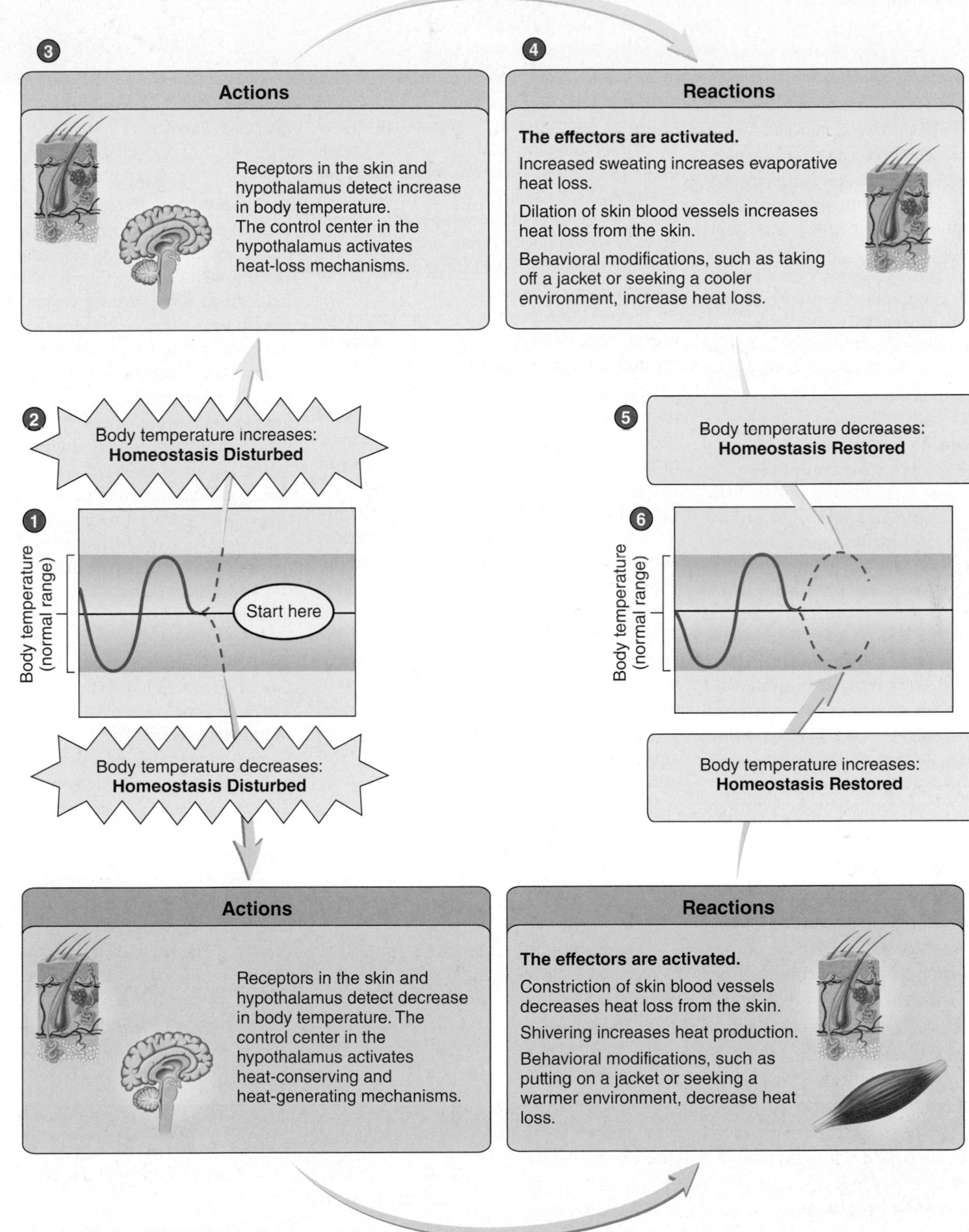
3
Actions
Receptors in the skin and hypothalamus detect increase in body temperature. The control center in the hypothalamus activates heat-loss mechanisms.
4
Reactions
The effectors are activated.
Increased sweating increases evaporative heat loss.
Dilation of skin blood vessels increases heat loss from the skin.
Behavioral modifications, such as taking off a jacket or seeking a cooler environment, increase heat loss.
2
Body temperature increases: Homeostasis Disturbed
5
Body temperature decreases: Homeostasis Restored
1
Body temperature (normal range)
Start here
6
Body temperature (normal range)
Body temperature decreases: Homeostasis Disturbed
Body temperature increases: Homeostasis Restored
Actions
Receptors in the skin and hypothalamus detect decrease in body temperature. The control center in the hypothalamus activates heat-conserving and heat-generating mechanisms.
Reactions
The effectors are activated.
Constriction of skin blood vessels decreases heat loss from the skin.
Shivering increases heat production.
Behavioral modifications, such as putting on a jacket or seeking a warmer environment, decrease heat loss.

HOMEOSTASIS FIGURE 25.18 Summary of Temperature Regulation

(1) Body temperature is within normal range. (2) Body temperature increases outside the normal range, which causes homeostasis to be disturbed. (3) Receptors in the skin and hypothalamus detect the increase in body temperature, and the control center in the hypothalamus responds to the change in body temperature. (4) The effectors are activated. Blood vessels in the skin dilate, and sweating increases to promote heat loss and evaporative cooling. (5) Body temperature decreases. (6) Body temperature returns to its normal range, and homeostasis is restored.

Clinical IMPACT — Hyperthermia and Hypothermia

Hyperthermia

Hyperthermia, elevated body temperature, develops when heat gain exceeds the body's ability to lose heat. Hyperthermia can result from exercise, exposure to hot environments, fever, or anesthesia.

Exercise increases body temperature because heat is a by-product of muscle activity (see chapter 9). Normally, while a person is exercising, vasodilation and increased sweating prevent harmful body temperature increases. However, in a hot, humid environment, the evaporation of sweat decreases, and exercise must be curtailed to prevent overheating.

Exposure to a hot environment normally activates heat-loss mechanisms, so that body temperature is maintained within the normal range, a negative-feedback mechanism. However, with prolonged exposure to a hot environment, the normal negative-feedback mechanisms are unable to keep body temperature from increasing above normal, resulting in **heat exhaustion.** Heavy sweating causes dehydration, decreased blood volume, decreased blood pressure, and increased heart rate. Heat exhaustion is characterized by wet, cool skin due to heavy sweating. In addition, the person usually feels weak, dizzy, and nauseated. Treatment includes moving to a cooler environment to allow heat loss, ceasing activity to reduce the heat produced by muscle metabolism, and restoring blood volume by drinking fluids.

Heat stroke is more severe than heat exhaustion because it results from a breakdown in the normal negative-feedback mechanisms of temperature regulation. If the temperature of the hypothalamus becomes too high, it no longer functions appropriately. Sweating stops, and the skin becomes dry and flushed. The person becomes confused, irritable, or even comatose. Treatment is the same as for heat exhaustion, except that efforts to promote heat loss from the skin should be increased—for example, by applying wet cloths to the skin or by immersing the person in a cool bath.

Fever is the development of a higher than normal body temperature due to invasion of the body by microorganisms or other foreign substances. Lymphocytes, neutrophils, and macrophages release chemicals called **pyrogens** (pī′rō-jenz), such as certain interleukins, interferons, and tissue necrosis factor. Pyrogens increase the synthesis of prostaglandins, which stimulate a rise in the temperature set point of the hypothalamus. Consequently, body temperature and metabolic rate increase. Physiologists believe fever is beneficial because it speeds the chemical reactions of the immune system (see chapter 22) and inhibits the growth of some microorganisms. However, body temperatures greater than 41°C (106°F) can be harmful. To lower body temperature, physicians prescribe aspirin, nonsteroidal anti-inflammatory drugs (NSAIDs), and acetaminophen, which act by inhibiting the synthesis of prostaglandins.

In people who have the inherited muscle disorder **malignant hyperthermia,** certain general anesthetics cause sustained, uncoordinated muscle contractions. Consequently, body temperature increases.

Therapeutic hyperthermia is an induced elevation in local or general body temperature sometimes used to treat tumors and infections.

Hypothermia

If heat loss exceeds the body's ability to produce heat, body temperature falls below normal. **Hypothermia** is a decrease in body temperature to 35°C (95°F) or below. Hypothermia usually results from prolonged exposure to a cold environment. At first, body temperature is maintained by normal negative-feedback mechanisms—that is, heat loss is decreased by constricting blood vessels in the skin, and heat production is increased by shivering. However, if body temperature decreases despite these mechanisms, hypothermia develops. The individual's thinking becomes sluggish, and movements are uncoordinated. Heart, respiratory, and metabolic rates decline, and death results unless body temperature is restored to normal. Treatment for hypothemia calls for rewarming the body a few degrees per hour.

Frostbite is damage to the skin and deeper tissues resulting from prolonged exposure to the cold. Specifically, damage results from direct cold injury to cells, injury from ice crystal formation, and reduced blood flow to affected tissues. The fingers, toes, ears, nose, and cheeks are most commonly affected. The consequences of frostbite can range from redness and discomfort to loss of the affected part due to death of the tissue. The best treatment is immersion in a warm-water bath. Patients should avoid rubbing the affected area or applying local dry heat.

Therapeutic hypothermia is sometimes used to slow the metabolic rate during surgical procedures, such as heart surgery. Due to the decreased metabolic rate, the tissues do not require as much oxygen as normal and are less likely to be damaged.

Learn to Predict ❮ Answer

From page 912

Although Sadie's suggested snack did contain a lot of calories, which she and David needed for their day at the park, the food choices were not ideal. Most of the calories in the cookies and grape soda were actually from simple sugars. Eating large amounts of simple sugars, such as Sadie's suggested snack, could result in large fluctuations in blood glucose levels. Though the children may initially have an increase in energy, they will most likely experience a drastic decrease in energy as well. Foods that include complex carbohydrates, such as those in the snack Sadie's mom suggested, have other nutrients, such as vitamins, many of which are necessary for normal metabolism. In addition, complex carbohydrates are digested and absorbed at a slower rate and do not contribute to drastic changes in blood glucose levels. Essentially, Sadie's mom selected food that would provide the children with energy and additional beneficial nutrients.

Answers to the rest of this chapter's Predict questions are in Appendix G.

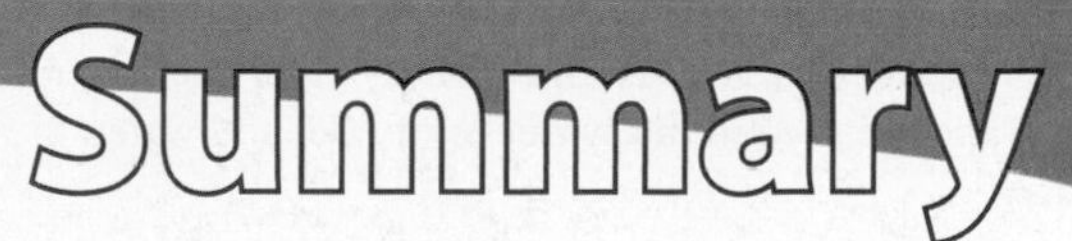
Summary

25.1 Nutrition (p. 913)

Nutrition is the ingestion and use of food, as well as the evaluation of food and drink requirements.

Nutrients

1. Nutrients are the chemicals used by the body: carbohydrates, lipids, proteins, vitamins, minerals, and water.
2. Essential nutrients must be ingested because the body either cannot manufacture them at all or cannot manufacture them in adequate amounts.

Kilocalories

1. A calorie (cal) is the heat (energy) necessary to raise the temperature of 1 g of water 1°C. A kilocalorie (kcal), or Calorie (Cal), is 1000 calories.
2. A gram of carbohydrate or protein yields 4 kcal, and a gram of lipids yields 9 kcal.

MyPlate

The MyPlate icon provides a visual reminder for making choices at mealtime, by selecting healthy foods from five food groups. Half the meal should be fruits and vegetables.

Carbohydrates

1. Carbohydrates are ingested as monosaccharides (glucose, fructose), disaccharides (sucrose, maltose, lactose), and polysaccharides (starch, glycogen, cellulose).
2. Polysaccharides and disaccharides are converted to glucose, which can be used for energy or stored as glycogen or lipids.
3. The Acceptable Macronutrient Distribution Range (AMDR) for carbohydrates is 45–65% of total kilocalories.

Lipids

1. Lipids are ingested as triglycerides (95%) or cholesterol and phospholipids (5%).
2. Monounsaturated fats and oils have one double bond, and polyunsaturated fats and oils have two or more double bonds.
3. Most unprocessed polyunsaturated oils occur in the *cis* form, whereas hydrogenated polyunsaturated oils are in the *trans* form.
4. Triglycerides are used for energy or stored in adipose tissue. Cholesterol forms other molecules, such as steroid hormones. Cholesterol and phospholipids are part of the plasma membrane.
5. The AMDR for lipids is 20–35%.

Proteins

1. Proteins are ingested and broken down into amino acids.
2. Proteins function in protection (antibodies), regulation (enzymes, hormones), structure (collagen), muscle contraction (actin and myosin), and transportation (hemoglobin, transport proteins); they also act as receptor molecules.
3. The AMDR for protein is 10–35% of total kilocalories.

Vitamins

1. Many vitamins function as coenzymes or as parts of coenzymes.
2. Most vitamins are not produced by the body and must be obtained in the diet. Some vitamins can be formed from provitamins.
3. Vitamins are classified as either fat-soluble or water-soluble.
4. Recommended Dietary Allowances (RDAs) are a guide for estimating the nutritional needs of groups of people based on their age, gender, and other factors.

Minerals

1. Minerals contribute to normal metabolism, add mechanical strength to bones and teeth, function as buffers, and are involved in osmotic balance.
2. The daily requirement for major minerals is 100 mg or more daily; for trace minerals, less than 100 mg daily is sufficient.

Daily Values for Nutrients

1. Daily Values are dietary references that can be used to plan a healthy diet.
2. Daily Values for vitamins and minerals are based on Reference Daily Intakes, which are generally the highest 1968 RDA values of age categories.
3. Daily Values are based on Daily Reference Values.
   - The Daily Reference Values for energy-producing nutrients (carbohydrates, total fat, saturated fat, and proteins) and dietary fiber are recommended percentages of the total kilocalories ingested daily for each nutrient.
   - The Daily Reference Values for total fats, saturated fats, cholesterol, and sodium are the uppermost limits considered desirable because of their link to diseases.
4. The % Daily Value is the percent of the recommended Daily Value of a nutrient found in one serving of a particular food.

25.2 Metabolism (p. 921)

1. Metabolism consists of catabolism and anabolism. Catabolism breaks down molecules and gives off energy. Anabolism builds up molecules and requires energy.
2. The energy in carbohydrates, lipids, and proteins is used to produce ATP through oxidation-reduction reactions.

25.3 Carbohydrate Metabolism (p. 922)

Glycolysis

Glycolysis is the breakdown of glucose into two pyruvic acid molecules. Also produced are two NADH molecules and two ATP molecules.

Lactic Acid Fermentation

1. Lactic acid fermentation is the breakdown of glucose in the absence of oxygen into two lactic acid. Two ATP molecules are also produced.
2. Lactic acid can be converted to glucose (Cori cycle) using aerobically produced ATP (oxygen deficit).

Aerobic Respiration

1. Aerobic respiration is the breakdown of glucose in the presence of oxygen to produce carbon dioxide, water, and 36 ATP molecules.
2. The first phase is glycolysis, which produces two ATP, two NADH, and two pyruvic acid molecules.
3. The second phase is the conversion of the two pyruvic acid molecules into two molecules of acetyl-CoA. These reactions also produce two NADH and two carbon dioxide molecules.
4. The third phase is the citric acid cycle, which produces two ATP, six NADH, two $FADH_2$, and four carbon dioxide molecules.

5. The fourth phase is the electron-transport chain. The high-energy electrons in NADH and $FADH_2$ enter the electron-transport chain and are used in the synthesis of ATP and water.

25.4 Lipid Metabolism (p. 930)

1. Triglycerides are broken down and released as free fatty acids.
2. Free fatty acids are taken up by cells and broken down by beta-oxidation into acetyl-CoA.
   - Acetyl-CoA can enter the citric acid cycle.
   - Acetyl-CoA can be converted into ketone bodies.

25.5 Protein Metabolism (p. 932)

1. New amino acids are formed by transamination, the transfer of an amine group to a keto acid.
2. Amino acids are used to synthesize proteins. If used for energy, ammonia is produced as a by-product of oxidative deamination. Ammonia is converted to urea and excreted.

25.6 Interconversion of Nutrient Molecules (p. 934)

1. Glycogenesis is the formation of glycogen from glucose.
2. Lipogenesis is the formation of lipids from glucose and amino acids.
3. Glycogenolysis is the breakdown of glycogen to glucose.
4. Gluconeogenesis is the formation of glucose from amino acids and glycerol.

25.7 Metabolic States (p. 935)

1. In the absorptive state, nutrients are used as energy or stored.
2. In the postabsorptive state, stored nutrients are used for energy.

25.8 Metabolic Rate (p. 937)

Metabolic rate is the total energy expenditure per unit of time, and it has three components.

Basal Metabolic Rate

Basal metabolic rate, the energy used at rest, is 60% of the metabolic rate.

Thermic Effect of Food

The energy used to digest and absorb food, called the thermic effect of food, is 10% of the metabolic rate.

Muscular Activity

Muscular energy, that used for muscle contraction, is 30% of the metabolic rate.

25.9 Body Temperature Regulation (p. 938)

1. Body temperature is maintained by balancing heat gain and heat loss.
   - Heat is produced through metabolism.
   - Heat is exchanged with the environment through radiation, conduction, convection, and evaporation.
2. The greater the temperature difference between the body and the environment, the greater the rate of heat exchange.
3. Body temperature is regulated by a set point in the hypothalamus.

REVIEW AND COMPREHENSION

1. Which of these statements concerning kilocalories is true?
   a. A kilocalorie is the amount of energy required to raise the temperature of 1 g of water 1°C.
   b. There are 9 kcal in a gram of protein.
   c. There are 4 kcal in a gram of lipids.
   d. A pound of body fat contains 3500 kcal.
2. What type of nutrient is recommended as the primary energy source in the diet?
   a. carbohydrates
   b. lipids
   c. proteins
   d. cellulose
3. A source of monounsaturated fats is
   a. fat associated with meat.
   b. egg yolks.
   c. whole milk.
   d. fish oil.
   e. olive oil.
4. A complete protein food
   a. provides the daily amount (grams) of protein recommended for a healthy diet.
   b. can be used to synthesize the nonessential amino acids.
   c. contains all 20 amino acids.
   d. includes beans, peas, and leafy green vegetables.
5. Concerning vitamins,
   a. most can be synthesized by the body.
   b. they are normally broken down before they can be used by the body.
   c. A, D, E, and K are water-soluble vitamins.
   d. many function as coenzymes.
6. Minerals
   a. are inorganic nutrients.
   b. compose about 4–5% of total body weight.
   c. act as buffers and osmotic regulators.
   d. are components of enzymes.
   e. All of these are correct.
7. Glycolysis
   a. is the breakdown of glucose to two pyruvic acid molecules.
   b. requires the input of two ATP molecules.
   c. produces two NADH molecules.
   d. does not require oxygen.
   e. All of these are correct.
8. Lactic acid fermentation _______________ oxygen and produces _______________ energy (ATP) for the cell than aerobic respiration.
   a. does not require, more
   b. does not require, less
   c. requires, more
   d. requires, less

9. The molecule that moves electrons from the citric acid cycle to the electron-transport chain is
   a. tRNA.
   b. mRNA.
   c. ADP.
   d. NADH.
   e. pyruvic acid.
10. The carbon dioxide you breathe out comes from
    a. glycolysis.
    b. the electron-transport chain.
    c. lactic acid fermentation.
    d. the food you eat.
11. Lipids are
    a. stored primarily as triglycerides.
    b. synthesized by beta-oxidation.
    c. broken down by oxidative deamination.
    d. All of these are correct.
12. Amino acids
    a. are classified as essential or nonessential.
    b. can be synthesized in a transamination reaction.
    c. can be used as a source of energy.
    d. can be converted to keto acids.
    e. All of these are correct.
13. Ammonia is
    a. a by-product of lipid metabolism.
    b. formed during ketogenesis.
    c. converted into urea in the liver.
    d. produced during lipogenesis.
    e. converted to keto acids.
14. The conversion of amino acids and glycerol into glucose is called
    a. gluconeogenesis.
    b. glycogenesis.
    c. glycogenolysis.
    d. ketogenesis.
15. Which of these events takes place during the absorptive state?
    a. Glycogen is converted into glucose.
    b. Glucose is converted into lipids.
    c. Ketones are produced.
    d. Proteins are converted into glucose.
16. Loss of heat resulting from loss of water from the body's surface is
    a. radiation.
    b. conduction.
    c. evaporation.
    d. convection.

Answers in Appendix E

CRITICAL THINKING

1. One serving of a food contains 2 g of saturated fat. What % Daily Value for saturated fat would appear on the food label for this food? (See the bottom of figure 25.2 for information needed to answer this question.)
2. An active teenage boy has a daily intake of 3000 kcal. What is the maximum amount (weight) of total fats he should consume, according to the Daily Values?
3. If the teenager in question 2 eats a serving of food that has a total fat content of 10 g/serving, what is his % Daily Value for total fat?
4. Why does a vegetarian usually have to be more careful about his or her diet than a person who eats meat?
5. Explain why a person suffering from copper deficiency feels tired all the time.
6. Why can some people lose weight on a 1200 kcal/day diet, whereas others cannot?
7. After learning that sweat evaporation results in loss of calories, an anatomy and physiology student enters a sauna to try to lose weight. He reasons that a liter (about a quart) of water weighs 1000 g, which is equivalent to 580,000 cal, or 580 kcal, of heat when lost as sweat. Therefore, instead of reducing his diet by 580 kcal/day, he believes that losing a liter of sweat every day in the sauna will cause him to lose about a pound of fat a week. Will this approach work? Explain.
8. Thyroid hormone increases the activity of the sodium-potassium pump. If a person produced excess amounts of thyroid hormone, how would basal metabolic rate, body weight, and body temperature be affected? How would the body attempt to compensate for the changes in body weight and temperature?
9. In some diseases, an infection causes a high fever, resulting in a crisis state. Once body temperature begins to return to normal, the person is on the way to recovery. If you were looking for symptoms in a person who had just passed through the crisis state, would you look for dry, pale skin or flushed, wet skin? Explain.

Answers in Appendix F

Visit this book's website at **www.mhhe.com/seeley10** for chapter quizzes, interactive learning exercises, and other study tools.

Mc Graw Hill
connect plus+
ANATOMY & PHYSIOLOGY

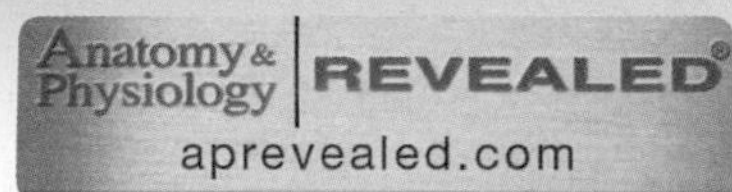
Anatomy & Physiology
REVEALED
aprevealed.com

26

Learn to Predict

Fifty-seven-year-old Sadie is living proof that a person can live with type 2 diabetes. Since being diagnosed with the condition 15 years ago, she has taken good care of herself, watching her diet and regularly monitoring her blood glucose and blood pressure at home. Therefore, she was immediately aware when her blood pressure began to rise. She also felt tired much of the time, and her face appeared puffy. Her physician confirmed the hypertension and detected generalized edema. After ordering several laboratory tests, he concluded that Sadie was suffering from chronic renal failure. With some information from chapters 19 and 20 and after reading chapter 26, explain the cause of chronic renal failure and predict the results of Sadie's blood test and urinalysis that led to this diagnosis. What is the probable prognosis for Sadie in the future?

Urinary System

It must have been fate. Sam met Dave when they played on the same softball team, and the two became great friends. A few years later, Dave was diagnosed with a fatal, progressive kidney disease. Like Sadie in this chapter's "Learn to Predict," Dave began feeling tired and suffering from generalized edema. Without a kidney transplant, Dave could expect to live only a few more years. Immediately, Sam volunteered to be tested as a possible kidney donor and, amazingly, he was a nearly perfect match. After months of testing and planning, one of Sam's kidneys was removed and placed into Dave's body. Within a week, Dave's new kidney was functioning at almost normal capacity. Meanwhile, Sam recovered quickly, and his spirits were high because his donation had saved his friend's life. Doctors assured him that his remaining kidney, containing more than a million nephrons, would be sufficient for his future needs. The kidneys are remarkable organs that perform life-sustaining functions as part of the urinary system.

Photo: Removal of a healthy kidney, as shown in this photo, is performed when the patient donates his or her kidney to a recipient who may otherwise die. The donor can expect to live a normal life with just one kidney.

Module 13
Urinary System

Anatomy & Physiology
REVEALED
aprevealed.com

26.1 Functions of the Urinary System

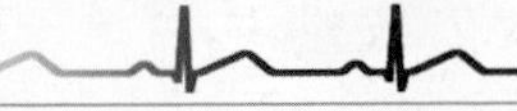

LEARNING OUTCOMES

After reading this section, you should be able to

A. List the organs of the urinary system.

B. Describe the main functions of the kidneys.

The **urinary system** consists of two kidneys; two ureters, which carry urine from the kidneys to the urinary bladder; a single, midline urinary bladder; and a single urethra, which carries urine from the bladder to the outside of the body (figure 26.1).

The kidneys are the major excretory organs of the body. The skin, liver, lungs, and intestines eliminate some waste products; however, if the kidneys fail to function, these other excretory organs cannot adequately compensate. The kidneys perform the following functions:

1. *Excretion.* The kidneys filter blood and produce a large volume of filtrate. Large molecules, such as proteins, and blood cells remain in the blood, whereas smaller molecules and ions enter the filtrate. As the filtrate flows through the kidneys, it is slowly modified until it is converted into urine. This conversion requires the reabsorption of most of the filtrate volume back into the blood, along with useful molecules and ions. Metabolic wastes, toxic molecules, and excess ions remain in a small volume of filtrate. Additional waste products are secreted into the filtrate, eventually forming urine.
2. *Regulation of blood volume and pressure.* The kidneys play a major role in controlling the extracellular fluid volume in the body by producing either a large volume of dilute urine or a small volume of concentrated urine, depending on the hydration level of the body. Consequently, the kidneys regulate blood volume and hence blood pressure.
3. *Regulation of the concentration of solutes in the blood.* The kidneys help regulate the concentration of primarily the major ions—Na^+, Cl^-, K^+, Ca^{2+}, HCO_3^-, and HPO_4^{2-}; they also regulate other solute concentration, such as urea.
4. *Regulation of extracellular fluid pH.* The kidneys secrete variable amounts of H^+ to help regulate the extracellular fluid acidity.
5. *Regulation of red blood cell synthesis.* The kidneys secrete the hormone erythropoietin, which stimulates the synthesis of red blood cells in red bone marrow (see chapter 19).
6. *Regulation of vitamin D synthesis.* The kidneys play an important role in controlling blood levels of Ca^{2+} by activating vitamin D (see chapter 6).

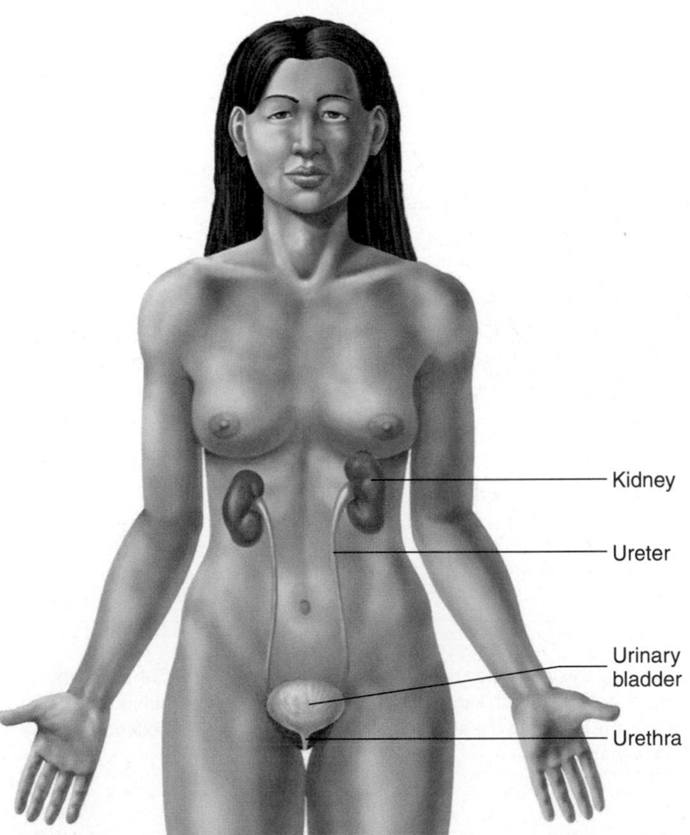
Kidney
Ureter
Urinary bladder
Urethra

FIGURE 26.1 APR **Urinary System**
The urinary system consists of two kidneys, two ureters, the urinary bladder, and the urethra.

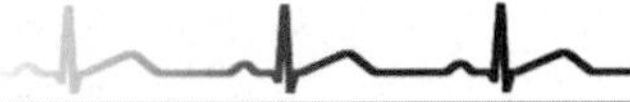

ASSESS YOUR PROGRESS

1. *Name the organs that make up the urinary system.*
2. *List the functions performed by the kidneys, and briefly describe each.*

26.2 Kidney Anatomy and Histology

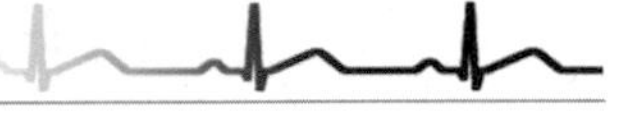

LEARNING OUTCOMES

After reading this section, you should be able to

A. Describe the location and external anatomy of the kidneys.

B. Describe the inner regions of the kidney.

C. Give the details of the nephron's structure and histology.

D. Explain the blood supply of the kidney.

Location and External Anatomy of the Kidneys

The **kidneys** are bean-shaped, and each is about the size of a tightly clenched fist. They lie behind the peritoneum on the posterior abdominal wall on each side of the vertebral column near the lateral borders of the psoas major muscles (figure 26.2). The kidneys extend from the level of the last thoracic (T12) to the third lumbar (L3) vertebrae, and the rib cage partially protects them. The liver is superior to the right kidney, causing the right kidney to be slightly lower than the left. Each kidney measures about 11 cm long, 5 cm wide, and 3 cm thick and weighs about 130 g. A **renal capsule,** a layer of fibrous connective tissue, surrounds each kidney. A dense layer of adipose tissue engulfs the renal capsule. This adipose tissue cushions the kidneys against mechanical shock. A thin layer of connective tissue, the **renal fascia,** surrounds the adipose tissue and helps anchor the kidneys and surrounding adipose tissue to the abdominal wall. Adipose tissue surrounds the renal fascia.

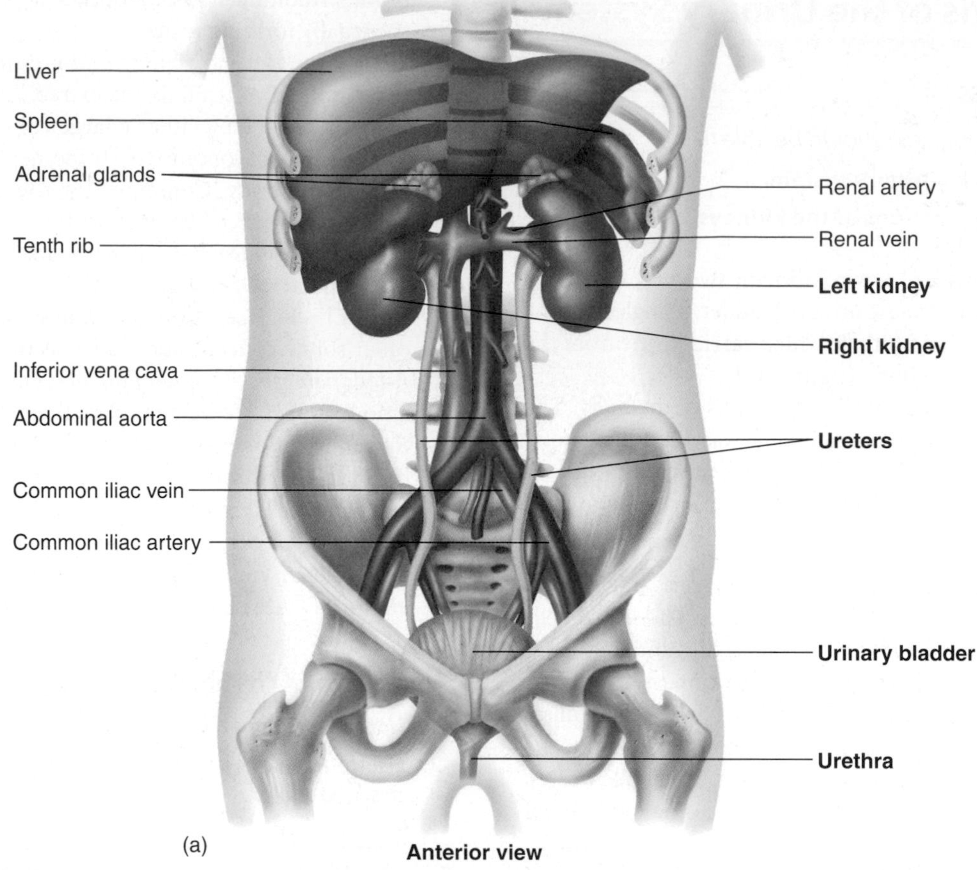
Liver
Spleen
Adrenal glands
Tenth rib
Inferior vena cava
Abdominal aorta
Common iliac vein
Common iliac artery
Renal artery
Renal vein
Left kidney
Right kidney
Ureters
Urinary bladder
Urethra
(a)
Anterior view

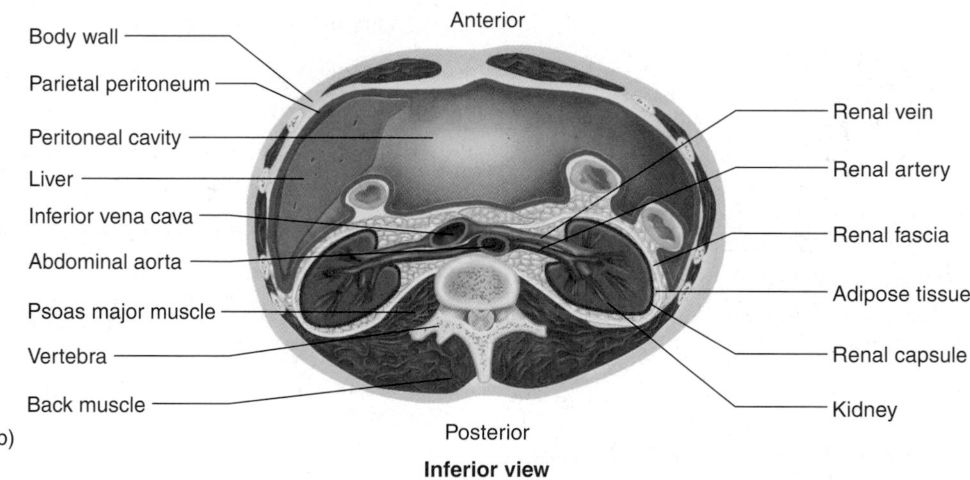
Anterior
Body wall
Parietal peritoneum
Peritoneal cavity
Liver
Inferior vena cava
Abdominal aorta
Psoas major muscle
Vertebra
Back muscle
Renal vein
Renal artery
Renal fascia
Adipose tissue
Renal capsule
Kidney
(b)
Posterior
Inferior view

FIGURE 26.2 APR **Anatomy of the Urinary System**

(*a*) The kidneys are located in the abdominal cavity, with the right kidney just below the liver and the left kidney below the spleen. A ureter extends from each kidney to the urinary bladder within the pelvic cavity. An adrenal gland is located at the superior pole of each kidney. (*b*) The kidneys are located behind the parietal peritoneum, surrounded by adipose tissue. A connective tissue layer, the renal fascia, anchors the kidney to the abdominal wall. The renal arteries extend from the abdominal aorta to each kidney, and the renal veins extend from the kidneys to the inferior vena cava.

The **hilum** (hī′lŭm) is a small area where the renal artery and nerves enter, and the renal vein and ureter exit, the kidney. It is located on the concave, medial side of the kidney. The hilum opens into the **renal sinus,** a cavity filled with adipose tissue and connective tissue. Structures that enter and leave the kidney pass through the renal sinus (figure 26.3).

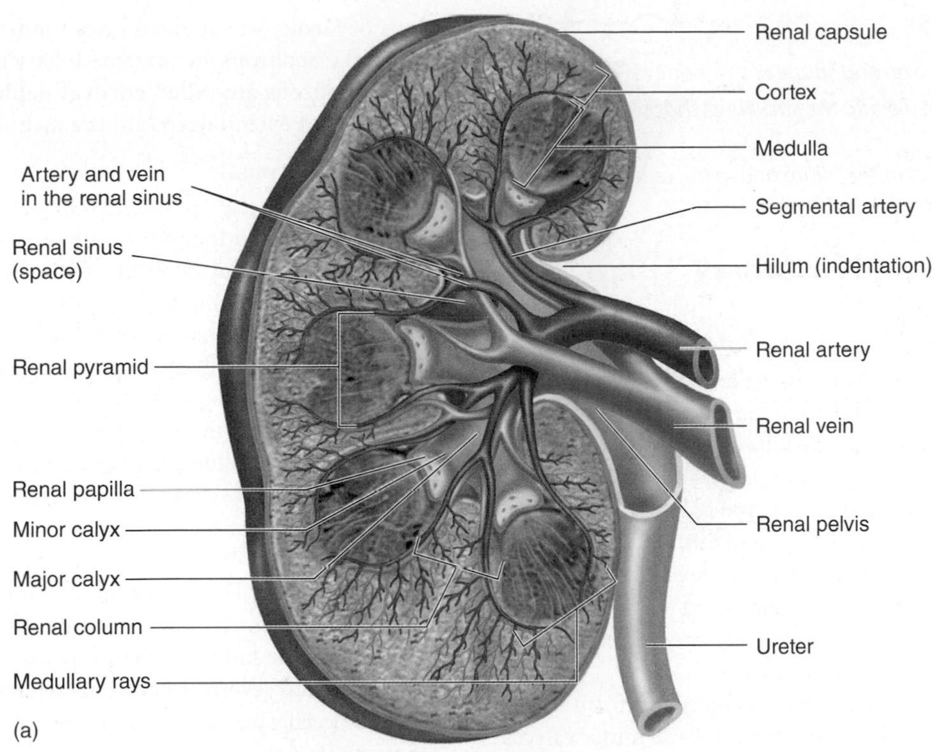
Renal capsule
Cortex
Medulla
Artery and vein in the renal sinus
Segmental artery
Renal sinus (space)
Hilum (indentation)
Renal artery
Renal pyramid
Renal vein
Renal papilla
Renal pelvis
Minor calyx
Major calyx
Renal column
Ureter
Medullary rays
(a)

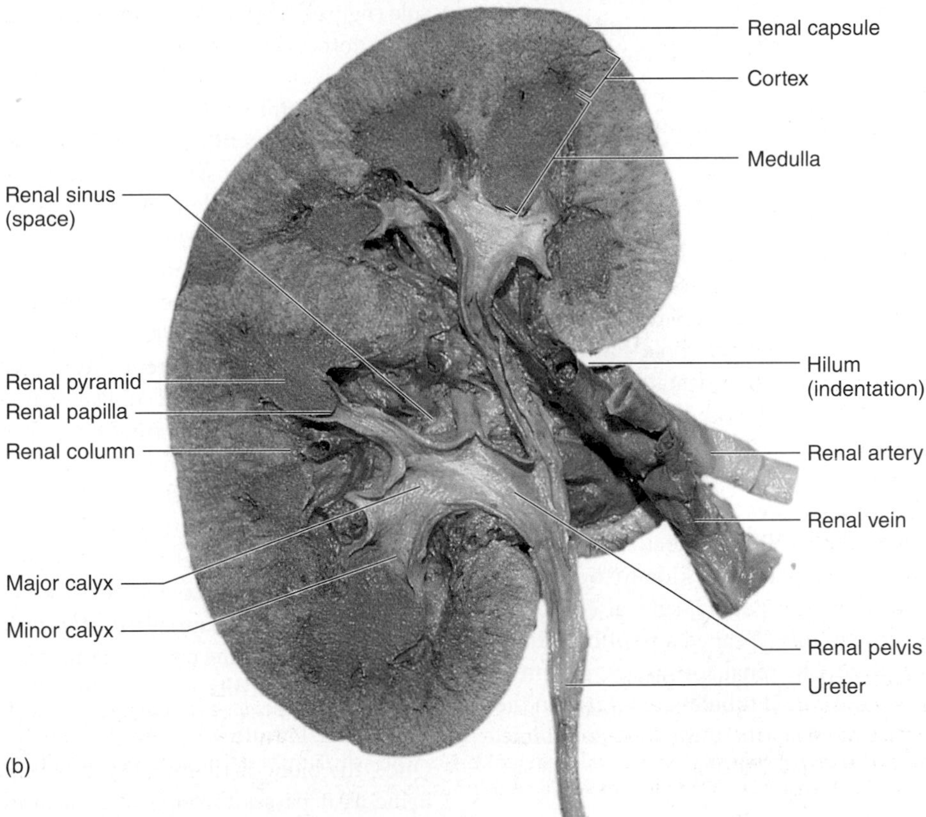
Renal capsule
Cortex
Medulla
Renal sinus (space)
Hilum (indentation)
Renal pyramid
Renal papilla
Renal column
Renal artery
Renal vein
Major calyx
Minor calyx
Renal pelvis
Ureter
(b)

FIGURE 26.3 APR **Frontal Section of the Kidney and Ureter**

(*a*) A frontal kidney section shows that the cortex forms the outer part of the kidney, and the medulla forms the inner part. A central cavity called the renal sinus contains the renal pelvis. The renal columns of the kidney project from the cortex into the medulla and separate the pyramids. (*b*) Photograph of a longitudinal section of a human kidney and ureter.

ASSESS YOUR PROGRESS 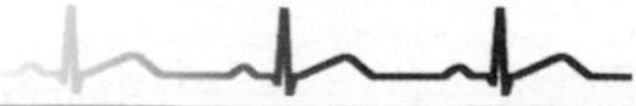

3. *Describe the location, size, and shape of the kidneys.*
4. *Describe the renal capsule and the structures that surround the kidney.*
5. *List the structures found at the hilum and in the renal sinus of a kidney.*

Internal Anatomy and Histology of the Kidneys

To fully appreciate the function of the kidneys, we must first understand their ultrastructure. The kidneys are organized into two major regions: an outer **cortex** and an inner **medulla** surrounding the renal sinus (figure 26.3). The medulla is composed of cone-shaped structures called renal pyramids. **Medullary rays** extend from the renal pyramids into the cortex. **Renal columns,** consisting of cortical tissue, project between the renal pyramids. The bases of the pyramids form the boundary between the cortex and the medulla. The tips of the pyramids, the **renal papillae,** point toward the renal sinus. **Minor calyces** (kal′i-sēz; sing. *calyx*) are funnel-shaped chambers into which the renal papillae extend.

The minor calyces of several pyramids merge to form larger funnels, the **major calyces.** Each kidney contains 8–20 minor calyces and 2 or 3 major calyces. The major calyces converge to form an enlarged chamber called the **renal pelvis,** which is surrounded by the renal sinus. The renal pelvis narrows into a small-diameter tube, the **ureter,** which exits the kidney at the hilum and connects to the urinary bladder. Urine formed within the kidney flows from the renal papillae into the minor calyces. From the minor calyces, urine flows into the major calyces, collects in the renal pelvis, and then leaves the kidney through the ureter.

Structure of a Nephron

The **nephron** (nef′ron) is the histological and functional unit of the kidney (figure 26.4). Each nephron consists of four basic components: a **renal corpuscle,** a proximal convoluted tubule, a loop of Henle (nephron loop), and a distal convoluted tubule. Each portion of a nephron plays a different role in urine formation. Generally speaking, the renal corpuscle filters the blood, the proximal convoluted tubule returns filtered substances to the blood, the loop of Henle helps conserve water and solutes, and the distal convoluted tubule rids the blood of additional wastes. The fluid in the distal convoluted tubule then empties into a collecting duct, which carries the newly formed urine from the cortex of the kidney toward the renal papilla. Near the tip of the renal papilla, several collecting ducts merge into a larger-diameter tubule called a **papillary duct,** which empties into a minor calyx. The renal corpuscles, proximal convoluted tubules, and distal convoluted tubules are located in the renal cortex, but the collecting ducts, parts of the loops of Henle, and the papillary ducts are in the renal medulla.

Types of Nephrons

There are approximately 1.3 million nephrons in each kidney. Most nephrons measure 50–55 mm in length. Nephrons whose renal corpuscles lie near the medulla are called **juxtamedullary** (juks′ta-med′ŭ-lār-ē; next to medulla) **nephrons.** They have long loops of Henle, which extend deep into the medulla. Only about 15% of the nephrons are juxtamedullary nephrons. The remainder of the nephrons are called **cortical nephrons,** and their loops of Henle do not extend deep into the medulla (figure 26.4).

The Renal Corpuscle

Each renal corpuscle consists of a Bowman capsule and a network of capillaries called the **glomerulus** (glō-măr′ū-lŭs), which is the filtration unit of the nephron (figure 26.5*a,b*). The wall of the Bowman capsule is indented to form a double-walled chamber. The glomerulus, which looks like a wad of yarn, fills the indentation. Fluid is filtered from the glomerulus into the Bowman capsule. The filtered fluid then flows into the proximal convoluted tubule, which carries it away from the Bowman capsule.

A Bowman capsule has an outer layer, called the **parietal layer,** and an inner layer, called the **visceral layer** (figure 26.5*b*). The parietal layer is constructed of simple squamous epithelial cells. The epithelial cells become cube-shaped at the beginning of the proximal convoluted tubule. The visceral layer is constructed of specialized cells called **podocytes,** which wrap around the glomerular capillaries.

Because the kidneys' main function is to filter the blood, the glomerulus has several unique characteristics that make these capillaries especially permeable. Numerous, windowlike openings, called **fenestrae** (fe-nes′trē), are in the endothelial cells of the glomerular capillaries. Gaps, called **filtration slits,** are between the cell processes of the podocytes that make up the visceral layer of the Bowman capsule (figure 26.5*c*). A basement membrane lies sandwiched between the endothelial cells of the glomerular capillaries and the podocytes of the Bowman capsule. Together, the capillary endothelium, the basement membrane, and the podocytes of the Bowman capsule form the kidney's **filtration membrane** (figure 26.5*d*), which performs the first major step in urine formation. Urine formation begins when fluid from the glomerular capillaries is filtered across the filtration membrane into the lumen, or space, inside the Bowman capsule.

An **afferent** (af′er-ent) **arteriole** supplies blood to the glomerulus, and an **efferent** (ef′er-ent) **arteriole** drains it (figure 26.5*a*). A layer of smooth muscle lines both the afferent and efferent arterioles. At the point where the afferent arteriole enters the renal corpuscle, the smooth muscle cells form a cufflike arrangement around the arteriole. These cells are called **juxtaglomerular cells.** Lying between the afferent and efferent arterioles adjacent to the renal corpuscle is part of the distal convoluted tubule of the nephron. Specialized tubule cells in this section are collectively called the **macula** (mak′ū-lă) **densa.** The juxtaglomerular cells of the afferent arteriole and the macula densa cells are in close contact with each other and together are called the **juxtaglomerular apparatus** (figure 26.5*b*). The juxtaglomerular apparatus secretes the enzyme renin and plays an important role in the regulation of filtrate formation and blood pressure.

The Renal Tubule

Once, the blood is filtered, the resulting fluid is modified to form urine as it passes through each section of the renal tubule. The **proximal convoluted tubule** measures approximately 14 mm long and 60 μm in diameter. Simple cuboidal epithelium makes up its wall. The cells rest on a basement membrane, which forms the outer surface of the tubule. Many microvilli project from the luminal (next to the filtrate) surface of the cells (figure 26.6*a,b*).

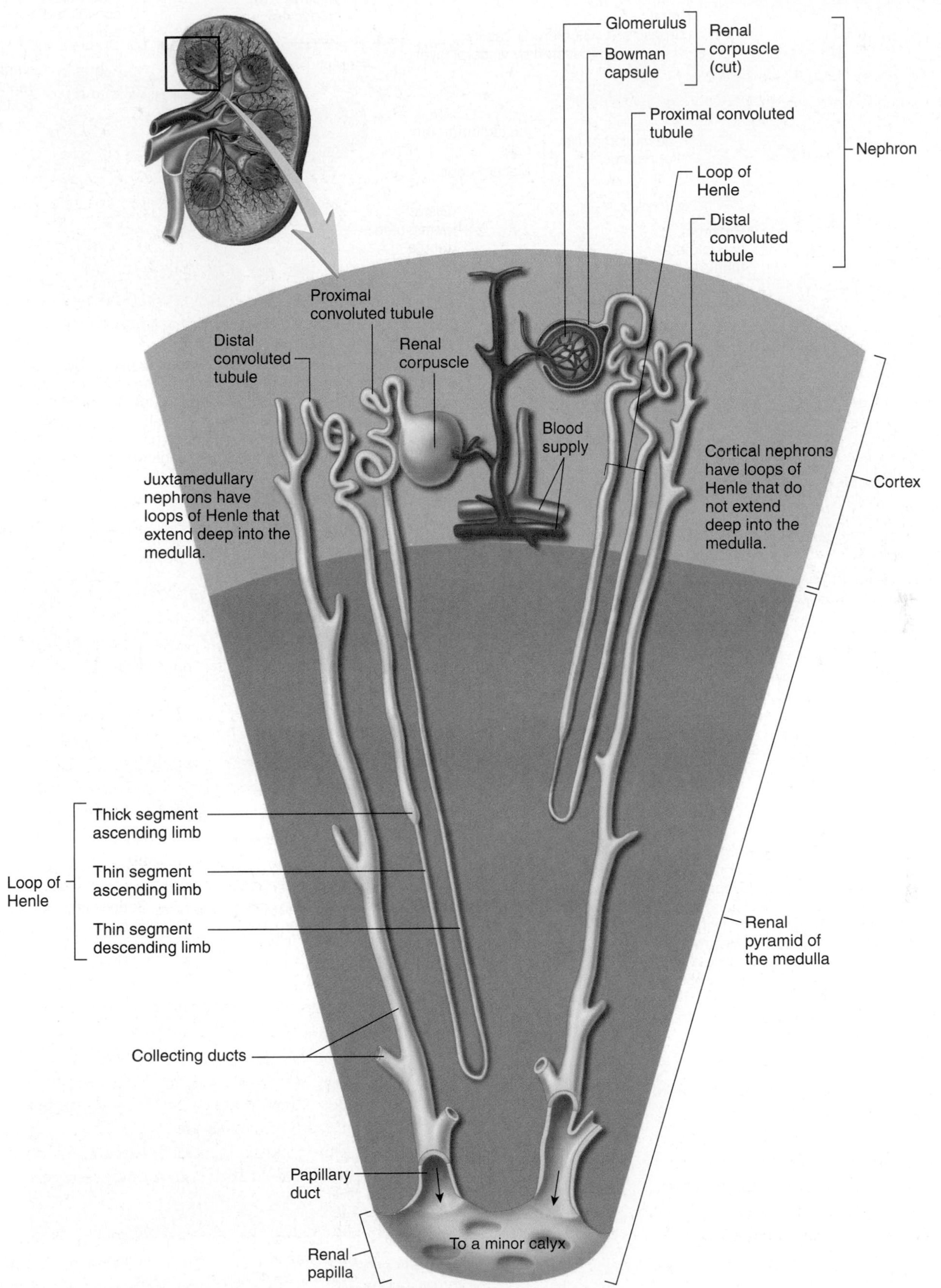
Glomerulus
Bowman capsule
Renal corpuscle (cut)
Proximal convoluted tubule
Loop of Henle
Distal convoluted tubule
Nephron
Proximal convoluted tubule
Distal convoluted tubule
Renal corpuscle
Blood supply
Juxtamedullary nephrons have loops of Henle that extend deep into the medulla.
Cortical nephrons have loops of Henle that do not extend deep into the medulla.
Cortex
Loop of Henle
Thick segment ascending limb
Thin segment ascending limb
Thin segment descending limb
Renal pyramid of the medulla
Collecting ducts
Papillary duct
Renal papilla
To a minor calyx

FIGURE 26.4 AP|R **Functional Unit of the Kidney—the Nephron**

A nephron consists of a renal corpuscle, proximal convoluted tubule, loop of Henle, and distal convoluted tubule. The distal convoluted tubule empties into a collecting duct. Juxtamedullary nephrons (those near the medulla of the kidney) have loops of Henle that extend deep into the medulla, whereas other nephrons do not. Collecting ducts undergo a transition to larger-diameter papillary ducts near the tip of the renal papilla. The papillary ducts empty into a minor calyx.

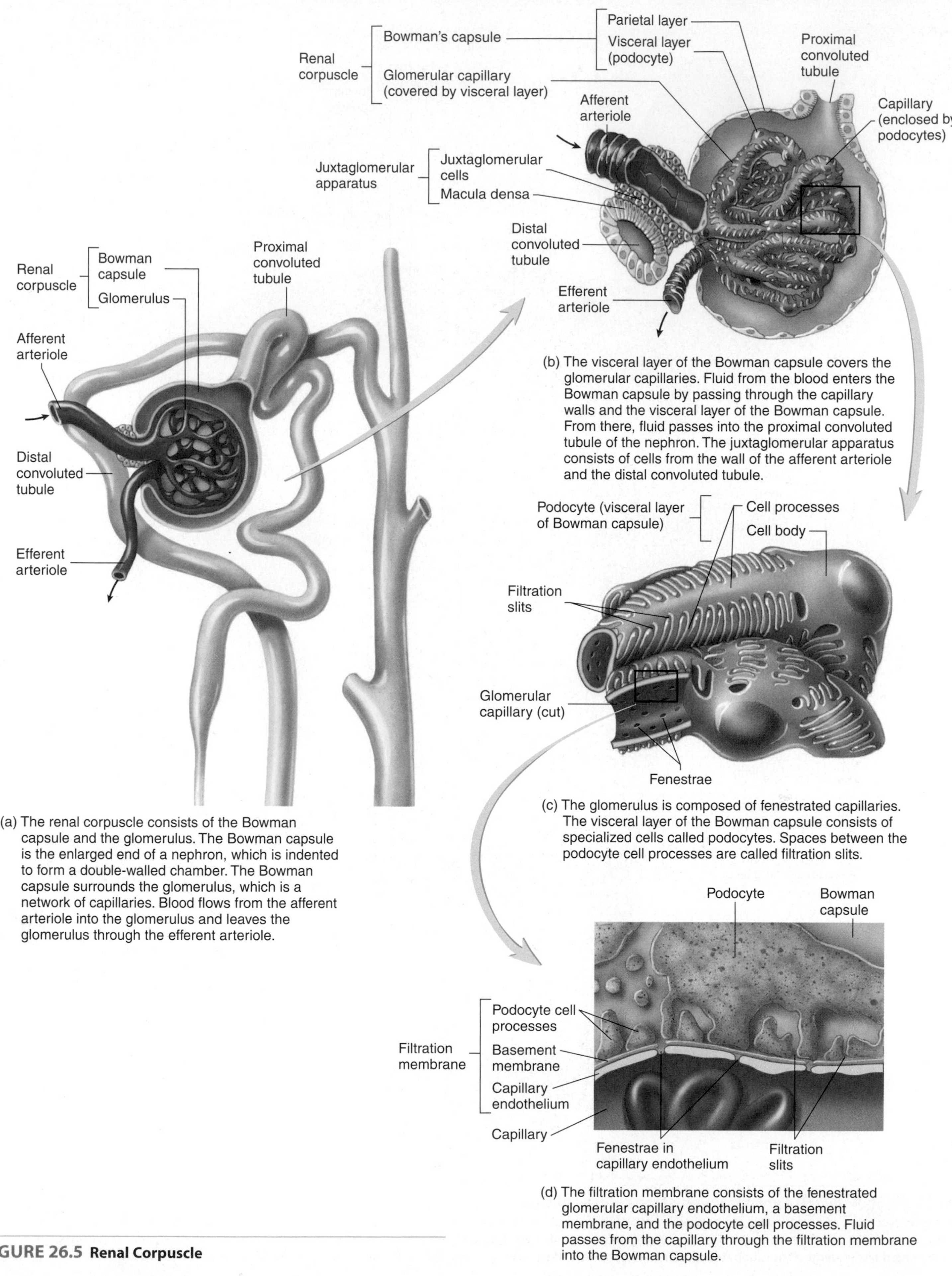

Renal corpuscle
Bowman capsule
Glomerulus
Proximal convoluted tubule
Afferent arteriole
Distal convoluted tubule
Efferent arteriole
Renal corpuscle
Bowman's capsule
Glomerular capillary (covered by visceral layer)
Parietal layer
Visceral layer (podocyte)
Proximal convoluted tubule
Afferent arteriole
Capillary (enclosed by podocytes)
Juxtaglomerular apparatus
Juxtaglomerular cells
Macula densa
Distal convoluted tubule
Efferent arteriole
Podocyte (visceral layer of Bowman capsule)
Cell processes
Cell body
Filtration slits
Glomerular capillary (cut)
Fenestrae
Podocyte
Bowman capsule
Filtration membrane
Podocyte cell processes
Basement membrane
Capillary endothelium
Capillary
Fenestrae in capillary endothelium
Filtration slits

(a) The renal corpuscle consists of the Bowman capsule and the glomerulus. The Bowman capsule is the enlarged end of a nephron, which is indented to form a double-walled chamber. The Bowman capsule surrounds the glomerulus, which is a network of capillaries. Blood flows from the afferent arteriole into the glomerulus and leaves the glomerulus through the efferent arteriole.

(b) The visceral layer of the Bowman capsule covers the glomerular capillaries. Fluid from the blood enters the Bowman capsule by passing through the capillary walls and the visceral layer of the Bowman capsule. From there, fluid passes into the proximal convoluted tubule of the nephron. The juxtaglomerular apparatus consists of cells from the wall of the afferent arteriole and the distal convoluted tubule.

(c) The glomerulus is composed of fenestrated capillaries. The visceral layer of the Bowman capsule consists of specialized cells called podocytes. Spaces between the podocyte cell processes are called filtration slits.

(d) The filtration membrane consists of the fenestrated glomerular capillary endothelium, a basement membrane, and the podocyte cell processes. Fluid passes from the capillary through the filtration membrane into the Bowman capsule.

FIGURE 26.5 Renal Corpuscle

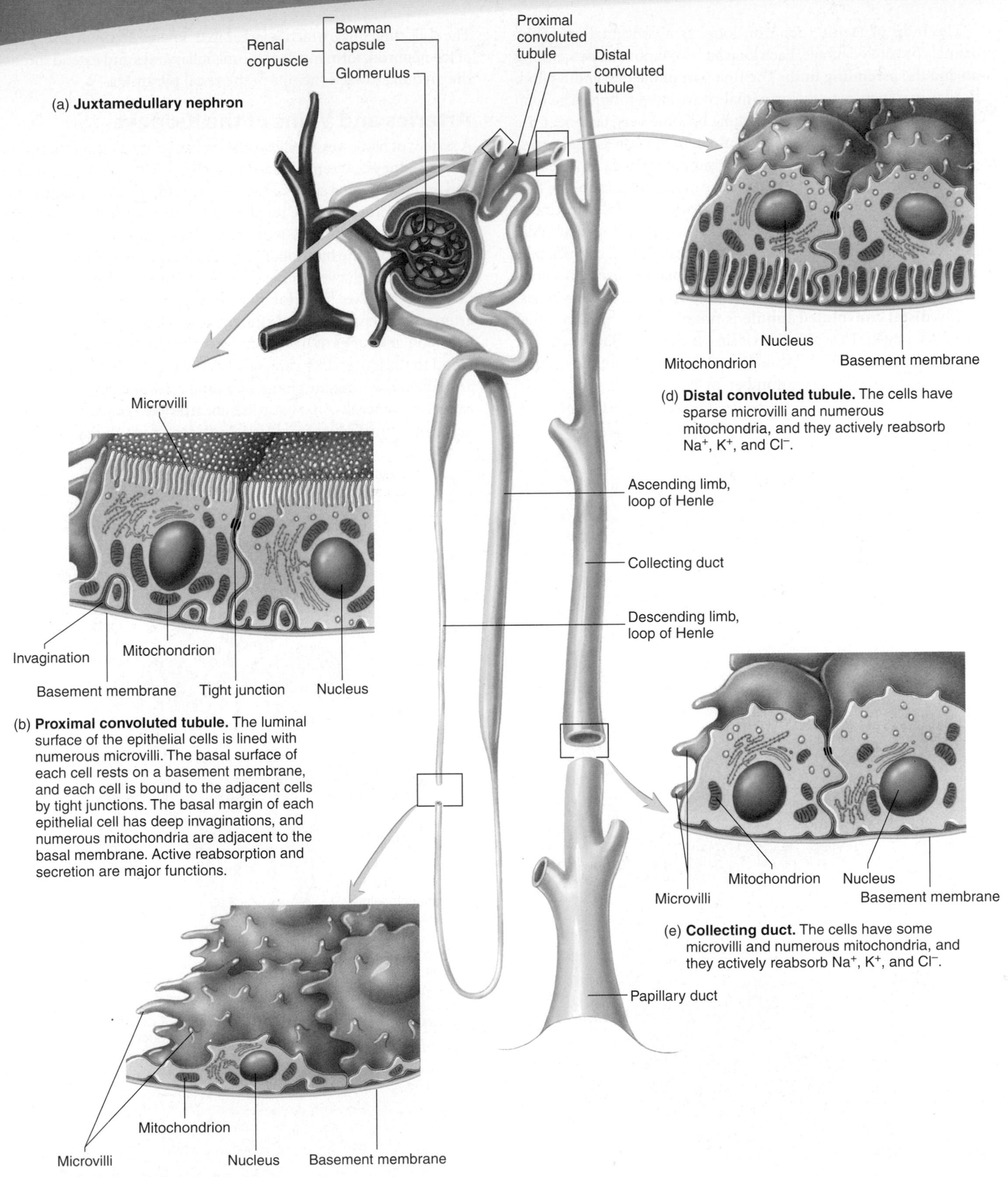

Renal corpuscle
Bowman capsule
Glomerulus
Proximal convoluted tubule
Distal convoluted tubule
(a) Juxtamedullary nephron
Microvilli
Invagination
Mitochondrion
Basement membrane
Tight junction
Nucleus
Nucleus
Mitochondrion
Basement membrane
Ascending limb, loop of Henle
Collecting duct
Descending limb, loop of Henle
Mitochondrion
Nucleus
Microvilli
Basement membrane
Papillary duct
Mitochondrion
Microvilli
Nucleus
Basement membrane

(b) **Proximal convoluted tubule.** The luminal surface of the epithelial cells is lined with numerous microvilli. The basal surface of each cell rests on a basement membrane, and each cell is bound to the adjacent cells by tight junctions. The basal margin of each epithelial cell has deep invaginations, and numerous mitochondria are adjacent to the basal membrane. Active reabsorption and secretion are major functions.

(c) **Descending limb of the loop of Henle.** The thin segment of the descending limb is composed of simple squamous epithelial cells that have microvilli and contain a relatively small number of mitochondria. Water easily diffuses from the thin segment into the interstitial fluid.

(d) **Distal convoluted tubule.** The cells have sparse microvilli and numerous mitochondria, and they actively reabsorb Na^+, K^+, and Cl^-.

(e) **Collecting duct.** The cells have some microvilli and numerous mitochondria, and they actively reabsorb Na^+, K^+, and Cl^-.

FIGURE 26.6 Histology of the Nephron

The **loop of Henle** (nephron loop) is a continuation of the proximal convoluted tubule. Each loop has two limbs: the **descending limb** and the **ascending limb.** The first part of the descending limb is similar in structure to the proximal convoluted tubule. The loop of Henle that extends into the medulla become very thin near the end of the loop (figure 26.6*a,c*). The lumen in the thin part narrows, and an abrupt transition occurs from simple cuboidal epithelium to simple squamous epithelium. Like the descending limb, the first part of the ascending limb is thin and made of simple squamous epithelium. Soon, however, it becomes thicker, and simple cuboidal epithelium replaces the simple squamous epithelium. The thick part of the ascending limb returns toward the renal corpuscle and ends by giving rise to the distal convoluted tubule near the macula densa.

The **distal convoluted tubule** is not as long as the proximal convoluted tubules. The epithelium is simple cuboidal, but the cells are smaller than the epithelial cells in the proximal convoluted tubules and do not possess a large number of microvilli (figure 26.6*d*). Several distal convoluted tubules connect to a single **collecting duct,** which is composed of simple cuboidal epithelium (figure 26.6*c*). The collecting duct, which is larger in diameter than other segments of the nephron, form much of the medullary rays and extend through the medulla toward the tips of the renal pyramids.

Arteries and Veins of the Kidneys

A system of blood vessels allows the exchange of materials that occurs in the kidneys. A **renal artery** branches off the abdominal aorta and enters the renal sinus of each kidney (figure 26.7*a*). **Segmental arteries** diverge from the renal artery to form **interlobar arteries,** which ascend within the renal columns toward the renal cortex. Branches from the interlobar arteries diverge near the base of each pyramid and arch over the bases of the pyramids to form the **arcuate** (ar′kū-āt) **arteries. Interlobular arteries** project from the arcuate arteries into the cortex, and afferent arterioles are derived from the interlobular arteries or their branches. The afferent arterioles supply blood to the glomerular capillaries of the renal corpuscles. Efferent arterioles arise from the glomerular capillaries and carry blood away from the glomeruli. After each efferent arteriole exits the glomerulus, it gives rise to a plexus of capillaries, called the **peritubular capillaries,**

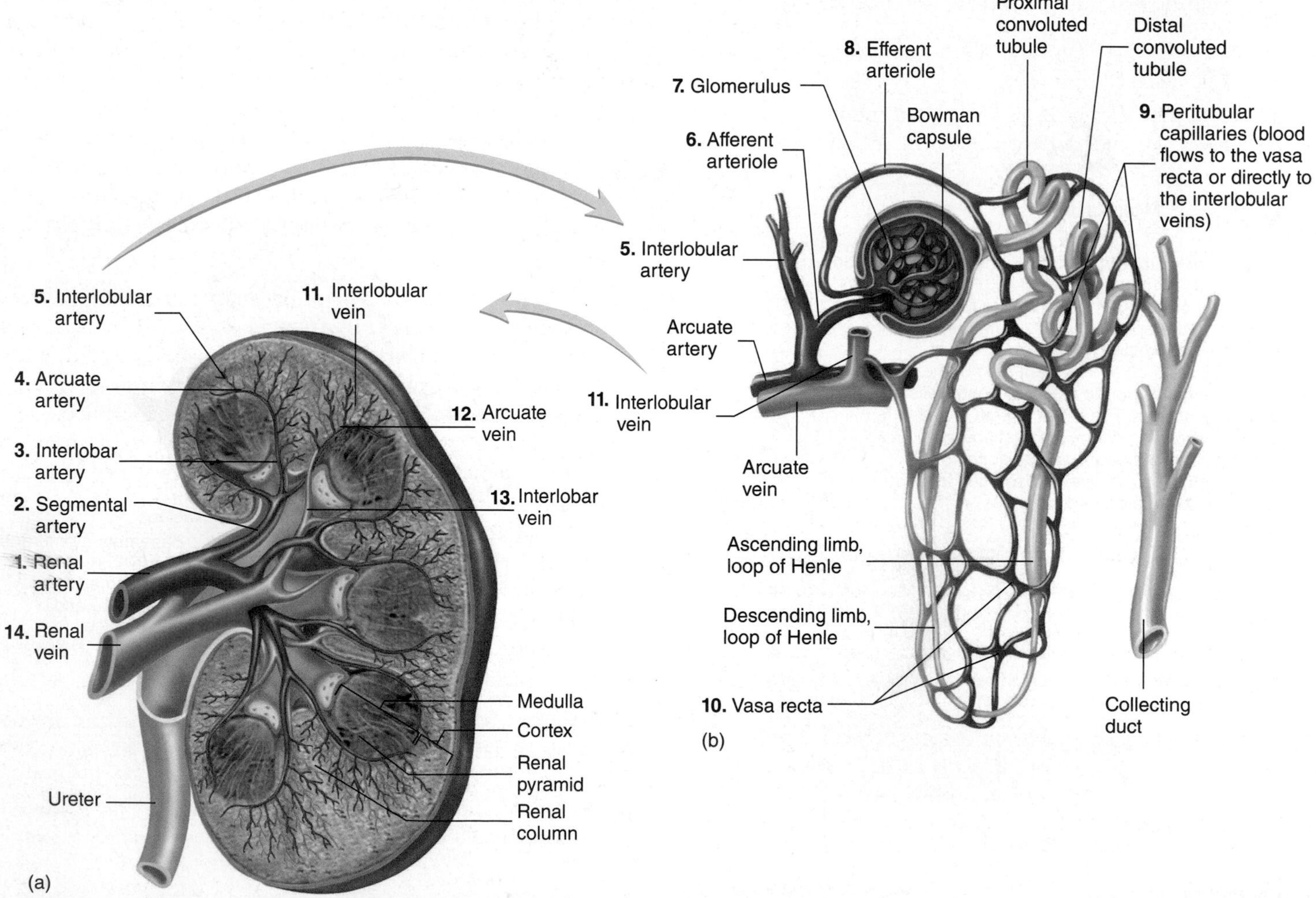

5. Interlobular artery
11. Interlobular vein
4. Arcuate artery
12. Arcuate vein
3. Interlobar artery
2. Segmental artery
13. Interlobar vein
1. Renal artery
14. Renal vein
Medulla
Cortex
Renal pyramid
Renal column
Ureter
(a)
Proximal convoluted tubule
Distal convoluted tubule
8. Efferent arteriole
7. Glomerulus
Bowman capsule
6. Afferent arteriole
9. Peritubular capillaries (blood flows to the vasa recta or directly to the interlobular veins)
5. Interlobular artery
Arcuate artery
11. Interlobular vein
Arcuate vein
Ascending limb, loop of Henle
Descending limb, loop of Henle
10. Vasa recta
Collecting duct
(b)

FIGURE 26.7 AP|R **Blood Flow Through the Kidney**
Numbers 1–14 show the sequence of blood flow through the kidney. (*a*) Blood flow through the larger arteries and veins of the kidney. (*b*) Blood flow through the arteries, capillaries, and veins that provide circulation to the nephrons.

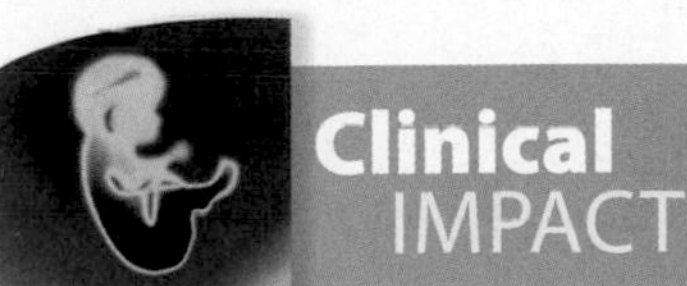
Clinical
IMPACT

Polycystic Kidney Disease

Polycystic kidney disease is the third leading cause of renal failure (after diabetes mellitus and high blood pressure). Approximately 90% of patients inherit the condition as an autosomal dominant trait. Consequently, if one parent carries an allele for this disorder, each child has a 50% chance of also having the disorder (see chapter 29). The gene for this condition is located on chromosome 16 and codes for a protein that may regulate cell-to-cell interactions.

In people affected by polycystic kidney disease, the kidneys are enlarged and often contain large, fluid-filled cysts varying in size from a few millimeters to centimeters. The cysts increase in number and enlarge as the person ages. Development of the cysts results from abnormal cell-to-cell interactions and causes excess proliferation of the epithelial cells that make up the kidney nephrons and collecting ducts.

Polycystic kidney disease is often detected using ultrasound techniques. The condition is usually diagnosed when patients are between 30 and 50 years of age. Approximately 50% of patients require hemodialysis (see Systems Pathology) by 70 years of age.

around the proximal and distal convoluted tubules. Associated with the juxtamedullary nephrons are specialized sets of peritubular capillaries called the **vasa recta** (vā′să rek′tă; figure 26.7*b*). The peritubular capillaries drain into **interlobular veins,** which in turn drain into the **arcuate veins.** The arcuate veins empty into the **interlobar veins,** which drain into the **renal vein.** The renal vein exits the kidney and connects to the inferior vena cava.

ASSESS YOUR PROGRESS

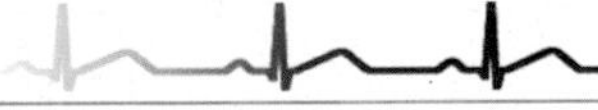

6. *What is the functional unit of the kidney? Name its parts.*
7. *Distinguish between cortical and juxtamedullary nephrons.*
8. *List the components of a renal corpuscle.*
9. *Describe the structure of the Bowman capsule, the glomerulus, and the filtration membrane.*
10. *Describe the structure of the afferent and efferent arterioles and the juxtaglomerular apparatus. What is the function of the juxtaglomerular apparatus?*
11. *Describe the structure and location of the proximal convoluted tubule, loop of Henle, distal convoluted tubule, collecting duct, and papillary duct.*
12. *Explain the blood supply for the kidney.*

26.3 Urine Production

LEARNING OUTCOMES

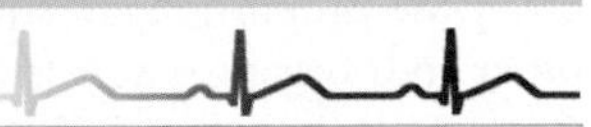

After reading this section, you should be able to

A. **Briefly describe the three processes necessary for urine formation.**
B. **Identify the principal factors that influence filtration and explain how they affect the rate of filtrate formation.**
C. **Explain how filtration is regulated.**
D. **Describe the role of the various regions of the kidney tubule in the process of reabsorption.**
E. **Explain how substances are able to move across the wall of the tubule.**
F. **Relate the types of substances that are moved during tubular secretion and explain how those substances are moved.**
G. **Describe the three mechanisms that explain the kidney's ability to concentrate urine.**

The primary function of the kidney is regulation of body fluid composition. The kidney is the organ that sorts the substances from the blood for either removal in the urine or return to the blood. Substances that are waste products, toxins, and excess materials are permanently removed from the body, whereas other substances need to be preserved to maintain homeostasis. The structural components that perform this sorting are the nephrons, the functional units of the kidney. If you have ever decided to organize your "junk" drawer in your desk or kitchen, you may realize just how difficult it is to quickly sort through all its contents. In fact, you may have found yourself simply emptying the drawer onto a table and then sorting the contents one by one as you place objects into a "save group" or a "throw away group." In a sense, the kidney uses the same approach when regulating blood composition. The "throw away" items end up in the urine, and the "save" items go back into the blood.

Scientists usually categorize urine formation into three major processes: filtration, tubular reabsorption, and tubular secretion (figure 26.8).

1. **Filtration** occurs when blood pressure nonselectively forces water and other small molecules out of glomerular capillaries and into the Bowman capsule, forming a fluid called filtrate.
2. **Tubular reabsorption** occurs when the nephron specifically returns water and some filtered substances to the blood. Most of the filtered water and useful solutes have been returned to the blood by the time the filtrate has been modified to urine, whereas the remaining waste or excess substances, and a small amount of water form urine. For certain solutes, there is a higher concentration in the urine compared with the plasma. This is a result of secretion and is explained more fully later in this chapter (table 26.1).
3. **Tubular secretion** occurs when the nephron cells transport solutes from the blood into the filtrate.

Therefore, urine consists of substances filtered directly from the blood and secreted directly from the blood into the nephron, minus any reabsorbed substances. The next few sections describe the overall process of urine formation. Later, the section "Urine Concentration Mechanism" explores the topic of solute movement in more detail.

Filtration

Filtration is a nonspecific process whereby materials are separated based on size or charge. A simple example of size filtration is demonstrated by a coffee maker. In this case, the driving force of filtration

FUNDAMENTAL **Figure**

Urine formation results from the following three processes:

1. **Filtration** — Filtration (*blue arrow*) is the movement of materials across the filtration membrane into the Bowman capsule to form filtrate.

2. **Tubular reabsorption** — Solutes are reabsorbed (*purple arrow*) across the wall of the nephron into the interstitial fluid by transport processes, such as active transport and cotransport.
   Water is reabsorbed (*orange arrow*) across the wall of the nephron by osmosis. Water and solutes pass from the interstitial fluid into the peritubular capillaries.

3. **Tubular secretion** — Solutes are secreted (*green arrow*) across the wall of the nephron into the filtrate.

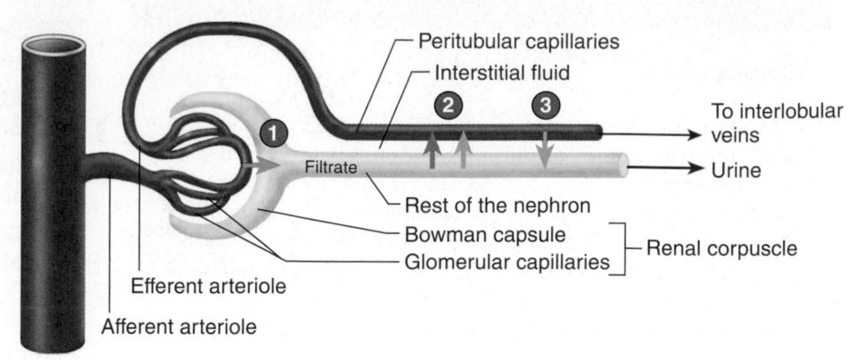
Peritubular capillaries
Interstitial fluid
2
3
1
To interlobular veins
Filtrate
Urine
Rest of the nephron
Bowman capsule
Glomerular capillaries
Renal corpuscle
Efferent arteriole
Afferent arteriole

PROCESS FIGURE 26.8 APR **Urine Formation**

TABLE 26.1 Concentrations of Major Solutes in Urine

Substance	Plasma	Filtrate	Net Movement of Solute*	Urine	Urine Concentration/ Plasma Concentration‡
Water (L)	180	180	178.6	1.4	——
Organic molecules (mg/100 mL)					
Protein	3900–5000	6–11	−100.0	0†	0
Glucose	100	100	−100.0	0	0
Urea	26	26	−11.4	1820	70
Uric acid	3	3	−2.7	42	14
Creatinine	1.1	1.1	0.5	196	180
Ions (mEq/L)					
Na^+	142	142	−141.0	128	0.9
K^+	5	5	−4.5	60	12.0
Cl^-	103	103	−101.9	134	1.3
HCO_3^-	28	28	−27.9	14	0.5

*In many cases, solute moves into and out of the nephron. Numbers indicate net movement. Negative numbers are net movement out of the filtrate, and positive numbers are net movement into the filtrate.

†Trace amounts of protein can be found in the urine. A value of zero is assumed here.

‡Represents solute added to urine via secretion from the interstitial fluid.

is gravity. The kidneys also demonstrate size filtration by filtering the blood, but in this case, the driving force of filtration is blood pressure. Thus, the first step in urine formation is a crude separation of some water and small solutes from blood cells and other large molecules, which occurs when blood flows through the glomerular capillaries. The resulting solution is collected by the Bowman capsule and is called filtrate. It is this filtrate that will be modified into urine.

The importance of filtration is indicated by the large percentage of total cardiac output, or blood, that is sent through the kidneys, given the proportionally small size of the kidneys compared with other body organs. The portion of the total cardiac output that flows through the kidneys is called the **renal fraction.** It varies from 12% to 30% of the cardiac output in healthy, resting adults, but it averages 21% (table 26.2).

TABLE 26.2 Calculation of Renal Flow Rates

	Amount per Minute (mL)	Calculation
Renal Blood Flow	1176	Amount of blood flowing through the kidneys per minute; equals cardiac output (5600 mL blood/min) times the percentage (21%; renal fraction) of cardiac output that enters the kidneys 5600 mL blood/min × 0.21 = 1176 mL blood/min
Renal Plasma Flow	650	Amount of plasma flowing through the kidneys per minute; equals renal blood flow times percentage of the blood that is plasma. Because the hematocrit is the percentage of the blood that consists of formed elements, the percentage of the blood that is plasma is 100 minus the hematocrit. Assuming a hematocrit of 45, the percentage of the blood that is plasma is 55% (100 − 45). Renal plasma flow is therefore 55% of renal blood flow. 1176 mL blood/min × 0.55 ≈ 650 mL plasma/min
Glomerular Filtration Rate (GFR)	125	Amount of plasma (filtrate) that enters the Bowman capsule per minute; equals renal plasma flow times the percentage (19%; filtration fraction) of the plasma that enters the renal capsule 650 mL plasma/min × 0.19 ≈ 125 mL filtrate/min
Urine	1	Nonreabsorbed filtrate that leaves the kidneys per minute; equals glomerular filtration rate times the percentage (0.8%) of the filtrate that is not reabsorbed into the blood 125 mL filtrate/min × 0.008 = 1 mL urine/min Milliliters of urine per minute can be converted to liters of urine per day by multiplying by 1.44. 1 mL urine/min × 1.44 = 1.4 L/day

Health professionals often use standard measures of blood flow through the kidney to determine whether the kidneys are functioning properly. One standard measure is the **renal blood flow rate.** Two pieces of information are used to calculate this number:

- Renal fraction, which is 21%
- Cardiac output, which is 5600 mL/min (see chapter 20)

Thus, renal blood flow rate is 5600 × 0.21 = 1176 mL/min.

Health professionals also use the **renal plasma flow rate,** which is equal to the renal blood flow rate multiplied by 55% (percentage of whole blood that is plasma; see chapter 19).

1176 mL/min × 0.55 = 646.8 mL plasma/min, or approximately 650 mL/min (0.650 L)

When the blood is filtered by the glomerulus, approximately 19% of the plasma is removed from the blood. This is called the **filtration fraction,** which is equal to 650 mL plasma/min × 0.19 = 123.5 mL plasma/min. Thus, approximately 125 milliliters (0.125 liters) of filtrate is produced each minute and is called the **glomerular filtration rate (GFR).** Approximately 180,000 milliliters (180 liters) of **filtrate** are produced daily. This enormous volume is equal to about 90 2-liter soft drink bottles per day. Because a healthy person produces only 1000–2000 milliliters (1–2 liters) of *urine* each day, the equivalent of one 2-liter soft drink bottle, it is readily apparent that not all of the filtrate becomes urine. In fact, about 99% of the filtrate volume is reabsorbed into the blood as it travels through the nephron, and less than 1% becomes urine. Although it may seem pointless to remove so much material from the blood only to return it right away, it is important that filtration remains continuous, so that waste products can be removed from the blood as quickly as possible.

Filtration Membrane

As mentioned earlier, the renal corpuscle is the site of blood filtration in the nephron. As such, it has several components that collectively form a filtration barrier, called the **filtration membrane.** The filtration membrane prevents blood cells and proteins from entering the lumen of the Bowman capsule on the basis of size and charge, but it allows other blood components to pass. The filtration membrane is many times more permeable than a typical capillary. Water and small solute molecules readily pass from the glomerular capillaries through this membrane into the Bowman capsule, but larger molecules do not. The filtration membrane is comprised of the following:

1. The fenestrae of the glomerular capillary
2. The basement membrane between the capillary wall and the visceral layer of the Bowman capsule
3. Podocytes of the visceral layer of the Bowman capsule (see figure 26.5*d*)

Together, these components prevent molecules larger than 7 nm in diameter or those having a molecular mass equal to or greater than 40,000 daltons from passing through. First, the fenestrae are about 7 nm in size and serve as the initial filter. Most plasma proteins are slightly larger than 7 nm in diameter and are retained in the glomerular capillaries. However, albumin, which has a diameter just slightly less than 7 nm, enters the filtrate only in small amounts. Therefore, the filtrate is not protein-free but, rather, contains about 0.03% protein. In addition, some protein hormones, such as thyrotropin-releasing hormone, oxytocin, and antidiuretic hormone, are small enough to pass through the filtration membrane. The proteins that are able to pass through the filtration membrane are actively reabsorbed by endocytosis and metabolized by the cells in the proximal convoluted tubule. Next, the basement membrane and the podocytes contain negatively charged glycoproteins, which repel negatively charged plasma proteins and prevent them from exiting the blood. Therefore, the combined effect of the filtration membrane components prevents most proteins from exiting the blood on the basis of size and charge, and only a small amount of protein is found in the urine of healthy people.

> **Predict 2**
>
> A hemoglobin molecule has a smaller diameter than an albumin molecule, but very little hemoglobin passes from the blood into the filtrate. Explain why. Under what circumstances do large amounts of hemoglobin enter the filtrate?

Filtration Pressure

The formation of filtrate is due to a pressure gradient in the renal corpuscle, called **filtration pressure** (figure 26.9). The filtration pressure is dependent on the combination of three different pressures:

- The **glomerular capillary pressure (GCP)** is an *outward pressure* from blood pressing on the capillary walls, or simply blood pressure. The GCP forces fluid and solutes out of the blood into the Bowman capsule. This glomerular capillary pressure is higher than that in other capillaries of the body because the diameter of the efferent arteriole is smaller than that of the afferent arteriole and glomerular capillaries. As you learned in chapter 21, when the diameter of a vessel decreases, the resistance to blood flow through the vessel is greater. The efferent arteriole has a smaller diameter than the afferent arteriole and glomerular capillary. Thus, as the blood flows from the larger-diameter afferent arteriole through the glomerular capillaries to the smaller-diameter efferent arteriole, the blood pressure increases in the glomerular capillaries. Consequently, filtrate is forced across the filtration membrane into the lumen of the Bowman capsule. The GCP is approximately 50 mm Hg compared with approximately 30 mm Hg at the arterial end of other capillary beds.
- The **capsular hydrostatic pressure (CHP)** is an *inward pressure* from the pressure of filtrate accumulation in the Bowman capsule. CHP is comparable to blood pressure in that, as blood pressing on the walls of a capillary creates blood pressure, filtrate pressing on the walls of the Bowman capsule creates CHP. The CHP equals about 10 mm Hg.
- The **blood colloid osmotic pressure (BCOP)** is also an *inward pressure* due to the osmotic force of plasma proteins within the glomerular capillaries. The presence of these proteins resists fluid movement from the glomerular capillary into the Bowman capsule. The BCOP is greater at the end of the glomerular capillary than at its beginning because, as fluid leaves the capillary and enters the Bowman capsule, there is a higher protein concentration in the glomerular capillary. The average BCOP is approximately 30 mm Hg.

To calculate filtration pressure, all three filtration pressures are summed, and we see that in a normal kidney GCP overrides CHP and BCOP, and the filtration pressure is a net *outward* pressure of approximately 10 mm Hg:

Filtration pressure		Glomerular capillary pressure		Capsular hydrostatic pressure		Blood colloid osmotic pressure
(10 mm Hg)	=	(50 mm Hg)	−	(10 mm Hg)	−	(30 mm Hg)

Normally, the filtrate does not exert an osmotic force on the plasma because its colloid osmotic pressure is very close to zero. This is because few proteins cross the filtration membrane. However, in a disease such as **glomerular nephritis,** the permeability of the filtration membrane increases, and more protein than normal enters the filtrate, increasing the colloid osmotic pressure of the filtrate. This results in elevated filtration pressure and an increase in the filtrate volume.

Regulation of Glomerular Filtration Rate

Autoregulation is the maintenance of a very stable glomerular filtration rate (GFR), despite wide fluctuations in systemic pressure from as low as 90 mm Hg to as high as 180 mm Hg. However, under severe conditions such as hemorrhage or dehydration, the mean

1 Glomerular capillary pressure (GCP), the blood pressure (50 mm Hg) within the glomerulus, moves fluid from the blood into the Bowman capsule.

2 Capsular hydrostatic pressure (CHP), the fluid pressure inside the Bowman capsule (10 mm Hg), moves fluid from the Bowman capsule into the blood.

3 Blood colloid osmotic pressure (BCOP), produced by the concentration of blood proteins in the glomerular capillaries (30 mm Hg), moves fluid from the Bowman capsule into the blood by osmosis.

4 Filtration pressure is equal to the glomerular capillary pressure minus the capsular hydrostatic and blood colloid osmotic pressures.

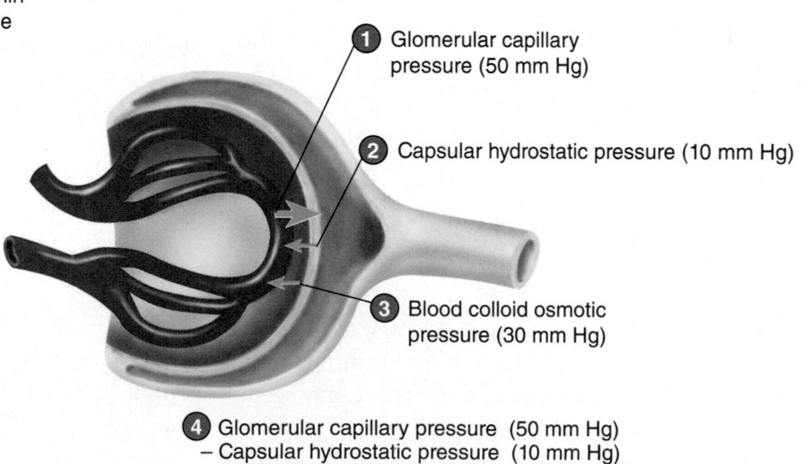
1 Glomerular capillary pressure (50 mm Hg)
2 Capsular hydrostatic pressure (10 mm Hg)
3 Blood colloid osmotic pressure (30 mm Hg)
4 Glomerular capillary pressure (50 mm Hg)
– Capsular hydrostatic pressure (10 mm Hg)
– Blood colloid osmotic pressure (30 mm Hg)
Filtration pressure = 50 – 10 – 30 = 10 mm Hg

PROCESS FIGURE 26.9 Filtration Pressure

Filtration pressure across the filtration membrane is equal to the glomerular capillary pressure (GCP) minus the blood colloid osmotic pressure (BCOP) in the glomerular capillary minus the capsular hydrostatic pressure (CHP) in the Bowman capsule.

arterial pressure can drop below 90 mm Hg, and the sympathetic nervous system causes a dramatic decrease in renal blood flow and GFR in an effort to maintain homeostatic blood pressure.

Mechanisms of Autoregulation

Autoregulation is achieved through two processes: the myogenic mechanism and tubuloglomerular feedback. As the name suggests, the myogenic mechanism is associated with the intrinsic properties of smooth muscle cells in the afferent (toward the glomerulus) and efferent (away from the glomerulus) arterioles. Smooth muscle cells seem to act as stretch receptors, and as blood pressure in the afferent arteriole increases, the walls of the vessel stretch. In response to the stretch, the smooth muscle constricts the afferent arteriole. On the other hand, when the blood pressure decreases, the walls of the afferent arteriole dilate. In this way, blood supply to the glomerulus, and thus GFR, fluctuates very little, even when the mean arterial pressure changes.

The tubuloglomerular feedback mechanism correlates filtrate flow past the macula densa of the juxtaglomerular apparatus to GFR. When the macula densa cells detect an increased flow rate, these cells send a signal to the juxtaglomerular cells of the afferent arteriole to constrict. Thus, glomerular filtration rate decreases due to a decreased glomerular capillary pressure.

Sympathetic Stimulation

Autoregulation maintains renal blood flow and filtrate formation at a relatively constant rate unless sympathetic stimulation is intense. Because norepinephrine-secreting sympathetic neurons innervate the blood vessels of the kidneys, sympathetic stimulation constricts the small arteries and afferent arterioles, thereby decreasing renal blood flow and filtrate formation. Intense sympathetic stimulation, as may occur during shock or intense exercise, decreases the rate of filtrate formation to only a few milliliters per minute; however, small changes in sympathetic stimulation have a minimal effect on renal blood flow and filtrate formation.

In response to severe stress or circulatory shock, renal blood flow can decrease to such low levels that the blood supply to the kidney is inadequate to maintain normal kidney metabolism. As a consequence, kidney tissues can be damaged and thus unable to perform their normal functions if blood flow is not reestablished. Therefore, shock should be treated quickly. On the other hand, reduced blood flow to the kidneys during stress or shock is consistent with homeostasis. Intense vasoconstriction maintains blood pressure at levels adequate to sustain blood flow to organs such as the heart and brain. A reduction in blood flow to organs such as the kidneys is only harmful if the lack of blood flow is prolonged.

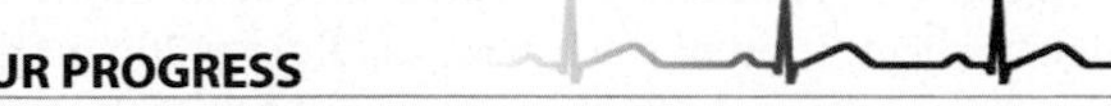

ASSESS YOUR PROGRESS

13. *Name the three general processes involved in producing urine.*
14. *Contrast the rates of renal blood flow, renal plasma flow, and glomerular filtration. How do they affect urine production?*
15. *Describe the filtration membrane. What substances do not pass through it?*
16. *What is filtration pressure? How does glomerular capillary pressure affect filtration pressure and the amount of urine produced?*
17. *How do systemic blood pressure and afferent arteriole diameter affect glomerular capillary pressure?*
18. *Describe autoregulation.*
19. *Explain the effect of sympathetic stimulation on the kidney and the GFR during rest, exercise, and shock.*

Tubular Reabsorption

Tubular reabsorption is the return of water and solutes filtered from the blood at the renal corpuscle to the blood. Nearly all (99%) of the water and solutes are rapidly returned to the blood via the renal tubules, and because of this, toxins are quickly removed from the circulation. However, the body would become overly dehydrated and deficient in important materials without proper tubular reabsorption. The filtrate leaves the lumen of the Bowman capsule and flows through the proximal convoluted tubule, the loop of Henle, and the distal convoluted tubule and then into the collecting ducts. As it passes through these structures, many of the substances in the filtrate are removed by one or more of several processes. These processes, such as simple and facilitated diffusion, active transport, symport, and osmosis, all result in tubular reabsorption. Inorganic salts, organic molecules, and about 99% of the filtrate volume leave the nephron and enter the interstitial fluid. Because the pressure is low in the peritubular capillaries, these substances enter the peritubular capillaries and flow through the renal veins to enter the general circulation (see figure 26.8).

Solutes reabsorbed from the lumen of the nephron to the interstitial fluid include amino acids, glucose, and fructose, as well as Na^+, K^+, Ca^{2+}, HCO_3^-, and Cl^-. A more complete list is provided in table 26.3 for each part of the nephron.

As the solutes in the nephron are reabsorbed, water follows the solutes by the process of osmosis (see chapter 3). The small volume of the filtrate (approximately 1%) that forms urine contains urea, uric acid, creatinine, K^+, and other substances. The regulation of solute reabsorption and the permeability characteristics of portions of the nephron allow for the production of a small volume of very concentrated urine or a large volume of very dilute urine. The following section describes the mechanisms responsible for reabsorption by the tubular wall of the nephron. The mechanisms that regulate urine concentration are then described in the section "Urine Concentration Mechanism."

Reabsorption in the Proximal Convoluted Tubule

The proximal convoluted tubule is responsible for the majority of reabsorption. The mechanisms underlying reabsorption can be better understood by considering the cells found there. These cells have an apical surface, which makes up the inside surface of the nephron; a basal surface, which forms the outer wall of the nephron; and lateral surfaces, which are bound to the surfaces of other cells of the nephron. Reabsorption of most solutes from the proximal convoluted tubule is linked to a steep Na^+ concentration gradient between the filtrate and the cytoplasm of the nephron cells. Active transport of Na^+ across the **basal membrane** of the nephron epithelial cells from the cytoplasm into the interstitial fluid creates a low concentration of Na^+ inside the cells (figure 26.10). At the basal membrane, the sodium-potassium pump moves Na^+ out of the cell

TABLE 26.3 Reabsorption of Major Solutes from the Nephron

Apical Membrane	Basal Membrane
Proximal Convoluted Tubule	
Substances Symported with Na^+	*Active Transport*
	Na^+ (exchanged for K^+)
K^+	*Facilitated diffusion*
Cl^-	K^+
Ca^{2+}	Cl^-
Mg^{2+}	Ca^{2+}
HCO_3^-	HCO_3^-
PO_4^{3-}	PO_4^{3-}
Amino acids	Amino acids
Glucose	Glucose
Fructose	Fructose
Galactose	Galactose
Lactate	Lactate
Succinate	Succinate
Citrate	Citrate
Diffusion Between Nephron Cells	
K^+	
Ca^{2+}	
Mg^{2+}	
Thick Ascending Limb of the Loop of Henle	
Substances Symported with Na^+	*Active Transport*
	Na^+ (exchanged for K^+)
K^+	*Facilitated diffusion*
Cl^-	K^+
	Cl^-
Diffusion Between Nephron Cells	
K^+	
Ca^{2+}	
Mg^{2+}	
Distal Convoluted Tubule and Collecting Duct	
Substances Symported with Na^+	*Active Transport*
	Na^+ (exchanged for K^+)
Cl^-	*Facilitated diffusion*
K^+	K^+
	Cl^-

and K^+ into the cell. Because the concentration of Na^+ in the lumen of the tubule is high, a large concentration gradient is present from the lumen of the nephron to the cytoplasm of the cells lining the nephron. This concentration gradient for Na^+ is responsible for the secondary active transport of many other solutes from the lumen of the nephron into the nephron cells (see chapter 3).

Carrier proteins that transport amino acids, glucose, and other solutes are located within the **apical membrane,** which separates the lumen of the nephron from the cytoplasm of epithelial cells. Each of these carrier proteins binds specifically to one of those substances to be transported and to Na^+. The concentration gradient for Na^+ provides the energy that moves both the Na^+ and the other molecules or ions from the lumen into the cell of the nephron. Once the symported molecules are inside the cell, they cross the basal membrane of the cell by facilitated diffusion or symport. The number of carrier proteins limits the rate at which a substance can be transported. For example, the high blood glucose in someone with untreated diabetes mellitus can lead to such high glucose levels in the filtrate that not all of it can be removed by the glucose transport proteins. The excess glucose remains in the filtrate and becomes part of the urine (see section 26.5).

Some solutes also diffuse from the lumen of the nephron into the interstitial fluid by moving *between* the cells. As other solutes are transported out of the lumen, through the proximal convoluted tubule cells, and into the interstitial fluid, water follows by osmosis. The reabsorption of water causes the concentration of solutes that remain in the lumen to increase. When the concentration of these solutes in the lumen becomes higher than in the interstitial fluid, these solutes will diffuse between the epithelial cells into the interstitial fluid. Examples of solutes that diffuse between nephron cells of the proximal convoluted tubule include K^+, Ca^{2+}, and Mg^{2+}. These solutes are reabsorbed by diffusion, even though the same ions are also sometimes reabsorbed by symport processes.

Reabsorption of solutes and water in the proximal convoluted tubule is extensive. As solute molecules are transported from the nephron to the interstitial fluid, water moves by osmosis in the same direction. By the time the filtrate has reached the end of the proximal convoluted tubule, its volume has been reduced by approximately 65%. Because the proximal convoluted tubule is permeable to water, the concentration of the filtrate there remains about the same as that of the interstitial fluid (300 mOsm/kg).

Reabsorption in the Loop of Henle

As the filtrate from the proximal convoluted tubule moves toward the loop of Henle, the wall of the nephron undergoes a histological change. As the loop of Henle descends into the medulla of the kidney, where the concentration of solutes in the interstitial fluid is very high, the wall transforms from simple cuboidal epithelial tissue to simple squamous epithelial tissue in the thin segment. Thus, the thin segment of the descending limb of the loop of Henle (figure 26.11*a*) is highly permeable to water and moderately permeable to urea, Na^+, and most other ions. As the filtrate passes through the thin segment of the descending limb, water moves out of the nephron by osmosis, and some solutes move into the nephron. By the time the filtrate has reached the end of the thin segment of the descending limb, the volume of the filtrate has been reduced by another 15%, and the concentration of the filtrate is equal to the high concentration of the interstitial fluid (1200 mOsm/L).

The thin segment of the ascending limb of the loop of Henle is permeable to solutes but impermeable to water (figure 26.11*b*). Therefore, although the gradient encourages water to leave the lumen of the thin segment of the ascending limb, no additional water exits. The ascending limb of the loop of Henle is surrounded by interstitial fluid, which becomes less concentrated toward the cortex. As the filtrate flows through the thin segment of the limb, solutes diffuse into the interstitial fluid, making the filtrate less concentrated.

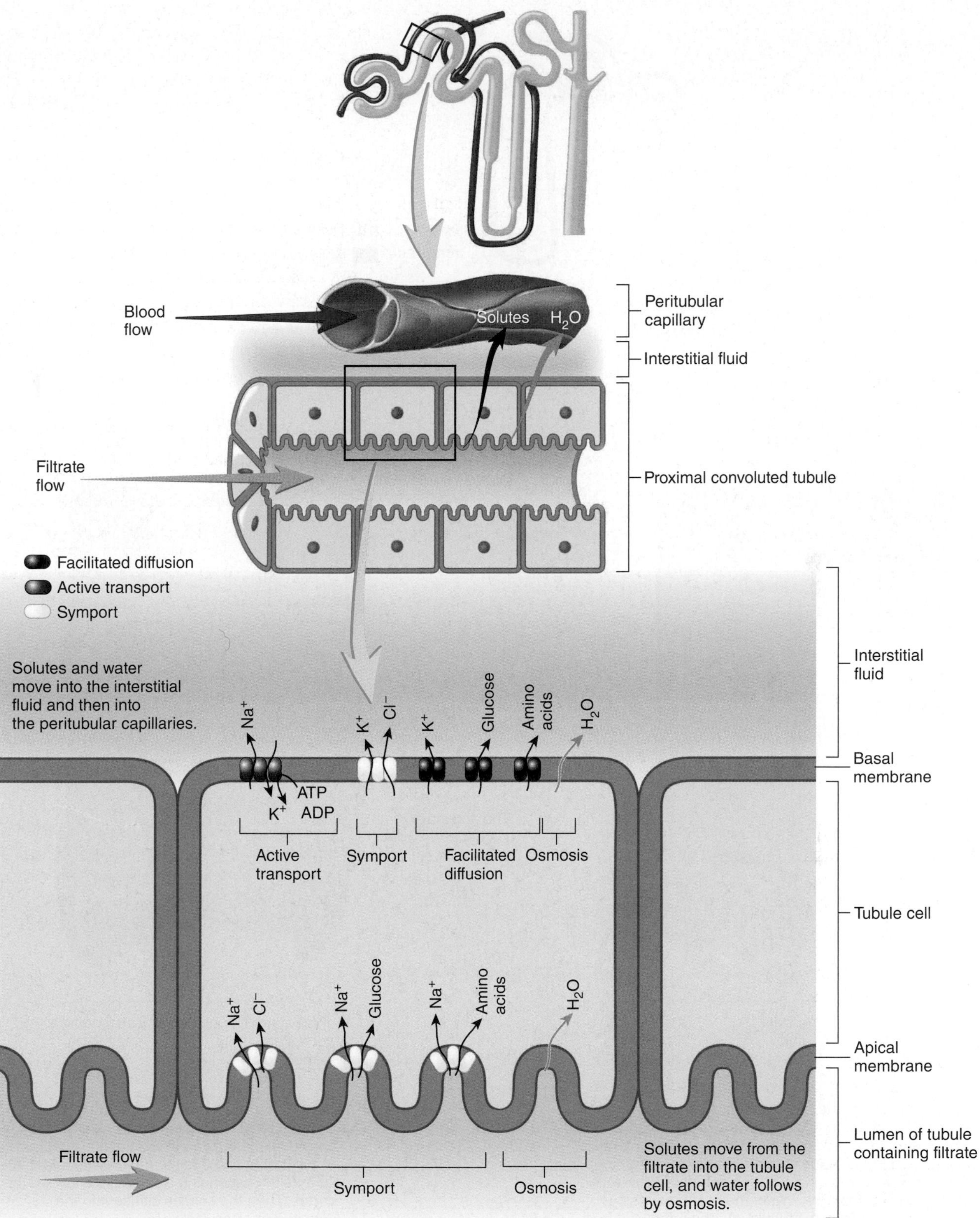

Blood flow
Solutes
H_2O
Peritubular capillary
Interstitial fluid
Filtrate flow
Proximal convoluted tubule
Facilitated diffusion
Active transport
Symport
Solutes and water move into the interstitial fluid and then into the peritubular capillaries.
Interstitial fluid
Na^+
K^+
Cl^-
K^+
Glucose
Amino acids
H_2O
ATP
K^+
ADP
Basal membrane
Active transport
Symport
Facilitated diffusion
Osmosis
Tubule cell
Na^+
Cl^-
Na^+
Glucose
Na^+
Amino acids
H_2O
Apical membrane
Filtrate flow
Symport
Osmosis
Solutes move from the filtrate into the tubule cell, and water follows by osmosis.
Lumen of tubule containing filtrate

FIGURE 26.10 Reabsorption of Solutes in the Proximal Convoluted Tubule

The symport of molecules and ions across the epithelial lining of the nephron depends on the active transport of Na^+, in exchange for K^+, across the basal membrane. Symport is the process by which carrier proteins move molecules or ions with Na^+ across the apical membrane. The Na^+ concentration gradient provides the energy for symport. Amino acids, glucose, K^+, Cl^-, and most other solutes are transported into the cells of the nephron with Na^+. Water enters and leaves the cell by osmosis. Glucose, amino acids, Na^+, Cl^-, and many other solutes leave the cells across the basal membrane by facilitated diffusion.

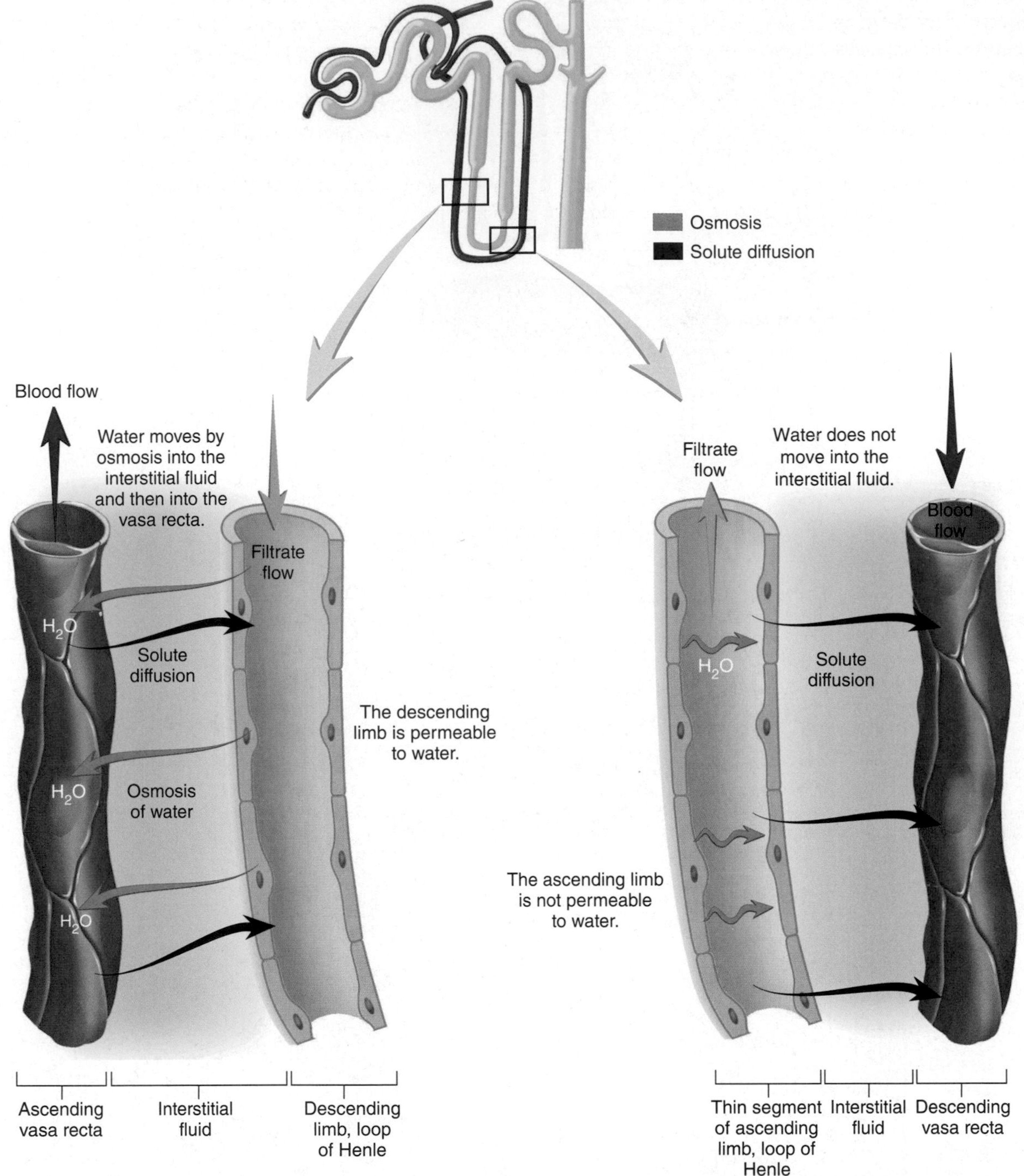

Osmosis
Solute diffusion
Blood flow
Water moves by osmosis into the interstitial fluid and then into the vasa recta.
Filtrate flow
H_2O
Solute diffusion
The descending limb is permeable to water.
H_2O
Osmosis of water
H_2O
Ascending vasa recta
Interstitial fluid
Descending limb, loop of Henle
Filtrate flow
Water does not move into the interstitial fluid.
Blood flow
H_2O
Solute diffusion
The ascending limb is not permeable to water.
Thin segment of ascending limb, loop of Henle
Interstitial fluid
Descending vasa recta

(a) The wall of the thin segment of the descending limb of the loop of Henle is permeable to water and, to a lesser extent, to solutes. The interstitial fluid in the medulla of the kidney and the blood in the vasa recta have a high solute concentration (high osmolality). Water therefore moves by osmosis from the tubule into the interstitial fluid and into the vasa recta. An additional 15% of the filtrate volume is reabsorbed. To a lesser extent, solutes diffuse from the vasa recta and interstitial fluid into the tubule.

(b) The thin segment of the ascending limb of the loop of Henle is not permeable to water but is permeable to solutes. The solutes diffuse out of the tubule and into the more dilute interstitial fluid as the ascending limb projects toward the cortex. Then the solutes diffuse into the descending vasa recta.

FIGURE 26.11 Reabsorption in the Loop of Henle: The Descending Limb and the Thin Segment of the Ascending Limb

Because the thick segment of the ascending limb is not freely permeable to water or solutes, solutes such as Na^+, K^+, and Cl^- must be actively transported from the thick segment of the ascending limb of the loop of Henle into the interstitial fluid. Symport is responsible for moving K^+ and Cl^- with Na^+ across the apical membrane of the ascending limb of the loop of Henle (figure 26.12). Once inside the cells of the ascending limb, Cl^- and K^+ exit the cells of the ascending limb via facilitated diffusion. The concentration of Na^+ in the lumen of the nephron is high, and the concentration inside the nephron cells is low. This concentration gradient is created by the active transport of the Na^+ out of the cell in exchange for K^+ across the basal membrane (figure 26.12).

As we follow the filtrate through the nephron, we see that it becomes very concentrated toward the tip of the loop of Henle, but the concentration of the filtrate is reduced to about 100 mOsm/kg by the time the fluid reaches the distal convoluted tubule. In contrast, the concentration of the interstitial fluid in the cortex is about 300 mOsm/kg. Thus, the filtrate entering the distal convoluted tubule is much more dilute (hypotonic) than the interstitial fluid surrounding it.

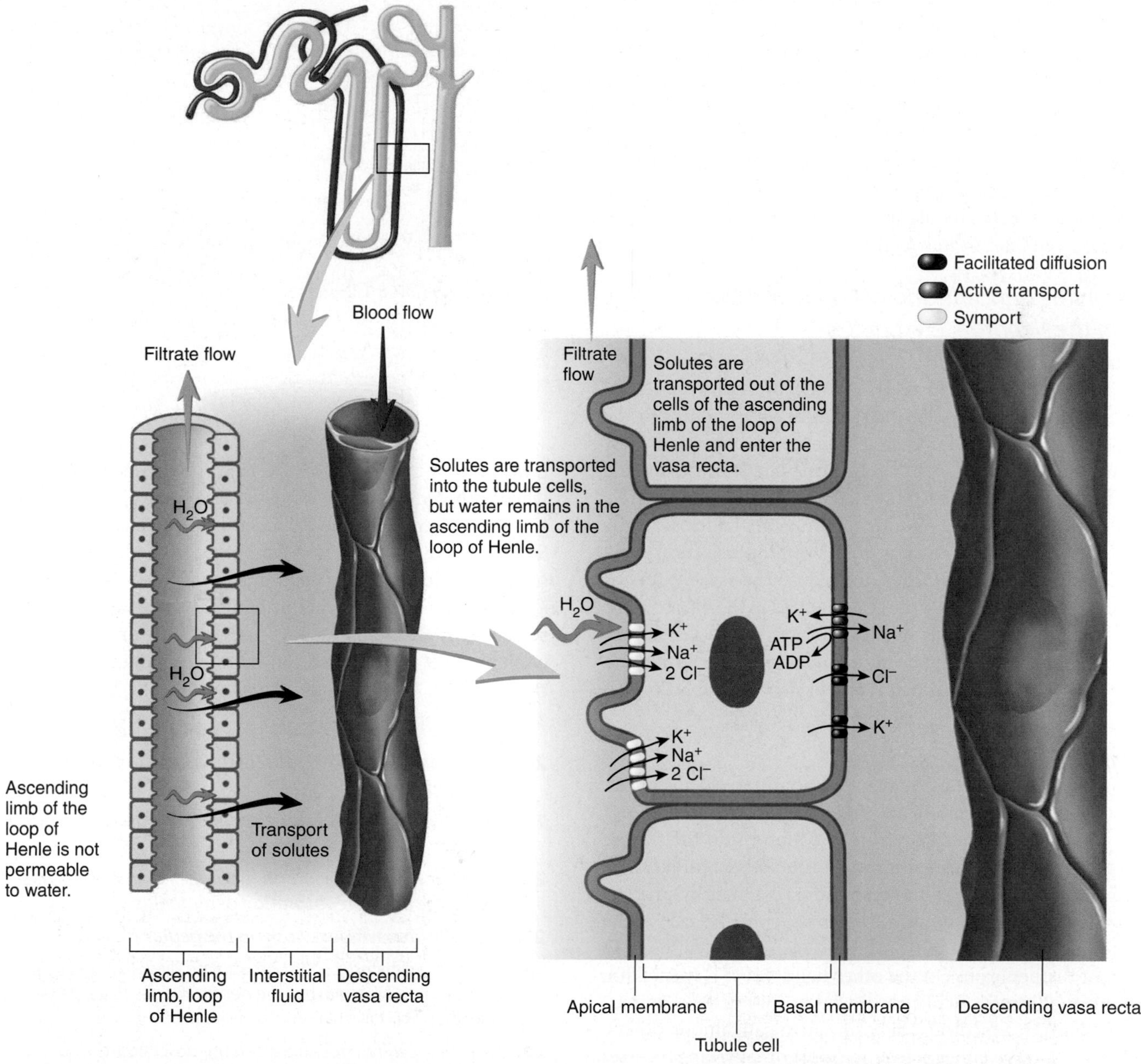
Facilitated diffusion
Active transport
Symport
Blood flow
Filtrate flow
Filtrate flow
Solutes are transported out of the cells of the ascending limb of the loop of Henle and enter the vasa recta.
Solutes are transported into the tubule cells, but water remains in the ascending limb of the loop of Henle.
H_2O
H_2O
H_2O
K^+
Na^+
2 Cl^-
K^+
Na^+
2 Cl^-
K^+
ATP
ADP
Na^+
Cl^-
K^+
Ascending limb of the loop of Henle is not permeable to water.
Transport of solutes
Ascending limb, loop of Henle
Interstitial fluid
Descending vasa recta
Apical membrane
Basal membrane
Descending vasa recta
Tubule cell

The wall of the ascending limb of the loop of Henle is not permeable to water. Sodium ions move across the wall of the basal membrane by active transport, establishing a concentration gradient for Na^+. Potassium ions and Cl^- are symported with Na^+ across the apical membrane, and ions pass by facilitated diffusion across the basal membrane of the tubule cells.

FIGURE 26.12 Reabsorption in the Thick Segment of the Ascending Limb of the Loop of Henle

Reabsorption in the Distal Convoluted Tubule and Collecting Duct

Some solutes (Na^+, Cl^-, H^+) are reabsorbed further along the nephron in the distal convoluted tubule or collecting duct. The reabsorption of these solutes is generally under hormonal control and depends on the current conditions of the body. Water permeability of the distal convoluted tubule and the collecting duct is not constant, but varies depending on hormone regulation (see section 26.4). Reabsorption of water is via osmosis across the wall of the distal convoluted tubule and the collecting duct when the hormone ADH is present (see chapter 18). The interstitial fluid surrounding the distal convoluted tubule and collecting duct is more concentrated than the filtrate, so the water moves toward the high solute concentration area. In these conditions, a small volume of concentrated urine is produced. ADH causes the tubule wall to become more permeable to water, a mechanism discussed in more detail later in this chapter. When ADH is absent, the distal convoluted tubule and collecting duct are not permeable to water and water stays in the filtrate. In this case, a large volume of dilute urine is produced.

The distal convoluted tubule also plays a major role in secretion, which is discussed later in this section.

Changes in the Concentration of Urea and Other Solutes in the Nephron

One of the nephron's major functions is to remove wastes from the body. For example, **urea** (ū-rē′ă), a protein breakdown product, enters the glomerular filtrate at the same concentration as in the plasma. Because renal tubules are not as permeable to urea as they are to water, as the volume of filtrate decreases in the nephron, the concentration of urea increases. Only 40–60% of the urea is passively reabsorbed in the nephron, although about 99% of the water is reabsorbed. In addition to urea, urate ions, creatinine, sulfates, phosphates, and nitrates are reabsorbed, but not to the same extent as water. They therefore become more concentrated in the filtrate as the volume of the filtrate becomes smaller. These substances are toxic if they build up in the body, so their accumulation in the filtrate and elimination in urine help maintain homeostasis (see table 26.1).

Tubular Secretion

Tubular secretion is the movement of nonfiltered substances, toxic by-products of metabolism, and drugs or molecules not normally produced by the body from the blood into the filtrate (table 26.4). As with tubular reabsorption, tubular secretion can be either active or passive. For example, ammonia is a toxic by-product of protein metabolism. It is produced when the epithelial cells of the nephron remove amino groups from amino acids, which diffuse into the lumen of the nephron. On the other hand, H^+, K^+, penicillin, and **para-aminohippuric acid** (par′ă-a-mī′nō-hi-pūr′ik; p-aminohippuric acid; **PAH;** a medical diagnostic chemical), among others, are actively secreted by either active transport or antiport processes into the nephron. An example of an antiport process in the kidney is that which moves H^+ from cells of the nephron into the nephron's lumen. Hydrogen ions bind the carrier proteins on the inside of the plasma membrane, and Na^+ binds to the carrier proteins on the outside of the plasma membrane. As Na^+ moves into the cell, H^+ moves out of the cell (figure 26.13). The secreted H^+ is produced when carbon dioxide and water react to form H^+ and HCO_3^-. The antiporters secrete H^+ into the nephron's lumen in exchange for Na^+. Sodium ions and HCO_3^- are symported across the basal membrane of the cell and enter the peritubular capillaries. Hydrogen ions are secreted into the proximal and distal convoluted tubules. This secretion of H^+ by the nephron plays a major role in regulating body fluid pH and is discussed in more detail in chapter 27.

TABLE 26.4 Secretion of Substances into the Nephron

Transport Process	Substance Transported
Proximal Convoluted Tubule	
Antiport	H^+
Active transport	Hydroxybenzoates
	Para-aminohippuric acid
	Neurotransmitters
	Dopamine
	Acetylcholine
	Epinephrine
	Bile pigments
	Uric acid
	Drugs and toxins
	Penicillin
	Atropine
	Morphine
	Saccharin
Diffusion	Ammonia
Distal Convoluted Tubule	
Active transport	K^+
Antiport	K^+
	H^+

ASSESS YOUR PROGRESS

20. *What is the direction of movement of substances in tubular reabsorption?*
21. *Describe what happens to most of the filtrate that enters the nephron tubule.*
22. *On what side of the nephron tubule cell does active transport take place during reabsorption of materials?*
23. *Describe how symport works in the nephron.*
24. *Name the substances that are moved by active and passive transport. In what part of the nephron does this movement take place?*
25. *Explain the differences between the descending limb and the ascending limb of the loop of Henle.*
26. *Where does tubular secretion take place? What is the direction of movement?*
27. *What substances are secreted? List the mechanisms by which these substances are transported.*

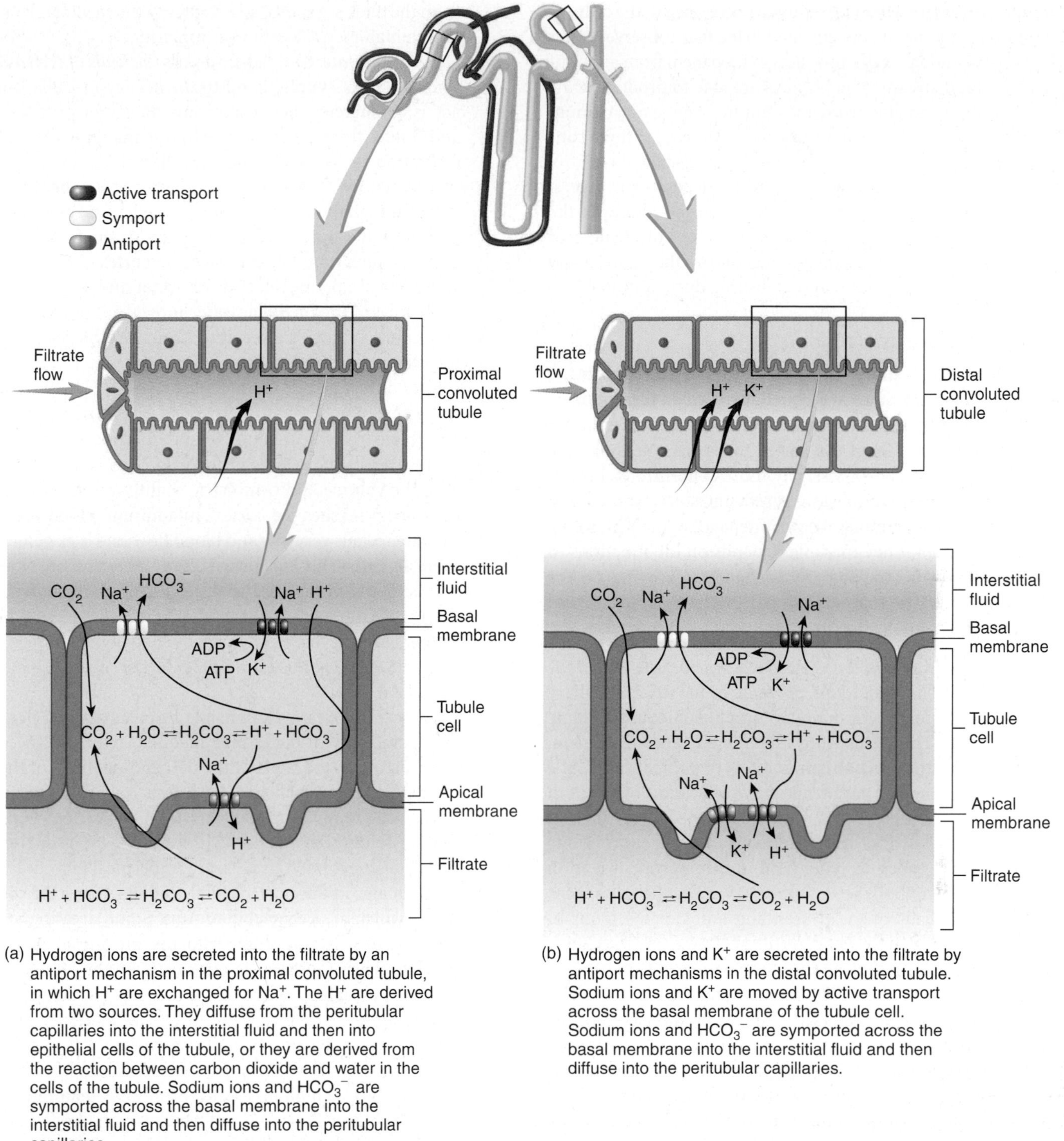

Active transport
Symport
Antiport
Filtrate flow
H^+
Proximal convoluted tubule
Filtrate flow
H^+ K^+
Distal convoluted tubule
Interstitial fluid
Basal membrane
Tubule cell
Apical membrane
Filtrate
CO_2 Na^+ HCO_3^-
Na^+ H^+
ADP
ATP K^+
$CO_2 + H_2O \rightleftharpoons H_2CO_3 \rightleftharpoons H^+ + HCO_3^-$
Na^+
H^+
$H^+ + HCO_3^- \rightleftharpoons H_2CO_3 \rightleftharpoons CO_2 + H_2O$
Na^+
K^+ H^+

(a) Hydrogen ions are secreted into the filtrate by an antiport mechanism in the proximal convoluted tubule, in which H^+ are exchanged for Na^+. The H^+ are derived from two sources. They diffuse from the peritubular capillaries into the interstitial fluid and then into epithelial cells of the tubule, or they are derived from the reaction between carbon dioxide and water in the cells of the tubule. Sodium ions and HCO_3^- are symported across the basal membrane into the interstitial fluid and then diffuse into the peritubular capillaries.

(b) Hydrogen ions and K^+ are secreted into the filtrate by antiport mechanisms in the distal convoluted tubule. Sodium ions and K^+ are moved by active transport across the basal membrane of the tubule cell. Sodium ions and HCO_3^- are symported across the basal membrane into the interstitial fluid and then diffuse into the peritubular capillaries.

FIGURE 26.13 Secretion of H^+ and K^+ into the Nephron

Urine Concentration Mechanism

As you have just read, the kidneys are remarkable at regulating blood composition. But how does the kidney move such a large volume of fluid from the blood into the filtrate and then back into the blood? The kidneys employ a unique mechanism, the countercurrent mechanism.

When a person drinks a large amount of liquid, the body must eliminate that excess fluid without losing dangerous amounts of electrolytes or other substances essential for maintaining homeostasis. The kidneys respond by producing a large volume of dilute urine. On the other hand, if a person did not drink enough water, producing an excessive amount of dilute urine would lead to rapid dehydration. Therefore, when water intake is restricted, the kidneys produce a small volume of concentrated urine that conserves water and contains just enough waste products to keep them from accumulating in the blood stream. The kidneys are able to produce urine with concentrations ranging from a minimum of 65 to a maximum of 1200 mOsm/kg while maintaining the extracellular fluid concentration very close to 300 mOsm/kg. The kidneys' ability to control the volume and concentration of the urine depends on several factors: the maintenance of a high concentration of solutes in the medulla, the countercurrent functions of the loops of Henle, and the antidiuretic hormone mechanism that controls the permeability of the distal convoluted tubules and collecting ducts to water.

Medullary Concentration Gradient

The kidney's ability to concentrate urine depends on maintaining a high concentration of solutes in the medullary region of the kidney. The interstitial fluid concentration is about 300 mOsm/kg in the cortical region. Moving from the cortex toward the medulla, the interstitial fluid becomes progressively more concentrated until it achieves a maximum concentration of 1200 mOsm/kg at the tip of the renal pyramid (see figure 26.16). Maintenance of the high solute concentration in the kidney medulla depends on the functions of the loops of Henle and the vasa recta (figure 26.14) and on the distribution of urea. The following major mechanisms create and maintain the high solute concentration in the renal medulla:

1. *Transport of solutes and diffusion of water across the wall of the loop of Henle.* The long loops of Henle of the juxtamedullary nephrons are essential for maintaining a high medullary solute concentration. They constitute a countercurrent mechanism. A **countercurrent mechanism** consists of parallel tubes (the loop of Henle and the vasa recta) in which fluid flows, but in opposite directions. Via this mechanism, substances such as water or solutes diffuse from the tubes carrying fluid in one direction to the tubes carrying fluid in the opposite direction so that the fluid in both sets of tubes has nearly the same composition. The walls of the descending limbs of the loops of Henle are permeable to water. As filtrate flows into the kidney medulla through the descending limbs, water diffuses out of the nephrons into the more concentrated interstitial fluid. The excess water that enters the interstitial fluid enters the vasa recta and is removed from the medulla (see function 2). Some solutes diffuse into the descending limb, but in relatively small amounts. By the time the filtrate reaches the tip of the loop of Henle, it is very concentrated. The walls of the thick and thin segments of the ascending limbs of the loops of Henle are impermeable to water. Consequently, solutes diffuse out of the thin segment of the ascending limb as it passes through progressively less concentrated interstitial fluid on its way back to the kidney cortex. Also, Na^+, K^+, and Cl^- are symported out of the thick segment of the ascending limb into the interstitial fluid. Thus, water enters the interstitial fluid from the descending limbs, and solutes enter the interstitial fluid from the ascending limbs (figure 26.14*a*). The solutes that diffuse from the thin segments, and those that are symported from the thick segments, add solutes to the medulla. This maintains a high medullary fluid osmolarity.
2. *Diffusion of water and solutes across the walls of the vasa recta.* The vasa recta supply blood to the kidney medulla, and they act as countercurrent mechanisms that remove excess water and solutes from the medulla without changing the high concentration of solutes in the medullary interstitial fluid. The vasa recta are a countercurrent mechanism because blood flows through them to the kidney medulla, and after the vessels turn near the tip of the renal pyramid, the blood flows the opposite direction, back toward the cortex. The walls of the vasa recta are permeable to both water and solutes. As blood flows toward the medulla, water moves out of the vasa recta, and some solutes diffuse into them. As blood flows back toward the cortex, water moves into the vasa recta, and some solutes diffuse out of them (figure 26.14*b*). The directions of diffusion are such that the vasa recta carry slightly more water and solute from the medulla than to it. Thus, the composition of the blood at both ends of the vasa recta is nearly the same, with the volume and osmolality slightly greater as the blood once again reaches the cortex. In addition, blood pressure in the vasa recta is very low and blood flow rate is extremely slow, even sluggish. This encourages ready diffusion of solutes into and back out of the vasa recta, ensuring the maintenance of the high medullary concentration gradient. Figure 26.14*c,d* further illustrate this mechanism by showing the close anatomical relationship of the loops of Henle, the collecting ducts, and the vasa recta.
3. *Urea cycling.* Urea is responsible for a substantial part of the high osmolality in the kidney medulla (figure 26.15). Due to their histology, the walls of the descending limbs of the loops of Henle are permeable to urea; thus, urea diffuses into the descending limbs from the interstitial fluid. However, due to their histology, the ascending limbs of the loops of Henle and the distal convoluted tubules are impermeable to urea, so the urea remains in the loop of Henle until it reaches the collecting ducts, which are permeable to urea. Some urea then diffuses out of the collecting ducts into the interstitial fluid of the medulla. Therefore, urea is recycled from the interstitial fluid into the descending limbs of the loops of Henle, through the ascending limbs, through the distal convoluted tubules, and into the collecting ducts. Most urea then diffuses from the collecting ducts back into the interstitial fluid of the medulla. Consequently, a high urea concentration is maintained in the medulla of the kidney. To summarize, several key events occur in the nephron to establish and maintain a high medullary solute concentration:
   a. Sodium ions and other solutes are actively transported into the interstitial fluid of the medulla, maintaining a high medullary osmolarity.
   b. Because blood flows sluggishly and there is low blood pressure in the vasa recta, solutes are not washed away from the medulla.
   c. Much urea returns to the medulla from the collecting duct, rather than exiting the urine.

(a)

1. Water diffuses out of the thin segment of the loop of Henle.
2. The filtrate concentration is 1200 mOsm.
3. Sodium and other solutes are actively transported out of the loop of Henle into the medulla.

(b)

1. Water diffuses out and solutes diffuse into the descending portion of the vasa recta.
2. The blood concentration is 1200 mOsm.
3. Water diffuses into and solutes diffuse out of the ascending portion of the vasa recta.
4. At the end of the vasa recta, the blood osmolality is only slightly greater than the osmolality of the blood at the beginning of the vasa recta.

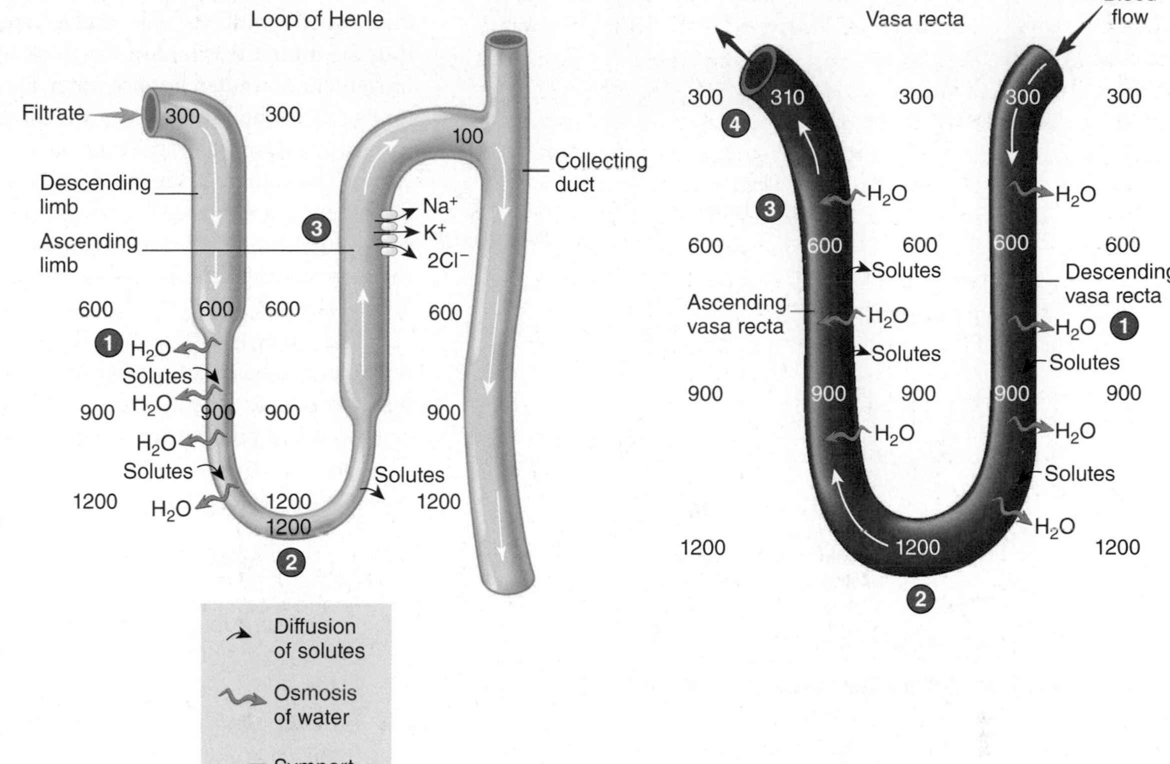

(a)
Loop of Henle
Filtrate
300
300
100
Collecting duct
Descending limb
Ascending limb
Na+
K+
2Cl−
600
600
600
600
H2O
Solutes
900
900
900
900
1200
1200
1200
1200
Diffusion of solutes
Osmosis of water
Symport of Na+, K+, and Cl−
(b)
Vasa recta
Blood flow
300
310
300
300
300
H2O
600
600
600
600
600
Solutes
Ascending vasa recta
Descending vasa recta
900
900
900
900
900
1200
1200
1200

(c)

1. Water moves out of the descending limb of the loop of Henle and enters the ascending vasa recta.
2. Solutes diffuse out of the acsending thin segment of the loop of Henle and enter the vasa recta.
3. Solutes transported out of the thick segment of the ascending limb of the loop of Henle enter the descending vasa recta. The vasa recta do not dilute the high medullary concentration.
4. The concentration of the filtrate is reduced to 100 mOsm/kg by the time it reaches the distal convoluted tubule.

(d)

1. Water and solutes, such as Na^+ and Cl^- leave the distal convoluted tubule and the collecting duct and enter the ascending vasa recta.

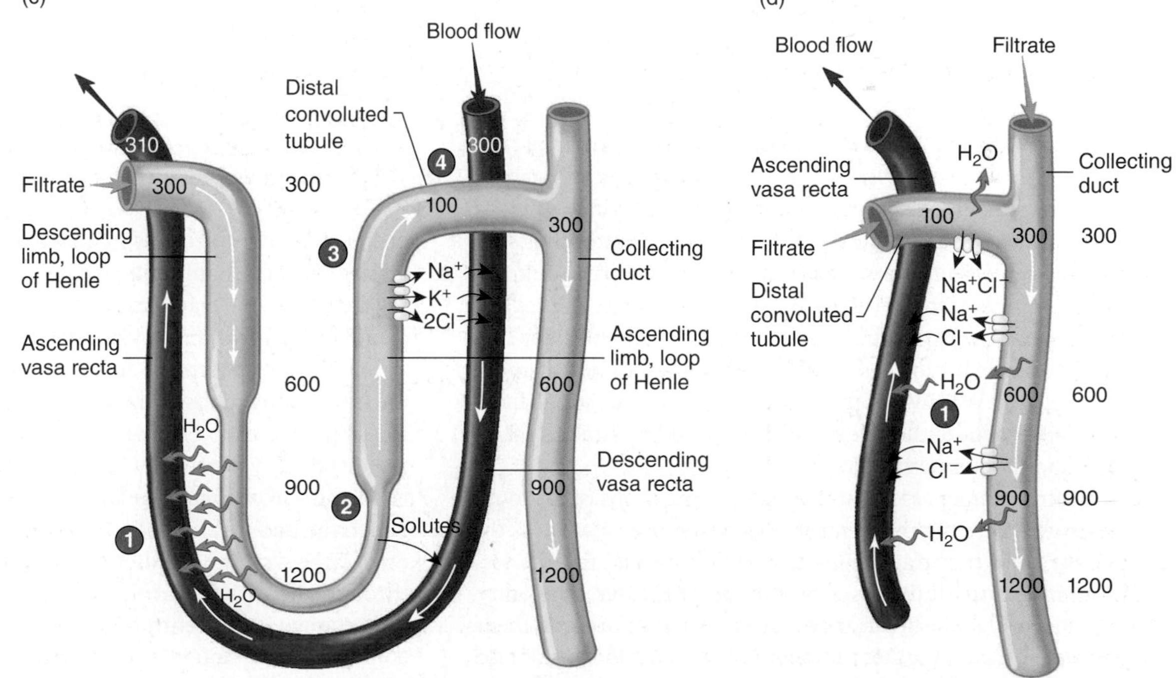

(c)
Blood flow
Distal convoluted tubule
310
300
Filtrate
300
300
100
300
Descending limb, loop of Henle
Na+
K+
2Cl−
Collecting duct
Ascending vasa recta
Ascending limb, loop of Henle
600
600
H2O
900
900
Descending vasa recta
Solutes
1200
1200
(d)
Blood flow
Filtrate
H2O
Ascending vasa recta
Collecting duct
100
300
300
Filtrate
Distal convoluted tubule
Na+Cl−
Na+
Cl−
600
600
900
900
1200
1200

FIGURE 26.14 Filtrate Concentration and the Medullary Concentration Gradient
The loop of Henle and the vasa recta function together to maintain a high concentration of solutes in the medulla of the kidney.

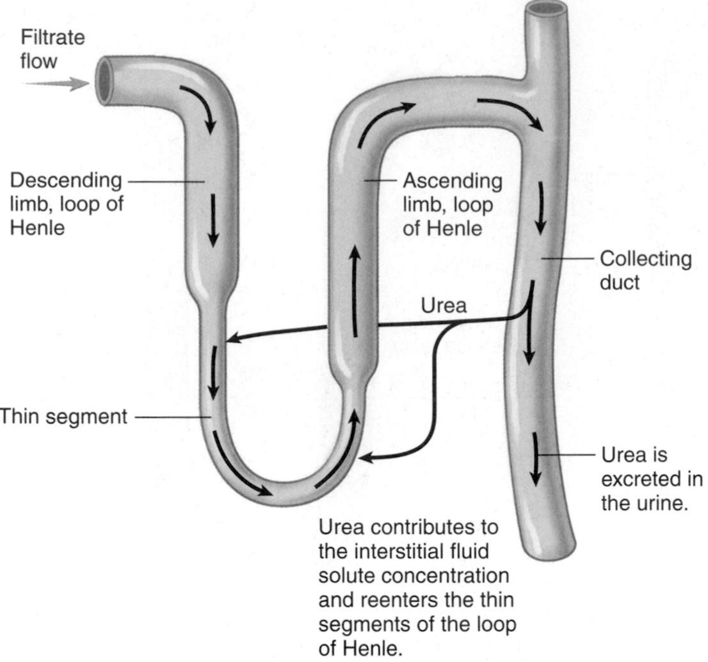

Filtrate flow
Descending limb, loop of Henle
Ascending limb, loop of Henle
Collecting duct
Urea
Thin segment
Urea is excreted in the urine.
Urea contributes to the interstitial fluid solute concentration and reenters the thin segments of the loop of Henle.

FIGURE 26.15 Medullary Concentration Gradient and Urea Cycling

The concentration of urea in the medulla of the kidney is high and contributes to the overall high concentration of solutes there. The wall of the collecting duct is permeable to urea. Urea diffuses out of the collecting duct into the interstitial fluid of the medulla. The wall of the descending limb of the loop of Henle is also permeable to urea. Urea diffuses from the interstitial fluid into the descending limb. Thus, a cycle is produced: Urea flows into the descending limb, through the ascending limb, through the distal convoluted tubule, through the collecting duct, out of the collecting duct, and back into the descending limb.

Summary of Changes in Filtrate Volume and Concentration

Following are the steps in urine formation from the proximal convoluted tubule to the collecting duct. In the average person, about 180 L of filtrate enter the proximal convoluted tubules daily. Glucose, amino acids, Na^+, Ca^{2+}, K^+, Cl^-, water, and other substances (see table 26.3) move from the lumens of the proximal convoluted tubules into the interstitial fluid. The excess solutes and water then enter the peritubular capillaries. Consequently, approximately 65% of the filtrate is reabsorbed as solutes and water move from the proximal convoluted tubules into the interstitial fluid. The osmolality of both the interstitial fluid and the filtrate is maintained at about 300 mOsm/kg.

The filtrate then passes into the descending limbs of the loops of Henle, which are highly permeable to water and solutes. As the descending limbs penetrate deep into the kidney medulla, the surrounding interstitial fluid has a progressively greater osmolality. Water diffuses out of the nephrons as solutes slowly diffuse into them. By the time the filtrate reaches the deepest part of the loops of Henle, its volume has been reduced by an additional 15% of the original volume, at least 80% of the filtrate volume has been reabsorbed, and its osmolality has increased to about 1200 mOsm/kg (figure 26.16).

After passing through the descending limbs of the loops of Henle, the filtrate enters the ascending limbs. Both the thin and thick segments are impermeable to water, but solutes diffuse out of the thin segment, and Na^+, Cl^-, and K^+ are symported from the filtrate into the interstitial fluid in the thick segments (figure 26.16). The movement of solutes, but not water, across the wall of the ascending limbs causes the osmolality of the filtrate to decrease from 1200 to about 100 mOsm/kg by the time the filtrate again reaches the kidney cortex. The volume of the filtrate does not change as it passes through the ascending limbs. As a result, the filtrate entering the distal convoluted tubules is dilute, compared with the concentration of the surrounding interstitial fluid, which has an osmolality of about 300 mOsm/kg.

The changes just described are *obligatory;* that is, they occur regardless of the concentration and volume of urine that the kidney finally produces. The mechanisms by which the kidney forms concentrated and dilute urine are described in section 26.4.

ASSESS YOUR PROGRESS

28. *List the major mechanisms that create and maintain the high solute concentration in the renal medulla.*
29. *Describe the roles of the loop of Henle, the vasa recta, and urea cycling in maintaining a high interstitial solute concentration in the kidney medulla.*
30. *Describe how the filtrate volume and concentration change as filtrate flows through the nephron and collecting ducts.*

26.4 Regulation of Urine Concentration and Volume

LEARNING OUTCOME

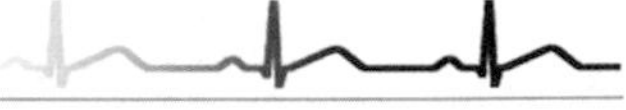

After reading this section, you should be able to

A. Explain how antidiuretic hormone, the renin-angiotensin-aldosterone hormone mechanism, and atrial natriuretic hormone influence the concentration and volume of urine.

Urine can be dilute or very concentrated, and it can be produced in large or small amounts. Urine concentration and volume are regulated by mechanisms that maintain the extracellular fluid osmolality and volume within narrow limits.

Filtrate reabsorption in the proximal convoluted tubules and the descending limbs of the loops of Henle is obligatory and therefore remains relatively constant. However, filtrate reabsorption in the distal convoluted tubules and collecting ducts is tightly regulated and can change dramatically, depending on the conditions to which the body is exposed. If homeostasis requires the elimination of a large volume of dilute urine, the dilute filtrate can pass through the distal convoluted tubules and collecting ducts with little change in concentration. On the other hand, if water must be conserved to maintain homeostasis, water is reabsorbed from the filtrate as it passes through the distal convoluted tubules and collecting ducts. This results in a small volume of very concentrated urine. The regulation of urine concentration and volume involves hormonal mechanisms, described next, as well as autoregulation and the sympathetic nervous system.

FUNDAMENTAL **Figure**

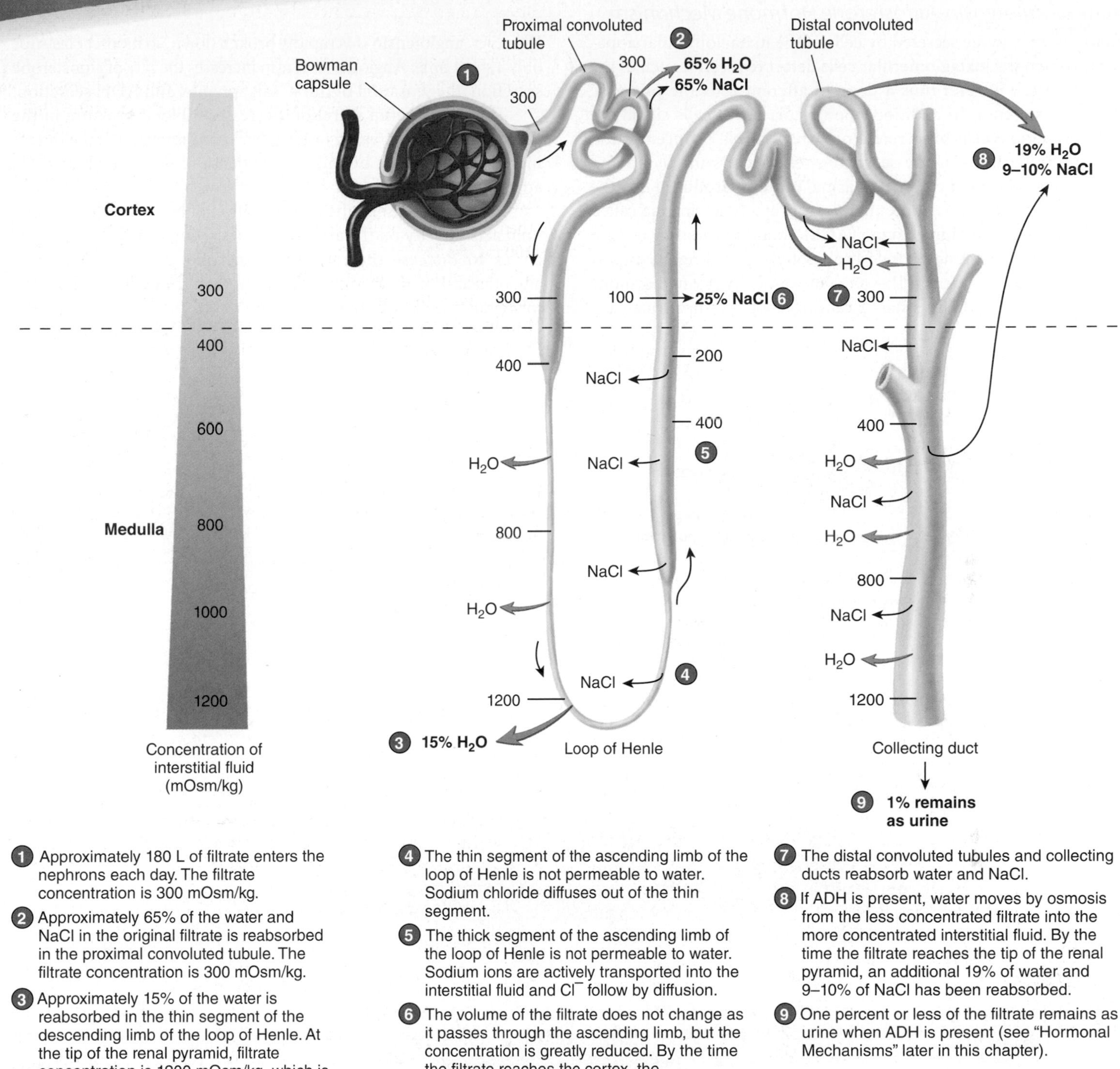

Proximal convoluted tubule
Distal convoluted tubule
Bowman capsule
1
2
300
300
65% H_2O
65% NaCl
8
19% H_2O
9–10% NaCl
Cortex
NaCl
H_2O
300
300
100
25% NaCl 6
7 300
400
400
200
NaCl
NaCl
600
400
400
H_2O
NaCl
5
H_2O
NaCl
Medulla
800
800
H_2O
NaCl
800
1000
H_2O
NaCl
NaCl
4
H_2O
1200
1200
1200
3 15% H_2O
Loop of Henle
Collecting duct
Concentration of interstitial fluid (mOsm/kg)
9 1% remains as urine

1 Approximately 180 L of filtrate enters the nephrons each day. The filtrate concentration is 300 mOsm/kg.

2 Approximately 65% of the water and NaCl in the original filtrate is reabsorbed in the proximal convoluted tubule. The filtrate concentration is 300 mOsm/kg.

3 Approximately 15% of the water is reabsorbed in the thin segment of the descending limb of the loop of Henle. At the tip of the renal pyramid, filtrate concentration is 1200 mOsm/kg, which is equal to the interstitial fluid concentration.

4 The thin segment of the ascending limb of the loop of Henle is not permeable to water. Sodium chloride diffuses out of the thin segment.

5 The thick segment of the ascending limb of the loop of Henle is not permeable to water. Sodium ions are actively transported into the interstitial fluid and Cl^- follow by diffusion.

6 The volume of the filtrate does not change as it passes through the ascending limb, but the concentration is greatly reduced. By the time the filtrate reaches the cortex, the concentration is 100 mOsm/kg, and an additional 25% of NaCl has been reabsorbed.

7 The distal convoluted tubules and collecting ducts reabsorb water and NaCl.

8 If ADH is present, water moves by osmosis from the less concentrated filtrate into the more concentrated interstitial fluid. By the time the filtrate reaches the tip of the renal pyramid, an additional 19% of water and 9–10% of NaCl has been reabsorbed.

9 One percent or less of the filtrate remains as urine when ADH is present (see "Hormonal Mechanisms" later in this chapter).

PROCESS FIGURE 26.16 Urine-Concentrating Mechanism

Hormonal Mechanisms

Two major hormonal mechanisms are involved in regulating urine concentration and volume: the renin-angiotensin-aldosterone hormone mechanism and the antidiuretic hormone (ADH) mechanism. Each mechanism is activated by different stimuli, but they work together to achieve homeostasis. The renin-angiotensin-aldosterone hormone mechanism is more sensitive to changes in blood pressure, and the ADH mechanism is more sensitive to changes in blood osmolality.

Renin-Angiotensin-Aldosterone Hormone Mechanism

Renin is an enzyme secreted by cells of the juxtaglomerular apparatus. When the juxtaglomerular cells detect reduced stretch of the afferent arteriole, and thus a drop in afferent arteriole pressure, they secrete renin. In addition, the macula densa cells signal the juxtaglomerular cells to secrete renin when the Na^+ concentration of the filtrate drops. Upon secretion, renin enters the blood and converts **angiotensinogen,** a plasma protein produced by the liver, to **angiotensin I.** Subsequently, a proteolytic enzyme called **angiotensin-converting enzyme (ACE),** found in capillary beds in organs such as the lungs, converts angiotensin I to **angiotensin II** (figure 26.17). Angiotensin II is a potent vasoconstricting hormone that increases peripheral resistance, causing blood pressure to increase. However, angiotensin II is rapidly broken down, so its effect lasts for only a short time. Angiotensin II also increases the rate of aldosterone secretion, the sensation of thirst, salt appetite, and ADH secretion.

The rate of renin secretion decreases if blood pressure in the afferent arteriole increases, or if the Na^+ concentration of the filtrate increases as it passes by the macula densa of the juxtaglomerular apparatuses.

A large decrease in the concentration of Na^+ in the interstitial fluid acts directly on the aldosterone-secreting cells of the adrenal cortex to increase the rate of aldosterone secretion. However, angiotensin II is much more important than the blood level of Na^+ for regulating aldosterone secretion.

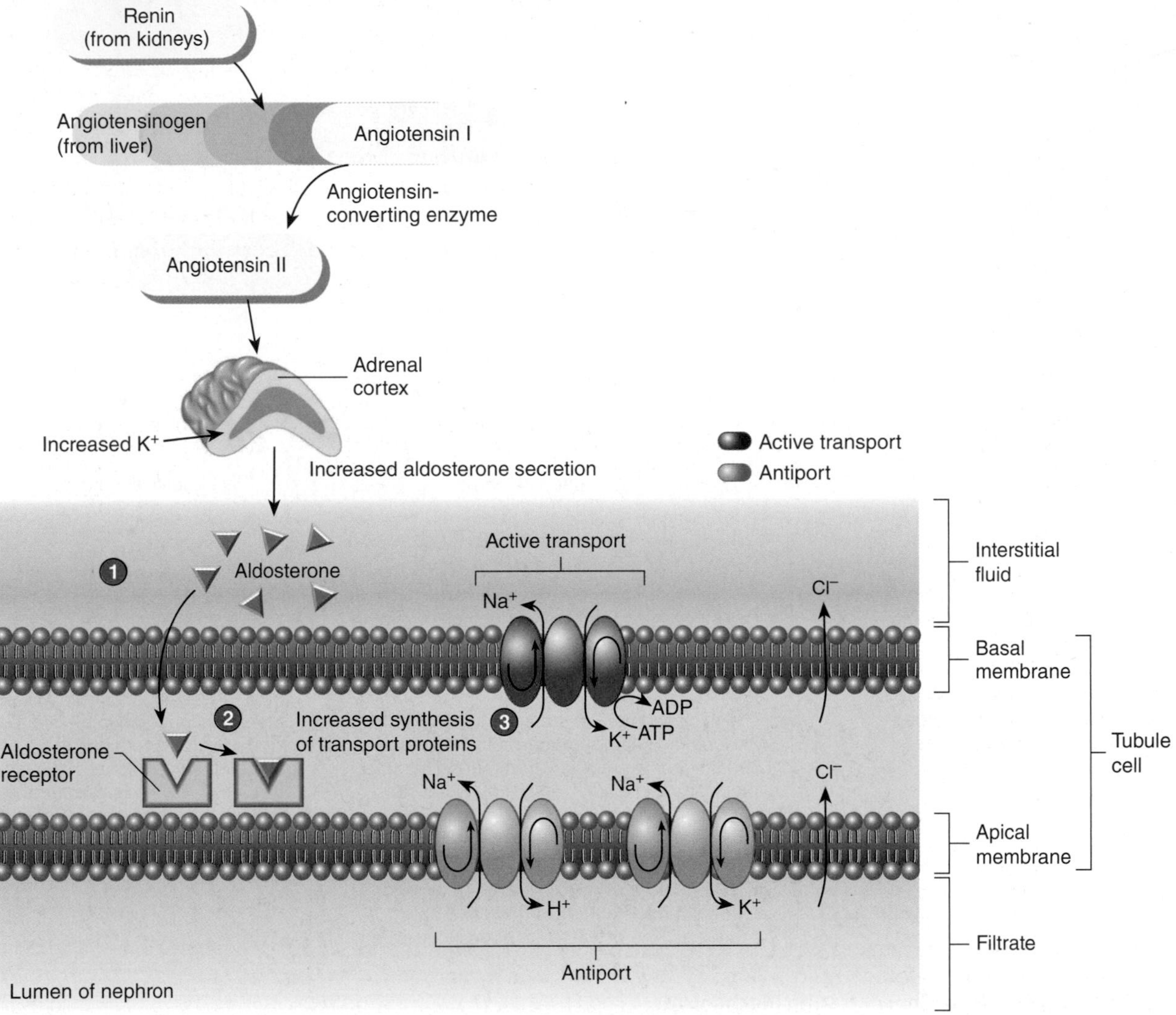
Renin
(from kidneys)
Angiotensinogen
(from liver)
Angiotensin I
Angiotensin-
converting enzyme
Angiotensin II
Adrenal
cortex
Increased K+
Increased aldosterone secretion
Active transport
Antiport
1
Aldosterone
Active transport
Na+
Cl−
Interstitial
fluid
Basal
membrane
Tubule
cell
2
Increased synthesis
of transport proteins
3
ADP
K+ ATP
Aldosterone
receptor
Na+
Na+
Cl−
Apical
membrane
H+
K+
Filtrate
Antiport
Lumen of nephron

1. Aldosterone secreted from the adrenal cortex enters cells of the distal convoluted tubule.
2. Aldosterone binds to nuclear receptors and increases the synthesis of transport proteins of the apical and basal membranes.
3. Newly synthesized transport proteins increase the rate at which Na^+ are absorbed and K^+ and H^+ are secreted. Chloride ions move with the Na^+ because they are attracted to the positive charge of Na^+.

PROCESS FIGURE 26.17 Effect of Aldosterone on the Distal Convoluted Tubule

Aldosterone, a steroid hormone secreted by the cortical cells of the adrenal glands (see chapter 18), passes through the bloodstream from the adrenal glands to the cells in the distal convoluted tubules and the collecting ducts. Aldosterone molecules diffuse through the plasma membranes and bind to receptor molecules within the cells. The combination of aldosterone molecules with their receptor molecules increases synthesis of the transport proteins that increase the transport of Na^+ across the basal and apical membranes of nephron cells. As a result, the rate of Na^+ reabsorption increases (figure 26.17).

Reduced secretion of aldosterone decreases the rate of Na^+ reabsorption. As a consequence, the concentration of Na^+ in the distal convoluted tubules and the collecting ducts remains high. Because the concentration of filtrate passing through the distal convoluted tubules and the collecting ducts has a greater-than-normal concentration of solutes, water's capacity to move by osmosis from the distal convoluted tubules and the collecting ducts is diminished, urine volume increases, and the urine has a greater concentration of Na^+.

Increases in blood K^+ levels act directly on the adrenal cortex to stimulate aldosterone secretion, whereas decreases in blood K^+ levels decrease aldosterone secretion (see chapter 27).

Predict 3

Drugs that increase urine volume are called diuretics. Some diuretics inhibit the active transport of Na^+ in the nephron. Explain how these diuretic drugs cause increased urine volume.

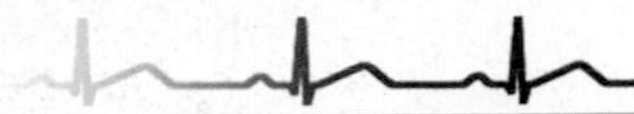

ASSESS YOUR PROGRESS

31. *What factors stimulate the release of renin? What will decrease the rate of renin secretion?*
32. *How is angiotensin II activated? What effects does it produce?*
33. *Where is aldosterone produced? What factors stimulate its secretion?*
34. *What are the effects of aldosterone on Na^+ and Cl^- transport? How does aldosterone affect urine concentration, urine volume, and blood pressure?*

Antidiuretic Hormone Mechanism

The distal convoluted tubules and collecting ducts remain relatively impermeable to water in the absence of **antidiuretic hormone (ADH),** also known as *vasopressin* (figure 26.18). ADH is secreted from the posterior pituitary. When little ADH is secreted, a large part of the 19% of the filtrate that is normally reabsorbed in the distal convoluted tubules and the collecting ducts becomes part of the urine. Insufficient ADH secretion results in a condition called **diabetes insipidus** (dī-ă-bē′tēz in-sip′i-dŭs); the word *diabetes* refers to the production of a large volume of urine, and the word *insipidus* means the urine is clear, tasteless, and dilute. People who secrete insufficient ADH often produce 10–20 L of urine per day and develop major problems, such as dehydration and ion imbalances. In contrast to diabetes insipidus, **diabetes mellitus** (me-lī′tŭs) refers to the production of a large volume of urine that contains a high concentration of glucose (*mellitus,* honeyed, sweet).

Clinical IMPACT — Diabetic Nephropathy and Renal Failure

Diabetic nephropathy (ne-frop′ă-thē) is a disease of the kidneys associated with diabetes mellitus, and it is the principal cause of chronic renal failure. This condition damages renal glomeruli and ultimately destroys functional nephrons through progressive scar tissue formation, mediated in part by an inflammatory response. The damaged glomeruli no longer filter the blood effectively, allowing proteins to pass through the filtration membrane and be excreted in the urine. The presence of protein in the urine of people who have type 2 diabetes strongly suggests significant diabetic nephropathy, which can lead to end-stage renal failure. About 1 in 14 Americans over age 30 have some degree of type 2 diabetes mellitus, and most hemodialysis patients have type 2 diabetes mellitus.

The development of diabetic nephropathy is complex. Although the mechanism is not completely understood, the level of angiotensin II is elevated in diabetes mellitus. This causes exaggerated efferent arteriole vasoconstriction and consequently increased glomerular capillary pressure. The increased glomerular capillary pressure damages the glomerular basement membrane, causing it to thicken and become more permeable. The glomerular basement membrane is also damaged by the production of glycoproteins called **advanced glycosylation end products (AGEs).** AGEs are produced when glucose forms irreversible cross-links with kidney and plasma proteins. The AGEs stimulate the secretion of growth factors from glomerular cells, which promote glomerular basement membrane thickening.

Because the glomerular basement membrane in patients with diabetes mellitus is more permeable than normal, plasma proteins cross the filtration membrane and enter the urine. The initial amount of protein entering the urine is small, a condition called microalbuminuria (mī′krō-al-boo-min-ū′rē-ă). However, as the number of functional nephrons in the kidney decreases, microalbuminuria eventually progresses to overt proteinuria (prō-tē-noo′rē-ă), the secretion of more than 300 mg albumin/day. By the time overt proteinuria has developed, which may take 10–15 years, the number of functional nephrons has decreased to less than 10% of normal, and the kidneys are no longer able to excrete adequate amounts of waste products. This condition is called **end-stage renal disease (ESRD).** In ESRD, renal failure has worsened to the point that kidney function is less than 10% of normal. Unless ESRD is treated by hemodialysis or kidney transplantation, the patient dies.

The use of **angiotensin-converting enzyme (ACE) inhibitors** slows or, in some cases, even halts the progression of proteinuria and end-stage renal disease. ACE inhibitors prevent the formation of angiotensin II; consequently, arterial blood pressure and glomerular capillary pressure remain within their normal ranges. When ACE inhibitors are used in combination with drugs called **angiotensin receptor blockers (ARBs),** which prevent angiotensin II molecules from binding to their receptors, proteinuria decreases up to 45%. People with type 2 diabetes who maintain their blood glucose within normal levels have a much lower incidence of diabetic nephropathy and ESRD.

FUNDAMENTAL **Figure**

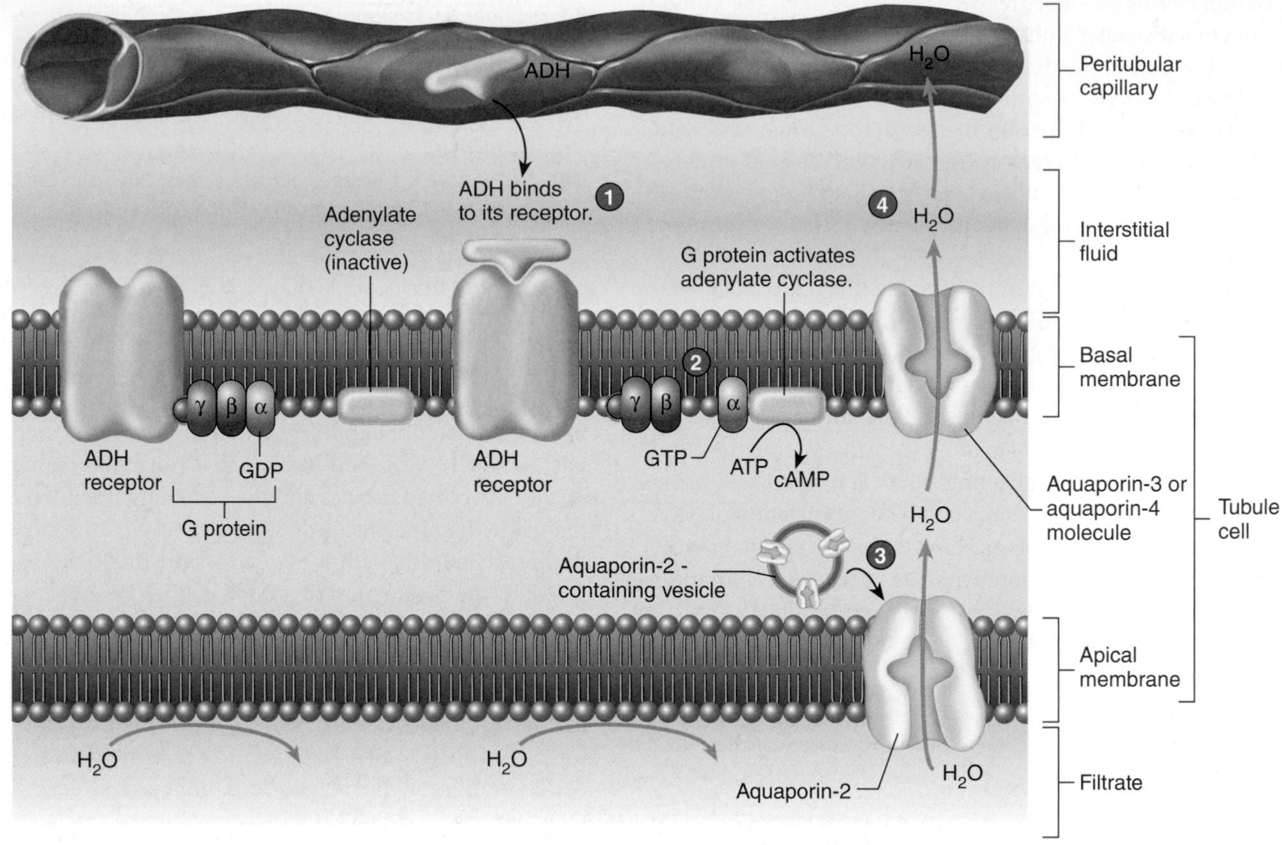
ADH
H_2O
Peritubular capillary
ADH binds to its receptor. 1
Adenylate cyclase (inactive)
4 H_2O
Interstitial fluid
G protein activates adenylate cyclase.
Basal membrane
2
γ β α
ADH receptor
GDP
G protein
ADH receptor
GTP
ATP
cAMP
Aquaporin-3 or aquaporin-4 molecule
Tubule cell
H_2O
Aquaporin-2 - containing vesicle
3
Apical membrane
H_2O
H_2O
Aquaporin-2
H_2O
Filtrate
cAMP increases aquaporin-2 to increase permeability of the membrane to H_2O.

1. ADH moves from the peritubular capillaries and binds to ADH receptors in the plasma membranes of the distal convoluted tubule cells and the collecting duct cells.
2. When ADH binds to its receptor, a G protein mechanism is activated, which in turn activates adenylate cyclase.
3. Adenylate cyclase increases the rate of cAMP synthesis. Cyclic AMP promotes the insertion of aquaporin-2 containing cytoplasmic vesicles into the apical membranes of the distal convoluted tubules and collecting ducts, thereby increasing their permeability to water. Water then moves by osmosis out of the distal convoluted tubules and collecting ducts into the tubule cells through the aquaporin-2 water channels.
4. Water exits the tubule cells and enters the interstitial fluid through aquaporin-3 and aquaporin-4 water channels in the basal membranes.

PROCESS FIGURE 26.18 Effect of Antidiuretic Hormone (ADH) on Nephron Water Movement

Neurons with cell bodies primarily in the supraoptic nuclei of the hypothalamus have axons that extend to the posterior pituitary gland. ADH is released into the blood from these neuron terminals. Cells called **osmoreceptor cells** in the supraoptic nuclei are very sensitive to even slight changes in the osmolality of the interstitial fluid. If the osmolality of the blood and interstitial fluid increases, these cells stimulate the ADH-secreting neurons. Action potentials are then propagated along the axons of the ADH-secreting neurons to the posterior pituitary gland, where the axons release ADH from their ends. Reduced osmolality of the interstitial fluid within the supraoptic nuclei inhibits ADH secretion from the posterior pituitary gland (see figure 18.6).

Baroreceptors that monitor blood pressure in the atria of the heart, large veins, carotid sinuses, and aortic arch also influence ADH secretion when the blood pressure increases or decreases by more than 5–10%. Decreases in blood pressure are detected by the baroreceptors, which consequently decrease the frequency of action potentials sent along the afferent pathways that ultimately extend to the supraoptic region of the hypothalamus. The result is increased ADH secretion.

When blood osmolality increases or when blood pressure declines significantly, ADH is secreted and acts on the kidneys to promote water reabsorption. Water reabsorption lowers blood osmolality. It also increases blood volume, which elevates blood pressure.

Conversely, when blood osmolality decreases or when blood pressure goes up, ADH secretion declines. The reduced ADH levels cause the kidneys to reabsorb less water and to produce a larger volume of dilute urine. The greater loss of water in the urine raises blood osmolality and lowers blood pressure. ADH secretion occurs in response to small changes in osmolality, whereas a substantial change in blood pressure is required to alter ADH secretion. Thus, ADH is more important in regulating blood osmolality than it is in regulating blood pressure.

Predict 4

Ethyl alcohol inhibits ADH secretion. Given this information, describe the mechanism by which alcoholic beverages affect urine production.

Formation of Concentrated Urine

Filtrate enters the distal convoluted tubules after passing through the loops of Henle and then passes through the collecting ducts. Near the ends of the distal convoluted tubules and in the collecting ducts, the walls of the tubules become very permeable to water, provided antidiuretic hormone (ADH) is present. Water then diffuses from the lumens of the nephrons into the more concentrated interstitial fluid.

ADH increases the permeability of the apical membranes of the distal convoluted tubules and collecting ducts to water by binding to membrane-bound receptors. This activates a G protein mechanism that increases cAMP synthesis inside these cells. Cyclic AMP promotes the insertion of aquaporins into the apical membrane (figure 26.18). **Aquaporins** are water channel proteins that increase the apical membrane's permeability to water. There are multiple forms of aquaporins. In cells of the distal convoluted tubules and collecting ducts, the basal membranes contain aquaporin molecules—aquaporin-3 and aquaporin-4—that are insensitive to ADH. These aquaporin molecules provide channels for water to exit from the collecting duct cells into the interstitial fluid. Aquaporin-2 molecules regulate water movement into the cells. In cells that have not been exposed to ADH, the aquaporin-2 molecules are found in the membranes of vesicles in the cytoplasm (see figure 3.3*b*). In response to ADH, the increased cAMP initiates the incorporation of the membranes of vesicles containing aquaporin-2 channels into the apical membrane. Thus, when ADH is present, water moves by osmosis out of the distal convoluted tubules and collecting ducts; conversely, when ADH is absent, water remains in the distal convoluted tubules and collecting ducts to become urine (figure 26.19). Abnormal aquaporin-2 genes can result in excessive urine production because these genes code for abnormal aquaporin-2 molecules that

Clinical IMPACT — Diuretics

Diuretics (dī-ū-ret′iks) are chemicals that increase the rate of urine formation. Although the definition is simple, a number of physiological mechanisms are involved.

Diuretics are used to treat hypertension, as well as several types of edema caused by congestive heart failure, cirrhosis of the liver, and other anomalies. However, treatment with diuretics can lead to complications, including dehydration and electrolyte imbalances.

The varying degree of diuretic chemical types is outlined in the following descriptions, along with their physiological mechanisms. The action of **carbonic anhydrase** (kar-bon′ik an-hī′drās) **inhibitors** reduces the rate of H^+ secretion and the reabsorption of bicarbonate ion (HCO_3^-). As H^+ is secreted into the nephron, it combines with HCO_3^- to form carbonic acid. Carbonic acid dissociates into water and carbon dioxide, which can diffuse across the wall of the nephron. Reduced H^+ secretion causes HCO_3^- to remain in the nephron. The HCO_3^- increases tubular osmotic pressure, causing osmotic diuresis. The diuretic effect is useful in treating conditions such as glaucoma and altitude sickness. However, with long-term use, carbonic anhydrase inhibitors tend to lose their diuretic effect.

Sodium ion reabsorption inhibitors include thiazide-type diuretics. They promote the loss of Na^+, Cl^-, and water in the urine. These diuretics are sometimes given to people who have hypertension. The increased loss of water in the urine lowers blood volume and thus blood pressure. Other inhibitors of Na^+ reabsorption, such as bumetanide, furosemide, and ethacrynic acid, specifically inhibit transport in the ascending limb of the loop of Henle. These diuretics are frequently used to treat congestive heart failure, cirrhosis of the liver, and renal disease. A possible side effect of these drugs is increased excretion of K^+ in the urine.

Certain **potassium-sparing diuretics** act on the distal convoluted tubules and the collecting ducts to reduce the exchange between Na^+ and K^+. Potassium-sparing diuretics are used to diminish the loss of K^+ in the urine, thereby preserving, or "sparing," these ions. Some potassium-sparing diuretic drugs act by competitive inhibition of aldosterone, whereas others inhibit the symporters for Na^+ in apical membranes of cells in the distal convoluted tubules and collecting ducts. Both types result in Na^+ diuresis and K^+ retention. Because K^+ depletion is one of the side effects of prolonged treatment with inhibitors of Na^+–Cl^- symporters in the ascending limb of the loop of Henle that are frequently used as diuretics, potassium-sparing diuretics are commonly used in combination with these Na^+–Cl^- symport inhibitors.

Osmotic diuretics freely pass into the filtrate and undergo limited reabsorption by the nephron. These diuretics increase urine volume by elevating the osmotic concentration of the filtrate, thus reducing the amount of water moving by osmosis out of the nephron. Urea, mannitol, and glycerine have been used as osmotic diuretics and can be effective in treating patients who have cerebral edema and edema in acute renal failure (see the Diseases and Disorders table, later in this chapter).

Xanthines (zan′thēnz), including caffeine and related substances, act as diuretics partly because they increase renal blood flow and the rate of glomerular filtrate formation. They also influence the nephron by decreasing Na^+ and Cl^- reabsorption.

Alcohol acts as a diuretic, although it is not used clinically for that purpose. It inhibits ADH secretion from the posterior pituitary and results in increased urine volume.

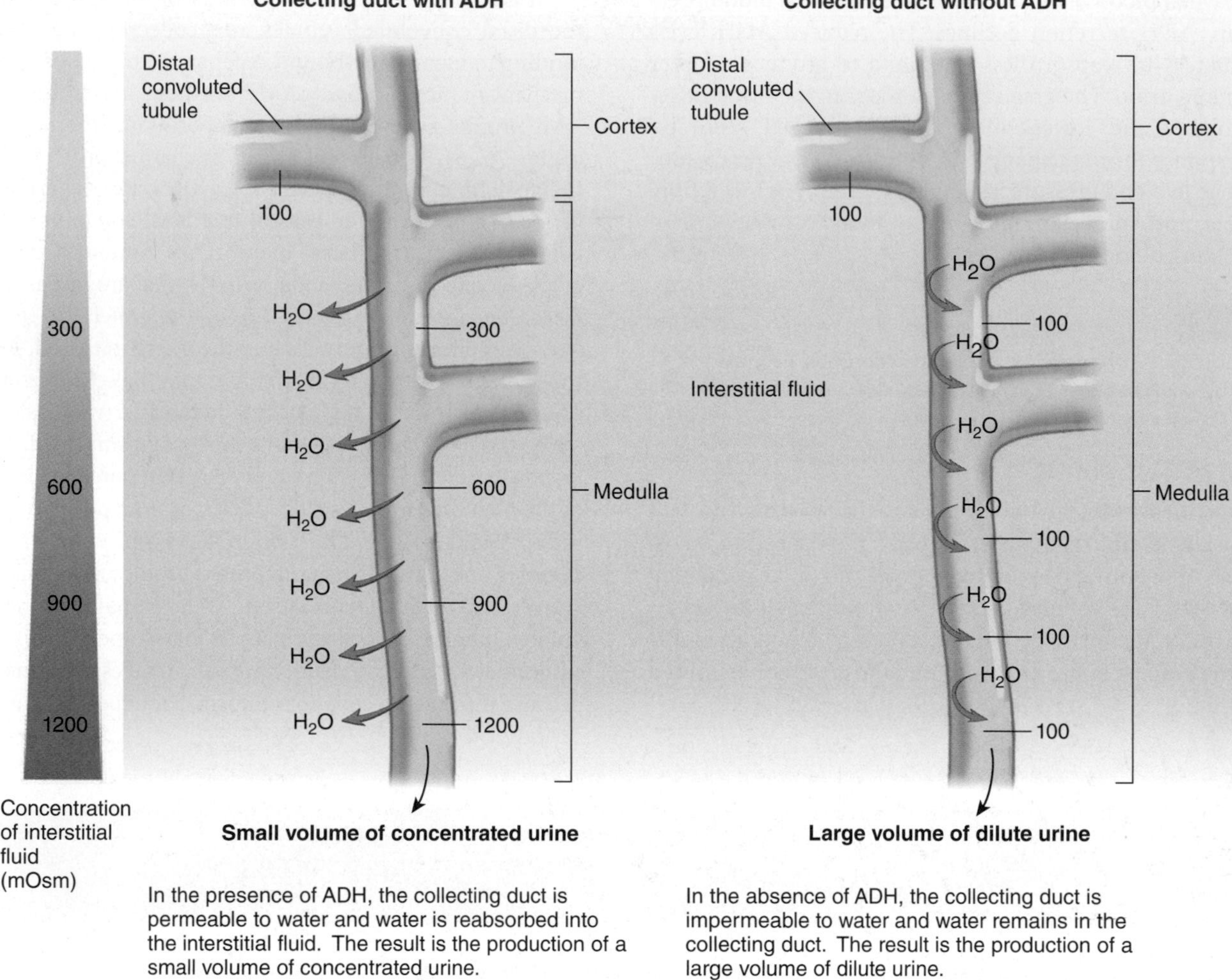
Collecting duct with ADH
Distal convoluted tubule
Cortex
100
H_2O
300
600
900
1200
Medulla
Concentration of interstitial fluid (mOsm)
Small volume of concentrated urine
In the presence of ADH, the collecting duct is permeable to water and water is reabsorbed into the interstitial fluid. The result is the production of a small volume of concentrated urine.
Collecting duct without ADH
Distal convoluted tubule
Cortex
100
Interstitial fluid
Medulla
Large volume of dilute urine
In the absence of ADH, the collecting duct is impermeable to water and water remains in the collecting duct. The result is the production of a large volume of dilute urine.

FIGURE 26.19 Effect of ADH on Urine Concentration and Volume

do not function normally. Thus, the number of functional aquaporins decreases, and water remains in the nephron.

The filtrate flows into the distal convoluted tubules and collecting ducts that pass through the kidney medulla with its high concentration of solutes. If ADH is present, water moves by osmosis from the distal convoluted tubules and the collecting ducts into the interstitial fluid. By the time the filtrate has reached the end of the collecting ducts, another 19% of the filtrate has been reabsorbed. Thus, 1% of the filtrate remains as urine, and 99% of the filtrate has been reabsorbed. The osmolality of the filtrate at the ends of the collecting ducts is approximately 1200 mOsm/kg (see figure 26.16).

In addition to the dramatic decrease in filtrate volume and the increase in filtrate osmolality, a marked alteration occurs in the filtrate composition. Waste products, such as creatinine and urea, and excess ions, such as K^+, H^+, phosphate, and sulfate, are at a much higher concentration in urine than in the original filtrate because water has been removed from the filtrate. Overall, the processes of reabsorption and secretion are selective so that, in the end, beneficial substances are retained in the body and toxic substances are eliminated.

Formation of Dilute Urine

If ADH is not present or if its concentration is reduced, the distal convoluted tubules and collecting ducts have a low permeability to water. Therefore, the amount of water moving by osmosis from the distal convoluted tubule and collecting duct is small. The concentration of the urine produced is less than 1200 mOsm/kg, and the volume is increased. The volume of this more dilute urine can be much larger than 1% of the filtrate formed each day. If no ADH is secreted, the osmolality of the urine may be close to the osmolality of the filtrate in the distal convoluted tubule, and the volume of urine may approach 20–30 L/day, which is the same volume as 10–15 2 L soda bottles per day (see figure 26.19*b*).

In a healthy person, even when the kidneys produce dilute urine, the concentration of waste products in the urine is large enough to maintain homeostasis. Again, as with the formation of concentrated urine, beneficial substances are retained, and both toxic substances and excess water are eliminated.

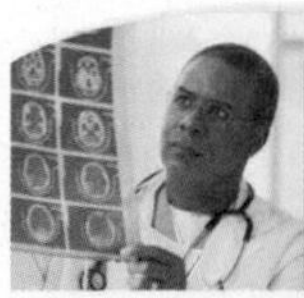

Case STUDY Diabetes Insipidus

Two infants were born to different families within the same week. Not long after the newborns arrived home from the hospital, their respective parents noticed that their diapers were excessively wet hour after hour throughout the day and night. In addition, both infants were irritable, had slight fevers, and had vomited, even though they had not eaten for several feedings. The parents took the babies to their pediatricians. Subsequently, blood tests indicated that both infants had high blood Na^+ levels. Following water deprivation tests, which monitor plasma levels of ADH, the physicians diagnosed nephrogenic diabetes insipidus. One infant was found to have an ADH receptor abnormality, whereas the other was diagnosed with an aquaporin-2 abnormality.

The term *diabetes* refers to a disease state characterized by polyuria, excess production of urine. There are two major causes of diabetes: (1) inadequate production of or response to insulin, called diabetes mellitus (see chapter 18), and (2) inadequate production of or response to ADH, called diabetes insipidus. Diabetes insipidus is a relatively rare disease that occurs in two varieties: **Central diabetes insipidus (CDI)** is caused by failure of ADH secretion, and **nephrogenic diabetes insipidus (NDI)** results when ADH secretion is normal but the ADH receptor, or the response to ADH, in the kidney is abnormal. Consequently, the G protein mechanism, which normally functions in the insertion of the aquaporin-2 water channel protein in the apical membranes, does not operate. In most cases, NDI results from an inherited condition that affects the function of the ADH receptor. NDI can also be acquired, but that usually happens later in life and can be due to several factors, including the use of certain prescription drugs or the existence of an underlying systemic disease.

Treatment of NDI includes ensuring a plentiful supply of water, following a low-sodium and sometimes a low-protein diet, and using thiazide diuretics (Na^+ reabsorption inhibitors) in combination with a potassium-sparing diuretic.

Clinical GENETICS Nephrogenic Diabetes Insipidus

There are three types of inherited nephrogenic diabetes insipidus (NDI). X-linked NDI, the most common form, affects more males than females. X-linked NDI is caused by a mutation in the V_2 ADH receptor gene on the X chromosome. Mutations in this gene result in defective ADH receptors, which prevent a normal response to ADH in the kidneys.

Autosomal recessive NDI is more rare than the X-linked form, and it affects males and females equally. This form of NDI requires both parents to be carriers for an abnormal aquaporin-2 gene. For children to have autosomal recessive NDI, they must inherit a recessive allele from each parent. In autosomal recessive NDI, there is a 25% chance that each child of heterozygous parents will have NDI.

Autosomal dominant NDI is the most rare form of NDI, and it affects males and females equally. With this form, only one parent must have a dominant allele for an abnormal aquaporin-2 gene, but that parent will also have symptoms of NDI. There is a 50% chance that each child will have NDI.

Predict 5

Use your knowledge of kidney physiology and figure 26.18 to answer the following questions.

a. Why did the two infants have high blood Na^+ levels and dilute urine?

b. Predict how the infants' plasma levels of ADH changed during the water deprivation test, given the diagnosis of NDI. (*Hint:* See "Formation of Concentrated Urine" earlier in this section.)

c. Why does an abnormal aquaporin-2 gene result in excessive urine production?

d. Predict plasma levels of ADH following a water deprivation test in an individual with central diabetes insipidus.

e. Why is treatment with a thiazide diuretic helpful to patients with NDI? (*Hint:* See Clinical Impact, "Diuretics.")

Predict 6

Amanda, an inexperienced runner, competed in her first marathon last spring in Phoenix, Arizona. During the run, the temperature reached 35°C (95°F) with 30% humidity. Amanda drank very little water during the race. When she finished 4½ hours later, she was dizzy and disoriented and had an increased heart rate. She was also very pale. Friends took her to a hospital, where the doctor diagnosed severe dehydration and prescribed IV fluids. Amanda did not urinate until nearly 12 hours later. Explain the physiological responses that resulted in her reduced urine production. (*Hint:* See the discussion of blood pressure regulation in chapter 21.)

Other Hormones

A polypeptide hormone called **atrial natriuretic** (nā′trē-ū-ret′ik) **hormone (ANH)** is secreted from cardiac muscle cells in the right atrium of the heart when blood volume in the right atrium increases and stretches the cardiac muscle cells (see chapter 21). Atrial natriuretic hormone inhibits Na^+ reabsorption in the kidney tubules. ANH also inhibits ADH secretion from the posterior pituitary gland. Consequently, increased ANH secretion increases the volume of urine produced and lowers blood volume and blood pressure. Atrial natriuretic hormone also dilates arteries and veins, which reduces peripheral resistance and lowers blood pressure. Thus, venous return and blood volume decrease in the right atrium.

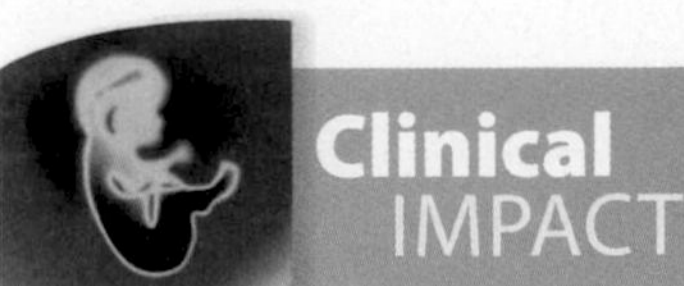
Clinical
IMPACT

Urine Concentration and Juxtamedullary Nephrons

Only the juxtamedullary nephrons have loops of Henle that descend deep into the medulla, but enough of them exist to maintain a high concentration of solutes in the interstitial fluid of the medulla. Not all of the nephrons need to have loops of Henle that descend into the medulla to concentrate urine effectively. The cortical nephrons function as the juxtamedullary nephrons do, but their loops of Henle are not as efficient at concentrating urine. However, because the filtrate from the cortical nephrons passes through the collecting ducts, water can diffuse out of the collecting ducts into the interstitial fluid. Thus, the filtrate becomes concentrated. Animals that concentrate urine more effectively than humans have a greater percentage of nephrons descending into the kidney medulla. For example, in desert mammals, many nephrons descend into the medulla, and the renal pyramids are longer than those in humans and most other mammals.

ASSESS YOUR PROGRESS

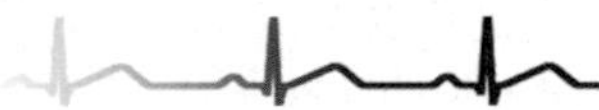

35. *Where is ADH produced? What factors stimulate an increase in ADH secretion?*
36. *How does ADH affect urine volume and concentration?*
37. *Describe how the presence of ADH causes the formation of a small volume of concentrated urine.*
38. *How does the absence of ADH cause the formation of a large volume of dilute urine?*
39. *Where is atrial natriuretic hormone produced, and how does it affect urine production?*

26.5 Plasma Clearance and Tubular Maximum

LEARNING OUTCOMES

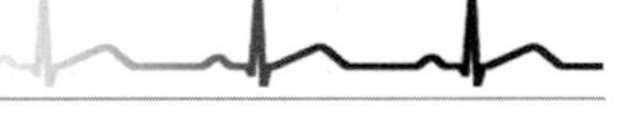

After reading this section, you should be able to

A. Define *plasma clearance* and show how it is calculated.

B. Describe why inulin is used to estimate GFR through plasma clearance.

C. Explain how plasma clearance is used to calculate renal plasma flow.

D. Define *tubular load* and *tubular maximum*.

A clinician who is concerned that a patient's kidney function is declining will measure the GFR by determining plasma clearance. **Plasma clearance** is a calculated value representing the volume of plasma that is cleared of a specific substance each minute. For example, if the clearance value is 100 mL/min for a substance, the substance is completely removed from 100 mL of plasma each minute. The plasma clearance can be calculated for any substance that enters the blood according to the following formula:

$$\text{Plasma clearance (mL/min)} = \text{Quantity of urine (mL/min)} \times \frac{\text{Concentration of substance in urine}}{\text{Concentration of substance in plasma}}$$

Plasma clearance can be used to estimate GFR if the appropriate substance is monitored (see table 26.2). Such a substance must have the following characteristics: (1) It must pass through the filtration membrane of the renal corpuscle as freely as water or other small molecules, (2) it must not be reabsorbed, (3) it must not be secreted into the nephron, and (4) it must not be either metabolized or produced in the kidneys. **Inulin** (in′ū-lin; not to be confused with the hormone insulin) is a nonphysiological polysaccharide that has these characteristics. As filtrate forms, it has the same concentration of inulin as plasma; however, as the filtrate flows through the nephron, all the inulin remains in the nephron to enter the urine. As a consequence, all the volume of plasma that becomes filtrate is cleared of inulin, and the plasma clearance for inulin is equal to the rate of glomerular filtrate formation.

GFR is reduced when a kidney fails. Therefore, measurement of the GFR can indicate the degree of kidney damage. The clearance value for urea and creatinine can also be used clinically. One advantage of using these substances is that they are naturally occurring metabolites, so foreign substances do not have to be injected. A high plasma concentration and a lower-than-normal clearance value for urea and creatinine indicate a reduced GFR and kidney failure. Creatinine clearance can also be used to monitor the progress of GFR changes in people experiencing kidney failure.

Plasma clearance can also be used to calculate renal plasma flow (see table 26.2). However, substances with the following characteristics must be used: (1) The substance must pass through the filtration membrane of the renal corpuscle, and (2) it must be secreted into the nephron at a sufficient rate that very little of it remains in the blood as the blood leaves the kidney. Para-aminohippuric acid meets these requirements. As blood flows through the kidney, essentially all the PAH is either filtered or secreted into the nephron. The clearance calculation for PAH is therefore a good estimate of the volume of plasma flowing through the kidney each minute. Also, if the hematocrit is known, the total volume of blood flowing through the kidney each minute can be calculated easily.

In addition, the concept of plasma clearance can be used to help determine how drugs or other substances are excreted by the kidney. A plasma clearance value greater than the inulin clearance value suggests that the substance is secreted by the nephron into the filtrate.

The **tubular load** of a substance is the total amount that passes through the filtration membrane into the nephrons each minute. Normally, glucose is almost completely reabsorbed from the nephron by active transport. However, the nephron's capacity to actively transport glucose across the epithelium of the nephron is limited. If the tubular load is greater than the nephron's capacity to reabsorb it, the excess glucose remains in the urine.

The maximum rate at which a substance can be actively reabsorbed is called the **tubular maximum** (figure 26.20). Each substance that is reabsorbed has its own tubular maximum, determined by the number of active transport carrier proteins and the rate at which they are able to transport molecules of the substance. For example, in people who have diabetes mellitus, the tubular load for glucose can exceed the tubular maximum by a substantial amount, thus allowing glucose to appear in the urine. Urine volume is also greater than normal because the glucose molecules in the filtrate increase the osmolality of the filtrate in the nephron and reduce the effectiveness of water reabsorption by osmosis.

Predict 7

A person is suspected of having chronic renal failure. To assess kidney function, urea clearance is measured and found to be very low. Explain what a very low urea clearance indicates for this patient. Compare that with the effect of chronic renal failure on the tendency for the blood K^+ level to be higher than normal and the blood Na^+ level to be lower than normal.

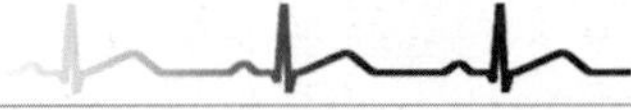

ASSESS YOUR PROGRESS

40. *What is plasma clearance and how is it calculated?*
41. *Explain why plasma clearance of inulin can be used to estimate GFR.*
42. *Describe how PAH is used to determine renal plasma flow.*
43. *Explain the significance of tubular load and tubular maximum.*

26.6 Urine Movement

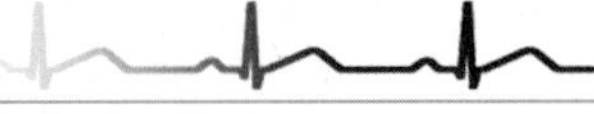

LEARNING OUTCOMES

After reading this section, you should be able to

A. **Describe the anatomy and histology of the ureters, urinary bladder, and urethra.**
B. **Explain the flow of urine from the nephron to the urinary bladder.**
C. **Discuss the micturition reflex.**

Anatomy and Histology of the Ureters and Urinary Bladder

The **ureters** are tubes through which urine flows from the kidneys to the urinary bladder. The ureters extend inferiorly and medially from the renal pelvis at the renal hilum of each kidney to the urinary bladder, which stores urine (figure 26.21; see figures 26.1 and 26.2). The **urinary bladder** is a hollow, muscular container that lies in the pelvic cavity just posterior to the symphysis pubis. The ureters enter on its posterolateral surface. In males, the urinary bladder is just anterior to the rectum; in females, it is just anterior to the vagina and inferior and anterior to the uterus. Its volume increases and decreases, depending on how much or how little urine is stored in it.

The **urethra,** which transports urine to the outside of the body, exits the urinary bladder inferiorly and anteriorly (figure 26.21, *center*). The triangular area of its wall between the two ureters posteriorly and the urethra anteriorly is called the **trigone** (trī′gōn). This region differs histologically from the rest of the urinary bladder wall and expands minimally during filling.

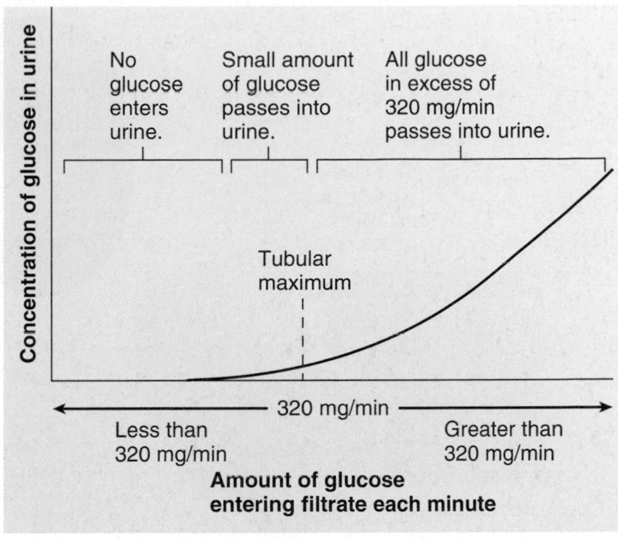
Concentration of glucose in urine
No glucose enters urine.
Small amount of glucose passes into urine.
All glucose in excess of 320 mg/min passes into urine.
Tubular maximum
320 mg/min
Less than 320 mg/min
Greater than 320 mg/min
Amount of glucose entering filtrate each minute

FIGURE 26.20 Tubular Maximum for Glucose
As the concentration of glucose increases in the filtrate, it reaches a point that exceeds the nephron's ability to actively reabsorb it. That concentration is called the tubular maximum. Beyond that concentration, the excess glucose enters the urine.

Transitional epithelium lines both the ureters and the urinary bladder. The rest of the walls of these structures consists of a lamina propria, a muscular coat, and a fibrous adventitia (figure 26.21*b,c*). The wall of the urinary bladder is much thicker than the wall of a ureter because it consists of layers of primarily smooth muscle, sometimes called the **detrusor** (dī-troo′ser) **muscle,** external to the epithelium. Contraction of this smooth muscle forces urine out of the urinary bladder. The epithelium itself ranges from four or five cells thick when the urinary bladder is empty to two or three cells thick when it is distended. Transitional epithelium is specialized, so that the cells slide past one another, and the number of cell layers decreases as the volume of the urinary bladder increases. The urethra is lined with stratified or pseudostratified columnar epithelium.

Where the urethra exits the urinary bladder, elastic connective tissue and smooth muscle keep urine from flowing out of the urinary bladder until the pressure in the urinary bladder is great enough to force urine to flow from it. In males, the elastic tissue and smooth muscle form an **internal urinary sphincter.** There is no functional internal urinary sphincter in females. In males, the internal urinary sphincter contracts to keep semen from entering the urinary bladder during sexual intercourse (see chapter 28). The **external urinary sphincter** is composed of skeletal muscle that surrounds the urethra as the urethra extends through the pelvic floor. This sphincter acts as a valve that controls the flow of urine through the urethra.

In males, the urethra extends to the end of the penis, where it opens to the outside (see chapter 28). The urethra is much shorter in females than in males, and it opens into the vestibule anterior to the vaginal opening.

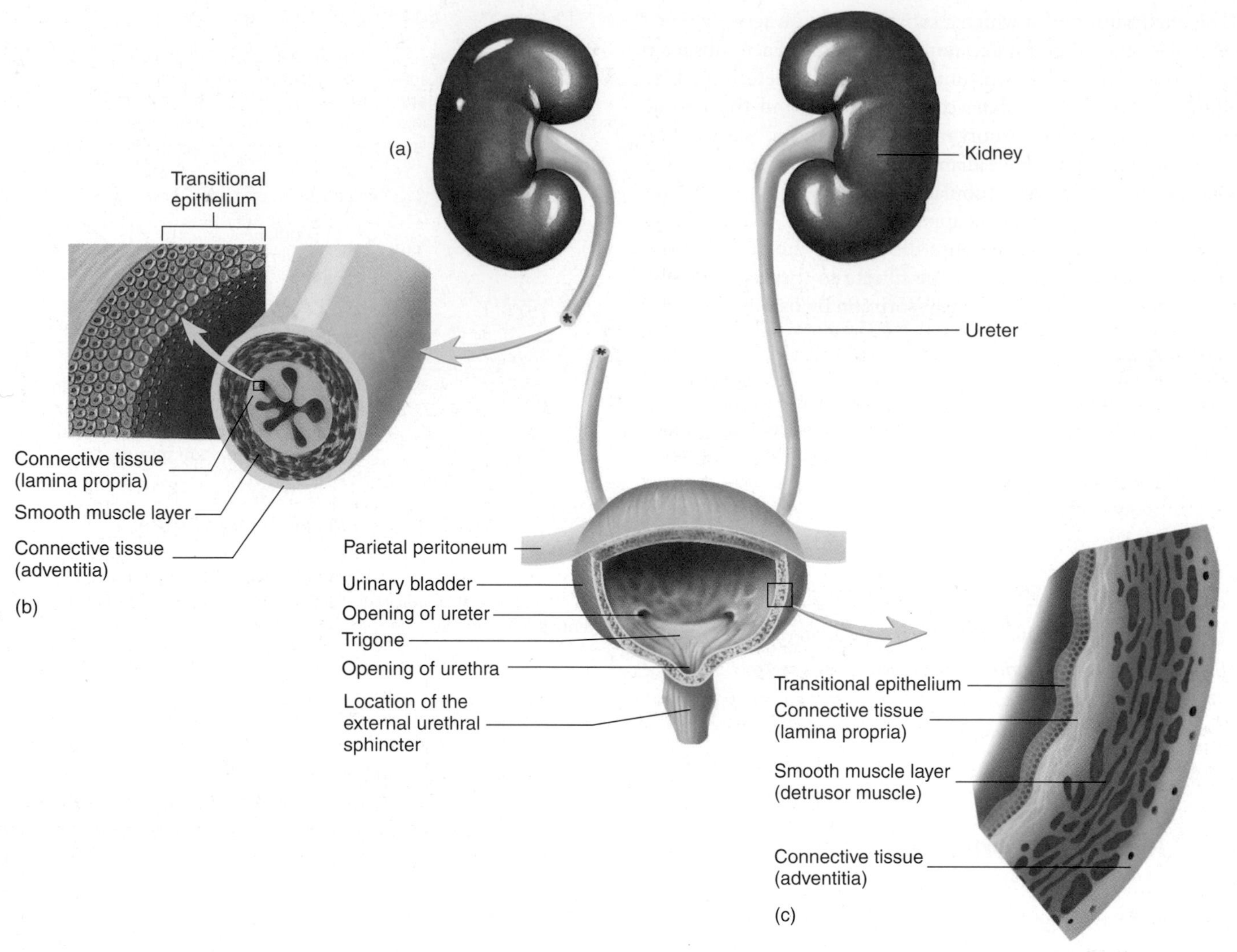
(a)
Kidney
Ureter
Transitional epithelium
Connective tissue (lamina propria)
Smooth muscle layer
Connective tissue (adventitia)
(b)
Parietal peritoneum
Urinary bladder
Opening of ureter
Trigone
Opening of urethra
Location of the external urethral sphincter
Transitional epithelium
Connective tissue (lamina propria)
Smooth muscle layer (detrusor muscle)
Connective tissue (adventitia)
(c)

FIGURE 26.21 Ureters and Urinary Bladder
(*a*) Ureters extend from the pelvis of the kidney to the urinary bladder. (*b*) The walls of the ureters and the urinary bladder are lined with transitional epithelium, which is surrounded by a connective tissue layer (lamina propria), smooth muscle layers, and a fibrous adventitia. (*c*) Section through the wall of the urinary bladder.

Predict 8

Cystitis (sis-ti′tis) is inflammation of the urinary bladder. It typically results from infection by bacteria from outside the body. Are males or females more prone to cystitis? Explain.

Urine Flow Through the Nephron and Ureters

Hydrostatic pressure averages 10 mm Hg in the Bowman capsule and nearly 0 mm Hg in the renal pelvis. This pressure gradient forces the filtrate to flow from the Bowman capsule through the nephron into the renal pelvis. Because the hydrostatic pressure is 0 mm Hg in the renal pelvis, no pressure gradient exists to force urine to flow through the ureters to the urinary bladder. However, the circular smooth muscle in the walls of the ureters undergoes peristaltic contractions, which force urine through the ureters. The peristaltic waves progress from the region of the renal pelvis to the urinary bladder. They occur from once every few seconds to once every 2–3 minutes. Parasympathetic stimulation increases their frequency, and sympathetic stimulation decreases it.

The peristaltic contractions of each ureter proceed at a velocity of approximately 3 cm/s and can generate pressures in excess of 50 mm Hg. Where the ureters penetrate the urinary bladder, they run obliquely through the trigone. Pressure inside the urinary bladder compresses that part of the ureter to prevent the backflow of urine.

When no urine is present in the urinary bladder, internal pressure is about 0 mm Hg; even when the urine volume is 100 mL, pressure rises to only 10 mm Hg. Pressure continues to rise slowly as volume increases to approximately 300 mL, but above volumes of 400 mL the pressure rises rapidly.

Micturition Reflex

The flow of urine from the kidney to the urinary bladder through the ureter is relatively continuous. The urinary bladder acts as a reservoir for urine until it can be eliminated relatively quickly at

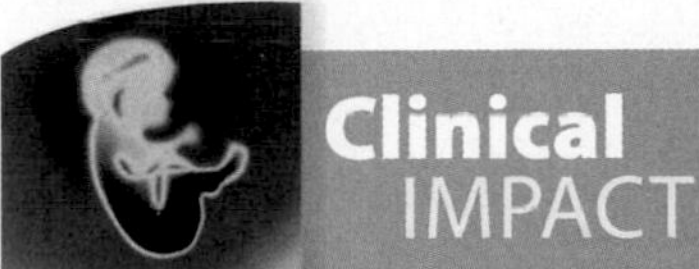
Clinical
IMPACT

Urinary Bladder Cancer

In the United States, urinary bladder cancer affects more than 60,000 new patients each year and is among the 10 most common cancers in men and women. However, with early detection (when the cancer is confined to the urinary bladder), the chance for survival is 94%, compared with 6% if the cancer has spread to other areas of the body. Unfortunately, early detection of urinary bladder cancer is especially challenging due to its rapid growth rate. Frequently, blood in the urine is a symptom but, because this symptom is also associated with other, less serious problems, it tends to be ignored.

Half the diagnosed cases of urinary bladder cancer can be attributed to cigarette smoking, even 10 years or more after a person has given up smoking. Other risk factors include being exposed to industrial aniline dyes and taking one of two specific prescription drugs, either phenacetin (an analgesic) or chlornaphazine (a drug used to treat polycythemia; see chapter 19).

Scientists are investigating ways to detect urinary bladder cancer early and noninvasively. Currently, urinary bladder cancer tests screen for abnormal cells, and the bladder is visually examined with a catheter in a process called **cystoscopy** (sís-tos′kŏ-pē). One test that holds promise is a urine test to measure the levels of an enzyme called telomerase, which is present in nearly all human cancer cells. Telomerase levels are measurable earlier in urinary bladder cancer than in many other cancers.

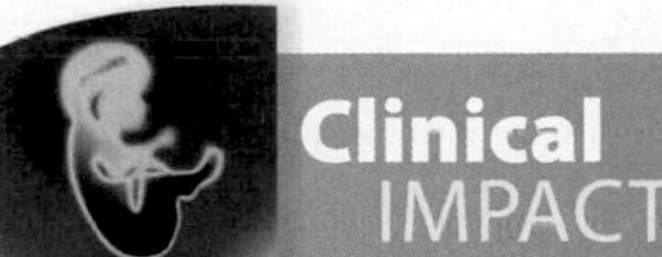
Clinical
IMPACT

Kidney Stones

Kidney stones are hard objects usually found in the pelvis of the kidney. They are typically 2–3 mm in diameter, with either a smooth or a jagged surface, but occasionally a large, branching kidney stone, called a **staghorn stone,** forms in the renal pelvis. About 1% of all autopsies reveal kidney stones, and many stones never cause symptoms. The symptoms associated with kidney stones occur when a stone passes into the ureter, resulting in referred pain down the back, side, and groin area. The ureter contracts around the stone, causing the stone to irritate the epithelium and produce bleeding, which appears as blood in the urine, a condition called **hematuria.** In addition to causing intense pain, kidney stones can block the ureter, cause ulceration in the ureter, and increase the probability of bacterial infection.

About 65% of all kidney stones are composed of calcium oxylate mixed with calcium phosphate, whereas another 15% are magnesium ammonium phosphate and 10% are uric acid or cystine; approximately 2.5% of each kidney stone is composed of mucoprotein.

The cause of kidney stones is usually obscure. Predisposing conditions include concentrated urine and an abnormally high calcium concentration in the urine, although the cause of the high calcium concentration is usually unknown. Magnesium ammonium phosphate stones are often found in people with recurrent kidney infections, and uric acid stones are common in people suffering from gout. Severe kidney stones must be surgically removed from the kidney. However, traditional surgical procedures have mainly been replaced by **lithotripsy** (lith′ō-trip-sē), in which kidney stones are pulverized using ultrasound or lasers.

an appropriate time and place. The urinary bladder can distend to accommodate a large volume of fluid; at its maximum volume, the urinary bladder can contain 1 L (about 1 quart) of urine, but discomfort becomes noticeable when urine volume exceeds approximately 500 mL. The urinary bladder's capacity to distend is due to three factors. First, the wall of the urinary bladder contains large folds, similar to those of the stomach, which unfold to enlarge the lumen. Second, the lining of the urinary bladder is transitional epithelium, which stretches. Third, the smooth muscle wall of the urinary bladder, with the exception of the trigone, also stretches to accommodate fluid. As urine enters, the urinary bladder lifts and expands superiorly to accommodate the fluid.

Elimination of urine from the urinary bladder is called **micturition** (mik-choo-rish′un). The **micturition reflex** is activated when the urinary bladder wall is stretched. Integration of the micturition reflex occurs in the sacral region of the spinal cord and is modified by centers in the pons and cerebrum.

Urine filling the urinary bladder stimulates stretch receptors, which produce action potentials. The action potentials are carried by sensory neurons to the sacral segments of the spinal cord through the pelvic nerves. In response, action potentials travel to the urinary bladder through parasympathetic fibers in the pelvic nerves (figure 26.22). The parasympathetic action potentials cause the smooth muscle of the urinary bladder (the detrusor muscle) to contract. In addition, decreased somatic motor action potentials cause the external urinary sphincter, which consists of skeletal muscle, to relax. Urine flows from the urinary bladder when the pressure there is great enough to force the urine through the urethra while the external urinary sphincter is relaxed. The micturition reflex normally produces a series of contractions of the urinary bladder.

Action potentials carried by sensory neurons from stretch receptors in the urinary bladder wall also ascend the spinal cord to a micturition center in the pons and to the cerebrum. The micturition reflex integrated in the spinal cord is automatic, but it is either stimulated or inhibited by descending action potentials sent to the sacral region of the spinal cord. For example, higher brain centers prevent micturition by sending action potentials from the cerebrum and pons through spinal pathways to inhibit the spinal micturition reflex. Consequently, parasympathetic stimulation of the urinary bladder is inhibited, and somatic motor neurons that keep the external urinary sphincter contracted are stimulated. The micturition reflex, integrated in the spinal cord, predominates in infants. The ability to inhibit micturition voluntarily develops at the age of 2–3 years; subsequently, the influence of the pons and cerebrum on the spinal micturition reflex predominates.

The slow increase in internal pressure helps explain why there is little urge to urinate when the urinary bladder contains less than 300 mL. As stated previously, though, the pressure in the urinary

1. Urine in the urinary bladder stretches the bladder wall.
2. Action potentials produced by stretch receptors are carried along pelvic nerves (*green line*) to the sacral region of the spinal cord.
3. Action potentials are carried by parasympathetic nerves (*red line*) to contract the smooth muscles of the urinary bladder.
4. Ascending pathways carry an increased frequency of action potentials up the spinal cord to the pons and cerebrum when the urinary bladder becomes stretched. This increases the conscious urge to urinate.
5. Descending pathways carry action potentials to the sacral region of the spinal cord to tonically inhibit the micturition reflex, preventing automatic urination when the bladder is full. Descending pathways facilitate the reflex when stretch of the urinary bladder produces the conscious urge to urinate. This reinforces the micturition reflex.
6. The brain voluntarily controls the external urethral sphincter through somatic motor nerves (*purple*), causing the sphincter to relax or constrict.

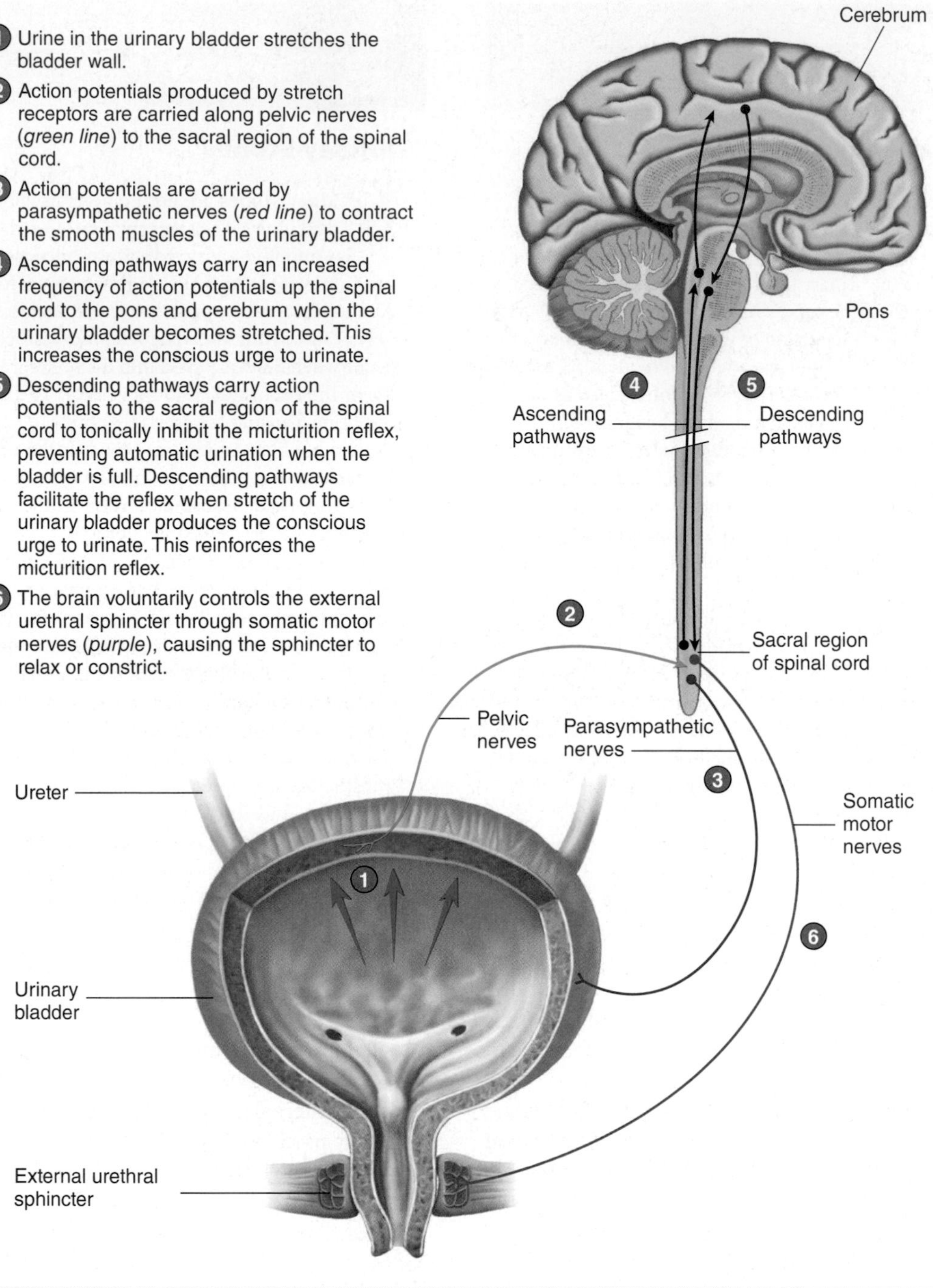

Cerebrum
Pons
4
5
Ascending pathways
Descending pathways
2
Sacral region of spinal cord
Pelvic nerves
Parasympathetic nerves
3
Somatic motor nerves
Ureter
1
6
Urinary bladder
External urethral sphincter

PROCESS FIGURE 26.22 AP|R **Micturition Reflex**

bladder increases rapidly once its volume exceeds approximately 400 mL. In addition, the frequency of action potentials conducted by the ascending spinal pathways to the pons and cerebrum also increases, resulting in a stronger urge to urinate.

Voluntary initiation of micturition requires an increase in action potentials sent from the cerebrum to facilitate the micturition reflex and to voluntarily relax the external urinary sphincter. In addition, voluntary contraction of the abdominal muscles increases abdominal pressure and thereby enhances the micturition reflex by increasing the pressure applied to the urinary bladder wall.

Normally, the urge to urinate results from stretch of the urinary bladder wall, but irritation of the urinary bladder or the urethra by a bacterial infection or some other condition can also initiate the urge to urinate, even if the urinary bladder is nearly empty.

ASSESS YOUR PROGRESS

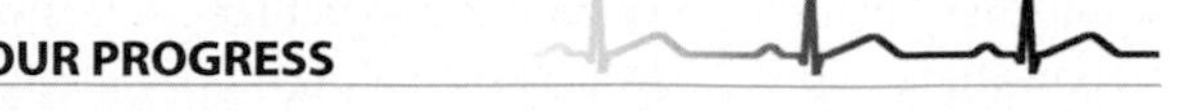

44. *What are the functions of the ureters, urinary bladder, and urethra? Describe their structure, including the epithelial lining of their inner surfaces.*

45. *What is the trigone?*
46. *What force moves urine through the nephron and ureters?*
47. *Explain the ability of the urinary bladder to distend.*
48. *Describe the micturition reflex. How is voluntary control of micturition accomplished?*

26.7 Effects of Aging on the Kidneys

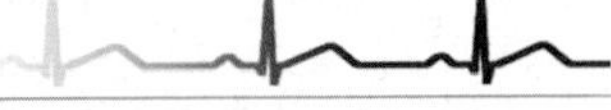

LEARNING OUTCOME

After reading this section, you will be able to

A. Describe the effects of aging on the kidneys.

Aging causes the kidneys to gradually decrease in size, beginning as early as age 20, becoming obvious by age 50, and continuing until death. The decrease in kidney size appears to be related to changes in the blood vessels of the kidney. The amount of blood flowing through the kidneys gradually decreases. Starting at age 20, there appears to be an approximately 10% decrease every 10 years. Small arteries, including the afferent and efferent arterioles, become irregular and twisted. Functional glomeruli are destroyed. By age 80, 40% of the glomeruli are not functioning. About 30% of the glomeruli that stop functioning no longer have a lumen through which blood flows. Other glomeruli thicken and assume a structure similar to that of arterioles. Some nephrons and collecting ducts become thicker, shorter, and more irregular in structure. The capacity to secrete and absorb declines, and whole nephrons stop functioning. The kidney's ability to concentrate urine gradually declines. Eventually, changes in the kidney increase the risk for dehydration because of the kidney's reduced ability to produce a concentrated urine. The ability to eliminate uric acid, urea, creatine, and toxins from the blood also decreases.

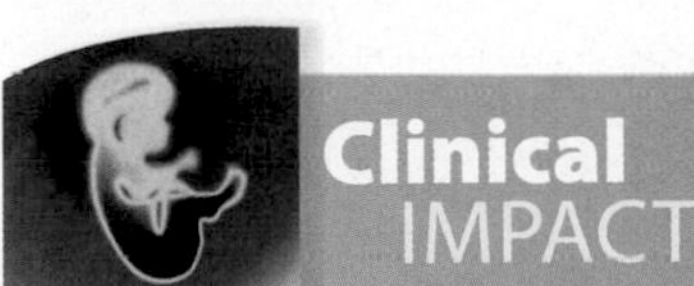

Clinical IMPACT

Automatic, Noncontracting, and Hyperexcitable Urinary Bladder

If the spinal cord is damaged above the sacral region, no micturition reflex exists for a time; however, if the urinary bladder is emptied frequently, the micturition reflex eventually regains the ability to cause the bladder to empty. Some time is generally required for the micturition reflex integrated within the spinal cord to begin to operate. Although a typical micturition reflex may exist, the person has no conscious control over its onset or duration. This condition is called **automatic bladder.**

Damage to the sacral region of the spinal cord or to the nerves that carry action potentials between the spinal cord and the urinary bladder can make the urinary bladder unable to contract even though the external urinary sphincter is relaxed. As a result, the micturition reflex cannot occur. The bladder fills to capacity, and urine is forced in a slow dribble through the external urinary sphincter. In elderly people and in patients with damage to the brainstem or spinal cord, inhibitory action potentials to the sacral region of the spinal cord can be lost. Without inhibition, the sacral centers are hyperexcitable, and even a small amount of urine in the bladder can elicit an uncontrollable micturition reflex.

An age-related loss of responsiveness to ADH and to aldosterone occurs. The kidney decreases renin secretion and has a reduced ability to participate in vitamin D synthesis, which contributes to Ca^{2+} deficiency, osteoporosis, and bone fractures.

Recall that one-third of one kidney is required to maintain homeostasis, and the additional kidney tissue beyond this constitutes

Diseases and Disorders

Urinary System

Kidney Disorders	Description
Inflammation of the Kidneys	
Glomerulonephritis (glō-mā r′ū-lō-nefrī′tis)	Inflammation of the filtration membrane within the renal corpuscle, causing an increase in the filtration membrane's permeability; plasma proteins and blood cells enter the filtrate, which increases urine volume due to increased osmotic concentration of the filtrate
Acute glomerulonephritis	Often occurs 1–3 weeks after a severe bacterial infection, such as "strep throat"; normally subsides after several days
Chronic glomerulonephritis	Long-term, progressive process whereby the filtration membrane thickens and is eventually replaced by connective tissue; the kidneys become nonfunctional
Pyelonephritis (pī′ē-lō-ne-frī′tis)	Often begins as a bacterial, usually *E. coli,* infection of the renal pelvis, which spreads to the rest of the kidney; the infection can destroy nephrons, corpuscles, and loops of Henle, dramatically reducing the kidney's ability to concentrate urine
Renal Failure	Can result from any condition that interferes with kidney function
Acute renal failure	Occurs when damage to the kidney is rapid and extensive; leads to accumulation of wastes in the blood; if renal failure is complete, death can occur in 1–2 weeks
Chronic renal failure	Caused by permanent damage to so many nephrons that the remaining nephrons are inadequate for normal kidney function; can result from chronic glomerulonephritis, trauma to the kidneys, tumors, or kidney stones

Go to: www.mhhe.com/seeley10 for additional information on these pathologies.

Systems PATHOLOGY | Acute Renal Failure

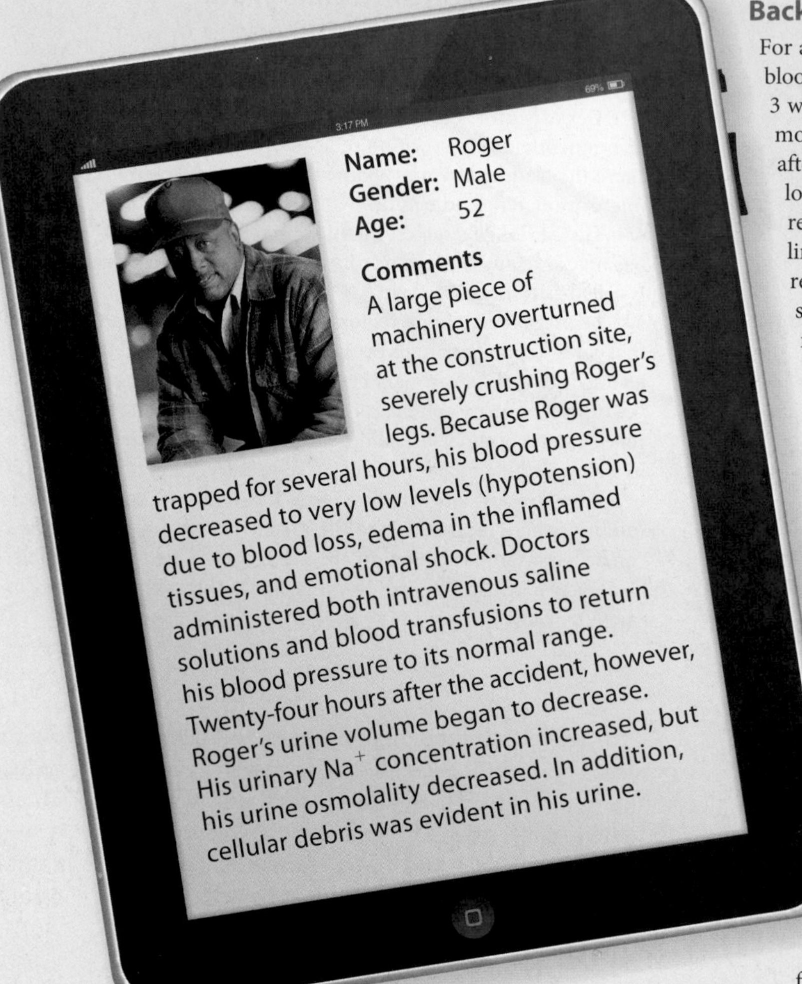

Name: Roger
Gender: Male
Age: 52
Comments
A large piece of machinery overturned at the construction site, severely crushing Roger's legs. Because Roger was trapped for several hours, his blood pressure decreased to very low levels (hypotension) due to blood loss, edema in the inflamed tissues, and emotional shock. Doctors administered both intravenous saline solutions and blood transfusions to return his blood pressure to its normal range. Twenty-four hours after the accident, however, Roger's urine volume began to decrease. His urinary Na+ concentration increased, but his urine osmolality decreased. In addition, cellular debris was evident in his urine.

Background Information

For approximately 7 days, Roger required renal dialysis to maintain his blood volume and ion concentrations within normal ranges. After about 3 weeks, his kidney function slowly began to improve, although many months passed before it was back to normal. In Roger's case, the events after 24 hours are consistent with acute renal failure caused by prolonged low blood pressure and lack of blood flow to the kidneys. The reduced blood flow was severe enough to cause damage to the epithelial lining of the kidney tubules. The period of reduced urine volume resulted from the tubular damage. Dead and damaged tubular cells sloughed off into the tubules and blocked them, so that filtrate could not flow through. In addition, the filtrate leaked from the blocked or partially blocked tubules back into the interstitial spaces and therefore back into the circulatory system. As a result, the amount of filtrate that became urine was markedly reduced.

Blood levels of urea and creatine usually increase due to reduced filtrate formation and reduced function of the tubular epithelium. A small amount of urine is produced that has a high Na^+ concentration, although the osmolality is usually close to the concentration of the body fluids. The kidney is not able to reabsorb Na^+, nor can it effectively concentrate urine.

Treatments for Renal Failure

Hemodialysis (hē′mō-dī-al′i-sis) is used when a person is suffering from severe acute or chronic kidney failure. The procedure substitutes for the excretory functions of the kidney. Hemodialysis is based on blood flow through tubes composed of a selectively permeable membrane. Blood is usually taken from an artery, passed through tubes of the dialysis machine, and then returned to a vein (figure 26A). On the outside of the dialysis tubes is a fluid, called dialysis fluid, which contains the same concentration of solutes as normal plasma, except for the metabolic waste products. As a consequence, the metabolic wastes diffuse from the blood to the dialysis fluid. The dialysis membrane

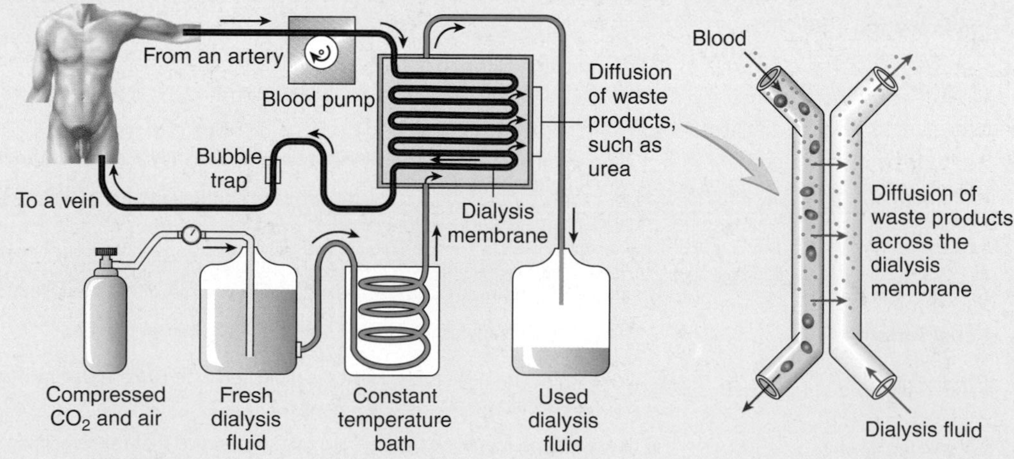

From an artery
Blood pump
Bubble trap
To a vein
Diffusion of waste products, such as urea
Dialysis membrane
Compressed CO_2 and air
Fresh dialysis fluid
Constant temperature bath
Used dialysis fluid
Blood
Diffusion of waste products across the dialysis membrane
Dialysis fluid

Figure 26A Hemodialysis

During hemodialysis, blood flows through a system of tubes composed of a selectively permeable membrane. Dialysis fluid, which has a composition similar to that of normal blood (except that the concentration of waste products is very low), flows in the opposite direction on the outside of the dialysis tubes. Waste products, such as urea, diffuse from the blood into the dialysis fluid. Other substances, such as Na^+, K^+, and glucose, can diffuse from the blood into the dialysis fluid if they are present in higher-than-normal concentrations, because these substances are present in the dialysis fluid at the same concentrations found in normal blood.

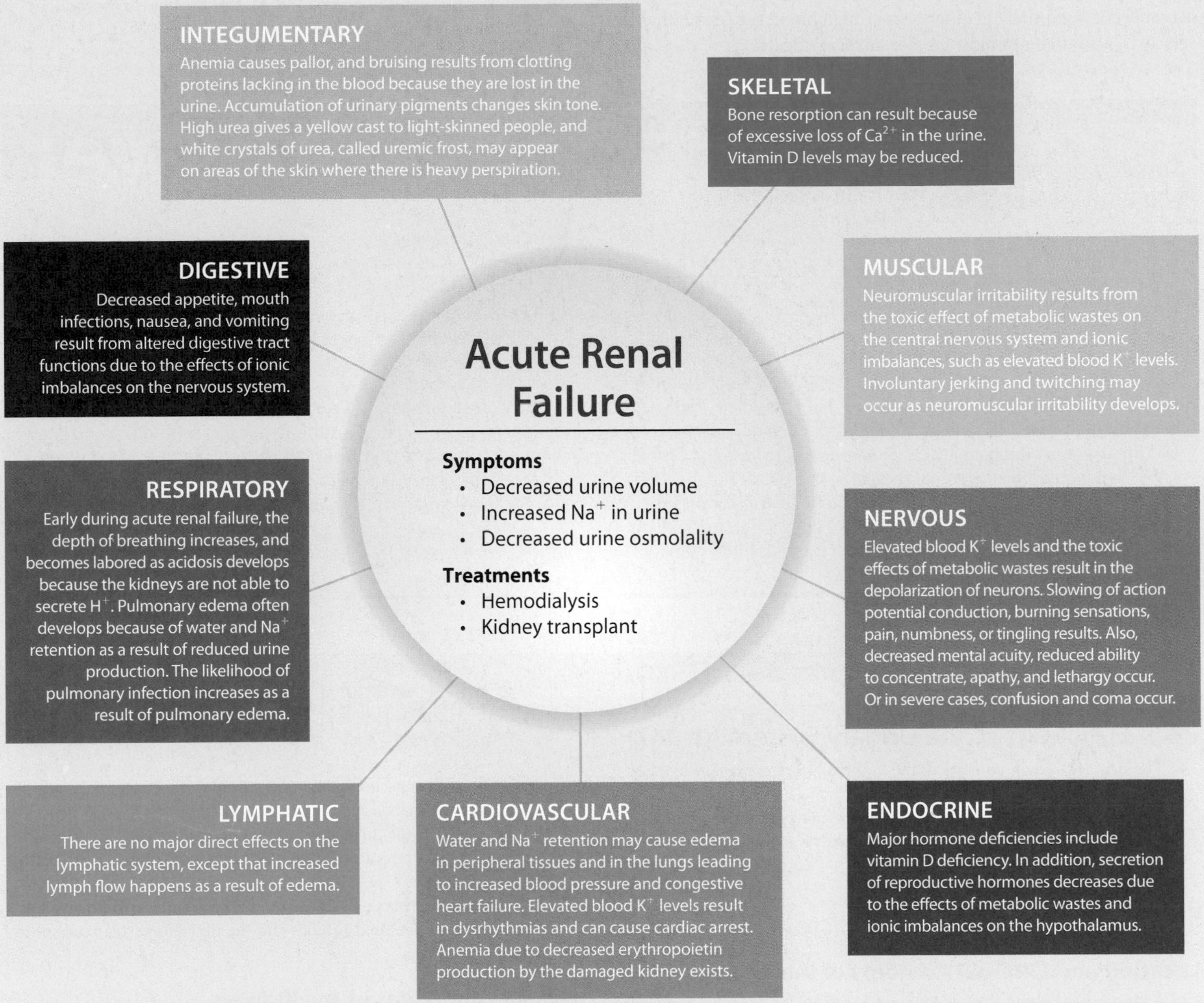

Acute Renal Failure

Symptoms
- Decreased urine volume
- Increased Na^+ in urine
- Decreased urine osmolality

Treatments
- Hemodialysis
- Kidney transplant

INTEGUMENTARY
Anemia causes pallor, and bruising results from clotting proteins lacking in the blood because they are lost in the urine. Accumulation of urinary pigments changes skin tone. High urea gives a yellow cast to light-skinned people, and white crystals of urea, called uremic frost, may appear on areas of the skin where there is heavy perspiration.

SKELETAL
Bone resorption can result because of excessive loss of Ca^{2+} in the urine. Vitamin D levels may be reduced.

MUSCULAR
Neuromuscular irritability results from the toxic effect of metabolic wastes on the central nervous system and ionic imbalances, such as elevated blood K^+ levels. Involuntary jerking and twitching may occur as neuromuscular irritability develops.

DIGESTIVE
Decreased appetite, mouth infections, nausea, and vomiting result from altered digestive tract functions due to the effects of ionic imbalances on the nervous system.

RESPIRATORY
Early during acute renal failure, the depth of breathing increases, and becomes labored as acidosis develops because the kidneys are not able to secrete H^+. Pulmonary edema often develops because of water and Na^+ retention as a result of reduced urine production. The likelihood of pulmonary infection increases as a result of pulmonary edema.

NERVOUS
Elevated blood K^+ levels and the toxic effects of metabolic wastes result in the depolarization of neurons. Slowing of action potential conduction, burning sensations, pain, numbness, or tingling results. Also, decreased mental acuity, reduced ability to concentrate, apathy, and lethargy occur. Or in severe cases, confusion and coma occur.

LYMPHATIC
There are no major direct effects on the lymphatic system, except that increased lymph flow happens as a result of edema.

CARDIOVASCULAR
Water and Na^+ retention may cause edema in peripheral tissues and in the lungs leading to increased blood pressure and congestive heart failure. Elevated blood K^+ levels result in dysrhythmias and can cause cardiac arrest. Anemia due to decreased erythropoietin production by the damaged kidney exists.

ENDOCRINE
Major hormone deficiencies include vitamin D deficiency. In addition, secretion of reproductive hormones decreases due to the effects of metabolic wastes and ionic imbalances on the hypothalamus.

has pores that are too small to allow plasma proteins to pass through them, and because the dialyis fluid contains the same beneficial solutes as the plasma, the net movement of these substances is zero. **Peritoneal** (per′i-tō-nē′ăl) **dialysis** is sometimes used to treat kidney failure. The principles by which peritoneal dialysis works are the same as for hemodialysis, but the dialysis fluid flows through a tube inserted into the peritoneal cavity. The visceral and parietal peritonea act as the dialysis membrane. Waste products diffuse from the blood vessels beneath the peritoneum, across the peritoneum, and into the dialysis fluid.

Kidney transplants are sometimes performed on people who have severe renal failure. Often, the donor has suffered an accidental death and had granted permission to have his or her kidneys used for transplantation. The major cause of kidney transplant failure is rejection by the recipient's immune system. Physicians therefore attempt to match the immune characteristics of the donor and recipient to reduce the tendency for rejection. Even with careful matching, recipients have to take medication for the rest of their lives to suppress their immune reactions. In most cases, the transplanted kidney functions well, and the tendency of the recipient's immune system to reject the transplanted kidney can be controlled.

Predict 9

Nine days after the accident, Roger began to appear pale, became dizzy on standing, and was very weak and lethargic. His hematocrit was elevated and his heart was arrhythmic. Explain these manifestations.

a reserve capacity. Therefore, the age-related changes in the kidney reduce the kidney's reserve capacity. As the functional kidney mass is reduced substantially in older people, high blood pressure, atherosclerosis, and diabetes have greater adverse effects.

ASSESS YOUR PROGRESS

49. *Discuss the effect of aging on the kidneys. Why do the kidneys gradually decrease in size?*

Answer

Learn to Predict

From page 946

We read in this chapter that chronic renal failure is caused by a decrease in the number of functional nephrons in the kidneys, which is common in type 2 diabetics, such as Sadie. Renal failure is most likely the result of damage to the glomerular basement membrane due to increased glomerular pressure and the production of advanced glycosylation end products, both of which are common side effects of type 2 diabetes.

Because Sadie's renal failure means her kidneys have a dramatically reduced filtration function, we would expect Sadie's blood tests to reveal high levels of glucose, K^+, and creatinine and low levels of Na^+. In addition, the creatinine clearance rate would be below normal, and there would be substantial protein in the urine. The increased blood glucose level results from a decrease in the tubular maximum for glucose reabsorption due to fewer functional nephrons. Similarly, the tubular maximum for Na^+ is decreased, and the kidneys' ability to secrete K^+ also decreases. Consequently, blood Na^+ levels decrease and blood K^+ levels increase. The low creatinine clearance rate is consistent with a decreased number of functional nephrons, and protein in the urine reflects the increased permeability of the filtration membrane in the remaining nephrons. The puffiness of Sadie's face indicates that fluid is being retained, which is consistent with the increase in blood pressure.

The consequences for Sadie are severe. She needs to take precautions to make it easier for the remaining nephrons to maintain homeostasis—for example, by carefully regulating her blood glucose and controlling the hypertension. If her condition continues to worsen, she may have to resort to dialysis and consider a kidney transplant.

Answers to the rest of this chapter's Predict questions are in Appendix G.

Summary

26.1 Functions of the Urinary System (p. 947)

1. The urinary system consists of the kidneys, ureters, urinary bladder, and urethra.
2. The urinary system eliminates wastes; regulates blood volume, ion concentration, and pH; and is involved with red blood cell and vitamin D production.

26.2 Kidney Anatomy and Histology (p. 947)

Location and External Anatomy of the Kidneys

1. A kidney lies behind the peritoneum on the posterior abdominal wall on each side of the vertebral column.
2. The renal capsule surrounds each kidney, and adipose tissue and the renal fascia engulf each kidney and anchor it to the abdominal wall.
3. Blood vessels and nerves enter and exit the kidney at the hilum, on the medial side of each kidney, which opens into the renal sinus, containing fat and connective tissue.

Internal Anatomy and Histology of the Kidneys

1. The two major regions of the kidney are the cortex and the medulla.
   - The renal columns extend toward the medulla between the renal pyramids.
   - The renal pyramids of the medulla project to the minor calyces.
2. The minor calyces open into the major calyces, which open into the renal pelvis. The renal pelvis leads to the ureter.
3. The functional unit of the kidney is the nephron. The parts of a nephron are the renal corpuscle, the proximal convoluted tubule, the loop of Henle, and the distal convoluted tubule.
   - The renal corpuscle consists of the Bowman capsule and the glomerulus. Materials leave the blood in the glomerulus and enter the Bowman capsule through the filtration membrane.
   - The nephron empties through the distal convoluted tubule into a collecting duct.
4. The juxtaglomerular apparatus consists of the macula densa (part of the distal convoluted tubule) and the juxtaglomerular cells of the afferent arteriole.

Arteries and Veins of the Kidneys

1. Arteries branch as follows: renal artery to segmental artery to interlobar artery to arcuate artery to interlobular artery to afferent arteriole.
2. Afferent arterioles supply the glomeruli.
3. Efferent arteries from the glomeruli supply the peritubular capillaries and vasa recta.
4. Veins form from the peritubular capillaries as follows: interlobular vein to arcuate vein to interlobar vein to renal vein.

26.3 Urine Production (p. 955)

Urine is produced by filtration, tubular reabsorption, and tubular secretion.

Filtration

1. The renal filtrate is plasma minus blood cells and blood proteins. Most (99%) of the filtrate is reabsorbed.
2. The filtration membrane is composed of a fenestrated endothelium, a basement membrane, and the slitlike pores formed by podocytes.

3. Filtration pressure is responsible for filtrate formation.
   - Filtration pressure is glomerular capillary pressure minus capsular hydrostatic pressure minus blood colloid osmotic pressure.
   - Filtration pressure changes are primarily caused by changes in glomerular capillary pressure.

Regulation of Glomerular Filtration Rate

1. Two important mechanisms regulating GFR are autoregulation and sympathetic stimulation.
2. Autoregulation dampens systemic blood pressure changes by altering afferent arteriole diameter.
3. Sympathetic stimulation decreases afferent arteriole diameter.

Tubular Reabsorption

1. Filtrate is reabsorbed by passive transport, including simple diffusion and facilitated diffusion. Filtrate is also reabsorbed through active transport and symport. Materials move from the nephron into the peritubular capillaries.
2. Specialization of tubule segments
   - The thin segment of the loop of Henle is specialized for passive transport.
   - The rest of the nephron and collecting ducts perform active transport, symport, and passive transport.
3. Substances transported
   - Active transport moves mainly Na^+ across the wall of the nephron. Other ions and molecules are moved primarily by symport.
   - Passive transport moves water, urea, and lipid-soluble, nonpolar compounds.

Tubular Secretion

1. Substances enter the proximal or distal convoluted tubules and the collecting ducts.
2. Hydrogen ions, K^+, and some substances not produced in the body are secreted by antiport mechanisms.

Urine Concentration Mechanism

1. The vasa recta, the loop of Henle, and the distribution of urea are responsible for the concentration gradient in the medulla. The concentration gradient is necessary for the production of concentrated urine.
2. Production of urine
   - In the proximal convoluted tubule, Na^+ and other substances are removed by active transport. Water follows passively, filtrate volume is reduced 65%, and the filtrate concentration is 300 mOsm/L.
   - In the descending limb of the loop of Henle, water exits passively and solute enters. The filtrate volume is reduced 15%, and the osmolality of the filtrate concentration is 1200 mOsm/kg.
   - In the ascending limb of the loop of Henle, Na^+, Cl^-, and K^+ are actively transported out of the filtrate, but water remains because this segment of the nephron is impermeable to water. The osmolality of the filtrate concentration is 100 mOsm/kg.

26.4 Regulation of Urine Concentration and Volume (p. 968)

Hormonal Mechanisms

1. Aldosterone, produced in the adrenal cortex, affects Na^+ and Cl^- transport in the nephron and collecting ducts.
   - A decrease in aldosterone results in less Na^+ reabsorption and an increase in urine concentration and volume. An increase in aldosterone results in greater Na^+ reabsorption and a decrease in urine concentration and volume.
   - Aldosterone production is stimulated by angiotensin II, increased blood K^+ concentration, and decreased blood Na^+ concentration.
2. Renin, produced by the kidneys, causes the production of angiotensin II.
   - Angiotensin II acts as a vasoconstrictor and stimulates aldosterone secretion, causing a decrease in urine production and an increase in blood volume.
   - Decreased blood pressure or decreased Na^+ concentration stimulates renin production.
3. ADH, secreted by the posterior pituitary, increases water permeability in the distal convoluted tubules and collecting ducts.
   - ADH decreases urine volume, increases blood volume, and thus increases blood pressure.
   - ADH release is stimulated by increased blood osmolality or decreased blood pressure.
   - Water movement out of the distal convoluted tubules and collecting ducts is regulated by ADH. If ADH is absent, water is not reabsorbed, and a dilute urine is produced. If ADH is present, water moves out, and a concentrated urine is produced.
4. Atrial natriuretic hormone, produced by the heart when blood pressure increases, inhibits ADH production and reduces the kidney's ability to concentrate urine.

26.5 Plasma Clearance and Tubular Maximum (p. 976)

1. Plasma clearance is the volume of plasma that is cleared of a specific substance each minute.
2. Tubular load is the total amount of a substance that enters the nephron each minute.
3. Tubular maximum is the fastest rate at which a substance is reabsorbed from the nephron.

26.6 Urine Movement (p. 977)

Anatomy and Histology of the Ureters and Urinary Bladder

1. Structure
   - The walls of the ureter and urinary bladder consist of the epithelium, the lamina propria, a muscular coat, and a fibrous adventitia.
   - The transitional epithelium permits changes in size.
2. Function
   - The ureters transport urine from the kidney to the urinary bladder.
   - The urinary bladder stores urine.

Urine Flow Through the Nephron and Ureters

1. Hydrostatic pressure forces urine through the nephron.
2. Peristalsis moves urine through the ureters.

Micturition Reflex

1. Stretch of the urinary bladder stimulates a reflex that causes the urinary bladder to contract and inhibits the urinary sphincters.
2. Higher brain centers can stimulate or inhibit the micturition reflex.

26.7 Effects of Aging on the Kidneys (p. 981)

1. The kidneys gradually decrease in size due to a decrease in renal blood flow.
2. The number of functional nephrons decreases.
3. Renin secretion and vitamin D synthesis decrease.
4. The nephron's ability to secrete and absorb declines.

REVIEW AND COMPREHENSION

1. Which of these is *not* a general function of the kidneys?
   a. regulation of blood volume
   b. regulation of solute concentration in the blood
   c. regulation of the pH of the extracellular fluid
   d. regulation of vitamin A synthesis
   e. regulation of red blood cell synthesis
2. The cortex of the kidney contains the
   a. hilum. c. adipose tissue. e. renal pelvis.
   b. glomeruli. d. renal pyramids.
3. Given these structures:
   (1) major calyx (3) renal papilla
   (2) minor calyx (4) renal pelvis

   Choose the arrangement that lists the structures in order as urine leaves the collecting duct and travels to the ureter.
   a. 1,4,2,3 b. 2,3,1,4 c. 3,2,1,4 d. 4,1,3,2 e. 4,3,2,1
4. Which of these structures contain(s) blood?
   a. glomerulus d. Bowman capsule
   b. vasa recta e. Both a and b are correct.
   c. distal convoluted tubule
5. The juxtaglomerular cells of the ________ and the macula densa cells of the ________ form the juxtaglomerular apparatus.
   a. afferent arteriole, proximal convoluted tubule
   b. afferent arteriole, distal convoluted tubule
   c. efferent arteriole, proximal convoluted tubule
   d. efferent arteriole, distal convoluted tubule
6. Given these blood vessels:
   (1) afferent arteriole (3) glomerulus
   (2) efferent arteriole (4) peritubular capillaries

   Choose the correct order as blood passes from an interlobular artery to an interlobular vein.
   a. 1,2,3,4 b. 1,3,2,4 c. 2,1,4,3 d. 3,2,4,1 e. 4,3,1,2
7. Which of these processes is responsible for kidney function?
   a. filtration c. reabsorption e. All of these are correct.
   b. secretion d. Both a and b are correct.
8. The amount of plasma that enters the Bowman capsule per minute is the
   a. GFR. c. renal fraction.
   b. renal plasma flow. d. renal blood flow.
9. If the glomerular capillary pressure is 40 mm Hg, the capsular hydrostatic pressure is 10 mm Hg, and the blood colloid osmotic pressure within the glomerulus is 30 mm Hg, the filtration pressure is
   a. −20 mm Hg. c. 20 mm Hg. e. 80 mm Hg.
   b. 0 mm Hg. d. 60 mm Hg.
10. Which of these conditions reduces filtration pressure in the glomerulus?
    a. elevated blood pressure
    b. constriction of the afferent arterioles
    c. decreased plasma protein in the glomerulus
    d. dilation of the afferent arterioles
    e. decreased capsular hydrostatic pressure
11. If blood pressure increases by 50 mm Hg,
    a. the afferent arterioles constrict.
    b. glomerular capillary pressure increases by 50 mm Hg.
    c. GFR increases dramatically.
    d. efferent arterioles constrict.
    e. All of these are correct.
12. Glucose is usually completely reabsorbed from the filtrate by the time the filtrate has reached the
    a. end of the proximal convoluted tubule.
    b. tip of the loop of Henle.
    c. end of the distal convoluted tubule.
    d. end of the collecting duct.
    e. Bowman capsule.
13. The greatest volume of water is reabsorbed from the nephron by the
    a. proximal convoluted tubule. c. distal convoluted tubule.
    b. loop of Henle. d. collecting duct.
14. Water leaves the nephron by
    a. active transport. d. facilitated diffusion.
    b. filtration into the capillary network. e. symport.
    c. osmosis.
15. Potassium ions enter the ________ by ________.
    a. proximal convoluted tubule, diffusion
    b. proximal convoluted tubule, active transport
    c. distal convoluted tubule, diffusion
    d. distal convoluted tubule, antiport
16. Reabsorption of most solute molecules from the proximal convoluted tubule is linked to the active transport of Na^+ across the
    a. apical membrane and out of the cell.
    b. apical membrane and into the cell.
    c. basal membrane and out of the cell.
    d. basal membrane and into the cell.
17. Which of these ions is used to symport amino acids, glucose, and other solutes through the apical membrane of nephron epithelial cells?
    a. K^+ b. Na^+ c. C^- d. Ca^{2+} e. Mg^{2+}
18. Which of the following contributes to the formation of a hyperosmotic environment in the medulla of the kidney?
    a. the effects of ADH on water permeability of the ascending limb of the loop of Henle
    b. the impermeability of the ascending limb of the loop of Henle to water
    c. the symport of Na^+, K^+, and Cl^- out of the ascending limb of the loop of Henle
    d. Both a and c are correct.
    e. Both b and c are correct.
19. At which of these sites is the osmolality of the filtrate at its lowest (lowest concentration)?
    a. glomerular capillary d. initial section of the distal convoluted tubule
    b. proximal convoluted tubule e. collecting duct
    c. tip of the loop of Henle
20. Increased aldosterone causes
    a. increased reabsorption of Na^+.
    b. decreased blood volume.
    c. decreased reabsorption of Cl^-.
    d. increased permeability of the distal convoluted tubule to water.
    e. increased volume of urine.
21. Juxtaglomerular cells are involved in the secretion of
    a. ADH. c. aldosterone.
    b. angiotensin. d. renin.
22. Angiotensin II
    a. causes vasoconstriction. d. increases the sensation of thirst.
    b. stimulates aldosterone secretion. e. All of these are correct.
    c. stimulates ADH secretion.

23. ADH governs the
    a. Na^+ pump of the proximal convoluted tubules.
    b. water permeability of the loop of Henle.
    c. Na^+ pump of the vasa recta.
    d. water permeability of the distal convoluted tubules and collecting ducts.
    e. Na^+ reabsorption in the proximal convoluted tubule.
24. A decrease in blood osmolality results in
    a. increased ADH secretion.
    b. increased permeability of the collecting ducts to water.
    c. decreased urine osmolality.
    d. decreased urine output.
    e. All of the these are correct.
25. The amount of a substance that passes through the filtration membrane into the nephrons per minutc is thc
    a. renal plasma flow.
    b. tubular load.
    c. plasma clearance.
    d. tubular maximum.
26. The urinary bladder
    a. is composed of skeletal muscle.
    b. is lined by simple columnar epithelium.
    c. is connected to the outside of the body by the ureter.
    d. is located in the pelvic cavity.
    e. has two urethras and one ureter attached to it.

Answers in Appendix E

CRITICAL THINKING

1. To relax after an anatomy and physiology examination, Mucho Gusto gocs to a local bistro and drinks 2 quarts of low-sodium beer. What effect does this beer have on urine concentration and volume? Explain the mechanisms involved.
2. Harry is doing yard work one hot summer day and refuses to drink anything until he is finished. He then drinks glass after glass of plain water. Assuming that he drinks enough water to rcplace all the water he lost as sweat, how does this much water affect urine concentration and volume? Explain the mechanisms involved.
3. A patient has the following symptoms: slight increase in extracellular fluid volume, large decrease in plasma sodium concentration, very concentrated urine, and cardiac fibrillation. An imbalance of what hormone is responsible for these symptoms? Are the symptoms caused by oversecretion or undersecretion of the hormone?
4. Propose several ways to decrease the GFR.
5. Design a kidney that can produce hyposmotic urine, which is less concentrated than plasma, or hyperosmotic urine, which is more concentrated than plasma, by the active transport of water instead of Na^+. Assume that the kidney's anatomical structure is the same as that in humans, but feel free to change anything else you choose.
6. If only a very small amount of urea, instead of its normal concentration, were present in the interstitial fluid of the kidney, how would the kidney's ability to concentrate urine be affected?
7. Some patients with hypertension are kept on a low-salt (low-sodium) diet. Propose an explanation for this therapy.
8. Research has shown that mammals with kidneys having relatively thicker medullas can produce more concentrated urine than humans. Explain why this is so.
9. Marvin Motormount was driving too fast on a remote mountain road at 3 a.m. when his car left the road and rolled down a steep hill. Marvin sustained numerous cuts and bruises. When medical help arrived 2 hours later, his systolic blood pressure was 70 mm Hg, and his pulse was weak (thready). Intravenous saline was administered immediately, and plasma and then whole blood were administered in the emergency room. After another hour, Marvin's blood pressure had returned to normal and he no longer appeared pale. While he was in the hospital, Marvin's urine volume decreased to less than 30 mL/h (<400 mL/day). A blood sample indicated elevated blood levels of urea, creatinine, and uric acid. He also exhibited hyperkalemia and some cardiac arrhythmia, and his arterial pH was <7.35 (below normal). Over the next few days, his red blood cell count decreased and he bruised easily. His jugular veins were distended, and there was some peripheral and pulmonary edema. From the following list, select the conditions that applied to Marvin at this time.
    1. hypoxic injury to the kidney
    2. increased reabsorption of wastes
    3. decreased H^+ secretion
    4. decreased K^+ secretion
    5. increased HCO_3^- reabsorption
    6. decreased erythropoietin secretion

    a. 1,2,3,4,5,6
    b. 2,3,4,5
    c. 1,2,3,4
    d. 1,2,3,4,6
    e. 1,2,6
10. Which of the following will help compensate for the low pH of the patient in question 9?
    a. increased respiration
    b. increased HCO_3^- reabsorption
    c. increased H^+ secretion
    d. All of these are correct.
    e. Both a and b are correct.
11. Renin-secreting tumors are usually found in the kidneys but rarely in other organs, such as the liver, lungs, pancreas, and ovaries. Predict the effects of renin-secreting tumors on blood K^+ levels, and explain the effects on action potential conduction in nerves and muscle tissues.
12. Even though mutations of aquaporin-3 and aquaporin-4 in the collecting duct have not been described in the literature, if mutations occurred that resulted in a reduced number of these aquaporins in the cells of the collecting ducts, how would urine volume and concentration be affected? Would ADH be an effective treatment?

Answers in Appendix F

Visit this book's website at **www.mhhe.com/seeley10** for chapter quizzes, interactive learning exercises, and other study tools.

Mc Graw Hill
connect plus+
ANATOMY & PHYSIOLOGY

Anatomy & Physiology
REVEALED
aprevealed.com

27

Learn to Predict

Satish and Kiran, two college students in search of adventure, embarked on an orienteering exercise—but it almost turned out to be their last adventure. Their compass malfunctioned, and they became lost in desert terrain during the hottest days of summer. One week later, they were discovered, near death, in a dry ravine. Rescuers estimated that the students had gone without water for 24 hours. After reading this chapter and recalling your study of the kidneys in chapter 26, explain the students' dehydration, the homeostatic mechanisms that would have tried to compensate for this condition, and the specific physiological cause of their illness.

Water, Electrolyte, and Acid-Base Balance

Maintaining appropriate levels of water in the body is crucial to achieving homeostasis. We are highlighting two extremes of water balance in our bodies: too little or too much. Here, we describe what happens when the body receives too much water; the "Learn to Predict" question describes what happens when the body receives too little water.

In 2007, a mother of three died from drinking too much water during a radio contest. The rules required the contestants to drink one 8 oz bottle of water every 15 minutes for almost 2 hours without urinating. The successful contestant would win a video-gaming system. By the time the contest was over, the woman had drunk almost 2 gallons of water and was complaining of an extreme headache. One hour later, her mother found her dead. Water intoxication, also known as hyperhydration or water poisoning, results from overconsumption of water that causes the levels of electrolytes in the blood to become overly diluted. Water can move into tissue cells via osmosis, causing malfunctions that lead to seizures, coma, and potentially death. In this chapter, you will learn how, under normal circumstances, the balance of electrolytes and body fluids is maintained within a very narrow range.

Photo: Water is not readily available in the desert. Hikers need to prepare for hot, dry conditions and pack plenty of drinking water.

Module 13
Urinary System

Anatomy & Physiology
REVEALED
aprevealed.com

27.1 Body Fluids

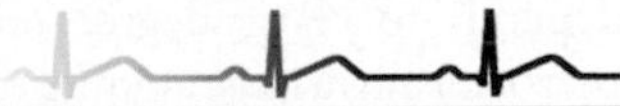

LEARNING OUTCOMES

After reading this section, you should be able to

A. Identify the major fluid compartments of the body and their subdivisions.

B. List the dominant cations and anions in the major fluid compartments.

C. Describe the causes of edema.

The adult human body is at least 50% water by weight (table 27.1). In infants, water makes up as much as 75% of body weight; the content is significantly less in children and slowly declines further as they get older. Because the water content of adipose tissue is relatively low, people with more adipose tissue have a smaller proportion of water in their bodies. For example, the relatively lower water content of adult females compared with that of adult males reflects the greater development of subcutaneous adipose tissue characteristic of women.

The body has two major fluid compartments, or areas: the intracellular (in-tră-sel′ū-lăr) fluid compartment and the extracellular (eks-tră-sel′ū-lăr) fluid compartment. The **intracellular fluid compartment** includes all the fluid in the trillions of body cells. The intracellular fluid from all cells has a similar composition, and it accounts for approximately 40% of total body weight.

The **extracellular fluid compartment** includes all the fluid outside the cells, constituting nearly 20% of total body weight. The extracellular fluid compartment can be divided into several subcompartments. The two major ones are interstitial fluid and plasma. **Interstitial** (in-ter-stish′ăl) **fluid** occupies the extracellular spaces outside the blood vessels, and **plasma** (plaz′mă) occupies the extracellular spaces within blood vessels. The other subcompartments of the extracellular compartment include lymph, cerebrospinal fluid, and synovial fluid. These constitute relatively small volumes.

Although the fluid composition in each subcompartment is different from that of the others, continuous and extensive water and ion exchange occurs between the subcompartments because the osmotic pressures of most fluid subcompartments are approximately equal. (Recall from chapter 2 that ions are charged particles, which are either positively charged [cations] or negatively charged [anions]. Also recall that the term **electrolyte** is used synonymously with *ion*.) On the other hand, large molecules, such as proteins, are much more restricted in their movement because they cannot penetrate the membranes that separate the fluid subcompartments (table 27.2). However, if the composition of one subcompartment is greatly altered, fluid exchange may not be equal. For example, edema occurs when fluid shifts from the plasma to the interstitial fluid.

TABLE 27.1 Approximate Volumes of Body Fluid Compartments*

			Extracellular Fluid		
	Total Body Water	Intracellular Fluid	Plasma	Interstitial Fluid	Total
Infants	75	45	4	26	30
Adult Males	60	40	5	15	20
Adult Females	50	35	5	10	15

*Expressed as percentage of body weight.

TABLE 27.2 Approximate Concentration of Major Solutes in Body Fluids*

Solute	Plasma	Interstitial Fluid	Intracellular Fluid†
Cations			
Sodium (Na^+)	153.2	145.1	12.0
Potassium (K^+)	4.3	4.1	150.0
Calcium (Ca^{2+})	3.8	3.4	4.0
Magnesium (Mg^{2+})	1.4	1.3	34.0
TOTAL	162.7	153.9	200.0
Anions			
Chloride (Cl^-)	111.5	118.0	4.0
Bicarbonate (HCO_3^-)	25.7	27.0	12.0
Phosphate (HPO_4^{2-} plus HPO_4^-)	2.2	2.3	40.0
Protein	17.0	0.0	54.0
Other	6.3	6.6	90.0
TOTAL	162.7	153.9	200.0

*Expressed as milliequivalents per liter (mEq/L).

†Data are from skeletal muscle.

Edema commonly results from an increase in the permeability of the capillary walls due to inflammation, allowing proteins to diffuse from the plasma into the interstitial fluid. Water moves in the same direction by osmosis. Edema can also result from a change in the hydrostatic pressure across capillary walls. Increased hydrostatic pressure in capillaries due to blocked veins or heart failure forces fluid from plasma into the interstitial spaces (see chapter 21).

ASSESS YOUR PROGRESS

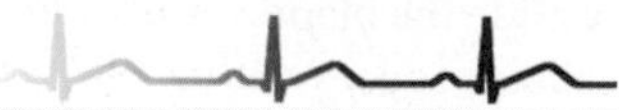

1. *What are the two major fluid compartments of the body? Name the subdivisions of the extracellular fluid compartment.*
2. *What cations and anions occur with the higher percentage in each major compartment?*
3. *Compare the osmotic concentration among most fluid compartments.*
4. *What factors contribute to edema?*

27.2 Regulation of Body Fluid Concentration and Volume

LEARNING OUTCOMES

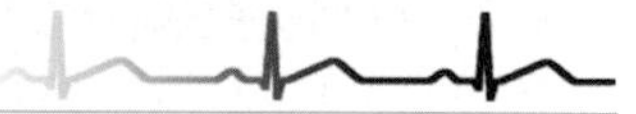

After reading this section, you should be able to

A. Explain how body water content is maintained.

B. Discuss how body fluid osmolality is achieved and held in homeostatic balance.

C. Explain the mechanisms that regulate extracellular fluid volume.

Regulation of Water Content

The body's water content is regulated so that its total volume remains constant. Thus, the volume of water taken into the body is equal to the volume lost each day. Changes in the water volume alter the osmolality of body fluids, blood pressure, and interstitial fluid pressure. The total volume of water entering the body each day is 1500–3000 mL. Most of that volume (90%) comes from ingested fluids, some comes from food, and a smaller amount, approximately 10%, is derived from the water produced during cellular metabolism (table 27.3; see figure 24.34).

TABLE 27.3 Summary of Water Intake and Loss

Sources of Water	Routes by Which Water Is Lost
Ingestion (90%)	Urine (61%)
Cellular metabolism (10%)	Evaporation (35%)
	Perspiration
	Insensible
	Sensible
	Respiratory passages
	Feces (4%)

The movement of water across the wall of the digestive tract depends on osmosis, and the volume of water entering the body depends, to a large degree, on the volume of water consumed. If a person drinks a large amount of dilute liquid—iced tea, for instance—the rate at which water enters the body fluids increases. If a person drinks a small volume of concentrated liquid, the rate decreases.

Regulation of Thirst

Although fluid consumption is heavily influenced by habit and by social settings, water ingestion does depend, at least in part, on thirst regulatory mechanisms. There are three principal sensors for thirst regulation: hypothalamic receptors, arterial baroreceptors, and juxtaglomerular apparatuses. The sensation of thirst results primarily from an increase in the osmolality of the extracellular fluids and from a reduction in plasma volume, which lowers blood pressure. Cells of the supraoptic nucleus within the hypothalamus can detect an increased extracellular fluid osmolality and initiate activity in neural circuits, resulting in a conscious sensation of thirst (figure 27.1).

Baroreceptors can also influence the sensation of thirst. When they detect a substantial decrease in blood pressure, action potentials are conducted to the brain along sensory neurons to influence the sensation of thirst. Low blood pressure associated with hemorrhagic shock, for example, is correlated with an intense sensation of thirst.

The role of the juxtaglomerular apparatus in thirst regulation centers around its ability to detect blood pressure. Renin is released from the juxtaglomerular apparatuses of the kidneys in response to reduced blood pressure in the kidney. Renin increases the formation of angiotensin II in the circulatory system (see chapters 21 and 26). Angiotensin II opposes a decrease in blood pressure by acting on the brain to stimulate the sensation of thirst, by acting on the adrenal cortex to increase aldosterone secretion, and by acting on blood vessel smooth muscle cells to increase vasoconstriction.

When people who are dehydrated drink water, they eventually consume a quantity sufficient to reduce the osmolality of the extracellular fluid to its normal value. They do not normally consume the water all at once. Instead, they drink intermittently until the proper osmolality of the extracellular fluid is established. The thirst sensation is temporarily reduced after a person drinks a small amount of liquid. At least two factors are responsible for this temporary interruption of the thirst sensation. First, when the oral mucosa becomes wet after it has been dry, sensory neurons conduct action potentials to the thirst center of the hypothalamus and temporarily decrease the sensation of thirst. Second, consumed fluid increases the digestive tract volume, and stretch of the digestive tract wall initiates sensory action potentials in stretch receptors. The sensory neurons conduct action potentials to the thirst center of the hypothalamus, where they temporarily suppress the sensation of thirst. Because the absorption of water from the digestive tract requires time, mechanisms that temporarily suppress thirst prevent a person from consuming extreme volumes of fluid that would exceed the amount required to reduce blood osmolality. Longer-term suppression of thirst results when extracellular fluid osmolality and blood pressure are within their normal ranges.

Learned behavior can be very important in avoiding periodic dehydration. Feeling thirsty promotes water consumption, and thirst

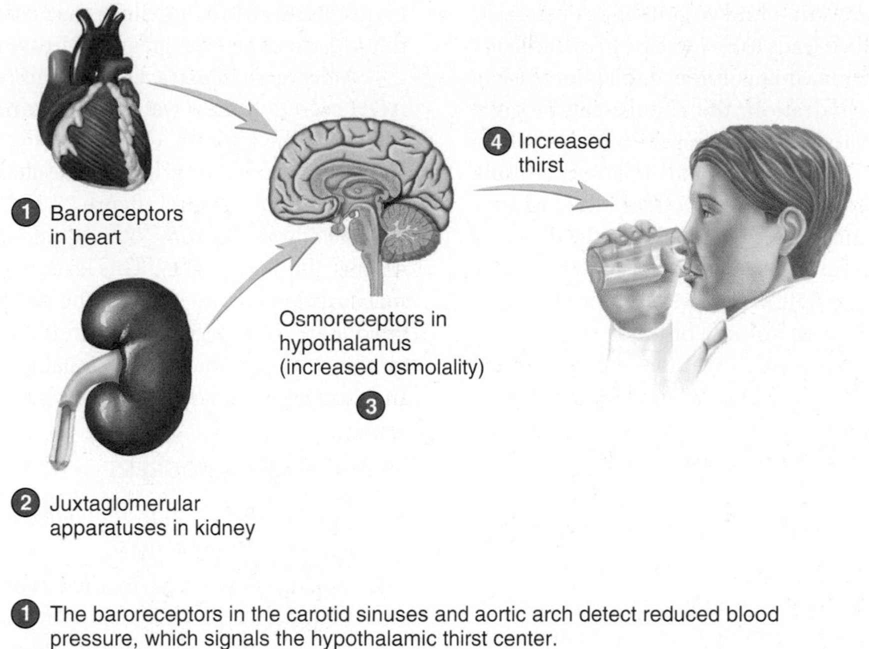
1 Baroreceptors in heart
2 Juxtaglomerular apparatuses in kidney
Osmoreceptors in hypothalamus (increased osmolality)
3
4 Increased thirst

1. The baroreceptors in the carotid sinuses and aortic arch detect reduced blood pressure, which signals the hypothalamic thirst center.
2. Simultaneously, the juxtaglomerular apparatuses detect low blood pressure, which activates the renin-angiotensin system to produce angiotensin II. Angiotensin II stimulates the hypothalamic thirst center.
3. Osmoreceptors in the hypothalamus shrink when blood osmolality goes up, triggering action potentials that stimulate thirst.
4. The combination of these inputs activates thirst and promotes water consumption.

PROCESS FIGURE 27.1 APR **Effect of Blood Osmolality and Blood Pressure on Thirst**

is relieved. Therefore, a healthy person usually drinks more than the minimum volume of fluid required to maintain homeostasis. The kidneys eliminate the excess water in urine.

Regulation of Water Loss

Water loss from the body occurs through three routes (table 27.3). The greatest amount of water, approximately 61%, is excreted through the urine. Approximately 35% of water loss occurs through evaporation from respiratory passages and the skin, which includes perspiration. Approximately 4% is lost in the feces.

The kidneys are the primary organs that regulate the composition and volume of body fluids by controlling the volume and concentration of water excreted in the form of urine (see chapter 26). The volume of water lost through the respiratory passages depends on the temperature and humidity of the air, body temperature, and the volume of air expired. Water lost through the diffusion and evaporation of water from the skin is called **insensible perspiration** (see chapter 25), and it regulates heat loss. For each degree that the body temperature rises above normal, an increased volume of 100–150 mL of water is lost each day in the form of insensible perspiration.

Sweat, or *sensible perspiration,* is secreted by the sweat glands (see chapters 5 and 25); in contrast to insensible perspiration, it contains solutes. Sweat resembles extracellular fluid in its composition, with sodium chloride as the major component, but it also contains some K^+, ammonia, and urea (table 27.4). The volume of sweat produced is determined primarily by neural mechanisms that regulate body temperature, although some sweat is produced as a result of sympathetic stimulation in response to stress. During exercise, elevated environmental temperature, or fever, the volume of sweat increases substantially and plays an important role in heat loss. Sweat losses of 8–10 L/day have been measured in outdoor workers in the summer. The volume of fluid lost as sweat is negligible for a person at rest in a cool environment.

Adequate fluid replacement during extensive sweating is important. Sweat is hyposmotic to plasma. The loss of a large volume of hyposmotic sweat causes a decrease in body fluid volume and an

TABLE 27.4 Composition of Sweat

Solute	Concentration (mM*)
Sodium	9.8–77.2
Potassium	3.9–9.2
Chloride	5.5–65.1
Ammonia	1.7–5.6
Urea	6.5–12.1

*1 mM is 1/1000 of a mole of solute in 1 liter of solution (see appendix C).

increase in body fluid concentration. Fluid volume is lost primarily from the extracellular space, which leads to an increase in extracellular fluid osmolality, a reduction in plasma volume, and an increase in hematocrit. During severe dehydration, the change can be great enough to cause blood viscosity to increase substantially, which increases the heart's workload enough that heart failure can result.

Relatively little water is lost by way of feces from the digestive tract. Although the total volume of fluid secreted into the digestive tract is large, nearly all the fluid is reabsorbed under normal conditions (see chapter 24). Exceptions are severe vomiting and diarrhea, which can result in a large volume of fluid loss.

The body can produce either a small volume of concentrated urine or a large volume of dilute urine, depending on the extracellular fluid osmolality and volume. These mechanisms operate to keep the total body water content within a narrow range of values.

Regulation of Extracellular Fluid Osmolality

Altering the water content of a solution changes its osmolality. Consider a solution in a pan on a stove. Adding water to the solution decreases its osmolality, or dilutes it. Boiling the solution in the pan removes water by evaporation, increasing its osmolality and concentrating it. Body fluid osmolality changes just as the water in the pan does but is maintained between 285 and 300 mOsm/kg.

An increase in the osmolality of the extracellular fluid triggers thirst and antidiuretic hormone (ADH) secretion. Consumed water is absorbed from the intestines and enters the extracellular fluid. ADH acts on the distal convoluted tubules and collecting ducts of the kidneys to increase the reabsorption of water from the filtrate. The increase in the amount of water entering the extracellular fluid decreases the osmolality (figures 27.2 and 27.3). The ADH and thirst mechanisms are sensitive to even small changes in extracellular fluid osmolality, and the response is fast (from minutes to a few hours).

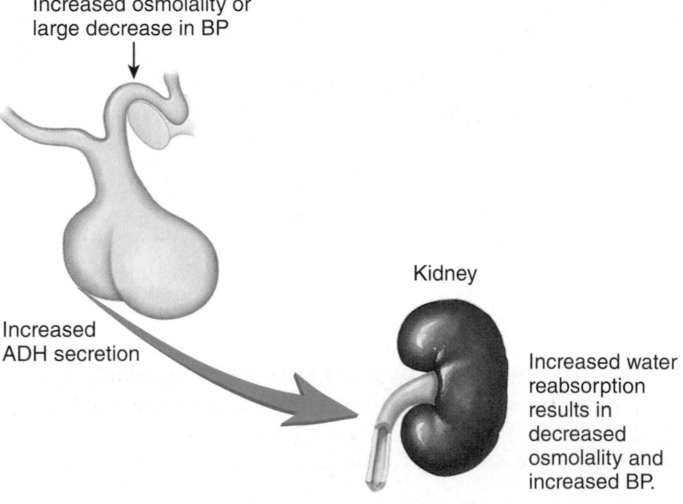

Increased osmolality or large decrease in BP
Kidney
Increased ADH secretion
Increased water reabsorption results in decreased osmolality and increased BP.

FIGURE 27.2 Effect of Blood Osmolality and Blood Pressure on Water Reabsorption in the Kidneys

Increased blood osmolality affects hypothalamic neurons, and decreased blood pressure (BP) affects baroreceptors in the aortic arch, carotid sinuses, and atrium. As a result of these stimuli, the rate of antidiuretic hormone (ADH) secretion from the posterior pituitary increases, which increases water reabsorption by the kidneys.

Larger increases in extracellular fluid osmolality, as occur during dehydration, cause an even greater increase in thirst and ADH secretion.

A decrease in extracellular fluid osmolality inhibits thirst and ADH secretion. Less water is consumed and reabsorbed from the filtrate in the kidneys. Consequently, more water is excreted as a large volume of dilute urine. The result is an increase in the osmolality of the extracellular fluid (figure 27.3). For example, drinking a large volume of water or some other dilute fluid results in reduced extracellular fluid osmolality. This leads to reduced ADH secretion, less reabsorption of water from the filtrate in the kidneys, and the production of a large volume of dilute urine. This response occurs quickly enough that the osmolality of the extracellular fluid is maintained within its normal range.

ASSESS YOUR PROGRESS

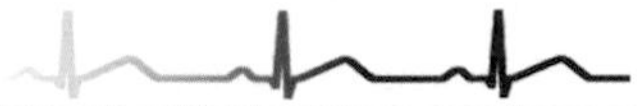

5. *List three factors that stimulate thirst. Name two factors that inhibit the sense of thirst.*
6. *Describe three routes for the loss of water from the body. Contrast insensible and sensible perspiration.*
7. *What are the primary organs that regulate the composition and volume of body fluids?*
8. *What two mechanisms are triggered by an increase in the osmolality of the extracellular fluid?*

Regulation of Extracellular Fluid Volume

The volume of extracellular fluid can change even if its osmolality is maintained within a narrow range. Sensory receptors that detect changes in blood pressure are important in regulating extracellular fluid volume. Carotid sinus and aortic arch baroreceptors monitor blood pressure in large arteries; receptors in the juxtaglomerular apparatuses monitor pressure changes in the afferent arterioles of the kidneys; and receptors in the walls of the atria of the heart and large veins monitor slight blood pressure changes that occur within them. These receptors activate neural mechanisms and three hormonal mechanisms that regulate extracellular fluid volume (figure 27.4):

1. *Neural mechanisms.* Neural mechanisms change the frequency of action potentials carried by sympathetic neurons to the afferent arterioles of the kidneys in response to changes in blood pressure. When baroreceptors detect an increase in arterial and venous blood pressure, the frequency of action potentials carried by sympathetic neurons to the afferent arterioles decreases. Consequently, the afferent arterioles dilate. This increases glomerular capillary pressure, resulting in an increase in the glomerular filtration rate (GFR), an increase in filtrate volume, and an increase in urine volume.

   When baroreceptors detect a decrease in arterial and venous blood pressure, the frequency of action potentials carried by sympathetic neurons to the afferent arterioles increases. Consequently, the afferent arterioles constrict. This decreases GFR, filtrate volume, and urine volume.
2. *Renin-angiotensin-aldosterone hormone mechanism.* The renin-angiotensin-aldosterone hormone mechanism responds to small changes in blood volume. Increased blood volume causes increased blood pressure. Juxtaglomerular cells detect a rise in

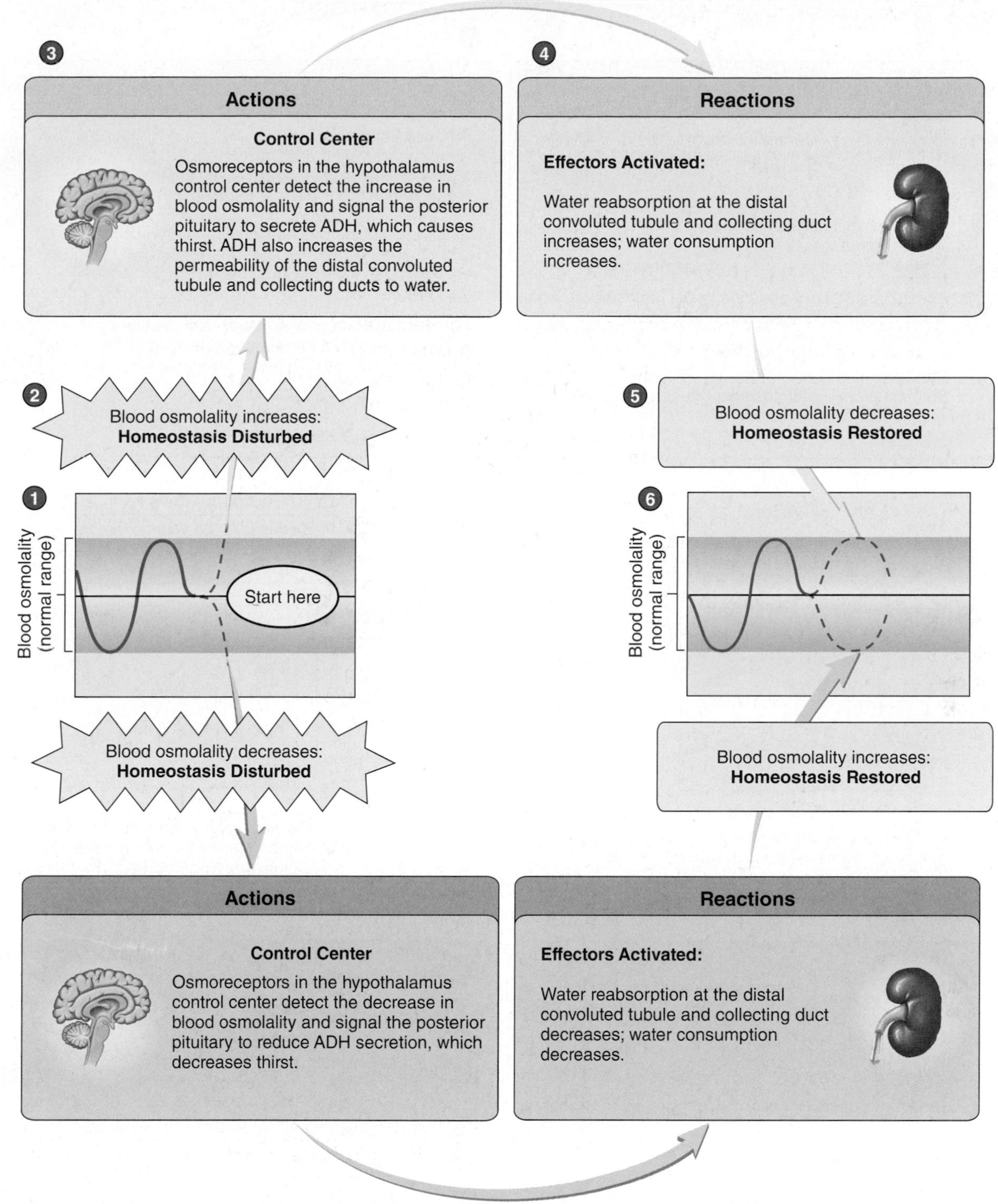

3
Actions
Control Center
Osmoreceptors in the hypothalamus control center detect the increase in blood osmolality and signal the posterior pituitary to secrete ADH, which causes thirst. ADH also increases the permeability of the distal convoluted tubule and collecting ducts to water.
4
Reactions
Effectors Activated:
Water reabsorption at the distal convoluted tubule and collecting duct increases; water consumption increases.
2
Blood osmolality increases:
Homeostasis Disturbed
5
Blood osmolality decreases:
Homeostasis Restored
1
Blood osmolality (normal range)
Start here
6
Blood osmolality (normal range)
Blood osmolality decreases:
Homeostasis Disturbed
Blood osmolality increases:
Homeostasis Restored
Actions
Control Center
Osmoreceptors in the hypothalamus control center detect the decrease in blood osmolality and signal the posterior pituitary to reduce ADH secretion, which decreases thirst.
Reactions
Effectors Activated:
Water reabsorption at the distal convoluted tubule and collecting duct decreases; water consumption decreases.

HOMEOSTASIS FIGURE 27.3 Blood Osmolality Regulation

(1) Blood osmolality is in the normal range. (2) Blood osmolality increases outside the normal range, which causes homeostasis to be disturbed. (3) The control center responds to the change in blood osmolality. (4) The control center causes ADH to be secreted, which increases water reabsorption at the distal convoluted tubule and the collecting duct. (5) These changes cause blood osmolality to decrease. (6) Blood osmolality returns to the normal range and homeostasis is restored. Observe the responses to a decrease in blood osmolality outside the normal range by following the *red arrows*.

blood pressure in the afferent arterioles and reduce renin secretion. Decreased renin secretion leads to a lower conversion rate of angiotensinogen to angiotensin I, which slows the conversion of angiotensin I to angiotensin II. With less angiotensin II, there is a decline in aldosterone secretion from the adrenal cortex. Lower aldosterone levels reduce the rate of Na^+ reabsorption from the distal convoluted tubules and collecting ducts. Consequently, more Na^+ remains in the filtrate, and less Na^+ is reabsorbed. The effect is to increase the osmolality of the filtrate, which lessens water reabsorption in the kidneys. The water remains, with the excess Na^+, in the filtrate. Thus, because the volume of urine produced by the kidneys goes up, the extracellular fluid volume is reduced, and blood pressure returns to the normal range (figure 27.4).

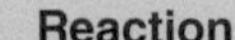

3

Actions

Blood vessels:
Sympathetic division baroreceptors detect increased blood volume, which causes vasodilation of renal arteries.

Heart:
Atrial cardiac muscle cells secrete ANH when blood volume increases.

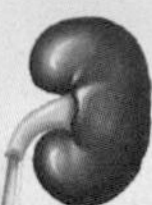

Kidney:
Juxtaglomerular apparatuses inhibit renin release when blood volume increases, which decreases aldosterone secretion.

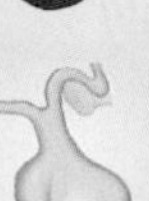

Pituitary:
Baroreceptors inhibit posterior pituitary ADH secretion when blood volume increases.

4

Reactions

Effectors Activated:

Increased renal blood flow increases the rate of filtrate formation, and more water is excreted in the urine.

Decreased aldosterone and increased ANH decreases Na^+ reabsorption into the distal convoluted tubule and collecting duct. More Na^+ in urine, which decreases blood volume.

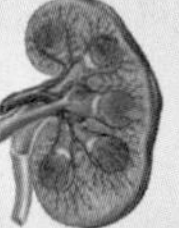

Decreased ADH decreases water reabsorption by the distal convoluted tubules and collecting ducts. Less water returns to the blood, and more water is excreted in the urine.

2 High blood volume induces elevated blood pressure: **Homeostasis Disturbed**

1

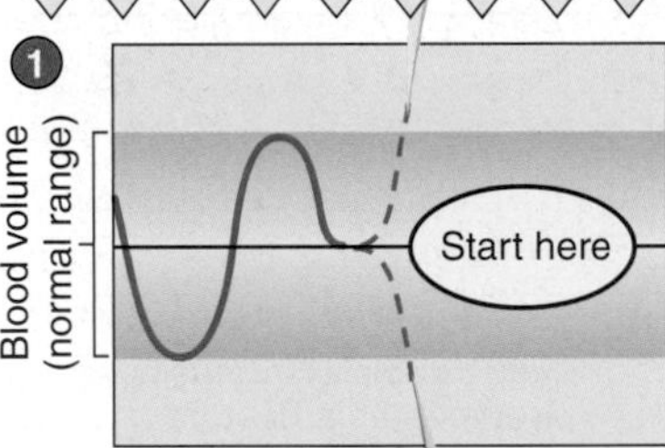

Blood volume (normal range)
Start here

Low blood volume induces lowered blood pressure: **Homeostasis Disturbed**

5 Reduced blood volume due to loss of water and Na^+ in the urine lowers blood pressure: **Homeostasis Restored**

6

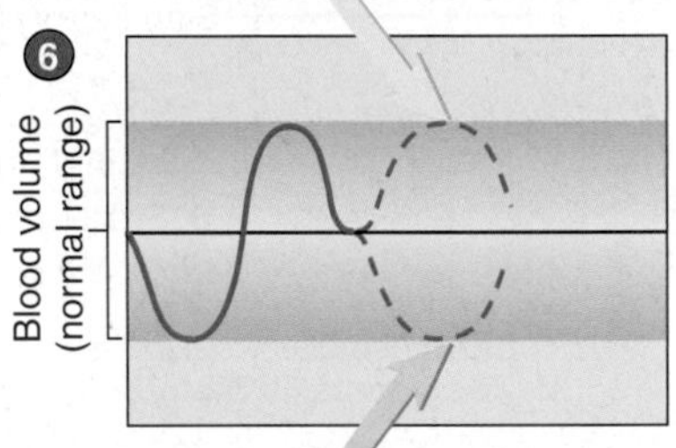

Blood volume (normal range)

Increased blood volume due to decreased Na^+ and water loss in the urine raises blood pressure: **Homeostasis Restored**

Actions

Blood vessels:
Sympathetic division baroreceptors detect decreased blood volume, which causes vasoconstriction of renal arteries.

Heart:
Atrial cardiac muscle cells do not secrete ANH when blood volume decreases.

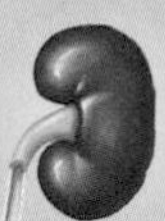

Kidney:
Juxtaglomerular apparatuses stimulate renin release when blood volume decreases, which increases aldosterone secretion.

Pituitary:
Baroreceptors stimulate posterior pituitary ADH secretion when blood volume decreases. Increased ADH also increases the sensation of thirst.

Reactions

Effectors Activated:

Decreased renal blood flow decreases filtrate formation, and less water is excreted in urine.

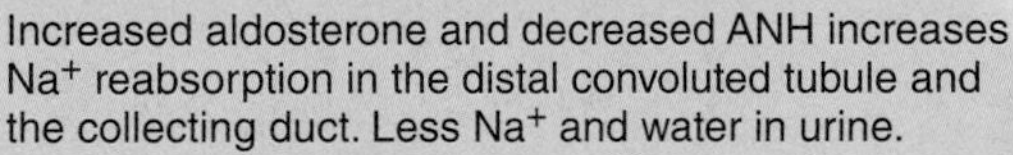

Increased aldosterone and decreased ANH increases Na^+ reabsorption in the distal convoluted tubule and the collecting duct. Less Na^+ and water in urine.

Increased ADH increases the permeability of the distal convoluted tubule and the collecting duct to water. Less water is excreted in urine.

HOMEOSTASIS FIGURE 27.4 Regulation of Blood Volume
(1) Blood volume is in the normal range. (2) Blood volume increases outside the normal range, which causes homeostasis to be disturbed. (3) The control centers respond to the change in blood volume. (4) The control centers cause ADH and aldosterone secretion to decrease, which reduces water reabsorption. The control centers also cause dilation of renal arteries, which increases urine production. The heart secretes ANH, which also increases urine production. (5) These changes cause blood volume to decrease. (6) Blood volume returns to the normal range and homeostasis is restored. Observe the responses to a decrease in blood volume outside the normal range by following the *red arrows*.

Whereas increased blood volume results in more urine, a decreased blood volume causes less urine to be produced. Once blood volume goes down, blood pressure decreases, and this triggers the juxtaglomerular cells to secrete renin, which increases angiotensin II. The increased angiotensin II causes an increase in the rate of aldosterone secretion from the adrenal cortex. Increased aldosterone increases the rate of Na^+ reabsorption, primarily from the distal convoluted tubules and collecting ducts. Consequently, less Na^+ remains in the filtrate, and more Na^+ is reabsorbed. The effect is to decrease the osmolality of the filtrate. This increases the kidneys' ability to reabsorb water and to increase extracellular fluid volume. Thus, the volume of urine produced by the kidneys decreases, and the extracellular fluid volume and blood pressure increase (figure 27.5; see figure 27.4).

3. *Atrial natriuretic hormone (ANH) mechanism.* In response to a larger-than-normal blood volume, the heart's atrial walls are stretched to a greater degree, which triggers ANH secretion. ANH reduces Na^+ reabsorption from the distal convoluted tubules and the collecting ducts. This increases the rate of Na^+ and water loss in the urine. Thus, increased ANH secretion decreases extracellular fluid volume and thus blood volume (figure 27.6; see figure 27.4).

   ANH does not appear to respond strongly to decreases in blood volume. However, a decrease in pressure in the atria of the heart inhibits the secretion of ANH. Without ANH, Na^+ reabsorption is inhibited in the distal convoluted tubules and collecting ducts. Therefore, the rate of Na^+ reabsorption increases, and water reabsorption increases. Thus, decreased ANH secretion is consistent with a reduced urine volume and an elevated extracellular fluid volume (see figure 27.4).

4. *Antidiuretic hormone (ADH) mechanism.* The ADH mechanism plays an important role in regulating extracellular fluid volume in response to large changes in blood pressure (of 5–10%). Increased blood pressure inhibits ADH secretion. As a result, the reabsorption of water from the lumens of the distal convoluted tubules and collecting ducts decreases, resulting in a larger volume of dilute urine. This response helps decrease extracellular fluid volume and blood pressure (see figures 27.2 and 27.4).

   Low blood pressure stimulates ADH secretion. Consequently, the reabsorption of water from the distal convoluted tubules and collecting ducts increases, resulting in a small volume of concentrated urine. This response helps increase extracellular fluid volume and blood pressure (see figures 27.2 and 27.4).

The mechanisms that maintain extracellular fluid concentration and volume function together. However, when the mechanisms do not function normally, extracellular fluid volume may increase even though the extracellular concentration of fluids does not change substantially. For example, excessive aldosterone secretion from an enlarged adrenal cortex increases Na^+ reabsorption by the kidneys, and the total volume of extracellular fluid increases. Certain mechanisms, such as the regulation of ADH secretion, keep the concentration of the body fluids constant. Even if elevated blood pressure causes edema, the osmolality of the extracellular fluid is maintained between 285 and 300 mOsm/kg. Similarly, in people suffering from heart failure, the resulting reduced blood pressure activates mechanisms, such as renin secretion, that increase blood pressure to its normal range. Consequently, aldosterone secretion increases, and the result is increased extracellular fluid volume and edema in the periphery, including the lungs.

FUNDAMENTAL **Figure**

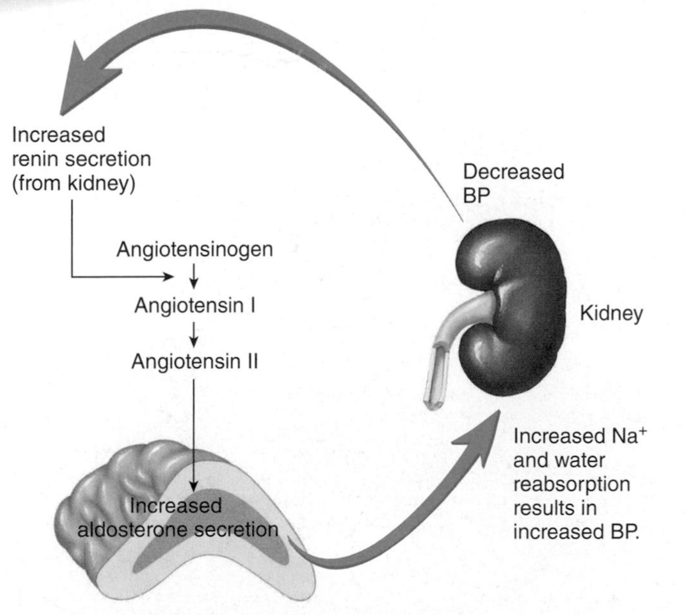

Increased renin secretion (from kidney)
Decreased BP
Angiotensinogen
Angiotensin I
Angiotensin II
Kidney
Increased Na+ and water reabsorption results in increased BP.
Increased aldosterone secretion

FIGURE 27.5 Effect of Blood Pressure on Na^+ and Water Reabsorption in the Kidneys

Low blood pressure (BP) stimulates renin secretion from the kidneys. Renin converts angiotensinogen to angiotensin I, which is converted to angiotensin II. Angiotensin II stimulates aldosterone secretion from the adrenal cortex. Aldosterone increases Na^+ and water reabsorption in the kidneys.

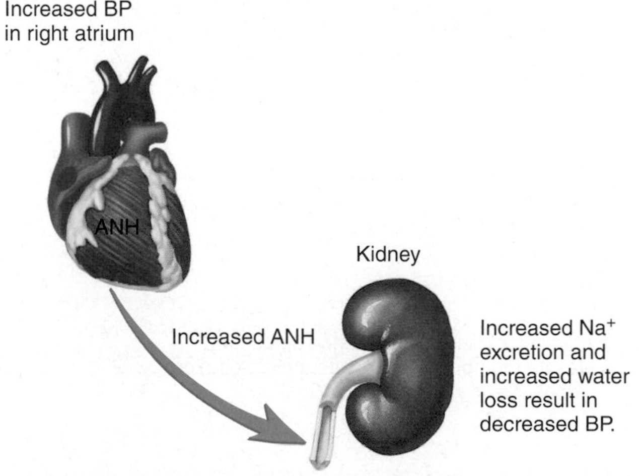
Increased BP
in right atrium
ANH
Kidney
Increased ANH
Increased Na+
excretion and
increased water
loss result in
decreased BP.

FIGURE 27.6 Effect of Blood Pressure in the Right Atrium on Na^+ and Water Excretion

Increased blood pressure (BP) in the right atrium of the heart causes increased secretion of atrial natriuretic hormone (ANH), which increases Na^+ excretion and water loss in urine.

ASSESS YOUR PROGRESS

9. *What sensory receptors are responsible for activating neural and hormonal mechanisms that regulate extracellular fluid volume?*
10. *What is the effect on sympathetic stimulation, afferent arterioles, GFR, filtrate volume, urine volume, and extracellular fluid volume when baroreceptors detect an increase in arterial and venous blood pressure?*
11. *Describe the response of the renin-angiotensin-aldosterone hormone mechanism to a decrease in blood pressure. How are extracellular fluid volume and urine volume affected?*
12. *What effect does atrial natriuretic hormone (ANH) have on extracellular fluid volume?*
13. *How does an increase in blood pressure affect the secretion of ADH? How does ADH affect extracellular fluid volume?*

27.3 Regulation of Intracellular Fluid Composition

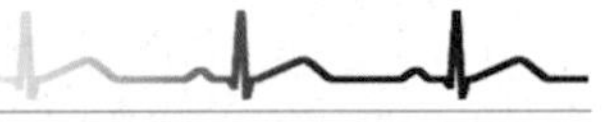

LEARNING OUTCOME

After reading this section, you should be able to

A. Describe how intracellular fluid composition is maintained.

The composition of intracellular fluid is markedly different from that of extracellular fluid. Plasma membranes, which separate the two compartments, are selectively permeable, meaning that they are relatively impermeable to proteins and other large molecules and have limited permeability to smaller molecules and ions. Consequently, most large molecules synthesized within cells, such as proteins, remain within the intracellular fluid. Some substances, such as electrolytes, are actively transported across the plasma membrane, and their concentrations in the intracellular fluid are determined by the transport processes and by the electrical charge difference across the plasma membrane (figure 27.7).

Water moves across the plasma membrane by osmosis. Thus, the net movement of water is affected by changes in the concentration of solutes in the extracellular and intracellular fluids. For example, as dehydration develops, the concentration of solutes in the extracellular fluid increases, allowing water to move by osmosis from the intracellular fluid into the extracellular fluid. If dehydration is severe, the cells can shrink and will function abnormally. If water intake increases after a period of dehydration, the concentration of solutes in the extracellular fluids decreases, which allows water to move back into the cells.

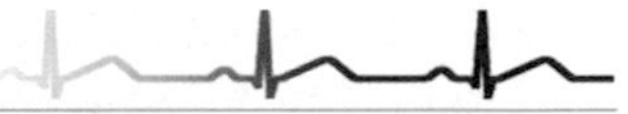

ASSESS YOUR PROGRESS

14. *What factors determine the composition of intracellular fluid?*
15. *What characteristic of the plasma membrane is responsible for maintaining the differences between intracellular and extracellular fluid?*

FUNDAMENTAL **Figure**

1 Large organic molecules, such as proteins, which cannot cross the plasma membrane, are synthesized inside cells and influence the concentration of solutes inside the cells.

2 The transport of ions, such as Na^+, K^+, and Ca^{2+}, across the plasma membrane influences the concentration of ions inside and outside the cell.

3 An electrical charge difference across the plasma membrane influences the distribution of ions inside and outside the cell.

4 The distribution of water inside and outside the cell is determined by osmosis.

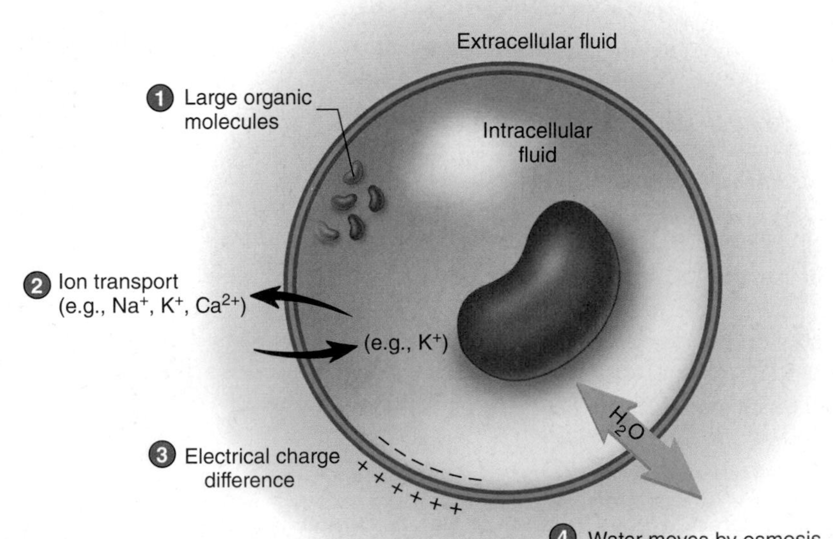
Extracellular fluid
1 Large organic molecules
Intracellular fluid
2 Ion transport (e.g., Na+, K+, Ca2+)
(e.g., K+)
H2O
3 Electrical charge difference
4 Water moves by osmosis.

PROCESS FIGURE 27.7 Regulation of Intracellular Fluid

27.4 Regulation of Specific Electrolytes in the Extracellular Fluid

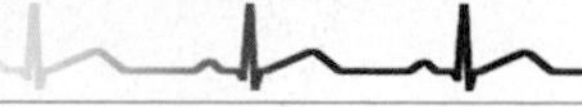

LEARNING OUTCOMES

After reading this section, you should be able to

A. Discuss how sodium ion concentration is achieved.
B. Describe the regulation of chloride ions.
C. Explain potassium ion homeostasis.
D. Demonstrate an understanding of calcium ion homeostasis.
E. Relate the mechanisms governing magnesium ion concentration.
F. Explain the regulation of phosphate ions.
G. Describe the causes and symptoms of abnormal plasma levels of selected electrolytes.

Electrolytes (ē-lek′trō-līts) are molecules or atoms with an electrical charge. The major extracellular ions are Na^+, Cl^-, K^+, Ca^{2+}, Mg^{2+}, and phosphate ions (Po_4^{3-}). Electrolytes are in the food and water we ingest. Organs—such as the kidneys and, to a lesser degree, the liver, skin, and lungs—remove them from the body. The concentrations of electrolytes in the extracellular fluid are regulated, so that they do not change unless the individual is growing, gaining weight, or losing weight. The regulation of each electrolyte involves the coordinated participation of several organ systems.

Regulation of Sodium Ions

Sodium ions are the dominant extracellular cations. Because of their abundance in the extracellular fluid, they exert substantial osmotic pressure. Approximately 90–95% of the osmotic pressure of the extracellular fluid is caused by Na^+ and the negative ions associated with it. In the United States, most people consume 20–30 times the amount of sodium chloride (salt) the body needs. Although less than 0.5 g is required to maintain homeostasis, the average individual ingests approximately 10–15 g of sodium chloride daily. Therefore, regulation of the body's Na^+ content depends primarily on the excretion of excess quantities of Na^+. On the other hand, when Na^+ intake is very low, the mechanisms for conserving Na^+ in the body take effect.

The kidneys are the major route by which Na^+ is excreted. Sodium ions readily pass from the glomerulus into the lumen of the Bowman capsule and are present in the same concentration in the filtrate as in the plasma. The concentration of Na^+ excreted in the urine is determined by the amount of Na^+ and water reabsorbed from filtrate in the nephron. If Na^+ reabsorption from the nephron decreases, large quantities are excreted in the urine. If Na^+ reabsorption from the nephron increases, only small quantities are excreted in the urine.

The rate of Na^+ transport in the proximal convoluted tubule is relatively constant, but the Na^+ transport mechanisms of the distal convoluted tubule and the collecting duct are under hormonal control. **Aldosterone** increases Na^+ reabsorption from the distal convoluted tubule and collecting duct. As little as 0.1 g of sodium is excreted in the urine each day in the presence of high blood levels of aldosterone. When aldosterone is absent, Na^+ reabsorption in the nephron is greatly reduced, and as much as 30–40 g of sodium can be lost in the urine daily.

Sodium ions are also excreted from the body in sweat. Normally, only a small quantity of Na^+ is lost each day in the form of sweat, but, as noted in section 27.2, the amount increases during heavy exercise in warm environments. The mechanisms that regulate sweating control the quantity of Na^+ excreted through the skin. As body temperature increases, thermoreceptor neurons within the hypothalamus respond by increasing the rate of sweat production. As the rate of sweat production increases, the quantity of Na^+ lost in the urine decreases to keep the extracellular concentration of Na^+ constant. Because of this mechanism, the loss of Na^+ in sweat is rarely physiologically significant.

The primary mechanisms that regulate Na^+ concentrations in the extracellular fluid do not directly monitor Na^+ levels but are sensitive to changes in extracellular fluid osmolality or blood pressure (table 27.5; see figure 27.4). The quantity of Na^+ in the body has a dramatic effect on extracellular osmotic pressure and extracellular fluid volume. For example, if the quantity of Na^+ increases, the osmolality of the extracellular fluid increases. Increased extracellular fluid osmolality stimulates ADH secretion, which increases water reabsorption in the kidneys and thirst. Consequently, water consumption, extracellular fluid volume, and water conservation increase. A decrease in the quantity of Na^+ in the body causes an opposite set of responses. ADH secretion decreases, which allows a large volume of dilute urine to be produced and decreases the sensation of thirst. As a result of these changes, extracellular osmolality increases. By regulating extracellular fluid osmolality and extracellular fluid volume, the concentration of Na^+ in the body fluids is maintained within a narrow range of values.

Elevated blood pressure under resting conditions increases Na^+ and water excretion (table 27.5; see figure 27.4). If blood pressure is low, the total Na^+ content of the body is usually also low. In response to low blood pressure, mechanisms such as the renin-angiotensin-aldosterone hormone mechanism are activated, increasing Na^+ concentration and water volume in the extracellular fluid (table 27.5; see figure 27.4).

Predict 2

In response to hemorrhagic shock, the kidneys produce a small volume of very concentrated urine. Explain how the rate of filtrate formation changes and how Na^+ transport is altered in the distal part of the nephron in response to hemorrhagic shock.

Cells in the walls of the atria synthesize ANH, which is secreted in response to elevated blood pressure within the right atrium. ANH acts on the kidneys to increase urine production by inhibiting the reabsorption of Na^+ (table 27.5; see figure 27.6). It also inhibits the effect of ADH on the distal convoluted tubules and collecting ducts and inhibits ADH secretion (see chapter 26, section 26.4).

Deviations from the normal concentration range for Na^+ in body fluids result in significant symptoms. A reduced plasma Na^+ concentration leads to **hyponatremia** (hī′pō-nă-trē′mē-ă); an elevated plasma Na^+ concentration results in **hypernatremia** (hī′per-nă-trē′mē-ă). The major causes and symptoms of each are listed in table 27.6.

TABLE 27.5 Homeostasis: Mechanisms Regulating Blood Sodium

Mechanism	Stimulus	Response to Stimulus	Effect of Response	Result
Response to Changes in Blood Osmolality				
Antidiuretic hormone (ADH); the most important regulator of blood osmolality	Increased blood osmolality (e.g., increased Na^+ concentration)	Increased ADH secretion from the posterior pituitary; mediated through cells in the hypothalamus	Increased water reabsorption in the kidney; production of a small volume of concentrated urine	Decreased blood osmolality as reabsorbed water dilutes the blood
	Decreased blood osmolality (e.g., decreased Na^+ concentration)	Decreased ADH secretion from the posterior pituitary; mediated through cells in the hypothalamus	Decreased water reabsorption in the kidney; production of a large volume of dilute urine	Increased blood osmolality as water is excreted from the blood into the urine
Response to Changes in Blood Pressure				
Renin-angiotensin-aldosterone hormone mechanism	Decreased blood pressure in the kidney's afferent arterioles	Increased renin release from the juxtaglomerular apparatuses; renin initiates the conversion of angiotensinogen to angiotensin; angiotensin I is converted to angiotensin II, which increases aldosterone secretion from the adrenal cortex	Increased Na^+ reabsorption in the kidney (because of increased aldosterone); increased water reabsorption as water follows the Na^+; decreased urine volume	Increased blood pressure as blood volume increases because of increased water reabsorption; blood osmolality is maintained because both Na^+ and water are reabsorbed*
	Increased blood pressure in the kidney's afferent arterioles	Decreased renin release from the juxtaglomerular apparatuses, resulting in reduced formation of angiotensin I; reduced angiotensin I leads to reduced angiotensin II, which causes a decrease in aldosterone secretion from the adrenal cortex	Decreased Na^+ reabsorption in the kidney (because of decreased aldosterone); decreased water reabsorption as less Na^+ is reabsorbed; increased urine volume	Decreased blood pressure as blood volume decreases because water is excreted in the urine; blood osmolality is maintained because both Na^+ and water are excreted in the urine*
Atrial natriuretic hormone (ANH)	Decreased blood pressure in the atria of the heart	Decreased ANH released from the atria	Increased Na^+ reabsorption in the kidney; increased water reabsorption as water follows the Na^+; decreased urinary volume	Increased blood pressure as blood volume increases because of increased water reabsorption; blood osmolality is maintained because both Na^+ and water are reabsorbed*
	Increased blood pressure in the atria of the heart	Increased ANH released from the atria	Decreased Na^+ reabsorption in the kidney; decreased water reabsorption as water is excreted with Na^+ in the urine; increased urinary volume	Decreased blood osmolality as blood volume decreases because water is excreted in the urine; blood osmolality is maintained because both Na^+ and water are excreted in the urine*
ADH—activated by significant decreases in blood pressure; normally regulates blood osmolality (see above)	Decreased arterial blood pressure	Increased ADH secretion from the posterior pituitary; mediated through baroreceptors	Increased water reabsorption in the kidney; production of a small volume of concentrated urine	Increased blood pressure resulting from increased blood volume; decreased blood osmolality
	Increased arterial blood pressure	Decreased ADH secretion from the posterior pituitary; mediated through baroreceptors	Decreased water reabsorption in the kidney; production of a large volume of dilute urine	Decreased blood pressure resulting from decreased blood volume; increased blood osmolality

Abbreviation: ADH = antidiuretic hormone.

*Assumes normal levels of ADH.

TABLE 27.6	Consequences of Abnormal Plasma Levels of Sodium Ions
HYPONATREMIA	
Causes	Inadequate dietary intake of sodium
	Extrarenal losses
	Dilution
	Hyperglycemia
Symptoms	Lethargy, confusion, apprehension, seizures, and coma
	When accompanied by reduced blood volume: reduced blood pressure, tachycardia, and decreased urine output
	When accompanied by increased blood volume: weight gain, edema, and distension of veins
HYPERNATREMIA	
Causes	High dietary sodium (rarely causes symptoms)
	Administration of hypertonic saline solutions
	Oversecretion of aldosterone
	Water loss
Symptoms	Thirst, fever, dry mucous membranes, and restlessness
	Most serious symptoms are convulsions and pulmonary edema
	When occurring with increased water volume: weight gain, edema, elevated blood pressure, and bounding pulse

ASSESS **YOUR PROGRESS**

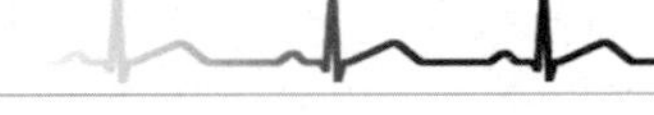

16. *Name the substance responsible for most of the osmotic pressure of the extracellular fluid.*
17. *How does aldosterone affect the concentration of Na^+ in the urine?*
18. *What role does sweating play in Na^+ balance?*
19. *How does increased blood pressure result in a loss of water and salt? What happens when blood pressure decreases?*
20. *What effect does ANH have on Na^+ and water loss in urine?*
21. *What are the causes of hypernatremia and hyponatremia? List the symptoms of both conditions.*

Regulation of Chloride Ions

The predominant anions in the extracellular fluid are **chloride ions.** The electrical attraction of anions and cations makes it difficult to separate these charged particles. Consequently, the regulatory mechanisms that influence the concentration of cations in the extracellular fluid also influence the concentration of anions. The mechanisms that regulate Na^+, K^+, and Ca^{2+} levels in the body are important in influencing Cl^- levels. Because Na^+ predominates, the mechanisms that regulate extracellular Na^+ concentration are the most important in regulating the extracellular Cl^- concentration.

ASSESS **YOUR PROGRESS**

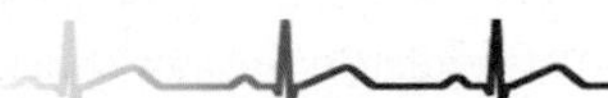

22. *What mechanisms regulate Cl^- concentrations?*

Regulation of Potassium Ions

Because the concentration gradient of **potassium ions** across the plasma membrane has a major influence on the resting membrane potential of electrically excitable cells, K^+ concentrations are tightly regulated. An increase in extracellular K^+ concentration leads to depolarization, and a decrease in extracellular K^+ concentration leads to hyperpolarization of the resting membrane potential. An abnormally low level of K^+ in the extracellular fluid is called **hypokalemia** (hī′pō-ka-lē′mē-ă); an abnormally high level of K^+ in the extracellular fluid is called **hyperkalemia** (hī′per-kă-lē′mē-ă). Major causes of hypokalemia and hyperkalemia and their symptoms are listed in table 27.7.

Potassium ions pass freely through the filtration membrane of the renal corpuscle. They are actively reabsorbed in the proximal convoluted tubules and actively secreted in the distal convoluted tubules and collecting ducts. Potassium ion secretion into the distal convoluted tubules and collecting ducts is highly regulated and primarily responsible for controlling the extracellular concentration of K^+.

Aldosterone plays a major role in regulating the concentration of K^+ in the extracellular fluid by increasing the rate of K^+ secretion in the distal convoluted tubules and collecting ducts. Aldosterone secretion from the adrenal cortex is stimulated by elevated K^+ blood levels (figure 27.8; see chapter 26). Aldosterone secretion is

TABLE 27.7	Consequences of Abnormal Concentrations of Potassium Ions
HYPOKALEMIA	
Causes	Alkalosis
	Insulin administration
	Reduced K^+ intake
	Increased renal loss
Symptoms	Decreased neuromuscular excitability
	Decreased smooth muscle tone
	Delayed ventricular depolarization
	Bradycardia
	Atrioventricular block
HYPERKALEMIA	
Causes	Loss of intracellular K^+ due to cell trauma or reduced permeability of plasma membrane
	Reduced renal excretion
Symptoms	
Mild	Increased neuromuscular irritability
	Intestinal cramping and diarrhea
	Rapid cardiac repolarization
Severe	Muscle weakness
	Loss of muscle tone and paralysis
	Reduced rate of cardiac action potential conduction

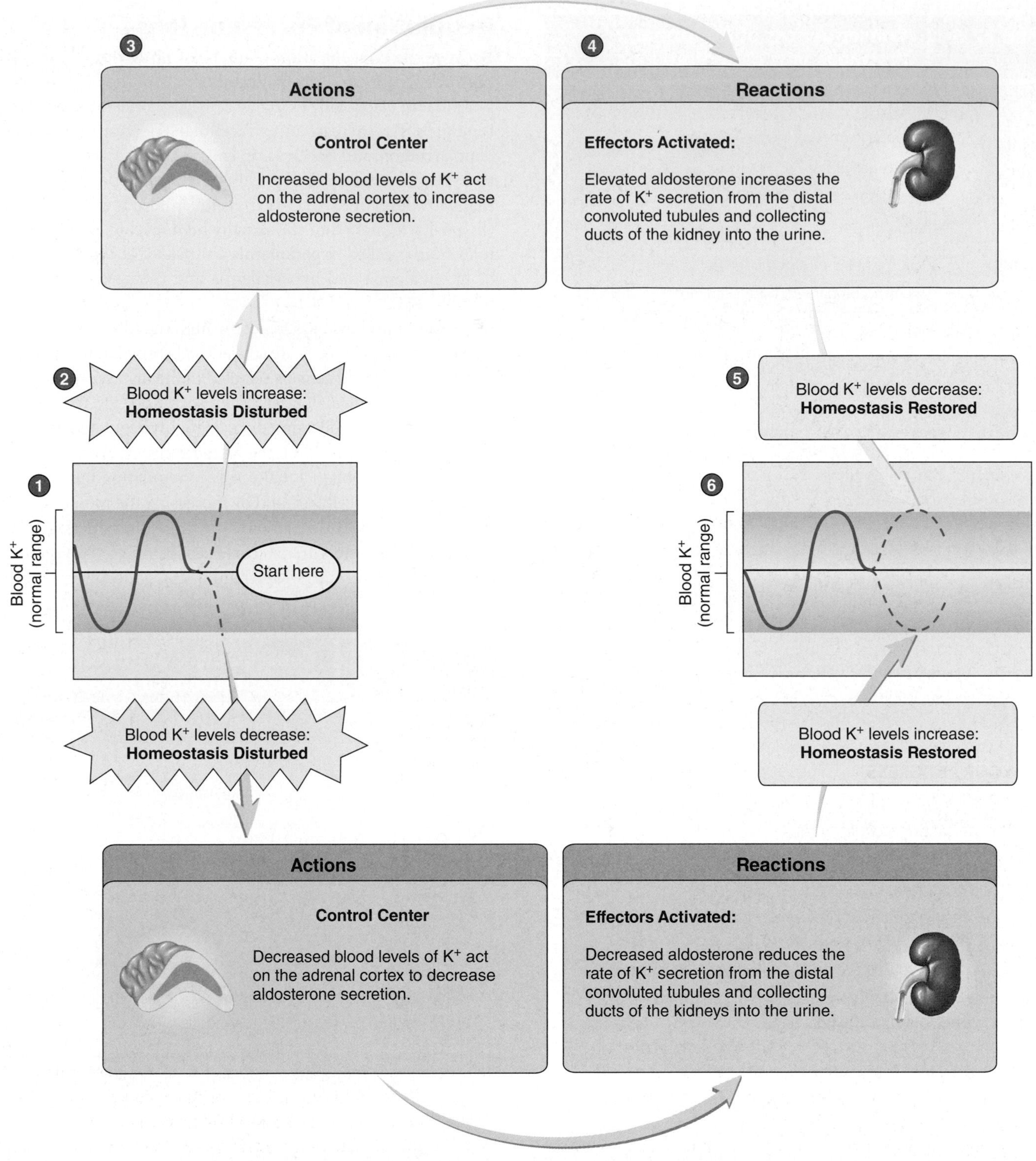

3
Actions
Control Center
Increased blood levels of K+ act on the adrenal cortex to increase aldosterone secretion.
4
Reactions
Effectors Activated:
Elevated aldosterone increases the rate of K+ secretion from the distal convoluted tubules and collecting ducts of the kidney into the urine.
2
Blood K+ levels increase:
Homeostasis Disturbed
5
Blood K+ levels decrease:
Homeostasis Restored
1
Blood K+ (normal range)
Start here
6
Blood K+ (normal range)
Blood K+ levels decrease:
Homeostasis Disturbed
Blood K+ levels increase:
Homeostasis Restored
Actions
Control Center
Decreased blood levels of K+ act on the adrenal cortex to decrease aldosterone secretion.
Reactions
Effectors Activated:
Decreased aldosterone reduces the rate of K+ secretion from the distal convoluted tubules and collecting ducts of the kidneys into the urine.

HOMEOSTASIS FIGURE 27.8 Summary of Blood K^+ Regulation

(1) Blood K^+ is in the normal range. (2) Blood K^+ increases outside the normal range, which causes homeostasis to be disturbed. (3) The control center responds to the change in blood K^+. (4) The control center causes aldosterone to be secreted, which increases K^+ secretion at the distal convoluted tubule and the collecting duct. (5) These changes cause blood K^+ to decrease. (6) Blood K^+ returns to the normal range and homeostasis is restored. Observe the responses to a decrease in blood K^+ outside the normal range by following the *red arrows*.

also stimulated in response to increased angiotensin II. Elevated aldosterone concentrations in the circulatory system increase K^+ secretion into the nephron, thereby lowering blood levels of K^+.

Circulatory system shock can result from plasma loss, dehydration, and tissue damage, as occurs in burn patients. This shock causes the extracellular K^+ to be more concentrated than normal, which stimulates aldosterone secretion from the adrenal cortex. Aldosterone is also secreted in response to decreased blood pressure, which stimulates the renin-angiotensin-aldosterone hormone mechanism. Homeostasis is reestablished as K^+ excretion increases. Also, increased Na^+ and water reabsorption stimulated by aldosterone results in an increase in extracellular fluid volume, which dilutes the K^+ in the body fluids. Blood pressure increases toward normal as water reabsorption increases and when vasoconstriction is stimulated by angiotensin II.

ASSESS YOUR PROGRESS

23. *What effect does an increase or a decrease in extracellular K^+ concentration have on resting membrane potential?*

24. *In what portion of the nephron is K^+ secreted?*

25. *How is the secretion of K^+ regulated?*

Regulation of Calcium Ions

As with other ions, the extracellular concentration of **calcium ions** is regulated within a narrow range. The normal concentration of Ca^{2+} in plasma is 9.4 mg/100 mL. **Hypocalcemia** (hī′pō-kal-sē′mē-ă) is a below-normal level of Ca^{2+} in the extracellular fluid, and **hypercalcemia** (hī′per-kal-sē′mē-ă) is an above-normal level of Ca^{2+} in the extracellular fluid. Major symptoms develop when the extracellular concentration of Ca^{2+} declines below 6 mg/100 mL or increases above 12 mg/100 mL. Changes in the extracellular concentration of Ca^{2+} markedly affect the electrical properties of excitable tissues. Hypocalcemia increases the plasma membrane's permeability to Na^+. As a result, nerve and muscle tissues undergo spontaneous action potential generation. Hypercalcemia decreases the plasma membrane's permeability to Na^+, preventing normal depolarization of nerve and muscle cells. High extracellular Ca^{2+} levels cause the deposition of calcium carbonate salts in soft tissues, resulting in irritation and inflammation of those tissues. Table 27.8 lists the major causes and symptoms of hypocalcemia and hypercalcemia.

The kidneys, digestive tract, and bones are important in maintaining extracellular Ca^{2+} levels (figure 27.9). Almost 99% of total body calcium is contained in bone. Calcium ion regulation partly involves the regulation of Ca^{2+} deposition into and reabsorption from bone (see chapter 6). However, long-term regulation of Ca^{2+} levels depends on maintaining a balance between Ca^{2+} absorption across the intestinal wall and Ca^{2+} excretion by the kidneys.

Parathyroid (par-ă-thī′royd) **hormone (PTH),** secreted by the parathyroid glands, increases extracellular Ca^{2+} levels and reduces extracellular phosphate levels. The rate of parathyroid hormone secretion is regulated by extracellular Ca^{2+} levels. Parathyroid cell receptors act as extracellular Ca^{2+} level sensors. Elevated Ca^{2+} levels inhibit parathyroid hormone secretion, and reduced levels stimulate it. Parathyroid hormone causes increased osteoclast activity, which results in the degradation of bone and the release of Ca^{2+} and phosphate ions into body fluids. Parathyroid hormone increases the rate of Ca^{2+} reabsorption from nephrons in the kidneys and increases the concentration of phosphate ions in the urine. It also increases the rate at which vitamin D is converted to 1,25-dihydroxycholecalciferol, or **active vitamin D.** Active vitamin D acts to increase Ca^{2+} absorption across the intestinal mucosa.

TABLE 27.8 Consequences of Abnormal Concentrations of Calcium Ions

HYPOCALCEMIA	
Causes	Nutritional deficiencies
	Vitamin D deficiency
	Decreased parathyroid hormone secretion
	Malabsorption of fats (reduces vitamin D absorption)
	Bone tumors that increase Ca^{2+} deposition
Symptoms	Confusion
	Muscle spasms
	Hyperreflexia
	Intestinal cramping
	Convulsions
	Tetany
	Inadequate respiratory movements
	Prolonged cardiac ventricular depolarization
HYPERCALCEMIA	
Causes	Excessive parathyroid hormone secretion
	Excess vitamin D
Symptoms	Fatigue
	Weakness
	Lethargy
	Anorexia
	Nausea
	Constipation
	Reduced cardiac ventricular depolarization
	Kidney stones

Lack of parathyroid hormone secretion results in a rapid decline in extracellular Ca^{2+} concentration. The causes of this decline are a reduced rate of Ca^{2+} absorption from the digestive tract, increased Ca^{2+} excretion by the kidneys, and reduced bone reabsorption. A lack of parathyroid hormone can result in death because of tetany of the respiratory muscles caused by hypocalcemia.

Vitamin D can be obtained from food or from vitamin D biosynthesis. Normally, vitamin D biosynthesis is adequate; however, because ultraviolet light is required for this process, prolonged lack of sun exposure reduces vitamin D blood levels.

Without vitamin D, the transport of Ca^{2+} across the wall of the digestive tract is negligible. This leads to inadequate Ca^{2+} absorption, even though the person may be consuming a large amount of this ion. Thus, Ca^{2+} absorption depends on getting adequate amounts of calcium and vitamin D.

Calcitonin (kal-si-tō′nin), which is secreted by the parafollicular cells of the thyroid gland, helps reduce extracellular Ca^{2+} levels.

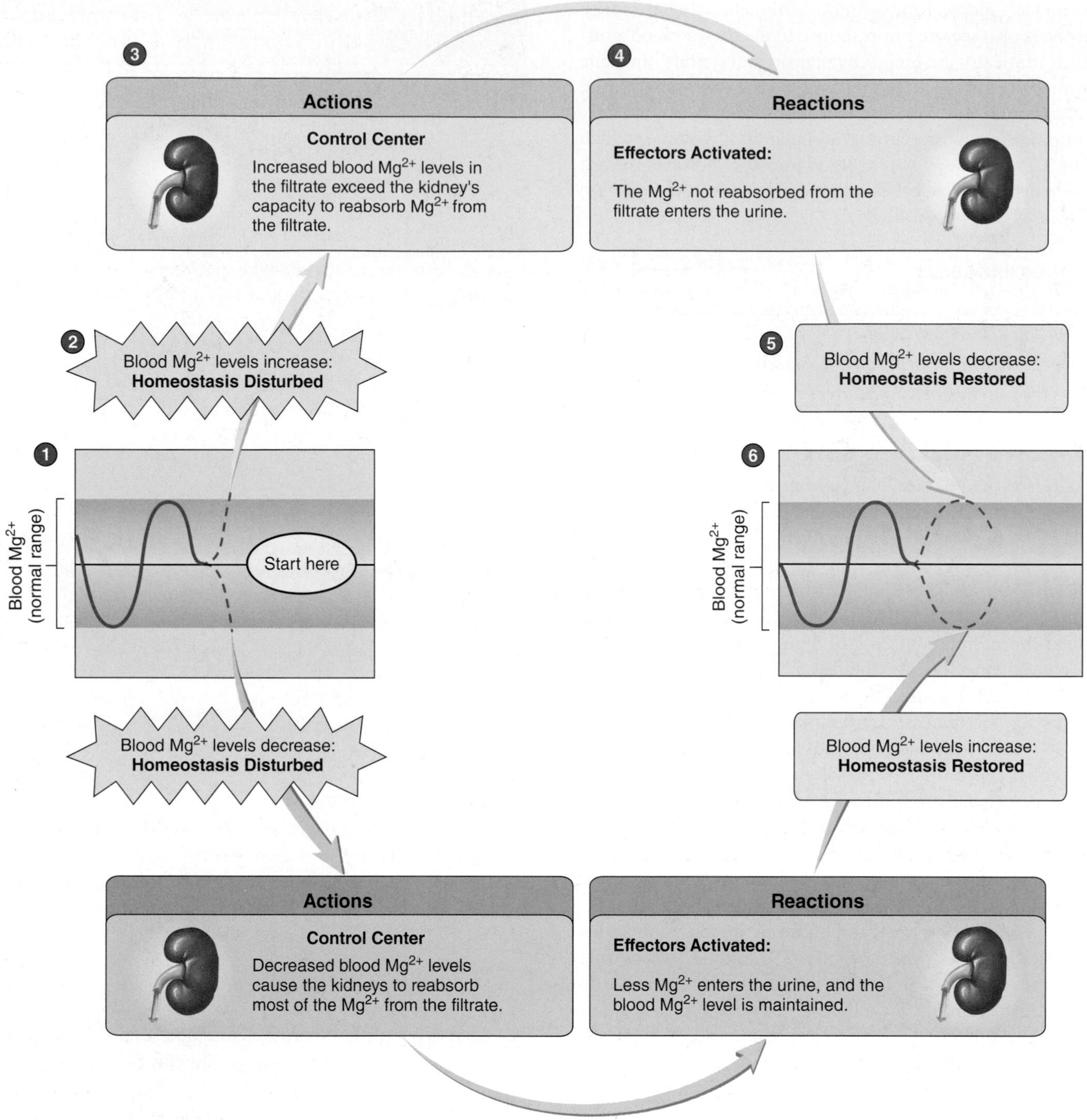

3 Actions
Control Center
Increased blood Mg^{2+} levels in the filtrate exceed the kidney's capacity to reabsorb Mg^{2+} from the filtrate.

4 Reactions
Effectors Activated:
The Mg^{2+} not reabsorbed from the filtrate enters the urine.

2 Blood Mg^{2+} levels increase:
Homeostasis Disturbed

5 Blood Mg^{2+} levels decrease:
Homeostasis Restored

1 Blood Mg^{2+} (normal range)
Start here

6 Blood Mg^{2+} (normal range)

Blood Mg^{2+} levels decrease:
Homeostasis Disturbed

Blood Mg^{2+} levels increase:
Homeostasis Restored

Actions
Control Center
Decreased blood Mg^{2+} levels cause the kidneys to reabsorb most of the Mg^{2+} from the filtrate.

Reactions
Effectors Activated:
Less Mg^{2+} enters the urine, and the blood Mg^{2+} level is maintained.

HOMEOSTASIS FIGURE 27.9 Summary of Blood Mg^{2+} Regulation

(1) Blood Mg^{2+} is in the normal range. (2) Blood Mg^{2+} increases outside the normal range, which causes homeostasis to be disturbed. (3) The control center responds to the change in blood Mg^{2+}. (4) The control center prevents Mg^{2+} reabsorption in the kidney. (5) These changes cause blood Mg^{2+} to decrease. (6) Blood Mg^{2+} returns to the normal range and homeostasis is restored. Observe the responses to a decrease in blood Mg^{2+} outside the normal range by following the *red arrows*.

However, greater-than-normal calcitonin levels in the blood do not reduce blood levels of Ca^{2+} to below-normal values. The major effect of calcitonin is in bone, where it inhibits osteoclasts. Thus, calcitonin prevents bone degeneration, which keeps blood Ca^{2+} levels from rising (see chapter 6). However, parathyroid hormone is more important than calcitonin for blood Ca^{2+} regulation.

ASSESS YOUR PROGRESS

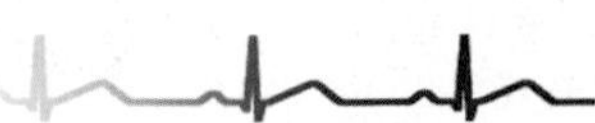

26. *Describe the effects of hypocalcemia and hypercalcemia on membrane potentials.*
27. *Explain the role of parathyroid hormone (PTH) in regulating extracellular Ca^{2+} concentration.*

28. *Discuss the role of vitamin D in regulating extracellular Ca^{2+} concentration.*

29. *Describe the role of calcitonin in regulating extracellular Ca^{2+} concentration.*

Regulation of Magnesium Ions

Most of the magnesium in the body is stored in the bones or intracellular fluid. Less than 1% of the total are ions in the extracellular fluid. Approximately one-half of those ions are bound to plasma proteins, and the rest are free. The free **magnesium ion (Mg^{2+})** concentration is 1.8–2.4 mEq/L. Magnesium ions are cofactors for intracellular enzymes, such as the sodium-potassium pump involved in actively transporting Na^+ out of and K^+ into cells. **Hypomagnesemia** is a below-normal blood level of magnesium, and **hypermagnesemia** is an above-normal blood level of magnesium. Low and high levels of plasma magnesium produce symptoms associated with the effect of magnesium on Na^+–K^+ active transport (table 27.9).

Free Mg^{2+} passes through the filtration membranes of the kidney into the filtrate. About 85–90% of those ions are reabsorbed from the filtrate, and only about 10–15% enter the urine. Of the Mg^{2+} reabsorbed, most is reabsorbed by the loop of Henle. The remainder is reabsorbed by the proximal convoluted tubule, distal convoluted tubule, and collecting duct.

The kidney's capacity to reabsorb Mg^{2+} is limited. If the level of free Mg^{2+} increases in the extracellular fluid, the excess Mg^{2+} remains in the filtrate, and the rate of Mg^{2+} loss in the urine increases. If the level of free Mg^{2+} decreases in the extracellular fluid, nearly all of the Mg^{2+} is reabsorbed, and the rate of Mg^{2+} loss in the urine decreases. The control of Mg^{2+} reabsorption is not clear, but a decreased extracellular concentration of Mg^{2+} causes an increased rate of reabsorption in the nephron (figure 27.9).

TABLE 27.9	Consequences of Abnormal Concentrations of Magnesium Ions
	HYPOMAGNESEMIA (rare)
Causes	Malnutrition
	Alcoholism
	Reduced magnesium intestinal absorption
	Renal tubular dysfunction
	Some diuretics
Symptoms	Irritability
	Muscle weakness
	Tetany
	Convulsions
	HYPERMAGNESEMIA (rare)
Causes	Renal failure
	Magnesium-containing antacids
Symptoms	Nausea
	Vomiting
	Muscle weakness
	Hypotension
	Bradycardia
	Reduced respiration

ASSESS YOUR PROGRESS 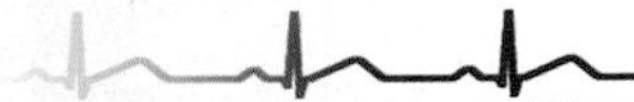

30. *In what part of the nephron is most Mg^{2+} reabsorbed?*

31. *What effect does a decreased extracellular concentration of Mg^{2+} have on its reabsorption in the nephron?*

Regulation of Phosphate Ions

About 85% of the phosphate in the body is in the form of calcium phosphate salts in bone (hydroxyapatite) and teeth. Most of the remaining phosphate is inside cells. Many of the phosphate ions are covalently bound to other organic molecules. Phosphate ions are bound to lipids (to form phospholipids), proteins, and carbohydrates, and they are important components of DNA, RNA, and ATP. Phosphates also play important roles in regulating enzyme activity, and phosphate ions dissolved in the intracellular fluid act as buffers (see section 27.5). The extracellular concentration of phosphate ions is between 1.7 and 2.6 mEq/L. Phosphate ions are in the form of $H_2PO_4^-$, HPO_4^{2-}, and PO_4^{3-}. The most common phosphate ion is HPO_4^{2-}.

The kidneys' capacity to reabsorb phosphate ions is limited. If the level of phosphate ions increases in the extracellular fluid, more phosphate is excreted in the urine. If the level of phosphate ions decreases in the extracellular fluid, less phosphate is excreted in the urine (figure 27.10).

Over time, a diet low in phosphate can increase the rate of phosphate reabsorption. Consequently, most of the phosphate that enters the filtrate is reabsorbed to maintain the extracellular phosphate concentration. Parathyroid hormone plays a significant role in regulating extracellular phosphate levels by promoting bone resorption, which releases Ca^{2+} and phosphate ions into the extracellular fluid. Thus, the kidneys do not need to conserve phosphate ions, and more are excreted in the urine. If phosphate levels in the extracellular fluid increase above normal levels, Ca^{2+} and phosphate ions precipitate as calcium phosphate salts in soft tissues.

Elevated blood levels of phosphate may occur with acute or chronic renal failure as a result of a critically reduced rate of filtrate formation by the kidneys. The rate of phosphate excretion is consequently reduced. Also, the chronic use of laxatives containing phosphates may cause elevated blood levels of phosphate. Symptoms of elevated phosphate levels are related to reduced blood Ca^{2+} levels because phosphate ions and Ca^{2+} precipitate out of solution and are deposited in the body's soft tissues. Prolonged elevation of blood levels of phosphate can result in calcium phosphate deposits in the joints and other tissues, such as the lungs and kidneys. A below-normal blood level of phosphate is called **hypophosphatemia** (hī′pō-fos-fă-tē′mē-ă), and an above-normal blood level of phosphate is called **hyperphosphatemia** (hī′per-fos-fă-tē′mē-ă). The consequences of increased and reduced plasma levels of phosphates are presented in table 27.10.

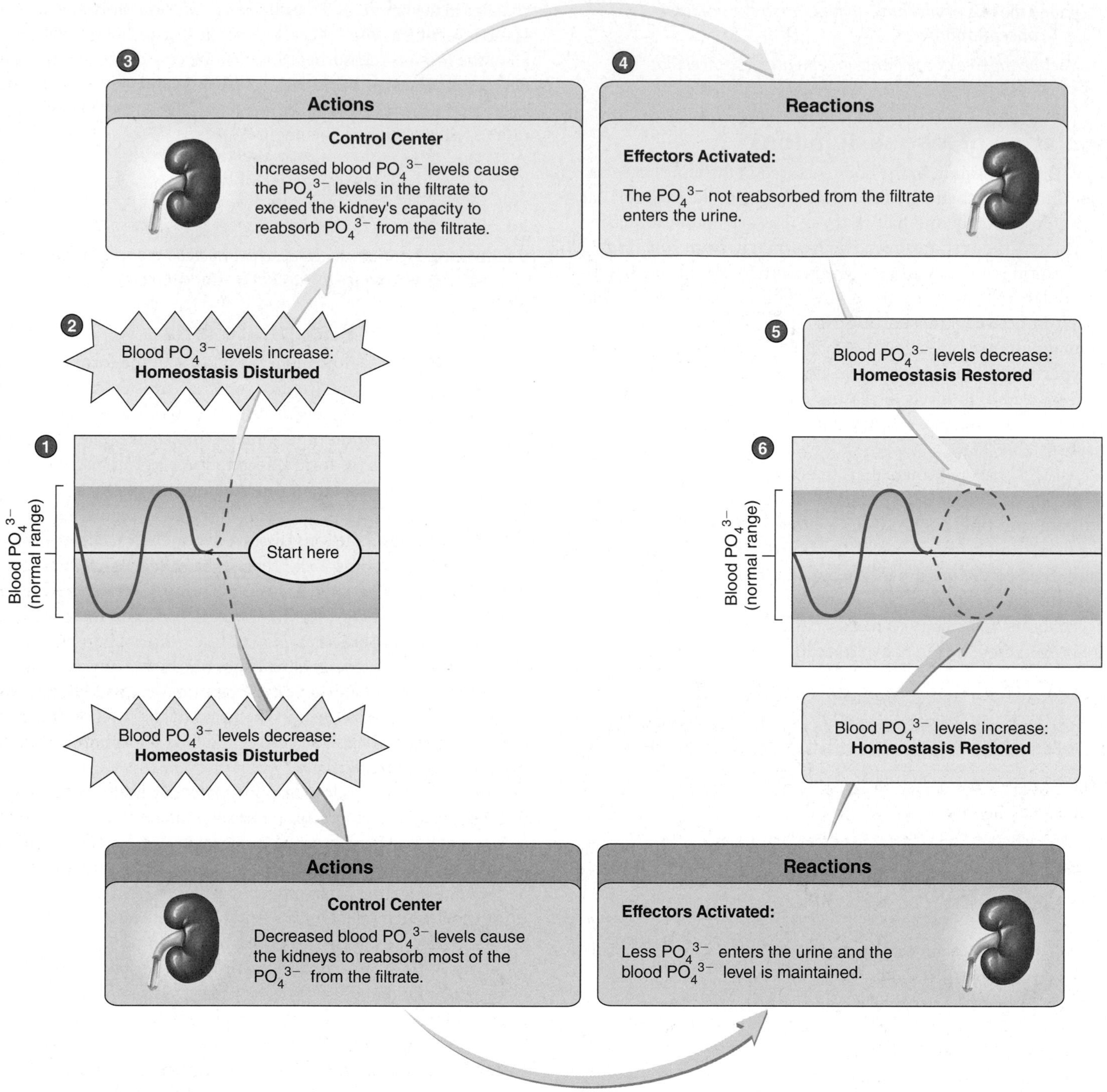

3

Actions

Control Center

Increased blood PO_4^{3-} levels cause the PO_4^{3-} levels in the filtrate to exceed the kidney's capacity to reabsorb PO_4^{3-} from the filtrate.

4

Reactions

Effectors Activated:

The PO_4^{3-} not reabsorbed from the filtrate enters the urine.

2

Blood PO_4^{3-} levels increase:
Homeostasis Disturbed

5

Blood PO_4^{3-} levels decrease:
Homeostasis Restored

1

Blood PO_4^{3-} (normal range)

Start here

6

Blood PO_4^{3-} (normal range)

Blood PO_4^{3-} levels decrease:
Homeostasis Disturbed

Blood PO_4^{3-} levels increase:
Homeostasis Restored

Actions

Control Center

Decreased blood PO_4^{3-} levels cause the kidneys to reabsorb most of the PO_4^{3-} from the filtrate.

Reactions

Effectors Activated:

Less PO_4^{3-} enters the urine and the blood PO_4^{3-} level is maintained.

HOMEOSTASIS FIGURE 27.10 Summary of Blood Phosphate Ion Regulation

(1) Blood PO_4^{3-} is in the normal range. (2) Blood PO_4^{3-} increases outside the normal range, which causes homeostasis to be disturbed. (3) The control center responds to the change in blood PO_4^{3-}. (4) The control center prevents PO_4^{3-} reabsorption in the kidney. (5) These changes cause blood PO_4^{3-} to decrease. (6) Blood PO_4^{3-} returns to the normal range and homeostasis is restored. Observe the responses to a decrease in blood PO_4^{3-} outside the normal range by following the *red arrows*.

ASSESS YOUR PROGRESS

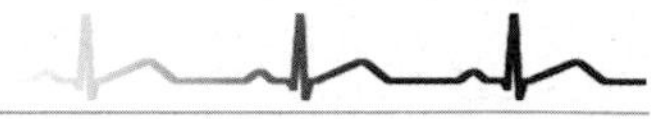

32. *Where is most of the phosphate in the body located? What is the most common phosphate ion?*

33. *Explain how the kidneys control plasma levels of phosphate ions.*

34. *How does an increased level of PTH affect tubular phosphate reabsorption?*

35. *What are the consequences of prolonged elevation of blood phosphate ions?*

TABLE 27.10	Consequences of Abnormal Concentrations of Phosphate Ions
	HYPOPHOSPHATEMIA
Causes	Reduced intestinal absorption due to vitamin D deficiency or alcohol abuse
	Hyperparathyroidism (elevated renal PO_4– excretion)
Symptoms	Reduced metabolic rate
	Reduced oxygen transport
	Reduced blood clotting
	Reduced white blood cell functions
	HYPERPHOSPHATEMIA
Causes	Renal failure
	Tissue destruction from chemotherapy
	Hyperparathyroidism (reduced renal PO_4– excretion)
Symptoms	Formation of calcium phosphate deposits in tissues of lungs, kidneys and joints
	Symptoms of reduced Ca^{2+} related to formation of deposits

27.5 Regulation of Acid-Base Balance

LEARNING OUTCOMES

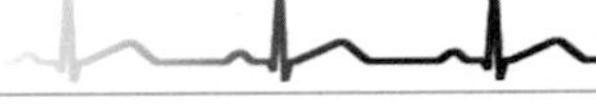

After reading this section, you should be able to

A. Demonstrate an understanding of acids and bases and their relationship with buffers.

B. Explain the actions of the three buffer systems of the body.

C. Describe the mechanisms of acid-base balance.

D. Explain the causes and effects of acid-base imbalances.

Hydrogen ions affect the activity of enzymes and interact with many electrically charged molecules. Consequently, most chemical reactions within the body are highly sensitive to the H^+ concentration of the fluid in which they occur. Maintaining the H^+ concentration within a narrow range of values is essential for normal metabolic reactions. The H^+ concentration is determined by acids and bases in the body. Recall from chapter 2 that the pH of a solution is a measure of H^+ concentration. However, the pH scale can be a little confusing because the relationship between the pH level and H^+ concentration is an inverse one. The lower the pH level, the greater the H^+ concentration. Likewise, the higher the pH level, the lower the H^+ concentration (see figure 2.13).

Acids and Bases

Acids release H^+ into a solution; bases remove H^+ from a solution. Acids and bases are classified as either strong or weak. A **strong acid** completely dissociates into its component ions; for example, hydrochloric acid dissociates into H^+ and Cl^- (figure 27.11). A **strong base** also completely dissociates into its component ions, which results in the removal of H^+ from a solution. For example, NaOH dissociates into Na^+ and OH^-. The OH^- can then react with H^+ to form H_2O, thus removing H^+ from the solution.

Like a strong acid, a **weak acid** releases H^+ into a solution; however, it does not completely dissociate into its component ions. The weak acid releases H^+ until equilibrium with the surrounding solution is reached, and then no more H^+ dissociates. Weak acids are common in living systems, and they play important roles in preventing large changes in body fluid pH. A **weak base**, such as NH_3^+, reduces the concentration of H^+ ions in a solution by binding to free H^+.

ASSESS YOUR PROGRESS

36. *Define* acid *and* base. *What is the relationship between pH and H^+ concentration?*

37. *Describe weak acids. Why are weak acids important in living systems?*

Mechanisms of Acid-Base Balance Regulation

The three major mechanisms that regulate H^+concentration are buffer systems, the respiratory system, and the kidneys. Buffers and other mechanisms of acid-base balance regulation work together to regulate acid-base balance (figure 27.12). Buffers almost instantaneously resist changes in the pH of body fluids, but the regulation of respiration and the function of the kidneys also play essential roles. The respiratory system responds within a few minutes to bring the pH of body fluids back toward its normal range. Still, the respiratory system's capacity to regulate pH is not as great as that of the kidneys, nor does the respiratory system have the same ability to return the pH to its precise range of normal values. In contrast, the kidneys respond more slowly, within hours to days, to alterations of body fluid pH, and their capacity to respond is substantial.

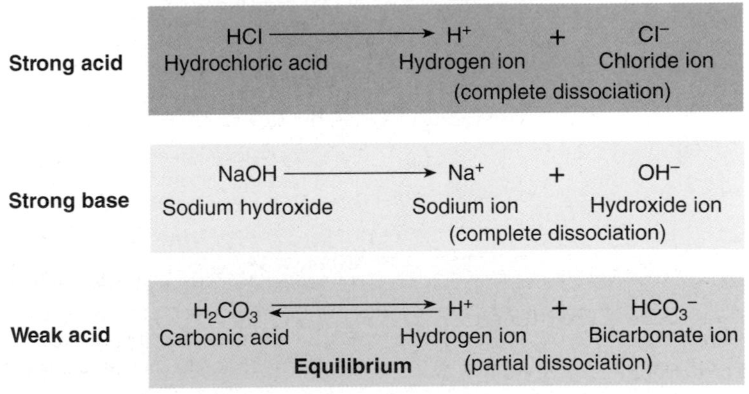

Strong acid
HCl → H+ + Cl–
Hydrochloric acid
Hydrogen ion
Chloride ion
(complete dissociation)
Strong base
NaOH → Na+ + OH–
Sodium hydroxide
Sodium ion
Hydroxide ion
(complete dissociation)
Weak acid
H2CO3 ⇌ H+ + HCO3–
Carbonic acid
Hydrogen ion
Bicarbonate ion
Equilibrium
(partial dissociation)

FIGURE 27.11 Comparison of Strong Acids and Bases with Weak Acids

Strong acids and bases completely dissociate when dissolved in water. Weak acids do not completely dissociate. Instead, weak acids partially dissociate, so that an equilibrium is established between the acid and the ions that are formed when the dissociation occurs.

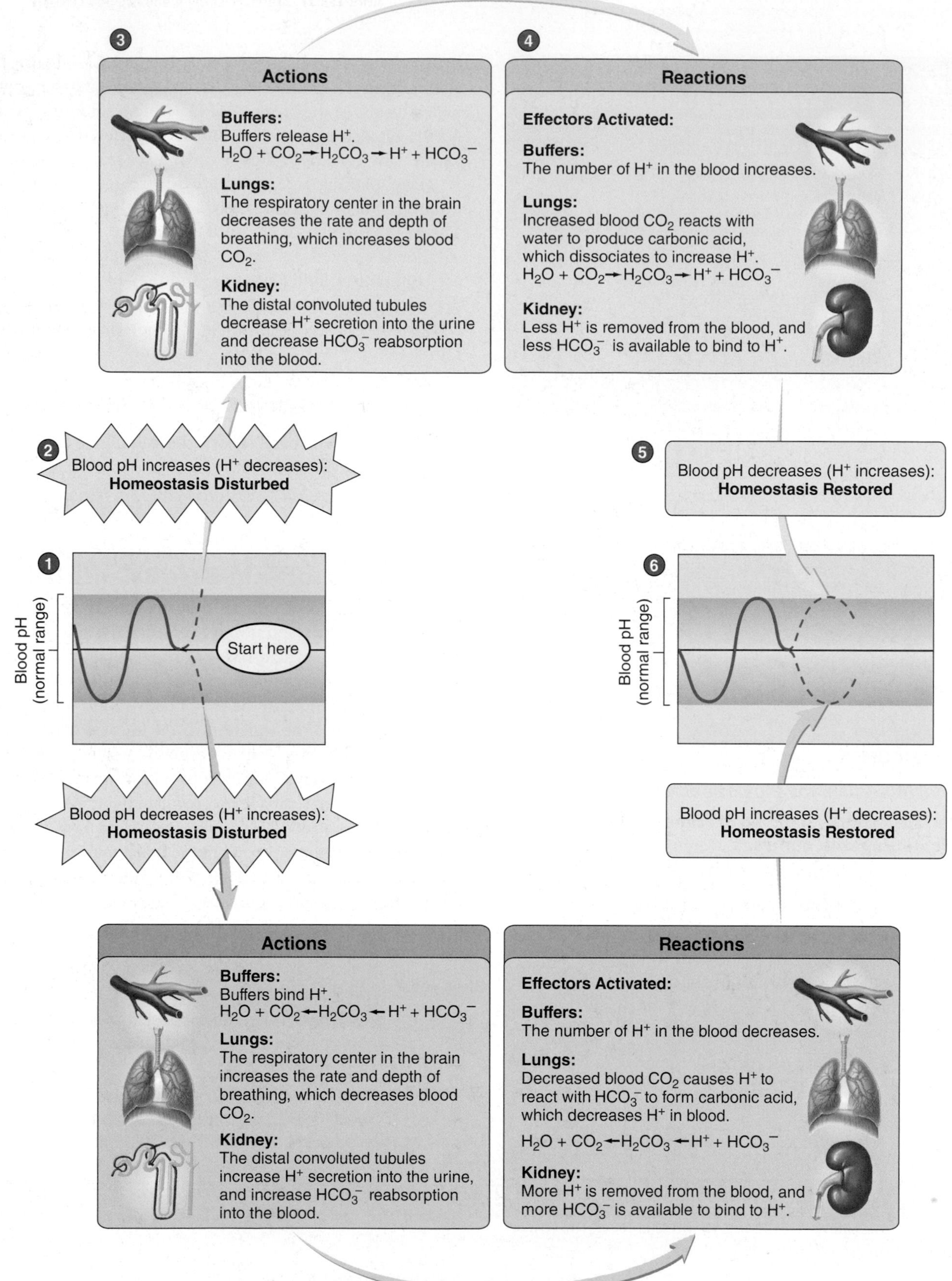
3
Actions
Buffers:
Buffers release H+.
H2O + CO2 → H2CO3 → H+ + HCO3−
Lungs:
The respiratory center in the brain decreases the rate and depth of breathing, which increases blood CO2.
Kidney:
The distal convoluted tubules decrease H+ secretion into the urine and decrease HCO3− reabsorption into the blood.
4
Reactions
Effectors Activated:
Buffers:
The number of H+ in the blood increases.
Lungs:
Increased blood CO2 reacts with water to produce carbonic acid, which dissociates to increase H+.
H2O + CO2 → H2CO3 → H+ + HCO3−
Kidney:
Less H+ is removed from the blood, and less HCO3− is available to bind to H+.
2
Blood pH increases (H+ decreases):
Homeostasis Disturbed
5
Blood pH decreases (H+ increases):
Homeostasis Restored
1
Blood pH (normal range)
Start here
6
Blood pH (normal range)
Blood pH decreases (H+ increases):
Homeostasis Disturbed
Blood pH increases (H+ decreases):
Homeostasis Restored
Actions
Buffers:
Buffers bind H+.
H2O + CO2 ← H2CO3 ← H+ + HCO3−
Lungs:
The respiratory center in the brain increases the rate and depth of breathing, which decreases blood CO2.
Kidney:
The distal convoluted tubules increase H+ secretion into the urine, and increase HCO3− reabsorption into the blood.
Reactions
Effectors Activated:
Buffers:
The number of H+ in the blood decreases.
Lungs:
Decreased blood CO2 causes H+ to react with HCO3− to form carbonic acid, which decreases H+ in blood.
H2O + CO2 ← H2CO3 ← H+ + HCO3−
Kidney:
More H+ is removed from the blood, and more HCO3− is available to bind to H+.

HOMEOSTASIS FIGURE 27.12 Summary of Acid-Base Balance Regulation

(1) Blood pH is in the normal range. (2) Blood pH increases outside the normal range, which causes homeostasis to be disturbed. (3) The blood pH control centers respond to the change in blood pH. (4) The control centers cause decreased H^+ secretion from the blood and increased carbonic acid production, which increases blood H^+ concentration. (5) These changes cause blood pH to decrease. (6) Blood pH returns to the normal range and homeostasis is restored. Observe the responses to a decrease in blood pH outside the normal range by following the *red arrows*.

Buffer Systems

Buffers (bŭf′erz; see chapter 2) resist changes in the pH of a solution. Buffers within body fluids stabilize the pH by chemically binding to excess H^+ when H^+ is added to a solution or by releasing H^+ when the H^+ concentration in a solution begins to fall.

Three important buffer systems function together to resist changes in the pH of body fluids: the carbonic acid/bicarbonate buffer system; the protein buffer system, such as hemoglobin and plasma proteins; and the phosphate buffer system (table 27.11).

Carbonic Acid/Bicarbonate Buffer System

Carbonic acid (H_2CO_3) is a weak acid formed when CO_2 and water mix. When it is dissolved in water, the following equilibrium is established:

$$H_2CO_3 \rightleftarrows HCO_3^- + H^+$$

The carbonic acid/bicarbonate buffer system depends on the equilibrium that is quickly established between H_2CO_3 and the H^+ and bicarbonate (HCO_3^-). When H^+ is added to this solution (by adding a small amount of a strong acid), a large proportion of the H^+ binds to HCO_3^- to form H_2CO_3, and only a small percentage remain as free H^+. Thus, the carbonic acid/bicarbonate buffer system resists a large decrease in pH when acidic substances are added to a solution containing H_2CO_3.

When H^+ is removed from a solution containing H_2CO_3 (by adding a small amount of a strong base), much of the H_2CO_3 forms HCO_3^- and H^+. Thus, a large increase in pH is resisted when basic substances are added to a solution containing H_2CO_3.

The carbonic acid/bicarbonate buffer system plays an important role in regulating the extracellular pH. It quickly responds to the addition of substances such as CO_2 or lactic acid produced by increased metabolism during exercise (see chapter 23) and to increased fatty acid and ketone body production during periods of elevated fat metabolism (see chapter 25). It also responds to the addition of basic substances, such as large amounts of $NaHCO_3$ consumed as an antacid. The carbonic acid/bicarbonate buffer system has a limited capacity to resist changes in pH, but it plays an essential role in the control of pH by both the respiratory system and the kidneys.

Protein Buffer System

Intracellular proteins and plasma proteins form a large pool of buffer molecules. They provide approximately three-fourths of the body's buffer capacity because of their high concentration. Hemoglobin in red blood cells is one of the most important intracellular proteins. Other intracellular molecules associated with nucleic acids, such as histone proteins, also act as buffers. The capacity of proteins to function as buffers is due to the functional groups of amino acids, such as carboxyl (–COOH) or amino ($-NH_2$) groups, which can act as weak acids and bases. Consequently, as the H^+ concentration increases, more H^+ binds to the functional groups; when the H^+ concentration decreases, H^+ is released from the functional groups (table 27.11).

Phosphate Buffer System

The concentration of phosphate and phosphate-containing molecules is low in the extracellular fluid, compared with that of the other buffer systems, but the phosphate buffer system is an important intracellular buffer system. Phosphate-containing molecules in solution, such as DNA, RNA, ATP, and phosphate ions, act as buffers. When the pH decreases, ions, such as HPO_4^-, bind H^+ to form $H_2PO_4^-$; however, when the pH becomes more basic, H_2PO_4 releases H^+ into solution. In this way, these two ions fluctuate between gaining and losing H^+ ions, which helps balance the pH (table 27.11).

ASSESS **YOUR PROGRESS** 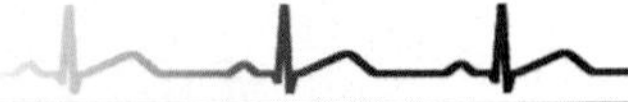

38. *What are the three major mechanisms that regulate H^+ concentration?*
39. *Compare the capacity and rate at which the respiratory system and kidneys control body fluid pH.*
40. *Define* buffer. *Describe how a buffer works when H^+ is added to a solution or when it is removed from a solution.*
41. *Name the three buffer systems of the body. Which of these systems provides the largest proportion of buffer capacity?*

Respiratory Regulation of Acid-Base Balance

The respiratory system regulates acid-base balance by influencing the carbonic acid/bicarbonate buffer system. Carbon dioxide (CO_2) reacts with water (H_2O) to form carbonic acid (H_2CO_3), which dissociates to form H^+ and HCO_3^- as follows:

$$H_2O + CO_2 \leftrightarrows H_2CO_3 \leftrightarrows H^+ + HCO_3^-$$

This reaction is in equilibrium. As CO_2 increases, CO_2 combines with H_2O. The higher the concentration of CO_2, the greater the amount of H_2CO_3 formed. Many H_2CO_3 molecules then dissociate to form H^+ and HCO_3^-. However, if CO_2 levels decline, the equilibrium shifts in the opposite direction, so that many H^+ and HCO_3^- ions combine to form H_2CO_3, which then forms CO_2 and H_2O. Thus, both H^+ and HCO_3^- decrease in the solution.

TABLE 27.11 Characteristics of Buffer Systems

Carbonic acid/ bicarbonate buffer system	Components of the carbonic acid/bicarbonate buffer system are not present in high enough concentrations in the extracellular fluid to constitute a powerful buffer system. However, the concentrations of the components of the buffer system are regulated. Therefore, it plays an exceptionally important role in controlling the pH of extracellular fluid.
Protein buffer system	Intracellular proteins and plasma proteins form a large pool of protein molecules that can act as buffer molecules. Because of their high concentration, they provide approximately three-fourths of the body's buffer capacity. Hemoglobin in red blood cells is an important intracellular protein. Other intracellular molecules, such as histone proteins and nucleic acids, also act as buffers.
Phosphate buffer system	Components of the phosphate buffer system are low in the extracellular fluids, compared with the other buffer systems, but it is an important intracellular buffer system.

FUNDAMENTAL **Figure**

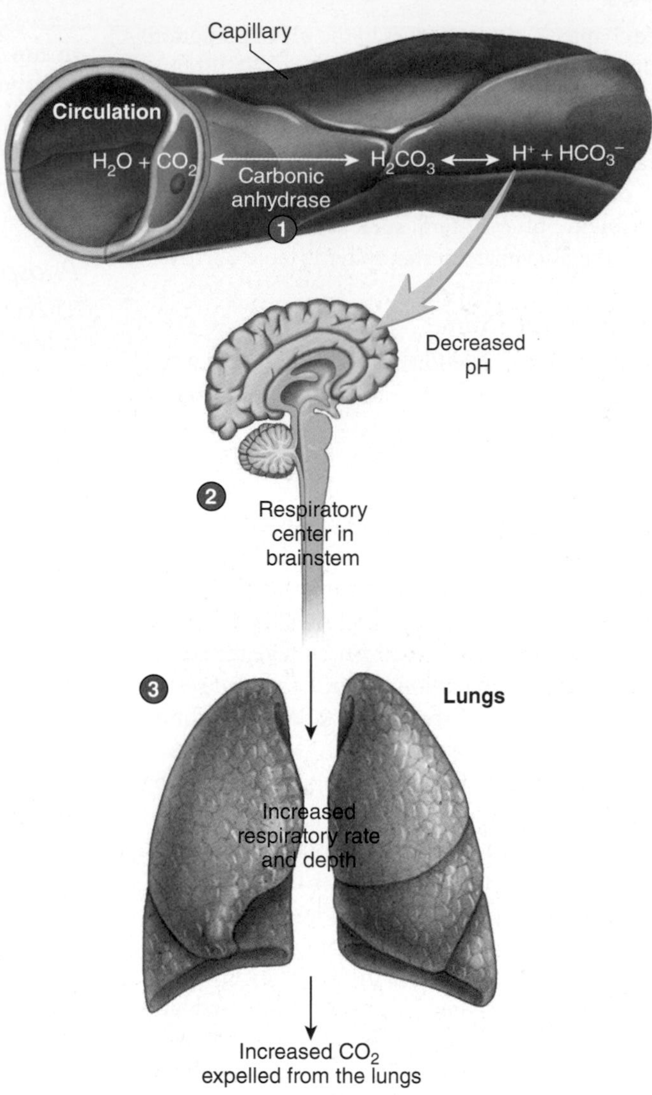

Capillary
Circulation
$H_2O + CO_2 \longleftrightarrow H_2CO_3 \longleftrightarrow H^+ + HCO_3^-$
Carbonic anhydrase
1
Decreased pH
2
Respiratory center in brainstem
3
Lungs
Increased respiratory rate and depth
Increased CO_2 expelled from the lungs

1. Carbon dioxide reacts with H_2O to form H_2CO_3. An enzyme, carbonic anhydrase, found in red blood cells and on the surface of blood vessel epithelium, catalyzed the reaction. Carbonic acid dissociates to form H^+ and HCO_3^-. An equilibrium is quickly established.

2. Decreased pH in the extracellular fluid stimulates the respiratory center and causes an increased rate and depth of breathing.

3. Increased rate and depth of breathing causes CO_2 to be expelled from the lungs, thus reducing the extracellular CO_2 levels. As CO_2 levels decrease,the extracellular concentration of H^+ decreases, and the extracellular fluid pH increases.

PROCESS FIGURE 27.13 Respiratory Regulation of Body Fluid Acid-Base Balance

The reaction between CO_2 and H_2O is catalyzed by an enzyme, **carbonic anhydrase,** which is found in a relatively high concentration in red blood cells and on the surface of capillary epithelial cells (figure 27.13). This enzyme does not influence equilibrium but accelerates the rate at which the reaction proceeds in either direction, so that equilibrium is achieved quickly.

Decreases in body fluid pH, regardless of the cause, stimulate neurons in the respiratory center in the brainstem and cause the rate and depth of ventilation to increase. The increased rate and depth of ventilation cause CO_2 to be eliminated from the body through the lungs at a greater rate, and the concentration of CO_2 in the body fluids decreases. As CO_2 levels decline, the carbonic acid/bicarbonate buffer system reacts. Hydrogen ions combine with HCO_3^- to form H_2CO_3, which then form CO_2 and H_2O. Consequently, the concentration of H^+ decreases toward its normal range as CO_2 exits through the respiratory system (figure 27.13).

Increases in body fluid pH, regardless of the cause, inhibit neurons in the respiratory center in the brainstem and cause the rate and depth of ventilation to decrease. The decreased rate and depth of ventilation cause less CO_2 to be eliminated from the body through the lungs. The concentration of CO_2 in the body fluids increases because CO_2 is continually produced as a by-product of metabolism in all tissues. Thus, the body fluid concentration of H_2CO_3 increases also. As H_2CO_3 increases, many H_2CO_3 molecules dissociate to form H^+ and HCO_3^-. This results in an elevated H^+ concentration, and the pH decreases toward its normal range.

Predict 3

Under stressful conditions, some people hyperventilate. What effect does the rapid rate of ventilation have on blood pH? Explain why breathing into a paper bag helps a hyperventilating person.

FUNDAMENTAL **Figure**

1. When the filtrate or blood pH decreases, H^+ combine with HCO_3^- to form carbonic acid that is converted into CO_2 and H_2O. The CO_2 diffuses into tubule cells.
2. In the tubule cells, CO_2 combines with H_2O to form H_2CO_3 that dissociates to form H^+ and HCO_3^-.
3. An antiport mechanism secretes H^+ into the filtrate in exchange for Na^+ from the filtrate. As a result, filtrate pH decreases.
4. Bicarbonate ions are symported with Na^+ into the interstitial fluid. They then diffuse into capillaries.
5. In capillaries, HCO_3^- combine with H^+. This decreases the H^+ concentration and increases blood pH.

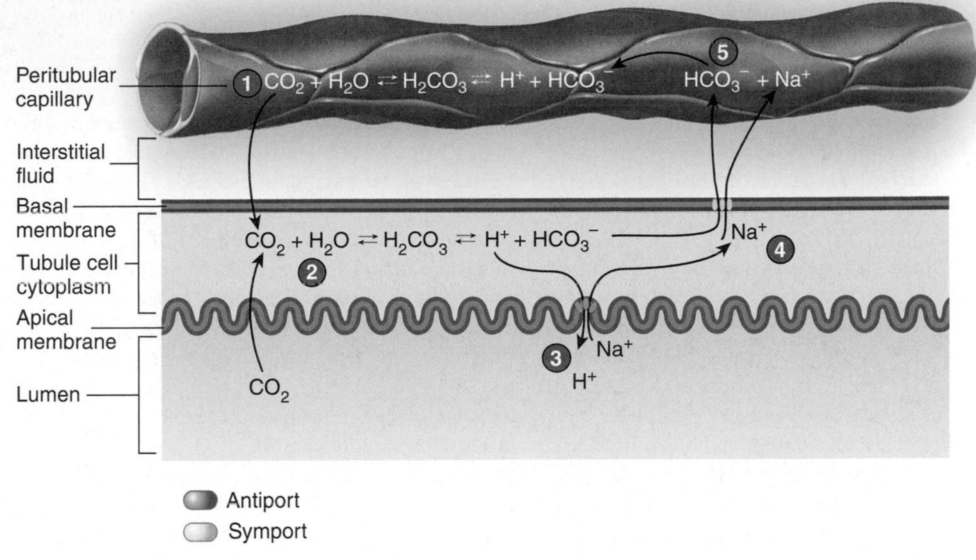

Peritubular capillary
Interstitial fluid
Basal membrane
Tubule cell cytoplasm
Apical membrane
Lumen
1 $CO_2 + H_2O \rightleftharpoons H_2CO_3 \rightleftharpoons H^+ + HCO_3^-$
5 $HCO_3^- + Na^+$
$CO_2 + H_2O \rightleftharpoons H_2CO_3 \rightleftharpoons H^+ + HCO_3^-$
2
Na^+
4
3
Na^+
H^+
CO_2
Antiport
Symport

PROCESS FIGURE 27.14 Kidney Regulation of Body Fluid Acid-Base Balance
As the extracellular pH decreases, the rate of H^+ secretion by the kidneys' tubule cells and HCO_3^- reabsorption increase.

Renal Regulation of Acid-Base Balance

Cells of the kidney tubules directly regulate acid-base balance by increasing or decreasing both the rate of H^+ secretion into the filtrate (figure 27.14) and the rate of HCO_3^- reabsorption. Carbonic anhydrase within nephron cells catalyzes the formation of H_2CO_3 from CO_2 and H_2O. The carbonic acid molecules dissociate to form H^+ and HCO_3^-. An antiport system then exchanges H^+ for Na^+ across the apical membrane of the cells. Thus, cells of the kidney tubules secrete H^+ into the filtrate and reabsorb Na^+. The Na^+ and HCO_3^- are symported across the basal membrane. After the Na^+ and HCO_3^- are symported from the kidney tubule cells, they diffuse into the peritubular capillaries. As a result, H^+ is secreted into the lumens of the kidney tubules, and HCO_3^- passes into the extracellular fluid.

The reabsorbed HCO_3^- combines with excess H^+ in the extracellular fluid to form H_2CO_3. This combination removes H^+ from the extracellular fluid and increases extracellular pH. The rate of H^+ secretion and HCO_3^- reabsorption increases when the pH of the body fluids decreases, and this process slows when the pH of the body fluids increases (figure 27.14).

Some of the H^+ secreted by cells of the kidney tubules into the filtrate combines with HCO_3^-, which enters the filtrate through the filtration membrane in the form of sodium bicarbonate ($NaHCO_3$). Within the kidney tubules, H^+ combines with HCO_3^- to form H_2CO_3, which then dissociates to form CO_2 and H_2O. The CO_2 diffuses from the filtrate into the cells of the kidney tubules, where it can react with H_2O to form H_2CO_3, which subsequently dissociates to form H^+ and HCO_3^- (figure 27.15). Once again, H^+ is antiported into the lumens of the kidney tubules in exchange for Na^+, whereas HCO_3^- enters the extracellular fluid. As a result, much of the HCO_3^- entering the filtrate reenters the extracellular fluid.

Hydrogen ions secreted into the nephron normally exceed the amount of HCO_3^- that enters the kidney tubules through the filtration membrane. Because the H^+ combines with HCO_3^-, almost all the HCO_3^- is reabsorbed from the kidney tubules (figure 27.15). Little HCO_3^- is excreted in the urine unless the pH of the body fluids becomes elevated.

The rate of H^+ secretion into the filtrate and the rate of HCO_3^- reabsorption into the extracellular fluid decrease if the pH of the body fluids increases. As a result, the amount of bicarbonate filtered into the kidney tubules exceeds the amount of secreted H^+, and the excess HCO_3^- passes into the urine. The excretion of excess HCO_3^- in the urine diminishes the amount of HCO_3^- in the extracellular fluid. This allows the amount of extracellular H^+ to increase; as a consequence, the pH of the body fluids decreases toward its normal range.

Although the kidney tubule cells respond directly to H^+, the hormone aldosterone can also alter the H^+ permeability of the kidney tubules. Aldosterone increases the rate of Na^+ reabsorption and K^+ secretion by the kidneys, but in high concentrations aldosterone also stimulates H^+ secretion. Elevated aldosterone levels, such as those occurring in patients with Cushing syndrome, can therefore elevate body fluid pH above normal (alkalosis). However, the major influence on the rate of H^+ secretion is the pH of the body fluids.

Predict 4

Predict the effect of aldosterone hyposecretion on body fluid pH.

The secretion of H^+ into the urine can decrease the filtrate pH to approximately 4.5. A filtrate pH below 4.5 inhibits the secretion of additional H^+. The H^+ that passes into the filtrate is more than the amount required to decrease the pH of an unbuffered solution below 4.5. Buffers in the filtrate combine with many of the secreted H^+.

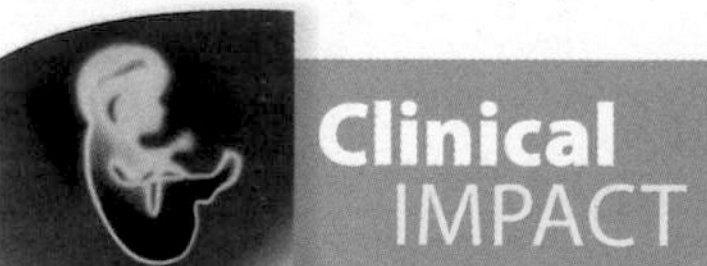
Clinical
IMPACT

Acidosis and Alkalosis

The normal pH of the body fluids is between 7.35 and 7.45. When the pH is below 7.35, the condition is called **acidosis** (as-i-dō′sis); when the pH is above 7.45, it is called **alkalosis** (al′kă-lō′sis).

Metabolism produces acidic products that lower the pH of the body fluids. For example, CO_2 is a by-product of metabolism that combines with water to form H_2CO_3. Likewise, anaerobic respiration produces lactic acid, protein metabolism produces phosphoric and sulfuric acids, and lipid metabolism produces fatty acids. These acidic substances must continuously be eliminated from the body to maintain homeostatic pH. Failure to eliminate the acidic products of metabolism results in acidosis. Excess elimination of the acidic products of metabolism results in alkalosis.

The major effect of acidosis is depression of the central nervous system. When blood pH falls below 7.35, the central nervous system malfunctions. The individual becomes disoriented, and possibly comatose as the condition worsens.

A major effect of alkalosis is hyperexcitability of the nervous system. Peripheral nerves are affected first, resulting in spontaneous nervous stimulation of muscles. Spasms, tetanic contractions, and possibly extreme nervousness or convulsions result. Severe alkalosis can cause death as a result of tetany of the respiratory muscles.

Although buffers help resist changes in the pH of body fluids, the respiratory system and the kidneys regulate the pH of the body fluids. Therefore, malfunctions in either the respiratory system or the kidneys can result in acidosis or alkalosis.

Acidosis and alkalosis are categorized according to the cause of the condition. **Respiratory acidosis** and **alkalosis** result from abnormalities in the respiratory system. **Metabolic acidosis** and **alkalosis** result from all causes other than abnormal respiratory functions.

Inadequate ventilation of the lungs causes respiratory acidosis (table 27A). The rate at which CO_2 is eliminated from the body fluids through the lungs falls. This increases the concentration of CO_2 in the body fluids. As CO_2 levels increase, CO_2 reacts with water to form H_2CO_3. Carbonic acid forms H^+ and HCO_3^-. The increase in H^+ concentration causes the pH of the body fluids to decrease. If the pH falls below 7.35, the symptoms of respiratory acidosis appear.

Buffers help resist a decrease in pH, and the kidneys help compensate for the failure of the lungs to prevent respiratory acidosis by increasing the rate at which they secrete H^+ into the filtrate and reabsorb HCO_3^-. However, the capacity of buffers to resist changes in pH can be exceeded, and a period of 1–2 days is required for the kidneys to become maximally functional. Thus, the kidneys are not effective if respiratory acidosis develops quickly. On the other hand, the kidneys are very effective if respiratory acidosis develops slowly, or if it lasts long enough for the kidneys to respond. For example, the kidneys cannot compensate for respiratory acidosis occurring in response to a severe asthma attack that begins quickly and subsides within hours. If, however, respiratory acidosis results from emphysema, which develops over a long time, the kidneys play a significant role in helping compensate.

Respiratory alkalosis results from hyperventilation of the lungs (table 27A). This increases the rate at which CO_2 is eliminated from the body fluids and results in a decreased concentration of CO_2 in the body fluids. As CO_2 levels decrease, H^+ reacts with HCO_3^- to form H_2CO_3. The H_2CO_3 forms H_2O and CO_2. The resulting decrease in the concentration of H^+ causes the pH of the body fluids to increase. If the pH increases above 7.45, the symptoms of respiratory alkalosis appear.

The kidneys help compensate for respiratory alkalosis by decreasing the rate of H^+ secretion into the filtrate and the rate of HCO_3^- reabsorption. If pH increases, the kidneys need 1–2 days to compensate. Thus, the kidneys are not effective if respiratory alkalosis develops quickly. However, they are very effective if respiratory alkalosis develops slowly. For example, the kidneys are not effective in compensating for respiratory alkalosis that occurs in response to hyperventilation triggered by emotions, which usually begins quickly and subsides within minutes or hours. However, if alkalosis results from staying at a high altitude over a 2- or 3-day period, the kidneys play a significant role in helping compensate.

Metabolic acidosis results from all conditions that decrease the pH of the body fluids below 7.35, with the exception of those resulting from altered function of the respiratory system (table 27A). As H^+ accumulates in the body fluids, buffers first resist a decline in pH. If the buffers cannot compensate for the increase in H^+, the respiratory center helps regulate body fluid pH. The reduced pH stimulates the respiratory center, which causes hyperventilation. During hyperventilation, CO_2 is eliminated at a greater rate. The elimination of CO_2 also eliminates excess H^+ and helps maintain the pH of the body fluids within a normal range.

If metabolic acidosis persists for many hours and if the kidneys are functional, the kidneys can also help compensate for metabolic acidosis by secreting H^+ at a greater rate and increasing the rate of HCO_3^- reabsorption. The symptoms of metabolic acidosis appear if the respiratory and renal systems are not able to maintain the pH of the body fluids within its normal range.

Metabolic alkalosis results from all conditions that increase the pH of the body fluids above 7.45, with the exception of those resulting from altered function of the respiratory system. As H^+ decreases in the body fluids, buffers first resist an increase in pH. If the buffers cannot compensate for the decrease in H^+, the respiratory center helps regulate body fluid pH. The increased pH inhibits respiration. Reduced respiration allows CO_2 to accumulate in the body fluids. Carbon dioxide reacts with water to produce H_2CO_3. If metabolic alkalosis persists for several hours and if the kidneys are functional, the kidneys reduce the rate of H^+ secretion to help reverse alkalosis (table 27A).

—Continued

TABLE 27A	Acidosis and Alkalosis
Acidosis	
Respiratory Acidosis	
Reduced elimination of CO_2 from the body fluids	
Asphyxia	
Hypoventilation (e.g., impaired respiratory center function due to trauma, tumor, shock, or renal failure)	
Advanced asthma	
Severe emphysema	
Metabolic Acidosis	
Elimination of large amounts of HCO_3^- resulting from mucous secretion (e.g., severe diarrhea and vomiting of lower intestinal contents)	
Direct reduction of body fluid pH as acid is absorbed (e.g., ingestion of acidic drugs, such as aspirin)	
Production of large amounts of fatty acids and other acidic metabolites, such as ketone bodies (e.g., untreated diabetes mellitus)	
Inadequate oxygen delivery to tissue, resulting in anaerobic respiration and lactic acid buildup (e.g., exercise, heart failure, or shock)	
Alkalosis	
Respiratory Alkalosis	
Reduced CO_2 levels in the extracellular fluid (e.g., hyperventilation due to emotions)	
Decreased atmospheric pressure, causing reduced oxygen levels, which stimulate the chemoreceptor reflex (e.g., cause hyperventilation at high altitudes)	
Metabolic Alkalosis	
Elimination of H^+ and reabsorption of HCO_3^- in the stomach or kidneys (e.g., severe vomiting or formation of acidic urine in response to excess aldosterone)	
Ingestion of alkaline substances (e.g., large amounts of sodium bicarbonate)	

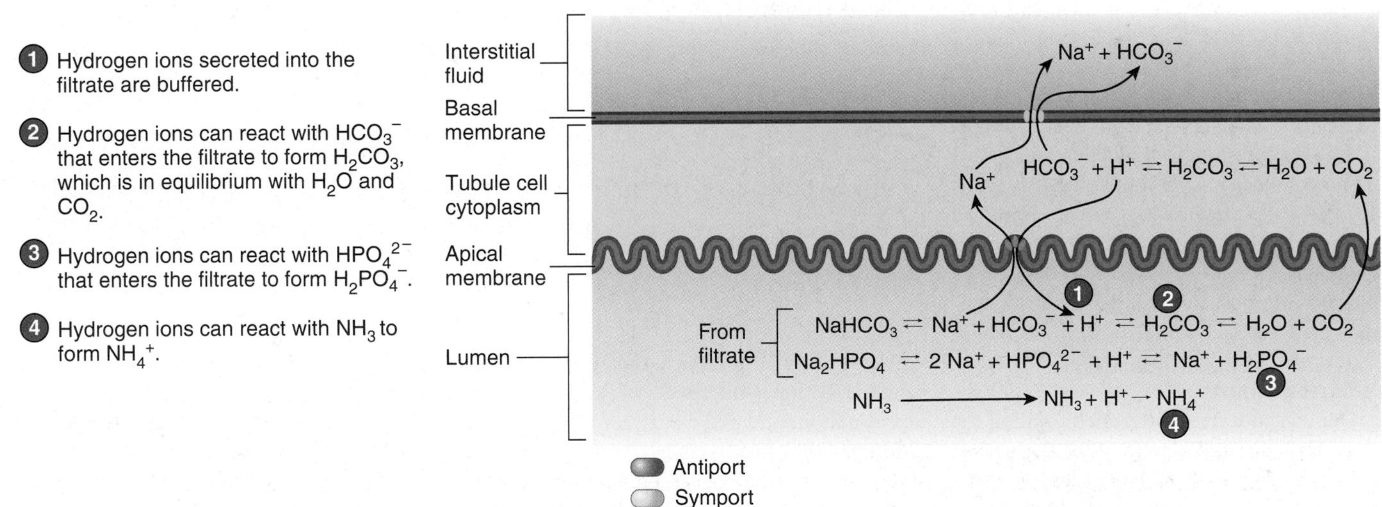
1 Hydrogen ions secreted into the filtrate are buffered.
2 Hydrogen ions can react with HCO_3^- that enters the filtrate to form H_2CO_3, which is in equilibrium with H_2O and CO_2.
3 Hydrogen ions can react with HPO_4^{2-} that enters the filtrate to form $H_2PO_4^-$.
4 Hydrogen ions can react with NH_3 to form NH_4^+.
Interstitial fluid
Basal membrane
Tubule cell cytoplasm
Apical membrane
Lumen
$Na^+ + HCO_3^-$
Na^+
$HCO_3^- + H^+ \rightleftharpoons H_2CO_3 \rightleftharpoons H_2O + CO_2$
From filtrate
$NaHCO_3 \rightleftharpoons Na^+ + HCO_3^- + H^+ \rightleftharpoons H_2CO_3 \rightleftharpoons H_2O + CO_2$
$Na_2HPO_4 \rightleftharpoons 2\,Na^+ + HPO_4^{2-} + H^+ \rightleftharpoons Na^+ + H_2PO_4^-$
$NH_3 \rightarrow NH_3 + H^+ \rightarrow NH_4^+$
Antiport
Symport

PROCESS FIGURE 27.15 Hydrogen Ion Buffering in the Filtrate
The secretion of H^+ into the filtrate decreases filtrate pH. As the concentration of H^+ increases in the filtrate, the ability of tubule cells to secrete additional H^+ becomes limited. Buffering the H^+ in the filtrate decreases its concentration and enables tubule cells to secrete additional H^+.

Bicarbonate ions, phosphate ions (HPO_4^{2-}), and **ammonia** (NH_3) in the filtrate act as buffers. Both HCO_3^- and HPO_4^{2-} enter the kidney tubules through the filtration membrane along with the rest of the filtrate, and NH_3 diffuses across the wall of nephron cells to enter the filtrate. These ions combine with H^+ secreted by the nephron (figure 27.16).

NH_3 is produced in the cells of the nephron when amino acids, such as glutamine, are deaminated. Subsequently, NH_3 diffuses from the nephron cells into the filtrate and combines with H^+ in the filtrate to form **ammonium ions (NH_4^+**; figure 27.15). The rate of NH_3 production increases when the pH of the body fluids has been depressed for 2–3 days, as occurs during prolonged respiratory or metabolic acidosis. The elevated ammonia production increases the buffering capacity of the filtrate, allowing the secretion of additional H^+ into the urine.

HCO_3^-, HPO_4^{2-}, and NH_3 constitute major buffers within the filtrate, but other weak acids, such as lactic acid in the filtrate, also combine with H^+ and increase the amount of H^+ that can be secreted into the filtrate.

ASSESS YOUR PROGRESS

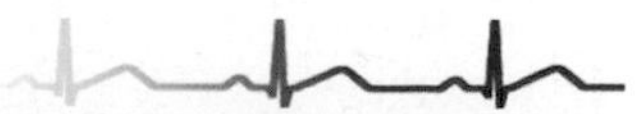

42. *What happens to blood pH when blood CO_2 levels go up or down? What causes this change?*
43. *What effect do increased CO_2 levels or decreased pH have on ventilation? How does this change in breathing affect blood pH?*
44. *Describe the process by which nephron cells move H^+ into the kidney tubule lumen and HCO_3^- into the extracellular fluid.*
45. *Name the factors that cause an increase and a decrease in H^+ secretion.*
46. *What is the purpose of buffers in the urine? Describe how the ammonia buffer system operates.*
47. *What is acidosis? What are the two types, and what causes each? How does the body compensate for each of the changes?*
48. *What is alkalosis? What are the two types, and what causes each? How does the body compensate for each of the changes?*

Case STUDY | Gastroenteritis

After work on a Friday night, Dan and some friends had a late dinner at a fast-food restaurant. Shortly afterward, Dan began feeling ill, so he went to his dorm room. By midnight, he was very nauseated and vomited repeatedly over the next several hours. The next morning, he was still vomiting at least once each hour, so he went to the student health center. At that point, it was about 12 hours since he had begun feeling nauseated. Each time he vomited, a significant volume of acidic gastric fluid was expelled. A physician examined him and ordered blood tests. The physician then diagnosed gastroenteritis caused by an unknown microorganism and prescribed medication for the infection. The next day, about 36 hours after Dan had begun to feel nauseated, he was reexamined.

Dan began to feel better by the end of the second day (about 48 hours after the nausea had begun). By 72 hours, he felt fairly normal again. After the nauseous feeling disappeared, Dan was very thirsty, and he drank a substantial amount of water and juice over several hours. Table 27B shows the results of Dan's physical examinations and blood analyses 12 hours and 36 hours after his illness began and compares them with normal values. After studying this table, explain the meaning of the various changes, as suggested in Predict 5.

TABLE 27B

	12 Hours	36 Hours	Normal
Body Weight	71 kg	68 kg	Not known
Blood Pressure	115/75 mm Hg	90/60 mm Hg	120/80 mm Hg
Heart Rate	77 beats/min	105 beats/min	72 beats/min
Plasma pH	7.48	7.5	7.35 to 7.45
Plasma HCO_3^-	32 mEq/L	36 mEq/L	22 to 28 mEq/L
Plasma P_{CO_2}	44 mm Hg	48 mm Hg	35 to 43 mm Hg
Skin Color	Pallor	Pallor	

Predict 5

a. Explain the changes in weight, blood pressure, heart rate, and skin color that Dan experienced between 12 and 36 hours.

b. Explain the changes in plasma pH and plasma HCO_3^- between 12 and 36 hours.

c. Explain the plasma P_{CO_2} at 12 and 36 hours.

d. Explain why all the parameters had returned to normal by 72 hours.

Answer

Learn to Predict ❮ From page 988

Satish and Kiran suffered from dehydration due to lack of water intake and increased water loss. In this chapter we learned that water is normally lost through urine, sensible and insensible perspiration, respiratory passages, and feces. The students' rate of water loss due to perspiration and through the respiratory passages increased dramatically in the hot, dry environment. As water loss continued, blood volume decreased, and plasma osmolality increased. The decrease in blood volume caused their renal arteries to constrict, reducing renal blood flow and the glomerular filtration rate; stimulated the renin-angiotensin-aldosterone hormone mechanism, increasing Na^+ reabsorption; and increased ADH secretion, reducing water loss in the urine. Their increased plasma osmolality also increased ADH secretion. ADH and aldosterone together maximize the kidneys' ability to conserve water; thus, their urine was very concentrated and its volume was small. However, Satish and Kiran continued to lose water by evaporation of perspiration and from the respiratory passages. Their blood osmolality continued to increase and plasma volume continued to decrease. As the plasma volume decreased, Satish and Kiran's hematocrit increased. Consequently, their blood became more viscous, and their hearts had to work harder to pump blood. We also learned in this chapter that increased plasma osmolality causes water to move from cells into the interstitial spaces and into the plasma. As a result, cells became dehydrated and malfunctioned. Satish and Kiran almost died because of the increased workload on their hearts and the decreased ability of their nerve and muscle cells to function.

Answers to the rest of this chapter's Predict questions are in Appendix G.

Summary

27.1 Body Fluids (p. 989)

1. Intracellular fluid is inside cells.
2. Extracellular fluid is outside cells and includes interstitial fluid and plasma.

27.2 Regulation of Body Fluid Concentration and Volume (p. 990)

Regulation of Water Content

1. Water crosses the wall of the digestive tract through osmosis.
2. An increase in extracellular osmolality or a decrease in blood pressure stimulates the sense of thirst.
3. Wetting of the oral mucosa or stretch of the digestive tract inhibits thirst.
4. Learned behavior plays a role in the amount of fluid ingested.
5. Water leaves the body by the following routes:
   - Through evaporation from the respiratory system and the skin (insensible perspiration and sweat)
   - Through the digestive tract (this amount is normally small, but vomiting or diarrhea can significantly increase it)
   - Through the kidneys, the primary regulator of water excretion; urine output can vary from a small amount of concentrated urine to a large amount of dilute urine

Regulation of Extracellular Fluid Osmolality

1. Increased water consumption and ADH secretion occur in response to increases in extracellular fluid osmolality. Decreased water consumption and ADH secretion occur in response to decreases in extracellular fluid osmolality.
2. Increased water consumption and ADH decrease extracellular fluid osmolality by increasing water absorption from the intestines and water reabsorption from the nephrons. Decreased water consumption and ADH increase extracellular fluid osmolality by decreasing absorption from the intestines and water reabsorption from the nephrons.

Regulation of Extracellular Fluid Volume

1. Increased extracellular fluid volume results in decreased aldosterone secretion, increased ANH secretion, decreased ADH secretion, and decreased sympathetic stimulation of afferent arterioles. These changes decrease Na^+ reabsorption and increase urine volume so as to decrease extracellular fluid volume.
2. Decreased extracellular fluid volume results in increased aldosterone secretion, decreased ANH secretion, increased ADH secretion, and increased sympathetic stimulation of the afferent arterioles. These changes increase Na^+ reabsorption and decrease urine volume so as to increase extracellular fluid volume.

27.3 Regulation of Intracellular Fluid Composition (p. 996)

1. Substances used or produced inside the cell and substances exchanged with the extracellular fluid determine the composition of intracellular fluid.
2. Intracellular fluid is different from extracellular fluid because the plasma membrane regulates the movement of materials.
3. The difference between intracellular and extracellular fluid concentrations determines water movement.

27.4 Regulation of Specific Electrolytes in the Extracellular Fluid (p. 997)

The intake and elimination of substances from the body and the exchange of substances between the extracellular and intracellular fluids determine extracellular fluid composition.

Regulation of Sodium Ions

1. Sodium is responsible for 90–95% of extracellular osmotic pressure.

2. The amount of Na^+ excreted in the kidneys is the difference between the amount of Na^+ that enters the nephron and the amount that is reabsorbed from the nephron.
   - The glomerular filtration rate determines the amount of Na^+ entering the nephron.
   - Aldosterone determines the amount of Na^+ reabsorbed.
3. Small quantities of Na^+ are excreted in sweat.
4. Increased blood osmolality leads to the production of a small volume of concentrated urine and to thirst. Decreased blood osmolality leads to the production of a large volume of dilute urine and to decreased thirst.
5. Increased blood pressure increases water and salt loss.
   - Elevated extracellular fluid volume reduces ADH secretion.
   - Renin secretion is inhibited, leading to reduced aldosterone production.

Regulation of Chloride Ions

Chloride ions are the dominant negatively charged ions in extracellular fluid.

Regulation of Potassium Ions

1. The extracellular concentration of K^+ affects resting membrane potentials.
2. The amount of K^+ excreted depends on the amount that enters with the glomerular filtrate, the amount actively reabsorbed by the nephron, and the amount secreted into the distal convoluted tubule.
3. Aldosterone increases the amount of K^+ secreted.

Regulation of Calcium Ions

1. Elevated extracellular Ca^{2+} levels prevent membrane depolarization. Decreased levels lead to spontaneous action potential generation.
2. Parathyroid hormone increases extracellular Ca^{2+} levels and decreases extracellular phosphate levels. It stimulates osteoclast activity, increases Ca^{2+} reabsorption from the kidneys, and stimulates active vitamin D production.
3. Vitamin D stimulates Ca^{2+} uptake in the intestines.
4. Calcitonin decreases extracellular Ca^{2+} levels.

Regulation of Magnesium Ions

The kidneys' capacity to reabsorb Mg^{2+} is limited, so excess Mg^{2+} is excreted in the urine, and decreased extracellular Mg^{2+} leads to increased Mg^{2+} reabsorption.

Regulation of Phosphate Ions

1. Under normal conditions, reabsorption of phosphate occurs at a maximum rate in the nephron.
2. An increase in plasma phosphate increases the amount of phosphate in the nephron beyond that which can be reabsorbed, and the excess is excreted in the urine.

27.5 Regulation of Acid-Base Balance (p. 1005)

Acids and Bases

Acids release H^+ into solution, and bases remove it.

Mechanisms of Acid-Base Balance Regulation

Buffers, the respiratory system, and the kidneys regulate acid-base balance.

Buffer Systems

1. A buffer resists changes in pH.
   - When H^+ is added to a solution, the buffer removes it.
   - When H^+ is removed from a solution, the buffer replaces it.
2. Three important buffers are carbonic acid/bicarbonate, proteins, and phosphate compounds.

Respiratory Regulation of Acid-Base Balance

Respiratory regulation of pH is achieved through the carbonic acid/bicarbonate buffer system.

1. As CO_2 levels increase, pH decreases.
2. As CO_2 levels decrease, pH increases.
3. Carbon dioxide levels and pH affect the respiratory centers. Hypoventilation increases blood CO_2 levels, and hyperventilation decreases blood CO_2 levels.

Renal Regulation of Acid-Base Balance

1. The secretion of H^+ into the filtrate and the reabsorption of HCO_3^- into extracellular fluid cause extracellular pH to increase.
   - Carbonic acid dissociates to form H^+ and HCO_3^- in nephron cells.
   - An antiport mechanism moves H^+ into the nephron lumen and Na^+ into the nephron cell.
   - Sodium ions and HCO_3^- diffuse into the extracellular fluid.
2. Bicarbonate ions in the filtrate are reabsorbed.
   - Bicarbonate ions combine with H^+ to form carbonic acid, which dissociates to form CO_2 and water.
   - Carbon dioxide diffuses into nephron cells and forms carbonic acid, which dissociates to form HCO_3^- and H^+.
   - Bicarbonate ions diffuse into the extracellular fluid, and H^+ is secreted into the filtrate.
3. The rate of H^+ secretion increases as body fluid pH decreases or as aldosterone levels increase.
4. Secretion of H^+ is inhibited when urine pH falls below 4.5.
   - Ammonia and phosphate buffers in the urine resist a drop in pH.
   - As the buffers absorb H^+, more H^+ is pumped into the urine.

REVIEW AND COMPREHENSION

1. The sensation of thirst increases when
   a. the levels of angiotensin II increase.
   b. the osmolality of the blood decreases.
   c. blood pressure increases.
   d. renin secretion decreases.
2. Insensible perspiration
   a. is lost through sweat glands.
   b. results in heat loss from the body.
   c. increases when ADH secretion increases.
   d. results in the loss of solutes, such as Na^+ and Cl^-.
3. The composition and volume of body fluids are regulated primarily by the
   a. skin.
   b. lungs.
   c. kidneys.
   d. heart.
   e. spleen.
4. Which of these conditions *decreases* extracellular fluid volume?
   a. constriction of afferent arterioles
   b. increased ADH secretion
   c. decreased ANH secretion
   d. decreased aldosterone secretion
   e. stimulation of sympathetic nerves to the kidneys

5. Which of these results in an increased blood Na^+ concentration?
   a. decrease in ADH secretion
   b. decrease in aldosterone secretion
   c. increase in ANH
   d. decrease in renin secretion
6. Which of these mechanisms is the most important for regulating blood osmolality?
   a. ADH
   b. renin-angiotensin-aldosterone
   c. ANH
   d. parathyroid hormone
7. A decrease in extracellular K^+
   a. produces depolarization of the plasma membrane.
   b. results when aldosterone levels increase.
   c. occurs when tissues are damaged (e.g., in burn patients).
   d. increases ANH secretion.
   e. increases PTH secretion.
8. Calcium ion concentration in the blood decreases when
   a. vitamin D levels are lower than normal.
   b. calcitonin secretion decreases.
   c. parathyroid hormone secretion increases.
   d. All of these are correct.
9. An acid
   a. is a solution that has a pH greater than 7.
   b. is a substance that releases H^+ into a solution.
   c. is considered weak if it completely dissociates in water.
   d. All of these are correct.
10. Buffers
    a. release H^+ when pH increases.
    b. resist changes in the pH of a solution.
    c. include the proteins of the blood.
    d. All of these are correct.
11. An increase in blood carbon dioxide levels is followed by a(n) ________ in H^+ and a(n) ________ in blood pH.
    a. increase, increase
    b. increase, decrease
    c. decrease, increase
    d. decrease, decrease
12. High levels of bicarbonate ions in the urine indicate
    a. a low level of H^+ secretion into the urine.
    b. that the kidneys are causing blood pH to increase.
    c. that urine pH is decreasing.
    d. All of these are correct.
13. High levels of ammonium ions in the urine indicate
    a. a high level of H^+ secretion into the urine.
    b. that the kidneys are causing blood pH to decrease.
    c. that urine pH is too basic.
    d. All of these are correct.
14. Blood plasma pH is normally
    a. slightly acidic.
    b. strongly acidic.
    c. slightly basic.
    d. strongly basic.
    e. neutral.
15. Acidosis
    a. increases neuron excitability.
    b. can produce tetany by affecting the peripheral nervous system.
    c. may lead to coma.
    d. may produce convulsions through the central nervous system.
16. Respiratory alkalosis is caused by ________ and can be compensated for by the production of a more ________ urine.
    a. hypoventilation, basic
    b. hypoventilation, acidic
    c. hyperventilation, acidic
    d. hyperventilation, basic

Answers in Appendix E

CRITICAL THINKING

1. In patients with diabetes mellitus, not enough insulin is produced; as a consequence, blood glucose levels increase. If blood glucose levels rise high enough, the kidneys are unable to absorb the glucose from the glomerular filtrate, and glucose "spills over" into the urine. What effect does this glucose have on urine concentration and volume? How does the body adjust to the excess glucose in the urine?
2. A patient suffering from a tumor in the hypothalamus produces excessive amounts of ADH, a condition called syndrome of inappropriate ADH (SIADH) production. For this patient, the excessive ADH production is chronic and has persisted for many months. A student nurse keeps a fluid intake-output record on the patient. She is surprised to find that fluid intake and urinary output are normal. What effect was she expecting? Can you explain why urinary output is normal?
3. Acetazolamide is a diuretic that blocks the activity of the enzyme carbonic anhydrase inside kidney tubule cells. This blockage prevents the formation of carbonic acid from carbon dioxide and water. Normally, carbonic acid dissociates to form H^+ and HCO_3^-, and the H^+ is exchanged for Na^+ from the urine. Blocking the formation of H^+ in the cells of the nephron tubule blocks sodium reabsorption, thus inhibiting water reabsorption and producing the diuretic effect. With this information in mind, what effect does acetazolamide have on blood pH, urine pH, and respiratory rate?
4. As part of a physiology experiment, an anatomy and physiology student is asked to breathe through a 3-foot-long glass tube. What effect does this action have on his blood pH, urine pH, and respiratory rate?
5. Harry ate a late meal at a cheap diner on the way home from work. A couple of hours later, he vomited three times, and then he consumed several packages of an antacid (mostly $NaHCO_3$) over the next several hours. By the evening of the next day, he was feeling better. During this ordeal, his blood pH did not deviate significantly. Select the mechanism(s) that helped maintain Harry's pH within a normal range.
   (1) increased respiratory rate
   (2) increased H^+ secretion by the nephron
   (3) decreased respiratory rate
   (4) decreased H^+ secretion by the nephron
   a. 1 b. 2 c. 1,2 d. 3 e. 3,4

Answers in Appendix F

28

Learn to Predict

Chase and Christina were studying for their last anatomy and physiology exam of the semester. After reviewing his notes on meiosis, Chase said, "Well, since meiosis is the same in males and females, I don't need to study the meiosis notes from the female reproductive system lecture." Christina quickly pointed out that he was wrong and that if he reviewed his notes he would see that meiosis is very different in males and females. With some information from previous chapters and after reading chapter 28, explain the major differences between meiosis in males and females.

Photo: Color-enhanced scanning electron micrograph of a human oocyte.

Reproductive System

"Do you know if it is a boy or a girl?" Expectant mothers answer this question continuously throughout their pregnancies. Gender is a common way that we classify people. Just think of all the times you have had to check a box for male or female while filling out a form. The reproductive system controls the development of the structural and functional differences between males and females, and it has profound effects on human behavior. Not only are male and female reproductive structures different, but they also influence structure and function in other areas of the body, such as the integumentary, muscular, and skeletal systems.

Although the differences between the male and female reproductive systems are striking, these two systems also have numerous similarities. Many reproductive organs of males and females are derived from the same embryological structures (see chapter 29), and some of the same hormones function in both males and females, even though these hormones act in very different ways.

Module 14
Reproductive System

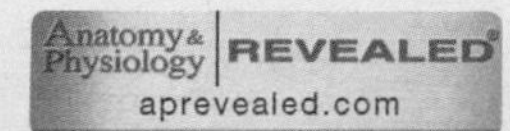
Anatomy & Physiology
REVEALED
aprevealed.com

28.1 Functions of the Reproductive System

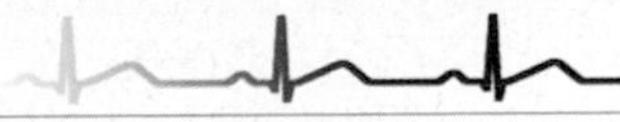

LEARNING OUTCOMES

After reading this section, you should be able to

A. List the functions of the reproductive system.

B. Distinguish among the functions that occur in males, females, and both.

Reproduction is an essential characteristic of living organisms, and functional male and female reproductive systems are necessary for humans to reproduce. In addition, even in people who do not reproduce, the reproductive system plays important roles. The reproductive system performs the following functions:

1. *Production of gametes.* Specialized organs of the reproductive system produce gametes: sperm cells in males and oocytes (eggs) in females.
2. *Fertilization.* The reproductive system enables fertilization of the oocyte by the sperm. The duct system in males nourishes sperm cells until they are mature and are deposited in the female reproductive tract by the penis. The female reproductive system receives the sperm cells from the male and transports them to the fertilization site.
3. *Development and nourishment of a new individual.* The female reproductive system nurtures a new individual in the uterus until birth and provides nourishment (milk) after birth.
4. *Production of reproductive hormones.* Hormones produced by the reproductive system control its development and the development of the gender-specific body form (table 28.1). These hormones are also essential for the normal function of the reproductive system and for reproductive behavior.

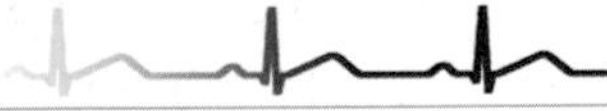

ASSESS YOUR PROGRESS

1. *What are the functions of the reproductive system?*
2. *What functions occur in both males and females, and what functions occur only in females?*

28.2 Anatomy of the Male Reproductive System

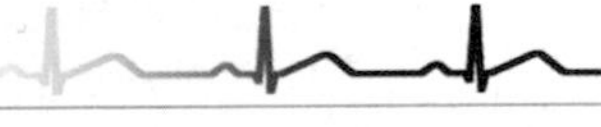

LEARNING OUTCOMES

After reading this section, you should be able to

A. Describe the scrotum and its role in regulating the temperature of the testes.

B. Describe the structure of the testes, the specialized cells of the testes, and the process of spermatogenesis.

C. List the ducts of the male reproductive system and explain their functions.

D. Describe the structure of the penis, seminal vesicles, prostate gland, and bulbourethral glands and explain their functions.

The male reproductive system consists of the testes (sing. *testis*), a series of ducts, accessory glands, and supporting structures. The ducts include the epididymides (sing. *epididymis*), the ducta deferentia (sing. *ductus deferens;* also *vas deferens*), and the urethra. Accessory glands include the seminal vesicles, the prostate gland, and the bulbourethral glands. Supporting structures include the scrotum and the penis (figure 28.1).

The testes and epididymides, in which the sperm cells develop, are located outside the body cavity in the scrotum. The ducta deferentia lead from the testes into the pelvis, where they join the ducts of the seminal vesicles to form the ampullae. Extensions of the ampullae, called the ejaculatory ducts, pass into the prostate and empty into the urethra within the prostate. The urethra, in turn, exits the pelvis and passes through the penis to the outside of the body.

Scrotum

The **scrotum** (skrō′tŭm) is a saclike structure that contains the testes. It is divided into two internal compartments by an incomplete connective tissue septum. Externally, the scrotum is marked in the midline by an irregular ridge, the **raphe** (rā′fē; a seam), which continues posteriorly to the anus and anteriorly onto the inferior surface of the penis. The outer layer of the scrotum includes the skin, a layer of superficial fascia consisting of loose connective tissue, and a layer of smooth muscle called the **dartos** (dar′tōs) **muscle.**

In cold temperatures, the dartos muscle contracts, causing the skin of the scrotum to become firm and wrinkled and reducing its overall size. At the same time, the **cremaster** (krē-mas′ter) **muscles** (see figure 28.6), which are extensions of abdominal muscles into the scrotum, contract and help pull the testes nearer the body, which helps keep the testes warm. When temperature increases due to a warmer environment or as a result of exercise or fever, the dartos and cremaster muscles relax, and the skin of the scrotum becomes loose and thin, allowing the testes to descend away from the body and keep cool. The response of the dartos and cremaster muscles is important because sperm cells are very temperature-sensitive and do not develop normally if the testes become too warm or too cool.

Perineum

The **perineum** (per′i-nē′ŭm) is the area between the thighs that is bounded by the symphysis pubis anteriorly, the coccyx posteriorly, and the ischial tuberosities laterally. The perineum is divided into two triangles by a set of muscles, the superficial transverse and deep transverse perineal muscles. These muscles run transversely between the two ischial tuberosities (see figure 10.19). The **urogenital** (ū′rō-jen′i-tăl) **triangle,** or *anterior triangle,* contains the base of the penis and the scrotum. The smaller **anal triangle,** or *posterior triangle,* contains the anal opening (figure 28.2).

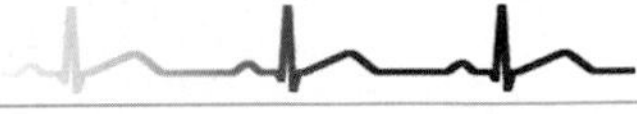

ASSESS YOUR PROGRESS

3. *List the structures of the male reproductive system.*
4. *Describe the structure of the scrotum.*
5. *Explain the role of the dartos and cremaster muscles in regulating the temperature of the testes.*
6. *Locate the boundaries of the perineum and the two triangles within it.*

TABLE 28.1 Major Reproductive Hormones and Their Effects

Hormone	Source	Target Tissue	Response
Males			
Gonadotropin-releasing hormone (GnRH)	Hypothalamus	Anterior pituitary	Stimulates secretion of LH and FSH
Luteinizing hormone (LH) (also called interstitial cell–stimulating hormone [ICSH] in males)	Anterior pituitary	Interstitial cells in the testes	Stimulates synthesis and secretion of testosterone
Follicle-stimulating hormone (FSH)	Anterior pituitary	Seminiferous tubules (sustentacular cells)	Supports spermatogenesis
Testosterone	Interstitial cells in the testes	Testes and body tissues	Supports spermatogenesis; stimulates development and maintenance of reproductive organs; causes development of secondary sex characteristics
		Anterior pituitary and hypothalamus	Inhibits GnRH, LH, and FSH secretion through negative feedback
Females			
Gonadotropin-releasing hormone (GnRH)	Hypothalamus	Anterior pituitary	Stimulates production of LH and FSH
Luteinizing hormone (LH)	Anterior pituitary	Ovaries	Causes follicles to complete maturation and undergo ovulation; causes ovulated follicle to become the corpus luteum
Follicle-stimulating hormone (FSH)	Anterior pituitary	Ovaries	Causes follicles to begin development
Prolactin	Anterior pituitary	Mammary glands	Stimulates milk secretion following childbirth
Estrogen	Follicles of ovaries	Uterus	Causes proliferation of endometrial cells
		Mammary glands	Causes development of mammary glands (especially duct systems)
		Anterior pituitary and hypothalamus	Has a positive-feedback effect before ovulation, resulting in increased LH and FSH secretion; has a negative-feedback effect, with progesterone, on the hypothalamus and anterior pituitary after ovulation, resulting in decreased LH and FSH secretion
		Other tissues	Causes development of secondary sex characteristics
Progesterone	Corpus luteum of ovaries	Uterus	Causes hypertrophy of endometrial cells and secretion of fluid from uterine glands; helps maintain pregnancy
		Mammary glands	Causes development of mammary glands (especially alveoli)
		Anterior pituitary	Has a negative-feedback effect, with estrogen, on the hypothalamus and anterior pituitary after ovulation, resulting in decreased LH and FSH secretion
		Other tissues	Causes development of secondary sex characteristics
Oxytocin*	Posterior pituitary	Uterus and mammary glands	Causes contraction of uterine smooth muscle during intercourse and childbirth; causes contraction of myoepithelial cells in the breast, resulting in milk letdown in lactating women
Human chorionic gonadotropin (hCG)	Placenta	Corpus luteum of ovaries	Maintains corpus luteum and increases its rate of progesterone secretion during the first one-third (first trimester) of pregnancy; increases testosterone production in testes of male fetuses

*Covered in chapter 29.

Testes

Testicular Histology

The **testes** (tes′tēz) are small, ovoid organs, each about 4–5 cm long, within the scrotum (see figure 28.1). They function as both exocrine and endocrine glands. Their major exocrine secretion is sperm cells, and their major endocrine secretion is the hormone testosterone.

The outer part of each testis is a thick, white capsule consisting mostly of fibrous connective tissue called the **tunica albuginea** (al-bū-jin′ē-ă). Extensions of the tunica albuginea enter the testis and form incomplete **septa** (sep′tă; figure 28.3*a*). The septa divide each testis into about 300–400 cone-shaped **lobules.** The lobules contain **seminiferous** (sem′i-nif′er-ŭs; seed carriers) **tubules,** in which sperm cells develop. Loose connective tissue surrounding the semi-

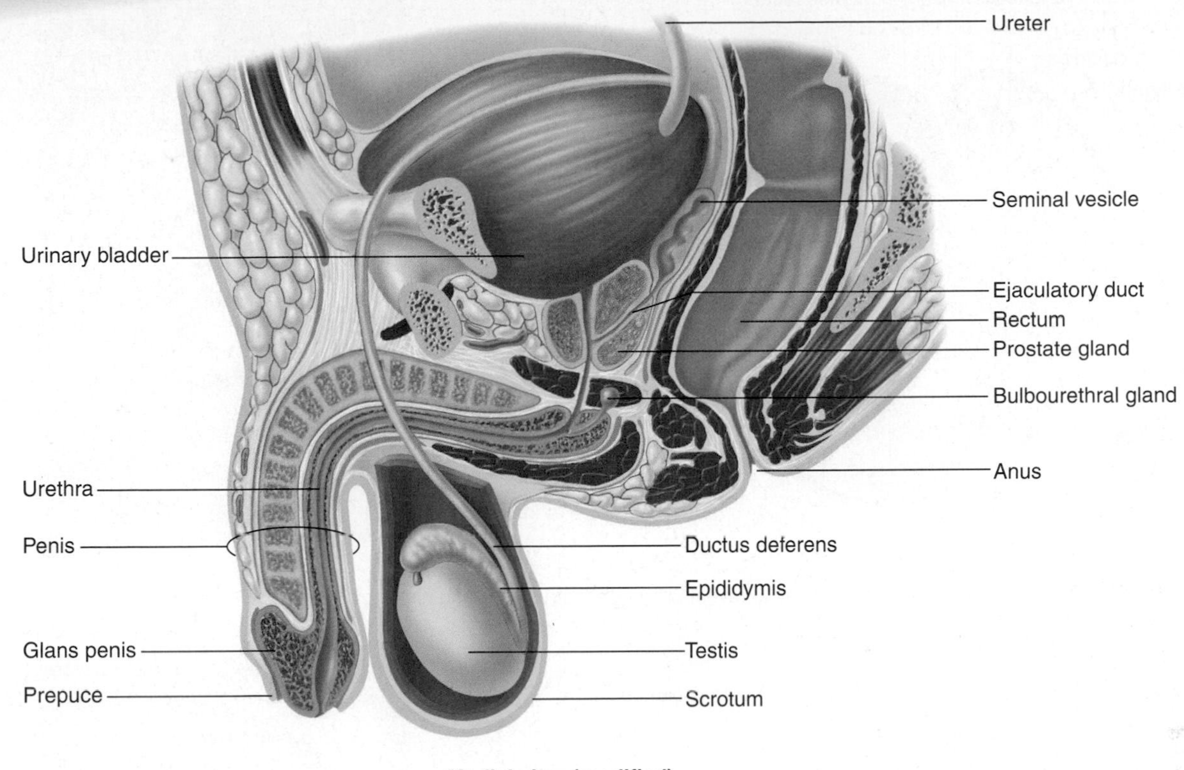
Ureter
Seminal vesicle
Urinary bladder
Ejaculatory duct
Rectum
Prostate gland
Bulbourethral gland
Anus
Urethra
Penis
Ductus deferens
Epididymis
Glans penis
Testis
Prepuce
Scrotum
Medial view (modified)

FIGURE 28.1 APR **Male Reproductive System**
Sagittal section of the male pelvis, showing the male reproductive structures. Some structures are drawn as a modified medial section to reveal the testis, epididymis, and seminal vesicles and to show the relationship of the ductus deferens to the ureter and urinary bladder.

niferous tubules contains clusters of endocrine cells called **interstitial cells,** or *Leydig cells* (figure 28.3*b*), which secrete testosterone.

The combined length of the seminiferous tubules in both testes is nearly half a mile. The seminiferous tubules empty into a set of short, straight tubules, the **tubuli recti,** which in turn empty into a tubular network called the **rete** (rē′tē; net) **testis.** The rete testis empties into 15–20 tubules called **efferent ductules** (dŭk′tools) that pierce the tunica albuginea to exit the testis. The efferent ductules have a ciliated pseudostratified columnar epithelium, which helps move sperm cells out of the testis.

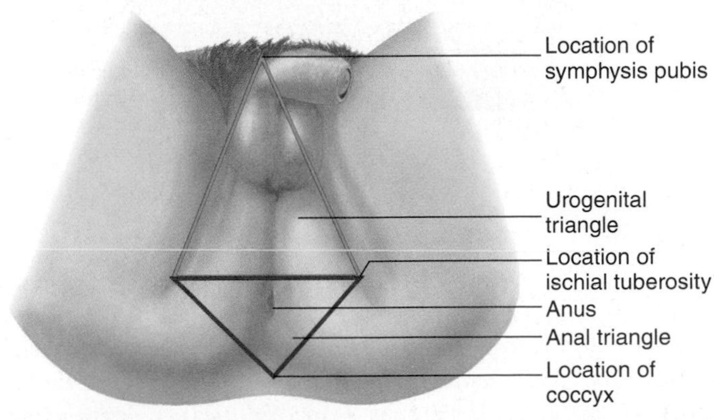
Location of symphysis pubis
Urogenital triangle
Location of ischial tuberosity
Anus
Anal triangle
Location of coccyx

FIGURE 28.2 Male Perineum
Inferior view of the male perineum.

Descent of the Testes

By approximately 8 weeks following fertilization, the testes have developed as retroperitoneal organs in the abdominopelvic cavity. Each testis is connected to a labioscrotal swelling, which becomes the scrotum, by a **gubernaculum** (goo′ber-nak′ū-lŭm), a fibromuscular cord (figure 28.4, *step 1;* see chapter 29). The testes descend toward the area where the **inguinal** (ing′gwi-năl) **canals** will form (figure 28.4, *step 2*). The gubernaculum extends through the inguinal canal, enlarging the canal. Between 7 and 9 months of development, the testes move through the inguinal canals into the scrotum (figure 28.4, *step 3*). As it moves into the scrotum, each testis is preceded by an outpocketing of the peritoneum called the **process vaginalis** (vaj′i-nă-lis). The superior part of each process vaginalis usually degenerates, and the inferior part remains as a small, closed sac called the **tunica** (too′ni-kă) **vaginalis** (figure 28.4, *step 4*). The tunica vaginalis is a serous membrane consisting of a layer of simple squamous epithelium resting on a basement membrane. The tunica vaginalis surrounds most of the testis in much the same way that the pericardium surrounds the heart, and the small amount of fluid it secretes allows the testes to move with little friction.

The testes have descended into the scrotum in approximately 79% of male infants delivered prior to 28 weeks of development. In male infants delivered after 28 weeks, greater than 97% show normal testes descent. By 9 months of age, 98.2% of male infants show normal testes descent.

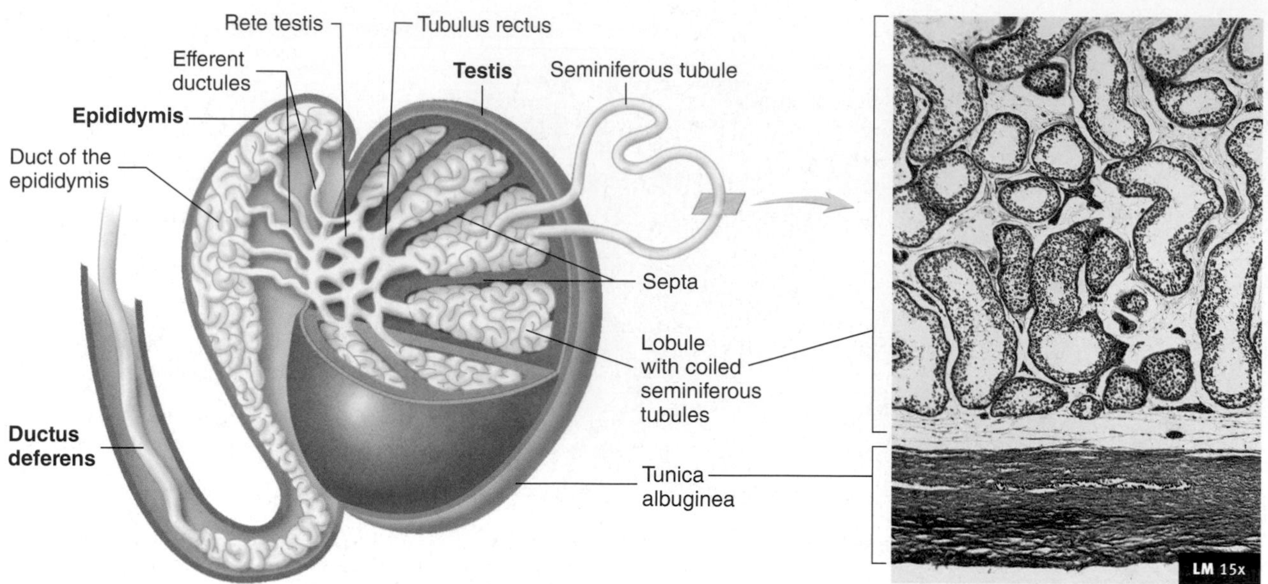

Rete testis
Tubulus rectus
Efferent ductules
Testis
Seminiferous tubule
Epididymis
Duct of the epididymis
Septa
Lobule with coiled seminiferous tubules
Ductus deferens
Tunica albuginea
LM 15x

(a) Gross anatomy of the testis, with a section cut away to reveal the internal structures and histology.

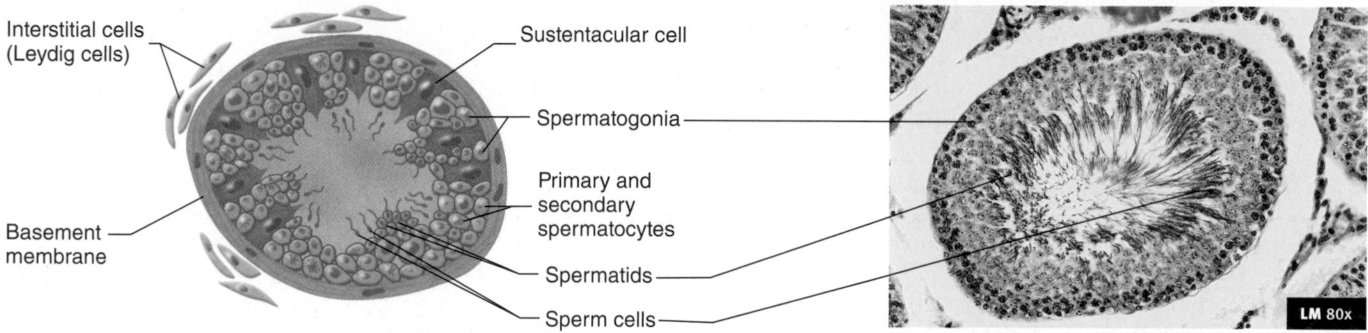

Interstitial cells (Leydig cells)
Sustentacular cell
Spermatogonia
Primary and secondary spermatocytes
Basement membrane
Spermatids
Sperm cells
LM 80x

(b) Cross section of a seminiferous tubule. Spermatogonia are near the periphery, and mature sperm cells are near the lumen of the seminiferous tubules.

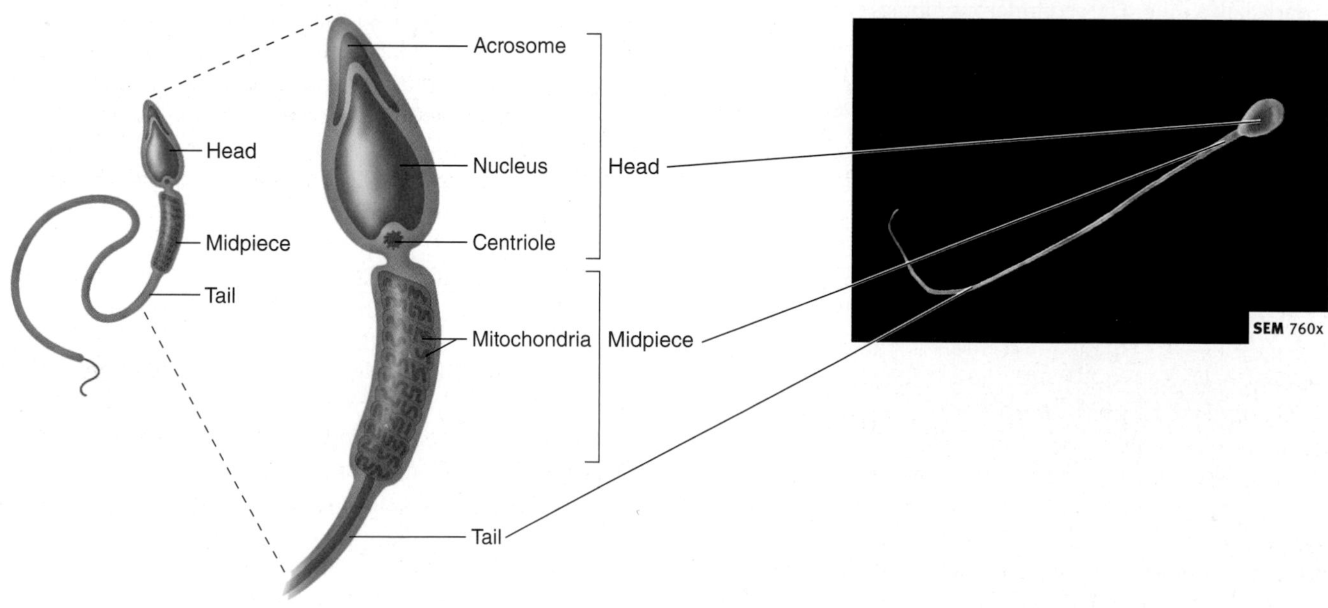

Acrosome
Head
Nucleus
Head
Midpiece
Centriole
Tail
Mitochondria
Midpiece
SEM 760x
Tail

(c) Head, midpiece, and tail of a mature sperm cell.

FIGURE 28.3 AP|R **Histology of the Testis**

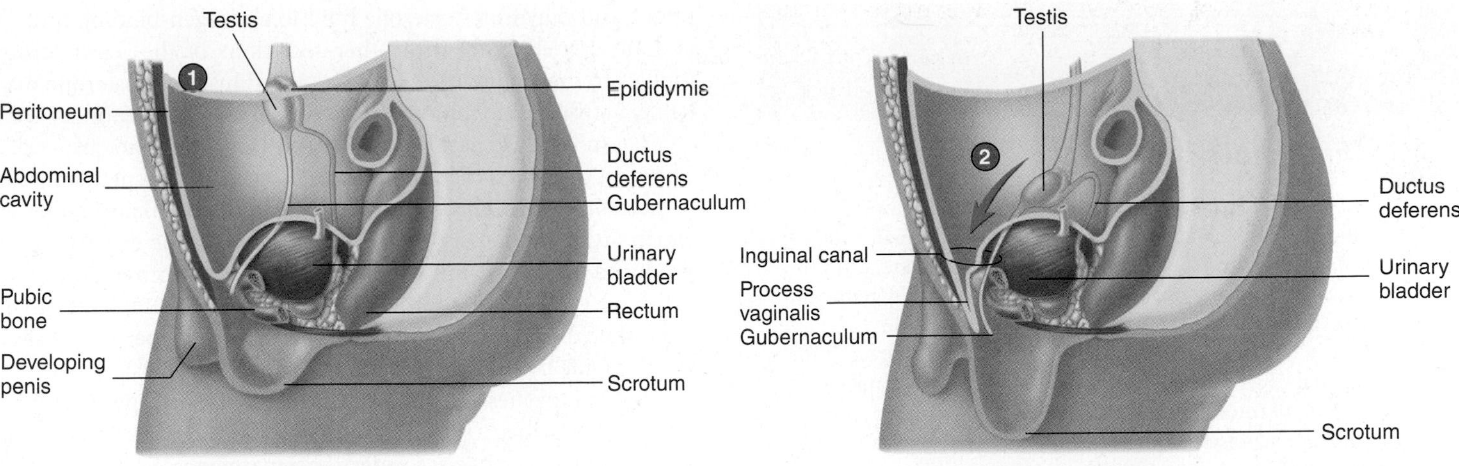
Testis
1
Peritoneum
Epididymis
Abdominal cavity
Ductus deferens
Gubernaculum
Urinary bladder
Pubic bone
Rectum
Developing penis
Scrotum
Approximately 8 weeks
Testis
2
Ductus deferens
Inguinal canal
Urinary bladder
Process vaginalis
Gubernaculum
Scrotum
Approximately 12 weeks

1 Each testis forms as a retroperitoneal structure near the level of the kidney.

2 The testis beneath the parietal peritoneum is descending toward the inguinal canal. The process vaginalis extends into the inguinal canal alongside the gubernaculum.

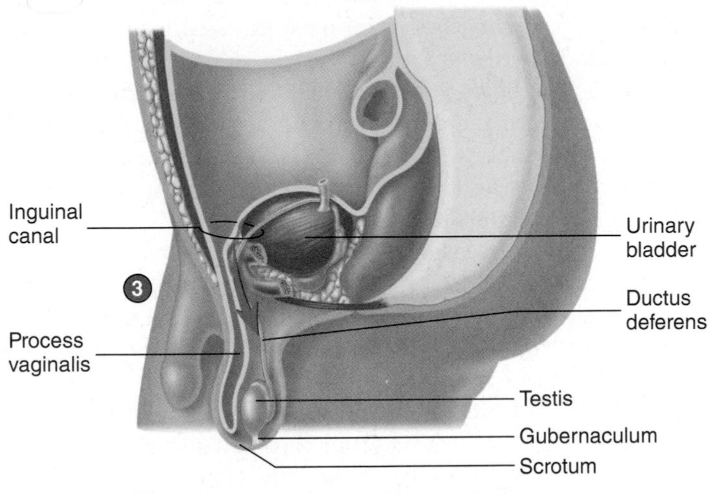
Inguinal canal
Urinary bladder
3
Ductus deferens
Process vaginalis
Testis
Gubernaculum
Scrotum
Between 7 and 9 months

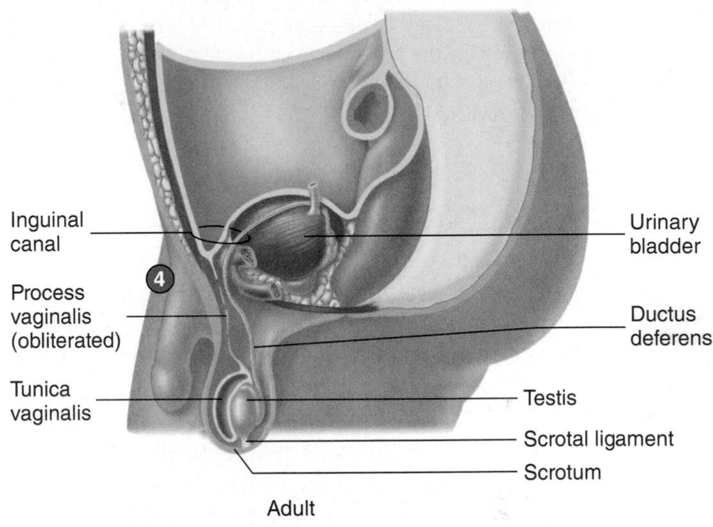
Inguinal canal
Urinary bladder
4
Process vaginalis (obliterated)
Ductus deferens
Tunica vaginalis
Testis
Scrotal ligament
Scrotum
Adult

3 The testis follows the process vaginalis and descends through the inguinal canal into the scrotum.

4 Between birth and adulthood, the process vaginalis is obliterated, and its inferior portion becomes the tunica vaginalis. The gubernaculum becomes the scrotal ligament.

PROCESS FIGURE 28.4 Descent of the Testes

The inguinal canals are bilateral, oblique passageways in the anterior abdominal wall. They originate at the **deep inguinal rings,** which open through the aponeuroses of the transversus abdominis muscles. The canals extend inferiorly and obliquely and end at the **superficial inguinal rings,** openings in the aponeuroses of the external abdominal oblique muscles. In females, the inguinal canals develop, but they are much smaller than in males, and the ovaries do not normally descend through them.

Cryptorchidism (krip-tōr′ki-dizm) is the failure of one or both of the testes to descend into the scrotum. Because the higher temperature of the abdominal cavity prevents normal sperm cell development, sterility can result if both testes are involved (see chapter 29). In addition, approximately 10% of testicular cancer cases occur in men with a history of cryptorchidism.

ASSESS YOUR PROGRESS

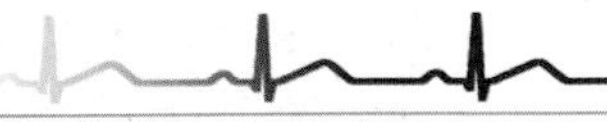

7. *Describe the covering and connective tissue of the testis.*
8. *Where are the seminiferous tubules and interstitial cells located? What are their functions?*
9. *List the ducts that will move sperm from the seminiferous tubules out of the testis.*
10. *When and how do the testes descend into the scrotum?*
11. *What is the function of the tunica vaginalis?*

Sperm Cell Development

Before puberty, the testes remain relatively simple and unchanged from the time of their initial development. The interstitial cells are

Clinical IMPACT

Inguinal Hernia

An **inguinal hernia** (her′nē-ă) is an abnormal opening in the abdominal wall in the inguinal region through which a structure, such as a portion of the small intestine, can protrude. Normally, the inguinal canals are closed, but they represent weak spots in the abdominal wall.

Inguinal hernias can be of two types—indirect or direct. If the deep inguinal ring remains open, or if it is weak and enlarges later in life, a loop of intestine can protrude into or even pass through the inguinal canal, resulting in an indirect inguinal hernia. A direct inguinal hernia results from a tear, or rupture, in a weakened area of the anterior abdominal wall near the inguinal canal, but not through the inguinal canal.

Inguinal hernias can be quite painful and even very dangerous, especially if a portion of the small intestine is compressed so that its blood supply is cut off. Fortunately, inguinal hernias can be repaired surgically. Because a male's inguinal canals are larger and weakened as a result of the testes passing through them on their way into the scrotum, males are much more prone to inguinal hernias than are females.

not particularly prominent during this period, and the seminiferous tubules lack a lumen and are not yet functional. At 12–14 years of age, the interstitial cells increase in number and size, a lumen develops in each seminiferous tubule, and sperm cell production begins. It takes approximately 74 days for sperm cells to be produced. For about 50 of those days, the sperm cells are in the seminiferous tubules.

Sperm cell development, or **spermatogenesis** (sper′mă′-tō-jen′ĕ-sis), occurs in the seminiferous tubules (figure 28.5*a*; see figure 28.3*b*). The seminiferous tubules contain two types of cells, germ cells and **sustentacular** (sŭs-ten-tak′ū-lăr) **cells,** or *Sertoli* (sĕr-tō′lē) *cells* (also sometimes referred to as *nurse cells*).

The germ cells are the ones that divide and differentiate to form sperm cells during spermatogenesis. The sustentacular cells are large cells that extend from the periphery to the lumen of the seminiferous tubule (figure 28.5*b;* see figure 28.3*b*). They nourish the germ cells and probably produce, together with the interstitial cells, a number of hormones, such as androgens, estrogens, and inhibins. In addition, tight junctions between the sustentacular cells form a **blood-testis barrier** between germ cells and sperm cells. The blood-testis barrier isolates the sperm cells from the immune system (figure 28.5*b*). This barrier is necessary because, as the sperm cells develop, they form surface antigens that could stimulate an immune response, resulting in their destruction.

Testosterone, produced by the interstitial cells, passes into the sustentacular cells and binds to receptors, enabling the sustentacular cells to function normally. In addition, testosterone is converted to two other steroids in the sustentacular cells: **dihydrotestosterone** (dī-hī′drō-tes-tos′ter-ōn) and **estradiol,** a specific type of estrogen. The sustentacular cells also secrete a protein called **androgen-binding** (an′drō-jen) **protein** into the seminiferous tubules. Testosterone and dihydrotestosterone bind to androgen-binding protein and are carried along with other secretions of the seminiferous tubules to the epididymis. Estradiol and dihydrotestosterone may be the active hormones that promote sperm cell formation.

Germ cells are partially embedded in the sustentacular cells. The most peripheral cells, those adjacent to the basement membrane of the seminiferous tubules, are **spermatogonia** (sper′mă-tō-gō′nē-ă), which divide by mitosis (figure 28.5*a*). Some of the daughter cells produced from these mitotic divisions remain spermatogonia and continue to produce additional spermatogonia. Other daughter cells differentiate to form **primary spermatocytes** (sper′mă-tō-sītz), which divide by meiosis (see Clinical Impact, "Meiosis").

Spermatogenesis begins when the primary spermatocytes divide. Each primary spermatocyte passes through the first meiotic division to become two **secondary spermatocytes.** Each secondary spermatocyte undergoes a second meiotic division to produce two even smaller cells called **spermatids** (sper′mă-tidz). Each spermatid contains one of each of the homologous pairs of chromosomes. Therefore, each sperm cell contains 22 autosomes and either an X or a Y chromosome. Each spermatid undergoes the last phase of spermatogenesis, called **spermiogenesis** (sper′mē-ō-jen′ĕ-sis), to form a mature **sperm cell,** or *spermatozoon* (sper′mă-tō-zō′on; pl. *spermatozoa,* sper′mă-tō-zō′ă; figure 28.5; see figure 28.3*c,d*). During spermiogenesis, each spermatid develops a head, a midpiece, and a tail, or flagellum. The nucleus of the sperm is located in the head. Just anterior to the nucleus is a vesicle called the **acrosome** (ak′rō-sōm), which contains enzymes necessary for the sperm cell to penetrate the female oocyte. The flagellum is similar to a cilium (see chapter 3), and microtubules within the flagellum move, propelling the sperm cell forward. The midpiece contains large numbers of mitochondria, which produce the ATP necessary for microtubule movement.

At the end of spermatogenesis, the developing sperm cells gather around the lumen of the seminiferous tubules, with their heads directed toward the surrounding sustentacular cells and their tails directed toward the center of the lumen (figure 28.5; see figure 28.3*b*). Finally, sperm cells are released into the lumen of the seminiferous tubules.

ASSESS YOUR PROGRESS 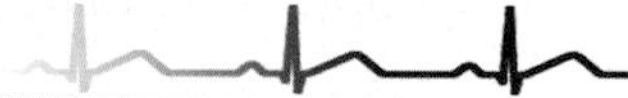

12. *Describe the role of germ cells, sustentacular cells, and the blood-testis barrier in the structure of the seminiferous tubules.*
13. *Describe the conversion of testosterone to other hormones in the sustentacular cells.*
14. *Where, specifically, are sperm cells produced in the testis?*
15. *Describe the role of meiosis in sperm cell formation.*
16. *List the parts of a mature sperm. What are their functions? Explain the events of spermiogenesis.*

Ducts

After being released into the seminiferous tubules, the sperm cells pass through the tubuli recti to the rete testis. From the rete testis, they pass through the efferent ductules, which leave the testis and enter the epididymis. The sperm cells then leave the epididymis, passing through the ductus deferens, ejaculatory duct, and urethra to the exterior of the body.

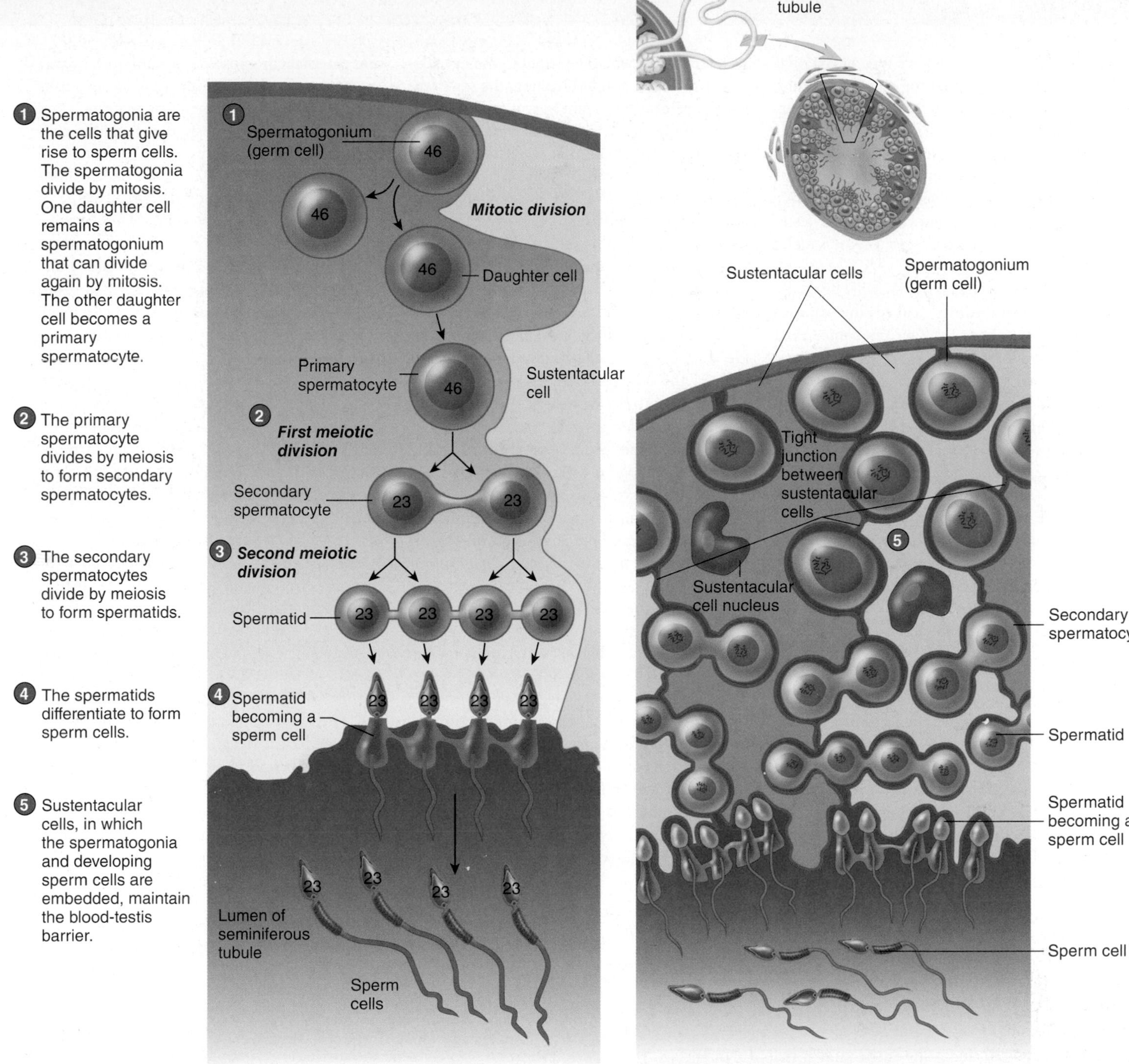

1 Spermatogonia are the cells that give rise to sperm cells. The spermatogonia divide by mitosis. One daughter cell remains a spermatogonium that can divide again by mitosis. The other daughter cell becomes a primary spermatocyte.
2 The primary spermatocyte divides by meiosis to form secondary spermatocytes.
3 The secondary spermatocytes divide by meiosis to form spermatids.
4 The spermatids differentiate to form sperm cells.
5 Sustentacular cells, in which the spermatogonia and developing sperm cells are embedded, maintain the blood-testis barrier.
1
Spermatogonium (germ cell)
46
46
Mitotic division
46
Daughter cell
Primary spermatocyte
46
Sustentacular cell
2
First meiotic division
Secondary spermatocyte
23
23
3
Second meiotic division
Spermatid
23
23
23
23
4
Spermatid becoming a sperm cell
23
23
23
23
23
23
23
23
Lumen of seminiferous tubule
Sperm cells
(a)
Seminiferous tubule
Sustentacular cells
Spermatogonium (germ cell)
Tight junction between sustentacular cells
5
Sustentacular cell nucleus
Secondary spermatocyte
Spermatid
Spermatid becoming a sperm cell
Sperm cell
(b)

PROCESS FIGURE 28.5 APR **Spermatogenesis**

(*a*) Meiosis during spermatogenesis. A section of a seminiferous tubule illustrating the process of meiosis and sperm cell formation. (*b*) The sustentacular cells extend from the periphery to the lumen of the seminiferous tubules shown in alternating dark and light shades for emphasis. The tight junctions that form between adjacent sustentacular cells form the blood-testis barrier. Spermatogonia are peripheral to the blood-testis barrier, and spermatocytes are central to it.

Clinical IMPACT — Meiosis

Meiosis (mī-ō′sis) is a type of cell division specialized for sexual reproduction. During meiosis, one cell undergoes two consecutive divisions to produce four genetically different daughter cells. Each of these daughter cells contains half as many chromosomes as the parent cell (figure 28A). In humans, meiosis occurs only in the testes and ovaries and produces sperm cells in males and oocytes in females.

In humans, the somatic cells normally have 46 chromosomes, called the **diploid** (dip′loyd) number (2*n*). Chromosomes exist in 23 **homologous** (hŏ-mol′ō-gŭs) **pairs**—22 autosomal pairs and 1 pair of sex chromosomes. The sex chromosome pair is composed of an X and a Y chromosome in males and two X chromosomes in females. One chromosome of each homologous pair is inherited from the male parent, and the other is inherited from the female parent. The chromosomes of each homologous pair are alike in size and shape and contain genes for the same traits.

Sperm cells and oocytes contain the **haploid** (hap′loyd) number (***n***) of chromosomes, which is half the diploid number, or 23. Each gamete contains one chromosome from each of the homologous pairs. Reduction in the number of chromosomes in sperm cells or oocytes to an *n* number is important. When a sperm cell and an oocyte fuse to form a fertilized egg, each provides an *n* number of chromosomes, which reestablishes a 2*n* number. If meiosis did not take place, the number of chromosomes in the fertilized oocyte would double each time fertilization occurred, and the extra chromosomal material would be lethal to the developing offspring. Upon fertilization, the sex of the baby is determined by the sperm cell. The baby is male if the oocyte is fertilized by a Y-carrying sperm cell or female if it is fertilized by an X-carrying sperm cell.

The two divisions of meiosis are called **meiosis I** and **meiosis II.** The stages of meiosis have the same names as the stages of mitosis—that is, prophase, metaphase, anaphase, and telophase—but distinct differences exist between mitosis and meiosis.

Before meiosis begins, all the chromosomes are duplicated. At the beginning of meiosis, each of the 46 chromosomes consists of two sister **chromatids** (krō′mă-tids) connected by a **centromere** (sen′trō-mēr; figure 28A, *step 1*). In prophase I, the chromosomes become visible, and the homologous pairs come together in a process called **synapsis** (si-nap′sis). Because each chromosome consists of two chromatids, the pairing of the homologous chromosomes brings two chromatids of each chromosome close together, an arrangement called a **tetrad** (figure 28A, *step 2*). Occasionally, part of a chromatid of one homologous chromosome breaks off and is exchanged with part of another chromatid from the other homologous chromosome. This exchange of genetic material between maternal and paternal chromosomes is called **crossing over** and may result in new gene combinations on the chromosomes.

During metaphase I, homologous pairs of chromosomes line up near the center of the cell (figure 28A, *step 3*). For each homologous pair, however, the orientation of the maternal and paternal chromosomes is random. The way the chromosomes align during synapsis results in the random assortment of maternal and paternal chromosomes in the daughter cells during meiosis. Crossing over and the random assortment of maternal and paternal chromosomes are responsible for the large degree of diversity in the genetic composition of sperm cells and oocytes produced by each individual.

During anaphase I, the homologous pairs are separated to each side of the cell (figure 28A, *step 4*). During telophase I, new nuclei form, and the cell completes division of the cytoplasm to form two cells (figure 28A, *step 5*). As a consequence, when meiosis I is complete, each daughter cell has 1 chromosome from each homologous pair. Since the chromosome number is reduced from a 2*n* number (46 chromosomes, or 23 pairs) to an *n* number (23 chromosomes, or 1 from each homologous pair) during the first meiotic division, this division is often called a **reduction division.**

At the end of meiosis I, each of the 23 chromosomes in the daughter cells still consists of two chromatids. The separation of the chromatids of the duplicated chromosomes occurs in meiosis II. The second meiotic division is similar to mitosis (figure 28A, *steps 6–9*). The duplicated chromosomes line up near the middle of the cell. Then the chromatids separate at the centromere, and each daughter cell receives one of the chromatids from each chromosome. When the centromere separates, each of the chromatids is called a chromosome. Consequently, each of the four daughter cells produced by meiosis contains 23 chromosomes.

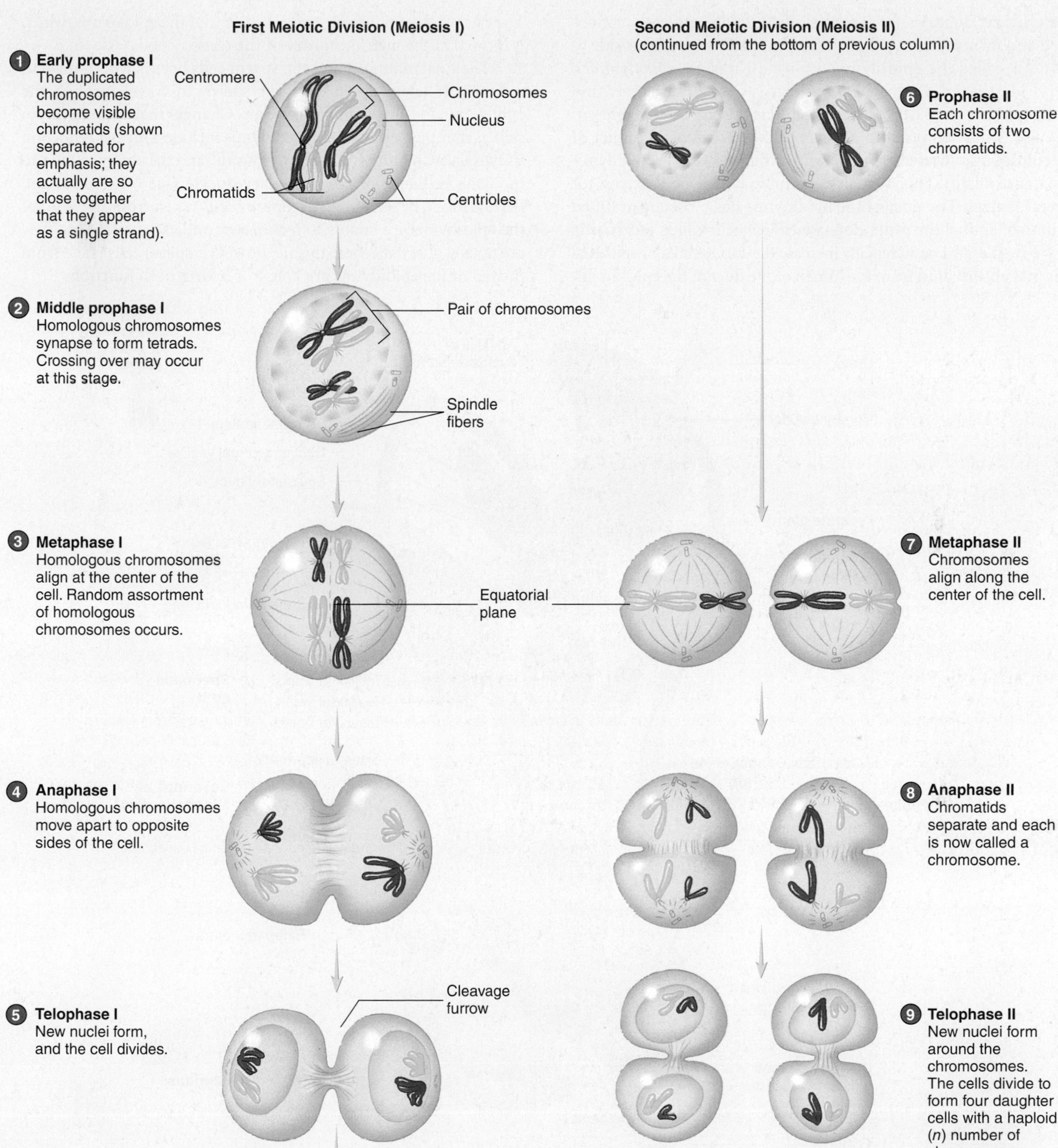

First Meiotic Division (Meiosis I)
Second Meiotic Division (Meiosis II)
(continued from the bottom of previous column)
1 Early prophase I
The duplicated chromosomes become visible chromatids (shown separated for emphasis; they actually are so close together that they appear as a single strand).
Centromere
Chromosomes
Nucleus
Chromatids
Centrioles
2 Middle prophase I
Homologous chromosomes synapse to form tetrads. Crossing over may occur at this stage.
Pair of chromosomes
Spindle fibers
3 Metaphase I
Homologous chromosomes align at the center of the cell. Random assortment of homologous chromosomes occurs.
Equatorial plane
4 Anaphase I
Homologous chromosomes move apart to opposite sides of the cell.
5 Telophase I
New nuclei form, and the cell divides.
Cleavage furrow
Prophase II (top of next column)
6 Prophase II
Each chromosome consists of two chromatids.
7 Metaphase II
Chromosomes align along the center of the cell.
8 Anaphase II
Chromatids separate and each is now called a chromosome.
9 Telophase II
New nuclei form around the chromosomes. The cells divide to form four daughter cells with a haploid (n) number of chromosomes.

PROCESS FIGURE 28A APR **Meiosis**

Epididymis

The **efferent ductules** from each testis become extremely convoluted and form a comma-shaped structure on the posterior side of the testis called the **epididymis** (ep-i-did′i-mis; pl. *epididymides,* ep-i-di-dim′i-dēz). Each epididymis consists of a head, a body, and a long tail (figure 28.6). The head contains the convoluted efferent ductules, which empty into a single convoluted tube, the **duct of the epididymis,** located primarily within the body of the epididymis (see figure 28.3*a*). This duct alone, if unraveled, would extend for several meters. The duct of the epididymis has a pseudostratified columnar epithelium with elongated microvilli called **stereocilia** (ster′ē-ō-sil′ē-ă). The stereocilia increase the surface area of epithelial cells that absorb fluid from the lumen of the duct of the epididymis. The duct of the epididymis ends at the tail of the epididymis, which is located at the inferior border of the testis.

The final maturation of the sperm cells occurs within the epididymis. It takes 12–16 days for sperm to travel through the epididymis and appear in the ejaculate. Changes that occur in sperm cells as they pass through the epididymis include a further reduction in cytoplasm, maturation of the acrosome, and the ability to bind to the zona pellucida of the secondary oocyte (see "Oogenesis and Fertilization" in section 28.4). Sperm cells taken from the head of the epididymis are unable to fertilize secondary oocytes, and they are not yet able to become motile; however, sperm cells taken from the tail of the epididymis are able to perform both functions.

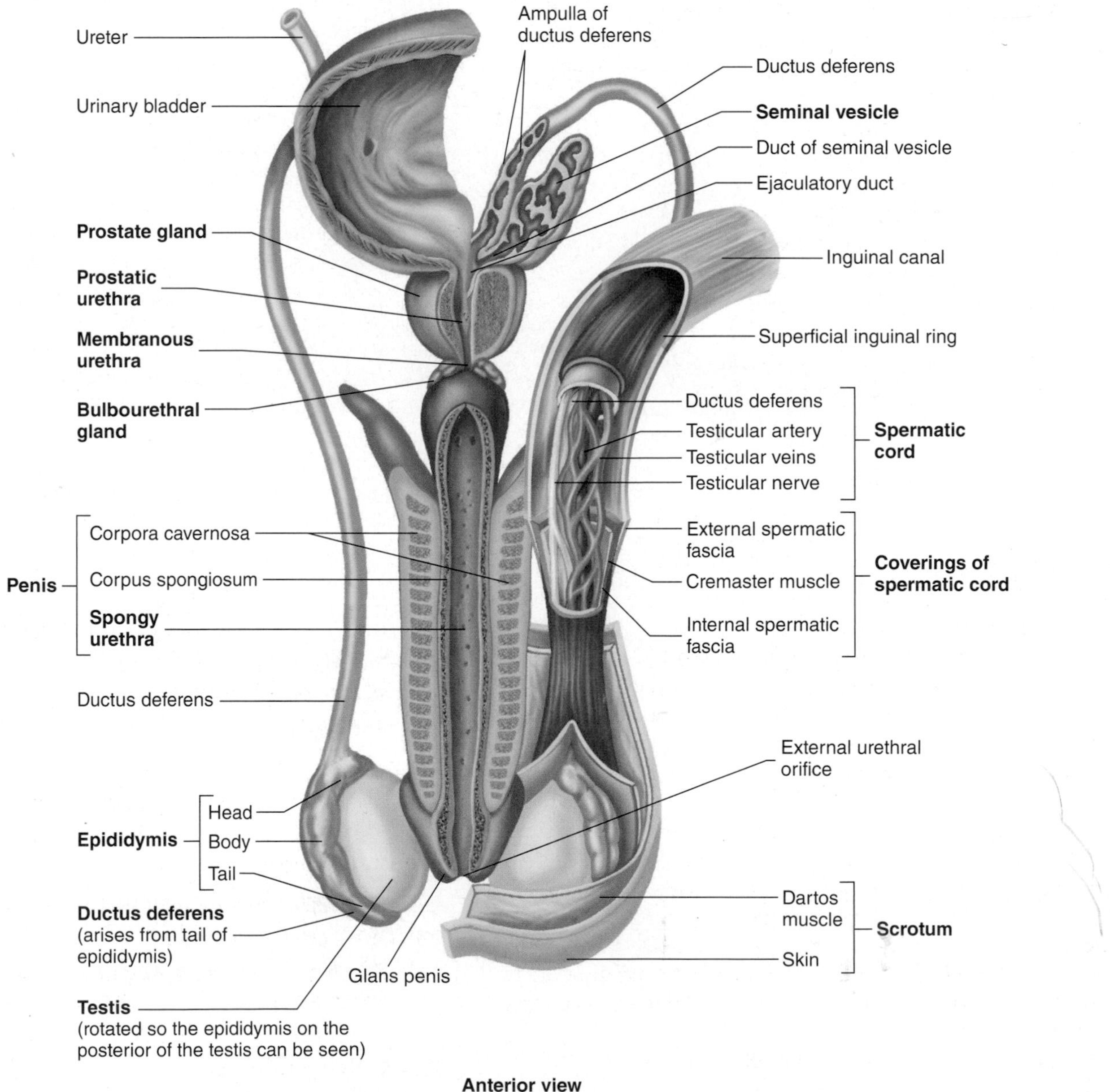
Ureter
Urinary bladder
Ampulla of ductus deferens
Ductus deferens
Seminal vesicle
Duct of seminal vesicle
Ejaculatory duct
Prostate gland
Prostatic urethra
Membranous urethra
Bulbourethral gland
Inguinal canal
Superficial inguinal ring
Ductus deferens
Testicular artery
Testicular veins
Testicular nerve
Spermatic cord
Penis
Corpora cavernosa
Corpus spongiosum
Spongy urethra
External spermatic fascia
Cremaster muscle
Internal spermatic fascia
Coverings of spermatic cord
Ductus deferens
External urethral orifice
Epididymis
Head
Body
Tail
Ductus deferens (arises from tail of epididymis)
Glans penis
Dartos muscle
Skin
Scrotum
Testis (rotated so the epididymis on the posterior of the testis can be seen)
Anterior view

FIGURE 28.6 AP|R **Male Reproductive Structures**
Testes, epididymis, ductus deferens, and glands of the male reproductive system. The urethra is cut open along its dorsal side.

Ductus Deferens and Ejaculatory Duct

The **ductus deferens** (pl. *ducta deferentia*), or *vas deferens,* emerges from the tail of the epididymis and ascends along the posterior side of the testis medial to the epididymis, where it associates with the blood vessels and nerves that supply the testis to form the **spermatic cord** (figure 28.6; see figures 28.1 and 28.3*a*). The spermatic cord consists of (1) the ductus deferens, (2) the testicular artery and venous plexus, (3) lymphatic vessels, (4) nerves, and (5) fibrous remnants of the process vaginalis. The coverings of the spermatic cord include the **external spermatic fascia** (fash′ē-ă); the cremaster muscle, an extension of the muscle fibers of the internal abdominal oblique muscle of the abdomen; and the **internal spermatic fascia** (figure 28.6).

The ductus deferens and the rest of the spermatic cord structures ascend and pass through the inguinal canal to enter the pelvic cavity (figure 28.6; see figure 28.1). The ductus deferens crosses the lateral and posterior walls of the pelvic cavity, travels over the ureter, and loops over the posterior surface of the urinary bladder to approach the prostate gland. The end of the ductus deferens enlarges to form an **ampulla** (am-pul′lă). The lumen of the ductus deferens is lined with pseudostratified columnar epithelium, which is surrounded by smooth muscle. Peristaltic contractions of this smooth muscle tissue help propel sperm cells through the ductus deferens.

Adjacent to the ampulla of each ductus deferens is a sac-shaped gland called the seminal vesicle. A short duct from the seminal vesicle joins the ampulla of the ductus deferens to form the **ejaculatory** (ē-jak′ū-lă-tōr-ē) **duct.** Each ejaculatory duct is approximately 2.5 cm long. These ducts project into the prostate gland and end by opening into the urethra (figure 28.6; see figure 28.1).

Urethra

The male **urethra** (ū-rē′thră) is about 20 cm long and extends from the urinary bladder to the distal end of the penis (figure 28.7; see figures 28.1 and 28.6). The urethra is a passageway for both urine and male reproductive fluids. The urethra is divided into three parts: the prostatic part, the membranous part, and the spongy part. The **prostatic** (pros-tat′ik) **urethra** is connected to the bladder and passes through the prostate gland. Fifteen to 30 small ducts from the prostate gland and the two ejaculatory ducts empty into the prostatic urethra. The **membranous urethra** is the shortest part of the urethra, extending from the prostate gland through the perineum. The **spongy urethra,** also called the *penile* (pē′nīl) *urethra,* is the longest part of the urethra; it extends from the membranous urethra through the length of the penis. In rare cases, the penis does not develop normally, and the urethra may open to the exterior along the inferior surface of the penis (see chapter 29). Stratified columnar epithelium lines most of the urethra, but transitional epithelium is in the prostatic urethra near the bladder, and stratified squamous epithelium is near the opening of the spongy urethra to the outside. Several minute, mucus-secreting **urethral glands** empty into the urethra.

Penis

The **penis** is the male organ of copulation, through which sperm cells are transferred from the male to the female. The penis contains three columns of erectile tissue (figure 28.7). Engorgement of this erectile tissue with blood causes the penis to enlarge and become firm, a process called **erection** (ē-rek′shŭn). Two of the erectile columns form the dorsum and sides of the penis and are called the **corpora cavernosa** (kōr′pōr-ă kav-er-nos′ă). The third column, the **corpus spongiosum** (kōr′pŭs spŭn′jē-ō′sŭm), forms the ventral portion of the penis. It expands to form a cap, the **glans penis,** over the distal end of the penis. The spongy urethra passes through the corpus spongiosum, penetrates the glans penis, and opens as the **external urethral orifice.**

At the base of the penis, the corpus spongiosum expands to form the **bulb of the penis,** and each corpus cavernosum expands to form the **crus** (kroos; pl. *crura,* kroo′ră) **of the penis.** Together, these structures constitute the **root of the penis,** and the crura attach the penis to the pelvic bones.

Skin is loosely attached to the connective tissue that surrounds the erectile columns in the shaft of the penis. The skin is firmly attached at the base of the glans penis, and a thinner layer of skin tightly covers the glans penis. The skin of the penis, especially the glans penis, is well supplied with sensory receptors. A loose fold of skin called the **prepuce** (prē′poos), or *foreskin,* covers the glans penis (see figure 28.1).

In many cultures, the prepuce is surgically removed shortly after birth, a procedure called **circumcision** (ser-kŭm-sizh′ŭn). There are no compelling medical reasons for circumcision. Uncircumcised males have a higher incidence of penile cancer, but the underlying causes seem related to chronic infections and poor hygiene. In the few cases in which the prepuce is "too tight" to be moved over the glans penis, circumcision may be necessary to avoid chronic infections and maintain normal circulation.

The primary nerves, arteries, and veins of the penis pass along its dorsal surface (figure 28.7*b*). Dorsal arteries, with dorsal nerves lateral to them, exist on each side of a single, midline dorsal vein. Additional deep arteries lie within the corpora cavernosa.

ASSESS YOUR PROGRESS

17. *Describe the structure and functions of the epididymis.*

18. *How do sperm move from the epididymis, and what changes occur in sperm cells while in the epididymis?*

19. *Describe the structure and functions of the ductus deferens.*

20. *List the components and coverings of the spermatic cord.*

21. *Relate the route by which the ductus deferens extends from the testis to the prostate gland.*

22. *What is the ejaculatory duct?*

23. *Distinguish among the three parts of the male urethra.*

24. *Describe the erectile tissue of the penis.*

25. *Describe the structures and locations of the glans penis, crus, bulb, and prepuce.*

Accessory Glands

Three exocrine glands secrete material into the ducts of the male reproductive tract. These glands are the seminal vesicles, the prostate gland, and the bulbourethral glands.

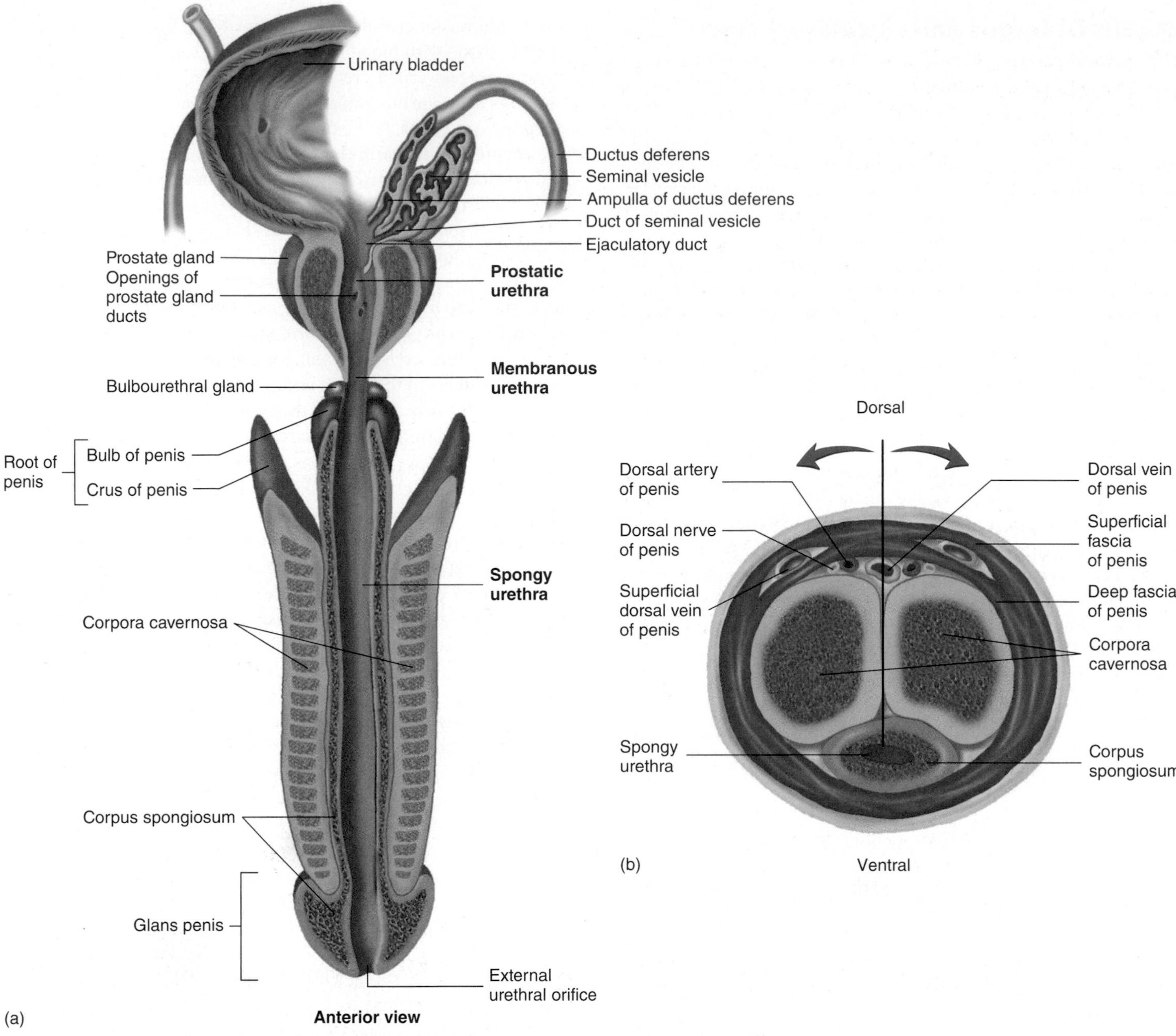

Urinary bladder
Ductus deferens
Seminal vesicle
Ampulla of ductus deferens
Duct of seminal vesicle
Ejaculatory duct
Prostate gland
Openings of prostate gland ducts
Prostatic urethra
Bulbourethral gland
Membranous urethra
Root of penis
Bulb of penis
Crus of penis
Spongy urethra
Corpora cavernosa
Corpus spongiosum
Glans penis
External urethral orifice
(a)
Anterior view
Dorsal
Dorsal artery of penis
Dorsal nerve of penis
Superficial dorsal vein of penis
Dorsal vein of penis
Superficial fascia of penis
Deep fascia of penis
Corpora cavernosa
Spongy urethra
Corpus spongiosum
(b)
Ventral

FIGURE 28.7 Penis
(*a*) Section through the spongy, or penile, urethra laid open and viewed from above. The prostate is also cut open to show the prostatic urethra. (*b*) Cross section of the penis, showing principal nerves, arteries, and veins along the dorsum of the penis. The *black line* and *blue arrows* depict the manner in which (*a*) is cut and laid open.

Seminal Vesicles

The **seminal vesicles** (sem′i-năl ves′i-klz) are sac-shaped glands located next to the ampullae of the ducta deferentia (see figure 28.6). Each gland is about 5 cm long and tapers into a short excretory duct that joins the ampulla of the ductus deferens to form the ejaculatory duct. The seminal vesicles have a capsule containing fibrous connective tissue and smooth muscle cells.

Prostate Gland

The **prostate** (pros′tāt; one standing before) **gland,** consisting of both glandular and muscular tissue, resembles a walnut in shape and size, being approximately 4 cm long and 2 cm wide. It is dorsal to the symphysis pubis at the base of the urinary bladder, where it surrounds the prostatic urethra and the two ejaculatory ducts (see figure 28.1). The prostate gland is composed of a fibrous connective tissue capsule containing distinct smooth muscle cells and numerous fibrous partitions, also containing smooth muscle, that radiate inward toward the urethra. Covering these muscular partitions is a layer of columnar epithelial cells that form saccular dilations into which the columnar cells secrete prostatic fluid. Fifteen to 30 small prostatic ducts carry these secretions into the prostatic urethra.

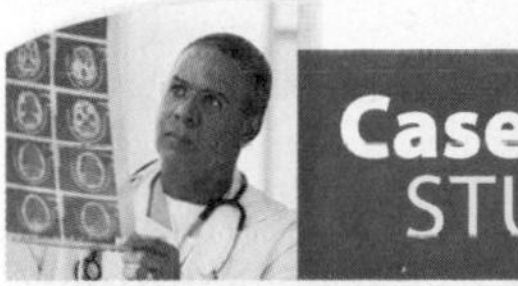

Case STUDY Prostate Cancer

Sixty-five-year-old Vern has a physical examination every year. Eight years ago, a test indicated that Vern's blood **prostate specific antigen (PSA)** levels were elevated compared with his previous tests. His physician reported moderate enlargement of the prostate gland but could detect no obvious tumorlike structures by a digital examination. Because of the increased PSA levels, needle biopsies of Vern's prostate gland were taken through his rectum. Suspicious cells consistent with prostate cancer were detected in one of the tissue samples. Vern's physician had the biopsy samples examined by another pathology laboratory, which did not confirm the first pathology report. As a consequence, Vern's physician explained that, since prostate cancer typically develops slowly, one option was to do nothing and continue having regular checkups.

Eight years later, a routine blood test showed another substantial increase in Vern's PSA levels, although, again, no tumor could be detected by a digital exam, and Vern had no symptoms, such as difficulty urinating. Needle biopsies of the prostate gland were taken again, and cancer cells were found in two of the six biopsy samples. The physician explained that, since the cancer had been discovered before it had metastasized, Vern's chances of surviving were high. Vern could choose to do nothing, have his prostate gland surgically removed, or treat the cancer with radiation therapy, hormonal therapy, or chemotherapy. Vern's physician indicated that doing nothing is a reasonable option for men who are significantly older than Vern because older men diagnosed with prostate cancer often die of other conditions before they succumb to prostate cancer.

Vern elected to have radiation therapy, which focuses radiation on the prostate gland to kill the cancer cells. Statistics indicate that surgery and radiation therapy have similar success rates for small, localized tumors like Vern's. The trauma of surgery and the higher probability of erectile dysfunction following surgery convinced Vern that radiation therapy was preferable. The physician explained that approximately 85% of patients like Vern are cancer-free 5 years after radiation treatments. Vern was glad that he had continued having annual physical examinations. Prostate cancer represents 29% of cancers in males in the United States and 14% of the deaths due to cancer. Only lung cancer causes more cancer deaths in men.

Predict 2

a. Given the elevated PSA levels and the suspicious cells appearing in the first biopsy, was it reasonable to conclude that cancer cells were not present in Vern's prostate gland, or was it an error? (*Hint:* See chapter 4.)
b. Explain why, following the first biopsy, the second pathology report did not confirm the results of the first pathology report.
c. Explain why cancer cells were present in only two of the six needle biopsy samples.
d. Since increased blood levels of PSA are a common diagnostic tool for prostate cancer, explain why Vern's PSA levels were elevated prior to the first biopsy.

Bulbourethral Glands

The **bulbourethral** (bŭl′bō-ū-rē′thrăl) **glands,** or *Cowper glands,* are a pair of small glands located near the membranous part of the urethra (see figures 28.1 and 28.6). In young males, each gland is about the size of a pea, but they decrease in size with age and are almost impossible to see in older men. Each bulbourethral gland is a compound mucous gland (see chapter 4). The small ducts of each gland unite to form a single duct, which then enters the spongy urethra at the base of the penis.

Semen

Collectively, the sperm cells and secretions from the male reproductive glands are called **semen** (sē′men). The seminal vesicles produce about 60% of the fluid, the prostate gland contributes about 30%, the testes contribute 5%, and the bulbourethral glands contribute 5%. **Emission** (ē-mish′ŭn) is the discharge of all these secretions from the ducta deferentia into the urethra. **Ejaculation** (ē-jak-ū-lā′shŭn) is the forceful expulsion of semen from the urethra caused by contraction of the urethra, the skeletal muscles in the pelvic floor, and the muscles at the base of the penis.

The bulbourethral glands and urethral mucous glands produce a mucous secretion just before ejaculation. This mucus lubricates the urethra, neutralizes the contents of the normally acidic spongy urethra, provides a small amount of lubrication during intercourse, and helps reduce vaginal acidity.

Testicular secretions include sperm cells, a small amount of fluid, and some metabolic by-products. The thick, mucuslike secretions of the seminal vesicles contain large amounts of fructose, citric acid, and other nutrients that nourish the sperm cells. The seminal vesicle secretions also contain fibrinogen, which is involved in a weak coagulation reaction of the semen immediately after ejaculation. In addition, seminal vesicle secretions contain prostaglandins that can cause uterine contractions to help transport sperm cells through the female reproductive tract.

The thin, milky secretions of the prostate have a basic pH. In combination with secretions of the seminal vesicles, bulbourethral glands, and urethral mucous glands, the prostatic secretions help neutralize the acidic urethra. The secretions of the prostate and seminal vesicles also help neutralize the acidic secretions of the testes and the vagina. In addition, the prostatic secretions are important in the transient coagulation of semen because they contain clotting factors that convert fibrinogen from the seminal vesicles to fibrin, resulting in coagulation. The coagulated material keeps the semen a sticky mass for a few minutes after ejaculation, and then fibrinolysin from the prostate causes the coagulum to dissolve, thereby releasing the sperm cells to make their way up the female reproductive tract.

Predict 3

Explain a possible function of the coagulation reaction.

Before ejaculation, the ductus deferens begins to contract rhythmically to propel sperm cells and testicular and epididymal secretions from the tail of the epididymis to the ampulla of the ductus deferens. Contractions of the ampullae, seminal vesicles, and ejaculatory ducts cause the sperm cells, along with testicular and epididymal secretions, to move into the prostatic urethra with the prostatic secretions. Secretions of the seminal vesicles then enter the prostatic urethra, where they mix with the other secretions.

Normal sperm cell counts in the semen range from 75 to 400 million sperm cells per milliliter of semen, and a normal ejaculation usually consists of about 2–5 mL of semen. The semen with the highest sperm count is expelled from the penis first because it contains the greater percentage of sperm-containing fluid from the epididymis. Sperm cells become motile after ejaculation once they are mixed with secretions of the male accessory glands and the female reproductive tract. The basic pH (an average of 7.5), nutrients, and removal of inhibitory substances from the surface of sperm cells appear to increase sperm cell motility. Enzymes carried in the acrosomal cap of each sperm cell help digest a path through the mucoid fluids of the female reproductive tract and through materials surrounding the oocyte. Once the acrosomal fluid is depleted from a sperm cell, the sperm cell is no longer capable of fertilization. As a result, most of the sperm cells (millions) are expended in moving the general group of sperm cells through the female reproductive tract.

ASSESS YOUR PROGRESS

26. *State where the seminal vesicles, prostate gland, and bulbourethral glands empty into the male reproductive duct system.*
27. *Define* emission *and* ejaculation.
28. *Describe the contributions to semen from the accessory glands.*
29. *What is the function of the secretions of each of the accessory glands?*

28.3 Physiology of Male Reproduction

LEARNING OUTCOMES

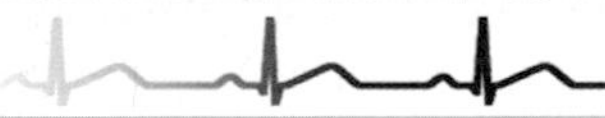

After reading this section, you should be able to

A. List the hormones that influence the male reproductive system and describe their functions.
B. Demonstrate an understanding of the changes that occur in males during puberty.
C. Explain the events that occur during the male sex act.

Normal function of the male reproductive system depends on both hormonal and neural mechanisms. Hormones are primarily responsible for the development of reproductive structures and the maintenance of their functional capacities, the development of secondary sex characteristics, and the control of sperm cell formation. Hormones also influence sexual behavior. Neural mechanisms are primarily involved in sexual behavior and control of the sexual act.

Regulation of Sex Hormone Secretion

Hormonal mechanisms that influence the male reproductive system involve the hypothalamus, the pituitary gland, and the testes (figure 28.8). A small peptide hormone called **gonadotropin-releasing hormone (GnRH)** is released from neurons in the hypothalamus. GnRH passes through the hypothalamohypophysial portal system to the anterior pituitary gland (see chapter 18). In response to GnRH, cells within the anterior pituitary gland secrete two hormones, referred to as **gonadotropins** (gō′nad-ō-trō′pinz, gon′ă-dō-trō′pinz) because they influence the function of the **gonads** (gō′nadz; testes or ovaries).

The two gonadotropins are **luteinizing hormone (LH)** and **follicle-stimulating hormone (FSH).** They are named for their functions in females, but they also have important functions in males. LH in males is sometimes called **interstitial cell–stimulating hormone (ICSH).** LH binds to the interstitial cells in the testes and causes them to increase their rate of testosterone synthesis and secretion. FSH binds primarily to sustentacular cells in the seminiferous tubules and promotes sperm cell development. Both gonadotropins bind to specific receptor molecules on the membranes of the cells they influence, and cyclic adenosine monophosphate (cAMP) is an important intracellular mediator in those cells.

For GnRH to stimulate the secretion of large quantities of LH and FSH, the anterior pituitary must be exposed to a series of brief increases and decreases in GnRH. Interestingly, GnRH can be produced synthetically; if administered in small amounts in frequent pulses or surges, it can be useful in treating male infertility. However, chronically elevated GnRH levels in the blood cause the anterior pituitary cells to become insensitive to stimulation by GnRH molecules, and little LH or FSH is secreted. Long-term administration of synthetic GnRH, therefore, can reduce sperm cell production, causing infertility.

Testosterone is the major male hormone secreted by the testes. It is classified as an **androgen** (Gr. *andros,* male human) because it stimulates the development of male reproductive structures (see chapter 29) and male secondary sex characteristics. The testes secrete other androgens, but they are produced in smaller concentrations and are less potent than testosterone. In addition, the testes secrete small amounts of estrogen and progesterone.

Testosterone has a major influence on many tissues. It plays an essential role in the embryonic development of reproductive structures, their further development during puberty, the development of secondary sex characteristics during puberty, the maintenance of sperm cell production, and the regulation of gonadotropin secretion. It also influences behavior.

Inside some target tissue cells, such as cells of the scrotum and penis, an enzyme converts testosterone to dihydrotestosterone. In these cells, dihydrotestosterone is the active hormone. If the enzyme is not active, these structures do not fully develop normally. In other target tissue cells, an enzyme converts testosterone to estrogen, and estrogen becomes the active hormone. Some brain cells convert testosterone to estrogen; in these cells, estrogen may be the active hormone responsible for certain aspects of male sexual behavior.

The sustentacular cells of the testes secrete a polypeptide hormone called **inhibin** (in-hib′in), which inhibits FSH secretion from the anterior pituitary.

1. GnRH from the hypothalamus stimulates the secretion of LH and FSH from the anterior pituitary.
2. LH stimulates testosterone secretion from the interstitial cells.
3. FSH stimulates sustentacular cells of the seminiferous tubules to increase spermatogenesis and to secrete inhibin.
4. Testosterone has a stimulatory effect on the sustentacular cells of the seminiferous tubules, as well as on the development of sex organs and secondary sex characteristics.
5. Testosterone has a negative-feedback effect on the hypothalamus and pituitary to reduce LH and FSH secretion.
6. Inhibin has a negative-feedback effect on the anterior pituitary to reduce FSH secretion.

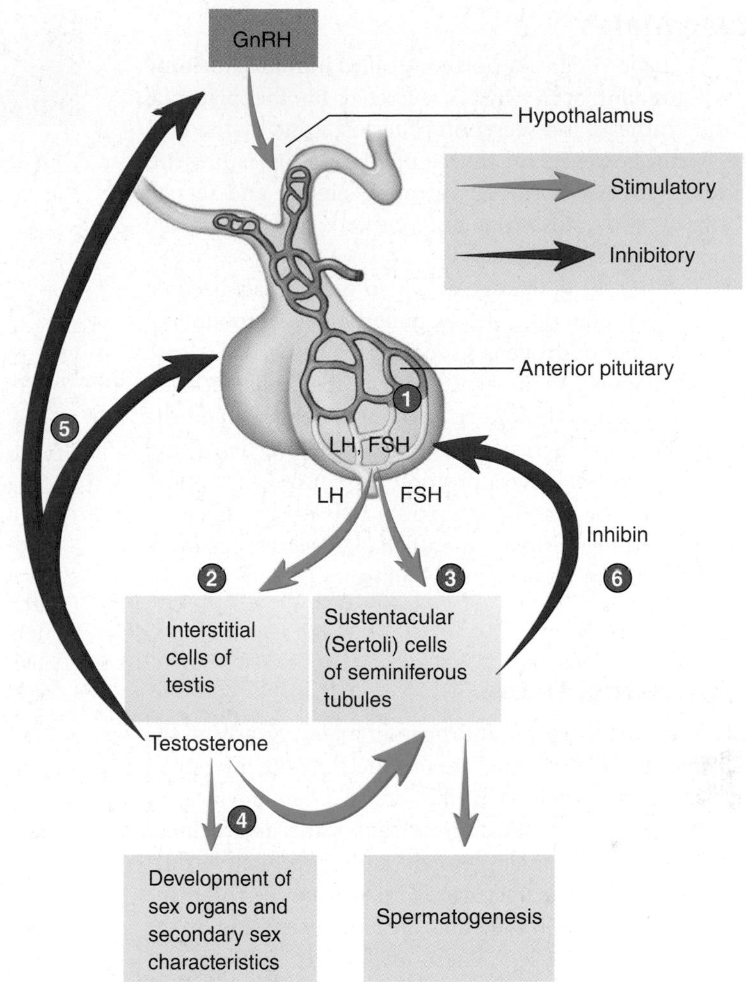
GnRH
Hypothalamus
Stimulatory
Inhibitory
Anterior pituitary
1
5
LH, FSH
LH
FSH
Inhibin
2
3
6
Interstitial cells of testis
Sustentacular (Sertoli) cells of seminiferous tubules
Testosterone
4
Development of sex organs and secondary sex characteristics
Spermatogenesis

PROCESS FIGURE 28.8 Regulation of Reproductive Hormone Secretion in Males

Clinical IMPACT — Male Infertility

Infertility (in-fer-til′i-tē) is the inability or the reduced ability to produce offspring. The most common cause of infertility in males is a low sperm cell count. A count of less than 20 million sperm cells per milliliter usually indicates infertility.

The sperm cell count can decrease because of damage to the testes as a result of trauma, radiation, cryptorchidism, or an infection, such as mumps. **Varicocele** (var′i-kō-sēl) is an abnormal dilation of a spermatic vein that results from incompetent or absent valves in spermatic veins, from thrombi, or from tumors. As a result, both testicular blood flow and spermatogenesis decrease. Reduced sperm cell counts can also result from inadequate secretion of LH and FSH, which can be caused by hypothyroidism, trauma to the hypothalamus, infarctions of the hypothalamus or anterior pituitary gland, or tumors. Decreased testosterone secretion reduces the sperm cell count as well. Some reports suggest that the average sperm cell count has decreased substantially since the end of World War II (1945), although there is some controversy about the accuracy of these reports. Researchers speculate that certain synthetic chemicals are responsible.

Even when the sperm cell count is normal, fertility can be reduced if sperm cell structure is abnormal, as occurs due to chromosomal abnormalities caused by genetic factors. Reduced sperm cell motility also results in infertility. A major cause of reduced sperm cell motility is the presence of antisperm antibodies, which are produced by the immune system and bind to sperm cells.

In cases of infertility due to low sperm count or reduced motility, fertility can sometimes be achieved by collecting several ejaculations and concentrating the sperm cells before inserting them into the female reproductive tract, a process called **artificial insemination** (in-sem-i-nā′shŭn).

Puberty in Males

Before birth, a gonadotropin-like hormone called **human chorionic** (kō-rē-on′ik) **gonadotropin (hCG),** secreted by the placenta, stimulates the synthesis and secretion of testosterone by the fetal testes. After birth, however, no source of stimulation is present, and the testes of the newborn baby atrophy slightly and secrete only small amounts of testosterone until puberty, which normally begins when a boy is 12–14 years old.

Puberty (pū′ber-tē) is the age at which individuals become capable of sexual reproduction. Before puberty, small amounts of testosterone and other androgens in males inhibit GnRH release from the hypothalamus. At puberty, the hypothalamus becomes much less sensitive to the inhibitory effect of androgens, and the rate of GnRH secretion increases, leading to increased LH and FSH release. Elevated FSH levels promote sperm cell formation, and elevated LH levels cause the interstitial cells to secrete larger amounts of testosterone. Testosterone still has a negative-feedback effect on GnRH secretion after puberty but is not capable of completely suppressing it.

Effects of Testosterone

Testosterone is by far the major androgen in males. Nearly all the androgens, including testosterone, are produced by the interstitial cells, with small amounts produced by the adrenal cortex and possibly by the sustentacular cells. Testosterone causes the enlargement and differentiation of the male genitals and reproductive duct system and is necessary for sperm cell formation. Testosterone also stimulates hair growth in the pubic area and extending up the linea alba, as well as on the legs, chest, axillary region, face, and back. It causes vellus hair to be converted to terminal hair, which is coarser and more pigmented.

Testosterone causes the texture of the skin to become rougher or coarser. The quantity of melanin in the skin also increases, making the skin darker. In addition, testosterone increases the rate of secretion from the sebaceous glands, especially on the face. Near puberty, the increased testosterone level and increased sebaceous gland secretion frequently cause acne (see chapter 5). Testosterone also causes hypertrophy of the larynx and reduced tension on the vocal folds, beginning near puberty. At first, the structural changes can make the voice difficult to control, but ultimately its normal masculine quality is achieved.

Testosterone stimulates metabolism so that males have a slightly higher metabolic rate than females. The red blood cell count increases by nearly 20% as a result of testosterone's effect on erythropoietin production. Testosterone also has a minor mineralocorticoid-like effect, causing Na^+ to be retained in the body and, consequently, an increased volume of body fluids. Testosterone promotes protein synthesis in most tissues; as a result, skeletal muscle mass increases at puberty. The average percentage of the body weight composed of skeletal muscle is greater for males than for females because of the effect of androgens.

Testosterone causes rapid bone growth and increases the deposition of Ca^{2+} in bone, resulting in increased height. However, the growth in height is limited because testosterone also stimulates ossification of the epiphyseal plates of long bones (see chapter 6).

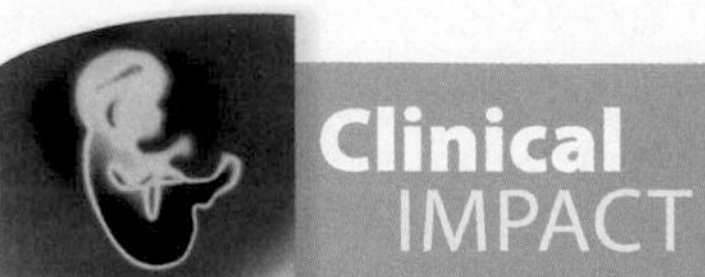

Clinical IMPACT

Anabolic Steroids

In an attempt to improve their performance, some athletes, especially those who depend on muscle strength, may either ingest or inject synthetic androgens, commonly called **anabolic steroids,** or simply *steroids.* These hormones have testosterone-like effects, such as stimulating the development of male secondary sex characteristics, but many anabolic steroids are structurally different from testosterone, and their effect on muscle is greater than their effect on the reproductive organs. However, when taken in large amounts, they can influence the reproductive system. Large doses of anabolic steroids have a negative-feedback effect on the hypothalamus and anterior pituitary, reducing GnRH, LH, and FSH levels. As a result, the testes can atrophy and sterility can develop. Other side effects of large doses of anabolic steroids include kidney and liver damage, heart attack, and stroke. In addition, anabolic steroids cause abrupt mood swings, usually toward intense anger and rage. Taking anabolic steroids is highly discouraged by medical professionals, violates the rules of most athletic organizations, and is illegal without a prescription.

Males who mature sexually at an earlier age grow rapidly but reach their maximum height earlier. Males who mature sexually at a later age do not exhibit a rapid period of growth, but they grow for a longer period and can become taller than those who mature sexually at an earlier age.

Male Sexual Behavior and the Male Sex Act

Testosterone is required to initiate and maintain male sexual behavior. Testosterone enters cells within the hypothalamus and the surrounding areas of the brain and influences their function, resulting in sexual behavior. However, male sexual behavior may depend, in part, on the conversion of testosterone to other steroids, such as estrogen, in cells of the brain.

The blood levels of testosterone remain relatively constant in a male from puberty until about 40 years of age. Thereafter, the levels slowly decline to about 20% of this value by 80 years of age, causing a slow decrease in sex drive and fertility.

The male sex act is a complex series of reflexes that result in erection of the penis, secretion of mucus into the urethra, emission, and ejaculation. Sensations that are normally interpreted as pleasurable occur during the male sex act and result in a climactic sensation, called **orgasm** (ōr′gazm), associated with ejaculation. After ejaculation, a phase called **resolution** occurs, which is characterized by a flaccid penis, an overall feeling of satisfaction, and the inability to achieve erection and a second ejaculation for a period that can range from many minutes to many hours or longer.

Sensory Action Potentials and Integration

Male sexual reflexes are initiated by a variety of sensory stimuli. Action potentials are conducted by sensory neurons from the genitals through the pudendal nerve to the sacral region of the spinal cord, where reflexes that result in the male sex act are integrated. Action potentials travel from the spinal cord to the cerebrum to produce conscious sexual sensations.

Rhythmic massage of the penis, especially the glans penis, produces extremely important sensory action potentials that initiate erection and ejaculation. In addition, sensory action potentials produced in surrounding tissues, such as the scrotum and the anal, perineal, and pubic regions, reinforce sexual sensations. Engorgement of the prostate and seminal vesicles with their secretions also causes sexual sensations. In some cases, mild irritation of the urethra, as may result from an infection, can cause sexual sensations.

Psychic stimuli, including sight, sound, odor, and thoughts, have a major effect on sexual reflexes. Thinking sexual thoughts or dreaming about erotic events tends to reinforce stimuli that trigger sexual reflexes, such as erection and ejaculation. Ejaculation while sleeping is a relatively common event in young males and is thought to be triggered by psychic stimuli associated with dreaming. Psychic stimuli can also inhibit the sex act, and thoughts that are not sexual tend to decrease the effectiveness of the male sex act.

Action potentials from the cerebrum that reinforce the sacral reflexes are not absolutely required for the culmination of the male sex act. The sex act can be performed by males who have suffered spinal cord injuries superior to the sacral region.

Erection, Emission, and Ejaculation

When erection occurs, the penis becomes enlarged and rigid. Erection is the first major component of the male sex act. Figure 28.9 illustrates the neural events that lead to an erection. First, action potentials travel from the spinal cord through the pudendal nerve to the arteries that supply blood to the erectile tissues. The nerve fibers release acetylcholine and **nitric oxide (NO)** as neurotransmitters. Acetylcholine binds to muscarinic receptors and activates a G protein mechanism that causes smooth muscle relaxation. Nitric oxide diffuses into the smooth muscle cells of blood vessels and combines with the enzyme guanylate cyclase, which converts guanosine triphosphate (GTP) to cyclic guanosine monophosphate (cGMP). The cGMP causes smooth muscle cells to relax and blood vessels to dilate (figure 28.9). At the same time, other arteries of the penis constrict to shunt blood to the erectile tissues. As a consequence, blood fills the sinusoids of the erectile tissue and compresses the veins. Because venous outflow is partially occluded, the blood pressure in the sinusoids causes inflation and rigidity of the erectile tissue. Nerve action potentials that result in erection come from parasympathetic centers (S2–S4) or sympathetic centers (T2–L1) in the spinal cord. Normally, the parasympathetic centers are more important but, in cases of damage to the sacral region of the spinal cord, erection can occur through the sympathetic system.

Parasympathetic action potentials also cause the mucous glands within the penile urethra and the bulbourethral glands at the base of the penis to secrete mucus.

As mentioned in section 28.2, emission is the accumulation of sperm cells and secretions of the prostate gland and seminal vesicles in the urethra. Sympathetic centers in the spinal cord (T12–L1), which are stimulated as the level of sexual tension increases, control emission. Sympathetic action potentials cause peristaltic contractions of the reproductive ducts and stimulate the seminal vesicles and the prostate gland to release their secretions. Consequently, semen accumulates in the prostatic urethra and produces sensory action potentials that pass through the pudendal nerves to the spinal cord.

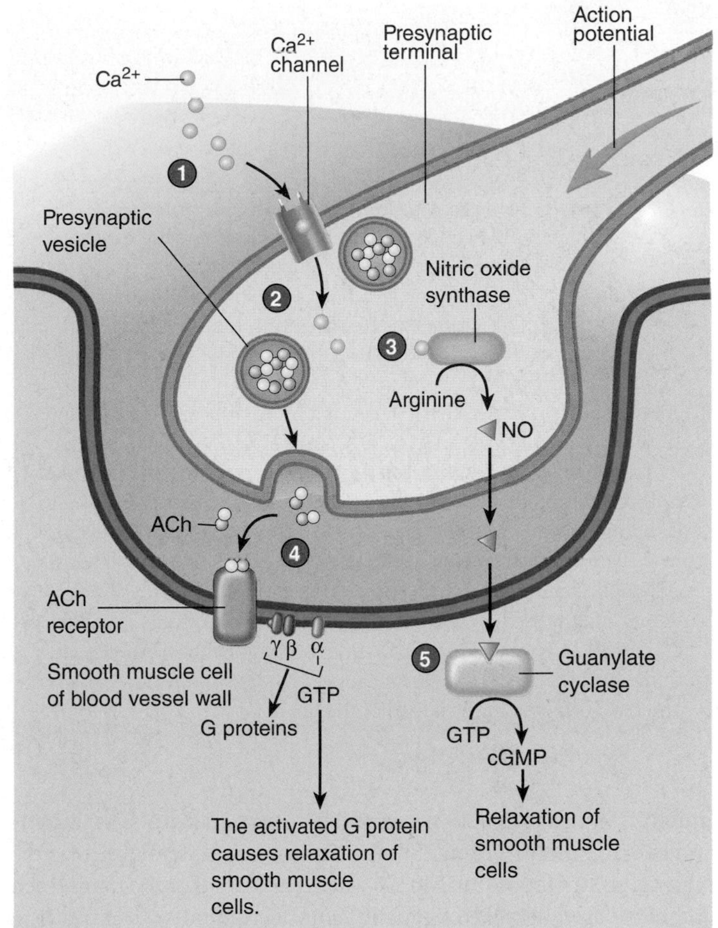

Ca^{2+}
Ca^{2+} channel
Presynaptic terminal
Action potential
1
Presynaptic vesicle
2
Nitric oxide synthase
3
Arginine
NO
ACh
4
ACh receptor
γ β α
Smooth muscle cell of blood vessel wall
GTP
G proteins
5
Guanylate cyclase
GTP
cGMP
The activated G protein causes relaxation of smooth muscle cells.
Relaxation of smooth muscle cells

1. Action potentials in parasympathetic neurons cause voltage-gated Ca^{2+} channels to open, and Ca^{2+} diffuse into the presynaptic terminals.
2. Calcium ions initiate the release of acetylcholine (ACh) from presynaptic vesicles.
3. Calcium ions also activate nitric oxide synthase, which promotes the synthesis of nitric oxide (NO) from arginine.
4. ACh binds to ACh receptors on the smooth muscle cells and activates a G protein mechanism. The activated G protein causes the relaxation of smooth muscle cells and erection of the penis.
5. NO binds to guanylate cyclase enzymes and activates them. The activated enzymes convert GTP to cGMP, which causes relaxation of the smooth muscle cells and erection of the penis.

PROCESS FIGURE 28.9 Neural Control of Erection

Predict 4

Mr. Grover suffers from the periodic inability to achieve an erection. His doctor could find no structural or physiological abnormalities, so he prescribed sildenafil. After taking the pills, Mr. Grover could sometimes, but not always, achieve an erection. Assuming no pathology affected Mr. Grover, explain his experiences.

Clinical IMPACT

Erectile Dysfunction

Failure to achieve erection, or **erectile dysfunction (ED),** sometimes called *impotence,* can be a major source of frustration for some men. ED can be due to reduced testosterone secretion resulting from hypothalamic, pituitary, or testicular complications. In other cases, ED is due to defective stimulation of the erectile tissue by nerve fibers or reduced response of the blood vessels to neural stimulation.

Some men can achieve erections by taking oral medication, such as sildenafil (Viagra), tadafil (Cialis), or vardenafil (Levitra), or by having specific drugs injected into the base of the penis, which increases blood flow into the erectile tissue of the penis, resulting in erection for many minutes. Sildenafil is a drug that blocks the activity of the enzyme that converts cGMP to GMP. Consequently, it allows cGMP to accumulate in smooth muscle cells in the arteries of erectile tissues and causes them to relax. This response is effective in enhancing erection in males. Sildenafil's action is not specific to the erectile tissue of the penis, however. It causes vasodilation in other tissues and can increase the workload of the heart.

Integration of these action potentials results in both sympathetic and somatic motor output. Sympathetic action potentials cause the internal sphincter of the urinary bladder to constrict, so that semen and urine are not mixed. Somatic motor action potentials travel to the skeletal muscles of the urogenital diaphragm and the base of the penis, causing several rhythmic contractions that force the semen out of the urethra, the process called ejaculation. In addition, muscle tension increases throughout the body.

ASSESS YOUR PROGRESS

30. *Where are GnRH, LH, FSH, and inhibin produced? What effects do they have on the male reproductive system?*
31. *Where is testosterone produced?*
32. *Explain the regulation of testosterone secretion.*
33. *What changes in hormone production occur at puberty?*
34. *Describe the effects of testosterone on the male body.*
35. *What effects do psychic, tactile, parasympathetic, and sympathetic stimulation have on the male sex act?*
36. *Describe the processes of erection, emission, ejaculation, orgasm, and resolution.*

28.4 Anatomy of the Female Reproductive System

LEARNING OUTCOMES

After reading this section, you should be able to

A. Name the organs of the female reproductive system and describe their functions.

B. Describe the anatomy and histology of the ovaries.

C. Discuss the development of the oocyte and the follicle and describe ovulation and fertilization.

D. Describe the structure of the uterine tubes, uterus, vagina, external genitalia, and mammary glands.

The female reproductive organs are the ovaries, the uterine tubes, the uterus, the vagina, the external genital organs, and the mammary glands. The internal reproductive organs are within the pelvis between the urinary bladder and the rectum (figure 28.10). The uterus and the vagina are in the midline, with the ovaries to each side of the uterus. A group of ligaments holds the internal reproductive organs in place. The most conspicuous is the **broad ligament,** an extension of the peritoneum that spreads out on both sides of the uterus and attaches to the ovaries and uterine tubes (figure 28.11).

Ovaries

The two **ovaries** (ō′var-ēz) are small organs about 2–3.5 cm long and 1–1.5 cm wide (figure 28.11). A peritoneal fold called the **mesovarium** (mez′ō-vā′rē-ŭm; mesentery of the ovary) attaches each ovary to the posterior surface of the broad ligament. Two other ligaments are associated with the ovary: the **suspensory ligament,** which extends from the mesovarium to the body wall, and the **ovarian ligament,** which attaches the ovary to the superior margin of the uterus. The ovarian arteries, veins, and nerves traverse the suspensory ligament and enter the ovary through the mesovarium.

Ovarian Histology

The visceral peritoneum, made up of simple cuboidal epithelium, covers the surface of the ovary, where it is called the **ovarian epithelium,** or *germinal epithelium.* Immediately below the epithelium is a capsule of dense fibrous connective tissue, the **tunica albuginea** (al-bū-jin′ē-ă). The **cortex** is the denser, outer part of the ovary, and the **medulla** is the looser, inner part of the ovary (figure 28.12). The connective tissue of the ovary is called the **stroma.** Blood vessels, lymphatic vessels, and nerves from the mesovarium enter the medulla. Numerous **ovarian follicles,** each of which contains an **oocyte** (ō′ō-sīt), are distributed throughout the stroma of the cortex.

Oogenesis and Fertilization

The formation of female gametes begins in the fetus. By the fourth month of development, the ovaries contain 5 million **oogonia** (ō-ō-gō′nē-ă; *oon,* egg + *gone,* generation), the cells from which oocytes develop (figure 28.13). By the time of birth, many of the oogonia have degenerated, and the remaining ones have begun meiosis. Also, oogonia can form after birth from stem cells, but the extent to which this occurs, and how long it occurs, is not clear. Meiosis stops, however, during the first meiotic division at a stage called prophase I (see Clinical Impact, "Meiosis," earlier in this chapter). The cell at this stage is called a **primary oocyte,** and at birth there are about 2 million of them. From birth to puberty, the number of primary oocytes decreases to around 300,000–400,000. Only about 400 primary oocytes will complete development and give rise to the secondary oocytes that are eventually released from the ovaries.

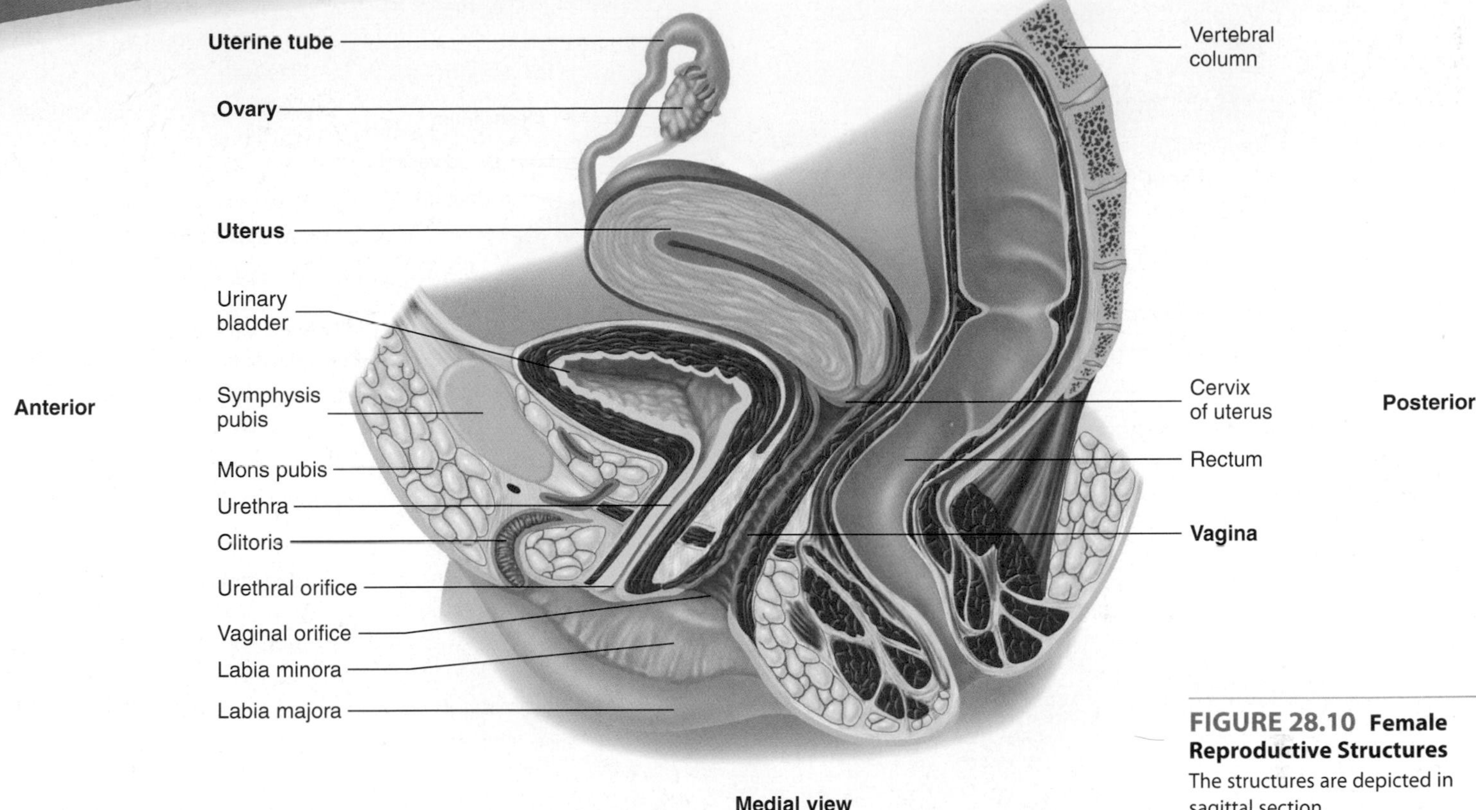
Uterine tube
Ovary
Uterus
Urinary bladder
Anterior
Symphysis pubis
Mons pubis
Urethra
Clitoris
Urethral orifice
Vaginal orifice
Labia minora
Labia majora
Vertebral column
Cervix of uterus
Posterior
Rectum
Vagina
Medial view

FIGURE 28.10 Female Reproductive Structures The structures are depicted in sagittal section.

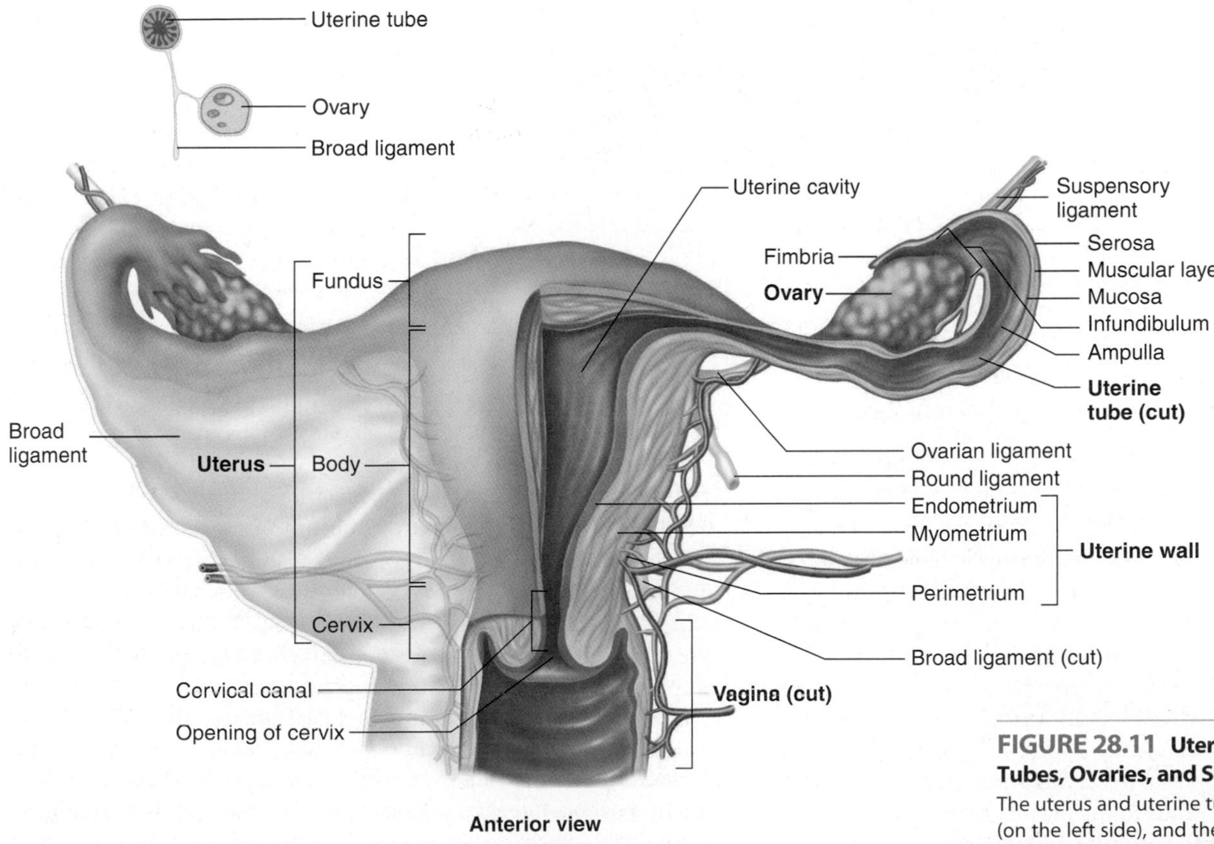

Uterine tube
Ovary
Broad ligament
Uterine cavity
Suspensory ligament
Fimbria
Ovary
Serosa
Muscular layer
Mucosa
Uterine tube wall
Infundibulum
Ampulla
Uterine tube (cut)
Fundus
Broad ligament
Uterus
Body
Ovarian ligament
Round ligament
Endometrium
Myometrium
Uterine wall
Perimetrium
Cervix
Broad ligament (cut)
Cervical canal
Vagina (cut)
Opening of cervix
Anterior view

FIGURE 28.11 Uterus, Vagina, Uterine Tubes, Ovaries, and Supporting Ligaments The uterus and uterine tubes are cut in section (on the left side), and the vagina is cut to show the internal anatomy. The inset (top, left) shows the relationships among the ovary, the uterine tube, and the ligaments that suspend them in the pelvic cavity.

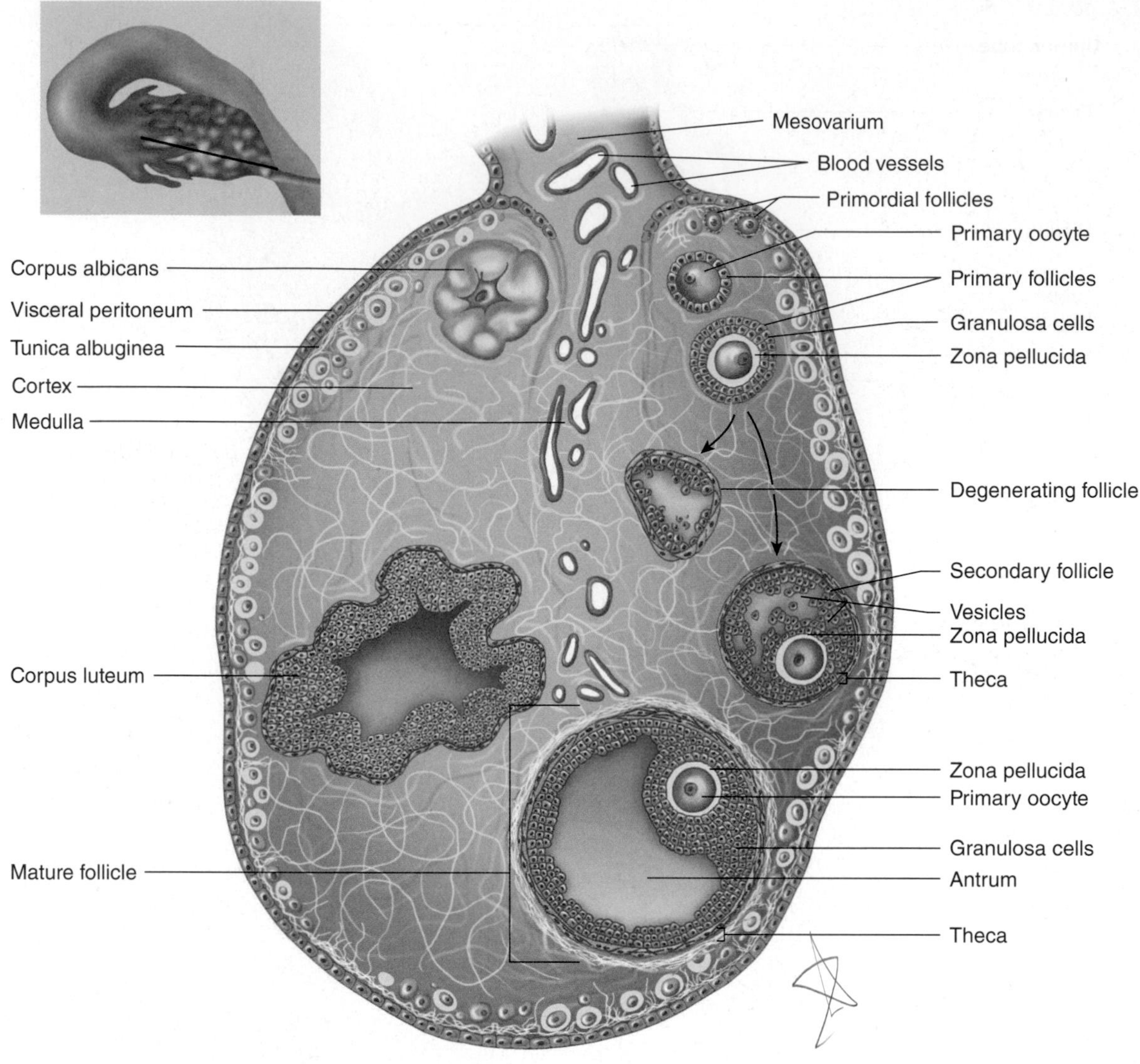
Mesovarium
Blood vessels
Primordial follicles
Primary oocyte
Primary follicles
Granulosa cells
Zona pellucida
Corpus albicans
Visceral peritoneum
Tunica albuginea
Cortex
Medulla
Degenerating follicle
Secondary follicle
Vesicles
Zona pellucida
Theca
Corpus luteum
Zona pellucida
Primary oocyte
Granulosa cells
Antrum
Mature follicle
Theca

FIGURE 28.12 APR **Histology of the Ovary**
The ovary is sectioned to illustrate its internal structure (the inset shows plane of section). Ovarian follicles from each major stage of development are shown.

Ovulation (ov′ū-lā′shŭn, ō′vū-lā′shŭn) is the release of a **secondary oocyte** from an ovary (figure 28.13, *step 8*). Just before ovulation, the primary oocyte completes the first meiotic division to produce a secondary oocyte and a **polar body.** Unlike meiosis in males, cytoplasm is not split evenly between the two cells. Most of the cytoplasm of the primary oocyte remains with the secondary oocyte. The cytoplasm contains organelles, such as mitochondria, and nutrients that increase the viability of the secondary oocyte. The polar body either degenerates or divides to form two polar bodies. Eventually, the polar bodies degenerate. The secondary oocyte begins the second meiotic division, but it stops in metaphase II.

After ovulation, the secondary oocyte may be fertilized by a sperm cell (figure 28.13, *step 9*). **Fertilization** (fer′til-i-zā-shŭn) begins when a sperm cell binds to the plasma membrane and penetrates the plasma membrane of a secondary oocyte. Subsequently, the secondary oocyte completes the second meiotic division to form two cells, each containing 23 chromosomes. One of these cells has very little cytoplasm and is another polar body that degenerates. In the other, larger cell, the 23 chromosomes from the sperm cell nucleus join with the 23 chromosomes from the oocyte to form a **zygote** (zī′gōt) and complete fertilization. The zygote has 23 pairs of chromosomes (a total of 46 chromosomes). All cells of the human body contain 23 pairs of chromosomes, except for the male and female gametes. The zygote divides by mitosis to form two cells, which divide to form four cells, and so on. Seven days after ovulation, the mass of cells may implant in or attach to the uterine wall. The implanted mass of cells continues to develop for approximately 9 months to form a new individual (see chapter 29).

Follicle Development

As discussed earlier, oogenesis begins when a female is in her mother's uterus. The primary oocytes present at birth are located in primordial follicles. A **primordial follicle** is a primary oocyte surrounded by a single layer of flat cells, called **granulosa cells** (figure 28.13).

FUNDAMENTAL **Figure**

1. Oogonia give rise to oocytes. Before birth, oogonia multiply by mitosis. During development of the fetus, many oogonia begin meiosis, but stop in prophase I and are now called primary oocytes. They remain in this state until puberty.
2. Before birth, the primary oocytes become surrounded by a single layer of granulosa cells, creating a primordial follicle. These are present until puberty.
3. After puberty, primordial follicles develop into primary follicles when the granulosa cells enlarge and increase in number.
4. Secondary follicles form when fluid-filled vesicles develop and thecal cells arise on the outside of the follicle.
5. Mature follicles form when the vesicles create a single antrum.
6. Just before ovulation, the primary oocyte completes meiosis I, creating a secondary oocyte and a nonviable polar body.
7. The secondary oocyte begins meiosis II, but stops at metaphase II.
8. During ovulation, the secondary oocyte is released from the ovary.
9. The secondary oocyte only completes meiosis II if it is fertilized by a sperm cell. The completion of meiosis II forms an oocyte and a second polar body. Fertilization is complete when the oocyte nucleus and the sperm cell nucleus unite, creating a zygote.
10. Following ovulation, the granulosa cells divide rapidly and enlarge to form the corpus luteum.
11. The corpus luteum degenerates to form a scar, or corpus albicans.

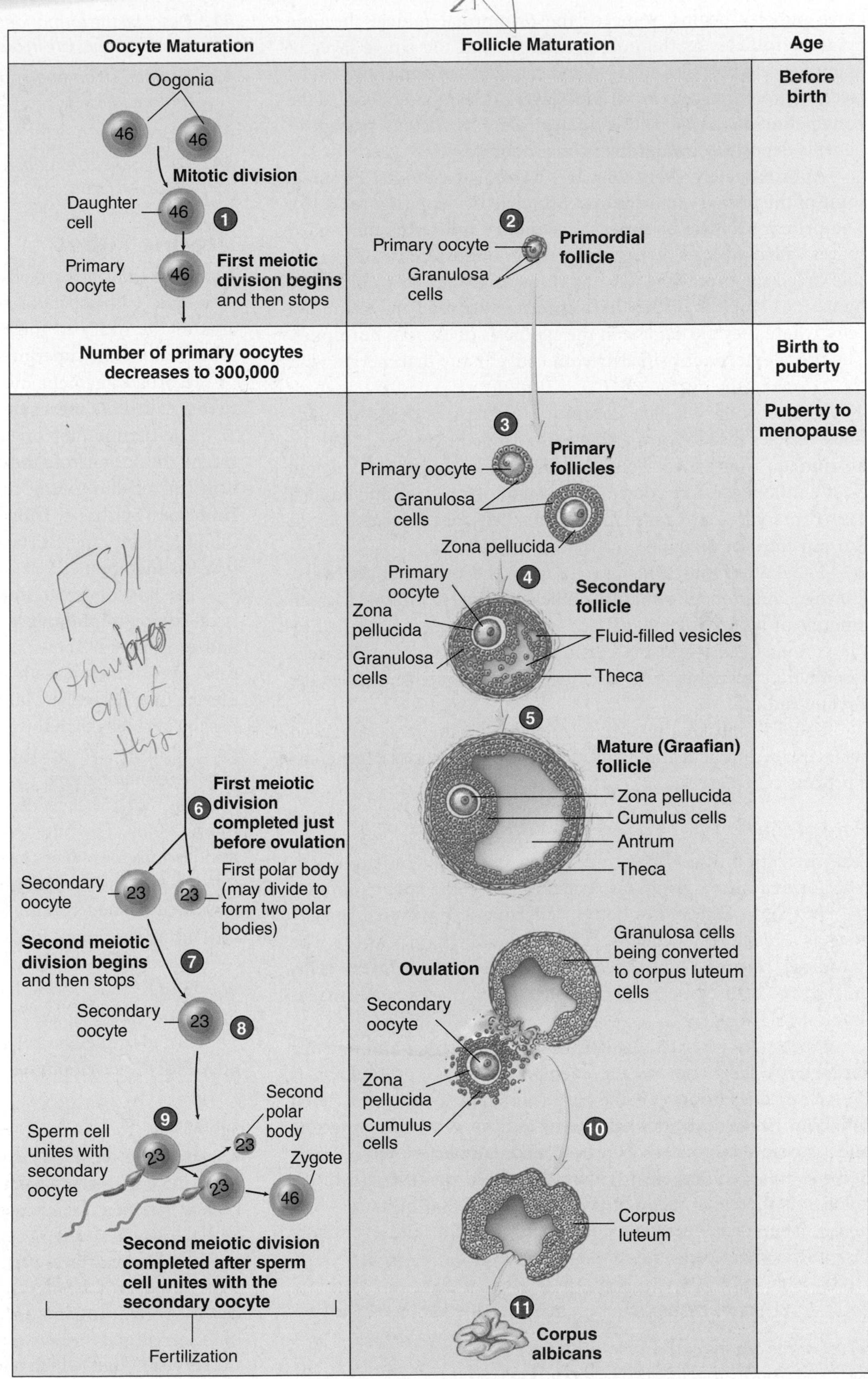

Oocyte Maturation
Follicle Maturation
Age
Before birth
Oogonia
46
46
Mitotic division
Daughter cell
46
1
Primary oocyte
46
First meiotic division begins and then stops
2
Primary oocyte
Primordial follicle
Granulosa cells
Birth to puberty
Number of primary oocytes decreases to 300,000
Puberty to menopause
3
Primary oocyte
Primary follicles
Granulosa cells
Zona pellucida
Primary oocyte
4
Secondary follicle
Zona pellucida
Fluid-filled vesicles
Granulosa cells
Theca
5
Mature (Graafian) follicle
Zona pellucida
Cumulus cells
Antrum
Theca
First meiotic division completed just before ovulation
6
Secondary oocyte
23
23
First polar body (may divide to form two polar bodies)
Second meiotic division begins and then stops
7
Secondary oocyte
23
8
Ovulation
Granulosa cells being converted to corpus luteum cells
Secondary oocyte
Zona pellucida
Cumulus cells
10
9
Sperm cell unites with secondary oocyte
23
Second polar body
23
Zygote
23
46
Second meiotic division completed after sperm cell unites with the secondary oocyte
Fertilization
Corpus luteum
11
Corpus albicans

PROCESS FIGURE 28.13 Maturation of the Oocyte and Follicle
The events leading to oocyte and follicle maturation are closely linked. Once these structures are mature, fertilization can result. The numbers written in the cells are total numbers of chromosomes.

Once puberty begins, some of the primordial follicles become **primary follicles** as the oocyte enlarges and the single layer of granulosa cells becomes enlarged and cuboidal. Subsequently, several layers of granulosa cells form, and a layer of clear material called the **zona pellucida** (zō′nă pellū′sid-dă; girdle + *pellucidus,* passage of light) is deposited around the primary oocyte.

Approximately every 28 days, hormonal changes stimulate some of the primary follicles to continue to develop (figure 28.13). The primary follicle becomes a **secondary follicle** as fluid-filled spaces called **vesicles** form among the granulosa cells, and a capsule called the **theca** (thē′kă; a box) forms around the follicle (see figure 28.12). Cells of the **theca interna** surround the granulosa cells, where they participate in the synthesis of ovarian hormones. The **theca externa** is primarily connective tissue that merges with the stroma of the ovary.

The secondary follicle continues to enlarge; when the fluid-filled vesicles fuse to form a single, fluid-filled chamber called the **antrum** (an′trŭm), the follicle is called a **mature follicle,** or *Graafian* (graf′ē-ăn) *follicle.* The oocyte is pushed off to one side and lies in a mass of granulosa cells called the **cumulus cells,** or *cumulus oophorus* (kū′mū-lŭs ō-of′ōr-ŭs).

The mature follicle forms a lump on the surface of the ovary. During ovulation, the mature follicle ruptures, forcing a small amount of blood, follicular fluid, and the oocyte, surrounded by the cumulus cells, into the peritoneal cavity. The cumulus cells resemble a crown radiating from the oocyte and are thus called the **corona radiata.**

Usually, only one mature follicle reaches the most advanced stages of development and is ovulated. The other follicles degenerate, a process called **atresia.**

Fate of the Follicle

After ovulation, the follicle still has an important function. It is transformed into a glandular structure called the **corpus luteum** (kōr′pŭs loo′tē-ŭm; yellow body), which has a convoluted appearance as a result of its collapse after ovulation (figure 28.13). The granulosa cells and the theca interna, now called **luteal cells,** enlarge and begin to secrete hormones—progesterone and smaller amounts of estrogen.

If pregnancy occurs, the corpus luteum enlarges and remains throughout pregnancy as the **corpus luteum of pregnancy.** If pregnancy does not occur, the corpus luteum remains functional for about 10–12 days and then begins to degenerate. Progesterone and estrogen secretion decreases, and connective tissue cells become enlarged and clear, giving the whole structure a whitish color; it is therefore called the **corpus albicans** (al′bī-kanz; white body). The corpus albicans continues to shrink and eventually disappears after several months or even years.

ASSESS YOUR PROGRESS

37. *List the organs of the female reproductive system.*

38. *Name and describe the ligaments that hold the ovaries in place.*

39. *Discuss the coverings and structure of the ovary.*

40. *Starting with oogonia, describe the formation of secondary oocytes by meiosis. What are polar bodies?*

41. *Describe the formation of a zygote. How many pairs of chromosomes are in a zygote, and where do they come from?*

42. *Distinguish among primordial, primary, secondary, and mature follicles.*

43. *Describe the process of ovulation.*

44. *What is the corpus luteum? What happens to it if fertilization occurs? If fertilization does not occur?*

Uterine Tubes

A **uterine tube,** also called a *fallopian* (fa-lō′pē-an) *tube* or *oviduct* (ō′vi-dŭkt), is associated with each ovary and extends from the area of the ovary to the uterus (see figure 28.11). Each tube is located along the superior margin of the broad ligament. The part of the broad ligament most directly associated with the uterine tube is called the **mesosalpinx** (mez′ō-sal′pinks).

The uterine tube opens directly into the peritoneal cavity to receive the oocyte from the ovary. It expands to form the **infundibulum** (in-fŭn-dib′ū-lŭm; funnel), and long, thin processes called **fimbriae** (fim′brē-ē; fringe) surround the opening of the infundibulum. The inner surfaces of the fimbriae consist of a ciliated mucous membrane.

The part of the uterine tube that is nearest the infundibulum is called the **ampulla.** It is the widest and longest part of the tube and accounts for about 7.5–8 cm of the total 10 cm length of the tube. Fertilization usually occurs in the ampulla. The part of the uterine tube nearest the uterus, the **isthmus,** is much narrower and has thicker walls than the ampulla. The **uterine part,** or *intramural part,* of the tube passes through the uterine wall and ends in a very small uterine opening.

The wall of each uterine tube consists of three layers (see figure 28.11). The outer **serosa** is formed by the peritoneum, the middle **muscular layer** consists of longitudinal and circular smooth muscle cells, and the inner **mucosa** consists of a mucous membrane of simple ciliated columnar epithelium. The mucosa is arranged into numerous longitudinal folds.

The mucosa of the uterine tubes provides nutrients for the oocyte or, if fertilization has occurred, for the developing embryonic mass (see chapter 29) as it passes through the uterine tube. The ciliated epithelium helps move the small amount of fluid and the oocyte, or the developing embryonic mass, through the uterine tubes.

Uterus

The **uterus** (ū′ter-ŭs) is the size and shape of a medium-sized pear—about 7.5 cm long and 5 cm wide (see figures 28.10 and 28.11). It is slightly flattened anteroposteriorly and is oriented in the pelvic cavity with the larger, rounded part, the **fundus** (fŭn′dŭs), directed superiorly and the narrower part, the **cervix** (ser′viks), directed inferiorly. The main part of the uterus, the **body,** is between the fundus and the cervix. A slight constriction called the **isthmus** marks the junction of the cervix and the body. Internally, the uterine cavity continues as the **cervical canal,** which opens through the **ostium** into the vagina.

The uterus is supported by the broad ligament, the **round ligaments** (see figure 28.11), and the **uterosacral ligaments.** The broad ligament is a peritoneal fold extending from the lateral margins of the uterus to the wall of the pelvis on either side. It also ensheathes

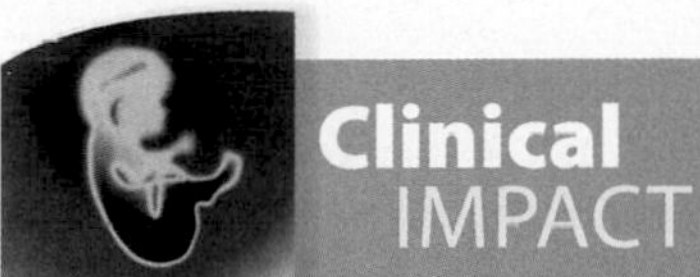
Clinical
IMPACT

Cervical Cancer

The National Cancer Institute estimated over 12,000 women will be diagnosed with cervical cancer in 2012 and that just over 4000 women will die from it. Fortunately, cervical cancer can be detected and treated. During its early stages, the cervical cells exhibit characteristic changes that are visible when viewed using a microscope. In a diagnostic test called a **Pap smear,** a physician inserts a swab through the vagina to obtain a sample of the epithelial cells from the cervix and the vaginal wall. These cells are smeared on a glass slide and later stained and examined microscopically for signs of cancer. Treatments for cervical cancer vary, depending on the stage of cancer development, but they include radiation therapy, chemotherapy, and surgery.

Preventive measures are available to guard against developing cervical cancer. An estimated 70% of all cervical cancer cases can be linked to infection by 2 of more than 100 types of human papillomavirus (HPV types 16 and 18). Therefore, the vaccine Gardasil has been developed to immunize women against HPV types 16 and 18, as well as against two other types (6 and 11), which are linked to 90% of genital warts cases. The Centers for Disease Control and Prevention (CDC) recommends vaccination of preteen girls and boys against HPV at ages 11 or 12 to protect against cervical cancer in females and genital warts in males and females. Vaccination is also recommended for teen girls and boys and women and men through age 26 who were not vaccinated when younger.

the ovaries and the uterine tubes. The round ligaments extend from the uterus through the inguinal canals to the labia majora of the external genitalia, and the uterosacral ligaments attach the lateral wall of the uterus to the sacrum. Normally, the uterus is *anteverted,* meaning that the body of the uterus is tipped slightly anteriorly. However, in some women, the uterus is retroverted, or tipped posteriorly.

In addition to the ligaments, skeletal muscles of the pelvic floor support the uterus inferiorly. If these muscles are weakened (e.g., in childbirth), the uterus can extend inferiorly into the vagina, a condition called a **prolapsed uterus.**

The uterine wall is composed of three layers: the perimetrium, the myometrium, and the endometrium (see figure 28.11). The **perimetrium** (per-i-mē′trē-ŭm), or *serous layer,* is the peritoneum that covers the uterus. The next layer, just deep to the perimetrium, is the **myometrium** (mī′ō-mē′trē-ŭm), or *muscular layer,* composed of a thick layer of smooth muscle. The myometrium accounts for the bulk of the uterine wall and is the thickest layer of smooth muscle in the body. In the cervix, the muscular layer contains less muscle and more dense connective tissue. The cervix is therefore more rigid and less contractile than the rest of the uterus. The innermost layer of the uterus is the **endometrium** (en′dō-mē′trē-ŭm), or *mucous membrane,* which consists of a simple columnar epithelial lining and a connective tissue layer called the lamina propria. Simple tubular glands, called spiral glands, are scattered about the lamina propria and open through the epithelium into the uterine cavity. The endometrium consists of two layers: A thin, deep **basal layer** is the deepest part of the lamina propria and is continuous with the myometrium, and a thicker, superficial **functional layer,** consisting of most of the lamina propria and the endothelium, lines the cavity itself. The functional layer is so named because it undergoes changes and sloughing during the female menstrual cycle. Small **spiral arteries** of the lamina propria supply blood to the functional layer of the endometrium. These blood vessels play an important role in the cyclic changes of the endometrium.

Columnar epithelial cells line the cervical canal, which contains **cervical mucous glands.** The mucus fills the cervical canal and acts as a barrier to substances that could pass from the vagina into the uterus. Near ovulation, the consistency of the mucus changes, easing the passage of sperm cells from the vagina into the uterus.

ASSESS YOUR PROGRESS

45. *Describe the structure of the uterine tube.*

46. *How are the uterine tubes involved in moving the oocyte or the zygote?*

47. *Name the parts of the uterus.*

48. *Describe the major ligaments holding the uterus in place.*

49. *Describe the layers of the uterine wall.*

Vagina

The **vagina** (vă-jī′nă) is the female organ of copulation, receiving the penis during intercourse. It also allows menstrual flow and childbirth. The vagina is a tube about 10 cm long that extends from the uterus to the outside of the body (see figure 28.11). Longitudinal ridges called **columns** extend the length of the anterior and posterior vaginal walls, and several transverse ridges called **rugae** (roo′gē) extend between the anterior and posterior columns. The superior, domed part of the vagina, the **fornix** (fōr′niks), is attached to the sides of the cervix, so that a part of the cervix extends into the vagina.

The wall of the vagina consists of an outer muscular layer and an inner mucous membrane. The muscular layer is smooth muscle that allows the vagina to increase in size to accommodate the penis during intercourse and to stretch greatly during childbirth. The mucous membrane is moist stratified squamous epithelium that forms a protective surface layer. The vaginal mucous membrane releases most of the lubricating secretions produced by the female during intercourse.

A thin mucous membrane called the **hymen** (hī′men) covers the **vaginal opening,** or *orifice.* Sometimes, the hymen completely closes the vaginal opening (a condition called imperforate hymen), and it must be removed to allow menstrual flow. More commonly, the hymen is perforated by one or several holes. The openings in the hymen are usually greatly enlarged during the first sexual intercourse. In addition, the hymen can be perforated earlier in a young woman's life, such as during strenuous physical exercise. Thus, the absence of an intact hymen does not necessarily indicate that a woman has had sexual intercourse, as was once thought.

ASSESS YOUR PROGRESS

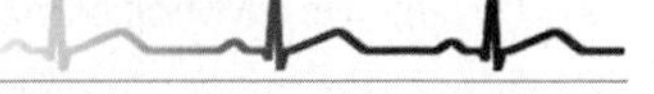

50. *What are the functions of the vagina?*

51. *Describe the layers of the vaginal wall. What are rugae and columns?*

52. *What is the hymen?*

External Genitalia

The external female genitalia, also referred to as the **vulva** (vŭl′vă), or *pudendum* (pū-den′dŭm), consist of the vestibule and its surrounding structures (figure 28.14). The **vestibule** (ves′ti-bool) is the space into which the vagina opens posteriorly and the urethra opens anteriorly. A pair of thin, longitudinal skin folds called the **labia** (lā′bē-ă; lips) **minora** (sing. *labium minus*) form a border on each side of the vestibule. A small, erectile structure called the **clitoris** (klit′ō-ris) is located in the anterior margin of the vestibule. Anteriorly, the two labia minora unite over the clitoris to form a fold of skin called the **prepuce.**

The clitoris is usually less than 2 cm in length and consists of a shaft and a distal glans. Well supplied with sensory receptors, it initiates and intensifies levels of sexual tension. The clitoris contains two erectile structures, the **corpora cavernosa,** each of which expands at the base end of the clitoris to form the **crus of the clitoris** and attaches the clitoris to the pelvic bones. The corpora cavernosa of the clitoris are comparable to the corpora cavernosa of the penis, and they become engorged with blood as a result of sexual excitement. In most women, this engorgement results in an increase in the diameter, but not the length, of the clitoris. With increased diameter, the clitoris makes better contact with the prepuce and surrounding tissues and is more easily stimulated.

Erectile tissue that corresponds to the corpus spongiosum of the male lies deep to and on the lateral margins of the vestibular floor on each side of the vaginal orifice. Each erectile body is called a **bulb of the vestibule.** Like other erectile tissue, it becomes engorged with blood and is more sensitive during sexual arousal. Expansion of the bulbs causes narrowing of the vaginal orifice and allows better contact of the vagina with the penis during intercourse.

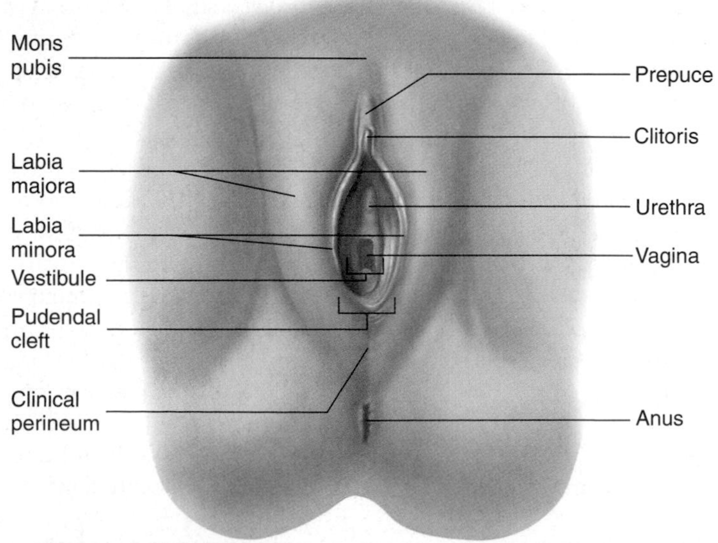
Mons pubis
Labia majora
Labia minora
Vestibule
Pudendal cleft
Clinical perineum
Prepuce
Clitoris
Urethra
Vagina
Anus

FIGURE 28.14 Female External Genitalia

On each side of the vestibule, between the vaginal opening and the labia minora, is an opening of the duct of the **greater vestibular gland.** Additional small mucous glands, the **lesser vestibular glands,** or *paraurethral glands,* are located near the clitoris and urethral opening. They produce a lubricating fluid that helps maintain the moistness of the vestibule.

Lateral to the labia minora are two prominent, rounded folds of skin called the **labia majora.** Subcutaneous adipose tissue is primarily responsible for the prominence of the labia majora. The two labia majora unite anteriorly in an elevation over the symphysis pubis called the **mons pubis** (monz pū′bis). The lateral surfaces of the labia majora and the surface of the mons pubis are covered with coarse hair. The medial surfaces are covered with numerous sebaceous and sweat glands. The space between the labia majora is called the **pudendal cleft.** Most of the time, the labia majora are in contact with each other across the midline, closing the pudendal cleft and concealing the deeper structures within the vestibule.

Perineum

The female **perineum** is divided into two triangles by the superficial and deep transverse perineal muscles (figure 28.15). The anterior urogenital triangle contains the external genitalia, and the posterior anal triangle contains the anal opening. The region between the vagina and the anus is the **clinical perineum.** The skin and muscle of this region can tear during childbirth. To prevent such tearing, an incision called an **episiotomy** (e-piz-ē-ot′ō-mē, e-pis-ē-ot′ō-mē) is sometimes made in the clinical perineum. This clean, straight incision is easier to repair than a tear would be. Alternatively, allowing the perineum to stretch slowly during delivery may prevent tearing, thereby making an episiotomy unnecessary.

ASSESS **YOUR PROGRESS** 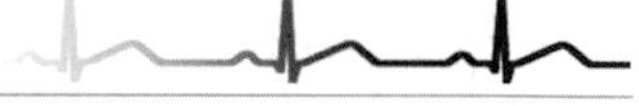

53. *What are the vulva and vestibule?*

54. *What erectile tissue is in the clitoris and bulb of the vestibule?*

55. *What is the function of the clitoris and bulb of the vestibule?*

56. *Describe the labia minora, the prepuce, the labia majora, the pudendal cleft, and the mons pubis.*

57. *Where are the greater and lesser vestibular glands located? What is their function?*

58. *Describe the perineum.*

59. *What is the clinical perineum? Explain the purpose of an episiotomy.*

Mammary Glands

The **mammary glands,** the organs of milk production, are located within the **breasts** (figure 28.16). The mammary glands are modified sweat glands. Externally, the breasts of both males and females have a raised **nipple** surrounded by a circular, pigmented **areola** (ă-rē′ō-lă). The areolae normally have a slightly bumpy surface caused by the presence of rudimentary mammary glands, called **areolar glands,** just below the surface. Secretions from these glands protect the nipple and the areola from chafing during nursing.

Before puberty, the general structure of the breasts is similar in both males and females. The breasts possess a rudimentary glandular system, which consists mainly of ducts with sparse alveoli.

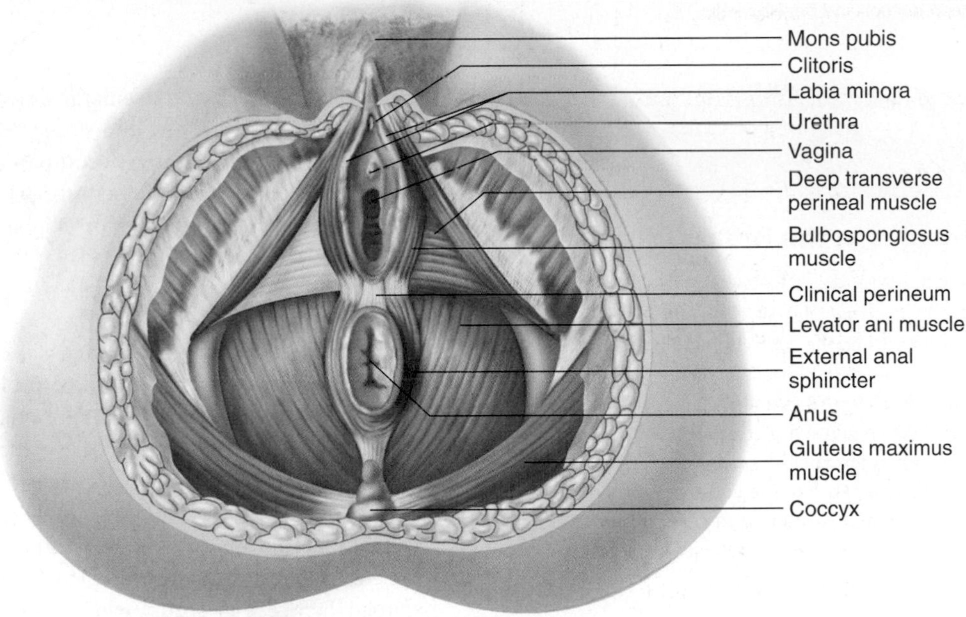
Mons pubis
Clitoris
Labia minora
Urethra
Vagina
Deep transverse perineal muscle
Bulbospongiosus muscle
Clinical perineum
Levator ani muscle
External anal sphincter
Anus
Gluteus maximus muscle
Coccyx

FIGURE 28.15 Inferior View of the Female Perineum

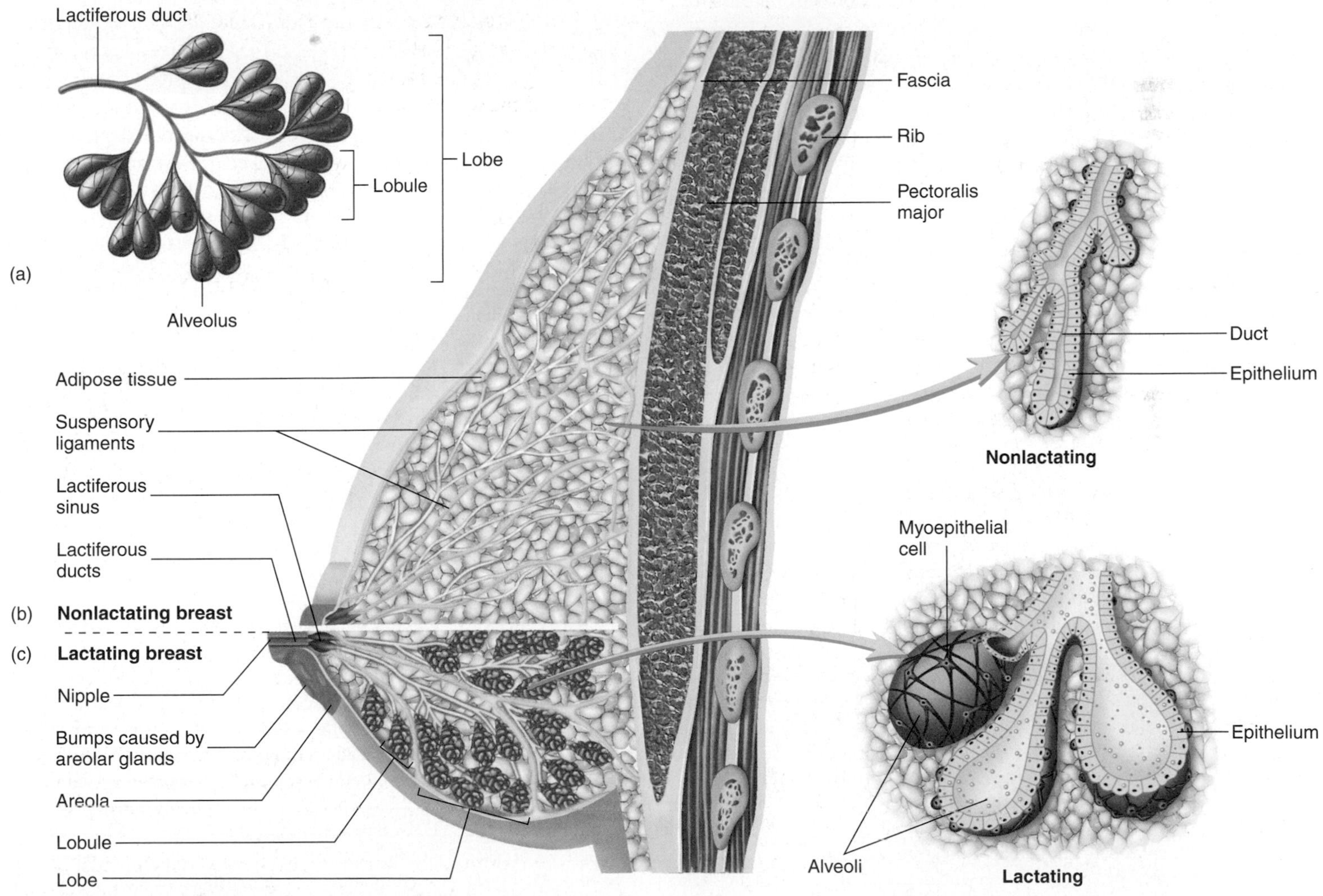

Lactiferous duct
Lobule
Lobe
(a)
Alveolus
Fascia
Rib
Pectoralis major
Adipose tissue
Suspensory ligaments
Lactiferous sinus
Lactiferous ducts
(b) **Nonlactating breast**
(c) **Lactating breast**
Nipple
Bumps caused by areolar glands
Areola
Lobule
Lobe
Duct
Epithelium
Nonlactating
Myoepithelial cell
Epithelium
Alveoli
Lactating

FIGURE 28.16 AP|R Anatomy of the Breast

(*a*) Lactiferous ducts divide to supply lobules, which form lobes. (*b*) In the nonlactating breast, only the duct system is present. (*c*) In the lactating breast, the ends of the mammary gland ducts have secretory sacs, called alveoli, that produce milk. Surrounding the alveoli are myoepithelial cells, which can contract, causing the milk to move out of the alveoli.

Clinical
IMPACT

Fibrocystic Changes in the Breast

Fibrocystic changes in the breast are benign and include the formation of fluid-filled cysts, hyperplasia (accelerated growth) of the duct system of the breast, and the deposition of fibrous connective tissue. Breast pain sometimes occurs, especially during the luteal phase of the menstrual cycle and continuing until menstruation. These changes occur in approximately 10% of women who are less than 21 years of age, 25% of women in their reproductive years, and 50% of women who are postmenopausal. The cause of the condition is not known.

Some evidence suggests that women with certain types of duct hyperplasia, along with a family history of breast cancer, have an increased likelihood of developing breast cancer.

The female breasts begin to enlarge during puberty, primarily under the influence of estrogen and progesterone. Increased sensitivity or pain in the breasts often accompanies this enlargement. Males often experience the same sensations during early puberty, and their breasts can even develop slight swellings; however, these symptoms usually disappear fairly quickly. On rare occasions, the breasts of a male become enlarged, a condition called **gynecomastia** (gī′ne-kō-mas′tē-ă).

Each adult female mammary gland usually consists of 15–20 glandular **lobes** covered by a considerable amount of adipose tissue. It is primarily this superficial adipose tissue that gives the breast its form. The lobes of each mammary gland form a conical mass, with the nipple located at the apex. Each lobe has a single **lactiferous** (lak-tif′er-ŭs; milk-producing) **duct,** which opens independently of other lactiferous ducts on the surface of the nipple. Just deep to the surface, each lactiferous duct enlarges to form a small, spindle-shaped **lactiferous sinus,** where milk accumulates during milk

Clinical GENETICS | Breast Cancer

Breast cancer is a serious, often fatal disease that most often occurs in women. It is the most common cancer in North American women. Greater than 75% of the breast cancer cases occur in women older than 50 years of age, and 85% involve cancer of the epithelium of the mammary gland ducts (ductal carcinoma).

The risk for breast cancer appears to be lower if a woman has had a child. Women who have never given birth are at a greater risk than those who have given birth. The younger a woman is when her first child is born, the lower her risk of developing breast cancer. However, the risk for breast cancer is not lowered in young women who become pregnant but do not carry to full term. It appears that differentiation of the breast epithelium caused by the first pregnancy decreases the risk for cancer by reducing the amount of undifferentiated epithelium in the breast.

Breast cancer development correlates with long-term exposure to high levels of estrogen. Since pregnancy delays the menstrual cycle and decreases estrogen secretion, this may explain its effect on breast cancer risk. In addition, decreased exposure of breast tissue to estrogen explains a reduced cancer risk because of early menopause or removal of the ovaries. Unfortunately, hormone replacement therapy for postmenopausal women correlates with an increased incidence of cancer; however, breast cancer incidence decreases with the use of drugs that block the effect of estrogen receptors.

A number of environmental factors have been associated with the development of breast cancer. Exposure to ionizing radiation, especially during adolescence and during pregnancy when epithelial cells are dividing more rapidly, is correlated with breast cancer. High fat intake and obesity are also linked to breast cancer, although the fish oil–derived omega-3 fatty acids may help prevent cancer. Breast cancer rates are low in Japanese women; however, Japanese women who immigrate to the United States and adopt a Western diet have breast cancer rates close to those of women in the United States. A number of environmental factors that mimic estrogen, such as components of plastics, fuels, pharmaceuticals, and chlorine-based chemicals (such as DDT, PCBs, and chlorofluorocarbons), may increase the incidence of breast cancer.

Between 5% and 10% of breast cancers are hereditary. Women who have inherited specific gene mutations have an increased risk for breast cancer. Therefore, a history of breast cancer in a close relative, such as a mother or sister, makes breast cancer more likely, and the risk increases further if more than one close relative has had breast cancer, especially if the cancer affected both breasts and if it occurred before menopause. Genetic tests can identify breast cancer genes. However, the presence of breast cancer genes does not mean that cancer will develop. Some women who are known to be at high risk because of their genetic makeup have frequent breast examinations, whereas others elect to have their breasts surgically removed before cancer develops.

Mutations in the *BRCA1* and *BRCA2* genes are responsible for about 30–40% of inherited breast cancers. The *BRCA* genes are tumor suppressor genes, which normally suppress cell division (see Clinical Genetics, "Genetic Changes in Cancer Cells," chapter 3). A person who has two mutated *BRCA* alleles is much more likely to develop breast cancer. Although a person with one normal and one mutated *BRCA* allele may have sufficient tumor suppressor activity, she has a greater risk of developing cancer because only one mutation in the remaining normal *BRCA* allele is necessary to eliminate tumor suppressor activity. Consequently, a person who has inherited two normal alleles is much less likely to develop breast cancer than a person who has inherited one or two mutated alleles.

Another tumor suppressor gene is called *p53*. The cells of almost 50% of all cancers have a mutated *p53* gene, and these cancers are more aggressive and more often fatal than cancers without a mutated *p53* gene. Approximately 20–40% of individuals with hereditary breast cancer have a mutated *p53* gene. *P53* is a regulatory gene that has been called the "guardian of the genome," since *p53* is activated when DNA is damaged. Once activated, *p53* arrests the cell cycle and initiates the repair of damaged DNA before the cell cycle continues. If the DNA damage is too extensive, *p53* causes cell death. Thus, the *p53* gene normally helps repair or eliminate cells that may become cancer cells. When *p53* is mutated, this protection against aberrant cell proliferation is reduced and there is a greatly increased risk for tumor formation.

ejection. The lactiferous duct supplying a lobe subdivides to form smaller ducts, each of which supplies a **lobule.** Within a lobule, the ducts branch and become even smaller. In milk-producing, or lactating, mammary glands, the ends of these small ducts expand to form secretory sacs called **alveoli. Myoepithelial cells** surround the alveoli and contract to expel milk (see chapter 29). In non-lactating mammary glands, only the duct system is present.

A group of **suspensory ligaments,** or *Cooper ligaments,* support and hold the breasts in place. These ligaments extend from the fascia over the pectoralis major muscles to the skin over the mammary glands and prevent the breasts from excessive sagging.

The nipples are very sensitive to tactile stimulation and contain smooth muscle cells that contract, causing the nipple to become erect in response to stimulation. These smooth muscle cells respond to stimuli such as touch, cold, and sexual arousal.

ASSESS YOUR PROGRESS

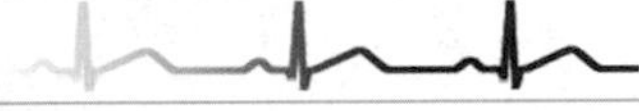

60. *Describe the anatomy of the mammary glands.*
61. *Trace the route taken by a drop of milk from its site of production to the outside of the body.*

28.5 Physiology of Female Reproduction

LEARNING OUTCOMES

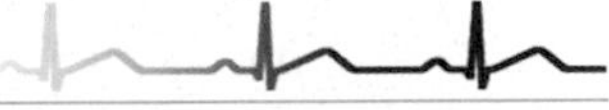

After reading this section, you should be able to

A. **Describe the changes that occur in females during puberty.**
B. **Explain the changes in the ovary and uterus during the menstrual and ovarian cycles.**
C. **List the hormones of the female reproductive system and explain their functions and how their secretion is regulated.**
D. **Explain the events that occur during the female sex act.**
E. **Describe the events that occur following fertilization of the oocyte and the process of implantation of the embryo.**
F. **Discuss menopause, including the changes that result from it.**

Female reproduction is under hormonal and nervous control. The development of the female reproductive organs and their normal function depend on a number of hormones. Estrogen and progesterone are the female reproductive hormones. The term *estrogen* actually refers to several hormones, including estradiol, estrone, and estriol. Estradiol is the primary estrogen in humans and the most prevalent in the circulation.

Puberty in Females

During puberty, females experience their first episode of menstrual bleeding, called **menarche** (me-nar′kē), between ages 11 and 16. The vagina, uterus, uterine tubes, and external genitalia begin to enlarge. Adipose tissue is deposited in the breasts and around the hips, causing them to enlarge and assume adult form. The ducts of the breasts develop, pubic and axillary hair grows, and the voice changes, although more subtly than in males. The development of sex drive is also associated with puberty.

Elevated rates of estrogen and progesterone secretion by the ovaries are primarily responsible for the changes associated with puberty. Before puberty, estrogen and progesterone are secreted in very small amounts, and LH and FSH levels remain very low. The low secretory rates are due to a lack of GnRH released from the hypothalamus. At puberty, not only are GnRH, LH, and FSH secreted in greater quantities than before puberty, but also the adult cyclic pattern of FSH and LH secretion is established.

ASSESS YOUR PROGRESS

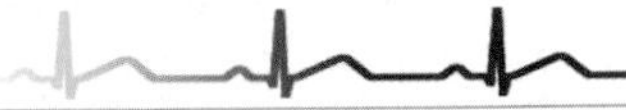

62. *Define* menarche. *Describe other physical changes that occur during female puberty.*
63. *What changes occur in LH, FSH, estrogen, and progesterone secretion during puberty?*

Menstrual Cycle

The term **menstrual** (men′stroo-ăl) **cycle** technically refers to the cyclic changes in sexually mature, nonpregnant females that culminate in menses. Typically, the menstrual cycle is about 28 days long, although it can be as short as 18 days in some women and as long as 40 days in others (figure 28.17). **Menses** (men′sēz) is a period of mild hemorrhage that occurs approximately once each month, during which the uterine epithelium is sloughed and expelled from the uterus. **Menstruation** (men-stroo-ā′shŭn) is the discharge of the blood and other elements of the uterine mucous membrane. Although the term *menstrual cycle* refers specifically to changes in the uterus, the term is often used to refer to all the cyclic events in the female reproductive system, including alterations in hormone secretion and changes in the ovaries (table 28.2).

The first day of menses is considered day 1 of the menstrual cycle, and menses typically lasts 4–5 days. Ovulation occurs on about day 14 of a 28-day menstrual cycle; however, the timing of ovulation varies from individual to individual and even within a single individual from one menstrual cycle to the next. The time between ovulation, on day 14, and the next menses is typically 14 days. The time between the first day of menses and the day of ovulation is more variable than the time between ovulation and the next menses. The time between the ending of menses and ovulation is called the **proliferative phase,** because of the rapid proliferation of the uterine mucosa, or the **follicular phase,** because of the rapid development of ovarian follicles. The period after ovulation and before the next menses is called the **secretory phase,** because of the maturation of and secretion by uterine glands, or the **luteal phase,** because of the existence of the corpus luteum.

Ovarian Cycle

The term **ovarian cycle** refers to the regular events that occur in the ovaries of sexually mature, nonpregnant women during the menstrual cycle. The hypothalamus and anterior pituitary release hormones that control these events. FSH from the anterior pituitary is primarily responsible for initiating the development of primary follicles, and as many as 25 follicles begin to mature during each

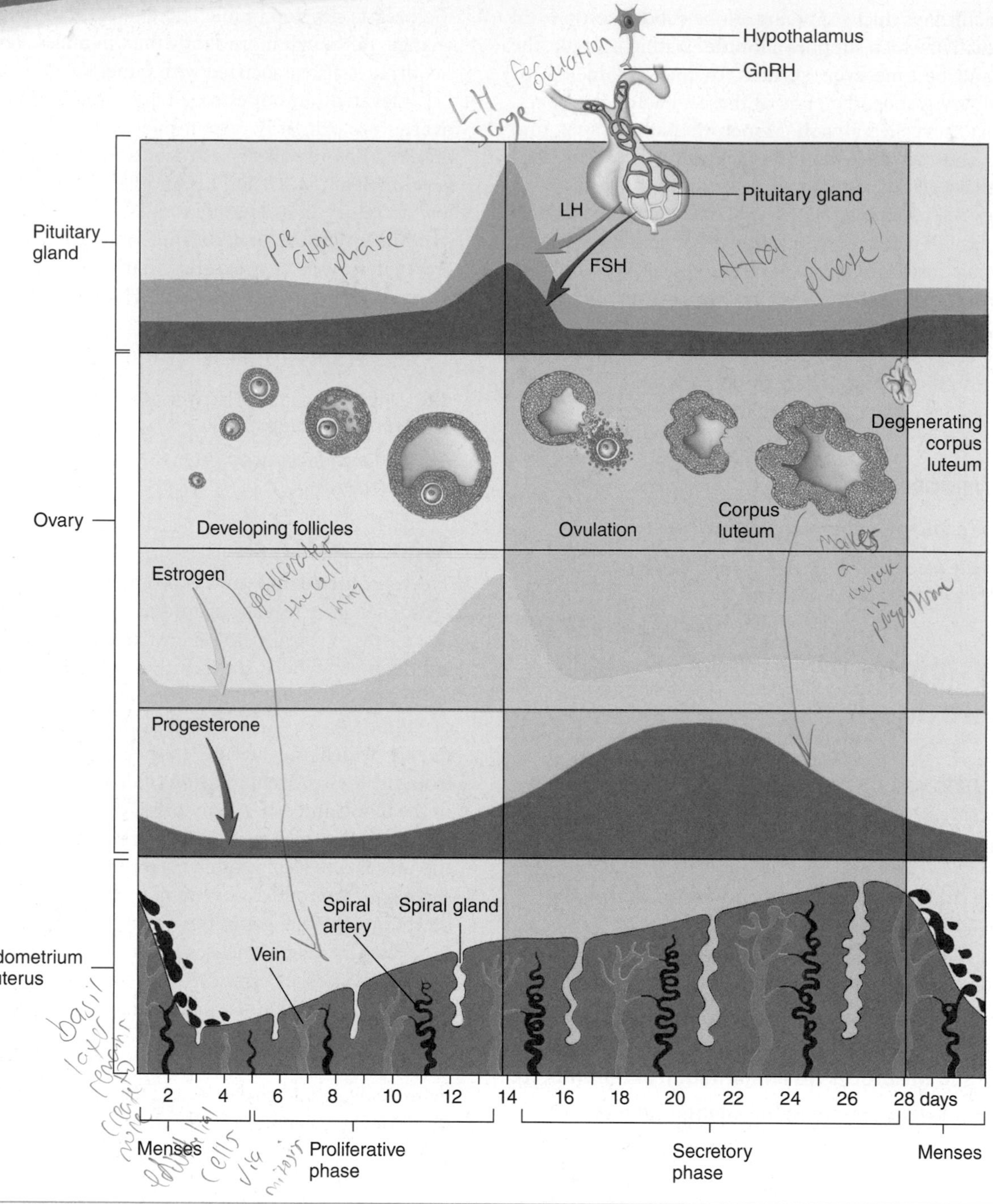

Hypothalamus
GnRH
Pituitary gland
LH
FSH
Pituitary gland
Ovary
Developing follicles
Ovulation
Corpus luteum
Degenerating corpus luteum
Estrogen
Progesterone
Endometrium of uterus
Spiral artery
Spiral gland
Vein
2
4
6
8
10
12
14
16
18
20
22
24
26
28 days
Menses
Proliferative phase
Secretory phase
Menses

FIGURE 28.17 Menstrual Cycle
This graph depicts the changes that occur in blood hormone levels, follicles, and the endometrium during the menstrual cycle.

menstrual cycle. However, normally only 1 is ovulated. The follicles that start to develop in response to FSH may not ovulate during the same menstrual cycle in which they begin to mature, but they may ovulate one or two cycles later. The remaining follicles degenerate. Larger, more mature follicles appear to secrete estrogen and other substances that have an inhibitory effect on other, less mature follicles.

Early in the menstrual cycle, the release of GnRH from the hypothalamus increases, as does the sensitivity of the anterior pituitary to GnRH. These changes stimulate the anterior pituitary to produce and release small amounts of FSH and LH. FSH and LH stimulate follicular growth and maturation. They also cause an increase in estrogen secretion by the developing follicles. FSH exerts its main effect on the granulosa cells, whereas LH exerts its initial effect on the theca interna cells and later on the granulosa cells.

LH stimulates the theca interna cells to produce androgens, which diffuse from these cells to the granulosa cells. FSH stimulates the granulosa cells to convert androgens to estrogen. In addition, FSH gradually increases LH receptors in the granulosa cells. Estrogen produced by the granulosa cells increases LH receptors in the theca interna cells. Consequently, theca interna cells and granulosa

TABLE 28.2 Hormone Secretion During the Menstrual Cycle

Menses	Proliferative Phase	Ovulation	Secretory Phase
Pituitary Hormones			
LH levels are low and remain low; FSH increases somewhat.	Near the end of the proliferative phase, LH and FSH levels begin to increase rapidly in response to increases in estrogen.	Increasing levels of LH trigger ovulation. Ovulation generally occurs after LH levels have reached their peak. FSH reaches a peak about the time of ovulation and initiates the development of follicles that may complete maturation during a later cycle.	LH and FSH levels decline to low levels following ovulation and remain low during the secretory phase in response to increases in estrogen and progesterone.
Developing Follicles			
FSH secreted during menses causes several follicles to begin to enlarge.	As several follicles continue to enlarge, they begin to secrete estrogen. In addition, many follicles degenerate. By the end of the proliferative phase, only one of the follicles has become a mature follicle that is capable of ovulation.	Normally, a single follicle reaches maturity and ovulates in response to LH. The oocyte and some cumulus cells are released during ovulation.	Following ovulation, the granulosa cells of the ovulated follicle change to luteal cells and begin secreting large amounts of progesterone and some estrogen.
Estrogen			
The ovarian follicles secrete very little estrogen.	Near the end of the proliferative phase, the enlarging follicles begin to secrete increasing amounts of estrogen. The estrogen causes the pituitary gland to secrete increasing quantities of LH and smaller quantities of FSH. The positive-feedback relationship between estrogen and LH results in rapidly increasing LH and estrogen levels several days prior to ovulation. The rapid increase in LH triggers ovulation.	Estrogen, secreted by developing follicles, reaches a peak at ovulation.	Following ovulation, estrogen levels decline. After the luteal cells have been established, smaller amounts of estrogen are secreted by the corpus luteum.
Progesterone			
The ovarian follicles secrete very little progesterone.	Progesterone levels are low during the proliferative phase.	Progesterone levels are low.	Following ovulation, progesterone levels increase due to progesterone secretion by the corpus luteum. Progesterone levels remain high throughout the secretory phase and fall rapidly just before menses unless pregnancy occurs.
Uterine Endometrium			
The endometrium of the uterus undergoes necrosis and is eliminated during menses. The necrosis is a result of decreasing progesterone concentrations near the end of the proliferative phase.	In response to estrogen, endometrial cells of the uterus undergo rapid cell division and proliferate rapidly. In addition, the number of progesterone receptors in the endometrial cells increases in response to estrogen.	Ovulation occurs over a short time. It signals the end of the proliferative phase, as estrogen levels decline, and the onset of the secretory phase, as progesterone levels begin to increase.	Progesterone causes the endometrial cells to enlarge and secrete a small amount of fluid. The endometrium continues to thicken throughout the secretory phase. Near the end of the secretory phase, declining progesterone levels allow the spiral arteries of the endometrium to constrict, causing ischemia, and the endometrium becomes necrotic unless pregnancy occurs.

cells cooperate to produce estrogen. Estrogen, in turn, increases receptors for LH in both theca interna cells and granulosa cells.

After LH receptors in the granulosa cells have increased, LH stimulates the granulosa cells to produce progesterone, which diffuses from the granulosa cells to the theca interna cells, where it is converted to androgens. These androgens are also converted to estrogen by the granulosa cells. Thus, the production of androgens by the theca interna cells increases, resulting in a gradual increase in estrogen secretion by granulosa cells throughout the follicular phase, even though only a small increase in LH secretion occurs. FSH levels actually decrease during the follicular phase because developing follicles produce inhibin, which has a negative-feedback effect on FSH secretion.

The gradual increase in estrogen levels, especially late in the follicular phase, begins to have a positive-feedback effect on LH and FSH release from the anterior pituitary. Consequently, as the estrogen level in the blood increases, it stimulates greater LH and FSH secretion. The sustained increase in estrogen is necessary for this positive-feedback effect. In response, LH and FSH secretion increases rapidly and in large amounts just before ovulation (figure 28.18). The increase in blood levels of both LH and FSH is called the **LH surge,** and the increase in FSH is called the **FSH surge.** The LH surge occurs several hours earlier and to a greater degree than the FSH surge, and the LH surge can last up to 24 hours.

The LH surge initiates ovulation and causes the ovulated follicle to become the corpus luteum. FSH can make the follicle more sensitive to the influence of LH by stimulating the synthesis of additional LH receptors in the follicles and by stimulating the development of follicles that may ovulate during later ovarian cycles.

The LH surge causes the primary oocyte to complete the first meiotic division just before or during the process of ovulation. Also, the LH surge triggers several events that are very much like inflammation in a mature follicle. These events result in ovulation. The follicle enlarges due to edema. Proteolytic enzymes break down the ovarian tissue around the follicle, causing the follicle to rupture, and the oocyte and some surrounding follicle cells are slowly extruded from the ovary.

Shortly after ovulation, the follicle's production of estrogen decreases. The remaining granulosa cells of the ovulated follicle are converted to corpus luteum cells and begin to secrete progesterone. After the corpus luteum forms, progesterone levels become much higher than before ovulation, and some estrogen is produced. The increased progesterone and estrogen have a negative-feedback effect on GnRH release from the hypothalamus. As a result, LH and FSH release from the anterior pituitary decreases. Estrogen and progesterone also cause the down-regulation of GnRH receptors in the anterior pituitary, and the anterior pituitary cells become

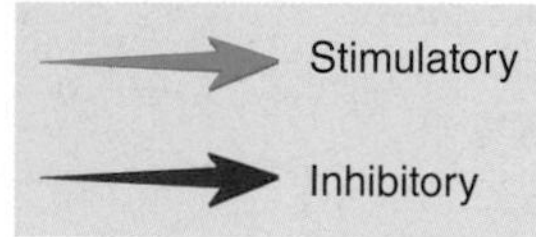
Stimulatory
Inhibitory

1 Before ovulation, FSH secretion increases, which stimulates follicles to develop and secrete estrogen.

2 Estrogen causes the endometrium to proliferate and the hypothalamus to increase LH secretion, which results in the LH and FSH surges prior to ovulation.

3 The LH surge causes a follicle to mature and ovulate. The corpus luteum develops and secretes progesterone and some estrogen.

4 The progesterone causes hypertrophy of the endometrium and has a negative-feedback effect on LH and FSH secretion. The corpus luteum continues to secrete progesterone for approximately 12 days after ovulation.

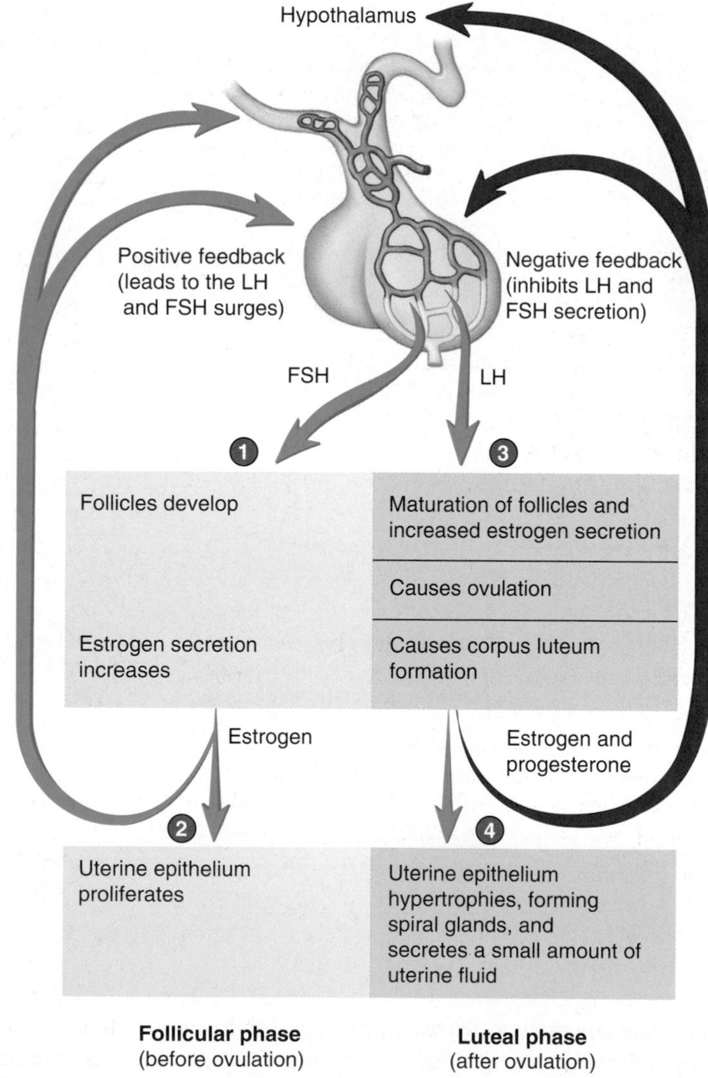
Hypothalamus
Positive feedback (leads to the LH and FSH surges)
Negative feedback (inhibits LH and FSH secretion)
FSH
LH
1
3
Follicles develop
Estrogen secretion increases
Maturation of follicles and increased estrogen secretion
Causes ovulation
Causes corpus luteum formation
Estrogen
Estrogen and progesterone
2
4
Uterine epithelium proliferates
Uterine epithelium hypertrophies, forming spiral glands, and secretes a small amount of uterine fluid
Follicular phase (before ovulation)
Luteal phase (after ovulation)

PROCESS FIGURE 28.18 Regulation of Hormone Secretion During the Menstrual Cycle
Hormone secretion is regulated from the anterior pituitary and the ovary before and after ovulation.

less sensitive to GnRH. Because of the decreased secretion of GnRH and decreased sensitivity of the anterior pituitary to GnRH, the rate of LH and FSH secretion declines to very low levels after ovulation (figure 28.18; see figure 28.17).

If the ovulated oocyte is fertilized, the outer layer of the developing embryo begins to secrete the LH-like substance **human chorionic gonadotropin (hCG),** which keeps the corpus luteum from degenerating. As a result, blood levels of estrogen and progesterone do not decrease, and menses does not occur. If fertilization does not occur, hCG is not produced. The cells of the corpus luteum begin to atrophy after day 25 or 26, and the blood levels of estrogen and progesterone decrease rapidly, resulting in menses.

Predict 5

Predict the effect on the ovarian cycle of administering a relatively large amount of estrogen and progesterone just before the preovulatory LH surge. Also predict the consequences of continually administering high concentrations of GnRH.

Uterine Cycle

The term **uterine cycle** refers to changes that occur primarily in the endometrium of the uterus during the menstrual cycle (figure 28.19; see figure 28.17). Other, more subtle changes also take place in the vagina and other structures during the menstrual cycle. Cyclic secretions of estrogen and progesterone are the primary cause of these changes.

The endometrium of the uterus begins to proliferate after menses. The remaining epithelial cells rapidly divide and replace the cells of the functional layer that were sloughed during the last menses. A relatively uniform layer of low cuboidal endometrial cells is produced. The cells later become columnar, and the layer of cells folds to form tubular **spiral glands.** Blood vessels called **spiral arteries** project through the delicate connective tissue that separates the individual spiral glands to supply nutrients to the endometrial cells. After ovulation, the endometrium becomes thicker, and the spiral glands develop to a greater extent and begin to secrete small amounts of a fluid rich in glycogen. Approximately 7 days after ovulation, or about day 21 of the menstrual cycle, the endometrium is prepared to receive a developing embryonic mass, if fertilization has occurred. If the developing embryonic mass arrives in the uterus too early or too late, the endometrium does not provide a suitable environment for it.

Estrogen causes the endometrial cells and, to a lesser degree, the myometrial cells to proliferate. It also makes the uterine tissue more sensitive to progesterone by stimulating the synthesis of progesterone receptor molecules within the uterine cells. After ovulation, progesterone from the corpus luteum binds to the progesterone receptors, resulting in cellular hypertrophy in the endometrium and myometrium and causing the endometrial cells to become secretory. Estrogen increases the tendency of the smooth muscle cells of the uterus to contract in response to stimuli, but progesterone inhibits smooth muscle contractions. When progesterone levels increase while estrogen levels are low, contractions of the uterine smooth muscle are reduced.

In menstrual cycles in which pregnancy does not occur, progesterone and estrogen levels decline to low levels as the corpus luteum degenerates. As a consequence, the uterine lining also begins to degenerate. The spiral arteries constrict in a rhythmic pattern for longer and longer periods as progesterone levels fall. As a result, all but the basal parts of the spiral glands become ischemic and then necrotic. As the cells become necrotic, they slough into the uterine lumen. The necrotic endometrium, mucous secretions, and a small amount of blood released from the spiral arteries make up the menstrual fluid. Decreases in progesterone levels and increases in

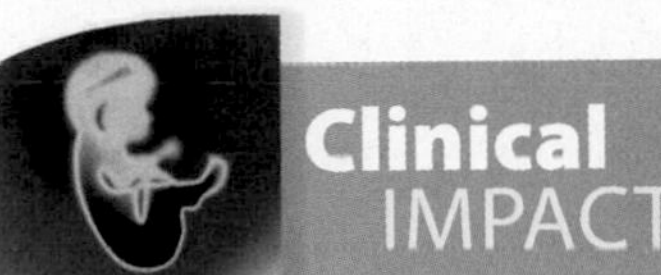

Clinical IMPACT

Menstrual Cramps and Amenorrhea

Menstrual cramps are the result of strong myometrial contractions that occur before and during menstruation. The cramps can result from excessive prostaglandin secretion. As the endometrium of the uterus sloughs off, it becomes inflamed, and prostaglandins are produced as part of the inflammatory process. Sloughing of the endometrium and uterine contractions are inhibited by progesterone but stimulated by estrogen. Many women can alleviate painful cramps by taking nonsteroidal anti-inflammatory drugs (NSAIDs), such as aspirin or ibuprofen, which inhibit prostaglandin biosynthesis, just before the onset of menstruation. These medications, however, are not effective in treating all painful menstruation, especially when the pain is not due to inflammation but to other conditions, such as tumors of the myometrium or obstruction of the cervical canal.

The absence of a menstrual cycle is called **amenorrhea** (ă-men-ō-rē′ă). If the pituitary gland does not function properly because of abnormal development, a female does not begin to menstruate at puberty. This condition is called **primary amenorrhea.** In contrast, if a female has had normal menstrual cycles and later stops menstruating, the condition is called **secondary amenorrhea.** One cause of secondary amenorrhea is anorexia, in which lack of food causes the hypothalamus to decrease GnRH secretion to levels so low that the menstrual cycle cannot occur. Many female athletes and ballet dancers who pursue rigorous training schedules have secondary amenorrhea. Physical stress coupled with inadequate food intake also result in very low GnRH secretion. Increased food intake, for anorexic women, and reduced training, for women who exercise intensely, generally restore normal hormone secretion and normal menstrual cycles.

Secondary amenorrhea can also result from a pituitary tumor that decreases FSH and LH secretion or from a lack of GnRH secretion from the hypothalamus due to head trauma or a tumor. In addition, secondary amenorrhea can occur due to a lack of normal hormone secretion from the ovaries, which can be caused by autoimmune diseases that attack the ovary or by polycystic ovarian disease, in which cysts in the ovary produce large amounts of androgens that are converted to estrogens by other body tissues. The increased estrogen prevents the normal cycle of FSH and LH secretion required for ovulation. Other hormone-secreting tumors of the ovary can also disrupt the normal menstrual cycle and result in amenorrhea.

Predict 6

Predict the effect on the endometrium of maintaining high progesterone levels in the circulatory system, including the period of time during which estrogen normally increases following menstruation.

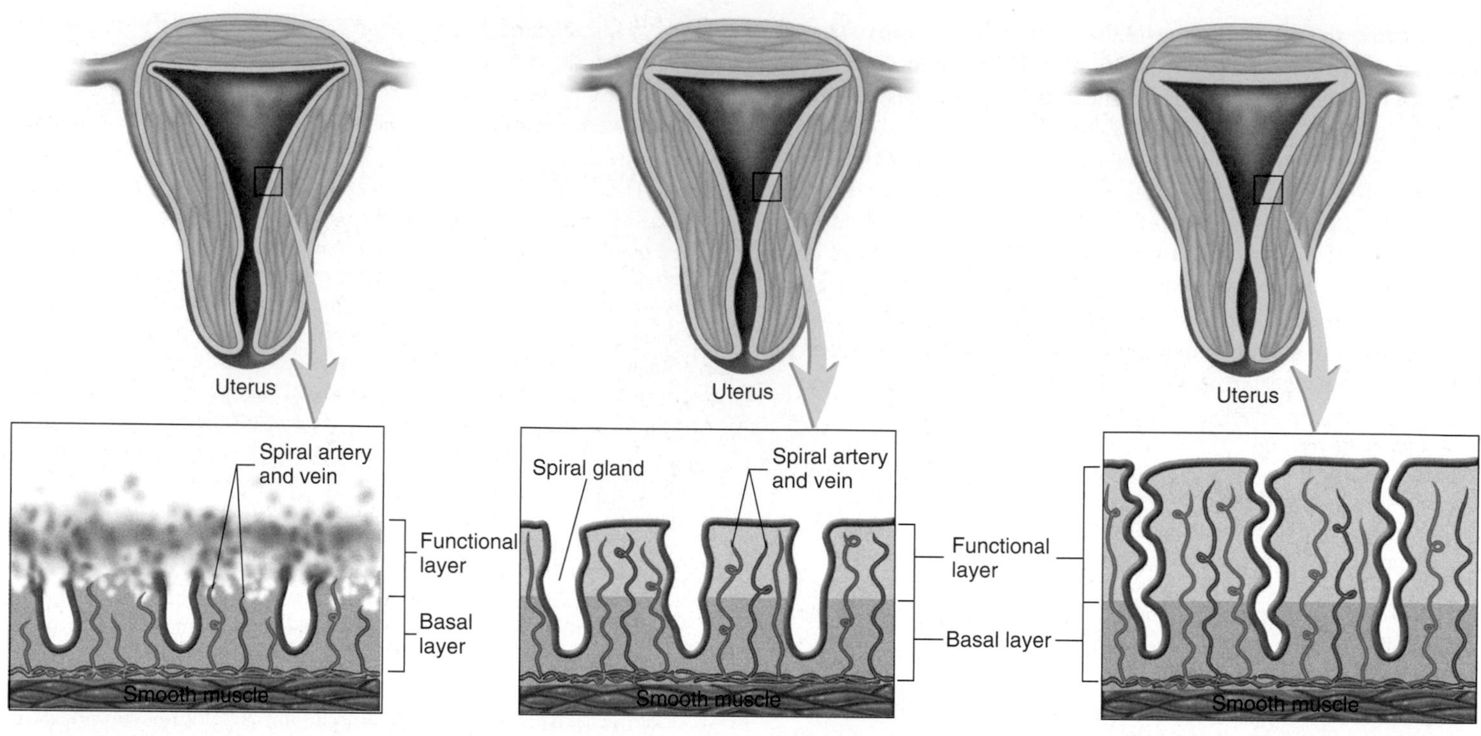
Uterus
Spiral artery and vein
Functional layer
Basal layer
Smooth muscle
Uterus
Spiral gland
Spiral artery and vein
Smooth muscle
Functional layer
Basal layer
Uterus
Smooth muscle

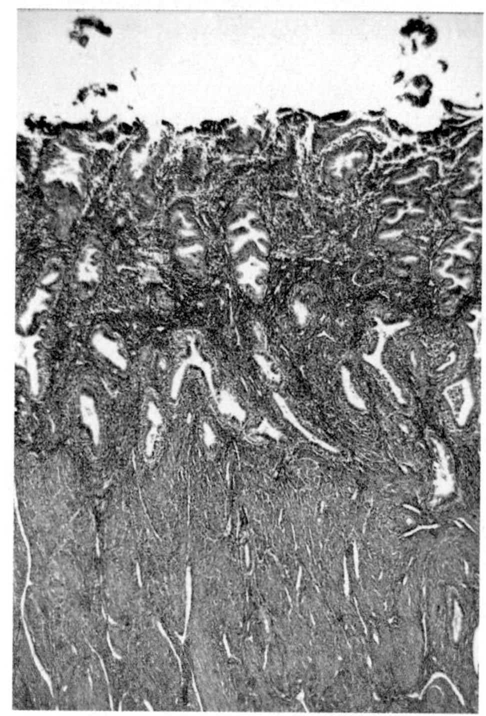

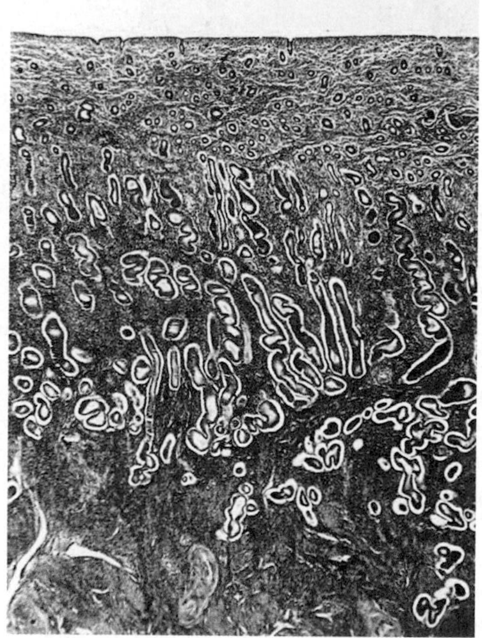

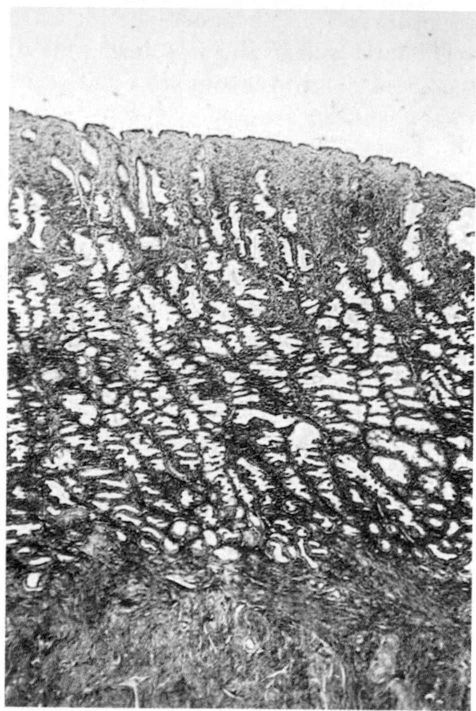

1 Menses
The apical portion of the endometrium is the functional layer. It sloughs off as the spiral arteries remain in a constricted state in response to low levels of progesterone, depriving the functional layer of an adequate blood supply. Functional layer tissue and some blood make up most of the menstrual fluid. The basal layer of the endometrium remains intact.

2 Proliferative phase
Epithelial cells of the basal layer of the endometrium proliferate in response to estrogen. As a result, the epithelial cells and loose connective tissue on which they rest form the tubular, spiral glands. The spiral arteries found in the loose connective tissue between the spiral glands nourish the functional layer.

3 Secretory phase
Epithelial cells of the basal and functional layers undergo hypertrophy in response to progesterone. As a result, the spiral glands become more elongated and more spiral. Consequently, the endometrial layer reaches its greatest thickness. The spiral arteries can be seen in the loose connective tissue between the spiral glands. The spiral arteries remain dilated due to the presence of progesterone.

PROCESS FIGURE 28.19 AP|R **Uterine Cycle**

inflammatory substances that stimulate myometrial smooth muscle cells cause uterine contractions, which expel the menstrual fluid from the uterus through the cervix and into the vagina.

ASSESS **YOUR PROGRESS**

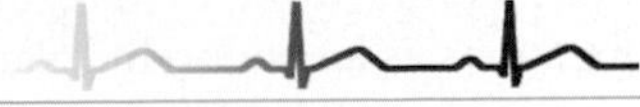

64. *What are the major events of the menstrual cycle?*

65. *What is the length of a typical menstrual cycle? What event marks the beginning of a cycle?*

66. *On which day does ovulation occur?*

67. *Where do the proliferative and secretory phases occur? What happens during these phases?*

68. *Where do the follicular and luteal phases occur? What happens during these phases?*

69. *What roles do FSH and LH play in the ovarian cycle?*

70. *Describe how the cyclic increase and decrease in FSH and LH is produced.*

71. *What is the importance of the LH surge and the FSH surge?*

72. *Where is hCG produced, and what effect does it have on the ovary?*

73. *What are the effects of estrogen and progesterone on the uterus?*

Female Sexual Behavior and the Female Sex Act

The female sex drive, like the sex drive in males, depends on hormones. The adrenal gland and other tissues, such as the liver, convert steroids, such as progesterone, to androgens. Androgens and possibly estrogens affect cells in the brain, especially in the hypothalamus, to influence sexual behavior. However, androgens and estrogen alone do not control sex drive. In other words, sex drive cannot be predictably increased simply by injecting these hormones into healthy women or men. Psychological factors also affect sexual behavior. For example, after removal of the ovaries or after menopause, many women report an increased sex drive because they no longer fear pregnancy.

The neural pathways, both sensory and motor, involved in controlling sexual responses are the same for males and females. Sensory action potentials are conducted from the genitals to the sacral region of the spinal cord, where reflexes that govern sexual responses are integrated. Ascending pathways, primarily the spinothalamic tracts (see chapter 14), conduct sensory information through the spinal cord to the brain, and descending pathways conduct action potentials back to the sacral region of the spinal cord. As a result, cerebral influences modulate the sacral reflexes. Motor action potentials are conducted from the spinal cord to the reproductive organs by both parasympathetic and sympathetic nerve fibers and to skeletal muscles by the somatic motor nerve fibers.

During sexual excitement, as a result of parasympathetic stimulation, erectile tissue within the clitoris and around the vaginal opening becomes engorged with blood. The nipples of the breast often become erect as well. The mucous glands within the vestibule, especially the vestibular glands, secrete small amounts of mucus. Large amounts of mucuslike fluid are also extruded into the vagina through its wall, although no well-developed mucous glands are within the vaginal wall. These secretions provide lubrication that allows for easy entry of the penis into the vagina and easy movement of the penis during intercourse. Tactile stimulation of the female's genitals that occurs during sexual intercourse, along with psychological stimuli, normally triggers an orgasm. The vaginal, uterine, and perineal muscles contract rhythmically, and muscle tension increases throughout much of the body. After the sex act, a period of resolution characterized by an overall sense of satisfaction and relaxation occurs. In contrast to males, females can be receptive to further stimulation and can experience successive orgasms. Although orgasm is a pleasurable component of sexual intercourse, it is not necessary for females to experience an orgasm for fertilization to occur.

Female Fertility and Pregnancy

After sperm cells are ejaculated into the vagina during sexual intercourse, they are transported through the cervix, the body of the uterus, and the uterine tubes to the ampulla (figure 28.20). The forces responsible for moving sperm cells through the female reproductive tract include the swimming ability of the sperm cells and possibly the muscular contractions of the uterus and the uterine tubes. During sexual intercourse, oxytocin is released from the posterior pituitary of the female, and the semen introduced into the vagina contains prostaglandins. Both of these hormones stimulate smooth muscle contractions in the uterus and uterine tubes.

While passing through the vagina, uterus, and uterine tubes, the sperm cells undergo **capacitation** (kă-pas′i-tā′shŭn), the removal of proteins and the modification of glycoproteins of the sperm cell plasma membranes. Following capacitation, as the sperm cells move through the female reproductive tract, some of them release

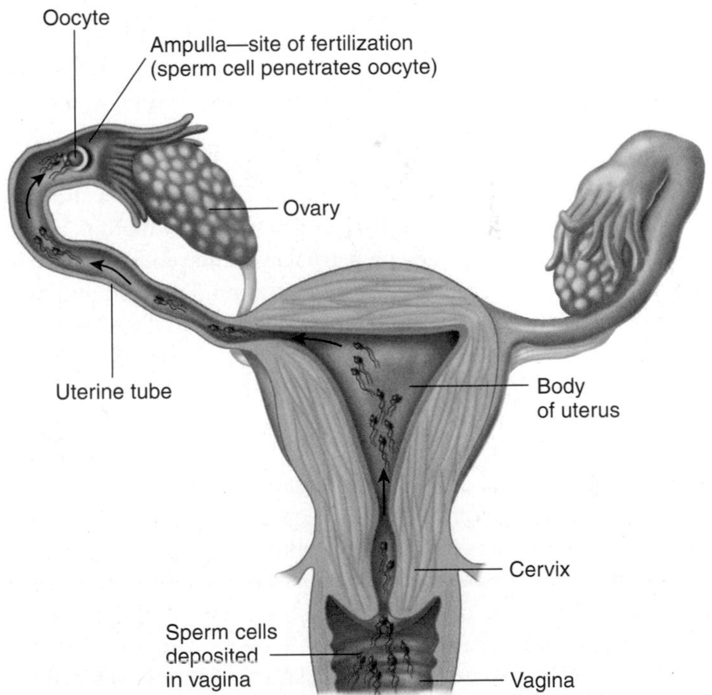
Oocyte
Ampulla—site of fertilization (sperm cell penetrates oocyte)
Ovary
Uterine tube
Body of uterus
Cervix
Sperm cells deposited in vagina
Vagina

FIGURE 28.20 Sperm Cell Movement
Sperm cells are deposited into the vagina as part of the semen when the male ejaculates. Sperm cells pass through the cervix, the body of the uterus, and the uterine tube. Fertilization normally occurs when the oocyte is in the upper one-third of the uterine tube (the ampulla).

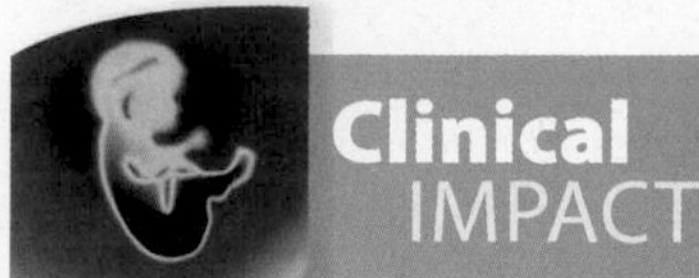
Clinical
IMPACT

Ectopic Pregnancy

If implantation occurs anywhere other than in the uterine cavity, an ectopic pregnancy results. The most common site of ectopic pregnancy is the uterine tube. Implantation in the uterine tube is eventually fatal to the fetus and can cause the tube to rupture. The possibility of hemorrhage makes ectopic pregnancy dangerous to the mother as well.

In rare cases, implantation occurs in the mesenteries of the abdominal cavity. Although the fetus may develop normally, the pregnancy is considered extremely high-risk due to the serious threat to the mother's life and that of the fetus. Severe hemorrhaging at or near the time of delivery is the most serious concern. Delivery by cesarean section, with special precautions to prevent uncontrolled bleeding, can result in a successful outcome for this type of ectopic pregnancy. However, maternal mortality rates for abdominal pregnancies are significantly higher than for uterine tube ectopic pregnancies.

acrosomal enzymes. These enzymes help the sperm cells penetrate the cervical mucus, cumulus mass, zona pellucida, and oocyte plasma membrane.

One sperm cell enters the secondary oocyte, and fertilization occurs (see chapter 29). The oocyte can be fertilized for up to 24 hours after ovulation, and some sperm cells remain viable in the female reproductive tract for up to 6 days, although most of them have degenerated after 24 hours. For fertilization to occur successfully, sexual intercourse must occur between 5 days before and 1 day after ovulation.

For the next several days following fertilization, a sequence of cell divisions occurs while the developing embryo passes through the uterine tube to the uterus. By 7 or 8 days after ovulation, which is day 21 or 22 of the average menstrual cycle, the endometrium of the uterus has been prepared for implantation. Estrogen and progesterone have caused it to reach its maximum thickness and secretory activity, and the developing embryo begins to implant. The outer layer of the developing embryo, the **trophoblast** (trof′ō-blast, trō′fō-blast), secretes proteolytic enzymes that digest the cells of the thickened endometrium (see chapter 29), and the developing embryo digests its way into the endometrium.

The trophoblast secretes hCG, which is transported in the blood to the ovary and causes the corpus luteum to remain functional. As a consequence, both estrogen and progesterone levels continue to increase rather than decrease. The secretion of hCG increases rapidly and reaches a peak about 8–9 weeks after fertilization. Subsequently, hCG levels in the circulatory system have declined to a lower level by 16 weeks and remain at a relatively constant level throughout the remainder of pregnancy. The detection of hCG in the urine is the basis for some pregnancy tests.

The estrogen and progesterone secreted by the corpus luteum are essential for the maintenance of pregnancy. After the **placenta** (plă-sen′tă) forms from the trophoblast and uterine tissue, however, it also begins to secrete estrogen and progesterone. After the first 3 months of pregnancy, the corpus luteum is no longer needed to maintain pregnancy; the placenta, in addition to its function of nutrient and waste exchange between the fetus and the mother, functions as an endocrine gland that secretes sufficient quantities of estrogen and progesterone to maintain pregnancy. Estrogen and progesterone levels increase in the woman's blood throughout pregnancy (figure 28.21).

Menopause

When a woman is 40–50 years old, menstrual cycles become less regular, and ovulation often does not occur consistently. Eventually, menstrual cycles stop completely. The cessation of menstrual cycles is called **menopause** (men′ō-pawz). The time from the onset of irregular cycles to their complete cessation, which is often 3 to 5 years, is called the **female climacteric** (klī-mak′ter-ik, klī-mak-ter′ik), or *perimenopause.*

Menopause is associated with changes in the ovaries. The number of follicles remaining in the ovaries of menopausal women is small. In addition, the follicles that remain become less sensitive to stimulation by LH and FSH; therefore, fewer mature follicles and corpora lutea are produced.

Older women experience gradual changes in response to the reduced amount of estrogen and progesterone produced by the ovaries (table 28.3). For example, some women experience sudden episodes of uncomfortable sweating (hot flashes), fatigue, anxiety, temporary decreases in sex drive, and occasionally severe emotional disturbances. Many of these symptoms can be treated successfully with hormone replacement therapy (HRT). HRT usually involves administering small amounts of estrogen, or estrogen in combination with progesterone, and then gradually decreasing the treatment over time. It appears that administering estrogen following menopause also helps prevent osteoporosis and may reduce colorectal cancer. However, although estrogen therapy has been successful, it prolongs the symptoms associated with menopause, in many women and some potential side effects are of concern, including an increased risk for breast, ovarian, and uterine cancer. In addition, HRT does not reduce the risk for heart disease for the first few years after the beginning of menopause. Some data indicate that the risk for heart attacks, strokes, and blood clots is also increased.

ASSESS YOUR PROGRESS

74. *Compare the female sex act with the male sex act.*
75. *Is an orgasm required for fertilization to occur?*
76. *Describe the transport of sperm cells through the female reproductive system.*
77. *What is capacitation of sperm cells?*
78. *Describe the events that follow fertilization.*
79. *Describe implantation of the embryo and formation of the placenta.*
80. *Differentiate between menopause and the female climacteric.*
81. *What causes the changes that lead to menopause?*

1. Human chorionic gonadotropin (hCG; *red line*) increases until it reaches a maximum concentration near the end of the first 3 months of pregnancy and then decreases to a low level thereafter.

2. Progesterone (*blue line*) continues to increase until it levels off near the end of pregnancy. Early in pregnancy, progesterone is produced by the corpus luteum in the ovary; by the second trimester, production shifts to the placenta.

3. Estrogen (*green line*) increases slowly throughout pregnancy but increases more rapidly as the end of pregnancy approaches. Early in pregnancy, estrogen is produced only in the ovary; by the second trimester, production shifts to the placenta.

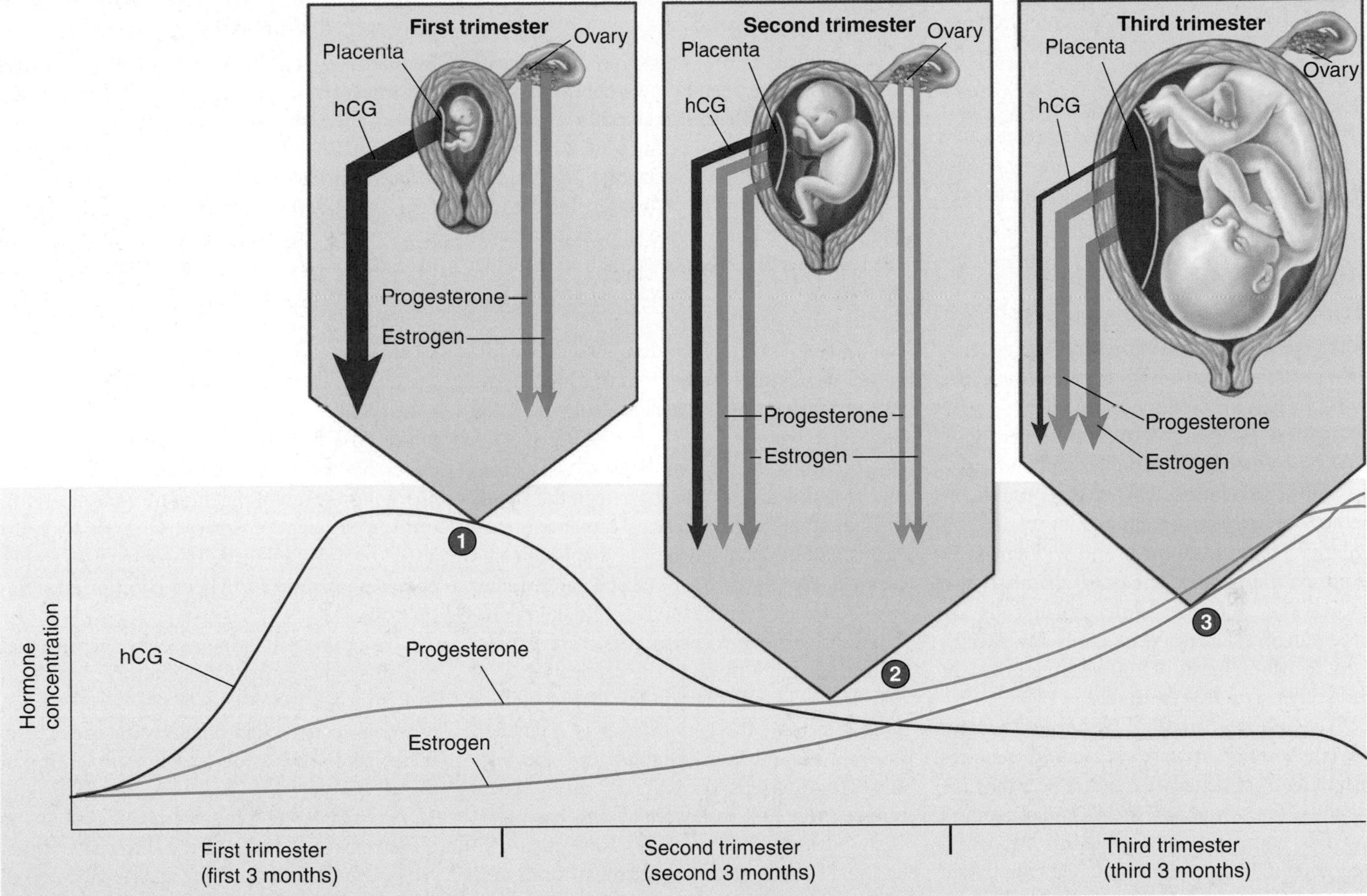
First trimester
Placenta
Ovary
hCG
Progesterone
Estrogen
Second trimester
Placenta
Ovary
hCG
Progesterone
Estrogen
Third trimester
Placenta
Ovary
hCG
Progesterone
Estrogen
1
2
3
Hormone concentration
hCG
Progesterone
Estrogen
First trimester (first 3 months)
Second trimester (second 3 months)
Third trimester (third 3 months)

PROCESS FIGURE 28.21 Changes in Hormone Concentration and Changes in Hormone Secretion During Pregnancy

TABLE 28.3 Possible Changes Caused by Decreased Ovarian Hormone Secretion in Postmenopausal Women

Affected Structures and Functions	Changes
Menstrual cycle	Five to 7 years before menopause, the cycle becomes less regular; finally, the number of cycles in which ovulation occurs decreases, and corpora lutea do not develop.
Uterine tubes	Little change occurs.
Uterus	Irregular menstruation is gradually followed by no menstruation; the chance of cystic glandular hypertrophy of the endometrium increases; the endometrium finally atrophies, and the uterus becomes smaller.
Vagina and external genitalia	The dermis and epithelial lining become thinner; the vulva becomes thinner and less elastic; the labia majora become smaller; pubic hair decreases; the vaginal epithelium produces less glycogen; vaginal pH increases; reduced secretion leads to dryness; the vagina is more easily inflamed and infected.
Skin	The epidermis becomes thinner; melanin synthesis increases.
Cardiovascular system	Hypertension and atherosclerosis occur more frequently.
Vasomotor instability	Hot flashes and increased sweating are correlated with the vasodilation of cutaneous blood vessels; hot flashes are not caused by abnormal FSH and LH secretion but are related to decreased estrogen levels.
Sex drive	Temporary changes, such as either decreases or increases in sex drive, are often associated with the onset of menopause.
Fertility	Fertility begins to decline approximately 10 years before the onset of menopause; by age 50, almost all oocytes and follicles have been lost; the loss is gradual, and no increased follicular degeneration is associated with the onset of menopause.

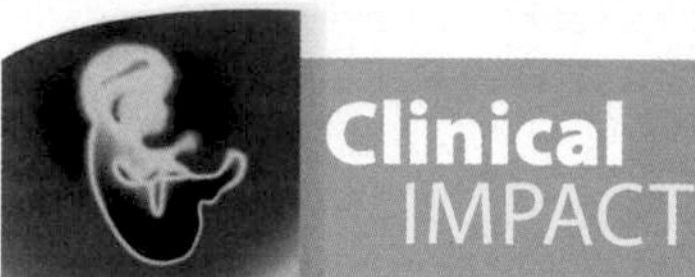

Clinical IMPACT

Control of Pregnancy

Many methods are used to prevent pregnancy (figure 28B), either by preventing fertilization or by preventing implantation of the developing embryo. Many of these techniques are quite effective when used perfectly and consistently. But most of these methods also have disadvantages, and the use of some of them is controversial.

Behavorial Methods

Abstinence, or refraining from sexual intercourse, is 100% effective in preventing pregnancy when practiced consistently. It is not an effective method when used only occasionally.

Coitus (kō′i-tŭs) **interruptus,** or *withdrawal,* is removal of the penis from the vagina just before ejaculation. This is a very unreliable method of preventing pregnancy because it requires perfect awareness and willingness to withdraw the penis at the correct time. It also ignores the fact that some sperm cells are present in preejaculatory emissions. Statistically, about 23 women out of 100 become pregnant while relying on this method.

The **calendar method,** also called *natural family planning* or *rhythm method,* requires abstaining from sexual intercourse near the time of ovulation. Although the calendar method provides some protection against becoming pregnant, it has a relatively high failure rate because of both the inability to predict the time of ovulation and the failure to abstain around that time. About 9 women out of 100 become pregnant while using the calendar method.

Continuous breastfeeding, or lactation (also known as *lactation amenorrhea,* or *LAM*), often stops the menstrual cycle for a few months after childbirth. Action potentials sent to the hypothalamus in response to suckling inhibit GnRH release from the hypothalamus, thus also inhibiting FSH and LH release from the anterior pituitary. Lactation, therefore, prevents the development of ovarian follicles and ovulation. This method is 98% effective as long as the baby is *exclusively* and frequently breastfed and the mother has not resumed menstruation. Supplemental feeding and introduction of solid food reduce the amount of suckling, the stimulus that reduces fertility. Furthermore, despite continuous lactation, the ovarian and uterine cycles eventually resume. Because ovulation occurs before menstruation, relying on lactation to prevent pregnancy is not consistently effective.

Barrier Methods

A **male condom** (kon′dom; figure *a*) is a sheath made of animal membrane, rubber, or latex. When placed over the erect penis, a condom acts as a barrier by collecting semen instead of allowing it to be released into the vagina. Condoms also provide protection against sexually transmitted diseases. Male condoms alone are 98% effective; when used with spermicide, they are 99% effective. A **vaginal condom,** or *female condom,* also acts as a barrier. A woman can place the vaginal condom into the vagina before sexual intercourse. Female condoms are 95% effective; using spermicide further increases their effectiveness.

Methods to prevent sperm cells from reaching the oocyte once they are in the vagina include a diaphragm, a cervical cap, spermicidal agents, and a vaginal sponge. The **diaphragm** and **cervical cap** (figure *b*) are flexible latex domes that are placed over the cervix within the vagina, where they prevent sperm cells from passing from the vagina through the cervical canal of the uterus. The diaphragm is a larger, shallow latex cup, and the cervical cap is a smaller, thimble-shaped cup. Diaphragms are 94% effective, whereas cervical cap effectiveness ranges from 71% in a women who has previously been pregnant to 86% in a woman who has never been pregnant. The most commonly used **spermicidal agents** (figure *c*) are foams, creams, and gels that kill the sperm cells. They are inserted into the vagina before sexual intercourse, often in conjunction with diaphragm or condom use. Alone, spermicidal agents are only about 85% effective.

Intrauterine (in′tră-yū′ter-in) **devices (IUDs;** figure *d*) are inserted into the uterus through the cervix. The two types of IUDs now available in the United States are the copper-containing ParaGard and the progestin hormone–coated Mirena. The ParaGard may be left in place for 12 years, whereas the Mirena may be left in place for 5 years. Both types of IUDs thicken cervical mucus, which bars sperm cells from entering the uterus. Some women stop ovulating when they have an IUD implanted. IUDs also alter the endometrium, which in theory may prevent implantation of an embryo. IUDs are 99.99% effective in preventing pregnancy.

Chemical Methods

Synthetic estrogen and progesterone in **oral contraceptives** (birth control pills; figure *e*) are among the most effective contraceptives, providing 99.9% effectiveness. The synthetic hormones may have more than one action, but they reduce LH and FSH release from the anterior pituitary. Estrogen and progesterone are present in high enough concentrations to have a negative-feedback effect on the pituitary gland, which prevents the large increase in LH and FSH secretion that triggers ovulation. Over the years, the dose of estrogen and progesterone in birth control pills has been reduced. The current lower dose has fewer side effects than earlier doses. For most women, the pill is effective and has a minimum frequency of complications, until at least age 35. However, the risk for heart attack or stroke increases in women who smoke or have a history of hypertension or coagulation disorders.

The **mini-pill** is an oral contraceptive that contains only synthetic progesterone. It reduces and thickens the mucus of the cervix, which prevents sperm cells from reaching the oocyte. It also prevents blastocysts from implanting in the uterus.

Progesterone-like chemicals, such as medroxyprogesterone (Depo-Provera), which are injected intramuscularly and slowly released into the circulatory system, can act as effective contraceptives. These injections can protect against pregnancy for approximately 1 month, depending on the amount injected, and are 99.9% effective. The **patch** (ortho Evra) is an adhesive skin patch containing synthetic estrogen and progesterone. It is worn on the lower abdomen, buttocks, or upper body and is 99.9% effective. The **vaginal contraceptive ring** (NuvaRing) is inserted into the vagina, where it releases synthetic estrogen and progesterone, and is 99.9% effective.

A drug called **mifepristone** (mif′pris-tōn), formerly *RU486,* blocks the action of progesterone, causing the endometrium of the uterus to slough off, as it does at the time of menstruation. It can therefore be used to induce menstruation and reduce the possibility of implantation when sexual intercourse has occurred near the time of ovulation. It can also be used to terminate pregnancies.

Chemical techniques to be used after intercourse have been developed, but they are only about 75% effective. **Morning-after pills,** similar in composition to birth control pills, are available. Alternatively doubling the number of birth control pills after sexual intercourse within 3 days and again after 12 more hours is sometimes recommended. The elevated estrogen and progesterone levels may inhibit the increase in LH that causes ovulation, alter the rate at which the fertilized oocyte is transported through the uterine tube to the uterus, or inhibit implantation. The precise effect of the elevated estrogen-like and progesterone-like substances depends on the stage of the menstrual cycle when they are taken.

Surgical Methods

Vasectomy (va-sek′to-mē; figure *f*) is a common method of rendering males permanently infertile without affecting the performance of the sex act. Vasectomy is a surgical procedure in which the ductus deferens from each testis is cut and tied off within the scrotal sac. This procedure prevents sperm cells from passing through the ductus deferens and becoming part of the ejaculate. Because such a small volume of ejaculate comes from the testis and epididymis, vasectomy has little effect on the volume of the ejaculated semen. The sequestered sperm cells are reabsorbed in the epididymis. Only 1–4 in 1000 surgeries of this type fail.

A common method of permanent birth control in females is **tubal ligation** (lī-gā′shŭn; figure *g*), in which the uterine tubes are tied and cut or clamped by means of an incision through the wall of the abdomen. This procedure closes off the pathway between the sperm cells and the oocyte. Commonly, tubal ligation is performed by **laparoscopy** (lap-ă-ros′kŏ-pē), in which an instrument is inserted into the abdomen through a small incision, so that only small openings need to be made to perform the operation.

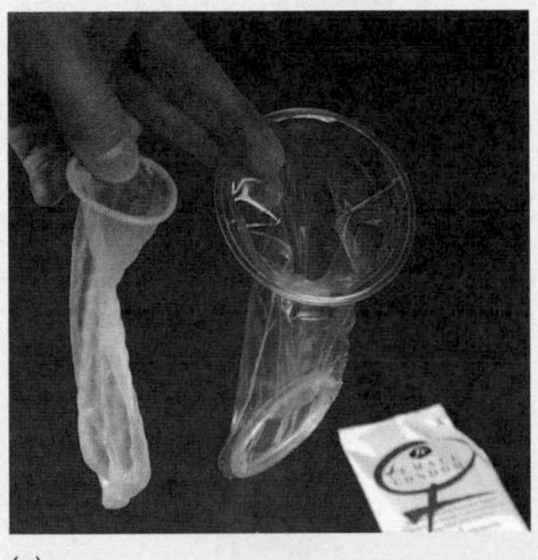

(a)

(b)

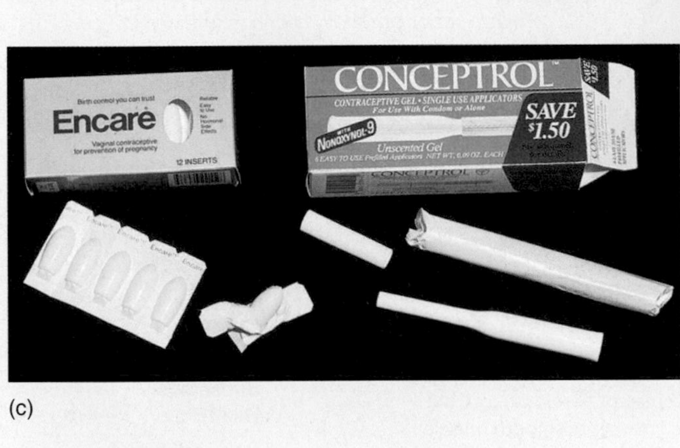
Encare
CONCEPTROL
SAVE $1.50
Unscented Gel
12 INSERTS

(c)

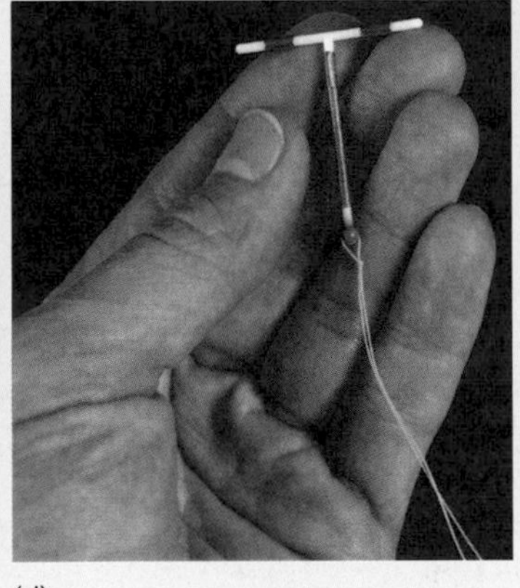

(d)

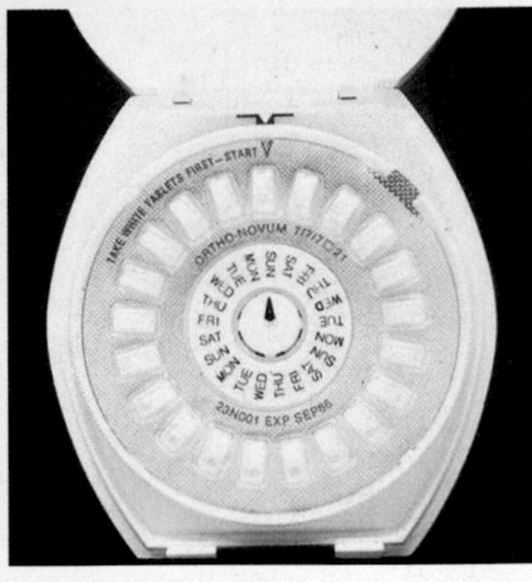
ORTHO-NOVUM 7/7/7 21

(e)

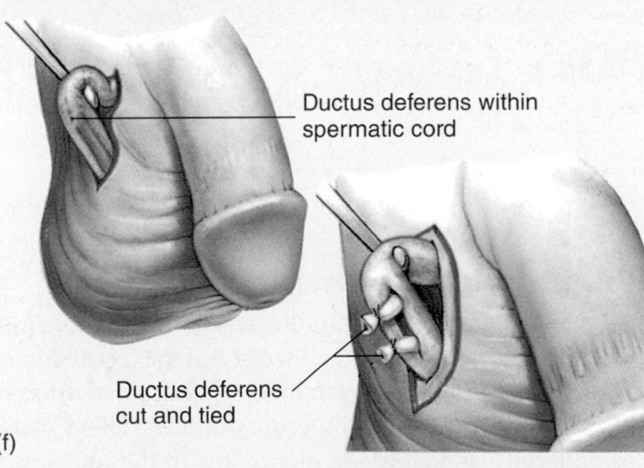
Ductus deferens within spermatic cord
Ductus deferens cut and tied

(f)

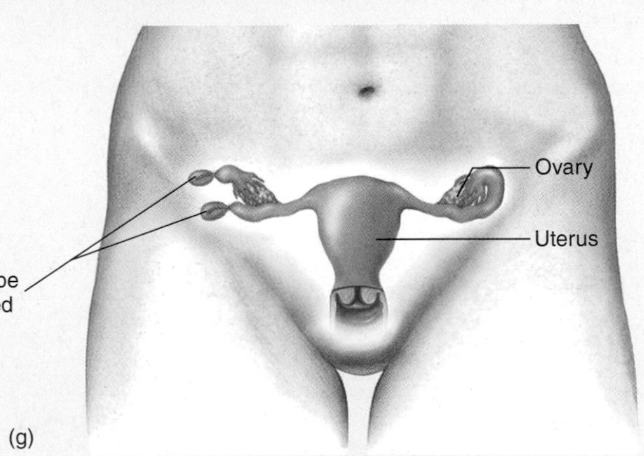
Ovary
Uterus
Uterine tube cut and tied

(g)

FIGURE 28B Contraceptive Devices and Techniques

(*a*) Condoms. (*b*) Cervical cap and diaphragm used with spermicidal gel. (*c*) Spermicidal gel. (*d*) Intrauterine device (IUD). (*e*) Oral contraceptives. (*f*) Vasectomy. (*g*) Tubal ligation.

Systems PATHOLOGY | Benign Uterine Tumors

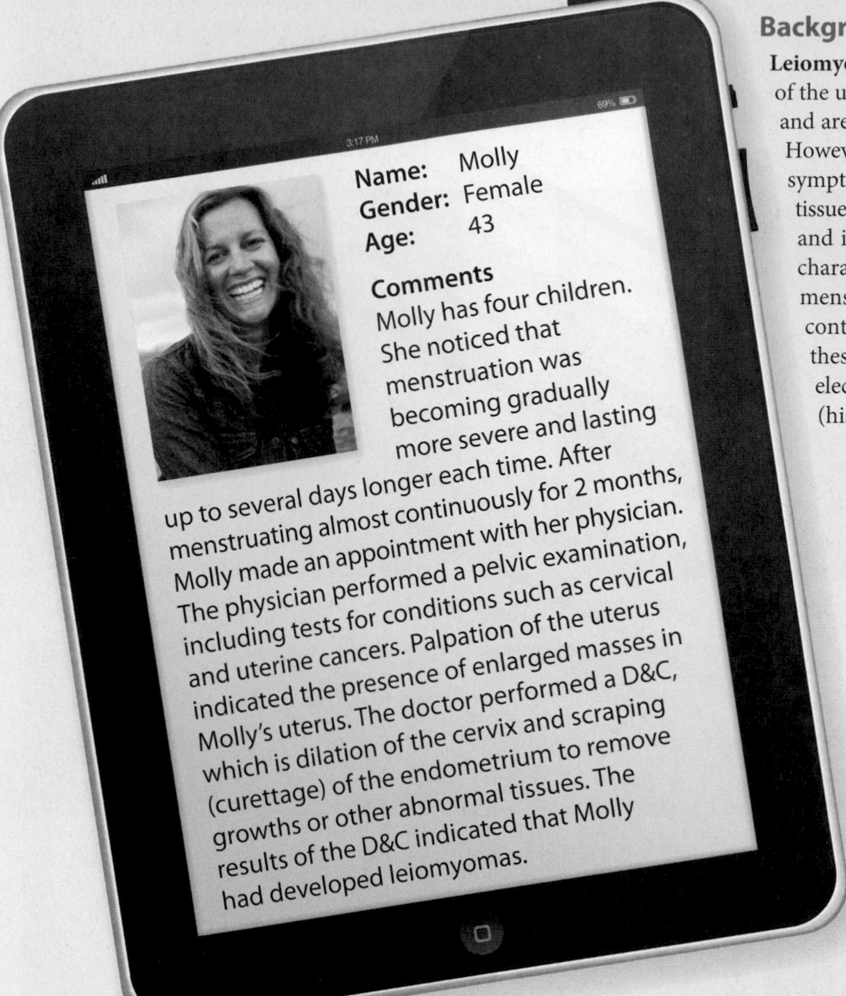
Name: Molly
Gender: Female
Age: 43
Comments
Molly has four children. She noticed that menstruation was becoming gradually more severe and lasting up to several days longer each time. After menstruating almost continuously for 2 months, Molly made an appointment with her physician. The physician performed a pelvic examination, including tests for conditions such as cervical and uterine cancers. Palpation of the uterus indicated the presence of enlarged masses in Molly's uterus. The doctor performed a D&C, which is dilation of the cervix and scraping (curettage) of the endometrium to remove growths or other abnormal tissues. The results of the D&C indicated that Molly had developed leiomyomas.

Background Information

Leiomyomas (lī′-ō-mī-ō′măs; figure 28C) are also called fibroid tumors of the uterus. They are one of the most common disorders of the uterus and are the most frequent tumor in women, affecting one of every four. However, three-fourths of the women with this condition experience no symptoms. The enlarged masses that originate from smooth muscle tissue compress the uterine lining (endometrium), resulting in ischemia and inflammation. The increased inflammation, which shares some characteristics with menstruation, results in frequent and severe menses, with associated abdominal cramping due to strong uterine contractions. Constant menstruation is a frequent manifestation of these tumors, and it is one of the most common reasons women elect to have the uterus removed, a procedure called a **hysterectomy** (his-ter-ek′tō-mē).

> **Predict 7**
>
> *When discussing her condition with her mother, Molly discovered that her mother had experienced frequent menses that were irregular and prolonged when she was in her late forties. Molly's mother did not have a hysterectomy, and in a few years the frequency of menstruation gradually began to subside. Explain.*

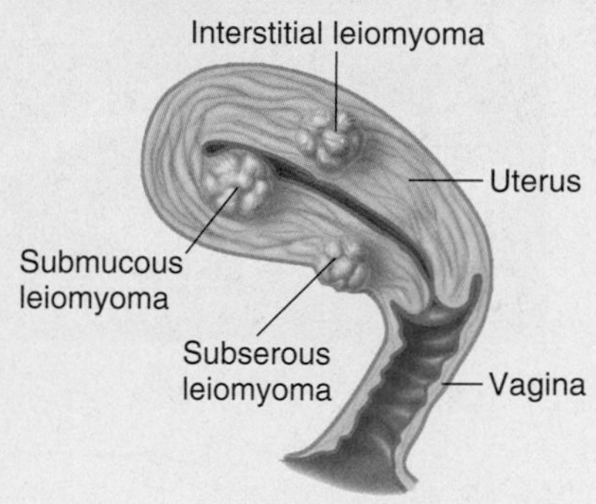
Interstitial leiomyoma
Uterus
Submucous leiomyoma
Subserous leiomyoma
Vagina
Medial view

Figure 28C Leiomyomas

Leiomyomas, or fibroid tumors, are enlarged masses of smooth muscle. They can be located near the mucosa (submucous), within the myometrium (interstitial), or near the serosa (subserous).

28.6 Effects of Aging on the Reproductive System

LEARNING OUTCOME

After reading this section, you should be able to

A. Describe the major age-related changes in the male and female reproductive systems.

Age-Related Changes in Males

Several age-related changes occur in the male reproductive system. In some but not all men, the size and weight of the testes decrease, with an associated decrease in the number of interstitial cells and a thinning of the wall of the seminiferous tubules. These changes may be secondary to a decrease in blood flow to the testes or to a gradual decrease in sex hormone production. In addition, the rate of sperm cell production decreases, and the number of abnormal sperm cells produced increases. However, sperm cell production does not stop, and it remains adequate for fertility for most men.

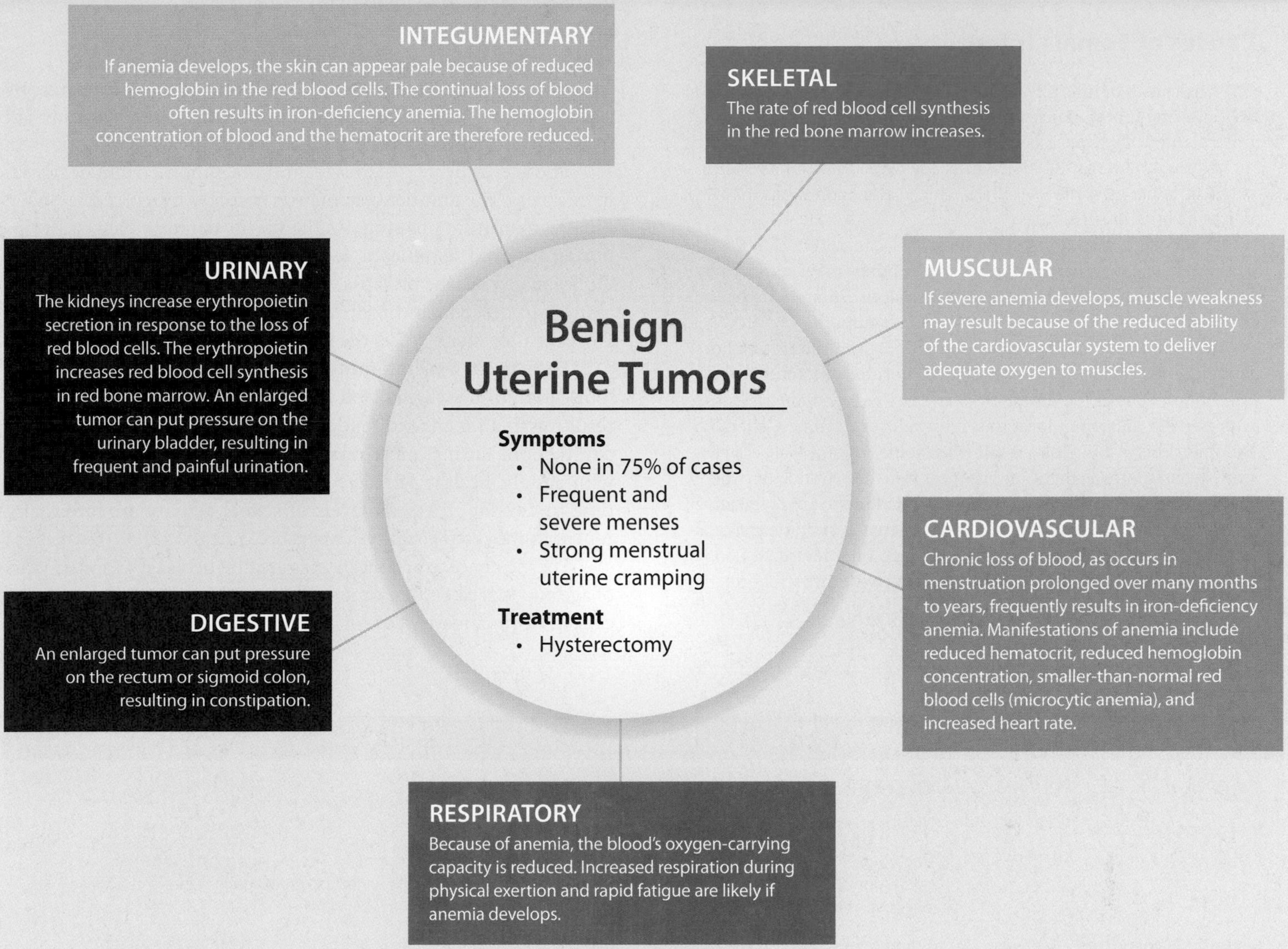

INTEGUMENTARY
If anemia develops, the skin can appear pale because of reduced hemoglobin in the red blood cells. The continual loss of blood often results in iron-deficiency anemia. The hemoglobin concentration of blood and the hematocrit are therefore reduced.

SKELETAL
The rate of red blood cell synthesis in the red bone marrow increases.

URINARY
The kidneys increase erythropoietin secretion in response to the loss of red blood cells. The erythropoietin increases red blood cell synthesis in red bone marrow. An enlarged tumor can put pressure on the urinary bladder, resulting in frequent and painful urination.

MUSCULAR
If severe anemia develops, muscle weakness may result because of the reduced ability of the cardiovascular system to deliver adequate oxygen to muscles.

Benign Uterine Tumors

Symptoms
- None in 75% of cases
- Frequent and severe menses
- Strong menstrual uterine cramping

Treatment
- Hysterectomy

CARDIOVASCULAR
Chronic loss of blood, as occurs in menstruation prolonged over many months to years, frequently results in iron-deficiency anemia. Manifestations of anemia include reduced hematocrit, reduced hemoglobin concentration, smaller-than-normal red blood cells (microcytic anemia), and increased heart rate.

DIGESTIVE
An enlarged tumor can put pressure on the rectum or sigmoid colon, resulting in constipation.

RESPIRATORY
Because of anemia, the blood's oxygen-carrying capacity is reduced. Increased respiration during physical exertion and rapid fatigue are likely if anemia develops.

Age-related changes become obvious in the prostate gland by age 40. By age 60, there is a clear decrease in blood flow to the prostate gland, an increased thickness in the epithelial cell lining of the prostate gland, and a decrease in the number of functional smooth muscle cells in the wall of the prostate. The changes in the prostate gland do not decrease fertility, but the incidence of benign prostatic hypertrophy (enlargement of the prostate) increases substantially and can lead to difficulty in urination due to compression of the prostatic urethra. A significant number (approximately 15%) of men older than 60 require medical treatment for benign prostatic hypertrophy; by age 80, this number is 50%. Before age 50, prostate cancer is rare. After 55, it is the third leading cause of death from cancer in men.

Erectile dysfunction increases in men with age. By age 60, approximately 15% of men have ED; by 80, 50% do. In addition, the amount of fibrous connective tissue in the erectile tissue of the penis increases, which generally decreases the speed of erection by age 60.

Although there is great variation among males, many exhibit decreases in the frequency of sexual activity and in sexual performance.

Clinical
IMPACT

Causes of Female Infertility

The causes of infertility in females include malfunctions of the uterine tubes, reduced hormone secretion from the pituitary gland or the ovaries, and interruption of implantation.

Uterine tube malfunction can occur when infections result in pelvic inflammatory disease (PID), which causes adhesions to form in one or both uterine tubes.

Inadequate secretion of LH and FSH can result in reduced ovulation. The insufficient amount of hormones may be caused by hypothyroidism, trauma to the hypothalamus, infarctions of the hypothalamus or anterior pituitary gland, or tumors.

Interruption of implantation may result from uterine tumors or conditions causing abnormal ovarian hormone secretion.

Endometriosis (en′dō-mē-trē-ō′sis), in which endometrial tissue is present in abnormal locations, also reduces fertility. Generally, endometriosis is thought to result when some endometrial cells pass from the uterus through the uterine tubes into the pelvic cavity, where they invade the peritoneum. Because the endometrium is sensitive to estrogen and progesterone, the areas where the endometrial cells have invaded periodically become inflamed. Endometriosis is one possible cause of abdominal pain associated with menstruation.

Psychological changes, age-related changes in the nervous system, and decreased blood flow explain some of the decline. The side effects of medications taken for other conditions are responsible for decreased sexual activity in many older men.

Age-Related Changes in Females

The most significant age-related change in females is menopause. By age 50, few viable follicles remain in the ovaries. As a result, the amount of estrogen and progesterone produced by the ovaries has decreased.

The uterus decreases in size, and the endometrium decreases in thickness. The time between menses becomes irregular and longer. Finally, menstruations stop. As the uterus decreases in size, it tips posteriorly and assumes a lower position in the pelvic cavity. Occasionally, uterine prolapse, in which the ligaments of the uterus allow it to descend and protrude into the vagina, occurs. Within 15 years after menopause, the uterus is 50% of its original size.

The vaginal wall becomes thinner and less elastic. There is less lubrication of the vagina, and the epithelial lining is more fragile, resulting in an increased tendency for vaginal infections. Vaginal contractions during intercourse decrease, and the vagina narrows with age. In healthy females, sexual excitement requires greater time to develop, the peak levels of sexual activity are lower, and return to the resting state occurs more quickly.

The incidence of breast cancer is greatest between 45 and 65 years of age and is greater for women who have a family history of breast cancer. Approximately 10% of all women will develop breast cancer.

Diseases and Disorders

TABLE 28.4 Reproductive System

Condition	Description
Infectious Diseases	
Pelvic inflammatory disease (PID)	Bacterial infection of the female pelvic organs; commonly caused by vaginal or uterine infection by the bacteria that cause gonorrhea or chlamydia; early symptoms include increased vaginal discharge and pelvic pain; antibiotics are effective; if untreated, can lead to sterility or be life-threatening
Sexually Transmitted Diseases	Commonly known as STDs; spread by intimate sexual contact
Nongonococcal urethritis (non-gon′ō-kok′ăl u-rē-thrī′tis)	Inflammation of the urethra that is not caused by gonorrhea; can be caused by trauma, insertion of a nonsterile catheter, or sexual contact; usually due to infection with the bacterium *Chlamydia trachomatis* (kla-mid′ē-ă tra-kō′mă-tis); may go unnoticed and result in pelvic inflammatory disease or sterility; antibiotics are effective treatment
Trichomoniasis (trik-ō-mō-nī′ă-sis)	Caused by *Trichomonas* (trik′ō-mō′nas), a protozoan commonly found in the vagina of women and in the urethra of men; results in a greenish-yellow discharge with a foul odor; more common in women than in men
Gonorrhea (gon-ō-rē′ă)	Caused by the bacterium *Neisseria gonorrhoeae* (nī-sē′rē-ă gon-ō-rē′ă), which attaches to the epithelial cells of the vagina or male urethra and causes pus to form; pain and discharge from the penis occur in men; asymptomatic in women in the early stages; can lead to sterility in men and pelvic inflammatory disease in women
Genital herpes (her′pēz)	Caused by herpes simplex 2 virus; characterized by lesions on the genitals that progress into blisterlike areas, making urination, sitting, and walking painful; antiviral drugs can be effective
Genital warts	Caused by a viral infection; very contagious; warts vary from separate, small growths to large, cauliflower-like clusters; lesions are not painful, but sexual intercourse with lesions is; treatments include topical medicines and surgery to remove the lesions
Syphilis (sif′i-lis)	Caused by the bacterium *Treponema pallidum* (trep-ō-nē′mă pal′i-dŭm); can be spread by sexual contact; multiple disease stages occur; children born to infected mothers may be developmentally delayed; antibiotics are effective
Acquired immunodeficiency syndrome (AIDS)	Caused by the human immunodeficiency virus (HIV), which ultimately destroys the immune system (see chapter 22); transmitted through intimate sexual contact or by allowing infected body fluids into the interior of another person

Go to: www.mhhe.com/seeley10 for additional information on these pathologies.

The most important measure to guard against death from breast cancer is early detection through breast self-exams and annual mammograms after age 40. The incidences of uterine cancer and cervical cancer increase between 50 and 65 years of age. Ovarian cancer increases in frequency in older women, and it is the second most common cancer of the reproductive system in older women. Annual medical checkups, including Pap smears, are important for early detection and treatment of these cancers.

ASSESS YOUR PROGRESS 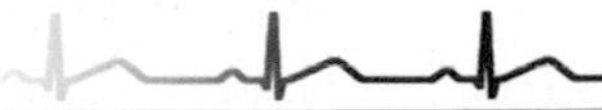

82. *List the major age-related changes that occur in the male reproductive system.*

83. *What are the age-related changes in the prostate gland?*

84. *List the major age-related changes that occur in the female reproductive system.*

85. *What are some of the long-term consequences of menopause?*

Answer

Learn to Predict ❮ From page 1016

In this chapter, we learned that meiosis is cell division that produces haploid cells. When comparing meiosis in males and females, we find that the processes differ in several ways: the stage in an individual's life when meiosis begins, the types of cells produced, the number of functional cells produced with each cell division, and the stage of life when meiosis ceases to occur.

In males, meiosis begins at puberty in the seminiferous tubules. Spermatogonia give rise to primary spermatocytes, which will undergo the process of meiosis. During this process, each primary spermatocyte eventually gives rise to four mature sperm cells. Males continue to produce sperm until death.

Meiosis in females is more complex. The process actually begins before a female is born. During fetal development, many of the oogonia in the ovaries degenerate. The remaining oogonia actually begin meiosis I and are called primary oocytes. At birth, the existing primary oocytes stop meiosis. After puberty and just before the ovulation of each oocyte, the primary oocyte that is ovulated completes the first meiotic division to produce one secondary oocyte and one polar body. The secondary oocyte begins the second meiotic division but will complete the process only if fertilized by a sperm cell. In the case of fertilization, the secondary oocyte divides to form two cells. One cell is another polar body and degenerates. In the other cell, the haploid sperm nucleus combines with the haploid oocyte nucleus to form a zygote. Thus, in females, each primary oocyte produces only one functional cell. For females, the process of meiosis stops at menopause.

Answers to the rest of this chapter's Predict questions are in Appendix G.

Summary

28.1 Functions of the Reproductive System (p. 1017)

The reproductive systems produce male and female gametes, enhance fertilization of an oocyte by a sperm cell, and produce sex hormones. In addition, the female reproductive system nurtures the new individual until birth.

28.2 Anatomy of the Male Reproductive System (p. 1017)

The male reproductive system includes the testes, ducts, accessory glands, and supporting structures.

Scrotum

1. The scrotum is a two-chambered sac that contains the testes.
2. The dartos and cremaster muscles help regulate testicular temperature.

Perineum

The perineum, the diamond-shaped area between the thighs, consists of a urogenital triangle and an anal triangle.

Testes

1. The tunica albuginea is the outer connective tissue capsule of the testes.
2. The testes are divided by septa into lobules that contain the seminiferous tubules and the interstitial cells.
3. The seminiferous tubules straighten to form the tubuli recti, which lead to the rete testis. The rete testis opens into the efferent ductules of the epididymis.
4. During development, the testes pass from the abdominal cavity through the inguinal canal to the scrotum.

Sperm Cell Development

1. Spermatogenesis begins in the seminiferous tubules at the time of puberty.
2. Spermatogonia divide (mitosis) to form primary spermatocytes.
3. Primary spermatocytes divide (first division of meiosis) to form secondary spermatocytes, which divide (second division of meiosis) to form spermatids.
4. Spermatids develop an acrosome and a flagellum to become sperm cells.
5. Sustentacular cells nourish the sperm cells, form a blood-testis barrier, and produce hormones.

Ducts

1. Efferent ductules extend from the testes to the head of the epididymis.
2. The epididymis, a coiled tube system, is located on the testis and is the site of sperm cell maturation. It consists of a head, a body, and a tail.
3. The ductus deferens passes from the epididymis into the abdominal cavity.
4. The end of the ductus deferens, called the ampulla, and the seminal vesicle join to form the ejaculatory duct.
5. The prostatic urethra extends from the urinary bladder and joins with the ejaculatory ducts to form the membranous urethra.
6. The membranous urethra extends through the urogenital diaphragm and becomes the spongy urethra, which continues through the penis.
7. The spermatic cord consists of the ductus deferens, blood and lymphatic vessels, nerves, and remnants of the process vaginalis. Coverings of the spermatic cord consist of the external spermatic fascia, cremaster muscle, and internal spermatic fascia.
8. The spermatic cord passes through the inguinal canal into the abdominal cavity.

Penis

1. The penis consists of erectile tissue.
   - The two corpora cavernosa form the dorsum and the sides of the penis.
   - The corpus spongiosum forms the ventral part and the glans penis.
2. The bulb of the penis and the crura form the root of the penis, and the crura attaches the penis to the pelvic bones.
3. The prepuce covers the glans penis.

Accessory Glands

1. The seminal vesicles empty into the ejaculatory ducts.
2. The prostate gland consists of glandular and muscular tissue and empties into the prostatic urethra.
3. The bulbourethral glands are compound mucous glands that empty into the spongy urethra.

Semen

1. Semen is a mixture of sperm cells and glandular secretions.
2. The bulbourethral glands and the urethral mucous glands produce mucus, which neutralizes the acidic pH of the urethra.
3. The testicular secretions contain sperm cells.
4. The seminal vesicle fluid contains fructose and fibrinogen.
5. The prostate secretions make the seminal fluid more pH-neutral. Clotting factors activate fibrinogen, and fibrinolysin breaks down fibrin.

28.3 Physiology of Male Reproduction (p. 1030)

Normal function of the male reproductive system depends on hormonal and neural mechanisms.

Regulation of Sex Hormone Secretion

1. GnRH is produced in the hypothalamus and released in surges.
2. GnRH stimulates LH and FSH release from the anterior pituitary.
   - LH stimulates the interstitial cells to produce testosterone.
   - FSH stimulates sperm cell formation.
3. Inhibin, produced by sustentacular cells, inhibits FSH secretion.

Puberty in Males

1. Before puberty, small amounts of testosterone inhibit GnRH release.
2. During puberty, testosterone does not completely suppress GnRH release, resulting in increased production of FSH, LH, and testosterone.

Effects of Testosterone

1. Interstitial cells, the adrenal cortex, and possibly the sustentacular cells produce testosterone.
2. Testosterone causes the development of male sex organs in the embryo and stimulates the descent of the testes.
3. Testosterone causes enlargement of the genitals and is necessary for sperm cell formation.
4. Other effects of testosterone
   - Hair growth stimulation (pubic area, axilla, and beard) and inhibition (male pattern baldness)
   - Enlargement of the larynx and deepening of the voice
   - Increased skin thickness and melanin and sebum production
   - Increased protein synthesis (muscle), bone growth, blood cell synthesis, and blood volume
   - Increased metabolic rate

Male Sexual Behavior and the Male Sex Act

1. Testosterone is required for normal sex drive.
2. Stimulation of the sex act can be tactile or psychological.
3. Afferent action potentials pass through the pudendal nerve to the sacral region of the spinal cord.
4. Parasympathetic stimulation
   - Erection is due to vasodilation of the blood vessels that supply the erectile tissue.
   - The glands of the urethra and the bulbourethral glands produce mucus.
5. Sympathetic stimulation causes erection, emission, and ejaculation.

28.4 Anatomy of the Female Reproductive System (p. 1034)

The female reproductive system includes the ovaries, uterine tubes, uterus, vagina, external genitals, and mammary glands.

Ovaries

1. The broad ligament, the mesovarium, the suspensory ligaments, and the ovarian ligaments hold the ovaries in place.
2. The peritoneum (ovarian epithelium) covers the surface of the ovaries.
3. The ovary has an outer capsule called the tunica albuginea and is divided internally into a cortex, which contains follicles, and a medulla, which receives blood and lymphatic vessels and nerves.
4. Oocyte development and fertilization
   - Oogonia proliferate and become primary oocytes that are in prophase I of meiosis.
   - Ovulation is the release of an oocyte from an ovary.
   - Prior to ovulation, a primary oocyte continues meiosis I and produces a secondary oocyte, which begins meiosis II, and a polar body, which either degenerates or divides to form two polar bodies.
   - Fertilization is the joining of a sperm cell and a secondary oocyte to form a zygote. A sperm cell enters a secondary oocyte, which then completes the second meiotic division and produces a polar body. A zygote is formed when the nuclei of the sperm cell and the oocyte fuse to form a diploid nucleus.
5. Follicle development
   - Primordial follicles are surrounded by a single layer of flat granulosa cells.
   - Primary follicles are primary oocytes surrounded by cuboidal granulosa cells.
   - The primary follicles become secondary follicles as granulosa cells increase in number and fluid begins to accumulate in the vesicles. The granulosa cells increase in number, and a theca forms around the secondary follicles.

- Mature follicles are enlarged secondary follicles at the surface of the ovary.

6. Ovulation occurs when the follicle swells and ruptures and the secondary oocyte is released from the ovary.
7. Fate of the follicle
   - The mature follicle becomes the corpus luteum.
   - If pregnancy occurs, the corpus luteum persists. If no pregnancy occurs, it becomes the corpus albicans.

Uterine Tubes

1. The mesosalpinx holds the uterine tubes.
2. The uterine tubes transport the oocyte or zygote from the ovary to the uterus.
3. Structures
   - The ovarian end of the uterine tube is expanded as the infundibulum. The opening of the infundibulum is the ostium, which is surrounded by fimbriae.
   - The infundibulum connects to the ampulla, which narrows to become the isthmus. The isthmus is the part of the uterine tube nearest the uterus.
4. The uterine tube consists of an outer serosa, a middle muscular layer, and an inner mucosa composed of simple ciliated columnar epithelium.
5. Movement of the oocyte
   - Cilia move the oocyte over the fimbriae surface into the infundibulum.
   - Peristaltic contractions and cilia move the oocyte within the uterine tube.
   - Fertilization occurs in the ampulla, where the zygote remains for several days.

Uterus

1. The uterus consists of the body, the isthmus, and the cervix. The uterine cavity and the cervical canal are the spaces formed by the uterus.
2. The uterus is held in place by the broad, round, and uterosacral ligaments.
3. The wall of the uterus consists of the perimetrium (serous membrane), the myometrium (smooth muscle), and the endometrium (mucous membrane).

Vagina

1. The vagina connects the uterus (cervix) to the outside of the body.
2. The vagina consists of a layer of smooth muscle and an inner lining of moist stratified squamous epithelium.
3. The vagina is folded into rugae and longitudinal folds.
4. The hymen covers the opening of the vagina.

External Genitalia

1. The vulva, or pudendum, comprises the external genitalia.
2. The vestibule is the space into which the vagina and the urethra open.
3. Erectile tissue
   - The two corpora cavernosa form the clitoris.
   - The corpora spongiosa form the bulbs of the vestibule.
4. The labia minora are folds that cover the vestibule and form the prepuce.
5. The greater and lesser vestibular glands produce a mucous fluid.
6. When closed, the labia majora cover the labia minora.
   - The pudendal cleft is a space between the labia majora.
   - The mons pubis is an elevated fat deposit superior to the labia majora.

Perineum

The clinical perineum is the region between the vagina and the anus.

Mammary Glands

1. The mammary glands are modified sweat glands located in the breasts.
   - The mammary glands consist of glandular lobes and adipose tissue.
   - The lobes consist of lobules that are divided into alveoli.
   - The lobes connect to the nipple through the lactiferous ducts.
   - The areola surrounds the nipple.
2. Suspensory ligaments support the breasts.

28.5 Physiology of Female Reproduction (p. 1043)

Puberty in Females

1. The first menstrual bleeding (menarche) occurs during puberty.
2. Puberty begins when GnRH levels increase.

Menstrual Cycle

1. Ovarian cycle
   - FSH initiates the development of the primary follicles.
   - The follicles secrete a substance that inhibits the development of other follicles.
   - LH stimulates ovulation and completion of the first meiotic division by the primary oocyte.
   - The LH surge stimulates the formation of the corpus luteum. If fertilization occurs, hCG stimulates the corpus luteum to persist. If fertilization does not occur, the corpus luteum becomes the corpus albicans.
2. A positive-feedback mechanism causes FSH and LH levels to increase near the time of ovulation.
   - Estrogen produced by the theca cells of the follicle stimulates GnRH secretion.
   - GnRH stimulates the production and release of FSH and LH, which stimulate more estrogen secretion, and so on.
   - Inhibition of GnRH levels causes FSH and LH levels to decrease after ovulation. Inhibition is due to the high levels of estrogen and progesterone produced by the corpus luteum.
3. Uterine cycle
   - Menses (from day 1 to day 4 or 5). The spiral arteries constrict, and endometrial cells die. The menstrual fluid is composed of sloughed cells, secretions, and blood.
   - Proliferative phase (from day 5 to day 14). Epithelial cells multiply and form glands, and the spiral arteries supply the glands.
   - Secretory phase (from day 15 to day 28). The endometrium becomes thicker, and the endometrial glands secrete.
   - Estrogen stimulates proliferation of the endometrium and synthesis of progesterone receptors.
   - Increased progesterone levels cause hypertrophy of the endometrium, stimulate gland secretion, and inhibit uterine contractions. Decreased progesterone levels cause the spiral arteries to constrict and start menses.

Female Sexual Behavior and the Female Sex Act

1. Female sex drive is partially influenced by androgens (produced by the adrenal gland) and other steroids (produced by the ovaries).
2. Parasympathetic effects
   - The erectile tissue of the clitoris and the bulbs of the vestibule become filled with blood.
   - The vestibular glands secrete mucus, and the vagina extrudes a mucuslike substance.

Female Fertility and Pregnancy

1. If fertilization is to occur, intercourse must take place between 5 days before and 1 day after ovulation.
2. Sperm cell transport to the ampulla depends on the ability of the sperm cells to swim and possibly on contractions of the uterus and the uterine tubes.

3. Implantation of the developing embryo into the uterine wall occurs when the uterus is most receptive.
4. Estrogen and progesterone, secreted first by the corpus luteum and later by the placenta, are essential for the maintenance of pregnancy.

Menopause

The female climacteric begins with irregular menstrual cycles and ends with menopause, the cessation of the menstrual cycle.

28.6 Effects of Aging on the Reproductive System (p. 1054)

Several age-related changes occur in the male and female reproductive systems.

Age-Related Changes in Males

1. Decreases occur in the size and weight of the testes, the number of interstitial cells, and the number of sperm produced, but sperm cell production is still adequate for fertilization. The wall of the seminiferous tubule becomes thin.
2. The prostate gland enlarges, and the incidence of prostate cancer increases.
3. Erectile dysfunction becomes more common, and sexual activity gradually decreases.

Age-Related Changes in Females

1. The most significant age-related change in females is menopause.
2. The uterus decreases in size, and the vaginal wall thins.
3. The incidence of breast, uterine, and ovarian cancer increases.

REVIEW AND COMPREHENSION

1. If an adult male walked into a swimming pool of cold water, which of these muscles would probably contract?
   a. cremaster muscle
   b. dartos muscle
   c. gubernaculum
   d. prepuce muscle
   e. Both a and b are correct.
2. Testosterone is produced in the
   a. interstitial cells.
   b. seminiferous tubules of the testes.
   c. anterior lobe of the pituitary gland.
   d. sperm cells.
3. Early in development (4 months after fertilization), the testes
   a. are found in the abdominal cavity.
   b. move through the inguinal canal.
   c. produce a membrane that becomes the scrotum.
   d. produce sperm cells.
   e. All of these are correct.
4. The site of spermatogenesis in the male is the
   a. ductus deferens.
   b. seminiferous tubules.
   c. epididymis.
   d. rete testis.
   e. efferent ductule.
5. The site of final maturation and storage of sperm cells before their ejaculation is the
   a. seminal vesicles.
   b. seminiferous tubules.
   c. glans penis.
   d. epididymis.
   e. sperm bank.
6. Given these structures:
   (1) ductus deferens
   (2) efferent ductule
   (3) epididymis
   (4) ejaculatory duct
   (5) rete testis

   Choose the arrangement that lists the structures in the order a sperm cell passes through them from the seminiferous tubules to the urethra.
   a. 2,3,5,4,1
   b. 2,5,3,4,1
   c. 3,2,4,1,5
   d. 3,4,2,1,5
   e. 5,2,3,1,4
7. Concerning the penis,
   a. the membranous urethra passes through the corpora cavernosa.
   b. the glans penis is formed by the corpus spongiosum.
   c. the penis contains four columns of erectile tissue.
   d. the crus of the penis is part of the corpus spongiosum.
   e. the bulb of the penis is covered by the prepuce.
8. Which of these glands is correctly matched with the function of its secretions?
   a. bulbourethral gland—neutralizes acidic contents of the urethra
   b. seminal vesicles—contain large amounts of fructose, which nourishes the sperm cells
   c. prostate gland—contains clotting factors that cause coagulation of the semen
   d. All of these are correct.
9. LH in the male stimulates
   a. the development of the seminiferous tubules.
   b. spermatogenesis.
   c. testosterone production.
   d. Both a and b are correct.
   e. All of these are correct.
10. Which of these factors causes a decrease in GnRH release?
    a. decreased inhibin
    b. increased testosterone
    c. decreased FSH
    d. decreased LH
11. In the male, before puberty
    a. FSH levels are higher than after puberty.
    b. LH levels are higher than after puberty.
    c. GnRH release is inhibited by testosterone.
    d. All of these are correct.
12. Testosterone
    a. stimulates the development of terminal hairs.
    b. decreases red blood cell count.
    c. prevents closure of the epiphyseal plate.
    d. decreases blood volume.
    e. All of these are correct.
13. Which of these events is consistent with erection of the penis?
    a. parasympathetic stimulation
    b. dilation of arterioles
    c. engorgement of sinusoids with blood
    d. occlusion of veins
    e. All of these are correct.

14. After ovulation, the mature follicle collapses, taking on a yellowish appearance to become the
    a. degenerating follicle.
    b. corpus luteum.
    c. corpus albicans.
    d. tunica albuginea.
    e. cumulus mass.
15. The ampulla of the uterine tube
    a. is the opening of the uterine tube into the uterus.
    b. has long, thin projections called the ostium.
    c. is connected to the isthmus of the uterine tube.
    d. is lined with simple cuboidal epithelium.
16. The layer of the uterus that undergoes the greatest change during the menstrual cycle is the
    a. perimetrium.
    b. hymen.
    c. endometrium.
    d. myometrium.
    e. broad ligament.
17. The vagina
    a. consists of skeletal muscle.
    b. has ridges called rugae.
    c. is lined with simple squamous epithelium.
    d. All of these are correct.
18. During sexual excitement, which of these structures fills with blood and causes the vaginal opening to narrow?
    a. bulbs of the vestibule
    b. clitoris
    c. mons pubis
    d. labia majora
    e. prepuce
19. Concerning the breasts,
    a. lactiferous ducts open on the areola.
    b. each lactiferous duct supplies an alveolus.
    c. they are attached to the pectoralis major muscles by suspensory ligaments.
    d. even before puberty, the female breast is quite different from the male breast.
20. The major secretory product of the mature follicle is
    a. estrogen.
    b. progesterone.
    c. LH.
    d. FSH.
    e. relaxin.
21. In the average adult female, ovulation occurs at day ________ of the menstrual cycle.
    a. 1
    b. 7
    c. 14
    d. 21
    e. 28
22. Which of these processes or phases in the monthly reproductive cycle of the human female occur at the same time?
    a. maximal LH secretion and menstruation
    b. early follicular development and the secretory phase of the uterus
    c. regression of the corpus luteum and an increase in ovarian progesterone secretion
    d. ovulation and menstruation
    e. proliferative stage of the uterus and increased estrogen production
23. During the proliferative phase of the menstrual cycle, one would normally expect
    a. the highest levels of estrogen that occur during the menstrual cycle.
    b. the mature follicle to be present in the ovary.
    c. an increase in the thickness of the endometrium.
    d. Both a and b are correct.
    e. All of these are correct.
24. The cause of menses in the menstrual cycle appears to be
    a. increased progesterone secretion from the ovary, which produces blood clotting.
    b. increased estrogen secretion from the ovary, which stimulates the muscles of the uterus to contract.
    c. decreased progesterone secretion by the ovary.
    d. decreased production of oxytocin, causing the muscles of the uterus to relax.
25. After fertilization, the successful development of a mature, full-term fetus depends on
    a. the release of human chorionic gonadotropin (hCG) by the developing placenta.
    b. the production of estrogen and progesterone by the placental tissues.
    c. maintenance of the corpus luteum for all 9 months.
    d. Both a and b are correct.
    e. All of these are correct.
26. A woman with a 28-day menstrual cycle is most likely to become pregnant as a result of intercourse on days
    a. 1–3.
    b. 5–8.
    c. 9–14.
    d. 15–20.
    e. 21–28.
27. Menopause
    a. develops when follicles become less responsive to FSH and LH.
    b. results from elevated estrogen levels in 40- to 50-year-old women.
    c. occurs because too many follicles develop during each cycle.
    d. results when follicles develop but contain no oocytes.
    e. occurs because FSH and LH levels decline.

Answers in Appendix E

CRITICAL THINKING

1. If an adult male were castrated (testes were removed), what would happen to the levels of GnRH, FSH, LH, and testosterone in his blood? What effect would these hormonal changes have on his sex characteristics and sexual behavior?
2. If a 9-year-old boy were castrated, what would happen to the levels of GnRH, FSH, LH, and testosterone in his blood? What effect would these hormonal changes have on his sex characteristics and sexual behavior as an adult?
3. Suppose you want to produce a birth control pill for men. On the basis of what you know about the male hormone system, what process should the pill affect? Discuss any possible side effects of the pill.
4. If the ovaries are removed from a postmenopausal woman, what happens to the levels of GnRH, FSH, LH, estrogen, and progesterone in her blood? What symptoms would you expect to observe?
5. If the ovaries are removed from a 20-year-old woman, what happens to the levels of GnRH, FSH, LH, estrogen, and progesterone in her blood?

What side effects would these hormonal changes have on her sex characteristics and sexual behavior?

6. A study divides healthy women into two groups (A and B). Both groups are composed of women who have been sexually active for at least 2 years and are not pregnant at the beginning of the experiment. The subjects weigh about the same amount, and none smokes cigarettes, although some drink alcohol occasionally. Group A women receive a placebo in the form of a sugar pill each morning of their menstrual cycles. Group B women receive a pill containing estrogen and progesterone each morning of their menstrual cycles. Then plasma LH levels are measured before, during, and after ovulation. The results are as follows:

Group	4 Days Before Ovulation	Day of Ovulation	4 Days After Ovulation
A	18 mg/100 mL	300 mg/100 mL	17 mg/100 mL
B	21 mg/100 mL	157 mg/100 mL	15 mg/100 mL

The number of pregnancies in group A is 37/100 women/year. The number of pregnancies in group B is 1.5/100 women/year. What conclusion can you reach on the basis of these data? Explain the mechanism involved.

7. A woman who is taking birth control pills that consist of only progesterone experiences the hot flashes of menopause. Explain why.

8. GnRH can be used to treat some women who want to have children but have not been able to get pregnant. Explain why it is critical to administer the correct concentration of GnRH at the right time during the menstrual cycle.

9. Dr. Procter has two patients, one of whom has elevated blood PSA. A digital exam reveals that both patient 1 and patient 2 have enlarged prostate glands. Patient 1's enlarged prostate has the same shape as a smaller prostate, except that it is larger than normal, with a smooth contour. Patient 2's prostate is enlarged and asymmetrical, with a rough contour. In what way are these patients' lives probably being affected by their enlarged prostates? Explain how the doctor was able to conclude that one of the patients is likely to have prostate cancer.

10. The left testis of a newborn baby failed to descend into his scrotal sac. For some reason, the condition was not treated, and the testis remained in that position until after puberty. Select the observation consistent with the fate of the left testis after puberty.
    (1) normal testosterone secretion
    (2) no interstitial cells in the testis
    (3) no sustentacular cells in the testis
    (4) no spermatogonia in the testis
    (5) increased number of interstitial cells
    (6) normal LH secretion
    a. 1
    b. 1,2,3,4
    c. 1,2,3,6
    d. 1,4,6
    e. 4,6

11. Norman had a stroke that decreased blood flow to his anterior pituitary gland. The condition had a sudden onset, and the manifestations lasted for about a week before collateral circulation developed and the manifestations disappeared. Which of the following are most consistent with this temporary interruption of anterior pituitary function?
    (1) increased testosterone levels in the blood
    (2) reduced sperm counts during the week Norman was in the hospital
    (3) decreased testosterone levels in the blood
    (4) normal sperm counts during the week Norman was in the hospital
    (5) increased LH secretion
    (6) decreased LH secretion
    a. 1,5
    b. 3,5
    c. 2,3,6
    d. 3,4,6
    e. 1,4,5

Answers in Appendix F

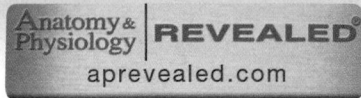

29

Development, Growth, Aging, and Genetics

Learn to Predict

After reading about the benefits of breast milk, Ming was determined to breastfeed her baby until he was a year old. Unfortunately, 2 weeks after her son's birth, Ming developed a serious urinary tract infection. Her physician prescribed an antibiotic and explained that she must not breastfeed her son while taking it. After reviewing chapter 28 and reading chapter 29, explain the effect on Ming if she stops breastfeeding for an extended time, and propose a strategy that would allow her to cease breastfeeding for awhile but resume after she has finished the antibiotic.

Photo: A woman and her newborn son.

The stages of life and associated activites are issues of great interest in today's society. We tend to view life stages very differently today than in the past. For example, in 1960, 20% of males and 12% of females graduating from high school attended college. Today, over half of all people 25 and older have attended some college. In addition, there are many more nontraditional college students than there were 30 years ago.

The life span is usually considered the period between birth and death; however, the 9 months before birth are a critical part of a person's existence. What happens in these 9 months profoundly affects the rest of a person's life. Although most people develop normally and are born without defects, approximately 3 out of every 100 people are born with a birth defect so severe that it requires medical attention during the first year of life. Later in life, many more people discover previously unknown problems, such as the tendency to develop asthma, certain brain disorders, or cancer. This chapter discusses the topics of development, growth, aging, and genetics.

Module 14
Reproductive System

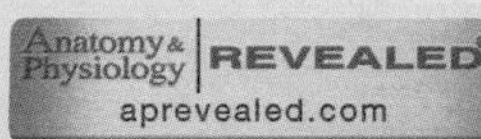
Anatomy & Physiology
REVEALED
aprevealed.com

29.1 Prenatal Development

LEARNING OUTCOMES

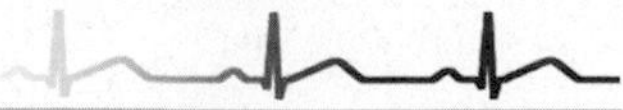

After reading this section, you should be able to

A. List the prenatal periods and state the major developmental events associated with each.

B. List the postnatal periods and state the age range for each.

C. Describe the process of fertilization.

D. Describe the blastocyst, the process of implantation, and placental formation.

E. List the three germ layers, describe their formation, and list the adult derivatives of each layer.

F. Discuss the formation of the neural tube and the neural crest cells.

G. Explain the formation of the digestive tract, the limbs, and the face.

H. Demonstrate knowledge of the formation of the major organ systems.

The human life span is divided into prenatal (before birth) and postnatal (after birth) periods. The prenatal period extends from conception until birth and is composed of three stages:

- The **germinal period** begins at fertilization and ends at 14 days (weeks 1 and 2 of development). This is the stage when the primitive germ layers form.
- The **embryonic period** comprises days 14–56 after fertilization (weeks 3–8 of development) and is the time when the major organ systems form. The developing human is called an **embryo** (em′brē-ō) at this stage.
- The **fetal period** extends from 56 days after fertilization to birth (the last 30 weeks of development). During this stage, the organ systems grow and become more mature, and the developing human is called a **fetus** (fē′tus).

Altogether, the prenatal period extends about 9 months. The medical community in general uses the mother's last **menstrual period (LMP)** to calculate the clinical age of the unborn child. Most embryologists, on the other hand, use **postovulatory age** to describe the timing of developmental events. Postovulatory age is used in this book. Because ovulation occurs about 14 days after LMP and fertilization occurs near the time of ovulation, postovulatory age is 14 days less than clinical age.

The first 9 months after conception is acutely important for the remainder of a person's life, but the prenatal period is only a small part of the total life span. The postnatal period is commonly divided into the following stages:

- *Neonatal period*: from birth to 1 month after birth
- *Infancy*: from 1 month to 1 or 2 years of age. The end of infancy is sometimes set at the time the child begins to walk.
- *Childhood*: from age 1 or 2 to puberty. During childhood, the individual develops considerably and forms many of the emotional characteristics that last throughout life.
- *Adolescence*: from puberty (age 11–14) to 20 years. Puberty usually occurs somewhat earlier in females (about 11–13 years) than in males (about 12–14 years). A period of rapid growth usually accompanies the onset of puberty.
- *Adult*: from age 20 to death. Full adult stature is usually achieved by age 17 or 18 in females and by 19 or 20 years in males. Adulthood is sometimes divided into three periods: young adult, age 20–40; middle age, age 40–65; and older adult, age 65 to death.

ASSESS YOUR PROGRESS

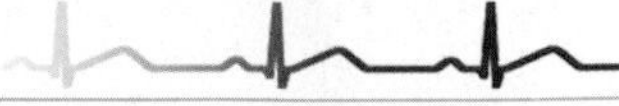

1. *Describe the three parts of the prenatal period, including the length of time for each part.*
2. *Distinguish between clinical age and postovulatory age.*
3. *List the stages of postnatal development, including the age range for each.*

Fertilization

Prenatal development begins at **fertilization,** when a sperm cell attaches to a secondary oocyte, and the contents of the sperm head enter the oocyte cytoplasm and join with the oocyte pronucleus (figure 29.1). Of the several hundred million sperm cells deposited in the vagina during sexual intercourse, only a few dozen reach the vicinity of the secondary oocyte in the ampulla of the uterine tube. The **corona radiata,** composed of cumulus cells expelled from the follicle with the oocyte during ovulation (see figure 28.13), acts as a barrier to those sperm cells that reach the oocyte. The flagella on the sperm cells propel them through the loose matrix between the follicular cells of the corona radiata. The **zona pellucida** is an extracellular membrane, comprised mostly of glycoproteins, that lies between the corona radiata and the oocyte. One particular zona pellucida glycoprotein, called **ZP3,** is a species-specific sperm cell receptor, to which molecules on the acrosomal cap of the sperm cell bind. This binding initiates the **acrosomal reaction,** which activates digestive enzymes in the acrosome, primarily hyaluronidase.

The first sperm cell through the zona pellucida attaches to a receptor molecule (integrin α6β1) on the surface of the oocyte plasma membrane and causes depolarization of the membrane within 2–3 seconds. This depolarization, called the **fast block to polyspermy,** prevents additional sperm cells from attaching to the oocyte plasma membrane. Depolarization also stimulates the intracellular release of Ca^{2+}, which in turn causes the exocytosis of water and other molecules from secretory vesicles, referred to as cortical granules, on the inner surface of the oocyte plasma membrane. The released fluid causes the oocyte to shrink and the zona pellucida to denature and expand away from the oocyte. The fluid-filled space between the oocyte plasma membrane and the zona pellucida is the **perivitelline space.** As the result of denaturation of the zona pellucida, ZP3 is inactivated, and no additional sperm cells can attach. This reaction is referred to as the **slow block to polyspermy.** Together, the fast block and the slow block ensure that the oocyte is fertilized by only one sperm cell.

The entrance of a sperm cell into the oocyte stimulates the female nucleus to undergo the second meiotic division, and the second polar body is formed. The nucleus that remains after the second meiotic division, called the **female pronucleus,** moves to the center of the oocyte, where it meets the **male pronucleus** of the sperm cell.

FUNDAMENTAL **Figure**

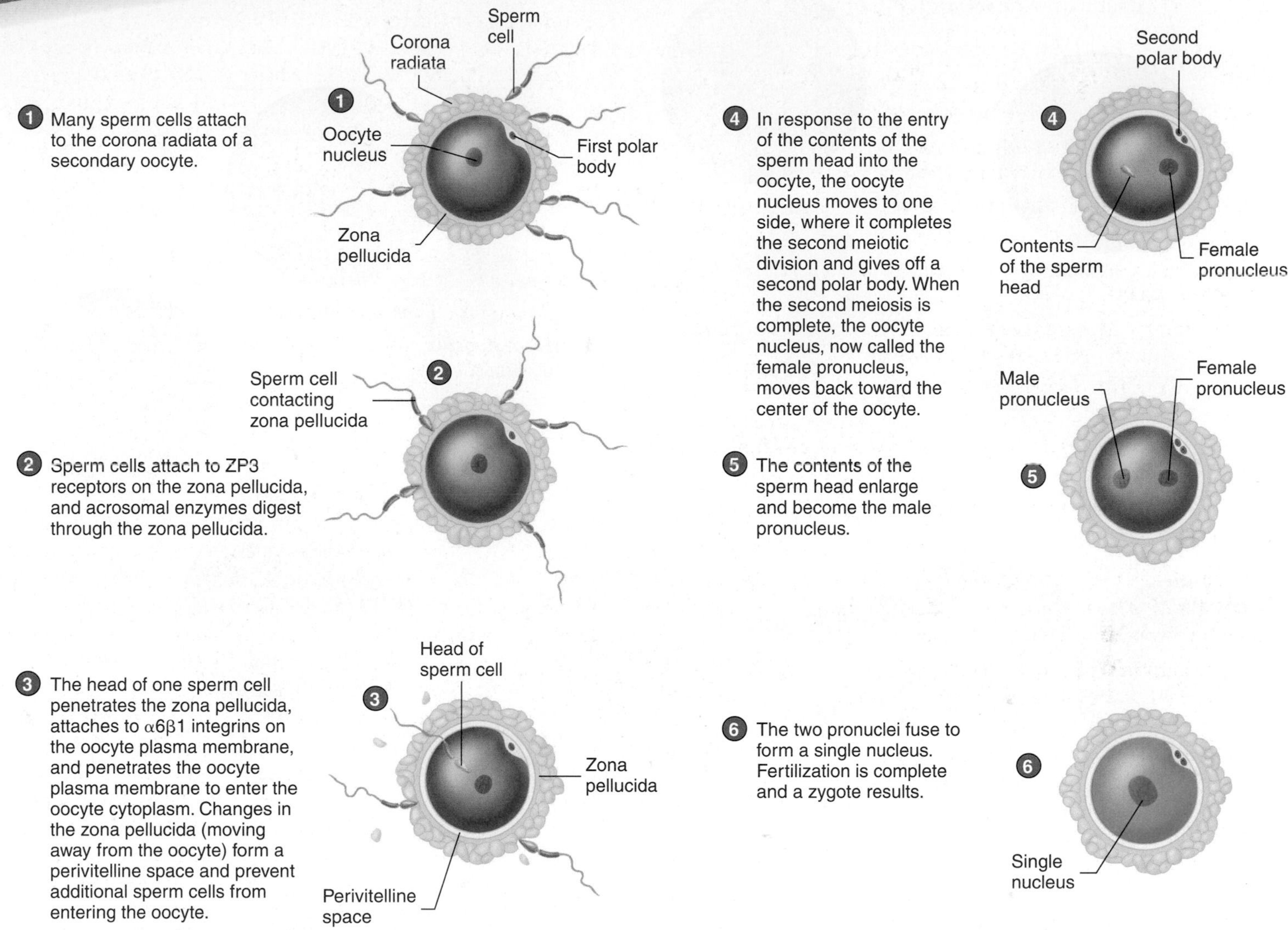

1 Many sperm cells attach to the corona radiata of a secondary oocyte.
1
Sperm cell
Corona radiata
Oocyte nucleus
First polar body
Zona pellucida
2 Sperm cells attach to ZP3 receptors on the zona pellucida, and acrosomal enzymes digest through the zona pellucida.
2
Sperm cell contacting zona pellucida
3 The head of one sperm cell penetrates the zona pellucida, attaches to α6β1 integrins on the oocyte plasma membrane, and penetrates the oocyte plasma membrane to enter the oocyte cytoplasm. Changes in the zona pellucida (moving away from the oocyte) form a perivitelline space and prevent additional sperm cells from entering the oocyte.
3
Head of sperm cell
Zona pellucida
Perivitelline space
4 In response to the entry of the contents of the sperm head into the oocyte, the oocyte nucleus moves to one side, where it completes the second meiotic division and gives off a second polar body. When the second meiosis is complete, the oocyte nucleus, now called the female pronucleus, moves back toward the center of the oocyte.
4
Second polar body
Contents of the sperm head
Female pronucleus
5 The contents of the sperm head enlarge and become the male pronucleus.
5
Male pronucleus
Female pronucleus
6 The two pronuclei fuse to form a single nucleus. Fertilization is complete and a zygote results.
6
Single nucleus

PROCESS FIGURE 29.1 AP|R **Fertilization**

Both the male and female pronuclei are haploid, each having one chromosome of each homologous pair (see chapter 28). Fusion of the pronuclei completes the process of fertilization and restores the diploid number of chromosomes. The product of fertilization is a single diploid cell, the **zygote** (zī′gōt; figure 29.2*a*).

Early Cell Division

About 18–36 hours after fertilization, the zygote divides to form two cells. Those two cells divide to form four cells, which divide to form eight cells, and so on (figure 29.2*b*–*d*). In the very early stages of development (days 1–4), the cells are said to be **totipotent** (tō-tīp′ō-tent; whole-powered), meaning that each cell has the potential to give rise to any tissue type necessary for development. However, the cells of the developing embryo soon undergo **differentiation,** or specialization. Once differentiation occurs, the dividing cells of the embryo are referred to as **pluripotent** (ploo-rip′ō-tent; multiple-powered), which means that any cell has the ability to develop into a wide range of tissues, but not into all the tissues necessary for development, as totipotent cells can. From this point on, the total number of embryonic cells can decrease, increase, or reorganize without affecting the normal development of the embryo.

Morula and Blastocyst

Once the dividing embryo consists of 12 or more cells, it is a solid sphere composed of numerous, smaller spheres and is called a **morula** (mōr′oo-lă, mōr′u-lă; mulberry; figure 29.2*d*). Four or 5 days after ovulation, the morula consists of about 32 cells. Near this time, a fluid-filled cavity called the **blastocele** (blas′tō-sēl) begins to form near the center of the cellular mass, resulting in a hollow sphere called a **blastocyst** (blas′tō-sist; figure 29.2*e*). A single layer of cells, the **trophoblast** (trof′ō-blast, trō′fō-blast; feeding layer), surrounds most of the blastocele. At one end of the blastocyst, however, the cells are several layers thick. The thickened area, called the **inner cell mass,** is the tissue from which the embryo proper develops.

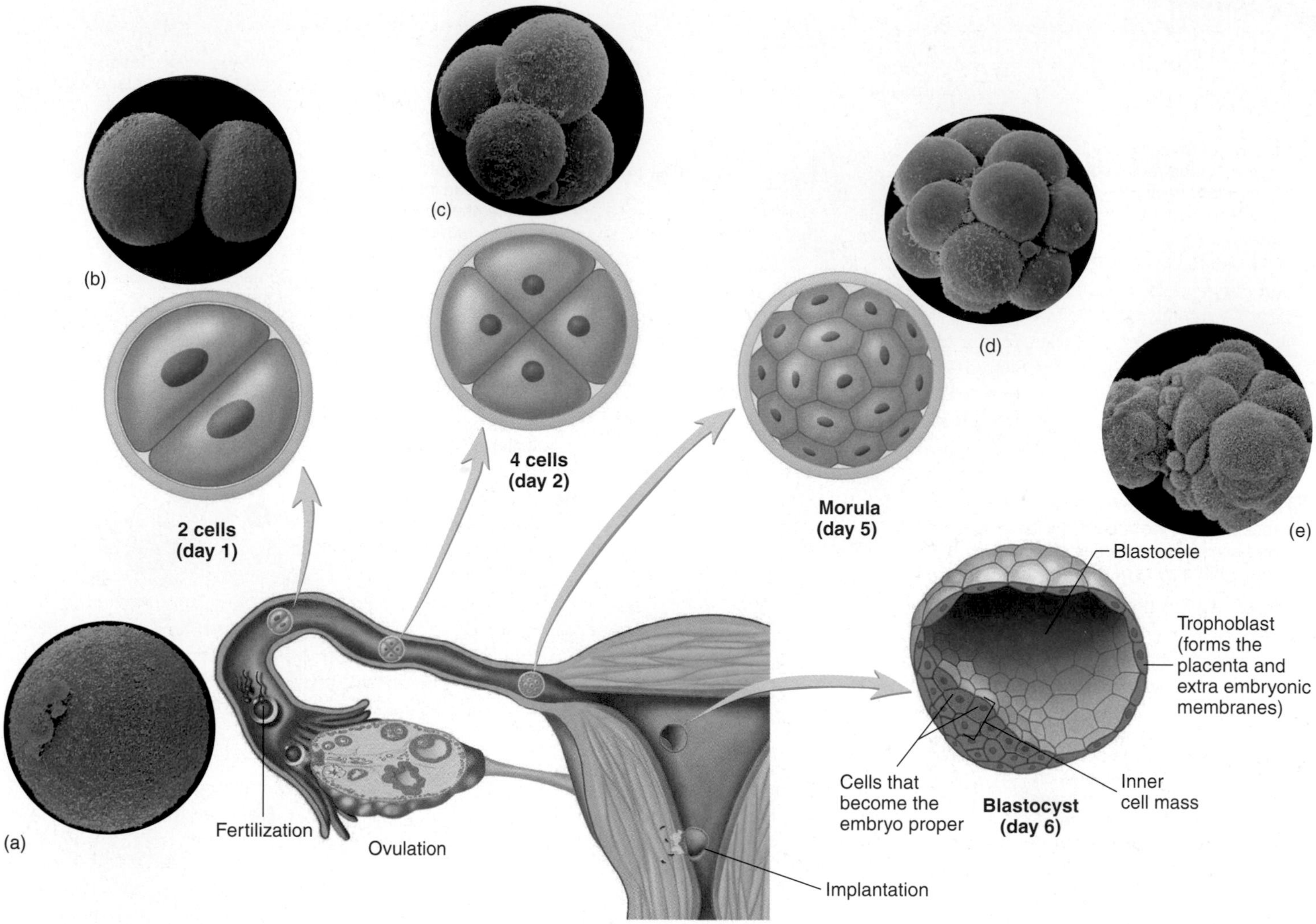

FIGURE 29.2 AP|R **Development of the Blastocyst and Implantation**

Successive cell divisions produce a multicellular morula by day 5, which becomes a hollow blastocyst on day 6. In the figure of the blastocyst, *green* cells are trophoblastic, and *orange* cells form the embryo proper. (*a*) Zygote (120 μm in diameter) with two polar bodies attached. (*b–d*) During the early cell divisions, the embryo divides into more and more cells, but the total size of the embryo remains relatively constant.

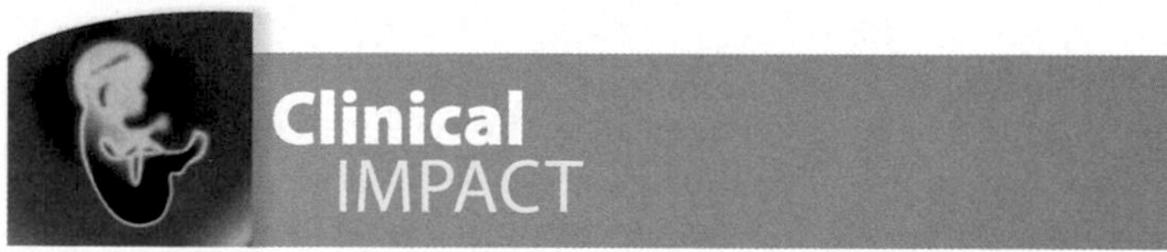
Clinical
IMPACT

Twins

In rare cases, following early cell divisions, a totipotent cell separates from the embryo and develops to form another individual, producing "identical twins," or **monozygotic twins.** Identical twins have identical genetic information in their cells. Other mechanisms that occur a little later in development can also cause identical twins.

Occasionally, a woman ovulates two or more secondary oocytes at the same time. Fertilization of two oocytes by different sperm cells results in "fraternal twins," or **dizygotic twins.** Dizygotic twins are not genetically identical. Multiple ovulations can occur naturally or result from the injection of drugs that stimulate gonadotropin release. These drugs are used to treat certain forms of infertility, and they sometimes result in multiple fetuses.

The pluripotent cells of the inner cell mass do not form all the tissues necessary for normal development. The placenta and the membranes (chorion and amnion) surrounding the embryo form from the trophoblast.

Implantation of the Blastocyst and Development of the Placenta

All the events of the early germinal phase, from the first cell division through formation of the blastocele and the inner cell mass, occur as the embryo moves from the site of fertilization in the ampulla of the uterine tube to the site of **implantation** (im-plan-tā′shun), which is the burrowing of the blastocyst into the uterine wall. About 7 days after fertilization, the blastocyst implants into the uterine wall, usually in the area of the uterine fundus.

As the blastocyst invades the uterine wall, two populations of trophoblast cells develop and form the embryonic portion of the **placenta** (figure 29.3), the organ of nutrient and waste exchange

Clinical
IMPACT

Stem Cell Research

During growth and development, many cells differentiate for a particular function, and many lose the ability to divide. However, some cells do not become fully specialized and retain the ability to undergo mitosis and differentiation. Such cells are called **stem cells.** Currently, stem cells are a hot topic for discussion, not only in the scientific community but also in the political arena. Stem cells have the potential to treat many diseases by replacing dysfunctional cells with normal cells. For example, bone marrow transplants have become a common treatment for certain types of leukemia. In this procedure, blood stem cells, found in the red bone marrow, are harvested from a donor and introduced into a leukemia patient. The stem cells provide a new source of normal blood cells for that individual. Investigations as to whether stem cells can be isolated and used to treat other diseases are underway.

Stem cells can be obtained from different sources, including adult tissues. Adult stem cells, however, produce only a limited number of cell types, depending on the tissue source of the stem cell. For example, adult liver stem cells can give rise to many cell types found in the liver, but they cannot give rise to other tissue cell types, such as blood cells. The most versatile stem cell is the totipotent zygote because it can give rise to all cell types, including placental tissue.

Also very versatile are the pluripotent cells of the inner cell mass, called **embryonic stem cells,** because they can give rise to all tissues, but they do not produce placental tissue. Embryonic stem cells were first isolated in 1998. Theoretically, embryonic stem cells have great potential for treating diseases. One problem, however, lies in the ability to artificially stimulate embryonic stem cells to differentiate into the right type of cell. Scientists are still in the early stages of determining the biochemical processes of how embryonic stem cells communicate to allow for the proper development of all tissues in multicellular organisms. Ideally, scientists would prefer to limit the number of cell types a particular stem cell can give rise to; otherwise, when these stem cells are grown in the lab, further work must be completed to isolate the correct cell type for a given treatment procedure.

Adult stem cells have had limited success in treating certain diseases. For example, stem cells obtained from adult blood can treat the complications associated with coronary heart disease, particularly the growth of new vessels around areas of blockage. However, it appears that adult stem cells are not effective for treating other diseases, such as type 1 diabetes mellitus and Parkinson disease. In these cases, embryonic stem cells may have a greater potential for providing a long-lasting cure.

Important ethical questions surround the use of embryonic stem cells. To harvest the embryonic cells, an embryo must be destroyed, which is unacceptable to opponents of the procedure, who argue that the zygote is a living human whose cells should not be destroyed. They believe that human life extends into the earliest stages of development. Conversely, proponents of embryonic stem cell use maintain that the definition of a human life does not extend to the zygote and that preserving the sanctity and quality of human life for people suffering from debilitating, often fatal, and currently incurable diseases is worth the price. Furthermore, proponents argue that the zygotes providing the embryonic stem cells were produced for some other reason, such as infertility treatment, and will more than likely be destroyed.

The debate will probably not end soon. Current research is attempting to obtain embryonic stem cells without needing to destroy embryos. Indeed, the recent development of **induced pluripotent stem (iPS)** cells has raised the exciting prospect that fully differentiated adult cells—for example, skin cells—can be reprogrammed to create stem cells similar to embryonic stem cells. Alternatively, the possibility of bypassing iPS cells by directly transdifferentiating, or reprogramming, readily accessible adult skin cells to generate specific cell types, such as neurons, affected by a disease may soon be possible. Whether iPS or transdifferentiated cells will be useful in treatment remains to be established, although in the near future the ability to create patient-specific cells will provide scientists with useful tools for researching disease mechanisms and possibly for screening new drugs.

between the embryo and the mother. The first proliferating population of individual trophoblast cells is called the **cytotrophoblast** (sī-tō-trof′ō-blast). The other trophoblast population is a nondividing syncytium, or multinucleated cell, called the **syncytiotrophoblast** (sin-sish′ē-ō-trō′fō-blast). The cytotrophoblast remains nearer the other embryonic tissues, and the syncytiotrophoblast invades the endometrium of the uterus. The syncytiotrophoblast is nonantigenic, meaning that, as it invades the maternal tissue, no immune reaction is triggered.

As the syncytiotrophoblast encounters maternal blood vessels, it surrounds them and digests the vessel wall, forming pools of maternal blood within cavities called **lacunae** (lă-koo′nē; figure 29.3, *step 3*). The lacunae are still connected to intact maternal vessels, so that blood circulates from the maternal vessels through the lacunae. Cords of cytotrophoblast surround the syncytiotrophoblast and lacunae (figure 29.3, *step 4*).

Branches, called **chorionic** (kō-rē-on′ik) **villi,** sprout from the cords of cytotrophoblasts and protrude into the lacunae, like fingers (figure 29.4*a*). The entire embryonic structure facing the maternal tissues is called the **chorion** (kō′rē-on). Embryonic mesoderm and blood vessels grow into the cords and villi as they protrude into the lacunae. In the mature placenta, the cytotrophoblast disappears, so that the embryonic blood supply is separated from the maternal blood supply by only the embryonic capillary wall, a basement membrane, and a thin layer of syncytiotrophoblast (figure 29.4*b*).

The site of implantation and the integrity of the placental attachment are important for preventing complications in a pregnancy. If the blastocyst implants near the cervix, a condition called **placenta previa** (prē′vē-ă) occurs. As the placenta grows, it may extend partially or completely across the internal cervical opening. As the fetus and placenta continue to grow and the uterus stretches, the region of the placenta over the cervical opening may tear, and hemorrhaging may occur. A second condition, **abruptio** (ab-rŭp′shē-ō) **placentae,** occurs when a normally positioned placenta tears away from the uterine wall. This condition is also accompanied by hemorrhaging. Both conditions can result in miscarriage and threaten the life of the mother.

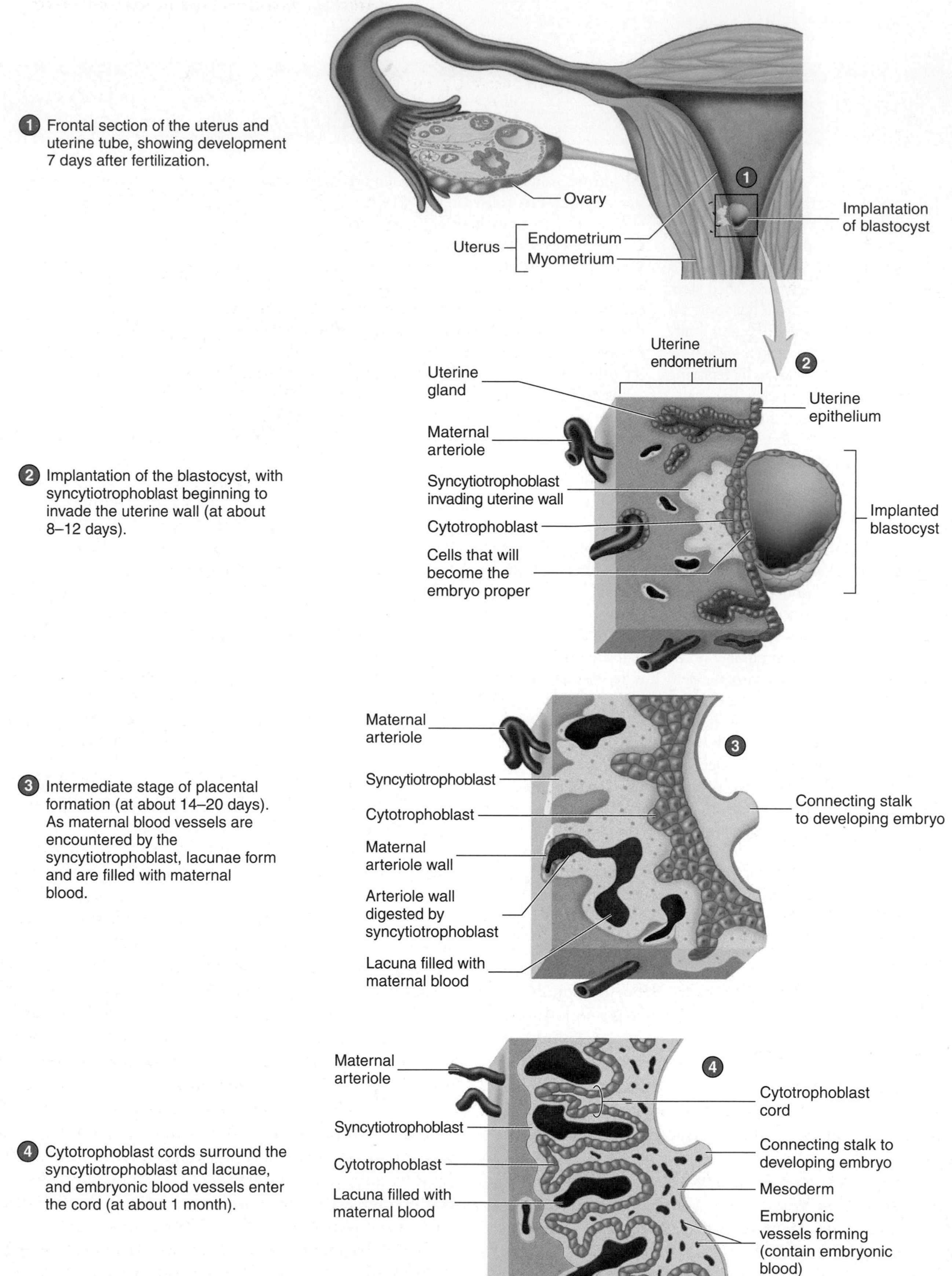

1 Frontal section of the uterus and uterine tube, showing development 7 days after fertilization.
Ovary
Uterus
Endometrium
Myometrium
Implantation of blastocyst
2 Implantation of the blastocyst, with syncytiotrophoblast beginning to invade the uterine wall (at about 8–12 days).
Uterine endometrium
Uterine gland
Uterine epithelium
Maternal arteriole
Syncytiotrophoblast invading uterine wall
Cytotrophoblast
Cells that will become the embryo proper
Implanted blastocyst
3 Intermediate stage of placental formation (at about 14–20 days). As maternal blood vessels are encountered by the syncytiotrophoblast, lacunae form and are filled with maternal blood.
Maternal arteriole
Syncytiotrophoblast
Cytotrophoblast
Maternal arteriole wall
Arteriole wall digested by syncytiotrophoblast
Lacuna filled with maternal blood
Connecting stalk to developing embryo
4 Cytotrophoblast cords surround the syncytiotrophoblast and lacunae, and embryonic blood vessels enter the cord (at about 1 month).
Maternal arteriole
Syncytiotrophoblast
Cytotrophoblast
Lacuna filled with maternal blood
Cytotrophoblast cord
Connecting stalk to developing embryo
Mesoderm
Embryonic vessels forming (contain embryonic blood)

PROCESS FIGURE 29.3 Implantation of the Blastocyst and Formation of the Placenta
As the blastocyst is implanted, syncytiotrophoblasts and cytotrophoblasts invade to form the placenta.

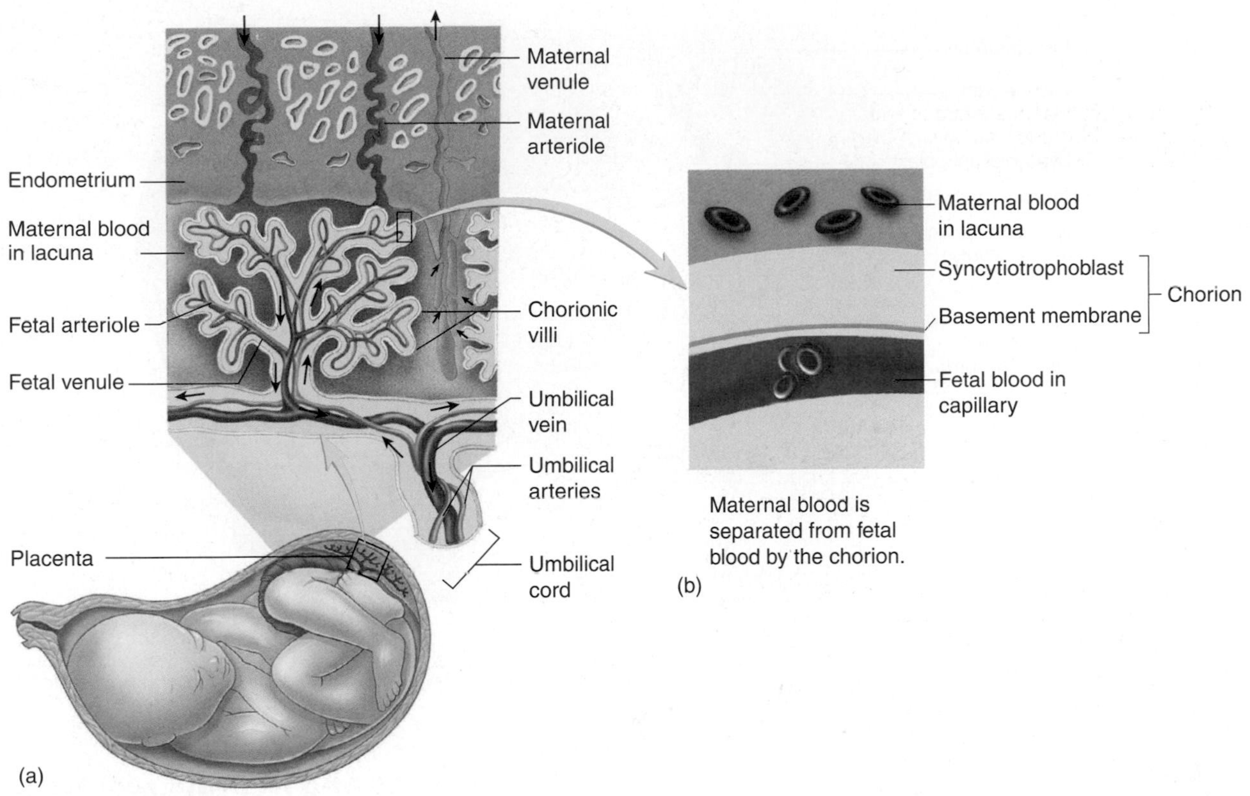
Maternal venule
Maternal arteriole
Endometrium
Maternal blood in lacuna
Fetal arteriole
Fetal venule
Chorionic villi
Umbilical vein
Umbilical arteries
Placenta
Umbilical cord
(a)
Maternal blood in lacuna
Syncytiotrophoblast
Basement membrane
Chorion
Fetal blood in capillary
Maternal blood is separated from fetal blood by the chorion.
(b)

FIGURE 29.4 Mature Placenta and Fetus
Fetal blood vessels and maternal blood vessels are in close contact, and nutrients are exchanged between fetal and maternal blood. However, fetal and maternal blood do not mix.

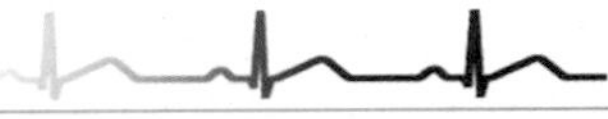

ASSESS YOUR PROGRESS

4. *Describe the events of fertilization. Where does fertilization occur?*
5. *Explain what occurs to prevent polyspermy.*
6. *What events occur during the first week after fertilization? Use the terms zygote, morula, blastocyste, blastocele, totipotent, and pluripotent in your explanation.*
7. *Describe the trophoblast and inner cell mass, and explain what develops from each.*
8. *Explain the process of implantation and the development of the placenta.*

Formation of the Germ Layers

After implantation, an **amniotic** (am-nē-ot′ik) **cavity** forms inside the inner cell mass and is surrounded by a layer of cells called the **amnion** (am′nē-on), or *amniotic sac.* Formation of the amniotic cavity causes part of the inner cell mass nearest the blastocele to separate as a flat disk of tissue called the **embryonic disk** (figure 29.5). This embryonic disk is composed of two layers of cells: an **epiblast** adjacent to the amniotic cavity and a **hypoblast** on the side of the disk opposite the amnion. The epiblast gives rise to the three germ layers, whereas the hypoblast gives rise to extraembryonic membranes. A third cavity, the **yolk sac,** forms inside the blastocele from the hypoblast. The amniotic sac and the yolk sac can be compared to two balloons pushed together, with the circular double layer where the two balloons meet representing the embryonic disk. The amniotic sac eventually enlarges to surround the developing embryo, providing a protective fluid environment, the "bag of waters," where the embryo forms.

About 13 or 14 days after fertilization, the embryonic disk becomes a slightly elongated oval structure. This phase of development, known as **gastrulation,** involves the movement of epiblast cells and results in the formation of three distinct germ layers that eventually give rise to the many body structures. Proliferating cells of the epiblast migrate toward the center and the caudal end of the disk, forming a thickened line called the **primitive streak.** Some epiblast cells migrate through the primitive streak. Some of these cells migrate toward and displace the hypoblast to form the **endoderm** (en′dō-derm; inside layer). Other cells emerge between the epiblast and the endoderm as the **mesoderm** (mez′o-derm; middle layer; figure 29.6). Those epiblast cells that do not migrate form the **ectoderm** (ek′tō-derm; outside layer). These three germ layers, the endoderm, mesoderm, and ectoderm, are the beginning of the embryo proper and all tissues of the adult can be traced to them (table 29.1). A cordlike structure called the **notochord** extends from the cephalic end of the primitive streak.

The development of the germ layers and the subsequent development of the organ systems is heavily dependent on cell communication. Some communication requires direct cell-to-cell contact, whereas other communication depends on diffusable molecules, such as **growth factors.** Two important families of growth factors are epidermal growth factors (EGF) and fibroblast growth factors (FGF).

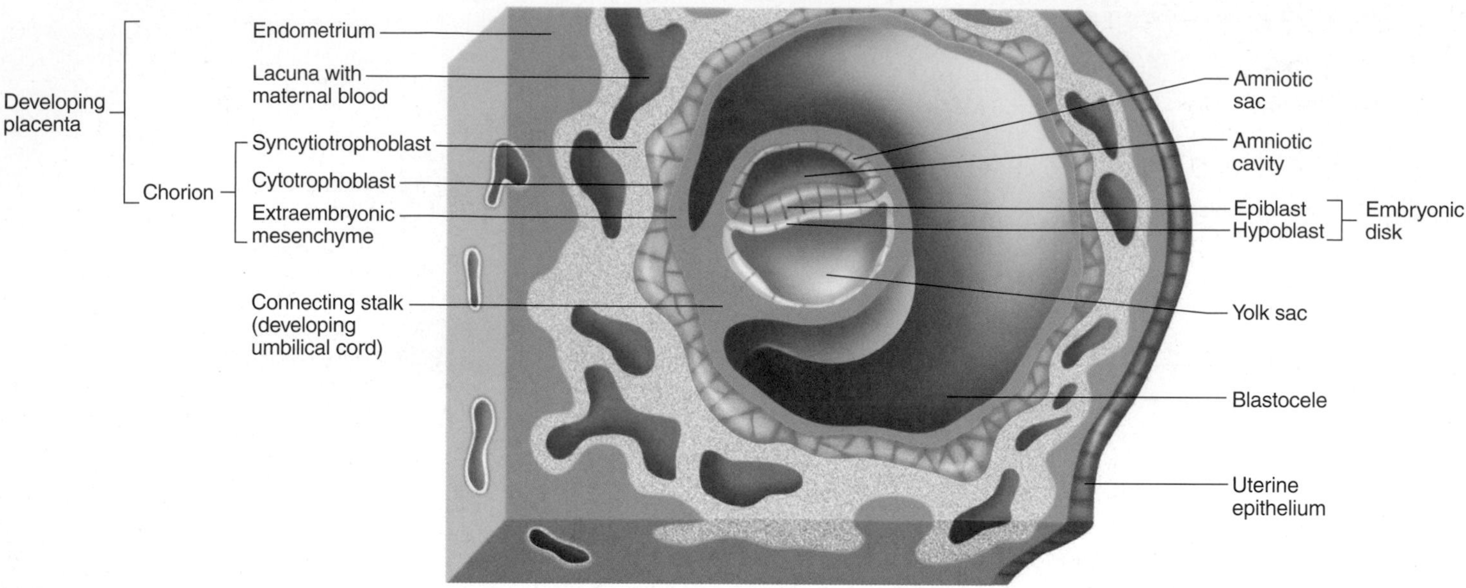
Developing placenta
Endometrium
Lacuna with maternal blood
Chorion
Syncytiotrophoblast
Cytotrophoblast
Extraembryonic mesenchyme
Connecting stalk (developing umbilical cord)
Amniotic sac
Amniotic cavity
Epiblast
Hypoblast
Embryonic disk
Yolk sac
Blastocele
Uterine epithelium

FIGURE 29.5 Embryonic Disk

The embryonic disk consists of epiblast (blue) and hypoblast (yellow), surrounded by the amniotic cavity and yolk sac. The connecting stalk, which attaches the embryo to the uterus, becomes part of the umbilical cord.

Predict 2

Occasionally, two primitive streaks form in one embryonic disk. Predict the result. What happens if the two primitive streaks are touching each other?

Neural Tube and Neural Crest Formation

The ectoderm near the cephalic end of the primitive streak is stimulated about 18 days after fertilization to form a thickened **neural plate.** The lateral edges of the plate begin to rise, like two ocean waves

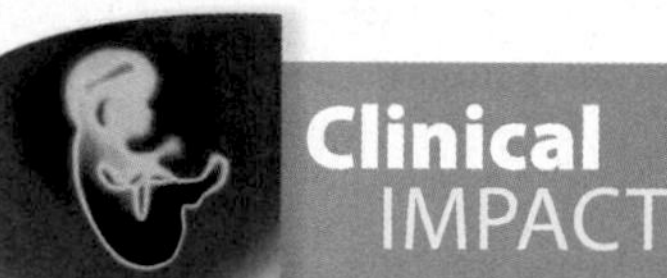

Clinical IMPACT — In Vitro Fertilization and Embryo Transfer

In a small number of women, normal pregnancy is not possible because of an anatomical or physiological condition. In 87% of these cases, the uterine tubes are incapable of allowing sperm cells to reach the oocyte or transporting the zygote to the uterus. Since 1978, two techniques—in vitro fertilization and embryo transfer—have made pregnancy possible in hundreds of such women. **In vitro fertilization (IVF)** involves removing secondary oocytes from a woman, placing the oocytes into a petri dish, and adding sperm cells to the dish, where fertilization and early development occur in vitro, which means "in glass." **Embryo transfer** involves removing the developing embryo from the petri dish and introducing the embryo into the uterus of a recipient woman.

For IVF and embryo transfer to be accomplished, a woman is first injected with a substance similar to luteinizing hormone (LH), which causes more than one follicle to ovulate at a time. Just before the follicles rupture, the secondary oocytes are surgically removed from the ovary. The oocytes are then incubated in a dish and maintained at body temperature for 6 hours when sperm cells are added to the dish. At that time, different techniques may be used to enhance sperm entry into the oocyte.

After 24–48 hours, when the embryos have divided to form two- to eight-cell masses, several of the embryos are transferred to the uterus. Typically, three embryos are introduced into the uterus at a time to increase the success rate, since only a small percentage of them are expected to survive. However, the rate of complications, such as multiple pregnancies, miscarriage, and prematurity, also increases with the greater the number of embryos transferred.

About one-third of transfers of three embryos end in multiple pregnancies. Of triplets born as a result of IVF, 64% require intensive care after birth, and 75% of quadruplets require intensive care, often for several weeks.

Prematurity from IVF pregnancies in the United Kingdom results in newborn mortality in 2.7% of cases, a rate three times that of natural pregnancies. As a result of the possible complications, no more than two or three embryos are transferred per IVF in the United Kingdom. In 2009, the birth of octuplets to a California woman with six other children led to serious discussions of the regulation of embryo transfers in the United States.

The success of IVF is dependent on many factors, including the reason for infertility, the mother's age, and the physician's expertise. The success rate for achieving pregnancy following IVF treatment has increased dramatically in recent years and is currently 31%. Of these pregnancies, 83% result in live births. By comparison, only 50% or fewer natural fertilizations result in successful delivery. The improved success rate is probably due to the fact that abnormal-appearing embryos are not used for IVF and more than one embryo is transferred.

FUNDAMENTAL **Figure**

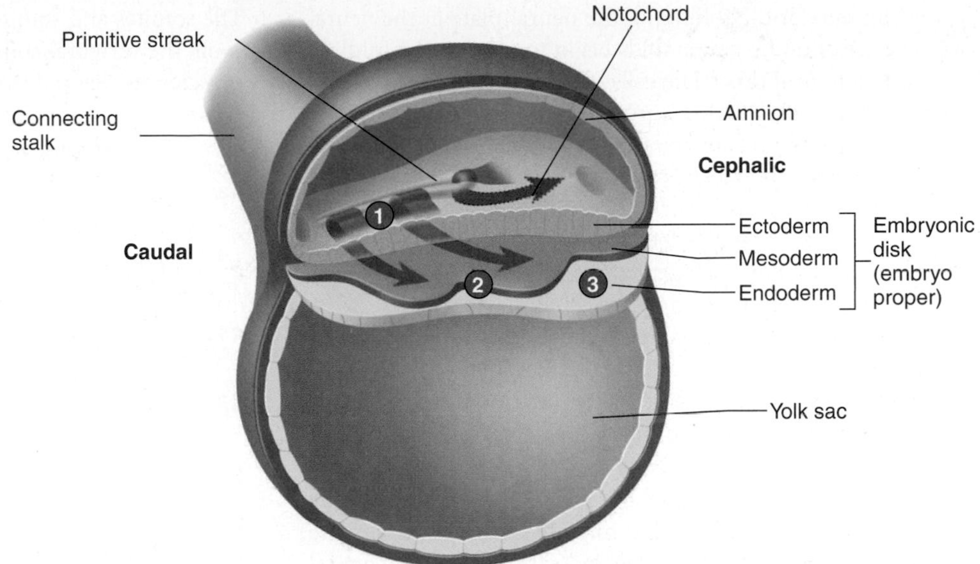

1 Cells in the epiblast move toward the primitive streak and migrate through the streak (*blue arrow tails*).
2 Cells of the epiblast that migrate through the primitive streak become endodermal and mesodermal cells (*red arrows*).
3 The mesoderm (*pink*) lies between the ectoderm (*blue*) and the endoderm (*yellow*).
Notochord
Primitive streak
Connecting stalk
Amnion
Cephalic
Caudal
Ectoderm
Mesoderm
Endoderm
Embryonic disk (embryo proper)
Yolk sac

PROCESS FIGURE 29.6 Embryonic Disk with a Primitive Streak
The head of the embryo will develop over the notochord.

TABLE 29.1 Germ Layer Derivatives

Ectoderm	Mesoderm
Epidermis of skin	Dermis of skin
Tooth enamel	Cardiovascular system
Lens and cornea of eye	Parenchyma of glands
Outer ear	Muscle
Nasal cavity	Bones (except facial)
Anterior pituitary	Microglia
Neuroectoderm	Kidneys
Brain and spinal cord	**Endoderm**
Somatic motor neurons	Lining of digestive tract
Preganglionic autonomic neurons	Lining of lungs
Neuroglia (except microglia)	Lining of hepatic, pancreatic, and other exocrine ducts
Posterior pituitary	Urinary bladder
Neural crest cells	Thymus
Melanocytes	Thyroid gland
Sensory neurons	Parathyroid glands
Postganglionic autonomic neurons	Tonsils
Adrenal medulla	
Facial bones	
Teeth (dentin, pulp, and cementum) and gingiva	
A few skeletal muscles in head	

coming together. These edges are called the **neural folds,** and between them lies a **neural groove** (figure 29.7). The underlying notochord stimulates the folding of the neural plate at the neural groove. The crests of the neural folds begin to meet in the midline and fuse into a **neural tube.** The cells of the neural tube are called **neuroectoderm** (table 29.1). Neuroectoderm becomes the brain, the spinal cord, and parts of the peripheral nervous system. The neural tube becomes completely closed by day 26. If the neural tube fails to close, major defects of the central nervous system can result (see Clinical Impact, "Neural Tube Defects," later in this chapter).

As the neural folds come together and fuse, a population of cells breaks away from the neuroectoderm all along the crests of the folds. In the body of the embryo, **neural crest cells** migrate along one of three distinct routes as they leave the neural folds. Those that migrate down along the side of the developing neural tube become autonomic ganglia neurons, adrenal medullary cells, or enteric nervous system neurons. Neural crest cells that migrate into the somites (see the next section, "Somite Formation") become sensory ganglia neurons. Those that migrate laterally between the somites and the ectoderm become melanocytes. In the head, neural crest cells contribute to the skull, the dentin of teeth, a few small skeletal muscles, and general connective tissue. Because neural crest cells in the head give rise to many of the same tissues as the mesoderm in the head and trunk, the general term **mesenchyme** (mez′en-kīm) is sometimes applied to cells of either neural crest or mesoderm origin.

Somite Formation

As the neural tube develops, the mesoderm immediately adjacent to the tube forms distinct segments called **somites** (sō′mītz; figure 29.7). In the head, the first few somites do not become clearly divided, but develop into indistinct, segmented structures called **somitomeres.** The somites and somitomeres eventually give rise to a part of the skull, the vertebral column, and skeletal muscle. Most of the head muscles are derived from the somitomeres.

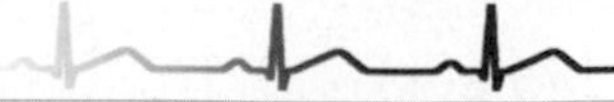

ASSESS YOUR PROGRESS

9. *Describe the process of gastrulation and the role of the primitive streak.*
10. *List the major body tissues, organs, and systems that originate from each of the three germ layers.*
11. *Describe the formation of the neural tube. What are the three routes taken by neural crest cells?*
12. *What is a somite? What do they develop into?*

Formation of the Gut and Body Cavities

At the same time the neural tube is forming, the embryo itself is becoming a tube along the upper part of the yolk sac. The **foregut** and **hindgut** develop as the cephalic and caudal ends of the yolk sac separate from the main yolk sac. This is the beginning of the digestive tract (figure 29.8*a*). The developing digestive tract pinches off from the yolk sac as a tube but remains attached to the yolk sac by a yolk stalk.

The ends of the foregut and hindgut (figure 29.8*b*) are in close relationship to the overlying ectoderm and form membranes called the oropharyngeal membrane and the cloacal membrane, respectively. The **oropharyngeal membrane** opens to form the mouth, and the **cloacal membrane** opens to form the urethra and anus. Thus, the digestive tract becomes a tube with openings at both ends.

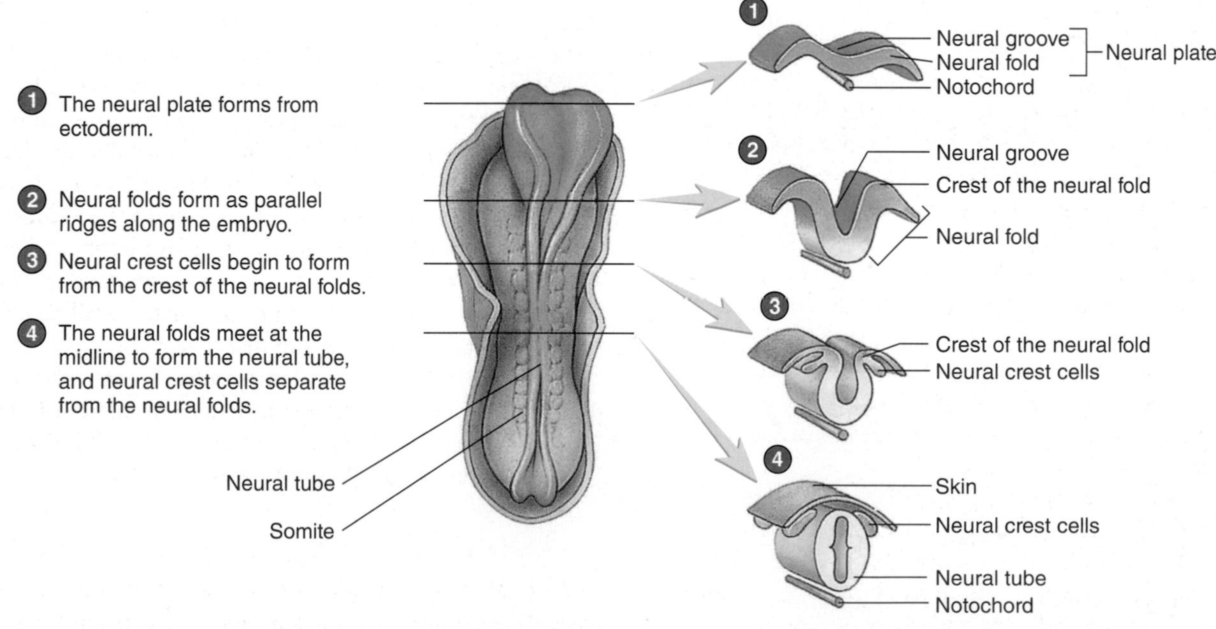
1 The neural plate forms from ectoderm.
2 Neural folds form as parallel ridges along the embryo.
3 Neural crest cells begin to form from the crest of the neural folds.
4 The neural folds meet at the midline to form the neural tube, and neural crest cells separate from the neural folds.
Neural tube
Somite
1
Neural groove
Neural fold
Neural plate
Notochord
2
Neural groove
Crest of the neural fold
Neural fold
3
Crest of the neural fold
Neural crest cells
4
Skin
Neural crest cells
Neural tube
Notochord

PROCESS FIGURE 29.7 Formation of the Neural Tube
The neural folds come together in the midline and fuse to form a neural tube. This fusion begins in the center and moves both cranially and caudally. Shown is an embryo at about 21 days after fertilization. The insets to the right show progressive closure of the neural tube at various levels of cross section.

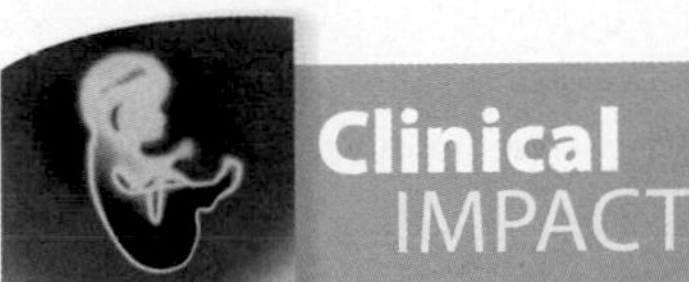
Clinical
IMPACT

Environmental Influences on Development

During the first 2 weeks of development, the embryo is quite resistant to environmental influences that may cause malformations. Factors that adversely affect the embryo at this age are more likely to kill it than cause malformations. However, between 2 weeks and the next 4–7 weeks (depending on the structure considered), the embryo is more sensitive to outside influences that cause malformations than at any other time.

A number of drugs and other environmental influences are known to affect the embryo and fetus during development. The two most common are alcohol and cigarette smoke. Alcohol consumption can result in **fetal alcohol syndrome,** which is primarily characterized by decreased mental function. Although excessive alcohol consumption, as occurs with alcoholism and binge drinking, is known to cause fetal alcohol syndrome, research results are inconsistent about the effects of lower levels of consumption. Most physicians recommend that alcohol not be consumed at all during pregnancy. Studies have shown that exposure to cigarette smoke throughout pregnancy can stunt the physical growth and mental development of the fetus.

Isotretinoin, more commonly known as Accutane®, is a drug used to treat severe acne; however, it has been shown to cause severe birth defects, including malformation of the ears, eyes, face, skull, heart, and brain. The risk of detrimental effects on the developing embryo and fetus is so great that the manufacturer of Accutane® requires both medical professionals who prescribe Accutane and their patients to participate in a pregnancy risk management program in order to obtain the medication. The program requires that woman prescribed Accutane® agree to use two forms of birth control and take monthly pregnancy tests while using the drug.

A considerable number of outpocketings, or **evaginations** (ē-vaj-i-nā′shŭnz), occur along the early digestive tract. The first to form is the allantois (figure 29.8*b*), part of which will form the urinary bladder. Other evaginations develop into structures such as the anterior pituitary, the thyroid gland, the lungs, the liver, the pancreas, and the urinary bladder. At the same time, solid bars of tissue known as **branchial arches** (figure 29.8*c*; see figure 29.9) form along the lateral sides of the head. The sides of the foregut expand as pockets between these branchial arches. The central expanded foregut is called the **pharynx,** and the pockets along both sides of the pharynx are called **pharyngeal pouches.** Adult derivatives of the pharyngeal pouches include the auditory tubes, tonsils, thymus, and parathyroids.

At about the same time the gut is developing, a series of isolated cavities starts to form within the embryo, thus beginning development of the **coelom** (sē′lom; figure 29.8), or body cavities. The most cranial group of cavities enlarges and fuses to form the **pericardial cavity.** Shortly thereafter, the coelomic cavity extends toward the caudal end of the embryo as the **pleural cavity** and the **peritoneal cavity.** Initially, these three cavities are continuous, but they eventually separate into three distinct adult cavities (see chapter 1).

Limb Bud Development

Arms and legs first appear as limb buds at about 28 days (figure 29.9). The **apical ectodermal ridge,** a specialized thickening of the ectoderm, develops on the lateral margin of each limb bud and stimulates its outgrowth. As the buds elongate, limb tissues are laid down in a proximal-to-distal sequence. For example, in the upper limb, the arm forms before the forearm, which in turn forms before the hand.

Development of the Face

The face develops by fusion of five embryonic structures: the **frontonasal process,** which forms the forehead, nose, and midportion of the upper jaw and lip; two **maxillary processes,** which form the lateral parts of the upper jaw and lip; and two **mandibular processes,** which form the lower jaw and lip (figure 29.10, *step 1*).

Nasal placodes (plak′ōdz), areas of thickening, which develop at the lateral margins of the frontonasal process, develop into the nose and the center of the upper jaw and lip (figure 29.10, *step 2*). As the brain enlarges and the face matures, the nasal placodes approach each other in the midline. The medial edges of the placodes fuse to form the midportion of the upper jaw and lip (figure 29.10, *steps 3 and 4*). This part of the frontonasal process is between the two maxillary processes, which are expanding toward the midline, and fuses with them to form the upper jaw and lip, known as the **primary palate** (figure 29.10, *step 4*). When the frontonasal and one or both maxillary processes fail to fuse, a **cleft lip** results. Because the frontonasal process is a midline structure that normally fuses with the two lateral maxillary processes during formation of the primary palate, cleft lips usually do not occur in the midline but to one side (or both sides). The cleft can vary in severity from a slight indentation in the lip to a fissure that extends from the mouth to the nares (nostril).

At about the same time the primary palate is forming, the lateral edges of the nasal placodes fuse with the maxillary processes to close off the groove extending from the mouth to the eye (figure 29.10, *steps 4 and 5*). On rare occasions, these structures fail to meet, resulting in a facial cleft extending from the mouth to the eye. The inferior margins of the maxillary processes fuse with the superior margins of the mandibular processes to decrease the size of the mouth.

All of the previously described fusions and the growth of the brain give the face a recognizably "human" appearance by about 50 days. The roof of the mouth, known as the **secondary palate,** begins as vertical shelves, which swing to a horizontal position and begin to fuse with each other at about 56 days of development. Fusion of the entire palate is not completed until about 90 days. If the secondary palate does not fuse, a midline fissure in the roof of the mouth, called a **cleft palate,** results. A cleft palate can range in severity from a slight cleft of the uvula (see figure 24.6) to a fissure extending the entire length of the palate. A cleft lip and cleft palate can occur together, forming a continuous fissure.

ASSESS YOUR PROGRESS

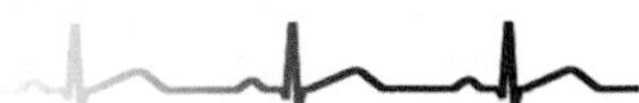

13. *Describe the formation of the gut and the body cavities.*
14. *How do limb buds develop? What does proximal-to-distal growth sequence mean?*
15. *Describe the process involved in forming the face. What clefts may result if these tissues fail to fuse?*

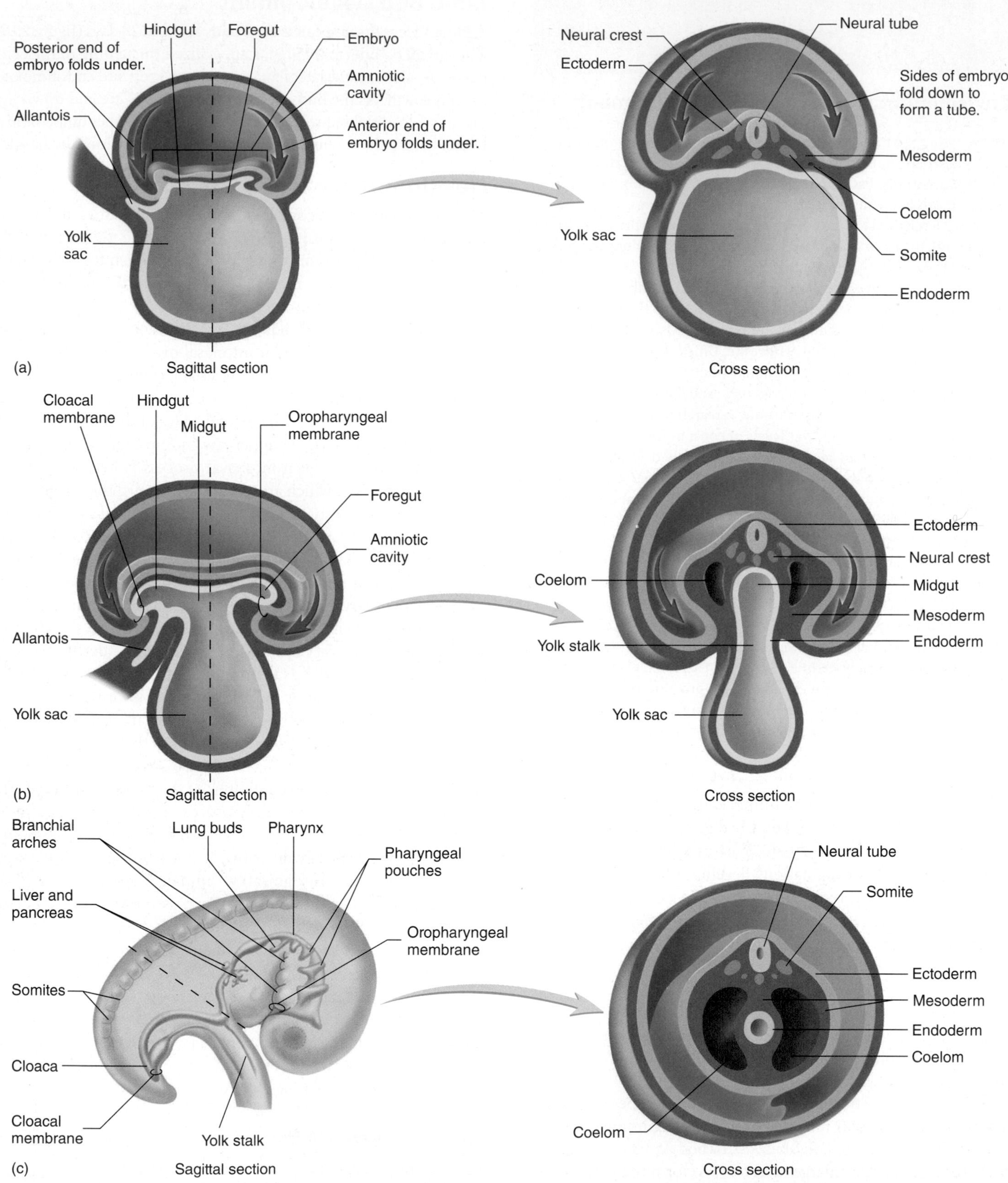

Hindgut
Foregut
Posterior end of embryo folds under.
Embryo
Amniotic cavity
Allantois
Anterior end of embryo folds under.
Yolk sac
(a)
Sagittal section
Neural crest
Neural tube
Ectoderm
Sides of embryo fold down to form a tube.
Mesoderm
Coelom
Yolk sac
Somite
Endoderm
Cross section
Cloacal membrane
Hindgut
Midgut
Oropharyngeal membrane
Foregut
Amniotic cavity
Allantois
Yolk sac
(b)
Sagittal section
Ectoderm
Neural crest
Coelom
Midgut
Mesoderm
Yolk stalk
Endoderm
Yolk sac
Cross section
Branchial arches
Lung buds
Pharynx
Pharyngeal pouches
Liver and pancreas
Oropharyngeal membrane
Somites
Cloaca
Cloacal membrane
Yolk stalk
(c)
Sagittal section
Neural tube
Somite
Ectoderm
Mesoderm
Endoderm
Coelom
Coelom
Cross section

FIGURE 29.8 Formation of the Digestive Tract

Blue arrows show the folding of the digestive tract into a tube. Dashed lines show the plane from which the cross sections were taken. (*a*) Twenty days after fertilization. (*b*) Twenty-five days after fertilization. (*c*) Thirty days after fertilization. Evaginations are identified along the pharynx and the liver and pancreas.

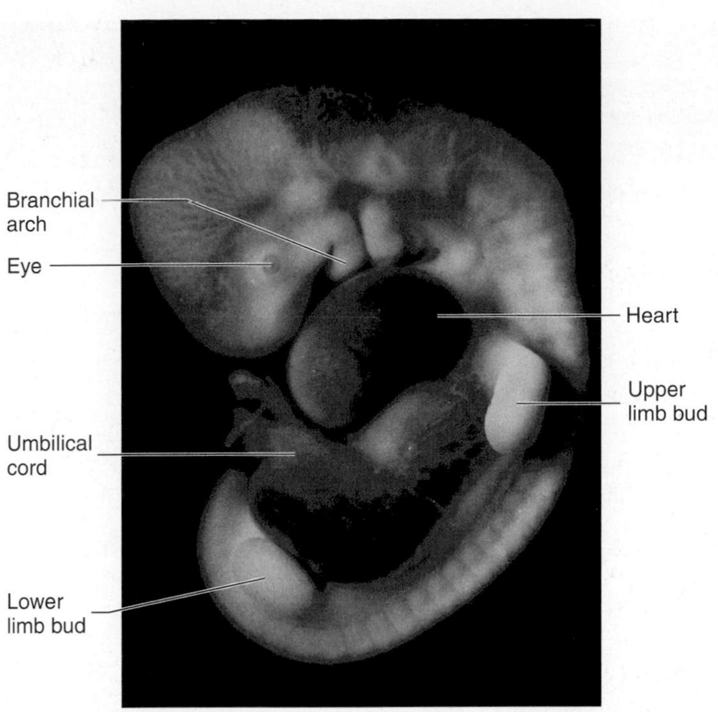
Branchial arch
Eye
Heart
Upper limb bud
Umbilical cord
Lower limb bud

FIGURE 29.9 Human Embryo 35 Days After Fertilization

Development of the Organ Systems

The major organ systems appear and begin to develop during the embryonic period. The period between 14 and 60 days is therefore called the period of **organogenesis** (table 29.2).

Skin

The **epidermis** of the skin is derived from ectoderm, and the **dermis** is derived from mesoderm or, in the case, of the face, from neural crest cells. Nails, hair, and glands develop from the epidermis (see chapter 5). Melanocytes and sensory receptors in the skin are derived from neural crest cells.

Skeleton

The skeleton develops from either mesoderm or the neural crest cells through intramembranous or endochondral bone formation (see chapter 6). The bones of the face develop from neural crest cells, whereas the rest of the skull, the vertebral column, and the ribs develop from somite- or somitomere-derived mesoderm. The appendicular skeleton develops from limb bud mesoderm.

Muscle

Myoblasts (mī′ō-blastz) are the early embryonic cells that give rise to skeletal muscle fibers. Myoblasts migrate from somites or somitomeres to sites of future muscle development, where they continue to divide and begin to fuse to form multinuclear cells called **myotubes.** Myotubes enlarge to become the muscle fibers of the skeletal muscles. Shortly after myotubes form, nerves grow into the area and innervate the developing muscle fibers. After the

1 **28 days after fertilization**
The face develops from five processes: frontonasal (*blue*), two maxillary (*yellow*), and two mandibular (*orange;* already fused).

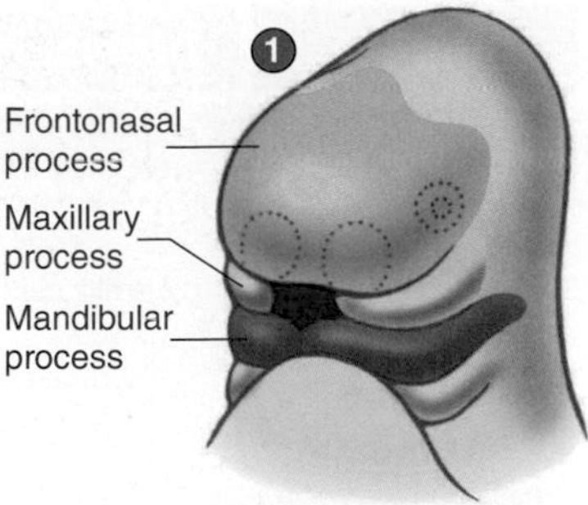
1
Frontonasal process
Maxillary process
Mandibular process

2 **33 days after fertilization**
Nasal placodes appear in the frontonasal process.

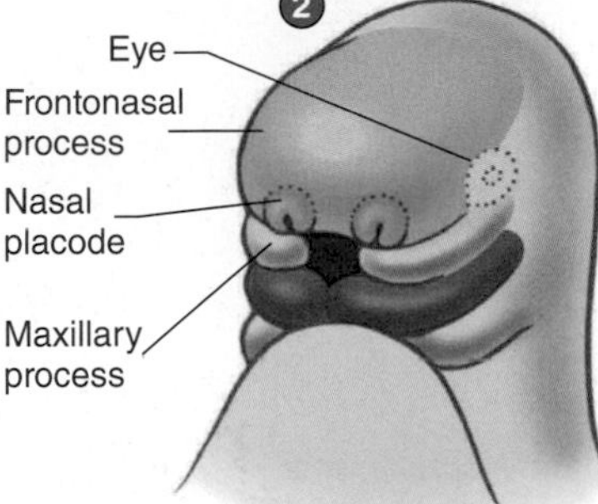
2
Eye
Frontonasal process
Nasal placode
Maxillary process

3 **40 days after fertilization**
Maxillary processes extend toward the midline. The nasal placodes also move toward the midline and fuse with the maxillary processes to form the jaw and lip.

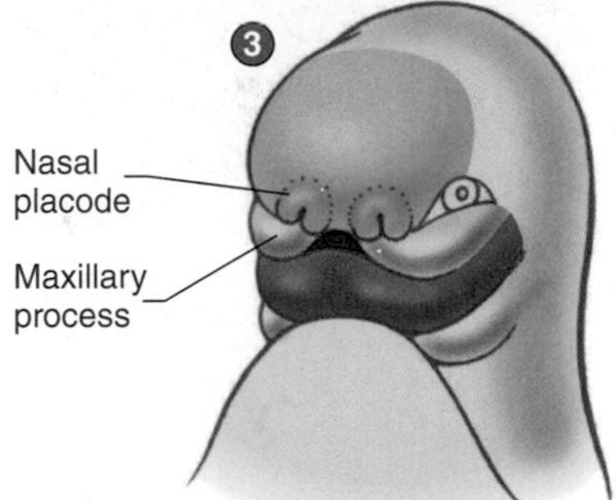
3
Nasal placode
Maxillary process

4 **48 days after fertilization**
Continued growth brings structures more toward the midline.

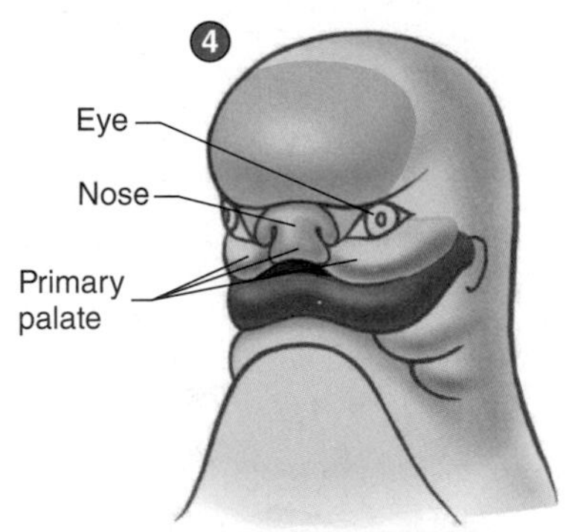
4
Eye
Nose
Primary palate

5 **14 weeks after fertilization**
Colors show the contributions of each process to the adult face.

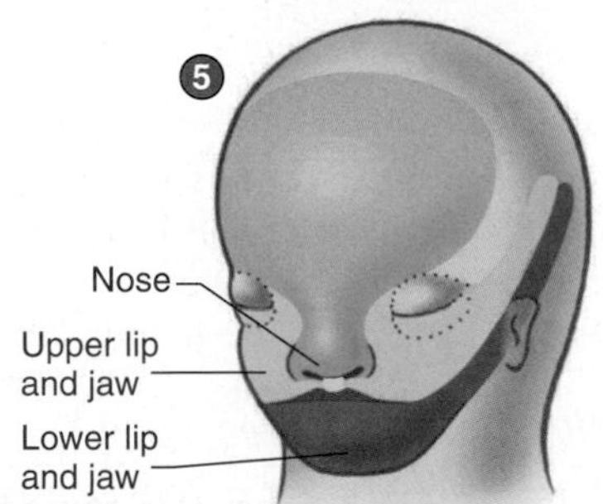
5
Nose
Upper lip and jaw
Lower lip and jaw

PROCESS FIGURE 29.10 Development of the Face

TABLE 29.2 Development of the Organ Systems

	Age (Days Since Fertilization)					
	1–5	6–10	11–15	16–20	21–25	26–30
General Features	Fertilization, morula, blastocyst	Blastocyst implants.	Primitive streak, three germ layers	Neural plate	Neural tube closed	Limb buds and other "buds" appear.
Integumentary System			Ectoderm, mesoderm			Melanocytes form from neural crest.
Skeletal System			Mesoderm		Neural crest (will form facial bones)	Limb buds
Muscular System			Mesoderm	Somites begin to form.		Somites are all present.
Nervous System			Ectoderm	Neural plate	Neural tube complete; neural crest forms; eyes and ears begin to form	Lens begins to form.
Endocrine System			Ectoderm, mesoderm, endoderm	Thyroid begins to develop.		Parathyroid glands and pancreas appear.
Cardiovascular System			Mesoderm	Blood islands form; two-tubed heart forms.	Single-tubed heart begins to beat.	Interatrial septum begins to form.
Lymphatic System			Mesoderm			Thymus appears.
Respiratory System			Mesoderm, endoderm		Diaphragm begins to form.	Trachea forms as single bud; lung buds (primary bronchi) form.
Digestive System			Mesoderm		Neural crest forms tooth dentin.	Liver and pancreas appear as buds.
			Endoderm		Foregut and hindgut form.	Tongue bud appears.
Urinary System			Mesoderm, endoderm		Pronephros develops; allantois appears.	Mesonephros appears.
Reproductive System			Mesoderm, endoderm		Primordial germ cells form on yolk sac.	Mesonephros appears; genital tubercle forms.

basic form of each muscle is established, muscle growth continues as the number of muscle fibers increases. The total number of muscle fibers is established before birth and remains relatively constant thereafter. Muscle enlargement after birth results from an increase in the size of individual fibers.

Nervous System

The nervous system is derived from the neural tube and the neural crest cells. Neural tube closure begins at about 21 days of development in the upper cervical region and proceeds into the head and down the spinal cord. Soon after the neural tube has closed at about 25 days of development, the part of the neural tube that becomes the brain begins to expand and develop a series of pouches (see figure 13.13). The central cavity of the neural tube becomes the ventricles of the brain and the central canal of the spinal cord.

Within the neural tube are the neuron cell bodies of somatic motor neurons and preganglionic neurons of the autonomic nervous system, which provide axons to the peripheral nervous system. Sensory neurons and postganglionic neurons of the autonomic nervous system are derived from neural crest cells.

Age (Days Since Fertilization)					
31–35	**36–40**	**41–45**	**46–50**	**51–55**	**56–60**
Hand and foot plates on limbs	Fingers and toes appear; lips form; embryo is 15 mm long.	External ear is forming; embryo is 20 mm long.	Embryo is 25 mm long.	Limbs elongate to adult proportions; embryo is 35 mm long.	Face is distinctly human in appearance.
Sensory receptors appear in skin.		Collagen fibers are clearly present in skin.		Extensive sensory endings are present in skin.	
Mesoderm condenses in areas of future bone.	Cartilage in site of future humerus	Cartilage in site of future ulna and radius	Cartilage in site of hand and fingers		Ossification begins in clavicle and then in other bones.
Muscle precursor cells enter limb buds.			Functional muscle		Nearly all muscles are appearing in an adult form.
Nerve processes enter limb buds.		External ear is forming; olfactory nerves begin to form.		Semicircular canals in inner ear are complete.	Eyelids form; cochlea in inner ear is complete.
Pituitary gland appears as evaginations from brain and mouth.	Gonadal ridges form; adrenal glands are forming.		Pineal gland appears.	Thyroid gland is in adult position, and its attachment to tongue is lost.	Anterior pituitary loses its connection to mouth.
Interventricular septum begins to form.		Interventricular septum is complete.	Interatrial septum is complete but still has opening until birth.		
Large lymphatic vessels form in neck.	Spleen appears.			Adult lymph pattern forms.	
Secondary bronchi to lobes form.	Tertiary bronchi to bronchopulmonary segments form.		Tracheal cartilage begins to form.		
Oropharyngeal membrane ruptures.		Secondary palate begins to form; tooth buds begin to form.			Secondary palate begins to fuse (fusion complete by 90 days).
Metanephros begins to develop.				Mesonephros degenerates.	Anal portion of cloacal membrane ruptures.
	Gonadal ridges form.	Primordial germ cells enter gonadal ridges.	Paramesonephric ducts appear.		Uterus is forming; external genitalia begin to differentiate in male and female.

Special Senses

The **olfactory bulbs** and **olfactory nerves** develop as an evagination from the telencephalon (see figure 13.15). The eyes develop as evaginations from the diencephalon. Each evagination elongates to form an **optic stalk,** and a bulb called the **optic vesicle** develops at the terminal end of each optic stalk. The optic vesicle reaches the side of the head and stimulates the overlying ectoderm to thicken into a **lens.** The sensory part of the ear appears as an ectodermal thickening, or placode, that invaginates and pinches off from the overlying ectoderm.

Endocrine System

An evagination from the floor of the diencephalon forms the **posterior pituitary gland.** The **anterior pituitary gland** develops from an evagination of ectoderm in the roof of the embryonic oral cavity and grows toward the floor of the brain. It eventually loses its connection with the oral cavity and becomes attached to the posterior pituitary gland (see chapter 18).

The **thyroid gland** originates as an evagination from the floor of the pharynx in the region of the developing tongue and moves into the lower neck, eventually losing its connection with the

Clinical IMPACT

Neural Tube Defects

Proper neural tube formation is necessary for the development of the nervous system; otherwise, several birth defects can occur. **Anencephaly** (an′en-sef′ă-lē; no brain) is a birth defect in which much of the brain fails to form. It results because the neural tube did not close in the region of the head. A baby born with anencephaly cannot survive.

Spina bifida (spī′nă bi′fi-dă; split spine) is a general term describing defects of the spinal cord, vertebral column, or both (figure 29A). Spina bifida can range from a simple defect, in which one or more vertebral spinous processes are split or missing but there is no clinical manifestation, to a severe defect that results in paralysis of the limbs or the bowels and bladder, depending on where the defect occurs.

The inclusion of **folic acid,** the B vitamin folate, in a woman's diet during the early stages of her pregnancy significantly reduces the risk for neural tube defects in her developing embryo.

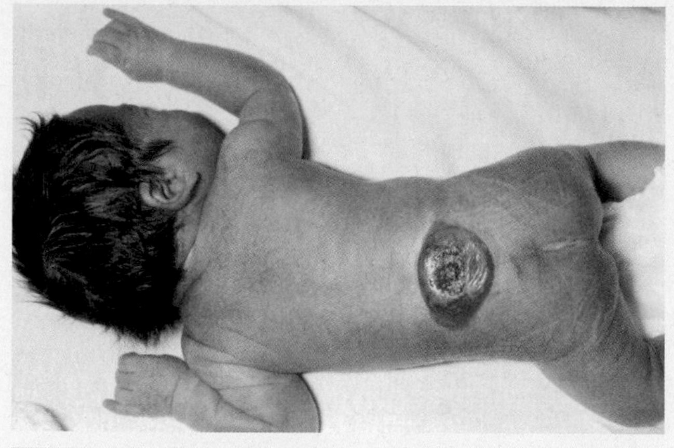

FIGURE 29A Spina Bifida

pharynx. The **parathyroid glands,** which are derived from the third and fourth pharyngeal pouches, migrate inferiorly and become associated with the thyroid gland.

The **adrenal medulla** arises from neural crest cells and consists of specialized postganglionic neurons of the sympathetic division of the autonomic nervous system (see chapter 16). The **adrenal cortex** is derived from mesoderm.

The **pancreas** originates as two evaginations from the duodenum, which come together to form a single gland (see figure 29.8*c*).

Cardiovascular System

The heart develops from two endothelial tubes (figure 29.11, *step 1*), which fuse into a single, midline heart tube (figure 29.11, *step 2*). Blood vessels and blood cells form from blood islands on the surface of the yolk sac and inside the embryo. **Blood islands** are small masses of mesoderm that become blood vessels on the outside and blood cells on the inside. These islands expand and fuse to form the blood vessels.

A series of dilations appears along the length of the primitive heart tube, and four major regions can be identified: the **sinus venosus,** the site where blood enters the heart; a single **atrium;** a single **ventricle;** and the **bulbus cordis,** where blood exits the heart (figure 29.11, *step 2*). The elongating heart, confined within the pericardium, becomes bent into a loop, the apex of which is the ventricle.

The major chambers of the heart, the atrium and the ventricle, expand rapidly. The right part of the sinus venosus becomes absorbed into the atrium, and the bulbus cordis is absorbed into the ventricle. The embryonic sinus venosus initiates contraction at one end of the tubular heart. Later in development, part of the sinus venosus becomes the sinoatrial node, the pacemaker in the adult heart.

The **interatrial septum,** which separates the two atria in the adult heart, is formed from two parts: the **septum primum** (primary septum) and the **septum secundum** (secondary septum). An opening in the interatrial septum called the **foramen ovale** (ō-val′ē) connects the two atria and allows blood to flow from the right to the left atrium in the embryo and fetus. An **interventricular septum** (figure 29.11, *steps 3–5*) develops, which divides the single ventricle into two chambers.

Respiratory System

The lungs develop as a single midline evagination from the foregut in the region of the future esophagus. This evagination branches to form two **lung buds** (figure 29.12, *step 1*). The lung buds elongate and branch, first forming the bronchi that project to the lobes of the lungs (figure 29.12, *step 2*) and then forming the bronchi that project to the bronchopulmonary segments of the lungs (figure 29.12, *step 3*). This branching continues (figure 29.12, *step 4*) until, by the end of the sixth month, about 17 generations of branching have occurred. Even after birth, some branching continues as the lungs grow larger and, in the adult, about 24 generations of branches have been established.

Urinary System

The kidneys develop from mesoderm located between the somites and the lateral part of the embryo. About 21 days after fertilization, the mesoderm in the cervical region differentiates into a structure called the **pronephros** (figure 29.13*a*), which consists of a duct and simple tubules connecting the duct to the open coelomic cavity. This type of kidney is the functional adult kidney in some lower chordates, but it is probably not functional in the human embryo and soon disappears.

The **mesonephros** (figure 29.13*a*) is a functional organ in the embryo. It consists of a caudal extension of the pronephric duct and a number of minute tubules, which are smaller and more complex than those of the pronephros. One end of each tubule opens into the mesonephric duct, and the other end forms a glomerulus (see chapter 26).

As the mesonephros is developing, the caudal end of the hindgut begins to enlarge to form the **cloaca** (klō-ā′kă; sewer), the common junction of the digestive, urinary, and genital systems (figure 29.13*b*). A **urorectal septum** divides the cloaca into two parts: a digestive part called the **rectum** and a urogenital part called the **urethra** (figure 29.13*c*). The cloaca is associated with two tubes:

1 **20 days after fertilization**
At this age, the heart consists of two parallel tubes that will fuse into a single, midline heart.

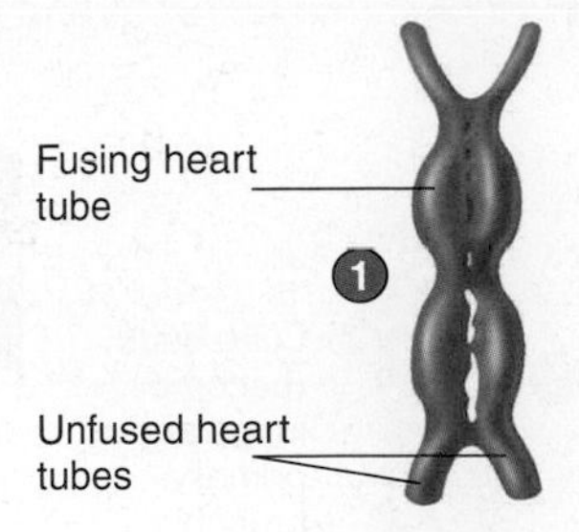
Fusing heart tube
1
Unfused heart tubes

2 **22 days after fertilization**
The two parallel tubes have fused to form one tube. This tube bends as it elongates (*blue arrows* suggest the direction of bending) within the confined space of the pericardium.

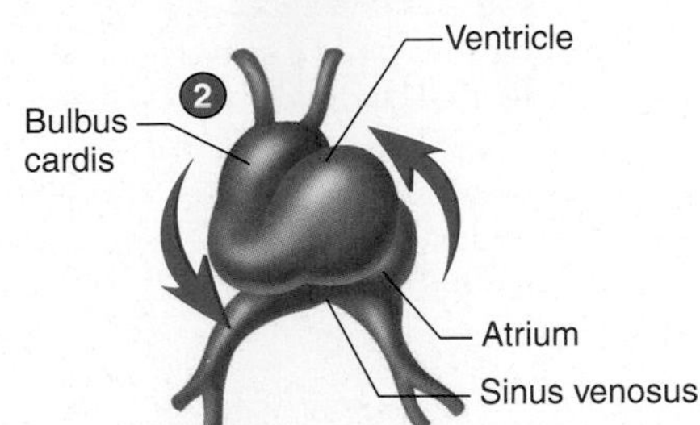
Ventricle
2
Bulbus cardis
Atrium
Sinus venosus

3 **31 days after fertilization**
The interatrial septum (septum primum, *green*) and the interventricular septum grow toward the center of the heart.

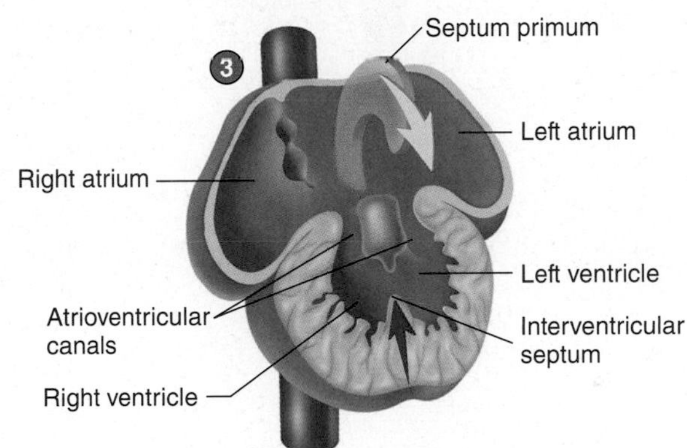
Septum primum
3
Left atrium
Right atrium
Left ventricle
Atrioventricular canals
Interventricular septum
Right ventricle

4 **35 days after fertilization**
The interventricular septum is nearly complete. A foramen opens in the septum primum (*green*) as the septum secundum begins to form (*blue*).

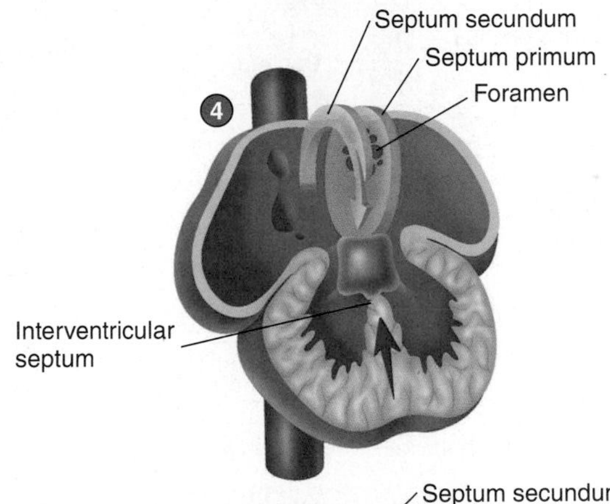
Septum secundum
Septum primum
Foramen
4
Interventricular septum

5 **Final embryonic condition of the interatrial septum**
Blood from the right atrium can flow through the foramen ovale into the left atrium. After birth, as blood begins to flow in the other direction, the septum primum is forced against the septum secundum, closing the foramen ovale.

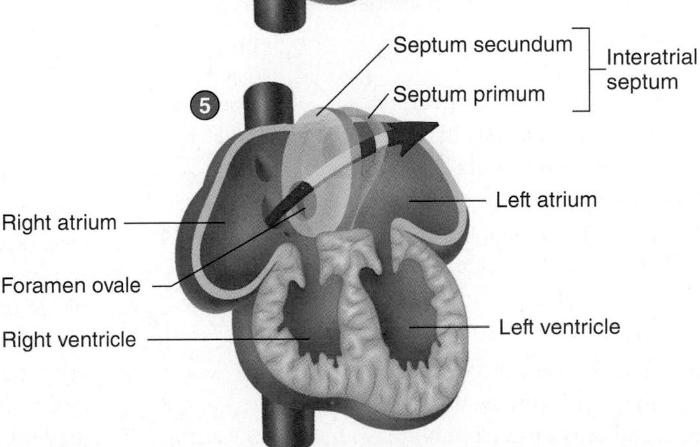
Septum secundum
Septum primum
Interatrial septum
5
Right atrium
Left atrium
Foramen ovale
Right ventricle
Left ventricle

PROCESS FIGURE 29.11 **Development of the Heart**

1 28 days after fertilization
A single bud forms and divides into two buds, which will become the lungs and primary bronchi.

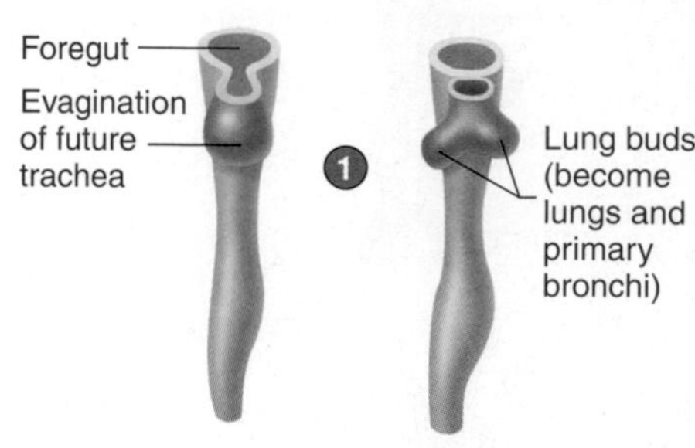
Foregut
Evagination of future trachea
1
Lung buds (become lungs and primary bronchi)

2 32 days after fertilization
Primary bronchi branch to form secondary bronchi, which supply the lung lobes.

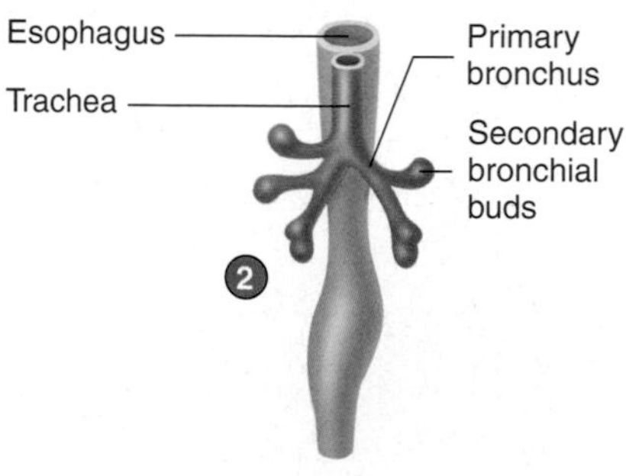
Esophagus
Trachea
Primary bronchus
Secondary bronchial buds
2

3 35 days after fertilization
Secondary bronchi branch to form tertiary bronchi, which supply the bronchopulmonary segments.

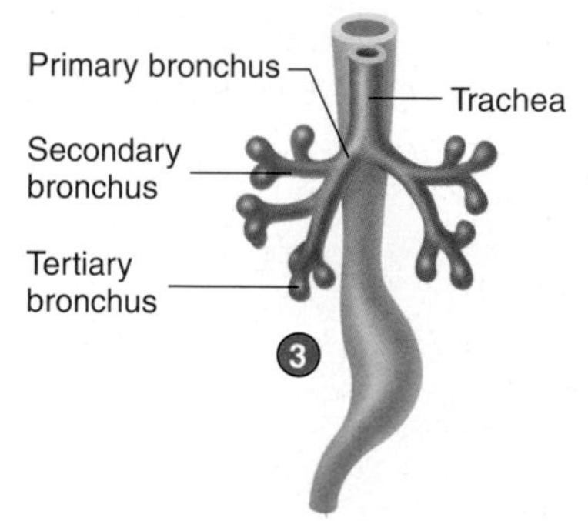
Primary bronchus
Trachea
Secondary bronchus
Tertiary bronchus
3

4 50 days after fertilization
The branching will continue to eventually form the extensive respiratory passages in the lungs.

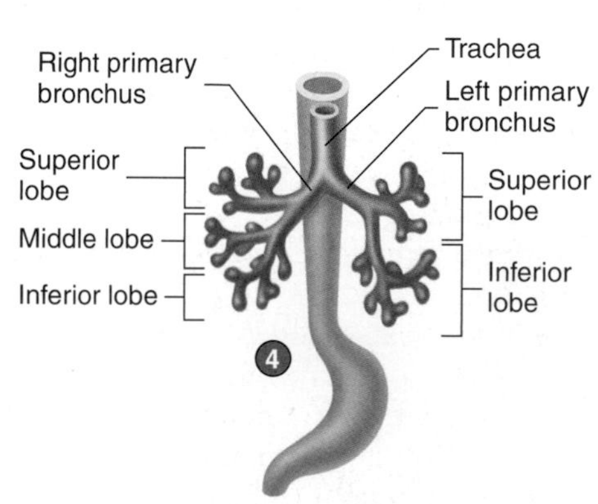
Right primary bronchus
Trachea
Left primary bronchus
Superior lobe
Middle lobe
Inferior lobe
Superior lobe
Inferior lobe
4

PROCESS FIGURE 29.12 Development of the Lungs

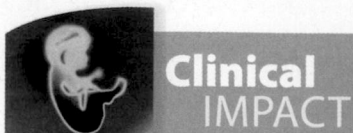

Clinical IMPACT

Heart Defects

If the septum secundum fails to grow far enough or if the foramen ovale becomes too large, an **atrial septal defect (ASD)** develops, allowing blood to flow from the left atrium to the right atrium in the newborn. If the interventricular septum does not grow enough to completely separate the ventricles, a **ventricular septal defect (VSD)** results. VSDs are more common than ASDs. Both ASDs and VSDs result in abnormal heart sounds, called a **heart murmur.** Blood passes through the ASD or the VSD from the left to the right side of the heart. The right side of the heart usually hypertrophies. In many cases, septal defects are not serious. However, in severe cases of VSD (Eisenmenger syndrome), increased pressure in the pulmonary blood vessels and decreased blood flow through the systemic blood vessels result in pulmonary edema, cyanosis (a bluish color to the skin due to deficient blood oxygenation), or heart failure.

the hindgut and the **allantois** (ă-lan′tō-is), a blind tube extending into the umbilical cord (see figure 29.8). The part of the allantois nearest the cloaca enlarges to form the urinary bladder, and the remainder, from the bladder to the umbilicus, forms the median umbilical ligament.

The mesonephric duct extends caudally as it develops and eventually joins the cloaca. At the point of junction, another tube, the **ureter,** begins to form. Its distal end enlarges and branches to form the duct system of the adult kidney, called the **metanephros** (last kidney), which takes over the function of the degenerating mesonephros (figure 29.13*d*).

Reproductive System

The male and female gonads appear as **gonadal ridges** along the ventral border of each mesonephros (figure 29.14*a*). **Primordial germ cells,** destined to become oocytes or sperm cells, form on the surface of the yolk sac, migrate into the embryo, and enter the gonadal ridge.

In the female, the ovaries descend from their original position high in the abdomen to a location within the pelvis. In the male, the testes descend even farther. As the testes reach the anteroinferior abdominal wall, two tunnels called the **inguinal canals** form through the abdominal musculature. The testes pass through these canals, leaving the abdominal cavity and coming to lie within the **scrotum** (see figure 28.4). Descent of the testes through the inguinal canals begins about 7 months after conception, and the testes enter the scrotum about 1 month before the infant is born. In approximately 3% of male children, one or both testes fail to enter the scrotum. This condition is called undescended testes, or **cryptorchidism** (krip-tōr′ki-dizm). Because testosterone is required for the testes to descend into the scrotum, cryptorchidism is often the result of inadequate testosterone secreted by the fetal testes. If neither testis descends and the defect is not corrected, the male will be infertile

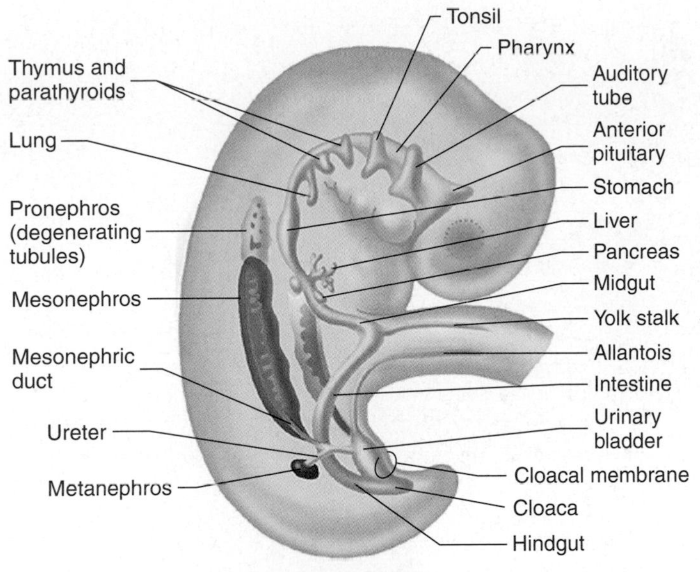
Tonsil
Pharynx
Thymus and parathyroids
Auditory tube
Anterior pituitary
Lung
Stomach
Pronephros (degenerating tubules)
Liver
Pancreas
Mesonephros
Midgut
Yolk stalk
Mesonephric duct
Allantois
Intestine
Ureter
Urinary bladder
Metanephros
Cloacal membrane
Cloaca
Hindgut

(a) The three parts of the developing kidney: pronephros, mesonephros, metanephros

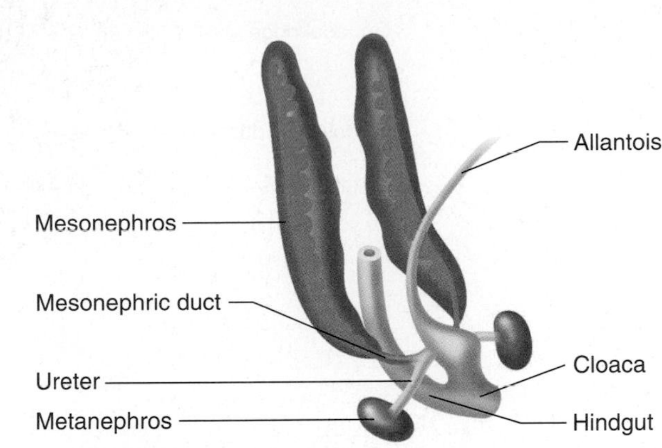
Allantois
Mesonephros
Mesonephric duct
Cloaca
Ureter
Metanephros
Hindgut

(b) The metanephros (adult kidney) enlarges as the mesonephros degenerates.

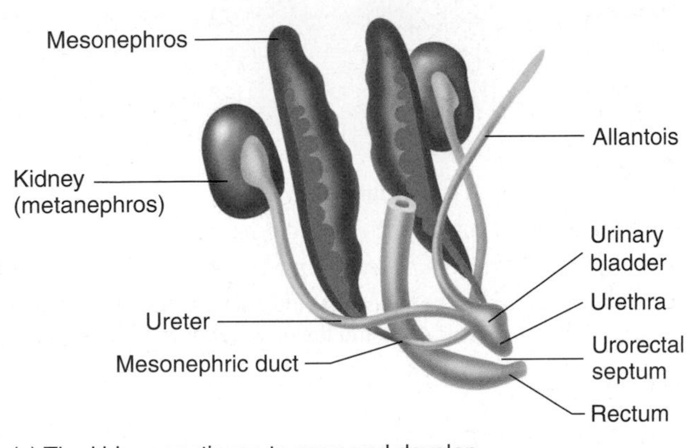
Mesonephros
Allantois
Kidney (metanephros)
Urinary bladder
Urethra
Ureter
Urorectal septum
Mesonephric duct
Rectum

(c) The kidney continues to grow and develop.

Kidney (metanephros)
Ureter
Urinary bladder
Urethra
Rectum

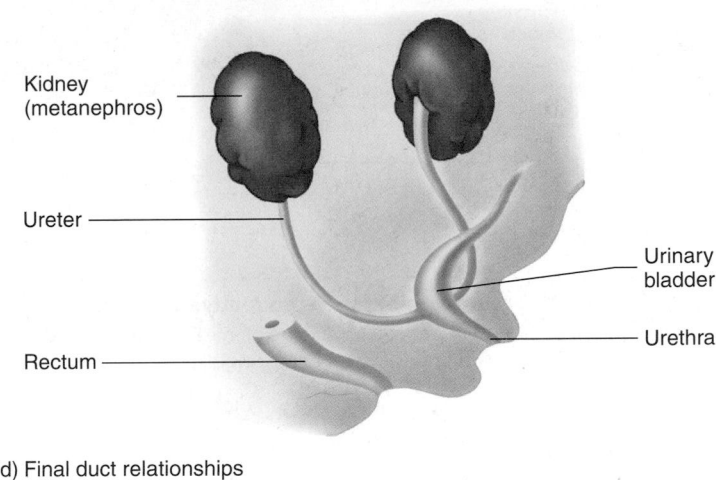

(d) Final duct relationships

FIGURE 29.13 Development of the Kidneys and Urinary Bladder

because the slightly higher temperature of the body cavity, compared with that of the scrotal sac, causes the spermatogonia to degenerate. Cryptorchidism is treated with hormone therapy, or it can be surgically corrected. Cryptorchidism is an important risk factor for testicular cancer.

Paramesonephric ducts (also called müllerian ducts) begin to develop just lateral to the **mesonephric ducts** (also called wolffian ducts) and grow inferiorly until they meet one another and enter the cloaca as a single, midline tube.

In male embryos, testosterone is secreted by the testes. It causes the mesonephric duct system to enlarge and differentiate to form the epididymis, ductus deferens, seminal vesicles, and prostate gland (figure 29.14*b*). Müllerian-inhibiting hormone, also secreted by the testes, causes the paramesonephric ducts to degenerate. In female embryos, neither testosterone nor müllerian-inhibiting hormone is secreted. As a result, the mesonephric duct system atrophies, and the paramesonephric duct system develops to form the uterine tubes, the uterus, and part of the vagina in females (figure 29.14*c*).

Like the other sex organs, the external genitalia begin as the same structures in the male and female and then diverge (figure 29.15). An enlargement called the **genital tubercle** develops in the groin of the embryo. **Genital folds** develop on each side of a **urethral groove,** and **labioscrotal swellings** develop lateral to the folds.

In the male, under the influence of dihydrotestosterone, derived from testosterone, the genital tubercle and the genital folds close over the urethral groove to form the penis. The testes move into the labioscrotal swellings, which become the scrotum of the male. If closure of the urethral groove does not proceed all the way to the end of the penis, a defect known as **hypospadias** (hī′pō-spā′dē-ăs) results.

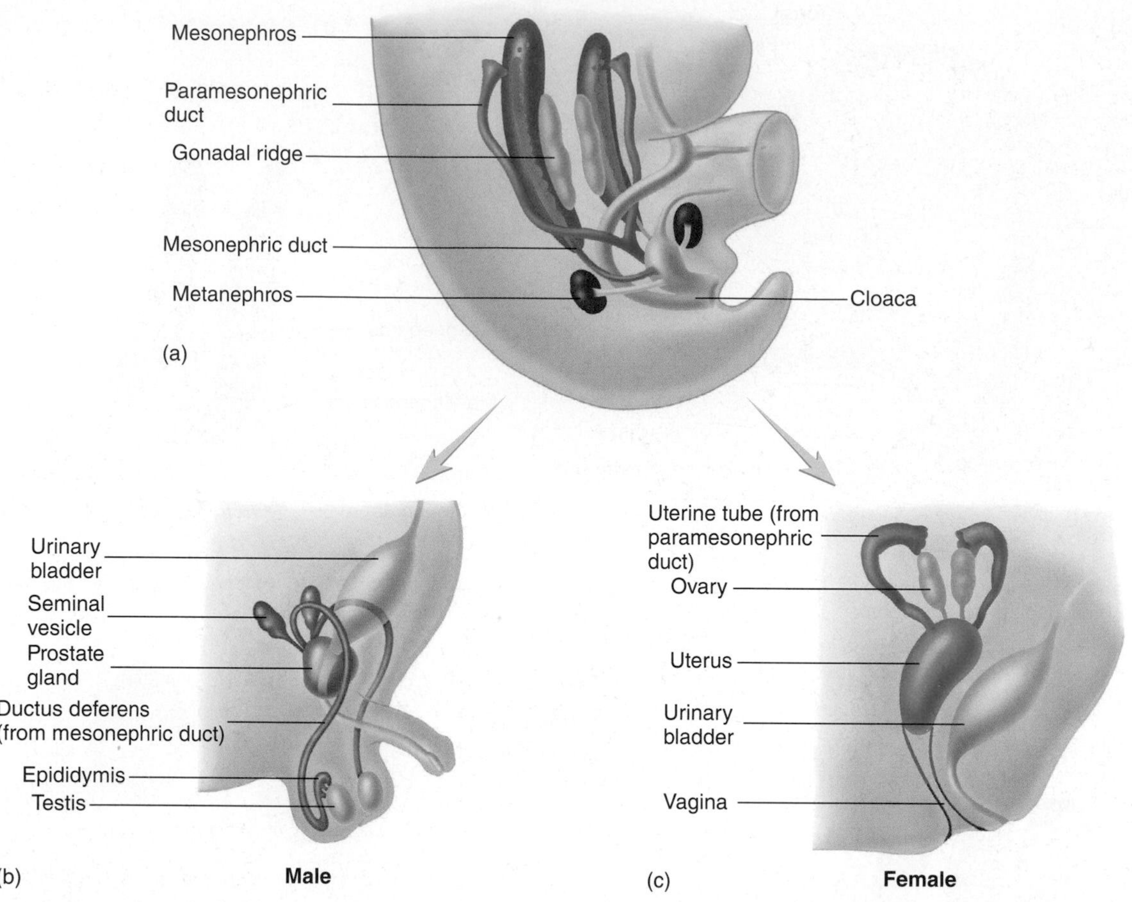
Mesonephros
Paramesonephric duct
Gonadal ridge
Mesonephric duct
Metanephros
Cloaca
(a)
Urinary bladder
Seminal vesicle
Prostate gland
Ductus deferens (from mesonephric duct)
Epididymis
Testis
(b)
Male
Uterine tube (from paramesonephric duct)
Ovary
Uterus
Urinary bladder
Vagina
(c)
Female

FIGURE 29.14 Differentiation of the Reproductive Systems
(*a*) Indifferent stage. (*b*) The male, under the influence of male hormones, develops a ductus deferens from the mesonephric duct, and the paramesonephric duct degenerates. (*c*) The female, without male hormones, develops a uterus and uterine tubes from the paramesonephric duct, and the mesonephros disappears.

In the female, in the absence of testosterone, the genital tubercle becomes the clitoris. The urethral groove disappears; genital folds do not fuse. As a result, the urethra opens somewhat posterior to the clitoris but anterior to the vaginal opening. The unfused genital folds become the labia minora, and the labioscrotal folds become the labia majora.

Predict 3

How would the failure to produce müllerian-inhibiting hormone affect the development of the internal reproductive system and external genitalia in a male embryo?

Growth of the Fetus

The embryo becomes a fetus approximately 60 days after fertilization (a 50-day-old embryo is shown in figure 29.16*a*). In the embryo, most of the organ systems are forming, whereas in the fetus the organs are present and continue to develop during the fetal period. Most morphological changes occur in the embryonic phase of development, whereas the fetal period is primarily a "growing phase."

The fetus grows from about 3 cm and 2.5 g at 60 days to 50 cm and 3300 g at term—more than a 15-fold increase in length and a 1300-fold increase in weight. The growth of the fetus also causes an increase in the size of the uterus (figure 29.17).

Fine, soft hair called **lanugo** (la-noo′gō) covers the fetus. A waxy coat of sloughed epithelial cells called the **vernix caseosa** (ver′niks kā-se-ō′sā) covers and protects the fetus from the somewhat toxic amniotic fluid formed by the accumulation of fetal waste products.

Subcutaneous adipose tissue that accumulates in the older fetus and newborn provides a nutrient reserve and helps insulate the baby. In addition, the subcutaneous adipose tissue aids the baby in sucking by strengthening and supporting the cheeks, so that negative pressure can be developed in the oral cavity.

Peak body growth occurs late in gestation, but growth of the placenta essentially stops at about 35 weeks, thus restricting further intrauterine growth.

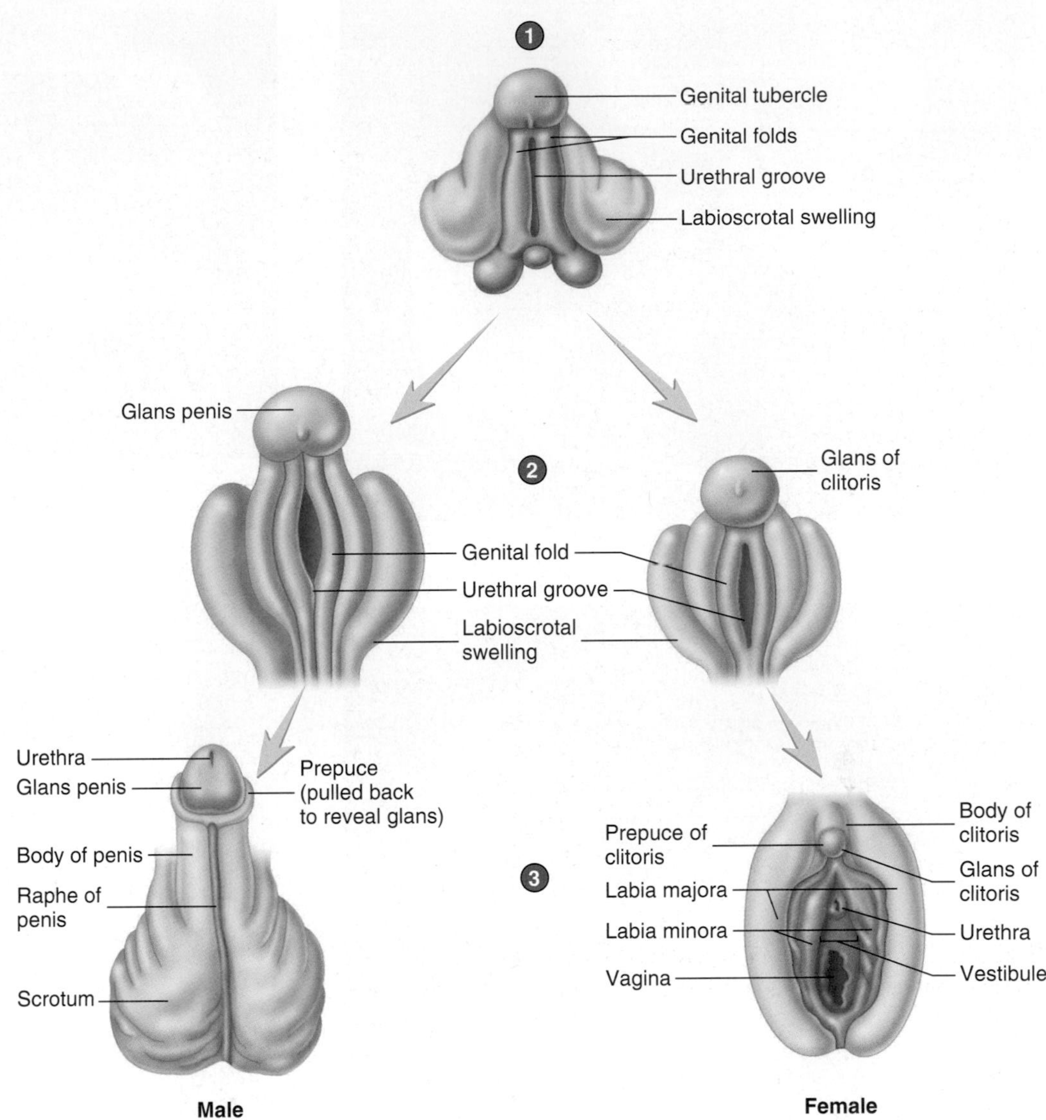
1
At approximately 5 weeks after fertilization, the genital tubercle, genital folds, and labioscrotal swellings are the same for the male and female.
Genital tubercle
Genital folds
Urethral groove
Labioscrotal swelling
2
By 10 weeks of development, the male penile structures are somewhat larger than those of the female clitoris. The urethral groove is still open in the male and female.
Glans penis
Glans of clitoris
Genital fold
Urethral groove
Labioscrotal swelling
3
Near term, the general adult condition is achieved. The labioscrotal swellings become the scrotum in the male and the labia majora in the female. The genital folds fuse in the male to form the body of the penis, whereas they form the labia minora in the female. The urethral groove in the male is closed by the raphe of the penis. It remains open in the female as the vestibule.
Urethra
Glans penis
Prepuce (pulled back to reveal glans)
Body of penis
Raphe of penis
Scrotum
Prepuce of clitoris
Labia majora
Labia minora
Vagina
Body of clitoris
Glans of clitoris
Urethra
Vestibule
Male
Female

PROCESS FIGURE 29.15 Development of the External Genitalia

At about 38 weeks of development, the fetus can survive outside the mother and is ready to be delivered. The average weight at this point is 3250 g for a female fetus and 3300 g for a male fetus.

ASSESS YOUR PROGRESS

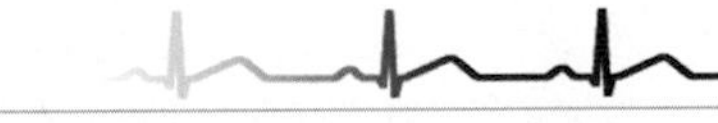

16. *Describe the formation of the following major organs and organ systems: skin, bones, skeletal muscles, nervous system, eyes, and respiratory system. When does each of these events occur?*
17. *Explain the formation of the following endocrine glands: anterior pituitary, posterior pituitary, thyroid, parathyroid, adrenal medulla, adrenal cortex, and pancreas.*
18. *Explain the process by which a one-chambered heart becomes a four-chambered heart.*
19. *Describe how the pronephros, mesonephros, and metanephros lead to the development of the kidneys.*
20. *Demonstrate your understanding of the effect hormones have on the development of the male and female reproductive systems.*
21. *Compare the male and female structures formed from each of the following: genital tubercle, genital folds, and labioscrotal swellings.*
22. *What major events distinguish between embryonic and fetal development?*

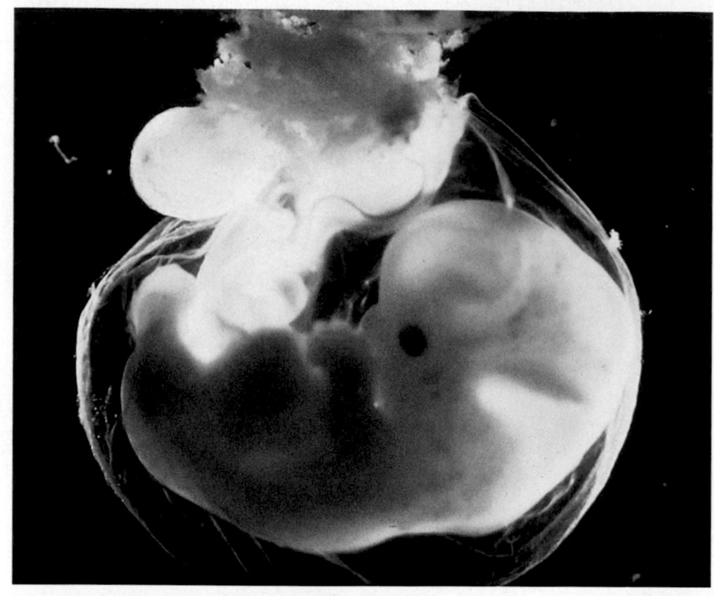

(a)

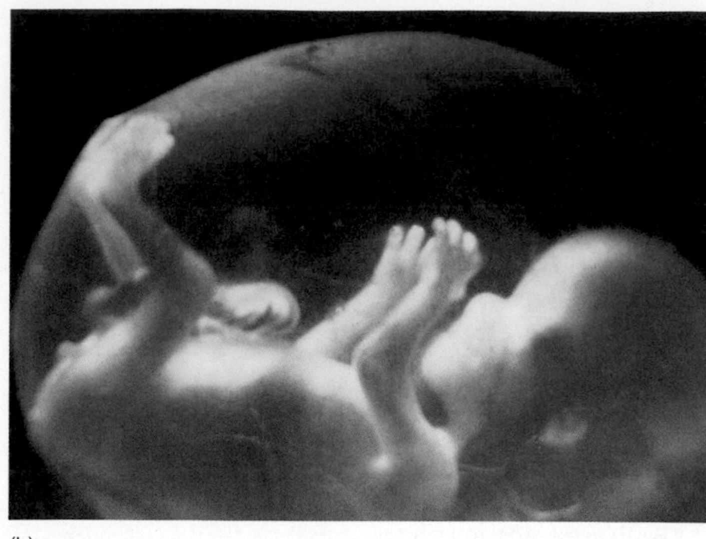

(b)

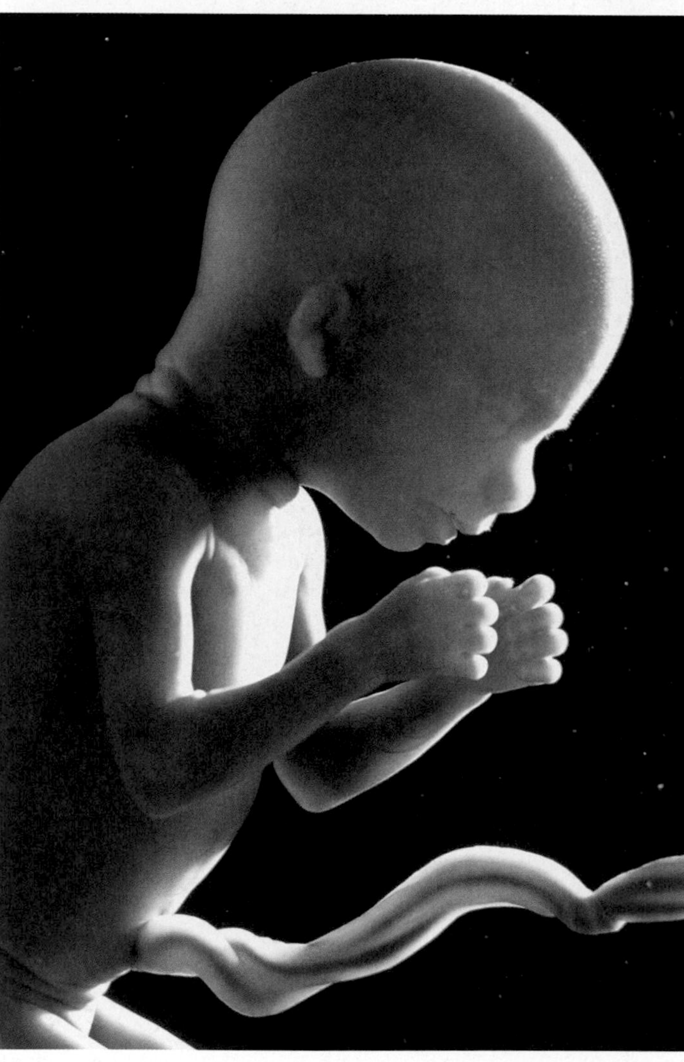

(c)

FIGURE 29.16 Embryos and Fetuses at Different Ages
(*a*) Fifty days after fertilization. (*b*) Three months after fertilization. (*c*) Four months after fertilization.

1 month

4 months

7 months

Umbilical cord
Mucus plug in cervical canal
Rectum
Vagina
Placenta
Urinary bladder
Symphysis pubis
Urethra

9 months

FIGURE 29.17 Enlargement of the Uterus During Fetal Development

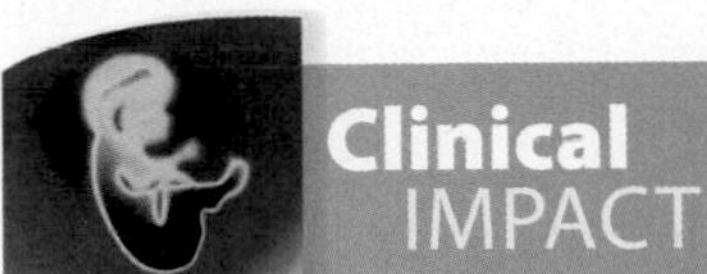

Clinical IMPACT — Fetal Monitoring

Monitoring of the fetus during pregnancy allows physicians to determine the likelihood of complications during delivery, as well as possible health conditions the fetus may have as a result of either genetics or a developmental error. **Amniocentesis** (am′nē-ō-sen-tē′sis) is the removal and analysis of amniotic fluid (figure 29B). As the fetus develops, it expels molecules of various types, as well as living cells, into the amniotic fluid. These molecules and cells can be collected and analyzed to detect a number of metabolic disorders. Furthermore, if the cells collected by amniocentesis are grown in culture, additional metabolic disorders can be detected. Chromosome analysis, called **karyotyping,** can also be performed on the cultured cells. Amniocentesis has been done as early as 10 weeks after fertilization, but the success rate in detecting disorders at that time is quite low. It is most commonly performed at 13–16 weeks after fertilization.

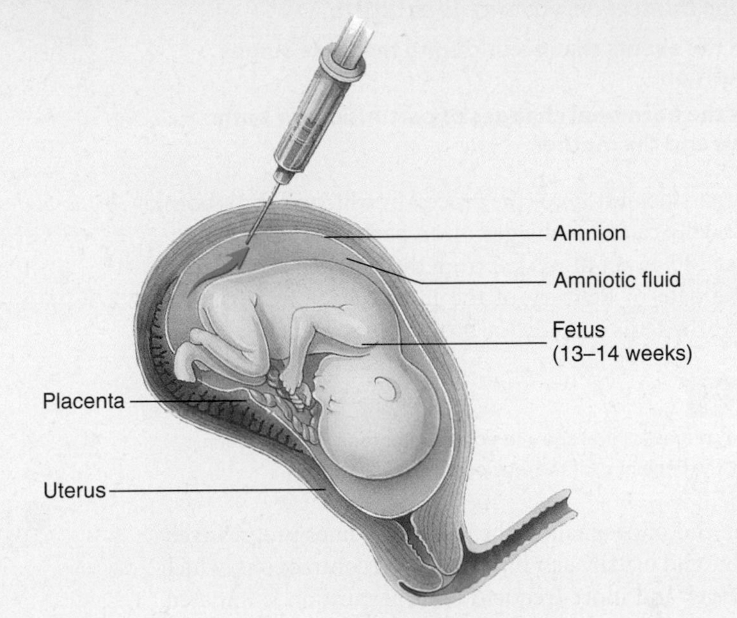
Amnion
Amniotic fluid
Fetus
(13–14 weeks)
Placenta
Uterus

FIGURE 29B Removal of Amniotic Fluid for Amniocentesis

Fetal tissue samples can also be obtained by **chorionic villus sampling,** in which a probe introduced into the uterine cavity through the cervix removes a small piece of the chorion. This technique has an advantage over amniocentesis in that it can be used as early as 8–10 weeks after fertilization. Furthermore, cells can be analyzed directly, as in karyotyping, rather than having to be cultured, as is required in amniocentesis.

One of the molecules normally produced by the fetus and released into the amniotic fluid is **α-fetoprotein.** This chemical shows up in the amniotic fluid if fetal tissues that are normally covered by skin are directly exposed to the amniotic fluid. This can happen with nervous tissue due to failure of the neural tube to close, or with abdominal tissues due to failure of the abdominal wall to fully form.

Some of the metabolic by-products from the fetus, such as α-fetoprotein and estriol, a weak form of estrogen produced in the placenta after 20 weeks of gestation, can enter the maternal blood. In some cases, the by-products are processed and passed to the maternal urine. The levels of these fetal products can then be measured in the mother's blood or urine to screen for possible conditions of the fetus. Further tests are required to verify suspected fetal conditions indicated by elevated levels of fetal metabolic by-products in the maternal blood or urine.

The fetus can be seen within the uterus by **ultrasound,** in which sound waves are bounced off the fetus like sonar and then analyzed and enhanced by computer. In another technique, called **fetoscopy,** a fiberoptic probe is introduced into the amniotic cavity and used to view the fetus. Because of the constantly increasing resolution in ultrasound technology and because it is noninvasive compared with fetoscopy, ultrasound is usually the preferred technique. However, some fetal defects cannot be adequately assessed by ultrasound and require fetoscopy. In that case, ultrasound is used to guide the fetoscope.

Ultrasound has not been found to pose any risk to the fetus or the mother. It is accomplished by placing a transducer on the mother's abdominal wall (transabdominal method) or by inserting the transducer into the woman's vagina (transvaginal method). The latter technique produces much higher resolution because fewer layers of tissue exist between the transducer and the uterine cavity. Transvaginal ultrasound can be used to identify the yolk sac of a developing embryo as early as 17 days after fertilization, and the embryo can be visualized at 25 days. Transabdominal ultrasound allows for fetal monitoring by 6–8 weeks after fertilization.

Fetal heart rate can be detected with an ultrasound stethoscope by the tenth week after fertilization and with a conventional stethoscope by 20 weeks. Fetal heart rate is most commonly monitored electronically, either indirectly by transducers on the mother's abdomen or by using a probe attached to the skin of the fetus. The normal fetal heart rate range is 110–160 bpm).

29.2 Parturition

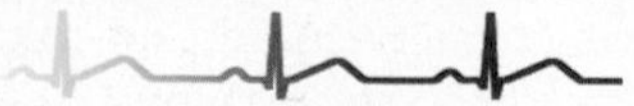

LEARNING OUTCOMES

After reading this section, you should be able to

A. Explain the events that occur during the three stages of parturition.

B. Discuss the hormonal changes of parturition for both the fetus and the mother.

Parturition (par-toor-ish′ŭn) is the process by which a baby is born. Physicians usually calculate the **gestation period,** or length of the pregnancy, as 280 days (40 weeks) from the last menstrual period (LMP) to the date of delivery of the infant. (In terms of postovulatory age, the fetus is ready for birth at 38 weeks.)

Predict 4

Compare and contrast clinical age and postovulatory age for fertilization, implantation, the beginning of the fetal period, and parturition.

Near the end of pregnancy, the uterus becomes progressively more excitable and usually exhibits occasional contractions, which become stronger and more frequent until parturition is initiated. The cervix gradually dilates, and strong uterine contractions help expel the fetus from the uterus through the vagina (figure 29.18).

Stages of Labor

Labor is the period during which the contractions occur that eventually expel the fetus and placenta from the uterus. It occurs as three stages:

1. *First stage.* The first stage of labor, often called the *dilation stage,* begins with the onset of regular uterine contractions and extends until the cervix has dilated to a diameter about the size of the

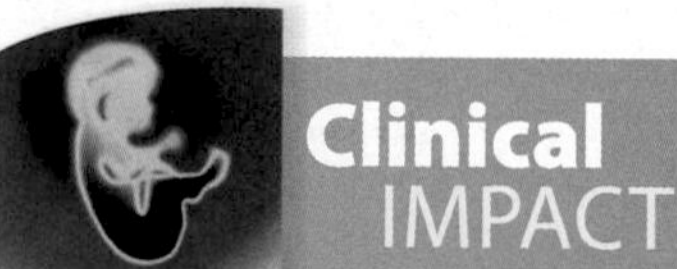

Clinical IMPACT

Prematurity

Occasionally, the fetus is delivered before it has sufficiently matured and is therefore considered **premature.** Prematurity is one of the most significant problems in pediatrics, the branch of medicine dealing with children, because many complications can result. The most significant complication is **infant respiratory distress syndrome,** which occurs because very young premature infants cannot produce **surfactant,** a mixture of phospholipids and protein that lines the inner surface of the lungs and allows them to expand as we breathe. Each year, 65,000 premature infants suffer from respiratory distress syndrome in the United States. Until recently, 10% of those infants died. Now, surfactant substitutes are being developed, and glucocorticoid administration can stimulate surfactant production. These therapies have cut the death rate in half, and more effective replacements are being investigated.

FUNDAMENTAL **Figure**

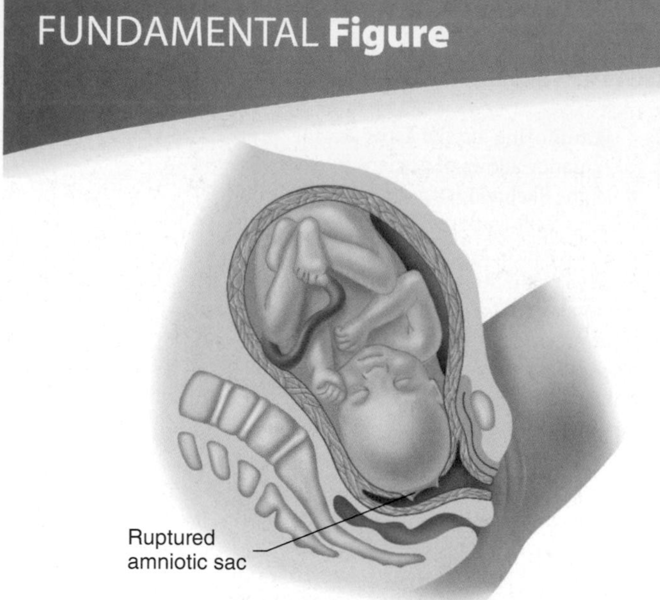
Ruptured
amniotic sac

① **First stage.** The cervix dilates, and the amniotic sac ruptures.

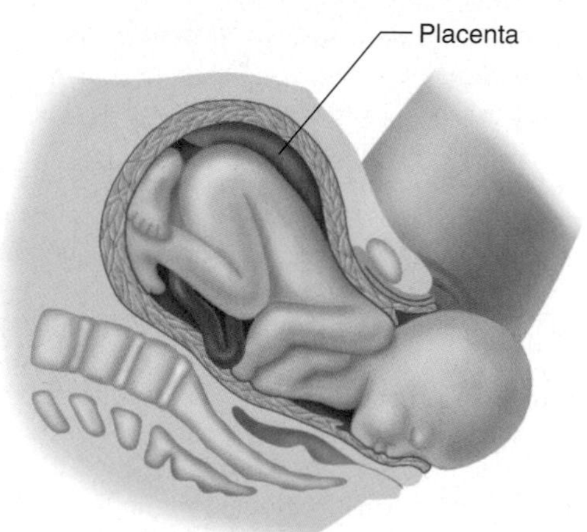
Placenta

② **Second stage.** The fetus is expelled from the uterus.

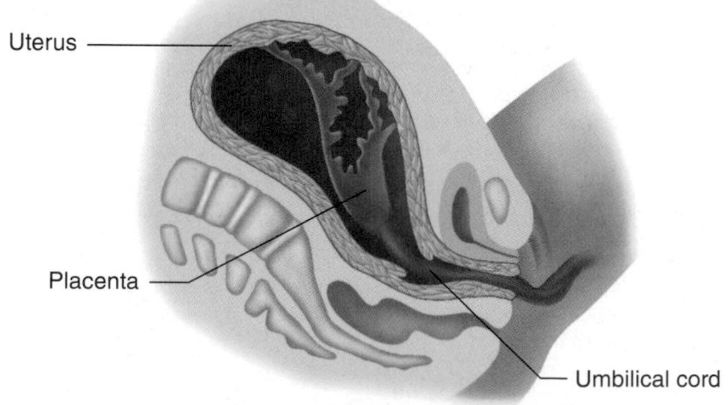
Uterus
Placenta
Umbilical cord

③ **Third stage.** The placenta is expelled.

PROCESS FIGURE 29.18 Parturition

fetus's head. This stage of labor commonly lasts 8–24 hours, but it may be as short as a few minutes, especially in women who have had more than one child. Normally during labor (95% of the time), the head of the fetus is in an inferior position within the woman's pelvis, so that the head acts as a wedge, forcing the cervix and vagina to open as the uterine contractions push against the fetus. During this stage of labor, the amniotic sac ruptures, releasing the amniotic fluid.

The **central tendon of the perineum** (see figure 10.19) is very important in supporting the uterus and vagina. Tearing or stretching of the tendon during childbirth may weaken the inferior support of these organs, and prolapse of the uterus may occur. **Prolapse** is a "sinking" of the uterus, so that the uterine cervix moves down into the vagina (first degree), moves down near the vaginal orifice (second degree), or protrudes through the vaginal orifice (third degree).

2. *Second stage.* The second stage of labor, often called the *expulsion stage,* lasts from the time of maximum cervical dilation until the baby exits the vagina. This stage may last from a minute to an hour or more. During this stage, contractions of the abdominal muscles assist the uterine contractions. The contractions generate enough pressure to compress blood vessels in the placenta that blood flow to the fetus is stopped. During periods of relaxation, blood flow to the placenta resumes. Occasionally, synthetic oxytocin (pitocin) is administered to women during labor to increase the force of the uterine contractions. However, caution must be exercised in using this drug, so that tetanic contractions, which would drastically reduce blood flow through the placenta, do not occur.
3. *Third stage.* During the third stage of labor, often called the *placental stage,* the placenta is expelled from the uterus. Uterine contractions cause the placenta to tear away from the wall of the uterus. Some bleeding from the uterine wall occurs as a consequence of the intimate contact between the placenta and the uterus, but the bleeding is normally limited because uterine smooth muscle contractions compress the blood vessels to the placenta.

After parturition, blood levels of estrogen and progesterone fall dramatically because the source of these hormones is gone once the placenta has been dislodged from the uterus. In addition, during the 4 or 5 weeks after parturition, the uterus becomes much smaller, although it remains somewhat larger than it was before pregnancy. The cells of the uterine lining become smaller, and many of them degenerate. A vaginal discharge composed of small amounts of blood and degenerating endometrium persists for 1 week or more after parturition.

Hormonal Stimulation of Parturition

The precise signal that triggers parturition is unknown, but many supporting factors have been identified (figure 29.19). Before parturition, the progesterone concentration in the maternal circulation

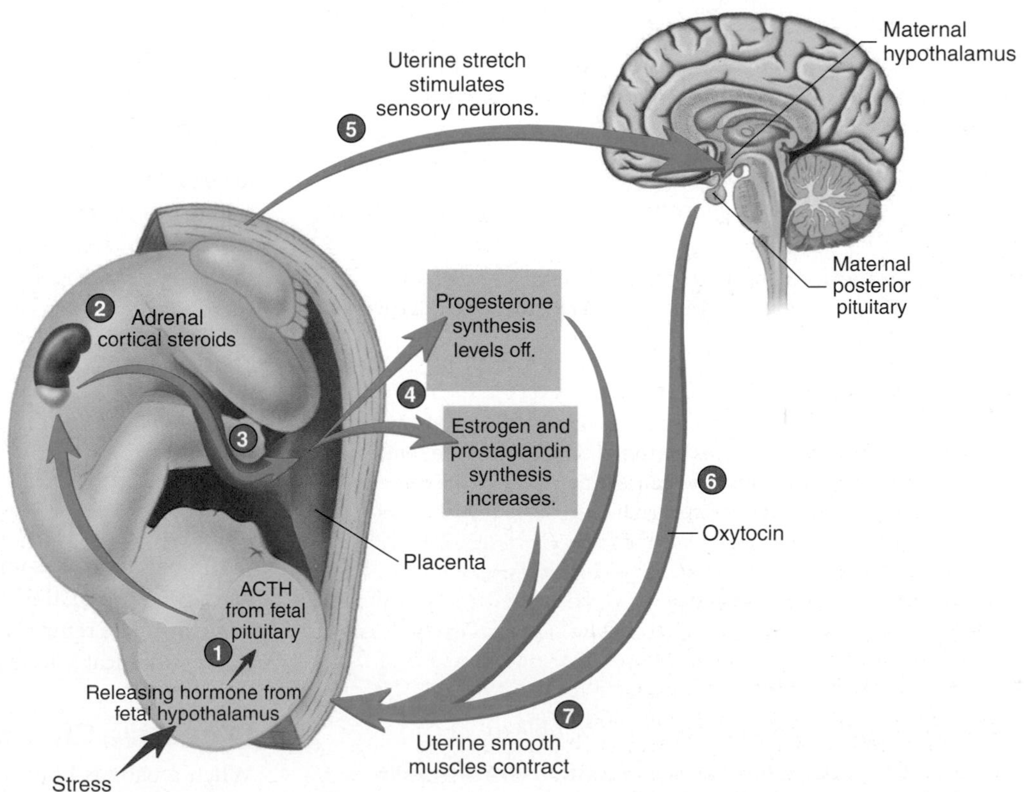

1 The fetal hypothalamus secretes a releasing hormone, corticotropin-releasing hormone (CRH), which stimulates adrenocorticotropic hormone (ACTH) secretion from the pituitary. The fetal pituitary secretes ACTH in greater amounts near parturition.
2 ACTH causes the fetal adrenal gland to secrete greater quantities of adrenal cortical steroids.
3 Adrenal cortical steroids travel in the umbilical blood to the placenta.
4 In the placenta, the adrenal cortical steroids cause progesterone synthesis to level off and estrogen and prostaglandin synthesis to increase, making the uterus more excitable.
5 The stretching of the uterus produces action potentials that are transmitted to the brain through ascending pathways.
6 Action potentials stimulate the secretion of oxytocin from the posterior pituitary.
7 Oxytocin causes the uterine smooth muscle to contract.
Uterine stretch stimulates sensory neurons.
Maternal hypothalamus
Maternal posterior pituitary
Adrenal cortical steroids
Progesterone synthesis levels off.
Estrogen and prostaglandin synthesis increases.
Oxytocin
Placenta
ACTH from fetal pituitary
Releasing hormone from fetal hypothalamus
Stress
Uterine smooth muscles contract

PROCESS FIGURE 29.19 Factors That Influence Parturition
Although the precise control of parturition is unknown, these changes appear to play a role.

is at its highest level and is exerting an inhibitory effect on uterine smooth muscle cells. Near the end of pregnancy, however, estrogen levels rapidly increase in the maternal circulation, and their excitatory influence overcomes the inhibitory influence of progesterone.

Before parturition, the adrenal glands of the fetus are greatly enlarged, and the rate of adrenocorticotropic hormone (ACTH) secretion by the fetus's anterior pituitary gland increases due to the stress of the confined space and limited oxygen supply created in the uterus by the growing fetus. ACTH causes the fetal adrenal cortex to produce glucocorticoids, which travel to the placenta, where they decrease the rate of progesterone secretion and increase the rate of estrogen synthesis. Also initiated is the synthesis of prostaglandins, which strongly stimulate uterine contractions.

During parturition, stretch of the uterine cervix initiates nervous reflexes that cause oxytocin to be released from the woman's posterior pituitary gland. Oxytocin stimulates uterine contractions, which move the fetus farther into the cervix, causing further stretch. Thus, by a positive-feedback mechanism, stretch stimulates oxytocin release and oxytocin causes further stretch until the time of delivery, when the cervix is no longer stretched.

Besides the effects already described, progesterone and estrogen also influence the secretion of oxytocin. Progesterone inhibits oxytocin release and decreases the number of oxytocin receptors. Therefore, decreased progesterone levels in the maternal circulation result in increased oxytocin secretion and more oxytocin receptors in the uterus. In addition, estrogen makes the uterus more sensitive to oxytocin stimulation by increasing the synthesis of oxytocin receptor sites. Estrogen may also increase the formation of gap junctions between myometrial cells, thereby enhancing the contractility of the uterus. Some evidence suggests that oxytocin also stimulates prostaglandin synthesis in the uterus. All these events support the development of strong uterine contractions.

ASSESS YOUR PROGRESS

23. *List the stages of labor; indicate when each stage begins and its approximate length.*

24. *Describe the hormonal changes that take place before and during delivery. How is stretch of the cervix involved in delivery?*

Predict 5

A woman is having an extremely prolonged labor. From her anatomy and physiology course, she remembers the role of calcium in muscle contraction and asks the doctor to give her a calcium injection to speed the delivery. Explain why the doctor would or would not do as she requests.

29.3 The Newborn

LEARNING OUTCOMES

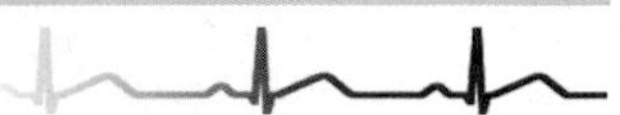

After reading this section, you should be able to

A. Discuss the respiratory, cardiovascular, and digestive changes that occur in the newborn.

B. Explain the significance of the Apgar score.

C. List the causes of congenital disorders.

The newborn baby, or **neonate,** immediately experiences several dramatic changes due to being separated from the maternal circulation and transferred from a fluid to a gaseous environment.

Respiratory and Cardiovascular Changes

The large, forced gasps of air taken in when the infant cries at the time of delivery help inflate the lungs. This initial inflation of the lungs causes important changes in the cardiovascular system (figure 29.20). Before birth, very little blood flows through the fetus's pulmonary arteries to the lungs and back to the heart through the pulmonary veins. As a result, the left atrium has little blood and very low blood pressure. Therefore, blood flows from the right atrium, through the foramen ovale, and into the left atrium.

Expansion of the lungs reduces the resistance to blood flow through the lungs, resulting in increased blood flow through the pulmonary arteries. Consequently, more blood flows from the right atrium to the right ventricle and into the pulmonary arteries, and less blood flows from the right atrium through the foramen ovale to the left atrium. In addition, more blood returns from the lungs through the pulmonary veins to the left atrium, which increases the pressure in the left atrium. The increased left atrial pressure and decreased right atrial pressure, resulting from decreased pulmonary resistance, force blood against the septum primum, causing the foramen ovale to close. This action functionally completes the separation of the heart into two pumps: the right side of the heart and the left side of the heart. The closed foramen ovale becomes the **fossa ovalis.**

Within 1 or 2 days after birth, the **ductus arteriosus,** which connects the pulmonary trunk to the aorta and allows blood to flow from the pulmonary trunk to the systemic circulation, closes off. This closure occurs because of the sphincterlike constriction of the artery and is probably stimulated by local changes in blood pressure and blood oxygen. Once closed, the ductus arteriosus is replaced by connective tissue and is known as the **ligamentum arteriosum.** If the ductus arteriosus does not close completely, it is said to be **patent.** This is a serious birth defect, resulting in marked elevation in pulmonary blood pressure because blood flows from the left ventricle to the aorta, through the ductus arteriosus, and to the pulmonary arteries. If not corrected, it can lead to irreversible degenerative changes in the heart and lungs.

Before birth, the fetal blood passes to the placenta through umbilical arteries from the internal iliac arteries and returns through an umbilical vein. The blood passes through the liver via the ductus venosus, which joins the inferior vena cava. At birth, when the umbilical cord is tied and cut, no more blood flows through the umbilical vein and arteries, and they degenerate. The remnant of the umbilical vein becomes the **ligamentum teres,** or *round ligament,* of the liver, and the ductus venosus becomes the **ligamentum venosum.** The remnants of the umbilical arteries become the cords of the umbilical arteries.

Digestive Changes

When a baby is born, it is suddenly separated from its source of nutrients, the maternal circulation. Because of this separation and the stress of birth and new life, the neonate usually loses 5–10% of its total body weight during the first few days of life.

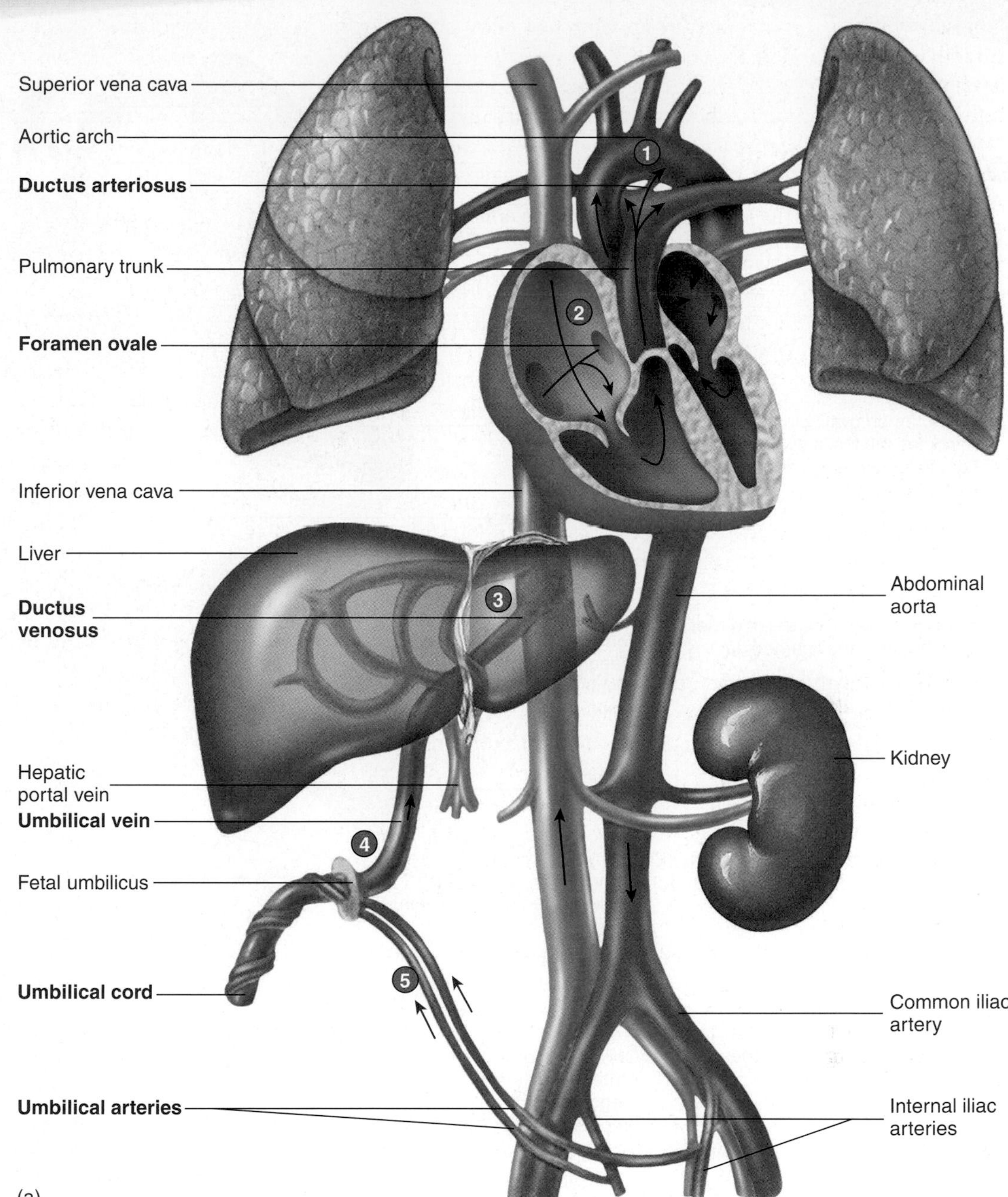

1 Blood bypasses the lungs by flowing from the pulmonary trunk through the ductus arteriosus to the aorta.
2 Blood also bypasses the lungs by flowing from the right to the left atrium through the foramen ovale.
3 Blood bypasses the liver sinusoids by flowing through the ductus venosus.
4 Oxygenated blood is returned to the fetus from the placenta by the umbilical vein.
5 Deoxygenated blood is carried from the fetus to the placenta through the umbilical arteries.
Superior vena cava
Aortic arch
Ductus arteriosus
Pulmonary trunk
Foramen ovale
Inferior vena cava
Liver
Ductus venosus
Hepatic portal vein
Umbilical vein
Fetal umbilicus
Umbilical cord
Umbilical arteries
Abdominal aorta
Kidney
Common iliac artery
Internal iliac arteries
(a)

PROCESS FIGURE 29.20 Circulatory Changes at Birth
(*a*) Circulatory conditions in the fetus.

Although the digestive system of the fetus becomes somewhat functional late in development, it is still very immature, compared with that of an adult. Late in gestation, the fetus swallows amniotic fluid from time to time. Shortly after birth, this swallowed fluid plus intestinal cells, intestinal mucus, and bile pass from the digestive tract as a greenish anal discharge called **meconium** (mē-kō′nē-ŭm).

The pH of the stomach at birth is nearly neutral because of the swallowed basic amniotic fluid. Within the first 8 hours of life, gastric acid secretion increases, causing the stomach pH to decrease. Maximum acidity is reached at 4–10 days, and the pH gradually increases for the next 10–30 days.

The neonatal liver is also functionally immature. It lacks adequate amounts of the enzyme required to produce bilirubin. This enzyme system usually develops within 2 weeks after birth in a healthy neonate; however, because it is not fully developed at birth, some full-term babies temporarily develop jaundice, characterized by elevated blood levels of bilirubin and a slightly yellowish cast to the skin or whites of the eye. Jaundice is also common in premature babies.

The newborn digestive system is capable of digesting lactose (milk sugar) from the time of birth. The pancreatic secretions are sufficiently mature for a milk diet, but the digestive system only gradually develops the ability to digest more solid foods over the

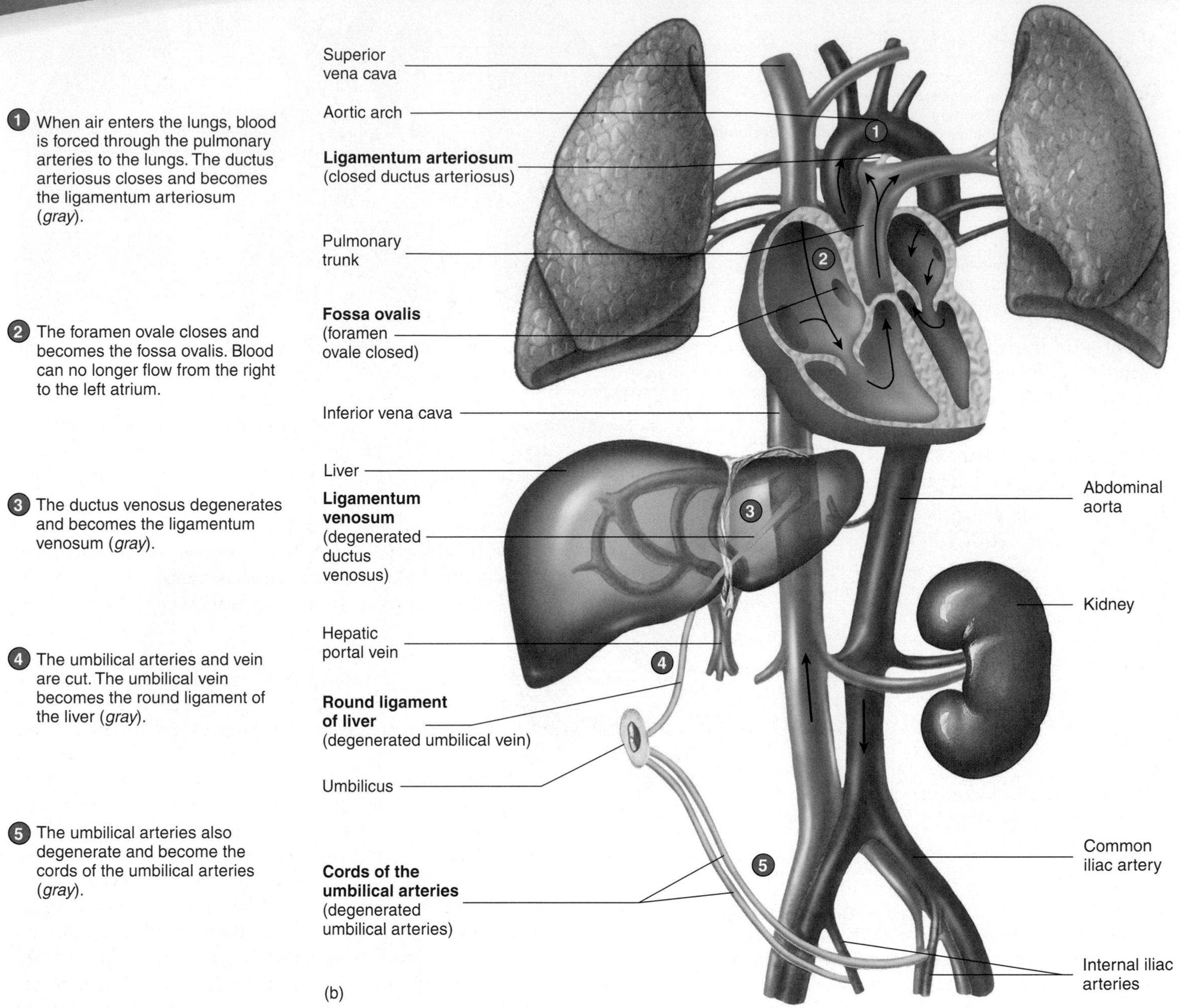

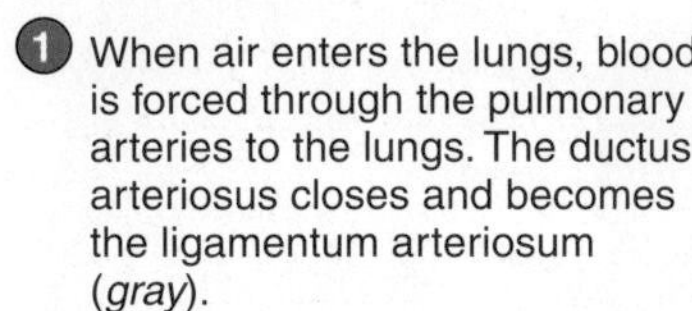
1 When air enters the lungs, blood is forced through the pulmonary arteries to the lungs. The ductus arteriosus closes and becomes the ligamentum arteriosum (gray).
2 The foramen ovale closes and becomes the fossa ovalis. Blood can no longer flow from the right to the left atrium.
3 The ductus venosus degenerates and becomes the ligamentum venosum (gray).
4 The umbilical arteries and vein are cut. The umbilical vein becomes the round ligament of the liver (gray).
5 The umbilical arteries also degenerate and become the cords of the umbilical arteries (gray).
Superior vena cava
Aortic arch
Ligamentum arteriosum (closed ductus arteriosus)
Pulmonary trunk
Fossa ovalis (foramen ovale closed)
Inferior vena cava
Liver
Ligamentum venosum (degenerated ductus venosus)
Hepatic portal vein
Round ligament of liver (degenerated umbilical vein)
Umbilicus
Cords of the umbilical arteries (degenerated umbilical arteries)
(b)
Abdominal aorta
Kidney
Common iliac artery
Internal iliac arteries

PROCESS FIGURE 29.20 (continued)
(*b*) Circulatory changes that occur at birth.

first year or two; therefore, new foods should be introduced gradually during the first 2 years. Parents are also advised to introduce only one new food at a time, so that, if an allergic reaction occurs, the cause is more easily determined.

Amylase secretion by the salivary glands and the pancreas remains low until after the first year. Lactase activity in the small intestine is high at birth but declines during infancy, although the levels still exceed those in adults. In many adults, lactase activity is lost, and an intolerance for milk develops (see chapter 24).

Apgar Scores

A newborn baby is usually evaluated soon after birth using a physiological assessment tool known as the **Apgar score.** The Apgar test, named for Virginia Apgar, the physician who developed it, evaluates the neonate's **a**ppearance, **p**ulse, **g**rimace, **a**ctivity, and **r**espiratory effort. Each of these characteristics is rated on a scale of 0–2, with 2 denoting normal function; 1, reduced function; and 0, seriously impaired function. The total Apgar score is the sum of the scores from the five characteristics, ranging from 0 to 10 (table 29.3). A total Apgar

TABLE 29.3 Apgar Rating Scales

Physiological Conditions	0	1	2
Appearance (skin color)	White or blue	Limbs blue, body pink	Pink
Pulse (rate)	No pulse	100 bpm	>100 bpm
Grimace (reflexive grimace initiated by stimulating the plantar surface of the foot)	No response	Facial grimaces, slight body movement	Facial grimaces, extensive body movement
Activity (muscle tone)	No movement, muscles flaccid	Limbs partially flexed, little movement, poor muscle tone	Active movement, good muscle tone
Respiratory effort (amount of breathing)	No breathing	Slow, irregular breathing	Good, regular breathing; strong cry

score of 8–10 at 1–5 minutes after birth is considered normal. Other systems that assess neonatal growth and development, including general external appearance and neurological development, also exist.

Congenital Disorders

The term *congenital* means "present at birth," and **congenital disorders** are abnormalities commonly referred to as birth defects. Of all congenital disorders, approximately 70% are of unknown cause, 15% have a known genetic cause, and the remaining 15% result from environmental factors or a combination of environmental and genetic factors. Environmental factors damage the fetus during development. Environmental agents that cause congenital disorders are called **teratogens** (ter′ă-tō-jenz). For example, fetal alcohol syndrome results when a pregnant woman drinks alcohol, which crosses the placenta and damages the fetus. The baby is born with a smaller-than-normal head, mental retardation, and possibly other defects. Researchers are working to identify various teratogens, so that women can avoid known teratogens and reduce the risk for congenital disorders.

ASSESS YOUR PROGRESS

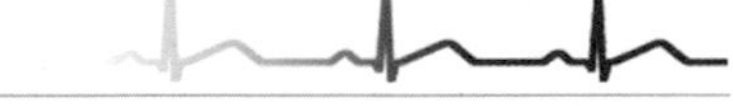

25. *What changes occur in the newborn's cardiovascular system shortly after birth? What do each of the following fetal structures become: foramen ovale, ductus arteriosus, umbilical vein, and ductus venosus?*

26. *What changes take place in the newborn's digestive system shortly after birth?*

27. *What does the Apgar score measure?*

28. *What are congenital disorders? What are some causes of these disorders?*

29. *What is a teratogen? Give an example.*

29.4 Lactation

LEARNING OUTCOMES

After reading this section, you should be able to

A. Describe the events of lactation.

B. Relate the roles of hormones in milk production and release.

Lactation is the production of milk by the mother's breasts (mammary glands; figure 29.21) following parturition.

During pregnancy, the high concentration and continuous presence of estrogens and progesterone cause the duct system and secretory units of the breasts to expand. The ducts grow and branch repeatedly to form an extensive network. Additional adipose tissue is deposited also; thus, the size of the breasts increases throughout pregnancy. Estrogen is primarily responsible for this breast growth, but normal development of the breast does not occur without the influence of several other hormones. Progesterone causes development of the breasts' secretory alveoli, which enlarge but do not usually secrete milk during pregnancy. The other necessary hormones are growth hormone, prolactin, thyroid hormones, glucocorticoids, and insulin. In addition, the placenta secretes a growth hormone–like substance (human somatotropin) and a prolactin-like substance (human placental lactogen), both of which support development of the breasts.

Prolactin, produced by the anterior pituitary gland, is responsible for milk production. Before parturition, high levels of estrogen stimulate increased prolactin production, but no milk is produced because high levels of estrogen and progesterone inhibit prolactin's effect on the mammary glands. After parturition, estrogen, progesterone, and prolactin levels decrease, allowing prolactin to stimulate milk production. Despite a decrease in the basal levels of prolactin, a reflex response produces surges of prolactin release. During suckling, mechanical stimulation of the breasts initiates nerve impulses that reach the hypothalamus, causing the secretion of **prolactin-releasing factor (PRF)** and inhibiting the release of **prolactin-inhibiting factor (PIF).** Consequently, prolactin levels temporarily increase and stimulate milk production.

For the first few days after birth, the mammary glands secrete **colostrum** (kō-los′trŭm), a high-protein material that contains many antibodies. Although colostrum is high in protein, it contains few lipids and less lactose than milk. Eventually, the breasts produce milk with a higher lipid and lactose content. Colostrum and milk not only provide nutrition but also contain antibodies (see chapter 22) that help protect the nursing baby from infections.

Because it takes time to produce milk, an increase in prolactin results in the production of milk to be used in the next nursing period. At the time of nursing, stored milk is released as a result of a reflex response. Mechanical stimulation of the breasts produces nerve impulses that cause the release of oxytocin from the posterior

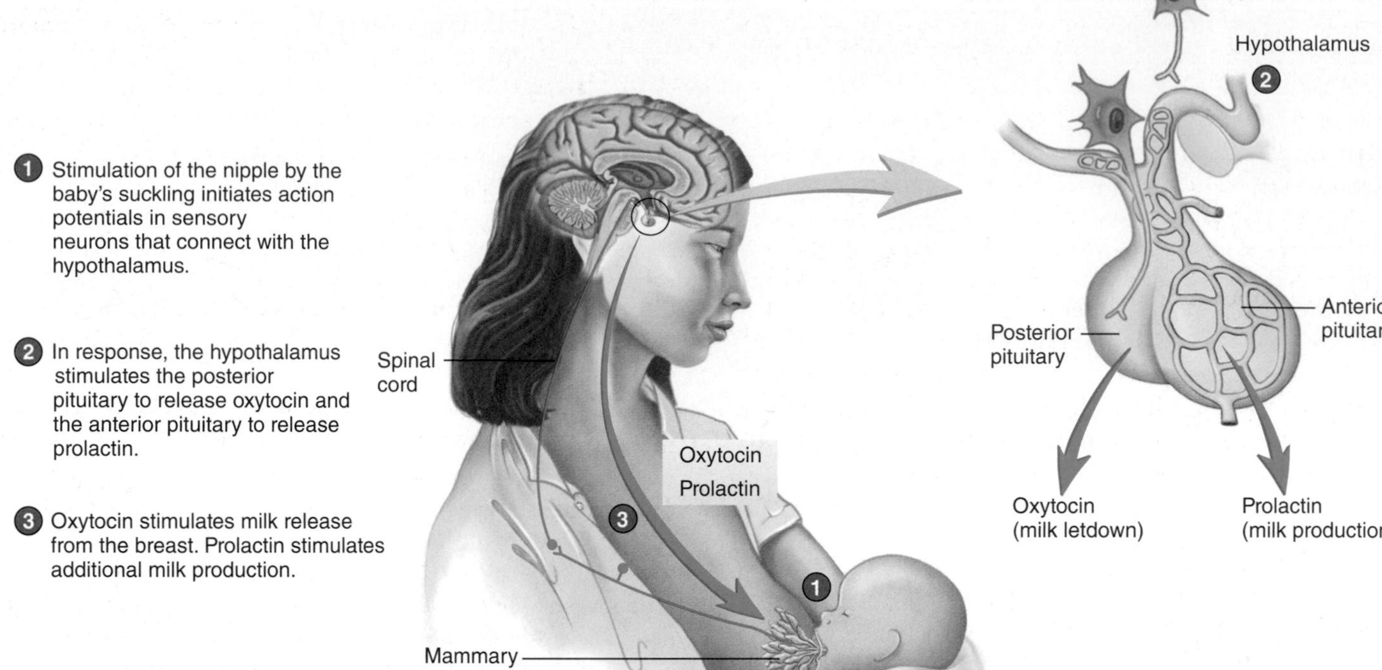

1 Stimulation of the nipple by the baby's suckling initiates action potentials in sensory neurons that connect with the hypothalamus.
2 In response, the hypothalamus stimulates the posterior pituitary to release oxytocin and the anterior pituitary to release prolactin.
3 Oxytocin stimulates milk release from the breast. Prolactin stimulates additional milk production.
Spinal cord
Oxytocin
Prolactin
Mammary gland
Hypothalamus
Posterior pituitary
Anterior pituitary
Oxytocin (milk letdown)
Prolactin (milk production)

PROCESS FIGURE 29.21 Hormonal Control of Lactation

pituitary, which stimulates cells surrounding the alveoli to contract. Milk is then released from the breasts, a process called **milk letdown** or *milk ejection*. In addition, higher brain centers can stimulate oxytocin release, and stimuli such as hearing an infant cry can result in milk letdown.

Repeated stimulation of prolactin release by suckling makes nursing (breastfeeding) possible for several years. If nursing stops, however, the ability to produce prolactin ceases, and milk production stops within a few days.

ASSESS YOUR PROGRESS

30. *What hormones are involved in preparing the breast for lactation?*
31. *Describe the events of milk production and milk letdown.*
32. *What is in colostrum?*

Predict 6

A first-time mother was breastfeeding her baby when she felt uterine cramps. Concerned, she consulted a nurse. What was the nurse's explanation?

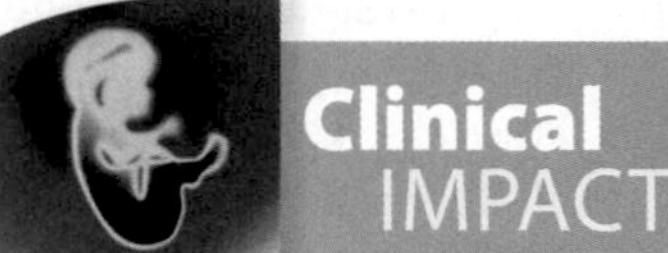

Clinical IMPACT

HIV and the Newborn

The human immunodeficiency virus (HIV) can be transmitted from a mother to her child in utero, during parturition, or during breastfeeding. HIV has been isolated from human breast milk and colostrum. In developed countries, HIV patients are advised not to breastfeed but to bottle-feed using formula instead. However, in underdeveloped countries, bottle-feeding is not yet a viable option. Evidence indicates that newborn babies given a single dose of an AIDS drug at birth have half the risk for HIV infection as long as they are not breastfed. Therefore, the current proposal for underdeveloped countries is that daily AIDS drug treatment be continued for newborns of HIV patients as long as the child is breastfeeding. In one study, treated infants had a 1% risk for HIV infection, compared with a 15% rate in untreated infants.

29.5 First Year After Birth

LEARNING OUTCOME

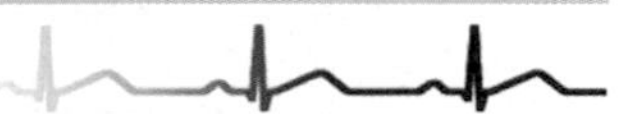

After reading this section, you should be able to

A. Describe the changes that occur during the first year after birth.

Many changes occur in an infant from birth until 1 year of age. The brain is still developing, and much of what the infant can do depends on how much brain development has occurred. It is estimated that the newborn's central nervous system contains the adult number of neurons, but subsequent brain growth and maturation add new neuroglia, myelin sheaths, and new connections between neurons that may remain throughout life.

During the first year of life, infants typically pass certain developmental milestones. However, the timing of these events varies considerably from child to child, and the following time frames are only rough estimates. By 6 weeks of age, the infant usually can hold up his head when placed in a prone position and begins to smile in response to people or objects. At 3 months, the infant exercises his limbs aimlessly, but he can control his arms and hands enough to voluntarily suck his thumb. The infant can follow a moving person with his eyes. At 4 months, the infant begins to raise himself by his arms. He can begin to grasp objects placed in his hand, coo and gurgle, roll from back to side, listen quietly to a person's voice or to music, hold his head erect, and play with his hands.

At 5 months, an infant can usually laugh, reach for objects, turn her head to follow an object, lift her head and shoulders, sit with support, and roll over. At 8 months, the infant recognizes familiar people, sits up without support, and reaches for specific objects. At 12 months, the infant may pull herself to a standing position and may be able to walk without support. She can pick up objects with her hands and examine them carefully. She can understand much of what is said and may say several words of her own.

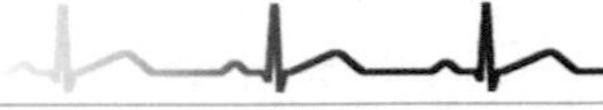

ASSESS YOUR PROGRESS

33. *List the major changes that occur during the first year of life. What do most of these activities depend on?*

29.6 Aging and Death

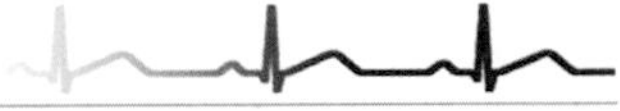

LEARNING OUTCOMES

After reading this section, you should be able to

A. Describe the process of aging.

B. Explain the events that occur at the time of death.

The development of a new human begins at fertilization, and so does the process of aging. Cells proliferate at an extremely rapid rate during early development, and then the process begins to slow as various cells become committed to specific body functions.

Many body cells, such as liver and skin cells, continue to proliferate throughout life, replacing dead or damaged tissue. However, many other cells, such as the neurons in the central nervous system, cease to proliferate once a certain number exist, and dead cells are not replaced. After the number of neurons reaches a peak, at about the time of birth, their numbers begin to decline. Neuronal loss is most rapid early in life and later decreases to a slower, steadier rate.

A natural, but as yet unexplained, decline occurs in mitochondrial DNA function with age. If this decline reaches a threshold, which apparently differs from tissue to tissue, the normal mitochondrial function is lost, and the tissue or organ may exhibit disease. In a small number of people, this mitochondrial degeneration occurs very early in life, resulting in premature aging.

The physical plasticity (i.e., the state of being soft and pliable) of young embryonic tissues results largely from the presence of large amounts of hyaluronate and relatively small amounts of collagen. Furthermore, the collagen and other related proteins that are present are not highly cross-linked; thus, the tissues are very flexible and elastic. Many of these proteins produced during development are permanent body components; as the person ages, more and more cross-links form between these protein molecules, thereby rendering the tissues more rigid and less elastic.

The tissues with the highest content of collagen and other related proteins are most severely affected by the collagen cross-linking and tissue rigidity associated with aging. One of the first structures to exhibit pathological changes as a result of this increased rigidity is the lens of the eye. Seeing close objects becomes more difficult with advancing age until most middle-aged people require reading glasses. Loss of elasticity also affects other tissues, including the joints, blood vessels, kidneys, lungs, and heart, and greatly reduces their ability to function.

Like nervous tissue, mature muscle cells do not normally proliferate after terminal differentiation occurs before birth. As a result, the total number of skeletal and cardiac muscle fibers declines with age. The strength of skeletal muscle reaches a peak between 20 and 30 years of life and declines steadily thereafter. Furthermore, like the collagen of connective tissue, the macromolecules of muscle undergo biochemical changes during aging and render the muscle tissue less functional. A good exercise program, however, can slow or even reverse this process (figure 29.22).

The decline in muscular function also contributes to a decline in cardiac function with advancing age. The heart loses elastic recoil ability and muscular contractility. As a result, total cardiac output declines, and less oxygen and fewer nutrients reach cells, such as neurons in the brain and chondrocytes in the joints, thereby contributing to the decline in these tissues. Reduced cardiac function also may result in decreased blood flow to the kidneys, which decreases their filtration ability. Degeneration of the connective tissues as a result of collagen cross-linking and other factors also decreases the filtration efficiency of the glomerular basement membrane.

Atherosclerosis (ath′er-ō-skler-ō′sis) is the deposit and subsequent hardening of a soft, gruel-like material, containing lipids, cholesterol, calcium, and other materials, in lesions of the tunica intima of large and medium-sized arteries. These deposits then become fibrotic and calcified, resulting in **arteriosclerosis** (ar-tēr′ē-ō-skler-ō′sis; hardening of the arteries). Arteriosclerosis interferes with normal blood flow and can result in a **thrombus,** the formation of a clot or plaque inside a vessel. A piece of the

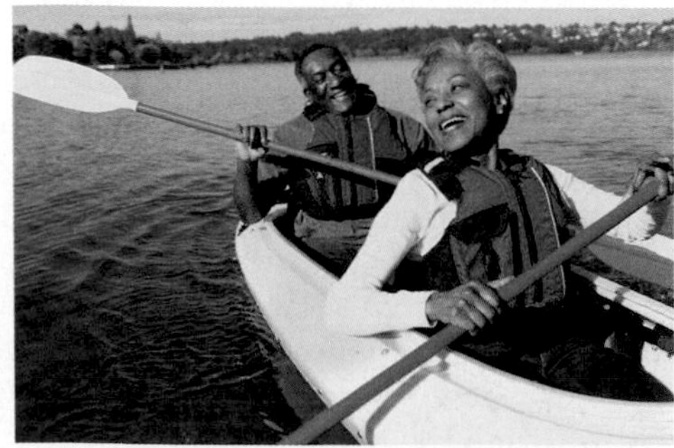

FIGURE 29.22 Active Older Adults
The population of older adults exhibits a range of variability. In some adults over 70 years, many systems are beginning to fail, whereas others can look forward to at least 10 more years of healthy living.

plaque, called an **embolus** (em′bō-lŭs) can break loose, float through the circulation, and lodge in smaller arteries, where it may cause myocardial infarctions or strokes. Atherosclerosis affects all middle-aged and older people to some extent and may even occur in young people. People with high blood cholesterol appear to be at increased risk of developing atherosclerosis. In addition to dietary influences, this condition seems to have a heritable component, and blood tests are available to screen people for high blood cholesterol.

Many other organs, including the liver, pancreas, stomach, and colon, undergo degenerative changes with age. The ingestion of harmful agents may accelerate such changes. For example, cigarette smoke induces degenerative changes in the lungs, and excessive alcohol consumption can cause sclerotic changes in the liver.

In addition to the previously described changes, cellular wear and tear contributes to aging. Progressive damage from many sources, including radiation and toxic substances, may result in irreversible cellular insults and may be one of the major factors leading to aging. Researchers have speculated that ingesting the antioxidant vitamins C and E in combination may help slow this part of aging by stimulating cell repair. Vitamin C also stimulates collagen production and may slow the loss of tissue plasticity associated with aging collagen.

According to the **free radical theory of aging,** free radicals, which are atoms or molecules with an unpaired electron, can react with and alter the structure of molecules that are critical for normal cell function. Alteration of these molecules can result in cell dysfunction, cancer, or other types of cellular damage. Free radicals are produced as a normal part of metabolism and introduced into the body from the environment through the air we breathe and the food we eat. The damage caused by free radicals may accumulate with age. Antioxidants, such as beta carotene (provitamin A), vitamin C, and vitamin E, can donate electrons to free radicals, without themselves becoming harmful. Thus, antioxidants may prevent the damage caused by the free radicals and may ward off age-related disorders, ranging from wrinkles to cancer. Experiments designed to test this hypothesis, however, have not consistently produced positive results.

As a result of poor diet, many people over age 50 do not get the minimum daily allotment of several vitamins and minerals. Feeling "bad" is not necessarily a part of aging but is mostly a result of poor nutrition and lack of exercise. Engaging in moderate exercise and avoiding overeating can prolong life. Moderate exercise can reduce the risk for heart attack by as much as 20%. It can also reduce the risk for stroke, high blood pressure, and some forms of cancer. Exercise can also increase a person's ability to reason and remember. Walking 30 minutes a day is recommended.

One characteristic of aging is an overall decrease in ATP production. This decrease is associated with a decline in oxidative phosphorylation, which has been shown in many cases to result from **mitochondrial DNA mutations.** Mitochondria lack the DNA repair mechanisms that exist in the nucleus for nuclear DNA, and thus mutations can accumulate in mitochondrial DNA.

Immune system changes may also contribute to the effects of aging. The aging immune system loses its ability to respond to outside antigens but becomes more sensitive to the body's own antigens. These autoimmune changes add to the degeneration of the tissues and may be responsible for conditions such as arthritis, chronic glomerular nephritis, and hyperthyroidism. In addition, T lymphocytes tend to lose their functional capacity with aging and cannot destroy abnormal cells as efficiently. This change may be one reason that certain types of cancer occur more readily in older people.

Genetic traits may also cause many of the changes associated with aging. As a general rule, animals with a very high metabolic rate have a shorter life span than those with a lower metabolic rate. In humans, a very small number of exceptional people have a slightly reduced normal body temperature, suggesting a lower metabolic rate. The same people often have an unusually long life span. This tendency appears to run in families and probably has some genetic basis. Studies of the general population suggest that, if your parents and grandparents have lived a long time, so will you.

Another piece of evidence suggesting a strong genetic component to aging comes from a disorder called **progeria** (prō-jĕr′ē-ă; premature aging). This genetic trait causes the degenerative changes of aging to begin occurring shortly after the first year of life, and the child may look like a very old person by age 7.

One of the greatest disadvantages of aging is the increasing inability to adjust to stress. Older people have a far more precarious homeostatic balance than younger people, and eventually the body encounters some stressor so great that it cannot recover, and death results. Death is usually not attributed to old age. Another problem, such as heart failure, renal failure, or stroke, is usually the cause.

Death was once defined as the loss of heartbeat and respiration. In recent years, however, more precise definitions of death have been developed because medical science has learned how to keep both the heart and lungs working artificially, and the heart can even be replaced by an artificial device. Modern definitions of death are based on the permanent cessation of life functions and the cessation of integrated tissue and organ function. The most widely accepted indication of death in humans is **brain death,** which is defined as irreparable brain damage manifested clinically by the absence of (1) response to stimulation, (2) spontaneous breathing and heartbeat, and (3) brainstem reflexes, in addition to an electroencephalogram that remains isoelectric ("flat") for at least 30 minutes.

ASSESS YOUR PROGRESS

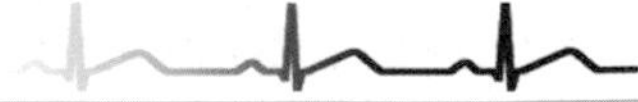

34. *How does the loss of cells that are not replaced affect the aging process? Give examples.*
35. *How does the loss of tissue plasticity affect the aging process? Give examples.*
36. *What tissue component is most affected by tissue rigidity?*
37. *Explain the free radical theory of aging.*
38. *How does aging affect the immune system?*
39. *What role does genetics play in aging?*
40. *What is the modern definition of death? How is death determined?*

29.7 Genetics

LEARNING OUTCOMES

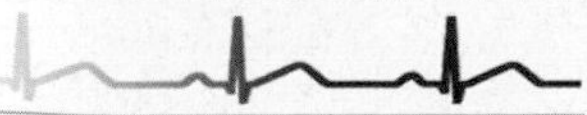

After reading this section, you should be able to

A. Define *genetics* and explain how chromosomes are related to inheritance.

B. Explain how Mendel's theories of genetics compare to modern concepts of genetics.

C. Define and give examples of *phenotypes* and *genotypes.*

D. Explain what a karyotype is.

E. Describe major patterns of inheritance.

F. Relate how mitosis differs from meiosis.

G. Explain how various genetic disorders can occur.

Genetics is the study of heredity—that is, the characteristics children inherit from their parents. Although the environment can influence gene expression, many of a person's physical characteristics, abilities, susceptibility to disease, and even life span are influenced by the genes inherited from his or her parents. Because many of the diseases caused by microorganisms are now preventable or treatable, diseases that have a genetic basis are receiving more attention.

An understanding of genetics is important for medical professionals. Knowing a patient's family medical history helps diagnose many diseases. The family medical history also allows a physician to determine the probability that a patient will develop certain diseases, such as heart disease or cancer, and to suggest preventive measures. Recent advances in genetics have shown how genes influence health and can provide new methods for treating certain diseases.

Some genetic diseases, such as hemophilia, have a simple pattern of inheritance, usually involving one or a few genes, and it is easy to predict whether a person will develop the disease. However, other genetic diseases have a more complex pattern of inheritance involving many more genes. **Mendelian genetics,** the study of how genetic traits are passed from parent to offspring, has been used to determine an individual's risk of developing certain genetic diseases.

With the completion of the Human Genome Project in 2003, more than 99% of the genes in human chromosomes, collectively known as the *human genome,* have been sequenced. The results of this project have provided a wealth of useful information, allowing researchers to better identify the genes associated with particular diseases, to understand the biochemical relationship between genes and diseases, and to create new treatments. The genetic approach to the diagnosis and management of disease is known as **genomic medicine.** Mendelian genetics has indeed come a long way since its early beginnings.

Mendelian Genetics

Gregor Mendel (1822–1884) was an Austrian monk who made revolutionary discoveries about inheritance patterns in pea plants. Because of his discoveries, Mendel is known as the "father of genetics." Mendel found that certain traits, such as purple versus white flowers, are transmitted from one generation of pea plants to another by means of discrete units he called "heritable factors." He concluded that each pea plant has two heritable factors for a characteristic, such as flower color. He proposed that, during the production of **gametes,** or sex cells, each gamete receives one of these factors. Then two gametes combine to produce the next generation, so that each member of the next generation of pea plants has two heritable factors for each characteristic. Mendel's "heritable factors" are now called genes. The genes an organism has for a given trait is called the **genotype** (jēn′ō-tīp). The expression of the genes as a trait is called the **phenotype** (fē′nō-tīp).

Environmental effects can influence gene expression to determine the phenotype of many traits. Even genetically identical twins can have different phenotypes because of environmental effects. For example, in addition to genes, height is affected by nutrition, and skin color by exposure to the sun. Knowledge of environmental influences can be used to improve our genetic potential and to prevent harmful effects. For example, a healthy diet can promote growth or help prevent diabetes, and refraining from smoking can reduce the risk of developing cancer.

Mendel proposed that genes occur in dominant and recessive forms. Alternate forms of genes are now called **alleles** (ă-lēlz′). By definition, the effects of a **dominant allele** for a trait mask the effects of the **recessive allele** for that trait. For example, in pea plants, the allele for purple flowers is dominant over the allele for white flowers. By convention, dominant alleles are indicated by uppercase letters, and recessive alleles are indicated by lowercase letters. For example, the letter *P* designates the dominant allele for purple flower color, and the letter *p* represents the recessive allele for white flower color. A plant with the *Pp* genotype has purple flowers because purple is dominant over white.

An organism is **homozygous** (hō-mō-zī′gŭs) for a trait if the two alleles for the trait are identical, whereas an organism is **heterozygous** (het′er-ō-zī′gŭs) if the two alleles for the trait are different. The possible genotypes and phenotypes for purple flower color are

Alleles	Genotype	Phenotype
PP	Homozygous dominant	Purple
Pp	Heterozygous	Purple
pp	Homozygous recessive	White

Note that the recessive trait is expressed only when it is not masked by the dominant trait.

The relationship between genotype and phenotype can be understood on the molecular level. For example, a recessive trait in humans called **albinism** (al′bi-nĭzm) results in a lack of normal coloring of the skin, hair, and eyes. Several human genes produce the enzymes that are necessary for the synthesis of melanin, the pigment responsible for skin, hair, and eye color (see chapter 5). An **albino,** a person with albinism, has light blonde or white hair and light-colored eyes and skin, with shades of pink, blue, and yellow (figure 29.23). The pink and blue colors result from blood seen in the eyes or through the skin. The yellow is from the natural accumulation of ingested yellow plant pigments in the skin (see chapter 5).

The normal alleles for the melanin-synthesizing enzymes produce normal, functional enzymes capable of catalyzing the steps

FIGURE 29.23 Albinism
An albino male with his normally pigmented father.

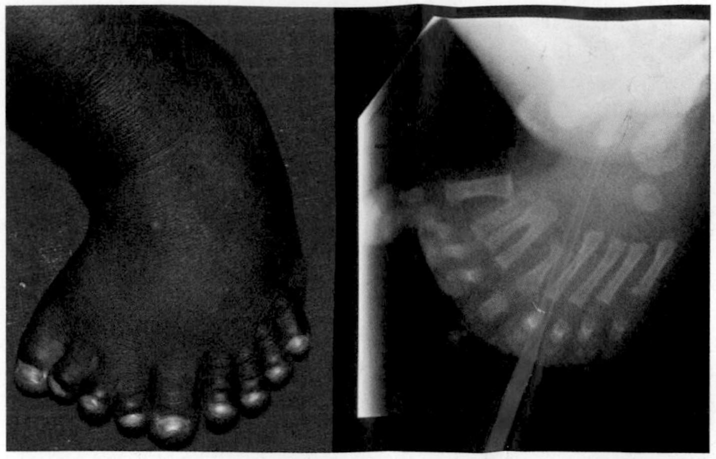

FIGURE 29.24 Polydactyly
Polydactyly, having extra fingers or toes, is determined by a dominant gene.

in melanin synthesis. When normal enzymes are produced, a person has normal pigmentation. But an abnormal allele for any one of the enzymes in the melanin pathway produces a defective enzyme, which blocks the pathway and stops the production of melanin. Type 1 albinism results from a defective enzyme that fails to initiate the synthesis of melanin from tyrosine. The normal allele of the gene for this enzyme, designated *A*, is dominant over the abnormal allele, designated *a*. The possible genotypes and phenotypes for albinism are

Alleles	Genotype	Phenotype
AA	Homozygous dominant	Normal pigmentation
Aa	Heterozygous	Normal pigmentation
aa	Homozygous recessive	Albino

A person with the genotype *AA* has the phenotype of normal pigmentation because the functional enzyme for melanin synthesis is produced by both alleles. A person with the genotype *Aa* also has the phenotype of normal pigmentation. Even though the *a* allele produces an abnormal, nonfunctional enzyme for melanin synthesis, the *A* allele produces sufficient amounts of the normal enzyme to keep pigmentation normal. A person with the genotype *aa* has the phenotype of albinism because only the abnormal, nonfunctional enzyme for melanin synthesis is produced.

Not all dominant traits are the normal condition, and not all recessive traits are abnormal. In some cases, the dominant trait is abnormal. For example, a person with **polydactyly** (pol-ē-dak′ti-lē) has extra fingers or toes (figure 29.24). One allele for polydactyly is dominant over the recessive, normal allele that results in the normal number of fingers or toes.

Predict 7

List all the possible genotypes and phenotypes for polydactyly. Use the letters *D* and *d* for the alleles.

Modern Concepts of Genetics

Although the existence of chromosomes was known in Mendel's time, the relationship between genes and chromosomes was not confirmed until T. H. Morgan was able to induce mutations (gene changes) in fruit flies in 1910. Researchers first isolated DNA as the genetic material in 1944, and James Watson and Francis Crick resolved the structure of DNA in 1953, which led to an understanding of the genetic code and protein synthesis (see chapter 3). The sequence of nucleotides of a human gene, specifically the gene that causes cystic fibrosis, was determined for the first time in 1989. The sequencing of more than 99% of the human genome was accomplished in 2003 (see Clinical Genetics, "The Human Genome Project," later in this chapter).

Chromosomes

In modern terminology, Mendel hypothesized that organisms have genes that control the expression of traits. Most genes, which are segments of DNA, occur in pairs of alleles. Chromosomes are made up of DNA and associated proteins found in the nuclei of somatic cells and gametes. **Somatic** (sō-mat′ik) **cells** are all the cells of the body except the gametes. Examples of somatic cells are epithelial cells, muscle cells, neurons, fibroblasts, lymphocytes, and macrophages. In males, the gametes are **sperm cells;** in females, the gametes are **oocytes** (egg cells; see chapter 28).

The normal number of chromosomes in a somatic cell is called the **diploid** (dip′loyd; twofold) number. The normal number of chromosomes in a gamete is the **haploid** (hap′loyd; single) number. In humans, the diploid number of chromosomes is 46. Chromosomes in diploid cells are paired, so humans have 23 pairs of chromosomes. The haploid number of chromosomes is 23.

Predict 8

Why does it make sense that the number of chromosomes in a gamete is half the number in a somatic cell?

The 23 pairs of chromosomes are divided into autosomal and sex chromosomes. Humans have 22 pairs of **autosomal** (aw-tō-sō′măl) **chromosomes,** which are all the chromosomes except the sex chromosomes, and 1 pair of **sex chromosomes,** which determines the sex of the individual. Sex chromosomes are denoted as **X** or **Y chromosomes.** A normal female has two X chromosomes (XX) in each somatic cell. One X chromosome of a female is derived from her mother; the other is derived from her father. A normal male has one X chromosome and one Y chromosome (XY) in each somatic cell. The X chromosome of a male is derived from his mother; the Y chromosome is derived from his father.

Genetic traits can be classified by the type of chromosome their alleles are located on and by whether their alleles are dominant or recessive. Thus, traits can be autosomal dominant or autosomal recessive, or they can be sex-linked dominant or sex-linked recessive. For example, albinism is an autosomal recessive trait.

A **karyotype** (kar′ē-ō-tīp) is a display of the chromosomes of a somatic cell during metaphase of mitosis. It is produced by photographing the cell's stained chromosomes through a microscope and arranging the photographed chromosomes in pairs (figure 29.25). For convenience, the autosomal chromosomes are numbered, from largest to smallest, 1 through 22. The sex chromosomes are denoted with an X or a Y. Note that a chromosome in a karyotype is a replicated mitotic chromosome—that is, it has two chromatids (figure 29.25). The chromatids of each chromosome are attached at the centromere, so that each replicated chromosome appears as a single structure.

Chromosome pairs are called **homologous** (hŏ-mŏl′ō-gŭs) **chromosomes,** and each member of the pair is considered a **homolog** (hŏm′ō-lŏg). One homolog is derived from a person's father, and the other is derived from a person's mother. A **genome** (jē′nōm, jĕ′nōm) consists of all the genes found in the haploid number of chromosomes from one parent. The combined genomes from both parents are responsible for all of a person's genetic traits. Each gene of the genome occupies a specific **locus,** or location, on a chromosome (figure 29.26). The locus on one chromosome may have one allele type, whereas the locus on its homolog may have another allele type.

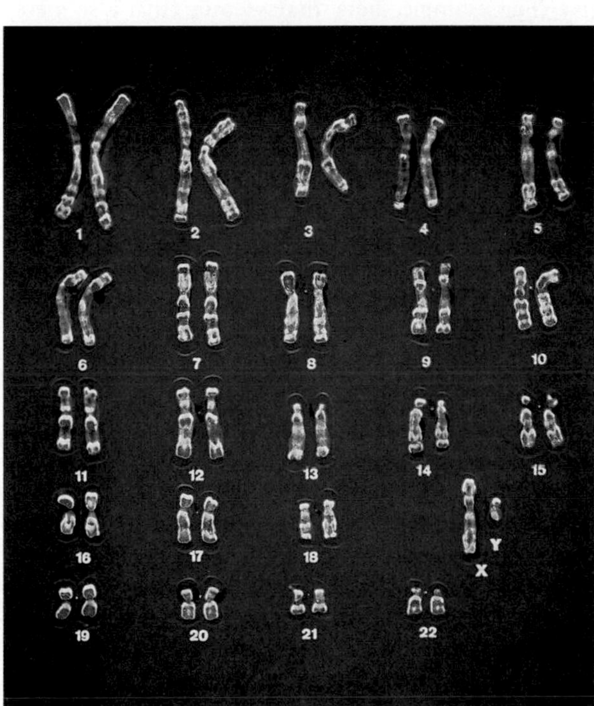
1
2
3
4
5
6
7
8
9
10
11
12
13
14
15
16
17
18
X
Y
19
20
21
22

FIGURE 29.25 Human Karyotype
The 23 pairs of chromosomes in a human cell consist of 22 pairs of autosomal chromosomes (numbered 1–22) and 1 pair of sex chromosomes. This is the karyotype of a male, as evidenced by the presence of an X and a Y sex chromosome. A female karyotype would have two X chromosomes.

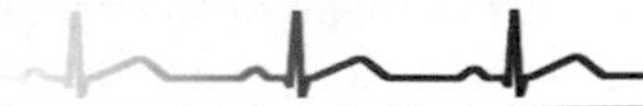

ASSESS YOUR PROGRESS

41. *What is genetics?*
42. *What is the difference between genotype and phenotype? Give an example of each.*
43. *What are alleles? If tall (T) plants are dominant over short (t) plants, what would the alleles and phenotypes be for homozygous dominant, heterozygous, and homozygous recessive?*
44. *How do the chromosomes in somatic cells and gametes differ from each other?*
45. *What are the number and type of chromosomes in the karyotype of a human somatic cell? How do the chromosomes of a male and a female differ?*
46. *What are homologous chromosomes?*

Multiple Alleles

Mendel believed that alleles came in two forms, dominant and recessive. We now know that alleles can exist in many forms, called **multiple alleles.** An individual has only two alleles for a given gene, one on each homologous chromosome. At the population level, however, many forms of an allele may exist.

Differences in alleles arise by mutation (see "Genetic Disorders," later in this section), in which the DNA nucleotide sequence is altered. Many alleles are possible in a population because any change in the nucleotide sequence of DNA, even in one base pair, potentially produces a different allele. A different form of an allele is called an **allelic variant,** a *mutated allele (gene),* or a *polymorphism* (many forms).

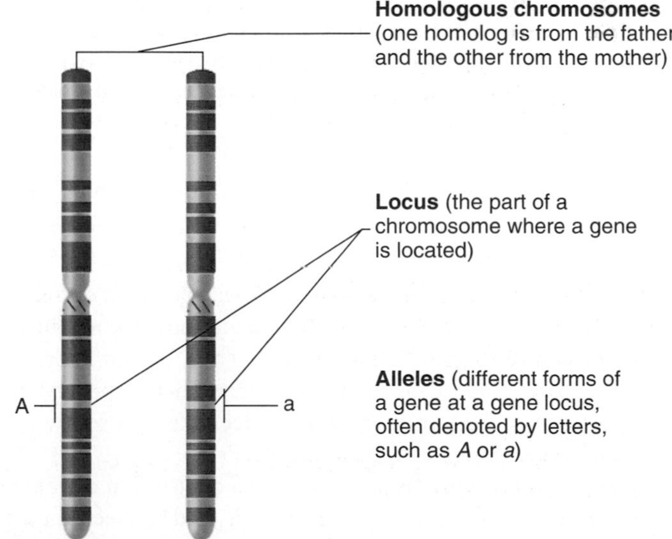
Homologous chromosomes (one homolog is from the father and the other from the mother)
Locus (the part of a chromosome where a gene is located)
A
a
Alleles (different forms of a gene at a gene locus, often denoted by letters, such as *A* or *a*)

FIGURE 29.26 Homologous Chromosomes
This drawing of a pair of unduplicated homologous chromosomes depicts them as compact bodies for simplicity. In the unduplicated state, chromosomes are dispersed within the nucleus as chromatin.

Allelic variants can result in either no effect on the phenotype or minor to major phenotypic changes. Allelic variants at a locus may encode for different sequences of amino acids in proteins. Recall from chapter 2 that the sequence of amino acids in a protein affects the shape of the protein, including the protein's domain, which is the functional part of the protein. An allelic variant that causes an amino acid change in a protein that does not significantly change protein shape may not affect the phenotype. However, the greater the effect of an amino acid change on a protein's shape, the greater the effect on the protein's function and the greater the effect on the phenotype. Millions of allelic variants have been identified. The significance of most of them is unknown.

The allelic variant present at a locus can affect a person's phenotype. For example, **phenylketonuria** (fen′il-kē′tō-noo′rē-ă) **(PKU)** is an autosomal recessive trait with multiple alleles. A gene on chromosome 12 encodes for an enzyme that converts the amino acid phenylalanine to the amino acid tyrosine. If phenylalanine is not converted to tyrosine, high levels of phenylalanine can accumulate, leading to brain cell damage. The more defective the enzyme, the greater the accumulation of phenylalanine and the greater the damage to the brain, in most cases. Over 400 disease-causing allelic variants of the PKU gene are known. The severity of PKU depends in part on which two allelic variants are present. The less severe of the two alleles determines the severity of the disorder. Symptoms of untreated PKU can vary from profound mental retardation to nearly normal mental abilities. Thus, even though PKU is an autosomal recessive trait, the expression of the recessive condition exhibits considerable variability because of multiple alleles. Furthermore, reducing phenylalanine in the diet can prevent the development of mental retardation.

Predict 9

Paula is a woman with PKU. In her case, the conversion of the amino acid phenylalanine to tyrosine is quite limited. Mental retardation, a common consequence of this condition, can be avoided if people who have PKU consume food containing very small amounts of phenylalanine. Paula is healthy even though she has PKU because from the time she was very small, her diet has been monitored so that it contains little phenylalanine. Paula is engaged to be married to a man named Marvin. No PKU has been identified in his family for many generations. What are the possible genotype(s) for Paula and Marvin's children?

Gene Dominance

In Mendel's experiments with pea plants' flower color, each dominant allele produced enough of its protein product to cause the maximum phenotypic response in the heterozygote. Thus, the homozygous dominant and the heterozygote had the same phenotype—purple flowers. This is called **complete dominance.** We now know that many other types of gene expression exist.

Complete dominance is at one end of a continuum of genetic expression. At the other end is codominance, and between them is incomplete dominance. In **codominance,** two alleles at the same locus are expressed, so that separate, distinguishable phenotypes occur at the same time. ABO blood types are an example of codominance. A gene with three alleles on chromosome 9 encodes for enzymes that add sugar molecules to certain carbohydrates found on the surface of red blood cells. The carbohydrates are part of glycoproteins (see figure 3.4), called A and B antigens (see chapter 19). These carbohydrates consist of different sugars joined together. Traditionally, in designating alleles for blood type, the A and B alleles are superscripted to a capital letter *I*, and the O allele is designated by the lowercase letter *i*. The I^A allele encodes for an enzyme that adds a particular sugar to the ends of the carbohydrate, producing the A antigen. The I^B allele encodes for a different enzyme that adds a different sugar to the ends of the carbohydrate, producing the B antigen. The *i* allele encodes for no functional enzyme and therefore is recessive to A and B. The possible genotypes and phenotypes for the ABO blood group are

Genotype	Phenotype
I^AI^A or I^Ai	Type A blood (A antigen only)
I^BI^B or I^Bi	Type B blood (B antigen only)
I^AI^B	Type AB blood (A and B antigens)
ii	Type O blood (neither A nor B antigen)

Type A blood results from the expression of only the I^A allele, type B blood from the expression of only the I^B allele, and type O blood from the expression of neither the I^A allele nor the I^B allele. Type AB blood illustrates codominance and results from the expression of both the I^A allele and the I^B allele at the same time.

In **incomplete dominance,** the dominant allele does not completely mask the effects of the recessive allele in the heterozygote. The heterozygote produces less of the protein product than the homozygous dominant and has phenotypic characteristics intermediate between the homozygous dominant and the homozygous recessive. For example, **beta thalassemia** (thăl-ă-sē′mē-ă) is a disorder of a gene on chromosome 11. It affects the synthesis of β-globulin polypeptide chains, which are part of the hemoglobin in red blood cells. Hemoglobin is a protein that transports oxygen. If normal amounts of the β-globulin polypeptide are produced, the β-globulin polypeptides join with other proteins to form hemoglobin. If lower than normal amounts of β-globulin polypeptide are synthesized, lower than normal amounts of hemoglobin are produced. In the homozygous dominant state, two normal alleles for β-globulin synthesis are present, normal amounts of hemoglobin are produced, and the phenotype is normal, meaning that normal oxygen transport is occurring. In the homozygous recessive condition, called **major thalassemia,** two abnormal alleles are present, and much lower than normal amounts of β-globulin polypeptide chains are synthesized. The result is severe **anemia** (ă-nē′mē-ă), which is a deficiency of hemoglobin in the blood. Symptoms include pallor, weakness, fatigue, and spleen enlargement. Blood transfusions are necessary to maintain hemoglobin levels. In the heterozygous condition, called **minor thalassemia,** one normal allele and one abnormal allele are present. The production of hemoglobin is intermediate between the normal phenotype and major thalassemia, and mild anemia results.

Polygenic Traits

In some cases, one gene (usually with two alleles) determines one phenotype. In other cases, many genes determine a phenotype; these so-called **polygenic traits** result from the interactions of many genes. Although all the genes contribute to the phenotype without being dominant or recessive to each other, each gene has its own characteristics, such as multiple alleles or incomplete dominance. Some of the genes may be more important than others for the expression of the phenotype, and the genes can be on different chromosomes. The result of all the gene interactions is a phenotype with a great amount of variability. Examples of polygenic traits are height, intelligence, eye color, and skin color.

Even though a complex combination of genes determines a polygenic trait, a defect in one of these genes can sometimes have a dramatic effect on phenotype. For example, even though many genes contribute to skin color, one defective gene can eliminate skin color completely, resulting in albinism.

Sex-Linked Traits

Sex-linked traits are traits affected by genes on the sex chromosomes; **X-linked traits** are affected by genes on the X chromosome, and **Y-linked traits** are affected by genes on the Y chromosome. Most sex-linked traits are X-linked, whereas only a few Y-linked traits exist, mainly because the Y chromosome is very small. An example of an X-linked recessive trait is **hemophilia A** (classic hemophilia), in which the body is unable to produce certain blood clotting factors (see chapter 19). Consequently, clotting is impaired, and persistent bleeding can occur either spontaneously or as a result of an injury. Traditionally, X-linked genes are designated by letters superscripted to the letter *X*. The possible genotypes and phenotypes for hemophilia A are

Genotype	Phenotype
X^HX^H or X^HX^h	Normal female
X^hX^h	Hemophiliac female
X^HY	Normal male
X^hY	Hemophiliac male

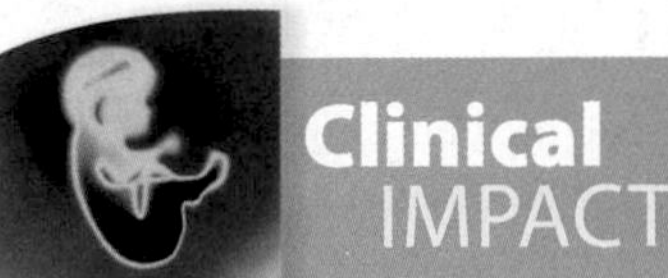

Clinical IMPACT

Sex Chromosome Abnormalities

A wide range of sex chromosome abnormalities exist. The presence of a Y chromosome makes a person genetically a male, and the absence of a Y chromosome makes a person genetically a female, regardless of the number of X chromosomes. Therefore, individuals with XO (Turner syndrome), XX, XXX, or XXXX karyotypes are females, and individuals with XY, XXY, XXXY, or XYY karyotypes are males. A YO condition is lethal because the genes on the X chromosome are necessary for survival. Secondary sex characteristics are usually underdeveloped in both the XO female and the XXY male (called Klinefelter syndrome), and additional X chromosomes (XXXX or XXXY) are often associated with some degree of mental retardation. Due to hormonal imbalances, the morphological features (phenotype) may be reversed from the genetic constitution (genotype). For example, a male with Klinefelter syndrome may display swelling of the breast.

Note that a female must have both recessive alleles (X^hX^h) to exhibit hemophilia, whereas a male, because he has only one X chromosome, has hemophilia if he has only one of the recessive alleles. For this reason, X-linked recessive traits are seen more frequently in males than in females.

ASSESS YOUR PROGRESS

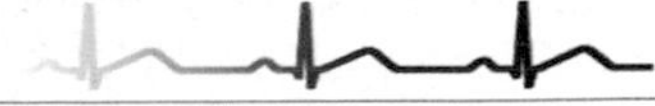

47. *Explain multiple allelism.*

48. *Distinguish among complete dominance, incomplete dominance, and codominance. Give examples of each.*

49. *What is the difference between multiple allelism and polygenic traits?*

50. *How are sex-linked traits inherited? Give an example.*

Predict 10

Wilma and Wally have one male child who has been diagnosed with Duchenne muscular dystrophy, an X-linked condition that results in severe atrophy of skeletal muscle. Neither parent has this condition. What is the probability that their next child will have Duchenne muscular dystrophy?

Meiosis and the Transmission of Genes

Gametes are haploid cells that are derived from diploid cells. In **meiosis** (mī-ō′sĭs), gametes are produced that have one homolog from each of the homologous pairs of chromosomes. Therefore, gametes have one-half the number of chromosomes and one-half as many alleles as the original diploid cells.

Before meiosis begins, DNA replication occurs, forming replicated chromosomes that each consist of two chromatids (figure 29.27). As a result of the first meiotic division, each daughter cell receives one replicated member of a homologous pair. No DNA replication takes place between the first and second meiotic divisions. During the second meiotic division, the chromatids of each chromosome separate. As a result of the second meiotic division, each cell receives one chromatid from each chromosome. That single chromatid is now referred to as a chromosome. Each gamete now has a haploid number of chromosomes. See chapter 28 for a more detailed discussion of gamete production.

Figure 29.27 shows the alleles for albinism on a pair of homologous chromosomes. On one homolog, the normal, dominant allele is indicated with an uppercase *A;* on the other homolog, the abnormal allele is indicated with a lowercase *a.* Figure 29.27 illustrates how the alleles on homologous chromosomes are distributed to gametes. In this case, the gametes have either an *A* or an *a* allele.

The probability of transmitting particular alleles to the next generation can be determined using a **Punnett square** if the genotypes of the parents are known (figure 29.28). The possible alleles in the gametes of one parent form the rows of the square, and the possible alleles in the gametes of the other parent form the columns. The body of the Punnett square shows all possible combinations of the parents' gametes. For example, suppose both parents are heterozygous (*Aa*) for albinism. The heterozygous parents can produce two types of gametes. One gamete has the *A* allele, and the other has the *a* allele. The probability that the child will be homozygous dominant (*AA*) is one out of four; heterozygous (*Aa*), two out of four; and homozygous recessive (*Aa*), one out of four.

1. A pair of unduplicated homologous chromosomes lie in the nucleus of a somatic cell. For simplicity, the chromosomes are shown as compact bodies. In the unduplicated state, chromosomes are dispersed within the nucleus as chromatin.
2. The dominant (*A*) and recessive (*a*) alleles for albinism are shown on the homologous chromosomes.
3. The DNA has replicated, and each replicated chromosome consists of two chromatids (see figure 3.40).
4. Each replicated homologous chromosome has two alleles because the DNA has replicated.
5. As a result of the first meiotic division, each new nucleus has a replicated homologous chromosome.
6. DNA is not replicated between the first and second meiotic divisions. As a result of the second meiotic division, the chromatids separate and are now called chromosomes.
7. Each homologous chromosome has an allele for albinism. Two types of gametes have been produced, those carrying the *A* allele and those carrying the *a* allele.

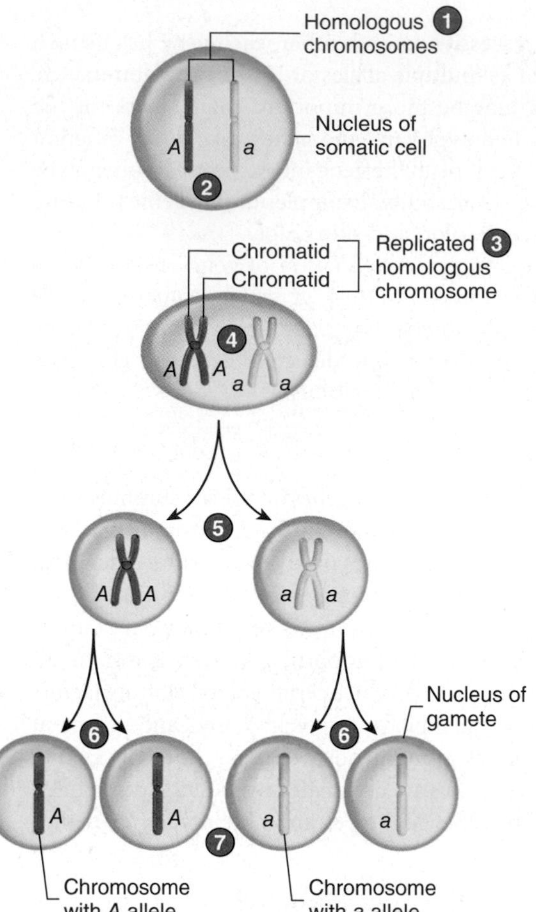
Homologous chromosomes 1
Nucleus of somatic cell
A
a
2
Chromatid
Chromatid
Replicated homologous chromosome 3
4
A
A
a
a
5
A
A
a
a
6
6
Nucleus of gamete
A
A
a
a
7
Chromosome with A allele
Chromosome with a allele

PROCESS FIGURE 29.27 APR **Meiosis and the Distribution of Alleles**

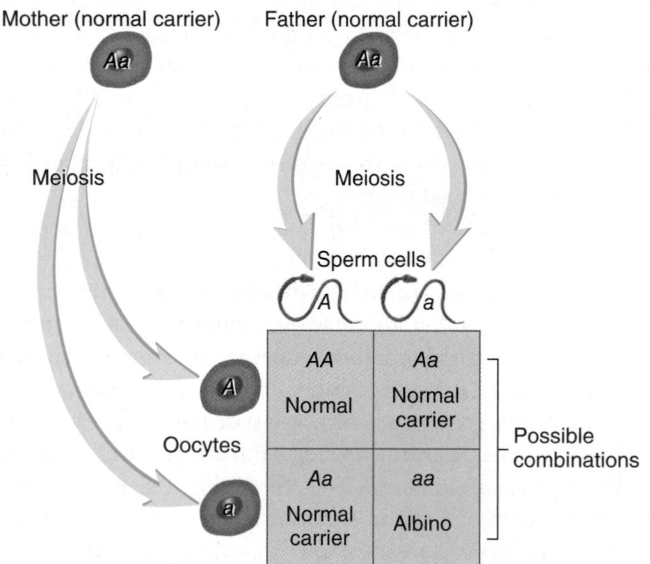
Mother (normal carrier)
Father (normal carrier)
Aa
Aa
Meiosis
Meiosis
Sperm cells
A
a
A
a
Oocytes
AA Normal
Aa Normal carrier
Aa Normal carrier
aa Albino
Possible combinations
Genotype and phenotype probabilities:
1/4 AA (normal) : 1/2 Aa (normal carrier) : 1/4 aa (albino)

FIGURE 29.28 Inheritance of a Recessive Trait: Albinism
In this Punnett square of a mating between two normal carriers, A represents the normal, pigmented condition, and a represents the recessive, unpigmented condition.

Albinism is an autosomal recessive trait. The homozygous dominant and heterozygous conditions result in normal pigmentation, whereas the homozygous recessive condition results in albinism. Both heterozygous parents have normal pigmentation, but there is a one out of four chance that any one of their children will be an albino. A **carrier** for a recessive trait is a person who is heterozygous for that trait, with one normal allele and one disorder-causing allele. The carrier does not exhibit the disorder but can pass it on to his or her children.

Genetic Disorders

A **genetic disorder** is a failure of structure, function, or both as a result of abnormalities in a person's genetic makeup—that is, his or her DNA. Humans may have a variety of genetic disorders (table 29.4), some of which were highlighted throughout this text. Genetic disorders determined by one gene are described by Mendelian characteristics—for example, as autosomal recessive. In some cases, this type of description is accurate, but in other cases the reality is more complex. For example, PKU is said to be autosomal recessive, but PKU has multiple alleles, which result in a spectrum of defects.

Genetic disorders involve either a single gene or an entire chromosome, as in the case of aneuploidy. Recall that a mutation is a change in a gene that usually involves a change in the number

TABLE **29.4** **Genetic Disorders**

Disorder	Description
Dominant Traits	
Achondroplasia	Dwarfism characterized by shortening of the upper and lower limbs
Huntington disease	Severe degeneration of the basal nuclei and frontal cerebral cortex; characterized by purposeless movements and mental deterioration; onset is usually between 40 and 50 years of age
Hypercholesterolemia	Elevated blood cholesterol levels that contribute to atherosclerosis and cardiovascular disease
Marfan syndrome	Abnormal connective tissue, resulting in increased height, elongated digits, and weakness in the aortic wall
Neurofibromatosis	Small, pigmented lesions (café-au-lait spots) in the skin and disfiguring tumors (noncancerous) caused by the proliferation of Schwann cells along nerves
Osteogenesis imperfecta	Abnormal collagen synthesis, resulting in brittle bones that break repeatedly
Recessive Traits	
Albinism	Lack of an enzyme necessary to produce the pigment melanin; characterized by lack of coloration in skin, hair, and eyes
Cystic fibrosis	Impaired transport of chloride ions across plasma membranes; results in excessive production of thick mucus, which blocks the respiratory and gastrointestinal tract; the most common fatal genetic disorder
Phenylketonuria	Lack of the enzyme necessary to convert the amino acid phenylalanine to the amino acid tyrosine; an accumulation of phenylalanine leads to mental retardation
Severe combined immune deficiency	Inability to form the white blood cells (B cells, T cells, and phagocytes) necessary for an immune system response
Sickle-cell disease	Inability to produce normal hemoglobin; results in abnormally shaped red blood cells that clog capillaries or rupture
Tay-Sachs disease	Lack of the enzyme necessary to break down certain lipids; an accumulation of lipids impairs action potential propagation, resulting in deterioration of mental and physical functions and death by 3–4 years of age
Thalassemia	Decreased rate of hemoglobin synthesis; results in anemia, enlarged spleen, increased cell numbers in red bone marrow, and congestive heart failure
Sex-Linked Traits	
Duchenne muscular dystrophy	Deletion or alteration of part of the X chromosome; results in progressive weakness and wasting of muscles
Hemophilia	Most commonly, failure to produce blood clotting factors, caused by a recessive gene; results in prolonged bleeding
Red-green color blindness	Most commonly, deficiency in functional green-sensitive cones, caused by a recessive gene; inability to distinguish between red and green colors
Chromosomal Disorders	
Down syndrome	Caused by having three chromosomes 21; results in mental retardation, short stature, and poor muscle tone
Klinefelter syndrome	Caused by two or more X chromosomes in a male (XXY); results in small testes, sterility, and development of femalelike breasts
Turner syndrome	Caused by having only one X chromosome; results in immature uterus, lack of ovaries, and short stature

or kinds of nucleotides composing the DNA (see chapter 2). Mutations are known to occur by chance (randomly without known cause), but they can also be caused by chemicals, radiation, and viruses. Agents that cause mutations are called **mutagens** (mū′tă-jenz). In most cases, the specific cause of a mutation cannot be determined. Once a mutation has occurred, however, the abnormal trait can be passed from one generation to the next.

A mutation involving a single nucleotide change is called a *point mutation*. Mutations that change the sequence or number of nucleotides involve changes in chromosome structure and therefore are called *structural mutations*. Structural mutations can involve deletions, the loss of part of a chromosome; duplications, the addition of a copied section of a chromosome; translocations, in which part of one chromosome becomes attached to another chromosome; and inversions, in which the DNA sequence of a section of a chromosome is reversed.

Furthermore, segregation errors that occur during meiosis result in changes in the chromosome number in the gametes. As the chromosomes separate during meiosis, the two members of a homologous pair may not segregate as they normally do. As a result, one of the daughter cells receives both chromosomes, and the other daughter cell receives none. This event is called **nondisjunction.** When the gametes are fertilized, the resulting zygote has either 47 chromosomes or 45 chromosomes, rather than the normal 46, a condition called **aneuploidy** (an′ū-ploy-dē). Aneuploidies include monosomies, in which a chromosome is missing, and trisomies, in which an extra chromosome is present. Aneuploidies are usually lethal and are one reason for a high rate of early embryo loss. **Down syndrome,** or *trisomy 21*, is a type of aneuploidy in which three chromosomes 21 are present. A **syndrome** (sĭn′drōm) is a set of signs and symptoms occurring together as the result of a single cause, such as a single mutation or one extra chromosome (a trisomy).

Clinical GENETICS | The Human Genome Project

The human genome consists of all the genes found in one homologous set of human chromosomes. Geneticists estimate that humans have 20,000–25,000 genes. A **genomic map** depicts the DNA nucleotide sequences of the genes and their locations on the chromosomes (figure 29C). Celera, a private corporation working on the **Human Genome Project,** announced on February 12, 2001, that it had completed the sequencing of the genome. However, researchers considered this a "working draft" of the genome, and a more complete draft was published in 2003. Online Mendelian Inheritance in Man (OMIM) is a database that contains information about many human genes and genetic disorders (www.ncbi.nlm.nih.gov/omim).

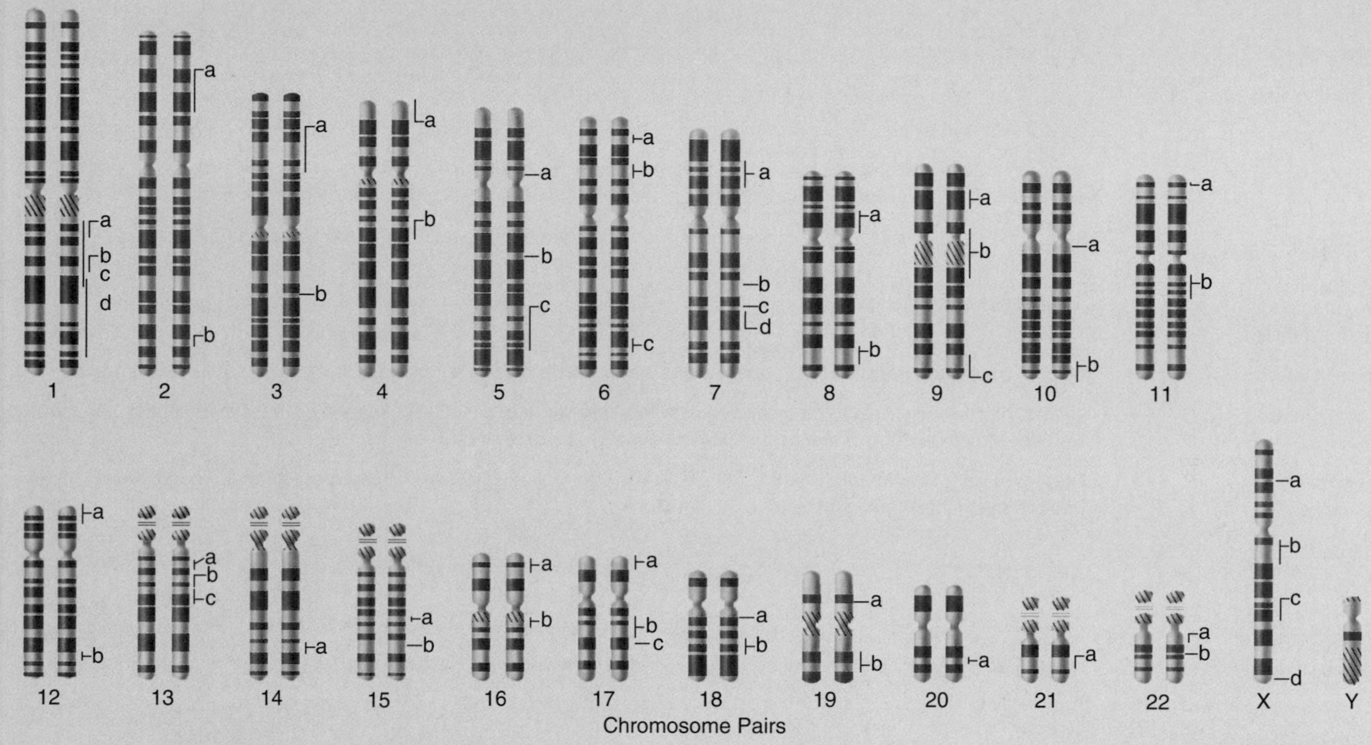
1
2
3
4
5
6
7
8
9
10
11
12
13
14
15
16
17
18
19
20
21
22
X
Y
Chromosome Pairs

1. a. Gaucher disease
   b. Prostate cancer
   c. Glaucoma
   d. Alzheimer disease*
2. a. Familial colon cancer*
   b. Waardenburg syndrome
3. a. Lung cancer
   b. Retinitis pigmentosa*
4. a. Huntington disease
   b. Parkinson disease
5. a. Cockayne syndrome
   b. Familial polyposis of the colon
   c. Asthma
6. a. Spinocerebellar ataxia
   b. Diabetes*
   c. Epilepsy*
7. a. Diabetes*
   b. Osteogenesis imperfecta
   c. Cystic fibrosis
   d. Obesity*
8. a. Werner syndrome
   b. Burkitt lymphoma
9. a. Malignant melanoma
   b. Friedreich ataxia
   c. Tuberous sclerosis
10. a. Multiple endocrine neoplasia, type 2
    b. Gyrate atrophy
11. a. Sickle-cell disease
    b. Multiple endocrine neoplasia
12. a. Zellweger syndrome
    b. Phenylketonuria (PKU)
13. a. Breast cancer*
    b. Retinoblastoma
    c. Wilson disease
14. a. Alzheimer disease*
15. a. Marfan syndrome
    b. Tay-Sachs disease
16. a. Polycystic kidney disease
    b. Crohn disease*
17. a. Tumor suppressor protein
    b. Breast cancer*
    c. Osteogenesis imperfecta
18. a. Amyloidosis
    b. Pancreatic cancer*
19. a. Familial hypercholesterolemia
    b. Myotonic dystrophy
20. a. Severe combined immunodeficiency
21. a. Amyotrophic lateral sclerosis*
22. a. DiGeorge syndrome
    b. Neurofibromatosis, type 2

X a. Duchenne muscular dystrophy
  b. Menkes syndrome
  c. X-linked severe combined immunodeficiency
  d. Factor VIII deficiency (hemophilia A)

*Gene responsible for only some cases.

FIGURE 29C Human Genomic Map
Representative genetic defects mapped to date. The bars and lines indicate the location of the genes listed for each chromosome.

—Continued

Armed with knowledge of the human genome and its effects on a person's physical, mental, and behavioral abilities, medicine and society are being transformed in many ways. Medicine, for example, is beginning to shift its emphasis from the curative to the preventive. The disorders or diseases a person is likely to develop may be prevented or their severity lessened. When prevention is not possible, knowledge of the enzymes or other molecules involved in a disorder may lead to new drugs and techniques that can compensate for the genetic disorder. Knowledge of the genes involved in a disorder may result in **gene therapy** or **genetic engineering** that repairs or replaces defective genes or gene products, resulting in cures or treatments for genetic disorders.

Despite the great promise of benefits from the Human Genome Project, the knowledge it yielded has raised a number of ethical and legal questions for society. Should a person's genomic information be public knowledge? Should persons with a genome that predisposes them to cancer or behavioral disorders be barred from certain types of employment or be refused medical insurance because they are at high risk? Can a person demand to know a prospective mate's genome? Should parents know the genome of their fetus and be allowed to make decisions regarding abortion based on this knowledge? Should the same genetic-engineering techniques that alter the genome to cure genetic disorders be used to create genomes that are deemed superior? Such questions raise the specter of genetic discrimination.

Down syndrome is an aneuploidy that is not always lethal. However, individuals with this genetic disorder exhibit physical and mental developmental problems, as well as an increased probability of developing certain cancers. These individuals also have a shortened life span.

Predict 11

Mr. and Mrs. Smith, both 40 years of age, are healthy with no known genetic conditions. Their newborn child has Down syndrome (trisomy 21). Explain the events that caused this condition.

ASSESS YOUR PROGRESS

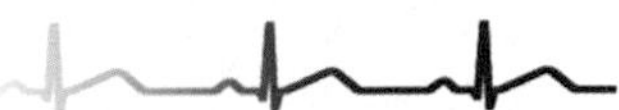

51. *What is meiosis? How does it differ from mitosis? What is the end result of meiosis?*
52. *What is a carrier?*
53. *What is a mutagen? Describe a point mutation, structural mutation, and nondisjunction.*
54. *What causes the genetic disorder Down syndrome?*

Answer

Learn to Predict

From page 1063

In this chapter, we learned that the production of breast milk depends on a number of hormones. Estrogen causes the duct system of the breast to develop and is mainly responsible for the enlargement of the breasts during pregnancy. Progesterone causes the development of the breasts' secretory alveoli, which enlarge but do not usually secrete milk during pregnancy. Other hormones necessary for development of the breast include growth hormone, prolactin, thyroid hormones, glucocorticoids, and insulin. Human somatotropin and human placental lactogen are secreted from the placenta. The high levels of estrogen and progesterone inhibit the effect of prolactin on milk secretion during pregnancy. After parturition, blood levels of estrogen, progesterone, and prolactin decrease.

Chapter 29 also explained that suckling stimulates the periodic pulses of prolactin and oxytocin. The pulses of prolactin, in the absence of high blood levels of estrogen and progesterone, cause the secretory units of the breasts to secrete milk. Oxytocin causes expulsion of the milk from the breast in response to suckling. Refraining from breastfeeding for a period of time results in no prolactin pulses. Consequently, the secretory units fail to secrete milk and quickly lose the capacity to do so. Ming can retain the capacity to breastfeed her baby by using a breast pump while she takes the prescribed antibiotic. The breast pump simulates suckling and stimulates the pulses of prolactin secretion. After finishing the antibiotic, Ming can then continue to breastfeed her baby.

Answers to the rest of this chapter's Predict questions are in Appendix G.

Summary

29.1 Prenatal Development (p. 1064)

1. Prenatal development is divided into three parts: the germinal period, during which the germ layers form; the embryonic period, when the organ systems form; and the fetal period, characterized by growth and maturation.
2. Postovulatory age is 14 days less than clinical age.
3. The stages of postnatal development are the neonatal period (birth to 1 month), infancy (1 month to 1–2 years), childhood (1–2 years to puberty), adolescence (puberty to 20 years), and adulthood (20 years to death).

Fertilization

Fertilization, the union of the oocyte and sperm, results in a zygote.

Early Cell Division

The cells of the early embryo are pluripotent (capable of making any cell of the body). In the very early stages of development, the cells are totipotent, meaning that each cell can give rise to any tissue necessary for development.

Morula and Blastocyst

The product of fertilization undergoes divisions until it becomes a mass called a morula and then a hollow ball of cells called a blastocyst.

Implantation of the Blastocyst and Development of the Placenta

The blastocyst implants into the uterus about 7 days after fertilization. The placenta is derived from the trophoblast of the blastocyst.

Formation of the Germ Layers

All body tissues are derived from three primary germ layers: endoderm, mesoderm, and ectoderm.

Neural Tube and Neural Crest Formation

The nervous system develops from a neural tube that forms in the ectodermal surface of the embryo and from neural crest cells derived from the developing neural tube.

Somite Formation

Segments called somites, which develop along the neural tube, give rise to the musculature, vertebral column, and ribs.

Formation of the Gut and Body Cavities

1. The digestive tract forms as the developing embryo closes off part of the yolk sac.
2. The coelom develops from small cavities that fuse within the embryo.

Limb Bud Development

The limbs develop from proximal to distal as outgrowths called limb buds.

Development of the Face

The face develops from the fusion of five major tissue processes.

Development of the Organ Systems

1. The epidermis of the skin develops from the ectoderm, and the dermis develops from the mesoderm or from neural crest cells. Melanocytes and sensory receptors develop from neural crest cells.
2. The skeletal system develops from mesoderm or neural crest cells.
3. Muscle develops from myoblasts, which migrate from somites.
4. The brain and spinal cord develop from the neural tube, and the peripheral nervous system develops from the neural tube and the neural crest cells.
5. The special senses develop mainly as neural tube or neural crest cell derivatives.
6. Many endocrine organs develop mainly as evaginations of the brain or digestive tract.
7. The heart develops as two tubes fuse into a single tube, which bends and develops septa to form four chambers.
8. The peripheral circulation develops from mesoderm as blood islands become hollow and fuse to form a network.
9. The lungs form as evaginations of the digestive tract. These evaginations undergo repeated branching.
10. The urinary system develops in three stages—pronephros, mesonephros, and metanephros—from the head to the tail of the embryo. The ducts join the allantois, part of which becomes the urinary bladder.
11. The reproductive systems develop in conjunction with the urinary system. The presence or absence of certain hormones is very important to sexual development.

Growth of the Fetus

1. The embryo becomes a fetus at 60 days.
2. The fetal period, from day 60 to birth, is a time of rapid growth.

29.2 Parturition (p. 1086)

The total length of gestation is 280 days (clinical age).

Stages of Labor

Uterine contractions force the baby out of the uterus during labor.

Hormonal Stimulation of Parturition

1. Increased estrogen and decreased progesterone help initiate parturition.
2. Fetal glucocorticoids act on the placenta to decrease progesterone synthesis and to increase estrogen and prostaglandin synthesis.
3. Stretch of the uterus and decreased progesterone levels stimulate oxytocin secretion, which stimulates uterine contraction.

29.3 The Newborn (p. 1088)

Shortly after birth, the newborn baby experiences several dramatic changes when it is separated from the maternal circulation and transferred from a fluid to a gaseous environment.

Respiratory and Cardiovascular Changes

1. The foramen ovale closes, separating the two atria.
2. The ductus arteriosus closes, and blood no longer flows between the pulmonary trunk and the aorta.
3. The umbilical vein and arteries degenerate.

Digestive Changes

1. Meconium is a mixture of cells from the digestive tract, amniotic fluid, bile, and mucus excreted by the newborn.
2. The stomach begins to secrete acid.

3. The liver does not form adult bilirubin for the first 2 weeks.
4. The neonate can digest lactose, but other foods must be gradually introduced.

Apgar Scores

1. *Apgar* represents appearance, pulse, grimace, activity, and respiratory effort.
2. Apgar and other methods are used to assess the physiological condition of the newborn.

Congenital Disorders

1. Congenital disorders are abnormalities present at birth.
2. Teratogens are environmental agents that cause some congenital disorders.

29.4 Lactation (p. 1091)

1. Estrogen, progesterone, and other hormones stimulate the growth of the breasts during pregnancy.
2. Suckling stimulates prolactin and oxytocin synthesis. Prolactin stimulates milk production, and oxytocin stimulates milk letdown.

29.5 First Year After Birth (p. 1092)

1. The number of neuron connections and neuroglia increases.
2. Motor skills develop gradually, especially head, eye, and hand movements.

29.6 Aging and Death (p. 1093)

1. Loss of cells that are not replaced contributes to aging.
   - A loss of neurons occurs.
   - Loss of muscle cells can affect skeletal and cardiac muscle function.
2. Loss of tissue plasticity results from cross-link formation between collagen molecules. The lens of the eye loses the ability to accommodate. Other organs, such as the joints, kidneys, lungs, and heart, also have reduced efficiency with advancing age.
3. The immune system loses the ability to act against foreign antigens and may attack self-antigens.
4. Many aging changes are probably genetic.
5. Death is the loss of brain functions.

29.7 Genetics (p. 1095)

1. Genetics is the study of heredity, the characteristics children inherit from their parents.
2. Genomic medicine uses an understanding of the biochemical relationship between genes and disease to diagnose and manage disease.

Mendelian Genetics

1. The genes an organism has for a given trait is called the genotype. The expression of the genes is called the phenotype.
2. Alleles are alternate forms of genes. A dominant allele masks the effects of a recessive allele for the same trait.
3. An organism homozygous for a trait has two identical alleles for the trait, whereas an organism heterozygous for a trait has two different alleles for the trait.

Modern Concepts of Genetics

1. Chromosomes
   - Somatic cells have a diploid number of chromosomes, whereas gametes have a haploid number. In humans, the diploid number is 46 and the haploid number is 23.
   - Humans have 22 pairs of autosomal chromosomes and 1 pair of sex chromosomes. Females have the sex chromosomes XX, and males have XY.
   - A karyotype is a display of the chromosomes of a somatic cell during metaphase of mitosis.
   - Chromosome pairs are called homologous chromosomes.
2. The genome consists of all the genes found in the haploid number of chromosomes from one parent.
3. Each gene occupies a specific locus, or location, on a chromosome.
4. Alleles exist in many different forms, called multiple alleles.
5. Gene dominance
   - In complete dominance, the dominant allele masks the effects of the recessive allele.
   - In codominance, two alleles at the same locus are expressed, so that separate, distinguishable phenotypes occur at the same time.
   - In incomplete dominance, the dominant allele does not completely mask the effects of the recessive allele.
6. Polygenic traits result from the interaction of many genes.
7. Sex-linked traits
   - Sex-linked traits are traits affected by genes on the sex chromosomes.
   - X-linked traits are affected by genes on the X chromosome, and Y-linked traits are affected by genes on the Y chromosome.
   - X-linked traits are seen more frequently in males than in females because males have only one X chromosome.

Meiosis and the Transmission of Genes

1. Meiosis results in the production of gametes.
2. A Punnett square can determine the probability of particular alleles being transmitted to the next generation.
3. A carrier for a recessive trait is heterozygous for the trait, having one normal allele and one disorder-causing allele.

Genetic Disorders

1. A mutation is a change in the number or kinds of nucleotides in DNA.
2. Some genetic disorders result from an abnormal distribution of chromosomes during gamete formation.

REVIEW AND COMPREHENSION

1. The major development of organ systems takes place in
   a. weeks 1 and 2 of development.
   b. weeks 3–8 of development.
   c. weeks 8–20 of development.
   d. the last 30 weeks of development.
2. Given these structures:
   (1) blastocyst (2) morula (3) zygote

   Choose the arrangement that lists the structures in the order in which they form during development.
   a. 1,2,3 c. 2,3,1 e. 3,2,1
   b. 1,3,2 d. 3,1,2
3. The embryo proper develops from the
   a. inner cell mass. c. blastocele.
   b. trophoblast. d. yolk sac.

4. The placenta
   a. develops from the trophoblast.
   b. allows maternal blood to mix with embryonic blood.
   c. invades the lacunae of the embryo.
   d. All of these are correct.
5. The embryonic disk
   a. forms between the amniotic cavity and the yolk sac.
   b. contains the primitive streak.
   c. becomes a three-layered structure.
   d. All of these are correct.
6. The brain develops from
   a. ectoderm.
   b. endoderm.
   c. mesoderm.
7. Most of the skeletal system develops from
   a. ectoderm.
   b. endoderm.
   c. mesoderm.
8. The somites give rise to the
   a. circulatory system.
   b. skeletal muscle.
   c. lungs.
   d. kidneys.
   e. brain.
9. The pericardial cavity forms from
   a. evagination of the early digestive tract.
   b. the neural tube.
   c. the coelom.
   d. the branchial arches.
   e. pharyngeal pouches.
10. The parts of the limbs develop
    a. in a proximal-to-distal sequence.
    b. in a distal-to-proximal sequence.
    c. at approximately the same time.
    d. before the primitive streak is formed.
11. Concerning development of the face,
    a. the face develops by the fusion of five embryonic structures.
    b. the maxillary processes normally meet at the midline to form the lip.
    c. the primary palate forms the roof of the mouth.
    d. clefts of the secondary palate normally occur to one side of the midline.
12. Concerning the development of the heart,
    a. the heart develops from a single tube, which results from the fusion of two tubes.
    b. the SA node develops in the wall of the sinus venosus.
    c. the foramen ovale lets blood flow from the right atrium to the left atrium.
    d. the bulbus cordis is absorbed into the ventricle.
    e. All of these are correct.
13. Given these structures:
    (1) mesonephros (2) metanephros (3) pronephros

    Choose the arrangement that lists the structures in the order in which they form during development.
    a. 1,2,3
    b. 1,3,2
    c. 2,3,1
    d. 3,1,2
    e. 3,2,1
14. A study of the early embryo indicates that the glans penis of the male develops from the same embryonic structure as which of these female structures?
    a. labia majora
    b. uterus
    c. clitoris
    d. vagina
    e. urinary bladder
15. Which hormones cause the differentiation of sex organs in the developing male fetus?
    a. FSH and LH
    b. LH and testosterone
    c. testosterone and dihydrotestosterone
    d. estrogen and progesterone
    e. GnRH and FSH
16. The onset of labor may be a result of
    a. increased estrogen secretion by the placenta.
    b. increased glucocorticoid secretion by the fetus.
    c. increased secretion of oxytocin.
    d. stretch of the uterus.
    e. All of these are correct.
17. Following birth,
    a. the ductus arteriosus closes.
    b. the pH of the stomach increases.
    c. the fossa ovalis becomes the foramen ovale.
    d. blood flow through the pulmonary arteries decreases.
    e. All of these events occur.
18. The hormone involved in milk production is
    a. oxytocin. c. estrogen. e. ACTH.
    b. prolactin. d. progesterone.
19. Which of these life stages is correctly matched with the time it occurs?
    a. neonate—birth to 1 month after birth
    b. infant—1 month to 6 months
    c. child—6 months to 5 years
    d. puberty—10–12 years
    e. middle age—20–40 years
20. Which of these occurs as we get older?
    a. Neurons replicate to replace lost neurons.
    b. Skeletal muscle cells replicate to replace lost muscle cells.
    c. Cross-links between collagen molecules increase.
    d. The immune system becomes less sensitive to the body's own antigens.
    e. Free radicals help prevent cancer.
21. A gene is
    a. the functional unit of heredity.
    b. a certain portion of a DNA molecule.
    c. a part of a chromosome.
    d. All of these are correct.
22. Which of these terms is correctly matched with its definition?
    a. autosome—an X or a Y chromosome
    b. phenotype—the genetic makeup of an individual
    c. allele—variant form of a gene occupying locus on a homologous chromosome
    d. heterozygous—having two identical genes for a trait
    e. recessive—a trait expressed when the genes are heterozygous
23. Which of these genotypes is heterozygous?
    a. *DD* c. *dd*
    b. *Dd* d. Both a and c are correct.

24. The AB blood type in the ABO blood group is an example of
    a. dominant versus recessive alleles.
    b. incomplete dominance.
    c. codominance.
    d. a polygenic trait.
    e. sex-linked inheritance.

25. Assume that a trait is determined by an X-linked dominant gene. If the mother exhibits the trait but the father does not, then their
    a. sons are more likely than their daughters to exhibit the trait.
    b. daughters are more likely than their sons to exhibit the trait.
    c. sons and daughters are equally likely to exhibit the trait.

Answers in Appendix E

CRITICAL THINKING

1. Triploidy is the presence of three sets of chromosomes in a cell. Although rare, triploidy does occur in humans. What failures during fertilization can lead to triploidy?
2. A physician tells a woman that she is pregnant and is 44 days past her LMP. Approximately how many days has the embryo been developing, and what developmental events are occurring?
3. A high fever can prevent neural tube closure. If a woman had a high fever approximately 35–45 days after her LMP, what kinds of congenital disorders might be seen in the developing embryo?
4. A woman goes into labor during the thirtieth week of her pregnancy. What are the effects of administering progesterone at this stage?
5. Three minutes after birth, a newborn has an Apgar score of 5 as follows: A, 0; P, 1; G, 1; A, 1; and R, 2. What are some possible causes for this low score? What should be done for this neonate?
6. When a woman breastfeeds, milk letdown can occur in the breast that is not being suckled. Explain how this response happens.
7. An 18-year-old woman consulted a physician because of her failure to initiate menses. She had experienced normal breast development at age 13, but no pubic or axillary hair had ever appeared. Her height was 5 feet 6 inches, weight 119 pounds, blood pressure 110/70, and pulse 60 beats/minute. Her physical examination confirmed the presence of well-developed breasts and the external genitalia of a normal female. However, the vagina ended in a blind pouch, and there was no evidence of a cervix, ovaries, or a uterus. Within the inguinal area, the examining physician could palpate a small, spherical mass on each side. Is this patient's genotype most likely XX, XO, or XY? Explain why the genotype you selected is consistent with the manifestations.
8. Select the conditions that would be present in a genetic male fetus having a mutation that causes the synthesis of an ineffective müllerian-inhibiting hormone.
   (1) Male internal reproductive structures develop.
   (2) Male internal reproductive structures do not develop.
   (3) Female internal reproductive structures develop.
   (4) Female internal reproductive structures do not develop.
   (5) Male external genitalia develop.
   (6) Female external genitalia develop.
   a. 2,4,6
   b. 1,3,5
   c. 1,2,6
   d. 2,3,5
9. The ability to roll the tongue to form a "tube" results from a dominant gene. Suppose that a woman and her son can roll their tongues, but her husband cannot. Is it possible to determine if the husband is the father of her son?
10. The ABO antigens are a group of molecules found on the surface of red blood cells. Can an individual who has blood type AB be a parent of a child with blood type O? Why or why not?

Answers in Appendix F

Visit this book's website at **www.mhhe.com/seeley10** for chapter quizzes, interactive learning exercises, and other study tools.

Mc Graw Hill
connect plus+
ANATOMY & PHYSIOLOGY

Anatomy & Physiology
REVEALED
aprevealed.com

Appendix A

PERIODIC TABLE OF THE ELEMENTS

9 **F** Fluorine 19.00	Atomic number (9) Atomic mass (19.00)

1 1A	2 2A	3 3B	4 4B	5 5B	6 6B	7 7B	8 8B	9 8B	10 8B	11 1B	12 2B	13 3A	14 4A	15 5A	16 6A	17 7A	18 8A
1 **H** Hydrogen 1.008																	2 **He** Helium 4.003
3 **Li** Lithium 6.941	4 **Be** Beryllium 9.012											5 **B** Boron 10.81	6 **C** Carbon 12.01	7 **N** Nitrogen 14.01	8 **O** Oxygen 16.00	9 **F** Fluorine 19.00	10 **Ne** Neon 20.18
11 **Na** Sodium 22.99	12 **Mg** Magnesium 24.31											13 **Al** Aluminum 26.98	14 **Si** Silicon 28.09	15 **P** Phosphorus 30.97	16 **S** Sulfur 32.07	17 **Cl** Chlorine 35.45	18 **Ar** Argon 39.95
19 **K** Potassium 39.10	20 **Ca** Calcium 40.08	21 **Sc** Scandium 44.96	22 **Ti** Titanium 47.88	23 **V** Vanadium 50.94	24 **Cr** Chromium 52.00	25 **Mn** Manganese 54.94	26 **Fe** Iron 55.85	27 **Co** Cobalt 58.93	28 **Ni** Nickel 58.69	29 **Cu** Copper 63.55	30 **Zn** Zinc 65.39	31 **Ga** Gallium 69.72	32 **Ge** Germanium 72.59	33 **As** Arsenic 74.92	34 **Se** Selenium 78.96	35 **Br** Bromine 79.90	36 **Kr** Krypton 83.80
37 **Rb** Rubidium 85.47	38 **Sr** Strontium 87.62	39 **Y** Yttrium 88.91	40 **Zr** Zirconium 91.22	41 **Nb** Niobium 92.91	42 **Mo** Molybdenum 95.94	43 **Tc** Technetium (98)	44 **Ru** Ruthenium 101.1	45 **Rh** Rhodium 102.9	46 **Pd** Palladium 106.4	47 **Ag** Silver 107.9	48 **Cd** Cadmium 112.4	49 **In** Indium 114.8	50 **Sn** Tin 118.7	51 **Sb** Antimony 121.8	52 **Te** Tellurium 127.6	53 **I** Iodine 126.9	54 **Xe** Xenon 131.3
55 **Cs** Cesium 132.9	56 **Ba** Barium 137.3	57 **La** Lanthanum 138.9	72 **Hf** Hafnium 178.5	73 **Ta** Tantalum 180.9	74 **W** Tungsten 183.9	75 **Re** Rhenium 186.2	76 **Os** Osmium 190.2	77 **Ir** Iridium 192.2	78 **Pt** Platinum 195.1	79 **Au** Gold 197.0	80 **Hg** Mercury 200.6	81 **Tl** Thallium 204.4	82 **Pb** Lead 207.2	83 **Bi** Bismuth 209.0	84 **Po** Polonium (210)	85 **At** Astatine (210)	86 **Rn** Radon (222)
87 **Fr** Francium (223)	88 **Ra** Radium (226)	89 **Ac** Actinium (227)	104 **Rf** Rutherfordium (257)	105 **Db** Dubnium (260)	106 **Sg** Seaborgium (263)	107 **Bh** Bohrium (262)	108 **Hs** Hassium (265)	109 **Mt** Meitnerium (266)	110 **Ds** Darmstadtium (269)	111 **Rg** Roentgenium (272)	112 **Cn** Copernicium	(113)	114 **Fl** Flerovium	(115)	116 **Lv** Livermorium	(117)	(118)

58 **Ce** Cerium 140.1	59 **Pr** Praseodymium 140.9	60 **Nd** Neodymium 144.2	61 **Pm** Promethium (147)	62 **Sm** Samarium 150.4	63 **Eu** Europium 152.0	64 **Gd** Gadolinium 157.3	65 **Tb** Terbium 158.9	66 **Dy** Dysprosium 162.5	67 **Ho** Holmium 164.9	68 **Er** Erbium 167.3	69 **Tm** Thulium 168.9	70 **Yb** Ytterbium 173.0	71 **Lu** Lutetium 175.0
90 **Th** Thorium 232.0	91 **Pa** Protactinium (231)	92 **U** Uranium 238.0	93 **Np** Neptunium (237)	94 **Pu** Plutonium (242)	95 **Am** Americium (243)	96 **Cm** Curium (247)	97 **Bk** Berkelium (247)	98 **Cf** Californium (249)	99 **Es** Einsteinium (254)	100 **Fm** Fermium (253)	101 **Md** Mendelevium (256)	102 **No** Nobelium (254)	103 **Lr** Lawrencium (257)

Legend
Metals
Metalloids
Nonmetals

The 1–18 group designation has been recommended by the International Union of Pure and Applied Chemistry (IUPAC) but is not yet in wide use.

The modern periodic table of the elements lists the known elements in order of their atomic masses. Each element has a box that contains the name of the element and its chemical symbol, atomic number, and atomic mass. The boxes are organized into a grid of horizontal rows, called periods, and vertical columns, called groups. Within a period, the elements are listed in order of increasing atomic number from left to right. Elements in a period have different chemical properties, whereas elements in a group have similar chemical properties.

The atomic number is the number of protons in an element. Each element has a unique number of protons and therefore a unique atomic number. There are 90 naturally occurring elements. Scientists have been able to create new elements by changing the number of protons in the nuclei of existing elements. Protons, neutrons, or electrons from one atom are accelerated to very high speeds and then smashed into the nucleus of another atom. The resulting changes in the nucleus produce a new element with a new atomic number. These artificially produced elements are usually unstable, and they quickly convert back to more stable elements. The synthetic elements are technetium (Tc, atomic number 43), promethium (Pm, atomic number 61), and all the elements with an atomic number of 93 or higher. An element with an atomic number of 116 has the highest number officially recognized by the International Union of Pure and Applied Chemistry (IUPAC), but elements with higher atomic numbers have reportedly been made. In addition, the IUPAC gives standard atomic weights to only four significant figures, atomic weights now give a range of values when previously there was just one given, and atomic weights are no longer given for compounds without naturally occurring isotopes.

Appendix B

SCIENTIFIC NOTATION

Very large numbers with many zeros, such as 1,000,000,000,000,000, or very small numbers, such as 0.0000000000000001, are very cumbersome to work with. Consequently, the numbers are expressed in a kind of mathematical shorthand known as scientific notation. Scientific notation has the following form:

$$M \times 10^n$$

where n specifies how many times the number M is raised to the power of 10. The exponent n has two meanings, depending on its sign. If n is positive, M is multiplied by 10 n times. For example, if $n = 2$ and $M = 1.2$, then

$$1.2 \times 10^2 = 1.2 \times 10 \times 10 = 120$$

In other words, if n is positive, the decimal point of M is moved to the right n times. In this case the decimal point of 1.2 is moved two places to the right.

$$1.20.$$

If n is negative, M is divided by 10 n times.

$$1.2 \times 10^{-2} = \frac{1.2}{(10 \times 10)} = \frac{1.2}{(100)} = 0.012$$

In other words, if n is negative, the decimal point of M is moved to the left n times. In this case, the decimal point of 1.2 is moved two places to the left.

$$0.01.2$$

If M is the number 1.0, it often is not expressed in scientific notation. For example, 1.0×10^2 is the same thing as 10^2, and 1.0×10^{-2} is the same thing as 10^{-2}.

Two common examples of the use of scientific notation in chemistry are Avogadro's number and pH. Avogadro's number, 6.023×10^{23}, is the number of atoms in 1 molar mass of an element. Thus

$$6.023 \times 10^{23} = 602{,}300{,}000{,}000{,}000{,}000{,}000{,}000$$

which is a very large number of atoms.

The pH scale is a measure of the concentration of hydrogen ions in a solution. A neutral solution has 10^{-7} moles of hydrogen ions per liter. In other words

$$10^{-7} = 0.0000001$$

which is a very small amount (1 ten-millionth of a gram) of hydrogen ions.

Appendix C

SOLUTION CONCENTRATIONS

Physiologists often express solution concentration in terms of percent, molarity, molality, and equivalents.

Percent

The weight-volume method of expressing percent concentrations states the weight of a solute in a given volume of solvent. For example, to prepare a 10% solution of sodium chloride, 10 g of sodium chloride is dissolved in a small amount of water (solvent) to form a salt solution. Then additional water is added to the salt solution to form 100 mL of salt solution. Note that the sodium chloride is dissolved in water and then diluted to the required volume. The sodium chloride is not dissolved directly in 100 mL of water.

Molarity

Molarity determines the number of moles of solute dissolved in a given volume of solvent. A 1 molar (1 M) solution is made by dissolving 1 mole (mol) of a substance in enough water to make 1 L of solution. For example, 1 mol of sodium chloride solution is made by dissolving 58.44 g of sodium chloride in enough water to make 1 L of solution. One mol of glucose solution is made by dissolving 180.2 g of glucose in enough water to make 1 L of solution. Both solutions have the same number (Avogadro's number) of formula units (NaCl) and molecules (glucose) in solution.

Osmoles, Osmolality, and Osmosis

An **osmole** is the measure of the number of particles in a solution. One osmole is the molecular mass, in grams, of a solute times the number of ions or particles into which it dissociates in solution. A milliosmole (mOsm) is 1/1000 of an osmole. The osmolality of a solution is the number of osmoles in a kilogram of solution. Water moves by osmosis from a solution with a lower osmolality to a solution with a higher osmolality. Thus, water moves by osmosis from a solution of 100 mOsm/kg toward a solution of 300 mOsm/kg.

Molality

Although 1 M solutions have the same number of solute molecules, they do not have the same number of solvent (water) molecules. Because 58.5 g of sodium chloride occupies less volume than 180 g of glucose, the sodium chloride solution has more water molecules. **Molality** is a method of calculating concentrations that takes into account the number of solute and solvent molecules. A 1 molal solution (1 *m*) is 1 mol of a substance dissolved in 1 kg of water. Thus, all 1-molal solutions have the same number of solvent molecules.

When sodium chloride, which is an ionic compound, is dissolved in water, it dissociates to form two ions, a sodium cation (Na^+) and a chloride anion (Cl^-). Glucose does not dissociate when dissolved in water, however, because it is a molecule. Thus, the sodium chloride solution contains twice as many particles as the glucose solution (one Na^+ and one Cl^- for each glucose molecule). To report the concentration of these substances in a way that reflects the number of particles in a given mass of solvent, the concept of **osmolality** is used. The osmolality of a solution is the molality of the solution times the number of particles into which the solute dissociates in 1 kg of solvent. Thus, 1 mol of sodium chloride in 1 kg of water is a 2-osmolal (osm) solution because sodium chloride dissociates to form two ions.

The osmolality of a solution is a reflection of the number, not the type, of particles in a solution. Thus, a 1 osm solution contains 1 osm of particles per kilogram of solvent, but the particles may be all one type or a complex mixture of different types.

The concentration of particles in body fluids is so low that the measurement milliosmole (mOsm), 1/1000 of an osmole, is used. Most body fluids have an osmotic concentration of approximately 300 mOsm and consist of many different ions and molecules. The osmotic concentration of body fluids is important because it influences the movement of water into or out of cells (see chapter 3).

Equivalents

Equivalents are a measure of the concentrations of ionized substances. One equivalent (Eq) is 1 mol of an ionized substance multiplied by the absolute value of its charge. For example, 1 mol of NaCl dissociates into 1 mol of Na^+ and 1 mol of Cl^-. Thus, there is 1 Eq of Na^+ (1 mol $\times$ 1) and 1 Eq of Cl^- (1 mol $\times$ 1). One mole of $CaCl_2$ dissociates into 1 mol of Ca^{2+} and 2 mol of Cl^-. Thus, there are 2 Eq of Ca^{2+} (1 mol $\times$ 2) and 2 Eq of Cl^- (2 mol $\times$ 1). In an electrically neutral solution, the equivalent concentration of positively charged ions is equal to the equivalent concentration of the negatively charged ions. One milliequivalent (mEq) is 1/1000 of an equivalent.

Appendix D

pH

Pure water weakly dissociates to form small numbers of hydrogen and hydroxide ions:

$$H_2O \leftrightarrow H^+ + OH^-$$

At 25°C, the concentration of both hydrogen ions and hydroxide ions is 10^{-7} mol/L. Any solution that has equal concentrations of hydrogen and hydroxide ions is considered **neutral.** A solution is an **acid** if it has a higher concentration of hydrogen ions than hydroxide ions, and a solution is a **base** if it has a lower concentration of hydrogen ions than hydroxide ions. In any aqueous solution (at 25°C), the hydrogen ion concentration $[H^+]$ times the hydroxide ion concentration $[OH^-]$ is a constant that is equal to 10^{-14}.

$$[H^+] \times [OH^-] = 10^{-14}$$

Consequently, as the hydrogen ion concentration decreases, the hydroxide ion concentration increases, and vice versa—for example,

	$[H^+]$	$[OH^-]$
Acidic solution	10^{-3}	10^{-11}
Neutral solution	10^{-7}	10^{-7}
Basic solution	10^{-12}	10^{-2}

Although the acidity or basicity of a solution could be expressed in terms of either hydrogen or hydroxide ion concentration, it is customary to use hydrogen ion concentration. The pH of a solution is defined as

$$pH = -\log_{10}(H^+)$$

Thus, a neutral solution with 10^{-7} mol of hydrogen ions per liter has a pH of 7.

$$\begin{aligned} pH &= -\log_{10}(H^+) \\ &= -\log_{10}(10^{-7}) \\ &= -(-7) \\ &= 7 \end{aligned}$$

In simple terms, to convert the hydrogen ion concentration to the pH scale, the exponent of the concentration (e.g., −7) is used, and it is changed from a negative to a positive number. Thus, an acidic solution with 10^{-3} mol of hydrogen ions/L has a pH of 3, whereas a basic solution with 10^{-12} hydrogen ions/L has a pH of 12.

Appendix E

ANSWERS TO REVIEW AND COMPREHENSION QUESTIONS

Chapter One
1. a; 2. b; 3. a; 4. c; 5. e; 6. a; 7. b; 8. c; 9. d; 10. c; 11. d; 12. a; 13. b; 14. b; 15. a; 16. c; 17. b; 18. a; 19. e

Chapter Two
1. e; 2. b; 3. b; 4. a; 5. d; 6. b; 7. e; 8. c; 9. e; 10. c; 11. d; 12. b; 13. c; 14. c; 15. d; 16. c; 17. d; 18. e; 19. b; 20. d; 21. b; 22. c; 23. d; 24. a; 25. e

Chapter Three
1. a; 2. e; 3. c; 4. e; 5. c; 6. e; 7. b; 8. a; 9. a; 10. d; 11. d; 12. b; 13. b; 14. c; 15. c; 16. c; 17. b; 18. e; 19. c; 20. a; 21. d; 22. c

Chapter Four
1. e; 2. c; 3. a; 4. b; 5. d; 6. c; 7. d; 8. b; 9. a; 10. b; 11. e; 12. a; 13. d; 14. b; 15. b; 16. d; 17. d; 18. a; 19. e; 20. b; 21. c; 22. c; 23. b; 24. c; 25. b; 26. d

Chapter Five
1. d; 2. e; 3. b; 4. a; 5. c; 6. d; 7. c; 8. d; 9. b; 10. a; 11. b; 12. b; 13. c; 14. c; 15. d; 16. b; 17. c; 18. b; 19. a; 20. b; 21. d; 22. c; 23. d; 24. c

Chapter Six
1. e; 2. d; 3. a; 4. b; 5. d; 6. c; 7. a; 8. b; 9. e; 10. c; 11. d; 12. d; 13. c; 14. b; 15. c; 16. e; 17. d; 18. a; 19. b; 20. c; 21. a; 22. e; 23. c; 24. e; 25. d

Chapter Seven
1. c; 2. c; 3. c; 4. d; 5. c; 6. a; 7. d; 8. a; 9. b; 10. a; 11. c; 12. a; 13. a; 14. d; 15. e; 16. b; 17. a; 18. c; 19. b; 20. c; 21. a

Chapter Eight
1. d; 2. d; 3. c; 4. d; 5. b; 6. e; 7. d; 8. c; 9. d; 10. e; 11. a; 12. b; 13. c; 14. d; 15. b; 16. b; 17. a; 18. c; 19. c

Chapter Nine
1. c; 2. e; 3. d; 4. c; 5. e; 6. b; 7. a; 8. b; 9. d; 10. b; 11. d; 12. c; 13. e; 14. d; 15. c; 16. c; 17. a; 18. b; 19. d; 20. a; 21. d; 22. c; 23. c; 24. c; 25. a

Chapter Ten
1. d; 2. b; 3. b; 4. c; 5. c; 6. d; 7. b; 8. c; 9. a; 10. a; 11. d; 12. a; 13. b; 14. d; 15. b; 16. d; 17. a; 18. c; 19. a; 20. e; 21. c; 22. b; 23. b; 24. d

Chapter Eleven
1. b; 2. c; 3. c; 4. b; 5. c; 6. a; 7. b; 8. d; 9. a; 10. c; 11. b; 12. a; 13. a; 14. b; 15. d; 16. c; 17. e; 18. b; 19. d; 20. d; 21. b; 22. b; 23. e; 24. e; 25. e; 26. a

Chapter Twelve
1. d; 2. c; 3. c; 4. b; 5. b; 6. d; 7. c; 8. c; 9. c; 10. d; 11. d; 12. c; 13. e; 14. c; 15. d

Chapter Thirteen
1. a; 2. c; 3. b; 4. e; 5. d; 6. b; 7. b; 8. b; 9. a; 10. e; 11. d; 12. d; 13. a; 14. c; 15. d; 16. b; 17. b; 18. a; 19. d; 20. b; 21. b; 22. e; 23. b; 24. c; 25. c; 26. b

Chapter Fourteen
1. c; 2. d; 3. e; 4. c; 5. a; 6. d; 7. b; 8. e; 9. a; 10. b; 11. b; 12. e; 13. c; 14. c; 15. b; 16. d; 17. a; 18. b; 19. b; 20. d; 21. b; 22. b; 23. d; 24. e; 25. d; 26. b; 27. b; 28. d; 29. e; 30. c

Chapter Fifteen
1. e; 2. c; 3. a; 4. b; 5. b; 6. a; 7. c; 8. d; 9. c; 10. d; 11. c; 12. c; 13. a; 14. d; 15. c; 16. d; 17. a; 18. d; 19. c; 20. a; 21. a; 22. b; 23. d; 24. a; 25. c

Chapter Sixteen
1. e; 2. d; 3. a; 4. e; 5. b; 6. d; 7. e; 8. d; 9. d; 10. d; 11. c; 12. a; 13. c; 14. e; 15. d; 16. c; 17. e

Chapter Seventeen
1. c; 2. b; 3. e; 4. d; 5. b; 6. a; 7. e; 8. b; 9. a; 10. e; 11. e; 12. e; 13. d; 14. c; 15. a; 16. c

Chapter Eighteen
1. e; 2. d; 3. e; 4. a; 5. c; 6. b; 7. b; 8. e; 9. c; 10. d; 11. e; 12. b; 13. c; 14. a; 15. a; 16. d; 17. c; 18. d; 19. e; 20. e; 21. b; 22. d; 23. a; 24. a; 25. b; 26. e; 27. c; 28. c; 29. e; 30. d

Chapter Nineteen
1. e; 2. c; 3. a; 4. d; 5. a; 6. d; 7. c; 8. d; 9. b; 10. a; 11. c; 12. b; 13. e; 14. e; 15. b; 16. a; 17. a; 18. d; 19. d; 20. c; 21. c

Chapter Twenty
1. e; 2. c; 3. a; 4. b; 5. c; 6. e; 7. a; 8. c; 9. d; 10. a; 11. b; 12. a; 13. a; 14. e; 15. b; 16. c; 17. d; 18. a; 19. a; 20. d; 21. c; 22. c; 23. c; 24. e; 25. c

Chapter Twenty-One
1. b; 2. a; 3. a; 4. d; 5. d; 6. c; 7. d; 8. a; 9. a; 10. d; 11. e; 12. c; 13. a; 14. b; 15. b; 16. a; 17. e; 18. d; 19. b; 20. b; 21. d; 22. d; 23. c; 24. b; 25. e

Chapter Twenty-Two
1. d; 2. b; 3. e; 4. e; 5. d; 6. b; 7. e; 8. b; 9. a; 10. d; 11. d; 12. d; 13. b; 14. e; 15. e; 16. a; 17. d; 18. c; 19. a; 20. b; 21. c; 22. a; 23. d; 24. d; 25. d

Chapter Twenty-Three
1. a; 2. e; 3. d; 4. c; 5. d; 6. b; 7. c; 8. b; 9. d; 10. a; 11. d; 12. c; 13. c; 14. c; 15. d; 16. b; 17. c; 18. b; 19. c; 20. a; 21. c

Chapter Twenty-Four
1. a; 2. d; 3. d; 4. a; 5. b; 6. a; 7. b; 8. a; 9. c; 10. d; 11. c; 12. e; 13. d; 14. d; 15. e; 16. b; 17. b; 18. a; 19. e; 20. e; 21. e; 22. a; 23. c; 24. a

Chapter Twenty-Five
1. d; 2. a; 3. e; 4. b; 5. d; 6. e; 7. e; 8. b; 9. d; 10. d; 11. a; 12. e; 13. c; 14. a; 15. b; 16. c

Chapter Twenty-Six
1. d; 2. b; 3. c; 4. e; 5. b; 6. b; 7. e; 8. a; 9. b; 10. b; 11. a; 12. a; 13. a; 14. c; 15. d; 16. c; 17. b; 18. e; 19. d; 20. a; 21. d; 22. e; 23. d; 24. c; 25. c; 26. d

Chapter Twenty-Seven
1. a; 2. b; 3. c; 4. d; 5. a; 6. a; 7. b; 8. a; 9. b; 10. d; 11. b; 12. a; 13. a; 14. c; 15. c; 16. d

Chapter Twenty-Eight
1. e; 2. a; 3. a; 4. b; 5. d; 6. e; 7. b; 8. d; 9. c; 10. b; 11. c; 12. a; 13. e; 14. b; 15. c; 16. c; 17. b; 18. a; 19. c; 20. a; 21. c; 22. e; 23. e; 24. c; 25. d; 26. c; 27. a

Chapter Twenty-Nine
1. b; 2. e; 3. a; 4. a; 5. d; 6. a; 7. c; 8. b; 9. c; 10. a; 11. a; 12. e; 13. d; 14. c; 15. c; 16. e; 17. a; 18. b; 19. a; 20. c; 21. d; 22. c; 23. b; 24. c; 25. c

Appendix F

ANSWERS TO CRITICAL THINKING QUESTIONS

Chapter 1

1. Student B is correct. Body temperature begins to rise as a result of exposure to the hot environment. Sweating eliminates heat from the body and lowers body temperature. Body temperature returning to its ideal normal value is an example of negative feedback. Student A probably thought it was positive feedback because sweating continued to increase. However, sweating is the response. The variable being regulated by sweating is body temperature.
2. Answer *e* is correct. Positive-feedback mechanisms continuously stimulate a response until the initial stimulus is removed and are sometimes harmful. The continually decreasing blood pressure is an example of a harmful positive feedback mechanism. Negative-feedback mechanisms result in a return to homeostasis. The elevated heart rate is a negative-feedback mechanism that attempts to return blood pressure to a normal value. In this case, the negative-feedback mechanism is inadequate to restore homeostasis, and medical intervention (a transfusion) is necessary.
3. When a boy is standing on his head, his nose is superior to his mouth. Directional terms refer to a person's body in the anatomical position, not to the body's current position.
4. The uterus is located in the pelvic cavity. The pelvic cavity, however, is surrounded by the bones of the pelvis and does not increase in size during pregnancy. Instead, as the fetus grows, the expanding uterus must move into the abdominal cavity, thereby crowding the abdominal organs and dramatically increasing the size of the abdominal cavity.
5. After the pole passes through the abdominal wall, it pierces the parietal peritoneum. In passing through the stomach, it penetrates the visceral peritoneum, the stomach itself, and the visceral peritoneum on the other side of the stomach. Because the diaphragm is lined inferiorly by parietal peritoneum and superiorly by parietal pleura, these are the next two membranes pierced. The pole then passes through the pleural space and visceral pleura to enter the lung.

Chapter 2

1. An atom of iron has 26 protons (the atomic number), 30 neutrons (the mass number minus the atomic number), and 26 electrons (because the number of electrons is equal to the number of protons). If an atom of iron loses 3 electrons, it has 3 more protons (positive charges) than electrons (negative charges). Therefore, the iron ion has an overall charge of +3, which is represented symbolically as Fe^{3+}.
2. The formation of free fatty acids and glycerol from a triglyceride is a decomposition reaction because a larger molecule breaks down into smaller molecules. All of the decomposition reactions in the body are collectively referred to as catabolism. This reaction can also be classified as a hydrolysis reaction because, as part of the reaction, a water molecule is split into hydrogen, which becomes part of the glycerol molecule, and hydroxide, which becomes part of a fatty acid molecule. Yes, the reverse anabolic reaction generates water.
3. The slight amount of heat functions as activation energy and starts a chemical reaction. The reaction releases a large amount of heat, causing the solution to become hot.
4. Heating (boiling) has destroyed the ability of the molecules in one or both of the solutions to function in the chemical reaction. This is called denaturation. There are two possibilities as to what is denatured: the reactants themselves or an enzyme that catalyzes the reaction.
5. Muscle contains proteins. To increase muscle mass, proteins must be synthesized from amino acids. The synthesis of molecules in living organisms requires the input of energy. That energy comes from the potential energy stored in the chemical bonds of food molecules, which is released during the decomposition of food molecules.
6. pH is a measure of H^+ concentration. If equal amounts of solutions A and B are mixed, the resulting H^+ concentration is the average value of the two solutions—that is, the pH is $(8 + 2) / 2 = 5$. A pH of 5 is acidic. This question illustrates an important point: The pH of a solution can be changed by adding a more acidic or more basic solution to it.
7. Rapid respiration before diving into the water causes blood levels of carbon dioxide to decrease. As a result, there is a slight increase in blood pH. Recall that carbon dioxide molecules react with water to produce carbonic acid, and some of the carbonic acid molecules dissociate to form hydrogen ions and bicarbonate ions. These reactions are reversible. As Ned hyperventilates, the blood carbon dioxide levels decrease, which causes some hydrogen ions to react with bicarbonate ions to produce carbonic acid. The carbonic acid then dissociates to form water and carbon dioxide, so there is a net loss of protons. While holding his breath under water, carbon dioxide levels increased in Ned's blood. The carbon dioxide molecules react with water to form carbonic acid, which then dissociates to form hydrogen ions and bicarbonate ions. As a result, there is a slight decrease in blood pH. However, the pH of the blood does not change dramatically, in part because of buffers in the blood.
8. As A and B are added to the solution, the enzyme E catalyzes the formation of C. However, when C binds to the active site of E, the ability of E to catalyze the formation of C is blocked. As more and more C is produced, the rate of formation of C is slowed. Because the reaction of C with E is reversible, there will always be some E that has a functional (not blocked) active site, and some A will therefore always combine with B.
9. Heating the substances might help because proteins can be denatured and can coagulate (as in frying an egg). Another possibility is to try dissolving the substances in water. Most lipids are insoluble in water, whereas many proteins either are soluble in water or form colloids with water.

Chapter 3

1. The cells within the wound swell with water and lyse when a hypotonic solution is introduced. This kills potentially metastatic cells that may still be present in the wound.

2. Water moves by osmosis from solution B into solution A. Because solution A is hyperosmotic to solution B, solution A has more solutes and less water than does solution B. Water therefore moves from solution B (with more water) to solution A (with less water).
3. Answer *b* is correct. At point A on the graph, the extracellular concentration is equal to the intracellular concentration. If movement were by simple diffusion or by facilitated diffusion, at this point the rate of movement would be zero. Because it is not zero, it is reasonable to conclude that the mechanism involved is active transport.
4. Answer *b* is correct. Because the solution is isotonic, no exchange of water occurs. Because the solution contains the same concentration of all substances except that it has no urea, only a net movement of urea occurs across the membrane.
5. It is obvious that the heart and leg muscles of people who are jogging require the formation of ATP as a source of energy. As the heart and leg muscles increase in size, more ATP is produced by a greater number of mitochondria. Mitochondria are the critical membrane-bound organelles that increased in number by dividing. The genetic information for some of the proteins in mitochondria comes from the DNA of the mitochondria, and the genetic information for other proteins comes from DNA from the nuclei of the muscle cells.
6. Because the drug inhibits mRNA synthesis, protein synthesis is stopped. If the cell releases proteins as they are synthesized, the rate of protein secretion dramatically decreases following the administration of the drug. On the other hand, if the cell releases stored proteins, the rate of secretion at first is normal and then gradually declines.

Chapter 4

1. The tissue is epithelial tissue because it is lining a free surface, and the epithelium is stratified because it consists of more than one layer. The types of stratified epithelium are stratified squamous, stratified cuboidal, stratified columnar, and transitional epithelium. The structure of the cells in the surface layers determines the tissue type. Flat cells in the surface layer indicate stratified squamous epithelium. Cuboidal cells in the surface layer indicate stratified cuboidal epithelium, and columnar cells in the surface layer point to stratified columnar epithelium. The surface cells of transitional epithelium are roughly cuboidal with cubelike or columnar cells beneath them. When transitional epithelium is stretched, the surface cells are still roughly cuboidal, but underlying layers can be somewhat flattened.
2. In general, epithelial cells undergo cell division (mitosis) in response to injury, and the newly produced cells replace the damaged cells. However, if the basement membrane is destroyed, nothing is present to provide scaffolding for the newly formed epithelial cells. Without the basement membrane, there is no effective way for the newly formed epithelial cells to repair a structure, such as a kidney tubule. Since the basement membranes appear mostly intact, the person is likely to survive, and the kidney will regain most of its ability to function.
3. Pseudostratified squamous epithelium has goblet cells that secrete mucus. The cilia move the mucus over the surface of the epithelium toward the upper portion of the trachea. Stratified squamous epithelium does not secrete abundant mucus, and it does not have cilia. Consequently, mucus secreted by the area of the trachea that is still lined by pseudostratified columnar epithelium is not moved over the portion of the trachea lined by stratified squamous epithelium. The mucus accumulates below the area of the trachea lined by stratified squamous epithelium, causing Willie's frequent cough.
4. Glands producing merocrine secretions do so with no loss of actual cellular material, whereas glands producing holocrine secretions shed entire cells. The cells rupture and die, and the entire cell becomes part of the secretion. You can chemically analyze the secretions for the types of molecules found in cellular organelles. For example, if phospholipids and proteins normally found in membranes are in the secretion, the secretion is a holocrine secretion. If the secretion is watery or contains products that are not found in membranes or organelles, it is a merocrine secretion.
5. The tissue described is dense, regularly arranged, collagenic connective tissue. Injury to this type of tissue affects structures made up of this type of connective tissue, which includes tendons. Damage to neck vertebrae can be ruled out because they are connected by ligaments containing abundant dense, regularly arranged elastic connective tissue. A ruptured intervertebral disk is not indicated because it would consist of dense, irregularly arranged collagenic connective tissue.
6. Histamine is one of the mediators of inflammation released in response to tissue damage. However, several other chemical mediators are also released. Antihistamines might reduce the inflammatory response somewhat, but they are not likely to have a major effect because of the other chemical mediators released at the same time. In certain types of inflammatory responses, such as allergic responses, histamines are released in large amounts. Under these conditions, antihistamines do reduce the inflammatory response.

Chapter 5

1. The stratum corneum, the outermost layer of the skin, consists of many rows of flat, dead epithelial cells. The many rows of cells, which are continuously shed and replaced, are responsible for the protective function of the integument. In infants, there are fewer rows of cells, resulting in skin that is more easily damaged than that of adults.
2. Melanocytes produce melanin, which protects underlying tissue from ultraviolet radiation. Therefore, we expect melanocytes to be present in the most superficial living tissue. Although the epidermis as a whole is the most superficial tissue of the body, not all regions of the epidermis are composed of living cells. Therefore, melanocytes are located in the most superficial living tissue layer, the stratum basale of the epidermis.
3. Alcohol is a solvent that dissolves lipids (see chapter 2). It removes the lipids from the skin, especially in the stratum corneum. The rate of water loss increases after soaking the hand in alcohol because of the removal of the lipids that normally prevent water loss.
4. Carotene, a yellow pigment from ingested plants, accumulates in lipids. The stratum corneum of a callus has more layers of cells than other, noncallused parts of the skin, and the cells in each layer are surrounded by lipids. The carotene in the lipids makes the callus appear yellow.
5. Yes, the skin (dermis) can be overstretched due to obesity or rapid growth.
6. Eyelashes have a short growth stage (30 days) and are therefore short. Fingernails grow continuously but are short because they are cut, broken off, or worn down.
7. The hair follicle, but not the hair, is surrounded with nerve endings that can detect movement and pulling of the hair. The hair is dead, keratinized epithelium, so cutting the hair is not painful.
8. Probably not, because, following removal of the nail from the nail fold, it may grow back into the nail fold and the ingrown toenail may recur. One solution is to remove the small part of the nail responsible for the ingrown toenail. Prior to this drastic approach, sterile gauze can be placed between the nail and the nail fold to force the nail away from the nail fold. After the nail fold is healed, the gauze can be removed.

9. Rickets is a disease of children resulting from inadequate vitamin D intake. Inadequate vitamin D leads to insufficient absorption of calcium from the small intestine, resulting in soft bones. If adequate vitamin D is ingested, rickets is prevented, whether a person is dark- or fair-skinned. However, if dietary vitamin D is inadequate, when the skin is exposed to ultraviolet light, 7-dehydrocholesterol is converted into cholecalciferol, which can be converted to vitamin D. Dark-skinned children are more susceptible to rickets because the additional melanin in their skin screens out the ultraviolet light and they produce less vitamin D.
10. Chemotherapeutic drugs decrease mitosis in the stratum basale and stratum spinosum of the epidermis. As a result, there is a decrease in the number of layers of epithelial cells in the other layers of the skin (the stratum granulosum, stratum lucidum, and stratum corneum) where the cells are apparent. The rate of mitosis in the nail matrix decreases, so the rate of nail growth and the rate of mitosis in the hair bulb also decrease. Hair growth can stop, and hairs already formed can fall out of the hair follicles. The number of melanocytes in the skin does not decrease substantially because they do not typically undergo mitosis. The rate of melanin synthesis may decrease. More importantly, though, the rate at which epithelial cells are formed decreases, and the number of layers of epithelial cells in the epidermis decreases, so the rate of pigmentation may decrease. Sebaceous glands depend on mitosis to provide the cells that contribute to the holocrine secretion they produce. The decreased rate of mitosis causes a reduced rate of sebum production.

Chapter 6

1. Normally, bone matrix and bone trabeculae are organized to be strongest along lines of stress. Random organization of the collagen fibers of bone matrix results in weaker bones. In addition, the reduced amount of trabecular bone makes the bone weaker. Fractures can occur when the weakened bone is subjected to stress.
2. Osteoporosis is depletion of bone matrix that results when more bone is destroyed than is formed. Because mechanical stress stimulates bone formation (osteoblast activity), running helps prevent osteoporosis in the bones being stressed. This includes the bones of the lower limbs and the spine.
3. The loss of bone density results because the bones are not bearing weight in the weightless environment. Therefore, osteoblasts are not sufficiently stimulated and bone reabsorption exceeds bone building. Bone loss can be slowed by stressing the bones using exercises against resistance, such as cycling.
4. Testosterone normally causes a growth spurt at puberty, followed by slower growth and closure of the epiphyseal plate. Without testosterone, growth is slower but proceeds longer, resulting in a taller than normal person.
5. Blood vessels in central canals run parallel to the long axis of the bone, and perforating canals run at approximately a right angle to the central canals. Thus, perforating canals connect to central canals, which allows blood vessels in the perforating canals to connect with blood vessels in the central canals. After a fracture, blood flow through the central canals stops back to the point where the blood vessels in the central canals connect to the blood vessels in the perforating canals. The regions of bone on both sides of the fracture associated with this lack of blood delivery die.
6. Hyperparathyroidism stimulates increased bone breakdown and can cause osteitis fibrosa cystica, a condition in which the bone is eaten away as Ca^{2+} is released from the bone. The result can be a deformed bone that is likely to fracture. Vitamin D therapy could help because vitamin D promotes an increase in blood Ca^{2+} and therefore increased deposition of Ca^{2+} in bone.

Chapter 7

1. An infection in the nasal cavity could spread to adjacent cavities and fossae, including the paranasal sinuses: (1) frontal, (2) maxillary, (3) ethmoidal, and (4) sphenoidal; (5) the orbit (through the nasolacrimal duct); (6) the cranial cavity (through the cribriform plate); and (7) the throat (through the posterior opening of the nasal cavity).
2. Falling on the top of the head could drive the occipital condyles into the superior articulating processes of the atlas, causing a fracture. An uppercut to the jaw would slightly lift the occipital condyles away from the superior articulating processes of the atlas and usually does not result in a fractured atlas. However, such a blow to the jaw can fracture the temporal bone where it articulates with the mandible.
3. Forceful rotation of the vertebral column is most likely to damage the articular processes, especially in the lumbar region, where the articular processes tend to prevent excessive rotation (the superior articular processes face medially and the inferior articular processes face laterally).
4. If the ulna and radius become fused, the radius can no longer rotate relative to the ulna. As a result, most of the rotation of the forearm and hand is lost.
5. The ischial tuberosity is the bony protuberance.
6. Women's hips are wider than men's. As the knees are positioned toward the midline, the slope of the femur from its proximal end toward its distal end is greater in women. As a result, more women than men tend to be knock-kneed.
7. The lateral malleolus extends farther distally than the medial malleolus does, thus making it more difficult to turn the foot laterally than to turn it medially. The styloid process of the radius extends farther distally than the styloid process of the ulna, thus making it more difficult to cock the wrist toward the thumb (laterally) than toward the little finger (medially).
8. Landing on the heels can fracture the calcaneus. A heavy object, such as a firefighter, landing on the dorsal surface of the foot could fracture the metatarsal bones or even the tarsal bones.

Chapter 8

1. If the sternocostal synchondrosis were to ossify, becoming a synostosis, there would no longer be any stretch through the costal cartilage, the thorax could not expand, and respiration would be severely hampered.
2. a. suture, little or no movement
   b. syndesmosis, some movement
   c. complex synovial joints: the humeroulnoradial joint is a hinge joint and the radioulnar joint is a pivot joint; all have considerable movement
3. a. flexion and supination
   b. flexion of the hip and extension of the knee
   c. abduction of the arm at the shoulder
   d. flexion of the knee and plantar flexion of the foot
4. The descriptions of Donnie's symptoms and the diagnosis of AS provide us with important information to answer these questions. First, Donnie's lower back is the site of pain and the discomfort is spreading up his vertebral column, but not to his legs. Also, we are told that AS causes fibrosis, ossification, and fusion of joints due to chronic inflammation of the joints, specifically at insertion points for ligaments, tendons, and the joint capsule. Because Donnie's pain and stiffness are in his back and affect his joints, we know from chapter 7 that Donnie's discomfort is specifically associated with the vertebrae and the sacrum. We learned in chapter 8 that the joints between vertebrae, or intervertebral joints, are cartilaginous. We can therefore conclude that AS primarily affects cartilaginous joints. The second part of the question asks us to

explain why chronic inflammation of the joints would lead to their fusion. Inflammation of the joints occurs at insertion sites for ligaments, tendons, and the joint capsule, all of which are near the ends of the articulating bones. Similar to the inflammation associated with bone repair described in chapter 6, chronic inflammation of the joints leads to collagen synthesis by fibroblasts. The collagen becomes organized into a fibrous scar similar to the granulation tissue that forms during bone repair. Since the inflammation is chronic, this fibrous scar continues to increase in size to the point that it covers the entire joint. The fibrous scar is later ossified, causing the bones of the joint to fuse.

Chapter 9

1. Botulism poisoning results from ingesting botulism toxin produced by the bacterium *Clostridium botulinum.* The toxin binds to presynaptic nerve terminals and prevents the release of acetylcholine. Thus, action potentials in nerves cannot produce action potentials in skeletal muscles, and skeletal muscles become paralyzed, which explains the difficulty in breathing and swallowing. Other reasonable explanations are that the toxin binds to and blocks the receptors for acetylcholine, that the toxin blocks the entry of Ca^{2+} into the presynaptic terminal and thus prevents acetylcholine release, and that the toxin specifically prevents the entry of ions through Na^+ channels of skeletal muscle fibers.
2. Muscular dystrophy results from gradual atrophy of skeletal muscle fibers and their replacement with connective tissue. Myasthenia gravis results from the degeneration of the receptors for acetylcholine on the postsynaptic membranes of skeletal muscle fibers. If an inhibitor of acetylcholinesterase is administered, the result should be an increase in the concentration of acetylcholine in the nerve muscle synapse. Thus, more acetylcholine is available to bind to acetylcholine receptors. In people suffering from myasthenia gravis, the increased concentration of acetylcholine in the synapse allows acetylcholine to bind a greater percentage of the acetylcholine receptors present and causes the muscle contractions to increase in strength. In people who have muscular dystrophy, the muscle contractions do not increase in strength because muscle atrophy is the cause of the weakness. The additional acetylcholine in the neuromuscular junction has no effect on the weakened muscle fibers.
3. Start with a subthreshold stimulus and increase the stimulus strength by very small increments. Apply the stimulus to the nerve of muscle A and muscle B. If the number of motor units is the same for both preparations, each time the stimulus strength is increased, the degree of tension produced by the muscles will also increase to the same degree in each muscle. If one muscle has more motor units than the other, the muscle with the greater number of motor units will exhibit a greater number of separate increases in tension, and the magnitude of the increases in tension will be smaller than those seen in the muscle with fewer motor units.
4. While the weight is being held steady, the cross-bridges are pulling the Z disks closer together, but the external load (the weight) is pulling the sarcomeres apart with equal force. Because the internal and external forces are equal, the cross-bridges are producing enough force to hold the weight steady, but not enough to shorten the muscle, so the sarcomeres remain the same length. When the person lowers the weight, the cross-bridges are producing less force than the weight. Thus, each time a cross-bridge detaches from actin, the thin filaments "slip" and the sarcomeres lengthen. When the person raises the weight, the cross-bridges are producing more force than the external load. Thus, the cross-bridges collectively are able to produce enough force to pull the Z disks closer together, and the sarcomeres get shorter.
5. The shape of an active tension curve for skeletal muscle can be seen in figure 9.21. In contrast, an active tension curve is much flatter for smooth muscle. That is, for each increase in the length of a muscle fiber, there is little change in the active tension produced by the smooth muscle fiber. Smooth muscle has the ability to increase in length without much increase in tension.
6. Both the 100-meter dash and weight lifting involve rapid and intense contractions of skeletal muscles that are completed quickly. These contractions depend on anaerobic respiration for a significant amount of the ATP produced. In contrast, the 10,000-meter run involves sustained muscular contractions that are not as rapid, but the slower contractions are repeated many times during the run. Aerobic respiration produces the majority of the ATP for the 10,000-meter run. Anaerobic respiration is associated with a decrease in creatine phosphate, an increase in creatine, an increase in lactic acid, and a decrease in glycogen, and the enzymes responsible for anaerobic respiration function more rapidly. Aerobic respiration is associated with increased enzyme activity in the mitochondria and an increase in carbon dioxide production. Oxygen is used more rapidly during aerobic respiration.
7. During the 100-meter race, Shorty depended on ATP produced by anaerobic respiration. That produced an oxygen deficit at the end of the run, which resulted in an elevated rate of respiration for a time. During the longer and slower run, most of the ATP for muscle contractions was produced by aerobic respiration, and very little oxygen deficit developed. Prolonged aerobic respiration is required to "pay back" the oxygen deficit. Therefore, Shorty's rate of respiration was prolonged after the 100-meter race but not after the longer but slower run.
8. High blood K^+ concentration also results in depolarization of smooth muscle plasma membranes. Depolarization of the smooth muscle plasma membrane results in increased muscle contractions and increased permeability of the plasma membrane to both Na^+ and Ca^{2+}, which causes further depolarization and an increase in the intracellular concentration of Ca^{2+}. These changes result in the production of action potentials and muscle contractions.
9. The muscles would contract. ATP would be available to bind to the myosin heads, thus allowing myosin molecules to be released from actin molecules. The cross-bridges would immediately re-form, and complete cross-bridge cycling would result in contraction of the muscle fibers. As long as Ca^{2+} were present at high concentrations in the sarcoplasm, contraction of the muscles would occur. If the sarcoplasmic reticulum were intact, ATP would be available to drive the active transport of Ca^{2+} into the sarcoplasmic reticulum. As the Ca^{2+} decreased in the sarcoplasm, relaxation would result. If the sarcoplasmic reticulum were not intact, however, and could not transport Ca^{2+} into the sarcoplasmic reticulum as fast as they leaked out, the muscle would remain contracted until it fatigued.
10. Hormones can bind to ligand-gated Ca^{2+} channels, and the channels, in response, open. Calcium ions diffuse into the cell and cause contraction. Only a small amount of depolarization results as Ca^{2+} diffuses into the cell; since Na^+ channels do not open, a large change in the resting membrane potential does not occur.
11. In experiment A, the students used anaerobic respiration as they started to run in place, but aerobic respiration also increased to meet most of their energy needs. When they stopped, their respiration rate was increased over resting levels because of the repayment of the oxygen deficit due to anaerobic respiration. In experiment B,

almost all of the students' respiration came from anaerobic respiration because the students held their breath while running in place. Consequently, the students had a much larger oxygen deficit. The students' respiratory rate and depth would be greater than in experiment A, or their respiration rates would be elevated for a longer period of time than in experiment A.

12. The color of the meat depends on the number of capillaries (blood is red) within the muscle and is based on its myoglobin content (myoglobin is also red). After cooking, the tissues look darker, not red, because the blood and myoglobin have been broken down by the heat. Thus, the dark meat is darker because it contains more capillaries and more myoglobin. This is consistent with slow-twitch muscle fibers, which are used for maintaining posture and for performing slow movements, such as walking. The white muscle, with fewer capillaries and lower myoglobin content, is consistent with fast-twitch muscle fibers, which are used for quick movements, such as running or flying.

Chapter 10

1.

Muscle	Action	Synergist	Antagonist
Longus capitis	Flexes neck	Rectus capitis anterior, longus colli	Most of the posterior neck muscles
Erector spinae	Extends vertebral column	Interspinales, multifidus, semispinalis thoracis	Most of the anterior abdominal muscles
Coraco-brachialis	Adducts arm	Latissimus dorsi, pectoralis major, teres major, teres minor	Deltoid, supra-spinatus
	Flexes arm	Deltoid (anterior), pectoralis major, biceps, brachii	Deltoid (posterior), latissimus dorsi, teres major, teres minor, infra-spinatus, sub-scapularis, triceps brachii

2. The brachioradialis originates on the humerus and inserts onto the distal end of the radius. The fulcrum of this lever system is the elbow joint. With a weight held in the hand, the pull, applied between the weight and the fulcrum, is a class III lever system. With the weight on the forearm, the weight is between the pull and the fulcrum and is a class II lever system. A greater weight can be lifted if placed on the forearm rather than in the hand, but weights placed on the forearm cannot be lifted as far.
3. The muscles that flex the head also oppose extension of the neck. In an accident causing hyperextension of the neck, these muscles can be stretched and torn. The muscles involved can include the sternocleidomastoid, longus capitis, rectus capitis anterior, and longus colli. Automobile headrests are designed so that, if adjusted correctly, the back of the head hits the headrest during a rear-end accident, thereby preventing hyperextension of the neck.
4. The only muscle that elevates the lower eyelid is the orbicularis oculi, which "closes the eye." With this muscle not functioning, the lower eyelid droops. The levator anguli oris, which elevates the angle of the mouth, was also affected, allowing the corner of the mouth to droop. The zygomaticus major may also have been affected, as it inserts onto the corner of the mouth (see figure 10.9).
5. The genioglossus muscle protrudes the tongue. If it becomes relaxed, or paralyzed, the tongue may fall back and obstruct the airway. This can be prevented or reversed by pulling forward and down on the mandible, thus opening the mouth. The genioglossus originates on the genu of the mandible. As the mandible is pulled down and forward, the genioglossus is pulled forward with the mandible, thus pulling the tongue forward also.
6. The rotator cuff muscles are the primary muscles holding the head of the humerus in the glenoid fossa, especially the supraspinatus. In fact, a torn rotator cuff, which usually involves a tear of the supraspinatus muscle, often results in dislocation of the shoulder.
7. Speedy has ruptured the calcaneal tendon, and the gastrocnemius and soleus muscles have retracted, causing the abnormal bulging of the calf muscles. Because the major plantar flexors are no longer connected to the calcaneus, the runner cannot plantar flex the foot, and the foot is abnormally dorsiflexed because the antagonists have been disconnected.
8. If you are sitting on a chair with the book open on the desk in front of you, very few muscles are required to turn the page. First the arm is extended slightly to push the forearm and hand along the edge of the book. This can be accomplished with the anterior portion of the deltoid muscle. The hand is then slightly supinated by means of the supinator muscle. The thumb and index finger are then flexed to grasp the page to be turned. This movement involves the flexor pollicis brevis and flexor digitorum superficialis to the index finger (the flexor digitorum profundus may also be involved). The page is then turned by extending the fingers (extensor digitorum), pronating the hand (pronator quadratus and pronator teres), and medially rotating the arm (pectoralis major, teres major, latissimus dorsi).

Chapter 11

1. Because the plasma membrane is much less permeable to Na^+ than to K^+, changes in the extracellular concentration of Na^+ affect the resting membrane potential less than do changes in the extracellular concentration of K^+. Therefore, increases in extracellular Na^+ have a minimal effect on the resting membrane potential. Because the membrane is much more permeable to Na^+ during the action potential, the elevated concentration of Na^+ in the extracellular fluid allows Na^+ to diffuse into the cell at a more rapid rate during the action potential, resulting in a greater degree of depolarization during the depolarization phase of the action potential.
2. Smooth muscle cells contract spontaneously in response to spontaneous depolarizations that produce action potentials. One way action potentials can be produced spontaneously is if membrane permeability to Na^+ spontaneously increases. As a result, some Na^+ enters the smooth muscle cells and causes a small depolarizing graded potential. The small depolarization can cause voltage-gated Na^+ channels to open, which results in further depolarization, thereby stimulating additional Na^+ voltage-gated ion channels to open. This positive-feedback cycle can continue until the plasma membrane is depolarized to its threshold level and an action potential is produced.
3. Action potential conduction along a myelinated nerve fiber is more energy-efficient because the action potential is propagated by saltatory conduction, which produces action potentials at the nodes of Ranvier. Compared with an unmyelinated nerve fiber, only a small portion of the myelinated neuron's membrane has action potentials. Thus, less Na^+ flows into the neuron (depolarization) and less K^+ flows out of the neuron (repolarization). Consequently, the sodium-potassium pump has to move fewer ions in order to restore ion concentrations. Because the sodium-potassium pump requires ATP, myelinated axons use less ATP than unmyelinated axons.
4. The inhibitory neuromodulator causes the postsynaptic neuron to become less sensitive to excitatory stimuli, probably by causing

hyperpolarization of the postsynaptic neuron. As a result, the excitatory neurotransmitter released from the excitatory neuron is less likely to produce postsynaptic action potentials.

5. With aging, the amount of myelin surrounding axons decreases, which decreases the speed of action potential propagation. Also, at synapses, action potentials in the presynaptic terminal take longer to cause the production of action potentials in the postsynaptic membrane. It is believed this results from a reduced release of neurotransmitter by the presynaptic terminal and a reduced number of receptors in the postsynaptic membrane.
6. Organophosphates inhibit acetylcholinesterase, thereby causing an increase in acetylcholine in the synaptic cleft, which leads to overproduction of action potentials, tetanus of muscles, and possible death from respiratory failure (see chapter 9). Curare is the best antidote because it blocks the effect of acetylcholine and counteracts the organophosphate. Too much curare, however, can cause flaccid paralysis of the respiratory muscles. Injecting acetylcholine would make the effect of the organophosphate worse. Potassium chloride causes depolarization of muscle fiber membranes, thereby making them more sensitive to acetylcholine.
7. If the motor neurons supplying skeletal muscle are innervated by both excitatory and inhibitory neurons, blocking the activity of the inhibitory neurons with strychnine results in overstimulation of the motor neurons by the excitatory neurons.
8. GABA binds to a ligand-gated receptor and opens Cl^- channels, causing hyperpolarization of the postsynaptic membrane. Alcohol enhances binding of GABA to its receptors. Therefore, alcohol enhances the hyperpolarization effect of GABA on postsynaptic membranes for synapses in which GABA is a neurotransmitter. Finally, since chronic alcohol consumption makes the GABA receptors less sensitive to both alcohol and GABA, the postsynaptic membranes of people who consume alcohol chronically are hyperpolarized less by a given amount of GABA than are those of other people. Therefore, the membrane is less hyperpolarized in the absence of alcohol, and this triggers withdrawal symptoms. Benzodiazepine enhances the effect of GABA on its receptors, thus also enhancing the hyperpolarizing effect of GABA in synapses in which GABA is a neurotransmitter.
9. When the neurotoxin binds to ligand-gated Na^+ channels in the postsynaptic membrane of a skeletal muscle fiber, they open, and Na^+ enters the cell, producing graded potentials. When the graded potentials reach threshold, an action potential is produced, stimulating the muscle fiber to contract. However, the neurotoxin tends to remain bound to the ligand-gated Na^+ channels, which prevents ACh from binding. Thus, the nervous system's ability to stimulate the muscle fiber decreases as more and more neurotoxin binds to ligand-gated Na^+ channels. Because the ligand-gated Na^+ channels with bound neurotoxin remain open, Na^+ continues to enter the muscle fiber, causing its resting membrane potential to depolarize. Eventually, the membrane becomes so depolarized that it is unresponsive to stimulation. Death from a cobra bite usually occurs because of paralysis of respiratory muscles.
10. A Na^+ channelopathy, in which Na^+ channels open more readily than normal or stay open longer than normal, could cause an increased production of action potentials. A Ca^{2+} channelopathy, in which more Ca^{2+} than normal enters the presynaptic terminal, could result in increased release of neurotransmitter from the presynaptic terminal and thus an increased production of action potentials.

Chapter 12

1. If the neuron with its cell body in the cerebrum is an inhibitory neuron and if it synapses with the motor neuron of a reflex arc, stimulation of the cerebral neuron can inhibit the reflex.
2. The phrenic nerve is cut in the thorax while the lung is being removed.
3. Pulling on the upper limb when it is raised over the head can damage the lower brachial plexus—in this case, the origin of the ulnar nerve. The ulnar nerve innervates muscles that abduct/adduct the fingers and flex the wrist.
4. The sciatic nerve has rootlets from L4 to S3. Depending on the rootlet compressed, pain can be felt in different locations.
5. a. obturator nerve
   b. femoral nerve
   c. sciatic (tibial) nerve
   d. obturator nerve
   e. obturator nerve, some from femoral nerve
6. The plaster cast is pressing against the common fibular (peroneal) nerve at the neck of the fibula. Tingling is expected along the lateral and anterior leg and the dorsum of the foot.
7. Lack of sensation in the lower limbs and inability to move them are consequences of complete transection (transverse cut) of the spinal cord. Because Cecil could still breathe on his own, the phrenic nerves, which innervate the diaphragm, must still be intact. The phrenic nerves originate from C3, C4, and C5, so the transection must have occurred below C5. Also, Cecil was able to move his upper limbs, and the upper arms are controlled by nerve fibers that originate from C3, C4, and C5. However, the hands and fingers (whose movement was somewhat impaired) are primarily controlled by nerve fibers that originate from C6, C7, C8, and T1. Thus, the transection is probably between C5 and C6 or between C6 and C7.

Chapter 13

1. If CSF does not drain properly, the fluid accumulates and exerts pressure on the brain (hydrocephalus). In the developing fetus, the ventricles enlarge because of the excess fluid pressure. The head also enlarges because the skull bones have not fused. However, the expansion of the head is not sufficient to relieve all the pressure exerted on the developing brain by the expanding ventricles. As a result, the cerebral cortex becomes proportionately thinner as it is compressed between the ventricles and the skull. In many cases, fewer gyri form in the cerebral cortex. Brain damage may or may not result, depending on the amount of excess CSF, the ventricular pressure generated, and the areas of the brain damaged by the pressure.
2. Enlargement of the lateral and third ventricles, without enlargement of the fourth ventricle, suggests a blockage between the third and fourth ventricles in the cerebral aqueduct. This defect, called aqueductal stenosis, is a common congenital problem.
3. Blood in the CSF taken through a spinal tap indicates the presence of blood in the subarachnoid space and suggests that the patient has a damaged blood vessel in the subarachnoid space.
4. I: Test vision; II: Have the person describe the smell of something placed under the nose; III: Test eye movement; IV: Test the ability to move the eye down and out; V: Test the sense of feeling in the face; VI: Test the ability to move the eye to the side; VII: Test the ability to taste an item on the front of the tongue and check for facial expression; VIII: Test the ability to hear; IX: Test the ability to swallow; X: Test the ability to swallow and check the uvula when the mouth is opened (the uvula will "point" away from the side where X is not working); XI: Test the ability to turn the head; XII: Have the person protrude his or her tongue (if XII is not functioning on one side, the tongue will "point" toward the damaged side).

5. The cerebellum, which is in charge of controlling coordinated muscle movement and maintaining muscle tone, was damaged in this patient.
6. The olfactory nerves (CN I) travel through the olfactory foramina within the cribriform plate of the ethmoid bone as they pass from the nasal epithelium to the olfactory bulbs (see table 13.5). In this case, the cribriform plate of the ethmoid bone was probably fractured, severing the connections of the olfactory nerves to the olfactory bulb and resulting in a loss of the sense of smell.
7. The abducens nerves supply the lateral rectus muscles, which are responsible for moving the eyes laterally (abducting the eyes). Damage to the abducens nerve on the left side reduced or eliminated Stanley's ability to abduct his left eye. The inability to move the left eye in concert with the right eye results in double vision.
8. Based on the source of pain, it appears that the maxillary branch of the trigeminal nerve on the right side is affected. This condition, known as trigeminal neuralgia, results in sharp bursts of pain that are often initiated by stimulation of a particular trigger area. Loss of tactile sensations from the affected area is likely to result. Because the main muscles of mastication are innervated by the mandibular branch of the trigeminal nerve, chewing is not greatly affected, and the maxillary branch of the trigeminal does not innervate the muscles of the pharynx and tongue that are critical for the swallowing process. Since Andy experiences no pain in the area of the nose or superior to the eyes, the ophthalmic branch of the trigeminal nerve does not appear to be involved.
9. It is likely that Afton experienced facial palsy (Bell palsy), which can be temporary. This condition may result from a stroke or a tumor, or it can be triggered by inflammation of the parotid gland, anesthesia, or exposure of the superficial branches of the nerve to cold (which is probably the cause in Afton's case). The loss of motor tone in the face is due to decreased innervation of the muscles of facial expression. Also, some muscles of the pharynx are affected, but these are not related to drooping of the face.

Chapter 14

1. The first sensations the woman perceives when she picks up an apple and bites into it are visual (special), tactile (general), and proprioceptive (general). The woman holds the apple in her hand and looks at it. The tactile sensations from mechanoreceptors in the hand tell her the apple is firm and smooth. The proprioceptive sensations originating in the joints of the hand tell her the size and shape of the apple. Visual input also tells her the size and color of the apple and that it has a smooth surface. As the woman bites into the apple and begins to chew, proprioceptive sensations from the teeth and jaws provide information on how widely she must open her mouth to accommodate the apple and how hard to bite down. Tactile sensations originating in the tongue and cheeks tell her the location of the bite of apple and its texture as it is moved about in the mouth. In chapter 15 you will learn that taste sensations (special, chemoreceptor) from the tongue indicate that the apple has both sweet and sour characteristics. Olfactory sensations (special) provide more specific information that the "fruity taste" is that of an apple.
2. a. The most likely explanation is that the olfactory neurons have adapted and no longer respond to the odor stimulus.
   b. The fact that people can hear the sound when they make a conscious effort indicates that the hair cells in the spiral organ have not adapted and are still able to detect the sound stimulus. Many action potentials arriving in the brain are prevented from causing conscious perception until we consciously "pay attention" to the stimulus. For example, you may not be paying attention to general conversations in a crowded room or hall until someone says your name. The sound of your name leaps out of the surrounding babble, and you are suddenly interested.
3. It is possible that the dorsal-column/medial-lemniscal system within the right side of the spinal cord was damaged. However, it is also possible that this system was damaged within the medulla oblongata, where neurons synapse and cross over to the left side of the brain, or within the tracts on the left side that ascend from the medulla oblongata to the thalamus. Another possibility is damage to the cerebral cortex on the left side. Additional information is needed to determine exactly where the injury is located.
4. The fibers of the lateral spinothalamic tract carry impulses for pain and temperature. A lesion in the area where these fibers decussate results in the bilateral loss of pain and temperature sensations only at the level of the lesion, but no loss of sensation below the lesion. This occurs because fibers decussating above or below the lesion, as well as tracts that pass lateral to the lesion, are unaffected. This disorder, called syringomyelia, is often caused by a spinal cord tumor.
5. The damaged tracts are the lateral corticospinal tract, controlling motor functions on the right side of the body, and the lateral spinothalamic tract for pain and temperature sensations from the left side of the body. Damage to these tracts in the right side of the spinal cord produces the observed symptoms because within the cord the lateral spinothalamic tract crosses over at the level of entry and is therefore located on the opposite side of the cord from its peripheral nerve endings, whereas the corticospinal tract lies on the same side of the cord as its target muscles.
6. Complete unilateral transection of the right side of the spinal cord would result in loss of motor function (lateral corticospinal tract), proprioception, and two-point discrimination (dorsal-column/medial-lemniscal system) on the same side of the body as the lesion, below the level of the lesion. Pain and temperature sensations (lateral spinothalamic tract) would be lost on the opposite side of the body from below the level of the lesion. These symptoms describe the Brown-Séquard syndrome. Light touch would not be greatly affected on either side because of the large number of collateral branches in the anterior spinothalamic tract.
7. Damage to the cerebellum can result in decreased muscle tone, balance impairment, a tendency to overshoot when reaching for or touching something, and an intention tremor. These symptoms are opposite those seen with basal ganglia dysfunction. Cerebellar dysfunction exhibits symptoms very similar to those seen in an inebriated person, and the same tests can be applied, such as having the person touch his or her nose or walk a straight line.
8. Memory for the 10 minutes prior to the accident was stored in short-term memory and was disrupted before it could be transferred to long-term memory. Anytime a person suffers a concussion, there is a possibility that postconcussion syndrome will develop later. Symptoms include muscle tension or migraine headaches, reduced alcohol tolerance, difficulty learning new things, reduction in creativity and motivation, fatigue, and personality changes. The syndrome may last a month to a year. Postconcussion syndrome may be the result of a slowly occurring subdural hematoma, which may be missed in an early examination.
9. The subdural hematoma is likely to be over the medial portion of the left side of the cerebral hemisphere in the area of the premotor cortex and expanding to the area of the primary motor cortex. The premotor area must be intact for a person to carry out complex, skilled, or learned movements, especially

those requiring manual dexterity. If blood is removed from the hematoma and if no more bleeding occurs, it is likely that Perry's motor movements will improve rapidly.

Chapter 15

1. The lens of the eye is biconvex and causes light rays to converge. If the lens is removed, the replacement lens should also cause light rays to converge. A biconvex lens or a lens with a single convex surface should work. Bifocals or trifocals can also be recommended to compensate for because of the loss of accommodation.
2. Carrots contain vitamin A (retinoic acid), which can be used to form retinal. Retinal and opsin combine to form rhodopsin, which is found in rods. Rhodopsin is necessary for rods to respond to low levels of light. Lack of vitamin A can result in lack of rhodopsin and night blindness.
3. This phenomenon is called a negative afterimage. While the man is staring at the clock, the darkest portion of the image (the black clock) causes dark adaptation in part of the retina—that is, part of the retina becomes more sensitive to light. At the same time, the lightest part of the image (the white wall) causes light adaptation in the rest of the retina, and that part of the retina becomes less sensitive to light. When the man looks at a white wall, the dark-adapted portion of the retina, which is more sensitive to light, produces more action potentials than does the light-adapted part of the retina. Consequently, he perceives a light clock against a darker background.
4. A lesion of the optic chiasm results in visual loss in both the right and left temporal fields, a condition called bitemporal hemianopsia, or tunnel vision. Tunnel vision can cause problems for normal functions, such as driving a car, because the peripheral vision is severely limited. This condition can also suggest a much more serious problem, such as a pituitary tumor just posterior to the optic chiasm.
5. Eyestrain, or eye fatigue, occurs primarily in the ciliary muscles. It occurs because close vision requires accommodation. Accommodation occurs as the ciliary muscles contract, releasing the tension of the suspensory ligaments and allowing the lens to become more rounded. Continued close vision requires the maintenance of accommodation, which requires that the ciliary muscles remain contracted for a long time, resulting in their fatigue.
6. The most likely area damaged is the spiral organ, where waves result in the production of action potentials. The action is much like ocean waves breaking on the shore during a violent storm, compared with those breaking in from a calm ocean. Specifically, damage likely occurs in the part of the spiral organ near the oval window because this part of the basilar membrane vibrates the most in response to high-frequency sounds.
7. Normally, as pressure changes, the auditory tubes open to allow equalization of pressure between the middle ear and the external environment. If this does not occur, the built-up pressure in the middle ear can rupture the tympanic membrane, or the pressure can be transmitted to the inner ear and cause sensorineural damage.
8. Normally, airborne sounds cause the tympanic membrane to vibrate, resulting in the movement of the middle ear ossicles and the production of waves in the perilymph of the scala vestibuli. Vibration of the skull bones can also cause vibration of the perilymph in the scala vestibuli.

Chapter 16

1. The sympathetic division of the ANS is responsible for dilation of the pupil. Preganglionic fibers from the upper thoracic region of the spinal cord pass through spinal nerves (T1 and T2), into the white rami communicantes, and into the sympathetic chain ganglia. The preganglionic fibers ascend the sympathetic chain and synapse with postganglionic neurons in the superior cervical sympathetic chain ganglia. The axons of the postganglionic neurons leave the sympathetic chain ganglia as small nerves that project to the iris of the eye.
2. Reduced salivary and lacrimal gland secretions can indicate damage to the facial nerves, which innervate the submandibular, sublingual, and lacrimal glands. The glossopharyngeal nerves innervate the parotid glands, but not the lacrimal glands.
3. a. pelvic splanchic nerves
   b. outflow of gray ramus
   c. vagus nerve
   d. oculomotor nerve
   e. pelvic splanchnic nerve
4. Inactivation of acetylcholinesterase results in a buildup of acetylcholine in synapses and an overstimulation of muscarinic receptors. One would expect mostly parasympathetic effects because the effects of acetylcholine are enhanced: blurred vision as a result of contraction of ciliary muscles, excess tear formation because of overstimulation of the lacrimal glands, and frequent or involuntary urination because of overstimulation of the urinary bladder. Pallor resulting from vasoconstriction in the skin is a sympathetic effect that would not be expected because skin blood vessels respond to norepinephrine. Muscle twitching or cramps of skeletal muscles might occur because they normally respond to acetylcholine. Atropine, a muscarinic blocking agent, can be used to treat exposure to malathion.
5. Epinephrine causes vasoconstriction and confines the drug to the site of administration. This increases the drug's duration of action locally and decreases its systemic effects. Vasoconstriction also reduces bleeding if a dry field (an area clear of blood on its surface) is required.
6. Because normal action potentials are produced, the drug does not act at the synapse between the preganglionic and postganglionic neurons. Because injected norepinephrine works, sympathetic receptors in the heart are functioning and are not affected by the drug. Therefore, the drug must somehow affect the postganglionic neurons. Possibly it inhibits neurotransmitter production or release from the postganglionic neurons.
7. Because cutting the white rami of T1–T4 does not affect the drug's action, sympathetic preganglionic neurons in the spinal cord and sympathetic centers in the brain can be ruled out as a site of action. Because cutting the vagus nerves eliminates the drug's effect, the drug cannot be acting at the synapse between the preganglionic neurons and the postganglionic neurons, or between the synapse of the postganglionic neuron and the effector of either division of the ANS. The drug must therefore excite parasympathetic centers in the brainstem, resulting in decreased heart rate.
8. a. Responses in a person who is extremely angry are primarily controlled by the sympathetic division of the ANS. These responses include increased heart rate and blood pressure, decreased blood flow to the internal organs, increased blood flow to skeletal muscles, decreased contractions of the intestinal smooth muscle, flushed skin in the face and neck region, and dilation of the pupils of the eyes.
   b. For a person who has just finished eating and is relaxing, parasympathetic reflexes are more important than sympathetic reflexes. The blood pressure and heart rate are at normal resting levels, the blood flow to the internal organs is greater, contractions of smooth muscle in the intestines are greater, and secretions that achieve digestion are more active. If the urinary bladder or the colon becomes distended, autonomic reflexes that result in urination or defecation can result. Blood flow to the skeletal muscles is reduced.

Chapter 17

1. Liver disease and kidney disease would increase the concentration of this hormone in the blood, and the concentration would remain high for a longer time. The liver modifies the hormone to cause it to be excreted by the kidneys more rapidly. In the case of liver disease, the hormone is not modified and excreted rapidly. Therefore, the concentration becomes higher than normal, and the high concentration remains for longer than normal. A similar result is seen if the kidneys are diseased and the hormone cannot be excreted rapidly.
2. Secretion of hormones is usually controlled by negative feedback. If a hormone controls the concentration of a substance in the circulatory system, the hormone is secreted in smaller amounts if the substance increases in the circulatory system. If a tumor begins to secrete the substance in large amounts, the presence of the substance has a negative-feedback effect on the secretion of the hormone, and the concentration of the hormone in the circulatory system is very low.
3. Usually, intracellular mediator mechanisms respond quickly, and the hormone's effect is brief. Nuclear receptor mechanisms usually take a long time (several hours) to respond, and their effects last much longer. If the hormone is large and water-soluble, it is probably functioning through an intracellular mediator mechanism; if the hormone is lipid-soluble, it is probably a nuclear receptor mechanism. If you have the ability to monitor the concentration of a suspected intracellular mediator and it increases in response to the hormone, or if you can inhibit the synthesis of an intracellular mediator and it prevents the target cells' response to the hormone, it is an intracellular mediator mechanism. If you can inhibit the synthesis of mRNA and this inhibits the action of the hormone, or if you can measure an increase in mRNA synthesis in response to the hormone, then the mechanism is a nuclear receptor mechanism.
4. When the hormone binds to its receptor, the α subunit of the G protein is released. However, GTP must bind to the α subunit before it can have its normal effect. If the α subunit cannot bind to GTP, the hormone has no effect on the target tissue.
5. Phosphodiesterase causes the conversion of cAMP to AMP, thus reducing the concentration of cAMP. Therefore, a drug that inhibits phosphodiesterase increases the amount of cAMP in cells where cAMP is produced. Thus, an inhibitor of phosphodiesterase increases a tissue's response to a hormone that has cAMP as an intracellular mediator.
6. Epinephrine's shorter half-life allows it to produce a short-lived response. The response to a potentially harmful situation is terminated shortly after the situation passes. If epinephrine had a longer half-life, the heart rate and the blood glucose level would be elevated for a long time, even if the harmful situation were very brief.
7. Because thyroid hormones are important in regulating the basal metabolic rate, their long half-life is an advantage. Thyroid hormones are secreted and have a prolonged effect without large fluctuations in the basal metabolic rate. If thyroid hormones had a short half-life, the basal metabolic rate could fluctuate with changes in the rate of secretion of thyroid hormones. Certainly, the rate of secretion of thyroid hormones would have to be controlled within narrow limits if it did have a short half-life.
8. Elevated GnRH levels in the blood as a result of the GnRH-secreting tumor would cause the down-regulation of GnRH receptors in the anterior pituitary. This would decrease the ability of GnRH to stimulate the anterior pituitary, and the rate of LH and FSH secretion by the anterior pituitary would decrease and remain decreased as long as the GnRH levels were chronically elevated. Therefore, the functions of the reproductive system controlled by LH and FSH would decrease.
9. Insulin levels normally change in order to maintain normal blood sugar levels, despite periodic fluctuations in sugar intake. A constant supply of insulin from a skin patch might result in insulin levels that are too low when blood sugar levels are high (after a meal) and might be too high when blood sugar levels are low (between meals). In addition, insulin is a protein hormone that would not readily diffuse through the lipid barrier of the skin (see chapter 5).

Chapter 18

1. The hypothalamohypophysial portal system allows neurohormones that function as releasing and inhibiting hormones, which are secreted by neurons in the hypothalamus, to be carried directly from the hypothalamus to the anterior pituitary gland. Consequently, the releasing and inhibiting hormones are not diluted or destroyed by the enzymes, which are abundant in the kidneys, liver, lungs, and general circulation, before they reach the anterior pituitary. Also, the time it takes for releasing and inhibiting hormones to reach the anterior pituitary is less than if they were secreted into the general circulation.
2. Polydipsia and polyuria are consistent with both diabetes mellitus and diabetes insipidus. Diabetes mellitus, however, is consistent with an increased urine osmolality because of the large amount of glucose lost in the urine. Diabetes insipidus is consistent with urine with a low specific gravity because little water is reabsorbed by the kidneys. Thus, urine has an osmolality close to that of the body fluids, and the rapid loss of dilute urine results in a decrease in blood pressure. Therefore, polyuria with a low specific gravity is consistent with diabetes insipidus but not with diabetes mellitus. The administration of ADH would reverse the symptoms of diabetes insipidus. Neither polydipsia nor polyuria results from a lack of glucagon or aldosterone.
3. The symptoms are consistent with acromegaly, which is a consequence of elevated GH secretion after the epiphyses have closed. Increased GH causes enlarged finger bones, the growth of bony ridges over the eyes, and increased growth of the jaw. The anterior pituitary tumor increases pressure at the base of the brain near the optic nerves as it enlarges. The pituitary rests in the sella turcica of the sphenoid bone; as it enlarges, pressure increases because the pituitary is nearly surrounded by rigid bone, and the brain is located just superior to the pituitary. As the anterior pituitary enlarges because of a tumor, it pushes superiorly, and pressure is applied to the ventral portion of the brain. In addition, the GH causes bone deposition on the inner surface of skull bones, which also increases the pressure inside the skull.
4. If hyperthyroidism results from a pituitary abnormality, laboratory tests should show elevated TSH levels in the circulatory system in addition to elevated T_3 and T_4 levels. If hyperthyroidism results from the production of a nonpituitary thyroid-stimulating substance, laboratory tests should also show elevated T_3 and T_4 levels, but TSH levels will be low because of the negative-feedback effects of T_3 and T_4 on the hypothalamus and pituitary gland.
5. It is likely that Julie's elevated ACTH levels are causing elevated blood levels of cortisol, which in turn are causing the observed symptoms. The elevated ACTH levels are probably due to a hormone-secreting tumor (adenoma) in the anterior pituitary gland. According to National Institutes of Health sources, pituitary adenomas cause 70% of Cushing syndrome cases, excluding those caused by glucocorticoid use. Most people with the disorder have a single adenoma. Cushing syndrome affects women five

times more often than men. The most widely used treatment is surgical removal of the tumor, known as transsphenoidal adenomectomy. Using a special microscope and fine instruments, the surgeon approaches the pituitary gland through a nostril or an opening made below the upper lip. Because this procedure is extremely delicate, patients are often referred to centers specializing in this type of surgery. The success rate of this procedure is more than 80% when performed by a surgeon with extensive experience. If surgery fails or produces only a temporary cure, the surgery can be repeated, often with good results. Radiation of the pituitary gland is another possible treatment.

6. The second student is correct. Low levels of vitamin D reduce calcium uptake in the digestive tract, which results in a decreased blood level of calcium ions. As blood calcium levels decrease, the rate of PTH secretion increases. Parathyroid hormone increases bone breakdown, which maintains blood calcium levels, even if vitamin D deficiency exists for a prolonged time. Osteomalacia results because of the increased bone reabsorption necessary to maintain normal blood calcium levels.
7. Because the person is diabetic and probably taking insulin, insulin shock is more likely than diabetic coma. To confirm the condition, however, a blood sample should be taken. If the condition is due to diabetic coma, the blood glucose levels will be elevated. If the condition is due to insulin shock, the blood glucose levels will be below normal. In the case of insulin shock, glucose can be administered intravenously. In the case of diabetic coma, insulin should be administered. An isotonic solution containing insulin can be administered to reduce the osmolality of the extracellular fluid.
8. Elevated epinephrine from the adrenal medulla promotes elevated blood pressure and increases the workload on the heart, increases the rate of metabolism, and results in increased sweating and nervousness. The risks for heart attack and stroke are increased. Elevated cortisol causes hyperglycemia and can lead to diabetes mellitus, a depressed immune system with increased susceptibility to infections, and the destruction of proteins, leading to tissue wasting.
9. We learn that Katie has androgen insensitivity syndrome. The name suggests that her tissues are not sensitive to androgens (malelike hormones). We learned that a tissue responds to hormones on the presence of specific receptors. With androgen insensitivity syndrome, the endocrine malfunction is not the production of the hormone, as in insulin-dependent (type I) diabetes mellitus, but the presence of the receptor in the target cells. Even though her cells produce plenty of the hormone testosterone, the target cells do not respond because the testosterone receptor has malfunctioned and does not recognize testosterone. Katie's feminine appearance is a direct result of the lack of a normal male body because her tissues were resistant to testosterone, which is responsible for growth of the male genitalia and other sex characteristics (see table 10.2 and chapter 19). Therefore, externally Katie looked female when she was born. Without testosterone working in her body or her brain and without normal social cues to associate with other males, she had always identified herself as female and was devastated to learn she could not have children. Her doctor and mother decided to wait until she was older to explain the entire condition to her.

Chapter 19

1. Because of the rapid destruction of the red blood cells, we would expect erythropoiesis to increase in an attempt to replace the lost red blood cells. The reticulocyte count would therefore be above normal. Jaundice is a symptom of hereditary hemolytic anemia because the destroyed red blood cells release hemoglobin, which is converted into bilirubin. Removal of the spleen cures the disease because the spleen is the major site of red blood cell destruction.
2. Blood doping increases the number of red blood cells in the blood, thereby increasing its oxygen-carrying capacity. The increased number of red blood cells also makes it more difficult for the blood to flow through the blood vessels, increasing the heart's workload.
3. The correct answer is *c*. Ben's red blood cells are of normal size. However, his reticulocyte level is low, which indicates a reduced rate of red blood cell synthesis. The low red blood cell count is also consistent with a reduced rate of red blood cell synthesis. Hemoglobin concentration and hematocrit are low, both of which are consistent with the reduced red blood cell count. The prothrombin time is longer than normal, the prothrombin count is low, and the platelet count is low. All these observations are consistent with aplastic anemia, in which the stem cells that produce blood cells in the red marrow are damaged. Chemicals such as benzene and chloramphenicol can also destroy red marrow cells and cause aplastic anemia.
4. Hypoventilation results in decreased blood oxygen levels, which stimulate erythropoiesis. Therefore, the number of red blood cells increases and produces secondary polycythemia.
5. Vitamin B_{12} and folic acid are necessary for blood cell division. Lack of these vitamins results in pernicious anemia. Iron is necessary for the production of hemoglobin. Lack of iron results in iron-deficiency anemia. Vitamin K is necessary for the production of many blood clotting factors. Lack of vitamin K can greatly increase blood clotting time, resulting in excessive bleeding.
6. Grace has hemolytic anemia. The RBC is lower than normal because the red blood cells are being destroyed faster than they are being replaced. With fewer red blood cells, hemoglobin and hematocrit are lower than normal. Bilirubin levels are above normal because of the breakdown of the hemoglobin released from the ruptured red blood cells.
7. The correct answer is *c*. As Pam went from a lower to a higher elevation, the barometric pressure decreased; therefore, the availability of oxygen in the air also decreased. Consequently, as Pam moved from sea level to a higher elevation, her kidneys secreted greater amounts of erythropoietin, and red blood cell synthesis increased. After about 4 days in Jackson Hole, Pam's blood should have increased numbers of reticulocytes and red blood cells. By day 6, Pam's red blood cell count had increased significantly and was still rising. Changes in oxygen levels should not affect platelet count.

Chapter 20

1. The walls of the ventricles are thicker than the walls of the atria because the ventricles must produce a greater pressure to pump blood into the arteries. Only a small pressure is required to pump blood from the atria into the ventricles during diastole. The wall of the left ventricle is thicker than the wall of the right ventricle because the left ventricle produces a much greater pressure to force blood through the aorta than the right ventricle produces to move blood through the pulmonary trunk and pulmonary arteries.
2. During systole, the cardiac muscle in the right and left ventricles contracts, which compresses the coronary arteries. During diastole, the cardiac muscle of the ventricles relaxes, and blood flow through the coronary arteries increases. The diastolic pressure is sufficient to cause blood to flow through the coronary arteries during diastole.
3. Contraction of the ventricles, beginning at the apex and moving toward the base of the

heart, forces blood out of the ventricles and toward their outflow vessels—the aorta and pulmonary trunks. The aorta and pulmonary trunks are located at the base of the heart.

4. Ectopic foci cause various regions of the heart to contract at different times. Ectopic action potentials in the ventricles cause the areas in which the action potentials develop to contract independently of all other areas of the ventricles. As a result, the coordinated contraction of ventricular cardiac muscle is interrupted, and pumping effectiveness is reduced. The normal filling and emptying of the ventricles during the cardiac cycle is also interrupted.
5. A drug that prolongs the plateau of cardiac muscle cell action potentials prolongs the time each action potential exists and increases the refractory period. Therefore, the drug slows the heart. A drug that shortens the plateau shortens the length of time each action potential exists and shortens the refractory period. Therefore, the drug can allow the heart rate to increase further.
6. Endurance-trained athletes have a decreased resting heart rate because their cardiac muscle undergoes hypertrophy in response to exercise. The hypertrophied cardiac muscle causes the stroke volume to increase substantially. The increased stroke volume is sufficient to maintain an adequate cardiac output and blood pressure even though the heart rate is slower.
7. The two heartbeats occurring closely together can be heard through the stethoscope because the heart valves open and close normally during each of the heartbeats even if they are close together. The second heartbeat, however, produces a greatly reduced stroke volume because there is not enough time for the ventricles to fill with blood between the first and second contractions. Thus, the preload is reduced. Because the preload is reduced, the second heartbeat has a greatly reduced stroke volume, which fails to produce a normal pulse. The pulse deficit results from the reduced stroke volume of the second of the two beats that are very close together.
8. Atrial contractions complete ventricular filling, but they are not primarily responsible for ventricular filling. Therefore, even if the atria are fibrillating, blood can still flow into the ventricles, and ventricular contractions can occur. As long as the ventricles contract rhythmically, the heart can pump an adequate amount of blood, even though the atria are not effective pumps. However, if the ventricles fail to contract forcefully and rhythmically, they cannot function as pumps. Thus, the stroke volume will become too low to maintain adequate blood flow to tissues. Therefore, atria transplants are not essential but ventricle transplants are.
9. An ECG measures the electrical activity of the heart and would not indicate a slight heart murmur. Heart murmurs are detected by listening to the heart sounds. The boy may have a heart murmur, but the mother does not understand the basis for making such a diagnosis.
10. When both common carotid arteries are clamped, the blood pressure within the internal carotid arteries drops dramatically. The decreased blood pressure is detected, and the baroreceptor reflex increases heart rate and stroke volume. The resulting increase in cardiac output causes the increase in blood pressure.
11. Venous return declines markedly in hemorrhagic shock because of the loss of blood volume. With decreased venous return, stroke volume decreases (Starling law of the heart). The decreased stroke volume results in a decreased cardiac output, which produces decreased blood pressure. In response to the decreased blood pressure, the baroreceptor reflex causes an increase in heart rate in an attempt to restore normal blood pressure. However, with inadequate venous return, the increased heart rate is not able to restore normal blood pressure.

Chapter 21

1. a. aorta, left coronary artery, circumflex artery, posterior interventricular artery or aorta, right coronary artery, posterior interventricular artery
   b. aorta, brachiocephalic artery, right common carotid artery, right internal carotid artery or aorta, left common carotid artery, left internal carotid artery
   c. aorta, brachiocephalic artery, right subclavian artery, right vertebral artery, basilar artery or aorta, left subclavian artery, left vertebral artery, basilar artery
   d. aorta, left or right common carotid artery, left or right external carotid artery
   e. aorta, left subclavian artery, axillary artery, brachial artery, radial or ulnar artery, deep or superficial palmar arch, digital artery (on the right: the brachiocephalic artery would be included)
   f. aorta, common iliac artery, external iliac artery, femoral artery, popliteal artery, anterior tibial artery
   g. aorta, celiac artery, common hepatic artery
   h. aorta, superior mesenteric artery, intestinal branches
   i. aorta, left or right internal iliac artery
2. a. great cardiac vein, coronary sinus or anterior cardiac vein
   b. transverse sinus, sigmoid sinus, internal jugular vein, brachiocephalic vein, superior vena cava
   c. retromandibular vein, external jugular vein, subclavian vein, brachiocephalic vein, superior vena cava
   d. *deep:* vein of hand, radial or ulnar vein, brachial vein, axillary vein, subclavian vein, brachiocephalic vein, superior vena cava
   superficial: vein of hand, radial or ulnar vein, cephalic or basilic vein, axillary vein, subclavian vein, brachiocephalic vein, superior vena cava
   e. *deep:* vein of foot, dorsalis veins of foot, anterior tibial vein, popliteal vein, femoral vein, external iliac vein, common iliac vein, inferior vena cava
   superficial: vein of foot, great saphenous vein, external iliac vein, common iliac vein, inferior vena cava; or vein of foot, small saphenous vein, popliteal vein, femoral vein, external iliac vein, common iliac vein, inferior vena cava
   f. gastric vein or gastroepiploic vein, hepatic portal vein, hepatic sinusoids, hepatic vein, inferior vena cava
   g. renal vein, inferior vena cava
   h. hemiazygous vein or accessory hemiazygous vein, azygous vein, superior vena cava
3. A superficial vessel is easiest, such as the right cephalic or basilic vein. The catheter is passed through the cephalic (or brachial) vein and the superior vena cava to the right atrium. Because the pulmonary veins are not readily accessible, dye is not normally placed directly into them. Instead, the dye is placed in the right atrium using the procedure just described. The dye passes from the right atrium into the right ventricle, the pulmonary arteries, the lungs, the pulmonary veins, and the left atrium. If the catheter has to be placed into the left atrium, it can be inserted through an artery, such as the femoral artery, and passed via the aorta to the left ventricle and then into the left atrium.
4. The resistance to blood flow is less in the venae cavae for two reasons: First, the diameter of one vena cava is greater than the diameter of the aorta; second, an increased diameter of a blood vessel reduces resistance to flow (Poiseuille's law). In addition, there are two venae cavae, the superior vena cava and the inferior vena cava, but only one aorta. The blood flow through the aorta and the venae cavae is about equal, but the velocity of blood flow is much higher in the aorta than it is in the venae cavae.
5. According to Laplace's law, as the diameter of a blood vessel increases, the force applied

to the vessel wall increases, even if the pressure remains constant. The increased connective tissue in the walls of the large blood vessels makes the wall of those vessels stronger and more capable of resisting the force applied to the wall.

6. The nursing student's diagnosis is incorrect. Blood pressure measurements are normally made in either the right or the left arm, both of which are close to the level of the heart. Blood pressure taken in the leg is influenced by pressure created by the pumping action of the heart, but the effect of gravity on the blood, as it flows into the leg, also influences the blood pressure in a substantial way. In this case, gravity increases blood pressure from about 120 mm Hg systolic to 200 mm Hg.
7. Decreased liver function includes a decrease in the synthesis of plasma proteins. Consequently, the concentration of plasma proteins decreases, and the colloid osmotic pressure of the blood decreases. Less water moves by osmosis into the capillaries at the venous ends, and edemas result.
8. Chemoreceptors in the medulla oblongata detect carbon dioxide and the pH of the blood. The normal blood levels of CO_2 and pH stimulate these chemoreceptors, which in turn stimulate the vasomotor center. The vasomotor center keeps blood vessels partially constricted under resting conditions. This basal level of activity is called vasomotor tone. Blowing off CO_2 reduces the blood levels of carbon dioxide and increases the pH of the body fluids. These changes reduce vasomotor tone and result in vasodilation. If a person hyperventilates and blows off CO_2, the stimulus to the vasomotor center decreases, which results in a decrease in vasomotor tone. The decrease in vasomotor tone results in a decrease in systemic blood pressure. If the blood pressure decreases enough, the blood flow to the brain decreases and can cause a sensation of dizziness or even loss of consciousness.
9. Epinephrine is secreted from the adrenal medulla in response to stressful stimuli. The responses it stimulates are consistent with increased physical activity. Vasoconstriction of the blood vessels in the skin shunts blood away from the skin to skeletal muscles. Vasodilation occurs in blood vessels of exercising skeletal muscles. Blood flow through the exercising skeletal muscles increases. Because epinephrine causes vasodilation of the blood vessels of cardiac muscle, blood flow through the cardiac muscle increases. This response is consistent with the increased work performed by the heart under conditions of increased physical activity.
10. Answer *b* is correct. Shortly after Dick was wounded, the loss of blood caused his blood pressure to decrease. Rapidly, the baroreceptor and adrenal medullary mechanisms were activated. His heart rate increased, and peripheral vasoconstriction increased as a result. There is no indication that the CNS ischemic response was activated, because it requires a severe decrease in blood flow to the brain. The renin-angiotensin-aldosterone mechanism is activated as the systemic blood pressure decreases. Angiotensin II is a potent vasoconstrictor that helps the baroreceptor mechanism increase peripheral resistance. The increase in aldosterone secretion increases Na^+ and, consequently, water reabsorption in the kidneys, but it requires approximately 24 hours to be active. The fluid shift mechanism operates, but it takes several hours to function. The ADH mechanism is activated in response to a substantial decrease in blood pressure; it helps maintain blood volume by reducing the amount of urine produced
11. The answer is *e*. It is possible that Mr.Wilson had a stroke, which activated the CNS ischemic response. The fact that he was unconscious is consistent with a stroke, and his rapid heart rate and high blood pressure are consistent with activation of the CNS ischemic response. A heart attack would not necessarily cause Mr. Wilson to be unconscious and would be associated with decreased blood pressure rather than increased blood pressure, especially if it caused unconsciousness. Shock would also be associated with a decrease in blood pressure.

Chapter 22

1. Elevating the limb reduces blood pressure in the limb, resulting in less fluid movement from the blood into the tissues (see chapter 21). Thus, the edema is reduced as fluid moves out of the tissues faster than it enters them. Massage moves lymph through the lymphatic vessels in the same fashion as the contraction of skeletal muscle does. The periodic application of pressure to lymphatic vessels forces lymph to flow toward the trunk of the body, but valves prevent flow in the reverse direction. The removal of lymph from the tissue helps relieve edema.
2. Normally, T cells are processed in the thymus and then migrate to other lymphatic tissues. Without the thymus, this processing is prevented. Because there are normally five T cells for every one B cell, the number of lymphocytes is greatly reduced. The loss of T cells results in an increased susceptibility to infections and an inability to reject grafts because of the loss of cell-mediated immunity. In addition, since helper T cells are involved with the activation of B cells, antibody-mediated immunity is also depressed.
3. That there is no immediate effect indicates a reservoir of T cells exists in the lymphatic tissue. As the reservoir is depleted over time, the number of lymphocytes decreases and cell-mediated immunity is depressed, the animal becomes more susceptible to infections, and the ability to reject grafts decreases. The ability to produce antibodies decreases because of the loss of helper T cells that are normally involved with the activation of B cells.
4. Injection B results in the greater amount of antibody production. At first, the antigen causes a primary response. A few weeks later, the slowly released antigen causes a secondary response, resulting in greatly increased production of antibodies. Injection A does not cause a secondary response because all of the antigen is eliminated by the primary response.
5. If the patient has already been vaccinated, the booster shot stimulates a memory (secondary) response and rapid production of antibodies against the toxin. If the patient has never been vaccinated, vaccinating now is not effective because there is not enough time for the patient to develop his or her own primary response. Therefore, antiserum is given to provide immediate, but temporary, protection. Sometimes both are given: The antiserum provides short-term protection, and the tetanus vaccine stimulates the patient's immune system to provide long-term protection. If the shots are given at the same location in the body, the antiserum (antibodies against the tetanus toxin) can cancel the effects of the tetanus vaccine (tetanus toxin is altered to be nonharmful).
6. The infant's antibody-mediated immunity is not functioning properly, whereas his cell-mediated immunity is working properly. This explains his susceptibility to extracellular bacterial infections and his resistance to intracellular viral infections. It took so long to become apparent because IgG from the mother crossed the placenta and provided the infant with protection. The infant began to get sick after these antibodies degraded.
7. (a) At the first location, an antibody-mediated response resulted in an immediate hypersensitivity reaction, which produced inflammation. Most likely, the response resulted from IgE antibodies. (b) At the second location, a cell-mediated response resulted in a delayed hypersensitivity reaction, which produced inflammation. This probably involved the release of cytokines and the lysis of cells. (c) At the

other locations, neither an antibody-mediated nor a cell-mediated response occurred.

8. The ointment was a good idea for the poison ivy, which causes a delayed hypersensitivity reaction—for example, too much inflammation. For the scrape, it was a bad idea because a normal amount of inflammation is beneficial and helps fight infection.
9. The correct answer is *b*. The immune response that occurred in the next several hours was primarily the innate immune response. Tissue was damaged, and mediators of inflammation were released. As a consequence, there were increases in capillary permeability, coagulation of blood, and chemotaxis of neutrophils; all are components of inflammation and the innate immune response. Increased mitosis of memory B cells is likely, but more than a few hours would be required, and it is most likely to occur in response to antigens. Bacteria and other antigens may enter the wound, and an adaptive immune response may result, but this, too, would take more than several hours to occur.
10. The correct answer is *a*. Upon first exposure to a type of antigen, the unprocessed antigens are ingested by macrophages, and the ingested antigens are broken down to fragments as they are processed. The processed antigen fragments are combined with MCH II molecules and transported to the cells' surfaces. A specific helper T cell recognizes the processed antigens bound to the MCH II molecules (the MHC II complex) at the macrophages' surfaces and binds to the MHC II complexes. Costimulation occurs, and interleukin-1 is released from the macrophages, while interleukin-2 is released from the T lymphocytes. As a result, the helper T cells undergo mitosis (see figure 22.17).

Chapter 23

1. We expect vital capacity to be greatest when standing because the abdominal organs move inferiorly, thereby allowing greater depression of the diaphragm and a greater inspiratory reserve volume.
2. The hose increases dead space and therefore decreases alveolar ventilation. Ima Diver has to compensate by increasing respiratory rate or tidal volume. If the hose is too long, she will not be able to compensate. Furthermore, with a long hose, air is simply moved back and forth in the hose, so that little exchange of air between the atmosphere and the lungs takes place. Another consideration is the effect of water pressure on the thorax, which decreases compliance and increases the work of ventilation. In fact, a few feet under water, there is enough pressure on the thorax to prevent the intake of air, even through a short hose connected to the atmosphere.
3. The increase in atmospheric pressure increases the partial pressure of oxygen. According to Henry's law, as the partial pressure of oxygen increases, the amount of oxygen dissolved in the body fluids increases. The increase in dissolved oxygen is detrimental to the gangrene bacteria. Because hemoglobin is already saturated with oxygen, the HBO treatment does not increase hemoglobin's ability to pick up oxygen in the lungs.
4. The patient's lungs expand because of the pressure generated by the rescuer's muscles of expiration. This fills the lungs with air that has a greater pressure than atmospheric pressure. Air flows out of the patient's lungs as a result of this pressure difference and because of the recoil of the thorax and lungs. Although the partial pressure of oxygen of the rescuer's expired air is less than atmospheric pressure, enough oxygen can be provided to sustain the patient. The lower partial pressure of oxygen can also activate the chemoreceptor reflex and stimulate the patient to breathe. In addition, the rescuer's partial pressure of carbon dioxide is higher than atmospheric pressure, and this can activate the chemosensitive area in the medulla.
5. The left side of the diaphragm moves superiorly. During inspiration, thoracic volume increases as the right side of the diaphragm moves inferiorly and the intercostal muscles move the ribs outward. Increased thoracic volume causes a decrease in pressure in the thoracic cavity. As a result, the pressure on the superior surface of the diaphragm is less than on the inferior surface. The paralyzed left side of the diaphragm moves superiorly because of this pressure difference.
6. At the end of expiration, the thoracic wall is not moving. Therefore, the forces causing the thoracic wall to move inward and outward must be equal. At the end of expiration, the lungs are adhering to the thoracic wall through the pleurae, and recoil of the lungs is pulling the thoracic wall inward. As a result of the pneumothorax, air enters the pleural cavity and the visceral and parietal pleurae separate from each other. The recoil of the lungs causes them to collapse. Without the inward force produced by lung recoil, the thoracic wall expands outward.
7. All else being equal (i.e., the thickness of the respiratory membrane, the diffusion coefficient of the gas, and the surface area of the respiratory membrane), diffusion is a function of the partial pressure difference of the gas across the respiratory membrane. The greater the difference in partial pressure, the greater the rate of diffusion. The greatest rate of oxygen diffusion should therefore occur at the end of inspiration, when the partial pressure of oxygen in the alveoli is at its highest. The greatest rate of carbon dioxide diffusion should occur at the end of inspiration, when the partial pressure of carbon dioxide in the alveoli is at its lowest.
8. EPO increases the number of red blood cells and consequently the blood viscosity. Thus, even though the blood has a higher oxygen-carrying capacity, it increases the workload on the heart and can cause a heart attack or stroke.
9. (a) Cutting the vagus nerves would eliminate the Hering-Breuer reflex and result in greater than normal inspiration. This would increase tidal volume. (b) Cutting the phrenic nerves would eliminate contraction of the diaphragm. Tidal volume would decrease drastically, and death would probably result. Cutting the intercostal nerves would eliminate raising of the ribs and sternum and decreased tidal volume, unless the diaphragm compensated.
10. While hyperventilating and making ready to leave your instructor behind, you might make the following arguments:
    - I would argue that hyperventilation increases the oxygen content of the air in the lungs; therefore, you would have more oxygen to use when holding your breath.
    - It is hemoglobin that is saturated. Hyperventilation increases the amount of oxygen dissolved in the blood plasma.
    - Hyperventilation decreases the amount of carbon dioxide in the blood. This makes it possible to hold your breath longer because of a decreased urge to take a breath.
    - Hyperventilation activates alveoli not in use because increasing alveolar oxygen and decreasing alveolar carbon dioxide cause lung arterioles to relax, thereby increasing blood flow through the lungs.
11. When Ima is hyperventilating, the stimulus for the hyperventilation is anxiety, which is more important than the carbon dioxide in controlling respiratory movements. As the blood levels of carbon dioxide decrease during hyperventilation, vasodilation occurs in the periphery. As a result, the systemic blood pressure decreases. The systemic blood pressure can decrease enough that blood flow to the brain decreases. Decreased blood flow to the brain results in a reduced oxygen level in the brain tissue, causing dizziness. Breathing into a paper bag raised Ima's

blood levels of carbon dioxide toward normal. Because the carbon dioxide did not rise above normal, it did not increase the urge to breathe. The more normal level of carbon dioxide prevented peripheral vasodilation. As Ima breathed into the paper bag, her anxiety likely subsided, allowing the normal regulation of respiration to resume.

Chapter 24

1. With the loss of the swallowing reflex, the vocal folds no longer occlude the glottis. Consequently, vomit can enter the larynx and block the respiratory tract.
2. Without adequate amounts of hydrochloric acid, the pH in the stomach is not low enough for the activation of pepsin. This loss of pepsin function would result in inadequate protein digestion. However, if the food is well chewed, proteolytic enzymes in the small intestine (e.g., trypsin, chymotrypsin) can still digest the protein. If the stomach secretion of intrinsic factor decreases, the absorption of vitamin B_{12} is hindered. Inadequate amounts of vitamin B_{12} can result in decreased red blood cell production (pernicious anemia).
3. Even though ulcers are usually caused by bacteria, overproduction of hydrochloric acid due to stress is a possible contributing factor. Reducing hydrochloric acid production is recommended. In addition to antibiotic therapy, commonly recommended treatments include relaxation, drugs that reduce stomach acid secretion, and antacids to neutralize the hydrochloric acid. Smaller meals are also advised because distension of the stomach stimulates acid production. In addition, proper diet is important, and patients are advised to avoid alcohol, caffeine, and large amounts of protein because they stimulate acid production. Stress also stimulates the sympathetic nervous system, which inhibits duodenal gland secretion. As a result, the duodenum has less mucous coating, making it more susceptible to gastric acid and enzymes. Relaxing after a meal helps decrease sympathetic activities and increase parasympathetic activities.
4. Blockage of the common bile duct can result in jaundice (due to an accumulation of bile pigments in the blood) and clay-colored stools (due to lack of bile pigments in the feces). Other symptoms include abdominal pain, nausea, and vomiting. In addition, fat absorption is impaired because of the absence of bile salts in the duodenum, and a loose, bulky stool results. Lack of fat absorption reduces the absorption of fat-soluble vitamins, such as vitamin K, impairing normal clotting function.
5. The patient is still able to defecate. Following a meal, the gastrocolic and duodenocolic reflexes initiate mass movement of the feces into the rectum. In the rectum, local reflexes and the defecation reflex (integrated in the sacral level of the cord and not requiring connections to high brain centers) cause defecation. However, the patient loses awareness of the need to defecate (due to loss of sensory input to the brain) and the ability to prevent defecation voluntarily via the external anal sphincter.
6. Cholera toxin irreversibly activates a G protein that causes persistent activation of the chloride channel. The activated channel allows excessive movements of chloride from the cells into the digestive tract. Water follows the osmotic gradient generated by the chloride, which leads to diarrhea. Conversely, mutations in the channel reduce the movement of chloride into the digestive tract. A thick mucus builds up on the surface of the epithelial cells, leading to some symptoms of cystic fibrosis.
7. Oral rehydration therapy relies on the principle of osmosis. Water follows solutes as they are absorbed across the intestinal epithelium. The combination of sodium and glucose is optimal, since the two molecules are cotransported by a symporter that is driven by a sodium gradient established by the Na^+–K^+ pump. Hence, the presence of sodium aids glucose absorption. Fructose is absorbed by a facilitated diffusion transporter that is not coupled to a sodium gradient.
8. Dietary changes are beneficial but alone would not likely be sufficient for FH patients. Only about 15% of the cholesterol in the body comes from diet; the remainder is synthesized, primarily in the liver. A diet reduced in total fat and cholesterol, along with exercise, would help lower serum LDL cholesterol levels. However, since FH patients no longer have the normal LDL receptor-mediated feedback loop, increased cellular synthesis of cholesterol would continue. Often statin drug therapy is necessary to restore cholesterol levels to normal.

Chapter 25

1. In figure 25.2, the Daily Value for saturated fat is listed as less than 20 g for a 2000 kcal/day diet. The % Daily Values appearing on food labels are based on a 2000 kcal/day diet. Therefore, the % Daily Value for saturated fat for one serving of this food would be 10% (2/20 = .10, or 10%).
2. According to the Daily Value guidelines, total fats should be no more than 30% of total kilocaloric intake. For a person consuming 3000 kcal/day, this is 900 kcal (3000 kcal × 0.30). There are 9 kcal in a gram of fat. Therefore, the maximum amount (weight) of fats the active teenage boy should consume is 100 g (900/9).
3. The % Daily Value is the amount of the nutrient in one serving divided by its Daily Value. Therefore, the % Daily Value is 10% (10/100 = .10, or 10%).
4. The protein in meat contains all the essential amino acids and is a complete protein food. Although plants contain proteins, a variety of plants must be consumed to ensure that all the essential amino acids are included in adequate amounts. Also, plants contain less protein per unit weight than meat, so a larger quantity of plants must be consumed to get the same amount of protein.
5. Copper is necessary for the proper functioning of the electron-transport chain. Inadequate copper in the diet results in reduced ATP production—that is, not enough energy.
6. A person loses weight when kilocalories used per day exceed kilocalories ingested per day. About 60% of the kilocalories used per day is due to basal metabolic rate. A person with a high basal metabolic rate loses weight faster than a person with a low basal metabolic rate, all else being equal. Another factor to consider is physical activity, which accounts for about 30% of kilocalories used per day. An active person loses more weight than a sedentary person does.
7. This approach will not work because he is not losing stored energy from adipose tissue. In the sauna, he gains heat, primarily by convection from the hot air and by radiation from the hot walls. The evaporating sweat is removing heat gained from the sauna. The loss of water will make him thirsty, and he will regain the lost weight from fluids he drinks and food he eats.
8. As ATP breakdown increased, more ATP would be produced to replace that used. Over an extended time, the ATP must be produced through aerobic respiration. Therefore, oxygen consumption and basal metabolic rate increase. The production of ATP requires the metabolism of carbohydrates, lipids, or proteins. As these molecules are used at a faster than normal rate, body weight decreases. Increased appetite and increased food consumption resist the loss in body weight. As ATP is produced and used, heat is released as a by-product. The heat raises body temperature, which is resisted by the dilation of blood vessels in the skin and by sweating.

9. During a fever, the body produces heat by shivering. The body also conserves heat by the constriction of blood vessels in the skin (producing pale skin) and by reduction in sweat loss (producing dry skin). When the fever breaks—that is, "the crisis is over"—heat is lost from the body to lower body temperature to normal. This is accomplished by the dilation of blood vessels in the skin (producing flushed skin) and increased sweat loss (producing wet skin).

Chapter 26

1. The beer consumed composes a large volume of hyposmotic fluid, which increases blood volume and causes blood osmolality to decrease. The increased blood volume is detected by baroreceptors, and the decreased blood osmolality is detected by osmoreceptors in the hypothalamus. The response to these stimuli is inhibition of ADH secretion. The alcohol in the beer also inhibits ADH secretion. The increased volume inhibits the renin-angiotensin-aldosterone hormone mechanism, which in turn inhibits aldosterone secretion. The changes in aldosterone, however, take much longer to influence kidney function than changes in ADH. As a result of these changes, a large volume of dilute urine is produced until the blood osmolality and blood volume return to normal.
2. The loss of sweat results in a loss of water and electrolytes. Replacing just the water restores blood volume and decreases blood osmolality. At first, the decreased osmolality inhibits ADH secretion, and dilute urine is produced. However, as blood volume decreases as a result of urine production, ADH secretion and the renin-angiotensin-aldosterone mechanisms are stimulated. Consequently, urine concentration increases, and only a small amount of urine is produced.
3. As aldosterone levels decrease, sodium reabsorption in the nephron decreases; consequently, plasma sodium levels decrease. The sodium is lost in the urine, and water follows the sodium by osmosis. Thus, a large amount of urine having a high concentration of sodium is produced. The loss of water reduces blood volume, which causes the low blood pressure. As aldosterone levels decrease, potassium secretion into the nephron also decreases, resulting in increased plasma potassium levels. The increased extracellular potassium causes the depolarization of nerve and muscle membranes, leading to tremors of skeletal muscles and cardiac arrhythmias, including fibrillation.
4. There are several ways to decrease the glomerular filtration rate:
   a. Decrease hydrostatic pressure in the glomerulus.
      1. Decrease systemic arterial blood pressure.
         a. Decrease extracellular fluid volume.
         b. Decrease peripheral resistance.
         c. Decrease cardiac output.
      2. Constrict or occlude the afferent arteriole.
      3. Relax the efferent arteriole.
   b. Increase glomerular capsule pressure.
   c. Increase the colloid osmotic pressure of the plasma.
   d. Decrease the permeability of the filtration barrier.
   e. Decrease the total area of the glomeruli available for filtration.
5. Assume that the ascending limb of the loop of Henle and the distal convoluted tubules are impermeable to sodium and other ions but actively pump out water. Other characteristics of the kidney are assumed to be unchanged. As the urine moves up the ascending limb, it becomes hyperosmotic, because sodium remains behind as water is pumped out. Assuming that the collecting ducts are impermeable to sodium, on reaching the collecting ducts the presence or absence of ADH determines the final concentration of the urine. If ADH is absent, little or no water is exchanged as the urine passes down the collecting ducts and a hyperosmotic urine is produced. On the other hand, if ADH is present, water moves from the interstitial fluid into the collecting ducts, thus diluting the urine and producing a hyposmotic urine.
6. Urea is partially responsible for the high osmolality of the interstitial fluid in the medulla of the kidney. Since a high osmolality of the interstitial fluid must exist for the kidney to produce a concentrated urine, a small amount of urea in the kidney results in the production of dilute urine.
7. A low-salt diet tends to reduce the osmolality of the blood. Consequently, ADH secretion is inhibited, producing dilute urine and thus eliminating water. This in turn reduces blood volume and blood pressure.
8. As the loops of Henle become longer, the mechanisms that increase the concentration of the interstitial fluid of the medulla become more efficient, thus raising the concentration of the interstitial fluid. The higher the concentration of interstitial fluid in the medulla of the kidney, the greater the concentration of the urine the kidney is able to produce.
9. Answer *d* is correct. For at least 2 hours before help arrived, Marvin suffered from hemorrhagic shock, which was responsible for hypoxic injury to his kidneys. Hypoxic injury is the consequence of ischemia due to prolonged shock, and it results in cells sloughing off and blocking the lumen of the nephron, so that the filtrate cannot flow through it. Marvin's urine volume decreased because cells of the nephron sloughed off and blocked the lumen of the nephron. The reabsorption of wastes was increased because the flow rate of the filtrate slowed and the wastes were able to pass through the wall of the nephron and reenter the circulatory system, rather than becoming part of the filtrate. This explains the increased blood volume and the distended neck veins. The high blood H^+ is due to decreased H^+ secretion; and the high blood K^+ is due to decreased K^+ secretion. There is no increase in blood HCO_3^- because the rate of HCO_3^- reabsorption is not increased. There is a decrease in erythropoietin secretion by the damaged kidney, which explains the reduced red blood cell count.
10. Answer *a* is correct. The respiratory system responds to below-normal pH by increasing the rate of respiration. However, Marvin's kidneys were not able to respond to the low pH because of the hypoxic injury. The hypoxic injury is responsible for the metabolic acidosis (lower-than-normal blood pH), and the damaged nephrons could not adequately increase HCO_3^- reabsorption or H^+ secretion.
11. High blood renin levels would result in increased blood aldosterone levels (see figure 26.17). The aldosterone increases the number of sodium potassium transport proteins in the basal membranes of cells in the distal convoluted tubules and collecting ducts. More Na^+ would be reabsorbed, and more K^+ would be excreted into the urine. Consequently, blood K^+ levels would be lower than normal. The low blood K^+ levels result in the hyperpolarization of membrane potentials. As a result, stimuli of a greater strength than normal are required to produce action potentials in nerve and muscle tissue.
12. Water that moves into the collecting duct cells would diffuse into the interstitial fluid at a reduced rate because of the reduced number of aquaporins. Urine volume would increase because of water retention in the distal convoluted tubule and collecting ducts. Urine concentration would decrease because of dilution by the water. ADH can cause an increase in the number of aquaporin-2 water channel proteins in the apical membrane of the distal convoluted tubules and collecting ducts, but not in aquaporin-3 or aquaporin-4 channels in the basal membranes. These channels determine the permeability of the basal

membranes to water. Therefore, ADH would not be an effective treatment. The net effect would be polyuria, which is similar to nephrogenic diabetes insipidus caused by an abnormal aquaporin-2 water channel protein.

Chapter 27

1. When excess glucose is not reabsorbed, it osmotically obligates water to remain in the nephron. This results in the production of a large amount of urine, called polyuria, with a consequent loss of water, salts, and glucose. The loss of water can be compensated for by increasing fluid intake. The intense thirst that stimulates increased fluid intake is called polydipsia. The loss of salts can be compensated for by increasing the salt intake. The high glucose levels in the blood increase the blood osmolality, thus stimulating the secretion of ADH. This increases the permeability of the distal convoluted tubule and collecting duct to water. Normally, this allows reabsorption of water from the collecting ducts and thus conserves water. However, if glucose levels in the urine are high enough, water loss increases even with high levels of ADH
2. When ADH levels first increase, the reabsorption of water increases, and urinary output is reduced. This also causes an increase in blood volume and therefore an increase in blood pressure. The increased blood pressure increases the glomerular filtration rate, which increases urinary output to normal levels. In addition, the increased blood volume inhibits the renin-angiotensin-aldosterone mechanism, inhibits aldosterone secretion, and stimulates natriuretic hormone secretion. These responses also increase urinary output.
3. Blocking H^+ secretion produces acidosis. Because H^+ is exchanged for Na^+, the Na^+ remains in the urine as sodium bicarbonate. This effectively prevents the reabsorption of HCO_3^- and produces an alkaline urine. The blood pH is reduced because H^+ is not being secreted as rapidly by the nephron. The respiratory rate increases because of the stimulatory effect of decreased blood pH on the respiratory center.
4. Breathing through the glass tube increases the dead air space and decreases the efficiency of gas exchange. Consequently, blood carbon dioxide levels increase and produce a decrease in blood pH. Compensatory responses include an increased respiratory rate and the production of acidic urine.
5. Answer *e* is correct. The rate of stomach secretion increased prior to the time Harry vomited. Gastric secretions include a high concentration of HCl. As the stomach secretes HCl, it secretes H^+ into the stomach and absorbs HCO_3^- into the blood. Also, after Harry took the antacid, which is mostly $NaHCO_3$, the HCO_3^- was absorbed and entered the circulatory system. A significant amount of HCO_3^- entering the circulatory system causes an increase in the blood pH. The increased pH affects the regulatory centers of the respiratory system and causes respiration to slow. Although the kidney's response is slower, by 24 hours later, the rate of H^+ secretion by the kidney slows, and HCO_3^- absorption slows in response to an increase in the blood pH. The slower rates of respiration, H^+ secretion, and HCO_3^- absorption by the kidneys help keep the blood pH from becoming higher than normal.

Chapter 28

1. Removing the testes would eliminate the major source of testosterone. Blood levels of testosterone would therefore decrease. Because testosterone has a negative-feedback effect on the hypothalamus and pituitary gland, GnRH, FSH, and LH secretion would increase, and the blood levels of these hormones would increase. An adult male's primary and secondary sex characteristics are already developed; however, removal of the testes would eliminate sperm production. Also, the lack of testosterone would cause a decrease in sex drive and muscular strength.
2. Prior to puberty, the levels of GnRH are very low because the hypothalamus is very sensitive to the inhibitory effects of testosterone. Since GnRH levels are low, so are FSH and LH levels. Loss of the testes and testosterone production would result in an increase in GnRH, FSH, and LH levels. Because little testosterone is produced, the boy would not develop sexually and would have no sex drive. Small amounts of androgens would be produced because the adrenal cortex produces some androgens. As an adult, he would be taller than normal, with thin bones and weak musculature. His voice would not deepen, and the normal masculine distribution of hair would not develop.
3. Ideally, the pill would inhibit spermatogenesis. Using the same approach as the birth control pill in females, the inhibition of FSH and LH secretion should work. Chronic administration of GnRH suppresses FSH and LH levels enough to cause infertility through down-regulation. However, lack of LH can also result in reduced testosterone levels and loss of sex drive. Some evidence indicates that administering testosterone in the proper amounts reduces FSH and LH secretion, thus leading to less sperm cell production while still maintaining normal sex drive. For a large percentage of males the technique results in a sperm concentration in the semen that is too low to allow fertilization. However, the technique is not sufficiently precise to be used as a standard birth control technique.
4. In a postmenopausal woman, the ovaries have stopped producing estrogen and progesterone. Without the negative-feedback effect of these hormones, the levels of GnRH, FSH, and LH increase. Removal of the nonfunctioning ovaries in a postmenopausal woman would not change the level of any of these hormones or produce any symptoms not already occurring due to the lack of ovarian function.
5. The removal of the ovaries from a 20-year-old woman eliminates the major site of estrogen and progesterone production, thereby causing an increase in GnRH, FSH, and LH levels due to lack of negative feedback. We would expect to see the symptoms of menopause, such as cessation of menstruation and reduction in the size of the uterus, vagina, and breasts. A temporary reduction in sex drive may also occur.
6. It is clear that estrogen and progesterone administration resulted in a large decrease in the amount of LH in the plasma the day of ovulation. The differences in plasma LH levels between the groups at other times are very small. The incidence of pregnancies suggests that the reduced plasma LH levels may result in no ovulation.
7. The progesterone inhibits GnRH in the hypothalamus. Consequently, the anterior pituitary is not stimulated to produce LH and FSH. Lack of LH prevents ovulation, and lack of FSH prevents development of the follicles. LH is also required for the maturation of follicles prior to ovulation. Without follicle development, inadequate estrogen is produced, which causes the hot flashes.
8. GnRH administered either before or after the normal time of ovulation does not result in ovulation because the anterior pituitary is less sensitive to the effect of GnRH during those times. Also, follicles in the ovary are not adequately developed. The concentration of GnRH must be carefully controlled because too little results in inadequate release of FSH and LH from the anterior pituitary. Too little FSH and LH fails to cause ovulation. Too much GnRH given at the proper time results in the maturation of more than one follicle and the release of an oocyte from more than one of the follicles. If the oocytes are fertilized, multiple embryos can result.
9. Because of their enlarged prostates, the patients are likely to have difficulty urinating.

An enlarged prostate gland compresses the prostatic urethra, slowing the emptying of the urinary bladder and enhancing the urgency to urinate. Dr. Procter conducted a digital exam by palpating the men's prostate glands through their rectums. The patient with the enlarged but smooth prostate is likely to have benign prostatic hypertrophy and low blood levels of PSA. Benign prostatic hypertrophy is general enlargement of the prostate without specific tumors. The patient with the enlarged but asymmetrical prostate gland is more likely to have elevated blood levels of PSA and prostate cancer. Lack of symmetry and a rough surface can be caused by tumors within the prostate gland.

10. The correct answer is *d*. If a testis fails to descend into the scrotal sac, the higher body temperature causes the spermatogonia to degenerate. Consequently, that testis produces no sperm cells. Sustentacular cells remain in the seminiferous tubules, and interstitial cells remain in the testis. Both testosterone secretion by the interstitial cells and LH secretion from the anterior pituitary gland remain normal.
11. The correct answer is *d*. Decreased blood flow to the anterior pituitary gland would result in decreased LH secretion, not increased LH secretion. The decreased secretion of LH from the anterior pituitary causes a decrease in the blood level of testosterone. Because it takes approximately 74 days to produce a sperm cell, the sperm count is likely to remain normal during the week when the blood LH and FSH levels are reduced.

Chapter 29

1. Triploidy can occur as a result of polyspermy, the fertilization of one oocyte with two sperm cells. Polyspermy is usually prevented by the fast and slow blocks to polyspermy, both of which depend on the depolarization of the oocyte membrane. If this depolarization does not occur, the zona pellucida does not degenerate, and other sperm cells can attach to the oocyte membrane, leading to polyspermy.
2. The approximate length of time the embryo has been developing, called the postovulatory age, is 14 days less than the time since the last menstrual period (LMP). In this case, the postovulatory age is 30 days (44 − 14). By this time, the neural tube has closed, the somites have formed, the digestive tract is developing, the limb buds have appeared, a tubular beating heart is present, and the lungs are developing. Because reproductive structures are still just forming, male and female embryos are indistinguishable at this age.
3. The fever occurred on days 21–31 of development, during the time of neural tube closure (days 18–25). If the fever prevented neural tube closure, the child could be born with anencephalus or spina bifida.
4. Progesterone reduces the release of oxytocin, the hormone that stimulates uterine contractions. Also, progesterone reduces the number of oxytocin receptors in the uterus. These effects reduce labor contractions and can prevent preterm delivery.
5. The Apgar score of 5 indicates appearance (A, 0) white or blue; pulse (P, 1) low; grimace (G, 1) slight; activity (A, 1) little movement and poor muscle tone; and respiration (R, 2) normal. The white or blue appearance (A, 0) is consistent with a poor circulation, as also indicated by reduced pulse (P, 1). The reduced heart rate, resulting in the low pulse, may indicate a circulatory system problem. The reduced reflexes and motor activity (G, 1; A, 1) can result from lack of oxygen in the muscles due to poor circulation. Because the infant has poor circulation despite normal respiration, clearing the airway (if obstructed) and administering oxygen are in order. This Apgar score can have several causes, and additional information would be necessary to determine the specific cause.
6. Suckling the breast stimulates the release of oxytocin from the neurohypophysis (posterior pituitary). Once the oxytocin is in the blood, it travels to both breasts and causes milk letdown.
7. The woman most likely has an XY genotype. The spherical structures in the inguinal area are testes, but a genetic abnormality caused the androgen receptors to be defective or absent. Consequently, reproductive structures could not respond to androgens secreted by the testes. The embryonic testes secreted müllerian-inhibiting hormone, and therefore the müllerian duct system degenerated. This explains the lack of a uterus and a cervix. Female external genitalia developed because of the absence of androgen receptors. Normal breast development occurred because of the small amount of estrogen produced by the adrenal glands and because some testosterone is converted to estrogen in peripheral tissues.
8. Answer *b* is correct. Müllerian-inhibiting hormone causes the duct system that gives rise to female internal reproductive structures to degenerate in male fetuses. If the hormone is not effective, female internal reproductive structures, such as oviducts and a uterus, develop, at least to some degree, along with internal male reproductive structures. Because the developing testes produce testosterone, male external reproductive structures develop.
9. It is not possible from this information alone. If tongue rolling is designated *T* and the inability to roll the tongue is designated *t*, then a *Tt* woman and a *tt* man have a 50% probability of having a *Tt* child and a 50% probability of having a *tt* child. Therefore, a man who cannot roll his tongue and a woman who can roll her tongue can have a child who can roll his or her tongue. This connection alone, however, is not sufficient to establish paternity.
10. A person with blood type AB has the genotype I^AI^B but no *i* allele. A person with blood type O has a genotype of *ii*. Because the AB parent has no *i* allele, an AB individual cannot be the parent of a child with blood type O.

Appendix G

ANSWERS TO PREDICT QUESTIONS

Chapter 1

2. This question is about the hierarchical levels of organization in the body. The question tells us that an organ, the pancreas, is the source of the health disorder. Next, the question explains that particular groups of cells (tissue) in this organ are responsible for the normal functioning of the body. Finally, the question tells us that the individual cells release the chemical insulin, which, in this case, is not being produced, as it should. It is the lack of this chemical that is disrupting the health of the individual. Considering the hierarchy of organization, chemical → cell → tissue → organ, an adjustment at any of these levels could theoretically solve the issue.

   Currently, the chemical level is where diabetes is most commonly corrected. Purified insulin is injected into the blood at certain times. However, there are also drugs that stimulate the pancreatic cells to increase insulin production or to make other cells more responsive to insulin. Next, the tissue level is an important area of research to determine whether groups of insulin-producing pancreatic cells (pancreatic islets; see chapter 18 opening photo and caption) can be isolated and implanted into the patient. Finally, the next higher level, the organ level, could be a level of correction by transplanting a new pancreas into a diabetes patient.

3. To answer this question, we must first recognize the components of the feedback systems that maintain homeostasis in the body: receptor, control center, and effector. Recall that any disruption in normal function (such as fainting) is a disruption in homeostasis.
   a. Normally, homeostasis of blood pressure is maintained in this kind of situation when receptors near the heart are stimulated by lower blood pressure upon standing (blood falls away from the brain due to gravity). These receptors send a signal to the brain control center, which stimulates the effector, the heart, to beat faster. The increased heart rate causes blood pressure to go back to normal, thus maintaining homeostasis.
   b. Because Molly's blood pressure dropped when she stood, the feedback system described in (a) was initiated and her heart rate increased. However, the elevated heart rate was not sufficient to prevent Molly from fainting. Too little blood was delivered to her brain, and she lost consciousness.
   c. When Molly fell to the floor, she returned to a lying-down position. This eliminated the pooling effect of the blood in the veins below the heart because of gravity. Therefore, blood return to the heart increased, blood pressure increased, blood flow to the brain increased, homeostasis of brain tissue was restored, and she regained consciousness.

4. Initially, we might think that Ashley's increased respiratory rate is indicative of positive feedback. However, we must first look at the underlying reason for her elevated breathing rate. If you have ever run a race, you know that, eventually, your breathing rate goes back to normal once you stop running. However, during the race your body needs more oxygen to be able to keep running. Because running requires oxygen, oxygen levels in the blood go down. The lowered oxygen level is detected by receptors, which communicate with the control center. The control center stimulates the effector, the diaphragm, to increase breathing rate. Once oxygen levels are returned to the set point, the breathing rate returns to normal. This is the essence of negative feedback: The response is stopped when the variable returns to the normal range.

5. The use of the anatomical directional terms can be compared to using the cartographic directions terms N, S, E, and W. As long as we know the reference points, and the corresponding terms, we can reference any part of the body. However, there are some terms for four-legged animals that are different from those for upright, two-legged humans, yet others are the same in both species. For example, in both humans and cats, *cephalic* means toward the head. But because the head of the cat points in the direction the cat walks, *anterior* also refers to the head, whereas in humans, *anterior* refers to the front of the body (e.g., face, stomach). In both humans and cats, *dorsal* means toward the back. But in cats, *superior* also means toward the back, whereas in humans, *superior* refers to the head. In humans, *posterior* is the other term for the rear (e.g. back of head, back).

Organism	Head	Back	Front of body
Cat	Cephalic/ anterior	Dorsal/ superior	Ventral/ inferior
Human	Cephalic/ superior	Dorsal/ posterior	Ventral/ anterior

6. In order to recognize which term to use here, we must first realize that directional terms are relative to the body. Therefore, it does not matter what position our body is in compared with the earth; body parts always have the same relationship to each other. Thus, the kneecap is always both proximal (closer to point of attachment to the body) and superior (closer to head) to the heel. It is also anterior to the heel because it is on the anterior side of the lower limb, whereas the heel is on the posterior side.

7. The first step is to define the abdominopelvic and peritoneal cavities. The abdominopelvic cavity is located inferiorly to the diaphragm and superiorly to the symphysis pubis. The peritoneal cavity is located between the visceral peritoneum, which covers organs in the abdominopelvic cavity, and the parietal peritoneum, which lines the wall of the abdominopelvic cavity. Thus, notice the bright white area in figure 1.16c. This is the peritoneal cavity containing only peritoneal fluid. Although the peritoneal cavity is around these organs, they are not within the peritoneal cavity. Second, looking at the right side of figure 1.16c again, notice there are organs behind the parietal peritoneum, but inside the abdominopelvic cavity. These

organs are also not within the peritoneal cavity and are considered retroperitoneal (e.g., the kidneys).

Chapter 2

2. The question asks us to differentiate between mass and weight. First, consider the definitions. Weight, in particular, is dependent on the force of gravity. Therefore, if an astronaut is in outer space, where the force of gravity from earth is nearly nonexistent, the astronaut is "weightless." However, the definition of mass is the amount of matter present in the object itself. Thus, no matter the location of an object, the mass remains constant.
3. The question asks us to predict the atomic structure of potassium. By definition, the atomic number (19) is the number of protons. Therefore, there are 19 protons in potassium. Since the number of electrons in an atom is the same as the number of protons, there are 19 electrons. To find the number of neutrons of any element, subtract the atomic number (19) from the mass number (39): $39 - 19 = 20$. Thus, there are 20 neutrons.
4. To answer this question, we must recall the relationship among CO_2, H_2O and H^+ in solution. Carbon dioxide readily combines with water resulting in the production of free H^+. Therefore, as the amount of CO_2 decreases, the reversible reaction will shift in the other direction to form CO_2. Similar to the trough of water example, if CO_2 levels decrease, it is like raising the right side of the trough, causing water to flow to the left. The reaction "flows" to the left: $CO_2 + H_2O \rightarrow H^+ + HCO_3^-$. In order for this to happen, free H^+ combines with HCO_3^-, decreasing its level in the blood. Section 2.3 explains that this decrease in H^+ levels changes the pH of the blood so that it becomes more basic.
5. To answer this question, it may be helpful to write out the formulas of the reactants and the products. Hydrogen gas is represented by H_2, whereas O_2 represents oxygen gas; the product, water, is represented by H_2O. The oxygen atom is more electronegative than hydrogen, and each hydrogen forms a polar covalent bond with the oxygen. Recall that a polar covalent bond occurs when the more electronegative atom can pull the electrons more strongly and the electrons associate with the oxygen more than they do the hydrogens. In a way, the oxygen has gained some negativity and its charge has been lowered. Thus, we say that the oxygen is reduced. On the other hand, the hydrogen atoms have partially lost their units of negativity, their electrons, and are more positive than they were before. In this case, we say the hydrogens are oxidized. The term *oxidized* is used because oxygen is an effective electron acceptor. A useful mnemonic device to remember these terms is OIL RIG (oxidation is loss [of electrons]; reduction is gain [of electrons]).
6. During exercise, our body is doing work by muscular contractions. Work involves converting one form of energy into another, and as we read in the previous section, this conversion is not 100% efficient. As a result, heat energy is released. When contracting our muscles, potential energy is converted to kinetic energy and heat energy. Thus, more heat is produced when exercising than when at rest, and our body temperature increases.
7. Recall that the definition of an acid is a proton donor. Thus, by losing an H^+, dihydrogen phosphate ion ($H_2PO_4^-$), it is functioning as the conjugate acid. When the H^+ is lost, monohydrogen phosphate ion (HPO_4^-) is formed and is the conjugate base. The definition of a base is a proton acceptor. Thus, if H^+ is added to the solution, it combines with HPO_4^- and forms $H_2PO_4^-$. In this way, the conjugate base, HPO_4^-, helps resist an increase in free H^+ by acting as an H^+ "sponge." If OH^- is added to the solution, it combines with any free H^+ to form water. Now, $H_2PO_4^-$ (the conjugate acid) can dissociate into H^+ and HPO_4^-, thus serving as a reservoir of H^+, and a decrease in H^+ is prevented, again resisting a change in pH. This is the essence of a buffer system: alternating between binding excess H^+ and donating H^+ when needed.

Chapter 3

2. First, we need to consider the normal process and identify the intracellular and extracellular areas involved. We are told that urea diffuses from liver cells, which is the intracellular region, to the blood, which is the extracellular region. Recall that diffusion is the net movement of molecules down their concentration gradient, so in this case from the area of higher urea concentration inside the cells to the area of lower urea concentration in the blood. The kidneys remove the urea from the blood; therefore, if the kidneys stopped functioning, the concentration of urea in the blood would increase. Eventually, this would eliminate the urea concentration gradient or even reverse it. Urea would remain in the cells and increase to toxic levels, which could damage or even kill the cells.
3. If the membrane is freely permeable, there is no barrier to the movement of solutes or water. The solutes and water would each move down their concentration gradients. Since the solute concentration is higher in the tube, the solutes diffuse from the tube to the beaker until equal amounts of solutes exist inside the tube and beaker (i.e., equilibrium). In a similar fashion, water, which is at a higher concentration in the beaker compared with the tube, will diffuse into the tube until equal amounts of water are inside the tube and beaker. As a result of the diffusion of the water and solutes, the solution concentrations inside the tube and beaker will be the same because they both contain the same amounts of solutes and water.
4. Remember that diffusion, whether simple or facilitated, is the movement of a substance down its concentration gradient. That means that the glucose concentration gradient between the extracellular fluid and the cytoplasm depends on the amount of glucose molecules only and is not affected by other molecules, even if they are similar. If glucose is converted to other molecules inside the cell, the concentration gradient is maintained, and the cell can continue to take up more glucose passively by facilitated diffusion.
5. To answer this question, we must first consider the functions involved in each cell described and identify the organelles that carry out these functions. Referring to table 3.1 would help quickly identify the organelles and functions. (a) Cells that synthesize and secrete proteins would require organelles involved in the manufacturing, packaging, and releasing of proteins from the cell. The rough endoplasmic reticulum, with the attached ribosomes, carries out the synthesis of proteins that will be released from the cell. The Golgi apparatus is also involved in the packaging of cellular materials that are secreted by packaging the proteins into secretory vesicles that move to the plasma membrane. (b) Active transport requires ATP to move materials across the plasma membrane, so we would expect the cell to have many mitochondria to produce ample ATP. (c) Recall that the organelle involved in the synthesis of lipids is the smooth endoplasmic reticulum. (d) Cells that ingest foreign substances by endocytosis would require the enzymes needed to break down those substances. We learned that lysosomes are vesicles of digestive enzymes that breakdown materials that have been brought into the cell.
6. The first step in answering this question is to review the pairing relationship between nucleotides of DNA and RNA. In DNA, the base pairing associations are A aligns with T, G aligns with C, C aligns with G, and T aligns with A. However, since T is not

found in RNA, the base pairing is slightly different in that A aligns with U instead of T, but all other base pairing patterns are the same. The first question asks for the sequence of the mRNA that is transcribed from the given DNA sequence. The answer is GCAUGCGGCUCUGCAGUUG. The second question asks for the sequence of the complementary strand of DNA. The answer is GCATGCGGCTCTGCAGTTG. Notice that the only difference in the first and second answer is that the Us have been replaced with Ts.

Chapter 4

2. a. The question asks about the relationship between form and function of tissues. First, consider the name of the tissue type: nonkeratinized, stratified epithelium. The term *stratified* means more than one layer of cells, whereas the term *simple* means a single layer of cells. In the digestive tract, a principle function is absorption, a process that would be hindered by the many layers of stratified epithelium. Stratified epithelium is more suited to areas where the layers would protect underlying tissues from abrasion. Cuboidal cells are specialized for secretion and absorption. These cells contain a large number of organelles that produce the secretions and transporters needed to support absorption.
   b. In this scenario, both tissue types are stratified, but one type lacks keratin. The protein keratin provides a tough layer that retards water movement. If keratin were absent from the epidermis, the body could not retain water effectively and would be more prone to damage from abrasion.
   c. In the mouth, because whole foods are sometimes very coarse, the tissue needs to be thick and tough like stratified squamous. If the mouth were lined with simple columnar, it would be severely damaged during chewing.
3. First, define a tight junction: an intercellular junction that forms a permeability barrier. The tight junctions impede simple diffusion in the space between cells. As a result, tight junctions force ions and water to move through the epithelial cells in order to move from one side of the epithelium to the other. The most likely type of epithelial tissue would be simple cuboidal or simple columnar, which can readily absorb or secrete materials. The NaCl may be moved via active transport or facilitated diffusion across the cell membrane. Once NaCl reaches a higher concentration on one side of the epithelial cell membrane, water will follow the NaCl via osmosis.
4. The question asks how the structure of a tissue's components contributes to its function. When the vertebrae flex, elastic ligaments attached to the vertebrae help them return to their normal, upright position. When a muscle contracts, the pull it exerts is transmitted along the length of its tendons. The tendons need to be very strong in that direction but not as strong in others. The collagen fibers, which are like microscopic ropes, are therefore all arranged in the same direction to maximize their strength. If tendons were elastic, muscle contraction tension would not move the bone effectively. Imagine trying to connect train cars end to end with rubber bands, rather than steel couplings. Movement of the train would be ineffective, as the engine would be able to move quite a distance before the next car would move, and so on.
5. The question tells us that (1) vitamin C is required for collagen formation, (2) scars are formed from collagen, and (3) there is a lack of vitamin C (scurvy). Therefore, if there is a lack of vitamin C, the density of collagen fibers in a scar may be reduced and the scar may not be as durable as a normal scar. Overall, there is slow and poor wound healing in a patient with scurvy.
6. First, consider the characteristics of hyaline cartilage that make it an effective tissue for ease of joint mobility. Hyaline cartilage provides a smooth surface, so that bones in joints can move easily. In contrast, dense irregular collagenous connective tissue is a fibrous meshwork that is noted for its strength and ability to withstand stretching—for example, in the dermis layer of the skin. When the smooth surface provided by hyaline cartilage is replaced by fibrous connective tissue, the smooth surface is replaced by a less smooth surface, and the movement of bones in joints is much more difficult. The increased friction helps increase the inflammation and pain that occur in the joints of people who have rheumatoid arthritis.
7. First, define inflammation. Inflammation produces five main symptoms: redness, heat, swelling, pain, and disturbance of function. These symptoms occur because of an increased blood flow to the area. Consequently, if the area is so badly damaged that blood vessels are destroyed, then without blood vessels, no inflammation occurs in the damaged site. However, blood vessels are still intact in the tissues surrounding the severely damaged area. Thus, the signs of inflammation appear around the periphery of the damaged tissues.

Chapter 5

2. In the description of the epidermis, the superficial layer of the skin, we learned that the keratinized cells are coated with lipids to prevent fluid loss. Recall from chapter 3 that substances that are lipid-soluble easily diffuse through lipid layers but water-soluble substances do not. By applying the same principles of diffusion across cell membranes to diffusion across the skin, we can predict that lipid-soluble substances diffuse easily but water-soluble substances do not.
3. From the description of the injury, we know that the hammer struck Bob's nail bed. It is apparent that the hit was hard enough to rupture small blood vessels deep to the nail matrix. Blood accumulated between the nail and the nail bed, causing the dark area. In chapter 4 we learned that inflammation is the response that occurs when tissues are damaged and a normal inflammatory event is edema, or swelling at the injury site. The accumulation of blood and edema increased the pressure deep to the nail body, which stimulated pain receptors. When a hole was drilled through the nail, the accumulated bloody fluid drained, reducing the pressure and, consequently, the pain. Since the nail matrix, which is proximal to the injury site, was not injured, the nail continued to grow over the next 2 months, until the injured area was pushed distally to the free edge of the nail.
4. a. In this chapter we learned that body temperature is affected by blood flow in the skin. When body temperature decreases, blood flow through the skin also decreases to reduce heat loss. Billy's ears and nose turned pale because of decreased blood flow. Constriction of blood vessels in Billy's skin reduced heat loss and helped maintain his body temperature.
   b. Recall that the redness of skin is due to increased blood flow through the skin, so Billy's ears and nose turned red because of increased blood flow. As skin temperature decreases and blood vessels constrict, tissue can be damaged by the lack of blood flow and ice crystal formation. Cold-induced vasodilation periodically increases blood flow and prevents or slows the rate of ice crystal formation. This strategy for maintaining tissue homeostasis is beneficial as long as body temperature can be maintained. Also, we learned that blood flow determines the distribution of heat to different regions of the body. The red appearance of the nose and ears indicates increased blood flow in these areas, and we can assume that this leads to warming of this tissue that is probably very cold due to the weather.
   c. The white color indicates the complete lack of blood flow through the skin of his

ears and nose. The tissue will most likely be damaged due to the absence of blood flow in these areas.

d. Remember that one of the major functions of the skin is to prevent microorganisms from entering the body. If frostbite destroys skin cells, this function can be compromised, resulting in infection.

5. We learned that one of the functions of the skin is to reduce water loss. Sam's burns resulted in severe damage to his skin, which most likely led to increased water loss at the injury site, causing dehydration and reduced urine production. We learned that Sam was administered large volumes of fluid to counteract his increased fluid loss. But how much fluid should be given? The amount of fluid given should match the amount that is lost, plus enough to keep the kidneys functioning properly. An adult receiving intravenous fluids should produce 30 to 50 mL of urine per hour, and children should produce 1 mL/kg of body weight per hour. By monitoring Sam's urine output, the nurse can determine if he is getting enough fluids. If his urine output is too low, more fluids can be given.

Chapter 6

2. First, let's consider the structure of cartilage. The book tells us that the perichondrium, which surrounds the cartilage, contains blood vessels, but the blood vessels do not enter the cartilage. The book also states that nutrients must diffuse through the matrix before reaching the chondrocytes. Logically, cells that aid in tissue repair would also enter cartilage more slowly than if blood vessels penetrated the cartilage. The question next asks whether the lack of perichondrium, blood vessels, and nerves would be advantageous for articular cartilage. When considering the function of articular cartilage, the absence of such structures makes sense. Articular cartilage provides a smooth, low-friction surface for two bones to move past each other easily. If there were solid structures within the joint, the effect of smooth movement would be lost. Imagine trying to ice skate on a rink whose ice had garden hoses frozen just under and along the surface. We would spend more time trying to get back to our feet than skating.
3. This question requires knowledge of cartilage histology, bone formation, and bone histology. Recall that, to acquire O_2 and nutrients, cartilage relies on diffusion though the matrix. As osteoclasts migrate into the developing bone structure and start to remove the cartilage, the cartilage becomes calcified, no more diffusion is possible, and the chondrocytes die. However, when we examine bone histology, we see that adjacent osteocytes are connected via cell processes. As the matrix is laid down, the area where the cell processes meet does not become covered in ossified matrix, forming canaliculi. Therefore, osteocytes continue receiving oxygen and nutrients through the canaliculi or from one osteocyte to another through cell processes.
4. To begin, define interstitial growth (from section 6.2): growth of cartilage through division of chondrocytes within the existing matrix. Because cartilage matrix is pliable, expansion of the existing matrix occurs as new matrix is added by the new chondrocytes. Remember that new bone growth is added to existing bone by appositional growth only. The definition of appositional growth is the addition of new osteoblasts and bone matrix on the surface of existing bone or cartilage. Next, consider the composition of bone matrix. It contains an inorganic component, calcium phosphate, which is not pliable. Therefore, if an existing osteocyte were to divide within its lacuna, the existing matrix would physically impede deposition of more matrix.
5. It is likely that Jill is still growing. Consequently, the epiphyseal plates in her long bones have not yet been converted to epiphyseal lines. If a break occurs in an epiphyseal plate, it can slow bone growth and interfere with bone elongation. As a result, the femur, and therefore her left leg, will be shorter than her right leg. Recovery is difficult because cartilage repairs slowly due to the fact that cartilage is much less vascular than bone.
6. To answer this question, first describe the physical difference between cartilage and bone: cartilage matrix is semisolid (pliable) when compared with bone matrix, which is solid (rigid). Next address what the advantage is of having cartilage covering articular surfaces. The advantage of covering articulations (joints) with cartilage is that it provides a flexible, smooth surface for bone movement. A major impetus for many joint replacement surgeries is degeneration of the articular cartilage, as movement of a joint without articular cartilage is quite painful (see chapter 8 Clinical Impact, "Joint Replacement").
7. The question tells us that Nellie's blood levels of estrogen are much higher than normal for a 12-year-old girl. The text explained that important growth stimulators are reproductive hormones, which usually promote a burst of growth at puberty (approximately 12 years for girls). Because Nellie's estrogen levels are higher than normal, she will most likely grow at a faster rate than she normally would have over the next 6 months. However, if the estrogen levels are not lowered back to normal, Nellie will probably be shorter at 18 than expected. Recall that, in addition, to stimulating a burst of growth at puberty, estrogen causes a closure of the epiphyseal plate and growth in bone length stops. Also, estrogen is more effective at this than testosterone, so Nellie may stop growing years before she would have with normal estrogen levels.
8. Hopefully, Betty will discourage her granddaughter from smoking, as it reduces estrogen levels. Lowered estrogen levels, as we learned while reading the background information, are a major contributor to osteoporosis. Additionally, Betty should encourage her granddaughter to get adequate calcium and vitamin D in her diet and to get some sunscreen-free sun exposure for vitamin D production (no more than 20 minutes a day). Betty and her granddaughter might consider taking a yoga class together as well. Exercise is an important source of increased bone density because it causes the muscles to put stress on the bones, which thickens the bones. Because the greater the density of bone before the onset of osteoporosis, the more tolerant it is to bone loss, most people (especially women in their 20s and 30s) need to make sure to get adequate calcium and exercise.
9. a. Henry's bone density is less than normal for a man his age. Less dense bone is more likely to break.

   b. Henry's eating habits have resulted in insufficient dietary intake of Ca^{2+} and vitamin D. Therefore, the absorption of Ca^{2+} from his intestine into his blood has been inadequate.

   c. We might expect Henry's blood Ca^{2+} to be low because of his diet. However, low blood Ca^{2+} levels stimulate increased PTH secretion. An increase in PTH maintains normal blood Ca^{2+} by increasing the number of osteoclasts, which break down bone and release Ca^{2+} into the blood. Thus, Henry's blood Ca^{2+} levels are maintained at the expense of his bones, which become less dense as more matrix than usual is broken down. Increased PTH levels also promote more Ca^{2+} reabsorption from the urine.

   d. Normally, exposure to sunlight activates a precursor molecule in the skin that eventually becomes activated vitamin D in the kidneys (see chapter 5). Henry produces few, if any, precursor molecules because of his nocturnal lifestyle. Therefore, Henry has low vitamin D levels and, so, has reduced absorption of Ca^{2+} from his small intestine.

e. Exercise is a major source of mechanical stress on bones, which increases osteoblast activity. Because Henry does not exercise, his osteoblasts have not been as active as they might have otherwise been, which has allowed the osteoclasts to dissolve his bone to a greater degree than normal. The overactivity of the osteoclasts partially accounts for Henry's lower bone density.

Chapter 7

2. After reading section 7.2, you probably realize that common structures, such as the "arm," the "hand," and the "nose," are actually a combination of multiple bones. Therefore, a "broken nose" could involve the nasal bones, the ethmoid bones, the vomer bone, the maxillae bones, or the inferior nasal conchae.
3. To answer this question, consider the function of the vertebrae and bones in general. The question asks why cervical vertebrae (in the neck) are smaller than lumbar vertebrae (in the lower back). Most bones are protective and weight bearing. The larger and thicker the bone, the more protection and weight-bearing functions it can provide. Because the lumbar vertebrae are much more inferior than the cervical, they bear more weight. Once again, we see "form follows function." Second, the question asks what the differences in vertebrae structure would be between an actively exercising person and a more sedentary person. Recall from chapter 6 (see section 6.7, "Bone Remodeling") that mechanical stress, such as that from exercise, applied to a bone stimulates osteoblasts to lay down new matrix. Thus, a person who regularly exercises will have a greater bone density in all the vertebrae than someone who never exercises.
4. In this scenario, Dr. Smart is able to diagnose a broken clavicle without x-rays. To determine how this is possible, we first need to know the normal position of the arm and the clavicle's role in attaining this. The text explains that the clavicle's job is to hold the upper limb away from the body. In addition, if we look at figure 7.22, it shows that the clavicle supports the scapula anteriorly. Therefore, when Sarah arrived at the emergency room, Dr. Smart probably saw that her shoulder was more inferior and anterior than normal and that her arm was resting against the side of her body and not being held away from the body, as it normally is.
5. In a dried skeleton, the joints between bones are now visible and all the long bones of the hand are obvious. In a live hand, the "knuckles" are actually where the phalanges arise. With the soft tissue intact, it appears that the metacarpals extend from the most proximal phalanx to the carpals. But if we look at figure 7.27, we see that the proximal phalanx is not tremendously shorter than the metacarpal bones, especially the third metacarpal and proximal phalanx. It is a common misconception that the metacarpals of the palm are part of the fingers.
6. The top of modern ski boots is placed high up the leg to make the weakest point of the fibula less susceptible to great strain during a fall. Modern ski boots are also designed to reduce ankle mobility, which increases comfort and performance.
7. To answer this question, we must first recognize the areas of the skeleton that do not have a thick layer of muscle or fat overlying them. Additionally, we must think about the regions of the body where pressure would occur in a bedridden individual: anywhere on the skull, the zygomatic bones, the acromion process, the scapula, the olecranon process, the coccyx, the greater trochanter, the lateral epicondyle of the femur, the patella, the lateral malleolus, the sacrum, and the anterior iliac spine.
8. The question addresses the form and function of the femur, especially in older people. Recall from chapter 6 that as people age bone density declines because the reproductive hormones decline. We learned that estrogen in women and testosterone in men help bones grow and stay dense. However, because older women's estrogen levels tend to be lower than older men's testosterone levels, their bones are even more fragile than men's. Additionally, the femoral neck is commonly injured in elderly people because it is the smallest portion of the femur, which supports the weight of the body. It also forms an angle between the pelvis and the shaft of the femur, so the downward force of gravity on the body places enormous pressure on this part of the femur. That pressure is usually resisted in younger people with strong bones, but not as much in the elderly.

Chapter 8

2. First, we must define a suture and its location. A suture is a seam between two skull bones. Second, we need to define a synostosis: an ossified joint, such that two bones have become one solid bone. Next, the question asks about the effect of skull bones fusing prematurely on a child's brain development. To address this, recall that, in a normal newborn, the sutures are more extensive than in an adult skull and are called fontanels (soft spots). The fontanels allow for expansion of the skull to accommodate brain growth. If a newborn's skull were completely solid, the brain's growth would be impeded, and developmental problems would arise in the child.
3. The first step is to determine the function of the articular cartilage. Look at figure 8.7 to find its location. There we can see the articular cartilage covers the ends of the bones, which meet at the joint. Depending on the specific joint, there may be significant force causing the bones to move toward each other; thus, the articular cartilage protects the ends of the bones and may endure significant mechanical stress. Next, describe the structure of the synovial membrane: It is a "thin, delicate membrane." It is a logical conclusion, then, that the synovial membrane would suffer damage if it covered the articular cartilage.
4. To begin, it is required to know what anatomical position is. Recall that, when a person is standing, the face is forward, the arms are to the sides, and the palms of the hands are facing forward. Next, it is necessary to know the different terms used to describe movements of limbs and body parts relative to others. Now, picture yourself in anatomical position: Your right arm would move laterally out to the side (abduction) and then your hand would move closer to your head, bending at the elbow (flexion). Flexion at the shoulder and elbow also works.
5. The first step is to locate the AC joint on figure 8.23. However, simply consider the name of the joint: acromio (acromion process) clavicular (clavicle) joint. Tearing the AC ligament disrupts a portion of the only bony attachment of the upper limb to the body. Looking at figure 8.23 we can see that a separated shoulder may also include tearing of the coracoclavicular ligament, further destabilizing the shoulder. If the connection of the scapula to the clavicle is broken, the scapula and humerus tend to be displaced inferiorly, and the proximal pivot point for the upper limb is disrupted, not allowing smooth rotation of the humerus in the glenoid cavity.
6. This question requires you to apply your knowledge of the knee ligaments to a clinical situation involving two tests of ligament integrity (anterior and posterior drawer tests). For Mr. Dent, the normal anterior drawer test result indicates that the anterior

cruciate ligament was not injured. However, the increased movement in the posterior direction indicates that his posterior cruciate ligament (PCL) was torn. The PCL connects the femur to the tibia at the back of the knee to limit the backward or posterior motion of the tibia. A PCL tear occurs most commonly when there is a strong direct blow to the front of the knee. The sudden force backwards can tear the PCL, especially if the knee is flexed (bent) close to a 90 degree angle. Common incidents that cause PCL damage are car accidents when the knee hits the dashboard and contact sports such as when a football player is tackled from the front below the knee.

Chapter 9

2. The first step to answering this question is to define the resting membrane potential and the distribution of ions across the cell membrane in a resting, electrically excitable cell. At rest, the concentration of K^+ is higher inside the cell than outside the cell. This means that a concentration gradient exists for K^+ from inside the plasma membrane to outside the plasma membrane. Therefore, if K^+ ion channels were opened, regardless of the stimulus that caused them to open, K^+ would diffuse out of the cell. Because K^+ is a positively charged ion and other, negatively charged particles, such as proteins, remain in the cell, the resting membrane potential would decrease or would become more negative.
3. The important concept here is the role of acetylcholine in skeletal muscle contraction. Acetylcholine is the neurotransmitter that first stimulates ligand-gated Na+ channels to open, thus depolarizing (exciting) the muscle fiber. Recall that a muscle fiber is also an electrically excitable cell and that, in order for it to generate its response (a contraction), it must fire an action potential. If insufficient acetylcholine were released from the presynaptic terminal of an axon to depolarize the skeletal muscle fiber membrane to threshold, an action potential would not be produced in the muscle fiber and the muscle could not contract. However, a single neuron may not release enough acetylcholine on its own; rather; it may release only enough to generate a subthreshold local potential. In that case a muscle contraction would require the release of acetylcholine from several neurons, and, if the local potentials were produced over a short period of time, they could summate (see chapter 11) and reach threshold, and an action potential would be produced.
4. To answer this question, we first need to understand what a motor unit is and how it works. In chapter 9 we learned that a motor unit is a motor neuron and all the muscle fibers it innervates. The pattern is that motor units with few muscle fibers perform more precise, delicate tasks (e.g., the motor units in our fingers), whereas motor units with a greater number of muscle fibers perform more gross movement tasks (e.g., the motor units in the gastrocnemius [calf] muscle). Therefore, upon recovery from poliomyelitis, muscle control decreases when reinnervation of muscle fibers occurs because the number of motor units in the muscle is decreased. The greater the number of motor units in a muscle, the greater the potential for fine gradations of muscle contraction as motor units are recruited. A smaller number of motor units means that gradations of muscle contraction are not as fine.
5. a. Before we can answer these questions, it is necessary to know what the function of acetylcholinesterase is. We can actually tell just by the name. Enzymes, which react with a certain chemical, are named by the chemical's name and the suffix *–ase.* Thus, acetyl cholinesterase reacts with acetylcholine by degrading it. If this enzyme cannot work, then acetylcholine is not broken down in the synaptic cleft. Instead, it accumulates there and continuously stimulates the muscle fiber. As a result, the muscle remains contracted until it fatigues. Death is caused by the victim's inability to breathe. Either the respiratory muscles are in spastic paralysis or they are so depleted of ATP that they cannot contract at all.
   b. Since the main interruption of organophosphate poison is disruption of acetylcholine break down and then continuous muscle stimulation, the primary goal is to either get rid of acetylcholine or prevent it from doing its job. One such drug for treating organophosphate poisoning is called curare. Curare binds to acetylcholine receptors and thus prevents acetylcholine from binding to them. Because curare does not activate the receptors, the muscles do not respond to nervous stimulation. In the event of too much curare the person suffers from flaccid paralysis and dies from suffocation because the respiratory muscles are not able to contract.
6. As a weight is lifted, the muscle contractions are concentric contractions. When a weight lifter lifts a heavy weight above the head, most of the muscle groups contract with a force while the muscle is shortening. Concentric contractions are a category of isotonic contractions in which tension in the muscle increases or remains about the same while the muscle shortens. While the weight is held above the head, the contractions are isometric contractions, because the length of the muscles does not change. While the weight is lowered, unless the weight lifter simply drops the weight, the length of the muscles increases as the weight is lowered for most of the muscle groups. Eccentric contractions are contractions in which tension is maintained in a muscle while the muscle increases in length. The major muscle groups are therefore contracting eccentrically while the weight is lowered. So Mary explained to the two students that they were each correct, since the weight lifter used all three types of contractions.
7. The first step to answer this question is to define glycogen and its role in exercise in the nondisease state. Glycogen is the stored form of glucose. Free, immediately usable glucose is in limited supply, and the body quickly relies on glycogen for sustained exercise. Without the ability to break down glycogen to glucose molecules, muscle depends on the uptake of glucose from outside the cells (e.g., from the blood) or from the metabolism of fatty acids. Consequently, a person's ability to carry out vigorous exercise, including anaerobic exercise, is reduced. Fatigue and lower exercise tolerance are characteristic of the condition, but exercise can be maintained at a slow pace.
8. This question is really asking about oxygen deficit and recovery oxygen consumption— in other words, how long a person needs to breathe heavily after exercise to return the body functions to baseline. The increase in breathing after running up the stairs is recovery oxygen consumption and is partially due to the oxygen deficit that occurs at the beginning of exercise. Because the duration and intensity of the exercise were similar in both boys, the only explanation for the difference in recovery oxygen consumption is their general physical fitness. Because of conditioning, Eric's muscles accumulated more mitochondria, were more vascular, and received more oxygen from his more efficient respiratory system. Thus, Eric's muscles produced more of his ATP aerobically. On the other hand, John was very out of shape, and so his muscles had fewer mitochondria and less blood supply and his respiratory system was not efficient. Therefore, John's muscles relied more on anaerobic respiration to produce

ATP. Anaerobic respiration is far less efficient than aerobic respiration and produces only 2 ATP per glucose molecule and lactic acid. When someone like John primarily depends on anaerobic respiration, there is a larger accumulated oxygen deficit during exercise and a larger recovery oxygen consumption after exercise. Therefore, after the exercise, John's metabolism switched to aerobic respiration to restore homeostasis. John's respiration rate was elevated longer than Eric's because John's homeostasis was disrupted more than Eric's, in part to metabolize the lactic acid produced during anaerobic metabolism.

9. Susan was unable to keep pace during the finishing sprint because her muscles were not trained for quick movements, such as sprinting. She had trained them only for aerobic, steady-state activity like that occurring during most of the race. As her coach, you should advise Susan to incorporate high-intensity sprinting activities into her training. Such activities will cause her fast-twitch muscle fibers to increase their performance capacity, which will increase her ability to sprint at the end of the race.
10. To answer this question, the first step is to refer to figure 9.25. We learned that a ligand is a chemical that, upon binding to its receptor, will stimulate a response. In figure 9.25, a smooth muscle cell is depicted, which shows that the ligand receptors are G protein–linked receptors. Binding of the ligand results in the opening of Ca^{2+} channels. Calcium ions diffuse into the cell and bind to calmodulin, which activates the enzyme myosin kinase to phosphorylate myosin heads and start the cross-bridge cycle. As long as intracellular Ca^{2+} levels are elevated, the cross-bridge cycle will take place. Because the cross-bridges release slowly, the contraction is sustained: Once a myosin head has been phosphorylated, it can form cross-bridges, move the actin, detach, and re-form cross-bridges again and again without breaking down new ATP molecules for each cycle.
11. DMD affects the muscles of respiration and causes deformity of the thoracic cavity. The reduced capacity of muscle tissue to contract is one factor that reduces the ability to breathe deeply or cough effectively. In addition, the thoracic cavity can become severely deformed because of the replacement of skeletal muscle with connective tissue. The deformity can reduce the ability to breathe deeply. DMD can also affect the muscle of the heart and cause heart failure. The boy with advanced DMD would have muscles that contained fewer muscle fibers and more fibrous connective tissue and fat. His posture would be much worse due to abnormal curvature of the spinal column (such as kyphoscoliosis) and more deformity of the thoracic cage. In general, his physical condition would be worse because of weakened heart muscle and difficulty breathing due to thoracic cage and vertebral column deformities. He would be much more likely to be wheelchair-bound due to severe weakness of his skeletal muscles.

Chapter 10

2. The question is asking about which muscles of the tongue allow people to "stick their tongue out," or protrude it. Refer to table 10.6, where the actions of tongue muscles are described. There, we find it is the extrinsic muscle the genioglossus that works to protrude the tongue. Further, the question tells us that Rachel's left mandible was broken and nerves were damaged. Recall from chapter 9 that skeletal muscles require a nerve impulse in order to contract. Therefore, we can conclude that the left genioglossus is unable to contract and the right is able to contract. So, when Rachel protrudes her tongue, it will only go to the left because it is only getting pushed out on the right side.
3. This question asks about the function of the eye muscles. Refer to table 10.8, where the actions of the lateral and medial rectus muscles of the eye are described. Notice that the lateral rectus "pulls" the eye laterally and the medial rectus "pulls" the eye medially. So, if the lateral rectus is weak and does not pull the eye laterally, the medial rectus will predominate, and the eye will deviate medially.
4. The first step is to remember that we learned in chapter 8 that the definition of abduction is movement away from the body's midline. Next, it is necessary to know the location of the supraspinatus muscle to answer the question. The supraspinatus rests in the supraspinous fossa of the scapula. Then refer to table 10.14, where the function of the supraspinatus is summarized. Also refer to figure 10.23, which focuses on the rotator cuff muscles, of which the supraspinatus is a member. There, you see that in extreme abduction the supraspinatus would be compressed against the acromion process of the scapula.
5. When doing chin-ups the elbow flexes as you pull your body up toward the bar and extends as you lower your body back down. Refer to table 10.16 for the muscles involved in flexing and extending the elbow. Notice that the biceps brachii both flexes the elbow and supinates the forearm; however, the brachialis only flexes the elbow. When the forearm is pronated, the biceps brachii cannot contract as forcefully when flexing the elbow because the distal end of the biceps brachii is partially medially rotated. Thus, chin-ups with the arm pronated are more difficult. Bodybuilders who want to build up the brachialis muscles do pronated chin-ups. When the forearm is supinated (palms toward you), both muscles can work optimally, making these the easier chin-ups.

Chapter 11

2. First, consider the basic function of the axon. At this point in the reading, we know that axons are a cellular projection from the neuron, sometimes called nerve fibers. We also know that the cell body is the site of protein synthesis for the entire cell. Therefore, if the distal portion of an axon is severed from the rest of the cell, it will die. There is no way for the distal axon to replenish the enzymes and other proteins essential for survival. On the other hand, any remaining portion of axon still attached to the cell body survives and, in many cases, grows to replace the severed portion.
3. Although this question seems to ask about the cause of death, it is really asking about what cellular changes nervous tissue would exhibit in the event of trauma. First recall the types of glial cells found in central nervous tissue, in particular any cells involved in tissue repair or protection. The primary cell type of the neuroglia with these functions is the microglia. They are defined as "cells that become mobile and phagocytic in response to inflammation" and are described as cells that "migrate to areas damaged by infection, trauma, or stroke." Second, the question tells us that the pathologist found evidence of a prior stroke. The pathologist recognized the prior stroke because of the large number of microglia in the brain tissue surrounding the blocked blood vessels. In addition, the pathologist knew at least one damaged area was several weeks old because it takes time for microglia to migrate to a damaged area. However, the second damaged area of the brain showed evidence of a reduced blood supply, but far fewer microglia. Recall from chapter 4 that cardinal signs of inflammation include redness and edema, both of which were evident in the area of the brain injured by the fall. Therefore, the pathologist could conclude that either the second stroke killed the man, and caused him to fall, or it caused him to fall and the trauma to the brain incurred during the fall caused his death.

4. The first step to answer this question is to define the basis for the resting membrane potential. Recall that the inside of the plasma membrane is more negative than the outside of the plasma membrane due to a higher concentration of negatively charged proteins inside the cell. The K^+ is also in higher concentration inside the cell and tends to "offset" the negative charge of the proteins. However, the K^+ leak channels allow K^+ to diffuse out of the cell, down the concentration gradient. Removal of K^+ removes positive charge inside the cell, and the membrane potential becomes more negative. Therefore, because Tissue A has more K^+ leak channels, more K^+ can leak out of Tissue A cells than in Tissue B cells, and the inside of Tissue A cells becomes more negative. The resting membrane potential is larger for Tissue A than for Tissue B; it is further from threshold.
5. First, consider the role Ca^{2+} plays in establishing the resting membrane potential. We learned that Ca^{2+} ions are important for keeping voltage-gated Na^+ channels closed. When voltage-gated Na^+ channels open, the cell depolarizes, or becomes more positively charged on the inside of the plasma membrane. Thus, if Ca^{2+} levels drop, the voltage-gated Na^+ channels open, Na^+ diffuses into the cell, making it more positive, and the resting membrane potential increases.
6. To answer this question, let us first describe the Na^+ concentration and its role in a normal excitable cell. We learned that, in a normal cell, the concentration of Na^+ is much higher outside the cell than inside the cell. As a result, when Na^+ channels open, Na^+ diffuses into the cell quickly, causing the changes in the membrane potential that result in an action potential. The movement of Na^+ into the cell is the result of its steep concentration gradient. Remember, also, that enough Na^+ must enter a cell for the membrane potential to reach threshold, opening the voltage-regulated Na^+ channels and causing an action potential. If the extracellular concentration of Na^+ were reduced, then the concentration gradient would also be reduced. The effect would be that, if the cell were stimulated, less Na^+ would enter the cell. The cell would not reach threshold, and an action potential would not occur.
7. To answer this question, we must first define threshold. Threshold is the membrane potential at which voltage-gated Na^+ channels first begin to open. A prolonged threshold stimulus will result in a single action potential at the end of each relative refractory period. However, a prolonged, stronger-than-threshold stimulus will produce a greater number of action potentials than a threshold stimulus of the same duration due to a higher Na^+ permeability in the axon. The stronger-than-threshold stimulus will even produce action potentials during the relative refractory period. Therefore, a neuron receiving a stronger stimulus will send a greater frequency of action potentials. Because of the all-or-none properties of neurons, one action potential has the same magnitude as another. Thus, in order for a neuron to communicate stimulus strength (e.g., warm water vs. scalding water), the neuron receiving the stronger stimulus sends a higher frequency of action potentials.
8. First, what is the absolute refractory period? Recall that it is the time during which a neuron cannot fire an action potential no matter how strong the stimulus. This is due to all the voltage-gated Na^+ channels already being open. Thus, if the axon is nonresponsive for 1 ms, action potentials can be generated no faster than every millisecond. Because there are 1000 ms in 1 second, the maximal frequency is 1000 action potentials/second.
9. The question asks us to compare signal transmission rates of electrical and chemical synapses. The first thing we need to do is consider the mechanism for each type of synapse. Electrical synapses require only ion flow from one neuron to the next. Chemical synapses rely on diffusion of the neurotransmitter across the synaptic cleft, binding to the receptor and then triggering a signal transduction pathway. Thus, electrical synapses take no more time to be transmitted than an individual action potential takes traveling from node to node in a myelinated axon, or down the axon in an unmyelinated axon. On the other hand, the chemical synapse is more complex and reliant on molecular conformation change, enzyme activity, and so on, all of which are slower processes.
10. The key piece of information in the question is the fact that neuron B releases both a neurotransmitter and a neuromodulator, which is excitatory (produces EPSPs). EPSPs depolarize neuron C and bring it closer to threshold than it is when stimulated by only neuron A. Neuron C does not require as much time for temporal summation to reach threshold to fire an action potential when stimulated by neuron B vs. neuron A. Therefore, neuron C fires more action potentials with stimulation by neuron B alone than with temporal summation of neuron A alone. In other words, the neurotransmitter from neuron B is more effective when released with the neuromodulator from neuron B than it is when released alone from neuron A.

Chapter 12

2. In chapter 11, we learned that a ganglion is a cluster of neuron cell bodies. Similarly, in this chapter we learned that the dorsal root ganglia are clusters of sensory cell bodies and the dorsal roots are bundles of sensory axons. The ganglia are larger in diameter than the roots because of the size difference between the cell bodies and axons. To answer the second question, refer to figure 12.4 to see the direction of action potential propagation in sensory axons and motor axons. Sensory axons carry action potentials from peripheral tissues to the central nervous system (CNS), which includes the brain and spinal cord. Motor axons carry action potentials from the CNS to peripheral tissues. Now, identify the types of axons in each structure listed. Recall that spinal nerves have both sensory and motor axons, so action potentials are propagated both to the spinal cord and away from the spinal cord. Dorsal roots contain only sensory axons, so action potentials are conducted to the spinal cord only. Finally, ventral roots contain only motor axons, so action potentials are conducted away from the spinal cord.
3. The body areas in figure 12.14 is color-coded and labeled for the nerve innervation. To answer this question, look at the left arm, forearm, and hand to see which nerves are indicated. In the figure, it is apparent that nerves C5–T1 are damaged. C5 and T1 innervate the left arm, C6 and T1 innervate the forearm, and C7 and C8 innervate the hand.
4. Recall that the phrenic nerve innervates the diaphragm, allowing for the contraction necessary for breathing. If the right phrenic nerve were damaged, then we would expect lack of muscle contraction in the right half of the diaphragm, affecting breathing. To answer the second part of the question, we need to consider the location of spinal cord injury to predict the effect it would have on the diaphragm. Remember that the phrenic nerve is part of the cervical plexus, which includes spinal nerves C1–C4. If the spinal cord were severed at the level of C2, the phrenic nerve would be damaged, and the contractions of the diaphragm would not occur, eliminating the person's ability to breath. Death would likely occur if medical assistance were not administered quickly. On the other hand, if the spinal cord were completely severed at the level of C6, the phrenic nerve would not be damaged, and the diaphragm would not be affected.
5. a. Looking at figure 12.14, we can see that nerves C7, C8, and T1 innervate the skin of the hand. Also after reviewing the description of the brachial plexus and viewing figures 12.19, 12.21, and 12.22, we can

conclude that the radial, ulnar, and median nerves innervate the skin of the hand.

b. Figures 12.19 and 12.21 indicate that the symptoms of pain, tingling, and numbness in the ring finger and little finger of the right hand that radiated down the posteromedial portion of the forearm and hand most likely involve damage to the ulnar nerve, although this does not explain the radiation of symptoms into the forearm and elbow.

c. When the physician carried out a careful examination of Sarah's right upper limb to map the extent of the pain and numbness, he most likely was trying to develop a differential diagnosis between the ulnar nerve damage and damage to the C8-T1 brachial plexus root. Cervical ribs compress the roots of the brachial plexus, which might be the cause of Sarah's symptoms. An x-ray of the neck can indicate the presence of extra rib, which most commonly affects the roots C8–T1.

d. Based on our answers for parts a and b, we can assume that the ulnar nerve is most likely affected, so the muscles innervated by the ulnar nerve may be affected. These muscles are listed in figure 12.21. However, the radial and median nerves are also associated with the C8–T1 root and may be affected; therefore, the muscles innervated by these might also be affected. These muscles are listed in figures 12.19 and 12.22.

6. Figure 12.25 indicates that the femoral nerve innervates several muscles involved in hip flexion and knee extension, both activities that were difficult for Carl. This indicates that the femoral nerve is involved. The sources of nerve fibers in the femoral nerve are L2, L3, and L4; therefore, the intervertebral disk involved compresses L2, L3, or L4 on the left side of the vertebral column. Recall that figure 12.14 includes a dermatomal map. We can see from the map that L3 innervates the dermatome of the medial thigh and is the most likely spinal nerve involved. Carl's motor movements were affected because the reduced control of action potentials from the femoral nerve to the muscles of the thigh caused muscle weakness. The referred pain results from compression of the spinal nerve that innervates the medial thigh and the knee. The compression stimulates action potentials in the nerves, and the pain is referred to the site of the sensory receptors for that nerve.

Chapter 13

2. We learned in this chapter that reflexes that maintain blood pressure are integrated by the medulla oblongata. In response to blood loss, the reflexes increase the heart rate. Similarly, the reflexes cause the constriction of blood vessels in the skin and viscera to increase blood volume and therefore blood pressure. The lack of blood flow through the skin results in pallor. Recall that respiratory reflexes are integrated in the medulla oblongata and the pons.

3. a. The Clinical Impact, "Traumatic Brain Injuries and Hematomas," explains that, often, when one part of the head suffers a heavy blow, the brain moves within the cranial cavity and hits the opposite side of the cranial cavity. In this case, the blow to the back of the head forced the brain anteriorly, and the frontal lobes struck the frontal bones with enough force to tear blood vessels between the brain and the dura. Subsequent bleeding from these vessels into the subdural space created the subdural hematoma. This type of injury is a countercoup brain injury because it occurs on the side of the brain opposite the point of traumatic impact.

   b. We can assume that the herniation of the medulla into the vertebral canal compressed the tissue of the medulla and therefore disrupted its function. Recall from its description that the medulla oblongata contains the centers for controlling respiration and heart rate. Damage to these areas of the brain interrupted the woman's ability to regulate her breathing rate and heart rate, which was most likely a major factor contributing to her death.

4. To answer this question, let us first review the functions of the oculomotor, trochlear, and abducens nerves, listed in table 13.5. Besides innervating the levator palpebrae superioris muscle, the oculomotor nerve innervates the four eye muscles that move the eyeball so that the gaze is directed superiorly, inferiorly, medially, or superolaterally. If the patient can move the eyes in these directions, the oculomotor nerve is not damaged. Similarly, the abducens nerve directs the gaze laterally, and the trochlear nerve directs the gaze inferolaterally. If the patient can move the eyes in these directions, the associated nerves are intact.

5. To answer this question we must first determine the function of the sternocleidomastoid muscle. The description in table 10.1 states that, when one sternocleidomastoid muscle contracts, it rotates the head and neck to the opposite side. If the innervation to one sternocleidomastoid muscle is eliminated due to accessory nerve injury, the opposite muscle is unopposed and turns the face toward the side of the injury. A person with wry neck whose head is turned to the left most likely has an injured left accessory nerve.

Chapter 14

2. First, let us review the sensory information carried by the spinothalamic tract. From table 14.3 we can see that these include pain, temperature, light touch, pressure, tickle, and itch. Next, we can also see from table 14.3 and figure 14.8 that crossover of axons of the spinothalamic tract occurs in the spinal cord. From this we can conclude that a lesion on one side of the spinal cord that interrupts the spinothalamic tract would eliminate the specific sensations carried by that tract below the level of the lesion, but on the opposite side of the body. Not all sensations associated with the spinothalamic tract will be eliminated. Pain and temperature sensation from the opposite side of the body below the lesion would be eliminated. There would be few, if any, clinical changes in detecting light touch because other tracts still carry this information.

3. To answer this question, we need to first identify the sensory pathway involved. Reviewing table 14.3, we can see that the dorsal-column/medial-leminiscal system carries the sensations of proprioception, fine touch (two-point discrimination), and vibration. We can also specifically identify the fasciculus gracilis tract, since Bill and Mary lost the sensations in the lower half of the body. The next step is to determine why Bill and Mary have similar symptoms but different injuries. Recall that the secondary neuron axons cross over along the sensory pathway, so that stimuli from one side of the body are conveyed to the primary sensory cortex on the opposite side. We can see that the crossover of axons for the fasciculus gracilis tract occurs in the medulla, which is part of the brainstem. Bill experienced a spinal cord injury that interrupted the fasciculus gracilis. Since the axons have not crossed over in the spinal cord, we can assume that the spinal cord injury was on the left side, the same side as the sensory receptors. Mary, on the other hand, experienced a brainstem injury. If the injury occurred below the medulla, the site of axon crossover, we would expect it to be on the left side, similar to what was seen in Bill's spinal cord injury. However, if the injury occurred above the medulla, we would assume that it affected the right side, since the secondary neuron axons cross over in the nucleus gracilis of the medulla.

4. Before answering this question, consider the activities carried out by the two sets of limbs. We use the lower limbs primarily for standing and walking, both of which are activities we do not focus our attention. We use our upper limbs for all kinds of activities, including writing, texting on our cellphones, or even playing video games. These activities require much more conscious effort compared with the activities of the lower limbs.
5. To answer the first part of the question, we must recall the types of pain and the types of structures associated with each type of pain. Constipation is causing pain associated with the colon, a deep structure. Remember that the deeper structures lack tactile receptors, so the type of pain experienced is diffuse. We also learned after reading about referred pain that the pain associated with deeper structures is usually felt in more superficial structures. Using figure 14A, we can predict that the man feels the pain around his navel, the common area for referred pain associated with the colon.
6. Recall that stimulating the reticular activating system (RAS) promotes consciousness. Also, remember that acoustic stimuli, such as a dripping faucet, stimulate the RAS. Luke could not sleep because his RAS was stimulated, promoting consciousness and preventing sleep.
7. The vagus nerve supplies the muscles of the larynx that aid in voice production; therefore, minor injury to this nerve in the neck can lead to hoarseness or changes in the voice. In our reading of "Motor Output and Reflexes Projecting Through the Brainstem," we also learned that the vagus nerve controls muscles of the pharynx, larynx, and soft palate associated with swallowing and speech. Damage to the vagus nerve could therefore affect these two activities.
8. After carefully reviewing Vern's symptoms, we can see that the stroke primarily affected muscle activities of his face, head, and neck on the right side. We can therefore conclude that the stroke occurred on the left side of his brain. After reviewing figure 14.13 we can predict the stroke specifically affected the lateral portion of the precentral gyrus of the cerebrum, where motor control of the lips, jaw, face, tongue, and pharynx are represented. After reading Clinical Impact, "Aphasia," we can conclude that Vern does not have expressive aphasia since he is able to understand language and can achieve speech that makes sense with words in proper sequence. In this case, the Wernicke area and the association areas around it appear to be unaffected. However, Vern does exhibit characteristics of expressive aphasia in that his speech is hesitant and distorted. This is usually the result of a lesion in the motor speech area, the Broca area.
9. a. We learned in chapter 11 that myelination speeds up action potential propagation along axons; therefore, demyelination would slow the speed of action potential propagation. We can assume that demyelination of the optic nerve slows the speed of action potential propagation of visual input to the cerebral cortex, affecting visual perception. In this case, demyelination resulted in blurred vision.
   b. It is possible that other neurons besides those of the optic nerve are affected by the demyelination. Demyelination of motor neurons in descending motor pathways would affect muscle activity and result in muscle weakness. Demyelination of sensory neurons in ascending sensory pathways would affect perception of stimuli and may cause the tingling sensations.
10. Recall that the right side of the body is monitored and controlled by the left side of the brain. The crossing over of axons from one side of the brainstem or spinal cord to the other side of the brainstem or spinal cord results in this pattern of regulation. Since Scott exhibited loss of skeletal muscle movement and sensations on his right side, we can conclude that the left side of the brainstem was more severely affected by the stroke. But why did Scott experience pain and temperature sensations on the left side of his face? At the level of the upper medulla oblongata, neither the motor nor the sensory pathways to the limbs have yet crossed over to the left side of the CNS. Most of the motor fibers cross at the inferior end of the medulla oblongata, whereas sensory pain and temperature axons cross over at the level where they enter the CNS. Loss of pain and temperature on the left side of the face indicates that the lesion occurred at the level where the sensory axons from the face had entered the CNS but had not yet crossed.

Chapter 15

2. Ernie's sensory neuropathy is the loss of taste sensation in the posterior one-third of the right half of his tongue. In reviewing "Neuronal Pathways for Taste" we learn that the glossopharyngeal nerves carry taste sensations from the posterior one-third of the tongue. Considering that only the right side of Ernie's tongue is affected, we can conclude that it is the right glossopharyngeal nerve that is damaged.
3. After reviewing the section "Lacrimal Apparatus" we learn that tears from the surface of the eyeball drain through the lacrimal caniculi into the lacrimal sac, which empties into the nasal cavity through the nasolacrimal duct. Our sense of smell is due to the presence of olfactory receptors in the nasal cavity. If medications are placed into the eyes, some may drain into the nasal cavity, which may stimulate the olfactory receptors of the olfactory organ. Our sense of taste is due to the presence of taste receptors in the mouth and pharynx. The ability to "taste" the medication is due to the fluid draining from the nasal cavity into the pharynx, stimulating taste receptors. Similarly, a person's nose "runs" when he or she cries because the excess tear production leads to more fluid drainage into the nasal cavity and therefore a "runny nose."
4. In this question, Max and his grandfather were looking at an object that was far away (the glacier) and an object that was close (the piece of ice). To answer this question, we need to consider three important factors for each scenario: accommodation, pupil constriction, and convergence. While Max and his grandfather were looking at the distant glacier, the ciliary muscles of their eyes were relaxed, and the suspensory ligaments of their ciliary bodies maintained elastic pressure on the lenses of the eyes, keeping the lenses relatively flat and allowing for distant vision. When they looked at the piece of ice that was close to them, accommodation occurred. The ciliary muscles contracted, pulling the choroids toward the lenses and reducing the tension on the suspensory ligaments. This allowed the lenses to assume a more spherical form because of their own elastic nature. More spherical lenses have more convex surfaces, causing greater refraction of light. Also, pupil constriction occurred when they were looking at the piece of ice. As the pupils constrict during close vision, the depth of focus is greater, and more light is required on the observed object. Convergence also occurs during close vision. As an object moves closer to the eye, the eyes must rotate medially so that the object is kept in focus on corresponding areas of each retina; otherwise, the object becomes blurry. The reason Max's grandfather had to reach for his glasses to view the piece of ice is because he suffers from presbyopia, which occurs because the lenses become sclerotic and less flexible as people age. Presbyopia is corrected by wearing reading glasses for close work and removing them when the person wants to see at a distance. Either bifocals or progressive lenses can also be used if a person also has myopia, or problems seeing distant objects clearly.

5. Recall that rod photoreceptors, which are distributed over most of the retina, are involved in both peripheral vision (out of the corner of the eye) and vision under conditions of very dim light. Stimulation of rods would explain why a person notices movements at night out of the corner of the eye. However, cone photoreceptors, which are concentrated in the fovea centralis, are involved in focusing directly on an object, and they do not function well in dim light. Thus, when the person tries to focus on the movement, the low light level is not enough to stimulate the cones, and the person cannot see what caused the movement.
6. Before answering this question, let us review figure 15.23 to see the normal visual fields. We can see that each visual field is divided into temporal and nasal portions. The visual pathway is then color-coded to correlate with each portion for the right and left side. In the case of lesion A, we can see that the temporal part of the left visual field (light green) and the nasal part of the right visual field (dark blue) are affected. The black areas correlate with the same areas in the visual field. Using the same strategy, we can see that lesion B would affect both the temporal part (light blue) and the nasal part (dark blue) of the right visual field, so the entire right visual field would be affected, and an oval representing it would be black. The left visual field would be normal, with the temporal (light green) and nasal (dark green) parts indicated. Following is an illustration of the right and left visual fields.

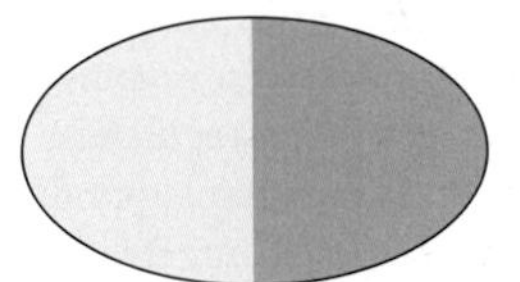

7. First, let us define *perfect pitch* and *tone deaf.* Perfect pitch is the ability to reproduce a pitch precisely just by being told its name or reading it on a sheet of music, with no other musical support, such as piano accompaniment. Tone deafness is the complete inability to recognize or reproduce musical pitches. Both of these characteristics can be due to variation in the structure of the basilar membrane, the inhibitory reflex from the superior olivary nucleus to the spiral organ, and the auditory cortex. Recall that the basilar membrane is structurally different from the base to the apex, allowing different segments to be sensitive to different frequencies (see figure 15.34). In individuals with perfect pitch, the structure of the basilar membrane may be such that tones are adequately spaced along the cochlear duct to facilitate clear separation of tones, whereas in tone-deaf individuals the spacing is inadequate to clearly separate tones.

   We also learned that the superior olivary nucleus of the medulla oblongata has an inhibitory effect that only allows action potentials from regions of the basilar membrane with maximum vibration to be conducted to the auditory cortex. This allows an individual to distinguish specific sound frequencies. For people with perfect pitch, the reflex from the superior olivary nucleus to the spiral organ may have a very narrow "window of function," whereas for individuals that are tone deaf, the reflex may not function well enough.

   Finally, at the auditory cortex, action potentials that originate in the spiral organ are translated to perceived sounds. Variation in translation accuracy would explain the difference between individuals with perfect pitch and those with tone deafness.
8. Recall that the sound wave amplitude determines the volume of a sound, whereas wave frequency determines the pitch of the sound. Loud sounds have sound waves with greater amplitude. This greater amplitude causes the basilar membrane to vibrate more violently over a wider range. The spreading of the wave in the basilar membrane to some extent counteracts the reflex from the superior olivary nucleus that is responsible for enabling a person to hear subtle tone differences.
9. We learned that the brain compares sensory input from the semicircular canals, eyes, and proprioceptors in the back and lower limbs and that conflicting information can cause motion sickness. If you close your eyes, one of these sources of information is eliminated and the brain has less conflicting input to compare, reducing the probability of motion sickness. We can perceive more motion in close objects than in distant objects, so looking at the horizon would also reduce the visual input of perceived motion to the brain and reduce the probability of motion sickness.

Chapter 16

2. This question is referring to the different types of cholinergic receptors: nicotinic receptors and muscarinic receptors. Remember that, even though these are cholinergic receptors to which acetylcholine normally binds, they were classified based on laboratory findings that nicotine binds to one type of cholinergic receptor and muscarine binds to the other type of cholinergic receptor. Also recall that all preganglionic neurons of the sympathetic and parasympathetic divisions and all postganlionic neurons of the parasympathetic division release acetylcholine. Also, the postganglionic neurons of the sympathetic division that innervate sweat glands also release acetylcholine. Figure 16.7 allows us to determine which of these synapses would be affected by nicotine and which of these synapses would be affected by muscarine. Nicotinic receptors are located within the autonomic ganglia in the membranes of postganglionic neurons of both the sympathetic and parasympathetic divisions. Consumption of nicotine would result in stimulation of the postganglionic neurons, and consequently, the stimulation of the effectors of both the sympathetic and parasympathetic divisions. Again, figure 16.7 illustrates that muscarinic receptors are located on the effectors of the parasympathetic division. After the consumption of muscarine, only the effectors that respond to acetylcholine would be affected. This includes all the effectors innervated by the parasympathetic division and the sweat glands, which are innervated by the sympathetic division.
3. Muscarinic receptors are associated with all effectors innervated by the parasympathetic division as well as the sweat glands. We can assume that bethanechol chloride binding to muscarinic receptors results in effects similar to those of parasympathetic stimulation of organs. Contraction of the urinary bladder is a normal response to the parasympathetic division, so we would expect the same result when bethanechol chloride is administered. Side effects of bethanechol chloride can be produced by the stimulation of muscarinic receptors elsewhere in the body. The stimulation of smooth muscle in the digestive tract can produce abdominal cramps, and stimulation of the air passageways can cause an asthmatic attack. Decreased tear production and salivation as well as dilation of the pupils are not expected side effects because parasympathetic stimulation causes increased tear production and salivation as well as constriction of the pupils. Recall that, even though the sympathetic division innervates sweat glands, these glands have muscarinic receptors, so we would expect increased sweating as a side effect of bethanecol chloride.
4. a. Dilation of the pupil is caused by the contraction of the dilator pupillae, which are the radial muscles of the iris.
   b. Recall from the discussion of the iris that the sympathetic division innervates and therefore controls the radial muscles

(dilator pupillae). The parasympathetic division innervates and therefore controls the circular muscles (sphincter pupillae).

c. A drug that mimics sympathetic stimulation, such as an adrenergic drug, could activate α1 receptor on the radial muscles of the iris and cause Sally's pupils to dilate. On the other hand, a drug that blocks parasympathetic stimulation, such as a muscarinic blocking agent, would prevent constriction of the pupil and therefore cause dilation.

d. Remember from chapter 15 that the ciliary muscles constrict, changing the shape of the lenses when viewing close objects. Blurred vision indicates the eyedrops are inhibiting ciliary muscle contraction. From table 16.3 we can see that ciliary muscle contraction is a parasympathetic effect, so we would predict that the eyedrops contain a muscarinic blocking agent, one that would affect parasympathetic effectors.

e. Notice that this scenario is essentially the opposite as that described in part c. Based on our answer for part c, we would expect that an adrenergic blocking agent that binds to α1 receptors and prevents sympathetic stimulation would cause the pupils to constrict. Similarly, a muscarinic agent could stimulate the circular muscles of the iris, causing the pupils to constrict.

f. Recall that sympathetic stimulation of blood vessels normally keeps them in a state of partial contraction. An adrenergic blocking agent that inhibits sympathetic stimulation allows the blood vessels in the conjunctiva of the eye to dilate, causing the appearance of blood-shot eyes.

5. Sweat glands are innervated by the sympathetic division of the autonomic nervous system, the effects of which are greater during times of physical activity. Prior to and during the first part of the race, there was little stimulation of Brad's sweat glands by the sympathetic division. However, Brad's body temperature began to rise because of the increased metabolism of his exercising skeletal muscles. The higher body temperature resulted in increased sympathetic stimulation of his sweat glands. The increased sweat production allowed for evaporative cooling, resulting in more and more heat loss from the body. Therefore, Brad's body temperature did not increase much during the race. After the race, the evaporation of sweat continued to cool Brad's skin quickly, and his body temperature began to fall. The slight decrease in body temperature caused the sympathetic stimulation of his sweat glands to decrease, so that the rate of sweating decreased. As the rate of sweating decreased, the rate at which heat was lost from his body also decreased. Consequently, Brad's body temperature changed only a small amount, and homeostasis was maintained both while he was exercising and after he stopped exercising.

6. Recall that the primary function of the autonomic nervous system is to maintain homeostasis. In both of the scenarios Sarah's blood pressure changes. In chapter 1 we learned that most homeostasis control mechanisms are negative-feedback mechanisms that reverse the change. Also, notice that both questions address sympathetic reflexes controlling blood vessels. Remember that blood vessels are only innervated by the sympathetic nervous system. When Sarah does a headstand, blood drains toward her head, and the blood pressure in the arteries of her chest and head increases. Sensory receptors detect the increase in blood pressure, so the frequency of action potentials in sensory neurons increases. The brain, in turn, activates sympathetic stimulation of the blood vessels, which causes them to dilate. Consequently, the blood pressure in the arteries of the neck and head does not increase dramatically. When Sarah stands quickly after crouching for a short time, blood tends to drain away from her head, and the blood pressure in the arteries of her chest and head decreases. Sensory receptors detect the decrease in blood pressure, so the frequency of action potentials in sensory neurons decreases. The brain, in turn, activates sympathetic reflexes that cause blood vessels to constrict. Consequently, the blood pressure in her neck and head does not fall dramatically.

Chapter 17

2. a. The major clue we are given is that Josie has low thyroid hormone levels, yet her TSH levels (the hormone that stimulates thyroid hormone release) are very high. Normally, thyroid hormone would negatively feedback at the anterior pituitary to inhibit TSH secretion once thyroid hormone levels had reached their set point. But since Josie's thyroid is unable to produce thyroid hormone, there has been a loss of negative feedback, and TSH never receives the signal to shut off, so the TSH levels rise above normal.

b. Usually, tumors create more cells than normal that contribute to the higher hormone levels. Therefore, if Josie had more thyroid hormone–secreting cells than normal, as with cancer, her thyroid hormone levels would be much higher than normal, not lower than normal as the tests indicated.

c. The thyroid gland requires iodine to produce thyroid hormones because iodine is an integral part of the hormone molecule. Thus, lack of iodine prevented the synthesis of thyroid hormones and the lack of negative feedback on the anterior pituitary gland. Thus, TSH levels increased, overstimulated the thyroid gland, and caused it to grow larger than normal.

d. By providing the thyroid gland with iodine, it was able to resume normal production of thyroid hormone. Once thyroid hormone levels reached their set point, they would negatively feedback at the anterior pituitary to inhibit TSH release, and its levels would return to normal.

e. Josie will probably need to continue taking thyroid hormone supplements for sometime. After a large portion of the thyroid gland was destroyed by 131I, its ability to secrete T3 and T4 would fall below their normal range of values unless Josie took supplemental thyroid hormone. Theoretically, her doctor could slowly reduce her supplement, which could cause her thyroid to grow to compensate for the reduced intake levels.

3. The question gives us two very important pieces of information: (1) Estrogen stimulates an increase in progesterone receptor number in the uterus and (2) progesterone action (binding to its receptors) is required for pregnancy. Thus, if too little estrogen is secreted, the up-regulation of receptors in the uterus for progesterone cannot occur. As a result, progesterone cannot prepare the uterus for the embryo to attach to its wall following ovulation, and pregnancy cannot occur. Because of the lack of up-regulation of progesterone receptors, the uterus cannot respond adequately to progesterone. If fewer than normal progesterone receptors are present, then a much larger than normal amount of progesterone is needed to produce its normal response.

4. The first piece of information we need to answer this question is given to us: The hormones that prevent asthma attacks work through cAMP mechanisms. The question then asks us to describe various cAMP-based mechanisms of hormone action. A drug can increase the cAMP concentration in a cell by stimulating its synthesis or by inhibiting its breakdown. A drug that binds to a receptor that increases adenylate cyclase activity increases cAMP synthesis. Because phosphodiesterase normally causes the breakdown of cAMP, an inhibitor of phosphodiesterase decreases the rate of cAMP breakdown and

causes cAMP to increase in the smooth muscle cells of the airway, producing relaxation.

5. This question highlights the idea of form follows function. Each class of hormones is best suited for a particular type of response. Nuclear receptor mechanisms result in the synthesis of new proteins that exist within the cell for a considerable amount of time. Thus, nuclear receptors are better adapted for mediating responses that last a relatively long time (i.e., for many minutes, hours, or longer). On the other hand, membrane-bound receptors that increase the synthesis of intracellular mediators, such as cAMP, normally activate enzymes already in the cytoplasm of the cell for shorter periods. The synthesis of cAMP occurs quickly, but the duration is short because cAMP is broken down quickly, and the activated enzymes are then deactivated. Therefore, membrane-bound receptor mechanisms are better adapted to short-term and rapid responses. In short, nuclear receptor mechanisms usually result in new proteins, whereas membrane-bound receptors usually activate already existing proteins.

Chapter 18

2. Secretions from the posterior pituitary are sometimes called neurohormones because the posterior pituitary is continuous with the brain and an extension of the nervous system. The neurons that secrete these hormones have their cell bodies in the hypothalamus of the brain, and their axons extend into the posterior pituitary. If the posterior pituitary is removed, the distal ends of the axons of the neurons that have their cell bodies in the hypothalamus are removed. The neurons survive, and after a few days, the proximal ends of the axons become capable of secreting the neuropeptides that the neurons normally secrete. Cells located entirely within the gland secrete the hormones of the anterior pituitary. They are released in response to neurohormones that travel through the hypothalamo-hypophysial portal system from the hypothalamus to the anterior pituitary. Removal of the anterior pituitary removes the cells that synthesize and secrete anterior pituitary hormones. Consequently, there is a permanent decrease in anterior pituitary hormones.
3. The beer Luke drank contained a substantial amount of water, as well as some alcohol. The water in the beer increased Luke's blood volume and reduced his blood osmolality, which caused ADH secretion from the posterior pituitary gland to decrease. In addition, the alcohol in the beer caused the rate of ADH secretion to decrease even further. The decreased ADH secretion caused the kidneys to increase the volume of dilute urine, which resulted in reduced blood volume with an increased osmolality. By morning, he was somewhat dehydrated, and that caused him to be thirsty.
4. Recall that growth hormone (GH) targets cartilage in the epiphyseal plate of long bones and stimulates cell division of the chondrocytes. Administering GH to young people before the growth of their long bones is complete causes their long bones to grow. While Mr. Hoops's son is actively growing, GH administration would cause him to grow taller. To accomplish this, GH has to be administered over a considerable length of time. However, there could also be unwanted changes consistent with acromegaly (oversecretion of GH). Other side effects, such as abnormal joint formation and diabetes mellitus, are also possible. In addition to undesirable changes in the skeleton, nerves frequently are compressed as a result of the proliferation of connective tissue. Because GH spares glucose usage, chronic hyperglycemia results, frequently leading to diabetes mellitus and severe atherosclerosis. Thus, Mr. Hoops's doctor would not prescribe GH.
5. Remember that hormones of the hypothalamus and anterior pituitary regulate thyroid hormones. TRH from the hypothalamus stimulates the secretion of TSH from the anterior pituitary. TSH travels to the thyroid to promote secretion of thyroid hormones. In response to the large amount of the abnormal TSH-like immunoglobulin (a protein), thyroid hormones are oversecreted by the thyroid gland (hyperthyroidism). Thus, Becky has a type of hyperthyroidism called Graves disease. Because production of the protein cannot be inhibited by thyroid hormones as TSH is, oversecretion of the thyroid gland is prolonged, and the symptoms associated with hyperthyroidism become obvious: her increased appetite and weight loss, as well as her other symptoms. Because TSH stimulates gland growth as well as hormone secretion, the thyroid gland enlarges. In addition, the increased thyroid hormones have a negative-feedback effect on the hypothalamus and pituitary gland. TRH secretion from the hypothalamus and TSH secretion from the anterior pituitary gland are inhibited.
6. Removal of the thyroid gland means that the tissue responsible for thyroid hormone (T_3 and T_4) secretion from thyroid follicles, and calcitonin secretion from parafollicular cells is gone. However, blood levels of Ca^{2+} remain within their normal range. Thyroid hormone does not affect Ca^{2+} levels and calcitonin is not essential for the maintenance of normal blood Ca^{2+} levels. However, recall that the four parathyroid glands are embedded in thyroid tissue on the posterior side of the thyroid gland. Although thyroid removal surgery is generally very safe, there is some risk of removing or damaging the parathyroid glands, especially in young children. Removal of the parathyroid gland eliminates PTH secretion. Without PTH, blood levels of Ca^{2+} fall. Recall from chapter 9 that, when the blood levels of Ca^{2+} fall below normal, the permeability of nerve and muscle cells to Na^+ increases. As a consequence, spontaneous action potentials are produced, causing tetanus of muscles. Death can result from tetany of respiratory muscles.
7. To answer this question, you must first realize that, because cortisone mimics the normal adrenal cortex hormone, cortisol, it will also act by negative feedback at the anterior pituitary. Cortisone inhibits ACTH secretion from the anterior pituitary. Second, it is important to understand that ACTH is required to prevent atrophy of the adrenal cortex. In the absence of ACTH, the adrenal cortex shrinks and may never recover to produce its normal secretions, even if ACTH secretion increases again.
8. a. Ethan's symptoms are consistent with those commonly found in people suffering from Cushing syndrome. Elevated blood glucose levels and reduced blood K^+ levels are also consistent with Cushing syndrome. Cortisol increases gluconeogenesis and decreases the use of glucose in metabolism. Therefore, blood glucose levels increase. Increased blood glucose levels stimulate insulin secretion, and insulin causes glucose to be taken up by adipose cells, where they convert glucose to fat. The decrease in muscle mass occurs because cortisol decreases protein synthesis and increases protein breakdown. Reduced blood levels of K^+ result from an increased rate of K^+ transport by the renal tubules and their elimination in the urine.
   b. Ethan's blood cortisol levels were very high, and his blood ACTH levels were very low. Blood CRH levels are not normally measured. ACTH from the anterior pituitary normally stimulates cortisol secretion from the adrenal glands. Because cortisol has a negative-feedback effect on the rate of ACTH secretion, and because the blood ACTH levels were low, the high blood levels of cortisol could not be due to elevated ACTH secretion from the anterior pituitary. A

possible cause of increased cortisol secretion is an adrenal gland tumor, which was confirmed by imaging techniques.

c. After the surgical removal of Ethan's left adrenal gland, the blood levels of cortisol decreased because the tumor was the source. As the blood cortisol levels decreased to normal, the rate of ACTH secretion from the anterior pituitary gland increased. Blood cortisol levels can be within the normal range of values if production of cortisol by the right adrenal gland compensates for the missing left adrenal gland. If cortisol production by the remaining adrenal gland is not adequate, additional cortisol will have to be administered.

d. If Ethan's high blood cortisol had been due to a hormone-secreting tumor of the anterior pituitary gland, blood levels of both ACTH and cortisol would have been elevated. There was no evidence that cancer was involved but, if cancer tissue was the source of ACTH, blood ACTH and cortisol levels would have increased.

9. Recall from chapter 16 that the parasympathetic nervous system is the "rest-and-digest" portion of the autonomic nervous system. Therefore, an increase in insulin secretion in response to parasympathetic stimulation and gastrointestinal hormones is consistent with the maintenance of homeostasis because parasympathetic stimulation and increased gastrointestinal hormones result from conditions such as eating a meal. Insulin levels therefore increase just before large amounts of glucose and amino acids enter the blood. The elevated insulin levels prevent a large increase in blood glucose and the loss of glucose in the urine.

10. It is important to understand that a large meal, high in carbohydrates, will increase blood glucose as the nutrients are transported from the small intestine into the blood. The body cells need to take up the glucose across their plasmal membranes either to utilize it immediately or to store it for later use. Cells take up glucose only in the presence of insulin. Glucose serves as the stimulus for insulin secretion and the inhibitor of glucagon secretion. Thus, after eating, insulin levels increase and glucagon levels decrease. Long before 12 hours has passed after a meal, blood glucose levels drop. Again, it is glucose that serves as the signal for hormone secretion. Now the reduction in blood glucose stimulates glucagon secretion and inhibits insulin secretion. Glucagon then promotes release of stored glucose into the blood to be utilized by cells until the next meal. In response to a meal high in protein but low in carbohydrates, insulin secretion is only slightly elevated, but glucagon secretion is increased. Insulin secretion is stimulated by the parasympathetic system and an increase in blood amino acid levels. Glucagon is stimulated by low blood glucose levels and by some amino acids. The lower insulin secretion causes some increase in the rate of glucose uptake and amino acid uptake, but the rate of uptake is not great enough to cause blood glucose levels to fall below normal values. Glucagon also causes glucose to be released from the liver. During exercise, sympathetic stimulation inhibits insulin secretion. As blood glucose levels decline, enhanced glucagon secretion occurs. The lower rate of insulin secretion reduces the rate at which tissues, such as skeletal muscle, take up glucose. Thus, during exercise, muscle depends on intracellular glycogen and fatty acids for energy, and blood glucose levels are maintained within the normal range of values. The actions of glucagon prevent glucose levels from dropping too much.

11. Recall from chapter 16 that the sympathetic nervous system is activated during intense exercise and inhibits insulin secretion. Blood glucose levels are not high because skeletal muscle tissue continues to take up some glucose and metabolize it. In chapter 9, we learned that skeletal muscles store glucose as glycogen, and muscle contraction depends on glucose and fatty acid metabolism. During a long run, muscle glycogen levels are depleted as the cells metabolize more and more glucose. The "kick" at the end of a race results from increased energy produced by anaerobic respiration, which uses glucose or glycogen as an energy source. However, because blood glucose levels and glycogen levels are low at the end of a long-distance race, the runners' source of energy is insufficient for the greatly increased muscle activity a sprint to the finish would require.

12. Removal of part of the thyroid gland reduces the amount of thyroid hormone secreted by the gland. Usually, enough thyroid tissue can be removed to reduce secretion of thyroid hormone to within the normal range. In addition, the remaining thyroid tissue normally does not enlarge enough to produce more thyroid hormone, although there are exceptions. The removal of the thyroid tissue does not remove the influence of the abnormal antibodies on the tissues behind the eyes. Thus, in many cases the condition's effect on the eyes is not improved.

Chapter 19

2. The fetus relies solely on the maternal blood supply for oxygen. Fetal hemoglobin must be more effective at binding oxygen than adult hemoglobin in the maternal blood, so that the fetal circulation can draw the needed oxygen away from the maternal circulation. If the hemoglobin of the maternal blood had an equal or higher oxygen affinity, the hemoglobin of fetal blood would not be able to draw away the required oxygen, and the fetus would die.

3. We learned in this chapter that reticulocytes are immature red blood cells released from the red bone marrow into the circulation. Consider that, by donating a unit of blood, Juan's red blood cell count dropped below normal, so we would expect his body to respond by producing more red blood cells. Erythropoietin secretion from the kidneys stimulates erythropoiesis, red blood cell production, in the red bone marrow to increase. As a result, we would expect Juan's reticulocyte count to increase during the week after he donated blood. This process continues at the increased rate until Juan's normal red blood cell count is reestablished.

4. To answer this question, we first need to determine the effect carbon monoxide has on red blood cells. Recall that carbon monoxide binds very readily to hemoglobin and does not tend to unbind. Therefore, carbon monoxide-bound hemoglobin in red blood cells can no longer transport oxygen. This essentially leads to a decrease in blood oxygen levels. We also learned that low blood oxygen levels stimulate red blood cell production by causing the release of erythropoietin, primarily by the kidneys. Erythropoietin stimulates red blood cell production in the red bone marrow. Thus, we can predict that the number of red blood cells in the person's blood would increase.

5. Referring to table 19.2 will help identify each of the white blood cells. It is also helpful to compare the size of each cell with red blood cells as well. In each figure, red blood cells are visible for easy comparison. (a) The cell in this figure is slightly larger than red blood cells and has a large, round nucleus with a small amount of cytoplasm. This cell is a lymphocyte. (b) The cell in this figure is larger than a red blood cell and has many bluish-purple granules. This cell is a basophil. (c) The cell in this figure is very large and has a kidney-shaped nucleus. It is a monocyte. (d) The cell in this figure has a three-lobed nucleus. It is most likely a neutrophil. (e) The cell in this figure has a bi-lobed nucleus and many bright red-stained granules. It is an eosinophil.

6. Platelets are normally inactive; however, they become activated at sites of tissue injury.

These are the areas where clot formation is needed to stop bleeding.

7. Before answering this question, consider that Cedric had two different drugs in his system that affect blood clotting. Both the aspirin and t-PA contributed to his rapid improvement. We learned in this chapter that aspirin inhibits thromboxane synthesis and consequently slows down the rate at which clotting occurs. Since Cedric had taken aspirin that morning, we can assume that his condition could have been worse if he had not done so. The t-PA increases the conversion of plasminogen to plasmin, which breaks down the clot. The t-PA acted quickly to remove the cardiac thrombosis before major damage to his heart occurred.
8. a. We learned that HDN is caused by an Rh incompatibility between a pregnant mother with Rh-negative blood and her Rh-positive child. If the mother is sensitized to the Rh antigen, she can produce anti-Rh antibodies that cross the placenta and cause agglutination and hemolysis of fetal red blood cells. A transfusion would replace the red blood cells lost by agglutination and hemolysis.
   b. Remember that an exchange transfusion is different than just a transfusion. In a transfusion, donor blood is introduced to the recipient, but in an exchange transfusion, the recipient's blood is replaced, meaning that as the donor blood is introduced into the system the recipient's blood is removed. An exchange transfusion not only increases the number of red blood cells but also decreases bilirubin and anti-Rh antibody levels by removing them in the withdrawn blood. The presence of fewer anti-Rh antibodies decreases the agglutination and lysis of red blood cells.
   c. Remember that erythropoietin stimulates red blood cell production. Even though new red blood cells were introduced during the exchange transfusion, administering erythropoietin will also increase Billy's red blood cell count.
   d. Erythropoietin levels increase as a result of low blood oxygen levels. This is directly related to the red blood cell count. Considering that Billy's red blood cell count was extremely low as a result of HDN, we would predict that his erythropoietin level would be higher than a fetus without HDN.
   e. A major physiological change that occurred after birth was that Billy was able to breathe on his own. The ability to oxygenate the blood using the lungs is greater than the ability to oxygenate the blood across the placenta. Billy's blood oxygen levels increased and his erythropoietin levels decreased. Thus, his production of red blood cells decreased and his anemia got worse.
   f. To treat HDN with an exchange transfusion, the donor's blood should be Rh-negative, even though the newborn is Rh-positive. Rh-negative red blood cells do not have Rh antigens. Therefore, any anti-Rh antibodies in the newborn's blood do not react with the transfused Rh-negative red blood cells. Eventually, all of the Rh-negative red blood cells die, and only Rh-positive red blood cells are produced by the newborn.
   g. Remember that blood types are genetically determined. Giving an Rh-postive newborn a transfusion of Rh-negative blood would not change the newborn's blood type. Even though Rh-negative blood is introduced to the body, over time all of the Rh-negative red blood cells die. The new red blood cells are produced from the bone marrow stem cells, which are genetically Rh-positive. Thus, only Rh-positive red blood cells are produced by the newborn.
9. After reading the question, we know to focus on blood tests that are associated with identification of bacterial infections. White blood cells defend the body against pathogens. A white blood differential cell count would be useful and would show an abnormally high neutrophil percentage because they increase during bacterial infections.

Chapter 20

2. The Clinical Impact, "Pericarditis and Cardiac Tamponade," explains that pericarditis is an inflammation of the serous pericardium. In Tony's case, the pericarditis is a result of a viral infection. Pericarditis can lead to fluid accumulation in the pericardial sac. Tony's pain results from inflammation and distention of the pericardial membranes. His pulse was weak because the fluid in the pericardial sac compressed both the heart and the veins delivering blood to the heart; consequently, the heart could not fill with as much blood as normal. With less blood entering the heart, Tony's stroke volume, or volume of blood pumped from the heart, also decreased, causing a weak pulse and lower blood pressure. Recall from chapter 16 that increased sympathethic stimulation of the heart occurs when blood pressure begins to decrease, resulting in tachycardia. The increased heart rate helps maintain blood pressure. The jugular veins, which carry blood toward the heart, are distended because the accumulated fluid in the pericardial sac compresses the heart, preventing complete filling of the heart and reducing the flow of blood from the jugular veins toward the heart. Tony's physician used a needle to remove excess fluid from the pericardial space. Reviewing figure 20.2, we can see that the needle is most likely inserted in the fifth or sixth intercostal space near the sternum and penetrates the following body layers: the skin, the subcutaneous tissue, the intercostal muscles, the fibrous pericardium, and the parietal pericardium.
3. The SA node is the pacemaker of the heart. Action potentials develop at a normal rate in the SA node and are conducted to the AV node and stimulate atrial contraction. In the AV node, the action potentials are delayed. The action potentials are rapidly conducted through the ventricular muscle and cause the ventricles to contract. If the delay is too long, action potentials are produced in the SA node again before action potential conduction through the AV node is complete. If more than one action potential at a time is produced in the SA node before conduction of action potentials through the ventricles is complete, the atria contract at a greater frequency than the ventricles. The atria may contract two or three times for each time the ventricles contract.
4. Before answering the question, we need to define tetanic contraction. Recall from chapter 9 that a tetanic contraction is a sustained contraction in which the frequency of stimulation of the muscle is so rapid that no relaxation occurs. The purpose of cardiac muscle contractions is to pump blood through the circulation by contracting and relaxing in a repeated cycle. Tetanic contractions in cardiac muscle would interrupt the pumping action produced by the cycle of contraction and relaxation, and the blood flow would cease. Tetanic skeletal muscle contractions are important to maintain posture or to hold a limb in a specific position.
5. In section 20.3, "Anatomy of the Heart," we learn that the left ventricle has the thicker wall. The pressure produced by the left ventricle is much higher than the pressure produced by the right ventricle. It is important for each ventricle to pump the same amount of blood because, with two connected circulation loops (pulmonary circuit and systemic circuit), the volume of blood flowing into one must equal the volume of blood flowing into the other, so that one does not become overfilled at the expense of the other. For example, if the right ventricle pumps less blood than the left ventricle, blood accumulates in the systemic blood

vessels. If the left ventricle pumps less blood than the right ventricle, blood accumulates in the pulmonary blood vessels.

6. The pumping action of the heart depends on coordinated cycles of contraction and relaxation of the chambers of the heart, allowing them to fill and empty in the proper sequence. Fibrillation destroys the coordinated contractions, and the cardiac muscle loses its function as a pump. Recall that the ventricles are the primary pumping chambers of the heart and the atria function more as blood reservoirs. The pumping action of the ventricles is necessary to maintain normal blood flow to the tissues of the body at all times. Ventricular fibrillation results in death because of the heart's inability to pump blood. The pumping action of the atria is most important during exercise; therefore, atrial fibrillation does not destroy the ventricles' ability to pump blood.
7. Figure 20.19 will be very useful in answering this question. From the figure we can see that ventricular diastole occurs between the second heart sound of one cardiac cycle and the first heart sound of the next cardiac cycle. The second heart sound is produced as the aortic semilunar valve closes. The aortic pressure decreases from 95 mm Hg to approximately 80 mm Hg when the left ventricle contracts again. This causes the bicuspid valve to close, producing the first heart sound of the next cardiac cycle. Now, let us consider the pressure changes in the left atrium and ventricle. When the aortic semilunar valve closes (second heart sound), the pressure in the left ventricle decreases rapidly to nearly 0 mm Hg. As soon as the pressure decreases below the pressure in the left atrium, the bicuspid valve opens, and blood flows into the left ventricle. Pressure remains low in the left atrium and the left ventricle, with pressure in the left atrium slightly greater than in the left ventricle. Finally, pressure increases a few mm Hg when the left atrium contracts, causing additional blood to flow from the left atrium into the left ventricle. When the left ventricle begins to contract, the pressure in the left ventricle increases but, as soon as the pressure is greater in the left ventricle than in the left atrium, the bicuspid valve closes, causing the first heart sound of the next cardiac cycle.
8. We learned in chapter 16 that sympathetic stimulation increases heart rate. If venous return remains constant, stroke volume decreases as the number of beats per minute increases. An increased heart rate and stroke volume means the heart is doing more work, which means the cardiac tissue requires more oxygen for cellular respiration. Dilation of the cardiac blood vessels is important because it allows for a greater blood supply to carry more oxygen to the cardiac tissue.
9. a. In the Clinical Impact, "Abnormal Heart Sounds," we learned that a stenosed valve has an abnormally narrow opening. The stenosed valve makes it more difficult for contractions of the left ventricle to force blood into the aorta. The stroke volume decreases because the left ventricle is unable to overcome the increased resistance to blood flow through the narrowed valve.
   b. Because of the increased resistance caused by the stenosed aortic semilunar valve, the left ventricle had to generate a greater force to cause blood to flow into the aorta. Just as skeletal muscle hypertrophies in response to the resistance of moving heavy weights, cardiac muscle hypertrophies in response to the greater resistance to blood flow through the stenosed valve.
   c. The mean arterial pressure is equal to cardiac output × peripheral resistance. Recall that cardiac output is equal to stroke volume × heart rate. We described earlier how a stenosed valve causes a decrease in stroke volume, which in turn causes a decrease in mean arterial pressure, unless there is a compensating increase in heart rate or peripheral resistance.
   d. When Norma rises to a standing position, gravity affects blood in the arteries and veins, and the blood tends to settle in the blood vessels below the heart. Consequently, there is decreased venous return to the heart, which results in decreased cardiac output (the Starling law of the heart). Decreased cardiac output results in decreased blood pressure, and therefore decreased blood flow to the brain, causing dizziness.
   e. The baroreceptor reflexes detect changes in blood pressure and lead to changes in heart rate and the force of contraction to maintain homeostatic blood pressure levels. We have already predicted that Norma's blood pressure decreases when she stands. This activates the baroreceptor reflexes. From figure 20.23, we see that her heart rate would increase. Although Norma has a normal baroreceptor reflex, the heart is not able to pump enough blood past the stenosed valve to increase stroke volume and maintain blood pressure.
   f. After reviewing Clinical Impact, "Angina, Infarctions, and the Treatment of Blocked Coronary Arteries," we know that angina pectoris results from inadequate blood delivery to heart muscle through the coronary circulation. Norma's stenosed valve has reduced blood pressure in the aorta, which reduces blood delivery to the body and to the coronary blood vessels. At rest, sufficient blood is delivered to the cardiac muscle, and there is no pain; however, when Norma starts exercising, the cardiac muscle requires more oxygen, which the blood cannot deliver through the stenosed valve. This results in an increase in anaerobic respiration and a decrease in pH, which is responsible for the pain of angina pectoris.
10. Rupture of the left ventricle can occur several days after a myocardial infarction. As the necrotic tissues are being removed by macrophages, the wall of the ventricle becomes thinner and may bulge during systole. If the wall of the ventricle becomes very thin before new connective tissue is deposited, it can rupture. If the left ventricle ruptures, blood flows from the left ventricle into the pericardial sac, resulting in cardiac tamponade. As blood fills the pericardial sac, it compresses the ventricle from the outside. As a consequence, the ventricle is not able to fill with blood, and its pumping ability is rapidly eliminated. Rupture of the left ventricular wall, therefore, quickly results in death.

Chapter 21

2. As stated in the question, atherosclerosis of vessels occurs when lipid deposits block the vessels, which results in reduced blood flow through the vessel. The tissues to which the blocked vessels supply blood will therefore have reduced oxygen and nutrients. Since the carotid artery supplies blood to the brain, we would expect atherosclerosis of these vessels to lead to reduced brain function, which may include confusion and loss of memory.
3. a. We learned in the section "Blood Flow and Poiseuille's Law" that a small change in the diameter of a vessel dramatically changes the resistance to flow and therefore the amount of blood that flows through it. Vasoconstriction of blood vessels in the skin in response to cold exposure causes a reduction in blood flow through the skin because the blood vessel diameter decreases and the resistance to blood flow increases. The skin will appear pale due to the lower blood flow.
   b. Just like the answer to part a, this question focuses on the relationship between blood vessel diameter and blood flow (Poiseuille's law). Vasodilation of blood

vessels in the skin in response to elevated body temperature would cause an increase in blood flow through the skin because the blood vessel diameter increases and the resistance to blood flow decreases. The skin will appear flushed due to the higher blood flow.

c. Poiseuille's law also states that the viscosity of blood determines resistance to blood flow. As viscosity of blood increases, blood flow decreases. Erythrocytosis results in an increase in red blood cell numbers, which also increases the viscosity of blood. We would therefore expect that erythrocytosis decreases blood flow.

4. Vasoconstriction of Richard's cutaneous blood vessels would cause a substantial increase in his systemic blood pressure. From Laplace's law, we learned that, as the diameter of a vessel increases, the force applied to the vessel wall increases. An aneurysm is a bulge that forms in an arterial wall that has weakened. The increase in Richard's blood pressure due to the sudden vasoconstriction of his cutaneous blood vessels would apply more force on the aneurysm. Because this is a weakened area of the artery wall, the increase in force applied to the aneurysm is more likely to rupture the wall. The rupture of an arterial wall is life threatening.

5. Arterial blood pressure can increase substantially without resulting in edema. Recall that the precapillary sphincters constrict to match blood flow into capillary beds with the metabolic needs of tissues. Thus, the capillary blood pressure does not change substantially, even if blood pressure increases to high levels. Blood pressure must increase above approximately 175 mm Hg before edema results. In contrast, a small increase in venous pressure leads to edema because there is not a sphincter muscle that protects the capillary from an increase in pressure due to resistance of blood flow out of the capillary bed. If venous pressure increases, the blood pressure in the capillaries must increase also to overcome the resistance to blood flow and maintain the normal movement of blood. The increase in capillary pressure changes the net filtration pressure, increasing the amount of fluid that flows from the capillary into the tissues and reducing the amount of interstitial fluid that enters the circulation, therefore resulting in edema.

6. a. The blocked vein in Harry's right leg caused edema and led to tissue ischemia. Edema developed inferior to the blocked vein. Blockage of the vein increased the capillary hydrostatic pressure in the capillary beds drained by the blocked vein. The increased capillary hydrostatic pressure increased the amount of fluid that flowed from the capillaries into the tissue spaces and reduced the amount of fluid that returned to the capillaries. Consequently, fluid accumulated in the tissue spaces and caused edema (see figure 21.36). The ischemia resulted in pain, much the way ischemia of the heart causes pain during myocardial infarctions (see chapter 20).

b. Emboli that originated in the posterior tibial vein would pass through the following parts of the circulatory system before lodging in the pulmonary arteries of the lungs: posterior tibial vein, popliteal vein, femoral vein, external iliac vein, common iliac vein, inferior vena cava, right atrium, right ventricle, pulmonary trunk, right or left pulmonary artery. Emboli lodge in branches of the pulmonary arteries and are most likely to lodge in the lungs because the pulmonary arteries branch many times before delivering blood to the pulmonary capillaries, and as they branch their diameters decrease. Even small emboli eventually lodge in the smaller branches of the pulmonary arteries. The other parts of the circulatory system through which the emboli pass have much larger diameters, so emboli can pass readily through them.

c. When emboli are large enough or numerous enough to block blood flow through a significant part of the lungs, resistance to blood flow through the lungs increases. The increased resistance increases the pulmonary venous pressure, which increases the afterload for the right ventricle. If the right ventricle is unable to overcome the increased afterload, failure of the right side of the heart can occur, and blood flow through the lungs is reduced.

d. First let us address the effect of pulmonary emboli on blood oxygen levels. Pulmonary emboli large enough to significantly reduce blood flow through the lungs reduce the lungs' ability to carry out gas exchange with blood, and blood oxygen levels decrease. Next, let us address the affect of pulmonary emboli on the left ventricle's ability to pump blood. Pulmonary emboli block normal blood flow through the lungs, thereby reducing the blood volume returning to the left ventricle. This would also cause a reduction in cardiac output. Recall that cardiac output also affects blood pressure, so we would expect hypotension, or a reduction in blood pressure, to occur. As blood pressure falls, the homeostatic mechanisms described in figure 21.40 are activated to increase blood pressure to normal levels. Manifestations of hypotension, such as increased heart rate, weak pulse, and pallor, would be present.

e. Heparin and coumadin are anticoagulants, discussed in chapter 19. They are prescribed to slow down the rate of blood clot formation. Heparin must be administered intravenously, but coumadin can be taken orally, which makes home use possible. Harry's prothrombin time will be checked periodically to ensure that enough anticoagulant is administered to prevent enlargement of the thrombus in the deep vein of his leg and to prevent additional emboli from forming. Remember that enzymes naturally found in our blood break down coagulated blood. In Harry's case the clots are removed because the slower rate of coagulation allows the clots to be broken down faster than they can form.

7. In chapter 16, we learned that, during times of rest, sympathetic stimulation decreases. We can predict that, while this athlete is relaxing, the sympathetic stimulation of the arteries in her skeletal muscles, the arteries in her digestive system, and the large veins decreases. As a result, vasoconstriction increases in the arteries of her muscles, and vasodilation occurs in the blood vessels of her digestive system and in the large veins. Blood flow decreases to her skeletal muscles, which are less active and require less oxygen and nutrients. Blood flow increases in her digestive system, which is more active during times of rest and relaxation and therefore has a higher oxygen and nutrient demand. In addition, blood accumulates in the large veins as they undergo vasodilation. Consequently, venous return to the heart decreases, which is consistent with the reduced cardiac output.

8. During a headstand, gravity acts on the blood, causing it to settle in the vessels of the head and chest. Blood pressure in the area of the aortic arch and carotid sinus baroreceptors increases. Figure 21.40 illustrates how the baroreceptor reflexes regulate changes in blood pressure. The increased blood pressure activates the baroreceptor reflexes, increasing parasympathetic stimulation of the heart and decreasing sympathetic stimulation. Thus, the heart rate decreases. As blood from the periphery runs down to the heart, venous return increases, causing stroke volume to increase (Starling law of the heart). The

activated baroreceptor reflexes also decrease vasomotor tone, and some peripheral vasodilation can occur.

9. The rapid loss of a large volume of blood activates the mechanisms responsible for maintaining blood pressure. Recall that blood pressure regulatory mechanisms include the baroreceptor reflexes, chemoreceptor reflexes, and hormonal mechanisms. In response to a dramatic drop in blood pressure due to the rapid loss of a large volume of blood, the baroreceptor mechanism, adrenal medullary mechanism, and chemoreceptor mechanism increase the heart rate and result in vasoconstriction of blood vessels, especially in the skin and viscera. Angiotensin II is produced quickly, and it causes vasoconstriction and stimulates aldosterone secretion. Aldosterone, which requires up to 24 hours to become maximally active, increases water reabsorption from the kidneys and reduces the loss of water in the form of urine. Blood volume is therefore increased. All of these mechanisms increase blood pressure back to its normal value.

   If blood is lost over several hours, the decrease in blood pressure is not as dramatic as when blood loss occurs quickly. Consequently, mechanisms that respond to a rapid and large decrease in blood pressure are stimulated to a lesser degree. These include the chemoreceptor reflex, the vasopressin mechanism, and the adrenal medullary mechanism. The baroreceptor reflexes are most sensitive to sudden decreases in blood pressure, but the baroreceptor reflexes are still sensitive to decreases in blood pressure that occur over a period of several hours. The baroreceptor reflexes that trigger vasoconstriction in response to the blood loss are substantial. The kidneys detect even small decreases in blood volume. Consequently, the renin-angiotensin-aldosterone mechanism is activated and remains active until the blood pressure is returned to its normal range of values. Aldosterone secretion increases and, though it requires several hours to become maximally active, it continues to stimulate water reabsorption by the kidneys, increasing blood volume until the blood pressure returns to its normal range of values.

Chapter 22

2. a. Cancer cells can break free from a tumor and spread, or metastasize, to other parts of the body by entering lymphatic or blood capillaries (see chapter 3). If the cancer cells enter the lymphatic capillaries, they are carried to the lymph nodes, which filter lymph. The first lymph nodes in which cancer cells are likely to become trapped are the sentinel lymph nodes.
   b. Test results showed that all of the sentinel lymph nodes contained cancer cells, indicating that cancer cells had spread into her lymphatic system. Cindy's cancerous tumor is in her left breast, and the axillary lymph nodes drain the superficial thorax and upper limb. Removal of these lymph nodes minimizes the risk of further metastasis.
   c. Recall from chapter 21 approximately one-tenth of the fluid entering tissues is normally removed by the lymphatic system. Removal of the lymph nodes and their attached lymphatic vessels disrupts the normal removal of fluid from tissues, resulting in lymphedema, the accumulation of fluid in the tissues.
   d. Remember that contraction of skeletal muscle and the thoracic pressure changes associated with breathing are both mechanisms for moving lymph through lymphatic vessels. Exercise increases skeletal muscle contractions and breathing. Consequently, more fluid can enter lymphatic capillaries, reducing edema.
   e. The external force applied to tissues by a compression bandage or garment reduces the movement of fluid from the circulation into tissues (see chapter 21) thereby reducing lymphedema.
   f. Recall that contraction of lymphatic vessels and the presence of valves along these vessels ensure that lymph moves through lymphatic vessels in one direction, from the peripheral tissues toward the venous system of the thorax. A compression pump mimics the normal contraction and relaxation of the lymphatic vessels, moving the lymph in a distal-to-proximal direction.
3. Remember that the combination of a MHC class I molecule and an antigen is necessary to activate T cells. MHC molecules are genetically determined. Thus, unless mouse B is essentially genetically identical to mouse A, the T cells from mouse A that are introduced to mouse B will not respond to the antigen. The T cells are MHC-restricted, meaning they must interact with the MHC proteins of mouse A as well as the antigen from virus X, to be activated and respond.
4. We learned that a B cell phagocytizes and processes antigens that combine with MHC molecules on the surface of cells. Helper T cells interact with the MHC-antigen complex to stimulate the B cell to divide. The daughter cells then produce antibodies. If the antigens are eliminated, the stimulus for B-cell proliferation and antibody production is then removed.
5. The first exposure to the disease-causing agent (antigen) evokes a primary immune response, which destroys the existing pathogens but also produces memory cells that can respond to future infections. As time passes, the antibodies produced during the primary immune response will degrade, and memory cells will die. If, before all of the memory cells are eliminated, a second exposure to the antigen occurs, a secondary response results, increasing the number of antibodies and memory cells again. The newly produced memory cells can provide immunity until the next exposure to the antigen.
6. We learned in this chapter that helper T cells are necessary for the effective activation of cytotoxic T cells. With the depression of helper T cell activity, cytotoxic T cell activity is also reduced.
7. First, recall that vaccines are a form of artificial active immunity caused by deliberately introducing an antigen into the body, stimulating a primary response by the immune system. The immune system responds to the vaccine by increasing the number of specific memory cells and antibodies for the particular disease. This provides long-lasting immunity without disease symptoms. The booster shot stimulates a memory (secondary) response, resulting in the formation of even more memory cells and antibodies. A booster shot improves the effectiveness of the immune system's ability to fight the types of infections for which a person is vaccinated.
8. After reading the Systems Pathology, we learned that SLE is an autoimmune disorder in which self-antigens activate immune responses. Often, this results in the formation of immune complexes and inflammation. Sometimes antibodies bind to antigens on plasma membranes, resulting in the rupture of the plasma membranes. Purpura results from bleeding into the skin. One cause of purpura is thrombocytopenia, a condition in which the number of platelets is greatly reduced, resulting in decreased platelet plug formation and blood clotting (see chapter 19). Considering that SLE is an autoimmune disorder, we can predict that the purpura is the result of the production of antibodies that destroy platelets, causing thrombocytopenia.

Chapter 23

2. It could be that Jake was fibbing when he claimed to be unafraid, trying to impress his girlfriend. But whether he was frightened or not, in order for his body to produce such a sound, certain muscles need to

contract. The muscles of expiration cause airflow from the lungs to the exterior through the larynx. As air rushes past the vocal cords, they vibrate. If air is explosively pushed past the vocal cords, they vibrate to a greater degree than when air is pushed past them to a lesser degree, and they produce a louder sound. As for the high pitch of the scream, it is the position of the vocal cords that is important. More tension results in a higher frequency of vibration of the vocal cords, producing a higher-pitched sound. How does the tension increase? The arytenoid cartilages are rotated medially, which moves the vocal cords medially and posteriorly. Finally, the epiglottis-controlling muscles cause it to tip anteriorly so that air can flow past and out of the mouth. For Jake, the result was a high-pitched scream that either endeared him to his girlfriend or left him single.

3. We learned that the cartilage rings of the trachea are incomplete, C-shaped rings and that the esophagus lies in the groove of those rings along the posterior side of the trachea. When a large mouthful of food is swallowed, it stretches the esophagus. Because of the groove along the trachea, movement of food is encouraged because the esophagus can expand. In addition, the trachea can remain open, so that breathing can resume immediately upon swallowing.
4. In order to calculate alveolar ventilation, you need to use the following formula: $VA = f(VT - VD)$ [f = respiratory rate; VT = tidal volume; VD = dead space]. $VA = 12(500 - 150) = 4200$ mL/min. If the tidal volume, dead space, and respiratory rate all increase during exercise, the new alveolar ventilation would be $VA = 24(4000 - 300) = 88{,}800$ mL/min. This means that, during exercise, the person gets 84,600 mL/min more air exchange, ultimately increasing the amount of O_2 acquired during exercise when the O_2 demand is higher.
5. The air the diver is breathing has a greater total pressure than atmospheric pressure at sea level. Consequently, the partial pressure of each gas in the air increases. According to Henry's law, as the partial pressure of a gas increases, the amount (concentration) of gas dissolved in the liquid (e.g., body fluids) with which the gas is in contact increases. When the diver suddenly ascends, the partial pressure of gases in the body returns toward sea level barometric pressure. As a result, the amount (concentration) of gas that can be dissolved in body fluids suddenly decreases. When the fluids can no longer hold all the gas, gas bubbles form.
6. At high altitudes, the atmospheric PO_2 decreases because of lower atmospheric pressure. The decreased atmospheric PO_2 results in a decrease in alveolar PO_2 and less oxygen diffusion into lung tissue. If the person's arterioles are especially sensitive to the lower oxygen levels, constriction of the arterioles reduces blood flow through the lungs, and the ability to oxygenate blood decreases.
7. Remember that the oxygen-hemoglobin dissociation curve normally shifts to the right in tissues. The shift of the curve to the left caused by CO reduces hemoglobin's ability to release oxygen to tissues, which contributes to the detrimental effects of CO poisoning. In the lungs, the shift to the left may slightly increase the hemoglobin's ability to pick up oxygen, but this effect is offset by its decreased ability to release oxygen to tissues.
8. A person who cannot synthesize BPG has mild erythrocytosis because his or her hemoglobin releases less oxygen to tissues. Consequently, increased erythropoietin will probably be released from the kidneys, and increased red blood cell production will occur in red bone marrow.
9. When CO_2 moves from fetal blood into maternal blood, it increases CO_2 levels inside maternal red blood cells. As a result, pH inside maternal red blood cells decreases, the affinity of maternal hemoglobin for oxygen decreases, and more oxygen is released. In other words, the maternal oxygen-hemoglobin curve shifts to the right. Simultaneously, the movement of CO_2 from fetal blood into maternal blood decreases CO_2 levels inside fetal red blood cells and increases pH inside fetal red blood cells. This means that the affinity of fetal hemoglobin for oxygen increases, and more oxygen will bind to fetal hemoglobin. In other words, the fetal oxygen-hemoglobin dissociation curve shifts to the left. This shifting of the maternal and fetal oxygen-hemoglobin curves is called the double Bohr effect. The double Bohr effect increases the delivery of oxygen from maternal blood to fetal blood because maternal hemoglobin releases more oxygen and fetal hemoglobin is more effective at picking up that oxygen.
10. First, we know that hyperventilation is rapid, shallow breathing. Thus, as less air is inhaled with each breath and the rate of breathing goes up, blood levels of CO_2 decline, which increases blood pH. Recall that, as CO_2 is lost, H^+ and HCO_3^- combine to form H_2CO_3, which in turn dissociates to form CO_2 and H_20. The lowered H^+ levels cause an increase in blood pH. Remember that the pH scale is an inverse scale where low H^+ concentration results in high pH. In addition, because CO_2 levels provide the urge to breathe, a person could theoretically hold his or her breath longer after hyperventilating. Holding one's breath increases blood CO_2 levels and decreases blood pH. On the other hand, holding your breath results in a decrease in pH because CO_2 accumulates in the blood. The CO_2 combines with H_2O to form H_2CO_3, which dissociates to form H^+ and HCO_3^-. The elevation in H^+ levels causes a decrease in blood pH (high H^+ concentration results in low pH).
11. To answer this question, we need to consider what types of stimuli affect the respiratory center. Look at figure 23.21. Notice that receptors for temperature (e.g., cold water), pain (e.g., very cold water), and touch (e.g., swatting on the buttocks) all stimulate the respiratory center. That is why we gasp when we jump into a cold swimming pool—it is involuntary—and babies were swatted on the buttocks—to stimulate their first breath of air.
12. Resistance to airflow is caused by a buildup of mucus and increased inflammatory damage to the air passageways. It becomes difficult to expel air, so FEV_1 decreases. As air is trapped in the alveoli, residual volume gradually increases. Because air exchange in the alveoli is less than normal, physiological dead space increases.

Chapter 24

2. Four. Recall that the greater omentum is a localized mesentery, which consists of serous membranes. Each single layer of the mesentery has two layers of simple squamous epithelium. Since the greater omentum is folded back on itself, that results in four layers of simple squamous epithelium.
3. First, realize that moving the soft palate can close the opening between the nasopharynx and the oropharynx. Normally, during swallowing, the soft palate is elevated, closing off the nasopharynx so that liquids and food bypass the nasal cavity, pass through the oropharynx, and enter the esophagus. If a person laughs suddenly while drinking a liquid, the liquid may be explosively expelled from the mouth and even the nose because the soft palate relaxes during laughing. If a person tries to swallow and speak at the same time, choking is most likely to occur. Speaking requires that the epiglottis be elevated, so that air can pass out of the larynx. When the epiglottis is raised, food or liquid can pass into the larynx and choke the person.
4. Alice's physician concluded that the inflammation of her larynx was due to the

reflux of gastric fluid into her esophagus while she was sleeping. In general, reflux is more likely after eating a meal and while lying down. Gastric acid secretion increases after a meal mainly because of the cephalic and gastric phases of gastric secretion, and gravity normally helps keep gastric fluid in the stomach. During the night, the gastric fluid moved through the esophagus and entered her larynx. An antacid was prescribed to neutralize the low pH of the gastric secretions. Two classes of drugs decrease H^+ secretion. Both histamine receptor antagonists and proton pump inhibitors reduce the movement of H^+ into the lumen of the stomach, thereby increasing the pH of the gastric fluid secreted by the stomach mucosa. The histamine receptor antagonist binds to histamine receptors and blocks the action of histamine, and the proton pump inhibitor reduces the H^+–K^+ exchange pump. Either type could have been prescribed for Alice. The smaller volume of gastric acid secreted by the stomach mucosa and the higher pH of the stomach secretions reduce the acid reflux that cause inflammation of her larynx.

5. In order to completely understand why Amos's doctor prescribed an antibiotic, read the Clinical Impact, "Peptic Ulcer," in this chapter. There, you will find that about 80% of peptic ulcers are due to a specific bacterial infection. Thus, based on the physician's recommendations, the cause of Amos's condition is a bacterial infection (*Helicobacter pylori*). The effect of the antacid is to increase the pH of the chyme that enters the small intestine from the stomach, which reduces the inflammation and pain. Amos cannot control the condition with antacids alone because the antacids do not eliminate the bacterial infection. Prolonged inflammation can cause perforation of the duodenum. The ulceration penetrates the mucosa, submucosa, muscularis, and serosa, allowing the intestinal contents to exit the duodenum. This condition can result in peritonitis and hemorrhage.
6. First, let us consider that acidic chyme in the small intestine is the stimulus. Stimuli are detected by receptors, and then control centers send a signal to initiate a response that will regulate homeostasis. In this case, the control center is the pancreas. In response to the acidic chyme, pancreatic secretin stimulates bicarbonate ion secretion from the pancreas (the effector), which neutralizes the acidic chyme. Thus, secretin prevents the acid levels in the chyme from becoming too high and keeps them in the normal range. The neutralization of the acidic chyme removes the stimulus for more secretin release, and bicarbonate ion is no longer secreted. Because the response was inhibited, this is an example of a negative-feedback system.
7. An enema is the introduction of fluid into the rectum, which causes it to distend. Recall that the defecation reflex is initiated by the movement of feces into the rectum and the subsequent stretch of the rectal wall. Therefore, because an enema stretches the rectal wall, it initiates the defecation reflex.
8. Recall that diarrhea is either increased stool frequency or increased volume, which can result in the abnormal loss of fluid and ions from the colon. This fluid loss from the colon affects the cardiovascular system in the same way that blood loss does. In either case, the result is hypovolemia, which causes a drop in blood pressure in a positive-feedback cycle. Eventually, heart failure results from insufficient blood flow to the heart itself.

Chapter 25

2. Most of the vitamins, with the exception of A, D, and niacin, are essential vitamins, meaning they cannot be produced by the body but must be obtained from the diet. Recall that, after a vitamin is destroyed, its function is lost. If the vitamins were broken down by digestion before being absorbed, they would not be functional, and vitamin deficiencies would occur.
3. To answer this question we need to first identify the Daily Value for carbohydrates, which is 300 g/day. The % Daily Value is then determined by dividing the amount in the serving of food (30 g) by the Daily Value (300 g). The % Daily Value for carbohydrates from one serving of this food product is 10% (30/300 = 0.10, or 10%).
4. The last step in the electron-transport chain is when the electrons are passed to oxygen to form water. If this step is blocked, the citric acid and electron-transport chain cannot function, so ATP will not be produced aerobically. Lactic acid fermentation alone cannot produce sufficient levels of ATP to maintain normal cellular activity, and death will occur.
5. We can see from table 25.1 that one cola or beer has about 145 kcal/serving. To determine the time it takes to burn these kcal, we divide the kcal/serving by the number of kcal used per hour. Watching TV uses 95 kcal/h, so 145/95 is equal to about 1.5 hours, or 1 hour and 30 minutes. Jogging at a pace of 6 mph uses 580 kcal/h, so 145/580 is equal to 0.25 h, or 15 minutes.
6. Recall that catabolism of food releases energy that can be used by the body for normal biological work, such as muscle contraction. However, about 40% of the total energy released is actually used for biological work. The remaining energy is lost as heat. Exercise increases the amount of biological work and therefore requires more energy in the form of ATP. As more ATP is produced to fuel the exercise, more heat is also generated as lost energy, thereby increasing body temperature. Shivering consists of small, rapid muscle contractions that produce heat in an effort to prevent a decrease in body temperature in the cold.

Chapter 26

2. Even though hemoglobin is a smaller molecule than albumin, it does not normally enter the filtrate because hemoglobin is contained within red blood cells, and these cells cannot pass through the filtration membrane. However, if red blood cells rupture, by a process called hemolysis, the hemoglobin is released into the plasma, and large amounts of hemoglobin enter the filtrate. Conditions that cause red blood cells to rupture in the circulatory system allow large amounts of hemoglobin to enter the urine.
3. In this chapter we learned that the reabsorption of water from the nephron is based on osmosis. If Na^+ and Cl^- are not actively transported out of the nephron, the concentration of ions inside the nephron stays elevated. The normal osmotic gradient that moves water out of the nephron is greatly reduced and the water stays in the nephron, which increases urine volume.
4. Inhibition of ADH secretion is one of alcohol's numerous effects on the body. Lack of ADH causes the distal convoluted tubules and the collecting ducts to be relatively impermeable to water. Therefore, the water cannot move by osmosis from the distal nephrons and collecting ducts but remains in the nephrons to become urine. In addition, because other fluids are normally consumed with the alcohol, the increased water intake also results in an increase in dilute urine.
5. a. Decreased water reabsorption in the distal convoluted tubules and collecting ducts results in increased water loss in urine and a large urine volume. Additionally, because the infants' kidneys were unable to reabsorb water from the filtrate, the infants' blood osmolality increased.
   b. The infants' plasma levels of ADH increased during the water deprivation test. The hallmark of NDI is normal secretion of ADH from the pituitary

followed by an abnormal response in the kidneys.

c. Normally, following exposure to ADH, the cells of the distal convoluted tubules and collecting ducts insert aquaporin-2 water channels into the apical membranes, which increases their permeability to water. However, in autosomal recessive and autosomal dominant NDI, the aquaporin-2 is dysfunctional, and water stays in the collecting duct to exit the body as part of urine.

d. Plasma levels of ADH will range from abnormally low to nondetectable because central diabetes insipidus results from a deficiency in ADH secretion.

e. Although it seems counterintuitive to treat a patient already suffering from polyuria with a drug that increases urine output, treatment with a Na^+ reabsorption inhibitor in conjunction with a low-sodium diet ultimately promotes homeostatic water balance in patients' cell fluids. By preventing plasma levels of sodium from becoming abnormally elevated, water is not prone to move by osmosis out of cells' cytoplasm into the extracellular fluid toward the blood plasma.

6. During the race, Amanda's blood volume decreased because of increased water loss through breathing and sweating. Because of Amanda's severe dehydration, her body immediately initiated the mechanism that maintains blood pressure. The vasoconstriction and reduced blood volume explain her pallor. Increased sympathetic stimulation caused the more rapid heart rate. The renal arteries also became vasoconstricted. Subsequently, the GFR decreased, and urine production was reduced. In addition, Amanda's ADH levels increased, which increased water reabsorption from the distal convoluted tubules and collecting ducts.

7. The question asks what is the relevance of urea clearance. Low urea clearance indicates that the amount of blood cleared of urea, a metabolic waste product, per minute is lower than normal. It is consistent with the reduced number of functional nephrons characteristic of advanced cases of renal failure. In addition, low urea clearance indicates that the GFR is reduced and blood levels of urea are increasing. A low blood Na^+ level is consistent with the reduced ability of the smaller number of functional nephrons to reabsorb Na^+ at the normal rate. The high blood K^+ level is consistent with the reduced ability of the smaller number of functional nephrons to secrete K^+ at the normal rate.

8. Recall that the female urethra is much shorter than the male urethra and is more accessible to bacteria from the external environment. For this reason, females are more susceptible to bladder infections than males.

9. After 9 days Roger's kidneys began to produce a large volume of urine with larger-than-normal Na^+ and K^+ concentrations. As a result, Roger became dehydrated by day 9. Dehydration results in reduced blood volume and blood pressure. His hematocrit was increased because the volume of his blood was decreased, but there was no decrease in the number of red blood cells. The percentage of the blood made of red blood cells therefore increased. The pale skin was the result of vasoconstriction, which was triggered by the reduced blood pressure. Dizziness resulted from reduced blood flow to the brain when Roger tried to stand and walk. He was lethargic in part because of reduced blood volume, but also because of low blood levels of K^+ and Na^+, caused by the loss of these ions in the urine. Low blood levels of Na^+ and K^+ alter the electrical activity of nerve and muscle cells and result in muscular weakness. The arrhythmia of his heart was due to low blood levels of K^+ and increased sympathetic stimulation, which was also triggered by low blood pressure.

Chapter 27

2. The first step is to define hemorrhagic shock. Recall from the chapter 21 Clinical Impact, "Shock," that hemorrhagic shock is due to excessive internal or external bleeding, which lowers blood volume. During hemorrhagic shock, blood pressure decreases and visceral blood vessels constrict. As a consequence, blood flow to the kidneys and the blood pressure in the glomeruli decrease dramatically. The total filtration pressure decreases, as does the amount of filtrate formed each minute. The rate at which Na^+ enters the nephron therefore decreases. In addition, the kidneys secrete large amounts of renin, which causes the formation of angiotensin I from angiotensinogen. Angiotensin I is converted to angiotensin II, which stimulates aldosterone secretion. Aldosterone increases the rate at which Na^+ is reabsorbed from the filtrate in the distal convoluted tubules and collecting ducts.

3. Remember that acidic (low) pH is due to an elevated concentration of H^+. Hydrogen ions in the blood are derived from the combination of CO_2 and water. When hyperventilating, the faster breathing rate results in a greater than normal rate of CO_2 loss from the circulatory system. Because CO_2 levels decrease, fewer H^+ are formed and the pH becomes more basic. Breathing into a paper bag corrects for the effects of hyperventilation because the person re-breathes air that has a higher concentration of CO_2. Carbon dioxide levels increase in the body, more H^+ is formed, and pH levels drop back into the normal range.

4. Aldosterone hyposecretion results in acidosis. Aldosterone increases the rate at which Na^+ is reabsorbed from nephrons, but it also increases the rate at which K^+ and H^+ are secreted. Hyposecretion of aldosterone decreases the rate at which H^+ is secreted by the nephrons and therefore can result in acidosis.

5. a. Each time Dan vomited, he lost a substantial amount of fluid. Because he was very nauseated, he did not consume any food or liquid. Water crosses the wall of the stomach by osmosis. Secretion of acid by the gastric glands of Dan's stomach produced the acidic gastric fluid, and vomiting eliminated the fluid from the stomach. As a result of the lost fluid volume, Dan's body weight decreased. His body weight before he became nauseated was not known. The 71 kg body weight at 12 hours is probably less than his normal body weight. By 36 hours, Dan had lost an additional 3 kg. Because of the decrease in the plasma volume, Dan's blood pressure began to fall. The mechanisms that regulate blood pressure and blood volume were activated (see figure 27.4). The decrease in blood pressure and the increase in heart rate at 36 hours were a result of the continued fluid loss as vomiting continued. Because of the baroreceptors reflex (see figure 21.40), there is increased sympathetic stimulation of the heart and peripheral blood vessels. As a result, heart rate and pallor increase due to vasoconstriction of the blood vessels of the skin. It is likely that his heart rate was increased somewhat by 12 hours, but it was clearly increased by 36 hours. Although not reported, the ADH mechanism caused his urine volume to decrease dramatically. Thirst would have increased if Dan had not been so nauseated. At 12 hours, the mechanisms that control blood pressure were adequate to maintain blood pressure within its normal range. By 36 hours, Dan's blood pressure had decreased in spite of the homeostatic mechanisms.

b. The gastric glands of the stomach secrete H^+. The rate of H^+ secretion had to increase in response to nausea and vomiting because Dan expelled a significant volume of acidic gastric fluid each time he vomited. The increased H^+ secretion caused metabolic acidosis (see Clinical Impact, "Acidosis and Alkalosis"). At the same time H^+ was secreted, HCO_3^- was absorbed into the circulatory system (see figure 24.12). Therefore, the plasma HCO_3^- levels and plasma pH were increased at both 12 and 36 hours. The secretion of H^+ into the stomach and the absorption of HCO_3^- into the blood caused the plasma pH to increase. The mechanisms that control blood pH were activated (see figure 27.13). However, they were not adequate to keep the plasma pH within its normal range. The plasma pH had increased at 12 hours and increased further at 36 hours.

c. The plasma PCO_2 levels were increased at both 12 and 36 hours. As the pH of the body fluids rose, stimulation of the respiratory centers decreased (see figure 23.22). Consequently, CO_2 elimination slowed, and plasma CO_2 levels increased.

d. After Dan stopped vomiting, the rate of fluid loss decreased. The rate of H^+ loss and HCO_3^- absorption slowed dramatically. He was also able to consume liquids. These changes and the mechanisms that regulate body fluid volume and plasma pH continued to function until homeostasis was reestablished.

Chapter 28

2. a. In chapter 3, we learned that cancer cells result from several mutations that accumulate over many generations of cell division. This may require several years to develop. Cells that had accumulated some mutations leading to cancer might have existed in Vern's prostate at the time of the first biopsy, but actual cancer cells might not have been present. Also, a biopsy removes a small sample of cells from the tissue. If only a few cancer cells were present, it is very likely that they were missed in the first biopsy.

b. Similar to the answer for part a, cancer cells result from the accumulation of mutations over generations of cell division. Pathologists examine biopsies and make conclusions based on their observations of the microscopic anatomy of the cells in a tissue sample. Cells that may become cancerous in the future may not possess the anatomical features the pathologist would use to classify them as cancerous. This was likely the case in the examination of the first biopsy. After 8 years had passed, some of these cells had accumulated mutations that altered their appearance, and the pathologist examining the second biopsy identified these cells.

c. We can assume that the six biopsy samples were of different regions of Vern's prostate gland. It appears that the cancerous tumor is small and the needle biopsy only passed through the tumor two of the six times.

d. It is likely that mutations that cause cells to secrete increased amounts of PSA accumulated in some cells of the prostate, but all of the mutations that cause cells to become cancerous may not have been present. In this case, elevated PSA levels indicate the increased risk of tumor formation.

3. Sperm cells are deposited into the female reproductive system and need to move quite a distance to come in contact with the oocyte (see figure 28.20). Coagulation may help keep the sperm cells together within the female reproductive tract, thereby increasing the likelihood of fertilization.

4. Erection is the result of neural stimulation of arteries that supply the erectile tissue. Erectile dysfunction (ED) is caused by either defective stimulation of the erectile tissue by nerve fibers or reduced response of the blood vessels to neural stimulation. Recall from Clinical Impact, "Erectile Dysfunction," that sildenafil does not stimulate erection, but instead increases the effectiveness of stimulation by enhancing the response of blood vessels to action potentials by slowly breaking down cGMP. As cGMP accumulates in the smooth muscle cells of the blood vessels, they relax, allowing blood to flow into the erectile tissue. Mr. Grover is sometimes able to achieve erections, especially after treatment, so it is reasonable to assume that he is unable to achieve erections at other times because of dysfunction in neural stimulation. There could be fewer action potentials reaching the erectile tissue due to nerve damage or decreased ability of the central nervous system to increase action potentials in the nerves. Another possibility is that Mr. Grover is unable to achieve minimal sexual excitement, which would decrease action potentials in the appropriate nerves.

5. The question addresses the time period in the menstrual cycle just before the LH surge, which promotes ovulation. Referring to figure 28.17, it is evident that estrogen and progesterone are normally at their lowest levels before the LH surge. In contrast, progesterone is at its highest level after ovulation and prevents further development of follicles. Therefore, administration of a large amount of progesterone and estrogen just before the preovulatory LH surge inhibits the release of GnRH, LH, and FSH. Consequently, ovulation does not occur. However, progesterone is the more potent hormone when it comes to inhibiting ovulation. Injections of a small amount of estrogen just before ovulation could stimulate GnRH, LH, and FSH secretion with little negative effect on ovulation. Continual administration of high concentrations of GnRH causes the anterior pituitary cells to become insensitive to GnRH. Thus, LH and FSH levels remain low and the ovarian cycle stops.

6. Recall that progesterone has an inhibitory effect on the hypothalamus and anterior pituitary (see figure 28.18). High progesterone levels after menses would therefore inhibit GnRH secretion from the hypothalamus and consequently FSH and LH secretion from the anterior pituitary. Without FSH and LH, the events of the ovarian cycle, including estrogen production, are inhibited. Remember that the proliferation of the endometrium is stimulated by estrogen. If estrogen levels do not increase following menses, the endometrium does not thicken. In addition, estrogen normally increases the synthesis of uterine progesterone receptors, so without estrogen the secretory response of the endometrium to the elevated progesterone would be inhibited.

7. Molly's mother could have had leiomyomas also, although, without direct data of medical examinations, one cannot be certain. If that was the cause of her irregular menstruations, they may have become less frequent as Molly's mother experienced menopause. During menopause, the uterus gradually becomes smaller, and eventually the cyclical changes in the endometrial lining cease. If the condition was relatively mild, the onset of menopause could explain the gradual disappearance of the irregular and prolonged menstruations. (Note: If the tumors are large, constant, and severe, menstruations are likely even if regular menstrual cycles stop due to menopause.)

Chapter 29

2. The primitive streak essentially forms the central axis of an embryo. If two primitive streaks formed in one embryonic disk, we would expect two different embryos, or essentially twins, to develop. If the two primitive streaks were touching each other, conjoined twins would develop. The degree to which the two primitive streaks were touching would determine the severity of the attachment.

3. Recall that the function of müllerian-inhibiting hormone is to prevent the development of the paramesonephric

duct system, which gives rise to the female internal reproductive system. The lack of müllerian-inhibiting hormone would therefore allow the paramesonephric duct system to develop, particularly the uterus and uterine tubes. Assuming that testosterone secretion is normal, the internal male reproductive system would also develop. In the presence of dihydrotestosterone, derived from testosterone, the external male genitalia would also develop.

4. Recall that clinical age is dependent on LMP (last menstrual period) of the mother and that developmental age begins at fertilization, which is assumed to occur on day 14 after LMP. Most of the times reported in the text are developmental age. To determine the clinical age, add 14 to the developmental age. The one exception is parturition, which is reported as clinical age. To determine the developmental age, subtract 14 from the clinical age. We can easily construct a table to compare the ages:

	Clinical Age	**Developmental Age**
Fertilization	14 days	0 days
Implantation	21 days	7 days
Beginning of fetal period	70 days	56 days
Parturition	280 days	266 days

5. The doctor would not do as the woman requested because elevated calcium levels could lead to tetanic uterine muscle contractions. Tetanic contractions might compress blood vessels and cut the blood supply to the fetus. Recall from chapter 27 that hypercalcemia can also result in heart arrythmias and skeletal muscle weakness. Instead of calcium, the doctor may administer oxytocin, which strengthens uterine contractions and is less likely to produce tetany.
6. Suckling causes a reflex release of oxytocin from the mother's posterior pituitary. Oxytocin causes expulsion of milk from the breast, but it also causes contraction of the uterus. Contraction of the uterus is responsible for the sensation of cramps in her abdomen.
7. Genotypes are the alleles a person has for a given trait, and the phenotype is the person's appearance. In the case of polydactyly, there are three possible genotypes: *DD* (homozygous dominant), *Dd* (heterozygous), and *dd* (homozygous recessive). Since polydactyly is a dominant trait, we would expect the individuals with genotype *DD* or *Dd* to exhibit polydactyly and the individuals with the genotype *dd* not to exhibit polydactyly.
8. When fertilization occurs, the two haploid gametes combine to form a new diploid zygote. If the gametes had the same number of chromosomes as somatic cells, the chromosome number would double with each generation.
9. To answer this question we must determine Paula's and Marvin's genotypes. Paula has PKU, which we know is an autosomal recessive trait. That means that she is homozygous recessive (*pp*). Marvin does not have PKU, so his genotype is either homozygous dominant (*PP*) or heterozygous (*Pp*). Since no PKU has appeared in Marvin's family for several generations, it is more likely that he is homozygous dominant. If Paula has two PKU alleles (*pp*) and Marvin has two normal alleles (*PP*), each of their children will receive one of the alleles from each parent. Therefore, all of their children will have one allele for PKU and one normal allele (*Pp*). Because PKU is a recessive disease, none of the children will have the disorder.
10. Duchenne muscular dystrophy is an X-linked condition. Remember that a male child receives an X chromosome from his mother and a Y chromosome from his father. We can therefore assume that Wilma is a "carrier" for Duchenne muscular dystrophy, meaning she is heterozygous for the condition. As a male, Wally has only one X chromosome in his cells, and his X chromosome must have the normal allele because Wally does not have Duchenne muscular dystrophy. Because each of their children will receive an X chromosome from Wilma, they may receive either an X chromosome with the normal allele or an X chromosome with the Duchenne muscular dystrophy allele. Males who receives a chromosome with a normal allele from Wilma and a Y chromosome from Wally will be normal. Males who receive an X chromosome with the muscular dystrophy allele from Wilma and the Y chromosome from Wally will have Duchenne muscular dystrophy. Therefore, the probability that their next child will have the condition is 0 if the child is female and 0.5 (or ½) if the child is male. However, the probability that each female child will be a carrier for the condition is 0.5 (or ½).
11. Down syndrome (trisomy 21) is an aneuploidy condition where the individual has three copies of chromosome 21. At fertilization, instead of the child receiving two chromosomes 21, one from each parent, the child has three chromosomes 21, two from one parent and one from the other. We learned in this chapter that this type of disorder is the result of nondisjunction, or the failure of chromosomes to segregate correctly during meiosis. One possible explanation for this condition is that nondisjunction occurred during oogenesis, producing an oocyte with two chromosomes 21. If this oocyte were fertilized by a sperm cell that had one chromosome 21, the resulting zygote would have three copies of chromosome 21. Nondisjunction during spermatogenesis and subsequent fertilization of a normal oocyte would also result in a zygote with three copies of chromosome 21. The probability of nondisjunction occurring during oogenesis increases in women after approximately age 35.

Glossary

Many of the words in this glossary and the text are followed by a simplified phonetic spelling showing pronunciation. The pronunciation key reflects standard clinical usage as presented in *Stedman's Medical Dictionary* (28th edition), a leading reference volume in the health sciences.

The phonetic system used is a basic one and has only a few conventions:

- Two diacritical marks are used; the macron (¯) for long vowels; and the breve (˘) for short vowels.
- Principal stressed syllables are followed by a prime (′); monosyllables do not have a stress mark.
- Other syllables are separated by hyphens.

The following pronunciation key provides examples and consonant sounds encountered in the phonetic system. No attempt has been made to accommodate the slurred sounds common in speech or regional variations in speech sounds. Note that a vowel with a breve (˘) is used for the indefinite vowel sound of the schwa (ə). Native pronunciation of foreign words is approximated as closely as possible.

Pronunciation Key

Vowels

ā	day, care, gauge	ĕ	taken, genesis
a	mat, damage	e	term, learn
ă	about, para	ī	pie
ĭ	pit, sieve, build	ū	unit, curable
ō	note, for so,	ŭ	cut
o	not, oncology, ought	ah	father
oo	food	aw	fall, cause, raw
ow	cow, out	ē	be, equal, ear
oy	troy, void		

Consonants

b	bad	n	no
ch	child	ng	ring
d	dog	p	pan
dh	this, smooth	r	rot
f	fit	s	so, miss
g	got	sh	should
h	hit	t	ten
j	jade	th	thin, with
k	kept	v	very
ks	tax	w	we
kw	quit	y	yes
l	law	z	zero
m	me	zh	azure, measure

In some words the initial sound is not that of the initial letter(s), or the initial letter(s) is not sounded or has a different sound, as in the following examples:

aerobe (ar′ob)	phthalein (thal′e-in)
eimuria (ime′re-a)	pneumonia (nu-mo′ne-a)
gnathic (nath′ik)	psychology (si-kol′o-je)
knuckle (nuk-l)	ptosis (to′sis)
oedipism (ed′i-pizm)	xanthoma (zan-tho′ma)

A

A band Length of the myosin myofilament in a sarcomere.

abdomen (ab-dō′men, ab′dō-men) Belly, between the thorax and the pelvis.

abduction (ab-dŭk′shŭn) [L., *abductio,* take away] Movement away from the midline.

absolute refractory period (ab′sō-loot rē-frak′ tōr-ē) Portion of the action potential during which the membrane is insensitive to all stimuli, regardless of their strength.

absorptive cell (ab-sōrp′tiv) Cell on the surface of villi of the small intestines and the luminal surface of the large intestine that is characterized by having microvilli; secretes digestive enzymes and absorbs digested materials on its free surface.

absorptive state Immediately after a meal when nutrients are being absorbed from the intestine into the circulatory system.

accommodation (ă-kom′ŏ-dā′shŭn) [L., *ac* + *commodo,* to adapt] Ability of electrically excitable tissues, such as nerve or muscle cells, to adjust to a constant stimulus so that the magnitude of the local potential decreases through time; also called adaptation.

acetabulum (as-ĕ-tab′ū-lŭm) [L., shallow vinegar vessel or cup] Cup-shaped depression on the external surface of the coxa.

acetylcholine (ACh) (as-e-til-kō′lēn) Neurotransmitter substance released from motor neurons, all preganglionic neurons of the parasympathetic and sympathetic divisions, all postganglionic neurons of the parasympathetic division, some postganglionic neurons of the sympathetic division, and some central nervous system neurons.

acetylcholinesterase (as′ē-til-kō-lin-es′ter-ās) Enzyme found in the synaptic cleft that causes the breakdown of acetylcholine to acetic acid and choline, thus limiting the stimulatory effect of acetylcholine.

Achilles tendon *See* calcaneal tendon.

acid (as′id) Molecule that is a proton donor; any substance that releases hydrogen ions (H^+).

acidic Solution containing more than 10^{-27} mol of hydrogen ions per liter; has a pH less than 7.

acinus; pl. **acini** (as′i-nŭs, as′ĭ-nī) [L., berry, grape] Grape-shaped secretory portion of a gland. The terms *acinus* and *alveolus* are sometimes used interchangeably. Some authorities differentiate the terms: Acini have a constricted opening into the excretory duct, whereas alveoli have an enlarged opening.

acromion (ă-krō′mē-on) [Gr., *akron,* extremity + omos, shoulder] Bone comprising the tip of the shoulder.

acrosome (ak′rō-sōm) [Gr., *akron,* extremity + *soma,* body] Cap on the head of the spermatozoon, with hydrolytic enzymes that help the spermatozoon penetrate the ovum.

actin filament (ak′tin) Thin myofilament within the sarcomere; composed of two F actin molecules, tropomyosin, and troponin molecules.

action potential [L., *potentia,* power, potency] Change in membrane potential in an excitable tissue that acts as an electric signal and is propagated in an all-or-none fashion.

activation energy (ak-ti-vā′shŭn) Energy that must be added to molecules to initiate a reaction.

active site Portion of an enzyme in which reactants are brought into close proximity and that plays a role in reducing activation energy of the reaction.

active tension Tension produced by the contraction of a muscle.

active transport Carrier-mediated process that requires ATP and can move substances against a concentration gradient.

adaptive immunity Immune status in which there is an ability to recognize, remember, and destroy a specific antigen.

adenohypophysis (ad′ĕ-nō-hī-pof′ĭ-sis) Portion of the hypophysis derived from the oral ectoderm; also called anterior pituitary.

adenosine diphosphate (ADP) (ă-den′ō-sēn) Adenosine, an organic base, with two phosphate groups attached to it. Adenosine diphosphate combines with a phosphate group to form adenosine triphosphate.

adenosine triphosphate (ATP) Adenosine, an organic base, with three phosphate groups attached to it. Energy stored in ATP is used in nearly all of the endergonic reactions in cells.

adenylate cyclase (ad′e-nil-āte sīklās) An enzyme acting on ATP to form 3′,5′-cyclic AMP plus pyrophosphate (two phosphate groups). A crucial step in the regulation and formation of the intracellular chemical signal 3′,5′-cyclic AMP.

adipocyte (ad′i-pō-sīt) Fat cell.

adipose (ad′i-pōs) [L., *adeps,* fat] Fat.

adrenal gland (ă-drē′năl) [L., *ad,* to + *ren,* kidney] Located near the superior pole of each kidney, it is composed of a cortex and a medulla. The adrenal medulla is a highly modified sympathetic ganglion that secretes the hormones epinephrine and norepinephrine; the cortex secretes aldosterone and cortisol as its major secretory products. Also called suprarenal gland.

adrenergic neuron (ad-rĕ-ner′jik) Nerve fiber that secretes norepinephrine (or epinephrine) as a neurotransmitter substance.

adrenergic receptor (ad-rĕ-ner′jik) Receptor molecule that binds to adrenergic agents, such as epinephrine and norepinephrine.

adrenocorticotropic hormone (ACTH) (ă-drē′nō-kōr′ti-kō-trō′pik) Hormone of the adenohypophysis that governs the nutrition and growth of the adrenal cortex, stimulates it to functional activity, and causes it to secrete cortisol.

adventitia (ad-ven-tish′ă) [L., *adventicius,* coming from abroad, foreign] Outermost covering of any organ or structure that is properly derived from outside the organ and does not form an integral part of the organ.

aerobic respiration (ār-ō′bik) Breakdown of glucose in the presence of oxygen to produce carbon dioxide, water, and approximately 38 ATPs; includes glycolysis, the citric acid cycle, and the electron-transport chain.

afferent arteriole (af′er-ent) Branch of an interlobular artery of the kidney that conveys blood to the glomerulus.

afferent division Nerve fibers that send impulses from the periphery to the central nervous system.

agglutination (ă-gloo-ti-nā-shŭn) [L., *ad,* to + *gluten,* glue] Process by which blood cells, bacteria, or other particles are caused to adhere to one another and form clumps.

agglutinin (ă-gloo′ti-nin) Antibody that binds to an antigen and causes agglutination.

agglutinogen (ă-gloo-tin′ō-jen) Antigen on surface of red blood cells that can stimulate the production of antibodies (agglutinins) that combine with the antigen and cause agglutination.

agranulocyte (ă-gran′ū-lō-sīt) Nongranular leukocyte (monocyte or lymphocyte).

ala; pl. **alae** (ā′lă, ā′lē) [L., a wing] Wing-shaped structure.

aldosterone (al-dos′ter-ōn) Steroid hormone produced by the zona glomerulosa of the adrenal cortex that facilitates potassium exchange for sodium in the distal renal tubule, causing sodium reabsorption and potassium and hydrogen secretion.

alkaline (al′kă-līn) Solution containing less than 10^{-7} mol of hydrogen ions per liter; has a pH greater than 7.0.

alkalosis (al-kă-lō′sis) Condition characterized by blood pH of 7.45 or above.

allantois (ă-lan′tō-is) Tube extending from the embryonic hindgut into the umbilical cord; forms the urinary bladder.

allele (ă-lēl′) [Gr., *allelon,* reciprocally] Any one of a series of two or more different genes that may occupy the same position or locus on a specific chromosome.

all-or-none principle When a stimulus is applied to a cell, an action potential is either produced or not. In muscle cells, the cell either contracts to the maximum extent possible (for a given condition) or does not contract.

alternative pathway Part of the nonspecific immune system for activation of complement.

alveolar duct (al-vē′ō-lăr) Part of the respiratory passages beyond a respiratory bronchiole; from it arise alveolar sacs and alveoli.

alveolar gland Gland in which the secretory unit has a saclike form and an obvious lumen.

alveolar sac Two or more alveoli that share an opening.

alveolus; pl. **alveoli** (al-vē′ō-lŭs, al-vē′ō-lī) Cavity. Examples include the sockets into which teeth fit, the endings of the respiratory system, and the terminal endings of secretory glands.

amino acid (ă-mē′nō) Class of organic acids that constitute the building blocks for proteins.

amplitude-modulated signal (am′pli-tood) Signal that varies in magnitude or intensity, such as with large versus small concentrations of hormones.

ampulla (am-pul′lă, am-pul′lē) [L., two-handled bottle] Saclike dilatation of a semicircular canal; contains the crista ampullaris. Wide portion of the uterine tube between the infundibulum and the isthmus.

amygdala (a-mig′da-la, -lē) [L. fr. Gr., *amygydale,* almond] Nucleus in the temporal lobe of the brain, amygdaloid nucleus; also called amydaloid nuclear complex.

amylase (am′il-ās) One of a group of starch-splitting enzymes that cleave starch, glycogen, and related polysaccharides.

anabolism (ă-nab′ō-lizm) [Gr., *anabole,* a raising up] All of the synthesis reactions that occur within the body; requires energy.

anaerobic respiration (an-ār-ō′bik) Breakdown of glucose in the absence of oxygen to produce lactic acid and two ATPs; consists of glycolysis and the reduction of pyruvic acid to lactic acid.

anal canal Terminal portion of the digestive tract.

anal triangle Posterior portion of the perineal region through which the anal canal opens.

analgesic (an-al-jē′zik) Compound capable of producing analgesia, without producing anesthesia or loss of consciousness, characterized by reduced response to painful stimuli.

anaphase (an′ă-fāz) Time during cell division when chromatids divide (or, in the case of first meiosis, when the chromosome pairs divide).

anastomoses (ă-nas′tō-mō′sez) Natural communication, direct or indirect, between two blood vessels or other tubular structures. An opening created by surgery, trauma, or disease between two or more normally separate spaces or organs.

anatomical dead air space Volume of the conducting airways from the external environment down to the terminal bronchioles.

androstenedione (an-drō-stēn-dī′ōn) Androgenic steroid of weaker potency than testosterone; secreted by the testis, ovary, and adrenal cortex.

anencephaly (an′en-sef′ă-lē) [Gr., *an* + *enkephalos,* no brain] Defective development of the brain and absence of the bones of the cranium. Only a rudimentary brainstem and a trace of basal ganglia are present.

aneurysm (an′ū-rizm) [Gr., *eurys,* wide] Dilated portion of an artery.

angiotensin I (an-jē-ō-ten′sin) Peptide derived when renin acts on angiotensinogen.

angiotensin II Peptide derived from angiotensin I; stimulates vasoconstriction and aldosterone secretion; also called active angiotensin.

anion (an′ī-on) Ion carrying a negative charge.

antagonist (an-tag′ŏ-nist) Muscle that works in opposition to another muscle.

anterior chamber Chamber of the eye between the cornea and the iris.

anterior interventricular sulcus Groove on the anterior surface of the heart, marking the location of the septum between the two ventricles.

anterior pituitary gland *See* adenohypophysis.

antibody (an′tē-bod-ē) Protein found in the plasma that is responsible for humoral immunity; binds specifically to antigen.

antibody-mediated immunity Immunity due to B cells and the production of antibodies.

anticoagulant (an′tē-kō-ag′ū-lant) Agent that prevents coagulation.

antidiuretic hormone (ADH) (an′tē-dī-ū-ret′ik) Hormone secreted from the neurohypophysis that acts on the kidney to reduce the output of urine; also called vasopressin because it causes vasoconstriction.

antigen (an′ti-jen) [anti(body) + Gr., *gen,* producing] Substance that induces a state of sensitivity or resistance to infection or toxic substances after a latent period; substance that stimulates the specific immune system; also called epitope.

antigenic determinant (an-ti-jen′ik) Specific part of an antigen that stimulates an immune system response by binding to receptors on the surface of lymphocytes; also called epitope.

antithrombin (an-tē-throm′bin) Substance that inhibits or prevents the effects of thrombin so that blood does not coagulate.

antrum (an′trŭm) [Gr., *antron,* a cave] Cavity of an ovarian follicle filled with fluid containing estrogen.

anulus fibrosus (an′ū-lŭs fī-brō′sus) [L., fibrous ring] Fibrous material forming the outer portion of an intervertebral disk.

anus (ā′nŭs) Lower opening of the digestive tract through which fecal matter is extruded.

aorta (ā-ōr′tă) [Gr., *aorte* from *aeiro,* to lift up] Large, elastic artery that is the main trunk of the systemic arterial system; carries blood from the left ventricle of the heart and passes through the thorax and abdomen.

aortic arch [L., bow] Curve between the ascending and descending portions of the aorta.

aortic body One of the smallest bilateral structures, similar to the carotid bodies, attached to a small branch of the aorta near its arch; contains chemoreceptors that respond primarily to decreases in blood oxygen; less sensitive to decreases in blood pH or increases in carbon dioxide.

apex (ā′peks) [L., summit or tip] Extremity of a conical or pyramidal structure. The apex of the heart is the rounded tip directed anteriorly and slightly inferiorly.

Apgar score Named for U.S. anesthesiologist Virginia Apgar (1909–1974). Evaluation of a newborn infant's physical status by assigning numerical values to each of five criteria: appearance (skin color), pulse (heart rate), grimace (response to stimulation), activity (muscle tone), and respiratory effort; a score of 10 indicates the best possible condition.

apical ectodermal ridge Layer of surface ectodermal cells at the lateral margin of the embryonic limb bud; stimulates growth of the limb.

apical foramen [L., aperture] Opening at the apex of the root of a tooth, gives passage to the nerve and blood vessels.

apocrine sweat gland (ap′ō-krin) [Gr., *apo,* away from + *krino,* to separate] Gland whose cells contribute cytoplasm to its secretion (e.g., mammary glands). Sweat glands that produce organic secretions traditionally are called apocrine. These sweat glands, however, are actually merocrine glands.

appendicular skeleton (ap′en-dik′ū-lăr) The portion of the skeleton consisting of the upper limbs and the lower limbs and their girdles.

appositional growth (ap-ō-zish′ŭn-al) [L., *ap* + *pono,* to put or place] To place one layer of bone, cartilage, or other connective tissue against an existing layer.

aqueous humor (ak′wē-ŭs, ā′kwē-ŭs) Watery, clear solution that fills the anterior and posterior chambers of the eye.

arachnoid (ă-rak′noyd) [Gr., *arachne,* spider, cobweb] Thin, cobweb-appearing meningeal layer surrounding the brain; the middle of the three layers.

arcuate artery (ar′kū-āt) Artery that originates from the interlobar arteries of the kidney and forms an arch between the cortex and medulla of the kidney.

areola (ă-rē′ō-lă, -lē) [L., area] Circular, pigmented area surrounding the nipple; its surface is dotted with little projections caused by the presence of the areolar glands beneath.

areolar gland (ă-rē′ō-lăr) Gland forming small, rounded projections from the surface of the areola of the mamma.

arrectores pilorum; pl. **arrector pili** (ă-rek-tō′rez pī-lōr′um, ă-rek′tōr pī′lī) [L., that which raises; hair] Smooth muscle attached to the hair follicle and dermis that raises the hair when it contracts.

arterial capillary (ar-tē′rē-ăl) Capillary opening from an arteriole or a metarteriole.

arteriole (ar-tēr′ē-ōl) Minute artery with all three tunics that transports blood to a capillary.

arteriosclerosis (ar-tēr′ē-ō-skler-ō′sis) [L., *arterio* + Gr., *sklerosis,* hardness] Hardening of the arteries.

arteriovenous anastomosis (ar-tēr′ē-ō-vē′nŭs ă-nas′tō-mō′sis) Vessel through which blood is shunted from an arteriole to a venule without passing through the capillaries.

artery (ar′ter-ē) Blood vessel that carries blood away from the heart.

articular cartilage (ar-tik′ū-lăr kar′ti-lij) Hyaline cartilage covering the ends of bones within a synovial joint.

articulation (ar-tik-ū-lā′shŭn) Place where two bones come together; also called joint.

arytenoid cartilages (ar-i-tē′noyd) Small, pyramidal laryngeal cartilages that articulate with the cricoid cartilage.

ascending aorta Part of the aorta from which the coronary arteries arise.

ascending colon (kō′lon) Portion of the colon between the small intestine and the right colic flexure.

asthma (az′mă) Condition of the lungs in which widespread narrowing of airways occurs, caused by contraction of smooth muscle, edema of the mucosa, and mucus in the lumen of the bronchi and bronchioles.

astrocyte (as′trō-sīt) [Gr., *astron,* star + *kytos,* a hollow, a cell] Star-shaped neuroglia cell involved with forming the blood-brain barrier.

atherosclerosis (ath′er-ō-skler-ō′sis) Arteriosclerosis characterized by irregularly distributed lipid deposits in the intima of large and medium-sized arteries.

atomic number (ă-tom′ik) Number of protons in each type of atom.

atrial diastole (ā′trē-ăl dī-as′tō-lē) Dilation of the heart's atria.

atrial natriuretic hormone (ANH) (ā′trē-ăl nā′trē-ū-ret′ik) Peptide released from the atria when atrial blood pressure is increased; lowers blood pressure by increasing the rate of urinary production, thus reducing blood volume.

atrial systole (ā′trē-ăl sis′tō-lē) Contraction of the atria.

atrioventricular (AV) bundle (ā′trē-ō-ven-trik′ū-lar) Bundle of modified cardiac muscle fibers that projects from the AV node through the interventricular septum.

atrioventricular (AV) node Small node of specialized cardiac muscle fibers that gives rise to the atrioventricular bundle of the conduction system of the heart.

atrioventricular valve One of two valves closing the openings between the atria and ventricles.

atrium; pl. **atria** (ā′trē-ŭm, ā′trē-ă) [L., entrance hall] One of two chambers of the heart into which veins carry blood.

auditory cortex (aw′di-tōr-ē kōr′teks) Portion of the cerebral cortex that is responsible for the conscious sensation of sound; in the dorsal portion of the temporal lobe within the lateral fissure and on the superolateral surface of the temporal lobe.

auditory ossicle (os′i-kl) Bone of the middle ear; includes the malleus, incus, and stapes.

auditory tube Auditory canal; extends from the middle ear to the nasopharynx.

auricle (aw′ri-kl) [L., *auris,* ear] Part of the external ear that protrudes from the side of the head; also called pinna. Small pouch projecting from the superior, anterior portion of each atrium of the heart.

auscultatory (aws-kŭltă-tō-rē) Relating to auscultation, listening to the sounds made by the various body structures as a diagnostic method.

autoimmune disease (aw-tō-i-mūn′ di-zēz′) Disease resulting from a specific immune system reaction against self-antigens.

autonomic ganglia (aw-tō-nom′ik gang′glē-ă) Ganglia containing the nerve cell bodies of the autoimmune division of the nervous system.

autonomic nervous system (ANS) Nervous system composed of nerve fibers that send impulses from the central nervous system to smooth muscle, cardiac muscle, and glands.

autophagia (aw-tō-fā′jē-ă) [Gr., *auto,* self + *phagein,* to eat] Segregation and disposal of organelles within a cell.

autoregulation (aw′tō-reg-ŭ-lā′shŭn) Maintenance of a relatively constant blood flow through a tissue despite relatively large changes in blood pressure; maintenance of a relatively constant glomerular filtration rate despite relatively large changes in blood pressure.

autorhythmic Spontaneous and periodic—for example, in smooth muscle, spontaneous (without nervous or hormonal stimulation) and periodic contractions.

autosome (aw′tō-sōm) [Gr., *auto,* self + *soma,* body] Any chromosome other than a sex chromosome; normally exist in pairs in somatic cells and singly in gametes.

axial skeleton (ak′sē-ăl) Skull, vertebral column, and rib cage.

axillary (ak′sil-ăr-ē) Relating to the axilla; the space below the shoulder joint, bounded by the pectoralis major anteriorly, the latissimus dorsi posteriorly, the serratus anterior medially, and the humerus laterally.

axolemma (ak′sō-lem′ă) [Gr., *axo* + *lemma,* husk] Plasma membrane of the axon.

axon (ak′son) [Gr., axis] Main central process of a neuron that normally conducts action potentials away from the neuron cell body.

axon hillock Area of origin of the axon from the nerve cell body.

axoplasm (ak′sō-plazm) Neuroplasm or cytoplasm of the axon.

B

baroreceptor (bar′ō-rē-sep′ter, bar′ō-rē-sep′tōr) Sensory nerve ending in the walls of the atria of the heart, venae cavae, aortic arch, and carotid sinuses; sensitive to stretching of the wall caused by increased blood pressure; also called pressoreceptor.

baroreceptor reflex Detects changes in blood pressure and produces changes in heart rate, heart force of contraction, and blood vessel diameter that return blood pressure to homeostatic levels.

basal ganglia (bā′săl gang′glē-ă) Nuclei at the base of the cerebrum involved in controlling motor functions.

base (bās) Molecule that is a proton acceptor; any substance that binds to hydrogen ions. Lower part or bottom of a structure; the base of the heart is the flat portion directed posteriorly and superiorly; veins and arteries project into and out of the base, respectively.

basement membrane (bās′ment mem′brān) Specialized extracellular material located at the base of epithelial cells and separating them from the underlying connective tissues.

basilar membrane (bas′i-lăr mem′brān) Wall of the membranous labyrinth bordering the scala tympani; supports the organ of Corti.

basophil (bā′sō-fil) [Gr., *basis,* baso + *phileo,* to love] White blood cell with granules that stain specifically with basic dyes; promotes inflammation.

B cell Type of lymphocyte responsible for antibody-mediated immunity.

belly (bel′ē) Largest portion of muscle between the origin and insertion.

beta-oxidation (bā′tă ok-si-dā′shŭn) Metabolism of fatty acids by removing a series of two-carbon units to form acetyl-CoA.

bicarbonate ion (bī-kar′bon-āt) Anion (HCO_3^-) remaining after the dissociation of carbonic acid.

bicuspid valve (bī-kŭs′pid) Valve closing the orifice between left atrium and left ventricle of the heart; also called mitral valve.

bile (bīl) Fluid secreted from the liver into the duodenum; consists of bile salts, bile pigments, bicarbonate ions, cholesterol, fats, fat-soluble hormones, and lecithin.

bile canaliculus (bīl kan′ă-lik′ū-lŭs) One of the intercellular channels approximately 1 μm or less in diameter that occurs between liver cells into which bile is secreted; empties into the hepatic ducts.

bile salt Organic salt secreted by the liver that functions as an emulsifying agent.

bilirubin (bil-i-roo′bin) [L., *bili* + *ruber,* red] Bile pigment derived from hemoglobin during the destruction of red blood cells.

biliverdin (bil-i-ver′din) Green bile pigment formed from the oxidation of bilirubin.

binocular vision (bin-ok′ū-lăr) [L., *bini,* paired + *oculus,* eye] Vision using two eyes at the same time; responsible for depth perception when the visual fields of the eyes overlap.

bipolar neuron (bī-pō′ler) One of the three categories of neurons consisting of a neuron with two processes—one dendrite and one axon—arising from opposite poles of the cell body.

blastocele (blas′tō-sēl) [Gr., *blastos,* germ + *koilos,* hollow] Cavity in the blastocyst.

blastocyst (blas′tō-sist) [Gr., *blastos,* germ + *kystis,* bladder] Stage of mammalian embryos that consists of the inner cell mass and a thin trophoblast layer enclosing the blastocele.

bleaching In response to light, retinal separates from opsin.

blind spot (blīnd) Point in the retina where the optic nerve penetrates the fibrous tunic; contains no rods or cones and therefore does not respond to light.

blood clot Coagulated phase of blood.

blood colloid osmotic pressure (BCOP) Osmotic pressure due to the concentration difference of proteins across a membrane that does not allow passage of the proteins.

blood groups Classification of blood based on the type of antigen found on the surface of red blood cells.

blood island Aggregation of mesodermal cells in the embryonic yolk sac that forms vascular endothelium and primitive blood cells.

blood pressure [L., *pressus,* to press] Tension of the blood within the blood vessels; commonly expressed in units of millimeters of mercury (mm Hg).

blood-brain barrier Permeability barrier controlling the passage of most large-molecular compounds from the blood to the cerebrospinal fluid and brain tissue; consists of capillary endothelium and may include the astrocytes.

blood-thymic barrier Layer of reticular cells that separates capillaries from thymic tissue in the cortex of the thymus gland; prevents large molecules from leaving the blood and entering the cortex.

Bohr effect Named for Danish physiologist Christian Bohr (1855–1911). Shift of the oxygen-hemoglobin dissociation curve to the right or left because of changes in blood pH. The definition sometimes is extended to include shifts caused by changes in blood carbon dioxide levels.

bony labyrinth (lab′i-rinth) Part of the inner ear; contains the membranous labyrinth that forms the cochlea, vestibule, and semicircular canals.

Boyle's law The pressure of a gas is equal to the volume of a container times the constant, K, for a given temperature. Assuming a constant temperature, the pressure of a gas is inversely proportional to its volume.

brachial (brā′kē-āl) [L., *brachium,* arm] Relating to the arm.

branchial arch Typically, six arches in vertebrates; in the lower vertebrates, they bear gills, but they appear transiently in the higher vertebrates and give rise to structures in the head and neck.

broad ligament Peritoneal fold passing from the lateral margin of the uterus to the wall of the pelvis on each side.

bronchiole (brong′kē-ōl) One of the finer subdivisions of the bronchial tubes, less than 1 mm in diameter; has no cartilage in its wall but does have relatively more smooth muscle and elastic fibers.

brush border Epithelial surface consisting of microvilli.

buffer (bŭf′er) Mixture of an acid and a base that reduces any changes in pH that would otherwise occur in a solution when acid or base is added to the solution.

bulb of the penis Expanded posterior part of the corpus spongiosum of the penis.

bulb of the vestibule Mass of erectile tissue on each side of the vagina.

bulbar conjunctiva (bŭl′bar kon-jŭnk-tī′vă) Conjunctiva that covers the surface of the eyeball.

bulbourethral gland (bŭl′bō-ū-rē′thrăl) One of two small compound glands that produce a mucoid secretion; it discharges through a small duct into the spongy urethra.

bulbus cordis (bŭl′bŭs) [L., plant bulb] End of the embryonic cardiac tube where blood leaves the heart; becomes part of the ventricle.

bursa; pl. **bursae** (ber′să, ber′sē) [L., purse] Closed sac or pocket containing synovial fluid, usually found in areas where friction occurs.

bursitis (ber-sī′tis) [L., *purse* + Gr., *ites,* inflammation] Inflammation of a bursa.

C

calcaneal tendon (kal-kā′nē-al) Common tendon of the gastrocnemius, soleus, and plantaris muscle that attaches to the calcaneus; also called Achilles tendon.

calcitonin (kal-si-tō′nin) Hormone released from parafollicular cells that acts on tissues to cause a decrease in blood levels of calcium ions.

calmodulin (kal-mod′ū-lin) [*calcium* + *modulate*] Protein receptor for Ca^{2+} that plays a role in many Ca^{2+}-regulated processes, such as smooth muscle contraction.

calorie (cal) (kal′ō-rē) [L., *calor,* heat] Unit of heat content or energy. The quantity of energy required to raise the temperature of 1 g of water 1°C.

calpain (kal′pān) Enzyme involved in changing the shape of dendrites; involved with long-term memory.

calyx; pl. **calyces** (kā′liks, kal′i-sēz) [Gr., cup of a flower] Flower-shaped or funnel-shaped structure; specifically, one of the branches or recesses of a renal pelvis into which the tips of the renal pyramids project.

cancer (kan′ser) Any of various types of malignant neoplasms, most of which invade surrounding tissues, may metastasize to several sites, and are likely to recur after attempted removal and to cause death of the patient unless adequately treated.

canine (kā′nīn) Referring to the cuspid tooth.

cannula (kan′ū-lă) [L., *canna,* reed] Tube; often inserted into an artery or a vein.

capacitation (kă-pas′i-tā′shŭn) [L., *capax,* capable of] Process whereby spermatozoa acquire the ability to fertilize ova; occurs in the female genital tract.

capitulum (kă-pit′ū-lŭm) [L., *caput,* head] Head-shaped structure.

carbaminohemoglobin (kar-bam′i-nō-hē-mō-glō′bin) Carbon dioxide bound to hemoglobin by means of a reactive amino group on the hemoglobin.

carbohydrate (kar-bō-hī′drāt) Monosaccharide (simple sugar) or the organic molecules composed of monosaccharides bound together by chemical bonds—for example, glycogen. For each carbon atom in the molecule, there are typically one oxygen molecule and two hydrogen molecules.

carbonic acid/bicarbonate buffer system One of the major buffer systems in the body; major components are carbonic acid and bicarbonate ions.

carbonic anhydrase Enzyme that catalyzes the reaction between carbon dioxide and water to form carbonic acid.

carcinoma (kar-si-nō′mă) Malignant neoplasm derived from epithelial tissue.

cardiac [Gr., *kardia,* heart] Related to the heart.

cardiac cycle [Gr., *kyklos,* circle] Complete round of cardiac systole and diastole.

cardiac nerve Nerve that extends from the sympathetic chain ganglia to the heart.

cardiac output (CO) Volume of blood pumped by the heart per minute; also called minute volume.

cardiac part Region of the stomach near the opening of the esophagus.

cardiac reserve [L., *re* + *servo,* to keep back, reserve] Work that the heart is able to perform beyond that required during ordinary circumstances of daily life.

carotid body (ka-rot′id) One of the small organs near the carotid sinuses; contains chemoreceptors that respond primarily to decreases in blood oxygen; less sensitive to decreases in blood pH or increases in carbon dioxide.

carotid sinus Enlargement of the internal carotid artery near the point where the internal carotid artery branches from the common carotid artery; contains baroreceptors.

carpal (kar′păl) [Gr., *karpos,* wrist] Bone of the wrist.

carrier Person in apparent health whose chromosomes contain a pathologic mutant gene, which may be transmitted to his or her children.

cartilage (kar′ti-lij) [L., *cartilage,* gristle] Firm, smooth, resilient, nonvascular connective tissue.

cartilaginous joint (kar-ti-laj′i-nŭs) Bones connected by cartilage; includes synchondroses and symphyses.

catabolism (kă-tab′ō-lizm) [Gr., *katabole,* a casting down] All of the decomposition reactions that occur in the body; releases energy.

catalyst (kat′ă-list) Substance that increases the rate at which a chemical reaction proceeds without being changed permanently.

cataract (kat′a-rakt) Complete or partial opacity of the lens of the eye.

cation (kat′ī-on) [Gr., *kation,* going down] Ion carrying a positive charge.

caveola; pl. **caveolae** (kav-ē-ō′lă, kav-ē-ō′lē) [L., small pocket] Shallow invagination in the membranes of smooth muscle cells that may perform a function similar to that of both the T tubules and sarcoplasmic reticulum of skeletal muscle.

cecum (sē′kŭm, sē′kă) [L., *caecus,* blind] Cul-de-sac forming the first part of the large intestine.

cell-mediated immunity Immunity due to the actions of T cells and null cells.

celom (sē′lom, sē-lō′mă) [Gr., *koilo* + *amma,* a hollow] Principal cavities of the trunk—for example, the pericardial, pleural, and peritoneal cavities. Separate in the adult, they are continuous in the embryo.

cementum (se-men′tŭm) [L., *caementum,* rough quarry stone] Layer of modified bone covering the dentin of the root and neck of a tooth; blends with the fibers of the periodontal membrane.

central nervous system (CNS) Major subdivision of the nervous system, consisting of the brain and spinal cord.

central vein Terminal branches of the hepatic veins that lie centrally in the hepatic lobules and receive blood from the liver sinusoids.

centrosome (sen′trō-sōm) Specialized zone of cytoplasm close to the nucleus and containing two centrioles.

cerebellum (ser-e-bel′ŭm) [L., little brain] Separate portion of the brain attached to the brainstem at the pons; important in maintaining muscle tone, balance, and coordination of movement.

cerebrospinal fluid (CSF) (ser′ĕ-brō-spī-năl) Fluid filling the ventricles and surrounding the brain and spinal cord.

ceruminous glands (sĕ-roo′mi-nŭs) Modified sebaceous glands in the external acoustic meatus that produce cerumen (earwax).

cervical canal (ser′vĭ-kal) Canal extending from the isthmus of the uterus to the opening of the uterus into the vagina.

cervix; pl. **cervices** (ser′viks, ser-vī′sēz) [L., neck] Lower part of the uterus extending from the isthmus of the uterus into the vagina.

chalazion (ka-lā′zē-on) Chronic inflammation of a meibomian gland; also called meibomian cyst.

cheek (chēk) Side of the face forming the lateral wall of the mouth.

chemical signal Molecule that binds to a macromolecule, such as receptors or enzymes, and alters their function; also called ligand.

chemoreceptor (kē′mō-rē-sep′tor) Sensory cell that is stimulated by a change in the concentration of chemicals to produce action potentials. Examples include taste receptors, olfactory receptors, and carotid bodies.

chemoreceptor reflex Chemoreceptors detect a decrease in blood oxygen, an increase in carbon dioxide, or a decrease in pH and produce an increased rate and depth of respiration and, by means of the vasomotor center, vasoconstriction.

chemosensitive area (kem-ō-sen′si-tiv, kē-mō-sen′si-tiv) Chemosensitive neurons in the medulla oblongata detect changes in blood, carbon dioxide, and pH.

chemotactic factor (kē-mō-tak′tik) Part of a microorganism or chemical released by tissues and cells that act as chemical signals to attract leukocytes.

chemotaxis (ke-mo-tak′sis) [Gr., *chemo* + *taxis,* orderly arrangement] Attraction of living protoplasm (cells) to chemical stimuli.

chief cell Cell of the parathyroid gland that secretes parathyroid hormone. Cell of a gastric gland that secretes pepsinogen.

chloride (klōr′īd) Compound containing chlorine—for example, salts of hydrochloric acid.

chloride shift Diffusion of chloride ions into red blood cells as bicarbonate ions diffuse out; maintains electrical neutrality inside and outside the red blood cells.

choana; pl. **choanae** (kō′an-ă, kō-ā′nē) *See* internal naris.

cholecystokinin (kō′lē-sis-tō-kī′nin) Hormone liberated by the upper intestinal mucosa on contact with gastric contents; stimulates the contraction of the gallbladder and the secretion of pancreatic juice high in digestive enzymes.

cholinergic neuron (kol-in-er′jik) Nerve fiber that secretes acetylcholine as a neurotransmitter substance.

chondroblast (kon′drō-blast) [Gr., *chondros,* gristle, cartilage + *blastos,* germ] Cartilage-producing cell.

chondrocyte (kon′drō-sīt) [Gr., *chondros,* gristle, cartilage + *kytos,* a cell] Mature cartilage cell.

chorda tympani; pl. **chordae** (kōr′dă tim′pan-ē, kōr′dē) Branch of the facial nerve that conveys taste sensation from the front two-thirds of the tongue.

chordae tendineae (kōr′dă ten′di-nē-ē) [L., cord] Tendinous strands running from the papillary muscles to the atrioventricular valves.

choroid (ko′royd) Portion of the vascular tunic associated with the sclera of the eye.

choroid plexus [Gr., *chorioeides,* membranelike] Specialized plexus located within the ventricles of the brain that secretes cerebrospinal fluid.

chromatid (krō′mă-tid) One-half of a chromosome; separates from its partner during cell division.

chromatin (krō′ma-tin) Colored material; the genetic material in the nucleus.

chromosome (krō′mō-sōm) Colored body in the nucleus, composed of DNA and proteins and containing the primary genetic information of the cell; 23 pairs in humans.

chronic pain Prolonged pain.

chylomicron (kī-lō-mī′kron) [Gr., *chylos,* juice + *micros,* small] Microscopic particle of lipid surrounded by protein; in chyle and blood.

chymotrypsin (kī-mō-trip′sin) Proteolytic enzyme formed in the small intestine from the pancreatic precursor chymotrypsinogen.

ciliary body (sil′ē-ar-ē) Structure continuous with the choroid layer at its anterior margin that contains smooth muscle cells; functions in accommodation.

ciliary gland Modified sweat gland that opens into the follicle of an eyelash, keeping it lubricated.

ciliary muscle Smooth muscle in the ciliary body of the eye.

ciliary process Portion of the ciliary body of the eye that attaches by suspensory ligaments to the lens.

ciliary ring Portion of the ciliary body of the eye that contains smooth muscle cells.

circumduction (ser-kŭm-dŭk′shŭn) [L., around + *ductus,* to draw] Movement in a circular motion.

circumferential lamellae (ser kŭm-fer en′shē-al lă-mel′ē) Lamellae covering the surface of and extending around compact bone inside the periosteum.

circumvallate papilla (ser-kŭm-val′āt pă-pil′ă) Type of papilla on the surface of the tongue surrounded by a groove.

cisterna; pl. **cisternae** (sis-ter′nă, sis-ter′nē) Interior space of the endoplasmic reticulum.

cisterna chyli (kīl′ē) [L., tank + Gr., *chylos,* juice] Enlarged inferior end of the thoracic duct that receives chyle from the intestine.

citric acid cycle (sit′rik) Series of chemical reactions in which citric acid is converted into oxaloacetic acid, carbon dioxide is formed, and energy is released. The oxaloacetic acid can combine with acetyl-CoA to form citric acid and restart the cycle. The energy released is used to form NADH, FADH, and ATP.

classical pathway Part of the specific immune system for activation of complement.

clavicle (klav′i-kl) The collarbone, between the sternum and scapula.

cleavage furrow (klēv′ij) Inward pinching of the plasma membrane that divides a cell into two halves, which separate from each other to form two new cells.

cleft palate (kleft) Failure of the embryonic palate to fuse along the midline, resulting in an opening through the roof of the mouth.

clinical age (klin′i-kl) Age of the developing fetus from the time of the mother's last menstrual period before pregnancy.

clinical perineum (klin′i-kl per′i-nē′ŭm) Portion of the perineum between the vaginal and anal openings.

clitoris (klit′ō-ris) Small, cylindrical, erectile body, rarely exceeding 2 cm in length, situated at the most anterior portion of the vulva and projecting beneath the prepuce.

cloaca (klō-ā′kă) [L., sewer] In early embryos, the endodermally lined chamber into which the hindgut and allantois empty.

cloning (klōn′ing) Growing a colony of genetically identical cells or organisms.

clot retraction Condensation of a clot into a denser, compact structure; caused by the elastic nature of fibrin.

coagulation (kō-ag-ū-lā′shŭn) Process of changing from liquid to solid, especially of blood; formation of a blood clot.

cochlear duct (kok′lē-ăr) Interior of the membranous labyrinth of the cochlea; also called cochlear canal or scala media.

cochlear nerve Nerve that carries sensory impulses from the organ of Corti to the vestibulocochlear nerve.

cochlear nucleus Neurons from the cochlear nerve synapse within the dorsal or ventral cochlear nucleus in the superior medulla oblongata.

codon (kō′don) Sequence of three nucleotides in mRNA or DNA that codes for a specific amino acid in a protein.

cofactor (kō′fak′ter, kō-fak′ tōr) Nonprotein component of an enzyme, such as coenzymes and inorganic ions essential for enzyme action.

collagen fibers (kol′lă-jen) [Gr., *koila,* glue + *gen,* producing] Ropelike protein of the extracellular matrix.

collateral ganglia (ko-lat′er-ăl gang′glē-ă) Sympathetic ganglia at the origin of large abdominal arteries; include the celiac, superior, and inferior mesenteric arteries; also called prevertebral ganglia.

collecting duct Straight tubule that extends from the cortex of the kidney to the tip of the renal pyramid. Filtrate from the distal convoluted tubes enters the collecting duct and is carried to the calyces.

colloid (kol′oyd) [Gr., *kolla,* glue + *eidos,* appearance] Atoms or molecules dispersed in a gaseous, liquid, or solid medium that resist separation from the liquid, gas, or solid.

colloidal solution (ko-loyd′ăl) Fine particles suspended in a liquid; resistant to sedimentation or filtration.

colon (kō′lon) Division of the large intestine that extends from the cecum to the rectum.

colostrum (kō-los′trŭm) Thin, white fluid; the first milk secreted by the breast at the termination of pregnancy; contains less fat and lactose than the milk secreted later.

columnar Shaped like a column.

commissure (kom′i-shŭr) [L., *commissura,* a joining together] Connection of nerve fibers between the cerebral hemispheres or from one side of the spinal cord to the other.

common bile duct Duct formed by the union of the common hepatic and cystic ducts; it empties into the small intestine.

common hepatic duct Part of the biliary duct system formed by the joining of the right and left hepatic ducts.

compact bone Bone that is denser and has fewer spaces than cancellous bone.

competition Similar molecules binding to the same carrier molecule or receptor site.

complement (kom′plě-ment) Group of serum proteins that stimulates phagocytosis and inflammation.

complement cascade Series of reactions in which each component activates the next component, resulting in activation of complement proteins.

compliance (kom-plī′ans) Change in volume (e.g., in lungs or blood vessels) caused by a given change in pressure.

compound (kom′pownd) Substance composed of two or more different types of atoms that are chemically combined.

concha; pl. **conchae** (kon′kă, kon′kē) [L., shell] Structure comparable to a shell in shape—for example, the three bony ridges on the lateral wall of the nasal cavity.

conduction (kon-dŭk′shŭn) [L., *con* + *ductus,* to lead, conduct] Transfer of energy, such as heat, from one point to another without evident movement in the conducting body.

cone (kōn) Photoreceptor in the retina of the eye; responsible for color vision.

congenital (kon-jen′i-tăl) [L., *congenitus,* born with] Occurring at birth; may be genetic or due to some influence (e.g., drugs) during development.

conjunctiva (kon-jŭnk-tī′vă) [L., *conjungo,* to bind together] Mucous membrane covering the anterior surface of the eyeball and lining the lids.

conjunctival fornix (kon-jŭnk-tī′văl fōr′niks) Area in which the palpebral and bulbar conjunctiva meet.

constant region Portion of an antibody that does not combine with an antigen and is the same in different antibodies.

continuous capillary [L., *capillaris,* relating to hair] Capillary in which pores are absent; is less permeable to large molecules than are other types of capillaries.

contraction phase (kon-trak′shŭn) One of the three phases of muscle contraction; the time during which tension is produced by the contraction of muscle.

convection (kon-vek′shŭn) [L., *con* + *vectus,* to carry or bring together] Transfer of heat in liquids or gases by movement of the heated particles.

coracoid (kōr′ă-koyd) [Gr., *korakodes,* crow's beak] Resembling a crow's beak—for example, a process on the scapula.

Cori cycle Named for Czech-U.S. biochemist and Nobel laureate Carl F. Cori (1896–1984). Lactic acid, produced by skeletal muscle, is carried in the blood to the liver, where it is aerobically converted into glucose. The glucose may return through the blood to skeletal muscle or may be stored as glycogen in the liver.

cornea (kōr′nē-ă) Transparent portion of the fibrous tunic that makes up the outer wall of the anterior portion of the eye.

corniculate cartilages (kōr-nik′ū-lāt) Conical nodules of elastic cartilage surmounting the apex of each arytenoid cartilage.

corona radiata Single layer of columnar cells derived from the cumulus mass, which anchor on the zona pellucida of the oocyte in a secondary follicle.

coronary (kōr′o-nār-ē) [L., *coronarius,* a crown] Resembling a crown; encircling.

coronary artery One of two arteries that arise from the base of the aorta and carry blood to the muscle of the heart.

coronary ligament Peritoneal reflection from the liver to the diaphragm at the margins of the bare area of the liver.

coronary sinus Short trunk that receives most of the veins of the heart and empties into the right atrium.

coronoid (kōr′ŏ-noyd) [Gr., *korone,* a crow] Shaped like a crow's beak—for example, a process on the mandible.

corpus; pl. **corpora** (kōr′pŭs, -pōr-ă) [L., *body*] Any body or mass; the main part of an organ.

corpus albicans (al′bi-kanz) Atrophied corpus luteum, leaving a connective tissue scar in the ovary.

corpus callosum (kăl-lō′sŭm) [L., *body* + *callous*] Largest commissure of the brain, connecting the cerebral hemispheres.

corpus cavernosum; pl. **corpora cavernosa** One of two parallel columns of erectile tissue forming the dorsal part of the body of the penis or the body of the clitoris.

corpus luteum (lū′tē-ŭm) Yellow endocrine body formed in the ovary in the site of a ruptured vesicular follicle immediately after ovulation; secretes progesterone and estrogen.

corpus luteum of pregnancy Large corpus luteum in the ovary of a pregnant female; secretes large amounts of progesterone and estrogen.

corpus spongiosum (spŭn′jē-ō′sŭm) Median column of erectile tissue located between and ventral to the two corpora cavernosa in the penis; posteriorly it forms the bulb of the penis, and anteriorly it terminates as the glans penis; it is traversed by the urethra. In the female, it forms the bulb of the vestibule.

corpus striatum (strī-ā′tŭm) [L., *corpus,* body + *striatus,* striated or furrowed] Caudate nucleus, putamen, and globus pallidus; so-named because of the striations caused by intermixing of gray and white matter, which result from the number of tracts crossing the anterior portion of the corpus striatum.

cortex; pl. **cortices** (kōr′teks, kōr′ti-sēz) [L., bark] Outer portion of an organ (e.g., adrenal cortex or cortex of the kidney).

corticotropin-releasing hormone (CRH) (kōr′ti-kō-trō′pin) Hormone from the hypothalamus that stimulates the anterior pituitary gland to release adrenocorticotropic hormone.

cortisol (kōr′ti-sol) Steroid hormone released by the zona fasciculata of the adrenal cortex; increases blood glucose and inhibits inflammation.

covalent bond (kō-vāl′ent) Chemical bond characterized by the sharing of electrons.

coxal bone (kok′să) Hipbone.

cranial nerve (krā′nē-ăl) Nerve that originates from a nucleus within the brain; there are 12 pairs of cranial nerves.

cranial vault Eight skull bones that surround and protect the brain; braincase.

craniosacral division (krā′nē-ō-sā′krăl) Parasympathetic division of the autonomic nervous system.

cranium (krā′nē-ŭm) [Gr., *kranion,* skull] Skull; in a more limited sense, the braincase.

cremaster muscle (krē-mas′ter) Extension of abdominal muscles originating from the internal oblique muscles; in the male, raises the testicles; in the female, envelops the round ligament of the uterus.

crenation (krē-nā′shŭn) [L., *crena,* notched] Denoting the outline of a shrunken cell.

cricoid cartilage (krī′koyd) Most inferior laryngeal cartilage.

cricothyrotomy (krī′kō-thī-rot′ō-mē) Incision through the skin and cricothyroid membrane for relief of respiratory obstruction.

crista, cristae (kris′tă, kris′tē) [L., crest] Shelflike infolding of the inner membrane of a mitochondrion.

crista ampullaris (kris′tă am-pul′ăr-ĭs) [L., crest] Elevation on the inner surface of the ampulla of each semicircular duct for dynamic or kinetic equilibrium.

critical closing pressure Pressure in a blood vessel below which the vessel collapses, occluding the lumen and preventing blood flow.

crown Part of a tooth that is covered with enamel.

cruciate (kroo′shē-āt) [L., *cruciatus,* cross] Resembling or shaped like a cross.

crus of the penis (krūs) Posterior portion of the corpus cavernosum of the penis attached to the ischiopubic ramus.

crypt (kript) Pitlike depression or tubular recess.

cryptorchidism (krip-tōr′ki-dizm) Failure of the testis to descend.

crystalline (kris′tă-lēn) Protein that fills the epithelial cells of the lens in the eye.

cuboidal Resembling a cube.

cumulus oophorus (ō-of′ōr-ŭs) [L., a heap] Mass of epithelial cells surrounding the oocyte; also called cumulus mass.

cuneiform cartilages (kū′nē-i-fōrm) Small rods of elastic cartilage above each corniculate cartilage in the larynx.

cupula; pl. cupulae (koo′poo-lă, kū′pū-lă, koo′poo-lē) [L., *cupa,* tub] Gelatinous mass that overlies the hair cells of the cristae ampullares of the semicircular ducts.

cuticle (kū′ti-kl) [L., *cutis,* skin] Outer, thin layer, usually horny—for example, the outer covering of hair or the growth of the stratum corneum onto the nail.

cystic duct (sis′tik) Duct leading from the gallbladder; joins the common hepatic duct to form the common bile duct.

cytokine (sī′tō-kīn) Protein or peptide secreted by a cell that regulates the activity of neighboring cells.

cytokinesis (sī′tō-ki-nē′sis) [Gr., *cyto,* cell + *kinsis,* movement] Division of the cytoplasm during cell division.

cytology (sī-tol′ō-jē) [Gr., *kytos,* a hollow (cell) + *logos,* study] Study of anatomy, physiology, pathology, and chemistry of the cell.

cytoplasm (sī′tō-plazm) Protoplasm of the cell surrounding the nucleus.

cytoplasmic inclusion (sī-tō-plaz′mik) Any foreign or other substance contained in the cytoplasm of a cell.

cytotoxic reaction (sī′tō-tok′sik) [Gr., *cyto,* cell + L., *toxic,* poison] Antibodies (IgG or IgM) combine with cells and activate complement, and cell lysis occurs.

cytotrophoblast (sī′tō-trof′ō-blast) Inner layer of the trophoblast, composed of individual cells.

D

Daily Values Dietary reference values useful for planning a healthy diet. The Daily Values are taken from the Reference Daily Intakes (RDIs) and the Daily Reference Values.

Daily Reference Values (DRVs) Recommended amounts in the diet for total fat, saturated fat, cholesterol, total carbohydrate, dietary fiber, sodium, potassium, and protein. The values for total fat, saturated fat, cholesterol, and sodium are the uppermost limits considered desirable because of their link to certain diseases.

Dalton's law Named for English chemist John Dalton (1766–1844). In a mixture of gases, the portion of the total pressure resulting from each type of gas is determined by the percentage of the total volume represented by each gas type.

dartos muscle (dar′tōs) Layer of smooth muscle in the skin of the scrotum; contracts in response to lower temperature and relaxes in response to higher temperature; raises and lowers testes in the scrotum.

deciduous tooth (dē-sid′ū-ŭs) Tooth of the first set of teeth; also called primary tooth.

decussate (dē′kŭ-sāt, dē-kŭs′āt) [L., *decusso,* X-shaped, from *decussis,* ten (X)] To cross.

deep inguinal ring (ing′gwi-năl) Opening in the transverse fascia through which the spermatic cord (or round ligament in the female) enters the inguinal canal.

defecation (def-ĕ-kā′shŭn) [L., *defaeco,* to remove the dregs, purify] Discharge of feces from the rectum.

defecation reflex Combination of local and central nervous system reflexes initiated by distension of the rectum and resulting in the movement of feces out of the lower colon.

deglutition (dē-gloo-tish′ŭn) [L., *de* + *glutio,* to swallow] Act of swallowing.

dendrite (den′drīt) [Gr., *dendrites,* tree] Branching processes of a neuron; receives stimuli and conducts potentials toward the cell body.

dendritic cell (den-drit′ik) Large cells with long, cytoplasmic extensions that are capable of taking up and concentrating antigens, leading to the activation of B or T lymphocytes.

dendritic spine Extension of nerve cell dendrites where axons form synapses with the dendrites; also called gemmule.

dental arch (den′tăl) [L., *arcus,* bow] Curved maxillary or mandibular arch in which the teeth are located.

dentin (den′tin) Bony material forming the mass of the tooth.

deoxyhemoglobin (dē-oks′ē-hē-mō-glō′bin) Hemoglobin without oxygen bound to it.

deoxyribonuclease (de-oks′ē-rī-bō-noo′klē-ās) Enzyme that splits DNA into its component nucleotides.

deoxyribonucleic acid (DNA) (dē-oks′ē-rī′bō-noo-klē′ic) Type of nucleic acid containing deoxyribose as the sugar component, found principally in the nuclei of cells; constitutes the genetic material of cells.

depolarization (dē-pō′lăr-i-zā′shŭn) **phase** Change in the electric charge difference across the plasma membrane that causes the difference to be smaller or closer to 0 mV; phase of the action potential in which the membrane potential moves toward zero, or becomes positive.

depression (dē-presh′ŭn) Movement of a structure in an inferior direction.

depth perception (per-sep′shun) Ability to distinguish between near and far objects and to judge their distance.

dermatome (der′mă-tōm) Area of skin supplied by a spinal nerve.

dermis (der′mis) [Gr., *derma,* skin] Dense irregular connective tissue that forms the deep layer of the skin.

descending aorta Part of the aorta, further divided into the thoracic aorta and abdominal aorta.

descending colon Part of the colon extending from the left colonic flexure to the sigmoid colon.

desmosome (dez′mō-sōm) [Gr., *desmos,* a band + *soma,* body] Point of adhesion between cells. Each contains a dense plate at the point of adhesion and a cementing extracellular material between the cells.

desquamate (des′kwă-māt) [L., *desquamo,* to scale off] Peeling or scaling off of the superficial cells of the stratum corneum.

diabetes insipidus (dī-ă-bē′tēz in-sip′ĭ-dŭs) Chronic excretion of large amounts of urine of low specific gravity accompanied by extreme thirst; results from inadequate output of antidiuretic hormone.

diabetes mellitus (me-lī′tŭs) Metabolic disease in which carbohydrate use is reduced and that of lipid and protein enhanced; caused by a deficiency of insulin or an inability to respond to insulin and is characterized, in more severe cases, by hyperglycemia, glycosuria, water and electrolyte loss, ketoacidosis, and coma.

diapedesis (dī′ă-pë-dē′sis) [Gr., *dia,* through + *pedesis,* a leaping] Passage of blood or any of its formed elements through the intact walls of blood vessels.

diaphragm (dī′ă-fram) Musculomembranous partition between the abdominal and thoracic cavities.

diaphysis (dī-af′i-sis) [Gr., growing between] Shaft of a long bone.

diastole (dī-as′tō-lē) [Gr., *diastole,* dilation] Relaxation of the heart chambers, during which they fill with blood; usually refers to ventricular relaxation.

diencephalon (dī-en-sef′ă-lon) [Gr., *dia,* through + *enkephalos,* brain] Second portion of the embryonic brain; in the inferior core of the adult cerebrum.

diffuse lymphatic tissue Dispersed lymphocytes and other cells with no clear boundary; found beneath mucous membranes, around lymph nodules, and within lymph nodes and spleen.

diffusion (di-fū′zhŭn) [L., *diffundo,* to pour in different directions] Tendency for solute molecules to move from an area of high concentration to an area of low concentration in solution; the product of the constant random motion of all atoms, molecules, or ions in a solution.

diffusion coefficient Measure of how easily a gas diffuses through a liquid or tissue.

digestive tract (di-jes′tiv, dī-jes′tiv) Mouth, oropharynx, esophagus, stomach, small intestine, and large intestine.

digit (dij′it) Finger, thumb, or toe.

dilator pupillae (dī′lā-tĕr pū-pil′ē) Radial smooth muscle cells of the iris diaphragm that cause the pupil of the eye to dilate.

diploid (dip′loyd) Normal number of chromosomes (in humans, 46 chromosomes) in somatic cells.

disaccharide (dī-sak′ă-rīd) Condensation product of two monosaccharides by the elimination of water.

dissociate (di-sō′sē-āt) [L., *dis* + *socio,* to disjoin, separate] Ionization in which ions are dissolved in water and the cations and anions are surrounded by water molecules.

distal convoluted tubule Convoluted tubule of the nephron that extends from the ascending limb of the loop of Henle and ends in a collecting duct.

distributing artery Medium-sized artery with a tunica media composed principally of smooth muscle; regulates blood flow to different regions of the body.

dominant (dom′i-nant) [L., *dominus,* a master] Gene that is expressed phenotypically to the exclusion of a contrasting recessive gene.

dorsal root (dōr′săl) Sensory (afferent) root of a spinal nerve.

dorsal root ganglion (gang′glē-on) Collection of sensory neuron cell bodies within the dorsal root of a spinal nerve; also called spinal ganglion.

ductus arteriosus (dŭk′tŭs ar-tēr′ē-ō-sŭs) Fetal vessel connecting the left pulmonary artery with the descending aorta.

ductus deferens (def′er-enz) Duct of the testicle, running from the epididymis to the ejaculatory duct; also called vas deferens.

duodenal gland (doo′ō-dē′năl, doo-od′ĕ-năl) Small gland that opens into the base of intestinal glands; secretes a mucoid alkaline substance.

duodenocolic reflex (doo-ō-dē′nō-kō-lik) Local reflex resulting in a mass movement of the contents of the colon; produced by stimuli in the duodenum.

duodenum (doo-ō-dē′nŭm, doo-od′ĕ-nŭm) [L., *duodeni*] First division of the small intestine; connects to the stomach.

dura mater (doo′ră mā′ter) [L., hard mother] Tough, fibrous membrane forming the outer covering of the brain and spinal cord.

E

eardrum (ēr′drŭm) Cellular membrane that separates the external from the middle ear; vibrates in response to sound waves; also called tympanic membrane.

ectoderm (ek′tō-derm) Outermost of the three germ layers of an embryo.

ectopic focus; pl. **foci** (ek-top′ik fōkŭs, fō′sī) Any pacemaker other than the sinus node of the heart; abnormal pacemaker; an ectopic pacemaker.

edema (e-dē′mă) [Gr., *oidema,* a swelling] Excessive accumulation of fluid within or around cells, usually causing swelling.

effector T cell (ē-fek′tŏr, ē-fek′tōr) Subset of T lymphocytes that is responsible for cell-mediated immunity.

efferent arteriole (ef′er-ent ar-tēr′ē-ōl) Vessel that carries blood from the glomerulus to the peritubular capillaries.

efferent division Nerve fibers that send impulses from the central nervous system to the periphery.

efferent ductule (ef′er-ent dŭk′tool) [L., *ductus,* duct] One of a number of small ducts leading from the testis to the head of the epididymis.

ejaculation (ē-jak-ū-lā′shŭn) Reflexive expulsion of semen from the penis.

ejaculatory duct (ē-jak′ū-lă-tōr-ē) Duct formed by the union of the ductus deferens and the excretory duct of the seminal vesicle; opens into the prostatic urethra.

ejection period (ē-jek′shŭn) Time in the cardiac cycle when the semilunar valves are open and blood is being ejected from the ventricles into the arterial system.

elastin (ĕ-las′tin) Yellow, elastic, fibrous mucoprotein that is the major connective tissue protein of elastic structures (e.g., large blood vessels and elastic ligaments).

electrocardiogram (ECG, EKG) (ē-lek-trō-kar′dē-ō-gram) [Gr., *elektron,* amber + *kardia,* heart + *gramma,* a drawing] Graphic record of the heart's electric currents obtained with an electrocardiograph.

electrolyte (ē-lek′trō-līt) [Gr., *electro* + *lytos,* soluble] Cation or anion in solution that conducts an electric current.

electron (ē-lek′tron) Negatively charged subatomic particle in an atom.

electron-transport chain Series of electron carriers in the inner mitochondrial membrane; they receive electrons from NADH and $FADH_2$, using the electrons in the formation of ATP and water.

element (el′ĕ-ment) [L., *elementum,* rudiment, beginning] Substance composed of atoms of only one kind.

elevation (el-ĕ-vā′shŭn) Movement of a structure in a superior direction.

embolism (em′bō-lizm) [Gr., *embolisma,* a piece of patch, literally something thrust in] Obstruction or occlusion of a vessel by a transported clot, a mass of bacteria, or other foreign material.

embolus; pl. **emboli** (em′bō-lŭs, em′bō-lī) [Gr., *embolos,* plug, wedge, or stopper] Plug, composed of a detached clot, a mass of bacteria, or another foreign body, occluding a blood vessel.

embryo (em′brē-ō) Developing human from the first to the eighth week of development.

embryonic disk (em-brē-on′ik) Point in the inner cell mass at which the embryo begins to be formed.

embryonic period From approximately the second to the eighth week of development, during which the major organ systems are organized.

emission (ē-mish′ŭn) [L., *emissio,* to send out] Discharge; accumulation of semen in the urethra prior to ejaculation. A nocturnal emission is a discharge of semen while asleep.

emmetropia (em-ĕ-trō′pē-ă) [Gr., *emmetros,* according to measure + *ops,* eye] In the eye, the state of refraction in which parallel rays are focused exactly on the retina; no accommodation is necessary.

emulsify (ē-mŭl′si-fī) To form an emulsion.

enamel (ē-nam′ĕl) Hard substance covering the exposed portion of the tooth.

endocardium; pl. **endocardia** (en-dō-kar′dē-ŭm, en-dō-kar′dē-ă) Innermost layer of the heart, including endothelium and connective tissue.

endocrine gland (en′dō-krin) [Gr., *endon,* inside + *krino,* to separate] Ductless gland that secretes a hormone internally, usually into the circulation.

endocytosis (en′dō-sī-tō′sis) Bulk uptake of material through the cell membrane.

endoderm (en′dō-derm) Innermost of the three germ layers of an embryo.

endolymph (en′dō-limf) [Gr., *endo* + L., *lympha,* clear fluid] Fluid found within the membranous labyrinth of the inner ear.

endometrium; pl. **endometria** (en′dō-mē′trē-ŭm, en′dō-mē′trē-ă) Mucous membrane composing the inner layer of the uterine wall; consists of a simple columnar epithelium and a lamina propria that contains simple tubular uterine glands.

endomysium (en′dō-miz′ē-ŭm, en′dō-mis′ē-ŭm) [Gr., *endo,* within + *mys,* muscle] Fine connective tissue sheath surrounding a muscle fiber.

endoneurium (en-dō-noo′rē-ŭm) [Gr., *endo,* within + *neuron,* nerve] Delicate connective tissue surrounding individual nerve fibers within a peripheral nerve.

endoplasmic reticulum; pl. **reticula** (en′dō-plas′mik re-tik′ū-lŭm, re-tik′ū-lă) Double-walled membranous network inside the cytoplasm; rough has ribosomes attached to the surface; smooth does not have ribosomes attached.

endorphin (en-dōr′fin) Opiate-like polypeptide found in the brain and other parts of the body; binds in the brain to the same receptors that bind exogenous opiates.

endosteum (en-dos′tē-ŭm) [Gr., *endo,* within + *osteon,* bone] Membranous lining of the medullary cavity and the cavities of spongy bone.

endothelium; pl. **endothelia** (en-dō-thē′lē-ŭm, en-dō-thē′lē-ă) [Gr., *endo,* within + *thele,* nipple] Layer of flat cells lining blood and lymphatic vessels and the chambers of the heart.

enkephalin (en-kef′ă-lin) Pentapeptide found in the brain; binds to specific receptor sites, some of which may be pain-related opiate receptors.

enteric nervous system (ENS) Complex network of neuron cell bodies and axons within the wall of the digestive tract; capable of controlling the digestive tract independently of the central nervous system through local reflexes.

enterokinase (en′tĕr-ō-kī′nās) Intestinal proteolytic enzyme that converts trypsinogen into trypsin.

enzyme (en′zīm) [Gr., *en,* in + *zyme,* leaven] Protein that acts as a catalyst.

eosinophil (ē-ō-sin′ō-fil) [Gr., *eos,* dawn + *philos,* fond] White blood cell that stains with acidic dyes; inhibits inflammation.

epicardium (ep-i-kar′dē-ŭm) [Gr., *epi,* on + *kardia,* heart] Serous membrane covering the surface of the heart; also called visceral pericardium.

epidermis (ep-i-derm′is) [Gr., *epi,* on + *derma,* skin] Outer portion of the skin formed of epithelial tissue that rests on or covers the dermis.

epididymis; pl. **epididymides** (ep-i-did′i-mis, -di-dim′i-dēz) [Gr., *epi,* on + *didymos,* twin] Elongated structure connected to the posterior surface of the testis, which consists of the head, body, and tail; site of storage and maturation of the spermatozoa.

epiglottis (ep-i-glot′is) [Gr., *epi,* on + *glottis,* mouth of the windpipe] Plate of elastic cartilage covered with mucous membrane; serves as a valve over the glottis of the larynx during swallowing.

epimysium (ep-i-mis′ē-ŭm) [Gr., *epi,* on + *mys,* muscle] Fibrous envelope surrounding a skeletal muscle.

epinephrine (ep′i-nef′rin) Hormone (amino acid derivative) similar in structure to the neurotransmitter norepinephrine; major hormone released from the adrenal medulla; increases cardiac output and blood glucose levels; also called adrenaline.

epineurium (ep-i-noo′rē-ŭm) [Gr., *epi,* on + *neuron,* nerve] Connective tissue sheath surrounding a nerve.

epiphyseal line (ep-i-fiz′ē-ăl) Dense plate of bone in a bone that is no longer growing, indicating the former site of the epiphyseal plate.

epiphyseal plate Site at which bone growth in length occurs; located between the epiphysis and diaphysis of a long bone; area of hyaline cartilage where cartilage growth is followed by endochondral ossification; also called metaphysis or growth plate.

epiphysis; pl. **epiphyses** (e-pif′i-sis, e-pif′i-sēz) [Gr., *epi,* on + *physis,* growth] Portion of a bone developed from a secondary ossification center and separated from the remainder of the bone by the epiphyseal plate.

epiploic appendage (ep′i-plō′ik) One of a number of little processes of peritoneum projecting from the serous coat of the large intestine except the rectum; they are generally distended with fat.

epithelium (ep-i-thē′lē-ŭm) [Gr., *epi,* on + *thele,* nipple] One of the four primary tissue types. *Nipple* refers to the tiny capillary-containing connective tissue in the lips, which is where the term was first used. The use of the term was later expanded to include all covering and lining surfaces of the body.

epitope (ep′i-tōp) [Gr., *epi,* on + *top,* place] *See* antigenic determinant.

eponychium (ep-ō-nik′ē-ŭm) [Gr., *epi,* on + *onyx,* nail] Outgrowth of the skin that covers the proximal and lateral borders of the nail; also called cuticle.

erection (ē rek′shŭn) [L., *erectio*, to set up] Condition of erectile tissue when filled with blood; tissue becomes hard and unyielding; especially refers to this state of the penis.

erythrocyte (ĕ-rith′rō-sīt) [Gr., *erythros*, red + *kytos*, cell] Red blood cell; biconcave disk containing hemoglobin.

erythropoiesis (ĕ-rith′rō-poy-ē′sis) [*erythrocyte* + Gr., *poiesis*, a making] Production of erythrocytes.

erythropoietin (EPO) (ĕ-rith′rō-poy′ĕ-tin) Protein that enhances erythropoiesis by stimulating the formation of proerythroblasts and the release of reticulocytes from bone marrow.

esophagus; pl. **esophagi** (ē-sof′ă-gŭs, ē-sof′ă-gī, ē-sof′ă-jī) [Gr., *oisophagos*, gullet] Portion of the digestive tract between the pharynx and stomach.

essential amino acid Amino acid, required by animals, that must be supplied in the diet.

estrogen (es′trō-jen) Substance that exerts biologic effects characteristic of estrogen hormone, such as stimulating female secondary sexual characteristics, growth, and maturation of long bones, and help controlling the menstrual cycle.

evagination (ē-vaj-i-nā′shŭn) [L., *e*, out + *vagina*, sheath] Protrusion of some part or organ from its normal position.

evaporation (ē-vap-ŏ-ra′shŭn) [L., *e*, out + *vaporare*, to emit vapor] Change from liquid to vapor form.

eversion (ē-ver′zhŭn) [L., *everto*, to overturn] Turning outward.

excitation-contraction coupling (ek-sī-tā′shŭn kon-trak′shŭn kŭp′ling) Stimulation of a muscle fiber produces an action potential that results in contraction of the muscle fiber.

excitatory postsynaptic potential (EPSP) (ek-sī′tă-tō-rē pōst-si-nap′tik pō-ten′shăl) Depolarization in the postsynaptic membrane that brings the membrane potential close to threshold.

exocrine gland (ek′sō-krin) [Gr., *exo*, outside + *krino*, to separate] Gland that secretes to a surface or outward through a duct.

exocytosis (ek′sō-sī-to′sis) Elimination of material from a cell through the formation of vacuoles.

expiratory reserve volume Maximum volume of air that can be expelled from the lungs after a normal expiration.

extension (eks-ten′shŭn) [L., *extensio*, to stretch out] To stretch out.

external anal sphincter Ring of striated muscular fibers surrounding the anus.

external auditory canal Short canal that opens to the exterior environment and terminates at the eardrum; part of the external ear.

external ear Portion of the ear that includes the auricle and external acoustic meatus; terminates at the eardrum.

external nose Nostril; anterior or external opening of the nasal cavity.

external spermatic fascia Outer fascial covering of the spermatic cord.

external urethral orifice Slitlike opening of the urethra in the glans penis.

external urinary sphincter Sphincter skeletal muscle around the base of the urethra external to the internal urinary sphincter.

exteroceptor (eks′ter-ō-sep′ter, eks′ter-ō-sep′tōr) [L., *exterus*, external + *receptor*, receiver] Sensory receptor in the skin or mucous membranes that responds to stimulation by external agents or forces.

extracellular (eks-tră-sel′ū-lăr) Outside the cell.

extracellular matrix; pl. **matrices** (eks-tră-sel′ū-lăr mā′triks, mā′tri-sēz) Nonliving chemical substances located between connective tissue cells.

extrinsic clotting pathway (eks-trin′sik) Series of chemical reactions resulting in clot formation; begins with chemicals (e.g., tissue thromboplastin) found outside the blood.

extrinsic muscle Muscle located outside the structure being moved.

eyebrow Short hairs on the bony ridge above the eyes.

eyelash Hair at the margins of the eyelids.

eyelid Movable fold of skin in front of the eyeball; also called palpebra.

F

F actin (ak′tin) Fibrous actin molecule that is composed of a series of globular actin molecules (G actin).

facilitated diffusion Carrier-mediated process that does not require ATP and moves substances into or out of cells from a high to a low concentration.

falciform ligament (fal′si-fōrm lig′ă-ment) Fold of peritoneum extending to the surface of the liver from the diaphragm and anterior abdominal wall.

fallopian tube (fa-lō′pē-an) *See* uterine tube.

false pelvis Portion of the pelvis superior to the pelvic brim; composed of the bone on the posterior and lateral sides and by muscle on the anterior side; also called greater pelvis.

falx cerebelli (falks ser-ĕ-bel′ī) Dural fold between the two cerebellar hemispheres.

falx cerebri (falx ser′ĕ-brī) Dural fold between the two cerebral hemispheres.

far point of vision Distance from the eye where accommodation is not needed to have the image focused on the retina.

fascia; pl. **fasciae** (fash′ē-ă, fash′ē-ē) [L., band or fillet] Loose areolar connective tissue found beneath the skin (hypodermis) or dense connective tissue that encloses and separates muscles.

fasciculus (fă-sik′ū-lŭs) [L., *fascis*, bundle] Band or bundle of nerve or muscle fibers bound together by connective tissue.

fat [A.S., *faet*] Greasy, soft-solid material found in animal tissues and many plants; composed of two types of molecules: glycerol and fatty acids.

fatigue (fă-tēg′) [L., *fatigo*, to tire] Period characterized by a reduced capacity to do work.

fat-soluble vitamin Vitamin, such as A, D, E, and K, that is soluble in lipids and absorbed from the intestine along with lipids.

fauces (faw′sēz) [L., throat] Space between the cavity of the mouth and the pharynx.

female climacteric Period of life occurring in women, encompassing termination of the reproductive period and characterized by endocrine, somatic, and transitory psychologic changes and ultimately menopause; also called perimenopause.

female pronucleus Nuclear material of the ovum after the ovum has been penetrated by the spermatozoon. Each pronucleus carries the haploid number of chromosomes.

fertilization (fer′til-i-zā′shŭn) Process that begins with the penetration of the secondary oocyte by the spermatozoon and is completed with the fusion of the male and female pronuclei.

fetus (fē′tus) Developing human following the embryonic period (after 8 weeks of development).

fibrin (fī′brin) Elastic filamentous protein derived from fibrinogen by the action of thrombin, which releases peptides from fibrinogen in coagulation of the blood.

fibroblast (fī′brō-blast) [L., *fibra*, fiber + Gr., *blastos*, germ] Spindle-shaped or stellate cells that form connective tissue.

fibrocyte (fī′brō-sīt) Mature cell of fibrous connective tissue.

fibrous joint (fī′brŭs) Bones connected by fibrous tissue with no joint cavity; includes sutures, syndesmoses, and gomphoses.

fibrous tunic Outer layer of the eye; composed of the sclera and the cornea.

filiform (fil′i-fōrm) Filament-shaped.

filtrate (fil′trāt) Liquid that has passed through a filter—for example, fluid that enters the nephron through the filtration membrane of the glomerulus.

filtration (fil-trā′shŭn) Movement, due to a pressure difference, of a liquid through a filter that prevents some or all of the substances in the liquid from passing through.

filtration fraction Fraction of the plasma entering the kidney that filters into Bowman's capsule. Normally, it is around 19%.

filtration membrane Membrane formed by the glomerular capillary endothelium, the basement membrane, and the podocytes of Bowman's capsule.

filtration pressure Pressure gradient that forces fluid from the glomerular capillary through the filtration membrane into Bowman's capsule; glomerular capillary pressure minus glomerular capsule pressure minus colloid osmotic pressure.

fimbria; pl. **fimbriae** (fim′brē-ă, fim′brē-ē) [L., fringe] Fringelike structure located at the ostium of the uterine tube.

first messenger *See* intercellular chemical signal.

fixator (fik-sā′ter) Muscle that stabilizes the origin of a prime mover.

flagellum; pl. **flagella** (flă-jel′ŭm, flă-jel′ă) [L., whip] Whiplike locomotory organelle of constant structural arrangement consisting of double peripheral microtubules and two single central microtubules.

flatus (flā′tŭs) [L., a blowing] Gas or air in the gastrointestinal tract that may be expelled through the anus.

flexion (flek′shŭn) [L., *flectus*] Bending.

focal point Point at which light rays cross after passing through a concave lens, such as the lens of the eye.

foliate (fō′lē-āt) Leaf-shaped.

follicle-stimulating hormone (FSH) (fol′i-kl) Hormone of the adenohypophysis that, in females, stimulates the Graafian follicles of the ovary and assists in follicular maturation and the secretion of estrogen; in males, FSH stimulates the epithelium of the seminiferous tubules and is partially responsible for inducing spermatogenesis.

follicular phase (fō-lik′ū-lăr) Time between the end of menses and ovulation, characterized by rapid division of endometrial cells and development of follicles in the ovary; also called proliferative phase.

foramen; pl. **foramina** (fō-rā′men, fō-ram′i-nă) Hole.

foramen ovale (o-val′ē) In the fetal heart, the oval opening in the septum secundum; the persistent part of septum primum acts as a valve for this interatrial communication during fetal life; postnatally, the septum primum becomes fused to the septum secundum to close the foramen ovale, forming the fossa ovale.

force That which produces a motion in the body; pull.

foregut Cephalic portion of the primitive digestive tube in the embryo.

foreskin *See* prepuce.

formed elements Cells (i.e., red and white blood cells) and cell fragments (i.e., platelets) of blood.

formula unit Relative number of cations and ions in an ionic compound.

fornix (fōr′niks) [L., arch, vault] Recess at the cervical end of the vagina. Recess deep to each eyelid where the palpebral and bulbar conjunctivae meet.

fovea centralis (fō′vē-ă) Depression in the middle of the macula where there are only cones and no blood vessels.

free energy Total amount of energy that can be liberated by the complete catabolism of food.

frenulum (fren′ū-lŭm) [L., *frenum,* bridle] Fold extending from the floor of the mouth to the midline of the undersurface of the tongue.

frequency-modulated signals Signals, all of which are identical in amplitude, that differ in their frequency—for example, strong stimuli may initiate a high frequency of action potentials and weak stimuli may initiate a low frequency of action potentials.

FSH surge Increase in plasma follicle-stimulating hormone (FSH) levels before ovulation.

fulcrum (F) (ful′krŭm) Pivot point.

fundus (fŭn′dŭs) [L., bottom] Bottom, or rounded end, of a hollow organ—for example, the fundus of the stomach or uterus.

fungiform (fŭn′ji-fōrm) Mushroom-shaped.

G

G actin (jē ak′tin) Globular protein molecules that, when bound together, form fibrous actin (F actin).

gallbladder (gawl′blad-er) Pear-shaped receptacle on the inferior surface of the liver; serves as a storage reservoir for bile.

gamete (gam′ēt) Ovum or spermatozoon.

gamma globulin (gam′ă glob′ū-lin) [L., *globulus,* globule] Plasma proteins that include the antibodies.

ganglion; pl. **ganglia** (gang′glē-on, gang′glē-ă) [Gr., swelling, or knot] Any group of nerve cell bodies in the peripheral nervous system.

gap junction Small channel between cells that allows the passage of ions and small molecules between cells; provides means of intercellular communication.

gastric gland (gas′trik) Gland located in the mucosa of the fundus and body of the stomach.

gastric inhibitory polypeptide Hormone secreted by the duodenum that inhibits gastric acid secretion.

gastric pit Small pit in the mucous membrane of the stomach, at the bottom of which are the mouths of the gastric glands that secrete mucus, hydrochloric acid, intrinsic factor, pepsinogen, and hormones.

gastrin (gas′trin) Hormone secreted in the mucosa of the stomach and duodenum that stimulates the secretion of hydrochloric acid by the parietal cells of the gastric glands.

gastrocolic reflex (gas′trō-kol′ik) Local reflex resulting in mass movement of the contents of the colon, which occurs after the entrance of food into the stomach.

gastroesophageal opening (gas′trō-ē-sof′ā-jē′āl) Opening of the esophagus into the stomach; also called cardiac opening.

gene (jēn) [Gr., *genos,* birth, descent] Functional unit of heredity. Each gene occupies a specific place, or locus, on a chromosome; it is capable of reproducing itself exactly at each cell division; it often is capable of directing the formation of an enzyme or another protein.

genetics (jĕ-net′iks) [Gr., *genesis,* origin or production] Branch of science that deals with heredity.

genital fold (jen′i-tāl) Paired longitudinal ridges developing in the embryo on each side of the urogenital orifice. In the male, they form part of the penis; in the female, they form the labia minora.

genital tubercle (jen′i-tāl) Median elevation just cephalic to the urogenital orifice of an embryo; gives rise to the penis of the male or the clitoris of the female.

genotype (jen′ō-tīp) [Gr., *genos,* birth, descent + *typos,* type] Genetic makeup of an individual.

germ cell (jerm) Spermatozoon or ovum.

germ layer One of three layers in the embryo (ectoderm, endoderm, or mesoderm) from which the four primary tissue types arise.

germinal center Lighter-staining center of a lymphatic nodule; area of rapid lymphocyte division.

gingiva (jin′ji-vă) Dense fibrous tissue, covered by mucous membrane, that covers the alveolar processes of the upper and lower jaws and surrounds the necks of the teeth.

girdle (ger′dl) Belt or zone; the bony region where the limbs attach to the body.

gland [L., *glans,* acorn] Secretory organ from which secretions may be released into the blood, into a cavity, or onto a surface.

glans penis [L., *glans,* acorn] Conical expansion of the corpus spongiosum that forms the head of the penis.

globin (glō′bin) Protein portion of hemoglobin.

glomerular capillary pressure (GCP) (glō-mār′ū-lăr) Blood pressure within the glomerulus.

glomerular filtration rate (GFR) Amount of plasma (filtrate) that filters into Bowman's capsules per minute.

glomerulus (glō-mār′ū-lŭs) [L., *glomus,* ball of yarn] Mass of capillary loops at the beginning of each nephron, nearly surrounded by Bowman's capsule.

glottis (glot′is) [Gr., aperture of the larynx] Vocal apparatus; includes vocal folds and the cleft between them.

glucocorticoid (gloo-kō-kōr′ti-koyd) Steroid hormone (e.g., cortisol) released by zonula fasciculata of the adrenal cortex; increases blood glucose and inhibits inflammation.

gluconeogenesis (gloo′kō-nē-ō-jen′ē-sis) [Gr., *glykys,* sweet + *neos,* new + *genesis,* production] Formation of glucose from noncarbohydrates, such as proteins (amino acids) or lipids (glycerol).

glycogenesis (glī′kō-jĕ-nō′-sis) Formation of glycogen from glucose molecules.

glycolysis (glī-kol′i-sis) [Gr., *glykys,* sweet + *lysis,* a loosening] Anaerobic process during which glucose is converted to pyruvic acid; net of two ATP molecules is produced during glycolysis.

goblet cell Mucus-producing epithelial cell that has its apical end distended with mucin.

Golgi apparatus (gol′jē) Named for Camillo Golgi, Italian histologist and Nobel laureate (1843–1926). Specialized endoplasmic reticulum that concentrates and packages materials for secretion from the cell.

Golgi tendon organ Proprioceptive nerve ending in a tendon.

gomphosis (gom-fō′sis) [Gr., *gomphos,* bolt, nail + *osis,* condition] Fibrous joint in which a peglike process fits into a hole.

gonad (gō′nad) [Gr., *gone,* seed] Organ that produces sex cells; testis of a male or ovary of a female.

gonadal ridge (gō-nad′āl) Elevation on the embryonic mesonephros; primordial germ cells become embedded in it, establishing it as a testis or an ovary.

gonadotropin (gō′nad-ō-trō′pin) Hormone capable of promoting gonadal growth and function. Two major gonadotropins are luteinizing hormone (LH) and follicle-stimulating hormone (FSH).

gonadotropin-releasing hormone (GnRH) Hypothalamic-releasing hormone that stimulates the secretion of gonadotropins (LH and FSH) from the adenohypophysis; also called luteinizing hormone–releasing hormone (LHRH).

granulocyte (gran′ū-lō-sīt) Mature granular white blood cell (neutrophil, basophil, or eosinophil).

granulosa cell (gran-ū-lō′să) Cell in the layer surrounding the primary follicle.

gray matter Collection of nerve cell bodies, their dendritic processes, and associated neuroglial cells within the central nervous system.

gray ramus communicans; pl. **rami communicantes** (rā′mŭs kō-mū′nī-kans, rā′mī kō-mū-nī-kan′tēz) Connection between a spinal nerve and a sympathetic chain ganglion through which unmyelinated postganglionic axons project.

greater omentum Peritoneal fold passing from the greater curvature of the stomach to the transverse colon, hanging like an apron in front of the intestines.

greater vestibular gland One of two mucus-secreting glands on each side of the lower part of the vagina. The equivalent of the bulbourethral glands in the male.

growth hormone Hormone that stimulates general growth of the individual; stimulates cellular amino acid uptake and protein synthesis; also called somatotropin.

gubernaculum (goo′ber-nak′ū-lŭm) [L., helm] Column of tissue that connects the fetal testis to the developing scrotum; involved in testicular descent.

gustatory (gŭs′tă-tōr-ē) Associated with the sense of taste.

gustatory hair Microvillus of gustatory cell in a taste bud.

gynecomastia (gī′nē-kō-mas′tē-ă) [Gr., *gyne,* woman + *mastos,* breast] Excessive development of the male mammary glands, which sometimes secrete milk.

H

H zone Area in the center of the A band in which there are no actin myofilaments; contains only myosin.

hair [A.S., hear] Columns of dead keratinized epithelial cells.

hair follicle Invagination of the epidermis into the dermis; contains the root of the hair and receives the ducts of sebaceous and apocrine glands.

Haldane effect Named for Scottish physiologist John S. Haldane (1860–1936). Hemoglobin that is not bound to carbon dioxide binds more readily to oxygen than hemoglobin that is bound to carbon dioxide.

half-life Time it takes for one-half of an administered substance to be lost through biologic processes.

haploid (hap′loyd) One set of chromosomes, in contrast to diploid; characteristic of gametes.

hapten (hap′ten) [Gr., *hapto,* to fasten] Small molecule that binds to a large molecule; together they stimulate the specific immune system.

hard palate Floor of the nasal cavity that separates the nasal cavity from the oral cavity; composed of the palatine processes of the maxillary bones and the horizontal plates of the palatine bones; also called bony palate.

haustra (haw′stră) [L., machine for drawing water] Sacs of the colon, caused by contraction of the taeniae coli, which are slightly shorter than the gut so that the latter is thrown into pouches.

haversian canal (ha-ver′shan) Named for seventeenth-century English anatomist Clopton Havers (1650–1702). Canal containing blood vessels, nerves, and loose connective tissue and running parallel to the long axis of the bone.

heart skeleton Fibrous connective tissue that provides a point of attachment for cardiac muscle cells, electrically insulates the atria from the ventricles, and forms the fibrous rings around the valves.

heat energy Energy that results from the random movement of atoms, ions, or molecules; the greater the amount of heat energy in an object, the higher the object's temperature.

helicotrema (hel′i-kō-trē′mă) [Gr., *helix,* spiral + *traema,* hole] Opening at the apex of the cochlea through which the scala vestibuli and the scala tympani of the cochlea connect.

helper T cell Subset of T lymphocytes that increases the activity of B cells and T cells.

hematocrit (hē′mă-tō-krit) [Gr., *hemato,* blood + *krin,* to separate] Percentage of blood volume occupied by erythrocytes.

hematoma (he-ma-to′ma) Localized mass of blood released from blood vessels but confined within an organ or a space; the blood is usually clotted.

heme (hēm) Oxygen-carrying, color-furnishing part of hemoglobin.

hemidesmosome (hem-ē-des′mō-sōm) Similar to half a desmosome, attaching epithelial cells to the basement membrane.

hemocytoblast (he′mo-si′to-blast) Blood stem cell derived from mesenchyme that can give rise to red and white blood cells and platelets.

hemoglobin (hē-mō-glō′bin) Red, respiratory protein of red blood cells; consists of 6% heme and 94% globin; transports oxygen and carbon dioxide.

hemolysis (hē-mol′i-sis) [Gr., *haima* + *lysis,* destruction] Destruction of red blood cells in such a manner that hemoglobin is released.

hemolytic disease of the newborn (HDN) Destruction of erythrocytes in the fetus or newborn caused by antibodies produced in an Rh-negative mother acting on the Rh-positive blood of the fetus or newborn.

hemopoiesis (hē′mō-poy-ē′sis) [Gr., *haima,* blood + *poiesis,* a making] Formation of the formed elements of blood—that is, red blood cells, white blood cells, and platelets; also called hematopoiesis.

hemopoietic tissue (hē′mō-poy-et′ik) [Gr., *haima,* blood + *poiesis,* to make] Blood-forming tissue.

hemostasis (hē′mō-stā-sis) Arrest of bleeding.

Henry's law Named for English chemist William Henry (1775–1837). The concentration of a gas dissolved in a liquid is equal to the partial pressure of the gas over the liquid times the solubility coefficient of the gas.

heparin (hep′ă-rin) Anticoagulant that prevents platelet agglutination and thus prevents thrombus formation.

hepatic artery (he-pa′tik) Branch of the aorta that delivers blood to the liver.

hepatic cord Plate of liver cells that radiates away from the central vein of a liver lobule.

hepatic portal system System of portal veins that carries blood from the intestines, stomach, spleen, and pancreas to the liver.

hepatic portal vein Portal vein formed by the superior mesenteric and splenic veins and entering the liver.

hepatic sinusoid (si′nŭ-soyd) Terminal blood vessel having an irregular and larger caliber than an ordinary capillary within the liver lobule.

hepatic vein Vein that drains the liver into the inferior vena cava.

hepatocyte (hep′ă-tō-sīt) Liver cell.

hepatopancreatic ampulla Dilation within the major duodenal papilla that normally receives both the common bile duct and the main pancreatic duct.

hepatopancreatic ampullar sphincter Smooth muscle sphincter of the hepatopancreatic ampulla; also called sphincter of Oddi.

Hering-Breuer reflex (her′ing broy′er) Named for German physiologist Heinrich Ewald Hering (1866–1948) and Austrian internist Josef Breuer (1842–1925). Sensory impulses from stretch receptors in the lungs arrest inspiration; expiration then occurs.

heterozygous (het′er-ō-zī′gŭs) [Gr., *heteros,* other + *zygon,* yoke] Having different allelic genes at one or more paired loci in homologous chromosomes.

hiatus (hī-ā′tŭs) [L., aperture, to yawn] Opening.

hilum (hī′lŭm) [L., small bit or trifle] Indented surface on many organs, serving as a point where nerves and vessels enter or leave.

hindgut Caudal or terminal part of the embryonic gut.

histamine (his′tă-mēn) Amine released by mast cells and basophils that promotes inflammation.

histology (his-tol′ō-jē) [Gr., *histo,* web (tissue) + *logos,* study] Science that deals with the microscopic structure of cells, tissues, and organs in relation to their function.

holocrine gland (hol′ō-krin) [Gr., *holos,* complete + *krino,* to separate] Gland whose secretion is formed by the disintegration of entire cells (e.g., sebaceous gland).

homeostasis (hō′mē-ō-stā′sis) [Gr., *homoio,* like + *stasis,* a standing] State of equilibrium in the body with respect to functions and composition of fluids and tissues.

homologous (hŏ-mol′ō-gŭs) [Gr., ratio or relation] Alike in structure or origin.

homozygous (hō-mō-zī′gŭs) [Gr., *homos,* the same + *zygon,* yoke] State of having identical allelic genes at one or more paired loci in homologous chromosomes.

hormone (hōr′mōn) [Gr., *hormon,* to set into motion] Substance secreted by endocrine tissues into the blood that acts on a target tissue to produce a specific response.

hormone receptor Protein or glycoprotein molecule of cells that specifically binds to hormones and produces a response.

horn Subdivision of gray matter in the spinal cord. The axons of sensory neurons synapse with neurons in the posterior horn, the cell bodies of motor neurons are in the anterior horn, and the cell bodies of autonomic neurons are in the lateral horn.

human chorionic gonadotropin (hCG) Hormone produced by the placenta; stimulates the secretion of testosterone by the fetus; during the first trimester, it stimulates ovarian secretion from the corpus luteum of the estrogen and progesterone required for the maintenance of the placenta. In a male fetus, it stimulates the secretion of testosterone by the fetal testis.

humoral immunity (hū′mōr-ăl) [L., *humor,* a fluid] Immunity due to antibodies in serum.

hyaline cartilage (hī′ă-lin) [Gr., *hyalos,* glass] Gelatinous, glossy cartilage tissue consisting of cartilage cells and their matrix; contains collagen, proteoglycans, and water.

hyaluronic acid (hī′ă-loo-ron′ik) Mucopolysaccharide made up of alternating β-(1,4)-linked residues of hyalobiuronic acid, forming a gelatinous material in the tissue spaces and acting as a lubricant and shock absorbant generally throughout the body.

hydrochloric acid (HCl) (hī-drō-klōr′ik) Acid of gastric juice.

hydrogen bond (hī′drō-jen) Hydrogen atoms bound covalently to either N or O atoms have a small positive charge that is weakly attracted to the small negative charge of other atoms, such as O or N; it can occur within a molecule or between different molecules.

hydrophilic (hi-dro-fil′ik) [Gr., *hydro,* water + *philos,* love] Denoting the property of attracting or associating with water molecules, possessed by polar molecules and ions; the opposite of hydrophobic.

hydrophobic (hi-dro-fob′ik) [Gr., *hydro,* water + *phobos,* fear] Lacking an attraction to water, possessed by nonpolar molecules; the opposite of hydrophilic.

hydroxyapatite (hī-drok′sē-ap′ă-tīt) Mineral with the empiric formula $3\ Ca_3(PO_4)_2 \cdot Ca(OH)_2$; the main mineral of bone and teeth.

hymen (hī′men) [Gr., membrane] Thin, membranous fold partly occluding the vaginal external orifice; normally disrupted by sexual intercourse or other mechanical phenomena.

hyoid bone (hī′oyd) (Gr., *hyoeides,* shaped like the Greek letter epsilon [ϵ]) U-shaped bone between the mandible and larynx.

hypercalcemia (hī′per-kal-sē′mē-ă) Abnormally high levels of calcium in the blood.

hypercapnia (hī′per-kap′nē-ă) Higher than normal levels of carbon dioxide in the blood or tissues.

hyperkalemia (hī′per-kă-lē′mē-ă) Greater than normal concentration of potassium ions in the circulating blood.

hypernatremia (hī′per-nă-trē′mē-ă) Abnormally high plasma concentration of sodium ions.

hyperosmotic (hī′per-oz-mot′īk) [Gr., *hyper,* above + *osmos,* an impulsion] Having a greater osmotic concentration or pressure than a reference solution.

hyperpolarization (hī′per-pō′lăr-i-zā′shŭn) Increase in the charge difference across the plasma membrane; causes the charge difference to move away from 0 mV.

hypertonic (hī-per-ton′ik) [Gr., *hyper,* above + *tonos,* tension] Solution that causes cells to shrink.

hypertrophy (hī-per′trō-fē) [Gr., *hyper,* above + *trophe,* nourishment] Increase in bulk or size; not due to an increase in number of individual elements.

hypocalcemia (hī-pō-kal-sē′mē-ă) Abnormally low levels of calcium in the blood.

hypocapnia (hī′pō-kap′nē-ă) Lower than normal levels of carbon dioxide in the blood or tissues.

hypodermis (hī′pō-der′mis) [Gr., *hypo,* under + *dermis,* skin] Loose areolar connective tissue found deep to the dermis that connects the skin to muscle or bone.

hypokalemia (hī′pō-ka-lē′mē-ă) Abnormally small concentration of potassium ions in the blood.

hyponatremia (hī′pō-nā-trē′mē-ă) Abnormally low plasma concentration of sodium ions.

hyponychium (hī-pō-nik′ē-ūm) [Gr., *hypo,* under + *onyx,* nail] Thickened portion of the stratum corneum under the free edge of the nail.

hypopolarization Change in the electric charge difference across the plasma membrane that causes the charge difference to be smaller or move closer to 0 mV.

hyposmotic (hī′pos-mot′ik) [Gr., *hypo,* under + *osmos,* an impulsion] Having a lower osmotic concentration or pressure than a reference solution.

hypospadias (hī′pō-spā′dē-ăs) [Gr., one having the orifice of the penis too low; *hypospao,* to draw away from under] Developmental anomaly in the wall of the urethra so that the canal is open for a greater or lesser distance on the undersurface of the penis; a similar defect in the female in which the urethra opens into the vagina.

hypothalamohypophysial portal system (hī′pō-thal′ă-mō-hī′pō-fiz′ē-ăl) Series of blood vessels that carries blood from the area of the hypothalamus to the anterior pituitary gland; originates from capillary beds in the hypothalamus and terminates as a capillary bed in the anterior pituitary gland.

hypothalamohypophyseal tract Nerve tract consisting of the axons of neurosecretory cells and extending from the hypothalamus into the posterior pituitary gland. Hormones produced in the neurosecretory cell bodies in the hypothalamus are transported through the hypothalamohypophyseal tract to the posterior pituitary gland, where they are stored for later release.

hypothalamus (hī′pō-thal′ă-mūs) [Gr., *hypo,* under + *thalamus,* bedroom] Important autonomic and neuroendocrine control center beneath the thalamus.

hypothenar (hī-pō-thē′nar) [Gr., *hypo,* under + *thenar,* palm of the hand] Fleshy mass of tissue on the medial side of the palm; contains muscles responsible for moving the little finger.

hypotonic (hī-pō-ton′ik) [Gr., *hypo,* under + *tonos,* tension] Solution that causes cells to swell.

I

I band Area between the ends of two adjacent myosin myofilaments within a myofibril; Z disk divides the I band into two equal parts.

ileocecal sphincter (il′ē-ō-sē′kăl) Thickening of circular smooth muscle between the ileum and the cecum, forming the ileocecal valve.

ileocecal valve Valve formed by the ileocecal sphincter between the ileum and the cecum.

ileum (il′ē-ŭm) [Gr., *eileo,* to roll up, twist] Third portion of the small intestine, extending from the jejunum to the ileocecal opening into the large intestine; the posterior inferior bone of the coxal bone.

immunity (i-mū′ni-tē) [L., *immunis,* free from service] Resistance to infectious disease and harmful substances.

immunization (im-mū′ni-zā′shun) Process by which a subject is rendered immune by deliberately introducing an antigen or antibody into the subject.

immunoglobulin (IG) (im′ū-nō-glob′ū-lin) Antibody found in the gamma globulin portion of plasma.

implantation (im-plan-tā′shŭn) Attachment of the blastocyst to the endometrium of the uterus, occurring 6 or 7 days after fertilization of the ovum.

impotence (im′pŏ-tens) Inability to accomplish the male sexual act; caused by psychic or physical factors; also called erectile dysfunction (ED).

incisor (in-sī′zŏr) [L., *incido,* to cut into] One of the anterior, cutting teeth.

incisura (in′sī-soo′ră) [L., a cutting into] Notch or indentation at the edge of any structure.

incus (ing′kus) [L., anvil] Middle of the three ossicles in the middle ear.

inferior colliculus (ko-lik′ū-lŭs) [L., *collis,* hill] One of two rounded eminences of the midbrain; involved with hearing.

inferior vena cava Vein that returns blood from the lower limbs and the greater part of the pelvic and abdominal organs to the right atrium.

inflammatory response (in-flam′ă-tōr-ē) Complex sequence of events involving chemicals and immune cells that results in the isolation and destruction of antigens and tissues near the antigens.

infundibulum (in-fŭn-dib′ū-lŭm) [L., funnel] Funnel-shaped structure or passage—for example, the infundibulum that attaches the hypophysis to the hypothalamus or the funnel-like expansion of the uterine tube near the ovary.

infusion Introduction of a fluid other than blood, such as a saline or glucose solution, into the blood.

inguinal canal (ing′gwi-năl) Passage through the lower abdominal wall that transmits the spermatic cord in the male and the round ligament in the female.

inhibin (in-hib′in) Polypeptide secreted from the testes that inhibits FSH secretion.

inhibitory neuron (in-hib′i-tōr-ē) Neuron that produces IPSPs and has an inhibitory influence.

inhibitory postsynaptic potential (IPSP) Hyperpolarization in the postsynaptic membrane, which causes the membrane potential to move away from threshold.

innate immunity (i′nāt, i-nāt′) Immune system response that is the same with each exposure to an antigen; there is no ability for the system to remember a previous exposure to the antigen.

inner cell mass Group of cells at one end of the blastocyst, part of which forms the body of the embryo.

inner ear Part of the ear that contains the sensory organs for hearing and balance; contains the bony and membranous labyrinth.

insensible perspiration [L., *per,* through + *spiro,* to breathe everywhere] Perspiration that evaporates before it is perceived as moisture on the skin; the term sometimes includes evaporation from the lungs.

insertion (in-ser′shŭn) More movable attachment point of a muscle; usually, the lateral or distal end of a muscle associated with the limbs; also called mobile end.

inspiratory capacity (in-spī′ră-tō-rē) Volume of air that can be inspired after a normal expiration; the sum of the tidal volume and the inspiratory reserve volume.

inspiratory reserve volume Maximum volume of air that can be inspired after a normal inspiration.

insulin (in′sŭ-lin) Protein hormone secreted from the pancreas that increases the uptake of glucose and amino acids by most tissues.

interatrial septum (in-ter-ā′trē-ăl) [L., *saeptum,* partition] Wall between the atria of the heart.

intercalated disk (in-ter′kă-lā-ted) Cell-to-cell attachment with gap junctions between cardiac muscle cells.

intercalated duct Minute duct of glands, such as the salivary gland, and the pancreas; leads from the acini to the interlobular ducts.

intercellular (in-ter-sel′ū-lăr) Between cells.

intercellular chemical signal Chemical that is released from cells and passes to other cells; acts as a signal that allows cells to communicate with each other; also called first messenger.

interferons (in-ter-fēr′onz) Proteins that prevent viral replication.

interlobar artery (in-ter-lō′bar) Branch of the segmental arteries of the kidney; runs between the renal pyramids and gives rise to the arcuate arteries.

interlobular artery (in-ter-lob′ū-lăr) Artery that passes between lobules of an organ; branches of the interlobar arteries of the kidney pass outward through the cortex from the arcuate arteries and supply the afferent arterioles.

interlobular duct Any duct leading from a lobule of a gland and formed by the junction of the intercalated ducts draining the acini.

interlobular vein Vein that parallels the interlobular arteries; in the kidney, it drains the peritubular capillary plexus, emptying into arcuate veins.

intermediate olfactory area Part of the olfactory cortex responsible for the modulation of olfactory sensations.

internal anal sphincter [Gr., *sphinkter,* band or lace] Smooth muscle ring at the upper end of the anal canal.

internal naris; pl. **nares** (nā′ris, nā′rēs) Opening from the nasal cavity into the nasopharynx.

internal spermatic fascia Inner connective tissue covering of the spermatic cord.

internal urinary sphincter Traditionally recognized as a sphincter composed of a thickening of the middle smooth muscle layer of the bladder around the urethral opening.

interphase (in′ter-fāz) Period between active cell divisions when DNA replication occurs.
interstitial (in-ter-stish′ăl) [L., *inter,* between + *sisto,* to stand] Space within tissue. Interstitial growth is growth from within.
interstitial cell Cell between the seminiferous tubules of the testes; secretes testosterone; also called Leydig cell.
interventricular septum (in-ter-ven-trik′ū-lăr) Wall between the ventricles of the heart.
intestinal gland (in-tes′ti-năl) Tubular gland in the mucous membrane of the small and large intestines; also called crypt.
intracellular (in-tră-sel′ū-lăr) Inside a cell.
intracellular mediator Molecule produced within a cell that binds to a macromolecule, such as receptors or enzymes inside that cell, that regulates their activities; also called second messenger.
intramural plexus (in′tră-mū′răl plek′sus) Combined submucosal and myenteric plexuses.
intrinsic clotting pathway (in-trin′sik) Series of chemical reactions resulting in clot formation that begins with chemicals (e.g., plasma factor XII) within the blood.
intrinsic factor Factor secreted by the parietal cells of gastric glands and required for adequate absorption of vitamin B_{12}.
intrinsic muscles Muscles located within the structure being moved.
intubation (in-too-ba′shun) Insertion of a tube into an opening, a canal, or a hollow organ.
inversion (in-ver′zhŭn) [L., *inverto,* to turn about] Turning inward.
ion (ī′on) [Gr., *ion,* going] Atom or group of atoms carrying a charge of electricity by virtue of having gained or lost one or more electrons.
ion channel Pore in the plasma membrane through which ions, such as sodium and potassium, move.
ionic bond (ī-on′ik) Chemical bond that is formed when one atom loses an electron and another accepts that electron.
iris (ī′ris) Specialized portion of the vascular tunic; the "colored" portion of the eye that can be seen through the cornea.
ischemia (is-kē′mē-ă) [Gr., *ischo,* to keep back + *haima,* blood] Reduced blood supply to an area of the body.
ischium (is′kē-ŭm) Superior bone of the coxal bone.
isomers (ī′sō-merz) [Gr., *isos,* equal + *meros,* part] Molecules having the same number and types of atoms but differing in their three-dimensional arrangement.
isometric contraction (ī-sō-met′rik) [Gr., *isos,* equal + *metron,* measure] Muscle contraction in which the length of the muscle does not change but the tension produced increases.
isosmotic (ī′sō-os-mot′ik) [Gr., *isos,* equal + *osmos,* an impulsion] Having the same osmotic concentration or pressure as a reference solution.
isotonic (ī′sō-ton′ik) [Gr., *isos,* equal + *tonos,* tension] Type of solution that causes cells to neither shrink nor swell.
isotope (ī′sō-tōp) [Gr., *isos,* equal + *topos,* part, place] Either of two or more atoms that have the same atomic number but a different number of neutrons.
isthmus (is′mŭs) Constriction connecting two larger parts of an organ, such as the constriction between the body and the cervix of the uterus or the portion of the uterine tube between the ampulla and the uterus.

J

jaundice (jawn′dis) [Fr., *jaune,* yellow] Yellowish staining of the integument, the sclerae, and the other tissues with bile pigments.
jejunum (jĕ-joo′nŭm) [L., *jejunus,* empty] Second portion of the small intestine; located between the duodenum and the ileum.
juxtaglomerular apparatus (jŭks′tă-glŏ-mer′ū-lăr) Complex consisting of juxtaglomerular cells of the afferent arteriole and macular densa cells of the distal convoluted tubule near the renal corpuscle; secretes renin.
juxtaglomerular cell Modified smooth muscle cell of the afferent arteriole located at the renal corpuscle; a component of the juxtaglomerular apparatus.
juxtamedullary nephron (jŭks′tă-med′ŭ-lăr-ē) Nephron located near the junction of the renal cortex and medulla.

K

karyotype (kar′ē-ō-tīp) Display of chromosomes arranged by pairs.
keratinization (ker′ă-tin-i-zā′shŭn) Production of keratin and changes in the chemical and structural character of epithelial cells as they move to the skin surface.
keratinized (ker′ă-ti-nīzd) [Gr., *keras,* horn] Having become a structure that contains keratin, a protein found in skin, hair, nails, and horns.
keratinocyte (ke-rat′i-nō-sīt) [Gr., *keras,* horn + *kytos,* cell] Epidermal cell that produces keratin.
keratohyalin (ker′ă-tō-hī′ă-lin) Nonmembrane-bound protein granule in the cytoplasm of stratum granulosum cells of the epidermis.
ketogenesis (kē-tō-jen′ĕ-sis) Production of ketone bodies, such as from acetyl-CoA.
ketone body (kē′tōn) One of a group of ketones, including acetoacetic acid, β-hydrobutyric acid, and acetone.
ketosis (kē-tō′sis) [*ketone* + *osis,* condition] Condition characterized by the enhanced production of ketone bodies, as in diabetes mellitus or starvation.
kidney (kid′nē) [A.S., *cwith,* womb, belly + *neere,* kidney] One of the two organs that excrete urine. The kidneys are bean-shaped organs approximately 11 cm long, 5 cm wide, and 3 cm thick lying on each side of the spinal column, posterior to the peritoneum, approximately opposite the twelfth thoracic and first three lumbar vertebrae.
kilocalorie (kcal) (kil′ō-kal-ō-rē) Quantity of energy required to raise the temperature of 1 kg of water 1°C; 1000 calories. Equal to one dietary calorie.
kinetic energy (ki-net′ik) Motion energy or energy that can do work.
kinetic labyrinth (lab′i-rinth) Part of the membranous labyrinth composed of the semicircular canals; detects dynamic or kinetic equilibrium, such as movement of the head.
kinetochore (ki-nē′tō-kōr, ki-net′o-) [Gr., *kinēto,* moving + Gr., *chōra,* space] Structural portion of the chromosome to which microtubules attach.
Korotkoff sounds (kō-rot′kof) Named for Russian physician Nikolai S. Korotkoff (1874–1920). Sounds heard over an artery when blood pressure is determined by the auscultatory method; caused by turbulent flow of blood.

L

labium majus; pl. **labia majora** (lā′bē-ŭm, lā′bē-ă) One of two rounded folds of skin surrounding the labia minora and vestibule; homolog of the scrotum in males.
labium minus; pl. **labia minora** One of two narrow longitudinal folds of mucous membrane enclosed by the labia majora and bounding the vestibule; anteriorly they unite to form the prepuce.
lacrimal apparatus (lak′ri-măl) Lacrimal, or tear, gland in the superolateral corner of the orbit of the eye and a duct system that extends from the eye to the nasal cavity.
lacrimal canaliculus Canal that carries excess tears away from the eye; located in the medial canthus and opening on the lacrimal papilla.
lacrimal gland Tear gland located in the superolateral corner of the orbit.
lacrimal papilla Small lump of tissue in the medial canthus or corner of the eye; the lacrimal canal opens within the lacrimal papilla.
lacrimal sac Enlargement in the lacrimal canal that leads into the nasolacrimal duct.
lactation (lak-tā′shŭn) [L., *lactatio,* suckle] Period after childbirth during which milk is formed in the breasts.
lacteal (lak′tē-ăl) Lymphatic vessel in the wall of the small intestine that carries chyle from the intestine and absorbs fat.
lactiferous duct (lak-tif′er-ŭs) One of 15–20 ducts that drain the lobes of the mammary gland and open onto the surface of the nipple.
lactiferous sinus Dilation of the lactiferous duct just before it enters the nipple.
lacuna; pl. **lacunae** (lă-koo′nă, -koo′nē) [L., *lacus,* a hollow, a lake] Small space or cavity; potential space within the matrix of bone or cartilage normally occupied by a cell that can be visualized only when the cell shrinks away from the matrix during fixation; space containing maternal blood within the placenta.
lag phase One of the three phases of muscle contraction; time between the application of the stimulus and the beginning of muscular contraction. Also called latent phase.
lamella; pl. **lamellae** (lă-mel′ă, lă-mel′ē) Thin sheet or layer of bone.
lamellated corpuscle (lam′ĕ-lāt-ed) Oval receptor found in the deep dermis or hypodermis (responsible for deep cutaneous pressure and vibration) and in tendons (responsible for proprioception); also called Pacinian corpuscle.
lamina; pl. **laminae** (lam′i-nă, lam′i-nē) [L., *lamina,* plate, leaf] Thin plate—for example, the thinner portion of the vertebral arch.
lamina propria (prō′prē-ă) Layer of connective tissue underlying the epithelium of a mucous membrane.
laminar flow (lam′i-nar) Relative motion of layers of a fluid along smooth, concentric, parallel paths.
Langerhans cell Named for German anatomist Paul Langerhans (1847–1888); dendritic cell found in the skin.

lanugo (lă-noo′gō) [L., *lana,* wool] Fine, soft, unpigmented fetal hair.

Laplace's law Named for French mathematician Pierre S. de Laplace (1749–1827); the force that stretches the wall of a blood vessel is proportional to the radius of the vessel times the blood pressure.

large intestine Portion of the digestive tract extending from the small intestine to the anus.

laryngitis (lar-in-jī′tis) Inflammation of the mucous membrane of the larynx.

laryngopharynx (lă-ring′gō-far-ingks) Part of the pharynx lying posterior to the larynx.

larynx; pl. **larynges** (lar′ingks, lă-rin′jēz) Organ of voice production located between the pharynx and the trachea; it consists of a framework of cartilages and elastic membranes housing the vocal folds and the muscles that control the position and tension of these elements.

last menstrual period (LMP) Beginning of the last menstruation before pregnancy; used clinically to time events during pregnancy.

lateral geniculate nucleus (je-nik′ū-lāt) Nucleus of the thalamus where fibers from the optic tract terminate.

lateral olfactory area (ol-fak′tŏ-rē) Part of the olfactory cortex involved in the conscious perception of olfactory stimuli.

lens Transparent biconvex structure lying between the iris and the vitreous humor.

lens fiber Epithelial cell that makes up the lens of the eye.

lesser omentum (ō-men′tŭm) [L., membrane that encloses the bowels] Peritoneal fold passing from the liver to the lesser curvature of the stomach and to the upper border of the duodenum for a distance of approximately 2 cm beyond the pylorus.

lesser vestibular gland (ves-tib′ū-lăr) Number of minute mucous glands opening on the surface of the vestibule between the openings of the vagina and urethra; also called paraurethral gland.

leukocyte (loo′kō-sīt) White blood cell.

leukocytosis (loo′kō-sī-tō′sis) Abnormally large number of white blood cells in the blood.

leukopenia (loo-kō-pē′nē-ă) Lower than normal number of white blood cells in the blood.

leukotriene (loo-kō-trī′ēn) Specific class of physiologically active fatty acid derivatives present in many tissues.

lever Rigid shaft capable of turning about a fulcrum or pivot point.

LH surge Increase in plasma luteinizing hormone (LH) levels before ovulation and responsible for initiating it.

ligamentum arteriosum (lig′ă-men′tŭm) Remains of the ductus arteriosus.

ligamentum venosum Remnant of the ductus venosus.

ligand *See* chemical signal.

ligand-gated ion channel Ion channel in a plasma membrane caused to either open or close by a ligand binding to a receptor.

limbic system (lim′bik) [L., *limbus,* border] Part of the brain involved with emotions and olfaction; includes the cingulate gyrus, hippocampus, habenular nuclei, parts of the basal ganglia, the hypothalamus (especially the mammillary bodies, the olfactory cortex, and various nerve tracts).

lingual tonsil (ling′gwăl) Collection of lymphoid tissue on the posterior portion of the dorsum of the tongue.

lipase (lip′ās) Any fat-splitting enzyme.

lipid (li′pid) [Gr., *lipos,* fat] Substance composed principally of carbon, oxygen, and hydrogen; contains a lower ratio of oxygen to carbon and is less polar than carbohydrates; generally soluble in nonpolar solvents.

lipid bilayer Double layer of lipid molecules forming the plasma membrane and other cellular membranes.

lipochrome (lip′ō-krōm) Lipid-containing pigment that is metabolically inert.

lipotropin (li-pō-trō′pin) One of the peptide hormones released from the adenohypophysis; increases lipolysis in fat cells.

liver (liv′er) Largest gland of the body, lying in the upper-right quadrant of the abdomen just inferior to the diaphragm; secretes bile and is of great importance in carbohydrate and protein metabolism and in detoxifying chemicals.

lobar bronchi Branch from a primary bronchus that conducts air to each lobe of the lungs. There are two branches in the left lung and three branches from the primary bronchus in the right lung. Also called secondary bronchus.

lobe (lōb) Rounded, projecting part, such as the lobe of a lung, the liver, or a gland.

lobule (lob′ūl) Small lobe or subdivision of a lobe, such as a lobule of the lung or a gland.

local inflammation Inflammation confined to a specific area of the body. Symptoms include redness, heat, swelling, pain, and loss of function.

local potential Depolarization that is not propagated and that is graded or proportional to the strength of the stimulus.

local reflex Reflex of the intramural plexus of the digestive tract that does not involve the brain or spinal cord.

locus; pl. **loci** (lō′kŭs, lō′sī) Place; usually a specific site.

loop of Henle Named for German anatomist Friedrich G. J. Henle (1809–1885). U-shaped part of the nephron extending from the proximal to the distal convoluted tubule and consisting of descending and ascending limbs. Some of the loops of Henle extend into the renal pyramids.

lower respiratory tract Larynx, trachea, and lungs.

lung recoil Decrease in the size of an expanded lung as a result of a decrease in the size (volume) of its alveoli; due to elastic recoil of elastic fibers surrounding alveoli and water surface tension of a thin film of water within alveoli.

lunula; pl. **lunulae** (loo′noo-lă, loo′noo-lē) [L., *luna,* moon] White, crescent-shaped portion of the nail matrix visible through the proximal end of the nail.

luteal phase (loo′tē-ăl) Portion of the menstrual cycle extending from the time of formation of the corpus luteum after ovulation to the time when menstrual flow begins; usually 14 days in length; also called secretory phase.

luteinizing hormone (LH) (loo′tē-ĭ-nīz-ing) In females, hormone stimulating the final maturation of the follicles and the secretion of progesterone by them, with their rupture releasing the ovum, and the conversion of the ruptured follicle into the corpus luteum; in males, stimulates the secretion of testosterone in the testes.

lymph (limf) [L., *lympha,* clear spring water] Clear or yellowish fluid derived from interstitial fluid and found in lymph vessels.

lymph node Encapsulated mass of lymph tissue found among lymph vessels.

lymphatic capillary Beginning of the lymphatic system of vessels; lined with flattened endothelium lacking a basement membrane.

lymphatic nodule Small accumulation of lymph tissue lacking a distinct boundary.

lymphatic sinus Channels in a lymph node crossed by a reticulum of cells and fibers.

lymphatic vessel One of the system of vessels carrying lymph from the lymph capillaries to the veins.

lymphedema (limf′e-dē′mă) Swelling of tissues resulting from the excessive accumulation of fluid caused by the removal, damage, or blockage of lymphatic vessels or lymph nodes; usually results in swelling of the arm or leg.

lymphoblast (lim′fō-blast) Cell that matures into a lymphocyte.

lymphocyte (lim′fō-sīt) Nongranulocytic white blood cell formed in lymphoid tissue.

lymphokine (lim′fō-kīn) Chemical produced by lymphocytes that activates macrophages, attracts neutrophils, and promotes inflammation.

lysis (lī′sis) [Gr., *lysis,* a loosening] Process by which a cell swells and ruptures.

lysosome (lī′sō-sōm) [Gr., *lysis,* loosening + *soma,* body] Membrane-bounded vesicle containing hydrolytic enzymes that function as intracellular digestive enzymes.

lysozyme (lī′sō-zīm) Enzyme that is destructive to the cell walls of certain bacteria; present in tears and some other fluids of the body.

M

M line Line in the center of the H zone made of delicate filaments that holds the myosin myofilaments in place in the sarcomere of muscle fibers.

macrophage (mak′rō-fāj) [Gr., *makros,* large + *phagein,* to eat] Any large, mononuclear phagocytic cell.

macula; pl. **maculae** (mak′ū-lă, mak′ū-lē) [L., a spot] Sensory structure in the utricle and saccule, consisting of hair cells and a gelatinous mass embedded with otoliths.

macula densa Cells of the distal convoluted tubule located at the renal corpuscle and forming part of the juxtaglomerular apparatus.

main bronchus; pl. **bronchi** (brong′kŭs, brong′kī) One of two tubes arising at the inferior end of the trachea; each primary bronchus extends into one of the lungs; also called primary bronchus.

major duodenal papilla Point of opening of the common bile duct and pancreatic duct into the duodenum.

major histocompatibility complex (MHC) molecules Genes that control the production of major histocompatibility complex proteins, which are glycoproteins found on the surfaces of cells. The major histocompatibility proteins serve as self-markers for the immune system and are used by antigen-presenting cells to present antigens to lymphocytes.

male pronucleus Nuclear material of the sperm cell after the ovum has been penetrated by the sperm cell.

malignant (mă-lig′nănt) Resistant to treatment; occurring in severe form and frequently fatal; having locally invasive and destructive growth and metastasis.

malleus; pl. **mallei** (mal′ē-ŭs, mal′ē-ī) [L., hammer] Largest of the three auditory ossicles; attached to the tympanic membrane.

mamillary bodies (mam′i-lār-ē) [L., breast- or nipple-shaped] Nipple-shaped structures at the base of the hypothalamus.

mamma; pl. **mammae** (mam′ă, mam′ē) Breast; the organ of milk secretion; one of two hemispheric projections of variable size situated in the subcutaneous layer over the pectoralis major muscle on each side of the chest; it is rudimentary in the male.

mammary ligaments (mam′ă-rē) Well-developed ligaments that extend from the overlying skin to the fibrous stroma of mammary gland; also called Cooper's ligaments.

manubrium; pl. **manubria** (mă-noo′brē-ŭm, mă-noo′bre-ă) [L., handle] Part of a bone representing the handle, such as the manubrium of the sternum representing the handle of a sword.

mass movement Forcible peristaltic movement of short duration, occurring only three or four times a day, which moves the contents of the large intestine.

mass number Number of protons plus the number of neutrons in each atom.

mastication (mas-ti-kā′shŭn) [L., *mastico*, to chew] Process of chewing.

mastication reflex Repetitive cycle of relaxation and contraction of the muscles of mastication.

mastoid (mas′toyd) [Gr., *mastos*, breast] Resembling a breast.

mastoid air cells Spaces within the mastoid process of the temporal bone connected to the middle ear by ducts.

mature follicle Ovarian follicle in which the oocyte attains its full size. The follicle contains a fluid-filled antrum and is surrounded by the theca interna and externa. Also called Graafian follicle.

maximal stimulus Stimulus resulting in a local potential just large enough to produce the maximum frequency of action potentials.

meatus (mē-ā′tŭs) [L., to go, pass] Passageway or tunnel.

mechanoreceptor (mek′ă-nō-rē-sep′tŏr) Sensory receptor that responds to mechanical pressures—for example, pressure receptors in the carotid sinus or touch receptors in the skin.

meconium (mē-kō′nē-ŭm) [Gr., *mekon*, poppy] First intestinal discharges of the newborn infant, greenish in color and consisting of epithelial cells, mucus, and bile.

medial olfactory area Part of the olfactory cortex responsible for the visceral and emotional reactions to odors.

medulla oblongata (me-dool′ă ob-long-gah′tă) Inferior portion of the brainstem that connects the spinal cord to the brain and contains autonomic centers controlling functions such as heart rate, respiration, and swallowing.

medullary cavity (med′ul-er-ē, med′oo-lār-ē) Large, marrow-filled cavity in the diaphysis of a long bone.

medullary ray Extension of the kidney medulla into the cortex, consisting of collecting ducts and loops of Henle.

megakaryoblast (meg-ă-kar′ē-ō-blast) [Gr., *mega* + *karyon*, nut, nucleus + *blastos*, germ] Cell that gives rise to platelets or thrombocytes.

meibomian cyst (mī-bō′mē-an) Named for German anatomist Hendrik Meibom (1638–1700). A chronic inflammation of a meibomian gland.

meibomian gland Sebaceous gland near the inner margins of the eyelid; secretes sebum that lubricates the eyelid and retains tears.

meiosis (mī-ō′sis) [Gr., a lessening] Cell division that results in the formation of gametes. Consists of two divisions, which result in one (female) or four (male) gametes, each of which contains one-half the number of chromosomes in the parent cell.

Meissner corpuscle (mīs′ner kōr′pŭs-l) Named for German histologist Georg Meissner (1829–1905). *See* tactile corpuscle.

melanin (mel′ă-nin) [Gr., *melas*, black] Group of related molecules responsible for skin, hair, and eye color. Most melanins are brown to black pigments; some are yellowish or reddish.

melanocyte (mel′ă-nō-sīt) [Gr., *melas*, black + *kytos*, cell] Cell found mainly in the stratum basale; produces the brown or black pigment melanin.

melanocyte-stimulating hormone (MSH) Peptide hormone secreted by the anterior pituitary; increases melanin production by melanocytes, making the skin darker in color.

melanosome (mel′ă-nō-sōm) [Gr., *melas*, black + *soma*, body] Membranous organelle containing the pigment melanin.

melatonin (mel-ă-tōn′in) Hormone (amino acid derivative) secreted by the pineal body; inhibits the secretion of gonadotropin-releasing hormone from the hypothalamus.

membrane attack complex (MAC) Channel through a plasma membrane produced by activated complement proteins, primarily complement protein C9; in nucleated cells, water enters the channel, causing lysis of cells.

membrane-bound receptor Receptor molecule, such as a hormone receptor, that is bound to the plasma membrane of the target cell.

membranous labyrinth (mem′bră-nŭs lab′i-rinth) Membranous structure within the inner ear consisting of the cochlea, vestibule, and semicircular canals.

membranous urethra (ū-rē′thră) Portion of the male urethra, approximately 1 cm in length, extending from the prostate gland to the beginning of the penile urethra.

memory cell Small lymphocyte that is derived from a B cell or T cell and that rapidly responds to a subsequent exposure to the same antigen.

menarche (me-nar′kē) [Gr., *mensis*, month + *arche*, beginning] Establishment of menstrual function; the time of the first menstrual period or flow.

meninx; pl. **meninges** (mē′ninks, mě-nin′jes) [Gr., membrane] Connective tissue membrane surrounding the brain.

meniscus; pl. **menisci** (me-nis′kus, me-nis′sī) Crescent-shaped intraarticular fibrocartilage found in certain joints, such as the crescent-shaped fibrocartilaginous structure of the knee.

menopause (men′ō-pawz) [Gr., *mensis*, month + *pausis*, cessation] Permanent cessation of the menstrual cycle.

menses (men′sēz) [L., *mensis*, month] Periodic hemorrhage from the uterine mucous membrane, occurring at approximately 28-day intervals.

menstrual cycle (men′stroo-ăl) Series of changes that occur in sexually mature, nonpregnant women and result in menses. Specifically refers to the uterine cycle but is often used to include both the uterine and ovarian cycles.

Merkel (tactile) disk (mer′kel) Named for German anatomist Friedrich Merkel (1845–1919).

merocrine gland (mer′ō-krin) [Gr., *meros*, part + *krino*, to separate] Gland that secretes products with no loss of cellular material—for example, water-producing sweat glands.

mesencephalon (mez-en-sef′ă-lon) [Gr., *mesos*, middle + *enkephalos*, brain] Midbrain in both the embryo and adult; consists of the cerebral peduncle and the corpora quadrigemini.

mesentery (mes′en-ter-ē) [Gr., *mesos*, middle + *enteron*, intestine] Double layer of peritoneum extending from the abdominal wall to the abdominal viscera, conveying to it its vessels and nerves.

mesoderm (mez′ō-derm) Middle of the three germ layers of an embryo.

mesonephros (mez′ō-nef′ros) One of three excretory organs appearing during embryonic development; forms caudal to the pronephros as the pronephros disappears. It is well developed and is functional for a time before the establishment of the metanephros, which gives rise to the kidney. It undergoes regression as an excretory organ, but its duct system is retained in the male as the efferent ductule and epididymis.

mesosalpinx (mez′ō-sal′pinks) [Gr., *mesos*, middle + *salpinx*, trumpet] Part of the broad ligament supporting the uterine tube.

mesothelium (mez-ō-thē′lē-ŭm) Single layer of flattened cells forming an epithelium that lines serous cavities, such as peritoneum, pleura, pericardium.

mesovarium (mez′ō-vā′rē-ŭm) Short peritoneal fold connecting the ovary with the broad ligament of the uterus.

messenger ribonucleic acid (mRNA) Type of RNA that moves out of the nucleus and into the cytoplasm, where it is used as a template to determine the structure of proteins.

metabolism (mě-tab′ō-lizm) [Gr., *metabole*, change] Sum of all the chemical reactions that take place in the body, consisting of anabolism and catabolism. *Cellular metabolism* refers specifically to the chemical reactions within cells.

metacarpal (met′ă-kar′păl) Relating to the fine bones of the hand between the carpus (wrist) and the phalanges.

metanephros (met-ă-nef′ros) Most caudally located of the three excretory organs appearing during embryonic development; becomes the permanent kidney of mammals. In mammalian embryos, it is formed caudal to the mesonephros and develops later as the mesonephros undergoes regression.

metaphase (met′ă-fās) Time during cell division when the chromosomes line up along the equator of the cell.

metarteriole (met′ar-tēr′ē-ōl) One of the small peripheral blood vessels that contain scattered groups of smooth muscle fibers in their walls; located between the arterioles and the true capillaries.

metastasis (mĕ-tas′tă-sis) Shifting of a disease or its local manifestations or the spread of a disease from one part of the body to another, as in a malignant neoplasm.

metatarsal (met′ă-tar′sal) [Gr., *meta,* after + *tarsos,* sole of the foot] Distal bone of the foot.

metencephalon (met′en-sef′ă-lon) [Gr., *meta,* after + *enkephalos,* brain] Second most posterior division of the embryonic brain; becomes the pons and cerebellum in the adult.

micelle (mi-sel′, mī-sel′) [L., *micella,* small morsel] Droplets of lipid surrounded by bile salts in the small intestine.

microfilament (mī-krō-fil′ă-ment) Small fibril forming bundles, sheets, or networks in the cytoplasm of cells; provides structure to the cytoplasm and mechanical support for microvilli and stereocilia.

microglia (mī-krog′lē-ă) [Gr., *micro* + *glia,* glue] Small neuroglial cells that become phagocytic and mobile in response to inflammation; considered to be macrophages within the central nervous system.

microtubule (mī-krō-too′būl) Hollow tube composed of tubulin, measuring approximately 25 nm in diameter and usually several micrometers long. Helps provide support to the cytoplasm of the cell and is a component of certain cell organelles, such as centrioles, spindle fibers, cilia, and flagella.

microvillus; pl. **microvilli** (mī′krō-vil′ŭs, mī′krō-vil′ī) Minute projection of the cell membrane that greatly increases the surface area.

micturition reflex (mik-choo-rish′ŭn) Contraction of the urinary bladder stimulated by stretching of the bladder wall; results in emptying of the bladder.

middle ear Air-filled space within the temporal bone; contains auditory ossicles; between the external and internal ear.

milk letdown Expulsion of milk from the alveoli of the mammary glands; stimulated by oxytocin.

mineral Inorganic nutrient necessary for normal metabolic functions.

mineralocorticoid (min′er-al-ō-kōr′ti-koyd) Steroid hormone (e.g., aldosterone) produced by the zona glomerulosa of the adrenal cortex; facilitates exchange of potassium for sodium in the distal renal tubule, causing sodium reabsorption and potassium and hydrogen ion secretion.

minor duodenal papilla Site of the opening of the accessory pancreatic duct into the duodenum.

minute ventilation Product of tidal volume times the respiratory rate.

mitochondrion; pl. **mitochondria** (mī-tō-kon′ drē-on, mī-tō-kon′drē-ă) [Gr., *mitos,* thread + *chandros,* granule] Small, spherical, rod-shaped or thin filamentous structure in the cytoplasm of cells that is a site of ATP production.

mitosis (mī-tō′sis) [Gr., thread] Cell division resulting in two daughter cells with exactly the same number and type of chromosomes as the mother cell.

modiolus (mō-dī′ō′lŭs) [L., nave of a wheel] Central core of spongy bone about which turns the spiral canal of the cochlea.

molar (mō′lăr) Tricuspid tooth; the three posterior teeth of each dental arch.

molecule (mol′ĕ-kūl) Substance composed of two or more atoms chemically combined to form a structure that behaves as an independent unit.

monoblast (mon′ō-blast) Cell that matures into a monocyte.

monocyte (mon′o-sīt) Type of white blood cell; large phagocytic white blood cell that moves from the blood into tissues and becomes a macrophage.

mononuclear phagocytic system (mon-ō-noo′klē-ăr fag-ō-sit′ik) Phagocytic cells, each with a single nucleus; derived from monocytes; also called reticuloendothelial system.

monosaccharide (mon-ō-sak′ă-rīd) Simple sugar carbohydrate that cannot form any simpler sugar by hydrolysis.

mons pubis (monz pū′bis) [L., mountain; the genitals] Prominence caused by a pad of fatty tissue over the symphysis pubis in the female.

morula (mōr′oo-lă, mōr′ū-lă) [L., *morus,* mulberry] Mass of 12 or more cells resulting from the early cleavage divisions of the zygote.

motor neuron Neuron that innervates skeletal, smooth, or cardiac muscle fibers.

motor unit Single neuron and the muscle fibers it innervates.

mucosa (mū-kō′să) [L., *mucosus* mucous] Mucous membrane consisting of epithelium and lamina propria. In the digestive tract, there is also a layer of smooth muscle.

mucous membrane (mū′kŭs) Thin sheet consisting of epithelium and connective tissue (lamina propria) that lines cavities that open to the outside of the body; many contain mucous glands that secrete mucus.

mucous neck cell One of the mucous-secreting cells in the neck of a gastric gland.

mucus (mū′kŭs) Viscous secretion produced by and covering mucous membranes; lubricates mucous membranes and traps foreign substances.

multiple motor unit summation Increased force of contraction of a muscle due to recruitment of motor units.

multiple-wave summation Increased force of contraction of a muscle due to increased frequency of stimulation.

multipolar neuron One of three categories of neurons consisting of a neuron cell body, an axon, and two or more dendrites.

muscarinic receptor (mŭs′kă-rin′ik) Class of cholinergic receptor that is specifically activated by muscarine in addition to acetylcholine.

muscle fiber Muscle cell.

muscle spindle Three to 10 specialized muscle fibers supplied by gamma motor neurons and wrapped in sensory nerve endings; detects stretch of the muscle and is involved in maintaining muscle tone.

muscle tone Relatively constant tension produced by a muscle for long periods as a result of asynchronous contraction of motor units.

muscle twitch Contraction of a whole muscle in response to a stimulus that causes an action potential in one or more muscle fibers.

muscular fatigue Fatigue due to a depletion of ATP within the muscle fibers.

muscularis (mŭs-kū-lā′ris) [Modern L., muscular] Muscular coat of a hollow organ or tubular structure.

muscularis mucosa Thin layer of smooth muscle found in most parts of the digestive tube; located outside the lamina propria and adjacent to the submucosa.

musculi pectinati (mŭs-kū-lī pek′tī-nă′tē) Prominent ridges of atrial myocardium located on the inner surface of much of the right atrium and both auricles.

mutation (mū-tā′shŭn) Change in the number or kinds of nucleotides in the DNA of a gene.

myelencephalon (mī′el-en-sef′ă-lon) [Gr., *myelos,* medulla, marrow + *enkephalos* brain] Most caudal portion of the embryonic brain; also called medulla oblongata.

myelin sheath (mī′ĕ-lin) Envelope surrounding most axons; formed by Schwann cell membranes being wrapped around the axon.

myelinated axon (mī′ĕ-li-nāt-ed ak′son) Nerve fiber having a myelin sheath.

myeloblast (mī′ĕ-lō-blast) Immature cell from which the different granulocytes develop.

myenteric plexus (mī′en-ter′ik) Plexus of unmelinated fibers and postganglionic autonomic cell bodies lying in the muscular coat of the esophagus, stomach, and intestines; communicates with the submucosal plexuses.

myoblast (mī′ō-blast) [Gr., *mys,* muscle + *blastos,* germ] Primitive, multinucleated cell with the potential to develop into a muscle fiber.

myofilament (mī-ō-fil′ă-ment) Extremely fine molecular thread helping form the myofibrils of muscle; thick myofilaments are formed of myosin, and thin myofilaments are formed of actin.

myometrium (mī′ō-mē′trē-ŭm) Muscular wall of the uterus; composed of smooth muscle; also called muscular layer.

myosin myofilament (mī′ō-sin mī-ō-fil′ă-ment) Thick myofilament of muscle fibrils; composed of myosin molecules.

N

nail (nāl) [A.S., naegel] Several layers of dead epithelial cells containing hard keratin on the ends of the digits.

nail bed Epithelial tissue resting on dermis under the nail between the nail matrix and hyponychium; contributes to the formation of the nail.

nail matrix Epithelial tissue resting on dermis under the proximal end of a nail; produces most of the nail.

nasal cavity (nā′zăl) Cavity between the external nares and the pharynx. It is divided into two chambers by the nasal septum and is bounded inferiorly by the hard and soft palates.

nasal septum Bony partition that separates the nasal cavity into left and right parts; composed of the vomer, the perpendicular plate of the ethmoid, and hyaline cartilage.

nasolacrimal duct (nā-zō-lak′ri-măl) Duct that leads from the lacrimal sac to the nasal cavity.

nasopharynx (nā-zō-far′ingks) Part of the pharynx that lies above the soft palate; anteriorly it opens into the nasal cavity.

near point of vision Closest point from the eye at which an object can be held without appearing blurred.

neck Slightly constricted part of a tooth, between the crown and the root.

neoplasm (nē′ō-plazm) Abnormal tissue that grows by cellular proliferation more rapidly than normal and continues to grow after the stimuli that initiated the new growth ceases.

nephron (nef′ron) [Gr., *nephros,* kidney] Functional unit of the kidney, consisting of the renal corpuscle, the proximal convoluted tubule, the loop of Henle, and the distal convoluted tubule.

nerve tract Bundles of parallel axons with their associated sheaths in the central nervous system.

neural crest (noor′ăl) Edge of the neural plate as it rises to meet at the midline to form the neural tube.

neural crest cells Cells derived from the crests of the forming neural tube in the embryo; together with the mesoderm, they form the mesenchyme of the embryo; they give rise to part of the skull, the teeth, melanocytes, sensory neurons, and autonomic neurons.

neural layer Portion of the retina containing rods and cones.

neural plate Region of the dorsal surface of the embryo that is transformed into the neural tube and neural crest.

neural tube Tube formed from the neuroectoderm by the closure of the neural groove; develops into the spinal cord and brain.

neuroectoderm (noor-ō-ek′tō-derm) Part of the ectoderm of an embryo giving rise to the brain and spinal cord.

neurohormone (noor′ō-hōr′mōn) Hormone secreted by a neuron.

neuromodulator Substance that influences the sensitivity of neurons to neurotransmitters but neither strongly stimulates nor strongly inhibits neurons by itself.

neuromuscular junction (noor-ō-mŭs′kū-lăr) Specialized synapse between a motor neuron and a muscle fiber.

neuron (noor′on) [Gr., nerve] Morphologic and functional unit of the nervous system, consisting of the nerve cell body, the dendrites, and the axon; also called nerve cell.

neuron cell body Enlarged portion of the neuron containing the nucleus and other organelles; also called nerve cell body.

neurotransmitter (noor′ō-trans-mit′er) [Gr., *neuro,* nerve + L., *transmitto,* to send across] Any specific chemical agent released by a presynaptic cell on excitation that crosses the synaptic cleft and stimulates or inhibits the postsynaptic cell.

neutral solution (noo′trăl) Solution, such as pure water, that has 10^{-7} mol of hydrogen ions per liter and an equal concentration of hydroxide ions; has a pH of 7.

neutron (noo′tron) [L., *neuter,* neither] Electrically neutral particle in the nuclei of atoms (except hydrogen).

neutrophil (noo′trō-fil) [L., *neuter,* neither + Gr., *philos,* fond] Type of white blood cell; small phagocytic white blood cell with a lobed nucleus and small granules in the cytoplasm.

nicotinic receptor (nik-ō-tin′ik) Class of cholinergic receptor molecule that is specifically activated by nicotine and by acetylcholine.

nipple (nip′l) Projection at the apex of the mamma, on the surface of which the lactiferous ducts open; surrounded by a circular pigmented area, the areola.

Nissl substance (nis′l) Named after German neurologist Franz Nissl (1860–1919). Areas in the neuron cell body containing rough endoplasmic reticulum.

nociceptor (nō-si-sep′ter) [L., *noceo,* to injure + *capio,* to take] Sensory receptor that detects painful or injurious stimuli; also called pain receptor.

nonelectrolyte (non-e-lek′tro-līt) [Gr., *electro* + *lytos,* soluble] Molecules that do not dissociate and do not conduct electricity.

norepinephrine (nōr′ep-i-nef′rin) Neurotransmitter substance released from most of the postganglionic neurons of the sympathetic division; hormone released from the adrenal cortex that increases cardiac output and blood glucose levels; also called noradrenaline.

nose, or **nasus** (nōz, nā′sŭs) Visible structure that forms a prominent feature of the face; nasal cavities.

notochord (nō′tō-kōrd) [Gr., *notor,* back + *chords,* cord] Small rod of tissue lying ventral to the neural tube. A characteristic of all vertebrates, in humans it becomes the nucleus pulposus of the intervertebral disks.

nuchal (noo′kăl) Back of the neck.

nuclear envelope (noo′klē-er) Double membrane structure surrounding and enclosing the nucleus.

nuclear pores Porelike openings in the nuclear envelope where the inner and outer membranes fuse.

nucleic acid (noo-klē′ik, noo-klā′ik) Polymer of nucleotides, consisting of DNA and RNA, forms a family of substances that comprise the genetic material of cells and control protein synthesis.

nucleolus; pl. **nucleoli** (noo-klē′ō-lŭs, noo-klē′ō-lī) Somewhat rounded, dense, well-defined nuclear body with no surrounding membrane; contains ribosomal RNA and protein.

nucleotide (noo′klē-ō-tīd) Basic building block of nucleic acids consisting of a sugar (either ribose or deoxyribose) and one of several types of organic bases.

nucleus; pl. **nuclei** (noo′klē-ŭs, noo′klē-ī) [L., inside of a thing] Cell organelle containing most of the genetic material of the cell; collection of nerve cell bodies within the central nervous system; center of an atom consisting of protons and neutrons.

nucleus pulposus (pŭl-pō′sŭs) [L., central pulp] Soft central portion of the intervertebral disk.

nutrient (noo′trē-ent) [L., *nutriens,* to nourish] Chemicals taken into the body that are used to produce energy, provide building blocks for new molecules, or function in other chemical reactions.

O

olecranon process (ō-lek′ră-non, ō′lē-krā′non) Process on the distal end of the ulna, forming the point of the elbow.

olfaction (ol-fak′shŭn) [L., *olfactus,* smell] Sense of smell.

olfactory area Extreme superior region of the nasal cavity.

olfactory bulb (ol-fak′tŏ-rē) Ganglion-like enlargement at the rostral end of the olfactory tract that lies over the cribriform plate; receives the olfactory nerves from the nasal cavity.

olfactory cortex Termination of the olfactory tract in the cerebral cortex within the lateral fissure of the cerebrum.

olfactory epithelium Epithelium of the olfactory recess containing olfactory receptors.

olfactory tract Nerve tract that projects from the olfactory bulb to the olfactory cortex.

oligodendrocyte (ol′i-gō-den′drō-sīt) Neuroglial cell that has cytoplasmic extensions that form myelin sheaths around axons in the central nervous system.

oncogene (ong′kō-jēn) Gene that can change or be activated to cause cancer.

oncology (ong-kol′ō-jē) Study of neoplasms.

oocyte (ō′ō-sīt) [Gr., *oon,* egg + *kytos,* a hollow (cell)] Immature ovum.

oogenesis (ō-ō-jen′ĕ-sis) Formation and development of a secondary oocyte or ovum.

oogonium (ō-ō-gō′nē-ŭm) [Gr., *oon,* egg + *gone,* generation] Primitive cell from which oocytes are derived by meiosis.

opposition Movement of the thumb and little finger toward each other; movement of the thumb toward any of the fingers.

opsin (op′sin) Protein portion of the rhodopsin molecule; a class of proteins that bind to retinal to form the visual pigments of the rods and cones of the eye.

opsonin (op′sŏ-nin) [Gr., *opsonein* to prepare food] Substance, such as antibody or complement, that enhances phagocytosis.

optic chiasm (op′tik kī′az′m) [Gr., two crossing lines; *chi* the letter χ] Point of crossing of the optic tracts.

optic disc Point at which axons of ganglion cells of the retina converge to form the optic nerve, which then penetrates through the fibrous tunic of the eye.

optic nerve Nerve carrying visual signals from the eye to the optic chiasm.

optic stalk Constricted proximal portion of the optic vesicle in the embryo; develops into the optic nerve.

optic tract Tract that extends from the optic chiasma to the lateral geniculate nucleus of the thalamus.

optic vesicle One of the paired evaginations from the walls of the embryonic forebrain from which the retina develops.

oral cavity (ōr′ăl) The mouth; consists of the space surrounded by the lips, cheeks, teeth, and palate; limited posteriorly by the fauces.

orbit (ōr′bit) Eye socket; formed by seven skull bones that surround and protect the eye.

organ of Corti (ōr′găn) Named for Italian anatomist Marquis Alfonso Corti (1822–1888). Spiral organ; rests on the basilar membrane and supports the hair cells that detect sounds.

organelle (or′gă-nel) [Gr., *organon,* tool] Specialized part of a cell with one or more specific individual functions.

orgasm (ōr′gazm) [Gr., *orgao,* to swell, be excited] Climax of the sexual act, associated with a pleasurable sensation.

origin (ōr′i-jin) Less movable attachment point of a muscle; usually the medial or proximal end of a muscle associated with the limbs; also called fixed end.

oropharynx (ōr′ō-far′ingks) Portion of the pharynx that lies posterior to the oral cavity; it is continuous above with the nasopharynx and below with the laryngopharynx.

oscillating circuit Neuronal circuit arranged in a circular fashion that allows action potentials produced in the circuit to keep stimulating the neurons of the circuit.

osmolality (os-mō-lal′i-tē) Osmotic concentration of a solution; the number of moles of solute in 1 kg of water times the number of particles into which the solute dissociates.

osmoreceptor cell (os-mō-rē-sep′ter, os′mō-rē-sep′tōr) [Gr., *osmos,* impulsion] Receptor in the central nervous system that responds to changes in the osmotic pressure of the blood.

osmosis (os-mō′sis) [Gr., *osmos,* thrusting or an impulsion] Diffusion of solvent (water) through a membrane from a less concentrated solution to a more concentrated solution.

osmotic pressure (os-mot′ik) Force required to prevent the movement of water across a selectively permeable membrane.

ossification (os′i-fi-kā′shŭn) [L., *os,* bone + *facio,* to make] Bone formation; also called osteogenesis.

osteoblast (os′tē-ō-blast) [Gr., *osteon,* bone + *blastos,* germ] Bone-forming cell.

osteoclast (os′tē-ō-klast) [Gr., *osteon,* bone + *klastos,* broken] Large, multinucleated cell that absorbs bone.

osteocyte (os′tē-ō-sīt) [Gr., *osteon,* bone + *kytos,* cell] Mature bone cell surrounded by bone matrix.

osteomalacia (os′tē-ō-mă-lā′shē-ă) Softening of bones due to calcium depletion; adult rickets.

osteon (os′tē-on) Central canal containing blood capillaries and the concentric lamellae around it; occurs in compact bone; also called haversian system.

osteoporosis (os′tē-ō-pō-rō′sis) [Gr., *osteon,* bone + *poros,* pore + *osis,* condition] Reduction in quantity of bone, resulting in porous bone.

ostium (os′tē-ŭm) [L., door, entrance, mouth] Small opening—for example, the opening of the uterine tube near the ovary or the opening of the uterus into the vagina.

otolith (ō′tō-lith) Crystalline particles of calcium carbonate and protein embedded in the maculae.

oval window (ō′văl) Membranous structure to which the stapes attaches; transmits vibrations to the inner ear.

ovarian cycle (ō-var′ē-an) Series of events that occur in a regular fashion in the ovaries of sexually mature, nonpregnant females; results in ovulation and the production of the hormones estrogen and progesterone.

ovarian epithelium Peritoneal covering of the ovary; also called germinal epithelium.

ovarian ligament Bundle of fibers passing to the uterus from the ovary.

ovary (ō′vă-rē) One of two female reproductive glands located in the pelvic cavity; produces the secondary oocyte, estrogen, and progesterone.

oviduct (ō′vi-dŭkt) *See* uterine tube.

ovulation (ov′ū-lā′shun) Release of an ovum, or secondary oocyte, from the vesicular follicle.

oxidation (ok-si-dā′shŭn) Loss of one or more electrons from a molecule.

oxidation-reduction reaction Reaction in which one molecule is oxidized and another is reduced.

oxidative deamination (ok-si-dā′tiv) Removal of the amine group of an amino acid to form a keto acid, ammonia, and NADH.

oxygen deficit (ok′sē-jen) Oxygen necessary for the synthesis of the ATP required to remove lactic acid produced by anaerobic respiration.

oxygen-hemoglobin dissociation curve Graph describing the relationship between the percentage of hemoglobin saturated with oxygen and a range of oxygen partial pressures.

oxyhemoglobin (ox′sē-hē-mō-glō′bin) Oxygenated hemoglobin.

P

P wave First complex of the electrocardiogram representing depolarization of the atria.

Pacinian corpuscle (pa-sin′ē-an) Named for Italian anatomist Filippo Pacini (1812–1883). *See* lamellated corpuscle.

palate (pal′ăt) [L., *palatum,* palate] Roof of the mouth.

palatine tonsil (pal′ă-tīn) One of two large, oval masses of lymphoid tissue embedded in the lateral wall of the oral pharynx.

palpebra; pl. **palpebrae** (pal-pē′bră, pal-pē′brē) [L., eyelid] Eyelid.

palpebral conjunctiva (pal-pē′brăl kon-jŭnk-tī′vă) Conjunctiva that covers the inner surface of the eyelids.

palpebral fissure Space between the upper and lower eyelids.

pancreas (pan′krē-as) [Gr., *pankreas,* sweetbread] Abdominal gland that secretes pancreatic juice into the intestine and insulin and glucagon from the pancreatic islets into the bloodstream.

pancreatic duct (pan-krē-at′ik) Excretory duct of the pancreas that extends through the gland from tail to head, where it empties into the duodenum at the greater duodenal papilla.

pancreatic islet Cellular mass varying from a few to hundreds of cells lying in the interstitial tissue of the pancreas; composed of different cell types that make up the endocrine portion of the pancreas and are the source of insulin and glucagon; also called islets of Langerhans.

pancreatic juice [L., *jus,* broth] External secretion of the pancreas; clear, alkaline fluid containing several enzymes.

papilla (pă-pil′ă) [L., nipple] Small, nipplelike process; projection of the dermis, containing blood vessels and nerves, into the epidermis; projections on the surface of the tongue.

papillary muscle (pap′i-lăr′ē) Nipplelike, conical projection of myocardium within the ventricle; the chordae tendineae are attached to the apex of the papillary muscle.

parafollicular cell (par-ă-fo-lik′ū-lăr) Endocrine cell in the thyroid gland; secretes the hormone calcitonin.

paramesonephric duct (par-ă-mes-ō-nef′rik) One of two embryonic tubes extending along the mesonephros and emptying into the cloaca; in the female, the duct forms the uterine tube, the uterus, and part of the vagina; in the male, it degenerates.

paranasal sinus (par-ă-nā′săl) Air-filled cavities within certain skull bones that connect to the nasal cavity; located in the frontal, maxillary, sphenoid, and ethmoid bones.

parasympathetic division (par-ă-sim-pa-thet′ik) Subdivision of the autonomic nervous system; characterized by having the cell bodies of its preganglionic neurons located in the brainstem and the sacral region of the spinal cord (craniosacral division); usually involved in activating vegetative functions, such as digestion, defecation, and urination.

parathyroid gland (par-ă-thī′royd) One of four glandular masses embedded in the posterior surface of the thyroid gland; secretes parathyroid hormone.

parathyroid hormone (PTH) Peptide hormone produced by the parathyroid gland; increases bone breakdown and blood calcium levels.

parietal (pă-rī′ĕ-tăl) [L., *paries,* wall] Relating to the wall of any cavity.

parietal cell Gastric gland cell that secretes hydrochloric acid.

parietal pericardium Serous membrane lining the fibrous portion of the pericardial sac.

parietal peritoneum Layer of peritoneum lining the abdominal walls.

parietal pleura Serous membrane that lines the different parts of the wall of the pleural cavity.

parotid gland (pă-rot′id) Largest of the salivary glands; situated anterior to each ear.

partial pressure Pressure exerted by a single gas in a mixture of gases.

passive tension Tension applied to a load by a muscle without contracting; produced when an external force stretches the muscle.

patella (pa-tel′ă) [L., *patina,* shallow disk] Kneecap.

pectoral girdle (pek′tŏ-răl) Site of attachment of the upper limb to the trunk; consists of the scapula and the clavicle; also called shoulder girdle.

pedicle (ped′ī-kl) [L., *pes,* feet] Stalk or base of a structure, such as the pedicle of the vertebral arch.

pelvic brim (pel′vik) Imaginary plane passing from the sacral promontory to the pubic crest.

pelvic girdle Site of attachment of the lower limb to the trunk; ring of bone formed by the sacrum and the coxal bones.

pelvic inlet Superior opening of the true pelvis.

pelvic outlet Inferior opening of the true pelvis.

pelvis (pel′vis) [L., basin] Any basin-shaped structure; cup-shaped ring of bone at the lower end of the trunk, formed from the ossa coxal bones, sacrum, and coccyx.

pennate (bipennate) (pen′āt) [L., *penna,* feather] Muscles with fasciculi arranged like the barbs of a feather along a common tendon.

pepsin (pep′sin) [Gr., *pepsis,* digestion] Principal digestive enzyme of the gastric juice, formed from pepsinogen; digests proteins into smaller peptide chains.

pepsinogen (pep-sin′ō-jen) [*pepsin* + Gr. *gen,* producing] Proenzyme formed and secreted by the chief cells of the gastric mucosa; the acidity of the gastric juice and pepsin itself converts pepsinogen into pepsin.

peptidase (pep′ti-dās) Enzyme capable of hydrolyzing one of the peptide links of a peptide.

peptide bond (pep′tīd) Chemical bond between amino acids.

Percent Daily Value (% Daily Value) Percent of the recommended daily value of a nutrient found in one serving of a particular food.

perforating canal Canal containing blood vessels and nerves and running through bone perpendicular to the haversian canals; also called Volkmann's canal.

periarterial lymphatic sheath (per′ē-ar-tē′rē-ăl) Dense accumulation of lymphocytes (white pulp) surrounding arteries within the spleen.

pericapillary cell One of the slender connective tissue cells in close relationship to the outside of the capillary wall; it is relatively undifferentiated and may become a fibroblast, macrophage, or smooth muscle cell.

pericardial cavity (per-i-kar′dē-ăl) Space within the mediastinum in which the heart is located.

pericardial fluid Viscous fluid contained within the pericardial cavity between the visceral and parietal pericardium; functions as a lubricant.

pericardium (per-i-kar′dē-ŭm) [Gr., *pericardion,* membrane around the heart] Membrane covering the heart; also called pericardial sac.

perichondrium (per-i-kon′drē-ŭm) [Gr., *peri,* around + *chondros,* cartilage] Double-layered connective tissue sheath surrounding cartilage.

perilymph (per′i-limf) [Gr., *peri,* around + L., *lympha,* clear fluid (lymph)] Fluid contained within the bony labyrinth of the inner ear.

perimetrium (per-i-mē′trē-ŭm) Outer serous coat of the uterus; also called serous layer.

perimysium (per-i-mis′ē-ŭm, per-i-miz′ē-ŭm) [Gr., *peri,* around + *mys,* muscle] Fibrous sheath enveloping a bundle of skeletal muscle fibers (muscle fascicle).

perineum (per′i-nē′ŭm) Area inferior to the pelvic diaphragm between the thighs; extends from the coccyx to the pubis.

perineurium (per-i-noo′rē-ŭm) [L., *peri,* around + Gr., *neuron,* nerve] Connective tissue sheath surrounding a nerve fascicle.

periodontal ligament (per′ē-ō-don′tăl) Connective tissue that surrounds the tooth root and attaches it to its bony socket.

periosteum (per-ē-os′tē-ŭm) [Gr., *peri,* around + *osteon,* bone] Thick, double-layered connective tissue sheath covering the entire surface of a bone, except the articular surface, which is covered with cartilage.

peripheral nervous system (PNS) (pē-rif′ē-răl) Major subdivision of the nervous system consisting of nerves and ganglia.

peripheral resistance (PR) Resistance to blood flow in all the blood vessels.

peristaltic wave (per-i-stal′tik) Contraction in a tube, such as the intestine, characterized by a wave of contraction in smooth muscle preceded by a wave of relaxation that moves along the tube.

peritubular capillary Capillary network located in the cortex of the kidney; associated with the distal and proximal convoluted tubules.

permanent tooth One of the 32 teeth belonging to the permanent dentition; also called secondary tooth.

peroneal (per-ō-nē′ăl) [Gr., *perone,* fibula] Associated with the fibula.

peroxisome (per-ok′si-sōm) Membrane-bounded body similar to a lysosome in appearance but often smaller and irregular in shape; contains enzymes that either decompose or synthesize hydrogen peroxide.

Peyer patch Named for Swiss anatomist Johann K. Peyer (1653–1712). Lymphatic nodule found in the lower half of the small intestine and the appendix.

phagocyte (fag′ō-sīt) Cell that ingests bacteria, foreign particles, and other cells.

phagocytosis (fag′ō-sī-tō′sis) [Gr., *phagein,* to eat + *kytos,* cell + *osis,* condition] Cells' ingestion of solid substances, such as other cells, bacteria, bits of necrosed tissue, and foreign particles.

phalange; pl. **phalanges** (fă-lanj′, fă-lan′jēz) [Gr., *phalanx,* line of soldiers] Bone of a finger or toe.

pharyngeal pouch (fă-rin′jē-ăl) Paired evagination of embryonic pharyngeal endoderm between the brachial arches that gives rise to the thymus, thyroid gland, tonsils, and parathyroid glands.

pharyngeal tonsil (fă-rin′jē-ăl) One of two collections of aggregated lymphoid nodules on the posterior wall of the nasopharynx.

pharynx (far′ingks) [Gr., *pharynx,* throat, the joint opening of the gullet and windpipe] Upper expanded portion of the digestive tube between the esophagus below and the oral and nasal cavities above and in front.

phenotype (fē′nō-tīp) [Gr., *phaino,* to display, show forth + *typos,* model] Characteristic observed in an individual due to the expression of his or her genotype.

phosphodiesterase (fos′fō-dī-es′ter-ās) Enzyme that splits phosphodiester bonds—that is, that breaks down cyclic AMP to AMP.

phospholipid (fos-fō-lip′id) Lipid with phosphorus, resulting in a molecule with a polar end and a nonpolar end; main component of the lipid bilayer.

phosphorylation (fos′fōr-i-lā′shŭn) Addition of phosphate to an organic compound.

photoreceptor (fō′tō-rē-sep′ter, fō′tō-rē-sep′tōr) [L., *photo,* light + *ceptus,* to receive] Sensory receptor that is sensitive to light—for example, rods and cones of the retina.

phrenic nerve (fren′ik) Nerve derived from spinal nerves C3–C5; supplies the diaphragm.

physiologic contracture (fiz-ē-ō-loj′ik kon-trak′chūr) Temporary inability of a muscle to either contract or relax because of a depletion of ATP so that active transport of calcium ions into the sarcoplasmic reticulum cannot occur.

physiological dead space Sum of anatomical dead air space plus the volume of any nonfunctional alveoli.

physiologic shunt Deoxygenated blood from the alveoli plus deoxygenated blood from the bronchi and bronchioles.

pia mater (pī′ă mā′ter, pē′a mah′ter) [L., tender mother] Delicate membrane forming the inner covering of the brain and spinal cord.

pigmented layer Pigmented portion of the retina.

pineal gland (pin′ē-ăl) [L., *pineus,* relating to pine trees] A small, pine cone–shaped structure that projects from the epiphysis of the diencephalon; produces melatonin; also called pineal body.

pinna (pin′ă) [L., *pinna* or *penna,* feather] *See* auricle.

pinocytosis (pin′ō-sī-tō′sis, pī′no-sī-tō′sis) [Gr., *pineo,* to drink + *kytos,* cell + *osis,* condition] Cell drinking; uptake of liquid by a cell.

pituitary gland Endocrine gland attached to the hypothalamus by the infundibulum; also called hypophysis.

plane (plān) [L., *planus,* flat] Flat surface; an imaginary surface formed by extension through any axis or two points—for example, a midsagittal plane, a coronal plane, and a transverse plane.

plasma (plaz′mă) [Gr., something formed] Fluid portion of blood.

plasma cell Cell derived from B cells; produces antibodies.

plasma clearance Volume of plasma per minute from which a substance can be completely removed by the kidneys.

plasmin (plaz′min) Enzyme derived from plasminogen; dissolves clots by converting fibrin into soluble products.

plateau phase Prolongation of the depolarization phase of a cardiac muscle cell membrane; results in a prolonged refractory period.

platelet (plāt′let) Irregularly shaped disk found in blood; contains granules in the central part and clear protoplasm peripherally but has no definite nucleus; also called thrombocyte.

platelet plug Accumulation of platelets that stick to each other and to connective tissue; prevents blood loss from damaged blood vessels.

pleural cavity (ploor′ăl) Potential space between the parietal and visceral layers of the pleura.

plexus; pl. **plexuses** (plek′sŭs, plek′sŭs-ez) [L., a braid] Intertwining of nerves or blood vessels.

plicae circulares (plī′kă, plī′sē) Numerous folds of the mucous membrane of the small intestine.

pluripotent (ploo-rip′ō-tent) [L., *pluris,* more + *potentia,* power] In development, a cell or group of cells that have not yet become fixed or determined as to what specific tissues they are going to become.

podocytes (pod′ō-sīts) [Gr., *pous, podos,* foot + *kytos,* a hollow (cell)] Epithelial cell of Bowman's capsule attached to the outer surface of the glomerular capillary basement membrane by cytoplasmic foot processes.

Poiseuille's law (pwah-zuh′yez) Named for French physiologist and physicist Jean Léonard Marie Poiseuille (1797–1869). The volume of a fluid passing per unit of time through a tube is directly proportional to the pressure difference between its ends and to the fourth power of the internal radius of the tube and inversely proportional to the tube's length and the viscosity of the fluid.

polar body (pō′lăr) One of the two small cells formed during oogenesis because of unequal division of the cytoplasm.

polar covalent bond Covalent bond in which atoms do not share their electrons equally.

polarization Development of differences in potential between two points in living tissues, as between the inside and outside of the plasma membrane.

polycythemia (pol′ē-sī-thē′mē-ă) Increase in red blood cell number above the normal.

polygenic (pol-ē-jen′ik) Relating to a hereditary disease or normal characteristic controlled by the interaction of genes at more than one locus.

polysaccharide (pol-ē-sak′ă-rīd) Carbohydrate containing a large number of monosaccharide molecules.

polyunsaturated fat Fatty acid that contains two or more double covalent bonds between its carbon atoms.

pons (ponz) [L., bridge] Portion of the brainstem between the medulla and midbrain.

popliteal (pop-lit′ē-ăl, pop-li-tē′ăl) [L., ham] Posterior region of the knee.

porta (pōr′tă) [L., gate] Fissure on the inferior surface of the liver where the portal vein, hepatic artery, hepatic nerve plexus, hepatic ducts, and lymphatic vessels enter or exit the liver.

portal system (pōr′tal) System of vessels in which blood, after passing through one capillary bed, is conveyed through a second capillary network.

portal triad Branches of the portal vein, hepatic artery, and hepatic duct bound together in the connective tissue that divides the liver into lobules.

postabsorptive state State following the absorptive state; blood glucose levels are maintained because of the conversion of other molecules to glucose.

posterior chamber (pos-tēr′ē-ŏr) Chamber of the eye between the iris and the lens.

posterior interventricular sulcus Groove on the diaphragmatic surface of the heart, marking the location of the septum between the two ventricles.

posterior pituitary gland Portion of the hypophysis derived from the brain. Major secretions include antidiuretic hormone and oxytocin. Also called neurohypophysis.

postganglionic neuron (pōst′gang-glē-on′ik) Autonomic neuron that has its cell body located within an autonomic ganglion and sends its axon to an effector organ.

postovulatory age Age of the developing fetus based on the assumption that fertilization occurs 14 days after the last menstrual period before the pregnancy.

postsynaptic (pōst-si-nap′tik) Relating to the membrane of a nerve, muscle, or gland that is in close association with a presynaptic terminal. The postsynaptic membrane has receptor molecules within it that bind to neurotransmitter molecules.

potential difference (pō-ten′shăl) Difference in electrical potential, measured as the charge difference across the plasma membrane.

potential energy [Gr., *en,* in + *ergon,* work] Energy in a chemical bond that is not being exerted or used to do work.

PQ interval Time elapsing between the beginning of the P wave and the beginning of the QRS complex in the electrocardiogram; also called PR interval.

precapillary sphincter (prē-kap′i-lār-ē sfingk′ter) Smooth muscle sphincter that regulates blood flow through a capillary.

preganglionic neuron (prē′gang-glē-on′ik) Autonomic neuron that has its cell body located within the central nervous system and sends its axon through a nerve to an autonomic ganglion, where it synapses with postganglionic neurons.

premolar (prē-mō′lăr) Bicuspid tooth.

prenatal (prē-nā′tal) [*pre* + L., *natus,* born] Preceding birth.

prepuce (prē′poos) In males, the free fold of skin that more or less completely covers the glans penis; the foreskin. In females, the external fold of the labia minora that covers the clitoris.

pressoreceptor (pres′ō-rē-sep′ter, pres′ō-rē-sep′tōr) *See* baroreceptor.

presynaptic terminal (prē′si-nap′tik) Enlarged axon terminal or terminal bouton.

primary palate In the early embryo, the structure that gives rise to the upper jaw and lips.

primary response Immune response that occurs as a result of the first exposure to an antigen.

primary spermatocyte (sper′mă-tō-sīt) Spermatocyte arising by a growth phase from a spermatogonium; gives rise to secondary spermatocytes after the first meiotic division.

prime mover Muscle that plays a major role in accomplishing a movement.

primitive streak (prim′i-tiv) Ectodermal ridge in the midline of the embryonic disk, from which arises the mesoderm by the inward and then lateral migration of cells.

primordial germ cell (prī-mōr′dē-ăl) Most primitive undifferentiated sex cell, found initially outside the gonad on the surface of the yolk sac.

PR interval *See* PQ interval.

process (pros′es, prō′ses) Projection on a bone.

processus vaginalis (prō-ses′ŭs vaj′i-năl-ŭs) Peritoneal outpocketing in the embryonic lower anterior abdominal wall that traverses the inguinal canal; in the male, it forms the tunica vaginalis testis and normally loses its connection with the peritoneal cavity.

proerythroblast Cell that matures into an erythrocyte.

progeria (prō-jēr′ē-ă) [Gr., *pro,* before + *ge* + *amras,* old age] Severe retardation of growth after the first year accompanied by a senile appearance and death at an early age.

progesterone (prō-jēs′tē-rōn) Steroid hormone secreted by the corpus luteum and one of the hormones secreted by the placenta.

prolactin (prō-lak′tin) Hormone of the adenohypophysis that stimulates the production of milk.

prolactin-inhibiting hormone (PIH) Neurohormone released from the hypothalamus that inhibits prolactin release from the adenohypophysis.

prolactin-releasing hormone (PRH) Neurohormone released from the hypothalamus that stimulates prolactin release from the adenohypophysis.

proliferative phase (prō-lif′er-ă-tiv) *See* follicular phase.

pronation (prō-nā′shŭn) [L., *pronare,* to bend forward] Rotation of the forearm so that the anterior surface is down (prone).

pronephros (prō-nef′ros) In the embryos of higher vertebrates, a series of tubules emptying into the celomic cavity. It is a temporary structure in the human embryo, followed by the mesonephros and still later by the metanephros, which gives rise to the kidney.

prophase (prō′fāz) First stage in cell division when chromatin strands condense to form chromosomes.

proprioception (prō-prē-ō-sep′shun) [L., *proprius,* one's own + *capio,* to take] Information about the position of the body and its various parts.

proprioceptor (prō′prē-ō-sep′ter) Sensory receptor associated with joints and tendons.

prostaglandin (pros′tă-glan′din) Class of physiologically active substances present in many tissues; among its effects are vasodilation, stimulation and contraction of uterine smooth muscle and the promotion of inflammation and pain.

prostate gland (pros′tāt) [Gr., *prostates,* one standing before] Gland that surrounds the beginning of the urethra in the male. The secretion of the gland is a milky fluid that is discharged by 20–30 excretory ducts into the prostatic urethra as part of the semen.

prostatic urethra (pros-tat′ik) Part of the male urethra, approximately 2.5 cm in length, that passes through the prostate gland.

protease (prō′tē-ās) Enzyme that breaks down proteins.

protein (prō′tēn, prō′tē-ĭn) [Gr., *proteios,* primary] Macromolecule consisting of long sequences of amino acids linked together by peptide bonds.

protein kinase (kin-āz) Class of enzymes that phosphorylates other proteins. Many of these kinases are responsive to other chemical signals (e.g., cAMP, cGMP, insulin, epidermal growth factor, calcium and calmodulin).

proteoglycan (prō′tē-ō-glī′kan) Macromolecule consisting of numerous polysaccharides attached to a common protein core.

prothrombin (prō-throm′bin) Glycoprotein present in blood that, in the presence of prothrombin activator, is converted to thrombin.

proton (prō′ton) [Gr., *protos,* first] Positively charged particle in the nuclei of atoms.

protraction (prō-trak′shŭn) [L., *protractus,* to draw forth] Movement forward or in the anterior direction.

provitamin (prō-vī′tă-min) Substance that may be converted into a vitamin.

proximal convoluted tubule (prok′si-măl) Part of the nephron that extends from the glomerulus to the descending limb of the loop of Henle.

pseudostratified epithelium Epithelium consisting of a single layer of cells but having the appearance of multiple layers.

pseudo-unipolar neuron One of the three categories of neurons consisting of a nerve cell body with a single axon projecting from it.

psychologic fatigue (sī-kō-loj′-ik) Fatigue caused by the central nervous system.

ptosis (tō′sis) [G., *ptosis,* a falling] Falling down of an organ—for example, the drooping of the upper eyelid.

puberty (pū′ber-tē) [L., *pubertas,* grown up] Series of events that transform a child into a sexually mature adult; involves an increase in the secretion of GnRH.

pubis (pū′bis) Anterior inferior bone of the coxal bone.

pudendal cleft (pū-den′dăl) Cleft between the labia majora.

pudendum (pū-den′dŭm) *See* vulva.

pulmonary artery (pŭl′mō-nār-ē) One of the arteries that extend from the pulmonary trunk to the right or left lungs.

pulmonary capacity Sum of two or more pulmonary volumes.

pulmonary trunk Large, elastic artery that carries blood from the right ventricle of the heart to the right and left pulmonary arteries.

pulmonary vein One of the veins that carry blood from the lungs to the left atrium of the heart.

pulp (pŭlp) [L., *pulpa,* flesh] Soft tissue within the pulp cavity of the tooth, consisting of connective tissue containing blood vessels, nerves, and lymphatics.

pulse pressure (pŭls) Difference between systolic and diastolic pressure.

pupil (pū′pĭl) Circular opening in the iris through which light enters the eye.

Purkinje fiber (pŭr-kĭn′jē) Named for Bohemian anatomist Johannes E. von Purkinje (1787–1869). Modified cardiac muscle cells found beneath the endocardium of the ventricles. Specialized to conduct action potentials.

pus (pŭs) Fluid product of inflammation; contains white blood cells, the debris of dead cells, and tissue elements liquefied by enzymes.

pyloric opening (pī-lōr′ik) Opening between the stomach and the superior part of the duodenum.

pyloric sphincter Thickening of the circular layer of the gastric musculature encircling the junction between the stomach and duodenum; also called pylorus.

pyrogen (pī′rō-jen) Chemical released by microorganisms, neutrophils, monocytes, and other cells that stimulates fever production by acting on the hypothalamus.

Q

QRS complex Principal deflection in the electrocardiogram, representing ventricular depolarization.

QT interval Time elapsing from the beginning of the QRS complex to the end of the T wave, representing the total duration of electrical activity of the ventricles.

R

radial pulse (rā′dē-ăl) Pulse detected in the radial artery.

radiation (rā′dē-ā′shŭn) [L., *radius,* ray, beam] Sending forth of light, short radiowaves, ultraviolet or x-rays, or any other rays for treatment or diagnosis or for other reasons; radiant heat.

radioactive isotope (rā′dē-ō-ak′tiv) Isotope with a nuclear composition that is unstable from which subatomic particles and electromagnetic waves are emitted.

ramus; pl. rami (rā′mŭs, rā′mī) [L., branch] One of the primary subdivisions of a nerve or blood vessel; the part of a bone that forms an angle with the main body of the bone.

raphe (rā′fē) [Gr., *rhaphe,* suture, seam] Central line running over the scrotum from the anus to the root of the penis.

receptor Structural protein or glycoprotein molecule on the cell surface or within the cytoplasm that binds to a specific factor (chemical signal).

recessive (rē-ses′iv) Gene that may not be expressed because of suppression by a contrasting dominant gene.

Recommended Dietary Allowances (RDAs) Guide for estimating the nutritional needs of groups of people based on their age, sex, and other factors; first established in 1941.

Recommended Daily Intake (RDI) Generally, the highest RDA value in each of four categories: infants, toddlers, people over 4 years of age, and pregnant or lactating women. The RDIs used to determine the Daily Values are based on the 1968 RDAs.

rectum (rek′tŭm) [L., *rectus,* straight] Portion of the digestive tract that extends from the sigmoid colon to the anal canal.

red marrow (mar′o) Soft, pulpy connective tissue filling the cavities of bones; consists of reticular fibers and the development stages of blood cells and platelets; gradually replaced by yellow marrow in long bones and the skull.

red pulp [L., *pulpa,* flesh] Reddish brown substance of the spleen consisting of venous sinuses and the tissues intervening between them, called pulp cords.

reduction (rē-dŭk′shŭn) Gain of one or more electrons by a molecule.

refraction (rē-frak-shŭn) Bending of a light ray when it passes from one medium into another of different density.

refractory period (rē-frak′tōr-ē) [Gr., *periodos,* a way around, a cycle] Period following effective stimulation during which excitable tissue, such as heart muscle, fails to respond to a stimulus of threshold intensity.

regeneration (rē′jen-er-ā′shŭn) Reproduction or reconstruction of a lost or injured part.

regulatory gene Gene involved with controlling the activity of structural genes.

relative refractory period Portion of the action potential following the absolute refractory period during which another action potential can be produced with a greater-than-threshold stimulus strength.

relaxation phase (rē-lak-sā′shŭn) Phase of muscle contraction following the contraction phase; the time from maximal tension production until tension decreases to its resting level.

relaxin Polypeptide hormone secreted by the corpus luteum and placenta during pregnancy; facilitates the birth process by causing a softening and lengthening of the pubic symphysis and cervix.

renal artery (rē′năl) Artery that originates from the aorta and delivers blood to the kidney.

renal blood flow rate Volume at which blood flows through the kidneys per minute; an average of approximately 1200 mL/min.

renal column Cortical substance separating the renal pyramids.

renal corpuscle Glomerulus and Bowman's capsule that encloses it.

renal fascia (fash′i-ă) Connective tissue surrounding the kidney that forms a sheath or capsule for the organ.

renal fat pad Fat layer that surrounds the kidney and functions as a shock-absorbing material.

renal fraction Portion of the cardiac output that flows through the kidneys; averages 21%.

renal pelvis Funnel-shaped expansion of the upper end of the ureter that receives the calyces.

renal pyramid One of a number of pyramidal masses seen on longitudinal section of the kidney; they contain part of the loops of Henle and the collecting tubules.

renin (rē′nin) Enzyme secreted by the juxtaglomerular apparatus that converts angiotensinogen to angiotensin I.

renin-angiotensin-aldosterone mechanism Renin, released from the kidneys in response to low blood pressure, converts angiotensinogen to angiotensin I. Angiotensin I is converted by angiotensin-converting enzyme to angiotensin II, which causes vasoconstriction, resulting in increased blood pressure. Angiotensin II also increases aldosterone secretion, which increases blood pressure by increasing blood volume.

repolarization phase (rē′pō-lăr-i-zā′shŭn) Phase of the action potential in which the membrane potential moves from its maximum degree of depolarization toward the value of the resting membrane potential.

reposition Return of a structure to its original position.

residual volume (rē-zid′ū-ăl) Volume of air remaining in the lungs after a maximum expiratory effort.

resolution (rez-ō-loo′shŭn) [L., *resolutio,* a slackening] Phase of the male sexual act after ejaculation during which the penis becomes flaccid; feeling of satisfaction; inability to achieve erection and second ejaculation. Last phase of the female sexual act, characterized by an overall sense of satisfaction and relaxation.

respiration (res-pi-rā′shŭn) [L., *respiratio,* to exhale, breathe] Process of life in which oxygen is used to oxidize organic fuel molecules, providing a source of energy, carbon dioxide, and water; movement of air into and out of the lungs, the exchange of gases with blood, the transportation of gases in the blood, and gas exchange between the blood and the tissues.

respiratory bronchiole (res′pi-ră-tōr-ē, rē-spīr′ă-tōr-ē) Smallest bronchiole (0.5 mm in diameter) that connects the terminal bronchiole to the alveolar duct.

respiratory membrane Membrane in the lungs across which gas exchange occurs with blood.

resting membrane potential (RMP) Electric charge difference inside a plasma membrane, measured relative to just outside the plasma membrane.

reticular (re-tik′ū-lăr) [L., *rete,* net] Relating to a fine network of cells or collagen fibers.

reticular cell Cell with processes making contact with those of other similar cells to form a cellular network; along with the network of reticular fibers, the reticular cells form the framework of bone marrow and lymphatic tissues.

reticulocyte (re-tik′ū-lō-sīt) Young red blood cell with a network of basophilic endoplasmic reticulum occurring in larger numbers during the process of active red blood cell synthesis.

reticuloendothelial system (re-tik′ū-lō-en-dō-thē′lē-ăl) *See* mononuclear phagocytic system.

retina (ret′i-nă) Nervous tunic of the eyeball.

retinaculum (ret-i-nak′ū-lŭm) [L., band, halter, to hold back] Dense regular connective tissue sheath holding down the tendons at the wrist, ankle, or other sites.

retraction (rē-trak′shŭn) [L., *retractio,* a drawing back] Movement in the posterior direction.

retroperitoneal (re′trō-per′i-tō-nē′ăl) Behind the peritoneum.

rhodopsin (rō-dop′sin) Light-sensitive substance found in the rods of the retina; composed of opsin loosely bound to retinal.

ribonuclease (rī-bō-nū′klē-ās) Enzyme that splits RNA into its component nucleotides.

ribonucleic acid (RNA) (rī′bō-noo-klē′ik) Nucleic acid containing ribose as the sugar component; found in all cells in both nuclei and cytoplasm; helps direct protein synthesis.

ribosomal RNA (rRNA) (rī′bō-sōm-ăl) RNA that is associated with certain proteins to form ribosomes.

ribosome (rī′bō-sōm) Small, spherical, cytoplasmic organelle where protein synthesis occurs.

right lymphatic duct Lymphatic duct that empties into the right subclavian vein; drains the right side of the head and neck, the right-upper thorax, and the right-upper limb.

rigor mortis (rig′er mōr′tĭs) Increased rigidity of muscle after death due to cross-bridge formation between actin and myosin as calcium ions leak from the sarcoplasmic reticulum.

rod Photoreceptor in the retina of the eye; responsible for noncolor vision in low-intensity light.

root Part below the neck of a tooth covered by cementum rather than enamel and attached by the periodontal ligament to the alveolar bone.

root of the penis Proximal attached part of the penis, including the two crura and the bulb.

rotation (rō-tā′shun) Movement of a structure about its axis.

rotator cuff (rō-tā′ter, rō-tā′tor) Four deep muscles that attach the humerus to the scapula.

round ligament Fibromuscular band that is attached to the uterus on each side in front of and below the opening of the uterine tube; it passes through the inguinal canal to the labium majus.

round ligament of the liver Remains of the umbilical vein.

round window Membranous structure separating the scala tympani of the inner ear from the middle ear.

Ruffini end organ (roo-fē′nēz) Named for Italian histologist Angelo Ruffini (1864–1929); receptor located deep in the dermis and responding to continuous touch or pressure.

ruga; pl. **rugae** (roo′gă, roo′gē) [L., a wrinkle] Fold or ridge; fold of the mucous membrane of the stomach when the organ is contracted; transverse ridge in the mucous membrane of the vagina.

S

saccule (sak′yūl) Part of the membranous labyrinth; contains a sensory structure, the macula, that detects static equilibrium.

salivary amylase (sal′i-vār-ē am′il-ās) Enzyme secreted in the saliva that breaks down starch to maltose and isomaltose.

salivary gland Gland that produces and secretes saliva into the oral cavity. The three major pairs of salivary glands are the parotid, submandibular, and sublingual glands.

salt Molecule consisting of a cation other than hydrogen and an anion other than hydroxide.

sarcolemma (sar′kō-lem′ă) [Gr., *sarco,* muscle + *lemma,* husk] Plasma membrane of a muscle fiber.

sarcoma (sar-kō′mă) Malignant neoplasm derived from connective tissue.

sarcomere (sar′kō-mēr) [Gr., *sarco,* muscle + *meros,* part] Part of a myofibril between adjacent Z disks.

sarcoplasm (sar′kō-plazm) [Gr., *sarco,* muscle + *plasma,* a thing formed] Cytoplasm of a muscle fiber, excluding the myofilaments.

sarcoplasmic reticulum (sar′kō-plaz′mik) [Gr., *sarco,* muscle + *plasma,* a thing formed + *reticulum,* net] Endoplasmic reticulum of muscle.

satellite cell (sat′ĕ-līt) Specialized cell that surrounds the cell bodies of neurons within ganglia.

saturated (satch′ŭ-rāt-ĕd) Fatty acid in which the carbon chain contains only single bonds between carbon atoms.

saturation (satch-ŭ-rā′shun) Point when all carrier molecules or enzymes are attached to substrate molecules and no more molecules can be transported or reacted.

scala tympani (skā′lă tim′pă-nī) [L., stairway] Division of the spiral canal of the cochlea lying below the spiral lamina and basilar membrane.

scala vestibuli (skā′lă ves-tib′ū-lī) Division of the cochlea lying above the spiral lamina and vestibular membrane.

scapula (skap′ū-lă) Bone forming the shoulder blade.

scar (skar) [Gr., *eschara,* scab] Fibrous tissue replacing normal tissue; also called cicatrix.

sciatic nerve (sī-at′ik) Tibial and common peroneal nerves bound together; also called ischiadic nerve.

sclera (sklēr′ă) White of the eye; white, opaque portion of the fibrous tunic of the eye.

scleral venus sinus Series of veins at the base of the cornea that drain excess aqueous humor from the eye.

scrotum; pl. **scrota, scrotums** (skrō′tŭm, skrō′tă, skrō′tŭmz) Musculocutaneous sac containing the testes.

sebaceous gland (sē-bā′shŭs) [L., *sebum,* tallow] Gland of the skin, usually associated with a hair follicle, that produces sebum.

second messenger *See* intracellular mediator.

secondary follicle Follicle in which the secondary oocyte is surrounded by granulosa cells at the periphery; contains fluid-filled antral spaces.

secondary oocyte (ō′ō-sīt) Oocyte in which the second meiotic division stops at metaphase II unless fertilization occurs.

secondary palate Roof of the mouth in the early embryo that gives rise to the hard and the soft palates.

secondary response Immune response that occurs when the immune system is exposed to an antigen against which it has already produced a primary response; also called memory response.

secondary spermatocyte (sper′mă-tō-sīt) Spermatocyte derived from a primary spermatocyte by the first meiotic division; each secondary spermatocyte gives rise by the second meiotic division to two spermatids.

secretin (se-krē′tin) Hormone formed by the epithelial cells of the duodenum; stimulates secretion of pancreatic juice high in bicarbonate ions.

secretion (se-krē′shŭn) Substance produced inside a cell and released from the cell.

secretory phase (se-krēt′ĕ-rē, sē′krĕ-tōr-ē) *See* luteal phase.

segmental artery One of five branches of the renal artery, each supplying a segment of the kidney.

segmental bronchi Extend from the secondary bronchus and conducts air to each lobule of the lungs.

self-antigen Antigen produced by the body that is capable of initiating an immune response against the body.

semen (sē′men) [L., seed (of plants, men, animals)] Penile ejaculate; thick, yellowish white, viscous fluid containing spermatozoa and secretions of the testes, seminal vesicles, prostate, and bulbourethral glands.

semicircular canal (sem′ē-sir′kū-lăr) Canal in the petrous portion of the temporal bone that contains sensory organs that detect kinetic or dynamic equilibrium. Three semicircular canals are within each inner ear.

seminal vesicle (sem′i-năl) One of two glandular structures that empty into the ejaculatory ducts; its secretion is one of the components of semen.

seminiferous tubule (sem′i-nif′er-ŭs) Tubule in the testis in which spermatozoa develop.

sensible perspiration (sen′si-bl pers-pi-rā′shŭn) Perspiration excreted by the sweat glands that appears as moisture on the skin; produced in large quantity when there is much humidity in the atmosphere.

septum primum (sep′tŭm prī′mŭm) First septum in the embryonic heart that arises on the wall of the originally single atrium of the heart and separates it into right and left chambers.

septum secundum (sek′ŭn-dŭm) Second of two major septal structures involved in the partitioning of the atrium, arising later than the septum primum and located to the right of it; it remains an incomplete partition until after birth, with its unclosed area constituting the foramen ovale.

serosa (se-rō′să) [L., *serosus,* serous] Outermost covering of an organ or a structure that lies in a body cavity.

serous fluid (ser′ŭs) Fluid similar to lymph that is produced by and covers serous membrane; it lubricates the serous membrane.

serous membrane Thin sheet composed of epithelial and connective tissues; it lines cavities that do not open to the outside of the body or contain glands but do secrete serous fluid.

serous pericardium Lining of the pericardial sac composed of a serous membrane.

Sertoli cell (ser-tō′lē) Named for Italian histologist Enrico Sertoli (1842–1910). Elongated cell in the wall of the seminiferous tubules to which spermatids are attached during spermatogenesis.

serum (sēr′ŭm) [L., whey] Fluid portion of blood after the removal of fibrin and blood cells.

sesamoid bone (ses′ă-moyd) [Gr., *sesamoceies,* like a sesame seed] Bone found within a tendon, such as the patella.

sex chromosomes Pair of chromosomes re-sponsible for sex determination, XX in female and XY in male.

sex-linked trait Characteristic resulting from the expression of a gene on a sex chromosome.

sigmoid colon (sig′moyd) Part of the colon between the descending colon and the rectum.

sigmoid mesocolon Fold of peritoneum attaching the sigmoid colon to the posterior abdominal wall.

simple epithelium Epithelium consisting of a single layer of cells.

sinoatrial (SA) node (si′nō-ā′trē-ăl) Mass of specialized cardiac muscle fibers; acts as the "pacemaker" of the cardiac conduction system.

sinus (sī′nŭs) [L., cavity] Hollow in a bone or other tissue; enlarged channel for blood or lymph.

sinus venosus End of the embryonic cardiac tube where blood enters the heart; becomes a portion of the right atrium, including the SA node.

sinusoid (si′nŭ-soyd) [L., sinus + Gr., *eidos,* resemblance] Terminal blood vessel having a larger diameter than an ordinary capillary.

sinusoidal capillary (si-nŭ-soy′dăl) Capillary with caliber of 10–20 μm or more; lined with a fenestrated type of endothelium.

small intestine [L., *intestinus,* entrails] Portion of the digestive tube between the stomach and the cecum; consists of the duodenum, jejunum, and ileum.

sodium-potassium (Na^+–K^+) pump Biochemical mechanism that uses energy derived from ATP to achieve the active transport of potassium ions opposite to that of sodium ions; also called sodium potassium ATP-ase.

soft palate Posterior muscular portion of the palate, forming an incomplete septum between the mouth and the oropharynx and between the oropharynx and the nasopharynx.

solute (sol′ūt, sō′loot) [L., *solutus,* dissolved] Dissolved substance in a solution.

solution (sō-loo′shŭn) [L., *solutio*] Homogenous mixture formed when a solute is dissolved in a solvent.

solvent (sol′vent) [L., *solvens,* to dissolve] Liquid that holds another substance in solution.

soma (sō′mă) [Gr., body] Neuron cell body or the enlarged portion of the neuron containing the nucleus and other organelles.

somatic (sō-mat′ik) [Gr., *somatikos,* bodily] Relating to the body; the cells of the body except the reproductive cells.

somatic nervous system Composed of nerve fibers that send impulses from the central nervous system to skeletal muscle.

somatomedin (sō′mă-tō-mē′din) Peptide synthesized in the liver capable of stimulating certain anabolic processes in bone and cartilage, such as synthesis of DNA, RNA, and protein.

somatotropin (sō′mă-tō-trō′pin) Protein hormone of the anterior pituitary gland; it promotes body growth, fat mobilization, and inhibition of glucose utilization.

somite (sō′mīt) [Gr., *soma*, body + *ite*] One of the paired segments consisting of cell masses formed in the early embryonic mesoderm on each side of the neural tube.

somitomere (sō′mīt-ō-mēr) Indistinct somite in the head region of the embryo.

spatial summation Summation of the local potentials in which two or more action potentials arrive simultaneously at two or more presynaptic terminals that synapse with a single neuron.

specific heat Heat required to raise the temperature of any substance 1°C compared with the heat required to raise the same volume of water 1°C.

speech Use of the voice in conveying ideas.

spermatic cord (sper-mat′ik) Cord formed by the ductus deferens and its associated structures; extends through the inguinal canal into the scrotum.

spermatid (sper′mă-tid) [Gr., *sperma*, seed + *id*] Cell derived from the secondary spermatocyte; gives rise to a spermatozoon.

spermatogenesis (sper′mă-tō-jen′ĕ-sis) Formation and development of the spermatozoon.

spermatogonium (sper′mă-tō-gō′nē-ŭm) [Gr., *sperma*, seed + *gone*, generation] Cell that divides by mitosis to form primary spermatocytes.

spermatozoon; pl. spermatozoa (sper′mă-tō-zō′on, sper′mă-to-zō′ă) [Gr., *sperma*, seed + *zoon*, animal] Male gamete or sex cell, composed of a head and a tail. The spermatozoon contains the genetic information transmitted by the male. Also called sperm cell.

sphenoid (sfē′noyd) [Gr., *shen*, wedge] Wedge-shaped.

sphincter pupillae (sfingk′ter pū-pil′ē) Circular smooth muscle fibers of the iris's diaphragm that constrict the pupil of the eye.

sphygmomanometer (sfig′mō-mă-nom′ĕ-ter) [Gr., *sphygmos*, pulse + *manos*, thin, scanty + *metron*, measure] Instrument for measuring blood pressure.

spinal nerve (spī′năl) One of 31 pairs of nerves formed by the joining of the dorsal and ventral roots that arise from the spinal cord.

spindle fiber (spin′dl) Specialized microtubule that develops from each centrosome and extends toward the chromosomes during cell division.

spiral artery (spī′răl) One of the corkscrewlike arteries in premenstrual endometrium; most obvious during the secretory phase of the uterine cycle.

spiral ganglion Cell bodies of sensory neurons that innervate hair cells of the organ of Corti are located in the spiral ganglion.

spiral lamina Attached to the modiolus and supports the basilar and vestibular membranes.

spiral ligament Attachment of the basilar membrane to the lateral wall of the bony labyrinth.

spiral organ Organ of Corti; rests on the basilar membrane and consists of the hair cells that detect sound; also called organ of Corti.

spiral tubular gland Well-developed simple or compound tubular gland; spiral in shape; within the endometrium of the uterus; prevalent in the secretory phase of the uterine cycle.

spirometer (spī-rom′ĕ-ter) [L., *spiro*, to breathe + Gr., *metron*, measure] Gasometer used for measuring the volume of respiratory gases; usually understood to consist of a counterbalanced cylindrical bell sealed by dipping into a circular trough of water.

spirometry (spī-rom′ĕ-trē) Making pulmonary measurements with a spirometer.

spleen (splēn) Large lymphatic organ in the upper part of the abdominal cavity on the left side between the stomach and diaphragm, composed of white and red pulp. It responds to foreign substances in the blood, destroys worn-out red blood cells, and is a storage site for blood cells.

spongy urethra Portion of the male urethra, approximately 15 cm in length, that traverses the corpus spongiosum of the penis.

squamous (skwā′mŭs) [L., *squama*, a scale] Scale-like, flat.

stapedius (stā-pē′dē-ŭs) Small skeletal muscles attached to the stapes.

stapes (stā′pēz) [L., stirrup] Smallest of the three auditory ossicles; attached to the oval window.

Starling law of the heart Named for English physiologist Ernest H. Starling (1866–1927). Force of contraction of cardiac muscle is a function of the length of its muscle fibers at the end of diastole; the greater the ventricular filling, the greater the stroke volume produced by the heart.

sternum (ster′nŭm) [L., *sternon*, chest] Breastbone.

steroid (stēr′oyd, ster′oyd) Large family of lipids, including some reproductive hormones, vitamins, and cholesterol.

stomach (stŭm′ŭk) Large sac between the esophagus and the small intestine, lying just beneath the diaphragm.

stratified epithelium (strat′i-fīd ep-i-thē′lē-ŭm) Epithelium consisting of more than one layer of cells.

stratum basale (strat-ŭm băh-săl′ē) [L., layer; basal] Basal, or deepest, layer of the epidermis; also called stratum germinativum.

stratum corneum (kōr′nē-ŭm) [L., layer + *corneus*, horny] Most superficial layer of the epidermis consisting of flat, keratinized, dead cells.

stratum granulosum (gran′ū-lō′sŭm) [L., layer; granulum, a small grain] Layer of cells in the epidermis filled with granules of keratohyalin.

stratum lucidum (lū′sid-ŭm) [L., layer + *lucidus*, clear] Clear layer of the epidermis found in thick skin between the stratum granulosum and the stratum corneum.

stratum spinosum (spī′nōs-ŭm) [L., layer + *spina*, spine] Layer of many-sided cells in the epidermis with intercellular connections (desmosomes) that give the cells a spiny appearance.

stria; pl. striae (strī′ă, strī′ē) [L., channel] Line or streak in the skin that is a different texture or color from the surrounding skin; also called stretch mark.

striated (strī′āt-ēd) [L., *striatus*, furrowed] Striped; marked by stripes or bands.

stroke volume (SV) [L., *volumen*, something rolled up, scroll, from *volvo*, to roll] Volume of blood pumped out of one ventricle of the heart in a single beat.

structural gene Gene that determines the structure of a specific protein or peptide.

sty (stī) Inflamed ciliary gland of the eye.

subcutaneous (sŭb′-koo-tā′nē-ŭs) [L., *sub*, under + *cutis*, skin] Under the skin; same tissue as the hypodermis.

sublingual gland (sŭb-ling′gwăl) One of two salivary glands in the floor of the mouth beneath the tongue.

submandibular gland (sŭb-man-dib′ū-lăr) One of two salivary glands in the neck, located in the space bounded by the two bellies of the digastric muscle and the angle of the mandible.

submucosa (sŭb-moo-kō′să) Layer of tissue beneath a mucous membrane.

submucosal plexus (sŭb-mū-kō′săl) [L., a braid] Gangliated plexus of unmyelinated nerve fibers in the intestinal submucosa.

substantia nigra (sŭb-stan′shē-ă nī′gră) [L., substance; black] Black nuclear mass in the midbrain; involved in coordinating movement and maintaining muscle tone.

subthreshold stimulus Stimulus resulting in a local potential so small that it does not reach threshold and produce an action potential.

sucrose (soo′krōs) Disaccharide composed of glucose and fructose; table sugar.

sulcus; pl. sulci (sool′kŭs, sŭl′sī) [L., furrow or ditch] Furrow or groove on the surface of the brain between the gyri; may also refer to a fissure.

superficial inguinal ring (ing′gwi-năl) Slitlike opening in the aponeurosis of the external oblique muscle of the abdominal wall through which the spermatic cord (round ligament in the female) emerges from the inguinal canal.

superior colliculus (ko-lik′ū-lŭs) [L., *collis*, hill] One of two rounded eminences of the midbrain; aids in coordination of eye movements.

superior vena cava (vē′nă cā′vă) Vein that returns blood from the head and neck, upper limbs, and thorax to the right atrium.

supination (soo′pi-nā′shŭn) [L., *supino*, to bend backward, place on back] Rotation of the forearm (when the forearm is parallel to the ground) so that the anterior surface is up (supine).

supramaximal stimulus Stimulus of greater magnitude than a maximal stimulus; however, the frequency of action potentials is not increased above that produced by a maximal stimulus.

suppressor T cell Subset of T lymphocytes that decreases the activity of B cells and T cells.

surfactant (ser-fak′tănt) Lipoproteins forming a monomolecular layer over pulmonary alveolar surfaces; stabilizes alveolar volume by reducing surface tension and the tendency for the alveoli to collapse.

suspension (sŭs-pen′shŭn) Liquid through which a solid is dispersed and from which the solid separates unless the liquid is kept in motion.

suspensory ligament (sŭs-pen′sŏ-rē) Band of peritoneum that extends from the ovary to the body wall; contains the ovarian vessels and nerves. Small ligament attached to the margin of the lens in the eye and the ciliary body to hold the lens in place.

suture (soo′choor) [L., *sutura*, a seam] Junction between flat bones of the skull.

sweat (swet) [A.S., *swat*] Perspiration; secretions produced by the sweat glands of the skin; also called sensible perspiration.

sweat gland Usually means structure that produces a watery secretion called sweat. Some sweat glands, however, produce viscous organic secretions. Also called sudoriferous gland.

sympathetic chain ganglion (sim-pă-thet′ik) Collection of sympathetic postganglionic neurons that are connected to each other to form a chain along both sides of the spinal cord; also called paravertebral ganglion.

sympathetic division Subdivision of the autonomic division of the nervous system characterized by having the cell bodies of its preganglionic neurons located in the thoracic and upper lumbar regions of the spinal cord (thoracolumbar division); usually involved in preparing the body for physical activity; also called thoracolumbar division.

symphysis; pl. **symphyses** (sim′fi-sis, sim′fă-sēz) [Gr., a growing together] Fibrocartilage joint between two bones.

synapse; pl. **synapses** (sin′aps, sĭ-naps′, sĭ-nap′sēz) [Gr., *syn,* together + *haptein,* to clasp] Functional membrane-to-membrane contact of a nerve cell with another nerve cell, muscle cell, gland cell, or sensory receptor; functions in the transmission of action potentials from one cell to another; also called neuromuscular junction.

synaptic cleft (si-nap′tik) Space between the presynaptic and the postsynaptic membranes.

synaptic fatigue Fatigue due to depletion of neurotransmitter vesicles in the presynaptic terminals.

synaptic vesicle Secretory vesicle in the presynaptic terminal containing neurotransmitter substances.

synchondrosis; pl. **synchondroses** (sin′kon-drō′sis, -sēz) [Gr., *syn,* together + *chondros,* cartilage + *osis,* condition] Union between two bones formed by hyaline cartilage.

syncytiotrophoblast (sin-sish′ē-ō-trō′fō-blast) Outer layer of the trophoblast composed of multinucleated cells.

syndesmosis; pl. **syndesmoses** (sin′dez-mō′sis, sin′dez-mō′sēz) [Gr., *syndeo,* to bind + *osis,* condition] Form of fibrous joint in which opposing surfaces that are some distance apart are united by ligaments.

synergist (sin′er-jist) Muscle that works with other muscles to cause a movement.

synovial (si-nō′vē-ăl) [Gr., *syn,* together + *oon,* egg] Relating to or containing synovia (a substance that serves as a lubricant in a joint, tendon sheath, or bursa).

synovial fluid Slippery fluid found inside synovial joints and bursae; produced by the synovial membranes.

systemic inflammation (sis-tem′ik) Inflammation that occurs in many areas of the body. In addition to symptoms of local inflammation, increased neutrophil numbers in the blood, fever, and shock can occur.

systole (sis′tō-lē) [Gr., *systole,* a contracting] Contraction of the heart chambers during which blood leaves the chambers; usually refers to ventricular contraction.

T

T cell Thymus-derived lymphocyte of immunologic importance; it is of long life and is responsible for cell-mediated immunity.

T wave Deflection in an electrocardiogram following the QRS complex, representing ventricular repolarization.

tactile corpuscle (tak′til) Oval receptor found in the papillae of the dermis; responsible for fine, discriminative touch; also called Meissner corpuscle.

tactile disk Cuplike receptor found in the epidermis; responsible for light touch and superficial pressure; also called Merkel's disk.

talus (tā′lŭs) [L., ankle bone, heel] Tarsal bone contributing to the ankle.

target tissue Tissue on which a hormone acts.

tarsal bone (tar′săl) [Gr., *tarsos,* sole of foot] One of seven ankle bones.

tarsal plate (tar′săl) Crescent-shaped layer of connective tissue that helps maintain the shape of the eyelid.

taste (tāst) Sensations created when a chemical stimulus is applied to the taste receptors in the tongue.

taste bud Sensory structure, mostly on the tongue, that functions as a taste receptor.

tectum (tek′tŭm) Roof of the midbrain.

tegmentum (teg-men′tŭm) Floor of the midbrain.

telencephalon (tel-en-sef′ă-lon) [Gr., *telos,* end + *enkephalos,* brain] Anterior division of the embryonic brain from which the cerebral hemispheres develop.

telophase (tel′ō-fāz) Time during cell division when the chromosomes are pulled by spindle fibers away from the cell equator and into the two halves of the dividing cell.

temporal summation (tem′pŏ-răl) Summation of the local potential that results when two or more action potentials arrive at a single synapse in rapid succession.

tendon (ten′dŏn) Band or cord of dense connective tissue that connects a muscle to a bone or another structure.

tensor tympani (ten′sōr tim′pa-nī) Small skeletal muscle attached to the malleus.

tentorium cerebelli (ten-tō′rē-ŭm ser′ĕ-bel′ī) Dural folds between the cerebrum and the cerebellum.

terminal bouton (bū-ton′) [Fr., button] Enlarged axon terminal or presynaptic terminal.

terminal cisterna (sis-ter′nă) [L., *terminus,* limit + *cista,* box] Enlarged end of the sarcoplasmic reticulum in the area of the T tubules.

terminal hair [L., *terminus,* a boundary, limit] Long, coarse, usually pigmented hair found in the scalp, eyebrows, and eyelids and replacing vellus hair.

terminal sulcus (sūl′kŭs) [L., furrow or ditch] V-shaped groove on the surface of the tongue at the posterior margin.

testis; pl. **testes** (tes′tis, tes′tēz) One of two male reproductive glands located in the scrotum; produces spermatozoa, testosterone, and inhibin.

testosterone (tes-tos′tĕ-rōn) Steroid hormone secreted primarily by the testes; aids in spermatogenesis, maintenance and development of male reproductive organs, secondary sexual characteristics, and sexual behavior.

tetraiodothyronine (T_4) (tet′ră-ī-ō′dō-thī′rō-nēn) One of the iodine-containing thyroid hormones; also called thyroxine.

thalamus (thal′ă-mŭs) [Gr., *thalamos,* bed, bedroom] Large mass of gray matter that forms the larger dorsal subdivision of the diencephalon.

theca (thē′kă) [Gr., *theke,* box] Sheath or capsule.

theca externa External fibrous layer of the theca of a vesicular follicle.

theca interna Inner vascular layer of the theca of the secondary and mature follicle; produces estrogen and contributes to the formation of the corpus luteum after ovulation.

thenar eminence (thē′nar) [Gr., palm of the hand] Fleshy mass of tissue at the base of the thumb; contains muscles responsible for thumb movements.

thick skin Skin in the palms, soles, and tips of the digits; has all five epidermal strata.

thin skin Skin over most of the body, usually without a stratum lucidum; has fewer layers of cells than thick skin.

thoracic cavity (thō-ras′ik) Space within the thoracic walls, bounded below by the diaphragm and above by the neck.

thoracic duct Largest lymph vessel in the body, beginning at the cisterna chyli and emptying into the left subclavian vein; drains the left side of the head and neck, the left-upper thorax, the left-upper limb, and the inferior half of the body.

thoracolumbar division (thōr′ă-kō-lŭm′bar) *See* sympathetic division.

thoroughfare channel Channel for blood through a capillary bed from an arteriole to a venule.

threshold potential (thresh′ōld) Value of the membrane potential at which an action potential is produced as a result of depolarization in response to a stimulus.

threshold stimulus Stimulus resulting in a local potential just large enough to reach threshold and produce an action potential.

thrombocyte (throm′bō-sīt) Platelet.

thrombocytopenia (throm′bō-sī′tō-pē′nē-ă) [*thrombocyte* + Gr., *penia,* poverty] Condition in which there is an abnormally small number of platelets in the blood.

thromboxane (throm-bok′sān) Specific class of physiologically active fatty acid derivatives present in many tissues.

thrombus; pl. **thrombi** (throm′bŭs, throm′bī) [Gr., *thrombos,* a clot] Clot in the cardiovascular system formed from constituents of blood; may be occlusive or attached to the vessel or heart wall without obstructing the lumen.

thymus (thī′mŭs) [Gr., *thymos,* sweetbread] Bilobed lymph organ located in the inferior neck and superior mediastinum; secretes the hormone thymosin.

thyroid cartilage (thī′royd) Largest laryngeal cartilage. It forms the laryngeal prominence, or Adam's apple.

thyroid gland [Gr., *thyreoeides,* shield] Endocrine gland located inferior to the larynx and consisting of two lobes connected by the isthmus; secretes the thyroid hormones triiodothyronine (T_3) and tetraiodothyronine (T_4).

thyroid-stimulating hormone (TSH) Glycoprotein hormone released from the hypothalamus; stimulates thyroid hormone secretion from the thyroid gland; also called thyrotropin.

thyrotropin (thī-rot′rō-pin, thī-rō-trō′pin) *See* thyroid-stimulating hormone (TSH).

thyroxine (thi-rok′sēn, thi-rok′sin) *See* tetraiodothyronine.

tidal volume (tī′dăl) Volume of air that is inspired or expired in a single breath during regular, quiet breathing.

tissue repair (tish′ū) Substitution of viable cells for damaged or dead cells by regeneration or replacement.

tolerance (tol′er-ăns) Failure of the specific immune system to respond to an antigen.

tongue (tŭng) Muscular organ occupying most of the oral cavity when the mouth is closed; major attachment is through its posterior portion.

tonicity (tō-nis′ĭ-tē) [Gr., *tonos,* tone] Osmotic pressure or tension of a solution, usually relative to that of blood; a state of continuous activity or tension caused by muscle contraction beyond the tension related to physical properties of muscle.

tonsil; pl. **tonsils** (ton′sil, ton′silz) [L., *tonsilla,* stake] Collection of lymphoid tissue; usually refers to a large collection of lymphatic tissue beneath the mucous membrane of the oral cavity and pharynx; lingual, pharyngeal, and palatine tonsils.

total lung capacity Volume of air contained in the lungs at the end of a maximum inspiration; equals vital capacity plus residual volume.

total tension Sum of active and passive tension.

trabecula; pl. **trabeculae** (tră-bek′ū-lă, tră-bek′ū-lē) [L., *trabs,* beam] One of the supporting bundles of fibers traversing the substance of a structure, usually derived from the capsule or one of the fibrous septa, such as trabeculae of lymph nodes, testes; a beam or plate of cancellous bone.

trachea (trā′kē-ă) [Gr., *tracheia arteria,* rough artery] Air tube extending from the larynx into the thorax, where it divides to form the bronchi; composed of 16–20 rings of hyaline cartilage.

tracheostomy (tra′ke-os′to-me) [*tracheo, trachea* + Gr., *stoma,* mouth] Operation to make an opening into the trachea; usually the opening is intended to be permanent, and a tube is inserted into the trachea to allow airflow.

tracheotomy (tra′ke-ot′o-me) [*tracheo, trachea* + Gr., *tome,* incision] Act of cutting into the trachea.

transcription (tran-skrip′shŭn) Process of forming RNA from a DNA template.

transfer RNA (tRNA) RNA that attaches to individual amino acids and transports them to the ribosomes, where they are connected to form a protein polypeptide chain.

transfusion (trans-fū′zhŭn) [L., *trans,* across + *fundo,* to pour from one vessel to another] Transfer of blood from one person to another.

transitional epithelium (tran-sish′ŭn-ăl) Stratified epithelium that may be either cuboidal or squamouslike, depending on the presence or absence of fluid in the organ (as in the urinary bladder).

translation (trans-lā′shŭn) Synthesis of polypeptide chains at the ribosome in response to information contained in mRNA molecules.

transverse colon (trans-vers′ kō′lon) Part of the colon between the right and left colic flexures.

transverse mesocolon (mez′ō-kō′lon) Fold of peritoneum attaching the transverse colon to the posterior abdominal wall.

transverse (T) tubule [L., *tubus,* tube] Tubule that extends from the sarcolemma to a myofibril of striated muscles.

treppe (trep′eh) [Ger., staircase] Series of successively stronger contractions that occur when a rested muscle fiber receives closely spaced stimuli of the same strength but with a sufficient stimulus interval to allow complete relaxation of the fiber between stimuli.

triacylglycerol (trī-as′il-glis′er-ol) *See* triglyceride.

triad (trī′ad) Two terminal cisternae and a T tubule between them.

tricuspid valve (trī-kŭs′pid) Valve closing the orifice between the right atrium and the right ventricle of the heart.

triglyceride (tri-glis′er-id) Three-carbon glycerol molecule with a fatty acid attached to each carbon; constitute approximately 95% of the fats in the human body. Also called triacylglycerol.

trigone (trī′gōn) [Gr., *trigonon,* triangle] Triangular, smooth area at the base of the bladder between the openings of the two ureters and that of the urethra.

triiodothyronine (T_3) (trī-ī′ō-dō-thī′rō-nēn) One of the iodine-containing thyroid hormones.

trochlea (trok′lē-ă) [L., pulley] Structure shaped like or serving as a pulley or spool.

trochlear nerve (trok′lē-ar) [L., *trochlea,* pulley] Cranial nerve IV, to the muscle (superior oblique) turning around a pulley.

trophoblast (trof′ō-blast) [Gr., *trophe,* nourishment + *blastos,* germ] Cell layer forming the outer layer of the blastocyst, which erodes the uterine mucosa during implantation; the trophoblast does not become part of the embryo but contributes to the formation of the placenta.

tropomyosin (trō-pō-mī′ō-sin) Fibrous protein found as a component of the actin myofilament.

troponin (trō′pō-nin) Globular protein component of the actin myofilament.

true pelvis Portion of the pelvis inferior to the pelvic brim.

true rib (ver-tĕ′brō-ster′năl) Rib that attaches by an independent costal cartilage directly to the sternum; also called vertebrosternal rib.

trypsin (trip′sin) Proteolytic enzyme formed in the small intestine from the inactive pancreatic precursor trypsinogen.

tubercle (too′ber-kl) Lump on a bone.

tubular load (too′bū-lăr) Amount of a substance per minute that crosses the filtration membrane into Bowman's capsule.

tubular maximum Maximum rate of secretion or reabsorption of a substance by the renal tubules.

tubular reabsorption Movement of materials, by means of diffusion, active transport, or symport, from the filtrate within a nephron to the blood.

tubular secretion Movement of materials, by means of active transport, from the blood into the filtrate of a nephron.

tumor (too′mŏr) Any swelling or growth; a neoplasm.

tunic (too′nik) [L., coat] One of the enveloping layers of a part; one of the coats of a blood vessel; one of the coats of the eye; one of the coats of the digestive tract.

tunica adventitia (too′ni-kă ad-ven-tish′ă) Outermost fibrous coat of a vessel or an organ that is derived from the surrounding connective tissue.

tunica albuginea (al-bū-jin′ē-ă) Dense, white, collagenous tunic surrounding a structure, such as the capsule around the testis.

tunica intima (in′ti-mă) Innermost coat of a blood vessel; consists of endothelium, a lamina propria, and an inner elastic membrane.

tunica media Middle, usually muscular, coat of an artery or another tubular structure.

turbulent flow Flow characterized by eddy currents exhibiting nonparallel blood flow.

tympanic membrane (tim-pan′ik) Eardrum; cellular membrane that separates the external from the middle ear; vibrates in response to sound waves.

U

unmyelinated axon (ŭn-mī′ĕ-li-nā-ted) Nerve fibers lacking a myelin sheath.

unsaturated (ŭn-sach′ŭr-āt-ed) Carbon chain of a fatty acid that possesses one or more double or triple bonds.

upper respiratory tract Nasal cavity, pharynx, and associated structures.

up-regulation Increase in the concentration of receptors in response to a signal.

ureter (ū-rē′ter, ū′rē-ter) [Gr., *oureter,* urinary canal] Tube conducting urine from the kidney to the urinary bladder.

urethral gland (ū-rē′thrăl) One of numerous mucous glands in the wall of the spongy urethra in the male.

urogenital triangle Anterior portion of the perineal region containing the openings of the urethra and vagina in the female and the urethra and root structures of the penis in the male.

uterine cycle (ū′ter-in, ū′ter-īn) Series of events that occur in a regular fashion in the uterus of sexually mature, nonpregnant females; prepares the uterine lining for implantation of the embryo.

uterine part Portion of the uterine tube that passes through the wall of the uterus.

uterine tube One of the tubes leading on each side from the uterus to the ovary; consists of the infundibulum, ampulla, isthmus, and uterine parts; also called fallopian tube or oviduct.

uterus (ū′ter-ŭs) Hollow, muscular organ in which the fertilized ovum develops into a fetus.

utricle (oo′tri-kl) Part of the membranous labyrinth; contains a sensory structure, the macula, that detects static equilibrium.

uvula (ū′vū-lă) [L., *uva,* grape] Small, grapelike appendage at the posterior margin of the soft palate.

V

vaccination (vak′si-nā′shŭn) Deliberate introduction of an antigen into a subject to stimulate the immune system and produce immunity to the antigen.

vaccine (vak′sēn, vak-sēn′) [L., *vaccinus,* relating to a cow] Preparation of killed microbes, altered microbes, or derivatives of microbes or microbial products intended to produce immunity. The method of administration is usually inoculation, but ingestion is preferred in some instances, and nasal spray is used occasionally.

vagina (vă-jī′nă) [L., sheath] Genital canal in the female, extending from the uterus to the vulva.

vapor pressure Partial pressure exerted by water vapor.

variable region Part of an antibody that combines with an antigen.

vas deferens (vas def′er-enz) *See* ductus deferens.

vasa recta (vā′să rek′tă) Specialized capillary that extends from the cortex of the kidney into the medulla and then back to the cortex.

vasa vasorum (vā′sor-ŭm) [L., vessel, dish] Small vessels distributed to the outer and middle coats of larger blood vessels.

vascular tunic (vas′kū-lăr) Middle layer of the eye; contains many blood vessels.

vasoconstriction (vā′sō-kon-strik′shŭn, vas′ō-kon-strik′shŭn) Decreased diameter of blood vessels.

vasodilation (vā′sō-dī-lā′shŭn) Increased diameter of blood vessels.

vasomotion (vā-sō-mō′shŭn) Periodic contraction and relaxation of the precapillary sphincter, resulting in cyclic blood flow through capillaries.

vasomotor center (vā-sō-mō′ter, vas-ō-mō′ter) Area within the medulla oblongata that regulates the diameter of blood vessels by way of the sympathetic nervous system.

vasomotor tone Relatively constant frequency of sympathetic impulses that keep blood vessels partially constricted in the periphery.

vasopressin (vā-sō-pres′in, vas-ō-pres′in) *See* antidiuretic hormone.

vellus hair (vel′ŭs) [L., fleece] Short, fine, usually unpigmented hair that covers the body except for the scalp, eyebrows, and eyelids. Much of the vellus is replaced at puberty by terminal hairs.

venous capillary (vē′nŭs) Capillary opening into a venule.

venous return Volume of blood returning to the heart.

venous sinus Endothelium-lined venous channel in the dura mater that receives cerebrospinal fluid from the arachnoid granulations.

ventilation (ven-ti-lā′shŭn) [L., *ventus,* the wind] Movement of gases into and out of the lungs.

ventral root (ven′trăl) Motor (efferent) root of a spinal nerve.

ventricle (ven′tri-kl) [L., *venter,* belly] Chamber of the heart that pumps blood into arteries (i.e., the left and right ventricles); in the brain, a fluid-filled cavity.

ventricular diastole (ven-trik′ū-lăr) Dilation of the heart ventricles.

ventricular systole Contraction of the ventricles.

venule (ven′ool, vē′nool) Minute vein, consisting of endothelium and a few scattered smooth muscles, that carries blood away from capillaries.

vermiform appendix (ver′mi-fōrm) [L., *vermis,* worm + *forma,* form; appendage] Wormlike sac extending from the blind end of the cecum.

vesicle (ves′i-kl) [L., *vesica,* bladder] Small sac containing a liquid or gas, such as a blister in the skin or an intracellular, membrane-bounded sac.

vestibular fold (ves-tib′ū-lăr) One of two folds of mucous membrane stretching across the laryngeal cavity from the angle of the thyroid cartilage to the arytenoid cartilage superior to the vocal cords; helps close the glottis; also called false vocal cord.

vestibular membrane Membrane separating the cochlear duct and the scala vestibuli.

vestibule (ves′ti-bool) [L., antechamber, entrance court] Anterior part of the nasal cavity just inside the external nares that is enclosed by cartilage; space between the lips and the alveolar processes and teeth; middle region of the inner ear containing the utricle and saccule; space behind the labia minora containing the openings of the vagina, urethra, and vestibular glands.

vestibulocochlear nerve (ves-tib′ū-lō-kok′lē-ăr) Nerve formed by the cochlear and vestibular nerves; extends to the brain.

villus; pl. villi (vil′ŭs, vil′ī) [L., shaggy hair (of beasts)] Projection of the mucous membrane of the intestine; leaf-shaped in the duodenum; becomes shorter, more finger-shaped, and sparser in the ileum.

visceral (vis′er-ăl) Relating to the internal organs.

visceral pericardium (per′i-kar′dē-ŭm) Serous membrane covering the surface of the heart; also called epicardium.

visceral peritoneum (per′i-tō-nē′ŭm) [Gr., *periteino,* to stretch over] Layer of peritoneum covering the abdominal organs.

visceral pleura (vis′er-ăl plūr′ă) Serous membrane investing the lungs and dipping into the fissures between the several lobes.

visceroreceptor (vis′er-ō-rē-sep′tŏr) Sensory receptor associated with the organs.

viscosity (vis-kos′i-tē) [L., *viscosus,* viscous] Resistance to flow or alteration of shape by any substance as a result of molecular cohesion.

visual cortex (vizh′oo-ăl) Area in the occipital lobe of the cerebral cortex that integrates visual information and produces the sensation of vision.

visual field Area of vision for each eye.

vital capacity (vīt-ăl) Greatest volume of air that can be exhaled from the lungs after a maximum inspiration.

vitamin (vīt′ă-min) [L., *vita,* life + amine] One of a group of organic substances present in minute amounts in natural foodstuffs that are essential to normal metabolism; insufficient amounts in the diet may cause deficiency diseases.

vitamin D Fat-soluble vitamin produced from precursor molecules in skin exposed to ultraviolet light; increases calcium and phosphate uptake from the intestines.

vitreous humor (vit′rē-ŭs) Transparent, jellylike material that fills the space between the lens and the retina.

Volkmann's canal Named for German surgeon Richard Volkmann (1830–1889). Canal in bone containing blood vessels; not surrounded by lamellae; runs perpendicular to the long axis of the bone and the haversian canals, interconnecting the latter with each other and the exterior circulation.

vulva (vŭl′vă) [L., wrapper or covering, seed covering, womb] External genitalia of the female, composed of the mons pubis, the labia majora and minora, the clitoris, the vestibule of the vagina and its glands, and the opening of the urethra and of the vagina. Also called pudendum.

W

water-soluble vitamin Vitamin, such as B complex and C, that is absorbed with water from the intestinal tract.

white matter Bundles of parallel axons with their associated sheath in the central nervous system.

white pulp Part of the spleen consisting of lymphatic nodules and diffuse lymphatic tissue; associated with arteries.

white ramus communicans; pl. rami communicantes (rā′mŭs kō-mū′nĭ-kans, rā′mī kō-mū-nī kan′tēz) Connection between a spinal nerve and a sympathetic chain ganglion through which myelinated preganglionic axons project.

wisdom tooth Third molar tooth on each side in each jaw.

X

xiphoid (zi′foyd) [Gr., *xiphos,* sword] Sword-shaped, with special reference to the sword tip; the inferior part of the sternum.

X-linked Gene located on an X chromosome.

Y

yellow marrow (mar′o) Connective tissue filling the cavities of bones; consists primarily of reticular fibers and fat cells; replaces red marrow in long bones and the skull.

Y-linked Gene located on a Y chromosome.

yolk sac (yōk, yōlk) Highly vascular layer surrounding the yolk of an embryo.

Z

Z disk Delicate, membranelike structure found at each end of a sarcomere to which actin myofilaments attach.

zona fasciculata (zō′nă fa-sik′ū-lă′tă) [L., *zone,* a girdle, one of the zones of the sphere] Middle layer of the adrenal cortex that secretes cortisol.

zona glomerulosa (glō-mār-ū-lōs-ă) Outer layer of the adrenal cortex that secretes aldosterone.

zona pellucida (pe-lū′sĭ-dă) Layer of viscous fluid surrounding the oocyte.

zona reticularis (rē-tik′ū-lar′is) Inner layer of the adrenal cortex that secretes androgens and estrogens.

zonula adherens (zō′nū-lă ad-her′enz) [L., a small zone; adhering] Small zone holding or adhering cells together.

zonula occludens (ō-klūd′enz) [L., occluding] Junction between cells in which the plasma membranes may be fused; occludes or blocks off the space between the cells.

zygomatic (zī-gō-mat′ik) [Gr., *zygon,* yoke] Yoking or joining; bony arch created by the junction of the zygomatic and temporal bones.

zygote (zī′gōt) [Gr., *zygotos,* yoked] Diploid cell resulting from the union of a sperm cell and an oocyte.

Credits

PHOTO CREDITS

Chapter 1

Opener: © Digital Vision/DV157/Getty Images RF; **1A:** © Omikron/Photo Researchers, Inc.; **1B:** © Bernard Benoit/Science Photo Library/Photo Researchers, Inc.; **1Ca:** © Scott Camazine; **1Cb:** © D. P. M. Ribotsky/Custom Medical Stock Photo; **1D:** © ISM/Phototake.com; **1E:** © Photodisc RF; **1F:** © McGill University/CNRI/Phototake.com; **1.1:** © Bart Harris/Corbis; **1.9, 1.10a-b, 1.12a:** © The McGraw-Hill Companies, Inc./Eric Wise, photographer; **1.12b-d:** © R. T Hutchings.

Chapter 2

Opener: The Protein Data Bank/RCSB (PDB ID: 2R9R). Long, S.B., Tao, X., Campbell, E.B., MacKinnon, R. Atomic structure of a voltage-dependent K+ channel in a lipid membrane-like environment. *Nature* 450: 376-382 (2007); **2.4c:** © Trent Stephens; **2.15c:** © Biophoto Associates/Photo Researchers, Inc.

Chapter 3

Opener: This water permeation through aquaporin image was made with VMD and is owned by Theoretical and Computational Biophysics Group, NIH Center for Macromolecular Modeling and Bioinformatics, at the Beckman Institute, University of Illinois at Urbana-Champaign; **3Aa:** © National High Magnetic Field Laboratory, The Florida State University; **3Ab:** © & Courtesy of Werner Franke and Ulrich Scheer; **3Ac:** © & Courtesy of Dr. Martin W. Goldberg; **3Ad:** Shahin, Schillers & Oberleithner, Institute of Physiology II, Medical Faculty, University of Muenster, Germany; **3.2b:** © J.J. Head/Carolina Biological Supply Company/Phototake.com; **3.13a-c:** © Dr. David Phillipps/Visuals Unlimited; **3.17b:** © Manfred Kage/kage-mikrokfotografie.com/Getty Images; **3.20b:** © Courtesy of Dr. Birgit H. Satir; **3.21b, 3.22b:** © Don Fawcett/Photo Researchers, Inc.; **3.22c:** © Bernard Gilula/Photo Researchers, Inc.; **3.23b:** © Dr. Donald Fawcett, A. Olins/Visuals Unlimited; **3.25b:** © J. David Robertson, from Charles Flickinger, Medical Cell Biology, Philadelphia; **3.26b:** © Robert Bollender/Don Fawcett/Visuals Unlimited; **3.29b:** © Don Fawcett/Visuals Unlimited; **3.30b:** © Biology Media/Photo Researchers, Inc.; **3.31b:** © E. de Harven/Photo Researchers, Inc.; **3.31c:** © Dr. Gopal Murti/Science Photo Library/Photo Researchers, Inc.; **3.33b:** © Courtesy Susumi Ito, Ph.D.; **3.41:** © Ed Reschke.

Chapter 4

Opener: Thomas Deerinck and Mark Ellisman, NCMIR, UCSD; **4.1b:** © Victor Eroschenko; **4.1c:** © Ed Reschke; **4.2a:** © Dr. Fred Hossler/Visuals Unlimited; **4.2b-c, 4.3a:** © Victor Eroschenko; **4.3b:** © The McGraw-Hill Companies, Inc./Dr. Alvin Telser, photographer; **4.3c:** © Dr. Richard Kessel/Visuals Unlimited; **4.4a, 4.4b** ***(both)*****, 4.7a-b:** © Victor Eroschenko; **4.8:** © Ed Reschke; **4.8b:** © Carolina Biological Supply/Phototake.com; **4.8c:** © The McGraw-Hill Companies, Inc./Dr. Alvin Telser, photographer; **4.9a1:** © Victor Eroschenko; **4.9a2:** © Ed Reschke/Peter Arnold, Inc./Getty Images; **4.9b1-2, 4.9c1:** © Victor Eroschenko; **4.9c2:** © Ed Reschke/Peter Arnold, Inc./Getty Images; **4.9d1-2:** © Ed Reschke; **4.10a:** © Carolina Biological Supply/Phototake.com; **4.10b-c, 4.11a:** © Victor Eroschenko; **4.11b:** © Trent Stephens; **4.12a-b, 4.14a-b:** © Ed Reschke; **4.14c:** © The McGraw-Hill Companies, Inc./Dennis Strete, photographer; **4.15a-b:** © Trent Stephens.

Chapter 5

Opener: © Jeff Greenberg/PhotoEdit, Inc.; **5.2a:** © Herve Conge/ISM/Phototake.com; **5.2b:** © The Bergman Collection; **Figure 5Aa:** © Dr. P. Marazzi/Photo Researchers, Inc.; **Figure 5Ab:** © Caliendo/Custom Medical Stock Photo; **Figure 5Ac:** © Thomas B. Habif; **p. 156:** © Image Source/Getty Images RF; **Figure 5D:** © James Stevenson/Science Photo Library/Photo Researchers, Inc.; **Figure 5E:** © Stan Levy/Photo Researchers, Inc.; **5.10a:** Courtesy of A. M. Kligman, Professor of Dermatology, University of Pennsylvania School of Medicine, Philadelphia, PA; **5.10b:** Edward Curtis photo from Library of Congress; Courtesy of A. M. Kligman, Professor of Dermatology, University of Pennsylvania School of Medicine, Philadelphia, PA.

Chapter 6

Opener: © AP Photo/The Decatur Daily/Dan Henry; **6.1:** © Ed Reschke; **6.2a-c:** © Trent Stephens; **6.3c:** © Bio-Photo Assocs/Photo Researchers, Inc.; **Figure 6A:** © Hilt & Cogburn, Manual of Orthopedics; **6.6:** © Robert Caladine/Visuals Unlimited; **6.7b:** © Trent Stephens; **6.11(1):** © Victor Eroschenko; **6.11(2):** © Dr. Richard Kessel/Visuals Unlimited; **6.11(3):** © Victor Eroschenko; **6.12:** © Visuals Unlimited; **6.14a:** © Ed Reschke/Peter Arnold, Inc./Getty Images; **6.14b:** © Bio-Photo Assocs/Photo Researchers, Inc.; **6.15:** © J. M. Booher; **6.17:** From: *Report of The Consultative Committee on The Primary School*, 1931; **6.18:** © Hulton Archive/Getty Images; **6.20a:** © Andrew F. Russo; **Figure 6C:** © Dr. Michael Klein/Peter Arnold, Inc./Getty Images; **Figure 6D:** © Princess Margaret Rose Orthopaedic Hospital/Science Photo Library/Photo Researchers, Inc.; **Figure 6E:** © Yoav Levy/Phototake.com.

Chapter 7

Opener: © Ingram Publishing RF; **7.5, 7.7:** © The McGraw-Hill Companies, Inc./Eric Wise, photographer; **7.10c-d:** © Jupiter Media/Alamy; **Figure 7A:** © Princess Margaret Rose Orthopaedic Hospital/Photo Researchers, Inc.; **7.14:** © The McGraw-Hill Companies, Inc./Eric Wise, photographer; **p. 213, 7.16h, 7.17c, 7.18c, 7.20c:** © Trent Stephens; **7.21:** © The McGraw-Hill Companies, Inc./Eric Wise, photographer; **7.23d:** © Trent Stephens; **7.26:** © The McGraw-Hill Companies, Inc./Eric Wise, photographer; **Figure 7D:** © Shutterstock.com; **7.31, 7.36:** © The McGraw-Hill Companies, Inc./Eric Wise, photographer.

Chapter 8

Opener: © Andrew F. Russo; **8.9, 8.10a-c, 8.11, 8.12:** © The McGraw-Hill Companies, Inc./Eric Wise, photographer; **8.13a:** © The McGraw-Hill Companies, Inc./Jill Braaten, Photographer; **8.13b, 8.14, 8.15, 8.16, 8.17, 8.128, 8.19, 8.20, 8.21:** © The McGraw-Hill Companies, Inc./Eric Wise, photographer; **8.25c:** © The McGraw-Hill Companies, Inc./Photo and Dissection by Christine Eckel; **8.26e:** © R. T. Hutchings; **Figure 8Ca:** © James Stevenson/Science Photo Library/Photo Researchers, Inc.; **Figure 8Cb:** © CNRI/Science Photo Library/Photo Researchers, Inc.

Chapter 9

Opener: © Dr. Fred Hossler/Visuals Unlimited, Inc.; **9.2:** © Ed Reschke; **9.5a, 9.6a:** © Don W. Fawcett/Photo Researchers, Inc.; **9.6b:** © Biophoto Associates/Photo Researchers, Inc.; **9.11c:** © Fred Hossler/Visuals Unlimited; **9.18b:** © Don Fawcett/Photo Researchers, Inc.; **9.23:** © Victor Eroschenko; **p. 302:** © Olga Donskaya/Shutterstock.com; **Figure 9A:** © Dr. Richard Kessel/Visuals Unlimited; **Figure 9B:** © Roberta Seidman; **Figure 9C:** © Andrew J. Kornberg.

Chapter 10

Opener: © Digital Vision, DV46/Getty Images RF; **10.4b, 10.5b:** © The McGraw-Hill Companies, Inc./Jill Braaten, Photographer; **10.8a-d:** © The McGraw-Hill Companies, Inc./Eric Wise, photographer; **10.17b:** © The McGraw-Hill Companies, Inc./Jill Braaten, photographer; **10.21b:** © Christine Eckel; **10.21c:** © The McGraw-Hill Companies, Inc./Jill Braaten, photographer; **10.22b:** © Christine Eckel; **10.22c:** © The McGraw-Hill Companies, Inc./Jill Braaten, photographer; **10.24b:** © The McGraw-Hill Companies, Inc./Rebecca Gray, photographer/Don Kincaid, dissections; **10.24c, 10.25d:** © The McGraw-Hill Companies, Inc./Jill Braaten, photographer; **10.26c:** © The McGraw-Hill Companies, Inc./Rebecca Gray, photographer/Don Kincaid, dissections;

10.26d, 10.27c: © The McGraw-Hill Companies, Inc./Jill Braaten, photographer; **10.28b:** © Christine Eckel; **10.28c, 10.29c, 10.32:** © The McGraw-Hill Companies, Inc./Jill Braaten, photographer; **10.34c:** © Christine Eckel; **10.34d, 10.35d:** © The McGraw-Hill Companies, Inc./Jill Braaten, photographer; **Figure 10A:** © Paul J. Sutton/DUOMO/PCN Photography.

Chapter 11

Opener: © Dennis Kunkel Microscopy, Inc.; **11.2a:** © Brand X Pictures/PunchStock RF; **11.2b:** © Royalty-Free/Corbis RF.

Chapter 12

Opener: © SPL/Photo Researchers, Inc.; **12.3b:** © Ed Reschke/Peter Arnold, Inc./Getty Images; **12.15b:** © The McGraw-Hill Companies, Inc./Rebecca Gray, photographer/Don Kincaid, dissections.

Chapter 13

Opener: © Royalty-Free/Corbis RF; **13.1:** © Courtesy of Branislav Vidic; **13.6a-c:** © The McGraw-Hill Companies, Inc./Rebecca Gray, photographer/Don Kincaid, dissections; **13.8a:** © R. T. Hutchings; **13.8b, 13.9b:** © The McGraw-Hill Companies, Inc./Rebecca Gray, photographer/Don Kincaid, dissections; **13.10b-c:** © Trent Stephens.

Chapter 14

Opener: Image by Tamily Weissman. The Brainbow mouse was produced by Livet J, Weissman TA, Kang H, Draft RW, Lu J, Bennis RA, Sanes JR, Lichtman JW. Nature (2007) 450:56-62.; **14.3:** © The McGraw-Hill Companies, Inc./Eric Wise, photographer; **14.20b:** © Marcus Raichle, MD, Washington University School of Medicine; **14.21a:** © PHANIE/Photo Researchers, Inc.; **p. 292:** © Digital Vision/Getty Images RF; **Figure 14Ba:** © Howard J. Radzyner/Phototake.com; **Figure 14Bb:** © CNRI/Science Photo Library/Photo Researchers, Inc.

Chapter 15

Opener: © Mireille Lavigne-Rebillard, INSERM Unit 583, Montpellier ("from the site Promenade around the cochlea (http://www.iurc.montp.inserm.fr/cric/audition/fran%E7ais/cochlea/fcochlea.htm)" by R. Pujol, S Blatrix and T. Pujol, CRIC, University of Montpellier/INSERM & www.neuroreille.com; **15.4f:** © Omikran/Photo Researchers, Inc.; **15.8:** © The McGraw-Hill Companies, Inc./Eric Wise, photographer; **15.12:** © The McGraw-Hill Companies, Inc./Rebecca Gray, photographer/Don Kincaid, dissections; **15.15a:** © A. L. Blum/Visuals Unlimited; **15.18b:** © Steve Gschmeissner/Photo Researchers, Inc.; **Figure 15Ba:** © Steve Allen/Getty Images RF; **Figure 15Bb:** © Prisma Bildagentur AG/Alamy; **15.23d:** © The McGraw-Hill Companies, Inc./Rebecca Gray, photographer/Don Kincaid, dissections; **15.25:** © The McGraw-Hill Companies, Inc./Eric Wise, photographer; **15.29a:** Courtesy of R. A. Jacobs and A. J. Hudspeth; **15.29b:** © Fred Hossler/Visuals Unlimited; **15.30b:** Courtesy of A. J. Hudspeth; **15.30c:** © David Corey, Harvard Medical School; **15.36d:** © Susumu Nishinag/Photo Researchers, Inc.; **15.37a-b:** © Trent Stephens; **15.38d:** © Dr. Richard Kessel & Dr. Randy Kardon/Tissues and Organs/Visuals Unlimited; **15.39b:** © Jerry Wachter/Photo Researchers, Inc.

Chapter 16

Opener: © Prof. P.M. Motta/Univ. "La Sapienza", Rome/Photo Researchers, Inc.

Chapter 17

Opener: © Quest/Science Photo Library/Photo Researchers, Inc.

Chapter 18

Opener, *left*: Courtesy of UAB Media Relations; **Opener,** *right*: Courtesy of Dr. Elizabeth Fenjves; **18.1c:** © Dr. John D. Cunningham/Visuals Unlimited; **18.1d:** © Science VU/Visuals Unlimited; **18.8c, 18.11b, 18.13b:** © Victor Eroschenko; **18.16:** © Bio-Photo Assocs/Photo Researchers, Inc.; **p. 630:** © Jon Feingersch Photography Inc./Image Source/VEER RF; **Figure 18A:** © Ken Greer/Visuals Unlimited.

Chapter 19

Opener: © Dennis Kunkel Microscopy, Inc.; **19.1:** © liquidlibrary/PictureQuest RF; **19.3a:** © National Cancer Institute/Science Photo Library/Photo Researchers, Inc.; **Figure 19A:** © Dr. Stanley Flegler/Visuals Unlimited; **19.7:** © Ed Reschke/Peter Arnold Inc./Getty Images; **19.8a-e:** © Victor Eroschenko; **19.10:** © Oliver Meckes/Photo Researchers, Inc.

Chapter 20

Opener: © Frieder Michler/Photo Researchers, Inc.; **20.5b:** © R. T. Hutchings; **20.8a-b:** © McMinn & Hutchings, Color Atlas of Human Anatomy/Mosby; **20.12b:** © Ed Reschke; **20.20:** © Terry Cockerham/Cynthia Alexander/Synapse Media Productions; **p. 700:** © Digital Vision/Getty Images RF; **Figure 20A** *(left)*: © Hank Morgan/Science Source/Photo Researchers, Inc.; **Figure 20A** *(right)*: © SPL/Photo Researchers, Inc.

Chapter 21

Opener: © Medicine and Health Care/EP017/Getty Images RF; **21.5:** © Carolina Biological Supply/Visuals Unlimited.

Chapter 22

Opener: © Dream Pictures/Vstock/Blend Images/Getty Images; **22.4:** © Victor Eroschenko; **22.6b:** © Trent Stephens; **22.7d:** © The McGraw-Hill Companies, Inc./APR; **22.8b:** © Trent Stephens; **p. 804:** © Image Source/Getty Images RF; **Figure 22A:** © ISM/Phototake.com.

Chapter 23

Opener: © David Phillips/Photo Researchers, Inc.; **23.2b:** © The McGraw-Hill Companies, Inc./Photo and dissection by Christine Eckel; **23.4b:** © CNRI/Phototake.com; **23.5a:** © John Cunningham/Visuals Unlimited; **23.5c** © Ed Reschke/Peter Arnold, Inc./Getty Images; **23.6b:** © Walter Reiter/Phototake.com; **23.7b:** © Carolina Biological Supply Company/Phototake.com; **p. 850:** © Thinkstock/Jupiterimages RF; **Figure 23A:** © Chris Cole/Riser/Getty Images; **Figure 23B** *(left)*: © Image Source/Getty Images RF; **Figure 23B** *(right)*: © Biodisc/Visuals Unlimited.

Chapter 24

Opener: © GJLP/Photo Researchers, Inc.; **24.5b-c:** © The McGraw-Hill Companies, Inc./Rebecca Gray, photographer/Don Kincaid, dissections; **24.9c:** © Ed Reschke; **24.11c:** © Victor Eroschenko; **24.16e:** © David M. Phillips/Visuals Unlimited; **24.25b:** © CNRI/Science Photo Library/Photo Researchers, Inc.; **24.26d:** © The McGraw-Hill Companies, Inc./Dr. Alvin Telser, photographer; **p. 904:** © Comstock Images RF; **Figure 24A:** © PictureNet/Corbis RF.

Chapter 25

Opener: U.S. Department of Agriculture; **25.17:** © M.M. Sweet/Flickr/Getty Images.

Chapter 26

Opener: © Hank Morgan/Photo Researchers, Inc.; **26.3b:** © The McGraw-Hill Companies, Inc./Rebecca Gray, photographer/Don Kincaid, dissections.

Chapter 27

Opener: © Brand X Pictures RF.

Chapter 28

Opener: © Dr. Yorgus Nikas/Wellcome Images; **28.3a:** © Biodisc/Visuals Unlimited; **28.3b:** © Ken Wagner/Phototake.com; **28.3c:** © Dennis Kunkel/Phototake.com; **28.19(1):** © Educational Images/Custom Medical Stock Photo; **28.19(2-3):** © Biophoto Associates/Photo Researchers, Inc.; **Figure 28Ba:** © M. Long/Visuals Unlimited; **Figure 28Bb:** © Ray Ellis/Photo Researchers, Inc.; **Figure 28Bc:** © The McGraw-Hill Companies, Inc.; **Figure 28Bd:** © Hank Morgan/Photo Researchers, Inc.; **Figure 28Be:** © The McGraw-Hill Companies, Inc./Bob Coyle photographer; **p. 1054:** © Brand X Pictures/PunchStock RF.

Chapter 29

Opener: © UFO RF/a.collectionRF/Getty Images RF; **29.2a-e:** © Dr. Yorgus Nikas/Jason Burns/Phototake.com; **29.2b:** © Dr. Yorgus Nikas/Jason Burns/Phototake.com; **29.9:** © John Giannicchi/Photo Researchers, Inc.; **Figure 29A:** © NMSB/Custom Medical Stock Photo; **29.16a-b:** © Petit Format/Nestle/Photo Researchers, Inc.; **29.16c:** © Tissuepix/Photo Researchers, Inc.; **29.22:** © Anderson Ross/Blend Images/Getty Images RF; **29.23:** © Norman Lightfoot/Photo Researchers, Inc; **29.24:** © Dr. M.A. Ansary/Photo Researchers, Inc.; **29.25:** © CNRI/Science Photo Library/Photo Researchers, Inc.

Index

Note: Page numbers followed by *f* or *t* indicate figures and tables, respectively.

A

B

E

G

H

I

M

N

O

P

Q

R

T

U

V

W

X

Y

Z

Selected Abbreviations

α alpha
ACE angiotensin-converting enzyme
acetyl-CoA acetyl coenzyme A
ACh acetylcholine
ADH antidiuretic hormone
ADP adenosine diphosphate
ANH atrial natriuretic hormone
ANS autonomic nervous system
apo E apolipoprotein E
ATP adenosine triphosphate
AV atrioventricular beta
$\bar{B}$COP blood colloid osmotic pressure
BMI body mass index
BMR basal metabolic rate
BP blood pressure
BPG 2,3-bisphosphoglycerate
bpm beats per minute
BUN blood urea nitrogen
$C_6H_{12}O_6$ glucose
$Ca_{10}(PO_4)_6(OH)_2$ hydroxyapatite
Ca^{2+} calcium ion
cal calorie
cAMP cyclic adenosine monophosphate
CBC complete blood count
cGMP cyclic guanosine monophosphate
CH_3COOH acetic acid
Cl^- chloride ion
CNS central nervous system
CO cardiac output
CO carbon monoxide
CO_2 carbon dioxide
—COOH carboxyl group
COX-1 cyclooxygenase-1
CP capsule pressure
CRH corticotrophin-releasing hormone
CSF cerebrospinal fluid
DAG diacylglycerol
DNA deoxyribonucleic acid
ECG or EKG electrocardiogram
EEG electroencephalogram
EGF epidermal growth factor
ENS enteric nervous system
EPSP excitatory postsynaptic potential
FAD flavin adenine dinucleotide
$FADH_2$ reduced flavin adenine diphosphate
Fe^{2+} iron ion
FEV_1 forced expiratory volume in one second
FGF fibroblast growth factor
FSH follicle-stimulating hormone gamma
$\bar{g}$ gram
GABA gamma-aminobutyric acid
GCP glomerular capillary pressure
GDP guanosine diphosphate
GFR glomerular filtration rate
GH growth hormone
GHIH growth hormone-inhibiting hormone
GHRH growth hormone-releasing hormone
GnRH gonadotropin-releasing hormone
GTP guanosine triphosphate
H^+ hydrogen ion
H_2CO_3 carbonic acid
H_2O water
H_2O_2 hydrogen peroxide
$H_2PO_4^-$ dihydrogen phosphate ion
HCG human chorionic gonadotropin
HCl hydrochloric acid
HCO_3^- bicarbonate ion
HDL high-density lipoprotein
Hg mercury
HIV human immunodeficiency virus
HLA human leukocyte antigen
HPO_4^{2-} monohydrogen phosphate ion
HR heart rate
I^- iodide ion
ICSH interstitial cell-stimulating hormone
IFP interstitial fluid pressure
Ig immunoglobulin
IP_3 inositol triphosphate
IPSP inhibitory postsynaptic potential
IU international units
K^+ potassium ion
kcal kilocalorie
kg kilogram
L liter
LDL low-density lipoprotein
LH luteinizing hormone
LHRH luteinizing hormone-releasing hormone
MAC membrane attack complex
MALT mucosa-associated lymphoid tissue
MAO monoamine oxidase
MAP mean arterial pressure
mEq milliequivalent
Mg^{2+} magnesium ion
MHC major histocompatibility complex
mOsm milliosmole
mRNA messenger ribonucleic acid
mV millivolt
Na^+ sodium ion
NaCl sodium chloride
NAD^+ nicotinamide adenine dinucleotide
NADH reduced nicotinamide adenine dinucleotide
$NaHCO_3$ sodium bicarbonate
NaOH sodium hydroxide
NFP net filtration pressure
$—NH_2$ amine group
NH_3 ammonia
NH_4^+ ammonium ion
NK cells natural killer cells
NO nitric oxide
O_2 oxygen
OH^- hydroxide ion
PAH *para*-aminohippuric acid
P_{ALV} alveolar pressure
P_B barometric air pressure
PGE prostaglandin E
PGF prostaglandin F
P_i inorganic phosphate
PIF prolactin-inhibiting factor
PIH prolactin-inhibiting hormone
PIP_2 phosphoinositol
PMNs polymorphonuclear neutrophils
PNS peripheral nervous system
PO_4^{3-} phosphate ion
P_{PL} pleural pressure
PR peripheral resistance
PRF prolactin-releasing factor
PRH prolactin-releasing hormone
PTH parathyroid hormone
RANKL receptor for activation of nuclear factor kappa B ligand
RAS reticular activating system
RBC red blood count
RDA recommended daily allowances
RDI reference daily intake
RhoGAM Rh_o (D) immune globulin
RMP resting membrane potential
RNA ribonucleic acid
SA sinoatrial
SV stroke volume
T_3 triiodothyronine
T_4 tetraiodothyronine
TF tissue factor
TGF-_ transforming growth factor beta
TMJ temporomandibular joint
TNF tumor necrosis factor
TRH thyroid-releasing hormone
TSH thyroid-stimulating hormone
V_A alveolar ventilation
VLDL very low-density lipoprotein
vWF von Willebrand factor
WBC white blood count

Prefixes, Suffixes, and Combining Forms

The ability to break down medical terms into separate components or to recognize a complete word depends on mastery of the combining forms (roots or stems) and the prefixes and suffixes that alter or modify their meanings. Common prefixes, suffixes, and combining forms are listed below in boldface type, followed by the meaning of each form and an example illustrating its use.

a-, an- without, lack of: *a*phasia (lack of speech), *an*aerobic (without oxygen)
ab- away from: *ab*ductor (leading away from)
-able capable: vi*able* (capable of living)
acou- hearing: *acou*stics (science of sound)
acr- extremity: *acro*megaly (large extremities)
ad- to, toward, near to: *ad*renal (near the kidney)
adeno- gland: *adeno*ma (glandular tumor)
-al expressing relationship: neur*al* (referring to nerves)
-algia pain: gastr*algia* (stomach pain)
angio- vessel: *angio*graphy (radiography of blood vessels)
ante- before, forward: *ante*cubital (before elbow)
anti- against, reversed: *anti*peristalsis (reversed peristalsis)
arthr- joint: *arthr*itis (inflammation of a joint)
-ary associated with: urin*ary* (associated with urine)
-asis condition, state of: homeost*asis* (state of staying the same)
auto- self: *auto*lysis (self breakdown)
bi- twice, double: *bi*cuspid (two cusps)
bio- live: *bio*logy (study of living)
-blast bud, germ: fibro*blast* (fiber-producing cell)
brady- slow: *brady*cardia (slow heart rate)
-c expressing relationship: cardia*c* (referring to heart)
carcin- cancer: *carcin*ogenic (causing cancer)
cardio- heart: *cardio*pathy (heart disease)
cata- down, according to: *cata*bolism (breaking down)
cephal- head: *cephal*ic (toward the head)
-cele hollow: blasto*cele* (hollow cavity inside a blastocyst)
cerebro- brain: *cerebro*spinal (referring to brain and spinal cord)
chol- bile: a*chol*ic (without bile)
cholecyst- gallbladder: *cholecyst*okinin (hormone causing the gallbladder to contract)
chondr- cartilage: *chondr*ocyte (cartilage cell)
-cide kill: bacteri*cide* (agent that kills bacteria)
circum- around, about: *circum*duction (circular movement)
-clast smash, break: osteo*clast* (cell that breaks down bone)
co-, com-, con- with, together: *co*enzyme (molecule that functions with an enzyme), *com*misure (coming together), *con*vergence (to incline together)
contra- against, opposite: *contra*lateral (opposite side)
crypto- hidden: *crypt*orchidism (undescended or hidden testes)
cysto- bladder, sac: *cysto*cele (hernia of a bladder)
-cyte-, cyto- cell: erythro*cyte* (red blood cell), *cyto*skeleton (supportive fibers inside a cell)
de- away from: *de*hydrate (remove water)
derm- skin: *derm*atology (study of the skin)
di- two: *di*ploid (two sets of chromosomes)
dia- through, apart, across: *dia*pedesis (ooze through)
dis- reversal, apart from: *dis*sect (cut apart)
-duct- leading, drawing: ab*duct* (lead away from)
-dynia pain: masto*dynia* (breast pain)
dys- difficult, bad: *dys*mentia (bad mind)
e- out, away from: *e*viscerate (take out viscera)
ec- out from: *ec*topic (out of place)
ecto- on outer side: *ecto*derm (outer skin)
-ectomy cut out: append*ectomy* (cut out the appendix)
-edem- swell: myo*edem*a (swelling of a muscle)
em-, en- in: *em*pyema (pus in), *en*cephalon (in the brain)
-emia blood: an*emia* (deficiency of blood)
endo- within: *endo*metrium (within the uterus)
entero- intestine: *enter*itis (inflammation of the intestine)
epi- upon, on: *epi*dermis (on the skin)
erythro- red: *erythro*cyte (red blood cell)
eu- well, good: *eu*phoria (well-being)
ex- out, away from: *ex*halation (breathe out)
exo- outside, on outer side: *exo*genous (originating outside)
extra- outside: *extra*cellular (outside the cell)
-ferent carry: af*ferent* (carrying to the central nervous system)
-form expressing resemblance: fusi*form* (resembling a fusion)
gastro- stomach: *gastro*dynia (stomach ache)
-genesis produce, origin: patho*genesis* (origin of disease)
gloss- tongue: hypo*gloss*al (under the tongue)
glyco- sugar, sweet: *glyco*lysis (breakdown of sugar)
-gram a drawing: myo*gram* (drawing of a muscle contraction)
-graph instrument that records: myo*graph* (instrument for measuring muscle contraction)
hem- blood: *hem*opoiesis (formation of blood)
hemi- half: *hemi*plegia (paralysis of half of the body)
hepato- liver: *hepa*titis (inflammation of the liver)
hetero- different, other: *hetero*zygous (different genes for a trait)
hist- tissue: *hist*ology (study of tissues)
homeo-, homo- same: *homeo*stasis (state of staying the same), *homo*logous (alike in structure or origin)
hydro- wet, water: *hydro*cephalus (fluid within the head)
hyper- over, above, excessive: *hyper*trophy (overgrowth)
hypo- under, below, deficient: *hypo*tension (low blood pressure)
-ia, -id expressing condition: neuralg*ia* (pain in nerve), flacc*id* (state of being weak)
-iatr- treat, cure: ped*iatr*ics (treatment of children)
-im not: *im*permeable (not permeable)
in- in, into: *in*jection (forcing fluid into)
infra- below, beneath: *infra*orbital (below the eye)
inter- between: *inter*costal (between the ribs)
intra- within: *intra*ocular (within the eye)
-ism condition, state of: dimorph*ism* (condition of two forms)